DONALD VOET

University of Pennsylvania

JUDITH G. VOET

Swarthmore College

BIOCHEMISTRY

SECOND EDITION

JOHN WILEY & SONS, INC.

New York   Chichester   Brisbane
Toronto   Singapore

Acquisitions Editor: *Nedah Rose*
Marketing Manager: *Catherine Faduska*
Designer: *Madelyn Lesure*
Manufacturing Manager: *Susan Stetzer*
Photo Researcher: *Hilary Newman*
Illustration Coordinator: *Edward Starr*
Illustrators: *J/B Woolsey Associates*
Production Management: *Pamela Kennedy Oborski and Suzanne Ingrao*
Cover and Part Opening Illustrations © *Irving Geis*
This book was set in 9.5/12 Times Roman by Progressive Information Technologies and printed and bound by Von Hoffman Press. The cover was printed by Lehigh Press Lithographers. Color separations by Lehigh Press Colortronics.

Cover Art: Two paintings of horse heart cytochrome *c* by Irving Geis in which the protein is illuminated by its single iron atom. On the front cover the hydrophilic side chains are drawn in green, and on the back cover the hydrophobic side chains are drawn in orange. The paintings are based on an X-ray structure by Richard Dickerson.

ISBN: 0-471-58651-X

Printed in the United States of America

10 9 8 7 6 5 4 3

To
Our parents, who encouraged us,
Our teachers, who enabled us, and
Our children, who put up with us.

Annual Supplement Subscription Notice

Every year the biochemical literature continues its phenomenal growth. It is therefore increasingly important for the student and teacher alike to keep up with the literature. BIOCHEMISTRY, Second Edition is updated with Annual Supplements prepared by Donald and Judith Voet as a guide for doing so.

Beginning in the Summer of 1996 and each year thereafter an Annual Supplement will be prepared. Each supplement is keyed to BIOCHEMISTRY in that new advances are organized in terms of the textbook sections in which they would logically fit. Since space limitations permit only the most cursory discussions of these topics, the interested reader should consult the pertinent "Additional References" provided at the end of each discussion. The Annual Supplements for the Second Edition of BIOCHEMISTRY will continue until the Third Edition of BIOCHEMISTRY is published.

If you purchased this product directly from John Wiley & Sons, Inc., we have already recorded your subscription for this update service. If, however, you purchased this product from a bookstore and wish to receive future updates, billed separately, with a 30-day examination review, please send your name, company name (if applicable), address, and the notation, "SUBSCRIPTION SERVICE FOR VOET/VOET: BIOCHEMISTRY, Second Edition" to:

> Supplement Department
> John Wiley & Sons, Inc.
> One Wiley Drive
> Somerset, NJ 08875
> 1-800-225-5945
> FAX: 908-302-2300

For customers outside the United States, please contact the Wiley office nearest you:

College Division
John Wiley & Sons Canada, Ltd.
22 Worcester Road
Etobicoke, Ontario M9W 1L1
CANADA
(416) 675-3580
1-800-263-1590
FAX 1-800-565-6802

Jacaranda Wiley Ltd.
33 Park Road
Milton, Queensland 4064
AUSTRALIA
(07) 369-9755
FAX: (07) 369-9155

Customer Services
John Wiley & Sons, Ltd.
Distribution Center
Southern Cross Trading Estate
1 Oldlands Way
Bognor Regis, West Sussex PO22 9SA
UNITED KINGDOM
(44)(243) 829121
FAX: (44)(243) 820250

John Wiley & Sons (SEA)
  Pte.Ltd.
37 Jalan Pemimpin
Block B #05-04
Union Industrial Building
SINGAPORE 2057
(65) 258-1147

Foreword

My own scientific career was a descent from higher to lower dimension, led by a desire to understand life. I went from animals to cells, from cells to bacteria, from bacteria to molecules, from molecules to electrons. The story had its irony, for molecules and electrons have no life at all. On my way life ran out between my fingers.

Albert Szent-Györgi, The Living State, *Academic Press, 1972*

Thirty years ago, concepts in biochemistry were visualized in textbooks, chiefly as a series of organic chemistry formulas in schematic line diagrams. It would not be surprising if students of that time believed that living organisms were indeed composed of lifeless molecules.

A sharp change in biochemistry teaching came in the late 1960s with elucidation of the first three-dimensional structures of large protein molecules such as myoglobin, hemoglobin and the enzyme, lysozyme. In those days, it might require thirty-five man-years to determine, by X-ray crystallography, the structure of a single protein molecule. Thus it would be many years before there were sufficient examples of three-dimensional protein structures to begin to understand how these complex molecular machines acted to control the myriad series of reactions in the living cell.

In 1990, the first edition of this book achieved a grand synthesis of biochemical science, incorporating the flood of new ideas and experiments of the 1970s and 1980s — the development of rapid sequencing of DNA and protein molecules, as well as a proliferation of new three-dimensional structures. With a clear understanding that the one-dimensional linear DNA sequence provided the information for the three-dimensional conformation of protein molecules, the next step was then manipulating the DNA sequence, producing "designer proteins" by gene splicing and site-directed mutagenesis. New techniques for the manipulation of DNA and proteins called for new techniques of visual presentation. The clear and concise writing and the profusion of vivid, colorful illustrations made the first edition a leader in the field of biochemistry publishing.

This second 1995 edition is again a grand synthesis of the continuing flood of new concepts, as well as new structures determined by X-ray crystallography and NMR spectroscopy. The chapter Eukaryotic Gene Expression takes the student to the cutting edge of nucleic acid research by explaining the control of DNA expression in embryogenesis as well as the pathological expression of cancer genes. The expression of DNA is regulated mainly by the control of DNA transcription to RNA. This is a complex process involving protein–protein interactions along with protein binding, and often bending the DNA molecule.

The final chapter, Molecular Physiology, anticipates twenty-first-century molecular medicine by uniting physiological function, cell biology, and anatomy with molecular interactions. Examples are given of the cascade of enzymes in blood clotting, and the cascades of hormones and protein kinases in signalling between cells.

The ambitious Human Genome project aims to sequence the three billion base-pairs of human DNA. There are some 4,000 genetic diseases presently known that are potential targets for gene therapy. The effect of genetic research on the practice of medicine can hardly be overestimated. This book stands as a valuable guide to the science of molecular medicine for the twenty-first century.

IRVING GEIS

New York City
December 1994

Preface

In the five years since the first edition of *Biochemistry* was published, the field of biochemistry has continued its phenomenal growth and at an ever-increasing pace. This expansion of our knowledge has been marked not so much by new paradigms, although there have been plenty of those, but by an enormous enrichment of almost every facet in the field. For example, the number of known protein and nucleic acid structures as determined by X-ray and NMR techniques has increased by over fivefold and, moreover, many of these structures have led to seminal advances in our understanding of a particular subfield. Likewise, the state of knowledge has exploded in such subdisciplines as eukaryotic and prokaryotic molecular biology, metabolic control, protein folding, electron transport, membrane transport, immunology, signal transduction, and so on. Indeed, these advances have affected our everyday lives in that they have changed the way that medicine is practiced, the way that we protect our own health, and the way in which food is produced.

We have reported many of these advances in the second edition of *Biochemistry* and have thereby substantially changed nearly every section in it. We have, nevertheless, largely maintained the pedagogical framework of the first edition. Consequently, the Preface to the First Edition applies equally well to the Second Edition.

The textbook is accompanied by the following ancillary materials:

- A *Solutions Manual,* containing detailed solutions for all of the text's end-of-chapter problems.

- A CD-ROM containing most of the illustrations in the text. With computerized projection equipment, these full-color images can be shown in any prearranged order to provide "slide shows" to accompany lectures. Alternatively, they can be used to print transparencies.

- A diskette containing KINEMAGES, computer animated color images, of selected proteins and nucleic acids that permit the student to manipulate these macromolecules in three dimensions.

In addition, we shall continue our previous practice of annually publishing *Supplements to Biochemistry* that summarize the highlights of the preceding year's biochemical advances. The first of these ~80-page *Supplements* will be available in June 1996 and can be obtained either individually or on a subscription basis.

Finally, we are particularly grateful to the many readers of the First Edition, students and teachers alike, who have taken the trouble to write us with suggestions on how to improve the textbook and to point out errors they have found. We earnestly hope that the readers of the Second Edition will continue this practice.

DONALD VOET
JUDITH G. VOET

Preface

to the First Edition

Biochemistry is a field of enormous fascination and utility, arising, no doubt, from our own self-interest. Human welfare, particularly its medical and nutritional aspects, has been vastly improved by our rapidly growing understanding of biochemistry. Indeed, scarcely a day passes without the report of a biomedical discovery that benefits a significant portion of humanity. Further advances in this expanding field of knowledge will no doubt lead to even more spectacular gains in our ability to understand nature and to control our destinies. It is therefore of utmost importance that individuals embarking on a career in biomedical sciences be well versed in biochemistry.

This textbook is a distillation of our experiences in teaching undergraduate and graduate students at the University of Pennsylvania and Swarthmore College and is intended to provide such students with a thorough grounding in biochemistry. In writing this text we have emphasized several themes. First, biochemistry is a body of knowledge compiled by people through experimentation. In presenting what is known, we therefore stress how we have come to know it. The extra effort the student must make in following such a treatment, we believe, is handsomely repaid since it engenders the critical attitudes required for success in any scientific endeavor. Although science is widely portrayed as an impersonal subject, it is, in fact, a discipline shaped through the often idiosyncratic efforts of individual scientists. We therefore identify some of the major contributors to biochemistry (the majority of whom are still professionally active) and, in many cases, consider the approaches they have taken to solve particular biochemical puzzles. The student should realize, however, that most of the work described could not have been done without the dedicated and often indispensable efforts of numerous coworkers.

The unity of life and its variation through evolution is a second dominant theme running through the text. Certainly one of the most striking characteristics of life on earth is its enormous variety and adaptability. Yet, biochemical research has amply demonstrated that all living things are closely related at the molecular level. As a consequence, the molecular differences among the various species have provided intriguing insights into how organisms have evolved from one another and have helped delineate the functionally significant portions of their molecular machinery.

A third major theme is that biological processes are organized into elaborate and interdependent control networks. Such systems permit organisms to maintain relatively constant internal environments, to respond rapidly to external stimuli, and to grow and differentiate. A fourth theme is that biochemistry has important medical consequences. We therefore frequently illustrate biochemical principles by examples of normal and abnormal human physiology.

We assume that students who use this text have had the equivalent of one year of college chemistry and at least one semester of organic chemistry so that they are familiar with both general chemistry and the basic principles and nomenclature of organic chemistry. We also assume that students have taken a one year college course in general biology in which elementary biochemical concepts were discussed. Students who lack these prerequisites are advised to consult the appropriate introductory textbooks in these subjects.

The text is organized into five parts:

I. **Introduction and Background:** An introductory chapter followed by chapters that review the properties of aqueous solutions and the elements of thermodynamics.

II. **Biomolecules:** A description of the structures and functions of proteins, carbohydrates, and lipids.

III. **Mechanisms of Enzyme Action:** An introduction to the properties, reaction kinetics, and catalytic mechanisms of enzymes.

IV. **Metabolism:** A discussion of how living things synthesize and degrade carbohydrates, lipids, amino acids, and nucleotides with emphasis on energy generation and consumption.

V. **The Expression and Transmission of Genetic Information:** An exposition of nucleic acid structures and both prokaryotic and eukaryotic molecular biology.

This organization permits us to cover the major areas of biochemistry in a logical and coherent fashion. Yet, modern biochemistry is a subject of such enormous scope that to maintain a relatively even depth of coverage throughout the text, we include more material than most one year biochemistry courses will cover in detail. This depth of coverage, we feel, is one of the strengths of this book; it permits the instructor to teach a course of his/her own design and yet provide the student with a resource on biochemical subjects not emphasized in the course.

The order in which the subject matter of the text is presented more or less parallels that of most biochemistry courses. However, several aspects of the text's organization deserve comment:

1. We present nucleic acid structures (Chapter 28) as part of molecular biology (Part V) rather than in our discussions of structural biochemistry (Part II) because nucleic acids are not mentioned in any substantive way until Part V. Instructors who, nevertheless, prefer to consider nucleic acid structures in a sequence different from that in the text can easily do so since Chapter 28 requires no familiarity with enzymology or metabolism.

2. We have split our presentation of thermodynamics among two chapters. Basic thermodynamic principles —enthalpy, entropy, free energy, and equilibrium— are discussed in Chapter 3 because these subjects are prerequisite for understanding structural biochemistry, enzyme mechanisms, and kinetics. Metabolic aspects of thermodynamics—the thermodynamics of phosphate compounds and oxidation-reduction reactions—are presented in Chapter 15 since knowledge of these subjects is not required until the chapters that follow.

3. Techniques of protein purification are described in a separate chapter (Chapter 5) that precedes the discussion of protein structure and function. We have chosen this order so that students will not feel that proteins are somehow "pulled out of a hat". Nevertheless, Chapter 5 has been written as a resource chapter to be consulted repeatedly as the need arises.

4. Chapter 9 describes the properties of hemoglobin in detail so as to illustrate concretely the preceding discussions of protein structure and function. This chapter introduces allosteric theory to explain the cooperative nature of hemoglobin oxygen binding. The subsequent extension of allosteric theory to enzymology (Chapter 12) is a relatively simple matter.

5. Concepts of metabolic control are presented in the chapters on glycolysis (Chapter 16) and glycogen metabolism (Chapter 17) through discussions of flux generation, allosteric regulation, substrate cycles, covalent enzyme modification, and cyclic cascades. We feel that these concepts are best understood when studied in metabolic context rather than as independent topics.

6. There is no separate chapter on coenzymes. These substances, we feel, are more logically studied in the context of the enzymatic reactions in which they participate.

7. Glycolysis (Chapter 16), glycogen metabolism (Chapter 17), the citric acid cycle (Chapter 19), and oxidative phosphorylation (Chapter 20) are detailed as models of general metabolic pathways with emphasis placed on many of the catalytic and control mechanisms of the enzymes involved. The principles illustrated in these chapters are reiterated in somewhat less detail in the other chapters of Part IV.

8. Consideration of membrane transport (Chapter 18) precedes that of mitochondrially based metabolic pathways including the citric acid cycle and oxidative phosphorylation. In this manner, the idea of the compartmentalization of biological processes can be easily assimilated.

9. Discussions of both the synthesis and the degradation of lipids has been placed in a single chapter (Chapter 23) as have the analogous discussions of amino acids (Chapter 24) and nucleotides (Chapter 26).

10. Energy metabolism is summarized and integrated in terms of organ specialization in Chapter 25, following the descriptions of carbohydrate, lipid, and amino acid metabolism.

11. The basic principles of both prokaryotic and eukaryotic molecular biology are introduced in sequential chapters on transcription (Chapter 29), translation (Chapter 30), and DNA replication, repair and recombination (Chapter 31). Viruses (Chapter 32) are then considered as paradigms of more complex cellular functions, followed by discussions of newly emerging concepts of eukaryotic gene expression (Chapter 33).

12. Chapter 34, the final chapter, is a series of minichapters that describe the biochemistry of a variety of well-characterized human physiological processes: blood clotting, the immune response, muscle contraction, hormonal communication, and neurotransmission.

The old adage that you learn a subject best by teaching it simply indicates that learning is an active rather than a passive process. The problems we provide at the end of each chapter are therefore designed to make students think rather than to merely regurgitate poorly assimilated and rapidly forgotten information. Few of the problems are trivial and some of them (particularly those marked with an asterisk) are quite difficult. Yet, successfully working out such problems can be one of the most rewarding aspects of the learning process. Only by thinking long and hard for themselves can students make a body of knowledge truly their own. The answers to the problems are worked out in detail in the *Solutions Manual* that accompanies this text. However, this manual can only be an effective learning tool if the student makes a serious effort to solve a problem before looking up its answer.

We have included lists of references at the end of every chapter to provide students with starting points for independent biochemical explorations. The enormity of the biochemical research literature precludes us from giving all but a few of the most important research reports. Rather, we list what we have found to be the most useful reviews and monographs on the various subjects covered in each chapter.

Biomedical research is advancing at such an astonishing pace that a seminal discovery often leads to the development of a mature subdiscipline within the period of a year or so. Consequently, a textbook on biochemistry can never be truly up to date. In order to alleviate this problem, we shall periodically bring out Supplements to this textbook that review the recent biochemical literature and list some

of its most important reviews and research reports. Nevertheless, students should be encouraged to peruse the current biochemical literature for only then will they acquire a feeling for the scope and excitement of modern biochemistry.

Finally, although we have made every effort to make this text error free, we are under no illusions that we have done so. We therefore request that readers provide us with their comments and criticisms.

DONALD VOET
JUDITH G. VOET

Acknowledgments

This textbook is the result of the dedicated effort of many individuals, several of whom deserve special mention:

Lindsay Ardwin, our Copy Editor, put the final polish on our manuscript and eliminated an enormous number of grammatical and typographical errors. Irving Geis provided us with his extraordinary molecular art, generously wrote the Foreword to this textbook, and gave freely of his wise counsel. Laura Ierardi cleverly combined text, figures, and tables in designing each of the textbook's pages. Suzanne Ingrao, our Production Editor, skillfully and patiently managed the production of the textbook. Madelyn Lesure designed the textbook's typography and its cover. Hilary Newman and Stella Kupferberg acquired many of the photographs in this textbook and kept track of all of them. Nedah Rose, our Editor, adroitly directed the entire project and kept our noses to the grindstone. Edward Starr and Ishaya Monokoff coordinated the illustration program. John and Bette Woolsey and their Associates refined and extended the remarkable collection of illustrations that they created for the First Edition of *Biochemistry*.

The atomic coordinates of many of the proteins and nucleic acids that we have drawn for use in this textbook were obtained from the Protein Data Bank at Brookhaven National Laboratory. We created these drawings using the molecular graphics programs RIBBONS by Mike Carson and INSIGHT II from BIOSYM Technologies. Many of the drawings generously contributed by others were made using either these programs or GRASP by Anthony Nicholls, Kim Sharp, and Barry Honig; MIDAS by Thomas Ferrin, Conrad Huang, Laurie Jarvis, and Robert Langridge; MOLSCRIPT by Per Kraulis; or O by Alwyn Jones.

We wish especially to thank those colleagues who reviewed this text, in both its current and its earlier versions:

Joseph Babitch, Texas Christian University
Robert Blankenshop, Arizona State University
Kenneth Brown, University of Texas at Arlington
Larry G. Butler, Purdue University
Carol Caparelli, Fox Chase Cancer Center
W. Scott Champney, East Tennessee State University
Glenn Cunningham, University of Central Florida
Eugene Davidson, Georgetown University
Don Dennis, University of Delaware
Walter A. Deutsch, Louisiana State University
William A. Eaton, National Institutes of Health
David Eisenberg, University of California at Los Angeles
David Fahrney, Colorado State University
Robert Fletterick, University of California at San Francisco
Scott Gilbert, Swarthmore College
James H. Hageman, New Mexico State University
Lowell Hager, University of Illinois, Urbana-Champaign
James H. Hammons, Swarthmore College
Edward Harris, Texas A & M University
Ralph A. Jacobson, California Polytechnic State University
Eileen Jaffe, University of Pennsylvania
Jan G. Jaworski, Miami University
William P. Jencks, Brandeis University
Mary Ellen Jones, University of North Carolina
Tokuji Kimura, Wayne State University
Barrie Kitto, University of Texas at Austin
Daniel J. Kosman, State University of New York at Buffalo
Thomas Laue, University of New Hampshire
Albert Light, Purdue University
Robert D. Lynch, University of Lowell
Sabeeha Merchant, University of California at Los Angeles
Ronald Montelaro, Louisiana State University
Scott Moore, Boston University
Harry F. Noller, University of California at Santa Cruz
John Ohlsson, University of Colorado
Alan R. Price, University of Michigan
Paul Price, University of California at San Diego
Thomas I. Pynadath, Kent State University
Ivan Rayment, University of Wisconsin
Frederick Rudolph, Rice University
Raghupathy Sarma, State University of New York at Stony Brook
Paul R. Schimmel, Massachusetts Institute of Technology
Thomas Schleich, University of California at Santa Cruz
Allen Scism, Central Missouri State University
Charles Shopsis, Adelphi University
Thomas Sneider, Colorado State University
Jochanan Stenish, Western Michigan University
Phyllis Strauss, Northeastern University
JoAnne Stubbe, Massachusetts Institute of Technology
William Sweeney, Hunter College
John Tooze, European Molecular Biology Organization
Francis Vella, University of Saskatchewan
Harold White, University of Delaware
William Widger, University of Houston
Ken Willeford, Mississippi State University
Lauren Williams, Georgia Institute of Technology
Jeffrey T. Wong, University of Toronto
James Zimmerman, Clemson University

Finally, we wish to thank Joel Sussman, Michal Harel, and their colleagues of the Weizmann Institute of Science, Israel, for their marvelous hospitality and stimulating conversations while we wrote much of the second edition of this textbook.

D.V.
J.G.V.

Brief Contents

Contents

PART
I

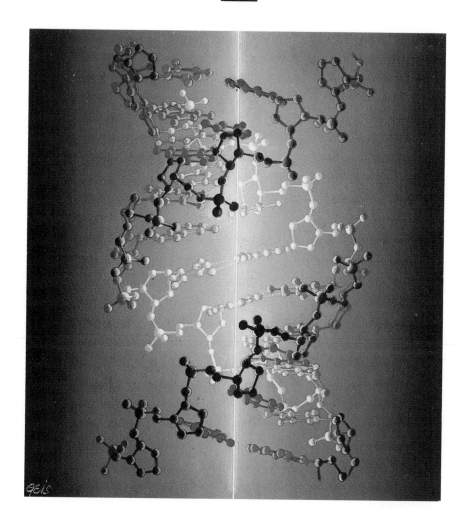

INTRODUCTION
AND
BACKGROUND

"Hot-wire" A-DNA illuminated by its helix axis.

1

Life

It is usually easy to decide whether or not something is alive. This is because living things share many common attributes, such as the capacity to extract energy from nutrients to drive their various functions, the power to actively respond to changes in their environment, and the ability to grow, to differentiate, and—perhaps most telling of all—to reproduce. Of course, a given organism may not have all of these traits. For example, mules, which are obviously alive, rarely reproduce. Conversely, inanimate matter may exhibit some lifelike properties. For instance, crystals may grow larger when immersed in a supersaturated solution of the crystalline material. Therefore, life, as are many other com- plex phenomena, is perhaps impossible to define in a pre- cise fashion. Norman Horowitz, however, has proposed a useful set of criteria for living systems: *Life possesses the properties of replication, catalysis, and mutability.* Much of this text is concerned with the manner in which living orga- nisms exhibit these properties.

Biochemistry is the study of life on the molecular level. The significance of such studies are greatly enhanced if they are related to the biology of the corresponding organisms or even communities of such organisms. This introductory chapter therefore begins with a synopsis of the biological realm. This is followed by an outline of biochemistry, a discussion of the origin of life, and finally, an introduction to the biochemical literature.

1. PROKARYOTES

It has long been recognized that life is based on morpholog- ical units known as **cells.** The formulation of this concept is generally attributed to an 1838 paper by Matthias Schlei- den and Theodor Schwann, but its origins may be traced to the seventeenth century observations of early microscopists such as Robert Hooke. There are two major classifications of cells: the **eukaryotes** (Greek: *eu,* good or true + *karyon,* kernel or nut), which have a membrane-enclosed **nucleus** encapsulating their **DNA (deoxyribonucleic acid);** and the **prokaryotes** (Greek: *pro,* before), which lack this organelle. Prokaryotes, which comprise the various types of bacteria, have relatively simple structures and are invariably unicellular (although they may form filaments or colonies of independent cells). Eukaryotes, which may be multicellular as well as unicellular, are vastly more complex than prokaryotes. (**Viruses,** which are much simpler entities than cells, are not classified as living because they lack the metabolic apparatus to reproduce outside their host cells. They are essentially large molecular aggregates.) This section is a discussion of prokaryotes. Eukaryotes are considered in the following section.

A. Form and Function

Prokaryotes are the most numerous and widespread organisms on earth. This is because their varied and often highly adaptable metabolisms suit them to an enormous variety of habitats. Besides inhabiting our familiar temperate and aerobic environment, certain types of bacteria may thrive in or even require conditions that are hostile to eukaryotes such as unusual chemical environments, high temperatures, and lack of oxygen. Moreover, the rapid reproductive rate of prokaryotes (optimally < 20 min per cell division for many species) permits them to take advantage of transiently favorable conditions, and conversely, the ability of many bacteria to form resistant **spores** allows them to survive adverse conditions.

Prokaryotes Have Relatively Simple Anatomies

Prokaryotes, which were first observed in 1683 by the inventor of the microscope, Antoni van Leeuwenhoek, have sizes that are mostly in the range 1 to 10 μm. They have one of three basic shapes (Fig. 1-1): spheroidal (**cocci**), rodlike (**bacilli**), and helically coiled (**spirilla**), but all have the same general design (Fig. 1-2). They are bounded, as are all cells, by an ~ 70-Å-thick **cell membrane (plasma membrane)** which consists of a lipid bilayer containing embed-

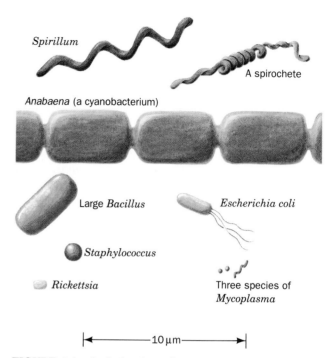

FIGURE 1-1. Scale drawings of some prokaryotic cells.

ded proteins that control the passage of molecules in and out of the cell and catalyze a variety of reactions. The cells of most prokaryotic species are surrounded by a rigid, 30- to 250-Å-thick polysaccharide **cell wall** that mainly functions to protect the cell from mechanical injury and to prevent it from bursting in media more osmotically dilute than its contents. Some bacteria further encase themselves in a gelatinous polysaccharide **capsule** that protects them from the defenses of higher organisms. Although prokaryotes lack the membranous subcellular organelles characteristic of

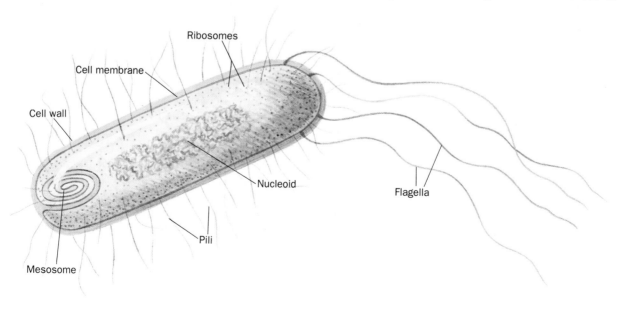

FIGURE 1-2. A schematic diagram of a prokaryotic cell.

eukaryotes (Section 1-2), their plasma membranes may be infolded to form multilayered structures known as **mesosomes.** The mesosomes are thought to serve as the site of DNA replication and other specialized enzymatic reactions.

The prokaryotic **cytoplasm** (cell contents) is by no means a homogeneous soup. Its single **chromosome** (DNA molecule, several copies of which may be present in a rapidly growing cell) is condensed to form a body known as a **nucleoid.** The cytoplasm also contains numerous species of **RNA (ribonucleic acid),** a variety of soluble **enzymes** (proteins that catalyze specific reactions), and many thousands of 250-Å-diameter particles known as **ribosomes,** which are the sites of protein synthesis.

Many bacterial cells bear one or more whiplike appendages known as **flagella,** which are used for locomotion. Certain bacteria also have filamentous projections named **pili,** some types of which function as conduits for DNA during sexual conjugation (a process in which DNA is transferred from one cell to another, Section 27-1D; prokaryotes usually reproduce by binary fission) or aid in the attachment of the bacterium to a host organism's cells.

The bacterium *Escherichia coli* (abbreviated *E. coli*) is the biologically most well-characterized organism as a result of its intensive biochemical and genetic study over the past 50 years. Indeed, much of the subject matter of this text deals with the biochemistry of *E. coli*. Cells of this normal inhabitant of the higher mammalian colon (Fig. 1-3) are typically 2-μm-long rods that are 1 μm in diameter and weigh $\sim 2 \times 10^{-12}$ g. Its DNA, which has a molecular mass of 2.5×10^9 **daltons (D),*** is thought to encode $\sim$ 3000 proteins (of which only $\sim$ 1000 have been identified), although not all of them are simultaneously present in a given cell. Altogether an *E. coli* contains some 3 to 6 thousand different types of molecules including proteins, nucleic acids, polysaccharides, lipids, and various small molecules and ions (Table 1-1).

Prokaryotes Employ a Wide Variety of Metabolic Energy Sources

The nutritional requirements of the prokaryotes are enormously varied. **Autotrophs** (Greek: *autos,* self + *trophikos,* to feed) can synthesize all their cellular constituents from simple molecules such as H_2O, CO_2, NH_3, and H_2S. Of course they need an energy source to do so as well as to power their other functions. **Chemolithotrophs** (Greek: *lithos,* stone) obtain their energy through the oxidation of inorganic compounds such as NH_3, H_2S, or even Fe^{2+}:

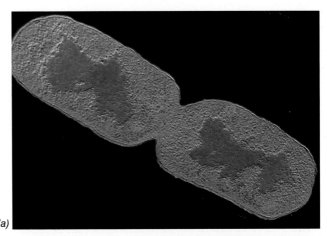

(a)

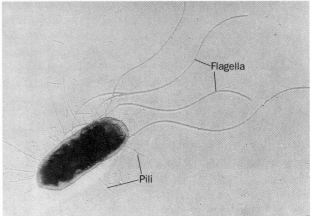

(b)

FIGURE 1-3. Electron micrographs of *E. coli* cells: *(a)* Stained to show internal structure. [CNRI.] *(b)* Stained to reveal flagella and pili. [Courtesy of Howard Berg, Harvard University.]

$$2NH_3 + 4 O_2 \longrightarrow 2HNO_3 + 2H_2O$$
$$H_2S + 2 O_2 \longrightarrow H_2SO_4$$
$$4FeCO_3 + O_2 + 6H_2O \longrightarrow 4Fe(OH)_3 + 4CO_2$$

Photoautotrophs do so via **photosynthesis** (Chapter 22), a process in which light energy powers the transfer of electrons from inorganic donors to CO_2 yielding **carbohydrates** $[(CH_2O)_n]$.

In the most widespread form of photosynthesis, the electron donor in the light-driven reaction sequence is H_2O.

$$nCO_2 + nH_2O \longrightarrow (CH_2O)_n + n O_2$$

This process is carried out by **cyanobacteria** (e.g., the green slimy organisms that grow on the walls of aquariums; cyanobacteria were formerly known as **bluegreen algae**), as well as by plants. This form of photosynthesis is thought to have generated the O_2 in the Earth's atmosphere. Some species of cyanobacteria have the ability to convert N_2 from the atmosphere to organic nitrogen compounds. This **nitrogen fixation** capacity gives them the simplest nutritional requirements of all organisms: With the exception of their need for small amounts of minerals, they can literally live on sunlight and air.

---

* The **molecular mass** of a particle may be expressed in units of daltons, which are defined as 1/12th the mass of a ^{12}C atom [atomic mass units (amu)]. Alternatively, this quantity may be expressed in terms of **molecular weight,** a dimensionless quantity defined as the ratio of the particle mass to 1/12th the mass of a ^{12}C atom and symbolized M_r (for relative molecular mass). In this text, we shall refer to the molecular mass of a particle rather than to its molecular weight.

TABLE 1-1. MOLECULAR COMPOSITION OF *E. COLI*

Component	Percentage by Weight
H_2O	70
Protein	15
Nucleic acids:	
DNA	1
RNA	6
Polysaccharides and precursors	3
Lipids and precursors	2
Other small organic molecules	1
Inorganic ions	1

Source: Watson, J.D., *Molecular Biology of the Gene* (3rd ed.), p. 69, Benjamin (1976).

In a more primitive form of photosynthesis, substances such as H_2, H_2S, thiosulfate, or organic compounds are the electron donors in light-driven reactions such as:

$$nCO_2 + 2nH_2S \longrightarrow (CH_2O)_n + nH_2O + 2nS$$

The **purple** and the **green photosynthetic bacteria** that carry out these processes occupy such oxygen-free habitats as shallow muddy ponds in which H_2S is generated by rotting organic matter.

Heterotrophs (Greek: *hetero,* other) obtain energy through the oxidation of organic compounds and hence are ultimately dependent on autotrophs for these substances. **Obligate aerobes** (which include animals) must utilize O_2, whereas **anaerobes** employ oxidizing agents such as sulfate (**sulfate-reducing bacteria**) or nitrate (**denitrifying bacteria**). Many organisms can partially metabolize various organic compounds in intramolecular oxidation–reduction processes known as **fermentation**. Facultative anaerobes such as *E. coli,* can grow either in the presence or the absence of O_2. **Obligate anaerobes**, in contrast, are poisoned by the presence of O_2. Their metabolisms are thought to resemble those of the earliest life forms (which arose some 3.5 billion years ago when the Earth's atmosphere lacked O_2; see Section 1-4B). At any rate, there are few organic compounds that cannot be metabolized by some prokaryotic organism.

B. *Prokaryotic Classification*

The traditional methods of **taxonomy** (the science of biological classification), which are based largely on the anatomical comparisons of both contemporary and fossil organisms, are essentially inapplicable to prokaryotes. This is because the relatively simple cell structures of prokaryotes, including those of ancient bacteria as revealed by their microfossil remnants, provide little indication of their phylogenetic relationships (**phylogenesis:** evolutionary development). Compounding this problem is the observation that prokaryotes exhibit little correlation between form and metabolic function. Moreover, the eukaryotic definition of a species as a population that can interbreed is meaningless for the asexually reproducing prokaryotes. Consequently, the conventional prokaryotic classification schemes are rather arbitrary and lack the implied evolutionary relationships of the eukaryotic classification scheme (Section 1-2B).

In the most widely used prokaryotic classification scheme, the **prokaryotae** (also known as **monera**) have two divisions: the cyanobacteria and the **bacteria**. The latter are further subdivided into 19 parts based on their various distinguishing characteristics, most notably cell structure, metabolic behavior, and staining properties.

A simpler classification scheme, which is based on cell wall properties, distinguishes three major types of prokaryotes: the **mycoplasma**, the **gram-positive bacteria,** and the **gram-negative bacteria.** Mycoplasma lack the rigid cell wall of other prokaryotes. They are the smallest of all living cells (as small as $0.12 \mu m$ in diameter, Fig. 1-1) and possess $\sim 20\%$ of the DNA of an *E. coli*. Presumably this quantity of genetic information approaches the minimum amount necessary to specify the essential metabolic machinery required for cellular life. Gram-positive and gram-negative bacteria are distinguished according to whether or not they take up **gram stain** (a procedure developed in 1884 by Christian Gram in which heat-fixed cells are successively treated with the dye crystal violet and iodine and then destained with either ethanol or acetone). Gram-positive bacteria possess a monolayered cell wall, whereas those of gram-negative bacteria, which include cyanobacteria, possess at least two structurally distinct layers (Section 10-3B).

The development, in recent years, of techniques for determining amino acid sequences in proteins (Section 6-1) and base sequences in nucleic acids (Section 28-6) has provided abundant indications as to the geneological relationships between organisms. Indeed, these techniques make it possible to place these relationships on a quantitative basis, and thus to construct a phylogenetically based classification system for prokaryotes.

By the analysis of ribosomal RNA sequences, Carl Woese demonstrated that a group of prokaryotes he named the **archaebacteria** seem as distantly related to the other prokaryotes, the **eubacteria** ("true" bacteria), as both these groups are to eukaryotes. The archaebacteria constitute three different kinds of unusual organisms: the **methanogens,** obligate anaerobes that produce methane (marsh gas) by the reduction of CO_2 with H_2; the **halobacteria**, which can live only in concentrated brine solutions ($>2M$ NaCl); and certain **thermoacidophiles,** organisms that inhabit acidic hot springs ($\sim 90°C$ and pH < 2). On the basis of a number of fundamental biochemical traits that differ among the archaebacteria, the eubacteria, and the eukaryotes, but that are common within each group, Woese proposed that these groups of organisms constitute the three primary kingdoms of life (rather than the traditional division into prokaryotes and eukaryotes). Moreover, the ob-

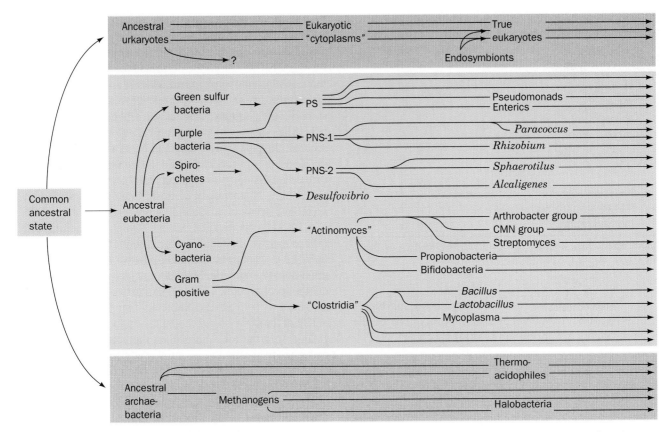

FIGURE 1-4. The major lines of prokaryotic descent. [After Fox, G.E., Stackebrandt, E., Hespell, R.B., Gibson, J., Maniloff, J., Dyer, T.A., Wolfe, R.S., Balch, W.E., Tanner, R.S., Magrum, L.J., Zablen, L.B., Blakemore, R., Gupta, R., Bonen, L., Lewis, B.J., Stahl, D.A., Leuhrsen, K.R., Chen, K.N., and Woese, C.R., *Science* **209**, 459 (1980).]

servation that these three kingdoms are genealogically equidistant suggests that all of them arose independently from some simple primordial life form (Fig. 1-4).

2. EUKARYOTES

Eukaryotic cells are generally 10 to 100 μm in diameter and thus have a thousand to a million times the volume of typical prokaryotes. It is not size, however, but a profusion of membrane-enclosed organelles, each with a specialized function, that best characterizes eukaryotic cells (Fig. 1-5). In fact, *eukaryotic structure and function is more complex than that of prokaryotes at all levels of organization*, from the molecular level on up.

Eukaryotes and prokaryotes have developed according to fundamentally different evolutionary strategies. Prokaryotes have exploited the advantages of simplicity and miniaturization: Their rapid growth rates permit them to occupy ecological niches in which there may be drastic fluctuations of the available nutrients. In contrast, the complexity of eukaryotes, which renders them larger and more slowly growing than prokaryotes, gives them the competitive advantage in stable environments with limited resources (Fig. 1-6). It is therefore erroneous to consider prokaryotes as

evolutionarily primitive with respect to eukaryotes. Both types of organisms are well adapted to their respective life styles.

The earliest known microfossils of eukaryotes date from ~ 1.4 billion years ago, some 2.1 billion years after life arose. This supports the classical notion that eukaryotes are descended from a highly developed prokaryote, possibly a mycoplasma. The differences between eukaryotes and modern prokaryotes, however, are so profound as to render this hypothesis improbable. Perhaps the early eukaryotes, which according to Woese's evidence evolved from a primordial life form, were relatively unsuccessful and hence rare. Only after they had developed some of the complex organelles described in the following section did they become common enough to generate significant fossil remains.

A. Cellular Architecture

Eukaryotic cells, like prokaryotes, are bounded by a plasma membrane. The large size of eukaryotic cells results in their surface-to-volume ratios being much smaller than those of prokaryotes (the surface area of an object increases as the square of its radius, whereas volume does so as the cube). This geometrical constraint, coupled with the fact that

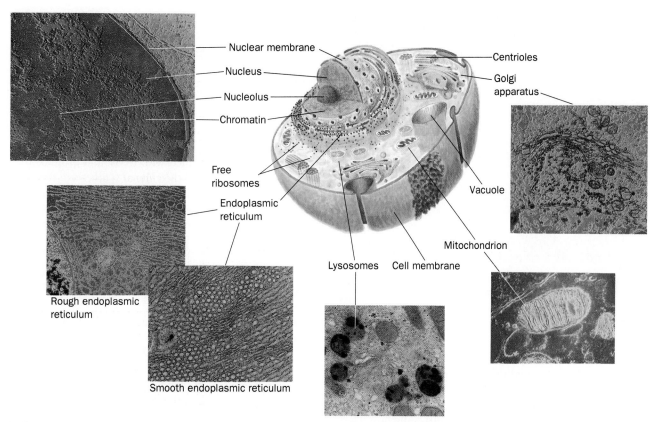

Nuclear membrane

Nucleus

Nucleolus

Chromatin

Free ribosomes

Endoplasmic reticulum

Rough endoplasmic reticulum

Smooth endoplasmic reticulum

Centrioles

Golgi apparatus

Vacuole

Mitochondrion

Lysosomes

Cell membrane

FIGURE 1-5. A schematic diagram of an animal cell accompanied by electron micrographs of its organelles. [Nucleus: Tektoff-Rhone-Merieux, CNRI; Rough endoplasmic reticulum and Golgi apparatus: Secchi-Lecaque/Roussel-UCLAF/CNRI; Smooth endoplasmic reticulum: David M. Phillips/Visuals Unlimited; Mitochondria: CNRI; Lysosome: Biophoto Associates/Photo Researchers, Inc.]

FIGURE 1-6. [Drawing by T.A. Bramley, in Carlile, M., *Trends Biochem. Sci.* 7, 128 (1982). Copyright © Elsevier Biomedical Press, 1982. Used by permission.]

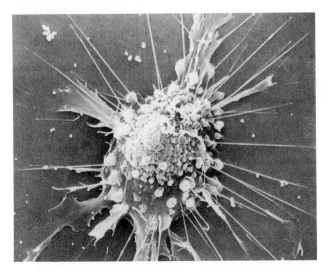

FIGURE 1-7. A scanning electron micrograph of a fibroblast. [Courtesy of Guenther Albrecht-Buehler, Northwestern University.]

many essential enzymes are membrane associated, partially rationalizes the large amounts of intracellular membranes in eukaryotes (the plasma membrane typically constitutes <10% of the membrane in a eukaryotic cell). Since all the matter that enters or leaves a cell must somehow pass through its plasma membrane, the surface areas of many eukaryotic cells are increased by numerous projections and/or invaginations (Fig. 1-7). Moreover, portions of the plasma membrane often bud inward, in a process known as **endocytosis,** so that the cell surrounds portions of the external medium. Thus eukaryotic cells can engulf and digest food particles such as bacteria, whereas prokaryotes are limited to the absorption of individual nutrient molecules. The reverse of endocytosis, a process termed **exocytosis,** is a common eukaryotic secretory mechanism.

The Nucleus Contains the Cell's DNA

The nucleus, the eukaryotic cell's most conspicuous organelle, is the repository of its genetic information. This information is encoded in the base sequences of DNA molecules that form the discrete number of chromosomes characteristic of each species. The chromosomes consist of **chromatin,** a complex of DNA and protein. The amount of genetic information carried by eukaryotes is enormous; for example, a human cell has over 700 times the DNA of *E. coli* (in the terms commonly associated with computer memories, the genetic complement in each human cell specifies over 700 megabytes of information—about 200 times the information content of this text). Within the nucleus, the genetic information encoded by the DNA is transcribed into molecules of RNA (Chapter 29) which, after extensive processing, are transported to the cytoplasm (in eukaroytes, the cell contents exclusive of the nucleus) where they direct the ribosomal synthesis of proteins (Chapter 30). The nuclear envelope consists of a double membrane that is perforated by numerous ~ 90-Å-wide

pores that regulate the flow of matter between the nucleus and the cytoplasm. These pores are of sufficient size to permit the passage of all but large molecular assemblies such as chromosomes and mature ribosomes.

The nucleus of most eukaryotic cells contains at least one dark-staining body known as the **nucleolus** which is the site of ribosomal assembly. It contains chromosomal segments bearing multiple copies of genes specifying ribosomal RNA. These genes are transcribed in the nucleolus and the resulting RNA is combined with ribosomal proteins that have been imported from their site of synthesis in the **cytosol** (the cytoplasm exclusive of its membrane-bound organelles). The resulting immature ribosomes are then exported to the cytosol where their assembly is completed. Thus protein synthesis can occur only in the cytosol.

The Endoplasmic Reticulum and the Golgi Apparatus Function To Modify Membrane-Bound and Secretory Proteins

The most extensive membrane in the cell, which was discovered in 1945 by Keith Porter, forms a labyrinthine compartment named the **endoplasmic reticulum.** A large portion of this organelle, which is called the **rough endoplasmic reticulum,** is studded with ribosomes that are engaged in the synthesis of proteins that are either membrane bound or destined for secretion. The **smooth endoplasmic reticulum,** which is devoid of ribosomes, is the site of lipid synthesis. Many of the products synthesized in the endoplasmic reticulum are eventually transported to the **Golgi apparatus** (named after Camillo Golgi, who first described it in 1898), a stack of flattened membranous sacs in which these products are further processed (Section 21-3B).

Mitochondria Are the Site of Oxidative Metabolism

The **mitochondria** (Greek: *mitos,* thread + *chondros,* granule) are the site of cellular **respiration** (aerobic metabolism) in almost all eukaryotes. These cytoplasmic organelles, which are large enough to have been discovered by nineteenth century cytologists, vary in their size and shape but are often ellipsoidal with dimensions of around $1.0 \times 2.0 \ \mu m$—much like a bacterium. A eukaryotic cell typically contains on the order of 2000 mitochondria, which occupy roughly one fifth of its total cell volume.

The mitochondrion, as the electron microscopic studies of George Palade and Fritjof Sjöstrand first revealed, has two membranes: a smooth outer membrane and a highly folded inner membrane whose invaginations are termed **cristae** (Latin: crests). Thus the mitochondrion contains two compartments, the **intermembrane space** and the internal **matrix space.** The enzymes that catalyze the reactions of respiration are components of either the gel-like **matrix** or the inner mitochondrial membrane. *These enzymes couple the energy-producing oxidation of nutrients to the energy-requiring synthesis of* **adenosine triphosphate** (ATP; Section 1-3B and Chapter 20). Adenosine triphosphate, after export to the rest of the cell, fuels its various energy-consuming processes.

Mitochondria are bacteria-like in more than size and shape. Their matrix space contains mitochondrion-specific DNA, RNA, and ribosomes that participate in the synthesis of several mitochondrial components. Moreover, they reproduce by binary fission, and the respiratory processes that they mediate bear a remarkable resemblance to those of modern aerobic bacteria. These observations led to the now widely accepted hypothesis championed by Lynn Margulis that mitochondria evolved from originally free-living aerobic bacteria, which formed a symbiotic relationship with a primordial anaerobic eukaryote. The eukaryote-supplied nutrients consumed by the bacteria were presumably repaid severalfold by the highly efficient oxidative metabolism that the bacteria conferred on the eukaryote. This hypothesis is corroborated by the observation that the amoeba *Pelomyxa pelustris,* one of the few eukaryotes that lack mitochondria, permanently harbors aerobic bacteria in such a symbiotic relationship.

Lysosomes and Peroxisomes Are Containers of Degradative Enzymes

Lysosomes, which were discovered in 1949 by Christian de Duve, are organelles bounded by a single membrane that are of variable size and morphology, although most have diameters in the range 0.1 to 0.8 μm. Lysosomes, which are essentially membranous bags containing a large variety of hydrolytic enzymes, function to digest materials ingested by endocytosis and to recycle cellular components (Section 30-6). Cytological investigations have revealed that lysosomes form by budding from the Golgi apparatus.

Peroxisomes (also known as **microbodies**) are membrane-enclosed organelles, typically 0.5 μm in diameter, that contain oxidative enzymes. They are so named because some peroxisomal reactions generate **hydrogen peroxide** (H_2O_2), a reactive substance that is either utilized in the enzymatic oxidation of other substances or degraded through a disproportionation reaction catalyzed by the enzyme **catalase:**

$$2H_2O_2 \longrightarrow 2H_2O + O_2$$

It is thought that peroxisomes function to protect sensitive cell components from oxidative attack by H_2O_2. Peroxisomes, like mitochondria, reproduce by fission and are therefore also thought to have descended from bacteria. Certain plants contain a specialized type of peroxisome, the **glyoxysome,** so named because it is the site of a series of reactions that are collectively termed the **glyoxylate pathway** (Section 21-2).

The Cytoskeleton Organizes the Cytosol

The cytosol, far from being a homogeneous solution, is a highly organized gel that can vary significantly in its composition throughout the cell. Much of its internal variability arises from the action of the **cytoskeleton,** an extensive array of filaments that gives the cell its shape and the ability to move and is responsible for the arrangement and internal motions of its organelles (Fig. 1-8).

The most conspicuous cytoskeletal components, the **microtubules,** are ~ 250-Å-diameter tubes that are composed of the protein **tubulin** (Section 34-3F). They form the supportive framework that guides the movements of organelles within a cell. For example, the **mitotic spindle** is an assembly of microtubules and associated proteins that participates in the separation of replicated chromosomes during

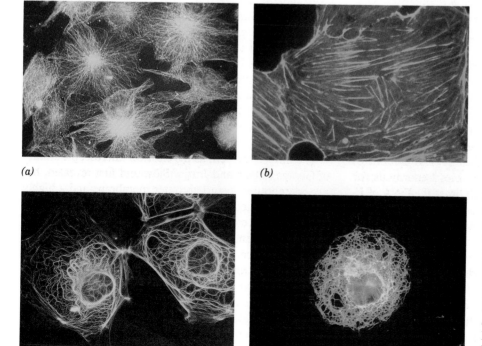

(a)

(b)

(c)

(d)

FIGURE 1-8. Immunofluorescence micrographs of cells that have been stained with fluorescently labeled antibodies raised against *(a)* tubulin, *(b)* actin, *(c)* keratin, and *(d)* **vimentin** (a protein constituent of a type of intermediate filament). [Parts *a* and *d*: K.G. Murti/Visuals Unlimited; Part *b*: M. Schliwa/Visuals Unlimited; Part *c*: Courtesy of Mary Osborn, Max-Planck Institut für Molecular Biologie, Germany.]

cell division. Microtubules are also major constitutents of **cilia,** the hairlike appendages extending from many cells, whose whiplike motions move the surrounding fluid past the cell or propel single cells through solution. Very long cilia, such as sperm tails, are termed **flagella** (prokaryotic flagella, which are composed of the protein **flagellin,** are quite different from and unrelated to those of eukaryotes). Mounting evidence suggests that cilia are also descended from free-living bacteria — perhaps spirochetes.

The **microfilaments** are ~ 90-Å in diameter fibers that consist of the protein **actin.** Microfilaments, as do microtubules, have a mechanically supportive function. Furthermore, through their interactions with the protein **myosin,** microfilaments form contractile assemblies that are responsible for many types of intracellular movements such as cytoplasmic streaming and the formation of cellular protuberances or invaginations. More conspicuously, however, actin and myosin are the major protein components of muscle (Section 34-3A).

The third major cytoskeletal component, the **intermediate filaments,** are protein fibers 100 to 150 Å in diameter. Their prominence in parts of the cell that are subject to mechanical stress suggests that they have a load-bearing function. For example, skin in higher animals contains an extensive network of intermediate filaments made of the protein **keratin** (Section 7-2A), which is largely responsible for the toughness of this protective outer covering. In contrast to the case with microtubules and microfilaments, the proteins forming intermediate filaments vary greatly in size and composition, both among the different cell types within a given organism and among the corresponding cell types in different organisms.

Plant Cells Are Enclosed by Rigid Cell Walls

Plant cells (Fig. 1-9) contain all of the previously described organelles. They also have several additional features, the most conspicuous of which is a rigid cell wall that lies outside the plasma membrane. These cell walls, whose

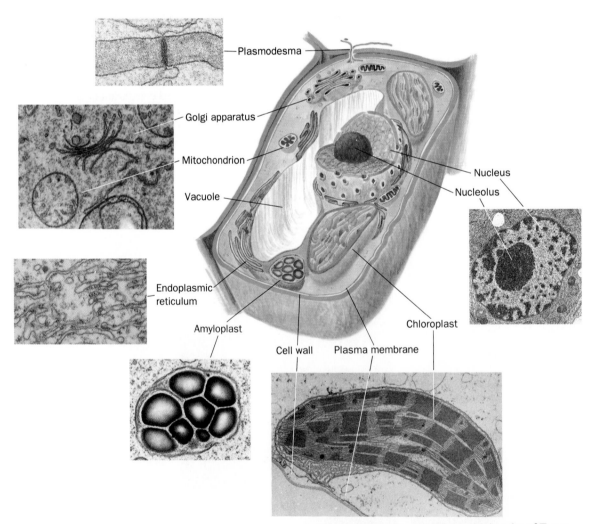

FIGURE 1-9. A drawing of a plant cell accompanied by electron micrographs of its organelles. [Plasmodesma: Courtesy of Hilton Mollenhauer, USDA; nucleus. Courtesy of Myron Ledbetter, Brookhaven National Laboratory; Golgi apparatus: Courtesy of W. Gordon Whaley, University of Texas; chloroplast: Courtesy of Lewis Shumway, College of Eastern Utah; amyloplast: Biophoto Associates; endoplasmic reticulum: Biophoto Associates/Photo Researchers, Inc.]

major component is the fibrous polysaccharide **cellulose** (Section 10-2C), account for the structural strength of plants.

A **vacuole** is a membrane-enclosed space filled with fluid. Although vacuoles occur in animal cells, they are most prominent in plant cells, where they typically occupy 90% of the volume of a mature cell. Vacuoles function as storage depots for nutrients, wastes, and specialized materials such as pigments. The relatively high concentration of solutes inside a plant vacuole causes it to take up water osmotically, thereby raising its internal pressure. This effect, combined with its cell walls' resistance to bursting, is largely responsible for the turgid rigidity of nonwoody plants.

Chloroplasts Are the Site of Photosynthesis in Plants

One of the definitive characteristics of plants is their ability to carry out photosynthesis. The site of photosynthesis is an organelle known as the **chloroplast** which, although generally several times larger than a mitochondrion, resembles it in that both organelles have an inner and an outer membrane. Furthermore, the chloroplast's inner membrane space, the **stroma,** is similar to the mitochondrial matrix in that it contains many soluble enzymes. However, the inner chloroplast membrane is not folded into cristae. Rather, the stroma encloses a third membrane system that forms interconnected stacks of disklike sacs called **thylakoids,** which

contain the photosynthetic pigment **chlorophyll.** The thylakoid uses chlorophyll-trapped light energy to generate ATP which is used in the stroma to drive biosynthetic reactions forming carbohydrates and other products (Chapter 22).

Chloroplasts, as do mitochondria, contain their own DNA, RNA and ribosomes, and they reproduce by fission. Apparently chloroplasts, much like mitochondria, evolved from an ancient cyanobacterium that took up symbiotic residence in an ancestral nonphotosynthetic eukaryote. In fact, several modern nonphotosynthetic eukaryotes have just such a symbiotic relationship with authentic cyanobacteria. Hence *most modern eukaryotes are genetic "mongrels" in that they simultaneously have nuclear, mitochondrial, peroxisomal, possibly ciliar, and—in the case of plants—chloroplast lines of descent.*

B. Phylogeny and Differentiation

One of the most remarkable characteristics of eukaryotes is their enormous morphological diversity, on both the cellular and organismal levels. Compare, for example, the architecture of the various human cells drawn in Fig. 1-10. Similarly, recall the great anatomical differences among say, an amoeba, an oak tree, and a human being.

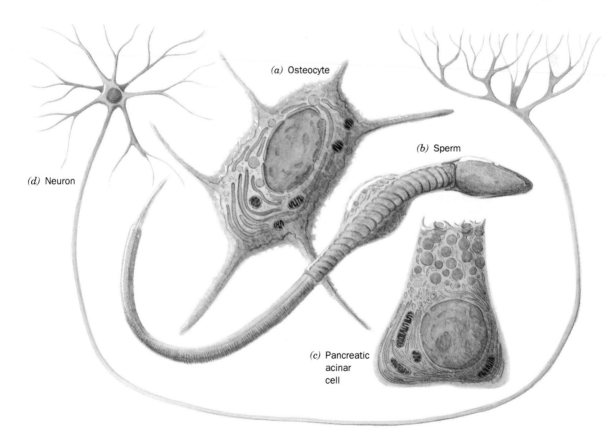

FIGURE 1-10. Drawings of some human cells: *(a)* an osteocyte (bone cell), *(b)* a sperm, *(c)* a pancreatic acinar cell (which secretes digestive enzymes), and *(d)* a neuron (nerve cell).

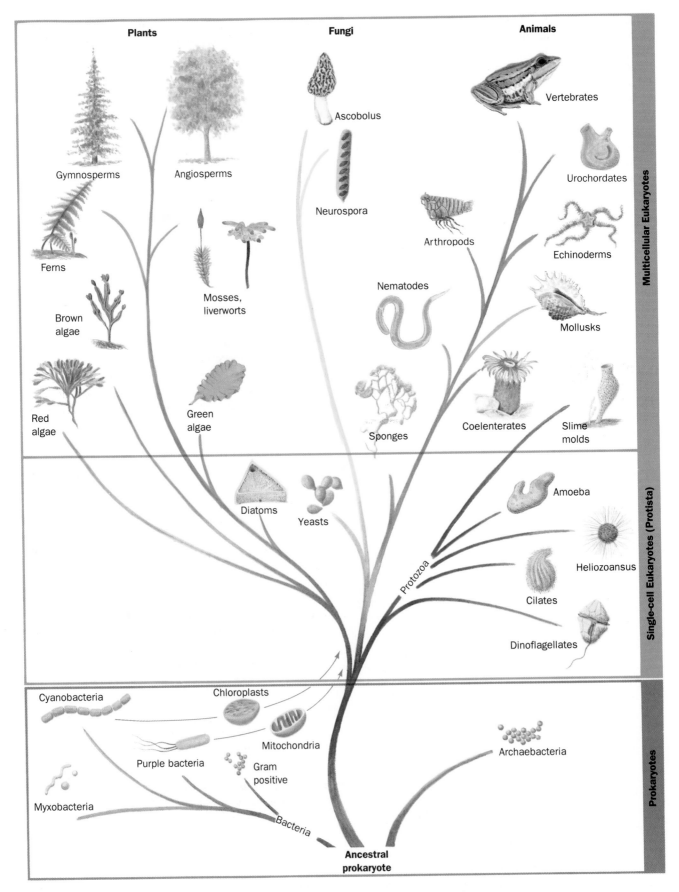

FIGURE 1-11. An evolutionary tree indicating the lines of descent of cellular life on Earth.

Taxonometric schemes based on gross morphology as well as on protein and nucleic acid sequences (Sections 6-1 and 28-6) indicate that eukaryotes may be classified into three kingdoms: **fungi, plantae** (plants), and **animalia** (animals). The relative structural simplicity of many unicellular eukaryotes, however, makes their classification under this scheme rather arbitrary. Consequently, these organisms are usually assigned a fourth eukaryotic kingdom, the **protista.** (Note that biological classification schemes are for the convenience of biologists; nature is rarely neatly categorized.) Figure 1-11 is a phylogenetic tree for eukaryotes.

Anatomical comparisons among living and fossil organisms indicate that the various kingdoms of multicellular organisms independently evolved from protista (Fig. 1-11).

The programs of growth, differentiation, and development followed by multicellular eukaryotes in their transformation from fertilized ova to adult organisms provide a remarkable indication of this evolutionary history. For example, all vertebrates exhibit gill-like pouches in their early embryonic stages, which presumably reflect their common fish ancestry (Fig. 1-12). Indeed, these early embryos are closely similar in size and anatomy even though their respective adult forms are vastly different in these characteristics. Such observations led Ernst Haeckel to formulate his famous (although overstated) dictum: *ontogeny recapitulates phylogeny* (ontogeny: biological development). The elucidation of the mechanism of cellular differentiation in eukaryotes is one of the major long-range goals of modern biochemistry.

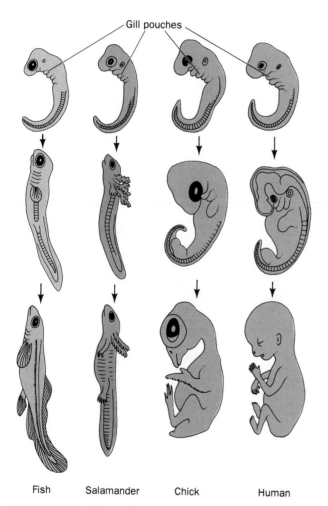

Gill pouches

Fish   Salamander   Chick   Human

FIGURE 1-12. The embryonic development of a fish, an amphibian (salamander), a bird (chick), and a mammal (human). At early stages they are closely similar, in both size and anatomy (the top drawings have about the same scale). Later they diverge in both these properties. [After Haeckel, E., *Anthropogenie oder Entwickelungsgeschichte des Menschen,* Engelmann (1874).]

3. BIOCHEMISTRY: A PROLOGUE

Biochemistry, as the name implies, is the chemistry of life. It therefore bridges the gap between chemistry, the study of the structures and interactions of atoms and molecules, and biology, the study of the structures and interactions of cells and organisms. Since living things are composed of inanimate molecules, *life, at its most basic level, is a biochemical phenomenon.*

Although living organisms, as we have seen, are enormously diverse in their macroscopic properties, there is a remarkable similarity in their biochemistry that provides a unifying theme with which to study them. For example, hereditary information is encoded and expressed in an almost identical manner in all cellular life. Moreover, the series of biochemical reactions, which are termed **metabolic pathways,** as well as the structures of the enzymes that catalyze them are, for many basic processes, nearly identical from organism to organism. This strongly suggests that all known life forms are descended from a single primordial ancestor in which these biochemical features first developed.

Although biochemistry is a highly diverse field, it is largely concerned with a limited number of interrelated issues. These are

1. What are the chemical and three-dimensional structures of biological molecules and assemblies, how do they form these structures, and how do their properties vary with them?

2. How do proteins work; that is, what are the molecular mechanisms of enzymatic catalysis, how do receptors recognize and bind specific molecules, and what are the intramolecular and intermolecular mechanisms by which receptors transmit information concerning their binding states?

3. How is genetic information expressed and how is it transmitted to future cell generations?

4. How are biological molecules and assemblies synthesized?

5. What are the control mechanisms that coordinate the myriads of biochemical reactions that take place in cells and in organisms?

6. How do cells and organisms grow, differentiate, and reproduce?

These issues are previewed in this section and further illuminated in later chapters. However, as will become obvious as you read further, in all cases, our knowledge, extensive as it is, is dwarfed by our ignorance.

A. Biological Structures

Living things are enormously complex. As indicated in Section 1-1A, even the relatively simple *E. coli* cell contains some 3 to 6 thousand different compounds, most of which are unique to *E. coli* (Fig. 1-13). Higher organisms have a correspondingly greater complexity. *Homo sapiens* (human beings), for example, may contain 100,000 different types of molecules, although only a minor fraction of them have been characterized. One might therefore suppose that to obtain a coherent biochemical understanding of any organism would be a hopelessly difficult task. This, however, is not the case. *Living things have an underlying regularity that derives from their being constructed in a hierarchical manner.* Anatomical and cytological studies have shown that multicellular organisms are organizations of organs, which are made of tissues consisting of cells, composed of subcellular organelles (e.g., Fig. 1-14). At this point in our hierarchical descent, we enter the biochemical realm since organelles consist of **supramolecular assemblies,** such as membranes or fibers, that are organized clusters of **macromolecules** (polymeric molecules with molecular masses from thousands of daltons on up).

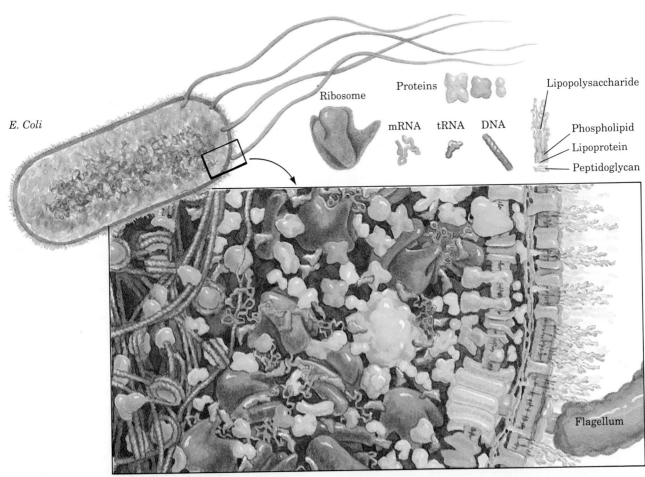

FIGURE 1-13. The simulated cross-section of an *E. coli* cell magnified around one millionfold. The right side of the drawing shows the multilayered cell wall and membrane, decorated on its exterior surface with lipopolysaccharides (Section 10-3B). A flagellum (*lower right*) is driven by a motor anchored in the inner membrane (Section 34-3G). The cytoplasm, which occupies the middle region of the drawing, is predominantly filled with ribosomes engaged in protein synthesis (Section 30-3). The left side of the drawing contains a dense tangle of DNA in complex with specific proteins. [After a drawing by David Goodsell, UCLA.]

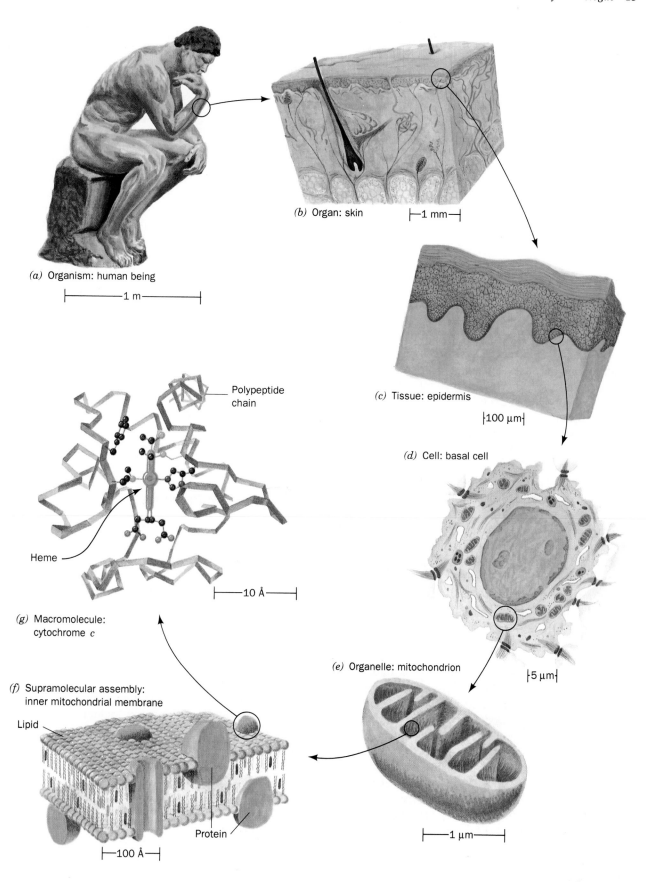

(a) Organism: human being

|——— 1 m ———|

(b) Organ: skin |— 1 mm —|

(c) Tissue: epidermis

|100 μm|

(d) Cell: basal cell

Polypeptide chain

Heme

|— 10 Å —|

(g) Macromolecule: cytochrome *c*

(f) Supramolecular assembly: inner mitochondrial membrane

Lipid

Protein

|— 100 Å —|

(e) Organelle: mitochondrion

|5 μm|

|——— 1 μm ———|

FIGURE 1-14. An example of the hierarchical organization of biological structures.

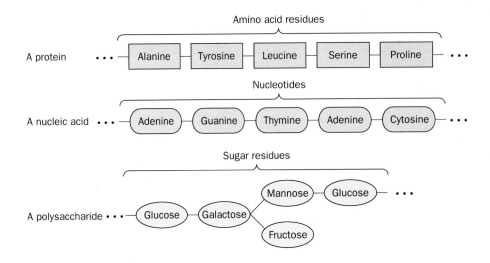

FIGURE 1-15. The polymeric organization of proteins, nucleic acids, and polysaccharides.

As Table 1-1 indicates, *E. coli*, and living things in general, contain only a few different types of macromolecules: **proteins** (Greek: *proteios*, of first importance), **nucleic acids**, and **polysaccharides** (Greek: *sakcharon*, sugar). *All of these substances have a modular construction; they consist of linked monomeric units that occupy the lowest level of our structural hierarchy.* Thus, as Fig. 1-15 indicates, proteins are polymers of amino acids (Section 4-1B), nucleic acids are polymers of nucleotides (Section 28-1), and polysaccharides are polymers of sugars (Section 10-2). **Lipids** (Greek: *lipos*, fat), the fourth major class of biological molecules, are too small to be classified as macromolecules but also have a modular construction (Section 11-1).

The task of the biochemist has been vastly simplified by the finding that *there are relatively few species of monomeric units that occur in each class of biological macromolecule.* Proteins are all synthesized from the same 20 species of **amino acids,** nucleic acids are made from 8 types of **nucleotides** (4 each in DNA and RNA), and there are ~ 8 commonly occurring types of **sugars** in polysaccharides. The great variation in properties observed among macromolecules of each type largely arises from the enormous number of ways its monomeric units can be arranged and, in many cases, derivatized.

One of the central questions in biochemistry is how biological structures are formed. As is explained in later chapters, the monomeric units of macromolecules are either directly acquired by the cell as nutrients or enzymatically synthesized from simpler substances. Macromolecules are synthesized from their monomeric precursors in complex enzymatically mediated processes.

Newly synthesized proteins spontaneously fold to assume their native conformations (Section 8-1A); that is, they undergo **self-assembly.** Apparently their amino acid sequences specify their three-dimensional structures. Likewise, the structures of other types of macromolecules are specified by the sequences of their monomeric units. The principle of self-assembly extends at least to the level of supramolecular assemblies. However, the way in which higher levels of biological structures are generated is largely unknown. The elucidation of the mechanisms of cellular and organismal growth and differentiation is a major area of biological research.

B. Metabolic Processes

There is a bewildering array of chemical reactions that simultaneously occur in any living cell. Yet, these reactions follow a pattern that organizes them into the coherent process we refer to as life. For instance, most biological reactions are members of a metabolic pathway; that is, they function as one of a sequence of reactions that produce one or more specific products. Moreover, one of the hallmarks of life is that the rates of its reactions are so tightly regulated that there is rarely an unsatisfied need for a reactant in a metabolic pathway or an unnecessary buildup of some product.

Metabolism has been traditionally (although not necessarily logically) divided into two major categories:

1. **Catabolism** or degradation, in which nutrients and cell constituents are broken down so as to salvage their components and/or to generate energy.

2. **Anabolism** or biosynthesis, in which biomolecules are synthesized from simpler components.

The energy required by anabolic processes is provided by catabolic processes largely in the form of **adenosine triphosphate (ATP).** For instance, such energy-generating processes as photosynthesis and the biological oxidation of nutrients produce ATP from **adenosine diphosphate (ADP)** and a phosphate ion.

Adenosine diphosphate (ADP)

Adenosine triphosphate (ATP)

Conversely, such energy-consuming processes as biosynthesis, the transport of molecules against a concentration gradient, and muscle contraction, are driven by the reverse of this reaction, the hydrolysis of ATP:

$$ATP + H_2O \rightleftharpoons ADP + HPO_4^{2-}$$

Thus, *anabolic and catabolic processes are coupled together through the mediation of the universal biological energy "currency," ATP.*

C. Expression and Transmission of Genetic Information

Deoxyribonucleic acid (DNA) is the cell's master repository of genetic information. This macromolecule, as is diagrammed in Fig. 1-16, consists of two strands of linked **nucleotides,** each of which is composed of a **deoxyribose** sugar residue, a phosphoryl group, and one of four bases: **adenine (A), thymine (T), guanine (G),** or **cytosine (C).** Genetic information is encoded in the sequence of these bases. Each DNA base is hydrogen bonded to a base on the opposite strand. However, A can only hydrogen bond with T, and G with C, so that the two strands are **complementary;**

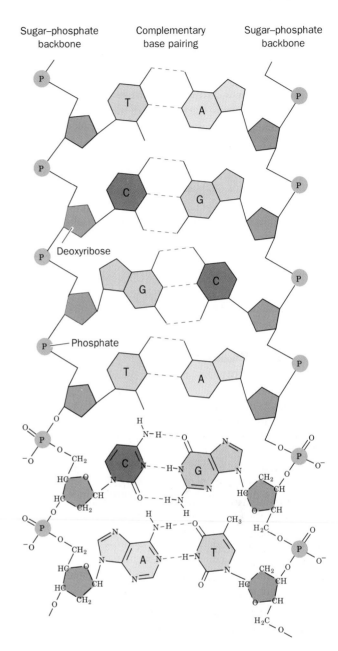

FIGURE 1-16. Double-stranded DNA. The two polynucleotide chains associate through complementary hydrogen bonding.

that is, the sequence of one strand implies the sequence of the other.

The division of a cell must be accompanied by the replication of its DNA. In this enzymatically mediated process, each DNA strand acts as a template for the formation of its complementary strand (Fig. 1-17; Section 31-1). Consequently, every progeny cell contains a complete DNA molecule (or set of DNA molecules), each of which consists of one parental strand and one daughter strand. Mutations arise when, through rare copying errors or damage to a parental strand, one or more wrong bases are incorporated into a daughter strand. Most mutations are either innocuous or deleterious. Occasionally, however, one results in a new characteristic that confers some sort of selective advantage on its recipient. Individuals with such mutations, according to the tenets of the Darwinian theory of evolution, have an increased probability of reproducing. New species arise through a progression of such mutations.

The expression of genetic information is a two-stage process. In the first stage, which is termed **transcription,** a DNA strand serves as a template for the synthesis of a complementary strand of ribonucleic acid (RNA; Section 29-2). This nucleic acid, which is generally single stranded, differs chemically from DNA (Fig. 1-16) only in that it has **ribose** sugar residues in place of DNA's deoxyribose and **uracil (U)** replacing DNA's thymine base.

Ribose   **Uracil**

In the second stage of genetic expression, which is known as **translation,** ribosomes enzymatically link together amino acids to form proteins (Section 30-3). The order in which the amino acids are linked together is prescribed by the RNA's sequence of bases. Consequently, since proteins are self-assembling, the genetic information encoded by DNA serves, through the intermediacy of RNA, to specify protein structure and function. Just which genes are expressed in a given cell under a particular set of circumstances is controlled by complex regulatory systems whose workings are understood only in outline.

4. THE ORIGIN OF LIFE

People have always pondered the riddle of their existence. Indeed, all known cultures, past and present, primitive and sophisticated, have some sort of a creation myth that rationalizes how life arose. Only in the modern era, however, has it been possible to consider the origin of life in terms of a scientific framework, that is, in a manner subject to experimental verification. One of the first to do so was Charles Darwin, the originator of the theory of evolution. In 1871, he wrote in a letter to a colleague:

> *It is often said that all the conditions for the first production of a living organism are now present, which could ever have been present. But if (and oh what a big if) we could conceive in some warm little pond, with all sorts of ammonia and phosphoric salts, light, heat, electricity, etc., present, that a protein compound was chemically formed ready to undergo*

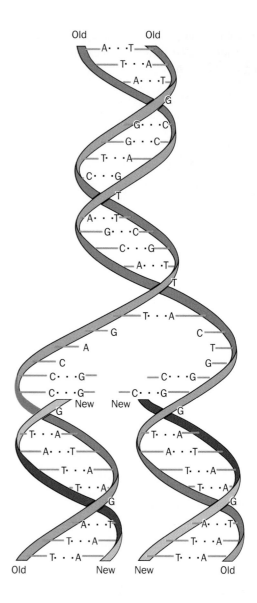

FIGURE 1-17.   Schematic diagram of DNA replication.

> *still more complex changes, at the present day such matter would be instantly devoured, or absorbed, which would not have been the case before living creatures were formed.*

Radioactive dating studies indicate that the earth formed some 4.6 billion years ago. Yet the earliest known fossil evidence of life, which was generated by organisms resembling modern bacteria, is ~ 3.5 billion years old, although there is evidence for biological carbon fixation as early as 3.8 billion years ago. Apparently the preceding "prebiotic era" left no direct record. *Clearly, then, we cannot hope to determine exactly how life arose. Through laboratory experimentation, however, we can at least demonstrate what sorts of abiotic chemical reactions may have led to the formation of a living system.* Moreover, we are not entirely without traces of prebiotic development. The underlying biochemical and genetic unity of modern organisms sug-

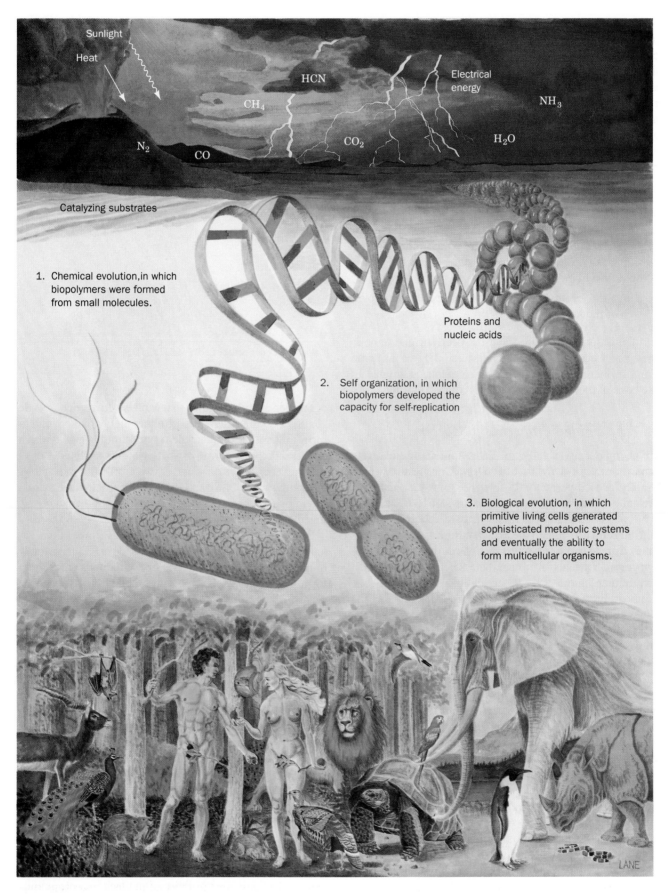

FIGURE 1-18. The three stages in the evolution of life.

gests that life as we know it arose but once (if life arose more than once, the other forms must have rapidly died out, possibly because they were "eaten" by the present form). Thus, by comparing the corresponding genetic messages of a wide variety of modern organisms it may be possible to derive reasonable models of the primordial messages from which they have descended.

It is generally accepted that the development of life occupied three stages (Fig. 1-18):

1. Chemical evolution, in which simple geologically occurring molecules reacted to form complex organic polymers.

2. The self-organization of collections of these polymers to form replicating entities. At some point in this process, the transition from a lifeless collection of reacting molecules to a living system occurred.

3. Biological evolution to ultimately form the complex web of modern life.

In this section, we outline what has been surmised about these processes. We precede this discussion by a consideration of why only carbon, of all the elements, is suitable as the basis of the complex chemistry required for life.

A. The Unique Properties of Carbon

Living matter, as Table 1-2 indicates, consists of a relatively small number of elements. C, H, O, N, P, and S, all of which readily form covalent bonds, comprise some 92% of the dry weight of living things (most organisms are ~ 70% water). The balance consists of elements that are mainly present as ions and for the most part occur only in trace quantities (they usually carry out their functions at the active sites of enzymes). Note, however, that there is no known biological requirement for 65 of the 90 naturally occurring elements. Conversely, with the exceptions of oxygen and calcium, the biologically most abundant elements are but minor constituents of the Earth's crust (the most abundant components of which are O, 47%; Si, 28%; Al, 7.9%; Fe, 4.5%; and Ca, 3.5%).

The predominance of carbon in living matter is no doubt a result of its tremendous chemical versatility compared with all the other elements. Carbon has the unique ability to form a virtually infinite number of compounds as a result of its capacity to make as many as four highly stable covalent bonds (including single, double, and triple bonds) combined with its ability to form covalently linked C—C chains of unlimited extent. Thus, of the over 13 million chemical compounds that are presently known, nearly 90% are organic (carbon-containing) substances. Let us examine the other elements in the periodic table to ascertain why they lack these combined properties.

Only five elements, B, C, N, Si, and P, have the capacity to make three or more bonds each and thus to form chains

TABLE 1-2. ELEMENTAL COMPOSITION OF THE HUMAN BODY

Element	Dry weight (%)	Elements present in trace amounts
C	61.7	B
N	11.0	F
O	9.3	Si
H	5.7	V
Ca	5.0	Cr
P	3.3	Mn
K	1.3	Fe
S	1.0	Co
Cl	0.7	Cu
Na	0.7	Zn
Mg	0.3	Se
		Mo
		Sn
		I

[Calculated from Frieden, E., *Sci. Am.* **227**(1): 54–55 (1972).]

of covalently linked atoms that can also have pendant side chains. The other elements are either metals, which tend to form ions rather than covalent bonds; noble gases, which are essentially chemically inert; or atoms such as H or O that can each make only one or two covalent bonds. However, although B, N, Si, and P can each participate in at least three covalent bonds, they are, for reasons indicated below, unsuitable as the basis of a complex chemistry.

Boron, having fewer valence electrons (three) than valence orbitals (four), is electron deficient. This severely limits the types and stabilities of compounds that B can form. Nitrogen has the opposite problem; its five valence electrons make it electron rich. The repulsions between the lone pairs of electrons on covalently bonded N atoms serve to greatly reduce the bond energy of an N—N bond ($171 \text{ kJ} \cdot \text{mol}^{-1}$ vs $348 \text{ kJ} \cdot \text{mol}^{-1}$ for a C—C single bond) relative to the unusually stable triple bond of the N_2 molecule ($946 \text{ kJ} \cdot \text{mol}^{-1}$). Even short chains of covalently bonded N atoms therefore tend to decompose, usually violently, to N_2. Silicon and carbon, being in the same column of the periodic table, might be expected to have similar chemistries. Silicon's large atomic radius, however, prevents two Si atoms from approaching each other closely enough to gain much orbital overlap. Consequently Si—Si bonds are weak ($177 \text{ kJ} \cdot \text{mol}^{-1}$) and the corresponding multiple bonds are rarely stable. Si—O bonds, in contrast, are so stable ($369 \text{ kJ} \cdot \text{mol}^{-1}$) that chains of alternating Si and O atoms are essentially inert (silicate minerals, whose frameworks consist of such bonds, form the Earth's crust). Indeed, science fiction writers have speculated that **silicones,** which are oily or rubbery organosilicon compounds with backbones of linked Si—O units; for example, **methyl silicones,**

$$\cdots \overset{\displaystyle CH_3}{\underset{\displaystyle CH_3}{Si}}-O-\overset{\displaystyle CH_3}{\underset{\displaystyle CH_3}{Si}}-O-\overset{\displaystyle CH_3}{\underset{\displaystyle CH_3}{Si}}-O-\overset{\displaystyle CH_3}{\underset{\displaystyle CH_3}{Si}}-O-\cdots$$

could form the chemical basis of extraterrestrial life forms. Yet, the very inertness of the Si—O bond makes this seem unlikely. Phosphorus, being below N in the periodic table, forms even less stable chains of covalently bonded atoms.

The foregoing does not imply that heteronuclear bonds are unstable. On the contrary, proteins contain C—N—C linkages, carbohydrates have C—O—C linkages, and nucleic acids possess C—O—P—O—C linkages. However, *these heteronuclear linkages are less stable than are C—C bonds. Indeed, they usually form the sites of chemical cleavage in the degradation of macromolecules and, conversely, are the bonds formed when monomer units are linked together to form macromolecules.* In the same vein, homonuclear linkages other than C—C bonds are so reactive that they are, with the exception of S—S bonds in proteins, extremely rare in biological systems.

B. Chemical Evolution

In the remainder of this section, we describe the most widely favored scenario for the origin of life. *Keep in mind, however, that there are valid scientific objections to this scenario as well as to the several others that have been seriously entertained so that we are far from certain as to how life arose.*

The solar system is thought to have formed by the gravitationally induced collapse of a large interstellar cloud of dust and gas. The major portion of this cloud, which was composed mostly of hydrogen and helium, condensed to form the sun. The rising temperature and pressure at the center of the proto-sun eventually ignited the self-sustaining thermonuclear reaction that has since served as the sun's energy source. The planets, which formed from smaller clumps of dust, were not massive enough to support such a process. In fact the smaller planets, including Earth, consist of mostly heavier elements because their masses are too small to gravitationally retain much H_2 and He.

The primordial Earth's atmosphere was quite different from what it is today. It could not have contained significant quantities of O_2, a highly reactive substance. Rather, in addition to the H_2O, N_2, and CO_2 that it presently has, the atmosphere probably contained smaller amounts of CO, CH_4, NH_3, SO_2, and possibly H_2, all molecules that have been spectroscopically detected in interstellar space. The chemical properties of such a gas mixture make it a **reducing atmosphere** in contrast to the Earth's present atmosphere, which is an **oxidizing atmosphere** (although recent contradictory evidence suggests that the primordial Earth had an oxidizing atmosphere).

In the 1920s, Alexander Oparin and J. B. S. Haldane independently suggested that *ultraviolet (UV) radiation from the sun [which is presently largely absorbed by an ozone (O_3) layer high in the atmosphere] or lightning discharges caused the molecules of the primordial reducing atmosphere to react to form simple organic compounds such as amino acids, nucleic acid bases, and sugars.* That this process is possible was first experimentally demonstrated in 1953 by Stanley Miller and Harold Urey who, in the apparatus diagrammed in Fig. 1-19, simulated effects of lightning storms in the primordial atmosphere by subjecting a refluxing mixture of H_2O, CH_4, NH_3, and H_2 to an electric

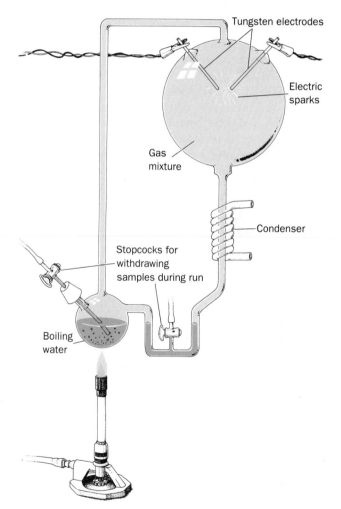

FIGURE 1-19. Apparatus for emulating the synthesis of organic compounds on the prebiotic Earth. A mixture of gases thought to resemble the primitive Earth's reducing atmosphere is subjected to an electric discharge, to simulate the effects of lightning, while the water in the flask is refluxed so that the newly formed compounds dissolve in the water and accumulate in the flask. [After Miller, S.L. and Orgel, L.E., *The Origins of Life on Earth*, p. 84, Prentice–Hall (1974).]

discharge for about a week. The resulting solution contained significant amounts of water-soluble organic compounds, the most abundant of which are listed in Table 1-3, together with a substantial quantity of insoluble tar (polymerized material). Several of the soluble compounds are amino acid components of proteins and many of the others, as we shall see, are also of biochemical significance. Similar experiments in which the reaction conditions, the gas mixture, and/or the energy source were varied have resulted in the synthesis of many other amino acids. This, together with the observation that carbonaceous meteorites contain many of the same amino acids, strongly suggests that these substances were present in significant quantities on the primordial Earth.

Nucleic acid bases can also be synthesized under supposed prebiotic conditions. In particular, adenine is formed by the condensation of HCN, a plentiful component of the prebiotic atmosphere, in a reaction catalyzed by NH_3 [note that the chemical formula of adenine is $(HCN)_5$]. The other bases have been synthesized in similar reactions involving HCN and H_2O. Sugars have been synthesized by the polymerization of formaldehyde (CH_2O) in reactions catalyzed by divalent cations, alumina, or clays. It is probably no accident that these compounds are the basic components of biological molecules. *They were apparently the most common organic substances in prebiotic times.*

The above described prebiotic reactions probably occurred over a period of hundreds of millions of years. Ultimately, it has been estimated, the oceans attained the organic consistency of a thin bouillon soup. Of course there must have been numerous places, such as tidal pools and shallow lakes, where the prebiotic soup became much more concentrated. In such environments its component organic molecules could have condensed to form, for example, polypeptides and polynucleotides (nucleic acids). Quite possibly these reactions were catalyzed by the adsorption of the reactants on minerals such as clays. If, however, life were to have formed, the rates of synthesis of these complex polymers would have had to be greater than their rates of hydrolysis. Therefore, the "pond" in which life arose may have been cold rather than warm, possibly even below 0°C (seawater freezes solidly only below −21°C), since hydrolysis reactions are greatly retarded at such low temperatures.

C. The Rise of Living Systems

Living systems have the ability to replicate themselves. The inherent complexity of such a process is such that no manmade device has even approached having this capacity. Clearly there is but an infinitesimal probability that a collection of molecules can simply gather at random to form a living entity (the likelihood of a living cell forming spontaneously from simple organic chemicals has been said to be comparable to that of a modern jet aircraft being assembled by a tornado passing through a junkyard). How then did life

TABLE 1-3. YIELDS FROM SPARKING A MIXTURE OF CH_4, NH_3, H_2O, AND H_2

Compound	Yield (%)
Glycine[a]	2.1
Glycolic acid	1.9
Sarcosine	0.25
Alanine[a]	1.7
Lactic acid	1.6
N-Methylalanine	0.07
α-Amino-*n*-butyric acid	0.34
α-Aminoisobutyric acid	0.007
α-Hydroxybutyric acid	0.34
β-Alanine	0.76
Succinic acid	0.27
Aspartic acid[a]	0.024
Glutamic acid[a]	0.051
Iminodiacetic acid	0.37
Iminoaceticpropionic acid	0.13
Formic acid	4.0
Acetic acid	0.51
Propionic acid	0.66
Urea	0.034
N-Methyl urea	0.051

[a] Amino acid constituent of proteins.

Source: Miller, S.J. and Orgel, L.E., *The Origins of Life on Earth,* p. 85, Prentice-Hall (1974).

arise? The answer, most probably, is that it was guided according to the Darwinian principle of the survival of the fittest as it applies at the molecular level.

Life Probably Arose through the Development of Self-Replicating RNA Molecules

The primordial self-replicating system is widely believed to have been a collection of nucleic acid molecules because such molecules, as we have seen in Section 1-3C, can direct the synthesis of molecules complementary to themselves. RNA, as does DNA, can direct the synthesis of a complementary strand. In fact, RNA serves as the hereditary material of many viruses (Chapter 32). The polymerization of the progeny molecules would, at first, have been a simple chemical process and hence could hardly be expected to be accurate. The early progeny molecules would therefore have been only approximately complementary to their parents. Nevertheless, repeated cycles of nucleic acid synthesis would eventually exhaust the supply of free nucleotides so that the synthesis rate of new nucleic acid molecules would be ultimately limited by the hydrolytic degradation rate of old ones. Suppose, in this process, a nucleic acid molecule randomly arose that, through folding, was more resistant to degradation than its cousins. The progeny of this molecule, or at least its more faithful copies, could then propagate at the expense of the nonresistant molecules; that is, the resistant molecules would have a Darwinian advantage over their fellows. Theoretical studies suggest that such a system of molecules would evolve so as to optimize its replication efficiency under its inherent physical and chemical limitations.

In the next stage of the evolution of life, it is thought the dominant nucleic acids evolved the capacity to influence the efficiency and accuracy of their own replication. This process occurs in living systems through the nucleic acid–directed ribosomal synthesis of enzymes that catalyze nucleic acid synthesis. How nucleic acid–directed protein synthesis could have occurred before ribosomes arose is unknown because nucleic acids are not known to interact selectively with particular amino acids. This difficulty exemplifies the major problem in tracing the pathway of prebiotic evolution. Suppose some sort of rudimentary nucleic acid–influenced system arose that increased the efficiency of nucleic acid replication. This system must have eventually been replaced, presumably with almost no trace of its existence, by the much more efficient ribosomal system. Our hypothetical nucleic acid synthesis system is therefore analogous to the scaffolding used in the construction of a building. After the building has been erected the scaffolding is removed, leaving no physical evidence that it ever was there. *Most of the statements in this section must therefore be taken as educated guesses.* Without having witnessed the event, it seems unlikely that we shall ever be certain of how life arose.

A plausible hypothesis for the evolution of self-replicating systems is that they initially consisted entirely of RNA, a scenario known as the "RNA world". This idea is based, in part, on the observation that certain species of RNA exhibit enzymelike catalytic properties (Section 29-4B). Moreover, since ribosomes are $\sim \frac{2}{3}$ RNA and only $\frac{1}{3}$ protein, it is plausible that the primordial ribosomes were entirely RNA. A cooperative relationship between RNA and protein might have arisen when these self-replicating protoribosomes evolved the ability to influence the synthesis of proteins that increased the efficiency and/or the accuracy of RNA synthesis. *From this point of view, RNA is the primary substance of life; the participation of DNA and proteins were later refinements that increased the Darwinian fitness of an already established self-replicating system.*

The types of systems that we have so far described were bounded only by the primordial "pond." A self-replicating system that developed a more efficient component would therefore have to share its benefits with all the "inhabitants" of the "pond," a situation that minimizes the improvement's selective advantage. Only through compartmentalization, that is, the generation of cells, could developing biological systems reap the benefits of any improvements that they might have acquired. Of course, cell formation would also hold together and protect any self-replicating system and therefore help it spread beyond its "pond" of origin. Indeed, the importance of compartmen-

talization is such that it may have preceded the development of self-replicating systems. The erection of cell boundaries is not without its price, however. Cells, as we shall see in later chapters, must expend much of their metabolic effort in selectively transporting substances across their cell membranes. How cell boundaries first arose, or even what they were made from, is presently unknown. However, one plausible theory holds that membranes first arose as empty vesicles whose exteriors could serve as attachment sites for such entities as enzymes and chromosomes in ways that facilitated their function. Evolution then flattened and folded these vesicles so that they enclosed their associated molecular assemblies, thereby defining the primordial cells.

Competition for Energy Resources Led to the Development of Metabolic Pathways, Photosynthesis, and Respiration

At this stage in their development, the entities we have been describing already fit Horowitz's criteria for life (exhibiting replication, catalysis, and mutability). The polymerization reactions through which these primitive organisms replicated were entirely dependent on the environment to supply the necessary monomeric units and the energy-rich compounds such as ATP or, more likely, just polyphosphates, that powered these reactions. As some of the essential components in the prebiotic soup became scarce, organisms developed the enzymatic systems that could synthesize these substances from simpler but more abundant precursors. As a consequence, energy-producing metabolic pathways arose. This latter development only postponed an "energy crisis," however, because these pathways consumed other preexisting energy-rich substances. The increasing scarcity of all such substances ultimately stimulated the development of photosynthesis to take advantage of a practically inexhaustible energy supply, the sun. Yet, this process, as we saw in Section 1-1A, consumes reducing agents such as H_2S. The eventual exhaustion of these substances led to the refinement of the photosynthetic process so that it used the ubiquitous H_2O as its reducing agent, thereby yielding O_2 as a byproduct. The recent discovery, in $\sim$ 3.5 billion-year-old rocks, of fossilized cyanobacteria-like microorganisms strongly suggests that oxygen-producing photosynthesis developed very early in the history of life.

The development of oxygen-producing photosynthesis led to yet another problem. The accumulation of the highly reactive O_2, which over the eons converted the reducing atmosphere of the prebiotic earth to the modern oxidizing atmosphere (21% O_2), eventually interfered with the existing metabolic apparatus, which had evolved to operate under reducing conditions. The O_2 accumulation therefore stimulated the development of metabolic refinements that protected organisms from oxidative damage. More importantly, it led to the evolution of a much more efficient form of energy metabolism than had previously been possible,

respiration (oxidative metabolism), which used the newly available O_2 as an oxidizing agent.

As previously outlined, the basic replicative and metabolic apparatus of modern organisms evolved quite early in the history of life on Earth. Indeed, many modern prokaryotes appear to resemble their very ancient ancestors. The rise of eukaryotes, as Section 1-2 indicates, occurred perhaps 2 billion years after prokaryotes had become firmly established. Multicellular organisms are a relatively recent evolutionary innovation, having not appeared, according to the fossil record, until ~ 700 million years ago.

5. THE BIOCHEMICAL LITERATURE

The biochemical literature contains the results of the work of tens of thousands of scientists extending over more than a century. Consequently a biochemistry text can report only selected highlights of this vast amount of information. Moreover, the tremendous rate at which biochemical knowledge is presently being acquired, which is perhaps greater than that of any other intellectual endeavor, guarantees that there will have been significant biochemical advances even in the year or so that it took to produce this text from its final draft. A serious student of biochemistry must therefore regularly read the biochemical literature to flesh out the details of subjects covered in (or omitted from) this text, as well as to keep abreast of new developments. This section provides a few suggestions on how to do so.

A. Conducting a Literature Search

The primary literature of biochemistry, those publications that report the results of biochemical research, is presently being generated at a rate of tens of thousands of papers per year appearing in over 200 periodicals. An individual can therefore only read this voluminous literature in a highly selective fashion. Indeed, most biochemists tend to "read" only those publications that are likely to contain reports pertaining to their interests. By "read" it is meant that they scan the tables of contents of these journals for the titles of articles that seem of sufficient interest to warrant further perusal (a convenient way of doing so is by using *Current Contents,* which is simply the collected tables of contents of recently published journals).

It is difficult to learn about a new subject by beginning with its primary literature. Instead, to obtain a general overview of a particular biochemical subject it is best to first peruse appropriate reviews and monographs (the update supplements to this textbook that we publish annually, and which are available from John Wiley & Sons, may also be useful in this endeavor). These usually present a synopsis of recent (at the time of their writing) developments in the area, often from the authors' particular point of view. There are more or less two types of reviews: Those that are essen-

TABLE 1-4.   SOME IMPORTANT BIOCHEMICAL REVIEW PUBLICATIONS

Accounts of Chemical Research

Advances in Enzymology and Related Areas of Molecular Biology

Advances in Protein Chemistry

Angewandte Chemie, International Edition in English[a]

Annual Review of Biochemistry

Annual Review of Biophysics and Biomolecular Structure[b]

Annual Review of Cell Biology

Annual Review of Genetics

Annual Review of Immunology

Annual Review of Medicine

Annual Review of Microbiology

Annual Review of Physiology

Annual Review of Plant Physiology and Plant Molecular Biology

Biochemical Journal[a]

BioEssays

Cell[a]

Chemtracts. Biochemistry and Molecular Biology[a]

Critical Reviews in Biochemistry and Molecular Biology

Critical Reviews in Eukaryotic Gene Expression

Current Biology

Current Opinion in Cell Biology

Current Opinion in Genetics and Development

Current Opinion in Structural Biology

Current Topics in Bioenergetics

Current Topics in Cell Regulation

Essays in Biochemistry

European Journal of Biochemistry[a]

FASEB Journal[a]

Harvey Lectures

Journal of Biological Chemistry[a]

Methods in Enzymology

Nature[a]

Progress in Biophysics and Molecular Biology

Progress in Nucleic Acid Research and Molecular Biology

Protein Science[a]

Quarterly Reviews of Biophysics

Science[a]

Scientific American

Structure[a]

Trends in Biochemical Sciences

Trends in Cell Biology

Trends in Genetics

[a] Periodicals that mainly publish research reports.

[b] Previously named *Annual Review of Biophysics and Biophysical Chemistry*

tially a compilation of facts and those that critically evaluate the data and attempt to place them in some larger context. The latter type of review is of course more valuable, particularly for a novice in the field. Most reviews are published in specialized books or journals, although many journals that publish research reports also occasionally print reviews. Table 1-4 provides a list of many of the important biochemical review publications.

Monographs and reviews relevant to a subject of interest are usually easy to find through the use of a library catalog and the subject indexes of the major review publications (the chapter-end references of this text may also be helpful in this respect). An important part of any review is its reference list. It usually lists previous reviews in the same or allied fields as well as indicating the most significant research reports in the area. Note the authors of these articles and the journals in which they tend to publish. When the most current reviews and research articles you have found tend to refer to the same group of earlier articles, you can be reasonably confident that your search for these earlier articles is largely complete. Finally, to familiarize yourself with the latest developments in the field, search the recent primary literature for the work of its most active research groups.

Biological Abstracts, Chemical Abstracts, and *Science Citation Index* are useful aids for locating references. These compendia list the articles in the many journals they cover by both author and subject (permuted title index). *Biological Abstracts* and *Chemical Abstracts* contain short English language abstracts of the articles listed (including many of foreign language articles). *Science Citation Index* lists all articles in a given year that cite a particular earlier paper, so it can be used to follow the developments in a field that build on a particular body of work.

Most academic libraries subscribe to computerized reference search services such as those of Chemical Abstracts (CAS Online), Current Contents, Medline, and Science Citation Index. If used properly, they can be highly efficient tools for locating specific information. Their disadvantage is their expense.

B. *Reading a Research Article*

Research reports more or less all have the same five-part format. They usually have a short abstract or summary located before (or, in some journals, after) the main body of the paper. The paper then continues (or begins) with an introduction, which often contains a short synopsis of the field, the motivation for the research reported, and a preview of its conclusions. The next section contains a descrip-

tion of the methods used to obtain the experimental data. This is followed by a presentation of the results of the investigation. Finally, there is a discussion section wherein the conclusions of the investigation are set forth and placed in the context of other work in the field. Most articles are so-called "full papers," which may be tens of pages long. However, many journals also contain "communications," which are usually only a page or two in length and are often published more quickly than are full papers.

It is by no means obvious how to read a scientific paper. Perhaps the worst way to do so is to read it from beginning to end as if it were some kind of a short story. In fact, most practicing scientists only occasionally read a research article in its entirety. It simply takes too long and is rarely productive. Rather, they scan selected parts of a paper and only dig deeper if it appears that to do so will be profitable. The following paragraph describes a reasonably efficient scheme for reading scientific papers. *This should be an active process in which the reader is constantly evaluating what is being read and relating it to his/her previous knowledge.* Moreover, the reader should maintain a healthy skepticism since there is a reasonable probability that any paper, particularly in its interpretation of experimental data and in its speculations, may be erroneous.

If the title of a paper indicates that it may be of interest then this should be confirmed by a reading of its abstract. For many papers, even those containing useful information, it is unnecessary to read further. If you choose to continue, it is probably best to do so by scanning the introduction so as to obtain an overview of the work reported. At this point most experienced scientists scan the conclusions section of the paper to gain a better understanding of what was found. If further effort seems warranted, they scan the results section to ascertain whether the experimental data support the conclusions. The methods section is usually not read in detail because it is often written in a condensed form that is only fully interpretable by an expert in the field. However, for such experts, the methods section may be the most valuable part of the paper. At this point, what to read next, if anything, is largely dictated by the remaining points of confusion. In many cases this confusion can only be eliminated by reading some of the references given in the paper. At any rate, unless you plan to repeat or extend some of the work described, it is rarely necessary to read an article in detail. To do so in a critical manner, you will find, takes several hours for a paper of even moderate size.

CHAPTER SUMMARY

Prokaryotes are single-celled organisms that lack a membrane-enclosed nucleus. Most prokaryotes have similar anatomies: a rigid cell wall surrounding a cell membrane that encloses the cytoplasm. The cell's single chromosome is condensed to form a nucleoid.

Escherichia coli, the biochemically most well-characterized organism, is a typical prokaryote. Prokaryotes have quite varied nutritional requirements. The chemolithotrophs metabolize inorganic substances. Photolithotrophs, such as cyanobacteria, carry out photosynthesis. Heterotrophs, which live by oxidizing organic substances, are classified as aerobes if they use oxygen in this process and as anaerobes if some other oxidizing agent serves as their terminal electron acceptor. Traditional prokaryotic classification schemes are rather arbitrary because of poor correlation between bacterial form and metabolism. Sequence comparisons of nucleic acids and proteins, however, have established that a class of bacteria, the archaebacteria, is as different from the other prokaryotes, the eubacteria, as these two groups are from eukaryotes.

Eukaryotic cells, which are far more complex than those of Prokaryotes, are characterized by having numerous membrane-enclosed organelles. The most conspicuous of these is the nucleus, which contains the cell's chromosomes, and the nucleolus, where ribosomes are assembled. The endoplasmic reticulum is the site of synthesis of lipids and of proteins that are destined for secretion. Further processing of these products occurs in the Golgi apparatus. The mitochondria, wherein oxidative metabolism occurs, are thought to have evolved from a symbiotic relationship between an aerobic bacterium and a primitive eukaryote. The chloroplast, the site of photosynthesis in plants, similarly evolved from a cyanobacterium. Other eukaryotic organelles include the lysosome, which functions as an intracellular digestive chamber, and the peroxisome, which contains a variety of oxidative enzymes including some that generate H_2O_2. The eukaryotic cytoplasm is pervaded by a cytoskeleton whose components include microtubules, which consist of tubulin; microfilaments, which are composed of actin; and intermediate filaments, which are made of different proteins in different types of cells. Eukaryotes have enormous morphological diversity on the cellular as well as on the organismal level. They have been classified into four kingdoms: protista, plantae, fungi, and animalia. The pattern of embryonic development in multicellular organisms partially mirrors their evolutionary history.

Organisms have a hierarchical structure that extends down to the submolecular level. They contain but three basic types of macromolecules; proteins, nucleic acids, and polysaccharides, as well as lipids, each of which are constructed from only a few different species of monomeric units. Macromolecules and supramolecular assemblies form their native biological structures through a process of self-assembly. The assembly mechanisms of higher biological structures is largely unknown. Metabolic processes are organized into a series of tightly regulated pathways. These are classified as catabolic or anabolic depending on whether they participate in degradative or biosynthetic processes. The common energy "currency" in all these processes is ATP, whose synthesis is the product of many catabolic pathways and whose hydrolysis drives most anabolic pathways. DNA, the cell's hereditary molecule, encodes genetic information in its sequence of bases. The complementary base sequences of its two strands permits them to act as templates for their own replication and for the synthesis of complementary strands of RNA. Ribosomes synthesize proteins by linking amino acids together in the order specified by the base sequences of RNAs.

Life is carbon based because only carbon, among all the elements in the periodic table, has a sufficiently complex chemistry together with the ability to form virtually infinite stable chains of covalently bonded atoms. Reactions among the molecules in the reducing atmosphere of the prebiotic Earth formed the simple organic precursors from which biological molecules developed. Eventually, in reactions that may have been catalyzed by minerals such as clays, polypeptides and polynucleotides formed. These evolved under the pressure of competition for the available monomeric units. Ultimately, a nucleic acid, most probably RNA, developed the capability of influencing its own replication by directing the synthesis of proteins that catalyze polynucleotide synthesis. This was followed by the development of cell membranes so as to form living entities. Subsequently, metabolic processes evolved to synthesize necessary intermediates from available precursors as well as the high-energy compounds required to power these reactions. Likewise, photosynthesis and respiration arose in response to environmental pressures brought about by the action of living organisms.

The sheer size and rate of increase of the biochemical literature requires that it be read to attain a thorough understanding of any aspect of biochemistry. The review literature provides an *entrée* into a given subspeciality. To remain current in any field, however, requires a regular perusal of its primary literature. This should be read in a critical but highly selective fashion.

REFERENCES

Prokaryotes and Eukaryotes

Attenborough, D., *Life on Earth,* Little, Brown (1980). [A beautifully illustrated exposition of evolutionary development.]

Becker, W.M., and Deamer, D.W., *The World of the Cell* (2nd ed.), Benjamin/Cummings (1991). [A highly readable cell biology text.]

Campbell, N.A., *Biology,* 3rd ed., Benjamin/Cummings (1993). [A comprehensive general biology text. There are several others available of similar content.]

de Duve, C., *A Guided Tour of the Living Cell,* Vols. 1 and 2, Scientific American Books (1984). [A fanciful but highly enlightening examination of the cell's inner workings.]

Dulbecco, R., *The Design of Life,* Yale University Press (1987). [An incisive introduction to modern concepts of biology and biochemistry.]

Fox, G.E., Stackebrandt, E., Hespell, R.B., Gibson, J., Maniloff, J., Dyer, T.A., Wolfe, R.S., Balch, W.E., Tanner, R.S., Magrum, L.J., Zablen, L.B., Blakemore, R., Gupta, R., Bonen, L., Lewis, B.J., Stahl, D.A., Leuhrsen, K.R., Chen, K.N., and Woese, C.R., The phylogeny of prokaryotes, *Science* **209**, 457–463 (1980). [A presentation of the evidence that archaebacteria, eubacteria, and eukaryotes form separate lines of descent.]

Fawcett, D.W., *The Cell,* Saunders (1981). [A collection of electron micrographs of cells and organelles.]

Frieden, E., The chemical elements of life, *Sci. Am.* **227**(1): 52–60 (1972).

Goodsell, D.S., *The Machinery of Life,* Springer-Verlag (1993); A look inside the living cell, *Am. Scientist* **80**, 457–465 (1992); and Inside a living cell, *Trends Biochem. Sci.* **16**, 203–206 (1991).

Holt, J.G., Krieg, N.R., Sneath, H.A., Staley, J.T., and Williams, S.T., *Bergey's Manual of Determinitive Bacteriology* (9th ed.), Williams and Wilkins (1994).

Holtzman, E. and Novikoff, A.B., *Cells and Organelles* (3rd ed.), Holt, Rinehart, & Winston (1984).

Lewin, R., *The Thread of Life,* Random House (1982). [A lavishly illustrated presentation of evolutionary biology.]

Margulis, L., Symbiosis and evolution, *Sci. Am.* **225**(2): 48–57 (1972). [A discussion of the evolution of mitochondria and chloroplasts.]

Margulis, L. and Schwartz, K.V., *Five Kingdoms. An Illustrated Guide to the Phyla of Life on Earth* (2nd ed.), Freeman (1987).

Sagan, D. and Margulis, L., *Garden of Microbial Delights,* Harcourt Brace Jovanovich (1988).

Stanier, R.Y., Ingrahan, J.L., Wheelis, M.L., and Painter, P.R., *The Microbial World* (5th ed.), Prentice–Hall (1986).

Origin of Life

Chyba, C. and Sagan, C., Endogenous production, exogenous delivery and impact-shock synthesis of organic molecules: an inventory for the origins of life, *Nature* **355**, 125–132 (1992). [Discusses the evidence that the heavy meteoritic bombardment of Earth before 3.5 billion years ago produced and/or delivered quantities of organic molecules comparible to those produced by other energy sources.]

de Duve, C., *Blueprint for a Cell. The Nature and Origin of Life,* Carolina Biological Supply Co. (1991).

Dickerson, R.E., Chemical evolution and the origin of life, *Sci. Am.* **239**(3): 70–86 (1978).

Dyson, F., *Origins of Life,* Cambridge University Press (1985). [A fascinating philosophical discourse on theories of life's origins by a respected theoretical physicist.]

Fraústo da Silva, J.R. and Williams, R.J.P., *The Biological Chemistry of the Elements,* Oxford (1991).

Knoll, A.H., The early evolution of eukaryotes: A geological perspective, *Science* **256**, 622–627 (1992).

Lamond, A.I. and Gibson, T.J., Catalytic RNA and the origin of genetic systems, *Trends Genet.* **6**, 145–149 (1990).

Orgel, L.E., Molecular replication, *Nature* **358**, 203–209 (1992).

Schopf, J.W., Microfossils of the early Archean Apex chert: New evidence of the antiquity of life, *Science* **260**, 640–646 (1993).

Schopf, J.W., The evolution of the earliest cells, *Sci. Am.* **239**(3): 110–138 (1978).

Schuster, P., Prebiotic evolution, *in* Gutfreund, H. (Ed.), *Biochemical Evolution, pp.* 15–87, Cambridge University Press (1981).

Sci. Am. **271**(4) (1994). [A special issue on "Life In the Universe".]

Shapiro, R., *Origins. A Skeptics Guide to the Creation of Life on Earth,* Summit Books (1986). [An incisive and entertaining critique of the reigning theories of the origin of life.]

Valentine, J.W., The evolution of multicellular plants and animals, *Sci. Am.* **239**(3): 140–158 (1978).

Watson, J.D., Hopkins, N.H., Roberts, J.W., Steitz, J.A., and Weiner, A.M., *Molecular Biology of the Gene* (4th ed.), Chapter 26, Benjamin/Cummings (1987).

Woese, C.R. and Pace, N.R., Probing RNA structure, function, and history by comparitive analysis, *in* Gesteland, R.F. and Atkins, J.F. (Eds.), *The RNA World, pp.* 91–117, Cold Spring Harbor Laboratories (1993).

PROBLEMS

It is very difficult to learn something well without somehow participating in it. The chapter-end problems are therefore an important part of this book. They contain few problems of the regurgitory type. Rather they are designed to make you think and to offer insights not discussed in the text. Their difficulties range from those that require only a few moments reflection to those that might take an hour or more of concentrated effort to work out. The more difficult problems are indicated by a leading asterisk (*). The answers to the problems are worked out in detail in the *Solutions Manual to Accompany Biochemistry* (2nd ed.) by Donald Voet and Judith G. Voet. You should, of course, make every effort to work out a problem before consulting the *Solutions Manual.*

1. Under optimal conditions for growth, an *E. coli* cell will divide around every 20 min. If no cells died, how long would it take a single *E. coli* cell, under optimal conditions in a 10-L culture flask, to reach its maximum cell density of 10^{10} cells·mL^{-1} (a "saturated" culture)? Assuming that optimum conditions could be maintained, how long would it take for the total volume of the cells alone to reach 1 km³? (Assume an *E. coli* cell to be a cylinder 2 μm long and 1 μm in diameter.)

2. Without looking them up, draw schematic diagrams of a bacterial cell and an animal cell. What are the functions of their various organelles? How many lines of descent might a typical animal cell have?

3. Compare the surface-to-volume ratios of a typical *E. coli* cell (its dimensions are given in Problem 1) and a spherical eukaryotic cell that is 20 μm in diameter. How does this difference affect the life styles of these two cell types? In order to improve their ability to absorb nutrients, the **brush border cells** of the intestinal epithelium have velvetlike patches of **microvilli** facing into the intestine. How does the surface-to-volume ratio of this eukaryotic cell change if 20% of its surface area is covered with cylindrical microvilli that are 0.1 μm in diameter, 1 μm in length, and occur on a square grid with 0.2- μm center-to-center spacing?

4. Many proteins in *E. coli* are normally present at concentrations of two molecules per cell. What is the molar concentration of such a protein (the dimensions of *E. coli* are given in Problem 1)? Conversely, how many glucose molecules does an *E. coli* cell contain if it has an internal glucose concentration of 1.0 m*M*?

5. The DNA of an *E. coli* chromosome measures 1.6 mm in length, when extended, and 20 Å in diameter. What fraction of an *E. coli* cell is occupied by its DNA (the dimensions of an *E. coli* are given in Problem 1)? A human cell has some 600 times the DNA of an *E. coli* cell and is typically spherical with a diameter of 20 μm. What fraction of such a human cell is occupied by its DNA?

*6. A new planet has been discovered that has approximately the same orbit about the sun as the Earth but is invisible from Earth because it is always on the opposite side of the sun. Interplanetary probes have already established that this planet has a significant atmosphere. The National Aeronautics and Space Administration is preparing to launch a new unmanned probe that will land on the surface of the planet. Outline a simple experiment for this lander that will test for the presence of life on the surface of this planet (assume that the life forms, if any, on the planet are likely to be microorganisms and therefore unable to walk up to the lander's video cameras and say "Hello").

7. It has been suggested that an all out nuclear war will so enshroud the Earth with clouds of dust and smoke that the entire surface of the planet will be quite dark and therefore intensely cold (well below 0°C) for several years (the so-called "nuclear winter"). In that case, it is thought, eukaryotic life would die out and bacteria would inherit the Earth. Why?

8. Green and purple photosynthetic bacteria are thought to resemble the first organisms that could carry out photosynthesis. Speculate on the composition of the Earth's atmosphere when these organisms first arose.

9. Explore your local biochemistry library (it may be disguised as a biology, chemistry, or medical library). Locate where the current periodicals, the bound periodicals, and the books are kept. Browse through the contents of a current major biochemistry journal, such as *Biochemistry, Cell,* or *Proceedings of the National Academy of Sciences,* and pick a title that interests you. Scan the corresponding paper and note its organization. Likewise, peruse one of the articles in the latest volume of *Annual Review of Biochemistry.*

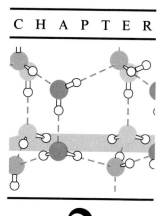

2

Aqueous Solutions

Life, as we know it, occurs, in aqueous solution. Indeed, terrestial life apparently arose in some primordial sea (Section 1-4B) and, as the fossil record indicates, did not venture onto dry land until comparatively recent times. Yet, even those organisms that did develop the capacity to live out of water still carry the ocean with them: The compositions of their intracellular and extracellular fluids are remarkably similar to that of seawater. This is true even of organisms that live in such unusual environments as saturated brine, acidic hot sulfur springs, and petroleum.

Water is so familiar, we generally consider it to be a rather bland fluid of simple character. It is, however, a chemically reactive liquid with such extraordinary physical properties that, if chemists had discovered it in recent times, it would undoubtedly have been classified as an exotic substance.

The properties of water are of profound biological significance. *The structures of the molecules on which life is based, proteins, nucleic acids, lipids, and complex carbohydrates, result directly from their interactions with their aqueous* *environment. The combination of solvent properties responsible for the intramolecular and intermolecular associations of these substances is peculiar to water; no other solvent even resembles water in this respect.* Although the hypothesis that life could be based on organic polymers other than proteins and nucleic acids seems plausible, it is all but inconceivable that the complex structural organization and chemistry of living systems could exist in other than an aqueous medium. Indeed, direct observations on the surface of Mars, the only other planet in the solar system with temperatures compatible with life, indicate that it is devoid both of water and of life.

Biological structures and processes can only be understood in terms of the physical and chemical properties of water. We therefore begin this chapter with a discussion of the molecular and solvent properties of water. In the following section we review its chemical behavior, that is, the nature of aqueous acids and bases.

1. PROPERTIES OF WATER

Water's peculiar physical and solvent properties stem largely from its extraordinary internal cohesiveness compared to that of almost any other liquid. In this section, we explore the physical basis of this phenomenon.

A. *Structure and Interactions*

The H_2O molecule has a bent geometry with an O—H bond distance of 0.958 Å and an H—O—H bond angle of

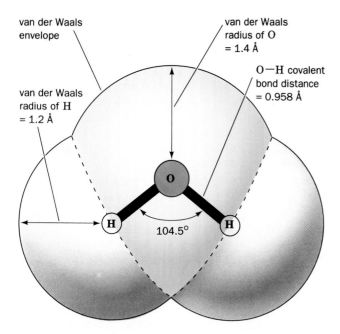

van der Waals envelope

van der Waals radius of H = 1.2 Å

van der Waals radius of O = 1.4 Å

O—H covalent bond distance = 0.958 Å

O

H   H

104.5°

FIGURE 2-1. The structure of the water molecule. The outline represents the van der Waals envelope of the molecule (where the attractive components of the van der Waals interactions balance the repulsive components). The skeletal model of the molecule indicates its covalent bonds.

104.5° (Fig. 2-1). The large electronegativity difference between H and O confers a 33% ionic character on the O—H bond as is indicated by water's dipole moment of 1.85 debye units. Water is clearly a highly polar molecule, a phenomenon with enormous implications for living systems.

Water Molecules Associate through Hydrogen Bonds

The electrostatic attractions between the dipoles of two water molecules tend to orient them such that the O—H bond on one water molecule points towards a lone-pair electron cloud on the oxygen atom of the other water molecule. This results in a directional intermolecular association known as a **hydrogen bond** (Fig. 2-2), an interaction that is crucial to both the properties of water itself and to its role as a biochemical solvent. In general, *a hydrogen bond may be represented as D—H · · · A, where D—H is a weakly*

acidic "donor group" such as N—H or O—H, and A is a lone-pair bearing and thus weakly basic "acceptor atom" such as N or O. The peculiar requirement of a hydrogen atom in the D—H · · · A interaction stems from the hydrogen atom's small size: Only a hydrogen nucleus can approach the lone-pair electron cloud of an acceptor atom closely enough to permit an electrostatic association of significant magnitude.

Hydrogen bonds are structurally characterized by an H · · · A distance that is at least 0.5 Å shorter than the calculated van der Waals distance (distance of closest approach between two nonbonded atoms) between these atoms. In water, for example, the O · · · H hydrogen bond distance is ~1.8 Å versus 2.6 Å for the corresponding van der Waals distance. The energy of a hydrogen bond (~20 $kJ \cdot mol^{-1}$ in H_2O) is small compared to covalent bond energies (for instance, 460 $kJ \cdot mol^{-1}$ for an O—H covalent bond). Nevertheless, most biological molecules have so many hydrogen bonding groups that hydrogen bonding is of paramount importance in determining their three-dimensional structures and their intermolecular associations. Hydrogen bonding is further discussed in Section 7-4B.

The Physical Properties of Ice and Liquid Water Largely Result from Intermolecular Hydrogen Bonding

The structure of ice provides a striking example of the cumulative strength of many hydrogen bonds. X-Ray and neutron diffraction studies have established that water molecules in ice are arranged in an unusually open structure. Each water molecule is tetrahedrally surrounded by four nearest neighbors to which it is hydrogen bonded (Fig. 2-3). In two of these hydrogen bonds, the central H_2O molecule is the "donor" and in the other two, it is the "acceptor." As a consequence of its open structure, water is one of the very few substances that expands on freezing (at 0°C, liquid water has a density of 1.00 $g \cdot mL^{-1}$, whereas ice has a density of 0.92 $g \cdot mL^{-1}$).

The expansion of water on freezing has overwhelming consequences for life on Earth. Suppose that water contracted upon freezing, that is, became more dense rather than less dense. Ice would then sink to the bottoms of lakes and oceans rather than float. This ice would be insulated from the sun so that oceans, with the exception of a thin surface layer of liquid in warm weather, would be permanently frozen solid (the water at great depths in even tropical oceans is close to 4°C, its temperature of maximum density). The reflection of sunlight by these frozen oceans and their cooling effect on the atmosphere would ensure that land temperatures would also be much colder than at present; that is, the Earth would have a permanent ice age. Furthermore, since life apparently evolved in the ocean, it seems unlikely that life could have developed at all if ice contracted upon freezing.

Although the melting of ice is indicative of the cooperative collapse of its hydrogen bonded structure, hydrogen bonds between water molecules persist in the liquid state. The heat of sublimation of ice at 0°C is 46.9 $kJ \cdot mol^{-1}$. Yet,

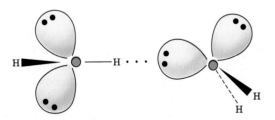

H   H · · ·   H   H

FIGURE 2-2. A hydrogen bond between two water molecules. The strength of this interaction is maximal when the O—H covalent bond points directly along a lone-pair electron cloud of the oxygen atom to which it is hydrogen bonded.

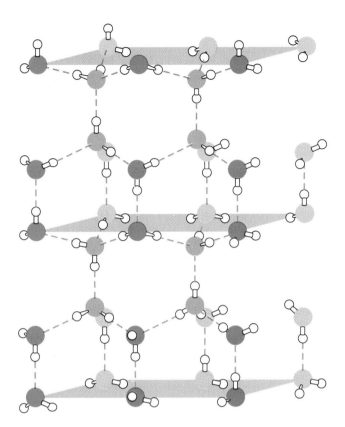

FIGURE 2-3. The structure of ice. The tetrahedral arrangement of the water molecules is a consequence of the roughly tetrahedral disposition of each oxygen atom's sp^3-hybridized bonding and lone pair orbitals (Fig. 2-2). Oxygen and hydrogen atoms are represented, respectively, by red and white spheres, and hydrogen bonds are indicated by dashed lines. Note the open structure that gives ice its low density relative to liquid water. [After Pauling, L., *The Nature of the Chemical Bond* (3rd ed.), p. 465, Cornell University Press (1960).]

only ~6 kJ · mol⁻¹ of this quantity can be attributed to the kinetic energy of gaseous water molecules. The remaining 41 kJ · mol⁻¹ must therefore represent the energy required to disrupt the hydrogen bonding interactions holding an ice crystal together. The heat of fusion of ice (6.0 kJ · mol⁻¹) is ~15% of the energy required to disrupt the ice structure. *Liquid water is therefore only ~15% less hydrogen bonded than ice at 0°C.* Indeed, the boiling point of water is 264°C higher than that of methane (CH_4), a substance with nearly the same molecular mass as H_2O but which is incapable of hydrogen bonding (in the absence of intermolecular associations, substances with equal molecular masses should have similar boiling points). This reflects the extraordinary internal cohesiveness of liquid water resulting from its intermolecular hydrogen bonding.

Liquid Water Has a Rapidly Fluctuating Structure

X-Ray scattering measurements of liquid water reveal a complex structure. Near 0°C, water exhibits an average nearest-neighbor O · · · O distance of 2.82 Å, which is slightly greater than the corresponding 2.76-Å distance in ice despite the greater density of the liquid. The X-ray data further indicate that each water molecule is surrounded by an average of 4.4 nearest neighbors, which strongly suggests that the short-range structure of liquid water is predominantly tetrahedral in character. This picture is corroborated by the additional intermolecular distances in liquid water of around 4.5 and 7.0 Å, which are near the expected second and third nearest-neighbor distances in an icelike tetrahedral structure. Liquid water, however, also exhibits a 3.5-Å intermolecular distance, which cannot be rationalized in terms of an icelike structure. These average distances, moreover, become less sharply defined as the temperature increases into the physiologically significant range, thereby signaling the thermal breakdown of the short-range water structure.

The structure of liquid water is not simply described. This is because each water molecule reorients about once every 10⁻¹² s, which makes the determination of water's instantaneous structure an experimentally and theoretically difficult problem (very few experimental techniques can make measurements over such short time spans). Indeed, only with the recent advent of molecular dynamics simulations, which calculate the trajectories of the individual atoms in a large collection of molecules over a period of time, have theoreticians felt that they had a reasonable understanding of liquid water on the molecular level.

For the most part, molecules in liquid water are each hydrogen bonded to four nearest neighbors as they are in ice. These hydrogen bonds are distorted, however, so that the networks of linked molecules are irregular and varied. For example, 4- to 7-membered rings of hydrogen bonded molecules commonly occur in liquid water, in contrast to the cyclohexane-like 6-membered rings characteristic of ice (Fig. 2-3). Moreover, these networks are continually breaking up and reforming over time periods on the order of 2 × 10⁻¹¹ s. *Liquid water therefore consists of a rapidly fluctuating, space-filling network of hydrogen bonded H_2O molecules that, over short distances, resembles that of ice.*

B. Water as a Solvent

Solubility depends on the ability of a solvent to interact with a solute more strongly than solute particles interact with each other. Water is said to be the "universal solvent." Although this statement cannot literally be true, water certainly dissolves more types of substances and in greater amounts than any other solvent. In particular, the polar character of water makes it an excellent solvent for polar and ionic materials, which are therefore said to be **hydrophilic** (Greek: *hydor*, water + *philos*, loving). On the other hand, nonpolar substances are virtually insoluble in water ("oil and water don't mix") and are consequently described as being **hydrophobic** (Greek: *phobos*, fear). Nonpolar substances, however, are soluble in nonpolar solvents such as CCl_4 or hexane. This information is summarized by another maxim, "like dissolves like."

Why do salts dissolve in water? Salts, such as NaCl or K_2HPO_4, are held together by ionic forces. The ions of a

TABLE 2-1. **Dielectric Constants and Permanent Molecular Dipole Moments of Some Common Solvents**

Substance	Dielectric Constant	Dipole Moment (debye)
Formamide	110.0	3.37
Water	78.5	1.85
Dimethyl sulfoxide	48.9	3.96
Methanol	32.6	1.66
Ethanol	24.3	1.68
Acetone	20.7	2.72
Ammonia	16.9	1.47
Chloroform	4.8	1.15
Diethyl ether	4.3	1.15
Benzene	2.3	0.00
Carbon tetrachloride	2.2	0.00
Hexane	1.9	0.00

Source: Brey, W.S., *Physical Chemistry and Its Biological Applications*, p. 26, Academic Press (1978).

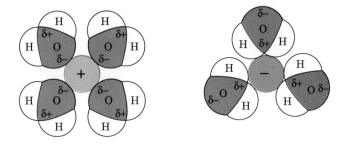

FIGURE 2-4.   Solvation of ions by oriented water molecules.

This arrangement greatly attenuates the coulombic forces between ions, which is why polar solvents have such high dielectric constants.

The orienting effect of ionic charges on dipolar molecules is opposed by thermal motions, which continually tend to randomly reorient all molecules. The dipoles in a solvated complex are therefore only partially oriented. The reason why the dielectric constant of water is so much greater than that of other liquids with comparable dipole moments is that liquid water's hydrogen bonded structure permits it to form oriented structures that resist thermal randomization, thereby more effectively distributing ionic charges.

The bond dipoles of uncharged polar molecules make them soluble in aqueous solutions for the same reasons that ionic substances are water soluble. The solubilities of polar and ionic substances are enhanced if they carry functional groups, such as hydroxyl ($-OH$), keto ($C=O$), carboxyl ($-CO_2H$ or $COOH$), or amino ($-NH_2$) groups, that can form hydrogen bonds with water as is illustrated in Fig. 2-5.

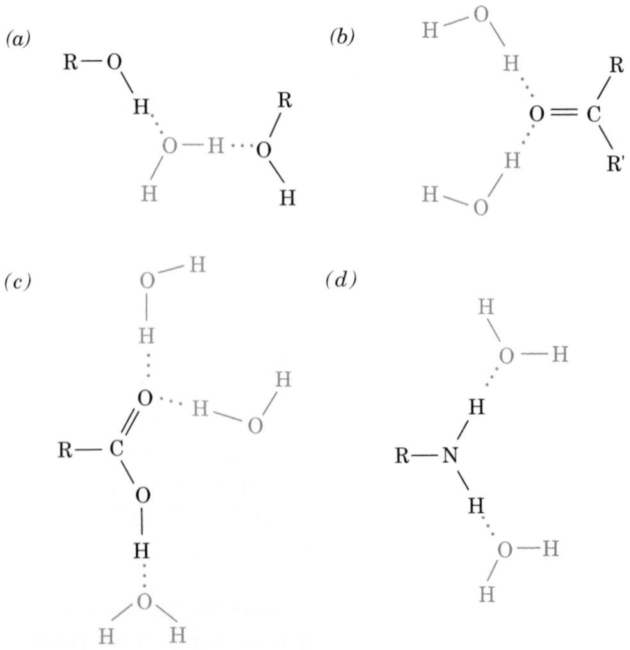

salt, as do any electrical charges, interact according to **Coulomb's law:**

$$F = \frac{kq_1q_2}{Dr^2} \qquad [2.1]$$

where F is the force between two electrical charges, q_1 and q_2, that are separated by the distance r, D is the **dielectric constant** of the medium between them, and k is a proportionality constant ($8.99 \times 10^9 \text{ J} \cdot \text{m} \cdot \text{C}^{-2}$). Thus, as the dielectric constant of a medium increases, the force between its embedded charges decreases; that is, the dielectric constant of a solvent is a measure of its ability to keep opposite charges apart. In a vacuum, D is unity and in air, it is only negligibly larger. The dielectric constants of several common solvents, together with their permanent molecular dipole moments, are listed in Table 2-1. Note that these quantities tend to increase together, although not in any regular way.

The dielectric constant of water is among the highest of any pure liquid, whereas those of nonpolar substances, such as hydrocarbons, are relatively small. The force between two ions separated by a given distance in nonpolar liquids such as hexane or benzene is therefore 30 to 40 times greater than that in water. Consequently, in nonpolar solvents (low D), ions of opposite charge attract each other so strongly that they coalesce to form a salt, whereas the much weaker forces between ions in water solution (high D) permit significant quantities of the ions to remain separated.

An ion immersed in a polar solvent attracts the oppositely charged ends of the solvent dipoles as is diagrammed in Fig. 2-4 for water. The ion is thereby surrounded by several concentric shells of oriented solvent molecules. Such ions are said to be **solvated** or, if water is the solvent, to be **hydrated.** The electric field produced by the solvent dipoles opposes that of the ion so that, in effect, the ionic charge is spread over the volume of the solvated complex.

FIGURE 2-5.   Hydrogen bonding between water and *(a)* hydroxyl groups, *(b)* keto groups, *(c)* carboxyl groups, and *(d)* amino groups.

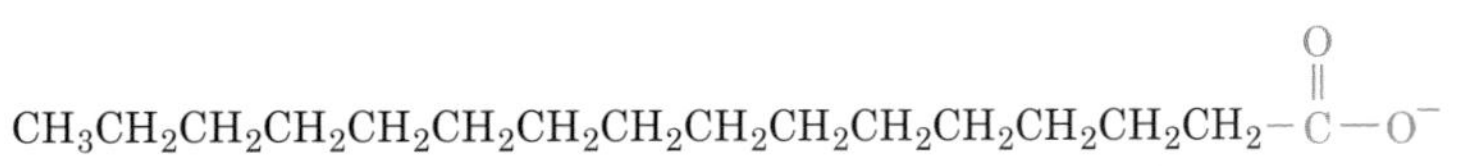

$$CH_3CH_2CH_2CH_2CH_2CH_2CH_2CH_2CH_2CH_2CH_2CH_2CH_2CH_2CH_2 - \overset{\overset{\textstyle O}{\|}}{C} - O^-$$

Palmitate ($C_{15}H_{31}COO^-$)

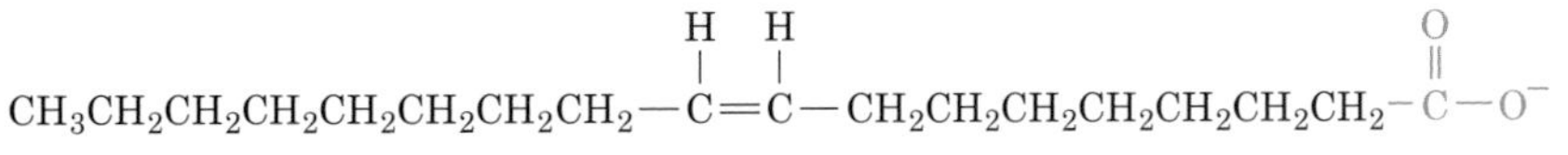

$$CH_3CH_2CH_2CH_2CH_2CH_2CH_2CH_2 - \overset{\overset{\textstyle H}{|}}{C} = \overset{\overset{\textstyle H}{|}}{C} - CH_2CH_2CH_2CH_2CH_2CH_2CH_2 - \overset{\overset{\textstyle O}{\|}}{C} - O^-$$

Oleate ($C_{17}H_{33}COO^-$)

FIGURE 2-6. Examples of fatty acid anions. They consist of a polar carboxylate group coupled to a long nonpolar hydrocarbon chain.

Indeed, water-soluble biomolecules such as proteins, nucleic acids, and carbohydrates bristle with just such groups. Nonpolar substances, in contrast, lack both hydrogen bonding donor and acceptor groups.

Amphiphiles Form Micelles and Bilayers

Most biological molecules have both polar (or ionically charged) and nonpolar segments and are therefore simultaneously hydrophilic and hydrophobic. Such molecules, for example, fatty acid ions (soap ions; Fig. 2-6), are said to be **amphiphilic** or, synonomously, **amphipathic** (Greek: *amphi,* both + *pathos,* passion). How do amphiphiles interact with an aqueous solvent? Water, of course, tends to hydrate the hydrophilic portion of an amphiphile, but it also tends to exclude its hydrophobic portion. Amphiphiles consequently tend to form water-dispersed structurally ordered aggregates. Such aggregates may take the form of **micelles,** which are globules of up to several thousand amphiphiles arranged with their hydrophilic groups at the globule surface so that they can interact with the aqueous solvent while the hydrophobic groups associate at the center so as to exclude solvent (Fig. 2-7a). Alternatively, the amphiphiles may arrange themselves to form bilayered sheets or vesicles (Fig. 2-7b) in which the polar groups face the aqueous phase.

The interactions stabilizing a micelle or bilayer are collectively described as **hydrophobic forces** or **hydrophobic interactions** to indicate that they result from the tendency of water to exclude hydrophobic groups. Hydrophobic interactions are relatively weak compared to hydrogen bonds and lack directionality. Nevertheless, hydrophobic interactions are of pivotal biological importance because, as we shall see in later chapters, they are largely responsible for the structural integrity of biological macromolecules (Sections 7-4C and 28-3D), as well as that of supramolecular aggregates such as membranes. Note that hydrophobic interactions are peculiar to an aqueous environment. Other polar solvents do not promote such associations.

C. Proton Mobility

When an electrical current is passed through an ionic solution, the ions migrate toward the electrode of opposite polarity at a rate proportional to the electrical field and inversely proportional to the frictional drag experienced by the ion as it moves through the solution. This latter quantity, as Table 2-2 indicates, varies with the size of the ion. Note, however, that the ionic mobilities of both H_3O^+ and OH^- are anomalously large compared to those of other ions. For H_3O^+ (the **hydronium ion,** which is abbreviated

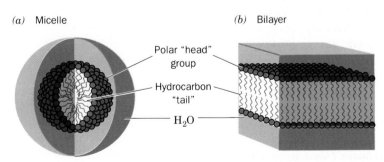

(a) Micelle   (b) Bilayer

Polar "head" group

Hydrocarbon "tail"

H_2O

FIGURE 2-7. The associations of amphipathic molecules in aqueous solutions. The polar "head" groups are hydrated, whereas the nonpolar "tails" aggregate so as to exclude the aqueous solution: (a) A spheroidal aggregate of amphipathic molecules known as a micelle. (b) An extended planar aggregate of amphipathic molecules called a bilayer. The bilayer may form a closed spheroidal shell, known as a vesicle, that encloses a small amount of aqueous solution.

TABLE 2-2. IONIC MOBILITIES^a IN H_2O AT 25°C

Ion	Mobility $\times 10^{-5}$ ($cm^2 \cdot V^{-1} \cdot s^{-1}$)
H_3O^+	362.4
Li^+	40.1
Na^+	51.9
K^+	76.1
NH_4^+	76.0
Mg^{2+}	55.0
Ca^{2+}	61.6
OH^-	197.6
Cl^-	76.3
Br^-	78.3
CH_3COO^-	40.9
SO_4^{2-}	79.8

^{a} Ionic mobility is the distance an ion moves in one second under the influence of an electric field of one volt per cm.

Source: Brey, W.S., *Physical Chemistry and Its Biological Applications,* p. 172, Academic Press (1978).

H^+; a bare proton has no stable existence in aqueous solution), this high migration rate results from the ability of protons to jump rapidly from one water molecule to another as is diagrammed in Fig. 2-8. Although a given hydronium ion can physically migrate through solution in the manner of say, a Na^+ ion, the rapidity of the proton-jump mechanism makes the H_3O^+ ion's effective ionic mobility much greater than it otherwise would be (the mean lifetime of a given H_3O^+ ion is 10^{-12} s at 25°C). The anomalously high ionic mobility of the OH^- ion is likewise accounted for

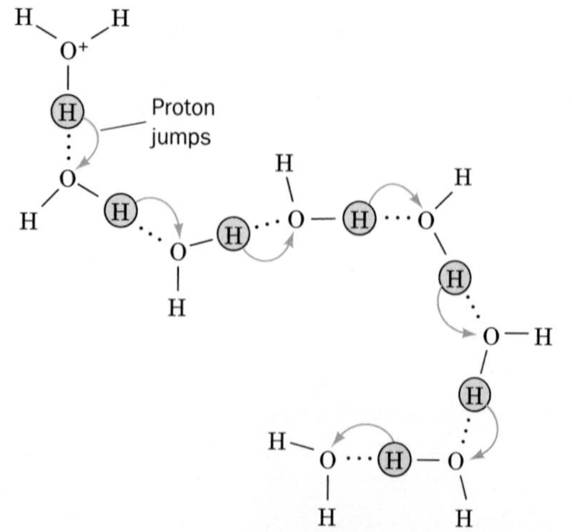

FIGURE 2-8. The mechanism of hydronium ion migration in aqueous solution via proton jumps. Proton jumps, which mostly occur at random, take place rapidly compared with direct molecular migration, thereby accounting for the observed high ionic mobilities of hydronium and hydroxyl ions in aqueous solutions.

by the proton-jump mechanism but, in this case, the apparent direction of ionic migration is opposite to the direction of proton jumping. Proton jumping is also responsible for the observation that *acid–base reactions are among the fastest reactions that take place in aqueous solutions* and it is probably of importance in biological proton-transfer reactions.

2. ACIDS, BASES, AND BUFFERS

Biological molecules, such as proteins and nucleic acids, bear numerous functional groups, such as carboxyl and amino groups, that can undergo acid–base reactions. Many properties of these molecules therefore vary with the acidities of the solutions in which they are immersed. In this section we discuss the nature of acid–base reactions and how acidities are controlled, both physiologically and in the laboratory.

A. Acid–Base Reactions

Acids and **bases,** in a definition coined in the 1880s by Svante Arrhenius, are, respectively, substances capable of donating protons and hydroxide ions. This definition is rather limited, because, for example, it does not account for the observation that NH_3, which lacks an OH^- group, exhibits basic properties. In a more general definition, which was formulated in 1923 by Johannes Brønsted and Thomas Lowry, *an acid is a substance that can donate protons (as in the Arrhenius definition) and a base is a substance that can accept protons.* Under this definition, in every acid–base reaction,

$$HA + H_2O \rightleftharpoons H_3O^+ + A^-$$

an acid (HA) reacts with a base (H_2O) to form the **conjugate base** of the acid (A^-) and the **conjugate acid** of the base (H_3O^+) (this reaction is usually abbreviated $HA \rightleftharpoons H^+ + A^-$ with the participation of H_2O implied). Accordingly, the acetate ion (CH_3COO^-) is the conjugate base of acetic acid (CH_3COOH) and the ammonium ion (NH_4^+) is the conjugate acid of ammonia (NH_3). (In a yet more general definition of acids and bases, Gilbert Lewis described an acid as a substance that can accept an electron pair and a base as a substance that can donate an electron pair. This definition, which is applicable to both aqueous and non-aqueous systems, is unnecessarily broad for describing most biochemical phenomena.)

The Strength of an Acid Is Specified by Its Dissociation Constant

The above acid dissociation reaction is characterized by its **equilibrium constant** which, for acid–base reactions, is known as a **dissociation constant,**

$$K = \frac{[H_3O^+][A^-]}{[HA][H_2O]} \qquad [2.2]$$

TABLE 2-3. DISSOCIATION CONSTANTS AND pK'S AT 25°C OF SOME ACIDS IN COMMON LABORATORY USE AS BIOCHEMICAL BUFFERS

Acid	K	pK
Oxalic acid	5.37×10^{-2}	1.27 (pK_1)
H_3PO_4	7.08×10^{-3}	2.15 (pK_1)
Citric acid	7.41×10^{-4}	3.13 (pK_1)
Formic acid	1.78×10^{-4}	3.75
Succinic acid	6.17×10^{-5}	4.21 (pK_1)
Oxalate$^-$	5.37×10^{-5}	4.27 (pK_2)
Acetic acid	1.74×10^{-5}	4.76
Citrate$^-$	1.74×10^{-5}	4.76 (pK_2)
Succinate$^-$	2.29×10^{-6}	5.64 (pK_2)
2-(N-Morpholino)ethane sulfonic acid (MES)	8.13×10^{-7}	6.09
Cacodylic acid	5.37×10^{-7}	6.27
H_2CO_3	4.47×10^{-7}	6.35 (pK_1)
Citrate^{2-}	3.98×10^{-7}	6.40 (pK_3)
N-(2-Acetamido)iminodiacetic acid (ADA)	2.69×10^{-7}	6.57
Piperazine-N,N'-bis(2-ethanesulfonic acid) (PIPES)	1.74×10^{-7}	6.76
N-(2-Acetamido)-2-aminoethanesulfonic acid (ACES)	1.58×10^{-7}	6.80
$H_2PO_4^-$	1.51×10^{-7}	6.82 (pK_2)
3-N-Morpholino)propanesulfonic acid (MOPS)	7.08×10^{-8}	7.15
N-2-Hydroxyethylpiperazine-N'-2-ethanesulfonic acid (HEPES)	3.39×10^{-8}	7.47
N-2-Hydroxyethylpiperazine-N'-3-propanesulfonic acid (HEPPS)	1.10×10^{-8}	7.96
N-[Tris(hydroxymethyl)methyl]glycine (Tricine)	8.91×10^{-9}	8.05
Tris(hydroxymethyl)aminomethane (TRIS)	8.32×10^{-9}	8.08
Glycylglycine	5.62×10^{-9}	8.25
N,N-Bis(2-hydroxyethyl)glycine (Bicine)	5.50×10^{-9}	8.26
Boric acid	5.75×10^{-10}	9.24
NH_4^+	5.62×10^{-10}	9.25
Glycine	1.66×10^{-10}	9.78
HCO_3^-	4.68×10^{-11}	10.33 (pK_2)
Piperidine	7.58×10^{-12}	11.12
HPO_4^{2-}	4.17×10^{-13}	12.38 (pK_3)

Source: Dawson, R.M.C., Elliott, D.C., Elliott, W.H., and Jones, K.M., *Data for Biochemical Research* (3rd ed.), pp. 424–425, Oxford Science Publications (1986) *and* Good, N.E., Winget, G.D., Winter, W., Connolly, T.N., Izawa, S., and Singh, R.M.M., *Biochemistry* **5**, 467 (1966).

a quantity that is a measure of the relative proton affinities of the HA/A$^-$ and H_3O^+/H_2O conjugate acid–base pairs. Here, as throughout the text, quantities in square brackets symbolize the molar concentrations of the indicated substances. Since in dilute aqueous solutions the water concentration is essentially constant with $[H_2O] = 1000$ g·L^{-1}/ 18.015 g·mol$^{-1} = 55.5M$, this term is customarily combined with the dissociation constant, which then takes the form

$$K_a = K[H_2O] = \frac{[H^+][A^-]}{[HA]} \qquad [2.3]$$

For brevity, however, we shall henceforth omit the subscript "a." The dissociation constants for acids useful in preparing biochemical solutions are listed in Table 2-3.

Acids may be classified according to their relative strengths, that is, according to their abilities to transfer a proton to water. Acids with dissociation constants smaller than that of H_3O^+ (which, by definition, is unity in aqueous solutions) are only partially ionized in aqueous solutions and are known are **weak acids** ($K < 1$). Conversely, **strong acids** have dissociation constants larger than that of H_3O^+ so that they are almost completely ionized in aqueous solutions ($K > 1$). The acids listed in Table 2-3 are all weak acids. However, many of the so-called "mineral acids," such as $HClO_4$, HNO_3, HCl, and H_2SO_4 (for the first ionization), are strong acids. Since strong acids rapidly transfer all their protons to H_2O, the strongest acid that can stably exist in aqueous solutions is H_3O^+. Likewise, there can be no stronger base in aqueous solutions than OH$^-$.

Water, being an acid, has a dissociation constant:

$$K = \frac{[H^+][OH^-]}{[H_2O]}$$

As above, the constant $[H_2O] = 55.5M$ can be incorporated into the dissociation constant to yield the expression for the ionization constant of water,

$$K_w = [H^+][OH^-] \qquad [2.4]$$

The value of K_w at 25°C is $10^{-14}M^2$. Pure water must contain equimolar amounts of H^+ and OH^- so that $[H^+] = [OH^-] = (K_w)^{1/2} = 10^{-7}M$. Since $[H^+]$ and $[OH^-]$ are reciprocally related by Eq. [2.4], if $[H^+]$ is greater than this value, $[OH^-]$ must be correspondingly less and vice versa. Solutions with $[H^+] = 10^{-7}M$ are said to be **neutral,** those with $[H^+] > 10^{-7}M$ are said to be **acidic,** and those with $[H^+] < 10^{-7}M$ are said to be **basic.** Most physiological solutions have hydrogen ion concentrations near neutrality. For example, human blood is normally slightly basic with $[H^+] = 4.0 \times 10^{-8}M$.

The values of $[H^+]$ for most solutions are inconveniently small and difficult to compare. A more practical quantity, which was devised in 1909 by Søren Sørensen, is known as the **pH:**

$$pH = -\log[H^+] \qquad [2.5]$$

The pH of pure water is 7.0, whereas acidic solutions have pH < 7.0 and basic solutions have pH > 7.0. For a $1M$ solution of a strong acid, pH = 0 and for a $1M$ solution of a strong base, pH = 14. Note that if two solutions differ in pH by one unit, they differ in $[H^+]$ by a factor of 10. The pH of a solution may be accurately and easily determined through electrochemical measurements with a device known as a **pH meter.**

The pH of a Solution Is Determined by the Relative Concentrations of Acids and Bases

The relationship between the pH of a solution and the concentrations of an acid and its conjugate base can be easily derived by rearranging Eq. [2.3]

$$[H^+] = K\left(\frac{[HA]}{[A^-]}\right)$$

and substituting it into Eq. [2.5]

$$pH = -\log K + \log\left(\frac{[A^-]}{[HA]}\right)$$

Defining $pK = -\log K$ in analogy with Eq. [2.5], we obtain the **Henderson–Hasselbalch equation:**

$$pH = pK + \log\left(\frac{[A^-]}{[HA]}\right) \qquad [2.6]$$

This equation indicates that *the pK of an acid is numerically equal to the pH of the solution when the molar concentrations of the acid and its conjugate base are equal.* Table 2-3 lists the pK values of several acids.

B. Buffers

A 0.01-mL droplet of $1M$ HCl added to 1 L of pure water will change the water's pH from 7 to 5, which represents a 100-fold increase in $[H^+]$. Yet, since the properties of biological substances vary significantly with small changes in pH, they require environments in which the pH is insensitive to additions of acids or bases. To understand how this is possible, let us consider the titration of a weak acid with a strong base.

Figure 2-9 shows how the pH values of 1-L solutions of $1M$ acetic acid, $H_2PO_4^-$, and ammonium ion (NH_4^+), vary with the quantity of OH^- added. Titration curves such as

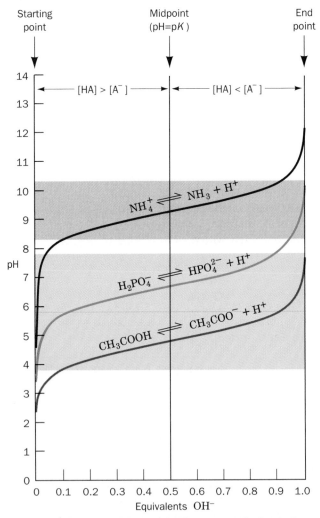

FIGURE 2-9. The acid–base titration curves of 1-L solutions of $1M$ acetic acid, $H_2PO_4^-$, and NH_4^+ by strong base. At the starting point of each titration, the acid form of the conjugate acid–base pair overwhelmingly predominates. At the midpoint of the titration, where pH = pK, the concentration of the acid is equal to that of its conjugate base. Finally, at the endpoint of the titration, where the equivalents of strong base added equal the equivalents of acid at the starting point, the conjugate base is in great excess over acid. The shaded bands indicate the pH ranges over which the corresponding solution can function effectively as a buffer.

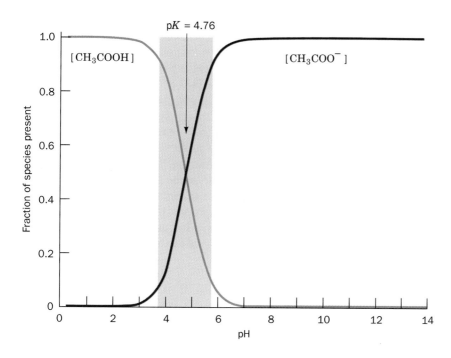

FIGURE 2-10. The distribution curves for acetic acid and acetate ion. The fraction of species present is given as the ratio of the concentration of CH_3COOH or CH_3COO^- to the total concentrations of these two species. The customarily accepted useful buffer range of $pK \pm 1$ is indicated by the shaded region.

those in Fig. 2-9, as well as distribution curves such as those in Fig. 2-10, may be calculated using the Henderson–Hasselbalch equation. Near the beginning of the titration, a significant fraction of the A^- present arises from the dissociation of HA. Similarly, near the endpoint, much of the HA derives from the reaction of A^- with H_2O. Throughout most of the titration, however, the OH^- added reacts essentially completely with the HA to form A^- so that

$$[A^-] = \frac{x}{V} \qquad [2.7]$$

where x represents the equivalents of OH^- added and V is the volume of the solution. Then, letting c_0 represent the equivalents of HA initially present,

$$[HA] = \frac{c_0 - x}{V} \qquad [2.8]$$

Incorporating these relationships into Eq. [2.6] yields

$$pH = pK + \log\left(\frac{x}{c_0 - x}\right) \qquad [2.9]$$

which accurately describes a titration curve except near its wings (these regions require more exact treatments that take into account the ionizations of water).

Several details about the titration curves in Fig. 2-9 should be noted:

1. The curves have a similar shape but are shifted vertically along the pH axis.

2. The pH at the **equivalence point** of each titration (where the equivalents of OH^- added equal the equivalents of HA initially present) is > 7 because of the reaction of A^- with H_2O to form $HA + OH^-$; similarly, each initial pH is < 7.

3. The pH at the midpoint of each titration is numerically equal to the pK of its corresponding acid; here, according to the Henderson–Hasselbalch equation, $[HA] = [A^-]$.

4. The slope of each titration curve is much less near its midpoint than it is near its wings. This indicates that *when $[HA] \approx [A^-]$, the pH of the solution is relatively insensitive to the addition of strong base or strong acid. Such a solution, which is known as an* **acid–base buffer,** *is resistant to pH changes because small amounts of added H^+ or OH^-, respectively, react with the A^- or HA present without greatly changing the value of $\log([A^-]/[HA])$.*

Buffers Stabilize a Solution's pH

The ability of a buffer to resist pH changes with added acid or base is directly proportional to the total concentration of the conjugate acid–base pair, $[HA] + [A^-]$. It is maximal when $pH = pK$ and decreases rapidly with a change in pH from that point. A good rule of thumb is that a weak acid is in its useful buffer range within 1 pH unit of its pK (the shaded regions of Figs. 2-9 and 2-10). Above this range, where the ratio $[A^-]/[HA] > 10$, the pH of the solution changes rapidly with added strong base. A buffer is similarly impotent with addition of strong acid when its pK exceeds the pH by more than a unit.

Biological fluids, both those found intracellularly and extracellularly, are heavily buffered. For example, the pH of the blood in healthy individuals is closely controlled at pH 7.4. The phosphate and carbonate ions that are components of most biological fluids are important in this respect because they have pK's in this range (Table 2-3). Moreover, many biological molecules, such as proteins, nucleic acids, and lipids, as well as numerous small organic molecules, bear multiple acid–base groups that are effective as buffer components in the physiological pH range.

The concept that the properties of biological molecules vary with the acidity of the solution in which they are dissolved was not fully appreciated before the beginning of the twentieth century so that the acidities of biochemical preparations made before that time were rarely controlled. Consequently these early biochemical experiments yielded poorly reproducible results. More recently, biochemical preparations have been routinely buffered to simulate the properties of naturally occurring biological fluids. Many of the weak acids listed in Table 2-3 are commonly used as buffers in biochemical preparations. In practice, the chosen weak acid and one of its soluble salts are dissolved in the (nearly equal) mole ratio necessary to provide the desired pH and, with the aid of a pH meter, the resulting solution is "fine tuned" by titration with strong acid or base.

C. Polyprotic Acids

Substances that bear more than one acid–base group, such as H_3PO_4 or H_2CO_3, as well as most biomolecules, are known as **polyprotic acids**. The titration curves of such substances, as is illustrated in Fig. 2-11 for H_3PO_4, are charac-

terized by multiple pK's, one for each ionization step. Exact calculations of the concentrations of the various ionic species present at a given pH is clearly a more complex task than for a **monoprotic acid.**

The pK's of two closely associated acid–base groups are not independent. The ionic charge resulting from a proton dissociation electrostatically inhibits further proton dissociation from the same molecule, thereby increasing the values of the corresponding pK's. This effect, according to Coulomb's law, decreases as the distance between the ionizing groups increases. For example, the pK's of **oxalic acid's** two adjacent carboxyl groups differ by 3 pH units (Table 2-3), whereas those of **succinic acid,** in which the carboxyl groups are separated by two methylene groups, differ by 1.4 units.

$$H-O-\overset{\overset{O}{\|}}{C}-\overset{\overset{O}{\|}}{C}-O-H \qquad H-O-\overset{\overset{O}{\|}}{C}-CH_2CH_2-\overset{\overset{O}{\|}}{C}-O-H$$

Oxalic acid   **Succinic acid**

Likewise, successive ionizations from the same center, such as in H_3PO_4 or H_2CO_3, have pK's that differ by 4 to 5 pH units. If the pK's for successive ionizations of a polyprotic acid differ by at least 3 pH units, it can be accurately as-

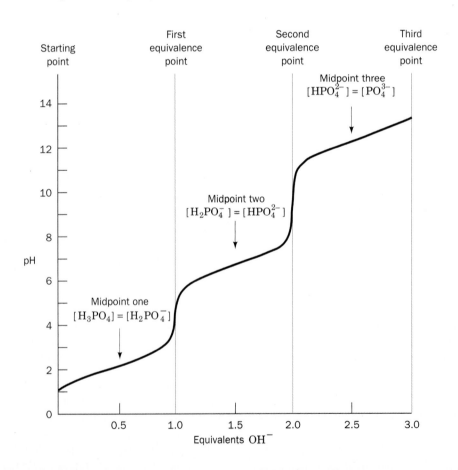

FIGURE 2-11. The titration curve of H_3PO_4. The two intermediate equivalence points occur at the steepest parts of the curve. Note the flatness of the curve near its starting points and endpoints in comparison with the curved ends of the titration curves in Fig. 2-9. This indicates that H_3PO_4 (pK_1 = 2.15) is verging on being a strong acid and PO_4^{3-} (pK_3 = 12.38) is verging on being a strong base.

sumed that, at a given pH, only the members of the conjugate acid–base pair characterized by the nearest pK are present in significant concentrations. This, of course, greatly simplifies the calculations for determining the concentrations of the various ionic species present.

Polyprotic Acids with Closely Spaced pK's Have Molecular Ionization Constants

If the pK's of a polyprotic acid differ by less than ~2 pH units, as is true in perhaps the majority of biomolecules, the ionization constants measured by titration are not true group ionization constants but rather, reflect the average ionization of the groups involved. The resulting ionization

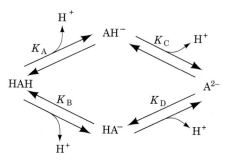

FIGURE 2-12. The ionization of an acid that has two nonequivalent protonation sites.

constants are therefore known as **molecular ionization constants.**

Consider the acid–base equilibria shown in Fig. 2-12 in which there are two nonequivalent protonation sites. Here, the quantities K_A, K_B, K_C, and K_D, the ionization constants for each group, are alternatively called **microscopic ionization constants.** The molecular ionization constant for the removal of the first proton from HAH is

$$K_1 = \frac{[H^+]([AH^-] + [HA^-])}{[HAH]} = K_A + K_B \quad [2.10]$$

Similarly, the molecular ionization constant K_2 for the removal of the second proton is

$$K_2 = \frac{[H^+][A^{2-}]}{[AH^-] + [HA^-]} = \frac{1}{(1/K_C) + (1/K_D)}$$
$$= \frac{K_C K_D}{(K_C + K_D)}$$

If $K_A \gg K_B$, then $K_1 \approx K_A$; that is, the first molecular ionization constant is equal to the microscopic ionization constant of the more acidic group. Likewise, if $K_D \gg K_C$, then $K_2 \approx K_C$, so that the second molecular ionization constant is the microscopic ionization constant of the less acidic group. If the ionization steps differ sufficiently in their pK's, the molecular ionization constants, as expected, become identical to the microscopic ionization constants.

CHAPTER SUMMARY

Water is an extraordinary substance, the properties of which are of great biological importance. A water molecule can simultaneously participate in as many as four hydrogen bonds: two as a donor and two as an acceptor. These hydrogen bonds are responsible for the open, low-density structure of ice. Much of this hydrogen bonded structure exists in the liquid phase, as is evidenced by the high boiling point of water compared to those of other liquids. Physical and theoretical evidence indicates that liquid water maintains a rapidly fluctuating, hydrogen bonded molecular structure that, over short ranges, resembles that of ice. The unique solvent properties of water derive from its polarity as well as its hydrogen bonding properties. In aqueous solutions, ionic and polar substances are surrounded by multiple concentric hydration shells of oriented water dipoles that act to attenuate the electrostatic interactions between the charges in the solution. The thermal randomization of the oriented water molecules is resisted by their hydrogen bonding associations, thereby accounting for the high dielectric constant of water. Nonpolar substances are essentially insoluble in water. However, amphipathic substances aggregate in aqueous solutions to form micelles and bilayers due to the combination of hydrophobic interactions among the nonpolar portions of these molecules and the hydrophilic interactions of their polar groups with the aqueous solvent. The H_3O^+ and OH^- ions have anoma-

lously large ionic mobilities in aqueous solutions because the migration of these ions through solution occurs largely via proton jumping from one H_2O molecule to another.

A Brønsted acid is a substance that can donate protons, whereas a Brønsted base can accept protons. Upon losing a proton, a Brønsted acid becomes its conjugate base. In an acid–base reaction, an acid donates its proton to a base. Water can react as an acid to form hydroxide ion, OH^-, or as a base to form hydronium ion, H_3O^+. The strength of an acid is indicated by the magnitude of its dissociation constant, K. Weak acids, which have a dissociation constant less than that of H_3O^+, are only partially dissociated in aqueous solution. Water has the dissociation constant $10^{-14}M$ at 25°C. A practical quantity for expressing the acidity of a solution is the pH = $-\log [H^+]$. The relationship between pH, pK, and the concentrations of the members of its conjugate acid–base pair is expressed by the Henderson–Hasselbalch equation. An acid–base buffer is a mixture of a weak acid with its conjugate base in a solution that has a pH near the pK of the acid. The ratio $[A^-]/[HA]$ in a buffer is not very sensitive to the addition of strong acids or bases so that the pH of a buffer is not greatly affected by these substances. Buffers are operationally effective only in the pH range of pK ± 1. Outside of this range, the pH of the solution changes rapidly with the addition of strong acid or base. Buffer capacity also depends on the total concentration of the conjugate acid–base pair. Biological fluids are generally buffered near neutrality. Many acids are polyprotic. However, unless the p*K*'s of their various ionizations differ by less than 2 or 3 pH units, pH calculations can effectively treat them as if they were a mixture of separate weak acids. For polyprotic acids with p*K*'s that differ by less than this amount, the observed molecular ionization constants are simply related to the microscopic ionization constants of the individual dissociating groups.

REFERENCES

Cooke, R. and Kuntz, I.D., The properties of water in biological systems, *Ann. Rev. Biophys. Bioeng.* 3, 95–126 (1974).

Edsall, J.T. and Wyman, J., *Biophysical Chemistry,* Vol. 1, Chapters 2, 8, and 9, Academic Press (1958). [Contains detailed treatments on the structure of water and on acid–base equilibria.]

Eisenberg, D. and Kauzman, W., *The Structure and Properties of Water,* Oxford University Press (1969). [A comprehensive monograph with a wealth of information.]

Franks, F., *Water,* The Royal Society of Chemistry (1993).

Montgomery, R. and Swenson, C.A., *Quantitative Problems in Biochemical Sciences* (2nd ed.), Chapters 7 and 8, Freeman (1978). [Contains additional problems.]

Segel, I.H., *Biochemical Calculations* (2nd ed.), Chapter 1, Wiley (1976). [An intermediate level discussion of acid–base equilibria with worked out problems.]

Stillinger, F.H., Water revisited, *Science* **209,** 451–457 (1980). [An outline of water structure on an elementary level.]

Tanford, C., *The Hydrophobic Effect: Formation of Micelles and Biological Membranes* (2nd ed.), Chapters 5 and 6, Wiley–Interscience (1980). [Discussion of the structure of water and of micelles.]

Westhof, E., *Water and Biological Macromolecules,* CRC Press (1993).

Zumdahl, S.S., *Chemistry* (3rd ed.), Chaps. 14 and 15, Heath (1993). [Discusses acid–base chemistry. Most other general chemistry textbooks contain similar information.]

PROBLEMS

1. Draw the hydrogen bonding pattern that water forms with acetamide (CH_3CONH_2) and pyridine (benzene with a CH group replaced by N).

2. Explain why the dielectric constants of the following pairs of liquids have the order given in Table 2-1: (a) carbon tetrachloride and chloroform; (b) ethanol and methanol; and (c) acetone and formamide.

3. "Inverted" micelles are made by dispersing amphipathic molecules in a nonpolar solvent, such as benzene, together with a small amount of water (counterions are also provided if the head groups are ionic). Draw the structure of an inverted micelle and describe the forces that stabilize it.

*4. Amphipathic molecules in aqueous solutions tend to concentrate at surfaces such as liquid–solid or liquid–gas interfaces. They are therefore said to be **surface-active molecules** or **surfactants**. Rationalize this behavior in terms of the properties of the amphiphiles and indicate the effect that surface-active molecules have on the surface tension of water (surface tension is a measure of the internal cohesion of a liquid as manifested by the force necessary to increase its surface area). Explain why surfactants such as soaps and detergents are effective in dispersing oily substances and oily dirt in aqueous solutions. Why do aqueous solutions of surfactants foam and why does the presence of oily substances reduce this foaming?

5. Indicate how hydrogen bonding forces and hydrophobic forces vary with the dielectric constant of the medium.

6. Using the data in Table 2-2, indicate the times it would take a K^+ and an H^+ ion to each move 1 cm in an electric field of 100 V·cm^{-1}.

7. Explain why the mobility of H^+ in ice is only about an order of magnitude less than that in liquid water, whereas the mobility of Na^+ in solid NaCl is zero.

8. Calculate the pH of: (a) 0.1*M* HCl; (b) 0.1*M* NaOH; (c) $3 \times 10^{-5}M$ HNO_3; (d) $5 \times 10^{-10}M$ $HClO_4$; and (e) $2 \times 10^{-8}M$ KOH.

9. The volume of a typical bacterial cell is on the order of 1.0 μm^3. At pH 7, how many hydrogen ions are contained inside a bacterial cell? A bacterial cell contains thousands of macromolecules, such as proteins and nucleic acids, that each bear multiple ionizable groups. What does your result indicate about the common notion that ionizable groups are continuously bathed with H^+ and OH^- ions?

10. Using the data in Table 2-3, calculate the concentrations of all molecular and ionic species and the pH in aqueous solutions that have the following formal compositions: (a) 0.01*M* acetic acid; (b) 0.25*M* ammonium chloride; (c) 0.05*M* acetic acid + 0.10*M* sodium acetate; and (d) 0.20*M* boric acid [$B(OH)_3$] + 0.05*M* sodium borate [$NaB(OH)_4$].

11. **Acid–base indicators** are weak acids that change color upon changing ionization states. When a small amount of an appropriately chosen indicator is added to a solution of an acid or base being titrated, the color change "indicates" the **endpoint** of the titration. **Phenolphthalein** is a commonly used acid–base indicator that, in aqueous solutions, changes from colorless to red-violet in the pH range between 8.2 and 10.0. Referring to Figs. 2-9 and 2-11, indicate the effectiveness of phenolphthalein for accurately detecting the endpoint of a titration with strong base of: (a) acetic acid; (b) NH_4Cl; and (c) H_3PO_4 (at each of its three equivalence points).

*12. The formal composition of an aqueous solution is $0.12M$ $K_2HPO_4 + 0.08M\ KH_2PO_4$. Using the data in Table 2-3, calculate the concentrations of all ionic and molecular species in the solution and the pH of the solution.

13. Distilled water in equilibrium with air contains dissolved carbon dioxide at a concentration of $1.0 \times 10^{-5}M$. Using the data in Table 2-3, calculate the pH of such a solution.

14. Calculate the formal concentrations of acetic acid and sodium acetate necessary to prepare a buffer solution of pH 5 that is $0.20M$ in total acetate. The pK of acetic acid is given in Table 2-3.

15. In order to purify a certain protein, you require $0.1M$ glycine buffer at pH 9.4. Unfortunately, your stockroom has run out of glycine. However, you manage to find two $0.1M$ glycine buffer solutions, one at pH 9.0 and the other at pH 10.0. What volumes of these two solutions must you mix in order to obtain 200 mL of your required buffer?

16. An enzymatic reaction takes place in a 10-mL solution that has a total citrate concentration of 120 mM and an initial pH of 7.00. During the reaction, 0.2 milliequivalents of acid are produced. Using the data in Table 2-3, calculate the final pH of the solution. What would the final pH of the solution be in the absence of the citrate buffer assuming that the other components of the solution have no significant buffering capacity and that the solution is initially at pH 7?

*17. A solution's **buffer capacity,** β, is defined as the ratio of an incremental amount of base added, in equivalents, to the corresponding pH change. This is the reciprocal of the slope of the titration curve, Eq. [2.9]. Derive the equation for β and show that it is maximal at pH = pK.

18. Using the data in Table 2-3, calculate the microscopic ionization constants for oxalic acid and for succinic acid. How do these values compare with their corresponding molecular ionization constants?

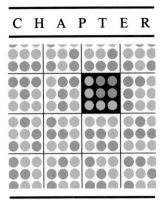

CHAPTER

3

Thermodynamic Principles: A Review

You can't win.
    First law of thermodynamics
You can't even break even.
    Second law of thermodynamics
You can't stay out of the game.
    Third law of thermodynamics

Living things require a continuous throughput of energy. For example, through photosynthesis, plants convert radiant energy from the sun, the primary energy source for life on Earth, to the chemical energy of carbohydrates and other organic substances. The plants, or the animals that eat them, then metabolize these substances to power such functions as the synthesis of biomolecules, the maintenance of concentration gradients, and the movement of muscles. These processes ultimately transform the energy to heat, which is dissipated to the environment. A considerable portion of the cellular biochemical apparatus must therefore be devoted to the acquisition and utilization of energy.

Thermodynamics (Greek: *therme,* heat + *dynamis,* power) is a marvelously elegant description of the relationships among the various forms of energy and how energy affects matter on the macroscopic as opposed to the molecular level; that is, it deals with amounts of matter large enough for their average properties, such as temperature and pressure, to be well defined. Indeed, the basic principles of thermodynamics were developed in the nineteenth century before the atomic theory of matter had been generally accepted.

With a knowledge of thermodynamics we can determine whether a physical process is possible. Thermodynamics is therefore essential for understanding why macromolecules fold to their native conformations, how metabolic pathways are designed, why molecules cross biological membranes, how muscles generate mechanical force, and so on. The list is endless. Yet, the reader should be cautioned that thermodynamics does not indicate the rates at which possible processes actually occur. For instance, although thermodynamics tells us that glucose and oxygen react with the release of copious amounts of energy, it does not indicate that this mixture is indefinitely stable at room temperature in the absence of the appropriate enzymes. The prediction of reaction rates requires, as we shall see in Section 13-1C, a mechanistic description of molecular processes. Yet, thermodynamics is also an indispensible guide in formulating such mechanistic models because such models must conform to thermodynamic principles.

Thermodynamics, as it applies to biochemistry, is most frequently concerned with describing the conditions under which processes occur *spontaneously* (by themselves). We shall consequently review the elements of thermodynamics that enable us to predict chemical and biochemical spontaneity: The first and second laws of thermodynamics, the concept of free energy, and the nature of processes at equilibrium. Familiarity with these principles is indispensible for understanding many of the succeeding discussions in this text. We shall, however, postpone consideration of the thermodynamic aspects of metabolism until Sections 15-4 through 15-6.

1. FIRST LAW OF THERMODYNAMICS: ENERGY IS CONSERVED

In thermodynamics, a **system** is defined as that part of the universe that is of interest, such as a reaction vessel or an organism; the rest of the universe is known as the **surroundings**. A system is said to be **open** or **closed** according to whether or not it can exchange matter and energy with its surroundings. Living organisms, which take up nutrients, release waste products, and generate work and heat, are examples of open systems; if an organism were sealed inside a perfectly insulated box, it would, together with the box, constitute a closed system.

A. Energy

The **first law of thermodynamics** is a mathematical statement of the law of conservation of energy: *Energy can be neither created nor destroyed.*

$$\Delta U = U_{final} - U_{initial} = q - w \qquad [3.1]$$

Here U is energy, q represents the **heat** absorbed *by* the system *from* the surroundings, and w is the **work** done *by* the system *on* the surroundings. Heat is a reflection of random molecular motion, whereas work, which is defined as force times the distance moved under its influence, is associated with organized motion. Force may assume many different forms, including the gravitational force exerted by one mass on another, the expansional force exerted by a gas, the tensional force exerted by a spring or muscle fiber, the electrical force of one charge on another, or the dissipa-

TABLE 3-1. THERMODYNAMIC UNITS AND CONSTANTS

Joule (J)
  $1 \text{ J} = 1 \text{ kg} \cdot \text{m}^2 \cdot \text{s}^{-2}$ $1 \text{ J} = 1 \text{ C} \cdot \text{V}$ (coulomb volt)
  $1 \text{ J} = 1 \text{ N} \cdot \text{m}$ (newton meter)
Calorie (cal)
  1 cal heats 1 g of H_2O from 14.5 to 15.5°C
  $1 \text{ cal} = 4.184 \text{ J}$
Large calorie (Cal)
  $1 \text{ Cal} = 1 \text{ kcal}$ $1 \text{ Cal} = 4184 \text{ J}$
Avogadro's number (N)
  $N = 6.0221 \times 10^{23}$ molecules $\cdot$ mol^{-1}
Coulomb (C)
  $1 \text{ C} = 6.241 \times 10^{18}$ electron charges
Faraday ($\mathscr{F}$)
  $1 \mathscr{F} = N$ electron charges
  $1 \mathscr{F} = 96,494 \text{ C} \cdot \text{mol}^{-1} = 96,494 \text{ J} \cdot \text{V}^{-1} \cdot \text{mol}^{-1}$
Kelvin temperature scale (K)
  0 K = absolute zero   273.15 K = 0°C
Boltzmann constant (k_B)
  $k_B = 1.3807 \times 10^{-23} \text{ J} \cdot \text{K}^{-1}$
Gas constant (R)
  $R = Nk_B$ $R = 1.9872 \text{ cal} \cdot \text{K}^{-1} \cdot \text{mol}^{-1}$
  $R = 8.3145 \text{ J} \cdot \text{K}^{-1} \cdot \text{mol}^{-1}$ $R = 0.08206 \text{ L} \cdot \text{atm} \cdot \text{K}^{-1} \cdot \text{mol}^{-1}$

tive forces of friction and viscosity. Processes in which the system releases heat, which by convention are assigned a negative q, are known as **exothermic processes** (Greek: *exo*, out of); those in which the system gains heat (positive q) are known as **endothermic processes** (Greek: *endon*, within). Under this convention, work done by the system against an external force is defined as a positive quantity.

The SI unit of energy, the **joule (J)**, is steadily replacing the **calorie (cal)** in modern scientific usage. The **large calorie (Cal**, with a capital C) is a unit favored by nutritionists. The relationships among these quantities and other units, as well as the values of constants that will be useful throughout this chapter, are collected in Table 3-1.

State Functions Are Independent of the Path a System Follows

Experiments have invariably demonstrated that the energy of a system depends only on its current properties or state, not on how it reached that state. For example, the state of a system composed of a particular gas sample is completely described by its pressure and temperature. The energy of this gas sample is a function only of these so-called **state functions** (quantities that depend only on the state of the system) and is therefore a state function itself. Consequently, there is no net change in energy ($\Delta U = 0$) for any process in which the system returns to its initial state (a **cyclic process**).

Neither heat nor work is separately a state function because each is dependent on the **path** followed by a system in changing from one state to another. For example, in the process of changing from an initial to a final state, a gas may do work by expanding against an external force, or do no work by following a path in which it encounters no external resistance. If Eq. [3.1] is to be obeyed, heat must also be path dependent. It is therefore meaningless to refer to the heat or work content of a system (in the same way that it is meaningless to refer to the number of one dollar bills and ten dollar bills in a bank account containing $85.00). To indicate this property, the heat or work produced during a change of state are never referred to as Δq or Δw but rather as just q or w.

B. Enthalpy

Any combination of only state functions must also be a state function. One such combination, which is known as the **enthalpy** (Greek: *enthalpein,* to warm in), is defined

$$H = U + PV \qquad [3.2]$$

where V is the volume of the system and P is its pressure. Enthalpy is a particularly convenient quantity with which to describe biological systems because *under constant pressure, a condition typical of most biochemical processes, the enthalpy change between the initial and final states of a process, ΔH, is the easily measured heat that it generates or absorbs.* To show this, let us divide work into two categories: pressure–volume (P–V) work, which is work performed by expansion against an external pressure ($P\Delta V$), and all other work (w'):

$$w = P\Delta V + w' \qquad [3.3]$$

Then, by combining Eqs. [3.2] and [3.3], we see that

$$\Delta H = \Delta U + P\Delta V = q_P - w + P\Delta V = q_P - w' \quad [3.4]$$

where q_P is the heat transferred at constant pressure. Thus if $w' = 0$, as is often true of chemical reactions, $\Delta H = q_P$. Moreover, the volume changes in most biochemical processes are negligible so that the differences between their ΔU and ΔH values are usually insignificant.

We are now in the position to understand the utility of state functions. For instance, suppose we wished to determine the enthalpy change resulting from the complete oxidation of 1 g of glucose to CO_2 and H_2O by muscle tissue. To make such a measurement directly would present enormous experimental difficulties. For one thing, the enthalpy changes resulting from the numerous metabolic reactions not involving glucose oxidation that normally occur in living muscle tissue would greatly interfere with our enthalpy measurement. Since enthalpy is a state function, however, we can measure glucose's enthalpy of combustion in any apparatus of our choosing, say, a constant pressure calorimeter rather than a muscle, and still obtain the same value. This, of course, is true whether or not we know the mechanism through which muscle converts glucose to CO_2 and H_2O, as long as we can establish that these substances actually are the final metabolic products. *In general, the change of enthalpy in any hypothetical reaction pathway can be determined from the enthalpy change in any other reaction pathway between the same reactants and products.*

We stated earlier in the chapter that thermodynamics serves to indicate whether a particular process occurs spontaneously. Yet, the first law of thermodynamics cannot, by itself, provide the basis for such an indication as the following example demonstrates. If two objects at different temperatures are brought into contact, we know that heat spontaneously flows from the hotter object to the colder one, never vice versa. Yet, either process is consistent with the first law of thermodynamics since the aggregate energy of the two objects is independent of their temperature distribution. Consequently, we must seek a criterion of spontaneity other than only conformity to the first law of thermodynamics.

2. SECOND LAW OF THERMODYNAMICS: THE UNIVERSE TENDS TOWARDS MAXIMUM DISORDER

When a swimmer falls into the water (a spontaneous process), the energy of the coherent motion of his body is converted to that of the chaotic thermal motion of the surrounding water molecules. The reverse process, the swimmer being ejected from still water by the sudden coherent motion of the surrounding water molecules, has never been witnessed even though such a phenomenon violates neither the first law of thermodynamics nor Newton's laws of motion. This is because *spontaneous processes are characterized by the conversion of order (in this case the coherent motion of the swimmers body) to chaos (here the random thermal motion of the water molecules).* The **second law of thermodynamics,** which expresses this phenomenon, therefore provides a criterion for determining whether a process is spontaneous. Note that thermodynamics says

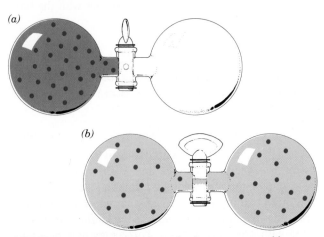

FIGURE 3-1. Two bulbs of equal volumes connected by a stopcock. In (*a*), a gas occupies the left bulb, the right bulb is evacuated, and the stopcock is closed. When the stopcock is opened (*b*), the gas molecules diffuse back and forth between the bulbs and eventually become distributed so that half of them occupy each bulb.

nothing about the rate of a process; that is the purview of **chemical kinetics** (Chapter 13). Thus a spontaneous process might proceed at only an infinitesimal rate.

A. Spontaneity and Disorder

The second law of thermodynamics states, in accordance with all experience, that *spontaneous processes occur in directions that increase the overall **disorder** of the universe*, that is, of the system and its surroundings. Disorder, in this context, is defined as the number of equivalent ways, W, of arranging the components of the universe. To illustrate this point, let us consider an isolated system consisting of two bulbs of equal volume containing a total of N identical molecules of ideal gas (Fig. 3-1). When the stopcock connecting the bulbs is open, there is an equal probability that a given molecule will occupy either bulb, so there are a total of 2^N equally probable ways that the N molecules may be distributed among the two bulbs. Since the gas molecules are indistinguishable from one another, there are only ($N +$ 1) different states of the system: those with 0, 1, 2, . . . , ($N - 1$), or N molecules in the left bulb. Probability theory indicates that the number of (indistinguishable) ways, W_L, of placing L of the N molecules in the left bulb is

$$W_L = \frac{N!}{L!(N-L)!}$$

The probability of such a state occurring is its fraction of the total number of possible states: $W_L/2^N$.

For any value of N, the state that is most probable, that is, the one with the highest value of W_L, is the one with half of the molecules in one bulb ($L = N/2$ for N even). As N becomes large, the probability that L is nearly equal to $N/2$ approaches unity: For instance, when $N = 10$ the probability that L is within 20% of $N/2$ (that is, 4, 5, or 6) is 0.66, whereas for $N = 50$ this probability (that L is in the range 20–30) is 0.88. For a chemically significant number of

molecules, say $N = 10^{23}$, the probability that the number of molecules in the left bulb differs from those in the right by as insignificant a ratio as 1 molecule in every 10 billion is 10^{-434}, which, for all intents and purposes, is zero. Therefore, the reason the number of molecules in each bulb of the system in Fig. 3-1b is always observed to be equal is not because of any law of motion; the energy of the system is the same for any arrangement of the molecules. *It is because the aggregate probability of all other states is so utterly insignificant* (Fig. 3-2). By the same token, the reason that our

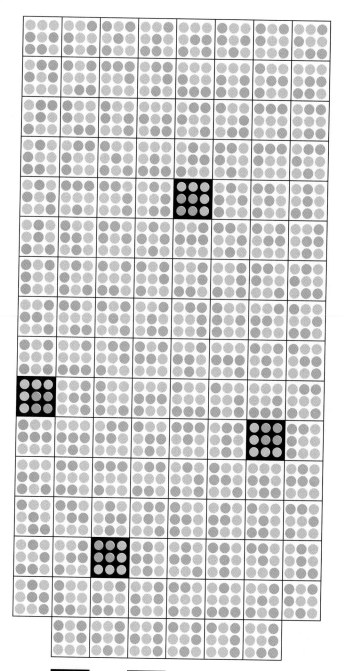

FIGURE 3-2. The improbability of even a small amount of order is illustrated by a simple "universe" consisting of a square array of 9 positions that collectively contain 4 identical "molecules" *(red dots)*. If the 4 molecules are arranged in a square, we shall call the arrangement a "crystal"; otherwise we shall call it a "gas." The total number of distinguishable arrangements of our 4 molecules in 9 positions is given by:

$$W = \frac{9 \cdot 8 \cdot 7 \cdot 6}{4 \cdot 3 \cdot 2 \cdot 1} = 126$$

Here, the numerator indicates that the first molecule may occupy any of the universe's 9 positions, the second molecule may occupy any of the 8 remaining unoccupied positions, and so on, whereas the denominator corrects for the number of indistinguishable arrangements of the 4 identical molecules. Of the 126 arrangements this universe can have, only 4 are crystals *(black squares)*. Thus, even in this simple universe, there is a more than 30-fold greater probability that it will contain a disordered gas, when arranged at random, than an ordered crystal. [Figure copyrighted© by Irving Geis.]

 A crystal   A gas

swimmer is never thrown out of the water or even noticeably disturbed by the chance coherent motion of the surrounding water molecules is that the probability of such an event is nil.

B. Entropy

In chemical systems, W, the number of equivalent ways of arranging a system in a particular state, is usually inconveniently immense. For example, when the above twin-bulb system contains N gas molecules, $W_{N/2} \approx 10^{N\ln 2}$ so that for $N = 10^{23}$, $W_{5 \times 10^{22}} \approx 10^{7 \times 10^{22}}$. In order to be able to deal with W more easily, we define, as did Ludwig Boltzmann in 1877, a quantity known as **entropy** (Greek: *en*, in + *trope*, turning):

$$S = k_B \ln W \qquad [3.5]$$

that increases with W but in a more manageable way. Here k_B is the **Boltzmann constant** (Table 3-1). For our twin-bulb system, $S = k_B N \ln 2$, so the entropy of the system in its most probable state is proportional to the number of gas molecules it contains. Note that *entropy is a state function because it depends only on the parameters that describe a state.*

The laws of random chance cause any system of reasonable size to spontaneously adopt its most probable arrangement, the one in which entropy is a maximum, simply because this state is so overwhelmingly probable. For example, assume that all N molecules of our twin-bulb system are initially placed in the left bulb (Fig. 3-1a; $W_N = 1$ and $S = 0$ since there is only one way of doing this). After the stopcock is opened, the molecules will randomly diffuse in and out of the right bulb until eventually they achieve their most probable (maximum entropy) state, that with half of the molecules in each bulb. The gas molecules will subsequently continue to diffuse back and forth between the bulbs but there will be no further macroscopic (net) change in the system. The system is therefore said to have reached **equilibrium.**

According to Eq. [3.5], the foregoing spontaneous expansion process causes the system's entropy to increase. In general, *for any constant energy process* ($\Delta U = 0$), *a spontaneous process is characterized by* $\Delta S > 0$. Since the energy of the universe is constant (energy can assume different forms but can be neither created nor destroyed), *any spontaneous process must cause the entropy of the universe to increase:*

$$\Delta S_{system} + \Delta S_{surroundings} = \Delta S_{universe} > 0 \qquad [3.6]$$

Equation [3.6] is the usual expression for the second law of thermodynamics. It is a statement of the general tendency of all spontaneous processes to disorder the universe; that is, *the entropy of the universe tends towards a maximum.*

The conclusions based on our twin-bulb apparatus may be applied to explain, for instance, why blood transports O_2 and CO_2 between the lungs and the tissues. Solutes in solution behave analogously to gases in that they tend to maintain a uniform concentration throughout their occupied volume because this is their most probable arrangement. In the lungs, where the concentration of O_2 is higher than that in the venous blood passing through them, more O_2 enters the blood than leaves it. On the other hand, in the tissues, where the O_2 concentration is lower than that in the arterial blood, there is net diffusion of O_2 from the blood to the tissues. The reverse situation holds for CO_2 transport since the CO_2 concentration is low in the lungs but high in the tissues. Keep in mind, however, that thermodynamics says nothing about the rates that O_2 and CO_2 are transported to and from the tissues. The rates of these processes depend on the physicochemical properties of the blood, the lungs, and the cardiovascular system.

Equation [3.6] does not imply that a particular system cannot increase its degree of order. As is explained in Section 3-3, however, *a system can only be ordered at the expense of disordering its surroundings to an even greater extent by the application of energy to the system.* For example, living organisms, which are organized from the molecular level upwards and are therefore particularly well ordered, achieve this order at the expense of disordering the nutrients they consume. Thus, *eating is as much a way of acquiring order as it is of gaining energy.*

A state of a system may constitute a distribution of more complicated quantities than those of gas molecules in a bulb or simple solute molecules in a solvent. For example, if our system consists of a protein molecule in aqueous solution, its various states differ, as we shall see, in the conformations of the protein's amino acid residues and in the distributions and orientations of its associated water molecules. The second law of thermodynamics applies here because a protein molecule in aqueous solution assumes its native conformation largely in response to the tendency of its surrounding water structure to be maximally disordered (Section 7-4C).

C. Measurement of Entropy

In chemical and biological systems, it is impractical, if not impossible, to determine the entropy of a system by counting the number of ways, W, it can assume its most probable state. An equivalent and more practical definition of entropy was proposed in 1864 by Rudolf Clausius: For spontaneous processes

$$\Delta S \geq \int_{initial}^{final} \frac{dq}{T} \qquad [3.7]$$

where T is the absolute temperature at which the change in heat occurs. The proof of the equivalence of our two definitions of entropy, which requires an elementary knowledge of statistical mechanics, can be found in many physical chemistry texts. It is evident, however, that any system becomes progressively disordered (its entropy increases) as its temperature rises (e.g., Fig. 3-3). The equality in Eq. [3.7]

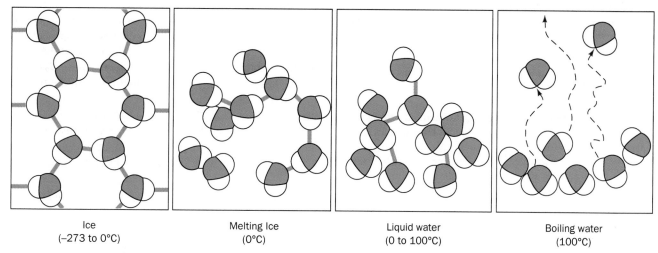

| Ice | Melting Ice | Liquid water | Boiling water |
| (−273 to 0°C) | (0°C) | (0 to 100°C) | (100°C) |

FIGURE 3-3. The structure of water, or any other substance, becomes increasingly disordered, that is, its entropy increases, as its temperature rises.

holds only for processes in which the system remains in equilibrium throughout the change; these are known as **reversible processes.**

For the constant temperature conditions typical of biological processes, Eq. [3.7] reduces to

$$\Delta S \geq \frac{q}{T} \qquad [3.8]$$

Thus the entropy change of a reversible process at constant temperature can be determined straightforwardly from measurements of the heat transferred and the temperature at which this occurs. However, since a process at equilibrium can only change at an infinitesimal rate (equilibrium processes are, by definition, unchanging), real processes can approach, but can never quite attain, reversibility. Consequently, *the universe's entropy change in any real process is always greater than its ideal (reversible) value.* This means that when a system departs from and then returns to its initial state via a real process, the entropy of the universe must increase even though the entropy of the system (a state function) does not change.

3. FREE ENERGY: THE INDICATOR OF SPONTANEITY

The disordering of the universe by spontaneous processes is an impractical criterion for spontaneity because it is rarely possible to monitor the entropy of the entire universe. Yet, the spontaneity of a process cannot be predicted from a knowledge of the system's entropy change alone. This is because exothermic processes ($\Delta H_{system} < 0$) may be spontaneous even though they are characterized by $\Delta S_{system} < 0$. For example, 2 mol of H_2 and 1 mol of O_2, when sparked, react in a decidedly exothermic reaction to form 2 mol of H_2O. Yet two water molecules, each of whose three atoms

are constrained to stay together, are more ordered than are the three diatomic molecules from which they formed. Similarly, under appropriate conditions, many **denatured** (unfolded) proteins will spontaneously fold to assume their highly ordered **native** (normally folded) conformations (Section 8-1A). What we really want, therefore, is a state function that predicts whether or not a given process is spontaneous. In this section, we consider such a function.

A. Gibbs Free Energy

The **Gibbs free energy:**

$$G = H - TS \qquad [3.9]$$

which was formulated by J. Willard Gibbs in 1878, is the required indicator of spontaneity for constant temperature and pressure processes. For systems that can only do pressure–volume work ($w' = 0$), combining Eqs. [3.4] and [3.9] while holding T and P constant yields

$$\Delta G = \Delta H - T\Delta S = q_P - T\Delta S \qquad [3.10]$$

But Eq. [3.8] indicates that $T\Delta S \geq q$ for spontaneous processes at constant T. Consequently, $\Delta G \leq 0$ *is the criterion of spontaneity we seek* for the constant T and P conditions that are typical of biochemical processes.

Spontaneous processes, that is, those with negative ΔG values, are said to be **exergonic** (Greek: *ergon*, work); they can be utilized to do work. Processes that are not spontaneous, those with positive ΔG values, are termed **endergonic;** they must be driven by the input of free energy (through mechanisms discussed in Section 3-4C). Processes at equilibrium, those in which the forward and backward reactions are exactly balanced, are characterized by $\Delta G = 0$. Note that the value of ΔG varies directly with temperature. This is why, for instance, the native structure of a protein, whose formation from its denatured form has both $\Delta H < 0$ and $\Delta S < 0$, predominates below the temperature at which

TABLE 3-2. THE VARIATION OF REACTION SPONTANEITY (SIGN OF ΔG) WITH THE SIGNS OF ΔH AND ΔS

ΔH	ΔS	$\Delta G = \Delta H - T\Delta S$
−	+	The reaction is both enthalpically favored (exothermic) and entropically favored. It is spontaneous (exergonic) at all temperatures.
−	−	The reaction is enthalpically favored but entropically opposed. It is spontaneous only at temperatures *below* $T = \Delta H/\Delta S$.
+	+	The reaction is enthalpically opposed (endothermic) but entropically favored. It is spontaneous only at temperatures *above* $T = \Delta H/\Delta S$.
+	−	The reaction is both enthalpically and entropically opposed. It is *un*spontaneous (endergonic) at all temperatures.

$\Delta H = T\Delta S$ (the **denaturation temperature**), whereas the denatured protein predominates above this temperature. The variation of the spontaneity of a process with the signs of ΔH and ΔS is summarized in Table 3-2.

B. Free Energy and Work

When a system at constant temperature and pressure does non-$P-V$ work, Eq. [3.10] must be expanded to

$$\Delta G = q_P - T\Delta S - w' \qquad [3.11]$$

or, because $T\Delta S \geq q_P$ (Eq. [3.8]),

$$\Delta G \leq -w'$$

so that

$$\Delta G \geq w' \qquad [3.12]$$

Since $P-V$ work is unimportant in biological systems, ΔG *for a biological process represents its maximum recoverable work*. The ΔG of a process is therefore indicative of the maximum charge separation it can establish, the maximum concentration gradient it can generate (Section 3-4A), the maximum muscular activity it can produce, and so on. In fact, for real processes, which can only approach reversibility, the inequality in Eq. [3.12] holds, so that *the work put into any system can never be fully recovered*. This is indicative of the inherent dissipative character of nature. Indeed, as we have seen, it is precisely this dissipative character that provides the overall driving force for any change.

It is important to reiterate that a large negative value of ΔG does not ensure a chemical reaction will proceed at a measurable rate. This depends on the detailed mechanism of the reaction, which is independent of the ΔG. For instance, most biological molecules, including proteins, nucleic acids, carbohydrates, and lipids, are thermodynamically unstable to hydrolysis but, nevertheless, spontaneously hydrolyze at biologically insignificant rates. Only with the introduction of the proper enzymes will the hydrolysis of these molecules proceed at a reasonable pace. Yet, a catalyst, which by definition is unchanged by a reaction, cannot affect the ΔG of a reaction. Consequently, *an enzyme can only accelerate the attainment of thermodynamic equilibrium; it cannot, for example, promote a reaction that has a positive ΔG.*

4. CHEMICAL EQUILIBRIA

The entropy (disorder) of a substance increases with its volume. For example, as we have seen for our twin-bulb apparatus (Fig. 3-1), a collection of gas molecules, in occupying all of the volume available to it, maximizes its entropy. Similarly, dissolved molecules become uniformly distributed throughout their solution volume. Entropy is therefore a function of concentration.

If entropy varies with concentration, so must free energy. Thus, as is shown in this section, the free energy change of a chemical reaction depends on the concentrations of both its reactants and its products. This phenomenon is of great biochemical significance because enzymatic reactions can operate in either direction depending on the relative concentrations of their reactants and products. Indeed, the directions of many enzymatically catalyzed reactions depend on the availability of their substrates and on the metabolic demand for their products (although most metabolic pathways operate unidirectionally; Section 15-6C).

A. Equilibrium Constants

The relationship between the concentration and the free energy of a substance A, which is derived in the appendix to this chapter, is approximately

$$\overline{G}_A - \overline{G}_A^\circ = RT \ln [A] \qquad [3.13]$$

where $\overline{G}_A$ is known equivalently as the **partial molar free energy** or the **chemical potential** of A (the bar indicates the quantity per mole), $\overline{G}_A^\circ$ is the partial molar free energy of A in its **standard state** (see Section 3-4B), R is the gas constant (Table 3-1), and [A] is the molar concentration of A. Thus for the general reaction,

$$a\text{A} + b\text{B} \rightleftharpoons c\text{C} + d\text{D}$$

since free energies are additive and the free energy change of a reaction is the sum of the free energies of the products less those of the reactants, the free energy change for this reaction is

$$\Delta G = c\overline{G}_C + d\overline{G}_D - a\overline{G}_A - b\overline{G}_B \qquad [3.14]$$

Substituting this relationship into Eq. [3.13] yields

$$\Delta G = \Delta G^\circ + RT \ln \left(\frac{[C]^c[D]^d}{[A]^a[B]^b} \right) \qquad [3.15]$$

where $\Delta G°$ is the free energy change of the reaction when all of its reactants and products are in their standard states. Thus the expression for the free energy change of a reaction consists of two parts: (1) a constant term whose value depends only on the reaction taking place, and (2) a variable term that depends on the concentrations of the reactants and the products, the stoichiometry of the reaction, and the temperature.

For a reaction at equilibrium, there is no *net* change because the free energy of the forward reaction exactly balances that of the backward reaction. Consequently, $\Delta G = 0$ so that Eq. [3.15] becomes

$$\Delta G° = -RT \ln K_{eq} \qquad [3.16]$$

where K_{eq} is the familiar **equilibrium constant** of the reaction:

$$K_{eq} = \frac{[C]_{eq}^c[D]_{eq}^d}{[A]_{eq}^a[B]_{eq}^b} = e^{-\Delta G°/RT} \qquad [3.17]$$

and the subscript "eq" in the concentration terms indicates their equilibrium values. (The equilibrium condition is usually clear from the context of the situation so that equilibrium concentrations are often expressed without this subscript.) *The equilibrium constant of a reaction may therefore be calculated from standard free energy data and vice versa.* Table 3-3 indicates the numerical relationship between $\Delta G°$ and K_{eq}. Note that a 10-fold variation of K_{eq} at $25°C$ corresponds to a 5.7-$kJ \cdot mol^{-1}$ change in $\Delta G°$, which is less than one half of the free energy of even a weak hydrogen bond.

Equations [3.15] through [3.17] indicate that when the reactants in a process are in excess of their equilibrium concentrations, the net reaction will proceed in the forward direction until the excess reactants have been converted to products and equilibrium is attained. Conversely, when products are in excess, the net reaction proceeds in the reverse reaction so as to convert products to reactants until the equilibrium concentration ratio is likewise achieved. Thus, as **Le Châtelier's principle** states, *any deviation from equilibrium stimulates a process that tends to restore the system to equilibrium. All closed systems must therefore inevitably reach equilibrium.* Living systems escape this thermodynamic *cul-de-sac* by being open systems (Section 15-6A).

The manner in which the equilibrium constant varies with temperature is seen by substituting Eq. [3.10] into Eq. [3.16] and rearranging:

$$\ln K_{eq} = \frac{-\Delta H°}{R}\left(\frac{1}{T}\right) + \frac{\Delta S°}{R} \qquad [3.18]$$

where $H°$ and $S°$ represent enthalpy and entropy in the standard state. If $\Delta H°$ and $\Delta S°$ are independent of temperature, as they often are to a reasonable approximation, a plot of $\ln K_{eq}$ versus $1/T$, known as a **van't Hoff plot,** yields a straight line of slope $-\Delta H°/R$ and intercept $\Delta S°/R$. This relationship permits the values of $\Delta H°$ and $\Delta S°$ to be deter-

TABLE 3-3. THE VARIATION OF K_{EQ} WITH $\Delta G°$ AT $25°C$.

K_{eq}	$\Delta G°$ $(kJ \cdot mol^{-1})$
10^6	-34.3
10^4	-22.8
10^2	-11.4
10^1	-5.7
10^0	0.0
10^{-1}	5.7
10^{-2}	11.4
10^{-4}	22.8
10^{-6}	34.3

mined from measurements of K_{eq} at two (or more) different temperatures. Calorimetric data, which until recently have been quite difficult to measure for biochemical processes, are therefore not required to obtain the values of $\Delta H°$ and $\Delta S°$. Consequently, most biochemical thermodynamic data have been obtained through the application of Eq. [3.18]. However, the recent development of the **scanning microcalorimeter** has made the direct measurement of ΔH (q_P) for biochemical processes a practical alternative. Indeed, a discrepancy between the values of $\Delta H°$ for a reaction as determined calorimetrically and from a van't Hoff plot suggests that the reaction occurs via one or more intermediate states in addition to the initial and final states implicit in the formulation of Eq. [3.18].

B. Standard Free Energy Changes

Since only free energy differences, ΔG, can be measured, not free energies themselves, it is necessary to refer these differences to some standard state in order to compare the free energies of different substances (likewise, we refer the elevations of geographic locations to sea level, which is arbitrarily assigned the height of zero). By convention, the free energy of all pure elements in their standard state: $25°C$, 1 atm, and in their most stable form (e.g., O_2 not O_3), is defined to be zero. The **free energy of formation** of any nonelemental substance, $\Delta G_f°$, is then defined as the change in free energy accompanying the formation of 1 mol of that substance, in its standard state, from its component elements in their standard states. The standard free energy change for any reaction can be calculated according to

$$\Delta G° = \sum \Delta G_f° \text{(products)} - \sum \Delta G_f° \text{(reactants)} \qquad [3.19]$$

Table 3-4 provides a list of standard free energies of formation, $\Delta G_f°$, for a selection of substances of biochemical significance.

Standard State Conventions in Biochemistry

The standard state convention commonly used in physical chemistry defines the standard state of a solute as that

TABLE 3-4. FREE ENERGIES OF FORMATION OF SOME COMPOUNDS OF BIOCHEMICAL INTEREST

Compound	$-\Delta G_f^\circ$ (kJ·mol^{-1})
Acetaldehyde	139.7
Acetate$^-$	369.2
Acetyl-CoA	374.1[a]
cis-Aconitate^{3-}	920.9
CO_2 (g)	394.4
CO_2 (aq)	386.2
HCO_3^-	587.1
Citrate^{3-}	1166.6
Dihydroxyacetone phosphate^{2-}	1293.2
Ethanol	181.5
Fructose	915.4
Fructose-6-phosphate^{2-}	1758.3
Fructose-1,6-biphosphate^{4-}	2600.8
Fumarate^{2-}	604.2
α-D-Glucose	917.2
Glucose-6-phosphate^{2-}	1760.2
Glyceraldehyde-3-phosphate^{2-}	1285.6
H^+	0.0
H_2 (g)	0.0
H_2O (l)	237.2
Isocitrate^{3-}	1160.0
α-Ketaglutarate^{2-}	798.0
Lactate$^-$	516.6
L-Malate^{2-}	845.1
OH^-	157.3
Oxaloacetate^{2-}	797.2
Phosphoenolpyruvate^{3-}	1269.5
2-Phosphoglycerate^{3-}	1285.6
3-Phosphoglycerate^{3-}	1515.7
Pyruvate$^-$	474.5
Succinate^{2-}	690.2
Succinyl-CoA	686.7[a]

[a] For formation from free elements + free CoA (coenzyme A).

Source: Metzler, D. E., *Biochemistry, The Chemical Reactions of Living Cells,* pp. 162–164, Academic Press (1977).

with unit **activity** at 25°C and 1 atm (activity is concentration corrected for nonideal behavior as is explained in the appendix to this chapter; for the dilute solutions typical of biochemical reactions, such corrections are small, so activities can be replaced by concentrations). However, because biochemical reactions usually occur in dilute aqueous solutions near neutral pH, a somewhat different standard state convention for biological systems has been adopted:

- Water's standard state is defined as that of the pure liquid, so that the activity of pure water is taken to be unity despite the fact that its concentration is 55.5*M*. In essence, the [H$_2$O] term is incorporated into the value of the equilibrium constant. This procedure simplifies the free energy expressions for reactions in dilute aqueous

solutions involving water as a reactant or product because the [H$_2$O] term can then be ignored.

- The hydrogen ion activity is defined as unity at the physiologically relevant pH of 7 rather than at the physical chemical standard state of pH 0, where many biological substances are unstable.

- The standard state of a substance that can undergo an acid–base reaction is defined in terms of the total concentration of its naturally occurring ion mixture at pH 7. In contrast, the physical chemistry convention refers to a pure species whether or not it actually exists at pH 0. The advantage of the biochemistry convention is that the total concentration of a substance with multiple ionization states, such as most biological molecules, is usually easier to measure than the concentration of one of its ionic species. Since the ionic composition of an acid or base varies with pH, however, the standard free energies calculated according to the biochemistry convention are valid only at pH 7.

Under the biochemistry convention, the standard free energy changes of substances are customarily symbolized by $\Delta G^{\circ\prime}$ in order to distinguish them from physical chemistry standard free energy changes, ΔG° (note that the value of ΔG for any process, being experimentally measurable, is independent of the chosen standard state; i.e., $\Delta G = \Delta G'$). Likewise, the biochemical equilibrium constant, which is defined by using $\Delta G^{\circ\prime}$ in place of ΔG° in Eq. [3.17], is represented by K_{eq}'.

The relationship between $\Delta G^{\circ\prime}$ and ΔG° is often a simple one. There are three general situations:

1. If the reacting species include neither H$_2$O nor H$^+$, the expressions for $\Delta G^{\circ\prime}$ and ΔG° coincide.

2. For a reaction in dilute aqueous solution that yields nH$_2$O molecules:

$$A + B \rightleftharpoons C + D + nH_2O$$

Equations [3.16] and [3.17] indicate that

$$\Delta G^\circ = -RT \ln K_{eq} = -RT \ln \left(\frac{[C][D][H_2O]^n}{[A][B]} \right)$$

Under the biochemistry convention, which defines the activity of pure water as unity,

$$\Delta G^{\circ\prime} = -RT \ln K_{eq}' = -RT \ln \left(\frac{[C][D]}{[A][B]} \right)$$

Therefore

$$\Delta G^{\circ\prime} = \Delta G^\circ + nRT \ln [H_2O] \qquad [3.20]$$

where [H$_2$O] = 55.5*M* (the concentration of water in aqueous solution), so that for a reaction at 25°C, which yields 1 mol of H$_2$O, $\Delta G^{\circ\prime} = \Delta G^\circ + 9.96$ kJ·mol^{-1}.

3. For a reaction involving hydrogen ions, such as

$$A + B \Longrightarrow C + HD$$
$$\Updownarrow K$$
$$D^- + H^+$$

where

$$K = \frac{[H^+][D^-]}{[HD]}$$

manipulations similar to those above lead to the relationship

$$\Delta G^{\circ\prime} = \Delta G^{\circ} - RT \ln (1 + [H^+]_0/K)$$
$$+ RT \ln [H^+]_0 \qquad [3.21]$$

where $[H^+]_0 = 10^{-7}M$, the only value of $[H^+]$ for which this equation is valid. Of course, if more than one ionizable species participates in the reaction and/or if any of them are polyprotic, Eq. [3.21] is correspondingly more complicated.

C. Coupled Reactions

The additivity of free energy changes allows an endergonic reaction to be driven by an exergonic reaction under the proper conditions. This phenomenon is the thermodynamic basis for the operation of metabolic pathways, since most of these reaction sequences are comprised of ender-

gonic as well as exergonic reactions. Consider the following two-step reaction process:

$$(1) \qquad A + B \Longrightarrow C + D \qquad \Delta G_1$$
$$(2) \qquad D + E \Longrightarrow F + G \qquad \Delta G_2$$

If $\Delta G_1 \geq 0$, Reaction (1) will not occur spontaneously. However, if ΔG_2 is sufficiently exergonic so that $\Delta G_1 + \Delta G_2 < 0$, then although the equilibrium concentration of D in Reaction (1) will be relatively small, it will be larger than that in Reaction (2). As Reaction (2) converts D to products, Reaction (1) will operate in the forward direction to replenish the equilibrium concentration of D. The highly exergonic Reaction (2) therefore *drives* the endergonic Reaction (1) and the two reactions are said to be **coupled** through their common intermediate, D. That these coupled reactions proceed spontaneously (although not necessarily at a finite rate) can also be seen by summing Reactions (1) and (2) to yield the overall reaction

$$(1 + 2) \qquad A + B + E \Longrightarrow C + F + G \qquad \Delta G_3$$

where $\Delta G_3 = \Delta G_1 + \Delta G_2 < 0$. *As long as the overall pathway (reaction sequence) is exergonic, it will operate in the forward direction.* Thus, the free energy of ATP hydrolysis, a highly exergonic process, is harnessed to drive many otherwise endergonic biological processes to completion (Section 15-4C).

APPENDIX: CONCENTRATION DEPENDENCE OF FREE ENERGY

To establish that the free energy of a substance is a function of its concentration, consider the free energy change of an ideal gas during a reversible pressure change at constant temperature ($w' = 0$, since an ideal gas is incapable of doing non-P–V work). Substituting Eqs. [3.1] and [3.2] into Eq. [3.9] and differentiating the result yields

$$dG = dq - dw + P\,dV + V\,dP - T\,dS \qquad [3.A1]$$

Upon substitution of the differentiated forms of Eqs. [3.3] and [3.8] in this expression, it reduces to

$$dG = V\,dP \qquad [3.A2]$$

The ideal gas equation is $PV = nRT$, where n is the number of moles of gas. Therefore

$$dG = nRT\frac{dP}{P} = nRT\,d\ln P \qquad [3.A3]$$

This gas phase result can be extended to the more biochemically relevant area of solution chemistry by application of

Henry's law for a solution containing the volatile solute A in equilibrium with the gas phase:

$$P_A = K_A X_A \qquad [3.A4]$$

Here P_A is the partial pressure of A when its mole fraction in the solution is X_A, and K_A is the **Henry's law constant** of A in the solvent being used. It is generally more convenient, however, to express the concentrations of the relatively dilute solutions of chemical and biological systems in terms of molarity rather than mole fractions. For a dilute solution

$$X_A \approx \frac{n_A}{n_{\text{solvent}}} = \frac{[A]}{[\text{solvent}]} \qquad [3.A5]$$

where the solvent concentration [solvent] is approximately constant. Thus

$$P_A \approx K'_A[A] \qquad [3.A6]$$

where $K'_A = K_A/[\text{solvent}]$. Substituting this expression into Eq. [3.A3] yields

$$dG_A = n_A RT\,d\,(\ln K'_A + \ln [A]) = n_A RT\,d\ln [A] \qquad [3.A7]$$

Free energy, as are energy and enthalpy, is a relative quantity that can only be defined with respect to some arbitrary standard state. The standard state is customarily taken to be $25\,°C$, 1-atm pressure, and, for the sake of mathematical simplicity, $[A] = 1$. The integration of Eq. [3.A7] from the standard state, $[A] = 1$, to the final state, $[A] = [A]$, results in

$$G_A - G_A^° = n_A RT \ln [A] \qquad [3.A8]$$

where $G_A^°$ is the free energy of A in the standard state and $[A]$ really represents the concentration ratio $[A]/1$. Since Henry's law is valid for real solutions only in the limit of infinite dilution, however, the standard state is defined as the entirely hypothetical state of $1\,M$ solute with the properties that it has at infinite dilution.

The free energy terms in Eq. [3.A8] may be converted from **extensive quantities** (those dependent on the amount of material) to **intensive quantities** (those independent of the amount of material) by dividing both sides of the equation by n_A. This yields

$$\overline{G}_A - \overline{G}_A^° = RT \ln [A] \qquad [3.A9]$$

Equation [3.A9] has the limitation that it refers to solutions that exactly follow Henry's law, although real solutions only do so in the limit of infinite dilution if the solute is, in fact, volatile. These difficulties can all be eliminated by replacing $[A]$ in Eq. [3.A9] by a quantity, a_A, known as the **activity** of A. This is defined

$$a_A = \gamma_A [A] \qquad [3.A10]$$

where γ_A is the **activity coefficient** of A. Equation [3.A9] thereby takes the form

$$\overline{G}_A - \overline{G}_A^° = RT \ln a_A \qquad [3.A11]$$

in which all departures from ideal behavior, including the provision that the system may perform non-P–V work, are incorporated in the activity coefficient, which is an experimentally measurable quantity. Ideal behavior is only approached at infinite dilution; that is, $\gamma_A \to 1$ as $[A] \to 0$. The standard state in Eq. [3.A11] is redefined as that of unit activity.

The concentrations of reactants and products in most biochemical reactions are usually so low (on the order of millimolar or less) that the activity coefficients of these various species are nearly unity. Consequently, the activities of most biochemical species under physiological conditions can be satisfactorily approximated by their molar concentrations:

$$\overline{G}_A - \overline{G}_A^° = RT \ln [A] \qquad [3.13]$$

CHAPTER SUMMARY

The first law of thermodynamics,

$$U = q - w \qquad [3.1]$$

where q is heat and w is work, is a statement of the law of conservation of energy. Energy is a state function because the energy of a system depends only on the state of the system. Enthalpy,

$$H = U + PV \qquad [3.2]$$

where P is pressure and V is volume, is a closely related state function that represents the heat at constant pressure under conditions where only pressure–volume work is possible. Entropy, which is also a state function, is defined

$$S = k_B \ln W \qquad [3.5]$$

where W, the disorder, is the number of equivalent ways the system can be arranged under the conditions governing it and k_B is the Boltzmann constant. The second law of thermodynamics states that the universe tends towards maximum disorder and hence $\Delta S_{universe} > 0$ for any real process. The Gibbs free energy of a system

$$G = H - TS \qquad [3.9]$$

decreases in a spontaneous, constant pressure process. In a process at equilibrium, the system suffers no net change so that $\Delta G = 0$. An ideal process, in which the system is always at equilibrium, is said to be reversible. All real processes are irreversible since processes at equilibrium can only occur at an infinitesimal rate. For a chemical reaction

$$a A + b B \rightleftharpoons c C + d D$$

the change in the Gibbs free energy is expressed

$$\Delta G = \Delta G^° + RT \ln \left(\frac{[C]^c [D]^d}{[A]^a [B]^b} \right) \qquad [3.15]$$

where $\Delta G^°$, the standard free energy change, is the free energy change at $25\,°C$, 1-atm pressure, and unit activities of reactants and products. The biochemical standard state, $\Delta G^{°\prime}$, is similarly defined but in dilute aqueous solution at pH 7 in which the activities of water and H^+ are both defined as unity. At equilibrium

$$\Delta G^{°\prime} = -RT \ln K_{eq}' = -RT \ln \left(\frac{[C]_{eq}^c [D]_{eq}^d}{[A]_{eq}^a [B]_{eq}^b} \right) \qquad [3.17a]$$

where K_{eq}' is the equilibrium constant under the biochemical convention. An endergonic reaction ($\Delta G > 0$) may be driven by an exergonic reaction ($\Delta G < 0$) if they are coupled and if the overall reaction is exergonic.

REFERENCES

Atkins, P.W., *The Second Law,* Scientific American Books (1984). [An insightful but nonmathematical exposition of the second law of thermodynamics.]

Atkins, P.W., *Physical Chemistry* (5th ed.), Chapters 1–10, Freeman (1994). [Most physical chemistry texts treat thermodynamics in some detail.]

Brey, W.S., *Physical Chemistry and Its Biological Applications,* Chapters 3 and 4, Academic Press (1978).

Dickerson, R.E., *Molecular Thermodynamics,* Benjamin (1969).

Edsall, J.T. and Gutfreund, H., *Biothermodynamics,* Wiley (1983).

Eisenberg, D. and Crothers, D., *Physical Chemistry with Applications to Life Sciences,* Chapters 1–5, Benjamin (1979).

Nash, L.K., *CHEMTHERMO: A Statistical Approach to Classical Chemical Thermodynamics,* Addison–Wesley (1971). [A delightfully written text on an elementary level.]

Segel, I.H., *Biochemical Calculations* (2nd ed.), Chapter 3, Wiley (1976). [Contains instructive problems accompanied by detailed solutions.]

Tinoco, I., Jr., Sauer, K., and Wang, J.C., *Physical Chemistry. Principles and Applications in Biological Sciences* (2nd ed.), Chapters 2–5, Prentice–Hall (1985).

van Holde, K.E., *Physical Biochemistry* (2nd ed.), Chapters 1–3, Prentice–Hall (1985). [The equivalence of the Boltzmann and Clausius formulations of the second law of thermodynamics is demonstrated on *pp.* 13–14.]

Wood, W.B., Wilson, J.H., Benbow, R.M., and Hood, L.E., *Biochemistry, A Problems Approach* (2nd ed.), Chapter 9, Benjamin/Cummings (1981). [A question and answer book.]

PROBLEMS

1. A common funeral litany is the Biblical verse: "Ashes to ashes, dust to dust." Why might a bereaved family of thermodynamicists be equally comforted by a recitation of the second law of thermodynamics?

2. How many flights of 4-m high stairs must an overweight person weighing 75 kg climb to atone for the indescretion of eating a 500-Cal hamburger? Assume that there is a 20% efficiency in converting nutritional energy to mechanical energy. The gravitational force of an object of mass m kg is $F = mg$, where the gravitational constant $g = 9.8$ m·s^{-2}.

3. In terms of thermodynamic concepts, why is it more difficult to park a car in a small space than it is to drive it out from such a space?

4. It has been said that an army of dedicated monkeys, typing at random, would eventually produce all of Shakespeare's works. How long, on average, would it take 1 million monkeys, each typing on a 46-key typewriter (space included but no shift key) at the rate of 1 key stroke per second, to type the phrase "to be or not to be"? How long, on average, would it take one monkey to do so at a computer if the computer would only accept the correct letter in the phrase and then would shift to its next letter (i.e., the computer knew what it wanted). What do these results indicate about the probability of order randomly arising from disorder vs order arising through a process of evolution?

5. Show that the transfer of heat from an object of higher temperature to one of lower temperature, but not the reverse process, obeys the second law of thermodynamics.

6. Carbon monoxide crystallizes with its CO molecules arranged in parallel rows. Since CO is a very nearly ellipsoidal molecule, in the absence of polarity effects, adjacent CO molecules could equally well line up in a head-to-tail or a head-to-head fashion. In a crystal consisting of 10^{23} CO molecules, what is the entropy of all the CO molecules being aligned head to tail?

7. The U.S. Patent Office has received, and continues to receive, numerous applications for perpetual motion machines. Perpetual motion machines have been classified as those of the first kind, which violate the first law of thermodynamics, and those of the second kind, which violate the second law of thermodynamics. The fallacy in a perpetual motion machine of the first kind is generally easy to detect. An example would be a motor-driven electrical generator that produces energy in excess of that input by the motor. The fallacy in a perpetual motion machine of the second type, however, is usually more subtle. Take, for example, a ship that uses heat energy extracted from the sea by a heat pump to boil water so as to power a steam engine that drives the ship as well as the heat pump. Show, in general terms, that such a propulsion system would violate the second law of thermodynamics.

8. Using the data in Table 3-4, calculate the values of $\Delta G°$ at 25°C for the following metabolic reactions:

   (a)　　$C_6H_{12}O_6 + 6\,O_2 \rightleftharpoons 6CO_2\,(aq) + 6H_2O\,(l)$
   　　　Glucose

   (b)　　$C_6H_{12}O_6 \rightleftharpoons 2(CH_3CH_2OH) + 2CO_2\,(aq)$
   　　　Glucose　　　　　**Ethanol**

   (c)　　$C_6H_{12}O_6 \rightleftharpoons 2(CH_3CHOCHOO^-) + 2H^+$
   　　　Glucose　　　　**Lactate**

[These reactions, respectively, constitute oxidative metabolism, alcoholic fermentation in yeast deprived of oxygen, and homolactic fermentation in skeletal muscle requiring energy faster than oxidative metabolism can supply it (Section 16-1B).]

*9. The native and denatured forms of a protein are generally in equilibrium as follows:

$$\text{Protein } \textit{(denatured)} \rightleftharpoons \text{protein } \textit{(native)}$$

For a certain solution of the protein **ribonuclease A,** in which the total protein concentration is $2.0 \times 10^{-3} M$, the concentrations of the denatured and native proteins at both 50 and 100°C are given in the following table:

Temperature (°C)	[Ribonuclease (denatured)] (M)	[Ribonuclease A (native)] (M)
50	5.1×10^{-6}	2.0×10^{-3}
100	2.8×10^{-4}	1.7×10^{-3}

(a) Determine $\Delta H°$ and $\Delta S°$ for the folding reaction assuming that these quantities are independent of temperature. (b) Calculate $\Delta G°$ for ribonuclease A folding at 25°C. Is this process spontaneous under standard state conditions at this temperature? (c) What is the denaturation temperature of ribonuclease A under standard state conditions?

*10. Using the data in Table 3-4, calculate $\Delta G_f°{}'$ for the following compounds at 25°C: (a) H_2O (*l*); (b) sucrose (sucrose + $H_2O \rightleftharpoons$ glucose + fructose: $\Delta G°{}' = -29.3$ kJ·mol^{-1}); and (c) ethyl acetate (ethyl acetate + $H_2O \rightleftharpoons$ ethanol + acetate$^-$ + H^+: $\Delta G°{}' = -19.7$ kJ·mol^{-1}; the pK of acetic acid is 4.76).

11. Calculate the equilibrium constants for the hydrolysis of the following compounds at pH 7 and 25°C: (a) phosphoenolpyruvate ($\Delta G°{}' = -61.9$ kJ·mol^{-1}); (b) pyrophosphate ($\Delta G°{}' = -33.5$ kJ·mol^{-1}); and (c) glucose-1-phosphate ($\Delta G°{}' = -20.9$ kJ·mol^{-1}).

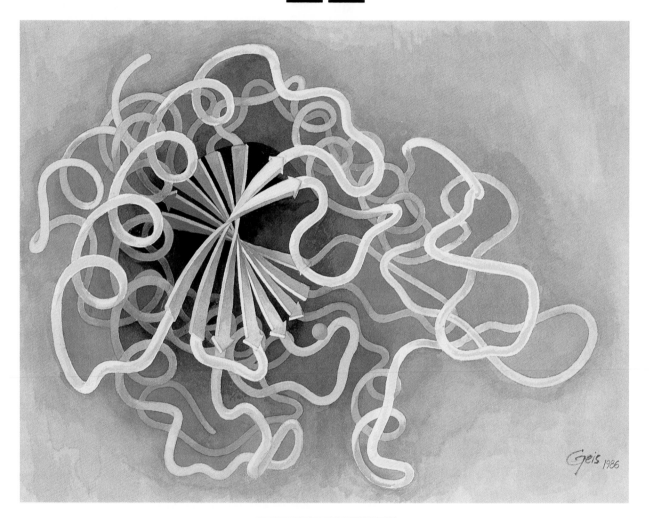

BIOMOLECULES

*The digestive enzyme bovine carboxypeptidase A
showing its central β sheet.*

$$\begin{array}{c} \text{COO}^- \\ | \\ \text{H}-\text{C}-\text{CH}_2- \\ | \\ \text{NH}_3^+ \end{array}$$

$$\begin{array}{c} \text{COO}^- \\ | \\ \text{H}-\text{C}-\text{CH}_2-\text{C} \\ | \\ \text{NH}_3^+ \end{array}$$

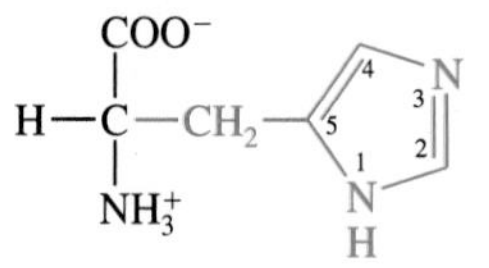

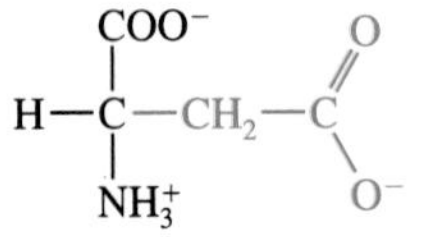

4

Amino Acids

1. Amino Acids of Proteins
    A. General Properties
    B. Peptide Bonds
    C. Classification and Characteristics
    D. Acid–Base Properties
    E. A Few Words on Nomenclature

2. Optical Activity
    A. An Operational Classification
    B. The Fischer Convention
    C. The Cahn–Ingold–Prelog System
    D. Chirality and Biochemistry

3. "Nonstandard" Amino Acids
    A. Amino Acid Derivatives in Proteins
    B. Specialized Roles of Amino Acids

It is hardly surprising that much of the early biochemical research was concerned with the study of proteins. Proteins form the class of biological macromolecules that have the most well-defined physicochemical properties and consequently they were generally easier to isolate and characterize than nucleic acids, polysaccharides, or lipids. Furthermore, proteins, particularly in the form of enzymes, have obvious biochemical functions. The central role that proteins play in biological processes has therefore been recognized since the earliest days of biochemistry. In contrast, the task of nucleic acids in the transmission and expression of genetic information was not realized until the late 1940s, the role of lipids in biological membranes was not appreciated until the 1960s, and the biological functions of polysaccharides are still somewhat mysterious.

In this chapter we study the properties of the monomeric units of proteins, the **amino acids.** It is from these substances that proteins are synthesized through processes that are discussed in Chapter 30. Amino acids are also energy metabolites and many of them are essential nutrients (Chapter 24). In addition, as we shall see, many amino acids and their derivatives are of biochemical importance in their own right (Section 4-3B).

1. THE AMINO ACIDS OF PROTEINS

The analyses of a vast number of proteins from almost every conceivable source have shown that *all proteins are composed of the 20 "standard" amino acids listed in Table 4-1.* These substances are known as **α-amino acids** because, with the exception of **proline,** they have a primary amino group and a carboxylic acid group substituent on the same carbon atom (Fig. 4-1; proline has a secondary amino group.)

A. General Properties

The pK values of the 20 "standard" α-amino acids of proteins are tabulated in Table 4-1. Here pK_1 and pK_2, respectively, refer to the α-carboxylic acid and α-amino groups, and pK_R refers to the side groups with acid–base properties. Table 4-1 indicates that the pK values of the α-carboxylic acid groups lie in a small range around 2.2 so that above pH 3.5 these groups are almost entirely in their carboxylate forms. The α-amino groups all have pK values near 9.4 and

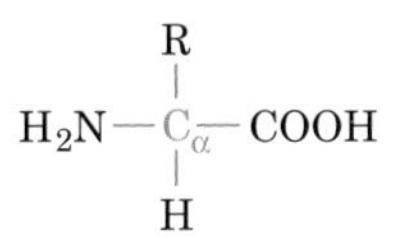

$$H_2N - \overset{\overset{\displaystyle R}{|}}{\underset{\underset{\displaystyle H}{|}}{C_\alpha}} - COOH$$

FIGURE 4-1. The general structural formula for α-amino acids. There are 20 different R groups in the commonly occurring amino acids (Table 4-1).

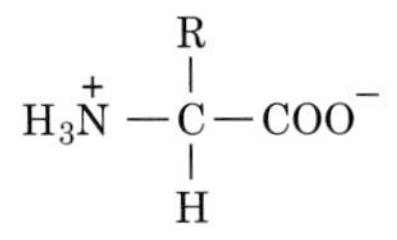

$$H_3\overset{+}{N} - \overset{\overset{\displaystyle R}{|}}{\underset{\underset{\displaystyle H}{|}}{C}} - COO^-$$

FIGURE 4-2. The zwitterionic form of the α-amino acids that occurs at physiological pH values.

are therefore almost entirely in their ammonium ion forms below pH 8.0. This leads to an important structural point: *In the physiological pH range, both the carboxylic acid and the amino groups of α-amino acids are completely ionized (Fig. 4-2).* An amino acid can therefore act as either an acid or a base. Substances with this property are said to be **amphoteric** and are referred to as **ampholytes** (*ampho*teric electro*lytes*). In Section 4-1D, we shall delve a bit deeper into the acid–base properties of the amino acids.

Molecules that bear charged groups of opposite polarity are known as **zwitterions** or **dipolar ions.** The zwitterionic character of the α-amino acids has been established by several methods including spectroscopic measurements and X-ray crystal structure determinations (in the solid state the α-amino acids are zwitterionic because the basic amine group abstracts a proton from the nearby acidic carboxylic acid group). Because amino acids are zwitterions, their physical properties are characteristic of ionic compounds. For instance, most α-amino acids have melting points near 300°C, whereas their nonionic derivatives usually melt around 100°C. Furthermore, amino acids, like other ionic compounds, are more soluble in polar solvents than in nonpolar solvents. Indeed, most α-amino acids are very soluble in water but are largely insoluble in most organic solvents.

B. Peptide Bonds

The α-amino acids polymerize, at least conceptually, through the elimination of a water molecule as is indicated in Fig. 4-3. The resulting CO—NH linkage is known as a **peptide bond.** Polymers composed of two, three, a few (3–10), and many **amino acid residues** (alternatively called **peptide units**) are known, respectively, as **dipeptides, tripeptides, oligopeptides,** and **polypeptides.** These substances, however, are often referred to simply as "peptides."

Proteins are molecules that consist of one or more polypeptide chains. These polypeptides range in length from ~40 to over 4000 amino acid residues and, since the average mass of an amino acid residue is ~110 D, have molecular masses that range from ~4 to over 440 kD.

*Polypeptides are **linear polymers;*** that is, each amino acid residue is linked to its neighbors in a head-to-tail fashion rather than forming branched chains. This observation reflects the underlying elegant simplicity of the way living systems construct these macromolecules for, as we shall see, the nucleic acids that encode the amino acid sequences of polypeptides are also linear polymers. This permits the direct correspondence between the monomer (nucleotide) sequence of a nucleic acid and the monomer (amino acid) sequence of the corresponding polypeptide without the added complication of specifying the positions and sequences of any branching chains.

With 20 different choices available for each amino acid residue in a polypeptide chain, it is easy to see that a huge number of different protein molecules can exist. For example, for dipeptides, each of the 20 different choices for the first amino acid residue can have 20 different choices for the second amino acid residue, for a total of $20^2 = 400$ distinct dipeptides. Similarly, for tripeptides, there are 20 possibilities for each of the 400 choices of dipeptides to yield a total of $20^3 = 8000$ different tripeptides. A relatively small protein molecule consists of a single polypeptide chain of 100 residues. There are $20^{100} = 1.27 \times 10^{130}$ possible unique polypeptide chains of this length, a quantity vastly greater than the estimated number of atoms in the universe (9×10^{78}). Clearly, nature can have made only a tiny fraction of the possible different protein molecules. Nevertheless, *the various organisms on Earth collectively synthesize an enormous number of different protein molecules whose great range of physicochemical characteristics stem largely from the varied properties of the 20 "standard" amino acids.*

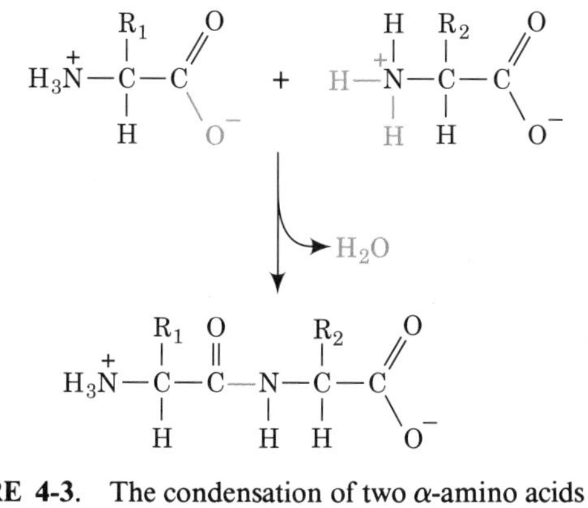

FIGURE 4-3. The condensation of two α-amino acids to form a dipeptide. The peptide bond is shown in red.

TABLE 4-1. Covalent Structures and Abbreviations of the "Standard" Amino Acids of Proteins, their Occurance, and the pK Values of their Ionizing Groups

Name, Three-letter Symbol, and One-letter Symbol	Structural Formula[a]	Residue Mass (D)[b]	Average Occurence in Proteins (%)[c]	pK_1 α-COOH[d]	pK_2 α-NH$_3^+$ [d]	pK_R Side chain[d]
Amino acids with nonpolar side chains						
Glycine Gly G		57.0	7.2	2.35	9.78	
Alanine Ala A		71.1	7.8	2.35	9.87	
Valine Val V		99.1	6.6	2.29	9.74	
Leucine Leu L		113.2	9.1	2.33	9.74	
Isoleucine Ile I		113.2	5.3	2.32	9.76	
Methionine Met M		131.2	2.2	2.13	9.28	
Proline Pro P		97.1	5.2	1.95	10.64	
Phenylalanine Phe F		147.2	3.9	2.20	9.31	
Tryptophan Trp W		186.2	1.4	2.46	9.41	

[a] The ionic forms shown are those predominating at pH 7.0 although residue mass is given for the neutral compound. The C$_\alpha$ atoms, as well as those atoms marked with an asterisk, are chiral centers with configurations as indicated according to Fischer projection formulas. The standard organic numbering system is provided for heterocycles.

[b] The residue masses are given for the neutral residues. For the molecular masses of the parent amino acids, add 18.0 D, the molecular mass of H$_2$O, to the residue masses. For side chain masses, subtract 56.0 D, the formula mass of a peptide group, from the residue masses.

[c] Calculated from a database of nonredundant proteins containing 300,688 residues as compiled by Doolittle, R. F. *in* Fasman, G, D. (Ed.), *Predictions of Protein Structure and the Principles of Protein Conformation,* Plenum Press (1989).

[d] *Source:* Dawson, R.M.C., Elliott, D.C., Elliott, W.H. and Jones, K.M., *Data for Biochemical Research* (3rd ed.), *pp.* 1–31, Oxford Science Publications (1986).

[e] The three- and one-letter symbols for asparagine *or* aspartic acid are Asx and B, whereas for glutamine *or* glutamic acid they are Glx and Z. The one-letter symbol for an undetermined or "nonstandard" amino acid is X.

Name, Three-letter Symbol, and One-letter Symbol	Structural Formula[a]	Residue Mass (D)[b]	Average Occurence in Proteins (%)[c]	pK_1 α-COOH[d]	pK_2 α-NH$_3^+$ [d]	pK_R Side chain[d]
Amino acids with uncharged polar side chains						
Serine Ser S	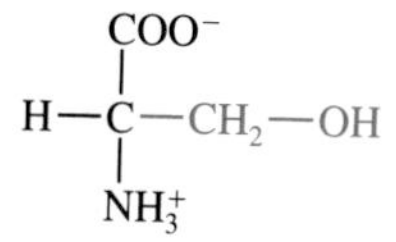	87.1	6.8	2.19	9.21	
Threonine Thr T	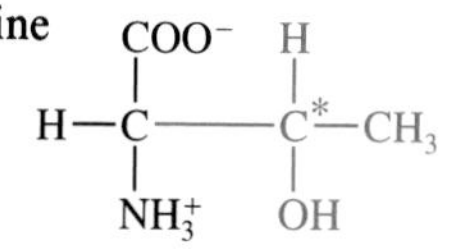	101.1	5.9	2.09	9.10	
Asparagine[e] Asn N	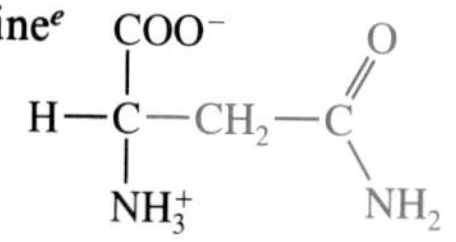	114.1	4.3	2.14	8.72	
Glutamine[e] Gln Q	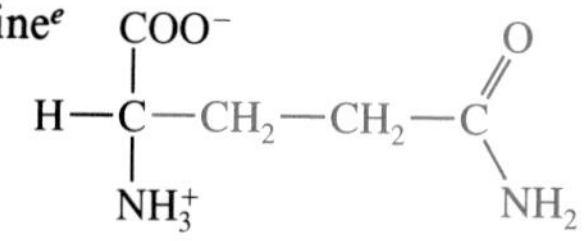	128.1	4.3	2.17	9.13	
Tyrosine Tyr Y	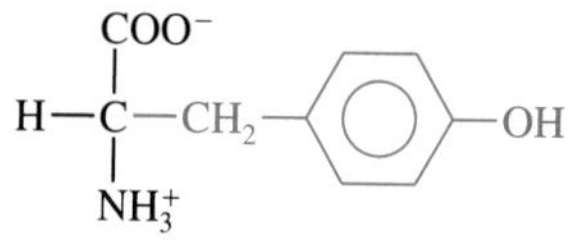	163.2	3.2	2.20	9.21	10.46 (phenol)
Cysteine Cys C	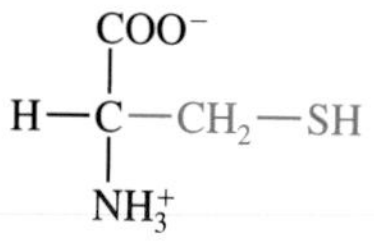	103.1	1.9	1.92	10.70	8.37 (sulfhydryl)
Amino acids with charged polar side chains						
Lysine Lys K	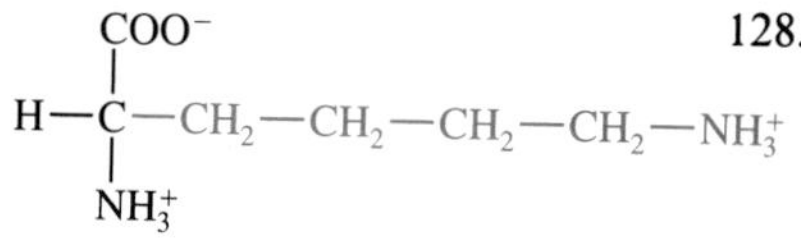	128.2	5.9	2.16	9.06	10.54 (ε-NH$_3^+$)
Arginine Arg R	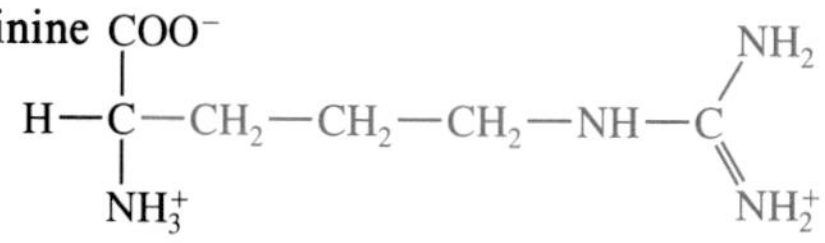	156.2	5.1	1.82	8.99	12.48 (guanidino)
Histidine His H	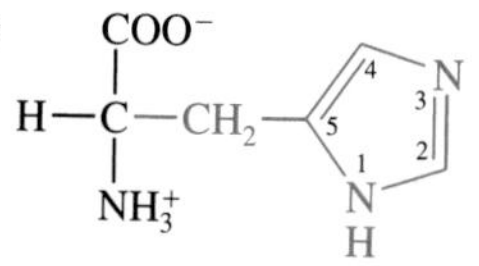	137.1	2.3	1.80	9.33	6.04 (imidazole)
Aspartic acid[e] Asp D	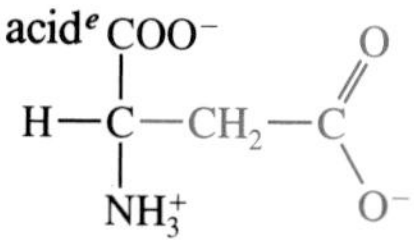	115.1	5.3	1.99	9.90	3.90 (β-COOH)
Glutamic acid[e] Glu E	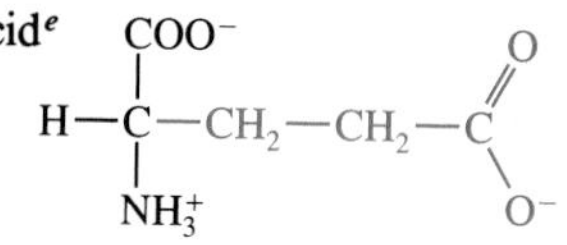	129.1	6.3	2.10	9.47	4.07 (γ-COOH)

C. Classification and Characteristics

The most common and perhaps the most useful way of classifying the 20 "standard" amino acids is according to the polarities of their side chains (**R groups**). This is because proteins fold to their native conformations largely in response to the tendency to remove their hydrophobic side chains from contact with water and to solvate their hydrophilic side chains (Chapters 7 and 8). According to this classification scheme, there are three major types of amino acids: (1) those with nonpolar R groups, (2) those with uncharged polar R groups, and (3) those with charged polar R groups.

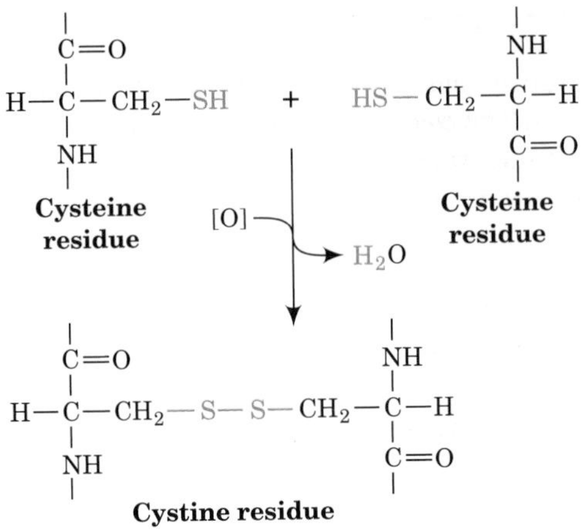

The Nonpolar Amino Acid Side Chains Have a Variety of Shapes and Sizes

Nine amino acids are classified as having nonpolar side chains. **Glycine** (which, when it was found to be a component of gelatin in 1820, was the first amino acid to be identified in protein hydrolyzates) has the smallest possible side chain, an H atom. **Alanine, valine, leucine,** and **isoleucine** have aliphatic hydrocarbon side chains ranging in size from a methyl group for alanine to isomeric butyl groups for leucine and isoleucine. **Methionine** has a thiol ether side chain that resembles an *n*-butyl group in many of its physical properties (C and S have nearly equal electronegativities and S is about the size of a methylene group). **Proline,** a cyclic secondary amino acid, has conformational constraints imposed by the cyclic nature of its pyrrolidine side group, which is unique among the "standard" 20 amino acids. **Phenylalanine,** with its phenyl moiety, and **tryptophan** with its indole group, contain aromatic side groups, which are characterized by bulk as well as nonpolarity.

Uncharged Polar Side Chains Have Hydroxyl, Amide, or Thiol Groups

Six amino acids are commonly classified as having uncharged polar side chains. **Serine** and **threonine** bear hydroxylic R groups of different sizes. **Asparagine** and **glutamine** have amide-bearing side chains of different sizes. **Tyrosine** has a phenolic group, which, together with the aromatic groups of phenylalanine and tryptophan, accounts for most of the UV absorbance and fluoresence exhibited by proteins. **Cysteine** has a thiol group that is unique among the 20 amino acids in that it often forms a disulfide bond to another cysteine residue (Fig. 4-4) through the oxidation of their thiol groups. This dimeric compound is referred to in the older biochemical literature as the amino acid **cystine.** The disulfide bond has great importance in protein structure: *It can join separate polypeptide chains or cross-link two cysteines in the same chain.* The confusing similarity between the names cysteine and cystine has led to the former occasionally being referred to as a **half-cystine** residue. However, the realization that cystine arises through the cross-linking of two cysteine residues

FIGURE 4-4.   The cystine residue consists of two disulfide-linked cysteine residues.

after polypeptide biosynthesis has occurred has caused the name cystine to become less commonly used.

Charged Polar Side Chains May Be Positively or Negatively Charged

Five amino acids have charged side chains. The basic amino acids are positively charged at physiological pH values; they are **lysine,** which has a butylammonium side chain, **arginine,** which bears a guanidino group, and **histidine,** which carries an imidazolium moiety. Of the 20 α-amino acids, only histidine, with $pK_R = 6.0$, ionizes within the physiological pH range. At pH 6.0, its imidazole side group is only 50% charged so that histidine is neutral at the basic end of the physiological pH range. As a consequence, histidine side chains often participate in the catalytic reactions of enzymes. The acidic amino acids, **aspartic acid** and **glutamic acid,** are negatively charged above pH 3; in their ionized state, they are often referred to as **aspartate** and **glutamate.** Asparagine and glutamine are, respectively, the amides of aspartic acid and glutamic acid.

The allocation of the 20 amino acids among the three different groups is, of course, rather arbitrary. For example, glycine and alanine, the smallest of the amino acids, and tryptophan, with its heterocyclic ring, might just as well be classified as uncharged polar amino acids. Similarly, tyrosine and cysteine, with their ionizable side chains, might also be thought of as charged polar amino acids, particularly at higher pH values, whereas asparagine and glutamine are nearly as polar as their corresponding carboxylates, aspartate and glutamate.

The 20 amino acids vary considerably in their physicochemical properties such as polarity, acidity, basicity, aromaticity, bulk, conformational flexibility, ability to cross-link, ability to hydrogen bond, and chemical reactivity. These several characteristics, many of which are interrelated, are largely responsible for proteins' great range of properties.

D. Acid - Base Properties

Amino acids and proteins have conspicuous acid-base properties. The α-amino acids have two or, for those with ionizable side groups, three acid-base groups. The titration curve of glycine, the simplest amino acid, is shown in Fig. 4-5. At low pH values, both acid-base groups of glycine are fully protonated so that it assumes the cationic form $^+H_3NCH_2COOH$. In the course of the titration with a strong base, such as NaOH, glycine loses two protons in the stepwise fashion characteristic of a polyprotic acid.

The pK values of glycine's two ionizable groups are sufficiently different so that the Henderson-Hasselbalch equation:

$$pH = pK + \log\left(\frac{[A^-]}{[HA]}\right) \qquad [2.6]$$

closely approximates each leg of its titration curve. Consequently, the pK for each ionization step is that of the midpoint of its corresponding leg of the titration curve (Sections 2-2A and C): At pH 2.35 the concentrations of the cationic form, $^+H_3NCH_2COOH$, and the zwitterionic form, $^+H_3NCH_2COO^-$, are equal; similarly, at pH 9.78 the concentrations of this zwitterionic form and the anionic form, $H_2NCH_2COO^-$, are equal. Note that *amino acids never assume the neutral form in aqueous solution.*

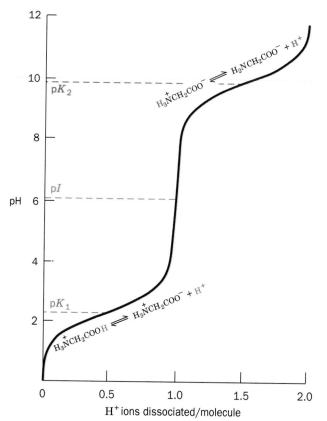

FIGURE 4-5. The titration curve of glycine. Other monoamino, monocarboxylic acids ionize in a similar fashion. [After Meister, A., *Biochemistry of Amino Acids* (2nd ed.), Vol. 1, *p.* 30, Academic Press (1965).]

The pH at which a molecule carries no net electric charge is known as its **isoelectric point, pI.** For the α-amino acids, the application of the Henderson-Hasselbalch equation indicates that, to a high degree of precision,

$$pI = \tfrac{1}{2}(pK_i + pK_j) \qquad [4.1]$$

where K_i and K_j are the dissociation constants of the two ionizations involving the neutral species. For monoamino, monocarboxylic acids such as glycine, K_i and K_j represent K_1 and K_2. However, for aspartic and glutamic acids, K_i and K_j are K_1 and K_R, whereas for arginine, histidine, and lysine, these quantities are K_R and K_2.

Acetic acid's pK (4.76), which is typical of aliphatic monocarboxylic acids, is ~2.4 pH units higher than the pK_1 of its α-amino derivative glycine. This large difference in pK values of the same functional group is caused, as is discussed in Section 2-2C, by the electrostatic influence of glycine's positively charged ammonium group; that is, its NH_3^+ group helps repel the proton from its COOH group. Conversely, glycine's carboxylate group increases the basicity of its amino group ($pK_2 = 9.78$) with respect to that of glycine methyl ester (pK = 7.75). However, the NH_3^+ groups of glycine and its esters are significantly more acidic than are aliphatic amines ($pK \approx 10.7$) because of the electron-withdrawing character of the carboxyl group.

The electronic influence of one functional group upon another is rapidly attenuated as the distance between the groups increases. Hence, the pK values of the α-carboxylate groups of amino acids and the side chain carboxylates of aspartic and glutamic acids form a series that is progressively closer in value to the pK of an aliphatic monocarboxylic acid. Likewise, the ionization constant of lysine's side chain amino group is indistinguishable from that of an aliphatic amine.

Proteins Have Complex Titration Curves

The titration curves of the α-amino acids with ionizable side chains, such as that of glutamic acid, exhibit the expected three pK values. However, the titration curves of polypeptides and proteins, an example of which is shown in Fig. 4-6, rarely provide any indication of individual pK values because of the large numbers of ionizable groups they represent (typically 30% of a protein's amino acid side chains are ionizable; Table 4-1). Furthermore, the covalent and three-dimensional structure of a protein may cause the pK of each ionizable group to shift by as much as several pH units from its value in the free α-amino acid as a result of the electrostatic influence of nearby charged groups, medium effects arising from the proximity of groups of low dielectric constant, and the effects of hydrogen bonding associations. The titration curve of a protein is also a function of the salt concentration, as is shown in Fig. 4-6, because the salt ions act electrostatically to shield the side chain charges from one another, thereby attenuating these charge-charge interactions.

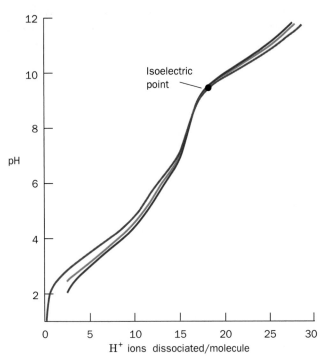

FIGURE 4-6. Titration curves of the enzyme ribonuclease A at 25°C. The concentration of KCl is 0.01*M* for the blue curve, 0.03*M* for the red curve, and 0.15*M* for the green curve. [After Tanford, C. and Hauenstein, J. D., *J. Am. Chem. Soc.* **78**, 5287 (1956).]

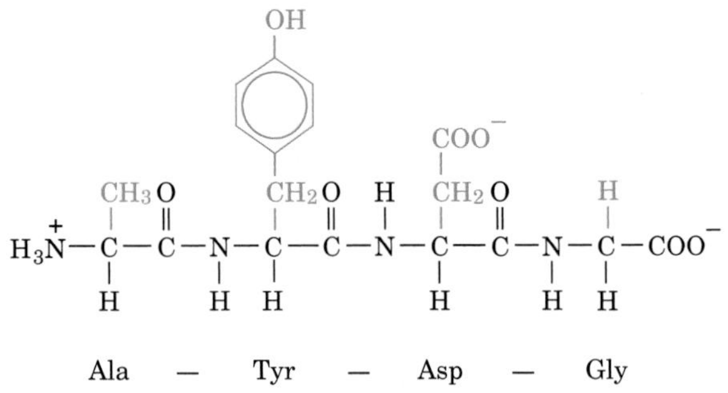

FIGURE 4-7. The tetrapeptide Ala-Tyr-Asp-Gly.

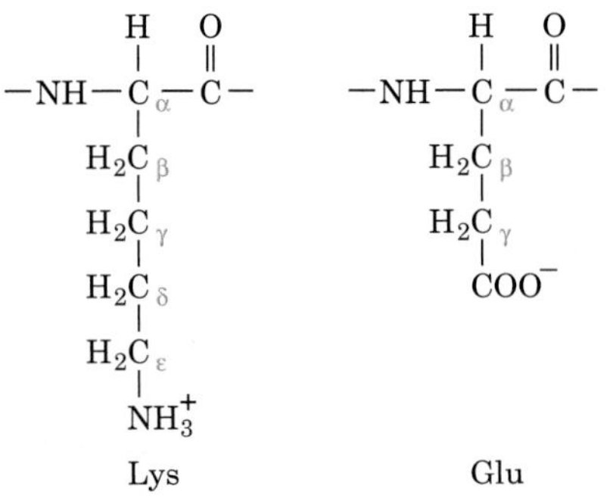

FIGURE 4-8. The Greek lettering scheme used to identify atoms in lysyl and glutamyl R groups.

E. A Few Words on Nomenclature

The three-letter abbreviations for the 20 amino acid residues are given in Table 4-1. It is worthwhile memorizing these symbols because they are widely used throughout the biochemical literature, including this text. These abbreviations are, in most cases, taken from the first three letters of the corresponding amino acid's name; they are conversationally pronounced as read.

The symbol **Glx** means Glu or Gln and, similarly, **Asx** means Asp or Asn. These ambiguous symbols stem from laboratory experience: Asn and Gln are easily hydrolyzed to aspartic acid and glutamic acid, respectively, under the acidic or basic conditions that are usually used to excise them from proteins (Section 6-1D). Therefore, without special precautions, we cannot determine whether a detected Glu was originally Glu or Gln, and likewise for Asp and Asn.

The one-letter symbols for the amino acids are also given in Table 4-1. This more compact code is often used when comparing the amino acid sequences of several similar proteins and hence should also be memorized by the serious student. Note that the one-letter symbols are usually the first letter of the amino acid residue's name. However, for those sets of residues that have the same first letter, this is only true of the most abundant residue of the set.

Amino acid residues in polypeptides are named by dropping the suffix **-ine** in the name of the amino acid and replacing it by **-yl**. Polypeptide chains are described by starting at the amino terminus (known as the **N-terminus**)

and sequentially naming each residue until the carboxyl terminus (the **C-terminus**) is reached. The amino acid at the C-terminus is given the name of its parent amino acid. Thus the compound shown in Fig. 4-7 is alanyltyrosylaspartylglycine. Of course such names for polypeptide chains of more than a few residues are extremely cumbersome. The use of abbreviations for amino acid residues partially relieves this problem. Thus the foregoing tetrapeptide is Ala-Tyr-Asp-Gly using the three-letter abbreviations and AYDG using the one-letter symbols. Note that these abbreviations are always written so that the N-terminus of the polypeptide chain is to the left and the C-terminus is to the right.

The various atoms of the amino acid side chains are often named in sequence with the Greek alphabet starting at the carbon atom adjacent to the peptide carbonyl group. Therefore, as Fig. 4-8 indicates, the Lys residue is said to have an ϵ-amino group and Glu has a γ-carboxyl group. Unfortunately, this labeling system is ambiguous for several amino acids. Consequently, standard numbering schemes for organic molecules are also employed. These are indicated in Table 4-1 for the heterocyclic side chains.

2. OPTICAL ACTIVITY

The amino acids as isolated by the mild hydrolysis of proteins are, with the exception of glycine, all **optically active;** that is, they rotate the plane of plane-polarized light (see below).

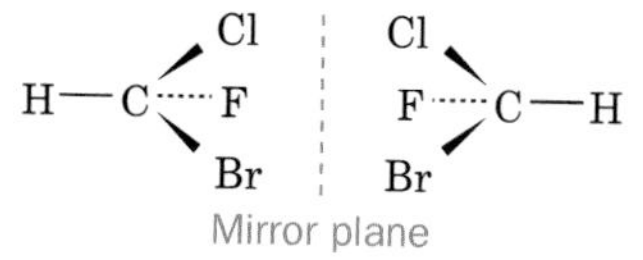

H—C⟋Cl⋯F⟍Br ┊ Cl⟍F⋯C—H⟋Br

Mirror plane

FIGURE 4-9. The two enantiomers of fluorochloro-bromomethane. The four substituents are tetrahedrally arranged about the central atom with the dotted lines indicating that a substituent lies behind the plane of the paper, a triangular line indicating that it lies above the plane of the paper, and a thin line indicating that it lies in the plane of the paper. The mirror plane relating the enantiomers is represented by a vertical dashed line.

Optically active molecules have an asymmetry such that they are not superimposable on their mirror image in the same way that a left hand is not superimposable on its mirror image, a right hand. This situation is characteristic of substances that contain tetrahedral carbon atoms that have four different substituents. The two molecules depicted in Fig. 4-9 are not superimposable since they are mirror images. The central atoms in such atomic constellations are known as **asymmetric centers** or **chiral centers** and are said to have the property of **chirality** (Greek: *cheir,* hand). The C_α atoms of all the amino acids, with the exception of glycine, are asymmetric centers. Glycine, which has two H atoms substituent to its C_α atom, is superimposable on its mirror image and is therefore not optically active.

Molecules that are nonsuperimposable mirror images are known as **enantiomers** of one another. Enantiomeric molecules are physically and chemically indistinguishable by most techniques. *Only when probed asymmetrically, for example, by plane-polarized light or by reactants that also contain chiral centers, can they be distinguished and/or differentially manipulated.*

There are three commonly used systems of nomenclature whereby a particular stereoisomer of an optically active molecule can be classified. These are explained in the following sections.

A. An Operational Classification

Molecules are classified as **dextrorotatory** (Greek: *dextro,* right) or **levorotatory** (Greek: *levo,* left) depending on whether they rotate the plane of plane-polarized light clockwise or counterclockwise from the point of view of the observer. This can be determined by an instrument known as a **polarimeter** (Fig. 4-10). A quantitative measure of the optical activity of the molecule is known as its **specific rotation:**

$$[\alpha]_D^{25} = \frac{\text{observed rotation (degree)}}{\text{optical path} \atop \text{length (dm)} \times {\text{concentration} \atop (g \cdot cm^{-3})}} \qquad [4.2]$$

where the superscript 25 refers to the temperature at which polarimeter measurements are customarily made (25°C) and the subscript D indicates the monochromatic light that is traditionally employed in polarimetry, the so-called D-line in the spectrum of sodium (589.3 nm). Dextrorotatory and levorotatory molecules are assigned positive and negative values of $[\alpha]_D^{25}$. Dextrorotatory molecules are therefore designated by the prefix (+) and their levorotatory enantiomers have the prefix (−). In an equivalent but archaic nomenclature, the lower case letters *d (dextro)* and *l (levo)* are used.

The sign and magnitude of a molecule's specific rotation depend on the structure of the molecule in a complicated and poorly understood manner. It is not yet possible to predict reliably the magnitude or even the sign of a given molecule's specific rotation. For example, proline, leucine,

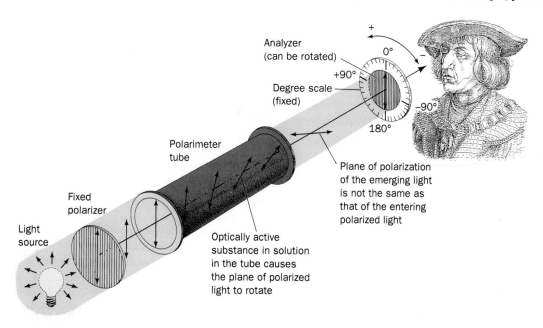

FIGURE 4-10. Schematic diagram of a polarimeter, a device used to measure optical rotation.

and arginine, which are isolated from proteins, have specific rotations in pure aqueous solutions of $-86.2°$, $-10.4°$, and $+12.5°$, respectively. Their enantiomers exhibit values of $[\alpha]_D^{25}$ of the same magnitude but of opposite signs. As might be expected from the acid–base nature of the amino acids, these quantities vary with the solution pH.

A problem with this operational classification system for optical isomers is that it provides no presently interpretable indication of the **absolute configuration** (spatial arrangement) of the chemical groups about a chiral center. Furthermore, a molecule with more than one asymmetric center may have an optical rotation that is not obviously related to the rotatory powers of the individual asymmetric centers. For this reason, the following relative classification scheme is more useful.

B. The Fischer Convention

In this system, the configuration of the groups about an asymmetric center is related to that of **glyceraldehyde,** a molecule with one asymmetric center. By a convention introduced by Emil Fischer in 1891, the $(+)$ and $(-)$ stereoisomers of glyceraldehyde are designated **D-glyceraldehyde** and **L-glyceraldehyde,** respectively (note the use of small upper case letters). With the realization that there was only a 50% chance that he was correct, Fischer assumed that the configurations of these molecules were those shown in Fig. 4-11. Fischer also proposed a convenient shorthand notation for these molecules, known as **Fischer projections,** which are also given in Fig. 4-11. In the Fischer convention, horizontal bonds extend above the plane of the paper and

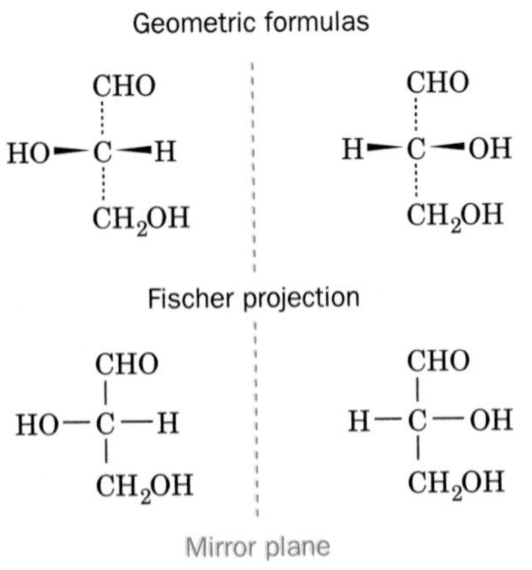

CHO   COO⁻

HO—C—H   H₃N⁺—C—H

CH₂OH   R

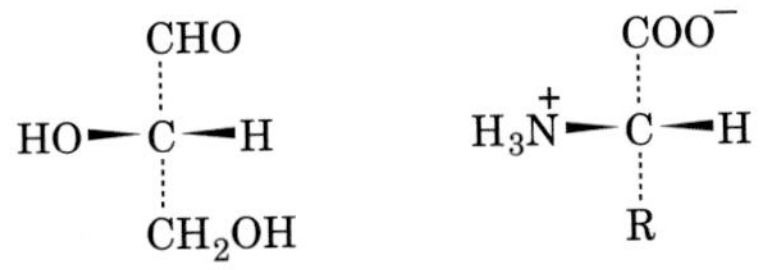

L-Glyceraldehyde   L-α-Amino Acid

FIGURE 4-12.   The configurations of L-glyceraldehyde and L-α-amino acids.

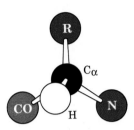

FIGURE 4-13.   The "CORN crib" mnemonic for the hand of L-amino acids. Looking at the C_α atom from its H atom substituent, its other substituents should read **CO—R—N** in the clockwise direction as shown. Here CO, R, and N, respectively, represent the carbonyl group, side chain, and main chain nitrogen atom. [After Richardson, J. S., *Adv. Protein Chem.* **34,** 171 (1981).]

vertical bonds extend below the plane of the paper as is explicitly indicated by the accompanying geometrical formulas.

The configuration of groups about a chiral center can be related to that of glyceraldehyde by chemically converting these groups to those of glyceraldehyde using reactions of known stereochemistry. For α-amino acids, the arrangement of the amino, carboxyl, R, and H groups about the C_α atom is related to that of the hydroxyl, aldehyde, CH₂OH, and H groups, respectively, of glyceraldehyde. In this way, L-glyceraldehyde and L-α-amino acids are said to have the same relative configurations (Fig. 4-12). Through the use of this method, the configurations of the α-amino acids can be described without reference to their specific rotations.

All α-amino acids derived from proteins have the L stereochemical configuration; that is, they all have the same relative configuration about their C_α atoms. In 1949, it was demonstrated by a then new technique in X-ray crystallography that Fischer's arbitrary choice was correct: The designation of the relative configuration of chiral centers is the same as their absolute configuration. The absolute configuration of L-α-amino acid residues may be easily remembered through the use of the "CORN crib" mnemonic that is diagrammed in Fig. 4-13.

Diastereomers Are Chemically and Physically Distinguishable

A molecule may have multiple asymmetric centers. For such molecules, the terms **stereoisomers** and **optical isomers** refer to molecules with different configurations about at least one of their chiral centers, but that are otherwise identical. The term enantiomer still refers to a mole-

Geometric formulas

CHO   CHO

HO—C—H   H—C—OH

CH₂OH   CH₂OH

Fischer projection

CHO   CHO

HO—C—H   H—C—OH

CH₂OH   CH₂OH

Mirror plane

L-Glyceraldehyde   D-Glyceraldehyde

FIGURE 4-11.   The Fischer convention configurations for naming the enantiomers of glyceraldehyde as represented by geometric formulas (*top*) and their corresponding Fischer projection formulas (*bottom*). Note that in Fischer projection, all horizontal bonds point above the page and all vertical bonds point below the page. The mirror planes relating the enantiomers are represented by a vertical dashed line. (Fischer projection formulas, as traditionally presented, omit the central C symbolizing the chiral carbon atom. The Fischer projection formulas in this text, however, will generally have a central C.)

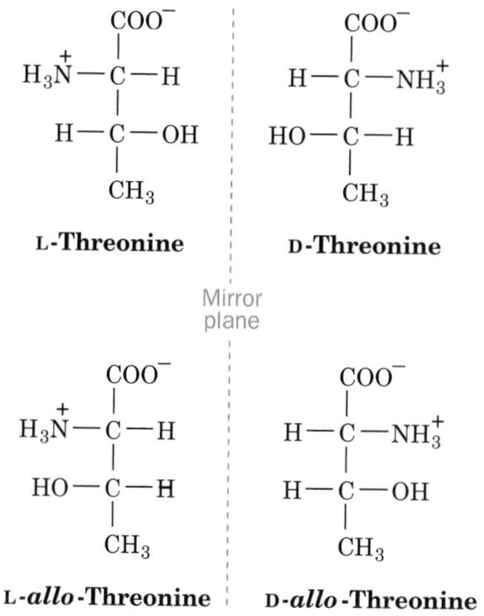

FIGURE 4-14. Fischer projections of threonine's four stereoisomers. The D and L forms are mirror images as are the D-*allo* and L-*allo* forms. D- and L-threonine are each diastereomers of both D-*allo*- and L-*allo*-threonine.

cule that is the mirror image of the one under consideration, that is, different in all its chiral centers. Since each asymmetric center in a chiral molecule can have two possible configurations, a molecule with n chiral centers has 2^n different possible stereoisomers and 2^{n-1} enantiomeric pairs. Threonine and isoleucine each have two chiral centers and hence $2^2 = 4$ possible stereoisomers. The forms of threonine and isoleucine that are isolated from proteins, which are by convention called the L forms, are indicated in Table 4-1. The mirror images of the L forms are the D forms. Their other two optical isomers are said to be **diastereomers** (or allo forms) of the enantiomeric D and L forms. The relative configurations of all four stereoisomers of threonine are given in Fig. 4-14. Note the following points:

1. The D-*allo* and L-*allo* forms are mirror images of each other, as are the D and L forms. Neither allo form is symmetrically related to either of the D or L forms.

2. In contrast to the case for enantiomeric pairs, diastereomers are physically and chemically distinguishable from one another by ordinary means such as melting points,

spectra, and chemical reactivity; that is, they are really different compounds in the usual sense.

A special case of diastereoisomerism occurs when the two asymmetric centers are chemically identical. Two of the four Fischer projections of the sort shown in Fig. 4-14 then represent the same molecule. This is because the two asymmetric centers in this molecule are mirror images of each other. Such a molecule is superimposible on its mirror image and is therefore optically inactive. This so-called meso form is said to be **internally compensated.** The three optical isomers of cystine are shown in Fig. 4-15 ,where it can be seen that the D and L isomers are mirror images of each other as before. Only L-cystine occurs in proteins.

C. The Cahn-Ingold-Prelog System

Despite its usefulness, the Fischer scheme is awkward and sometimes ambiguous for molecules with more than one asymmetric center. For this reason, the following absolute nomenclature scheme was formulated in 1956 by Robert Cahn, Christopher Ingold, and Vladimir Prelog. In this system, the four groups surrounding a chiral center are ranked according to a specific although arbitrary priority scheme: *Atoms of higher atomic number bonded to a chiral center are ranked above those of lower atomic number.* For example, the oxygen atom of an OH group takes precedence over the carbon atom of a CH_3 group that is bonded to the same chiral C atom. If any of the first substituent atoms are of the same element, the priority of these groups is established from the atomic numbers of the second, third, etc., atoms outward from the asymmetric center. Hence a CH_2OH group takes precedence over a CH_3 group. There are other rules (given in the references and in many organic chemistry texts) for assigning priority ratings to substituents with multiple bonds or differing isotopes. The order of priority of some common functional groups is

$$SH > OH > NH_2 > COOH > CHO$$
$$> CH_2OH > C_6H_5 > CH_3 > {}^2H > {}^1H$$

Note that each of the groups substituent to a chiral center must have a different priority rating; otherwise the center could not be asymmetric.

The prioritized groups are assigned the letters W, X, Y, Z such that their order of priority rating is $W > X > Y > Z$.

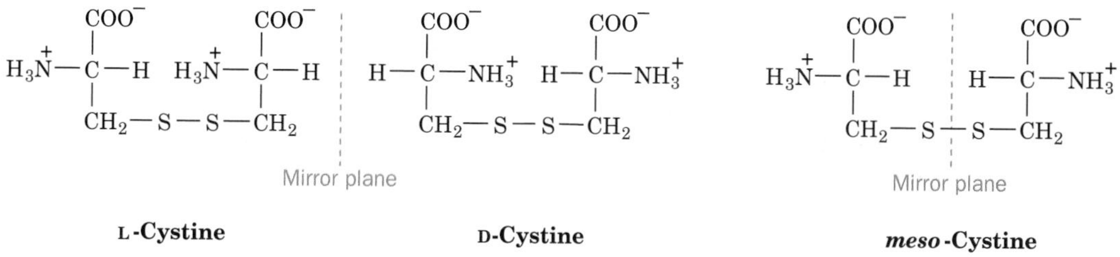

FIGURE 4-15. The three stereoisomers of cystine. The D and L forms are related by mirror symmetry, whereas the meso form has internal mirror symmetry and therefore lacks optical activity.

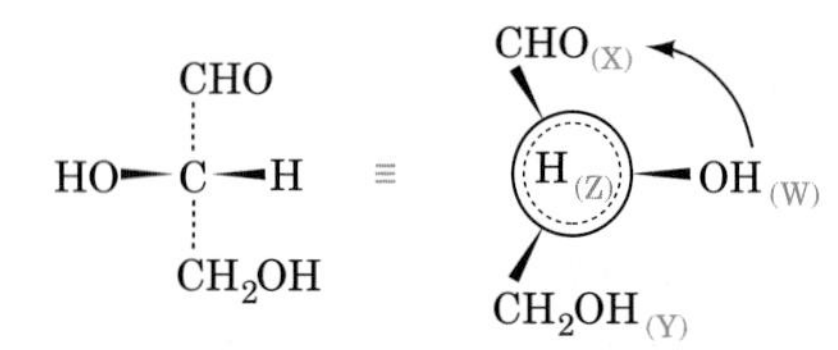

L-Glyceraldehyde   *(S)*-**Glyceraldehyde**

FIGURE 4-16. The structural formula of L-glyceraldehyde and its equivalent (*RS*)-system representation indicating that it is (*S*)-glyceraldehyde. In the latter drawing, the chiral C atom is represented by the large circle, and the H atom, which is located behind the plane of the paper, is represented by the smaller concentric dashed circle.

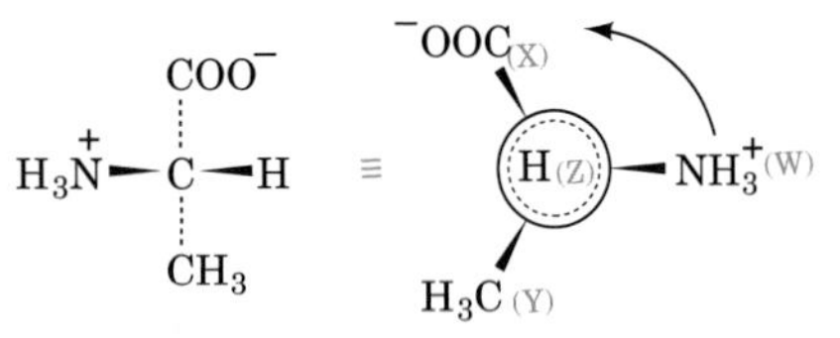

L-Alanine   *(S)*-**Alanine**

FIGURE 4-17. The structural formula of L-alanine and its equivalent (*RS*)-system representation indicating that it is (*S*)-alanine.

To establish the configuration of the chiral center, it is viewed from the asymmetric center towards the Z group (lowest priority). *If the order of the groups* $W \rightarrow X \rightarrow Y$ *as seen from this direction is clockwise, then the configuration of the asymmetric center is designated (R)* (Latin: *rectus*, right). *If the order of* $W \rightarrow X \rightarrow Y$ *is counterclockwise, the asymmetric center is designated (S)* (Latin: *sinistrus*, left). L-Glyceraldehyde is therefore designated (*S*)-glyceraldehyde (Fig. 4-16) and similarly, L-alanine is (*S*)-alanine (Fig. 4-17). In fact, all the L-amino acids from proteins are (*S*)-amino acids, with the exception of L-cysteine, which is (*R*)-cysteine.

A major advantage of this so-called **Cahn–Ingold–Prelog** or **(*RS*) system** is that the chiralities of compounds with multiple asymmetric centers can be unambiguously described. Thus, in the (*RS*) system, L-threonine is (2*S*,3*R*)-threonine, whereas L-isoleucine is (2*S*,3*S*)-isoleucine (Fig. 4-18).

Prochiral Centers Have Distinguishable Substituents

Two chemically identical substituents to an otherwise chiral tetrahedral center are geometrically distinct; that is, the center has no rotational symmetry so that it can be unambiguously assigned left and right sides. Consider, for example, the substituents to the C1 atom of ethanol (the CH$_2$ group; Fig. 4-19*a*). If one of the H atoms was converted to another group (not CH$_3$ or OH), C1 would be a chiral center. The two H atoms are therefore said to be **prochiral.** If we arbitrarily assign the H atoms the subscripts *a* and *b* (Fig. 4-19), then H$_b$ is said to be **pro-R** because in sighting

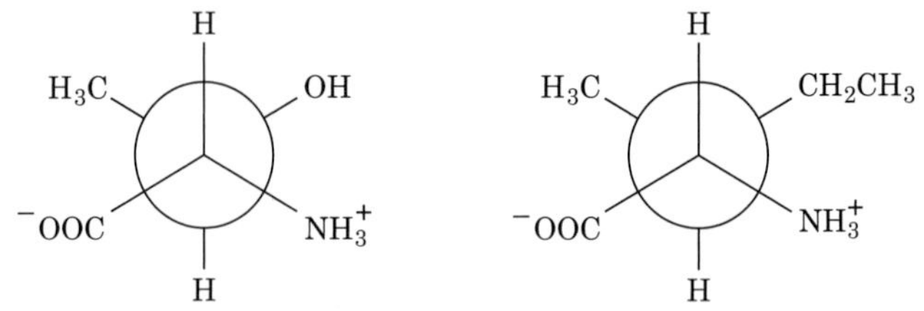

(2*S*, 3*R*)-Threonine   **(2*S*, 3*S*)-Isoleucine**

FIGURE 4-18.   Newman projection diagrams of the stereoisomers of threonine and isoleucine derived from proteins. Here the C$_\alpha$—C$_\beta$ bond is viewed end on. The nearer atom, C$_\alpha$, is represented by the confluence of the three bonds to its substituents, whereas the more distant atom, C$_\beta$, is represented by a circle from which its three substituents project.

from C1 towards H$_a$ (as if it was the Z group of a chiral center), the order of priority of the other substituents decreases in a clockwise direction (Fig. 4-19*b*). Similarly, H$_a$ is said to be **pro-S** (Fig. 4-19*c*).

Planar objects with no rotational symmetry also have the property of prochirality. For example, in many enzymatic reactions, stereospecific addition to a trigonal carbon atom occurs from a particular side of that carbon atom to yield a chiral center (Section 12-2A). If a trigonal carbon is facing the viewer such that the order of priority of its substituents decreases in a clockwise manner (Fig. 4-20*a*), that face is designated as the **re face** (after *rectus*). The opposite face is designated as the **si face** (after *sinistrus*) since the priorities of its substituents decrease in the counterclockwise direction (Fig. 4-20*b*). Comparison of Figs. 4-19*b* and 4-20*a* indicates that an H atom adding to the *re* side of acetaldehyde atom C1 occupies the *pro-R* position of the resulting tetrahedral center. Conversely, a *pro-S* H atom is generated by *si* side addition to this trigonal center (Figs. 4-19*c* and 4-20*b*).

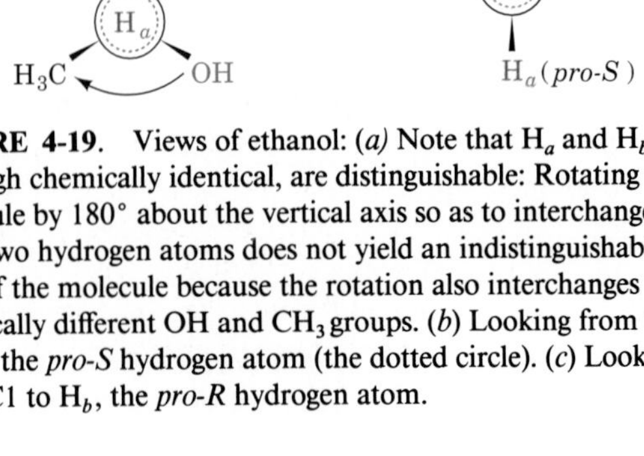

FIGURE 4-19. Views of ethanol: (*a*) Note that H$_a$ and H$_b$, although chemically identical, are distinguishable: Rotating the molecule by 180° about the vertical axis so as to interchange these two hydrogen atoms does not yield an indistinguishable view of the molecule because the rotation also interchanges the chemically different OH and CH$_3$ groups. (*b*) Looking from C1 to H$_a$, the *pro-S* hydrogen atom (the dotted circle). (*c*) Looking from C1 to H$_b$, the *pro-R* hydrogen atom.

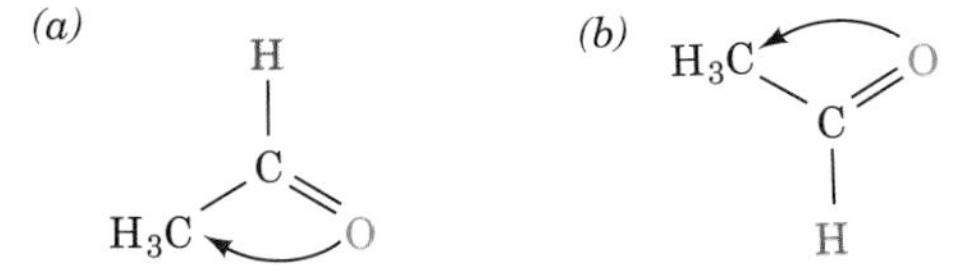

FIGURE 4-20. Views of acetaldehyde onto (*a*) its *re* face, and (*b*) its *si* face.

Closely related compounds, which have the same configurational representation under the Fischer DL convention, may have different representations under the (*RS*) system. Consequently, we shall use the Fischer convention in most cases. The (*RS*) system, however, is indispensible for describing prochirality and stereospecific reactions, so we shall find it invaluable for describing enzymatic reactions.

D. Chirality and Biochemistry

The ordinary chemical synthesis of chiral molecules produces **racemic** mixtures of these molecules (equal amounts of each member of an enantiomeric pair) because ordinary chemical and physical processes have no stereochemical bias. Consequently, there are equal probabilities for an asymmetric center of either hand to be produced in any such process. In order to obtain a product with net optical activity, a chiral process must be employed. This usually takes the form of using chiral reagents, although, at least in principle, the use of any asymmetric influence such as light that is plane polarized in one direction can produce a net asymmetry in a reaction product.

One of the most striking characteristics of life is its production of optically active molecules. *The biosynthesis of a substance possessing asymmetric centers almost invariably produces a pure stereoisomer.* The fact that the amino acid residues of proteins all have the L configuration is just one example of this phenomenon. This observation has prompted the suggestion that a simple diagnostic test for the past or present existence of extraterrestrial life, be it on moon rocks or in meteorites that have fallen to earth, would be the detection of net optical activity in these materials. Any such finding would suggest that the asymmetric molecules thereby detected had been biosynthetically produced. Thus, even though α-amino acids have been extracted from carbonaceous meteorites, the observation that they come in racemic mixtures suggests that they are of chemical rather than biological origin.

One of the enigmas of the origin of life is why terrestrial life is based on certain chiral molecules rather than their enantiomers; that is, on L-amino acids, for example, rather than D-amino acids. Arguments that physical effects such as polarized light might have promoted significant net asymmetry in prebiotically synthesized molecules (Section 1-4B) have not been convincing. Perhaps L-amino acid–based life forms arose at random and simply "ate" any D-amino acid–based life forms.

3. "NONSTANDARD" AMINO ACIDS

The 20 common amino acids are by no means the only amino acids that occur in biological systems. "Nonstandard" amino acid residues are often important constituents of proteins and biologically active polypeptides. Many amino acids, however, are not constituents of proteins. Together with their derivatives, they play a variety of biologically important roles.

A. Amino Acid Derivatives in Proteins

The "universal" genetic code, which is nearly identical in all known life forms (Section 30-1), specifies only the 20 "standard" amino acids of Table 4-1. Nevertheless, many other amino acids, a selection of which is given in Fig. 4-21, are components of certain proteins. *In all known cases but one (Section 30-2D), however, these unusual amino acids result from the specific modification of an amino acid residue after the polypeptide chain has been synthesized.* Among the most prominent of these modified amino acid residues are **4-hydroxyproline** and **5-hydroxylysine.** Both of these amino acid residues are important structural constituents of the fibrous protein **collagen,** the most abundant protein in mammals (Section 7-2C). Amino acids of proteins that form complexes with nucleic acids are often modified. For example, ribosomal proteins (Section 30-3A) and the chromosomal proteins known as **histones** (Section 33-1A) may be specifically methylated, acetylated, and/or phosphorylated. Several of these derivatized amino acid residues are presented in Fig. 4-21. **N-Formylmethionine** is initially the N-terminal residue of all prokaryotic proteins, but is usually removed as part of the protein maturation process (Section 30-3C). γ-**Carboxyglutamic acid** is a constituent of several proteins involved in blood clotting (Section 34-1B). Note that in most cases, these modifications are important, if not essential, for the function of the protein.

D-Amino acid residues are components of many of the relatively short (< 20 residues) bacterial polypeptides that are enzymatically rather than ribosomally synthesized. These polypeptides are perhaps most widely distributed as constituents of bacterial cell walls (Section 10-3B), which D-amino acids render less susceptible to attack by the **peptidases** (enzymes that hydrolyze peptide bonds) that many organisms employ to digest bacterial cell walls. Likewise, D-amino acids are components of many bacterially produced peptide antibiotics including **valinomycin, gramicidin A** (Section 18-2C), and **actinomycin D** (Section 29-2D). D-Amino acid residues are also functionally essential components of several ribosomally synthesized polypeptides of eukaryotic as well as prokaryotic origin. These D-amino acid residues are posttranslationally formed, most probably through the enzymatically mediated inversion of the preexisting L-amino acid residues.

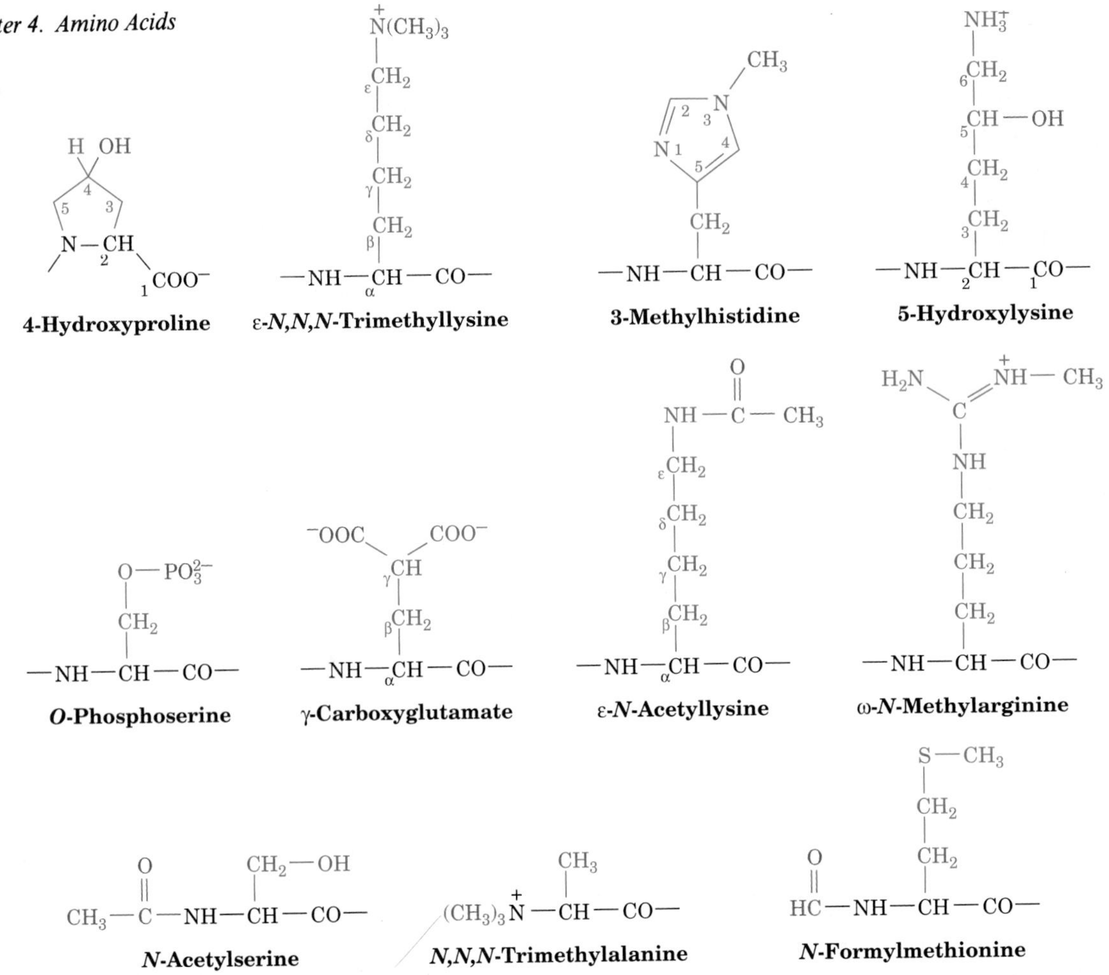

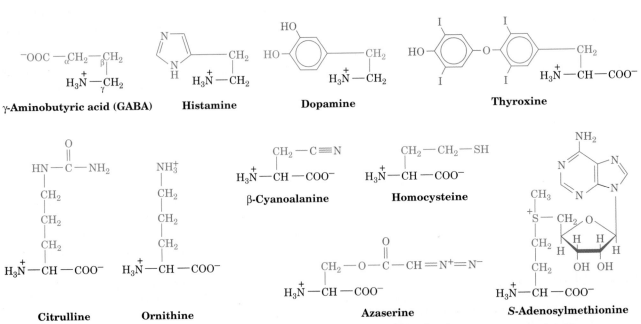

FIGURE 4-21. Some uncommon amino acid residues that are components of certain proteins. All of these residues are modified from one of the 20 "standard" amino acids after polypeptide chain biosynthesis. Those amino acid residues that are derivatized at their N_α position occur at the N-termini of proteins.

FIGURE 4-22. Some biologically produced derivatives of "standard" amino acids and amino acids that are not components of proteins.

B. *Specialized Roles of Amino Acids*

Besides their role in proteins, amino acids and their derivatives have many biologically important functions. A few examples of these substances are shown in Fig. 4-22. This alternative use of amino acids is an example of the biological opportunism that we shall repeatedly encounter: *Nature tends to adapt materials and processes that are already present to new functions.*

Amino acids and their derivatives often function as chemical messengers in the communications between cells. For example, glycine, γ-aminobutyric acid (GABA; a glutamate decarboxylation product), and **dopamine** (a tyrosine product) are neurotransmitters (substances released by nerve cells to alter the behavior of their neighbors; Section 34-4C); **histamine** (the decarboxylation product of histidine) is a potent local mediator of allergic reactions; and **thyroxine** (a tyrosine product) is an iodine-containing thy-

roid hormone that generally stimulates vertebrate metabolism (Section 34-4A).

Certain amino acids are important intermediates in various metabolic processes. Among them are **citrulline** and **ornithine,** intermediates in urea biosynthesis (Section 24-2B); **homocysteine,** an intermediate in amino acid metabolism (Section 24-3E); and **S-adenosylmethionine,** a biological methylating reagent (Section 24-3E).

Nature's diversity is remarkable. About 250 different amino acids have been found in various plants and fungi. For the most part, their biological roles are obscure although the fact that many are toxic suggests that they have a protective function. Indeed, some of them, such as **azaserine,** are medically useful antibiotics. Many of these amino acids are simple derivatives of the 20 "standard" amino acids although some of them, including azaserine and β-cyanoalanine (Fig. 4-22), have unusual structures.

CHAPTER SUMMARY

Proteins are linear polymers that are synthesized from the same 20 "standard" α-amino acids through their condensation to form peptide bonds. These amino acids all have a carboxyl group with a pK near 2.2 and an amino substituent with a pK near 9.4 attached to the same carbon atom, the C_α atom. The α-amino acids are zwitterionic compounds, $^+H_3N-CHR-COO^-$, in the physiological pH range. The various amino acids are usually classified according to the polarities of their side chains, R, which are also substituent to the C_α atom. Glycine, alanine, valine, leucine, isoleucine, methionine, proline (which is really a secondary amino acid), phenylalanine, and tryptophan are nonpolar amino acids; serine, threonine, asparagine, glutamine, tyrosine, and cysteine are uncharged and polar; and lysine, arginine, histidine, aspartic acid, and glutamic acid are charged and polar. The side chains of many of these amino acids bear acid–base groups and hence the properties of the proteins containing them are pH dependent.

The C_α atoms of all α-amino acids except glycine each bear four different substituents and are therefore chiral centers. According to the Fischer convention, which relates the configuration of D- or L-glyceraldehyde to that of the asymmetric center of interest, all the amino acids of proteins have the L configuration; that is, they all have the same absolute configuration about their C_α atom. According to the Cahn–Ingold–Prelog (RS) system of chirality nomenclature, they are, with the exception of cysteine, all (S)-amino acids. The side chains of threonine and isoleucine also contain chiral centers. A prochiral center has no rotational symmetry and hence its substituents, in the case of a central atom, or its faces, in the case of a planar molecule, are distinguishable.

Amino acid residues other than the 20 from which proteins are synthesized also have important biological functions. These "nonstandard" residues result from the specific chemical modifications of amino acid residues in preexisting proteins. Amino acids and their derivatives also have independent biological roles such as neurotransmitters, metabolic intermediates, and poisons.

REFERENCES

History

Vickery, H.B. and Schmidt, C.L.A., The history of the discovery of amino acids, *Chem. Rev.* **9,** 169–318 (1931).

Vickery, H.B., The history of the discovery of the amino acids. A review of amino acids discovered since 1931 as components of native proteins, *Adv. Protein Chem.* **26,** 81–171 (1972).

Properties of Amino Acids

Barrett, G.C. (Ed.), *Chemistry and Biochemistry of Amino Acids,* Chapman & Hall (1985).

Cohn, E.J. and Edsall, J.T., *Proteins, Amino Acids and Peptides as*

Ions and Dipolar Ions, Academic Press (1943). [A classic work in its field.]

Davies, J.S. (Ed.), *Amino Acids and Peptides,* Chapman & Hall (1985). [A "sourcebook" on amino acids.]

Edsall, J.T. and Wyman, J., *Biophysical Chemistry,* Vol. 1, Academic Press (1958). [A detailed treatment of the physical chemistry of amino acids.]

Jakubke, H.-D. and Jeschkeit, H., *Amino Acids, Peptides and Proteins,* translated into English by Cotterrell, G. P., Wiley (1977).

Meister, A., *Biochemistry of the Amino Acids* (2nd ed.), Vol. 1, Academic Press (1965). [A compendium of information on amino acid properties.]

Optical Activity

Cahn, R.S., An introduction to the sequence rule, *J. Chem. Ed.* **41,** 116–125 (1964). [A presentation of the Cahn–Ingold–Prelog system of nomenclature.]

Huheey, J.E., A novel method for assigning *R,S* labels to enantiomers, *J. Chem. Ed.,* **63,** 598–600 (1986).

Mislow, K., *Introduction to Stereochemistry,* Benjamin (1966).

Solomons, T.W.G., *Organic Chemistry* (5th ed.), Chapter 5, Wiley (1992). [A discussion of chirality. Most other organic chemistry textbooks contain similar material.]

"Nonstandard" Amino Acids

Amino Acids and Peptides, The Royal Society of Chemistry. [An annual series containing literature reviews on amino acids.]

Fowden, L., Lea, P.J., and Bell, E.A., The non-protein amino acids of plants, *Adv. Enzymol.* **50,** 117–175 (1979).

Fowden, L., Lewis, D., and Tristram, H., Toxic amino acids: their action as antimetabolites, *Adv. Enzymol.* **29,** 89–163 (1968).

Kleinkauf, H. and Döhren, H., Nonribosomal polypeptide formation on multifunctional proteins, *Trends Biochem. Sci.* **8,** 281–283 (1993).

Mor, A., Amiche, M., and Nicholas, P., Enter a new post-transcriptional modification: D-amino acids in gene-encoded peptides, *Trends Biochem. Sci.* **17,** 481–485 (1992).

Thompson, J. and Donkersloot, J.A., N-(Carboxyalkyl)amino acids: Occurence, synthesis, and functions, *Annu. Rev. Biochem.* **61,** 517–557 (1992).

PROBLEMS

1. Name the 20 standard amino acids without looking them up. Give their three-letter and one-letter symbols. Identify the two standard amino acids that are isomers and the two others that, although not isomeric, have essentially the same molecular mass for the neutral molecules.

2. Draw the following oligopeptides in their predominant ionic forms at pH 7: (a) Phe-Met-Arg, (b) tryptophanyllysylaspartic acid, and (c) Gln-Ile-His-Thr.

3. How many different pentapeptides are there that contain one residue each of Gly, Asp, Tyr, Cys, and Leu?

4. Draw the structures of the following two oligopeptides with their cysteine residues cross-linked by a disulfide bond: Val-Cys; Ser-Cys-Pro.

*5. What are the concentrations of the various ionic species in a $0.1M$ solution of lysine at pH 4, 7, and 10?

6. Derive Eq. [4.1] for a monoamino, monocarboxylic acid (use the Henderson–Hasselbalch equation).

*7. The **isoionic point** of a compound is defined as the pH of a pure water solution of the compound. What is the isoionic point of a $0.1M$ solution of glycine?

8. Normal human hemoglobin has an isoelectric point of 6.87. A mutant variety of hemoglobin, known as **sickle-cell hemoglobin,** has an isoelectric point of 7.09. The titration curve of hemoglobin indicates that, in this pH range, 13 groups change ionization states per unit change in pH. Calculate the difference in ionic charge between molecules of normal and sickle-cell hemoglobin.

9. Indicate whether the following familiar objects are chiral, prochiral, or nonchiral.

   (a) A glove
   (b) A tennis ball
   (c) A good pair of scissors
   (d) A screw
   (e) This page
   (f) A toilet paper roll
   (g) A snowflake
   (h) A spiral staircase
   (i) A flight of normal stairs
   (j) A paper clip
   (k) A shoe
   (l) A pair of glasses

10. Draw four equivalent Fischer projection formulas for L-alanine (see Figs. 4-11 and 4-12).

*11. (a) Draw the structural formula and the Fischer projection formula of (*S*)-3-methylhexane. (b) Draw all the stereoisomers of 2,3-dichlorobutane. Name them according to the (*RS*) system and indicate which of them has the meso form.

12. Identify and name the prochiral centers or faces of the following molecules:

   (a) Acetone
   (b) Propene
   (c) Glycine
   (d) Alanine
   (e) Lysine
   (f) 3-Methylpyridine

C H A P T E R

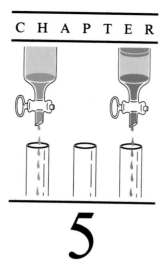

5

Techniques of Protein Purification

A major portion of most biochemical investigations involves the purification of the materials under consideration because these substances must be relatively free of contaminants if they are to be properly characterized. This is often a formidable task because a typical cell contains thousands of different substances, many of which closely resemble other cellular constituents in their physical and chemical properties. Furthermore, the material of interest may be unstable and exist in vanishingly small amounts. Typically, a substance that comprises < 0.1% of a tissue's dry weight must be brought to ~98% purity. Purification problems of this magnitude would be considered unreasonably difficult by most synthetic chemists. It is therefore hardly surprising that our understanding of biochemical processes has by and large paralleled our ability to purify biological materials.

This chapter presents an overview of the most commonly used techniques for the isolation, the purification, and, to some extent, the characterization of proteins as well as other types of biological molecules. These methods are the basic tools of biochemistry whose operation dominates the day-to-day efforts of the practicing biochemist. Furthermore, many of these techniques are routinely used in clinical applications. Indeed, *a basic comprehension of the methods described here is necessary for an appreciation of the significance and the limitations of much of the information presented in this text*. This chapter should therefore be taken as reference material to be consulted repeatedly as the need arises while reading other chapters. Techniques that are specific for the purification of biological molecules other than proteins are described in the appropriate chapters.

1. PROTEIN ISOLATION

Proteins constitute a major fraction of the mass of all organisms. A particular protein, such as **hemoglobin** in red blood cells, may be the dominant substance present in a tissue. Alternatively, a protein such as the *lac* **repressor** of *E. coli* (Section 29-3B) may normally have a population of only a few molecules per cell. Similar techniques are used for the isolation and purification of both proteins, although, in general, the lower the initial concentration of a substance, the more effort is required to isolate it in pure form.

In this section we discuss the care and handling of proteins and outline the general strategy for their purification. For many proteins, the isolation and purification procedure is an exacting task requiring days of effort to obtain only a few milligrams or less of the desired product. However, as we shall see, modern analytical techniques have achieved such a high degree of sensitivity that this small amount of material is often sufficient to characterize a protein extensively. You should note that the techniques described in this chapter are applicable to the separations of most types of biological molecules.

A. Selection of a Protein Source

Proteins with identical functions generally occur in a variety of organisms. For example, most of the enzymes that mediate basic metabolic processes or that are involved in the expression and transmission of genetic information are common to all cellular life. Of course, there is usually considerable variation in the properties of a particular protein from various sources. In fact, different variants of a given protein may occur in different tissues from the same organism or even in different compartments in the same cell. Therefore, if flexibility of choice is possible, the isolation of a protein may be greatly simplified by a judicious choice of the protein source. This choice should be based on such criteria as the ease of obtaining sufficient quantities of the tissue from which the protein is to be isolated, the amount of the chosen protein in that tissue, and any properties peculiar to the specific protein chosen that would aid in its stabilization and isolation. Tissues from domesticated animals such as chickens, cows, pigs, or rats are often chosen. Alternative sources might be easily obtainable microorganisms such as *E. coli* or **baker's yeast (*Saccharomyces cerevisiae*).** We shall see, however, that proteins from a vast variety of organisms have been studied.

The recent development of **molecular cloning techniques** (Section 28-8) has generated an entirely new protein production method. Almost any protein-encoding gene can be isolated from its parent organism, specifically altered (genetically engineered) if desired, and expressed at high levels (overproduced) in a conveniently grown organism such as *E. coli* or yeast. Indeed, the cloned protein may constitute up to 40% of the overproducer's total cell protein. This high level of protein production generally renders the cloned protein far easier to isolate than it would be from its parent organism (in which it may normally occur in vanishingly small amounts).

B. Methods of Solubilization

The first step in the isolation of a protein, or any other biological molecule, is to get it into solution. In some cases, such as with blood serum proteins, nature has already done so. However, a protein must usually be liberated from the cells that contain it. The method of choice for this procedure depends on the mechanical characteristics of the source tissue as well as on the location of the required protein in the cell.

If the protein of interest is located in the cytosol of the cell, its liberation requires only the breaking open (**lysis**) of the cell. In the simplest and gentlest method of doing so, which is known as **osmotic lysis,** the cells are suspended in a **hypotonic solution;** that is, a solution in which the total molar concentration of solutes is less than that inside the cell in its normal physiological state. Under the influence of osmotic forces, water diffuses into the more concentrated intracellular solution, thereby causing the cells to swell and burst. This method works well with animal cells, but with cells that have a cell wall, such as bacteria or plant cells, it is usually ineffective. The use of an enzyme, such as **lysozyme,** which chemically degrades bacterial cell walls (Section 14-2), is sometimes effective with such cells. Detergents or organic solvents such as acetone or toluene are also useful in lysing cells but care must be exercised in their use as they may denature the protein of interest (Section 7-4E).

Many cells require some sort of mechanical disruption process to break them open. This may include grinding with sand or alumina, the use of a high-speed blender (similar to the familiar kitchen appliance), a homogenizer (an implement for crushing tissue between a closely fitting piston and sleeve), a French press (a device that shears open cells by squirting them at high pressure through a small orifice), or sonication (breaking open cells through the use of ultrasonic vibrations). Once the cells have been broken open, the crude **lysate** may be filtered or centrifuged to remove the particulate cell debris, thereby leaving the protein of interest in the supernatant solution.

If the required protein is a component of subcellular assemblies such as membranes or mitochondria, a considerable purification of the protein can be effected by first separating the subcellular assembly from the rest of the cellular material. This is usually accomplished by **differential centrifugation,** a process in which the cell lysate is centrifuged at a speed that removes only the cell components denser than the desired organelle followed by centrifugation at a speed that spins down the component of interest. The required protein is then usually separated from the purified subcellular component by extraction with concentrated salt solutions or, in the case of proteins tightly bound to membranes, with the use of detergent solutions or organic solvents, such as butanol, that solubilize lipids.

C. Stabilization of Proteins

Once a protein has been removed from its natural environment, it becomes exposed to many agents that can irreversibly damage it. These influences must be carefully controlled at all stages of a purification process or the yield of the desired protein may be greatly reduced or even eliminated.

The structural integrity of many proteins is sensitive to pH as a consequence of their numerous acid–base groups. To prevent damage to biological materials due to variations in pH, they are routinely dissolved in buffer solutions effective in the pH range over which the material is stable.

Proteins are easily **denatured** (destroyed) by high temperatures. Although the thermal stabilities of proteins vary widely, many of them slowly denature above 25°C. Therefore, the purification of proteins is normally carried out at temperatures near 0°C. However, there are numerous proteins that require lower temperatures, some even lower than −100°C, for stability. Conversely, some **cold-labile** proteins become unstable below characteristic temperatures.

The thermal stability characteristics of a protein can sometimes be used to advantage in its purification. A heat-stable protein in a crude mixture can be greatly purified by briefly heating the mixture so as to denature and precipitate most of the contaminating proteins without affecting the desired protein.

Cells contain **proteases** (enzymes that cleave the peptide bonds of proteins) and other degradative enzymes that, upon lysis, are liberated into solution along with the protein of interest. Care must be taken that the protein is not damaged by these enzymes. Degradative enzymes may often be rendered inactive at pH's and temperatures that are not harmful to the protein of interest. Alternatively, these enzymes can often be specifically inhibited by chemical agents without affecting the desired protein. Of course, as the purification of a protein progresses, more and more of these degradative enzymes are eliminated.

Some proteins are more resistant than others to proteolytic degradation. The purification of a protein that is particularly resistant to proteases may be effected by maintaining conditions in a crude protein mixture under which the proteolytic enzymes present are active. This so-called **autolysis** technique simplifies the purification of the resistant protein because it is generally far easier to remove selectively the degradation products of contaminating proteins than it is the intact proteins.

Many proteins are denatured by contact with the air–water interface and, at low concentrations, a significant fraction of the protein present may be lost by adsorption to surfaces. Hence, a protein solution should be handled so as to minimize frothing and kept relatively concentrated. There are, of course, other factors to which a protein may be sensitive, including the oxidation of cysteine residues to form disulfide bonds; heavy metal contaminants, which may irreversibly bind to the protein; and the salt concentration and polarity of the solution, which must be kept within the stability range of the protein. Finally, many microorganisms consider proteins to be delicious, so proteins should be stored under conditions that inhibit the growth of microorganisms.

D. Assay of Proteins

To purify any substance, some means must be found for quantitatively detecting its presence. A protein rarely comprises more than a few percent by weight of its tissue of origin and is usually present in much smaller amounts. Yet, much of the material from which it is being extricated closely resembles the protein of interest. Accordingly, an assay must be specific for the protein being purified and highly sensitive to its presence. Furthermore, the assay must be convenient to use because it is done repeatedly at every stage of the purification process.

Among the most straightforward of protein assays are those for enzymes that catalyze reactions with readily detectable products. Perhaps such a product has a characteristic spectroscopic absorption or fluorescence that can be monitored. Alternatively, the enzymatic reaction may consume or generate acid so that the enzyme can be assayed by acid–base titrations. If an enzymatic reaction product is not easily quantitated, its presence may still be revealed by further chemical treatment to yield a more readily observable product. Often, this takes the form of a **coupled enzymatic reaction,** in which the product of the enzyme being assayed is converted, by an added enzyme, to an observable substance.

Proteins that are not enzymes may be assayed through their ability to bind specific substances or the observation of their biological effects. For example, receptor proteins are often assayed by incubating them with a radioactive molecule that they specifically bind, passing the mixture through a protein-retaining filter, and then measuring the amount of radioactivity bound to the filter (Section 34-4B). The presence of a hormone may be revealed by its effect on some standard tissue sample or on a whole organism. The latter type of assays are ususally rather lengthy procedures because the response elicited by the assay may take days to develop. In addition, their reproducibility is often less than satisfactory because of the complex behavior of living systems. Such assays are therefore used only when no alternative procedure is available.

Immunochemical Techniques Can Readily Detect Small Quantities of Specific Proteins

Immunochemical procedures provide protein assay techniques of high sensitivity and discrimination. These methods employ **antibodies,** proteins produced by an animal's immune system in response to the introduction of a foreign protein and which specifically bind to this foreign protein (antibodies and the immune system are discussed in Section 34-2).

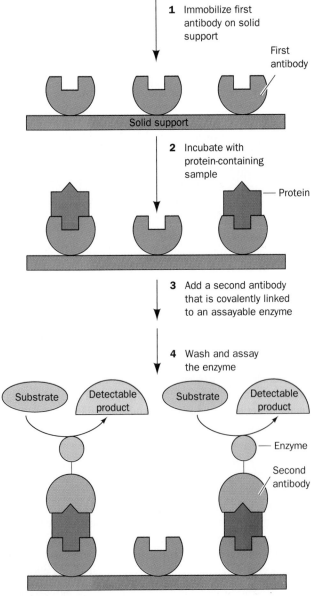

1 Immobilize first antibody on solid support

First antibody

Solid support

2 Incubate with protein-containing sample

Protein

3 Add a second antibody that is covalently linked to an assayable enzyme

4 Wash and assay the enzyme

Substrate   Detectable product   Substrate   Detectable product

Enzyme

Second antibody

FIGURE 5-1. An enzyme-linked immunosorbant assay (ELISA).

Antibodies extracted from the blood serum of an animal that has been immunized against a particular protein are the products of many different antibody-producing cells. They therefore form a heterogeneous mixture of molecules, which vary in their exact specificities and binding affinities for their target protein. Antibody-producing cells normally die after a few cell divisions, so one of them cannot be cultured (cloned) to produce a single species of antibody in useful quantities. Such **monoclonal antibodies** may be obtained, however, by fusing a cell producing the desired antibody with a cell of an immune system cancer known as a **myeloma** (Section 34-2B). The resulting **hybridoma** cell has an unlimited capacity to divide and, when raised in cell culture, produces large quantities of the monoclonal antibody.

A protein can be directly detected through its precipitation by its corresponding antibodies or, in a so-called **radioimmunoassay,** it can be indirectly detected by determining the degree with which it competes with a radioactively labeled standard for binding to the antibody (Section 34-4A). In an **enzyme-linked immunosorbent assay** (ELISA; Fig. 5-1):

1. Antibody against the protein of interest is immobilized on an inert solid such as polystyrene.

2. The solution being assayed for the protein is applied to the antibody-coated surface under conditions which the antibody binds the protein and the unbound protein is then washed away.

3. The resulting protein–antibody complex is further reacted with a second protein-specific antibody to which an easily assayed enzyme has been covalently linked.

4. After washing away any unbound antibody-linked enzyme, the enzyme in the immobilized antibody–protein–antibody–enzyme complex is assayed thereby indicating the amount of the protein present.

Both radioimmunoassays and ELISAs are widely used to detect small amounts of specific proteins and other biological substances in both laboratory and clinical applications. For example, a commonly available pregnancy test, which is reliably positive within a few days post-conception, uses an ELISA to detect the placental hormone **chorionic gonadotropin** (Section 34-4A) in the mother's urine.

E. General Strategy of Protein Purification

The fact that proteins are well-defined substances was not widely accepted until after 1926, when James Sumner first crystallized an enzyme, jack bean **urease.** Before that, it was thought that the high molecular masses of proteins resulted from a colloidal aggregation of rather ill-defined and mysterious substances of lower molecular mass. Once it was realized that it was possible, in principle, to purify proteins, work to do so began in earnest.

In the first half of the twentieth century, the protein purification methods available were extremely crude by today's standards. Protein purification was an arduous task that was as much an art as a science. Usually, the development of a satisfactory purification procedure for a given protein was a matter of years of labor ultimately involving huge quantities of starting material. Nevertheless, by 1940, ~20 enzymes had been obtained in pure form.

Since then, many thousands of proteins have been purified and characterized to varying extents. Modern techniques of separation have such a high degree of discrimination that one can now obtain, in quantity, a series of proteins with such similar properties that only a few years ago their mixture was thought to be a pure substance. Nevertheless, the development of an efficient procedure for the purification of a given protein may still be an intellectually challenging and time-consuming task.

Proteins are purified by fractionation procedures. In a series of independent steps, the various physicochemical properties of the protein of interest are utilized to separate it progressively from other substances. The idea here is not necessarily to minimize the loss of the desired protein, but to eliminate selectively the other components of the mixture so that only the required substance remains.

It may not be philosophically possible to prove that a substance is pure. However, *the operational criterion for establishing purity takes the form of the method of exhaustion: the demonstration, by all available methods, that the sample of interest consists of only one component.* Therefore, as new separation techniques are devised, standards of purity may have to be revised. Experience has shown that when a sample of material previously thought to be pure is subjected to a new separation technique, it is occasionally found to be a mixture of several components.

The characteristics of proteins and other biomolecules that are utilized in the various separation procedures are solubility, ionic charge, molecular size, adsorption properties, and binding specificity for other biological molecules. Some of the procedures we shall discuss and the protein characteristics they depend upon are as follows:

Characteristic	Procedure
Charge:	1. Ion exchange chromatography
	2. Electrophoresis
	3. Isoelectric focusing
Polarity:	1. Adsorption chromatography
	2. Paper chromatography
	3. Reverse-phase chromatography
	4. Hydrophobic interaction chromatography
Size:	1. Dialysis and ultrafiltration
	2. Gel electrophoresis
	3. Gel filtration chromatography
	4. Ultracentrifugation
Specificity:	1. Affinity chromatography

In the remainder of this chapter, we discuss these separation procedures.

2. SOLUBILITIES OF PROTEINS

A protein's multiple acid–base groups make its solubility properties dependent on the concentration of dissolved salts, the polarity of the solvent, the pH, and the temperature. Different proteins vary greatly in their solubilities under a given set of conditions: Certain proteins precipitate from solution under conditions in which others remain quite soluble. This effect is routinely used as the basis for protein purification.

A. *Effects of Salt Concentrations*

The solubility of a protein in aqueous solution is a sensitive function of the concentrations of dissolved salts (Figs. 5-2 through 5-4). The salt concentration in Figs. 5-2 and 5-3 is expressed in terms of the **ionic strength**, I, which is defined

$$I = \tfrac{1}{2} \sum c_i Z_i^2 \qquad [5.1]$$

where c_i is the molar concentration of the ith ionic species and Z_i is its ionic charge. The use of this parameter to account for the effects of ionic charges results from theoretical considerations of ionic solutions. However, as Fig. 5-3 indicates, a protein's solubility at a given ionic strength varies with the types of ions in solution. The order of effectiveness of these various ions in influencing protein solubility is quite similar for different proteins and is apparently mainly due to the ions' size and hydration.

The solubility of a protein at low ionic strength generally increases with the salt concentration (left side of Fig. 5-3 and

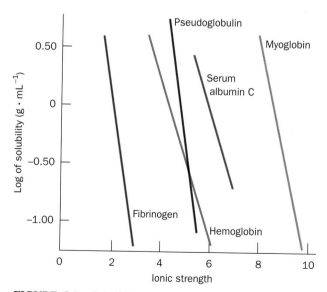

FIGURE 5-2. Solubilities of several proteins in ammonium sulfate solutions. [After Cohn, E.J. and Edsall, J.T., *Proteins, Amino Acids and Peptides*, p. 602, Academic Press (1943).]

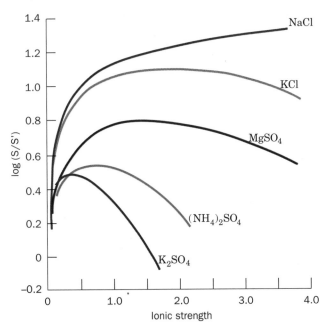

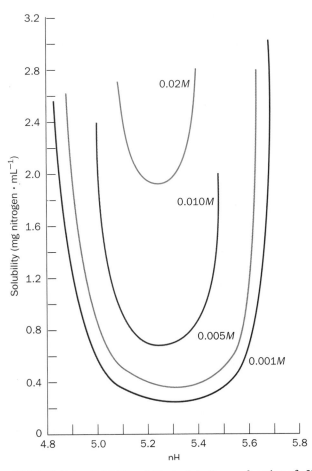

FIGURE 5-3. Solubility of carboxy-hemoglobin at its isoelectric point as a function of ionic strength and ion type. Here S and S′ are, respectively, the solubilities of the protein in the salt solution and in pure water. The logarithm of their ratios is plotted so that the solubility curves can be placed on a common scale. [After Green, A.A., *J. Biol. Chem.* **95,** 47 (1932).]

the different curves of Fig. 5-4). The explanation of this **salting in** phenomenon is that as the salt concentration of the protein solution increases, the additional counterions more effectively shield the protein molecules' multiple ionic charges and thereby increase the protein's solubility.

At high ionic strengths, the solubilities of proteins, as well as those of most other substances, decrease. This effect, known as **salting out,** is primarily a result of the competition between the added salt ions and the other dissolved solutes for molecules of solvation. At high salt concentrations, so many of the added ions are solvated that the amount of bulk solvent available becomes insufficient to dissolve other solutes. In thermodynamic terms, the solvent's activity (effective concentration; Appendix to Chapter 3) is decreased. Hence, solute–solute interactions become stronger than solute–solvent interactions and the solute precipitates.

Salting out is the basis of one of the most commonly used protein purification procedures. Figure 5-2 shows that the solubilities of different proteins vary widely as a function of salt concentration. For example, at an ionic strength of 3, fibrinogen is much less soluble than the other proteins in Fig. 5-2. *By adjusting the salt concentration in a solution containing a mixture of proteins to just below the precipitation point of the protein to be purified, many unwanted proteins can be eliminated from the solution. Then, after the precipitate is removed by filtration or centrifugation, the salt concentration of the remaining solution is increased so as to precipitate the desired protein.* In this manner, a significant purification and concentration of large quantities of pro-

FIGURE 5-4. Solubility of β-lactoglobulin as a function of pH at several NaCl concentrations. [After Fox, S. and Foster, J.S., *Introduction to Protein Chemistry, p.* 242, Wiley (1957).]

tein can be conveniently effected. Consequently, salting out is often the initial step in protein purification procedures. Ammonium sulfate is the most commonly used reagent for salting out proteins because its high solubility (3.9M in water at 0°C) permits the achievement of solutions with high ionic strengths (up to 23.4 in water at 0°C).

Certain ions, notably I^-, ClO_4^-, SCN^-, Li^+, Mg^{2+}, Ca^{2+}, and Ba^{2+}, increase the solubilities of proteins rather than salting them out. These ions also tend to denature proteins (Section 7-4E). Conversely, ions that decrease the solubilities of proteins stabilize their native structures so that proteins that have been salted out are not denatured.

B. Effects of Organic Solvents

Water-miscible organic solvents, such as acetone and ethanol, are generally good protein precipitants because their low dielectric constants lower the solvating power of their aqueous solutions for dissolved ions such as proteins. The different solubilities of proteins in these mixed solvents form the basis of a useful fractionation technique. This procedure is normally used near 0°C or less because, at higher temperatures, organic solvents tend to denature proteins. The lowering of the dielectric constant by organic solvents also magnifies the differences in the salting out behavior of proteins so that these two techniques can be effectively combined. Some water-miscible organic solvents, however, such as dimethyl sulfoxide (DMSO) or *N,N*-dimethylformamide (DMF), are rather good protein solvents because of their relatively high dielectric constants.

C. Effects of pH

Proteins generally bear numerous ionizable groups which have a variety of p*K*'s. At a pH characteristic for each protein, the positive charges on the molecule exactly balance its negative charges. At this pH, the protein's **isoelectric point,** p*I* (Section 4-1D), the protein molecule carries no net charge and is therefore immobile in an electric field.

Figure 5-4 indicates that the solubility of the protein *β*-lactoglobulin is a minimum near its p*I* of 5.2 in dilute NaCl solutions and increases more or less symmetrically about the p*I* with changes in pH. This solubility behavior, which is shared by most proteins, is easily explained. Physicochemical considerations suggest that the solubility properties of uncharged molecules are insensitive to the salt concentration. To a first approximation, therefore, a protein at its isoelectric point should not be subject to salting in. Conversely, as the pH is varied from a protein's p*I*, that is, as the protein's net charge increases, it should be increasingly subject to salting in because the electrostatic interactions between neighboring molecules that promote aggregation and precipitation should likewise increase. Hence, *in solutions of moderate salt concentrations, the solubility of a protein as a function of pH is expected to be at a minimum at the protein's pI and to increase about this point with respect to pH.*

Proteins vary in their amino acid compositions and therefore, as Table 5-1 indicates, in their p*I*'s. This phenomenon is the basis of a protein purification procedure known as **isoelectric precipitation** in which the pH of a protein mixture is adjusted to the p*I* of the protein to be isolated so as to selectively minimize its solubility. In practice, this technique is combined with salting out so that the protein being purified is usually salted out near its p*I*.

D. Crystallization

Once a protein has been brought to a reasonable state of purity, it may be possible to crystallize it. This is usually done by bringing the protein solution just past its saturation point with the types of precipitating agents discussed above. Upon standing for a time (as little as a few minutes, as much as several months), often while the concentration of the precipitating agent is being slowly increased, the protein may precipitate from the solution in crystalline form. It may be necessary to attempt the crystallization under different solution conditions and with various precipitating agents before crystals are obtained. The crystals may range

TABLE 5-1. **THE ISOELECTRIC POINTS OF SEVERAL COMMON PROTEINS**

Protein	Isoelectric pH
Pepsin	<1.0
Ovalbumin (hen)	4.6
Serum albumin (human)	4.9
Tropomyosin	5.1
Insulin (bovine)	5.4
Fibrinogen (human)	5.8
γ-Globulin (human)	6.6
Collagen	6.6
Myoglobin (horse)	7.0
Hemoglobin (human)	7.1
Ribonuclease A (bovine)	7.8
Cytochrome *c* (horse)	10.6
Histone (bovine)	10.8
Lysozyme (hen)	11.0
Salmine (salmon)	12.1

FIGURE 5-5. Protein crystals: (*a*) azurin from *Pseudomonas aeruginosa*, (*b*) flavodoxin from *Desulfovibrio vulgaris*, (*c*) rubredoxin from *Clostridium pasteurianum*, (*d*) azidomet myohemerythrin from the marine worm *Siphonosoma funafuti*, (*e*) lamprey hemoglobin, and (*f*) bacteriochlorophyll *a* protein from *Prosthecochloris aestuarii*. These proteins are colored because of their associated chromophores (light-absorbing groups); proteins are colorless in the absence of such bound groups. [Parts *a*–*c* courtesy of Larry Sieker, University of Washington; Parts *d* and *e* courtesy of Wayne Hendrikson, Columbia University; and Part *f* courtesy of John Olsen, Brookhaven National Laboratories and Brian Matthews, University of Oregon.]

in size from microscopic to 1 mm or more across. Crystals of the latter size, which generally require great care to grow, may be suitable for X-ray crystallographic analysis (Section 7-3A). Several such crystals are shown in Fig. 5-5.

3. CHROMATOGRAPHIC SEPARATIONS

In 1903, the Russian botanist Mikhail Tswett described the separations of plant leaf pigments in solution through the use of solid adsorbents. He named this process **chromatography** (Greek: *chroma,* color + *graphein,* to write), presumably resulting from the colored bands that formed in the adsorbents as the components of the pigment mixtures separated from one another (and possibly because Tswett means color in Russian).

Modern separation methods rely heavily on chromatographic procedures. In all of them, a mixture of substances to be fractionated is dissolved in a liquid or gaseous fluid known as the **mobile phase.** The resultant solution is percolated through a column consisting of a porous solid matrix known as the **stationary phase,** which in certain types of chromatography may be associated with a bound liquid. The interactions of the individual solutes with the stationary phase act to retard their progress through the matrix in a manner that varies with the properties of each solute. If the mixture being fractionated starts its journey through the column in a narrow band, the different retarding forces on each component that cause them to migrate at different rates will eventually cause the mixture to separate into bands of pure substances.

The power of chromatography derives from the continuous nature of the separation processes. A single purification step (or "plate" as it is often termed in analogy with distillation processes) may have very little tendency to separate a mixture into its components. However, since this process is applied in a continuous fashion so that it is, in effect, repeated hundreds or even hundreds of thousands of times, the segregation of the mixture into its components ultimately occurs. The separated components can then be collected into separate fractions for analysis and/or further fractionation.

The various chromatographic methods are classified according to their mobile and stationary phases. For example, in gas–liquid chromatography the mobile and stationary phases are gaseous and liquid, respectively, whereas in liquid–liquid chromatography they are immiscible liquids, one of which is bound to an inert solid support. Chromatographic methods may be further classified according to the nature of the dominant interaction between the stationary phase and the substances being separated. For example, if the retarding force is ionic in character, the separation technique is referred to as **ion exchange chromatography,** whereas if it is a result of the adsorption of the solutes onto a solid stationary phase, it is known as **adsorption chromatography.**

As has been previously mentioned, a cell contains huge numbers of different components, many of which closely resemble one another in their various properties. Therefore, the isolation procedures for most biological substances incorporate a number of independent chromatographic steps in order to purify the substance of interest according to several criteria. In this section, the most commonly used of these chromatographic procedures are described.

A. Ion Exchange Chromatography

In the process of **ion exchange,** *ions that are electrostatically bound to an insoluble and chemically inert matrix are reversibly replaced by ions in solution:*

$$R^+A^- + B^- \rightleftharpoons R^+B^- + A^-$$

Here, R^+A^- is an **anion exchanger** in the A^- form and B^- represents anions in solution. **Cation exchangers** similarly bear negatively charged groups that reversibly bind cations. Polyanions and polycations therefore bind to anion and cation exchangers, respectively. However, proteins and other **polyelectrolytes** (polyionic polymers) that bear both positive and negative charges can bind to both cation and anion exchangers depending on their net charge. *The affinity with which a particular polyelectrolyte binds to a given ion exchanger depends on the identities and concentrations of the other ions in solution because of the competition among these various ions for the binding sites on the ion exchanger. The binding affinities of polyelectrolytes bearing acid–base groups are also highly pH dependent because of the variation of their net charges with pH.* These principles are used to great advantage in isolating biological molecules by **ion exchange chromatography** (Fig. 5-6), as described below.

In purifying a given protein (or some other polyelectrolyte), the pH and the salt concentration of the buffer solution in which the protein is dissolved are chosen so as to immobilize the desired protein on the selected ion exchanger. The impure protein solution is applied to a column in which the ion exchanger has been packed and the column is washed with this buffer solution.

Various proteins bind to the ion exchanger with different affinities. As the column is washed, a process known as **elution,** *those proteins with relatively low affinities for the ion exchanger move through the column faster than the proteins that bind to the ion exchanger with higher affinities.* This occurs because the progress of a given protein through the column is retarded relative to that of the solvent due to interactions between the protein molecules and the ion exchanger. The greater the binding affinity of a protein for the ion exchanger, the more it will be retarded. Thus, proteins that bind tightly to the ion exchanger can be eluted by changing the elution buffer to one with a higher salt concentration (and/or a different pH), a process called **stepwise elution.** With the use of a fraction collector, purification of a protein can be effected by selecting only those fractions of the column effluent that contain the desired protein.

Gradient Elution Improves Chromatographic Separations

The purification process can be further improved by washing the protein-loaded column using the method of

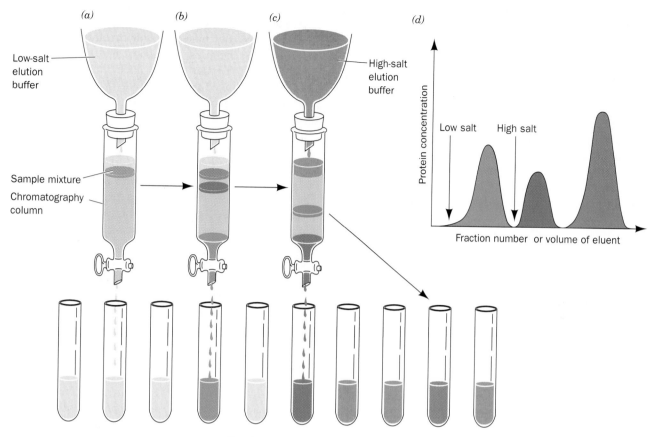

Fractions sequentially collected

FIGURE 5-6. A schematic diagram illustrating the separation of several proteins by ion exchange chromatography using stepwise elution. Here the tan region of the column represents the ion exchanger and the colored bands represent the various proteins. (*a*) The protein mixture is bound to the topmost portion of the ion exchanger in the chromatography column. (*b*) As the elution progresses, the various proteins separate into discrete bands as a consequence of their different mobilities on the ion exchanger under the prevailing solution conditions. Here the first band of protein has passed through the column and been isolated as a separate fraction, whereas the remaining, less mobile, bands remain near the top of the column. (*c*) The salt concentration in the elution buffer is increased to elute the remaining bands. (*d*) The elution diagram of the protein mixture from the column.

gradient elution. Here the salt concentration and/or pH is continuously varied as the column is eluted so as to release sequentially the various proteins that are bound to the ion exchanger. This procedure generally leads to a better separation of proteins than does elution of the column by a single solution or stepwise elution.

Many different types of elution gradients have been successfully employed in purifying biological molecules. The most widely used of these is the **linear gradient** in which the concentration of the eluant solution varies linearly with the volume of solution passed. A simple device for generating such a gradient is illustrated in Fig. 5-7. Here the solute

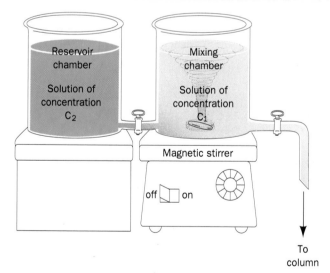

FIGURE 5-7. A device for generating a linear concentration gradient. Two connected open chambers, which have identical cross-sectional areas, are initially filled with equal volumes of solutions of different concentrations. As the solution of concentration c_1 drains out of the mixing chamber, it is partially replaced by a solution of concentration c_2 from the reservoir chamber. The concentration of the solution in the mixing chamber varies linearly from its initial concentration, c_1, to the final concentration, c_2, as is expressed by Eq. [5.2]

concentration, c, in the solution being withdrawn from the mixing chamber, is expressed by

$$c = c_2 - (c_2 - c_1)f \qquad [5.2]$$

where c_1 is the solution's initial concentration in the mixing chamber, c_2 is its concentration in the reservoir chamber, and f is the remaining fraction of the combined volumes of the solutions initially present in both reservoirs. Linear gradients of increasing salt concentration are probably more commonly used than all other means of column elution. However, gradients of different shapes can be generated by using two or more chambers of different cross-sectional areas or programmed mixing devices.

Several Types of Ion Exchangers Are Available

Ion exchangers consist of charged groups covalently attached to a support matrix. The chemical nature of the charged groups determines the types of ions that bind to the ion exchanger and the strength with which they bind. The chemical and mechanical properties of the support matrix govern the flow characteristics, ion accessibility, and stability of the ion exchanger.

Several classes of materials, colloquially referred to as **resins,** are in general use as support matrices for ion exchangers in protein purification, including cellulose (Fig. 5-8) and cross-linked polyacrylamide or polydextran gels (see Section 5-3C). Table 5-2 contains descriptions of some commercially available ion exchangers in common use.

Cellulosic ion exchangers are among the materials most commonly employed to separate biological molecules. The cellulose, which is derived from wood or cotton, is lightly derivatized with ionic groups to form the ion exchanger. The most often used cellulosic anion exchanger is **diethylaminoethyl (DEAE)-cellulose,** whereas **carboxymethyl (CM)-cellulose** is the most popular cellulosic cation exchanger (Fig. 5-8).

Gel-type ion exchangers can have the same sorts of charged groups as do cellulosic ion exchangers. The advantage of gel-type ion exchangers is that they combine the separation properties of gel filtration (Section 5-3C) with

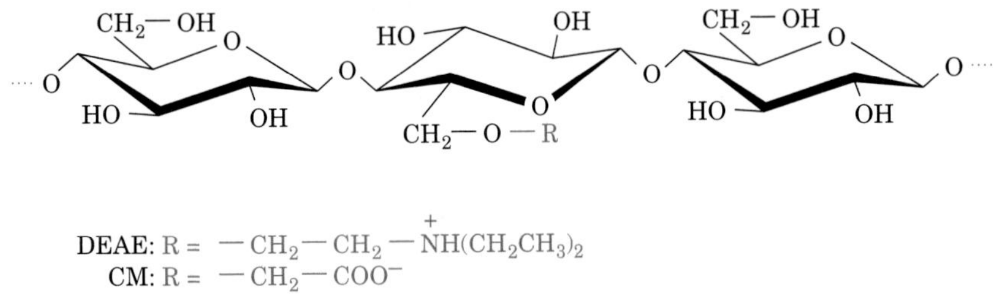

DEAE: $R = -CH_2-CH_2-\overset{+}{N}H(CH_2CH_3)_2$
CM: $R = -CH_2-COO^-$

FIGURE 5-8. Schematic diagram of cellulose-based ion exchangers.

TABLE 5-2.   SOME COMMONLY USED ION EXCHANGERS

Name[a]	Type	Ionizable group	Remarks
DEAE-Cellulose	Basic	Diethylaminoethyl $-CH_2CH_2N(C_2H_5)_2$	Used in fractionation of acidic and neutral proteins.
ECTEOLA-Cellulose	Basic	Mixed amines	Used for chromatography of nucleic acids.
CM-cellulose	Acidic	Carboxymethyl $-CH_2COOH$	Used for fractionation of basic and neutral proteins.
P-cellulose	Strongly and weakly acidic	$-OPO_3H_2$	Dibasic; binds basic proteins strongly.
DEAE-Sephadex	Basic cross-linked dextran gel	$-CH_2CH_2N(C_2H_5)_2$	Combined chromatography and gel filtration of acidic and neutral proteins.
CM-Sephadex	Acidic cross-linked dextran gel	$-CH_2COOH$	Combined chromatography and gel filtration of basic and neutral proteins.
Bio-Gel CM 100	Acidic cross-linked polyacrylamide gel	$-CH_2COOH$	Combined chromatography and gel filtration of basic and neutral proteins.

[a] Sephadex gels are manufactured by Pharmacia Fine Chemicals AB, Uppsala, Sweden and Bio-Gels are manufactured by BioRad Laboratories, Richmond, California, USA.

those of ion exchange. Because of their high degree of substitution of charged groups, which results from their porous structures, these gels have a higher loading capacity than do cellulosic ion exchangers.

One disadvantage of cellulosic and gel-type matrices is that they are easily compressed (usually by the high pressures resulting from attempts to increase the eluant flow rate) thereby greatly reducing eluant flow. This problem has been alleviated in recent years by the development of non-compressible matrices such as derivatized silica or coated glass beads. Such materials allow very high flow rates and pressures, even when they are very finely powdered, and hence permit more effective chromatographic separations (see HPLC in Section 5-3E).

B. Paper Chromatography

Paper chromatography, developed in 1941 by Archer Martin and Richard Synge, has played an indispensable role in biochemical analysis due to its ability to efficiently separate small molecules such as amino acids and oligopeptides and its requirement for only the simplest of equipment. Although paper chromatography has been largely supplanted by the more modern techniques discussed in this chapter, we briefly describe it here because of its historical importance and because many of its principles and ancillary techniques are directly applicable to these more modern techniques.

In paper chromatography (Fig. 5-9), a few drops of solution containing a mixture of the components to be separated are applied (spotted) ~2 cm above one end of a strip of

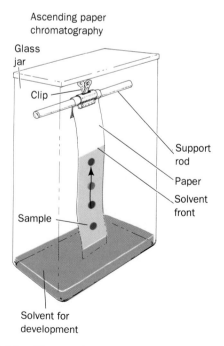

FIGURE 5-9. The experimental arrangement for chromatography.

filter paper. After drying, that end of the paper is dipped into a solvent mixture consisting of aqueous and organic components; for example, water/butanol/acetic acid in 4:5:1 ratio, 77% aqueous ethanol, 6:7:7 water/*t*-amyl alcohol/pyridine, or 1:6:3 water/*n*-propanol/concentrated NH₄OH. The paper should also be in contact with the equilibrium vapors of the solvent. The solvent soaks into the paper by capillary action because of the fibrous nature of the paper. The aqueous component of the solvent binds to the cellulose of the paper and thereby forms a stationary gellike phase with it. The organic component of the solvent continues migrating, thus forming the mobile phase.

The rates of migration of the various substances being separated are governed by their relative solubilities in the polar stationary phase and the nonpolar mobile phase. In a single step of the separation process, a given solute is distributed between the mobile and stationary phases according to its **partition coefficient,** an equilibrium constant defined as

$$K_p = \frac{\text{concentration in stationary phase}}{\text{concentration in mobile phase}} \qquad [5.3]$$

The molecules are therefore separated according to their polarities, with nonpolar molecules moving faster than polar ones.

After the solvent front has migrated an appropriate distance, the **chromatogram** is removed from the solvent and dried. The separated materials, if not colored, may be detected by several means:

1. Radioactively labeled materials may be located by a variety of radiation detection methods.

2. Materials that are fluorescent or quench the normal fluorescence of the paper can be seen under ultraviolet (UV) light.

3. Materials may be visualized by spraying the chromatogram with a reagent solution that forms a colored product upon reaction with the substance under investigation. For example, α-amino acids and other primary amines react with **ninhydrin** to form an intensely purple compound (Fig. 5-10). Secondary amines such as proline also react with ninhydrin but form a yellow compound.

The migration rate of a substance may be expressed according to the ratio

$$R_f = \frac{\text{distance traveled by substance}}{\text{distance traveled by solvent front}} \qquad [5.4]$$

For a given solvent system and paper type, each substance has a characteristic R_f value.

Paper chromatography can be used as a preparative technique for purifying small amounts of materials. A solution containing the substance of interest is applied in a line (streaked) across the bottom of a sheet of filter paper and the entire sheet is chromatographed as has been described. The substance of interest is located on the chromatogram by

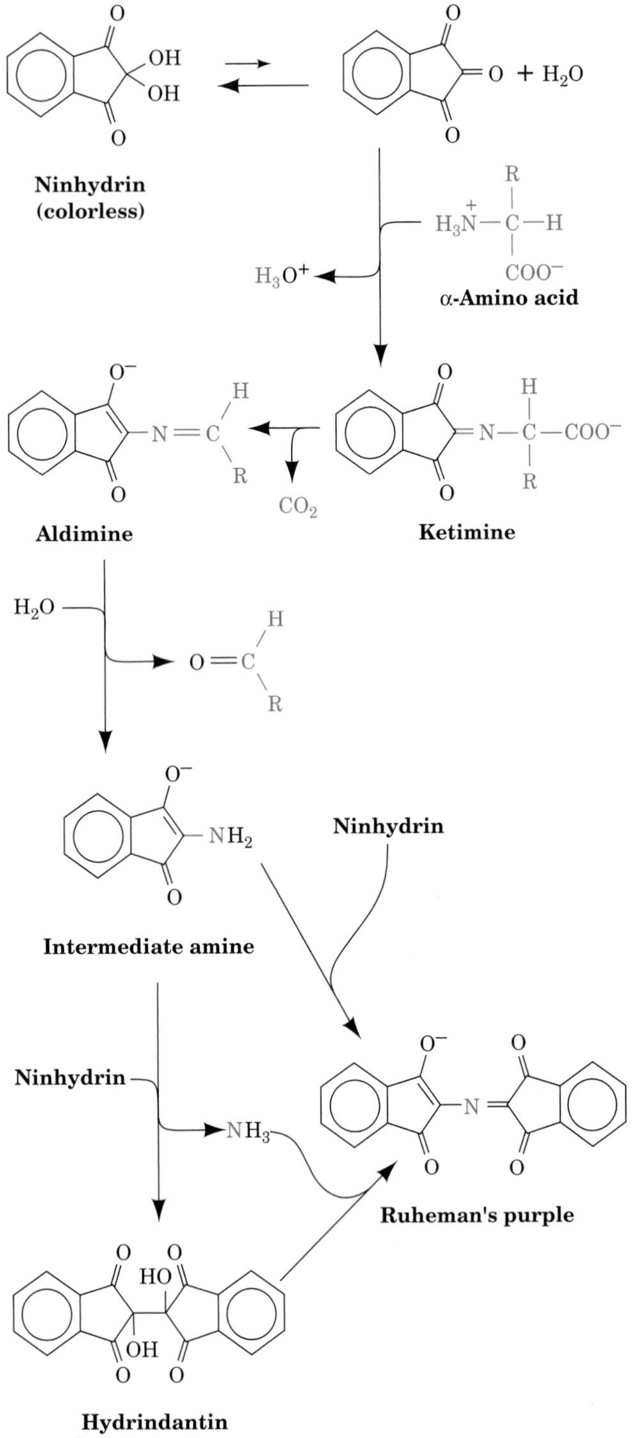

FIGURE 5-10.   The reaction of ninhydrin with α-amino acids. In the initial reaction, the α-amino acid forms a Schiff base (a ketimine) with the ninhydrin, which is subsequently oxidatively decarboxylated to form an aldimine. This hydrolyzes to form an intermediate amine which, in turn, can react directly or indirectly with a second molecule of ninhydrin to form the intensely colored **Ruheman's purple.** Note that only the nitrogen atom of this pigment arises from the α-amino acid. Primary amines and peptides also react with ninhydrin to form Ruheman's purple, but in these cases, a proton rather than CO_2 is lost to form the aldimine. Ninhydrin preparations used in quantitative analysis usually contain **hydrindantin** to assure maximum color development.

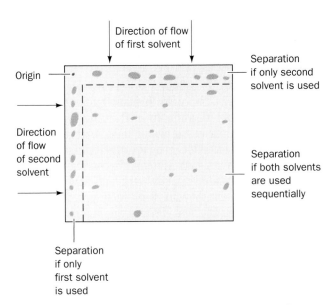

FIGURE 5-11.   Two-dimensional paper chromatography.

applying an appropriate detection technique to a small strip that has been cut out from the chromatogram along the direction of solvent migration. Finally, the band of purified substance is cut out and the purified substance recovered by eluting it from the paper with a suitable solvent.

A complex mixture that is incompletely separated in a single paper chromatogram can often be fully resolved by **two-dimensional paper chromatography** (Fig. 5-11). In this technique, a chromatogram is made as previously described except that the sample is spotted onto one corner of a sheet of filter paper and the chromatogram is run parallel to an edge of the paper. After the chromatography has been completed and the paper dried, the chromatogram is rotated 90° and is chromatographed parallel to the second edge using another solvent system. Since each compound migrates at a characteristic rate in a given solvent system, the second chromatographic step should greatly enhance the separation of the mixture into its components.

C. Gel Filtration Chromatography

*In **gel filtration chromatography**, which is also called **size exclusion** or **molecular sieve chromatography**, molecules are separated according to their size and shape.* The stationary phase in this technique consists of beads of a hydrated, spongelike material containing pores that span a relatively narrow size range of molecular dimensions. If an aqueous solution containing molecules of various sizes is passed through a column containing such "molecular sieves," the molecules that are too large to pass through the pores are excluded from the solvent volume inside the gel beads. These larger molecules therefore traverse the column more rapidly, that is, in a smaller eluent volume, than the molecules that pass through the pores (Fig. 5-12).

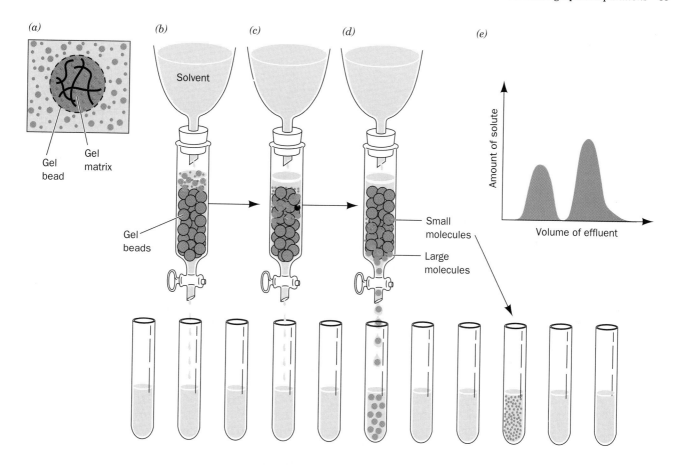

FIGURE 5-12. A schematic illustration of gel filtration chromatography. (*a*) A gel bead, whose periphery is represented by a dashed line, consists of a gel matrix (*wavy solid lines*) that encloses an internal solvent space. Smaller molecules (*small red dots*) can freely enter the internal solvent space of the gel bead from the external solvent space. However, larger molecules (*large blue dots*) are too large to penetrate the gel pores. (*b*) The sample solution begins to enter the gel column (in which the gel beads are now represented by brown spheres). (*c*) The

smaller molecules can penetrate the gel and consequently migrate through the column more slowly than the larger molecules that are excluded from the gel. (*d*) The larger molecules emerge from the column to be collected separately from the smaller molecules, which require additional solvent for elution from the column. (*e*) The elution diagram of the chromatogram indicating the complete separation of the two components with the larger component eluting first.

The molecular mass of the smallest molecule unable to penetrate the pores of a given gel is said to be the gel's **exclusion limit.** This quantity is to some extent a function of molecular shape because elongated molecules, as a consequence of their higher radius of hydration, are less likely to penetrate a given gel pore than spherical molecules of the same molecular volume.

The behavior of a molecule on a particular gel column can be quantitatively characterized. If V_x is the volume occupied by the gel beads and V_0, the **void volume,** is the volume of the solvent space surrounding the beads, then V_t, the total **bed volume** of the column, is simply their sum:

$$V_t = V_x + V_0 \qquad [5.5]$$

V_0 is typically ~35% of V_t.

The **elution volume** of a given solute, V_e, is the volume of solvent required to elute the solute from the column after it has first contacted the gel. The void volume of a column is

easily measured as the elution volume of a solute whose molecular mass is larger than the exclusion limit of the gel. The behavior of a particular solute on a given gel is therefore characterized by the ratio V_e/V_0, the **relative elution volume,** a quantity that is independent of the size of the particular column used.

Molecules with molecular masses ranging below the exclusion limit of a gel will elute from the gel in the order of their molecular masses, with the largest eluting first. This is because the pore sizes in any gel vary over a limited range so that larger molecules have less of the gel's interior volume available to them than smaller molecules do. This effect is the basis of gel filtration chromatography.

Gel Filtration Chromatography Can Be Used to Estimate Molecular Masses

There is a linear relationship between the relative elution volume of a substance and the logarithm of its molecular

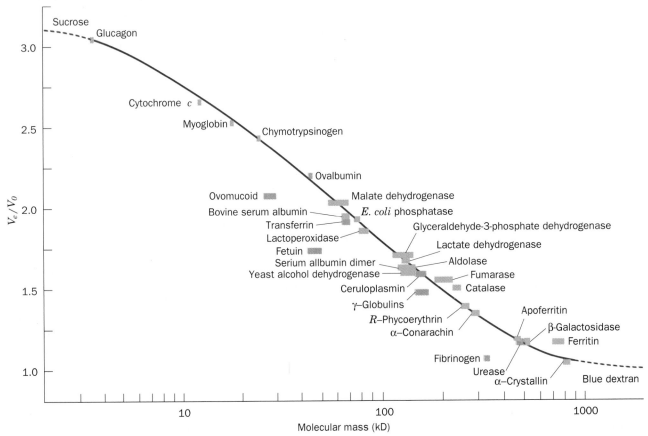

FIGURE 5-13. A plot of relative elution volume versus the logarithm of molecular mass for a variety of proteins on a cross-linked dextran column (Sephadex G-200) at pH 7.5. Orange bars represent glycoproteins (proteins with attached carbohydrate groups). [After Andrews, P., *Biochem. J.,* **96,** 597 (1965).]

mass over a considerable molecular mass range (Fig. 5-13). If a plot such as Fig. 5-13 is made for a particular gel filtration column using macromolecules of known molecular masses, *the molecular mass of an unknown substance can be estimated from its position on the plot. The precision of this technique is limited by the accuracy of the underlying assumption that the known and unknown macromolecules have identical shapes.* Nevertheless, gel filtration chromatography is often used to estimate molecular masses because it can be applied to quite impure samples (providing that the molecule of interest can be identified) and because it can be rapidly carried out using simple equipment.

Most Gels Are Made from Dextran, Agarose, or Polyacrylamide

The most commonly used materials for making chromatographic gels are **dextran** (a high molecular mass polymer of glucose produced by the bacterium *Leuconostoc mesenteroides*), **agarose** (a linear polymer of alternating D-galactose and 3,6-anhydro-L-galactose from red algae), and **polyacrylamide** (see Section 5-4B). The properties of several gels that are commonly employed in separating biological molecules are listed in Table 5-3. The porosity of dextran-based gels, sold under the trade name Sephadex, is

TABLE 5-3. Some Commonly Used Gel Filtration Materials

Name[a]	Type	Fractionation Range (kD)
Sephadex G-10	Dextran	0.05–0.7
Sephadex G-25	Dextran	1–5
Sephadex G-50	Dextran	1–30
Sephadex G-100	Dextran	4–150
Sephadex G-200	Dextran	5–600
Bio-Gel P-2	Polyacrylamide	0.1–1.8
Bio-Gel P-6	Polyacrylamide	1–6
Bio-Gel P-10	Polyacrylamide	1.5–20
Bio-Gel P-30	Polyacrylamide	2.4–40
Bio-Gel P-100	Polyacrylamide	5–100
Bio-Gel P-300	Polyacrylamide	60–400
Sepharose 6B	Agarose	10–4,000
Sepharose 4B	Agarose	60–20,000
Sepharose 2B	Agarose	70–40,000

[a] Sephadex and Sepharose gels are products of Pharmacia Fine Chemicals AB.; Bio-Gel gels are manufactured by BioRad Laboratories.

controlled by the molecular mass of the dextran used and the introduction of glyceryl ether units that cross-link the hydroxyl groups of the polyglucose chains. The several classes of Sephadex that are available have exclusion limits between 0.7 and 800 kD. The pore size in polyacrylamide gels is similarly controlled by the extent of cross-linking of neighboring polyacrylamide molecules (Section 5-4B). They are commercially available under the trade name of Bio-Gel and have exclusion limits between 0.2 and 400 kD. Very large molecules and supramolecular assemblies can be separated using agarose gels, which have exclusion limits ranging up to 40,000 kD.

Gel filtration is often used to "desalt" a protein solution. For example, an ammonium sulfate-precipitated protein can be easily freed of ammonium sulfate by dissolving the protein precipitate in a minimum volume of suitable buffer and applying this solution to a column of gel with an exclusion limit less than the molecular mass of the protein. Upon elution of the column with buffer, the protein will precede the ammonium sulfate through the column.

Dextran and polyacrylamide gels can be derivatized with ionizable groups such as DEAE and CM to form ion exchange gels (Section 5-3A). Substances that are chromatographed on these gels are therefore subject to separation according to their ionic charges as well as their sizes and shapes.

Dialysis Is a Form of Molecular Filtration

Dialysis is a process that separates molecules according to size through the use of semipermeable membranes containing pores of less than macromolecular dimensions. These pores allow small molecules, such as those of solvents, salts, and small metabolites, to diffuse across the membrane but block the passage of larger molecules. **Cello-phane** (cellulose acetate) is the most commonly used dialysis material, although many other substances such as **nitrocellulose** and **collodion** are similarly employed.

Dialysis (which is not considered to be a form of chromatography) is routinely used to change the solvent in which macromolecules are dissolved. A macromolecular solution is sealed inside a dialysis bag (usually made by knotting dialysis membrane tubing at both ends), which is immersed in a relatively large volume of the new solvent (Fig. 5-14a). After several hours of stirring, the solutions will have equilibrated but with the macromolecules remaining inside the dialysis bag (Fig. 5-14b). The process can be repeated several times to replace one solvent system completely by another.

Dialysis can be used to concentrate a macromolecular solution by packing a filled dialysis bag in a polymeric desiccant, such as **polyethylene glycol** [$HOCH_2(CH_2$—O—$CH_2)_nCH_2OH$], which cannot penetrate the membrane. Concentration is effected as water diffuses across the membrane to be absorbed by the polymer. A related technique that is used to concentrate macromolecular solutions is known as **ultrafiltration.** Here a macromolecular solution is forced, under pressure, through a semipermeable membranous disk or bag. Solvent and small solutes pass through the membrane, leaving behind a more concentrated macromolecular solution. Since ultrafiltration membranes with different pore sizes are available, ultrafiltration can also be used to separate different sized macromolecules.

D. Affinity Chromatography

A striking characteristic of many proteins is their ability to bind specific molecules tightly but noncovalently. This property can be used to purify such proteins by **affinity chromatography** (Fig. 5-15). In this technique, a molecule, known as a **ligand,** which specifically binds to the protein of interest, is covalently attached to an inert and porous matrix. *When an impure protein solution is passed through this chromatographic material, the desired protein binds to the immobilized ligand, whereas other substances are washed through the column with the buffer. The desired protein can then be recovered in highly purified form by changing the elution conditions such that the protein is released from the chromatographic matrix.* The great advantage of affinity chromatography is its ability to exploit the desired protein's unique biochemical properties rather than the small differences in physicochemical properties between proteins that other chromatographic methods must utilize.

The chromatographic matrix in affinity chromatography must be chemically inert, have high porosity, and have large numbers of functional groups capable of forming covalent linkages to ligands. Of the few materials available that meet these criteria, agarose, which has numerous free hydroxyl groups is by far the most widely used. If the ligand has a

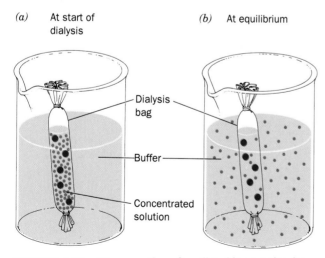

FIGURE 5-14. The separation of small and large molecules by dialysis. (*a*) Only small molecules can diffuse through the pores in the bag. (*b*) At equilibrium the concentrations of small molecules are nearly the same inside and outside the bag, whereas the macromolecules remain in the bag.

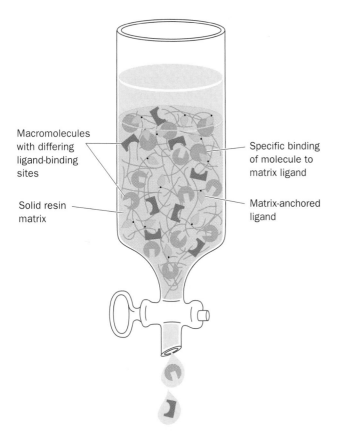

FIGURE 5-15. The separation of macromolecules by affinity chromatography. The cutout squares, semicircles, and triangles are schematic representations of ligand-binding sites on macromolecules. Only those ligand-binding sites represented by the orange circles with triangle cutouts specifically bind to the chromatographic matrix-anchored ligands (*yellow*).

Macromolecules with differing ligand-binding sites

Specific binding of molecule to matrix ligand

Solid resin matrix

Matrix-anchored ligand

The ligand used in the affinity chromatography isolation of a particular protein must have an affinity high enough to immobilize the protein on the agarose gel but not so high as to prevent its subsequent release. If the ligand is a substrate for an enzyme being isolated, the chromatography conditions must be such that the enzyme does not function catalytically or the ligand will be destroyed.

After a protein has been bound to an affinity chromatography column and washed free of impurities, it must be released from the column. One method of doing so is to elute the column with a solution of a compound that has higher affinity for the protein-binding site than the bound ligand. Another is to alter the solution conditions such that the protein–ligand complex is no longer stable, for example, by changes in pH, ionic strength, and/or temperature. However, care must be taken that the solution conditions are not so inhospitable to the protein being isolated that it is irreversibly damaged. An example of protein purification by affinity chromatography is shown in Fig. 5-18.

Affinity chromatography has been used to isolate such substances as enzymes, antibodies, transport proteins, hormone receptors, membranes, and even whole cells. For instance, the protein hormone **insulin** (Section 25-2) has been covalently attached to agarose and used to isolate **insulin**

primary amino group that is not essential for its binding to the protein of interest, the ligand can be covalently linked to the agarose in a two-step process (Fig. 5-16):

1. Agarose is reacted with **cyanogen bromide** to form an "activated" but stable intermediate (which is commercially available).

2. Ligand reacts with the activated agarose to form covalently bound product.

Many proteins are unable to bind their cyanogen bromide–coupled ligands due to steric interference with the agarose matrix. This problem is alleviated by attaching the ligand to the agarose by a flexible "spacer" group. This is conveniently done through the use of commercially available activated resins. One such resin is "epoxy-activated" agarose, in which a spacer group (containing, e.g., a chain of 12 atoms) links the resin to a reactive epoxy group. The epoxy group can react with many of the nucleophilic groups on ligands, thereby permitting the ligand of choice to be covalently linked to the agarose via a tether of defined length (Figure 5-17).

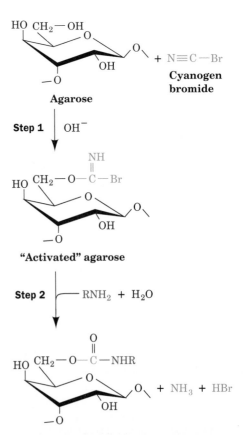

FIGURE 5-16. The formation of cyanogen bromide–activated agarose (*top*) and its reaction with a primary amine to form a covalently attached ligand for affinity chromatography (*bottom*).

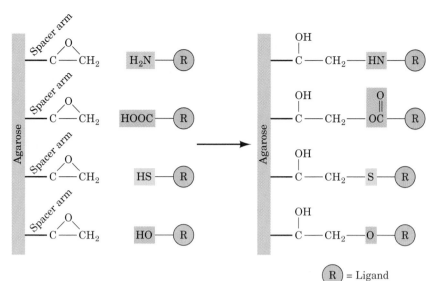

FIGURE 5-17. Examples of the various types of nucleophilic groups that can be covalently attached to epoxy-activated agarose via reaction with its epoxide groups.

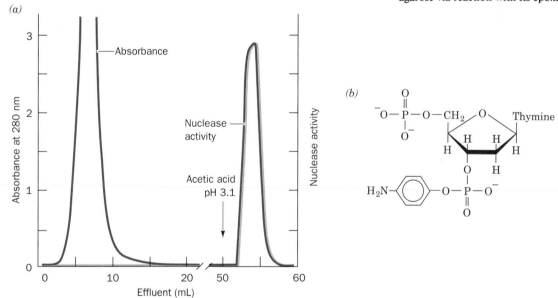

FIGURE 5-18. (a) The purification of **staphylococcal nuclease** (a DNA-hydrolyzing enzyme) by affinity chromatography on a nuclease-specific agarose column. The compound shown in Part b whose diphosphothymidine moiety specifically binds to the enzyme, was covalently linked to cyanogen bromide–activated agarose. An 0.8 × 5-cm column was equilibrated with 0.05M borate buffer, pH 8.0, containing 0.01M CaCl₂.

Approximately 40 mg of partially purified material was applied to the column in 3.2 mL of the same buffer. After 50 mL of buffer had been passed through the column at a flow rate of 70 mL·h⁻¹, 0.1M acetic acid was added to elute the enzyme. All of the original enzymatic activity, comprising 8.2 mg of pure nuclease, was recovered. [After Cuatrecasas, P., Wilchek, M., and Anfinsen, C. B., *Proc. Natl. Acad. Sci.* **61,** 636 (1968).]

receptor protein, a cell-surface protein whose other properties were unknown and which is present in only very small amounts. The separation power of affinity chromatography for a specific protein is often far greater than that of other chromatographic techniques (e.g., Table 5-4). Indeed, the replacement of many chromatographic steps in a tried-and-true protein isolation protocol by a single affinity chromatographic step often results in purer protein in higher yield.

Immunoaffinity Chromatography Employs the Binding Specificity of Monoclonal Antibodies

A melding of immunochemistry with affinity chromatography has generated a powerful method for purifying biological molecules. Cross-linking monoclonal antibodies (Section 5-1D) to a suitable column material yields a substance that will bind only the protein against which the antibody has been raised. Such **immunoaffinity chromatography** can achieve a 10,000-fold purification in a single

TABLE 5-4. **PURIFICATION OF RAT LIVER GLUCOKINASE**

Stage	Specific Activity (nkat·g⁻¹)[a]	Yield (%)	Fold[b] Purification
Scheme A: A "traditional" chromatographic procedure			
1. Liver supernatant	0.17	100	1
2. (NH₄)₂SO₄ precipitate	c	c	c
3. DEAE-Sephadex chromatography by stepwise elution with KCl	4.9	52	29
4. DEAE-Sephadex chromatography by linear gradient elution with KCl	23	45	140
5. DEAE-cellulose chromatography by linear gradient elution with KCl	44	33	260
6. Concentration by stepwise KCl elution from DEAE-Sephadex	80	15	480
7. Bio-Gel P-225 chromatography	130	15	780
Scheme B: An affinity chromatography procedure			
1. Liver supernatant	0.092	100	1
2. DEAE-cellulose chromatography by stepwise elution with KCl	20.1	104	220
3. Affinity chromatography[d]	**420**	**83**	**4500**

[a] A **katal** (abbreviation **kat**) is the amount of enzyme that catalyzes the transformation of 1 mol of substrate per second under standard conditions. One nanokatal (nkat) is 10^{-9} kat.

[b] Calculated from specific activity; the first step is arbitrarily assigned unity.

[c] The activity could not be accurately measured as this stage because of uncertainty in correcting for contamination by other enzymes.

[d] The affinity chromatography material was made by linking glucosamine (an inhibitor of glucokinase) through a 6-aminohexanoyl spacer arm to NCBr activated agarose.

Source: Cornish-Bowden, A., *Fundamentals of Enzyme Kinetics, p.* 48, Butterworths (1979) as adapted from Parry, M.J. and Walker, D.G., *Biochem. J.* **99,** 266 (1966) for Scheme A and from Holroyde, M.J., Allen, B.M., Storer, A.C., Warsey, A.S., Chesher, J.M.E., Trayer, I.P., Cornish-Bowden, A., and Walker, D.G., *Biochem. J.* **153,** 363 (1976) for Scheme B.

step. Disadvantages of immunoaffinity chromatography include the technical difficulty of producing monoclonal antibodies and the harsh conditions that are often required to elute the bound protein.

E. Other Chromatographic Techniques

A number of other chromatographic techniques are of biochemical value. These are briefly discussed below.

Adsorption Chromatography Separates Nonpolar Substances

In **adsorption chromatography** (the original chromatographic method), molecules are physically adsorbed on the surface of insoluble substances such as **alumina** (Al_2O_3), charcoal, **diatomaceous earth** (also called **kieselguhr,** the silicaceous fossils of unicellular organisms known as diatoms), finely powdered sucrose, or **silica gel** (silicic acid), through van der Waals and hydrogen bonding associations. The molecules are then eluted from the column by a pure solvent such as chloroform, hexane, or ethyl ether or by a mixture of such solvents. The separation process is based on the partition of the various substances between the polar column material and the nonpolar solvent. This procedure is most often used to separate nonpolar molecules rather than proteins.

Hydroxyapatite Chromatography Separates Proteins

Proteins are adsorbed by gels of crystalline **hydroxyapatite,** an insoluble form of calcium phosphate with empirical formula $Ca_5(PO_4)_3OH$. The separation of the proteins occurs upon gradient elution of the column with phosphate buffer (the presence of other anions is unimportant). The physicochemical basis of this fractionation procedure is not fully understood but apparently involves the adsorption of anions to the Ca^{2+} sites and cations to the PO_4^{3-} sites of the hydroxyapatite crystalline lattice.

Thin Layer Chromatography Is Used to Separate Organic Molecules

In **thin layer chromatography (TLC),** a thin (~0.25 mm) coating of a solid material spread on a glass or plastic plate is utilized in a manner similar to that of the paper in paper chromatography. In the case of TLC, however, the chromatographic material can be a variety of substances such as ion exchangers, gel filtration agents, and physical adsorbents. According to the choice of solvent for the mobile phase, the separation may be based on adsorption, partition, gel filtration, or ion exchange processes, or some combination of these. The advantages of thin layer chromatography in convenience, rapidity, and high resolution have led to its routine use in the analysis of organic molecules.

Reverse-Phase Chromatography Separates Nonpolar Substances Including Denatured Proteins

Reverse-phase chromatography (RPC) is a form of liquid–liquid partition chromatography in which the polar character of the phases is reversed relative to that of paper chromatography: The stationary phase consists of a nonpolar liquid immobilized on a relatively inert solid and the mobile phase is a more polar liquid. Reverse-phase chromatography was first developed to separate mixtures of nonpolar substances such as lipids but has also been found to be effective in separating polar substances such as oligonucleotides and proteins, provided that they have exposed nonpolar areas. Although nonpolar side chains tend to inhabit the water-free interiors of native proteins (Section 7-3B), denaturation results in the exposure of these side chains to the solvent. Even when still in the native state, a significant fraction of these hydrophobic groups are at least partially exposed to the solvent at the protein surface. Consequently, under suitable conditions, proteins hydrophobically interact with the nonpolar groups on an immobilized matrix. The hydrophobic interactions in RPC are strong, so the eluting mobile phase must be highly nonpolar (containing high concentrations of organic solvents such as acetonitrile) to dislodge adsorbed substances from the stationary phase. RPC therefore usually denatures proteins.

Hydrophobic Interaction Chromatography Separates Native Proteins on the Basis of Surface Hydrophobicity

Hydrophobic interactions form the basis not only of RPC but of **hydrophobic interaction chromatography (HIC)**. However, whereas the stationary phase in RPC is strongly hydrophobic in character, resulting in protein denaturation, in HIC it is a hydrophilic substance, such as an agarose gel, that is only lightly substituted with hydrophobic groups, usually octyl or phenyl residues. The resulting hydrophobic interactions in HIC are therefore relatively weak, so proteins maintain their native structures. The eluants in HIC, whose gradients must progressively reduce these weak hydrophobic interactions, are aqueous buffers with, for example, decreasing salt concentrations (hydrophobic interactions are strengthened by increased ionic strength; Section 5-2A), increasing concentrations of detergents, or increasing pH. Thus, HIC separates native proteins according to their degree of surface hydrophobicity, a criterion that differs from those on which other types of chromatography are based.

HPLC Has Permitted Greatly Improved Separations

In **high-performance liquid chromatography (HPLC)**, a separation may be based on adsorption, ion exchange, size exclusion, HIC, or RPC as previously described. The separations are greatly improved, however, through the use of high-resolution columns and the column retention times are much reduced. The narrow and relatively long columns are packed with a noncompressible matrix of fine glass or plastic beads coated with a thin layer of the stationary phase. Alternatively, the matrix may consist of **silica**

```
   OH      OH      OH
   |       |       |
  Si — O — Si — O — Si
   |       |       |
   O — Si — O — Si — O
   |       |       |
  Si — O — Si — O — Si
```
Silica

whose available hydroxyl groups can be derivatized with many of the commonly used functional groups of ion exchange chromatography, RPC, HIC, or affinity chromatography. The mobile phase is one of the solvent systems previously discussed including gradient elutions with binary or even ternary mixtures. In the case of HPLC, however, the mobile phase is forced through the tightly packed column at pressures of up to 5000 psi (pounds per square inch) leading to greatly reduced analysis times. The elutants are detected as they leave the column by such methods as UV absorption, refractive index, or fluorescence measurements. The advantages of HPLC are

1. Its high resolution, which permits the routine purification of mixtures that have defied separation by other techniques.
2. Its speed, which permits most separations to be accomplished in significantly < 1 h.
3. Its high sensitivity, which, in favorable cases, permits the quantitative estimation of less than picomole quantities of materials.
4. Its capacity for automation.

Thus, few biochemistry laboratories now function without access to an HPLC system. HPLC is also often utilized in the clinical analyses of body fluids because it can rapidly, routinely, and automatically yield reliable quantitative estimates of nanogram quantities of biological materials such as vitamins, steroids, lipids, and drug metabolites.

4. ELECTROPHORESIS

Electrophoresis, the migration of ions in an electric field, is widely used for the analytical separation of biological molecules. The laws of electrostatics state that the electrical force, $F_{electric}$, on an ion with charge q in an electric field of strength E is expressed by

$$F_{electric} = qE \qquad [5.6]$$

The resulting electrophoretic migration of the ion through the solution is opposed by a frictional force

$$F_{friction} = vf \qquad [5.7]$$

where v is the rate of migration (velocity) of the ion and f is its **frictional coefficient.** *The frictional coefficient is a measure of the drag that the solution exerts on the moving ion and is dependent on the size, shape, and state of solvation of the ion as well as on the viscosity of the solution.* In a constant electric field, the forces on the ion balance each other:

$$qE = vf \qquad [5.8]$$

so that each ion moves with a constant characteristic velocity. An ion's **electrophoretic mobility,** μ, is defined

$$\mu = \frac{v}{E} = \frac{q}{f} \qquad [5.9]$$

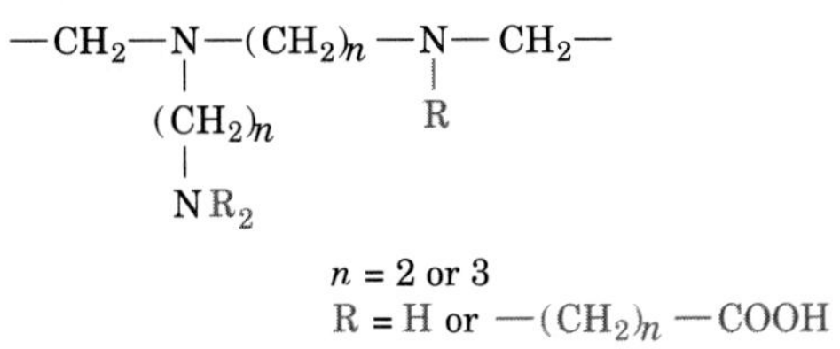

$$-CH_2-N-(CH_2)_n-N-CH_2-$$

$$\underset{\underset{NR_2}{|}}{(CH_2)_n} \qquad R$$

$$n = 2 \text{ or } 3$$
$$R = H \text{ or } -(CH_2)_n-COOH$$

FIGURE 5-29. The general formula of the polyampholytes used in isoelectric focusing.

*electrophoresed through a solution having a stable **pH gradient** in which the pH smoothly increases from anode to cathode, each protein will migrate to the position in the pH gradient corresponding to its isoelectric point.* If a protein molecule diffuses away from this position, its net charge will change as it moves into a region of different pH and the resulting electrophoretic forces will move it back to its isoelectric position. Each species of protein is thereby "focused" into a narrow band about its isoelectric point that may be as thin as 0.01 pH unit.

A pH gradient produced by mixing two different buffers together in continuously varying ratios is unstable in an electric field because the buffer ions migrate to the electrode of opposite polarity. Rather, the pH gradient in isoelectric focusing is formed by a mixture of low molecular mass (300–600 D) oligomers bearing aliphatic amino and carboxylic acid groups (Fig. 5-29) that have a range of isoelectric points. Under the influence of an electric field, a solution of these **polyampholytes** will segregate according to their isoelectric points such that the most acidic gather at the anode with the progressively more basic positioning themselves ever closer to the cathode. The pH gradient, which is maintained by the electric field, arises from the buffering action of these polyampholytes.

The pH gradient can be stabilized against convection by preparing it in a sucrose concentration gradient in which the sucrose concentration and therefore the density of the viscous solution increases from top to bottom. Alternatively, convection may be eliminated through use of a lightly cross-linked polyacrylamide gel in the form of a tube or a thin slab. The gel often contains urea

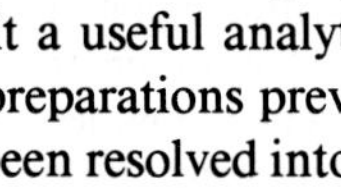

$$\overset{\overset{\textstyle O}{\|}}{H_2N-C-NH_2}$$

Urea

at a concentration of ~6*M*. This powerful protein denaturing agent, unlike SDS, is uncharged and hence cannot directly affect the charge of a protein.

The fact that isoelectric focusing separates proteins into sharp bands makes it a useful analytical and preparative tool. Many protein preparations previously thought to be homogeneous have been resolved into several components by isoelectric focusing. Isoelectric focusing can be com-

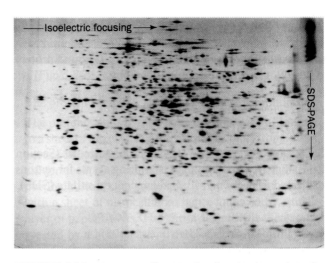

FIGURE 5-30. An autoradiogram showing the separation of *E. coli* proteins using isoelectric focusing in one dimension and SDS–PAGE in the second dimension. A 10-μg sample of proteins from *E. coli* that had been labeled with [¹⁴C]amino acids were subjected to isoelectric focusing in a 2.5 × 130-mm tube of urea-containing polyacrylamide gel. The gel was then extruded from its tube, placed at one edge of an SDS–polyacrylamide slab gel, and subjected to electrophoresis. Over 1000 spots were counted on the original autoradiogram, which resulted from an 825-h exposure. [Courtesy of Patrick O'Farrell, University of California at San Francisco.]

bined with electrophoresis in an extremely powerful two-dimensional separation technique (Fig. 5-30).

E. Capillary Electrophoresis

Although gel electrophoresis in its various forms is a common and highly effective method for separating charged molecules, it typically requires several hours for a run and is difficult to quantitate and automate. These disadvantages are largely overcome through the use of capillary electrophoresis (CE), a technique in which electrophoresis is carried out in very thin (1- to 10-μm inner diameter) capillary tubes made of quartz, glass, or plastic. Such narrow capillaries rapidly dissipate heat and hence permit the use of high electric fields (typically 100 to 300 V·cm⁻¹, about 10 times that of most other electrophoretic techniques) which reduces separation times to a few minutes. These rapid separations, in turn, minimize band broadening caused by diffusion, thereby yielding extremely sharp separations. Capillaries can be filled with buffer (as in moving boundary electrophoresis, but here the capillary's narrow bore all but eliminates convective mixing), SDS–polyacrylamide gel (separation according to molecular mass; Section 5-4C), or polyampholytes (isoelectric focusing; Section 5-4D). These CE techniques have extremely high resolution and can be automated in much the same way as is HPLC, that is, with automatic sample loading and on-line sample detection. Since CE can only separate small amounts of material, it is largely limited to use as an analytical technique.

5. ULTRACENTRIFUGATION

If a container of sand and water is shaken and then allowed to stand quietly, the sand will rapidly sediment to the bottom of the container due to the influence of the Earth's gravity (an acceleration $g = 9.81$ m·s^{-2}). Yet macromolecules in solution, which experience the same gravitational field, do not exhibit any perceptible sedimentation because their random thermal (Brownian) motion keeps them uniformly distributed throughout the solution. *Only when they are subjected to enormous accelerations will the sedimentation behavior of macromolecules begin to resemble that of sand grains.*

The ultracentrifuge, which was developed around 1923 by the Swedish biochemist The Svedberg, can attain rotational speeds as high as 80,000 rpm (revolutions per minute) so as to generate centrifugal fields in excess of 600,000g. Using this instrument, Svedberg first demonstrated that proteins are macromolecules with homogeneous compositions and that many proteins are composed of subunits. More recently, ultracentrifugation has become an indispensable tool for the isolation of proteins, nucleic acids, and subcellular particles. In this section we outline the theory and practice of ultracentrifugation.

A. Sedimentation

The rate at which a particle sediments in the ultracentrifuge is related to its mass. The force, $F_{sedimentation}$, acting to sediment a particle of mass m that is located a distance r from a point about which it is revolving with angular velocity ω (in

rad · s^{-1}), is the centrifugal force ($m\omega^2 r$) on the particle less the buoyant force ($V_p\rho\omega^2 r$) exerted by the solution:

$$F_{sedimentation} = m\omega^2 r - V_p\rho\omega^2 r \qquad [5.10]$$

Here V_p is the particle volume and ρ is the density of the solution. However, the motion of the particle through the solution, as we have seen in our study of electrophoresis, is opposed by the frictional force:

$$F_{friction} = vf \qquad [5.7]$$

where $v = dr/dt$ is the rate of migration of the sedimenting particle and f is its frictional coefficient. The particle's frictional coefficient can be determined from measurements of its rate of diffusion.

Under the influence of gravitational (centrifugal) force, the particle accelerates until the forces on it exactly balance:

$$m\omega^2 r - V_p\rho\omega^2 r = vf \qquad [5.11]$$

The mass of 1 mol of particles, M, is

$$M = mN \qquad [5.12]$$

where N is Avogadro's number (6.022×10^{23}). Thus, a particle's volume, V_p, may be expressed in terms of its molar mass:

$$V_p = \overline{V}m = \frac{\overline{V}M}{N} \qquad [5.13]$$

where $\overline{V}$, the particle's **partial specific volume**, is the volume change when 1 g (dry weight) of particles is dissolved in an infinite volume of the solute. For most proteins dissolved in pure water at 20°C, $\overline{V}$ is near 0.73 cm^3·g^{-1} (Table 5-5). Indeed, for proteins of known amino acid composi-

TABLE 5-5. PHYSICAL CONSTANTS OF SOME PROTEINS

Protein	Molecular Mass (kD)	Partial Specific Volume, $\overline{V}_{20,w}$ (cm^3·g^{-1})	Sedimentation Coefficient, $s_{20,w}$ (S)	Frictional Ratio f/f_0
Lipase (milk)	6.7	0.714	1.14	1.190
Ribonuclease A (bovine pancreas)	12.6	0.707	2.00	1.066
Cytochrome c (bovine heart)	13.4	0.728	1.71	1.190
Myoglobin (horse heart)	16.9	0.741	2.04	1.105
α-Chymotrypsin (bovine pancreas)	21.6	0.736	2.40	1.130
Crotoxin (rattlesnake)	29.9	0.704	3.14	1.221
Concanavalin B (jack bean)	42.5	0.730	3.50	1.247
Diphtheria toxin	70.4	0.736	4.60	1.296
Cytochrome oxidase (*P. aeruginosa*)	89.8	0.730	5.80	1.240
Lactate dehydrogenase H (chicken)	150	0.740	7.31	1.330
Catalase (horse liver)	222	0.715	11.20	1.246
Fibrinogen (human)	340	0.725	7.63	2.336
Hemocyanin (squid)	612	0.724	19.50	1.358
Glutamate dehydrogenase (bovine liver)	1015	0.750	26.60	1.250
Turnip yellow mosaic virus protein	3013	0.740	48.80	1.470

Source: Smith, M.H., *in* Sober, H.A. (Ed.), *Handbook of Biochemistry and Molecular Biology* (2nd ed.), *p.* C-10, CRC Press (1970).

tion, $\overline{V}$ is closely approximated by the sum of the partial specific volumes of its component amino acid residues, thereby indicating that the atoms in proteins are closely packed (Section 7-3B).

A Particle may be Characterized by Its Sedimentation Rate

Substituting Eqs. [5.12] and [5.13] into Eq. [5.11] yields

$$vf = \frac{M(1 - \overline{V}\rho)\omega^2 r}{N} \qquad [5.14]$$

Now define the **sedimentation coefficient,** s, as

$$s = \frac{v}{\omega^2 r} = \frac{1}{\omega^2}\left(\frac{d\ln r}{dt}\right) = \frac{M(1 - \overline{V}\rho)}{Nf} \qquad [5.15]$$

The sedimentation coefficient, a quantity that is analogous to the electrophoretic mobility (Eq. [5.9]) in that it is a velocity per unit force, is usually expressed in units of 10^{-13} s, which are known as **Svedbergs (S).** For the sake of uniformity, the sedimentation coefficient is customarily

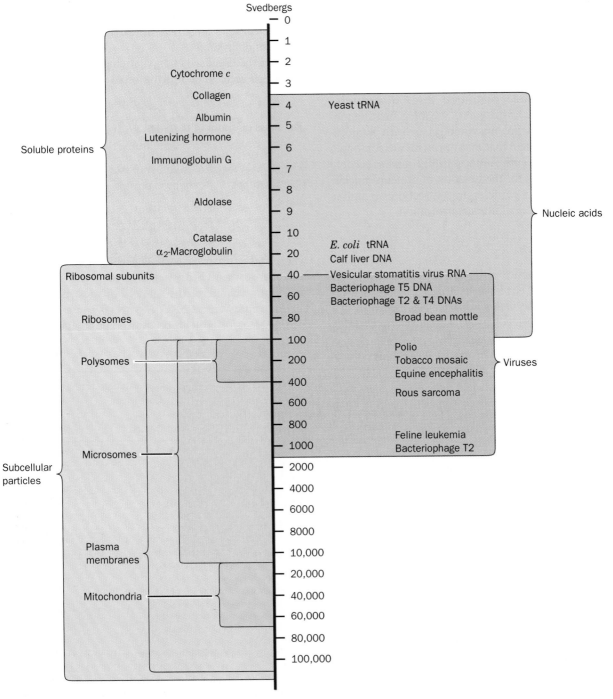

FIGURE 5-31. The sedimentation coefficients in Svedbergs for some biological materials. [After a diagram supplied by Beckman Instruments, Inc.]

corrected to the value that would be obtained at 20°C in a solvent with the density and viscosity of pure water. This is symbolized $s_{20,w}$. Table 5-5 and Fig. 5-31 indicate the values of $s_{20,w}$ in Svedbergs for a variety of biological materials.

Equation [5.15] indicates that a particle's mass, $m = M/N$, can be determined from the measurement of its sedimentation coefficient, s, if its frictional coefficient, f, is known. Indeed, before about 1970, most macromolecular mass determinations were made using the **analytical ultracentrifuge**, a device in which the sedimentation rates of molecules under centrifugation can be optically measured (the masses of macromolecules are too high to be accurately determined by such classical physical techniques as melting point depression or osmotic pressure measurements). Although the advent of much simpler molecular mass determination methods, such as gel filtration chromatography (Section 5-3C) and SDS–PAGE (Section 5-4C), had caused analytical ultracentrifugation to largely fade from use, recently developed instrumentation has led to a resurgence in the use of analytical ultracentrifugational measurements. They are particularly useful in characterizing systems of associating macromolecules.

The Frictional Ratio Is Indicative of Molecular Solvation and Shape

For an unsolvated spherical particle of radius r_p, the frictional coefficient is determined according to the **Stokes equation:**

$$f = 6\pi\eta r_p \qquad [5.16]$$

where η is the **viscosity** of the solution. Solvation increases the frictional coefficient of a particle by increasing its effective or **hydrodynamic volume**. Furthermore, f is minimal when the particle is a sphere. This is because a nonspherical particle has a larger surface area than a sphere of equal volume and therefore must, on the average, present a greater surface area towards the direction of movement than a sphere.

The frictional coefficient, f, of a particle of known mass and partial specific volume can be ultracentrifugationally determined using Eq. [5.15]. The effective or **Stokes radius** of a particle in solution can be calculated by solving Eq. [5.16] for r_p, given the experimentally determined values of f and η. Conversely, the minimal frictional coefficient of a particle, f_0, can be calculated from the mass and the partial specific volume of the particle by assuming it to be spherical ($V_p = \frac{4}{3}\pi r_p^3$) and unsolvated:

$$f_0 = 6\pi\eta\left(\frac{3M\overline{V}}{4\pi N}\right)^{1/3} \qquad [5.17]$$

If the **frictional ratio**, f/f_0, of a particle is much greater than unity, it must be concluded that the particle is highly solvated and/or significantly elongated. The frictional ratios of a selection of proteins are presented in Table 5-5. The "globular" proteins, which are known from structural studies to be relatively compact and spheroidal (Section 7-3B), have frictional ratios ranging up to ~1.5. Fibrous molecules

such as DNA and the blood clotting protein **fibrinogen** (Section 34-1A) have larger frictional ratios. Upon denaturation, the frictional coefficients of globular proteins increase by as much as twofold because denatured proteins assume flexible and fluctuating **random coil** conformations in which all parts of the molecule are in contact with solvent (Section 7-1D).

B. Preparative Ultracentrifugation

Preparative ultracentrifuges, which as their name implies, are designed for sample preparation, differ from analytical ultracentrifuges in that they lack sample observation facilities. Preparative rotors contain cylindrical sample tubes whose axes may be parallel, at an angle, or perpendicular to the rotor's axis of rotation, depending on the particular application (Fig. 5-32).

In the derivation of Eq. [5.15], it was assumed that sedimentation occurred through a homogeneous medium. Sedimentation may be carried out in a solution of an inert substance, however, such as sucrose or CsCl, in which the concentration, and therefore the density, of the solution increases from the top to the bottom of the centrifuge tube. The use of such **density gradients** greatly enhances the resolving power of the ultracentrifuge. Two applications of density gradients are widely employed: (1) **zonal ultracentrifugation** and (2) **equilibrium density gradient ultracentrifugation.**

Zonal Ultracentrifugation Separates Particles According to Their Sedimentation Coefficients

In zonal ultracentrifugation, a macromolecular solution is carefully layered on top of a density gradient prepared by

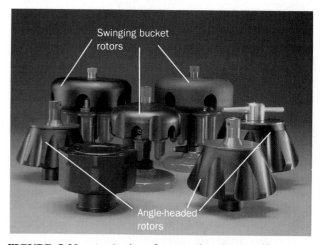

FIGURE 5-32. A selection of preparative ultracentrifuge rotors. The sample tubes of the swinging bucket rotors (*rear*) are hinged so that they swing from the vertical to the horizontal position as the rotor starts spinning, whereas the sample tubes of the other rotors have a fixed angle relative to the rotation axis. [Courtesy of Beckman Instruments, Inc.]

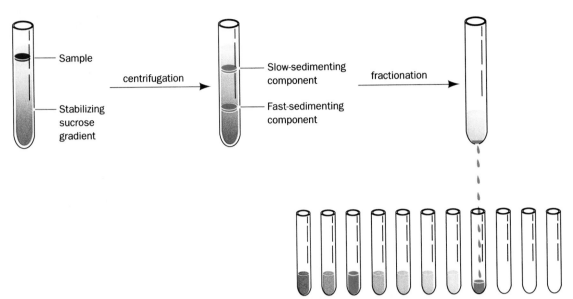

FIGURE 5-33. A diagrammatic representation of zonal ultracentrifugation. The sample is layered onto a sucrose gradient (*left*). Under centrifugation (*middle*), each particle sediments at a rate that depends largely on its mass. After the end of the run, the centrifugation tube is punctured and the separated particles (zones) are collected (*right*).

use of a device resembling that diagrammed in Fig. 5-7. The purpose of the density gradient is to allow smooth passage of the various macromolecular zones by damping out convective mixing of the solution. Sucrose, which forms a syrupy and biochemically benign solution, is commonly used to form a density gradient for zonal ultracentrifugation. The density gradient is normally rather shallow because the maximum density of the solution must be less than that of the least dense macromolecule of interest. Nevertheless, consideration of Eq. [5.15] indicates that the sedimentation rate of a macromolecule is a more sensitive function of molecular size than density. Consequently, *zonal ultracentrifugation separates similarly shaped macromolecules largely on the basis of their molecular masses.*

During centrifugation, each species of macromolecule moves through the gradient at a rate largely determined by its sedimentation coefficient and therefore travels as a zone that can be separated from other such zones as is diagrammed in Fig. 5-33. After centrifugation, fractionation is commonly effected by puncturing the bottom of the celluloid centrifuge tube with a needle, allowing its contents to drip out, and collecting the individual zones for subsequent analysis.

Equilibrium Density Gradient Ultracentrifugation Separates Particles According to Their Densities

In **equilibrium density gradient ultracentrifugation** [alternatively, **isopycnic ultracentrifugation**; (Greek: *isos*, equal + *pyknos*, dense)], the sample is dissolved in a relatively concentrated solution of a dense, fast-diffusing (and therefore low molecular mass) substance, such as CsCl or Cs_2SO_4, and is spun at high speed until the solution achieves equilibrium. *The high centrifugal field causes the low molecular mass solute to form a steep density gradient (Fig. 5-34) in which the sample components band at positions where their densities are equal to that of the solution;*

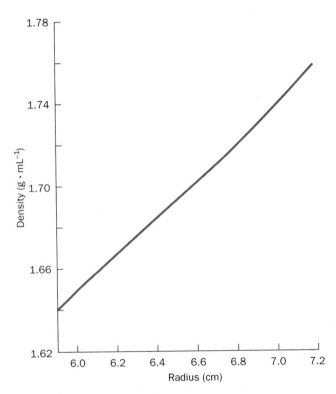

FIGURE 5-34. The equilibrium density distribution of a CsCl solution in an ultracentrifuge spinning at 39,460 rpm. The initial density of the solution was 1.7 g·mL^{-1}. [After Ifft, J.B., Voet, D.H., and Vinograd, J., *J. Phys. Chem.* **65,** 1138 (1961).]

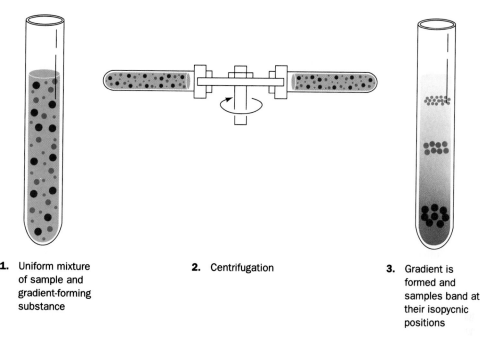

1. Uniform mixture of sample and gradient-forming substance

2. Centrifugation

3. Gradient is formed and samples band at their isopycnic positions

FIGURE 5-35. Isopycnic ultracentrifugation. The centrifugation of a uniform mixture of a macromolecular sample dissolved in a solution of a dense, fast-diffusing solute such as CsCl (*left*). At equilibrium in a centrifugational field, the solute forms a density gradient in which the macromolecules migrate to their positions of buoyant density (*right*).

that is, where $(1 - \overline{V}\rho)$ in Eq. [5.15] is zero (Fig. 5-35). These bands are collected as separate fractions when the sample tube is drained as described above. The salt concentration in the fractions and hence the solution density is easily determined with an **Abbé refractometer,** an optical instrument that measures the refractive index of a solution. *The equilibrium density gradient technique is often the method of choice for separating mixtures whose components have a range of densities.* These substances include nucleic acids, viruses, and certain subcellular organelles such as ribosomes. However, isopycnic ultracentrifugation is rather ineffective for the fractionation of protein mixtures because most proteins have similar densities (high salt concentrations also salt out or possibly denature proteins).

CHAPTER SUMMARY

Macromolecules in cells are solubilized by disrupting the cells by various chemical or mechanical means such as detergents or blenders. Partial purification by differential centrifugation is used after cell lysis to remove cell debris or to isolate a desired subcellular component. When out of the protective environment of the cell, proteins and other macromolecules must be treated so as to prevent their destruction by such influences as extremes of pH and temperature, enzymatic and chemical degradation, and rough mechanical handling. The state of purity of a substance being isolated must be monitored throughout the purification procedure by a specific assay.

Proteins are conveniently purified on a large scale by fractional precipitation, in which protein solubilities are varied by changing the salt concentration or pH.

Ion exchange chromatography employs support materials such as cellulose or cross-linked polyacrylamide gels. Separations are based on differential electrostatic interactions between charged groups on the ion exchange materials and those on the substances being separated. In paper chromatography, compounds are separated by partition between a moving nonpolar solvent phase and a stationary aqueous phase that is bound to the paper fibers. Separation may be enhanced by repeating the chromatographic separation in a second dimension using a different solvent system. Molecules may be located by the use of specific stains, such as ninhydrin for amino acids and polypeptides, or through radioactive labeling. In gel filtration chromatography, molecules are separated according to their size and shape through the use of cross-linked dextran, polyacrylamide, or agarose beads that have pores of molecular dimensions. A calibrated gel filtration column can be used to estimate the molecular masses of macromolecules. Affinity chromatography separates biomolecules according to their unique biochemical abilities to bind other molecules specifically. High-performance liquid chromatography (HPLC) utilizes any of the foregoing separation techniques but uses high-resolution chromatographic materials, high solvent pressures, and automatic solvent mixing and monitoring systems so as to obtain much greater degrees of separation than are achieved with the more conventional chromatographic procedures. Adsorption chromatography, thin layer chromatography (TLC), reverse-phase chromatography, and hydrophobic interaction chromatography also have valuable biochemical applications.

In electrophoresis, charged molecules are separated according to their rates of migration in an electric field on a solid support such as paper, cellulose acetate, cross-linked polyacrylamide gel, or agarose. Paper electrophoresis may be combined with paper chromatography in the two-dimensional technique of fingerprinting. Gel electrophoresis employs a cross-linked polyacrylamide or agarose gel support, so that molecules are separated according to size by gel filtration as well as according to charge. The separated molecules may be visualized by means of stains, autoradiography, or immunoblotting. The anionic detergent sodium dodecyl sulfate (SDS) denatures proteins and uniformly coats them so as to give most proteins a similar charge density and shape. SDS–PAGE may be used to estimate macromolecular masses. In isoelectric focusing, macromolecules are immersed in a stable pH gradient and subjected to an electric field that causes them to migrate to their isoelectric positions. In capillary electrophoresis, the use of thin capillary tubes and high electric fields permits rapid and highly resolved separations of small amounts of material.

In ultracentrifugation, molecules are separated by subjecting them to gravitational fields large enough to counteract diffusional forces. Molecules may be separated and their molecular masses estimated from their rates of sedimentation through a solvent or a preformed gradient of an inert low molecular mass material such as sucrose. Alternately, molecules may be separated according to their buoyant densities in a solution with a density gradient of a dense, fast-diffusing substance such as CsCl. The deviation of a molecule's frictional ratio from unity is indicative of its degrees of solvation and elongation.

REFERENCES

General

Boyer, R.F., *Modern Experimental Biochemistry* (2nd ed.), Benjamin/Cummings (1993).

Creighton, T.E. (Ed.), *Protein Structure. A Practical Approach,* IRL Press (1989). [Chapters 1–3 describe various electrophoretic methods. Chapter 4 is on the immunological detection of proteins.]

Deutscher, M.P. (Ed.), *Guide to Protein Purification, Methods Enzymol.* **182** (1990). [A compendium of useful explanations and recipes.]

Freifelder, D., *Physical Biochemistry* (2nd ed.), Freeman (1983). [A textbook on the techniques used in biophysical analysis.]

Harris, E.L.V. and Angal, S., *Protein Purification Methods. A Practical Approach,* IRL Press (1989) and *Protein Purification Applications. A Practical Approach*, IRL Press (1989). [Two companion volumes.]

Janson, J.C. and Rydén, L. (Eds.), *Protein Purification,* VCH Publishers (1989). [Contains detailed discussions of a variety of chromatographic and electrophoretic separation techniques.]

Robyt, J.F. and White, B.J., *Biochemical Techniques,* Brooks/Cole (1987).

Scopes, R., *Protein Purification: Principles and Practice* (3rd ed.), Springer–Verlag (1994).

Tinoco, I., Jr., Sauer, K., and Wang, J.C., *Physical Chemistry. Principles and Applications in Biological Sciences* (2nd ed.), Chapter 6, Prentice–Hall (1985).

Wilson, K.J., Micro-level protein and peptide separations, *Trends Biochem. Sci.* **14,** 252–255 (1989). [Discusses HPLC and PAGE techniques, including CE.]

Solubility and Crystallization

Arakawa, T. and Timasheff, S.N., Theory of protein solubility, *Methods Enzymol.* **114,** 49–77 (1985).

Ducruix, A. and Giegé. R., *Crystallization of Nucleic Acids and Proteins. A Practical Approach,* IRL Press (1992).

Edsall, J.T. and Wyman, J., *Biophysical Chemistry,* Vol. 1, Academic Press (1958). [A classical, detailed treatise on the acid–base and electrostatic properties of amino acids and proteins.]

McPherson, A., Crystallization of macromolecules: general principles, *Methods Enzymol.* **114,** 112–120 (1985).

Chromatography

Ackers, G.K., Molecular sieve methods of analysis, *in* Neurath, H. and Hill, R.L. (Eds.), *The Proteins* (3rd ed.), Vol. 1, *pp.* 1–94, Academic Press (1975).

Dean, P.D.G., Johnson, W.S., and Middle, F.A. (Eds.), *Affinity Chromatography. A Practical Approach,* IRL Press (1985).

Fallon, H., Booth, R.F.G., and Bell, L.D., Applications of HPLC in biochemistry, *in* Burdon, R.H. and van Knippenberg, P.H. (Eds.), *Laboratory Techniques In Biochemistry and Molecular Biology,* Vol. 17, Elsevier (1987).

Fischer, L., Gel filtration chromatography (2nd ed.), *in* Work, T.S. and Burdon, R.H. (Eds.), *Laboratory Techniques in Biochemistry and Molecular Biology,* Vol. 1, Part II, North–Holland Biomedical Press (1980).

Oliver, R.W.A. (Ed.), *HPLC of Macromolecules. A Practical Approach,* IRL Press (1989).

Petersen, E.A., Cellulosic ion exchangers, *in* Work, T.S. and Work, E. (Eds.), *Laboratory Techniques in Biochemistry and Molecular Biology,* Vol. 2, Part II, North–Holland (1980).

Schott, H., *Affinity Chromatography,* Dekker (1984).

Electrophoresis

Cantor, C.R. and Schimmel, P.R., *Biophysical Chemistry,* Chapter 12, Freeman (1980).

Celis, J.E. and Bravo, R. (Eds.), *Two-Dimensional Gel Electrophoresis of Proteins,* Academic Press (1984).

Grossman, P.D. and Colburn, J.C. (Eds.), *Capillary Electrophoresis: Theory and Practice,* Academic Press (1992).

Hames, B.D. and Rickwood, D. (Eds.), *Gel Electrophoresis of Proteins. A Practical Approach* (2nd. ed.), IRL Press (1990).

Harrington, M.G., Gudeman, D., Zewart, T., Yun, M. and Hood, L., Analytical and micropreparative two-dimensional electrophoresis of proteins, *Methods* **3**, 98–108 (1991).

Karger, B.L., High-performance capillary electrophoresis, *Nature* **339**, 641–642 (1989) *and* High-performance capillary electrophoresis in the biological sciences, *J. Chromatog.* **492**, 585–614 (1989).

Li, S.F.Y., *Capillary Electrophoresis,* Elsevier (1992). [A detailed exposition.]

Righetti, P.G., Immobilized pH gradients: theory and methodology, *in* Burdon, R.H. and van Knippenberg (Eds.), *Laboratory Techniques in Biochemistry and Molecular Biology,* Vol. 20, Elsevier (1990). [Discusses isoelectric focusing.]

Strahler, J.R. and Hanash, S.M., Immobilized pH gradients: Analytical and preparative use, *Methods* **3**, 109–114 (1991).

Weber, K. and Osborn, M., Proteins and sodium dodecyl sulfate: Molecular weight on polyacrylamide and related procedures, *in* Neurath, H. and Hill, R.L. (Eds.), *The Proteins* (3rd ed.), Vol. 1, *pp.* 179–223, Academic Press (1975). [An article by the inventors of this classical method of molecular mass determination.]

Ultracentrifugation

Cantor, C.R. and Schimmel, P.R., *Biophysical Chemistry,* Chapters 10 and 11, Freeman (1980).

Harding, S.E., Rowe, A.J., and Horton, J.C. (Eds.), *Analytical Ultracentrifugation in Biochemistry and Polymer Science,* Royal Society of Chemistry (1992).

Hinton, R. and Dobrata, M., Density gradient ultracentrifugation, *in* Work, T.S. and Work, E. (Eds.), *Laboratory Techniques in Biochemistry and Molecular Biology,* Vol. 6, Part I, North–Holland (1978).

Schachman, H.K., *Ultracentrifugation in Biochemistry,* Academic Press (1959). [A classic treatise on ultracentrifugation.]

van Holde, K.E., Sedimentation analyses of proteins, *in* Neurath, H. and Hill, R.L. (Eds.), *The Proteins* (3rd ed.), Vol. 1, *pp.* 225–291, Academic Press (1975).

PROBLEMS

1. What are the ionic strengths of $1.0 M$ solutions of NaCl, $(NH_4)_2SO_4$, and K_3PO_4? In which of these solutions would a protein be expected to be most soluble; least soluble?

2. An **isotonic saline solution** (one that has the same salt concentration as blood) is 0.9% NaCl. What is its ionic strength?

3. In what order will the following amino acids be eluted from a column of P-cellulose ion exchange resin by a buffer at pH 6: arginine, aspartic acid, histidine, and leucine?

4. In what order will the following proteins be eluted from a CM-cellulose ion exchange column by an increasing salt gradient at pH 7: fibrinogen, hemoglobin, lysozyme, pepsin, and ribonuclease A (see Table 5-1)?

5. What is the order of the R_f values of the following amino acids in their paper chromatography with a water/butanol/acetic acid solvent system in which the pH of the aqueous phase is 4.5: alanine, aspartic acid, lysine, glutamic acid, phenylalanine, and valine?

6. What is the order of elution of the following proteins from a Sephadex G-50 column: catalase, α-chymotrypsin, concanavalin B, lipase, and myoglobin (see Table 5-5)?

7. Estimate the molecular mass of an unknown protein that elutes from a Sephadex G-50 column between cytochrome *c* and ribonuclease A (see Table 5-5).

8. A gel-chromatography column of Bio-Gel P-30 with a bed volume of 100 mL is poured. The elution volume of the protein hexokinase (96 kD) on this column is 34 mL. That of an unknown protein is 50 mL. What is the void volume of the column, the volume occupied by the gel, and the relative elution volume of the unknown protein?

9. What chromatographic method would be suitable for separating the following pairs of substances? (a) Ala-Phe-Lys, Ala-Ala-Lys; (b) lysozyme, ribonuclease A (see Table 5-1); and (c) hemoglobin, myoglobin (see Table 5-1).

10. The neurotransmitter γ-aminobutyric acid is thought to bind to specific receptor proteins in nerve tissue. Design a procedure for the partial purification of such a receptor protein.

11. A mixture of amino acids consisting of arginine, cysteine, glutamic acid, histidine, leucine, and serine is applied to a strip of paper and subjected to electrophoresis using a buffer at pH 7.5. What are the directions of migration of these amino acids and what are their relative mobilities?

*12. Sketch the appearance of a fingerprint of the following tripeptides: Asn-Arg-Lys, Asn-Leu-Phe, Asn-His-Phe, Asp-Leu-Phe, and Val-Leu-Phe. Assume the paper chromatographic step is carried out using a water/butanol/acetic acid solvent system (pH 4.5) and the electrophoretic step takes place in a buffer at pH 6.5.

13. What is the molecular mass of a protein that has a relative electrophoretic mobility of 0.5 in an SDS–polyacrylamide gel such as that of Fig. 5-28.

14. Explain why the molecular mass of fibrinogen is significantly overestimated when measured using a calibrated gel filtration column (Fig. 5-13) but can be determined with reasonable accuracy from its electrophoretic mobility on SDS–polyacrylamide gel (see Table 5-5).

15. What would be the relative arrangement of the following proteins after they had been subjected to isoelectric focusing: insulin, cytochrome *c*, histone, myoglobin, and ribonuclease A (see Table 5-1)?

16. Calculate the centrifugal acceleration, in gravities (*g*'s), on a particle located 6.5 cm from the axis of rotation of an ultracentrifuge rotating at 60,000 rpm (1 $g = 9.81$ m·s^{-2}).

17. In a dilute buffer solution at 20°C, rabbit muscle aldolase has a frictional coefficient of 8.74×10^{-8} g·s^{-1}, a sedimentation coefficient of 7.35 S, and a partial specific volume of 0.742 cm·g^{-1}. Calculate the molecular mass of aldolase assuming the density of the solution to be 0.998 g·cm^{-3}.

*18. The sedimentation coefficient of a protein was measured by observing its sedimentation at 20°C in an ultracentrifuge spinning at 35,000 rpm.

Time, t (min)	Distance of Boundary from Center of Rotation, r (cm)
4	5.944
6	5.966
8	5.987
10	6.009
12	6.032

The density of the solution is 1.030 g·cm^{-3}, the partial specific volume of the protein is 0.725 cm^3·g^{-1}, and its frictional coefficient is 3.72×10^{-8} g·s^{-1}. Calculate the protein's sedimentation coefficient, in Svedbergs, and its molecular mass.

D	T	L	M	E	Y	L	E	N	P	K	K	Y	I	P	G
D	T	L	M	E	Y	L	E	N	P	K	K	Y	I	P	G
E	T	L	M	E	Y	L	E	N	P	K	K	Y	I	P	G
D	T	L	M	E	Y	L	E	N	P	K	K	Y	I	P	G
D	T	L	M	E	Y	L	E	N	P	K	K	Y	I	P	G
D	T	L	M	E	Y	L	E	N	P	K	K	Y	I	P	G
E	T	L	R	I	Y	L	E	N	P	K	K	Y	I	P	G
D	T	L	F	E	Y	L	E	N	P	K	K	Y	I	P	G
D	T	L	F	E	Y	L	E	N	P	K	K	Y	I	P	G
D	T	L	F	E	Y	L	E	N	P	K	K	Y	I	P	G
D	T	L	F	E	Y	L	E	N	P	K	K	Y	I	P	G

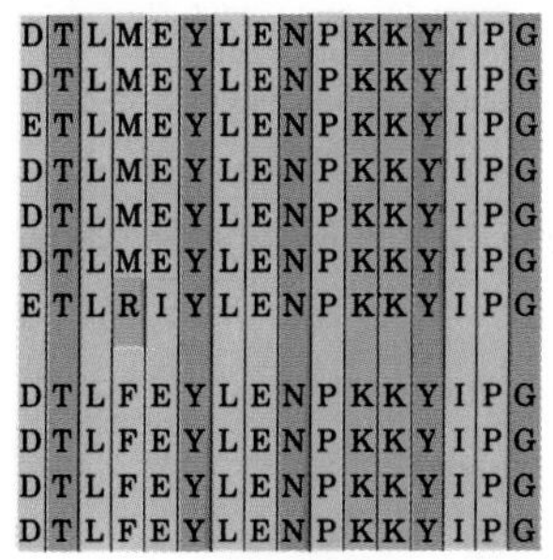

6

Covalent Structures of Proteins

1. Primary Structure Determination

Proteins are at the center of the action in biological processes. They function as enzymes, which catalyze the complex set of chemical reactions that are collectively referred to as life. Proteins serve as regulators of these reactions, both directly as components of enzymes and indirectly in the form of chemical messengers, known as hormones, as well as the receptors for those hormones. They act to transport and store biologically important substances such as metal ions, O_2, glucose, lipids, and many other molecules. In the form of muscle fibers and other contractile assemblies, proteins generate the coordinated mechanical motion of numerous biological processes, including the separation of chromosomes during cell division and the movement of your eyes as you read this page. Proteins, such as **rhodopsin** in the retina of your eye, acquire sensory information that is processed through the action of nerve cell proteins. The proteins of the immune system, such as the **immunoglobulins,** form an essential biological defense system in higher animals. Proteins are major active elements in, as well as products of, the expression of genetic information; the nucleic acids are, for the most part, information banks upon which proteins act. However, proteins also have important passive roles such as that of **collagen,** which provides bones, tendons, and ligaments with their characteristic tensile strength. Clearly, there is considerable validity to the old cliché that proteins are the "building blocks" of life.

Protein function can be understood only in terms of protein structure, that is, the three-dimensional relationships between a protein's component atoms. The structural descriptions of proteins, as well as those of other polymeric

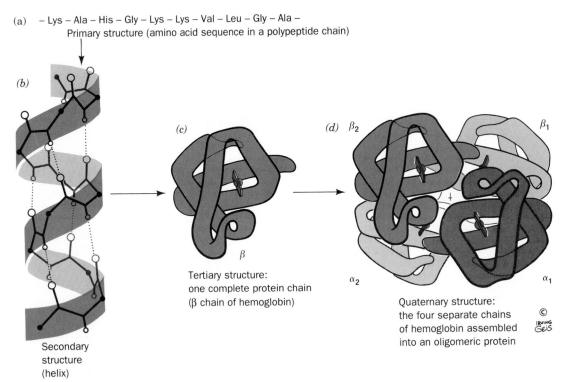

(a)   – Lys – Ala – His – Gly – Lys – Lys – Val – Leu – Gly – Ala –
Primary structure (amino acid sequence in a polypeptide chain)

(b)

Secondary
structure
(helix)

(c)

Tertiary structure:
one complete protein chain
(β chain of hemoglobin)

(d)   β₂ β₁

α₂ α₁

Quaternary structure:
the four separate chains
of hemoglobin assembled
into an oligomeric protein

© IRVING GEIS

FIGURE 6-1.   The structural hierarchy in proteins: (*a*) primary structure, (*b*) secondary structure, (*c*) tertiary structure, and (*d*) quaternary structure. [Figure copyrighted © by Irving Geis.]

materials, have been traditionally described in terms of four levels of organization (Fig. 6-1):

1. A protein's **primary structure (1° structure)** is the amino acid sequence of its polypeptide chain(s).

2. **Secondary (2°) structure** is the local spatial arrangement of a polypeptide's backbone atoms without regard to the conformations of its side chains.

3. **Tertiary (3°) structure** refers to the three-dimensional structure of an entire polypeptide. The distinction between secondary and tertiary structures is, of necessity, somewhat vague; in practice, the term secondary structure alludes to easily characterized structural entities such as helices.

4. Many proteins are composed of two or more polypeptide chains, loosely referred to as **subunits,** which associate through noncovalent interactions and, in some cases, disulfide bonds. A protein's **quaternary (4°) structure** refers to the spatial arrangement of its subunits.

In this, the first of four chapters on protein structure, we discuss the 1° structures of proteins: how they are elucidated and their biological and evolutionary significance. We also survey methods of chemically synthesizing polypeptide chains. The 2°, 3°, and 4° structures of proteins, which as we shall see are a consequence of their 1° structures, are treated in Chapter 7. In Chapter 8 we take up

protein folding, dynamics, and structural evolution, and in Chapter 9 we analyze hemoglobin as a paradigm of protein structure and function.

1. PRIMARY STRUCTURE DETERMINATION

The first determination of the complete amino acid sequence of a protein, that of the bovine polypeptide hormone **insulin** by Frederick Sanger in 1953, was of enormous biochemical significance in that it definitively established that proteins have unique covalent structures. Since that time, the amino acid sequences of several thousand proteins have been elucidated. This extensive information has been of central importance in the formulation of modern concepts of biochemistry for several reasons:

1. The knowledge of a protein's amino acid sequence is essential for an understanding of its molecular mechanism of action as well as being prerequisite for the elucidation of its X-ray and nuclear magnetic resonance (NMR) structures (Section 7-3A).

2. Sequence comparisons among analogous proteins from the same individual, from members of the same species, and from members of related species have yielded im-

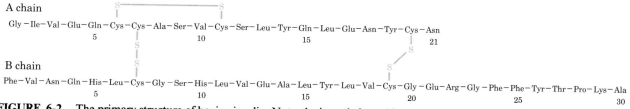

FIGURE 6-2. The primary structure of bovine insulin. Note the intrachain and interchain disulfide bond linkages.

portant insights into how proteins function and have indicated the evolutionary relationships among the proteins and the organisms that produce them. These analyses, as we shall see in Section 6-3, complement and extend analogous taxonometric studies based on anatomical comparisons.

3. Amino acid sequence analyses have important clinical applications because many inherited diseases are caused by mutations leading to an amino acid change in a protein. Recognition of this fact has led to the development of valuable diagnostic tests for many such diseases and, in several cases, to symptom-relieving therapy.

The elucidation of the 51-residue primary structure of insulin (Fig. 6-2) was the labor of many scientists over the period of a decade that altogether utilized ~100 g of protein. Procedures for primary structure determination have since been so refined and automated that proteins of similar size can be sequenced by an experienced technician in a few days using only a few micrograms of protein. The sequencing of the 1021-residue enzyme *β*-galactosidase in 1978 signaled that the sequence analysis of almost any protein could be reasonably attempted. Despite these technical advances, the basic procedure for primary structure determination using the techniques of protein chemistry is that developed by Sanger. The procedure consists of three conceptual parts, each of which can be broken down into several laboratory steps.

1. **Prepare the protein for sequencing:**
   a. Determine the number of chemically different polypeptide chains (subunits) in the protein.
   b. Cleave the protein's disulfide bonds.
   c. Separate and purify the unique subunits.
   d. Determine the subunits' amino acid compositions.

2. **Sequence the polypeptide chains:**
   a. Fragment the individual subunits at specific points to yield peptides small enough to be sequenced directly.
   b. Separate and purify the fragments.
   c. Determine the amino acid sequence of each peptide fragment.

   d. Repeat Step 2(a) with a fragmentation process of different specificity so that the subunit is cleaved at different peptide bonds from before. Separate these peptide fragments as in Step 2(b) and determine their amino acid sequences as in Step 2(c).

3. **Organize the completed structure:**
   a. Span the cleavage points between one set of peptide fragments by the other. By comparison, the sequences of these sets of polypeptides can be arranged in the order that they occur in the subunit, thereby establishing its amino acid sequence.
   b. Elucidate the positions of the disulfide bonds, if any, between and within the subunits.

We discuss these various steps in the following sections.

A. End Group Analysis: How Many Different Types of Subunits?

Each polypeptide chain (if it is not chemically blocked or circular) has an N-terminal residue and a C-terminal residue. By identifying these **end groups,** we can establish the number of chemically distinct polypeptides in a protein. For example, insulin has equal amounts of the N-terminal residues Phe and Gly, which indicates that it has equal numbers of two nonidentical polypeptide chains.

N-Terminus Identification

There are several effective methods by which a polypeptide's end groups may be identified. **1-Dimethylaminonaphthalene-5-sulfonyl chloride (dansyl chloride)** reacts with primary amines (including the *ε*-amino group of Lys) to yield dansylated polypeptides (Fig. 6-3). Acid hydrolysis (Section 6-1D) liberates the N-terminal residue as a **dansyl-amino acid** which exhibits such intense yellow fluorescence that it can be chromatographically identified from as little as 100 picomol of material [1 picomol (pmol) = 10^{-12} mol].

In the most useful method of N-terminal residue identification, the **Edman degradation** (named after its inventor, Pehr Edman), **phenylisothiocyanate (PITC, Edman's reagent)** reacts with the N-terminal amino groups of proteins under mildly alkaline conditions to form their **phenylthio-**

In order to expose all disulfide groups to the reducing agent, the reaction is usually carried out under conditions that denature the protein (disrupt its native conformation; see below). The resulting free sulfhydryl groups are alkylated, usually by treatment with **iodoacetic acid,**

$$Cys — CH_2 — SH \quad + \quad ICH_2COO^-$$

$$\textbf{Cysteine} \qquad\qquad \textbf{Iodoacetate}$$

$$Cys — CH_2 — S — CH_2COO^- \quad + \quad HI$$

$$\textbf{S-Carboxymethylcysteine}$$

to prevent the reformation of disulfide bonds through oxidation by O_2. S-Alkyl derivatives are stable in air and under the conditions used for the subsequent cleavage of peptide bonds.

C. Separation, Purification, and Characterization of the Polypeptide Chains

A protein's nonidentical polypeptides must be separated and purified in preparation for their amino acid sequence determination. Subunit dissociation, as well as denaturation, occurs under acidic or basic conditions, at low salt concentrations, at elevated temperatures, or through the use of denaturing agents such as urea, its iminium analog **guanidinium ion,**

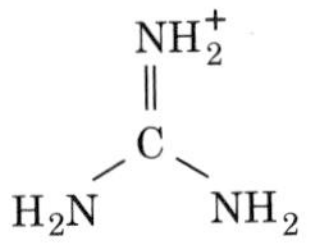

$$\begin{array}{c} NH_2^+ \\ \| \\ C \\ H_2N \quad\quad NH_2 \end{array}$$

$$\textbf{Guanidinium ion}$$

or detergents such as sodium dodecyl sulfate (SDS; Section 5-4C). The dissociated subunits can then be separated by methods described in Chapter 5 that capitalize on small differences in polypeptide size and polarity. Ion exchange and gel filtration chromatography are most often used.

It is, of course, desirable to know the number of residues in the polypeptide to be sequenced, which can be estimated from its molecular mass (~110 D/residue). Molecular mass can be measured with an accuracy of no better than 5 to 10% by the usual laboratory techniques of gel filtration chromatography or SDS–PAGE (Sections 5-3C and 4C). In recent years, however, **mass spectrometry (MS)** has provided a faster and far more accurate means to determine the molecular masses of macromolecules. A major difficulty of this method for large organic molecules had been the need to vaporize them without thermal decomposition (mass spectrometry requires the molecules being analyzed to be in the gas phase). The development of three techniques has eliminated this problem: (1) **plasma desorption,** in which the high-energy particles arising from the fission decay of ^{252}Cf (Californium) nuclei pass through a thin sheet of Al foil on which the sample has been deposited; (2) **matrix-assisted laser desorption,** in which the sample is embedded in a solid matrix of low molecular mass UV-absorbing organic molecules and irradiated with intense short pulses of UV laser light; and (3) **electrospray,** in which a dilute acidic solution of the macromolecule is sprayed from a metal syringe needle maintained at +5000 V, forming fine highly charged droplets from which the solvent rapidly evaporates. Each of these techniques injects the intact macromolecules into the gas phase as positively charged ions which are then directed into the mass spectrometer. The molecular masses

TABLE 6-2. SPECIFICITIES OF VARIOUS ENDOPEPTIDASES

Enzyme	Source	Specificity	Comments
Trypsin	Bovine pancreas	R_{n-1} = positively charged residues: Arg, Lys; $R_n \neq$ Pro	Highly specific
Chymotrypsin	Bovine pancreas	R_{n-1} = bulky hydrophobic residues: Phe, Trp, Tyr; $R_n \neq$ Pro	Cleaves more slowly for R_{n-1} = Asn, His, Met, Leu
Elastase	Bovine pancreas	R_{n-1} = small neutral residues: Ala, Gly, Ser, Val; $R_n \neq$ Pro	
Thermolysin	*Bacillus thermoproteolyticus*	R_n = Ile, Met, Phe, Trp, Tyr, Val; $R_{n-1} \neq$ Pro	Occasionally cleaves at R_n = Ala, Asp, His, Thr; heat stable
Pepsin	Bovine gastric mucosa	R_n = Leu, Phe, Trp, Tyr; $R_{n-1} \neq$ Pro	Also others; quite nonspecific; pH optimum = 2
Endopeptidase V8	*Staphylococcus aureus*	R_{n-1} = Glu	

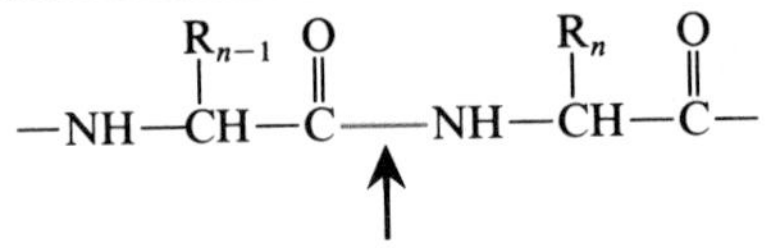

of picomolar amounts of >100 kD polypeptides can thereby be determined with accuracies of ~0.01%.

D. Amino Acid Composition

Before we begin the actual sequencing of a polypeptide chain, it is desirable to know its amino acid composition, that is, the number of each type of amino acid residue present. *The amino acid composition of a subunit is determined by its complete hydrolysis followed by the quantitative analysis of the liberated amino acids.* Polypeptide hydrolysis can be accomplished by either chemical (acid or base) or enzymatic means, although none of these methods alone is fully satisfactory. For acid-catalyzed hydrolysis, the polypeptide is dissolved in $6N$ HCl, sealed in an evacuated tube to prevent the air oxidation of the sulfur-containing amino acids, and heated at 100 to 120°C for 10 to 100 h. The long hydrolysis times are required for the complete liberation of the aliphatic amino acids Val, Leu, and Ile. Unfortunately, not all side chains are impervious to these harsh conditions. Ser, Thr, and Tyr are partially degraded, although by following their disappearance as a function of hydrolysis time correction factors for these losses can be established. A more serious problem is that acid hydrolysis largely destroys the Trp residues. Moreover, Gln and Asn are converted to Glu and Asp plus NH_4^+ so that only the amounts of Asx (= Asp + Asn), Glx (= Glu + Gln), and NH_4^+ (= Asn + Gln) can be independently measured after acid hydrolysis.

Base-catalyzed hydrolysis of polypeptides is carried out in 2 to $4N$ NaOH at 100°C for 4 to 8 h. This treatment is even more problematic because it causes the decomposition of Cys, Ser, Thr, and Arg and partially deaminates and racemizes the other amino acids. Hence alkaline hydrolysis is principally used to measure Trp content.

The complete enzymatic digestion of a polypeptide requires mixtures of peptidases because individual peptidases do not cleave all peptide bonds. Tables 6-1 and 6-2 indicate the specificities of the exopeptidases and **endopeptidases** (enzymes that catalyze the hydrolysis of internal peptide bonds) commonly used for this purpose. **Pronase,** a mixture of relatively nonspecific proteases from *Streptomyces griseus,* is also often used to effect complete proteolysis. The amount of enzyme used is limited to ~1% by weight of the polypeptide to be hydrolyzed because proteolytic enzymes, being proteins themselves, are self-degrading so that they will, if used too generously, significantly contaminate the final digest. Enzymatic digestion is most often used for determining the amounts of Trp, Asn, and Gln in a polypeptide, which are destroyed by the harsher chemical methods.

Amino Acid Analysis Has Been Automated

The amino acid content of a polypeptide hydrolysate can be quantitatively determined through the use of an automated **amino acid analyzer.** Such an instrument separates amino acids by ion exchange chromatography, a technique pioneered by William Stein and Stanford Moore, or by reverse-phase chromatography using HPLC (Section 5-3E). The amino acids are pre- or post-column derivatized by treatment with either dansyl chloride, Edman's reagent, or **o-phthalaldehyde (OPA)** + 2-mercaptoethanol. The latter reagents react with amino acids to form highly fluorescent adducts:

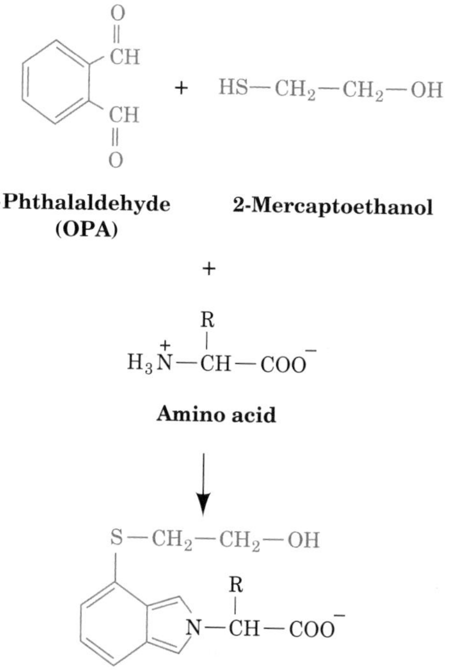

The amino acids are then identified according to their characteristic elution volumes (retention times on HPLC; Fig. 6-6) and quantitatively estimated from their fluorescence intensities (UV absorbances for PTC-amino acids). With modern amino acid analyzers, the complete analysis of a protein digest can be performed in <1 h with a sensitivity that can detect as little as 1 pmol of each amino acid.

The Amino Acid Compositions of Proteins Are Indicative of Their Structures

The amino acid analysis of a vast number of proteins indicates that they have considerable variation with respect to their amino acid compositions. Leu, Ala, Gly, Ser, Val, and Glu are the most common amino acid residues (>6% abundance), whereas His, Met, Cys, and Trp occur least frequently (<3% abundance; Table 4-1). Indeed, many proteins lack one or more amino acids. The ratio of polar to nonpolar residues is generally >1 for globular proteins and tends to decrease with increasing protein size. This is because, as we shall see in Chapters 7 and 8, globular proteins have a hydrophobic core and a hydrophilic exterior; that is, they have a micellelike structure. Nonpolar residues predominate in membrane-bound proteins, however, because these proteins, being immersed in a nonpolar environment (Section 11-3A), must also have a hydrophobic exterior.

F. Separation and Purification of the Peptide Fragments

Once again we must employ separation techniques, this time to isolate the peptide fragments of specific cleavage operations for subsequent sequence determinations. The nonpolar residues of peptide fragments are not excluded from the aqueous environment as they are in native proteins (Chapters 7 and 8). Consequently, many peptide fragments aggregate, precipitate, and/or strongly adsorb to chromatographic materials, which often results in unacceptable peptide losses. Until around 1980, the trial-and-error development of methods that could satisfactorily separate a mixture of peptide fragments constituted the major technical challenge of a protein sequence determination, as well as its most time-consuming step. Such methods involved the use of denaturants, such as urea and SDS, to solubilize the peptide fragments, and the selection of chromatographic materials and conditions that would reduce their adsorptive losses. The advent of reverse-phase chromatography by HPLC (Section 5-3E), however, has largely reduced the separation of peptide fragments to a routine procedure.

G. Sequence Determination

Once the manageably sized peptide fragments that were formed through specific cleavage reactions have been isolated, their amino acid sequences can be determined. *This is usually accomplished through repeated cycles of the Edman degradation* (Section 6-1A). An automated device for doing so, a **sequenator,** was first developed by Edman and Geoffrey Begg. In modern instruments, the peptide sample is dried onto a disk of glass fiber paper which is impregnated with a polymeric quaternary ammonium salt, **polybrene,** that immobilizes the peptide but is readily penetrated by Edman reagents. Accurately measured quantities of reagents, delivered as vapors in a stream of argon

(thereby minimizing peptide loss), are then added to the reaction cell at programmed intervals. The thiazolinone-amino acids are automatically removed, converted to the corresponding PTH-amino acids (Fig. 6-4), and identified chromatographically. Such instruments are capable of processing up to one residue per hour.

Usually, a peptide's 40 to 60 N-terminal residues can be identified (100 or more with the most advanced systems) before the cumulative effects of incomplete reactions, side reactions, and peptide loss make further amino acid identification unreliable. Since less than a picomole of a PTH amino acid can be detected and identified by a reverse-phase HPLC system equipped with a UV detector, sequence analysis can be carried out on as little as 5 to 10 pmol (<0.1 μg—an invisibly small amount) of a peptide.

H. Ordering the Peptide Fragments

With the peptide fragments individually sequenced, what remains is to elucidate the order in which they are connected in the original polypeptide. *We do so by comparing the amino acid sequences of one set of peptide fragments with those of a second set whose specific cleavage sites overlap those of the first set (Fig. 6-7).* The overlapping peptide segments must be of sufficient length to identify each cleavage site uniquely, but as there are 20 possibilities for each amino acid residue, an overlap of only a few residues is usually enough.

I. Assignment of Disulfide Bond Positions

The final step in an amino acid sequence analysis is to determine the positions (if any) of the disulfide bonds. This is done by cleaving a sample of the native protein, with its disulfide bonds intact, so as to yield pairs of peptide fragments, each containing a single Cys, that are linked by a disulfide bond.

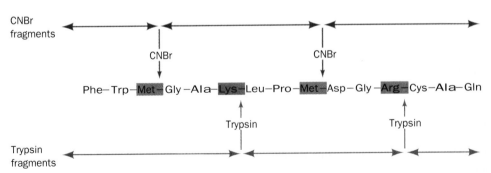

FIGURE 6-7. The amino acid sequence of a polypeptide chain is determined by comparing the sequences of two sets of mutually overlapping peptide fragments. In this example, the two sets of peptide fragments are generated by cleaving the polypeptide after all its Arg and Lys residues with trypsin and, in a separate reaction, after all its Met residues by treatment with cyanogen bromide. The order of the first two tryptic

peptides is established, for example, by the observation that the Gly-Ala-Lys-Leu-Pro-Met cyanogen bromide peptide has its N- and C-terminal sequences in common with the C- and N-termini, respectively, of the two tryptic peptides. In this manner the order of the peptide fragments in their parent polypeptide chain is established.

Such peptide fragments may be identified through **diagonal electrophoresis** (Fig. 6-8). In this technique, the partial peptide digest is electrophoretically separated in two dimensions by identical procedures. After the first separation, the electrophoretogram is exposed to performic acid vapor, which oxidizes the disulfide linkages to cysteic acid (Section 6-1B). After the second separation, those peptide fragments that were not modified by the performic acid treatment will be located along the diagonal of the electrophoretogram because their rates of migration are the same in both directions. Those polypeptides that were originally joined by a disulfide bond will each be decomposed to two peptides that have different migration rates from before and therefore lie off the diagonal of the electrophoretogram.

After the isolation of a disulfide-linked polypeptide fragment, the disulfide bond is cleaved and alkylated (Section 6-1B) and the sequences of the two polypeptides are determined (Fig. 6-9). The various pairs of such polypeptide fragments are identified by the comparison of their sequences with that of the protein, thereby establishing the locations of the disulfide bonds.

J. Peptide Sequencing by Mass Spectrometry

Mass spectrometry (MS) is emerging as a practical technique for sequencing polypeptides of up to ~25 residues. This is done through an ionization technique named **fast atom bombardment (FAB)** in concert with a **tandem mass spectrometer** (two mass spectrometers coupled in series). In FAB, the macromolecule of interest, dissolved in a low-volatility solvent such as glycerol, is bombarded with a beam of Ar or Xe atoms in the ionization chamber of the tandem instrument's first mass spectrometer. This process results in the ejection of $(M + H)^+$ ions from the glycerol solution where M represents the compound of interest such as a polypeptide. The first mass spectrometer functions to select the polypeptide ion of interest from other polypeptide ions

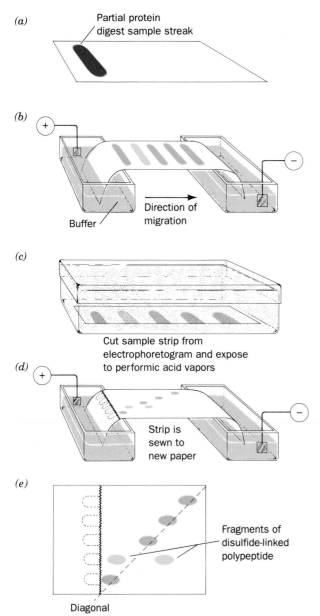

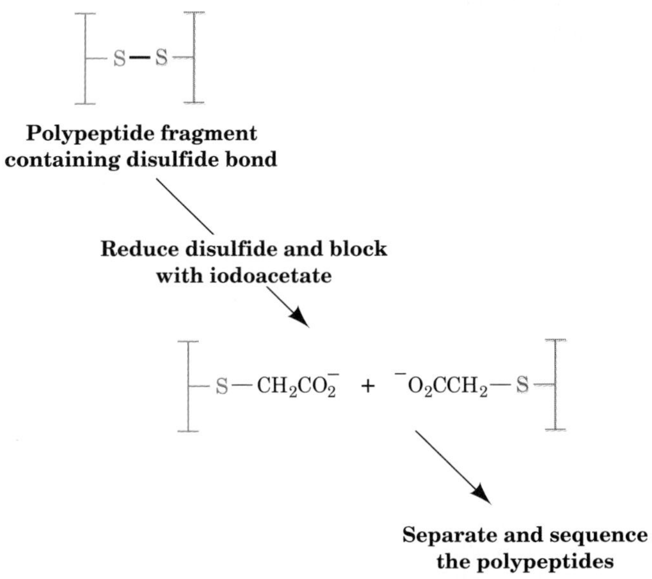

FIGURE 6-8. Diagonal electrophoresis: (*a*) A partial protein digest with its disulfide bonds intact is streaked onto a sheet of paper and (*b*) subjected to electrophoresis. (*c*) A guide strip is cut from the edge of the electrophoretogram and exposed to performic acid vapor that oxidizes each —S—S— linkage to two —SO₃H groups. (*d*) The guide strip is sewn to a second sheet of paper and subjected to electrophoresis in the perpendicular direction under the same conditions as before, followed by staining. (*e*) Peptides not on the diagonal of the second electrophoretogram, such as the two derived from the yellow fragment, contain cysteic acid residues and hence were originally linked by at least one disulfide bond. The parent peptide fragment can be located on the first electrophoretogram and eluted for further analysis. In an alternative process, the partial digest is separated by gel electrophoresis, reductively cleaved, and then identically separated in the second dimension.

FIGURE 6-9. A protein's disulfide bond positions are determined by identifying pairs of disulfide-linked peptide fragments.

and chemical contaminants that may be present. The selected polypeptide ion then passes into a collision cell, where it is made to collide with chemically inert atoms such as helium. The energy thereby imparted to the polypeptide ion causes it to fragment from either end in known ways to yield a series of increasingly smaller ions. The molecular masses of these fragments are then determined by the second mass spectrometer with an accuracy of better than ±0.5 D.

By comparing the molecular masses of successively larger sets of fragments, the molecular masses and therefore the identities of the corresponding amino acid residues can be determined. The sequence of the entire polypeptide can thus be elucidated. Computerization of this comparison process has reduced the time required to sequence an entire polypeptide to only a few minutes as compared to the hour or so required for just one cycle of Edman degradation. Moreover, the sequences of several polypeptides in a mixture can be determined, even in the presence of contaminants, by sequentially selecting the corresponding polypeptide ions in the first mass spectrometer of the tandem instrument. Hence, in separating and purifying the polypeptide fragments of a protein digest in preparation for their sequencing, less effort need be expended for mass spectrometric as compared to Edman techniques. Finally, mass spectrometry can be used to sequence peptides with blocked N-termini (a common eukaryotic posttranslational modification that prevents Edman degradation) and to characterize other posttranslational modifications such as phosphorylations (Section 4-3A) and glycosylations (Section 10-3C). The major disadvantages of mass spectrometric polypeptide sequence determination are the high expense of mass spectrometers and their lesser sensitivities compared to those of advanced sequencers using Edman degradation.

K. Peptide Mapping

The sequence determination of a protein is an exacting and time-consuming process. Once the primary structure of a protein has been elucidated, however, that of a nearly identical protein, such as one arising from a closely related species, a mutation, or a chemical modification, can be more easily determined. This can be done through the combined paper chromatography and paper electrophoresis (Section 5-4A) of partial protein digests, a technique synonomously known as **fingerprinting** or **peptide mapping**. The peptide fragments incorporating the amino acid variations migrate to different positions on their fingerprint (peptide map) than do the corresponding peptides of the original protein (Fig. 6-10). The variant peptides can then be eluted from the fingerprint and sequenced to establish the differences between the primary structures of the original and variant proteins. An alternative form of peptide mapping separates the peptide fragments by isoelectric focusing followed by SDS–PAGE to yield a two-dimensional gel (Section 5-4D).

L. Nucleic Acid Sequencing

The amino acid sequences of proteins are specified by the base sequences of nucleic acids (Section 30-1) so that, with a knowledge of the genetic code, a protein's primary struc-

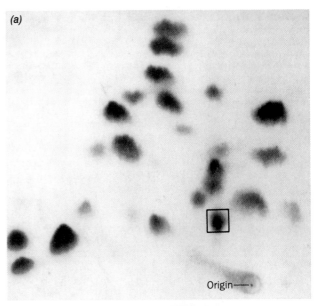

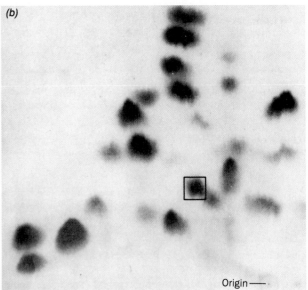

FIGURE 6-10. A comparison of the ninhydrin-stained fingerprints of trypsin-digested (*a*) hemoglobin A (HbA) and (*b*) hemoglobin S (HbS). The peptides that differ in these two forms of hemoglobin are boxed. These peptides constitute the eight N-terminal residues of the β subunit of hemoglobin. Their amino acid sequences are

Hemoglobin A	Val-	His-	Leu-	Thr-	Pro-	**Glu-**	Glu-	Lys
Hemoglobin S	Val-	His-	Leu-	Thr-	Pro-	**Val-**	Glu-	Lys
	β1	2	3	4	5	**6**	7	8

[Courtesy of Corrado Baglioni, State University of New York at Albany.]

ture can be inferred from that of a corresponding nucleic acid. Techniques for sequencing nucleic acids initially lagged far behind those for proteins, but by the late 1970s, DNA sequencing methods had advanced to the point that it became easier to sequence a DNA segment than the protein it specified (nucleic acid sequencing is the subject of Section 28-6). Although protein primary structures are now routinely inferred from DNA sequences, direct protein sequencing remains an indispensable biochemical tool for several important reasons:

1. Disulfide bonds can be located only by protein sequencing.

2. Many proteins are modified after their biosynthesis by the excision of certain residues and by the specific derivatization of others (Section 30-5). The identities of these modifications, which are often essential for the protein's biological function, can be determined only by directly sequencing the protein.

3. It is often difficult to identify and isolate a nucleic acid that encodes the protein of interest. Indeed, one of the most effective ways of doing so is to determine the amino acid sequence of at least a portion of the protein, infer the base sequence of the DNA segment that encodes this polypeptide segment, chemically synthesize this DNA, and use it to identify and isolate the gene(s) containing its base sequence through techniques discussed in Sections 28-8C and D.

4. A common error in DNA sequencing is the inadvertent insertion or deletion of a single nucleotide. The genetic code consists of consecutive triplets of nucleotides, each specifying a single amino acid residue (Section 30-1B). The erroneous addition or deletion of a nucleotide from a gene sequence therefore changes the gene's apparent reading frame and thus changes the predictions for all the amino acid residues past the point of error. Double checking the predicted amino acid sequence by directly sequencing a series of oligopeptides scattered throughout the protein readily detects such errors.

5. The "standard" genetic code is not universal: Those of mitochondria and certain protozoa are slightly different (Section 30-1E). In addition, in certain species of protozoa, the RNA transcripts are "edited"; that is, their sequences are altered before they are translated (Section 29-4A). These genetic code anomalies were discovered by comparing the amino acid sequences of proteins and the base sequences of their corresponding genes. If there are other genetic code anomalies, they will no doubt be discovered in a like manner.

2. PROTEIN MODIFICATION

A common strategy for identifying the residues of a protein essential for its biological function is to treat the protein with reagents that react with specific types of side chains. A protein will probably be inactivated by such a **group-specific reagent** if it chemically alters one of the protein's essential residues (although derivatizing a nonessential residue in a way that structurally alters the protein may also inactivate it). If a protein has been sequenced, it is a usually simple matter to identify the modified protein's altered residues through fingerprinting. Even without the knowledge of a protein's primary structure, the comparison of the fingerprints of the modified and unmodified proteins can yield valuable information concerning the altered residues. Some of the most useful group-specific reagents and their products are listed in Table 6-3.

TABLE 6-3.   GROUP-SPECIFIC REAGENTS FOR AMINO ACID RESIDUE MODIFICATIONS

Side Chain	Reagent	Product	Other Reactive Groups
Lys	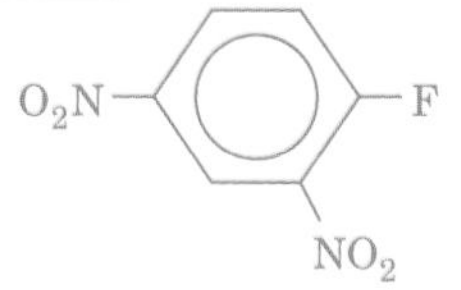 O_2N——⬡——F, NO_2 **1-Fluoro-2,4-dinitrobenzene (FDNB)**	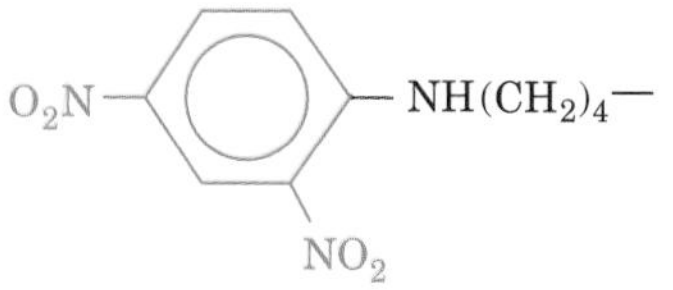 O_2N——⬡——$NH(CH_2)_4$—, NO_2 **Dinitrophenylated (DNP)-Lys**	Cys, His, Tyr, N-terminal amine
	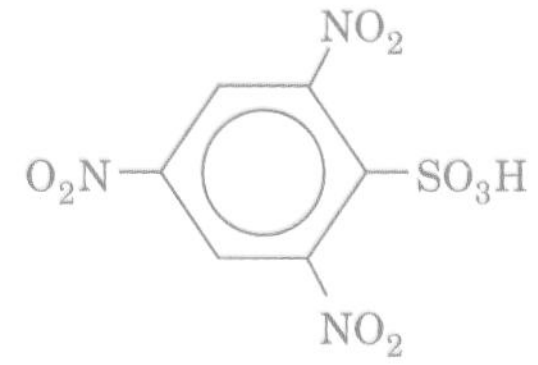 NO_2, O_2N——⬡——SO_3H, NO_2 **Trinitrobenzene sulfonic acid**	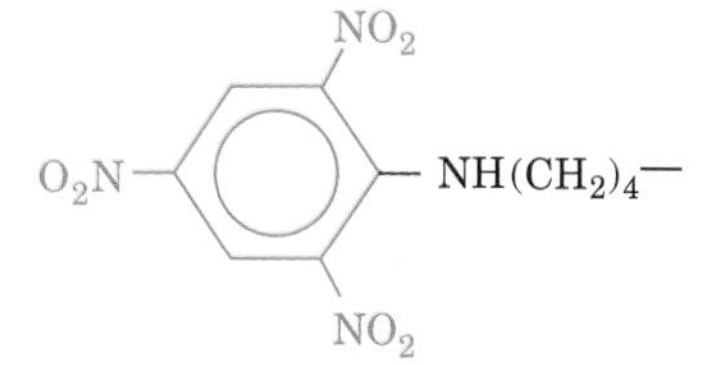 NO_2, O_2N——⬡——$NH(CH_2)_4$—, NO_2 **Trinitrophenyl-Lys**	N-terminal amine

TABLE 6-3. (CONT.)

Side Chain	Reagent	Product	Other Reactive Groups
Lys *(continued)*	$F_3C-\overset{\overset{\displaystyle O}{\|\|}}{C}-S-CH_2-CH_3$ **Ethylthiotrifluoroacetate**	$F_3C-\overset{\overset{\displaystyle O}{\|\|}}{C}-NH(CH_2)_4-$ **Trifluoroacetyl-Lys**	N-terminal amine
	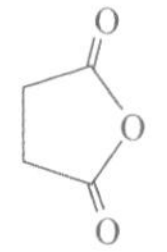 **Succinic anhydride**	$^-OOC-CH_2-CH_2-\overset{\overset{\displaystyle O}{\|\|}}{C}-NH(CH_2)_4-$ **Succinyl-Lys**	N-terminal amine
Arg	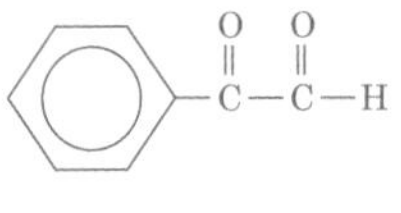 **Phenylglyoxal**	Arg reacts with two phenyl-glyoxal molecules to form a product of unknown structure	
	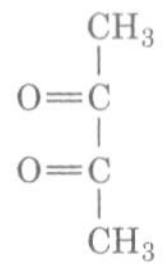 $\overset{\overset{\displaystyle CH_3}{\|}}{\underset{\underset{\displaystyle CH_3}{\|}}{\overset{O=C}{O=C}}}$ **2,3-Butanedione**	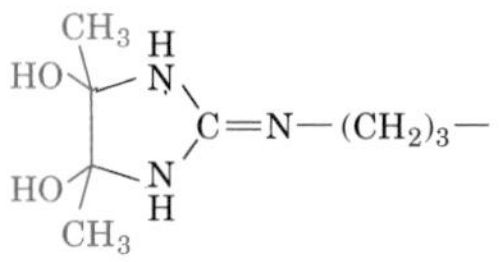 $C=N-(CH_2)_3-$ (cyclic product with CH_3, HO, N, H groups)	
Cys	ICH_2-COO^- **Iodoacetate**	$^-OOC-CH_2-S-CH_2-$ *S*-**Carboxymethyl-Cys**	Asp, Glu, His, Lys, Met
	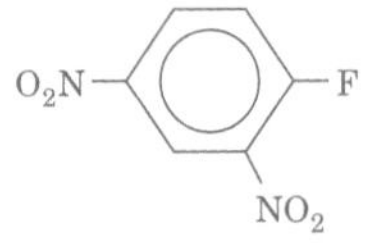 O_2N—(ring)—F, NO_2 **1-Fluoro-2,4-dinitrobenzene (FDNB)**	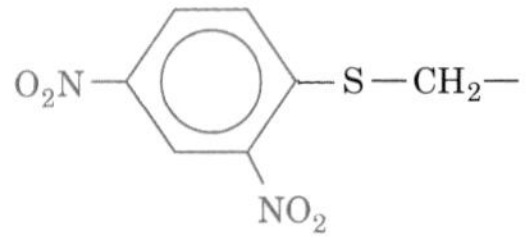 O_2N—(ring)—$S-CH_2-$, NO_2 **Dinitrophenylated (DNP)-Cys**	Lys, His, Tyr
	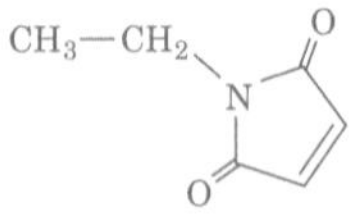 **N-Ethylmaleimide**	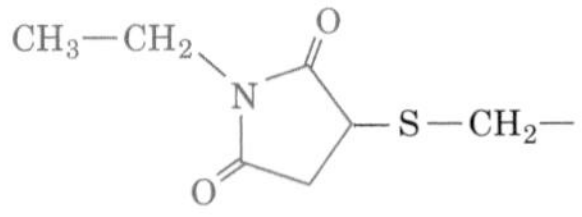 CH_3-CH_2 ... $S-CH_2-$	Specific at slightly acidic pH's
	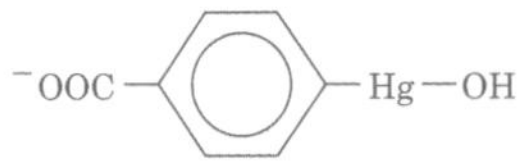 ^-OOC—(ring)—$Hg-OH$ **p-Hydroxymercuribenzoate**	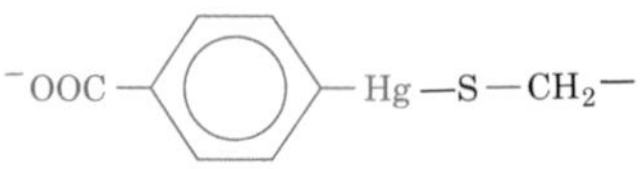 ^-OOC—(ring)—$Hg-S-CH_2-$	
	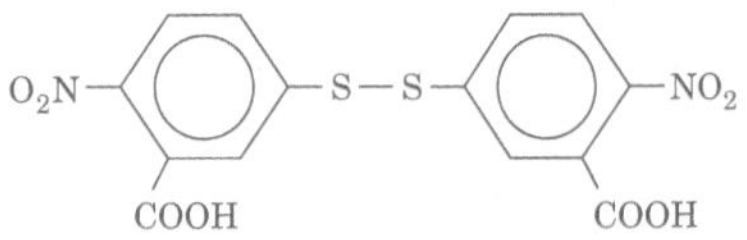 O_2N—(ring)—$S-S$—(ring)—NO_2, $COOH$, $COOH$ **5,5'-Dithiobis(2-nitrobenzoic acid) (DTNB)**	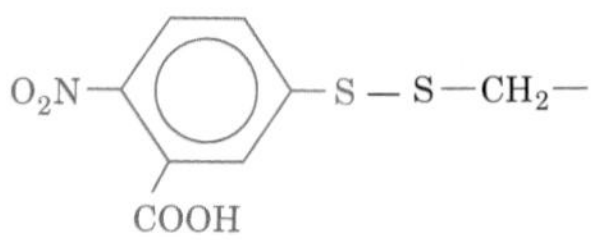 O_2N—(ring)—$S-S-CH_2-$, $COOH$	
	$H-\overset{\overset{\displaystyle O}{\|\|}}{C}-O-O-H$ **Performic acid**	$^-O_3S-CH_2-$ **Cysteic acid**	Met, Cys-S-S-Cys

TABLE 6-3. (CONT.)

Side Chain	Reagent	Product	Other Reactive Groups
Cys-S-S-Cys	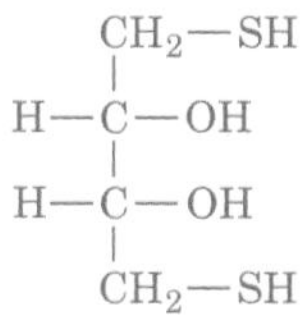 HS—CH$_2$—CH$_2$—OH **2-Mercaptoethanol** 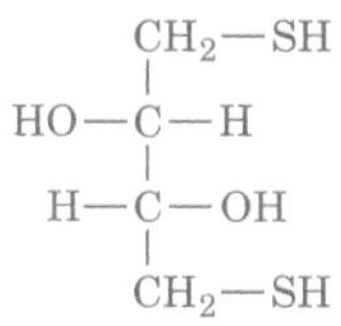 CH$_2$—SH HO—C—H H—C—OH CH$_2$—SH **Dithiothreitol** CH$_2$—SH H—C—OH H—C—OH CH$_2$—SH **Dithioerythritol**	HS—CH$_2$— **Cys-SH**	
	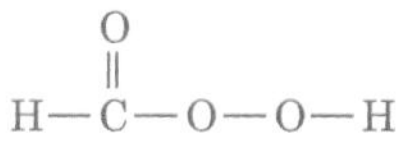 O ‖ H—C—O—O—H **Performic acid**	$^-$O$_3$S—CH$_2$— **Cysteic acid**	Cys, Met
Met	N≡C—Br **Cyanogen bromide**	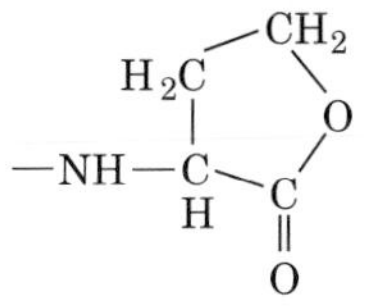 H$_2$C—CH$_2$ ... —NH—C—O ... C=O **Peptidyl homoserine lactone**	
	ICH$_2$—COO$^-$ **Iodoacetate**	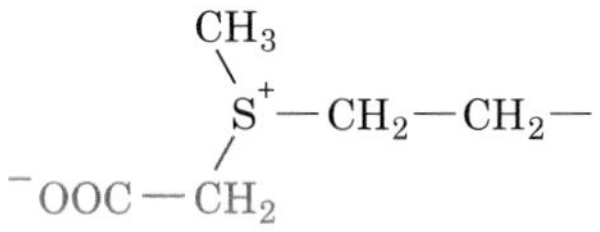 CH$_3$ S$^+$—CH$_2$—CH$_2$— $^-$OOC—CH$_2$ **S-Carboxymethyl-Met**	Asp, Cys, Glu, His, Lys
	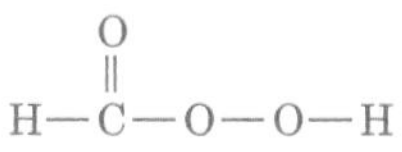 O ‖ H—C—O—O—H **Performic acid**	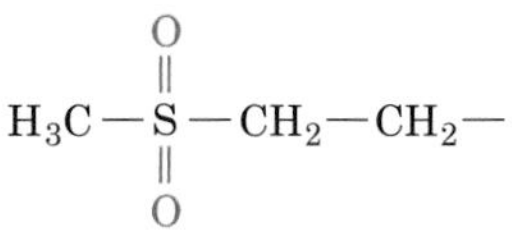 O ‖ H$_3$C—S—CH$_2$—CH$_2$— ‖ O **Methionine sulfone**	Cys, Cys-S-S-Cys
Asp, Glu	CH$_2$=N$^+$=N$^-$ **Diazomethane**	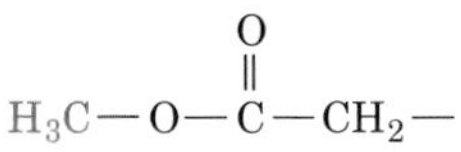 O ‖ H$_3$C—O—C—CH$_2$— **Methyl ester**	

TABLE 6-4. AMINO ACID SEQUENCES OF CYTOCHROMES *C* FROM 38 SPECIES

Group	Species	Sequence (positions −9 to 42)
Mammals	Man, chimpanzee	a GDVEKGKKIFIMKCSQCHTVEKGGKHKTGPNLHGLFGRKTGQA
	Rhesus monkey	a GDVEKGKKIFIMKCSQCHTVEKGGKHKTGPNLHGLFGRKTGQA
	Horse	a GDVEKGKKIFVQKCAQCHTVEKGGKHKTGPNLHGLFGRKTGQA
	Donkey	a GDVEKGKKIFVQKCAQCHTVEKGGKHKTGPNLHGLFGRKTGQA
	Cow, pig, sheep	a GDVEKGKKIFVQKCAQCHTVEKGGKHKTGPNLHGLFGRKTGQA
	Dog	a GDVEKGKKIFVQKCAQCHTVEKGGKHKTGPNLHGLFGRKTGQA
	Rabbit	a GDVEKGKKIFVQKCAQCHTVEKGGKHKTGPNLHGLFGRKTGQA
	California gray whale	a GDVEKGKKIFVQKCAQCHTVEKGGKHKTGPNLHGLFGRKTGQA
	Great gray kangaroo	a GDVEKGKKIFVQKCAQCHTVEKGGKHKTGPNINGIFGRKTGQA
Other vertebrates	Chicken, turkey	a GDIEKGKKIFVQKCSQCHTVEKGGKHKTGPNLHGLFGRKTGQA
	Pigeon	a GDIEKGKKIFVQKCSQCHTVEKGGKHKTGPNLHGLFGRKTGQA
	Pekin duck	a GDVEKGKKIFVQKCSQCHTVEKGGKHKTGPNLNGLIGRKTGQA
	Snapping turtle	a GDVEKGKKIFVQKCAQCHTVEKGGKHKTGPNLHGLFGRKTGQA
	Rattlesnake	a GDVEKGKKIFTMKCSQCHTVEKGGKHKTGPNLHGLFGRKTGQA
	Bullfrog	a GDVEKGKKIFVQKCAQCHTCEKGGKHKVGPNLYGLIGRKTGQA
	Tuna	a GDVAKGKKTFVQKCAQCHTVENGGKHKVGPNLWGLFGRKTGQA
	Dogfish	a GDVEKGKKVFVQKCAQCHTVENGGKHKTGPNLSGLFGRKTGQA
Insects	*Samia cynthia* (a moth)	h GVPAGNAENGKKIFVQRCAQCHTVEAGGKHKVGPNLHGFYGRKTGQA
	Tobacco hornworm moth	h GVPAGNADNGKKIFVQRCAQCHTVEAGGKHKVGPNLHGFFGRKTGQA
	Screwworm fly	h GVPAGDVEKGKKIFVQRCAQCHTVEAGGKHKVGPNLHGLFGRKTGQA
	Drosophila (fruit fly)	h GVPAGDVEKGKKLFVQRCAQCHTVEAGGKHKVGPNLHGLIGRKTGQA
Fungi	Baker's yeast	h TEFKAGSAKKGATLFKTRCLQCHTVEKGGPHKVGPNLHGIFGRHSGQA
	Candida krusei (a yeast)	h PAPFEQGSAKKGATLFKTRCAQCHTIEAGGPHKVGPNLHGIFSRHSGQA
	Neurospora crassa (a mold)	h GFSAGDSKKGANLFKTRCAQCHTLEEGGGNKIGPALHGLFGRKTGSV
Higher plants	Wheat germ	a ASFSEAPPGNPDAGAKIFKTKCAQCHTVDAGAGHKQGPNLHGLFGRQSGTT
	Buckwheat seed	a ATFSEAPPGNIKSGEKIFKTKCAQCHTVEKGAGHKQGPNLNGLFGRQSGTT
	Sunflower seed	a ASFAEAPPGDPTTGAKIFKTKCAQCHTVEKGAGHKQGPNLNGLFGRQSGTT
	Mung bean	a ASFBEAPPGBSKSGEKIFKTKCAQCHTVDKGAGHKQGPNLNGLFGRQSGTT
	Cauliflower	a ASFBEAPPGBSKAGEKIFKTKCAQCHTVDKGAGHKQGPNLNGLFGRQSGTT
	Pumpkin	a ASFBEAPPGBSKAGEKIFKTKCAQCHTVDKGAGHKQGPNLNGLFGRQSGTT
	Sesame seed	a ASFBEAPPGBVKSGEKIFKTKCAQCHTVEKGAGHKQGPNLNGLFGRQSGTT
	Castor bean	a ASFBEAPPGBVKAGEKIFKTKCAQCHTVEKGAGHKQGPNLNGLFGRQSGTT
	Cottonseed	a ASFZEAPPGBAKAGEKIFKTKCAQCHTVDKGAGHKQGPNLNGLFGRQSGTT
	Abutilon seed	a ASFZEAPPGBAKAGEKIFKTKCAQCHTVEKGAGHKQGPNLNGLFGRQSGTT

Number of different amino acids:
1 3 5 5 5 1 3 3 4 1 4 3 2 1 3 1 1 1 1 4 2 4 1 2 3 2 1 4 1 1 2 1 5 1 3 3 2 1 3 2 1 3 3

a The amino acid side chains have been shaded according to their polarity characteristics so that an invariant or conservatively substituted residue is identified by a vertical band of a single color. The letter a at the beginning of the chain indicates that the N-terminal amino group is acetylated; an h indicates that the acetyl group is absent.

Source: After Dickerson, R.E., *Sci. Am.* **226**(4): 58–72 (1972), with corrections from Dickerson, R.E., and Timkovich, R., *in* Boyer, P.D. (Ed.), *The Enzymes* (3rd ed.), Vol. 11, *pp.* 421–422, Academic Press (1975). Table copyrighted © by Irving Geis.

It is believed that the electron-transport chain took its present form between 1.5 and 2 billion years ago as organisms evolved the ability to respire (Section 1-4C). Since that time, the components of this multienzyme system have changed very little, as is evidenced by the observation that the cytochrome *c* from any eukaryotic organism, say a pigeon, will react *in vitro* (in the test tube) with the cytochrome oxidase from any other eukaryote, for instance, wheat. Indeed, hybrid cytochromes *c* consisting of covalently linked fragments from such distantly related species as horse and yeast (prepared by methods discussed in Section 6-4B) exhibit biological activity.

Protein Sequence Comparisons Yield Taxonometric Insights

Emanuel Margoliash, Emil Smith, and others have elucidated the amino acid sequences of the cytochromes *c* from nearly 100 widely diverse eukaryotic species ranging in complexity from yeast to humans. The sequences from 38 of these organisms are arranged in Table 6-4 so as to maximize the similarities between vertically aligned residues. The various residues in this table have been shaded according to their physical properties in order to illuminate the conservative character of the amino acid substitutions. Inspection of Table 6-4 indicates that cytochrome *c* is an

```
     45        50        55        60        65        70        75        80        85        90        95       100   104
P G Y S Y T A A N K N K G I I W G E D T L M E Y L E N P K K Y I P G T K M I F V G I K K K E E R A D L I A Y L K K A T N E
P G Y S Y T A A N K N K G I I W G E D T L M E Y L E N P K K Y I P G T K M I F V G I K K K E E R A D L I A Y L K K A A N E
P G F T Y T D A N K N K G I T W K E E T L M E Y L E N P K K Y I P G T K M I F A G I K K K T E R E D L I A Y L K K A T N E
P G F S Y T D A N K N K G I T W K E E T L M E Y L E N P K K Y I P G T K M I F A G I K K K T E R E D L I A Y L K K A T N E
P G F S Y T D A N K N K G I T W G E E T L M E Y L E N P K K Y I P G T K M I F A G I K K K G E R E D L I A Y L K K A T N E
V G F S Y T D A N K N K G I T W G E D T L M E Y L E N P K K Y I P G T K M I F A G I K K K D E R A D L I A Y L K K A T K E
V G F S Y T D A N K N K G I T W G E E T L M E Y L E N P K K Y I P G T K M I F A G I K K K G E R A D L I A Y L K K A T N E
P G F T Y T D A N K N K G I I W G E D T L M E Y L E N P K K Y I P G T K M I F A G I K K K G E R A D L I A Y L K K A T N E

E G F S Y T D A N K N K G I T W G E D T L M E Y L E N P K K Y I P G T K M I F A G I K K K S E R V D L I A Y L K D A T S K
E G F S Y T D A N K N K G I T W G E D T L M E Y L E N P K K Y I P G T K M I F A G I K K K A E R A D L I A Y L K Q A T A K
E G F S Y T D A N K N K G I T W G E D T L M E Y L E N P K K Y I P G T K M I F A G I K K K S E R A D L I A Y L K D A T A K
E G F S Y T E A N K N K G I T W G E E T L M E Y L E N P K K Y I P G T K M I F A G I K K K A E R A D L I A Y L K D A T S K
V G Y S Y T A A N K N K G I I W G D D T L M E Y L E N P K K Y I P G T K M V F T G L S K K K E R T N L I A Y L K E K T A A
A G F S Y T D A N K N K G I T W G E D T L M E Y L E N P K K Y I P G T K M I F A G I K K K G E R Q D L I A Y L K S A C S K
E G Y S Y T D A N K S K G I V W N N D T L M E Y L E N P K K Y I P G T K M I F A G I K K K G E R Q D L V A Y L K S A T S -
Q G F S Y T D A N K S K G I T W Q Q E T L R I Y L E N P K K Y I P G T K M I F A G L K K K S E R Q D L I A Y L K K T A A S

P G F S Y S N A N K A K G I T W G D D T L F E Y L E N P K K Y I P G T K M V F A G L K K A N E R A D L I A Y L K E S T K -
P G F S Y S N A N K A K G I T W Q D D T L F E Y L E N P K K Y I P G T K M V F A G L K K A N E R A D L I A Y L K Q A T K -
A G F A Y T N A N K A K G I T W Q D D T L F E Y L E N P K K Y I P G T K M I F A G L K K P N E R G D L I A Y L K S A T K -
A G F A Y T N A N K A K G I T W Q D D T L F E Y L E N P K K Y I P G T K M I F A G L K K P N E R G D L I A Y L K S A T K -

Q G Y S Y T D A N I K K N V L W D E N N M S E Y L T N P X K Y I P G T K M A F G G L K K E K D R N D L I T Y L K K A C E -
Q G Y S Y T D A N K R A G V E W A E P T M S D Y L E N P X K Y I P G T K M A F G G L K K A K D R N D L V T Y M L E A S K -
D G Y A Y T D A N K Q K G I T W D E N T L F E Y L E N P X K Y I P G T K M A F G G L K K D K D R N D I I T F M K E A T A -

A G Y S Y S A A N K N K A V E W E E N T L Y D Y L L N P X K Y I P G T K M V F P G L X K P Q D R A D L I A Y L K K A T S S
A G Y S Y S A A N K N K A V T W G E D T L Y E Y L L N P X K Y I P G T K M V F P G L X K P Q E R A D L I A Y L K D S T E -
A G Y S Y S A A N K N M A V I W E E N T L Y D Y L E N P X K Y I P G T K M V F P G L X K P Q E R A D L I A Y L K T S T A -
A G Y S Y S T A N K N M A V I W E E K T L Y D Y L E N P X K Y I P G T K M V F P G L X K P Q D R A D L I A Y L K E S T A -
A G Y S Y S A A N K N K A V E W E E K T L Y D Y L E N P X K Y I P G T K M V F P G L X K P Q D R A D L I A Y L K E A T A -
P G Y S Y S A A N K N R A V E W E E K T L Y D Y L E N P X K Y I P G T K M V F P G L X K P Q D R A D L I A Y L K E A T A -
P G Y S Y S A A N K N M A V I W G E N T L Y D Y L E N P X K Y I P G T K M V F P G L X K P Q E R A D L I A Y L K E A T A -
A G Y S Y S A A N K N M A V Q W G E N T L Y A Y L E N P X K Y I P G T K M V F P G L X K P Q D R A D L I A Y L K E A T A -
A G Y S Y S A A N K N M A V Q W G E N T L Y D Y L E N P X K Y I P G T K M V F P G L X K P Q D R A D L I A Y L K E S T A -
P G Y S Y S A A N K N M A V N W G E N T L Y D Y L E N P X K Y I P G T K M V F P G L X K P Q D R A D L I A Y L K E S T A -

6 1 2 3 1 2 5 1 1 2 6 4 3 2 7 1 7 4 5 2 2 5 4 1 1 3 1 1 1 1 1 1 1 1 1 1 1 3 1 5 1 2 2 1 6 9 2 1 7 2 2 2 2 2 2 6 4 4 5 4
```

Hydrophilic, acidic: [D] Asp [E] Glu

Hydrophilic, basic: [H] His [K] Lys [R] Arg [X] TrimethylLys

Polar, uncharged: [B] Asn or Asp [G] Gly [N] Asn [Q] Gln
[S] Ser [T] Thr [W] Trp [Y] Tyr [Z] Gln or Glu

Hydrophobic: [A] Ala [C] Cys [F] Phe [I] Ile [L] Leu
[M] Met [P] Pro [V] Val

evolutionary conservative protein. A total of 38 of its 105 residues (23 in all that have been sequenced) are invariant and most of the remaining residues are conservatively substituted. In contrast, there are eight positions that each accommodate six or more different residues and, accordingly, are described as being hypervariable.

The clear biochemical role of certain residues makes it easy to surmise why they are invariant. For instance, His 18

and Met 80 form ligands to the redox-active Fe atom of cytochrome *c*; the substitution of any other residues in these positions inactivates this protein. However, the biochemical significance of most of the invariant and conservatively substituted residues of cytochrome *c* can only be profitably assessed in terms of the protein's three-dimensional structure and is therefore deferred until Section 8-3A. In what follows, we consider what insights can be

TABLE 6-5. Amino Acid Difference Matrix for 26 Species of Cytochrome *C*[a]

Average differences

	Man, chimpanzee	Rhesus monkey	Horse	Donkey	Pig, cow, sheep	Dog	Gray whale	Rabbit	Kangaroo	Chicken, turkey	Penguin	Pekin duck	Rattlesnake	Snapping turtle	Bullfrog	Tuna fish	Screw worm fly	Silkworm moth	Wheat	Neurospora crassa	Baker's yeast	Candida krusei
Man, chimpanzee	0																					
Rhesus monkey	1	0																				
Horse	12	11	0																			
Donkey	11	10	1	0																		
Pig, cow, sheep	10	9	3	2	0																	
Dog	11	10	6	5	3	0																
Gray whale	10	9	5	4	2	3	0															
Rabbit	9	8	6	5	4	5	2	0														
Kangaroo	10	11	7	8	6	7	6	6	0													
Chicken, turkey	13	12	11	10	9	10	9	8	12	0												
Penguin	13	12	12	11	10	10	9	8	10	2	0											
Pekin duck	11	10	10	9	8	8	7	6	10	3	3	0										
Rattlesnake	14	15	22	21	20	21	19	18	21	19	20	17	0									
Snapping turtle	15	14	11	10	9	9	8	9	11	8	8	7	22	0								
Bullfrog	18	17	14	13	11	12	11	11	13	11	12	11	24	10	0							
Tuna fish	21	21	19	18	17	18	17	17	18	17	18	17	26	18	15	0						
Screw worm fly	27	26	22	22	22	21	22	21	24	23	24	22	29	24	22	24	0					
Silkworm moth	31	30	29	28	27	25	27	26	28	28	27	31	28	29	32	14	0					
Wheat	43	43	46	45	45	44	44	44	47	46	46	46	46	48	49	45	45	0				
Neurospora crassa	48	47	46	46	46	46	46	49	47	48	46	47	49	49	48	41	47	54	0			
Baker's yeast	45	45	46	45	45	45	45	46	46	45	46	47	49	47	47	45	47	47	41	0		
Candida krusei	51	51	51	50	50	49	50	50	51	50	51	51	53	51	48	47	47	50	42	27	0	

Average differences (values along upper diagonal): 10.0, 5.1, 9.9, 14.3, 12.6, 18.5, 25.9, 47.0

[a] Each table entry indicates the number of amino acid differences between the cytochromes *c* of the species noted to the left of and above that entry.

[Table copyrighted © by Irving Geis.]

gleaned solely from the comparisons of the amino acid sequences of related proteins. The conclusions we draw are surprisingly far reaching.

The easiest way to compare the evolutionary differences between two homologous proteins is simply to count the amino acid differences between them (more realistically, we should infer the minimum number of DNA base changes to convert one protein to the other but, because of the infrequency with which mutations are accepted, counting amino acid differences yields similar information). Table 6-5 is a tabulation of the amino acid sequence differences among 22 of the cytochromes *c* listed in Table 6-4. It has been boxed off to emphasize the relationships among groups of similar species. The order of these differences largely parallels that expected from classical taxonomy. Thus primate cytochromes *c* more nearly resemble those of other mammals than they do, for example, those of insects (8–12 differences for mammals vs 26–31 for insects). Similarly, the cytochromes *c* of fungi differ as much from those of mammals (45–51 differences) as they do from those of insects (41–47) or higher plants (47–54).

By computer analysis of data such as those in Table 6-5, *a kind of family tree, known as a **phylogenetic tree**, can be constructed which indicates the ancestral relationships* among the organisms that produced the proteins. That for cytochrome *c* is sketched in Fig. 6-14. Similar trees have been derived for other proteins. Each branch point of a tree indicates the probable existence of a common ancestor for all the organisms above it. The relative evolutionary distances between neighboring branch points are expressed as the number of amino acid differences per 100 residues of the protein (*Percentage of Accepted point Mutations*, or **PAM units**). This furnishes a quantitative measure of the degree of relatedness of the various species that macroscopic taxonomy cannot provide. Note that the evolutionary distances of modern cytochromes *c* from the lowest branch point on their tree are all approximately equal. Evidently, the cytochromes *c* of the so-called lower forms of life have evolved to the same extent as those of the higher forms.

Proteins Evolve at Characteristic Rates

The evolutionary distances between various species can be plotted against the time when, according to radiodated fossil records, the species diverged. For cytochrome *c*, this plot is essentially linear, thereby indicating that cytochrome *c* has accumulated mutations at a constant rate over the geological time scale (Fig. 6-15). This is also true for the other three proteins whose rates of evolution are plotted in Fig. 6-15. Each has its characteristic rate of change, known as a **unit evolutionary period,** which is defined as the time required for the amino acid sequence of a protein to change by 1% after two species have diverged. For cytochrome *c*, the unit evolutionary period is 20.0 million years. Compare this with the much less variant **histone H4** (600 million years) and the more variant hemoglobin (5.8 million years) and **fibrinopeptides** (1.1 million years).

The foregoing information does not imply that the rates of mutation of the DNAs specifying these proteins differ, but rather that *the rate that mutations are accepted into a protein depends on the extent that amino acid changes affect its function.* Cytochrome *c*, for example, is a rather small protein that, in carrying out its biological function, must interact with large protein complexes over much of its surface area. Any mutational change to cytochrome *c* will, most likely, affect these interactions unless, of course, the complexes simultaneously mutate to accommodate the change, a very unlikely occurrence. This accounts for the evolutionary stability of cytochrome *c*. Histone H4 is a protein that binds to DNA in eukaryotic chromosomes (Section 33-1A). Its central role in packaging the genetic archives evidently makes it extremely intolerant of any mutational changes. Indeed, histone H4 is so well adapted to its function that the histones H4 from peas and cows, species that diverged 1.2 billion years ago, differ by only two conservative changes in their 102 amino acids. Hemoglobin, like cytochrome *c*, is an intricate molecular machine (Section 9-2). It functions as a free floating molecule, however, so that its surface groups are usually more tolerant of change than are those of cytochrome *c* (although not in the

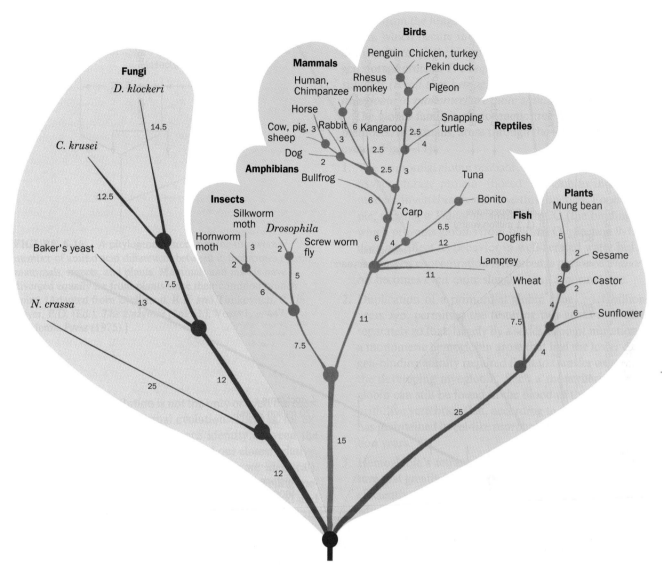

FIGURE 6-14. A phylogenetic tree of cytochrome *c* that was generated by the computer-aided analysis of difference data such as that in Table 6-5. Each branch point indicates the existence of an organism deduced to be ancestral to the species connected above it. The numbers beside each branch indicate the inferred differences, in PAM units, between the cytochromes *c* of its flanking branch points or species. [After Dayhoff, M.O., Park, C.M., and McLaughlin, P.J., *in* Dayhoff, M.O. (Ed.), *Atlas of Protein Sequence and Structure, p. 8*, National Biomedical Research Foundation (1972).]

case of HbS; Section 9-3B). This accounts for hemoglobin's greater rate of evolution. The fibrinopeptides are polypeptides of ~20 residues that are proteolytically cleaved from the vertebrate protein **fibrinogen** when it is converted to **fibrin** in the blood clotting process (Section 34-1A). Once they have been excised, the fibrinopeptides are discarded, so there is relatively little selective pressure on them to maintain their amino acid sequence and thus their rate of variation is high. If it is assumed that the fibrinopeptides are evolving at random, then the foregoing unit evolutionary periods indicate that in hemoglobin only 1.1/5.8 = 1/5 of the random amino acid changes are acceptable, that is, innocuous, whereas this quantity is 1/18 for cytochrome *c* and 1/550 for histone H4.

Mutational Rates Are Constant in Time

Amino acid substitutions in a protein mostly result from single base changes in the gene specifying the protein (Section 30-1). If such **point mutations** mainly occur as a consequence of errors in the DNA replication process, then the rate at which a given protein accumulates mutations would be constant with respect to numbers of generations. If, however, the mutational process results from the random chemical degradation of DNA, then the mutation rate would be constant with absolute time. To choose between these alternative hypotheses, let us compare the rate of cytochrome *c* divergence in insects with that in mammals.

Insects have shorter generation times than mammals. Therefore, if DNA replication were the major source of

benzyl ester link from the polypeptide's C-terminus to the support resin may be cleaved by treatment with liquid HF:

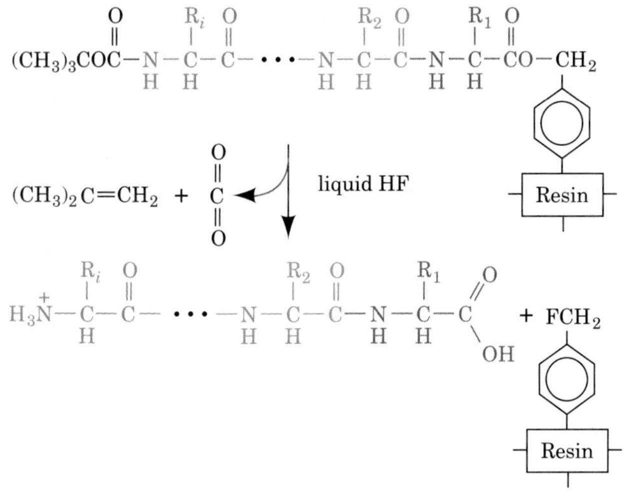

The Boc group linked to the polypeptide's N-terminus, as well as the benzyl groups protecting its side chains, are also removed by this treatment.

B. Problems and Prospects

The steps just outlined seem simple enough, but they are not as straightforward as we have implied. A major difficulty with the entire procedure is its low cumulative yield. Let us examine the reasons for this. To synthesize a polypeptide chain with n peptide bonds requires at least $2n$ reaction steps — one for coupling and one for deblocking each residue. If a protein-sized polypeptide is to be synthesized in reasonable yield, then each reaction step must be essentially quantitative; anything less greatly reduces the yield of final product. For example, in the synthesis of a 101-residue polypeptide chain, in which each reaction step occurs with an admirable 98% yield through 200 reaction steps, the overall yield is only $0.98^{200} \times 100 = 2\%$. Therefore, although oligopeptides are now routinely made, the synthesis of large polypeptides requires almost fanatical attention to chemical detail.

An ancillary problem is that the newly liberated synthetic polypeptide must be purified. This may be a difficult task because a significant level of incomplete reactions and/or side reactions at every stage of the solid phase synthesis will result in almost a continuum of closely related products for large polypeptides. The use of reverse-phase HPLC techniques (Section 5-3E), however, greatly facilitates this purification process.

Using the automated solid phase technique, Merrifield synthesized the nonapeptide hormone **bradykinin** in 85% yield.

$$Arg-Pro-Pro-Gly-Phe-Ser-Pro-Phe-Arg$$

Bradykinin

However, it was only recently, through steady progress in improving reaction yields (to >99.5% on average) and eliminating side reactions, that it became possible to synthesize ~100-residue polypeptides of reasonable quality. Thus, Stephen Kent synthesized the 99-residue **HIV protease** [an enzyme that is essential for the maturation of **human immunodeficiency virus (HIV, the AIDS virus)**] in such high yield and purity that, after being renatured (folded to its native conformation; Section 8-1A), it exhibited full biological activity. Indeed, this synthetic protein was crystallized and its X-ray structure was shown to be identical to that of biologically synthesized HIV protease. Kent also synthesized HIV protease from D-amino acids and experimentally verified, for the first time, that such a protein has the opposite chirality of its biologically produced counterpart. Moreover, this D-amino acid protease catalyzes the cleavage of its target polypeptide made from D-amino acids but not that made from L-amino acids as does naturally occurring HIV protease.

Genetic manipulations have, since the early 1980s, enabled the biosynthesis of proteins that have one or more amino acid residues which differ in a specified manner from those of the natural protein. These "genetic engineering" methods are already in routine use for the production of large quantities of specifically altered proteins for medical, industrial, and agricultural purposes. We shall have much more to say about these methods in Section 28-8. However, polypeptides containing "nonstandard" amino acid residues or even nonamino acid linkages must be synthetically produced since biologically generated polypeptides are made from only the 20 "standard" amino acids.

CHAPTER SUMMARY

The initial step in the amino acid sequence determination of a protein is to ascertain its content of chemically different polypeptides by end group analysis. The protein's disulfide bonds are then chemically cleaved, the different polypeptides are separated and purified, and their amino acid compositions are determined. Next, the purified polypeptides are specifically cleaved, by enzymatic or chemical means, to smaller peptides that are separated, purified, and then sequenced by (automated) Edman degradation or mass spectrometric methods. Repetition of this process, using a cleavage method of different specificity, generates overlapping peptides whose amino acid sequences, when compared to those of the first group of peptide fragments, indicate their order in the parent polypeptide. The primary structure determination is completed by establishing the positions of the disulfide bonds. This requires the degradation of the protein with its disulfide bonds intact. Then, by sequencing the pairs of disulfide-linked peptide fragments, their positions in the intact protein can be deduced. Once a primary structure is known, its minor variants, which may arise from mutations or chemical modifications, can be easily analyzed by peptide mapping.

Inactivation of a protein by treatment with group-specific reagents may serve to identify its essential residues. An enzyme's active site residues may be unusually reactive.

Sickle-cell anemia is a molecular disease of individuals who are homozygous for a gene specifying an altered β chain of hemoglobin. Fingerprinting and sequencing studies have identified this alteration as arising from a point mutation that changes Glu $\beta 6 \rightarrow$ Val. In the heterozygous state, the sickle-cell trait confers resistance to malaria without causing deleterious effects. This accounts for its high incidence in populations living in malarial regions. The cytochromes c from many eukaryotic species contain many amino acid residues that are invariant or conservatively substituted. Hence, this protein is well adapted to its function. The amino acid differences between the various cytochromes c have permitted the generation of their phylogenetic tree, which closely parallels that determined by classical taxonometry. The number of sequence differences between homologous proteins from related species plotted against the time when, according to the fossil record, these species diverged from a common ancestor, reveals that acceptable point mutations in proteins occur at a constant rate. Proteins whose functions are relatively intolerant to sequence changes evolve more slowly than those that are more tolerant to such changes. Phylogenetic analysis of the globin family—myoglobin and the α and β chains of hemoglobin—reveals that these proteins arose through gene duplication. In this process, the original function of the protein is maintained, while the duplicated copy evolves a new function. Many, if not most, proteins have evolved through gene duplication.

The strategy of polypeptide chemical synthesis involves coupling amino acids, one at a time, to the N-terminus of a growing polypeptide chain. The α-amino group of each amino acid must be chemically protected during the coupling reaction and then unblocked before the next coupling step. Reactive side chains must also be chemically protected but then unblocked at the conclusion of the synthesis. The difficulty in recovering the intermediate product of each of the many steps of such a synthesis has been eliminated by the development of solid phase synthesis techniques. These methods have led to the synthesis of numerous biologically active polypeptides and have recently become capable of synthesizing small proteins in useful amounts.

REFERENCES

General

Creighton, T.E., *Proteins* (2nd ed.), Chapters 1 and 3, Freeman (1993).

Wood, W.B., Wilson, J.H., Benbow, R.M., and Hood, L.E., *Biochemistry, A Problems Approach* (2nd ed.), Chapter 3, Benjamin/Cummings (1981). [A question-and-answer approach to biochemical instruction.]

Primary Structure Determination

Allen, G., *Sequencing of Proteins and Peptides, in* Burdon, R.M. and van Knippenberg, P.H. (Eds.), *Laboratory Techniques in Biochemistry and Molecular Biology,* Vol. 9 (2nd revised ed.), Elsevier (1989).

Bhown, A.S. (Ed.), *Protein/Peptide Sequence Analysis: Current Methodologies,* CRC Press (1988).

Biemann, K., Mass spectrometry of peptides and proteins, *Annu. Rev. Biochem.* **61,** 977–1010 (1992).

Bogusky, M.S., Ostell, J., and States, D.J., Molecular sequence data-bases and their uses, *in* Rees, A.R., Sternberg, M.J.E., and Wetzel, R. (Eds.), *Protein Engineering. A Practical Approach, pp.* 57–88, IRL Press (1992).

Chait, B.T. and Kent, S.B.H., Weighing naked proteins: Practical, high accuracy mass measurement of peptides and proteins, *Science* **257,** 1885–1894 (1992).

Creighton, T.E. (Ed.), *Protein Structure. A Practical Approach,* IRL Press (1989). [Chapters 5, 6, and 7 are respectively concerned with peptide mapping, determining the numbers of specific types of residues in a protein, and identifying the disulfide bonds in a protein.]

Fenselau, C., Beyond gene sequencing: Analysis of protein structure with mass spectrometry, *Annu. Rev. Biophys. Biophys. Chem.* **20,** 205–220 (1991).

Findlay, J.B.C. and Geisow, M.J. (Eds.), *Protein Sequencing, A Practical Approach,* IRL Press (1989).

Hunkapiller, M.W., Strickler, J.E., and Wilson, K.J., Contemporary methodology for protein structure determination, *Science* **226,** 304–311 (1984).

Matsudaira, P.T. (Ed.), *A Practical Guide to Protein and Peptide Purification and Microsequencing,* Academic Press (1989).

McCloskey, J.A. (Ed.), *Mass Spectrometry, Methods Enzymol.* **193** (1990). [Contains a section on the mass spectrometry of proteins and peptides.]

Sanger, F., Sequences, sequences, and sequences, *Annu. Rev. Biochem.* **57,** 1–28 (1988). [A scientific autobiography that provides a glimpse of the early difficulties in sequencing proteins.]

Senko, M.W. and McLafferty, F.W., Mass spectrometry of macromolecules: Has its time now come? *Annu. Rev. Biophys. Biomol. Struct.* **23,** 763–785 (1994).

Shively, J.E., Paxton, R.J., and Lee, T.D., Highlights of protein structural analysis, *Trends Biochem. Sci.* **14,** 246–252 (1989).

Protein Modification

Balaram, P., Non-standard amino acids in peptide design and protein engineering, *Curr. Opin. Struct. Biol.* **2,** 845–851 (1992).

Imoto, I. and Yamada, H., Chemical modification, *in* Creighton, T.E. (Ed.), *Protein Function. A Practical Approach, pp.* 247–277, IRL Press (1989).

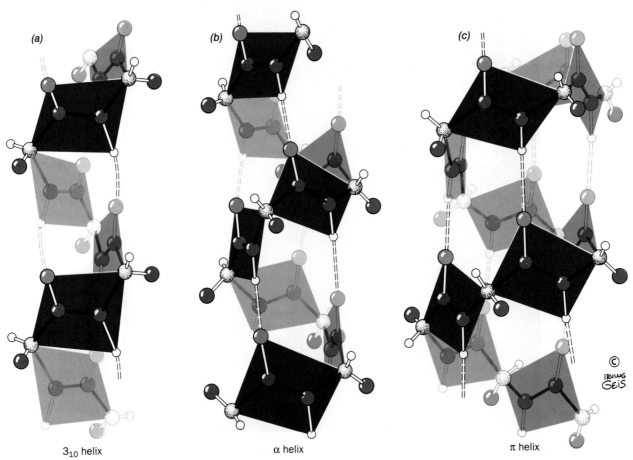

3₁₀ helix   α helix   π helix

FIGURE 7-14. Two polypeptide helices that occasionally occur in proteins compared with the commonly occurring α helix. (*a*) The 3₁₀ helix is characterized by 3.0 peptide units per turn and a pitch of 6.0 Å, which makes it thinner and more elongated than the α helix. (*b*) The α helix has 3.6 peptide units per turn and a pitch of 5.4 Å (also see Fig. 7-11). (*c*) The π helix, with 4.4 peptide units per turn and a pitch of 5.2 Å, is wider and shorter than the α helix. The peptide planes are indicated. [Figure copyrighted © by Irving Geis.]

closed by the hydrogen bond. With this notation, an α helix is a 3.6₁₃ helix.

The right-handed 3₁₀ helix (Fig. 7-14*a*), which has a pitch of 6.0 Å, is thinner and rises more steeply than does the α helix (Fig. 7-14*b*). Its torsion angles place it in a mildly forbidden zone of the Ramachandran diagram that is rather near the position of the α helix (Fig. 7-7) and its R groups experience some steric interference. This explains why the 3₁₀ helix is only occasionally observed in proteins, and then mostly in short segments that are frequently distorted from the ideal 3₁₀ conformation. The 3₁₀ helix most often occurs as a single-turn transition between one end of an α helix and the adjoining portion of a polypeptide chain.

The **π helix** (4.4₁₆ helix), which also has a mildly forbidden conformation (Fig. 7-7), has only rarely been observed and then only as segments of longer helices. This is probably because its comparatively wide and flat conformation (Fig. 7-14*c*) results in an axial hole that is too small to admit water molecules but yet too wide to allow van der Waals associations across the helix axis; this greatly reduces its

stability relative to more closely packed conformations. The 2.2₇ ribbon, which as Fig. 7-7 indicates, has strongly forbidden conformation angles, has never been observed.

Certain synthetic homopolypeptides assume conformations that are models for helices in particular proteins. **Polyproline** is unable to assume any common secondary structure due to the conformational constraints imposed by its cyclic pyrrolidine side chains. Furthermore, the lack of a hydrogen substituent on its backbone nitrogen precludes any polyproline conformation from being knit together by hydrogen bonding. Nevertheless, under the proper conditions, polyproline precipitates from solution as a left-handed helix of all-trans peptides that has 3.0 residues per helical turn and a pitch of 9.4 Å (Fig. 7-15). This rather extended conformation permits the Pro side chains to avoid each other. Curiously, **polyglycine,** the least conformationally constrained polypeptide, precipitates from solution as a helix whose parameters are essentially identical to those of polyproline, the most conformationally constrained polypeptide (although the polyglycine helix may be either right

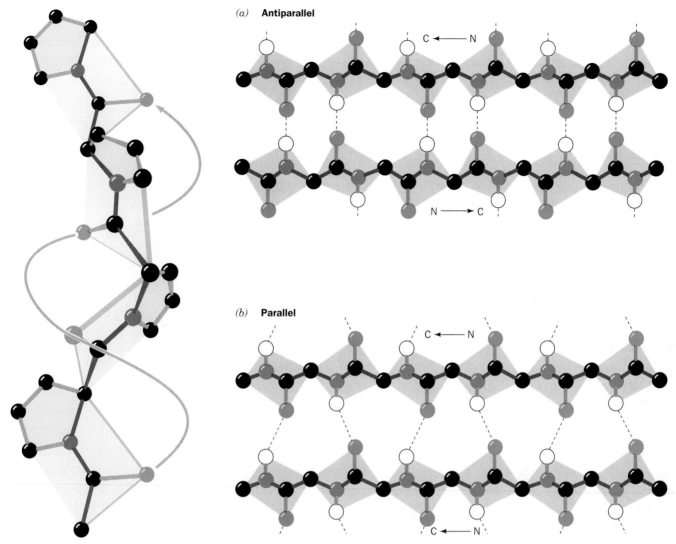

(a) **Antiparallel**

C ← N

N → C

(b) **Parallel**

C ← N

C ← N

FIGURE 7-15. The **polyproline II** helix. Polyglycine forms a nearly identical helix **(polyglycine II).** [Figure copyrighted © by Irving Geis.]

FIGURE 7-16. The hydrogen bonding associations in β pleated sheets. Side chains are omitted for clarity. (*a*) The antiparallel β pleated sheet. (*b*) The parallel β pleated sheet. [Figure copyrighted © by Irving Geis.]

or left handed because Gly is nonchiral). The structures of the polyglycine and polyproline helices are of biological significance because they form the basic structural motif of collagen, a structural protein that contains a remarkably high proportion of both Gly and Pro (Section 7-2C).

C. Beta Structures

In 1951, the year that they proposed the α helix, Pauling and Corey also postulated the existence of a different polypeptide secondary structure, the β **pleated sheet.** As with the α helix, the β pleated sheet's conformation has repeating ϕ and ψ angles that fall in the allowed region of the Ramachandran diagram (Fig. 7-7) and utilizes the full hydrogen

bonding capacity of the polypeptide backbone. *In β pleated sheets, however, hydrogen bonding occurs between neighboring polypeptide chains rather than within one as in α helices.*

β Pleated sheets come in two varieties:

1. The **antiparallel β pleated sheet,** in which neighboring hydrogen bonded polypeptide chains run in opposite directions (Fig. 7-16*a*).

2. The **parallel β pleated sheet,** in which the hydrogen bonded chains extend in the same direction (Fig. 7-16*b*).

The conformations in which these β **structures** are optimally hydrogen bonded vary somewhat from that of a fully extended polypeptide ($\phi = \psi = \pm 180°$) as indicated in Fig. 7-7. They therefore have a rippled or pleated edge-on appearance (Fig. 7-17), which accounts for the appellation "pleated sheet." In this conformation, successive side chains of a polypeptide chain extend to opposite sides of the pleated sheet with a two-residue repeat distance of 7.0 Å.

β Sheets are common structural motifs in proteins. In globular proteins, they consist of from 2 to as many as 15 polypeptide strands, the average being 6 strands, which have an aggregate width of ~25 Å. The polypeptide chains in a β sheet are known to be up to 15 residues long, with the average being 6 residues that have a length of ~21 Å. A 6-stranded antiparallel β sheet, for example, occurs in the jack bean protein **concanavalin A** (Fig. 7-18).

Parallel β sheets of less than five strands are rare. This observation suggests that parallel β sheets are less stable than antiparallel β sheets, possibly because the hydrogen bonds of parallel sheets are distorted in comparison to those of the antiparallel sheets (Fig. 7-16). Mixed parallel–antiparallel β sheets are common but occur with only ~40% of the frequency that would be expected for the random mixing of strand directions.

The β pleated sheets in globular proteins invariably exhibit a pronounced right-handed twist when viewed along

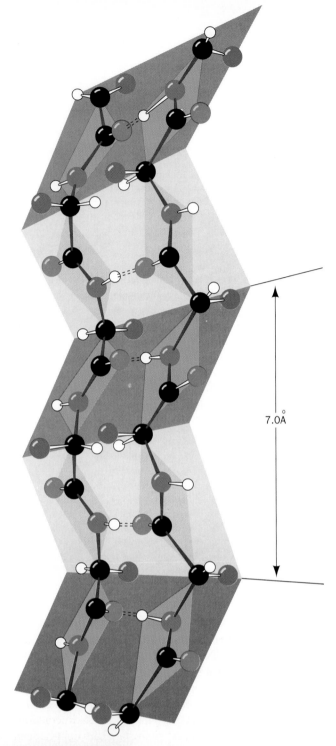

7.0Å

FIGURE 7-17. A two-stranded β antiparallel pleated sheet drawn so as to emphasize its pleated appearance. Dashed lines indicate hydrogen bonds. Note that the R groups (*purple balls*) on each polypeptide chain alternately extend to opposite sides of the sheet and that they are in register on adjacent chains. [Figure copyrighted © by Irving Geis.]

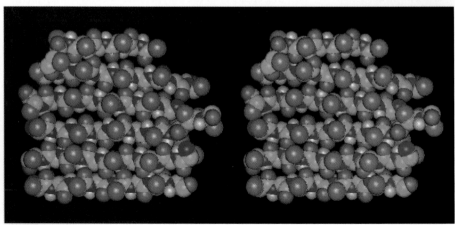

FIGURE 7-18. A stereo, space-filling representation of the six-stranded antiparallel β pleated sheet in jack bean concanavalin A as determined by X-ray crystal structure analysis. In the main chain, carbon atoms are green, nitrogen atoms are blue, oxygen atoms are red, and hydrogen atoms are white. R groups are represented by large purple balls. Instructions for viewing stereo drawings are given in the appendix to this chapter.

(a)

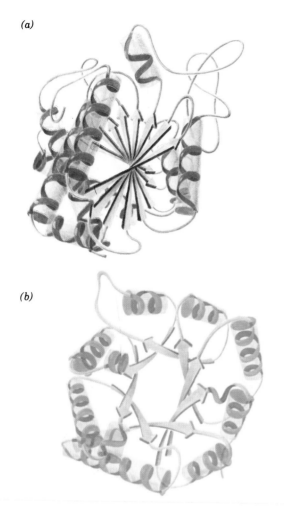

(b)

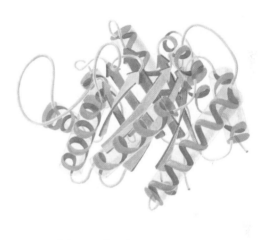

FIGURE 7-19. Polypeptide chain folding in proteins illustrating the right-handed twist of β sheets. Here the polypeptide backbones are represented by ribbons with α helices shown as coils and the strands of β sheets indicated by arrows pointing towards the C-terminus. Side chains are not shown. (*a*) Bovine carboxypeptidase A, a 307-residue protein, contains an eight-stranded mixed β sheet that forms a saddle-shaped curved surface with a right-handed twist. (*b*) **Triose phosphate isomerase,** a 247-residue chicken muscle enzyme, contains an eight-stranded parallel β sheet that forms a cylindrical structure known as a **β barrel** [here viewed from the top (*left*) and from the side (*right*)]. Note that the crossover connections between successive strands of the β barrel, which each consist predominantly of an α helix, are outside the β barrel and have a right-handed helical sense. [After drawings by Jane Richardson, Duke University.]

their polypeptide strands (e.g., Fig. 7-19). Such twisted β sheets are important architectural features of globular proteins since β sheets often form their central cores (Fig. 7-19). Conformational energy calculations indicate that a β sheet's right-handed twist is a consequence of nonbonded interactions between the chiral L-amino acid residues in the sheet's extended polypeptide chains. These interactions tend to give the polypeptide chains a slight right-handed helical twist (Fig. 7-19) which distorts and hence weakens the β sheet's interchain hydrogen bonds. A particular β sheet's geometry is thus the result of a compromise between optimizing the conformational energies of its polypeptide chains and preserving its hydrogen bonds.

The **topology** (connectivity) of the polypeptide strands in a β sheet can quite complex; the connecting links of these assemblies often consist of long runs of polypeptide chain which frequently contain helices (e.g., Fig. 7-19). The link connecting two consecutive antiparallel strands is topologically equivalent to a simple hairpin turn (Fig. 7-20a). However, tandem parallel strands must be linked by a cross-over connection that is out of the plane of the β sheet. Such crossover connections almost always have a right-handed

(a)   (b)

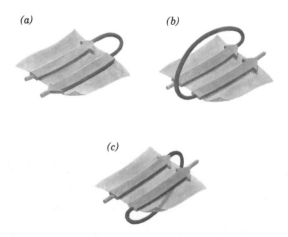

(c)

FIGURE 7-20. The connections between adjacent polypeptide strands in β pleated sheets: (*a*) The hairpin connection between antiparallel strands is topologically in the plane of the sheet. (*b*) A right-handed crossover connection between successive strands of a parallel β sheet. Nearly all such crossover connections in proteins have this chirality (see, e.g., Fig. 7-19b). (*c*) A left-handed cross-over connection between parallel β sheet strands. Connections with this chirality are rare. [After Richardson, J.S., *Adv. Protein Chem.* **34,** 290, 295 (1981).]

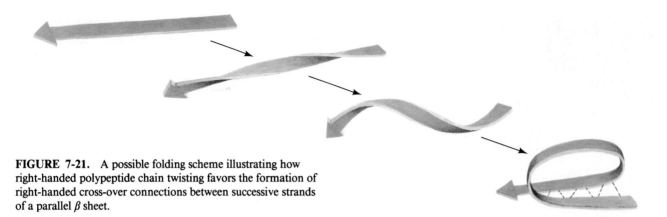

FIGURE 7-21. A possible folding scheme illustrating how right-handed polypeptide chain twisting favors the formation of right-handed cross-over connections between successive strands of a parallel β sheet.

helical sense (Fig. 7-20*b*), which is thought to better fit the β sheets' inherent right-handed twist (Fig. 7-21).

D. Nonrepetitive Structures

Regular secondary structures—helices and β sheets—comprise around half of the average globular protein. The protein's remaining polypeptide segments are said to have a **coil** or **loop conformation.** That is not to say, however, that these nonrepetitive secondary structures are any less ordered than are helices or β sheets; they are simply irregular and hence more difficult to describe. You should therefore not confuse the term coil conformation with the term **random coil,** which refers to the totally disordered and rapidly fluctuating set of conformations assumed by denatured proteins and other polymers in solution.

Globular proteins consist largely of approximately straight runs of secondary structure joined by stretches of polypeptide that abruptly change direction. Such **reverse turns** or **β bends** (so named because they often connect successive strands of antiparallel β sheets) almost always occur at protein surfaces; indeed, they partially define these surfaces. Most reverse turns involve four successive amino acid residues more or less arranged in one of two ways, Type I and Type II, that differ by a 180° flip of the peptide unit linking residues 2 and 3 (Fig. 7-22). Both types of β bends are stabilized by a hydrogen bond although deviations from these ideal conformations often disrupt this hydrogen bond. Type I β bends may be considered to be distorted sections of 3_{10} helix. In Type II β bends, the oxygen atom of residue 2 crowds the C_β atom of residue 3, which is therefore usually Gly. Residue 2 of either type of β bend is often Pro since it can facilely assume the required conformation.

Almost all proteins of >60 residues contain one or more loops of 6 to 16 residues that are not components of helices or β sheets and whose end-to-end distances are <10 Å. Such

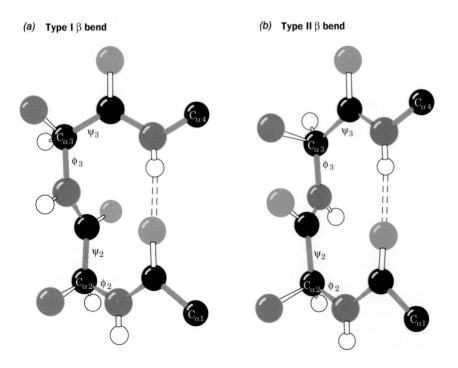

(a) **Type I β bend**

(b) **Type II β bend**

FIGURE 7-22. Reverse turns in polypeptide chains: (*a*) A Type I β bend, which has the following torsion angles:

$$\phi_2 = -60°, \quad \psi_2 = -30°,$$
$$\phi_3 = -90°, \quad \psi_3 = 0°.$$

(*b*) A Type II β bend, which has the following torsion angles:

$$\phi_2 = -60°, \quad \psi_2 = 120°,$$
$$\phi_3 = 90°, \quad \psi_3 = 0°.$$

Variations from these ideal conformation angles by as much as 30° are common. Hydrogen bonds are represented by dashed lines. [Figure copyrighted © by Irving Geis.]

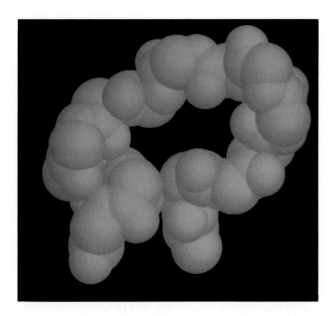

FIGURE 7-23. Space-filling representation of an Ω loop comprising residues 40 to 54 of cytochrome *c*. Only backbone atoms are shown; the addition of side chains would fill in the loop. [Courtesy of George Rose, Washington University School of Medicine.]

Ω loops (so named because they have the necked-in shape of the Greek upper case letter omega; Fig. 7-23), which may contain reverse turns, are compact globular entities because their side chains tend to fill in their internal cavities. Since Ω loops are almost invariably located on the protein surface, they may have an important role in biological recognition processes.

Many proteins have regions that are truly disordered. Extended, charged surface groups such as Lys side chains or the N- or C-termini of polypeptide chains are good examples: They often wave around in solution because there are few forces to hold them in place (Section 7-4). Sometimes entire polypeptide chain segments are disordered. Such segments may have functional roles, such as the binding of a specific molecule, so they may be disordered in one state of the protein (molecule absent) and ordered in another (molecule bound). This is one mechanism whereby a protein can interact flexibly with another molecule in the performance of its biological function.

2. FIBROUS PROTEINS

Fibrous proteins are highly elongated molecules whose secondary structures are their dominant structural motifs. Many fibrous proteins, such as those of skin, tendon, and bone, function as structural materials that have a protective, connective, or supportive role in living organisms. Others, such as muscle and ciliary proteins, have motive functions. In this section, we shall discuss structure–

function relationships in four common and well-characterized fibrous proteins: keratin, silk fibroin, collagen, and elastin (muscle and ciliary proteins are considered in Section 34-3). The structural simplicity of these proteins relative to those of globular proteins (Section 7-3) makes them particularly amenable to understanding how their structures suit them to their biological roles.

Fibrous molecules rarely crystallize and hence are usually not subject to structural determination by single-crystal X-ray structure analysis (Section 7-3A). Rather than crystallizing, they associate as fibers in which their long molecular axes are more or less parallel to the fiber axis but in which they lack specific orientation in other directions. The X-ray diffraction pattern of such a fiber, Fig. 7-24, for example, contains little information, far less than would be obtained if the fibrous protein could be made to crystallize. Consequently, the structures of fibrous proteins are not known in great detail. Nevertheless, the original X-ray studies of proteins were carried out in the early 1930s by William Astbury on such easily available protein fibers as wool and tendon. Since the first X-ray crystal structure of a protein was not determined until the late 1950s, these fiber studies constituted the first tentative steps in the elucidation of the structural principles governing proteins and formed much of the experimental basis for Pauling's formulation of the α helix and β pleated sheet.

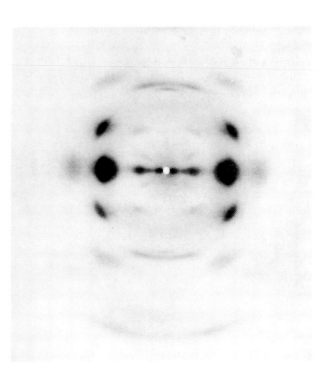

FIGURE 7-24. The X-ray diffraction photograph of a fiber of *Bombyx mori* silk obtained by shining a collimated beam of monochromatic X-rays through the silk fiber and recording the diffracted X-rays on a sheet of photographic film placed behind the fiber. The photograph has only a few spots and thus contains little structural information. [From March, R.E., Corey, R.B., and Pauling, L., *Biochim. Biophys. Acta* **16,** 5 (1955).]

A. α Keratin—A Helix of Helices

Keratin is a mechanically durable and chemically unreactive protein occurring in all higher vertebrates. It is the principal component of their horny outer epidermal layer, comprising up to 85% of the cellular protein, and its related appendages such as hair, horn, nails, and feathers. Keratins have been classified as either **α keratins,** which occur in mammals, or **β keratins,** which occur in birds and reptiles. Mammals have around 30 keratin variants which are expressed in a tissue-specific manner and which are classified as belonging to families of relatively acidic (Type I) and relatively basic (Type II) polypeptides. Keratin filaments, which form the intermediate filaments of skin cells (Section 1-2A), must contain at least one member of each type.

Electron microscopic studies indicate that hair, which is composed mainly of α keratin, consists of a hierarchy of structures (Figs. 7-25 and 7-26). A typical hair is ~20 μm in diameter and is constructed from dead cells, each of which contains packed **macrofibrils** (~2000 Å in diameter) that are oriented parallel to the hair fiber (Fig. 7-25). The macrofibrils are constructed from **microfibrils** (~80 Å wide) that are cemented together by an amorphous protein matrix of high sulfur content.

Moving to the molecular level, the X-ray diffraction pattern of α keratin resembles that expected for an α helix (hence the name α keratin). Yet, α keratin exhibits a 5.1-Å spacing rather than the 5.4-Å distance corresponding to the pitch of the α helix. This observation, together with a variety of physical and chemical evidence, suggests that *α keratin polypeptides form closely associated pairs of α helices in which each pair is composed of a Type I and a Type II*

keratin chain twisted in parallel into a left-handed coil (Fig. 7-26a). The normal 5.4-Å repeat distance of each α helix in the pair is thereby tilted with respect to the axis of this assembly, yielding the observed 5.1-Å spacing. This assembly is said to have a **coiled coil** structure because each α helix axis itself follows a helical path.

The conformation of α keratin's coiled coil is a consequence of its primary structure: The central ~310-residue segment of each polypeptide chain has a 7-residue pseudorepeat, *a-b-c-d-e-f-g,* with nonpolar residues predominating at positions *a* and *d.* Since an α helix has 3.6 residues per turn, α keratin's *a* and *d* residues line up on one side of the α helix to form a hydrophobic strip that promotes its lengthwise association with a similar strip on another such α helix (Fig. 7-27; hydrophobic residues, as we shall see in Section 7-4C, have a strong tendency to associate). Indeed, the slight discrepancy between the 3.6 residues per turn of a normal α helix and the ~3.5 residue repeat of α keratin's hydrophobic strip is responsible for the coiled coil's coil. The resulting 18° inclination of the α helices relative to one another permits their contacting side chains to interdigitate efficiently, thereby greatly increasing their favorable interactions. Coiled coils, as we shall see, occur in several globular proteins as well as in other intermediate filament proteins.

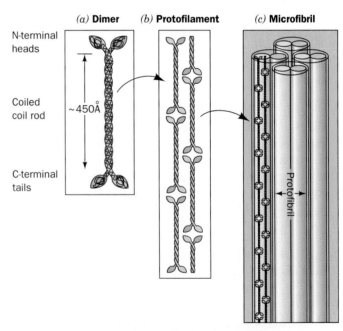

FIGURE 7-26. The structure of α keratin. (*a*) The central ~310 residues of one polypeptide chain each of Types I and II α keratins associate in a dimeric coiled coil. The conformations of the polypeptides' globular N- and C-terminal domains are unknown. (*b*) Protofilaments are formed from two staggered and antiparallel rows of associated head-to-tail coiled coils. (*c*) The protofilaments dimerize to form a protofibril, four of which form a microfibril. The structures of these latter assemblies are poorly characterized.

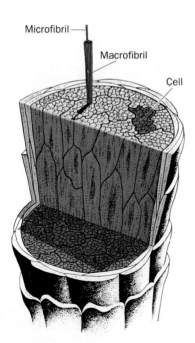

FIGURE 7-25.   The macroscopic organization of hair. [Figure copyrighted © by Irving Geis.]

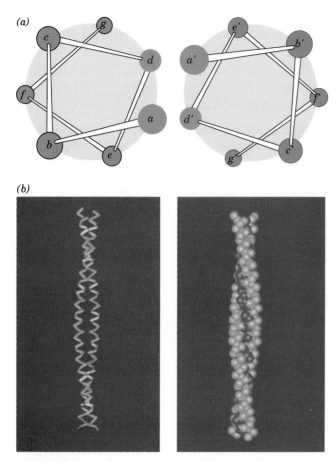

FIGURE 7-27. The two-stranded coiled coil. (*a*) View down the coil axis showing the interactions between the nonpolar edges of the α helices. The α helices have the pseudorepeating heptameric sequence *a-b-c-d-e-f-g* in which residues *a* and *d* are predominantly nonpolar. [After McLachlan, A.D. and Stewart, M., *J. Mol. Biol.* **98**, 295 (1975).] (*b*) Side view in which the polypeptide backbone is represented in skeletal (*left*) and space-filling (*right*) forms. Note the interlocking of the contacting nonpolar side chains (shown as red spheres) in the space-filling model. [Courtesy of Carolyn Cohen, Brandeis University.]

The higher order substructure of α keratin is poorly understood. The N- and C-terminal domains of each polypeptide probably have a flexible conformation and facilitate the assembly of the coiled coils into ~30-Å-wide protofilaments. These are thought to consist of two staggered antiparallel rows of head-to-tail aligned coiled coils (Fig. 7-26*b*). Two such protofilaments are thought to comprise a ~50-Å-wide protofibril, four of which, in turn, form a microfibril (Fig. 7-26*c*).

α Keratin is rich in Cys residues which form disulfide bonds that cross-link adjacent polypeptide chains. This accounts for its insolubility and resistance to stretching, two of α keratin's most important biological properties. The α keratins are classified as "hard" or "soft" according to whether they have a high or low sulfur content. Hard keratins, such as those of hair, horn, and nail, are less pliable than soft keratins, such as those of skin and callus, because the disulfide bonds resist any forces tending to deform

them. The disulfide bonds can be reductively cleaved with mercaptans (Section 6-1B). Hair so treated can be curled and set in a "permanent wave" by application of an oxidizing agent which reestablishes the disulfide bonds in the new "curled" conformation. Although the insolubility of α keratin prevents most animals from digesting it, the clothes moth larva, which has a high concentration of mercaptans in its digestive tract, can do so to the chagrin of owners of woolen clothing.

The springiness of hair and wool fibers is a consequence of the coiled coil's tendency to untwist when stretched and to recover its original conformation when the external force is relaxed. After some of its disulfide bonds have been cleaved, however, an α keratin fiber can be stretched to over twice its original length by the application of moist heat. In this process, as X-ray analysis indicates, the α helical structure extends with concomitant rearrangement of its hydrogen bonds to form a β pleated sheet. β Keratin, such as that of feathers, exhibits a similar X-ray pattern in its native state (hence the β sheet).

Keratin Defects Result in a Loss of Skin Integrity

The inherited skin diseases **epidermolysis bullosa simplex (EBS)** and **epidermolytic hyperkeratosis (EHK)** are characterized by skin blistering arising from the rupture of epidermal basal cells (Fig. 1-14*d*) and suprabasal cells, respectively, as caused by mechanical stresses that normally would be harmless. Symptomatic variations in these conditions range from severely incapacitating, particularly in early childhood, to barely noticeable. In families afflicted with EBS, sequence abnormalities may be present in either keratin 14 or keratin 5, the dominant Types I and II keratins in basal skin cells. EHK is similarly caused by defects in keratins 1 or 10, the dominant Types I and II keratins in suprabasal cells (which arise through the differentiation of basal cells, a process in which the synthesis of keratins 14 and 5 is switched off and that of keratins 1 and 10 is turned on). These defects evidently interfere with normal filament formation thereby demonstrating the function of the keratin cytoskeleton in maintaining the mechanical integrity of the skin.

B. Silk Fibroin — A β Pleated Sheet

Insects and arachnids (spiders) produce **silk** to fabricate structures such as cocoons, webs, nests, and egg stalks. The silk is stored in aqueous solution in the gland that produces it but, during spinning, is converted to a water-insoluble form. Most silks consist of the fibrous protein **fibroin** and a gummy amorphous protein named **sericin** that cements the fibroin fibers together. An adult moth emerging from its sealed cocoon secretes a protease (**cocoonase**) that digests sericin, thereby enabling the moth to push the fibroin filaments aside and escape from the cocoon. In the preparation of silk cloth, which consists only of fibroin, the sericin is removed by treatment with boiling soap solution.

FIGURE 7-28. The domestic silkworm *Bombyx mori* in the process of constructing its cocoon. [Hans Pfletschinger/Peter Arnold, Inc.]

Silk fibroin from the cultivated larvae (silkworms) of the moth *Bombyx mori* (Fig. 7-28) exhibits an X-ray diffraction pattern (Fig. 7-24) indicating that *its polypeptide chains form antiparallel β pleated sheets in which the chains extend parallel to the fiber axis.* Sequence studies have shown that long stretches of the chain are comprised of the six-residue repeat

(-Gly-Ser-Gly-Ala-Gly-Ala-)$_n$

This sequence forms β sheets with its Gly side chains extending from one surface and its Ser and Ala side chains extending from the other surface (as in Fig. 7-17). The β sheets thereby stack to form a microcrystalline array such that layers of contacting Gly side chains from neighboring sheets alternate with layers of contacting Ser and Ala side chains (Fig. 7-29). This structure, in part, accounts for silk's mechanical properties. *Silk fibers are strong but only slightly extensible because appreciable stretching would require breaking the covalent bonds of its nearly fully extended polypeptide chains.* Yet, the fibers are flexible because neighboring β sheets associate only through relatively weak van der Waals forces.

Although large segments of silk fibroin have the repeating hexameric amino acid sequence, it also has regions in which bulky residues such as Tyr, Val, Arg, and Asp occur. These residues distort and therefore disorder the microcrystalline array drawn in Fig. 7-29. Silk fibers are consequently composed of alternating crystalline and amorphous regions. The amorphous regions are largely responsible for the extensibility of the silk fibers. Silks from different species have different proportions of bulky side groups and therefore different mechanical properties. The greater a fibroin's proportion of bulky groups, the less its elasticity (the ability to resist deformation and to recover its original shape when

the deforming forces are removed) and the greater its extensibility. Hence, *the mechanical properties of silk fibroin can be understood in terms of its structure which, in turn, depends on its amino acid sequence.*

C. Collagen — A Triple Helical Cable

Collagen occurs in all multicellular animals and is the most abundant protein of vertebrates. It is an extracellular protein that is organized into insoluble fibers of great tensile strength. This suits collagen to its role as the major stress-bearing component of **connective tissues** such as bone, teeth, cartilage, tendon, ligament, and the fibrous matrices of skin and blood vessels. Collagen occurs in virtually every tissue.

A single molecule of Type I collagen has a molecular mass of ~285 kD, a width of ~14 Å, and a length of ~3000 Å. It is composed of three polypeptide chains. Mammals have at least 30 genetically distinct polypeptide chains comprising 16 collagen variants that occur in different tissues of the same individual. The most prominant of these are listed in Table 7-2.

Collagen has a distinctive amino acid composition: Nearly one third of its residues are Gly; another 15 to 30% of its residues are Pro and 4-hydroxyproline (Hyp).

4-Hydroxyprolyl residue (Hyp)   **3-Hydroxyprolyl residue**

5-Hydroxylysyl residue (Hyl)

3-Hydroxyproline and **5-hydroxylysine (Hyl)** also occur in collagen but in smaller amounts. Radioactive labeling experiments have established that these nonstandard hydroxylated amino acids are not incorporated into collagen during polypeptide synthesis: If ^{14}C-labeled 4-hydroxyproline is administered to a rat, the collagen synthesized is not radioactive, whereas radioactive collagen is produced if the

(a)

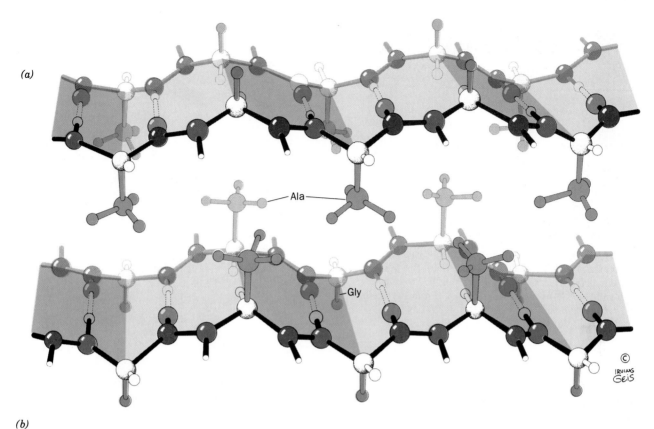

Ala

Gly

© IRVING GEIS

(b)

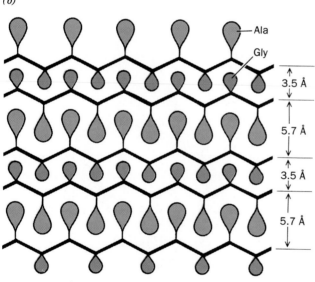

Ala

Gly

3.5 Å

5.7 Å

3.5 Å

5.7 Å

FIGURE 7-29. The three-dimensional architecture of silk fibroin. (*a*) Silk's alternating Gly and Ala (or Ser) residues extend to opposite sides of a given β sheet so that the Ala side chains extending from one β sheet efficiently nestle between those of the neighboring sheet and likewise for the Gly side chains. (*b*) Gly side chains from neighboring β sheets are in contact as are those of Ala and Ser. The intersheet spacings consequently have the alternating values 3.5 and 5.7 Å. [Figure copyrighted © by Irving Geis.]

rat is fed ^{14}C-labeled proline. The hydroxylated residues appear after the collagen polypeptides are synthesized, when certain Pro residues are converted to Hyp in a reaction catalyzed by the enzyme **prolyl hydroxylase.**

Hyp confers stability upon collagen, probably through intramolecular hydrogen bonds that may involve bridging water molecules. If, for example, collagen is synthesized under conditions that inactivate prolyl hydroxylase, it loses its native conformation (denatures) at 24°C, whereas normal collagen denatures at 39°C (denatured collagen is known as **gelatin**). Prolyl hydroxylase requires **ascorbic acid (vitamin C)**

TABLE 7-2. THE MOST ABUNDANT TYPES OF COLLAGEN

Type	Chain Composition	Distribution
I	$[\alpha 1(I)]_2 \alpha 2(I)$	Skin, bone, tendon, blood vessels, cornea
II	$[\alpha 1(II)]_3$	Cartilage, intervertebral disk
III	$[\alpha 1(III)]_3$	Blood vessels, fetal skin

Source: Eyre, D.R., *Science* **207**, 1316 (1980).

Ascorbic acid (vitamin C)

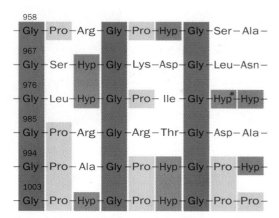

FIGURE 7-30. The amino acid sequence at the C-terminal end of the triple helical region of the bovine α1(I) collagen chain. Note the repeating triplets Gly-X-Y, where X is often Pro and Y is often Hyp. Here Gly is shaded in purple, Pro in tan, and Hyp and Hyp* (3-hydroxyPro) in brown. [From Bornstein, P. and Traub, W., *in* Neurath, H. and Hill. R.L. (Eds.), *The Proteins* (3rd ed.), Vol. 4, *p.* 483, Academic Press (1979).]

to maintain its enzymatic activity. In the vitamin C deficiency disease **scurvy,** the collagen synthesized cannot form fibers properly. This results in the skin lesions, blood vessel fragility, and poor wound healing that are symptomatic of scurvy.

The amino acid sequence of bovine collagen a1(I), which is similar to that of other collagens, consists of monotonously repeating triplets of sequence Gly-X-Y over a continuous 1011-residue stretch of its 1042-residue polypeptide chain (Fig. 7-30). Here X is often Pro and Y is often Hyp. The restriction of Hyp to the Y position stems from the specificity of prolyl hydroxylase. 5-Hydroxylysine is similarly restricted to the Y position.

The high Gly, Pro, and Hyp content of collagen suggests that its polypeptide backbone conformation resembles those of the polyglycine II and polyproline II helices (Fig. 7-15). X-Ray and model building studies indicate that *collagen's three polypeptide chains, which individually resemble polyproline II helices, are parallel and wind around each other with a gentle, right-handed, ropelike twist to form a triple-helical structure (Fig. 7-31).* Every third residue of each polypeptide chain passes through the center of the triple helix, which is so crowded that only a Gly side chain can fit there. This crowding explains the absolute requirement for a Gly at every third position of a collagen polypeptide chain (Fig. 7-30). It also requires that the three polypeptide chains be staggered so that the Gly, X, and Y residues from the three chains occur at similar levels (Fig. 7-32). The staggered peptide groups are oriented such that the N—H of each Gly makes a strong hydrogen bond with the carbonyl oxygen of an X residue on a neighboring chain. The bulky and relatively inflexible Pro and Hyp residues confer rigidity on the entire assembly.

Collagen's well-packed, rigid, triple helical structure is responsible for its characteristic tensile strength. As with the twisted fibers of a rope, the extended and twisted polypep-

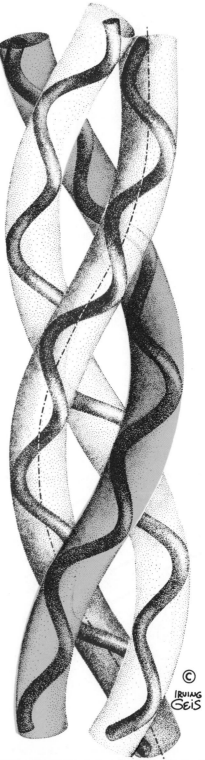

FIGURE 7-31. The triple helix of collagen, indicating how the left-handed polypeptide helices are twisted together to form a right-handed superhelical structure. Ropes and cables are similarly constructed from hierarchies of fiber bundles that are alternately twisted in opposite directions. An individual polypeptide helix has 3.3 residues per turn and a pitch of 10.0 Å (in contrast to polyproline II's 3.0 residues per turn and pitch of 9.4 Å; Fig. 7-15). The collagen triple helix has 10 Gly-X-Y units per turn and a pitch of 86.1 Å. [Figure copyrighted © by Irving Geis.]

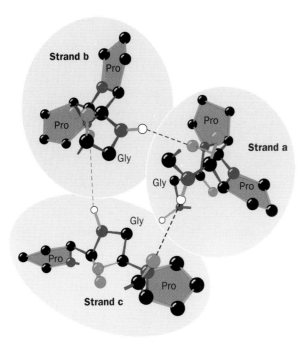

FIGURE 7-32. A projection down the triple helix axis of the collagen-like polymer (Gly-Pro-Pro)$_n$ as viewed from its carboxyl end. The residues in each chain, Gly-X-Y, are vertically staggered such that a Gly, an X, and a Y residue from different chains are on the same level along the helix axis. The dashed lines represent hydrogen bonds between each Gly N—H group and the oxygen of the succeeding X residue on a neighboring chain. Every third residue on each chain must be Gly because there is no room near the helix axis for the side chain of any other residue. The bulky pyrrolidine side chains (*brown*) of the Pro residues are situated on the periphery of the triple helix where they are sterically unhindered. [After Yonath, A. and Traub, W., *J. Mol. Biol.* **43**, 461 (1969).]

FIGURE 7-33. An electron micrograph of collagen fibrils from skin. [Courtesy of Jerome Gross, Massachusetts General Hospital.]

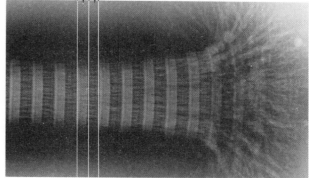

FIGURE 7-34. The banded appearance of collagen fibrils in the electron microscope arises from the schematically represented staggered arrangement of collagen molecules (*above*) that results in a periodically indented surface. *D*, the distance between cross striations, is ~680 Å, so the length of a 3000-Å-long collagen molecule is 4.4*D*. [Courtesy of Karl A. Piez, Collagen Corporation.]

tide chains of collagen convert a longitudinal tensional force to a more easily supported lateral compressional force on the almost incompressible triple helix. This occurs because the oppositely twisted directions of collagen's polypeptide chains and triple helix (Fig. 7-31) prevent the twists from being pulled out under tension (note that successive levels of fiber bundles in ropes and cables are likewise oppositely twisted). The successive helical hierarchies in other fibrous proteins exhibit similar alternations of twist directions, for example, keratin (Section 7-2A) and muscle (Section 34-3A).

Collagen Is Organized into Fibrils

Types I, II, III, V, and XI collagens form distinctive banded fibrils (Fig. 7-33) that are mostly, if not entirely, composed of several different types of collagens. These fibrils have a periodicity of 680 Å and a diameter of 100 to 2000 Å depending on the types of collagen they contain and their tissue of origin (the other collagen types form different sorts of aggregates such as networks; we will not discuss them further). Computerized model building studies indicate that collagen molecules are laterally organized in a precisely staggered array (Fig. 7-34). The darker portions of the banded structures correspond to the 400-Å "holes" on the surface of the fibril between head-to-tail aligned collagen molecules. Structural and energetic considerations suggest that the conformations of individual collagen molecules, much like those of individual α helices and β sheets, are but marginally stable. The driving force for the assembly of collagen molecules into a fibril is apparently provided by the added hydrophobic interactions within the fibrils in a manner analogous to the packing of secondary structural elements to form a globular protein (Section 7-3B).

Collagen contains covalently attached carbohydrates in amounts that range from ~0.4 to 12% by weight, depending on the collagen's tissue of origin. The carbohydrates, which consist mostly of glucose, galactose, and their disaccharides, are covalently attached to collagen at its 5-hydroxylysyl residues by specific enzymes.

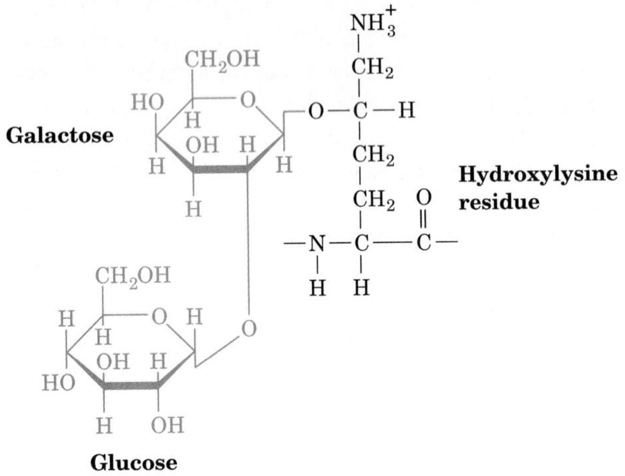

Although the function of carbohydrates in collagen is unknown, the observation that they are located in the "hole" regions of the collagen fibrils suggests that they are involved in directing fibril assembly.

Collagen Fibrils Are Covalently Cross-Linked

Collagen's insolubility in solvents that disrupt hydrogen bonding and ionic interactions is explained by the observation that it is both intramolecularly and intermolecularly covalently cross-linked. The cross-links cannot be disulfide bonds, as in keratin, because collagen is almost devoid of Cys residues. Rather, they are derived from Lys and His side chains in reactions such as those in Fig. 7-35. **Lysyl**

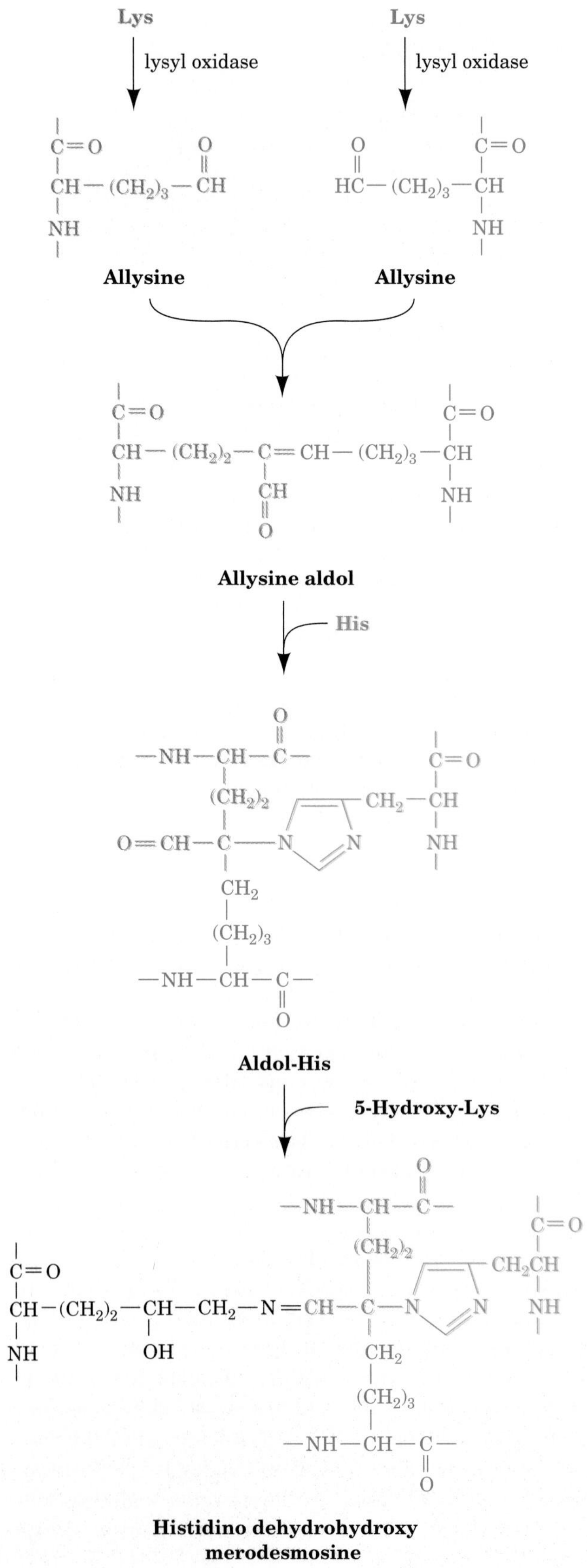

FIGURE 7-35. A biosynthetic pathway for cross-linking Lys, 5-hydroxylysyl, and His side chains in collagen. The first step in the reaction is the lysyl oxidase–catalyzed oxidative deamination of Lys to form the aldehyde allysine. Two such aldehydes then undergo an aldol condensation to form **allysine aldol.** This product can react with His to form **aldol histidine.** This, in turn, can react with 5-hydroxylysine to form a Schiff base (an imine bond), thereby cross-linking four side chains.

TABLE 7-3. THE ARRANGEMENT OF COLLAGEN FIBRILS IN VARIOUS TISSUES

Tissue	Arrangement
Tendon	Parallel bundles
Skin	Sheets of fibrils layered at many angles
Cartilage	No distinct arrangement
Cornea	Planar sheets stacked crossways so as to minimize light scatter

oxidase, a Cu-containing enzyme that converts Lys residues to those of the aldehyde **allysine,** is the only enzyme implicated in this cross-linking process. Up to four side chains can be covalently bonded to each other. The cross-links do not form at random but, instead, tend to occur near the N- and C-termini of the collagen molecules.

The importance of cross-linking to the normal functioning of collagen is demonstrated by the disease **lathyrism** which occurs in humans and other animals as a result of the regular ingestion of seeds from the sweet pea *Lathyrus odoratus.* The symptoms of this condition are serious abnormalities of the bones, joints, and large blood vessels which are caused by an increased fragility of the collagen fibers. The causative agent of lathyrism, **β-aminopropionitrile,**

$$N\equiv C-CH_2-CH_2-NH_3^+$$
β-Aminopropionitrile

inactivates lysyl oxidase by covalently binding to its active site. This results in markedly reduced cross-linking in the collagen of lathrytic animals.

The degree of cross-linking of the collagen from a particular tissue increases with the age of the animal. This is why meat from older animals is tougher than that from younger animals. In fact, individual molecules of collagen (called **tropocollagen**) can only be extracted from the tissues of very young animals. Collagen cross-linking is not the central cause of aging, however, as is demonstrated by the observation that lathyrogenic agents do not slow the aging process.

The collagen fibrils in various tissues are organized in ways that largely reflect the functions of the tissues (Table 7-3). Thus tendons (the "cables" connecting muscles to bones), skin (a tear-resistant outer fabric), and cartilage (which has a load-bearing function) must support stress in predominantly one, two, and three dimensions, respectively, and their component collagen fibrils are arrayed accordingly. How collagen fibrils are laid down in these arrangements is unknown. However, some of the factors guiding collagen molecule assembly are discussed in Sections 30-5A and B.

Collagen Defects Are Responsible for a Variety of Human Diseases

Several rare heritable disorders of collagen are known. Mutations of Type I collagen, which constitutes the major structural protein in most human tissues, usually result in **osteogenesis imperfecta** (brittle bone disease). The severity of this disease varies with the nature and position of the mutation: Even a single amino acid change can have lethal consequences. Mutations may affect the structure of the collagen molecule or how it forms fibrils. These mutations tend to be dominant because they affect either the folding of the triple helix or fibril formation even when normal chains are also involved. All known amino acid changes within Type I collagen's triple helical region result in abnormalities, indicating that the structural integrity of this region is essential for proper collagen function.

Many collagen disorders are characterized by deficiencies in the amount of a particular collagen type synthesized, or by abnormal activities of collagen-processing enzymes such as lysyl hydroxylase or lysyl oxidase. One group of at least 10 different collagen deficiency diseases, the **Ehlers–Danlos syndromes,** are all characterized by hyperextensibility of the joints and skin. The "India-rubber man" of circus fame had an Ehlers–Danlos syndrome. Many degenerative diseases exhibit collagen abnormalities in the affected tissues. Examples of such tissues are the cartilage in **osteoarthritis** and the fibrous **atherosclerotic plaques** in human arteries.

D. Elastin—A Nonrepetitive Coil

Elastin is a protein with rubberlike elastic properties whose fibers can stretch to several times their normal length. It is the principle protein component of the elastic yellow connective tissue that occurs in the lungs, the walls of large blood vessels such as the aorta, and elastic ligaments such as those in the neck. The inelastic white connective tissue of tendons contains only a small amount of elastin. The hyperextensibility of the joints and skin characteristic of certain collagen deficiency diseases results from the loss of rigidity ordinarily conferred by collagen coupled with the normal presence of elastin.

Elastin, like silk fibroin and collagen, has a distinctive amino acid composition. It consists predominantly of small, nonpolar residues: It is one-third Gly, over one-third Ala + Val, and is rich in Pro. However, it contains little hydroxyproline, no hydroxylysine, and few polar residues.

Elastin forms a three-dimensional network of fibers that exhibit no recognizable periodicity in the electron microscope. Furthermore, according to X-ray analyses, the fibers are devoid of regular secondary structure.

The covalent cross-links in elastin are formed by allysine aldol, which also occurs in collagen (Fig. 7-35), and the compounds **lysinonorleucine, desmosine,** and **isodesmosine.**

Desmosine

Lysinonorleucine

Isodesmosine

Lysinonorleucine, which likewise occurs in collagen, results from the reduction of the Schiff base (imine bond) formed by the condensation of a Lys side chain with that of allysine (Fig. 7-35). Desmosine and isodesmosine are unique to elastin and are responsible for its yellow color; they result from the condensation of three allysine and one lysine side chains. The primary structure of elastin consists of alternating hydrophobic segments, thought to be responsible for the protein's elastic properties, and Lys-rich segments that contain the cross-links. The Lys residues usually occur in pairs, which suggests that a given desmosine or isodesmosine serves to cross-link two rather than four elastin polypeptides.

3. GLOBULAR PROTEINS

Globular proteins comprise a highly diverse group of substances that, in their native state, exist as compact spheroidal molecules. Enzymes are globular proteins as are transport and receptor proteins. In this section we consider the tertiary structures of globular proteins. However, since

most of our detailed structural knowledge of proteins, and thus to a large extent their function, has resulted from X-ray crystal structure determinations of globular proteins and, more recently, from their nuclear magnetic resonance (NMR) structure determinations, we begin this section with a discussion of the capabilities and limitations of these powerful techniques.

A. Interpretation of Protein X-Ray and NMR Structures

X-Ray crystallography is a technique that directly images molecules. X-Rays must be used to do so because, according to optical principles, the uncertainty in locating an object is approximately equal to the wavelength of the radiation used to observe it (both X-ray wavelengths and covalent bond distances are ~1.5 Å; individual molecules cannot be seen in a light microscope because visible light has a minimum wavelength of 4000 Å). There is, however, no such thing as an X-ray microscope because there are no X-ray lenses. Rather, a crystal of the molecule to be visualized is exposed to a collimated beam of monochromatic X-rays and the consequent diffraction pattern is recorded on photographic film (Fig. 7-36) or by a radiation counter. The intensities of the diffraction maxima (darkness of the spots on the film) are then used to construct mathematically the three-dimensional image of the crystal structure through methods that are outside the scope of this text. In

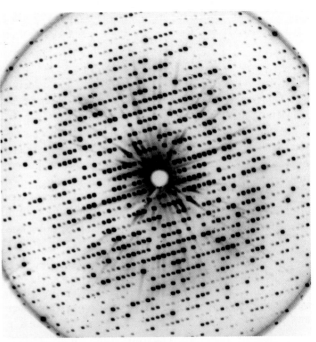

FIGURE 7-36. An X-ray diffraction photograph of a single crystal of sperm whale myoglobin. The intensity of each diffraction maximum (the darkness of each spot) is a function of the myoglobin crystal's electron density. The photograph contains only a small fraction of the total diffraction information available from a myoglobin crystal. [Courtesy of John Kendrew, Cambridge University.]

(a)

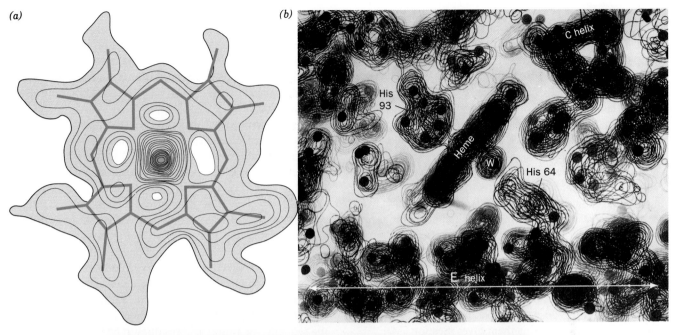

(b)

(c)

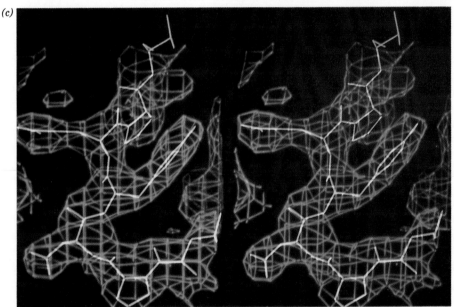

FIGURE 7-37. Electron density maps of proteins. (*a*) A section through the 2.0-Å-resolution electron density map of sperm whale myoglobin, which contains the heme group (*red*). The large peak at the center of the map represents the electron-dense Fe atom. [After Kendrew, J.C., Dickerson, R.E., Strandberg, B.E., Hart, R.G., Davies, D.R., Phillips, D.C., and Shore, V.C., *Nature* **185**, 434 (1960).] (*b*) A portion of the 2.4-Å-resolution electron density map of myoglobin constructed from a stack of contoured transparencies. Dots have been placed at the positions deduced for the nonhydrogen atoms. The heme group is seen edge-on together with its two associated His residues and a water molecule, W. An α helix, the so-called E helix (Fig. 7-12), extends across the bottom of the map. Another α helix, the C helix, extends into the plane of the paper on the upper right. Note the hole along its axis. [Courtesy of John Kendrew, Cambridge University, U.K.] (*c*) A portion of the 3.0-Å-resolution electron density map of a human rhinovirus (the cause of the common cold, Section 32-2C) contoured in three dimensions on a graphics computer and shown in stereo. Only a single contour level (*orange*) is shown, together with an atomic model of the corresponding polypeptide segment (*white*). Instructions for viewing stereo diagrams are given in the appendix to this chapter. [Courtesy of Michael Rossmann, Edward Arnold, and Gerrit Vriend, Purdue University.]

what follows, we discuss some of the special problems associated with interpreting the X-ray crystal structures of proteins.

X-Rays interact almost exclusively with the electrons in matter, not the nuclei. An X-ray structure is therefore an image of the electron density of the object under study. Such **electron density maps** may be presented as a series of parallel sections through the object. On each section, the electron density is represented by contours (Fig. 7-37*a*) in the same way that altitude is represented by the contours on a topographic map. A stack of such sections, drawn on transparencies, yields a three-dimensional electron density map (Fig. 7-37b). Modern structural analysis, however, is often carried out with the aid of graphics computers, on which electron density maps are contoured in three dimensions (Fig. 7-37c).

Protein Crystal Structures Exhibit Less Than Atomic Resolution

The molecules in protein crystals, as in other crystalline substances, are arranged in regularly repeating three-dimensional lattices. Protein crystals, however, differ from those of most small organic and inorganic molecules in being highly hydrated; they are typically 40 to 60% water by volume. The aqueous solvent of crystallization is necessary for the structural integrity of the protein crystals as J. D. Bernal and Dorothy Crowfoot Hodgkin first noted in 1934 when they carried out the original X-ray studies of protein crystals. This is because water is required for the structural integrity of native proteins themselves (Section 7-4).

The large solvent content of protein crystals gives them a soft, jellylike consistency so that their molecules lack the rigid order characteristic of crystals of small molecules such as NaCl or glycine. The molecules in a protein crystal are typically disordered by a few angstroms so that the corresponding electron density map lacks information concerning structural details of smaller size. The crystal is therefore said to have a resolution limit of that size. Protein crystals typically have resolution limits in the range 2 to 3.5 Å, although a few are better ordered (have higher resolution, that is, a lesser resolution limit) and many are less ordered (have lower resolution).

Since an electron density map of a protein must be interpreted in terms of its atomic positions, the accuracy and even the feasibility of a crystal structure analysis depends on the crystal's resolution limit. Figure 7-38 indicates how the quality (degree of focus) of an electron density map varies with its resolution limit. At 6-Å resolution, the presence of a molecule the size of diketopiperazine is difficult to

discern. At 2.0-Å resolution, its individual atoms cannot yet be distinguished, although its molecular shape has become reasonably evident. At 1.5-Å resolution, which roughly corresponds to a bond distance, individual atoms become partially resolved. At 1.1-Å resolution, atoms are clearly visible.

Most protein crystal structures are too poorly resolved for their electron density maps to reveal clearly the positions of individual atoms (e.g., Fig. 7-37). Nevertheless, the distinctive shape of the polypeptide backbone usually permits it to be traced, which, in turn, allows the positions and orientations of its side chains to be deduced (e.g., Fig. 7-37c). Yet, side chains of comparable size and shape, such as those of Leu, Ile, Thr, and Val, cannot be differentiated with a reasonable degree of confidence (hydrogen atoms, having but one electron, are not visible in protein X-ray structures), so that a protein structure cannot be elucidated from its electron density map alone. Rather, the primary structure of the protein must be known, thereby permitting the sequence of amino acid residues to be fitted, by eye, to its electron density map. Mathematical refinement can then reduce the errors in the crystal structure's atomic positions to around 0.1 Å (in contrast, the errors in small molecule X-ray structure determinations may be as little as 0.001 Å).

Most Crystalline Proteins Maintain Their Native Conformations

What is the relationship between the structure of a protein in a crystal and that in solution where most proteins normally function? Several lines of evidence indicate that *crystalline proteins assume very nearly the same structures that they have in solution:*

1. A protein molecule in a crystal is essentially in solution because it is bathed by solvent of crystallization over all of its surface except for the few, generally small patches that contact neighboring protein molecules. In fact, the 40 to 60% water content of typical protein crystals is similar to that of many cells (e.g., see Fig 1-13).

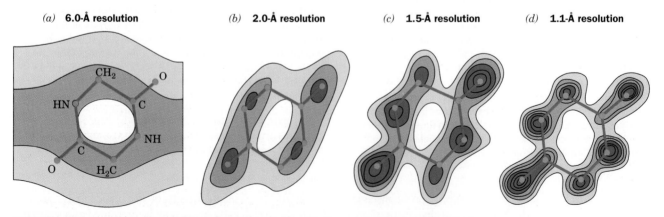

(a) **6.0-Å resolution**   *(b)* **2.0-Å resolution**   *(c)* **1.5-Å resolution**   *(d)* **1.1-Å resolution**

FIGURE 7-38.   A section through the electron density map of diketopiperazine calculated at the indicated resolution levels. Hydrogen atoms are not apparent in this map because of their low electron density. [After Hodgkin, D.C., *Nature* **188**, 445 (1960).]

2. A protein may crystallize in one of several forms or "habits," depending on crystallization conditions, that differ in how the protein molecules are arranged in space relative to each other. In the numerous cases in which different crystal forms of the same protein have been independently analyzed, the molecules have virtually identical conformations. Similarly, in the several cases that both the X-ray crystal structure and the solution NMR structure of the same protein have been determined, the two structures are, for the most part, identical to within experimental error (see below). Evidently, crystal packing forces do not greatly perturb the structures of protein molecules.

3. The most compelling evidence that crystalline proteins have biologically relevant structures, however, is the observation that many enzymes are catalytically active in the crystalline state. The catalytic activity of an enzyme is very sensitive to the relative orientations of the groups involved in binding and catalysis (Chapter 14). Active crystalline enzymes must therefore have conformations that closely resemble their solution conformations.

Protein Structure Determination by 2D-NMR

The determination of the three-dimensional structures of small globular proteins in aqueous solution has become possible, since the mid 1980s, through the development of **two-dimensional (2D) NMR spectroscopy** (and, more recently, of 3D and 4D techniques), in large part by Kurt Wüthrich. Such NMR measurements, whose description is beyond the scope of this text, yield the interatomic distances between specific protons that are <5 Å apart in a protein of known sequence that has no more than ~200 residues. The interproton distances may be either through space, as determined by nuclear Overhauser effect spectroscopy (NOESY, Fig. 7-39a), or through bonds, as determined by correlated spectroscopy (COSY). These dis-

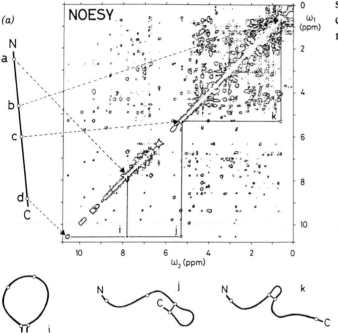

(a)

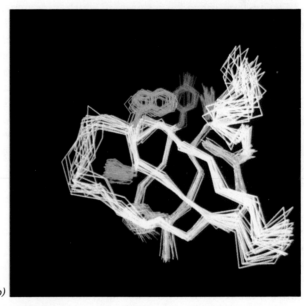

(b)

FIGURE 7-39. The 2D proton NMR structures of proteins. (*a*) A NOESY spectrum of a protein presented as a contour plot with two frequency axes, ω_1 and ω_2. The conventional 1D-NMR spectrum of the protein, which occurs along the diagonal of the plot ($\omega_1 = \omega_2$), is too crowded with peaks to be directly interpretable (even a small protein has hundreds of protons). The off-diagonal peaks, the so-called cross peaks, each arise from the interaction of two protons that are <5 Å apart in space and whose 1D-NMR peaks are located where the horizontal and vertical lines through the cross peak intersect the diagonal [a **nuclear Overhauser effect (NOE)**]. For example, the line to the left of the spectrum represents the extended polypeptide chain with its N- and C-terminal ends identified by the letters N and C and with the positions of four protons, a to d, represented by small circles. The dashed arrows indicate the diagonal NMR peaks to which these protons give rise. Cross peaks, such as *i, j,* and *k,* which are each located at the intersections of the horizontal and vertical lines through two diagonal peaks, are indicative of an NOE between the corresponding two protons, indicating that they are <5 Å apart. These distance relationships are schematically indicated by the three circular structures drawn below the spectrum. Note that the assignment of a distance relationship between two protons in a polypeptide requires that the NMR peaks to which they give rise and their positions in the polypeptide be known, which requires that the polypeptide's amino acid sequence has been previously determined. [After Wüthrich, K., *Science* **243**, 45 (1989).] (*b*) The NMR structure of a 64-residue polypeptide comprising the **Src protein SH3 domain** (Section 34-4B). The drawing represents 20 superimposed structures that are consistent with the 2D- and 3D-NMR spectra of the protein (each calculated from a different, randomly generated starting structure). The polypeptide backbone, as represented by its connected C_α atoms, is white and its Phe, Tyr, and Trp side chains are yellow, red, and blue, respectively. It can be seen that the polypeptide backbone folds into two 3-stranded antiparallel β sheets that form a sandwich. [Courtesy of Stuart Schreiber, Harvard University.]

tances, together with known geometric constraints such as covalent bond distances and angles, group planarity, chirality, and van der Waals radii, are used to compute the protein's three-dimensional structure. However, since interproton distance measurements are imprecise, they are insufficient to imply a unique structure. Rather, they are consistent with an ensemble of closely related structures. Consequently, an NMR structure of a protein (or any other macromolecule with a well-defined structure) is often presented as a representative sample of structures that are consistent with the constraints (e.g., Figure 7-39*b*). The "tightness" of a bundle of such structures is indicative both of the accuracy with which the structure is known, which in the most favorable cases is roughly comparible to that of an X-ray crystal structure with a resolution of 2 to 2.5 Å, and of the conformational fluctuations that the protein undergoes (Section 8-2).

In most of the several cases in which both the NMR and X-ray crystal structures of a particular protein have been determined, the two structures are in good agreement. There are, however, a few instances in which there are real differences between the corresponding X-ray and NMR structures. These, for the most part, involve surface residues that, in the crystal, participate in intermolecular contacts and are thereby perturbed from their solution conformations. NMR methods, besides providing mutual crosschecks with X-ray techniques, can determine the structures of proteins and other macromolecules that fail to crystallize. Moreover, since NMR can probe motions over

time scales spanning 10 orders of magnitude, it can be used to study protein folding and dynamics (Chapter 8).

Protein Molecular Structures Are Most Effectively Illustrated in Simplified Form

The several hundred nonhydrogen atoms of even a small protein makes understanding the detailed structure of a protein a considerable effort. The most instructive method of studying a protein structure is the hands-on examination of its skeletal (ball-and-stick) model. Unfortunately, such models are rarely available and photographs of them are too cluttered to be of much use. A practical alternative is a computer-generated stereo diagram in which the polypeptide backbone is represented only by its C_α atoms and only a few key side chains are included (Fig. 7-40). Another possibility is an artistic rendering of a protein model that has been simplified and slightly distorted to improve its visual clarity (Fig. 7-41). A further level of abstraction may be obtained by representing the protein in a cartoon form that emphasizes its secondary structure (Fig. 7-42; also see Fig. 7-19). Computer-generated drawings of space-filling models, such as Figs. 7-12 and 7-18, may also be employed to illustrate certain features of protein structures.

B. Tertiary Structure

The **tertiary structure (3° structure)** of a protein is its three-dimensional arrangement; that is, the folding of its 2° structural elements, together with the spatial dispositions of its

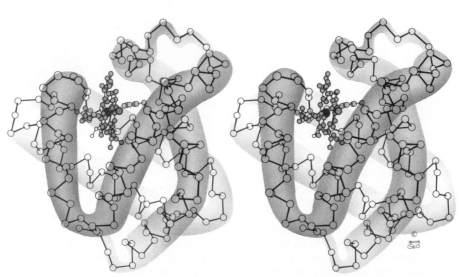

FIGURE 7-40.   A computer-drawn stereo diagram of sperm whale myoglobin in which the C_α atoms are represented by balls and the peptide groups linking them are represented by solid bonds. The 153-residue polypeptide chain is folded into eight α helices (highlighted here by hand-drawn envelopes), connected by short polypeptide links. The protein's bound heme group (*purple*) in complex with an O_2 molecule (*orange sphere*) is shown together with its two closely associated His side chains (*light blue*). Hydrogen atoms have been omitted for the sake of clarity. Instructions for viewing stereo diagrams are given in the appendix to this chapter. [Figure copyrighted © by Irving Geis.]

side chains. The first protein X-ray structure, that of sperm whale myoglobin, was elucidated in the late 1950s by John Kendrew and coworkers. Its polypeptide chain follows such a tortuous, wormlike path (Figs. 7-40 through 7-42), that these investigators were moved to indicate their disappoint-

ment at its lack of regularity. In the intervening years, well over 500 protein structures have been reported. Each of them is a unique, highly complicated entity. Nevertheless, their tertiary structures have several outstanding features in common as we shall see below.

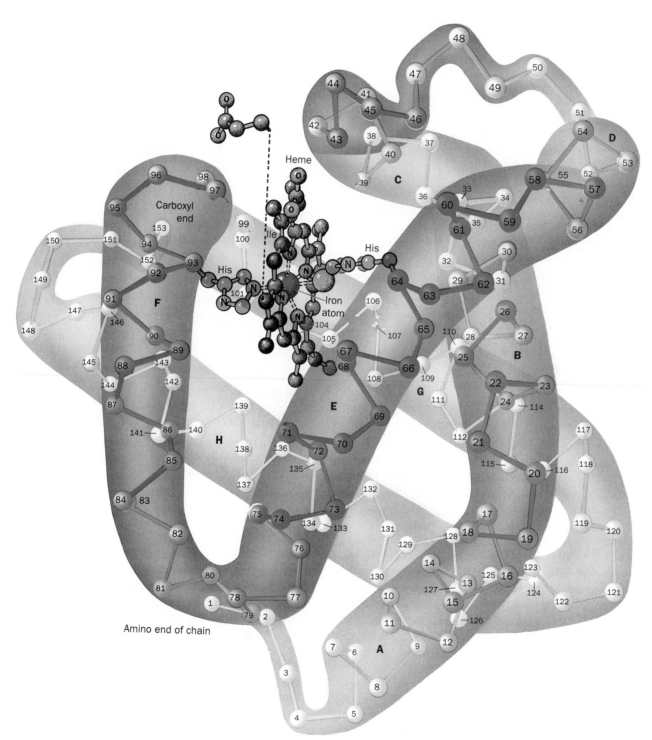

FIGURE 7-41. An artist's rendering of sperm whale myoglobin analogous to Fig. 7-40. One of the heme group's propionic acid side chains has been displaced for clarity. The

amino acid residues are consecutively numbered, starting from the N-terminus, and the eight helices are likewise designated A through H. [Figure copyrighted © by Irving Geis.]

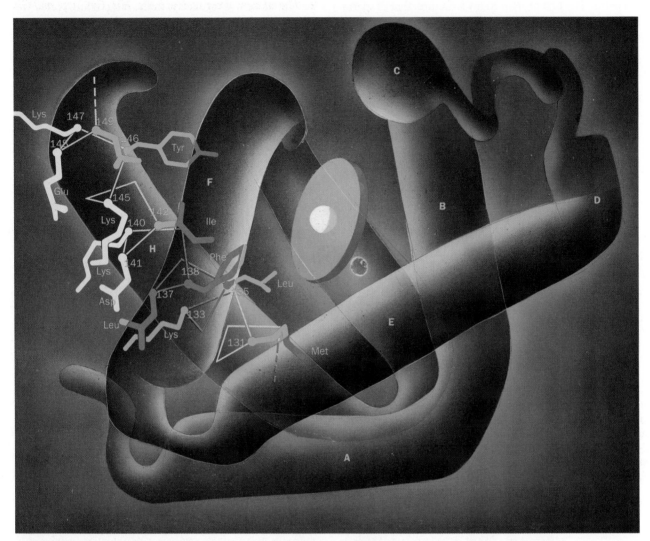

FIGURE 7-42. A cartoon of sperm whale myoglobin, oriented similarly to Figs. 7-40 and 7-41, which emphasizes the protein's α helical secondary structure (*cylinders*). The pink disk with its central white sphere represents the protein's associated heme group with its bound iron atom. Many of the H helix side chains are shown with polar and nonpolar groups, respectively, colored blue and red. [Figure copyrighted © by Irving Geis.]

Globular Proteins May Contain Both α Helices and β Sheets

The major types of secondary structural elements, α helices and β pleated sheets, commonly occur in globular proteins but in varying proportions and combinations. Some proteins, such as myoglobin, consist only of α helices spanned by short connecting links that have a coil conformation (Fig. 7-42). Others, such as concanavalin A, have a large proportion of β sheets but are devoid of α helices (Fig. 7-43). Most proteins, however, have significant amounts of both types of secondary structure (on average, ~31% α helix and ~28% β sheet). Human **carbonic anhydrase** (Fig. 7-44) as well as carboxypeptidase and triose phosphate isomerase (Fig. 7-19) are examples of such proteins.

Side Chain Location Varies with Polarity

The primary structures of globular proteins generally lack the repeating or pseudorepeating sequences that are responsible for the regular conformations of fibrous proteins. The amino acid side chains in globular proteins are, nevertheless, spatially distributed according to their polarities:

1. *The nonpolar residues Val, Leu, Ile, Met, and Phe largely occur in the interior of a protein, out of contact with the aqueous solvent.* The hydrophobic interactions that promote this distribution, which are largely responsible for the three-dimensional structures of native proteins, are further discussed in Section 7-4C.

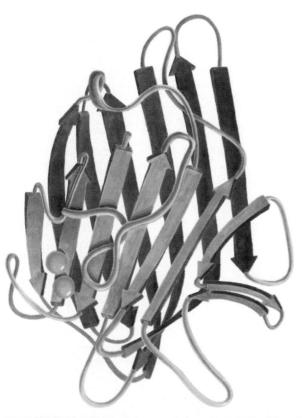

2. *The charged polar residues Arg, His, Lys, Asp, and Glu are largely located on the surface of a protein in contact with the aqueous solvent.* This is because the immersion of an ion in the virtually anhydrous interior of a protein results in the uncompensated loss of much of its hydration energy. In the instances that these groups occur in the interior of a protein, they often have a specific chemical function such as promoting catalysis or participating in metal ion binding (e.g., the metal ion–liganding His residues in Figs. 7-41 and 7-44).

3. The uncharged polar groups Ser, Thr, Asn, Gln, Tyr, and Trp, are usually on the protein surface but frequently occur in the interior of the molecule. In the latter case, these residues are almost always hydrogen bonded to other groups in the protein. In fact, *nearly all buried hydrogen bond donors form hydrogen bonds with buried acceptor groups;* in a sense, the formation of a hydrogen bond "neutralizes" the polarity of a hydrogen bonding group.

FIGURE 7-43. The jack bean protein concanavalin A largely consists of extensive regions of antiparallel β pleated sheet, here represented by arrows pointing towards the polypeptide chain's C-terminus. The balls represent protein-bound metal ions. The back sheet is shown in a space-filling representation in Fig. 7-18. [After a drawing by Jane Richardson, Duke University.]

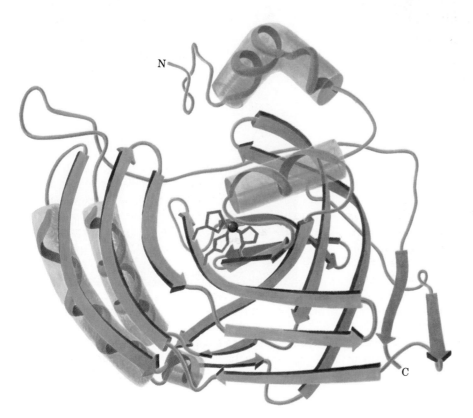

FIGURE 7-44. Human carbonic anhydrase in which α helices are represented as cylinders and each strand of β sheet is drawn as an arrow pointing towards the polypeptide's C-terminus. The gray ball in the middle represents a Zn^{2+} ion that is coordinated by three His side chains (*blue*). Note that the C-terminus is tucked through the plane of a surrounding loop of polypeptide chain so that carbonic anhydrase is one of the rare native proteins in which a polypeptide chain forms a knot. [After Kannan, K.K., Liljas, A., Waara, I., Bergsten, P.-C., Lovgren, S., Strandberg, B., Bengtsson, J., Carlbom, U., Friedborg, K., Jarup, L., and Petef, M., *Cold Spring Harbor Symp. Quant. Biol.* **36,** 221 (1971).]

(a)

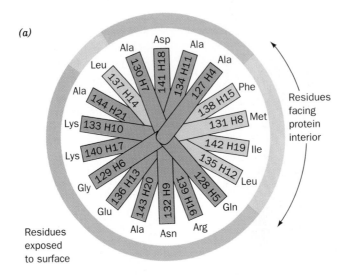

Residues facing protein interior

Residues exposed to surface

(b)

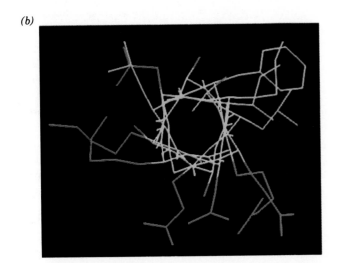

(c)

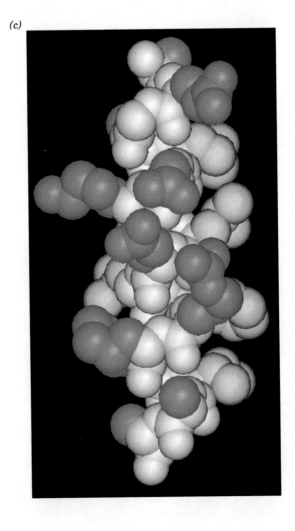

FIGURE 7-45. The H helix of sperm whale myoglobin. (*a*) A **helical wheel** representation in which side chain positions about the α helix are projected down the helix axis onto a plane. Here each residue is identified both according to its sequence in the polypeptide chain, and according to its position in the H helix. The residues lining the side of the helix facing the protein's interior regions are all nonpolar. The other residues, except Leu 137, which contacts the protein segment linking helices E and F (Figs. 7-40 and 7-41), are exposed to the solvent and are all more or less polar. (*b*) A skeletal model, viewed as in Part *a*, in which the main chain is white, nonpolar side chains are yellow, and polar side chains are purple. (*c*) A space-filling model, viewed from the bottom of the page in Parts *a* and *b* and colored as in Part *b*. Compare these diagrams with the drawing of the H helix in Fig. 7-42.

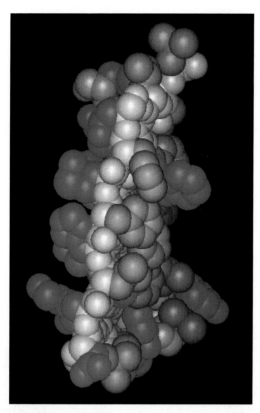

FIGURE 7-46. A space-filling model of an antiparallel β sheet from concanavilin A in side view with the interior of the protein (the surface of a second antiparallel β sheet; see Fig. 7-43) to the right and the exterior to the left. The main chain is white, nonpolar side chains are brown, and polar side chains are purple.

This side chain distribution is apparent in Figs. 7-42 and 7-45, which show the surface and interior exposures of the amino acid side chains of myoglobin's H helix. This arrangement is likewise seen on the covers of this textbook, which show the distributions of polar *(front cover)* and nonpolar *(back cover)* residues of cytochrome *c,* as well as in Fig. 7-46, which shows one of the antiparallel β pleated sheets of concanavilin A.

Globular Protein Cores Are Efficiently Arranged With Their Side Chains in Relaxed Conformations

Globular proteins are quite compact; there is very little space inside them so that water is largely excluded from their interiors. The micellelike arrangement of their side chains (polar groups outside, nonpolar groups inside) has led to their description as "oil drops with polar coats." This generalization, although picturesque, lacks precision. The **packing density** (ratio of the volume enclosed by the van der Waals envelopes of the atoms in a region to the total volume of the region) of the internal regions of globular proteins averages ~0.75, which is in the same range as that of molecular crystals of small organic molecules. In comparison, equal-sized close-packed spheres have a packing density of 0.74, whereas organic liquids (oil drops) have packing densities that are mostly between 0.60 and 0.70. *The interior of a protein is therefore more like a molecular crystal than an*

oil drop; that is, it is efficiently packed. The ability of most hydrogen bonding donors to find acceptors under such constrained conditions is explained by the observation that most hydrogen bonding partners reside on residues that are close in sequence (which is, in turn, explained by the facts that backbone N—H groups comprise the majority of the hydrogen bonding donors in proteins and that most protein residues are members of secondary structural elements).

The bonds of protein side chains, including those occupying protein cores, almost invariably have low-energy staggered torsion angles (Fig. 7-5*b*). Evidently, interior side chains adopt relaxed conformations despite their profusion of intramolecular interactions (Section 7-4).

Large Polypeptides Form Domains

Polypeptide chains that consist of more than ~200 residues usually fold into two or more globular clusters known as **domains,** which give these proteins a bi- or multilobal appearance. Most domains consist of 100 to 200 amino acid residues and have an average diameter of ~25 Å. Each subunit of **glyceraldehyde-3-phosphate dehydrogenase,** for example, has two distinct domains (Fig. 7-47). A polypeptide chain wanders back and forth within a domain but neighboring domains are usually connected by one, or less commonly two, polypeptide segments. *Domains are therefore structurally independent units that each have the characteristics of a small globular protein.* Indeed, limited proteolysis of a multidomain protein often liberates its domains without greatly altering their structures. Nevertheless, the domain structure of a protein is not always obvious since its domains may make such extensive contacts with each other that the protein appears to be a single globular entity.

An inspection of the various protein structures diagrammed in this chapter reveals that domains consist of two or more layers of secondary structural elements. The reason for this is clear: At least two such layers are required to seal off a domain's hydrophobic core from the aqueous environment.

Domains often have a specific function such as the binding of a small molecule. In Fig. 7-47, for example, **nicotinamide adenine dinucleotide (NAD⁺)** binds to the first domain of glyceraldehyde-3-phosphate dehydrogenase. Small molecule–binding sites in multidomain proteins often occur in the clefts between domains; that is, the small molecules are bound by groups from two domains. This arrangement arises, in part, from the need for a flexible interaction between the protein and the small molecule that the relatively pliant covalent connection between the domains can provide.

Supersecondary Structures Have Structural and Functional Roles

Certain groupings of secondary structural elements, named **supersecondary structures** or **motifs,** occur in many unrelated globular proteins:

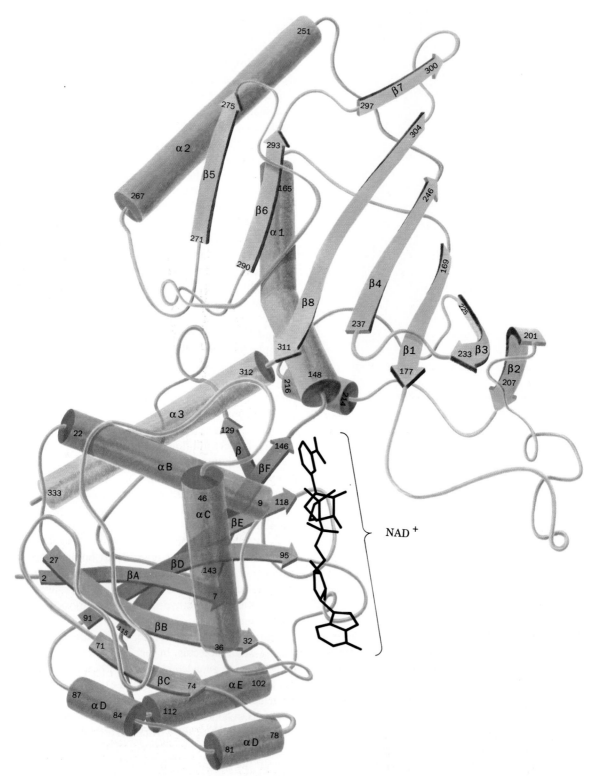

FIGURE 7-47.   One subunit of the enzyme glyceraldehyde-3-phosphate dehydrogenase from *Bacillus stearothermophilus.* The polypeptide folds into two distinct domains. The first domain (*red,* residues 1–146) binds NAD+ (*black*) near the C-terminal ends of its parallel *β* strands, and the second domain (*green*) binds glyceraldehyde-3-phosphate (not shown). [After Biesecker, G., Harris, J.I., Thierry, J.C., Walker, J.E., and Wonacott, A., *Nature* **266,** 331 (1977).]

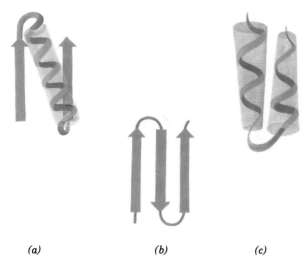

FIGURE 7-48. Schematic diagrams of (*a*) a $\beta\alpha\beta$ motif, (*b*) a β hairpin motif, and (*c*) an $\alpha\alpha$ motif.

1. The most common form of supersecondary structure is the **$\beta\alpha\beta$ motif,** in which the usually right-handed cross-over connection between two consecutive parallel strands of a β sheet consists of an α helix (Fig. 7-48*a*).

2. Another common supersecondary structure, the **β hairpin motif,** consists of an antiparallel β sheet formed by sequential segments of polypeptide chain that are connected by relatively tight reverse turns (Fig. 7-48*b*).

3. In an **$\alpha\alpha$ motif,** two successive antiparallel α helices pack against each other with their axes inclined so as to permit energetically favorable intermeshing of their contacting side chains (Fig. 7-48*c*). Such associations stabilize the coiled coil conformation of α keratin (Section 7-2A).

4. Extended β sheets often roll up to form **β barrels** (e.g., Fig. 7-19*b*). Three different β barrel topologies (the ways in which the strands and their interconnections are arranged) have been named in analogy with geometric motifs found on Native American and Greek weaving and pottery (Fig. 7-49).

Supersecondary structures may have functional as well as structural significance. A $\beta\alpha\beta\alpha\beta$ unit, for example, in which the β strands form a parallel sheet with right-handed α helical crossover connections (two overlapping $\beta\alpha\beta$ units), was shown by Michael Rossmann to form a nucleotide-binding site in many enzymes. In most proteins that bind dinucleotides, two such $\beta\alpha\beta\alpha\beta$ units combine to form a motif alternatively known as a **dinucleotide-binding fold** or a **Rossmann fold** (Fig. 7-50). In some cases, the second α helix in a $\beta\alpha\beta\alpha\beta$ unit is replaced by a length of nonhelical polypeptide. This occurs, for example, between the βE and βF strands of glyceraldehyde-3-phosphate dehydrogenase (Fig. 7-47).

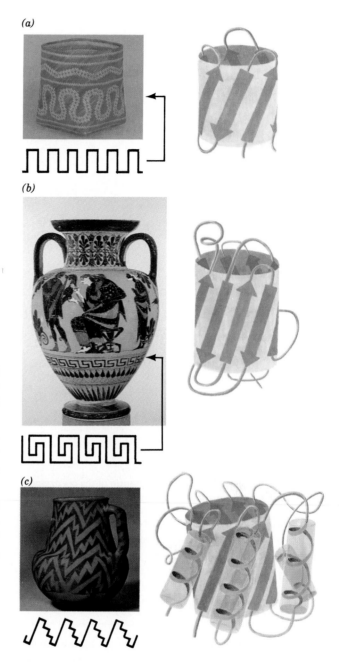

FIGURE 7-49. Comparisons of the backbone folding patterns of protein β barrels (*right*) with geometric motifs commonly used to decorate Native American and Greek weaving and pottery (*left*). (*a*) Native American polychrome cane basket and the polypeptide backbone of **rubredoxin** from *Clostridium pasteurianum* showing its linked β meanders. [Museum of the American Indian, Heye Foundation.] (*b*) Red figured Greek amphora with its Greek key border area showing Cassandra and Ajax (about 450 B.C.) and the polypeptide backbone of human **prealbumin** with its "Greek key" pattern. [The Metropolitan Museum of Art, Fletcher Fund, 1956.] (*c*) Early Anasazi redware pitcher from New Mexico and the polypeptide backbone of chicken muscle triose phosphate isomerase showing its "lightning" pattern of overlapping $\beta\alpha\beta$ units. This so-called α/β barrel is also diagrammed in Fig. 7-19*b*. [Museum of the American Indian, Heye Foundation.] [After Richardson, J.S., *Nature* **268,** 498 (1977).]

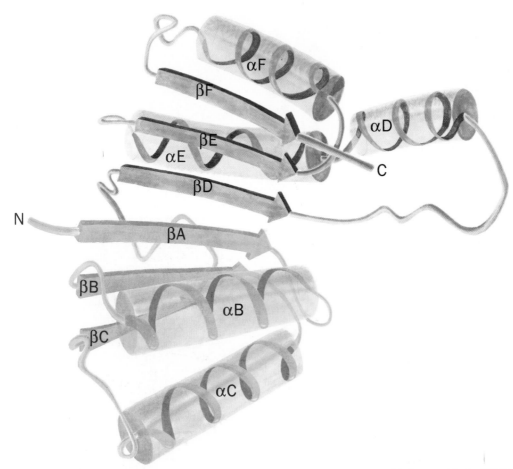

FIGURE 7-50. An idealized representation of the coenzyme-binding domain from various dehydrogenases. This domain consists of two structurally similar *βαβαβ* units, drawn here with one yellow and the other blue, each of which binds a nucleotide portion of NAD⁺ so as to form a dinucleotide-binding or Rossmann fold. Compare this figure with the NAD⁺-binding domain of glyceraldehyde-3-phosphate dehydrogenase (Fig. 7-47). [After Rossmann, M.G., Liljas, A., Brändén, C.-I., and Banaszak, L.J., *in* Boyer, P.D. (Ed.), *The Enzymes,* Vol. 11 (3rd ed.), *p.* 68, Academic Press (1975).]

4. PROTEIN STABILITY

Incredible as it may seem, thermodynamic measurements indicate that *native proteins are only marginally stable entities under physiological conditions.* The free energy required to denature them is ~0.4 kJ·mol⁻¹ of amino acid residues so that 100-residue proteins are typically stable by only around 40 kJ·mol⁻¹. In contrast, the energy required to break a typical hydrogen bond is ~20 kJ·mol⁻¹. The various noncovalent influences to which proteins are subject—electrostatic interactions (both attractive and repulsive), hydrogen bonding (both intramolecular and to water), and hydrophobic forces—each have energetic magnitudes that may total thousands of kilojoules per mole over an entire protein molecule. Consequently, *a protein structure is the result of a delicate balance among powerful countervailing forces.* In this section we discuss the nature of these forces and end by considering protein denaturation: that is, how these forces can be disrupted.

A. Electrostatic Forces

Molecules are collections of electrically charged particles and hence, to a reasonable degree of approximation, their interactions are determined by the laws of classical electrostatics (more exact calculations require the application of quantum mechanics). The energy of association, U, of two electric charges, q_1 and q_2, that are separated by the distance r, is found by integrating the expression for Coulomb's law, Eq. [2.1], to determine the work necessary to separate these charges by an infinite distance:

$$U = \frac{kq_1q_2}{Dr} \qquad [7.1]$$

Here $k = 9.0 \times 10^9$ J·m·C⁻² and D is the dielectric constant of the medium in which the charges are immersed (recall that $D = 1$ for a vacuum and, for the most part, increases with the polarity of the medium; Table 2-1). The dielectric constant of a molecule-sized region is difficult to estimate. For the interior of a protein, it is usually taken to be in the range 3 to 5 in analogy with the measured dielectric constants of substances that have similar polarities such as benzene and diethyl ether.

Ionic Interactions Are Strong but Do Not Greatly Stabilize Proteins

The association of two ionic protein groups of opposite charge is known as an **ion pair** or **salt bridge.** According to Eq. [7.1], the energy of a typical ion pair, say the carboxyl group of Glu and the ammonium group of Lys, whose charge centers are separated by 4.0 Å in a medium of dielectric constant 4, is -86 kJ $\cdot$ mol^{-1} (one electronic charge = 1.60×10^{-19} C). Free ions in aqueous solution are highly solvated, however, so that the free energy of solvation of two separated ions is about equal to the free energy of formation of their unsolvated ion pairs. *Ion pairs therefore contribute little stability towards a protein's native structure.* This accounts for the observations that although ~75% of charged residues occur in ion pairs, very few ion pairs are buried (unsolvated) and that ion pairs that are exposed to the aqueous solvent tend to be but poorly conserved among homologous proteins.

Dipole–Dipole Interactions Are Weak but Significantly Stabilize Protein Structures

The noncovalent associations between electrically neutral molecules, collectively known as **van der Waals forces,** arise from electrostatic interactions among permanent and/or induced dipoles. These forces are responsible for numerous interactions of varying strengths between nonbonded neighboring atoms. (The hydrogen bond, a special class of dipolar interaction, is considered separately in Section 7-4B.)

Interactions among permanent dipoles are important structural determinants in proteins because many of their groups, such as the carbonyl and amide groups of the peptide backbone, have permanent dipole moments. These interactions are generally much weaker than the charge–charge interactions of ion pairs. Two carbonyl groups, for example, each with dipoles of 4.2×10^{-30} C·m (1.3 debye units) that are oriented in an optimal head-to-tail arrangement (Fig. 7-51*a*) and separated by 5 Å in a medium of dielectric constant 4, have a calculated attractive energy of only -9.3 kJ·mol^{-1}. Furthermore, these energies vary with r^{-3} so they rapidly attenuate with distance. In α helices, however, the dipolar amide and carbonyl groups of the polypeptide backbone all point in the same direction (Fig. 7-11) so that their interactions are associative and tend to be additive (these groups, of course, also form hydrogen bonds but here we are concerned with their residual electric fields). The carbonyl groups all have their oxygen atoms pointing towards the C terminal end of the α helix, giving it a significant dipole moment that is positive towards the N terminus and negative towards the C terminus. Consequently, *in*

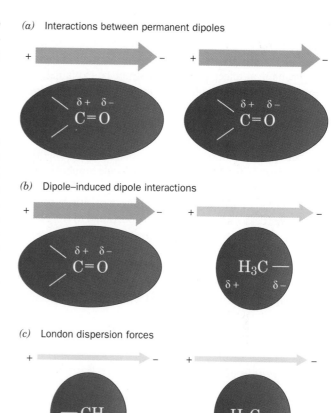

FIGURE 7-51. Dipole–dipole interactions. The strength of each dipole is represented by the thickness of the accompanying arrow. (*a*) Interactions between permanent dipoles. These interactions, here represented by carbonyl groups lined up head to tail, may be attractive, as shown here, or repulsive, depending on the relative orientations of the dipoles. (*b*) Dipole–induced dipole interactions. A permanent dipole (here shown as a carbonyl group) induces a dipole in a nearby group (here represented by a methyl group) by electrostatically distorting its electron distribution (*shading*). This always results in an attractive interaction. (*c*) London dispersion forces. The instantaneous charge imbalance (*shading*) resulting from the motions of the electrons in a molecule (*left*) induce a dipole in a nearby group (*right*); that is, the motions of the electrons in neighboring groups are correlated. This always results in an attractive interaction.

the low dielectric constant core of a protein, dipole–dipole interactions significantly influence protein folding.

A permanent dipole also induces a dipole moment on a neighboring group so as to form an attractive interaction (Fig. 7-51*b*). Such dipole–induced dipole interactions are generally much weaker than are dipole–dipole interactions.

Although nonpolar molecules are nearly electrically neutral, at any instant they have a small dipole moment resulting from the rapid fluctuating motion of their electrons. This transient dipole moment polarizes the electrons in a neighboring group, thereby giving rise to a dipole moment (Fig. 7-51c) such that, near their van der Waals contact distances, the groups are attracted to one another (a quantum mechanical effect that really cannot be explained in terms of only classical physics). These so-called **London dispersion forces** are extremely weak. The 8.2-kJ·mol^{-1} heat of vaporization of CH_4, for example, indicates that the attractive interaction of a nonbonded $H \cdots H$ contact between neighboring CH_4 molecules is roughly -0.3 kJ·mol^{-1} (in the liquid, a CH_4 molecule touches its 12 nearest neighbors with ~2 $H \cdots H$ contacts each).

London forces are only significant for contacting groups because their association energy is proportional to r^{-6}. Nevertheless, *the great numbers of interatomic contacts in proteins makes London forces a major influence in determining their conformations.* London forces also provide much of the binding energy in the sterically complementary interactions between proteins and the molecules that they specifically bind.

B. Hydrogen Bonding Forces

Hydrogen bonds (D—H$\cdots$A), as we discussed in Section 2-1A, are predominantly electrostatic interactions between a weakly acidic donor group (D—H) and an acceptor atom (A) that bears a lone pair of electrons. In biological systems, D and A can both be the highly electronegative N and O atoms and occasionally S atoms. Hydrogen bonds, which have association energies in the range -12 to -30 kJ·mol^{-1}, are much more directional than are van der Waals forces although less so than are covalent bonds. The D$\cdots$A distance is normally in the range 2.7 to 3.1 Å. Hydrogen bonds tend to be linear with the D—H bond pointing along the acceptor's lone pair orbital. Large deviations from this ideal geometry are not unusual, however. For example, in the hydrogen bonds of both α helices (Fig. 7-11) and antiparallel β pleated sheets (Fig. 7-16a), the N—H bonds point approximately along the C=O bonds rather than along an O lone pair orbital, and in parallel β pleated sheets (Fig. 7-16b), the hydrogen bonds depart significantly from linearity. Indeed, many of the hydrogen bonds in proteins are members of networks in which each donor is hydrogen bonded to multiple acceptors and each acceptor is hydrogen bonded to multiple donors.

The internal hydrogen bonding groups of a protein are arranged such that nearly all possible hydrogen bonds are formed (Section 7-3B). Clearly, hydrogen bonding has a major influence on the structures of proteins. An unfolded protein, however, makes all its hydrogen bonds with the water molecules of the aqueous solvent (water, it will be recalled, is a strong hydrogen bonding donor and acceptor). The free energy of stabilization that internal hydrogen bonds confer upon a native protein is therefore equal to the difference in the free energy of hydrogen bonding between the native protein and the unfolded protein. Since the various hydrogen bonds in question, to a first approximation, all have the same free energy, *internal hydrogen bonding cannot significantly stabilize, and, indeed, may even slightly destabilize, the structure of a native protein relative to its unfolded state.*

Despite the foregoing, *the internal hydrogen bonds of a protein provide a structural basis for its native folding pattern:* If a protein folded in a way that prevented some of its internal hydrogen bonds from forming, their free energy would be lost and such conformations would be less stable than those that are fully hydrogen bonded. Indeed, the formation of α helices and β sheets efficiently satisfies the polypeptide backbone's hydrogen bonding requirements. This argument also applies to the van der Waals forces discussed in the previous section.

C. Hydrophobic Forces

*The **hydrophobic effect** is the name given to those influences that cause nonpolar substances to minimize their contacts with water, and amphipathic molecules, such as soaps and detergents, to form micelles in aqueous solutions (Section 2-1B).* Since native proteins form a sort of intramolecular micelle in which their nonpolar side chains are largely out of contact with the aqueous solvent, *hydrophobic interactions must be an important determinant of protein structures.*

The hydrophobic effect derives from the special properties of water as a solvent, only one of which is its high dielectric constant. In fact, other polar solvents, such as dimethylsulfoxide (DMSO) and *N,N*-dimethylformamide (DMF), tend to denature proteins. The thermodynamic data of Table 7-4 provide considerable insight as to the origin of the hydrophobic effect because the transfer of a hydrocarbon from water to a nonpolar solvent resembles the transfer of a nonpolar side chain from the exterior of a protein in aqueous solution to its interior. The isothermal Gibbs free energy changes ($\Delta G = \Delta H - T\Delta S$) for the transfer of a hydrocarbon from an aqueous solution to a nonpolar solvent is negative in all cases, which indicates, as we know to be the case, that such transfers are spontaneous processes (oil and water do not mix). What is perhaps unexpected is that these transfer processes are endothermic (positive ΔH) for aliphatic compounds and athermic ($\Delta H = 0$) for aromatic compounds; that is, *it is enthalpically more or equally favorable for nonpolar molecules to dissolve in water than in nonpolar media.* In contrast, the entropy component of the unitary free energy change, $-T\Delta S_u$ (see footnote *a* to Table 7-4), is large and negative in all cases. Clearly, *the transfer of a hydrocarbon from an aqueous medium to a nonpolar medium is entropically driven. The same is true of the transfer of a nonpolar protein group from an aqueous environment to the protein's nonpolar interior.*

TABLE 7-4. **THERMODYNAMIC CHANGES FOR TRANSFERRING HYDROCARBONS FROM WATER TO NONPOLAR SOLVENTS AT 25°C**[a]

Process	ΔH (kJ·mol^{-1})	$-T\Delta S_u$ (kJ·mol^{-1})	ΔG_u (kJ·mol^{-1})
CH$_4$ in H$_2$O $\rightleftharpoons$ CH$_4$ in C$_6$H$_6$	11.7	−22.6	−10.9
CH$_4$ in H$_2$O $\rightleftharpoons$ CH$_4$ in CCl$_4$	10.5	−22.6	−12.1
C$_2$H$_6$ in H$_2$O $\rightleftharpoons$ C$_2$H$_6$ in benzene	9.2	−25.1	−15.9
C$_2$H$_4$ in H$_2$O $\rightleftharpoons$ C$_2$H$_4$ in benzene	6.7	−18.8	−12.1
C$_2$H$_2$ in H$_2$O $\rightleftharpoons$ C$_2$H$_2$ in benzene	0.8	−8.8	−8.0
Benzene in H$_2$O $\rightleftharpoons$ liquid benzene[b]	0.0	−17.2	−17.2
Toluene in H$_2$O $\rightleftharpoons$ liquid toluene[b]	0.0	−20.0	−20.0

[a] ΔG_u, the **unitary Gibbs free energy change,** is the Gibbs free energy change, ΔG, corrected for its concentration dependence so that it reflects only the inherent properties of the substance in question and its interaction with solvent. This relationship, according to Equation [3.13], is

$$\Delta G_u = \Delta G - nRT \ln \frac{[A_f]}{[A_i]}$$

where $[A_i]$ and $[A_f]$ are the initial and final concentrations of the substance under consideration, respectively, and n is the number of moles of that substance. Since the second term in this equation is a purely entropic term (concentrating a substance increases its order), ΔS_u, the **unitary entropy change,** is expressed

$$\Delta S_u = \Delta S + nR \ln \frac{[A_f]}{[A_i]}$$

[b] Data measured at 18°C.

Source: Kauzmann, W., *Adv. Protein Chem.* **14**, 39 (1959).

What is the physical mechanism whereby nonpolar entities are excluded from aqueous solution? Recall that entropy is a measure of the order of a system; it decreases with increasing order (Section 3-2). Thus the decrease in entropy when a nonpolar molecule or side chain is solvated by water (the reverse of the foregoing process) must be due to an ordering process. This is an experimental observation, not a theoretical conclusion. The magnitudes of the entropy changes are too large to be attributed only to changes in the conformations of the hydrocarbons; rather, as Henry Frank and Marjorie Evans pointed out in 1945, *these entropy changes mainly arise from some sort of ordering of the water structure.*

Liquid water has a highly ordered and extensively hydrogen bonded structure (Section 2-1A). The insinuation of a nonpolar group into this structure disrupts it: A nonpolar group can neither accept nor donate hydrogen bonds, so the water molecules at the surface of the cavity occupied by the nonpolar group cannot hydrogen bond to other molecules in their usual fashion. In order to recover the lost hydrogen bonding energy, these surface waters must orient themselves so as to form a hydrogen bonded network enclosing the cavity (Fig. 7-52). This orientation constitutes an ordering of the water structure since the number of ways that

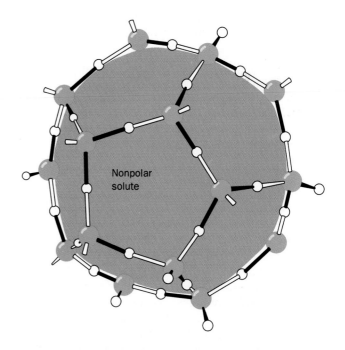

FIGURE 7-52. The orientational preference of water molecules next to a nonpolar solute. In order to maximize their hydrogen bonding energy, these water molecules tend to straddle the inert solute such that two or three of their tetrahedral directions are tangential to its surface. This permits them to form hydrogen bonds with neighboring water molecules lining the nonpolar surface. This ordering of water molecules extends several layers of water molecules beyond the first hydration shell of the nonpolar solute.

water molecules can form hydrogen bonds about the surface of a nonpolar group is less than the number of ways that they can hydrogen bond in bulk water.

Unfortunately, the complexity of liquid water's basic structure (Section 2-1A) has not yet allowed a detailed structural description of this ordering process. One model that has been proposed is that water forms quasi-crystalline hydrogen bonded cages about the nonpolar groups similar to those of **clathrates** (Fig. 7-53). The magnitudes of the entropy changes that result when nonpolar substances are dissolved in water, however, indicate that the resulting water structures can only be slightly more ordered than bulk water. They also must be quite different from that of ordinary ice, because, for instance, the solvation of nonpolar groups by water causes a large decrease in water volume (e.g., the transfer of CH_4 from hexane to water shrinks the water solution by 22.7 mL · mol^{-1} of CH_4), whereas the freezing of water results in a 1.6-mL · mol^{-1} expansion.

The unfavorable free energy of hydration of a nonpolar substance caused by its ordering of the surrounding water molecules has the net result that *the nonpolar substance is excluded from the aqueous phase.* This is because the surface area of a cavity containing an aggregate of nonpolar molecules is less than the sum of the surface areas of the cavities that each of these molecules would individually occupy. The aggregation of the nonpolar groups thereby minimizes the surface area of the cavity and therefore the entropy loss of the entire system. In a sense, the nonpolar groups are squeezed out of the aqueous phase by the hydrophobic interactions. Thermodynamic measurements indicate that the free energy change of removing a —CH_2— group from an aqueous solution is about −3 kJ·mol^{-1}. Although this is a relatively small amount of free energy, *in molecular assemblies involving large numbers of nonpolar contacts, hydrophobic interactions are a potent force.*

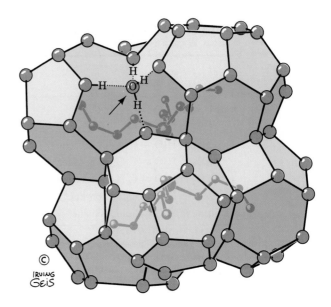

FIGURE 7-53. The structure of the clathrate (n-C_4H_9)$_3$S$^+$F$^-$· 23H$_2$O. Clathrates are crystalline complexes of nonpolar compounds with water (usually formed at low temperatures and high pressures) in which the nonpolar molecules are enclosed, as shown, by a polyhedral cage of tetrahedrally hydrogen bonded water molecules (here represented by only their oxygen atoms). The hydrogen bonding interactions of one such water molecule (*arrow*) are shown in detail. [Figure copyrighted © by Irving Geis.]

Walter Kauzmann pointed out in the 1950s that *hydrophobic forces are a major influence in causing proteins to fold into their native conformations.* Figure 7-54 indicates that the amino acid side chain **hydropathies** (indexes of combined hydrophobic and hydrophilic tendencies; Table 7-5) are, in fact, good predictors of which portions of a polypeptide chain are inside a protein, out of contact with

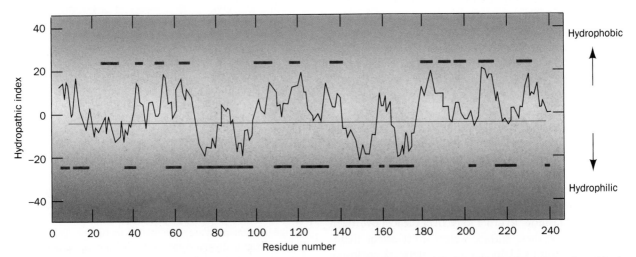

FIGURE 7-54. The hydropathic index (sum of the hydropathies of nine consecutive residues; see Table 7-5) versus the residue sequence number for bovine **chymotrypsinogen.** A large positive hydropathic index is indicative of a hydrophobic region of the polypeptide chain, whereas a large negative value is indicative of a hydrophilic region. The bars above the midpoint line denote the protein's interior regions, as determined by X-ray crystallography, and the bars below the midpoint line indicate the protein's exterior regions. [After Kyte, J. and Doolittle, R.F., *J. Mol. Biol.* **157,** 111 (1982).]

TABLE 7-5. HYDROPATHY SCALE FOR AMINO ACID SIDE CHAINS

Side Chain	Hydropathy
Ile	4.5
Val	4.2
Leu	3.8
Phe	2.8
Cys	2.5
Met	1.9
Ala	1.8
Gly	−0.4
Thr	−0.7
Ser	−0.8
Trp	−0.9
Tyr	−1.3
Pro	−1.6
His	−3.2
Glu	−3.5
Gln	−3.5
Asp	−3.5
Asn	−3.5
Lys	−3.9
Arg	−4.5

Source: Kyte, J. and Doolitle, R.F., *J. Mol. Biol.* **157**, 110 (1982).

the aqueous solvent, and which portions are outside, in contact with the aqueous solvent. In proteins, the effects of hydrophobic forces are often termed **hydrophobic bonding**, presumably to indicate the specific nature of protein folding under the influence of the hydrophobic effect. You should keep in mind, however, that hydrophobic bonding does not generate the directionally specific interactions usually associated with the term "bond."

D. Disulfide Bonds

Since disulfide bonds form as a protein folds to its native conformation (Section 8-1B), they function to stabilize its three-dimensional structure. The relatively reducing chemical character of the cytoplasm, however, greatly diminishes the stability of intracellular disulfide bonds. In fact, almost all proteins with disulfide bonds are secreted to more oxidized extracellular destinations where their disulfide bonds are effective in stabilizing protein structures [secreted proteins fold to their native conformations—and hence form their disulfide bonds—in the endoplasmic reticulum (Section 11-4B) which, unlike other cell compartments, has an oxidizing environment]. Apparently, the relative "hostility" of extracellular environments towards proteins (e.g., uncontrolled temperatures and pH's) requires the additional structural stability conferred by disulfide bonds.

E. Protein Denaturation

The low conformational stabilities of native proteins make them easily susceptible to denaturation by altering the balance of the weak nonbonding forces that maintain the na-

tive conformation. When a protein in solution is heated, its conformationally sensitive properties, such as optical rotation (Section 4-2A), viscosity, and UV absorption, change abruptly over a narrow temperature range (Fig. 7-55). *Such a nearly discontinuous change indicates that the native protein structure unfolds in a cooperative manner: Any partial unfolding of the structure destabilizes the remaining structure, which must simultaneously collapse to the random coil.* The temperature at the midpoint of this process is known as the protein's **melting temperature,** T_m, in analogy with the melting of a solid. Most proteins have T_m values well below 100°C. Among the exceptions to this generalization, however, are the proteins of **thermophilic bacteria**, organisms that inhabit hot springs with temperatures approaching 100°C. Interestingly, the X-ray structures of these heat-stable proteins are but subtly different from those of their normally stable homologs.

In addition to high temperatures, proteins are denatured by a variety of other conditions and substances:

1. pH variations alter the ionization states of amino acid side chains (Table 4-1), which changes protein charge distributions and hydrogen bonding requirements.

2. Detergents, some of which significantly perturb protein structures at concentrations as low as $10^{-6}M$, hydrophobically associate with the nonpolar residues of a protein, thereby interfering with the hydrophobic interactions responsible for the protein's native structure.

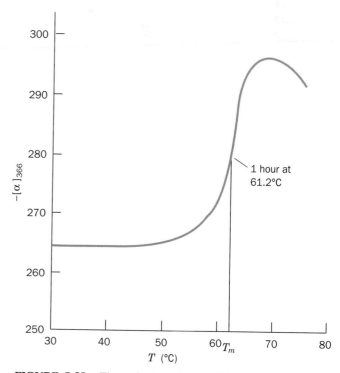

FIGURE 7-55. The optical rotation, at 366 nm, as a function of temperature, of bovine pancreatic ribonuclease A (RNase A) in 0.15*M* KCl and 0.013*M* sodium cacodylate buffer, pH 7. The melting temperature, T_m, is defined as the midpoint of the transition. [After von Hippel, P.H. and Wong, K.Y., *J. Biol. Chem.* **10**, 3911 (1965).]

3. High concentrations of water-soluble organic substances, such as aliphatic alcohols, interfere with the hydrophobic forces stabilizing protein structures through their own hydrophobic interactions with water. Organic substances with several hydroxyl groups, such as ethylene glycol or sucrose,

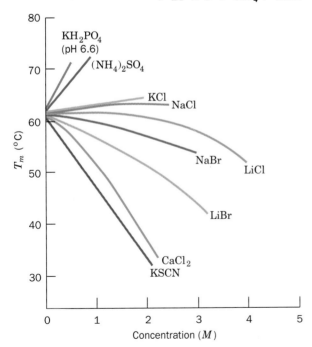

Ethylene glycol   **Sucrose**

however, are relatively poor denaturants because their hydrogen bonding ability renders them less disruptive of water structure.

The influence of salts is more variable. Figure 7-56 shows the effects of a number of salts on the T_m of bovine pancreatic **ribonuclease A (RNase A).** Some salts, such as $(NH_4)_2SO_4$ and KH_2PO_4, stabilize the native protein structure (raise its T_m); others, such as KCl and NaCl, have little effect; and yet others, such as KSCN and LiBr, destabilize it. The order of effectiveness of the various ions in stabilizing a protein, which is largely independent of the identity of the protein, parallels their capacity to salt out proteins (Section 5-2A). This order is known as the **Hofmeister series:**

Anions $SO_4^{2-} > H_2PO_4^- > CH_3COO^- > Cl^-$
$$> Br^- > I^- > ClO_4^- > SCN^-$$

Cations $NH_4^+, Cs^+, K^+, Na^+ > Li^+$
$$> Mg^{2+} > Ca^{2+} > Ba^{2+}$$

The ions in the Hofmeister series that tend to denature proteins, I^-, ClO_4^-, SCN^-, Li^+, Mg^{2+}, Ca^{2+}, and Ba^{2+}, are said to be **chaotropic.** This list should also include the guanidinium ion (Gu^+) and the nonionic urea,

$$H_2N - \underset{\underset{NH_2^+}{||}}{C} - NH_2 \qquad H_2N - \underset{\underset{O}{||}}{C} - NH_2$$

Guanidinium ion   **Urea**

which, in concentrations in the range 5 to 10M, are the most commonly used protein denaturants. The effect of the various ions on proteins is largely cumulative: GuSCN is a much more potent denaturant than the often used GuCl, whereas Gu_2SO_4 stabilizes protein structures.

Chaotropic agents increase the solubility of nonpolar substances in water. Consequently, their effectiveness as denaturing agents stems from their ability to disrupt hydrophobic interactions although the manner in which they do so is not well understood. Conversely, those substances listed that stabilize proteins strengthen hydrophobic forces, thus increasing the tendency of water to expel proteins. This accounts for the correlation between the abilities of an ion to stabilize proteins and to salt them out.

5. QUATERNARY STRUCTURE

Proteins, because of their multiple polar and nonpolar groups, stick to almost anything; anything, that is, but other proteins. This is because the forces of evolution have arranged the surface groups of proteins so as to prevent their association under physiological conditions. If this were not the case, their resulting nonspecific aggregation would render proteins functionally useless (recall, e.g., the consequences of sickle-cell anemia; Section 6-3A). In his pioneering ultracentrifugational studies on proteins, however, The Svedberg discovered that some proteins are composed of more than one polypeptide chain. Subsequent studies established that this is, in fact, true of most proteins, including nearly all those with molecular masses >100 kD. Furthermore, these polypeptide **subunits** associate in a geometrically specific manner. The spatial arrangement of these subunits is known as a protein's **quaternary structure (4° structure).**

There are several reasons that multisubunit proteins are so common. In large assemblies of proteins, such as collagen fibrils, the advantages of subunit construction over the synthesis of one huge polypeptide chain are analogous to those of using prefabricated components in constructing a building. Defects can be repaired by simply replacing the flawed subunit, the site of subunit manufacture can be different from the site of assembly into the final product, and the only genetic information necessary to specify the entire edifice is that specifying its few different self-assembling subunits. In the case of enzymes, increasing a protein's size

FIGURE 7-56. The melting temperature of RNase A as a function of the concentrations of various salts. All solutions also contain 0.15M KCl and 0.013M sodium cacodylate buffer, pH 7. [After von Hippel, P.J. and Wong, K.Y., *J. Biol. Chem.* **10,** 3913 (1965).]

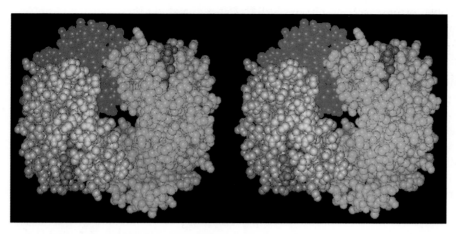

FIGURE 7-57. A stereo, space-filling drawing showing the quaternary structure of hemoglobin. The α_1, α_2, β_1, and β_2 subunits are colored yellow, green, light blue, and purple, respectively. Heme groups are red. The protein is viewed along its molecular twofold rotation axis which relates the $\alpha_1\beta_1$ protomer to the $\alpha_2\beta_2$ protomer. Instructions for viewing stereo drawings are given in the appendix to this chapter.

tends to better fix the three-dimensional positions of the groups forming the enzyme's active site. Increasing the size of an enzyme through the association of identical subunits is more efficient, in this regard, than increasing the length of its polypeptide chain since each subunit has an active site. More importantly, however, the subunit construction of many enzymes provides the structural basis for the regulation of their activities. Mechanisms for this indispensable function are discussed in Sections 9-4 and 12-4.

In this section we discuss how the subunits of multisubunit proteins associate, what sorts of symmetries they have, and how their stoichiometries may be determined.

A. *Subunit Interactions*

A multisubunit protein may consist of identical or nonidentical polypeptide chains. Recall that hemoglobin, for example, has the subunit composition $\alpha_2\beta_2$. We shall refer to proteins with identical subunits as **oligomers** and to these identical subunits as **protomers**. A protomer may therefore consist of one polypeptide chain or several unlike polypeptide chains. In this sense, hemoglobin is a **dimer** (oligomer of two protomers) of $\alpha\beta$ protomers (Fig. 7-57).

The contact regions between subunits closely resemble the interior of a single subunit protein. They contain closely packed nonpolar side chains, hydrogen bonds involving the polypeptide backbones and their side chains, and, in some cases, interchain disulfide bonds.

B. *Symmetry in Proteins*

In the vast majority of oligomeric proteins, the protomers are symmetrically arranged; that is, the protomers occupy geometrically equivalent positions in the oligomer. This implies that each protomer has exhausted its capacity to bind to other protomers; otherwise, higher oligomers would form. As a result of this limited binding capacity, protomers pack about a single point to form a closed shell. Proteins cannot have inversion or mirror symmetry, however, because such symmetry operations convert chiral L-residues to D-residues. Thus, *proteins can only have rotational symmetry.*

Various types of rotational symmetry occur in proteins:

1. Cyclic symmetry

In the simplest type of rotational symmetry, **cyclic symmetry,** subunits are related (brought to coincidence) by a single axis of rotation (Fig. 7-58a). Objects with 2, 3, . . . , or *n*-fold rotational axes are said to have C_2, C_3, . . . , or C_n symmetry, respectively. An oligomer

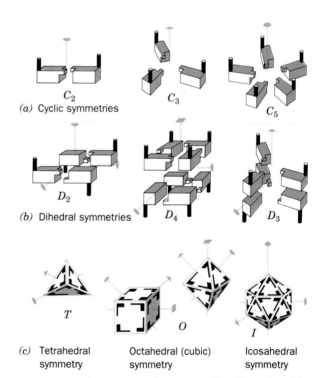

FIGURE 7-58. Some possible symmetries of proteins with identical protomers. The lenticular shape, the triangle, the square, and the pentagon at the ends of the dashed lines indicate, respectively, the unique twofold, threefold, fourfold, and fivefold rotational axes of the objects shown. (*a*) Assemblies with the cyclic symmetries C_2, C_3, and C_5. (*b*) Assemblies with the dihedral symmetries D_2, D_4, and D_3. In these objects, a twofold axis is perpendicular to the vertical two-, four-, and threefold axes. (*c*) Assemblies with *T*, *O*, and *I* symmetry. Note that the tetrahedron has some but not all of the symmetry elements of the cube, and that the cube and the octahedron have the same symmetry. [Figure copyrighted © by Irving Geis.]

with C_n symmetry consists of n protomers that are related by $(360/n)°$ rotations. C_2 symmetry is the most common symmetry in proteins; higher cyclic symmetries are relatively rare.

A common mode of association between protomers related by a twofold rotation axis is the continuation of a β sheet across subunit boundaries. In such cases, the twofold axis is perpendicular to the β sheet so that two symmetry equivalent strands hydrogen bond in an antiparallel fashion. In this manner, the sandwich of two four-stranded β sheets of the **prealbumin** protomer is extended across a twofold axis to form a sandwich of two eight-stranded β sheets (Fig. 7-59).

2. Dihedral symmetry

Dihedral symmetry (D_n), a more complicated type of rotational symmetry, is generated when an n-fold rotation axis and a twofold rotation axis intersect at right angles (Fig. 7-58b). An oligomer with D_n symmetry consists of $2n$ protomers. The D_2 symmetry is, by far, the most common type of dihedral symmetry in proteins. Under the proper conditions, many oligomers with D_n symmetry will dissociate into two oligomers, each with C_n symmetry (and which were related by the twofold rotation axis in the D_n oligomer). These, in turn, dissociate to their component protomers under more stringent dissociating conditions.

3. Other rotational symmetries

The only other types of rotationally symmetric objects are those that have the rotational symmetries of a tetrahedron (T), a cube or octahedron (O), or an icosahedron (I), and have 12, 24, and 60 equivalent positions, respectively (Fig. 7-58c). The subunit arrangements in the protein coats of the so-called spherical viruses are based on icosahedral symmetry (Section 32-2A).

Under favorable conditions electron microscopy can provide dramatic indications of oligomeric symmetry.

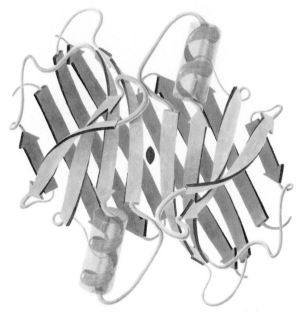

FIGURE 7-59.   A prealbumin dimer viewed down its twofold axis (*red symbol*). Each protomer consists of a sandwich of two four-stranded β sheets. Note how both of these β sheets are continued in an antiparallel fashion in the other protomer to form a sandwich of two eight-stranded β sheets. Two of these dimers associate back to back in the native protein to form a tetramer with D_2 symmetry. [After a drawing by Jane Richardson, Duke University.]

Electron microscopy studies suggest, for example, that the 600 kD *E. coli* **glutamine synthetase** has D_6 symmetry (Fig. 7-60). Unfortunately, since this technique has insufficient resolution to reveal the relative orientations of the protein subunits (i.e., the directions of the arrows in the interpretive drawing of Fig. 7-60), such symmetry assignments must be taken as tentative; only X-ray crystal structure analysis can unambiguously establish the geometric relationships

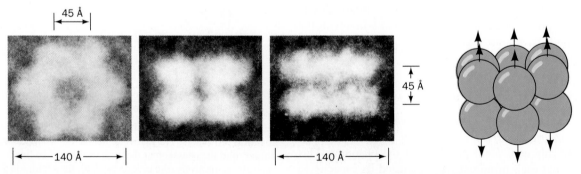

FIGURE 7-60.   Sets of five superimposed electron micrographs (to enhance real detail) of *E. coli* glutamine synthetase molecules in their three characteristic orientations. The mean dimensions are indicated. When the oligomeric molecule rests on its face, it appears to be a hexagonal ring of subunits (*left*). Molecules on edge, however, show two layers of subunits as four spots when viewed exactly between the subunits (*middle*), or as two parallel streaks when viewed in other directions (*right*). This suggests, as the accompanying drawing indicates, that the enzyme molecule has 12 identical subunits organized with D_6 symmetry into two hexagons that are stacked with their subunits in apposition. [Courtesy of Earl Stadtman, NIH.]

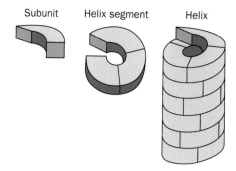

FIGURE 7-61. A helical structure composed of a single kind of subunit.

among protein subunits. In the case of glutamine synthetase, however, X-ray studies have confirmed that it indeed has D_6 symmetry (Section 24-5A).

Helical Symmetry

Some protein oligomers have **helical symmetry** (Fig. 7-61). The chemically identical subunits in a helix are not strictly equivalent because, for instance, those at the end of the helix have a different environment than those in the middle. Nevertheless, the surroundings of all subunits in a long helix, except those near its ends, are sufficiently similar that the subunits are said to be **quasi-equivalent**. The subunits of many structural proteins, for example, those of muscle (Section 34-3A), assemble into fibers with helical symmetry.

C. Determination of Subunit Composition

The number of different types of subunits in an oligomeric protein may be determined by end group analysis (Section 6-1A). In principle, the subunit composition of a protein may be determined by comparing its molecular mass with those of its component subunits. In practice, however, experimental difficulties, such as the partial dissociation of a supposedly intact protein and uncertainties in molecular mass determinations, often provide erroneous results.

Hydridization Yields Quaternary Structural Information

An alternative procedure may be used if two chemically different and therefore separable species of the protein are available. The species may be proteins with slightly different 1° structures from different organisms or, as is often the case, variants of a protein that occur in the same organism. The two different oligomeric proteins are purified, mixed together, dissociated to their component subunits by exposure to mildly denaturing conditions (e.g., by changing the pH or adding urea), and then allowed to reassemble (e.g., by restoring the pH or dialyzing out the urea). If the native proteins are *n*-mers, S_n and S'_n, this procedure will yield $(n + 1)$ species of **hybrid molecules** with the mixed subunit compositions $S_n, S_{n-1}S', S_{n-2}S'_2, \ldots, S'_n$, which can be

analyzed, for example, by electrophoresis. For instance, vertebrates possess two varieties of the enzyme **lactate dehydrogenase (LDH):** the M type, which predominates in skeletal muscle, and the H type, which predominates in heart tissue. Hybridization of these oligomers, in this case by repeated freezing and thawing, yields five **isozymes** (isoenzymes; catalytically and structurally similar enzymes from the same organism) of LDH that have the subunit compositions M_4, M_3H, M_2H_2, MH_3, and H_4 (Fig. 7-62). This demonstrates that LDH is a tetramer.

In a related method, a protein subunit may be labeled, for example, by succinylation,

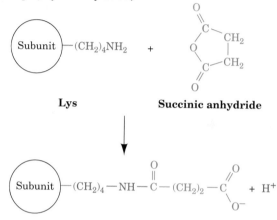

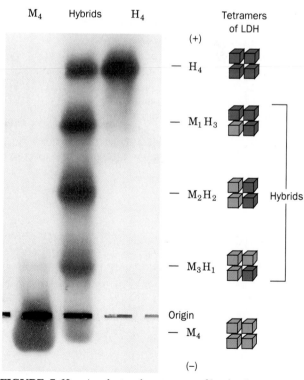

FIGURE 7-62. An electrophoretogram of bovine lactate dehydrogenase. The M and H forms of LDH (*outer lanes*) have different electrophoretic mobilities. Upon hybridization of these oligomers, five electrophoretically distinct isozymes are formed (*center lane*), which indicates that LDH is a tetramer. [Courtesy of Clement Markert, North Carolina State University at Raleigh.]

which alters the electrophoretic mobility of a protein by changing its ionic charge. John Gerhart and Howard Schachman used this technique to determine the geometric distribution of subunits in *E. coli* **aspartate transcarbamoylase (ATCase).** ATCase has two types of subunits, the catalytic subunit, c, and the regulatory subunit, r (their enzymatic roles are discussed in Section 12-4). Molecular mass measurements (c = 33 kD, r = 17 kD, and ATCase = 300 kD) indicate that ATCase has the subunit composition c_6r_6. This was corroborated by preliminary X-ray studies which established that ATCase has D_3 symmetry (recall that a protein of D_3 symmetry must have six protomers; Fig. 7-58*b*). Treatment with organic mercurials such as ***para*-hydroxymercuribenzoate,** which reacts with Cys sulfhydryl groups

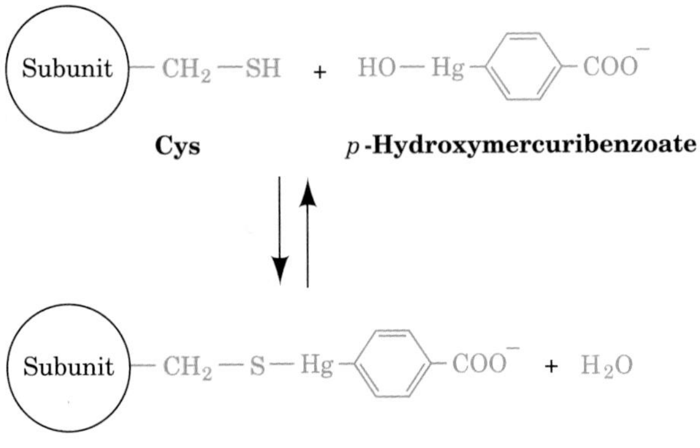

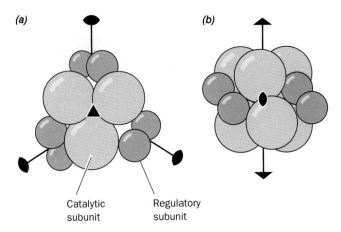

Catalytic subunit   Regulatory subunit

FIGURE 7-63. The quaternary structure of *E. coli* aspartate transcarbamoylase as established by X-ray structure analysis. Catalytic subunits and regulatory subunits are represented, respectively, by large orange spheres and small purple spheres. The molecule has D_3 symmetry. (*a*) View along the threefold axis (*triangle*). (*b*) View along a twofold axis (*lenticular shapes*). [After Kantrowitz, E. R., Pastra-Landis, S.C., and Lipscomb, W.N., *Trends Biochem. Sci.* **5**, 150 (1980).]

causes ATCase to dissociate according to the reaction

$$c_6r_6 \rightarrow 2c_3 + 3r_2$$

The catalytic trimers, c_3, were isolated and succinylated to form c_3^s. When these were mixed with unmodified catalytic trimers and excess regulatory dimers, r_2, under conditions that ATCase reforms, only three products could be electrophoretically distinguished: c_6r_6, $c_3c_3^sr_6$, and $c_6^sr_6$. This indicates that the catalytic subunits were not exchanged be-

FIGURE 7-64. Dimethylsuberimidate and glutaraldehyde are bifunctional reagents that react to covalently cross-link two Lys residues.

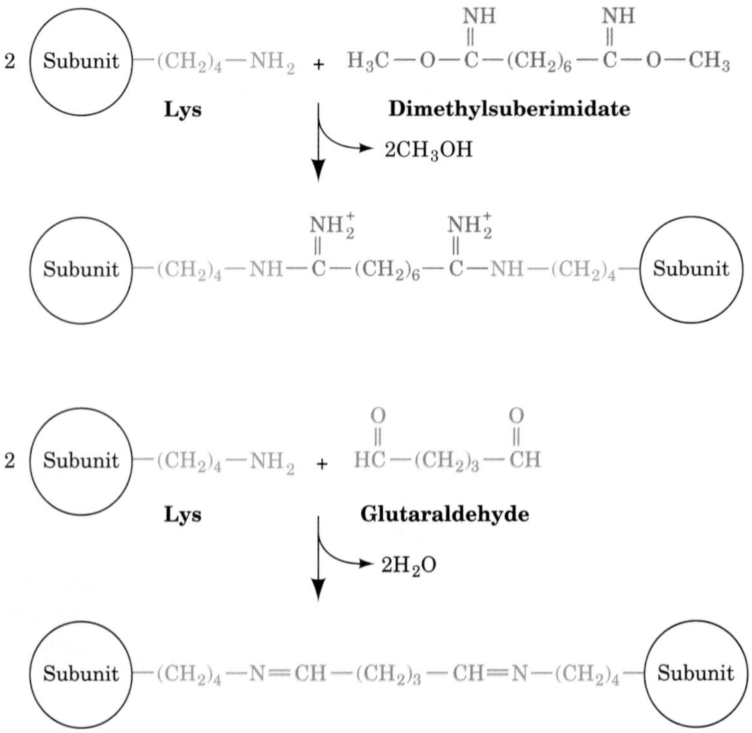

tween catalytic trimers in ATCase; for example, no $c_4c_2^sr_6$ was formed. The c_3 trimers must therefore be separate entities in the enzyme. Similar studies using succinylated regulatory dimers, r_2^s, established that regulatory dimers likewise maintain their integrity in ATCase. Accordingly, the subunit composition of ATCase is more realistically represented as $(c_3)_2(r_2)_3$. This result was later confirmed by the X-ray crystal structure of ATCase (Fig. 7-63).

Cross-Linking Agents Stabilize Oligomers

Another method for 4° structure analysis, which is especially useful for oligomeric proteins that decompose easily, employs **cross-linking agents** such as **dimethylsuberimidate** or **glutaraldehyde** (Fig. 7-64). If carried out at sufficiently low protein concentrations to eliminate intermolecular reactions, cross-linking reactions will covalently join only the subunits in a molecule that are no further apart than the length of the cross-link (assuming, of course, that the proper amino acid residues are present). The molecular mass of a cross-linked protein therefore places a lower limit on its number of subunits. Such studies can also provide some indication of the distance between subunits, particularly if a series of cross-linking agents with different lengths is employed.

APPENDIX: VIEWING STEREO PICTURES

Although we live in a three-dimensional world, the images that we see have been projected onto the two-dimensional plane of our retinas. Depth perception therefore involves binocular vision: The slightly different views perceived by each eye are synthesized by the brain into a single three-dimensional impression.

Two-dimensional pictures of complex three-dimensional objects are difficult to interpret because most of the information concerning the third dimension is suppressed. This information can be recovered by presenting each eye with the image only it would see if the three-dimensional object were actually being viewed. A **stereo pair** therefore consists of two images, one for each eye. Corresponding points of stereo pairs are generally separated by ~6 cm, the average distance between human eyes. Stereo drawings are usually computer generated because of the required precision of the geometric relationship between the members of a stereo pair.

In viewing a stereo picture, one must overcome the visual habits of a lifetime because each eye must see its corresponding view independently. Viewers are commercially available to aid in this endeavor. However, with some training and practice, equivalent results can be obtained without their use.

To train yourself to view stereo pictures, you should become aware that each eye sees a separate image. Hold your finger up about a foot (30 cm) before your eyes while fixing your gaze on some object beyond it. You may realize that you are seeing two images of your finger. If, after some concentration, you are still aware of only one image, try blinking your eyes alternately to ascertain which of your eyes is seeing the image you perceive. Perhaps alternately covering and uncovering this dominant eye while staring past your finger will help you become aware of the independent workings of your eyes.

The principle involved in seeing a stereo picture is to visually fuse the left member of the stereo pair seen by the left eye with the right member seen by the right eye. To do this, sit comfortably at a desk, center your eyes about a foot over a stereo drawing such as Fig. 7-65 and stare through it at a point about a foot below the drawing. Try to visually fuse the central members of the four out-of-focus images you see. When you have succeeded, your visual system will "lock onto" it and this fused central image will appear three dimensional. Ignore the outer images. You may have to slightly turn the book, which should be held perfectly flat, or your head in order to bring the two images to the same level. It may help to place the book near the edge of a desk,

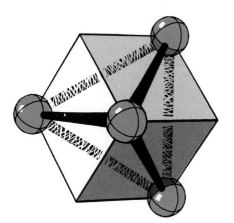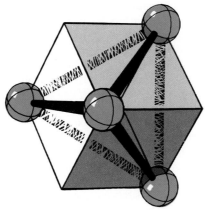

FIGURE 7-65. A stereo drawing of a tetrahedron inscribed in a cube. When properly viewed, the apex of the tetrahedron should appear to be pointing towards the viewer.

center your finger about a foot below the drawing, and fixate on your finger while concentrating on the stereo pair. Another trick is to hold your flattened hand or an index card between your eyes so that the left eye sees only the left half of the stereo pair and the right eye sees only the right half and then fuse the two images you see.

The final step in viewing a stereo picture is to focus on the image while maintaining fusion. This may not be easy because our ingrained tendency is to focus on the point at which our gaze converges. It may help to move your head

closer to or further from the picture. Most people (including the authors) require a fair amount of practice to become proficient at seeing stereo without a viewer. However, the three-dimensional information provided by stereo pictures, not to mention their esthetic appeal, makes it worth the effort. In any case, the few stereo figures used in this text have been selected for their visual clarity without the use of stereo; stereo will simply enhance their impression of depth.

CHAPTER SUMMARY

The peptide group is constrained by resonance effects to a planar, trans conformation. Steric interactions further limit the conformations of the polypeptide backbone by restricting the torsion angles, ϕ and ψ, of each peptide group to three small regions of the Ramachandran diagram. The α helix, whose conformation angles fall within the allowed regions of the Ramachandram diagram, is held together by hydrogen bonds. The 3_{10} helix, which is more tightly coiled than the α helix, lies in a mildly forbidden region of the Ramachandran diagram. Its infrequent occurrences are most often as single-turn terminations of α helices. In the parallel and antiparallel β pleated sheets, two or more almost fully extended polypeptide chains associate such that neighboring chains are hydrogen bonded. These β sheets have a right-handed curl when viewed along their polypeptide chains. The polypeptide chain often reverses its direction through a β bend. Other arrangements of the polypeptide chain, which are collectively known as coil conformations, are more difficult to describe but are no less ordered than are α or β structures.

The mechanical properties of fibrous proteins can often be correlated with their structures. Keratin, the principal component of hair, horn, and nails, forms protofibrils that consist of two pairs of α helices in which the members of each pair are twisted together into a left-handed coil. The pliability of keratin decreases as the content of disulfide cross-links between the protofibrils increases. Silk fibroin forms flexible but inextensible fibers of great strength. It exists as a semicrystalline array of antiparallel β sheets in which layers of Gly side chains alternate with layers of Ala and Ser side chains. Collagen is the major protein component of connective tissue. Its every third residue is Gly and many of the others are Pro and Hyp. This permits collagen to form a ropelike triple helical structure that has great tensile strength. Collagen molecules aggregate in a staggered array to form fibrils that are covalently cross-linked by groups derived from their His and Lys side chains. Elastin, which has elastic properties, forms a three-dimensional network of fibers that exhibit no regular structure. Its polypeptide strands are cross-linked in a manner similar to that in collagen.

The accuracies of protein X-ray structure determinations are limited by crystalline disorder to resolutions that are mostly in the range 2.0 to 3.5 Å. This requires that a protein's structure be determined by fitting its primary structure to its electron density map. Several lines of evidence indicate that protein crystal structures are nearly identical to their solution structures. The structures of small proteins may also be determined in solution by

2D-NMR techniques which, for the most part, yield results similar to those of X-ray crystal structures. A globular protein's 3° structure is the arrangement of its various elements of 2° structure together with the spatial dispositions of its side chains. Its amino acid residues tend to segregate according to residue polarity. Nonpolar residues preferentially occur in the interior of a protein out of contact with the aqueous solvent, whereas charged polar residues are located on its surface. Uncharged polar residues may occur at either location but, if they are internal, they form hydrogen bonds with other protein groups. The interior of a protein molecule resembles a crystal of an organic molecule in its packing efficiency. Larger proteins often fold into two or more domains that may have functionally and structurally independent properties. Certain groupings of secondary structural elements, known as supersecondary structures, repeatedly occur as components of globular proteins. They may have functional as well as structural significance.

Proteins have marginally stable native structures that form as a result of a fine balance among the various noncovalent forces to which they are subject: ionic and dipolar interactions, hydrogen bonding, and hydrophobic forces. Ionic interactions are relatively weak in aqueous solutions due to the solvating effects of water. The various interactions among permanent and induced dipoles, which are collectively referred to as van der Waals forces, are even weaker and are effective only at short range. Nevertheless, because of their large numbers, they cumulatively have an important influence on protein structures. Hydrogen bonding forces are far more directional than are other noncovalent forces. They add little stability to a protein structure, however, because the hydrogen bonds that native proteins form internally are no stronger than those that unfolded proteins form with water. Yet, a protein can only fold stably in ways that almost all of its possible internal hydrogen bonds are formed so that hydrogen bonding is important in specifying the native structure of a protein. Hydrophobic forces arise from the unfavorable ordering of water structure that results from the hydration of nonpolar groups. By folding such that its nonpolar groups are out of contact with the aqueous solvent, a protein minimizes these unfavorable interactions. The fact that most protein denaturants interfere with the hydrophobic effect demonstrates the importance of hydrophobic forces in stabilizing native protein structures. Disulfide bonds often stabilize the native structures of extracellular proteins.

Many proteins consist of noncovalently linked aggregates of subunits in which the subunits may or may not be identical. Most oligomeric proteins are rotationally symmetric. The protomers in many fibrous proteins are related by helical symmetry. The subunit structures of proteins may be elucidated by a variety of techniques, including the hybridization of subunits with their naturally occurring or derivatized variants, cross-linking studies, electron microscopy, and X-ray crystal structure analysis.

REFERENCES

General

Branden, C. and Tooze, J., *Introduction to Protein Structure,* Garland Publishing (1991).

Cantor, C.R. and Schimmel, P.R., *Biophysical Chemistry,* Chapters 2 and 5, Freeman (1980).

Creighton, T.E., *Proteins* (2nd ed.), Chapters 4–6, Freeman (1993).

Dickerson, R.E. and Geis, I., *The Structure and Action of Proteins,* Benjamin/Cummings (1969). [A marvelously illustrated exposition of the fundamentals of protein structure.]

Perutz, M., *Protein Structure. New Approaches to Disease and Therapy,* Freeman (1992). [A series of short articles on the structures of a variety of proteins and their biomedical implications.]

Schultz, G.E. and Schirmer, R.H., *Principles of Protein Structure,* Chapters 2–5 and 7, Springer–Verlag (1979). [An advanced text.]

Wood, W.B., Wilson, J.H., Benbow, R.M., and Hood, L.E., *Biochemistry. A Problems Approach* (2nd ed.), Chapters 4 and 5, Benjamin/Cummings (1981). [A question-and-answer approach to learning protein structures.]

Secondary Structure

Leszczynski, J.F. and Rose, G.D., Loops in globular proteins: a novel category of secondary structure, *Science* **234,** 849–855 (1986).

Toniolo, C. and Benedetti, E.,The polypeptide 3_{10}-helix, *Trends Biochem. Sci.* **16,** 350–353 (1991).

Rose, G.D., Gierasch, L.M., and Smith, J.A., Turns in polypeptides and proteins, *Adv. Protein Chem.* **37,** 1–109 (1985).

Salemme, F.R., Structural properties of protein β-sheets, *Prog. Biophys. Mol. Biol.* **42,** 95–133 (1983).

Fibrous Proteins

Bornstein, P. and Traub, W., The chemistry and biology of collagen, *in* Neurath, H. and Hill, R.L. (Eds.), *The Proteins* (3rd ed.), Vol. 4, *pp.* 412–632, Academic Press (1979).

Byers, P.H., Disorders of collagen synthesis and structure, *in* Scriver, C.R., Beaudet, A.L., Sly, W.S., and Valle, D. (Eds.), *The Metabolic Basis of Inherited Disease* (6th ed.), *pp.* 2805–2842, McGraw–Hill (1989).

Engel, J. and Prokop, D.J., The zipper-like folding of collagen triple helices and the effects of mutations that disrupt the zipper, *Annu. Rev. Biophys. Biophys. Chem.* **20,** 137–152 (1991).

Eyre, D.R., Paz, M.A., and Gallop, P.M., Cross-linking in collagen and elastin, *Annu. Rev. Biochem.* **53,** 717–748 (1984).

Fuchs, E. and Coulombe, P.A., Of mice and men: Genetic skin diseases of keratin, *Cell* **69,** 899–902 (1992).

Jones, E.Y. and Miller, A., Analysis of structural design features in collagen, *J. Mol. Biol.* **218,** 209–219 (1991).

Kaplan, D., Adams, W.W., Farmer, B., and Viney, C., *Silk Polymers,* Am. Chem. Soc. (1994).

Martin, G.R., Timpl, R., Müller, P.K., and Kühn, K., The genetically distinct collagens, *Trends Biochem. Sci.* **10,** 285–287 (1985).

Nimni, M.E. (Ed.), *Collagen,* Vol. I, CRC Press (1988). [Contains articles on collagen biochemistry.]

Prockop, D.J., Mutations in collagen genes as a cause of connective-tissue disease, *New Engl. J. Med.* **326,** 540–546 (1992).

Robert, L. and Hornebeck, W. (Eds.), *Elastin and Elastases,* Vol. 1, CRC Press (1989).

Steinert, P.M. and Parry, D.A.D., Intermediate filaments, *Annu. Rev. Cell Biol.* **1,** 41–65 (1985). [Discusses the structure of a keratin.]

van der Rest, M. and Bruckner, P., Collagens: diversity at the molecular and supramolecular levels, *Curr. Opin. Struct. Biol.* **3,** 430–436 (1993).

Vuorio, E. and de Crombrugghe, B., The family of collagen genes, *Annu. Rev. Biochem.* **59,** 837–872 (1990). [Discusses the various types of collagens.]

Globular Proteins

Chothia, C., Principles that determine the structures of proteins, *Annu. Rev. Biochem.* **53,** 537–572 (1984).

Clore, G.M. and Gronenborn, A.M., Two-, three-, and four-dimensional NMR methods for obtaining larger and more precise three-dimensional structures of proteins in solution, *Annu. Rev. Biophys. Biophys. Chem.* **20,** 29–63 (1991).

Cohen, C. and Parry, D.A.D., α-Helical coiled coils and bundles: How to design an α-helical protein, *Proteins* **7,** 1–15 (1990).

Glusker, J.P., Lewis, M., and Rossi, M., *Crystal Structure Analysis for Chemists and Biologists,* VCH Publishers (1994).

Lesk, A.M., *Protein Architecture. A Practical Approach,* IRL Press (1991).

McRee, D.E., *Practical Protein Cystallography,* Academic Press (1993).

Rees, A.R., Sternberg, M.J.E., and Wetzel, R., *Protein Engineering. A Practical Approach,* IRL Press (1992). [Chapters 1 and 2 contain synopses of protein structure determination by X-ray crystallographic and NMR methods.]

Rhodes, G., *Crystallography Made Clear: A Guide for Users of Macromolecular Models,* Academic Press (1993).

Richards, F.M., Areas, volumes, packing, and protein structure, *Annu. Rev. Biophys. Bioeng.* **6,** 151–176 (1977).

Richardson, J.S. and Richardson, D.C., Principles and patterns of protein conformation, *in* Fasman, G.D. (Ed.), *Prediction of Protein Structure and the Principles of Protein Conformation, pp.* 1–

98, Plenum Press (1989). [A comprehensive account of protein conformations based on X-ray structures.]

Richardson, J.S., The anatomy and taxonomy of protein structures, *Adv. Protein Chem.* **34**, 168–339 (1981). [A detailed discussion of the structural principles governing globular proteins accompanied by an extensive collection of their cartoon representations.]

Thornton, J.M., Protein structures: The end point of the folding pathway, *in* Creighton, T.E. (Ed.), *Protein Folding, pp.* 59–81, Freeman (1992).

Wagner, G., Hyberts, S.G., and Havel, T.F., NMR structure determination in solution. A critique and comparison with X-ray crystallography, *Annu. Rev. Biophys. Biomol. Struct.* **21**, 167–198 (1992).

Wüthrich, K., Protein structure determination in solution by nuclear magnetic resonance spectroscopy, *Science* **243**, 45–50 (1989).

Wyckoff, H.W., Hirs, C.H.W., and Timasheff, S.N. (Eds.), *Diffraction Methods for Biological Macromolecules,* Parts A and B, *Methods Enzymol.* **114, 115** (1985). [A series of articles on the theory and practice of X-ray crystallography.]

Protein Stability

Alber, T., Stabilization energies of protein conformation, *in* Fasman, G.D. (Ed.), *Prediction of Protein Structure and the Principles of Protein Conformation, pp.* 161–192, Plenum Press (1989).

Burley, S.K. and Petsko, G.A., Weakly polar interactions in proteins, *Adv. Protein Chem.* **39**, 125–189 (1988).

Creighton, T.E., Stability of folded proteins, *Curr. Opin. Struct. Biol.* **1**, 5–16 (1991).

Edsall, J.T. and McKenzie, H.A., Water and proteins, *Adv. Biophys.* **16**, 51–183 (1983).

Eigenbrot, C. and Kossiakoff, A.A., Structural consequences of mutation, *Curr. Opin. Biotech.* **3**, 333–337 (1992).

Fersht, A.R. and Serrano, L., Principles of protein stability derived from protein engineering experiments, *Curr. Opin. Struct. Biol.* **3**, 75–83 (1993). [Discusses how the roles of specific side chains in proteins can be quantitatively determined by mutationally changing them and calorimetrically measuring the stabilities of the resulting proteins.]

Harvey, S.C., Treatment of electrostatic effects in macromolecular modeling, *Proteins* **5**, 78–92 (1989).

Jeffrey, G.A. and Saenger, W., *Hydrogen Bonding in Biological Structures,* Springer–Verlag (1991).

Kauzmann, W., Some factors in the interpretation of protein denaturation, *Adv. Protein Chem.* **14**, 1–63 (1958). [A classic review that first pointed out the importance of hydrophobic bonding in stabilizing proteins.]

Matthews, B.W., Structural and genetic analysis of protein stability, *Annu. Rev. Biochem.* **62**, 139–160 (1993).

Ramachandran, G.N. and Sasisekharan, V., Conformation of polypeptides and proteins, *Adv. Protein Chem.* **23**, 283–437 (1968).

Richards, F.M., Folded and unfolded proteins: An introduction, *in* Creighton, T.E. (Ed.), *Protein Folding, pp.* 1–58, Freeman (1992).

Saenger, W., Structure and dynamics of water surrounding biomolecules, *Annu. Rev. Biophys. Biophys. Chem.* **16**, 93–114 (1987).

Schellman, J.A., The thermodynamic stability of proteins, *Annu. Rev. Biophys. Biophys. Chem.* **16**, 115–137 (1987).

Stickle, D.F., Presta, L.G., Dill, K.A., and Rose, G.D., Hydrogen bonding in globular proteins, *J. Mol. Biol.* **226**, 1143–1159 (1992).

Tanford, C., *The Hydrophobic Effect: Formation of Micelles and Biological Membranes* (2nd ed.), Chapters. 2–4 and 13, Wiley (1980).

Teeter, M.M., Water-protein interactions: Theory and experiment, *Annu. Rev. Biophys. Biophys. Chem.* **20**, 577–600 (1991).

Yang, A.-S. and Honig, B., Electrostatic effects on protein stability, *Curr. Opin. Struct. Biol.* **2**, 40–45 (1992).

Quaternary Structure

Eisenstein, E. and Schachman, H.K., Determining the roles of subunits in protein function, *in* Creighton, T.E. (Ed.), *Protein Function. A Practical Approach, pp.* 135–176, IRL Press (1989).

Klotz, I.M., Darnell, D.W., and Langerman, N.R., Quaternary structure of proteins, *in* Neurath, H. and Hill, R.L. (Eds.), *The Proteins* (3rd ed.), Vol. 1, *pp.* 226–411, Academic Press (1975).

Matthews, B.W. and Bernhard, S.A., Structure and symmetry in oligomeric proteins, *Annu. Rev. Biophys. Bioeng.* **6**, 257–317 (1973).

PROBLEMS

1. What is the length of an α helical section of a polypeptide chain of 20 residues? What is its length when it is fully extended (all trans)?

*2. From an examination of Figs. 7-7 and 7-8, it is apparent that the polypeptide conformation angle ϕ is more constrained than is ψ. By referring to Fig. 7-4, or better yet, by examining a molecular model, indicate the sources of the steric interference that limit the allowed values of ϕ when $\psi = 180°$.

3. For a polypeptide chain made of γ-amino acids, state the nomenclature of the helix analogous to the 3_{10} helix of α-amino acids. Assume the helix has a pitch of 9.9 Å and a rise per residue of 3.2 Å.

*4. Table 7-6 gives the torsion angles, ϕ and ψ, of hen egg white lysozyme for residues 24–73 of this 129-residue protein. (a) What is the secondary structure of residues 26–35? (b) What is the secondary structure of residues 42–53? (c) What is the

TABLE 7-6. TORSION ANGLES (ϕ, ψ) FOR RESIDUES 24 TO 73 OF HEN EGG WHITE LYSOZYME

Residue Number	Amino Acid	ϕ (deg)	ψ (deg)	Residue Number	Amino Acid	ϕ (deg)	ψ (deg)
24	Ser	−60	147	49	Gly	95	−75
25	Leu	−49	−32	50	Ser	−18	138
26	Gly	−67	−34	51	Thr	−131	157
27	Asn	−58	−49	52	Asp	−115	130
28	Trp	−66	−32	53	Tyr	−126	146
29	Val	−82	−36	54	xxx	67	−179
30	Cys	−69	−44	55	Ile	−42	−37
31	Ala	−61	−44	56	Leu	−107	14
32	Ala	−72	−29	57	Gln	35	54
33	Lys	−66	−65	58	Ile	−72	133
34	Phe	−67	−23	59	Asn	−76	153
35	Glu	−81	−51	60	Ser	−93	−3
36	Ser	−126	−8	61	Arg	−83	−19
37	Asn	68	27	62	Trp	−133	−37
38	Phe	79	6	63	Trp	−91	−32
39	Asn	−100	109	64	Cys	−151	143
40	Thr	−70	−18	65	Asn	−85	140
41	Glu	−84	−36	66	Asp	133	8
42	Ala	−30	142	67	Gly	73	−8
43	Thr	−142	150	68	Arg	−135	17
44	Asn	−154	121	69	Thr	−122	83
45	Arg	−91	136	70	Pro	−39	−43
46	Asn	−110	174	71	Gly	−61	−11
47	Thr	−66	−20	72	Ser	−45	122
48	Asp	−96	36	73	Arg	−124	146

Source: Imoto, T., Johnson, L.N., North, A.C.T., Phillips, D.C., and Rupley, J.A., *in* Boyer, P.D. (Ed.), *The Enzymes* (3rd ed.), Vol. 7, *pp.* 693–695, Academic Press (1972).

probable identity of residue 54? (d) What is the secondary structure of residues 69–71? (e) What additional information besides the torsion angles, ϕ and ψ, of each of its residues are required to define the three-dimensional structure of a protein?

5. Hair splits most easily along its fiber axis, whereas fingernails tend to split across the finger rather than along it. What are the directions of the keratin fibrils in hair and in fingernails? Explain your reasoning.

6. What structural features are responsible for the observations that α keratin fibers can stretch to over twice their normal length, whereas silk is nearly inextensible?

7. What is the growth rate, in turns per second, of the α helices in a hair that is growing 15 cm·year^{-1}?

8. Can polyproline form a collagenlike triple helix? Explain.

9. As Mother Nature's chief engineer, you have been asked to design a five-turn α helix that is destined to have half its circumference immersed in the interior of a protein. Indicate the helical wheel projection of your prototype α helix and its amino acid sequence (see Fig. 7-45).

10. β-Aminopropionitrile is effective in reducing excessive scar tissue formation after an injury (although its use is contraindicated by side effects). What is the mechanism of action of this lathyrogen?

11. Proteins have been classified as α, β, α/β, or $\alpha + \beta$ proteins depending on whether their tertiary structures, respectively, consist of mostly α helices, mostly β sheets, alternating α helices and β sheets, or some α helices and β sheets that tend to aggregate together rather than alternate along the polypeptide chain. By inspection, classify the following proteins according to this nomenclature and, where possible, identify their supersecondary structures: carboxypeptidase A (Fig. 7-19a), triose phosphate isomerase (Fig. 7-19b), myoglobin (Fig. 7-42), concanavalin A (Fig. 7-43), carbonic anhydrase (Fig. 7-44), glyceraldehyde-3-phosphate dehydrogenase (Fig. 7-47), and prealbumin (Fig. 7-59).

12. The coat protein of tomato bushy stunt virus consists of 180 chemically identical subunits, each of which is composed of ~386 amino acid residues. The probability that a wrong amino acid residue will be biosynthetically incorporated in a polypeptide chain is 1 part in 3000 per residue. Calculate the average number of coat protein subunits that would have to be synthesized in order to produce a perfect viral coat. What would this number be if the viral coat were a single polypeptide chain with the same number of residues that it actually has?

13. State the rotational symmetry of the following objects: (a) a starfish, (b) a square pyramid, (c) a rectangular box, and (d) a trigonal bipyramid.

disulfide bonds have formed is proinsulin converted to the two-chained active hormone by the specific proteolytic excision of an internal 33-residue segment known as its C chain. Nevertheless, two sets of observations suggest that the C chain does not direct the folding of the A and B chains but, rather, simply holds them together while they form their native disulfide bonds: (1) Under proper renaturing conditions, native insulin is obtained from scrambled insulin in 25 to 30% yield, which increases to 75% when the A and B chains are chemically cross-linked; and (2) sequence comparisons of proinsulins from a variety of species indicate that mutations are accepted into the C chain at a rate which is eight times that for the A and B chains.

Protein Folding Is Directed Mainly by Internal Residues

Numerous protein modification studies have been aimed at determining the role of various classes of amino acid residues in protein folding. In one particularly revealing study, the free primary amino groups of RNase A (Lys residues and the N-terminus) were derivatized with 8-residue chains of poly-DL-alanine. Intriguingly, these large, water-soluble poly-Ala chains could be simultaneously coupled to RNase's 11 free amino groups without significantly altering the protein's native conformation or its ability to refold. Since these free amino groups are all located on the exterior of RNase A, this observation suggests that *it is largely a protein's internal residues that direct its folding to the native conformation.* Similar conclusions have been reached from studies of protein structure and evolution (Section 8-3): Mutations that change surface residues are accepted more frequently and are less likely to affect protein conformations than are changes of internal residues. It is therefore not surprising that the perturbation of protein folding by limited concentrations of denaturing agents indicates that *protein folding is driven by hydrophobic forces.*

Globular proteins have packing densities comparable to those of organic crystals (Section 7-3B) because the side chains in a protein's interior fit together with exquisite complementarity. To ascertain whether this phenomenon is an important determinant of protein structure, Eaton Lattman and George Rose analyzed 67 globular proteins of known structure for the existence of preferred interactions between side chains. They found none, thereby indicating that, at least in globular proteins, *the native fold determines the packing but packing does not determine the native fold.* This view is corroborated by the widespread occurrence of protein families whose members assume the same fold even though they may be so distantly related as to have little sequence similarity (e.g., Section 8-3A). Evidently, *there are a large number of ways in which a protein's internal residues can pack together efficiently.*

Helices and Sheets May Predominate in Proteins Simply Because They Fill Space Efficiently

Why do proteins contain such a high proportion (~60%, on average) of α helices and β sheets? Hydrophobic interactions, although the dominant influence responsible for the compact nonpolar cores of proteins, lack the specificity to restrict polypeptides to particular conformations. Similarly, the observation that polypeptide segments in the coil conformation are no less hydrogen bonded than helices and sheets suggests that the conformations available to polypeptides are not greatly limited by their hydrogen bonding requirements. Rather, as Ken Dill has shown, it appears that helices and sheets form largely as a consequence of steric constraints in compact polymers. Exhaustive simulations of the conformations which simple flexible chains (such as a string of pearls) can assume indicate that the proportion of helices and sheets increases dramatically with a chain's level of compaction (number of intrachain contacts); that is, helices and sheets are particularly compact entities. Thus, most ways to compact a chain involve the formation of helices and sheets. In native proteins, such elements of secondary structure are fine tuned to form α helices and β sheets by short-range forces such as hydrogen bonding, ion pairing, and van der Waals interactions. It is probably these less dominant but more specific forces that "select" the unique native structure of a protein from among its relatively small number of hydrophobically generated compact conformations (recall that most hydrogen bonds in proteins link residues that are close together in sequence; Section 7-3B).

B. Folding Pathways

How does a protein fold to its native conformation? One might guess that this process occurs through the protein's random exploration of all the conformations available to it until it eventually "stumbles" onto the correct one. A "back-of-the-envelope" calculation first made by Cyrus Levinthal, however, convincingly demonstrates that this cannot possibly be the case: Assume that the $2n$ torsional angles, ϕ and ψ, of an n-residue protein each have three stable conformations. This yields $3^{2n} \approx 10^n$ possible conformations for the protein, which is a gross underestimate, if only because the side chains are ignored. If a protein can explore new conformations at the rate that single bonds can reorient, it can find $\sim 10^{13}$ conformations per second, which is, no doubt, an overestimate. We can then calculate the time, t, in seconds, required for a protein to explore all the conformations available to it:

$$t = \frac{10^n}{10^{13}} \qquad [8.1]$$

For a rather small protein of $n = 100$ residues, $t = 10^{87}$ s, which is immensely more than the apparent age of the universe (20 billion years $= 6 \times 10^{17}$ s).

It would obviously take even the smallest protein an absurdly long time to explore all its possible conformations. Yet, many proteins fold to their native conformation in less than a few seconds. They therefore must fold by some sort of ordered set of pathways in which the approach to the native state is accompanied by sharply increasing conformational stability (decreasing free energy). An analogy to

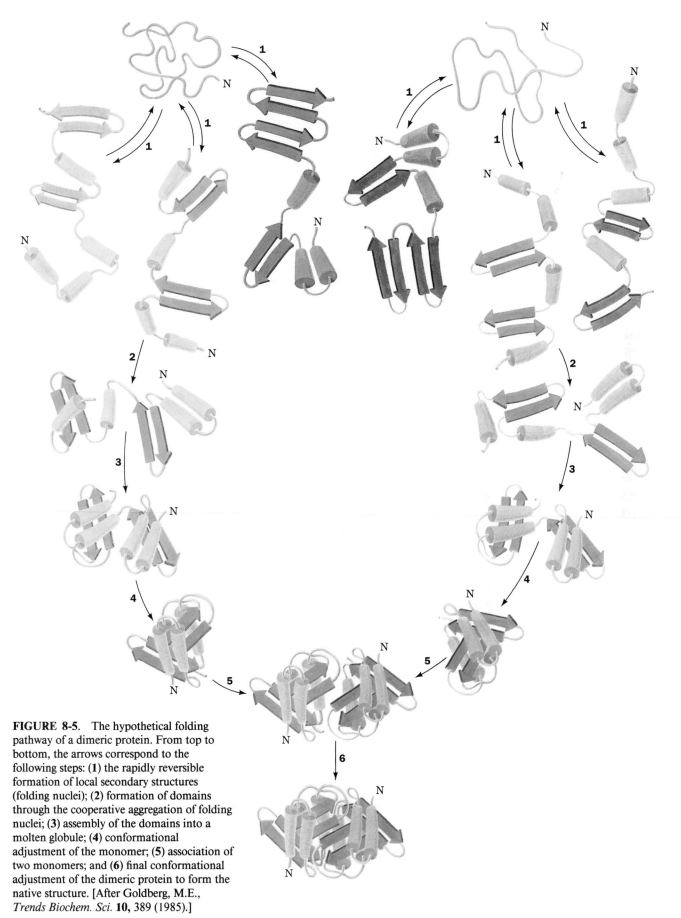

FIGURE 8-5. The hypothetical folding pathway of a dimeric protein. From top to bottom, the arrows correspond to the following steps: (**1**) the rapidly reversible formation of local secondary structures (folding nuclei); (**2**) formation of domains through the cooperative aggregation of folding nuclei; (**3**) assembly of the domains into a molten globule; (**4**) conformational adjustment of the monomer; (**5**) association of two monomers; and (**6**) final conformational adjustment of the dimeric protein to form the native structure. [After Goldberg, M.E., *Trends Biochem. Sci.* **10**, 389 (1985).]

this situation is a ball rolling down a hill to a valley: If the topography of the hill (conformational free energy map) is rather flat and featureless, the ball (polypeptide chain) could take many different, mostly indirect pathways (conformational sequences) in rolling to the valley (folding to its native state). If, however, the hill (conformational free energy map) slopes sharply into a canyon that leads to the valley (native state), then the ball (polypeptide chain) would follow a more or less direct pathway in reaching it.

Protein folding is thought to occur via a multistage process (Fig. 8-5):

1. The folding of a random coil polypeptide begins with the random formation of short stretches of 2° structure, such as α helices and β turns, that can act as **nuclei** (scaffolding) for the stabilization of additional ordered regions of the protein. These nuclei are probably small enough (8–15 residues) to "flicker" in and out of existence in under a millisecond.

2. Nuclei with the proper nativelike structure probably grow by the diffusion, random collision, and adhesion of two or more such nuclei. The stabilities of these ordered regions increase with size so that once they have randomly reached a certain minimum size, they spontaneously grow in a cooperative fashion until they form a nativelike domain. The existence of such a hierarchical folding process, in which folding units condense to form larger folding units, etc., is corroborated by the observation that domains are composed of subdomains which, in turn, may consist of sub-subdomains, etc. Perhaps the usual upper limit of ~200 residues in a domain reflects the special requirement that it must rapidly fold to its native conformation in an ordered sequence.

3. In multidomain proteins, the domains come together to form a so-called **molten globule**, an entity that has extensive 2° structure but disordered 3° structure such that its hydrophobic side chains remain at least partially exposed to solvent.

4. Through a series of relatively small conformational adjustments the molten globule rearranges to a more compact 3° structure, the native conformation of a single-subunit protein.

5. In a multisubunit protein, the requisite number of subunits assemble in a nativelike 4° structure.

6. Finally, a further series of slight conformational adjustments yields the native protein structure.

BPTI Folds to Its Native Conformation via an Ordered Pathway

The most convincing experimental evidence favoring the above protein folding scheme comes from Thomas Creighton's renaturation studies of **bovine pancreatic trypsin inhibitor (BPTI)**. This 58-residue monomeric protein has three disulfide bonds (Fig. 8-6); it binds to and inactivates trypsin in the pancreas, thereby protecting that secre-

tory organ from self-digestion (Section 14-3). Creighton monitored the folding of BPTI by identifying the order of disulfide bond formation as reduced and unfolded BPTI was oxidized and renatured. *Each distinguishable disulfide-bonded species represents a subset of the conformations that the BPTI polypeptide can assume, so that by following the time course of the appearance of these various species, the approximate conformational path taken by the renaturing protein was deduced.*

The investigation began with fully reduced BPTI, which is completely unfolded (random coil) under physiological conditions. Disulfide bond formation and subsequent renaturation was induced by reacting the protein with oxidized dithiothreitol (Section 6-1B). As the protein renatured, the intermediates formed were irreversibly trapped by blocking their remaining free Cys-SH groups with iodoacetate (Section 6-1B), thereby adding a negative charge to the protein for each alkylated Cys-SH group. These stabilized intermediates were then separated by ion exchange chromatography and the positions of their disulfide bonds determined by diagonal electrophoresis (Section 6-1I).

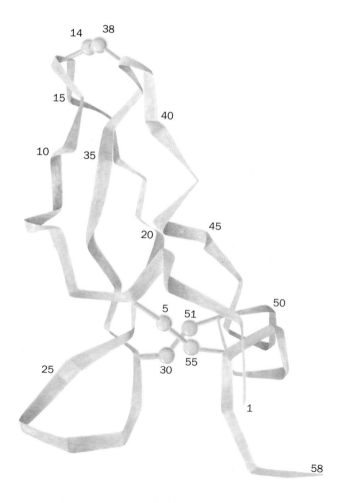

FIGURE 8-6. The polypeptide backbone and disulfide bonds of native BPTI. [After a drawing by Michael Levitt, *in* Creighton, T.E., *J. Mol. Biol.* **95**, 168 (1975).]

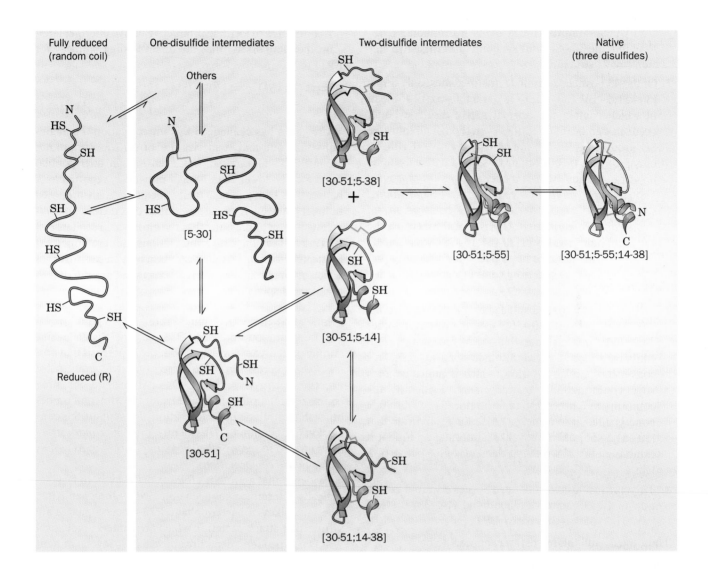

FIGURE 8-7. The proposed renaturation pathway of BPTI showing the conformations of its polypeptide backbone as deduced from disulfide trapping experiments and NMR measurements (note that these views of the protein differ from that in Fig. 8-6 by a slight rotation about the vertical axis). The sequence numbers of the Cys residues involved in each disulfide bond are given in brackets below the diagram representing each folding intermediate. The two one-disulfide intermediates, [5–30] and [30–51], are in rapid equilibrium. The "+" between intermediates [30–51;5–38] and [30–51;5–14] indicates that both are formed directly from the one-disulfide intermediates, that both convert directly to [30–51;5–55], and that either or both are intermediates in the rearrangement of [30–51;14–38] to [30–51;5–55]. [After Creighton, T. E., *Biochem. J.* **270,** 12 (1990).]

These experiments indicate that *BPTI follows a limited but indirect set of pathways in folding to its native structure* (Fig. 8-7):

1. The six Cys residues of the fully reduced BPTI (R), are equally likely to participate in forming the initial disulfide bond. Yet, after the molecule has equilibrated through a series of rapid internal disulfide interchange reactions, only 2 of the 15 possible one-disulfide intermediates, [5–30] (where the numbers indicate the Cys residues participating in a disulfide bond) and [30–51], exist in significant quantities. (An intermediate's relative abundance at equilibrium is indicative of its thermodynamic stability relative to other intermediates; Section 3-4A.) Of these, only [30–51] has a disulfide bond that occurs in the native protein and only this species reacts in significant amounts with sulfhydryl reagents to form a second disulfide bond. A variety of conformational studies indicate that [30–51] is a conformationally fluctuating molecule that, much of the

time, forms the native protein's β sheet and C-terminal α helix, the two major two elements of BPTI's 2° structure, which comprise much of its hydrophobic core. Note that the 30–51 disulfide bond links these two elements of 2° structure.

2. Of the 45 possible 2-disulfide intermediates, only three, [30–51;5–14], [30–51;5–38], and [30–51;14–38], all of which contain the original native disulfide bond, occur in significant quantities even though all four free SH groups of [30–51] are equally reactive. Of these, only [30–51;14–38] contains a second native disulfide bond but, curiously, *it is a deadend species that must first convert to either of the non-native disulfide-linked species, [30–51;5–14] or [30–51;5–38], in order to fold to the native conformation.* This is probably because the conformational rigidity of [30–51;14–38], which NMR measurements indicate has a nativelike structure, prevents Cys 5 and Cys 55 from coming together to form a disulfide bond.

3. Before any additional disulfide bond formation occurs, [30–51;5–14] and [30–51;5–38] convert to another 2-disulfide intermediate, [30–51;5–55], which also contains a second native disulfide bond (5–55). This conversion is relatively slow, which is indicative of a large conformational rearrangement. Evidently, the conformation necessary to form the 5–55 disulfide bond is difficult to achieve. NMR studies indicate that [30–51;5–55] has a nativelike conformation in that it exhibits all the 2° structural elements of the native protein.

4. This conclusion is corroborated by the very rapid formation of BPTI's third disulfide bond (14–38) to yield [30–51;5–55;14–38], native BPTI (N).

Thus, the overall folding pathway appears to be:

$$R \rightleftharpoons [30\text{–}51] \rightleftharpoons [30\text{–}51;5\text{–}14] \text{ or } [30\text{–}51;5\text{–}38]$$
$$\rightleftharpoons [30\text{–}51;5\text{–}55] \rightleftharpoons N$$

If, in fact, proteins generally achieve their native conformations via such indirect folding pathways, this would vastly complicate efforts to understand the folding process. However, Peter Kim has reinvestigated the folding pathway of BPTI by trapping the intermediates in acid, which inhibits the formation of the thiolate ion required in the disulfide interchange reaction (Fig. 8-3). He isolated and characterized these acid-stabilized folding intermediates by reverse-phase HPLC techniques. His data revealed that all well-populated folding intermediates have only native disulfide bonds (i.e., non-native species such as [30–51;5–14] and [30–51;5–38] were detected in only small amounts). Kim therefore concluded that the above-described slow conformational rearrangement must be accompanied by one native disulfide bond being replaced by another, that is, [30–51;14–38] $\rightleftharpoons$ [30–51;5–55]. This implies that non-native species, such as [30–51;5–38], which must transiently

occur via the intramolecular disulfide interchange reactions through which native intermediates interconvert, are not committed to forming productive native intermediates. Kim therefore asserts that the direct folding pathway for BPTI is:

$$R \rightleftharpoons [30\text{–}51] \rightleftharpoons [30\text{–}51;14\text{–}38] \rightleftharpoons$$
$$[30\text{–}51;5\text{–}55] \rightleftharpoons N$$

Creighton has challenged this interpretation: He maintains that non-native species are of kinetic importance along BPTI's folding pathway even though their populations may be lower than he had previously reported and, hence, that Kim's data actually corroborate his proposed indirect pathway. Kim has responded that there is no compelling reason to presume that the amino acid sequence of BPTI specifies non-native interactions that are important for protein folding. It is the resolution of such controversies which often sparks significant advances in our understanding of the disputed phenomena.

Primary Structures Determine Protein Folding Pathways

Renaturation studies of BPTI and other disulfide-containing proteins are consistent with the hypothesis that *protein folding occurs through a largely ordered, although not necessarily direct, pathway.* Observations on the time course of protein folding made using experimental probes other than disulfide bond formation also support this conclusion (see Section 8-2). Evidently, *protein primary structures evolved to specify efficient folding pathways as well as stable native conformations.*

Evidence that a protein's folding pathway is, in fact, genetically determined has been obtained by Jonathan King. The tail spikes of **bacteriophage P22** (bacteriophages are viruses that attack bacteria), which are trimers of identical 76-kD polypeptides, denature above 80°C. Several mutant varieties of this protein fail to renature at 39°C. Yet, at 30°C, the mutant proteins fold to structures whose properties, including their T_m values, are indistinguishable from that of the nonmutant tail spike protein. The amino acid changes causing these **temperature-sensitive mutations** apparently act to destabilize intermediate states in the folding process but do not affect the native protein's stability. This observation suggests that *a protein's amino acid sequence dictates its native structure by specifying the series of steps comprising its folding pathway.*

C. The Roles of Folding Accessory Proteins

Most unfolded proteins renature *in vitro* (in the laboratory) over periods ranging from minutes to days and, quite often, with low efficiency, that is, with a large fraction of the polypeptide chains assuming quasi-stable non-native conformations and/or forming nonspecific aggregates. *In vivo* (in the cell), however, polypeptides efficiently fold to their native conformations as they are being synthesized, a process that normally requires only a few minutes. This is because all cells contain three types of accessory proteins that function to assist polypeptides in folding to their native conformations and in assembling to their 4° structures: protein disulfide isomerases, **peptidyl prolyl cis–trans isomerases,** and **molecular chaperones.** We discuss these essential proteins below.

Protein Disulfide Isomerase Facilitates Disulfide Interchange Reactions

   Protein disulfide isomerase (PDI), which we encountered in Section 8-1A, is a homodimeric eukaryotic enzyme of 486-residue subunits. It catalyzes disulfide interchange reactions, thereby facilitating the shuffling of the disulfide bonds in proteins until they achieve their native pairing. Curiously, PDI is also the β subunit of the $\alpha_2\beta_2$ heterotetramer prolyl hydroxylase, the enzyme that hydroxylates the Pro residues of collagen (Section 7-2C). The significance of this observation is unknown.

   Although the X-ray structure of PDI is unknown, that of the oxidized form of its *E. coli* counterpart, **DsbA protein** (the product of the *dsbA* gene), has been determined by John Kuriyan and Jennifer Martin (Fig. 8-8). This 189-residue monomeric protein is folded into two domains, one of which structurally resembles the ubiquitous disulfide-containing redox protein **thioredoxin** (Section 26-4A). This domain contains the protein's only two Cys residues which, in the protein's oxidized state, form a disulfide bond. Although disulfide bonds in native proteins are usually buried and frequently occur in hydrophobic environments, one of

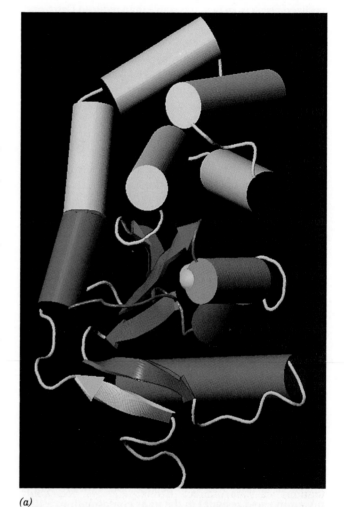

(a)

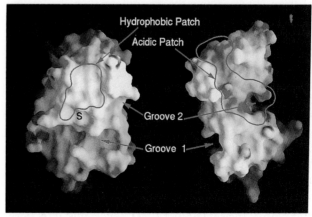

(b)

FIGURE 8-8. The X-ray structure of *E. coli* DsbA protein, a protein disulfide isomerase. (*a*) A ribbon diagram in which those parts of the protein that structurally resemble thioredoxin are purple or green and the other parts are white. The green helix contains the protein's disulfide-linked sulfur atoms (*yellow spheres*) at its N-terminus. (*b*) The molecular surface and charge distribution. The left view is oriented similarly to that in (*a*) and the right view is related to that on the left by a 90° rotation about the vertical axis. The surface is colored according to charge: Negatively charged groups (Asp and Glu) are red, positively charged groups (Arg and Lys) are blue, and uncharged groups are white. The one surface-exposed S atom in DsbA's disulfide bond is denoted by "S." Structural features that may be important for polypeptide binding are indicated. Part of Groove 1 near the disulfide group is lined with hydrophobic residues. [Courtesy of John Kuriyan, The Rockefeller University, New York, New York, and Jennifer Martin, University of Queensland, Australia.]

DsbA's S atoms is exposed on the protein surface, where it is flanked on one side by a long and deep groove lined with hydrophobic residues and on the opposite side by a surface-exposed patch of hydrophobic residues (Fig. 8-8*b*). Thus, DsbA's active site disulfide bond is centered in a region that appears capable of binding unfolded polypeptide segments. Moreover, even though disulfide bonds almost always stabilize proteins (Section 7-4D) and are usually unreactive, oxidized DsbA is, in fact, less stable than reduced DsbA and has a highly reactive, that is, strongly oxidizing, disulfide bond. It has therefore been proposed that DsbA facilitates proper disulfide bond formation by binding partially folded polypeptide chains before they have formed any disulfide bonds. Since sequence alignments indicate that PDI contains two thioredoxin-like domains, it is likely that PDI functions similarly to DsbA.

Some proteins have, in essence, a built-in PDI activity. For example, BPTI is synthesized with an N-terminal, 13-residue, Cys-containing segment, a so-called **propeptide,** that is eventually enzymatically excised to form the mature protein. **Pro-BPTI** (BPTI with its attached propeptide) folds, *in vitro,* to its native conformation with a rate and yield substantially greater than that of BPTI itself. Replacing the propeptide's single Cys residue with Ala abolishes this effect. Moreover, a single Cys residue tethered to the C-terminus of BPTI with a flexible linker of repeating Ser-Gly-Gly residues exhibits essentially the same effects on BPTI folding. Evidently, the propeptide facilitates folding by providing a tethered, solvent-accessible, thiol-disulfide reagent in a manner similar to that of PDI (which, *in vitro,* may work synergistically with Pro-BPTI's propeptide to further increase the rate and efficiency of BPTI folding).

Peptidyl Prolyl Cis–Trans Isomerases Facilitate the Formation of the Cis Peptide Bonds Preceding Pro Residues

Although polypeptides are probably biosynthesized with almost all of their Xaa–Pro peptide bonds (where Xaa is any amino acid residue) in the trans conformation, ~10% of these bonds assume the cis conformation in globular proteins because, as we have seen in Section 7-1A, the energy difference between their cis and trans conformations is relatively small. **Peptidyl prolyl cis–trans isomerases (PPIs;** alternatively known as **rotamases)** catalyze the otherwise slow interconversion of Xaa–Pro peptide bonds between their cis and trans conformations, thereby accelerating the folding of Pro-containing polypeptides. Two structurally unrelated families of PPIs, collectively named the **immunophilins,** have been characterized: The **cyclophilins** (so named because they are inhibited by the immunosuppressive drug **cyclosporin A,** a fungally produced 11-residue cyclic peptide) and the family for which **FK506 binding protein (FKBP)** is prototypic (**FK506** is a fungally produced macrocyclic lactone that is also an immunosup-

pressive drug; medicinal chemists tend to identify the often huge numbers of related drug candidates they deal with by serial numbers rather than by trivial names).

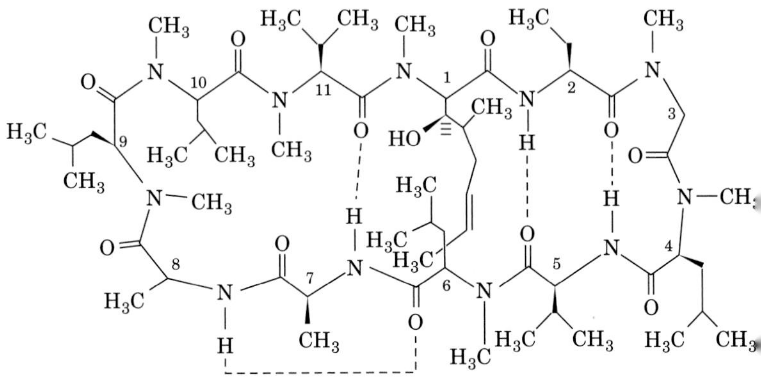

Cyclosporin A

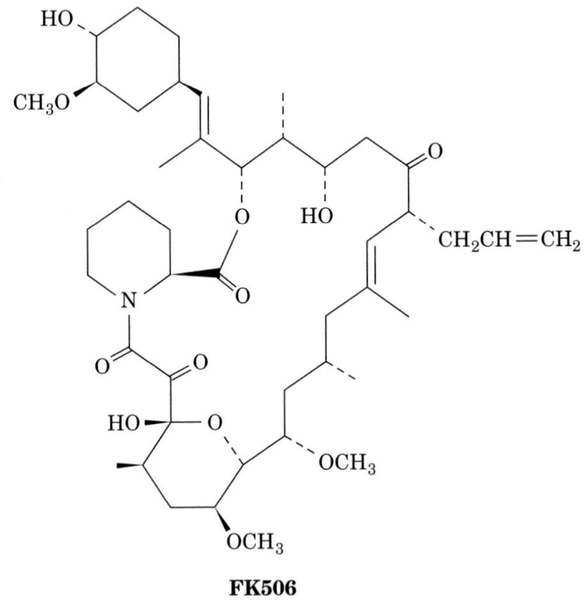

FK506

The X-ray and NMR structures of the cyclophilin–cyclosporin A and FKBP–FK506 complexes suggest that these proteins promote the cis–trans isomerization of an Xaa–Pro peptide bond by preferentially binding it in a twisted (nonplanar) conformation (such a mode of catalysis, known as transition state binding, is discussed in Section 14-1F).

Cyclosporin A and FK506 are highly effective agents for the treatment of autoimmune disorders and for preventing organ-transplant rejection. Indeed, until the advent of cyclosporin A in the early 1980s, the long-term survival of a transplanted organ was a rare occurrence. The more recently discovered FK506 is an even more potent immunosuppressant.

The immunosuppressive properties of both cyclosporin A and FK506 stem from the abilities of their respective complexes with cyclophillin and FKBP to prevent the expression of genes involved in the activation of *T* lymphocytes (the immune system cells responsible for **cellular immunity;** the immune response is discussed in Section 34-2) by interfering with these cells' intracellular signaling pathways. Enigmatically, there is no obvious relationship between the immunophilins' immunosuppressive properties and rotamase activities: Both cyclosporin A and FK506 are effective immunosuppressants at concentrations far below those of the cyclophilin and FKBP in cells; and mutational changes that destroy cyclophilin's rotamase activity do not eliminate its ability to bind cyclosporin A or for the resulting complex to interfere with *T* lymphcyte signaling. Perhaps the cyclophilin–cyclosporin A and FKBP–FK506 complexes mimic the effects of naturally occurring substances that function to modulate the immune response.

Molecular Chaperones Prevent the Improper Folding and Aggregation of Proteins

Unfolded proteins contain numerous solvent-exposed hydrophobic regions and therefore have a great tendency to form both intramolecular and intermolecular aggregates. **Molecular chaperones** are proteins that function to prevent or reverse such improper associations, particularly in multidomain and multisubunit proteins. They do so by binding to an unfolded or aggregated polypeptide's solvent-exposed hydrophobic surfaces and subsequently releasing them, often repeatedly, in a manner that facilitates their proper folding and/or 4° assembly. This idea is consistent with the X-ray structure, determined by Scott Hultgren, of the *E. coli* chaperone protein **PapD** in complex with a conserved 19-residue segment of the pilus proteins whose folding PapD facilitates (Fig. 8-9). PapD exhibits a wide crevice containing several conserved, solvent-exposed hydrophobic residues that evidently functions to specifically bind PapD's target polypeptide segments.

The mechanisms by which molecular chaperones carry out their functions are poorly understood. However, many of them are **ATPases** (enzymes that catalyze ATP hydrolysis), which bind to unfolded polypeptides and apparently apply the free energy of ATP hydrolysis to effect their release in a favorable manner. Whatever the case, it appears, as John Ellis has pointed out, that molecular chaperones function analogously to their human counterparts: *They inhibit inappropriate interactions between potentially complementary surfaces and disrupt unsuitable liasons so as to facilitate more favorable associations.*

The molecular chaperones comprise several unrelated classes of proteins including:

1. The **heat shock proteins 70 (Hsp70),** so named because the rate of synthesis of these ~70-kD monomers greatly

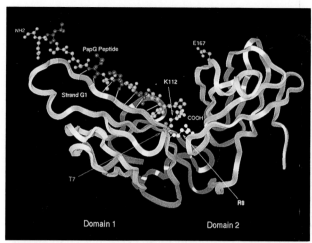

FIGURE 8-9. The X-ray structure of the *E. coli* PapD protein, a molecular chaperone that functions to facilitate the proper assembly of pili (Section 1-1A) but that is not a component of them. It is shown here in its complex with a 19-residue polypeptide that corresponds to the conserved C-terminal segment of pilus subunits. PapD protein consists of two structurally similar domains oriented so as to give the protein the shape of a boomerang. The peptide is represented by a ball-and-stick model whose conserved residues are highlighted in red but which is otherwise light blue. It binds to the G1 β strand of PapD in an extended conformation, thereby augmenting the protein's β sheet by one strand (peptide–protein hydrogen bonds are represented by white lines). The peptide's terminal carboxylate group (*yellow*) is anchored in PapD's interdomain cleft via hydrogen bonds to the protein's invariant Arg 8 (R8, *white*) and Lys 112 (K112; *orange*) residues, whose mutation abolishes PapD's ability to bind pilus subunits. The conserved hydrophobic residues of the G1 β strand are highlighted in green on the PapD ribbon (*violet*). This crystal structure is thought to provide a "snapshot" of a PapD–pilus subunit interaction. [Courtesy of Scott Hultgren, Washington University, St. Louis, Missouri.]

increases at elevated temperatures, are highly conserved proteins in both prokaryotes and eukaryotes. They function, in part, to reverse the denaturation and aggregation of proteins, processes that are accelerated at elevated temperatures. For example, certain Hsp70 proteins bind to nascent (not yet fully synthesized) polypeptide chains as they emerge from the ribosome.

2. The **chaperonins,** large, multisubunit, cagelike proteins that are universal components of bacteria, mitochondria, chloroplasts, and possibly eukaryotes. Two families of chaperonins are known: (1) The **Hsp60** proteins (**GroEL** in *E. coli* and **Cpn60** in chloroplasts), which consist of 14 identical ~60-kD subunits arranged to form two apposed rings of 7 subunits, each surrounding a central cavity large enough to enclose a 90-kD globular

protein (Fig. 8-10); and (2) the **Hsp10 proteins (GroES** in *E. coli* and **Cpn10** in chloroplasts), which form single heptameric rings of identical 10-kD proteins. Ulrich Hartl has demonstrated that GroEL and GroES act in concert in an ATP-driven process (Fig. 8-11) to enclose unfolded proteins in a protected environment that prevents their nonspecific aggregation while they spontaneously fold to their native conformations. This mechanism is corroborated by the observation that chaperonins do not increase the rate of protein folding but, rather, increase the yield of correctly folded product.

3. The **nucleoplasmins,** acidic nuclear proteins whose presence is required for the proper *in vivo* assembly of **nucleosomes** (particles in which eukaryotic DNA is packaged) from their component DNA and histones (Section 33-1B).

The Concept of Self-Assembly Must Take Accessory Proteins Into Account

Many proteins can fold/assemble to their native conformations in the absence of accessory proteins, albeit with low efficiency. Moreover, accessory proteins are not components of the native proteins whose folding/assembly they facilitate. Hence, accessory proteins must mediate the proper folding/assembly of a polypeptide to a conformation/complex governed solely by the polypeptide's amino acid sequence. Nevertheless, the concept that proteins are self-assembling entities must be modified to incorporate the effects of accessory proteins.

D. Prediction of Protein Structures

Since the primary structure of a protein specifies its three-dimensional structure, it should be possible, at least in principle, to predict the native structure of a protein from a knowledge of only its amino acid sequence. This might be done using theoretical methods based on physicochemical principles, or by empirical methods in which predictive schemes are distilled from the analyses of known protein structures. Theoretical methods, which usually attempt to determine the minimum energy conformation of a protein, are mathematically quite sophisticated and require extensive computations. The enormous difficulty in making such calculations sufficiently accurate and yet computationally tractable has, so far, limited their success. Nevertheless, an understanding of how and why proteins fold to their native structures must ultimately be based on such theoretical methods.

Empirical methods are usually much easier to apply than are theoretical methods and have had remarkable success in secondary structure prediction. Clearly, certain amino acid sequences limit the conformations available to a polypeptide chain in an easily understood manner. For example, a Pro residue cannot fit into the interior portions of a regular α helix or β sheet because its pyrrolidine ring would fill the

FIGURE 8-10. An electron micrograph–derived 3D image of the **Hsp60** chaperonin from the photosynthetic bacterium *Rhodobacter sphaeroides*. Hsp60 consists of 14 identical ~ 60-kD subunits arranged to form two apposed rings of 7 subunits, each surrounding a central cavity. The image of Hsp60, which is viewed with its 7-fold axis tipped towards the viewer, reveals that each subunit consists of two major domains, one in contact with the opposing heptameric ring, and the other at the end of the cylindrical protein molecule. The spherical density occupying the protein's central cavity is thought to represent a bound polypeptide. The cavity presumably provides a protected microenvironment in which a polypeptide can progressively fold itself (see Fig. 8-11). [Courtesy of Helen Saibil and Steve Wood, Birbeck College, London.]

space normally occupied by part of an abutting segment of chain and because it lacks the backbone N—H group with which to contribute a hydrogen bond. Likewise, steric interactions between several sequential amino acid residues with side chains branched at C_β (for instance, Ile and Thr) will destabilize an α helix. Furthermore, there are more subtle effects that may not be apparent without a detailed analysis of known protein structures. In what follows, we shall examine two schemes for predicting secondary structures empirically.

The Chou and Fasman Scheme

The first empirical structure prediction scheme we shall consider, which was developed by Peter Chou and Gerald Fasman, is among the most reliable of those available, and yet, is easy to apply. Its use requires two definitions. The frequency, f_α, with which a given residue occurs in an α helix in a set of protein structures is defined as

$$f_\alpha = \frac{n_\alpha}{n} \qquad [8.2]$$

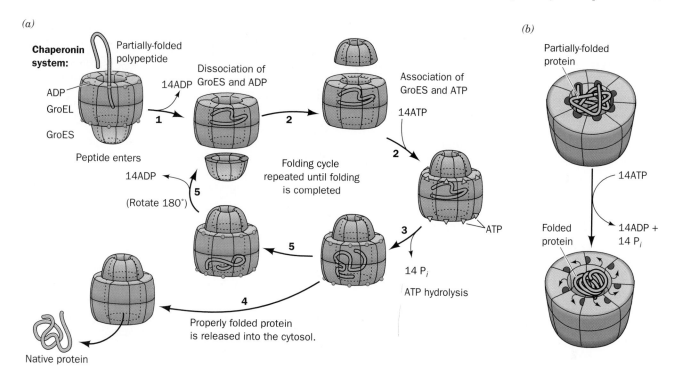

FIGURE 8-11. (*a*) The reaction cycle of the *E. coli* chaperonins GroEL and GroES in protein folding: (**1**) GroEL in asymmetric complex with one heptameric ring of GroES and 14 ADP's (one per GroEL subunit) binds an unfolded polypeptide in its central cavity in a process that releases all 14 ADP's and the bound GroES. (**2**) GroEL binds 14 ATPs, thereby weakening the interaction between GroEL and the unfolded polypeptide and causing the rebinding of GroES to the opposite face of GroEL. (**3**) All 14 ATPS are simultaneously hydrolyzed, thereby releasing the bound polypeptide within GroEL. This permits the polypeptide, which is probably in its molten globule state, to fold in a protected microenvironment, out of contact with other partially folded polypeptides with which it would otherwise aggregate. (**4**) If the polypeptide has folded to its native conformation, it is released from GroEL.

(**5**) If, however, the polypeptide has failed to fully attain its native fold, it remains bound to GroEL and reenters the reaction cycle at step 2 (note that in doing so, the diagrammed GroEL turns over by 180°). GroES, which binds but does not hydrolyze ATP, presumably facilitates the binding of the ATPs to GroEL, coordinates their simultaneous hydrolysis, and prevents the escape of a partially folded polypeptide from the GroEL cavity. (*b*) A model for the ATP-dependent release of an unfolded polypeptide from its multiple attachment sites in GroEL. ATP binding and hydrolysis conformationally mask the hydrophobic sites of GroEL (*darker areas*) that bind to the unfolded polypeptide, thereby permitting it to fold in an isolated environment. [After Hartl, F.-U., Hlodan, R., and Langer, T., *Trends Biochem. Sci.* **19**, 23 (1994).]

where n_α is the number of amino acid residues of the given type that occur in α helices and n is the total number of residues of this type in the set. The propensity of a particular amino acid residue to occur in an α helix is defined as

$$P_\alpha = \frac{f_\alpha}{\langle f_\alpha \rangle} \qquad [8.3]$$

where $\langle f_\alpha \rangle$ is the average value of f_α for all 20 residues. Accordingly, a value of $P_\alpha > 1$ indicates that a residue occurs with greater than average frequency in an α helix. The propensity, P_β, of a residue to occur in a β sheet is similarly defined.

Table 8-1 contains a list of α and β propensities based on the analysis of 29 X-ray structures. In accordance with its value of a given propensity, a residue is classified as a strong former (*H*), former (*h*), weak former (*I*), indifferent former

(*i*), breaker (*b*), or strong breaker (*B*) of that secondary structure. Using these data, Chou and Fasman formulated the following empirical rules to predict the secondary structures of proteins:

1. A cluster of four helix-forming residues (H_α or h_α, with I_α counting as one half h_α) out of six contiguous residues will nucleate a helix. The helix segment propagates in both directions until the average value of P_α for a tetrapeptide segment falls below 1.00. A Pro residue, however, can occur only at the N-terminus of an α helix.

2. A cluster of three β sheet formers (H_β or h_β) out of five contiguous residues nucleates a sheet. The sheet is propagated in both directions until the average value of P_β for a tetrapeptide segment falls below 1.00.

3. For regions containing both α- and β-forming sequences, the overlapping region is predicted to be helical

TABLE 8-1. PROPENSITIES AND CLASSIFICATIONS OF AMINO ACID RESIDUES FOR α HELICAL AND β SHEET CONFORMATIONS

Residue	P_α	Helix Classification	P_β	Sheet Classification
Ala	1.42	H_α	0.83	i_β
Arg	0.98	i_α	0.93	i_β
Asn	0.67	b_α	0.89	i_β
Asp	1.01	I_α	0.54	B_β
Cys	0.70	i_α	1.19	h_β
Gln	1.11	h_α	1.10	h_β
Glu	1.51	H_α	0.37	B_β
Gly	0.57	B_α	0.75	b_β
His	1.00	I_α	0.87	h_β
Ile	1.08	h_α	1.60	H_β
Leu	1.21	H_α	1.30	h_β
Lys	1.16	h_α	0.74	b_β
Met	1.45	H_α	1.05	h_β
Phe	1.13	h_α	1.38	h_β
Pro	0.57	B_α	0.55	B_β
Ser	0.77	i_α	0.75	b_β
Thr	0.83	i_α	1.19	h_β
Trp	1.08	h_α	1.37	h_β
Tyr	0.69	b_α	1.47	H_β
Val	1.06	h_α	1.70	H_β

Source: Chou, P.Y. and Fasman, G.D., *Annu. Rev. Biochem.* **47**, 258 (1978).

if its average value of P_α is greater than its average value of P_β; otherwise a sheet conformation is assumed.

These rather simple empirical rules predict the α helix and β sheet strand positions in a protein with an average reliability of ~50% and, in the most favorable cases, ~80% (Fig. 8-12; note that since proteins consist, on average, of ~31% α helix and ~28% β sheet, random predictions of these secondary structures would average ~30% correct).

Reverse Turns Are Characterized by a Minimum in Hydrophobicity Along a Polypeptide Chain

The positions of reverse turns can also be predicted by the Chou and Fasman method. However, since a reverse turn usually consists of four consecutive residues, each with a different conformation (Section 7-1D), their prediction algorithm is necessarily more cumbersome than those for sheets and helices.

Rose has proposed a simpler empirical method for predicting reverse turns. Reverse turns nearly always occur on the surface of a protein and, in part, define that surface. Since the core of a protein consists of hydrophobic groups and its surface is relatively hydrophilic, reverse turns occur at positions along a polypeptide chain where the hydropathy (Table 7-5) is a minimum. Using this method for

partitioning a polypeptide chain, we can deduce the positions of most reverse turns by inspection (Fig. 8-12). Since this method often predicts reverse turns to occur in helical regions (helices are all turns), it should be applied only to regions that are not predicted to be helical.

Physical Basis of α Helix Propensity

Why do amino acid residues have such different propensities for forming α helices? Brian Matthews has, in part, answered this question through structural and thermodynamic analyses of the enzyme **lysozyme** from **bacteriophage T4** (a bacterial virus) in which Ser 44, a solvent-exposed residue in the middle of a 12-residue (3.3 turn) α helix, was mutagenically replaced, in turn, by all 19 other amino acids. The X-ray structures of 13 of these variant proteins revealed that, with the exception of Pro, the substitutions caused no significant distortion to the α helix backbone and, hence, that differences in α helix propensities are unlikely to arise from strain. However, for 17 of the amino acids (all but Pro, Gly, and Ala), the stability of the α helix increases with the amount of side chain hydrophobic surface that is buried (brought out of contact with the solvent) when residue 44 is transferred from a fully extended state to an α helix. The low α helix propensity of Pro is due to the strain generated by its presence in an α helix, and that of Gly arises from the entropy cost associated with restricting this most conformationally flexible of residues to an α helical conformation (compare Figs. 7-7 and 7-9) and its lack of hydrophobic stabilization. The high α helix propensity of Ala, however, is caused by its lack of a γ substituent (possessed by all residues but Gly and Ala) and hence the absence of the entropy cost associated with conformationally restricting such a group within an α helix together with its small amount of hydrophobic stabilization.

Secondary Structures Are Partially Dictated by Tertiary Structures

Secondary structures are greatly influenced by tertiary interactions. For instance, in a sample of 62 unrelated proteins of known structure comprising some 10,000 amino acid residues, the longest segments with identical amino acid sequences are pentapeptides. In 6 of the 25 such pentapeptide pairs in the sample, the pentapeptide is part of an α helix in one protein and part of a β strand in another. The inability of sophisticated secondary structure prediction schemes to improve significantly on the reliability of simpler schemes is therefore explained by the failure of all of them to take tertiary interactions into account.

Rose has discovered a class of tertiary interactions that appear to be important for helix formation. A backbone C=O group in an α helix accepts a hydrogen bond from the backbone N—H group located four residues further along the polypeptide chain (Section 7-1B). Consequently, an α helix's four N-terminal N—H groups and four C-terminal C=O groups (which account for over half the hydrogen bonds in an average length α helix of ~11 resi-

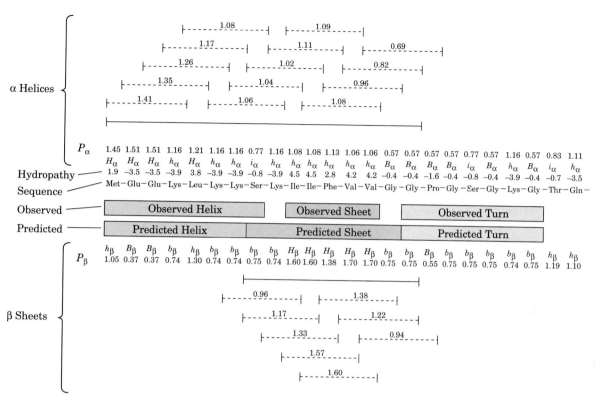

FIGURE 8-12. The prediction of α helices and β sheets by the method of Chou and Fasman and the prediction of reverse turns by the method of Rose for the N-terminal 24 residues of adenylate kinase. The helix and sheet propensities and classifications are taken from Table 8-1. The solid lines indicate all hexapeptide sequences that can nucleate an α helix (*top*) and all pentapeptide sequences that can nucleate a β sheet (*bottom*) as is explained in the text. The average helix and sheet propensities for each tetrapeptide segment in the helix and sheet regions are given above the corresponding dashed lines. Twelve of the 15 residues are observed to have their predicted secondary structures (*middle*) so that the prediction accuracy, in this case, is 80%. Reverse turns are predicted to occur in sequences in which the hydropathy (Table 7-5) is a minimum and which do not occur in helical regions. The region that matches this criterion is observed to have a reverse turn. [After Schultz, G.E. and Schirmer, R.H., *Principles of Protein Structure, p.* 121, Springer–Verlag (1979).]

dues) lack intrahelical hydrogen bonding partners. Most α helices in known protein X-ray structures are flanked by residues whose side chains, as conformational analysis indicates, can form hydrogen bonds to these terminal groups. This explains, for example, the observed high preference for an α helix to be preceded by Asn and Pro and to be succeeded by Gly (Asn's side chain C=O and Pro's backbone C=O are, respectively, readily positioned to accept hydrogen bonds from succeeding backbone N—H groups; a Gly at the C-terminus of an α helix, but no other residue, can assume a conformation in which its backbone N—H donates a hydrogen bond to the backbone C=O group three residues down the helix—a 3_{10} helical conformation). The observation that not all of these possible hydrogen bonds occur in native proteins led Rose to postulate that these interactions often function to nucleate helix formation. Such transient interactions may explain the existence of temperature-sensitive folding mutants (Section 8-1B), and why identical pentapeptides occur in both α helices and β sheets.

Theoretical calculations of tertiary structures that minimize the conformational energies of polypeptide chains have, so far, yielded results that are not significantly better than random models. The major problem is that polypeptide chains have astronomical numbers of non-native low-energy conformations so that it is presently not feasible, even with the fastest available computers, to determine a polypeptide's lowest energy conformation. Conformational energy calculations have been useful, however, in predicting, for example, how a protein of known structure alters its conformation in response to an amino acid residue change and how an enzyme conformationally adjusts to the binding of its substrate. The extrapolation of this approach to predicting the structure of a protein whose sequence is homologous to those of known structures has yielded encouraging results.

2. PROTEIN DYNAMICS

The fact that X-ray studies yield time-averaged "snapshots" of proteins may leave the false impression that proteins have fixed and rigid structures. In fact, as is becoming

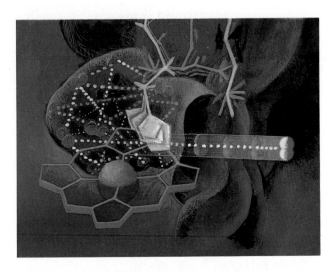

FIGURE 8-13. An artist's conception of the "breathing" motions in myoglobin that permit the escape of its bound O_2 molecule (*double red spheres*). The dotted lines trace a trajectory an O_2 molecule might take in worming its way through the rapidly fluctuating protein before finally escaping. O_2 binding presumably resembles the reverse of this process. [Figure copyrighted © by Irving Geis.]

increasing clear, *proteins are flexible and rapidly fluctuating molecules whose structural mobilities have functional significance.* For example, X-ray studies indicate that the heme groups of myoglobin and hemoglobin are so surrounded by protein that there is no clear path for O_2 to approach or escape from its binding pocket. Yet, we know that myoglobin and hemoglobin readily bind and release O_2. These proteins must therefore undergo conformational fluctuations, **breathing motions,** that permit O_2 reasonably free access to their heme groups (Fig. 8-13). The three-dimensional structures of myoglobin and hemoglobin undoubtedly evolved the flexibility to facilitate the diffusion of O_2 to its binding pocket.

The intramolecular motions of proteins have been classified into three broad categories according to their coherence:

1. **Atomic fluctuations,** such as the vibrations of individual bonds, which have time periods ranging from 10^{-15} to 10^{-11} s and spatial displacements between 0.01 and 1 Å.

2. **Collective motions,** in which groups of covalently linked atoms, which vary in size from amino acid side chains to entire domains, move as units with time periods ranging from 10^{-12} to 10^{-3} s and spatial displacements between 0.01 and > 5 Å. Such motions may occur frequently or infrequently compared with their characteristic time period.

3. **Triggered conformational changes,** in which groups of atoms varying in size from individual side chains to

complete subunits move in response to specific stimuli such as the binding of a small molecule, for example, the binding of O_2 to hemoglobin (Sections 9-2A and C). Triggered conformational changes occur over time spans ranging from 10^{-9} to 10^3 s and result in atomic displacements between 0.5 and > 10 Å.

In this section, we discuss how these various motions are characterized and their structural and functional significance. We shall mainly be concerned with atomic fluctuations and collective motions; triggered conformational changes are considered in later chapters in connection with specific proteins.

Proteins Have Mobile Structures

X-Ray crystallographic analysis is a powerful technique for the analysis of motion in proteins; it reveals not only the average positions of the atoms in a crystal, but also their mean-square displacements from these positions. X-Ray analysis indicates, for example, that myoglobin has a rigid core surrounding its heme group and that the regions toward the periphery of the molecule have a more mobile character. A similar analysis of lysozyme has produced the intriguing observation that the enzyme's active site cleft, which undergoes an ~1-Å closure upon binding substrate (Section 14-2A), is among the regions of the protein with the greatest mobility.

Theoretical considerations of the internal motions in proteins by Martin Karplus indicate that *a protein's native structure really consists of a large collection of conformational substates that have essentially equal stabilities.* These substates, which each have slightly different atomic arrangements, randomly interconvert at rates that increase with temperature. Computer simulations of protein internal motions (e.g., Fig. 8-14) suggest that the interior of a protein typically has a fluidlike character for structural displacements of up to ~2 Å, that is, over excursions that are somewhat larger than a bond distance.

Gregory Petsko and Dagmar Ringe have demonstrated the functional significance of the internal motions in proteins. Both experimental and theoretical evidence indicates that below around 220 K ($-53°$C), collective motions in proteins are arrested, leaving atomic fluctuations as the dominant intramolecular motions. X-Ray studies have shown that, at 228 K, the enzyme RNase A, in its crystalline form, readily binds an unreactive substrate analog (protein crystals generally contain large water-filled cavities through which small molecules readily diffuse; at low temperatures, the water is prevented from freezing by the addition of an antifreeze such as methanol). Yet, when the same experiment is performed at 212 K, the substrate analog does not bind to the enzyme, even after 6 days exposure. Likewise, at 228 K, substrate-free solvent washes bound substrate analog out of the crystal within minutes but, if the temperature is lowered to 212 K before this occurs, the substrate analog

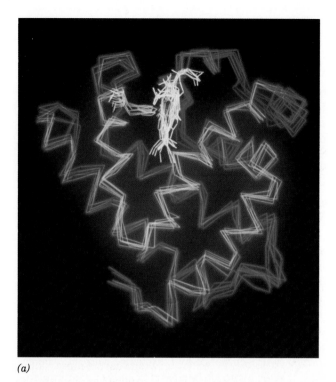

(a)

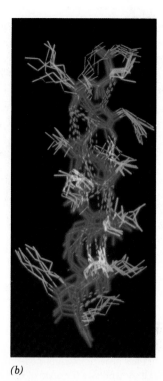

(b)

FIGURE 8-14. The internal motions of myoglobin as simulated by computerized molecular dynamics calculations. Several "snapshots" of the molecule calculated at intervals of 5×10^{-12} s are superimposed. (*a*) The C_α backbone and the heme group. The backbone is shown in blue, the heme in yellow, and the proximal His residue in orange. (*b*) An α helix. The backbone is shown in blue, the side chains in green, and the helix hydrogen bonds as dashed orange lines. Note that the helices tend to move in a coherent fashion so as to retain their shape. [Courtesy of Martin Karplus, Harvard University.]

remains bound to the crystalline enzyme for at least 2 days. Evidently, the freezing out of its collective motions below 220 K renders RNase A too rigid to bind or release substrate.

Protein Core Mobility Is Revealed by Aromatic Ring Flipping

The rate at which an internal Phe or Tyr ring in a protein undergoes 180° "flips" about its C_β—C_γ bonds is indicative of the protein core's rigidity. This is because these bulky asymmetric groups, in the close packed interior of a protein, can move only when the surrounding groups move aside transiently (although note that these rings have the shape of flattened ellipsoids rather than thin disks). The rate that a particular aromatic ring flips is best inferred from an analysis of its NMR spectrum (infrequent motions such as ring flipping are not detected by X-ray crystallography since this technique reveals the average structure of a protein). NMR measurements indicate that the ring flipping rate varies from over 10^6 s^{-1} to one of immobility (<1 s^{-1}) depending on both the protein and the location of the aromatic ring within the protein. Thus, at 4°C, four of BPTI's eight Phe and Tyr rings flip at rates $> 5 \times 10^4$ s^{-1}, whereas the remaining four rings flip at rates ranging between 30 and <1 s^{-1}. These ring-flipping rates sharply increase with temperature, as expected.

Infrequent Motions Can Be Detected through Hydrogen Exchange

Conformational changes occurring over time spans of more than several seconds can be chemically characterized through **hydrogen exchange studies.** Weakly acidic protons, such as those of amine and hydroxyl groups (X—H), exchange with those of water as can be demonstrated with the use of tritiated water [HTO; tritium (T or ^{3}H) is a radio-active isotope of ^{1}H].

$$X—H + HTO \rightleftharpoons X—T + H_2O$$

Under physiological conditions, most small organic molecules, such as amino acids and dipeptides, completely exchange such protons in times ranging from milliseconds to seconds.

Proteins bear numerous exchangeable protons such as those of its backbone amide groups. Indeed, hydrogen exchange studies of native proteins indicate that these protons exchange at rates that vary from milliseconds to many years

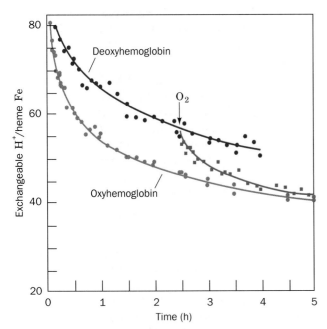

FIGURE 8-15. The hydrogen–tritium "exchange-out" curve for hemoglobin that has been preequilibrated with tritiated water. The vertical axis expresses the ratio of exchangeable protons to heme Fe atoms. Exchange-out was initiated by replacing the protein's tritiated water solvent with untritiated water through rapid gel filtration (Section 5-3C). As the exchange-out proceeded, additional gel filtration separations were performed and the amount of tritium remaining bound to the protein measured. At the arrow, O_2 was added to exchanging deoxyhemoglobin (hemoglobin lacking bound O_2). The changing slopes of these curves indicate that the hydrogen exchange rates of the ~ 80 exchangeable protons of each hemoglobin subunit vary by factors of many decades and that O_2 binding increases the exchange rates for ~ 10 of these protons (the structural changes that O_2 binding induces in hemoglobin are discussed in Section 9-2). [After Englander, S.W. and Mauel, C., *J. Biol. Chem.* **247,** 2389 (1972).]

(Fig. 8-15). Yet protein interiors, as we have seen, are largely excluded from contact with their surrounding aqueous solvent (Section 7-3B), and, moreover, protons cannot exchange with solvent while they are engaged in hydrogen bonding. The observation that the internal protons of a protein do, in fact, exchange with solvent, must therefore be a consequence of transient local unfolding or "breathing" that physically and chemically exposes these exchangeable protons to the solvent. Hence, *the rate at which a particular proton undergoes hydrogen exchange is a reflection of the conformational mobility of its surroundings.* This hypothesis is corroborated by the observation that the hydrogen exchange rates of proteins decrease as their denaturation temperatures increase and that these exchange rates are sensitive to the proteins' conformational states (Fig. 8-15).

Protein Folding Can Be Monitored By Pulsed H/D Exchange Methods

Deuterium (D or ^2H), a stable isotope of ^1H, has an NMR spectrum in an entirely different frequency range from that of ^1H and, hence, in a method devised by Walter Englander and Robert Baldwin, the exchange of D for ^1H can be readily followed by NMR spectroscopy. The advent of 2D-NMR techniques (Section 7-3B) has led to the development of **pulsed H/D exchange** methods for following the time course of protein folding. The protein of interest, usually with its native disulfide bonds intact, is denatured by guanidinium chloride or urea in D_2O solution such that all of the protein's peptide nitrogen atoms become deuterated (ND). Folding then is initiated in a rapid mixing device by diluting the denaturant solution with ^1H_2O while the pH is simultaneously lowered so as to arrest hydrogen exchange (near neutrality, hydrogen exchange reactions are catalyzed by OH^- and, therefore, their rates are highly pH dependent). After a preset folding time, t_f, the pH is rapidly increased (the so-called labeling pulse) to initiate hydrogen exchange. Peptide nitrogen atoms whose D atoms have not formed hydrogen bonds by t_f exchange with ^1H, whereas those that are hydrogen bonded at t_f, and hence unavailable for hydrogen exchange, remain deuterated. After a short time (10 to 40 ms), the labeling pulse is terminated by rapidly lowering the pH. Folding is then allowed to go to completion and the H/D ratio at each exchangeable site is determined by 2D NMR. By repeating the analysis for several values of t_f, the time course of hydrogen bond formation at each peptide group can be determined.

The few small proteins whose folding has so far been investigated by pulsed H/D exchange exhibit remarkably similar folding patterns. Within tens of milliseconds, a folding intermediate forms that contains nonexchanging NH groups in regions corresponding to most of the native protein's secondary structure. This intermediate appears to have the molten globule conformation discussed in Section 8-1B. Over the next second or so, the protein gradually rigidifies to yield the native conformation via what may be several different pathways that are thought to differ by the cis–trans isomerization states of their Xaa–Pro peptide bonds. In any case, proteins appear to more or less follow the model represented by Fig. 8-5 in folding to their native conformations.

3. STRUCTURAL EVOLUTION

Proteins, as we discussed in Section 6-3, evolve through point mutations and gene duplications. Over eons, through processes of natural selection and/or neutral drift, homologous proteins thereby diverge in character and develop new functions. How these primary structure changes affect function, of course, depends on the protein's three-dimensional structure. In this section, we explore the effects of evolutionary change on protein structures.

A. Structures of Cytochromes c

The *c*-type cytochromes are small globular proteins that contain a covalently bound heme group (iron–protoporphyrin IX; Fig. 8-16). The X-ray structures of horse, tuna, and bonito cytochromes *c* (see the cover illustrations), which were elucidated by Richard Dickerson, are quite similar and thus permit the structural significance of cytochrome *c*'s amino acid sequences (Section 6-3B) to be assessed.

The internal residues of cytochrome *c*, particularly those lining its heme pocket, tend to be invariant or conservatively substituted, whereas surface positions have greater variability. This observation is, in part, an indication of the more exacting packing requirements of a protein's internal regions compared to those of its surface (Section 7-3B). Certain invariant or highly conserved residues (Table 6-4) have specific structural and/or functional roles in cytochrome *c*:

1. The invariant Cys 14, Cys 17, His 18, and Met 80 residues form covalent bonds with the heme group (Fig. 8-16).

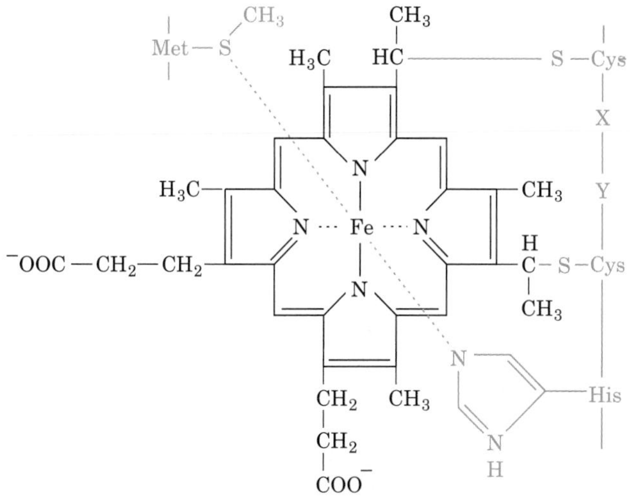

FIGURE 8-16. The molecular formula of iron–protoporphyrin IX (heme). In *c*-type cytochromes, the heme is covalently bound to the protein (*red*) by two thioether bonds linking what were the heme vinyl groups to two Cys residues that occur in the sequence Cys-X-Y-Cys-His. Here X and Y symbolize other amino acids. A fifth and sixth ligand to the Fe atom, both normal to the heme plane, are formed by a nitrogen of the His side chain in this sequence and the sulfur of a Met residue that is further along the polypeptide chain. The iron atom, which is thereby octahedrally liganded, can stably assume either the Fe(II) or the Fe(III) oxidation state. Heme also occurs in myoglobin and hemoglobin but without the thioether bonds or the Met ligand.

2. The nine invariant or highly conserved Gly residues occupy close-fitting positions in which larger side chains would significantly alter the protein's three-dimensional structure.

3. The highly conserved Lys residues 8, 13, 25, 27, 72, 73, 79, 86, and 87 are distributed in a ring around the exposed edge of the otherwise buried heme group. There is considerable evidence that this unusual constellation of positive charges specifically associates with complementary sets of negative charges on the physiological reaction partners of cytochrome *c*, cytochrome *c* reductase, and cytochrome *c* oxidase (Section 20-2C).

Prokaryotic *c*-Type Cytochromes Are Structurally Related to Cytochrome *c*

Although cytochrome *c* occurs only in eukaryotes, similar proteins known as **c-type cytochromes** are common in prokaryotes, where they function to transfer electrons at analogous positions in a variety of respiratory and photosynthetic electron-transport chains. Unlike the eukaryotic proteins, however, the prokaryotic *c*-type cytochromes exhibit considerable sequence variability among species. For example, the more than 30 bacterial *c*-type cytochromes whose primary structures are known have from 82 to 134 amino acid residues, whereas eukaryotic cytochromes *c* have a narrower range—between 103 and 112 residues. The primary structures of several representative *c*-type cytochromes have few obvious similarities (Fig. 8-17). Yet their X-ray structures closely resemble each other, particularly in their chain folding and side group packing in the regions surrounding the heme group (Fig. 8-18). Furthermore, most of them have aromatic rings in analogous positions and orientations relative to their heme groups as well as similar distributions of positively charged Lys residues about the perimeters of their heme crevices. The major structural differences between these *c*-type cytochromes stem from various loops of polypeptide chain that are located on their surfaces.

The proper alignments of analogous *c*-type cytochrome residues (thin lines in Fig. 8-17) could not have been made on the basis of only their primary structures: These proteins have diverged so far that their three-dimensional structures were essential guides for this task. Three-dimensional structures are evidently more indicative of the similarities among these distantly related proteins than are primary structures. *It is the essential structural and functional elements of proteins, rather than their amino acid residues, that are conserved during evolutionary change.*

B. Gene Duplication

Gene duplication may promote the evolution of new functions through structural evolution. In over half of the multidomain proteins of known structure, two of the domains are structurally quite similar. Consider, for example, the

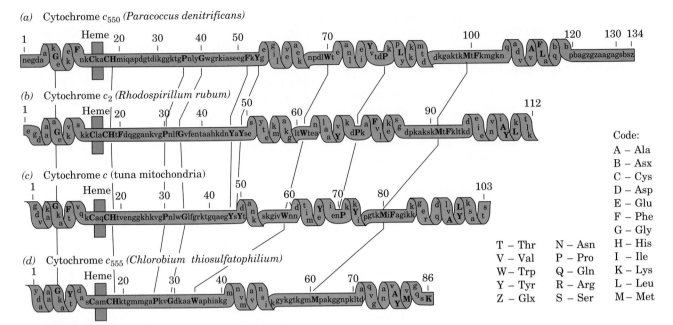

FIGURE 8-17. The primary structures of some representative *c*-type cytochromes: (*a*) Cytochrome c_{550} (the subscript indicates the protein's peak absorption wavelength in visible light in nanometers, nm) from *Paracoccus denitrificans,* a respiring bacterium that can use nitrate as an oxidant. (*b*) Cytochrome c_2 (the subscript has only historical significance) from *Rhodospirillum rubrum,* a purple photosynthetic bacterium. (*c*) Cytochrome *c* from tuna mitochondria. (*d*) Cytochrome c_{555} from *Chlorobium thiosulfatophilum,* a green photosynthetic bacterium that utilizes H_2S as a hydrogen source. Thin lines connect structurally significant or otherwise invariant residues (*capitalized*). Helical regions are indicated to facilitate structural comparisons with Fig. 8-18. [After Salemme, F.R., *Annu. Rev. Biochem.* **46**, 307 (1977).]

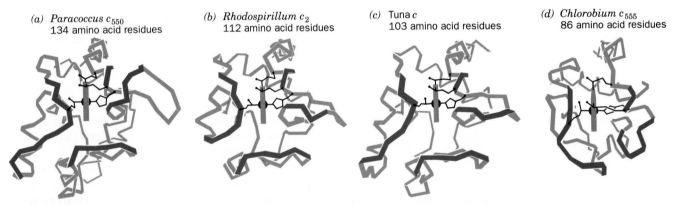

FIGURE 8-18. The three-dimensional structures of the *c*-type cytochromes whose primary structures are displayed in Fig. 8-17. The polypeptide backbones (*blue*) are shown in analogous orientations such that their heme groups (*red*) are viewed edge on. The Cys, Met, and His side chains that covalently link the heme to the protein are also shown. (*a*) Cytochrome c_{550} from *P. denitrificans.* (*b*) Cytochrome c_2 from *Rs. rubrum.* (*c*) Tuna cytochrome *c.* (*d*) Cytochrome c_{555} from *C. thiosulfatophilum.* [Figures copyrighted © by Irving Geis.]

two domains of the bovine liver enzyme **rhodanese** (Fig. 8-19). It seems highly improbable that its two complex but conformationally similar domains could have independently evolved their present structures (a process known as **convergent evolution**). More likely, they arose through the duplication of the gene specifying an ancestral domain followed by the fusion of the resulting two genes to yield a single gene specifying a polypeptide that folds into two similar domains. The differences between the two domains is therefore due to their **divergent evolution.** The same argument can be made about the PapD protein with its two β-barrel domains (Fig. 8-9).

Structurally similar domains occur in proteins whose other domains bear no resemblance to one another. The

Domain 1

Domain 2

FIGURE 8-19. The two structurally similar domains of rhodanese. [After drawings provided by Jane Richardson, Duke University.]

redox enzymes known as dehydrogenases, for example, each consist of two domains: a coenzyme-binding domain (Fig. 7-50) that is structurally similar in all the dehydrogenases, and a dissimilar substrate-binding domain that determines the specificity and mode of action of each enzyme. Indeed, in some dehydrogenases, such as glyceraldehyde-3-phosphate dehydrogenase (Fig. 7-47), the coenzyme-binding domain occurs at the N-terminal end of the polypeptide chain, whereas in others it occurs at the C-terminal end. Each of these dehydrogenases must have arisen by the fusion of the gene specifying an ancestral coenzyme-binding domain with a gene coding for a proto-substrate-binding domain. This must have happened very early in evolutionary history, perhaps in the precellular stage (Section 1-4C), because there are no significant sequence similarities among these coenzyme-binding domains. Evidently, a domain is as much a unit of evolution as it is a unit of structure. *By genetically combining these structural modules in various ways, nature can develop new functions far more rapidly than it can do so by the evolution of completely new structures through point mutations.*

CHAPTER SUMMARY

Under renaturing conditions, many proteins fold to their native structures in a matter of seconds. Proteins must therefore fold in an ordered manner rather than go through a random search of all their possible conformations. This has been confirmed in the case of BPTI, in which the native disulfide bonds reform in a largely specific but possibly indirect sequence. Even though it is clear that a protein's primary structure dictates its three-dimensional structure, many proteins, nevertheless, require the assistance of accessory proteins such as protein disulfide isomerase, peptidyl prolyl cis–trans isomerases, and molecular chaperones to fold/assemble to their native structures. The prediction of protein secondary structures from only amino acid sequences has been reasonably successful using empirical techniques. Techniques for predicting the tertiary structures of proteins, however, are still at a rudimentary stage of development.

Proteins are flexible and fluctuating molecules whose group motions have characteristic periods ranging from 10^{-15} to over 10^3 s.

X-Ray analysis, which reveals the average atomic mobilities in a protein, indicates that proteins tend to be more mobile at their peripheries than in their interiors. Theoretical analysis of protein mobilities suggests that the native protein structures each consist of a large number of closely related and rapidly interconverting conformational substates of nearly equal stabilities. Without this flexibility, enzymes would be nonfunctional. The rates of aromatic ring flipping, as revealed by NMR measurements, indicate that internal group mobilities within proteins vary both with the protein and with the position within the protein. Hydrogen exchange studies demonstrate that proteins have a great variety of infrequently occurring internal motions. The exchange of internal protons with solvent probably results from the transient local unfolding of the protein. Pulsed H/D exchange studies provide a useful probe of protein folding pathways.

The X-ray structures of eukaryotic cytochromes *c* demonstrate that internal residues and those having specific structural and functional roles tend to be conserved during evolution. Prokaryotic c-type cytochromes from a variety of organisms structurally resemble each other and those of eukaryotes even though there are few similarities among their amino acid sequences. This indicates that the three-dimensional structures of proteins rather than their amino acid sequences are conserved during evolutionary change. The structural similarities between the domains in many multidomain proteins indicates that these proteins arose through the duplication of a gene specifying an ancestral domain followed by their fusion. Similarly, the structural resemblance between the coenzyme-binding domains of dehydrogenases suggests that these proteins arose by duplication of a primordial coenzyme-binding domain followed by its fusion with a gene specifying a proto-substrate-binding domain. In this manner, proteins with new functions can evolve much faster than by a series of point mutations.

REFERENCES

Protein Folding

Accessory Folding Proteins, Adv. Prot.. Chem. **44** (1993). [Contains authoritative articles on protein disulfide isomerase, peptidyl prolyl cis–trans isomerase, and several types of molecular chaperones.]

Anfinsen, C.B., Principles that govern the folding of protein chains, *Science* **181**, 223–230 (1973). [A Nobel laureate explains how he got his prize.]

Behe, M., Lattman, E.E., and Rose, G.D., The protein folding problem: The native fold determines the packing but does packing determine the native fold? *Proc. Natl. Acad. Sci.* **88**, 4195–4199 (1991).

Blaber, M., Zhang, X., and Matthews, B.W., Structural basis of amino acid α helix propensity, *Science* **260**, 1637–1640 (1993).

Chothia, C. and Finkelstein, A.V., The classification and origins of protein folding patterns, *Annu. Rev. Biochem.* **59**, 1007–1039 (1990).

Chou, P.Y. and Fasman, G.D., Empirical predictions of protein structure, *Annu. Rev. Biochem.* **47**, 251–276 (1978); *and* Prediction of the secondary structure of proteins from their amino acid sequence, *Adv. Enzymol.* **47**, 45–148 (1978). [Expositions of one of the most widely used methods for the prediction of protein secondary structures.]

Creighton, T.E., *Proteins* (2nd ed.), Chapter 7, Freeman (1993).

Creighton, T.E. (Ed.), *Protein Folding*, Freeman (1992). [A series of authoritative reviews.]

Creighton, T.E., The disulfide folding pathway of BPTI *and* Weissman, J.S. and Kim, P.S., Response, *Science* **256**, 111–114 (1992). [A critique of Weissman and Kim's conclusions (see Weissman and Kim reference below) and their reply.]

Creighton, T.E., Protein folding, *Biochem. J.* **270**, 1–16 (1990).

Dill, K.A., Folding proteins: finding a needle in a haystack, *Curr. Opin. Struct. Biol.* **3**, 99–103 (1993); *and* Dominant forces in protein folding, *Biochemistry* **29**, 7133–7155 (1990).

Dill, K.A.. and Shortle, D., Denatured states of proteins, *Annu. Rev. Biochem.* **60**, 795–825 (1991).

Ellis, R.J. and van der Vies, S.M., Molecular chaperones, *Annu. Rev. Biochem.* **60**, 321–347 (1991).

Fasman, G.D. (Ed.), *Prediction of Protein Structure and the Principles of Protein Conformation*, Plenum Press (1989). [Contains a wealth of articles on protein folding and protein structure prediction.]

Frauenfelder, H., Parak, F., and Young, R.D., Conformational substates in proteins, *Annu. Rev. Biophys. Biophys. Chem.* **17**, 451–479 (1988).

Gething, M.-J. and Sambrook, J., Protein folding in the cell, *Nature* **355**, 33–45 (1992). [A detailed review on protein folding accessory proteins.]

Goldenberg, D.P., Native and non-native intermediates in the BPTI folding pathway, *Trends Biochem. Sci.* **17**, 257–261, 339 (1992).

Goldenberg, D.P., Genetic studies of protein stability and mechanisms of folding, *Annu. Rev. Biophys. Biophys. Chem.* **17**, 481–507 (1988).

Goldenberg, D.P. and King, J., Trimeric intermediates in the *in vivo* folding and subunit assembly of the tail spike endorhamnosidase of bacteriophage P22, *Proc. Natl. Acad. Sci.* **79**, 3403–3407 (1982). [Evidence that protein folding pathways are genetically controlled.]

Hartl, F.-U., Hlodan, R., and Langer, T., Molecular chaperones in protein folding: the art of avoiding sticky situations, *Trends Biochem. Sci.* **19**, 20–25 (1994).

Hendrick, J.P. and Hartl, F.-U., Molecular chaperone functions of heat shock proteins, *Annu. Rev. Biochem.* **62**, 349–384 (1993); *and* Hartl, F.U. and Martin, J., Protein folding in the cell: The role of molecular chaperones Hsp70 and Hsp60, *Annu. Rev. Biophys. Biomol. Struct.* **21**, 293–322 (1992).

Jaenicke, R., Role of accessory proteins in protein folding, *Curr. Opin. Struct. Biol.* **3**, 104–112 (1993).

Kabsch, W. and Sander, C., On the use of sequence homologies to predict protein structure: identical pentapeptides can have completely different conformations, *Proc. Natl. Acad. Sci.* **81**, 1075–1078 (1984).

Kim, P.S. and Baldwin, R.L., Intermediates in the folding reactions of small proteins, *Annu. Rev. Biochem.* **59**, 631–660 (1990).

Kuehn, M.J., Ogg, D.J., Kihlber, J., Slonim, L.N., Flemmer, K., Bergfors, T., and Hultgren, S.A., Structural basis of pilus subunit recognition by the PapD chaperone, *Science* **262**, 1234–1241 (1993).

Lattman, E.E. and Rose, G.D., Protein folding—what's the question? *Proc. Natl. Acad. Sci.* **90**, 439–441 (1993).

Martin, J.L., Bardwell, J.C.A., and Kuriyan, J., Crystal structure of the DsbA protein required for disulfide bond formation in vivo, *Nature* **365**, 464–468 (1993).

Matthews, C.R., Pathways of protein folding, *Annu. Rev. Biochem.* **62**, 653–684 (1993).

Richards, F.M., The protein folding problem, *Sci. Am.* **264**(1): 54–63 (1991).

Richardson, J.S. and Richardson, D.C., Amino acid preferences for specific locations at the ends of α helices, *Science* **240**, 1648–1652 (1988).

Rose, G.D., Prediction of chain turns in globular proteins on a hydrophobic basis, *Nature* **272**, 586–590 (1978).

Rose, G.D. and Wolfenden, R., Hydrogen bonding, hydrophobicity, packing, and protein folding, *Annu. Rev. Biophys. Biomol. Struct.* **22**, 381–415 (1993).

Saibil, H. and Wood, S., Chaperonins, *Curr. Opin. Struct. Biol.* **3**, 207–213 (1993).

Sali, A., Overington, J.P., Johnson, M.S., and Blundell, T.L., From comparisons of protein sequences and structures to protein modeling and design, *Trends Biochem. Sci.* **15**, 235–240 (1990).

Schreiber, S.L., Chemistry and biology of immunophilins and their immunosuppressive ligands, *Science* **251**, 238–287 (1991).

Schultz, G.E., A critical evaluation of methods for prediction of protein secondary structures, *Annu. Rev. Biophys. Biophys. Chem.* **17**, 1–21 (1988).

Schultz, G.E. and Schirmer, R.H., *Principles of Protein Structure,* Chapters 6 and 8, Springer–Verlag (1979).

Sternberg, M.J.E., Secondary structure prediction, *Curr. Opin. Struct. Biol.* **2**, 237-241 (1992).

Walsh, C.T., Zydowsky, L.D., and McKeon, F.D., Cyclosporin A, the cyclophilin class of peptidylprolyl isomerases, and blockade of T cell signal transduction, *J. Biol. Chem.* **267**, 13115–13118 (1992).

Wang, C.-C. and Tsou, C.-L., The insulin A and B chains contain sufficient structural information to form the native molecule, *Trends Biochem. Sci.* **16**, 279-281 (1991).

Weissman, J.S. and Kim, P.S., Reexamination of the folding of BPTI: Predominance of native intermediates, *Science* **253**, 1386–1393 (1991); *and* Kinetic role of nonnative species in the folding of bovine pancreatic trypsin inhibitor, *Proc. Natl. Acad. Sci.* **89**, 9900–9904 (1992).

Weissman, J.S. and Kim, P.S., The pro region of BPTI facilitates folding, *Cell* **71**, 841–851 (1992).

Protein Dynamics

Baldwin, R.L., Pulsed H/D-exchange studies of folding intermediates, *Curr. Opin. Struct. Biol.* **3**, 84–91 (1993).

Chothia, C. and Lesk, A.M., Helix movements in proteins, *Trends Biochem. Sci.* **10**, 116–118 (1985).

Englander, S.W. and Mayne, L., Protein folding studied using hydrogen-exchange labeling and two-dimensional NMR, *Annu. Rev. Biophys. Biomol. Struct.* **21**, 243–265 (1992).

Huber, R., Flexibility and rigidity of proteins and protein-pigment complexes, *Angew. Chem. Int. Ed. Engl.* **27**, 79–88 (1988).

Karplus, M. and McCammon, J.A., The dynamics of proteins, *Sci. Am.* **254**(4): 42–51 (1986).

Karplus, M. and Petsko, G.A., Molecular dynamics simulations in biology, *Nature* **347**, 631–639 (1990).

McCammon, J.A. and Harvey, S.C., *Dynamics of Proteins and Nucleic Acids,* Cambridge University Press (1987).

Rasmussen, B.F., Stock, A.M., Ringe, D., and Petsko, G.A., Crystalline ribonuclease A loses function below the dynamical transition at 220 K, *Nature* **357**, 423–424 (1992).

Ringe, D. and Petsko, G.A., Mapping protein dynamics by X-ray diffraction, *Prog. Biophys. Mol. Biol.* **45**, 197–235 (1985).

Rogero, J.R., Englander, J.J., and Englander, S.W., Measurement and identification of breathing units in hemoglobin by hydrogen exchange, *in* Sarma R.H. (Ed.), *Biomolecular Stereodynamics,* Vol. 2, *pp.* 287–298, Adenine Press (1981).

Woodward, C.K., Hydrogen exchange rates and protein folding, *Curr. Opin. Struct. Biol.* **4**, 112–116 (1994).

Structural Evolution

Bajaj, M. and Blundell, T., Evolution and the tertiary structure of proteins, *Annu. Rev. Biophys. Bioeng.* **13**, 453–492 (1983).

Dickerson, R.E., The structure and history of an ancient protein, *Sci. Am.* **226**(4): 58–72 (1972); *and* Cytochrome *c* and the evolution of energy metabolism, *Sci. Am.* **242**(3): 137–149 (1980).

Dickerson, R.E., The cytochromes *c*: An exercise in scientific serendipity, *in* Sigman, D.S. and Brazier, M.A. (Eds.), *The Evolution of Protein Structure and Function, pp.* 172–202, Academic Press (1980).

Dickerson, R.E., Timkovitch, R., and Almassy, R.J., The cytochrome fold and the evolution of bacterial energy metabolism, *J. Mol. Biol.* **100**, 473–491 (1976).

Eventhoff, W. and Rossmann, M., The structures of dehydrogenases, *Trends Biochem. Sci.* **1**, 227–230 (1976).

Mathews, F.S., The structure, function and evolution of proteins, *Prog. Biophys. Mol. Biol.* **45**, 1–56 (1985).

Moore, G.R. and Pettigrew, G.W., *Cytochromes c,* Springer-Verlag (1990). [Chapters 4 and 5 discuss the structures of *c*-type cytochromes.]

Rossmann, M.G. and Argos, P., The taxonomy of protein structure, *J. Mol. Biol.* **109**, 99–129 (1977).

Salemme, R., Structure and function of cytochromes *c*, *Annu. Rev. Biochem.* **46**, 299–329 (1977).

Schultz, G.E. and Schirmer, R.H., *Principles of Protein Structure,* Chapter 9, Springer–Verlag (1979).

PROBLEMS

1. How long should it take the polypeptide backbone of a 6-residue folding nucleus to explore all its possible conformations? Repeat the calculation for 10-, 15-, and 20-residue folding nuclei. Explain why folding nuclei are thought to be no larger than 15 residues?

*2. Consider a protein with 10 Cys residues. Upon air oxidation, what fraction of the denatured and reduced protein will randomly reform the native set of disulfide bonds if: (a) The native protein has five disulfide bonds? (b) The native protein has three disulfide bonds?

3. Why are β sheets more commonly found in the hydrophobic interiors of proteins rather than on their surfaces so as to be in contact with the aqueous solvent?

4. Under physiological conditions, polylysine assumes a random coil conformation. Under what conditions might it form an α helix?

5. Explain why Pro residues can occupy the N-terminal turn of an α helix.

*6. Predict the secondary structure of the C peptide of proinsulin (Fig. 8-4). Is it likely to have a supersecondary structure?

7. As Mother Nature's chief engineer, now certified as a master helix builder, you are asked to repeat Problem 9 in Chapter 7 with the stipulation that the α helix really be helical. Use Table 8-1.

8. Explain why β sheets are unlikely to form folding nuclei.

9. Folding nuclei are thought to be predominantly stabilized by hydrophobic forces. Why aren't hydrogen bonding forces equally effective?

10. Indicate the probable effects of the following mutational changes on the structure of a protein. Explain your reasoning. (a) Changing a Leu to a Phe, (b) changing a Lys to a Glu, (c) changing a Val to a Thr, (d) changing a Gly to an Ala, and (e) changing a Met to a Pro.

11. Explain why Trp rings are usually completely immobile in proteins that have rapidly flipping Phe and Tyr rings.

*12. Discuss the merits of the hypothesis that the coenzyme-binding domains of the dehydrogenases arose by convergent evolution.

C H A P T E R

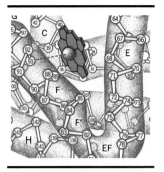

9

Hemoglobin: Protein Function in Microcosm

1. Hemoglobin Function

The existence of hemoglobin, the red blood pigment, is evident to every child who scrapes a knee. Its brilliant red color, widespread occurrence, and ease of isolation have made it an object of inquiry since ancient times. Indeed, the early history of protein chemistry is essentially that of hemoglobin. The observation of crystalline hemoglobin was first reported in 1849, and by 1909 a photographic atlas of hemoglobin crystals from 109 species had been published. In contrast, it was not until 1926 that crystals of an enzyme, those of jack bean **urease,** were first reported. Hemoglobin was one of the first proteins to have its molecular mass accurately determined, the first protein to be characterized by ultracentrifugation, the first to be associated with a specific physiological function (that of oxygen transport) and, in sickle-cell anemia, the first in which a point mutation was demonstrated to cause a single amino acid change (Section 6-3A). Theories formulated to account for the cooperative binding of oxygen to hemoglobin (Section 9-4) have also been successful in explaining the control of enzyme activity. The first protein X-ray structures to be elucidated were those of hemoglobin and myoglobin. This central role in the development of protein chemistry together with its enzymelike O_2-binding properties have caused hemoglobin to be dubbed an "honorary enzyme."

Hemoglobin is not just a simple oxygen tank. Rather, it is a sophisticated oxygen delivery system that provides the proper amount of oxygen to the tissues under a wide variety of circumstances. In this chapter, we discuss hemoglobin's properties, structure, and mechanism of action, both to understand the workings of this physiologically essential molecule and to illustrate the principles of protein structure that we have developed in the preceding chapters. We also consider the properties of abnormal hemoglobins and their relationship to human disease. Finally, we discuss theories of cooperative interactions among proteins, both to better understand the properties of hemoglobin and to set the stage for our later consideration of how enzyme action is regulated.

1. HEMOGLOBIN FUNCTION

Hemoglobin **(Hb),** as we have seen in Chapters 6 and 7, is a tetrameric protein, $\alpha_2\beta_2$ (alternatively, a dimer of $\alpha\beta$ protomers). The α and β subunits are structurally and evolutionarily related to each other and to myoglobin **(Mb),** the monomeric oxygen-binding protein of muscle (Section 6-3C).

Hemoglobin transports oxygen from the lungs, gills, or skin of an animal to its capillaries for use in respiration. Very small organisms do not require such a protein because their respiratory needs are satisfied by the simple passive diffusion of O_2 through their bodies. However, since the transport rate of a diffusing substance varies inversely with the square of the distance it must diffuse, the O_2 diffusion rate through tissue thicker than ~1 mm is too slow to support life. The evolution of organisms as large and complex as annelids (e.g., earthworms) therefore required the development of circulatory systems that actively transport O_2 and nutrients to the tissues. The blood of such organisms must contain an oxygen transporter such as Hb because the solubility of O_2 in **blood plasma** (the fluid component of blood) is too low (~$10^{-4}M$ under physiological conditions) to carry sufficient O_2 for metabolic needs. In contrast, whole blood, which normally contains ~150 g of $Hb \cdot L^{-1}$, can carry O_2 at concentrations as high as $0.01M$, about the same as in air. [Although many invertebrate species have hemoglobin-based oxygen transport systems, others produce one of two alternative types of O_2-binding proteins: (1) **hemocyanin,** a Cu-containing protein that is blue in complex with oxygen and colorless otherwise; or (2) **hemerythrin,** a nonheme Fe-containing protein that is burgundy colored in complex with oxygen and colorless otherwise. Antarctic icefish, the only adult vertebrates that lack hemoglobin—their blood is colorless—are viable because of their reduced need for O_2 at low temperatures combined with the relatively high aqueous solubility of O_2 at the $-1.9°C$ temperature of their environment (recall that the solubilities of gases increase with decreasing temperature.]

Although Mb was originally assumed to function only to store oxygen, it is now apparent that *its major physiological role is to facilitate oxygen transport in rapidly respiring muscle.* The rate that O_2 can diffuse from the capillaries to the tissues, and thus the level of respiration, is limited by the oxygen's low solubility in aqueous solution. Mb increases the effective solubility of O_2 in muscle, the most rapidly respiring tissue under conditions of high exertion. Thus, in rapidly respiring muscle, Mb functions as a kind of molecu-

lar bucket brigade to facilitate O_2 diffusion. The O_2 storage function of Mb is probably only significant in aquatic mammals such as seals and whales, which have Mb concentrations in their muscles around 10-fold greater than those in terrestrial mammals.

In this section, we begin our discussions of hemoglobin by considering its chemical and physical properties and how they relate to its physiological function. Hemoglobin structure and the mechanisms by which it carries out these physiological functions are discussed in Section 9-2.

A. Heme

*Mb and each of the four subunits of Hb noncovalently bind a single **heme** group (Fig. 9-1; spelled "haem" in British English).* This is the same group that occurs in the cytochromes (Section 8-3A) and in certain redox enzymes such as **catalase.** Heme is responsible for the characteristic red color of blood and is the site at which each **globin** monomer binds one molecule of O_2 (globins are the heme-free proteins of Hb and Mb). The heterocyclic ring system of heme is a **porphyrin** derivative; it consists of four **pyrrole** rings (lettered A–D in Fig. 9-1) linked by methene bridges. The porphyrin in heme, with its particular arrangement of four methyl, two propionate, and two vinyl substituents, is known as **protoporphyrin IX.** Heme, then, is protoporphyrin IX with a centrally bound iron atom. *In Hb and Mb,*

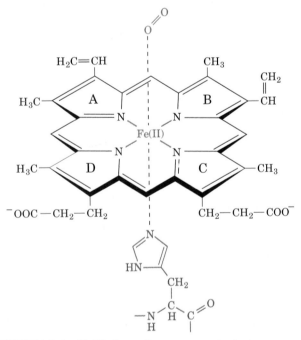

FIGURE 9-1. Fe(II)–heme (ferroprotoporphyrin IX) shown liganded to His and O_2 as it is in oxygenated myoglobin and oxygenated hemoglobin. Note that the heme is a conjugated system so that, although two of its Fe—N bonds are coordinate covalent bonds (bonds in which the bonding electron pair is formally contributed by only one of the atoms forming the bond), all of the Fe—N bonds are equivalent. The pyrrole ring lettering scheme is shown.

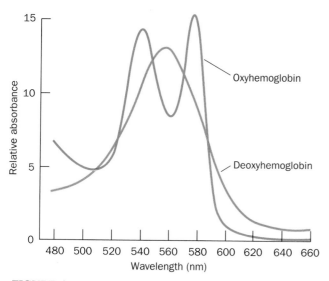

FIGURE 9-2. The visible absorption spectra of oxygenated and deoxygenated hemoglobins.

the iron atom normally remains in the Fe(II) (ferrous) oxidation state whether or not the heme is oxygenated (binds O_2).

The Fe atom in deoxygenated Hb and Mb is 5-coordinated by a square pyramid of N atoms: four from the porphyrin and one from a His side chain of the protein. Upon oxygenation, the O_2 binds to the Fe(II) on the opposite side of the porphyrin ring from the His ligand so that the Fe(II) is octahedrally coordinated; that is, the ligands occupy the six corners of an octahedron centered on the Fe atom (Fig. 9-1). *Oxygenation changes the electronic state of the Fe(II)–heme as is indicated by the color change of blood from the dark purplish hue characteristic of venous blood to the brilliant scarlet color of arterial blood and blood from a cut finger (Fig. 9-2).*

Certain small molecules, such as CO, NO, and H_2S, coordinate to the sixth liganding position of the Fe(II) in Hb and Mb with much greater affinity than does O_2. This, together with their similar binding to the hemes of cytochromes, accounts for the highly toxic properties of these substances.

The Fe(II) of Hb or Mb can be oxidized to Fe(III) to form **methemoglobin (metHb)** or **metmyoglobin (metMb).** MetHb does not bind O_2; its Fe(III) is already octahedrally coordinated with an H_2O molecule in the sixth liganding position. The brown color of dried blood and old meat is that of metHb and metMb. Erythrocytes (red blood cells) contain the enzyme **methemoglobin reductase,** which converts the small amount of metHb that spontaneously forms back to the Fe(II) form.

B. Oxygen Binding

The binding of O_2 to myoglobin is described by a simple equilibrium reaction

$$Mb + O_2 \rightleftharpoons MbO_2$$

with dissociation constant

$$K = \frac{[Mb][O_2]}{[MbO_2]} \qquad [9.1]$$

(biochemists usually express equilibria in terms of dissociation constants, the reciprocals of the more chemically traditional association constants). The O_2 dissociation of Mb may be characterized by its **fractional saturation,** Y_{O_2}, defined as the fraction of O_2-binding sites occupied by O_2.

$$Y_{O_2} = \frac{[MbO_2]}{[Mb] + [MbO_2]} = \frac{[O_2]}{K + [O_2]} \qquad [9.2]$$

Since O_2 is a gas, its concentration is conveniently expressed by its partial pressure, pO_2 (also called the **oxygen tension**). Equation [9.2] may therefore be expressed:

$$Y_{O_2} = \frac{pO_2}{K + pO_2} \qquad [9.3]$$

Now define p_{50} as the value of pO_2 when $Y_{O_2} = 0.50$, that is, when half of myoglobin's O_2-binding sites are occupied. Substituting this value into equation [9.3] and solving for K yields $K = p_{50}$. Hence our expression for the fractional saturation of Mb finally becomes:

$$Y_{O_2} = \frac{pO_2}{p_{50} + pO_2} \qquad [9.4]$$

Hemoglobin Cooperatively Binds O_2

Myoglobin's O_2-dissociation curve (Fig. 9-3) closely follows the hyperbolic curve described by Eq. [9.4]; its p_{50} is 2.8 torr (1 torr = 1 mm Hg at $0°C$ = 0.133 kPa; 760 torr = 1 atm). Mb therefore gives up little of its bound O_2 over the normal physiological range of pO_2 in blood (100 torr in

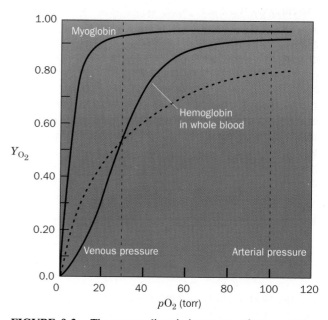

FIGURE 9-3. The oxygen dissociation curves of Mb and Hb in whole blood. The normal sea level values of human arterial and venous pO_2 values are indicated. The dashed line is a hyperbolic O_2 dissociation curve with the same p_{50} as Hb (26 torr).

arterial blood and 30 torr in venous blood); for example, $Y_{O_2} = 0.97$ at $pO_2 = 100$ torr and 0.91 at 30 torr. In contrast, hemoglobin's O_2-dissociation curve (Fig. 9-3), which has a **sigmoidal** shape (S shape) that Eq. [9.4] does not describe, indicates that the amount of O_2 bound by Hb changes significantly over the normal physiological range of pO_2 in blood; for example, $Y_{O_2} = 0.95$ at 100 torr and 0.55 at 30 torr in whole blood for a difference in Y_{O_2} of 0.40. Mb therefore binds O_2 under conditions that Hb releases it. Thus, the two proteins form a sophisticated O_2 transport system that delivers O_2 from lung to muscle (where pO_2 may be <20 torr). Hemoglobin's sigmoidal O_2-dissociation curve is of great physiological importance; *it permits the blood to deliver much more O_2 to the tissues than it could if Hb had a hyperbolic O_2-dissociation curve with the same p_{50} (26 torr; dashed curve in Fig. 9-3). Such a hyperbolic curve has $Y_{O_2} = 0.79$ at 100 torr and 0.54 at 30 torr for a difference in Y_{O_2} of only 0.25.*

A sigmoidal dissociation curve is diagnostic of a **cooperative interaction** between a protein's small molecule binding sites; that is, the binding of one small molecule affects the binding of others. In this case, the binding of O_2 increases the affinity of Hb for binding additional O_2. The structural mechanism of hemoglobin cooperativity is described in Section 9-2C.

The Hill Equation Phenomenologically Describes Hemoglobin's O_2-Binding Curve

The earliest attempt to analyze hemoglobin's sigmoidal O_2-dissociation curve was formulated by Archibald Hill in 1910. We shall follow his analysis in general form because it is useful for characterizing the cooperative behavior of oligomeric enzymes as well as that of hemoglobin.

Consider a protein E consisting of n subunits that can each bind a molecule S, which, in analogy with the substituents of metal ion complexes, is known as a **ligand**. Assume that the ligand binds with infinite cooperativity,

$$E + nS \rightleftharpoons ES_n$$

that is, the protein either has all or none of its ligand-binding sites occupied so that there are no observable intermediates ES_1, ES_2, etc. The dissociation constant for this reaction is

$$K = \frac{[E][S]^n}{[ES_n]} \qquad [9.5]$$

and, as before, its fractional saturation is expressed:

$$Y_S = \frac{n[ES_n]}{n([E] + [ES_n])} \qquad [9.6]$$

Combining Eqs. [9.5] and [9.6] yields

$$Y_S = \frac{[E][S]^n/K}{[E](1 + [S]^n/K)}$$

which upon algebraic rearrangement and cancellation of terms becomes the **Hill equation:**

$$Y_S = \frac{[S]^n}{K + [S]^n} \qquad [9.7]$$

which, in a manner analogous to Eq. [9.4], describes the degree of saturation of a multisubunit protein as a function of ligand concentration.

Infinite ligand-binding cooperativity (n equal to the number of protein subunits), as assumed in deriving the Hill equation, is a physical impossibility. Nevertheless, n may be taken to be a nonintegral parameter related to the degree of cooperativity among interacting ligand-binding sites rather than the number of subunits per protein. The Hill equation then becomes a useful empirical curve-fitting relationship rather than an indicator of a particular model of ligand binding. *The quantity n, the **Hill constant**, increases with the degree of cooperativity of a reaction and thereby provides a convenient, although simplistic, characterization of a ligand-binding reaction.* If $n = 1$, Eq. [9.7] describes a hyperbola, as do Eqs. [9.3] and [9.4] for Mb, and the ligand-binding reaction is said to be **noncooperative**. A reaction with $n > 1$ is described as **positively cooperative**: Ligand binding increases the affinity of E for further ligand binding (cooperativity is infinite in the limit that n is equal to the number of ligand-binding sites in E). Conversely, if $n < 1$, the reaction is termed **negatively cooperative**: Ligand binding reduces the affinity of E for subsequent ligand binding.

Hill Equation Parameters May Be Graphically Evaluated

The Hill constant, n, and the dissociation constant, K, that best describe a saturation curve can be graphically determined by rearranging Eq. [9.7] as follows:

$$\frac{Y_S}{1 - Y_S} = \frac{\dfrac{[S]^n}{K + [S]^n}}{1 - \dfrac{[S]^n}{K + [S]^n}} = \frac{[S]^n}{K}$$

and then taking the log of both sides to yield a linear equation:

$$\log\left(\frac{Y_S}{1 - Y_S}\right) = n \log[S] - \log K \qquad [9.8]$$

The linear plot of $\log[Y_S/(1 - Y_S)]$ versus $\log[S]$, the **Hill plot**, has a slope of n and an intercept on the $\log[S]$ axis of $(\log K)/n$ (recall that the linear equation $y = mx + b$ describes a line with a slope of m and an x intercept of $-b/m$).

For Hb, if we substitute pO_2 for $[S]$ as was done for Mb, the Hill equation becomes:

$$Y_{O_2} = \frac{(pO_2)^n}{K + (pO_2)^n} \qquad [9.9]$$

As in Eq. [9.4], let us define p_{50} as the value of pO_2 at $Y_{O_2} = 0.50$. Then, substituting this value into Eq. [9.9],

$$0.50 = \frac{(p_{50})^n}{K + (p_{50})^n}$$

so that

$$K = (p_{50})^n \qquad [9.10]$$

Substituting this result back into Eq. [9.9] yields

$$Y_{O_2} = \frac{(pO_2)^n}{(p_{50})^n + (pO_2)^n} \qquad [9.11]$$

(*Note:* Eq. [9.4] is a special case of Eq. [9.11] with $n = 1$.) Equation [9.8] for the Hill plot of Hb therefore takes the form

$$\log\left(\frac{Y_{O_2}}{1 - Y_{O_2}}\right) = n \log pO_2 - n \log p_{50} \qquad [9.12]$$

so that *this plot has a slope of n and an intercept on the log pO_2 axis of p_{50}.*

Figure 9-4 shows the Hill plots for Mb and Hb. For Mb it is linear with a slope of 1, as expected. Although Hb does not bind O_2 in a single step as is assumed in deriving the Hill equation, its Hill plot is essentially linear for values of Y_{O_2} between 0.1 and 0.9. Its maximum slope, which occurs near $pO_2 = p_{50}$ [$Y_{O_2} = 0.5$; $Y_{O_2}/(1 - Y_{O_2}) = 1$], is normally taken to be the Hill constant. For normal human Hb, the Hill constant is between 2.8 and 3.0; that is, hemoglobin oxygen binding is highly, but not infinitely, cooperative. Many abnormal hemoglobins exhibit smaller Hill constants (Section 9-3A), indicating that they have a less than normal degree of cooperativity. At Y_{O_2} values near 0, when few Hb

molecules have bound even one O_2 molecule, the Hill plot of Hb assumes a slope of 1 (Fig. 9-4, lower asymptote) because the Hb subunits independently compete for O_2 as do molecules of Mb. At Y_{O_2} values near 1, when at least three of each of hemoglobin's four O_2-binding sites are occupied, the Hill plot also assumes a slope of 1 (Fig. 9-4, upper asymptote) because the few remaining unoccupied sites are on different molecules and therefore bind O_2 independently.

Extrapolating the lower asymptote in Fig. 9-4 to the horizontal axis indicates, according to Eq. [9.11], that $p_{50} = 30$ torr for binding the first O_2 to Hb. Likewise, extrapolating the upper asymptote yields $p_{50} = 0.3$ torr for binding hemoglobin's fourth O_2. Thus *the fourth O_2 to bind to Hb does so with 100-fold greater affinity than the first.* This difference, as we shall see in Section 9-2C, is entirely due to the influence of the globin on the O_2 affinity of heme. It corresponds to a free energy difference of 11.4 kJ·mol⁻¹ between binding the first and binding the last O_2 to Hb (Section 3-4A).

More sophisticated mathematical models than the Hill equation have been developed for analyzing the cooperative binding of ligands to proteins. They are examined in Section 9-4.

Globin Prevents Oxyheme from Autooxidizing

Globin not only modulates the O_2-binding affinity of heme, but makes reversible O_2 binding possible. Fe(II)–heme by itself is incapable of binding O_2 reversibly. Rather, in the presence of O_2, it autooxidizes irreversibly to the Fe(III) form through the intermediate formation of a complex consisting of an O_2 bridging the Fe atoms of two hemes. This reaction can be inhibited by derivatizing the heme with bulky groups that sterically prevent the close

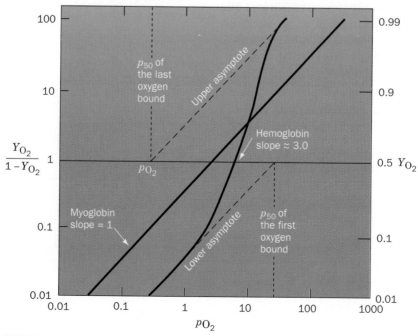

FIGURE 9-4. Hill plots for Mb and purified Hb. Note that this is a log–log plot. At $pO_2 = p_{50}$, the value of $Y_{O_2}/(1 - Y_{O_2}) = 1$.

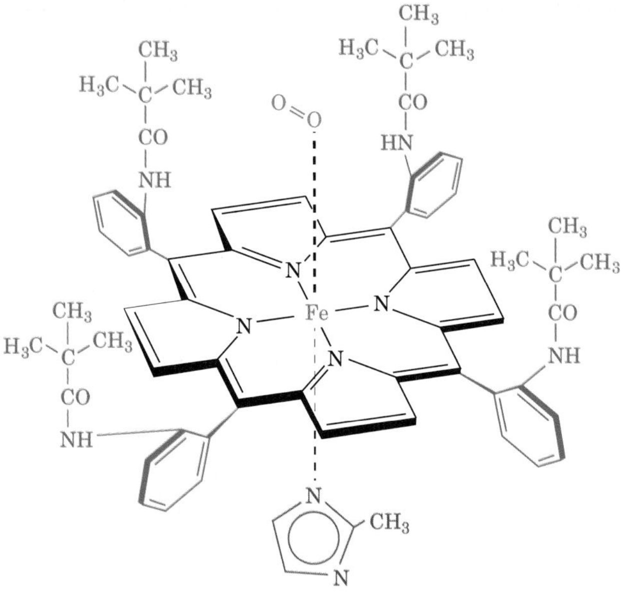

FIGURE 9-5. A picket-fence Fe(II)–porphyrin complex with bound O_2. [After Collman, J.P., Brauman, J.I., Rose, E., and Suslick, K.S., *Proc. Natl. Acad. Sci.* **75**, 1053 (1978).]

face-to-face approach of two hemes. Such **picket-fence** Fe(II)–porphyrin complexes (Fig. 9-5), which James Collman first synthesized, bind O_2 reversibly. The backside of this porphyrin is unhindered and is complexed with a substituted imidazole in a manner similar to that in Mb and Hb. In fact, the O_2 affinity of the picket-fence complex is similar to that of Mb. Thus, the globins of Mb and Hb function to prevent the autooxidation of oxyheme by surrounding it, rather like a hamburger bun surrounds a hamburger, so that only its propionate side chains are exposed to the aqueous solvent (Section 9-2B).

C. Carbon Dioxide Transport and the Bohr Effect

In addition to being an O_2 carrier, *Hb plays an important role in the transport of CO_2 by the blood.* When Hb (but not Mb) binds O_2 at physiological pH's, it undergoes a conformational change (Section 9-2B) that makes it a slightly stronger acid. It therefore releases protons on binding O_2:

$$Hb(O_2)_n H_x + O_2 \rightleftharpoons Hb(O_2)_{n+1} + xH^+$$

where $n = 0, 1, 2,$ or 3 and $x \approx 0.6$ under physiological conditions. Conversely, *increasing the pH, that is, removing protons, stimulates Hb to bind O_2* (Fig. 9-6). This phenomenon, whose molecular basis is discussed in Section 9-2E, is known as the **Bohr effect** after Christian Bohr (the father of the pioneering atomic physicist Niels Bohr) who first reported it in 1904.

The Bohr Effect Facilitates O_2 Transport

The ~0.8 molecules of CO_2 formed per molecule of O_2 consumed by respiration diffuse from the tissues to the capillaries largely as dissolved CO_2 as a result of the slowness of the reaction forming bicarbonate:

$$CO_2 + H_2O \rightleftharpoons H^+ + HCO_3^-$$

This reaction, however, is catalyzed in the erythrocyte by carbonic anhydrase (Fig. 7-44). Accordingly, most of the CO_2 in the blood is carried in the form of bicarbonate (in the absence of carbonic anhydrase, the hydration of CO_2 would equilibrate 100-fold more slowly, so bubbles of the only slightly soluble CO_2 would form in the blood and tissues).

In the capillaries, where pO_2 is low, the H^+ generated by bicarbonate formation is taken up by Hb, which is thereby induced to unload its bound O_2. This H^+ uptake, moreover, facilitates CO_2 transport by stimulating bicarbonate formation. Conversely, in the lungs, where pO_2 is high, O_2 binding by Hb releases the Bohr protons, which drive off the CO_2. These reactions are closely matched, so they cause very little change in blood pH.

The Bohr effect provides a mechanism whereby additional O_2 can be supplied to highly active muscles. Such muscles generate acid (Section 16-3A) so fast that they lower the pH of the blood passing through them from 7.4 to 7.2. At pH 7.2, Hb releases ~10% more O_2 at the < 20 torr pO_2 in these muscles than it does at pH 7.4 (Fig. 9-6).

CO_2 and Cl^- Modulate Hemoglobin's O_2 Affinity

CO_2 modulates O_2 binding directly and by combining reversibly with the N-terminal amino groups of blood proteins to form **carbamates**:

$$R-NH_2 + CO_2 \rightleftharpoons R-NH-COO^- + H^+$$

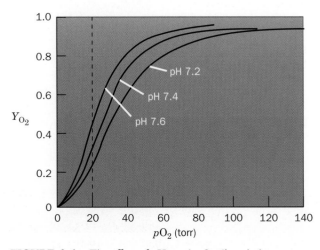

FIGURE 9-6. The effect of pH on the O_2-dissociation curve of Hb: The Bohr effect. The vertical dashed line indicates the pO_2 in actively respiring muscle tissue. [After Benesch, R.E. and Benesch, R., *Adv. Protein Chem.* **28**, 212 (1974).]

The conformation of deoxygenated Hb (**deoxyHb**), as we shall see in Section 9-2B, is significantly different from that of oxygenated Hb (**oxyHb**). Consequently, deoxyHb binds more CO_2 as carbamate than does oxyHb. CO_2, like H^+, is therefore a modulator of hemoglobin's O_2 affinity: A high CO_2 concentration, as occurs in the capillaries, stimulates Hb to release its bound O_2. Note the complexity of this Hb–O_2–CO_2–H^+ equilibrium: The protons released by carbamate formation are, in part, taken up through the Bohr effect, thereby increasing the amount of O_2 that Hb would otherwise release. Although the difference in CO_2 binding between the oxy and deoxy states of hemoglobin accounts for only ~5% of the total blood CO_2, it is nevertheless responsible for around one half of the CO_2 transported by blood. This is because only ~10% of the total blood CO_2 turns over in each circulatory cycle.

Cl^- is also bound more tightly to deoxyHb than to oxyHb (Section 9-2E). Accordingly, hemoglobin's O_2 affinity also varies with $[Cl^-]$. HCO_3^- freely permeates the erythrocyte membrane (Section 11-3C) so that once formed, it equilibrates with the surrounding plasma. The need for charge neutrality on both sides of the membrane, however, requires that Cl^-, which also freely permeates the membrane, replace the HCO_3^- that leaves the erythrocyte (the erythrocyte membrane is impermeable to cations). Consequently, $[Cl^-]$ in the erythrocyte is greater in the venous blood than it is in the arterial blood. *Cl^- is therefore also a modulator of hemoglobin's O_2 affinity.*

D. Effect of BPG on O_2 Binding

Purified (stripped) hemoglobin has a much greater O_2 affinity than does hemoglobin in whole blood (Fig. 9-7). This observation led Joseph Barcroft, in 1921, to speculate that blood contains some other substance that complexes with Hb so as to reduce its O_2 affinity. In 1967, Reinhold and Ruth Benesch demonstrated that this substance is **D-2,3-bisphosphoglycerate (BPG)**

D-2,3-Bisphosphoglycerate (BPG)

[previously known as **2,3-diphosphoglycerate (DPG)**]. BPG binds tightly to deoxyHb in a 1:1 mole ratio ($K = 1.5 \times 10^{-5}M$) but only weakly to oxyHb. The presence of BPG therefore decreases hemoglobin's oxygen affinity by keeping it in the deoxy conformation; for example, the p_{50} of stripped hemoglobin is increased from 12 to 22 torr by 4.7 mM BPG, its normal concentration in erythrocytes

(similar to that of Hb). Organic polyphosphates, such as **inositol hexaphosphate (IHP)** and **adenosine triphosphate (ATP),**

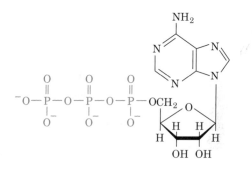

Inositol hexaphosphate (IHP)

Adenosine triphosphate (ATP)

also have this effect on Hb. In fact, in birds, IHP functionally replaces BPG and ATP does so in fish and most amphibians. The ~2 mM ATP normally present in mammalian erythrocytes is prevented from binding to Hb by its complexation with Mg^{2+}.

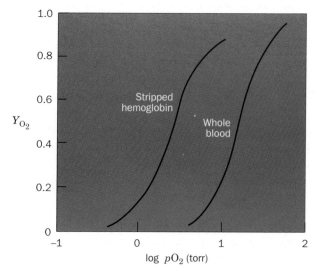

FIGURE 9-7. Comparison of the O_2-dissociation curves of "stripped" Hb and whole blood in 0.01M NaCl at pH 7.0. [After Benesch, R.E. and Benesch, R., *Adv. Protein Chem.* **28,** 217 (1974).]

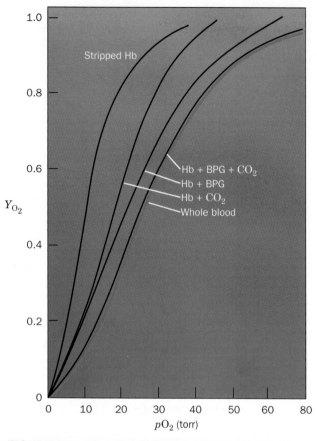

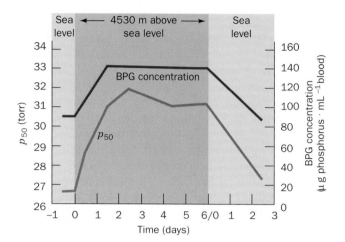

FIGURE 9-9. The effect of high-altitude exposure on the p_{50} and the BPG concentration of blood in sea level–adapted individuals. The region on the right marked "Sea level" indicates the effects of exposure to sea level on high altitude–adapted individuals. [After Lenfant, C., Torrance, J.D., English, E., Finch, C.A., Reynafarje, C., Ramos, J., and Faura, J., *J. Clin. Invest.* **47,** 2653 (1968).]

FIGURE 9-8. The effects of BPG and CO_2, both separately and combined, on hemoglobin's O_2-dissociation curve compared with that of whole blood (*red curve*). In the Hb solutions, which were $0.1M$ KCl and pH 7.22, $pCO_2 = 40$ torr and the BPG concentration was 1.2 times that of Hb. The blood had $pCO_2 = 40$ torr and a plasma pH of 7.40, which corresponds to a pH of 7.22 inside the erythrocyte. [After Kilmartin, J.V. and Rossi-Bernardi, L., *Physiol. Rev.* **53,** 884 (1973).]

BPG has an indispensable physiological function: In arterial blood, where pO_2 is ~100 torr, Hb is ~95% saturated with O_2, but in venous blood, where pO_2 is ~30 torr, it is only ~55% saturated (Fig. 9-3). Consequently, in passing through the capillaries, Hb unloads ~40% of its O_2. *In the absence of BPG, little of this O_2 is released since hemoglobin's O_2 affinity is increased, thus shifting the O_2-dissociation curve significantly towards lower pO_2 (Fig. 9-8, left).*

CO_2 and BPG independently modulate hemoglobin's O_2 affinity. Figure 9-8 indicates that stripped Hb can be made to have the same oxygen-dissociation curve as the Hb in whole blood by adding CO_2 and BPG in the concentrations found in erythrocytes (the pH and [Cl$^-$] are also the same). Hence, *the presence of these four substances in whole blood*

—*BPG, CO_2, H^+, and Cl^-—accounts for the O_2-binding properties of Hb.*

Increased BPG Levels Are Partially Responsible for High-Altitude Adaptation

High-altitude adaptation is a complex physiological process that involves an increase in the amount of hemoglobin per erythrocyte and in the number of erythrocytes. It normally requires several weeks to complete. Yet, as is clear to anyone who has climbed to high altitude, even a 1-day stay there results in a noticeable degree of adaptation. This effect results from a rapid increase in the erythrocyte BPG concentration (Fig. 9-9; BPG, which cannot pass through the erythrocyte membrane, is synthesized in the erythrocyte; Section 16-2H). The consequent decrease in O_2-binding affinity, as indicated by its elevated p_{50}, increases the amount of O_2 that hemoglobin unloads in the capillaries (Fig. 9-10). Similar increases in BPG concentration occur in individuals suffering from disorders that limit the oxygenation of the blood (**hypoxia**), such as various anemias and cardiopulmonary insufficiency.

Fetal Hemoglobin Has a Low BPG Affinity

The effects of BPG also help supply the fetus with oxygen. A fetus obtains its O_2 from the maternal circulation via the placenta. This process is facilitated because fetal hemoglobin (HbF) has a higher O_2 affinity than does maternal hemoglobin (HbA; recall that HbF has the subunit compo-

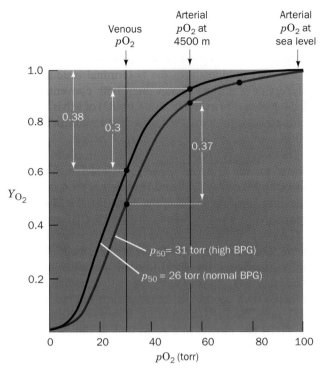

FIGURE 9-10. The O_2-dissociation curves of blood adapted to sea level (*black curve*) and to high altitude (*red curve*). Between the sea level arterial and venous pO_2 values of 100 and 30 torr, respectively, Hb normally unloads 38% of the O_2 it can maximally carry. However, when the arterial pO_2 drops to 55 torr, as it does at an altitude of 4500 m, this difference is reduced to 30% in nonadapted blood. High-altitude adaptation increases the BPG concentration in erythrocytes, which shifts the O_2-dissociation curve of Hb to the right. The amount of O_2 that Hb delivers to the tissues is thereby restored to 37% of its maximum load.

sition $\alpha_2\gamma_2$, in which the γ subunit is a variant of HbA's β subunit; Section 6-3C). BPG occurs in about the same concentrations in adult and fetal erythrocytes but binds more tightly to deoxyHbA than to deoxyHbF; this accounts for HbF's greater O_2 affinity. In the next section we shall develop the structural rationale for the effect of BPG and for the other aspects of O_2 binding.

2. STRUCTURE AND MECHANISM

The determination of the first protein X-ray structures, those of sperm whale myoglobin by John Kendrew in 1959 and of human deoxyhemoglobin and horse methemoglo-

bin by Max Perutz shortly thereafter, ushered in a revolution in biochemical thinking that has reshaped our understanding of the chemistry of life. Before the advent of protein crystallography, macromolecular structures, if they were considered at all, were thought of as having a rather hazy existence of uncertain biological significance. However, as the elucidation of macromolecular structures has continued at an ever quickening pace, it has become clear that *life is based on the interactions of complex, structurally well-defined macromolecules.*

The story of hemoglobin's structural determination is a tale of enormous optimism and tenacity. Perutz began this study in 1937 at Cambridge University as a graduate student of J. D. Bernal (who, with Dorothy Crowfoot Hodgkin, had taken the first X-ray diffraction photographs of hydrated protein crystals in 1934). In 1937, the X-ray crystal structure determination of even the smallest molecule required many months of hand computation and the largest structure yet determined was that of the dye phthalocyanin, which has 40 nonhydrogen atoms. Since hemoglobin has ~4500 nonhydrogen atoms, it must have seemed to Perutz's colleagues that he was pursuing an impossible goal. Nevertheless, the laboratory director, Lawrence Bragg (who in 1912, with his father William Bragg, had determined the first X-ray structure, that of NaCl), realized the tremendous biological significance of determining a protein structure and supported the project.

It was not until 1953 that Perutz finally hit upon the method that would permit him to solve the X-ray structure of hemoglobin, that of isomorphous replacement. Kendrew, a colleague of Perutz, used this technique to solve the X-ray structure of sperm whale myoglobin, first at low resolution in 1957, and then at high resolution in 1959. Hemoglobin's greater complexity delayed its low-resolution structural determination until 1959, and it was not until 1968, over 30 years after he had begun the project, that Perutz and his associates obtained the high-resolution X-ray structure of horse methemoglobin. Those of human and horse deoxyhemoglobins followed shortly thereafter. Since then, the X-ray structures of hemoglobins from numerous different species, from mutational variants, and with different bound ligands have been elucidated. This, together with many often ingenious physicochemical investigations, has made hemoglobin the most intensively studied, and perhaps the best understood, of proteins.

In this section, we examine the molecular structures of myoglobin and hemoglobin and consider the structural basis of hemoglobin's oxygen-binding cooperativity, the Bohr effect and BPG binding.

A. *Structure of Myoglobin*

Myoglobin consists of eight helices (labeled A–H) that are linked by short polypeptide segments to form an ellipsoidal molecule of approximate dimensions $44 \times 44 \times 25$ Å (Fig.

(a)

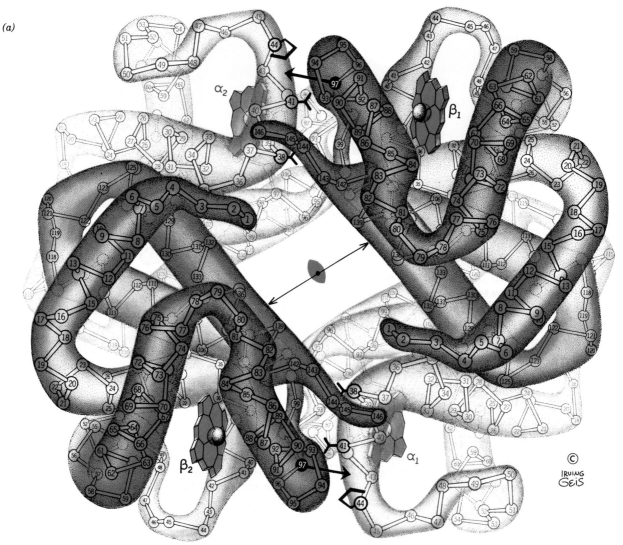

FIGURE 9-13.   The structures of (*a*) deoxyHb and (*b*) oxyHb (*Opposite*), as viewed down their exact twofold axes. The C_α atoms, numbered from each N-terminus, and the heme groups are shown. The Hb tetramer contains a solvent-filled central channel paralleling its twofold axis whose flanking β chains draw closer together upon oxygenation (compare the lengths of double-headed arrows). In the deoxy state, His FG4(97)β (*small single-headed arrow*) fits between Thr C6(41)α and Pro CD2(44)α (*lower right* and *upper left*). The relative movements of the two $\alpha\beta$ protomers upon oxygenation (*large gray arrows in Part b*) shift His FG4(97)β to a new position between Thr C3(38)α and Thr C6(41)α. See Fig. 7-57 for a similarly viewed space-filling model of deoxyHb. [Figure copyrighted © by Irving Geis.]

9-2C). In contrast, contacts between like subunits, $\alpha_1-\alpha_2$ and $\beta_1-\beta_2$, are few and largely polar in character. This is because like subunits face each other across an ~20-Å-diameter solvent-filled channel that parallels the 50-Å length of the exact twofold axis (Figs. 7-57 and 9-13).

Oxy- and Deoxyhemoglobins Have Different Quaternary Structures

Oxygenation causes such extensive quaternary structural changes to Hb that oxy- and deoxyHb have different crystalline forms; indeed, crystals of deoxyHb shatter upon exposure to O_2. The crystal structures of hemoglobin's oxy and deoxy forms therefore had to be determined independently. *The quaternary structural change preserves hemoglobin's exact twofold symmetry and takes place entirely across its $\alpha_1-\beta_2$ (and $\alpha_2-\beta_1$) interface.* The $\alpha_1-\beta_1$ (and $\alpha_2-\beta_2$) contact is unchanged, presumably as a result of its more extensive close associations. This contact provides a convenient frame of reference from which the oxy and deoxy conformations may be compared. Viewed in this way, oxygenation rotates the $\alpha_1\beta_1$ dimer ~15° with respect to the $\alpha_2\beta_2$ dimer (Fig. 9-14) so that some atoms at the α_1-

(b)

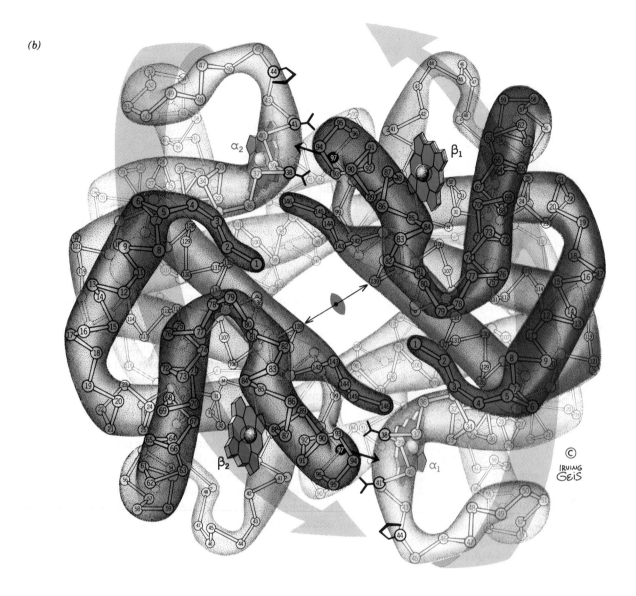

β_2 interface shift by as much as 6 Å relative to each other (compare Fig. 9-13*a* and *b*).

The quaternary conformation of deoxyHb is named the **T state.** That of oxyHb, which is essentially independent of the ligand used to induce it (e.g., O_2, met, CO, CN^-, and NO hemoglobins all have the same quaternary structure), is called the **R state.** Similarly, the tertiary conformational states for the deoxy and liganded subunits are designated as the **t** and **r states,** respectively. The structural differences between these quaternary and these tertiary conformations are described in the following section in terms of hemoglo-

bin's O_2-binding mechanism.

C. Mechanism of Oxygen-Binding Cooperativity

The positive cooperativity of O_2 binding to Hb arises from the effect of the ligand-binding state of one heme on the ligand-binding affinity of another. Yet, the distances of 25 to 37 Å between the hemes in an Hb molecule are too large for these heme–heme interactions to be electronic in character. Rather, *they are mechanically transmitted by the pro-*

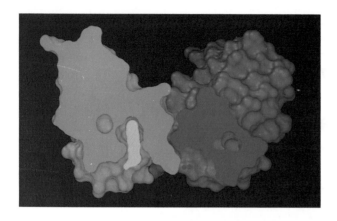

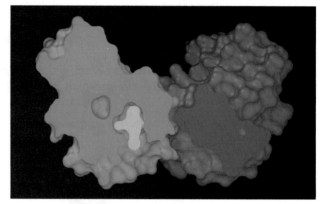

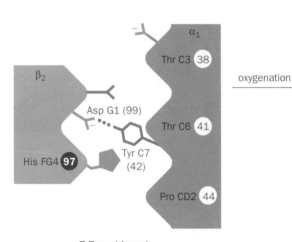

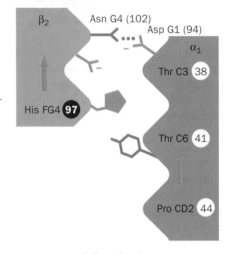

T Form (deoxy)   **R Form (oxy)**

FIGURE 9-17. Surface drawings of the $\alpha_1\beta_2$ dimer of (*a*) T-state Hb, and (*b*) R-state Hb, that have been sectioned perpendicular to the protein's exact twofold axis so as to show its α_1C–β_2FG interface. Each drawing is accompanied by a corresponding schematic diagram of the α_1–β_2 contact. Upon a T→R transformation, this contact snaps from one position to the other with no stable intermediate. The subunits are joined by different sets of hydrogen bonds in the two quarternary states. Figures 9-13, 9-14 and 9-19 provide additional structural views of these interactions. [Courtesy of Michael Pique, Research Institute of Scripps Clinic, San Diego, California.]

associations throughout the change (Fig. 9-19). *These side chains therefore act as flexible joints or hinges about which the α_1 and β_2 subunits pivot during the quaternary change.*

The T State Is Stabilized by a Network of Salt Bridges That Must Break To Form the R State

The R state is stabilized by ligand binding. But in the absence of ligand, why is the T state more stable than the R state? In the electron density maps of R-state Hb, the C-terminal residues of each subunit (Arg 141α and His 146β) appear as a blur, which suggests that these residues are free to wave about in solution. Maps of the T form, however, show these residues firmly anchored in place via several intersubunit and intrasubunit salt bridges which evidently help stabilize the T state (Figs. 9-18 and 9-19). *The structural changes accompanying the T→R transition tear away these salt bridges in a process driven by the Fe—O_2 bond's energy of formation.*

Hemoglobin's O_2-Binding Cooperativity Derives from the T→R Conformational Shift

The hemoglobin molecule resembles a finely tooled mechanism that has very little slop. The binding of O_2 requires a series of tightly coordinated movements:

1. The Fe(II) of any subunit cannot move into its heme plane without the reorientation of its proximal His so as to prevent this residue from bumping into the porphyrin ring.

2. The proximal His is so tightly packed by its surrounding groups that it cannot reorient unless this movement is accompanied by the previously described translation of the F helix across the heme plane.

3. The F helix translation is only possible in concert with the quaternary shift that steps the α_1C–β_2FG contact one turn along the α_1C helix.

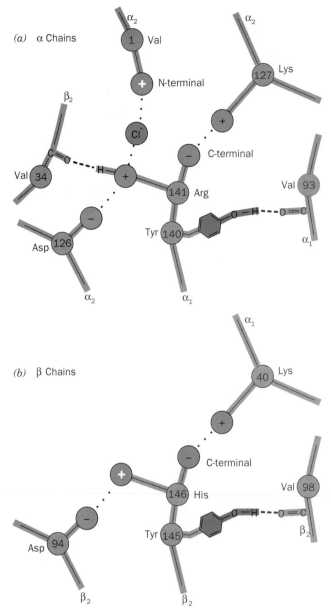

(a) α Chains

(b) β Chains

FIGURE 9-18. Networks of salt bridges and hydrogen bonds in deoxyHb involving the last two residues of *(a)* the α chains, and *(b)* the β chains of deoxyHb. All of these bonds are ruptured in the T→R transition. The two groups that participate in the Bohr effect by becoming partially deprotonated in the R state are indicated by white plus signs. [Figure copyrighted © by Irving Geis.]

4. The inflexibility of the $\alpha_1-\beta_1$ and $\alpha_2-\beta_2$ interfaces requires that this shift simultaneously occur at both the $\alpha_1-\beta_2$ and the $\alpha_2-\beta_1$ interfaces.

Consequently, *no one subunit or dimer can greatly change its conformation independently of the others. Indeed, the two stable positions of the $\alpha_1C-\beta_2FG$ contact limit the Hb molecule to only two quaternary forms, R and T.*

We are now in a position to structurally rationalize hemoglobin's O_2-binding cooperativity. Any deoxyHb subunit binding O_2 is constrained to remain in the t state by the T conformation of the tetramer. However, *the t state has reduced O_2 affinity, most probably because its Fe—O_2 bond is stretched beyond its normal length by the steric repulsions between the heme and the O_2, and in the β subunits, by the need to move Val E11 out of the O_2-binding site.* As more O_2 is bound to the Hb tetramer, this strain, which derives from the Fe—O_2 bond energy, accumulates in the liganded subunits until it is of sufficient strength to snap the molecule into the R conformation. All the subunits are thereby converted to the r state whether or not they are liganded. *Unliganded subunits in the r state have an increased O_2 affinity because they are already in the O_2-binding conformation.* This accounts for the high O_2 affinity of nearly saturated Hb.

Hemoglobin's Sigmoidal O_2-Binding Curve Is a Composite of Its Hyperbolic R- and T-State Curves

The relative stabilities of the T and R states, as indicated by their free energies, vary with fractional saturation (Fig. 9-20*a*). In the absence of ligand, the T state is more stable than the R state, and vice versa when all ligand-binding sites are occupied. The formation of Fe—O_2 bonds causes the free energy of both the T and the R states to decrease (become more stable) with oxygenation, although the rate of this decrease is smaller for the T state as a result of the strain that liganding imposes on t-state subunits. The $R \rightleftharpoons T$ transformation is, of course, an equilibrium process so that Hb molecules, at intermediate levels of fractional saturation (1, 2, or 3 bound O_2 molecules), continually interconvert between the R and the T states.

The O_2-binding curve of Hb can be understood as a composite of those of its R and T states (Fig. 9-20*b*). For pure states, such as R or T, these curves are hyperbolic because ligand binding at one protomer is unaffected by the state of other protomers in the absence of a quaternary structural change. At low pO_2's, Hb follows the low-affinity T-state curve and at high pO_2's, it follows the high-affinity R-state curve. At intermediate pO_2's, Hb exhibits an O_2 affinity that changes from T-like to R-like as pO_2 increases. The switchover results in the sigmoidal shape of hemoglobin's O_2-binding curve.

D. Testing the Perutz Mechanism

The Perutz mechanism is a description of the dynamic behavior of Hb that is largely based on the static structures of its R and T end states. Accordingly, without the direct demonstration that Hb actually follows the postulated pathway in changing conformational states, the Perutz mechanism must be taken as being at least partially conjectural. Unfortunately, the physical methods that can follow dynamic changes in proteins are, as yet, incapable of providing de-

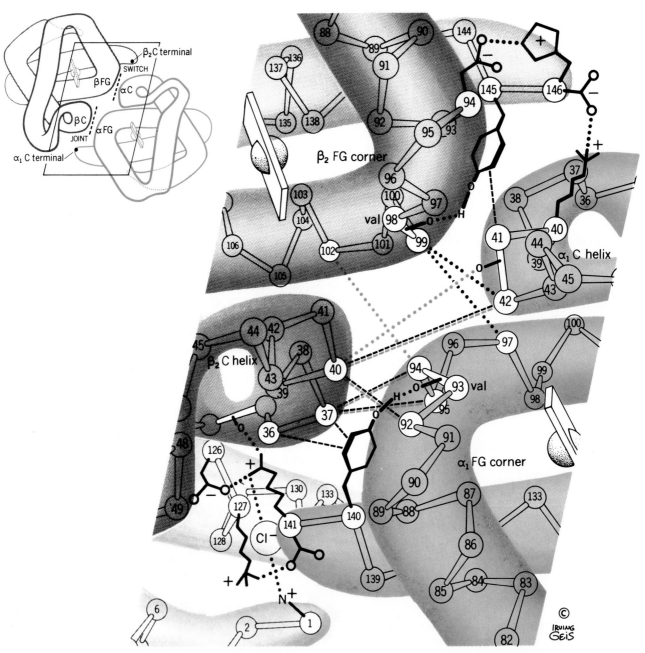

FIGURE 9-19. The hemoglobin $\alpha_1\beta_2$ interface as viewed perpendicularly to Fig. 9-13. The boxed area on the left is shown in greater detail on the right. Hydrogen bonds and salt bridges are represented by dotted lines, black for deoxyHb and blue for oxyHb, whereas van der Waals contacts are likewise indicated by dashed lines. Note that the α_1C–β_2FG interface (the "switch" region) undergoes significant readjustment in the T→R transition, whereas the pseudosymmetrically related α_1FG–β_2C interface (the "flexible joint") only undergoes small reorientations. Also note that the T-state salt bridges involving the C-terminal residues [Arg 141α (*below*) and His 146β (*above*)] are ruptured by the T→R transition. [Figure copyrighted © by Irving Geis.]

tailed descriptions of these changes. Nevertheless, certain aspects of the Perutz mechanism are supported by static measurements as is described below and in Section 9-3.

C-Terminal Salt Bridges Are Required to Maintain the T State

The proposed function of the C-terminal salt bridges in stabilizing the T state has been corroborated by chemically

modifying human Hb. Removal of the C-terminal Arg 141α (by treating isolated α chains with carboxypeptidase followed by reconstitution) drastically reduces the cooperativity of O_2 binding (Hill constant = 1.7; reduced from its normal value of 2.8). It is abolished by the further removal of the other C-terminal residue, His 146β (Hill constant ≈ 1.0). Apparently, in the absence of its C-terminal salt bridges, the T form of Hb is unstable. Indeed, human deoxy-

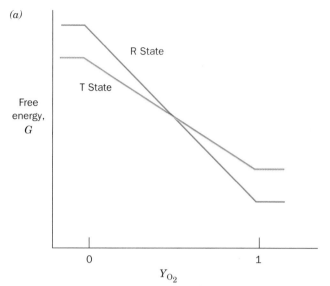

(a)

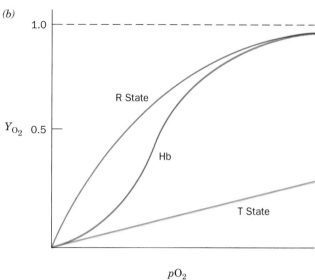

(b)

FIGURE 9-20. (*a*) The variation of the free energies of hemoglobin's T and R states with their fractional saturation, Y_{O_2}. In the absence of O_2, the T state is more stable and, when saturated with O_2, the R state is more stable. The free energy of both states is reduced with increasing oxygenation as a consequence of O_2 liganding. The Fe(II)—O_2 bonding is more

exergonic in the R state than it is in the T state, however, so that the relative stabilities of these two states reverse order at intermediate levels of oxygenation. (*b*) The sigmoid O_2-binding curve of Hb (*purple*) is a composite of its hyperbolic R-state (*red*) and T-state (*blue*) binding curves: It is more T-like at lower pO_2 values and more R-like at higher pO_2 values.

Hb, with its C-terminal residues removed, crystallizes in a form very similar to that of normal human oxyHb.

Fe—O_2 Bond Tension Has Been Spectroscopically Demonstrated

If movement of the Fe into the heme plane upon oxygenation is mechanically coupled via the proximal His to the T→R transformation, then conversely, forcing oxyHb into the T form must exert a tension on the Fe, through the proximal His, that tends to pull the Fe out of the heme plane. Perutz demonstrated the existence of this tension as follows. IHP's six phosphate groups cause it to bind to deoxyHb with much greater affinity than does BPG (the structural basis of BPG binding to Hb is discussed in Section 9-2F); the presence of IHP therefore tends to force Hb into the T state. Conversely, nitric oxide (NO) binds to Hb far more strongly than does O_2 and thereby tends to force Hb into the R state. Spectroscopic analysis indicates the consequences of simultaneously binding both NO and IHP to Hb:

1. The NO, as expected, pulls the Fe into the plane of the heme.

2. The IHP forces the Hb molecule into the T state which, through the "gears and levers" coupling the 4° and 3° conformational changes, pulls the proximal His in the opposite direction, away from the Fe.

The bond between the proximal His and the Fe lacks the strength to withstand these two opposing "irresistible" forces; it simply breaks. The spectroscopic observation of this phenomenon therefore confirms the existence of the heme – protein tension predicted by the Perutz mechanism.

E. Origin of the Bohr Effect

The Bohr effect, hemoglobin's release of H$^+$ on binding O_2, is also observed when Hb binds other ligands. *It arises from pK changes of several groups caused by changes in their local environments that accompany hemoglobin's T→R transition.* The groups involved include the N-terminal amino groups of the α subunits and the C-terminal His of the β subunits. These have been identified through chemical and structural studies and their quantitative contributions to the Bohr effect have been estimated.

Reaction of the α subunits of Hb with **cyanate** results in the specific **carbamoylation** of its N-terminal amino group (Fig. 9-21). When such carbamoylated α subunits are mixed with normal β subunits, the resulting reconstituted Hb lacks 20 to 30% of the normal Bohr effect. The reason for this is seen on comparing the structure of deoxyHb with that of carbamoylated deoxyHb. In deoxyHb, a Cl$^-$ ion

$$R-NH_2 \; + \; \overset{O}{\underset{\|}{C}}=N^- \quad \xrightarrow{\; H^+ \;} \quad R-NH-\overset{O}{\underset{\|}{C}}-NH_2$$

Terminal amino group	Cyanate	Carbamoylated terminal amino group

FIGURE 9-21. The reaction of cyanate with the unprotonated (nucleophilic) forms of primary amino groups. At physiological pH's, N-terminal amino groups, which have pK's near 8.0, readily react with cyanate. Lys ε-amino groups (pK $\approx$ 10.8), however, are fully protonated under these conditions and are therefore unreactive.

binds between the N-terminal amino group of Val 1α_2 and the guanidino group of Arg 141α_1 (the C-terminal residue; Fig. 9-18a). This Cl$^-$ is absent in carbamoylated deoxyHb. It is also absent in normal R-state Hb because its C-terminal residues are not held in place by salt bridges (the origin of the preferential binding of Cl$^-$ to deoxyHb; Section 9-1C). N-Terminal amino groups of polypeptides normally have pK's near 8.0. On deoxyHb α subunits, however, the N-terminal amino group is electrostatically influenced by its closely associated Cl$^-$ to increase its positive charge by binding protons more tightly, that is, to increase its pK. Since at the pH of blood (7.4) N-terminal amino groups are normally only partially charged, this pK shift causes them to bind significantly more protons in the T state than in the R state.

The Hb β chain also contributes to the Bohr effect. Removal of its C-terminal residue, His 146β, reduces the Bohr effect by 40%. In normal deoxyHb, the imidazole ring of His 146β associates with the carboxylate of Asp 94β on the same subunit (Figs. 9-18b and 9-19) to form a salt bridge that is absent in the R state. Proton NMR measurements indicate that formation of this salt bridge increases the pK of the imidazole group from 7.1 to 8.0. This effect more than accounts for His 146β's share of the Bohr effect.

About 30 to 40% of the Bohr effect remains unaccounted for. It no doubt arises from small contributions of many of the residues whose environments are altered upon hemoglobin's R→T transition. A variety of evidence suggests that His 122α, His 143β, and Lys 82β are among these residues.

F. Structural Basis of BPG Binding

BPG decreases the oxygen-binding affinity of Hb by preferentially binding to its deoxy state (Section 9-1D). The binding of the physiologically quadruply charged BPG to deoxyHb is weakened by high salt concentrations, which suggests that this association is ionic in character. This explanation is corroborated by the X-ray structure of a BPG–deoxyHb complex which indicates that BPG binds in the central cavity of deoxyHb on its twofold axis (Fig. 9-22). The anionic groups of BPG are within hydrogen bonding and salt bridging distances of the cationic Lys EF6(82), His H21(143), His NA2(2), and N-terminal amino groups of both β subunits (Fig. 9-22). The T→R transformation brings the two β H helices together, which narrows the central cavity (compare Fig. 9-13a and b) and expels the BPG. It also widens the distance between the β N-terminal amino groups from 16 to 20 Å, which prevents their simultaneous hydrogen bonding with BPG's phosphate groups. BPG therefore stabilizes the T conformation of Hb by cross-linking its β subunits. This shifts the T ⇌ R equilibrium towards the T state, which lowers hemoglobin's O$_2$ affinity.

The structure of the BPG–deoxyHb complex also indicates why fetal hemoglobin (HbF) has a reduced affinity for BPG relative to HbA (Section 9-1D). The cationic His H21(143)β of HbA is changed to an uncharged Ser residue in HbF's β-like γ subunit, thereby eliminating a pair of ionic

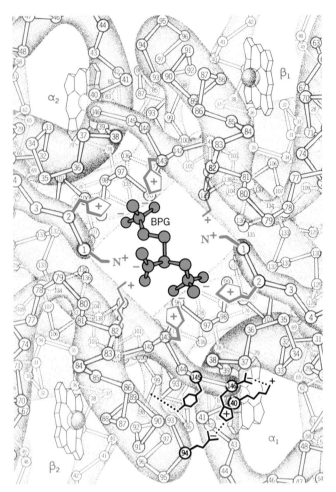

FIGURE 9-22. The binding of BPG to deoxyHb as viewed down the molecule's exact twofold axis (the same view as in Fig. 9-13a). BPG (*red*), with its five anionic groups, binds in the central cavity of deoxyHb, where it is surrounded by a ring of eight cationic side chains (*blue*) extending from the two β subunits. In the R state, the central cavity is too narrow to contain BPG (Fig. 9-13b). The arrangement of salt bridges and hydrogen bonds between the α_1 and β_2 subunits that partially stabilizes the T state (Figs. 9-18b and 9-19) is indicated on the lower right. [Figure copyrighted © by Irving Geis.]

interactions stabilizing the BPG–deoxyHb complex (Fig. 9-22).

G. Role of the Distal Histidine Residue

O$_2$ binding paradoxically protects the heme iron from autooxidation: The rate of Mb oxidation decreases as the partial pressure of O$_2$ increases. This is because heme iron oxidation is catalyzed by protons that are reduced by the heme iron and that in turn reduce O$_2$ in the solvent to **superoxide ion** (O$_2^-\cdot$). Bound O$_2$ evidently shields the Fe from the attacking protons.

The replacement, using genetic engineering techniques, of the distal His residue in Mb by any other residue reduces Mb's oxygen affinity and increases its rate of autooxidation.

Asp, a proton source, at this position increases the rate of Mb autooxidation by 350-fold, the largest increase of all residue replacements, whereas Phe, Met, and Arg provide only 50-fold accelerations, the smallest observed increases. However, the imidazole ring of the distal His, which has a pK of 5.5 and is therefore neutral at neutral pH and whose unprotonated N_ε atom faces the heme pocket (Fig. 9-12), acts as a proton trap, thereby protecting the Fe from protons. Thus, to quote Perutz, "Evolution is a brilliant chemist."

3. ABNORMAL HEMOGLOBINS

Mutant hemoglobins have provided a unique opportunity to study structure–function relationships in proteins because Hb is the only protein of known structure that has a large number of well-characterized naturally occurring variants. The examination of individuals with physiological disabilities, together with the routine electrophoretic screening of human blood samples, has led to the discovery of nearly 500 variant hemoglobins, ~95% of which result from single amino acid substitutions in a globin polypeptide chain. In this section, we consider the nature of these **hemoglobinopathies.** Hemoglobin diseases characterized by defective globin synthesis, the **thalassemias,** are the subject of Section 33-2G. It should be noted that ~300,000 individuals with serious hemoglobin disorders are born every year and that ~5% of the world's population are carriers of an inherited variant hemoglobin.

A. Molecular Pathology of Hemoglobin

The physiological effect of an amino acid substitution on Hb can, in most cases, be understood in terms of its molecular location:

1. Changes in surface residues

Changes of surface residues are usually innocuous because most of these residues have no specific functional role [although sickle-cell Hb (HbS) is a glaring exception to this generalization; Section 9-3B]. For example, **HbE** [Glu B8(26)$\beta \rightarrow$ Lys], the most common human Hb mutant after HbS (possessed by up to 10% of the populace in parts of Southeast Asia), has no clinical manifestations in either heterozygotes or homozygotes. About half of the known Hb mutations are of this type and have been discovered only accidentally or through surveys of large populations.

2. Changes in internally located residues

Changing an internal residue often destabilizes the Hb molecule. The degradation products of these hemoglobins, particularly those of heme, form granular precipitates (known as **Heinz bodies**) that are hydrophobically adsorbed to the erythrocyte cell membrane. The membrane's permeability is thereby increased, causing premature cell lysis. Carriers of unstable hemoglobins therefore suffer from **hemolytic anemia** of varying degrees of severity.

The structure of Hb is so delicately balanced that small structural changes may render it nonfunctional. This can occur through the weakening of the heme–globin association or as a consequence of other conformational changes. For instance, the heme group is easily dislodged from its closely fitting hydrophobic binding pocket. This occurs in **Hb Hammersmith** (Hb variants are often named after the locality of their discovery) in which Phe CD1(42)β, an invariant residue that wedges the heme into its pocket (see Figs. 9-12 and 9-15), is replaced by Ser. The resulting gap permits water to enter the heme pocket, which causes the hydrophobic heme to drop out easily (Phe CD1 and the proximal His F8 are the only invariant residues among all known hemoglobins). Similarly, in **Hb Bristol,** the substitution of Asp for Val E11(67)β, which partially occludes the O_2 pocket, places a polar group in contact with the heme. This weakens the binding of the heme to the protein, probably by facilitating the access of water to the subunit's otherwise hydrophobic interior.

Hb may also be destabilized by the disruption of elements of its 2°, 3°, and/or 4° structures. The instability of **Hb Bibba** results from the substitution of a helix-breaking Pro for Leu H19(136)α. Likewise, the instability of **Hb Savannah** is caused by the substitution of Val for the highly conserved Gly B6(24)β, which is located on the B helix where it crosses the E helix with insufficient clearance for side chains larger than an H atom (Fig. 9-13, and Fig. 9-11 where Gly B6 is residue 25). The $\alpha_1-\beta_1$ contact, which does not significantly dissociate under physiological conditions, may do so upon structural alteration. This occurs in **Hb Philly** in which Tyr C1(35)α, which participates in the hydrogen bonded network that helps knit together the $\alpha_1-\beta_1$ interface, is replaced by Phe.

3. Changes Stabilizing Methemoglobin

Changes at the O_2-binding site that stabilize the heme in the Fe(III) oxidation state eliminate the binding of O_2 to the defective subunits. Such methemoglobins are designated **HbM** and individuals carrying them are said to have **methemoglobinemia.** These individuals usually have bluish skin, a condition known as **cyanosis,** which results from the presence of deoxyHb in their arterial blood.

All known methemoglobins arise from substitutions that provide the Fe atom with an anionic oxygen atom ligand. In **Hb Boston,** the substitution of Tyr for His E7(58)α (the distal His, which protects the heme from oxidation; Section 9-2G) results in the formation of a 5-coordinate Fe(III) complex with the phenolate ion of the mutant Tyr E7 displacing the imidazole ring of His F8(87) as the apical ligand (Fig. 9-23a). In **Hb Milwaukee,** the γ-carboxyl group of the Glu that replaces Val E11(67)β forms an ion pair with a 5-coordinate Fe(III) complex (Fig. 9-23b). Both the phenolate and glutamate ions in these methemoglobins so stabilize the Fe(III) oxi-

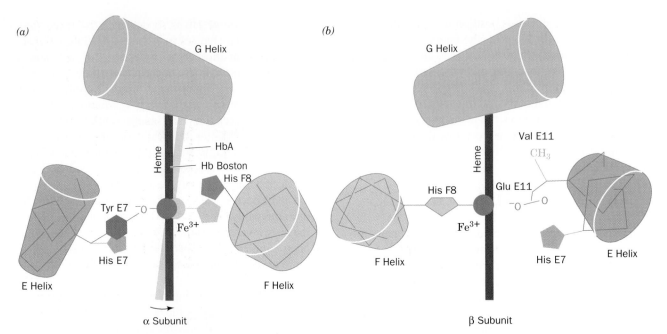

FIGURE 9-23. Mutations stabilizing the Fe(III) oxidation state of heme: (*a*) Alterations in the heme pocket of the α subunit on changing from deoxyHbA to Hb Boston [His E7(58)α→Tyr]. The phenolate ion of the mutant Tyr becomes the fifth ligand of the Fe atom, thereby displacing the proximal His [F8(87)α]. [After Pulsinelli, P.D., Perutz, M.F., and Nagel, R.L., *Proc. Natl. Acad. Sci.* **70**, 3872 (1973).] (*b*) The structure of the heme pocket of the β subunit in Hb Milwaukee [Val E11(67)β→Glu]. Here the mutant Glu residue's carboxyl group forms an ion pair with the heme iron atom so as to stabilize its Fe(III) state. [From Perutz, M.F., Pulsinelli, P.D., and Ranney, H.M., *Nature* **237**, 260 (1972).]

dation state that methemoglobin reductase is ineffective in converting them to the Fe(II) form.

Individuals with HbM are alarmingly cyanotic and have blood that is chocolate brown, even when their normal subunits are oxygenated. In northern Japan, this condition is named "black mouth" and has been known for centuries; it is caused by the presence of **HbM Iwate** [His F8(87)α→Tyr]. Methemoglobins have Hill constants of ~1.2. This indicates a reduced cooperativity in comparison with HbA even though HbM, which can bind only two oxygen molecules, can have a maximum Hill constant of 2 (the unmutated β chains remain functional). Surprisingly, heterozygotes with HbM, which have an average of one functional α subunit per Hb molecule, have no apparent physical disabilities. Evidently, the amount of O_2 released in their capillaries is within normal limits. Homozygotes of HbM, however, are unknown; this condition is, no doubt, lethal.

4. Changes at the α_1-β_2 contact

Changes at the α_1-β_2 contact often interfere with hemoglobin's quaternary structural changes. Most such hemoglobins have an increased O_2 affinity so that they release less than normal amounts of O_2 in the tissues. Individuals with such defects compensate for it by increasing the concentration of erythrocytes in their blood. This condition, which is named **polycythemia,** often gives them a ruddy complexion. Some amino acid substitutions at the α_1-β_2 interface instead result in a reduced O_2 affinity. Individuals carrying such hemoglobins are cyanotic.

Amino acid substitutions at the α_1-β_2 contact may change the relative stabilities of hemoglobin's R and T forms, thereby altering its O_2 affinity. For example, the replacement of Asp G1(99)β by His in **Hb Yakima** eliminates the hydrogen bond at the α_1-β_2 contact that stabilizes the T form of Hb (Fig. 9-17a). The interloping imidazole ring also acts as a wedge that pushes the subunits apart and displaces them toward the R state. This change shifts the T→R equilibrium almost entirely to the R state, which results in Hb Yakima having an increased O_2 affinity (p_{50} = 12 torr under physiological conditions vs 26 torr for HbA) and a total lack of cooperativity (Hill constant = 1.0). In contrast, the replacement of Asn G4(102)β by Thr in **Hb Kansas** eliminates the hydrogen bond in the α_1-β_2 contact that stabilizes the R state (Fig. 9-17b) so that this Hb variant remains in the T state upon binding O_2. Hb Kansas therefore has a low O_2 affinity (p_{50} = 70 torr) and a low cooperativity (Hill constant = 1.3).

B. Molecular Basis of Sickle-Cell Anemia

Most harmful Hb variants occur in only a few individuals, in many of whom the mutation apparently originated. However, ~10% of American blacks and as many as 25% of African blacks are heterozygotes for **sickle-cell hemoglobin (HbS).** HbS arises, as we have seen (Section 6-3A), from the substitution of a hydrophobic Val residue for the hydrophilic surface residue Glu A3(6)β (Fig. 9-13). The prevalence of HbS results from the protection it affords heterozygotes against malaria. However, homozygotes for HbS, of which there are some 50,000 in the United States, are severely afflicted by hemolytic anemia together with painful, debilitating, and sometimes fatal blood flow blockages caused by the irregularly shaped and inflexible erythrocytes characteristic of the disease.

HbS Fibers Are Stabilized by Intermolecular Contacts Involving Val $\beta6$ and Other Residues

The sickling of HbS-containing erythrocytes (Fig. 6-11b) results from the aggregation (polymerization) of deoxyHbS into rigid fibers that extend throughout the length of the cell (Fig. 9-24). Electron microscopy indicates that these fibers are ~220-Å-diameter elliptical rods consisting of 14 hexagonally packed and helically twisting strands of deoxyHbS molecules that associate in parallel pairs (Figs. 9-25 and 9-26a).

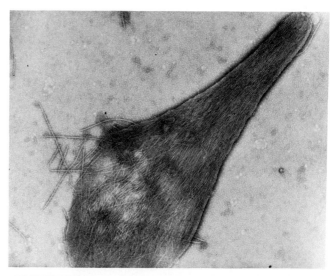

FIGURE 9-24. An electron micrograph of deoxyHbS fibers spilling out of a ruptured erythrocyte. [Courtesy of Robert Josephs, University of Chicago.]

(a)

(b)

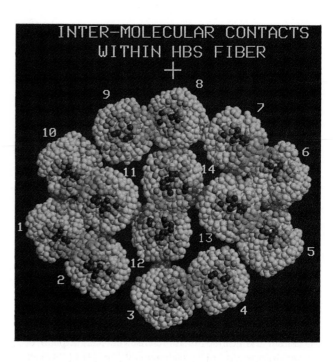

FIGURE 9-25. 220 Å in diameter fibers of deoxyHbS: (*a*) An electron micrograph of a negatively stained fiber. The accompanying cutaway interpretive drawing indicates the relationship between the inner and outer strands; the spheres represent individual HbS molecules. The fiber has a layer repeat distance of 64 Å and a moderate twist such that it repeats every 350 Å along the fiber axis. [Courtesy of Stuart Edelstein, University of Geneva.] (*b*) A model, viewed in cross-section, of the HbS fiber based on the crystal structure of HbS and three-dimensional reconstructions of electron micrographs of HbS fibers. The residues in the 14 HbS molecules are represented by spheres centered on the C_α positions. The residues making inter–double strand, intra–double strand lateral, and intra–double strand axial contacts are colored red, green and blue, respectively, with lighter and darker toned residues making intermolecular contacts <8 Å and <5 Å, respectively. The α and β chain residues outside the contact regions are colored white. [Courtesy of Stanley Watowich, Leon Gross, and Robert Josephs, University of Chicago.]

(a)

(b)

(c)

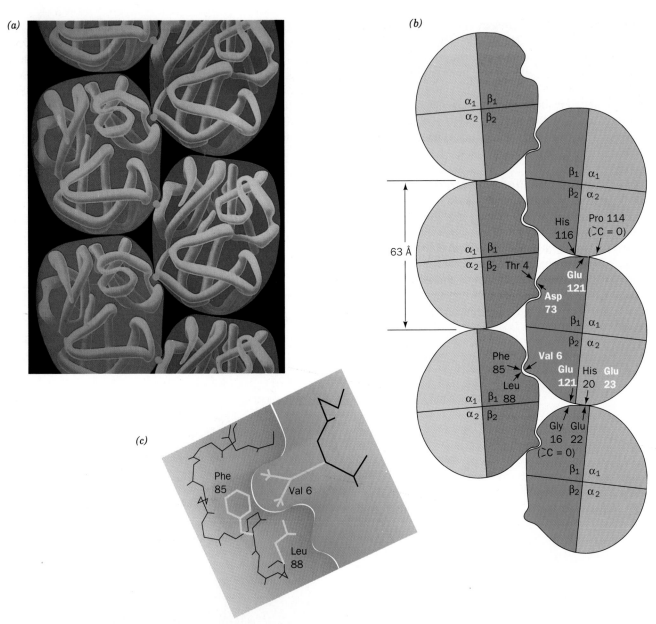

FIGURE 9-26. The structure of the deoxyHbS fiber. (*a*) The arrangement of the deoxyHbS molecules in the fiber. [Figure copyrighted © by Irving Geis.] (*b*) A schematic diagram indicating the intermolecular contacts in the crystal structure of deoxyHbS. The white-lettered residues are implicated in forming these contacts. Note that the only intermolecular association in which the mutant residue Val 6β participates involves subunit β₂; Val 6 of subunit β₁ is free. [After Wishner, B.C., Ward, K.B., Lattman, E.E., and Love, W.E., *J. Mol. Biol.* **98,** 192 (1975).] (*c*) The mutant Val 6β₂ fits neatly into a hydrophobic pocket formed mainly by Phe 85 and Leu 88 of an adjacent β₁ subunit. This pocket, which is located between helices E and F at the periphery of the heme pocket, is absent in oxyHb and is too hydrophobic to contain the normally occurring Glu 6β side chain. [Figure copyrighted © by Irving Geis.]

The structural relationship among the HbS molecules in the pairs of parallel HbS strands has been established by the X-ray structure analysis of deoxyHbS crystals. When this crystal structure was first determined, it was unclear whether the intermolecular contacts in the crystal resembled those in the fiber. However, the subsequent observation that HbS fibers slowly convert to these crystals with little change in their overall X-ray diffraction pattern indi- cates that the fibers structurally resemble the crystals. The crystal structure of deoxyHbS consists of double filaments of HbS molecules whose several different intermolecular contacts are diagrammed in Fig. 9-26*b*. Only one of the two Val 6β's per Hb molecule contacts a neighboring molecule. In this contact, the mutant Val side chain occupies a hydro- phobic surface pocket on an adjacent molecule's β subunit (Fig. 9-26*c*). This pocket is absent in oxyHb. Other contacts involve residues that also occur in HbA, including Asp 73β and Glu 23α (Fig. 9-26b). The observation that deoxyHbA does not aggregate into fibers, however, even at very high concentrations, indicates that *the contact involving Val 6β is essential for fiber formation*. This conclusion is corroborated by the observation that a genetically engineed human Hb in which Glu 6β is replaced by Ile (which differs from Val by an additional CH_2 group and is therefore even more hydrophobic) has half the solubility of HbS in 1.8*M* phosphate.

The importance of the other intermolecular contacts to the structural integrity of HbS fibers has been demonstrated by studying the effects of other mutant hemoglobins on HbS gelation (polymerization). For example, the doubly mutated **Hb Harlem** (Glu 6β→Val + Asp 73 β→Asn) requires a higher concentration to gel than does HbS (Glu 6β→Val); similarly, mixtures of HbS and **Hb Korle-Bu** (Asp 73β→Asn) gel less readily than equivalent mixtures of HbS and HbA. These observations suggest that Asp 73β occupies an important intermolecular contact site in HbS fibers (Fig. 9-26b). Likewise, the observation that hybrid tetramers consisting of α subunits from **Hb Memphis** (Glu 23α→Gln) and β subunits from HbS gel less readily than does HbS indicates that Glu 23α also participates in the polymerization of HbS fibers (Fig. 9-26b). The other white-lettered residues in Fig. 9-26b have been similarly implicated in sickling interactions.

The Initiation of HbS Gelation Is a Complex Process

The gelation of HbS, both in solution and within the red cell, follows an unusual time course. A solution of HbS can be brought to conditions under which it will gel by lowering the pO_2, raising the HbS concentration, and/or raising the temperature. *Upon achieving gelation conditions, there is a reproducible delay that varies according to conditions from milliseconds to days: During this time, no HbS fibers can be detected. Only after the delay do fibers first appear and gelation is then completed in about half the delay time* (Fig. 9-27a).

William Eaton and James Hofrichter discovered that the delay time, t_d, has a concentration dependence described by

$$\frac{1}{t_d} = k\left(\frac{c_t}{c_s}\right)^n \qquad [9.13]$$

where c_t is the total HbS concentration prior to gelation, c_s is the solubility of HbS measured after gelation is complete, and k and n are constants. Graphical analysis of the data indicates that $k \approx 10^{-7}$ s^{-1} and that n is between 30 and 50 (Fig. 9-27b). This is a remarkable result: *No other known solution process even approaches a 30th power concentration dependence.*

A two-stage process accounts for Eq. [9.13]:

1. At first, HbS molecules sequentially aggregate to form a **nucleus** consisting of *m* HbS molecules (Fig. 9-28a):

$$HbS \rightleftharpoons (HbS)_2 \rightleftharpoons (HbS)_3 \rightleftharpoons \cdots \rightleftharpoons (HbS)_m \rightarrow Growth$$

Prenuclear aggregates are unstable and easily decompose, but once a nucleus has formed it assumes a stable structure that rapidly elongates to form an HbS fiber.

2. Once a fiber has formed, it can nucleate the growth of other fibers (Fig. 9-28b). These newly formed fibers, in turn, nucleate the growth of yet other fibers, etc., so that this latter process is autocatalytic.

The initial **homogeneous nucleation** process (taking place in solution) accounts for the very high concentration depen-

(a)

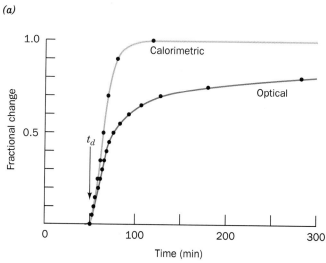

(b)

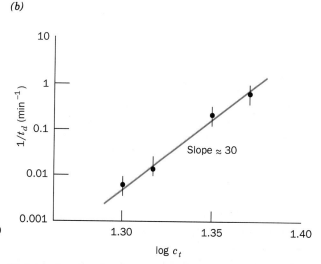

FIGURE 9-27. The time course of deoxyHbS gelation. (a) The extent of gelation as monitored calorimetrically (*yellow*) and optically (*purple*). Gelation of the 0.233 g·mL^{-1} deoxyHbS solution was initiated by rapidly increasing the temperature from 0°, where HbS is soluble, to 20°C; t_d is the delay time.

(b) A log–log plot showing the concentration dependence of $1/t_d$ for the gelation of deoxyHbS at 30°C. The slope of this line is ~30. [After Hofrichter, J., Ross, P.D., and Eaton, W.A., *Proc. Natl. Acad. Sci.* **71**, 4865, 4867 (1974).]

(a)   Homogeneous nucleation

(b)   Heterogeneous nucleation

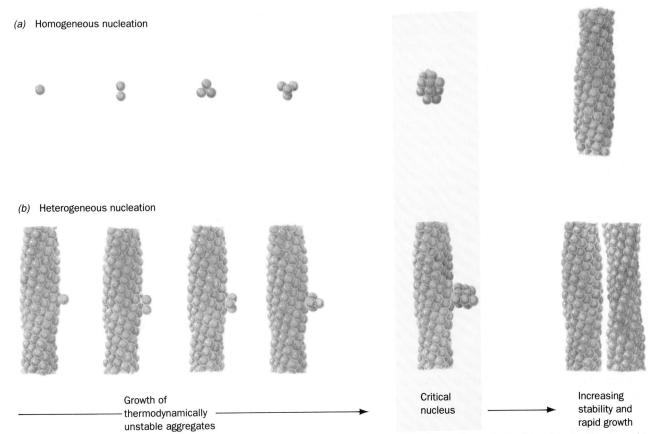

Growth of thermodynamically unstable aggregates →   Critical nucleus   →   Increasing stability and rapid growth

FIGURE 9-28.   The double nucleation mechanism for deoxyHbS gelation: (*a*) The initial aggregation of HbS molecules (*circles*) occurs very slowly because this process is thermodynamically unfavorable and hence the intermediates tend to decompose rather than grow. However, once an aggregate reaches a certain size, the **critical nucleus,** its further growth becomes thermodynamically favorable, leading to rapid fiber formation. (*b*) Each fiber, in turn, can nucleate the growth of other fibers, leading to the explosive appearance of polymer. [After Ferrone, F.A., Hofrichter, J., and Eaton, W.A., *J. Mol. Biol.* **183,** 614 (1985).]

dence in Eq. [9.13], whereas the secondary **heterogeneous nucleation** process (taking place on a surface—that of a fiber in this case) is responsible for the rapid onset of gelation (Fig. 9-27*a*).

The foregoing kinetic hypothesis suggests why sickle-cell anemia is characterized by episodic "crises" caused by blood flow blockages. HbS fibers dissolve essentially instantaneously upon oxygenation so that none are present in arterial blood. Erythrocytes take from 0.5 to 2 s to pass through the capillaries, where deoxygenation renders HbS insoluble. If the delay time, t_d, for sickling is greater than this transit time, no blood flow blockage occurs (although sickling that occurs in the veins damages the erythrocyte membrane). However, Eq. [9.13] indicates that small increases in HbS concentration, c_t, and/or small decreases in HbS solubility, c_s, caused by conditions known to trigger sickle-cell crises, such as dehydration, O_2 deprivation, and fever, result in significant decreases of t_d. Once a blockage occurs, the resulting lack of O_2 and slow down of blood flow in the area compound the situation.

The kinetic hypothesis of sickling has profound clinical implications for the treatment of sickle-cell anemia. Heterozygotes of HbS, whose blood usually contains ~60%

HbA and 40% HbS, rarely show any symptoms of sickling. The t_d for the gelation of their Hb is ~10^6-fold greater than that of homozygotes. Accordingly, a treatment of sickle-cell anemia that increases t_d by this amount, which corresponds to decreasing the ratio c_t/c_s by a factor of ~1.6, would relieve the symptoms of this disease. Three different therapeutic strategies to increase t_d, and thus inhibit HbS gelation, are under investigation:

1. The disruption of intermolecular interactions, thus increasing c_s. Of particular interest are synthetic oligopeptides that have been designed with the aid of the X-ray structure of HbS to bind stereospecifically to its intermolecular contact regions.

2. The use of agents that increase hemoglobin's O_2 affinity, thus decreasing c_t. For example, the administration of cyanate carbamoylates the N-terminal amino groups of Hb (Fig. 9-21). This treatment eliminates some of the salt bridges that stabilize the T state (Section 9-2E) and thereby increases the O_2 affinity of Hb. Although cyanate is an effective *in vitro* antisickling agent, its clinical use has been discontinued because of toxic side effects, cataract formation and peripheral nervous system dam-

age, that probably result from the carbamoylation of proteins other than Hb.

3. Lowering the HbS concentration (c_i) in erythrocytes by increasing erythrocyte volume. Agents that alter erythrocyte membrane permeability so as to permit the influx of water have promise in this regard.

Replacing HbS with other Hb molecules is also a promising possibility. Homozygotes for HbS with high levels of HbF in their blood, for example, have a relatively mild form of sickle-cell anemia. This observation has prompted the search for agents that can "switch on" the synthesis of HbF γ subunits in preference to that of mutant HbS β subunits. The use of vasodilators (substances that dilate blood vessels) so as to reduce the entrapment of sickled erythrocytes in the capillaries may also relieve the symptoms of sickle-cell disease.

4. ALLOSTERIC REGULATION

One of the outstanding characteristics of life is the high degree of control exercised in almost all of its processes. Through a great variety of regulatory mechanisms, the exploration of which constitutes a significant portion of this text, an organism is able to respond to changes in its environment, maintain intra- and intercellular communications, and execute an orderly program of growth and development. Regulation is exerted at every organizational level in living systems, from the control of rates of reactions on the molecular level, through the control of expression of genetic information on the cellular level, to the control of behavior on the organismal level. It is therefore not surprising that many, if not most, diseases are caused by aberrations in biological control processes.

Our exploration of the structure and function of hemoglobin continues with a theoretical discussion of the regulation of ligand binding to proteins through **allosteric interactions** (Greek: *allos*, other + *stereos*, solid or space). These cooperative interactions occur when the binding of one ligand at a specific site is influenced by the binding of another ligand, known as an **effector** or **modulator**, at a different (allosteric) site on the protein. If the ligands are identical, this is known as a **homotropic effect,** whereas if they are different, it is described as a **heterotropic effect.** These effects are termed **positive** or **negative** depending on whether the effector increases or decreases the protein's ligand-binding affinity.

Hemoglobin, as we have seen, exhibits both homotropic and heterotropic effects. The binding of O_2 to Hb results in a positive homotropic effect since it increases hemoglobin's O_2 affinity. In contrast, BPG, CO_2, H^+, and Cl^- are negative heterotropic effectors of O_2 binding to Hb because they decrease its affinity for O_2 (negative) and are chemically different from O_2 (heterotropic). The O_2 affinity of Hb, as we have seen, depends on its quaternary structure. *In gen-*

eral, allosteric effects result from interactions among subunits of oligomeric proteins.

Even though hemoglobin catalyzes no chemical reaction, it binds ligands in the same manner as do enzymes. Since an enzyme cannot catalyze a reaction until after it has bound its **substrate** (the molecule undergoing reaction), the enzyme's catalytic rate varies with its substrate-binding affinity. Consequently, the cooperative binding of O_2 to Hb is taken as a model for the allosteric regulation of enzyme activity. Indeed, in this section, we shall consider several models of allosteric regulation that, for the most part, were formulated to explain the O_2-binding properties of Hb. Following this, we shall compare these models with the realities of Hb behavior.

A. The Adair Equation

The derivation of the Hill equation (Section 9-1B) is predicated on the assumption of all-or-none O_2 binding. The observation of partially oxygenated Hb molecules, however, led Gilbert Adair, in 1924, to propose that the binding of ligands to proteins occurs sequentially with dissociation constants that are not necessarily equal. The expression for the saturation function under this model is straightforwardly derived.

For a protein such as Hb with four ligand-binding sites, the reaction sequence is

$$E + S \rightleftharpoons ES \qquad k_1 = 4K_1$$
$$ES + S \rightleftharpoons ES_2 \qquad k_2 = \tfrac{3}{2}K_2$$
$$ES_2 + S \rightleftharpoons ES_3 \qquad k_3 = \tfrac{2}{3}K_3$$
$$ES_3 + S \rightleftharpoons ES_4 \qquad k_4 = \tfrac{1}{4}K_4$$

where the K_i are the **macroscopic** or **apparent dissociation constants** for binding the ith ligand to the protein,

$$K_i = \frac{[ES_{i-1}][S]}{[ES_i]} \qquad [9.14]$$

and the k_i are the **microscopic** or **intrinsic dissociation constants,** that is, the individual dissociation constants for the ligand-binding sites. The intrinsic dissociation constants are equal to the apparent dissociation constants multiplied by **statistical factors,** 4, $\tfrac{3}{2}$, $\tfrac{2}{3}$, and $\tfrac{1}{4}$, that account for the number of ligand-binding sites on the protein molecule. The statistical factor 4 derives from the fact that a tetrameric protein E bears four sites that can bind ligand to form ES (that is, the concentration of ligand-binding sites is 4[E]) but only one site from which ES can dissociate ligand to form E (that is, the concentration of bound ligand is 1[E]); the statistical factor $\tfrac{3}{2}$ is a result of there being three remaining sites on ES that can bind ligand to form ES_2 and two sites from which ES_2 can dissociate ligand to form ES; etc. In general, for a protein with n equivalent binding sites:

$$k_i = \frac{(n - i + 1)[ES_{i-1}][S]}{i[ES_i]} = \left(\frac{n - i + 1}{i}\right) K_i \quad [9.15]$$

since $(n - i + 1)[ES_{i-1}]$ is the concentration of free ligand-

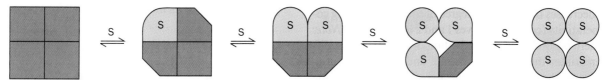

FIGURE 9-35.   The sequential binding of ligand in the sequential model of allosterism. Ligand-binding progressively induces conformational changes in the subunits, with the greatest changes occurring in those subunits that have bound ligand. The coupling between subunits is not necessarily of sufficient strength to maintain the symmetry of the oligomer as it is in the symmetry model.

coupling, however, conformational changes occur sequentially as more and more ligand is bound (Fig. 9-35). Thus, *the essence of the sequential model is that a protein's ligand-binding affinity varies with its number of bound ligands, whereas in the symmetry model this affinity depends only on the protein's quaternary state.*

The degree of coupling between oligomer subunits depends on how these subunits are arranged, that is, on the protein's symmetry. Consequently, in the sequential model, the fractional saturation has a different algebraic form for each oligomeric symmetry. The form of the Adair equation (Eq. [9.17] for a tetramer) similarly depends on the number of subunits in the protein. In fact, the sequential model of allosterism may be considered an extension of the Adair model that provides a physical rationalization for the values of its microscopic dissociation constants, k_i.

D. Hemoglobin Cooperativity

Hemoglobin's fractional saturation curve is closely approximated by both the symmetry model and the sequential model (Fig. 9-36). Clearly such curves cannot by themselves be used to differentiate between these two models, if, in fact, either is correct. It is of interest, however, to compare these models with the mechanistic model of Hb we developed in Section 9-2C.

Hb, of course, is not composed of identical subunits as the symmetry model demands. At least to a first approximation, however, the functional differences of hemoglobin's closely related α and β subunits may be ignored (although their structural differences are essential to the molecular mechanism of Hb cooperativity). *To this approximation, Hb largely follows the symmetry model, although it also exhibits some features of the sequential model* (although to what extent is still a matter of debate). The quaternary T→R conformation change is concerted as the symmetry model requires. Yet, ligand binding to the T state does cause small tertiary structural changes as the induced-fit model predicts. This phenomenon is evident in the X-ray structure of human hemoglobin whose α subunits are fully oxygenated and whose β subunits are oxygen free. This partially liganded hemoglobin remains in the T state but its α subunit Fe's are 0.15 Å closer to the still domed porphyrins than they are in deoxyHb (25% of the total distance moved in the T→R transition). *Such tertiary structural*

changes are undoubtedly responsible for the buildup of strain that eventually triggers the T→R transition.

The symmetry and sequential models are often indispensable guides in rationalizing the allosteric behavior of a protein. If, however, the experience with Hb is any guide, a protein's actual allosteric mechanism is likely to be more complex than either of these idealities. Indeed, recent thermodynamic studies by Gary Ackers of cooperativity among hemoglobin's various possible liganding states, together with determinations of their quaternary structures, indicate that hemoglobin's allosteric mechanism follows a set of pathways that have a heretofore unrecognized symmetry. A Hb tetramer undergoes a quaternary T→R switch only when at least one liganding site on each of its component $\alpha\beta$ dimers is occupied. Cooperativity arises from both concerted quaternary switching (as called for by the symmetry model) and from sequential modulation of ligand binding within each quaternary state through ligand-induced alterations in tertiary structure (in accord with the sequential model).

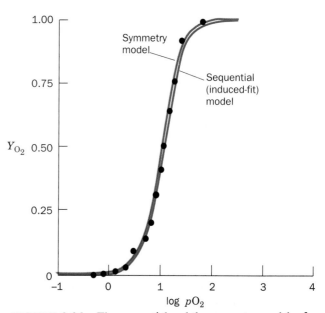

FIGURE 9-36.   The sequential and the symmetry models of allosterism can provide equally good fits to the measured O$_2$-dissociation curve of Hb. [After Koshland, D.E., Jr., Némethy, G., and Filmer, D., *Biochemistry* **5**, 382 (1966).]

APPENDIX: DERIVATIONS OF SYMMETRY MODEL EQUATIONS

A. Homotropic Interactions — Equation [9.22]

The fractional saturation Y_S for ligand binding is expressed:

$$Y_S = \frac{([R_1] + 2[R_2] + \cdots + n[R_n]) + ([T_1] + 2[T_2] + \cdots + n[T_n])}{n\{([R_0] + [R_1] + \cdots + [R_n]) + ([T_0] + [T_1] + \cdots + [T_n])\}}$$

[9.21]

Defining $\alpha = [S]/k_R$ and $c = k_R/k_T$, and using Eq. [9.20] to substitute $[R_{n-1}]$ for $[R_n]$, $[R_{n-2}]$ for $[R_{n-1}]$, etc., the terms enclosed by the first parentheses in the numerator of Eq. [9.21] are reduced to

$$[R_0]\left\{n\alpha + \frac{2n(n-1)\alpha^2}{2} + \cdots + \frac{n\,n!\,\alpha^n}{n!}\right\}$$

$$= [R_0]\alpha n\left\{1 + \frac{2(n-1)\alpha}{2} + \cdots + \frac{n(n-1)!\alpha^{n-1}}{n(n-1)!}\right\}$$

$$= [R_0]\alpha n(1+\alpha)^{n-1}$$

and similarly, the terms in the first parentheses of the denominator of Eq. [9.21] become

$$[R_0]\left\{1 + n\alpha + \cdots + \frac{n!\,\alpha^n}{n!}\right\} = [R_0](1+\alpha)^n$$

Likewise, the terms in the second parentheses of the numerator and the denominator of Eq. [9.21] assume the respective forms

$$[T_0]([S]/k_T)n(1 + [S]/k_T)^{n-1} = L[R_0]c\alpha n(1+c\alpha)^{n-1}$$

and

$$[T_0](1 + [S]/k_T)^n = L[R](1+c\alpha)^n$$

Accordingly,

$$Y_S = \frac{[R_0]\alpha n(1+\alpha)^{n-1} + L[R_0]c\alpha n(1+c\alpha)^{n-1}}{n\{[R_0](1+\alpha)^n + L[R_0](1+c\alpha)^n\}}$$

which, upon cancellation of terms, yields the equation describing the symmetry model for homotropic interactions:

$$Y_S = \frac{\alpha(1+\alpha)^{n-1} + Lc\alpha(1+c\alpha)^{n-1}}{(1+\alpha)^n + L(1+c\alpha)^n}$$

[9.22]

B. Heterotropic Interactions — Equation [9.23]

For an oligomer that binds activator A and substrate S to only its R state, and inhibitor I to only its T state, the fractional saturation for substrate, Y_S, the fraction of substrate-binding sites occupied by substrate, is expressed:

$$Y_S = \frac{\sum\limits_{i=1}^{n} \sum\limits_{j=0}^{n} i[R_{i,j}]}{n\left(\sum\limits_{i=0}^{n} \sum\limits_{j=0}^{n} [R_{i,j}] + \sum\limits_{k=0}^{n} [T_k]\right)}$$

Here the subscripts i, j, and k indicate the respective numbers of S, A, and I molecules that are bound to one oligomer; that is, $R_{i,j} \equiv RS_iA_j$ and $T_k \equiv TI_k$. Then defining $\alpha = [S]/k_R$ and following the foregoing derivation of Eq. [9.22]:

$$Y_S = \frac{\left(\sum\limits_{j=0}^{n} [R_{0,j}]\right)\alpha n(1+\alpha)^{n-1}}{n\left\{\left(\sum\limits_{j=0}^{n} [R_{0,j}]\right)(1+\alpha)^n + \sum\limits_{k=0}^{n} [T_k]\right\}} = \frac{\alpha(1+\alpha)^{n-1}}{(1+\alpha)^n + L'}$$

where

$$L' = \sum_{k=0}^{n} [T_k] \Bigg/ \sum_{j=0}^{n} [R_{0,j}]$$

In analogy with the definition of α, we define $\beta = [I]/k_I$ and $\gamma = [A]/k_A$, and again follow the derivation of Eq. [9.22] to obtain:

$$\sum_{k=0}^{n} [T_k] = [T_0](1+\beta)^n$$

and

$$\sum_{j=0}^{n} [R_{0,j}] = [R_{0,0}](1+\gamma)^n$$

so that

$$L' = \frac{L(1+\beta)^n}{(1+\gamma)^n}$$

The symmetry model equation extended to include heterotropic effects is therefore expressed:

$$Y_S = \frac{\alpha(1+\alpha)^{n-1}}{(1+\alpha)^n + \dfrac{L(1+\beta)^n}{(1+\gamma)^n}}$$

[9.23]

CHAPTER SUMMARY

The heme group in myoglobin and in each subunit of hemoglobin reversibly binds O_2. In deoxyHb, the Fe(II) is 5-coordinated to the four pyrrole nitrogen atoms of the protoporphyrin IX and to the protein's proximal His. Upon oxygenation, O_2 becomes the sixth ligand of Fe(II). Mb has a hyperbolic fractional saturation curve (Hill constant, $n = 1$). However, that of Hb is sigmoidal ($n \approx 2.8$) as

a consequence of its cooperative O_2 binding: Hb binds its fourth O_2 with 100-fold greater affinity than its first O_2. The variation of O_2 affinity with pH, the Bohr effect, causes Hb to release O_2 in the tissues in response to the binding of protons liberated by the hydration of CO_2 to HCO_3^-. Hb facilitates the transport of CO_2, both directly, by binding CO_2 as N-terminal carbamate, and indirectly, by increasing the concentration of HCO_3^- through the Bohr effect. The presence of BPG in erythrocytes, which only binds to deoxy-Hb, further modulates the O_2 affinity of Hb. Short-term high-altitude adaptation results from an increase of BPG concentration in the erythrocytes, which increases the amount of O_2 delivered to the tissues by decreasing hemoglobin's O_2 affinity.

The α and β subunits of Hb consist mostly of seven or eight consecutive helices arranged to form a hydrophobic pocket that almost completely envelops the heme. Oxygen binding moves the Fe(II) from a position 0.6 Å out of the heme plane on the side of the proximal His to the center of the heme, thereby relieving the steric interference that would otherwise occur between the bound O_2 and the porphyrin. The Fe(II) pulls the attached proximal His after it in a motion that can only occur if its imidazole ring reorients so as to avoid collision with the heme. In the T→R conformational transition, the symmetry equivalent $\alpha_1 C - \beta_2 FG$ and $\alpha_2 C - \beta_1 FG$ contacts simultaneously shift between two stable positions. Intermediate positions are sterically prevented so that these contacts act as a two-position conformational switch. The Perutz mechanism of O_2 binding proposes that the low O_2 affinity of the T state arises from strain that prevents the Fe(II) from moving into the heme plane to form a strong Fe—O_2 bond. This strain is relieved by the concerted 4° shift of the Hb molecule to the high O_2 affinity R state. The quaternary shift is opposed by a network of salt bridges in the T state that involve the C-terminal carboxyl groups and that are ruptured in the R state. The stability of the R state relative to the T state increases with the degree of oxygenation as a result of the strain of binding O_2 in the T state. The existence of this strain has been demonstrated through the breakage of the Fe(II)—proximal His bond upon hemoglobin's simultaneous binding of IHP, a tight-binding BPG analog that forces Hb into the T state, and NO, a strong ligand that forces Hb into the R state. The Bohr effect results from increases in the pK's of the α N-terminal amino group and His 146β on forming the T-state salt bridges. Lys 82β also participates in the Bohr effect. BPG binding occurs in the central cavity of T-state Hb through several salt bridges. The distal His residue protects deoxyHb from autooxidation by taking up the protons that would otherwise catalyze the oxidation of the heme Fe.

Almost 500 mutant varieties of Hb are known. About one half of them are innocuous because they result in surface residue changes. However, alterations of internal residues often disrupt the structure of Hb, which causes hemolytic anemia. Changes at the O_2-binding site that stabilize the Fe(III) state eliminate O_2 binding to these subunits, which results in cyanosis. Mutations affecting subunit interfaces may stabilize either the R state or the T state, which, respectively, increase and decrease hemoglobin's O_2 affinity. Sickle-cell anemia is caused by the homozygous Hb mutant Glu 6β→Val, which promotes the gelation of the resulting deoxyHbS to form rigid 14-strand fibers that deform erythrocytes. Under gelation conditions, fiber growth occurs via a two-stage nucleation mechanism, resulting in a delay time that varies with the 30th to 50th power of the initial HbS concentration. Agents that increase this delay time to longer than the transit times of erythrocytes through the capillaries should therefore prevent sickling and thus relieve the symptoms of sickle-cell anemia.

The Adair equation rationalizes the O_2-binding cooperativity of Hb by assigning a separate dissociation constant to each O_2 bound. Positive cooperativity results if these constants decrease sequentially. However, the Adair equation offers no physical insight as to why this occurs. The symmetry model proposes that symmetrical oligomers can exist in one of two conformational states, R and T, that differ in ligand-binding affinity. Ligand binding to the high-affinity state forces the oligomer to assume this conformation, which facilitates the binding of additional ligand. This homotropic model is extended to heterotropic effects by postulating that activator and substrate can bind only to the R state and inhibitor can bind only to the T state. The binding of activator forces the oligomer into the R state, which facilitates the binding of substrate and additional activator. The binding of inhibitor, however, forces the oligomer into the T state, which prevents substrate and activator binding. The sequential model postulates that an induced fit between ligand and substrate confers conformational strain on the protein that alters its affinity for binding other ligands without requiring the oligomer to maintain its symmetry. The Perutz mechanism for O_2 binding to Hb is largely consistent with the symmetry model but exhibits some elements of the sequential model.

REFERENCES

General

Antonini, E., Rossi-Bernardi, L., and Chiancone, E. (Eds.), *Methods Enzymol.* **76** (1981). [A collection of articles on techniques of purification and characterization of hemoglobins.]

Bunn, H.F. and Forget, B.G., *Hemoglobin: Molecular, Genetic and Clinical Aspects,* Saunders (1986). [A valuable compendium on normal and abnormal hemoglobins.]

Dickerson, R.E. and Geis, I., *Hemoglobin,* Benjamin/Cummings (1983). [A beautifully written and lavishly illustrated treatise on the structure, function, and evolution of hemoglobin.]

Everse, J., Vandegriff, K.K. and Winslow, R.M., Hemoglobins, Parts B and C, *Meth. Enzymol.* **231** *and* **232** (1994).

Judson, H.F., *The Eighth Day of Creation,* Chapters 9 and 10, Simon & Schuster (1979). [Includes a fascinating historical account of how our present perception of hemoglobin structure and function came about.]

Perutz, M.F., Hemoglobin structure and respiratory transport, *Sci. Am.* **239**(6): 92–125 (1978). [An instructive discussion of the structure of hemoglobin and its mechanism of oxygen transport.]

Stamatoyannopoulos, G., Nienhuis, A.W., Leder, P., and Majerus, P.W. (Eds.), *The Molecular Basis of Blood Diseases,* Chapters 5 and 6, Saunders (1987). [Authoritative discussions of hemoglobin structure and function and on sickle-cell disease.]

Structures of Myoglobin, Hemoglobin, and Model Compounds

Collman, J.P., Synthetic models for oxygen-binding hemoproteins, *Acc. Chem. Res.* **10**, 265–272 (1977).

Fermi, G., Perutz, M.F., Shaanan, B., and Fourme, R., The crystal structure of human deoxyhaemoglobin at 1.74 Å, *J. Mol. Biol.* **175**, 159–174 (1984).

Jameson, G.B., Molinaro, F.S., Ibers, J.A., Collman, J.P., Brauman, J.I., Rose, E., and Suslick, K.S., Models for the active site of oxygen-binding hemoproteins. Dioxygen binding properties and the structures of (2-methylimidazole)-*meso*-tetra($\alpha,\alpha,\alpha,\alpha$-$o$-pivalamidophenyl)porphinato iron(II)-ethanol and its dioxygen adduct, *J. Am. Chem. Soc.* **102**, 3224–3237 (1980).

Liddington, R., Derewenda, Z., Dodson, G., and Harris, D., Structure of the liganded T state of haemoglobin identifies the origin of cooperative oxygen binding, *Nature* **331**, 725–728 (1988).

Phillips, S.E.V. and Schoenborn, B.P., Neutron diffraction reveals oxygen–histidine hydrogen bond in oxymyoglobin, *Nature* **292**, 81–82 (1982).

Phillips, S.E.V., Structure and refinement of oxymyoglobin at 1.6 Å resolution, *J. Mol. Biol.* **142**, 531–554 (1980).

Shaanan, B., Structure of human oxyhaemoglobin at 2.1 Å resolution, *J. Mol. Biol.* **171**, 31–59 (1983).

Takano, T., Structure of myoglobin refined at 2.0 Å resolution, *J. Mol. Biol.* **110**, 537–568, 569–584 (1977).

Mechanism of Hemoglobin Oxygen Binding

Baldwin, J. and Chothia, C., Haemoglobin: the structural changes related to ligand binding and its allosteric mechanism, *J. Mol. Biol.* **129**, 175–220 (1979). [The exposition of a detailed mechanism of O_2 binding to Hb based on the structures of oxy and deoxyHb.]

Baldwin, J., Structure and cooperativity of haemoglobin, *Trends Biochem. Sci.* **5**, 224–228 (1980). [A review of the previous reference.]

Gelin, B.R., Lee, A.W.-N., and Karplus, M., Haemoglobin tertiary structural change on ligand binding, *J. Mol. Biol.* **171**, 489–559 (1983). [A theoretical study of the dynamics of O_2 binding to Hb.]

Perutz, M.F., Myoglobin and haemoglobin: Role of distal residues in reactions with haem ligands, *Trends Biochem. Sci.* **14**, 42–44 (1989).

Perutz, M.F., Mechanisms of cooperativity and allosteric regulation in proteins, *Quart. Rev. Biophys.* **22**, 139–236 (1989). [Contains a detailed structural description of allosterism in hemoglobin.]

Perutz, M.F., Regulation of oxygen affinity of hemoglobin, *Annu. Rev. Biochem.* **48**, 327–386 (1979). [A detailed examination of the Perutz mechanism in light of structural and spectroscopic data.]

Perutz, M.F., Stereochemistry of cooperative effects in haemoglobin, *Nature* **228**, 726–734 (1970). [A landmark paper in which the Perutz mechanism was first proposed. Although many of its details have since been modified, the basic model remains intact.]

Bohr Effect and BPG Binding

Arnone, A., X-ray diffraction study of binding of 2,3-diphosphoglycerate to human deoxyhaemoglobin, *Nature* **237**, 146–149 (1972).

Arnone, A., X-ray studies of the interaction of CO_2 with human deoxyhaemoglobin, *Nature* **247**, 143–145 (1974).

Benesch, R.E. and Benesch, R., The mechanism of interaction of red cell organic phosphates with hemoglobin, *Adv. Protein Chem.* **28**, 211–237 (1974).

Kilmartin, J.V. and Rossi-Bernardi, L., Interactions of hemoglobin with hydrogen ion, carbon dioxide and organic phosphates, *Physiol. Rev.* **53**, 836–890 (1973).

Lenfant, C., Torrance, J., English, E., Finch, C.A., Reynafarje, C., Ramos, J., and Faura, J., Effect of altitude on oxygen binding by hemoglobin and on organic phosphate levels, *J. Clin. Invest.* **47**, 2652–2656 (1968).

Perutz, M.F., Kilmartin, J.V., Nishikura, K., Fogg, J.H., and Butler, P.J.G., Identification of residues contributing to the Bohr effect of human haemoglobin, *J. Mol. Biol.* **138**, 649–670 (1980).

Abnormal Hemoglobins

Baudin-Chich, V., Pagnier, J., Marden, M., Bohn, B., Lacaze, N., Kister, J., Schaad, O., Edelstein, S.J., and Poyart, C., Enhanced polymerization of recombinant human deoxyhemoglobin β6 Glu→Ile, *Proc. Natl. Acad. Sci.* **87**, 1845–1849 (1990).

Eaton, W.A. and Hofrichter, J., Sickle cell hemoglobin polymerization, *Adv. Prot. Chem.* **40**, 63–279 (1990). [An authoritative and exhaustive review of HbS polymerization.]

Manning, J.M., Covalent inhibitors of the gelation of sickle cell hemoglobin and their effects on function, *Adv. Enzymol. Rel. Areas Mol. Biol.* **64**, 55–91 (1991).

Noguchi, C.T. and Schechter, A.N., Sickle hemoglobin polymerization in solution and in cells, *Annu. Rev. Biophys. Biophys. Chem.* **14**, 239–263 (1985).

Padlan, E.A. and Love, W.E., Refined crystal structure of deoxyhemoglobin S, *J. Biol. Chem.* **260**, 8280–8291 (1985).

Perutz, M., *Protein Structure. New Approaches to Disease and Therapy,* Chapter 6, Freeman (1992).

Perutz, M.F. and Lehmann, H., Molecular pathology of human haemoglobin, *Nature* **219**, 902–909 (1968). [A ground-breaking study correlating the clinical symptoms and inferred structural alterations of numerous mutant hemoglobins.]

Watowich, S.J., Gross, L.J., and Josephs, R., Intermolecular contacts within sickle hemoglobin fibers, *J. Mol. Biol.* **209**, 821–828 (1989).

Weatherall, D.J., Clegg, J.B., Higgs, D.R., and Wood, W.G., The hemoglobinopathies, *in* Scriver, C.R., Beaudet, A.L., Sly, W.S., and Valle, D. (Eds.), *The Metabolic Basis of Inherited Disease* (6th ed.), *pp.* 2281–2339, McGraw–Hill (1989). [A detailed review.]

Allosteric Regulation

Ackers, G.K. and Hazzard, J.H., Transduction of binding energy into hemoglobin cooperativity, *Trends Biochem. Sci.* **18**, 385–390 (1993); *and* Ackers, G.K., Doyle, M.L., Myers, D., and Daugherty, M.A., Molecular code for cooperativity in hemoglobin, *Science* **255**, 54–63 (1992). [Presents thermodynamic arguments that both of Hb's αβ dimers must be ligated for quaternary switching to occur.]

Baldwin, J.M., A model of co-operative oxygen binding to haemoglobin, *Br. Med. Bull.* **32**, 213–218 (1976).

Cantor, C.R. and Schimmel, P.R., *Biophysical Chemistry,* Chapter 17, Freeman (1980).

Fersht, A., *Enzyme Structure and Mechanism* (2nd ed.), Chapter 10, Freeman (1985).

Kuby, S.A., *A Study of Enzymes,* Vol. I, Chapter 7, CRC Press (1991).

Koshland, D.E., Jr., Némethy, G., and Filmer, D., Comparison of experimental binding data and theoretical models in proteins containing subunits, *Biochemistry* **5**, 365–385 (1966). [The formulation of the sequential model of allosteric regulation.]

Monod, J., Wyman, J., and Changeux, J.-P., On the nature of allosteric transitions: a plausible model, *J. Mol. Biol.* **12**, 88–118 (1965). [The exposition of the symmetry model of allosteric regulation.]

Riggs, A. Hemoglobins, *Curr. Opin. Struct. Biol.* **1**, 915–921 (1991).

PROBLEMS

1. The urge to breathe in humans results from a high blood CO_2 content; there are no direct physiological sensors of blood pO_2. Skindivers often **hyperventilate** (breathe rapidly and deeply for several minutes) just before making a protracted dive in the belief that they will thereby increase the O_2 content of their blood. This belief results from the fact that hyperventilation represses the breathing urge by expelling significant quantities of CO_2 from the blood. In light of what you know about the properties of hemoglobin, is hyperventilation a useful procedure? Is it safe? Explain.

2. Explain why n in the Hill equation can never be larger than the number of ligand-binding sites on the protein.

*3. In the Bohr effect, protonation of the N-terminal amino groups of hemoglobin's α chains is responsible for ~30% of the 0.6 mol of H^+ that combine with Hb upon the release of 1 mol of O_2 at pH 7.4. Assuming that this group has pK 7.0 in oxyHb, what is its pK in deoxyHb?

4. As one of the favorites to win the La Paz, Bolivia marathon, you have trained there for the several weeks it requires to become adapted to its 3700-m altitude. A manufacturer of running equipment who sponsors an opponent has invited you for the weekend to a prerace party at a beach house near Lima, Peru with the assurance that you will be flown back to La Paz at least a day before the race. Is this a token of his respect for you or an underhanded attempt to handicap you in the race? Explain (see Fig. 9-9).

5. In active muscles, the pO_2 may be 10 torr at the cell surface and 1 torr at the mitochondia (the organelles where oxidative metabolism occurs). How does myoglobin ($p_{50} = 2.8$ torr) facilitate the diffusion of O_2 through these cells? Active muscles consume O_2 much faster than do other tissues. Would myoglobin also be an effective O_2-transport protein in other tissues? Explain.

6. Erythrocytes that have been stored for over a week in standard acid–citrate–dextrose medium become depleted in BPG. Discuss the merits of using fresh versus week-old blood in blood transfusions.

7. The following fractional saturation data have been measured for a certain blood sample:

pO_2	Y_{O_2}	pO_2	Y_{O_2}
20	0.14	60	0.59
30	0.26	70	0.66
40	0.39	80	0.72
50	0.50	90	0.76

What are the Hill constant and the p_{50} of this blood sample? Are they normal?

8. An anemic individual, whose blood has only half the normal Hb content, may appear to be in good health. Yet, a normal individual is incapacitated by exposure to sufficient carbon monoxide to occupy half his heme sites (pCO of 1 torr for ~1 h; CO binds to Hb with 200 times greater affinity than does O_2). Explain.

*9. The X-ray structure of Hb Rainier (Tyr $145\beta \rightarrow$ Cys) indicates that the mutant Cys residue forms a disulfide bond with Cys 93β of the same subunit. This holds the β subunit's C-terminal residue in a quite different orientation than it assumes in HbA. How would the following quantities for Hb Rainier compare with those of HbA? Explain. (a) The oxygen affinity, (b) the Bohr effect, (c) the Hill constant, and (d) the BPG affinity.

10. The crocodile, which can remain under water without breathing for up to 1 h, drowns its air-breathing prey and then dines at its leisure. An adaptation that aids the crocodile in doing so is that it can utilize virtually 100% of the O_2 in its blood, whereas humans, for example, can extract only ~65% of the O_2 in their blood. Crocodile Hb does not bind BPG. However, crocodile deoxyHb preferentially binds HCO_3^-. How does this help the crocodile obtain its dinner?

11. The gelation time of an equimolar mixture of HbA and HbS is less than that of a solution of only HbS in the same concentration that it has in the mixture. What does this observation imply about the participation of HbA in the gelation of HbS?

12. The severely anemic condition of homozygotes for HbS results in an elevated BPG content in their erythrocytes. Discuss whether or not this is a beneficial effect.

13. As organizer of an expedition that plans to climb several very high mountains, it is your responsibility to choose its members. Each of the applicants for one of the positions on the team is a heterozygote for one of the following abnormal hemoglobins: (1) HbS, (2) **Hb Hyde Park** [His F8(92)$\beta \rightarrow$ Tyr], (3) **Hb Riverdale–Bronx** [Gly B6(24)$\beta \rightarrow$ Arg], (4) **Hb Memphis** [Glu B4(23)$\alpha \rightarrow$ Gln], and (5) **Hb Cowtown** [His HC3(146)$\beta \rightarrow$ Leu]. Assuming that all of these candidates are equal in ability at low altitudes, which one would you choose for the position? Explain your reasoning.

14. Show that the Adair equation for a tetramer reduces to the Hill equation for $k_1 \approx k_2 \approx k_3 \gg k_4$ and to a hyperbolic relationship for $k_1 = k_2 = k_3 = k_4$.

15. Derive the equilibrium constant for the reaction $R_2 \rightleftharpoons T_2$ for a symmetry model n-mer in terms of the parameters L, c, and α.

*16. Derive the equation for the fraction of protein molecules in the R state, $\overline{R}$, for the homotropic symmetry model in terms of the parameters n, L, c, and α. Plot this function versus α for $n = 4$, $L = 1000$, and $c = 0$ and discuss its physical significance.

17. In the symmetry model of allosterism, why must an inhibitor (which causes a negative heterotropic effect with the substrate) undergo a positive homotropic effect?

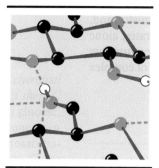

10

Sugars and Polysaccharides

Carbohydrates or **saccharides** (Greek: *sakcharon,* sugar) are essential components of all living organisms and are, in fact, the most abundant class of biological molecules. The name carbohydrate, which literally means "carbon hydrate," stems from their chemical composition, which is roughly $(C \cdot H_2O)_n$, where $n \geq 3$. The basic units of carbohydrates are known as **monosaccharides.** Many of these compounds are synthesized from simpler substances in a process named gluconeogenesis (Section 21-1). Others (and ultimately nearly all biological molecules) are the products of photosynthesis (Section 22-3), the light-powered combination of CO_2 and H_2O through which plants and certain

bacteria form "carbon hydrates." The metabolic breakdown of monosaccharides (Chapters 16 and 19) provides most of the energy used to power biological processes. Monosaccharides are also principal components of nucleic acids (Section 28-1), as well as important elements of complex lipids (Section 11-1D).

Oligosaccharides consist of a few covalently linked monosaccharide units. They are often associated with proteins **(glycoproteins)** and lipids **(glycolipids)** in which they have both structural and regulatory functions. **Polysaccharides** consist of many covalently linked monosaccharide units and have molecular masses ranging well into the millions of daltons. They have indispensable structural functions in all types of organisms but are most conspicuous in plants because **cellulose,** their principal structural material, comprises up to 80% of their dry weight. Polysaccharides such as **starch** in plants and **glycogen** in animals serve as important nutritional reservoirs.

The elucidation of the structures and functions of carbohydrates has lagged well behind those of proteins and nucleic acids. This can be attributed to several factors. Carbohydrate compounds are often heterogeneous, both in size and in composition, which greatly complicates their physical and chemical characterization. They are not subject to the types of genetic analysis that have been invaluable in the study of proteins and nucleic acids because saccharide sequences are not genetically specified but are built up through the sequential actions of specific enzymes (Section 21-3B). Furthermore, it has been difficult to establish assays for the biological activities of polysaccharides because of their largely passive roles. Nevertheless, it is abundantly

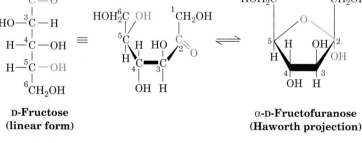

FIGURE 10-3. The reactions of alcohols with (*a*) aldehydes to form hemiacetals and (*b*) ketones to form hemiketals.

(Fig. 10-3). The hydroxyl and either the aldehyde or the ketone functions of monosaccharides can likewise react intramolecularly to form cyclic hemiacetals and hemiketals (Fig. 10-4). The configurations of the substituents to each carbon atom of these sugar rings are conveniently represented by their **Haworth projection formulas.**

A sugar with a 6-membered ring is known as a **pyranose** in analogy with **pyran,** the simplest compound containing such a ring. Similarly, sugars with 5-membered rings are designated **furanoses** in analogy with **furan.** The cyclic forms of glucose and fructose with 6- and 5-membered

rings are therefore known as **glucopyranose** and **fructofuranose,** respectively.

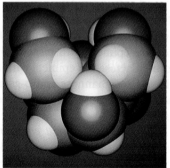

Pyran   **Furan**

Cyclic Sugars Have Two Anomeric Forms

The Greek letters preceding the names in Fig. 10-4 still need to be explained. The cyclization of a monosaccharide renders the former carbonyl carbon asymmetric. The resulting pair of diastereomers are known as **anomers** and the hemiacetal or hemiketal carbon is referred to as the **anomeric** carbon. In the α anomer, the OH substituent to the anomeric carbon is on the opposite side of the sugar ring from the CH_2OH group at the chiral center that designates the D or L configuration (C5 in hexoses). The other anomer is known as the β form (Fig. 10-5).

The two anomers of D-glucose, as any pair of diastereomers, have different physical and chemical properties. For example, the values of the specific optical rotation, $[\alpha]_D^{20}$, for α-D-glucose and β-D-glucose are, respectively, $+112.2°$ and $+18.7°$. When either of these pure substances is dissolved in water, however, the specific optical rotation

FIGURE 10-4. The reactions of (*a*) D-glucose in its linear form to yield the cyclic hemiacetal α-D-glucopyranose, and (*b*) D-fructose in its linear form to yield the hemiketal α-D-

fructofuranose. The cyclic sugars are shown as both Haworth projections and space-filling models. [Courtesy of Robert Stodola, Fox Chase Cancer Center.]

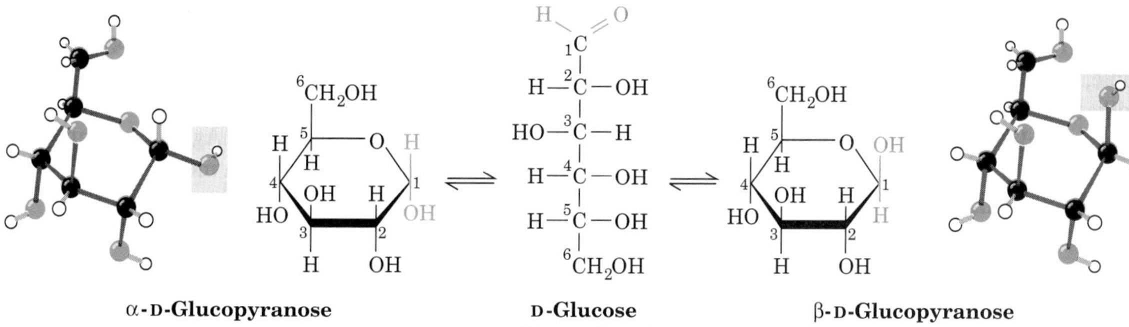

α-D-**Glucopyranose**   D-**Glucose**   β-D-**Glucopyranose**
(linear form)

FIGURE 10-5. The anomeric monosaccharides α-D- glucopyranose and β-D-glucopyranose, drawn as both Haworth projections and ball-and-stick models. These pyranose sugars interconvert through the linear form of D-glucose and differ only by the configurations about their anomeric carbon atoms, C1.

of the solution slowly changes until it reaches an equilibrium value of $[\alpha]_D^{20} = +52.7°$. This phenomenon is known as **mutarotation;** in glucose, it results from the formation of an equilibrium mixture consisting of 63.6% of the β anomer and 36.4% of the α anomer (the optical rotations of separate molecules in solution are independent of each other so that the optical rotation of a solution is the weighted average of the optical rotations of its components). The interconversion between these anomers occurs via the linear form of glucose (Fig. 10-5). Yet, since the linear forms of these monosaccharides are normally present in only minute amounts, these carbohydrates are accurately described as cyclic polyhydroxy hemiacetals or hemiketals.

Sugars Are Conformationally Variable

Hexoses and pentoses may each assume pyranose or furanose forms. The equilibrium composition of a particular monosaccharide depends somewhat on conditions but mostly on the identity of the monosaccharide. For instance, NMR measurements indicate that whereas glucose almost exclusively assumes its pyranose form in aqueous solutions, fructose is 67% pyranose and 33% furanose, and ribose is 75% pyranose and 25% furanose (although in polysaccharides, glucose, fructose, and ribose residues are exclusively in their respective pyranose, furanose, and furanose forms). Although, in principle, hexoses and larger sugars can form rings of seven or more atoms, such rings are rarely observed because of the greater stabilities of the 5- and 6-membered rings that these sugars can also form. The internal strain of 3- and 4-membered sugar rings makes them unstable with respect to linear forms.

The use of Haworth formulas may lead to the erroneous impression that furanose and pyranose rings are planar. This cannot be the case, however, because all of the atoms in these rings are tetrahedally (sp³) hybridized. The pyranose ring, like the cyclohexane ring, may assume a **boat** or a **chair** conformation (Fig. 10-6). The relative stabilities of these various conformations depends on the stereochemical interactions between the substituents on the ring. The boat conformer crowds the substituents on its "bow" and

"stern" and eclipses those along its sides so that in cyclohexane it is ~25 kJ · mol⁻¹ less stable than the chair conformer. The ring substituents on the chair conformer (Fig. 10-6b) fall into two geometrical classes: the rather close-fitting **axial** groups that extend parallel to the ring's threefold rotational axis, and the staggered, and therefore largely unencumbered, **equatorial** groups. Since the axial and equatorial groups on a cyclohexane ring are conformationally interconvertible, a given ring has two alternative chair forms (Fig. 10-7); the one that predominates usually has the lesser

(a)   Steric   (b)   Symmetry
      crowding   axis

Boat   **Chair**

FIGURE 10-6. The conformations of the cyclohexane ring. (a) In the boat conformation, substituents at the "bow" and "stern" (red) are sterically crowded, whereas those along its sides (green) are eclipsed. (b) In the chair conformation, the substituents that extend parallel to the ring's threefold rotation axis are designated axial [a] and those that extend roughly outward from this symmetry axis are designated equatorial [e]. The equatorial substituents about the ring are staggered so that they alternately extend above and below the mean plane of the ring.

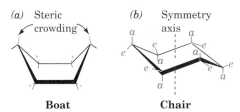

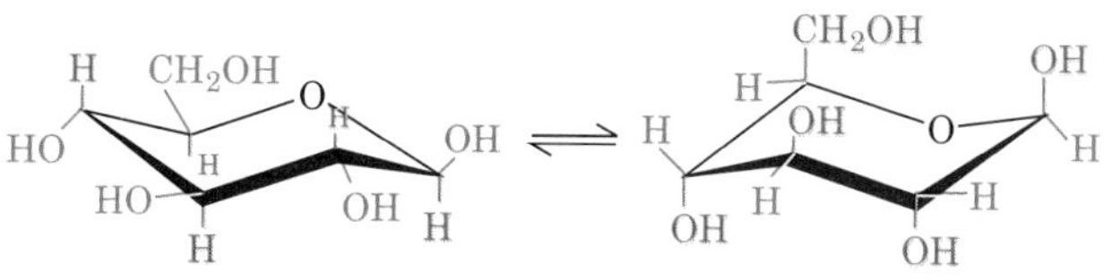

FIGURE 10-7. The two alternative chair conformations of β-D-glucopyranose. In the conformation on the left, which predominates, the relatively bulky OH and CH₂OH substituents all occupy equatorial positions, whereas in that on the right (drawn in ball-and-stick form in Fig. 10-5, right) they occupy the more crowded axial positions.

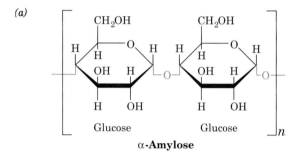

FIGURE 10-16. Chitin is a $\beta(1 \rightarrow 4)$-linked homopolymer of *N*-acetyl-D-glucosamine.

lose. Chitin is a homopolymer of $\beta(1 \rightarrow 4)$-linked *N*-acetyl-D-glucosamine residues (Fig. 10-16). It differs chemically from cellulose only in that each C2-OH group is replaced by an acetamido function. X-ray analysis indicates that chitin and cellulose have similar structures.

D. Storage Polysaccharides: Starch and Glycogen

Starch Is a Food Reserve in Plants and a Major Nutrient for Animals

Starch is a mixture of glucans that plants synthesize as their principal food reserve. It is deposited in the cytoplasm of plant cells as insoluble granules composed of **α-amylose** and **amylopectin.** α-Amylose is a linear polymer of several thousand glucose residues linked by $\alpha(1 \rightarrow 4)$ bonds (Fig. 10-17a). Note that although α-amylose is an isomer of cellulose, it has very different structural properties. This is because cellulose's β-glycosidic linkages cause each successive glucose residue to flip 180° with respect to the preceding residue so that the polymer assumes an easily packed fully extended conformation (Fig. 10-15). In contrast, α-amylose's α-glycosidic bonds cause it to adopt an irregularly aggregating helically coiled conformation (Fig. 10-17b).

Amylopectin consists mainly of $\alpha(1 \rightarrow 4)$-linked glucose residues but is a branched molecule with $\alpha(1 \rightarrow 6)$ branch points every 24 to 30 glucose residues on average (Fig. 10-18). Amylopectin molecules contain up to 10^6 glucose residues, which makes them among the largest molecules occurring in nature. Storage of glucose as starch greatly reduces the large intracellular osmotic pressures that would result from its storage in monomeric form because osmotic pressure is proportional to the number of solute molecules in a given volume.

Starch Digestion Occurs in Stages

The digestion of starch, the main carbohydrate source in the human diet, begins in the mouth. Saliva contains **α-amylase,** which randomly hydrolyzes all the $\alpha(1 \rightarrow 4)$ glucosidic bonds of starch except its outermost bonds and those next to branches. By the time thoroughly chewed food

reaches the stomach, where the acidity inactivates α-amylase, the average chain length of starch has been reduced from several thousand to fewer than eight glucose units. Starch digestion continues in the small intestine under the influence of pancreatic α-amylase, which is similar to the salivary enzyme. This enzyme degrades starch to a mixture of the disaccharide maltose, the trisaccharide **maltotriose,** which contains three $\alpha(1 \rightarrow 4)$-linked glucose residues, and oligosaccharides known as **dextrins** that contain the $\alpha(1 \rightarrow 6)$ branches. These oligosaccharides are hydrolyzed to their component monosaccharides by specific enzymes contained in the brush border membranes of the intestinal mucosa: an **α-glucosidase,** which removes one glucose residue at a time from oligosaccharides, an **α-dextrinase** or **debranching enzyme,** which hydrolyzes $\alpha(1 \rightarrow 6)$ and $\alpha(1 \rightarrow 4)$ bonds, a **sucrase,** and, at least in infants, a lactase.

(a)

(b)

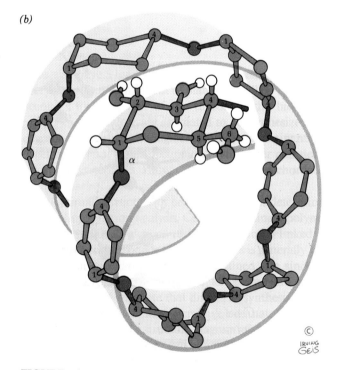

FIGURE 10-17. α-Amylose. (*a*) Its D-glucose residues are linked by $\alpha(1 \rightarrow 4)$ bonds (*red*). Here *n* is several thousand. (*b*) This regularly repeating polymer forms a left-handed helix. Note the great differences in structure and properties that result from changing α-amylose's $\alpha(1 \rightarrow 4)$ linkages to the $\beta(1 \rightarrow 4)$ linkages of cellulose (Fig. 10-15). [Figure copyrighted © by Irving Geis.]

(a)

Branch

CH$_2$OH   CH$_2$OH

$\alpha(1 \longrightarrow 6)$ branch point

Main chain

CH$_2$OH   CH$_2$OH   CH$_2$ CH$_2$OH

Amylopectin

(b)

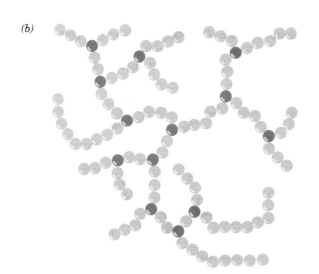

FIGURE 10-18. Amylopectin. (*a*) Its primary structure near one of its $\alpha(1 \rightarrow 6)$ branch points (*red*). (*b*) Its bushlike structure with glucose residues at branch points indicated in red. The actual distance between branch points averages 24 to 30 glucose residues. Glycogen has a similar structure but is branched every 8 to 12 residues.

The resulting monosaccharides are absorbed by the intestine and transported to the bloodstream.

Glycogen Is "Animal Starch"

Glycogen, the storage polysaccharide of animals, is present in all cells but is most prevalent in skeletal muscle and liver, where it occurs as cytoplasmic granules (Fig. 10-19). The primary structure of glycogen resembles that of amylopectin but glycogen is more highly branched, with branch points occurring every 8 to 12 glucose residues. Glycogen's degree of polymerization is nevertheless similar to that of amylopectin. In the cell, glycogen is degraded for metabolic use by **glycogen phosphorylase,** which phosphorolytically cleaves glycogen's $\alpha(1 \rightarrow 4)$ bonds sequentially inward from its nonreducing ends to yield **glucose-1-phosphate.** Glycogen's highly branched structure, which has many nonreducing ends, permits the rapid mobilization of glucose in times of metabolic need. The $\alpha(1 \rightarrow 6)$ branches of glycogen are cleaved by a debranching enzyme. These enzymes play an important role in glucose metabolism and are discussed further in Section 17-1.

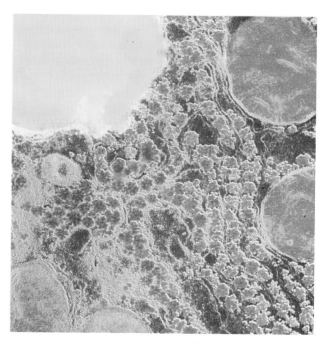

FIGURE 10-19. A photomicrograph showing the glycogen granules (*pink*) in the cytoplasm of a liver cell (the greenish object is a mitochondrion and the yellow object is a fat globule). Note that these granules tend to aggregate. The glycogen content of liver may reach as high as 10% of its net weight. [CNRI.]

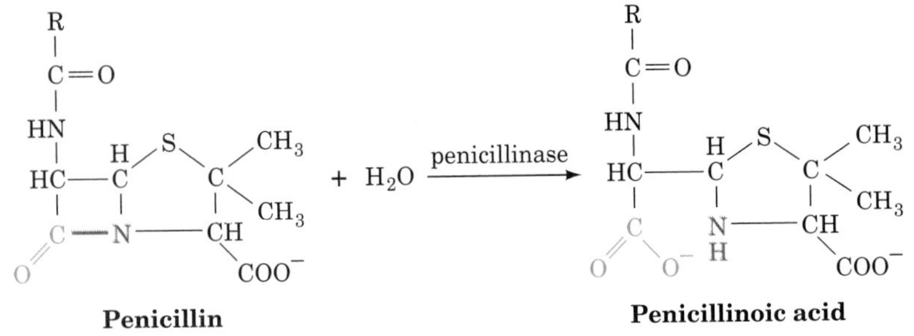

FIGURE 10-26. Penicillinase inactivates penicillin by catalyzing the hydrolysis of its β-lactam ring to form **penicillinoic acid.**

and characterizing penicillin arising from its instability, led to the passage of over a decade before penicillin was ready for routine clinical use. Penicillin specifically binds to and inactivates enzymes that function to cross-link the peptidoglycan strands of bacterial cell walls. Since cell wall expansion also requires the action of enzymes that degrade cell walls, *exposure of growing bacteria to penicillin results in their lysis;* that is, penicillin disrupts the normal balance between cell wall biosynthesis and degradation. However, since no human enzyme binds penicillin, it is of low human toxicity, a therapeutic necessity.

> Penicillin-treated bacteria that are kept in a hypertonic medium remain intact, even though they have no cell wall. Such bacteria, which are called **protoplasts** or **spheroplasts,** are spherical and extremely fragile because they are encased by only their plasma membranes. Protoplasts immediately lyse upon transfer to a normal medium.

Most bacteria that are resistant to penicillin secrete **penicillinase,** which inactivates penicillin by cleaving the amide bond of its β-lactam ring (Fig. 10-26). However, the observation that penicillinase activity varies with the nature of penicillin's R group has prompted the semisynthesis of penicillins, such as **ampicillin** (Fig. 10-25), which are clinically effective against penicillin-resistant strains of bacteria.

Bacterial Cell Walls Are Studded with Antigenic Groups

The surfaces of gram-positive bacteria are covered by **teichoic acids** (Greek: *teichos,* city walls), which account for up to 50% of the dry weight of their cell walls. Teichoic acids are polymers of glycerol or ribitol linked by phosphodiester bridges (Fig. 10-27). The hydroxyl groups of this sugar–phosphate chain are substituted by D-Ala residues and saccharides such as glucose or NAG. Teichoic acids are anchored to the peptidoglycans via phosphodiester bonds to the C6 OH groups of their NAG residues. They often terminate in **lipopolysaccharides** (lipids that contain polysaccharides; Section 11-1).

The outer membranes of gram-negative bacteria (Fig. 10-23b) are composed of complex lipopolysaccharides, proteins, and phospholipids that are organized in a complicated manner. The **periplasmic space,** an aqueous compartment that lies between the plasma membrane and the peptidoglycan cell wall, contains proteins that transport sugars and other nutrients. The outer membrane functions as a barrier to exclude harmful substances (such as gram

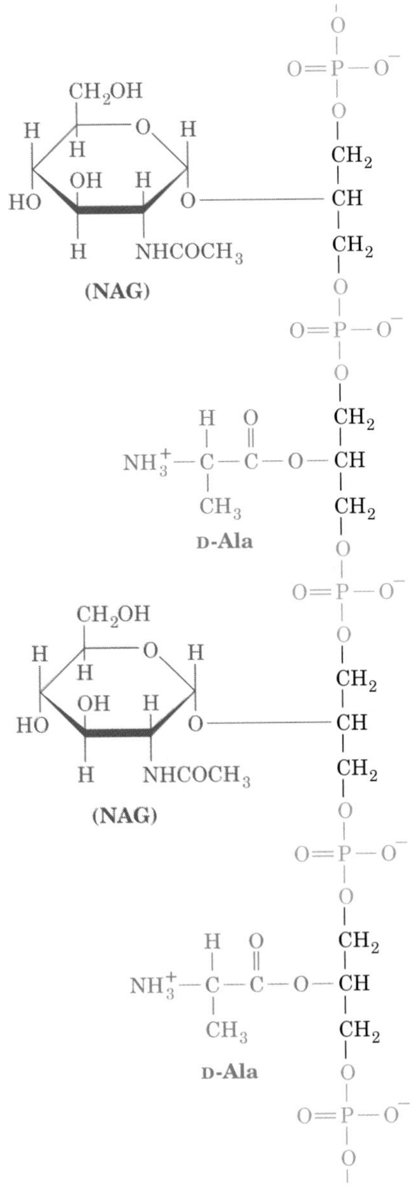

FIGURE 10-27. A segment of a teichoic acid with a glycerol phosphate backbone that bears alternating residues of D-Ala and NAG.

stain). This accounts for the observation that gram-negative bacteria are less affected by lysozyme and penicillin, as well as by other antibiotics, than are gram-positive bacteria.

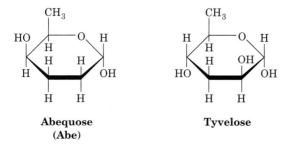

**2-Keto-3-deoxyoctanoate
(KDO)**

L-Glycero-D-manoheptose

**Abequose
(Abe)**

Tyvelose

FIGURE 10-28. Some of the unusual monoaccharides that occur in the O-antigens of gram-negative bacteria. These sugars rarely occur in other organisms.

The outer surfaces of gram-negative bacteria are coated with complex and often unusual polysaccharides known as **O-antigens** that uniquely mark each bacterial strain (Fig. 10-28). The observation that mutant strains of pathogenic bacteria lacking O-antigens are nonpathogenic suggests that O-antigens participate in the recognition of host cells. O-Antigens, as their name implies, are also the means by which a host's immunological defense system recognizes invading bacteria as foreign (Section 34-2A). As part of the ongoing biological warfare between pathogen and host, O-antigens are subject to rapid mutational alteration so as to generate new bacterial strains that the host does not initially recognize (the mutations are in the genes specifying the enzymes that synthesize the O-antigens).

C. Glycoprotein Structure and Function

Glycoprotein Carbohydrate Chains Are Highly Diverse

Almost all the secreted and membrane-associated proteins of eukaryotic cells are glycosylated. Indeed, protein glycosylation is more abundant than all other types of post-translational modifications combined. Oligosaccharides form two types of direct attachments to these proteins: **N-linked** and **O-linked**. Sequence analyses of glycoproteins have led to the following generalizations about these attachments.

1. *In N-glycosidic (N-linked) attachments, an NAG is invariably β-linked to the amide nitrogen of an Asn in the sequence Asn-X-Ser or Asn-X-Thr, where X is any amino acid residue except possibly Pro or Asp (Fig. 10-29a). The oligosaccharides in these linkages usually have a*

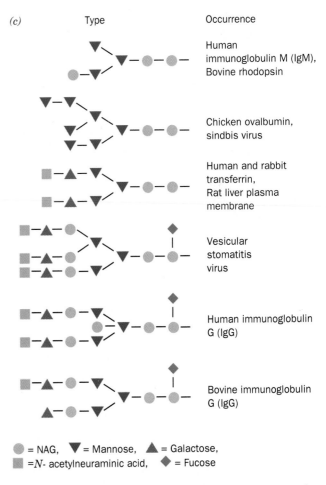

FIGURE 10-29. *N*-Linked oligosaccharides. (*a*) All *N*-glycosidic protein attachments occur through a *β*-*N*-acetylglucosamino–Asn bond in which the Asn occurs in the sequence Asn-X-Ser/Thr (*red*) where X is any amino acid. (*b*) *N*-Linked oligosaccharides usually have the branched (mannose)₃(NAG)₂ core shown. (*c*) Some examples of *N*-linked oligosaccharides. [After Sharon, N. and Lis, H., *Chem. Eng. News* **59**(13): 28 (1981).]

distinctive **core** (innermost sequence; Fig. 10-29*b*) whose peripheral mannose residues are linked to either mannose or NAG residues. These latter residues may, in turn, be linked to yet other sugar residues so that an enormous diversity of *N*-linked oligosaccharides is possible. Some examples of such oligosaccharides are shown in Fig. 10-29*c*.

2. The most common *O*-glycosidic (*O*-linked) attachment involves the disaccharide core *β-galactosyl-(1 → 3)-α-N-acetylgalactosamine* α linked to the OH group of either Ser or Thr (Fig. 10-30*a*). Less commonly, galactose, mannose, or xylose form α-*O*-glycosides with Ser or Thr (Fig. 10-30*b*). Galactose also forms *O*-glycosidic bonds to the 5-hydroxylysyl residues of collagen (Section 7-2C). However, there seem to be few, if any, additional generalizations that can be made about *O*-glycosidically linked oligosaccharides. They vary in size from a single galactose residue in collagen to the chains of up to 1000 disaccharide units in proteoglycans.

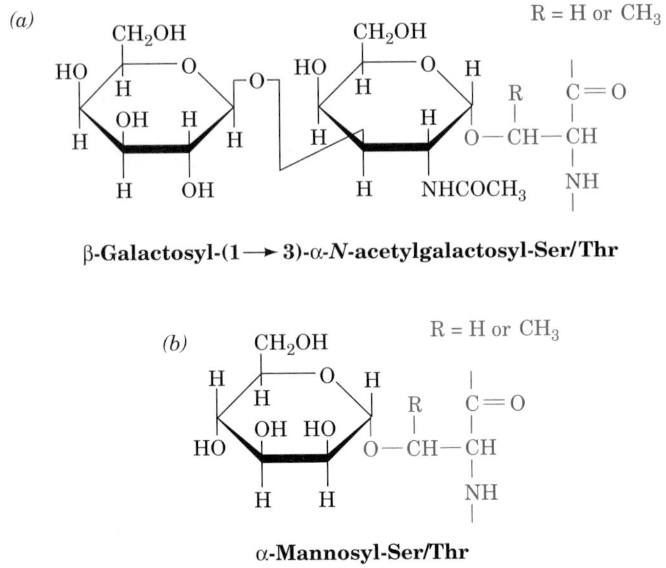

Oligosaccharides tend to attach to proteins at sequences that form *β* bends. Taken with their hydrophilic character, this observation suggests that *oligosaccharides extend from the surfaces of proteins rather than participating in their internal structures.* Indeed, the relatively few glycoprotein X-ray structures that have yet been reported, for example, those of **immunoglobulin G** (Section 34-2B) and the influenza virus **hemagglutinin** (Section 32-4B), are consistent with this hypothesis. This accounts for the observation that the protein structures of glycoproteins are unaffected by the removal of their associated oligosaccharides.

Both experimental and theoretical studies indicate that oligosaccharides have mobile and rapidly fluctuating conformations. Thus, representations in which oligosaccharides are shown as having fixed three-dimensional structures do not tell the whole story.

N-Linked Glycoproteins Exhibit Numerous Glycoforms

*Cells tend to synthesize a large repertoire of a given N-linked glycoprotein, in which each variant species (**glycoform**) differs somewhat in the sequences, locations, and numbers of its covalently attached oligosaccharides.* For example, one of the simplest glycoproteins, bovine pancreatic **ribonuclease B (RNase B),** differs from the well-characterized and carbohydrate-free enzyme RNase A (Section 8-1A) only by the attachment of a single *N*-glycosidically linked oligosaccharide chain. The oligosaccharide has the core sequence diagrammed in Fig. 10-31 with considerable microheterogeneity in the position of a sixth mannose residue. Nevertheless, the oligosaccharide does not appear to affect the conformation, substrate specificity, or catalytic properties of RNase A. In contrast, human **granulocyte–macrophage colony-stimulating factor (GM-CSF),** a 127-amino acid residue protein growth factor that promotes the development, activation, and survival of the white blood cells known as **granulocytes** and **macrophages,** is variably glycosylated at two *N*-linked sites and five *O*-linked sites.

(a)

β-Galactosyl-(1 ⟶ 3)-α-*N*-acetylgalactosyl-Ser/Thr

(b)

α-Mannosyl-Ser/Thr

FIGURE 10-30. Some common *O*-glycosidic attachments of oligosaccharides to glycoproteins (*red*).

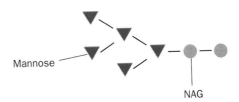

FIGURE 10-31. The microheterogeneous *N*-linked oligosaccharide of RNase B has the (mannose)$_5$(NAG)$_2$ core shown. A sixth mannose residue occurs at various positions on this core.

Through the generation of mutant varieties of GM-CSF that lack one or both of these *N*-glycosylation sites, it was found that the lifetime of GM-CSF in the bloodstream increases with its level of glycosylation. However, GM-CSF that is produced by *E. coli* and hence is unglycosylated (bacteria do not glycosylate the proteins they synthesize) has a 20-fold higher specific biological activity than does the naturally occurring glycoprotein.

As the foregoing examples suggest, no generalization can be made about the effects of glycosylation on protein properties; they must be experimentally determined on a case-by-case basis. Nevertheless, it is becoming increasingly evident that glycosylation can affect protein properties in many ways, including protein folding, physical stability, specific bioactivity, rate of clearance from the bloodstream, and protease resistance. Thus, the species-specific and tissue-specific distribution of glycoforms that each cell synthesizes endows it with a characteristic spectrum of biological properties.

O-Linked Glycoproteins Often Serve Protective Functions

O-Linked polysaccharides tend not to be uniformly distributed along polypeptide chains. Rather, they are clustered into heavily glycosylated (65 to 85% carbohydrate by weight) segments in which glycosylated Ser and Thr residues comprise 25 to 40% of the sequence. The carbohydrates' hydrophilic and steric interactions cause these heavily glycosylated regions, which are also rich in Pro and other helix-breaking residues, to assume extended conformations. For example, **mucins,** the protein components of **mucus,** are exceedingly large (~ 10^7 D) *O*-linked glycoproteins whose carbohydrate chains are often sulfated and hence mutually repelling. Mucins therefore consist of stiff chains with random coil conformations that occupy time-averaged volumes approximating those of small bacteria. Consequently, mucins, at their physiological concentrations, form intertangled networks that comprise the viscoelastic gels which protect and lubricate the mucous membranes that secrete them.

Eukaryotic cells, as we shall see in Section 11-3D, have a thick and fuzzy coating of glycoproteins and **glycolipids** named the **glycocalyx** that prevents the close approach of macromolecules and other cells. How, then, can cells interact? Many cell-surface proteins, such as the receptors for various macromolecules, have relatively short and presumably stiff *O*-glycosylated regions that link these glycoproteins' membrane-bound domains to their functional domains. This arrangement is thought to extend the functional domains in a lollipop-like manner above the cell's densely packed glycocalyx, thereby permitting the functional domain to interact with extracellular macromolecules that cannot penetrate the glycocalyx.

Oligosaccharide Markers Mediate a Variety of Intercellular Interactions

Glycoproteins are important constituents of plasma membranes (Section 11-3). The location of their carbohydrate moieties can be determined by electron microscopy. The glycoproteins are labeled with lectins that have been conjugated (covalently cross-linked) to **ferritin,** an iron-transporting protein that is readily visible in the electron microscope because of its electron-dense iron hydroxide core. Such experiments, with lectins of different specificities and with a variety of cell types, have demonstrated that *the carbohydrate groups of membrane-bound glycoproteins are, for the most part, located on the external surface of the cell membrane.* Thus, the viability of cultured cells from multicellular organisms that have any of a large number of glycosylation mutations and the infrequent viability of whole organisms that bear such mutations indicate that oligosaccharides are important for intercellular communications but not for intracellular "housekeeping" functions.

A further indication that oligosaccharides function as biological markers is the observation that the carbohydrate content of a glycoprotein often governs its metabolic fate. For example, the excision of sialic acid residues from certain radioactively labeled blood plasma glycoproteins by treatment with **sialidase** greatly increases the rate at which these glycoproteins are removed from the circulation. The glycoproteins are taken up and degraded by the liver in a process that depends on the recognition by liver cell receptors of sugar residues such as galactose and mannose which are exposed by the sialic acid excision. A diverse series of receptors, each specific for a particular type of sugar residue, participates in removing any particular glycoprotein from the blood. A variety of glycoforms for a given glycoprotein therefore probably ensures that it has a range of lifetimes in the blood. *Similar "ticketing" mechanisms probably govern the compartmentation and degradation of glycoproteins within cells.*

The observation that cancerous cells are more susceptible to agglutination by lectins than are normal cells led to the discovery that *there are significant differences between the cell-surface carbohydrate distributions of cancerous and noncancerous cells* (Fig. 10-32). Normal cells stop growing when they touch each other, a phenomenon known as **con-**

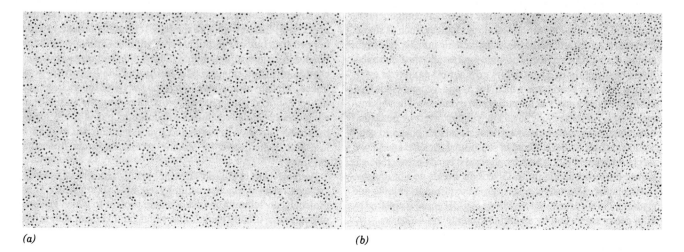

(a) *(b)*

FIGURE 10-32. The surfaces of (*a*) a normal mouse cell, and (*b*) a cancerous cell as seen in the electron microscope. Both cells were incubated with the ferritin-labeled lectin concanavalin A. The lectin is evenly dispersed on the normal cell but is aggregated into clusters on the cancerous cell. [Courtesy of Garth Nicholson, University of Texas M. D. Anderson Cancer Center.]

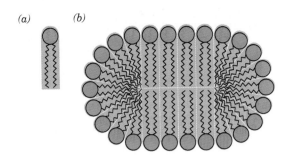

(a)   *(b)*

FIGURE 11-12. The cylindrical van der Waals envelope of phospholipids (*a*) causes them to form extended disklike micelles (*b*) that are better described as lipid bilayers.

diameters ~1000 Å are made by injecting an ethanolic solution of phospholipid into water or by dissolving phospholipid in a detergent solution and then dialyzing out the detergent. Once formed, liposomes are quite stable and, in fact, may be separated from the solution in which they reside by dialysis, gel filtration chromatography, or centrifugation. Liposomes with differing internal and external environments can therefore be readily prepared. *Biological membranes consist of lipid bilayers with which proteins are associated (Section 11-3A).* Liposomes composed of synthetic lipids and/or lipids extracted from biological sources (e.g., lecithin from egg yolks) have therefore been extensively studied as models for biological membranes.

The number of molecules in such micelles depends on the amphiphile, but for many substances, it is on the order of several hundred. For a given amphiphile, these numbers span a narrow range: Less would expose the hydrophobic core of the micelle to water, whereas more would give the micelle an energetically unfavorable hollow center (Fig. 11-11*c*). Of course, a large micelle could flatten out to eliminate this hollow center but the resulting decrease of curvature at the flattened surfaces would also generate empty spaces (Fig. 11-11*d*).

Glycerophospholipids and Sphingolipids Tend to Form Bilayers

The two hydrocarbon tails of glycerophospholipids and sphingolipids give these amphiphiles more or less rectangular shapes (Fig. 11-12*a*). The steric requirements of packing such molecules together yields large disklike micelles (Fig. 11-12*b*) that are really extended bimolecular leaflets. The existence of such **lipid bilayers** was first proposed in 1925 by E. Gorter and F. Grendel, on the basis of their observation that lipids extracted from erythrocytes covered twice the area when spread as a monolayer at the air–water interface (Fig. 11-10) than in the erythrocyte plasma membrane (the erythrocyte's only membrane). Lipid bilayers have thicknesses of ~60 Å, as measured by electron microscopy and X-ray diffraction techniques, the value expected when their hydrocarbon tails are more or less extended. We shall see below that *lipid bilayers form the structural basis of biological membranes.*

B. Liposomes

A suspension of phospholipids in water forms multilamellar vesicles that have an onionlike arrangement of lipid bilayers (Fig. 11-13*a*). Upon **sonication** (agitation by ultrasonic vibrations), these structures rearrange to form **liposomes**—closed, self-sealing, solvent-filled vesicles that are bounded by only a single bilayer (Fig. 11-13*b*). They usually have diameters of several hundred Å and, in a given preparation, are rather uniform in size. Liposomes with

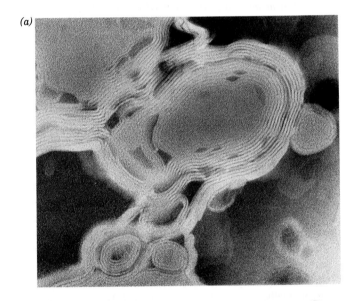

(a)

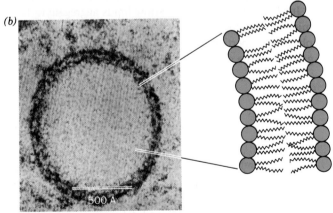

(b)

500 Å

FIGURE 11-13. (*a*) An electron micrograph of a multilamellar phospholipid vesicle in which each layer is a lipid bilayer. [Courtesy of Alec D. Bangham, Institute of Animal Physiology, U.K.] (*b*) An electron micrograph of a liposome. Its wall, as the accompanying diagram indicates, consists of a bilayer. [Courtesy of Walter Stoekenius, University of California at San Francisco.]

Lipid Bilayers Are Impermeable to Most Polar Substances

Since biological membranes form cell and organelle boundaries, it is important to determine their ability to partition two aqueous compartments. The permeability of a lipid bilayer to a given substance may be determined by forming liposomes in a solution containing the substance, changing the external aqueous solution, and then measuring the rate at which the substance of interest appears in the new external solution. It has been found in this way that *lipid bilayers are extraordinarily impermeable to ionic and polar substances and that the permeabilities of such substances increase with their solubilities in nonpolar solvents.* This suggests that to penetrate a lipid bilayer, a solute molecule must shed its hydration shell and become solvated by the bilayer's hydrocarbon core. Such a process is highly unfavorable for polar molecules so that even the ~30-Å thickness of a lipid bilayer's hydrocarbon core forms an effective barrier for polar substances. However, measurements using tritiated water indicate that lipid bilayers are appreciably permeable to water. Despite the polarity of water, its small molecular size makes it significantly soluble in the hydrocarbon core of lipid bilayers and therefore able to permeate them.

> The stability of liposomes and their impermeability to many substances makes them promising vehicles for the delivery of therapeutic agents, such as drugs and enzymes, to particular tissues. Liposomes are absorbed by many cells through fusion with their plasma membranes. If methods can be developed for targeting liposomes to specific cell populations, then drugs could be directed towards particular tissues through liposome encapsulation. Encouraging progress towards this goal has been made (Fig. 11-14).

Lipid Bilayers Are Two-Dimensional Fluids

The transfer of a lipid molecule across a bilayer (Fig. 11-15a), a process termed **transverse diffusion** or a **flip-flop,** is an extremely rare event. This is because a flip-flop requires the polar head group of the lipid to pass through the hydrocarbon core of the bilayer. The flip-flop rates of phospholipids, as measured by several techniques, are characterized by half-times that are minimally several days.

In contrast to their low flip-flop rates, *lipids are highly mobile in the plane of the bilayer* (**lateral diffusion,** Fig. 11-15b). The X-ray diffraction patterns of bilayers at physiological temperatures have a diffuse band, centered at a spacing of 4.6 Å, whose width is a measure of the distribution of lateral spacings between the hydrocarbon chains in the bilayer plane. This band, which resembles one in the X-ray diffraction patterns of liquid paraffins, is indicative that *the bilayer is a two-dimensional fluid in which the hydrocarbon chains undergo rapid fluxional (continuously changing) motions involving rotations about their C—C bonds.*

The lateral diffusion rate of lipid molecules can be quantitatively determined from the rate of **fluorescence photo-**

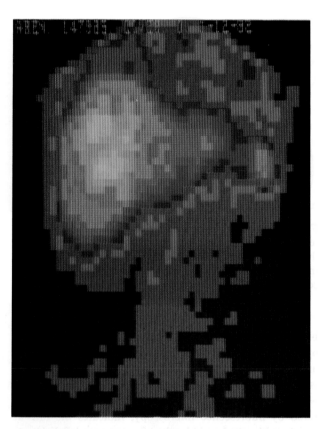

FIGURE 11-14. γ-Ray scan of a human subject taken 24 h after the intravenous injection of liposomes labeled with a γ-ray emitter. Liposomal uptake (*blue and white squares*) preferentially occurs in the liver (*left*) and the spleen (*right*). [Courtesy of Gabriel Lopez-Berestein, University of Texas M. D. Anderson Cancer Center.]

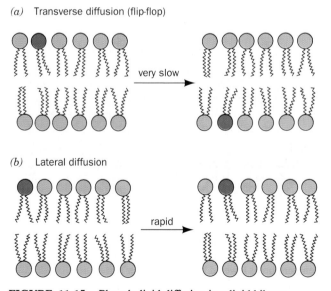

FIGURE 11-15. Phospholipid diffusion in a lipid bilayer. (a) Transverse diffusion (flip-flops) occurs through the transfer of a phospholipid molecule from one bilayer leaflet to the other. (b) Lateral diffusion occurs through the pairwise exchange of neighboring phospholipid molecules in the same bilayer leaflet.

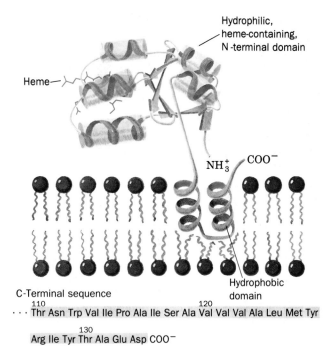

Hydrophilic,
heme-containing,
N-terminal domain

Heme

NH_3^+ COO^-

C-Terminal sequence
110   120
··· Thr Asn Trp Val Ile Pro Ala Ile Ser Ala Val Val Val Ala Leu Met Tyr

Hydrophobic
domain

130
Arg Ile Tyr Thr Ala Glu Asp COO^-

FIGURE 11-24. Liver cytochrome b_5 in association with a membrane. The protein's enzymatically active N-terminal domain (*purple*), whose X-ray structure has been determined, is anchored in the membrane by a hydrophobic and presumably α helical C-terminal segment (*brown*) that begins and ends with hydrophilic segments (*purple*). The amino acid sequence of the horse enzyme indicates that this hydrophobic anchor consists of a 13-residue segment ending 9 residues from the polypeptide's C-terminus (*below*). [Ribbon diagram of the N-terminal domain after a drawing by Jane Richardson, Duke University. Amino acid sequence from Ozols, J. and Gerard, C., *J. Biol. Chem.* **253**, 8549 (1977).]

due N-terminal fragment, and an ~50-residue C-terminal fragment that remains embedded in the membrane (Fig. 11-24). *The asymmetric orientation of integral membrane proteins in the membrane is maintained by their infinitesimal flip-flop rates (even slower than those of lipids), which result from the greater sizes of the membrane protein "head groups" in comparison to those of lipids.* The origin of this asymmetry is discussed in Section 11-4.

Only a handful of integral membrane proteins have yet been crystallized—and then only in the presence of deter-

gents. As a consequence, the three-dimensional structures of very few such proteins are known in detail. In the remainder of this subsection, we discuss the structures of three integral membrane proteins, **bacteriorhodopsin,** the bacterial **photosynthetic reaction center,** and **porins.**

Bacteriorhodopsin Contains a Bundle of Seven Hydrophobic Helical Rods

One of the structurally best characterized integral membrane proteins is **bacteriorhodopsin** from the halophilic (salt loving) bacterium *Halobacter halobium* that inhabits such salty places as the Dead Sea (it grows best in 4.3*M* NaCl and is nonviable below 2.0*M* NaCl; seawater contains 0.6*M* NaCl). Under low O_2 conditions, its cell membrane develops ~0.5-μm wide patches of **purple membrane** whose only protein component is bacteriorhodopsin. This 247-residue protein is a light-driven proton pump; it generates a proton concentration gradient across the membrane that powers the synthesis of ATP (by a mechanism discussed in Section 20-3B). Bacteriorhodopsin's light-absorbing element, **retinal,** is covalently bound to its Lys 216 (Fig. 11-25). This **chromophore** (light-absorbing group), which is responsible for the membrane's purple color, is also the light-sensitive element in vision.

The purple membrane, which is 75% protein and 25% lipid, has an unusual structure compared to most other membranes (Section 11-3B): Its bacteriorhodopsin molecules are arranged in a highly ordered two-dimensional array (a two-dimensional crystal). This permitted Richard Henderson and Nigel Unwin, through **electron crystallography** (a technique they devised, resembling X-ray crystallography, in which the electron beam of an electron microscope is used to elicit diffraction from two-dimensional crystals), to determine the structure of bacteriorhodopsin to near-atomic resolution (3.5 Å in directions parallel to the membrane but only ~10 Å in the perpendicular direction).

Bacteriorhodopsin consists largely of a bundle of seven ~25-residue α helical rods that span the lipid bilayer in directions almost perpendicular to the bilayer plane (Fig. 11-26). The ~20-Å spaces between the protein molecules in the purple membrane are occupied by this bilayer. Adjacent α helices, which are largely hydrophobic in character, are connected in a head-to-tail fashion by short polypeptide loops (Fig. 11-26). This arrangement places the protein's charged residues near the surfaces of the membrane in con-

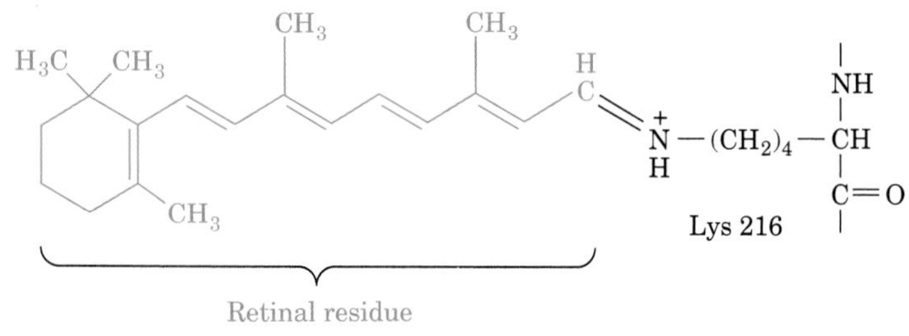

FIGURE 11-25. Retinal, the prosthetic group of bacteriorhodopsin, forms a Schiff base with Lys 216 of the protein. A similar linkage occurs in **rhodopsin,** the photoreceptor of the eye.

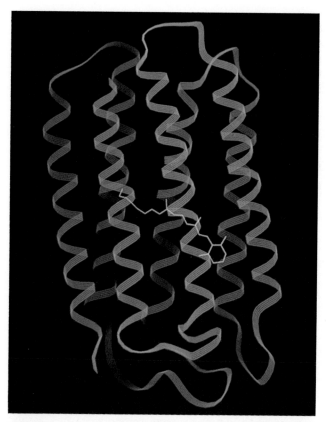 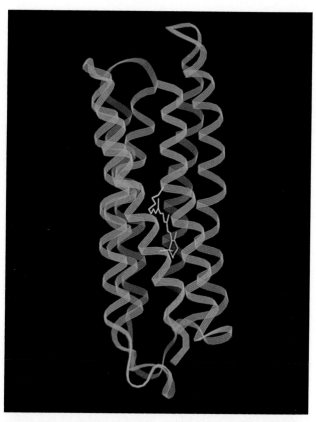

FIGURE 11-26.   A ribbon diagram of bacteriorhodopsin as viewed in two perpendicular directions from within the membrane plane. The polypeptide backbone is blue and the bound retinal is yellow. The conformations of some of the loop regions are tentative. [Courtesy of Richard Henderson, MRC Laboratory of Molecular Biology, Cambridge, England.]

tact with the aqueous solvent. The internal charged residues in this model line the center of the helix bundle so as to form a hydrophilic channel that undoubtedly facilitates the passage of protons. Other membrane pumps and channels (Chapter 18) probably have similar structures.

The Photosynthetic Reaction Center Contains Eleven Transmembrane Helices

The primary photochemical process of photosynthesis in purple photosynthetic bacteria is mediated by the so-called **photosynthetic reaction center** (Section 22-2B), a transmembrane protein consisting of at least three nonidentical ~300-residue subunits that collectively bind four **chlorophyll** molecules, four other chromophores, and a nonheme Fe atom. The 1187-residue photosynthetic reaction center of *Rhodopseudomonas (Rps.) viridis,* whose X-ray structure was determined in 1984 by Hartmut Michel, Johann Deisenhofer, and Robert Huber, was the first transmembrane protein to be described in atomic detail (Fig. 11-27). The protein's membrane-spanning portion consists of 11 α helices that form a 45-Å long cylinder with the expected hydrophobic surface.

Hydrophobic forces, as we have seen in Section 7-4, are the dominant interactions stabilizing the three-dimen-

sional structures of water-soluble globular proteins. However, since the transmembrane regions of integral membrane proteins are immersed in nonpolar environments, what stabilizes their structures? Analysis of the photosynthetic reaction center indicates that *its transmembrane region has a hydrophobic organization opposite to that of water-soluble proteins: The membrane-exposed residues of the photosynthetic reaction center are more hydrophobic, on average, than its interior residues, even though these interior residues have average hydrophobicities comparable to those of water-soluble proteins.* Evidently, the structures of transmembrane and water-soluble proteins are stabilized by similar forces.

Porins Are Channel-Forming Proteins that Contain Transmembrane β Barrels

The outer membranes of gram-negative bacteria (Section 10-3B) protect them from hostile environments but must nevertheless be permeable to small polar solutes such as nutrients. These outer membranes consequently contain embedded channel-forming proteins called **porins,** which are usually trimers of identical 30- to 50-kD subunits that permit the passage of less than ~600-D solutes. Porins also occur in eukaryotes in the outer membranes of mitochon-

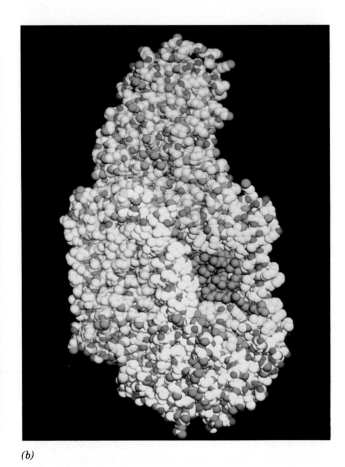

(a)

(b)

FIGURE 11-27. The X-ray structure of the photosynthetic reaction center of *Rps. viridis.* (*a*) A ribbon diagram in which only the C$_\alpha$ backbone and the prosthetic groups (*yellow*) are shown. The H, M, and L subunits (*pink, blue,* and *orange,* respectively) collectively have 11 transmembrane helices. The 4-heme *c*-type cytochrome (*green*), which does not occur in all species of photosynthetic bacteria, is bound to the external face of the complex. The position that the transmembrane protein is thought to occupy in the lipid bilayer is indicated schematically. [Based on an X-ray structure by Johann Deisenhofer, Robert Huber, and Hartmut Michel, Max-Planck-Institut für Biochemie, Germany.] (*b*) A space-filling model in which nitrogen atoms are blue, oxygens are red, sulfurs are yellow, and the carbon atoms of the H, M, L, and cytochrome subunits are tinted pink, blue, orange, and green, respectively. Exposed portions of prosthetic groups are brown. Note how few polar groups (nitrogens and oxygens) are externally exposed in the portion of the protein that is immersed in the nonpolar region of the lipid bilayer. [From Deisenhofer, J. and Michel, H., *Les Prix Nobel* (1989).]

dria and chloroplasts (thereby providing a further indication that these organelles are descended from bacteria; Section 1-2A).

The X-ray structures of three porins have recently been elucidated: A *Rhodobacter (Rb.) capsulatus* porin, determined by Georg Schulz, and the *E. coli* **OmpF** and **PhoE** porins, determined by Johan Jansonius. The 340- and 330-residue OmpF and PhoE porins share 63% sequence identity but have little sequence similarity with the 301-res-idue *Rb. capsulatus* porin. Nevertheless, all three porins have closely similar structures. Each monomer of these trimeric proteins predominantly consists of a 16-stranded antiparallel β barrel which forms a solvent-accessible pore along the barrel axis that has a length of ~55 Å and a minimum diameter of ~7 Å (Fig. 11-28). In the OmpF and PhoE porins, the N- and C-termini associate via a salt bridge in the 16th β strand, thereby forming a pseudocyclic structure (Fig. 11-28*a*). Note that a β barrel fully satisfies the polypeptide backbone's hydrogen bonding potential, as does an α helix. As expected, the side chains at the protein's membrane-exposed surface are nonpolar, thereby forming a ~25-Å-high hydrophobic band encircling the trimer (Fig 11-28c). In contrast, the side chains at the solvent-exposed surface of the protein, including those lining the walls of the aqueous channel, are polar. Possible mechanisms for solute selectivity by these porins are discussed in Section 18-2D.

B. Fluid Mosaic Model of Membrane Structure

The demonstrated fluidity of artificial lipid bilayers suggests that biological membranes have similar properties. This idea was proposed in 1972 by S. Jonathan Singer and Garth Nicholson in their unifying theory of membrane structure known as the **fluid mosaic model.** The theory postulates that integral proteins resemble "icebergs" floating in a two-dimensional lipid "sea" (Fig. 11-21) and that these proteins freely diffuse laterally in the lipid matrix unless their movements are restricted by associations with other cell components.

The Fluid Mosaic Model Has Been Verified Experimentally

The validity of the fluid mosaic model has been established in several ways. Perhaps the most vivid is an experiment by Michael Edidin (Fig. 11-29). Cultured mouse cells were fused with human cells by treatment with **Sendai virus** to yield a hybrid cell known as a **heterokaryon.** The mouse cells were labeled with mouse protein-specific antibodies to which a green-fluorescing dye had been covalently linked **(immunofluorescence).** The proteins on the human cells were similarly labeled with a red-fluorescing marker. Upon cell fusion, the mouse and human proteins, as seen under the fluorescence microscope, were segregated on the two

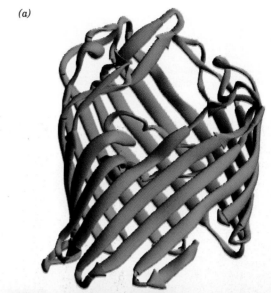

(a)

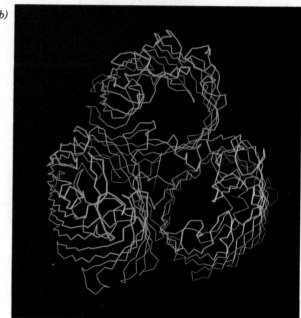

(b)

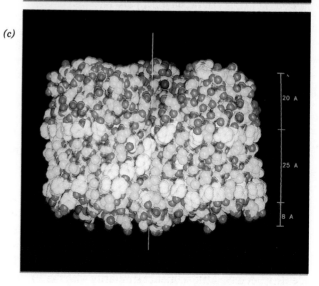

(c)

FIGURE 11-28. The X-ray crystal structure of the *E. coli* OmpF porin. (*a*) A ribbon diagram of the monomer. The C-terminal strand of this 16-stranded antiparallel β barrel is continued by its N-terminal segment, thereby forming a pseudocontinuous strand. [Based on an X-ray structure by Johan Jansonius.] (*b*) The C$_\alpha$ backbone of the trimer viewed ~30° from its three-fold axis of symmetry showing the pore through each subunit. The subunits are differently colored. It can be seen at the interface between the blue and green subunits that the strands in adjacent β sheets are essentially perpendicular to each other. (*c*) A space-filling model of the trimer viewed perpendicular to its three-fold axis (*vertical green line*). N atoms are blue, O atoms are red, and C atoms are yellow, except those in the side chains of aromatic residues, which are white. The aromatic groups appear to delimit a ~25-Å-high hydrophobic band (*scale at right*) that is immersed in the nonpolar portion of the bacterial outer membrane (with the cell's exterior at the tops of Parts *a* and *c*). Compare this hydrophobic band with that in Fig. 11-27b. [Parts *b* and *c* courtesy of Tilman Schirmer and Johan Jansonius, University of Basel, Basel, Switzerland.]

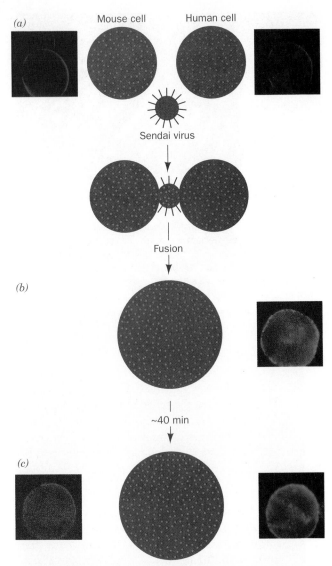

(a) Mouse cell   Human cell

Sendai virus

Fusion

(b)

~40 min

(c)

FIGURE 11-29. Sendai virus-induced fusion of a mouse cell with a human cell and the subsequent intermingling of their cell-surface components as visualized by immunofluorescence. Human and mouse antigens are labeled with red and green fluorescent markers, respectively. (*a*) The membrane-encapsulated Sendai virus specifically binds to cell-surface receptors on both types of cells and subsequently fuses to their cell membranes. (*b*) This results in the formation of a cytoplasmic bridge between the cells that expands so as to form the heterokaryon. (*c*) After 40 min, the red and green markers are fully intermingled. The photomicrographs were taken through filters that allowed only red or green light to reach the camera; that in Part (*b*) is a double exposure. [Immunofluorescence photomicrographs courtesy of Michael Edidin, The Johns Hopkins University.]

halves of the heterokaryon. After 40 min at 37°C, however, these proteins had thoroughly intermingled. The addition of substances that inhibit metabolism or protein synthesis did not slow this process, but lowering the temperature below 15°C did. These observations indicate that the mix-

ing process is independent of both metabolic energy and the insertion into the membrane of newly synthesized proteins. Rather, it is a result of the diffusion of existing proteins throughout the fluid membrane, a process that slows as the temperature is lowered.

Fluorescence photobleaching recovery measurements (Fig. 11-16) indicate that membrane proteins vary in their lateral diffusion rates. Some 30 to 90% of these proteins are freely mobile; they diffuse at rates only an order of magnitude or so slower than those of the much smaller lipids so that they typically take from 10 to 60 min to diffuse the 20-μm length of a eukaryotic cell. Other proteins diffuse more slowly, and some, because of submembrane attachments, are essentially immobile.

The distribution of proteins in membranes may be visualized through electron microscopy using the **freeze-fracture** and **freeze-etch** *techniques.* In the freeze-fracture procedure, which was devised by Daniel Branton, a membrane specimen is rapidly frozen to near liquid nitrogen temperatures (−196°C). This immobilizes the sample and thereby minimizes its disturbance by subsequent manipulations. The specimen is then fractured with a cold microtome knife, which often splits the bilayer into monolayers (Fig. 11-30). Since the exposed membrane itself would be destroyed by an electron beam, its metallic replica is made by coating the membrane with a thin layer of carbon, shadowing it (covering it by evaporative deposition under high

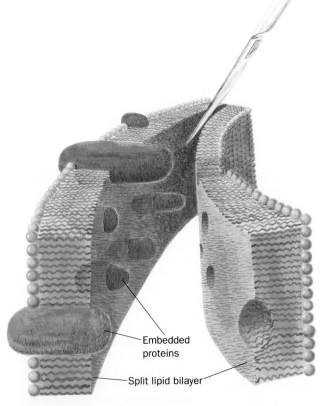

Embedded proteins

Split lipid bilayer

FIGURE 11-30. A membrane that has been split by freeze fracture, as is schematically diagrammed, exposes the interior of the lipid bilayer and its embedded proteins.

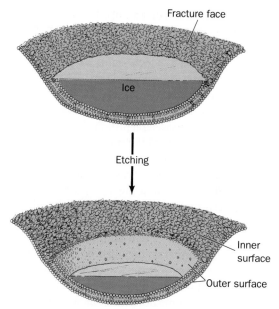

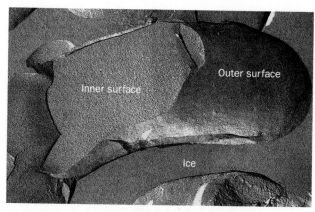

FIGURE 11-32. A freeze-etch electron micrograph of a human erythrocyte plasma membrane. The exposed interior face of the membrane is studded with numerous globular particles that are integral membrane proteins (see Fig. 11-30). The outer surface of the membrane appears smoother than the inner surface because proteins do not project very far beyond the outer membrane surface. [Courtesy of Vincent Marchesi, Yale University.]

FIGURE 11-31. In the freeze-etch procedure, the ice that encases a freeze-fractured membrane (*top*) is partially sublimed away so as to expose the outer membrane surface (*bottom*) for electron microscopy.

vacuum) with platinum, and removing the organic matter by treatment with acid. Such a metallic replica can be examined by electron microscopy. In the freeze-etch procedure, the external surface of the membrane adjacent to the cleaved area revealed by freeze fracture may also be visualized by first subliming (etching) away, at −100°C, some of the ice in which it is encased (Fig. 11-31).

Freeze-etch electron micrographs of most biological membranes show an inner fracture face that is studded with embedded 50- to 85-Å-diameter globular particles (Fig. 11-32) that appear to be distributed randomly. These particles correspond to membrane proteins, as is demonstrated by their disappearance when the membrane is treated with proteases before its freeze fracture. This is further corroborated by the observation that the myelin membrane, which has a low protein content, as well as liposomes composed of

only lipids, have smooth inner fracture faces. Outer membrane surfaces also have a relatively smooth appearance (Fig. 11-32) because integral proteins do not protrude very far beyond them. The distributions of individual external proteins may be visualized by staining procedures, such as the use of ferritin-labeled antibodies, to yield electron micrographs similar in appearance to Fig. 10-32.

Membrane Lipids Are Asymmetrically Distributed

The distribution of lipids in a membrane has been established through the use of phospholipid-hydrolyzing enzymes known as **phospholipases**. Phospholipases cannot pass through membranes so that only phospholipids on the external surfaces of intact cells are susceptible to their action. Such studies indicate that *the lipids in biological membranes, like the proteins, are asymmetrically distributed* (e.g., Fig. 11-33). Carbohydrates, as we have seen (Section 10-3C), are located exclusively on the external surfaces of plasma membranes.

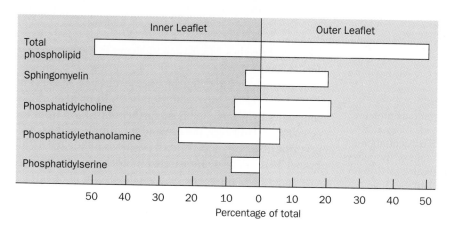

FIGURE 11-33. The asymmetric distribution of phospholipids in human erythrocyte membranes, expressed in mol %. [After Rothman, J.E. and Lenard, J., *Science* **194**, 1744 (1977).]

C. *The Erythrocyte Membrane*

The erythrocyte membrane's relative simplicity, availability, and ease of isolation have made it the most extensively studied and best understood biological membrane. It is therefore a model for the more complex membranes of other cell types. A mature mammalian erythrocyte is devoid of organelles and carries out few metabolic processes; it is essentially a membranous bag of hemoglobin. Erythrocyte membranes can therefore be obtained by osmotic lysis, which causes the cell contents to leak out. The resultant membranous particles are known as erythrocyte **ghosts** because, upon return to physiological conditions, they reseal to form particles that retain their original shape. Indeed, by transferring sealed ghosts to another medium, their contents can be made to differ from the external solution.

Erythrocyte Membranes Contain a Variety of Proteins

The erythrocyte membrane has a more or less typical plasma membrane composition of about one-half protein, somewhat less lipid, and the remainder carbohydrate (Table 11-4). *Its proteins may be separated by SDS–polyacrylamide gel electrophoresis (Section 5-4C) after first solubilizing the membrane in a 1% SDS solution.* The resulting electrophoretogram for a human erythrocyte membrane exhibits seven major and many minor bands when stained with Coomassie brilliant blue (Fig. 11-34). If the electrophoretogram is instead treated with **periodic acid–Schiff's reagent (PAS),** which stains carbohydrates, four so-called PAS bands become evident. The polypeptides corresponding to bands 1, 2, 4.1, 4.2, 5, and 6 are readily extracted from the membrane by changes in ionic strength or pH and hence are peripheral proteins. These proteins are located on the inner side of the membrane, as is indicated by the observation that they are not altered by the incubation of intact erythrocytes or sealed ghosts with proteolytic enzymes or membrane-impermeable protein labeling reagents. These proteins are altered, however, if "leaky" ghosts are so treated.

In contrast, bands 3, 7, and all four PAS bands correspond to integral proteins; they can be released from the membrane only by extraction with detergents or organic solvents. Of these, band 3 and PAS bands 1 and 2 correspond to transmembrane proteins as indicated by their different labeling patterns when intact cells are treated with membrane-impermeable protein-labeling reagents and when these reagents are introduced inside sealed ghosts.

The transport of CO_2 in blood (Section 9-1C) requires that the erythrocyte membrane be permeable to HCO_3^- and Cl^- (the maintenance of electroneutrality requires that for every HCO_3^- to enter a cell, a Cl^- or some other anion must leave the cell; Section 9-1C). The rapid transport of these and other anions across the erythrocyte membrane is mediated by a specific **anion channel** of which there are ~1 million/cell (comprising >30% of the membrane protein). Band 3 protein (929 residues and 5–8% carbohydrate) specifically reacts with anionic protein-labeling reagents that block the anion channel, thereby indicating that the anion channel is composed of band 3 protein. Furthermore, cross-linking studies with bifunctional reagents (Section 7-5C) demonstrate that the anion channel is at least a dimer. Hemoglobin and the glycolytic (glucose metabolizing) enzymes **aldolase, phosphofructokinase (PFK),** and the band 6 protein **glyceraldehyde-3-phosphate dehydrogenase (GAPDH;** Section 16-2) all specifically and reversibly bind to band 3 protein on the cytoplasmic side of the membrane. The functional significance of this observation is unknown.

The Erythrocyte's Membrane Skeleton Is Responsible for Its Shape

A normal erythrocyte's biconcave disklike shape (Fig. 6-11*a*) assures the rapid diffusion of O_2 to its hemoglobin molecules by placing them no further than 1 μm from the cell surface. However, the rim and the dimple regions of an erythrocyte do not occupy fixed positions on the cell membrane. This can be demonstrated by anchoring an erythrocyte to a microscope slide by a small portion of its surface and inducing the cell to move laterally with a gentle flow of

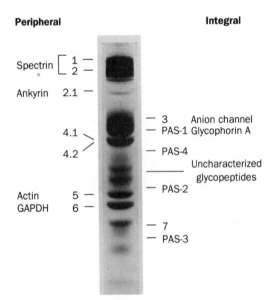

Peripheral   **Integral**

Spectrin $\begin{bmatrix} 1 \\ 2 \end{bmatrix}$

Ankyrin 2.1

— 3 Anion channel
— PAS-1 Glycophorin A

4.1
4.2
— PAS-4

Uncharacterized glycopeptides

— PAS-2

Actin 5
GAPDH 6

— 7
— PAS-3

FIGURE 11-34. An SDS–polyacrylamide gel electrophoretogram of human erythrocyte membrane proteins as stained by Coomassie brilliant blue. The bands designated 4.1 and 4.2 are not separated with the 1% SDS concentration used. The minor bands are not labeled for the sake of simplicity. The positions of the four sialoglycoproteins revealed by PAS staining are indicated. [Courtesy of Vincent Marchesi, Yale University.]

isotonic buffer. A point originally on the rim of the erythrocyte will move across the dimple to the rim on the opposite side of the cell from where it began. Evidently, the membrane rolls across the cell while maintaining its shape, much like the tread of a tractor. This remarkable mechanical property of the erythrocyte membrane results from the presence of a submembranous network of proteins that function as a membrane "skeleton." Indeed, this property is partially duplicated by a mechanical model consisting of a geodesic sphere (a spheroidal cage) that is freely jointed at the intersections of its struts but constrained from collapsing much beyond a flat surface. When placed inside an evacuated plastic bag, this cage also assumes a biconcave disklike shape.

The fluidity and flexibility imparted to an erythrocyte by its membrane skeleton has important physiological consequences. A slurry of solid particles of a size and concentration equal to that of red cells in blood has the flow characteristics approximating that of sand. Consequently, in order for blood to flow at all, much less for its erythrocytes to squeeze through capillary blood vessels smaller in diameter than they are, erythrocyte membranes, with their membrane skeletons, must be fluidlike and easily deformable.

The protein **spectrin**, so called because it was discovered in erythrocyte ghosts, accounts for ~ 75% of the erythrocyte membrane skeleton. It is composed of two similar polypeptide chains, band 1 (α subunit; 280 kD) and band 2 (β subunit; 246 kD), which sequence analysis indicates each consist of repeating 106-residue segments that are predicted to fold into triple-stranded α helical coiled coils (Figs. 11-35a and b). Electron microscopy indicates that these large polypeptides are loosely intertwined to form a flexible wormlike $\alpha\beta$ dimer that is ~ 1000 Å long (Fig. 11-35c). Two such heterodimers further associate in a head-to-head manner to form an $(\alpha\beta)_2$ tetramer. These tetramers, of which there are ~ 100,000/cell, are cross-linked at both ends by attachments to bands 4.1 and 5 to form a dense and irregular protein meshwork that underlies the erythrocyte plasma membrane (Fig. 11-35c and d). Band 5, a globular protein that forms filamentous oligomers, has been identified as **actin,** a common cytoskeletal element in other cells (Section 1-2A) and a major component of muscle (Section 34-3A). Spectrin also associates with band 2.1, an 1880-residue monomer known as **ankyrin,** which, in turn, binds to band 3, the anion channel protein. This attachment anchors the membrane skeleton to the membrane. Indeed, upon solubilization of spectrin and actin by low ionic strength solutions, erythrocyte ghosts lose their biconcave shape: Their integral proteins, which normally occupy fixed positions in the membrane plane, become laterally mobile. Ankyrin's N-terminal segment consists almost entirely of 22 tandem 33-residue repeats, which are postulated to form binding sites for various integral membrane proteins. Immunochemical studies have revealed spectrinlike, ankyrinlike, and band 4.1-like proteins in a variety of tissues.

Hereditary Spherocytosis and Elliptocytosis Arise from Erythrocyte Membrane Skeleton Defects

Individuals with **hereditary spherocytosis** have spheroidal erythrocytes that are relatively fragile and inflexible. These individuals suffer from hemolytic anemia because the spleen, a labyrinthine organ with narrow passages that normally filters out aged erythrocytes (which lose flexibility towards the end of their ~ 120-day lifetime), prematurely removes spherocytotic erythrocytes. The hemolytic anemia may be alleviated by the spleen's surgical removal. However, the primary defects in spherocytotic cells are reduced synthesis of spectrin, the production of an abnormal spectrin that binds band 4.1 protein with reduced affinity, or the absence of band 4.1 protein.

Hereditary elliptocytosis (having elongated or elliptical red cells; also known as **hereditary ovalcytosis**), a condition that is common in certain areas of Southeast Asia and Melanesia, confers resistance to malaria in heterozygotes (but apparently is lethal in homozygotes). This condition arises from defects in the erythrocyte anion channel. A common such defect consists of a 9-residue deletion that inactivates this transmembrane protein. The consequent reduced capacity of red cells to import phosphate or sulfate ions may inhibit the intraerythrocytotic growth of rapidly developing malarial parasites.

The camel, the renowned "ship of the desert," provides a striking example of adaptation involving the erythrocyte membrane. This remarkable animal is still active after loss of 30% of its body weight in water and, when thus dehydrated, can drink sufficient water in a few minutes to become fully rehydrated. The rapid uptake of such a large amount of water by the blood, which must deliver it to the cells, would lyse the erythrocytes of most animals. Yet, camel erythrocytes, which have the shape of flattened ellipsoids rather than biconcave disks, are resistant to osmotic lysis. Camel spectrin binds to its membrane with particular tenacity, but upon spectrin removal, which requires a strong denaturing agent such as guanidinium chloride, camel erythrocytes assume a spherical shape.

D. Blood Groups

The outer surfaces of erythrocytes and other eukaryotic cells are covered with complex carbohydrates that are components of plasma membrane glycoproteins and glycolipids. They form a thick, fuzzy cell coating, the **glycocalyx** (Fig. 11-36), which contains numerous identity markers that function in various recognition processes. The human erythrocyte has some 100 known **blood group determinants** that comprise 15 genetically distinct blood group systems. Of these, only two—the **ABO blood group system** (discovered in 1900 by Karl Landsteiner) and the **rhesus (Rh) blood group system**—have major clinical importance. The

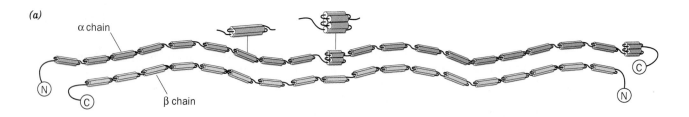

(a)

α chain

β chain

N

C

N

C

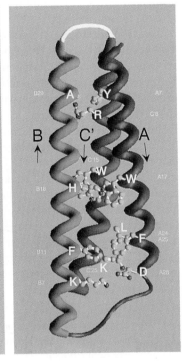

(b)

C

A'

B'

B

A

C'

A

Y

R

B

C'

A

W

W

H

L

F

F

K

D

K

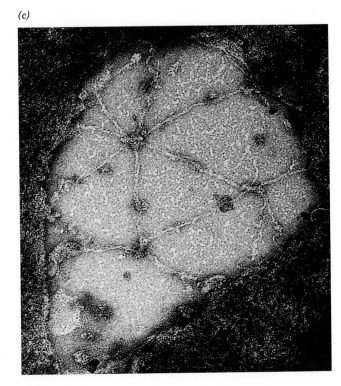

(c)

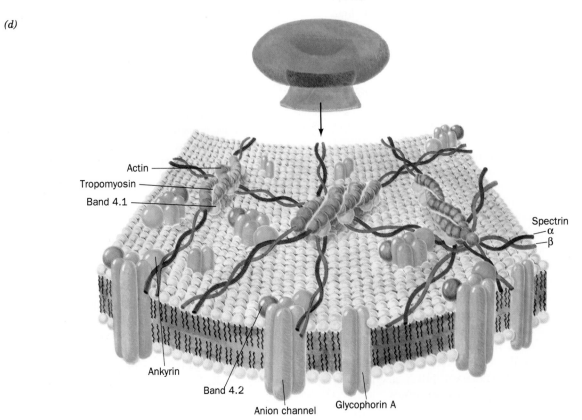

(d)

Actin

Tropomyosin

Band 4.1

Ankyrin

Band 4.2

Anion channel

Glycophorin A

Spectrin

α

β

302

FIGURE 11-35 (*Opposite*). The human erythrocyte membrane skeleton. (*a*) The structure of an $\alpha\beta$ dimer of spectrin. Both of these antiparallel polypeptides contain multiple 106-residue repeats, which are thought to form triple-helical bundles that are flexibly connected by nonhelical segments. Two of these heterodimers join, head to head, to form an $(\alpha\beta)_2$ heterotetramer. [After Speicher, D.W. and Marchesi, V., *Nature* **311**, 177 (1984).] (*b*) The X-ray structure of a 106-residue segment of *Drosophila* α-spectrin that forms a three-helix bundle. The left panel shows the protein as it occurs in the crystal structure: Two identical polypeptides, one shown in red hues and the other in yellow and green, associate as a dimer that consists of two symmetry related bundles of three covalently noncontiguous helices. Since the C and C' helices are chemically identical, the right panel shows the structure of a covalently contiguous three-helix bundle thought to be representative of those in spectrin, which was modeled simply by exchanging and bending the white polypeptide segments in the crystal structure. Some of the side chains that pack between the α helices are shown in ball-and-stick form with C yellow, N blue, and O red. [Courtesy of Stephen Harrison and Daniel Branton, Harvard University.] (*c*) An electron micrograph of an erythrocyte membrane skeleton that has been stretched to an area 9 to 10 times greater than that of the native membrane. Stretching makes it possible to obtain clear images of the membrane skeleton, which in its native state is so densely packed and irregularly flexed that it is difficult to pick out individual molecules and to ascertain how they are interconnected. Note the predominantly hexagonal network composed of spectrin tetramers cross-linked by junctions containing actin and band 4.1. [Courtesy of Daniel Branton, Harvard University.] (*d*) A model of the erythrocyte membrane skeleton. The so-called junctional complex, which is magnified in this drawing, contains actin, **tropomyosin** (which, in muscle, also associates with actin; Section 34-3A), and band 4.1 protein, as well as **adducin, dematin,** and **tropomodulin** (not shown). [After Goodman, S.R., Krebs, K.E., Whitfield, C.F., Riederer, B.M., and Zagen, I.S., *CRC Crit. Rev. Biochem.* **23**, 196 (1988).]

FIGURE 11-36. The erythrocyte glycocalyx as revealed by electron microscopy using special staining techniques. It is up to 1400 Å thick and composed of closely packed, 12- to 25-Å diameter oligosaccharide filaments. [Courtesy of Harrison Latta, UCLA.]

various blood groups are identified by means of suitable antibodies or by specific plant lectins.

> Knowledge of blood group substances, and of their inheritance according to simple Mendelian laws, has been useful for legal and historical as well as medical purposes. The use of blood types in disproving paternity has even become the stuff of soap operas. Similarly, the analysis of tissue dust from the mummy of Tutankhamen, an Egyptian Pharaoh who reigned from 1334 to 1325 B.C., has indicated his probable relationship to Smenkhkare, another eighteenth dynasty Pharaoh.

ABO Blood Group Substances Are Carbohydrates

The ABO system consists of three blood group substances, the A, B, and H antigens, which are components of erythrocyte surface sphingoglycolipids. [Antigens are characteristic constellations of chemical groups that elicit the production of specific antibodies when injected into an animal (Section 34-2A). Each antibody molecule can specifically bind to at least two of its corresponding antigen molecules, thereby cross-linking them.] Individuals with type A cells have A antigens on their cell surfaces and carry anti-B antibodies in their serum; those with type B cells, which bear B antigens, carry anti-A antibodies; those with type AB cells, which bear both A and B antigens, carry neither anti-A nor anti-B antibodies; and type O individuals, whose cells bear neither antigen, carry both anti-A and anti-B antibodies. Consequently, the transfusion of type A blood into a type B individual, for example, causes an anti-A antibody–A antigen reaction, which agglutinates (clumps together) the transfused erythrocytes, resulting in an often fatal blockage of blood vessels. The H antigen is discussed below.

The ABO blood group substances are not confined to erythrocytes but also occur in the plasma membranes of many tissues as glycolipids of considerable diversity. In fact, in the ~80% of the population known as secretors, these antigens are secreted as *O*-linked components of glycoproteins into various body fluids including saliva, milk, seminal fluid, gastric juice, and urine. These diverse molecules, which are 85% carbohydrate by weight and have molecular masses ranging into thousands of kD, consist of multiple oligosaccharides attached to a polypeptide chain.

The A, B, and H antigens differ only in the sugar residues at their nonreducing ends (Table 11-5). The H antigen occurs in type O individuals; it is also the precursor oligosaccharide of A and B antigens. Type A individuals have a 303-residue glycosyltransferase that specifically adds an *N*-acetylgalactosamine residue to the terminal position of the H antigen, whereas in type B individuals, this enzyme, which differs by four amino acid residues from that of type A individuals, instead adds a galactose residue. In type O individuals, the enzyme is inactive because its synthesis terminates after its 115th residue.

TABLE 11-5. STRUCTURES OF THE A, B, AND H ANTIGENIC DETERMINANTS IN ERYTHROCYTES

Type	Antigen
H	Galβ(1 → 4)GlcNAc $\cdots$ ↑ 1,2 L-Fuc α
A	GalNAc α(1 → 3)Galβ(1 → 4)GlcNAc $\cdots$ ↑ 1,2 L-Fuc α
B	Galα(1 → 3)Galβ(1 → 4)GlcNAc $\cdots$ ↑ 1,2 L-Fuc α

Abbreviations: Gal = Galactose, GalNAc = *N*-acetylgalactosamine, GlcNAc = *N*-acetylglucosamine, L-Fuc = L-fucose.

MN Blood Groups Arise from Glycophorin *A* Variants

The **MN blood group system** constitutes another well-characterized set of human blood group determinants. Antigens of this system occur only in the erythrocyte membrane as part of the transmembrane glycoprotein, glycophorin *A* (Fig. 11-22). This protein is also the site of the influenza virus receptor (Section 32-4A), as well as a receptor for erythrocyte invasion by the malarial parasite *Plasmodium falciparum* (Section 6-3A). The PAS band 1 protein (Fig. 11-34) is a dimer of glycophorin *A*, which is formed through an SDS-resistant association between hydrophobic sections of the polypeptide chains; this dimer is presumably the protein's native form. The PAS band 2 protein is the monomeric form of glycophorin *A*.

Treatment of erythrocytes or glycophorin *A* with neuraminidase abolishes their reactivity to anti-M or anti-N antibodies as well as destroying their influenza virus receptor

activity and reducing the invasion of *P. falciparum*. Thus, sialic acid (*N*-acetylneuraminic acid), which is cleaved by neuraminidase, forms part of the MN antigenic determinants. However, there are no differences between the oligosaccharides of M- and N-specific glycophorins *A*. Rather, these proteins differ in their amino acid sequence. Glycophorin *A*^{M} has a Ser at position 1 (the N-terminal residue) and a Gly at position 5, whereas these residues are Leu and Glu in glycophorin *A*^{N} (Fig. 11-22). The erythrocytes of heterozygotes with both M and N antigenicity bear both these glycophorin *A* variants.

It has been suggested that glycophorin *A*'s numerous negatively charged sialic acid residues prevent erythrocytes, which are closely packed in the blood stream, from adhering to one another. Yet, individuals who genetically lack glycophorin *A* suffer no apparent ill effects. It nevertheless seems unlikely that the million or so glycophorin *A* molecules per erythrocyte exist solely for the convenience of invading parasites.

E. Gap Junctions

Most eukaryotic cells are in metabolic as well as physical contact with neighboring cells. This contact is brought about by tubular particles, named **gap junctions,** that join discrete regions of neighboring plasma membranes much like hollow rivets (Fig. 11-37). Indeed, these intercellular channels are so widespread that many whole organs are continuous from within. Thus gap junctions are important intercellular communication channels. For example, the synchronized contraction of heart muscle is brought about by flows of ions through gap junctions (heart muscle is not innervated as is skeletal muscle). Likewise, gap junctions serve as conduits for some of the substances that mediate embryonic development; blocking gap junctions with antibodies that bind to them causes developmental abnormalities in species as diverse as hydra, frogs, and mice. Gap junctions also function to nourish cells that are distant from the blood supply, such as bone and lens cells.

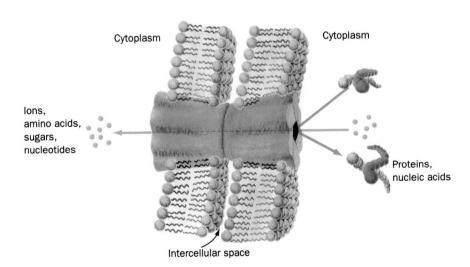

FIGURE 11-37. Gap junctions between adjacent cells consist of two apposed plasma membrane-embedded hexagonal studs that bridge the gap between the cells. Small molecules and ions, but not macromolecules, can pass between cells via the gap junction's central channel.

Mammalian gap junction channels are 16 to 20 Å in diameter, which Werner Loewenstein established by microinjecting single cells with fluorescent molecules of various sizes and observing with a fluorescence microscope whether the fluorescent probe passed into neighboring cells. The molecules and ions that can pass freely between neighboring cells are limited in molecular mass to a maximum of ~ 1200 D; macromolecules such as proteins and nucleic acids cannot leave a cell via this route.

The diameter of a gap junction channel varies with Ca^{2+} concentration: The channels are fully open when the Ca^{2+} level is $< 10^{-7}M$ and narrow as the Ca^{2+} concentration increases until, above $5 \times 10^{-5}M$, they close. This shutter system is thought to protect communities of interconnected cells from the otherwise catastrophic damage that would result from the death of any of their members. Cells generally maintain very low cytosolic Ca^{2+} concentrations ($< 10^{-7}M$) by actively pumping Ca^{2+} out of the cell as well as into their mitochondria and endoplasmic reticulum (Section 18-3B; Ca^{2+} is an important intracellular messenger whose cytosolic concentration is precisely regulated). Ca^{2+} floods back into leaky or metabolically depressed cells, thereby inducing closure of their gap junctions and sealing them off from their neighbors.

Gap Junction Channels Are Formed by Hexagons of Subunits

Purified gap junctions consist of a single type of ~ 32-kD protein subunits known as **connexins**. A single gap junction consists of two apposed hexagonal rings of these subunits, one from each of the adjoining plasma membranes (Fig. 11-37). Freeze-etch electron microscopy indicates that membrane-bound gap junctions form rafts of hexagonally

packed doughnut-shaped particles of 80 to 90 Å in diameter (Fig. 11-38). This crystal-like packing permitted Unwin to determine their 18-Å resolution structure through electron crystallography in a manner similar to that used with bacteriorhodopsin (Section 11-3A). Connexin subunits are rods that are 25 Å in diameter and 75 Å long. In the absence of Ca^{2+}, they are inclined with respect to the sixfold axis so as to form a central channel that runs the length of the gap junction (Fig. 11-39, *left*). In 0.05 mM Ca^{2+}, however, the rods are nearly perpendicular to the plane of the junction and their central channel is closed at its cytoplasmic end (Fig. 11-39, *right*). Unwin proposed that this closure is achieved by relatively slight tilting and twisting motions of the connexin subunits centered at their bases, which, through the lever arms of their 75- Å lengths, results in large (~ 9 Å) radial displacements at their cytoplasmic ends. However, X-ray studies by Lee Makowski suggest that the cytoplasmic domains of gap junctions contain gates that close through more localized motions. The resolution of this discrepancy is the object of ongoing research.

4. MEMBRANE ASSEMBLY AND PROTEIN TARGETING

As cells grow and divide, they synthesize new membranes. How are such asymmetric membranes generated? One way in which this might occur is through self-assembly. Indeed, when the detergent used to disperse a biological membrane is removed, liposomes form in which functional integral proteins are embedded. In most cases, however, these model membranes are symmetrical, both in their lipid distribution between the inner and outer leaflets of the bilayer and in the positions and orientations of their proteins. An alternative hypothesis of membrane assembly is that *it occurs on the scaffolding of preexisting membranes; that is, membranes are generated by expansion of old ones rather than by creation of new ones.* In this section we shall see that this is, in fact, the case.

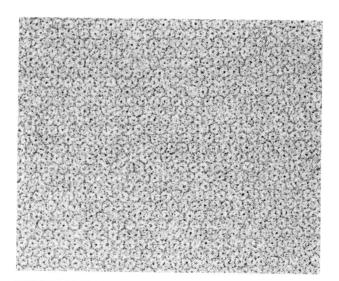

FIGURE 11-38. An electron micrograph of a gap junction–containing membrane. The gap junctions are arranged in a hexagonal lattice with a repeat distance of 80 to 85 Å. Note the densely stained central hole in each gap junction. [From Unwin, P.T.N. and Zampighi, G., *Nature* **283**, 546 (1980).]

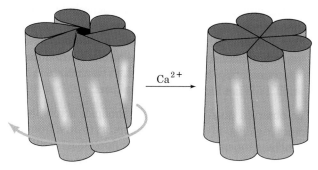

FIGURE 11-39. A model for the Ca^{2+}-induced closure of the gap junction central channel. [After Unwin, P.T.N. and Zampighi, G., *Nature* **283**, 549 (1980).]

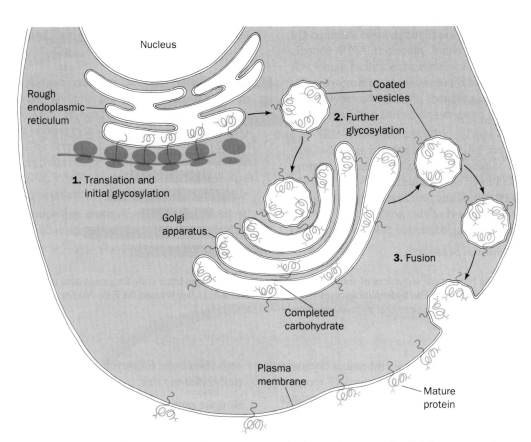

FIGURE 11-44. The posttranslational processing of integral membrane proteins. **(1)** During their ribosomal synthesis, their glycosylation is initiated in the lumen of the endoplasmic reticulum. **(2)** After ribosomal synthesis is completed, coated vesicles containing the protein bud off from the endoplasmic reticulum and move to the Golgi apparatus where protein processing is completed. **(3)** Later, coated vesicles containing the mature protein bud off from the Golgi apparatus and fuse to the membrane for which the protein is targeted, here shown as the plasma membrane.

processing occurs (Fig. 11-44). Proteins transit from one end of the Golgi stack to the other while being modified (mostly glycosylated) in a stepwise manner, a process that is described in Section 21-3B. Proteins are carried from compartment to compartment by specialized membranous transport vesicles. In the final Golgi compartment, the processed proteins are sorted and sent to their final cellular destinations.

Membrane, Secretory, and Lysosomal Proteins Are Transported in Coated Vesicles

The vehicles in which proteins are transported between the RER, the Golgi apparatus, and their final destinations are known as **coated vesicles** (Fig. 11-45). This is because these membranous sacs are encased on their outer (cytosolic) face by a polyhedral framework of the nonglycosylated protein **clathrin,** which is believed to act as a flexible scaffolding in promoting vesicle formation. A vesicle buds off from its membrane of origin and later fuses to its target membrane. *This process preserves the orientation of the transmembrane protein (Fig. 11-46), so that the lumens of the ER and Golgi apparatus are topologically equivalent to the outside of the cell. This explains why the carbohydrate*

moieties of integral membrane glycoproteins occur only on the external surfaces of plasma membranes.

The vesicles mediating intra-Golgi transport have a second type of coat which consists of four major proteins of 160, 110, 98, and 61 kD designated α-, β-, γ-, and δ-COPs (for coat proteins). Relatively little is yet known about the functions of these proteins.

ER-Resident Proteins Have the C-Terminal Sequence KDEL

Most soluble ER-resident proteins in mammals have the C-terminal sequence Lys-Asp-Glu-Leu (KDEL using the one letter amino acid symbols; Table 4-1) whose alteration results in the secretion of the resulting protein (the corresponding yeast proteins end in HDEL). By what means are these "KDEL proteins" selectively retained in the ER? Since KDEL proteins freely diffuse within the ER, it seems unlikely that they are immobilized by a membrane-bound receptor within the ER. Rather, it has been shown that KDEL proteins, as do secretory and lysosomal proteins, readily leave the ER via membranous vesicles but that KDEL proteins are promptly retrieved from some later compartment and returned to the ER. Thus, genetically

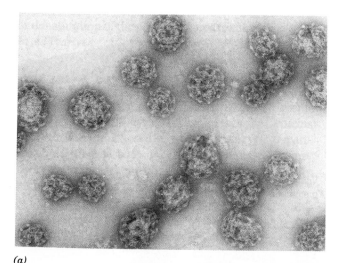

(a)

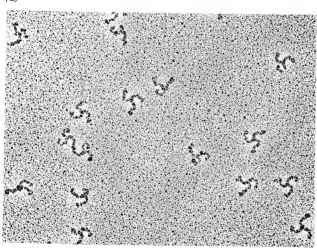

(b)

appending KDEL to the lysosomal protease **cathepsin D** causes it to accumulate in the ER, but it nevertheless acquires an *N*-acetylglucosaminyl-1-phosphate group, a modification that is made in an early Golgi compartment. Presumably, an unknown environmental feature, such as Ca^{2+} concentration or pH, stimulates a membrane-bound receptor in a post-ER compartment to bind the KDEL signal and the resulting complex is returned to the ER in a specialized vesicle. In fact, such receptors have been identified in yeast and humans. However, the observation that former KDEL proteins whose KDEL sequences have been deleted are, nevertheless, secreted relatively slowly suggests that there are mechanisms for retaining these proteins in the ER by actively withholding them from the bulk flow of proteins through the secretory pathway.

Bacterial Membrane Proteins Are Also Preceded by Signal Peptides

The signal hypothesis also applies to bacteria. Proteins that traverse the bacterial plasma membrane have leading signal peptides similar to those of eukaryotes and are synthesized by membrane-bound ribosomes. The importance of signal peptides in directing a protein to its cellular destination was demonstrated through the use of genetically engineered *E. coli.* **Maltose-binding protein,** which is involved in the uptake of the disaccharide maltose, is normally secreted across the plasma membrane into the **periplasmic compartment** (the space between the plasma membrane and the cell wall in gram-negative bacteria; Fig. 10-23*b*). Mutations that change even one hydrophobic residue of maltose-binding protein's signal peptide to a charged resi-

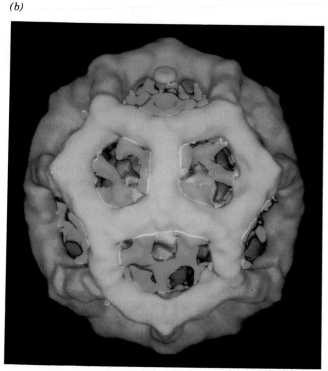

(c)

FIGURE 11-45. Coated vesicles are membranous sacs encased in polyhedral frameworks of clathrin and its associated proteins. Clathrin cages, which form the structural skeletons of coated vesicles, can be reversibly dissociated to flexible three-legged protein complexes known as **triskelions**. (*a*) An electron micrograph of coated vesicles. [Courtesy of Barbara Pearse, Medical Research Council, U.K.] (*b*) Electron micrograph of triskelions. These protein complexes consist of three so-called heavy chains (91 kD) and three light chains (23-27 kD), each of which binds to a heavy chain near the vertex of the triskelion. The variable orientations of their legs is indicative of their flexibility. [Courtesy of Daniel Branton, Harvard University.] (*c*) A three-dimensional map, generated from electron micrographs, of a clathrin coat. The polyhedral clathrin coat is shown in orange, the clathrin terminal domains are green, and an inner shell of accessory proteins is blue. Each vertex of the polyhedron is the center of a triskelion and its edges, which are ~150 Å in length, are formed by the overlapping legs of adjoining triskelions. Such frameworks, which consist of 12 pentagons and a variable number of hexagons (for geometric reasons explained in Section 32-2A), are the most parsimonious way of enclosing spheroidal objects in polyhedral cages. The accessory proteins (**adaptins**) are thought to bind the membrane-spanning receptors for the specific proteins that the coated vesicle sequesters. [Courtesy of Barbara Pearse, Medical Research Council, U.K.]

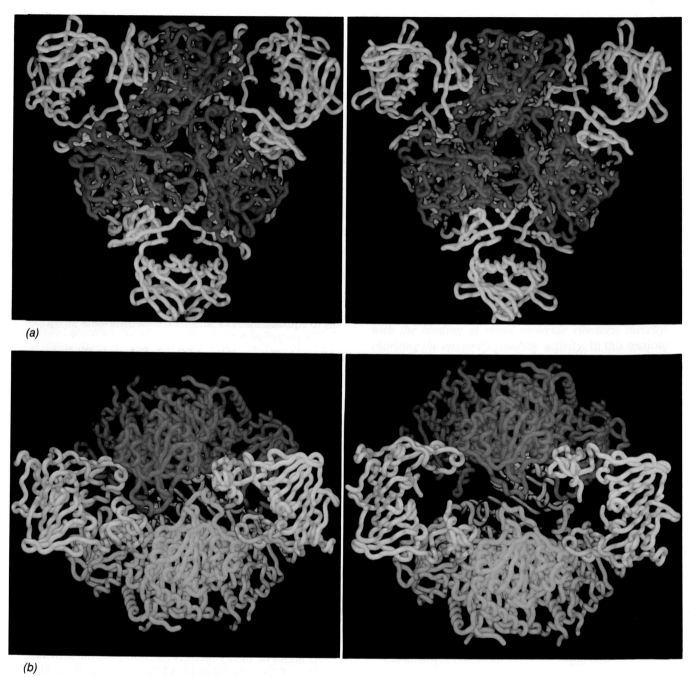

(a)

(b)

FIGURE 12-7. The X-ray structure of ATCase. The polypeptide backbones of T-state ATCase *(left)* and R-state ATCase *(right)* as viewed (a) along the protein's molecular threefold axis of symmetry and (b) along a molecular twofold axis of symmetry. The regulatory dimers (*yellow*) join the upper catalytic trimer (*red*) to the lower catalytic trimer (*blue*). [Courtesy of Michael Pique, The Scripps Research Institute, La Jolla, California.]

binds tightly to R-state but not to T-state ATCase (the use of unreactive substrate analogs is common in the study of enzyme mechanisms because they form stable complexes that are amenable to structural study rather than rapidly reacting to form products as do true substrates).

The X-ray structures of the T-state ATCase–CTP complex and the R-state ATCase–PALA complex reveal that the T → R transition maintains the protein's D_3 symmetry. The comparison of these two structures (Fig. 12-7) indi-

cates that in the T → R transition, the enzyme's catalytic trimers separate along the molecular threefold axis by ~11 Å and reorient about this axis relative to each other by 12° such that these trimers assume a more nearly eclipsed configuration. In addition, the regulatory dimers rotate clockwise by 15° about their twofold axes and separate by ~4 Å along the threefold axis. Such large quaternary shifts are reminiscent of those in hemoglobin (Section 9-2B).

ATCase's substrates, carbamoyl phosphate and aspartate, each bind to a separate domain of the catalytic subunit (Fig. 12-8). The binding of PALA to the enzyme, which presumably mimics the binding of both substrates, induces active site closure in a manner that would bring them together so as to promote their reaction. The resulting atomic shifts, up to 8 Å for some residues (Fig. 12-8), trigger ATCase's T → R quaternary shift. Indeed, *ATCases's tertiary and quaternary shifts are so tightly coupled through extensive intersubunit contacts (see below) that they cannot occur independently (Fig. 12-9)*. The binding of substrate to one catalytic subunit therefore increases the substrate-binding affinity and catalytic activity of the other catalytic subunits and hence accounts for the enzyme's positively cooperative substrate binding—much as occurs in hemoglobin (Section 9-2C). Thus, low levels of PALA actually activate ATCase by promoting its T → R transition: ATCase has such high affinity for this unreactive bisubstrate analog that the binding of one molecule of PALA converts all six of its catalytic subunits to the R state. Evidently, *ATCase closely follows the symmetry model of allosterism (Section 9-4B)*.

The Structural Basis of Allosterism in ATCase

What are the interactions that stabilize the T and R states of ATCase and why must their interconversion be concerted? The region of the protein that undergoes the most profound conformational rearrangement upon the T → R transition is a flexible loop composed of residues 230 to 250 in the catalytic (*c*) subunit, the so-called 240s loop [the

symmetry-related red and blue loops at the centers of Fig. 12-7*b* that lie side by side in the T state (*left*) but are vertically apposed in the R state (*right*)]. In the T state, each 240s loop forms two intersubunit hydrogen bonds with the vertically opposite *c* subunit (Fig. 12-7*b*, *left*), together with an intrasubunit hydrogen bond. Domain closure as a consequence of substrate binding (Figs. 12-8 and 12-9) ruptures these hydrogen bonds and replaces them, in the R state, with new intrachain hydrogen bonds. The consequent reorientation of the 240s loop is thought to be largely responsible for the quarternary shift to the R state (see below). Since the Glu 239 carboxyl group is the acceptor in all of the above T-state interchain and R-state intrachain hydrogen bonds, this hypothesis is corroborated by the observation that the mutation of Glu 239 to Gln converts ATCase to an enzyme that is devoid of both homotropic and heterotropic effects and that has a quaternary structure midway between those of the R and T states.

The structural basis for heterotropic effects in ATCase is gradually being unveiled. Both the inhibitor CTP and the activator ATP bind to the same site on the outer edge of the regulatory (*r*) subunit, about 60 Å away from the nearest catalytic site. CTP binds preferentially to the T state, increasing its stability, while ATP binds preferentially to the R state, increasing its stability. The binding of these effectors to their less favored states also has structural consequences. When CTP binds to R-state ATCase, it reorients several residues at the nucleotide binding site, which in-

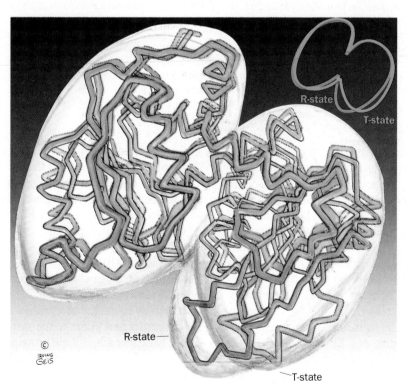

FIGURE 12-8. The comparison of the polypeptide backbones of the ATCase catalytic subunit in the T state *(orange)* and the R state *(blue)*. The subunit consists of two domains, with the one on the left containing the carbamoyl phosphate binding site and that on the right forming the aspartic acid binding site. The

T ⇌ R transition brings the two domains together such that their two bound substrates can react to form product. [Figure copyrighted © by Irving Geis and Eric Gouaux. X-Ray structure by William Lipscomb, Harvard University.]

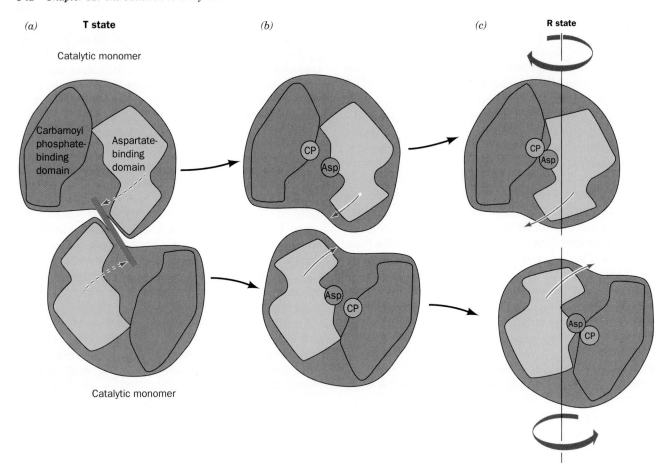

FIGURE 12-9. A schematic diagram indicating the tertiary and quarternary conformational changes in two vertically interacting catalytic ATCase subunits. (*a*) In the absence of bound substrate the protein is held in the T state because the motions that bring together the two domains of each subunit (*dashed arrows*) are prevented by steric interference (*purple bar*) between the contacting aspartic acid-binding domains. (*b*) The binding of carbamoyl phosphate (CP) followed by aspartic acid (Asp) to their respective binding sites causes the subunits to move apart and rotate with respect to each other so as to permit the T$\rightleftharpoons$R transition. (*c*) In the R state, the two domains of each subunit come together so as to promote the reaction of their bound substrates to form products. [Figure copyrighted © by Irving Geis.]

duces a contraction in the length of the regulatory dimer (r_2). This distortion, through the interactions of residues at the r–c interface, causes the catalytic trimers (c_3) to come together by 0.5 Å (become more T-like, that is, less active, which presumably destabilizes the R state). This, in turn, reorients key residues in the enzyme's active sites, thereby decreasing the enzyme's catalytic activity. ATP has essentially opposite effects when binding to the T-state enzyme: It causes the catalytic trimers to move apart by 0.4 Å (become more R-like, that is, more active, which presumably destabilizes the T state), thereby reorienting key residues in the enzyme's active sites so as to increase the enzyme's catalytic activity. The binding of CTP to T-state ATCase does not further compress the catalytic trimers but, nevertheless, perturbs active site residues in a way that further stabilizes the T state. Although the X-ray structure of ATP complexed to R-state ATCase has not yet been reported, it is expected that ATP binding perturbs the R state in a manner analogous but opposite to the binding of CTP to T-state ATCase.

Allosteric Transitions In Other Enzymes Resemble Those of Hemoglobin and ATCase

Allosteric enzymes are widely distributed in nature and tend to occupy key regulatory positions in metabolic pathways. Three such enzymes, in addition to hemoglobin and ATCase, have had their X-ray structures determined in both their R and T states: **phosphofructokinase** (Sections 16-2C and 16-4B), **fructose-1,6-bisphosphatase** (Section 16-4B), and **glycogen phosphorylase** (Section 17-1A). In all five proteins, quaternary changes, through which binding and catalytic effects are communicated among active sites, are concerted and preserve the symmetry of the protein. This is because each of these proteins has two sets of alternative contacts which are stabilized largely by hydrogen bonds that mostly involve side chains of opposite charge. In all five proteins, the quarternary shifts are primarily rotations of subunits relative to one another with only small translations. Secondary structures are largely preserved in T → R transitions, which is probably important for mechanically transmitting heterotropic effects over the tens of

A necessary in these proteins. The ubiquity of these structural features among allosteric proteins of known structures suggests that the regulatory mechanisms of other allosteric enzymes will, by and large, follow this model.

5. A PRIMER OF ENZYME NOMENCLATURE

Enzymes, as we have seen throughout the text so far, are commonly named by appending the suffix *-ase* to the name of the enzyme's substrate or to a phrase describing the enzyme's catalytic action. Thus urease catalyzes the hydrolysis of urea and alcohol dehydrogenase catalyzes the oxidation of alcohols to their corresponding aldehydes. Since there were at first no systematic rules for naming enzymes, this practice occasionally resulted in two different names being used for the same enzyme or, conversely, in the same name being used for two different enzymes. Moreover, many enzymes, such as catalase, which mediates the dismutation of H_2O_2 to H_2O and O_2, were given names that provide no clue as to their function; even such atrocities as "old yellow enzyme" have crept into use. In an effort to eliminate this confusion and to provide rules for rationally naming the rapidly growing number of newly discovered enzymes, a scheme for the systematic functional classification and nomenclature of enzymes was adopted by the International Union of Biochemistry and Molecular Biology (IUBMB).

Enzymes are classified and named according to the nature of the chemical reactions they catalyze. There are six major classes of reactions that enzymes catalyze (Table 12-3), as well as subclasses and sub-subclasses within these classes. Each enzyme is assigned two names and a four-number classification. Its **recommended name** is convenient for everyday use and is often an enzyme's previously used trivial name. Its **systematic name** is used when ambiguity

TABLE 12-3. ENZYME CLASSIFICATION ACCORDING TO REACTION TYPE

Classification	Type of Reaction Catalyzed
1. Oxidoreductases	Oxidation–reduction reactions
2. Transferases	Transfer of functional groups
3. Hydrolases	Hydrolysis reactions
4. Lyases	Group elimination to form double bonds
5. Isomerases	Isomerization
6. Ligases	Bond formation coupled with ATP hydrolysis

must be minimized; it is the name of its substrate(s) followed by a word ending in *-ase* specifying the type of reaction the enzyme catalyzes according to its major group classification. For example, the most recent version of *Enzyme Nomenclature* (see references) indicates that the enzyme whose recommended name is carboxypeptidase A (Section 6-1A) has the systematic name **peptidyl-L-amino acid hydrolase** and the **classification number** EC 3.4.17.1. Here "EC" stands for Enzyme Commission, the first number (3) indicates the enzyme's major class (hydrolases; Table 12-3), the second number (4) denotes its subclass [acting on peptide bonds (peptidases)], the third number (17) designates its subsubclass (metallocarboxypeptidases; carboxypeptidase A has a bound Zn^{2+} ion that is essential for its catalytic activity), and the fourth number (1) is the enzyme's arbitrarily assigned serial number in its subsubclass. As another example, the enzyme with the recommended name alcohol dehydrogenase (Section 12-2A) has the systematic name **alcohol:NAD$^+$ oxidoreductase** and the classification number EC 1.1.1.1. In this text, as in general biochemical terminology, we shall most often use the recommended names of enzymes but, when ambiguity must be minimized, we shall refer to an enzyme's systematic name.

CHAPTER SUMMARY

Enzymes specifically bind their substrates through geometrically and physically complementary interactions. This permits enzymes to be absolutely stereospecific, both in binding substrates and in catalyzing reactions. Enzymes vary in the more stringent requirement of geometric specificity. Some are highly specific for the identity of their substrates, whereas others can bind a wide range of substrates and catalyze a variety of related types of reactions.

Enzymatic reactions involving oxidation–reduction reactions and many types of group-transfer processes are mediated by coenzymes. Many vitamins are coenzyme precursors.

Enzymatic activity may be regulated by the allosteric alteration of substrate-binding affinity. For example, the rate of the reaction catalyzed by *E. coli* ATCase is subject to positive homotropic control by substrates, heterotropic inhibition by CTP, and hetero-

tropic activation by ATP. ATCase has the subunit composition c_6r_6. Its isolated catalytic trimers are catalytically active but not subject to allosteric control. The regulatory dimers bind ATP and CTP. Substrate binding induces a tertiary conformational shift in the catalytic subunits, which increases the subunit's substrate-binding affinity and catalytic efficiency. This tertiary shift is strongly coupled to ATCase's large quarternary T → R conformational shift, thereby accounting for the enzyme's allosteric properties. Other allosteric enzymes appear to operate in a similar manner.

Enzymes are systematically classified according to their recommended name, their systematic name, and their classification number, which is indicative of the type of reaction catalyzed by the enzyme.

REFERENCES

General

Dixon, M. and Webb, E.C., *Enzymes* (3rd ed.), Academic Press (1979). [A treatise on enzymes.]

History

Friedmann, H.C. (Ed.), *Enzymes,* Hutchinson Ross (1981). [A compendium of classic enzymological papers published between 1761 and 1974; with commentary.]

Fruton, J.S., *Molecules and Life, pp.* 22–86, Wiley (1972).

Schlenk, F., Early research on fermentation—a story of missed opportunities, *Trends Biochem. Sci.* **10,** 252–254 (1985).

Substrate Specificity

Creighton, D.J. and Murthy, N.S.R.K., Stereochemistry of enzyme-catalyzed reactions at carbon, *in* Sigman, D.S. and Boyer, P.D. (Eds.), *The Enzymes* (3rd ed.), Vol. 19, *pp.* 323–421, Academic Press (1990). [Section II discusses the stereochemistry of reactions catalyzed by nicotinamide-dependent dehydrogenases.]

Fersht, A., *Enzyme Structure and Mechanism* (2nd ed.), Freeman (1985).

Popják, G., Specificity of enzyme reactions, *in* Boyer, P.D. (Ed.), *The Enzymes* (3rd ed.), Vol. 2, *pp.* 217–279, Academic Press (1970).

Weinhold, E.G., Glasfeld, A., Ellington, A.D., and Benner, S.A., Structural determinants of stereospecificity in yeast alcohol dehydrogenase, *Proc Natl. Acad. Sci.* **88,** 8420–8424 (1991).

Regulation of Enzyme Activity

Allewell, N.M., *Eschericia coli* aspartate transcarbamoylase: structure, energetics, and catalytic and regulatory mechanisms, *Annu. Rev. Biophys. Biophys. Chem.* **18,** 71–92 (1989).

Evans, P.R., Structural aspects of allostery, *Curr. Opin. Struct. Biol.* **1,** 773–779 (1991).

Gouaux, J.E., Stevens, R.C., Ke, H., and Lipscomb, W.N., Crystal structure of the Glu-289 → Gln mutant of aspartate carbamoyltransferase at 3.1-Å resolution: An intermediate quarternary structure, *Proc. Natl. Acad. Sci.* **86,** 8212–8216 (1989).

Kantrowitz, E.R. and Lipscomb, W.N., *Eschericia coli* aspartate transcarbamylase: the molecular basis for a concerted allosteric transition, *Trends Biochem. Sci.* **15,** 53–59 (1990).

Lipscomb, W.N., Structure and function of allosteric enzymes, *Chemtracts–Biochem. Mol. Biol.* **2,** 1–15 (1991).

Schachman, H.K., Can a simple model account for the allosteric transition of aspartate transcarbamoylase? *J. Biol. Chem.* **263,** 18583–18586 (1988).

Stevens, R.C. and Lipscomb, W.N., A molecular mechanism for pyrimidine and purine nucleotide control of aspartate transcarbamoylase, *Proc. Natl. Acad. Sci.* **89,** 5281–5285 (1992).

Zhang, Y. and Kantrowitz, E.R., Probing the regulatory site of *Escherichia coli* aspartate transcarbamoylase by site specific mutagenesis, *Biochemistry* **31,** 792–798 (1992).

Enzyme Nomenclature

Dixon, M. and Webb, E.C., *Enzymes* (3rd ed.), Academic Press (1979). [Chapter V contains the rules for enzyme classification, and the Table of Enzymes provides their systematic names and classification numbers.]

Enzyme Nomenclature, Academic Press (1992). [Recommendations of the Nomenclature Committee of the IUBMB on the nomenclature and classification of enzymes.]

Tipton, K.F., The naming of parts, *Trends Biochem. Sci.* **18,** 113–115 (1993). [A discussion of the advantages of a consistent naming scheme for enzymes and the difficulties of formulating one.]

PROBLEMS

1. Indicate the products of the YADH reaction with normal acetaldehyde and NADH in D_2O solution.

2. Indicate the product(s) of the YADH-catalyzed oxidation of the chiral methanol derivative (*R*)-TDHCOH.

3. The enzyme **fumarase** catalyzes the hydration of the double bond of **fumarate:**

Fumarate		**L-Malate**

Predict the action of fumarase on **maleate,** the cis isomer of fumarate. Explain.

4. Write a balanced equation for the chymotrypsin-catalyzed reaction between an ester and an amino acid.

5. Hominy grits, a regional delicacy of the Southern United States, is made from corn that has been soaked in a weak lye (NaOH) solution. What is the function of this unusual treatment?

6. Which of the curves in Fig. 12-5 exhibits the greatest cooperativity? Explain.

7. What are the advantages of having the final product of a multistep metabolic pathway inhibit the enzyme that catalyzes the first step?

8. Using the references, find the systematic names and classification numbers for the enzymes whose recommended names are catalase, aspartate carbamoyltransferase (aspartate transcarbamoylase), and trypsin.

C H A P T E R

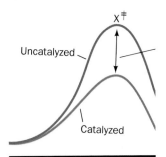

13

Rates of Enzymatic Reactions

Kinetics is the study of the rates at which chemical reactions occur. A major purpose of such a study is to gain an understanding of a reaction mechanism, that is, a detailed description of the various steps in a reaction process and the sequence with which they occur. Thermodynamics, as we saw in Chapter 3, tells us whether a given process can occur spontaneously but provides little indication as to the nature or even the existence of its component steps. In contrast, *the rate of a reaction and how this rate changes in response to different conditions is intimately related to the path followed by the reaction and is therefore indicative of its reaction mechanism.*

In this chapter, we take up the study of **enzyme kinetics,** a subject that is of enormous practical importance in biochemistry because:

1. It is through kinetic studies that the binding affinities of substrates and inhibitors to an enzyme can be determined and that the maximum catalytic rate of an enzyme can be established.

2. By observing how the rate of an enzymatic reaction varies with the reaction conditions and combining this information with that obtained from chemical and structural studies of the enzyme, the enzyme's catalytic mechanism may be elucidated.

3. Most enzymes, as we shall see in later chapters, function as members of metabolic pathways. The study of the kinetics of an enzymatic reaction leads to an understanding of that enzyme's role in an overall metabolic process.

4. Under the proper conditions, the rate of an enzymatically catalyzed reaction is proportional to the amount of the enzyme present and therefore most enzyme assays (measurements of the amount of enzyme present) are based on kinetic studies of the enzyme. Measurements of enzymatically catalyzed reaction rates are therefore among the most commonly employed procedures in biochemical and clinical analyses.

We begin our consideration of enzyme kinetics by reviewing chemical kinetics because enzyme kinetics is based on this formalism. Following that, we derive the basic equations of enzyme kinetics, describe the effects of inhibitors on enzymes, and consider how the rates of enzymatic reactions vary with pH. We end by outlining the kinetics of complex enzymatic reactions.

Kinetics is, by and large, a mathematical subject. Although the derivations of kinetic equations are occasionally rather detailed, the level of mathematical skills it requires should not challenge anyone who has studied elementary calculus. Nevertheless, to prevent mathematical detail from obscuring the underlying enzymological principles, the derivations of all but the most important kinetic equations have been collected in the appendix to this chapter. Those who wish to cultivate a deeper understanding of enzyme kinetics are urged to consult this appendix.

1. CHEMICAL KINETICS

Enzyme kinetics is a branch of chemical kinetics and, as such, shares much of the same formalism. In this section we shall therefore review the principles of chemical kinetics so that, in later sections, we can apply them to enzymatically catalyzed reactions.

A. Elementary Reactions

A reaction of overall stoichiometry

$$A \longrightarrow P$$

may actually occur through a sequence of **elementary reactions** (simple molecular processes) such as

$$A \longrightarrow I_1 \longrightarrow I_2 \longrightarrow P$$

Here A represents reactants, P products, and I_1 and I_2 symbolize **intermediates** in the reaction. *The characterization of the elementary reactions comprising an overall reaction process constitutes its mechanistic description.*

Rate Equations

At constant temperature, elementary reaction rates vary with reactant concentration in a simple manner. Consider the general elementary reaction:

$$a\text{A} + b\text{B} + \cdots + z\text{Z} \longrightarrow \text{P}$$

The rate of this process is proportional to the frequency with which the reacting molecules simultaneously come together,

that is, to the products of the concentrations of the reactants. This is expressed by the following **rate equation**

$$\text{Rate} = k[\text{A}]^a[\text{B}]^b \cdots [\text{Z}]^z \qquad [13.1]$$

where k is a proportionality constant known as a **rate constant**. The **order** of a reaction is defined as $(a + b + \cdots + z)$, the sum of the exponents in the rate equation. *For an elementary reaction, the order corresponds to the **molecularity** of the reaction, the number of molecules that must simultaneously collide in the elementary reaction.* Thus the elementary reaction A→P is an example of a **first-order** or **unimolecular** reaction, whereas the elementary reactions 2A→P and A + B→P are examples of **second-order** or **bimolecular** reactions. Unimolecular and bimolecular reactions are common. **Termolecular** reactions are unusual and fourth- and higher order reactions are unknown. This is because the simultaneous collision of three molecules is a rare event; that of four or more molecules essentially never occurs.

B. Rates of Reactions

We can experimentally determine the order of a reaction by measuring [A] or [P] as a function of time; that is,

$$v = -\frac{d[\text{A}]}{dt} = \frac{d[\text{P}]}{dt} \qquad [13.2]$$

where v is the instantaneous rate or **velocity** of the reaction. For the first-order reaction A→P:

$$v = -\frac{d[\text{A}]}{dt} = k[\text{A}] \qquad [13.3a]$$

For second-order reactions such as 2A→P:

$$v = -\frac{d[\text{A}]}{dt} = k[\text{A}]^2 \qquad [13.3b]$$

whereas for A + B→P, a second-order reaction that is first order in [A] and first order in [B],

$$v = -\frac{d[\text{A}]}{dt} = -\frac{d[\text{B}]}{dt} = k[\text{A}][\text{B}] \qquad [13.3c]$$

The rate constants of first- and second-order reactions must have different units. In terms of units, Eq. [13.3a] is expressed $M \cdot s^{-1} = kM$. Therefore, k must have units of reciprocal seconds (s^{-1}) in order for Eq. [13.3a] to balance. Similarly, for second-order reactions, $M \cdot s^{-1} = kM^2$ so that k has the units $M^{-1} s^{-1}$.

The order of a specific reaction can be determined by measuring the reactant or product concentrations as a function of time and comparing the fit of these data to equations describing this behavior for reactions of various orders. To do this we must first derive these equations.

First-Order Rate Equation

The equation for [A] as a function of time for a first-order reaction, A→P, is obtained by rearranging Eq. [13.3a]

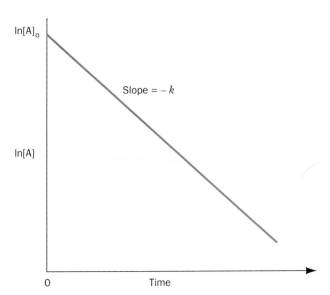

FIGURE 13-1. A plot of ln [A] versus time for a first-order reaction illustrating the graphical determination of the rate constant k using Eq. [13.4a].

$$\frac{d[A]}{[A]} \equiv d \ln [A] = -k \, dt$$

and integrating it from $[A]_o$, the initial concentration of A, to $[A]$, the concentration of A at time t:

$$\int_{[A]_o}^{[A]} d \ln [A] = -k \int_0^t dt$$

This results in

$$\ln [A] = \ln [A]_o - kt \qquad [13.4a]$$

or, by taking the antilogs of both sides,

$$[A] = [A]_o e^{-kt} \qquad [13.4b]$$

Equation [13.4a] is a linear equation in terms of the variables ln [A] and t as is diagrammed in Fig. 13-1. Therefore, if a reaction is first order, a plot of ln [A] versus t will yield a straight line whose slope is $-k$, the negative of the first-order rate constant, and whose intercept on the ln [A] axis is ln $[A]_o$.

Substances that are inherently unstable, such as radioactive nuclei, decompose through first-order reactions (first-order processes are not just confined to chemical reactions). One of the hallmarks of a first-order reaction is that *the time for half of the reactant initially present to decompose, its* **half-time** *or* **half-life,** $t_{1/2}$, *is a constant and hence independent of the initial concentration of the reactant.* This is easily demonstrated by substituting the relationship $[A] = [A]_o/2$ when $t = t_{1/2}$ into Eq. [13.4a] and rearranging:

$$\ln \left(\frac{[A]_o/2}{[A]_o} \right) = -kt_{1/2}$$

Thus

$$t_{1/2} = \frac{\ln 2}{k} = \frac{0.693}{k} \qquad [13.5]$$

In order to appreciate the course of a first-order reaction, let us consider the decomposition of ^{32}P, a radioactive isotope that is widely used in biochemical research. It has a half-life of 14 days. Thus, after 2 weeks, one half of the ^{32}P initially present in a given sample will have decomposed; after another 2 weeks, one half of the remainder or three quarters of the original sample will have decomposed; *etc.* The long-term storage of waste ^{32}P therefore presents little problem, since after 1 year (26 half-lives), only 1 part in $2^{26} = 67$ million of the original sample will remain. How much will remain after 2 years? In contrast, ^{14}C, another commonly employed radioactive tracer, has a half-life of 5715 years: Only a small fraction of a given quantity of ^{14}C will decompose over the course of a human lifetime.

Second-Order Rate Equation for One Reactant

In a second-order reaction with one type of reactant, 2A→P, the variation of [A] with time is quite different from that in a first-order reaction. Rearranging Eq. [13.3b] and integrating it over the same limits used for the first-order reaction yields:

$$\int_{[A]_o}^{[A]} -\frac{d[A]}{[A]^2} = k \int_0^t dt$$

so that

$$\frac{1}{[A]} = \frac{1}{[A]_o} + kt \qquad [13.6]$$

Equation [13.6] is a linear equation in terms of the variables $1/[A]$ and t. Consequently, Eqs. [13.4a] and [13.6] may be used to distinguish a first-order from a second-order reaction by plotting ln [A] versus t and $1/[A]$ versus t and observing which, if any, of these plots is a straight line.

Figure 13-2 compares the different shapes of the progress curves describing the disappearance of A in first- and sec-

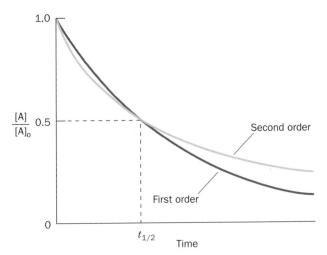

FIGURE 13-2. A comparison of the progress curves for first- and second-order reactions that have the same value of $t_{1/2}$. [After Tinoco, I., Jr., Sauer, K., and Wang., J. C., *Physical Chemistry. Principles and Applications in Biological Sciences* (2nd ed.), *p.* 291, Prentice–Hall (1985).]

ond-order reactions having the same half-times. Note that before the first half-time, the second-order progress curve descends more steeply than the first-order curve, but after this time the first-order progress curve is the more rapidly decreasing of the two. The half-time for a second-order reaction is expressed $t_{1/2} = 1/k[A]_o$ and therefore, in contrast to a first-order reaction, is dependent on the initial reactant concentration.

C. Transition State Theory

The goal of kinetic theory is to describe reaction rates in terms of the physical properties of the reacting molecules. A theoretical framework for doing so, that explicitly considers the structures of the reacting molecules and how they collide, was developed in the 1930s, principally by Henry Eyring. This view of reaction processes, known as **transition state theory** or **absolute rate theory,** is the foundation of much of modern kinetics and has provided an extraordinarily productive framework for understanding how enzymes catalyze reactions.

The Transition State

Consider a bimolecular elementary reaction involving three atoms A, B, and C:

$$A—B + C \longrightarrow A + B—C$$

Clearly atom C must approach the diatomic molecule A—B so that, at some point in the reaction, a high-energy (unstable) complex represented as $A \cdots B \cdots C$ exists in which the A—B covalent bond is in the process of breaking while the B—C bond is in the process of forming.

Let us consider the simplest example of this reaction: That of a hydrogen atom with diatomic hydrogen (H_2) to yield a new H_2 molecule and a different hydrogen atom:

$$H_A—H_B + H_C \longrightarrow H_A + H_B—H_C$$

The potential energy of this triatomic system as a function of the relative positions of its component atoms is plotted in Figs. 13-3a and b. Its shape is of two long and deep valleys parallel to the coordinate axes with sheer walls rising towards the axes and less steep ones rising towards a plateau where both coordinates are large (the region of point b). The two valleys are joined by a pass or saddle near the origin of the diagram (point c). The minimum energy configuration is that of an H_2 molecule and an isolated atom, that is, with one coordinate large and the other at the H_2 covalent bond distance [near points a (the reactants) and d (the products)]. During a collision, the reactants generally approach one another with little deviation from the minimum energy reaction pathway (line a—c—d) because other trajectories would require much greater energy. As the atom and molecule come together, they increasingly repel one another (have increasing potential energy) and therefore usually fly apart. *If, however, the system has sufficient kinetic energy to continue its coalescence, it will cause the covalent bond of*

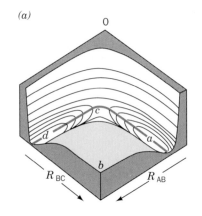

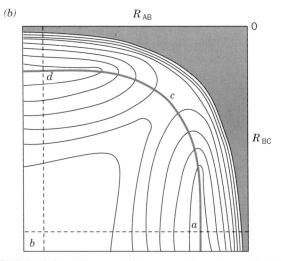

FIGURE 13-3. The potential energy of the colinear $H + H_2$ system as a function of its internuclear distances, R_{AB} and R_{BC}. The reaction is represented as: (a) A perspective drawing. (b) The corresponding contour diagram. The points a and d are approaching potential energy minima, b is approaching a maximum, and c is a saddle point. [After Frost, A. A. and Pearson, R. G., *Kinetics and Mechanism* (2nd ed.), p. 80, Wiley (1961).]

*the H_2 molecule to weaken until ultimately, if the system reaches the saddle point (point c), there is an equal probability that either the reaction will occur or that the system will decompose back to its reactants. Therefore, at this saddle point, the system is said to be at its **transition state** and hence to be an **activated complex.** Moreover, since the concentration of the activated complex is small, *the decomposition of the activated complex is postulated to be the rate-determining process of this reaction.**

The minimum energy pathway of a reaction is known as its **reaction coordinate.** Figure 13-4a, which is called a **transition state diagram** or a **reaction coordinate diagram,** shows the potential energy of the $H + H_2$ system along the reaction coordinate (line a—c—d in Fig. 13-3). It can be seen that the transition state is the point of highest energy on the reaction coordinate. If the atoms in the triatomic system are of different types, as is diagrammed in Fig. 13-4b, the tran-

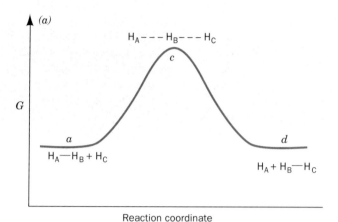

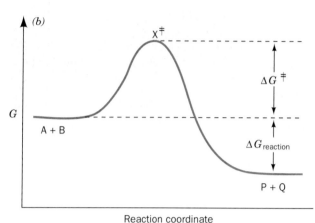

FIGURE 13-4. Transition state diagrams: (*a*) For the H + H$_2$ reaction. This is a section taken along the *a—c—d* line in Fig. 13-3*b*. (*b*) For a spontaneous reaction, that is, one in which the free energy decreases.

sition state diagram is no longer symmetrical because there is an energy difference between reactants and products.

Thermodynamics of the Transition State

The realization that the attainment of the transition state is the central requirement in any reaction process led to a detailed understanding of reaction mechanisms. For example, consider a bimolecular reaction that proceeds along the following pathway:

$$A + B \underset{}{\overset{K^{\ddagger}}{\rightleftharpoons}} X^{\ddagger} \overset{k'}{\longrightarrow} P + Q$$

where X‡ represents the activated complex. Therefore, considering the preceding discussion,

$$\frac{d[P]}{dt} = k[A][B] = k'[X^{\ddagger}] \qquad [13.7]$$

where k is the ordinary rate constant of the elementary reaction and k' is the rate constant for the decomposition of X‡ to products.

In contrast to stable molecules, such as A and P, which occur at energy minima, the activated complex occurs at an energy maximum and is therefore only metastable (like a ball balanced on a pin). Transition state theory nevertheless assumes that X‡ is in rapid equilibrium with the reactants; that is,

$$K^{\ddagger} = \frac{[X^{\ddagger}]}{[A][B]} \qquad [13.8]$$

where $K^{\ddagger}$ is an equilibrium constant. *This central assumption of transition state theory permits the powerful formalism of thermodynamics to be applied to the theory of reaction rates.*

If $K^{\ddagger}$ is an equilibrium constant it can be expressed as:

$$-RT \ln K^{\ddagger} = \Delta G^{\ddagger} \qquad [13.9]$$

where $\Delta G^{\ddagger}$ is the Gibbs free energy of the activated complex less that of the reactants (Fig. 13-4*b*), T is the absolute

temperature, and R (= 8.3145 J·K^{-1} mol^{-1}) is the gas constant (this relationship between equilibrium constants and free energy is derived in Section 3-4A). Then combining Eqs. [13.7] through [13.9] yields

$$\frac{d[P]}{dt} = k'e^{-\Delta G^{\ddagger}/RT}[A][B] \qquad [13.10]$$

This equation indicates that the rate of a reaction depends not only on the concentrations of its reactants but also decreases exponentially with $\Delta G^{\ddagger}$. *Thus, the larger the difference between the free energy of the transition state and that of the reactants, that is, the less stable the transition state, the slower the reaction proceeds.*

In order to continue, we must now evaluate k', the rate of passage of the activated complex over the maximum in the transition state diagram (sometimes referred to as the **activation barrier** or the **kinetic barrier** of the reaction). This transition state model permits us to do so (although the following derivation is by no means rigorous). The activated complex is held together by a bond that is associated with the reaction coordinate and that is assumed to be so weak that it flies apart during its first vibrational excursion. Therefore, k' is expressed

$$k' = \kappa\nu \qquad [13.11]$$

where ν is the vibrational frequency of the bond that breaks as the activated complex decomposes to products and κ, the **transmission coefficient,** is the probability that the breakdown of the activated complex, X‡, will be in the direction of product formation rather than back to reactants. For most spontaneous reactions in solution, κ is between 0.5 and 1.0; for the colinear H + H$_2$ reaction, we saw that it is 0.5.

We have nearly finished our job of evaluating k'. All that remains is to determine the value of ν. Planck's law states that

$$\nu = \varepsilon/h \qquad [13.12]$$

where, in this case, ε is the average energy of the vibration that leads to the decomposition of X^+, and $h \, (= 6.6261 \times 10^{-34} \, J \cdot s)$ is Planck's constant. Statistical mechanics tells us that at temperature T, the classical energy of an oscillator is

$$\varepsilon = k_B T \qquad [13.13]$$

where $k_B \, (= 1.3807 \times 10^{-23} \, J \cdot K^{-1})$ is a constant of nature known as the **Boltzmann constant** and $k_B T$ is essentially the available thermal energy. Combining Eqs. [13.11] through [13.13]

$$k' = \frac{\kappa k_B T}{h} \qquad [13.14]$$

Then assuming, as is done for most reactions, that $\kappa = 1$ (κ can rarely be calculated with any confidence), the combination of Eqs. [13.7] and [13.10] with [13.15] yields the expression for the rate constant k of our elementary reaction:

$$k = \frac{k_B T}{h} \, e^{-\Delta G^+/RT} \qquad [13.15]$$

This equation indicates that *the rate of reaction decreases as its free energy of activation, ΔG^+, increases.* Conversely, as the temperature rises, so that there is increased thermal energy available to drive the reacting complex over the activation barrier, the reaction speeds up. (Of course, enzymes, being proteins, are subject to thermal denaturation so that the rate of an enzymatically catalyzed reaction falls precipitously with increasing temperature once the enzyme's denaturation temperature has been surpassed.) Keep in mind, however, that transition state theory is an ideal model; real systems behave in a more complicated, although qualitatively similar, manner.

Multistep Reactions Have Rate-Determining Steps

Since chemical reactions commonly consist of several elementary reaction steps, let us consider how transition state theory treats such reactions. For a multistep reaction such as,

$$A \xrightarrow{\ k_1\ } I \xrightarrow{\ k_2\ } P$$

where I is an intermediate of the reaction, there is an activated complex for each elementary reaction step; the shape of the transition state diagram for such a reaction reflects the relative rates of the elementary reactions involved. For this reaction, if the first reaction step is slower than the second reaction step ($k_1 < k_2$), then the activation barrier of the first step must be higher than that of the second step, and conversely if the second reaction step is the slower (Fig. 13-5). Since the rate of formation of product P can only be as fast as the slowest elementary reaction, *if one reaction step of an overall reaction is much slower than the other, the slow step acts as a "bottleneck" and is therefore said to be the* **rate-determining step** *of the reaction.*

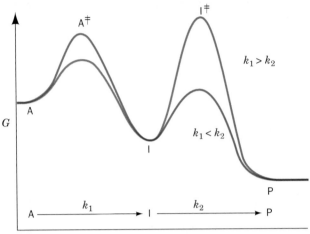

FIGURE 13-5.   The transition state diagram for the two-step overall reaction $A \rightarrow I \rightarrow P$. For $k_1 < k_2$ (*green curve*), the first step is rate determining, whereas if $k_1 > k_2$ (*red curve*), the second step is rate determining.

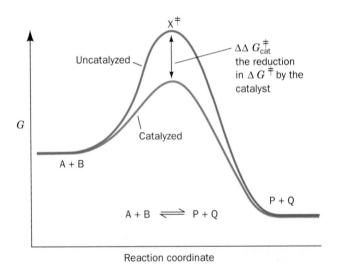

FIGURE 13-6.   A schematic diagram illustrating the effect of a catalyst on the transition state diagram of a reaction. Here $\Delta\Delta G^+ = \Delta G_{uncat}^+ - \Delta G_{cat}^+$.

Catalysis Reduces ΔG^+

Biochemistry is, of course, mainly concerned with enzyme-catalyzed reactions. *Catalysts act by lowering the activation barrier for the reaction being catalyzed* (Fig. 13-6). If a catalyst lowers the activation barrier of a reaction by $\Delta\Delta G_{cat}^+$ then, according to Eq. [13.15], the rate of the reaction is enhanced by the factor $e^{\Delta\Delta G_{cat}^+/RT}$. Thus, a 10-fold rate enhancement requires that $\Delta\Delta G_{cat}^+ = 5.71 \, kJ \cdot mol^{-1}$, less than half the energy of a typical hydrogen bond; a million-fold rate acceleration occurs when $\Delta\Delta G_{cat}^+ = 34.25 \, kJ \cdot mol^{-1}$, a small fraction of the energy of most covalent bonds. The rate enhancement is therefore a sensitive function of $\Delta\Delta G^{\ddagger}_{cat}$.

Note that the kinetic barrier is lowered by the same amount for both the forward and the reverse reactions (Fig. 13-6). Consequently, a catalyst equally accelerates the forward and the reverse reactions so that the equilibrium constant for the reaction remains unchanged. The chemical mechanisms through which enzymes lower the activation barriers of reactions are the subject of Section 14-1. There we shall see that the most potent such mechanism often involves the enzymatic binding of the transition state of the catalyzed reaction in preference to the substrate.

2. ENZYME KINETICS

The chemical reactions of life are mediated by enzymes. These remarkable catalysts, as we saw in Chapter 12, are individually highly specific for particular reactions. Yet, collectively they are extremely versatile in that the several thousand enzymes now known carry out such diverse reactions as hydrolysis, polymerization, functional group transfer, oxidation–reduction, dehydration, and isomerization, to mention only the most common classes of enzymatically mediated reactions. Enzymes are not passive surfaces on which reactions take place but rather, are complex molecular machines that operate through a great diversity of mechanisms. For instance, some enzymes act on only a single substrate molecule; others act on two or more different substrate molecules whose order of binding may or may not be obligatory. Some enzymes form covalently bound intermediate complexes with their substrates; others do not.

Kinetic measurements of enzymatically catalyzed reactions are among the most powerful techniques for elucidating the catalytic mechanisms of enzymes. The remainder of this chapter is therefore largely concerned with the development of the kinetic tools that are most useful in the determination of enzymatic mechanisms. We begin, in this section, with a presentation of the basic theory of enzyme kinetics.

A. The Michaelis–Menten Equation

The study of enzyme kinetics began in 1902 when Adrian Brown reported an investigation of the rate of hydrolysis of sucrose as catalyzed by the yeast enzyme **invertase** (now known as *β*-**fructofuranosidase**):

$$Sucrose + H_2O \longrightarrow glucose + fructose$$

Brown demonstrated that when the sucrose concentration is much higher than that of the enzyme, the reaction rate becomes independent of the sucrose concentration; that is, the rate is **zero order** with respect to sucrose. He therefore

proposed that the overall reaction is composed of two elementary reactions in which the substrate forms a complex with the enzyme that subsequently decomposes to products and enzyme:

$$E + S \underset{k_{-1}}{\overset{k_1}{\rightleftharpoons}} ES \xrightarrow{k_2} P + E$$

Here E, S, ES, and P symbolize the enzyme, substrate, **enzyme–substrate complex,** and products, respectively (for enzymes composed of multiple identical subunits, E refers to active sites rather than enzyme molecules). According to this model, *when the substrate concentration becomes high enough to entirely convert the enzyme to the ES form, the second step of the reaction becomes rate limiting and the overall reaction rate becomes insensitive to further increases in substrate concentration.*

The general expression for the **velocity** (rate) of this reaction is

$$v = \frac{d[P]}{dt} = k_2[ES] \qquad [13.16]$$

The overall rate of production of [ES] is the difference between the rates of the elementary reactions leading to its appearance and those resulting in its disappearance:

$$\frac{d[ES]}{dt} = k_1[E][S] - k_{-1}[ES] - k_2[ES] \qquad [13.17]$$

This equation cannot be explicitly integrated, however, without simplifying assumptions. Two possibilities are

1. Assumption of equilibrium

In 1913, Leonor Michaelis and Maude Menten, building upon earlier work by Victor Henri, assumed that $k_{-1} \gg k_2$, so that the first step of the reaction achieves equilibrium.

$$K_S = \frac{k_{-1}}{k_1} = \frac{[E][S]}{[ES]} \qquad [13.18]$$

Here K_S is the dissociation constant of the first step in the enzymatic reaction. With this assumption, Eq. [13.17] can be integrated. Although this assumption is not often correct, in recognition of the importance of this pioneering work, the noncovalently bound enzyme–substrate complex ES is known as the **Michaelis complex.**

2. Assumption of steady state

Figure 13-7 illustrates the progress curves of the various participants in the preceding reaction model under the physiologically common condition that substrate is in great excess over enzyme. With the exception of the initial stage of the reaction, the so-called **transient phase,** which is usually over within milliseconds of mixing the enzyme and substrate, [ES] remains approximately constant until the substrate is nearly exhausted. Hence, the rate of synthesis of ES must equal its rate of consumption over most of the course of the reaction; that is, [ES]

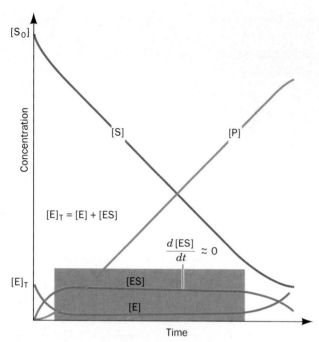

FIGURE 13-7. The progress curves for the components of a simple Michaelis–Menten reaction. Note that with the exception of the transient phase of the reaction, which occurs before the shaded block, the slopes of the progress curves for [E] and [ES] are essentially zero so long as [S] > > [E]$_T$ (within the shaded block). [After Segel, I. H., *Enzyme Kinetics, p. 27,* Wiley (1975).]

maintains a **steady state.** One can therefore assume with a reasonable degree of accuracy that [ES] is constant; that is,

$$\frac{d[ES]}{dt} = 0 \qquad [13.19]$$

This so-called **steady-state assumption** was first proposed in 1925 by G. E. Briggs and James B. S. Haldane.

In order to be of use, kinetic expressions for overall reactions must be formulated in terms of experimentally measurable quantities. The quantities [ES] and [E] are not, in general, directly measurable but the total enzyme concentration

$$[E]_T = [E] + [ES] \qquad [13.20]$$

is usually readily determined. The rate equation for our enzymatic reaction is then derived as follows. Combining Eq. [13.17] with the steady state assumption, Eq. [13.19], and the conservation condition, Eq. [13.20], yields:

$$k_1([E]_T - [ES])[S] = (k_{-1} + k_2)[ES]$$

which upon rearrangement becomes

$$[ES](k_{-1} + k_2 + k_1[S]) = k_1[E]_T[S]$$

Dividing both sides by k_1 and solving for [ES],

$$[ES] = \frac{[E]_T [S]}{K_M + [S]}$$

where K_M, which is known as the **Michaelis constant,** is defined

$$K_M = \frac{k_{-1} + k_2}{k_1} \qquad [13.21]$$

The meaning of this important constant is discussed below.

The **initial velocity** of the reaction from Eq. [13.16] can then be expressed in terms of the experimentally measureable quantities [E]$_T$ and [S].

$$v_o = \left(\frac{d[P]}{dt}\right)_{t=0} = k_2[ES] = \frac{k_2[E]_T[S]}{K_M + [S]} \qquad [13.22]$$

The use of the initial velocity (operationally taken as the velocity measured before more than ~10% of the substrate has been converted to product) rather than just the velocity minimizes such complicating factors as the effects of reversible reactions, inhibition of the enzyme by product, and progressive inactivation of the enzyme.

The **maximal velocity** of a reaction, V_{max}, occurs at high substrate concentrations when the enzyme is **saturated,** that is, when it is entirely in the ES form:

$$V_{max} = k_2[E]_T \qquad [13.23]$$

Therefore, combining Eqs. [13.22] and [13.23], we obtain

$$v_o = \frac{V_{max}[S]}{K_M + [S]} \qquad [13.24]$$

*This expression, the **Michaelis–Menten equation,** is the basic equation of enzyme kinetics.* It describes a rectangular hyperbola such as is plotted in Fig. 13-8 (although this curve is rotated by 45° and translated to the origin with respect to the examples of hyperbolas seen in most elementary algebra texts). The saturation function for oxygen binding to myoglobin, Eq. [9.4], has the same functional form.

Significance of the Michaelis Constant

The Michaelis constant, K_M, has a simple operational definition. At the substrate concentration where [S] = K_M,

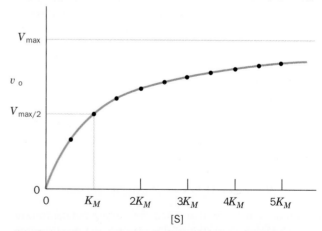

FIGURE 13-8. A plot of the initial velocity v_o of a simple Michaelis–Menten reaction versus the substrate concentration [S]. Points are plotted in 0.5-K_M intervals of substrate concentration between 0.5 K_M and 5 K_M.

TABLE 13-1. THE VALUES OF K_M, k_{CAT}, AND k_{CAT}/K_M FOR SOME ENZYMES AND SUBSTRATES

Enzyme	Substrate	K_M (M)	k_{cat} (s^{-1})	k_{cat}/K_M (M^{-1} s^{-1})
Acetylcholinesterase	Acetylcholine	9.5×10^{-5}	1.4×10^{4}	1.5×10^{8}
Carbonic anhydrase	CO_2	1.2×10^{-2}	1.0×10^{6}	8.3×10^{7}
	HCO_3^-	2.6×10^{-2}	4.0×10^{5}	1.5×10^{7}
Catalase	H_2O_2	2.5×10^{-2}	1.0×10^{7}	4.0×10^{8}
Chymotrypsin	*N*-Acetylglycine ethyl ester	4.4×10^{-1}	5.1×10^{-2}	1.2×10^{-1}
	N-Acetylvaline ethyl ester	8.8×10^{-2}	1.7×10^{-1}	1.9
	N-Acetyltyrosine ethyl ester	6.6×10^{-4}	1.9×10^{2}	2.9×10^{5}
Fumarase	Fumarate	5.0×10^{-6}	8.0×10^{2}	1.6×10^{8}
	Malate	2.5×10^{-5}	9.0×10^{2}	3.6×10^{7}
Urease	Urea	2.5×10^{-2}	1.0×10^{4}	4.0×10^{5}

Eq. [13.24] yields $v_o = V_{max}/2$ so that K_M is the substrate concentration at which the reaction velocity is half-maximal. Therefore, if an enzyme has a small value of K_M, it achieves maximal catalytic efficiency at low substrate concentrations.

The magnitude of K_M varies widely with the identity of the enzyme and the nature of the substrate (Table 13-1). It is also a function of temperature and pH (see Section 13-4). The Michaelis constant (Eq. [13.21]) can be expressed as

$$K_M = \frac{k_{-1}}{k_1} + \frac{k_2}{k_1} = K_S + \frac{k_2}{k_1} \qquad [13.25]$$

Since K_S is the dissociation constant of the Michaelis complex, as K_S decreases, the enzyme's affinity for substrate increases. K_M is therefore also a measure of the affinity of the enzyme for its substrate providing k_2/k_1 is small compared with K_S, that is, $k_2 < k_{-1}$.

B. Analysis of Kinetic Data

There are several methods for determining the values of the parameters of the Michaelis–Menten equation. At very

high values of [S], the initial velocity v_o asymptotically approaches V_{max}. In practice, however, it is very difficult to assess V_{max} accurately from direct plots of v_o versus [S] such as Fig. 13-8. Even at such high substrate concentrations as [S] = 10 K_M, Eq. [13.24] indicates that v_o is only 91% of V_{max} so that the extrapolated value of the asymptote will almost certainly be underestimated.

A better method for determining the values of V_{max} and K_M, which was formulated by Hans Lineweaver and Dean Burk, uses the reciprocal of Eq. [13.24]:

$$\frac{1}{v_o} = \left(\frac{K_M}{V_{max}}\right)\frac{1}{[S]} + \frac{1}{V_{max}} \qquad [13.26]$$

This is a linear equation in $1/v_o$ and $1/[S]$. If these quantities are plotted, the so-called **Lineweaver–Burk** or **double-reciprocal plot,** the slope of the line is K_M/V_{max}, the $1/v_o$ intercept is $1/V_{max}$, and the extrapolated $1/[S]$ intercept is $-1/K_M$ (Fig. 13-9). A disadvantage of this plot is that most experimental measurements involve relatively high [S] and are therefore crowded onto the left side of the graph. Furthermore, for small values of [S], small errors in v_o lead to large errors in $1/v_o$ and hence to large errors in K_M and V_{max}.

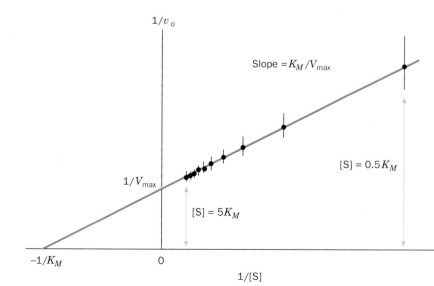

FIGURE 13-9. A double reciprocal (Lineweaver–Burk) plot with error bars of ± 0.05 V_{max}. The indicated points are the same as those in Fig. 13-8. Note the large effect of small errors at small [S] (large $1/[S]$) and the crowding together of points at large [S].

Several other types of plots, each with its advantages and disadvantages, have been formulated for the determination of V_{max} and K_M from kinetic data. With the advent of conveniently available computers, however, kinetic data are commonly analyzed by mathematically sophisticated statistical treatments. Nevertheless, Lineweaver–Burk plots are valuable for the visual presentation of kinetic data as well as being useful in the analysis of kinetic data from enzymes requiring more than one substrate (Section 13-5C).

k_{cat}/K_M Is a Measure of Catalytic Efficiency

An enzyme's kinetic parameters provide a measure of its catalytic efficiency. We may define the **catalytic constant** of an enzyme as

$$k_{cat} = \frac{V_{max}}{[E]_T} \qquad [13.27]$$

This quantity is also known as the **turnover number** of an enzyme because it is the number of reaction processes (turnovers) that each active site catalyzes per unit time. The turnover numbers for a selection of enzymes are given in Table 13-1. Note that these quantities vary by over eight orders of magnitude depending on the identity of the enzyme as well as that of its substrate. Equation [13.23] indicates that for the Michaelis–Menten model, $k_{cat} = k_2$. For enzymes with more complicated mechanisms, k_{cat} may be a function of several rate constants.

When $[S] \ll K_M$, very little ES is formed. Consequently, $[E] \approx [E]_T$, so that Eq. [13.22] reduces to a second-order rate equation:

$$v_o \approx \left(\frac{k_2}{K_M}\right)[E]_T[S] \approx \left(\frac{k_{cat}}{K_M}\right)[E][S] \qquad [13.28]$$

k_{cat}/K_M is the apparent second-order rate constant of the enzymatic reaction; the rate of the reaction varies directly with how often enzyme and substrate encounter one another in solution. *The quantity k_{cat}/K_M is therefore a measure of an enzyme's catalytic efficiency.*

Some Enzymes Have Attained Catalytic Perfection

Is there an upper limit on enzymatic catalytic efficiency? From Eq. [13.21] we find

$$\frac{k_{cat}}{K_M} = \frac{k_2}{K_M} = \frac{k_1 k_2}{k_{-1} + k_2} \qquad [13.29]$$

This ratio is maximal when $k_2 \gg k_{-1}$; that is, when the formation of product from the Michaelis complex, ES, is fast compared to its decomposition back to substrate and enzyme. Then $k_{cat}/K_M = k_1$, the second-order rate constant for the formation of ES. The term k_1, of course, can be no greater than the frequency with which enzyme and substrate molecules collide with each other in solution. This **diffusion-controlled limit** is in the range of 10^8 to $10^9 M^{-1} s^{-1}$. Thus, enzymes with such values of k_{cat}/K_M must catalyze a reaction almost every time they encounter a substrate molecule. Table 13-1 indicates that several enzymes, namely, catalase, acetylcholinesterase, fumarase, and possibly carbonic anhydrase, have achieved this state of virtual catalytic perfection.

Since the active site of an enzyme generally occupies only a small fraction of its total surface area, how can any enzyme catalyze a reaction every time it encounters a substrate molecule? Although the answer to this question is yet unclear, structural and theoretical evidence is accumulating suggesting that the arrangements of charged groups on the surfaces of enzymes serve to electrostatically guide polar substrates to their enzymes' active sites.

C. Reversible Reactions

The Michaelis–Menten model implicitly assumes that enzymatic reverse reactions may be neglected. Yet, many enzymatic reactions are highly reversible (have a small free energy of reaction) and therefore have products that back react to form substrates at a significant rate. In this section we therefore relax the Michaelis–Menten restriction of no back reaction and, by doing so, discover some interesting and important kinetic principles.

The One-Intermediate Model

Modification of the Michaelis–Menten model to incorporate a back reaction yields the following reaction scheme:

$$E + S \underset{k_{-1}}{\overset{k_1}{\rightleftharpoons}} ES \underset{k_{-2}}{\overset{k_2}{\rightleftharpoons}} P + E$$

(Here ES might just as well be called EP because this model does not specify the nature of the intermediate complex.) The equation describing the kinetic behavior of this model, which is derived in Appendix A of this chapter, is expressed

$$v = \frac{\dfrac{V_{max}^f[S]}{K_M^S} - \dfrac{V_{max}^r[P]}{K_M^P}}{1 + \dfrac{[S]}{K_M^S} + \dfrac{[P]}{K_M^P}} \qquad [13.30]$$

where

$$V_{max}^f = k_2[E]_T \qquad V_{max}^r = k_{-1}[E]_T$$
$$K_M^S = \frac{k_{-1} + k_2}{k_1} \qquad K_M^P = \frac{k_{-1} + k_2}{k_{-2}}$$

and

$$[E]_T = [E] + [ES]$$

This is essentially a Michaelis–Menten equation that works backwards as well as forwards. Indeed, at [P] = 0, that is, when $v = v_o$, this equation becomes the Michaelis–Menten equation.

The Haldane Relationship

At equilibrium, $v = 0$ so Eq. [13.30] can be solved to yield

$$K_{eq} = \frac{[P]}{[S]} = \frac{V^f_{max} K^P_M}{V^r_{max} K^S_M} \qquad [13.31]$$

which is known as the **Haldane relationship.** This relationship demonstrates that *the kinetic parameters of a reversible enzymatically catalyzed reaction are not independent of one another. Rather, they are related by the equilibrium constant for the overall reaction which, of course, is independent of the presence of the enzyme.*

Kinetic Data Cannot Unambiguously Establish a Reaction Mechanism

An enzyme that forms a reversible complex with its substrate should likewise form one with its product; that is, have a mechanism such as:

$$E + S \underset{k_{-1}}{\overset{k_1}{\rightleftharpoons}} ES \underset{k_{-2}}{\overset{k_2}{\rightleftharpoons}} EP \underset{k_{-3}}{\overset{k_3}{\rightleftharpoons}} P + E$$

The equation describing the kinetic behavior of this two-intermediate model, whose derivation is analogous to that described in Appendix A for the one-intermediate model, has a form identical to that of Eq. [13.30]. However, its parameters V^f_{max}, V^r_{max}, K^S_M, and K^P_M are defined in terms of the six kinetic constants of the two-intermediate model rather than the four of the one-intermediate model. In fact, the steady state rate equations for reversible reactions with three or more intermediates also have this same form but with yet different definitions of the four parameters.

The values of V^f_{max}, V^r_{max}, K^S_M, and K^P_M in Eq. [13.30] can be determined by suitable manipulations of the initial substrate and product concentrations under steady state conditions. This, however, will not yield the values of the rate constants for our two-intermediate model because there are six such constants and only four equations describing their relationships. Moreover, steady state kinetic measurements are incapable of distinguishing the number of intermediates in a reversible enzymatic reaction because the form of Eq. [13.30] does not change with this number of intermediates.

The functional identities of the equations describing these reaction schemes may be understood in terms of an analogy between our *n*-intermediate reversible reaction model and a "black box" containing a system of water pipes with one inlet and one drain:

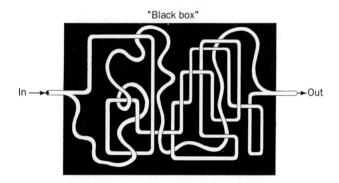

At steady state, that is, after the pipes have filled with water, one can measure the relationship between input pressure and output flow. However, such measurements yield no information concerning the detailed construction of the plumbing connecting the inlet to the drain. This would require additional information such as opening the black box and tracing the pipes. *Likewise, steady state kinetic measurements can provide a phenomenological description of enzymatic behavior, but the nature of the intermediates remains indeterminate. Rather, these intermediates must be detected and characterized by independent means such as by spectroscopic analysis.*

The foregoing discussion brings to light a central principle of kinetic analysis: *The steady state kinetic analysis of a reaction cannot unambiguously establish its mechanism.* This is because no matter how simple, elegant, or rational a mechanism one postulates that fully accounts for kinetic data, there are an infinite number of alternate mechanisms, perhaps complicated, awkward, and seemingly irrational, that can account for these kinetic data equally well. Usually it is the simpler and more elegant mechanism that turns out to be correct, but this is not always the case. *If, however, kinetic data are not compatible with a given mechanism, then the mechanism must be rejected.* Therefore, although kinetics cannot be used to establish a mechanism unambiguously without confirming data, such as the physical demonstration of an intermediate's existence, the steady state kinetic analysis of a reaction is of great value because it can be used to eliminate proposed mechanisms.

3. INHIBITION

Many substances alter the activity of an enzyme by combining with it in a way that influences the binding of substrate and/or its turnover number. Substances that reduce an enzyme's activity in this way are known as **inhibitors.**

Many inhibitors are substances that structurally resemble their enzyme's substrate but either do not react or react very slowly compared to substrate. Such inhibitors are commonly used to probe the chemical and conformational nature of a substrate-binding site as part of an effort to elucidate the enzyme's catalytic mechanism. In addition, many enzyme inhibitors are effective chemotherapeutic agents since an "unnatural" substrate analog can block the action of a specific enzyme. For example, **methotrexate** (also called **amethopterin**) chemically resembles **dihydrofolate**. Methotrexate binds tightly to the enzyme **dihydrofolate reductase**, thereby preventing it from carrying out its normal function, the reduction of dihydofolate to **tetrahydrofolate**, an essential cofactor in the biosynthesis of the DNA precursor **thymidylic acid** (Section 26-4B).

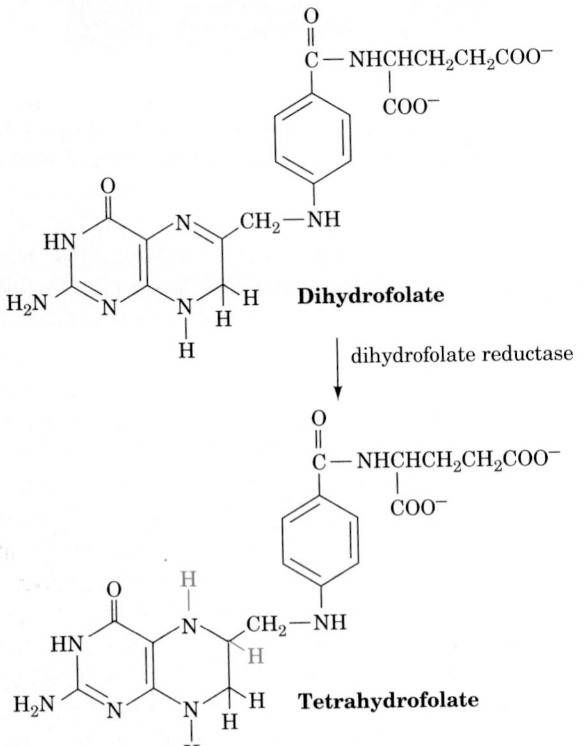

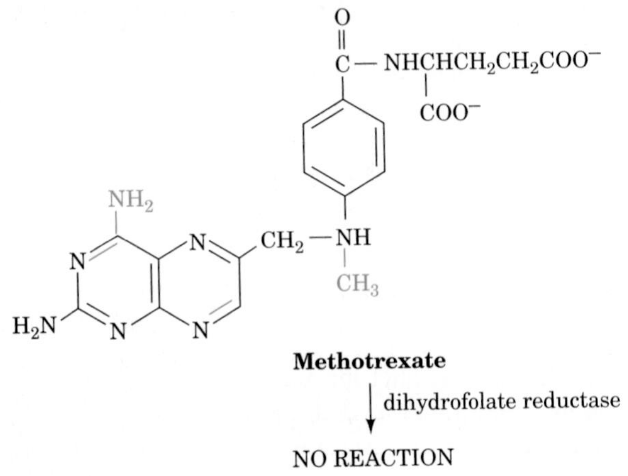

Rapidly dividing cells, such as cancer cells, which are actively engaged in DNA synthesis, are far more susceptible to methotrexate than are slower growing cells such as those of most normal mammalian tissues. Hence, methotrexate, when administered in proper dosage, kills cancer cells without fatally poisoning the host.

There are various mechanisms through which enzyme inhibitors can act. In this section, we discuss several of the simplest such mechanisms and their effects on the kinetic behavior of enzymes that follow the Michaelis–Menten model.

A. Competitive Inhibition

A substance that competes directly with a normal substrate for an enzymatic-binding site is known as a **competitive inhibitor.** Such an inhibitor usually resembles the substrate to the extent that it specifically binds to the active site but differs from it so as to be unreactive. Thus methotrexate is a competitive inhibitor of dihydrofolate reductase. Similarly, **succinate dehydrogenase,** a citric acid cycle enzyme that functions to convert **succinate** to **fumarate** (Section 19-3F), is competitively inhibited by **malonate,** which structurally resembles succinate but cannot be dehydrogenated.

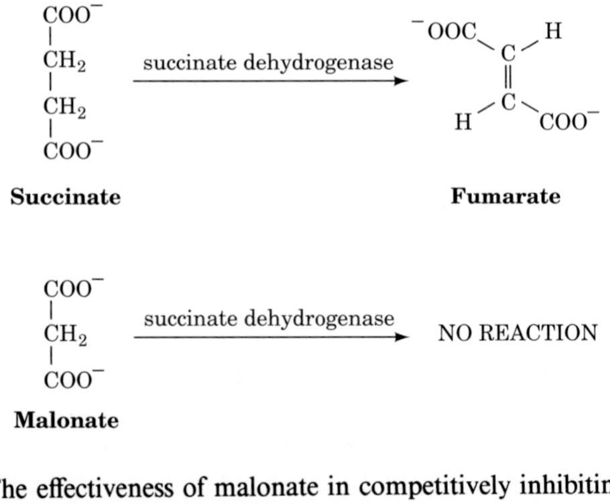

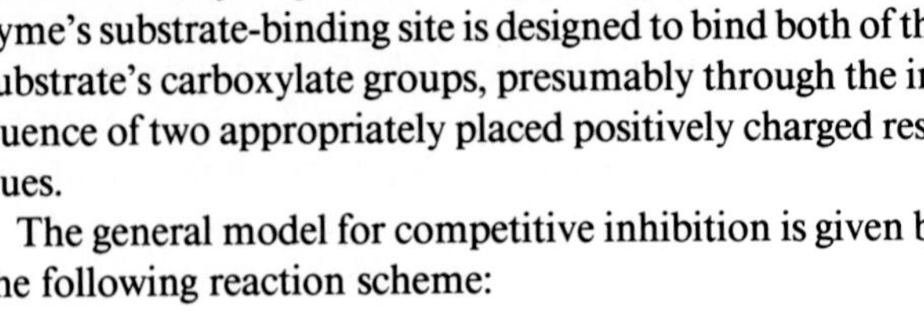

The effectiveness of malonate in competitively inhibiting succinate dehydrogenase strongly suggests that the enzyme's substrate-binding site is designed to bind both of the substrate's carboxylate groups, presumably through the influence of two appropriately placed positively charged residues.

The general model for competitive inhibition is given by the following reaction scheme:

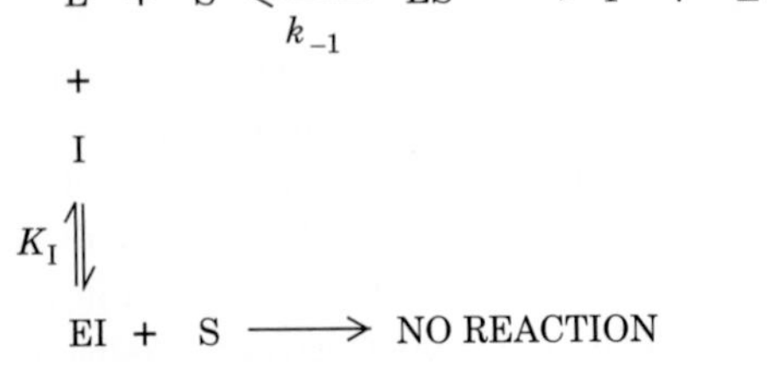

$$E + S \underset{k_{-1}}{\overset{k_1}{\rightleftharpoons}} ES \overset{k_2}{\longrightarrow} P + E$$

$$+$$

$$I$$

$$K_I \Updownarrow$$

$$EI + S \longrightarrow \text{NO REACTION}$$

Here it is assumed that I, the inhibitor, binds reversibly to the enzyme and is in rapid equilibrium with it so that

$$K_I = \frac{[E][I]}{[EI]} \qquad [13.32]$$

and EI, the enzyme–inhibitor complex, is catalytically inactive. *A competitive inhibitor therefore acts by reducing the concentration of free enzyme available for substrate binding.*

Our goal, as before, is to express v_o in terms of measurable quantities; in this case $[E]_T$, $[S]$, and $[I]$. We begin, as in the derivation of the Michaelis–Menten equation, with the expression for the conservation condition, which must now take into account the existence of EI.

$$[E]_T = [E] + [EI] + [ES] \qquad [13.33]$$

The enzyme concentration can be expressed in terms of $[ES]$ by rearranging Eq. [13.17] under the steady state condition:

$$[E] = \frac{K_M[ES]}{[S]} \qquad [13.34]$$

That of the enzyme–inhibitor complex is found by rearranging Eq. [13.32] and substituting Eq. [13.34] into it

$$[EI] = \frac{[E][I]}{K_I} = \frac{K_M[ES][I]}{[S]K_I} \qquad [13.35]$$

Substituting the latter two results into Eq. [13.33] yields

$$[E]_T = [ES]\left\{\frac{K_M}{[S]}\left(1 + \frac{[I]}{K_I}\right) + 1\right\}$$

which can be solved for $[ES]$ by rearranging it to

$$[ES] = \frac{[E]_T[S]}{K_M\left(1 + \frac{[I]}{K_I}\right) + [S]}$$

so that, according to Eq. [13.22], the initial velocity is expressed

$$v_o = k_2[ES] = \frac{k_2[E]_T[S]}{K_M\left(1 + \frac{[I]}{K_I}\right) + [S]} \qquad [13.36]$$

Then defining

$$\alpha = \left(1 + \frac{[I]}{K_I}\right) \qquad [13.37]$$

and $V_{max} = k_2[E]_T$ as in Eq. [13.23],

$$v_o = \frac{V_{max}[S]}{\alpha K_M + [S]} \qquad [13.38]$$

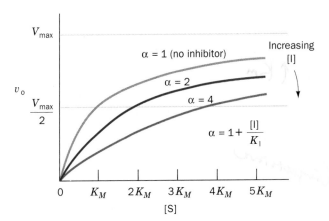

FIGURE 13-10. A plot of the initial velocity v_o of a simple Michaelis–Menten reaction versus the substrate concentration $[S]$ in the presence of different concentrations of a competitive inhibitor.

This is the Michaelis–Menten equation with K_M modulated by α, a function of the inhibitor concentration (which, according to Eq. [13.37], must always be ≥ 1). The value of $[S]$ at $v_o = V_{max}/2$ is therefore αK_M.

Figure 13-10 shows the hyperbolic plot of Eq. [13.38] for various values of α. Note that as $[S] \to \infty$, $v_o \to V_{max}$ for any value of α. The larger the value of α, however, the greater $[S]$ must be to approach V_{max}. Thus, the inhibitor does not affect the turnover number of the enzyme. Rather, the presence of I has the effect of making $[S]$ appear more dilute than it actually is, or alternatively, making K_M appear larger than it really is. Conversely, increasing $[S]$ shifts the substrate-binding equilibrium towards ES. Hence, there is true competition between I and S for the enzyme's substrate-binding site; their binding is mutually exclusive.

Recasting Eq. [13.38] in the double-reciprocal form yields

$$\frac{1}{v_o} = \left(\frac{\alpha K_M}{V_{max}}\right)\frac{1}{[S]} + \frac{1}{V_{max}} \qquad [13.39]$$

A plot of this equation is linear and has a slope of $\alpha K_M/V_{max}$, a $1/[S]$ intercept of $-1/\alpha K_M$, and a $1/v_o$ intercept of $1/V_{max}$ (Fig. 13-11). *The double-reciprocal plots for a competitive inhibitor at various concentrations of I intersect at $1/V_{max}$ on the $1/v_o$ axis; this is diagnostic for competitive inhibition as compared with other types of inhibition (Sections 13-3B and C).*

By determining the values of α at different inhibitor concentrations, the value of K_I can be found from Eq. [13.37]. In this way, competitive inhibitors can be used to probe the

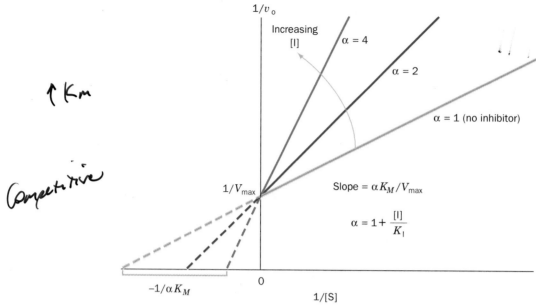

FIGURE 13-11. A Lineweaver–Burk plot of the competitively inhibited Michaelis–Menten enzyme described by Fig. 13-10. Note that all lines intersect on the $1/v_o$ axis at $1/V_{max}$.

structural nature of an active site. For example, to ascertain the importance of the various segments of an ATP molecule

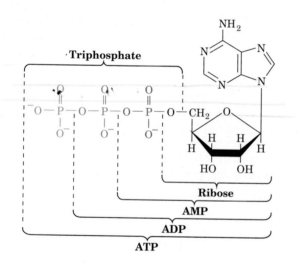

for binding to the active site of an ATP-requiring enzyme, one might determine the K_I, say, for ADP, AMP (adenosine monophosphate), ribose, triphosphate ion, *etc.* Since many of these ATP components are catalytically inactive, inhibition studies are the most convenient means of monitoring their binding to the enzyme.

If the inhibitor binds irreversibly to the enzyme, the inhibitor is classified as an **inactivator** as is any agent that somehow inactivates the enzyme. Inactivators truly reduce the effective level of $[E]_T$ at all values of $[S]$. Reagents that

modify specific amino acid residues, such as those listed in Table 6-3, can act in this manner.

B. Uncompetitive Inhibition

In **uncompetitive inhibition,** the inhibitor binds directly to the enzyme–substrate complex but not to the free enzyme:

$$E \;+\; S \;\underset{k_{-1}}{\overset{k_1}{\rightleftharpoons}}\; ES \;\overset{k_2}{\longrightarrow}\; P \;+\; E$$
$$+$$
$$I$$
$$K_I' \Big\Updownarrow$$
$$ESI \longrightarrow NO\ REACTION$$

The inhibitor-binding step, which has the dissociation constant

$$K_I' = \frac{[ES][I]}{[ESI]} \qquad [13.40]$$

is assumed to be at equilibrium. The binding of the uncompetitive inhibitor, which need not resemble the substrate, is envisioned to cause structural distortion of the active site, thereby rendering the enzyme catalytically inactive. (If the inhibitor binds to enzyme alone, it does so without affecting its affinity for substrate.)

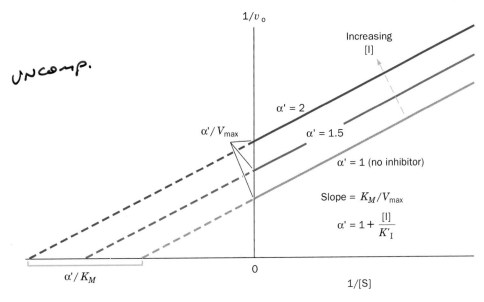

FIGURE 13-12. A Lineweaver–Burk plot of a simple Michaelis–Menten enzyme in the presence of uncompetitive inhibitor. Note that all lines have identical slopes of K_M/V_{max}.

The Michaelis–Menten equation for uncompetitive inhibition, which is derived in Appendix B of this chapter, is

$$v_o = \frac{V_{max}[S]}{K_M + \alpha'[S]} \qquad [13.41]$$

where

$$\alpha' = 1 + \frac{[I]}{K_I'} \qquad [13.42]$$

Inspection of this equation indicates that *at high values of [S], v_o asymptotically approaches V_{max}/α' so that, in contrast to competitive inhibition, the effects of uncompetitive inhibition on V_{max} are not reversed by increasing the substrate concentration.* However, at low substrate concentrations, that is, when $[S] \ll K_M$, the effect of an uncompetitive inhibitor becomes negligible, again the opposite behavior of a competitive inhibitor.

When cast in the double-reciprocal form, Eq. [13.41] becomes

$$\frac{1}{v_o} = \left(\frac{K_M}{V_{max}}\right)\frac{1}{[S]} + \frac{\alpha'}{V_{max}} \qquad [13.43]$$

The Lineweaver–Burk plot for uncompetitive inhibition is linear with slope K_M/V_{max} as in the uninhibited reaction, and with $1/v_o$ and $1/[S]$ intercepts of α'/V_{max} and $-\alpha'/K_M$, respectively. *A series of Lineweaver–Burk plots at various uncompetitive inhibitor concentrations consists of a family of parallel lines (Fig. 13-12). This is diagnostic for uncompetitive inhibition.*

Uncompetitive inhibition requires that the inhibitor affect the catalytic function of the enzyme but not its substrate binding. For single-substrate enzymes it is difficult to conceive of how this could happen with the exception of small inhibitors such as protons (see Section 13-4) or metal ions. As we discuss in Section 13-5C, however, uncompetitive inhibition is important for multisubstrate enzymes.

C. Mixed Inhibition ≈ NON comP

If both the enzyme and the enzyme–substrate complex bind inhibitor, the following model results:

$$
\begin{array}{ccccc}
E & + & S & \underset{k_{-1}}{\overset{k_1}{\rightleftharpoons}} & ES & \overset{k_2}{\longrightarrow} & P & + & E \\
+ & & & & + \\
I & & & & I \\
K_I \Updownarrow & & & & K_I' \Updownarrow \\
EI & & & & ESI \longrightarrow NO\ REACTION
\end{array}
$$

Both of the inhibitor-binding steps are assumed to be at equilibrium but with different dissociation constants:

$$K_I = \frac{[E][I]}{[EI]} \quad \text{and} \quad K_I' = \frac{[ES][I]}{[ESI]} \qquad [13.44]$$

This phenomenon is alternatively known as **mixed inhibition** or **noncompetitive inhibition**. Presumably a mixed inhibitor binds to enzyme sites that participate in both substrate binding and catalysis.

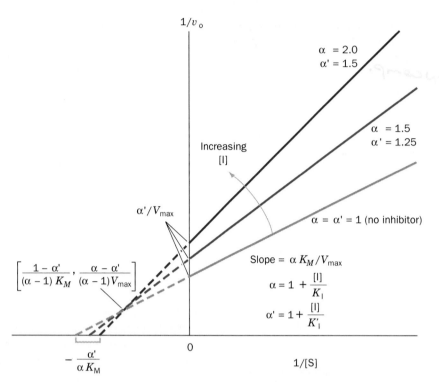

FIGURE 13-13. A Lineweaver–Burk plot of a simple Michaelis–Menten enzyme in the presence of a mixed inhibitor. Note that the lines all intersect to the left of the $1/v_o$ axis. The coordinates of this intersection point are given in brackets. Note that when $K_I = K'_I$, $\alpha = \alpha'$ and the lines interesect on the $1/[S]$ axis at $-1/K_M$.

The Michaelis–Menten equation for mixed inhibition, which is derived in Appendix C of this chapter, is

$$v_o = \frac{V_{max}[S]}{\alpha K_M + \alpha'[S]} \qquad [13.45]$$

where α and α' are defined in Eqs. [13.37] and [13.42], respectively. It can be seen from Eq. [13.45] that the name mixed inhibition arises from the fact that the denominator has the factor α multiplying K_M as in competitive inhibition, (Eq. [13.38]) and the factor α' multiplying [S] as in uncompetitive inhibition (Eq. [13.41]). Mixed inhibitors are therefore effective at both high and low substrate concentrations.

The Lineweaver–Burk equation for mixed inhibition is

$$\frac{1}{v_o} = \left(\frac{\alpha K_M}{V_{max}}\right)\frac{1}{[S]} + \frac{\alpha'}{V_{max}} \qquad [13.46]$$

The plot of this equation consists of lines that have slope $\alpha K_M/V_{max}$ with a $1/v_o$ intercept of α'/V_{max} and a $1/[S]$ intercept of $-\alpha'/\alpha K_M$ (Fig. 13-13). Algebraic manipulation of Eq. [13.46] for different values of [I] reveals that this equation describes a family of lines that intersect to the left of the $1/v_o$ axis (Fig. 13-13); for the special case in which $K_I = K'_I$ ($\alpha = \alpha'$), the intersection is, in addition, on the $1/[S]$ axis.

Table 13-2 provides a summary of the preceding results concerning the inhibition of simple Michaelis–Menten en-

TABLE 13-2. The Effects of Inhibitors on the Parameters of the Michaelis–Menten Equation[a]

Type of Inhibition	V^{app}_{max}	K^{app}_M
None	V_{max}	K_M
Competitive	V_{max}	αK_M
Uncompetitive	V_{max}/α'	K_M/α'
Mixed	V_{max}/α'	$\alpha K_M/\alpha'$

[a] $\alpha = 1 + \dfrac{[I]}{K_I}$ and $\alpha' = 1 + \dfrac{[I]}{K'_I}$

zymes. The quantities K^{app}_M and V^{app}_{max} are the "apparent" values of K_M and V_{max} that would actually be observed in the presence of inhibitor for the Michaelis–Menten equation describing the inhibited enzymes.

4. EFFECTS OF pH

Enzymes, being proteins, have properties that are quite pH sensitive. Most proteins, in fact, are active only within a narrow pH range, typically 5 to 9. This is a result of the effects of pH on a combination of factors: (1) the binding of substrate to enzyme, (2) the catalytic activity of the enzyme, (3) the ionization of substrate, and (4) the variation of protein structure (usually significant only at extremes of pH).

pH Dependence of Simple Michaelis–Menten Enzymes

The initial rates for many enzymatic reactions exhibit bell-shaped curves as a function of pH (e.g., Fig. 13-14). These curves reflect the ionizations of certain amino acid residues that must be in a specific ionization state for enzyme activity. The following model can account for such pH effects.

$$
\begin{array}{ccc}
E^- & & ES^- \\
K_{E2} \big\Vert H^+ & & K_{ES2} \big\Vert H^+ \\
EH + S \underset{k_{-1}}{\overset{k_1}{\rightleftharpoons}} ESH & \overset{k_2}{\longrightarrow} & P + EH \\
K_{E1} \big\Vert H^+ & & K_{ES1} \big\Vert H^+ \\
EH_2^+ & & ESH_2^+
\end{array}
$$

In this expansion of the simple one substrate–no back reaction mechanism, it is assumed that only EH and ESH are catalytically active.

The Michaelis–Menten equation for this model, which is derived in Appendix D, is

$$v_o = \frac{V'_{max}[S]}{K'_M + [S]} \qquad [13.47]$$

Here the apparent Michaelis–Menten parameters are defined

$$V'_{max} = V_{max}/f_2 \qquad \text{and} \qquad K'_M = K_M(f_1/f_2)$$

where

$$f_1 = \frac{[H^+]}{K_{E1}} + 1 + \frac{K_{E2}}{[H^+]}$$

$$f_2 = \frac{[H^+]}{K_{ES1}} + 1 + \frac{K_{ES2}}{[H^+]}$$

and V_{max} and K_M refer to the active forms of the enzyme, EH and ESH. Note that at any given pH, Eq. [13.47] behaves as a simple Michaelis–Menten equation, but because of the pH dependence of f_1 and f_2, v_o varies with pH in a bell-shaped manner (e.g., Fig. 13-14).

Evaluation of Ionization Constants

The ionization constants of enzymes that obey Eq. [13.47] can be evaluated by the analysis of the curves of log V'_{max} versus pH, which provides values of K_{ES1} and K_{ES2} (Fig. 13-15a), and of log (V'_{max}/K'_M) versus pH, which yields K_{E1} and K_{E2} (Fig. 13-15b). This, of course, entails the determination of the enzyme's Michaelis–Menten parameters at each of a series of different pH's.

The measured pK's often provide valuable clues as to the identities of the amino acid residues essential for enzymatic

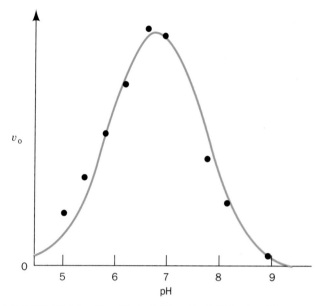

FIGURE 13-14. The effect of pH on the initial rate of the reaction catalyzed by the enzyme fumarase. [After Tanford, C., *Physical Chemistry of Macromolecules, p. 647*, Wiley (1961).]

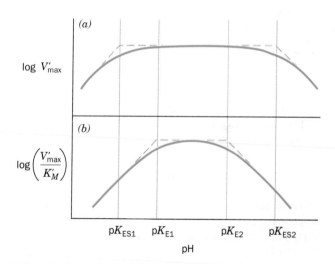

FIGURE 13-15. The pH dependence of (a) log V'_{max} and (b) log (V'_{max}/K'_M) illustrating how the values of the molecular ionization constants can be determined by graphical extrapolation.

activity. For example, a measured pK of ~4 suggests that an Asp or Glu residue is essential to the enzyme. Similarly, pK's of ~6 or ~10 suggest the participation of a His or a Lys residue, respectively. However, a given acid–base group may vary by as much as several pH units from its expected value as a consequence of the electrostatic influence of nearby charged groups, as well as of the proximity of regions of low polarity. For example, the carboxylate group of a Glu residue forming a salt bridge with a Lys residue is stabilized by the nearby positive charge and therefore has a lower pK than it would otherwise have; that is, it is more difficult to protonate. Conversely, a carboxylate group immersed in a region of low polarity is less acidic than normal because it attracts protons more strongly than if it were in a region of higher polarity. The identification of a kinetically characterized pK with a particular amino acid residue must therefore be verified by other types of measurements such as the use of group-specific reagents to inactivate a putative essential residue (Section 6-2).

5. BISUBSTRATE REACTIONS

We have heretofore been concerned with reactions involving enzymes that require only a single substrate. Yet, enzymatic reactions involving two substrates and yielding two products

$$A \ + \ B \ \overset{E}{\rightleftharpoons} \ P \ + \ Q$$

account for ~60% of known biochemical reactions. Almost all of these so-called **bisubstrate reactions** are either **transferase** reactions in which the enzyme catalyzes the transfer of a specific functional group, X, from one of the substrates to the other:

$$P - X \ + \ B \ \overset{E}{\rightleftharpoons} \ P \ + \ B - X$$

or oxidation–reduction reactions in which reducing equivalents are transferred between the two substrates. For example, the hydrolysis of a peptide bond by trypsin (Section 6-1E) is the transfer of the peptide carbonyl group from the peptide nitrogen atom to water (Fig. 13-16a). Similarly, in the alcohol dehydrogenase reaction (Section 12-2A), a hydride ion is formally transferred from ethanol to NAD$^+$ (Fig. 13-16b). Although such bisubstrate reactions could, in principle, occur through a vast variety of mechanisms, only a few types are commonly observed.

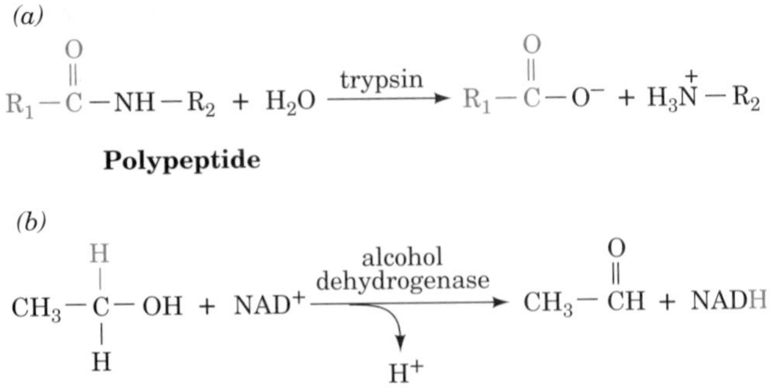

A. Terminology

We shall follow the nomenclature system introduced by W. W. Cleland for representing enzymatic reactions:

1. Substrates are designated by the letters A, B, C, and D *in the order that they add to the enzyme.*

2. Products are designated P, Q, R, and S *in the order that they leave the enzyme.*

3. Stable enzyme forms are designated E, F, and G with E being the free enzyme, if such distinctions can be made. A stable enzyme form is defined as one that by itself is incapable of converting to another stable enzyme form (see below).

4. The number of reactants and products in a given reaction are specified, in order, by the terms **Uni** (one), **Bi** (two), **Ter** (three), and **Quad** (four). A reaction requiring

FIGURE 13-16. Some bisubstrate reactions: (*a*) In the peptide hydrolysis reaction catalyzed by trypsin, the peptide carbonyl group, with its pendent polypeptide chain, is transferred from the peptide nitrogen atom to a water molecule. (*b*) In the alcohol dehydrogenase reaction, a hydride ion is formally transferred from ethanol to NAD$^+$.

one substrate and yielding three products is designated a Uni Ter reaction. In this section, we shall be concerned with reactions that require two substrates and yield two products, that is, Bi Bi reactions. Keep in mind, however, that there are numerous examples of even more complex reactions.

Types of Bi Bi Reactions

Enzyme-catalyzed group-transfer reactions fall under two major mechanistic classifications:

1. Sequential Reactions

*Reactions in which all substrates must combine with the enzyme before a reaction can occur and products be released are known as **Sequential reactions**.* In such reactions, the group being transferred, X, is directly passed from A (= P—X) to B yielding P and Q (= B—X). Hence, such reactions are also called **single displacement reactions**.

Sequential reactions can be subclassified into those with a compulsory order of substrate addition to the enzyme, which are said to have an **Ordered mechanism,** and those with no preference for the order of substrate addition, which are described as having a **Random mechanism.** In the Ordered mechanism, the binding of the first substrate is apparently required for the enzyme to form the binding site for the second substrate, whereas for the Random mechanism, both binding sites are present on the free enzyme.

Let us describe enzymatic reactions using Cleland's shorthand notation. The enzyme is represented by a horizontal line and successive additions of substrates and release of products are denoted by vertical arrows. Enzyme forms are placed under the line and rate constants, if given, are to the left of the arrow or on top of the line for forward reactions. An **Ordered Bi Bi** reaction is represented:

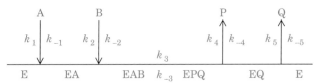

where A and B are said to be the **leading** and **following** substrates, respectively. Here, only minimal details are given concerning the interconversions of intermediate enzyme forms because, as we have seen for reversible single-substrate enzymes, steady state kinetic measurements provide no information concerning the number of intermediates in a given reaction step. Many NAD^+- and $NADP^+$-requiring dehydrogenases follow an Ordered Bi Bi mechanism in which the coenzyme is the leading reactant.

A **Random Bi Bi** reaction is diagrammed:

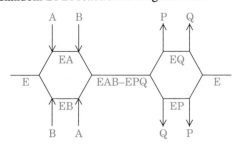

Some dehydrogenases and kinases operate through Random Bi Bi mechanisms.

2. Ping Pong Reactions

*Mechanisms in which one or more products are released before all substrates have been added are known as **Ping Pong reactions**.* The **Ping Pong Bi Bi** reaction is represented by

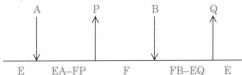

In it, a functional group X of the first substrate A ($=P-X$) is displaced from the substrate by the enzyme E to yield the first product P and a stable enzyme form F ($=E-X$) in which X is tightly (often covalently) bound to the enzyme (Ping). In the second stage of the reaction, X is displaced from the enzyme by the second substrate B to yield the second product Q ($=B-X$), thereby regenerating the original form of the enzyme, E (Pong). Such reactions are therefore also known as **double-displacement reactions**. *Note that in Ping Pong Bi Bi reactions, the substrates A and B do not encounter one another on the surface of the enzyme.* Many enzymes, including chymotrypsin (Section 14-3), transaminases, and some flavoenzymes, react with Ping Pong mechanisms.

B. Rate Equations

Steady state kinetic measurements can be used to distinguish among the foregoing bisubstrate mechanisms. In order to do so, one must first derive their rate equations. This can be done in much the same manner as for single-substrate enzymes, that is, solving a set of simultaneous linear equations consisting of an equation expressing the steady state condition for each kinetically distinct enzyme complex plus one equation representing the conservation condition for the enzyme. This, of course, is a more complex undertaking for bisubstrate enzymes than it is for single-substrate enzymes.

The rate equations for the above described bisubstrate mechanisms in the absence of products are given below in double reciprocal form.

Ordered Bi Bi

$$\frac{1}{v_o} = \frac{1}{V_{max}} + \frac{K_M^A}{V_{max}[A]} + \frac{K_M^B}{V_{max}[B]} + \frac{K_S^A K_M^B}{V_{max}[A][B]} \quad [13.48]$$

Rapid Equilibrium Random Bi Bi

The rate equation for the general Random Bi Bi reaction is quite complicated. However, in the special case that both substrates are in rapid and independent equilibrium with the enzyme; that is, the EAB–EPQ interconversion is rate determining, the initial rate equation reduces to the following relatively simple form. This mechanism is known as the

Rapid Equilibrium Random Bi Bi mechanism:

$$\frac{1}{v_o} = \frac{1}{V_{max}} + \frac{K_S^A K_M^B}{V_{max} K_S^B [A]} + \frac{K_M^B}{V_{max}[B]} + \frac{K_S^A K_M^B}{V_{max}[A][B]} \quad [13.49]$$

Ping Pong Bi Bi

$$\frac{1}{v_o} = \frac{K_M^A}{V_{max}[A]} + \frac{K_M^B}{V_{max}[B]} + \frac{1}{V_{max}} \quad [13.50]$$

Physical Significance of the Bisubstrate Kinetic Parameters

The kinetic parameters in the equations describing bisubstrate reactions have meanings similar to those for single-substrate reactions. V_{max} is the maximal velocity of the enzyme obtained when both A and B are present at saturating concentrations, K_M^A and K_M^B are the respective concentrations of A and B necessary to achieve $\frac{1}{2}V_{max}$ in the presence of a saturating concentration of the other, and K_S^A and K_S^B are the respective dissociation constants of A and B from the enzyme, E.

C. Differentiating Bisubstrate Mechanisms

One can discriminate between Ping Pong and Sequential mechanisms from their contrasting properties in linear plots such as those of the Lineweaver–Burk type.

(a)

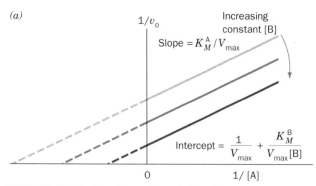

(b)

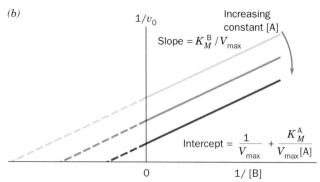

FIGURE 13-17. Double-reciprocal plots for an enzymatic reaction with a Ping Pong Bi Bi mechanism. (*a*) Plots of $1/v_o$ versus $1/[A]$ at various constant concentrations of B. (*b*) Plots of $1/v_o$ versus $1/[B]$ at various constant concentrations of A.

Diagnostic Plot for Ping Pong Bi Bi Reactions

A plot of $1/v_o$ versus $1/[A]$ at constant [B] for Eq. [13.50] yields a straight line of slope K_M^A/V_{max} and an intercept on the $1/v_o$ axis equal to the last two terms in Eq. [13.50]. Since the slope is independent of [B], such plots for different values of [B] yield a family of parallel lines (Fig. 13-17). A plot of $1/v_o$ versus $1/[B]$ for different values of [A] likewise yields a family of parallel lines. *Such parallel lines are diagnostic for a Ping Pong mechanism.*

Diagnostic Plot for Sequential Bi Bi Reactions

The equations representing the Ordered Bi Bi mechanism (Eq. [13.48]) and the Rapid Equilibrium Random Bi Bi mechanism (Eq. [13.49]) have identical functional dependence on [A] and [B].

Equation [13.48] can be rearranged to

$$\frac{1}{v_o} = \frac{K_M^A}{V_{max}}\left(1 + \frac{K_S^A K_M^B}{K_M^A[B]}\right)\frac{1}{[A]} + \frac{1}{V_{max}}\left(1 + \frac{K_M^B}{[B]}\right) \quad [13.51]$$

Thus plotting $1/v_o$ versus $1/[A]$ for constant [B] yields a

linear plot with a slope equal to the coefficient of $1/[A]$ and an intercept on the $1/v_o$ axis equal to the second term of Eq. [13.51] (Fig. 13-18*a*). Alternatively, Eq. [13.48] can be rearranged to

$$\frac{1}{v_o} = \frac{K_M^B}{V_{max}}\left(1 + \frac{K_S^A}{[A]}\right)\frac{1}{[B]} + \frac{1}{V_{max}}\left(1 + \frac{K_M^A}{[A]}\right) \quad [13.52]$$

which yields a linear plot of $1/v_o$ versus $1/[B]$ for constant [A] with a slope equal to the coefficient of $1/[B]$ and an intercept on the $1/v_o$ axis equal to the second term of Eq. [13.52] (Fig. 13-18*b*). *The characteristic feature of these plots, which is indicative of a Sequential mechanism, is that the lines intersect to the left of the $1/v_o$ axis.*

Differentiating Random and Ordered Sequential Mechanisms

The Ordered Bi Bi mechanism may be experimentally distinguished from the Random Bi Bi mechanism through **product inhibition studies.** If only one product of the reaction, P or Q, is added to the reaction mixture, the reverse reaction still cannot occur. Nevertheless, by binding to the

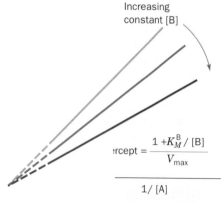

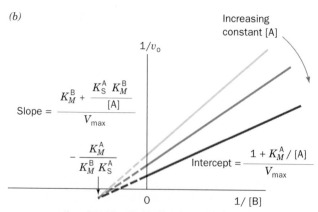

... s of an enzymatic ...m. (*a*) Plots of $1/v_o$ ...tions of B. (*b*) Plots of ...entrations of A. The

corresponding plots for Rapid Equilibrium Random Bi Bi reactions have identical appearances; their lines all intersect to the left of the $1/v_o$ axis.

TABLE 13-3. PATTERNS OF PRODUCT INHIBITION FOR SEQUENTIAL BISUBSTRATE MECHANISMS

Mechanism	Product Inhibitor	A Variable	B Variable
Ordered Bi Bi	P	Mixed	Mixed
	Q	Competitive	Mixed
Rapid Equilibrium Random Bi Bi	P	Competitive	Competitive
	Q	Competitive	Competitive

enzyme, this product will inhibit the forward reaction. For an Ordered Bi Bi reaction, Q (= B—X, the second product to be released) directly competes with A (= P—X, the leading substrate) for binding to E and hence is a competitive inhibitor of A when [B] is fixed (the presence of X in Q = B—X interferes with the binding of A = P—X). However, since B combines with EA, not E, Q is a mixed inhibitor of B when [A] is fixed (Q interferes with both the binding of B to enzyme and with the catalysis of the reaction). Similarly, P, which combines only with EQ, is a mixed inhibitor of A when [B] is held constant and of B when [A] is held constant. In contrast, in a Rapid Equilibrium Bi Bi reaction, since both products as well as both substrates can combine directly with E, both P and Q are competitive inhibitors of A when [B] is constant and of B when [A] is constant. These product inhibition patterns are summarized in Table 13-3.

D. Isotope Exchange

Mechanistic conclusions based on kinetic analyses alone are fraught with uncertainties and are easily confounded by inaccurate experimental data. A particular mechanism for an enzyme is therefore greatly corroborated if the mechanism can be shown to conform to experimental criteria other than kinetic analysis.

*Sequential (single-displacement) and Ping Pong (double-displacement) bisubstrate mechanisms may be differentiated through the use of **isotope exchange** studies.* Double-displacement reactions are capable of exchanging an isotope from the first product P back to the first substrate A in the absence of the second substrate. Consider an overall Ping Pong reaction catalyzed by the bisubstrate enzyme E

$$P—X + B \xrightleftharpoons{E} P + B—X$$

in which, as usual, A = P—X, Q = B—X, and X is the group that is transferred from one substrate to the other in the course of the reaction. Only the first step of the reaction can take place in the absence of B. If a small amount of isotopically labeled P, denoted P*, is added to this reaction mixture then, in the reverse reaction, P*—X will form:

Forward reaction $E + P—X \longrightarrow E—X + P$
Reverse reaction $E—X + P* \longrightarrow E + P*—X$

that is, isotopic exchange will occur.

In contrast, let us consider the first step of a Sequential reaction. Here a noncovalent enzyme–substrate complex forms:

$$E + P—X \rightleftharpoons E \cdot P—X$$

Addition of P* cannot result in an exchange reaction because no covalent bonds are broken in the formation of E·P—X, that is, there is no P released from the enzyme to exchange with P*. The demonstration of isotopic exchange for a bisubstrate enzyme is therefore convincing evidence favoring a Ping Pong mechanism.

Isotope Exchange in Sucrose Phosphorylase and Maltose Phosphorylase

The enzymes **sucrose phosphorylase** and **maltose phosphorylase** provide two clearcut examples of enzymatically catalyzed isotopic exchange reactions. Sucrose phosphorylase catalyzes the overall reaction

Glucose—fructose + phosphate
Sucrose

$\Updownarrow$ E

Glucose-1-phosphate + fructose

If the enzyme is incubated with sucrose and isotopically labeled fructose in the absence of phosphate, it is observed that the label passes into the sucrose:

Glucose— fructose + fructose*
Sucrose

$\Updownarrow$ E

Glucose—fructose* + fructose

For the reverse reaction, if the enzyme is incubated with glucose-1-phosphate and ^{32}P-labeled phosphate, this label exchanges into the glucose-1-phosphate:

Glucose-1-phosphate + phosphate*

$\Updownarrow$ E

Glucose-1-phosphate* + phosphate

These observations indicate that a tight glucosyl–enzyme complex is formed with the release of fructose, thereby establishing that the sucrose phosphorylase reaction occurs via a Ping Pong mechanism. This finding has been conclusively corroborated by the isolation and characterization of this glucosyl–enzyme complex.

The enzyme **maltose phosphorylase** catalyzes a similar overall reaction:

$$\text{Glucose} - \text{glucose} \; + \; \text{phosphate}$$
$$\textbf{Maltose}$$
$$\Updownarrow \text{E}$$
$$\text{Glucose-1-phosphate} \; + \; \text{glucose}$$

In contrast to sucrose phosphorylase, however, it does not catalyze isotopic exchange between glucose-1-phosphate and [^{32}P]phosphate or between maltose and [^{14}C]glucose. Likewise, a glucosyl–enzyme complex has not been detected. This evidence is consistent with maltose phosphorylase having a sequential mechanism.

Appendix: Derivations of Michaelis–Menten Equation Variants

A. The Michaelis–Menten Equation for Reversible Reactions—Equation [13.30]

The conservation condition for the reversible reaction with one intermediate (Section 13-2C) is

$$[\text{E}]_\text{T} = [\text{E}] + [\text{ES}] \qquad [13.\text{A}1]$$

The steady state condition is

$$\frac{d[\text{ES}]}{dt} = k_1[\text{E}][\text{S}] + k_{-2}[\text{E}][\text{P}] - (k_{-1} + k_2)[\text{ES}] = 0$$
$$[13.\text{A}2]$$

so that

$$[\text{E}] = \left(\frac{k_{-1} + k_2}{k_1[\text{S}] + k_{-2}[\text{P}]}\right)[\text{ES}] \qquad [13.\text{A}3]$$

Substituting this result into Eq. [13.A1] yields

$$[\text{E}]_\text{T} = \left(\frac{k_{-1} + k_2}{k_1[\text{S}] + k_{-2}[\text{P}]} + 1\right)[\text{ES}] \qquad [13.\text{A}4]$$

The velocity of the reaction is expressed

$$v = -\frac{d[\text{S}]}{dt} = k_1[\text{E}][\text{S}] - k_{-1}[\text{ES}] \qquad [13.\text{A}5]$$

which can be combined with Eq. [13.A3] to give

$$v = \left(\frac{k_1[\text{S}](k_{-1} + k_2)}{k_1[\text{S}] + k_{-2}[\text{P}]} - k_{-1}\right)[\text{ES}] \qquad [13.\text{A}6]$$

which, in turn, is combined with Eq. [13.A4] to yield

$$v = \left(\frac{k_1 k_2[\text{S}] - k_{-1}k_{-2}[\text{P}]}{k_{-1} + k_2 + k_1[\text{S}] + k_{-2}[\text{P}]}\right)[\text{E}]_\text{T} \qquad [13.\text{A}7]$$

Dividing the numerator and denominator of this equation by $(k_{-1} + k_2)$ results in

$$v = \left(\frac{k_2\left(\dfrac{k_1}{k_{-1} + k_2}\right)[\text{S}] - k_{-1}\left(\dfrac{k_{-2}}{k_{-1} + k_2}\right)[\text{P}]}{1 + \left(\dfrac{k_1}{k_{-1} + k_2}\right)[\text{S}] + \left(\dfrac{k_{-2}}{k_{-1} + k_2}\right)[\text{P}]}\right)[\text{E}]_\text{T}$$
$$[13.\text{A}8]$$

Then, if we define the following parameters analogously with the constants of the Michaelis–Menten equation (Eqs. [13.23] and [13.21]),

$$V^f_\text{max} = k_2[\text{E}]_\text{T} \qquad V^r_\text{max} = k_{-1}[\text{E}]_\text{T}$$
$$K^\text{S}_M = \frac{k_{-1} + k_2}{k_1} \qquad K^\text{P}_M = \frac{k_{-1} + k_2}{k_{-2}}$$

we obtain the Michaelis–Menten equation for a reversible, one-intermediate reaction:

$$v = \frac{\dfrac{V^f_\text{max}[\text{S}]}{K^\text{S}_M} - \dfrac{V^r_\text{max}[\text{P}]}{K^\text{P}_M}}{1 + \dfrac{[\text{S}]}{K^\text{S}_M} + \dfrac{[\text{P}]}{K^\text{P}_M}} \qquad [13.30]$$

B. Michaelis–Menten Equation for Uncompetitive Inhibition—Equation [13.41]

For uncompetitive inhibition (Section 13-3B), the inhibitor binds to the Michaelis complex with dissociation constant

$$K'_\text{I} = \frac{[\text{ES}][\text{I}]}{[\text{ESI}]} \qquad [13.\text{A}9]$$

The conservation condition is

$$[\text{E}]_\text{T} = [\text{E}] + [\text{ES}] + [\text{ESI}] \qquad [13.\text{A}10]$$

Substituting in Eqs. [13.34] and [13.A9] yields

$$[\text{E}]_\text{T} = [\text{ES}]\left(\frac{K_M}{[\text{S}]} + 1 + \frac{[\text{I}]}{K'_\text{I}}\right) \qquad [13.\text{A}11]$$

Defining α' similarly to Eq. [13.37] as

$$\alpha' = 1 + \frac{[\text{I}]}{K'_\text{I}} \qquad [13.\text{A}12]$$

and v_0 and V_max as in Eqs. [13.22] and [13.23], respectively,

$$v_0 = k_2[\text{ES}] = \frac{V_\text{max}}{\dfrac{K_M}{[\text{S}]} + \alpha'} \qquad [13.\text{A}13]$$

which upon rearrangement yields the Michaelis–Menten equation for uncompetitive inhibition:

$$v_o = \frac{V_{max}[S]}{K_M + \alpha'[S]} \qquad [13.41]$$

C. The Michaelis–Menten Equation for Mixed Inhibition—Equation [13.45]

In mixed inhibition (Section 13-3C), the inhibitor-binding steps have different dissociation constants:

$$K_I = \frac{[E][I]}{[EI]} \quad \text{and} \quad K_I' = \frac{[ES][I]}{[ESI]} \qquad [13.A14]$$

(Here, for the sake of mathematical simplicity, we are making the thermodynamically unsupportable assumption that EI does not react with S to form ESI. Inclusion of this reaction requires a more complex derivation than that given here but leads to results that are substantially the same.) The conservation condition for this reaction scheme is

$$[E]_T = [E] + [EI] + [ES] + [ESI] \qquad [13.A15]$$

so that substituting in Eqs. [13.A14]

$$[E]_T = [E]\left(1 + \frac{[I]}{K_I}\right) + [ES]\left(1 + \frac{[I]}{K_I'}\right) \qquad [13.A16]$$

Defining α and α' as in Eqs. [13.38] and [13.A12], respectively, Eq. [13.A16] becomes

$$[E]_T = [E]\alpha + [ES]\alpha' \qquad [13.A17]$$

Then substituting in Eq. [13.34]

$$[E]_T = [ES]\left(\frac{\alpha K_M}{[S]} + \alpha'\right) \qquad [13.A18]$$

Defining v_o and V_{max} as in Eqs. [13.22] and [13.23] results in the Michaelis–Menten equation for mixed inhibition:

$$v_o = \frac{V_{max}[S]}{\alpha K_M + \alpha'[S]} \qquad [13.45]$$

D. The Michaelis–Menten Equation for Ionizable Enzymes—Equation [13.47]

In the model presented in Section 13-4 to account for the effect of pH on enzymes, the dissociation constants for the ionizations are

$$K_{E2} = \frac{[H^+][E^-]}{[EH]} \qquad K_{ES2} = \frac{[H^+][ES^-]}{[ESH]}$$

$$\qquad\qquad\qquad\qquad\qquad\qquad\qquad [13.A19]$$

$$K_{E1} = \frac{[H^+][EH]}{[EH_2^+]} \qquad K_{ES1} = \frac{[H^+][ESH]}{[ESH_2^+]}$$

Protonation and deprotonation are among the fastest

known reactions so that, with the exception of the few enzymes with extremely high turnover numbers, it can be reasonably assumed that all acid–base reactions are at equilibrium. The conservation condition is

$$[E]_T = [ESH]_T + [EH]_T \qquad [13.A20]$$

where $[E]_T$ is the total enzyme present in any form,

$$[EH]_T = [EH_2^+] + [EH] + [E^-]$$

$$= [EH]\left(\frac{[H^+]}{K_{E1}} + 1 + \frac{K_{E2}}{[H^+]}\right) = [EH]f_1 \quad [13.A21]$$

and

$$[ESH]_T = [ESH_2^+] + [ESH] + [ES^-]$$

$$= [ESH]\left(\frac{[H^+]}{K_{ES1}} + 1 + \frac{K_{ES2}}{[H^+]}\right) = [ESH]f_2 \quad [13.A22]$$

Then making the steady state assumption

$$\frac{d[ESH]}{dt} = k_1[EH][S] - (k_{-1} + k_2)[ESH] = 0 \qquad [13.A23]$$

and solving for [EH]

$$[EH] = \frac{(k_{-1} + k_2)[ESH]}{k_1[S]} = \frac{K_M[ESH]}{[S]} \qquad [13.A24]$$

Therefore, from Eq. [13.A21],

$$[EH]_T = \frac{K_M[ESH]f_1}{[S]} \qquad [13.A25]$$

which, together with Eqs. [13.A20] and [13.A22], yields

$$[E]_T = [ESH]\left(\frac{K_M f_1}{[S]} + f_2\right) \qquad [13.A26]$$

As in the simple Michaelis–Menten derivation, the initial rate is

$$v_o = k_2[ESH] = \frac{k_2[E]_T}{\left(\frac{K_M f_1}{[S]}\right) + f_2} = \frac{(k_2/f_2)[E]_T[S]}{K_M(f_1/f_2) + [S]} \quad [13.A27]$$

Then defining the "apparent" values of K_M and $V_{max} = k_2[E]_T$ at a given pH:

$$K_M' = K_M(f_1/f_2) \qquad [13.A28]$$

and

$$V_{max}' = V_{max}/f_2 \qquad [13.A29]$$

the Michaelis–Menten equation modified to account for pH effects is

$$v_o = \frac{V_{max}'[S]}{K_M' + [S]} \qquad [13.47]$$

CHAPTER SUMMARY

Complicated reaction processes occur through a series of elementary reaction steps defined as having a molecularity equal to the number of molecules that simultaneously collide to form products. The order of a reaction can be determined from the characteristic functional form of its progress curve. Transition state theory postulates that the rate of a reaction depends on the free energy of formation of its activated complex. This complex, which occurs at the free energy maximum of the reaction coordinate, is poised between reactants and products and is therefore also known as the transition state. Transition state theory explains that catalysis results from the reduction of the free energy difference between the reactants and the transition state.

In the simplest enzymatic mechanism, the enzyme and substrate reversibly combine to form an enzyme–substrate complex known as the Michaelis complex, which may irreversibly decompose to form product and the regenerated enzyme. The rate of product formation is expressed by the Michaelis–Menten equation, which is derived under the assumption that the concentration of the Michaelis complex is constant, that is, at a steady state. The Michaelis–Menten equation, which has the functional form of a rectangular hyperbola, has two parameters: $V_{\max}$, the maximal rate of the reaction, which occurs when the substrate concentration is saturating, and K_M, the Michaelis constant, which has the value of the substrate concentration at the half-maximal reaction rate. These parameters may be graphically determined using the Lineweaver–Burk plot. Physically more realistic models of enzyme mechanisms than the Michaelis–Menten model assume the enzymatic reaction to be reversible and to have one or more intermediates. The functional form of the equations describing the reaction rates for these models is independent of their number of intermediates so that the models cannot be differentiated using only steady state kinetic measurements.

Enzymes may be inhibited by competitive inhibitors, which compete with the substrate for the enzymatic binding site. The effect of a competitive inhibitor may be reversed by increasing the substrate concentration. An uncompetitive inhibitor inactivates a Michaelis complex upon binding to it. The maximal rate of an uncompetitively inhibited enzyme is a function of inhibitor concentration and therefore the effect of an uncompetitive inhibitor cannot be reversed by increasing substrate concentration. In mixed inhibition, the inhibitor binds to both the enzyme and the enzyme–substrate complex to form a complex that is catalytically inactive. The rate equation describing this situation has characteristics of both competitive and uncompetitive reactions.

The rate of an enzymatic reaction is a function of hydrogen ion concentration. At any pH, the rate of a simple enzymatic reaction can be described by the Michaelis–Menten equation. However, its parameters $V_{\max}$ and K_M vary with pH. By the evaluation of kinetic rate curves as a function of pH, the pK's of an enzyme's ionizable binding and catalytic groups can be determined, which may help identify these groups.

The majority of enzymatic reactions are bisubstrate reactions in which two substrates react to form two products. Bisubstrate reactions may have Ordered or Random Sequential mechanisms or Ping Pong Bi Bi mechanisms, among others. The initial rate equations for any of these mechanisms involve five parameters, which are analogous to either Michaelis–Menten equation parameters or equilibrium constants. The various bisubstrate mechanisms may be experimentally differentiated according to the forms of their double-reciprocal plots and from the nature of their product inhibition patterns. Isotope exchange reactions provide an additional, nonkinetic method of differentiating bisubstrate mechanisms.

REFERENCES

Chemical Kinetics

Atkins, P.W., *Physical Chemistry* (5th. ed.), Chapters 25–27, Freeman (1994). [Most physical chemistry textbooks have similar coverage.]

Frost, A.A. and Pearson, R.G., *Kinetics and Mechanism* (2nd ed.), Wiley (1961). [A good introduction to chemical kinetics.]

Hammes, G.G., *Principles of Chemical Kinetics,* Academic Press (1978).

Laidler, K.J., *Theories of Chemical Reaction Rates,* McGraw–Hill (1969).

Enzyme Kinetics

Cleland, W.W., Steady state kinetics, *in* Boyer, P.D. (Ed.), *The Enzymes* (3rd ed.), Vol. 2, *pp.* 1–65, Academic Press (1970); *and* Steady-state kinetics, *in* Sigman, D.S. and Boyer, P.D. (Eds.), *The Enzymes* (3rd. ed.), Vol. 19, *pp.* 99–158, Academic Press (1990).

Cleland, W.W., Determining the mechanism of enzyme-catalyzed reactions by kinetic studies, *Adv. Enzymol.* **45,** 273 (1977).

Cornish-Bowden, A. and Wharton, C.W., *Enzyme Kinetics,* IRL Press (1988).

Cornish-Bowden, A., *Fundamentals of Enzyme Kinetics,* Butterworths (1979). [A lucid and detailed account of enzyme kinetics.]

Dixon, M. and Webb, E.C., *Enzymes* (3rd ed.), Chapter IV, Academic Press (1979). [An almost exhaustive treatment of enzyme kinetics.]

Fersht, A., *Enzyme Structure and Mechanism* (2nd ed.), Chapters 2–7, Freeman (1985).

Hammes, G.G., *Enzyme Catalysis and Regulation,* Chapter 3, Academic Press (1982).

Knowles, J.R., The intrinsic pK_a-values of functional groups in enzymes: Improper deductions from the pH-dependence of steady state parameters, *CRC Crit. Rev. Biochem.* **4,** 165 (1976).

Kuby, S.A., *A Study of Enzymes,* Vol. I, CRC Press (1991). [Contains several chapters on enzyme kinetics.]

Piszkiewicz, D., *Kinetics of Chemical and Enzyme Catalyzed Reactions,* Oxford University Press (1977). [A highly readable discussion of enzyme kinetics.]

Purich, D.L. (Ed.), Enzyme kinetics and mechanisms, *Methods Enzymol.* **63** and **64** (1979). [A collection of articles on advanced topics.]

Segel, I.H., *Enzyme Kinetics,* Wiley (1975). [A detailed and understandable treatise providing full explanations of many aspects of enzyme kinetics.]

Tinoco, I., Jr., Sauer, K., and Wang, J.C., *Physical Chemistry. Principles and Applications for Biological Sciences* (2nd ed.), Chapters 7 and 8, Prentice–Hall (1985).

Wood, W.B., Wilson, J.H., Benbow, R.M., and Hood, L.E., *Biochemistry. A Problems Approach* (2nd ed.), Chapter 8, Benjamin/Cummings (1981). [Contains instructive problems on enzyme kinetics with answers worked out in detail.]

PROBLEMS

1. The hydrolysis of sucrose:

$$\text{Sucrose} + H_2O \longrightarrow \text{glucose} + \text{fructose}$$

takes the following time course.

Time (min)	[Sucrose] (M)
0	0.5011
30	0.4511
60	0.4038
90	0.3626
130	0.3148
180	0.2674

Determine the first-order rate constant and the half-life of the reaction. Why does this bimolecular reaction follow a first-order rate law? How long will it take to hydrolyze 99% of the sucrose initially present? How long will it take if the amount of sucrose initially present is twice that given in the table?

2. By what factor will a reaction at 25°C be accelerated if a catalyst reduces the free energy of its activated complex by $1 \text{ kJ} \cdot \text{mol}^{-1}$; by $10 \text{ kJ} \cdot \text{mol}^{-1}$?

3. For a Michaelis–Menten reaction, $k_1 = 5 \times 10^7 M^{-1} \text{ s}^{-1}$, $k_{-1} = 2 \times 10^4 \text{ s}^{-1}$, and $k_2 = 4 \times 10^2 \text{ s}^{-1}$. Calculate K_S and K_M for this reaction. Does substrate binding achieve equilibrium or the steady state?

*4. The following table indicates the rates at which a substrate reacts as catalyzed by an enzyme that follows the Michaelis–Menten mechanism: (1) in the absence of inhibitor; (2) and (3) in the presence of 10 mM concentration, respectively, of each of two inhibitors. Assume $[E]_T$ is the same for all reactions.

[S] (mM)	(1) v_0 ($\mu M \cdot s^{-1}$)	(2) v_0 ($\mu M \cdot s^{-1}$)	(3) v_0 ($\mu M \cdot s^{-1}$)
1	2.5	1.17	0.77
2	4.0	2.10	1.25
5	6.3	4.00	2.00
10	7.6	5.7	2.50
20	9.0	7.2	2.86

(a) Determine K_M and V_{max} for the enzyme. For each inhibitor determine the type of inhibition and K_I and/or K_I'. What additional information would be required to calculate the turnover number of the enzyme? (b) For [S] = 5 mM, what fraction of the enzyme molecules have a bound substrate in the absence of inhibitor, in the presence of 10-mM inhibitor of type (2), and in the presence of 10-mM inhibitor of type (3)?

*5. Ethanol in the body is oxidized to acetaldehyde by liver alcohol dehydrogenase (LADH). Other alcohols are also oxidized by LADH. For example, methanol, which is mildly intoxicating, is oxidized by LADH to the quite toxic product formaldehyde. The toxic effects of ingesting methanol (a component of many commercial solvents) can be reduced by administering ethanol. The ethanol acts as a competitive inhibitor of the methanol by displacing it from LADH. This provides sufficient time for the methanol to be harmlessly excreted by the kidneys. If an individual has ingested 100 mL of methanol (a lethal dose), how much 100 proof whiskey (50% ethanol by volume) must he imbibe to reduce the activity of his LADH towards methanol to 5% of its original value? The adult human body contains ~40 L of aqueous fluids throughout which ingested alcohols are rapidly and uniformly mixed. The densities of ethanol and methanol are both 0.79 g·cm^{-3}. Assume the K_M values of LADH for ethanol and methanol to be $1.0 \times 10^{-3}M$ and $1.0 \ 10^{-2}M$, respectively, and that $K_I = K_M$ for ethanol.

6. The K_M of a Michaelis–Menten enzyme for a substrate is $1.0 \times 10^{-4}M$. At a substrate concentration of $0.2M$, $v_0 = 43 \ \mu M \cdot \text{min}^{-1}$ for a certain enzyme concentration. However, with a substrate concentration of $0.02M$, v_0 has the same value. (a) Using numerical calculations, show that this observation is accurate. (b) What is the best range of [S] for measuring K_M?

7. Why are uncompetitive and mixed inhibitors generally considered to be more effective *in vivo* (in a living organism) than competitive inhibitors?

8. Explain why an exact fit to a kinetic model of the experimental parameters describing a reaction does not prove that the reaction follows the model.

9. An enzyme that follows the model for pH effects presented in Section 13-4 has $pK_{ES1} = 4$ and $pK_{ES2} = 8$. What is the pH at which V_{max}' is a maximum for this enzyme? What fraction of V_{max} does V_{max}' achieve at this pH?

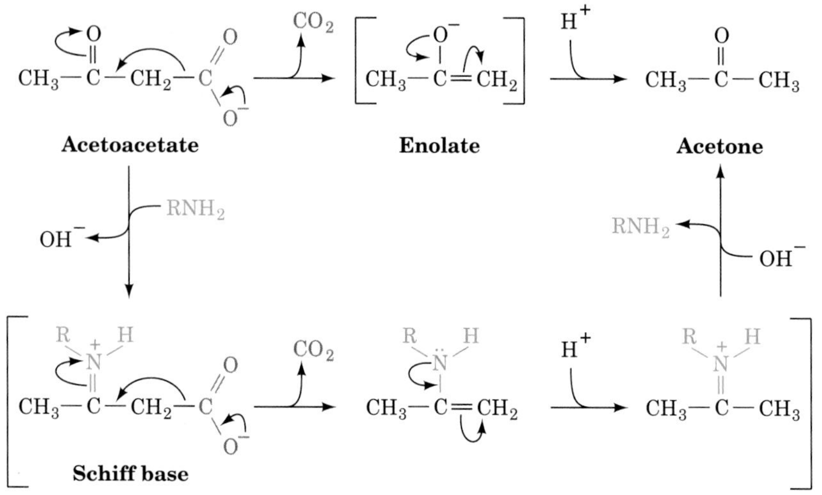

FIGURE 14-4. The uncatalyzed reaction mechanism for the decarboxylation of acetoacetate (*top*) and the reaction mechanism as catalyzed by primary amines (*bottom*).

α-amino group. The covalent intermediate, in this case, has been isolated through $NaBH_4$ reduction of its imine bond to an amine, thereby irreversibly inhibiting the enzyme. Other enzyme functional groups that participate in covalent catalysis include the imidazole moiety of His, the thiol group of Cys, the carboxyl function of Asp, and the hydroxyl group of Ser. In addition, several coenzymes, most notably **thiamine pyrophosphate** (Section 16-3B) and **pyridoxal phosphate** (Section 24-1A), function in association with their apoenzymes mainly as covalent catalysts.

C. Metal Ion Catalysis

Nearly one third of all known enzymes require the presence of metal ions for catalytic activity. There are two classes of metal ion-requiring enzymes that are distinguished by the strengths of their ion–protein interactions:

1. *Metalloenzymes* contain tightly bound metal ions, most commonly transition metal ions such as Fe^{2+}, Fe^{3+}, Cu^{2+}, Zn^{2+}, Mn^{2+}, or Co^{3+}.

2. *Metal-activated enzymes* loosely bind metal ions from solution, usually the alkali and alkaline earth metal ions Na^+, K^+, Mg^{2+}, or Ca^{2+}.

Metal ions participate in the catalytic process in three major ways:

1. By binding to substrates so as to orient them properly for reaction.

2. By mediating oxidation–reduction reactions through reversible changes in the metal ion's oxidation state.

3. By electrostatically stabilizing or shielding negative charges.

In this section we shall be mainly concerned with the third aspect of metal ion catalysis. The other forms of enzyme-mediated metal ion catalysis are considered in later chapters in conjunction with discussions of specific enzyme mechanisms.

Metal Ions Promote Catalysis through Charge Stabilization

In many metal ion–catalyzed reactions, the metal ion acts in much the same way as a proton to neutralize negative charge, that is, it acts as a Lewis acid. Yet, *metal ions are often much more effective catalysts than protons because metal ions can be present in high concentrations at neutral pH's and can have charges $> +1$.* Metal ions have therefore been dubbed "superacids."

The decarboxylation of **dimethyloxaloacetate,** as catalyzed by metal ions such as Cu^{2+} and Ni^{2+}, is a nonenzymatic example of catalysis by a metal ion:

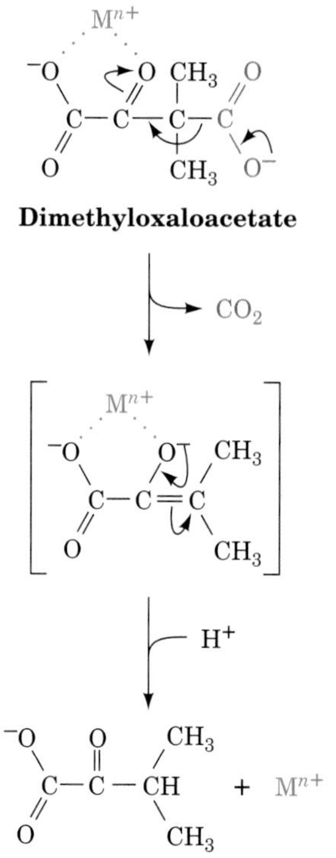

Here the metal ion (M^{n+}), which is chelated by the dimethyloxaloacetate, electrostatically stabilizes the developing enolate ion of the transition state. This mechanism is supported by the observation that acetoacetate, which cannot form such a chelate, is not subject to metal ion–catalyzed decarboxylation. Most enzymes that decarboxylate oxaloacetate require a metal ion for activity.

Metal Ions Promote Nucleophilic Catalysis via Water Ionization

A metal ion's charge makes its bound water molecules more acidic than free H_2O and therefore a source of OH^- ions even below neutral pH's. For example, the water molecule of $(NH_3)_5Co^{3+}(H_2O)$ ionizes according to the reaction:

$$(NH_3)_5Co^{3+}(H_2O) \rightleftharpoons (NH_3)_5Co^{3+}(OH^-) + H^+$$

with a pK of 6.6, which is some 9 pH units below the pK of free H_2O. *The resulting metal ion–bound hydroxyl group is a potent nucleophile.*

An excellent example of this phenomenon occurs in the catalytic mechanism of **carbonic anhydrase** (Section 9-1C), a widely occurring enzyme that catalyzes the reaction:

$$CO_2 + H_2O \rightleftharpoons HCO_3^- + H^+$$

Carbonic anhydrase contains an essential Zn^{2+} ion that is implicated in the enzyme's catalytic mechanism as follows:

1. The crystal structure of human carbonic anhydrase (Fig. 7-44) reveals that its Zn^{2+} lies at the bottom of a 15-Å-deep active site cleft, where it is tetrahedrally coordinated by three evolutionarily invariant His side chains and a H_2O molecule. This Zn^{2+}-polarized H_2O ionizes

in a process facilitated through general base catalysis most probably by His 64 (Fig. 14-5). Although His 64 is too far away from the Zn^{2+}-bound water to directly remove its proton, these entities are linked by two intervening water molecules to form a hydrogen bonded network that is thought to act as a proton shuttle.

2. The resulting Zn^{2+}-bound OH^- nucleophilically attacks the nearby enzymatically bound CO_2, thereby converting it to HCO_3^-.

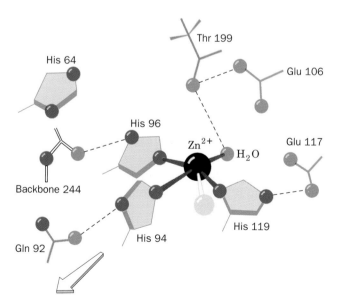

Im = imidazole

3. The catalytic site is then regenerated by the binding and ionization of another H_2O to the Zn^{2+}, possibly before the departure of the HCO_3^- ion, so as to transiently form a 5-coordinated Zn^{2+} complex.

Metal Ions Promote Reactions through Charge Shielding

Another important enzymatic function of metal ions is **charge shielding.** For example, the actual substrates of **kinases** (phosphoryl-transfer enzymes utilizing ATP) are Mg^{2+}–ATP complexes such as

$$\text{Adenine—Ribose—O—}\overset{\displaystyle O}{\underset{\displaystyle O}{\overset{\|}{\underset{\|}{P}}}}\text{—O—}\overset{\displaystyle O}{\underset{\displaystyle O}{\overset{\|}{\underset{\|}{P}}}}\text{—O—}\overset{\displaystyle O^-}{\underset{\displaystyle O}{\overset{|}{\underset{\|}{P}}}}\text{—O}^-$$

rather than just ATP. Here the Mg^{2+} ion's role, in addition to its orienting effect, is to shield electrostatically the negative charges of the phosphate groups. Otherwise, these charges would tend to repel the electron pairs of attacking nucleophiles, especially those with anionic character.

D. Electrostatic Catalysis

The binding of substrate generally excludes water from an enzyme's active site. The local dielectric constant of the active site therefore resembles that in an organic solvent, where electrostatic interactions are much stronger than

FIGURE 14-5. The active site of human carbonic anhydrase. The light grey ligand to the Zn^{2+} indicates the probable fifth Zn^{2+} coordination site. The arrow points towards the opening of the active site cavity. [After Sheridan, R. P. and Allen, L. C., *J. Am. Chem. Soc.* **103**, 1545 (1981).]

they are in aqueous solutions (Section 7-4A). The charge distribution in a medium of low dielectric constant can greatly influence chemical reactivity. Thus, as we have seen, the pK's of amino acid side chains in proteins may vary by several units from their nominal values (Table 4-1) because of the proximity of charged groups.

Although experimental evidence and theoretical analyses on the subject are still sparse, *there are mounting indications that the charge distributions about the active sites of enzymes are arranged so as to stabilize the transition states of the catalyzed reactions.* Such a mode of rate enhancement, which resembles the form of metal ion catalysis discussed above, is termed **electrostatic catalysis.** Moreover, in several enzymes, *these charge distributions apparently serve to guide polar substrates towards their binding sites so that the rates of these enzymatic reactions are greater than their apparent diffusion-controlled limits (Section 13-2B).*

E. Catalysis through Proximity and Orientation Effects

Although enzymes employ catalytic mechanisms that resemble those of organic model reactions, they are far more catalytically efficient than these models. Such efficiency must arise from the specific physical conditions at enzyme catalytic sites that promote the corresponding chemical reactions. The most obvious effects are **proximity** and **orientation:** *Reactants must come together with the proper spatial relationship for a reaction to occur.* For example, in the bimolecular reaction of imidazole with *p*-nitrophenylacetate,

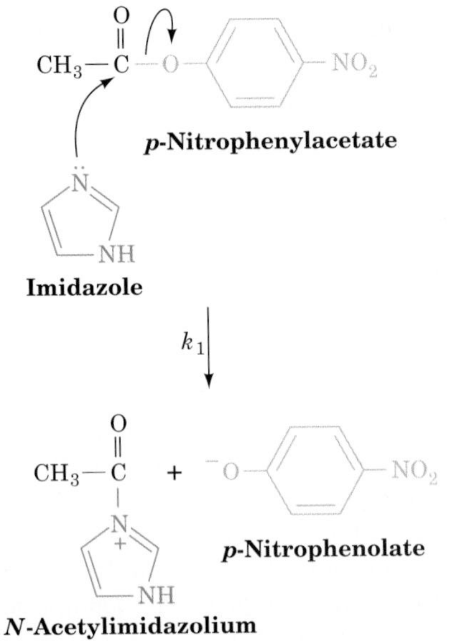

p-Nitrophenylacetate

Imidazole

k_1

p-Nitrophenolate

N-Acetylimidazolium

the progress of the reaction is conveniently monitored by the appearance of the intensely yellow *p*-**nitrophenolate** ion:

$$\frac{d\,[p\text{-}NO_2\phi O^-]}{dt} = k_1[\text{imidazole}][p\text{-}NO_2\phi Ac]$$

$$= k_1'[p\text{-}NO_2\phi Ac] \qquad [14.4]$$

Here k_1', the pseudo-first-order rate constant, is 0.0018 s^{-1} when [imidazole] $= 1M$ (ϕ = phenyl). However, for the intramolecular reaction

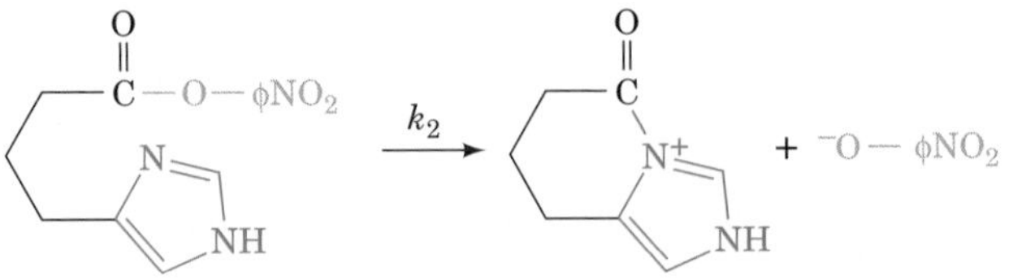

the first-order rate constant $k_2 = 0.043$ s^{-1}; that is, $k_2 = 24\,k_1'$. Thus, when the $1M$ imidazole catalyst is covalently attached to the reactant, it is 24-fold more effective than when it is free in solution; that is, *the imidazole group in the intramolecular reaction behaves as if its concentration is 24M.* This rate enhancement has contributions from both proximity and orientation.

Proximity Alone Contributes Relatively Little to Catalysis

Let us make a rough calculation as to how the rate of a reaction is affected purely by the proximity of its reacting groups. Following Daniel Koshland's treatment, we shall make several reasonable assumptions:

1. Reactant species, that is, functional groups, are about the size of water molecules.

2. Each reactant species in solution has 12 nearest-neighbor molecules, as do packed spheres of identical size.

3. Chemical reactions occur only between reactants that are in contact.

4. The reactant concentration in solution is low enough so that the probability of any reactant species being in simultaneous contact with more than one other reactant molecule is negligible.

Then the reaction:

$$A + B \xrightarrow{k_1} A\!-\!B$$

obeys the second-order rate equation

$$v = \frac{d[A\!-\!B]}{dt} = k_1[A][B] = k_2[A,B]_{pairs} \qquad [14.5]$$

where $[A,B]_{pairs}$ is the concentration of contacting molecules of A and B. The value of this quantity is

$$[A,B]_{pairs} = \frac{12[A][B]}{55.5M} \qquad [14.6]$$

since there are 12 ways that A can be in contact with B, and $[A]/55.5M$ is the fraction of sites occupied by A in water solution ([H$_2$O] $= 55.5M$ in dilute aqueous solutions) and hence the probability that a molecule of B will be next to one of A. Combining Eqs. [14.5] and [14.6] yields

$$v = k_1\!\left(\frac{55.5}{12}\right)[A,B]_{pairs} = 4.6k_1[A,B]_{pairs} \qquad [14.7]$$

Thus, in the absence of other effects, this model predicts that for the intramolecular reaction,

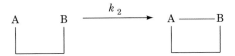

$k_2 = 4.6k_1$, which is a rather small rate enhancement. Factors that will increase this value other than proximity alone clearly must be considered.

Arresting Reactants' Relative Motions and Properly Orienting Them Can Result in Large Catalytic Rate Enhancements

The foregoing theory is, of course, quite simple. For example, it does not take into account the motions of the reacting groups with respect to one another. Yet, in the transition state complex, the reacting groups have little relative motion. In fact, as Thomas Bruice demonstrated, the rates of intramolecular reactions are greatly increased by arresting a molecule's internal motions (Table 14-1). Thus, when an enzyme brings two molecules together in a bimolecular reaction, as William Jencks pointed out, not only does it increase their proximity, but it freezes out their relative translational and rotational motions (decreases their entropy), thereby enhancing their reactivity. Table 14-1 indicates that such rate enhancements can be enormous.

Another effect that we have neglected in our treatment of proximity is that of orientation. Molecules are not equally reactive in all directions as Koshland's simple theory assumes. Rather, *they react most readily only if they have the proper relative orientation (Fig. 14-6)*. For example, in an S_N2 (bimolecular nucleophilic substitution) reaction, the incoming nucleophile optimally attacks its target along the direction opposite to that of the bond to the leaving group (backside attack). The approaches of reacting atoms along a trajectory that deviates by as little as 10° from this optimum direction results in a significantly reduced reactivity. It has been estimated that *properly orientating substrates can increase reaction rates by a factor of up to ~100*. In a related phenomenon, a molecule may be maximally reactive only when it assumes a conformation that aligns its various orbitals in a way that minimizes the electronic energy of

TABLE 14-1. RELATIVE RATES OF ANHYDRIDE FORMATION FOR ESTERS POSESSING DIFFERENT DEGREES OF MOTIONAL FREEDOM IN THE REACTION:

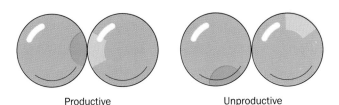

Reactants[a]	Relative Rate Constant
$CH_3COO\phi Br$ + CH_3COO^-	1.0
	$\sim 1 \times 10^3$
	$\sim 2.2 \times 10^5$
	$\sim 5 \times 10^7$

[a] Curved arrows indicate rotational degrees of freedom.
Source: Bruice, T.C., *Annu. Rev. Biochem.* **45**, 353 (1976).

its transition state, an effect termed **stereoelectronic assistance.**

Enzymes, as we shall see in Sections 14-2 and 14-3, bind substrates in a manner that both immobilizes them and aligns them so as to optimize their reactivities. The free energy required to do so is derived from the specific binding energy of substrate to enzyme.

F. Catalysis by Preferential Transition State Binding

The rate enhancements effected by enzymes are often greater than can be reasonably accounted for by the catalytic mechanisms so far discussed. However, we have not yet considered one of the most important mechanisms of enzymatic catalysis: *The binding of the transition state to an enzyme with greater affinity than the corresponding substrates or products.* When taken together with the previously described catalytic mechanisms, preferential transition state binding rationalizes the observed rates of enzymatic reactions.

The original concept of transition state binding proposed that enzymes mechanically strained their substrates towards the transition state geometry through binding sites into which undistorted substrates did not properly fit. This so-called **rack mechanism** (in analogy with the medieval torture device) was based on the extensive evidence for the

FIGURE 14-6. Molecules are susceptible to chemical attack over only limited regions of their surfaces (represented by the colored areas). Without the proper relative orientation (*left*), reactions do not occur (*right*).

role of strain in promoting organic reactions. For example, the rate of the reaction,

is 315 times faster when R is CH_3 rather than when it is H because of the greater steric repulsions between the CH_3 groups and the reacting groups. Similarly, ring opening reactions are considerably more facile for strained rings such as cyclopropane than for unstrained rings such as cyclohexane. In either process, *the strained reactant more closely resembles the transition state of the reaction than does the corresponding unstrained reactant.* Thus, as was first suggested by Linus Pauling and further amplified by Richard Wolfenden and Gustav Lienhard, *interactions that preferentially bind the transition state increase its concentration and therefore proportionally increase the reaction rate.*

Let us quantitate this statement by considering the kinetic consequences of preferentially binding the transition state of an enzymatically catalyzed reaction involving a single substrate. The substrate S may react to form product P either spontaneously or through enzymatic catalysis:

$$S \xrightarrow{k_N} P$$

$$ES \xrightarrow{k_E} EP$$

Here k_E and k_N are the first-order rate constants for the catalyzed and uncatalyzed reactions, respectively. The relationships between the various states of these two reaction pathways are indicated in the following scheme:

$$E + S \underset{K_N^\ddagger}{\rightleftharpoons} S^\ddagger + E \longrightarrow P + E$$

$$\Big\Updownarrow K_R \quad\quad \Big\Updownarrow K_T \quad\quad \Big\Updownarrow$$

$$ES \underset{K_E^\ddagger}{\rightleftharpoons} ES^\ddagger \longrightarrow EP$$

where

$$K_R = \frac{[ES]}{[E][S]} \qquad\qquad K_T = \frac{[ES^+]}{[E][S^+]}$$

$$K_N^\ddagger = \frac{[E][S^+]}{[E][S]} \quad\text{and}\quad K_E^\ddagger = \frac{[ES^+]}{[ES]}$$

are all association constants. Consequently,

$$\frac{K_T}{K_R} = \frac{[S][ES^+]}{[S^+][ES]} = \frac{K_E^\ddagger}{K_N^\ddagger} \qquad [14.8]$$

According to transition state theory, Eqs. [13.7] and [13.14], the rate of the uncatalyzed reaction can be expressed

$$v_N = k_N[S] = \left(\frac{\kappa k_B T}{h}\right)[S^+] = \left(\frac{\kappa k_B T}{h}\right)K_N^\ddagger[S] \quad [14.9]$$

Similarly, the rate of the enzymatically catalyzed reaction is

$$v_E = k_E[ES] = \left(\frac{\kappa k_B T}{h}\right)[ES^+] = \left(\frac{\kappa k_B T}{h}\right)K_E^\ddagger[ES] \quad [14.10]$$

Therefore, combining Eqs. [14.8] to [14.10],

$$\frac{k_E}{k_N} = \frac{K_E^\ddagger}{K_N^\ddagger} = \frac{K_T}{K_R} \qquad [14.11]$$

This equation indicates that *the more tightly an enzyme binds its reaction's transition state* (K_T) *relative to the substrate* (K_R), *the greater the rate of the catalyzed reaction* (k_E) *relative to that of the uncatalyzed reaction* (k_N); *that is, catalysis results from the preferential binding and therefore the stabilization of the transition state* $(S^\ddagger)$ *relative to that of the substrate* (S) *(Fig. 14-7).*

According to Eq. [13.15], the ratio of the rates of the catalyzed versus the uncatalyzed reaction is expressed

$$\frac{k_E}{k_N} = \exp{[(\Delta G_N^\ddagger - \Delta G_E^\ddagger)/RT]} \qquad [14.12]$$

A rate enhancement factor of 10^6 therefore requires that an enzyme bind its transition state complex with 10^6-fold higher affinity than its substrate which corresponds to a $34.2\,\text{kJ} \cdot \text{mol}^{-1}$ stabilization at $25°C$. This is roughly the free energy of two hydrogen bonds. Consequently, *the enzymatic binding of a transition state* $(ES^\ddagger)$ *by two hydrogen bonds that cannot form in the Michaelis complex* (ES) *should result in a rate enhancement of* $\sim 10^6$ *based on this effect alone.*

It is commonly observed that the specificity of an enzyme is manifested by its turnover number (k_{cat}) rather than by its expressed substrate-binding affinity. In other words, an enzyme binds poor substrates, which have a low reaction rate, as well as or even better than good ones, which have a high reaction rate. Such enzymes apparently use a good substrate's intrinsic binding energy to stabilize the correspond-

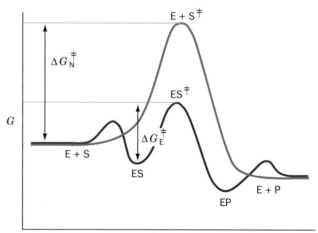

FIGURE 14-7. Reaction coordinate diagrams for a hypothetical enzymatically catalyzed reaction involving a single substrate (*blue*), and the corresponding uncatalyzed reaction (*red*).

ing transition state; that is, *a good substrate does not necessarily bind to its enzyme with high affinity, but does so upon activation to the transition state.*

Transition State Analogs Are Competitive Inhibitors

If an enzyme preferentially binds its transition state, then it can be expected that **transition state analogs,** *stable molecules that resemble* $S^\ddagger$ *or one of its components, are potent competitive inhibitors of the enzyme.* For example, the reaction catalyzed by **proline racemase** from *Clostridium sticklandii* is thought to occur via a planar transition state:

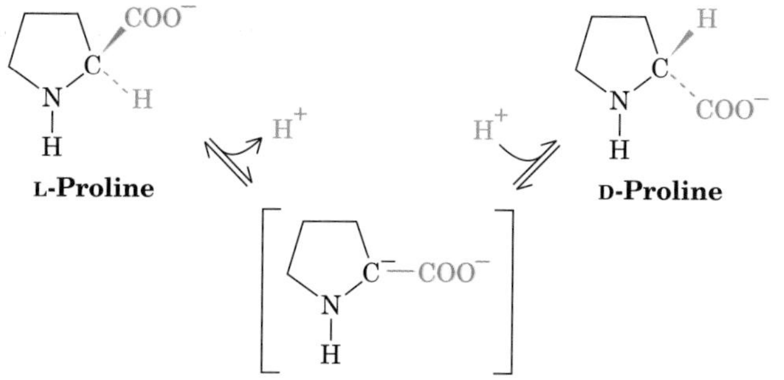

L-Proline

D-Proline

Planar transition state

Proline racemase is competitively inhibited by the planar analogs of proline, **pyrrole-2-carboxylate** and **Δ-1-pyrroline-2-carboxylate,**

Pyrrole-2-carboxylate   **Δ-1-Pyrroline-2-carboxylate**

both of which bind to the enzyme with 160-fold greater affinity than does proline. These compounds are therefore thought to be analogs of the transition state in the proline racemase reaction. In contrast, **tetrahydrofuran-2-carboxylate,**

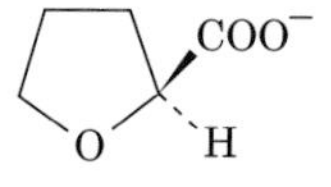

Tetrahydrofuran-2-carboxylate

which more closely resembles the tetrahedral structure of proline, is not nearly as good an inhibitor as these compounds. A 160-fold increase in binding affinity corresponds, according to Eq. [14.12], to a 12.6 $kJ \cdot mol^{-1}$ increase in the free energy of binding. This quantity presumably reflects the additional binding affinity that proline racemase has for proline's planar transition state over that of the undistorted molecule.

Hundreds of transition state analogs for various enzymatic reactions have been reported. Some are naturally occurring antibiotics. Others were designed to investigate the mechanisms of particular enzymes and/or to act as specific enzymatic inhibitors for therapeutic or agricultural use. Indeed, *the theory that enzymes bind transition states with higher affinity than substrates has led to a rational basis for drug design based on the understanding of specific enzyme reaction mechanisms.*

2. LYSOZYME

In the remainder of this chapter we shall investigate the catalytic mechanisms of several well-characterized enzymes. In doing so, we shall see how enzymes apply the catalytic principles described in Section 14-1. You should note that *the great catalytic efficiency of enzymes arises from their simultaneous use of several of these catalytic mechanisms.*

Lysozyme is an enzyme that destroys bacterial cell walls. It does so, as we saw in Section 10-3B, by hydrolyzing the $\beta(1{\rightarrow}4)$ glycosidic linkages from **N-acetylmuramic acid (NAM)** to **N-acetylglucosamine (NAG)** in the alternating NAM–NAG polysaccharide component of cell wall peptidoglycans (Fig. 14-8). It likewise hydrolyzes $\beta(1{\rightarrow}4)$-linked poly(NAG) (chitin), a cell wall component of most fungi. Lysozyme occurs widely in the cells and secretions of vertebrates, where it may function as a bacteriocidal agent. However, the observation that few pathogenic bacteria are susceptible to lysozyme alone has prompted the suggestion that this enzyme mainly helps dispose of bacteria after they have been killed by other means.

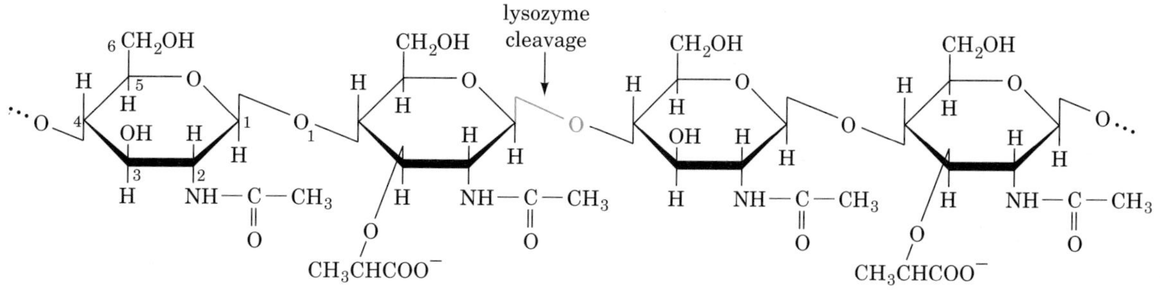

FIGURE 14-8. The alternating NAG–NAM polysaccharide component of bacterial cell walls, showing the position of the lysozyme cleavage site.

Hen egg white (HEW) lysozyme is the most widely studied species of lysozyme and is one of the mechanistically best understood enzymes. It is a rather small protein (14.6 kD) whose single polypeptide chain consists of 129 amino acid residues and is internally cross-linked by four disulfide bonds (Fig. 14-9). HEW lysozyme catalyzes the hydrolysis of its substrate at a rate that is ~10^8-fold greater than that of the uncatalyzed reaction.

A. Enzyme Structure

The elucidation of an enzyme's mechanism of action requires a knowledge of the structure of its enzyme–substrate complex. This is because, even if the active site residues have been identified through chemical and physical means, their three-dimensional arrangements relative to the substrate as well as to each other must be known for an understanding of how the enzyme works. However, an enzyme binds its good substrates only transiently before it catalyzes a reaction and releases the products. Consequently, *most of our knowledge of enzyme–substrate complexes derives from X-ray studies of enzymes in complex with inhibitors or poor substrates* that remain stably bound to the enzyme for the day or more required to measure a protein crystal's X-ray diffraction intensities (although techniques for measuring X-ray intensities in less than 1 s have been developed recently). The large solvent-filled channels that occupy much of the volume of most protein crystals (Section 7-3A) often permit the formation of enzyme–inhibitor complexes by the diffusion of inhibitor molecules into crystals of the native protein.

The X-ray structure of HEW lysozyme, which was elucidated by David Phillips in 1965, was the second structure of a protein and the first of an enzyme to be determined at high resolution. The protein molecule is roughly ellipsoidal in shape with dimensions $30 \times 30 \times 45$ Å (Fig. 14-10). *Its most striking feature is a prominent cleft, the substrate-binding site, that traverses one face of the molecule.* The polypeptide chain forms five helical segments as well as a three-stranded antiparallel β sheet that comprises much of one wall of the binding cleft (Fig. 14-10b). As expected, most of the nonpolar side chains are in the interior of the molecule, out of contact with the aqueous solvent.

The Nature of the Binding Site

NAG oligosaccharides of less than five residues are but very slowly hydrolyzed by HEW lysozyme (Table 14-2) although these substrate analogs bind to the enzyme's active site and are thus its competitive inhibitors. The X-ray structure of the (NAG)$_3$–lysozyme complex reveals that (NAG)$_3$ is bound on the right side of the enzymatic binding cleft as drawn in Fig. 14-10a for substrate residues A, B, and C. This inhibitor associates with the enzyme through strong hydrogen bonding interactions, some of which involve the acetamido groups of residues A and C, as well as through close-fitting hydrophobic contacts. In an example of in-

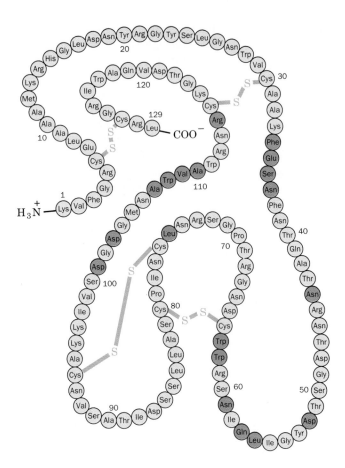

FIGURE 14-9. The primary structure of HEW lysozyme. The amino acid residues that line the substrate-binding pocket are shown in dark purple.

FIGURE 14-10 (*Opposite*). The X-ray structure of HEW lysozyme. (*a*) The polypeptide chain is shown with a bound (NAG)$_6$ substrate (*green*). The positions of the backbone C$_\alpha$ atoms are indicated together with those of the side chains that line the substrate-binding site and form disulfide bonds. The substrate's sugar rings are designated A, at its nonreducing end (*right*), through F, at its reducing end (*left*). Lysozyme catalyzes the hydrolysis of the glycosidic bond between residues D and E. Rings A, B, and C are observed in the X-ray structure of the complex of (NAG)$_3$ with lysozyme; the positions of rings D, E, and F were inferred from model building studies. [Figure copyrighted © by Irving Geis.] (*b*) A ribbon diagram of lysozyme highlighting the protein's secondary structure and indicating the positions of its catalytically important side chains. (*c*) A computer-generated model showing the protein's molecular envelope (*purple*) and C$_\alpha$ backbone (*blue*). The side chains of the catalytic residues, Asp 52 (*above*) and Glu 35 (*below*), are colored yellow. Note the enzyme's prominent substrate-binding cleft. [Courtesy of Arthur Olson, The Scripps Research Institute, La Jolla, California.] Parts *a*, *b* and *c* have approximately the same orientation.

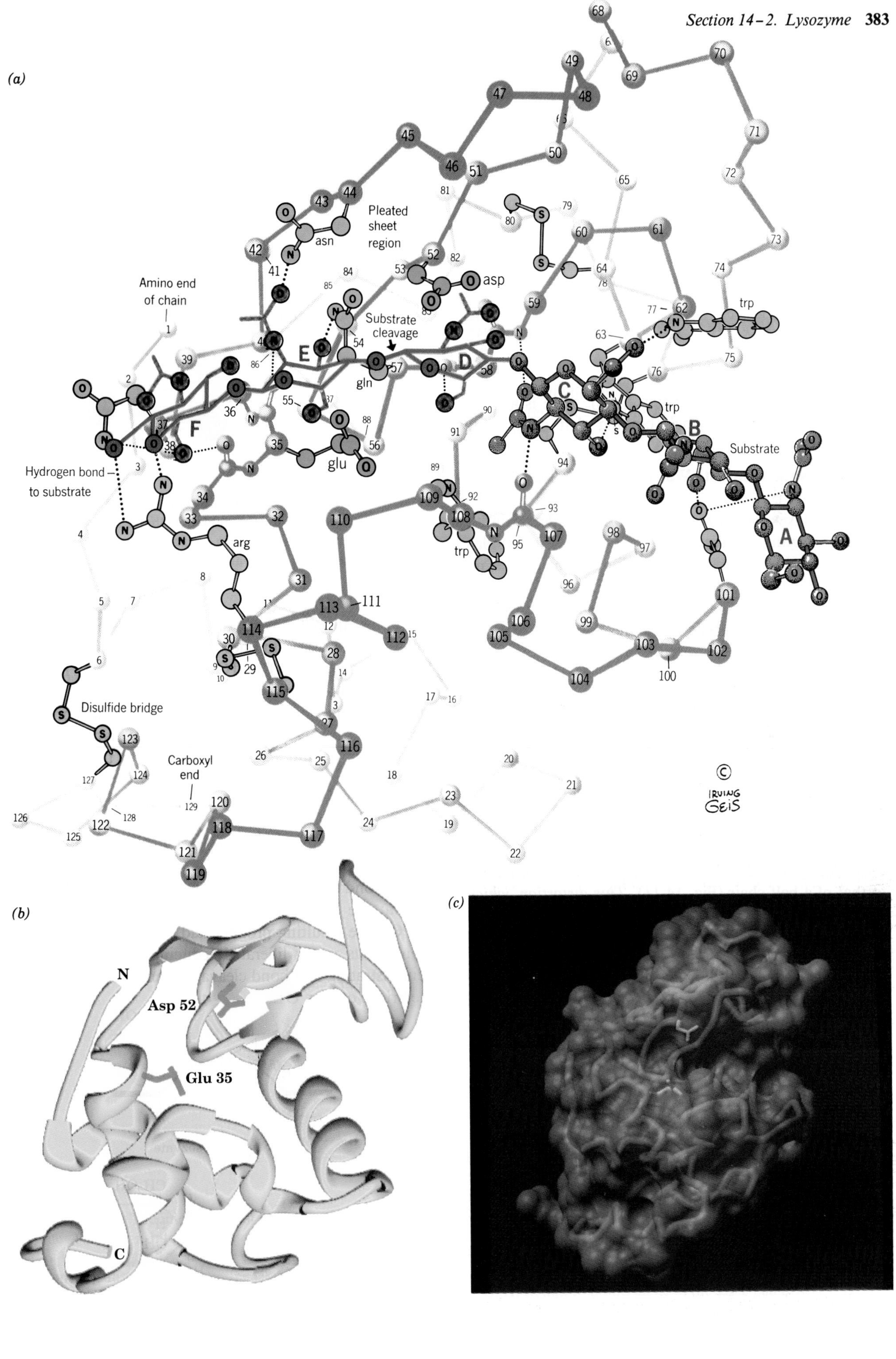

(a)

Amino end
of chain

Pleated
sheet
region

asn

Substrate
cleavage

Hydrogen bond
to substrate

Disulfide bridge

Carboxyl
end

gln

glu

arg

trp

trp

trp

asp

Substrate

E

D

C

B

A

F

© IRVING GEIS

(b)

N

Asp 52

Glu 35

C

(c)

properties are the side chains of Glu 35 and Asp 52, residues that are invariant in the family of lysozymes of which HEW lysozyme is the prototype. These side chains, which are disposed to either side of the $\beta(1\rightarrow4)$ glycosidic linkage to be cleaved (Fig. 14-10*a* and *b*), have markedly different environments. Asp 52 is surrounded by several conserved polar residues with which it forms a complex hydrogen bonded network. Asp 52 is therefore predicted to have a normal pK; that is, it should be unprotonated and hence negatively charged throughout the 3 to 8 pH range in which lysozyme is active. In contrast, *the carboxyl group of Glu 35 is nestled in a predominantly nonpolar pocket where, as we discussed in Section 14-1D, it is likely to remain protonated at unusually high pH's for carboxyl groups.* Indeed, neutron diffraction studies, which provide similar information to X-ray diffraction studies but also reveal the positions of hydrogen atoms, indicate that Glu 35 is protonated at physiological pH's. The closest approaches between the carboxyl O atoms of both Asp 52 and Glu 35 and the C1—O1 bond of NAG residue D are each ~3 Å, which makes them the prime candidates for electrostatic and acid catalysts, respectively.

The Phillips Mechanism

With much of the foregoing information, Phillips postulated the following enzymatic mechanism for lysozyme (Fig. 14-14):

1. Lysozyme attaches to a bacterial cell wall by binding to a hexasaccharide unit. In the process, *residue D is distorted towards the half-chair conformation* in response to the unfavorable contacts that its —C6H$_2$OH group would otherwise make with the protein.

2. *Glu 35 transfers its proton to the O1 of the D ring, the only polar group in its vicinity (general acid catalysis). The C1—O1 bond is thereby cleaved, generating a resonance-stabilized oxonium ion at C1.*

3. The ionized carboxyl group of Asp 52 acts to *stabilize the developing oxonium ion through charge–charge interactions (electrostatic catalysis).* This carboxylate group cannot form a covalent bond with the substrate because the ~3-Å distance between C1 and a carboxyl O atom of Asp 52 is much greater than the ~1.5 Å length of a C—O covalent bond [i.e., the reaction occurs via an S$_N$1 mechanism (unimolecular nucleophilic substitution), not an S$_N$2 mechanism involving the transient formation of a C—O bond to the enzyme]. The bond cleavage reaction is facilitated by the strain in the D ring that distorts it to the planar half-chair conformation. This is a result of the oxonium ion's required planarity; that is, *the initial binding conformation of the D ring resembles that of the reaction's transition state (transition state binding catalysis; Fig. 14-15).*

4. At this point, the enzyme releases the hydrolyzed E ring with its attached polysaccharide, yielding a **glycosyl–enzyme intermediate.** This oxonium ion subsequently

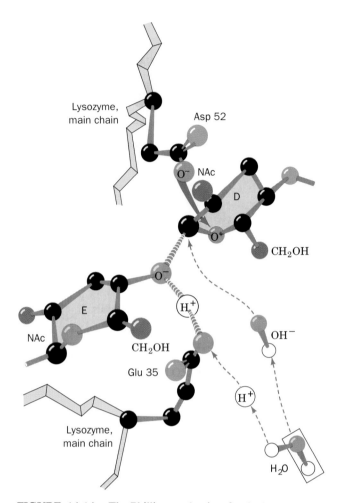

FIGURE 14-14. The Phillips mechanism for the lysozyme reaction. The cleavage of the glycosidic bond between the substrate D and E rings occurs through protonation of the bridge oxygen atom by Glu 35. The resulting D-ring oxonium ion is stabilized by the proximity of the Asp 52 carboxylate group and the enzyme-induced distortion of the D ring. Once the E ring is released, H$_2$O from solution provides both an OH$^-$ that combines with the oxonium ion and an H$^+$ that reprotonates Glu 35. NAc represents the *N*-acetylamino substituent at C2 of each glucose ring.

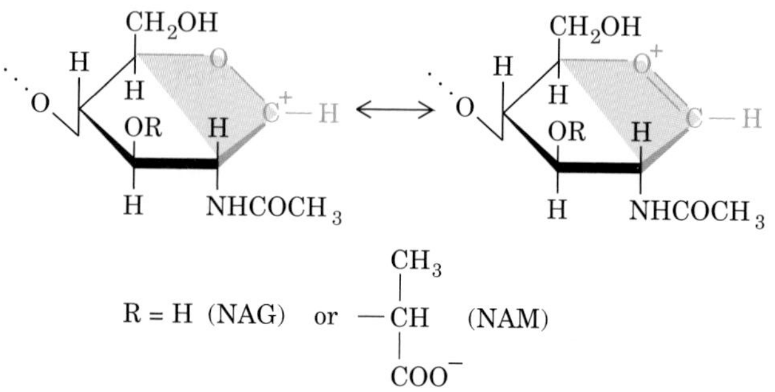

FIGURE 14-15. The oxonium ion transition state of the D ring in the lysozyme reaction is stabilized by resonance. This requires that atoms C1, C2, C5, and O5 be coplanar (*shading*); that is, the hexose ring must assume the half-chair conformation.

adds H₂O from solution in a reversal of the preceding steps to form product and to reprotonate Glu 35. *The reaction's retention of configuration is dictated by the shielding of one of the oxonium ion's faces by the enzymatic cleft.* The enzyme then releases the D-ring product with its attached saccharide, thereby completing the catalytic cycle.

C. Testing the Phillips Mechanism

The Phillips mechanism was formulated largely on the basis of structural investigations of lysozyme and a knowledge of the mechanism of nonenzymatic acetal hydrolysis. A variety of evidence has since been gathered that bears on the validity of this mechanism.

Identification of the Catalytic Residues

Lysozyme's catalytically important groups have been experimentally identified through the use of group-specific reagents (Section 6-2):

Asp 52. The ethylating agent **triethoxonium fluoroborate** reacts with carboxylic acids as follows:

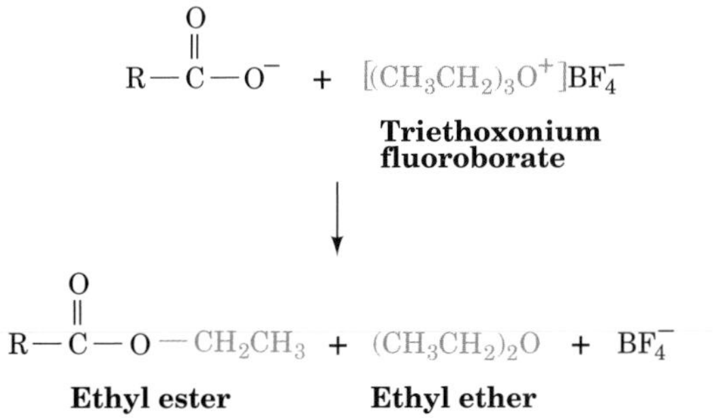

$$R-\overset{\overset{\text{O}}{\|}}{C}-O^- \ + \ [(CH_3CH_2)_3O^+]BF_4^-$$

Triethoxonium fluoroborate

$$\downarrow$$

$$R-\overset{\overset{\text{O}}{\|}}{C}-O-CH_2CH_3 \ + \ (CH_3CH_2)_2O \ + \ BF_4^-$$

Ethyl ester   **Ethyl ether**

At pH 4.5, this reagent reacts only with the β carboxyl group of lysozyme's Asp 52. The resulting monoesterified enzyme binds substrate in a normal manner but is catalytically inactive. Similarly, the mutagenesis of Asp 52 to Asn, which has a polarity comparable to that of Asp, yields an enzyme with no more than 5% of wild-type lysozyme's catalytic activity even though this mutation causes a ~2-fold increase in the enzyme's affinity for substrate. *Asp 52 must therefore be essential for enzymatic activity.*

Glu 35. In the native enzyme, Glu 35 is in van der Waals contact with Trp 108 (Fig. 14-10*a*). The reaction of lysozyme with I₂ specifically oxidizes Trp 108 (and none of the enzyme's other five Trp residues) to form a modified enzyme that is totally inactive. The X-ray structure of this modified enzyme indicates that its only chemical change is the formation of a covalent bond between Trp 108 and Glu 35 (Fig. 14-16). Lysozyme's only conformational change upon I₂ oxidation is a reorientation of Glu 35's side chain that does not significantly affect the enzyme's substrate-binding affinity. Furthermore, the mutagenesis of Glu 35 to Gln yields a protein with no detectable catalytic activity

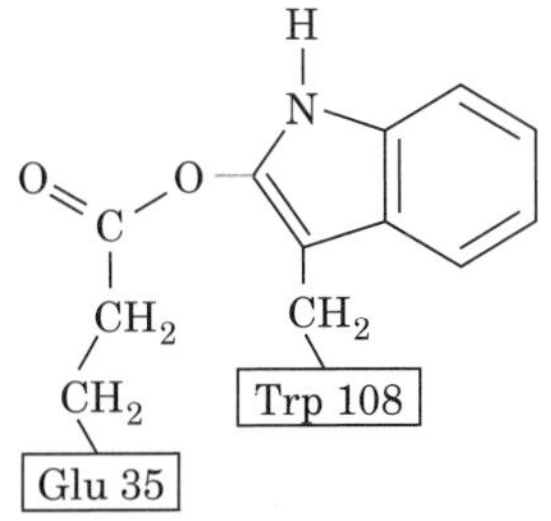

FIGURE 14-16. The I₂ oxidation of lysozyme results in the formation of a covalent bond between the side chains of Glu 35 and Trp 108.

(<0.1% of wild-type) although it has only a ~1.5-fold decrease in substrate affinity. *Glu 35 must therefore also be essential for lysozyme's catalytic activity.*

Verification of Asp 52 and Glu 35 Involvement. The analysis of lysozyme's pH rate profile (Section 13-4) indicates that this enzyme has two catalytically important ionizable groups whose p*K*'s are 3.5 ± 0.2 and 6.3 ± 0.2. The latter ionization, although abnormally high for a carboxyl group, is attributed to Glu 35 because of this ionization's disappearance upon I₂ oxidation of lysozyme. The only other reasonable possibility for an ionization with this p*K* is lysozyme's sole His residue but this residue is located far from the active site. The absence of the ionization with p*K* 3.5 in the triethoxonium fluoroborate–treated enzyme demonstrates that this ionization is due to Asp 52. This observation corroborates the mechanistic postulate that lysozyme is catalytically active only when Asp 52 is ionized and Glu 35 is not.

Noninvolvement of Other Amino Acid Residues. Lysozyme's other carboxyl groups besides Glu 35 and Asp 52 do not participate in the catalytic process, as was demonstrated by reacting lysozyme with carboxyl-specific reagents in the presence of substrate. This treatment yields an almost fully active enzyme in which all carboxyl groups but Glu 35 and Asp 52 are derivatized. Other group-specific reagents that modify, for instance, His, Lys, Met, or Tyr residues but induce no major protein structure disruptions cause little change in lysozyme's catalytic efficiency.

Role of Strain

Many of the mechanistic investigations of lysozyme have had the elusive goal of establishing the catalytic role of strain. Not all of these studies, as we shall see, have supported the Phillips mechanism, thereby stimulating a series of investigations that have only recently settled this issue. In the remainder of this section, we discuss the highlights of these studies to illustratrate how scientific models evolve.

Measurements of the binding equilibria of various oligosaccharides to lysozyme indicate that all saccharide residues except that binding to the D subsite contribute energetically towards the binding of substrate to lysozyme; binding NAM in the D subsite requires a free energy input of 12

TABLE 14-3. BINDING FREE ENERGIES OF
HEW LYSOZOME SUBSITES

Site	Bound Saccharide	Binding Free Energy $(kJ \cdot mol^{-1})$
A	NAG	−7.5
B	NAM	−12.3
C	NAG	−23.8
D	**NAM**	**+12.1**
E	NAG	−7.1
F	NAM	−7.1

Source: Chipman, D.M. and Sharon, N., *Science* **165**, 459 (1969).

$kJ \cdot mol^{-1}$ *(Table 14-3). The Phillips mechanism explains this observation as being indicative of the energy penalty of straining the D ring from its preferred chair conformation towards the half-chair form.*

As we have discussed in Section 14-1F, an enzyme that catalyzes a reaction by the preferential binding of its transition state has a greater binding affinity for an inhibitor that has the transition state geometry (transition state analog) than it does for its substrate. The δ-lactone analog of $(NAG)_4$ (Fig. 14-17) is a transition state analog of lysozyme since *this compound's lactone ring has the half-chair conformation that geometrically resembles the proposed oxonium ion transition state of the substrate's D ring.* X-Ray studies indicate, in accordance with prediction, that this inhibitor binds to lysozyme's A—B—C—D subsites such that the lactone ring occupies the D subsite in a half-chairlike conformation.

Despite the foregoing, *the role of substrate distortion in lysozyme catalysis has been questioned.* Theoretical studies by Michael Levitt and Arieh Warshel on substrate binding by lysozyme suggested that the protein is too flexible to mechanically distort the D ring of a bound substrate. Rather, these calculations impled that transition state stabilization occurs through the displacement by substrate of several tightly bound water molecules from the D subsite. The resulting desolvation of the Asp 52 carboxylate group would significantly enhance its capacity to electrostatically stabilize the transition state oxonium ion. This study therefore concluded that "electrostatic strain" rather than steric strain is the more important factor in stabilizing lysozyme's transition state.

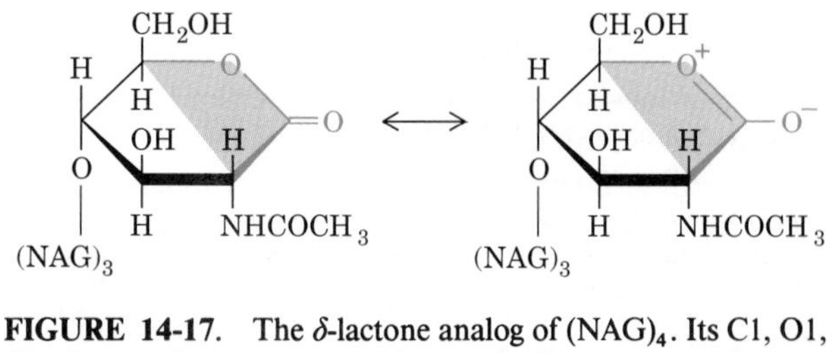

FIGURE 14-17. The δ-lactone analog of $(NAG)_4$. Its C1, O1, C2, C5, and O5 atoms are coplanar (*shading*) because of resonance as is the D ring in the transition state of the lysozyme reaction (compare with Fig. 14-15).

In an effort to obtain further experimental information bearing on the Phillips strain mechanism, Nathan Sharon and David Chipman determined the D subsite-binding affinities of several saccharides by comparing the lysozyme-binding affinities of various substrate analogs. The NAG lactone inhibitor binds to the D subsite with 9.2 $kJ \cdot mol^{-1}$ greater affinity than does NAG. This quantity corresponds, according to Eq. [13.15], to no more than an ~40-fold rate enhancement of the lysozyme reaction as a result of strain (recall that the difference in binding energy between a transition state analog and a substrate is indicative of the enzyme's rate enhancement arising from the preferential binding of the transition state complex). Such an enhancement is hardly a major portion of lysozyme's ~10^8-fold rate enhancement (accounting for only ~20% of the reaction's $\Delta\Delta G_{cat}^{\ddagger}$; Section 13-1C). Moreover, an **N-acetyl-xylosamine (XylNAc)** residue,

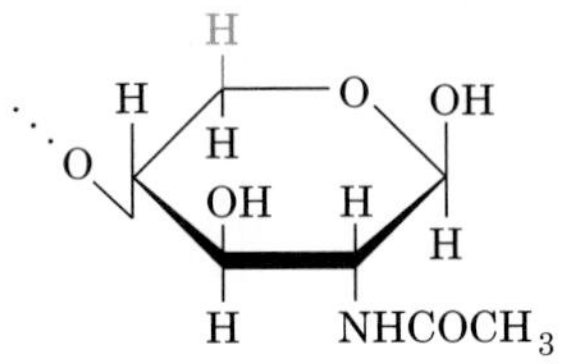

N-Acetylxylosamine residue

which lacks the sterically hindered —C6H₂OH group of NAM and NAG, has only marginally greater binding affinity for the D subsite (-3.8 $kJ \cdot mol^{-1}$) than does NAG (-2.5 $kJ \cdot mol^{-1}$). Yet, recall the Phillips mechanism postulates that it is the unfavorable contacts made by this —C6H₂OH group that promotes D-ring distortion. Nevertheless, lysozyme does not hydrolyze saccharides with XylNAc in the D subsite.

The apparent inconsistancies among the foregoing experimental observations have been largely rationalized by Michael James' particularly accurate (refined 1.5 Å resolution) X-ray crystal structure determination of lysozyme in complex with NAM–NAG–NAM. This trisaccharide binds, as expected, to the B, C, and D subsites of lysozyme. *The NAM in the D subsite, in agreement with the Phillips mechanism, is distorted to the half-chair conformation with its —C6H₂OH group in a nearly axial position due to steric clashes that would otherwise occur with the acetamido group of the C subsite NAG* [although, contrary to the original Phillips mechanism (Section 14-2A), Glu 35 and Trp 108 are too far away from the —C6H₂OH group to contribute to this distortion]. This strained conformation is stabilized by a strong hydrogen bond between the D ring O6 and the backbone NH of Val 109 (transition state stabilization). Indeed, the mutation of Val 109 to Pro, which lacks the NH group to make such a hydrogren bond, inactivates the enzyme. Lysozyme's lack of hydrolytic activity when XylNAc occupies its D subsite is likewise explained by the absence of this hydrogen bond and the consequent lesser stability of the XylNAc ring's half-chair transition state.

The unexpectedly small free energy differences in binding NAG, NAG lactone, and XylNAc to the D subsite are explained by the observation that undistorted NAG and XylNAc can be modeled into the D subsite as it occurs in the lysozyme·NAM-NAG-NAM complex. NAM's bulky lactyl side chain prevents it from binding to the D subsite in this manner.

The closest distance that an Asp 52 carboxylate O atom can approach C1 of the D site NAM without significantly disrupting the protein structure now appears to be 2.3 Å rather than the 3.0 Å surmised by the original Phillips mechanism. This distance is, nevertheless, ~0.8 Å longer than a C—O covalent bond. Moreover, to form such a bond, the Asp 52 carboxylate group would have to approach the NAM C1 atom from an unfavorable direction (stereoelectronic hindrance). Hence, it is still all but certain that the lysozyme reaction occurs via an electrostatically stabilized oxonium ion intermediate (Figure 14-14) rather than a covalent intermediate. Finally, a carboxyl O atom of Glu 35 forms a strong hydrogen bond with the O1 atom of the D site NAM which, no doubt, facilitates the proton transfer step in the Phillips mechanism. Thus, the Phillips mechanism appears to be substantially correct.

3. SERINE PROTEASES

Our next example of enzymatic mechanisms is a diverse group of proteolytic enzymes known as the **serine proteases** (Table 14-4). These enzymes are so named because they have a common catalytic mechanism characterized by the possession of a peculiarly reactive Ser residue that is essential for their enzymatic activity. The serine proteases are the most thoroughly understood family of enzymes as a result of their extensive examination over a more than 40-year period by kinetic, chemical, and physical techniques. In this section, we mainly study the best characterized serine proteases, **chymotrypsin, trypsin,** and **elastase.** We also consider how these three enzymes, which are synthesized in inactive forms, are physiologically activated.

A. Kinetics and Catalytic Groups

Chymotrypsin, trypsin, and elastase are digestive enzymes that are synthesized by the pancreatic acinar cells (Fig. 1-10c) and secreted, via the pancreatic duct, into the duodenum (the small intestine's upper loop). All of these enzymes catalyze the hydrolysis of peptide (amide) bonds but with different specificities for the side chains flanking the scissile (to be cleaved) peptide bond (recall that chymotrypsin is specific for a bulky hydrophobic residue preceding the scissile peptide bond, trypsin is specific for a positively charged residue, and elastase is specific for a small neutral residue; Table 6-2). Together, they form a potent digestive team.

Ester Hydrolysis as a Kinetic Model

That chymotrypsin can act as an esterase as well as a protease is not particularly surprising since the chemical mechanisms of ester and amide hydrolysis are almost identical. The study of chymotrypsin's esterase activity has led to important insights concerning this enzyme's catalytic mechanism. Kinetic measurements by Brian Hartley of the chymotrypsin-catalyzed hydrolysis of *p*-nitrophenylacetate

p-Nitrophenylacetate

Acetate   *p*-Nitrophenolate

indicated that the reaction occurs in two phases (Fig. 14-18):

1. The "burst phase," in which the highly colored *p*-nitrophenolate ion is rapidly formed in amounts stoichiometric with the quantity of active enzyme present.

TABLE 14-4. A Selection of Serine Proteases

Enzyme	Source	Function
Trypsin	Pancreas	Digestion of proteins
Chymotrypsin	Pancreas	Digestion of proteins
Elastase	Pancreas	Digestion of proteins
Thrombin	Vertebrate serum	Blood clotting
Plasmin	Vertebrate serum	Dissolution of blood clots
Kallikrein	Blood and tissues	Control of blood flow
Complement C1	Serum	Cell lysis in the immune response
Acrosomal protease	Sperm acrosome	Penetration of ovum
Lysosomal protease	Animal cells	Cell protein turnover
Cocoonase	Moth larvae	Dissolution of cocoon after metamorphosis
α-Lytic protease	*Bacillus sorangium*	Possibly digestion
Proteases A and B	*Streptomyces griseus*	Possibly digestion
Subtilisin	*Bacillus subtilus*	Possibly digestion

Source: Stroud, R.M., *Sci. Am.* **231**(1): 86 (1974).

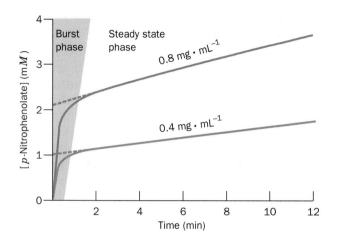

FIGURE 14-18. The time course of *p*-nitrophenylacetate hydrolysis as catalyzed by two different concentrations of chymotrypsin. The enzyme rapidly binds substrate and releases the first product, *p*-nitrophenolate ion, but the second product, acetate ion, is released more slowly. Consequently, the rate of *p*-nitrophenolate generation begins rapidly (burst phase) but slows as acyl–enzyme complex accumulates until the rate of *p*-nitrophenolate generation approaches that of acetate release (steady state). The extrapolation of the steady state curve to zero time (*dashed lines*) indicates the initial concentration of active enzyme. [After Hartley, B.S. and Kilby, B.A., *Biochem. J.* **56,** 294 (1954).]

2. The "steady state phase," in which *p*-nitrophenolate is generated at a reduced but constant rate that is independent of substrate concentration.

These observations have been interpreted in terms of a two-stage reaction sequence in which the enzyme (1) rapidly reacts with the *p*-nitrophenylacetate to release *p*-nitrophenylate ion forming a covalent acyl–enzyme intermediate that (2) is slowly hydrolyzed to release acetate:

Chymotrypsin evidently follows a Ping Pong Bi Bi mechanism (Section 13-5A). Chymotrypsin-catalyzed amide hydrolysis has been shown to follow a reaction pathway similar to that of ester hydrolysis but with the first step of the reaction, enzyme acylation, being rate determining rather than the deacylation step.

Identification of the Catalytic Residues

Chymotrypsin's catalytically important groups have been identified by chemical labeling studies. These are described below.

Ser 195. A diagnostic test for the presence of the **active Ser** of serine proteases is its reaction with **diisopropylphosphofluoridate (DIPF):**

which irreversibly inactivates the enzyme. Other Ser residues, including those on the same protein, do not react with DIPF. *DIPF reacts only with Ser 195 of chymotrypsin, thereby demonstrating that this residue is the enzyme's active Ser.*

The use of DIPF as an enzyme inactivating agent came about through the discovery that organophosphorus compounds such as DIPF are potent nerve poisons. The neurotoxicity of DIPF arises from its ability to inactivate **acetylcholinesterase,** a serine esterase that catalyzes the hydrolysis of **acetylcholine.**

Acetylcholine is a **neurotransmitter:** It transmits nerve impulses across the **synapses** (junctions) between certain types of nerve cells (Section 34-4C). The inactivation of acetyl-cholinesterase prevents the otherwise rapid hydrolysis of the acetylcholine released by a nerve impulse and thereby interferes with the regular sequence of nerve impulses. DIPF is of such great toxicity to humans that it has been used militarily as a nerve gas. Related compounds, such as **parathion** and **malathion,**

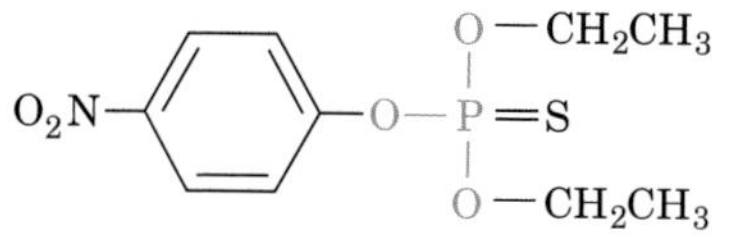

Parathion

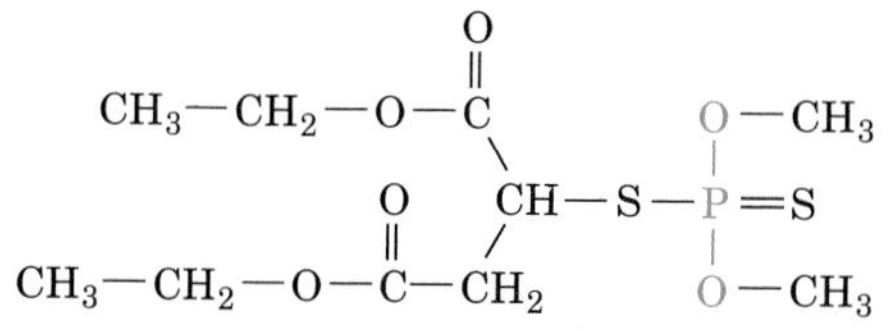

Malathion

are useful insecticides because they are far more toxic to insects than to mammals.

His 57. A second catalytically important residue was discovered through **affinity labeling.** In this technique, a substrate analog bearing a reactive group specifically binds at the enzyme's active site, where it reacts to form a stable covalent bond with a nearby susceptible group (these reactive substrate analogs have therefore been described as the "Trojan horses" of biochemistry). The affinity labeled groups can subsequently be identified by fingerprinting (Section 6-1J). Chymotrypsin specifically binds **tosyl-L-phenylalanine chloromethyl ketone (TPCK),**

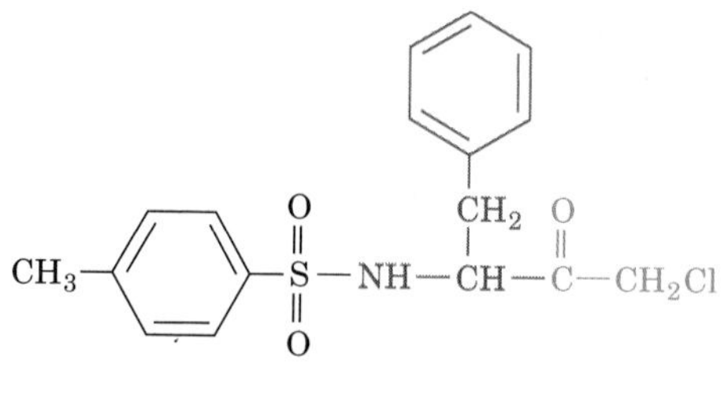

because of its resemblance to a Phe residue (one of chymotrypsin's preferred residues; Table 6-2). Active site–bound TPCK's chloromethyl ketone group is a strong alkylating agent; it reacts only with His 57 (Fig. 14-19), thereby inactivating the enzyme. The TPCK reaction is inhibited by **β-phenylpropionate,**

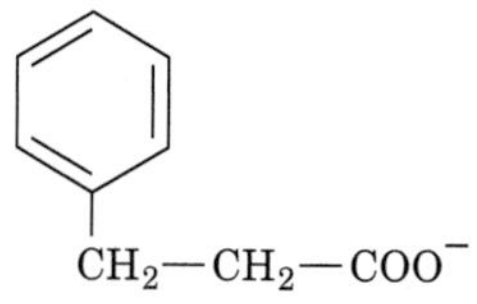

β-Phenylpropionate

a competitive inhibitor of chymotrypsin that presumably competes with TPCK for its enzymatic binding site. More-

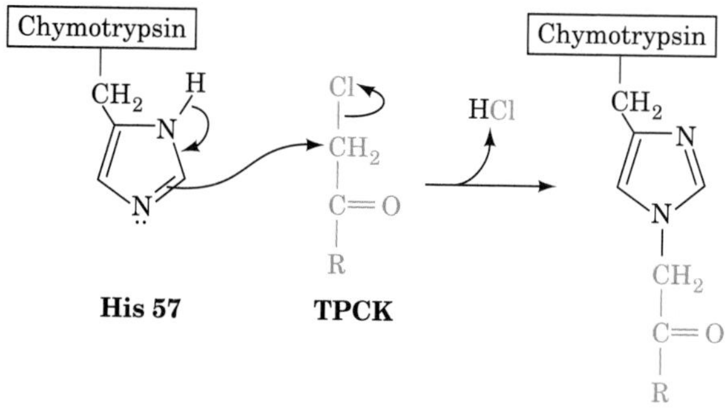

FIGURE 14-19. The reaction of TPCK with His 57 of chymotrypsin.

over, the TPCK reaction does not occur in 8M urea, a denaturing reagent, or with DIP–chymotrypsin, in which the active site is blocked. These observations establish that *His 57 is an essential active site residue of chymotrypsin.*

B. X-Ray Structures

Bovine chymotrypsin, bovine trypsin, and porcine elastase are strikingly homologous: The primary structures of these ~240-residue monomeric enzymes are ~40% identical and their internal sequences are even more alike (in comparison, the α and β chains of human hemoglobin have a 44% sequence identity). Furthermore, *all of these enzymes have an active Ser and a catalytically essential His as well as similar kinetic mechanisms.* It therefore came as no surprise when their X-ray structures all proved to be closely related.

To most conveniently compare the structures of these three digestive enzymes, they have been assigned the same amino acid residue numbering scheme. Bovine chymotrypsin is synthesized as an inactive 245-residue precursor named **chymotrypsinogen** that is proteolytically converted to chymotrypsin (Section 14-3E). In what follows, the numbering of the amino acid residues in chymotrypsin, trypsin, and elastase will be that of the corresponding residues in bovine chymotrypsinogen.

The X-ray structure of bovine chymotrypsin was elucidated in 1967 by David Blow. This was followed by the determination of the structures of bovine trypsin (Fig. 14-20) by Robert Stroud and Richard Dickerson, and porcine elastase by David Shotton and Herman Watson. Each of these proteins is folded into two domains, both of which have extensive regions of antiparallel β-sheets in a barrel-like arrangement but contain little helix. *The catalytically essential His 57 and Ser 195 are located at the substrate-binding site together with the invariant (in all serine proteases) Asp 102, which is buried in a solvent-inaccessible pocket. These three residues form a hydrogen bonded constellation referred to as the **catalytic triad** (Figs. 14-20 and 14-21).*

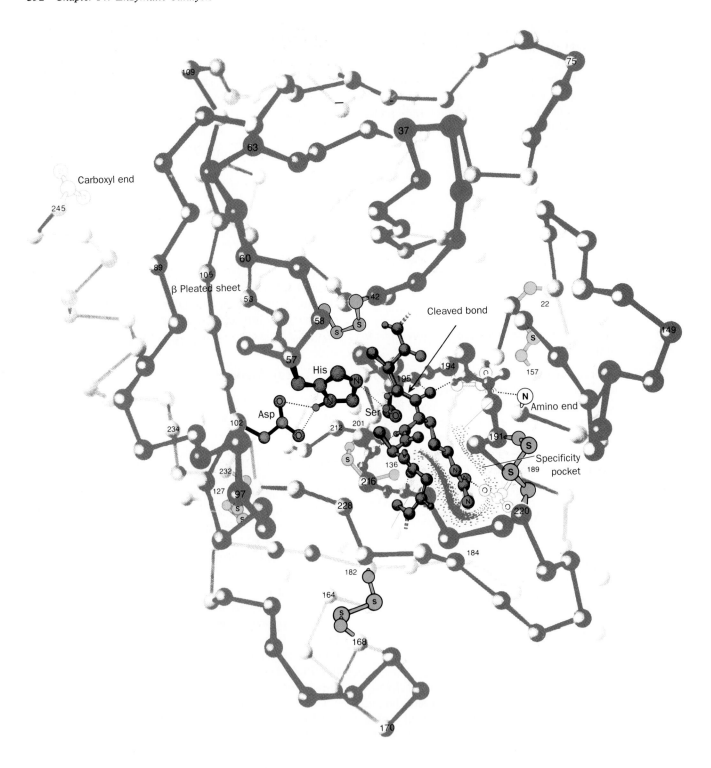

FIGURE 14-20. The X-ray structure of bovine trypsin. (*a*) A drawing of the enzyme in complex with a polypeptide substrate (*green*) that has its Arg side chain occupying the enzyme's specificity pocket (*stippling*).The C$_\alpha$ backbone of the enzyme is shown together with its disulfide bonds and the side chains of the catalytic triad, Ser 195, His 57, and Asp 102. The active sites of chymotrypsin and elastase contain almost identically arranged catalytic triads. [Figure copyrighted © by Irving Geis.]

(*b*) A ribbon diagram of trypsin highlighting its secondary structure and indicating the arrangement of its catalytic triad. (*c*) A computer-generated drawing showing the surface of trypsin (*blue*) superimposed on its polypeptide backbone (*purple*). The side chains of the catalytic triad are shown in green. [Courtesy of Arthur Olson, The Scripps Research Institute, La Jolla, California.] Parts *a*, *b*, and *c* have approximately the same orientation.

(b)

(c)

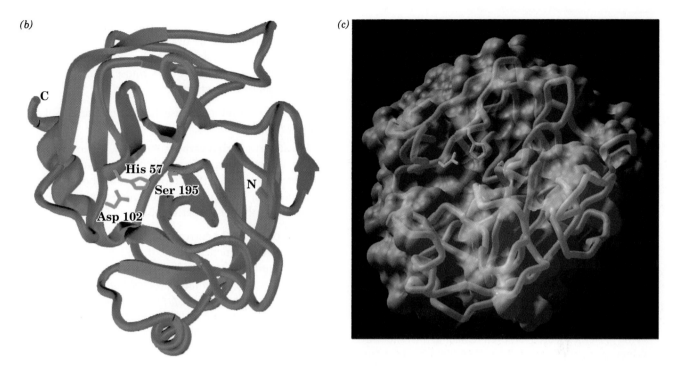

FIGURE 14-20. *(Continued)*

Substrate Specificities Can Be Only Partially Rationalized

The X-ray structures of the above three enzymes suggest the basis for their differing substrate specificities (Table 6-2):

1. In chymotrypsin, the bulky aromatic side chain of the preferred Phe, Trp, or Tyr residue that contributes the carbonyl group of the scissile peptide fits snugly into a slitlike hydrophobic pocket located near the catalytic groups.

2. In trypsin, the residue corresponding to chymotrypsin Ser 189, which lies at the back of the binding pocket, is the anionic residue Asp. The cationic side chains of trypsin's preferred residues, Arg or Lys, can therefore form ion pairs with this Asp residue. The rest of chymotryp-

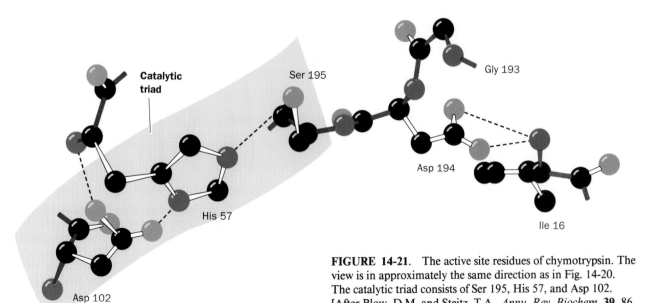

FIGURE 14-21. The active site residues of chymotrypsin. The view is in approximately the same direction as in Fig. 14-20. The catalytic triad consists of Ser 195, His 57, and Asp 102. [After Blow, D.M. and Steitz, T.A., *Annu. Rev. Biochem.* **39**, 86 (1970).]

sin's specificity pocket is preserved in trypsin so that it can accommodate the bulky side chains of Arg and Lys.

3. Elastase is so named because it rapidly hydrolyzes the otherwise nearly indigestible Ala, Gly, and Val-rich protein elastin (elastin is discussed in Section 7-2D). Elastase's binding pocket is largely occluded by the side chains of a Val and a Thr residue that replace two Gly's lining the specificity pocket in both chymotrypsin and trypsin. Consequently elastase, whose substrate-binding pocket is better described as a depression, specifically cleaves peptide bonds after small neutral residues, particularly Ala. In contrast, chymotrypsin and trypsin hydrolyze such peptide bonds extremely slowly because these small substrates cannot be sufficiently immobilized on the enzyme surface for efficient catalysis to occur (Section 14-1E).

Despite the foregoing, the mutagenic change of trypsin's Asp 189 to Ser by William Rutter did not switch its specificity to that of chymotrypsin but instead yielded a poor, nonspecific protease. Moreover, even replacing the other three residues in trypsin's specificity pocket that differ in chymotrypsin, with those of chymotrypsin, fails to yield a significantly improved enzyme. However, trypsin is converted to a reasonably active chymotrypsin-like enzyme when, in addition to the foregoing, its two surface loops that connect the walls of the specificity pocket (residues 185-188 and 221-225) are replaced by those of chymotrypsin. Curiously, these loops, whose sequences are largely conserved in each enzyme, are not structural components of either the specificity pocket or the extended substrate binding site in chymotrypsin or in trypsin (Figure 14-20a). Thus, the structural basis of the difference in specificity between chymotrypsin and trypsin remains, at least in part, enigmatic. These results therefore highlight an important caveat for genetic engineers: *Enzymes are so exquisitely tailored to their functions that they often respond to mutagenic tinkering in unexpected ways.*

Evolutionary Relationships among Serine Proteases

We have seen that sequence and structural homologies among proteins reveal their evolutionary relationships (Sections 6-3 and 8-3). *The great similarities among chymotrypsin, trypsin, and elastase indicate that these proteins evolved through gene duplications of an ancestral serine protease followed by the divergent evolution of the resulting enzymes (Section 6-3C).*

Several serine proteases from various sources provide further insights into the evolutionary relationships among the serine proteases. ***Streptomyces griseus* protease A (SGPA)** is a bacterial serine protease of chymotryptic specificity that exhibits extensive structural similarity, although only ~20% sequence identity, with the pancreatic serine proteases. The primordial trypsin gene evidently arose before the divergence of prokaryotes and eukaryotes.

There are two known serine proteases whose primary and tertiary structures bear no discernable relationship to each other or to chymotrypsin but which, nevertheless, contain catalytic triads at their active sites whose structures closely resemble that of chymotrypsin:

1. **Subtilisin,** an endopeptidase that was originally isolated from *Bacillus subtilus.*

2. Wheat germ **serine carboxypeptidase II,** an exopeptidase whose structure is surprisingly similar to that of carboxypeptidase A (Figure 7-19a) even though the latter protease has an entirely different catalytic mechanism from that of the serine proteases (see Problem 3).

Since the orders of the corresponding active site residues in the amino acid sequences of the three types of serine proteases are quite different (Fig. 14-22), *it seems highly improbable that they could have evolved from a common ancestor serine protease. These proteins apparently constitute a remarkable example of **convergent evolution:** Nature seems to have independently discovered the same catalytic mechanism at least three times.*

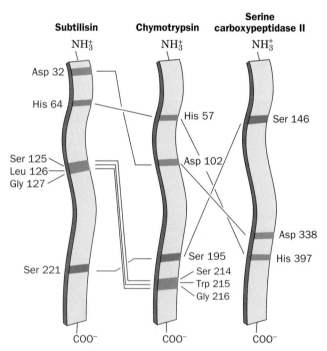

FIGURE 14-22. A diagram indicating the relative positions of the active site residues in the primary structures of subtilisin (*left*), chymotrypsin (*middle*), and serine carboxypeptidase II (*right*). The catalytic triad consists of Ser 221, His 64, and Asp 32 in subtilisin and of Ser 146, His 397, and Asp 338 in serine carboxypeptidase II. The peptide backbones of Ser 214, Trp 215, and Gly 216 in chymotrypsin, and their counterparts in subtilisin, participate in substrate-binding interactions. [After Robertus, J.D., Alden, R.A., Birktoft, J.J., Kraut, J., Powers, J.C., and Wilcox, P.E., *Biochemistry* **11,** 2449 (1972).]

C. Catalytic Mechanism

The extensive active site homologies among the various serine proteases indicates that they all have the same catalytic mechanism. On the basis of considerable chemical and structural data gathered in many laboratories, the following catalytic mechanism has been formulated for the serine proteases, here given in terms of chymotrypsin (Fig. 14-23):

1. After chymotrypsin has bound substrate to form the Michaelis complex, *Ser 195, in the reaction's rate-determining step, nucleophilically attacks the scissile peptide's carbonyl group to form a transition state complex known as the **tetrahedral intermediate** (covalent catalysis).* X-Ray studies indicate that Ser 195 is ideally positioned to carry out this nucleophilic attack (proximity and orientation effects). The imidazole ring of His 57 takes up the liberated proton, thereby forming an imidazolium ion (general base catalysis). This process is aided by the polarizing effect of the unsolvated carboxylate ion of Asp 102, which is hydrogen bonded to His 57 (electrostatic catalysis; see Section 14-3D). Indeed, the mutagenic replacement of trypsin's Asp 102 by

Asn leaves the enzyme's K_M substantially unchanged at neutral pH but reduces its k_{cat} to <0.05% of its wild-type value. Neutron diffraction studies have demonstrated that *Asp 102 remains a carboxylate ion rather than abstracting a proton from the imidazolium ion to form an uncharged carboxylic acid group.* The tetrahedral intermediate has a well-defined, although transient, existence. We shall see that *much of chymotrypsin's catalytic power derives from its preferential binding of this transition state (transition state binding catalysis).*

2. The tetrahedral intermediate decomposes to the **acyl–enzyme intermediate** under the driving force of proton donation from N3 of His 57 (general acid catalysis). The amine leaving group (R'NH₂, the new N-terminal portion of the cleaved polypeptide chain) is released from the enzyme and replaced by water from the solvent. The acyl–enzyme intermediate is extremely unstable to hydrolytic cleavage because of the enzyme's catalytic properties (see below). Despite this instability, the X-ray structure of elastase's acyl–enzyme intermediate has been reported. It was trapped at −55°C, at which tempera-

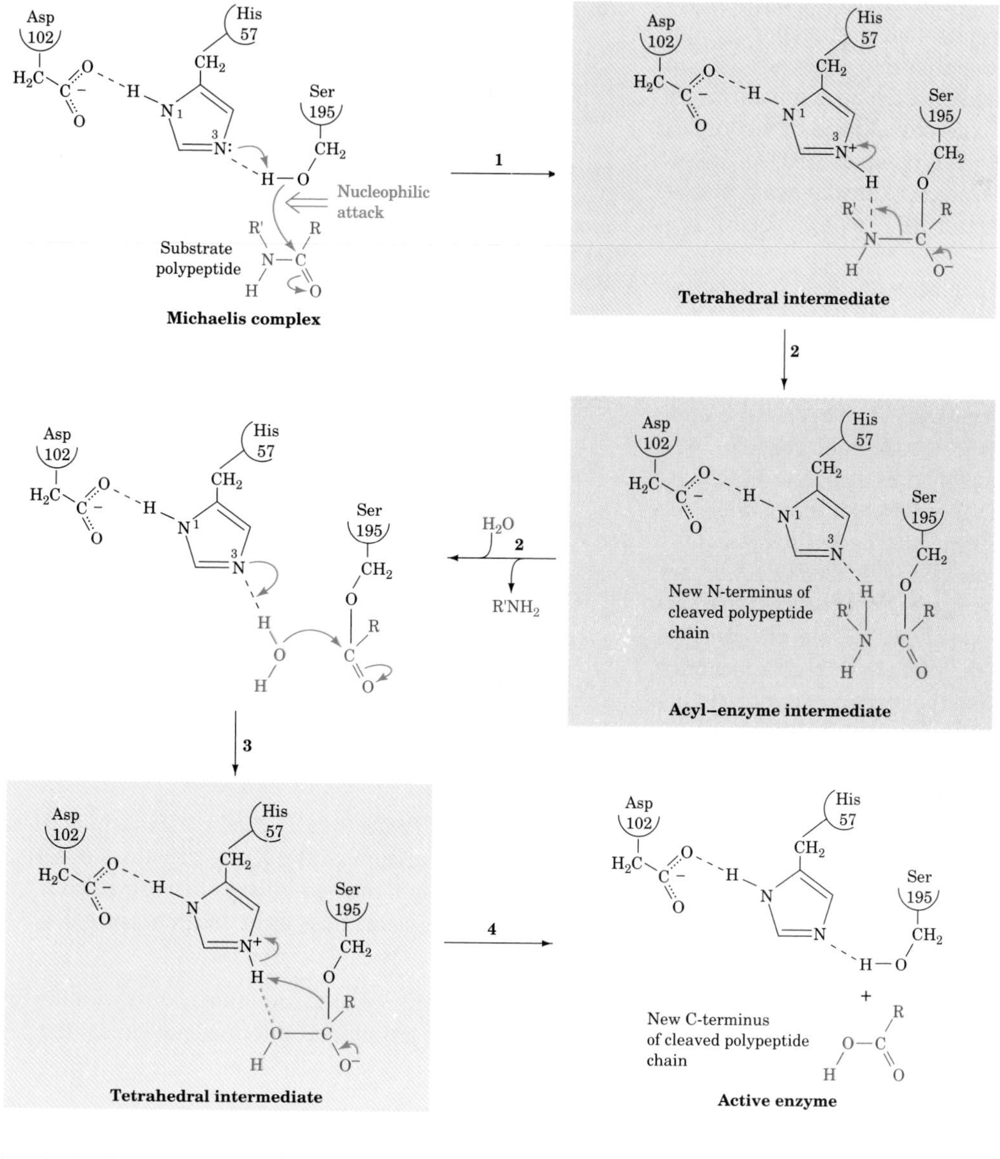

FIGURE 14-23. The catalytic mechanism of the serine proteases. The reaction involves **(1)** the nucleophilic attack of the active site Ser on the carbonyl carbon atom of the scissile peptide bond to form the tetrahedral intermediate; **(2)** the decomposition of the tetrahedral intermediate to the acyl–enzyme intermediate through general acid catalysis by the active site Asp-polarized His, followed by loss of the amine product and its replacement by a water molecule; **(3)** the reversal of Step 2 to form a second tetrahedral intermediate, and **(4)** the reversal of Step 1 to yield the reaction's carboxyl product and the active enzyme.

ture the rate of the enzymatic reaction is slowed by many orders of magnitude.

3 & 4. The acyl–enzyme intermediate is deacylated by what is essentially the reversal of the previous steps followed by the release of the resulting carboxylate product (the new C-terminal portion of the cleaved polypeptide chain), thereby regenerating the active enzyme. In this process, water is the attacking nucleophile and Ser 195 is the leaving group.

D. Testing the Catalytic Mechanism

The formulation of the foregoing model for catalysis by serine proteases has prompted numerous investigations of its validity. In this section we discuss several of the most revealing of these studies.

The Tetrahedral Intermediate Is Mimicked in a Complex of Trypsin with Trypsin Inhibitor

Perhaps the most convincing structural evidence for the existence of the tetrahedral intermediate was provided by Robert Huber in an X-ray study of the complex between **bovine pancreatic trypsin inhibitor (BPTI)** and trypsin. The 58-residue protein BPTI, whose folding pathway we examined in Section 8-1B, binds to and inactivates trypsin; this interaction prevents any trypsin that is prematurely activated in the pancreas from digesting that organ (see Section 14-3E). BPTI binds to the active site region of trypsin across a tightly packed interface that is cross-linked by a complex network of hydrogen bonds. This complex's $10^{13}M^{-1}$ association constant, among the largest of any known protein–protein interaction, emphasizes BPTI's physiological importance.

The portion of BPTI in contact with the trypsin active site resembles bound substrate. The side chain of BPTI Lys 15I (here "I" differentiates BPTI residues from trypsin residues) occupies the trypsin specificity pocket (Fig. 14-24a) and the peptide bond between Lys 15I and Ala 16I is positioned as if it were the scissile peptide bond (Fig. 14-24b). What is most remarkable about this structure is that *its active site complex assumes a conformation well along the reaction coordinate towards the tetrahedral intermediate: The side chain oxygen of trypsin Ser 195, the active Ser, is in closer-than-van der Waals contact (2.6 Å) with the pyramidally distorted carbonyl carbon of BPTI's "scissile" peptide.* Despite this close contact, the proteolytic reaction cannot proceed past this point along the reaction coordinate because of the rigidity of the active site complex and because it is so tightly sealed that the leaving group cannot leave and water cannot enter the reaction site.

Protease inhibitors are common in nature, where they have protective and regulatory functions. For example, certain plants release protease inhibitors in response to insect bites, thereby causing the offending insect to starve by inactivating its digestive enzymes. Protease inhibitors constitute ~10% of the nearly 200 proteins of blood serum. For instance, **α_1-proteinase inhibitor,** which is secreted by the liver,

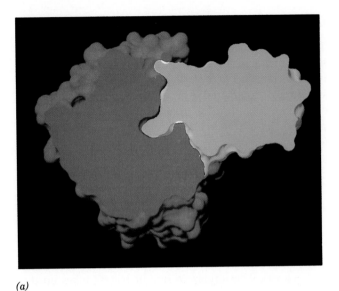

(a)

(b)

FIGURE 14-24. The trypsin–BPTI complex. (*a*) The X-ray structure shown as a computer-generated cutaway drawing indicating how trypsin (*red*) binds BPTI (*green*). The green protrusion extending into the red cavity near the center of the figure represents the Lys 15I side chain occupying trypsin's specificity pocket. Note the close complementary fit of these two proteins. [Courtesy of Michael Connolly, New York University.] (*b*) Trypsin Ser 195, the active Ser, is in closer-than-van der Waals contact with the carbonyl carbon of BPTI's scissile peptide, which is pyramidally distorted towards Ser 195. The normal proteolytic reaction is apparently arrested somewhere along the reaction coordinate between the Michaelis complex and the tetrahedral intermediate.

inhibits **leukocyte elastase** (leukocytes are a type of white blood cell; the action of leukocyte elastase is thought to be part of the inflammatory process). Pathological variants of α_1-proteinase inhibitor with reduced activity are associated with **pulmonary emphysema,** a degenerative disease of the lungs resulting from the hydrolysis of its elastic fibers. Smokers also suffer from reduced activity of their α_1-proteinase inhibitor because of the oxidation of its active site Met residue. Full activity of this inhibitor is not regained until several hours after smoking.

Serine Proteases Preferentially Bind the Transition State

Detailed comparisons of the X-ray structures of several serine protease–inhibitor complexes have revealed a further structural basis for catalysis in these enzymes (Fig. 14-25):

1. The conformational distortion that occurs with the formation of the tetrahedral intermediate causes the car-

(a) *(b)*

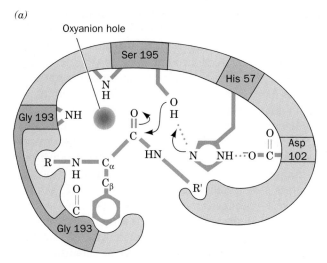

FIGURE 14-25. Transition state stabilization in the serine proteases: (*a*) In the Michaelis complex, the trigonal carbonyl carbon of the scissile peptide is conformationally constrained from binding in the oxyanion hole (*upper left*). (*b*) In the tetrahedral intermediate, the carbonyl oxygen of the scissile peptide (the oxyanion) has entered the oxyanion hole thereby hydrogen bonding to the backbone NH groups of Gly 193 and

Ser 195. The consequent conformational distortion permits the NH group of the peptide preceding the scissile peptide to form an otherwise unsatisfied hydrogen bond to Gly 193. Serine proteases therefore preferentially bind the tetrahedral intermediate. [After Robertus, J.D., Kraut, J., Alden, R.A., and Birktoft, J.J., *Biochemistry* **11**, 4302 (1972).]

bonyl oxygen of the scissile peptide to move deeper into the active site so as to occupy a previously unoccupied position — the **oxyanion hole.**

2. *There it forms two hydrogen bonds with the enzyme that cannot form when the carbonyl group is in its normal trigonal conformation.* These two enzymatic hydrogen bond donors were first noted by Joseph Kraut to occupy corresponding positions in chymotrypsin and subtilisin. He proposed the existence of the oxyanion hole based on the premise that convergent evolution had made the active sites of these unrelated enzymes functionally identical.

3. The tetrahedral distortion, moreover, permits the formation of an otherwise unsatisfied hydrogen bond between the enzyme and the backbone NH group of the residue preceding the scissile peptide. Consequently, *the enzyme binds the tetrahedral intermediate in preference to either the Michaelis complex or the acyl–enzyme intermediate.*

It is this phenomenon that is responsible for much of the catalytic efficiency of serine proteases. In fact, the reason that DIPF is such an effective inhibitor of serine proteases is because its tetrahedral phosphate group makes this compound a transition state analog of the enzyme.

The Water Molecule Attacking the Acyl–Enzyme Intermediate Has Been Directly Observed

A water molecule that seems poised to hydrolyze the acyl–enzyme intermediate of trypsin has been visualized by Robert Sweet through the use of a recently developed method for rapidly (in less than one second) recording an

X-ray diffraction data set. A crystal of trypsin was acylated by reacting it with *p*-nitrophenyl **guanidinobenzoate** to form *p*-guanidinobenzoyl–trypsin,

_p_-Nitrophenyl guanidinobenzoate

_p_-Nitrophenol

_p_-Guanidinobenzoyl–trypsin

a poorly reactive acyl–enzyme intermediate that is stable for days at pH 5.5 but which is hydrolyzed within hours at pH 8.5. Within three minutes after initiating the deacylation reaction by rapidly changing the pH of the acylated crystal from 5.5 to 8.5, a water molecule appeared in a previously unoccupied space above the plane of the ester carboxyl group. This water molecule is positioned along an unobstructed route to nucleophilically attack and hence hydrolyze the acyl–enzyme intermediate. However, when trypsin hydrolyzes a normal substrate, the polypeptide acyl group would occupy the space taken up by this water molecule. Nevertheless, structural evidence suggests that such a polypeptide is so loosely bound to the enzyme that it could be transiently displaced by an attacking water molecule.

The Role of the Catalytic Triad

The earlier literature postulated that the Asp 102-polarized His 57 side chain directly abstracts a proton from Ser 195, thereby converting its weakly nucleophilic —CH₂OH group to a highly nucleophilic alkoxide ion, —CH₂O⁻.

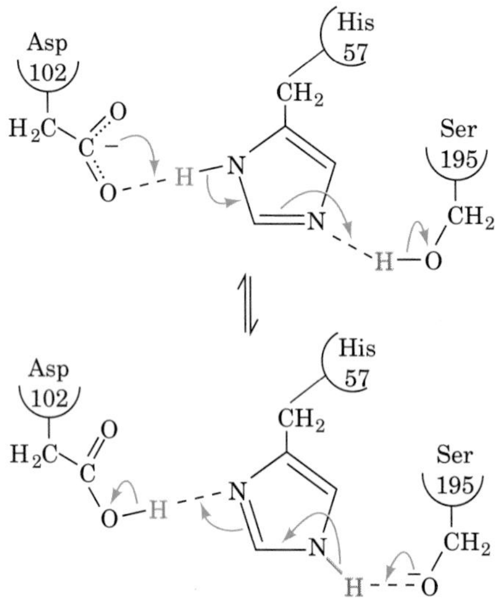

"Charge relay system"

In the process, the anionic charge of Asp 102 was thought to be transferred, via a tautomeric shift of His 57, to Ser 195. The catalytic triad was therefore originally named the **charge relay system.** It is now realized, however, that such a mechanism is implausible because an alkoxide ion ($pK \geq 15$) has far greater proton affinity than does His 57 ($pK \approx 7$, as measured by NMR techniques). How, then can Asp 102 nucleophilically activate Ser 195?

A possible solution to this conundrum has been pointed out by W.W. Cleland and Maurice Kreevoy and, independently, by John Gerlt and Paul Gassman. Proton transfers between hydrogen bonded groups (D—H · · · A) only occur at physiologically reasonable rates when the pK of the proton donor is no more than 2 or 3 pH units greater than that of the protonated form of the proton acceptor (the height of the kinetic barrier, $\Delta G^{\ddagger}$, for the protonation of an acceptor by a more basic donor increases with the difference between the pK's of the donor and acceptor). However, when the pK's of hydrogen bonding donor (D) and acceptor (A) groups are nearly equal, the distinction between them breaks down: *The hydrogen atom becomes more or less equally shared between them* (D · · · H · · · A). Such **low-barrier hydrogen bonds** are unusually strong; they have, as studies of model compounds indicate, association free energies as high as −40 to −80 kJ·mol⁻¹ vs the −12 to −30 kJ·mol⁻¹ for normal hydrogen bonds (the energy of the normally covalent D—H bond is subsumed into the low-barrier hydrogen bonding system). Thus, an effective enzymatic "strategy" would be

to convert a weak hydrogen bond in the Michaelis complex to a strong hydrogen bond in the transition state, thereby facilitating proton transfer while applying the difference in the free energy between the normal and low-barrier hydrogen bonds to preferentially binding the transition state. Indeed, as Perry Frey has shown, the NMR spectrum of the proton linking His 57 to Asp 102 in chymotrypsin is consistent with the formation of a low-barrier hydrogen bond in the transition state (see Fig. 14-25b; the pK's of protonated His 57 and Asp 102 are nearly equal in the anhydrous environment of the active site complex), which presumably promotes proton transfer from Ser 195 to His 57 as in the charge relay mechanism.

Despite the foregoing, blocking the action of the catalytic triad through the specific methylation of His 57 by treating chymotrypsin with **methyl-*p*-nitrobenzene sulfonate,**

Methyl-*p*-nitrobenzene sulfonate

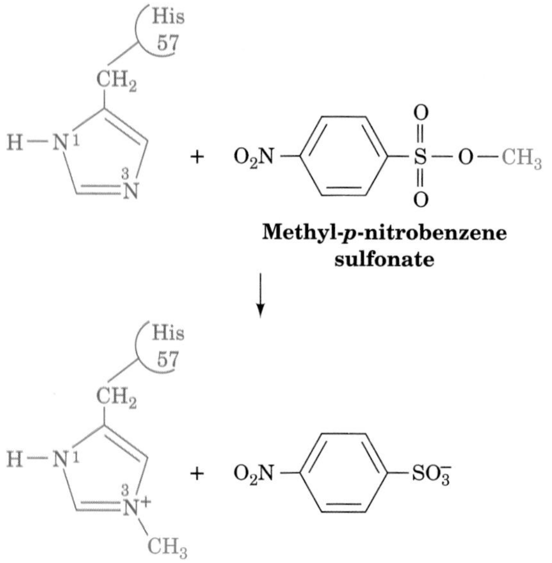

yields an enzyme that is a reasonably good catalyst: It enhances the rate of proteolysis by as much as a factor of 2 × 10⁶ over the uncatalyzed reaction, whereas the native enzyme has a rate enhancement factor of ~10¹⁰. Similarly, the mutation of Ser 195, His 57, or even all three residues of the catalytic triad, yields enzymes that enhance proteolysis rates by ~5 × 10⁴-fold over that of the uncatalyzed reaction. Evidently, the catalytic triad provides a nucleophile and is an alternate source and sink of protons (general acid–base catalysis). However, a large portion of chymotrypsin's rate enhancement must be attributed to its preferential binding of the catalyzed reaction's transition state.

E. Zymogens

Proteolytic enzymes are usually biosynthesized as somewhat larger inactive precursors known as **zymogens** (enzyme precursors, in general, are known as **proenzymes**). In the case of digestive enzymes, the reason for this is clear: If these enzymes were synthesized in their active forms, they would digest the tissues that synthesized them. Indeed,

acute pancreatitis, a painful and sometimes fatal condition that can be precipitated by pancreatic trauma, is characterized by the premature activation of the digestive enzymes synthesized by this gland.

Serine Proteases Are Autocatalytically Activated

Trypsin, chymotrypsin, and elastase are activated according to the following pathways:

Trypsin. The activation of **trypsinogen,** the zymogen of trypsin, occurs as a two-stage process when trypsinogen enters the duodenum from the pancreas. **Enteropeptidase** (originally named **enterokinase**), a serine protease that is secreted under hormonal control by the duodenal mucosa, specifically hydrolyzes trypsinogen's Lys 15—Ile 16 peptide bond, thereby excising its N-terminal hexapeptide (Fig. 14-26). This yields the active enzyme, which has Ile 16 at its N-terminus. Since this activating cleavage occurs at a trypsin-sensitive site (recall that trypsin cleaves after Arg and Lys residues), the small amount of trypsin produced by enteropeptidase also catalyzes activation, generating more trypsin, etc.; that is, trypsinogen activation is autocatalytic.

Chymotrypsin. Chymotrypsinogen is activated by the specific tryptic cleavage of its Arg 15—Ile 16 peptide bond, to form π-chymotrypsin (Fig. 14-27). π-Chymotrypsin subsequently undergoes autolysis (self-digestion) to specifically excise two dipeptides, Ser 14–Arg 15 and Thr 147–Asn 148, thereby yielding the equally active enzyme α-chymotrypsin (heretofore and hereafter referred to as chymotrypsin). The biochemical significance of this latter process, if any, is unknown.

$$\overset{+}{H_3N}-\overset{10}{Val}-(Asp)_4-\overset{15}{Lys}-\overset{16}{Ile}-Val-\cdots$$

Trypsinogen

enteropeptidase or trypsin

$$\overset{+}{H_3N}-Val-(Asp)_4-Lys \ + \ Ile-Val-\cdots$$

Trypsin

FIGURE 14-26. The activation of trypsinogen to form trypsin occurs by proteolytic excision of the N-terminal hexapeptide as catalyzed by either enteropeptidase or trypsin. Chymotrypsinogen residue numbering is used here; that is, Val 10 is actually trypsinogen's N-terminus and Ile 16 is trypsin's N-terminus.

Elastase. Proelastase, the zymogen of elastase, is activated similarly to trypsinogen by a single tryptic cleavage that excises a short N-terminal polypeptide.

Biochemical "Strategies" That Prevent Premature Zymogen Activation

Trypsin activates pancreatic **procarboxypeptidases A** and **B** and **prophospholipase A₂** (the action of phospholipase A_2 is outlined in Section 23-1) as well as the pancreatic serine proteases. Premature trypsin activation can consequently trigger a series of events that lead to pancreatic self-digestion. Nature has therefore evolved an elaborate defense against such inappropriate trypsin activation. We have already seen (Section 14-3D) that pancreatic trypsin

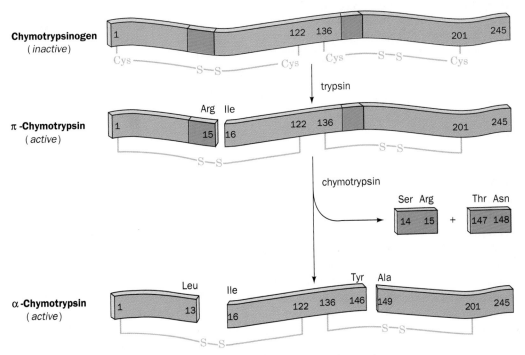

FIGURE 14-27. The activation of chymotrypsinogen by proteolytic cleavage. Both π- and α-chymotrypsin are enzymatically active.

inhibitor binds essentially irreversibly to any trypsin formed in the pancreas so as to inactivate it. Furthermore, the trypsin-catalyzed activation of trypsinogen (Fig. 14-26) occurs quite slowly, presumably because the unusually large negative charge of its highly evolutionarily conserved N-terminal hexapeptide repels the Asp at the back of trypsin's specificity pocket. Finally, pancreatic zymogens are stored in intracellular vesicles called **zymogen granules** whose membranous walls are thought to be resistant to enzymatic degradation.

Zymogens Have Distorted Active Sites

Since the zymogens of trypsin, chymotrypsin, and elastase have all their catalytic residues, why aren't they enzymatically active? Comparisons of the X-ray structures of trypsinogen with that of trypsin and of chymotrypsinogen with that of chymotrypsin show that upon activation, the newly liberated N-terminal Ile 16 residue moves from the surface of the protein to an internal position, where its free cationic amino group forms an ion pair with the invariant anionic Asp 194 (Fig. 14-21). Aside from this change, however, the structures of these zymogens closely resemble those of their corresponding active enzymes. Surprisingly, this resemblance includes their catalytic triads, an observation which led to the discovery that these zymogens are actually enzymatically active, albeit at a very low level. Careful comparisons of the corresponding enzyme and zymogen structures, however, revealed the reason for this low activity: *The zymogens' specificity pockets and oxyanion holes are improperly formed such that, for example, the amide NH of chymotrypsin's Gly 193 points in the wrong direction to form a hydrogen bond with the tetrahedral intermediate (see Fig. 14-25).* Hence, the zymogens' very low enzymatic activity arises from their reduced ability to bind substrate productively and to stabilize the tetrahedral intermediate. These observations provide further structural evidence favoring the role of transition state binding in the catalytic mechanism of serine proteases.

4. GLUTATHIONE REDUCTASE

Lysozyme and the serine proteases all catalyze hydrolytic reactions. In contrast, the third enzyme that we shall consider in mechanistic detail, **glutathione reductase,** catalyzes an oxidation–reduction reaction. Such reactions are extremely important in metabolic processes. We have chosen to study gluthathione reductase, which sequentially catalyzes several electron-transfer processes, because it is one of the few such enzymes in which the pathway of electron flow has been well characterized.

Glutathione reductase is a nearly ubiquitous enzyme that catalyzes the NADPH-dependent reduction of **glutathione disulfide (GSSG) to glutathione (GSH):**

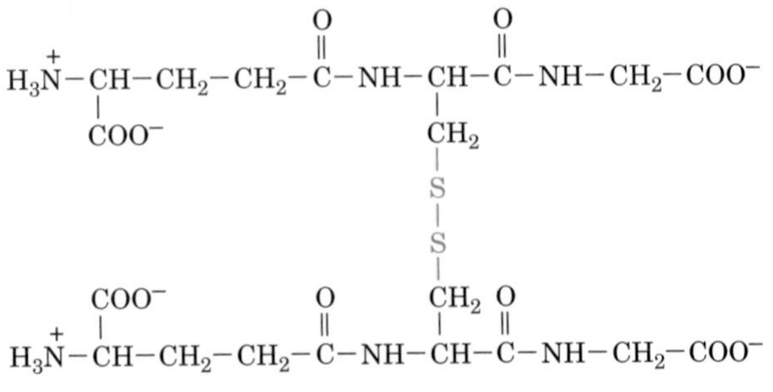

Glutathione disulfide (GSSG)

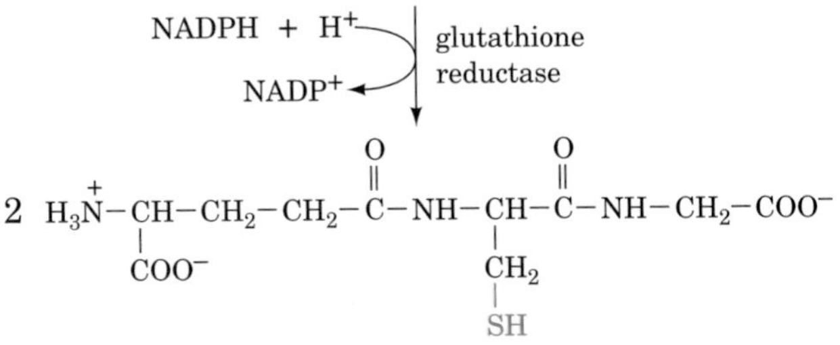

Glutathione (GSH)
(γ-L-Glutamyl-L-cysteinyl glycine)

(the structures of $NADP^+$ and NADPH are indicated in Fig. 12-2). This process normally produces a GSH:GSSG ratio of over 100:1, which permits GSH to function as an intracellular reducing agent (the thermodynamics of oxidation–reduction reactions is discussed in Section 15-5). For example, the inactivation of proteins (P) that have free SH groups through the spontaneous oxidative formation of mixed disulfides

$$P—SH + P'—SH + \tfrac{1}{2}O_2 \rightleftharpoons P—S—S—P' + H_2O$$

is reversed through disulfide interchange with GSH.

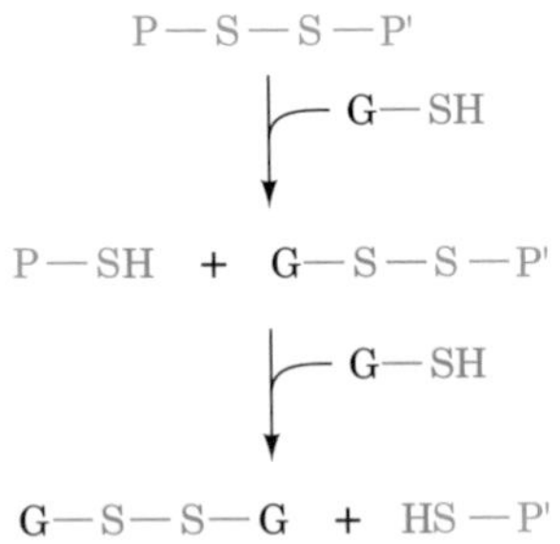

GSH also acts as a coenzyme in several enzymatically catalyzed reductions and plays an important role in the transport of amino acids into certain cells (Section 24-4C).

FAD Is an Essential Redox Coenzyme

Glutathione reductase contains the electron-transfer prosthetic group **flavin adenine dinucleotide (FAD;** Fig. 14-28). **Flavins** (substances that contain the **isoalloxazine** ring) can undergo two sequential one-electron transfers (Fig. 14-28), or a simultaneous two-electron transfer that bypasses the semiquinone state. The glutathione reductase reaction involves the simultaneous transfer of two electrons so that,

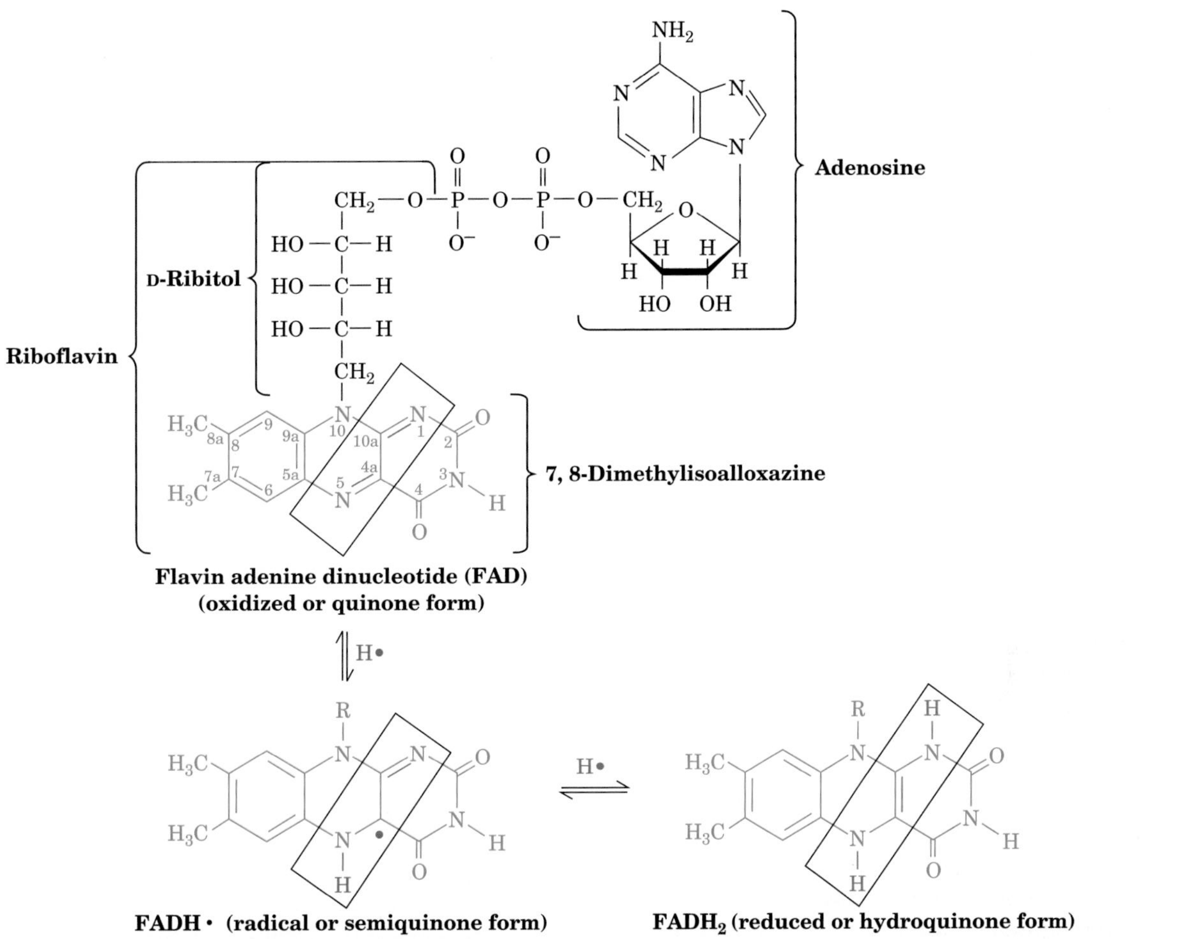

Flavin adenine dinucleotide (FAD)
(oxidized or quinone form)

FADH · (radical or semiquinone form) **FADH$_2$ (reduced or hydroquinone form)**

FIGURE 14-28. The molecular formula and reactions of the coenzyme flavin adenine dinucleotide (FAD). The term "flavin" is synonymous with the isoalloxazine ring system. The D-ribitol residue is derived from the alcohol of the sugar D-ribose. FAD may be half-reduced to the stable radical FADH· or fully reduced to FADH$_2$ (*boxes*). Consequently, different FAD-containing enzymes cycle between different oxidation states of FAD. FAD is usually tightly bound to its enzymes so that this coenzyme normally is a prosthetic group rather than a cosubstrate as is the case, for example, with NAD$^+$.

in this case, the FAD never assumes its radical form. The oxidation state of the flavin in a **flavoprotein** (flavin-containing protein) is readily established from its characteristic UV–visible spectrum: FAD is an intense yellow, whereas FADH$_2$ is pale yellow.

Humans and other higher animals are unable to synthesize the isoalloxazine component of flavins, so they must obtain this substance from their diets, for example, in the form of **riboflavin (vitamin B$_2$)** (Fig. 14-28). Riboflavin deficiency is quite rare in humans, in part because of the tight binding of flavin prosthetic groups to their apoenzymes. The symptoms of riboflavin deficiency, which are associated with general malnutrition or bizarre diets, include an inflamed tongue, lesions in the corners of the mouth, and dermatitis.

Glutathione Reductase Catalyzes a Two-Stage Reaction

Glutathione reductase from human erythrocytes is a dimer of identical 478-residue subunits that are covalently linked by an intersubunit disulfide bond. In the absence of GSSG, the enzyme catalyzes the first stage of a two-stage reaction:

First stage: $E + NADPH + H^+ \rightleftharpoons EH_2 + NADP^+$

where E represents fully oxidized glutathione reductase and EH$_2$ is a stable two-electron reduced intermediate whose chemical nature we shall presently discuss. Upon subsequent addition of GSSG, EH$_2$ reacts to form products and complete the catalytic cycle.

Second stage: $EH_2 + GSSG \rightleftharpoons E + 2GSH$

The glutathione reductase reaction is more complex than these overall reactions suggest. Vincent Massey and Charles Williams demonstrated that *oxidized glutathione reductase (E) contains a "redox-active" disulfide bond, which in EH$_2$ has accepted an electron pair through bond cleavage to form a dithiol.*

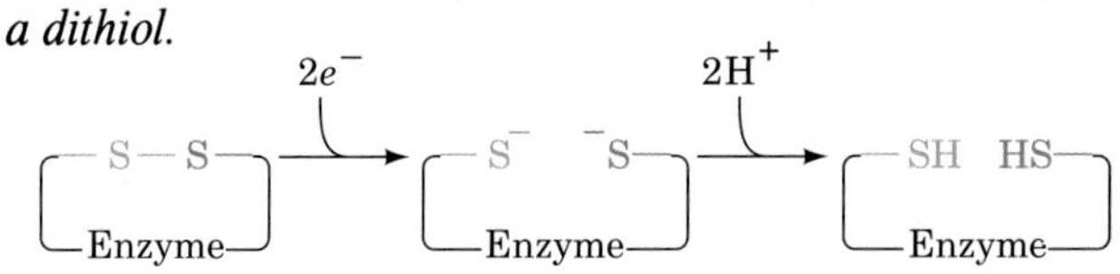

Through the use of arsenite, an ion that specifically complexes vicinal (adjacent) dithiols but not disulfide bonds,

they obtained the following information:

1. The spectrum of oxidized glutathione reductase (E) is unaffected by arsenite.

2. When NADPH reacts with the enzyme in the presence of arsenite, the reduced glutathione reductase (EH_2) produced binds arsenite to form an enzymatically inactive species.

3. The spectrum of the inactive EH_2 indicates that its FAD prosthetic group is fully oxidized.

Glutathione reductase must therefore have a second electron acceptor besides FAD; arsenite's known specificity suggests that the acceptor is a disulfide.

Glutathione reductase's amino acid sequence indicates that its redox-active disulfide bond forms between Cys 58 and Cys 63, which occur on a highly conserved segment of the enzyme's polypeptide chain. Thus, in the first stage of the glutathione reductase reaction, NADPH reduces the enzyme's redox-active disulfide and, in the second stage, this reduced disulfide reduces GSSG to 2GSH.

The X-Ray Structure of Glutathione Reductase

The X-ray structure of human erythrocyte glutathione reductase, which was determined by Georg Schulz and Heiner Schirmer, indicates that the protein's monomer units each contain considerable secondary structure and are folded into five domains (Fig. 14-29a,b). The active site involves at least four of these domains and the GSSG-binding site spans both subunits (Fig. 14-29c,c). Electrons are passed from NADPH to FAD, thereby reducing the flavin, which in turn reduces the redox-active disulfide and ultimately GSSG. The flavin is almost completely buried in the protein, which prevents the surrounding solution from interfering with the electron-transfer process ($FADH_2$, but neither NADPH nor thiols, is rapidly oxidized by O_2).

The arrangement of the groups in glutathione reductase's catalytic center and its reaction sequence is diagrammed in Fig. 14-30. In the absence of NADPH, the phenol side chain of Tyr 197 covers the nicotinamide-binding pocket so as to shield the flavin from contact with the solution (Fig. 14-30a). This side chain moves aside in the presence of

NADPH, thereby permitting the nicotinamide ring to bind parallel to and in van der Waals contact with the flavin ring of the fully oxidized enzyme E (Fig. 14-31). The H_S substituent to reduced nicotinamide's prochiral C4 atom (that facing the flavin), the H atom which is known to be lost in the glutathione reductase reaction, lies near flavin atom N5, the position through which electrons often enter a flavin ring upon its reduction. This positioning is particularly significant in view of the catalytic mechanism for glutathione reductase described below.

Catalytic Mechanism

Protein X-ray structures neither reveal H atom positions nor indicate pathways of electron transfer. The electron-transfer pathway in the glutathione reductase reaction has nevertheless been inferred from the X-ray structures of a series of stable enzymatic reaction intermediates as augmented with a variety of enzymological data (Fig. 14-30):

First Stage of the Catalytic Cycle: NADPH Reduction

1. NADPH binds to the oxidized enzyme, E, and immediately reduces the flavin, yielding $NADP^+$. The resulting reduced flavin anion ($FADH^-$) has but transient existence. The redox-active disulfide bond linking Cys 58 and Cys 63 lies between the GSSG-binding site and the flavin ring on the opposite side of the flavin from the nicotinamide pocket (Fig. 14-29c). Spectroscopic studies as well as the chemistry of model compounds suggest that rapid electron transfer between the flavin and the redox-active disulfide bond occurs through the transient formation of a covalent bond between the sulfur of Cys 63 and flavin atom C4a. This releases Cys 58, which acquires a solvent proton yielding a thiol group (Fig. 14-32; Step 1). The covalent Cys 63–flavin adduct rapidly collapses to a **charge-transfer complex** (a noncovalent interaction in which an electron pair is partially transferred from a donor, in this case the Cys 63 thiolate ion, to an acceptor, in this case the oxidized flavin ring; Fig. 14-32, Step 2).

2. The release of $NADP^+$ yields reduced glutathione reductase, EH_2. This model is corroborated by the X-ray structure of EH_2, which indicates that the redox-active disulfide bridge has opened such that the resulting Cys 63 thiol group is in contact with the flavin ring near its C4a position (Fig. 14-31). The red color of EH_2 crystals is indicative of this charge-transfer complex (the yellow color of crystals of oxidized glutathione reductase, E, is indicative of oxidized FAD).

Second Stage of the Catalytic Cycle: Glutathione Reduction

3. The second stage of the glutathione reductase reaction begins with the binding of GSSG to EH_2.

4. S_{58} (the Cys 58 thiol) nucleophilically attacks S_I (the S on the GS unit nearest Cys 58), yielding a mixed disulfide

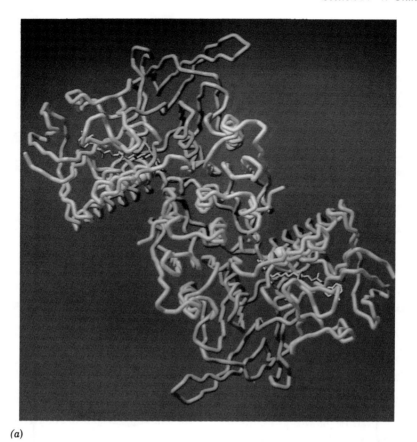

(a)

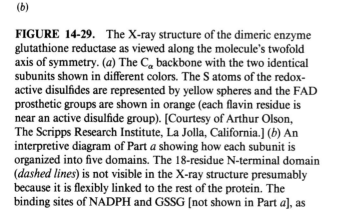

(b)

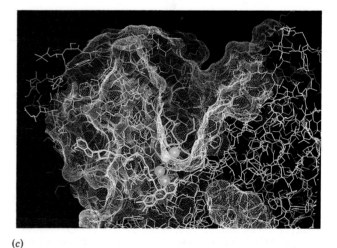

(c)

FIGURE 14-29.   The X-ray structure of the dimeric enzyme glutathione reductase as viewed along the molecule's twofold axis of symmetry. (*a*) The C_α backbone with the two identical subunits shown in different colors. The S atoms of the redox-active disulfides are represented by yellow spheres and the FAD prosthetic groups are shown in orange (each flavin residue is near an active disulfide group). [Courtesy of Arthur Olson, The Scripps Research Institute, La Jolla, California.] (*b*) An interpretive diagram of Part *a* showing how each subunit is organized into five domains. The 18-residue N-terminal domain (*dashed lines*) is not visible in the X-ray structure presumably because it is flexibly linked to the rest of the protein. The binding sites of NADPH and GSSG [not shown in Part *a*], as well as those of FAD, are indicated. The two subunits are covalently linked by a disulfide bridge across the molecular twofold axis. [After Pai, E.F. and Schulz, G.E., *J. Biol. Chem.* **258**, 1753 (1983).] (*c*) The active site region of glutathione reductase showing the FAD and GSSR positions. Acidic residues (Asp and Glu) are red, basic residues (Arg, Lys, and His) are blue, and all other residues are white. The dot surface, which is colored according to the nearest residue, represents the protein's solvent-accessible surface. The FAD has thick yellow bonds, the GSSR has thick green bonds, and the redox-active sulfur atoms of both the enzyme (*below*) and substrate (*above*) are represented by green spheres. [Courtesy of John Kuriyan, The Rockefeller University.]

treated pathogens); **NADH peroxidase,** a hydroperoxidase from *Streptococcus faecalis* 10C1 that lacks a redox-active disulfide but, instead, has a single Cys residue that, in the oxidized enzyme, is converted to a **sulfenic acid** (Cys—SOH) residue; and **thioredoxin reductase,** which functions to maintain the redox-active protein **thioredoxin** in its reduced state (Section 26-4A).

All of the above enzymes have sequence identities in the range 20 to 35%. Not surprisingly, therefore, they have similar tertiary and quaternary structures as well as catalytic mechanisms. In an exception to this generalization, however, thioredoxin reductase has a quaternary structure unlike that of other disulfide oxidoreductases of known structure. Thus, in glutathione reductase, for example, a given active site consists of residues from both subunits of the dimeric enzyme, whereas each active site of thioredoxin reductase, also a dimeric enzyme, consists of residues from only one subunit. Moreover, the redox-active disulfides of these two enzymes lie on opposite sides of the flavin ring. Since these enzymes have many common structural features but, nevertheless, have quite different arrangements of active site residues, it appears that they have diverged from a common ancestral nucleotide-binding protein (divergent evolution) and later independently acquired their disulfide reductase activities (convergent evolution).

In this chapter we have discussed the various mechanisms that enzymes utilize in catalyzing reactions and have shown in detail how these mechanisms are applied in three of the best understood systems. Throughout the text we shall continue our discussion of catalytic mechanisms, using the principles we have set out here as they are understood to apply to the various enzymes we encounter along the way.

CHAPTER SUMMARY

Most enzymatic mechanisms of catalysis have ample precedent in organic catalytic reactions. Acid–base catalyzed reactions occur, respectively, through the donation or abstraction of a proton to or from a reactant so as to stabilize the reaction's transition state complex. Enzymes often employ ionizable amino acid side chains as general acid–base catalysts. Covalent catalysis involves nucleophilic attack of the catalyst on the substrate to transiently form a covalent bond followed by the electrophilic stabilization of a developing negative charge in the reaction's transition state. Various protein side chains as well as certain coenzymes can act as covalent catalysts. Metal ions, which are common enzymatic components, catalyze reactions by stabilizing developing negative charges in a manner resembling general acid catalysis. Metal ion–bound water molecules are potent sources of OH^- ions at neutral pH's. Metal ions also facilitate enzymatic reactions through the charge shielding of bound substrates. The arrangement of charged groups about an enzymatic active site of low dielectric constant in a manner that stabilizes the transition state complex results in the electrostatic catalysis of the enzymatic reaction. Enzymes catalyze reactions by bringing their substrates into close proximity in reactive orientations. The enzymatic binding of the substrates in a bimolecular reaction arrests their relative motions resulting in a rate enhancement. The preferential enzymatic binding of the transition state of a catalyzed reaction over the substrate is an important rate enhancement mechanism. Transition state analogs are potent competitive inhibitors because they bind to the enzyme more tightly than does the corresponding substrate.

Lysozyme catalyzes the hydrolysis of $\beta(1 \rightarrow 4)$-linked poly(NAG–NAM), the bacterial cell wall polysaccharide, as well as that of poly(NAG). According to the Phillips mechanism, lysozyme binds a hexasaccharide so as to distort its D ring towards the half-chair conformation of the planar oxonium ion transition state. This is followed by cleavage of the C1—O1 bond between the D and E rings as promoted by proton donation from Glu 35. Finally, the resulting oxonium ion transition state is electrostatically stabilized by the nearby carboxyl group of Asp 52 so that the E ring can be replaced by OH^- to form the hydrolyzed product. The roles of Glu 35 and Asp 52 in lysozyme catalysis have been verified through chemical modification and mutagenisis studies. Similarly, structural and binding studies indicate that strain is of major catalytic importance in lysozyme.

Serine proteases constitute a widespread class of proteolytic enzymes that are characterized by the possession of a reactive Ser residue. The pancreatically synthesized digestive enzymes trypsin, chymotrypsin, and elastase are sequentially and structurally related but have different side chain specificities for their substrates. All have the same catalytic triad, Asp 102, His 57, and Ser 195, at their active sites. Subtilisin and serine carboxypeptidase II are unrelated serine proteases that have essentially the same active site geometry as do the pancreatic enzymes. Catalysis in serine proteases is initiated by the nucleophilic attack of the active Ser on the carbonyl carbon atom of the scissile peptide to form the tetrahedral intermediate transition state, a process that is facilitated by the formation of a low-barrier hydrogen bond between Asp 102 and His 57. The tetrahedral intermediate, which is stabilized by its preferential binding to the enzyme's active site, then decomposes to the acyl–enzyme intermediate under the impetus of proton donation from the Asp 102-polarized His 57. After the replacement of the leaving group by solvent H_2O, the catalytic process is reversed to yield the second product and the regenerated enzyme. The Asp 102–His 57 couple therefore functions in the reaction as a proton shuttle, a process that is facilitated by the formation of a low-barrier hydrogen bond in the transition state. The active Ser is not unusually reactive but is ideally situated to nucleophilically attack the activated scissile peptide. The X-ray structure of the trypsin–BPTI complex indicates the existence of the tetrahedral intermediate. The pancreatic serine proteases are synthesized as zymogens to prevent pancreatic self-digestion. Trypsinogen is activated by a single proteolytic cleavage by enteropeptidase. The resulting trypsin similarly activates trypsinogen as well as chymotrypsinogen, proelastase, and other pancreatic digestive enzymes. Trypsinogen's catalytic triad is structurally intact. The zymogen's low catalytic activity arises from a distortion of its specificity pocket and oxyanion hole so that it is unable to productively bind substrate or preferentially bind the tetrahedral intermediate.

Glutathione reductase is a nearly ubiquitous FAD-containing enzyme that catalyzes the NADPH reduction of GSSG to GSH. The first stage of the reaction is the formation of a stable two-electron reduced intermediate, EH_2. The electron pair is passed from the reduced nicotinamide ring, through the parallel flavin ring, to the enzyme's redox-active disulfide. The latter stage occurs via the transient formation of a covalent bond between the S atom of Cys

63 and flavin C4a. In EH_2, the Cys 63 thiol group forms a charge-transfer complex with the oxidized flavin ring. In the second step of the glutathione reductase reaction, EH_2 reacts with GSSG, through disulfide interchange reactions, to regenerate the oxidized enzyme and 2GSH. Glutathione reductase is a member of the family of disulfide oxidoreductases that catalyze a variety of different oxidation–reduction reactions.

REFERENCES

General

Bender, M.L., Bergeron, R.J., and Komiyama, M., *The Bioorganic Chemistry of Enzymatic Catalysis,* Wiley (1984).

Fersht, A., *Enzyme Structure and Mechanism* (2nd ed.), Freeman (1985).

Jencks, W.P., *Catalysis in Chemistry and Enzymology,* McGraw-Hill (1969). [A classic and, in many ways, still current work.]

Walsh, C., *Enzymatic Reaction Mechanisms,* Freeman (1979). [A compendium of enzymatic reactions.]

Catalytic Mechanisms

Atkins, W.M. and Sligar, S.G., Protein engineering for studying enzyme catalytic mechanism, *Curr. Opin. Struct. Biol.* **1,** 611–616 (1991).

Bruice, T.C., Proximity effects and enzyme catalysis, *in* Boyer, P.D. (Ed.), *The Enzymes* (3rd ed.), Vol. 2, *pp.* 217–279, Academic Press (1970).

Bruice, T.C., Some pertinent aspects of mechanism as determined with small molecules, *Annu. Rev. Biochem.* **45,** 331–373 (1976).

Christianson, D.W., Structural biology of zinc, *Adv. Protein Chem.* **42,** 281–355 (1991).

Glusker, J.P., Structural aspects of metal liganding to functional groups in proteins, *Adv. Protein Chem.* **42,** 1–76 (1991).

Jencks, W.P., Binding energy, specificity, and enzymatic catalysis: the Circe effect, *Adv. Enzymol.* **43,** 219–410 (1975).

Hackney, D.D., Binding energy and catalysis, *in* Sigman, D.S. and Boyer, P.D. (Eds.), *The Enzymes* (3rd ed.), Vol. 19, *pp.* 1–36, Academic Press (1990).

Kraut, J., How do enzymes work? *Science* **242,** 533–540 (1988).

Lolis, E. and Petsko, G.A., Transition-state analogues in protein crystallography: Probes of the structural source of enzyme catalysis, *Annu. Rev. Biochem.* **59,** 597–630 (1990).

Mooser, G., Glycosidases and glycosyltransferases, *in* Sigman, D.S. (Ed.), *The Enzymes* (3rd ed.), Vol. 20, *pp.* 187–233, Academic Press (1992). [Section II discusses lysozyme.]

Navia, M.A. and Murcko, M.A., Use of structural information in drug design, *Curr. Opin. Struct. Biol.* **2,** 202–210 (1992).

Page, M.I., Entropy, binding energy, and enzyme catalysis, *Angew. Chem. Int. Ed. Engl.* **16,** 449–459 (1977).

Villafranca, J.J. and Nowak, T., Metal ions at enzyme active sites, *in* Sigman, D.S. (Ed.), *The Enzymes* (3rd ed.), Vol. 20, *pp.* 63–94, Academic Press (1992).

Warshel, A., Computer simulations of enzymatic reactions, *Curr. Opin. Struct. Biol.* **2,** 230–236 (1992).

Williams, R.J.P., Are enzymes mechanical devices? *Trends Biochem. Sci.* **18,** 115–117 (1993). [Argues that the mechanical aspects of enzymes have recieved insufficient consideration.]

Wolfenden, R., Analogue approaches to the structure of the transition state in enzyme reactions, *Acc. Chem. Res.* **5,** 10–18 (1972).

Lysozyme

Beddell, C.R., Blake, C.C.F., and Oatley, S.J., An X-ray study of the structure and binding properties of iodine-inactivated lysozyme, *J. Mol. Biol.* **97,** 643–654 (1975).

Blake, C.C.F., Johnson, L.N., Mair, G.A., North, A.C.T., Phillips, D.C., and Sarma, V.R., Crystallographic studies of the activity of hen egg-white lysozyme, *Proc. R. Soc. London Ser. B* **167,** 378–388 (1967).

Chipman, D.M. and Sharon, N., Mechanism of lysozyme action, *Science* **165,** 454–465 (1969).

Ford, L.O., Johnson, L.N., Machin, P.A., Phillips, D.C., and Tijan, R., Crystal structure of a lysozyme–tetrasaccharide lactone complex, *J. Mol. Biol.* **88,** 349–371 (1974).

Imoto, T., Johnson, L.N., North, A.C.T., Phillips, D.C., and Rupley, J.A., Vertebrate lysozymes, *in* Boyer, P.D. (Ed.), *The Enzymes* (3rd ed.), Vol. 7, *pp.* 665–868, Academic Press (1972). [An exhaustive review.]

Johnson, L.N., Cheetham, J., McLaughlin, P.J., Acharya, K.R., Barford, D., and Phillips, D.C., Protein–oligosaccharide interactions: Lysozyme, phosphorylase, amylases, *Curr. Top. Microbiol. Immunol.* **139,** 81–134 (1988).

Kelly, J.A., Sielecki, A.R., Sykes, B.D., James, M.N.G., and Phillips, D.C., X-Ray crystallography of the binding of the bacterial cell wall trisaccharide NAM–NAG–NAM to lysozyme, *Nature* **282,** 875–878 (1979).

Kirby, A.J., Mechanism and stereoelectronic effects in the lysozyme reaction, *CRC Crit. Rev. Biochem.* **22,** 283–315 (1987).

McKenzie, H.A. and White, F.H., Jr., Lysozyme and α-lactalbumin: Structure function and interrelationships, *Adv. Protein Chem.* **41,** 173-315 (1991).

Parsons, S.M. and Raftery, M.A., The identification of aspartic acid 52 as being critical to lysozyme activity, *Biochemistry* **8,** 4199–4205 (1969).

Phillips, D.C., The three-dimensional structure of an enzyme molecule, *Sci. Am.* **215**(5): 75–80 (1966). [A marvelously illustrated article on the structure and mechanism of lysozyme.]

Secemski, I.I., Lehrer, S.S., and Lienhard, G.E., A transition state analogue for lysozyme, *J. Biol. Chem.* **247**, 4740–4748 (1972). [Binding studies on the lactone derivative of (NAG)$_4$.]

Schindler, M., Assaf, Y., Sharon, N., and Chipman, D.M., Mechanism of lysozyme catalysis: role of ground-state strain in subsite D in hen egg-white and human lysozymes, *Biochemistry* **16**, 423–431 (1977).

Strynadka, N.C.J. and James, M.N.G., Lysozyme revisited: Crystallographic evidence for distortion of an *N*-acetylmuramic acid residue bound in site D, *J. Mol. Biol.* **220**, 401–424 (1991).

Warshel, A. and Levitt, M., Theoretical studies of enzymatic reactions; dielectric, electrostatic and steric stabilization of the carbonium ion in the reaction of lysozyme, *J. Mol. Biol.* **103**, 227–249 (1976). [Theoretical indications that lysozyme catalysis occurs through electrostatic rather than steric strain.]

Serine Proteases

Cleland, W.W. and Kreevoy, M.M., Low-barrier hydrogen bonds and enzymic catalysis, *Science* **264**, 1887–1890 (1994); *and* Gerlt, J.A. and Gassman, P.G., Understanding the rates of certain enzyme-catalyzed reactions: Proton abstraction from carbon acids, acyl-transfer reactions, and displacement of phosphodiesters, *Biochemistry* **32**, 11943–11952 (1993).

Corey, D.R. and Craik, C.S., An investigation into the minimum requirements for peptide hydrolysis by mutation of the catalytic triad of trypsin, *J. Am. Chem. Soc.* **114**, 1784–1790 (1992).

Frey, P.A., Whitt, S.A., and Tobin, J.B., A low-barrier hydrogen bond in the catalytic triad of serine proteases, *Science* **264**, 1927–1930 (1994).

Hedstrom, L., Szilagyi, L., and Rutter, W.J., Converting trypsin to chymotrypsin: The role of surface loops, *Science* **255**, 1249–1253 (1992).

James, M.N.G., Sielecki, A.R., Brayer, G.D., Delbaere, L.T.J., and Bauer, C.A., Structure of product and inhibitor complexes of *Streptomyces griseus* protease A at 1.8 Å resolution, *J. Mol. Biol.* **144**, 45–88 (1980).

Kossiakoff, A.A., Catalytic properties of trypsin, *in* Jurnak, F.A. and McPherson, A., *Biological Macromolecules and Assemblies,* Vol. 3, *pp.* 369–412, Wiley (1987).

Laskowski, M., Jr., and Kato, I., Protein inhibitors of proteinases, *Annu. Rev. Biochem.* **49**, 593–626 (1980).

Liao, D.-I. and Remington, S.J., Structure of wheat serine carboxypeptidase II at 3.5-Å resolution, *J. Biol. Chem.* **265**, 6528–6531 (1990).

Neurath, H., Evolution of proteolytic enzymes, *Science* **224**, 350–357 (1984).

Phillips, M.A. and Fletterick, R.J., Proteases, *Curr. Opin. Struct. Biol.* **2**, 713–720 (1992).

Singer, P.T., Smalås, A., Carty, R.P., Mangel, W.F., and Sweet, R.M., The hydrolytic water molecule in trypsin, revealed by time-resolved Laue crystallography, *Science* **259**, 669–673 (1993).

Sprang, S., Standing, T., Fletterick, R.J., Stroud, R.M., Finer-Moore, J., Xuong, N.-H., Hamlin, R., Rutter, W.J., and Craik, C.S., The three-dimensional structure of Asn^{102} mutant of trypsin: role of Asp^{102} in serine protease catalysis, *Science* **237**, 905–909 (1987); *and* Craik, C.S., Roczniak, S., Largman, C., and Rutter, W.J., The catalytic role of the active site aspartic acid in serine proteases, *Science* **237**, 909–913 (1987).

Steitz, T.A. and Shulman, R.G., Crystallographic and NMR studies of the serine proteases, *Annu. Rev. Biophys. Bioeng.* **11**, 419–444 (1982).

Stroud, R.M., A family of protein-cutting proteins, *Sci. Am.* **231**(1): 74–88 (1974).

Stroud, R.M., Kossiakoff, A.A., and Chambers, J.L., Mechanism of zymogen activation, *Annu. Rev. Biophys. Bioeng.* **6**, 177–193 (1977).

Glutathione Reductase and Related Enzymes

Ghisla, S. and Massey, V., Mechanisms of flavoprotein-catalyzed reactions, *Eur. J. Biochem.* **181**, 1–17 (1989).

Karplus, P.A. and Schulz, G.E., Substrate binding and catalysis by glutathione reductase derived from refined enzyme: Substrate crystal structures at 2Å resolution, *J. Mol. Biol.* **210**, 163–180 (1989).

Karplus, P.A. and Schultz, G.E., Refined structure of glutathione reductase at 1.54 Å resolution, *J. Mol. Biol.* **195**, 701–729 (1987).

Matthews, F.S., New flavoenzymes, *Curr. Opin. Struct. Biol.* **1**, 954–967 (1991).

Pai, E.F., Variations on a theme: the family of FAD-dependent NAD(P)H-(disulfide)-oxidoreductases, *Curr. Opin. Struct. Biol.* **1**, 796–803 (1991).

Schulz, G.E. and Pai, E.F., The catalytic mechanism of glutathione reductase as derived from X-ray diffraction analyses of reaction intermediates, *J. Biol. Chem.* **258**, 1752–1757 (1983).

Thieme, R., Pai, E.F., Schirmer, R.H., and Schulz, G.E., Three-dimensional structure of glutathione reductase at 2 Å resolution, *J. Mol. Biol.* **152**, 763–782 (1981).

PROBLEMS

1. Explain why γ-pyridone is not nearly as effective a catalyst for glucose mutarotation as is α-pyridone. What about β-pyridone?

2. RNA is rapidly hydrolyzed in alkaline solution to yield a mixture of nucleotides whose phosphate groups are bonded to either the 2′ or the 3′ positions of the ribose residues. DNA, which lacks RNA's 2′ OH groups, is resistant to alkaline degradation. Explain.

3. Carboxypeptidase A, a Zn^{2+}-containing enzyme, hydrolyzes the C-terminal peptide bonds of polypeptides (Section 6-1A).

In the enzyme–substrate complex, the Zn^{2+} ion is coordinated to three enzyme side chains, the carbonyl oxygen of the scissile peptide bond, and a water molecule. A plausible model for the enzyme's reaction mechanism that is consistent with X-ray and enzymological data is diagrammed in Fig. 14-33. What are the roles of the Zn^{2+} ion and Glu 270 in this mechanism?

4. In the following lactonization reaction,

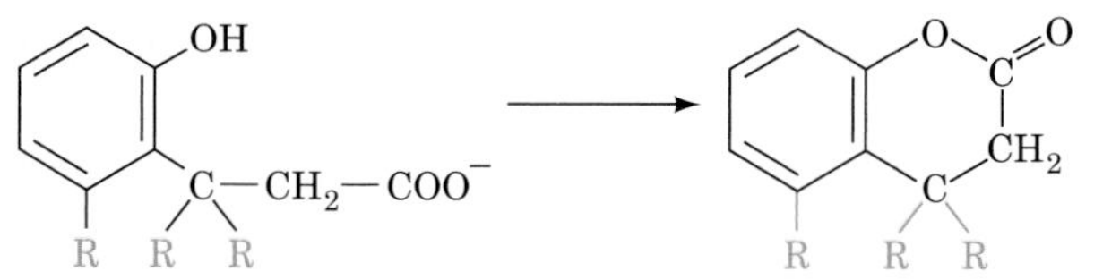

the relative reaction rate when $R = CH_3$ is 3.4×10^{11} times that when $R = H$. Explain.

*5. Derive the analog of Eq. [14.11] for an enzyme that catalyzes the reaction:

$$A + B \longrightarrow P$$

Assume the enzyme must bind A before it can bind B.

$$E + A + B \rightleftharpoons EA + B \rightleftharpoons EAB \longrightarrow EP$$

6. Explain, in thermodynamic terms, why an "enzyme" that stabilizes its Michaelis complex as much as its transition state does not catalyze a reaction.

7. Suggest a transition state analog for proline racemase that differs from those discussed in the text. Justify your suggestion.

8. Wolfenden has stated that it is meaningless to distinguish between the "binding sites" and the "catalytic sites" of enzymes. Explain.

9. Explain why oxalate ($^-OOCCOO^-$) is an inhibitor of oxaloacetate decarboxylase.

10. In light of the information given in this chapter, why are enzymes such large molecules? Why are active sites almost always located in clefts or depressions in enzymes rather than on protrusions?

11. Predict the effects on lysozyme catalysis of changing Phe 34, Ser 36, and Trp 108 to Arg, assuming that this change does not significantly alter the structure of the protein.

*12. The incubation of $(NAG)_4$ with lysozyme results in the slow formation of $(NAG)_6$ and $(NAG)_2$. Propose a mechanism for this reaction. What aspect of the Phillips mechanism is established by this reaction?

13. Why does the following $\beta(1 \rightarrow 4)$-linked tetrasaccharide

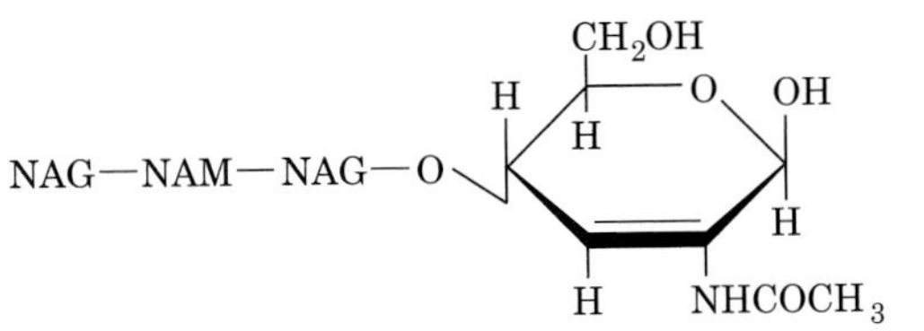

bind to lysozyme with 100-fold greater affinity than does NAG–NAM–NAG–NAM?

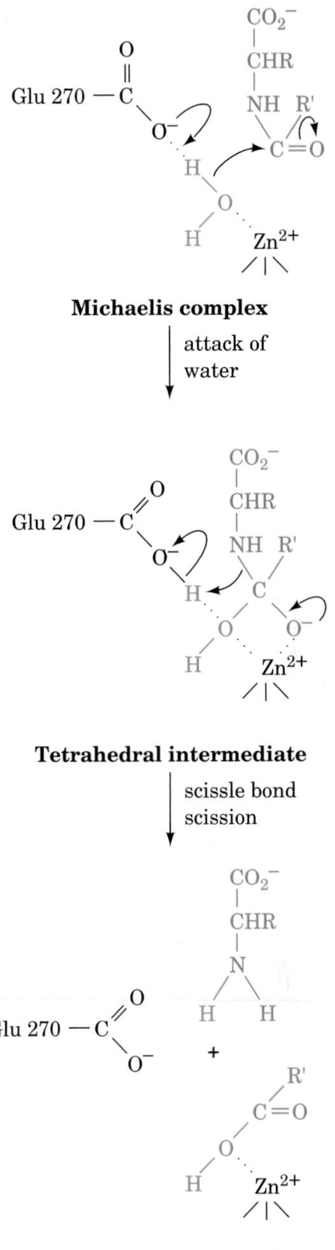

Michaelis complex

attack of water

Tetrahedral intermediate

scissle bond scission

Enzyme-product complex

FIGURE 14-33. The mechanism of carboxypeptidase A.

14. A major difficulty in investigating the properties of the pancreatic serine proteases is that these enzymes, being proteins themselves, are self-digesting. This problem is less severe, however, for solutions of chymotrypsin than it is for solutions of trypsin or elastase. Explain.

15. The comparison of the active site geometries of chymotrypsin and subtilisin under the assumption that their similarities have catalytic significance has led to greater mechanistic understanding of both these enzymes. Discuss the validity of this strategy.

16. Benzamidine ($K_I = 1.8 \times 10^{-5}M$) and **leupeptin** ($K_I = 3.8 \times 10^{-7}M$)

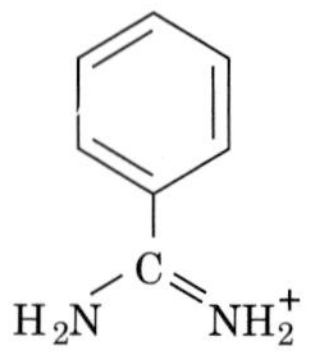

Benzamidine 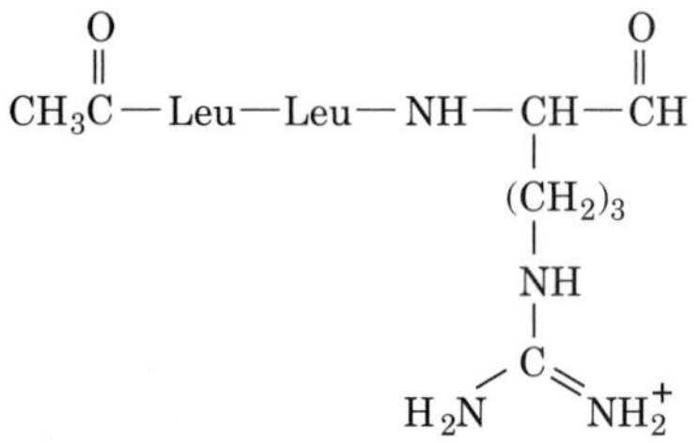   **Leupeptin**

are both specific competitive inhibitors of trypsin. Explain their mechanisms of inhibition. Design leupeptin analogs that inhibit chymotrypsin and elastase.

17. Trigonal boronic acid derivatives have a high tendency to form tetrahedral adducts. **2-Phenylethyl boronic acid**

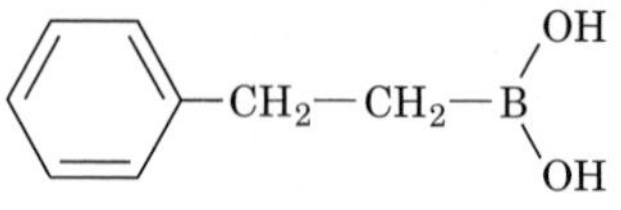

2–Phenylethyl boronic acid

is an inhibitor of subtilisin and chymotrypsin. Indicate the structure of these enzyme–inhibitor complexes.

18. Tofu (bean curd), a high-protein soybean product that is widely consumed in the Far East, is prepared in such a way as to remove the trypsin inhibitor present in soybeans. Explain the reason(s) for this treatment.

19. Explain why mutating all three residues of trypsin's catalytic triad has essentially no greater effect on the enzyme's catalytic rate enhancement than mutating only Ser 195.

20. Explain why chymotrypsin is not self-activating as is trypsin.

21. The reaction of glutathione reductase with an excess of NADPH in the presence of arsenite yields a nonphysiological four-electron reduced form of the enzyme. What is the chemical nature of this catalytically inactive species?

22. Two-electron reduced glutathione reductase (EH$_2$), but not the oxidized enzyme (E), reacts with iodoacetate (ICH$_2$COO$^-$) to yield an inactive enzyme. Explain.

PART

IV

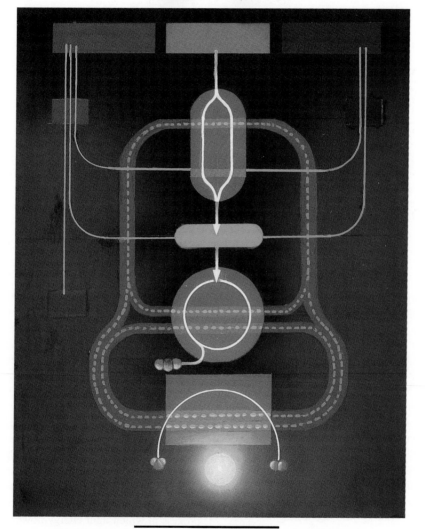

METABOLISM

Schematic diagram of the major pathways of energy metabolism.

15

Introduction to Metabolism

Living organisms are not at equilibrium. Rather, they require a continuous influx of free energy to maintain order in a universe bent on maximizing disorder. **Metabolism** is the overall process through which living systems acquire and utilize the free energy they need to carry out their various functions. *They do so by coupling the exergonic reactions of nutrient oxidation to the endergonic processes required to maintain the living state* such as the performance of mechanical work, the active transport of molecules against concentration gradients, and the biosynthesis of complex molecules. How do living things acquire this necessary free energy? And what is the nature of the energy-coupling process? **Phototrophs** (plants and certain bacteria) acquire free energy from the sun through **photosynthesis,** a process in which light energy powers the endergonic reaction of CO_2 and H_2O to form carbohydrates and O_2 (Chapter 22). **Chemotrophs** obtain their free energy by oxidizing organic compounds (carbohydrates, lipids, proteins) obtained from other organisms, ultimately phototrophs. This free energy is most often coupled to endergonic reactions through the intermediate synthesis of "high-energy" phosphate compounds such as **adenosine triphosphate (ATP;** Section 15-4). In addition to being completely oxidized, nutrients are broken down in a series of metabolic reactions to common intermediates that are used as precursors in the synthesis of other biological molecules.

A remarkable property of living systems is that, despite the complexity of their internal processes, they maintain a steady state. This is strikingly demonstrated by the observation that, over a 40-year time span, a normal human adult

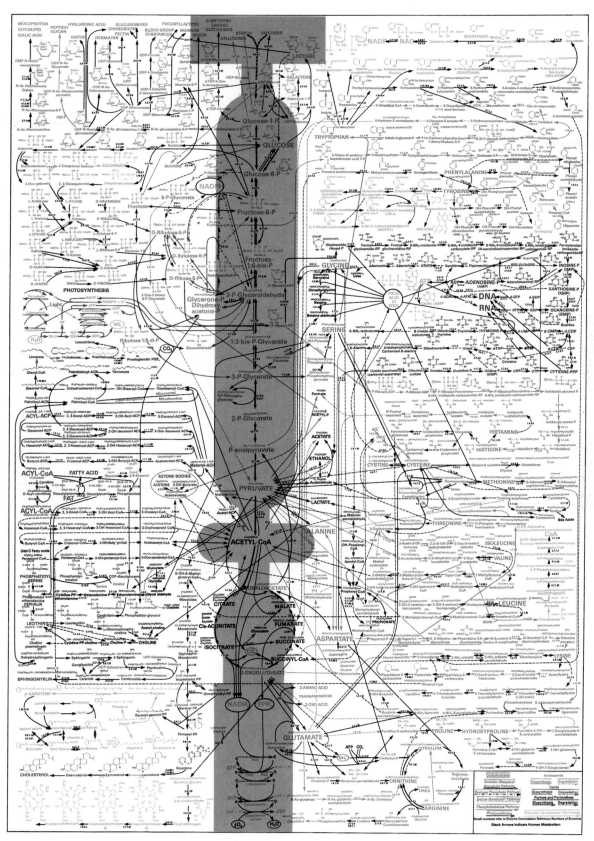

FIGURE 15-1. A map of the major metabolic pathways in a typical cell. The main pathways of glucose metabolism are shaded. [Designed by D. E. Nicholson. Published by BDH Ltd., Poole 2, Dorset, England.]

consumes literally tons of nutrients and imbibes over 20,000 L of water, but does so without significant weight change. This steady state is maintained by a sophisticated set of metabolic controls. In this introductory chapter to metabolism, we outline the general characteristics of metabolic pathways and their regulation, study the main types of chemical reactions that comprise these pathways, and consider the experimental techniques that have been most useful in their elucidation. We then discuss the free energy changes associated with reactions of phosphate compounds and oxidation–reduction reactions. Finally, we consider the thermodynamic nature of biological processes; that is, what properties of life are responsible for its self-sustaining character.

1. METABOLIC PATHWAYS

Metabolic pathways are series of consecutive enzymatic reactions that produce specific products. Their reactants, intermediates, and products are referred to as **metabolites.** Since an organism utilizes many metabolites, it has many metabolic pathways. Figure 15-1 shows a metabolic map for a typical cell with many of its interconnected pathways. Each reaction on the map is catalyzed by a distinct enzyme, of which there are more than 2000 known. At first glance, this network seems hopelessly complex. Yet, by focusing on its major areas in the following chapters, for example, the main pathways of glucose oxidation (the shaded areas of Fig. 15-1), we shall become familiar with its most important avenues and their interrelationships.

The reaction pathways that comprise metabolism are often divided into two categories: Those involved in degradation **(catabolism)** and those involved in biosynthesis **(anabolism).** In catabolic pathways, complex metabolites are exergonically broken down to simpler products. The free energy released during these processes is conserved by the synthesis of ATP from ADP and phosphate or by the reduction of the coenzyme $NADP^+$ to NADPH (Fig. 12-2). ATP and NADPH are the major free energy sources for anabolic pathways (Fig. 15-2).

A striking characteristic of degradative metabolism is that *it converts a large number of diverse substances (carbohydrates, lipids, and proteins) to common intermediates.* These intermediates are then further metabolized in a central oxidative pathway that terminates in a few end products. Figure 15-3 outlines the breakdown of various foodstuffs, first to their monomeric units, and then to the common intermediate, **acetyl-coenzyme A (acetyl-CoA)** (Section 19-2; structural formula in Fig. 19-2). This is followed by the oxidation of the acetyl group to CO_2 and H_2O by the sequential actions of the **citric acid cycle** (Chapter 19), the **electron-transport chain,** and **oxidative phosphorylation** (Chapter 20).

Biosynthesis carries out the opposite process. *Relatively few metabolites, mainly pyruvate, acetyl-CoA, and the citric*

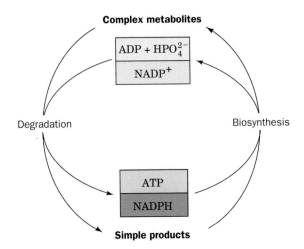

FIGURE 15-2. ATP and NADPH are the sources of free energy for biosynthetic reactions. They are generated through the degradation of complex metabolites.

acid cycle intermediates, serve as starting materials for a host of varied biosynthetic products. In the next several chapters we discuss many degradative and biosynthetic pathways in detail. For now, let us consider some general characteristics of these processes.

Four principal characteristics of metabolic pathways stem from their function of generating products for use by the cell:

1. Metabolic pathways are irreversible
They are highly exergonic (have large negative free energy changes), so their reactions go to completion. This characteristic provides the pathway with direction. Consequently, *if two metabolites are metabolically interconvertible, the pathway from the first to the second must differ from the pathway from the second back to the first:*

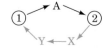

The reason for this difference is that if the route from the first metabolite to the second is exergonic, free energy must be supplied in order to bring it "back up the hill." This requires a different pathway for at least some of the reaction steps. *The existence of independent interconversion routes, as we shall see, is an important property of metabolic pathways because it allows independent control of the rates of the two processes.* If metabolite 2 is required by the cell, it is necessary to "turn off" the pathway from 2 to 1 while "turning on" the pathway from 1 to 2. Such independent control would be impossible without different pathways.

2. Every metabolic pathway has a first committed step
Although metabolic pathways are irreversible, most of their component reactions function close to equilibrium. Early in each pathway, however, there is generally

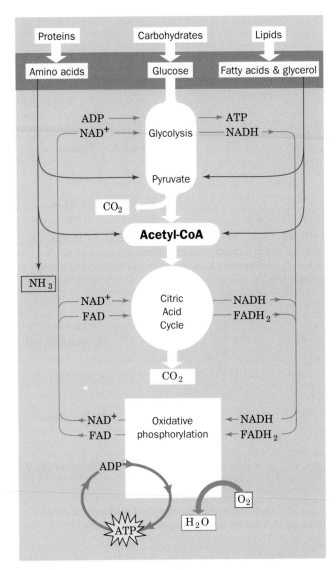

FIGURE 15-3. Complex metabolites such as carbohydrates, proteins, and lipids are degraded first to their monomeric units, chiefly glucose, amino acids, fatty acids, and glycerol, and then to the common intermediate, acetyl coenzyme A (acetyl-CoA). The acetyl group is then oxidized to CO_2 via the citric acid cycle with concomitant reduction of NAD^+ and FAD. Reoxidation of these latter coenzymes by O_2 via the electron-transport chain and oxidative phosphorylation yields H_2O and ATP.

an irreversible (exergonic) reaction that "commits" the intermediate it produces to continue down the pathway.

3. All metabolic pathways are regulated

In order to exert control on the flux of metabolites through a metabolic pathway, it is necessary to regulate its rate-limiting step. The first committed step, being irreversible, functions too slowly to permit its substrates and products to equilibrate. Since most of the other reactions in a pathway function close to equilibrium, the first committed step is often its rate-limiting step. Most metabolic pathways are therefore controlled by regulat-

ing the enzymes that catalyze their first committed steps. This is the most efficient way to exert control because it prevents the unnecessary synthesis of metabolites further along the pathway when they are not required. Specific aspects of such flux control are discussed in Section 16-4A.

4. Metabolic pathways in eukaryotic cells occur in specific cellular locations

The synthesis of metabolites in specific membrane-bounded subcellular compartments makes their transport between these compartments a vital component of eukaryotic metabolism. Biological membranes are selectively permeable to metabolites because of the presence in membranes of specific transport proteins. For example, ATP is generated in the mitochondria but much of it is utilized in the cytosol. The transport protein that facilitates the passage of ATP through the mitochondrial membrane is discussed in Section 18-4C, along with the characteristics of membrane transport processes in general. The synthesis and utilization of acetyl-CoA is also compartmentalized. This metabolic intermediate is utilized in the cytosolic synthesis of fatty acids, but is synthesized in mitochondria. Yet there is no transport protein for acetyl-CoA in the mitochondrial membrane. How cells solve this fundamental problem is discussed in Section 23-4D.

2. ORGANIC REACTION MECHANISMS

Almost all of the reactions that occur in metabolic pathways are enzymatically catalyzed organic reactions. Section 14-1 details the various mechanisms enzymes have at that their disposal for catalyzing reactions: Acid – base catalysis, covalent catalysis, metal ion catalysis, electrostatic catalysis, proximity and orientation effects, and transition state binding. Yet, few enzymes alter the chemical mechanisms of these reactions, so *much can be learned about enzymatic mechanisms from the study of nonenzymatic model reactions.* We therefore begin our study of metabolic reactions by outlining the types of reactions we shall encounter and the mechanisms by which they have been observed to proceed in nonenzymatic systems.

Christopher Walsh has classified biochemical reactions into four categories: (1) **group-transfer reactions;** (2) **oxidations and reductions;** (3) **eliminations, isomerizations, and rearrangements;** and (4) **reactions that make or break carbon–carbon bonds.** Much is known about the mechanisms of these reactions and about the enzymes that catalyze them. The discussions in the next several chapters focus on these mechanisms as they apply to specific metabolic interconversions. In this section we outline the four reaction categories and discuss how our knowledge of their reaction mechanisms derives from the study of model organic reactions. We begin by briefly reviewing the chemical logic used in analyzing these reactions.

A. Chemical Logic

A covalent bond consists of an electron pair shared between two atoms. In breaking such a bond, the electron pair can either remain with one of the atoms (**heterolytic bond cleavage**) or separate such that one electron accompanies each of the atoms (**homolytic bond cleavage**) (Fig. 15-4). Homolytic bond cleavage, which usually produces unstable radicals, occurs mostly in oxidation–reduction reactions. Heterolytic C—H bond cleavage involves either carbanion and proton (H^+) formation or carbocation (carbonium ion) and hydride ion (H^-) formation. Since hydride ions are highly reactive species and carbon atoms are slightly more electronegative than hydrogen atoms, bond cleavage in which the electron pair remains with the carbon atom is the predominant mode of C—H bond breaking in biochemical systems. Hydride ion abstraction occurs only if the hydride is transferred directly to an acceptor such as NAD^+ or $NADP^+$.

Compounds participating in reactions involving heterolytic bond cleavage and bond formation are categorized into two broad classes: electron rich and electron deficient. Electron-rich compounds, which are called **nucleophiles** (nucleus lovers), are negatively charged or contain unshared electron pairs that easily form covalent bonds with electron-deficient centers. Biologically important nucleophilic groups include amino, hydroxyl, imidazole, and sulfhydryl functions (Fig. 15-5). The nucleophilic forms of these groups are also their basic forms. Indeed, nucleophilicity and basicity are closely related properties (Section 14-1B): A compound acts as a base when it forms a covalent bond with H^+, whereas it acts as a nucleophile when it forms a covalent bond with an electron-deficient center other than H^+, usually an electron-deficient carbon atom.

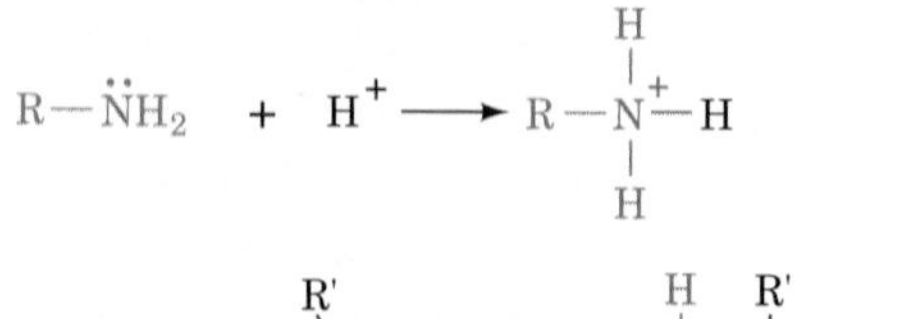

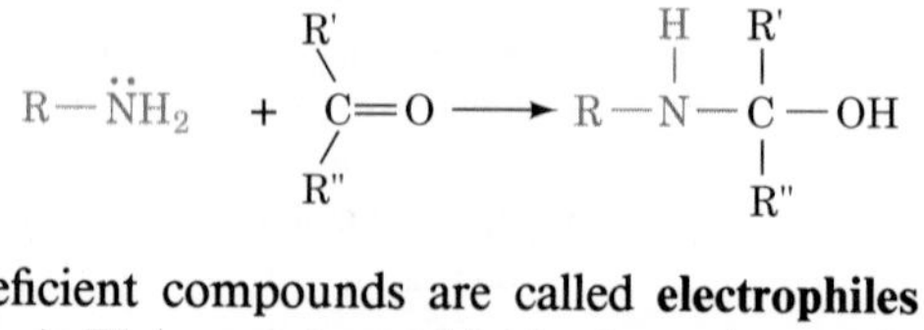

Electron-deficient compounds are called **electrophiles** (electron lovers). They may be positively charged, contain an unfilled valence electron shell, or contain an electronegative atom. The most common electrophiles in biochemical systems are H^+, metal ions, the carbon atoms of carbonyl groups, and cationic imines (Fig. 15-6).

Reactions are best understood if the electron pair rearrangements involved in going from reactants to products can be traced. In illustrating these rearrangements we shall use the **curved arrow convention** in which the movement of

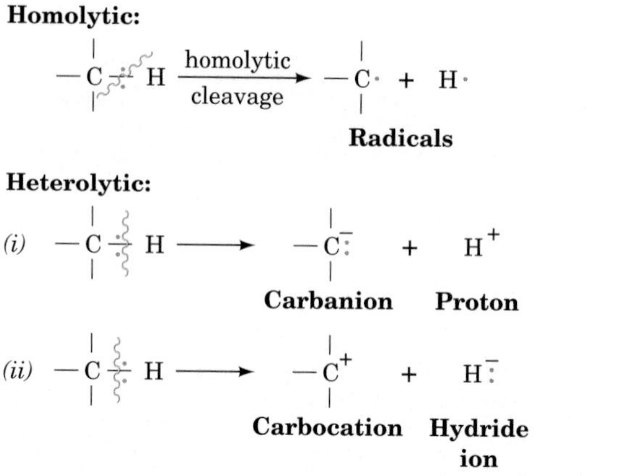

FIGURE 15-4. Modes of C—H bond breaking. Homolytic cleavage yields radicals, whereas heterolytic cleavage yields either (*i*) a carbanion and a proton or (*ii*) a carbocation and a hydride ion.

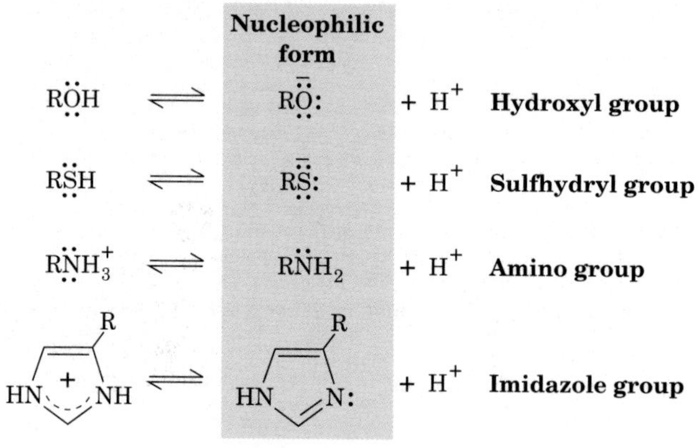

FIGURE 15-5. Biologically important nucleophilic groups. Nucleophiles are the conjugate bases of weak acids.

an electron pair is symbolized by a curved arrow emanating from the electron pair and pointing to the electron-deficient center attracting the electron pair. For example, imine formation, a biochemically important reaction between an amine and an aldehyde or ketone, is represented:

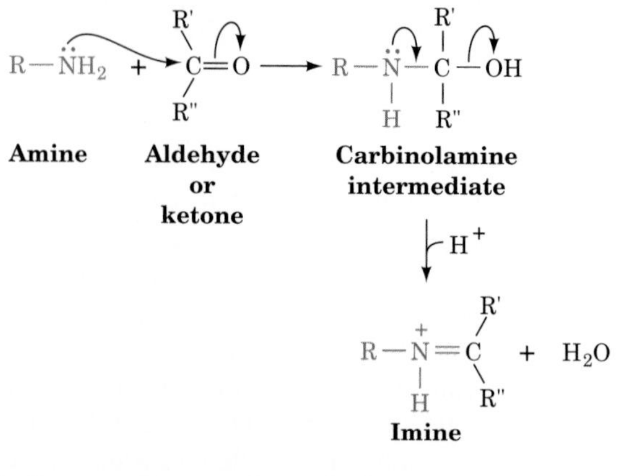

In the first reaction step, the amine's unshared electron pair adds to the electron-deficient carbonyl carbon atom while

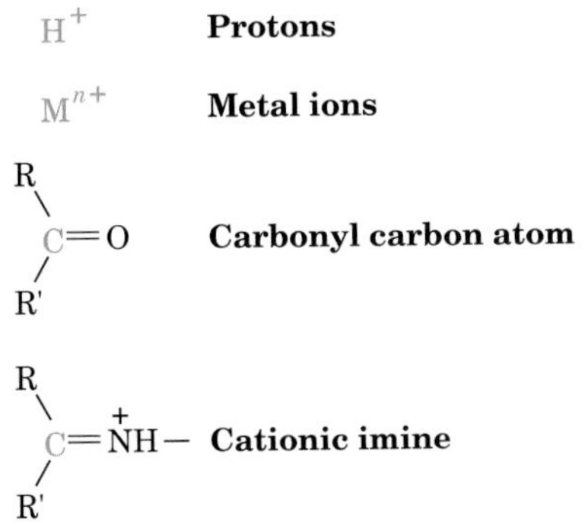

FIGURE 15-6. Biologically important electrophiles. Electrophiles contain an electron-deficient atom *(red)*.

one electron pair from its C=O double bond transfers to the oxygen atom. In the second step, the unshared electron pair on the nitrogen atom adds to the electron-deficient carbon atom with the elimination of water. *At all times, the rules of chemical reason apply to the system:* For example, there are never five bonds to a carbon atom or two bonds to a hydrogen atom.

B. Group-Transfer Reactions

The group transfers that occur in biochemical systems involve the transfer of an electrophilic group from one nucleophile to another.

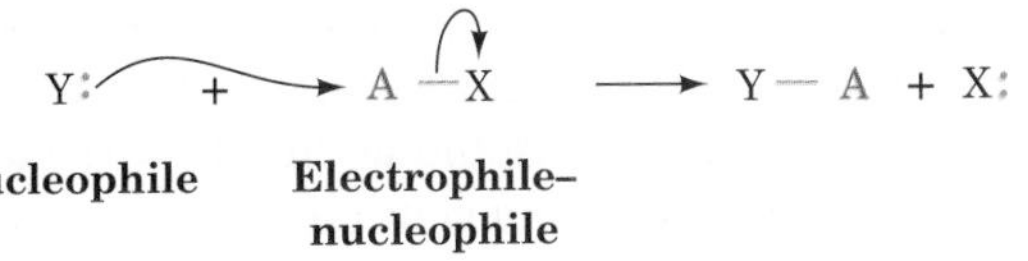

They could equally well be called nucleophilic substitution reactions. The most commonly transferred groups in biochemical reactions are acyl groups, phosphoryl groups, and glycosyl groups (Fig. 15-7):

1. **Acyl group transfer** from one nucleophile to another almost invariably involves the addition of a nucleophile to the acyl carbonyl carbon atom so as to form a tetrahedral intermediate (Fig. 15-7a). Peptide bond hydrolysis, as catalyzed, for example, by chymotrypsin (Section 14-3C), is a familiar example of such a reaction.

2. **Phosphoryl group transfer** proceeds via the addition of a nucleophile to a phosphoryl phosphorus atom to yield a

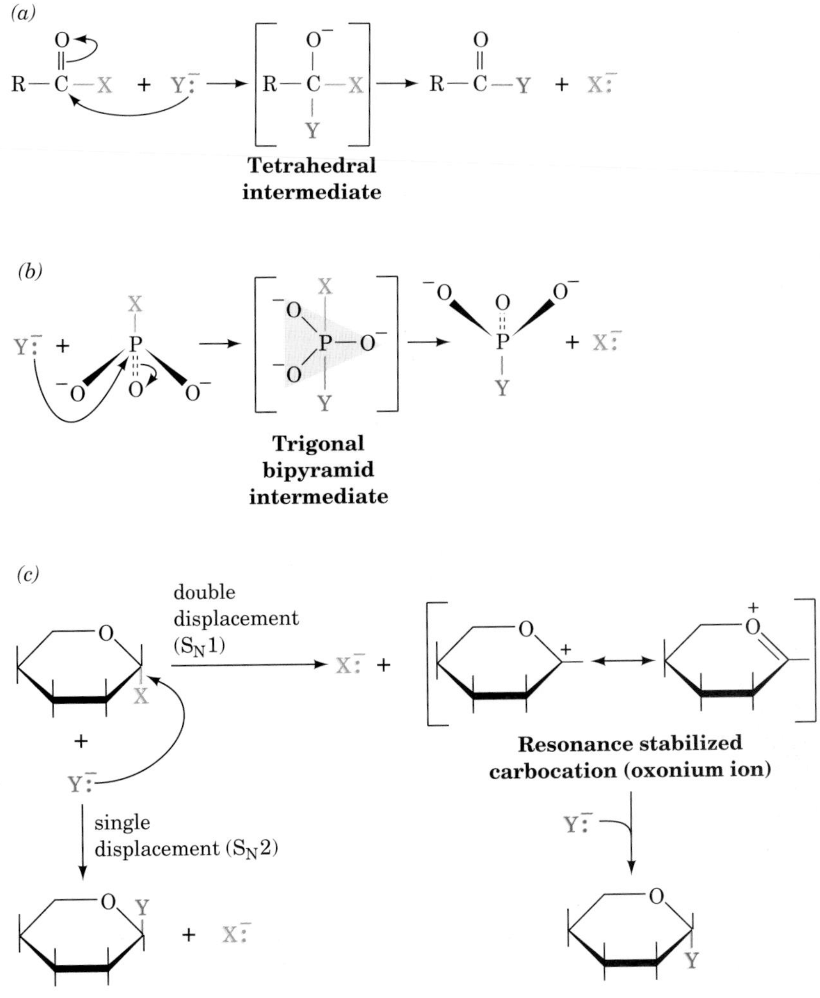

FIGURE 15-7. Types of metabolic group-transfer reactions: (*a*) Acyl group transfer involves addition of a nucleophile (Y) to the electrophilic carbon atom of an acyl compound to form a tetrahedral intermediate. The original acyl carrier (X) is then expelled to form a new acyl compound. (*b*) Phosphoryl group transfer involves addition of a nucleophile (Y) to the electrophilic phosphorus atom of a tetrahedral phosphoryl group. This yields a trigonal bipyramidal intermediate whose apical positions are occupied by the leaving group (X) and the attacking group (Y). Elimination of the leaving group (X) to complete the transfer reaction results in the phosphoryl group's inversion of configuration. (*c*) Glycosyl group transfer involves the substitution of one nucleophilic group for another at C1 of a sugar ring. This reaction usually occurs via a double displacement mechanism in which the elimination of the original glycosyl carrier (X) is accompanied by the intermediate formation of a resonance-stabilized carbocation (oxonuim ion) followed by the addition of the adding nucleophile (Y). The reaction also may occur via a single displacement mechanism in which Y directly displaces X with inversion of configuration.

trigonal bipyramidal intermediate whose apexes are occupied by the adding and leaving groups (Fig. 15-7b). The overall reaction results in the tetrahedral phosphoryl group's inversion of configuration. Indeed, chiral phosphoryl compounds have been shown to undergo just such an inversion. For example, Jeremy Knowles has synthesized ATP made chiral at its γ-phosphoryl group by isotopic substitution and demonstrated that this group is inverted upon its transfer to glucose in the reaction catalyzed by **hexokinase** (Fig. 15-8).

3. **Glycosyl group transfer** involves the substitution of one nucleophilic group for another at C1 of a sugar ring (Fig. 15-7c). This is the central carbon atom of an acetal. Chemical models of acetal reactions generally proceed via acid-catalyzed cleavage of the first bond to form a resonance-stabilized carbocation at C1 (an oxonium ion). The lysozyme-catalyzed hydrolysis of bacterial cell wall polysaccharides (Section 14-2B) is such a reaction.

C. Oxidations and Reductions

Oxidation–reduction (redox) reactions involve the loss or gain of electrons. The thermodynamics of these reactions is discussed in Section 15-5. Many of the redox reactions that occur in metabolic pathways involve C—H bond cleavage with the ultimate loss of two bonding electrons by the car-

bon atom. These electrons are transferred to an electron acceptor such as NAD^+ (Fig. 12-2). Whether these reactions involve homolytic or heterolytic bond cleavage has not always been rigorously established. In most instances heterolytic cleavage is assumed when radical species are not observed. It is useful, however, to visualize redox C—H bond cleavage reactions as hydride transfers as diagrammed below for the oxidation of an alcohol by NAD^+:

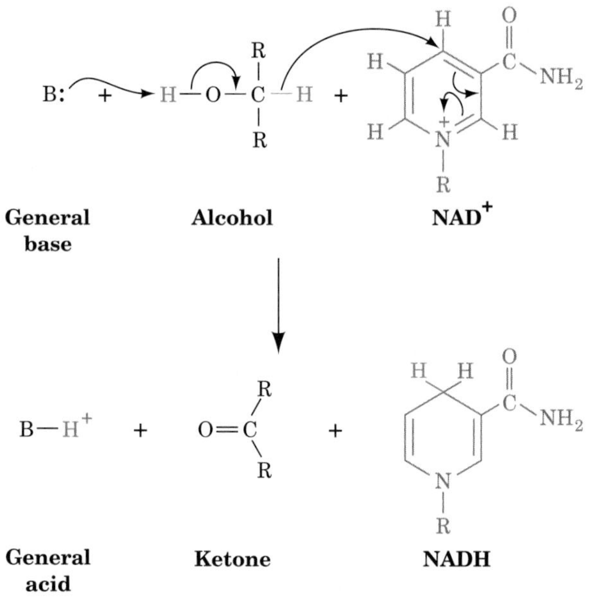

The terminal acceptor for the electron pairs removed from metabolites by their oxidation is, for aerobic organisms, molecular oxygen (O_2). Recall that this molecule is a ground state diradical species whose unpaired electrons have parallel spins. The rules of electron pairing (the Pauli exclusion principle) therefore dictate that O_2 can only accept unpaired electrons; that is, electrons must be transferred to O_2 one at a time (in contrast to redox processes in which electrons are transferred in pairs). Electrons that are removed from metabolites as pairs must therefore be passed to O_2 via the electron-transport chain one at a time. This is accomplished through the use of conjugated coenzymes that have stable radical oxidation states and can therefore undergo both $1e^-$ and $2e^-$ redox reactions. One such coenzyme is FAD (whose structure and oxidation states are indicated in Fig. 14-28).

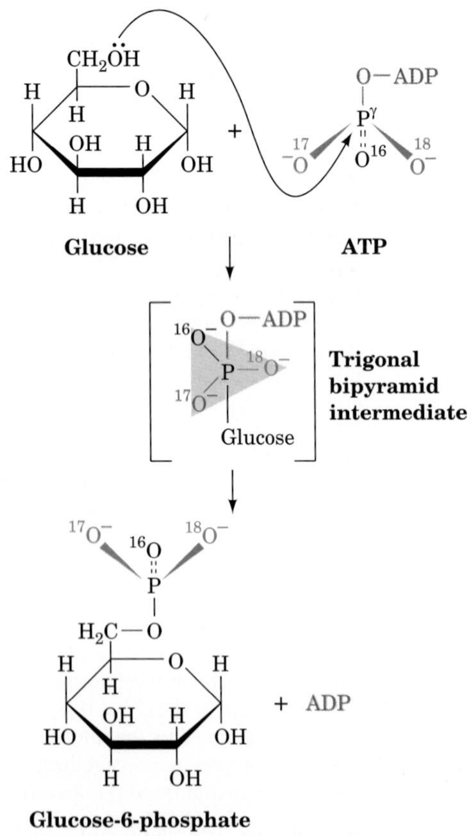

D. Eliminations, Isomerizations, and Rearrangements

Elimination Reactions Form Carbon–Carbon Double Bonds

Elimination reactions result in the formation of a double bond between two previously single-bonded saturated centers. The substances eliminated may be H_2O, NH_3, an

FIGURE 15-8. In the phosphoryl-transfer reaction catalyzed by hexokinase, the γ-phosphoryl group of ATP made chiral by isotopic substitution undergoes inversion of configuration.

alcohol (ROH), or a primary amine (RNH_2). The dehydration of an alcohol, for example, is an elimination reaction:

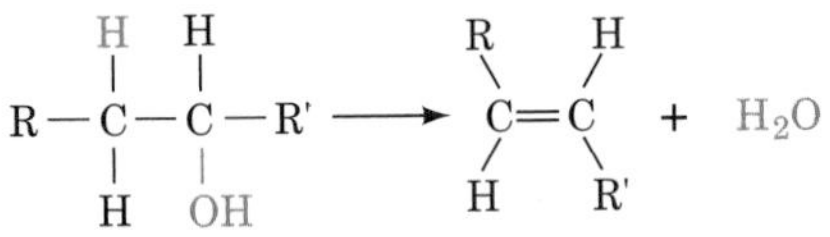

Bond breaking and bond making in this reaction may proceed via one of three mechanisms (Fig. 15-9a): (1) concerted; (2) stepwise with the C—O bond breaking first to form a carbocation; or (3) stepwise with the C—H bond breaking first to form a carbanion.

Enzymes catalyze dehydration reactions by either of two simple mechanisms: (1) protonation of the OH group by an acidic group (acid catalysis), or (2) abstraction of the proton by a basic group (base catalysis). Moreover, in a stepwise reaction, the charged intermediate may be stabilized by an oppositely charged active site group (electrostatic catalysis). The glycolytic enzyme **enolase** (Section 16-2I) and the citric acid cycle enzyme **fumarase** (Section 19-3G) catalyze such dehydration reactions.

Elimination reactions may take one of two possible stereochemical courses (Fig. 15-9b): (1) trans (anti) eliminations, the most prevalent biochemical mechanism, and

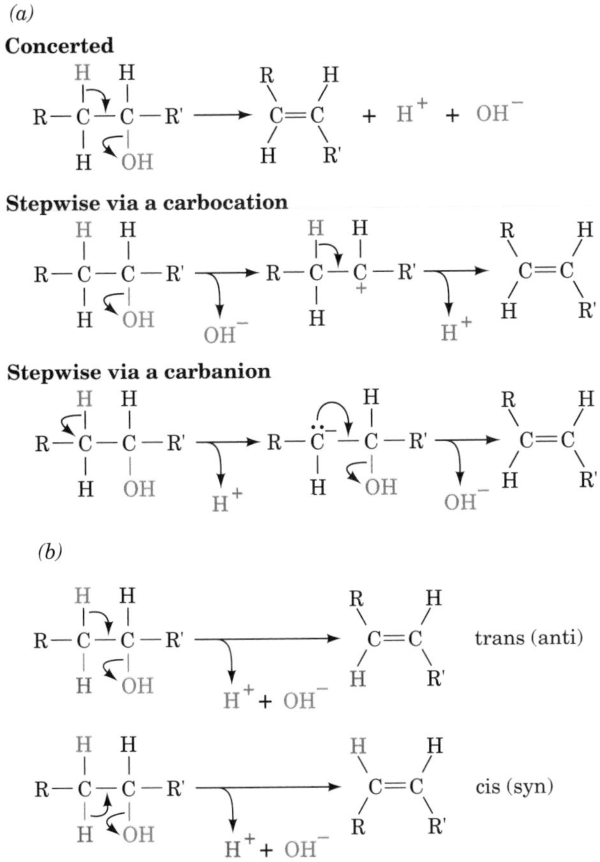

(a)

Concerted

Stepwise via a carbocation

Stepwise via a carbanion

(b)

FIGURE 15-9. Possible elimination reaction mechanisms using dehydration as an example. Reactions may be (*a*) either concerted, stepwise via a carbocation intermediate, or stepwise via a carbanion intermediate; and may occur (*b*) with either trans (anti) or cis (syn) stereochemistry.

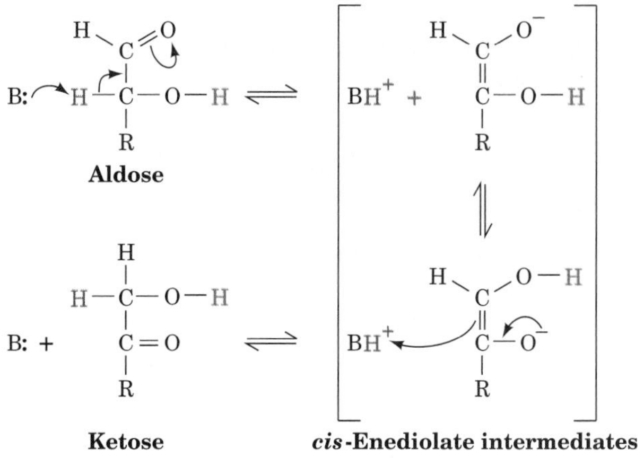

FIGURE 15-10. The mechanism of aldose–ketose isomerization. The reaction occurs with acid–base catalysis and proceeds via *cis*-enediolate intermediates.

(2) cis (syn) eliminations, which are biochemically less common.

Biochemical Isomerizations Involve Intramolecular Hydrogen Atom Shifts

Biochemical **isomerization reactions** involve the intramolecular shift of a hydrogen atom so as to change the location of a double bond. In such a process, a proton is removed from one carbon atom and added to another. The metabolically most prevalent isomerization reaction is the **aldose–ketose interconversion,** a base-catalyzed reaction that occurs via **enediolate anion** intermediates (Fig. 15-10). The glycolytic enzyme **phosphoglucose isomerase** catalyzes such a reaction (Section 16-2B).

Racemization is an isomerization reaction in which a hydrogen atom shifts its stereochemical position at a molecule's only chiral center so as to invert that chiral center. Such an isomerization is called an **epimerization** in a molecule with more than one chiral center.

Rearrangements Produce Altered Carbon Skeletons

Rearrangement reactions break and reform C—C bonds so as to rearrange a molecule's carbon skeleton. There are few such metabolic reactions. One is the conversion of L-methylmalonyl-CoA to **succinyl-CoA** by **methylmalonyl-CoA mutase,** an enzyme whose prosthetic group is a **vitamin B_{12}** derivative:

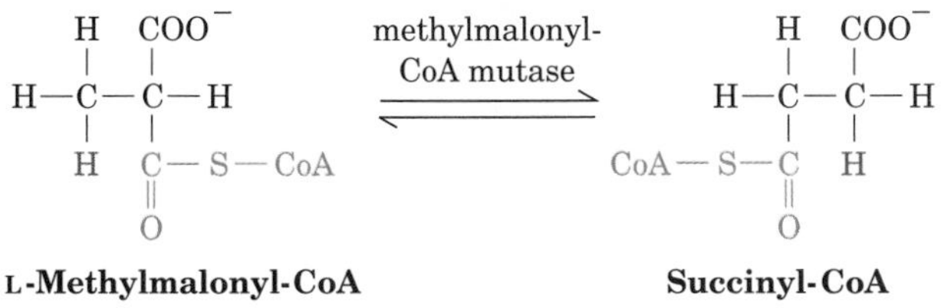

L-Methylmalonyl-CoA   **Succinyl-CoA**

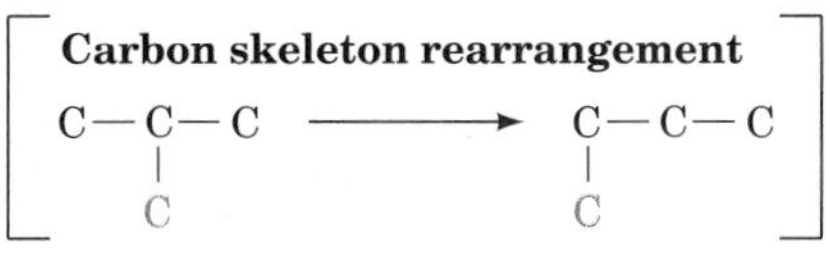

This reaction is involved in the oxidation of fatty acids with an odd number of carbon atoms (Section 23-2E) and several amino acids (Section 24-3E).

E. Reactions That Make and Break Carbon–Carbon Bonds

Reactions that make and break carbon–carbon bonds form the basis of both degradative and biosynthetic metabolism. The breakdown of glucose to CO_2 involves five such cleav-

ages, whereas its synthesis involves the reverse process. Such reactions, considered from the synthetic direction, involve addition of a nucleophilic carbanion to an electrophilic carbon atom. The most common electrophilic carbon atoms in such reactions are the sp^2-hybridized carbonyl carbon atoms of aldehydes, ketones, esters, and CO_2.

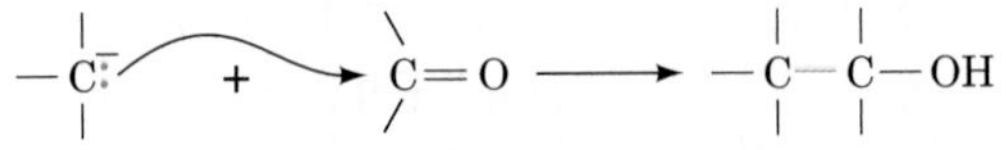

Stabilized carbanions must be generated to add to these

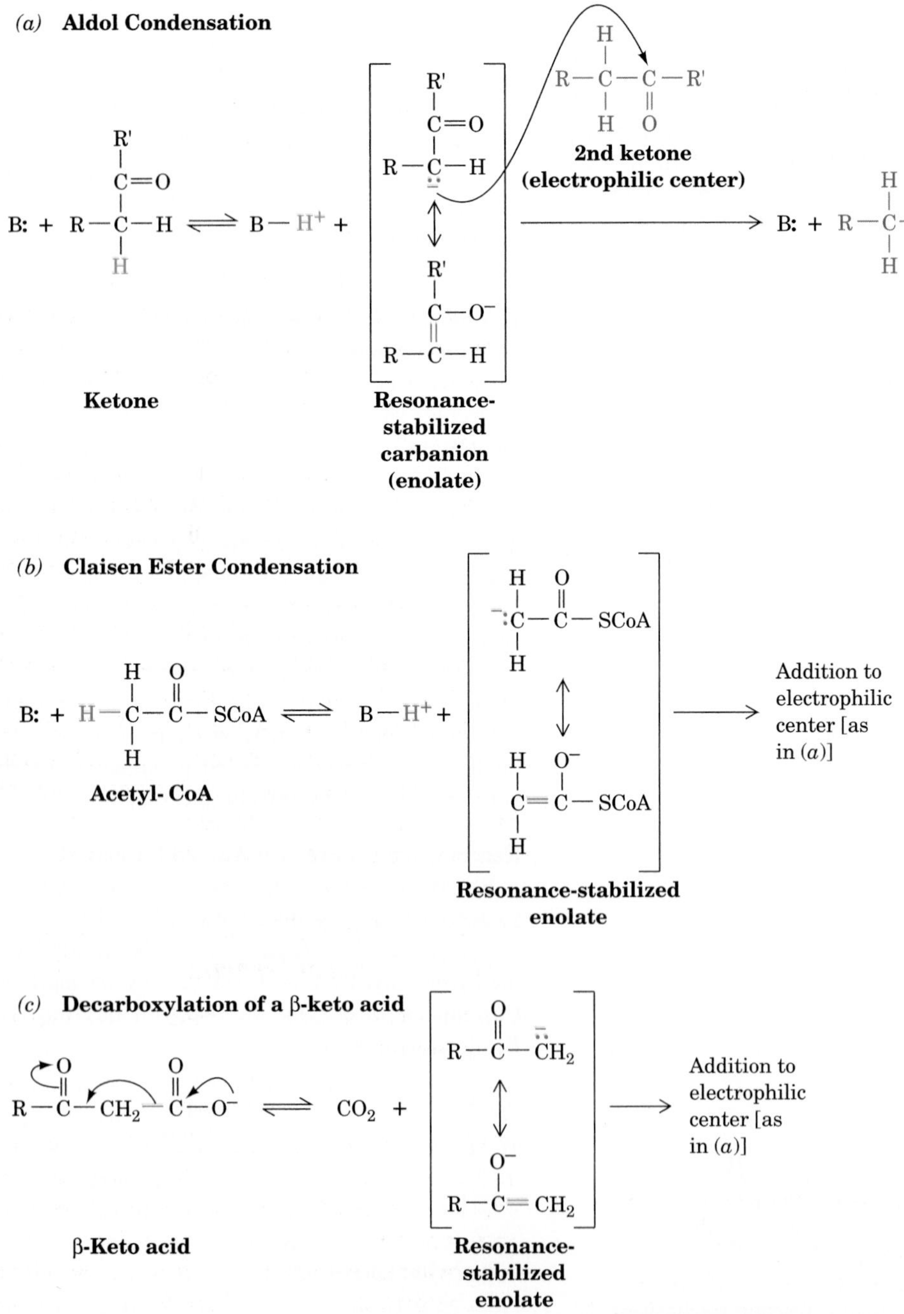

FIGURE 15-11. Examples of C—C bond formation and cleavage reactions: (*a*) aldol condensation, (*b*) Claisen ester condensation, and (*c*) decarboxylation of a β-keto acid. All three types of reaction involve generation of a resonance-stabilized carbanion followed by addition of this carbanion to an electrophilic center.

electrophilic centers. Three examples are the **aldol conden-sation** (catalyzed, e.g., by **aldolase;** Section 16-2D), **Claisen ester condensation** (citrate synthase; Section 19-3A), and the decarboxylation of a β-keto acid (isocitrate dehydrogen-ase; Section 19-3C, and **fatty acid synthase;** Section 23-4C). In nonenzymatic systems, both the aldol condensation and Claisen ester condensation involve the base-catalyzed gen-eration of a carbanion α to a carbonyl group (Fig. 15-11a and b). The carbonyl group is electron withdrawing and thereby provides resonance stabilization by forming an en-olate (Fig. 15-12a). The enolate may be further stabilized by neutralizing its negative charge. Enzymes do so through hydrogen bonding or protonation (Fig. 15-12b), conversion of the carbonyl group to a protonated Schiff base (covalent catalysis; Fig. 15-12c), or by its coordination to a metal ion (metal ion catalysis; Fig. 15-12d). Decarboxylation of a β-keto acid does not require base catalysis for the generation of the resonance-stabilized carbanion; the highly exergonic formation of CO_2 provides its driving force (Fig. 15-11c).

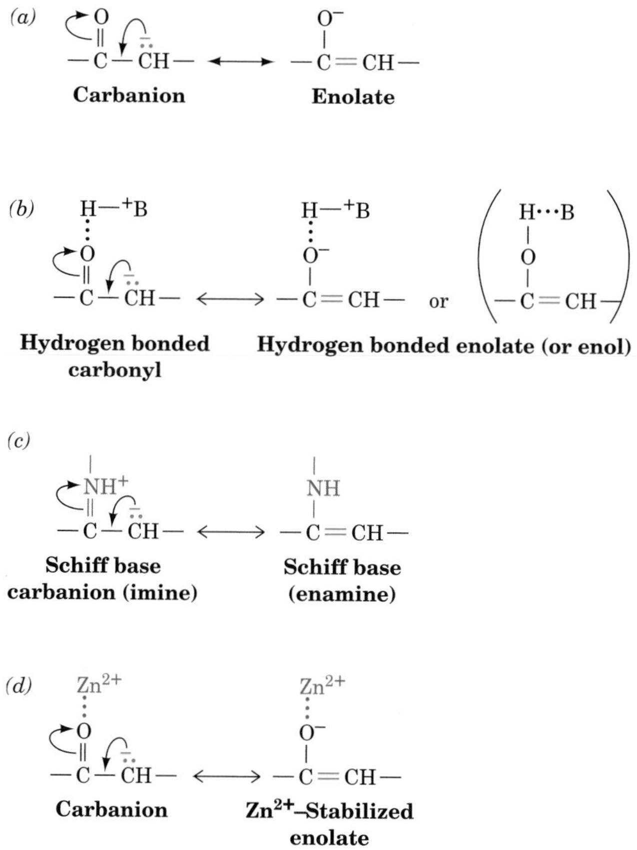

FIGURE 15-12. The stabilization of carbanions: (*a*) Carbanions adjacent to carbonyl groups are stabilized by the formation of enolates. (*b*) Carbanions adjacent to carbonyl groups hydrogen bonded to general acids are stabilized electrostatically or by charge neutralization. (*c*) Carbanions adjacent to protonated imines (Schiff bases) are stabilized by the formation of enamines. (*d*) Metal ions stabilize carbanions adjacent to carbonyl groups by the electrostatic stabilization of the enolate.

3. EXPERIMENTAL APPROACHES TO THE STUDY OF METABOLISM

A metabolic pathway can be understood at several levels:

1. In terms of the sequence of reactions by which a specific nutrient is converted to end products, and the energetics of these conversions.

2. In terms of the mechanisms by which each intermediate is converted to its successor. Such an analysis requires the isolation and characterization of the specific en-zymes that catalyze each reaction.

3. In terms of the control mechanisms that regulate the flow of metabolites through the pathway. An exquisitely complex network of regulatory processes renders meta-bolic pathways remarkably sensitive to the needs of the organism; the output of a pathway is generally only as great as required.

As you might well imagine, the elucidation of a metabolic pathway on all of these levels is a complex process, involv-ing contributions from a variety of disciplines. Most of the techniques used to do so involve somehow perturbing the system and observing the perturbation's effect on growth or on the production of metabolic intermediates. One such technique is the use of metabolic inhibitors that block metabolic pathways at specific enzymatic steps. Another is the study of genetic abnormalities that interrupt specific metabolic pathways. Techniques have also been developed for the dissection of organisms into their component organs, tissues, cells, and subcellular organelles, and for the purification and identification of metabolites as well as the enzymes that catalyze their interconversions. The use of isotopic tracers to follow the paths of specific atoms and molecules through the metabolic maze has become routine. New techniques utilizing NMR technology are able nonin-vasively to trace metabolites as they react *in vivo*. This sec-tion outlines the use of these various techniques.

A. Metabolic Inhibitors, Growth Studies, and Biochemical Genetics

Pathway Intermediates Accumulate in the Presence of Metabolic Inhibitors

The first metabolic pathway to be completely traced was the conversion of glucose to ethanol in yeast by a process known as **glycolysis** (Section 16-1A). In the course of these studies, certain substances, called **metabolic inhibitors,** were found to block the pathway at specific points, thereby causing preceding intermediates to build up. For instance, iodoacetate causes yeast extracts to accumulate fructose-1,6-bisphosphate, whereas fluoride causes the buildup of two phosphate esters, 3-phosphoglycerate and 2-phospho-glycerate. The isolation and characterization of these inter-mediates was vital to the elucidation of the glycolytic pathway: Chemical intuition combined with this information led to the prediction of the pathway's intervening steps. Each of the proposed reactions was eventually shown to occur *in vitro* (in the "test tube") as catalyzed by a purified enzyme.

Genetic Defects Also Cause Metabolic Intermediates to Accumulate

Archibald Garrod's realization, in the early 1900s, that human genetic diseases are the consequence of deficiencies in specific enzymes also contributed to the elucidation of metabolic pathways. For example, upon the ingestion of either phenylalanine or tyrosine, individuals with the largely harmless inherited condition known as **alcaptonuria,** but not normal subjects, excrete **homogentisic acid** in their urine (Sections 24-3H and 27-1C). This is because the liver of alcaptonurics lacks an enzyme that catalyzes the breakdown of homogentisic acid. Another genetic disease, **phenylketonuria** (Section 24-3H), results in the accumulation of **phenylpyruvate** in the urine (which, if untreated, causes severe mental retardation in infants). Ingested

phenylalanine and phenylpyruvate appear as phenylpyruvate in the urine of affected subjects, whereas tyrosine is metabolized normally. The effects of these two abnormalities suggested the pathway for phenylalanine metabolism diagrammed in Fig. 15-13. However, the supposition that phenylpyruvate but not tyrosine occurs on the normal pathway of phenylalanine metabolism because phenylpyruvate accumulates in the urine of phenylketonurics has proved incorrect. This indicates the pitfalls of relying solely on metabolic blocks and the consequent buildup of intermediates as indicators of a metabolic pathway. In this case, phenylpyruvate formation was later shown to arise from a normally minor pathway that becomes significant only when the phenylalanine concentration is abnormally high, as it is in phenylketonurics.

Metabolic Blocks Can Be Generated by the Genetic Manipulation of Microorganisms

Early metabolic studies led to the astounding discovery that *the basic metabolic pathways in most organisms are essentially identical.* This metabolic uniformity has greatly facilitated the study of metabolic reactions. Thus, although a mutation that inactivates or deletes an enzyme in a pathway of interest may be unknown in higher organisms, it can be readily generated in rapidly reproducing microorganisms through the use of **mutagens** (chemical agents that induce genetic changes; Section 30-1A), X-rays, or, more recently, through genetic engineering techniques (Section 28-8). Desired mutants are identified by their requirement of the pathway's end product for growth. For example, George Beadle and Edward Tatum proposed a pathway of arginine biosynthesis in the mold *Neurospora crassa* based on their analysis of three arginine-requiring **auxotrophic**

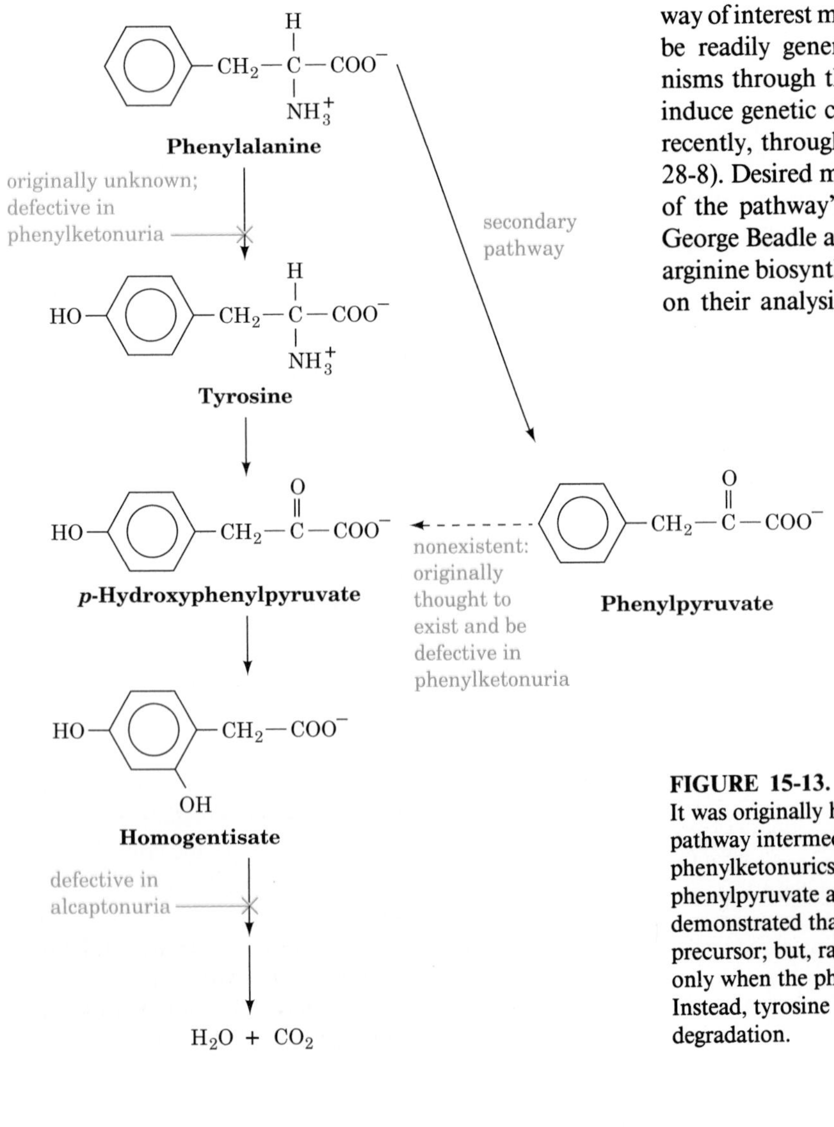

FIGURE 15-13. The pathway for phenylalanine degradation. It was originally hypothesized that phenylpyruvate was a pathway intermediate based on the observation that phenylketonurics excrete ingested phenylalanine and phenylpyruvate as phenylpyruvate. Further studies, however, demonstrated that phenylpyruvate is not a homogentisate precursor; but, rather, phenylpyruvate production is significant only when the phenylalanine concentration is abnormally high. Instead, tyrosine is the normal product of phenylalanine degradation.

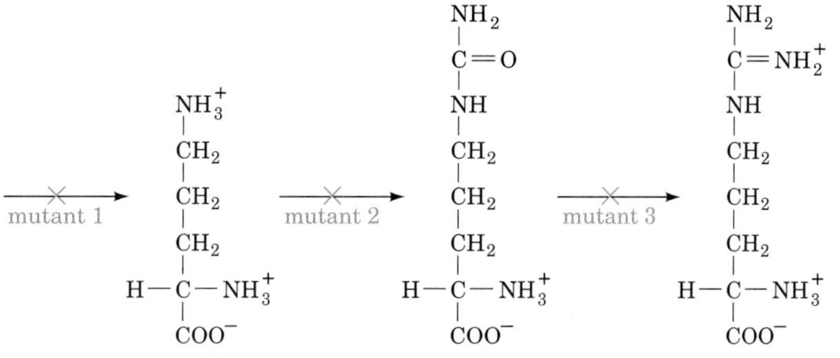

$$NH_3^+$$
$$|$$
$$CH_2$$
$$|$$
$$CH_2$$
$$|$$
$$CH_2$$
$$|$$
$$H - C - NH_3^+$$
$$|$$
$$COO^-$$
Ornithine

$$NH_2$$
$$|$$
$$C = O$$
$$|$$
$$NH$$
$$|$$
$$CH_2$$
$$|$$
$$CH_2$$
$$|$$
$$CH_2$$
$$|$$
$$H - C - NH_3^+$$
$$|$$
$$COO^-$$
Citrulline

$$NH_2$$
$$|$$
$$C = NH_2^+$$
$$|$$
$$NH$$
$$|$$
$$CH_2$$
$$|$$
$$CH_2$$
$$|$$
$$CH_2$$
$$|$$
$$H - C - NH_3^+$$
$$|$$
$$COO^-$$
Arginine

mutant 1 → mutant 2 → mutant 3 →

FIGURE 15-14. The pathway of arginine biosynthesis indicating the positions of genetic blocks. All of these mutants grow in the presence of arginine, but mutant 1 also grows in the presence of the (nonstandard) α-amino acids **citrulline** or **ornithine** and mutant 2 grows in the presence of citrulline. This is because in mutant 1, an enzyme leading to the production of ornithine is absent but enzymes further along the pathway are normal. In mutant 2, the enzyme catalyzing citrulline production is defective, whereas in mutant 3 an enzyme involved in the conversion of citrulline to arginine is lacking.

mutants (mutants requiring a specific nutrient for growth), which were isolated after X-irradiation (Fig. 15-14). This landmark study also conclusively demonstrated that enzymes are specified by genes (Section 27-1C).

Genetic Manipulations Have Been Developed for the Expression of Foreign Genes in Higher Organisms

The recently developed techniques for expressing foreign genes in multicellular host organisms such as mice and pigs **(transgenic animals;** Fig. 27-24 and Section 28-8F) constitutes a major breakthrough in genetic manipulation. In addition to creating metabolic blocks, *it is now possible to express genes in tissues where they were not originally present.* An example of this is the introduction of a gene encoding **creatine kinase** (Section 15-4C) into mouse liver. This enzyme, which is normally present in many tissues, including brain and muscle but not liver, catalyzes the formation of **phosphocreatine** (Section 15-4C), a substance that functions to generate ATP rapidly when it is in short supply. The expression of creatine kinase in liver causes the liver to synthesize phosphocreatine when the animal is fed creatine, as demonstrated by localized *in vivo* NMR techniques (Figure 15-15; NMR is discussed below). The presence of phosphocreatine in transgenic mouse livers protects the animals against the sharp drop in [ATP] ordinarily caused by fructose overload (Section 16-5A). This genetic manipulation technique is being used to study mechanisms of metabolic control *in vivo*.

Metabolic pathways are regulated both by controlling the activities of regulatory enzymes (Sections 16-4 and 17-3) and by controlling their concentrations at the level of gene expression (Sections 29-3, 30-4, and 33-3). The important question of how hormones and diet control metabolic processes at the level of gene expression is being addressed through the use of transgenic animals. **Reporter genes** (genes whose products are easily detected) are placed under the influence of genetic regulatory elements **(promoters)**

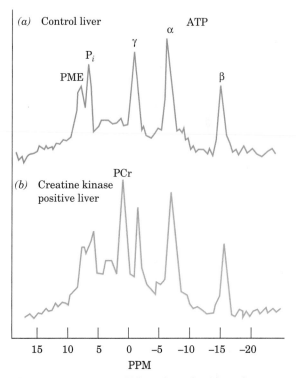

FIGURE 15-15. The expression of creatine kinase in transgenic mouse liver as demonstrated by localized *in vivo* ^{31}P NMR. (*a*) The spectrum of a normal mouse liver after the mouse had been fed a diet supplemented with 2% creatine. The peaks corresponding to inorganic phosphate (P_i), the α, β, and γ phosporyl groups of ATP, and phosphomonoesters (PME) are labeled. (*b*) The spectrum of the liver of a mouse transgenic for creatine kinase that had been fed a diet supplemented with 2% creatine. The phosphocreatine peak is labeled ΓCr. [After Koretsky, A.P., Brosnan, M.J., Chen, L., Chen, J., and Van Dyke, T.A., *Proc. Natl. Acad. Sci.* **87**, 3114 (1990)].

that control the expression of specific regulatory enzymes and the resulting composite gene is expressed in animals. The transgenic animals can then be treated with specific hormones and/or diets and the production of the reporter gene product measured. For example, the promoter for the enzyme **phosphoenolpyruvate carboxykinase (PEPCK)** was attached to the structural gene encoding **growth hormone (GH).** PEPCK, an important regulatory enzyme in **gluconeogenesis** (the synthesis of glucose from noncarbohydrate precursors; Section 21-1), is normally present in liver and kidneys but not in blood. GH, however, is secreted into the blood and its presence there can be readily quantitated by an ELISA (Section 5-1D). Mice transgenic for PEPCK/GH were fed either a high-carbohydrate/low-protein diet or a high-protein/low carbohydrate diet, which are known to decrease and increase PEPCK activity, respectively. GH in high concentrations was detected only in the serum of PEPCK/GH mice on a high-protein diet, thereby indicating that the GH was synthesized under the same dietary control as that of the PEPCK expressed by the normal gene. Thus, the activity of PEPCK in PEPCK/GH mice can be continuously monitored, albeit indirectly through serum GH assay (the direct measurement of PEPCK in mouse liver or kidney requires the sacrifice of the animal and hence can be done only once). Such use of reporter genes promises to be of great value in the study of the genetic control of metabolism.

B. Isotopes in Biochemistry

The specific labeling of metabolites such that their interconversions can be traced is an indispensable technique for elucidating metabolic pathways. Franz Knoop formulated this technique in 1904 to study fatty acid oxidation. He fed dogs fatty acids chemically labeled with phenyl groups and isolated the phenyl-substituted end products from their urine. From the differences in these products when the phenyl-substituted starting material contained odd and even numbers of carbon atoms he deduced that fatty acids are degraded in C_2 units (Section 23-2).

Isotopes Specifically Label Molecules Without Altering Their Chemical Properties

Chemical labeling has the disadvantage that the chemical properties of labeled metabolites differ from those of normal metabolites. This problem is eliminated by labeling molecules of interest with **isotopes** (atoms with the same number of protons but a different number of neutrons in their nuclei). Recall that the chemical properties of an element are a consequence of its electron configuration, which, in turn, is determined by its atomic number, not its atomic mass. The metabolic fate of a specific atom in a metabolite can therefore be elucidated by isotopically labeling that position and following its progress through the metabolic pathway of interest. The advent of isotopic labeling and tracing techniques in the 1940s therefore revolu-

tionized the study of metabolism. (**Isotope effects,** which are changes in reaction rates arising from the mass differences between isotopes, are in most instances, negligible. Where they are significant, most noticeably between hydrogen and its isotopes deuterium and tritium, they have been used to gain insight into enzymatic reaction mechanisms.)

NMR Can Be Used To Study Metabolism in Whole Animals

Nuclear magnetic resonance (NMR) detects specific isotopes due to their characteristic nuclear spins. Among the isotopes that NMR can detect are ^1H, ^{13}C, and ^{31}P. Since the NMR spectrum of a particular nucleus varies with its immediate environment, it is possible to identify the peaks corresponding to specific atoms even in relatively complex mixtures.

The development of magnets large enough to accommodate animals and humans, and to localize spectra to specific organs, has made it possible to study metabolic pathways noninvasively by NMR techniques. Thus, ^{31}P NMR can be used to study energy metabolism in muscle by monitoring the levels of ATP, ADP, inorganic phosphate, and phosphocreatine (Figure 15-15). Indeed, a ^{31}P NMR system has been patented to measure the muscular metabolic efficiency and maximum power of race horses while they are walking or running on a motor-driven treadmill in order to identify promising animals and to evaluate the efficacy of their training and nutritional programs.

Isotopically labeling specific atoms of metabolites with ^{13}C (which is only 1.10% naturally abundant) permits the metabolic progress of the labeled atoms to be followed by ^{13}C NMR. Figure 15-16 shows *in vivo* ^{13}C NMR spectra of a rat liver before and after an injection of D-[1-^{13}C]glucose. The ^{13}C can be seen entering the liver and then being converted to glycogen (the storage form of glucose; Chapter 17). ^1H NMR techniques are now being used to determine the *in vivo* levels of a variety of metabolites in tissues such as brain and muscle.

The Detection of Radioactive Isotopes

All elements have isotopes. For example, the atomic mass of naturally occurring Cl is 35.45 D because, at least on earth, it is a mixture of 55% ^{35}Cl and 45% ^{36}Cl (other isotopes of Cl are present in only trace amounts). Stable isotopes are generally identified and quantitated by mass spectrometry or NMR techniques. Many isotopes, however, are unstable; they undergo **radioactive decay,** a process that involves the emission from the radioactive nuclei of subatomic particles such as helium nuclei (**α particles),** electrons (**β particles),** and/or photons (**γ radiation).** Radioactive nuclei emit radiation with characteristic energies. For example, ^3H, ^{14}C, and ^{32}P all emit β particles but with respective energies of 0.018, 0.155, and 1.71 MeV. The radiation from ^{32}P is therefore highly penetrating, whereas that from ^3H and ^{14}C is not. (^3H and ^{14}C, as all radioactive

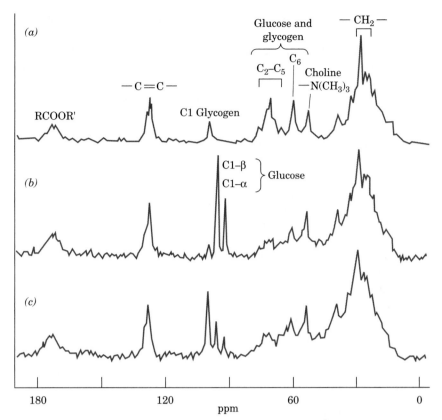

FIGURE 15-16. The conversion of [1-^{13}C]-glucose to glycogen as observed by localized *in vivo* ^{13}C NMR. (*a*) The natural abundance ^{13}C NMR spectrum of the liver of a live rat. Note the resonance corresponding to C1 of glycogen. (*b*) The ^{13}C NMR spectrum of the liver of the same rat ~5 min after it was intravenously injected with 100 mg of [1-^{13}C]glucose (90% enriched). The resonances of the C1 atom of both the α and β anomers of glucose are clearly distinguishable from each other and from the resonance of the C1 atom of glycogen. (*c*) The ^{13}C NMR spectrum of the liver of the same rat ~30 minutes after the [1-^{13}C]glucose injection. The C1 resonances of both the α and β glucose anomers are much reduced while the C1 resonance of glycogen has increased. [After Reo, N.V., Siegfried, B.A., and Acherman, J.J.H., *J. Biol. Chem.* **259**, 13665 (1984)].

isotopes, must, nevertheless, be handled with great caution because they can cause genetic damage upon ingestion.)

Radiation can be detected by a variety of techniques. Those most commonly used in biochemical investigations are **proportional counting** (known in its simplest form as **Geiger counting**), **liquid scintillation counting,** and **autoradiography.** Proportional counters electronically detect the ionizations in a gas caused by the passage of radiation. Moreover, they can also discriminate between particles of different energies and thus simultaneously determine the amounts of two or more different isotopes present.

Although proportional counters are quite simple to use, the radiation from two of the most widely used isotopes in biochemical analysis, ^{3}H and ^{14}C, have insufficient penetrating power to enter a proportional counter's detection chamber with reasonable efficiency. This limitation is circumvented through liquid scintillation counting. In this technique, a radioactive sample is dissolved or suspended in a solution containing fluorescent substances that emit a pulse of light when struck by radiation. The light is detected electronically so that the number of light pulses can be counted. The emitting nucleus can also be identified be-

cause the intensity of a light pulse is proportional to the radiation energy (the number of fluorescent molecules excited by a radioactive particle is proportional to the particle's energy).

In autoradiography, radiation is detected by its blackening of photographic film. The radioactive sample is laid on, or in some cases mixed with, the photographic emulsion and, after sufficient exposure time (from minutes to months), the film is developed. Autoradiography is widely used to locate radioactive substances in polyacrylamide gels (e.g., Fig. 5-30).

Radioactive Isotopes Have Characteristic Half-Lives

Radioactive decay is a random process whose rate for a given isotope depends only on the number of radioactive atoms present. It is therefore a simple first-order process whose half-life, $t_{1/2}$, is a function only of the rate constant, k, for the decay process (Section 13-1B):

$$t_{1/2} = \frac{\ln 2}{k} = \frac{0.693}{k} \qquad [13.5]$$

Because k is different for each radioactive isotope, each has

TABLE 15-1.   Some Trace Isotopes of Biochemical Importance

Nucleus	Stable Isotopes	
	Natural Abundance (%)	
^2H	0.015	
^{13}C	1.10	
^{15}N	0.37	
^{18}O	0.20	

Nucleus	Radioactive Isotopes	
	Radiation Type	**Half-Life**
^3H	β	12.32 years
^{14}C	β	5715 years
^{22}Na	β^+, γ	2.60 years
^{32}P	β	14.28 days
^{35}S	β	87.2 days
^{45}Ca	β	162.7 days
^{60}Co	β, γ	5.271 years
^{125}I	γ	59.4 days
^{131}I	β, γ	8.04 days

Source: Holden, N.E., *in* Lide, D.R. (Ed.), *Handbook of Chemistry and Physics* (75th ed.), pp. **11**-35 to 139, CRC Press (1994).

a characteristic half-life. The properties of some isotopes in common biochemical use are listed in Table 15-1.

Isotopes Are Indispensable for Establishing the Metabolic Origins of Complex Metabolites and Precursor–Product Relationships

The metabolic origins of complex molecules such as heme, cholesterol, and phospholipids may be determined by administering isotopically labeled starting materials to animals and isolating the resulting products. One of the early advances in metabolic understanding resulting from the use of isotopic tracers was the demonstration, by David Shemin and David Rittenberg in 1945, that the nitrogen atoms of heme are derived from glycine rather than from ammonia, glutamic acid, proline, or leucine (Section 24-4A). They showed this by feeding rats these ^{15}N-labeled nutrients, isolating the heme in their blood, and analyzing it for ^{15}N content. Only when the rats were fed [^{15}N]glycine did the heme contain ^{15}N (Fig. 15-17). This technique was also used to demonstrate that all of cholesterol's carbon atoms are derived from acetyl-CoA (Section 23-6A).

Isotopic tracers are also useful in establishing the order of appearance of metabolic intermediates, their so-called **precursor–product relationships.** An example of such an

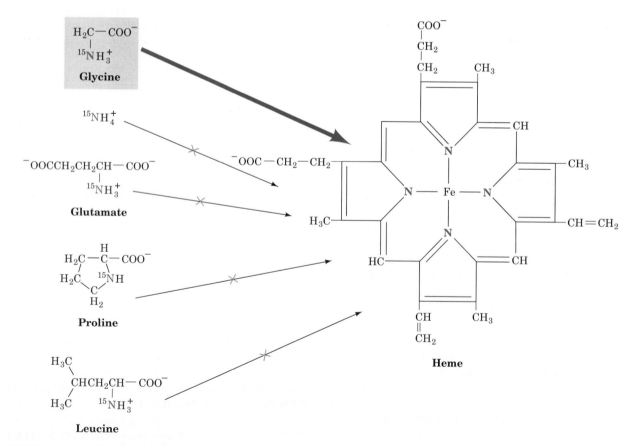

FIGURE 15-17.   The metabolic origin of the nitrogen atoms in heme. Only [^{15}N]glycine, of many ^{15}N-labeled metabolites, is an ^{15}N-labeled heme precursor.

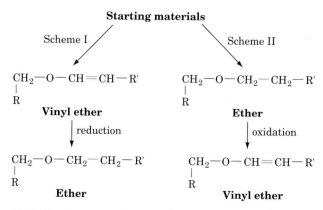

Starting materials

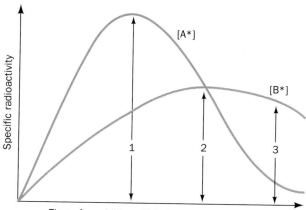

FIGURE 15-18. Two possible pathways for the biosynthesis of ether- and vinyl ether–containing phospholipids. **(I)** The vinyl ether is the precursor and the ether is the product. **(II)** The ether is the precursor and the vinyl ether is the product.

FIGURE 15-19. The flow of a pulse of radioactivity from precursor (A*, *orange*) to product (B*, *purple*). At point 1, product radioactivity is increasing and is less than that of its precursor; at point 2, product radioactivity is maximal and is equal to that of its precursor; and at point 3, product radioactivity is decreasing and is greater than that of its precursor.

analysis concerns the biosynthesis of the complex phospholipids called **plasmalogens** and **alkylacylglycerophospholipids** (Section 23-8A). Alkylacylglycerophospholipids are ethers, whereas the closely related plasmalogens are vinyl ethers. Their similar structures brings up the interesting question of their biosynthetic relationship: Which is the precursor and which is the product? Two possible modes of synthesis can be envisioned (Fig. 15-18):

I. The starting material is converted to the vinyl ether (plasmalogen), which is then reduced to yield the ether (alkylacylglycerophospholipid). Accordingly, the vinyl ether would be the precursor and the ether the product.

II. The ether is formed first and then oxidized to yield the vinyl ether. The ether would then be the precursor and the vinyl ether the product.

Precursor–product relationships can be most easily sorted out through the use of radioactive tracers. A pulse of the labeled starting material is administered to an organism and the specific radioactivities of the resulting metabolic products are followed with time (Fig. 15-19):

$$\text{Starting material*} \rightarrow A^* \rightarrow B^* \rightarrow \text{later products*}$$

(here the * represents the radioactive label). Metabolic pathways, as we shall see in Section 15-6B, normally operate in a steady state; that is, the throughput of metabolites in each of its reaction steps is equal. Moreover, the rates of most metabolic reactions are first order for a given substrate. Making these assumptions, we note that the rate of change of B's radioactivity, [B*], is equal to the rate of passage of label from A* to B* less the rate of passage of label from B* to the pathway's next product:

$$\frac{d[B^*]}{dt} = k[A^*] - k[B^*] = k([A^*] - [B^*]) \quad [15.1]$$

where k is the pseudo-first-order rate constant for both the conversion of A to B and the conversion of B to its product,

and t is time. Inspection of this equation indicates the criteria that must be met to establish that A is the precursor of B (Fig. 15-19):

1. Before the radioactivity of the product [B*] is maximal, $d[B^*]/dt > 0$, so $[A^*] > [B^*]$; that is, *while the radioactivity of a product is rising, it should be less than that of its precursor.*

2. When [B*] is maximal, $d[B^*]/dt = 0$, so $[A^*] = [B^*]$; that is, *when the radioactivity of a product is at its peak, it should be equal to that of its precursor.* This result also implies that *the radioactivity of a product peaks after that of its precursor.*

3. After [B*] begins to decrease, $d[B^*]/dt < 0$, so $[A^*] < [B^*]$; that is, *after the radioactivity of a product has peaked, it should remain greater than that of its precursor.*

Such a determination of the precursor–product relationship between alkylacylglycerophospholipid and plasmalogen, using ^{14}C-labeled starting materials, indicated that the ether is the precursor and the vinyl ether is the product (Fig. 15-18, Scheme II).

C. Isolated Organs, Cells, and Subcellular Organelles

In addition to understanding the chemistry and catalytic events that occur at each step of a metabolic pathway, it is important to learn where a given pathway occurs within an organism. Early workers studied metabolism in whole animals. For example, the role of the pancreas in diabetes was established by Frederick Banting and George Best in 1921

TABLE 15-2. **METABOLIC FUNCTIONS OF EUKARYOTIC ORGANELLES**

Organelle	Function
Mitochondrion	Citric acid cycle, oxidative phosphorylation, fatty acid oxidation, amino acid breakdown
Cytosol	Glycolysis, pentose phosphate pathway, fatty acid biosynthesis, many reactions of gluconeogenesis
Lysosomes	Enzymatic digestion of cell components and ingested matter
Nucleus	DNA replication and transcription, RNA processing
Golgi apparatus	Posttranslational processing of membrane and secretory proteins; formation of plasma membrane and secretory vesicles
Rough endoplasmic reticulum	Synthesis of membrane-bound and secretory proteins
Smooth endoplasmic reticulum	Lipid and steroid biosynthesis
Peroxisomes (glyoxisomes in plants)	Oxidative reactions catalyzed by amino acid oxidases and catalase; glyoxylate cycle reactions in plants

by surgically removing that organ from dogs and observing that these animals then developed the disease.

The metabolic products produced by a particular organ can be studied by **organ perfusion** or in **tissue slices.** In organ perfusion, a specific organ is surgically removed from an animal and the organ's arteries and veins are connected to an artificial circulatory system. The composition of the material entering the organ can thereby be controlled and its metabolic products monitored. Metabolic processes can be similarly studied in slices of tissue thin enough to be nourished by free diffusion in an appropriate nutrient solution. Otto Warburg pioneered the tissue slice technique in the early twentieth century through his studies of respiration, in which he used a manometer to measure the changes in gas volume above tissue slices as a consequence of their O_2 consumption.

A given organ or tissue generally contains several cell types. **Cell sorters** are devices that can separate cells according to type once they have been treated with the enzymes trypsin and collagenase to destroy the intercellular matrix that binds them into a tissue. This technique allows further localization of metabolic function. A single cell type may also be grown in **tissue culture** for study. Although culturing cells often results in their loss of differentiated function, techniques have been developed for maintaining several cell types that still express their original characteristics.

Metabolic pathways in eukaryotes are compartmentalized in various subcellular organelles as Table 15-2 indicates (these organelles are described in Section 1-2A). For example, oxidative phosphorylation occurs in the mitochondrion, whereas glycolysis and fatty acid biosynthesis

occur in the cytosol. Such observations are made by breaking cells open and fractionating their components by differential centrifugation (Section 5-1B), possibly followed by zonal ultracentrifugation through a sucrose density gradient or by equilibrium density gradient ultracentrifugation in a CsCl density gradient, which, respectively, separate particles according to their size and density (Section 5-5B). The cell fractions are then analyzed for biochemical function.

4. THERMODYNAMICS OF PHOSPHATE COMPOUNDS

The endergonic processes that maintain the living state are driven by the exergonic reactions of nutrient oxidation. This coupling is most often mediated through the syntheses of a few types of "high-energy" intermediates whose exergonic consumption drives endergonic processes. These intermediates therefore form a sort of universal free energy "currency" through which free energy–producing reactions "pay for" the free energy–consuming processes in biological systems.

Adenosine triphosphate (ATP; Fig. 15-20), which occurs in all known life forms, is the "high-energy" intermediate that constitutes the most common cellular energy currency. Its central role in energy metabolism was first recognized in 1941 by Fritz Lipmann and Herman Kalckar. ATP consists of an **adenosine** moiety to which three **phosphoryl groups** ($-PO_3^{2-}$) are sequentially linked via a **phosphoester bond** followed by two **phosphoanhydride bonds. Adenosine diphosphate (ADP)** and **5′-adenosine monophosphate (AMP)** are similarly constituted but with only two and one phosphoryl units, respectively.

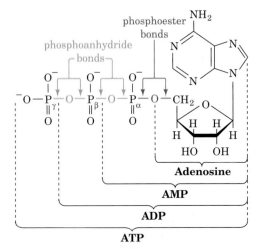

FIGURE 15-20. The structure of ATP indicating its relationship to ADP, AMP, and adenosine. The phosphoryl groups, starting with that on AMP, are referred to as the α, β, and γ phosphates. Note the differences between phosphoester and phosphoanhydride bonds.

In this section we consider the nature of phosphoryl-transfer reactions, discuss why some of them are so exergonic, and outline how the cell consumes and regenerates ATP.

A. Phosphoryl-Transfer Reactions

Phosphoryl-transfer reactions,

$$R_1—O—PO_3^{2-} + R_2—OH \rightleftharpoons R_1—OH + R_2—O—PO_3^{2-}$$

are of enormous metabolic significance. Some of the most important reactions of this type involve the synthesis and hydrolysis of ATP:

$$ATP + H_2O \rightleftharpoons ADP + P_i$$
$$ATP + H_2O \rightleftharpoons AMP + PP_i$$

where P_i and PP_i, respectively, represent **orthophosphate** (PO_4^{3-}) and **pyrophosphate** ($P_2O_7^{4-}$) in any of their ionization states. *These highly exergonic reactions are coupled to numerous endergonic biochemical processes so as to drive them to completion. Conversely, ATP is regenerated by coupling its formation to a more highly exergonic metabolic process* (the thermodynamics of coupled reactions is discussed in Section 3-4C).

To illustrate these concepts, let us consider two examples of phosphoryl-transfer reactions. The initial step in the metabolism of glucose is its conversion to glucose-6-phosphate (Section 16-2A). Yet, the direct reaction of glucose and P_i is thermodynamically unfavorable (Fig. 15-21a). In biological systems, however, this reaction is coupled to the exergonic hydrolysis of ATP, so the overall reaction is thermodynamically favorable. ATP can be similarly regenerated by coupling its synthesis from ADP and P_i to the even more exergonic hydrolysis of **phosphoenolpyruvate** (Fig. 15-21b; Section 16-2J).

The bioenergetic utility of phosphoryl-transfer reactions stems from their kinetic stability to hydrolysis combined with their capacity to transmit relatively large amounts of free energy. The $\Delta G°'$ values of hydrolysis of several phosphorylated compounds of biochemical importance are tabulated in Table 15-3. The negatives of these values are often referred to as **phosphate group-transfer potentials;** they are a measure of the tendency of phosphorylated compounds to transfer their phosphoryl groups to water. Note that ATP has an intermediate phosphate group-transfer potential. Under standard conditions, the compounds above ATP in Table 15-3 can spontaneously transfer a phosphoryl group to ADP to form ATP, which can, in turn, spontaneously transfer a phosphoryl group to the hydrolysis products (ROH form) of the compounds below it.

ΔG of ATP Hydrolysis Varies with pH, Divalent Metal Ion Concentration, and Ionic Strength

The ΔG of a reaction varies with the total concentrations of its reactants and products and thus with their ionic states

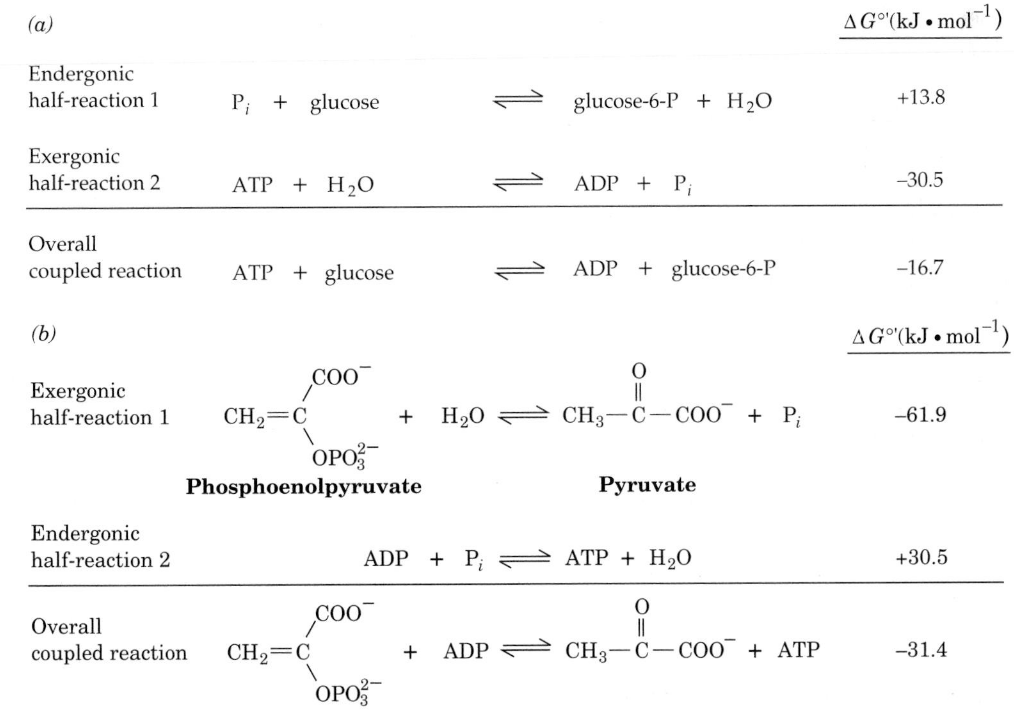

FIGURE 15-21. Some overall coupled reactions involving ATP: *(a)* The phosphorylation of glucose to form glucose-6-phosphate and ADP. *(b)* The phosphorylation of ADP by phosphoenolpyruvate to form ATP and pyruvate. Each reaction has been conceptually decomposed into a direct phosphorylation step (half-reaction 1) and a step in which ATP is hydrolyzed (half-reaction 2). Both half-reactions proceed in the direction in which the overall reaction is exergonic ($\Delta G < 0$).

TABLE 15-3. STANDARD FREE ENERGIES OF PHOSPHATE HYDROLYSIS OF SOME COMPOUNDS OF BIOLOGICAL INTEREST

Compound	$\Delta G^{\circ\prime}$ ($kJ \cdot mol^{-1}$)
Phosphoenol pyruvate	-61.9
1,3-Biphosphoglycerate	-49.4
Acetyl phosphate	-43.1
Phosphocreatine	-43.1
PP_i	-33.5
ATP ($\longrightarrow$ AMP + PP_i)	-32.2
ATP ($\longrightarrow$ ADP + P_i)	-30.5
Glucose-1-phosphate	-20.9
Fructose-6-phosphate	-13.8
Glucose-6-phosphate	-13.8
Glycerol-3-phosphate	-9.2

Source: Jencks, W.P., *in* Fasman, G.D. (Ed.), *Handbook of Biochemistry and Molecular Biology* (3rd ed.), Physical and Chemical Data, Vol. I, *pp.* 296–304, CRC Press (1976).

(Eq. [3.15]). The ΔG's of hydrolysis of phosphorylated compounds are therefore highly dependent on pH, divalent metal ion concentration (divalent metal ions such as Mg^{2+} have high phosphate-binding affinities), and ionic strength. Reasonable estimates of the intracellular values of these quantities as well as of [ATP], [ADP], and [P_i] (which are generally on the order of millimolar) indicate that ATP hydrolysis under physiological conditions has $\Delta G \approx -50$ $kJ \cdot mol^{-1}$ rather than the -30.5-$kJ \cdot mol^{-1}$ value of its $\Delta G^{\circ\prime}$. Nevertheless, for the sake of consistency in comparing reactions, we shall usually refer to the latter value.

The above situation for ATP is not unique. It is important to keep in mind that *within a given cell, the concentrations of most substances vary both with location and time. Indeed, the concentrations of many ions, coenzymes, and metabolites commonly vary by several orders of magnitude across membranous organelle boundaries.* Unfortunately, it is usually quite difficult to obtain an accurate measurement of the concentration of any particular chemical species in a specific cellular compartment. The ΔG's for most *in vivo* reactions are therefore little more than estimates.

B. Rationalizing the "Energy" in "High-Energy" Compounds

Bonds whose hydrolysis proceeds with large negative values of $\Delta G^{\circ\prime}$ (customarily more negative than -25 $kJ \cdot mol^{-1}$) are often referred to as **"high-energy" bonds** or **"energy-rich" bonds** and are frequently symbolized by the squiggle ($\sim$). Thus ATP may be represented as AR—P$\sim$P$\sim$P, where A, R, and P symbolize adenyl, ribosyl, and phosphoryl groups, respectively. Yet, the phosphoester bond joining the adenosyl group of ATP to its α-phosphoryl group appears to be not greatly different in electronic character from the so-called "high-energy" bonds bridging its β and γ phosphoryl groups. In fact, none of these bonds has any unusual properties, so the term "high-energy" bond is

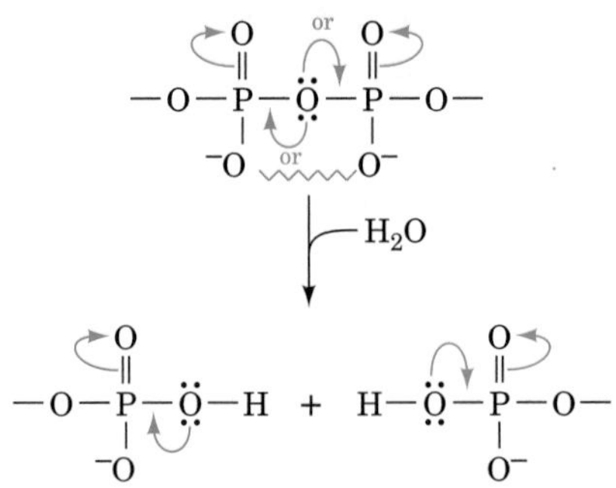

FIGURE 15-22. The competing resonances (*curved arrows* from central O) and charge–charge repulsions (*zigzag line*) between the phosphoryl groups of a phosphoanhydride decrease its stability relative to its hydrolysis products.

somewhat of a misnomer. (In any case, it should not be confused with the term "bond energy," which is defined as the energy required to break, not hydrolyze, a covalent bond.) Why then, should the phosphoryl-transfer reactions of ATP be so exergonic? The answer comes from the comparison of the stabilities of the reactants and products of these reactions.

Several different factors appear to be responsible for the "high-energy" character of phosphoanhydride bonds such as those in ATP (Fig. 15-22):

1. The resonance stabilization of a phosphoanhydride bond is less than that of its hydrolysis products. This is because a phosphoanhydride's two strongly electron-withdrawing phosphoryl groups must compete for the π electrons of its bridging oxygen atom, whereas this competition is absent in the hydrolysis products. In other words, the electronic requirements of the phosphoryl groups are less satisfied in a phosphoanhydride than in its hydrolysis products.

2. Of perhaps greater importance is the destabilizing effect of the electrostatic repulsions between the charged groups of a phosphoanhydride in comparison to that of its hydrolysis products. In the physiological pH range, ATP has three to four negative charges whose mutual electrostatic repulsions are partially relieved by ATP hydrolysis.

3. Another destabilizing influence, which is difficult to assess, is the smaller solvation energy of a phosphoanhydride in comparison to that of its hydrolysis products. Some estimates suggest that this factor provides the dominant thermodynamic driving force for the hydrolysis of phosphoanhydrides.

A further property of ATP that suits it to its role as an energy intermediate stems from the relative kinetic stability of phosphoanhydride bonds to hydrolysis. Most types of anhydrides are rapidly hydrolyzed in aqueous solution. Phosphoanhydride bonds, however, have unusually large free energies of activation. Consequently, ATP is reasonably stable under physiological conditions but is readily hydrolyzed in enzymatically mediated reactions.

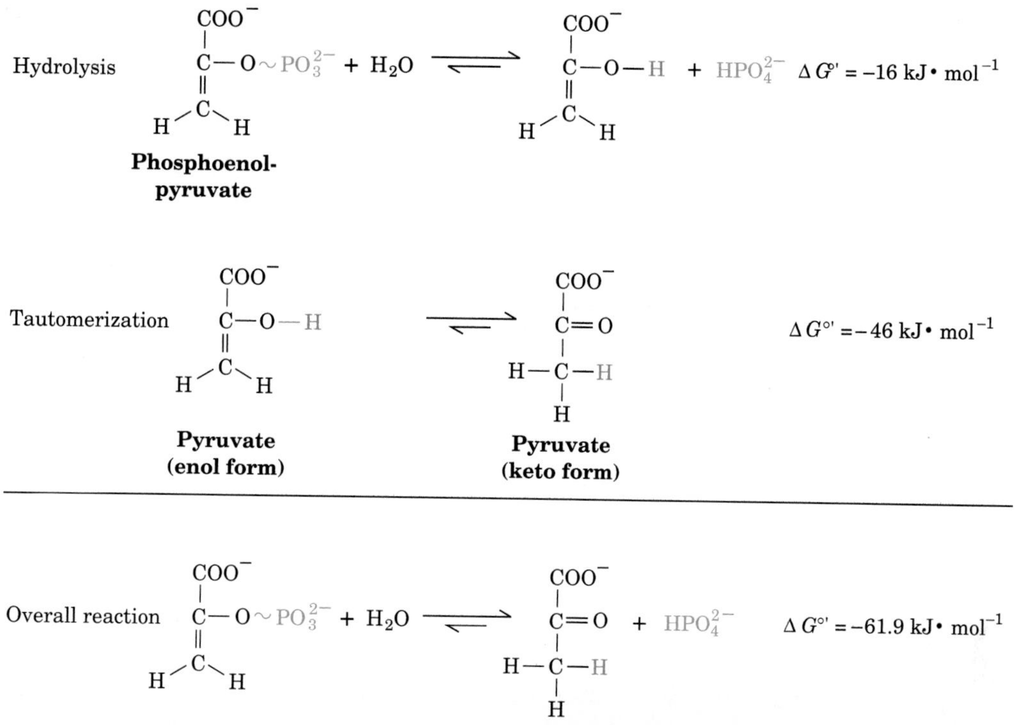

Hydrolysis

$$\underset{\substack{\text{Phosphoenol-}\\\text{pyruvate}}}{\overset{\displaystyle COO^-}{\underset{\displaystyle \underset{H}{\overset{\displaystyle\parallel}{C}}{}_{H}}{\overset{\displaystyle |}{C}}-O\sim PO_3^{2-}}} + H_2O \;\rightleftharpoons\; \underset{\substack{}}{\overset{\displaystyle COO^-}{\underset{\displaystyle \underset{H}{\overset{\displaystyle\parallel}{C}}{}_{H}}{\overset{\displaystyle |}{C}}-O-H}} \;+\; HPO_4^{2-} \qquad \Delta G' = -16 \text{ kJ} \cdot \text{mol}^{-1}$$

Tautomerization

$$\underset{\substack{\textbf{Pyruvate}\\\textbf{(enol form)}}}{\overset{\displaystyle COO^-}{\underset{\displaystyle \underset{H}{\overset{\displaystyle\parallel}{C}}{}_{H}}{\overset{\displaystyle |}{C}}-O-H}} \;\rightleftharpoons\; \underset{\substack{\textbf{Pyruvate}\\\textbf{(keto form)}}}{\overset{\displaystyle COO^-}{\underset{\displaystyle \underset{H}{\overset{\displaystyle |}{C}}{}_{H}}{\overset{\displaystyle |}{\underset{\displaystyle}{C}}=O}}} \qquad \Delta G^{\circ\prime} = -46 \text{ kJ} \cdot \text{mol}^{-1}$$

Overall reaction

$$\underset{\substack{}}{\overset{\displaystyle COO^-}{\underset{\displaystyle \underset{H}{\overset{\displaystyle\parallel}{C}}{}_{H}}{\overset{\displaystyle |}{C}}-O\sim PO_3^{2-}}} + H_2O \;\rightleftharpoons\; \underset{\substack{}}{\overset{\displaystyle COO^-}{\underset{\displaystyle \underset{H}{\overset{\displaystyle |}{C}}{}_{H}}{\overset{\displaystyle |}{\underset{\displaystyle}{C}}=O}}} \;+\; HPO_4^{2-} \qquad \Delta G^{\circ\prime} = -61.9 \text{ kJ} \cdot \text{mol}^{-1}$$

FIGURE 15-23. The hydrolysis of phosphoenolpyruvate. The reaction is broken down into two steps, hydrolysis and tautomerization.

Other "High-Energy" Compounds

The compounds in Table 15-3 with phosphate group-transfer protentials significantly greater than that of ATP have additional destabilizing influences:

1. Acyl phosphates

The hydrolysis of **acyl phosphates** (mixed phosphoric–carboxylic anhydrides), such as **acetyl phosphate** and **1,3-bisphosphoglycerate,**

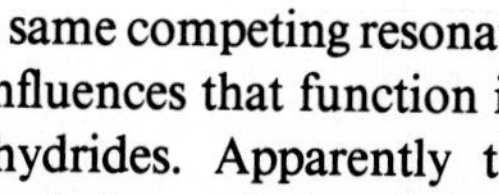

$$\underset{\textbf{Acetyl phosphate}}{CH_3 - \overset{\displaystyle O}{\overset{\displaystyle\parallel}{C}} \sim OPO_3^{2-}}$$

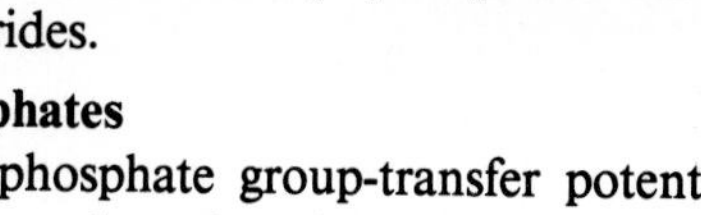

$$\underset{\textbf{1,3-Bisphosphoglycerate}}{^{-2}O_3POCH_2 - \overset{\displaystyle OH}{\underset{\displaystyle |}{CH}} - \overset{\displaystyle O}{\overset{\displaystyle\parallel}{C}} \sim OPO_3^{2-}}$$

is driven by the same competing resonance and differential solvation influences that function in the hydrolysis of phosphoanhydrides. Apparently these effects are more pronounced for acyl phosphates than for phosphoanhydrides.

2. Enol phosphates

The high phosphate group-transfer potential of **enol phosphates** such as phosphoenolpyruvate (Fig. 15-21*b*), derives from their **enol** hydrolysis product being less stable than its **keto** tautomer. Consider the hydrolysis reaction of an enol phosphate as occurring in two steps

(Fig. 15-23). The hydrolysis step is subject to the driving forces discussed above. *It is therefore the highly exergonic enol–keto conversion that provides phosphoenolpyruvate with the added thermodynamic impetus to phosphorylate ADP to form ATP.*

3. Phosphoguanidines

The high phosphate group-transfer potentials of **phosphoguanidines,** such as **phosphocreatine** and **phosphoarginine,** largely result from the competing resonances in their **guanidino** group, which are even more pronounced than they are in the phosphate group of phosphoanhydrides (Fig. 15-24). Consequently, phosphocreatine can phosphorylate ADP (see Section 15-4C).

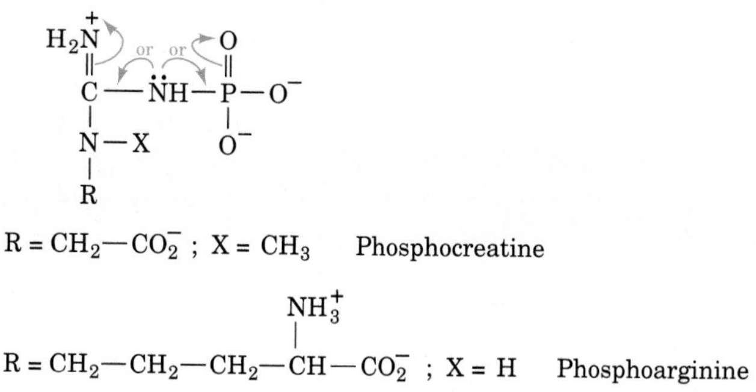

$$R = CH_2 - CO_2^- \;;\; X = CH_3 \qquad \text{Phosphocreatine}$$

$$R = CH_2 - CH_2 - CH_2 - \underset{\displaystyle \overset{\displaystyle NH_3^+}{\overset{\displaystyle |}{}}}{CH} - CO_2^- \;;\; X = H \qquad \text{Phosphoarginine}$$

FIGURE 15-24. The competing resonances in phosphoguanidines.

Compounds such as **glucose-6-phosphate** or **glycerol-3-phosphate**,

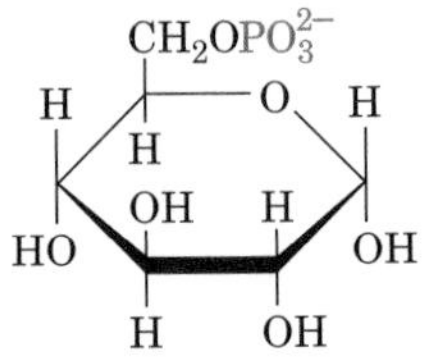

α-D-**Glucose-6-phosphate** 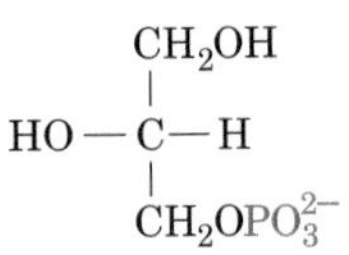   L-**Glycerol-3-phosphate**

which are below ATP in Table 15-3, have no significantly different resonance stabilization or charge separation in comparison with their hydrolysis products. Their free energies of hydrolysis are therefore much less than those of the preceding "high-energy" compounds.

C. *The Role of ATP*

As Table 15-3 indicates, *in the thermodynamic hierarchy of phosphoryl-transfer agents, ATP occupies the middle rank.* This enables ATP to serve as an energy conduit between "high-energy" phosphate donors and "low-energy" phosphate acceptors (Fig. 15-25). Let us examine the general biochemical scheme of how this occurs.

In general, the highly exergonic phosphoryl-transfer reactions of nutrient degradation are coupled to the formation of ATP from ADP and P_i through the auspices of various enzymes known as **kinases;** these enzymes catalyze the transfer of phosphoryl groups between ATP and other molecules. Consider the two reactions in Fig. 15-21*b.* If carried out independently, these reactions would not influence each other. In the cell, however, the enzyme **pyruvate kinase** couples the two reactions by catalyzing the transfer of the phosphoryl group of phosphoenolpyruvate directly to ADP to result in an overall exergonic reaction.

Consumption of ATP

In its role as the universal energy currency of living systems, ATP is consumed in a variety of ways:

1. Early stages of nutrient breakdown

The exergonic hydrolysis of ATP to ADP may be enzymatically coupled to an endergonic phosphorylation reaction to form "low-energy" phosphate compounds. We have seen one example of this in the hexokinase-catalyzed formation of glucose-6-phosphate (Fig. 15-21*a*).

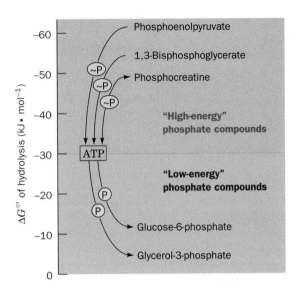

FIGURE 15-25.   The flow of phosphoryl groups from high-energy phosphate donors, via the ATP–ADP system, to low-energy phosphate acceptors.

Another example is the **phosphofructokinase**-catalyzed phosphorylation of **fructose-6-phosphate** to form **fructose-1,6-bisphosphate** (Fig. 15-26). Both of these reactions occur in the first stage of glycolysis (Section 16-2).

2. Interconversion of nucleoside triphosphates

Many biosynthetic processes, such as the synthesis of proteins and nucleic acids, require nucleoside triphosphates other than ATP. These include the ribonucleoside triphosphates CTP, GTP, and UTP (Section 1-3C) which, together with ATP, are utilized, for example, in the biosynthesis of RNA (Section 29-2) and the deoxyribonucleoside triphosphate DNA precursors dATP, dCTP, dGTP, and dTTP (Section 31-1A). All these **nucleoside triphosphates (NTPs)** are synthesized from ATP and the corresponding **nucleoside diphosphate (NDP)** in reactions catalyzed by the nonspecific enzyme **nucleoside diphosphate kinase:**

$$\text{ATP} + \text{NDP} \rightleftharpoons \text{ADP} + \text{NTP}$$

The $\Delta G°'$ values for these reactions are nearly zero as might be expected from the structural similarities among the NTPs. These reactions are driven by the depletion of the NTPs through their exergonic hydrolysis

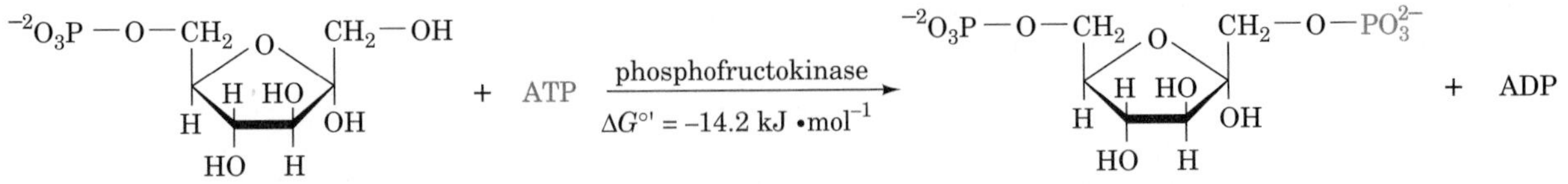

Fructose-6-phosphate

+ ATP $\xrightarrow[\Delta G°' = -14.2 \text{ kJ} \cdot \text{mol}^{-1}]{\text{phosphofructokinase}}$

Fructose-1,6-bisphosphate

+ ADP

FIGURE 15-26.   The phosphorylation of fructose-6-phosphate by ATP to form fructose-1,6-bisphosphate and ADP.

in the biosynthetic reactions in which they participate (Section 3-4C).

3. Physiological processes

The hydrolysis of ATP to ADP and P_i energizes many essential endergonic physiological processes such as muscle contraction and the transport of molecules and ions against concentration gradients. In general, these processes result from conformational changes in proteins (enzymes) that occur in response to their binding of ATP. This is followed by the exergonic hydrolysis of ATP and release of ADP and P_i, thereby causing these processes to be unidirectional (irreversible).

4. Additional phosphoanhydride cleavage in highly endergonic reactions

Although many reactions involving ATP yield ADP and P_i (**orthophosphate cleavage**), others yield AMP and PP_i (**pyrophosphate cleavage**). In these latter cases, the PP_i is rapidly hydrolyzed to $2P_i$ by **inorganic pyrophosphatase** ($\Delta G°' = -33.5$ kJ·mol^{-1}) so that *the pyrophosphate cleavage of ATP ultimately results in the hydrolysis of two "high-energy" phosphoanhydride bonds.* The first step in the oxidation of fatty acids (Fig. 15-27 and Section 23-2A) provides an example of this process. Pyrophosphate cleavage alone is insufficiently exergonic to drive the fatty acid activation reaction to completion.

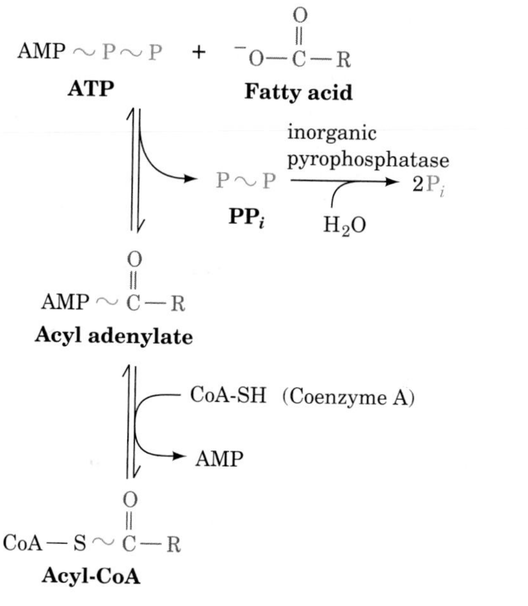

FIGURE 15-27. The activation of a fatty acid involves its conversion to a thioester. In the first reaction step, the fatty acid is **adenylylated** by reaction with ATP. In the second step, coenzyme A (CoA-SH), a complicated organic molecule bearing a sulfhydryl group (Fig. 19-2), displaces the AMP moiety to form an acyl-CoA adduct. For these two steps, $\Delta G°' = +4.6$ kJ·mol^{-1} so that, even under physiological conditions, the reaction is unfavorable. However, the hydrolysis of the product PP_i, which has $\Delta G°' = -33.5$ kJ·mol^{-1}, drives the activation reaction to completion ($\Delta G°' = -33.5 + 4.6 = -28.9$ kJ·mol^{-1}).

This reaction is made irreversible, however, by the additional thermodynamic impetus of PP_i hydrolysis. Nucleic acid biosynthesis from the appropriate NTPs also releases PP_i (Sections 29-2 and 31-1A). The free energy changes of these vital reactions are around zero, so the subsequent hydrolysis of PP_i is essential for the synthesis of nucleic acids.

Formation of ATP

To complete its intermediary metabolic function, ATP must be replenished. This is accomplished through three types of processes:

1. Substrate-level phosphorylation

ATP may be formed, as is indicated in Fig. 15-21b, from phosphoenolpyruvate by direct transfer of a phosphoryl group from a "high-energy" compound to ADP. Such reactions, which are referred to as **substrate-level phosphorylations,** most commonly occur in the early stages of carbohydrate metabolism (Section 16-2).

2. Oxidative phosphorylation and photophosphorylation

Both oxidative metabolism and photosynthesis act to generate a proton (H^+) concentration gradient across a membrane (Sections 20-3 and 22-2D). Discharge of this gradient is enzymatically coupled to formation of ATP from ADP and P_i (the reverse of ATP hydrolysis). In oxidative metabolism, this process is called **oxidative phosphorylation,** whereas in photosynthesis it is termed **photophosphorylation.** Most of the ATP produced by respiring and photosynthesizing organisms is generated in this manner.

3. Adenylate kinase reaction

The AMP resulting from pyrophosphate cleavage reactions of ATP is converted to ADP in a reaction catalyzed by the enzyme **adenylate kinase:**

$$AMP + ATP \rightleftharpoons 2ADP$$

The ADP is subsequently converted to ATP through substrate-level phosphorylation, oxidative phosphorylation, or photophosphorylation.

Rate of ATP Turnover

The cellular role of ATP is that of a free energy transmitter rather than a free energy reservoir. The amount of ATP in a cell is typically only enough to supply its free energy needs for a minute or two. Hence, ATP is continually being hydrolyzed and regenerated. Indeed, ^{32}P-labeling experiments indicate that the metabolic half-life of an ATP molecule varies from seconds to minutes depending on the cell type and its metabolic activity. For instance, brain cells have only a few seconds supply of ATP (which, in part, accounts for the rapid deterioration of brain tissue by oxygen deprivation). *An average person at rest consumes and regenerates ATP at a rate of ~3 mol (1.5 kg)·h^{-1} and as much as an order of magnitude faster during strenuous activity.*

Phosphocreatine Provides a "High-Energy" Reservoir for ATP Formation

Muscle and nerve cells, which have a high ATP turnover, have a free energy reservoir that functions to regenerate ATP rapidly. In vertebrates, phosphocreatine (Fig. 15-24) functions in this capacity. It is synthesized by the reversible phosphorylation of creatine by ATP as catalyzed by **creatine kinase:**

$$ATP + creatine \rightleftharpoons phosphocreatine + ADP$$
$$\Delta G°' = +12.6 \text{ kJ} \cdot \text{mol}^{-1}$$

Note that this reaction is endergonic under standard conditions; however, the intracellular concentrations of its reactants and products are such that it operates close to equilibrium ($\Delta G \approx 0$). Accordingly, when the cell is in a resting state, so that [ATP] is relatively high, the reaction proceeds with net synthesis of phosphocreatine, whereas at times of high metabolic activity, when [ATP] is low, the equilibrium shifts so as to yield net synthesis of ATP. *Phosphocreatine thereby acts as an ATP "buffer" in cells that contain creatine kinase.* A resting vertebrate skeletal muscle normally has sufficient phosphocreatine to supply its free energy needs for several minutes (but for only a few seconds at maximum exertion). In the muscles of some invertebrates, such as lobsters, phosphoarginine performs the same function. These phosphoguanidines are collectively named **phosphagens.**

5. OXIDATION–REDUCTION REACTIONS

Oxidation–reduction reactions, processes involving the transfer of electrons, are of immense biochemical significance; living things derive most of their free energy from them. In photosynthesis (Chapter 22), CO_2 is **reduced** (gains electrons) and H_2O is **oxidized** (loses electrons) to yield carbohydrates and O_2 in an otherwise endergonic process that is powered by light energy. In aerobic metabolism, which is carried out by all eukaryotes and many prokaryotes, the overall photosynthetic reaction is essentially reversed so as to harvest the free energy of oxidation of carbohydrates and other organic compounds in the form of ATP (Chapter 20). Anaerobic metabolism generates ATP, although in lower yields, through intramolecular oxidation–reductions of various organic molecules; for example, glycolysis (Chapter 16), or in certain anaerobic bacteria, through the use of non-O_2 oxidizing agents such as sulfate or nitrate. In this section we outline the thermodynamics of oxidation–reduction reactions in order to understand the quantitative aspects of these crucial biological processes.

A. The Nernst Equation

Oxidation–reduction reactions (also known as **redox** or **oxido reduction reactions**) resemble other types of chemical reactions in that they involve group transfer. For instance, hydrolysis transfers a functional group to water. In oxidation–reduction reactions, the "groups" transferred are electrons, which are passed from an **electron donor (reductant** or **reducing agent)** to an **electron acceptor (oxidant** or **oxidizing agent).** For example, in the reaction

$$Fe^{3+} + Cu^+ \rightleftharpoons Fe^{2+} + Cu^{2+}$$

Cu^+, the reductant, is oxidized to Cu^{2+} while Fe^{3+}, the oxidant, is reduced to Fe^{2+}.

Redox reactions may be divided into two **half-reactions** or **redox couples,** such as

$$Fe^{3+} + e^- \rightleftharpoons Fe^{2+} \text{ (reduction)}$$
$$Cu^+ \rightleftharpoons Cu^{2+} + e^- \text{(oxidation)}$$

whose sum is the above whole reaction. These half-reactions occur during oxidative metabolism in the vital mitochondrial electron transfer mediated by **cytochrome *c* oxidase** (Section 20-2C). Note that for electrons to be transferred, both half-reactions must occur simultaneously. In fact, the electrons are the two half-reactions' common intermediate.

Electrochemical Cells

A half-reaction consists of an electron donor and its conjugate electron acceptor; in the oxidation half-reaction shown above, Cu^+ is the electron donor and Cu^{2+} is its conjugate electron acceptor. Together these constitute a **conjugate redox pair** analogous to the conjugate acid–base pair (HA and A^-) of a Brønsted acid (Section 2-2A). An important difference between redox pairs and acid–base pairs, however, is that *the two half-reactions of a redox reaction, each consisting of a conjugate redox pair, may be physically separated so as to form an **electrochemical cell** (Fig. 15-28). In such a device, each half-reaction takes place in its

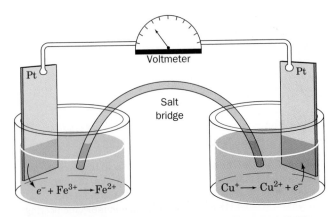

FIGURE 15-28. An example of an electrochemical cell. The half-cell undergoing oxidation (here $Cu^+ \rightarrow Cu^{2+} + e^-$) passes the liberated electrons through the wire to the half-cell undergoing reduction (here $e^- + Fe^{3+} \rightarrow Fe^{2+}$). Electroneutrality in the two half-cells is maintained by the transfer of ions through the electrolyte-containing salt bridge.

separate **half-cell,** and electrons are passed between half-cells as an electric current in the wire connecting their two electrodes. A salt bridge is necessary to complete the electrical circuit by providing a conduit for ions to migrate in the maintenance of electrical neutrality.

The free energy of an oxidation–reduction reaction is particularly easy to determine through a simple measurement of the voltage difference between its two half-cells. Consider the general redox reaction:

$$A_{ox}^{n+} + B_{red} \rightleftharpoons A_{red} + B_{ox}^{n+}$$

in which n electrons per mole of reactants are transferred from reductant (B_{red}) to oxidant (A_{ox}^{n+}). The free energy of this reaction is expressed, according to Eq. [3.15], as

$$\Delta G = \Delta G° + RT \ln \left(\frac{[A_{red}][B_{ox}^{n+}]}{[A_{ox}^{n+}][B_{red}]} \right) \qquad [15.2]$$

Equation [3.12] indicates that, under reversible conditions,

$$\Delta G = - w' = - w_{el} \qquad [15.3]$$

where w', the non-pressure–volume work, is, in this case, w_{el}, the electrical work required to transfer the n moles of electrons through the electric potential difference $\Delta\mathscr{E}$. This, according to the laws of electrostatics, is

$$w_{el} = n\mathscr{F}\Delta\mathscr{E} \qquad [15.4]$$

where $\mathscr{F}$, the **faraday,** is the electrical charge of 1 mol of electrons (1 $\mathscr{F}$ = 96,494 C·mol^{-1} = 96,494 J·V^{-1} mol^{-1}). Thus, substituting Eq. [15.4] into Eq. [15.3],

$$\Delta G = -n\mathscr{F}\Delta\mathscr{E} \qquad [15.5]$$

Combining Eqs. [15.2] and [15.5], and making the analogous substitution for $\Delta G°$, yields the **Nernst equation:**

$$\Delta\mathscr{E} = \Delta\mathscr{E}° - \frac{RT}{n\mathscr{F}} \ln \left(\frac{[A_{red}][B_{ox}^{n+}]}{[A_{ox}^{n+}][B_{red}]} \right) \qquad [15.6]$$

which was originally formulated in 1881 by Walther Nernst. Here $\Delta\mathscr{E}$, the **electromotive force (emf)** or **redox potential,** may be described as the "electron pressure" that the electrochemical cell exerts. The quantity $\Delta\mathscr{E}°$, the redox potential when all components are in their standard states, is called the **standard redox potential.** If these standard states refer to biochemical standard states (Section 3-4B), then $\Delta\mathscr{E}°$ is replaced by $\Delta\mathscr{E}°'$. Note that a postive $\Delta\mathscr{E}$ in Eq. [15.5] results in a negative ΔG; in other words, *a positive $\Delta\mathscr{E}$ is indicative of a spontaneous reaction, one that can do work.*

B. Measurements of Redox Potentials

The free energy change of a redox reaction may be determined, as Eq. [15.5] indicates, by simply measuring its redox potential with a voltmeter (Fig. 15-28). Consequently, voltage measurements are commonly employed to characterize the sequence of reactions comprising a meta-

bolic electron-transport pathway (such as mediates, e.g., oxidative metabolism; Chapter 20).

Any redox reaction can be divided into its component half-reactions:

$$A_{ox}^{n+} + ne^- \rightleftharpoons A_{red}$$
$$B_{ox}^{n+} + ne^- \rightleftharpoons B_{red}$$

where, by convention, both half-reactions are written as reductions. These half-reactions can be assigned **reduction potentials,** $\mathscr{E}_A$ and $\mathscr{E}_B$, in accordance with the Nernst equation:

$$\mathscr{E}_A = \mathscr{E}_A° - \frac{RT}{n\mathscr{F}} \ln \left(\frac{[A_{red}]}{[A_{ox}^{n+}]} \right) \qquad [15.7a]$$

$$\mathscr{E}_B = \mathscr{E}_B° - \frac{RT}{n\mathscr{F}} \ln \left(\frac{[B_{red}]}{[B_{ox}^{n+}]} \right) \qquad [15.7b]$$

For the redox reaction of any two half-reactions:

$$\Delta\mathscr{E}° = \mathscr{E}°_{(e- \text{ acceptor})} - \mathscr{E}°_{(e- \text{ donor})} \qquad [15.8]$$

Thus, when the reaction proceeds with A as the electron acceptor and B as the electron donor, $\Delta\mathscr{E}° = \mathscr{E}_A° - \mathscr{E}_B°$ and similarly for $\Delta\mathscr{E}$.

Reduction potentials, like free energies, must be defined with respect to some arbitrary standard. By convention, standard reduction potentials are defined with respect to the standard hydrogen half-reaction

$$2H^+ + 2e^- \rightleftharpoons H_2 \,(g)$$

in which H^+ at pH 0, 25°C, and 1 atm is in equilibrium with H_2 (g) that is in contact with a Pt electrode. This half-cell is arbitrarily assigned a standard reduction potential $\mathscr{E}°$ of 0 V (1 V = 1 J·C^{-1}). For the biochemical convention, we likewise define the standard (pH = 0) hydrogen half-reaction as having $\mathscr{E}' = 0$ so that the hydrogen half-cell at the biochemical standard state (pH = 7) has $\mathscr{E}°' = -0.421$ V (Table 15-4). When $\Delta\mathscr{E}$ is positive, ΔG is negative (Eq. [15.5]), indicating a spontaneous process. In combining two half-reactions under standard conditions, the direction of spontaneity therefore involves the reduction of the redox couple with the more positive standard reduction potential. In other words, *the more positive the standard reduction potential, the greater the tendency for the redox couple's oxidized form to accept electrons and thus become reduced.*

Biochemical Half-Reactions Are Physiologically Significant

The biochemical standard reduction potentials ($\mathscr{E}°'$) of some biochemically important half-reactions are listed in Table 15-4. The oxidized form of a redox couple with a large positive standard reduction potential has a high affinity for electrons and is a strong electron acceptor (oxidizing agent), whereas its conjugate reductant is a weak electron donor (reducing agent). For example, O_2 is the strongest oxidizing agent in Table 15-4, whereas H_2O, which tightly holds its electrons, is the table's weakest reducing agent.

**TABLE 15-4. STANDARD REDUCTION POTENTIALS OF SOME BIOCHEMICALLY
IMPORTANT HALF-REACTIONS**

Half Reaction	$\mathscr{E}°'$ (V)
$\frac{1}{2}O_2 + 2H^+ + 2e^- \rightleftharpoons H_2O$	0.815
$SO_4^{2-} + 2H^+ + 2e^- \rightleftharpoons SO_3^{2-} + H_2O$	0.48
$NO_3^- + 2H^+ + 2e^- \rightleftharpoons NO_2^- + H_2O$	0.42
Cytochrome a_3 $(Fe^{3+}) + e^- \rightleftharpoons$ cytochrome a_3 (Fe^{2+})	0.385
$O_2(g) + 2H^+ + 2e^- \rightleftharpoons H_2O_2$	0.295
Cytochrome a $(Fe^{3+}) + e^- \rightleftharpoons$ cytochrome a (Fe^{2+})	0.29
Cytochrome c $(Fe^{3+}) + e^- \rightleftharpoons$ cytochrome c (Fe^{2+})	0.235
Cytochrome c_1 $(Fe^{3+}) + e^- \rightleftharpoons$ cytochrome c_1 (Fe^{2+})	0.22
Cytochrome b $(Fe^{3+}) + e^- \rightleftharpoons$ cytochrome b (Fe^{2+}) (*mitochondrial*)	0.077
Ubiquinone $+ 2H^+ + 2e^- \rightleftharpoons$ ubiquinol	0.045
Fumarate$^-$ $+ 2H^+ + 2e^- \rightleftharpoons$ succinate$^-$	0.031
FAD $+ 2H^+ + 2e^- \rightleftharpoons FADH_2$ (*in flavoproteins*)	~0.
Oxaloacetate$^-$ $+ 2H^+ + 2e^- \rightleftharpoons$ malate$^-$	-0.166
Pyruvate$^-$ $+ 2H^+ + 2e^- \rightleftharpoons$ lactate$^-$	-0.185
Acetaldehyde $+ 2H^+ + 2e^- \rightleftharpoons$ ethanol	-0.197
FAD $+ 2H^+ + 2e^- \rightleftharpoons FADH_2$ (*free coenzyme*)	-0.219
$S + 2H^+ + 2e^- \rightleftharpoons H_2S$	-0.23
Lipoic acid $+ 2H^+ + 2e^- \rightleftharpoons$ dihydrolipoic acid	-0.29
$NAD^+ + H^+ + 2e^- \rightleftharpoons NADH$	-0.315
$NADP^+ + H^+ + 2e^- \rightleftharpoons NADPH$	-0.320
Cystine $+ 2H^+ + 2e^- \rightleftharpoons$ 2 cysteine	-0.340
Acetoacetate$^-$ $+ 2H^+ + 2e^- \rightleftharpoons \beta$-hydroxybutyrate$^-$	-0.346
$H^+ + e^- \rightleftharpoons \frac{1}{2}H_2$	-0.421
Acetate$^-$ $+ 3H^+ + 2e^- \rightleftharpoons$ acetaldehyde $+ H_2O$	-0.581

Source: Mostly from Loach, P.A., *In* Fasman, G.D. (Ed.), *Handbook of Biochemistry and Molecular Biology* (3rd ed.), *Physical and Chemical Data,* Vol. I, *pp.* 123–130, CRC Press (1976).

The converse is true of half-reactions with large negative standard reduction potentials. Since electrons spontaneously flow from low to high reduction potentials, they are transferred, under standard conditions, from the reduced products in any half-reaction in Table 15-4 to the oxidized reactants of any half-reaction above it (although this may not occur at a measurable rate in the absence of a suitable enzyme). Note that Fe^{3+} ions of the various cytochromes tabulated in Table 15-4 have significantly different redox potentials. This indicates that *the protein components of redox enzymes play active roles in electron-transfer reactions by modulating the redox potentials of their bound redox-active centers.*

Electron-transfer reactions are of great biological importance. For example, in the mitochondrial electron-transport chain (Section 20-2), the primary source of ATP in eukaryotes, electrons are passed from NADH (Fig. 12-2) along a series of electron acceptors of increasing reduction potential (many of which are listed in Table 15-4), to O_2. ATP is generated from ADP and P_i by coupling its synthesis to this free energy cascade. *NADH thereby functions as an energy-rich electron-transfer coenzyme.* In fact, the oxidation of one NADH to NAD^+ supplies sufficient free energy to generate three ATPs. The NAD^+/NADH redox couple functions as the electron acceptor in many exergonic metabolite oxidations. In serving as the electron donor in ATP

synthesis, it fulfills its cyclic role as a free energy conduit in a manner analogous to ATP. The metabolic roles of redox coenzymes are further discussed in succeeding chapters.

C. Concentration Cells

A concentration gradient has a lower entropy (greater order) than the corresponding uniformly mixed solution and therefore requires the input of free energy for its formation. Consequently, discharge of a concentration gradient is an exergonic process that may be harnessed to drive an endergonic reaction. For example, discharge of a proton concentration gradient (generated by the reactions of the electron-transport chain) across the inner mitochondrial membrane, drives the enzymatic synthesis of ATP from ADP and P_i (Section 20-3). Likewise, nerve impulses, which require electrical energy, are transmitted through the discharge of $[Na^+]$ and $[K^+]$ gradients that nerve cells generate across their cell membranes (Section 34-4C). Quantitation of the free energy contained in a concentration gradient is accomplished by use of the concepts of electrochemical cells.

The reduction potential and free energy of a half-cell vary with the concentrations of its reactants. An electrochemical cell may therefore be constructed from two half-cells that contain the same chemical species but at different concentrations. The overall reaction for such an electrochemical cell may be represented

$$A_{ox}^{n+}(\text{half-cell 1}) + A_{red}(\text{half-cell 2}) \rightleftharpoons$$
$$A_{ox}^{n+}(\text{half-cell 2}) + A_{red}(\text{half-cell 1}) \qquad [15.9]$$

and, according to the Nerst equation, since $\Delta\mathscr{E}°$ vanishes when the same reaction occurs in both cells

$$\Delta\mathscr{E} = \frac{RT}{n\mathscr{F}} \ln\left(\frac{[A_{ox}^{n+}(\text{half-cell 2})][A_{red}(\text{half-cell 1})]}{[A_{ox}^{n+}(\text{half-cell 1})][A_{red}(\text{half-cell 2})]}\right)$$

Such **concentration cells** are capable of generating electrical work until they reach equilibrium. This occurs when the concentrations in the half-cells become equal ($K_{eq} = 1$). The reaction, in effect, constitutes a mixing of the two half-cells; the free energy generated is a reflection of the entropy of this mixing. The thermodynamics of concentration gradients as they apply to membrane transport is discussed in Section 18-1.

6. THERMODYNAMICS OF LIFE

One of the last refuges of **vitalism,** the doctrine that biological processes are not bound by the physical laws that govern inanimate objects, was the belief that living things can somehow evade the laws of thermodynamics. This view was partially refuted by elaborate calorimetric measurements on living animals that are entirely consistent with the energy conservation predictions of the first law of thermodynamics. However, the experimental verification of the second law of thermodynamics in living systems is more difficult. It has not been possible to measure the entropy of living matter since the heat, q_p, of a reaction at a constant T and P is only equal to $T\Delta S$ if the reaction is carried out reversibly (Eq. [3.8]). Obviously, the dismantling of a living organism to its component molecules for such a measurement would invariably result in its irreversible death. Consequently, the present experimentally verified state of knowledge is that the entropy of living matter is less than that of the products to which it decays.

In this section we consider the special aspects of the thermodynamics of living systems. Knowledge of these matters, which is by no means complete, has enhanced our understanding of how metabolic pathways are regulated, how cells respond to stimuli, and how organisms grow and change with time.

A. Living Systems Cannot Be at Equilibrium

Classical or **equilibrium thermodynamics** (Chapter 3) applies largely to reversible processes in closed systems. The fate of any closed system, as we discussed in Section 3-4A, is that it must inevitably reach equilibrium. For example, if its reactants are in excess, the forward reaction will proceed faster than the reverse reaction until equilibrium is attained ($\Delta G = 0$). In contrast, open systems may remain in a non-equilibrium state as long as they are able to acquire free energy from their surroundings in the form of reactants, heat, or work. While classical thermodynamics provides invaluable information concerning open systems by indicating whether a given process can occur spontaneously, further thermodynamic analysis of open systems requires the application of the relatively recently elucidated principles of **nonequilibrium** or **irreversible thermodynamics.** In contrast to classical thermodynamics, this theory explicitly takes time into account.

Living organisms are open systems and therefore can never be at equilibrium. As indicated above, they continuously ingest high-enthalpy, low-entropy nutrients, which they convert to low-enthalpy, high-entropy waste products. The free energy resulting from this process is used to do work and to produce the high degree of organization characteristic of life. If this process is interrupted, the organism ultimately reaches equilibrium, which for living things is synonymous with death. For example, one theory of aging holds that senescence results from the random but inevitable accumulation in cells of genetic defects that interfere with and ultimately disrupt the proper functioning of living processes. [The theory does not, however, explain how single-celled organisms or the germ cells of multicellular organisms (sperm and ova), which are in effect immortal, are able to escape this so-called **error catastrophe.**]

Living systems must maintain a nonequilibrium state for several reasons:

1. Only a nonequilibrium process can perform useful work.

2. The intricate regulatory functions characteristic of life require a nonequilibrium state because a process at equilibrium cannot be directed (similarly, a ship that is dead in the water will not respond to its rudder).

3. The complex cellular and molecular systems that conduct biological processes can be maintained only in the nonequilibrium state. Living systems are inherently unstable because they are degraded by the very biochemical reactions to which they give rise. Their regeneration, which must occur almost simultaneously with their degradation, requires the continuous influx of free energy. For example, the ATP-generating consumption of glucose (Section 16-2), as has been previously mentioned, occurs with the initial consumption of ATP through its reactions with glucose to form glucose-6-phosphate and with fructose-6-phosphate to form fructose-1,6-bisphosphate. Consequently, if metabolism is suspended long enough to exhaust the available ATP supply, glucose metabolism cannot be resumed. Life therefore differs in a fundamental way from a complex machine such as a computer. Both require a throughput of free energy to be active. However, the function of the machine is based on a static structure so that the machine can be repeatedly switched on and off. Life, in contrast, is based on a self-destructing but self-renewing process which, once interrupted, cannot be reinitiated.

B. Nonequilibrium Thermodynamics and the Steady State

In a nonequilibrium process, something (such as matter, electrical charge, or heat) must flow, that is, change its spatial distribution. In classical mechanics, the acceleration of mass occurs in response to force. *Similarly, flow in a thermodynamic system occurs in response to a thermodynamic force (driving force), which results from the system's nonequilibrium state.* For example, the flow of matter in diffusion is motivated by the thermodynamic force of a concentration gradient; the migration of electrical charge (electric current) occurs in response to a gradient in an electric field (a voltage difference); the transport of heat results from a temperature gradient; and a chemical reaction results from a difference in chemical potential. Such flows are said to be **conjugate** to their thermodynamic force.

A thermodynamic force may also promote a **nonconjugate flow** under the proper conditions. For example, a gradient in the concentration of matter can give rise to an electric current (a concentration cell), heat (such as occurs upon mixing H_2O and HCl), or a chemical reaction (the mitochondrial production of ATP through the dissipation of a proton gradient). Similarly, a gradient in electrical potential can motivate a flow of matter (electrophoresis), heat (resistive heating), or a chemical reaction (the charging of a battery). When a thermodynamic force stimulates a nonconjugate flow, the process is called **energy transduction.**

Living Things Maintain the Steady State

*Living systems are, for the most part, characterized by being in a **steady state**.* By this it is meant that all flows in the system are constant so that the system does not change with time. Some environmental steady state processes are schematically illustrated in Fig. 15-29. Ilya Prigogine, a pioneer in the development of irreversible thermodynamics, has shown that a steady state system produces the maximum amount of useful work for a given energy expenditure under the prevailing conditions. *The steady state of an open system is therefore its state of maximum thermodynamic efficiency.* Furthermore, in analogy with Le Châtelier's principle, slight perturbations from the steady state give rise to changes in flows that counteract these perturbations so as to return the system to the steady state. *The steady state of an open system is therefore analogous to the equilibrium state of a closed system; both are stable states.*

In the following chapters it will be seen that many biological regulatory mechanisms function to maintain a steady state. For example, the flow of reaction intermediates through a metabolic pathway is often inhibited by an excess of final product and stimulated by an excess of starting material through the allosteric regulation of its key enzymes (Section 12-4). Living things have apparently evolved so as to take maximum thermodynamic advantage of their environment.

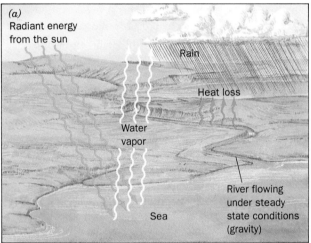

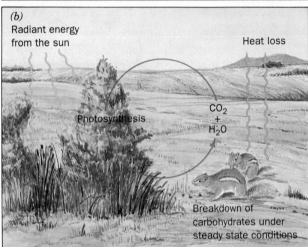

FIGURE 15-29. Two examples of open systems in a steady state. (*a*) A constant flow of water in the river occurs under the influence of the force of gravity. The water level in the reservoir is maintained by rain, the major source of which is the evaporation of seawater. Hence the entire cycle is ultimately powered by the sun. (*b*) The steady state of the biosphere is similarly maintained by the sun. Plants harness the sun's radiant energy to synthesize carbohydrates from CO_2 and H_2O. The eventual metabolism of the carbohydrates by the plants or by the animals that eat them results in the release of their stored free energy and the return of the CO_2 and H_2O to the environment to complete the cycle.

C. Thermodynamics of Metabolic Control

Enzymes Selectively Catalyze Required Reactions

Biological reactions are highly specific; only reactions that lie on metabolic pathways take place at significant rates despite the many other thermodynamically favorable reactions that are also possible. As an example, let us consider the reactions of ATP, glucose, and water. Two thermodynamically favorable reactions that ATP can undergo are phosphoryl transfer to form ADP and glucose-6-phosphate, and hydrolysis to form ADP and P_i (Fig. 15-21*a*). The free energy profiles of these reactions are diagrammed in Fig. 15-30. ATP hydrolysis is thermodynamically favored over

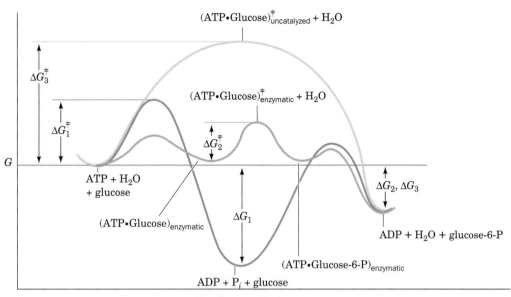

FIGURE 15-30. Reaction coordinate diagrams for (1) the reaction of ATP and water (*purple curve*), and the reaction of ATP and glucose (2) in the presence (*orange curve*), and (3) in the absence (*yellow curve*) of an appropriate enzyme. Although the hydrolysis of ATP is a more exergonic reaction than the phosphorylation of glucose (ΔG_1 more negative than ΔG_2), the latter reaction is predominant in the presence of a suitable enzyme because it is kinetically favored ($\Delta G_2^{\ddagger} < \Delta G_1^{\ddagger}$).

the phosphoryl transfer to glucose. However, their relative rates are determined by their free energies of activation to their transition states ($\Delta G^{\ddagger}$ values; Section 13-1C) and the relative concentrations of glucose and water. The larger $\Delta G^{\ddagger}$, the slower the reaction. In the absence of enzymes, $\Delta G^{\ddagger}$ for the phosphoryl-transfer reaction is greater than that for hydrolysis, so the hydrolysis reaction predominates (although neither reaction occurs at a biologically significant rate).

The free energy barriers of both of the nonenzymatic reactions are far higher than that of the enzyme-catalyzed phosphoryl transfer to glucose. Hence enzymatic formation of glucose-6-phosphate is kinetically favored over the nonenzymatic hydrolysis of ATP. *It is the role of an enzyme, in this case hexokinase, to selectively reduce the free energy of activation of a chemically coupled reaction so that it approaches equilibrium faster than the more thermodynamically favored uncoupled reaction.*

Many Enzymatic Reactions Are Near Equilibrium

Although metabolism as a whole is a nonequilibrium process, many of its component reactions function close to equilibrium. The reaction of ATP and creatine to form phosphocreatine (Section 15-4C) is an example of such a reaction. The ratio [creatine]/[phosphocreatine] depends on [ATP] because creatine kinase, the enzyme catalyzing this reaction, has sufficient activity to equilibrate the reaction rapidly. The net rate of such an equilibrium reaction is effectively regulated by varying the concentrations of its reactants and/or products.

Pathway Throughput Is Controlled by Enzymes Operating Far from Equilibrium

Other biological reactions function far from equilibrium. For example, the phosphofructokinase reaction (Fig. 15-26) has an equilibrium constant of $K'_{eq} = 300$ but under physiological conditions in rat heart muscle has the mass action ratio [fructose-1,6-bisphosphate][ADP]/[fructose-6-phosphate][ATP] = 0.03, which corresponds to $\Delta G = -25.7$ kJ·mol^{-1} (Eq. [3.15]). This situation arises from a buildup of reactants because there is insufficient phosphofructokinase activity to equilibrate the reaction. Changes in substrate concentrations therefore have relatively little effect on the rate of the phosphofructokinase reaction; the enzyme is essentially saturated. Only changes in the activity of the enzyme, through allosteric interactions, for example, can significantly alter this rate. An enzyme, such as phosphofructokinase, is therefore analogous to a dam on a river. It controls substrate **flux** (rate of flow) by varying its activity (allosterically or by other means), much as a dam controls the flow of a river by varying the opening of its flood gates.

Understanding of how reactant flux in a metabolic pathway is controlled requires knowledge of which reactions are functioning near equilibrium and which are far from it. Most enzymes in a metabolic pathway operate near equilibrium and therefore have net rates that vary with their substrate concentrations. However, as we shall see in the following chapters (particularly Section 16-4), *certain allosteric enzymes, which are strategically located in a metabolic pathway, operate far from equilibrium. The relative insensitivity of the rates of the reactions catalyzed by such "flux-generating" enzymes to variations in the concentrations of their substrates permits the establishment of a steady state flux of metabolites through the pathway. This* situation, as we have seen, maximizes the pathway's thermodynamic efficiency and allows the flux to be allosterically controlled.

CHAPTER SUMMARY

Metabolic pathways are series of consecutive enzymatic reactions that produce specific products for use by an organism. The free energy released by degradation (catabolism) is, through the intermediacy of ATP and NADPH, used to drive the endergonic processes of biosynthesis (anabolism). Carbohydrates, lipids, and proteins are all converted to the common intermediate acetyl-CoA, whose acetyl group is then converted to CO_2 and H_2O through the action of the citric acid cycle and oxidative phosphorylation. A relatively few metabolites serve as starting materials for a host of biosynthetic products. Metabolic pathways have four principal characteristics: (1) metabolic pathways are irreversible so if two metabolites are interconvertible, the synthetic route from the first to the second must differ from the route from the second to the first; (2) every metabolic pathway has an exergonic first committed step; (3) all metabolic pathways are regulated, usually at the first committed step; and (4) metabolic pathways in eukaryotes occur in specific subcellular compartments.

Almost all metabolic reactions fall into four categories: (1) group-transfer reactions; (2) oxidation–reduction reactions; (3) eliminations, isomerizations, and rearrangements; and (4) reactions that make or break carbon–carbon bonds. Most of these reactions involve heterolytic bond cleavage or formation occurring through addition of nucleophiles to electrophilic carbon atoms. Group-transfer reactions therefore involve transfer of an electrophilic group from one nucleophile to another. The main electrophilic groups transferred are acyl groups, phosphoryl groups, and glycosyl groups. The most common nucleophiles are amino, hydroxyl, imidazole, and sulfhydryl groups. Oxidation–reduction reactions involve loss or gain of electrons. Oxidation at carbon usually involves C—H bond cleavage, with the ultimate loss by C of the two bonding electrons through their transfer to an electron acceptor such as NAD^+. The terminal electron acceptor in aerobes is O_2. Elimination reactions are those in which a C=C double bond is created from two saturated carbon centers with the loss of H_2O, NH_3, ROH, or RNH_2. Dehydration reactions are the most common eliminations. Isomerizations involve shifts of double bonds within molecules. Rearrangements are biochemically uncommon reactions in which intramolecular C—C bonds are broken and reformed to produce new carbon skeletons. Reactions that make and break C—C bonds form the basis of both degradative and biosynthetic metabolism. In the synthetic direction, these reactions involve addition of a nucleophilic carbanion to an electrophilic carbon atom. The most common electrophilic carbon atom is the carbonyl carbon, whereas carbanions are usually generated by removal of a proton from a carbon atom adjacent to a carbonyl group or by decarboxylation of a β-keto acid.

Experimental approaches employed in elucidating metabolic pathways include the use of metabolic inhibitors, growth studies, and biochemical genetics. Metabolic inhibitors block pathways at specific enzymatic steps. Identification of the resulting intermediates indicates the course of the pathway. Mutations, which occur naturally in genetic diseases or can be induced in microorganisms by mutagens, X-rays, or genetic engineering, may also result in the absence or inactivity of an enzyme. Modern genetic techniques make it possible to express foreign genes in higher organisms (transgenic animals) and study the effects of these genes on metabolism. When isotopic labels are incorporated into metabolites and allowed to enter a metabolic system, their paths may be traced from the distribution of label in the intermediates. NMR is a noninvasive technique that may be used to detect and study metabolites *in vivo*. Studies on isolated organs, tissue slices, cells, and subcellular organelles have contributed enormously to our knowledge of the localization of metabolic pathways.

Free energy is supplied to endergonic metabolic processes by the ATP produced via exergonic metabolic processes. ATP's -30.5 $kJ \cdot mol^{-1}$ $\Delta G°'$ of hydrolysis, is intermediate between those of "high-energy" metabolites such as phosphoenol pyruvate and "low-energy" metabolites such as glucose-6-phosphate. The "high-energy" phosphoryl groups are enzymatically transferred to ADP and the resulting ATP, in a separate reaction, phosphorylates "low-energy" compounds. ATP may also undergo pyrophosphate cleavage to yield PP_i, whose subsequent hydrolysis adds further thermodynamic impetus to the reaction. ATP is present in too short a supply to act as an energy reservoir. This function, in vertebrate nerve and muscle cells, is carried out by phosphocreatine, which under low-ATP conditions, readily transfers its phosphoryl group to ADP to form ATP.

The half-reactions of redox reactions may be physically separated to form an electrochemical cell. The reduction potential for the reduction of A by B,

$$A_{ox}^{n+} + B_{red} \rightleftharpoons A_{red} + B_{ox}^{n+}$$

in which n electrons are transferred, is given by the Nernst equation

$$\Delta \mathscr{E} = \Delta \mathscr{E}° - \frac{RT}{n\mathscr{F}} \ln \left(\frac{[A_{red}][B_{ox}^{n+}]}{[A_{ox}^{n+}][B_{red}]} \right)$$

The change in reduction potential of such a reaction is related to the reduction potentials of its component half-reactions, $\mathscr{E}_A$ and $\mathscr{E}_B$, by

$$\Delta \mathscr{E} = \mathscr{E}_A - \mathscr{E}_B$$

If $\mathscr{E}_A > \mathscr{E}_B$, then A_{ox}^{n+} has a greater electron affinity than does B_{ox}^{n+}. The reduction potential scale is defined by arbitrarily setting the reduction potential of the standard hydrogen half-cell to zero. Redox reactions are of great metabolic importance. For example, the oxidation of NADH yields three ATPs through the mediation of the electron-transport chain.

Living organisms are open systems and therefore cannot be at equilibrium. They must continuously dissipate free energy in order to carry out their various functions and to preserve their highly ordered structures. The study of nonequilibrium thermodynamics has indicated that the steady state, which living processes maintain, is the state of maximum efficiency under the constraints governing open systems. Control mechanisms that regulate biological processes preserve the steady state by regulating the activities of enzymes that are strategically located in metabolic pathways.

REFERENCES

Metabolic Studies

Beadle, G.W., Biochemical genetics, *Chem. Rev.* **37**, 15–96 (1945). [A classical review summarizing the "one gene–one enzyme" hypothesis.]

Cerdan, S. and Seelig, J., NMR studies of metabolism, *Annu. Rev. Biophys. Biophys. Chem.* **19**, 43–67 (1990).

Cooper, T.G., *The Tools of Biochemistry,* Chapter 3, Wiley–Interscience (1977). [A presentation of radiochemical techniques.]

Freifelder, D., *Biophysical Chemistry* (2nd ed.), Chapters 5 and 6, Freeman (1982). [A discussion of the principles of radioactive counting and autoradiography.]

Fruton, J.S. and Simmons, S., *General Biochemistry,* Chapter 16, Wiley (1958). [Outlines the classical methods for the study of intermediate metabolism.]

Goodridge, A.G., The new metabolism: molecular genetics in the analysis of metabolic regulation, *FASEB J.,* **4**, 3099–3110 (1990).

Hevesy, G., Historical sketch of the biological application of tracer elements, *Cold Spring Harbor Symp. Quant. Biol.* **13**, 129–150 (1948).

Jeffrey, F.M.H., Rajagopal, A., Malloy, C.R., and Sherry, A.D., ^{13}C -NMR: a simple yet comprehensive method for analysis of intermediary metabolism, *Trends Biochem. Sci.* **16**, 5–10 (1991).

Koretsky, A.P., Investigation of cell physiology in the animal using transgenic technology, *Am. J. Physiol.,* **262**, C261–C275 (1992).

Koretsky, A.P. and Williams, D.S., Application of localized *in vivo* NMR to whole organ physiology in the animal, *Annu. Rev. Physiol.,* **54**, 799–826 (1992).

McGrane, M.M., Yun, J.S., Patel, Y.M., and Hanson, R.W., Metabolic control of gene expression: *in vivo* studies with transgenic mice, *Trends Biochem. Sci.,* **17**, 40–44 (1992).

Shemin, D. and Rittenberg, D., The biological utilization of glycine for the synthesis of the protoporphyrin of hemoglobin, *J. Biol. Chem.* **166**, 621–625 (1946).

Suckling, K.E. and Suckling, C.J., *Biological Chemistry,* Cambridge University Press (1980). [Presents the organic chemistry of biochemical reactions.]

Walsh, C., *Enzymatic Reaction Mechanisms,* Chapter 1, Freeman (1979). [A discussion of the types of biochemical reactions.]

Westheimer, F.H., Why nature chose phosphates, *Science* **235**, 1173–1178 (1987).

Bioenergetics

Alberty, R.A., Standard Gibbs free energy, enthalpy and entropy changes as a function of pH and pMg for reactions involving adenosine phosphates, *J. Biol. Chem.* **244**, 3290–3302 (1969).

Caplan, S.R., Nonequilibrium thermodynamics and its application to bioenergetics, *Curr. Top. Bioenerg.* **4**, 1–79 (1971).

Crabtree, B. and Taylor, D.J., Thermodynamics and metabolism, in Jones, M.N. (Ed.), *Biochemical Thermodynamics,* pp. 333–378, Elsevier (1979).

Dickerson, R.E., *Molecular Thermodynamics,* Chapter 7, Benjamin (1969). [An interesting chapter on the thermodynamics of life.]

Henley, H.J.M., An introduction to nonequilibrium thermodynamics, *J. Chem. Ed.* **41**, 647–655 (1964).

Katchelsky, A. and Curran, P.F., *Nonequilibrium Thermodynamics in Biophysics,* Harvard University Press (1965).

Lehninger, A.L., *Bioenergetics* (2nd ed.), Benjamin (1972). [An introductory work by one of the field's originators.]

Morowitz, H.J., *Foundations of Bioenergetics,* Academic Press (1978).

PROBLEMS

1. Glycolysis (glucose breakdown) has the overall stoichiometry:

$$\text{Glucose} + 2ADP + 2P_i + 2NAD^+ \longrightarrow$$
$$2\text{pyruvate} + 2ATP + 2NADH + 2H^+ + 2H_2O$$

whereas that of gluconeogenesis (glucose synthesis) is

$$2\text{Pyruvate} + 4ATP + 2NADH + 2H^+ + 4H_2O \longrightarrow$$
$$\text{glucose} + 4ADP + 4P_i + 2NAD^+$$

What is the overall stoichiometry of the glycolytic breakdown of 1 mol of glucose followed by its gluconeogenic synthesis? Explain why it is necessary that the pathways of these two processes be independently controlled and why they must differ by at least one reaction.

2. It has been postulated that a trigonal bipyrimidal pentacovalent phosphorus intermediate can undergo a vibrational deformation process known as **pseudorotation** in which its apical ligands exchange with two of its equatorial ligands via a tetragonal pyrimidal transition state:

Trigonal bipyramid   **Trigonal bipyramid**
[X and Y apical]   **[O2 and O3 apical]**

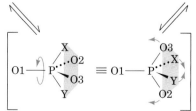

Tetragonal pyramidal
transition state

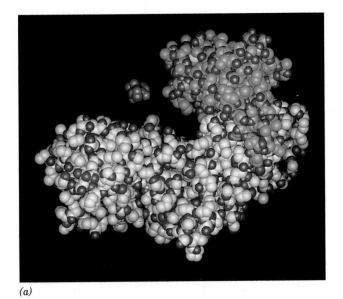

(a)

(b)

FIGURE 16-5. A space-filling model of a subunit of (*a*) yeast hexokinase and (*b*) its complex with glucose (*purple*). Note the prominant bilobal appearance of the free enzyme (the C atoms in the small lobe are shaded green, whereas those in the large lobe are light gray; the N and O atoms are blue and red). In the enzyme–substrate complex these lobes have swung together so as to engulf the substrate. [Based on an X-ray structure by Thomas Steitz, Yale University.]

vation that **xylose,** which differs from glucose only by the lack of the —$C6H_2OH$ group,

α- D-**Xylose**

greatly enhances the rate of ATP hydrolysis by hexokinase (presumably xylose induces the activating conformational change while water occupies the binding site of the missing hydroxymethyl group). Clearly, *this substrate-induced conformational change in hexokinase is responsible for the enzyme's specificity.* In addition, the active site polarity is reduced by exclusion of water, thereby expediting the nucleophilic reaction process. Other kinases have the same deeply clefted structure as hexokinase (Section 16-2G) and undergo conformational changes upon binding their substrates. This suggests that all kinases have similar mechanisms for maintaining specificity.

B. Phosphoglucose Isomerase

Reaction 2 of glycolysis is the conversion of G6P to **fructose-6-phosphate (F6P)** by **phosphoglucose isomerase** (PGI; also called **glucose-6-phosphate isomerase).** This is the isomerization of an aldose to a ketose:

Glucose 6-phosphate (G6P)

phosphoglucose isomerase (PGI)

Fructose 6-phosphate (F6P)

Since G6P and F6P both exist predominantly in their cyclic forms (Fig. 10-4 shows these structures for the unphosphorylated sugars), the reaction requires ring opening, followed by isomerization, and subsequent ring closure. The determination of the enzyme's pH dependence led to a hypothesis for amino acid side chain participation in the catalytic

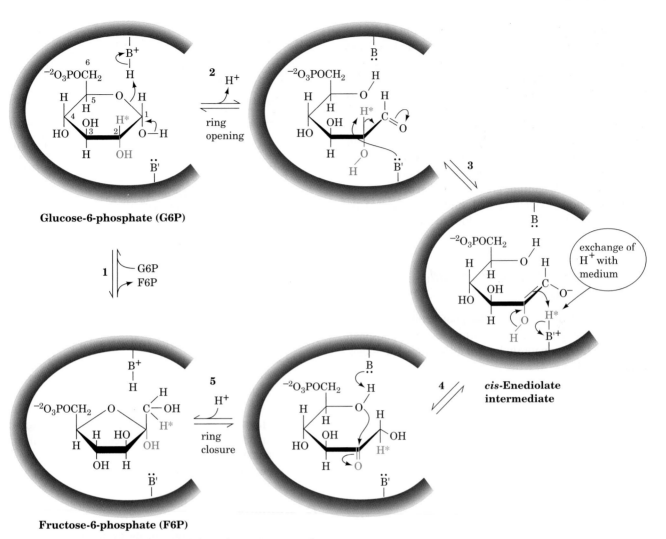

FIGURE 16-6. The reaction mechanism of phosphoglucose isomerase. The active site catalytic residues (BH⁺ and B′) are thought to be Lys and Glu, respectively.

mechanism. The catalytic rate exhibits a bell-shaped pH dependence curve with characteristic pK's of 6.7 and 9.3, which suggests the catalytic participation both of a His and a Lys (Section 13-4). Indeed, comparison of the amino acid sequences of PGI from several different organisms reveals that both His and Lys are conserved during evolution. However, a Glu residue is also conserved, and as we have seen in lysozyme (Section 14-2C), Glu can have an unusually high pK under certain environmental conditions. Glu may well account for the observed pK of 6.7. In fact, the X-ray structure of PGI reveals that a Glu is properly positioned at the enzyme's active site to act as a general base.

A proposed reaction mechanism for the PGI reaction involves general acid–base catalysis by the enzyme (Fig. 16-6):

Step 1 Substrate binding.

Step 2 An acid, presumably the Lys ε-amino group, catalyzes ring opening.

Step 3 A base, presumably the carboxylate group of Glu, abstracts the acidic proton from C2 to form a *cis*-enediolate intermediate (this proton is acidic because it is α to a carbonyl group).

Step 4 The proton is replaced on C1 in an overall proton transfer. Protons abstracted by bases are labile and exchange rapidly with solvent protons. Nevertheless, Irwin Rose confirmed this step by demonstrating that 2-[³H]G6P is occasionally converted to 1-[³H]F6P by intramolecular proton transfer before the ³H has a chance to exchange with the medium.

Step 5 Ring closure to form the product, which is subsequently released to yield free enzyme, thereby completing the catalytic cycle.

PGI, like most enzymes, catalyzes reactions with nearly absolute stereospecificity. To appreciate this, let us compare the proposed enzymatic reaction mechanism with that in

structures resemble that of the proposed enediol or ene-diolate intermediate:

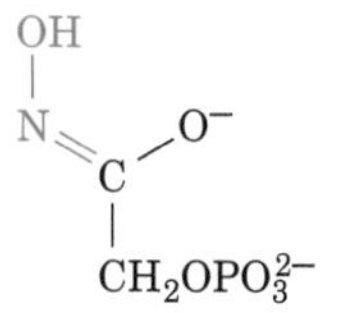

Phosphoglyco-hydroxamate

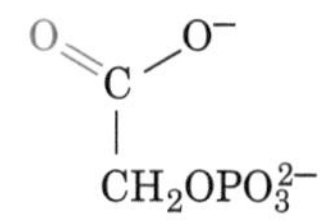

2-Phosphoglycolate

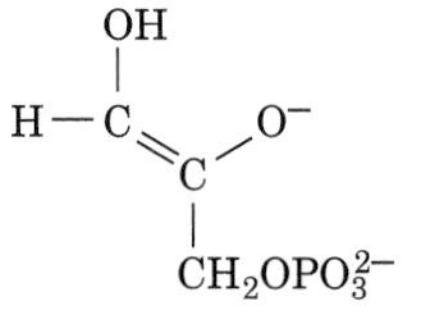

Proposed enediolate intermediate

Since enzymes catalyze reactions by binding the transition state complex more tightly than the substrate (Section 14-1F), phosphoglycohydroxamate and 2-phosphoglycolate should bind more tightly to TIM than to substrate. In fact, phosphoglycohydroxamate and 2-phosphoglycolate bind 155- and 100-fold more tightly to TIM than do either GAP or DHAP.

Glu 165 Functions as a General Base

The pH dependence of the TIM reaction is a bell-shaped curve with pK's of 6.5 and 9.5. The similarity of these pK's to the corresponding quantities of the phosphoglucose isomerase reaction suggests the participation of both an acid and a base in the TIM reaction. However, pH studies alone, as we have seen, are difficult to interpret in terms of specific amino acid residues since the active site environment may alter the pK of an acidic or basic group.

Affinity labeling reagents have been employed in an effort to identify the base at the active site of TIM. Both **bromohydroxyacetone phosphate** and **glycidol phosphate**

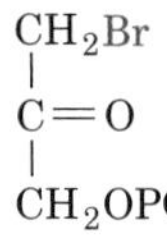

Bromohydroxyacetone phosphate

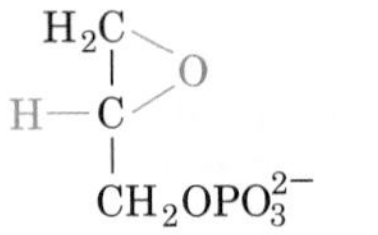

Glycidol phosphate

inactivate TIM by forming esters of Glu 165, whose carboxylate group, X-ray studies indicate, is ideally situated to abstract the C2 proton from the substrate (general base catalysis). In fact, the mutagenic replacement of Glu 165 by Asp, which X-ray studies show withdraws the carboxylate group only ~1 Å further away from the substrate than its position in the wild-type enzyme, reduces TIM's catalytic power ~1000-fold. Note that Glu 165's pK is drastically altered from the 4.1 value of the free amino acid to the

observed 6.5 value. This provides yet another striking example of the effect of the environment on the properties of amino acid side chains.

The TIM Reaction Probably Occurs via Concerted General Acid–Base Catalysis Involving Low Barrier Hydrogen Bonds

The X-ray structure of yeast TIM in complex with phosphoglycohydroxamate indicates that His 95 is hydrogen bonded to and hence is properly positioned to protonate the carbonyl oxygen atom of GAP (general acid catalysis).

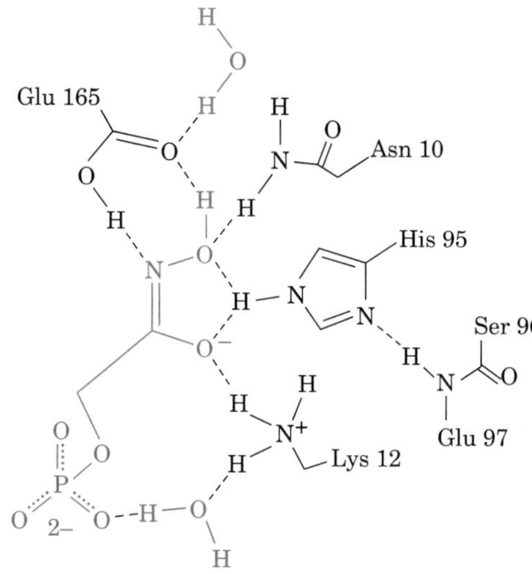

However, NMR studies indicate that His 95 is in its neutral imidazole form rather than its protonated imidazolium form. How can an imidazole N3—H group, which has a highly basic pK of ~14, protonate a carbonyl oxygen atom that, when protonated, has a very acidic pK of <0? Likewise, how can the Glu 165 carboxylate group (pK 6.5) abstract the C2 proton (pK ~17) from GAP? A plausible answer is that these proton shifts are facilitated by the formation of low-barrier hydrogen bonds. These unusually strong associations (-40 to -80 kJ·mol^{-1}, vs -12 to -30 kJ·mol^{-1} for normal hydrogen bonds), as we have seen in the case of the serine protease catalytic triad (Section 16-3D), form when the pK's of the hydrogen bonding donor and acceptor groups are nearly equal. They can be important contributors to rate enhancement if they form in the transition state of an enzymatically catalyzed reaction.

In converting GAP to the enediol (or enediolate) intermediate (Fig. 16-10, *left*), the pK of the protonated form of its carbonyl oxygen, which becomes a hydroxyl group, increases to ~14, which closely matches that of neutral His 95. The resulting low-barrier hydrogen bond between this hydroxyl group and His 95 permits the neutral imidazole side chain to protonate the oxygen atom. Likewise, as the carbonyl oxygen is protonated, the pK of the C2—H proton decreases to ~7, close to the pK of the Glu 165 carboxylate. It therefore appears that the reaction occurs via si-

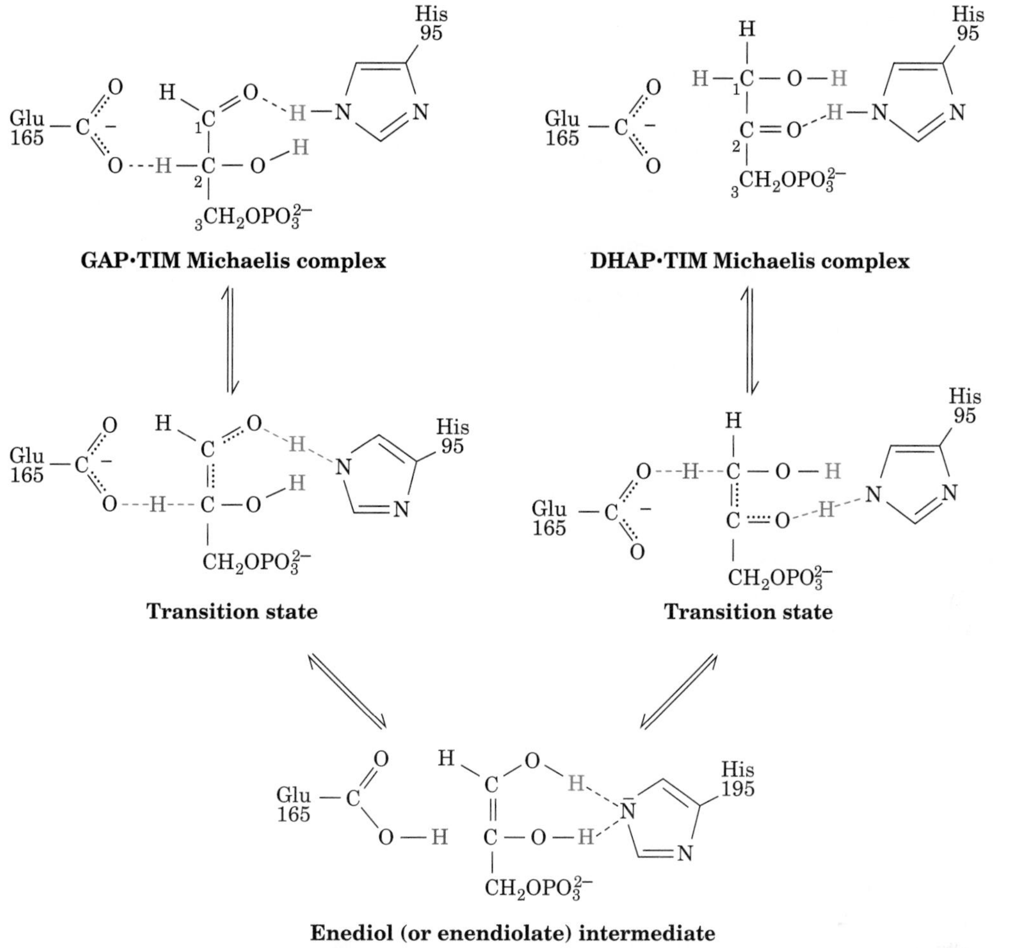

GAP·TIM Michaelis complex

DHAP·TIM Michaelis complex

Transition state

Transition state

Enediol (or enendiolate) intermediate

FIGURE 16-10. The proposed enzymatic mechanism of the TIM reaction. The reaction proceeds via the concerted abstraction of the C2—H proton of GAP by the carboxylate group of Glu 165 and the protonation of the GAP carbonyl oxygen atom by the imidazole group of His 95. The pK's of the corresponding donor and acceptor groups participating in each proton transfer process become nearly equal in the transition state and hence form low-barrier hydrogen bonds (*red dashed lines*) which act to stabilize the transition state. The resulting enediol (or possibly the electrostatically stabilized enediolate) intermediate then reacts in a similar fashion, with the carboxyl group of Glu 165 protonating C1 while the deprotonated N3 atom of His 95 abstracts the proton on the 2-hydroxyl group to yield DHAP.

multaneous proton abstraction by Glu 165 and protonation by His 95 (concerted general acid–base catalysis). The low-barrier hydrogen bonds postulated to form in the transition state, but not in the Michaelis complex, between Glu 165 and C2—H and between His 95 and the carbonyl oxygen atom are thought to provide some of the transition state stabilization necessary to catalyze the reaction. The positively charged side chain of Lys 12, which is probably responsible for the 9.5 pK observed in TIM's pH rate profile, is thought to electrostatically stabilize the negatively charged transition state. The conversion of the enediol(ate) intermediate to DHAP is likewise facilitated by the formation of transition state low-barrier hydrogen bonds (Fig. 16-10, *right*).

A Flexible Loop Both Preferentially Binds and Protects the Enediol Intermediate

The comparison of the structure of the TIM·phosphoglycohydroxamate complex with that of TIM alone reveals that when substrate binds to TIM, a conserved 10-residue loop closes over the active site like a hinged lid in a movement that involves main chain shifts of >7 Å (Fig. 16-11). A four-residue segment of this loop makes a hydrogen bond with the phosphate group of the substrate. Mutagenic excision of these four residues does not significantly distort the protein so that substrate binding is not greatly impaired. The catalytic power of the mutant enzyme is, nevertheless, reduced 10^5-fold and it only weakly binds phosphoglycohydroxamate. Evidently, loop

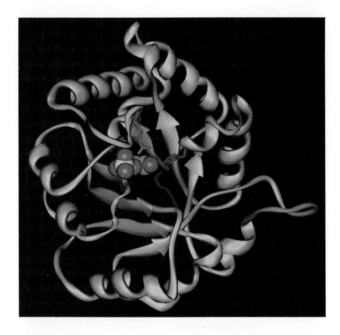

FIGURE 16-11. A ribbon diagram of yeast TIM in complex with its transition state analog 2-phosphoglycolate. A single 248-residue subunit of this homodimeric enzyme is viewed roughly along the axis of its α/β barrel. The enzyme's flexible loop, residues 168 through 177, is light blue and the side chains of Lys 12, His 95, and Glu 165 are dark blue, purple, and red, respectively. The 2-phosphoglycolate is represented by a space-filling model colored according to atom type (C, green; O, red; P, yellow). [Based on an X-ray structure determined by Gregory Petsko, Brandeis University.]

closure preferentially stabilizes the enzymatic reaction's enediol-like transition state.

Loop closure in the TIM reaction also provides a striking example of the stereoelectronic control that enzymes can exert on a reaction (Section 14-1E). In solution, the enediol intermediate readily breaks down with the elimination of the phosphate at C3 to form the toxic compound **methylglyoxal** (Fig. 16-12a). On the enzyme's surface, however, this reaction is prevented because the phosphate group is held by the flexible loop in the plane of the enediol, a position that disfavors phosphate elimination. In order for this elimination to occur, the C—O bond to the phosphate group must lie, as shown in Fig. 16-12a, in the plane perpendicular to that of the enediol. This is because, if the phosphate group were to be eliminated while this C—O bond was in the plane of the enediol as diagrammed in Fig. 16-12b, the CH₂ group of the resulting enol product would be twisted 90° out of the plane of the rest of the molecule. Such a conformation is energetically prohibitive because it prevents the formation of the enol's double bond by eliminating the overlap between its component p orbitals. In the mutant enzyme lacking the flexible loop, the enediol is able to escape: ~85% of the enediol intermediate is released into solution, where it rapidly decomposes to methylglyoxal and P_i. Thus, flexible loop closure also ensures that substrate is efficiently transformed to product.

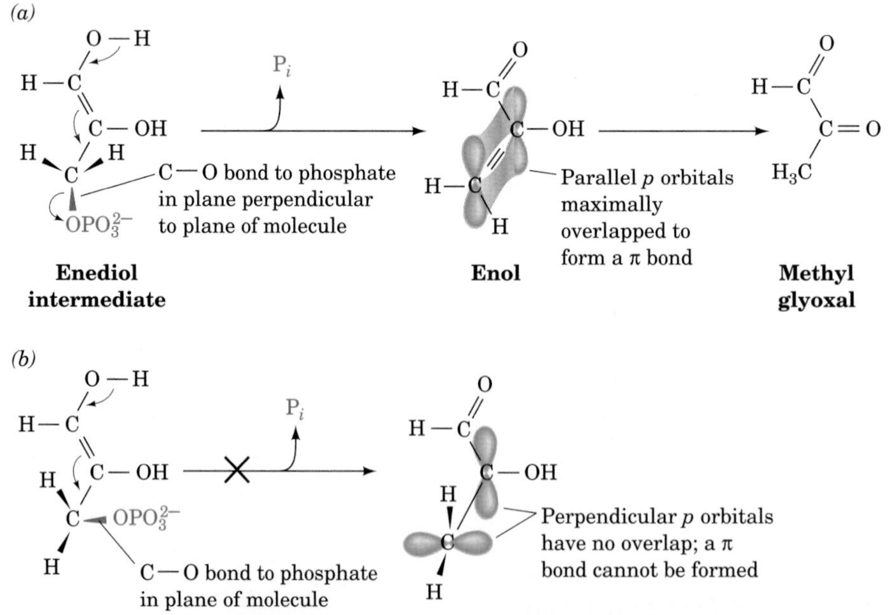

FIGURE 16-12. The spontaneous decomposition of the enediol intermediate in the TIM reaction to form methylglyoxal through the elimination of a phosphate group. (*a*) This reaction can occur only when the C—O bond to the phosphate group lies in a plane that is nearly perpendicular to that of the enediol so as to permit the formation of a double bond in the intermediate enol product. (*b*) When the C—O bond to the phosphate group lies in a plane that is nearly parallel to that of the enediol, the p orbitals on the resulting intermediate product would be perpendicular to each other and hence lack the overlap necessary to form a π bond, that is, a double bond. The resulting unsatisfied bonding capacity greatly increases the energy of the reaction intermediate and hence makes the reaction highly unfavorable.

TIM Is a Perfect Enzyme

TIM, as Jeremy Knowles demonstrated, has achieved catalytic perfection in that the rate of bimolecular reaction between enzyme and substrate is diffusion controlled; that is, product formation occurs as rapidly as enzyme and substrate can collide in solution so that any increase in TIM's catalytic efficiency would not increase the reaction rate (Section 13-2B). Because of the high interconversion efficiency of GAP and DHAP, these two metabolites are maintained in equilibrium: $K = [GAP]/[DHAP] = 4.73 \times 10^{-2}$; that is, $[DHAP] \gg [GAP]$ at equilibrium. However, *as GAP is utilized in the succeeding reaction of the glycolytic pathway, more DHAP is converted to GAP so that these compounds maintain their equilibrium ratio.* One common pathway therefore accounts for the metabolism of both products of the aldolase reaction.

α/β Barrel Enzymes May Have Evolved by Divergent Evolution

TIM was the first protein known to contain an α/β **barrel,** a folding motif in which eight β strands, alternating in sequence with eight α helices, are arranged such that the β strands form an eight-stranded parallel β sheet that wraps in a cylinder and that is surrounded by the eight α helices (Figs. 16-11, 7-19b and 7-49c). This particularly striking supersecondary structure (which is also called a **TIM barrel**) has since been found in over 20 different proteins. Essentially all of these α/β barrel-containing proteins are enzymes (including four glycolytic enzymes: aldolase, TIM, enolase, and pyruvate kinase), even though only around half of the ~400 different proteins of known X-ray structure are enzymes (and hence ~10% of these enzymes contain α/β barrels). Intriguingly, the active sites of all known α/β barrel enzymes are located in the mouth of the barrel at the end that contains the C-terminal ends of the β strands, although there is no obvious structural rationale for this. Thus, despite the fact that few of these proteins exhibit significant mutual sequence homology, it has been postulated that all of them have evolved from a common ancestor (divergent evolution). However, it has also been argued that the α/β barrel is a particularly stable arrangement which nature has independently discovered on several occasions (convergent evolution). Convincing evidence supporting either view has not been forthcoming, so the debate as to the evolutionary relationships among α/β barrel proteins continues.

Let us now take stock of where we are in our travels down the glycolytic pathway. At this point, the glucose, which has been transformed into two GAPs, has completed the preparatory stage of glycolysis. This process has required the expenditure of two ATPs. However, this investment has resulted in the conversion of one glucose to two C_3 units, each of which has a phosphoryl group that, with a little chemical artistry, can be converted to a "high-energy" compound (Section 15-14B) whose free energy of hydroly-

sis can be coupled to ATP synthesis. *This energy investment will be doubly repaid in the final stage of glycolysis in which the two phosphorylated C_3 units are transformed to two pyruvates with the coupled synthesis of four ATPs per glucose.*

F. Glyceraldehyde-3-Phosphate Dehydrogenase: First "High-Energy" Intermediate Formation

Reaction 6 of glycolysis involves the oxidation and phosphorylation of GAP by NAD^+ and P_i as catalyzed by **glyceraldehyde-3-phosphate dehydrogenase (GAPDH;** Fig. 7-46). This is the first instance of the chemical artistry alluded to above. *In this reaction, aldehyde oxidation, an exergonic reaction, drives the synthesis of the acyl phosphate 1,3-bisphosphoglycerate (1,3-BPG; previously called 1,3-diphosphoglycerate).* Recall that acyl phosphates are compounds with high phosphate group-transfer potential (Section 15-4B).

Glyceraldehyde 3-phosphate (GAP)

glyceraldehyde 3-phosphate dehydrogenase (GAPDH)

1,3-Bisphosphoglycerate (1,3-BPG)

Mechanistic Studies

Several key enzymological experiments have contributed to the elucidation of the GAPDH reaction mechanism:

1. GAPDH is inactivated by alkylation with stoichiometric amounts of iodoacetate. The presence of **carboxymethylcysteine** in the hydrolysate of the resulting alkylated enzyme (Fig. 16-13a) suggests that GAPDH has an active site Cys sulfhydryl group.

2. GAPDH quantitatively transfers ^3H from C1 of GAP to NAD^+ (Fig. 16-13b), thereby establishing that this reaction occurs via direct hydride transfer.

3. GAPDH catalyzes exchange of ^{32}P between $[^{32}P]P_i$ and the product analog **acetyl phosphate** (Fig. 16-13c). Such isotope exchange reactions are indicative of an acyl–enzyme intermediate (Section 13-5D).

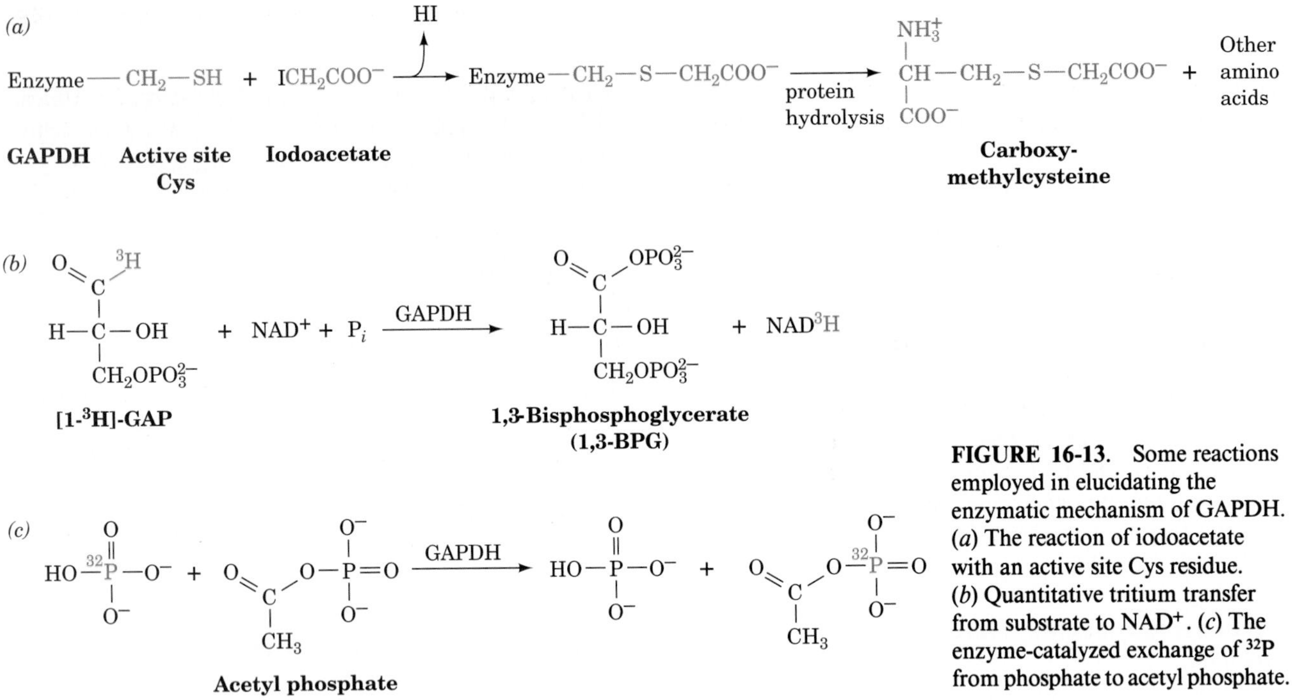

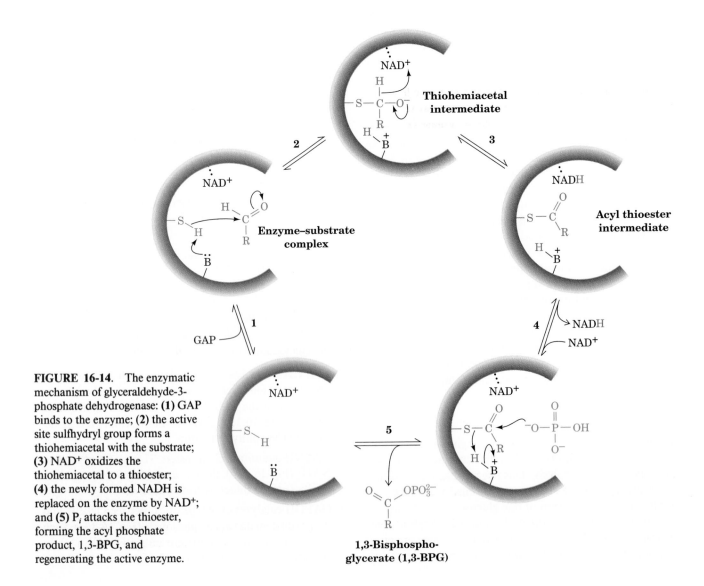

FIGURE 16-13. Some reactions employed in elucidating the enzymatic mechanism of GAPDH. (*a*) The reaction of iodoacetate with an active site Cys residue. (*b*) Quantitative tritium transfer from substrate to NAD$^+$. (*c*) The enzyme-catalyzed exchange of ^{32}P from phosphate to acetyl phosphate.

FIGURE 16-14. The enzymatic mechanism of glyceraldehyde-3-phosphate dehydrogenase: **(1)** GAP binds to the enzyme; **(2)** the active site sulfhydryl group forms a thiohemiacetal with the substrate; **(3)** NAD$^+$ oxidizes the thiohemiacetal to a thioester; **(4)** the newly formed NADH is replaced on the enzyme by NAD$^+$; and **(5)** P$_i$ attacks the thioester, forming the acyl phosphate product, 1,3-BPG, and regenerating the active enzyme.

David Trentham has proposed a mechanism for GAPDH based on this information and the results of kinetic studies (Fig. 16-14):

Step 1 GAP binds to the enzyme.

Step 2 The essential sulfhydryl group, acting as a nucleophile, attacks the aldehyde to form a **thiohemiacetal.**

Step 3 The thiohemiacetal undergoes oxidation to an **acyl thioester** by direct transfer of a hydride to NAD$^+$. This intermediate, which has been isolated, has a high group-transfer potential. *The energy of aldehyde oxidation has not been dissipated but has been conserved through the synthesis of the thioester and the reduction of NAD$^+$ to NADH.*

Step 4 Another molecule of NAD$^+$ replaces NADH.

Step 5 The thioester intermediate undergoes nucleophilic attack by P_i to regenerate free enzyme and form 1,3-BPG. This "high-energy" mixed anhydride generates ATP from ADP in the next reaction of glycolysis.

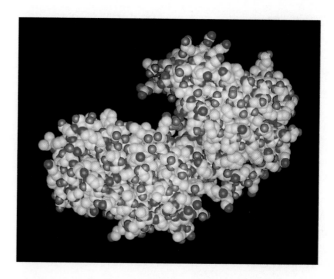

FIGURE 16-15. A space-filling model of yeast phosphoglycerate kinase showing its deeply clefted bilobal structure. The substrate-binding site is at the bottom of the cleft as marked by the P atom (*purple*) of 3PG. Compare this structure with that of hexokinase (Fig. 16-5a). [Based on an X-ray structure by Herman Watson, University of Bristol, U.K.]

G. Phosphoglycerate Kinase: First ATP Generation

Reaction 7 of the glycolytic pathway results in the first formation of ATP together with **3-phosphoglycerate (3PG)** in a reaction catalyzed by **phosphoglycerate kinase (PGK):**

1,3-Bisphosphoglycerate (1,3-BPG)

Mg^{2+} ⇅ phosphoglycerate kinase (PGK)

3-Phosphoglycerate (3PG)

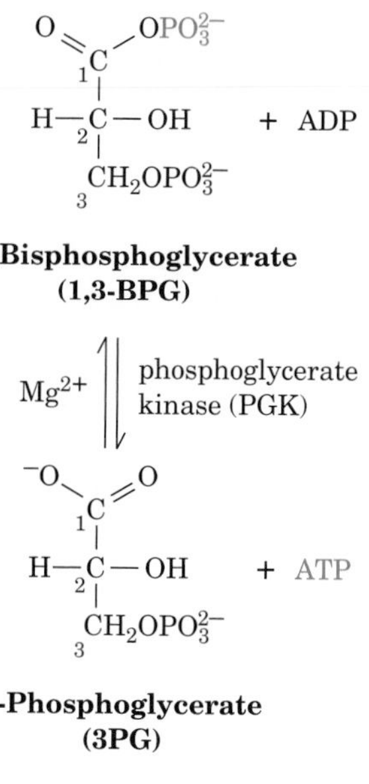

(*Note:* The name "kinase" is given to any enzyme that transfers a phosphoryl group between ATP and a metabolite. Nothing is implied as to the exergonic direction of transfer.)

PGK (Fig. 16-15) is conspicuously bilobal in appearance. The Mg^{2+}–ADP binding site is located on one domain, ~10 Å from the 1,3-BPG binding site, which is on the other

domain. Physical measurements suggest that, *upon substrate binding, the two domains of PGK swing together so as to permit the substrates to react in a water-free environment as occurs with hexokinase (Section 16-2A).* Indeed, the appearance of PGK is remarkably similar to that of hexokinase (Fig. 16-5a), even though the structures of these proteins are otherwise unrelated.

Figure 16-16 indicates a reaction mechanism for PGK that is consistent with its observed sequential kinetics. The terminal phosphoryl oxygen of ADP nucleophilically at-

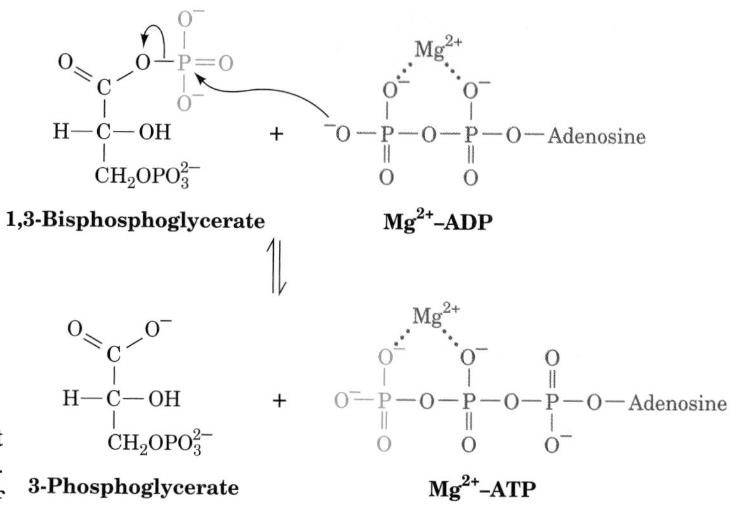

1,3-Bisphosphoglycerate   **Mg^{2+}–ADP**

3-Phosphoglycerate   **Mg^{2+}–ATP**

FIGURE 16-16. The mechanism of the PGK reaction. The Mg^{2+} positions are shown as examples; their actual binding sites are unknown.

tacks the C1 phosphorus atom of 1,3-BPG to form the reaction product.

The energetics of the overall GAPDH–PGK reaction pair are:

$$GAP + P_i + NAD^+ \longrightarrow 1,3\text{-BPG} + NADH$$
$$\Delta G^{\circ\prime} = +6.7 \text{ kJ} \cdot \text{mol}^{-1}$$

$$1,3\text{-BPG} + ADP \longrightarrow 3PG + ATP$$
$$\Delta G^{\circ\prime} = -18.8 \text{ kJ} \cdot \text{mol}^{-1}$$

$$GAP + P_i + NAD^+ + ADP \longrightarrow 3PG + NADH + ATP$$
$$\Delta G^{\circ\prime} = -12.1 \text{ kJ} \cdot \text{mol}^{-1}$$

Although the GAPDH reaction is endergonic, the strongly exergonic nature of the transfer of a phosphoryl group from 1,3-BPG to ADP makes the overall synthesis of NADH and ATP from GAP, P_i, NAD^+, and ADP favorable.

H. Phosphoglycerate Mutase

In Reaction 8 of glycolysis, 3PG is converted to **2-phosphoglycerate (2PG)** by **phosphoglycerate mutase (PGM)**:

3-Phosphoglycerate (3PG) ⇌ (phosphoglycerate mutase (PGM)) **2-Phosphoglycerate (2PG)**

A **mutase** catalyzes the transfer of a functional group from one position to another on a molecule. This reaction is necessary preparation for the next reaction in glycolysis, which generates a "high-energy" phosphoryl compound for use in ATP synthesis.

Reaction Mechanism of PGM

At first sight, the reaction catalyzed by PGM appears to be a simple intramolecular phosphoryl transfer. This is not the case, however. *The active enzyme has a phosphoryl group at its active site, which it transfers to the substrate to form a bisphospho intermediate. This intermediate then rephosphorylates the enzyme to form the product and regenerate the active phosphoenzyme.* The following experimental data permitted the elucidation of PGM's enzymatic mechanism:

1. Catalytic amounts of **2,3-bisphosphoglycerate (2,3-BPG;** previously known as **2,3-diphosphoglycerate)**

2,3-Bisphosphoglycerate (2,3-BPG)

are required for enzymatic activity; that is, 2,3-BPG acts as a reaction primer.

2. Incubation of the enzyme with catalytic amounts of ^{32}P-labeled 2,3-BPG yields a ^{32}P-labeled enzyme. Zelda Rose demonstrated that this was a result of the phosphorylation of a His residue:

Enzyme—CH_2 ... PO_3^{2-}

Phospho His residue

3. The enzyme's X-ray structure shows His at the active site (Fig. 16-17). In the active enzyme, His 8 is phosphorylated.

These data are consistent with a mechanism in which the active enzyme contains a phospho-His residue at the active site (Fig. 16-18):

Step 1   3PG binds to the phosphoenzyme in which His 8 is phosphorylated.

Step 2   This phosphoryl group is transferred to the substrate, resulting in an intermediate 2,3-BPG · enzyme complex.

Steps 3 & 4   The complex decomposes to form the product 2PG with regeneration of the phosphoenzyme.

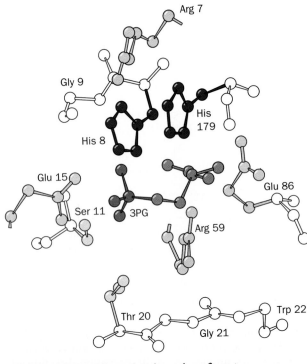

FIGURE 16-17. The active site region of yeast phosphoglycerate mutase (dephospho form) showing the substrate, 3-phosphoglycerate, and some of the side chains that approach it. His 8 is phosphorylated in the active enzyme. [After Winn, S.I., Watson, H.I., Harkins, R.N., and Fothergill, L.A., *Phil. Trans. R. Soc. London Ser. B* **293**, 126 (1981).]

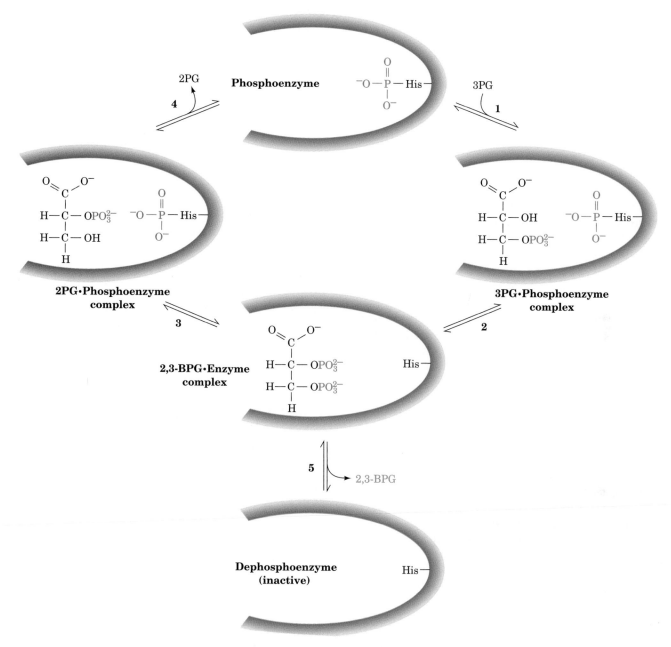

FIGURE 16-18. A proposed reaction mechanism for phosphoglycerate mutase. The active form of the enzyme contains a phospho-His residue at the active site. (**1**) Formation of an enzyme–substrate complex; (**2**) transfer of the enzyme-bound phosphoryl group to the substrate; (**3**) rephosphorylation of the enzyme by the other phosphoryl group of the substrate; and (**4**) release of product regenerating the active phospho-enzyme. (**5**) Occasionally 2,3-BPG dissociates from the enzyme, leaving it in an inactive, dephospho form that must be rephosphorylated by the reverse reaction.

The phosphoryl group on 3PG therefore ends up on the C2 of the next 3PG to undergo reaction.

Occasionally, 2,3-BPG dissociates from the enzyme (Fig. 16-18; Step 5) leaving it in an inactive form. Trace amounts of 2,3-BPG must therefore always be available to regenerate the active phosphoenzyme by the reverse reaction.

Glycolysis Influences Oxygen Transport

2,3-BPG specifically binds to deoxyhemoglobin and thereby alters the oxygen affinity of hemoglobin (Section 9-1D). The concentration of 2,3-BPG in erythrocytes is much higher (~5 mM) than the trace amounts required for its use as a primer of PGM. Erythrocytes synthesize and degrade 2,3-BPG by a detour from the glycolytic pathway diagrammed in Fig. 16-19. **Bisphosphoglycerate mutase** catalyzes the transfer of a phosphoryl group from C1 to C2 of 1,3-BPG. The resulting 2,3-BPG is hydrolyzed to 3PG by **2,3-bisphosphoglycerate phosphatase.** The rate of glycolysis affects the oxygen affinity of hemoglobin through the mediation of 2,3-BPG. Consequently, inherited defects of

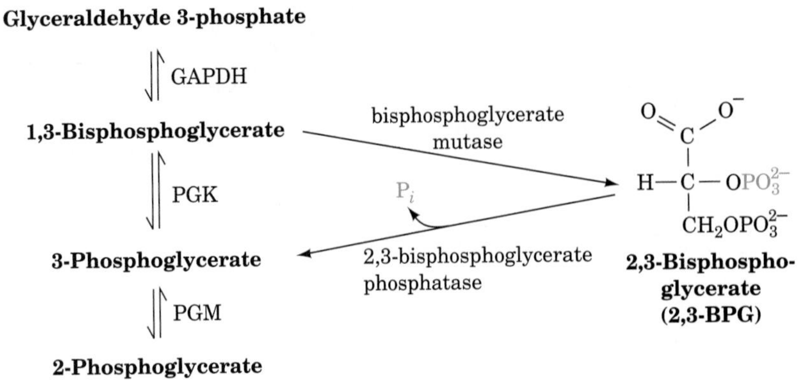

Glyceraldehyde 3-phosphate

$\Updownarrow$ GAPDH

1,3-Bisphosphoglycerate → bisphosphoglycerate
mutase

$\Updownarrow$ PGK P_i

3-Phosphoglycerate ← 2,3-bisphosphoglycerate
phosphatase

$\Updownarrow$ PGM

2-Phosphoglycerate

$$\underset{\text{O}}{\overset{\text{O}}{\diagdown}}\underset{\text{C}}{\diagup}\text{O}^-$$

H—C—OPO_3^{2-}

$CH_2OPO_3^{2-}$

**2,3-Bisphospho-
glycerate
(2,3-BPG)**

FIGURE 16-19. The pathway for the synthesis and degradation of 2,3-BPG in erythrocytes is a detour from the glycolytic pathway.

glycolysis in erythrocytes alter the capacity of the blood to transport oxygen (Fig. 16-20). For example, the concentration of glycolytic intermediates in hexokinase-deficient erythrocytes is less than normal because hexokinase catalyzes the first reaction of glycolysis. This results in a diminished 2,3-BPG concentration and therefore in increased hemoglobin oxygen affinity. Conversely, pyruvate kinase deficiency decreases hemoglobin oxygen affinity through the increase of 2,3-BPG resulting from the blockade of the last reaction in glycolysis.

I. Enolase: Second "High-Energy" Intermediate Formation

In Reaction 9 of glycolysis, 2PG is dehydrated to **phosphoenolpyruvate (PEP)** in a reaction catalyzed by **enolase:**

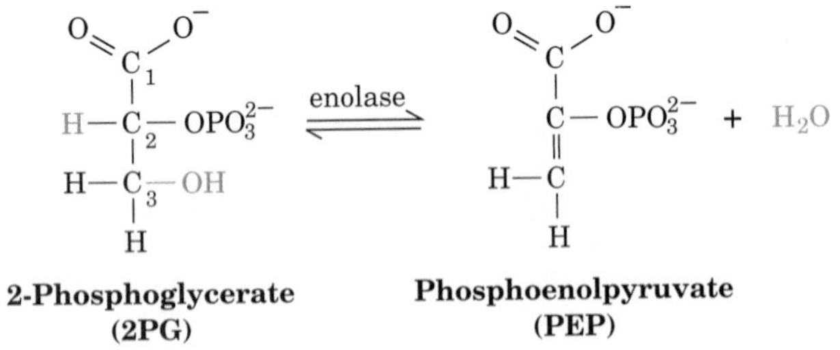

$$\underset{\text{O}}{\overset{\text{O}}{\diagdown}}\underset{\overset{|}{\text{C}}_1}{\diagup}\text{O}^-$$

H—$\overset{|}{\underset{|}{C}}_2$—$OPO_3^{2-}$ $\xrightarrow{\text{enolase}}$

H—$\overset{|}{\underset{|}{C}}_3$—OH
H

**2-Phosphoglycerate
(2PG)**

$$\underset{\text{O}}{\overset{\text{O}}{\diagdown}}\underset{\text{C}}{\diagup}\text{O}^-$$

C—OPO_3^{2-} + H_2O

H—C
H

**Phosphoenolpyruvate
(PEP)**

The enzyme forms a complex with a divalent cation such as Mg^{2+} before the substrate is bound. As is mentioned in Section 16-1A, fluoride ion inhibits glycolysis with the accumulation of 2PG and 3PG. It does so by strongly inhibiting enolase in the presence of P_i. F^- and P_i form a tightly bound complex with the Mg^{2+} at the enzyme's active site, blocking substrate binding and thereby inactivating the enzyme. Enolase's substrate, 2PG, therefore builds up and, as it does so, is equilibrated with 3PG by PGM.

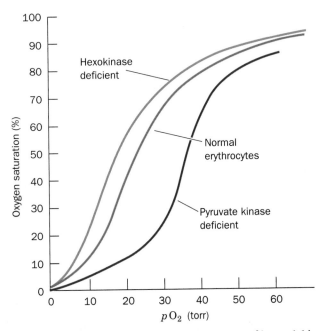

FIGURE 16-20. The oxygen-saturation curves of hemoglobin in normal erythrocytes (*red curve*) and those from patients with hexokinase deficiency (*green*) and with pyruvate kinase deficiency (*purple*). [After Delivoria-Papadopoulos, M., Oski, F. A., and Gottlieb, A. J., *Science* **165**, 601 (1969).]

Catalytic Mechanism of Enolase

The dehydration (elimination of H_2O) catalyzed by enolase might occur in one of three ways (Fig. 15-9*a*): (1) the C3—OH group can leave first, generating a carbocation at C3; (2) the C2 proton can leave first, generating a carbanion at C2; or (3) the reaction can be concerted. Isotope exchange studies by Paul Boyer demonstrated that the C2 proton of 2PG exchanges with solvent 12 times faster than the rate of PEP formation. However, the C3 oxygen ex-

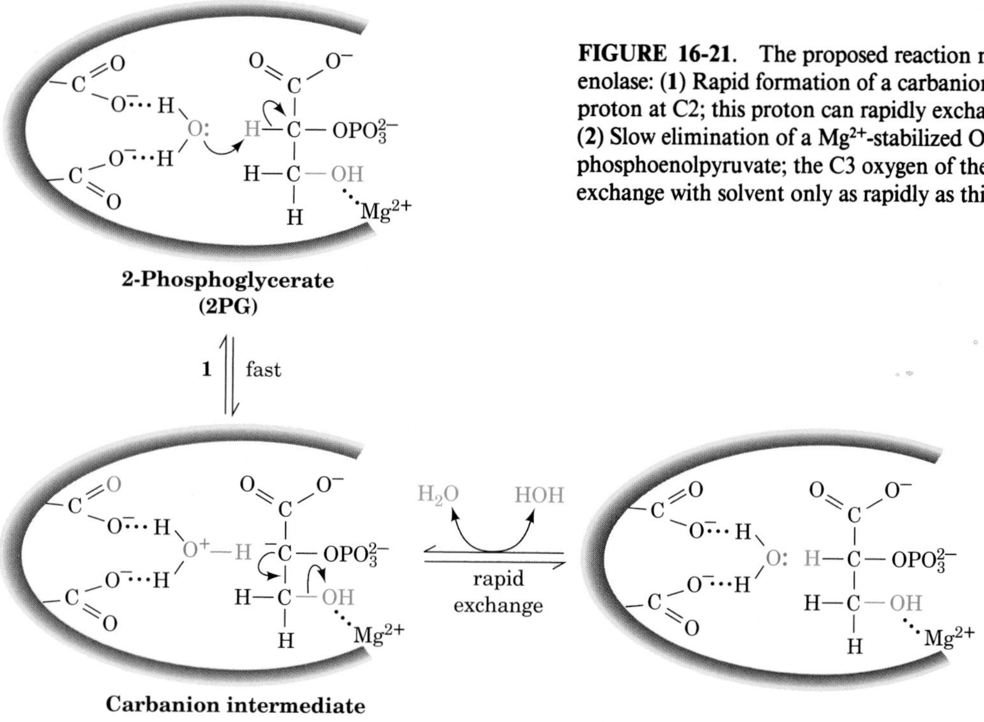

2-Phosphoglycerate
(2PG)

1 ⇅ fast

Carbanion intermediate

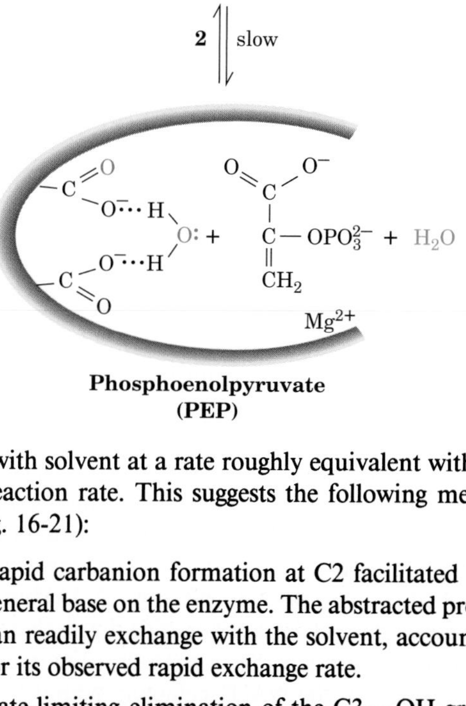

2 ⇅ slow

Phosphoenolpyruvate
(PEP)

H₂O → HOH

rapid
exchange

FIGURE 16-21. The proposed reaction mechanism of enolase: (1) Rapid formation of a carbanion by removal of a proton at C2; this proton can rapidly exchange with the solvent. (2) Slow elimination of a Mg^{2+}-stabilized OH group to form phosphoenolpyruvate; the C3 oxygen of the substrate can exchange with solvent only as rapidly as this step occurs.

the carboxyl groups of Glu 168 and Glu 211 and lies within 2.6 Å of the proton to be abstracted.

J. Pyruvate Kinase: Second ATP Generation

In Reaction 10 of glycolysis, the final reaction, **pyruvate kinase (PK)** couples the free energy of PEP hydrolysis to the synthesis of ATP to form **pyruvate:**

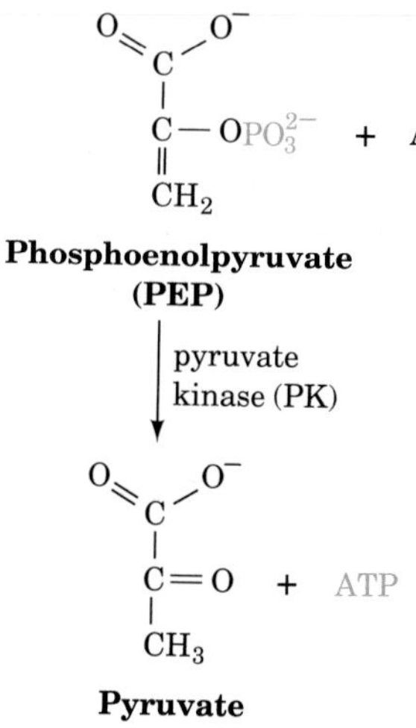

Phosphoenolpyruvate
(PEP)

pyruvate
kinase (PK)

Pyruvate

+ ADP + H⁺

+ ATP

changes with solvent at a rate roughly equivalent with the overall reaction rate. This suggests the following mechanism (Fig. 16-21):

Step 1 Rapid carbanion formation at C2 facilitated by a general base on the enzyme. The abstracted proton can readily exchange with the solvent, accounting for its observed rapid exchange rate.

Step 2 Rate-limiting elimination of the C3—OH group. This is consistent with the slow rate of exchange of this hydroxyl group with solvent.

The enolase reaction (Fig. 16-21) is of mechanistic interest because it involves the abstraction of a relatively nonacidic proton, followed by the elimination of an OH⁻ ion, which is a poor leaving group. The X-ray structure of enolase in complex with Mg^{2+} and 2PG shows that the hydroxyl group of 2PG participates in coordinating the Mg^{2+}. Such coordination makes the OH⁻ ion a much better leaving group. The base used to abstract the proton from C2 is thought to be a water molecule that is hydrogen bonded to

Catalytic Mechanism of PK

The PK reaction, which requires the participation of both monovalent (K^+) and divalent (Mg^{2+}) cations, occurs as follows (Fig. 16-22):

Step 1 A β-phosphoryl oxygen of ADP nucleophilically attacks the PEP phosphorus atom, thereby displacing **enol pyruvate** and forming ATP. This reaction conserves the free energy of PEP hydrolysis.

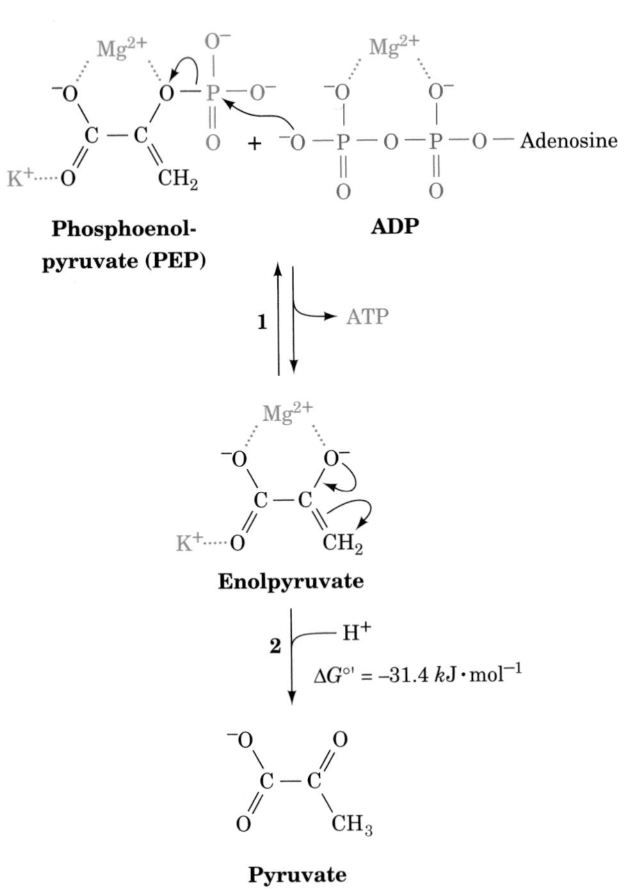

FIGURE 16-22. The mechanism of the reaction catalyzed by pyruvate kinase: (**1**) Nucleophilic attack of an ADP β-phosphoryl oxygen atom on the phosphorus atom of PEP to form ATP and phosphoenolpyruvate; and (**2**) tautomerization of enolpyruvate to pyruvate.

Step 2 Enol pyruvate converts to pyruvate. This enol–keto tautomerization is sufficiently exergonic to drive the coupled endergonic synthesis of ATP (Section 15-4C).

We can now see the "logic" of the enolase reaction. The standard free energy of hydrolysis of 2PG is only $\Delta G°' = -17.6$ kJ·mol^{-1}, which is insufficient to drive ATP synthesis ($\Delta G°' = 30.5$ kJ·mol^{-1} for ATP synthesis from ADP and P$_i$). The dehydration of 2PG results in the formation of a "high-energy" compound capable of such synthesis [the standard free energy of hydrolysis of PEP is -61.9 kJ·mol^{-1} (Fig. 15-23)]. In other words, PEP is a "high-energy" compound, 2PG is not.

3. FERMENTATION: THE ANAEROBIC FATE OF PYRUVATE

For glycolysis to continue, NAD$^+$, which cells have in limited quantities, must be recycled after its reduction to NADH by GAPDH (Fig. 16-3; Reaction 6). In the presence of oxygen, the reducing equivalents of NADH are passed into the mitochondria for reoxidation (Chapter 20). Under anaerobic conditions, on the other hand, the NAD$^+$ is replenished by the reduction of pyruvate in an extension of the glycolytic pathway. Two processes for the anaerobic replenishment of NAD$^+$ are homolactic and alcoholic fermentation, which occur in muscle and yeast, respectively.

A. Homolactic Fermentation

In muscle, particularly during vigorous activity when the demand for ATP is high and oxygen has been depleted, **lactate dehydrogenase (LDH)** catalyzes the oxidation of NADH by pyruvate to yield NAD$^+$ and **lactate**. This reaction is often classified as Reaction 11 of glycolysis:

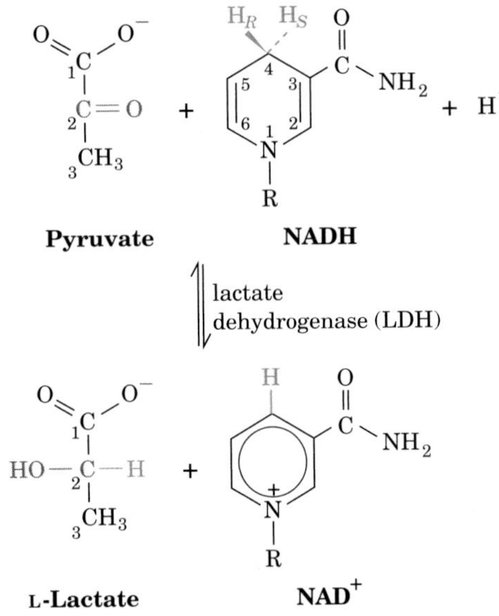

LDH, as do other NAD$^+$-requiring enzymes, catalyzes its reaction with absolute stereospecificity: The *pro-R* (A-side) hydrogen at C4 of NADH is stereospecifically transferred to the *re* face of pyruvate at C2 to form L- (or *S*-) lactate. This regenerates NAD$^+$ for participation in the GAPDH reaction. The hydride transfer to pyruvate is from the same face of the nicotinamide ring as that to acetaldehyde in the alcohol dehydrogenase reaction (Section 12-2A) but from the opposite (*si*) face of the nicotinamide ring as that to GAP in the GAPDH reaction (Section 16-2F). Figure 16-23 is a superposition of the NAD$^+$ cofactors bound to GAPDH and LDH in their respective crystal structures. The coenzyme conformations in the two enzymes are similar except that the orientations of the nicotinamide rings differ by ~180°.

Mammals have two different types of LDH subunits, the M type and the H type, which together form five tetrameric isozymes: M_4, M_3H, M_2H_2, MH_3, and H_4 (Section 7-5C). Although these hybrid forms occur in most tissues, the H-type subunit predominates in aerobic tissues such as heart muscle, while the M-type subunit predominates in tissues that are subject to anaerobic conditions such as skeletal muscle and liver. H_4 LDH has a low K_M for pyruvate and is allosterically inhibited by high levels of this metabolite,

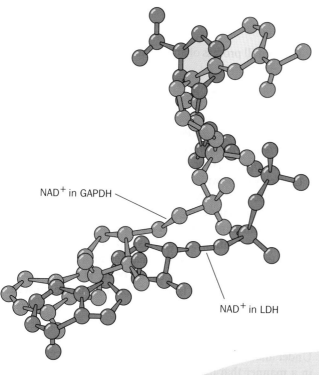

FIGURE 16-23. The superposition of the NAD⁺ molecules in the crystal structures of LDH (*red*) and GAPDH (*green*). The nicotinamide rings (*shaded*) of the two coenzymes face in opposite directions. [After Rossmann, M.G., Liljas, A., Bränden, C.-I., and Banaszak, L.J., *in* Boyer, P.D. (Ed.), *The Enzymes* (3rd ed.), Vol. 11, *p.* 85, Academic Press (1975).]

whereas the M_4 isozyme has a higher K_M for pyruvate and is not inhibited by it. The other isozymes have intermediate properties that vary with the ratio of their two types of subunits. It has therefore been proposed, although not without disagreement, that H-type LDH is better adapted to function in the oxidation of lactate to pyruvate, whereas M-type LDH is more suited to catalyze the reverse reaction.

The X-ray structure of dogfish M_4 LDH was elucidated by Michael Rossmann. The complex of LDH with a synthetic adduct of NAD⁺ and pyruvate (Fig. 16-24) suggests a

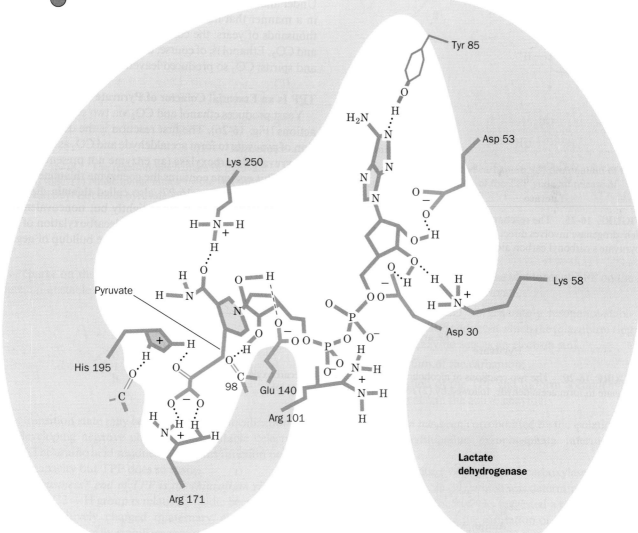

FIGURE 16-24. The binding of a synthetic covalent NAD⁺– pyruvate adduct in the active site of lactate dehydrogenase. The pyruvate residue is drawn in green and the NAD⁺ is brown.

[After Holbrook, J.J., Liljas, A., Steindel, S.J., and Rossmann, M.G., *in* Boyer, P.D. (Ed.), *The Enzymes* (3rd ed.), Vol. 11, *p.* 240, Academic Press (1975).]

cept this proton but its pK is too low to do so efficiently. It is therefore proposed that ylid formation on the enzyme's surface involves conversion of the aminopyrimidine to its imino tautomeric form in a reaction catalyzed by Glu 51 (Fig. 16-30). The imine, in turn, accepts a proton from C2, thereby forming the ylid, followed by tautomerization back to the amino form. The participation of N1′ and the 4′-amino group of the aminopyrimidine are supported by experiments showing that TPP analogs missing either of these functionalities are catalytically inactive.

Beriberi Is a Thiamine Deficiency Disease

The ability of TPP's thiazolium ring to add to carbonyl groups and act as an "electron sink" makes it the coenzyme most utilized in α-keto acid decarboxylations. TPP is also involved in decarboxylation reactions that we shall encounter in other metabolic pathways. Consequently, thiamine (**vitamin B$_1$**), which is neither synthesized nor stored in significant amounts by the tissues of most vertebrates, is required in their diets. Its deficiency in humans results in an ultimately fatal condition known as **beriberi** that is characterized by neurological disturbances causing pain, paralysis and atrophy (wasting) of the limbs, and/or cardiac failure resulting in edema (the accumulation of fluid in tissues and body cavities). Beriberi was particularly prevalent in the rice-consuming areas of the Orient because of the custom of polishing this staple grain to remove its coarse but thiamine-containing outer layers. Beriberi frequently develops in chronic alcoholics as a consequence of their penchant for drinking but not eating.

FIGURE 16-29. A portion of the X-ray structure of pyruvate decarboxylase from *Saccharomyces uvarum* (brewer's yeast) in complex with its TPP cofactor. The enzyme's identical 563-residue subunits form a tightly associated dimer, two of which associate loosely to form a tetramer. The TPP and the side chain of Glu 51 are shown in skeletal form with C green, N blue, O red, S yellow, and P orange. The TPP binds in a cavity situated between the dimer's two subunits (*light blue and purple*) where it hydrogen bonds to Glu 51. [Based on an X-ray structure by William Furey and Martin Sax, Veterans Administration Medical Center and University of Pittsburgh, Pittsburgh, Pennsylvania.]

Reduction of Acetaldehyde and Regeneration of NAD$^+$

The acetaldehyde formed by the decarboxylation of pyruvate is reduced to ethanol by NADH in a reaction catalyzed by **alcohol dehydrogenase (ADH)**. Each subunit of the tetrameric yeast ADH (YADH) binds one NADH and one Zn^{2+} ion. The Zn^{2+} ion functions to polarize the carbonyl group of acetaldehyde (Fig. 16-31), so as to stabilize the developing negative charge in the transition state of the reaction (the role of metal ions in enzymes is discussed in Section 14-1C). This facilitates the transfer of NADH's *pro-R* hydrogen (the same atom that LDH transfers) to acetaldehyde's *re* face, forming ethanol with the transferred hydrogen in the *pro-R* position (Section 12-2A).

Both homolactic and alcoholic fermentation have the same function: the anaerobic regeneration of NAD$^+$ for

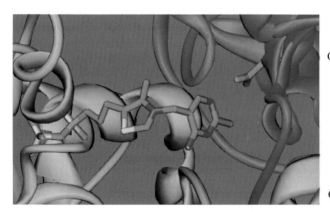

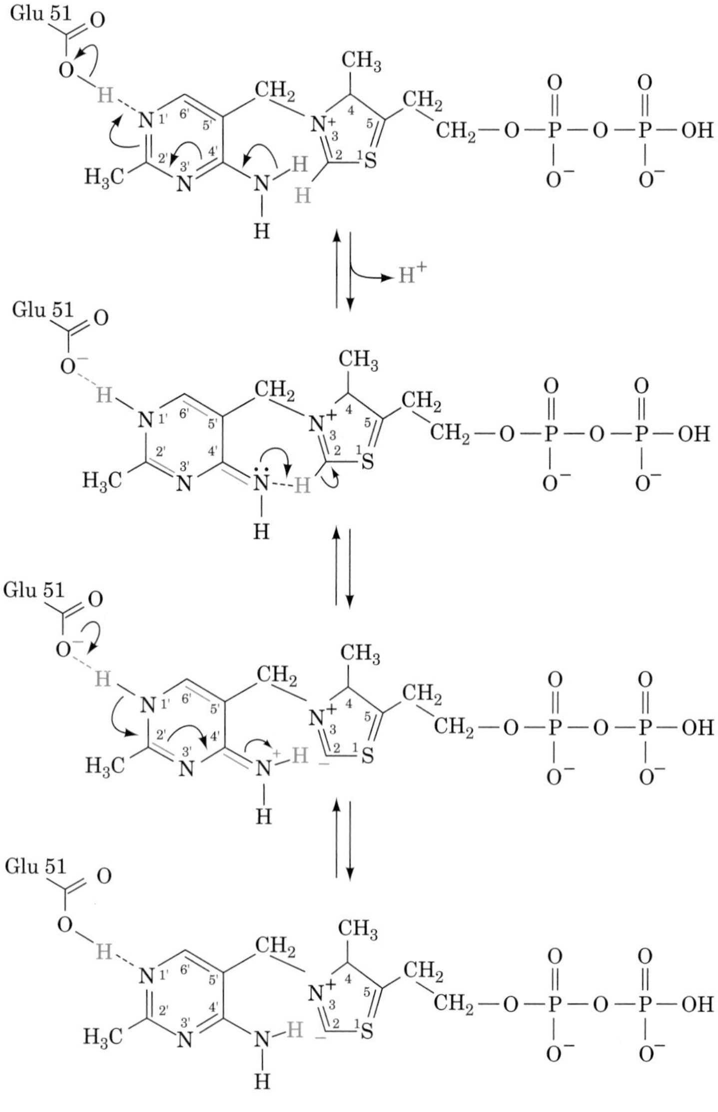

FIGURE 16-30. The formation of the active ylid form of TPP in the pyruvate decarboxylase reaction showing the participation of TPP's aminopyrimidine ring with general acid catalysis by Glu 51.

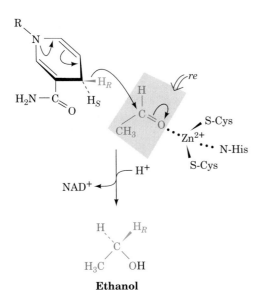

Ethanol

FIGURE 16-31. The reaction mechanism of alcohol dehydrogenase involves direct hydride transfer of the *pro-R* hydrogen of NADH to the *re* face of acetaldehyde.

continued glycolysis. Their main difference is in their metabolic products.

Mammalian liver ADH (**LADH**) functions to metabolize the alcohols anaerobically produced by intestinal flora as well as those from external sources (the direction of the ADH reaction varies with the relative concentrations of ethanol and acetaldehyde). Each subunit of this dimeric enzyme binds one NAD^+ and two Zn^{2+} ions, although only one of these ions participates directly in catalysis. There is significant amino acid sequence similarity between YADH and LADH, so it is commonly assumed that both enzymes have the same general mechanism.

C. Energetics of Fermentation

Thermodynamics permits us to dissect the process of fermentation into its component parts and to account for the free energy changes that occur. This enables us to calculate the efficiency with which the free energy of degradation of glucose is utilized in the synthesis of ATP. The overall reaction of homolactic fermentation is

$$\text{Glucose} \longrightarrow 2\text{lactate} + 2H^+$$
$$\Delta G^{\circ\prime}(\text{pH } 7) = -196 \text{ kJ} \cdot \text{mol}^{-1} \text{ of glucose}$$

($\Delta G^{\circ\prime}$ is calculated from the data in Table 3-4 using Eqs. [3.19] and [3.21] adapted for $2H^+$ ions.) For alcoholic fermentation, the overall reaction is

$$\text{Glucose} \longrightarrow 2CO_2 + 2\text{ethanol}$$
$$\Delta G^{\circ\prime} = -235 \text{ kJ} \cdot \text{mol}^{-1} \text{ of glucose}$$

Each of these reactions is coupled to the net formation of two ATPs, which requires $\Delta G^{\circ\prime} = +61 \text{ kJ} \cdot \text{mol}^{-1}$ of glucose consumed (Table 15-3). Dividing the $\Delta G^{\circ\prime}$ of ATP forma-

tion by that of lactate formation indicates that homolactic fermentation is 31% "efficient"; that is, 31% of the free energy released by this process under standard biochemical conditions is sequestered in the form of ATP. The rest is dissipated as heat, thereby making the process irreversible. Likewise, alcoholic fermentation is 26% efficient under biochemical standard state conditions. Actually, *under physiological conditions, where the concentrations of reactants and products differ from those of the standard state, these reactions have a free energy efficiency of >50%.*

Glycolysis Is Used for Rapid ATP Production

Anaerobic fermentation utilizes glucose in a profligate manner compared to oxidative phosphorylation: Fermentation results in the production of 2 ATPs per glucose, whereas oxidative phosphorylation yields 38 ATPs per glucose (Chapter 20). This accounts for Pasteur's observation that yeast consumes far more sugar when growing anaerobically than when growing aerobically (the **Pasteur effect;** Section 20-4C). However, *the rate of ATP production by anaerobic glycolysis can be up to 100 times faster than that of oxidative phosphorylation. Consequently, when tissues such as muscle are rapidly consuming ATP, they regenerate it almost entirely by anaerobic glycolysis.* (Homolactic fermentation does not really "waste" glucose since the lactate so produced is aerobically reconverted to glucose by the liver; Section 21-1C.)

Skeletal muscles consist of both **slow-twitch** (Type I) and **fast-twitch** (Type II) **fibers.** Fast-twitch fibers, so called because they predominate in muscles capable of short bursts of rapid activity, are nearly devoid of mitochondria so that they must obtain nearly all of their ATP through anaerobic glycolysis, for which they have a particularly large capacity. Muscles designed to contract slowly and steadily, in contrast, are enriched in slow-twitch fibers that are rich in mitochondria and obtain most of their ATP through oxidative phosphorylation. (Fast- and slow-twitch fibers were originally known as white and red fibers, respectively, because otherwise pale colored muscle tissue, when enriched with mitochondria, takes on the red color characteristic of their heme-containing cytochromes. However, fiber color has been shown to be an imperfect indicator of muscle physiology.)

In a familiar example, the flight muscles of migratory birds such as ducks and geese, which need a continuous energy supply, are rich in slow-twitch fibers and therefore such birds have dark breast meat. In contrast, the flight muscles of less ambitious fliers, such as chickens and turkeys, which are used only for short bursts (often to escape danger), consist mainly of fast-twitch fibers that form white meat. In humans, the muscles of sprinters are relatively rich in fast-twitch fibers, whereas distance runners have a greater proportion of slow-twitch fibers (although their muscles have the same color). World class distance runners have a remarkably high capacity to generate ATP aerobically. This was demonstrated by the noninvasive [31]P

NMR monitoring of the ATP, P_i, phosphocreatine, and pH levels in their exercising but untrained forearm muscles. These observations suggest that the muscles of these athletes are better endowed genetically for endurance exercise than those of "normal" individuals.

4. CONTROL OF METABOLIC FLUX

Living organisms, as we saw in Section 15-6, are thermodynamically open systems that tend to maintain a steady state rather than reaching equilibrium (death for living things). Thus the *flux (rate of flow) of intermediates through a metabolic pathway is constant; that is, the rates of synthesis and breakdown of each pathway intermediate maintain it at a constant concentration.* Such a state, it will be recalled, is one of maximum thermodynamic efficiency (Section 15-6B).

The concentrations of intermediates and the level of metabolic flux at which a pathway is maintained varies with the needs of the organism through a highly responsive system of precise controls. Such pathways are analogous to rivers that have been dammed to provide a means of generating electricity. Although water is continually flowing in and out of the lake formed by the dam, a relatively constant water level is maintained. The rate of water outflow from the lake is precisely controlled at the dam and is varied in response to the need for electrical power. In this section, we examine the mechanisms by which metabolic pathways in general, and the glycolytic pathway in particular, are controlled in response to biological energy needs.

A. Flux Generation

Since a metabolic pathway is a series of enzyme-catalyzed reactions, it is easiest to describe the flux of metabolites through the pathway by considering its reaction steps individually. The flux of metabolites, J, through each reaction step is the rate of the forward reaction, v_f, less that of the reverse reaction, v_r:

$$J = v_f - v_r \qquad [16.1]$$

At equilibrium, by definition, there is no flux ($J = 0$), although v_f and v_r may be quite large. At the other extreme, in reactions that are far from equilibrium, $v_f \gg v_r$, so that the flux is essentially equal to the rate of the forward reaction, $J \approx v_f$. *The flux throughout a steady state pathway is constant and is set (generated) by the pathway's rate-determining step (or steps).* Consequently, *control of flux through a metabolic pathway requires:* (1) *that the flux through this* **flux-generating step** *vary in response to the organism's metabolic requirements, and* (2) *that this change in flux be communicated throughout the pathway.* We begin our discussion with the second process, the communication of flux

changes in the rate-determining step of a pathway to the other enzymes of the pathway.

The Rates of Enzymatic Reactions Respond to Changes in Flux

Let us consider how a constant flux is maintained throughout a metabolic pathway by analyzing the response of an enzyme-catalyzed reaction to a change in the flux of the reaction preceding it. In the following steady state pathway:

$$S \xrightarrow[\text{rate-determining}]{J} A \underset{v_r}{\overset{v_f}{\rightleftarrows}} B \xrightarrow{J} P$$
$$\text{step}$$

the flux, J, through the reaction $A \rightleftharpoons B$, which must be identical to the flux through the rate-determining step, is expressed by Eq. [16.1]. If the flux of the rate-determining step increases by the amount ΔJ, the increase must be communicated to the next reaction step in the pathway by an increase in v_f (Δv_f) in order to reestablish the steady state. Qualitatively, we can see that this occurs because an increase in J causes an increase in [A], which in turn causes an increase in v_f. The amount of increase in [A] (Δ[A]) that causes v_f to increase the appropriate amount (Δv_f) is described as follows.

$$\Delta J = \Delta v_f \qquad [16.2]$$

Dividing Eq. [16.2] by J, multiplying the right side by v_f/v_f, and substituting in Eq. [16.1] yields

$$\frac{\Delta J}{J} = \frac{\Delta v_f}{v_f}\frac{v_f}{J} = \frac{\Delta v_f}{v_f}\frac{v_f}{(v_f - v_r)} \qquad [16.3]$$

which relates $\Delta J/J$, the fractional change in flux through the rate-determining step, and $\Delta v_f/v_f$, the fractional change in v_f, the forward rate of the next reaction in the pathway.

In Section 13-2A, we discussed the relationship between substrate concentration and the rates of an enzymatic reaction as expressed by the Michaelis–Menten equation:

$$v_f = \frac{V_{max}^f[A]}{K_M + [A]} \qquad [13.24a]$$

In the simplest and physiologically most common situation, [A] $\ll K_M$ so that

$$v_f = \frac{V_{max}^f[A]}{K_M} \qquad [16.4]$$

and

$$\Delta v_f = \frac{V_{max}^f \Delta[A]}{K_M} \qquad [16.5]$$

Hence,

$$\frac{\Delta v_f}{v_f} = \frac{\Delta[A]}{[A]} \qquad [16.6]$$

that is, the fractional change in reaction rate is equal to the fractional change in substrate concentration. Then, by substituting Eq. [16.6] into Eq. [16.3], we find that

$$\frac{\Delta J}{J} = \frac{\Delta[A]}{[A]} \frac{v_f}{(v_f - v_r)} \qquad [16.7]$$

This equation relates the fractional change in flux through a metabolic pathway's rate-determining step to the fractional change in substrate concentration necessary to communicate that change to the following reaction steps. *The quantity $v_f/(v_f - v_r)$ is a measure of the sensitivity of a reaction's fractional change in flux to its fractional change in substrate concentration.* This quantity is also a measure of the reversibility of the reaction, that is, how close it is to equilibrium:

1. In an irreversible reaction, v_r approaches 0 and $v_f/(v_f - v_r)$ approaches 1. The reaction therefore requires a nearly equal fractional increase in its substrate concentration in order to respond to a fractional increase in flux.

2. As a reaction approaches equilibrium, v_r approaches v_f and $v_f/(v_f - v_r)$ approaches infinity. The reaction's response to a fractional increase in flux therefore requires a much smaller fractional increase in its substrate concentration.

Consequently, *the ability of a reaction to communicate a change in flux increases as the reaction approaches equilibrium.* A series of sequential reactions that are all near equilibrium therefore have the same flux.

The Flux through a Pathway Is Controlled at Its Rate-Determining Step

The metabolic flux through an entire pathway is determined by its rate-determining step (or steps) which, by definition, is much slower than the following reaction step(s). The product(s) of the rate-determining step is therefore removed before it can equilibrate with reactant so that the rate-determining step functions far from equilibrium and has a large negative free energy change. In an analogous manner, the flow of a river can only be controlled by a dam, which creates a difference in water levels between its upstream and downstream sides; this is a situation that also has a large negative free energy change, in this case resulting from the hydrostatic pressure head. Yet, as we have just seen, the fractional change in flux, $\Delta J/J$, of a nonequilibrium reaction ($v_f \gg v_r$), is only directly proportional to the fractional change in its substrate concentration, $\Delta[A]/[A]$; that is, its substrate concentration must double (in the absence of other controlling effects) in order to double the reaction flux (Eq. [16.7]). Yet, some pathway fluxes vary by factors that are much greater than can be explained by changes in substrate concentrations. For example, glycolytic fluxes are known to vary by factors of 100 or more, whereas variations of substrate concentrations over such a large range are unknown. Consequently, although changes in substrate concentration can communicate a change in flux at the rate-determining step to the other (near equilib-

rium; $v_f \approx v_r$) reaction steps of the pathway, there must be other mechanisms that control the flux of the rate-determining step.

The flux through the rate-determining step of a pathway may be altered by several mechanisms:

1. **Allosteric control:** Many enzymes are allosterically regulated (Section 12-4) by effectors that are often substrates, products, or coenzymes in the pathway but not necessarily of the enzyme in question (feedback regulation). One such enzyme is PFK, an important glycolytic control enzyme (Section 16-4B).

2. **Covalent modification (enzymatic interconversion):** Many enzymes that control pathway fluxes have specific sites that may be enzymatically phosphorylated and dephosphorylated or covalently modified in some other way. Such enzymatic modification processes, which are themselves subject to control, greatly alter the activities of the modified enzymes. This flux control mechanism is discussed in Section 17-3.

3. **Substrate cycles:** If v_f and v_r in Eq. [16.7] represent the rates of two opposing nonequilibrium reactions that are catalyzed by different enzymes, v_f and v_r may be independently varied. The flux through such a substrate cycle, as we shall see in the next section, is more sensitive to the concentrations of allosteric effectors than is the flux through a single unopposed nonequilibrium reaction.

4. **Genetic control:** Enzyme concentrations, and hence enzyme activities, may be altered by protein synthesis in response to metabolic needs. Genetic control of enzyme concentrations is a major concern of Part V of this text.

Mechanisms 1 to 3 can respond rapidly (within seconds or minutes) to external stimuli and are therefore classified as "short-term" control mechanisms. Mechanism 4 responds more slowly to changing conditions (within hours or days in higher organisms) and is therefore referred to as a "long-term" control mechanism.

B. *Control of Glycolysis in Muscle*

Elucidation of the flux control mechanisms of a given pathway involves the determination of the pathway's regulatory enzymes controlling the rate-determining steps together with the identification of the modulators of these enzymes and their mechanism(s) of modulation. A hypothesis may then be formulated that can be tested *in vivo*. A common procedure for establishing control mechanisms involves three steps.

1. Identification of the rate-determining step(s) of the pathway. One way to do so is to measure the *in vivo* ΔG's of all the reactions in the pathway to determine how close to equilibrium they function. Those that operate far from equilibrium are potential control points; the enzymes catalyzing them may be regulated by one or

more of the mechanisms listed above. Another way of establishing the rate-determining step(s) of a pathway is to measure the effect of a known inhibitor on a specific reaction step and on the flux through the pathway as a whole. The ratio of the fractional change in the activity of the inhibited enzyme to the fractional change in the total flux will vary between 0 and 1. The closer the ratio is to 1, the more control is exerted by that particular enzyme on the total flux through the pathway.

2. *In vitro* identification of allosteric modifiers of the enzymes catalyzing the rate-determining reactions. The mechanisms by which these compounds act are determined from their effects on the enzyme's kinetics. From this information, a model of the allosteric control mechanisms for the pathway may be formulated.

3. Measurement of the *in vivo* levels of the proposed regulators under various conditions to establish whether these concentration changes are consistent with the proposed control mechanism.

Free Energy Changes in the Reactions of Glycolysis

Let us examine the thermodynamics of glycolysis with an eye towards understanding its control mechanisms. This must be done separately for each type of tissue in question because different tissues control glycolysis in different ways. We shall confine ourselves to muscle tissue. First we establish the pathway's possible control points through the identification of its nonequilibrium reactions. Table 16-1 lists the standard free energy change $(\Delta G^{\circ\prime})$ and the actual physiological free energy change (ΔG) associated with each reaction in the pathway. It is important to realize that the free energy changes associated with the reactions under standard conditions may differ dramatically from those in effect under physiological conditions. For example, the $\Delta G^{\circ\prime}$ for aldolase is $+22.8$ kJ·mol^{-1}, whereas under physiological conditions in heart muscle it is close to zero, indicating that the *in vivo* activity of aldolase is sufficient to equilibrate its substrates and products. The same is true of the GAPDH + PGK reaction series.

Only three reactions, those catalyzed by hexokinase, phosphofructokinase, and pyruvate kinase, function with large negative free energy changes in heart muscle under physiological conditions. These nonequilibrium reactions of glycolysis are the candidates for the flux-control points. The other glycolytic reactions function near equilibrium: Their forward and reverse rates are much faster than the actual flux through the pathway. Consequently, these equilibrium reactions are very sensitive to changes in the concentration of pathway intermediates and rapidly communicate any changes in flux generated at the rate-determining step(s) throughout the rest of the pathway.

Phosphofructokinase Is the Major Flux-Controlling Enzyme of Glycolysis in Muscle

In vitro kinetic studies of hexokinase, phosphofructokinase, and pyruvate kinase indicate that each is controlled

TABLE 16-1. STANDARD FREE ENERGY CHANGES ($\Delta G^{\circ\prime}$), AND PHYSIOLOGICAL FREE ENERGY CHANGES (ΔG) IN HEART MUSCLE, OF THE REACTIONS OF GLYCOLYSIS[a]

Reaction	Enzyme	$\Delta G^{\circ\prime}$ (kJ·mol^{-1})	ΔG (kJ·mol^{-1})
1	Hexokinase	−20.9	−27.2
2	PGI	+2.2	−1.4
3	PFK	−17.2	−25.9
4	Aldolase	+22.8	−5.9
5	TIM	+7.9	+4.4
6 + 7	GADPH + PGK	−16.7	−1.1
8	PGM	+4.7	−0.6
9	Enolase	−3.2	−2.4
10	PK	−23.0	−13.9

[a] Calculated from data in Newsholme, E.A., and Start, C., *Regulation in Metabolism, p. 97*, Wiley (1973).

by a variety of compounds, some of which are listed in Table 16-2. Yet, when the G6P source for glycolysis is glycogen, rather than glucose, as is often the case in skeletal muscle (Section 17-1), the hexokinase reaction is not required. *PFK, an elaborately regulated enzyme functioning far from equilibrium, evidently is the major control point for glycolysis in muscle under most conditions.*

PFK (Fig. 16-32a) is a tetrameric enzyme with two conformational states, R and T, that are in equilibrium. ATP is both a substrate and an allosteric inhibitor of PFK. Each subunit has two binding sites for ATP: a substrate site and an inhibitor site. The substrate site binds ATP equally well in either conformation, but the inhibitor site binds ATP almost exclusively in the T state. The other substrate of PFK, F6P, preferentially binds to the R state. Consequently, at high concentrations, ATP acts as a heterotropic allosteric inhibitor of PFK by binding to the T state, thereby shifting the T ⇌ R equilibrium in favor of the T state and thus decreasing PFK's affinity for F6P (this is similar to the action of 2,3-BPG in decreasing the affinity of hemoglobin for O$_2$; Section 9-2F). In graphical terms, at high concentrations of ATP, the hyperbolic (noncooperative) curve of PFK activity versus [F6P] is converted to the sigmoidal (cooperative) curve characteristic of allosteric enzymes

TABLE 16-2. SOME EFFECTORS OF THE NONEQUILIBRIUM ENZYMES OF GLYCOLYSIS

Enzyme	Inhibitors	Activators[a]
Hexokinase	G6P	
PFK	ATP, citrate, PEP	ADP, AMP, cAMP, FBP, F2,6P, F6P, NH$_4^+$, P$_i$
PK (muscle)	ATP	

[a] The activators for PFK are better described as deinhibitors of ATP because they reverse the effect of inhibitory concentrations of ATP.

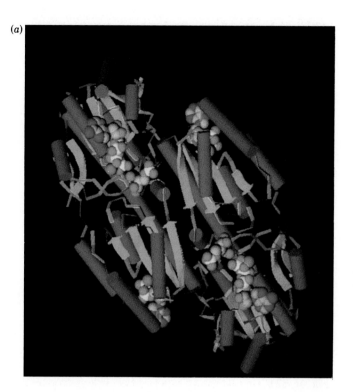

(a)

FIGURE 16-32. The X-ray structure of PFK from *Bacillus stearothermophilus*. (*a*) A ribbon diagram showing two subunits of the tetrameric molecule (related by a twofold axis perpendicular to the page through the center of the figure). Each of the subunits in the protein is in association with its substrates F6P (*near the center of each subunit*) and ATP·Mg^{2+} (*lower right and upper left;* the green balls represent Mg^{2+}), together with the activator ADP·Mg^{2+} (*top right and lower left, in the rear*). [Courtesy of Arthur Lesk, Cambridge University and EMBL. X-Ray structure determined by Phillip Evans.] (*b*) A superposition of those segments of the T state (*blue*) and R-state (*red*) enzymes that undergo a large conformational rearrangement upon the T→R allosteric transition (indicated by the arrows). Residues of the R state structure are marked by a prime. Also shown are bound ligands: the nonphysiological inhibitor 2-phosphoglycolate (**PGC;** a PEP analog) for the T state, and the cooperative substrate F6P and the activator ADP for the R state. [After Schirmer, T. and Evans, P. R., *Nature,* **343,** 142 (1990).]

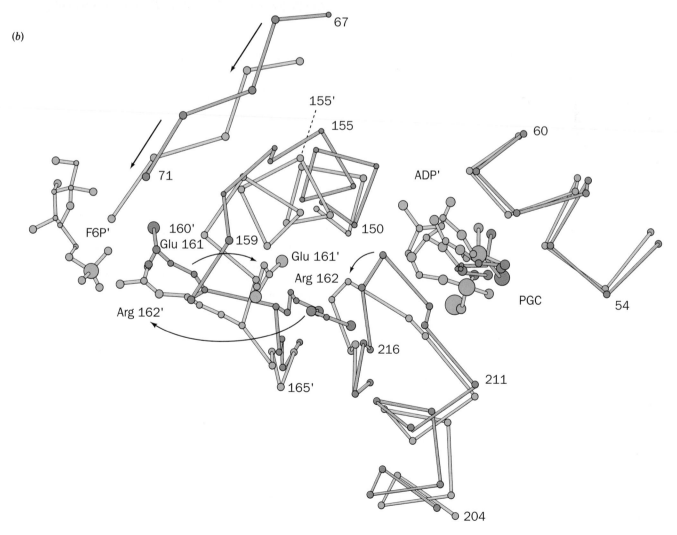

(b)

(Fig. 16-33; cooperative and noncooperative processes are discussed in Section 9-1B). For example, when [F6P] = 0.5 m*M* (the dashed line in Fig. 16-33), the enzyme is maximally active, but in the presence of 1 m*M* ATP, the activity drops to 15% of its original level (a nearly sevenfold decrease). [Actually, the most potent allosteric effector of PFK is **fructose-2,6-bisphosphate (F2,6P)**. We discuss the role of F2,6P in controlling PFK activity when we study the mechanism by which the liver maintains blood glucose concentrations (Section 17-3F).]

Structural Basis for PFK's Allosteric Change in F6P Affinity

The X-ray structures of PFK from several organisms have been determined for both the R and the T states by Phillip Evans. The R state of PFK is homotropically stabilized by the binding of its substrate fructose-6-phosphate (F6P). In the R state of *Bacillus stearothermophilus* PFK, the side chain of Arg 162 forms a salt bridge with the phosphoryl group of an F6P bound in an active site of another subunit (Fig. 16-32*b*). However, Arg 162 is located at the end of a helical turn that unwinds upon transition to the T state. The positively charged side chain of Arg 162 thereby swings away and is replaced by the negatively charged side chain of Glu 161. As a consequence, the doubly negative phosphoryl group of F6P has a greatly diminished affinity for the T-state enzyme. The unwinding of this helical turn, which is obligatory for the R→T transition, is prevented by the binding of the activator ADP to its effector site on PFK. Evidently, the same conformational shift is responsible for both the homotropic and the heterotropic allosteric effects in PFK.

AMP Overcomes the ATP Inhibition of PFK

Direct allosteric regulation of PFK by ATP may superficially appear to be the means by which glycolytic flux is controlled. After all, when [ATP] is high as a result of low metabolic demand, PFK is inhibited and flux through the pathway is low; conversely, when [ATP] is low, flux through the pathway is high and ATP is synthesized to replenish the pool. Consideration of the physiological variation in ATP concentration, however, indicates that the situation must be more complex. The metabolic flux through glycolysis may vary by 100-fold or more, depending on the metabolic demand for ATP. However, measurements of [ATP] *in vivo* at various levels of metabolic activity indicate that [ATP] varies <10% between rest and vigorous exertion. Yet, *there is no known allosteric mechanism that can account for a 100-fold change in flux of a nonequilibrium reaction with only 10% change in effector concentration.* Thus, some other mechanism, or mechanisms, must be responsible for controlling glycolytic flux.

The inhibition of PFK by ATP is relieved by AMP. This results from AMP's preferential binding to the R state of PFK. If a PFK solution containing 1 m*M* ATP and 0.5 m*M* F6P is brought to 0.1 m*M* in AMP, the activity of PFK rises

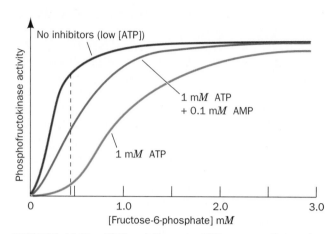

FIGURE 16-33. PFK activity versus F6P concentration under various conditions: blue, no inhibitors (low, noninhibitory [ATP]); green, 1 m*M* ATP (inhibitory); and red, 1 m*M* ATP + 0.1 m*M* AMP. [After data from Mansour, T. E. and Ahlfors, C. E., *J. Biol. Chem.* **243**, 2523–2533 (1968).]

from 15 to 50% of its maximal activity, a threefold increase (Fig. 16-33).

[ATP] decreases by only 10% in going from a resting state to one of vigorous activity because it is buffered by the action of two enzymes: creatine kinase (Section 15-4C) and, of particular importance to this discussion, **adenylate kinase** (also known as **myokinase**). Adenylate kinase catalyzes the reaction

$$2ADP \rightleftharpoons ATP + AMP \qquad K = \frac{[ATP][AMP]}{[ADP]^2} = 0.44$$

which rapidly equilibrates the ADP resulting from ATP hydrolysis in muscle contraction with ATP and AMP.

In muscle, [ATP] is ~50 times [AMP] and ~10 times [ADP] so that, *as a result of the adenylate kinase reaction, a 10% decrease in [ATP] will cause over a fourfold increase in [AMP] (see Problem 11 in this chapter).* Consequently, a metabolic signal consisting of a decrease in [ATP] too small to relieve PFK inhibition is amplified significantly by the adenylate kinase reaction, which increases [AMP] by an amount sufficient to produce a much larger increase in PFK activity.

Substrate Cycling Can Increase Flux Sensitivity

Even though a mechanism exists for amplifying the effect of a small change in [ATP] by producing a larger change in [AMP], a fourfold increase in [AMP] would allosterically increase the activity of PFK by only ~10-fold, an amount insufficient to account for the observed 100-fold increase in glycolytic flux. Small changes in effector concentration (and therefore v_f) can only cause relatively large changes in the flux through a reaction ($v_f - v_r$) if the reaction is functioning close to equilibrium. The reason for this high sensitivity is that for such reactions, the term $v_f/(v_f - v_r)$ in Eq. [16.7] is large, that is, the reverse reaction contributes significantly to the value of the net flux. This is not the case for the PFK reaction.

Such equilibrium-like conditions may be imposed on a nonequilibrium reaction if a second enzyme catalyzes the regeneration of substrate from product in a thermodynamically favorable manner. Then v_r is no longer negligible compared to v_f (although increasing v_r forces the forward reaction even further out of equilibrium than it would otherwise be). This situation requires that the forward process (formation of FBP from F6P) and reverse process (breakdown of FBP to F6P) be accomplished by different reactions since the laws of thermodynamics would otherwise be violated. In the following paragraphs, we discuss the nature of such **substrate cycles.**

Under physiological conditions, the reaction catalyzed by PFK:

Fructose-6-phosphate + ATP $\longrightarrow$
$$\text{fructose-1,6-bisphosphate} + \text{ADP}$$

is highly exergonic ($\Delta G = -25.9$ kJ·mol^{-1}, Table 16-1). Consequently, the back reaction has a negligible rate compared to the forward reaction. **Fructose-1,6-bisphosphatase (FBPase),** however, which is present in many mammalian tissues (and which is an essential enzyme in gluconeogenesis; Section 21-1), catalyzes the exergonic hydrolysis of FBP ($\Delta G = -8.6$ kJ·mol^{-1}):

Fructose-1,6-bisphosphate + H$_2$O $\longrightarrow$
$$\text{fructose-6-phosphate} + \text{P}_i$$

Note that the combined reactions catalyzed by PFK and FBPase result in net ATP hydrolysis:

$$\text{ATP} + \text{H}_2\text{O} \Longleftrightarrow \text{ADP} + \text{P}_i$$

Such a set of opposing reactions is known as a substrate cycle because it cycles a substrate to an intermediate and back again. When this set of reactions was discovered, it was referred to as a **futile cycle** since its net result seemed to be the useless consumption of ATP. In fact, when it was found that the PFK activators AMP and F2,6P allosterically inhibit FBPase it was suggested that only one of these enzymes was functional in a cell under any given set of conditions. It was subsequently demonstrated, however, that both enzymes often function simultaneously at significant rates.

Substrate Cycling Can Account for Glycolytic Flux Variation

Eric Newsholme has proposed that substrate cycles are not at all "futile," but rather, have a regulatory function. The *in vivo* activities of enzymes and concentrations of metabolites are extremely difficult to measure so that their values are rarely known accurately. However, let us make the physiologically reasonable assumption that a fourfold increase in [AMP], resulting from the adenylate kinase reaction, causes PFK activity (v_f) to increase from 10 to 90%

of its maximum and FBPase activity (v_r) to decrease from 90 to 10% of its maximum. The maximum activity of muscle PFK is known from *in vitro* studies to be ~10-fold greater than that of muscle FBPase. Hence, if we assign full activity of PFK to be 100 arbitrary units, then full activity of FBPase is 10 such units. The flux through the PFK reaction in glycolysis under conditions of low [AMP] is

$$J_{low} = v_f(low) - v_r(low) = 10 - 9 = 1$$

where v_f is catalyzed by PFK and v_r by FBPase. The flux under conditions of high [AMP] is

$$J_{high} = v_f(high) - v_r(high) = 90 - 1 = 89$$

Substrate cycling could therefore amplify the effect of changes in [AMP] on the net rate of phosphorylation of F6P. Without the substrate cycle, a fourfold increase in [AMP] increases the net flux by about ninefold, whereas with the cycle the same increase in [AMP] causes a $J_{high}/J_{low} = 89/1 \approx 90$-fold increase in net flux. Consequently, under the above assumptions, *a 10% change in [ATP] could stimulate a 90-fold change in flux through the glycolytic pathway by a combination of the adenylate kinase reaction and substrate cycles.*

Physiological Impact of Substrate Cycling

Substrate cycling, if it has a regulatory function, does not increase the maximum flux through a pathway. On the contrary, it functions to decrease its minimum flux. In a sense, the substrate is put into a "holding pattern." In the case described above, *the cycling of substrate is the energetic "price" that a muscle must pay to be able to change rapidly from a resting state, in which substrate cycling is maximal, to one of sustained high activity.* However, the rate of substrate cycling may itself be under hormonal or nervous control so as to increase the sensitivity of the metabolic system under conditions when high activity (fight or flight) is anticipated (we address the involvement of hormones in metabolic regulation in Sections 17-3E and F).

In some tissues, substrate cycles function to produce heat. For example, many insects require a thoracic temperature of 30°C to be able to fly. Yet bumblebees are capable of flight at ambient temperatures as low as 10°C. Bumblebee flight muscle FBPase has a maximal activity similar to that of its PFK (10-fold greater than our example for mammalian muscle); furthermore, unlike all other known muscle FBPases, it is not inhibited by AMP. This permits the FBPase and PFK of bumblebee flight muscle to be highly active simultaneously so as to generate heat. Since the maximal rate of FBP cycling possible in bumblebee flight muscle generates only 10 to 15% of the required heat, however, other mechanisms of thermogenesis must also be operative. Nevertheless, FBP cycling is probably signficant because, unlike bumblebees, honeybees, which have no FBPase activity in their flight muscles, cannot fly when the temperature is low.

combined with X-ray structural studies. The glycolytic enzymes exhibit stereospecificity in the reactions that they catalyze. In at least two kinases, phosphoryl transfer from substrate to water is prevented by substrate-induced conformational changes that form the active site and exclude water from it.

The NAD$^+$ consumed in the formation of 1,3-BPG must be regenerated if glycolysis is to continue. In the presence of O_2, NAD$^+$ is regenerated by oxidative phosphorylation in the mitochondria. Under anaerobic conditions in muscle, pyruvate is reduced by NADH, yielding lactate and NAD$^+$ in a reaction catalyzed by lactate dehydrogenase. In many muscles, particularly during strenuous activity, the process of homolactic fermentation is a major free energy source. In anaerobic yeast, NAD$^+$ is regenerated by alcoholic fermentation in two reactions. First pyruvate is decarboxylated to acetaldehyde by pyruvate decarboxylase, an enzyme that requires thiamine pyrophosphate as a cofactor. The acetaldehyde is then reduced by NADH to form ethanol and NAD$^+$ in a reaction catalyzed by alcohol dehydrogenase.

The flux through a reaction that is close to equilibrium is very sensitive to changes in substrate concentration. Hence, the steady state flux through a metabolic pathway can only be controlled by a nonequilibrium reaction. Nonequilibrium reactions are regulated by allosteric interactions, substrate cycles, covalent modification, and genetic (long-term) control mechanisms. In muscle glycolysis, phosphofructokinase (PFK) catalyzes the flux-generating step. Although PFK is inhibited by high concentrations of one of its substrates, ATP, the 10% variation of [ATP] over the range of metabolic activity has insufficient influence on PFK activity to account for the observed 100-fold range in glycolytic flux. [AMP] has a four fold variation in response to the 10% variation of [ATP] through the action of adenylate kinase. Although AMP relieves the ATP inhibition of PFK, its concentration variation is also insufficient to account for the observed glycolytic flux range. However, the product of the PFK reaction, fructose-1,6-bisphosphate, is hydrolyzed to F6P by FBPase, which is inhibited by AMP. The substrate cycle catalyzed by these two enzymes confers, at least in principle, the necessary sensitivity of the glycolytic flux to variations in [AMP]. Substrate cycling is an important source of nonshivering thermogenesis.

Digestion of carbohydrates yields glucose as the primary product. Other prominent products are fructose, galactose, and mannose. These monosaccharides are metabolized through their conversion to glycolytic intermediates.

REFERENCES

General

Fersht, A., *Enzyme Structure and Mechanism* (2nd ed.), Freeman (1985).

Fruton, J.S., *Molecules and Life: Historical Essays on the Interplay of Chemistry and Biology,* Wiley–Interscience (1974). [Includes a detailed historical account of the elucidation of fermentation.]

Saier, M.H., Jr., *Enzymes in Metabolic Pathways,* Chapter 5, Harper & Row (1987).

Walsh, C., *Enzymatic Reaction Mechanisms,* Freeman (1979).

Enzymes of Glycolysis

The Enzymes of Glycolysis: Structure, Activity and Evolution, *Phil. Trans. R. Soc. London Ser. B* **293**, 1–214 (1981). [A collection of authoritative discussions on the enzymes of glycolysis. Also available from the Royal Society as a bound volume.]

Anderson, C.M., Zucker, F.H., and Steitz, T.A., Space-filling models of kinase clefts and conformation changes, *Science* **204**, 375–380 (1979). [The examination of computer-generated space-filling models demonstrates that kinases have deep active site clefts that close upon binding substrate.]

Bennett, W.S., Jr. and Steitz, T.A., Glucose-induced conformational change in yeast hexokinase, *Proc. Natl. Acad. Sci.* **75**, 4848–4852 (1978).

Biesecker, G., Harris, J.I., Thierry, J.C., Walker, J.E., and Wonacott, A.J., Sequence and structure of D-glyceraldehyde-3-phosphate dehydrogenase from *Bacillus stearothermophilus, Nature* **266**, 328–333 (1977).

Boyer, P.D. (Ed.), *The Enzymes* (3rd ed.), Vols. 5–9 and 13, Academic Press (1972–1976). [Contains detailed reviews of the various glycolytic enzymes.]

Cleland, W.W. and Kreevoy, M.M., Low-barrier hydrogen bonds and enzymic catalysis, *Science* **264**, 1887–1890 (1994); *and* Gerlt, J.A. and Gassman, P.G., Understanding the rates of certain enzyme-catalyzed reactions: Proton abstraction from carbon acids, acyl-transfer reactions, and displacement of phosphodiesters, *Biochemistry* **32**, 11943–11952 (1993).

Davenport, R C., Bash, P.A., Seaton, B.A., Karplus, M., Petsko, G.A., and Ringe, D., Structure of the triosephosphate isomerase–phosphoglycohydroxamate complex: An analogue of the intermediate on the reaction pathway, *Biochemistry* **30**, 5821–5826 (1991); *and* Lolis, E. and Petsko, G.A., Crystallographic analysis of the complex between triosephosphate isomerase and 2-phosphoglycolate at 2.5-Å resolution: Implications for catalysis, *Biochemistry* **29**, 6619–6625 (1990).

Delivoria-Papadopoulos, M., Oska, F.A., and Gottlieb, A.J., Oxygen–hemoglobin dissociation curves: effect of inherited enzyme defects of the red cell, *Science* **165**, 601–602 (1969).

Evans, P.R. and Hudson, P.J., Structure and control of phosphofructokinase from *Bacillus stearothermophilus, Nature* **279**, 500–504 (1979).

Farber, G.K., An α/β-barrel full of evolutionary trouble, *Curr. Opin. Struct. Biol.* **3**, 409–412 (1993); *and* Farber, G.K. and Petsko, G.A., The evolution of a/β barrel enzymes, *Trends Biochem. Sci.* **15**, 228–234 (1990).

Gamblin, S.J., Davies, G.J., Grimes, J M., Jackson, R.M., Littlechild, J.A., and Watson, H.C., Activity and specificity of human aldolases, *J. Mol. Biol.* **219**, 573–576 (1991). [The X-ray structure of aldolase extended to 2.0-Å resolution.]; *and* Sygusch, J., Beaudry, D., and Allaire, M., Molecular architecture of rabbit skeletal muscle aldolase at 2.7-Å resolution, *Proc. Natl. Acad. Sci.* **84**, 7846–7850 (1987).

Harlos, K., Vas, M., and Blake, C.C.F., Crystal structure of the binary complex of pig muscle phosphoglycerate kinase and its substrate 3-phospho-D-glycerate, *Proteins* **12**, 133–144 (1992);

and Banks, R.D., Blake, C.C.F., Evans, P.R., Rice, D.W., Hardy, G.W., Merritt, M., and Phillips, A.W., Sequence, structure and activity of phosphoglycerate kinase: A possible hinge bending enzyme, *Nature* **279**, 773–777 (1979).

Joseph, D., Petsko, G.A., and Karplus, M., Anatomy of a conformational change: Hinged "lid" motion of the triosephosphate isomerase loop, *Science* **249**, 1425–1428 (1990).

Knowles, J.R., Enzyme catalysis: not different, just better, *Nature* **350**, 121–124 (1991). [A lucid discussion of TIM's catalytic mechanism.]

Kuby, S.A (Ed.)., *A Study of Enzymes,* Vol. II, CRC Press (1991). [Chapters 17, 18, 19, and 20 discuss the mechanisms of adenylate kinase, PFK, PGI and TIM, and aldolase, respectively. Chapter 4 discusses thiamine-dependent reaction mechanisms.]

Leboida, L. and Stec, B., Mechanism of enolase: The crystal structure of enolase-Mg^{2+}-2-phosphoglycerate/phosphoenolpyruvate complex at 2.2-Å resolution, *Biochemistry* **30**, 2817–2822 (1991).

Littlechild, J.A. and Watson, H.C., A data-based reaction mechanism for type I fructose bisphosphate aldolase, *Trends Biochem. Sci.* **18**, 36–39 (1993).

Marsh, J.J. and Lebherz, H.G., Fructose-bisphosphate aldolases: An evolutionary history, *Trends Biochem. Sci.* **17**, 110–113 (1992).

Maurer, P.J. and Nowak, T., Fluoride inhibition of yeast enolase. 1. Formation of ligand complexes, *Biochemistry* **20**, 6894–6900 (1981); *and* Nowak, T. and Maurer, P.J., Fluoride inhibition of yeast enolase. 2. Structural and kinetic properties of ligand complexes detemined by nuclear relaxation rate studies, *Biochemistry* **20**, 6901–6911 (1981).

Muirhead, H., Pyruvate kinase, *in* Jurnak, F.A. and McPherson, A. (Eds.), *Biological Macromolecules and Assemblies,* Vol. 3., *pp.* 141–186, Wiley (1987).

Muirhead, H. and Watson, H. Glycolytic enzymes: from hexose to pyruvate, *Curr. Opin. Struct. Biol.,* **2**, 870–876 (1992).

Seeholzer, S.H., Phosphoglucose isomerase: A ketol isomerase with aldol C2-epimerase activity, *Proc. Natl. Acad. Sci.* **90**, 1237–1241 (1993).

Takahashi, I., Takahashi, Y., and Hori, K., Site-directed mutagenesis of human aldolase isozymes: The role of Cys 72 and Cys 338 residues of aldolase A and of the carboxy-terminal Tyr residue of aldolases A and B., *J. Biochem.* **105**, 281–286 (1989).

Enzymes of Anaerobic Fermentation

Boyer, P.D. (Ed.), *The Enzymes* (3rd ed.), Vol. 11, Academic Press (1975). [Contains authoritative reviews on alcohol dehydrogenase, lactate dehydrogenase, and the evolutionary and structural relationships among the dehydrogenases.]

Dyda, F., Furey, W., Swaminathan, S., Sax, M., Farrenkopf, B. and Jordan, F., Catalytic centers in the thiamin diphosphate dependent enzyme pyruvate decarboxylase at 2.4-Å resolution, *Biochemistry* **32**, 6165-6170 (1993).

Golbik, R., Neef, H., Hübner, G., König, S., Seliger, B., Meshalkina, L., Kochetov, G.A., and Schellenberger, A., Function of the aminopyrimidine part in thiamine pyrophosphate enzymes, *Bioorganic Chemistry* **19**, 10-17 (1991).

Park, J.H., Brown, R.L., Park, C.R., Cohn, M., and Chance, B., Energy metabolism in the untrained muscle of elite runners as observed by ^{31}P magnetic resonance spectroscopy: evidence suggesting a genetic endowment for endurance exercise. *Proc. Natl. Acad. Sci.* **85**, 8780–8785 (1988).

Control of Metabolic Flux

Boscá, L. and Corredor, C., Is phosphofructokinase the rate-limiting step of glycolysis? *Trends Biochem. Sci.* **9**, 372–373 (1984).

Crabtree, B. and Newsolme, E.A., A systematic approach to describing and analyzing metabolic control systems, *Trends Biochem. Sci.* **12**, 4–12 (1987).

Fell, D.A., Metabolic control analysis: a survey of its theoretical and experimental development, *Biochem. J.* **286**, 313–330 (1992).

Kacser, H. and Porteous, J.W., Control of metabolism: what do we have to measure?, *Trends Biochem. Sci.* **12**, 5–14 (1987).

Katz, J. and Rognstad, R., Futile cycling in glucose metabolism, *Trends Biochem. Sci.* **3**, 171–174 (1978).

Lardy, H. and Schrago, E., Biochemical aspects of obesity, *Annu. Rev. Biochem.* **59**, 689–710 (1990).

Newsholme, E.A., Challiss, R.A.J., and Crabtree, B., Substrate cycles: their role in improving sensitivity in metabolic control, *Trends Biochem. Sci.* **9**, 277–280 (1984).

Perutz, M.F., Mechanism of cooperativity and allosteric regulation in proteins, *Q. Rev. Biophys.* **22**, 139–236 (1989). [Section 6 discusses PFK.]

Pettigrew, D.W. and Frieden, C., Rabbit muscle phosphofructokinase. A model for regulatory kinetic behavior, *J. Biol. Chem.* **254**, 1896–1901 (1979).

Schirmer, T. and Evans, P.R., Structural basis of the allosteric behaviour of phosphofructokinase, *Nature* **343**, 140–145 (1990).

Metabolism of Hexoses Other Than Glucose

Scriver, C.R., Beaudet, A.L., Sly, W.S., and Valle, D. (Eds.), *The Metabolic Basis of Inherited Disease* (6th ed.), McGraw–Hill, New York (1989). [Chapters 11 and 13 discuss fructose and galactose metabolism and their genetic disorders.]

PROBLEMS

1. Write out the reactions of the glycolytic pathway from glucose to lactate using structural formulas for all intermediates. Learn the names of these intermediates and the enzymes that catalyze the reactions.

2. $\Delta G°'$ for the aldolase reaction is $+22.8$ kJ·mol^{-1}. In the cell, at 37°C, the mass action ratio [DHAP]/[GAP] = 5.5. Calculate the equilibrium ratio of [FBP]/[GAP] when [GAP] is (a) $2 \times 10^{-5} M$ and (b) $10^{-3} M$.

However, the PP_i formed is hydrolyzed in a highly exergonic reaction by the omnipresent enzyme **inorganic pyrophosphatase.** The overall reaction for the formation of UDPG is therefore also highly exergonic:

	$\Delta G^{\circ\prime}(kJ \cdot mol^{-1})$
G1P + UTP $\rightleftharpoons$ UDPG + PP_i	~ 0
H_2O + PP_i $\longrightarrow$ $2P_i$	-33.5
Overall G1P + UTP $\longrightarrow$ UDPG + $2P_i$	-33.5

The cleavage of a nucleoside triphosphate to form PP_i is a common biosynthetic strategy. The free energy of PP_i hydrolysis can then be utilized together with the free energy of nucleoside triphosphate hydrolysis to drive an otherwise endergonic reaction to completion (Section 15-4C).

B. Glycogen Synthase

In the next step of glycogen synthesis, the glycogen synthase reaction, the glucosyl unit of UDPG is transferred to the C4—OH group on one of glycogen's nonreducing ends to form an $\alpha(1\rightarrow4)$-glycosidic bond (Fig. 17-7). The glycogen synthase reaction, like those of glycogen phosphorylase and lysozyme, is thought to involve a glucosyl oxonium ion transition state since it is also inhibited by 1,5-gluconolactone, an analog that mimics the oxonium ion's half-chair geometry.

The $\Delta G^{\circ\prime}$ for the glycogen synthase reaction is -13.4 $kJ \cdot mol^{-1}$, making the overall reaction spontaneous under the same conditions that glycogen breakdown by glycogen phosphorylase is also spontaneous. The rates of both reactions may then be independently controlled. There is, however, an energetic price for doing so. In this case, *for each molecule of G1P that is converted to glycogen and then regenerated, one molecule of UTP is hydrolyzed to UDP and P_i. The cyclic synthesis and breakdown of glycogen is therefore not a perpetual motion "machine" but, rather, is an "engine" that is powered by UTP hydrolysis.* The UTP is replenished through a phosphate-transfer reaction mediated by **nucleoside diphosphate kinase** (Section 26-3B):

$$UDP + ATP \rightleftharpoons UTP + ADP$$

so that UTP hydrolysis is energetically equivalent to ATP hydrolysis.

Glycogen synthase cannot simply link together two glucose residues; it can only extend an already existing $\alpha(1\rightarrow4)$-linked glucan chain. How, then, is glycogen synthesis initiated? The answer is that the first step in glycogen synthesis is the attachment of a glucose residue to the Tyr 194 OH group of a protein named **glycogenin** by a **tyrosine glucosyltransferase.** Glycogenin then autocatalytically extends the glucan chain by up to seven additional UDP–glucose–supplied residues, forming a "primer" for the initiation of glycogen synthesis. Only at this point does glycogen synthase commence glycogen synthesis, which it initiates on the "primer" while tightly complexed to glycogenin. However, these proteins dissociate after the growing glycogen granule has reached some minimum size. Analysis of glycogen granules for glycogenin and glycogen synthase shows that they are present in a 1 : 1 ratio. This implies

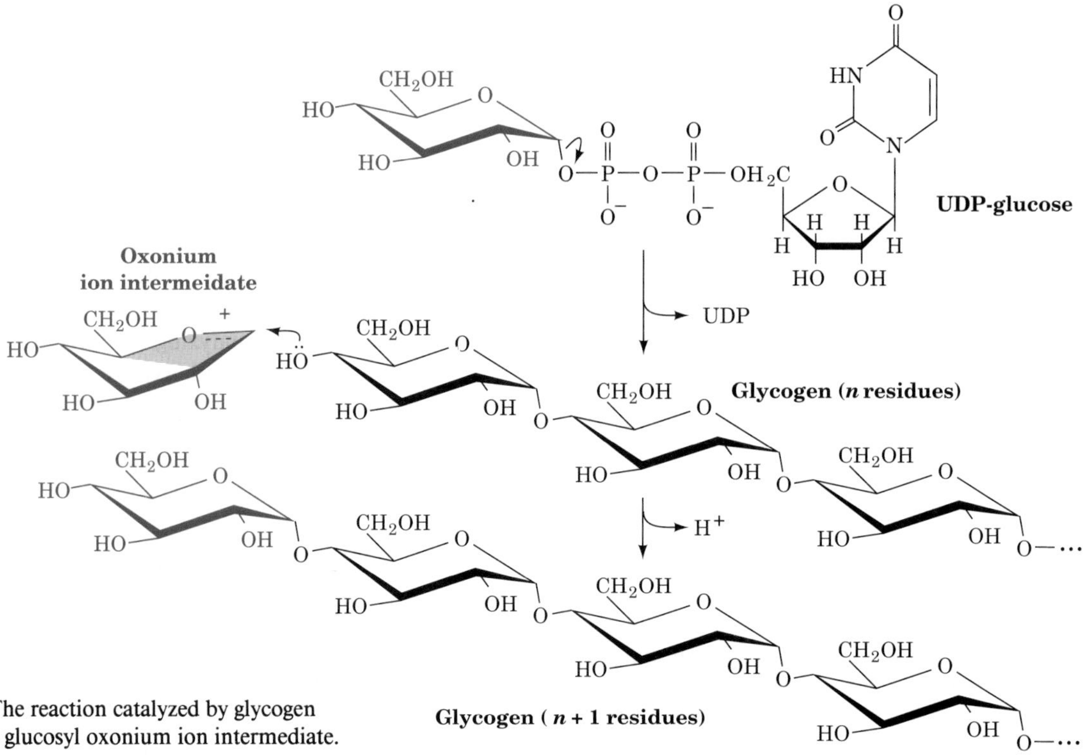

FIGURE 17-7. The reaction catalyzed by glycogen synthase involves a glucosyl oxonium ion intermediate.

that each glycogen granule, which contains a single glycogen molecule, has but one molecule each of glycogenin and glycogen synthase.

C. Glycogen Branching

Glycogen synthase catalyzes only $\alpha(1\rightarrow4)$-linkage formation to yield α-amylose. Branching to form glycogen is accomplished by a separate enzyme, **amylo-(1,4→1,6)-transglycosylase (branching enzyme),** which is distinct from glycogen debranching enzyme. Branches are created by the transfer of terminal chain segments consisting of ~7 glucosyl residues to the C6—OH groups of glucose residues on the same or another glycogen chain (Fig. 17-8). Each transferred segment must come from a chain of at least 11 residues and the new branch point must be at least 4 residues away from other branch points.

Debranching (Section 17-1C) involves breaking and reforming $\alpha(1\rightarrow4)$-glycosidic bonds and only the hydrolysis of $\alpha(1\rightarrow6)$-glycosidic bonds; branching, on the other hand, involves breaking $\alpha(1\rightarrow4)$-glycosidic bonds and reforming $\alpha(1\rightarrow6)$ linkages. The fact that glycogen debranching requires two types of reactions and branching requires only one is explained by the thermodynamics of the system. The free energy of hydrolysis of an $\alpha(1\rightarrow4)$-glycosydic bond is -15.5 kJ·mol^{-1}, whereas that of an $\alpha(1\rightarrow6)$-glycosydic

bond is only -7.1 kJ·mol^{-1}. Consequently, the hydrolysis of an $\alpha(1\rightarrow4)$-glycosidic bond drives the synthesis of an $\alpha(1\rightarrow6)$-glycosidic bond, but the reverse reaction is endergonic.

3. CONTROL OF GLYCOGEN METABOLISM

We have just seen that both glycogen synthesis and breakdown are exergonic under the same physiological conditions. If both pathways operate simultaneously, however, all that is achieved is wasteful hydrolysis of UTP. This situation is similar to that of the phosphofructokinase–fructose bisphosphatase substrate cycle (Section 16-4B). Glycogen phosphorylase and glycogen synthase therefore must be under stringent control such that glycogen is either synthesized or utilized according to cellular needs. The astonishing mechanism of this control is the next topic of our discussion. It involves not only allosteric regulation and substrate cycles but enzyme-catalyzed covalent modification of both glycogen synthase and glycogen phosphorylase. The covalent modification reactions are themselves ultimately under hormonal control through an enzymatic cascade.

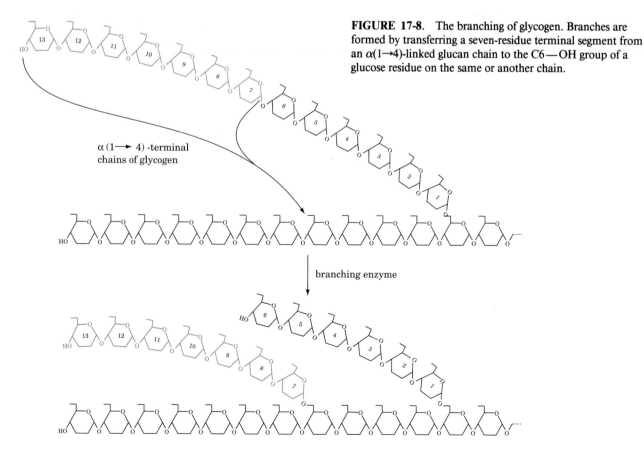

FIGURE 17-8. The branching of glycogen. Branches are formed by transferring a seven-residue terminal segment from an $\alpha(1\rightarrow4)$-linked glucan chain to the C6—OH group of a glucose residue on the same or another chain.

$\alpha(1\longrightarrow4)$-terminal chains of glycogen

branching enzyme

A. Direct Allosteric Control of Glycogen Phosphorylase and Glycogen Synthase

As we saw in Section 16-4A, the net flux of reactants, J, through a step in a metabolic pathway is the difference between the forward and reverse reaction velocities, v_f and v_r. The variation in the flux through any step in a pathway with a change in substrate concentration maximizes as that reaction step approaches equilibrium ($v_f \approx v_r$; Eq. [16.7]). The flux through a near-equilibrium reaction is therefore all but uncontrollable. As we have seen for the case of PFK and FBPase, however, *precise flux control of a pathway is possible when an enzyme functioning far from equilibrium is opposed by a separately controlled enzyme. Then, v_f and v_r vary independently. In fact, under these circumstances, even the flux direction is controlled if v_r can be made larger than v_f.* Exactly this situation occurs in glycogen metabolism through the opposition of the glycogen phosphorylase and glycogen synthase reactions. The rates of both of these reactions are under allosteric control by effectors that include ATP, G6P, and AMP. In muscle, glycogen phosphorylase is activated by AMP and inhibited by ATP and G6P (Fig. 17-9, *left*). Glycogen synthase, on the other hand, is activated by G6P. When there is high demand for ATP (low [ATP], low [G6P], and high [AMP]), glycogen phosphorylase is stimulated and glycogen synthase is inhibited, so flux

through this pathway favors glycogen breakdown. When [ATP] and [G6P] are high, the reverse is true and glycogen synthesis is favored.

The structural differences between the active (R) and inactive (T) conformations of glycogen phosphorylase (Fig. 17-10a) are now fairly well understood in terms of the symmetry model of allosterism (Section 9-4B). The T-state enzyme has a buried active site and hence a low affinity for its substrates, whereas the R-state enzyme has an accessible catalytic site and a high-affinity phosphate binding site.

AMP promotes phosphorylase's T (*inactive*)→R (*active*) conformational shift by binding to the R state of the enzyme at its allosteric effector site (Fig. 17-9, *left*). In doing so, AMP's adenine, ribose, and phosphate groups bind to separate segments of the polypeptide chain so as to link the active site, the subunit interface, and the N-terminal region (Fig. 17-10b), the latter having undergone a large conformational shift (36 Å for Ser 14) from its position in the T-state enzyme (Fig. 17-10a). AMP binding also causes glycogen phosphorylase's tower helices (Figs. 17-2 and 17-10) to tilt and pull apart so as to achieve a more favorable packing. These tertiary movements trigger a concerted T→R transition, which largely consists of a ~10° relative rotation of the two subunits about an axis at the subunit interface that is perpendicular to the dimer's twofold axis of symmetry. The enzyme's two-fold symmetry is thereby preserved in ac-

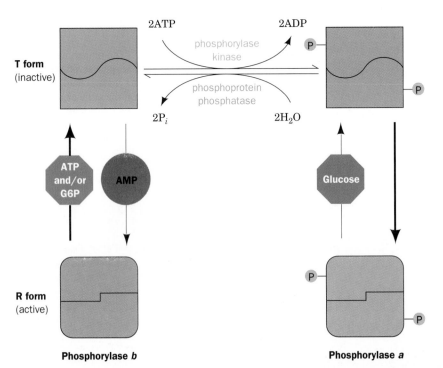

FIGURE 17-9. The control of glycogen phosphorylase activity. The enzyme may assume the enzymatically inactive T conformation (*above*) or the catalytically active R form (*below*). The conformation of phosphorylase *b* is allosterically controlled by effectors such as AMP, ATP, and G6P and is mostly in the T state under physiological conditions. In contrast, the modified form of the enzyme, phosphorylase *a*, is largely unresponsive

to these effectors and is mostly in the R state unless there is a high level of glucose. Under usual physiological conditions, the enzymatic activity of glycogen phosphorylase is essentially determined by its rates of modification and demodification. Note that only the T-form enzyme is subject to phosphorylation and dephosphorylation, so effector binding influences the rates of these modification/demodification events.

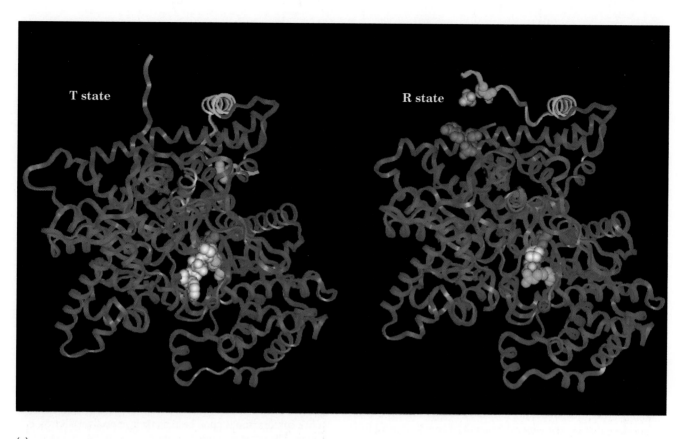

(a)

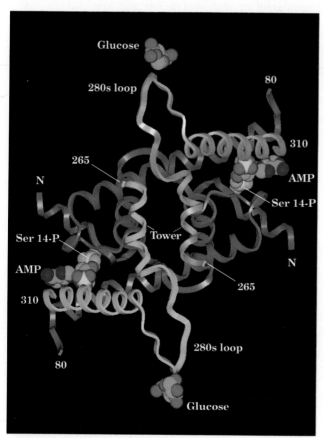

(b)

FIGURE 17-10. Conformational changes in glycogen phosphorylase. (*a*) A ribbon diagram of one subunit of the dimeric enzyme glycogen phosphorylase *b* shown (*left*) in the T state in the absence of allosteric effectors and (*right*) in the R state with bound AMP. The view is of the lower (*orange*) subunit in Fig. 17-2*b* as seen from the top of the page. The tower helix is dark blue, the N-terminal helix is light blue, and the N-terminal residues that change conformation upon AMP binding are green. Of the groups that are shown in space-filling representation, Ser 14, the phosphorylation site, is light green; AMP is orange; the active site PLP is red; the Arg 569 side chain, which reorients in the T → R transition so as to interact with the substrate phosphate, is light blue; residues 282 to 284 of the 280s loop, which in the R state are mostly disordered and hence not seen, are white; and the phosphates, both at the active site and at the R state Ser 14 phosphorylation site, are yellow. (*b*) The portion of the glycogen phosphorylase *a* dimer in the vicinity of the dimer interface showing the position of the Ser 14 phosphate group, the AMP bound in the allosteric effector site, and the active site–bound glucose molecule. The view is along the molecular 2-fold axis and hence is similar to that in Fig. 17-2*b*. Residues 6 to 80 and 265 to 310 are, respectively, light blue and darker blue in one subunit and pink and purple in the other. The AMP and glucose are shown in space-filling representation with C green, N dark blue, O red, and P yellow. The Ser 14 phosphate group is also shown in space-filling representation with O orange and P white. [X-Ray coordinates courtesy of Stephen Sprang, University of Texas Southwest Medical Center.]

cordance with the symmetry model of allosterism. The movement of the tower helices also displaces and disorders a loop (the 280s loop, residues 282–286), which covers the T-state active site so as to prevent substrate access. It also causes the Arg 569 side chain, which is located in the active site near the PLP phosphoryl group and the P_i-binding site, to rotate in a way that increases the enzyme's binding affinity for its anionic P_i substrate (Figure 17-10a).

Curiously, ATP also binds to the allosteric effector site, but in the T state so that it inhibits rather than promotes the T→R conformational shift. This is because, as structural analysis indicates, the β and γ phosphate groups of ATP bind to the enzyme so as to displace its ribose and α phosphate groups relative to those of AMP and thus destabilize the R state. The inhibitory action of ATP on phosphorylase is therefore simply understood: It competes with AMP for binding to phosphorylase and, in doing so, prevents the relative motions of the three polypeptide segments required for phosphorylase activation.

The above allosteric interactions are superimposed on an even more sophisticated control system involving covalent modifications (phosphorylation/dephosphorylation) of glycogen phosphorylase and glycogen synthase. These modifications alter the structures of the enzymes so as to change their responses to allosteric regulators. We shall therefore discuss the general concept of covalent modification and how it increases the sensitivity of a metabolic system to effector concentration changes. We subsequently consider the functions of such modifications in glycogen metabolism. Only then will we be in a position to take up the detailed consideration of allosteric regulation in glycogen metabolism.

B. Covalent Modification of Enzymes by Cyclic Cascades: Effector "Signal" Amplification

*Glycogen synthase and glycogen phosphorylase can each be enzymatically interconverted between two forms with different kinetic and allosteric properties through a complex series of reactions known as a **cyclic cascade**. The interconversion of these different enzyme forms involves distinct, enzyme-catalyzed **covalent modification** and **demodification reactions**.*

Compared with other regulatory enzymes, enzymatically interconvertible enzyme systems:

1. Can respond to a greater number of allosteric stimuli.

2. Exhibit greater flexibility in their control patterns.

3. Possess enormous amplification potential in their responses to variations in effector concentrations.

This is because *the enzymes that modify and demodify a target enzyme are themselves under allosteric control. It is therefore possible for a small change in concentration of an allosteric effector of a modifying enzyme to cause a large*

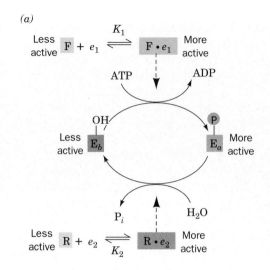

(a)

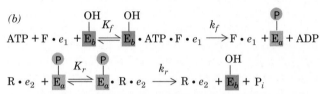

(b)

FIGURE 17-11. (*a*) A monocyclic enzyme cascade. F and R are, respectively, the modifying and demodifying enzymes. These are allosterically converted from their inactive to their active conformations upon binding their respective effectors, e_1 and e_2. The target enzyme, E, is more active in the modified form (E_a) and less active in the unmodified form (E_b). Dashed arrows symbolize catalysis of the indicated reactions. (*b*) Chemical equations for the interconversion of the target enzyme's unmodified and modified forms E_b and E_a.

change in the concentration of an active, modified target enzyme. Such a cyclic cascade is diagrammed in Fig. 17-11.

Description of a General Cyclic Cascade

Figure 17-11a shows a general scheme for a cyclic cascade where, by convention, the more active target enzyme form has the subscript a and the less active form has the subscript b. Here, modification, in this case, phosphorylation, activates the enzyme. Note that the modifying enzymes, F and R, are active only when they have bound their respective allosteric effectors e_1 and e_2. The kinetic mechanisms for the interconversion of the unmodified and modified forms of the target enzyme, E_b and E_a, are indicated in Fig. 17-11b.

In the steady state, the fraction of E in the active form, $[E_a]/[E]_T$ (where $[E]_T = [E_a] + [E_b]$ is the total enzyme concentration) determines the rate of the reaction catalyzed by E. This fraction is a function of the total concentrations of the modifying enzymes, $[F]_T$ and $[R]_T$, the concentrations of their allosteric effectors, e_1 and e_2, the dissociation constants of these effectors, K_1 and K_2, and the substrate dissociation constants, K_f and K_r, of the target enzymes, as well as the rate constants, k_f and k_r, for the interconversions themselves (Fig. 17-11). This relationship is obviously quite complex. Nevertheless, it can be shown that, in a cyclic cascade, a relatively small change in the concentration of e_1, the allosteric effector of the modifying enzyme F, can result in a much larger change in $[E_a]/[E]_T$, the fraction of E in the active form. In other words, *the cascade functions to amplify the sensitivity of the system to an allosteric effector.*

We have so far considered the covalent modification of only one enzyme, a **monocyclic cascade.** Imagine a **bicyclic cascade** involving the covalent modification of one of the modifying enzymes (F), as well as the metabolic target enzyme (E) (Fig. 17-12). As you might expect, the amplification potential of a "signal," e_1, as well as the control flexibility of such a system, is enormous.

The activities of both glycogen phosphorylase and glycogen synthase are controlled by bicyclic cascades. Let us now examine the enzymatic interconversions involved in these bicyclic cascades. We shall specifically focus on the covalent modifications of glycogen phosphorylase and glycogen synthase, the structural effects of these covalent modifications, and how these structural changes affect the interactions of their allosteric effectors. We shall then consider the cyclic cascades as a whole, studying the various modification enzymes involved and their "ultimate" allosteric effectors. Finally, we shall see how the various cyclic cascades of glycogen metabolism function in different physiological situations.

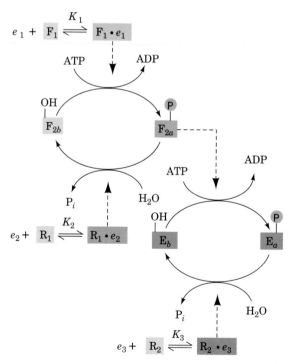

FIGURE 17-12. A bicyclic enzyme cascade. See the legend of Fig. 17-11 for symbol definitions. In a bicyclic cascade, one of the modifying enzymes (F_2) is also subject to chemical modification. It is active in the modified state (F_{2a}) and inactive in the unmodified state (F_{2b}).

C. Glycogen Phosphorylase Bicyclic Cascade

In 1938, Carl and Gerti Cori found that glycogen phosphorylase exists in two forms, the *b* form that requires AMP for activity, and the *a* form that is active without AMP. It nevertheless took 20 years for the development of the protein chemistry techniques through which Edwin Krebs and Edmund Fischer demonstrated, in 1959, that phosphorylases *a* and *b* correspond to forms of the protein in which a specific residue, Ser 14, is enzymatically phosphorylated or dephosphorylated, respectively.

Glycogen Phosphorylase: The Cascade's Target Enzyme

The activity of glycogen phosphorylase is allosterically controlled, as we saw, through AMP activation and ATP, G6P, and glucose inhibition (Section 17-3A). Superimposed upon this allosteric control is control by enzymatic interconversion through a bicyclic cascade involving the actions of three enzymes (Figs. 17-12 and 17-13, *left*):

1. **Phosphorylase kinase,** which specifically phosphorylates Ser 14 of glycogen phosphorylase *b* (Fig. 17-12, enzyme F_2).

2. **cAMP-dependent protein kinase,** which phosphorylates and thereby activates phosphorylase kinase (Fig. 17-12, enzyme F_1).

3. **Phosphoprotein phosphatase-1,** which dephosphorylates and thereby deactivates both glycogen phosphorylase *a* and phosphorylase kinase (Fig. 17-12, enzymes R_1 and R_2).

In an interconvertible enzyme system, the "modified" form of the enzyme bears the prefix *m* and the "original" (unmodified) form bears the prefix *o*, whereas the enzyme's most active and least active forms are identified by the suffixes *a* and *b*, respectively. In this case, *o*-phosphorylase *b* (unmodified, least active) is the form under allosteric control by AMP, ATP, and G6P (Fig. 17-9, *left*). Phosphorylation to yield *m*-phosphorylase *a* (modified, most active) all but removes the effects of these allosteric modulators. In terms of the symmetry model of allosterism (Section 9-4B), *the phosphorylation of Ser 14 shifts the enzyme's T (inactive) $\rightleftharpoons$ R (active) equilibrium in favor of the R state (Fig. 17-9, right).* Indeed, *phosphorylase a's Ser 14-phosphoryl group is analogous to an allosteric activator:* It forms ion pairs with two Arg side chains on the opposite subunit, thereby knitting the subunits together in much the same way as does AMP when it binds tightly to a site between the subunits (Fig. 17-10*b*).

In the resting cell, the concentrations of ATP and G6P are high enough to inhibit phosphorylase *b. The level of phosphorylase activity is therefore largely determined by the fraction of the enzyme present as phosphorylase a.* The steady state fraction of phosphorylated enzyme (E_a) depends on the relative activities of phosphorylase kinase (F_2), cAMP-dependent protein kinase (F_1), and phosphoprotein phosphatase-1 (R_1 and R_2). This interrelationship

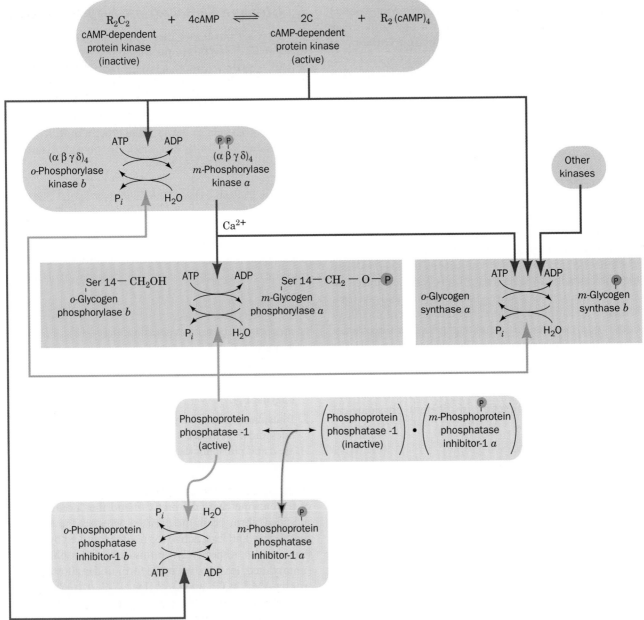

FIGURE 17-13. A schematic diagram of the major enzymatic modification/demodification systems involved in the control of glycogen metabolism in muscle. Modification systems are shaded in purple, demodification systems are shaded in yellow, and the target enzymes are shaded in green. Colored arrows indicate facilitation of a modification or demodification reaction. Note that glycogen phosphorylase activity is controlled by a bicyclic enzyme cascade (*left*) and glycogen synthase activity is controlled by both a bicyclic and a monocyclic enzyme cascade (*right*). By convention, the modified form of the enzyme bears the prefix *m* and the "original" (unmodified) form bears the prefix *o*. The most active and least active forms of the enzymes are identified by the suffixes *a* and *b*, respectively. Further control of phosphoprotein phosphatase-1 covalent modification is diagrammed in Fig. 17-18.

is remarkably elaborate for glycogen phosphorylase. Let us consider the actions of these enzymes.

cAMP-Dependent Protein Kinase: A Crucial Regulatory Link

Phosphorylase kinase, which converts phosphorylase b to phosphorylase a, is itself subject to covalent modification (Fig. 17-13). *For phosphorylase kinase to be fully active, Ca^{2+} must be present (see below) and the protein must be phosphorylated.*

In both the glycogen phosphorylase and glycogen synthase cascades, *the primary intracellular signal, e_1, is **adenosine-3′,5′-cyclic monophosphate (cAMP).*** The cAMP concentration in a cell is a function of the ratio of its rate of synthesis from ATP by **adenylate cyclase,** and its rate of breakdown to AMP by a specific **phosphodiesterase.**

ATP

PP$_i$ ← adenylate cyclase

3',5'-Cyclic AMP (cAMP)

H$_2$O → phosphodiesterase

AMP

Adenylate cyclase is, in turn, activated by certain hormones (Section 17-3E).

*cAMP is absolutely required for the activity of **cAMP-dependent protein kinase (cAPK),** an enzyme that phosphorylates specific Ser and/or Thr residues of numerous cellular proteins, including phosphorylase kinase and glycogen synthase.* These proteins all contain cAPK's consensus recognition sequence, Arg-Arg-X-Ser/Thr-Y, where Ser/Thr is the phosphorylation site, X is any small residue, and Y is a large hydrophobic residue. In the absence of cAMP, cAPK

is an inactive tetramer consisting of two regulatory and two catalytic subunits, R$_2$C$_2$. The cAMP binds to the regulatory subunits so as to cause the dissociation of active catalytic monomers (Fig. 17-13; *top*). *The intracellular concentration of cAMP therefore determines the fraction of cAPK in its active form and thus the rate at which it phosphorylates its substrates.* In fact, in all known eukaryotic cases, the physiological effects of cAMP are exerted through the activation of specific protein kinases.

The X-ray structure of the 350-residue C subunit of mouse cAPK in complex with Mg–ATP and a 20-residue inhibitor peptide has been determined by Susan Taylor and Janusz Sowadski (Fig. 17-14), and that of a similar complex

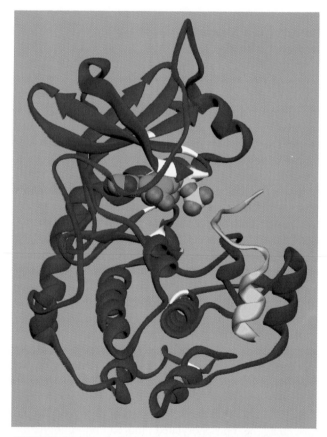

FIGURE 17-14. A ribbon diagram of the C subunit of mouse cAMP-dependent protein kinase (cAPK) in complex with ATP and a 20-residue segment of a naturally occurring protein kinase inhibitor. The protein's main chain is purple with the 11 residues that are highly conserved in all known protein kinases colored white. The polypeptide inhibitor is dark yellow and its pseudo–target sequence, Arg-Arg-Asn-Ala-Ile, is light blue (note that the enzyme's true target sequence is Arg-Arg-X-Ser/Thr-Y, where X is a small residue, Y is a large hydrophobic residue, and Ser/Thr, which the polypeptide inhibitor lacks, is the residue that the enzyme phosphorylates). The ATP is shown in space-filling representation colored according to atom type (C, green; N, blue; O, red; and P, yellow). Note that the inhibitor's pseudo–target sequence is in close proximity to the ATP's γ phosphate group, the group that the enzyme transfers.

of the porcine heart enzyme was determined by Robert Huber. The C subunit, in a manner resembling other kinases of known structure (e.g., Figs. 16-5 and 16-12), is bilobal with a deep cleft between the lobes that is occupied by the Mg–ATP and the segment of the inhibitor peptide that includes the above 5-residue consensus sequence. This cleft must therefore contain cAPK's catalytic site.

Protein kinases play key roles in the signaling pathways by which many hormones, growth factors, neurotransmitters, and toxins affect the functions of their target cells (Sections 33-4C and 34-4B), as well as in controlling metabolic pathways. Indeed, ~10% of the proteins in mammalian cells are phosphorylated. The >100 different protein kinases that have been sequenced share a conserved catalytic core corresponding to residues 40–280 of cAPK's C subunit.

Phosphorylase Kinase: Coordination of Enzyme Activation with [Ca²⁺]

Phosphorylase kinase is activated by Ca²⁺ concentrations as low as $10^{-7}M$ as well as by covalent modification. This 1200-kD enzyme consists of four nonidentical subunits that form the active oligomer $(\alpha\beta\gamma\delta)_4$. The isolated γ subunit has full catalytic activity (ability to convert phosphorylase *b* to phosphorylase *a*), whereas the α, β, and δ subunits are inhibitors of the catalytic reaction. The δ subunit, which is known as **calmodulin (CaM),** confers Ca²⁺ sensitivity on the complex. When Ca²⁺ binds to calmodulin's four Ca²⁺-binding sites, this ubiquitous eukaryotic regulatory protein undergoes an extensive conformational change (see below) that activates phosphorylase kinase. Glycogen phosphorylase therefore becomes phosphorylated and the rate of glycogen breakdown increases. *The physiological significance of this Ca²⁺ activation process is that muscle contraction is triggered by a transient increase in the level of cytosolic Ca²⁺ through its release from intracellular reservoirs by nerve impulses (Section 34-3C). The rate of glycogen breakdown is therefore linked to the rate of muscle contraction, an important regulatory link because glycogen breakdown in muscle provides fuel for glycolysis which, in turn, generates the ATP required for muscle contraction.*

Sites on both the α and β subunits of phosphorylase kinase may be phosphorylated. Indeed, phosphorylation causes the enzyme to become activated at much lower Ca²⁺ concentrations and full enzyme activity is obtained in the presence of Ca²⁺ only when both these subunits are phosphorylated. Since the phosphorylation of phosphorylase kinase ultimately occurs in response to the presence of certain hormones (Section 17-3E), whereas Ca²⁺ release occurs in response to nerve impulses, these two types of signals act synergistically on muscle cells to stimulate glycogenolysis.

Calmodulin: A Ca²⁺-Activated Switch

Calmodulin is a ubiquitous eukaryotic Ca²⁺-binding protein that participates in numerous cellular regulatory processes. The X-ray structure of this highly conserved

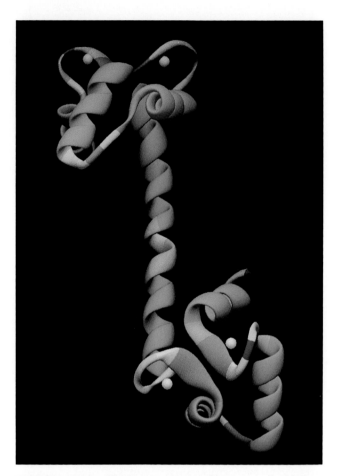

FIGURE 17-15. The X-ray structure of rat testes calmodulin. This monomeric 148-residue protein contains two remarkably similar globular domains separated by a seven-turn α helix. The residues are color coded according to their backbone conformation angles (ϕ and ψ; Fig. 7-7): light blue, α helical angles; green, β sheet angles; yellow, between helix and sheet; and purple, left-handed helix. The Gly residues are white and the N-terminus is dark blue. The two Ca²⁺-binding sites in each domain are represented by white spheres. [Courtesy of Mike Carson, University of Alabama at Birmingham. X-Ray structure determined by Charles Bugg, University of Alabama at Birmingham.]

148-residue protein, which was determined by Charles Bugg, has a curious dumbbell-like shape in which CaM's two globular domains are connected by a seven-turn α helix (Fig. 17-15). CaM has two high-affinity Ca²⁺-binding sites on each of its globular domains, both of which are formed by nearly superimposable helix–loop–helix motifs known as **EF hands** (Fig. 17-16) that also occur in several other Ca²⁺-sensing proteins of known structure. The Ca²⁺ ion in each of these sites is octahedrally coordinated by oxygen atoms from the backbone and side chains of the loop as well as from a protein-associated water molecule.

The binding of Ca²⁺ to either domain of CaM induces a conformational change in that domain, which exposes an otherwise buried Met-rich hydrophobic patch. This patch,

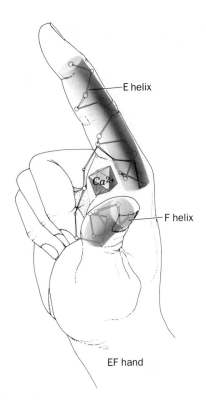

FIGURE 17-16. The Ca²⁺-binding sites in many proteins that function to sense the level of Ca²⁺ are formed by helix–loop–helix motifs named EF hands. [After Kretsinger, R.H., *Annu. Rev. Biochem.* **45**, 241 (1976).]

in turn, binds with high affinity to the CaM-binding domain of the phosphorylase kinase γ subunit, as well as to the CaM-binding domains of many other Ca²⁺-regulated proteins, and in doing so modulates the activities of these proteins. These CaM-binding domains have little mutual sequence homology but are all basic amphiphilic α helices. In fact, ~20-residue segments of these helices, as well as synthetic amphiphilic helices composed of only Leu, Lys, and Trp residues, bind Ca²⁺–CaM as tightly as the target proteins themselves.

Despite uncomplexed CaM's extended appearance in its X-ray structure (Fig. 17-15), a variety of studies indicate that both of its globular domains can simultaneously bind to a single target helix. Evidently, CaM's central α helix serves as a flexible tether rather than as a rigid spacer, a property that probably further increases the range of target sequences to which CaM can bind. An NMR structure (Fig. 17-17), determined by Marius Clore, Angela Gronenborn, and Ad Bax, of (Ca²⁺)₄–CaM in complex with its 26-residue CaM-binding target polypeptide of skeletal muscle **myosin light chain kinase (MLCK;** a homolog of the cAPK C subunit, which phosphorylates and thereby activates the light chains of the muscle protein **myosin;** Section 34-3D) confirms this idea, as does the closely similar X-ray structure of a related complex by Florante Quiocho. [Indeed, the extended conformation of CaM's central helix in Fig. 17-15 may well be an artifact arising from crystal packing forces

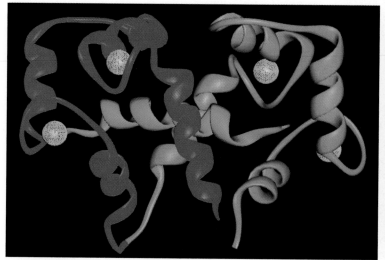
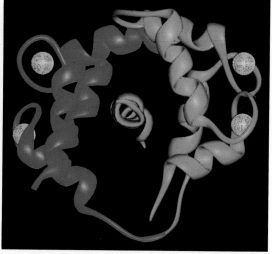

FIGURE 17-17. Ribbon diagrams showing the NMR structure of (Ca²⁺)₄–CaM from the fruit fly *Drosophila melanogaster* in complex with its 26-residue target polypeptide from rabbit skeletal muscle myosin light chain kinase. The N-terminal domain of CaM is blue, its C-terminal domain is red, the target polypeptide is green, and the Ca²⁺ ions are represented by light blue spheres. (*a*) A view of the complex in which the N-terminus of the target polypeptide is on the right, and (*b*) the perpendicular view as seen from the right side of Part *a*. In both views, the pseudo–twofold axis relating the N- and C-terminal domains of CaM is approximately vertical. Note how the middle segment of the long central helix in uncomplexed CaM (Fig.

17-15) has unwound and bent (bottom loop in *b*) such that CaM forms a globular protein that largely encloses the helical target polypeptide within a hydrophobic tunnel in a manner resembling two hands holding a rope (the target polypeptide assumes the random coil conformation in solution). However, the conformations of CaM's two globular domains are essentially unchanged by the complexation. Evidently, CaM's bound Ca²⁺ ions serve to organize and stabilize the target binding conformations of its globular domains. [Based on an NMR structure by Marius Clore, Angela Gronenborn, and Ad Bax, National Institutes of Health.]

considering that this helix's central two turns contact no other portion of the protein and hence are maximally solvent-exposed (almost all other known α helices are at least partially buried in a protein) and that a polypeptide with the sequence of this helix assumes a random coil conformation in aqueous solution.]

How does Ca^{2+}–CaM activate its target protein kinases? MLCK contains a C-terminal segment whose sequence resembles that of MLCK's target polypeptide on the light chain of myosin but lacks a phosphorylation site. A model of MLCK, based on the X-ray structure of the 30% identical C subunit of cAPK, strongly suggests that this pseudosubstrate peptide autoinactivates MLCK by binding to its active site. Indeed, the excision of MLCK's pseudosubstrate peptide by limited proteolysis permanently activates this enzyme. MLCK's CaM-binding target polypeptide overlaps its pseudosubstrate sequence. Thus, *the binding of Ca²⁺–CaM to this target polypeptide extracts the pseudosubstrate from MLCK's active site, thereby activating this enzyme.*

Ca^{2+}–CaM's other target proteins, including the phosphorylase kinase γ subunit, are presumably activated in the same way. cAPK's R (regulatory) subunit contains a similar pseudosubstrate sequence adjacent to its two tandem cAMP-binding domains. In this case, however, it appears that the autoinhibitory pseudosubstrate peptide is allosterically ejected from the C subunit's active site by the binding

of cAMP to the R subunit (which lacks a Ca^{2+}–CaM binding site).

Phosphoprotein Phosphatase-1

The steady state phosphorylation levels of most enzymes involved in cyclic cascades are maintained by the opposition of kinase-catalyzed phosphorylations and the hydrolytic dephosphorylations catalyzed by phosphoprotein phosphatase-1. This enzyme hydrolyzes the phosphoryl groups from *m*-glycogen phosphorylase *a*, both α and β subunits of phosphorylase kinase and, as is discussed below, two other proteins involved in glycogen metabolism.

Phosphoprotein phosphatase-1 in muscle is only active when it is bound to glycogen through its glycogen-binding **G subunit.** The activity of protein phosphatase-1 and its affinity for the G subunit are regulated by phosphorylation of the G subunit at two separate sites. Phosphorylation of site 1 by an **insulin-stimulated protein kinase** (Fig. 17-18) activates phosphoprotein phosphatase-1, whereas phosphorylation of site 2 by cAMP-dependent protein kinase (which can also phosphorylate site 1) causes the enzyme to be released into the cytoplasm where it cannot dephosphorylate the glycogen-bound enzymes of glycogen metabolism. In the cytosol, protein phosphatase-1 is also inhibited by its binding to the protein **phosphoprotein phosphatase inhibitor 1 (inhibitor-1).** This latter protein provides yet another example of control by enzymatic interconversion: It too is

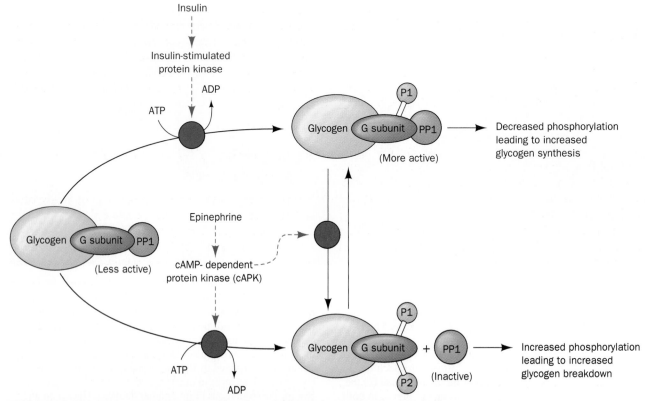

FIGURE 17-18.   The antagonistic effects of insulin and epinephrine on glycogen metabolism in muscle occur through their effects on the glycogen-bound G subunit of phosphoprotein phosphatase-1 **(PP1).** Green dots and dashed arrows indicate activation.

modified by cAMP-dependent protein kinase and demodified by phosphoprotein phosphatase-1 (Fig. 17-13, *bottom*), although, in this case, a Thr, not a Ser, is phosphorylated/dephosphorylated. The protein is a functional inhibitor only when it is phosphorylated. *The concentration of cAMP therefore controls the fraction of an enzyme in its phosphorylated form, not only by increasing the rate at which it is phosphorylated, but also by decreasing the rate at which it is dephosphorylated. In the case of glycogen phosphorylase, an increase in [cAMP] results not only in an increase in this enzyme's rate of activation, but also in a decrease in its rate of deactivation.*

The activity of phosphoprotein phosphatase-1 in liver is controlled by its binding to *m*-phosphorylase *a*. Among the major conformational changes that phosphorylase undergoes in converting from the T to the R state is the movement of the Ser 14-phosphoryl group from the surface of the T-state (inactive) enzyme to a position buried a few angstroms beneath the protein's surface at the dimer interface in the R-state (active) enzyme (Fig. 17-10). *Both the R and T forms of phosphorylase a strongly bind phosphoprotein phosphatase-1, but only in the T state enzyme is the Ser 14-phosphoryl group accessible for hydrolysis, which converts phosphorylase a to phosphorylase b. Consequently, when phosphorylase a is in its active R form, it effectively removes phosphoprotein phosphatase-1 from circulation.* However, under the conditions that phosphorylase *a* converts to the T state (Section 17-3G), phosphoprotein phosphatase-1 hydrolyzes the now exposed Ser 14-phosphoryl group, thereby converting *m*-phosphorylase *a* to *o*-phosphorylase *b*, which has only a low affinity for binding phosphoprotein phosphatase-1. One effect of phosphorylase *a* demodification, therefore, is to relieve the inhibition of phosphoprotein phosphatase-1 by releasing it and thus allowing it to excise the phosphoryl groups of other susceptible phosphoproteins. Since glycogen phosphorylase is in ~10-fold greater concentration than is phosphoprotein phosphatase-1, this release only occurs when more than ~90% of the glycogen phosphorylase is in the *o*-phosphorylase *b* form. Glycogen synthase is among the proteins that are dephosphorylated by phosphoprotein phosphatase-1 when it is released from phosphorylase. In contrast to phosphorylase, dephosphorylation activates glycogen synthase. This enzyme is involved in its own bicyclic cascade whose properties we shall now examine.

D. Glycogen Synthase Bicyclic Cascade

Like glycogen phosphorylase, glycogen synthase exists in two enzymatically interconvertible forms:

1. The modified (*m*; phosphorylated) form that is inactive under physiological conditions (the *b* form).

2. The original (*o*; dephosphorylated) form that is active (the *a* form).

m-Glycogen synthase *b* is under allosteric control; it is strongly inhibited by physiological concentrations of ATP, ADP, and P_i. This inhibition may be overcome by G6P but only at nonphysiological concentrations above 10 mM. Since the physiological concentration of G6P in muscle tissue is only 0.2 to 0.4 mM, the modified enzyme is almost totally inactive *in vivo*. The activity of the unmodified enzyme is essentially independent of these effectors, so the cell's glycogen synthase activity varies with the fraction of the enzyme in its unmodified form.

The mechanistic details of the interconversion of modified and unmodified forms of glycogen synthase are particularly complex and are therefore not as well understood as those of glycogen phosphorylase. It has been clearly established that the fraction of unmodified glycogen synthase is, in part, controlled by a bicyclic cascade involving phosphorylase kinase and phosphoprotein phosphatase-1, enzymes that are also involved in the glycogen phosphorylase bicyclic cascade (Fig. 17-13, *right*). However, 6 other protein kinases are known to at least partially deactivate human muscle glycogen synthase by phosphorylating this homotetramer at 1 or more of 9 Ser residues on its 737-residue subunits. These enzymes include cAMP-dependent protein kinase, so glycogen synthase deactivation may also be considered to occur via a monocyclic cascade; **calmodulin-dependent protein kinase,** which is activated by the presence of Ca^{2+}; **protein kinase C,** which responds to the extracellular presence of certain hormones via a mechanism described in Sections 34-4B and 17-3G; and **glycogen synthase kinase-3,** whose mode of activation is unknown. Why glycogen synthase deactivation is so elaborately regulated compared to its activation or the activation/deactivation of glycogen phosphorylase is unclear, although, whatever the reasons, it closely monitors the organism's metabolic state.

E. Integration of Glycogen Metabolism Control Mechanisms

Whether there is net synthesis or degradation of glycogen and at what rate depends on the relative balance of the active forms of glycogen synthase and glycogen phosphorylase. This, in turn, largely depends on the rates of the phosphorylation and dephosphorylation reactions of the two bicyclic cascades. These cascades, one controlling the rate of glycogen breakdown and the other controlling the rate of glycogen synthesis, are intimately related. They are linked by cAMP-dependent protein kinase and phosphorylase kinase which, through phosphorylation, activate phosphorylase as they inactivate glycogen synthase (Fig. 17-13). The cascades are also linked by phosphoprotein phosphatase-1, which in liver is inhibited by phosphorylase *a* and therefore unable to activate (dephosphorylate) glycogen synthase unless it first inactivates (also by dephosphorylation) phosphorylase *a*.

Hormones Trigger Glycogen Metabolism through the Intermediacy of Second Messengers

Glycogen metabolism in the liver is ultimately controlled by the polypeptide hormone **glucagon**.

$\overset{+}{H_3N}$ - His - Ser - Glu-Gly - Thr- Phe -Thr - Ser - Asp- Tyr- 10

Ser - Lys - Tyr- Leu- Asp- Ser - Arg - Arg - Ala - Gln- 20

Asp - Phe - Val - Gln - Trp - Leu- Met- Asn - Thr - COO^- 29

Glucagon

In muscles and various tissues, control is exerted by **insulin** (Fig. 6-2) and by the adrenal hormones **epinephrine (adrenalin)** and **norepinephrine (noradrenalin)**.

OH
OH
HO—C—H
CH₂
⁺NH₂
X   X = CH₃   **Epinephrine**
          X = H   **Norepinephrine**

Hormonal stimulation of cells at their plasma membranes occurs through the mediation of transmembrane proteins called **receptors.** *Different cell types have different complements of receptors and thus respond to different sets of hormones. This response involves the release inside the cell of molecules known as* **second messengers,** *that is, intracellular mediators of the externally received hormonal message.* Different receptors act to release different second messengers. Indeed, cAMP was identified by Earl Sutherland as the the first known instance of a second messenger through his demonstration that glucagon and epinephrine act at cell surfaces to stimulate adenylate cyclase to increase [cAMP] [the mechanism of adenylate cyclase activation, as well as a discussion of other second messengers, including Ca^{2+}, **inositol-1,4,5-triphosphate (IP₃)** and **diacylglycerol (DG)**, is elaborated in Section 34-4B]. Following this discovery, it was realized that cAMP, which is present in all forms of life, is an essential control element in many biological processes.

When hormonal stimulation by glucagon or epinephrine increases the intracellular cAMP concentration, the cAMP-dependent protein kinase activity increases, increasing the rates of phosphorylation of many proteins and decreasing their dephosphorylation rates as well. A decrease in dephosphorylation rates, as previously noted, increases the phosphorylation level of phosphoprotein phosphatase inhibitor, which in turn inhibits phosphoprotein phosphatase-1. An increase in the concentration of phosphorylase *a* also contributes to the inhibition of phosphoprotein phosphatase-1.

Because of the amplifying properties of the cyclic cascades, a small change in [cAMP] results in a large change in the fraction of enzymes in their phosphorylated forms. When a large fraction of the glycogen metabolism enzymes are present in their phosphorylated forms, the metabolic flux is in the direction of glycogen breakdown since glycogen phosphorylase is active and glycogen synthase is inactive. When [cAMP] decreases, phosphorylation rates decrease, dephosphorylation rates increase, and the fraction of enzymes in their dephospho forms increases. The resultant activation of glycogen synthase and the inhibition of glycogen phosphorylase causes a change in the flux direction towards net glycogen synthesis.

F. Maintenance of Blood Glucose Levels

An important function of the liver is to maintain the blood concentration of glucose, the brain's primary fuel source, at ~5 m*M*. When blood [glucose] decreases beneath this level, usually during exercise or well after meals have been digested, the liver releases glucose into the bloodstream. The process is mediated by the hormone glucagon as follows:

1. Low concentrations of blood glucose cause the pancreatic α cells to secrete glucagon into the bloodstream.

2. Glucagon receptors on liver cell surfaces respond to the presence of glucagon by activating adenylate cyclase, thereby increasing the [cAMP] inside these cells.

3. The [cAMP] increase, as described above, triggers an increase in the rate of glycogen breakdown, leading to increased intracellular [G6P].

4. G6P, in contrast to glucose, cannot pass through the cell membrane. However, in liver, which does not employ glucose as a major energy source, the enzyme **glucose-6-phosphatase** hydrolyzes G6P:

$$G6P + H_2O \longrightarrow glucose + P_i$$

The resulting glucose enters the bloodstream, thereby increasing the blood glucose concentration. Muscle and brain cells, however, lack glucose-6-phosphatase so that they retain their G6P.

How does this delicately balanced system respond to an increase in blood [glucose]? When blood sugar is high, normally immediately after meals have been digested, glucagon levels decrease and insulin is released from the pancreatic β cells. *The rate of glucose transport across many cell membranes increases in response to insulin (Section 18-2D), [cAMP] decreases, and glycogen metabolism therefore shifts from glycogen breakdown to glycogen synthesis.* The mechanism of insulin action is quite complex and has only recently begun to be understood (Section 34-4B), but one of its target enzymes seems to be protein phosphatase-1.

In muscle, insulin and epinephrine have antagonistic effects on glycogen metabolism. Epinephrine promotes glycogenolysis by activating the cAMP-dependent phosphorylation cascade, which stimulates glycogen breakdown

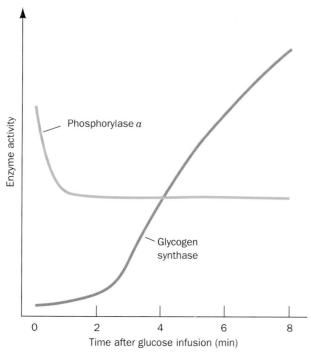

FIGURE 17-19. The enzymatic activities of phosphorylase *a* and glycogen synthase in mouse liver in response to an infusion of glucose. Phosphorylase *a* is rapidly inactivated and, somewhat later, glycogen synthase is activated. [After Stalmans, W., De Wulf, H., Hue, L., and Hers, H.-G., *Eur. J. Biochem.* **41,** 129 (1974).]

while inhibiting glycogen synthesis. Insulin, as we saw in Section 17-3C, activates insulin-stimulated protein kinase to phosphorylate site 1 on the glycogen-binding G subunit of protein phosphatase-1 so as to activate this protein and thus dephosphorylate the enzymes of glycogen metabolism (Fig. 17-18). The storage of glucose as glycogen is thereby stimulated through the inhibition of glycogen breakdown and the stimulation of glycogen synthesis.

In liver, it is thought that glucose itself may be the messenger to which the glycogen metabolism system responds. *Glucose inhibits phosphorylase a by binding only to the active site of the enzyme's inactive T state, but in a manner different from that of substrate.* The presence of glucose therefore shifts phosphorylase *a*'s T ⇌ R equilibrium towards the T state (Fig. 17-9, *right*). This conformational shift, as we saw in Section 17-3C, exposes the Ser 14-phosphoryl group to phosphoprotein phosphatase-1, resulting in the demodification of phosphorylase *a*. An increase in glucose concentration therefore promotes inactivation of glycogen phosphorylase *a* through the enzyme's conversion to phosphorylase *b* (Fig. 17-19; i.e., phosphorylase *a* acts as a glucose receptor). The concomitant release of phosphoprotein phosphatase-1 (recall that it specifically binds to phosphorylase *a*), moreover, results in the activation (dephosphorylation) of *m*-glycogen synthase *b*. Above a glucose concentration of 7 m*M*, this process reverses the flux

of glycogen metabolism. The liver can thereby store the excess glucose as glycogen.

Glucokinase Forms G6P at a Rate Proportional to the Glucose Concentration

The liver's function in "buffering" the blood [glucose] is made possible because this organ contains a variant of hexokinase (the first glycolytic enzyme) known as **glucokinase** (also called **hexokinase D** or **hexokinase IV**). The hexokinase in most cells obeys Michaelis–Menten kinetics, has a high glucose affinity ($K_M < 0.1$ m*M*; the value of [glucose] at which the enzyme achieves half maximal velocity; Section 13-2A), and is inhibited by its reaction product, glucose-6-phosphate (G6P). Glucokinase, in contrast, has a much lower glucose affinity (reaching half of its maximal velocity at ~5 m*M*) and displays sigmoidal kinetics with a Hill constant (Section 9-1B) of 1.5, so *its activity increases rapidly with the blood [glucose] over the normal physiological range (Fig. 17-20; see Problem 4, this chapter). Glucokinase, moreover, is not inhibited by physiological concentrations of G6P.* Consequently, the higher the blood [glucose], the faster the liver converts glucose to G6P (liver cells, unlike most cells, are freely permeable to glucose; their glucose transport rate is unresponsive to insulin). Thus at low blood [glucose], the liver does not compete with other tissues for the available glucose supply, whereas at high blood [glucose], when the glucose needs of these tissues are met, the liver converts the excess glucose to glycogen. (Note that glucokinase is a monomeric enzyme so that its sigmoidal rate increase with [glucose] is a puzzling observation in light

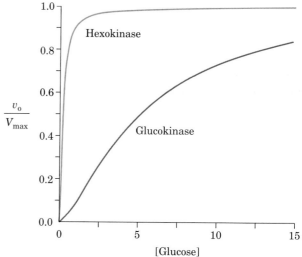

FIGURE 17-20. A comparison of the relative enzymatic activities of hexokinase and glucokinase over the physiological blood glucose range. The affinity of glucokinase for glucose ($K_{0.5} = 5$ m*M*) is much lower than that of hexokinase ($K_M = 0.1$ m*M*) and exhibits sigmoid rather than hyperbolic variation with [glucose]. [The glucokinase curve was generated using the Hill equation (Eq. [9.7]) with $K = 10$ m*M* and $n = 1.5$ as obtained from Cardenas, M. L., Rabajille, E., and Niemeyer, H., *Eur. J. Biochem,* **145**, 163–171 (1984).]

sides of a membrane generates a chemical potential difference:

$$\Delta \overline{G}_A = \overline{G}_A(in) - \overline{G}_A(out) = RT \ln \left(\frac{[A]_{in}}{[A]_{out}} \right) \quad [18.2]$$

Consequently, if the concentration of A outside the membrane is greater than that inside, $\Delta \overline{G}_A$ for the transfer of A from outside to inside will be negative and the spontaneous net flow of A will be inward. If, however, [A] is greater inside than outside, $\Delta \overline{G}_A$ is positive and an inward net flow of A can only occur if an exergonic process, such as ATP hydrolysis, is coupled to it to make the overall free energy change negative.

Membrane Potentials Arise from Transmembrane Concentration Differences of Ionic Substances

The permeabilities of biological membranes to ions such as H^+, Na^+, K^+, Cl^-, and Ca^{2+} are controlled by specific membrane-embedded transport systems that we shall discuss in later sections. *The resulting charge differences across a biological membrane generate an electric potential difference*, $\Delta \Psi = \Psi(in) - \Psi(out)$, where $\Delta \Psi$ is termed the **membrane potential**. Consequently, if A is ionic, Eq. [18.2] must be amended to include the electrical work required to transfer a mole of A across the membrane from outside to inside:

$$\Delta \overline{G}_A = RT \ln \left(\frac{[A]_{in}}{[A]_{out}} \right) + Z_A \mathscr{F} \Delta \Psi \quad [18.3]$$

where Z_A is the ionic charge of A; $\mathscr{F}$, the Faraday constant, is the charge on a mole of electrons ($96,494 \, C \cdot mol^{-1}$); and, $\overline{G}_A$ is now termed the **electrochemical potential** of A.

Membrane potentials in living cells can be measured directly with microelectrodes. $\Delta \Psi$ values of $-100 \, mV$ (inside negative) are not uncommon (note that $1 \, V = 1 \, J \cdot C^{-1}$). Thus the last term of Eq. [18.3] is often significant for ionic substances.

2. KINETICS AND MECHANISMS OF TRANSPORT

Thermodynamics indicates whether a given transport process will be spontaneous but, as we saw for chemical and enzymatic reactions, provides no indication of the rates of these processes. Kinetic analyses of transport processes together with mechanistic studies have nevertheless permitted these processes to be characterized. There are two types of transport processes: **Nonmediated transport** and **mediated transport**. Nonmediated transport occurs through simple diffusion. In contrast, *mediated transport occurs through the action of specific carrier proteins* that are variously called **carriers**, **permeases**, **porters**, **translocases**, **translocators**, and **transporters**. Mediated transport is fur-

ther classified into two categories depending on the thermodynamics of the system:

1. **Passive-mediated transport** or **facilitated diffusion** in which specific molecules flow from high concentration to low concentration so as to equilibrate their concentration gradients.

2. **Active transport** in which specific molecules are transported from low concentration to high concentration, that is, against their concentration gradients. Such an endergonic process must be coupled to a sufficiently exergonic process to make it favorable.

In this section, we consider the nature of nonmediated transport and then compare it to passive-mediated transport as exemplified by the erythrocyte glucose transporter, ionophores, and porins. Active transport is examined in succeeding sections.

A. Nonmediated Transport

The driving force for the nonmediated flow of a substance A through a medium is A's electrochemical potential gradient. This relationship is expressed by the **Nernst–Planck equation**:

$$J_A = -[A] U_A \, (d\overline{G}_A/dx) \quad [18.4]$$

where J_A is the flux (rate of passage per unit area) of A, x is distance, $d\overline{G}_A/dx$ is the electrochemical potential gradient of A, and U_A is its **mobility** (velocity per unit force) in the medium. If we assume, for simplicity, that A is an uncharged molecule so that $\overline{G}_A$ is given by Eq. [18.1], the Nernst–Planck equation reduces to

$$J_A = -D_A \, (d[A]/dx) \quad [18.5]$$

where $D_A \equiv RTU_A$ is the **diffusion coefficient** of A in the medium of interest. This is **Fick's first law of diffusion**, which states that *a substance diffuses in the direction that eliminates its concentration gradient, $d[A]/dx$, at a rate proportional to the magnitude of this gradient.*

For a membrane of thickness x, Eq. [18.5] is approximated by

$$J_A = \frac{D_A}{x}([A]_{out} - [A]_{in}) = P_A([A]_{out} - [A]_{in}) \quad [18.6]$$

where D_A is the diffusion coefficient of A inside the membrane and $P_A = D_A/x$ is termed the membrane's **permeability coefficient** for A. The permeability coefficient is indicative of the solute's tendency to transfer from the aqueous solvent to the membrane's nonpolar core. It should therefore vary with the ratio of the solute's solubility in a nonpolar solvent resembling the membrane's core (e.g., olive oil) to that in water, a quantity known as the solute's **partition coefficient** between the two solvents. Indeed, the fluxes of

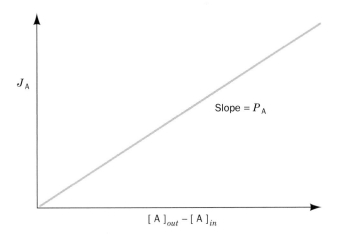

FIGURE 18-1. The linear relationship between diffusional flux (J_A) and ($[A]_{out} - [A]_{in}$) across a semipermeable membrane; see Eq. [18.6].

many nonelectrolytes across erythrocyte membranes vary linearly with their concentration differences across the membrane as predicted by Eq. [18.6] (Fig. 18-1). Moreover, their permeability coefficients, as obtained from the slopes of plots such as Fig. 18-1, correlate rather well with their measured partition coefficients between nonpolar solvents and water (Fig. 18-2).

B. Kinetics of Mediated Transport: Glucose Transport into Erythrocytes

Despite the success of the foregoing model in predicting the rates at which many molecules pass through membranes, there are numerous combinations of solutes and membranes that do not obey Eq. [18.6]. The flux in such a system is not linear with the solute concentration difference

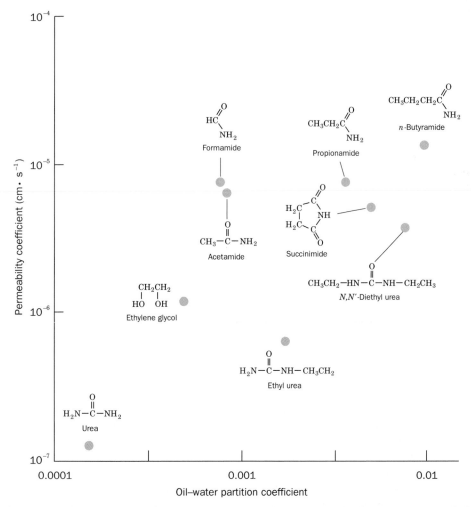

FIGURE 18-2. The permeability coefficients of various organic molecules in plasma membranes from the alga *Nitella mucronata* versus their partition coefficients between olive oil and water (a measure of a molecule's polarity). This more or less linear log–log plot indicates that the rate-limiting step for the nonmediated entry of a molecule into a cell is its passage through the membrane's hydrophobic core. [Based on data from Collander, R., *Physiol. Plant.* **7**, 433–434 (1954).]

across the corresponding membrane (Fig. 18-3) and, furthermore, the solute's permeability coefficient is much larger than is expected on the basis of its partition coefficient. Such behavior indicates that *these solutes are conveyed across membranes in complex with carrier molecules; that is, they undergo mediated transport.*

The system that transports glucose across the erythrocyte membrane provides a well-characterized example of passive-mediated transport: It invariably transports glucose down its concentration gradient but not at the rate predicted by Eq. [18.6]. Indeed, the **erythrocyte glucose transporter** exhibits four characteristics that differentiate mediated from nonmediated transport: (1) *speed and specificity*, (2) *saturation kinetics*, (3) *susceptibility to competitive inhibition*, and (4) *susceptibility to chemical inactivation.* In the following paragraphs we shall see how the erythrocyte glucose transporter exhibits these qualities.

Speed and Specificity

Table 18-1 indicates that the permeability coefficients of D-glucose and D-mannitol in synthetic bilayers, and that of D-mannitol in the erythrocyte membrane, are in reasonable agreement with the values calculated from the diffusion and partition coefficients of these sugars in olive oil. However, the experimentally determined permeability coefficient for D-glucose in the erythrocyte membrane is four orders of magnitude greater than its predicted value. *The erythrocyte membrane must therefore contain a system that rapidly transports glucose and that can distinguish D-glucose from D-mannitol.*

Saturation Kinetics

The concentration dependence of glucose transport indicates that its flux obeys the relationship:

$$J_A = \frac{J_{max}[A]}{K_M + [A]}$$ [18.7]

This **saturation function** has a familiar hyperbolic form (Fig. 18-3). We have seen it in the equation describing the binding of O_2 to myoglobin (Eq. [9.2]) and in the Michaelis–Menten equation describing the rates of enzymatic reactions (Eq. [13.24]). Here, as before, K_M may be

TABLE 18-1. PERMEABILITY COEFFICIENTS OF NATURAL AND SYNTHETIC MEMBRANES TO D-GLUCOSE AND D-MANNITOL AT 25°C

Membrane Preparation	Permeability Coefficient (cm · s⁻¹)	
	D-Glucose	**D-Mannitol**
Synthetic lipid bilayer	2.4×10^{-10}	4.4×10^{-11}
Calculated nonmediated diffusion	4×10^{-9}	3×10^{-9}
Intact human erythrocyte	2.0×10^{-4}	5×10^{-9}

Source: Jung, C.Y., *in* Surgenor, D. (Ed.), *The Red Blood Cell,* Vol. 2, p. 709, Academic Press (1975).

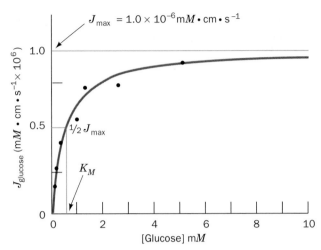

FIGURE 18-3. The variation of glucose flux into human erythrocytes with the external glucose concentration at 5°C. The black dots are experimentally determined data points, and the solid green line is computed from Eq. [18.7] with $J_{max} = 1.0 \times 10^{-6}$ mM·cm·s⁻¹ and $K_M = 0.5$ mM. The nonmediated glucose flux increases linearly with [glucose] (Fig. 18-1) but would not visibly depart from the baseline on the scale of this drawing. [Based on data from Stein, W.D., *Movement of Molecules across Membranes*, p. 134, Academic Press (1967).]

defined operationally as the concentration of glucose when the transport flux is half of its maximal rate, $J_{max}/2$. *This observation of **saturation kinetics** for glucose transport was the first evidence that a specific saturable number of sites on the membrane were involved in the transport of any substance.*

The transport process can be described by a simple four-step kinetic scheme involving binding, transport, dissociation, and recovery (Fig. 18-4). Its binding and dissociation steps are analogous to the recognition of a substrate and the release of product by an enzyme. The mechanisms of transport and recovery are under active investigation and are discussed in Section 18-2D.

Susceptibility to Competitive Inhibition

Many compounds structurally similar to D-glucose inhibit glucose transport. A double-reciprocal plot (Section 13-2B) for the flux of glucose into erythrocytes in the presence or absence of 6-*O*-benzyl-D-galactose (Fig. 18-5) shows behavior typical of competitive inhibition of glucose transport (competitive inhibition of enzymes is discussed in Section 13-3A). *Susceptibility to competitive inhibition indicates that there is a limited number of sites available for mediated transport.*

Susceptibility to Chemical Inactivation

Treatment of erythrocytes with $HgCl_2$, which reacts with protein sulfhydryl groups (Section 6-2) and thus inactivates many enzymes, causes the rapid, saturatable flux of glucose to disappear so that its permeability constant approaches that of mannitol. *The erythrocyte glucose transport sys-*

tem's susceptibility to such protein-modifying agents indicates that it, in fact, is a protein.

All of the above observations indicate that *glucose transport across the erythrocyte membrane is mediated by a limited number of protein carriers.* Before we discuss the mechanism of this transport system, however, we shall examine some simpler models of facilitated diffusion.

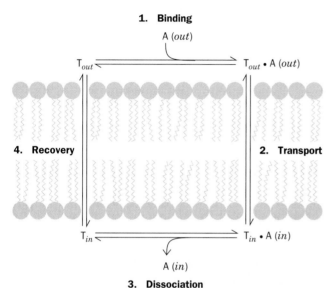

1. Binding

A (*out*)

$T_{out} \rightleftarrows T_{out} \cdot A\ (out)$

4. Recovery   **2. Transport**

$T_{in} \rightleftarrows T_{in} \cdot A\ (in)$

A (*in*)

3. Dissociation

FIGURE 18-4. A general kinetic scheme for membrane transport involving four steps: binding, transport, dissociation, and recovery. T is the transport protein whose binding site for solute A is located on either the inner or the outer side of the membrane at any one time.

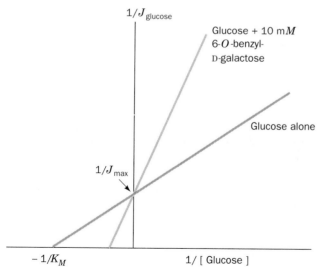

$1/J_{glucose}$

Glucose + 10 m*M* 6-*O*-benzyl-D-galactose

Glucose alone

$1/J_{max}$

$-1/K_M$

$1/$ [Glucose]

FIGURE 18-5. Double-reciprocal plots for the net flux of glucose into erythrocytes in the presence and absence of 6-*O*-benzyl-D-galactose. The pattern is that of competitive inhibition. [After Barnett, J.E.G., Holman, G.D., Chalkley, R.A., and Munday, K.A., *Biochem. J.* **145**, 422 (1975).]

C. Ionophores and Porins

Our understanding of mediated transport has been enhanced by the study of **ionophores,** substances that vastly increase the permeability of membranes to particular ions.

Ionophores May Be Carriers or Channel Formers

Ionophores are organic molecules of diverse types, many of which are antibiotics of bacterial origin. Cells and organelles actively maintain concentration gradients of various ions across their membranes (Section 18-3A). The antibiotic properties of ionophores arise from their tendency to discharge these vital concentration gradients.

There are two types of ionophores:

1. *Carriers, which increase the permeabilities of membranes to their selected ion by binding it, diffusing through the membrane, and releasing the ion on the other side (Fig. 18-6a). For net transport to occur, the uncomplexed ionophore must then return to the original side of the membrane ready to repeat the process. Carriers therefore share the common property that their ionic complexes are soluble in nonpolar solvents.*

2. *Channel formers, which form transmembrane channels or pores through which their selected ions can diffuse (Fig. 18-6b).*

Both types of ionophores transport ions at a remarkable rate. For example, a single molecule of the carrier antibiotic **valinomycin** transports up to 10^4 K$^+$ ions/s across a membrane. Channel formers have an even greater ion throughput; for example, each membrane channel composed of the antibiotic **gramicidin A** permits the passage of over 10^7 K$^+$ ions/s. Clearly, the presence of either type of ionophore, even in small amounts, greatly increases the permeability of a membrane towards the specific ions transported. However, *since ionophores passively permit ions to diffuse across*

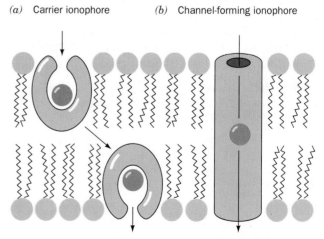

(*a*) Carrier ionophore   (*b*) Channel-forming ionophore

FIGURE 18-6. The ion transport modes of ionophores: (*a*) Carrier ionophores transport ions by diffusing through the lipid bilayer. (*b*) Channel-forming ionophores span the membrane with a channel through which ions can diffuse.

a membrane in either direction, their effect can only be to equilibrate the concentrations of their selected ions across the membrane.

Carriers and channel formers are easily distinguished experimentally through differences in the temperature dependence of their action. Carriers depend on their ability to diffuse freely across the membrane. Consequently, cooling a membrane below its transition temperature (the temperature below which it becomes a gel-like solid; Section 11-2B) essentially eliminates its ionic permeability in the presence of carriers. In contrast, membrane permeability in the presence of channel formers is rather insensitive to temperature because, once in place, channel formers need not move to mediate ion transport.

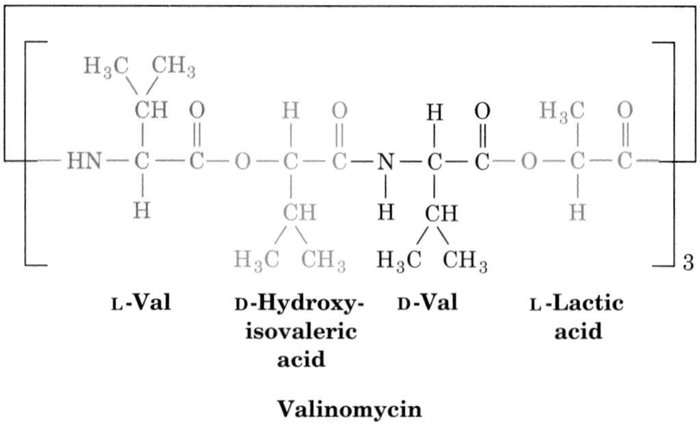

The K⁺–Valinomycin Complex Has a Polar Interior and a Hydrophobic Exterior

Valinomycin, which is perhaps the best characterized carrier ionophore, specifically binds K^+ and the biologically unimportant Rb^+. It is a **cyclic depsipeptide** that has both D- and L-amino acid residues (Fig. 18-7; a depsipeptide contains ester linkages as well as peptide bonds). The X-ray structure of valinomycin's K^+ complex (Fig. 18-8a) indicates that the K^+ is octahedrally coordinated by the carbonyl groups of its 6 Val residues, which also form its ester linkages. The cyclic, intramolecularly hydrogen bonded valinomycin backbone follows a zigzag path that surrounds the K^+ coordination shell with a sinuous molecular bracelet. *Its methyl and isopropyl side chains project outward*

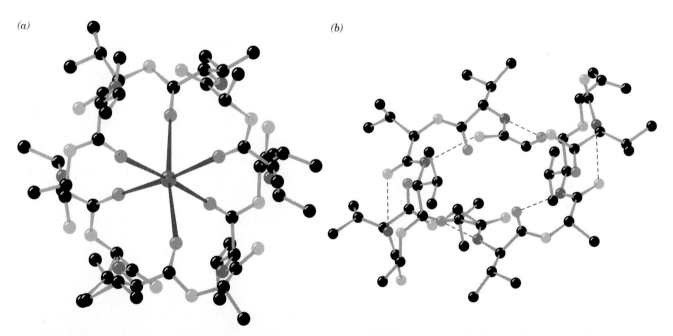

ʟ-Val   ᴅ-Hydroxy-   ᴅ-Val   ʟ-Lactic
         isovaleric   acid
         acid

Valinomycin

FIGURE 18-7. Valinomycin is a cyclic depsipeptide (has both ester and amide bonds) that contains both ᴅ- and ʟ-amino acids.

from the bracelet to provide the spheroidal complex with a hydrophobic exterior that makes it soluble in nonpolar solvents and in the hydrophobic cores of lipid bilayers. Uncomplexed valinomycin (Fig. 18-8b) has a more open conformation than its K^+ complex, which presumably facilitates the rapid binding of K^+.

K^+ (ionic radius, $r = 1.33$ Å) and Rb^+ ($r = 1.49$ Å) fit snugly into valinomycin's coordination site. However, the rigidity of the valinomycin complex makes this site too large to accommodate Na^+ ($r = 0.95$ Å) or Li^+ ($r = 0.60$ Å) properly; that is, valinomycin's six carbonyl oxygen atoms cannot simultaneously coordinate these ions. Complexes of

(a)

(b)

FIGURE 18-8. Valinomycin X-ray structures. (*a*) The K^+ complex. The six oxygen atoms that octahedrally complex the K^+ ion are darker red than the other oxygen atoms. [After Neupert-Laves, K. and Dobler, M., *Helv. Chim. Acta* **58**, 439

(1975).] (*b*) Uncomplexed valinomycin. [After Smith, G.D., Duax, W.L., Langs, D.A., DeTitta, G.T., Edmonds, R.C., Rohrer, D.C., and Weeks, C M., *J. Am. Chem. Soc.* **97**, 7242 (1975).] Hydrogen atoms are not shown.

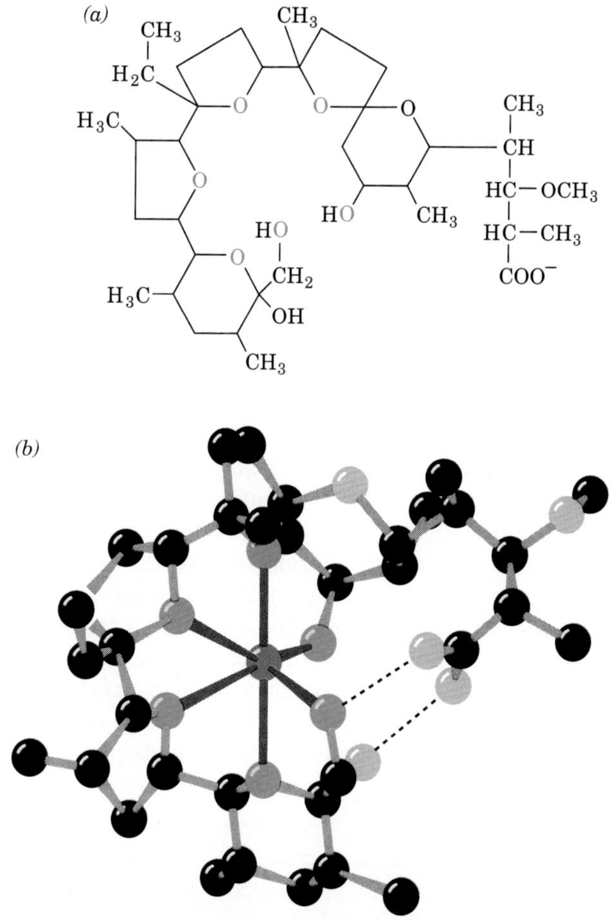

(a)

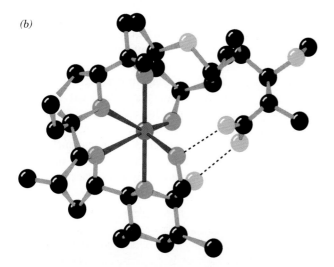

(b)

FIGURE 18-9. Monensin. (*a*) The structural formula with the six oxygen atoms that octahedrally complex Na⁺ indicated in red. (*b*) The X-ray structure of the Na⁺ complex (hydrogen atoms not shown). [After Duax, W.L., Smith, G.D., and Strong, P.D., *J. Am. Chem. Soc.* **102,** 6728 (1980).]

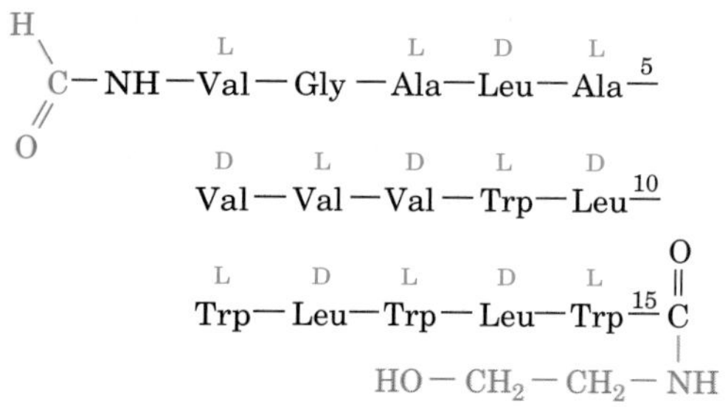

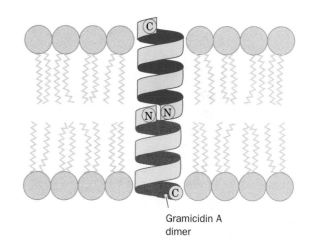

Gramicidin A

FIGURE 18-10. Gramicidin A consists of 15 alternating D- and L-amino acid residues and is blocked at both its N- and C-termini.

FIGURE 18-11. A schematic diagram of the transmembrane channel formed by two molecules of gramicidin A. The molecules presumably dimerize by a hydrogen bonding association between their *N*-formyl ends (N).

these ions with water are therefore energetically more favorable than their complexes with valinomycin. This accounts for valinomycin's 10,000-fold greater binding affinity for K⁺ over Na⁺. Indeed, no known substance discriminates so acutely between Na⁺ and K⁺.

The Na⁺-binding ionophore **monensin** (Fig. 18-9*a*), a linear polyether carboxylic acid, is chemically different from valinomycin. Nevertheless, X-ray analysis reveals that monensin's Na⁺ complex has the same general features as valinomycin's K⁺ complex in that monensin octahedrally coordinates Na⁺ so as to wrap it in a nonpolar jacket (Fig. 18-9*b*). Other carrier ionophores have similar characteristics.

Gramicidin A Forms Helical Transmembrane Channels

Gramicidin A is a channel-forming ionophore from *Bacillus brevis* that permits the passage of protons and alkali cations but is blocked by Ca²⁺. It is a 15-residue linear

polypeptide of alternating L and D residues that is chemically blocked at its amino terminus by formylation and at its carboxyl terminus by an amide bond with ethanolamine (Fig. 18-10). Note that all of its residues are hydrophobic as is expected for a small transmembrane polypeptide. NMR and X-ray crystallographic evidence indicate that *gramicidin A dimerizes in a head-to-head fashion to form a transmembrane channel* (Fig. 18-11). This is corroborated by the observation that two gramicidin A molecules with their N-terminal amino groups covalently cross-linked form a functional ion channel.

The gramicidin A channel cannot be α helical because α helices lack a central channel and cannot consist of alternating L and D residues. Dan Urry has proposed that gramicidin A forms a novel helix that he named the *β* **helix** because it resembles a rolled up parallel *β* sheet. Successive backbone N—H groups in this model alternately point up and down the helix to hydrogen bond with backbone carbonyl groups. An NMR structure, by Timothy Cross, of gramicidin A in a lipid bilayer supports this model and shows that the helix is right handed with 6 to 7 residues per turn (Fig. 18-12). As a consequence of its alternating D and L residues, the side chains of the β helix festoon its periphery to form the channel's required hydrophobic exterior (recall that in a β sheet of all L-amino acid residues, the side chains

alternately extend to opposite sides of the sheet; Section 7-1C). The polar backbone groups thus line the central channel and thereby facilitate the passage of ions. The four Trp side chains in the C-terminal half of each polypeptide chain are oriented with their polar N—H groups directed towards the bilayer surface, thereby orienting the dimeric helix perpendicular to the bilayer. Indeed, the replacement of these Trp residues by Phe significantly reduces the channel's conductivity without altering its backbone conformation.

Porin Structures Explain Their Ion Selectivities

The only channel-forming proteins whose X-ray structures are presently known are the bacterial **porins**, trimeric transmembrane proteins, each of whose identical subunits consists mainly of a 16-stranded antiparallel β barrel that forms a solvent-accessible channel along its barrel axis (Section 11-3A). In the *E. coli* OmpF porin (Figs. 11-28 and 18-13), this ~50-Å-long channel is constricted near its center to an elliptical pore that has a minimum cross-section of 7 × 11 Å. Consequently, solutes of more than ~600 D are too large to pass through a porin channel. The OmpF porin is weakly cation selective, whereas the 63% identical *E. coli* PhoE porin is weakly anion selective. This

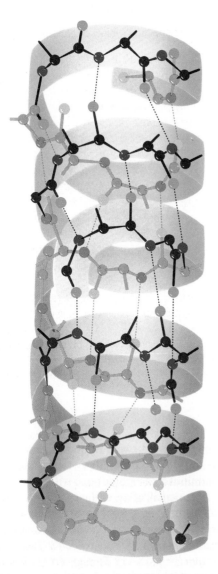

FIGURE 18-12.   A model for the gramicidin A transmembrane channel. Each gramicidin A molecule of the head-to-head dimer forms a right-handed helix in which the hydrogen bonding pattern resembles that in a parallel β pleated sheet. The novel hydrogen bonding arrangement of this 26-Å-long "β helix" is made possible because the alternating D and L configurations of the amino acid residues permit both successive NH and successive CO groups to point in opposite directions along the helix. The helix's bore diameter of 4 Å is sufficient to permit passage of alkali metal cations. [After Dobler, M., *Ionophores and Their Structures, p.* 215, Wiley–Interscience (1981). Based on a model proposed by D.W. Urry.]

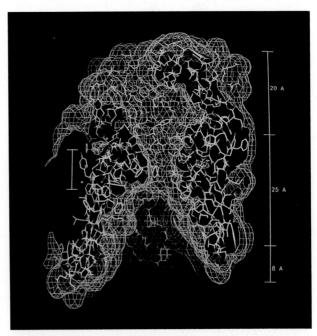

FIGURE 18-13.   A longitudinal section through the pore of the OmpF porin subunit as seen in its X-ray structure. The meshwork indicates the protein's solvent-accessible surface (the part of the protein that is in contact with water). The narrowest part of the pore, which is ~9 Å in length, is indicated by the short vertical bar on the left. The vertical scale on the right is the same as that in Fig. 11-28*c* and, hence, reference to this figure indicates the position of the hydrophobic band on the exterior of the protein. [Courtesy of Tilman Schirmer and Johan Jansonius, University of Basel, Basel, Switzerland.]

difference in ion selectivity is largely explained by the presence of Lys 131 in PhoE, the homolog of Gly 131 in OmpF. The side chain of Lys 131 protrudes into the channel so as to extend a row of basic side chains lining one wall of the constriction zone. The mutagenic replacement of Lys 131 with Glu largely abolishes the anion selectivity of PhoE, whereas replacement of any of several Lys residues lining the mouth of the channel by Glu has a lesser effect on ion selectivity. Evidently, Lys 131 electrostatically facilitates the passage of anions, probably by providing an anion binding site, whereas it is likely that the Lys side chains at the mouth of the channel function to attract anions into the channel.

D. *Mechanism of Passive-Mediated Glucose Transport*

Integral membrane proteins either are exposed only at one surface of the membrane or, in the case of transmembrane proteins, are oriented in only one direction with respect to the membrane (Section 11-3A). Since protein flip-flop rates are negligible, *the mobile carrier model that describes the mechanism of ionophores such as valinomycin (Fig. 18-6a) is not applicable to protein-mediated transport; rather, some sort of channel or pore mechanism appears likely.*

Glucose Transport Occurs via a Gated Pore Mechanism

The erythrocyte glucose transporter is a 55-kD glycoprotein which, according to sequence hydropathy analysis (Sections 7-4C and 11-3A), has four major domains (Fig. 18-14): (1) a bundle of 12 membrane-spanning α helices that are thought to form a hydrophobic cylinder surrounding a hydrophilic channel through which the glucose is transported; (2) a large, highly charged, cytoplasmic domain located between helices 6 and 7; (3) a smaller, carbohydrate-bearing, external domain located between helices 1 and 2; and (4) a relatively large C-terminal domain which is also cytoplasmic. The glucose transporter accounts for 2% of erythrocyte membrane proteins and runs as band 4.5 in SDS–PAGE gels of erythrocyte membranes (Section 11-3C; it is not visible on the gel depicted in Fig. 11-34 because the heterogeneity of its oligosaccharide component makes the protein band diffuse).

Two observations support the hypothesis that the erythrocyte glucose transporter is asymmetrically disposed in the membrane (Fig. 18-14b):

1. **Galactose oxidase** oxidizes the galactose units of the glucose transporter's carbohydrate moiety only when the oxidase is outside the erythrocyte. The transporter's galactose units must therefore be located on the erythrocyte cell surface.

2. Trypsin only disrupts glucose transport when it acts from within an erythrocyte ghost (erythrocytes made devoid of cytoplasm). The transporter's trypsin-sensi-

tive amino acid residues must therefore be located only on the cytoplasmic surface of the erythrocyte plasma membrane.

The glucose-binding sites on the two sides of the erythrocyte membrane have different steric requirements as well. John Barnett showed that addition of a propyl group to glucose C1 prevents glucose binding to the outer surface of the membrane, whereas addition of a propyl group to C6 prevents binding to the inner surface. He therefore proposed that this transmembrane protein has two alternate conformations: one with the glucose site facing the external cell surface, requiring O1 contact, and leaving O6 free, and the other with the glucose site facing the internal cell surface, requiring O6 contact and leaving O1 free (Fig. 18-15). *Transport apparently takes place by binding glucose to the protein on one face of the membrane, followed by a conformational change that closes the first site while exposing the other.* Glucose can then dissociate from the protein having been translocated across the membrane. The transport cycle is completed by the reversion of the glucose transporter to its initial conformation in the absence of bound glucose. Since this cycle can occur in either direction, the direction of net glucose transport is from high to low glucose concentrations. The glucose transporter thereby provides a means of equilibrating the glucose concentration across the erythrocyte membrane without any accompanying leakage of small molecules or ions.

The mechanism of glucose transport across erythrocyte membranes is a general one, often referred to as a **gated pore**. Indeed, *all known transport proteins appear to be asymmetrically situated transmembrane proteins that alternate between two conformational states in which the ligand-binding sites are exposed, in turn, to alternate sides of the membrane.*

Eukaryotes Express a Variety of Glucose Transporters

The erythrocyte glucose transporter, known also as **GLUT1**, has a highly conserved amino acid sequence (98% sequence identity between humans and rats), which suggests that all segments of this protein are functionally significant. GLUT1 is expressed in most tissues, although in liver and muscle, tissues that are highly active in glucose transport, it is present in only tiny amounts. At least four other glucose transporters, **GLUT2** through **GLUT5**, are 40 to 65% identical to GLUT1 but have different tissue distributions. For example, GLUT2 is prominant in pancreatic β cells (which secrete insulin in response to increased [glucose]; Section 17-3F) and liver (where its defects result in Type I glycogen storage disease; Section 17-4), whereas **GLUT4** occurs mainly in muscle and fat cells. Note that the tissue distributions of these glucose transporters correlate with the response of these tissues to insulin: Liver is unresponsive to insulin (liver functions, in part, to maintain the level of blood glucose; Section 17-3F), whereas muscle and fat cells take up glucose when stimulated by insulin.

2. E_2—P has an outward-facing high-affinity K^+-binding site ($K_M = 0.05M$, well below the extracellular $[K^+]$), and hydrolyzes to form $P_i + E_2$ only when K^+ is bound.

An Ordered Sequential Kinetic Reaction Mechanism Accounts for the Coupling of Active Transport with ATP Hydrolysis

The (Na^+-K^+)–ATPase is thought to operate in accordance with the following ordered sequential reaction scheme (Fig. 18-20):

1. $E_1 \cdot 3Na^+$, which acquired its Na^+ inside the cell, binds ATP to yield the ternary complex $E_1 \cdot ATP \cdot 3Na^+$.

2. The ternary complex reacts to form the "high-energy" aspartyl phosphate intermediate $E_1 \sim P \cdot 3Na^+$.

3. This "high-energy" intermediate relaxes to its "low-energy" conformation, E_2—$P \cdot 3Na^+$, and releases its bound Na^+ outside the cell, that is, Na^+ is transported through the membrane.

4. E_2—P binds $2K^+$ from outside the cell to form E_2—$P \cdot 2K^+$.

5. The phosphate group is hydrolyzed yielding $E_2 \cdot 2K^+$.

6. $E_2 \cdot 2K^+$ changes conformation, releases its $2K^+$ inside the cell, and replaces it with $3Na^+$, thereby completing the transport cycle.

The enzyme is thought to have only one set of cation-binding sites, which apparently changes both its orientation and its specificity during the course of the transport cycle.

The obligatory order of the reaction requires that ATP can be hydrolyzed only as Na^+ is transported "uphill." Conversely, Na^+ can be transported "downhill" only if ATP is concomitantly synthesized. Consequently, although each of the above reaction steps is, in fact, individually reversible, the cycle, as is diagrammed in Fig. 18-20, circulates only in the clockwise direction under normal physiological conditions; that is, ATP hydrolysis and ion transport are coupled processes. Note that the **vectorial** (unidirectional) nature of the reaction cycle results from the alternation of some of the steps of the exergonic ATP hydrolysis reaction (Steps 1+2 and Step 5) with some of the steps of the endergonic ion transport process (Steps 3+4 and Step 6). Thus, neither reaction can go to completion unless the other one also does.

Mutual Destabilization Accounts for the Rate of Na^+ and K^+ Transport

The above ordered kinetic mechanism accounts only for the coupling of active transport with ATP hydrolysis. *In order to maintain a reasonable rate of transport, the free energies of all its intermediates must be roughly equal. If some intermediates were much more stable than others, the stable intermediates would accumulate, thereby severely reducing the overall transport rate.* For example, in order for Na^+ to be transported out of the cell, uphill, its binding to E_1 must be strong on the inside and weak to E_2 on the outside. Strong binding means greater stability and a potential bottleneck. This difficulty is counteracted by the phosphorylation of $E_1 \cdot 3Na^+$ and its subsequent conformational change to yield the low Na^+ affinity E_2—P (Steps 2 and 3, Fig. 18-20). Likewise, the strong binding of K^+ to E_2—P on the outside is attenuated by its dephosphorylation and conformational change to yield the low K^+ affinity E_1 (Steps 5 and 6, Fig. 18-20). It is these mutual destabilizations that permit Na^+ and K^+ to be transported at a rapid rate.

Cardiac Glycosides Specifically Inhibit the (Na^+-K^+)–ATPase

Study of the (Na^+-K^+)–ATPase has been greatly facilitated by the use of **cardiac glycosides** (also called **cardio-**

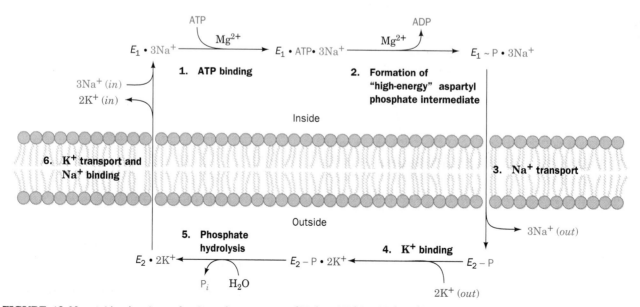

FIGURE 18-20.   A kinetic scheme for the active transport of Na^+ and K^+ by (Na^+-K^+)–ATPase.

(a)

(b)

Digitoxin (digitalin)

Ouabain

FIGURE 18-21. (*a*) The leaves of the purple foxglove plant are the source of the heart muscle stimulant digitalis. [Derek Fell.] (*b*) Digitoxin (digitalin), the major component of digitalis; and ouabain, a cardiac glycoside isolated from the East African Ouabio tree, are among the most commonly prescribed cardiac drugs.

tonic steroids), natural products that increase the intensity of heart muscle contraction. Indeed, **digitalis,** an extract of purple foxglove leaves (Fig. 18-21*a*), which contains a mixture of cardiac glycosides including **digitoxin (digitalin;** Fig. 18-21*b*), has been used to treat congestive heart failure for centuries. The cardiac glycoside **ouabain** (pronounced wabane; Fig. 18-21*b*), a product of the East African Ouabio tree, has been long used as an arrow poison. These two steroids, which are still among the most commonly prescribed cardiac drugs, inhibit the (Na^+-K^+)-ATPase by binding strongly to an externally exposed portion of the enzyme (the drugs are ineffective when injected inside cells) so as to block Step 5 in Fig. 18-20. The resultant increase in intracellular $[Na^+]$ stimulates the cardiac (Na^+-Ca^{2+}) antiport system, which pumps Na^+ out of and Ca^{2+} into the cell (Section 20-1B). The increased cytosolic $[Ca^{2+}]$ boosts the $[Ca^{2+}]$ in other cellular organelles, principally the sarcoplasmic reticulum. Thus, the release of Ca^{2+} to trigger muscle contraction (Section 34-3C) produces a larger than normal increase in cytosolic $[Ca^{2+}]$, thereby intensifying the force of cardiac muscle contraction. Ouabain, which was once thought to be produced only by plants, has recently been discovered also to be an animal hormone that is secreted by the adrenal cortex and functions to regulate cell $[Na^+]$ and overall body salt and water balance.

B. Ca^{2+}–ATPase

Ca^{2+} often acts as a second messenger in a manner similar to cAMP. Transient increases in cytosolic $[Ca^{2+}]$ trigger numerous cellular responses, including muscle contraction (Section 34-3C), release of neurotransmitters (Section 34-4C), and, as we have seen, glycogen breakdown (Section 17-3C). Moreover, Ca^{2+} is an important activator of oxidative metabolism (Section 19-4).

The use of phosphate as a basic energy currency requires cells to maintain a low internal $[Ca^{2+}]$ because, for example, $Ca_3(PO_4)_2$ has a maximum aqueous solubility of 65 μM. Thus, the $[Ca^{2+}]$ in the cytosol (~ 0.1 μM) is four orders of magnitude less than it is in the extracellular spaces (~ 1500 μM). This large concentration gradient is maintained by the active transport of Ca^{2+} across the plasma membrane, the endoplasmic reticulum (the sarcoplasmic reticulum in muscle), and the mitochondrial inner membrane. We discuss the mitochondrial system in Section 20-1B. Plasma membrane and endoplasmic reticulum each contain a P-type Ca^{2+}–**ATPase** (Ca^{2+} **pump**) that actively pumps Ca^{2+} out of the cytosol at the expense of ATP hydrolysis. Their kinetic mechanisms (18-22) are very similar to that of the (Na^+-K^+)-ATPase (Fig. 18-20).

Calmodulin Regulates the Plasma Membrane Ca^{2+} Pump

For a cell to maintain its proper physiological state, it must regulate the activities of its ion pumps precisely. *The regulation of the Ca^{2+} pump in the plasma membrane is controlled by the level of Ca^{2+} through the mediation of*

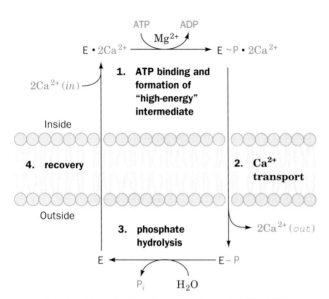

FIGURE 18-22. The kinetic mechanism of Ca^{2+}–ATPase. Here *(in)* refers to the cytosol and *(out)* refers to the outside of the cell for plasma membrane Ca^{2+}–ATPase or the lumen of the endoplasmic reticulum (sarcoplasmic reticulum) for the Ca^{2+}–ATPase of that membrane.

calmodulin (CaM). This ubiquitous eukaryotic Ca^{2+}-binding protein participates in numerous cellular regulatory processes including, as we have seen, the control of glycogen metabolism (Section 17-3C).

Ca^{2+}–Calmodulin activates the Ca^{2+}–ATPase of plasma membranes. The activation, as deduced from the study of the isolated ATPase, results in a decrease in its K_M for Ca^{2+} from 20 to 0.5 μM. Ca^{2+}–CaM activates the Ca^{2+} pump by binding to an inhibitory polypeptide segment of the pump in a manner similar to the way in which Ca^{2+}–CaM activates its target protein kinases (Section 17-3C). Evidence supporting this mechanism comes from proteolytically excising the Ca^{2+} pump's CaM-binding polypeptide, yielding a truncated pump that is active even in the absence of CaM. Synthetic peptides corresponding to this CaM-binding domain not only bind Ca^{2+}–CaM but inhibit the truncated pump by increasing its K_M for Ca^{2+} and decreasing its V_{max}. This suggests that, in the absence of Ca^{2+}–CaM, the CaM-binding domain of the pump interacts with the rest of the protein so as to inhibit its activity. When the Ca^{2+} concentration increases, Ca^{2+}–CaM forms and binds to the CaM-binding domain of the pump in a way that causes it to dissociate from the rest of the pump, thereby relieving the inhibition.

Now we can see how Ca^{2+} regulates its own cytoplasmic concentration: At Ca^{2+} levels below calmodulin's ~1-μM dissociation constant for Ca^{2+}, the Ca^{2+}–ATPase is relatively inactive due to autoinhibition by its CaM-binding domain. If, however, the $[Ca^{2+}]$ rises to this level, Ca^{2+} binds to calmodulin which, in turn, binds to the CaM-bind-

ing domain so as to relieve the inhibition, thereby activating the Ca^{2+} pump:

$$Ca^{2+} + CaM \rightleftharpoons$$
$$Ca^{2+}-CaM^* + pump \;(inactive) \rightleftharpoons$$
$$Ca^{2+}-CaM^* \cdot pump \;(active)$$

(CaM* indicates activated calmodulin). This interaction decreases the pump's K_M for Ca^{2+} to below the ambient $[Ca^{2+}]$, thereby causing Ca^{2+} to be pumped out of the cytosol. When the $[Ca^{2+}]$ decreases sufficiently, Ca^{2+} dissociates from calmodulin and this series of events reverses itself, thereby inactivating the pump. The entire system is therefore analogous to a basement sump pump that is automatically activated by a float when the water reaches a preset level.

C. $(H^+ - K^+)$ - ATPase of Gastric Mucosa

Parietal cells of the mammalian gastric mucosa secrete HCl at a concentration of $0.15M$ (pH 0.8). Since the cytosolic pH of these cells is 7.4, this represents a pH difference of 6.6 units, the largest known in eukaryotic cells. The secreted protons are derived from the intracellular hydration of CO_2 by carbonic anhydrase:

$$CO_2 + H_2O \rightleftharpoons HCO_3^- + H^+$$

The secretion of H^+ involves the participation of an $(H^+ - K^+)$ - ATPase, an electroneutral antiport with structure and properties similar to that of $(Na^+ - K^+)$ - ATPase. Like the related $(Na^+ - K^+)$ - and Ca^{2+} - ATPases, it is phosphorylated during the transport process. In this case, however, the K^+, which enters the cell as H^+ is pumped out, is subsequently externalized by its electroneutral cotransport with Cl^-. HCl is therefore the overall transported product.

For many years, effective treatment of peptic ulcers, which was a frequently fatal condition caused by the attack of stomach acid on the gastric mucosa, often required the surgical removal of the affected portions of the stomach. The discovery, by James Black, of **cimetidine,**

Cimetidine

Histamine

which inhibits stomach acid secretion, has almost entirely eliminated the need for this dangerous and debilitating surgery. The (H^+-K^+)–ATPase of the gastric mucosa is activated by histamine stimulation of a cell-surface receptor in a process mediated by cAMP. Cimetidine (trade name: Tagamet) and its analogs, which competitively inhibit the binding of histamine to this receptor, are presently among the most commonly prescribed drugs in the United States.

D. Group Translocation

Group translocation is a variation of ATP-driven active transport that most bacteria use to import certain sugars. It is required for many bacterial processes, both useful and harmful (to humans), such as those that produce cheese, soy sauce, and dental cavities. *It differs from active transport in that the molecules transported are simultaneously modified chemically.* The most extensively studied example of group translocation is the **phosphoenolpyruvate-dependent phosphotransferase system (PTS)** of *E. coli* discovered by Saul Roseman in 1964. Phosphoenolpyruvate (PEP) is the phosphoryl donor for this system (recall that PEP is the "high-energy" phosphoryl donor for ATP synthesis in the pyruvate kinase reaction of glycolysis; Section 16-2J). *The PTS simultaneously transports and phosphorylates sugars. Since the cell membrane is impermeable to sugar phosphates, once they enter the cell, they remain there.* Some of the PTS-transported sugars are listed in Table 18-2.

TABLE 18-2. SOME OF THE SUGARS TRANSPORTED BY THE *E. COLI* PEP-DEPENDENT PHOSPHOTRANSFERASE SYSTEM (PTS)

Glucose	Galactitol
Fructose	Mannitol
Mannose	Sorbitol
N-Acetylglucosamine	Xylitol

The PTS system involves two soluble cytoplasmic proteins, **Enzyme I (E I)** and **HPr** (for histidine-containing phosphocarrier protein), which participate in the transport of all sugars (Fig. 18-23). In addition, for each sugar the system transports, there is a specific transmembrane transport protein **E II**, and in some cases, a sugar-specific protein **E III** that is membrane bound for some sugars and cytoplasmic for others. Glucose transport, for example, requires the participation of E II^{glc} and E III^{glc}.

Glucose transport, which resembles that of other sugars, involves the transfer of a phosphoryl group from PEP to glucose with net inversion of configuration about the phosphorus atom. Since each phosphoryl transfer involves inversion (Section 15-2B), an odd number of transfers must be involved. Four phosphorylated protein intermediates have been identified, indicative of five phosphoryl transfers:

$$PEP \rightarrow E\ I \rightarrow HPr \rightarrow E\ III^{glc} \rightarrow E\ II^{glc} \rightarrow glucose$$

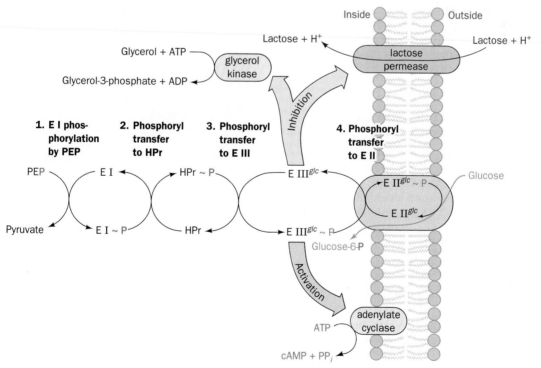

FIGURE 18-23. Transport of glucose by the PEP-dependent phosphotransferase system (PTS). HPr and E I are cytoplasmic proteins common to all sugars transported. E II^{glc} and E III^{glc} are proteins specific for glucose. E III^{glc} inhibits non-PTS transport proteins such as the lactose permease (Section 18-4B) and enzymes such as glycerol kinase. Adenylate cyclase is activated by the presence of E III^{glc}~P (or possibly inhibited by the presence of E III^{glc}).

The transport process occurs as follows (Fig. 18-23):

1. PEP phosphorylates E I at N3 of His 189 to form a reactive phosphohistidine adduct.

Phosphohistidine

2. The phosphoryl group is transferred to N1 of His 15 on HPr. His is apparently a favored phosphoryl group acceptor in phosphoryl transfer reactions. It also participates in the phosphoglycerate mutase reaction of glycolysis (Section 16-2H).

3. HPr~P continues the phosphoryl-transfer chain by phosphorylating E III^{glc}, again on a His residue, His 90.

4. The fourth phosphoryl transfer is to Cys 421 of E II^{glc}.

5. The phosphoryl group is finally transferred from E II^{glc} to glucose, which, in the process, is transported across the membrane by E II^{glc}. Glucose is released into the cytoplasm only after it has been phosphorylated to glucose-6-phosphate (G6P).

Thus the transport of glucose is driven by its indirect, exergonic phosphorylation by PEP. The PTS is an energy-efficient system since only one ATP-equivalent is required to both transport and phosphorylate glucose. When the active transport and phosphorylation steps occur separately, as they do in many cells, two ATPs are hydrolyzed per glucose processed.

Bacterial Sugar Transport Is Genetically Regulated

The PTS is more complex than the other transport systems we have encountered, probably because it is part of a complicated regulatory system governing sugar transport. When any of the sugars transported by the PTS is abundant, the active transport of sugars which enter the cell via other transport systems is inhibited. This inhibition, called **catabolite repression,** is mediated through the cAMP concentration (Section 29-3C). cAMP activates the transcription of genes that encode various sugar transport proteins, including **lactose permease** (Section 18-4B). The presence of glucose results in a decrease in [cAMP] which, in turn, represses the synthesis of these other sugar transport proteins. Direct inhibition of the sugar transport proteins themselves, as well as of certain enzymes, also occurs.

The mechanism for control of [cAMP] is thought to reside in E III^{glc}, which is transiently phosphorylated in Step 3 of the PTS transport process (Fig. 18-23). When glucose is plentiful, this enzyme is present mostly in its dephospho form since E III^{glc}~P rapidly transfers its phosphoryl group through to glucose. Under these conditions, adenylate cyclase is inactive, although whether dephospho E III^{glc} inhibits this enzyme or E III^{glc}~P activates it is unclear. However, dephospho E III^{glc} binds to and inhibits many non-PTS transporters and enzymes that participate in the the metabolism of sugars other than glucose (the metabolite of choice for many bacteria), including lactose permease and **glycerol kinase** (Section 16-5). In the absence of glucose, E III^{glc} is converted to E III^{glc}~P, thereby relieving the inhibition of non-PTS transporters. In addition, adenylate cyclase is activated to produce cAMP, which, in turn, induces the increased production of some of the non-PTS transporters and enzymes that E III^{glc} inhibits. This is a form of energy conservation for the cell. Why synthesize the proteins required for the transport and metabolism of all sugars when the metabolism of only one sugar at a time will do?

FIGURE 18-24. The X-ray structure of *E. coli* E III^{glc} (*yellow,* a 168-residue monomer) in complex with one of its regulatory targets, glycerol kinase (*blue,* a tetramer of identical 501-residue subunits). The two proteins associate, in part, by tetrahedrally coordinating a Zn^{2+} ion via the side chains of His 75 and His 90 of E III^{glc}, a carboxylate oxygen from Glu 478 of glycerol kinase, and a water molecule. These groups are shown in ball-and-stick form with C grey, N blue, O red, and Zn^{2+} white. The Zn^{2+}-mediated interaction between E III^{glc} and glycerol kinase inactivates glycerol kinase, presumably through an induced fit mechanism. The phosphorylation of E III^{glc} at His 90 disrupts this interaction, thereby reversing the inhibition of glycerol kinase. [Courtesy of James Remington, University of Oregon.]

The X-Ray Structure of E III^{glc} in Complex with Glycerol Kinase

The X-ray structures of E III^{glc}, both alone and in complex with one of its regulatory targets, glycerol kinase, which were determined by James Remington and Roseman, have revealed how E III^{glc} inhibits at least some of its targets and why E III^{glc}~P does not do so. E III^{glc} contains two His residues, His 75 and His 90, that are required for phosphoryl transfer, although only His 90 is necessary for E III^{glc} to accept a phosphate from HPr. The X-ray structure of *E. coli* E III^{glc} alone reveals that these two His residues lie in close proximity (their N3 atoms are 3.3 Å apart) in a depression on the surface of the protein that is surrounded by a remarkable ~18-Å in diameter hydrophobic ring consisting of 11 Phe, Val, and Ile side chains.

The X-ray structure of E III^{glc} in complex with glycerol kinase (Fig. 18-24) confirms that this hydrophobic gasket is indeed the site of interaction between the two proteins and reveals how the phosphorylation of His 90 disrupts this interaction. The two active site His residues, which are completely buried within the hydrophobic interaction surface, coordinate with a previously unanticipated Zn^{2+} ion, which is additionally coordinated to Glu 478 of glycerol kinase and a water molecule. The phosphorylation of E III^{glc} His 90 to yield E III^{glc}~P no doubt disrupts this intermolecular interaction, thereby releasing glycerol kinase and reversing its inhibition.

4. ION GRADIENT–DRIVEN ACTIVE TRANSPORT

Systems such as the (Na$^+$–K$^+$)–ATPase discussed above utilize the free energy of ATP hydrolysis to generate electrochemical potential gradients across membranes. Conversely, *the free energy stored in an electrochemical potential gradient may be harnessed to power various endergonic physiological processes.* Indeed, ATP synthesis by mitochondria and chloroplasts is powered by the dissipation of proton gradients generated through electron transport and photosynthesis (Sections 20-3C and 22-2D). In this section we discuss active transport processes that are driven by the dissipation of ion gradients. We consider three examples: intestinal uptake of glucose by the **Na$^+$–glucose symport,** uptake of lactose by *E. coli* **lactose permease,** *and the mitochondrial **ADP–ATP transporter.***

A. Na$^+$–Glucose Symport

Nutritionally derived glucose is actively concentrated in **brush border cells** of the intestinal epithelium by a Na$^+$-dependent symport (Fig. 18-25). It is transported from these cells to the circulatory system via a passive-mediated glucose uniport located on the capillary side of the cell and

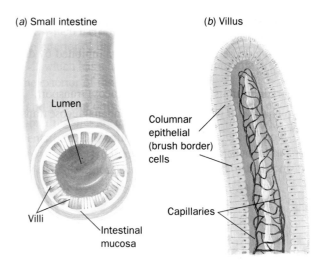

(a) Small intestine   *(b)* Villus

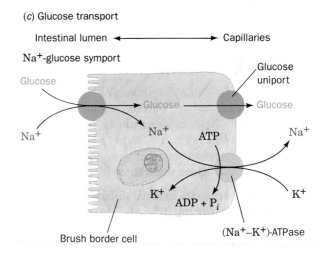

(c) Glucose transport

FIGURE 18-25. Glucose transport in the intestinal epithelium. The brushlike villi lining the small intestine greatly increase its surface area, thereby facilitating the absorption of nutrients. The brush border cells from which the villi are formed actively concentrate glucose from the intestinal lumen in symport with Na$^+$, a process that is driven by the (Na$^+$–K$^+$)–ATPase, which is located on the capillary side of the the cell and functions to maintain a low internal [Na$^+$]. The glucose is exported to the bloodstream via a separate passive-mediated uniport system like that in the erythrocyte.

which is similar to that of the erythrocyte membrane (Section 18-2B). Note that *although the immediate energy source for glucose transport from the intestine is the Na$^+$ gradient, it is really the free energy of ATP hydrolysis that powers this process through the maintenance of the Na$^+$ gradient by the (Na$^+$–K$^+$)–ATPase.* Nevertheless, since glucose enhances Na$^+$ resorption, which in turn enhances water resorption, glucose is often fed to individuals suffering from salt and water losses resulting from diarrhea.

cates, in agreement with the above electron microscopic observations, that E$_2$ consists of eight tightly associated trimers arranged at the corners of a cube (Fig. 19-9). The resulting hollow cagelike structure contains channels large enough to allow substrates to diffuse in and out. In fact, the

X-ray structures of the ternary complex of the catalytic domain with coenzyme A and dihydrolipoic acid reveal that these substrates bind in extended conformations at opposite ends of a 30-Å-long channel that is located at the interface between two of the subunits in a trimer. This ar-

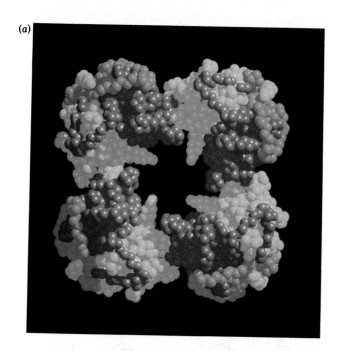

(a)

FIGURE 19-9. The X-ray structure of the *A. vinelandii* dihydrolipoyl transacetylase (E$_2$) catalytic domain. (*a*) A space-filling drawing (each residue is represented by a sphere centered on its C$_\alpha$ atom) of 8 trimers (24 identical subunits) arranged at the corners of a cube as viewed along one of the cube's 4-fold axes (only the forward half of the complex is visible). The edge length of the cube is ~125 Å. Note that the subunits in a trimer are extensively associated, but that the interactions between contacting trimers are relatively tenuous. [Courtesy of Wim Hol, University of Washington.] (*b*) Ribbon diagram of a trimer as viewed along its 3-fold axis (along the cube's body diagonal) from outside the complex. Coenzyme A (*purple*) and lipoamide (*light blue*), in skeletal form, are shown bound in the active site of the red subunit. Note how the N-terminal "elbow" of each subunit extends over a neighboring subunit; its deletion greatly destabilizes the complex. [Based on an X-ray structure by Wim Hol, University of Washington.]

(b)

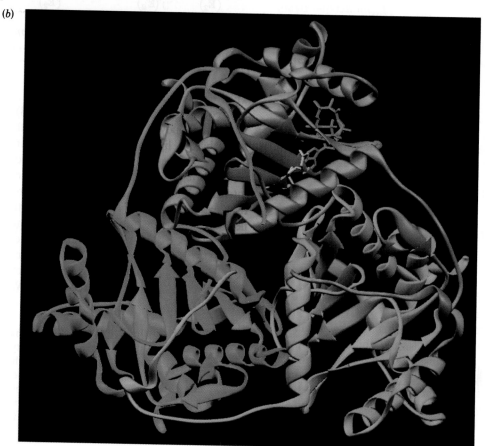

rangement requires CoA to approach its binding site from inside the cube. The flexible lipoyl-containing domains of E_2 are thought to protrude from the central core so as to interact with lipoyl-containing domains of neighboring E_2 subunits, as well as with E_1 and E_3.

In mammals and yeast the already complicated structure of the pyruvate dehydrogenase multienzyme complex has a further level of complexity in that it contains additional subunits: About six copies of the so-called **protein X** and one to three copies of **pyruvate dehydrogenase kinase** and **pyruvate dehydrogenase phosphatase**. Protein X contains a lipoyllysine-containing domain similar to E_2 and can accept an acetyl group, but its C-terminal domain has no catalytic activity, and the removal of its lipoyllysine domain does not diminish catalytic activity. It's main role appears to be to aid in the binding of E_3, since limited proteolysis of protein X decreases E_3 binding ability. The kinase and phosphatase are important in the regulation of the catalytic activity of the complex (Section 19-2B).

Arsenic Compounds Are Poisonous because They Sequester Lipoamide

Arsenic has been known to be a poison since ancient times. As(III) compounds, such as **arsenite** (AsO_3^{3-}) and organic arsenicals, are toxic because of their ability to covalently bind sulfhydryl compounds (Section 14-4). This is particularly true of sulfhydryls such as lipoamide that can form bidentate adducts.

Arsenite   **Dihydro-lipoamide**

Organic arsenical

The resultant inactivation of lipoamide-containing enzymes, especially pyruvate dehydrogenase and α-ketoglutarate dehydrogenase (Section 19-3D), brings respiration to a halt.

Organic arsenicals are more toxic to microorganisms than to humans, apparently because of differences in the sensitivities of their various enzymes to these compounds. This differential toxicity is the basis for the early twentieth century use of these organic arsenicals in the treatment of syphilis (now superseded by penicillin) and trypanosomiasis (typanosomes are parasitic protozoa). These compounds were really the first antibiotics, although, not surprisingly, they had severe side effects.

Arsenic is often suspected as a poison in untimely deaths. In fact, it is thought that Napoleon Bonaparte died from arsenic poisoning while in exile on the island of St. Helena. This suspicion, and the chemical analyses it sparked, makes a fascinating chemical anecdote. The finding that a lock of Napoleon's hair indeed contained high levels of arsenic strongly supports the notion that arsenic poisoning was the cause of his death. Was it murder or environmental pollution? Arsenic-containing dyes were used in wallpaper at the time and, it was eventually determined, in damp weather fungi convert the arsenic to a volatile compound. Samples of the wallpaper in Napoleon's room have survived and their analysis indicates that they contain arsenic. Napoleon's arsenic poisoning may therefore have been unintentional.

Retrospective detective work also suggests that Charles Darwin was a victim of chronic arsenic poisoning. For most of his life after he returned from his epic voyage, Darwin complained of numerous ailments, including eczema, vertigo, headaches, arthritis, gout, palpitations, and nausea, all symptoms of arsenic poisoning. Fowler's solution, a common nineteenth century tonic, contained 10 mg of arsenite·mL^{-1}. Many patients, quite possibly Darwin himself, took this "medication" for years.

B. Control of Pyruvate Dehydrogenase

The pyruvate dehydrogenase multienzyme complex regulates the entrance of acetyl units derived from carbohydrate sources into the citric acid cycle. The decarboxylation of pyruvate by E_1 is irreversible and, since there are no other pathways in mammals for the synthesis of acetyl-CoA from pyruvate, it is crucial that the reaction be carefully controlled. Two regulatory systems are employed:

1. Product inhibition by NADH and acetyl-CoA (Fig. 19-10*a*)

2. Covalent modification by phosphorylation/dephosphorylation of the pyruvate dehydrogenase (E_1) subunit (Fig. 19-10*b*; enzymatic regulation by covalent modification is discussed in Section 17-3B).

Control by Product Inhibition

NADH and acetyl-CoA compete with NAD$^+$ and CoA for binding sites on their respective enzymes. They also drive the reversible transacetylase (E_2) and dihydrolipoyl dehydrogenase (E_3) reactions backwards (Fig. 19-10a). High ratios of [NADH]/[NAD$^+$] and [acetyl-CoA]/[CoA] therefore maintain E_2 in the acetylated form, incapable of accepting the hydroxyethyl group from the TPP on E_1. This, in turn, ties up the TPP on the E_1 subunit in its hydroxyethyl form, decreasing the rate of pyruvate decarboxylation.

Control by Phosphorylation/Dephosphorylation

Control by phosphorylation/dephosphorylation occurs only in eukaryotic enzyme complexes. These complexes

(a)   **Product inhibition**

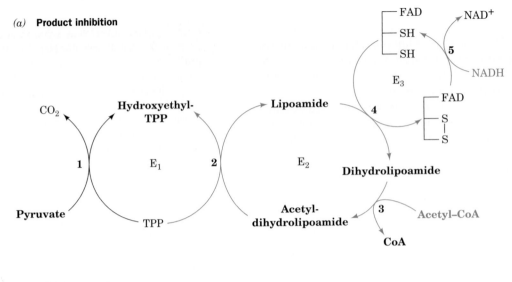

(b)   **Covalent modification**

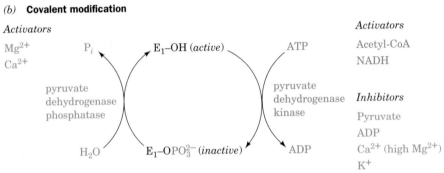

FIGURE 19-10.   The factors controlling the activity of the pyruvate dehydrogenase multienzyme complex. (*a*) Product inhibition. NADH and acetyl-CoA, respectively, compete with NAD^+ and CoA in Reactions **3** and **5** of the pyruvate dehydrogenase reaction sequence. When the relative concentrations of NADH and acetyl-CoA are high, the reversible reactions catalyzed by E_2 and E_3 are driven backwards (*red arrows*), thereby inhibiting further formation of acetyl-CoA.

(*b*) Covalent modification in the eukaryotic complex. Pyruvate dehydrogenase (E_1) is inactivated by the specific phosphorylation of one of its Ser residues in a reaction catalyzed by pyruvate dehydrogenase kinase (*right*). This phosphoryl group is hydrolyzed through the action of pyruvate dehydrogenase phosphatase (*left*), thereby reactivating E_1. The activators and inhibitors of the kinase are listed on the right and the activators of the phosphatase are listed on the left.

contain pyruvate dehydrogenase kinase and pyruvate dehydrogenase phosphatase bound to the dihydrolipoyl transacetylase core. *The kinase inactivates the pyruvate dehydrogenase (E_1) subunit by catalyzing the phosphorylation of a specific dehydrogenase Ser residue by ATP (Fig. 19-10b). Hydrolysis of this phospho-Ser residue by the phosphatase reactivates the complex.*

The products of the reaction, NADH and acetyl-CoA, in addition to their direct effects on the pyruvate dehydrogenase multienzyme complex, activate pyruvate dehydrogenase kinase. The resultant phosphorylation inactivates the complex just as the products themselves inhibit it. Insulin is involved in the control of this system through its indirect activation of pyruvate dehydrogenase phosphatase. Recall that insulin activates glycogen synthesis as well by activating phosphoprotein phosphatase (Section 17-3C). Insulin, in response to increases in blood [glucose] is now seen as promoting the synthesis of acetyl-CoA as well as glycogen.

As we shall see in Section 23-4, acetyl-CoA is the precursor to fatty acids in addition to being the fuel for the citric acid cycle. Various other activators and inhibitors regulate the pyruvate dehydrogenase system (Fig. 19-10*b*), but in contrast to the glycogen metabolism control system (Section 17-3), it is unaffected by cAMP.

3. ENZYMES OF THE CITRIC ACID CYCLE

In this section we discuss the reaction mechanisms of the eight citric acid cycle enzymes. Our knowledge of these mechanisms rests on an enormous amount of experimental work; as we progress, we shall pause to examine some of these experimental details. Consideration of how this cycle is regulated and its relationship to cellular metabolism are the subjects of the following sections.

A. Citrate Synthase

Citrate synthase (originally named **citrate condensing enzyme***) catalyzes the condensation of acetyl-CoA and oxaloacetate (Reaction 1 of Fig. 19-1).* This initial reaction of the citric acid cycle is the point at which carbon atoms are "fed into the furnace" as acetyl-CoA. The citrate synthase reaction proceeds with an ordered sequential kinetic mechanism (Section 13-5B), oxaloacetate adding to the enzyme before acetyl-CoA.

The X-ray structure of the free dimeric enzyme shows it to adopt an "open form" in which the two domains of each subunit form a deep cleft that contains the oxaloacetate-binding site (Fig. 19-11a). Upon binding oxaloacetate, however, the smaller domain undergoes a remarkable 18° rotation relative to the larger domain, which closes the cleft (Fig. 19-11b).

The X-ray structures of two inhibitors of citrate synthase in ternary complex with the enzyme and oxaloacetate have also been determined. **Acetonyl-CoA**, an inhibitory analog of acetyl-CoA in the ground state, and **carboxymethyl-CoA**, a proposed transition state analog (see below),

$$CoAS - C \overset{O}{\underset{CH_3}{\big\backslash}}$$

Acetyl-CoA

$$CoAS - C \overset{OH}{\underset{CH_2}{\big\backslash}}$$

Proposed enol intermediate

$$CoAS - CH_2 - C \overset{O}{\underset{CH_3}{\big\backslash}}$$

Acetonyl-CoA (ground-state analog)

$$CoAS - CH_2 - C \overset{OH}{\underset{O}{\big\backslash}}$$

Carboxymethyl-CoA (transition state analog)

bind to the enzyme in its "closed" form, thereby identifying the acetyl-CoA binding site. The existence of the "open" and "closed" forms explains the enzyme's ordered sequential kinetic behavior: *The conformational change induced by oxaloacetate binding generates the acetyl-CoA binding site while sealing off the solvent's access to the bound oxaloacetate.* This is a classic example of the induced-fit model of substrate binding (Section 9-4C). Hexokinase exhibits similar behavior (Section 16-2A).

The citrate synthase reaction is a mixed aldol–Claisen ester condensation, subject to general acid–base catalysis and the intermediate participation of the enol form of acetyl-CoA. The X-ray structure of the enzyme in ternary complex with oxaloacetate and carboxymethyl-CoA reveals that three of its ionizable side chains are properly oriented to play catalytic roles: His 274, Asp 375, and His 320. The N1 atoms in both of these His side chains are hydrogen bonded to two backbone NH groups indicating that these N1 atoms are not protonated. These observations

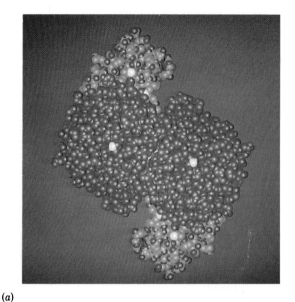

(a)

(b)

FIGURE 19-11. A space-filling drawing of citrate synthase in (a) the open conformation and (b) the closed, substrate-binding conformation. The C atoms of the small domain in each subunit of the enzyme are green and those of the large domain are purple. N, O, and S atoms in both domains are blue, red, and yellow. The view is along the homodimeric protein's twofold rotation axis. The large conformational shift between the open and closed forms entails relative interatomic movements of up to 15 Å. [Courtesy of Anne Dallas, University of Pennsylvania; and Helen Berman, Fox Chase Cancer Center. Based on X-Ray structures determined by James Remington and Robert Huber, Max-Planck-Institut für Biochemie, Germany.]

have led James Remington to propose the following three-step mechanism (Fig. 19-12):

1. The enol of acetyl-CoA is generated in the rate-limiting step of the reaction with the catalytic participation of

ergy changes for the eight citric acid cycle enzymes and estimates of the physiological free energy changes for the reactions in heart muscle or liver tissue. We can see that three of the enzymes are likely to function far from equilibrium under physiological conditions (negative ΔG): Citrate synthase, NAD$^+$-dependent isocitrate dehydrogenase, and α-ketoglutarate dehydrogenase. We shall therefore focus our discussion on how these enzymes are regulated (Fig. 19-18).

The Citric Acid Cycle Is Largely Regulated by Substrate Availability, Product Inhibition, and Inhibition by Other Cycle Intermediates

In heart muscle, where the citric acid cycle functions mainly to generate ATP for use in muscle contraction, the enzymes of the cycle almost always act as a functional unit with their metabolic flux proportional to the rate of cellular oxygen consumption. *Since oxygen consumption, NADH reoxidation, and ATP production are tightly coupled (Section 20-5), the citric acid cycle must be regulated by feedback mechanisms that coordinate its NADH production with energy expenditure.* Unlike the rate-limiting enzymes of glycolysis and glycogen metabolism, which utilize elaborate systems of allosteric control, substrate cycles, and covalent modification as flux control mechanisms, the regulatory enzymes of the citric acid cycle seem to be controlled almost entirely in three simple ways: (1) substrate availability, (2) product inhibition, and (3) competitive feedback inhibition by intermediates further along the cycle. We shall encounter several examples of these straightforward mechanisms in the following discussion.

Perhaps the most crucial regulators of the citric acid cycle are its substrates, acetyl-CoA and oxaloacetate, and its product NADH. Both acetyl-CoA and oxaloacetate are present in mitochondria at concentrations that do not saturate citrate synthase. The metabolic flux through the enzyme therefore varies with substrate concentration and is subject to control by substrate availability. The production of acetyl-CoA from pyruvate is regulated by the activity of pyruvate dehydrogenase (Section 19-2B). Oxaloacetate is in equilibrium with malate, its concentration fluctuating with the [NADH]/[NAD$^+$] ratio according to the equilibrium expression

$$K = \frac{\text{[oxaloacetate][NADH]}}{\text{[malate][NAD}^+\text{]}}$$

In the transition from low to high work and respiration rates, mitochondrial [NADH] decreases. The consequent increase in [oxaloacetate] stimulates the citrate synthase reaction, which controls the rate of citrate formation.

The observation that [citrate] invariably falls as the work load increases indicates that the rate of citrate removal increases more than its rate of formation. The rate of citrate removal is governed by NAD$^+$-dependent isocitrate dehydrogenase (aconitase functions close to equilibrium), which is strongly inhibited *in vitro* by NADH (product inhibition).

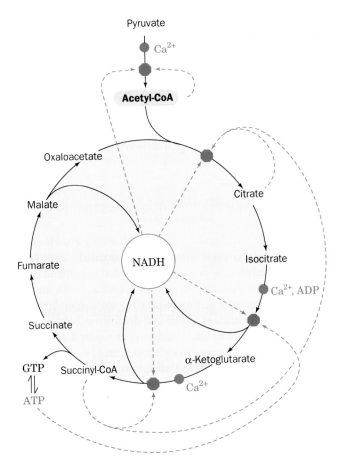

FIGURE 19-18. A diagram of the citric acid cycle and the pyruvate dehydrogenase reaction, indicating their points of inhibition (*red octagons*) and the pathway intermediates that function as inhibitors (*dashed red arrows*). ADP and Ca^{2+} (*green dots*) are activators.

Citrate synthase is also inhibited by NADH. Evidently, NAD$^+$-dependent isocitrate dehydrogenase is more sensitive to [NADH] changes than citrate synthase.

The decrease in [citrate] that occurs upon transition from low to high work and respiration rates results in a domino effect:

1. Citrate is a competitive inhibitor of oxaloacetate for citrate synthase (product inhibition); the fall in [citrate] caused by increased isocitrate dehydrogenase activity increases the rate of citrate formation.

2. α-Ketoglutarate dehydrogenase is also strongly inhibited by its products, NADH and succinyl-CoA. Its activity therefore increases when [NADH] decreases.

3. Succinyl-CoA also competes with acetyl-CoA in the citrate synthase reaction (competitive feedback inhibition).

This interlocking system serves to keep the citric acid cycle coordinately regulated.

ADP, ATP, and Ca²⁺ Are Allosteric Regulators of Citric Acid Cycle Enzymes

In vitro studies on the enzymes of the citric acid cycle have identified a few allosteric activators and inhibitors. Increased workload is accompanied by increased [ADP] resulting from the consequent increased rate of ATP hydrolysis. ADP acts as an allosteric activator of isocitrate dehydrogenase by decreasing its apparent K_M for isocitrate. ATP, which builds up when muscle is at rest, inhibits this enzyme.

Ca²⁺, among its many biological functions, is an essential metabolic regulator. It stimulates glycogen breakdown (Section 17-3C), triggers muscle contraction (Section 34-3C), and mediates many hormonal signals as a second messenger (Section 34-4B). Ca²⁺ also plays an important role in the regulation of the citric acid cycle (Fig. 19-18). It activates pyruvate dehydrogenase phosphatase, which in turn activates the pyruvate dehydrogenase complex to produce acetyl-CoA. In addition, Ca²⁺ activates both isocitrate dehydrogenase and α-ketoglutarate dehydrogenase. Thus, the same signal stimulates muscle contraction and the production of the ATP to fuel it.

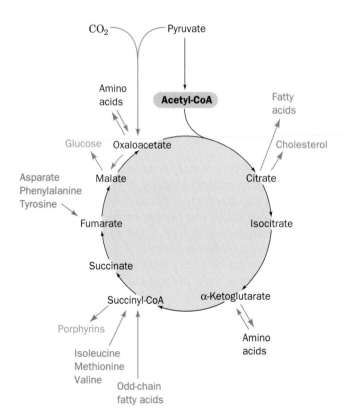

FIGURE 19-19. A diagram of the citric acid cycle, indicating the positions at which intermediates are drawn off for use in anabolic pathways (*red arrows*) and the points where anaplerotic reactions replenish depleted cycle intermediates (*green arrows*). Reactions involving amino acid transamination and deamination are reversible, so their direction varies with metabolic demand.

In the liver, the role of the citric acid cycle is more complex than in heart muscle. The liver synthesizes many substances required by the body including glucose, fatty acids, cholesterol, amino acids, and porphyrins. Reactions of the citric acid cycle play a part in many of these biosynthetic pathways in addition to their role in energy metabolism. In the next section, we discuss the contribution of the citric acid cycle to these processes.

A Bacterial Isocitrate Dehydrogenase Is Regulated by Phosphorylation

Escherichia coli isocitrate dehydrogenase is a dimer of identical 416-residue subunits that is inactivated by phosphorylation of its Ser 113, an active site residue. In contrast, most other enzymes that are known to be subject to covalent modification/demodification, for example, glycogen phosphorylase (Section 17-3), are phosphorylated at allosteric sites. In the case of isocitrate dehydrogenase, phosphorylation renders the enzyme unable to bind its substrate, isocitrate.

Comparison of the X-ray structures, determined by Daniel Koshland and Robert Stroud, of isocitrate dehydrogenase alone, in its phosphorylated form, and with bound isocitrate reveal only small conformational differences, suggesting that electrostatic repulsions between the anionic isocitrate and Ser phosphate groups prevent the enzyme from binding substrate. Evidently, phosphorylation can regulate enzyme activity by directly interfering with active site ligand binding as well as by inducing a conformational change from an allosteric site.

5. THE AMPHIBOLIC NATURE OF THE CITRIC ACID CYCLE

Ordinarily one thinks of a metabolic pathway as being either catabolic with the release (and conservation) of free energy, or anabolic with a requirement for free energy. The citric acid cycle is, of course, catabolic because it involves degradation and is a major free energy conservation system in most organisms. Cycle intermediates are only required in catalytic amounts to maintain the degradative function of the cycle. However, several biosynthetic pathways utilize citric acid cycle intermediates as starting materials (anabolism). The citric acid cycle is therefore **amphibolic** (both anabolic and catabolic).

All of the biosynthetic pathways that utilize citric acid cycle intermediates also require free energy. Consequently, the catabolic function of the cycle cannot be interrupted; *cycle intermediates that have been siphoned off must be replaced.* Although the mechanistic aspects of the enzymes involved in the pathways that utilize and replenish citric acid cycle intermediates are discussed in subsequent chapters, it is useful to briefly mention these metabolic interconnections here (Fig. 19-19).

Pathways That Utilize Citric Acid Cycle Intermediates

1. **Glucose biosynthesis (gluconeogenesis;** Section 21-1), which occurs in the cytosol, utilizes oxaloacetate. Oxaloacetate is not transported across the mitochondrial membrane, but malate is. Malate that has been transported across the mitochondrial membrane is converted to oxaloacetate in the cytosol for gluconeogenesis.

2. **Lipid biosynthesis,** which includes **fatty acid biosynthesis** (Section 23-4) and **cholesterol biosynthesis** (Section 23-6A), is a cytosolic process that requires acetyl-CoA. Acetyl-CoA is generated in the mitochondrion and is not transported across the inner mitochondrial membrane. Cytosolic acetyl-CoA is therefore generated by the breakdown of citrate, which can cross the inner mitochondrial membrane, in a reaction catalyzed by **ATP-citrate lyase** (Section 23-4D):

    ATP + citrate + CoA $\rightleftharpoons$
    $\qquad$ ADP + P$_i$ + oxaloacetate + acetyl-CoA

3. **Amino acid biosynthesis** utilizes citric acid cycle intermediates in two ways. α-Ketoglutarate is used to synthesize glutamate in a reductive amination reaction involving either NAD$^+$ or NADP$^+$ catalyzed by **glutamate dehydrogenase** (Section 24-1):

    α-Ketoglutarate + NAD(P)H + NH$_4^+$ $\rightleftharpoons$
    $\qquad$ glutamate + NAD(P)$^+$ + H$_2$O

    α-Ketoglutarate and oxaloacetate are also used to synthesize glutamate and aspartate in transamination reactions (Section 24-1):

    α-Ketoglutarate + alanine $\rightleftharpoons$ glutamate + pyruvate

    and

    $\qquad$ Oxaloacetate + alanine $\rightleftharpoons$ aspartate + pyruvate

4. **Porphyrin biosynthesis** (Section 24-4A) utilizes succinyl-CoA as a starting material.

Reactions That Replenish Citric Acid Cycle Intermediates
Reactions that replenish citric acid cycle intermediates are called **anaplerotic reactions** (filling up, Greek: *ana,*

up + *plerotikos,* to fill). The main reaction of this type is catalyzed by **pyruvate carboxylase,** which produces oxaloacetate (Section 21-1A):

Pyruvate + CO$_2$ + ATP + H$_2$O $\rightleftharpoons$
$\qquad$ oxaloacetate + ADP + P$_i$

This enzyme "senses" the need for more citric acid cycle intermediates through its activator, acetyl-CoA. Any decrease in the rate of the cycle caused by insufficient oxaloacetate or other cycle intermediates results in an increased level of acetyl-CoA because of its underutilization. This activates pyruvate carboxylase, which replenishes oxaloacetate, increasing the rate of the cycle. Of course, if the citric acid cycle is inhibited at some other step, by high NADH concentration, for example, increased oxaloacetate concentration will not activate the cycle. The excess oxaloacetate instead equilibrates with malate, which is transported out of the mitochondria for use in gluconeogenesis.

Degradative pathways generate citric acid cycle intermediates:

1. **Oxidation of odd-chain fatty acids** (Section 23-2E) leads to the production of succinyl-CoA.

2. **Breakdown of the amino acids isoleucine, methionine, and valine** (Section 24-3E) also leads to the production of succinyl-CoA.

3. **Transamination and deamination of amino acids** leads to the production of α-ketoglutarate and oxaloacetate. These reactions are reversible and, depending on metabolic demand, serve to remove or replenish these citric acid cycle intermediates.

The citric acid cycle is truly at the center of metabolism (see Fig. 15-1). Its reduced products, NADH and FADH$_2$, are reoxidized by the electron-transport chain during oxidative phosphorylation and the free energy released is coupled to the biosynthesis of ATP. Citric acid cycle intermediates are utilized in the biosynthesis of many vital cellular constituents. In the next few chapters we shall explore the interrelationships of these pathways in more detail.

CHAPTER SUMMARY

The citric acid cycle, the common mode of oxidative metabolism in most organisms, is mediated by eight enzymes that collectively convert 1 acetyl-CoA to 2 CO$_2$ molecules so as to yield 3 NADHs, 1 FADH$_2$, and 1 GTP (or ATP). The NADH and FADH$_2$ are oxidized by O$_2$ in the electron-transport chain with the concomitant synthesis of 11 ATPs.

Pyruvate, the end product of glycolysis under aerobic conditions, is converted to acetyl-CoA by the pyruvate dehydrogenase

multienzyme complex, a symmetrical cluster of three enzymes: Pyruvate dehydrogenase, dihydrolipoyl transacetylase, and dihydrolipoyl dehydrogenase. The pyruvate dehydrogenase subunit catalyzes the conversion of pyruvate to CO$_2$ and a hydroxyethyl-TPP intermediate. The latter is channeled to dihydrolipoyl transacetylase, which oxidizes the hydroxyethyl group to acetate and transfers it to CoA to form acetyl-CoA. The lipoamide prosthetic group, which is reduced to the dihydro form in the process, is reoxidized by dihydrolipoamide dehydrogenase in a reaction involving bound FAD that reduces NAD^+ to NADH. Dihydrolipoamide transacetylase is inactivated by the formation of a covalent adduct between lipoamide and As(III) compounds. The activity of the pyruvate dehydrogenase complex varies with the [NADH]/[NAD$^+$] and [acetyl-CoA]/[CoA] ratios. In eukaryotes, the pyruvate dehydrogenase subunit is also inactivated by phosphorylation of a specific Ser residue and is reactivated by its removal. These modifications are mediated, respectively, by pyruvate dehydrogenase kinase and pyruvate dehydrogenase phosphatase, which are components of the multienzyme complex and respond to the levels of metabolic intermediates such as NADH and acetyl-CoA.

Citrate is formed by the condensation of acetyl-CoA and oxaloacetate by citrate synthase. The citrate is dehydrated to *cis*-aconitate and then rehydrated to isocitrate in a stereospecific reaction catalyzed by aconitase. This enzyme is specifically inhibited by (2*R*,3*R*)-fluorocitrate, which is enzymatically synthesized from fluoroacetate and oxaloacetate. Isocitrate is oxidatively decarboxylated to α-ketoglutarate by isocitrate dehydrogenase, which produces NADH and CO_2. The α-ketoglutarate, in turn, is oxidatively decarboxylated by α-ketoglutarate dehydrogenase, a multienzyme

complex homologous to the pyruvate dehydrogenase multienzyme complex. This reaction generates the second NADH and CO_2. The resulting succinyl-CoA is converted to succinate with the generation of GTP (ATP in plants and bacteria) by succinyl-CoA synthetase. The succinate is stereospecifically dehydrogenated to fumarate by succinate dehydrogenase in a reaction that generates $FADH_2$. The final two reactions of the citric acid cycle, which are catalyzed by fumarase and malate dehydrogenase, in turn hydrate fumarate to *S*-malate and oxidize this alcohol to its corresponding ketone, oxaloacetate, with concomitant production of the pathway's third and final NADH.

The enzymes of the citric acid cycle act as a functional unit that keeps pace with the metabolic demands of the cell. The flux-controlling enzymes appear to be citrate synthase, isocitrate dehydrogenase, and α-ketoglutarate dehydrogenase. Their activities are controlled by substrate availability, product inhibition, inhibition by cycle intermediates, and activation by Ca^{2+}.

Several anabolic pathways utilize citric acid cycle intermediates as starting materials. These essential substances are replaced by anaplerotic reactions of which the major one is synthesis of oxaloacetate from pyruvate and CO_2 by pyruvate carboxylase.

REFERENCES

General

Cunningham, E.B., *Biochemistry: Mechanisms of Metabolism,* Chapter 10, McGraw–Hill (1978).

Holmes, F.L., *Hans Krebs:* Vol. 1: *The Formation of a Scientific Life, 1900–1933; and* Vol. 2: *Architect of Intermediary Metabolism, 1933–1937,* Oxford University Press (1991 and 1993). [The biography of the discoverer of the citric acid cycle through the time of its discovery.]

Kornberg, H.L., Tricarboxylic acid cycles, *BioEssays* **7,** 236–238 (1987). [An historical synopsis of the intellectual background leading to the discovery of the citric acid cycle.]

Krebs, H.A., The history of the tricarboxylic acid cycle, *Perspect. Biol. Med.* **14,** 154–170 (1970).

Enzyme Mechanisms

Angelides, J.K. and Hammes, G.G., Mechanism of action of the pyruvate dehydrogenase multienzyme complex from *E. coli, Proc. Natl. Acad. Sci.* **75,** 4877–4880 (1978).

Cleland, W.W. and Kreevoy, M.M., Low-barrier hydrogen bonds and enzymic catalysis, *Science* **264,** 1887–1890 (1994).

Karpusas, M., Branchaud, B., and Remington, S.J., Proposed mechanism for the condensation reaction of citrate synthase: 1.9-Å structure of the ternary complex with oxaloacetate and carboxymethyl coenzyme A, *Biochemistry* **29,** 2213–2219 (1990).

Lauble, H., Kennedy, M.C., Beinert, H., and Stout, D.C., Crystal structures of aconitase with isocitrate and nitroisocitrate bound, *Biochemistry* **31,** 2735–2748 (1992).

Mattevi, A., de Kok, A., and Perham, R.N., The pyruvate dehydrogenase multienzyme complex, *Curr. Opin. Struct. Biol.* **2,** 877-887 (1992).

Mattevi, A., Obmolova, G., Schulze, E., Kalk, K.H., Westphal,

A.H., de Kok, A., and Hol, W.G.J., Atomic structure of the cubic core of the pyruvate dehydrogenase multienzyme complex, *Science* **255,** 1544–1550 (1992).

Mattevi, A., Obmolova, G., Sokatch, J.R., Betzel, C., and Hol, W.G.J., The refined crystal structure of *Pseudomonas putida* lipoamide dehydrogenase complexed with NAD^+ at 2.45 Å resolution, *Proteins* **13,** 336–351 (1992).

Perham, R.N., Domains, motifs, and linkers in 2-oxo acid dehydrogenase multienzyme complexes: a paradigm in the design of a multifunctional protein, *Biochemistry* **30,** 8051–8512 (1991). [An authoritative review on this family of multienzyme complexes.]

Porter, D.J.T. and Bright, H.J., 3-Carbanionic substrate analogues bind very tightly to fumarase and aspartase, *J. Biol. Chem.* **255,** 4772–4780 (1980).

Reed, L.J. and Hackert, M.L., Structure-function relationships in dihydrolipoamide acyltransferases, *J. Biol. Chem.* **265,** 8971–8974 (1990).

Remington, S.J., Structure and mechanism of citrate synthase, *Curr. Top. Cell Reg.* **33,** 202–229 (1992); *and* Mechanisms of citrate synthase and related enzymes (triose phosphate isomerase and mandelate racemase), *Curr. Opin. Struct. Biol.* **2,** 730–735 (1992).

Srere, P.A., The enzymology of the formation and breakdown of citrate, *Adv. Enzymol.* **43,** 57–101 (1975).

Walsh, C., *Enzymatic Reaction Mechanisms,* Freeman (1979). [Contains discussions of the mechanisms of various citric acid cycle enzymes.]

Wegenknecht, T., Grassucci, R., Radke, G.A., and Roche, T.E., Cryoelectron microscopy of mammalian pyruvate dehydrogenase complex, *J. Biol. Chem.* **266,** 24650–24656 (1991).

Weiland, O.H., The mammalian pyruvate dehydrogenase complex: structure and regulation, *Rev. Physiol. Biochem. Pharmacol.* **96,** 123–170 (1983).

Wilkinson, K.D. and Williams, C.H., Jr., Evidence for multiple electronic forms of two-electron-reduced lipoamide dehydrogenase from *E. coli, J. Biol. Chem.* **254**, 852–862 (1979).

Wolodk, W.T., Fraser, M.E., James, M.N.G., and Bridger, W.A., The crystal structure of succinyl-CoA synthetase from *Eschericia coli* at 2.5Å resolution, *J. Biol. Chem.* **269**, 10883–10890 (1994).

Zheng, L., Kennedy, M.C., Beinert, H., and Zalkin, H. Mutational analysis of active site residues in pig heart aconitase, *J. Biol. Chem.* **267**, 7895–7903 (1992).

Metabolic Poisons

Committee on the Medical and Biological Effects of Environmental Pollutants, Subcommittee on Arsenic, *Arsenic,* National Research Council, National Academy of Sciences (1977).

Gibble, G.W., Fluoroacetate toxicity, *J. Chem. Ed.* **50**, 460–462 (1973).

Jones, D.E.H. and Ledingham, K.W.D., Arsenic in Napoleons's wallpaper, *Nature* **299**, 626–627 (1982).

Winslow, J.H., *Darwin's Victorian Malady,* American Philosophical Society (1971).

Control Mechanisms

Denton, R.M. and Halestrap, A.P., Regulation of pyruvate metabolism in mammalian tissues, *Essays Biochem.* **15**, 37–77 (1979).

Hansford, R.G., Control of mitochondrial substrate oxidation, *Curr. Top. Bioenerg.* **10**, 217–278 (1980).

Hurley, J.H., Dean, A.M., Sohl, J.L., Koshland, D.E., Jr., and Stroud, R.M., Regulation of an enzyme by phosphorylation at the active site, *Science* **249**, 1012–1016 (1990).

Reed, L.J., Damuni, Z., and Merryfield, M.L., Regulation of mammalian pyruvate and branched- chain α-keto-acid dehydrogenase complexes by phosphorylation and dephosphorylation, *Curr. Top. Cell. Reg.* **27**, 41–49 (1985).

Stroud, R.M., Mechanisms of biological control by phosphorylation, *Curr. Opin. Struct. Biol.* **1**, 826-835 (1991). [Reviews, among other things, the inactivation of isocitrate dehydrogenase by phosphorylation.]

Williamson, J.R., Mitochondrial function in the heart, *Annu. Rev. Physiol.* **41**, 485–506 (1979); *and* Mitochondrial metabolism and cell regulation, *in* Packer, L. and Gómez-Puyon, A. (Eds.), *Mitochondria, pp.* 79–107, Academic Press (1976).

PROBLEMS

1. Trace the course of the radioactive label in 2-[^{14}C]glucose through glycolysis and the citric acid cycle. At what point(s) in the cycle will the radioactivity be released as $^{14}CO_2$? How many turns of the cycle will be required for complete conversion of the radioactivity to CO_2? Repeat this problem for pyruvate that is ^{14}C-labeled at its methyl group.

2. Given the following information, calculate the physiological ΔG of the isocitrate dehydrogenase reaction at 25°C and pH 7.0: [NAD$^+$]/[NADH] = 8; [α-ketoglutarate] = 0.1 mM; [isocitrate] = 0.02 mM; assume standard conditions for CO_2 ($\Delta G°'$ is given in Table 19-2). Is this reaction a likely site for metabolic control? Explain.

3. The oxidation of acetyl-CoA to two molecules of CO_2 involves the transfer of four electron pairs to redox coenzymes. In which of the cycle's reactions do these electron transfers occur? Identify the redox coenzyme in each case. For each reaction, draw the structural formulas of the reactants, intermediates, and products and show, using curved arrows, how the electrons are transferred.

4. The citrate synthase reaction has been proposed to proceed via the formation of the enol form of acetyl-CoA. How, then, would you account for the observation that ^{3}H is not incorporated into acetyl-CoA when acetyl-CoA is incubated with citrate synthase in ^3H_2O?

5. Malonate is a competitive inhibitor of succinate in the succinate dehydrogenase reaction. Sketch the graphs that would be obtained upon plotting $1/v$ versus 1/[succinate] at three different malonate concentrations. Label the lines for low, medium, and high [malonate].

6. Krebs found that malonate inhibition of the citric acid cycle could be overcome by raising the oxaloacetate concentration. Explain the mechanism of this process in light of your findings in Problem 5.

7. Which of the following metabolites undergo net oxidation by the citric acid cycle: (a) α-ketoglutarate, (b) succinate, (c) citrate, and (d) acetyl-CoA?

8. Although there is no net synthesis of intermediates by the citric acid cycle, citric acid cycle intermediates are used in biosynthetic reactions such as the synthesis of porphyrins from succinyl-CoA. Give a reaction for the net synthesis of succinyl-CoA from pyruvate.

9. Oxaloacetate and α-ketoglutarate are precursors of the amino acids aspartate and glutamate as well as being catalytic intermediates in the citric acid cycle. Describe the net synthesis of α-ketoglutarate from pyruvate in which no citric acid cycle intermediates are depleted.

10. Lipoic acid is bound to enzymes that catalyze oxidative decarboxylation of α-keto acids. (a) What is the chemical mode of attachment of lipoic acid to enzymes? (b) Using chemical structures, show how lipoic acid participates in the oxidative decarboxylation of α-keto acids.

11. **British anti-lewisite (BAL),** which was designed to counter the effects of the arsenical war gas **lewisite,** is useful in treating arsenic poisoning. Explain.

$$CH_2 — SH$$
$$|$$
$$CH — SH \qquad Cl — CH = CH — AsCl_2$$
$$|$$
$$CH_2 — OH$$

British anti-lewisite   **Lewisite**
(BAL)

C H A P T E R

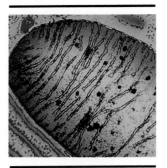

20

Electron Transport and Oxidative Phosphorylation

In 1789, Armand Séguin and Antoine Lavoisier (the father of modern chemistry) wrote:

> . . . *in general, respiration is nothing but a slow combustion of carbon and hydrogen, which is entirely similar to that which occurs in a lamp or lighted candle, and that, from this point of view, animals that respire are true combustible bodies that burn and consume themselves.*

Lavoisier had by this time demonstrated that living animals consume oxygen and generate carbon dioxide. It was not until the early twentieth century, however, after the rise of enzymology, that it was established, largely through the work of Otto Warburg, that biological oxidations are catalyzed by intracellular enzymes. As we have seen, glucose is completely oxidized to CO_2 through the enzymatic reactions of glycolysis and the citric acid cycle. In this chapter we shall examine the fate of the electrons that are removed from glucose by this oxidation process.

The complete oxidation of glucose by molecular oxygen is described by the following redox equation:

$$C_6H_{12}O_6 + 6\,O_2 \longrightarrow 6CO_2 + 6H_2O$$
$$\Delta G°' = -2823 \text{ kJ} \cdot \text{mol}^{-1}$$

To see more clearly the transfer of electrons, let us break this equation down into two half-reactions. In the first reaction the glucose carbon atoms are oxidized:

$$C_6H_{12}O_6 + 6H_2O \longrightarrow 6CO_2 + 24H^+ + 24e^-$$

and in the second, molecular oxygen is reduced:

$$6\,O_2 + 24H^+ + 24e^- \longrightarrow 12H_2O$$

In living systems, the electron-transfer process connecting these half-reactions occurs through a multistep pathway that harnesses the liberated free energy to form ATP.

The 12 electron pairs involved in glucose oxidation are not transferred directly to O_2. Rather, as we have seen, *they are transferred to the coenzymes NAD^+ and FAD to form 10 NADH + 2 $FADH_2$ (Fig. 20-1)* in the reactions catalyzed by the glycolytic enzyme glyceraldehyde-3-phosphate dehydrogenase (Section 16-2F), pyruvate dehydrogenase (Section 19-2A), and the citric acid cycle enzymes isocitrate dehydrogenase, α-ketoglutarate dehydrogenase, succinate dehydrogenase (the only FAD reduction), and malate dehydrogenase (Section 19-3). *The electrons then pass into the electron-transport chain where, through reoxidation of NADH and $FADH_2$, they participate in the sequential oxidation–reduction of over 10 redox centers before reducing O_2 to H_2O. In this process, protons are expelled from the mitochondrion. The free energy stored in the resulting pH gradient drives the synthesis of ATP from ADP and P_i through* **oxidative phosphorylation.** Reoxidation of each NADH results in the synthesis of 3 ATPs, and reoxidation of $FADH_2$ yields 2 ATPs for a total of 38 ATPs for each glucose completely oxidized to CO_2 and H_2O (including the 2 ATPs made in glycolysis and the 2 ATPs made in the citric acid cycle).

In this chapter we explore the mechanisms of electron transport and oxidative phosphorylation and their regulation. We begin with a discussion of mitochondrial structure and transport systems.

1. THE MITOCHONDRION

The mitochondrion (Section 1-2A) is the site of eukaryotic oxidative metabolism. It contains, as Albert Lehninger and Eugene Kennedy demonstrated in 1948, the enzymes that mediate this process, including pyruvate dehydrogenase, the citric acid cycle enzymes, the enzymes catalyzing fatty acid oxidation (Section 23-2C), and the enzymes and redox proteins involved in electron transport and oxidative phosphorylation. It is therefore with good reason that the mitochondrion is described as the cell's "power plant."

A. Mitochondrial Anatomy

Mitochondria vary considerably in size and shape depending on their source and metabolic state. They are typically ellipsoids of ~0.5-μm diameter and 1-μm length (about the size of a bacterium; Fig. 20-2). The mitochondrion is bounded by a smooth outer membrane and contains an extensively invaginated inner membrane. The number of invaginations, called **cristae,** varies with the respiratory activity of the particular type of cell. This is because the proteins mediating electron transport and oxidative phosphorylation are bound to the inner mitochondrial membrane so that the respiration rate varies with membrane surface area. Liver, for example, which has a relatively low respiration rate, contains mitochondria with relatively few cristae,

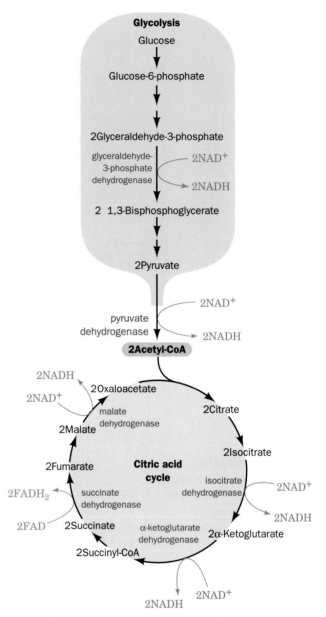

FIGURE 20-1. The sites of electron transfer that form NADH and $FADH_2$ in glycolysis and the citric acid cycle.

whereas those of heart muscle contain many. Nevertheless, the aggregate area of the inner mitochondrial membranes in a liver cell is ~15-fold greater than that of its plasma membrane. The inner mitochondrial compartment consists of a gel-like substance of <50% water, named the **matrix,** which contains remarkably high concentrations of the soluble enzymes of oxidative metabolism (e.g., citric acid cycle enzymes), as well as substrates, nucleotide cofactors, and inorganic ions. The matrix also contains the mitochondrial genetic machinery—DNA, RNA, and ribosomes—that generates several (but by no means all) mitochondrial proteins (Section 1-2A).

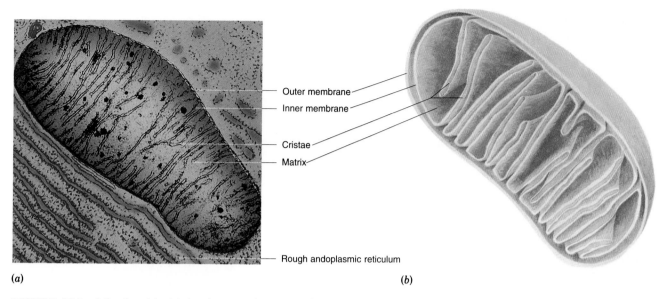

(a)

(b)

FIGURE 20-2. Mitochondria. (a) An electron micrograph of an animal mitochondrion. [Secchi-Lacaque/Roussel-UCLAF/CNRI.] (b) Cutaway diagram of a mitochondrion.

The Inner Mitochondrial Membrane Compartmentalizes Metabolic Functions

The outer mitochondrial membrane contains **porin,** a protein that forms nonspecific pores that permit free diffusion of up to 10-kD molecules (the X-ray structures of bacterial porins are discussed in Section 11-3A). The inner membrane, which is ~75% protein by mass, is considerably richer in proteins than the outer membrane (Fig. 20-3). It is freely permeable only to O_2, CO_2, and H_2O and contains, in addition to respiratory chain proteins, numerous transport proteins that control the passage of metabolites such as ATP, ADP, pyruvate, Ca^{2+}, and phosphate (see below). *This controlled impermeability of the inner mitochondrial membrane to most ions, metabolites, and low molecular mass compounds permits the generation of ionic gradients across this barrier and results in the compartmentalization of metabolic functions between cytosol and mitochondria.*

B. Mitochondrial Transport Systems

The inner mitochondrial membrane is impermeable to most hydrophilic substances. It must therefore contain specific transport systems to permit the following processes:

1. Glycolytically produced cytosolic NADH must gain access to the electron-transport chain for aerobic oxidation.

2. Mitochondrially produced metabolites such as oxaloacetate and acetyl-CoA, the respective precursors for cytosolic glucose and fatty acid biosynthesis, must reach their metabolic destinations.

3. Mitochondrially produced ATP must reach the cytosol, where most ATP-utilizing reactions take place, whereas

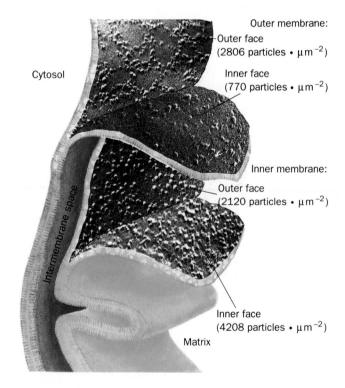

FIGURE 20-3. Freeze-fracture and freeze-etch electron micrographs of the inner and outer mitochondrial membranes. The inner membrane contains about twice the density of embedded particles as does the outer membrane. [Courtesy of Lester Packer, University of California at Berkeley.]

ADP and P_i, the substrates for oxidative phosphorylation, must enter the mitochondrion.

We have already studied the ADP–ATP translocator and its dependence on $\Delta\Psi$, the electric potential difference across the mitochondrial membrane (Section 18-4C). The export of oxaloacetate and acetyl-CoA from the mitochondrion are, respectively, discussed in Sections 21-1A and 23-4D. In the remainder of this section we examine the mitochondrial transport systems for P_i and Ca^{2+} and shuttle systems for NADH.

P_i Transport

ATP is generated from ADP + P_i in the mitochondrion but is utilized in the cytosol. The P_i produced is returned to the mitochondrion via an electroneutral P_i—H^+ symport that is driven by ΔpH. The electrochemical potential gradient generated by the redox-driven proton pumps of electron transport (Section 20-3B) is therefore responsible for maintaining high mitochondrial ADP and P_i concentrations in addition to providing the free energy for ATP synthesis.

Ca^{2+} Transport

Since Ca^{2+}, like cAMP, functions as a second messenger (Section 17-3C), its concentrations in the various cellular compartments must be precisely controlled. The mitochondrion, endoplasmic reticulum, and extracellular spaces act as Ca^{2+} storage tanks. We studied the Ca^{2+}-ATPases of the plasma membrane, endoplasmic reticulum, and sarcoplasmic reticulum in Section 18-3B. Here we consider the mitochondrial Ca^{2+}-transport systems.

Mitochondrial inner membrane systems separately mediate the influx and the efflux of Ca^{2+} (Fig. 20-4). The Ca^{2+} influx is driven by the inner mitochondrial membrane's membrane potential ($\Delta\Psi$, negative inside), which attracts positively charged ions. The rate of influx varies with the external $[Ca^{2+}]$ because the K_M for Ca^{2+} transport by this system is greater than the cytosolic Ca^{2+} concentration. In heart, brain, and skeletal muscle mitochondria especially, Ca^{2+} efflux is independently driven by the Na^+ gradient across the inner mitochondrial membrane. Ca^{2+} exits the matrix only in exchange for Na^+ so that this system is an antiport. This exchange process normally operates at its maximal velocity. *Mitochondria (as well as endoplasmic and sarcoplasmic reticulum) therefore can act as a "buffer" for cytosolic Ca^{2+} (Fig. 20-5):* If cytosolic $[Ca^{2+}]$ rises, the rate of mitochondrial Ca^{2+} influx increases while that of Ca^{2+} efflux remains constant, causing the mitochondrial $[Ca^{2+}]$ to increase while the cytosolic $[Ca^{2+}]$ decreases to its original level (its set-point). Conversely, a decrease in cytosolic $[Ca^{2+}]$ reduces the influx rate, causing net efflux of $[Ca^{2+}]$ and an increase of cytosolic $[Ca^{2+}]$ back to the set-point.

Oxidation carried out by the citric acid cycle in the mitochondrial matrix is controlled by the matrix $[Ca^{2+}]$ (Section 19-4). It is interesting to note, therefore, that in response to increases in cytosolic $[Ca^{2+}]$ caused by increased muscle activity, the matrix $[Ca^{2+}]$ increases, thereby activating the enzymes of the citric acid cycle. This leads to an increase in [NADH], whose reoxidation by oxidative phosphorylation (as we study in this chapter), generates the ATP needed for this increased muscle activity.

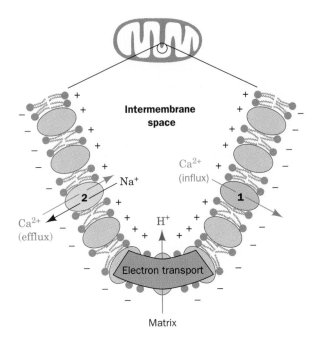

FIGURE 20-4. The two mitochondrial Ca^{2+}-transport systems. System 1 mediates Ca^{2+} influx to the matrix in response to the membrane potential (negative inside). System 2 mediates Ca^{2+} efflux in exchange for Na^+.

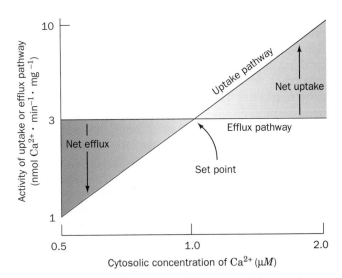

FIGURE 20-5. The regulation of cytosolic $[Ca^{2+}]$. The efflux pathway operates at a constant rate independent of $[Ca^{2+}]$, whereas the activity of the influx pathway varies with $[Ca^{2+}]$. At the set-point, the activities of the two pathways are equal and there is no net Ca^{2+} flux. An increase in cytosolic $[Ca^{2+}]$ results in net influx, and a decrease in cytosolic $[Ca^{2+}]$ results in net efflux. Both effects lead to the restoration of the cytosolic $[Ca^{2+}]$. [After Nicholls, D., *Trends Biochem. Sci.* **6,** 37 (1981).]

Cytoplasmic Shuttle Systems "Transport" NADH across the Inner Mitochondrial Membrane

Although most of the NADH generated by glucose oxidation is formed in the mitochondrial matrix via the citric acid cycle, that generated by glycolysis occurs in the cytosol.

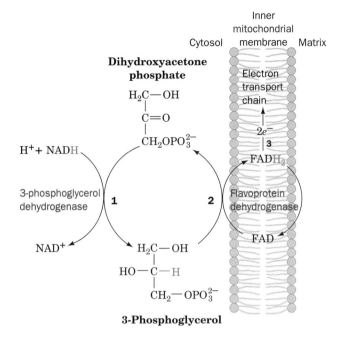

Dihydroxyacetone phosphate

FIGURE 20-6. The glycerophosphate shuttle. The electrons of cytosolic NADH are transported to the mitochondrial electron-transport chain in three steps (shown in red as hydride transfers): **(1)** Cytosolic oxidation of NADH by dihydroxyacetone phosphate catalyzed by 3-phosphoglycerol dehydrogenase. **(2)** Oxidation of 3-phosphoglycerol by flavoprotein dehydrogenase with reduction of FAD to $FADH_2$. **(3)** Reoxidation of $FADH_2$ with passage of electrons into the electron-transport chain.

Yet, the inner mitochondrial membrane lacks an NADH transport protein. *Only the electrons from cytosolic NADH are transported into the mitochondrion by one of several ingenious "shuttle" systems.* In the **glycerophosphate shuttle** (Fig. 20-6) of insect flight muscle (the tissue with the largest known sustained power output—about the same power-to-weight ratio as a small automobile engine), **3-phosphoglycerol dehydrogenase** catalyzes the oxidation of cytosolic NADH by dihydroxyacetone phosphate to yield NAD^+, which reenters glycolysis. The electrons of the resulting **3-phosphoglycerol** are transferred to **flavoprotein dehydrogenase** to form $FADH_2$. This enzyme, which is situated on the inner mitochondrial membrane's outer surface, supplies electrons to the electron-transport chain in a manner similar to that of succinate dehydrogenase (Section 20-2C). *The glycerophosphate shuttle therefore results in the synthesis of 2 ATPs for every cytoplasmic NADH reoxidized.*

The **malate–aspartate shuttle** (Fig. 20-7) of mammalian systems is more complex but more energy efficient than the glycerophosphate shuttle. Mitochondrial NAD^+ is reduced by cytosolic NADH through the intermediate reduction

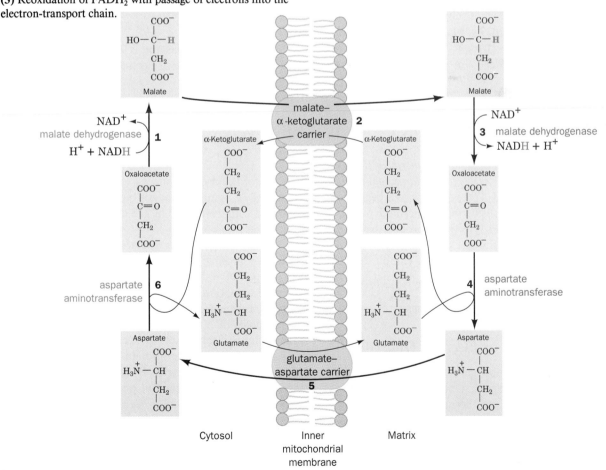

FIGURE 20-7. The malate–aspartate shuttle. The electrons of cytosolic NADH are transported to mitochondrial NADH (shown in red as hydride transfers) in Steps **1** to **3**. Steps **4** to **6** then serve to regenerate cytosolic oxaloacetate.

and subsequent regeneration of oxaloacetate. This process occurs in two phases of three reactions each:

Phase A (transport of electrons into the matrix):

1. NADH is reoxidized by cytosolic oxaloacetate through the action of cytosolic malate dehydrogenase.

2. The **malate–α-ketoglutarate carrier** transports the malate formed in Reaction 1 into the mitochondrial matrix in exchange for α-ketoglutarate.

3. In the mitochondrial matrix, malate is reoxidized to oxaloacetate by mitochondrial malate dehydrogenase (Section 19-3H), thereby reducing mitochondrial NAD⁺ to NADH.

Phase B (regeneration of cytosolic oxaloacetate):

4. A transaminase (Section 24-1A) converts mitochondrial oxaloacetate to aspartate with concomitant conversion of glutamate to α-ketoglutarate.

5. Aspartate is transported from the matrix to the cytosol by the **glutamate–aspartate carrier** in exchange for cytosolic glutamate.

6. Cytosolic aspartate is converted to oxaloacetate by a transaminase in conjunction with α-ketoglutarate conversion to glutamate.

The electrons of cytosolic NADH are thereby transferred to mitochondrial NADH, which is subject to reoxidation via the electron-transport chain. *The malate–aspartate shuttle yields three ATPs for every cytosolic NADH, one more than the glycerophosphate shuttle.*

2. ELECTRON TRANSPORT

In the electron-transport process, the free energy of electron transfer from NADH and FADH₂ to O₂ via protein-bound redox centers is coupled to ATP synthesis. We begin our study of this process by considering its thermodynamics. We then examine the path of electrons through the redox centers of the system and discuss experiments used to unravel this path. Finally, we study the four complexes that comprise the electron-transport chain. In the next section we discuss how the free energy released by the electron-transport process is coupled to ATP synthesis.

A. Thermodynamics of Electron Transport

We can estimate the thermodynamic efficiency of electron transport through knowledge of standard reduction potentials. As we have seen in our thermodynamic considerations of oxidation–reduction reactions (Section 15-5), an oxidized substrate's electron affinity increases with its standard reduction potential, $\mathscr{E}°'$ [the voltage generated by the reaction of the half-cell under standard biochemical conditions (1M reactants and products with [H⁺] defined as 1 at

pH 7) relative to the standard hydrogen electrode; Table 15-4 lists the standard reduction potentials of several half-reactions of biochemical interest]. The standard reduction potential difference, $\Delta\mathscr{E}°'$, for a redox reaction involving any two half-reactions is therefore expressed:

$$\Delta\mathscr{E}°' = \mathscr{E}°'_{(e^- \text{ acceptor})} - \mathscr{E}°'_{(e^- \text{ donor})}$$

NADH Oxidation Is a Highly Exergonic Reaction

The half-reactions for O₂ oxidation of NADH are (Table 15-4):

$$\text{NAD}^+ + \text{H}^+ + 2e^- \rightleftharpoons \text{NADH} \qquad \mathscr{E}°' = -0.315 \text{ V}$$

and

$$\tfrac{1}{2}\text{O}_2 + 2\text{H}^+ + 2e^- \rightleftharpoons \text{H}_2\text{O} \qquad \mathscr{E}°' = 0.815 \text{ V}$$

Since the $\text{O}_2/\text{H}_2\text{O}$ half-reaction has the greater standard reduction potential and therefore the higher electron affinity, the NADH half-reaction is reversed so that NADH is the electron donor in this couple and O₂ the electron acceptor. The overall reaction is

$$\tfrac{1}{2}\text{O}_2 + \text{NADH} + \text{H}^+ \rightleftharpoons \text{H}_2\text{O} + \text{NAD}^+$$

so that

$$\Delta\mathscr{E}°' = 0.815 - (-0.315) = 1.130 \text{ V}$$

The standard free energy change for the reaction can then be calculated from Eq. [15.7]:

$$\Delta G°' = -n\mathscr{F}\Delta\mathscr{E}°'$$

where $\mathscr{F}$, the Faraday constant, is 96,494 C·mol⁻¹ of electrons and n is the number of electrons transferred per mole of reactants. Thus, since 1 V = 1 J·C⁻¹, for NADH oxidation:

$$\Delta G°' = -2 \frac{\text{mol } e^-}{\text{mol reactant}} \times 96,494 \frac{\text{C}}{\text{mol } e^-} \times 1.13 \text{ J·C}^{-1}$$
$$= -218 \text{ kJ·mol}^{-1}$$

In other words, the oxidation of 1 mol of NADH by O₂ (the transfer of $2e^-$) under standard biochemical conditions is associated with the release of 218 kJ of free energy.

Electron Transport Is Thermodynamically Efficient

The standard free energy required to synthesize 1 mol of ATP from ADP + P$_i$ is 30.5 kJ·mol⁻¹. The standard free energy of oxidation of NADH by O₂, if coupled to ATP synthesis, is therefore sufficient to drive the formation of several moles of ATP. This coupling, as we shall see, is achieved by an electron-transport chain in which electrons are passed through three protein complexes containing redox centers with progressively greater electron affinity (increasing standard reduction potentials) instead of directly to O₂. *This allows the large overall free energy change to be broken up into three smaller packets, each of which is coupled with ATP synthesis in a process called* **oxidative phosphorylation.** *Oxidation of one NADH therefore results in the synthesis of three ATPs.* (Oxidation of FADH₂, whose entrance into the electron-transport chain is regulated by a fourth protein complex, is similarly coupled to the synthesis of two ATPs.) The thermodynamic efficiency of oxidative phosphorylation is therefore 3×30.5 kJ·mol^{-1} × 100/218 kJ·mol^{-1} = 42% under standard biochemical conditions. However, under physiological conditions in active mitochondria (where the reactant and product concentrations as well as the pH deviate from standard conditions), this thermodynamic efficiency is thought to be ~70%. In comparison, the energy efficiency of a typical automobile engine is < 30%.

B. The Sequence of Electron Transport

The free energy necessary to generate ATP is extracted from the oxidation of NADH and FADH$_2$ by the electron-transport chain, a series of four protein complexes through which electrons pass from lower to higher standard reduction potentials (Fig. 20-8). Electrons are carried from **Complexes I** and **II** to **Complex III** by **coenzyme Q** (**CoQ** or **ubiquinone;** so named because of its ubiquity in respiring organisms), and from Complex III to **Complex IV** by the peripheral membrane protein **cytochrome c** (Sections 6-3B and 8-3A).

Complex I catalyzes oxidation of NADH by CoQ:

$$NADH + CoQ\ (oxidized) \longrightarrow NAD^+ + CoQ\ (reduced)$$
$$\Delta\mathscr{E}°' = 0.360\ V \qquad \Delta G°' = -69.5\ kJ\cdot mol^{-1}$$

Complex III catalyzes oxidation of CoQ(reduced) by cytochrome c.

$$CoQ\ (reduced) + cytochrome\ c\ (oxidized) \longrightarrow$$
$$CoQ\ (oxidized) + cytochrome\ c\ (reduced)$$
$$\Delta\mathscr{E}°' = 0.190\ V \qquad \Delta G°' = -36.7\ kJ\cdot mol^{-1}$$

Complex IV catalyzes oxidation of cytochrome c (reduced) by O$_2$, the terminal electron acceptor of the electron-transport process.

$$Cytochrome\ c\ (reduced) + \tfrac{1}{2}O_2 \longrightarrow$$
$$cytochrome\ c\ (oxidized) + H_2O$$
$$\Delta\mathscr{E}°' = 0.580\ V \qquad \Delta G°' = -112\ kJ\cdot mol^{-1}$$

The changes in standard reduction potential of an electron pair as it successively traverses Complexes I, III, and IV corresponds, at each stage, to sufficient free energy to power the synthesis of an ATP molecule.

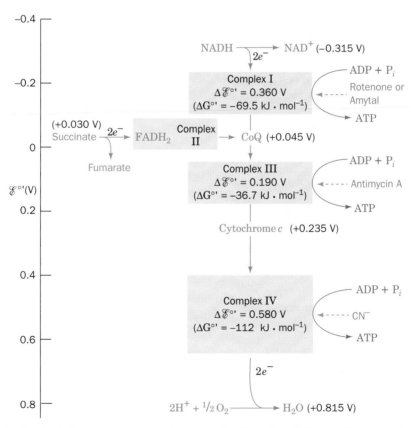

FIGURE 20-8. The mitochondrial electron-transport chain. The standard reduction potentials of its most mobile components (*green*) are indicated, as are the points where sufficient free energy is harvested to synthesize ATP (*blue*) and the sites of action of several respiratory inhibitors (*red*). [Note that Complexes I, III, and IV do not directly synthesize ATP but, rather, sequester the free energy necessary to do so by pumping protons outside the mitochondrion to form a proton gradient; Section 20-3]

Complex II catalyzes the oxidation of FADH$_2$ by CoQ.

$$FADH_2 + CoQ \ (oxidized) \longrightarrow FAD + CoQ \ (reduced)$$
$$\Delta \mathscr{E}°' = 0.015 \ V \qquad \Delta G°' = -2.9 \ kJ \cdot mol^{-1}$$

This redox reaction does not release sufficient free energy to synthesize ATP; it functions only to inject the electrons from FADH$_2$ into the electron-transport chain.

The Workings of the Electron-Transport Chain Have Been Elucidated through the Use of Inhibitors

Our understanding of the sequence of events in electron transport is largely based on the use of specific inhibitors.

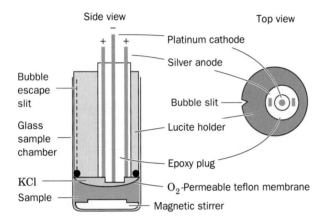

FIGURE 20-9. The oxygen electrode consists of an Ag/AgCl reference electrode and a Pt electrode, both immersed in a KCl solution and in contact with the sample chamber through an O$_2$-permeable Teflon membrane. O$_2$ is reduced to H$_2$O at the Pt electrode, thereby generating a voltage with respect to the Ag/AgCl electrode that is proportional to the O$_2$ concentration in the sealed sample chamber. [After Cooper, T.G., *The Tools of Biochemistry, p.* 69, Wiley (1977).]

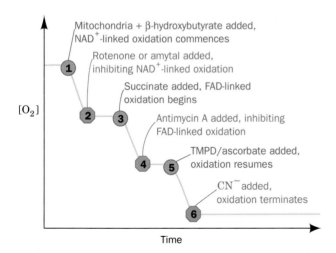

FIGURE 20-10. Idealized oxygen electrode trace of a mitochondrial suspension containing excess ADP and P$_i$. At the numbered points, the indicated reagents are injected into the sample chamber and the resulting changes in [O$_2$] are recorded. The numbers refer to the discussion in the text. [After Nicholls, D.G., *Bioenergetics, p.* 110, Academic Press (1982).]

This sequence has been corroborated by measurements of the standard reduction potentials of the redox components of each of the complexes as well as by determining the stoichiometry of the electron transport and coupled ATP synthesis.

The rate at which O$_2$ is consumed by a suspension of mitochondria is a sensitive measure of the functioning of the electron-transport chain. It is conveniently measured with an **oxygen electrode** (Fig. 20-9). Compounds that inhibit electron transport, as judged by their effect on O$_2$ disappearance in such an experimental system, have been invaluable experimental probes in tracing the path of electrons through the electron-transport chain and in determining the points of entry of electrons from various substrates. Among the most useful such substances are **rotenone** (a plant toxin used by Amazonian Indians to poison fish and which is also used as an insecticide), **amytal** (a barbiturate), **antimycin A** (an antibiotic), and **cyanide**.

The following experiment illustrates the use of these inhibitors:

A buffered solution containing excess ADP and P$_i$ is equilibrated in the reaction vessel of an oxygen electrode. Reagents are then injected into the chamber and the O$_2$ consumption recorded (Fig. 20-10):

1. Mitochondria and **β-hydroxybutyrate** are injected into the chamber. Mitochondria mediate the NAD^+-linked oxidation of β-hydroxybutyrate (Section 23-3).

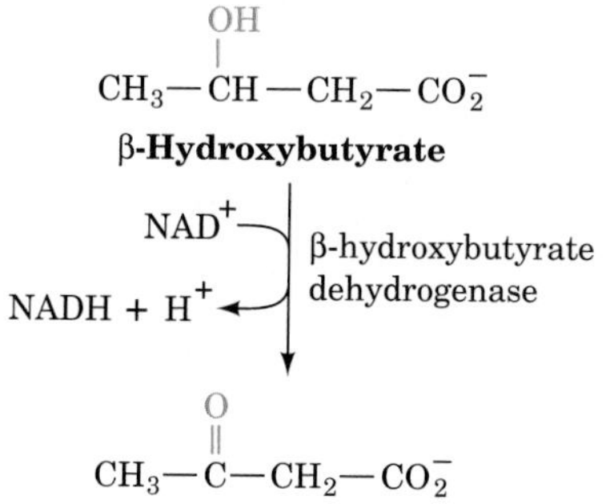

$$
\begin{array}{c}
OH \\
| \\
CH_3 - CH - CH_2 - CO_2^-
\end{array}
$$

β-Hydroxybutyrate

$$NAD^+ \qquad NADH + H^+$$
β-hydroxybutyrate dehydrogenase

$$
\begin{array}{c}
O \\
\parallel \\
CH_3 - C - CH_2 - CO_2^-
\end{array}
$$

Acetoacetate

As the resulting NADH is oxidized by the electron-transport chain with O_2 as the terminal electron acceptor, the O_2 concentration in the reaction mixture decreases.

2. Addition of rotenone or amytal completely stops the β-hydroxybutyrate oxidation.

3. Addition of succinate, which undergoes FAD-linked oxidation, causes the $[O_2]$ to resume its decrease. Electrons from $FADH_2$ are therefore still able to reduce O_2 in the presence of rotenone; that is, *electrons from $FADH_2$ enter the electron-transport chain after the rotenone-blocked step.*

4. Addition of antimycin A inhibits electron transport from $FADH_2$.

5. Although NADH and $FADH_2$ are the electron-transport chain's two physiological electron donors, nonphysiological reducing agents can also be used to probe the flow of electrons. **Tetramethyl-*p*-phenylenediamine (TMPD)** is an ascorbate-reducible redox carrier that transfers electrons directly to cytochrome *c*.

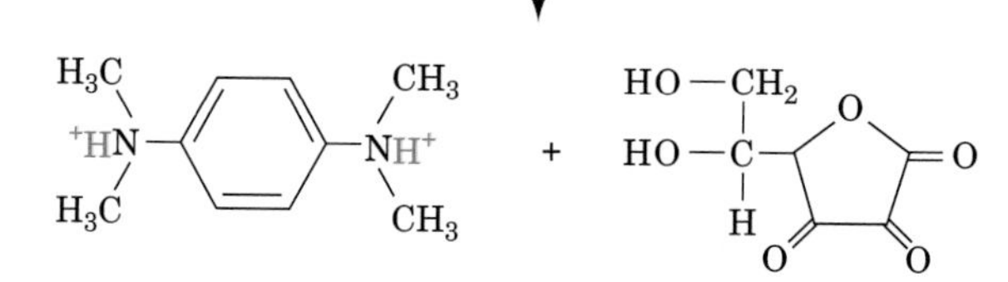

Tetramethyl-*p*-phenylenediamine (TMPD), oxidized form   **Ascorbic acid**

TMPD, reduced form   **Dehydroascorbic acid**

Addition of TMPD and ascorbate to the antimycin A-inhibited reaction mixture results in resumption of oxygen consumption; *evidently there is a third point for electrons to enter the electron-transport chain.*

6. The addition of CN^- completely inhibits oxidation of all three electron donors, indicating that it blocks the electron-transport chain after the third point of entry of electrons.

Experiments such as these established the order of electron flow through the electron-transport chain complexes and the positions blocked by various electron-transport inhibitors (Fig. 20-8). This order was confirmed and extended by observations that the standard reduction potentials of the redox carriers forming the electron-transport chain complexes are very close to the standard reduction potentials of their electron donor substrates (Table 20-1). *The three jumps in reduction potential between NADH, CoQ, cytochrome c, and O_2 are each of sufficient magnitude to drive ATP synthesis.* Indeed, these redox potential jumps correspond to the points of inhibition of rotenone (or amytal), antimycin A, and CN^-.

TABLE 20-1. REDUCTION POTENTIALS OF ELECTRON-TRANSPORT CHAIN COMPONENTS IN RESTING MITOCHONDRIA.

Component	$\mathscr{E}°'$ (V)
NADH	−0.315
Complex I (NADH–CoQ reductase; 850 kD, 26 subunits):	
FMN	?
(Fe–S)N-1a	−0.380
(Fe–S)N-1b	−0.250
(Fe–S)N-2	−0.030
(Fe–S)N-3,4	−0.245
(Fe–S)N-5,6	−0.270
Succinate	0.030
Complex II (succinate–CoQ reductase; 127 kD, 5 subunits):	
FAD	−0.040
(Fe–S)S-1	−0.030
(Fe–S)S-2	−0.245
(Fe–S)S-3	0.060
Cytochrome b_{560}	−0.080
Coenzyme Q	0.045
Complex III (CoQ–cytochrome *c* reductase; 280 kD, 10 subunits):	
Cytochrome b_H (b_{562})	0.030
Cytochrome b_L (b_{566})	−0.030
(Fe–S)	0.280
Cytochrome c_1	0.215
Cytochrome *c*	0.235
Complex IV (cytochrome *c* oxidase; ~200 kD, 6–13 subunits):	
Cytochrome *a*	0.210
Cu_A	0.245
Cu_B	0.340
Cytochrome a_3	0.385
O_2	0.815

Source: Wilson, D.F., Erecińska, M. and Dutton, P.L., *Annu. Rev. Biophys. Bioeng.* **3**, 205 and 208 (1974); *and* Wilson, D.F., *In* Bittar, E.E. (Ed.), *Membrane Structure and Function*, Vol. 1, p. 160, Wiley (1980).

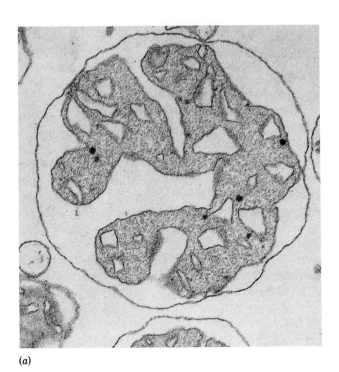

(a)

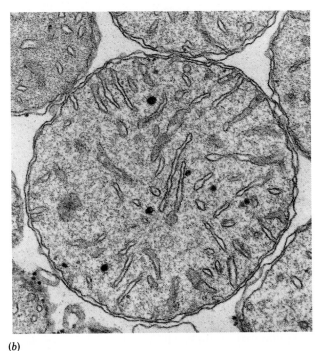

(b)

FIGURE 20-11. Electron micrographs of mouse liver mitochondria in (*a*) the actively respiring state and (*b*) the resting state. The cristae in actively respiring mitochondria are

far more condensed than they are in resting mitochondria. [Courtesy of Charles Hackenbrock, University of North Carolina Medical School.]

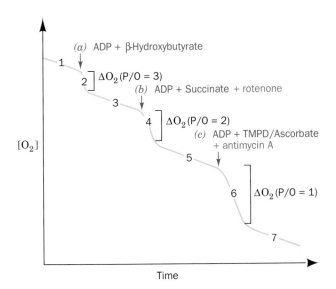

FIGURE 20-12. Stoichiometry of coupled oxidation and phosphorylation (the P/O ratio) with different electron donors. Mitochondria are incubated in excess phosphate buffer in the sample chamber of an oxygen electrode. (*a*) 90 μmol of ADP and excess β-hydroxybutyrate are added. Respiration continues until all the ADP is phosphorylated. ΔO_2 in Region 2 is 15 μmol, corresponding to 30 μmol of NADH oxidized; P/O = 90/30 = 3. (*b*) 90 μmol of ADP and excess succinate are added together with rotenone to inhibit electron transfer from NADH. ΔO_2 in Region 4 is 22.5 μmol, corresponding to 45 μmol of FADH$_2$ oxidized; P/O = 90/45 = 2. (*c*) 90 μmol of ADP and excess TMPD/ascorbate are added with antimycin A to inhibit electron transfer from FADH$_2$. ΔO_2 in Region 6 is 45-μmol, corresponding to 90 μmol of ascorbate oxidized; P/O = 90/90 = 1.

Phosphorylation and Oxidation Are Rigidly Coupled

The foregoing thermodynamic studies suggest that oxidation of NADH, FADH$_2$, and ascorbate by O$_2$ are associated with the synthesis of three, two, and one ATPs, respectively. This stoichiometry, called the **P/O ratio,** has been confirmed experimentally through measurements of O$_2$ uptake by resting and active mitochondria. An example of a typical experiment used to determine P/O ratio is as follows: A suspension of mitochondria (isolated by differential centrifugation after cell disruption; Section 5-1B) containing an excess of P$_i$ but no ADP is incubated in an oxygen electrode reaction chamber. *Oxidation and phosphorylation are closely coupled in well-functioning mitochondria, so electron transport can occur only if ADP is being phosphorylated (Section 20-3).* Indeed, mitochondrial metabolism is so tightly regulated that even the appearances of actively respiring and resting mitochondria are greatly different (Fig. 20-11). Since no ADP is present in the reaction mixture, the mitochondria are resting and the O$_2$ consumption rate is minimal (Fig. 20-12; Region 1). The system is then manipulated as follows:

(a) 90 μmol of ADP and an excess of β-**hydroxybutyrate** (an NAD$^+$-linked substrate) are added. The mitochondria immediately enter the active state and the rate of oxygen consumption increases (Fig. 20-12; Region 2) and is maintained at this elevated level until all the ADP is phosphorylated. The mitochondria then return to the resting state (Fig. 20-12; Region 3). Phosphorylation of 90 μmol of ADP under these conditions consumes 15 μmol of O$_2$. Since the oxidation of NADH by

O_2 consumes twice as many moles of NADH as of O_2, the P/O ratio for NADH reoxidation at Region 2 is 90 μmol of ADP/(2 × 15 μmol of O_2) = 3; that is, *3 mol of ADP are phosphorylated per mole of NADH oxidized.*

(b) The experiment is continued by inhibiting electron transfer from NADH by rotenone and adding an additional 90 μmol of ADP (Fig. 20-12; Region 4), this time together with an excess of the FAD-linked substrate succinate. Oxygen consumption again continues until all the ADP is phosphorylated, and the system again returns to the resting state (Fig. 20-12; Region 5). Calculation of the P/O ratio for $FADH_2$ oxidation yields the value 2; that is, *2 mol of ADP are phosphorylated per mole of $FADH_2$ oxidized.*

(c) In the same manner, *the oxidation of ascorbate/TMPD yields a P/O ratio of 1 (Fig. 20-12; Regions 6 and 7).*

These conclusions agree with the inhibitor studies indicating that there are three entry points for electrons into the electron-transport chain and with the standard reduction potential measurements exhibiting three potential jumps, each sufficient to provide the free energy for ATP synthesis (Fig. 20-8).

The P/O Ratios May Be Subject to Revision

Measurements of P/O ratios are subject to systematic experimental errors for which it is difficult to correct, such as inaccuracies in the measurement of the oxygen concentration, the presence of AMP, and proton leakage from mitochondria. Thus, the widely accepted P/O values of 3, 2, and 1 associated with NADH-, $FADH_2$-, and ascorbate/TMPD-linked oxidation may well be in error. Indeed, measurements by Peter Hinkle have yielded values close to 2.5, 1.5, and 1 for these quantities (we shall see in Section 20-3 that P/O ratios need not have integer values because the number of protons pumped out of the mitochondrion by any component of the electron transport chain may not be an integer multiple of the number of protons required to

synthesize ATP from ADP + P_i). If these values are correct, then the number of ATPs that are synthesized per molecule of glucose oxidized is 2.5 ATP/NADH × 10 NADH/glucose + 1.5 ATP/$FADH_2$ × 2 $FADH_2$/glucose + 2 ATP/glucose from the citric acid cycle + 2 ATP/glucose from glycolysis = 32 ATP/glucose rather than the conventional value of 38 ATP/glucose implied by P/O ratios of 3, 2, and 1. However, since there is significant disagreement as to the validity of the revised P/O ratios, we shall, for the sake of consistency, use the more established values of 3, 2, and 1 throughout this textbook. You should nevertheless keep in mind that these values are disputed.

How the free energy of electron transport is actually coupled to ATP synthesis, a subject of active research, is discussed in Section 20-3. We first examine the structures of the four respiratory complexes in order to understand how they are related to the function of the electron-transport chain. Keep in mind, however, that as in most areas of biochemistry, this field is under intense investigation and much of the information we need for a complete understanding of these relationships has yet to be uncovered.

C. Components of the Electron-Transport Chain

Many of the proteins embedded in the inner mitochondrial membrane are organized into the four respiratory complexes of the electron-transport chain. Each complex consists of several protein components that are associated with a variety of redox-active prosthetic groups with successively increasing reduction potentials (Table 20-1). The complexes are all laterally mobile within the inner mitochondrial membrane; they do not appear to form any stable higher structures. Indeed, they are not present in equimolar ratios. In the following paragraphs, we examine their structures and the agents that transfer electrons between them. Their relationships are summarized in Fig. 20-13.

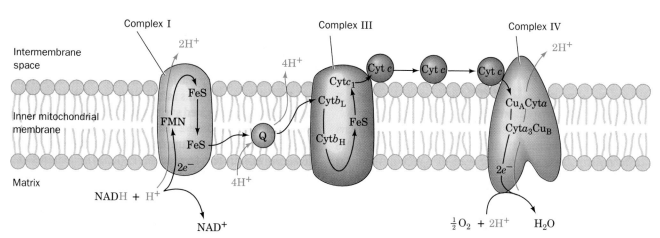

FIGURE 20-13. A diagram of the mitochondrial electron-transport chain indicating the pathway of electron transfer (*black*) and proton pumping (*red*). Electrons are transferred between Complexes I and III by the membrane-soluble CoQ and between Complexes III and IV by the peripheral membrane protein cytochrome *c*. Complex II (not shown) transfers electrons from succinate to CoQ.

The standard redox potential for electron transfer from succinate to CoQ (Fig. 20-8) is insufficient to provide the free energy necessary to drive ATP synthesis. The complex is, nevertheless, important because it allows these relatively high-potential electrons to enter the electron-transport chain.

3. Complex III (Coenzyme Q – Cytochrome *c* Reductase)

Complex III passes electrons from reduced CoQ to cytochrome c. It contains two **b-cytochromes**, one **cytochrome c_1**, and one [2Fe–2S] cluster (Table 20-1).

Cytochromes: Electron-Transport Heme Proteins

Cytochromes, whose function was elucidated in 1925 by David Keilin, are redox-active proteins that occur in all organisms except a few types of obligate anaerobes. These proteins contain heme groups that reversibly alternate between their Fe(II) and Fe(III) oxidation states during electron transport.

The heme groups of the reduced [Fe(II)] cytochromes have prominent visible absorption spectra consisting of three peaks: The α, β, and γ **(Soret)** bands (Fig. 20-16a). The wavelength of the α peak, which varies characteristically with the particular reduced cytochrome species (it is absent in oxidized cytochromes), is useful in differentiating the various cytochromes. Accordingly, the spectra of mitochondrial membranes (Fig. 20-16b) indicate that they contain three cytochrome species, **cytochromes *a*, *b*, and *c*.**

Within each group of cytochromes, different heme group environments may be characterized by slightly different α

peak wavelengths. For example, Complex III has two *b*-type cytochrome hemes: That absorbing maximally at 562 nm is referred to as b_H (for high potential; formerly called b_K) or b_{562}, whereas that absorbing maximally at 566 nm is referred to as b_L (for low potential; formerly called b_T) or b_{566}. (The second type of nomenclature is a recent adoption that identifies a cytochrome with the wavelength, in nanometers, at which its α band absorbance is maximal. Previously, cytochromes were identified nondescriptively with either numbers or letters.) Complex II had previously been noted to contain cytochrome b_{560}.

Each group of cytochromes contains a differently substituted porphyrin ring (Fig. 20-17a) coordinated with the redox-active iron atom. The *b*-type cytochromes contain **protoporphyrin IX,** which also occurs in hemoglobin (Section 9-1A). The heme group of *c*-type cytochromes differs from protoporphyrin IX in that its vinyl groups have added Cys sulfhydryls across their double bonds to form thioether linkages to the protein. Heme *a* contains a long hydrophobic tail of isoprene units attached to the porphyrin, as well as a formyl group in place of a methyl substituent. The axial ligands of the heme iron also vary with the cytochrome type. In cytochromes *a* and *b*, both ligands are His residues, whereas in cytochrome *c*, one is His and the other is Met (Fig. 20-17b).

Complex III is arranged asymmetrically in the inner mitochondrial membrane as judged from various chemical labeling studies. Both **cytochrome c_1** and the nonheme iron protein [often called the **Rieske iron–sulfur protein (ISP)** after its discoverer, John Rieske] are located on the mem-

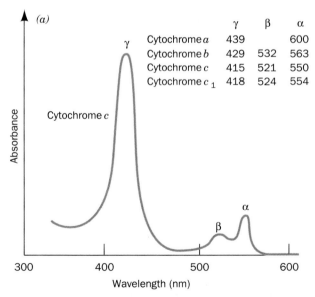

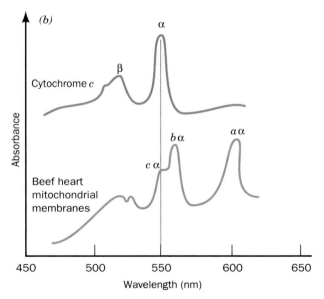

FIGURE 20-16. Visible absorption spectra of cytochromes. (*a*) Absorption spectrum of reduced cytochrome *c* showing its characteristic α, β, and γ (Soret) absorption bands. The absorption maxima for cytochromes *a, b, c,* and c_1 are listed. (*b*) The three separate α bands in the visible absorption spectra of beef heart mitochondrial membranes (*below*) indicate the presence of cytochromes *a, b,* and *c*. The spectrum of purified cytochrome *c* (*above*) is provided for reference. [After Nicholls, D.G. and Ferguson, S.J., *Bioenergetics* **2,** *p.* 113, Academic Press (1992).]

The table shown within figure (a):

	γ	β	α
Cytochrome *a*	439		600
Cytochrome *b*	429	532	563
Cytochrome *c*	415	521	550
Cytochrome *c*$_1$	418	524	554

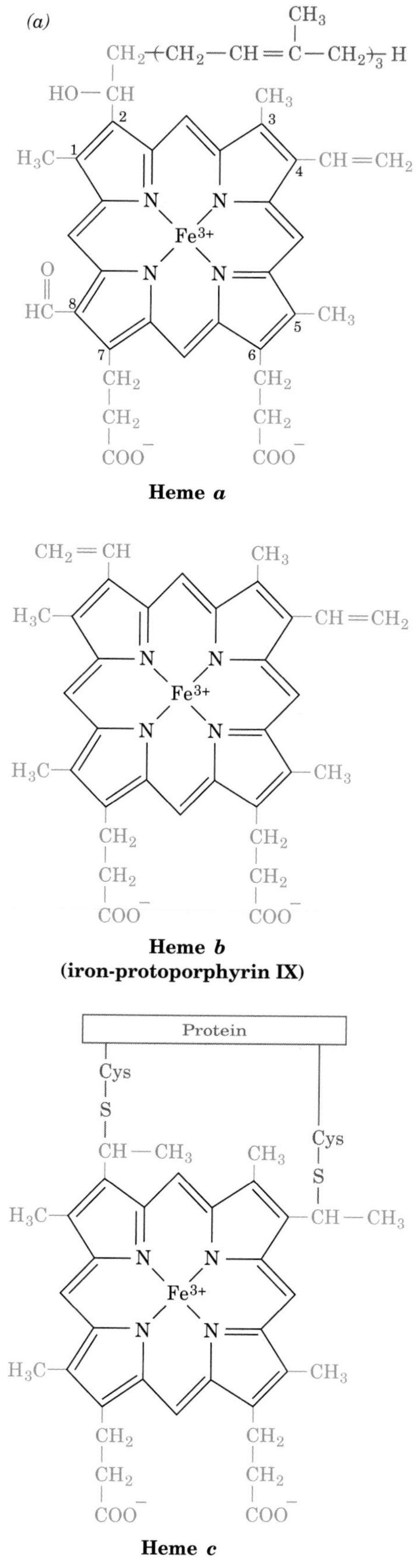

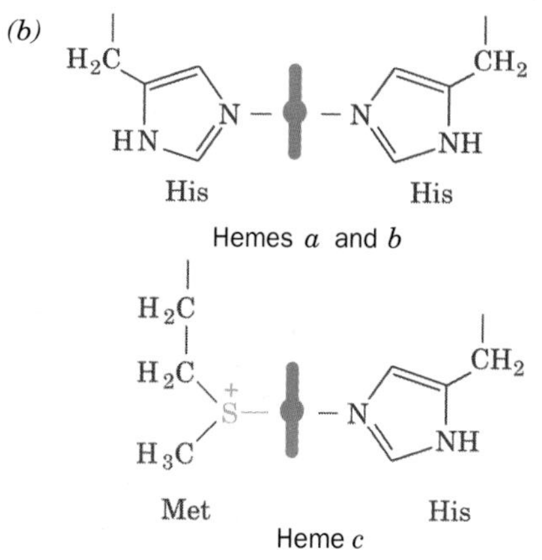

(a)

Heme *a*

Heme *b*
(iron-protoporphyrin IX)

Heme *c*

(b)

Hemes *a* and *b*

Met   His
Heme *c*

FIGURE 20-17. The (a) chemical structures and (b) axial liganding of the heme groups contained in cytochromes *a, b,* and *c.*

brane's outer surface, whereas cytochrome *b* is a transmembrane protein. Cytochrome *b* is a particularly interesting protein because it contains both *b*-type cytochrome hemes, b_H and b_L, associated with a single polypeptide chain.

Since only a few X-ray structures of integral membrane proteins are known (Section 11-3A), investigators have attempted to predict the gross structures of such proteins from their amino acid sequences. The amino acid sequence of cytochrome *b*, which is encoded by mitochondrial DNA, has been deduced from the base sequence of this DNA (Section 30-1E). The polypeptide chains from various species are 380 to 385 amino acids long and exhibit considerable sequence homology. They apparently have eight >20-residue stretches of predominantly hydrophobic residues that are predicted to form stable helices that can span the membrane bilayer. Therefore, it is likely that the polypeptide chain of cytochrome *b* spans the membrane eight times (Fig. 20-18). The two heme groups are postulated to be coordinated to four invariant His residues, located on Helices B and D at positions 82 and 183 on the cytoplasmic side, and 96 and 197 on the matrix side. The heme group closer to the cytoplasmic side of the membrane is thought to be b_L (b_{566}), whereas that closer to the matrix side is thought to be b_H (b_{562}). The route of electrons through the complex is discussed in Section 20-3B, along with the mechanism by which Complex III preserves the free energy of electron transfer from $CoQH_2$ to cytochrome *c* for ATP synthesis.

4. Cytochrome *c*

Cytochrome *c* is a peripheral membrane protein of known crystal structure (Fig. 8-18*c* and cover illustrations) that is loosely bound to the outer surface of the inner mitochondrial membrane. *It alternately binds to cytochrome c_1 of Complex III and to cytochrome c oxidase (Complex IV) and thereby functions to shuttle electrons between them.*

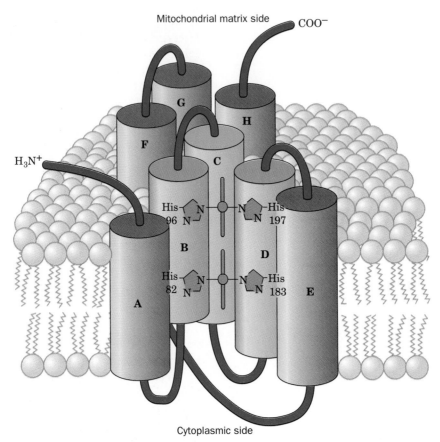

FIGURE 20-18. Predicted secondary structure of mitochondrial cytochrome *b* based on its amino acid sequences in several species. The polypeptide has eight stretches of nonpolar amino acids that can form helices long enough to traverse the membrane (represented as cylinders). The invariant His residues 82 and 183 as well as 96 and 197, form the axial ligands to the protein's two heme groups, thereby bridging helices B and D. [After Esposti, M.D., De Vries, S., Crimi, M., Ghelli, A., Patarnello, T., and Meyer, A., *Biochem. Biophys. Acta* **1143**, 266 (1993)]

Cytochrome *c*'s binding site contains several invariant Lys residues that lie in a ring around the exposed edge of its otherwise buried heme group (Fig. 20-19). This binding site has been identified by **differential labeling:** Treatment of cytochrome *c* with acetic anhydride (which acetylates Lys residues) in the presence and absence of cytochrome c_1 demonstrated that cytochrome c_1 completely shields these cytochrome *c* Lys residues. The reactivities of other cytochrome *c* Lys residues that are distant from the exposed heme edge are unaffected by complex formation. Nearly identical results were obtained when cytochrome c_1 was replaced by cytochrome *c* oxidase. Evidently, both these proteins have negatively charged sites that are complementary to the ring of positively charged Lys residues on cytochrome *c* (see below).

The Influence of Protein Structure on the Rate of Electron Transfer

Reduced hemes are highly reactive entities; they can transfer electrons over distances of 10 to 20 Å at physiologically significant rates. Hence cytochromes, in a sense, have the opposite function of enzymes: Instead of persuading unreactive substrates to react, they must prevent their hemes from transferring electrons nonspecifically to other cellular components. This, no doubt, is why these hemes are almost entirely enveloped by protein. However, cytochromes must also provide a path for electron transfer to an appropriate partner.

The mechanism of electron transfer within and between proteins and the role of protein structure in these processes are areas of active theoretical and experimental research. For example, Leslie Dutton has shown that for electron transfer within proteins, the experimentally measured electron transfer rates depend only upon the distance between the electron donor and the electron acceptor, falling off exponentially with the distance between them with a ~10-fold decrease in rate for each 1.7-Å increase in distance. Evidently, the rate of intraprotein electron transfer is largely independent of the structure of the protein in which the redox centers are embedded. However, since electron transfer occurs far more efficiently through bonds than through space, protein structure appears to be an important determinant of the rate of electron transfer between proteins.

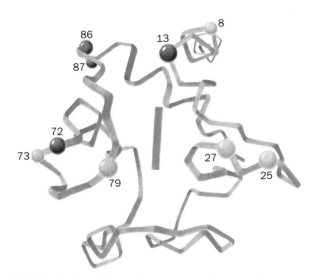

FIGURE 20-19. Ribbon diagram of cytochrome *c* showing the Lys residues involved in intermolecular complex formation with cytochrome *c* oxidase or reductase as inferred from chemical modification studies. Dark and light blue balls, respectively, mark the positions of Lys residues whose ε-amino groups are strongly and less strongly protected by cytochrome *c* oxidase and reductase against acetylation by acetic anhydride. Note that these Lys residues form a ring around the heme (*solid bar*) on one face of the protein. [After Mathews, F.S., *Prog. Biophys. Mol. Biol.* **45,** 45 (1986).]

The Complex of Cytochrome *c* with Cytochrome *c* Peroxidase Provides a Structural Model for Electron Transfer

The only redox partner of cytochrome *c* whose atomic structure has been determined is yeast **cytochrome *c* peroxidase (CCP),** a heme-containing, 296-residue, monomeric protein. It catalyzes the two-electron reduction of organic hydroperoxides (ROOH) in a three-step reaction cycle that results in the oxidation of two molecules of cytochrome *c*:

$$CCP + ROOH + 2H^+ \longrightarrow CCP(I)^{2+} + ROH + H_2O$$
$$CCP(I)^{2+} + cyt\ c\ (Fe^{2+}) \longrightarrow CCP(II)^+ + cyt\ c\ (Fe^{3+})$$
$$CCP(II)^+ + cyt\ c\ (Fe^{2+}) \longrightarrow CCP + cyt\ c\ (Fe^{3+})$$

Here CCP(I)$^{2+}$ represents a $2e^-$ oxidized state and CCP(II)$^+$ represents a $1e^-$ oxidized state of CCP.

The individually determined X-ray structures of CCP and yeast cytochrome *c* (which closely resembles that of the highly homologous tuna cytochrome *c*; Section 8-3A) fit together with remarkable precision such that the ring of Lys residues on cytochrome *c* is juxtaposed with a ring of negative charges surrounding the CCP heme crevice. The heme groups of the two proteins in this hypothetical model are parallel to each other and to the aromatic rings of His 181 of CCP and the invariant Phe 82 of cytochrome *c*. This led Thomas Poulos and Joseph Kraut to propose that this pi system forms the pathway for electron transfer from one heme to the other through the ~18 Å of intervening protein.

However, subsequent mutagenesis experiments have demonstrated that neither His 181 nor Phe 82 are required for efficient electron transfer from cytochrome *c* to CCP.

More recently, Kraut has determined the X-ray structure of a 1:1 complex of yeast cytochrome *c* with yeast CCP (Fig. 20-20). In this structure, the two hemes are, in fact, inclined to each other by ~60° and cytochrome *c* residues Lys 73 and Lys 87 form ion pairs with CCP residues Glu 290 and Asp 34, respectively. Electrons are now postulated to be transferred via what is essentially the shortest straight line route between the hemes, which runs from cytochrome *c*'s exposed heme edge through the CCP polypeptide backbone sigma bonds of Ala 194, Ala 193, Gly 192, and Trp 191, and finally through the indole ring of Trp 191, which is in van der Waals contact with and perpendicular to the CCP heme (Fig. 20-20). The significance of this electron-transfer pathway is supported by the observations that Trp 191 is the site of an intermediate radical during CCP reduction and that its replacement by Phe all but eliminates the electron

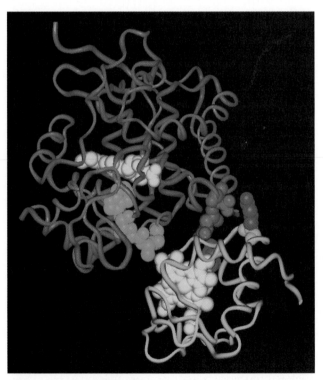

FIGURE 20-20. The X-ray structure of the yeast cytochrome *c*–CCP complex, indicating the proposed electron-transfer pathway. The cytochrome *c* (*light blue*) and CCP (*purple*) are represented by their polypetide backbones. The groups that comprise the proposed electron-transfer pathway are shown in space-filling form with the hemes yellow and CCP residues Ala 194, Ala 193, Gly 192, and Trp 191 (with its indole ring nearly perpendicular to the CCP heme) green. The ion pairing interactions between Lys 87 and Lys 73 of cytochrome c (*blue*) and Asp 34 and Glu 290 of CCP (*red*) are also shown in space-filling form. [Based on an X-ray structure by Joseph Kraut, University of California at San Diego.]

transfer. Moreover, the X-ray structure of the 1:1 complex of horse heart cytochrome *c* with CCP closely resembles that of the yeast cytochrome *c*-containing complex, even though the two complexes were crystallized under different conditions and are differently arranged within their respective crystals. Thus it appears that electron transfer occurs through sigma bonds rather than through a pi bonding system.

Despite the foregoing, Brian Hoffman has demonstrated, through spectroscopic techniques, that CCP has two binding sites for cytochrome *c*, one that has high affinity for cytochrome *c* but mediates electron transfer with low efficiency, and another that has low affinity for cytochrome *c* but mediates electron transfer with high efficiency. Presumably, the two binding sites on CCP each mediate only one of the two electron transfer reactions from cytochrome *c* to CCP. It seems likely that the high-affinity complex is that observed in the CCP–cytochrome *c* X-ray structure, whereas the low-affinity complex may well resemble the original Poulos–Kraut model.

5. Complex IV (Cytochrome *c* Oxidase)

Cytochrome c oxidase catalyzes the one-electron oxidations of four consecutive reduced cytochrome c molecules and the concomitant four-electron reduction of one O_2 molecule:

$$4\text{Cytochrome } c^{2+} + 4\text{H}^+ + \text{O}_2 \longrightarrow$$
$$4\text{cytochrome } c^{3+} + 2\text{H}_2\text{O}$$

Mammalian Complex IV is an ~200-kD transmembrane protein composed of 6 to 13 subunits (Fig. 20-21) whose largest and most hydrophobic Subunits, I, II, and III, are encoded by mitochondrial DNA. These subunits collectively have 18 hydrophobic segments that probably form membrane-spanning helices similar to those of cytochrome *b* (Fig. 20-18). Complex IV exists in membranes as a dimer. Subunits I and II of this complex contain all four of its redox-active centers: two *a*-type hemes (*a* and a_3; Fig. 20-17) that alternate between their Fe^{2+} and Fe^{3+} oxidation states, and two Cu atoms that alternate between their +1 and +2 oxidation states (Table 20-1). Heme *a* and the Cu atom designated Cu_A (alternatively, Cu_a) are of low potential (~0.24 V), whereas heme a_3 and Cu_B (alternatively, Cu_{a3}) are of higher potential (~0.34 V; Table 20-1). A variety of spectroscopic evidence indicates that Cu_A is liganded to Subunit II through two Cys and two His residues and is part of the cytochrome *c*-binding site. Hemes *a* and a_3 and Cu_B are located on Subunit I. Heme a_3 and Cu_B are bridged, most probably by an S atom, to form a binuclear complex that comprises the O_2-binding site (Fig. 20-22*a*).

Cytochrome *c* oxidase's interaction site with cytochrome *c* presumably has several Asp and/or Glu residues that interact with the above-discussed ring of Lys residues surrounding cytochrome *c*'s heme crevice. Indeed, differential labeling of cytochrome *c* oxidase's carboxyl groups in the presence and absence of cytochrome *c*, demonstrated

(*a*)

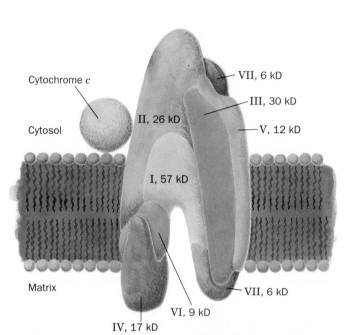

(*b*)

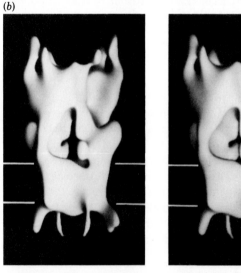

FIGURE 20-21. Cytochrome *c* oxidase. (*a*) The structure and orientation of cytochrome *c* oxidase in relation to the inner mitochondrial membrane. Only one monomer unit of the dimeric protein complex is shown. The overall Y shape of the monomer was elucidated by electron microscopy combined with image reconstruction techniques. Spatial relationships among the subunits were deduced from the binding of specific antibodies together with the results of cross-linking studies. The complex, evidently, contains two differently located copies of Subunit VII. The subunit molecular masses are indicated. (*b*) A stereo view of the 3-dimensional structure of bovine heart cytochrome *c* oxidase as determined by electron crystallography to 20-Å resolution. The estimated position of the membrane bilayer is represented by horizontal lines. The bulk of the dimeric protein complex is located on the cytosolic side of the membrane (*above*). Note the cleft between the two monomers, the probable cytochrome *c* binding site. Directions for viewing stereo images are given in the Appendix to Chapter 7. [Courtesy of Richard Henderson, Cambridge University, U.K.]

(a)

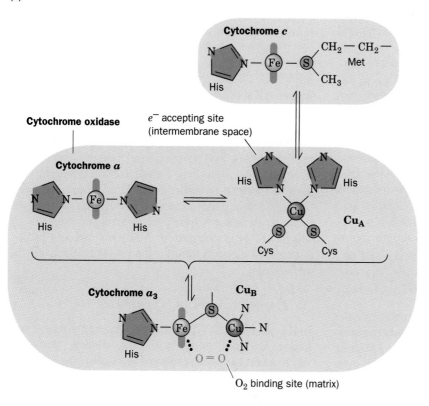

(b)

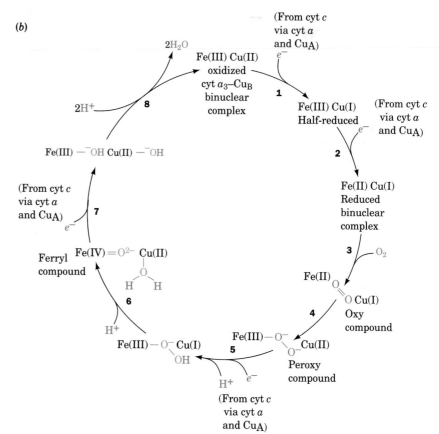

FIGURE 20-22. The cytochrome *c* oxidase reaction. (*a*) The flow of electrons from cytochrome *c* through the four redox-active centers of cytochrome *c* oxidase to O_2. The most probable ligands of the heme Fe and Cu ions are indicated, although they have not been determined unambiguously in every case. O_2 binds to the cytochrome a_3–Cu_B binuclear complex where it is reduced to $2H_2O$ by the stepwise transfer of four electrons from cytochrome *a* and Cu_A together with the acquisition of four protons. The entire reaction is extremely fast; it goes to completion in ~1 ms at room temperature. (*b*) The reaction sequence for the reduction of O_2 by the cytochrome a_3–Cu_B binuclear complex of cytochrome *c* oxidase. The numbered steps are explained in the text. [Modified from Varotsis, C., Zhang, Y., Appelman, E.H., and Babcock, G.T. *Proc. Natl. Acad. Sci.,* **90**, 240 (1993).]

that cytochrome c shields the invariant residues Asp 112, Glu 114, and Glu 198 of cytochrome c oxidase Subunit II. Glu 198 is located between the two Cys residues of Subunit II that are thought to ligand Cu_A. This observation supports the spectroscopic evidence that places the cytochrome c-binding site on Subunit II in close proximity to Cu_A. Cross-linking studies have additionally shown that the cytochrome c surface opposite the electron-transferring site interacts with Subunit III, suggesting the existence of a cytochrome c-binding cleft formed by Subunits II and III. Indeed, the low-resolution structure of the cytochrome c oxidase dimer (Figure 20-21b), as determined by electron crystallography, reveals a cavity between its two monomers that is proposed to form the cytochrome c binding site. Moreover, this transmembrane protein exhibits a convoluted shape that extends well above the membrane on its cytosolic side.

Reaction Sequence for the Reduction of O_2 by Cytochrome c Oxidase

The reduction of O_2 to 2 H_2O by cytochrome c oxidase takes place on the cytochrome a_3–Cu_B binuclear complex (Fig. 20-22a). This reaction, elucidated largely by Mårten Wikström and Gerald Babcock, involves four consecutive one-electron transfers from the Cu_A and cytochrome a sites and occurs as follows (Fig. 20-22b):

1. & 2. The binuclear Fe(III)$_{a3}$–Cu(II)$_B$ complex is reduced, by two one-electron transfers from cytochrome c via cytochrome a and Cu_A, to its Fe(II)$_{a3}$–Cu(I)$_B$ form.

3. O_2 binds to this reduced binuclear complex so as to bridge its Fe(II)$_{a3}$ and Cu(I)$_B$ atoms.

4. Internal electron redistribution rapidly yields the stable peroxy adduct Fe(III)—O^-—O^- Cu(II).

5. A further one-electron transfer together with the acquisition of a proton converts the adduct to Fe(III)—O^-—OH Cu(I).

6. The acquisition of a second proton and an electronic rearrangement results in Fe(IV)=O^{2-} H_2O—Cu(II) [where Fe(IV) is said to have the **ferryl** oxidation state].

7. The fourth one-electron transfer together with a proton rearrangement then yields Fe(III)—OH^- $^-$HO—Cu(II).

8. Finally, the acquisition of two more protons yields 2H_2O together with the Fe(III)$_{a3}$–Cu(II)$_B$ complex, thereby completing the cycle.

3. OXIDATIVE PHOSPHORYLATION

The endergonic synthesis of ATP from ADP and P$_i$ in mitochondria, which, as we shall see, is catalyzed by **proton-translocating ATP synthase (Complex V),** is driven by the electron-transport process. Yet, since Complex V is physically distinct from the proteins mediating electron transport (Complexes I–IV), *the free energy released by electron transport must be conserved in a form that ATP synthase can utilize.* Such energy conservation is referred to as **energy coupling** or **energy transduction.**

The physical characterization of energy coupling has proved to be surprisingly elusive; many sensible and often ingenious ideas have failed to withstand the test of experimental scrutiny. In this section we first examine some of the hypotheses that have been formulated to explain the coupling of electron transport and ATP synthesis. We shall then explore the coupling mechanism that has garnered the most experimental support, analyze the mechanism by which ATP is synthesized by ATP synthase, and finally, discuss how electron transport and ATP synthesis can be uncoupled.

A. Energy Coupling Hypotheses

In the more than 50 years that electron transport and oxidative phosphorylation have been studied, numerous mechanisms have been proposed to explain how these processes are coupled. In the following paragraphs, we examine the mechanisms that have received the greatest experimental attention:

1. The chemical coupling hypothesis

In 1953, Edward Slater formulated the **chemical-coupling hypothesis,** in which he proposed that electron transport yielded reactive intermediates whose subsequent breakdown drove oxidative phosphorylation. We have seen, for example, that such a mechanism is responsible for ATP synthesis in glycolysis (Sections 16-2F and G). Thus, the exergonic oxidation of glyceraldehyde 3-phosphate by NAD$^+$ yields 1,3-bisphosphoglycerate, a reactive ("high-energy") acyl phosphate whose phosphoryl group is then transferred to ADP to form ATP in the phosphoglycerate kinase reaction. The difficulty with such a mechanism for oxidative phosphorylation, which has largely caused it to be abandoned, is that despite intensive efforts in numerous laboratories over many years, no appropriate reactive intermediates have been identified.

2. The conformational-coupling hypothesis

The **conformational-coupling hypothesis,** which Paul Boyer formulated in 1964, proposes that electron transport causes proteins of the inner mitochondrial membrane to assume "activated" or "energized" conformational states. These proteins are somehow associated with ATP synthase such that their relaxation back to the deactivated conformation drives ATP synthesis. As with the chemical-coupling hypothesis, the conformational-coupling hypothesis has found little experimental support. However, conformational coupling of a different

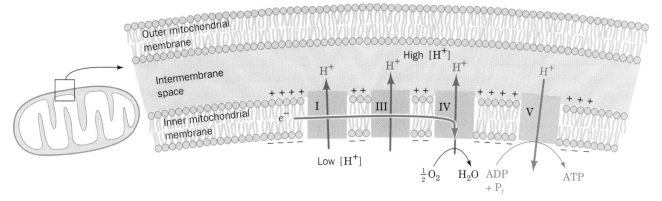

FIGURE 20-23. The coupling of electron transport (*green arrow*) and ATP synthesis by the generation of a proton electrochemical gradient across the inner mitochondrial membrane. H⁺ is pumped out of the mitochondrion during electron transport (*blue arrows*) and its exergonic return powers the synthesis of ATP (*red arrows*).

sort appears to be involved in ATP synthesis (Section 20-3C).

3. The chemiosmotic hypothesis

The chemiosmotic hypothesis, proposed in 1961 by Peter Mitchell, has spurred considerable controversy, as well as research, and now appears to be the model most consistent with the experimental evidence. It postulates that *the free energy of electron transport is conserved by pumping H⁺ from the mitochondrial matrix to the intermembrane space so as to create an electrochemical H⁺ gradient across the inner mitochondrial membrane. The electrochemical potential of this gradient is harnessed to synthesize ATP (Fig. 20-23).*

Several key observations are explained by the chemiosmotic hypothesis:

(a) Oxidative phosphorylation requires an intact inner mitochondrial membrane.

(b) The inner mitochondrial membrane is impermeable to ions such as H⁺, OH⁻, K⁺, and Cl⁻, whose free diffusion would discharge an electrochemical gradient.

(c) Electron transport results in the transport of H⁺ out of intact mitochondria, thereby creating a measureable electrochemical gradient across the inner mitochondrial membrane.

(d) Compounds that increase the permeability of the inner mitochondrial membrane to protons, and thereby dissipate the electrochemical gradient, allow electron transport (from NADH and succinate oxidation) to continue but inhibit ATP synthesis; that is, they "uncouple" electron transport from oxidative phosphorylation. Conversely, increasing the acidity outside the inner mitochondrial membrane stimulates ATP synthesis.

In the remainder of this section we examine the mechanisms through which electron transport can result in proton translocation and how an electrochemical gradient can interact with ATP synthase to drive ATP synthesis.

B. Proton Gradient Generation

Electron transport, as we shall see, causes Complexes I, III, and IV to transport protons across the inner mitochondrial membrane from the matrix, a region of low [H⁺] and negative electrical potential, to the intermembrane space (which is in contact with the cytosol), a region of high [H⁺] and positive electrical potential (Fig. 20-13). The free energy sequestered by the resulting electrochemical gradient [which, in analogy to the term electromotive force (emf), is called **proton-motive force (pmf)**] *powers ATP synthesis.*

Proton Pumping Is an Endergonic Process

The free energy change of transporting a proton out of the mitochondrion against an electrochemical gradient is expressed by Eq. [18.3] which, in terms of pH, is

$$\Delta G = 2.3RT\,[\text{pH}(in) - \text{pH}(out)] + Z\mathscr{F}\Delta\Psi \qquad [20.1]$$

where Z is the charge on the proton (including sign), $\mathscr{F}$ is the faraday constant, and $\Delta\Psi$ is the membrane potential. The sign convention for $\Delta\Psi$ is that when an ion is transported from negative to positive, $\Delta\Psi$ is positive. Since pH(*out*) is less than pH(*in*), the export of protons from the mitochondrial matrix (against the proton gradient) is an endergonic process. In addition, *proton transport out of the matrix makes the inner membrane's internal surface more negative than its external surface.* Outward transport of a positive ion is consequently associated with a positive $\Delta\Psi$ and an increase in free energy (endergonic process), whereas the outward transport of a negative ion yields the opposite result. Clearly, it is always necessary to describe membrane polarity when specifying a membrane potential.

The measured membrane potential across the inner membrane of a liver mitochondrion, for example, is 0.168 V (inside negative; which corresponds to an ~210,000-

$V \cdot cm^{-1}$ electric field across its ~80-Å thickness). The pH of its matrix is 0.75 units higher than that of its intermembrane space. ΔG for proton transport out of this mitochondrial matrix is therefore $21.5 \text{ kJ} \cdot mol^{-1}$.

The Passage of About Three Protons Is Required to Synthesize One ATP

An ATP molecule's estimated physiological free energy of synthesis, around $+40$ to $+50 \text{ kJ} \cdot mol^{-1}$, is too large to be driven by the passage of a single proton back into the mitochondrial matrix; at least two protons are required. This number is difficult to measure precisely, in part because transported protons tend to leak back across the mitochondrial membrane. However, most estimates indicate that around three protons are passed per ATP synthesized.

Two Mechanisms of Proton Transport Have Been Proposed

Three of the four electron-transport complexes, Complexes I, III, and IV, are involved in proton translocation. Two mechanisms have been entertained that would couple the free energy of electron transport with the active transport of protons — the **redox loop mechanism** and the **proton pump mechanism.**

1. The Redox Loop Mechanism

This mechanism, proposed by Mitchell, requires that the redox centers of the respiratory chain (FMN, CoQ, cytochromes, and iron–sulfur clusters) be so arranged in the membrane that reduction would involve a redox center simultaneously accepting e^- and H^+ from the matrix side of the membrane. Reoxidation of this redox center by the next center in the chain would involve release of H^+ on the cytosolic side of the membrane together with the transfer of

electrons back to the matrix side (Fig. 20-24a). Electron flow from one center to the next would therefore yield net translocation of H^+ and the creation of an electrochemical gradient ($\Delta\Psi$ and ΔpH).

The redox loop mechanism requires that the first redox carrier contain more hydrogen atoms in its reduced state than in its oxidized state and that the second redox carrier have no difference in its hydrogen atom content between its reduced and oxidized states. Are these requirements met in the electron-transport chain? Some of the redox carriers, FMN and CoQ, in fact, contain more hydrogen atoms in their reduced state than in their oxidized state and thus can qualify as proton carriers as well as electron carriers. If these centers were spatially alternated with pure electron carriers (cytochromes and iron–sulfur clusters), such a mechanism could well be accommodated (Fig. 20–24b).

The main difficulty with the redox loop mechanism involves the deficiency of $(H^+ + e^-)$ carriers that can alternate with pure e^- carriers. While there are as many as 15 pure e^- carriers (up to 8 iron–sulfur proteins, 5 cytochromes, and 2 Cu atoms), only two $(H^+ + e^-)$ carriers are known. The fact that there are 3 complexes with standard reduction potential changes large enough to provide free energy for ATP synthesis suggests the need for at least 3 proton-transport sites. This problem is emphasized in Fig. 20-24b by showing X as an unknown $(H^+ + e^-)$ carrier.

In an attempt to solve this problem, Mitchell has described a way coenzyme Q might be involved twice in proton translocation in Complex III (CoQ–cytochrome c reductase). In the so-called **Q cycle,** which has since garnered considerable experimental support as the mechanism of electron and proton transport in Complex III, $CoQH_2$ undergoes a two-cycle reoxidation involving the semiquinone, $CoQ^{\overline{\cdot}}$, as a stable intermediate (Fig. 20-25):

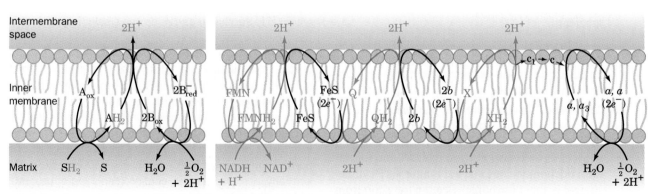

(a) Single redox loop   *(b)* Redox loop mechanism

FIGURE 20-24. The redox loop mechanism. *(a)* A single redox loop for electron transport–linked H^+ translocation. *(b)* The proposed redox loop mechanism of electron transport–linked proton translocation in mitochondria. FMN and CoQ function as $(H^+ + e^-)$ carriers, whereas the iron–sulfur clusters and the cytochromes function as pure e^- carriers. These components are so arranged as to require that electron transport be accompanied by H^+ translocation. Note that a mystery $(H^+ + e^-)$ carrier, X, is included to account for a third H^+ translocation site.

FIGURE 20-25. The Q cycle, an electron-transport cycle in Complex III proposed to account for H^+ translocation during the transport of electrons from cytochrome b to cytochrome c: The overall cycle is actually two cycles, the first requiring reactions 1 through 7 and the second requiring reactions 1 through 6 and 8. **(1)** Coenzyme QH_2 is supplied by Complex I on the matrix side of the membrane. **(2)** QH_2 diffuses to the cytosolic side of the membrane. **(3)** QH_2 reduces the Reiske iron–sulfur protein (ISP) forming $CoQ^{\cdot-}$ semiquinone and releasing $2H^+$. The ISP goes on to reduce cytochrome c_1. **(4)** $CoQ^{\cdot-}$ reduces cytochrome b_L to form coenzyme Q. **(5)** Q diffuses to the matrix side. **(6)** Cytochrome b_L reduces cytochrome b_H. **(7, cycle 1 only)** Q is reduced to $CoQ^{\cdot-}$ by cytochrome b_H. **(8, cycle 2 only)** $CoQ^{\cdot-}$ is reduced to $CoQH_2$ by cytochrome b_H. [After Trumpower, B.L., *J. Biol. Chem.* **265,** 11410 (1990).]

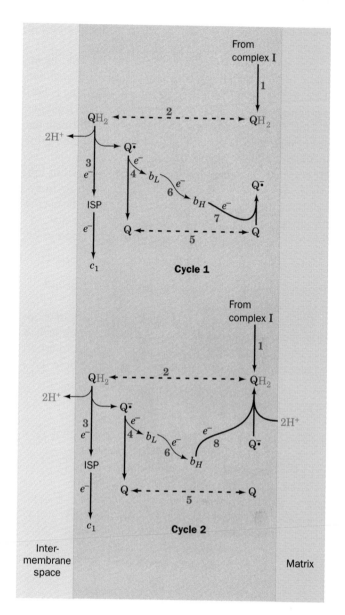

Cycle 1

$$CoQH_2 + cyt\ c_1(Fe^{3+}) \longrightarrow$$
$$CoQ^{\cdot-} + cyt\ c_1(Fe^{2+}) + 2H^+(cytosolic)$$

Cycle 2

$$CoQH_2 + CoQ^{\cdot-} + cyt\ c_1(Fe^{3+}) + 2H^+(mitochondrial) \longrightarrow$$
$$CoQ + CoQH_2 + cyt\ c_1(Fe^{2+}) + 2H^+(cytosolic)$$

Overall 2 e^- transfer from $CoQH_2$ to cytochrome c_1

$$CoQH_2 + 2cyt\ c_1(Fe^{3+}) + 2H^+(mitochondrial) \longrightarrow$$
$$CoQ + 2cyt\ c_1(Fe^{2+}) + 4H^+(cytosolic)$$

Thus, half the electrons liberated by the oxidation of $CoQH_2$ to CoQ are used to reduce Q to QH_2. *The result of this sequence in Complex III is the transport of two protons for each electron transferred from Complex I via $CoQH_2$ to cytochrome c_1.* The Q cycle cannot operate in Complex IV (cytochrome c oxidase), however, because Complex IV contains no ($H^+ + e^-$) carriers even though it pumps protons from the matrix to the cytosol during electron transport.

2. The Proton Pump Mechanism

The proton pump mechanism does not require that the redox centers themselves be hydrogen carriers. In this model, *the transfer of electrons results in conformational changes to the complex. The translocation of protons occurs as a result of the influence of these conformational changes on the pK's of amino acid side chains and their alternate exposure to the internal and external side of the membrane (Fig. 20-26).* We have already seen that conformation can influence pK. The Bohr effect in hemoglobin, for example, results from conformational changes induced by O_2 binding, which induces pK changes in protein acid–base groups (Section 9-2E). If such a protein were located in a membrane and if, in addition to pK changes, the conformational changes altered the side of the membrane to which the affected amino acid side chains were exposed, the result

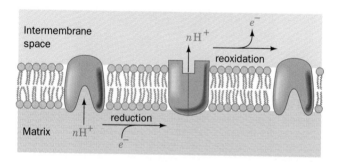

FIGURE 20-26. The proton pump mechanism of electron transport–linked proton translocation. At each H^+ translocation site, n protons bind to amino acid side chains on the matrix side of the membrane. Reduction causes a conformational change that decreases the pK's of these side chains and exposes them to the cytosolic side of the membrane where the protons dissociate. Reoxidation results in a conformational change that restores the pump to its original conformation.

would be H$^+$ transport and the system would be a proton pump.

One documented proton pump is the intrinsic membrane protein **bacteriorhodopsin** of *Halobacter halobium*. This protein, which has seven membrane-spanning helical segments forming a polar channel (Section 11-3A), obtains the free energy required for pumping protons through the absorption of light by its retinaldehyde–Schiff base prosthetic group. The proposed mechanism for light-induced H$^+$ transport in bacteriorhodopsin is diagrammed in Fig. 20-27. A similar mechanism is thought to operate in cytochrome *c* oxidase, with oxidation–reduction causing the pK-altering conformational change. The particular cytochrome *c* oxidase groups involved in this change are as yet unidentified.

Proton-Transport Mechanisms May Be Distinguished Experimentally

If the electrochemical H$^+$ gradient across the mitochondrial membrane were generated by a proton pump such as bacteriorhodopsin, the stoichiometry of proton transport would depend on the number of groups whose pK's

changed in going from the oxidized to the reduced conformation and the magnitudes of these pK changes. The redox loop mechanism, in contrast, requires that one proton be translocated for each electron transported in Complex I and two protons transported for each electron in Complex III if the Q cycle is employed. [There are no (H$^+$ + *e*$^-$) carriers in Complex IV so the redox loop mechanism cannot operate there.] It might therefore be possible to distinguish between the redox loop and proton pump mechanisms if the actual ratio of protons translocated to electrons transported were known. Of course, since three complexes translocate protons, there is also the possibility that different mechanisms may operate at different points in the electron-transport chain. This complicated situation is currently receiving much research attention.

C. Mechanism of ATP Synthesis

The free energy of the electrochemical proton gradient across the mitochondrial membrane is harnessed in the synthesis of ATP by **proton-translocating ATP synthase** *(also known as* **proton pumping ATPase** *and* **F$_1$F$_0$–ATPase**). In the following subsections we discuss the location and structure of this ATP synthase and the mechanism by which it harnesses proton flux to drive ATP synthesis. As with many other aspects of oxidative phosphorylation, this mechanism is by no means well established and is an area of active investigation.

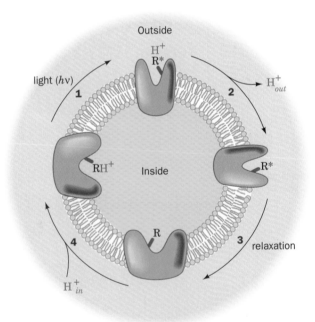

FIGURE 20-27. The proton pump of bacteriorhodopsin: (1) In its resting state, the retinaldehyde–Schiff base prosthetic group (R) is protonated and faces the inside of the membrane. Upon the absorption of light (*h*v), the pump undergoes a conformational change that causes R*H$^+$ to face the outside of the membrane (R* is the light-excited form of R). (2) The conformational change lowers the pK of R* compared to R, which causes H$^+$ to dissociate from R* on the outside of the membrane. (3) The pump relaxes back to its resting state so that R again faces the inside of the membrane. (4) The pK increase accompanying relaxation causes R to be protonated by an H$^+$ from the inside of the membrane. The pump has thereby recovered its original state, with the only change to the system being the translocation of a proton from the inside to the outside of the membrane.

TABLE 20-2. COMPONENTS OF MITOCHONDRIAL PROTON-TRANSLOCATING SYNTHASE (F$_1$F$_0$-ATPASE)[a]

Component	Subunit Composition	Function
F$_1$	$\alpha_3\beta_3\gamma\delta\varepsilon$	β contains the ATP synthase site; δ forms the gate coupling the F$_0$ proton channel with F$_1$
F$_0$	4–5 Types of subunit including 6–10 copies of DCCD-binding proteolipid	DCCD-binding proteolipid oligomer forms the proton channel
Stalk	One copy each of OSCP and F$_6$	Required to bind F$_0$ to F$_1$
Associated polypeptides	IF$_1$	Inhibits ATP hydrolysis; binds to the F$_1$ β subunit
	F$_B$	

[a] Total molecular mass = 450 kD.

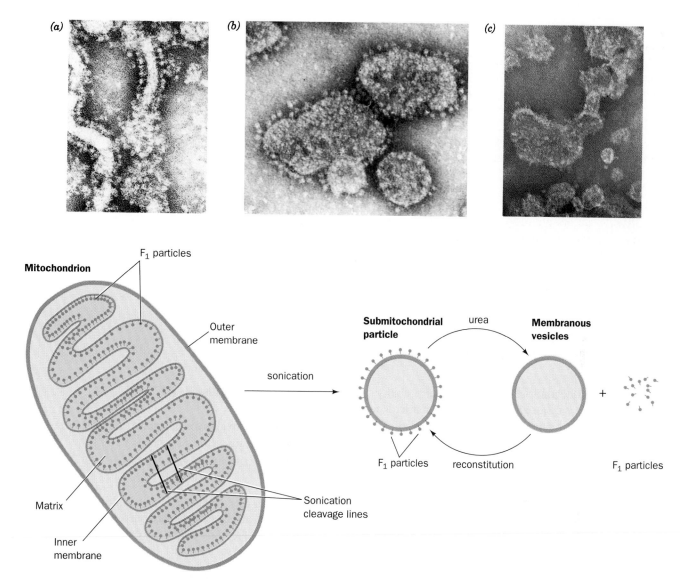

FIGURE 20-28. Electron micrographs and their interpretive drawings (*below*) of the mitochondrial membrane at various stages of dissection: (*a*) Cristae from intact mitochondria showing their F_1 "lollipops" projecting into the matrix. [From Parsons, D. F., *Science* **140**, 985 (1963). Copyright © 1963 American Association for the Advancement of Science. Used by permission.] (*b*) Submitochondrial particles, showing their outwardly projecting F_1 lollipops. Submitochondrial particles are prepared by the sonication (ultrasonic disruption) of inner mitochondrial membranes. [Courtesy of Peter Hinkle, Cornell University.] (*c*) Submitochondrial particles after treatment with urea. [Courtesy of Efraim Racker, Cornell University.]

Proton-Translocating ATP Synthase Is a Multisubunit Transmembrane Protein

Proton-translocating ATP synthase is the most complex structure in the inner mitochondrial membrane. It contains two major substructures and several different subunits (Table 20-2). Electron micrographs of mitochondria show lollipop-shaped structures studding the matrix surface of the inner mitochondrial membrane (Fig. 20-28*a*). Similar entities have been observed to line the inner surface of the bacterial plasma membrane and in chloroplasts (Section 22-2D). Sonication of the inner mitochondrial membrane yields sealed vesicles, **submitochondrial particles,** from which the "lollipops" project (Fig. 20-28*b*) and which can carry out ATP synthesis.

Efraim Racker discovered that the proton-translocating ATP synthase from submitochondrial particles is comprised of two functional units, F_0 and F_1. F_0 is a water-insoluble transmembrane protein composed of as many as 10 to 12 different types of subunits (although only 3 in *E. coli*) that contains a channel for proton translocation. F_1 is a water-soluble peripheral membrane protein, composed of five types of subunits, that is easily dissociated from F_0 by treatment with urea. Solubilized F_1 can hydrolyze ATP but cannot synthesize it (hence the name ATPase). Submitochondrial particles from which F_1 has been removed by urea treatment no longer exhibit the lollipops in their electron micrographs (Fig. 20-28*c*) and lack the ability to synthesize ATP. If, however, F_1 is added back to these F_0-con-

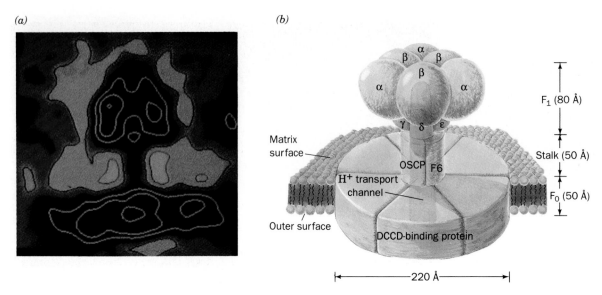

(a)   *(b)*

FIGURE 20-29. (*a*) A cryoelectron micrograph of *E. coli* proton-translocating ATP synthase (F_1F_0-ATPase) with (*b*) an interpretive drawing indicating the positions of the component subunits in the mitochondrial complex. [Electron micrograph courtesy of Edward Gogol and Roderick Capaldi, University of Oregon.]

taining submitochondrial particles, their ability to synthesize ATP is restored and their electron micrographs again exhibit the lollipops. Thus *the lollipops are the F_1 particles.* Electron micrographs of F_1F_0 particles clearly show their dumbbell-shaped structure in which F_0 and F_1 are connected by a 50-Å stalk (Fig. 20-29). This stalk contains at least two proteins, **oligomycin-sensitivity-conferring protein (OSCP)** and **coupling factor 6 (F_6)**.

F_1's catalytic site for ATP synthesis is contained on the β subunit of this $\alpha_3\beta_3\gamma\delta\varepsilon$ nonomer. The δ subunit is required for binding of F_1 to F_0. **Oligomycin,** a *Streptomyces*-produced antibiotic,

Oligomycin B

inhibits ATP synthase by binding to a subunit of F_0 (not OSCP) so as to interfere with H^+ transport through F_0.

Dicyclohexylcarbodiimide (DCCD),

Dicyclohexylcarbodiimide (DCCD)

a lipid-soluble carboxyl reagent, also inhibits proton transport through F_0 by reacting (analogously with EDAC; Table 6-3) with a single Glu residue on one of the subunits of mammalian F_0 (Asp in *E. coli*). Reaction with DCCD usually implies that a carboxylic acid group is located in a lipid environment; that is, "buried" in the membrane. Mammalian F_0 contains six copies of this **DCCD-binding protein** (also known as **DCCD-binding proteolipid**), which are thought to associate like staves of a barrel so as to form the polar H^+-transport channel that contains the buried Glu residues (Fig. 20-29).

The X-ray structure of the F_1 subunit of the F_1F_0-ATPase from bovine heart mitochondria has been determined by John Walker and Andrew Leslie. This 3440-residue (371-kD) protein, the largest asymmetric particle whose X-ray structure has yet been determined, consists of an 80-Å-high and 100-Å-wide spheroid that is mounted on a 30-Å-long stem (Fig. 20-30*a*). F_1's α and β subunits, which are 20% identical in sequence and have nearly identical folds, are arranged alternately, like the segments of an orange, about the upper portion of a 90-Å-long α helix formed by the C-terminal segment of the γ subunit. The C-terminus of this helix protrudes into a 15-Å-deep dimple that is centrally located at the top of the spheroid (Fig. 20-30*b*). The lower half of the helix forms a bent left-handed antiparallel coiled coil with the N-terminal segment of the γ subunit. This

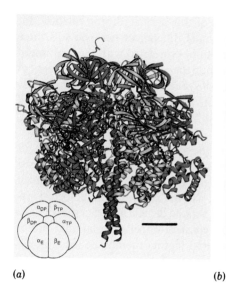

(a)

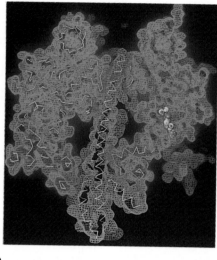

(b)

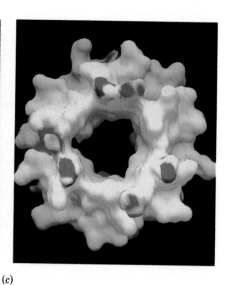

(c)

FIGURE 20-30. The X-ray structure of F$_1$-ATPase from bovine heart mitochondria. (*a*) A ribbon diagram in which the α, β, and γ subunits are red, yellow, and blue, respectively, and the nucelotides are black in ball-and-stick representation. The inset drawing indicates the orientation of these subunits in this view. The bar is 20 Å long. (*b*) Cross-section through the electron density map of the protein in which the density for the α and β subunits is blue and that for the γ subunit is orange. The superimposed C$_\alpha$ backbones for these subunits are yellow and a bound ADPNP is represented in space-filling form (C yellow, N blue, O red). Note the large central cavity surrounding the γ subunit between the two regions where it contacts the $\alpha_3\beta_3$ assembly. (*c*) The surface of the inner portion of the $\alpha_3\beta_3$ assembly through which the C-terminal helix of the γ subunit penetrates as viewed from the top of Parts *a* and *b*. The surface is colored according to its electrical potential with positive potentials blue and negative potentials red. Note the absence of charge on the inner surface of this sleeve. The portion of the γ subunit's C-terminal helix that contacts this sleeve is similarly devoid of charge. [From Abrahams, J.P., Leslie, A.G.W., Lutter, R., and Walker, J.E., *Nature* **370**, 623 and 627 (1994).]

coiled coil is almost certainly a part of the ~ 50-Å-long stalk seen in electron micrographs of the F$_1$F$_0$-ATPase such as Fig. 20-29*a*. Three large segments of the γ subunit as well as the entire δ and ε subunits are not visible in this X-ray structure.

The cylical arrangement and structural similarities of F$_1$'s α and β subunits gives it both pseudo-threefold and pseudo-sixfold rotational symmetry. Nevertheless, the protein is asymmetric. This is in part due to the presence of the γ subunit, but also because each of the α and β subunits takes up a somewhat different conformation. Thus, one β subunit (designated β_{TP}) binds a molecule of the nonhydro-lyzable ATP analog **adenosine-5'-(β,γ-imido)triphosphate (ADPNP or AMP-PNP),**

$$^-O-\overset{\displaystyle O}{\underset{\displaystyle O^-}{\overset{\displaystyle \|}{P}}}-NH-\overset{\displaystyle O}{\underset{\displaystyle O^-}{\overset{\displaystyle \|}{P}}}-O-\overset{\displaystyle O}{\underset{\displaystyle O^-}{\overset{\displaystyle \|}{P}}}-O-CH_2$$

Adenosine-5'-(β, γ-imido)triphosphate (ADPNP)

the second (β_{DP}) binds ADP, and the third (β_E) has an empty and distorted binding site. The α subunits, however, all bind ADPNP, although they also differ conformation-ally from one another. The ADPNP and ADP binding sites each lie at a radius of ~ 20 Å near an interface between adjacent α and β subunits and, in fact, all incorporate a few residues from the adjacent subunit.

The Binding Change Mechanism: Proton-Translocating ATP Synthase Is Driven by Conformational Changes

The mechanism of ATP synthesis by proton-translo-cating ATP synthase can be conceptually broken down into three phases:

1. Translocation of protons carried out by F$_0$.

2. Catalysis of formation of the phosphoanhydride bond of ATP carried out by F$_1$.

3. Coupling of the dissipation of the proton gradient with ATP synthesis, which requires interaction of F$_1$ and F$_0$.

The available evidence supports a mechanism for ATP formation, proposed by Boyer, that resembles the confor-mational coupling hypothesis of oxidative phosphorylation (Section 20-3A). However, the conformational changes in the ATP synthase that power ATP formation are generated by proton translocation rather than by direct electron transfer as proposed in the original formulation of the con-formational coupling hypothesis.

F$_1$ is proposed to have three interacting catalytic pro-tomers, each in a different conformational state: one that binds substrates and products loosely (L state), one that binds them tightly (T state), and one that does not bind them at all (open or O state). The free energy released on proton translocation is harnessed to interconvert these three states. The phosphoanhydride bond of ATP is synthesized only in the T state and ATP is released only in the O state. The reaction involves three steps (Fig. 20-31):

1. Binding of ADP and P_i to the "loose" (L) binding site.

2. A free energy-driven conformational change that converts the L site to a "tight" (T)-binding site that catalyzes the formation of ATP. This step also involves conformational changes of the other two subunits that convert the ATP-containing T site to an "open" (O) site and convert the O site to an L site.

3. ATP is synthesized at the T site on one subunit while ATP dissociates from the O site on another subunit. The free energy supplied by the proton flow primarily facilitates the release of the newly synthesized ATP from the enzyme, that is, it drives the T → O transition, thereby disrupting the enzyme-ATP interactions that had previously promoted the spontaneous formation of ATP from ADP + P_i in the T site.

How is the free energy of proton transfer coupled to the synthesis of ATP? Boyer had proposed that *the binding changes are driven by the rotation of the catalytic assembly, $\alpha_3\beta_3$, with respect to other portions of the F_1F_0-ATPase* (although more as a way to rationalize the single reaction pathway exhibited by F_1's three catalytic sites in terms of its known asymmetry). This hypothesis is supported by the X-ray structure of F_1. Thus, the closely fitting nearly circular arrangement of the α and β subunits' inner surface about the γ subunit's helical C-terminus is reminiscent of a cylindrical bearing rotating in a sleeve (Figs. 20-30*b and c*). Indeed, the contacting hydrophobic surfaces in this assembly are devoid of the hydrogen bonding and ionic interactions that would interfere with their free rotation (Fig. 20-30*c*); that is, the bearing and sleeve appear to be "lubricated." Moreover, the central cavity in the $\alpha_3\beta_3$ assembly (Fig. 20-30*b*) would permit the passage of the γ subunit's N-terminal

helix within the core of this particle during rotation. Finally, the conformational differences between F_1's three catalytic sites appear to be correlated with the position of the γ subunit. Apparently the γ subunit, which is thought to rotate within the fixed $\alpha_3\beta_3$ assembly, acts as a molecular cam shaft in linking the proton gradient-driven rotational motor to the conformational changes in the catalytic sites of F_1.

The description of the molecular motor responsible for the putative rotational motion of the γ subunit must at least await the structural determination F_0. However, such rotating systems are not without precedent in biological systems: The motion of bacterial flagella has been shown to be driven by a proton gradient-impelled rotational motor mounted in the bacterial plasma membrane (Section 34-3F).

The "tight coupling" between electron transport and ATP synthesis in the mitochondrion depends on the impermeability of the inner mitochondrial membrane. This impermeability allows an electrochemical gradient to be established across this membrane during the H^+ translocation associated with electron transport. The only way for H^+ to reenter the matrix is through the F_0 portion of the proton-translocating ATP synthase. The electrochemical gradient therefore builds until the free energy required to transport H^+ balances the free energy of electron transport. Electron transport must then cease. ATP synthesis, by dissipating the electrochemical gradient, allows electron transport to continue.

D. Uncoupling of Oxidative Phosphorylation

Electron transport (the oxidation of NADH and $FADH_2$ by O_2) and oxidative phosphorylation (the synthesis of ATP) are normally tightly coupled. In the resting state, when oxidative phosphorylation is minimal, the electrochemical gradient across the inner mitochondrial membrane builds up to the extent that it prevents further proton pumping and therefore inhibits electron transport. Over the years,

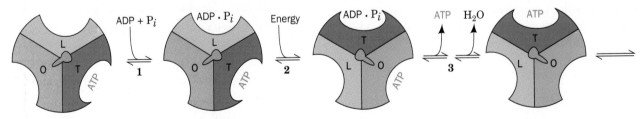

FIGURE 20-31. The energy-dependent binding change mechanism for ATP synthesis by proton-translocating ATP synthase. F_1 has three chemically identical but conformationally distinct interacting $\alpha\beta$ protomers: O, the open conformation, has very low affinity for ligands and is catalytically inactive; L has loose binding for ligands and is catalytically inactive; T has tight binding for ligands and is catalytically active. ATP synthesis occurs in three steps: **(1)** Binding of ADP and P_i to site L. **(2)** Energy-dependent conformational change converting binding site L to T, T to O, and O to L. **(3)** Synthesis of ATP at site T and release of ATP from site O. The enzyme returns to its initial state after two more passes of this reaction sequence. The energy that drives the conformational change is apparently transmitted to the catalytic $\alpha_3\beta_3$ assembly via the rotation of the $\gamma\delta\epsilon$ assembly, here represented by the centrally located asymmetric object *(green)*. [After Cross, R. L., *Annu. Rev. Biochem.* **50,** 687 (1980).]

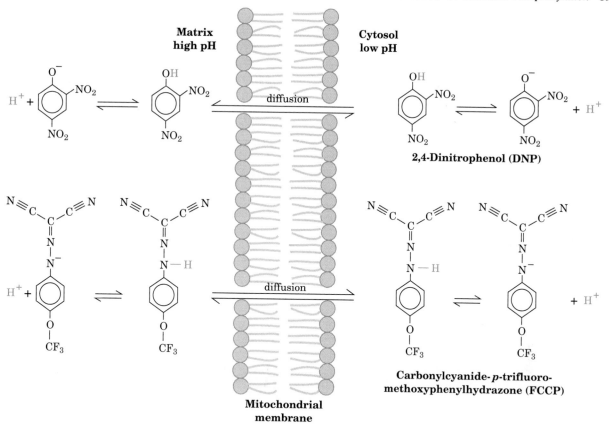

FIGURE 20-32. The proton-transporting ionophores DNP and FCCP uncouple oxidative phosphorylation from electron transport by discharging the electrochemical proton gradient generated by electron transport.

however, many compounds, including **2,4-dinitrophenol (DNP)** and **carbonylcyanide-*p*-trifluoromethoxyphenylhydrazone (FCCP),** have been found to "uncouple" these processes. The chemiosmotic hypothesis has provided a rationale for understanding the mechanism by which these uncouplers act.

The presence in the inner mitochondrial membrane of an agent that increases its permeability to H^+ uncouples oxidative phosphorylation from electron transport by providing a route for the dissipation of the proton electrochemical gradient that does not require ATP synthesis. Uncoupling therefore allows electron transport to proceed unchecked even when ATP synthesis is inhibited. DNP and FCCP are lipophilic weak acids that therefore readily pass through membranes in their neutral, protonated state. In a pH gradient, they bind protons on the acidic side of the membrane, diffuse through, and release them on the alkaline side, thereby dissipating the gradient (Fig. 20-32). Thus, *such uncouplers are proton-transporting ionophores* (Section 18-2C).

Even before the mechanism of uncoupling was known, it was recognized that metabolic rates were increased by such compounds. Studies at Stanford University in the early part of the twentieth century documented an increase in respiration and weight loss caused by DNP. The compound was

even used as a "diet pill" for several years. In the words of Efraim Racker *(A New Look at Mechanisms in Bioenergetics, p. 155)*:

> *In spite of warnings from the Stanford scientists, some enterprising physicians started to administer dinitrophenol to obese patients without proper precautions. The results were striking. Unfortunately in some cases the treatment eliminated not only the fat but also the patients, and several fatalities were reported in the Journal of the American Medical Association in 1929. This discouraged physicians for a while . . .*

Hormonally Controlled Uncoupling in Brown Adipose Tissue Functions to Generate Heat

The dissipation of an electrochemical H^+ gradient, which is generated by electron transport and uncoupled from ATP synthesis, produces heat. Heat generation is the physiological function of **brown adipose tissue (brown fat).** This tissue is unlike typical (white) adipose tissue in that, besides containing large amounts of triacylglycerols, it contains numerous mitochondria whose cytochromes cause its brown color. Newborn mammals that lack fur, such as humans, as well as hibernating mammals, contain brown fat in their neck and upper back that functions in **nonshivering thermogenesis,** that is, as a "biological heating pad." (The ATP hydrolysis that occurs during the muscle contractions of shivering—or any other movement—also produces heat. Nonshivering thermogenesis through substrate cycling is discussed in Section 16-4B.)

The mechanism of heat generation in brown fat involves the regulated uncoupling of oxidative phosphorylation in their mitochondria. These mitochondria contain **uncoupling protein (UCP;** also called **thermogenin),** a protein dimer of 32-kD subunits that is absent in the mitochondria of other tissues, which acts as a channel to control the permeability of the inner mitochondrial membrane to protons. In cold-adapted animals, UCP constitutes up to 15% of brown fat inner mitochondrial membrane proteins. The flow of protons through this channel protein is inhibited by physiological concentrations of purine nucleotides (ADP,

ATP, GDP, GTP) but this inhibition can be overcome by free fatty acids. The components of this system interact under hormonal control.

Thermogenesis in brown fat mitochondria is activated by free fatty acids. These counteract the inhibitory effects of purine nucleotides, thereby stimulating the flux through the proton channel and uncoupling electron transport from oxidative phosphorylation. *The concentration of fatty acids in brown adipose tissue is controlled by the hormone norepinephrine (noradrenaline)*

Norepinephrine

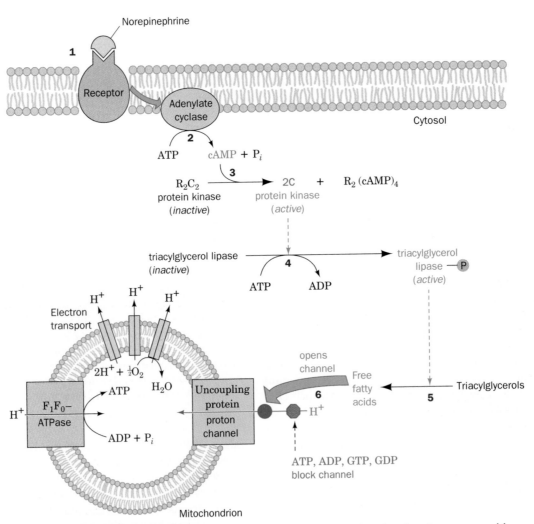

FIGURE 20-33. The mechanism of hormonally induced uncoupling of oxidative phosphorylation in brown fat mitochondria: **(1)** Norepinephrine binds to its cell-surface receptor. **(2)** The norepinephrine–receptor complex stimulates adenylate cyclase, thereby causing cAMP levels to rise. **(3)** cAMP binding activates cAMP-dependent protein kinase (cAPK). **(4)** cAPK phosphorylates hormone-sensitive triacylglycerol lipase, thereby activating it. **(5)** Triacylglycerols are hydrolyzed, yielding free fatty acids. **(6)** Free fatty acids overcome the purine nucleotide block of the proton channel formed by the uncoupling protein, allowing H^+ to enter the mitochondrion uncoupled from ATP synthesis.

with cAMP acting as a second messenger (Section 17-3). Under norepinephrine stimulation (Fig. 20-33), the adenylate cyclase component of the norepinephrine receptor system synthesizes cAMP as described in Section 34-4B. The cAMP, in turn, allosterically activates cAMP-dependent protein kinase (cAPK), which activates **hormone-sensitive triacylglycerol lipase** by phosphorylating it (Section 23-5). Finally, the activated lipase hydrolyzes triacylglycerols to yield the free fatty acids that open the proton channel.

4. CONTROL OF ATP PRODUCTION

An adult woman requires some 1500 to 1800 kcal (6300–7500 kJ) of metabolic energy per day. This corresponds to the free energy of hydrolysis of over 200 mol of ATP to ADP and P_i. Yet the total amount of ATP present in the body at any one time is <0.1 mol; obviously, this sparse supply of ATP must be continually recycled. As we have seen, when carbohydrates serve as the energy supply and aerobic conditions prevail, this recycling involves glycogenolysis, glycolysis, the citric acid cycle, and oxidative phosphorylation.

Of course the need for ATP is not constant. There is a 100-fold change in ATP utilization between sleep and vigorous activity. *The activities of the pathways that produce ATP are under strict coordinated control so that ATP is never produced more rapidly than necessary.* We have already discussed the control mechanisms of glycolysis, glycogenolysis, and the citric acid cycle (Sections 16-4, 17-3, and 19-4). In this section we discuss the mechanisms through which oxidative phosphorylation is controlled and observe how all four systems are synchronized to produce ATP at precisely the rate required at any particular moment.

A. Control of Oxidative Phosphorylation

In our discussion of the control of glycolysis, we saw that most of the reactions in a metabolic pathway function close to equilibrium. *The few irreversible reactions constitute the potential control points of the pathway and usually are catalyzed by regulatory enzymes that are under allosteric control.* In the case of oxidative phosphorylation, the pathway from NADH to cytochrome c functions near equilibrium $(\Delta G' \approx 0)$:

$$\tfrac{1}{2}\text{NADH} + \text{cytochrome } c^{3+} + \text{ADP} + P_i \rightleftharpoons$$
$$\tfrac{1}{2}\text{NAD}^+ + \text{cytochrome } c^{2+} + \text{ATP}$$

for which

$$K_{eq} = \left(\frac{[\text{NAD}^+]}{[\text{NADH}]}\right)^{\tfrac{1}{2}} \frac{[c^{2+}]}{[c^{3+}]} \frac{[\text{ATP}]}{[\text{ADP}][P_i]} \qquad [20.2]$$

This pathway is therefore readily reversed by the addition of ATP. *In the cytochrome c oxidase reaction, however, the terminal step of the electron-transport chain is irreversible and is therefore a prime candidate as the control site of the pathway.* Cytochrome c oxidase, in contrast to most regulatory enzyme systems, appears to be controlled exclusively by the availability of one of its substrates, reduced cytochrome c (c^{2+}). Since this substrate is in equilibrium with the rest of the coupled oxidative phosphorylation system, (Eq. [20.2]), its concentration ultimately depends on the intramitochondrial [NADH]/[NAD⁺] ratio and the **ATP mass action ratio** $([\text{ATP}]/[\text{ADP}][P_i])$. By rearranging Eq. [20.2], the ratio of reduced to oxidized cytochrome c is expressed

$$\frac{[c^{2+}]}{[c^{3+}]} = \left(\frac{[\text{NADH}]}{[\text{NAD}^+]}\right)^{\tfrac{1}{2}} \left(\frac{[\text{ADP}][P_i]}{[\text{ATP}]}\right) K_{eq} \qquad [20.3]$$

Consequently, the higher the [NADH]/[NAD⁺] ratio and the lower the ATP mass action ratio, the higher $[c^{2+}]$ (reduced cytochrome c) and thus the higher the cytochrome c oxidase activity.

How is this system affected by changes in physical activity? In an individual at rest, ATP hydrolysis to ADP and P_i is minimal and the ATP mass action ratio is high; the concentration of reduced cytochrome c is therefore low and oxidative phosphorylation is minimal. Increased activity results in hydrolysis of ATP to ADP and P_i, thereby decreasing the ATP mass action ratio and increasing the concentration of reduced cytochrome c. This results in an increase in the electron-transport rate and its coupled phosphorylation. Such control of oxidative phosphorylation by the ATP mass action ratio is called **acceptor control** because the rate of oxidative phosphorylation increases with the concentration of ADP, the phosphoryl group acceptor.

The compartmentalization of the cell into mitochondria, where ATP is synthesized, and cytosol, where ATP is utilized, presents an interesting control problem: Is it the ATP mass action ratio in the cytosol or in the mitochondrial matrix that ultimately controls oxidative phosphorylation? Clearly the ATP mass action ratio that exerts direct control must be that of the mitochondrial matrix where ATP is synthesized. However, the inner mitochondrial membrane, which is impermeable to adenine nucleotides and P_i, depends on specific transport systems to maintain communication between the two compartments (Section 18-4C). This organization makes it possible for the transport of adenine nucleotides or P_i to be the rate-limiting step in oxidative phosphorylation. Martin Klingenberg has proposed just such a control function for the ADP–ATP translocator. David Wilson and Maria Erecińska assert that there is equilibration of ATP, ADP, and P_i between cytosol and mitochondria and that the cytosolic ATP mass action ratio ultimately controls mitochondrial oxidative phosphorylation. As in most areas of investigation, this sort of controversy is what spurs research.

B. Coordinated Control of ATP Production

Glycolysis, the citric acid cycle, and oxidative phosphorylation constitute the major pathways for cellular ATP production. Control of oxidative phosphorylation by the ATP mass action ratio depends, of course, on an adequate supply of electrons to fuel the electron-transport chain. This aspect of the system's control is, in turn, dependent on the [NADH]/[NAD$^+$] ratio (Eq. [20.3]), which is maintained high by the combined action of glycolysis and the citric acid cycle in converting 10 molecules of NAD$^+$ to NADH per molecule of glucose oxidized (Fig. 20-1). It is clear, therefore, that coordinated control is necessary for the three processes. This is provided by the regulation of each of the control points of glycolysis (phosphofructokinase; PFK) and the citric acid cycle (pyruvate dehydrogenase, citrate synthase, isocitrate dehydrogenase, and α-ketoglutarate dehydrogenase) by adenine nucleotides or NADH or both as well as by certain metabolites (Fig. 20-34).

Citrate Inhibits Glycolysis

The main control points of glycolysis and the citric acid cycle are regulated by several effectors besides adenine nucleotides or NADH (Fig. 20-34). This is an extremely complex system with complex demands. Its many effectors, which are involved in various aspects of metabolism, increase its regulatory sensitivity. One particularly interesting regulatory effect is the inhibition of PFK by citrate. When demands for ATP decrease, [ATP] increases and [ADP] decreases. The citric acid cycle slows down at its isocitrate dehydrogenase (activated by ADP) and α-ketoglutarate dehydrogenase (inhibited by ATP) steps, thereby causing the citrate concentration to build up. Citrate can leave the mitochondrion via a specific transport system and, *once in the cytosol, acts to restrain further carbohydrate breakdown by inhibiting PFK.*

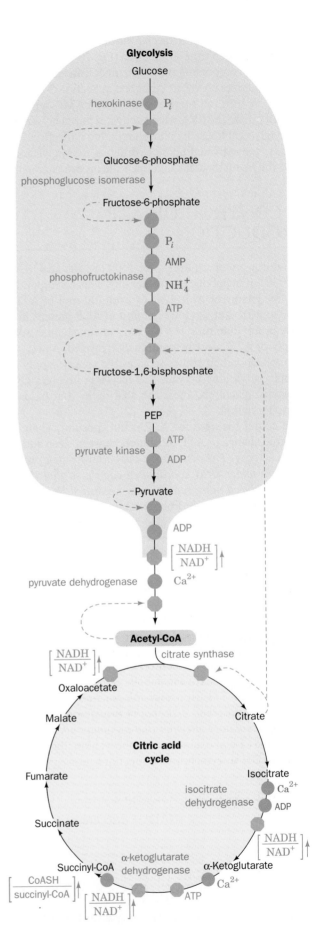

FIGURE 20-34. A schematic diagram representing the coordinated control of glycolysis and the citric acid cycle by ATP, ADP, AMP, P$_i$, Ca^{2+}, and the [NADH]/[NAD$^+$] ratio (the vertical arrows indicate increases in this ratio). Here a green dot signifies activation and a red octagon represents inhibition. [After Newsholme, E.A. and Leech, A.R., *Biochemistry for the Medical Sciences,* pp. 316, 320, Wiley (1983).]

C. Physiological Implications of Aerobic versus Anaerobic Metabolism

In 1861, Louis Pasteur observed that *when yeast are exposed to aerobic conditions, their glucose consumption and ethanol production drop precipitously* (the **Pasteur effect;** alcoholic fermentation in yeast to produce ATP, CO_2, and ethanol are discussed in Section 16-3B). An analogous effect is observed in mammalian muscle; the concentration of lactic acid, the anaerobic product of muscle glycolysis, drops dramatically when cells switch to aerobic metabolism. (This situation is of opposite concern in yeast and muscle since alcohol production is the desired pathway in yeast, at least for winemakers, whereas lactic acid accumulation in muscle—or more precisely, the associated decrease in pH—leads to soreness and fatigue.)

Aerobic ATP Production Is Far More Efficient Than Anaerobic ATP Production

One reason for the decrease in glucose consumption on switching from anaerobic to aerobic metabolism is clear from an examination of the stoichiometries of anaerobic and aerobic breakdown of glucose ($C_6H_{12}O_6$).

Anaerobic glycolysis:

$$C_6H_{12}O_6 + 2ADP + 2P_i \longrightarrow$$
$$2\text{lactate} + 2H^+ + 2H_2O + 2ATP$$

Aerobic metabolism of glucose:

$$C_6H_{12}O_6 + 38ADP + 38P_i + 6\,O_2 \longrightarrow$$
$$6CO_2 + 44H_2O + 38ATP$$

(3 ATPs for each of the 10 NADHs generated per glucose oxidized, 2 ATPs for each of the 2 FADH$_2$s generated, 2 ATPs produced in glycolysis, and $2GTP \rightleftharpoons 2ATP$ produced in the citric acid cycle.) Thus *aerobic metabolism is 19 times more efficient than anaerobic glycolysis in producing ATP.* The switch to aerobic metabolism therefore rapidly increases the ATP mass action ratio. As the ATP mass action ratio increases, the rate of electron transport decreases, which has the effect of increasing the [NADH]/[NAD$^+$] ratio. The increases in [ATP] and [NADH] inhibit their target enzymes in the citric acid cycle and in the glycolytic pathway. *The activity of PFK, the citrate- and adenine nucleotide–regulated rate-controlling enzyme of glycolysis, decreases manyfold on switching from anaerobic to aerobic metabolism. This accounts for the dramatic decrease in glycolysis.*

Anaerobic Glycolysis Has Advantages as Well as Limitations

Animals can sustain anaerobic glycolysis for only short periods of time. This is because PFK, which cannot function effectively much below pH 7, is inhibited by the acidification arising from lactic acid production. Despite this limitation and the low efficiency of glycolytic ATP production, *the enzymes of glycolysis are present in such great concentrations that when they are not inhibited, ATP can be produced much more rapidly than through oxidative phosphorylation.*

The different characteristics of aerobic and anaerobic metabolism permit us to understand certain aspects of cancer cell metabolism and cardiovascular disease.

Cancer Cell Metabolism

As Warburg first noted in 1926, certain cancer cells produce more lactic acid under aerobic conditions than do normal cells. This is because the glycolytic pathway in these cells produces pyruvate more rapidly than the citric acid cycle can accommodate. How can this happen given the interlocking controls on the system? One explanation is that these controls have broken down in cancer cells. Another is that their ATP utilization occurs at rates too rapid to be replenished by oxidative phosphorylation. This would alter the ratios of adenine nucleotides so as to relieve the inhibition of PFK. Attempts to understand the metabolic differences between cancer cells and normal cells may one day provide a clue to the treatment of certain forms of this devastating disease.

Cardiovascular Disease

Oxygen deprivation of certain tissues resulting in cardiovascular disease is of major medical concern. For example, two of the most common causes of human death, **myocardial infarction** (heart attack) and **stroke,** are caused by interruption of the blood (O_2) supply to a portion of the heart or the brain, respectively. It seems obvious why this should result in a cessation of cellular activity but why does it cause cell death?

In the absence of O_2, a cell, which must then rely only on glycolysis for ATP production, rapidly depletes its stores of phosphocreatine (a source of rapid ATP production; Section 15-4C) and glycogen. As the rate of ATP production falls below the level required by membrane ion pumps for the maintenance of proper intracellular ionic concentrations, the osmotic balance of the system is disrupted so that the cell and its membrane-enveloped organelles begin to swell. The resulting overstretched membranes become permeable, thereby leaking their enclosed contents. [In fact, a useful diagnostic criterion for myocardial infarction is the presence in the blood of heart-specific enzymes, such as the H-type isozyme of lactate dehydrogenase (Section 7-5C), which leak out of necrotic (dead) heart tissue.] Moreover, the decreased intracellular pH that accompanies anaerobic glycolysis (because of lactic acid production) permits the released lysosomal enzymes (which are active only at acidic pH's) to degrade the cell contents. Thus, the cessation of metabolic activity results in irreversible cell damage. Rapidly respiring tissues, such as those of heart and brain, are particularly susceptible to such damage.

CHAPTER SUMMARY

Oxidative phosphorylation is the process through which the NADH and FADH$_2$ produced by nutrient oxidation are oxidized with the concomitant formation of ATP. The process takes place in the mitochondrion, an ellipsoidal organelle that is bounded by a permeable outer membrane and contains an impermeable and highly invaginated inner membrane that encloses the matrix. Enzymes of oxidative phosphorylation are embedded in the inner mitochondrial membrane. P$_i$ is imported into the mitochondrion by a specific transport protein. Ca^{2+} import and Ca^{2+} export proteins operate to maintain a constant cytosolic [Ca^{2+}]. NADH's electrons are imported into the mitochondrion by one of several shuttle systems such as the glycerophosphate shuttle or the malate–aspartate shuttle.

The standard free energy change for the oxidation of NAD$^+$ by O$_2$ is $\Delta G°' = -218$ kJ$\cdot$mol^{-1}, whereas that for the synthesis of ATP from ADP and P$_i$ is $\Delta G°' = 30.5$ kJ$\cdot$mol^{-1}. Consequently, the molar free energy of oxidation of NADH by O$_2$ is sufficient to power the synthesis of several moles of ATP under standard conditions. The electrons generated by oxidation of NADH and FADH$_2$ pass through four protein complexes, the electron-transport chain, with the coupled synthesis of ATP. Complexes I, III, and IV participate in the oxidation of NADH producing three ATPs per NADH, whereas FADH$_2$ oxidation, which involves Complexes II, III, and IV, produces only two ATPs per FADH$_2$. Thus, the ratio of moles of ATP produced per mole of coenzyme oxidized by O$_2$, the P/O ratio, is three for NADH oxidation and two for FADH$_2$ oxidation (although some measurements indicate that these quantities are 2.5 and 1.5). The route taken by electrons through the electron-transport chain was elucidated, in part, through the use of electron-transport inhibitors. Rotenone and amytal inhibit Complex I, antimycin A inhibits Complex III, and CN$^-$ inhibits Complex IV. Also involved were measurements of the reduction potentials of the electron-carrying prosthetic groups contained in the electron-transport complexes. Complex I contains FMN and six to seven iron–sulfur clusters in a 26-subunit membrane protein complex. This complex passes electrons from NADH to CoQ, a nonpolar small molecule that diffuses freely within the membrane. Complex II contains the citric acid cycle enzyme succinate dehydrogenase and also passes electrons to CoQ, in this case from succinate through FAD and an iron–sulfur cluster. CoQ passes electrons to Complex III, which contains two b-type cytochromes, one iron–sulfur cluster, and cytochrome c_1. Electrons from cytochrome c_1 of Complex III are passed to Cu$_A$ and cytochrome a of Complex IV (cytochrome c oxidase) via the peripheral membrane protein cytochrome c. Complex IV passes electrons through Cu$_A$, cytochrome a, Cu$_B$, and cytochrome a_3 to O$_2$ which, in a four-electron process, is reduced to H$_2$O.

The mechanism by which the free energy released by the electron-transport chain is stored and utilized in ATP synthesis is best described by the chemiosmotic hypothesis. This hypothesis states that the free energy released by electron transport is conserved by the generation of an electrochemical proton gradient across the inner mitochondrial membrane (outside positive and acidic), which is harnessed to synthesize ATP. The proton gradient is created and maintained by the obligatory outward translocation of H$^+$ across the inner mitochondrial membrane as electrons travel through Complexes I, III, and IV. Transport of 2e$^-$ through one of these complexes creates a proton gradient sufficient for synthesis of one ATP. The mechanism of this H$^+$ translocation is a subject of active research. The energy stored in the electrochemical proton gradient is utilized by proton-translocating ATP synthase (proton-pumping ATPase, F$_1$F$_0$-ATPase) in the synthesis of ATP by coupling this process to the exergonic transport of H$^+$ back into the mitochondrial matrix. Proton-translocating ATPase contains two oligomeric components: F$_1$, a peripheral membrane protein that appears as "lollipops" in electron micrographs of the inner mitochondrial membrane, and F$_0$, an integral membrane protein that contains the proton channel. The conformational changes that promote the synthesis ATP from ADP + P$_i$ apparently arise through the proton gradient-driven rotation of the γ subunit relative to the catalytic $\alpha_3\beta_3$ assembly. Compounds such as 2,4-dinitrophenol are uncouplers of oxidative phosphorylation because they carry H$^+$ across the mitochondrial membrane, thereby dissipating the proton gradient and allowing electron transport to continue without concomitant ATP synthesis. Brown fat mitochondria contain a regulated uncoupling system that, under hormonal control, generates heat instead of ATP.

Under aerobic conditions, the rate of ATP synthesis by oxidative phosphorylation is regulated, in a phenomenon known as acceptor control, by the ATP mass action ratio. ATP synthesis is tightly coupled to the oxidation of NADH and FADH$_2$ by the electron-transport chain. Glycolysis and the citric acid cycle are coordinately controlled so as to produce NADH and FADH$_2$ only at a rate required to meet the system's demand for ATP.

REFERENCES

Historical Overview

Ernster, L. and Schatz, G., Mitochondria: a historical review, *J. Cell Biol.* **91**, 227s–255s (1981).

Fruton, J.S., *Molecules and Life*, pp. 262–396, Wiley–Interscience (1972).

Krebs, H., *Otto Warburg. Cell Physiologist, Biochemist, and Eccentric*, Clarendon Press (1981). [A biography of one of the pioneers in the biochemical study of respiration, by a distinguished student.]

Racker, E., *A New Look At Mechanisms in Bioenergetics*, Academic Press (1976). [A fascinating personal account by one of the outstanding contributors to the field.]

General

Ernster, L. (Ed.), *Bioenergetics*, Elsevier (1984).

Harold, F.M., *The Vital Force: A Study of Bioenergetics*, Chapter 7, Freeman (1986).

Hatefi, Y., The mitochondrial electron transport chain and oxidative phosphorylation system, *Annu. Rev. Biochem.* **54**, 1015–1069 (1985).

Hinkle, P.C. and McCarty, R.E., How cells make ATP, *Sci. Am.* **238**(3): 104–123 (1978).

Martonosi, A.N. (Ed.), *The Enzymes of Biological Membranes* (2nd ed.), Vol. 4, *Bioenergetics of Electron and Proton Transport,* Plenum Press (1985).

Newsholme, E. and Leech, T., *The Runner,* Fitness Books (1983). [A delightful book on the physiology and biochemistry of running.]

Nicholls, D.G. and Ferguson, S.J., *Bioenergetics 2,* Academic Press (1992). [An authoritative monograph devoted almost entirely to the mechanism of oxidative phosphorylation and the techniques used to elucidate it.]

Mitochondrial Structure

Wallace, D.C., Mitochondrial genetics: A paradigm for aging and degenerative diseases, *Science* **256**, 628–632 (1992).

Electron Transport

Babcock, G.T. and Wikström, M., Oxygen activation and the conservation of energy in cell respiration, *Nature* **356**, 301–309 (1992). [Provides a detailed description of the probable chemical mechanism for O_2 activation and reduction by cytochrome c oxidase (Complex IV) together with a discussion of how proton pumping is coupled to this process.]

Barber, J., Further evidence for the common ancestry of cytochrome b–c complexes, *Trends Biochem. Sci.* **9**, 209–211 (1984).

Beratan, D.N., Onuchic, J.N., Winkler, J.R., and Gray, H.B., Electron-tunneling pathways in proteins, *Science* **258**, 1740–1741 (1992).

Calhoun, M.W., Thomas, J.W., and Gennis, R.B., The cytochrome superfamily of redox-driven proton pumps, *Trends Biochem. Sci.* **19**, 325–330 (1994).

Capaldi, R.A., Structure and function of cytochrome c oxidase, *Annu. Rev. Biochem.* **59**, 569–596 (1990); *and* Structural features of the mitochondrial electron-transfer chain, *Curr. Opin. Struct. Biol.* **1**, 562–568 (1991).

Chan, S.I. and Li, P.M., Cytochrome c oxidase: Understanding nature's design of a proton pump, *Biochemistry* **29**, 1–12 (1990).

Esposti, M.D., De Vries, S., Crimi, M., Ghelli, A., Patarnello, T., and Meyer, A., Mitochondrial cytochrome b: evolution and structure of the protein, *Biochim. Biophys. Acta* **1143**, 244–271 (1993).

Hinkle, P.C., Kumar, M.A., Resetar, A., and Harris, D.L., Mechanistic stoichiometry of mitochondrial oxidative phosphorylation, *Biochemistry* **30**, 3576–3582 (1991). [Describes measurements of the P/O ratios indicating that their values are 2.5, 1.5, and 1.]

Lee, C.P. (Ed.), *Curr. Top. Bioenerg.* **15** (1987). [Contains articles on the structures of the components of the electron-transport chain.]

Mathews, F.S., The structure, function and evolution of cytochromes, *Prog. Biophys. Mol. Biol.* **45**, 1–56 (1985).

Moore, G.R. and Pettigrew, G.W., *Cytochomes c. Evolutionary, Structural and Physicochemical Aspects,* Springer-Verlag (1990).

Moser, C.C., Keske, J.M., Warncke, K., Farid, R.S., and Dutton, L.S., Nature of biological electron transfer, *Nature* **355**, 796–802 (1992).

Pelletier, H. and Kraut, J., Crystal structure of a complex between electron transfer partners, cytochrome c peroxidase and cytochrome c, *Science* **258**, 1748–1755 (1992).

Poulos, T.L. and Kraut, J., A hypothetical model of the cytochrome c peroxidase · cytochrome c electron transfer complex, *J. Biol. Chem.* **255**, 10322–10330 (1980).

Sareste, M., Structural features of cytochrome oxidase, *Q. Rev. Biophys.* **23**, 331–366 (1990).

Scott, R.A., X-Ray absorption spectroscopic investigations of cytochrome c oxidase structure and function, *Annu. Rev. Biophys. Biophys. Chem.* **18**, 137–158 (1989).

Smith, H.T., Ahmed, A.J., and Millett, F., Electrostatic interaction of cytochrome c with cytochrome c_1 and cytochrome oxidase, *J. Biol. Chem.* **256**, 4984–4990 (1981).

Stemp, E.D.A. and Hoffman, B.M., Cytochrome c peroxidase binds two molecules of cytochrome c: Evidence for a low-affinity, electron-transfer-active site on cytochrome c peroxidase, *Biochemistry* **32**, 10848–10865 (1993); *and* Zhou, J.S. and Hoffman, B.M., Stern-Volmer in reverse: 2:1 stoichiometry of the cytochrome c–cytochrome c peroxidase electron-transfer complex, *Science* **265**, 1693–1696 (1994).

Trumpower, B.L., The protonmotive Q cycle, *J. Biol. Chem.* **265**, 11409–11412 (1990).

Valpuesta, J.M., Henderson, R., and Frey, T.G., Electron-cryomicroscopic analysis of crystalline cytochrome oxidase, *J. Mol. Biol.* **214**, 237–251 (1991).

Varotsis, C., Zhang, Y., Appelman, E.H., and Babcock, G.T., Resolution of the reaction sequence during the reduction of O_2 by cytochrome oxidase, *Proc. Natl. Acad. Sci.* **90**, 237–241 (1993).

Vervoort, J., Electron-transfering proteins, *Curr. Opin. Struct. Biol.* **1**, 889–894 (1991).

Walker, J.E., The NADH:ubiquinone oxidoreductase (complex I) of respiratory chains, *Q. Rev. Biophys.* **25**, 253–324 (1992). [An exhaustive review.]

Oxidative Phosphorylation

Abrahams, J.P., Leslie, A.G.W., Lutter, R., and Walker, J.E., Structure at 2.8 Å resolution of F_1-ATPase from bovine heart mitochondria, *Nature* **370**, 621–628 (1994).

Boyer, P.D., The binding change mechanism for ATP synthase – Some probabilities and possibilities, *Biochim. Biophys. Acta* **1140**, 215–250 (1993).

Capaldi, R.A., Aggeler, R., Turina, P., and Wilkens, S., Coupling between catalytic sites and the proton channel in F_1F_0-type ATPase, *Trends. Biochem. Sci.* **19**, 284–289 (1994).

Engelbrecht, S. and Junge, W., Subunit δ of H^+-ATPases: at the interface between proton flow and ATP synthesis, *Biochim. Biophys. Acta* **1015**, 379–390 (1990).

Khorana, H.G., Two light-transducing membrane proteins: bacteriorhodopsin and the mammalian rhodopsin, *Proc. Natl. Acad. Sci.* **90**, 1166–1171 (1993).

Klingenberg, M., Mechanism and evolution of the uncoupling protein of brown adipose tissue, *Trends Biochem. Sci.* **15**, 108–112 (1990).

Lehninger, A.L., Reynafarje, B., Alexandre, A., and Villalobo, A., Respiration-coupled H$^+$ ejection by mitochondria, *Ann. N.Y. Acad. Sci.* **341**, 585–592 (1980).

Mitchell, P., Vectorial chemistry and the molecular mechanics of chemiosmotic coupling: power transmission by proticity, *Biochem. Soc. Trans.* **4**, 398–430 (1976).

Nelson, N. and Taiz, L., The evolution of H$^+$-ATPases, *Trends Biochem. Sci.* **14**, 113-116 (1989).

Nicholls, D.G. and Rial, E., Brown fat mitochondria, *Trends Biochem. Sci.* **9**, 489–491 (1984).

Pedersen, P.L. and Amzel, L.M., ATP synthases, *J. Biol. Chem.* **268**, 9937–9940 (1993).

Penefsky, H.S. and Cross, R.L., Structure and mechanism of F$_0$F$_1$-type ATP synthases and ATPases, *Adv. Enzymol.* **64**, 173-215 (1991).

Wikström, M., Identification of the electron transfers in cytochrome oxidase that are coupled to proton pumping, *Nature* **338**, 776–778 (1989).

Control of ATP Production

Brown, G.C., Control of respiration and ATP synthesis in mammalian mitochondria and cells, *Biochem. J.* **284**, 1–13 (1992).

Erecińska, M. and Wilson, D.F., Regulation of cellular energy metabolism, *J. Membr. Biol.* **70**, 1–14 (1982).

Harris, D.A. and Das, A.M., Control of mitochondrial ATP synthesis in the heart, *Biochem. J.* **280**, 561–573 (1991).

Klingenberg, M., The ADP,ATP shuttle of the mitochondrion, *Trends Biochem. Sci.* **4**, 249–252 (1979).

PROBLEMS

1. Rank the following redox-active coenzymes and prosthetic groups of the electron-transport chain in order of increasing affinity for electrons: cytochrome *a*, CoQ, FAD, cytochrome *c*, NAD$^+$.

2. Why is the oxidation of succinate to fumarate only associated with the production of two ATPs during oxidative phosphorylation, whereas the oxidation of malate to oxaloacetate is associated with the production of three ATPs?

3. What is the thermodynamic efficiency of oxidizing FADH$_2$ so as to synthesize two ATPs under standard biochemical conditions?

4. Sublethal cyanide poisoning may be reversed by the administration of nitrites. These substances oxidize hemoglobin, which has a relatively low affinity for CN$^-$, to methemoglobin, which has a relatively high affinity for CN$^-$. Why is this treatment effective?

5. Match the compound with it's behavior: (1) rotenone, (2) dinitrophenol, and (3) antimycin A. (a) Inhibits oxidative phosphorylation when the substrate is pyruvate but not when the substrate is succinate. (b) Inhibits oxidative phosphorylation when the substrate is either pyruvate or succinate. (c) Allows pyruvate to be oxidized by mitochondria even in the absence of ADP.

6. **Nigericin** is an ionophore (Section 18-2C) that exchanges K$^+$ for H$^+$ across membranes. Explain how the treatment of functioning mitochondria with nigericin uncouples electron transport from oxidative phosphorylation. Does valinomycin, an ionophore that transports K$^+$ but not H$^+$, do the same? Explain.

7. The difference in pH between the internal and external surfaces of the inner mitochondrial membrane is 1.4 pH units (external side acidic). If the membrane potential is assumed to be 0.06 V (inside negative) what is the free energy released upon transporting 1 mol of protons back across the membrane? How many protons must be transported to provide enough free energy for the synthesis of 1 mol of ATP (assume standard biochemical conditions)?

8. Explain why: (a) Submitochondrial particles from which F$_1$ has been removed are permeable to protons. (b) Addition of oligomycin to F$_1$-depleted submitochondrial particles decreases this permeability severalfold.

9. Oligomycin and cyanide both inhibit oxidative phosphorylation when the substrate is either pyruvate or succinate. Dinitrophenol can be used to distinguish between these inhibitors. Explain.

10. For the oxidation of a given amount of glucose, does nonshivering thermogenesis by brown fat or shivering thermogenesis by muscle produce more heat?

11. How does atractyloside affect mitochondrial respiration? (*Hint:* see Section 18-4C.)

12. Certain unscrupulous operators offer, for a fee, to freeze recently deceased individuals in liquid nitrogen until medical science can cure the disease from which they died. What is the biochemical fallacy of this procedure?

21

Other Pathways of Carbohydrate Metabolism

chapter, we examine several other carbohydrate metabolism pathways of importance:

1. **Gluconeogenesis,** through which noncarbohydrate precursors such as lactate, pyruvate, glycerol, and amino acids are converted to glucose.

2. The **glyoxylate pathway,** through which plants convert acetyl-CoA to glucose.

3. Oligosaccharide and glycoprotein biosynthesis, through which oligosaccharides are synthesized and added to specific amino acid residues of proteins.

4. The **pentose phosphate pathway,** an alternate pathway of glucose degradation, which generates **NADPH,** the source of reducing equivalents in reductive biosynthesis, and **ribose-5-phosphate,** the sugar precursor of the nucleic acids.

This chapter completes our study of carbohydrate metabolism in animals; photosynthesis, which occurs only in plants and certain bacteria, is the subject of Chapter 22.

Heretofore, we have dealt with many aspects of carbohydrate metabolism. We have seen how the free energy of glucose oxidation is sequestered in ATP through glycolysis, the citric acid cycle, and oxidative phosphorylation. We have also studied the mechanism by which glucose is stored as glycogen for future use and how glycogen metabolism is controlled in response to the needs of the organism. In this

1. GLUCONEOGENESIS

Glucose occupies a central role in metabolism, both as a fuel and as a precursor of essential structural carbohydrates and other biomolecules. The brain and red blood cells are

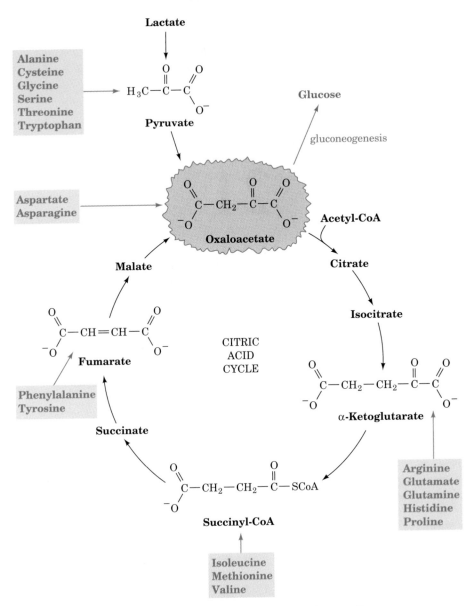

FIGURE 21-1. The pathways converting lactate, pyruvate, and citric acid cycle intermediates to oxaloacetate. The carbon skeletons of all amino acids but leucine and lysine may be, at least in part, converted to oxaloacetate and thus to glucose by these reactions.

almost completely dependent on glucose as an energy source. Yet the liver's capacity to store glycogen is only sufficient to supply the brain with glucose for about half a day under fasting or starvation conditions. Thus, *when fasting, most of the body's glucose needs must be met by gluconeogenesis (literally, new glucose synthesis), the biosynthesis of glucose from noncarbohydrate precursors.* Indeed, isotopic labeling studies determining the source of glucose in the blood during a fast showed that gluconeogenesis is responsible for 64% of total glucose production over the first 22 hours of the fast and account for almost all the glucose production by 46 hours. Thus, gluconeogenesis provides a substantial fraction of the glucose produced in

fasting humans, even after a few hours. Gluconeogenesis occurs in liver and, to a smaller extent, in kidney.

The noncarbohydrate precursors that can be converted to glucose include the glycolysis products lactate and pyruvate, citric acid cycle intermediates, and the carbon skeletons of most amino acids. First, however, all these substances must be converted to oxaloacetate, the starting material for gluconeogenesis (Fig. 21-1). The only amino acids that cannot be converted to oxaloacetate in animals are leucine and lysine because their breakdown yields only acetyl-CoA (Section 24-3F). *There is no pathway in animals for the net conversion of acetyl-CoA to oxaloacetate.* Likewise, fatty acids cannot serve as glucose precursors in animals because most

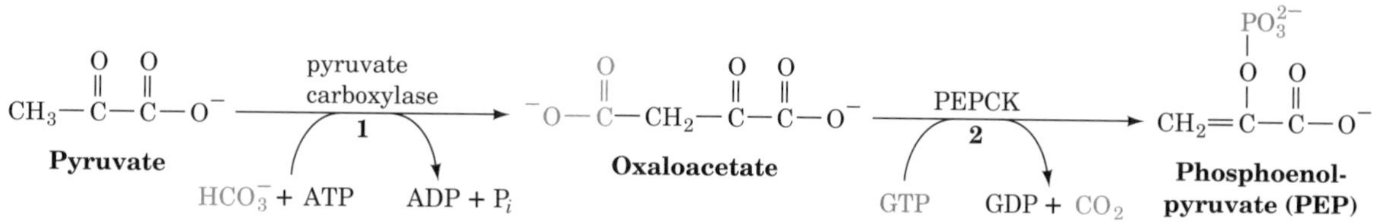

FIGURE 21-2. The conversion of pyruvate to oxaloacetate and then to phosphoenolpyruvate involves **(1)** pyruvate carboxylase and **(2)** PEP carboxykinase (PEPCK).

fatty acids are degraded completely to acetyl-CoA (Section 23-2C). Unlike animals, however, plants do contain a pathway for the conversion of acetyl-CoA to oxaloacetate, the **glyoxylate cycle** (Section 21-2), so that lipids can serve as a plant cell's only carbon source. Glycerol, a triacylglycerol breakdown product, is converted to glucose via synthesis of the glycolytic intermediate dihydroxyacetone phosphate as described in Section 23-1.

A. The Gluconeogenesis Pathway

Gluconeogenesis utilizes glycolytic enzymes. Yet, three of these enzymes, hexokinase, phosphofructokinase (PFK), and pyruvate kinase, catalyze reactions with large negative free energy changes in the direction of glycolysis. These reactions must therefore be replaced in gluconeogenesis by reactions that make glucose synthesis thermodynamically favorable. Here, as in glycogen metabolism (Section 17-1D), we see the recurrent theme that *biosynthetic and degradative pathways differ in at least one reaction. This not only permits both directions to be thermodynamically favorable under the same physiological conditions but allows the pathways to be independently controlled so that one direction can be activated while the other is inhibited.*

Pyruvate Is Converted to Oxaloacetate before Conversion to Phosphoenolpyruvate

The formation of phosphoenolpyruvate (PEP) from pyruvate, the reverse of the pyruvate kinase reaction, is endergonic and therefore requires free energy input. This is accomplished by first converting the pyruvate to oxaloacetate. Oxaloacetate is a "high-energy" intermediate whose exergonic decarboxylation provides the free energy necessary for PEP synthesis. The process requires the participation of two enzymes (Fig. 21-2):

1. **Pyruvate carboxylase** catalyzes the ATP-driven formation of oxaloacetate from pyruvate and HCO_3^-.

2. **PEP carboxykinase (PEPCK)** converts oxaloacetate to PEP in a reaction that uses GTP as a phosphorylating agent.

Pyruvate Carboxylase Has a Biotin Prosthetic Group

Pyruvate carboxylase, discovered in 1959 by Merton Utter, is a tetrameric protein of identical ~120-kD sub-

units, each of which has a **biotin** prosthetic group. *Biotin (Fig. 21-3a) functions as a CO_2 carrier by forming a carboxyl substituent at its **ureido group** (Fig. 21-3b).* Biotin is covalently bound to the enzyme by an amide linkage between the carboxyl group of its valerate side chain and the ε-amino group of an enzyme Lys residue to form a **biocytin** (alternatively, **biotinyllysine**) residue (Fig. 21-3b). The biotin ring system is therefore at the end of a 14Å-long flexible arm, much like that of the lipoic acid prosthetic group in the pyruvate dehydrogenase complex (Section 19-2A).

Biotin, which was first identified in 1935 as a growth factor in yeast, is an essential human nutrient. Its nutritional deficiency is rare, however, because it occurs in many

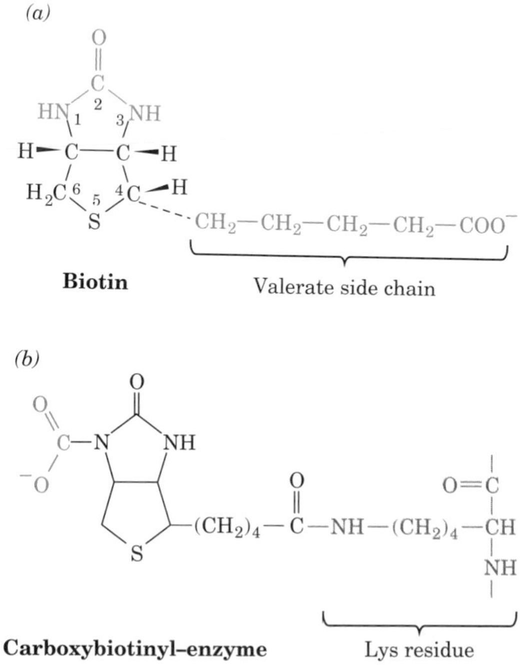

FIGURE 21-3. (*a*) Biotin consists of an imidazoline ring that is cis fused to a tetrahydrothiophene ring bearing a valerate side chain. The chirality at each of its three asymmetric centers is indicated. Positions 1, 2, and 3 constitute a ureido group. (*b*) Carboxybiotinyl–enzyme: N1 of the biotin ureido group is the carboxylation site. Biotin is covalently attached to carboxylases by an amide linkage between its valeryl carboxyl group and an ε-amino group of an enzyme Lys side chain.

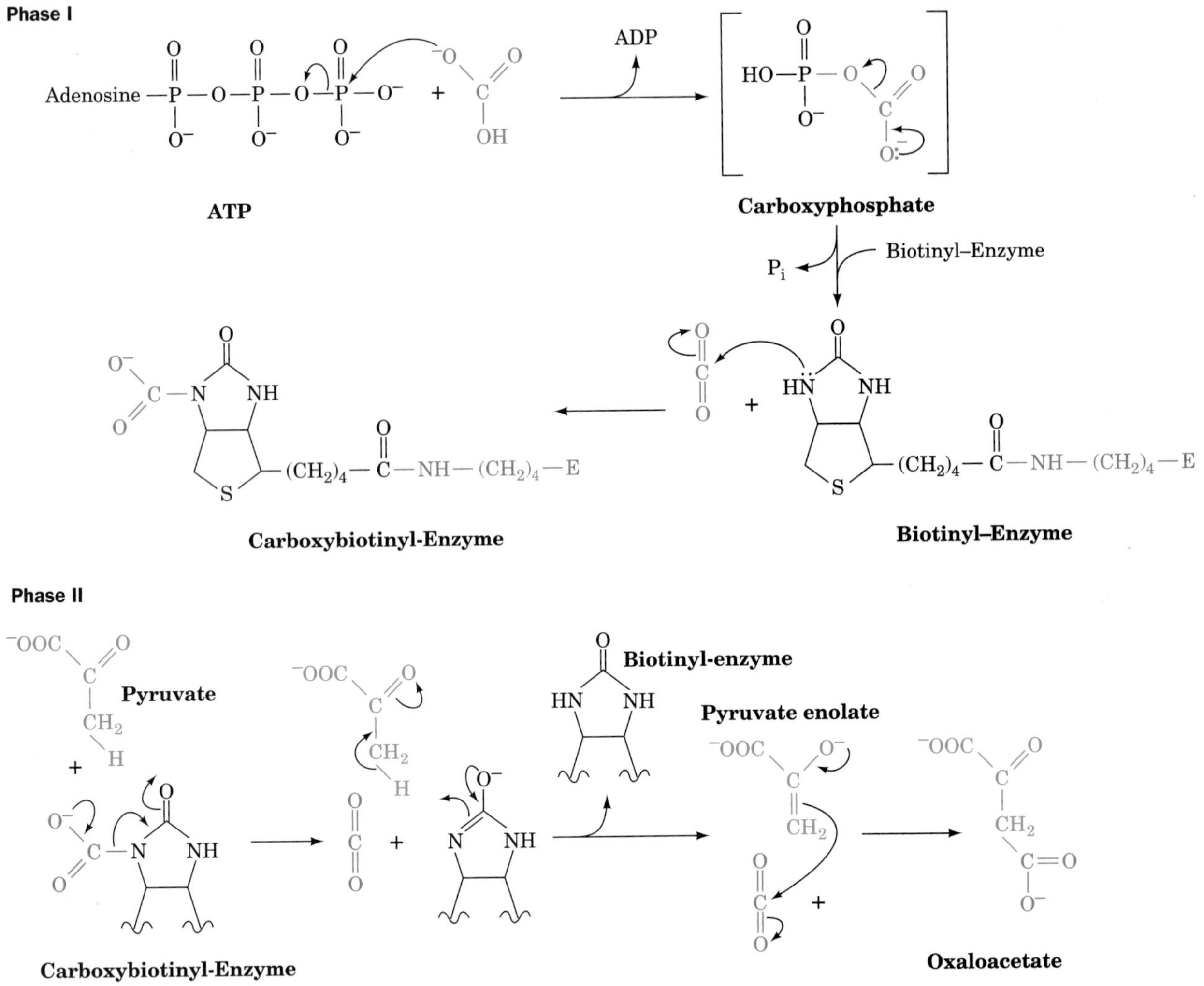

Phase I

Phase II

FIGURE 21-4. The two-phase reaction mechanism of pyruvate carboxylase. **Phase I** is a three-step reaction in which carboxyphosphate is formed from bicarbonate and ATP, followed by the generation of CO_2 on the enzyme, which then carboxylates biotin. **Phase II** is a three-step reaction in which CO_2 is produced at the active site via the elimination of the biotinyl enzyme, which accepts a proton from pyruvate to generate pyruvate enolate. This, in turn, nucleophilically attacks the CO_2, yielding oxaloacetate. [After Knowles, J. R., *Annu. Rev. Biochem.* **58**, 217 (1989).]

foods and is synthesized by intestinal bacteria. Human biotin deficiency almost always results from the consumption of large amounts of raw eggs. This is because egg whites contain a protein, **avidin,** that binds biotin so tightly (dissociation constant, $K = 10^{-15}M$) as to prevent its intestinal absorption (cooked eggs do not cause this problem because cooking denatures avidin). The presence of avidin in eggs is thought to inhibit the growth of microorganisms in this highly nutritious environment.

The Pyruvate Carboxylase Reaction

The pyruvate carboxylase reaction occurs in two phases (Fig. 21-4):

Phase I Biotin is carboxylated at its N1′ atom by bicarbonate ion in a three-step reaction in which the hydrolysis of ATP to ADP + P_i functions, via the intermediate formation of **carboxyphosphate,** to dehydrate bicarbonate. This yields free CO_2, which has sufficient free energy to carboxylate biotin. The resulting carboxyl group is activated relative to bicarbonate ($\Delta G^{\circ\prime}$ for its cleavage is -19.7 kJ·mol^{-1}) and can therefore be transferred without further free energy input.

Phase II The activated carboxyl group is transferred from carboxybiotin to pyruvate in a three-step reaction to form oxaloacetate.

These two reaction phases occur on different subsites of the same enzyme; the 14-Å arm of biocytin serves to transfer the biotin ring between the two sites.

Acetyl-CoA Regulates Pyruvate Carboxylase

Oxaloacetate synthesis is an anaplerotic (filling up) reaction that increases citric acid cycle activity (Section 19-4). Accumulation of the citric acid cycle substrate acetyl-CoA therefore signals the need for more oxaloacetate. Indeed, acetyl-CoA is a powerful allosteric activator of pyruvate carboxylase; the enzyme is all but inactive without bound acetyl-CoA. *If, however, the citric acid cycle is inhibited (by ATP and NADH whose presence in high concentrations indicates a satisfied demand for oxidative phosphorylation; Section 19-4), oxaloacetate instead undergoes gluconeogenesis.*

PEP Carboxykinase

PEPCK, a monomeric 74-kD enzyme, catalyzes the GTP-driven decarboxylation of oxaloacetate to form PEP and GDP (Fig. 21-5). Note that the CO_2 that carboxylates pyruvate to yield oxaloacetate is eliminated in the formation of PEP. Oxaloacetate may therefore be considered as "activated" pyruvate with CO_2 and biotin facilitating the activation at the expense of ATP hydrolysis. Acetyl-CoA is

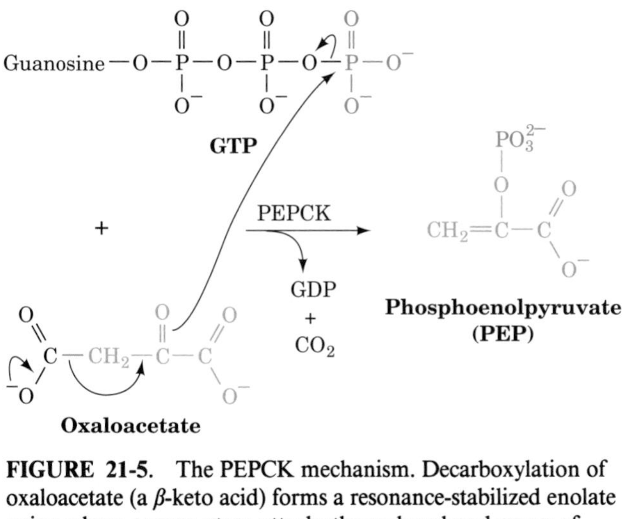

FIGURE 21-5. The PEPCK mechanism. Decarboxylation of oxaloacetate (a β-keto acid) forms a resonance-stabilized enolate anion whose oxygen atom attacks the γ-phosphoryl group of GTP forming PEP and GDP.

similarly activated for fatty acid biosynthesis through such a carboxylation–decarboxylation process (Section 23-4B).

Gluconeogenesis Requires Metabolite Transport between Mitochondria and Cytosol

The generation of oxaloacetate from pyruvate or citric acid cycle intermediates occurs only in the mitochondrion, whereas the enzymes that convert PEP to glucose are cytosolic. The cellular location of PEPCK varies with the species. In mouse and rat liver it is located almost exclusively in the cytosol, in pigeon and rabbit liver it is mitochondrial, and in guinea pig and humans it is more or less equally distributed between both compartments. In order for gluconeogenesis to occur, either oxaloacetate must leave the mitochondrion for conversion to PEP, or the PEP formed there must enter the cytosol.

PEP is transported across the mitochondrial membrane by specific membrane transport proteins. There is, however, no such transport system for oxaloacetate. It must first be converted either to aspartate (Fig. 21-6, Route 1) or to malate (Fig. 21-6, Route 2), for which mitochondrial transport systems exist (Section 20-1B). The difference between these two routes involves the transport of NADH reducing equivalents. The **malate dehydrogenase** route (Route 2) results in the transport of reducing equivalents from the mitochondrion to the cytosol, since it utilizes mitochondrial NADH and produces cytosolic NADH. The **aspartate aminotransferase** route (Route 1) does not involve NADH. Cytosolic NADH is required for gluconeogenesis so, under most conditions, the route through malate is a necessity. If the gluconeogenic precursor is, however, lactate (Section 21-1C), its oxidation to pyruvate generates cytosolic NADH so that either transport route may then be used. Of course, as we have seen, during oxidative metabolism the two routes may also alternate (with Route 2 reversed) to form the malate–aspartate shuttle, which transports NADH reducing equivalents into the mitochondrion (Section 20-1B).

Hydrolytic Reactions Bypass PFK and Hexokinase

The opposing pathways of gluconeogenesis and glycolysis utilize many of the same enzymes (Fig. 21-7). However, the free energy change is highly unfavorable in the gluconeogenic direction at two other points in the pathway in addition to the pyruvate kinase reaction: the PFK reaction and the hexokinase reaction. At these points, instead of generating ATP by reversing the glycolytic reactions, FBP and G6P are hydrolyzed, releasing P_i in exergonic processes catalyzed by **fructose-1,6-bisphosphatase** (**FBPase**) and **glucose-6-phosphatase**, respectively. *Glucose-6-phosphatase is unique to liver and kidney, permitting them to supply glucose to other tissues.*

Because of the presence of separate gluconeogenic enzymes at the three irreversible steps in the glycolytic conversion of glucose to pyruvate, both glycolysis and gluconeogenesis are rendered thermodynamically favorable. This is accomplished at the expense of the free energy of hydrolysis

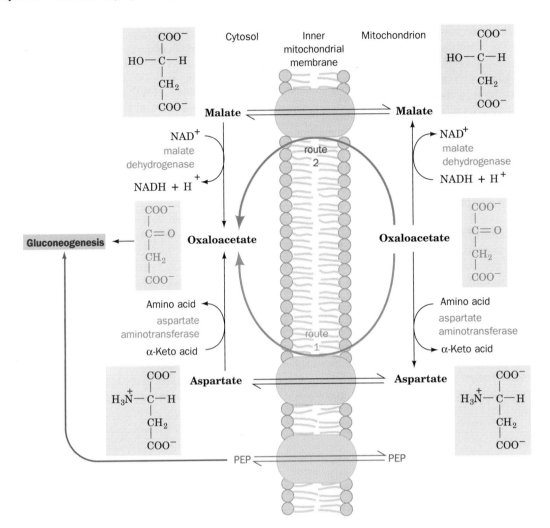

FIGURE 21-6.   The transport of PEP and oxaloacetate from the mitochondrion to the cytosol. PEP is directly transported between these compartments. Oxaloacetate, however, must first be converted to either aspartate through the action of **aspartate aminotransferase** (Route 1) or to malate by malate dehydrogenase (Route 2). Route 2 involves the mitochondrial oxidation of NADH followed by the cytosolic reduction of NAD^+ and therefore also transfers NADH reducing equivalents from the mitochondrion to the cytosol.

of two molecules each of ATP and GTP per molecule of glucose synthesized by gluconeogenesis in addition to that which would be consumed by the direct reversal of glycolysis.

Glycolysis:

$$Glucose + 2NAD^+ + 2ADP + 2P_i \longrightarrow$$
$$2pyruvate + 2NADH + 4H^+ + 2ATP + 2H_2O$$

Gluconeogenesis:

$$2Pyruvate + 2NADH + 4H^+ + \mathbf{4ATP} + \mathbf{2GTP} + 6H_2O$$
$$\longrightarrow glucose + 2NAD^+ + 4ADP + 2GDP + 6P_i$$

Overall:

$$2ATP + 2GTP + 4H_2O \longrightarrow 2ADP + 2GDP + 4P_i$$

Such free energy losses in a cyclic process are thermody-namically inescapable. They are the energetic price that must be paid to maintain independent regulation of the two pathways.

B. Regulation of Gluconeogenesis

If both glycolysis and gluconeogenesis were to proceed in an uncontrolled manner, the net effect would be a futile cycle wastefully hydrolyzing ATP and GTP. This does not occur. Rather, *these pathways are reciprocally regulated so as to meet the needs of the organism*. In the fed state, when the blood glucose level is high, the liver is geared toward fuel conservation: Glycogen is synthesized and the glycolytic pathway and pyruvate dehydrogenase are activated, break-ing down glucose to acetyl-CoA for fatty acid biosynthesis and fat storage. In the fasted state, however, the liver main-

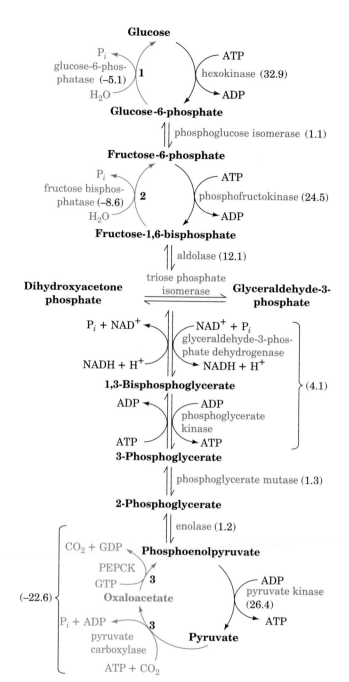

FIGURE 21-7. The pathways of gluconeogenesis and glycolysis. The three numbered steps, which are catalyzed by different enzymes in gluconeogenesis, have red arrows. The ΔG's for the reactions in the direction of gluconeogenesis under physiological conditions in liver are given in parentheses in $kJ \cdot mol^{-1}$. [ΔG's obtained from Newsholme, E.A. and Leech, A.R., *Biochemistry for the Medical Sciences*, p. 448, Wiley (1983).]

tains the blood glucose level both by stimulating glycogen breakdown and by reversing the flux through glycolysis toward gluconeogenesis (using mainly protein degradation products and involving the **glucose–alanine cycle**; Section 24-1A).

Glycolysis and Gluconeogenesis Are Controlled by Allosteric Interactions and Covalent Modifications

The rate and direction of glycolysis and gluconeogenesis are controlled at the points in these pathways where the forward and reverse directions can be independently regulated: the reactions catalyzed by (1) hexokinase/glucose-6-phosphatase, (2) PFK/FBPase, and (3) pyruvate kinase/pyruvate carboxylase–PEPCK (Fig. 21-7). Table 21-1 lists these regulatory enzymes and their regulators. The dominant mechanisms are allosteric interactions and cAMP-dependent covalent modifications (phosphorylation/dephosphorylation; Section 17-3). cAMP-dependent covalent modification renders this system sensitive to control by glucagon and other hormones that alter cAMP levels.

One of the most important allosteric effectors involved in the regulation of glycolysis and gluconeogenesis is fructose 2,6-bisphosphate (F2,6P), which activates PFK and inhibits FBPase (Section 17-3F). The concentration of F2,6P is controlled by its rates of synthesis and breakdown by phosphofructokinase-2 (PFK-2) and fructose bisphosphatase-2 (FBPase-2), respectively. Control of the activities of PFK-2 and FBPase-2 is therefore an important aspect of gluconeogenic regulation even though these enzymes do not catalyze reactions of the pathway. PFK-2 and FBPase-2 activities, which occur on separate domains of the same bifunctional enzyme, are subject to allosteric regulation as well as con-

TABLE 21-1. Rᴇɢᴜʟᴀᴛᴏʀs ᴏꜰ Gʟᴜᴄᴏɴᴇᴏɢᴇɴɪᴄ Eɴᴢʏᴍᴇ Aᴄᴛɪᴠɪᴛʏ

Enzyme	Allosteric Inhibitors	Allosteric Activators	Enzyme Phosphorylation	Protein Synthesis
PFK	ATP, citrate	AMP, F2,6P		
FBPase	AMP, F2,6P			
PK	Alanine	F1,6P	Inactivates	
Pyruvate carboxylase		Acetyl-CoA		
PEPCK				Stimulated by glucagon
PFK-2	Citrate	AMP, F6P, P_i	Inactivates	
FBPase-2	F6P	Glycerol-3-P	Activates	

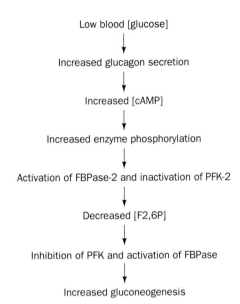

Low blood [glucose]

↓

Increased glucagon secretion

↓

Increased [cAMP]

↓

Increased enzyme phosphorylation

↓

Activation of FBPase-2 and inactivation of PFK-2

↓

Decreased [F2,6P]

↓

Inhibition of PFK and activation of FBPase

↓

Increased gluconeogenesis

FIGURE 21-8. Hormonal regulation of [F2,6P] activates gluconeogenesis in liver in response to low blood [glucose].

trol by covalent modifications (Table 21-1). Low levels of blood glucose result in hormonal activation of gluconeogenesis through regulation of [F2,6P] (Fig. 21-8).

Activation of gluconeogenesis in liver also involves inhibition of glycolysis at the level of pyruvate kinase. *Liver pyruvate kinase is inhibited, both allosterically by alanine (a pyruvate precursor; Section 24-1A) and by phosphorylation.* Glycogen breakdown, in contrast, is stimulated by phosphorylation (Section 17-3C). Both pathways then flow towards G6P, which is converted to glucose for export to muscle and brain. Muscle pyruvate kinase, an isozyme of the liver enzyme, is not subject to these controls. Indeed, such controls would be counterproductive in muscle since this tissue lacks the ability to synthesize glucose via gluconeogenesis.

C. The Cori Cycle

Muscle contraction is powered by hydrolysis of ATP, which is then regenerated through oxidative phosphorylation in the mitochondria of slow-twitch (red) muscle fibers and by glycolysis yielding lactate in fast-twitch (white) muscle fibers. Slow-twitch fibers also produce lactate when ATP demand exceeds oxidative flux. The lactate is transferred, via the bloodstream, to the liver, where it is reconverted to pyruvate by lactate dehydrogenase and then to glucose by gluconeogenesis. Thus, through the intermediacy of the bloodstream, liver and muscle participate in a metabolic cycle known as the **Cori cycle** (Fig. 21-9) in honor of Carl and Gerti Cori who first described it. This is the same ATP-consuming glycolysis/gluconeogenesis "futile cycle" we discussed above. Here, however, instead of occurring in the same cell, the two pathways occur in different organs. Liver ATP is used to resynthesize glucose from lactate produced

in muscle. The resynthesized glucose is returned to the muscle, where it is stored as glycogen and used, on demand, to generate ATP for muscle contraction. The ATP utilized by the liver for this process is regenerated by oxidative phosphorylation. After vigorous exertion, it often takes at least 30 min for the oxygen consumption rate to return to its resting level, a phenomenon known as **oxygen debt.**

2. THE GLYOXYLATE PATHWAY

Plants, but not animals, possess enzymes that mediate the net conversion of acetyl-CoA to oxaloacetate. This is accomplished via the **glyoxylate pathway** (Fig. 21-10), a route involving enzymes of both the mitochondrion and the **glyoxysome** (a membranous plant organelle; Section 1-2A). Mitochondrial oxaloacetate is converted to aspartate by aspartate aminotransferase and transported to the glyoxysome, where it is reconverted to oxaloacetate (Fig. 21-10, Reactions 1). The oxaloacetate is then condensed with acetyl-CoA to form citrate, which is isomerized to isocitrate as in the citric acid cycle (Fig. 21-10, Reactions 2 and 3). Glyoxysomal **isocitrate lyase** then cleaves isocitrate to succinate and **glyoxylate** (hence, the pathway's name; Fig. 21-10, Reaction 4). Succinate is transported to the mitochondrion where it enters the citric acid cycle for conversion back to oxaloacetate, completing the cycle. *The glyoxylate pathway therefore results in the net conversion of acetyl-CoA to glyoxylate instead of to two molecules of CO_2 as occurs in the citric acid cycle.*

Glyoxylate is converted to oxaloacetate in two reactions (Fig. 21-10):

Reaction 5. Malate synthase, a glyoxysomal enzyme, condenses glyoxylate with a second molecule of acetyl-CoA to form malate.

Reaction 6. Cytosolic malate dehydrogenase catalyzes the oxidation of malate to oxaloacetate by NAD^+.

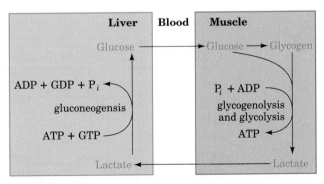

FIGURE 21-9. The Cori cycle. Lactate produced by muscle glycolysis is transported by the bloodstream to the liver, where it is converted to glucose by gluconeogenesis. The bloodstream carries the glucose back to the muscles, where it may be stored as glycogen.

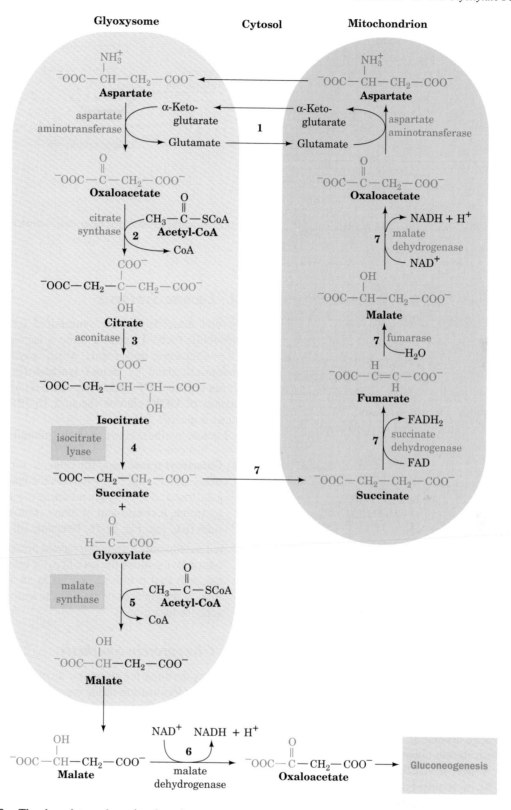

FIGURE 21-10. The glyoxylate pathway involves the participation of both mitochondrial and glyoxysomal enzymes. Isocitrate lyase and malate synthase, enzymes unique to plant glyoxysomes, are boxed. The pathway results in the net conversion of two acetyl-CoA to oxaloacetate. **(1)** Mitochondrial oxaloacetate is converted to aspartate, transported to the glyoxysome, and reconverted to oxaloacetate. **(2)** Oxaloacetate is condensed with acetyl-CoA to form citrate. **(3)** Aconitase catalyzes the conversion of citrate to isocitrate. **(4)** Isocitrate lyase catalyzes the cleavage of isocitrate to succinate and glyoxylate. **(5)** Malate synthase catalyzes the condensation of glyoxylate with acetyl-CoA to form malate. **(6)** After transport to the cytosol, malate dehydrogenase catalyzes the oxidation of malate to oxaloacetate, which can then be used in gluconeogenesis. **(7)** Succinate is transported to the mitochondrion, where it is reconverted to oxaloacetate via the citric acid cycle.

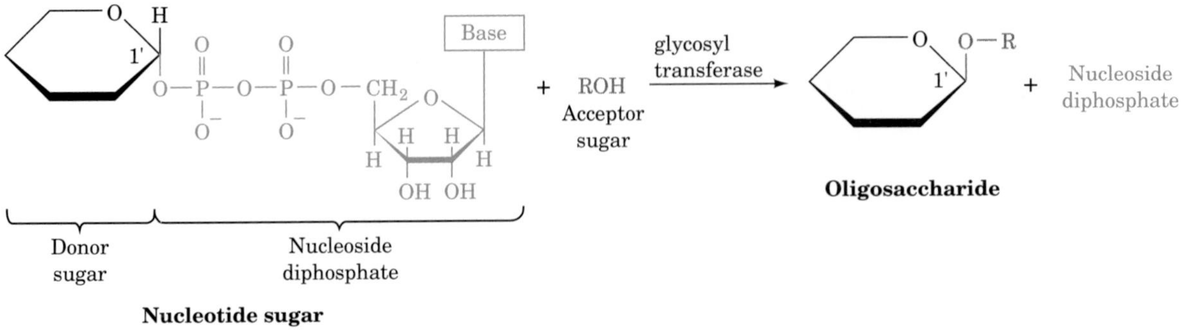

FIGURE 21-11. Nucleotide sugars are glycosyl donors in oligosaccharide biosynthesis catalyzed by glycosyl transferases.

The overall reaction of the glyoxylate cycle is therefore the formation of oxaloacetate from two molecules of acetyl-CoA.

$$\text{2Acetyl-CoA} + \text{2NAD}^+ + \text{FAD} \longrightarrow$$
$$\text{oxaloacetate} + \text{2CoA} + \text{2NADH} + \text{FADH}_2 + \text{2H}^+$$

Isocitrate lyase and malate synthase, the only enzymes of the glyoxylate pathway unique to plants, enable germinating seeds to convert their stored triacylglycerols, through acetyl-CoA, to glucose.

3. BIOSYNTHESIS OF OLIGOSACCHARIDES AND GLYCOPROTEINS

Oligosaccharides consist of monosaccharide units joined together by glycosidic bonds (linkages between C1, the anomeric carbon, of one unit and an OH group of a second unit; Section 10-1C). About 80 different kinds of naturally occurring glycosidic linkages are known, most of which involve mannose, N-acetylglucosamine, N-acetylmuramic acid, glucose, fucose (6-deoxygalactose), galactose, N-acetylneuraminic acid (sialic acid), and N-acetylgalactosamine (Section 10-1). Glycosidic linkages also occur to lipids (glycosphingolipids; Section 11-1D), and proteins (glycoproteins; Section 10-3C).

Glycosidic bond formation requires free energy input under physiological conditions ($\Delta G^{\circ\prime} = 16 \text{ kJ} \cdot \text{mol}^{-1}$). This free energy, as we have seen in the case of glycogen synthesis (Section 17-2B), is acquired through the conversion of monosaccharide units to nucleotide sugars. A nucleotide at a sugar's anomeric carbon atom is a good leaving group and thereby facilitates formation of a glycosidic bond to a second sugar unit via reactions catalyzed by **glycosyl transferases** (Fig. 21-11). The nucleotides that participate in monosaccharide transfers are UDP, GDP, and CMP; a given sugar is associated with only one of these nucleotides (Table 21-2).

A. *Lactose Synthesis*

Several disaccharides are synthesized for future use as metabolic fuels. Typical of these is lactose [β-galactosyl-$(1 \rightarrow 4)$-glucose; milk sugar], which is synthesized in the mammary gland by **lactose synthase** (Fig. 21-12). The donor sugar is UDP–galactose, which is formed by epimerization of UDP–glucose (Section 16-5B). The acceptor sugar is glucose.

Lactose synthase consists of two subunits:

1. **Galactosyl transferase,** the catalytic subunit, occurs in many tissues, where it catalyzes the reaction of UDP–galactose and N-acetylglucosamine to yield N-acetyllactosamine, a constituent of many complex oligosaccharides (see, e.g., Fig. 21-16, Reaction 10).

2. **α-Lactalbumin,** a mammary gland protein with no catalytic activity, alters the specificity of galactosyl transferase such that it utilizes glucose as an acceptor, rather than N-acetylglucosamine, to form lactose instead of N-acetyllactosamine.

B. *Glycoprotein Synthesis*

Proteins destined for secretion, incorporation into membranes, or localization inside membranous organelles con-

TABLE 21-2. SUGAR NUCLEOTIDES AND THEIR CORRESPONDING MONOSACCHARIDES IN GLYCOSYL TRANSFERASE REACTIONS

UDP	GDP	CMP
N-Acetylgalactosamine	Fucose	Sialic acid
N-Acetylglucosamine	Mannose	
N-Acetylmuramic acid		
Galactose		
Glucose		
Glucuronic acid		
Xylose		

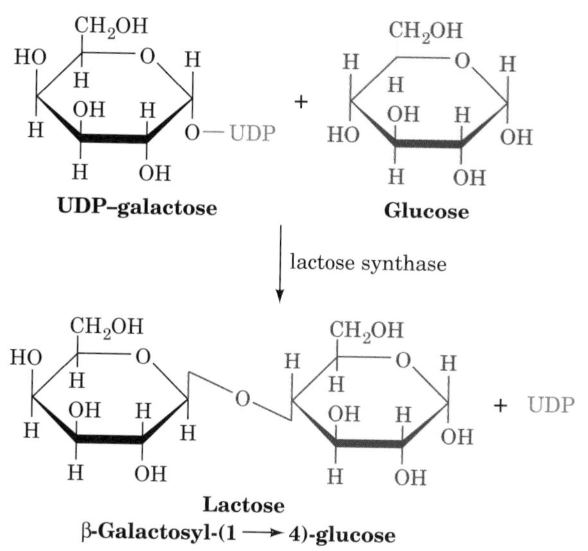

UDP–galactose + **Glucose**

lactose synthase

Lactose
β-**Galactosyl-(1 ⟶ 4)-glucose**

FIGURE 21-12. Lactose synthase catalyzes the formation of lactose from UDP–galactose and glucose.

tain carbohydrates and are therefore classified as glycoproteins. *Glycosylation and oligosaccharide processing play an indispensable role in the sorting and the distribution of these proteins to their proper cellular destinations.* Their polypeptide components are ribosomally synthesized and processed by addition and modification of oligosaccharides.

The oligosaccharide portions of glycoproteins, as we have seen in Sections 10-3C and 11-5A, are classified into three groups:

1. *N*-**Linked oligosaccharides,** which are attached to their polypeptide chain by a β-*N*-glycosidic bond to an Asn residue in the sequence Asn-X-Ser or Asn-X-Thr, where X is any amino acid residue except Pro or perhaps Asp (Fig. 21-13a).

2. *O*-**Linked oligosaccharides,** which are attached to their polypeptide chain through an α-*O*-glycosidic bond to Ser or Thr (Fig. 21-13b) or, only in collagens, to 5-hydroxylysine residues (Fig. 21-13c).

3. **Glycosylphosphatidylinositol(GPI)-membrane anchors,** which are attached to their polypeptide chain through an amide bond between mannose-6-phosphoethanolamine and the C-terminal carboxyl group (Fig. 21-13d).

We shall consider the synthesis of these three types of oligosaccharides separately.

N-Linked Glycoproteins Are Synthesized in Four Stages

N-Linked glycoproteins are formed in the endoplasmic reticulum and further processed in the Golgi apparatus. Synthesis of their carbohydrate moieties occurs in four stages:

1. Synthesis of a lipid-linked oligosaccharide precursor.

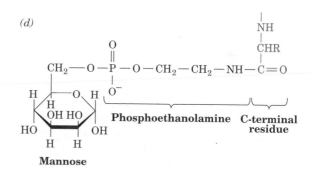

(a) **Asn**

(b) **Ser (Thr)**

(c) **5-Hydroxylysine**

(d) **Mannose** — **Phosphoethanolamine** **C-terminal residue**

FIGURE 21-13. Types of saccharide–polypeptide linkages in glycoproteins. (a) *N*-Linked glycosidic bond to an Asn residue in the sequence Asn-X-Ser/Thr. (b) *O*-Linked glycosidic bond to a Ser (or Thr) residue. (c) *O*-Linked glycosidic bond to a 5-hydroxylysine residue in collagen. (d) Amide bond between the C-terminal amino acid of a protein and the phosphoethanolamine bridge to the 6 position of mannose in the glycophosphoinositol (GPI)-anchor.

2. Transfer of this precursor to the NH_2 group of an Asn residue on a growing polypeptide.

3. Removal of some of the precursor's sugar units.

4. Addition of sugar residues to the remaining core oligosaccharide.

We shall discuss these stages in order.

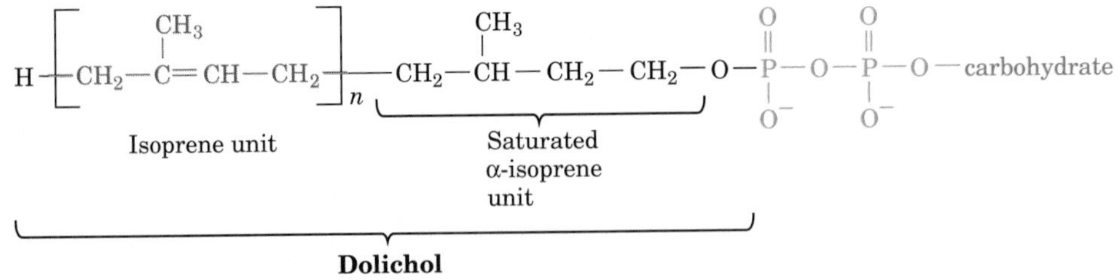

FIGURE 21-14. The carbohydrate precursors of *N*-linked glycosides are synthesized as dolichol pyrophosphate glycosides. Dolichols are long-chain polyisoprenols ($n = 14-24$) in which the α-isoprene unit is saturated.

N-Linked Oligosaccharides Are Constructed on Dolichol Carriers

N-Linked oligosaccharides are initially synthesized as lipid-linked precursors. The lipid component in this process is **dolichol,** a long-chain polyisoprenol of 14 to 24 isoprene units (17–21 units in animals and 14–24 units in fungi and plants; isoprene units are C_5 units with the carbon skeleton of isoprene; Section 23-6A), which is linked to the oligosaccharide precursor via a pyrophosphate bridge (Fig. 21-14). Dolichol apparently anchors the growing oligosaccharide to the endoplasmic reticulum membrane. Involvement of lipid-linked oligosaccharides in *N*-linked glycoprotein synthesis was first demonstrated in 1972 by Armando Parodi and Luis Leloir, who showed that, when a lipid-linked oligosaccharide containing [^{14}C]glucose is incubated with rat liver microsomes (vesicular fragments of isolated endoplasmic reticulum), the radioactivity becomes associated with protein.

N-Linked Glycoproteins Have a Common Oligosaccharide Core

The pathway of dolichol-PP-oligosaccharide synthesis involves stepwise addition of monosaccharide units to the growing glycolipid by specific glycosyl transferases to form a common "core" structure. Each monosaccharide unit is added by a unique glycosyl transferase (Fig. 21-15). For example, in Reaction 2 of Fig. 21-15, five mannosyl units are added through the action of five different mannosyl transferases, each with a different oligosaccharide-acceptor specificity. The oligosaccharide core, the product of Reaction 9 in Fig. 21-15, has the composition (*N*-acetylglucosamine)$_2$(mannose)$_9$(glucose)$_3$.

Although nucleotide sugars are the most common monosaccharide donors in glycosyl transferase reactions, *several mannosyl and glucosyl residues are transferred to the growing dolichol-PP-oligosaccharide from their corresponding dolichol-P derivatives.* The requirement for **dolichol-P-mannose** was discovered by Stuart Kornfeld, who found that mutant mouse lymphoma cells (lymphoma is a type of cancer) unable to synthesize the normal lipid-linked oligosaccharides formed a defective, smaller glycolipid. These cells contain all the requisite glycosyl transferases but are unable to synthesize dolichol-P-mannose (Reaction 4 in Fig. 21-15 is blocked). When this substance is supplied to the mutant cells, mannosyl units are added to the defective dolichol-PP-oligosaccharide.

Dolichol-PP-Oligosaccharide Synthesis Involves Topographical Changes of the Intermediates

Reactions 1, 2, 4, and 7 of Figure 21-15 all occur on the cytoplasmic side of the endoplasmic reticulum (ER) membrane. This was determined by using "right-side-out" rough ER vesicles and showing that various membrane impermeant reagents can disrupt one or another of these reactions. Reactions 6, 9, and 10 occur in the lumen of the ER as judged by the inability of concanavalin A, a carbohydrate binding protein, to react with the products of these reactions until the membrane is permeabilized. The (mannose)$_5$(*N*-acetylglucosamine)$_2$-PP-dolichol product of Step 2, the dolichol-P-mannose product of Reaction 4, and the dolichol-P-glucose product of Step 7 must therefore be translocated across the ER membrane (Reaction 3, 5, and 8) such that they extend from its lumenal surface in order for the synthesis of *N*-linked oligosaccharides to continue. The mechanisms of these various translocation processes are unknown.

Asparagine-Linked Oligosaccharides Are Cotranslationally Added to Proteins

Vesicular-stomatitis virus (VSV), which infects cattle, producing influenza-like symptoms, provides an excellent model system for studying *N*-linked glycoprotein processing. The VSV coat consists of host-cell membrane in which a single viral glycoprotein, the **VSV G protein,** is embedded. Since a viral infection almost totally usurps an infected cell's protein synthesizing machinery, a VSV-infected cell's Golgi apparatus, which normally contains hundreds of different types of glycoproteins, contains virtually no other glycoprotein but G protein. Consequently, the maturation of the G protein is relatively easy to follow.

Study of VSV-infected cells indicated that the *transfer of the lipid-linked oligosaccharide to a polypeptide chain occurs while the polypeptide chain is still being synthesized.* The G protein is *N*-glycosylated by **membrane-bound oligosaccharide-transferring enzyme,** which recognizes the amino acid sequence Asn-X-Ser/Thr (Fig. 21-15, Reaction 10; Fig. 21-16, Reaction 1). Yet, only about one third of the

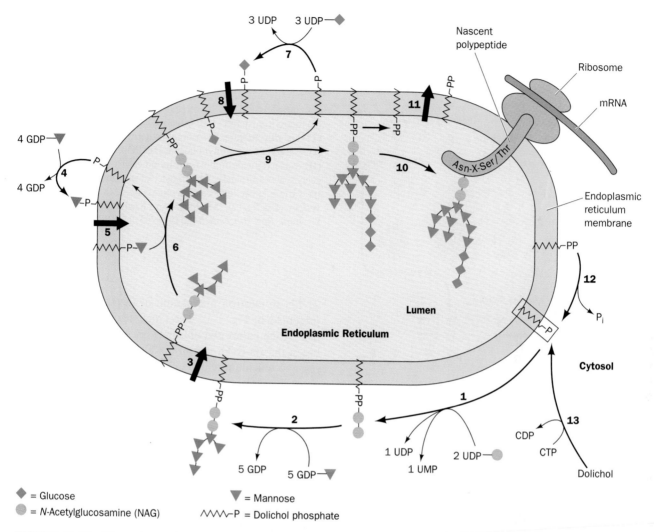

FIGURE 21-15. The pathway of dolichol-PP-oligosaccharide synthesis: **(1)** Addition of *N*-acetylglucosamine-1-P and a second *N*-acetylglucosamine to dolichol-P. **(2)** Addition of five mannosyl residues from GDP–mannose in reactions catalyzed by five different mannosyl transferases. **(3)** Membrane translocation of dolichol-PP-(*N*-acetylglucosamine)₂(mannose)₅ to the lumen of the endoplasmic reticulum. **(4)** Cytosolic synthesis of dolichol-P-mannose from GDP-mannose and dolichol-P. **(5)** Membrane translocation of dolichol-P-mannose to the lumen of the endoplasmic reticulum. **(6)** Addition of four mannosyl residues from dolichol-P-mannose in reactions

catalyzed by four different mannosyl transferases. **(7)** Cytosolic synthesis of dolichol-P-glucose from UDPG and dolichol-P. **(8)** Membrane translocation of dolichol-P-glucose to the lumen of the endoplasmic reticulum. **(9)** Addition of three glucosyl residues from dolichol-P-glucose. **(10)** Transfer of the oligosaccharide from dolichol-PP to the polypeptide chain at an Asn residue in the sequence Asn-X-Ser/Thr, releasing dolichol-PP. **(11)** Translocation of dolichol-PP to the cytoplasmic surface of the endoplasmic reticulum membrane. **(12)** Hydrolysis of dolichol-PP to dolichol-P. **(13)** Dolichol-P can also be formed by phosphorylation of dolichol by CTP.

Asn-X-Ser/Thr sites of eukaryotic proteins are actually *N*-glycosylated. The application of structure prediction algorithms (Section 8-1C) to the amino acid sequences flanking known *N*-glycosylation sites, together with glycosylation studies of model polypeptides, suggests that these sites occur at β turns or loops in which the Asn peptide N—H group is hydrogen bonded to the Ser/Thr hydroxyl O atom. This explains why Pro cannot occupy the X position; it would prevent Asn-X-Ser/Thr from assuming the putative required hydrogen bonded conformation.

Glycoprotein Processing Begins in the Endoplasmic Reticulum and Is Completed in the Golgi Apparatus

Processing of primary glycoproteins begins in the endoplasmic reticulum by the enzymatic trimming (removal) of their three glucose residues (Fig. 21-16, Reactions 2 and 3) and one of their mannose residues (Fig. 21-16, Reaction 4). The glycoproteins are then transported, in membranous vesicles, to the Golgi apparatus for further processing.

The Golgi apparatus consists of a stack of 4 to 6 or more (depending on the species) flattened membranous sacs (Fig.

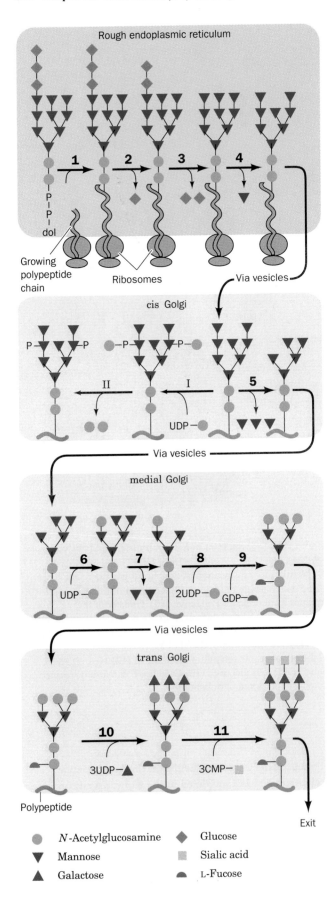

Rough endoplasmic reticulum

Growing polypeptide chain

Ribosomes

Via vesicles

cis Golgi

Via vesicles

medial Golgi

Via vesicles

trans Golgi

Polypeptide

Exit

- ● *N*-Acetylglucosamine
- ▼ Mannose
- ▲ Galactose
- ◆ Glucose
- ■ Sialic acid
- ◖ L-Fucose

FIGURE 21-16. Schematic pathway of oligosaccharide processing on newly synthesized vesicular-stomatitis virus glycoprotein. The reactions are catalyzed by: **(1)** membrane-bound oligosaccharide-transferring enzyme, **(2)** α-glucosidase I, **(3)** α-glucosidase II, **(4)** ER α-1,2-mannosidase, **(5)** Golgi α-mannosidase I, **(6)** *N*-acetylglucosaminyltransferase I, **(7)** Golgi α-mannosidase II, **(8)** *N*-acetylglucosaminyltransferase II, **(9)** fucosyltransferase, **(10)** galactosyltransferase, and **(11)** sialyltransferase. Lysosomal proteins are modified by: **(I)** *N*-acetylglucosaminyl phosphotransferase and **(II)** *N*-acetylglucosamine-1-phosphodiester α-*N*-acetylglucosaminidase. The transfer of intermediates between the various subcellular compartments occurs via membranous vesicles (Fig. 21-17). [Modified from Kornfeld, R. and Kornfeld, S., *Annu. Rev. Biochem.* **54**, 640 (1985).]

21-17; an electron micrograph of the Golgi apparatus is presented in Fig. 1-5). The Golgi stack has two distinct faces, each comprised of a network of interconnected tubules: the **cis Golgi network,** which is opposite the endoplasmic reticulum and is the port through which proteins enter the Golgi apparatus; and the **trans Golgi network,** though which processed proteins exit. The intervening Golgi stack contains at least three different types of sacs, the **cis, medial,** and **trans cisternae,** each of which, as shown by James Rothman and Stuart Kornfeld, contains different sets of glycoprotein processing enzymes. As a glycoprotein traverses the Golgi stack, from the cis to the medial to the trans cisternae, mannose residues are trimmed and *N*-acetylglucosamine, galactose, fucose, and sialic acid residues are added to complete its processing (Fig. 21-16; Reactions 5–11). The glycoproteins are then sorted in the trans Golgi network for transfer to their respective cellular destinations. The glycoproteins are transported between these various locations in membranous vesicles (Section 11-3F).

There is enormous diversity among the different oligosaccharides of *N*-linked glycoproteins as is indicated, for example, in Fig. 10-29c. Indeed, *even glycoproteins with a given polypeptide chain exhibit considerable microheterogeneity* (Section 10-3C), presumably as a consequence of incomplete glycosylation and lack of absolute specificity on the part of glycosyl transferases and glycosylases.

The processing of all *N*-linked oligosaccharides appears to be identical through Reaction 4 of Fig. 21-16 so that all of them have a common (*N*-acetylglucosamine)$_2$(mannose)$_3$ core (five "noncore" mannose residues are subsequently trimmed from VSV glycoprotein in Reactions 5 and 7). The diversity of the *N*-linked oligosaccharides therefore arises through divergence from this sequence after Reaction 7. The resulting oligosaccharides are classified into three groups:

1. **High-mannose oligosaccharides** (Fig. 21-18a), which contain 2 to 9 mannose residues appended to the common pentasaccharide core (red residues in Fig. 21-18).

FIGURE 21-17. Most proteins destined for secretion or insertion into a membrane are synthesized by ribosomes (*blue dots*) attached to the rough endoplasmic reticulum (rough ER; *top*). As they are synthesized, the proteins (*red dots*) are either injected into the lumen of the endoplasmic reticulum or inserted into its membrane (Section 11-4B). After initial processing, the proteins are encapsulated in vesicles formed from endoplasmic reticulum membrane, which subsequently fuse with the cis Golgi network. The proteins are progressively processed (Fig. 21-16), according to their cellular destinations, in the cis, medial, and trans cisternae of the Golgi, between which they are transported by other membranous vesicles. Finally, in the trans Golgi network (*bottom*), the completed glycoproteins are sorted for delivery to their final destinations, for example, lysosomes, the plasma membrane, or secretory granules, to which they are transported by yet other vesicles.

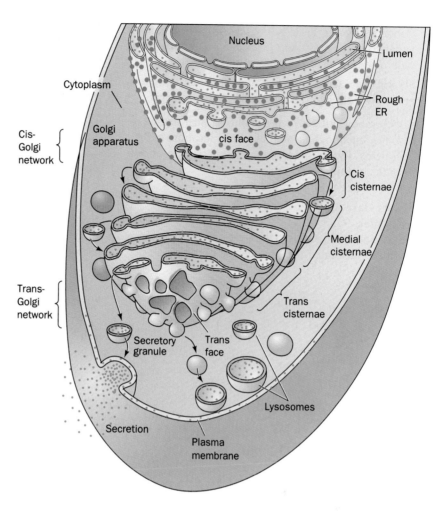

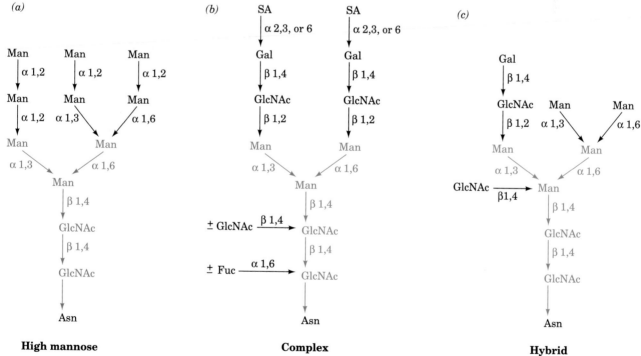

High mannose **Complex** **Hybrid**

FIGURE 21-18. Typical primary structures of (*a*) high-mannose, (*b*) complex, and (*c*) hybrid *N*-linked oligosaccharides. The pentasaccharide core common to all *N*-linked oligosaccharides is indicated in red. [After Kornfield, R. and Kornfield, S., *Annu. Rev. Biochem.* **54,** 633 (1985).]

2. **Complex oligosaccharides** (Fig. 21-18*b*), which contain variable numbers of *N*-acetyllactosamine units as well as sialic acid and/or fucose residues linked to the core.

3. **Hybrid oligosaccharides** (Fig. 21-18*c*), which contain elements of both high-mannose and complex chains.

It is unclear how different types of oligosaccharides are related to the functions and/or final cellular locations of their glycoproteins. Lysosomal glycoproteins, however, appear to be of the high-mannose variety.

Inhibitors Have Aided the Study of *N*-Linked Glycosylation

Elucidation of the events in the glycosylation process has been greatly facilitated through the use of inhibitors that block specific glycosylation enzymes. Two of the most useful are the antibiotics **tunicamycin** (Fig. 21-19*a*), a hydrophobic analog of UDP–*N*-acetylglucosamine, and **bacitracin** (Fig. 21-20), a cyclic polypeptide. Both were discovered because of their ability to inhibit bacterial cell wall biosynthesis, a process that also involves the participation of lipid-linked oligosaccharides. Tunicamycin blocks the formation of dolichol-PP-oligosaccharides by inhibiting the synthesis of dolichol-PP-*N*-acetylglucosamine from dolichol-P and UDP–*N*-acetylglucosamine (Fig. 21-15, Reaction 1). Tunicamycin resembles an adduct of these reactants (Fig. 21-19*b*) and, in fact, binds to the enzyme with a dissociation constant of $7 \times 10^{-9}M$.

Bacitracin forms a complex with dolichol-PP that inhibits its dephosphorylation (Fig. 21-15, Reaction 12), thereby

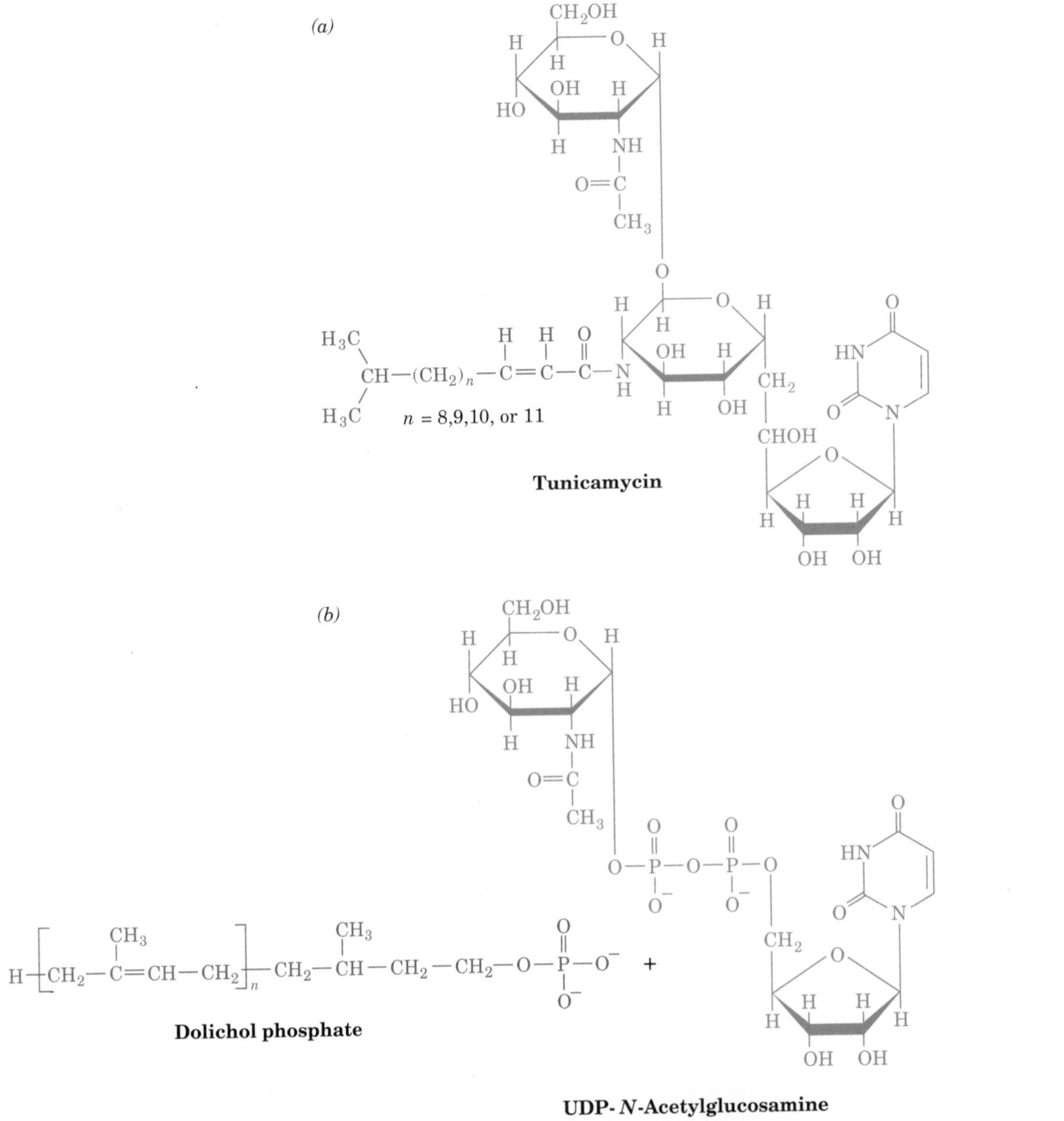

FIGURE 21-19. Comparison of the chemical structures of *(a)* tunicamycin and *(b)* dolichol-P + UDP-*N*-acetylglucosamine.

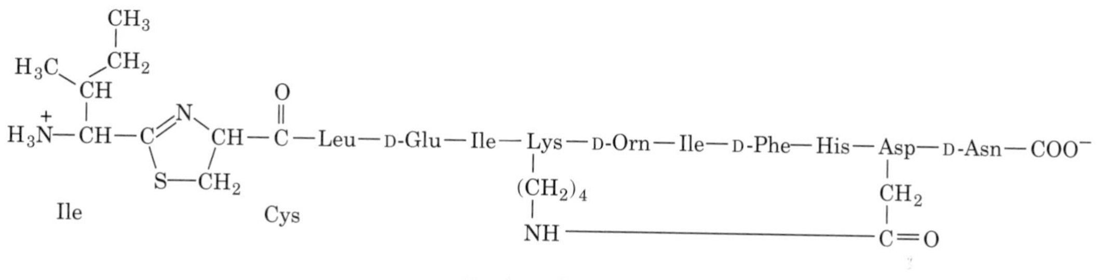

Bacitracin

FIGURE 21-20. The chemical structure of bacitracin. Note that this dodecapeptide has four D-amino acid residues and two unusual intrachain linkages. "Orn" represents the nonstandard amino acid residue ornithine (Fig. 4-22).

preventing glycoprotein synthesis from lipid-linked oligosaccharide precursors. Bacitracin is clinically useful because it destroys bacterial cell walls but does not affect animal cells because it cannot cross cell membranes (bacterial cell wall biosynthesis is an extracellular process).

O-Linked Oligosaccharides Are Posttranslationally Formed

The study of the biosynthesis of **mucin**, an *O*-linked glycoprotein secreted by the submaxillary salivary gland, indicates that *O*-linked oligosaccharides are synthesized in the Golgi apparatus by serial addition of monosaccharide units to a completed polypeptide chain (Fig. 21-21). Synthesis starts with the transfer of *N*-acetylgalactosamine (GalNAc) from UDP-GalNAc to a Ser or Thr residue on the polypeptide by **GalNAc transferase.** In contrast to *N*-linked oligosaccharides, which are transferred to an Asn in a specific amino acid sequence, the *O*-glycosylated Ser and Thr residues are not members of any common sequence. Rather, it is thought that the location of glycosylation sites is specified only by the secondary or tertiary structure of the polypeptide. Glycosylation continues with stepwise addition of galactose, sialic acid, *N*-acetylglucosamine, and fucose by the corresponding glycosyl transferases.

Oligosaccharides on Glycoproteins Act as Recognition Sites

Glycoproteins synthesized in the endoplasmic reticulum and processed in the Golgi apparatus are targeted for secretion, insertion into cell membranes, or incorporation into cellular organelles such as lysosomes (Fig. 21-17). This suggests that *oligosaccharides serve as recognition markers for this sorting process.* For example, the study of I-cell disease (Section 11-4C) demonstrated that in glycoprotein enzymes destined for the lysosome, a mannose residue is converted to mannose-6-phosphate (M6P) in the cis cisternae of the Golgi. The process involves two enzymes (Fig. 21-16, Reactions I and II), which are thought to recognize lysosomal protein precursors by various structural features on these proteins rather than a specific amino acid sequence. In the trans cisternae, M6P-bearing glycoproteins are

sorted into lysosome-bound coated vesicles through their specific binding to one of two M6P receptors, one of which is a 275-kD membrane glycoprotein called the **M6P/IGF-II receptor** (because it has been found that this M6P receptor and the **insulinlike growth factor II receptor** are the same protein). Individuals with I-cell disease lack the en-

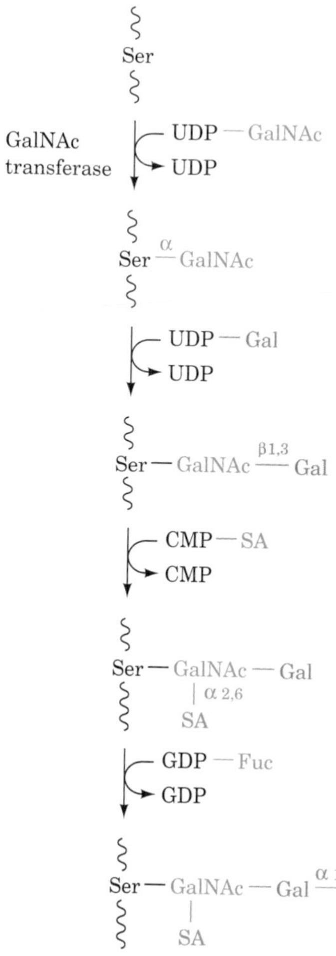

FIGURE 21-21. The proposed synthesis pathway for the carbohydrate moiety of an *O*-linked oligosaccharide chain of canine submaxillary mucin. SA = sialic acid.

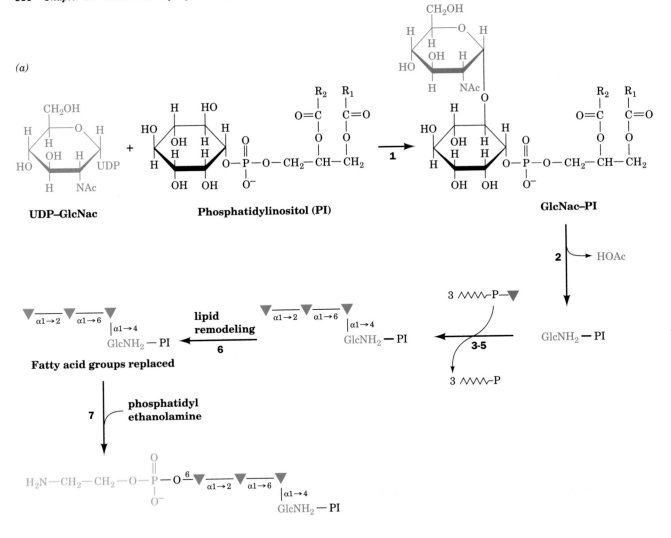

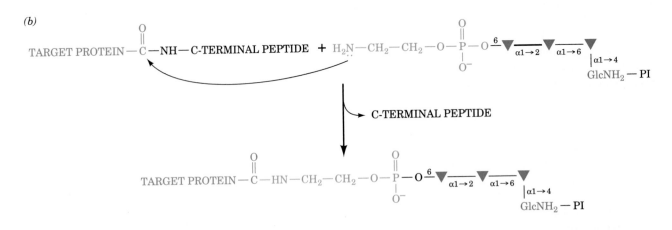

∿∿–P = Dolichol   ▼ = Mannose

FIGURE 21-22. (*a*) The pathway of synthesis of the tetrasaccharide core of glycophosphatidylinositol (GPI). (**1**) Addition of GlcNAc to phosphatidylinositol (PI). (**2**) Deacetylation of GlcNAc. (**3-5**) Addition of three mannosyl residues from dolichol-P-mannose catalyzed by three separate glycosylases. (**6**) Lipid remodeling (replacement of the fatty acyl groups on PI). (**7**) Transfer of phosphoethanolamine from phosphatidyl ethanolamine to the 6-hydroxyl group of the terminal mannose residue of the core tetrasaccharide. (*b*) Transamidation of the target protein resulting in a C-terminal amide link to the GPI anchor.

zyme catalyzing mannose phosphorylation (Fig. 21-16, Reaction I), resulting in the secretion of the normally lysosome-resident enzymes.

ABO blood group antigens (Section 11-3D) are *O*-linked glycoproteins. Their characteristic oligosaccharides are components of both cell-surface lipids and of proteins that occur in various secretions such as saliva. These oligosaccharides form antibody recognition sites.

Glycoproteins are believed to mediate cell–cell recognition. For example, an *O*-linked oligosaccharide on a glycoprotein that coats the mouse ovum surface (zona pellucida) acts as the sperm receptor. Even when this oligosaccharide is separated from its protein, it retains the ability to bind mouse sperm.

GPI-Linked Proteins

Glycosylphosphatidylinositol (GPI) groups function to anchor a wide variety of proteins to the exterior surface of the eukaryotic plasma membrane, providing an alternative to transmembrane polypeptide domains (Section 11-5A; Fig. 11-49). This anchoring results from transamidation of a preformed GPI glycolipid within 1 min of the synthesis and transfer of a target protein to the endoplasmic reticulum. The core GPI structure is synthesized on the lumenal side of the endoplasmic reticulum membrane from phosphatidylinositol, UDP–*N*-acetylglucosamine (UDPGlcNAc), Dolichol-P-mannose (dol-P-man; Fig. 21-14) and phosphatidylethanolamine (Table 11-2) as shown in Figure 21-22*a*. This core is modified with a variety of additional sugar residues, depending on the species and the protein to which it is attached. There is considerable diversity in the fatty acid residues of GPI-anchors due to the extensive lipid remodeling that occurs during anchor synthesis. Target proteins become anchored to the membrane surface when the amino group of the GPI phosphoethanolamine nucleophilically attacks a specific amino acyl group of the protein near its C-terminus, resulting in a transamidation that releases a 20- to 30-residue hydrophobic C-terminal peptide (Fig. 21-22*b*). Since GPI groups are appended to proteins on the lumenal surface of the RER, GPI-anchored proteins occur on the exterior surface of the plasma membrane (Fig. 11-46).

4. THE PENTOSE PHOSPHATE PATHWAY

ATP is the cell's "energy currency"; its exergonic hydrolysis is coupled to many otherwise endergonic cell functions. *Cells have a second currency, reducing power.* Many endergonic reactions, notably the reductive biosynthesis of fatty acids (Section 23-4) and cholesterol (Section 23-6A), as well as photosynthesis (Section 22-3A), require NADPH in addition to ATP. Despite their close chemical resemblance, *NADPH and NADH are not metabolically interchangeable*

(recall that these coenzymes differ only by a phosphate group at the 2′-OH group of NADPH's adenosine moiety; Fig. 12-2). Whereas NADH participates in utilizing the free energy of metabolite oxidation to synthesize ATP (oxidative phosphorylation), *NADPH is involved in utilizing the free energy of metabolite oxidation for otherwise endergonic reductive biosynthesis.* This differentiation is possible because the dehydrogenase enzymes involved in oxidative and reductive metabolism exhibit a high degree of specificity towards their respective coenzymes. Indeed, cells normally maintain their [NAD$^+$]/[NADH] ratio near 1000, which favors metabolite oxidation, while keeping their [NADP$^+$]/[NADPH] ratio near 0.01, which favors metabolite reduction.

NADPH is generated by the oxidation of G6P via an alternative pathway to glycolysis, the **pentose phosphate pathway** *[also called the* **hexose monophosphate (HMP) shunt** *or the* **phosphogluconate pathway;** *Fig. 21-23].* The first evidence of this pathway's existence was obtained in the 1930s by Otto Warburg, who discovered NADP$^+$ through his studies on the oxidation of G6P to 6-phosphogluconate. Further indications came from the observation that tissues continue to respire in the presence of high concentrations of fluoride ion which, it will be recalled, blocks glycolysis by inhibiting enolase (Section 16-2I). It was not until the 1950s, however, that the pentose phosphate pathway was elucidated by Frank Dickens, Bernard Horecker, Fritz Lipmann, and Efraim Racker. Tissues most heavily involved in fatty acid and cholesterol biosynthesis (liver, mammary gland, adipose tissue, and adrenal cortex), are rich in pentose phosphate pathway enzymes. Indeed, some 30% of the glucose oxidation in liver occurs via the pentose phosphate pathway.

The overall reaction of the pentose phosphate pathway is

$$3G6P + 6NADP^+ + 3H_2O \rightleftharpoons$$
$$6NADPH + 6H^+ + 3CO_2 + 2F6P + GAP$$

However, the pathway may be considered to have three stages:

1. Oxidative reactions (Fig. 21-23, Reactions 1–3), which yield NADPH and **ribulose-5-phosphate (Ru5P).**

$$3G6P + 6NADP^+ + 3H_2O \longrightarrow$$
$$6NADPH + 6H^+ + 3CO_2 + 3Ru5P$$

2. Isomerization and epimerization reactions (Fig. 21-23, Reactions 4 and 5), which transform Ru5P either to **ribose-5-phosphate (R5P)** or to **xylulose-5-phosphate (Xu5P).**

$$3Ru5P \rightleftharpoons R5P + 2Xu5P$$

3. A series of C—C bond cleavage and formation reactions (Fig. 21-23, Reactions 6–8) that convert two molecules of Xu5P and one molecule of R5P to two molecules of

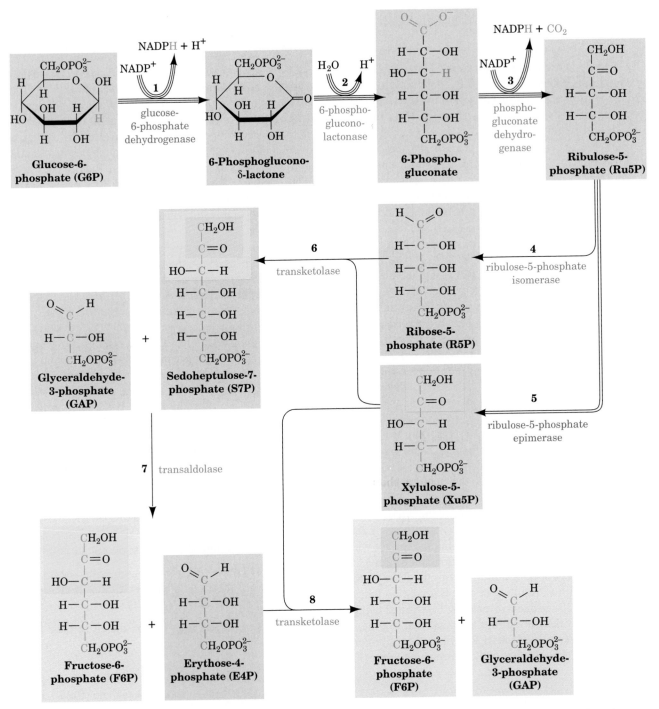

FIGURE 21-23. The pentose phosphate pathway. The number of lines in an arrow represents the number of molecules reacting in one turn of the pathway so as to convert three G6Ps to three CO_2s, two F6Ps, and one GAP. For the sake of clarity, sugars from Reaction 3 onward are shown in their linear forms. The carbon skeleton of R5P and the atoms derived from it are drawn in red and those from Xu5P are drawn in green. The C_2 units transferred by transketolase are shaded in green and the C_3 units transferred by transaldolase are shaded in blue.

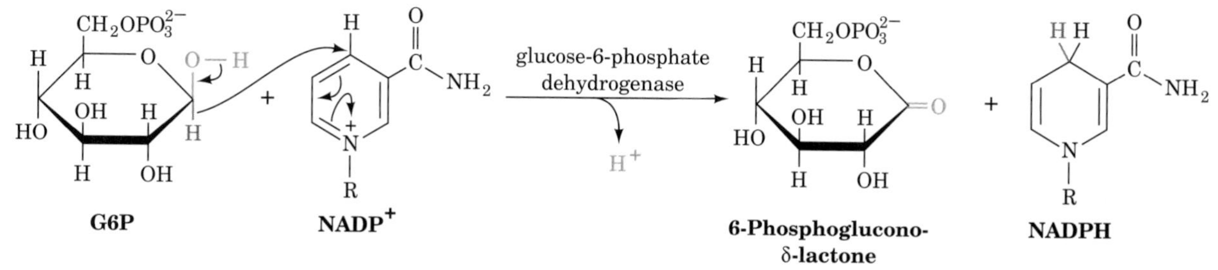

FIGURE 21-24. The glucose-6-phosphate dehydrogenase reaction.

fructose-6-phosphate (F6P) and one of glyceraldehyde-3-phosphate (GAP).

$$R5P + 2Xu5P \rightleftharpoons 2F6P + GAP$$

The reactions of Stages 2 and 3 are freely reversible so that the products of the pathway vary with the needs of the cell. For example, when R5P is required for nucleotide biosynthesis, Stage 3 produces less F6P and GAP. In this section, we discuss the three stages of the pentose phosphate pathway and how this pathway is controlled. We close by considering the consequences of one of its abnormalities.

A. Oxidative Reactions of NADPH Production

Only the first three reactions of the pentose phosphate pathway are involved in NADPH production.

1. **Glucose-6-phosphate dehydrogenase (G6PD)** catalyzes net transfer of a hydride ion to NADP$^+$ from C1 of G6P to form **6-phosphoglucono-δ-lactone** (Fig. 21-24). G6P, a cyclic hemiacetal with C1 in the aldehyde oxidation state, is thereby oxidized to a cyclic ester (lactone). The enzyme is specific for NADP$^+$ and is strongly inhibited by NADPH.

2. **6-Phosphogluconolactonase** increases the rate of hydrolysis of 6-phosphoglucono-δ-lactone to **6-phosphogluconate** (the nonenzymatic reaction occurs at a significant rate), the substrate of the next oxidative enzyme in the pathway.

3. **Phosphogluconate dehydrogenase** catalyzes the oxidative decarboxylation of 6-phosphogluconate, a β-hydroxy acid, to Ru5P and CO$_2$ (Fig. 21-25). The reaction is similar to that catalyzed by the citric acid cycle enzyme isocitrate dehydrogenase (Section 19-3C).

Formation of Ru5P completes the oxidative portion of the pentose phosphate pathway. *It generates two molecules of NADPH for each molecule of G6P that enters the pathway.* The product Ru5P must subsequently be converted to R5P or Xu5P for further use.

B. Isomerization and Epimerization of Ribulose-5-Phosphate

Ru5P is converted to R5P by **ribulose-5-phosphate isomerase** *(Fig. 21-23, Reaction 4) and to Xu5P by* **ribulose-5-phosphate epimerase** *(Fig. 21-23, Reaction 5).* These isomerization and epimerization reactions, as discussed in

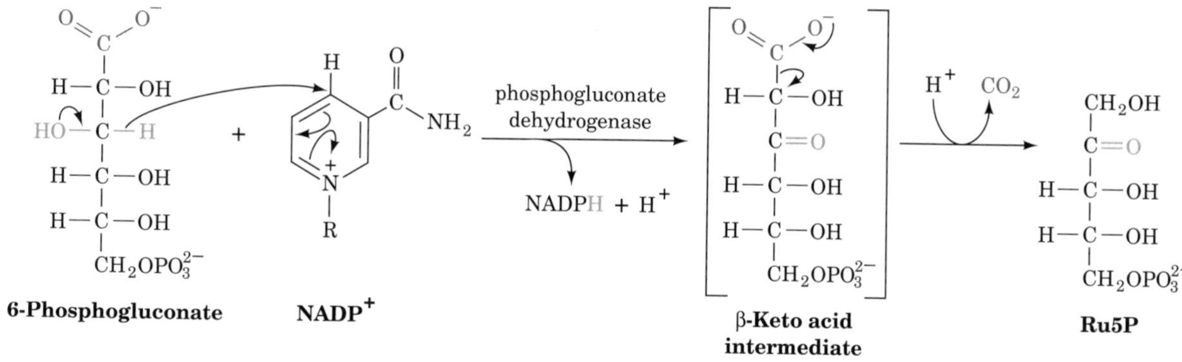

FIGURE 21-25. The phosphogluconate dehydrogenase reaction. Oxidation of the OH group forms an easily decarboxylated β-keto acid (although the proposed intermediate has not been isolated).

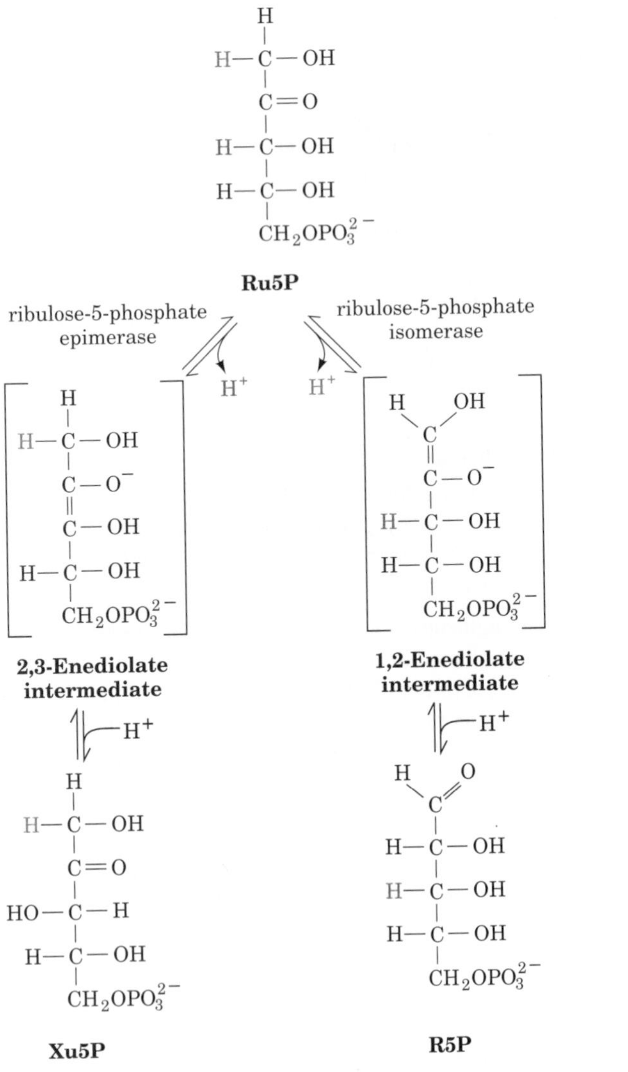

FIGURE 21-26. The ribulose-5-phosphate isomerase and ribulose-5-phosphate epimerase reactions both involve enediolate intermediates. In the isomerase reaction (*right*), a base on the enzyme removes a proton from C1 of Ru5P to form a 1,2-enediolate and then adds a proton at C2 to form R5P. In the epimerase reaction (*left*), a base on the enzyme removes a C3 proton to form a 2,3-enediolate. A proton is then added to the same carbon atom but with inversion of configuration to yield Xu5P.

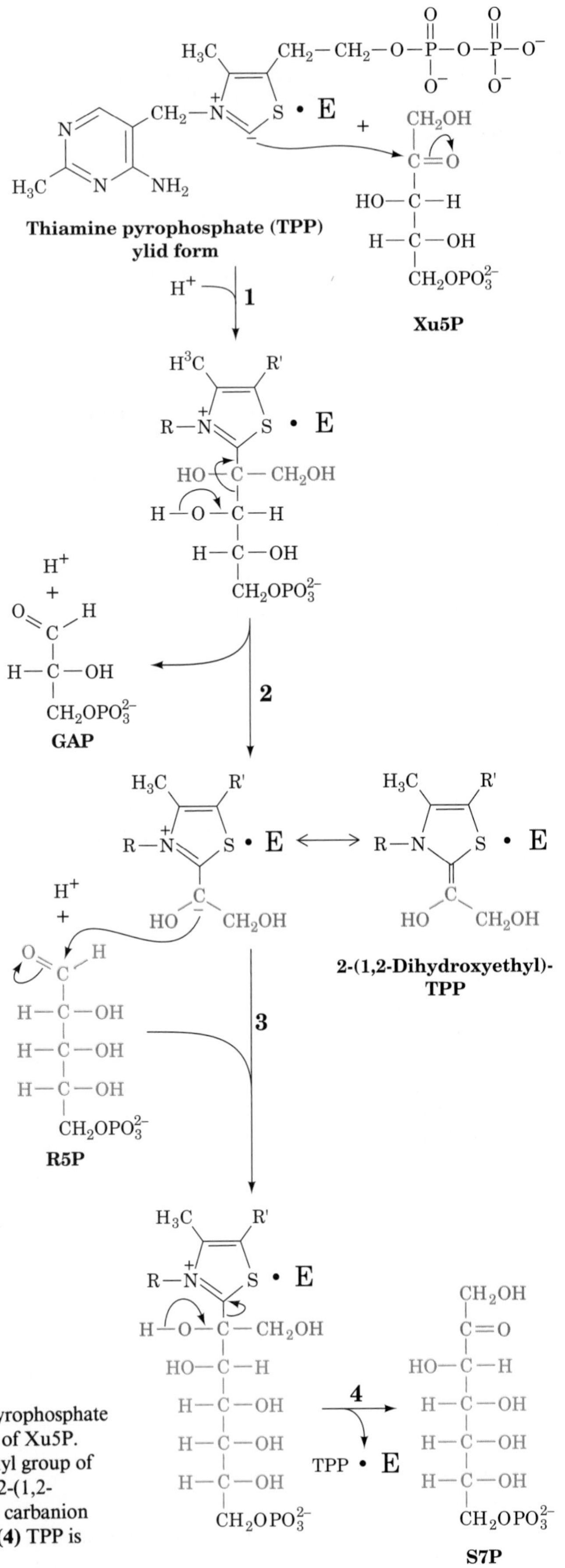

FIGURE 21-27. Transketolase utilizes the coenzyme thiamine pyrophosphate to stabilize the carbanion formed on cleavage of the C2—C3 bond of Xu5P. The reaction occurs as follows: **(1)** The TPP ylid attacks the carbonyl group of Xu5P. **(2)** C2—C3 bond cleavage yields GAP and enzyme-bound 2-(1,2-dihydroxyethyl)-TPP, a resonance stabilized carbanion. **(3)** The C2 carbanion attacks the aldehyde carbon of R5P forming an S7P–TPP adduct. **(4)** TPP is eliminated yielding S7P and the regenerated TPP–enzyme.

Section 15-2B, are both thought to occur via enediolate intermediates (Fig. 21-26).

R5P is an essential precursor in the biosynthesis of nucleotides (Sections 26-2, 3, and 6). If, however, more R5P is formed than the cell needs, the excess, along with Xu5P, is converted to the glycolytic intermediates F6P and GAP as is described below.

C. Carbon—Carbon Bond Cleavage and Formation Reactions

The conversion of three C_5 sugars to two C_6 sugars and one C_3 sugar involves a remarkable "juggling act" catalyzed by two enzymes, **transaldolase** and **transketolase**. As we discussed in Section 15-2E, enzymatic reactions that make or break carbon–carbon bonds usually have mechanisms that involve generation of a stabilized carbanion and its addition to an electrophilic center such as an aldehyde. This is the dominant theme of both the transaldolase and the transketolase reactions.

Transketolase Catalyzes the Transfer of C_2 Units

Transketolase, which has a thiamine pyrophosphate cofactor (TPP; Section 16-3B), catalyzes the transfer of a C_2 unit from Xu5P to R5P, yielding GAP and **sedoheptulose-7-phosphate (S7P)** (Fig. 21-23, Reaction 6). The reaction involves the intermediate formation of a covalent adduct between Xu5P and TPP (Fig. 21-27). The X-ray structure of this dimeric enzyme shows that the TPP binds in a deep cleft between the subunits such that residues from both subunits participate in its binding, just as in pyruvate decarboxylase (another TPP-requiring enzyme; Figure 16-29). In fact the structures are so similar that it is likely that they diverged from a common ancestor.

Transaldolase Catalyzes the Transfer of C_3 Units

Transaldolase catalyzes the transfer of a C_3 unit from S7P to GAP yielding **erythrose-4-phosphate (E4P)** and F6P (Fig. 21-23, Reaction 7). The reaction occurs by aldol cleavage (Section 16-2D), which begins with the formation of a Schiff base between an ε-amino group of an essential enzyme Lys residue and the carbonyl group of S7P (Fig. 21-28).

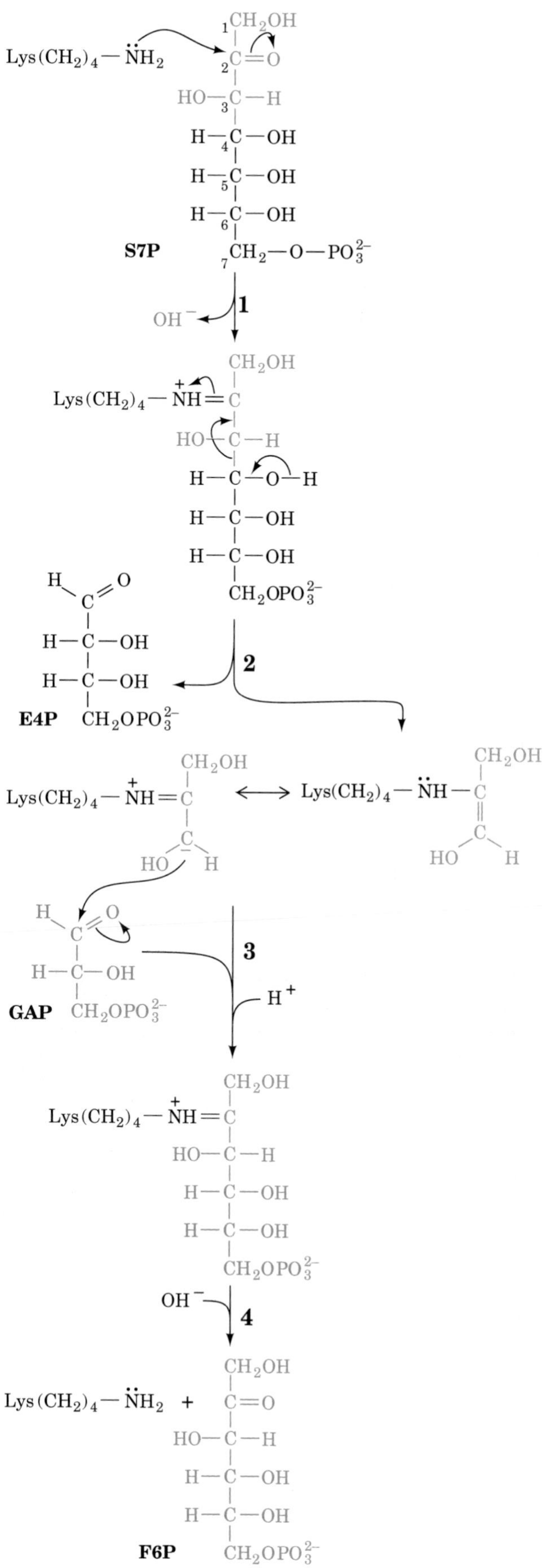

FIGURE 21-28. Transaldolase contains an essential Lys residue that forms a Schiff base with S7P to facilitate an aldol cleavage reaction. The reaction occurs as follows: (1) The ε-amino group of an essential Lys residue forms a Schiff base with the carbonyl group of S7P. (2) A Schiff base–stabilized C3 carbanion is formed in an aldol cleavage reaction between C3 and C4 that eliminates E4P. (3) The enzyme-bound resonance-stabilized carbanion adds to the carbonyl C atom of GAP, forming F6P linked to the enzyme via a Schiff base. (4) The Schiff base hydrolyzes, regenerating active enzyme and releasing F6P.

A Second Transketolase Reaction Yields GAP and a Second F6P Molecule

In a second transketolase reaction, a C_2 unit is transferred from a second molecule of Xu5P to E4P to form GAP and another molecule of F6P (Fig. 21-23, Reaction 8). The third phase of the pentose phosphate pathway thus transforms two molecules of Xu5P and one of R5P to two molecules of F6P and one molecule of GAP. These carbon skeleton transformations (Fig. 21-23, Reactions 6–8) are summarized in Fig. 21-29.

$$(6) \quad C_5 + C_5 \rightleftharpoons C_7 + C_3$$

$$(7) \quad C_7 + C_3 \rightleftharpoons C_6 + C_4$$

$$(8) \quad \underline{C_5 + C_4 \rightleftharpoons C_6 + C_3}$$

$$(\text{Sum}) \quad 3C_5 \rightleftharpoons 2C_6 + C_3$$

FIGURE 21-29. The carbon—carbon bond formations and cleavages that convert three C_5 sugars to two C_6 and one C_3 sugar in the pentose phosphate pathway. The number to the left of each reaction is keyed to the corresponding reaction in Fig. 21-23.

D. Control of the Pentose Phosphate Pathway

The principal products of the pentose phosphate pathway are R5P and NADPH. The transaldolase and transketolase reactions serve to convert excess R5P to glycolytic intermediates when the metabolic need for NADPH exceeds that of R5P in nucleotide biosynthesis. The resulting GAP and F6P can be consumed through glycolysis and oxidative phosphorylation or recycled by gluconeogenesis to form G6P. *In the latter case, 1 molecule of G6P can be converted, via 6 cycles of the pentose phosphate pathway and gluconeogenesis, to 6 CO$_2$ molecules with the concomitant generation of 12 NADPH molecules.* When the need for R5P outstrips that for NADPH, F6P and GAP can be diverted from the glycolytic pathway for use in the synthesis of R5P by reversal of the transaldolase and transketolase reactions.

Flux through the pentose phosphate pathway and thus the rate of NADPH production is controlled by the rate of the glucose-6-phosphate dehydrogenase reaction (Fig. 21-23, Reaction 1). The activity of this enzyme, which catalyzes the pentose phosphate pathway's first committed step ($\Delta G = -17.6$ kJ · mol^{-1} in liver), is regulated by the NADP$^+$ concentration (substrate availability). When the cell consumes NADPH, the NADP$^+$ concentration rises, increasing the rate of the glucose-6-phosphate dehydrogenase reaction, thereby stimulating NADPH regeneration.

E. Glucose-6-Phosphate Dehydrogenase Deficiency

NADPH is required for several reductive processes in addition to biosynthesis. For example, erythrocyte membrane integrity requires a plentiful supply of reduced glutathione (GSH), a Cys-containing tripeptide (Section 14-4). A major function of GSH in the erythrocyte is to eliminate H_2O_2 and organic hydroperoxides. H_2O_2, a toxic product of various oxidative processes, reacts with double bonds in the fatty acid residues of the erythrocyte cell membrane to form organic hydroperoxides. These, in turn, react to cleave fatty acid C—C bonds, thereby damaging the membrane. In erythrocytes, the unchecked buildup of peroxides results in premature cell lysis. Peroxides are eliminated through the action of **glutathione peroxidase**, one of the handful of en-

zymes with a selenium cofactor, yielding glutathione disulfide (GSSG).

$$2GSH + R-O-O-H \xrightarrow{\text{glutathione peroxidase}} GSSG + ROH + H_2O$$
Organic hydroperoxide

GSH is subsequently regenerated by the NADPH reduction of GSSG catalyzed by glutathione reductase (Section 14-4).

$$GSSG + NADPH + H^+ \xrightarrow{\text{glutathione reductase}} 2GSH + NADP^+$$

A steady supply of NADPH is therefore vital for erythrocyte integrity.

Primaquine Causes Hemolytic Anemia in Glucose-6-Phosphate Dehydrogenase Mutants

A genetic defect, common in African, Asian, and Mediterranean populations, results in severe hemolytic anemia upon infection or the administration of certain drugs including the antimalarial agent **primaquine.**

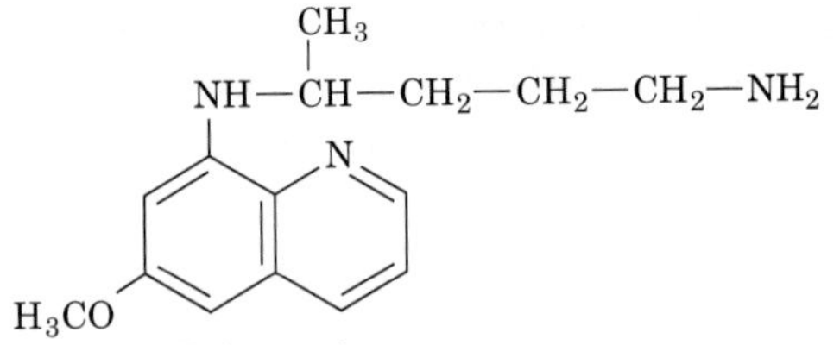

$$\begin{array}{c} CH_3 \\ | \\ NH-CH-CH_2-CH_2-CH_2-NH_2 \end{array}$$

Primaquine

Similar effects, which go by the name of **favism,** occur when individuals bearing this trait eat **fava beans (broad beans,** *Vicia faba*), a staple Middle Eastern vegetable that contains small quantities of toxic glycosides. This trait has been traced to an altered gene for glucose-6-phosphate dehydrogenase (G6PD). Under most conditions, mutant erythrocytes have sufficient enzyme activity for normal function. Agents such as primaquine and fava beans, however, stimulate peroxide formation, thereby increasing the demand for NADPH to a level that mutant cells cannot meet.

The major reason for low enzymatic activity in affected cells appears to be an accelerated rate of breakdown of the mutant enzyme (protein degradation is discussed in Section 30-6). This explains why patients with G6PD deficiency react to primaquine with hemolytic anemia but recover within a week despite continued primaquine treatment. Mature erythrocytes lack a nucleus and protein synthesizing machinery and therefore cannot synthesize new enzyme molecules to replace degraded ones (they likewise cannot synthesize new membrane components, which is why they are so sensitive to membrane damage in the first place). The initial primaquine treatments result in the lysis of old red blood cells whose defective G6PD has been largely degraded. Lysis products stimulate the release of young cells that contain more enzyme and are therefore better able to cope with primaquine stress.

It is estimated that ~400 million people are deficient in G6PD, which makes this condition the most common human enzymopathy. Indeed, ~400 G6PD variants have been reported (although only ~35 have yet been confirmed by DNA sequencing), several of which occur with high incidence. For example, the so-called type A⁻ deficiency, which exhibits ~10% of the normal G6PD activity, has an incidence of 11% among black Americans. This, together with the high prevalence of defective G6PD in malarial areas of the world, suggests that such mutations confer resistance to the malarial parasite, *Plasmodium falciparum*. Indeed, erythrocytes with G6PD deficiency appear to be less suitable hosts for plasmodia than normal cells because the parasite requires the products of the pentose phosphate pathway and/or because the erythrocyte is lysed before the parasite has had a chance to mature. Thus, like the sickle-cell trait (Section 6-3A), *a defective G6PD confers a selective advantage on individuals living where malaria is endemic.*

Curiously, however, only females who are heterozygous for this sex-linked trait are resistant to malaria. This is because plasmodia eventually adapt to living in G6PD-deficient erythrocytes. Thus, G6PD deficiency has little anti-malarial effect in males (whose cells each contain only one X chromosome; Section 27-1B) or homozygous females (whose cells each contain two X chromosomes). However, since, in human females, one of the two X chromosomes in each cell is inactivated essentially at random (Section 33-3A), half of the erythrocytes in heterozygotes for G6PD deficiency are G6PD deficient and the rest are normal, a phenomenon that apparently interferes with the ability of plasmodia to adapt to G6DP deficiency.

CHAPTER SUMMARY

Lactate, pyruvate, citric acid cycle intermediates, and many amino acids may be converted, by gluconeogenesis, to glucose via the formation of oxaloacetate. For this to occur, the three irreversible steps of glycolysis must be bypassed. The pyruvate kinase reaction is bypassed by converting pyruvate to oxaloacetate in an ATP-driven reaction catalyzed by the biotinyl–enzyme pyruvate carboxylase. The oxaloacetate is subsequently phosphorylated by GTP and decarboxylated to PEP in a reaction catalyzed by PEPCK. For this to happen in species in which PEPCK is a cytosolic enzyme, the oxaloacetate must be transported from the mitochondrion to the cytosol via its interim conversion to either malate or aspartate. Conversion to malate concomitantly transports reducing equivalents to the cytosol in the form of NADH. The two other irreversible steps of glycolysis, the PFK reaction and the hexokinase reaction, are bypassed by simply hydrolyzing their products, FBP and G6P, by FBPase and glucose-6-phosphatase, respectively. A glucose molecule may therefore be synthesized from pyruvate at the expense of four ATPs more than is generated by the reverse process. Glycolysis and gluconeogenesis are reciprocally regulated so as to consume glucose when the demand for ATP is high and synthesize it when the demand is low. The control points in these processes are at pyruvate kinase/pyruvate carboxylase–PEPCK, PFK/FBPase, and hexokinase/glucose-6-phosphatase. Regulation of these enzymes is exerted largely through allosteric interactions and cAMP-dependent enzyme modifications. Muscle, which is incapable of gluconeogenesis, transfers much of the lactate it produces to the liver via the blood for conversion to glucose and return to the muscle. This Cori cycle shifts the metabolic burden of oxidative ATP generation for gluconeogenesis from muscle to liver.

Animals cannot convert fatty acids to glucose because they lack the enzymes necessary to synthesize oxaloacetate from acetyl-CoA. Plants, however, can do so via the glyoxylate cycle, a glyoxysomal process that converts two molecules of acetyl-CoA to one molecule of oxaloacetate via the intermediate formation of glyoxylate.

Glycosidic bonds are formed by transfer of the monosaccharide unit of a sugar nucleotide to a second sugar unit. Such reactions occur in the synthesis of disaccharides such as lactose and of the carbohydrate components of glycoproteins. In *N*-linked glycoproteins, the carbohydrate component is attached to the protein via an *N*-glycosidic bond to an Asn residue in the sequence Asn-X-Ser/Thr. In *O*-linked glycoproteins, the carbohydrate attachment is an *O*-glycosidic bond to Ser or Thr, or in collagens, to 5-hydroxylysine. In GPI-anchored proteins the carbohydrate is linked to the protein through an intermediary phosphoethanolamine bridge which forms an amide bond to the protein's C-terminal amino acid residue. Synthesis of *N*-linked oligosaccharides begins in the endoplasmic reticulum with the multistep formation of a lipid-linked precursor consisting of dolichol pyrophosphate bonded to a common 14-residue core oligosaccharide. The carbohydrate is then transferred to an Asn residue of a growing polypeptide chain. The immature *N*-linked glycoprotein is subsequently transferred, via a membranous vesicle, to the cis cisternae of the Golgi apparatus. Processing is completed by trimming of mannose residues followed by attachment of a variety of other monosaccharides as catalyzed by specific enzymes in the cis, medial, and trans Golgi cisternae. Completed *N*-linked glycoproteins are sorted in the trans Golgi cisternae according to the identities of their carbohydrate components for transport, via membranous vesicles, to their final cellular destinations. Three major types of *N*-linked oligosaccharides have been identified, high mannose, complex, and hybrid oligosaccharides, all of which contain a common pentasaccharide core. Studies of glycoprotein formation have been facilitated by the use of antibiotics, such as tunicamycin and bacitracin, which inhibit specific enzymes involved in the synthesis of these oligosaccharides. *O*-Linked oligosaccharides are synthesized in the Golgi apparatus by sequential attachments of specific monosaccharide units to certain Ser or Thr residues. Carbohydrate components of glycoproteins are thought to act as recognition markers for the transport of glycoproteins to their proper cellular destinations and for cell–cell and antibody recognition. The GPI-anchor is appended to proteins on the lumenal surface of the endoplasmic reticulum, thereby targeting GPI-anchored proteins to the external surface of the plasma membrane.

The cell uses NAD^+ in oxidative reactions, and employs NADPH in reductive biosynthesis. NADPH is synthesized by the pentose phosphate pathway, an alternate mode of glucose oxidation. This pathway also synthesizes R5P for use in nucleotide biosynthesis. The first three reactions of the pentose phosphate pathway involve oxidation of G6P to Ru5P with release of CO_2 and formation of two NADPH molecules. This is followed by reactions that either isomerize Ru5P to R5P or epimerize it to Xu5P. Each molecule of R5P not required for nucleotide biosynthesis, together with two Xu5P, is converted to two molecules of F6P and one molecule of GAP via the sequential actions of transketolase, transaldolase, and, again, transketolase. The products of the pentose phosphate pathway depend on the needs of the cell. The F6P and GAP may be metabolized through glycolysis and the citric acid cycle or recycled via gluconeogenesis. If NADPH is in excess, the latter portion of the pentose phosphate pathway may be reversed to synthesize R5P from glycolytic intermediates. The pentose phosphate pathway is controlled at its first committed step, the glucose-6-phosphate dehydrogenase reaction, by the $NADP^+$ concentration. A genetic deficiency in glucose-6-phosphate dehydrogenase leads to hemolytic anemia on administration of the antimalarial drug primaquine. This deficiency, which results from the accelerated degradation of the mutant enzyme, provides resistance against malaria to females heterozygotic for this trait.

REFERENCES

Gluconeogenesis

Hers, H.G. and Hue, L., Gluconeogenesis and related aspects of glycolysis, *Annu. Rev. Biochem.* **52**, 617–653 (1983).

Hue, L., The role of futile cycles in the regulation of carbohydrate metabolism in the liver, *Adv. Enzymol.* **52**, 247–330 (1981).

Knowles, J.R., The mechanism of biotin-dependent enzymes, *Annu. Rev. Biochem.* **58**, 195–221 (1989).

Krauss-Friedman, N., *Hormonal Control of Gluconeogenesis*, Vols. I and II, CRC Press (1986).

Pilkis, S.J., Mahgrabi, M.R., and Claus, T.H., Hormonal regulation of hepatic gluconeogenesis and glycolysis, *Annu. Rev. Biochem.* **57**, 755–783 (1988).

Rothman, D.L., Magnusson, I., Katz, L.D., Shulman, R.G., and Shulman, G.I., Quantitation of hepatic gluconeogenesis in fasting humans with ¹³C NMR, *Science* **254**, 573–576 (1991).

The Glyoxylate Pathway

Tolbert, N.E., Metabolic pathways in peroxisomes and glyoxysomes, *Annu. Rev. Biochem.* **50**, 133–157 (1981).

Oligosaccharide Biosynthesis

Abeijon, C. and Hirschberg, C.B., Topography of glycosylation reactions in the endoplasmic reticulum, *Trends Biochem. Sci.* **17**, 32–36 (1992).

Elbein, A.D., Inhibitors of the biosynthesis and processing of N-linked oligosaccharide chains, *Annu. Rev. Biochem.* **56**, 497–534 (1987).

Elbein, A.D., The role of lipid-linked saccharides in the biosynthesis of complex carbohydrates, *Annu. Rev. Plant Physiol.* **30**, 239–272 (1979).

Englund, P.T., The structure and biosynthesis of glycosyl phosphatidylinositol protein anchors, *Annu. Rev. Biochem.* **62**, 65–100 (1993).

Ferguson, M.A.J., Glycosyl-phosphatidylinositol membrane anchors: The tale of a tail, *Biochem. Soc. Trans.* **20**, 243–256 (1992).

Florman, H.M. and Wasserman, P.M., *O*-Linked oligosaccharides of mouse egg ZP3 account for its sperm receptor activity, *Cell* **41**, 313–324 (1985).

Hauri, H.-P. and Schweizer, A., The endoplasmic reticulum–Golgi intermediate compartment, *Curr. Opin. Cell Biol.* **4**, 600–608 (1992).

Hirschberg, C.B. and Snider, M.D., Topography of glycosylation in the rough endoplasmic reticulum and the Golgi apparatus, *Annu. Rev. Biochem.* **56**, 63–87 (1987).

Keller, R.K., Boon, D.Y., and Crum, F.C., *N*-Acetylglucosamine-1-phosphate transferase from hen oviduct: solubilization, characterization and inhibition by tunicamycin, *Biochemistry* **18**, 3946–3952 (1979).

Kornfeld, R. and Kornfeld, S., Assembly of asparagine-linked oligosaccharides, *Annu. Rev. Biochem.* **54**, 631–664 (1985).

Lodish, H.F., Transport of secretory and membrane glycoproteins from the rough endoplasmic reticulum to the Golgi, *J. Biol. Chem.* **263**, 2107–2110 (1988).

Mellman, I. and Simons, K., The Golgi complex: In vitro veritas, *Cell* **68**, 829–840 (1992).

Parodi, A.J. and Leloir, L.F., The role of lipid intermediates in the glycosylation of proteins in the eucaryotic cell, *Biochim. Biophys. Acta* **559**, 1–37 (1979).

Pfeffer, S.R. and Rothman, J.E., Biosynthetic protein transport and sorting by the endoplasmic reticulum and Golgi, *Annu. Rev. Biochem.* **56**, 829–852 (1987).

Presper, K.A. and Heath, E.C., Glycosylated lipid intermediates involved in glycoprotein biosynthesis, *in* Boyer, P.D. (Ed.), *The Enzymes* (3rd ed.), Vol. 16, pp. 449–488, Academic Press (1983).

Rothman, J.E., The compartmental organization of the Golgi apparatus, *Sci. Am.* **253**(3): 74–89 (1985).

Schachter, H., Enzymes associated with glycosylation, *Curr. Opin. Struct. Biol.* **1**, 755–765 (1991).

Schwartz, R.T. and Datema, R., Inhibitors of trimming: new tools in glycoprotein research, *Trends Biochem. Sci.* **9**, 32–34 (1984).

Shaper, J.H. and Shaper, N.L., Enzymes associated with glycosylation, *Curr. Opin. Struct. Biol.* **2**, 701–709 (1992).

Tartakoff, A.M. and Singh, N., How to make a glycoinositol phospholipid anchor, *Trends Biochem. Sci.* **17**, 470–473 (1992).

von Figura, K. and Hasilik, A., Lysosomal enzymes and their receptors, *Annu. Rev. Biochem.* **55**, 167–193 (1986).

The Pentose Phosphate Pathway

Adams, M.J., Ellis, G.H., Gover, S., Naylor, C.E., and Phillips, C., Crystallographic study of coenzyme, coenzyme analogue and substrate binding in 6-phosphogluconate dehydrogenase: implications for NADP specificity and enzyme mechanism, *Structure* **2**, 651–668 (1994).

Beutler, E., The molecular biology of G6PD variants and other red cell enzyme defects, *Annu. Rev. Med.* **43**, 47–59 (1992).

Lindqvist, Y. and Schneider, G. Thiamin diphosphate dependent enzymes: transketolase, pyruvate oxidase and pyruvate decarboxylase, *Curr. Opin. Struct. Biol.* **3**, 896–901 (1993); *and* Muller, Y.A., Lindqvist, Y., Furey, W., Schulz, G.E., Jordan, F., and Schneider, G., A thiamin diphosphate binding fold revealed by comparison of the crystal structures of transketolase, pyruvate oxidase and pyruvate decarboxylase, *Structure.* **1**, 95–103 (1993).

Luzzato, L. and Mehta, A., Glucose-6-phosphate dehydrogenase deficiency, *in* Scriver, C.R., Beudet, A., Sly, W.S., and Valle, D. (Eds.), *The Metabolic Basis of Inherited Disease* (6th ed.), *pp.* 2237–2265, McGraw-Hill (1989).

Vulliamy, T., Mason, P., and Luzzatto, L., The molecular basis of glucose-6-phosphate dehydrogenase deficiency, *Trends Genet.* **8**, 138–143 (1992).

Wood, T., *The Pentose Phosphate Pathway,* Academic Press (1985).

PROBLEMS

1. Compare the relative energetic efficiencies, in ATPs per mole of glucose oxidized, of glucose oxidation via glycolysis + the citric acid cycle vs glucose oxidation via the pentose phosphate pathway + gluconeogenesis. Assume that NADH and NADPH are each energetically equivalent to three ATP's.

2. Although animals cannot synthesize glucose from acetyl-CoA, if a rat is fed ^{14}C-labeled acetate, some of the label will appear in the glycogen extracted from its muscles. Explain.

3. Substances that inhibit specific trimming steps in the processing of *N*-linked glycoproteins have been useful tools in elucidating the pathway of this process. Explain.

4. Through clever genetic engineering you have developed an unregulatable enzyme that can interchangably use NAD^+ or $NADP^+$ in a redox reaction. What would be the physiological consequence(s) on an organism of having such an enzyme?

5. What is the free energy change of the reaction

$$NADH + NADP^+ \rightleftharpoons NAD^+ + NADPH$$

under physiological conditions? Assume that $\Delta G^{\circ\prime} = 0$ for this reaction and that $T = 37°C$.

6. If G6P is ^{14}C-labeled at its C2 position, what is the distribution of the radioactive label in the products of the pentose phosphate pathway after one turnover of the pathway? What is the distribution of the label after passage of these products through gluconeogenesis followed by a second round of the pentose phosphate pathway?

7. The relative metabolic activities in an organism of glycolysis + the citric acid cycle vs the pentose phosphate pathway + gluconeogenesis can be measured by comparing the rates of $^{14}CO_2$ generation upon administration of glucose labeled with ^{14}C at C1 with that of glucose labeled at C6. Explain.

8. In light of the finding that an otherwise benign or even advantageous mutation leads to abnormal primaquine sensitivity combined with the fact that human beings have enormous genetic complexity, comment on the possibility of developing drugs that exhibit no atypical side effects in any individual.

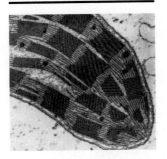

C H A P T E R

22

Photosynthesis

Life on Earth depends on the sun. *Plants and cyanobacteria chemically sequester light energy through photosynthesis, a light-driven process in which CO_2 is "fixed" to yield carbohydrates (CH_2O).*

$$CO_2 + H_2O \xrightarrow{\text{light}} (CH_2O) + O_2$$

This process, in which both CO_2 and H_2O are reduced to yield carbohydrate and O_2, is essentially the reverse of oxidative carbohydrate metabolism. Photosynthetically produced carbohydrates therefore serve as an energy source for the organism that produced them as well as for nonphotosynthetic organisms that directly or indirectly consume photosynthetic organisms. In fact, even modern industry is highly dependent on the products of photosynthesis because coal, oil, and gas (the so-called fossil fuels) are thought to be the remains of ancient organisms. It is estimated that photosynthesis annually fixes $\sim 10^{11}$ tons of carbon, which represents the storage of over 10^{18} kJ of energy. Moreover,

photosynthesis, over the eons, has produced the O_2 in the Earth's atmosphere.

The notion that plants obtain nourishment from such insubstantial things as light and air took nearly two centuries to evolve. In 1648, the Flemish physician Jean-Baptiste von Helmont reported that growing a potted willow tree from a shoot caused an insignificant change in the weight of the soil in which the tree had been rooted. Although another century was to pass before the law of conservation of matter was formulated, van Helmont attributed the tree's weight gain to the water it had taken up. This idea was extended in 1727 by Stephen Hales who proposed that plants extract some of their matter from the air.

The first indication that plants produce oxygen was found by the English clergyman and pioneering chemist Joseph Priestley who reported:

> *Finding that candles burn very well in air in which plants had grown a long time, and having some reason to think, that there was something attending vegetation, which restored air that had been injured by respiration, I thought it was possible that the same process might also restore the air that had been injured by the burning of candles. Accordingly, on the 17th of August, 1771, I put a sprig of mint into a quantity of air, in which a wax candle had burned out, and found that, on the 27th of the same month, another candle burned perfectly well in it.*

Although Priestley later discovered oxygen, which he named "dephlogisticated air," it was Antoine Lavoisier who elucidated its role in combustion and respiration. Nevertheless, Priestley's work inspired the Dutch physician Jan Ingen-Housz, who in 1779 demonstrated that the "purify-

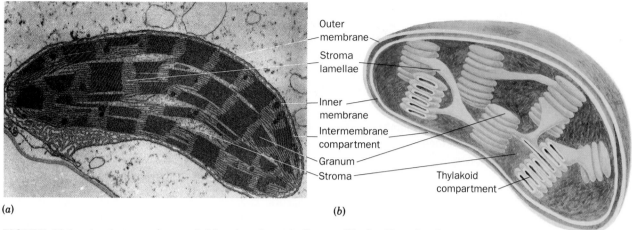

FIGURE 22-1. An electron micrograph (*a*) and a schematic diagram (*b*) of a chloroplast from corn. [Electron micrograph courtesy of Lester Shumway, College of Eastern Utah.]

ing" power of plants resides in the influence of sunlight upon their green parts. In 1782, the Swiss pastor Jean Senebier showed that CO_2, which he called "fixed air," is taken up during photosynthesis. His compatriot, Théodore de Saussure found, in 1804, that the combined weights of the organic matter produced by plants and the oxygen they evolve is greater than the weight of the CO_2 they consume. He therefore concluded that water, the only other substance he added to his system, was also necessary for photosynthesis. The final ingredient in the overall photosynthetic recipe was established in 1842 by the German physiologist Robert Mayer, one of the formulators of the first law of thermodynamics, who concluded that plants convert light energy to chemical energy.

1. CHLOROPLASTS

*The site of photosynthesis in eukaryotes (algae and higher plants) is the **chloroplast** (Section 1-2A), a member of the membranous subcellular organelles peculiar to plants known as **plastids**.* The first indication that chloroplasts have a photosynthetic function was Theodor Englemann's observation, in 1882, that small, motile, O_2-seeking bacteria congregate at the surface of the alga *Spirogyra,* overlying its single chloroplast, but only while the chloroplast is illuminated. Chloroplasts must therefore be the site of light-induced O_2 evolution, that is, photosynthesis. Chloroplasts, of which there are 1 to 1000 per cell, vary considerably in size and shape but are typically ~5-μm long ellipsoids. Like mitochondria, which they resemble in many ways, chloroplasts have a highly permeable outer membrane and a nearly impermeable inner membrane separated by a narrow intermembrane space (Fig. 22-1). The inner membrane encloses the **stroma,** a concentrated solution of enzymes that also contains the DNA, RNA, and ribosomes involved in the synthesis of several chloroplast proteins—

much like the mitochondrial matrix. The stroma, in turn, surrounds a third membranous compartment, the **thylakoid** (Greek: *thylakos,* a sac or pouch). The thylakoid is probably a single highly folded vesicle, although in most organisms it appears to consist of stacks of disklike sacs named **grana,** which are interconnected by unstacked **stroma lamellae.** A chloroplast usually contains 10 to 100 grana. Thylakoid membranes arise from invaginations in the inner membrane of developing chloroplasts and therefore resemble mitochondrial cristae.

The lipids of the thylakoid membrane have a distinctive composition. They consist of only ~10% phospholipids; the majority, ~80%, are uncharged **mono-** and **digalactosyl diacylglycerols,** and the remaining ~10% are the sulfolipids **sulfoquinovosyl diacylglycerols.**

$$X = OH \qquad \textbf{Galactosyl diacylglycerol}$$

$$X = \qquad \textbf{Digalactosyl diacylglycerol}$$

$$X = SO_3^- \qquad \textbf{Sulfoquinovosyl diacylglycerol}$$

The acyl chains of these lipids have a high degree of unsaturation, which gives the thylakoid membrane a highly fluid character.

Photosynthesis occurs in two distinct phases:

1. The **light reactions,** which use light energy to generate NADPH and ATP.

2. The **dark reactions,** actually light-independent reactions, which use NADPH and ATP to drive the synthesis of carbohydrate from CO_2 and H_2O.

The light reactions occur in the thylakoid membrane and involve processes that resemble mitochondrial electron transport and oxidative phosphorylation (Sections 20-2 and 3). In photosynthetic prokaryotes, which lack chloroplasts, the light reactions take place in the cell's plasma (inner) membrane or in highly invaginated structures derived from it called **chromatophores** (e.g., Fig. 22-2; recall that chloroplasts evolved from cyanobacteria that assumed a symbiotic relationship with a nonphotosynthetic eukaryote; Section 1-2A). In eukaryotes, the dark reactions occur in the stroma through a cyclic series of enzyme-catalyzed reactions. In the following sections, we consider the light and dark reactions in detail.

2. LIGHT REACTIONS

In the first decades of this century, it was generally assumed that light, as absorbed by photosynthetic pigments, directly reduced CO_2 which, in turn, combined with water to form carbohydrate. In this view, CO_2 is the source of the O_2 generated by photosynthesis. In 1931, however, Cornelis van Niel showed that green photosynthetic bacteria, anaerobes that use H_2S in photosynthesis, generate sulfur:

$$CO_2 + 2H_2S \xrightarrow{\text{light}} (CH_2O) + 2S + H_2O$$

The chemical similarity between H_2S and H_2O led van Niel to propose that the general photosynthetic reaction is

$$CO_2 + 2H_2A \xrightarrow{\text{light}} (CH_2O) + 2A + H_2O$$

where H_2A is H_2O in green plants and cyanobacteria and H_2S in photosynthetic sulfur bacteria. This suggests that photosynthesis is a two-stage process in which light energy is harnessed to oxidize H_2A (the light reactions):

$$2H_2A \xrightarrow{\text{light}} 2A + 4[H]$$

and the resulting reducing agent [H] subsequently reduces CO_2 (the dark reactions):

$$4[H] + CO_2 \longrightarrow (CH_2O) + H_2O$$

Thus, in aerobic photosynthesis, H_2O, not CO_2, is photolyzed (split by light).

The validity of van Niel's hypothesis was established unequivocally by two experiments. In 1937, Robert Hill discovered that when isolated chloroplasts that lack CO_2 are illuminated in the presence of an artificial electron acceptor such as ferricyanide [$Fe(CN)_6^{3-}$], O_2 is evolved with con-

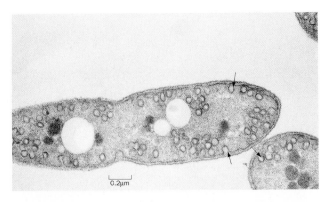

FIGURE 22-2. An electron micrograph of a section through the purple photosynthetic bacterium *Rhodobacter sphaeroides.* Its plasma membrane invaginates to form externally connected tubules known as chromatophores (*arrows;* seen here in circular cross-section) that are the sites of photosynthesis. [Courtesy of Gerald A. Peters, Virginia Commonwealth University.]

comitant reduction of the acceptor [to ferrocyanide, $Fe(CN)_6^{2-}$, in our example]. This so-called **Hill reaction** demonstrates that CO_2 does not participate directly in the O_2-producing reaction. It was discovered eventually that the natural photosynthetic electron acceptor is $NADP^+$ (Section 22-2C) whose reduction product, NADPH, is utilized in the dark reactions to reduce CO_2 to carbohydrate (Section 22-3A). In 1941, when the oxygen isotope ^{18}O became available, Samuel Ruben and Martin Kamen directly demonstrated that the source of the O_2 formed in photosynthesis is H_2O:

$$H_2{}^{18}O + CO_2 \xrightarrow{\text{light}} (CH_2O) + {}^{18}O_2$$

This section is a discussion of the major aspects of the light reactions.

A. Absorption of Light

The principal photoreceptor in photosynthesis is **chlorophyll.** This cyclic tetrapyrrole, like the heme group of globins and cytochromes (Sections 9-1A and 20-2C), is derived biosynthetically from protoporphyrin IX. Chlorophyll, however, differs from heme in four major respects (Fig. 22-3):

1. Its central metal ion is Mg^{2+} rather than Fe(II) or Fe(III).

2. It has a cyclopentanone ring, Ring V, fused to pyrrole Ring III.

3. Pyrrole Ring IV is partially reduced in **chlorophyll *a*** (**Chl *a***) and **chlorophyll *b* (Chl *b*),** the two major chlorophyll varieties in eukaryotes and cyanobacteria, whereas in **bacteriochlorophyll *a* (BChl *a*)** and **bacteriochlorophyll *b* (BChl *b*),** the principal chlorophylls of photosynthetic bacteria, Rings II and IV, are partially reduced.

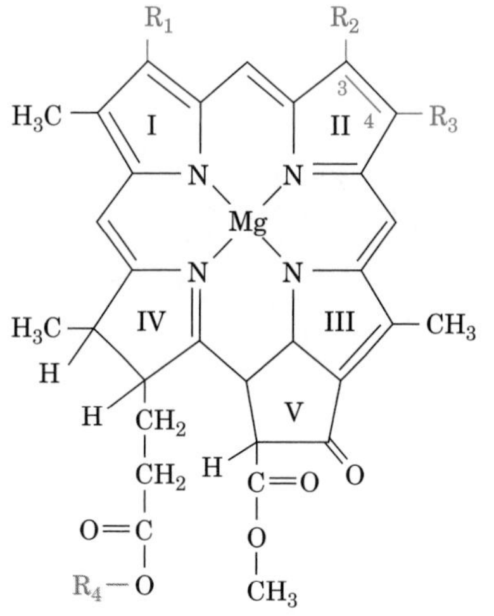

Chlorophyll 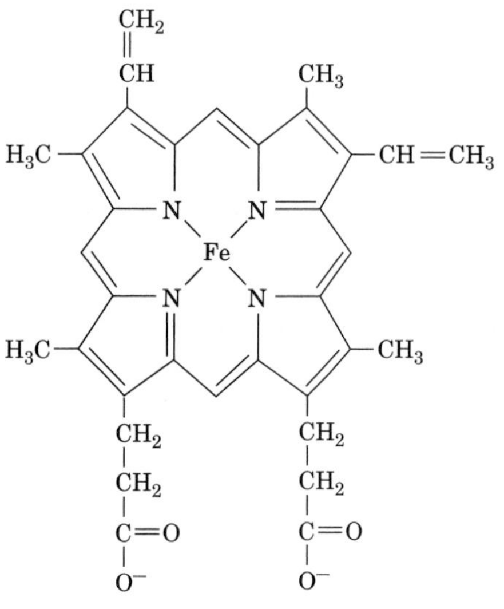   **Iron–protoporphyrin IX**

	R_1	R_2	R_3	R_4
Chlorophyll a	$-CH=CH_2$	$-CH_3$	$-CH_2-CH_3$	P
Chlorophyll b	$-CH=CH_2$	$\overset{\overset{O}{\|\|}}{-C}-H$	$-CH_2-CH_3$	P
Bacteriochlorophyll a	$\overset{\overset{O}{\|\|}}{-C}-CH_3$	$-CH_3{}^a$	$-CH_2-CH_3{}^a$	P or G
Bacteriochlorophyll b	$\overset{\overset{O}{\|\|}}{-C}-CH_3$	$-CH_3{}^a$	$=CH-CH_3{}^a$	P

^{a} No double bond between positions C3 and C4.

$$P = -CH_2$$

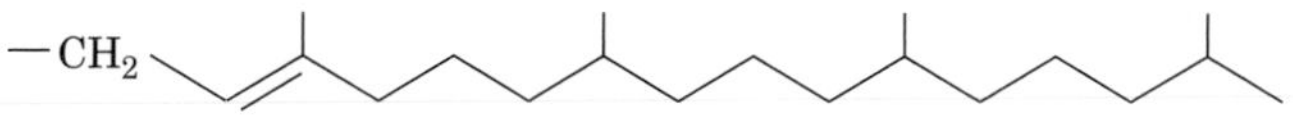

Phytyl side chain

$$G = -CH_2$$

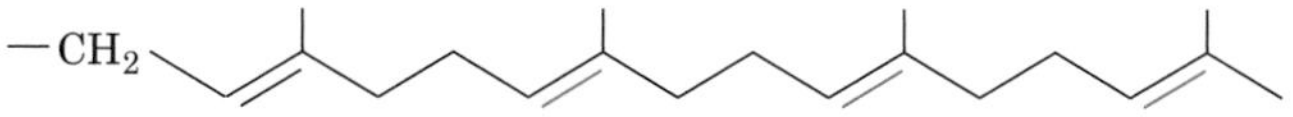

Geranylgeranyl side chain

FIGURE 22-3. The molecular formulas of chlorophylls a and b and bacteriochlorophylls a and b compared to that of iron protoporphyrin IX (heme). The isoprenoid phytyl and geranylgeranyl tails presumably increase the chlorophylls' solublity in nonpolar media.

4. The propionyl side chain of Ring IV is esterified to a tetraisoprenoid alcohol. In Chls a and b as well as in BChl b it is **phytol** but in BChl a it is either phytol or **geranylgeraniol** depending on the bacterial species.

Chl b has a formyl group in place of the methyl substitutent to Ring II of Chl a. Similarly, BChl a and BChl b have different substituents to atom C4.

Light and Matter Interact in Complex Ways

   As photosynthesis is a light-driven process, it is worth-while reviewing how light and matter interact. Electromag-

netic radiation is propagated as discrete **quanta (photons)** whose energy E is given by **Plank's law:**

$$E = h\nu = \frac{hc}{\lambda} \qquad [22.1]$$

where h is **Planck's constant** (6.626×10^{-34} J·s), c is the speed of light (2.998×10^8 m·s^{-1} in a vacuum), ν is the frequency of the radiation, and λ is its wavelength (visible light ranges in wavelength from 400 to 700 nm). Thus red light with $\lambda = 700$ nm has an energy of 171 kJ·einstein^{-1} (an **einstein** is a mole of photons).

Molecules, like atoms, have numerous electronic quantum states of differing energies. Moreover, because molecules contain more than one nucleus, each of their electronic states has an associated series of vibrational and rotational substates that are closely spaced in energy (Fig. 22-4). Absorption of light by a molecule usually occurs through the promotion of an electron from its ground (lowest energy) state molecular orbital to one of higher energy. However, *a given molecule can only absorb photons of certain wavelengths because, as is required by the law of conservation of energy, the energy difference between the two states must exactly match the energy of the absorbed photon.*

The amount of light absorbed by a substance at a given wavelength is described by the **Beer–Lambert** law:

$$A = \log \frac{I_0}{I} = \varepsilon c l \qquad [22.2]$$

where A is the absorbance, I_0 and I are, respectively, the intensities of incident and transmitted light, c is the molar concentration of the sample, ℓ is the length of the light path through the sample in cm, and ε is the molecule's **molar extinction coefficient.** Consequently, a plot of A versus λ for a given molecule, its **absorption spectrum** (Fig. 22-5), is indicative of its electronic structure.

The various chlorophylls are highly conjugated molecules (Fig. 22-3). It is just such molecules that strongly absorb visible light (the spectral band in which the solar radiation reaching the Earth's surface is of peak intensity; Fig. 22-5). In fact, the peak molar extinction coefficients of the various chlorophylls, over $10^5 \ M^{-1} \cdot cm^{-1}$, are among the highest known for organic molecules. Yet, the relatively

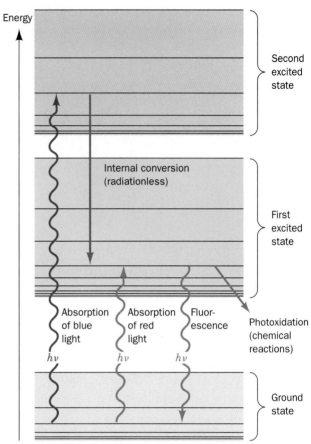

FIGURE 22-4. An energy diagram schematically indicating the electronic states of chlorophyll and their most important modes of interconversion. The wiggly arrows represent the absorption of photons or their fluorescent emission. Excitation energy may also be dissipated in radiationless processes such as internal conversion (heat production) or chemical reactions.

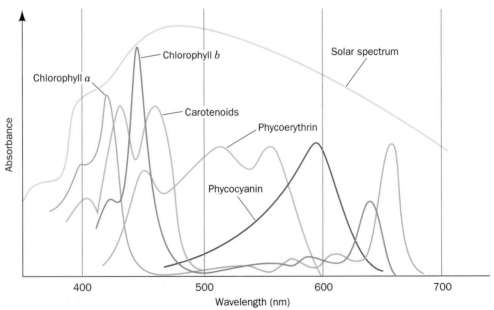

FIGURE 22-5. The absorption spectra of various photosynthetic pigments. The chlorophylls have two absorption bands, one in the red and one in the blue. Phycoerythrin absorbs blue and green light, whereas phycocyanin absorbs yellow light. Together, these pigments absorb most of the visible light in the solar spectrum.

small chemical differences among the various chlorophylls greatly affect their absorption spectra. These spectral differences, as we shall see, are functionally significant.

An electronically excited molecule can dissipate its excitation energy in many ways. Those modes with the greatest photosynthetic significance are (Fig. 22-4):

1. **Internal conversion,** a common mode of decay in which electronic energy is converted to the kinetic energy of molecular motion, that is, to heat. This process occurs very rapidly, being complete in $<10^{-11}$ s. Many molecules relax in this manner to their ground states. Chlorophyll molecules, however, usually relax only to their lowest excited states. Therefore, *the photosynthetically applicable excitation energy of a chlorophyll molecule that has absorbed a photon in its short-wavelength band, which corresponds to its second excited state, is no different than if it had absorbed a photon in its less energetic long-wavelength band.*

2. **Fluorescence,** in which an electronically excited molecule decays to its ground state by emitting a photon. Such a process requires $\sim10^{-8}$ s so it occurs much more slowly than internal conversion. Consequently, a fluorescently emitted photon generally has a longer wavelength (lower energy) than that initially absorbed. Fluorescence accounts for the dissipation of only 3 to 6% of the light energy absorbed by living plants. However, chlorophyll in solution, where of course the photosynthetic uptake of this energy cannot occur, has an intense red fluorescence.

3. **Exciton transfer** (also known as **resonance energy transfer**), in which an excited molecule directly transfers its excitation energy to nearby unexcited molecules with similar electronic properties. This process occurs through interactions between the molecular orbitals of the participating molecules in a manner analogous to the interactions between mechanically coupled pendulums of similar frequencies. An exciton (excitation) may be serially transferred between members of a group of molecules or, if their electronic coupling is strong enough, the entire group may act as a single excited "supermolecule." We shall see that *exciton transfer is of particular importance in funneling light energy to photosynthetic reaction centers.*

4. **Photooxidation,** in which a light-excited donor molecule is oxidized by transferring an electron to an acceptor molecule, which is thereby reduced. This process occurs because the transferred electron is less tightly bound to the donor in its excited state than it is to the ground state. In photosynthesis, excited chlorophyll (Chl*) is such a donor. *The energy of the absorbed photon is thereby chemically transferred to the photosynthetic reaction system.* Photooxidized chlorophyll, Chl$^+$, a cationic free radical, eventually returns to its ground state by oxidizing some other molecule.

Light Absorbed by Antenna Chlorophylls and Accesory Pigments Is Transferred to Photosynthetic Reaction Centers

The primary reactions of photosynthesis, as is explained in Sections 22-2B and C, take place at **photosynthetic reaction centers.** Yet, *photosynthetic organelles contain far more chlorophyll molecules than reaction centers.* This was demonstrated in 1932 by Robert Emerson and William Arnold in their studies of O_2 production by the green alga *Chlorella* (a favorite experimental subject), which had been exposed to repeated brief (10-μs) flashes of light. The amount of O_2 generated per flash was maximal when the interval between flashes was at least 20 ms. Evidently, this is the time required for a single turnover of the photosynthetic reaction cycle. Emerson and Arnold then measured the variation of O_2 yield with flash intensity when the flash interval was the optimal 20 ms. With weak flashes, the O_2 increased linearly with flash intensity such that about one molecule of O_2 was generated per eight photons absorbed (Fig. 22-6). With increasing flash intensity the efficiency of this process fell off, no doubt because the number of photons began to approach the number of photochemical units. What was unanticipated, however, was that each flash of saturating intensity produced only one molecule of O_2 per ~2400 molecules of chlorophyll present. Since at least eight photons must be sequentially absorbed to liberate one O_2 molecule (Section 22-2C), these results suggest that the photosynthetic apparatus contains $\sim2400/8 = 300$ chlorophyll molecules per photosynthetic reaction center.

With such a great excess of chlorophyll molecules per reaction center, it seems unlikely that all participate directly in photochemical reactions. Rather, as subsequent experiments have shown, *most chlorophylls function to gather light; that is, they act as light-harvesting antennas.* These **antenna chlorophylls** pass the energy of an absorbed photon, by exciton transfer, from molecule to molecule until

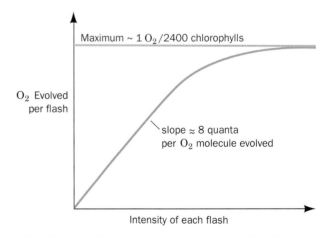

FIGURE 22-6. The amount of O_2 evolved by *Chlorella* algae versus the intensity of light flashes that are separated by dark intervals >20 ms.

the excitation reaches a photosynthetic reaction center (Fig. 22-7a). There, the excitation is trapped because reaction center chlorophylls, although chemically identical to antenna chlorophylls, have slightly lower excited state energies because of their different environments (Fig. 22-7b).

Transfer of energy from the antenna system to a reaction center occurs in $<10^{-10}$ s with an efficiency of $>90\%$. This high efficiency depends on the chlorophyll molecules having appropriate spacings and relative orientations. Even in bright sunlight, a reaction center intercepts only ~1 photon per second, a metabolically insignificant rate, and hence, these **light-harvesting complexes (LHCs)** serve an essential function.

Although the structures of the LHCs are not well understood, it appears that they consist of arrays of membrane-bound hydrophobic proteins that each contain several pigment molecules. For example, an antenna protein from the green photosynthetic bacterium *Prosthecochloris aestuarii,* whose X-ray structure was determined by Brian Matthews (Fig. 22-8a), is a trimer of identical 357-residue subunits, each of which contains 7 BChl *a* molecules and largely consists of 15 strands of β sheet wrapped around a chlorophyll-containing core. This protein therefore has been described as a "string bag" for holding pigment molecules. The porphyrin rings are not in van der Waals contact and exhibit a complex pattern of relative orientations, an arrangement that presumably optimizes the efficiency of exciton transfer throughout the LHC.

Most LHCs contain organized arrays of other light-absorbing substances besides chlorophyll. These **accessory pigments** function to fill in the absorption spectra of the antenna complexes in spectral regions where chlorophylls do not absorb strongly (Fig. 22-5). **Carotenoids,** which are linear polyenes such as β-**carotene,**

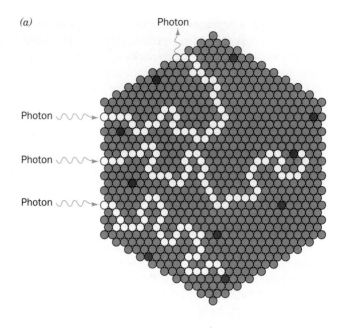

(a)

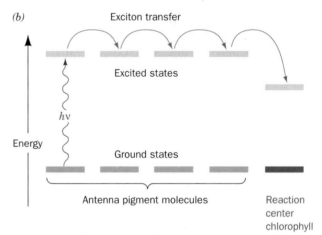

(b)

FIGURE 22-7. The flow of energy through a photosynthetic antenna complex. (*a*) The excitation resulting from photon absorption randomly migrates by exciton transfer among the molecules of the antenna complex (*light green circles*) until it is either trapped by a reaction center chlorophyll (*dark green circles*) or, less frequently, fluorescently reemitted. (*b*) The excitation is trapped by the reaction center chlorophyll because its lowest excited state has a lower energy than those of the antenna pigment molecules.

β-Carotene

are components of all green plants and many photosynthetic bacteria and are therefore the most common accessory pigments (they are largely responsible for the brilliant fall colors of deciduous trees as well as for the orange color of carrots, after which they are named). For instance, **LHC-II,** the most abundant membrane protein in the chloroplasts of green plants, is a 232-residue transmembrane protein that binds at least 7 Chl *a*'s, 5 Chl *b*'s and 2 carotenoids (Fig. 22-8b), thereby accounting for around half the chlorophyll in the biosphere. LHC-II's carotenoids serve an additional function besides that of light-gathering antennas: Through electronic interactions, they prevent their associated light-excited chlorophyll molecules from transferring this excitation to O_2, which would otherwise yield a highly reactive and hence destructive form of O_2.

Water-dwelling photosynthetic organisms, which are responsible for nearly half of the photosynthesis on Earth,

(a)

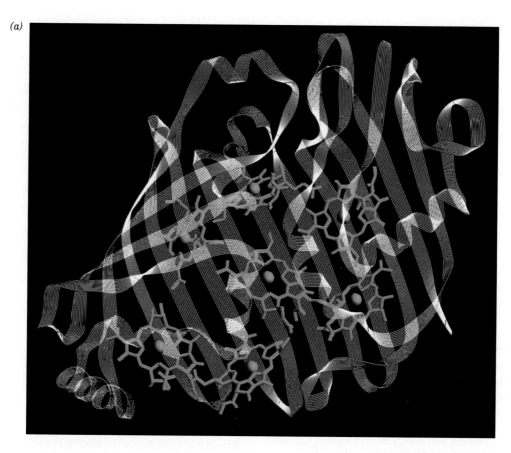

(b)

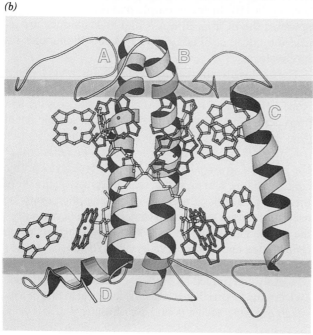

FIGURE 22-8. The structures of two antenna proteins. (a) The X-ray structure of a subunit of the homotrimeric bacterio-chlorophyll *a* protein from *P. aestuarii*. The polypeptide backbone of one subunit (*white ribbon*) consists largely of a 15-stranded antiparallel β sheet. Its seven bound BChl *a* molecules are shown in green with their Mg^{2+} ions drawn as spheres and their phytyl side chains omitted for clarity. [Based on an X-ray structure by Brian Matthews, University of Oregon.] (b) The structure of a subunit of the trimeric protein LHC-II from pea chloroplasts as determined by the electron crystallography of two-dimensional crystals. This highly conserved transmembrane protein's seven Chl *a*, five Chl *b*, and two carotenoid molecules are dark green, light green, and yellow, respectively, and its chlorophyll-bound Mg^{2+} ions are pink. The blue bands indicate the approximate boundaries of the thylakoid membrane in which the protein is normally embedded with its upper side, as shown, facing the stroma. Note that much of the protein, including its A and B helices and most of its pigment molecules, exhibits approximate two-fold symmetry with the pseudo-two-fold axis perpendicular to the plane of the membrane. [Courtesy of Werner Kühlbrandt, European Molecular Biology Laboratory, Heidelberg, Germany.]

additionally contain other types of accessory pigments. This is because light outside the wavelengths 450 to 550 nm (blue and green light) is absorbed almost completely by passage through more than 10 m of water. In red algae and cyanobacteria, Chl *a* therefore is replaced as an antenna pigment by a series of linear tetrapyrroles, notably the red

phycoerythrobilin and the blue **phycocyanobilin** (spectra in Fig. 22-5).

Phycoerythrobilin

The lowest excited states of these various **bilins** have higher energies than those of the chlorophylls, thereby facilitating energy transfer to the photosynthetic reaction centers. The bilins are covalently linked to **phycobiliproteins** which are, in turn, arranged in organized high molecular mass particles called **phycobilisomes**. The phycobilisomes are bound to the outer faces of photosynthetic membranes so as to funnel excitation energy to reaction centers over long distances with >90% efficiency.

B. Electron Transport in Photosynthetic Bacteria

Photosynthesis is a process in which electrons from excited chlorophyll molecules are passed through a series of acceptors that convert electronic energy to chemical energy. Thus two questions arise: (1) What is the mechanism of energy transduction; and (2) How do photooxidized chlorophyll molecules regain their lost electrons? We shall see that photosynthetic bacteria solve these problems somewhat differently from cyanobacteria and plants. We first discuss these mechanisms in photosynthetic bacteria, where they are simpler and better understood. Electron transport in cyanobacteria and plants is the subject of Section 22-2C.

The Photosynthetic Reaction Center Is a Transmembrane Protein Containing a Variety of Chromophores

The first indication that chlorophyll undergoes direct photooxidation during photosynthesis was obtained by Louis Duysens in 1952. He observed that illumination of membrane preparations from the **purple photosynthetic bacterium** *Rhodospirillum (Rs.) rubrum* caused a slight (~2%) bleaching of their absorbance at 870 nm, which returned to their original levels in the dark. Duysens suggested that this bleaching is caused by photooxidation of a bacteriochlorophyll complex that he named **P870** (P for pigment and 870 nm is the position of the major longwave absorption band of BChl *a*; photosynthetic bacteria tend to inhabit murky stagnant ponds so that they require a near

infrared–absorbing species of chlorophyll). The ability to detect the presence of P870 eventually led to the purification and characterization of the photosynthetic reaction centers to which it is bound.

Reaction center particles from several species of purple photosynthetic bacteria have similar compositions. That from *Rhodopseudomonas (Rps.) viridis* consists of three hydrophobic subunits: H (258 residues), L (273 residues), and M (323 residues). The L and M subunits of this membrane-spanning particle collectively bind four molecules of BChl *b* (which maximally absorbs light at 960 nm), two molecules of **bacteriopheophytin *b*** (BPheo *b*; BChl *b* in which the Mg^{2+} is replaced by two protons), one nonheme/non-Fe–S Fe(II) ion, one molecule of the redox coenzyme ubiquinone (Section 20-2B), and one molecule of the related **menaquinone**

Menaquinone

(**vitamin K_2**, a substance required for proper blood clotting; Section 34-1B). In many purple photosynthetic bacteria, however, the BChl *b*, BPheo *b*, and menaquinone are replaced by BChl *a*, BPheo *a*, and a second ubiquinone, respectively.

The photosynthetic reaction center of *Rps. viridis*, whose X-ray structure was determined by Johann Deisenhofer, Robert Huber, and Hartmut Michel in 1984, was the first transmembrane protein to be described in atomic detail (Fig. 11-27). *The protein's membrane-spanning portion consists of 11 α helices that form a 45-Å-long cylinder with the expected hydrophobic surface.* A *c*-type cytochrome containing four hemes, which is an integral constituent of the reaction center complex in only some photosynthetic bacteria, binds to the reaction center on the external side of the plasma membrane. In fact, the photosynthetic reaction center from another bacterial species, *Rhodobacter (Rb.) sphaeroides*, whose X-ray structure (Figure 22-9) was independently determined by Marianne Schiffer and by Douglas Rees and George Feher, is nearly identical to that of *Rps. viridis* but lacks such a bound cytochrome.

Two BChl *b* Molecules Form a "Special Pair"

The most striking aspect of the reaction center is that its chromophoric prosthetic groups are arranged with nearly perfect twofold symmetry (Fig. 22-10*a*). This symmetry arises because the L and M subunits, with which these prosthetic groups are exclusively associated, have homologous sequences and similar folds. Two of the BChl *b* molecules,

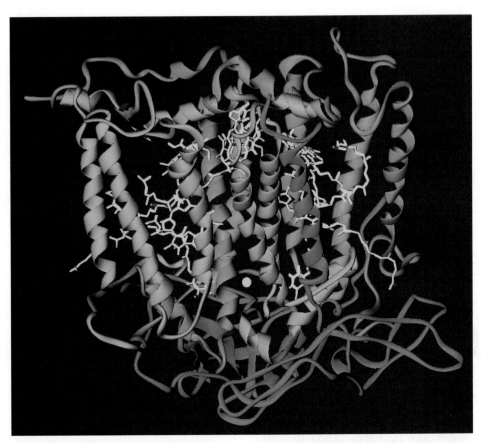

FIGURE 22-9. A ribbon diagram of the photosynthetic reaction center from *Rb. sphaeroides* as viewed from within the plane of the plasma membrane such that the cytoplasm is below. The H, M, and L subunits (281, 307, and 260 residues) are pink, blue, and orange, respectively. The prosthetic groups are yellow and are shown in skeletal form with the exception of the Fe(II) atom, which is represented by a sphere. The 11 largely vertical helices that form the central portion of the protein constitute its transmembrane region. Compare this structure with that of the photosynthetic reaction center from *Rps. viridis* (Fig. 11-27), whose H, M, and L subunits are 39, 50, and 59% identical to those of *Rb. sphaeroides*. Note that the *Rb. sphaeroides* protein lacks the four-heme *c*-type cytochrome (*green* in Fig. 11-27) on its periplasmic surface and that the Q_A prosthetic group, whose quinone ring lies to the right of the Fe(II), is ubiquinone in *Rb. sphaeroides* but menaquinone in *Rps. viridis*. [Based on an X-ray structure by Marianne Schiffer, Argonne National Laboratory.]

the so-called **"special pair,"** are closely associated; they are nearly parallel and have an Mg—Mg distance of ~7 Å. The "special pair" occupies a predominantly hydrophobic region of the protein and each of its Mg^{2+} ions has a His side chain as a fifth ligand [much like the Fe(II) in deoxyhemoglobin]. Each member of the "special pair" is in contact with another His-liganded BChl *b* molecule which, in turn, is associated with a BPheo *b* molecule. The menaquinone is in close association with the L subunit BPheo *b* (Fig. 22-10*a*, *right*), whereas the ubiquinone, which is but loosely bound to the protein, associates with the M subunit BPheo *b* (Fig. 22-10*a*, *left*). These various chromophores are closely associated with a number of protein aromatic rings, which are therefore also thought to participate in the electron-transfer process described below. The Fe(II) is posi-

tioned between the menaquinone and ubiquinone rings and is octahedrally liganded by four His side chains and the two carboxyl oxygen atoms of a Glu side chain. Curiously, the two symmetry related groups of chromophores are not functionally equivalent; electrons, as we shall see, are almost exclusively transferred through the L subunit (the right sides of Figs. 22-9 and 22-10). This effect is generally attributed to subtle structural differences between the L and M subunits.

The Electronic States of Molecules Undergoing Fast Reactions Can Be Monitored by EPR and Laser Spectroscopy Techniques

The turnover time of a photosynthetic reaction cycle, as we have seen, is only a few milliseconds. Its sequence of

FIGURE 22-10. The sequence of excitations in the bacterial photosynthetic reaction center of *Rps. viridis.* The reaction center chromophores are shown in the same view as in Fig. 11-27, which resembles that in Fig. 22-9. Note that their rings, but not their aliphatic side chains, are arranged with close to twofold symmetry. (*a*) At zero time, a photon is absorbed by the "special pair" of BChl *a* molecules, thereby collectively raising them to an excited state [in each step, the excited molecule(s) is shown in red]. (*b*) Within 3 ps, an excited electron has passed to the BPheo *a* of the L subunit (right arm of the system) without becoming closely associated with the accessory BChl *a.* The special pair is thereby left with a positive charge. (*c*) Some 200 ps later, the excited electron has transferred to the menaquinone (Q_A, which is ubiquinone in *Rb. sphaeroides*). (*d*) Within the next 100 μs, the "special pair" has been reduced (via an electron transport chain discussed in the text) thereby eliminating its positive charge while the excited electron migrates to the ubiquinone (Q_B). After a second such electron has been transferred to Q_B, it picks up two protons from solution and exchanges with the membrane-bound ubiquinone pool.

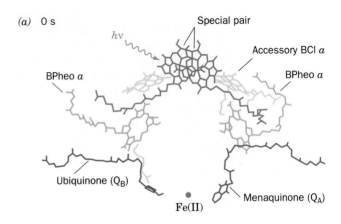

(*a*) 0 s

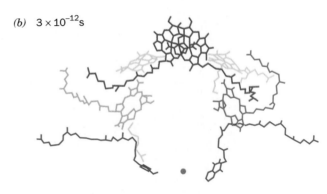

(*b*) 3×10^{-12} s

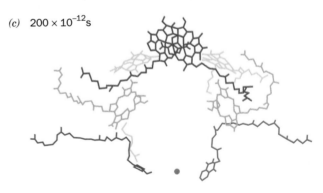

(*c*) 200×10^{-12} s

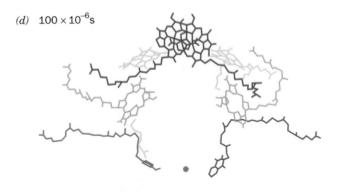

(*d*) 100×10^{-6} s

reactions can therefore only be traced by measurements that can follow extremely rapid electronic changes in molecules. Two techniques are well suited to this task:

1. **Electron paramagnetic resonance (EPR) spectroscopy** [also called **electron spin resonance (ESR) spectroscopy**], which detects the spins of unpaired electrons in a manner analogous to the detection of nuclear spins in NMR spectroscopy. A molecular species with unpaired electrons, such as an organic radical or a transition metal ion, has a characteristic EPR spectrum because its unpaired electrons interact with the magnetic fields generated by the nuclei and the other electrons of the molecule. Paramagnetic species as short lived as 10^{-11} s can exhibit definitive EPR spectra.

2. **Optical spectroscopy** using pulsed lasers. Laser flashes shorter than 10 femtosecond (fs, 10^{-15} s) have been generated. By monitoring the bleaching (disappearance) of certain absorption bands and the emergence of others, laser spectroscopy can track the time course of a fast reaction process.

Photon Absorption Rapidly Photooxidizes the "Special Pair"

The sequence of photochemical events mediated by the photosynthetic reaction center is diagrammed in Fig. 22-10:

(**a**) The primary photochemical event of bacterial photosynthesis is absorption of a photon by the special pair (P870 or **P960** depending on whether it consists of BChl *a* or *b;* here, for argument's sake, we assume it to be P870). This event is nearly instantaneous; it occupies the ~3-fs oscillation time of a light wave. EPR measurements established that P870 is, in fact, a pair

of BChl *a* molecules and indicated that the excited electron is delocalized over both of them.

(**b**) P870*, the excited state of P870, has but a fleeting existence. Laser spectroscopy has demonstrated that within ~3 picoseconds (ps; 10^{-12} s) after its formation, P870* has transferred an electron to the BPheo a on the right in Fig. 22-10b to yield P870$^+$ BPheo a^-. In forming this radical pair, the transferred electron must pass near but seems not to reduce the intervening BChl a (which is therefore termed an accessory chlorophyll), although its position strongly suggests that it has an important role in conveying electrons.

(c) By some 200 ps later, the electron has further migrated to the menaquinone (or, in many species, the second ubiquinone), designated Q_A, to form the anionic semiquinone radical Q_A^-. All these electron transfers, as dia-grammed in Fig. 22-11, are to progressively lower energy states, which makes this process all but irreversible.

Rapid removal of the excited electron from the vicinity of P870$^+$ is an essential feature of the photosynthetic reaction center; this prevents back reactions that would return the electron to P870$^+$ so as to provide the time required for the wasteful internal conversion of its excitation energy to heat. In fact, *this sequence of electron transfers is so efficient that its overall **quantum yield** (ratio of molecules reacted to photons absorbed) is virtually 100%.* No man-made device has yet approached this level of efficiency.

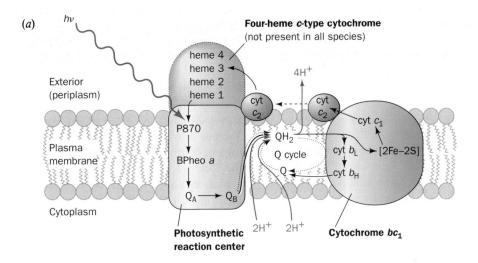

(a)

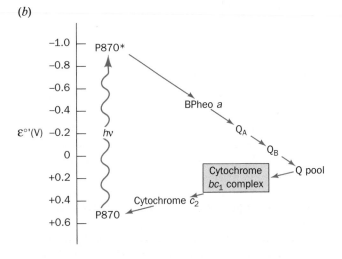

(b)

FIGURE 22-11. The photosynthetic electron transport system of purple photosynthetic bacteria. (*a*) A schematic diagram indicating the arrangement of the system components in the bacterial plasma membrane and the flows of electrons (*black arrows*) and protons (*blue arrows*) that photon (*h*ν) absorption promotes through them. The system contains two protein complexes, the photosynthetic reaction center and cytochrome bc_1. Two electrons liberated from P870 by the consecutive absorption of two photons are taken up by ubiquinone (Q_B) together with two protons from the cytoplasm to yield ubiquinol (QH_2). The QH_2 is released from the photosynthetic reaction center and diffuses (*dotted arrows*) through the membrane to cytochrome bc_1 which, in a two-electron reaction, oxidizes it to ubiquinone with the concomitant liberation of its two protons to the external medium. One of the two electrons is passed, via the [2Fe–2S] cluster and cytochrome c_1, to cytochrome c_2, a peripheral membrane protein that then diffuses across the external surface of the membrane so as to return the electron to P870 of the photosynthetic reaction center. The second electron from QH_2 passes, via a Q cycle, through hemes b_L and b_H of cytochrome bc_1 and then contributes to the reduction of a molecule of ubiquinone (Q) with the concomitant uptake of two more cytoplasmic protons (two rounds of a Q cycle are required for the reduction of one molecule of Q to QH_2; Fig. 20-25). The resulting QH_2 diffuses back to cytochrome bc_1. There it is again oxidized, with the liberation of its two protons to the exterior and the return of one of its two electrons, via cytochrome c_2, to P870, thereby completing the electrical circuit. Note that in every turn of a Q cycle, half the electrons liberated by the oxidation of QH_2 to Q are used to reduce Q to QH_2 so that, after a large number of turns, an electron that enters the Q cycle, on average, passes through it twice before being returned to P870. Thus, the net result of the absorption of two photons by the photosynthetic reaction center is the translocation of four H$^+$ from the cytoplasm to the external medium. (*b*) The approximate standard reduction potentials of the photosynthetic electron transport system's various components.

Electrons Are Returned to the Photooxidized Special Pair via an Electron-Transport Chain

The remainder of the photosynthetic electron-transport process occurs on a much slower time scale. Within $\sim$100 μs after its formation, Q_A^-, which occupies a hydrophobic pocket in the protein, transfers its excited electron to the more solvent-exposed ubiquinone, Q_B, to form Q_B^- (Fig. 22-10d). The nonheme Fe(II) is not reduced in this process and, in fact, its removal only slightly affects the electron transfer rate, so that the Fe(II) probably functions to fine tune the reaction center's electronic character. Q_A never becomes fully reduced; it shuttles between its oxidized and semiquinone forms. Moreover, the lifetime of Q_A^- is so short that it never becomes protonated. In contrast, once the reaction center again becomes excited, it transfers a second electron to Q_B^- to form the fully reduced Q_B^{2-}. This anionic quinol takes up two protons from the solution on the cytoplasmic side of the plasma membrane to form Q_BH_2. Thus Q_B *is a molecular transducer that converts two light-driven one-electron excitations to a two-electron chemical reduction.*

The electrons taken up by Q_BH_2 are eventually returned to P870$^+$ via a complex electron-transport chain (Fig. 22-11). The details of this process are more species dependent than the preceding and are not as well understood. The available redox carriers include a membrane-bound pool of ubiquinone molecules, **cytochrome bc_1**, and **cytochrome c_2**. Cytochrome bc_1 is a transmembrane protein complex composed of a [2Fe–2S] cluster–containing subunit; a heme c-containing cytochrome c_1; a cytochrome b that contains two functionally inequivalent heme b's, b_H and b_L (H and L for high and low potential); and, in some species, a fourth subunit. Note that cytochrome bc_1 is strikingly similar to the proton-translocating Complex III of mitochondria (Section 20-2C). The electron transport pathway leads from Q_BH_2 on the cytoplasmic side of the plasma membrane, through the ubiquinone pool, with which Q_BH_2 exchanges, to cytochrome bc_1, and then to cytochrome c_2 on the external (periplasmic) side of the plasma membrane. The reduced cytochrome c_2, which, as its name implies, closely resembles mitochondrial cytochrome c, diffuses along the external membrane surface until it reacts with the membrane-spanning reaction center to transfer an electron to P870$^+$ (the structures of several c-type cytochromes, including that of cytochrome c_2 from *Rs. rubrum*, are diagrammed in Fig. 8-18). In *Rps. viridis*, the four-heme c-type cytochrome bound to the reaction center complex on the external side of the plasma membrane (Fig. 11-27) is interposed between cytochrome c_2 and P870$^+$. Note that one of this c-type cytochrome's hemes is positioned to reduce the photooxidized special pair. The reaction center is thereby prepared to absorb another photon.

Photosynthetic Electron Transport Drives the Formation of a Proton Gradient

Since electron transport in purple photosynthetic bacteria is a cyclic process (Fig. 22-11), *it results in no net oxidation–reduction. Rather, it functions to translocate the cytoplasmic protons acquired by Q_BH_2 across the plasma membrane, thereby making the cell alkaline relative to its environment.* The mechanism of this process is essentially identical to that of proton transport in mitochondrial Complex III (Section 20-3A); that is, in addition to the translocation of the two H$^+$ resulting from the two-electron reduction of Q_B to QH$_2$, a Q cycle mediated by cytochrome bc_1 translocates two H$^+$ for a total of four H$^+$ translocated per two photons absorbed (Fig. 22-11a; also see Fig. 20-25). *Synthesis of ATP, a process known as **photophosphorylation**, is driven by the dissipation of the resulting pH gradient in a manner that closely resembles ATP synthesis in oxidative phosphorylation (Section 20-3C).* The mechanism of photophosphorylation is further discussed in Section 22-2D.

Photosynthetic bacteria use photophosphorylation-generated ATP to drive their various endergonic processes. However, unlike cyanobacteria and plants, which generate their required reducing equivalents by light-driven oxidation of H$_2$O (see below), photosynthetic bacteria must obtain their reducing equivalents from the environment. Various substances, such as H$_2$S, S, $S_2O_3^{2-}$, H$_2$, and many organic compounds, function in this capacity depending on the bacterial species.

Modern photosynthetic bacteria are thought to resemble the original photosynthetic organisms. These presumably arose very early in the history of cellular life when environmentally supplied sources of high-energy compounds were dwindling but reducing agents were still plentiful (Section 1-4C). During this era, photosynthetic bacteria were no doubt the dominant form of life. However, their very success eventually caused them to exhaust the available reductive resources. The ancestors of modern cyanobacteria adapted to this situation by evolving a photosynthetic system with sufficient electromotive force to abstract electrons from H$_2$O. The gradual accumulation of the resulting toxic waste product, O$_2$, forced photosynthetic bacteria, which cannot photosynthesize in the presence of O$_2$ (although some species have evolved the ability to respire), into the narrow ecological niches to which they are presently confined.

C. Two-Center Electron Transport

Plants and cyanobacteria use the reducing power generated by light-driven oxidation of H$_2$O to produce NADPH. The component half-reactions of this process, together with their standard reduction potentials, are

$$O_2 + 4e^- + 4H^+ \rightleftharpoons 2H_2O \qquad \mathscr{E}°' = +0.815 \text{ V}$$

and

$$NADP^+ + H^+ + 2e^- \rightleftharpoons NADPH \qquad \mathscr{E}°' = -0.320 \text{ V}$$

Hence, the overall four-electron reaction and its standard redox potential is

$$2NADP^+ + 2H_2O \rightleftharpoons 2NADPH + O_2 + 2H^+$$
$$\Delta\mathscr{E}°' = -1.135 \text{ V}$$

This latter quantity corresponds (Eq. [15.5]) to a standard free energy change of $\Delta G°' = 438$ kJ·mol^{-1}, which Eq. [22.1] indicates is the energy of an einstein of 223-nm photons (UV light). Clearly, *even if photosynthesis were 100% efficient, which it is not, it would require more than one photon of visible light to generate a molecule of O$_2$. In fact, experimental measurements indicate that algae minimally require 8 to 10 photons of visible light to produce one molecule of O$_2$*. In the following subsections, we discuss how plants and cyanobacteria manage this multiphoton process.

Photosynthetic O$_2$ Production Requires Two Sequential Photosystems

Two seminal observations led to the elucidation of the basic mechanism of photosynthesis in plants:

1. The quantum yield for O$_2$ evolution by *Chlorella pyrenoidosa* varies little with the wavelength of the illuminating light between 400 and 675 nm but decreases precipitously above 680 nm (Fig. 22-12, lower curve). This phenomenon, the "red drop," was unexpected because Chl *a* absorbs such far-red light (Fig. 22-5).

2. Shorter wavelength light, such as yellow-green light, enhances the photosynthetic efficiency of 700-nm light well in excess of the energy content of the shorter-wavelength light; that is, *the rate of O$_2$ evolution by both lights is greater than the sum of the rates for each light acting alone (Fig. 22-12, upper curve)*. Moreover, this enhancement still occurs if the yellow-green light is switched off several seconds before the red light is turned on and vice versa.

These observations clearly indicate that two processes are involved. They are explained by a mechanistic model, the **Z-scheme,** which postulates that *O$_2$-producing photosynthesis occurs through the actions of two photosynthetic reaction centers that are connected essentially in series (Fig. 22-13):*

1. **Photosystem I (PSI)** generates a strong reductant capable of reducing NADP$^+$ and, concomitantly, a weak oxidant.

2. **Photosystem II (PSII)** generates a strong oxidant capable of oxidizing H$_2$O and, concomitantly, a weak reductant.

The weak reductant reduces the weak oxidant so that *PSI and PSII form a two-stage electron "energizer." Both photosystems must therefore function for photosynthesis (electron transfer from H$_2$O to NADP$^+$ forming O$_2$ and NADPH) to occur.*

The red drop is explained in terms of the Z-scheme by the observation that PSII is only poorly activated by 680-nm light. In the presence of only this far-red light, PSI is activated but is unable to obtain more than a few of the electrons it is capable of energizing. Yellow-green light, however, efficiently stimulates PSII to supply these electrons. The observation that the far-red and yellow-green lights can

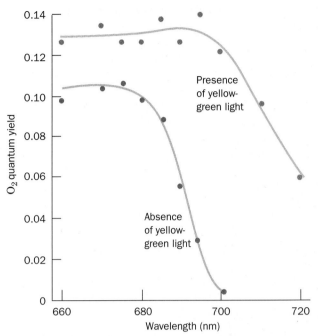

FIGURE 22-12. The quantum yield for O$_2$ production by *Chlorella* algae as a function of the wavelength of the incident light in the absence (*lower curve*) and the presence (*upper curve*) of supplementary yellow-green light. The upper curve has been corrected for the amount of O$_2$ production stimulated by the supplementary light alone. Note that the lower curve falls off precipitously above 680 nm (the red drop). However, the supplementary light greatly increases the quantum yield in the wavelength range above 680 nm (far-red) in which the algae absorb light. [After Emerson, R., Chalmers, R., and Cederstrand, C., *Proc. Natl. Acad. Sci.* **49,** 137 (1957).]

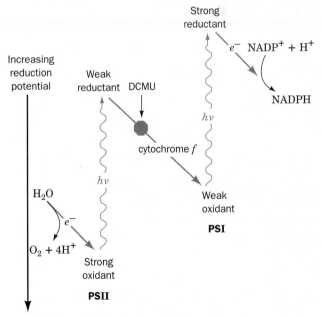

FIGURE 22-13. The Z-scheme for photosynthesis in plants and cyanobacteria. Two photosystems, PSI and PSII, function to drive electrons from H$_2$O to NADPH. The reduction potential increases downwards so that electron flow occurs spontaneously in this direction. The herbicide DCMU (see text) blocks photosynthetic electron transport from PSII to cytochrome *f*.

be alternated indicates that both photosystems remain activated for a time after the light is switched off.

The validity of the Z-scheme was established as follows. The oxidation state of **cytochrome *f***, a *c*-type cytochrome of the electron-transport chain connecting PSI and PSII (see below), can be spectroscopically monitored. Illumination of algae with 680-nm (far-red) light results in the oxidation of cytochrome *f* (Fig. 22-14). However, the additional imposition of a 562-nm (yellow-green) light results in this protein's partial rereduction. In the presence of the herbicide **3-(3,4-dichlorophenyl)-1,1-dimethylurea (DCMU),**

3-(3,4-Dichlorophenyl)-1,1-dimethylurea (DCMU)

which abolishes photosynthetic oxygen production, 680-nm light still oxidizes cytochrome *f* but simultaneous

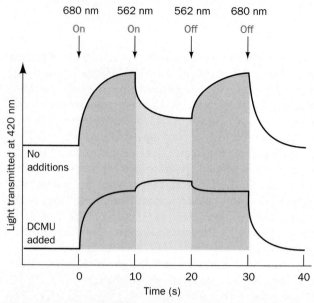

FIGURE 22-14. The oxidation state of cytochrome *f* in *Porphyridium cruentum* algae as monitored by a weak beam of 420 nm (blue-violet) light. An increase in the transmitted light signals the oxidation of cytochrome *f*. In the upper curve, strong light at 680 nm (far-red) causes the oxidation of the cytochrome *f* but the superposition of 562-nm (yellow-green) light causes its partial re-reduction. In the lower curve, the presence of the herbicide DCMU, which inhibits photosynthetic electron transport, causes 562-nm light to further oxidize, rather than reduce, the cytochrome *f*.

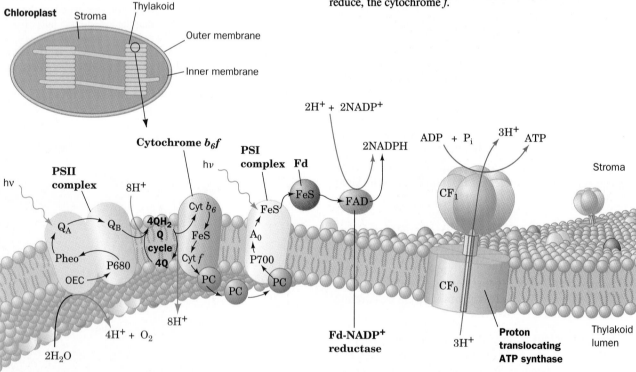

FIGURE 22-15. A schematic representation of the thylakoid membrane showing the components of its electron-transport chain. The system consists of three protein complexes: PSII, the cytochrome b_6f complex, and PSI, which are electrically "connected" by the diffusion of the electron carriers plastoquinol (Q) and plastocyanin (PC). Light-driven transport of electrons (*black arrows*) from H_2O to $NADP^+$ forming NADPH motivates the transport of protons (*red arrows*) into the thylakoid space (Fd is ferredoxin). Additional protons are split off from water by the oxygen-evolving complex (OEC) yielding O_2. The resulting proton gradient powers the synthesis of ATP by the CF_1CF_0 proton translocating ATP synthase [F_1 and F_0 are chloroplast (C) analogs of mitochondrial F_1 and F_0]. The membrane also contains light-harvesting complexes whose component chlorophylls and other chromophores transfer their excitations to PSI and PSII. [After Ort, D.R. and Good, N.E., *Trends Biochem. Sci.* **13,** 469 (1988).]

562-nm light only oxidizes it further. The explanation for these effects is that 680-nm light, which efficiently activates only PSI, causes it to withdraw electrons from (oxidize) cytochrome *f*. The 562-nm light also activates PSII, which thereby transfers electrons to (reduces) cytochrome *f*. DCMU blocks electron flow from PSII to cytochrome *f* (Fig. 22-13), so an increased intensity of light, whatever its wavelength, only serves to activate PSI further.

O₂-Producing Photosynthesis Is Mediated by Three Transmembrane Protein Complexes Linked by Mobile Electron Carriers

The pathway of electron transport in the chloroplast has been traced in broad outline but not in the same detail as in the simpler photosynthetic bacterial systems. *The components involved in the electron transport from H₂O to NADPH are largely organized into three thylakoid membrane-bound particles (Fig. 22-15): (1) PSII, (2)* **cytochrome b₆ f complex**, *and (3) PSI.* As in oxidative phosphorylation, electrons are transferred between these complexes via mobile electron carriers. The ubiquinone analog **plastoquinone (Q)**, via its reduction to **plastoquinol (QH₂)**,

links PSII to the cytochrome $b_6 f$ complex which, in turn, interacts with PSI through the mobile protein **plastocyanin (PC)**. In what follows, we trace the electron pathway through this chloroplast system, insofar as it is known, from H_2O to $NADP^+$ (Fig. 22-16).

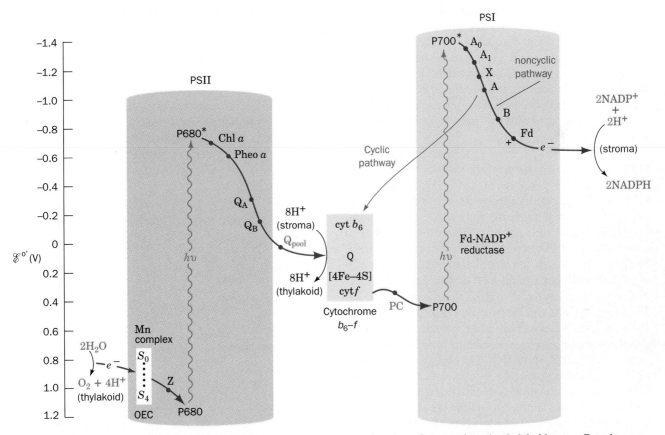

FIGURE 22-16. A detailed diagram of the Z-scheme of photosynthesis. Electrons ejected from P680 by the absorption of photons are replaced with electrons abstracted from H_2O by an Mn complex (OEC), thereby forming O_2 and four H^+. Each ejected electron is passed through a chain of electron carriers to a pool of plastoquinone molecules (Q). The resulting plastoquinol, in turn, reduces the cytochrome $b_6 f$ particle (*yellow box*) that transfers electrons with the concomitant translocation of protons into the thylakoid space. Cytochrome $b_6 f$ then transfers the electrons, via a poorly characterized pathway, to plastocyanin (PC). The plastocyanin regenerates photooxidized P700. The electron ejected from P700, through the intermediacy of a chain of electron carriers, reduces $NADP^+$ to NADPH in noncyclic electron transport. Alternatively, the electron may be returned to the cytochrome $b_6 f$ complex in a cyclic process that translocates protons into the thylakoid space.

O₂ Is Generated in a Five-Stage Water-Splitting Reaction Mediated by an Mn-Containing Protein Complex

The oxidation of two molecules of H_2O to form one molecule of O_2 requires four electrons. Since transfer of a single electron from H_2O to $NADP^+$ requires two photochemical events, this accounts for the observed minimum of 8 to 10 photons absorbed per molecule of O_2 produced.

Must the four electrons necessary to produce a given O_2 molecule be removed by a single photosystem or can they be extracted by several different photosystems? Pierre Joliet and Bessel Kok answered this question by analyzing the rate at which dark-adapted chloroplasts produce O_2 when exposed to a series of short flashes. O_2 was evolved with a peculiar oscillatory pattern (Fig. 22-17). There is virtually no O_2 evolved by the first two flashes. The third flash results in the maximum O_2 yield. Thereafter, the amount of O_2 produced peaks with every fourth flash until the oscillations damp out to a steady state. This periodicity indicates that each O_2-evolving center cycles through five different states, S_0 through S_4 (Fig. 22-18). Each of the transitions between S_0 and S_4 is a photon-driven redox reaction; that from S_4 to S_0 results in the release of O_2. Thus, *each O_2 molecule must be produced by a single photosystem.* The observation that O_2 evolution peaks at the third rather than the fourth flash indicates that the oxygen-evolving center's resting state is predominantly S_1 rather than S_0. The oscillations gradually damp out because a small fraction of the reaction centers fail to be excited or become doubly excited by a given flash of light so that the reaction centers eventually lose synchrony. The five reaction steps release a total of four water-derived protons into the inner thylakoid space in a stepwise manner (Fig. 22-18).

Since the S states function to abstract electrons from H_2O, their standard reduction potentials must average more than the 0.815-V value of the O_2/H_2O half-reaction. PSII has the remarkable capacity of stabilizing these highly reactive intermediates for extended periods (typically minutes) in close proximity to water. We are just beginning to understand how this occurs. PSII contains four protein-bound Mn ions which, upon excitation of chloroplasts with short flashes of light, exhibit EPR signals that have a four-flash periodicity similar to that of O_2 production (Fig. 22-17). These Mn ions, together with 2 or 3 Ca^{2+} ions and 4 or 5 Cl^- ions, form a catalytically active complex, the **oxygen-evolving complex (OEC)**, which binds two H_2O molecules so as to facilitate O_2 formation. The OEC cycles through a series of oxidation states [the S states, which appear to involve various combinations of Mn(III) and Mn(IV)] while abstracting protons and electrons from the H_2O molecules, and finally releases O_2 into the inner thylakoid space. In one of several models that have been proposed which are more or less consistent with the results of EPR and other experimental measurements, the S_0, S_1, and S_2 states are cubane-like Mn_4O_4 complexes and the S_3 and S_4 states are adamantane-like Mn_4O_6 complexes (Fig. 22-19). In the $S_4 \rightarrow S_0$ transition, the adamantane-like structure changes to the cubane-like structure with the release of O_2.

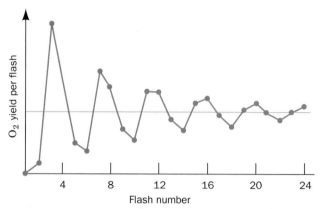

FIGURE 22-17. The O_2 yield per flash in dark-adapted spinach chloroplasts. Note that the yield peaks on the third flash and then on every fourth flash thereafter until the curve eventually damps out to its average value. [After Forbush, B., Kok, B., and McGloin, M. P., *Photochem. Photobiol.* **14**, 309 (1971).]

The next link in the PSII electron transport chain is a substance known as Z (Fig. 22-16), which relays electrons from the Mn–protein water-splitting complex to the reaction center of PSII. The existence of Z is signaled by a transient EPR spectrum of illuminated chloroplasts that parallels the S-state transitions. The change in this spectrum upon feeding deuterated tyrosine to cyanobacteria under conditions such that they incorporate this amino acid in their proteins indicates that Z^+ is a tyrosine radical (EPR spectra reflect the nuclear spins of the atoms with which the unpaired electrons interact).

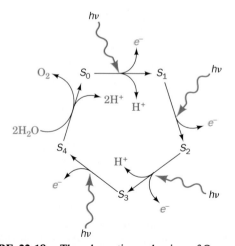

FIGURE 22-18. The schematic mechanism of O_2 generation in chloroplasts. Four electrons are stripped, one at a time in light-driven reactions ($S_0 \rightarrow S_4$), from two bound H_2O molecules. In the recovery step ($S_4 \rightarrow S_0$), which is light independent, O_2 is released and two more H_2O molecules are bound. Three of these five steps release protons into the thylakoid space.

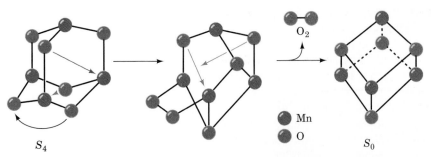

FIGURE 22-19. A proposed structure of PSII's water-splitting Mn complex (OEC). In the $S_4 \rightarrow S_0$ transition, the complex changes from the adamantane-like Mn_4O_6 complex to the cubane-like Mn_4O_4 complex with the release of O_2 (adamantane and cubane are saturated hydrocarbons with C atoms in the positions occupied by the Mn and O atoms in these complexes). The Mn complex retains the cubane-like structure through its S_1 and S_2 states, switching back to the adamantane-like structure in its S_3 state. [After Brudvig, G.W. and Crabtree, R.H., *Proc. Natl. Acad. Sci.* **83**, 4586 (1986).]

The PSII Reaction Center Resembles That of Photosynthetic Bacteria

PSII reaction center's photon-absorbing species is named **P680,** after the wavelength of its absorption maximum. Spectroscopic analysis of P680 indicates that it consists of Chl *a* but it has not been definitively established whether it is a "special pair" of Chl *a* molecules, similar to P870 of purple photosynthetic bacteria (Section 22-2B), or a monomer. The $P680^+$ formed by light excitation, which is among the most powerful biological oxidants known, abstracts electrons from H_2O via the intermediacy of Z and the *S* states.

The chain of electron carriers on the reducing side of P680 bears a remarkable resemblance to the bacterial photosynthetic reaction center (Section 22-2B) even though the two systems operate over different ranges of reduction potentials (compare Figs. 22-11b and 22-16). Indeed, *the two sets of proteins have similar amino acid sequences, indicating that they arose from a common ancestor.* A single electron is transferred, as diagrammed in the central portion of Fig. 22-16, from P680* to a molecule of **pheophytin *a* (Pheo *a;* Chl *a* with its Mg^{2+} replaced by two protons), probably via a Chl *a* molecule, and then to a plastoquinone–Fe(II) complex designated Q_A. Subsequently, two electrons are transferred, one at a time, to a second plastoquinone molecule, Q_B, which takes up two protons at the stromal surface of the thylakoid membrane. The resulting plastoquinol, Q_BH_2, then exchanges with a membrane-bound pool of plastoquinone molecules. DCMU, as well as many other commonly used herbicides, competes with plastoquinone for the Q_B-binding site on PSII, which explains how they inhibit photosynthesis.

Electron Transport through the Cytochrome b_6f Complex Generates a Proton Gradient

From the plastoquinone pool, electrons pass through the **cytochrome b_6f complex.** This integral membrane assembly, which closely resembles cytochrome bc_1, its bacterial counterpart (Section 22-2B), as well as Complex III of the mitochondrial electron-transport chain (Section 20-2C), contains one molecule of **cytochrome *f,*** (*f* for *feuille,* French for leaf), one two-heme–containing cytochrome b_6, one [2Fe–2S] iron–sulfur protein, and one bound plastoquinol. *The cytochrome b_6f complex transports protons as well as electrons from the outside to the inside of the thylakoid membrane.* This proton translocation probably occurs through a Q cycle (Section 20-3B and Fig. 22-11a) in which plastoquinone is the $(H^+ + e^-)$ carrier. However, the roles of the various available e^- carriers in this process have not been sorted out. The Q cycle mechanism predicts that two protons are translocated across the thylakoid membrane for every electron transported but the experimental difficulties of measuring this ratio have precluded its unambiguous determination. It is, nevertheless, clear that *electron transport, via the cytochrome b_6f complex, generates much of the electrochemical proton gradient that drives the synthesis of ATP (see below).*

The 285-residue cytochrome *f* from turnip, the largest of the four polypeptides in the cytochrome b_6f complex, contains a single transmembrane segment near its C-terminus (residues 251–270, which presumably forms an α helix) oriented such that the protein's N-terminal 250 residues extend into the thylakoid lumen. The X-ray structure of the 252-residue N-terminal segment of cytochrome *f*, determined by Janet Smith, reveals an elongated two-domain protein that is dominated by β sheets (Fig. 22-20a) and thus has an entirely different fold from those of other *c*-type cytochromes of known structure (e.g., Figs. 8-18 and 11-27a). Cytochrome *f*'s single heme *c* group is nevertheless covalently linked to the larger domain of the protein via the two Cys residues in a Cys-X-Y-Cys-His sequence that is characteristic of *c*-type cytochromes and whose His residue forms one of the Fe(III)'s two axial ligands (Fig. 8-16). Intriguingly, however, the second axial ligand is the protein's N-terminal amino group, a group that has previously not been observed to be a heme ligand.

Plastocyanin Transports Electrons from Cytochrome b_6f to PSI

Electron transfer between cytochrome *f*, the terminal electron carrier of the cytochrome b_6f complex, and PSI is mediated by **plastocyanin (PC),** a peripheral membrane

(a)

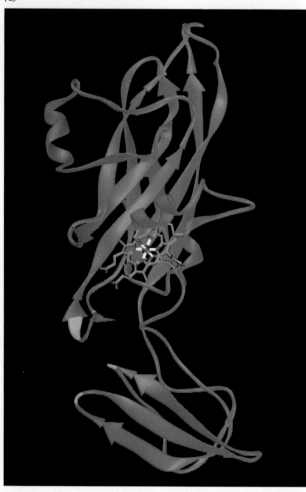

(b)

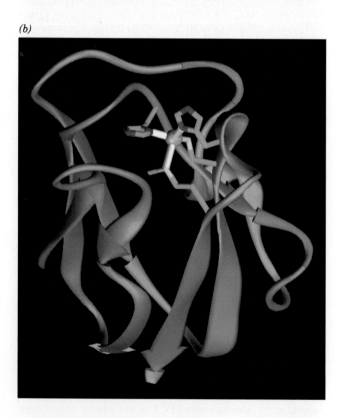

FIGURE 22-20. (*a*) A ribbon diagram of turnip cytochrome *f*. The heme group and the groups that covalently link it to the protein (Cys 21, Cys 24, His 25, and the N-terminal amino group) are shown in stick form with their C, N, O, and S atoms colored green, dark blue, red, and yellow; the heme's Fe atom is represented by a brown sphere. The five Lys and Arg residues that form a positively charged patch on the small domain's surface are light blue. [Based on an X-ray structure by Janet Smith, Purdue University.] (*b*) A ribbon diagram of plastocyanin (PC) from poplar leaves. This 99-residue monomeric protein, a member of the family of **blue copper proteins,** folds into a β sandwich. Its Cu atom *(orange sphere),* which alternates between its Cu(I) and Cu(II) oxidation states, is tetrahedrally liganded by the side chains of His 37, Cys 84, His 87, and Met 92, which are shown in stick form with their C, N, and S atoms green, blue, and yellow. Six conserved Asp and Glu residues that form a negatively charged patch on the protein's surface are red. [Based on an X-ray structure by Mitchell Guss and Hans Freeman, University of Sydney, Australia.]

protein located on the thylakoid luminal surface (Fig. 22-15). The Cu-containing redox center of this mobile 10.5-kD monomer cycles between its Cu(I) and Cu(II) oxidation states. The X-ray structure of PC from poplar leaves, determined by Hans Freeman, shows that the Cu atom is coordinated with distorted tetrahedral geometry by a Cys, a Met, and two His residues (Fig. 22-20*b*). Cu(II) complexes with four ligands normally adopt a square planar coordination geometry, whereas those of Cu(I) are generally tetrahedral. Evidently, the strain of Cu(II)'s protein-imposed tetrahedral coordination in PC promotes its reduction to Cu(I). This hypothesis accounts for PC's high standard reduction potential (0.370 V) compared to that of the normal Cu(II)/Cu(I) half-reaction (0.158 V). This is an example of how proteins modulate the reduction potentials of their redox centers so as to match them to their function—in the case of plastocyanin, the efficient transfer of electrons from the cytochrome b_6f complex to PSI.

The structures of cytochrome *f* and PC suggest how these proteins associate. Cytochrome *f*'s Lys 187, a member of a conserved group of five positively charged residues on the surface of the protein's small domain, can be crosslinked to Asp 44 on PC, which similarly occupies a conserved negatively charged surface patch. Quite possibly, the two proteins associate through electrostatic interactions, much like cytochrome *c* is thought to interact with its redox partners in the mitochondrial electron transport chain, cytochrome *c* reductase and cytochrome *c* oxidase (Section 20-2C).

PSI Has Both Similarities to and Major Differences with PSII and the Bacterial Photosynthetic Reaction Center

PSI consists of two large, ~45% identical (~83 kD) protein subunits named A and B, five small ones (8–16 kD) named I, J, K, L, and M, numerous chlorophyll *a* molecules

(~100 in cyanobacteria and ~200 in eukaryotes), 12 to 16 β-carotene molecules, three [4Fe–4S] clusters (Section 20-2C), and two molecules of **phylloquinone (vitamin K$_1$;** note that it has the same phytyl side chain as chlorophylls; Fig. 22-3).

Phylloquinone

Although PSI does not appear to be related to PSII or to the photosynthetic reaction center of purple photosynthetic bacteria, it is related to that of **green sulfur bacteria,** a second class of photosynthetic bacteria.

The photon-absorbing center of PSI, **P700,** probably consists of a dimer of Chl a molecules. Photooxidation of P700 yields P700$^+$, a weak oxidant that subsequently accepts an electron directly from plastocyanin. On the reducing side of P700, the analysis of light-induced EPR changes indicates that the electron passes through a chain of electron carriers of increasing reduction potential (right side of Fig. 22-16). The first of these carriers, designated A$_0$, appears to be a Chl a monomer, whereas the second carrier, A$_1$, is probably a phylloquinone. The electron finally proceeds through three [4Fe–4S] clusters (Section 20-2C) designated F$_X$, F$_A$, and F$_B$. In contrast, the terminal electron carriers of the other photosystems are all quinones.

The low (6-Å) resolution X-ray structure of PSI from the thermophilic cyanobacterium *Synechococcus* sp. has been determined by Wolfram Saenger (Fig. 22-21). The protein is a symmetric trimeric complex, each of whose monomeric units (which are comprised of the above seven subunits) consist of a catalytic domain and a smaller domain that joins the monomeric units together to form the trimer. Altogether, 28 helices, the three [4Fe–4S] clusters, one phylloquinone, and 45 Chl a molecules can be identified with reasonable confidence in the monomeric unit. Most of the helices are more or less parallel to the trimer's threefold axis (which is perpendicular to the membrane in which PSI is embedded). Eight of these helices appear to be related to another eight by a local twofold axis and hence are probably components of PSI's two large subunits, A and B. This local twofold axis passes through one of the [4Fe–4S] clusters, thereby identifying it as F$_X$, the cluster that had previously been shown to be coordinated by two Cys residues each from the A and B subunit. Two of the Chl a molecules, which are parallel, 9 Å apart, and close to the local twofold

axis, are therefore assumed to be P700, PSI's photon-absorbing center, which apparently resembles the "special pair" in the bacterial photosynthetic reaction center. Interestingly, P700 is flanked by two symmetry-related Chl a molecules in a manner reminiscent of the arrangement of the accessory BChl a molecules relative to the special pair in the bacterial photosynthetic reaction center (Section 22-2B). As is true of the accessory BChl a, there is no spectroscopic evidence that these Chl a molecules participate in electron transport. P700 is located near the tip of a 10-Å-deep indentation on the luminal surface of PS1 that, it is proposed, may be the docking site for plastoquinone.

Further along the electron-transport chain, electron density that is consistent with the Chl a designated A$_0$ and the phylloquinone designated A$_1$ has been tentatively identified. These are followed by F$_X$ and then F$_A$ and F$_B$. Since F$_A$ and F$_B$ are both known to be liganded to subunit C, the low-resolution structure provides no indication as to which of them is which.

PSI-Activated Electrons May Reduce NADP$^+$ or Motivate Proton Gradient Formation

Electrons ejected from PSI may follow either of two alternative pathways:

1. Most electrons follow a noncyclic pathway by passing to an 11-kD, [2Fe–2S]-containing soluble ferredoxin, Fd, that is located in the stroma. Reduced Fd, in turn, reduces NADP$^+$ in a reaction mediated by the 314-residue, monomeric, FAD-containing **ferredoxin-NADP$^+$ reductase (FNR,** Fig. 22-22), to yield the final product of the chloroplast light reaction, NADPH. Two reduced Fd molecules, in turn, deliver one electron each to the FAD of FNR, which thereby sequentially assumes the neutral semiquinone and fully reduced states before transferring the electrons to the NADP$^+$ in a two-electron reduction.

2. Some electrons are returned from PSI, via cytochrome b_6, to the plastoquinone pool, thereby traversing a cyclic pathway that translocates protons across the thylakoid membrane (Fig. 22-16). This accounts for the observation that chloroplasts absorb more than eight photons per O_2 molecule evolved. Note that the cyclic pathway is independent of the action of PSII and hence does not result in the evolution of O_2. PSI, in this way, functionally resembles the photosynthetic bacterial system. It therefore came as a surprise when it was discovered that PSII, but not PSI, is related genetically to bacterial photosystems.

The cyclic electron flow presumably functions to increase the amount of ATP produced relative to that of NADPH and thus permits the cell to adjust the relative amounts of these two substances produced according to its needs. However, the mechanism that apportions electrons between the cyclic and noncyclic pathways is unknown.

(a)

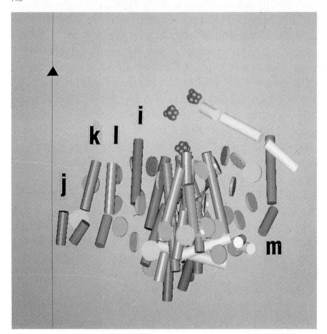

(b)

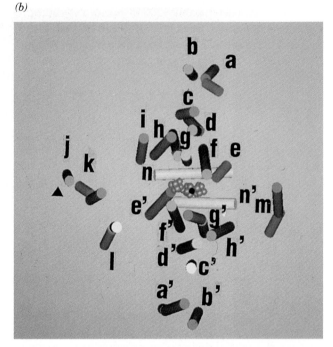

(c)

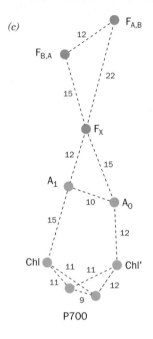

FIGURE 22-21. The 6-Å-resolution X-ray structure of a subunit of PSI from *Synechococcus* sp. (*a*) A view from within the thylakoid membrane with the stromal side above. The transmembrane helices are represented by blue cylinders, horizontal helices by white cylinders, antenna chlorophylls by green disks, electron carriers by yellow disks, and [4Fe–4S] clusters by clusters of red spheres. The vertical line and triangle (*left*) represent the trimeric protein's threefold axis of symmetry. (*b*) Same as Part *a* but rotated by 90° about the horizontal axis such that the stromal side is towards the viewer. For clarity, the chlorophylls and nontransmembrane helices (except for *n* and *n'*) have been omitted. The approximate twofold axis relating helices *a–h* and *a'–h'* in subunits A and B is indicated by a filled circle. (*c*) The arrangements of the centers of the electron carriers as viewed approximately as in Part *a*. The distances between centers are given in Å. [Courtesy of Norbert Krauss and Wolfram Saenger, Freie Universität Berlin, Germany.]

PSI and PSII Occupy Different Parts of the Thylakoid Membrane

Freeze-fracture electron microscopy (Section 11-3B) has revealed that the protein complexes of the thylakoid membrane have characteristic distributions (Fig. 22-23):

1. PSI occurs mainly in the unstacked stroma lamellae, in contact with the stroma, where it has access to $NADP^+$.

2. PSII is located almost exclusively between the closely stacked grana, out of direct contact with the stroma.

3. Cytochrome b_6f is uniformly distributed throughout the membrane.

The high mobilities of plastoquinone and plastocyanin, the electron carriers that shuttle electrons between these particles, permits photosynthesis to proceed at a reasonable rate.

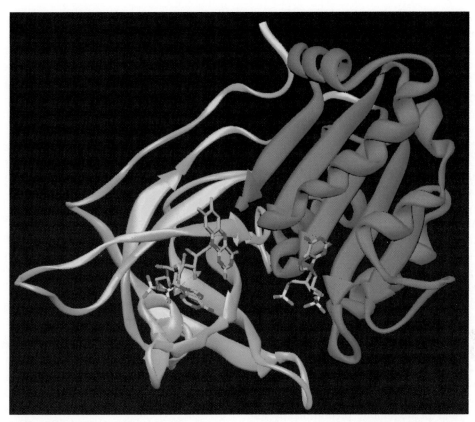

FIGURE 22-22. A ribbon drawing of ferredoxin-NADP⁺ reductase from spinach chloroplasts in complex with FAD and **2′-phospho-AMP.** The protein folds into two domains: Residues 1 to 161 (*gold*), which largely form the FAD binding site, fold into an antiparallel β barrel, whereas residues 162 to 314 (*purple*), which provide the bulk of the NADP⁺ binding site, form a dinucleotide-binding fold (Section 7-3B). The FAD and 2′-phospho-AMP are shown in ball-and-stick form with their C, N, O, and P atoms colored green, blue, red, and yellow, respectively. The 2′-phospho-AMP, an NADP⁺ component, marks a portion of the of NADP⁺ binding site. The cleft between the domains that faces the viewer appears likely to be the ferredoxin binding site. [Based on an X-ray structure by Andrew Karplus, Cornell University, and Jon Herriott, University of Washington.]

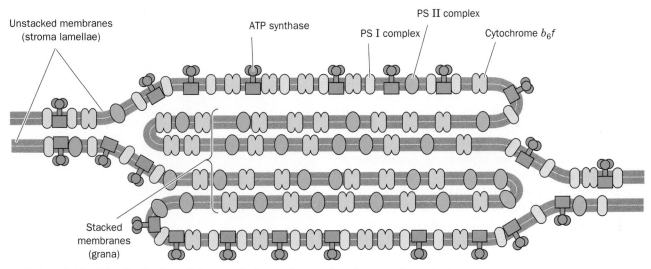

FIGURE 22-23. The distribution of photosynthetic protein complexes between the stacked (grana) and the unstacked (stroma exposed) regions of the thylakoid membrane. [After Anderson, J. M. and Anderson, B., *Trends Biochem. Sci.* **7,** 291 (1982).]

What function is served by the segregation of PSI and PSII? If these two photosystems were in close proximity, the higher excitation energy of PSII (P680 vs P700) would cause it to pass a large fraction of its absorbed photons to PSI via exciton transfer; that is, PSII would act as a light-harvesting antenna for PSI (Fig. 22-7*b*). The separation of these particles by around 100 Å eliminates this difficulty.

The physical separation of PSI and PSII also permits the chloroplast to respond to changes in illumination. The relative amounts of light absorbed by the two photosystems vary with how the light-harvesting complexes (LHCs) are distributed between the stacked and unstacked portions of the thylakoid membrane. Under high illumination (normally direct sunlight, which contains a high proportion of short-wavelength blue light), all else being equal, PSII absorbs more light than PSI. PSI is then unable to take up electrons as fast as PSII can supply them, so the plastoquinone is predominantly in its reduced state. The reduced plastoquinone activates a protein kinase to phosphorylate specific Thr residues of the LHCs, which, in response, migrate to the unstacked regions of the thylakoid membrane where they bind to PSI. A greater fraction of the incident light is thereby funneled to PSI. Under low illumination (normally shady light, which contains a high proportion of long-wavelength red light), PSI takes up electrons faster than PSII can provide them so that plastoquinone predominantly assumes its oxidized form. The LHCs are consequently dephosphorylated and migrate to the stacked portions of the thylakoid membrane, where they drive PSII. The chloroplast therefore maintains the balance between its two photosystems by a light-activated feedback mechanism.

D. Photophosphorylation

Chloroplasts generate ATP in much the same way as mitochondria, that is, by coupling the dissipation of a proton gradient to the enzymatic synthesis of ATP (Section 20-3C). This was clearly demonstrated by the imposition of an artificially produced pH gradient across the thylakoid membrane. Chloroplasts were soaked, in the dark, for several hours in a succinic acid solution of pH 4 so as to bring the thylakoid space to this pH (the thylakoid membrane is permeable to un-ionized succinic acid). The abrupt transfer of these chloroplasts to an $ADP + P_i$-containing buffer at pH 8 resulted in an impressive burst of ATP synthesis: About 100 ATPs were synthesized per molecule of cytochrome f present. Moreover, the amount of ATP synthesized was unaffected by the presence of electron-transport inhibitors such as DCMU. This, together with the observations that photophosphorylation requires an intact thylakoid membrane and that proton translocators such as 2,4-dinitrophenol (Section 20-3D) uncouple photophosphorylation from light-driven electron transport, provide convincing evidence favoring Peter Mitchell's chemiosmotic hypothesis (Section 20-3A).

Chloroplast Proton-Translocating ATP Synthase Resembles That of Mitochondria

Electron micrographs of thylakoid membrane stromal surfaces and bacterial plasma membrane inner surfaces reveal lollipop-shaped structures (Fig. 22-24). These closely resemble the F_1 units of the proton-translocating ATP synthase studding the matrix surfaces of inner mitochondrial membranes (Fig. 20-28*a*). In fact, the chloroplast ATP synthase, which is termed **CF$_1$CF$_0$ complex** (C for chloroplast), has remarkably similar properties to the mitochondrial F_1F_0 complex (Section 20-3C). For example,

1. Both F_0 and CF_0 units are hydrophobic transmembrane proteins that contain a proton translocating channel.

2. Both F_1 and CF_1 are hydrophilic peripheral membrane proteins of subunit composition $\alpha_3\beta_3\gamma\delta\varepsilon$, of which β is a reversible ATPase, and γ forms the gate controlling proton flow from $(C)F_0$ to $(C)F_1$.

3. Both ATP synthases are inhibited by oligomycin and by dicyclohexylcarbodiimide (DCCD).

Clearly, proton-translocating ATP synthases must have evolved very early in the history of cellular life. Note, however, that whereas chloroplast ATP synthase translocates protons out of the thylakoid space (Fig. 22-15), mitochondrial ATP synthase conducts them into the matrix space (Section 20-3A). Chloroplast ATP synthase is located in the unstacked portions of the thylakoid membrane, in contact

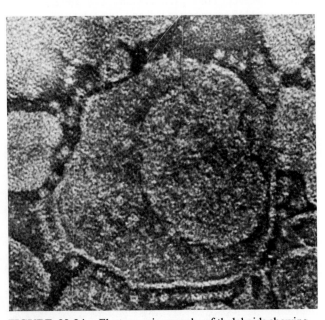

FIGURE 22-24. Electron micrographs of thylakoids showing the CF_1 "lollipops" of their ATP synthases projecting from their stromal surfaces. Compare this with Fig. 20-28*a* and *b*. [Courtesy of Peter Hinkle, Cornell University.]

with the stroma, where there is room for the bulky CF_1 globule and access to ADP (Fig. 22-23).

Photosynthesis with Noncyclic Electron Transport Produces around 1.25 ATPs per Absorbed Photon

At saturating light intensities, chloroplasts generate proton gradients of ~3.5 pH units across their thylakoid membranes. This, as we have seen (Figs. 22-15 and 22-16), arises from two sources:

1. The evolution of a molecule of O_2 from two H_2O molecules releases four protons into the thylakoid space. These protons should be considered as being supplied from the stroma by the protons and H atoms taken up in the synthesis of NADPH.

2. The transport of the liberated four electrons through the cytochrome $b_6 f$ complex occurs with the translocation of what is estimated to be eight protons from the stroma to the thylakoid space.

Altogether ~12 protons are translocated per molecule of O_2 produced by noncyclic electron transport.

The thylakoid membrane, in contrast to the inner mitochondrial membrane, is permeable to ions such as Mg^{2+} and Cl^-. Translocation of protons and electrons across the thylakoid membrane is consequently accompanied by the passage of these ions so as to maintain electrical neutrality (Mg^{2+} out and Cl^- in). This all but eliminates the membrane potential, $\Delta\Psi$ (Eq. [20.1]). *The electrochemical gradient in chloroplasts is therefore almost entirely a result of the pH gradient.*

Chloroplast ATP synthase, according to most estimates, produces one ATP for every three protons it transports out of the thylakoid space. Noncyclic electron transport in chloroplasts therefore results in the production of $\sim\frac{12}{3} = 4$ molecules of ATP per molecule of O_2 evolved (although this quantity is subject to revision) or around half an ATP per photon absorbed. Cyclic electron transport is a more productive ATP generator since it yields two thirds of an ATP (two protons) per absorbed photon. The noncyclic process, of course, also yields NADPH, each molecule of which has the free energy to produce three ATPs (Section 20-2A), for a total of six more ATP equivalents per O_2 produced. Consequently, the energetic efficiency of the noncyclic process is $\frac{4}{8} + \frac{6}{8} = 1.25$ ATP equivalents per absorbed photon.

3. DARK REACTIONS

In the previous section we saw how light energy is harnessed to generate ATP and NADPH. In this section we discuss how these products are used to synthesize carbohydrates and other substances from CO_2.

A. The Calvin Cycle

The metabolic pathway by which plants incorporate CO_2 into carbohydrates was elucidated between 1946 and 1953 by Melvin Calvin, James Bassham, and Andrew Benson. They did so by tracing the metabolic fate of the radioactive label from $^{14}CO_2$ as it passed through a series of photosynthetic intermediates. The basic experimental strategy they used was to expose growing cultures of algae, such as *Chlorella*, to $^{14}CO_2$ for varying times and under differing illumination conditions and then to drop the cells into boiling alcohol so as to disrupt them while preserving their labeling pattern. The radioactive products were subsequently separated and identified (an often difficult task) through the use of the then recently developed technique of two-dimensional paper chromatography (Section 5-3B) coupled with autoradiography. The overall pathway, diagrammed in Fig. 22-25, is known as the **Calvin cycle** or the **reductive pentose phosphate cycle**.

Some of Calvin's earliest experiments indicated that algae exposed to $^{14}CO_2$ for a minute or more had synthesized a complex mixture of labeled metabolic products, including sugars and amino acids. By inactivating the algae within 5 s of their exposure to $^{14}CO_2$, however, it was shown that *the first stable radioactively labeled compound formed is 3-phosphoglycerate (3PG), which is initially labeled only in its carboxyl group.* This result immediately suggested, in analogy with most biochemical experience, that the 3PG was formed by the carboxylation of a C_2 compound. Yet, the failure to find any such precursor eventually forced this hypothesis to be abandoned. The actual carboxylation reaction was discovered through an experiment in which illuminated algae had been exposed to $^{14}CO_2$ for ~10 min so that the levels of their labeled photosynthetic intermediates had reached a steady state. The CO_2 was then withdrawn. As expected, the carboxylation product, 3PG, decreased in concentration (Fig. 22-26) because it was depleted by reactions further along the pathway. The concentration of **ribulose-5-phosphate (Ru5P)**,

$$
\begin{array}{c}
CH_2OH \\
| \\
C = O \\
| \\
H - C - OH \\
| \\
H - C - OH \\
| \\
CH_2OPO_3^{2-}
\end{array}
$$

Ribulose-5-phosphate (Ru5P)

however, simultaneously increased. Evidently, Ru5P is the Calvin cycle's carboxylation substrate. If so, the resulting C_6 carboxylation product must split into two C_3 compounds, one of which is 3PG (Fig. 22-25, Reaction 2). A consideration of the oxidation states of Ru5P and CO_2 indicates that,

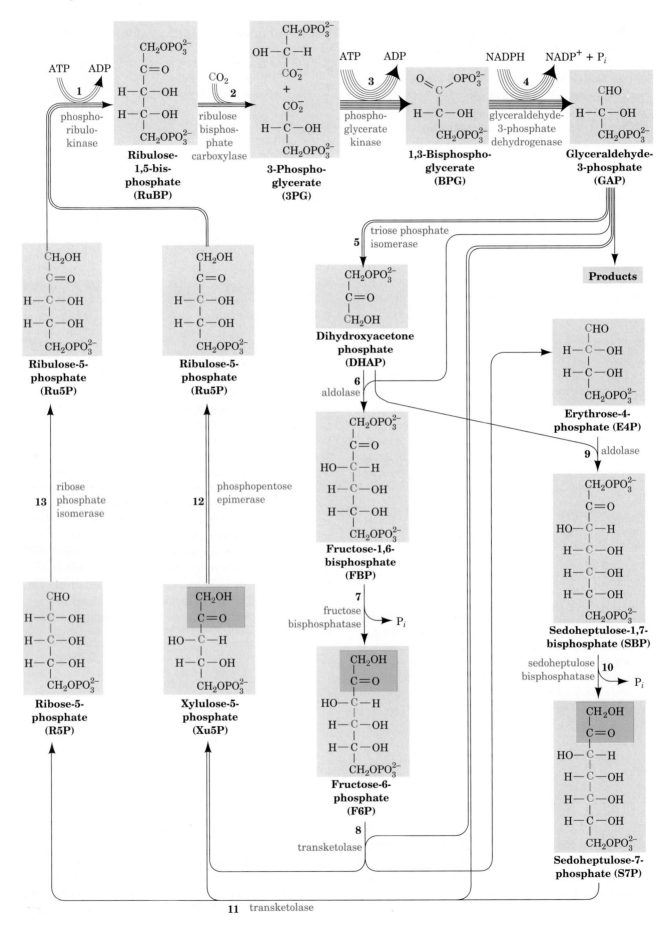

FIGURE 22-25 (*Opposite*). The Calvin cycle. The number of lines in an arrow indicates the number of molecules reacting in that step for a single turn of the cycle that converts three CO_2 molecules to one GAP molecule. For the sake of clarity, the sugars are all shown in their linear forms, although the hexoses and heptoses predominantly exist in their cyclic forms (Section 10-1B). The ^{14}C-labeling patterns generated in one turn of the cycle through the use of $^{14}CO_2$ are indicated in red. Note that two of the Ru5Ps are labeled only at C3, whereas the third Ru5P is equally labeled at C1, C2, and C3.

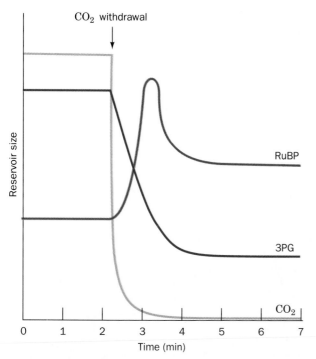

FIGURE 22-26. The time course of the levels of 3PG (*purple curve*) and RuBP (*green curve*) in steady state $^{14}CO_2$-labeled, illuminated algae during a period in which the CO_2 (*orange curve*) is abruptly withdrawn. In the absence of CO_2, the 3PG concentration rapidly decreases because it is taken up by the reactions of the Calvin cycle but cannot be replenished by them. Conversely, the RuBP concentration transiently increases as it is synthesized from the residual pool of Calvin cycle intermediates but, in the absence of CO_2, cannot be used for their regeneration.

in fact, both C_3 compounds must be 3PG and that the carboxylation reaction requires no external redox source.

While the search for the carboxylation substrate was going on, several other photosynthetic intermediates had been identified and, through chemical degradation studies, their labeling patterns had been elucidated. For example, the hexose fructose-1,6-bisphosphate (FBP) is initially labeled only at its C3 and C4 positions (Fig. 22-25) but later becomes labeled to a lesser degree at its other atoms. Similarly, a series of tetrose, pentose, hexose, and heptose phos-

phates were isolated that had the identities and initial labeling patterns indicated in Fig. 22-25. A consideration of the flow of the labeled atoms through these various intermediates led, in what is a milestone of metabolic biochemistry, to the deduction of the Calvin cycle as is diagrammed in Fig. 22-25. The existence of many of its postulated reactions was eventually confirmed by *in vitro* studies using purified enzymes.

The Calvin Cycle Generates GAP from CO_2 via a Two-Stage Process

The Calvin cycle may be considered to have two stages:

Stage 1 The production phase (top line of Fig. 22-25), in which three molecules of Ru5P react with three molecules of CO_2 to yield six molecules of glyceraldehyde-3-phosphate (GAP) at the expense of nine ATP and six NADPH molecules. *The cyclic nature of the pathway makes this process equivalent to the synthesis of one GAP from three CO_2 molecules.* Indeed, at this point, one GAP can be bled off from the cycle for use in biosynthesis (see Stage 2).

Stage 2 The recovery phase (bottom lines of Fig. 22-25), in which the carbon atoms of the remaining five GAPs are shuffled in a remarkable series of reactions, similar to those of the pentose phosphate pathway (Section 21-4), to reform the three Ru5Ps with which the cycle began. Indeed, the elucidation of the pentose phosphate pathway at about the same time that the Calvin cycle was being worked out provided much of the biochemical evidence in support of the Calvin cycle. This stage can be conceptually decomposed into four sets of reactions (with the numbers keyed to the corresponding reactions in Fig. 22-25):

6. $C_3 + C_3 \longrightarrow C_6$

8. $C_3 + C_6 \longrightarrow C_4 + C_5$

9. $C_3 + C_4 \longrightarrow C_7$

11. $C_3 + C_7 \longrightarrow C_5 + C_5$

The overall stoichiometry for this process is therefore

$$5C_3 \longrightarrow 3C_5$$

Note that this stage of the Calvin cycle occurs without further input of free energy (ATP) or reducing power (NADPH).

Most Calvin Cycle Reactions Also Occur in Other Metabolic Pathways

The types of reactions that comprise the Calvin cycle are all familiar (Section 21-4), with the exception of the carboxylation reaction. This first stage of the Calvin cycle begins with the phosphorylation of Ru5P by **phosphoribulokinase** to form **ribulose-1,5-bisphosphate (RuBP)**. Following the carboxylation step, which is discussed below, the resulting 3PG is converted first to 1,3-bisphosphoglycerate (BPG) and then to GAP. This latter sequence is the reverse of two consecutive glycolytic reactions (Sections 16-2G and F) except that the Calvin cycle reaction involves NADPH rather than NADH.

The second stage of the Calvin cycle begins with the reverse of a familiar glycolytic reaction, the isomerization of GAP to dihydroxyacetone phosphate (DHAP) by triose phosphate isomerase (Section 16-2E). Following this, DHAP is directed along two analogous paths (Fig. 22-25): Reactions 6–8 or Reactions 9–11. Reactions 6 and 9 are aldolase-catalyzed aldol condensations in which DHAP is linked to an aldehyde (aldolase is specific for DHAP but accepts a variety of aldehydes). Reaction 6 is also the reverse of a glycolytic reaction (Section 16-2D). Reactions 7

and 10 are phosphate hydrolysis reactions that are catalyzed, respectively, by fructose bisphosphatase (FBPase, which we previously encountered in our discussion of glycolytic futile cycles and gluconeogenesis; Sections 16-4B and 21–1A), and **sedoheptulose bisphosphatase (SBPase)**. The remaining Calvin cycle reactions are catalyzed by enzymes that also participate in the pentose phosphate pathway. In Reactions 8 and 11, both catalyzed by **transketolase**, a C_2 keto unit (shaded in green in Fig. 22-25) is transferred from a ketose to GAP to form **xylulose-5-phosphate (Xu5P)** and leave the aldoses **erythrose-4-phosphate (E4P)** in Reaction 8 and **ribose-5-phosphate (R5P)** in Reaction 11. The E4P produced by Reaction 8 feeds into Reaction 9. The Xu5Ps produced by Reactions 8 and 11 are converted to Ru5P by **phosphopentose epimerase** in Reaction 12. The R5P from Reaction 11 is also converted to Ru5P by **ribose phosphate isomerase** in Reaction 13, thereby completing a turn of the Calvin cycle. Thus only 3 of the 11 Calvin cycle enzymes, phosphoribulokinase, the

(a)

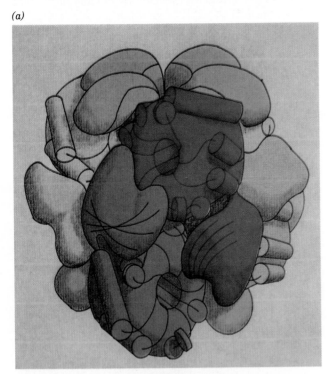

(b)

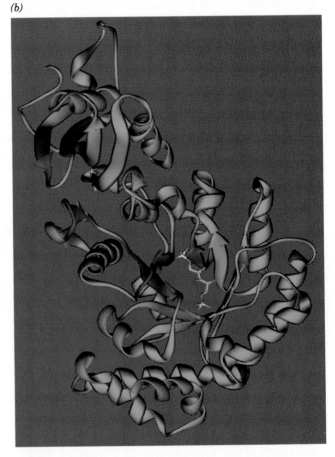

FIGURE 22-27. The X-ray structure of tobacco RuBP carboxylase. (*a*) A drawing showing the quaternary structure of this L_8S_8 protein. The S subunits (six visible) are blue and each of the bilobal L subunits (four visible) is a different color. The protein, which has D_4 symmetry (the symmetry of a square prism; Section 7-5B), is viewed here along one of its twofold axes (which relates the red and dark green L subunits). The fourfold rotation axis, which is nearly vertical in this view, relates the members of the S_4 tetramer visible on the top of the complex. [Courtesy of David Eisenberg, UCLA.] (*b*) A ribbon diagram of an L subunit in complex with the transition state inhibitor **2-carboxyarabinitol-1,5-bisphosphate (CABP;** see text)

as drawn in skeletal form with its C, O, and P atoms light blue, red, and yellow, respectively. The subunit is in approximately the same orientation as the dark green L subunit in Part *a*. Note that the CABP is bound in the mouth of the enzyme's α/β barrel. [Based on an X-ray structure by David Eisenberg.]

carboxylation enzyme **ribulose bisphosphate carboxylase,** and SBPase, have no equivalents in animal tissues.

RuBP Carboxylase Catalyzes CO₂ Fixation in an Exergonic Process

The enzyme that catalyzes CO_2 fixation, ribulose bisphosphate carboxylase **(RuBP carboxylase),** is arguably the world's most important enzyme since nearly all life on Earth ultimately depends on its action. This protein, presumably as a consequence of its low catalytic efficiency ($k_{cat} = \sim 3 \ s^{-1}$), comprises up to 50% of leaf proteins and is therefore the most abundant protein in the biosphere (it is estimated to be synthesized at the rate of $\sim 4 \times 10^9$ tons/y which fixes $\sim 10^{11}$ tons of CO_2/y; in comparison crude oil is consumed at the rate of $\sim 3 \times 10^9$ tons/y). RuBP carboxylase from higher plants and most photosynthetic microorganisms consists of eight large (L) subunits (477 residues in tobacco leaves) encoded by chloroplast DNA, and eight small (S) subunits (123 residues) specified by a nuclear gene (the RuBP carboxylase from certain photosynthetic bacteria is an L_2 dimer whose L subunit has 28% sequence identity and is structurally similar to that of the L_8S_8 enzyme). X-Ray studies by Carl-Ivar Brändén and by David Eisenberg demonstrated that the L_8S_8 enzyme has the symmetry of a square prism (Fig. 22-27a). The L subunit contains the enzyme's catalytic site as is demonstrated by its enzymatic activity in the absence of the S subunit. It consists of two domains (Fig. 22-27b): Residues 1-150 form a mixed five-stranded β sheet and residues 151-475 fold into an α/β barrel (Fig. 7-19b) which, as do all known α/β barrel enzymes (Section 16-2E), contains the enzyme's active site at the mouth of the barrel near the C-terminus of its β strands. The function of the S subunit is unknown; attempts to show that it has a regulatory role, in analogy with other enzymes, have been unsuccessful.

The accepted mechanism of RuBP carboxylase, which was largely formulated by Calvin, is indicated in Fig. 22-28. Abstraction of the C3 proton of RuBP, the reaction's rate-determining step, generates an enediolate that nucleophilically attacks CO_2 (not HCO_3^-). The resulting β-keto acid is rapidly attacked at its C3 position by H_2O to yield an adduct that splits, by a reaction similar to aldol cleavage, to yield the two product 3PG molecules. Evidence favoring this mechanism is

1. The C3 proton of enzyme-bound RuBP exchanges with solvent, an observation compatible with the existence of the enediolate intermediate.

2. The C2 and C3 oxygen atoms remain attached to their respective C atoms, which eliminates mechanisms in-

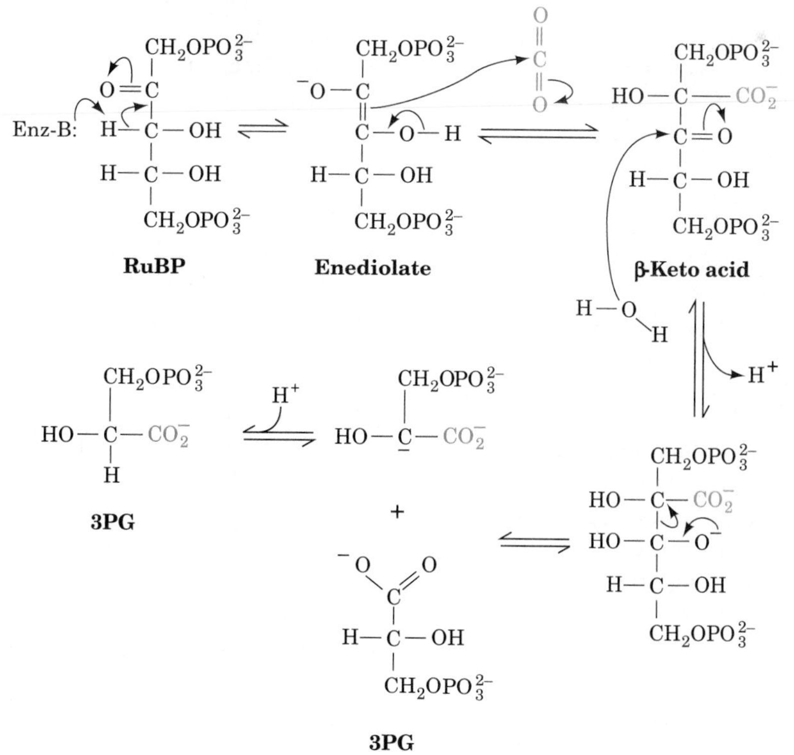

FIGURE 22-28. The probable reaction mechanism of the carboxylation reaction catalyzed by RuBP carboxylase. The reaction proceeds via an enediolate intermediate that nucleophilically attacks CO_2 to form a β-keto acid. This intermediate reacts with water to yield two molecules of 3PG.

volving a covalent adduct such as a Schiff base between RuBP and the enzyme.

3. The trapping of the proposed β-keto acid intermediate by borohydride reduction, and the tight enzymatic binding of its analogs such as **2-carboxyarabinitol-1-phosphate (CA1P)** and **2-carboxyarabinitol-1,5-bisphosphate (CABP),**

$$
\begin{array}{cc}
\text{CH}_2\text{OPO}_3^{2-} & \text{CH}_2\text{OPO}_3^{2-} \\
| & | \\
\text{HO}-\text{C}-\text{CO}_2^- & \text{HO}-\text{C}-\text{CO}_2^- \\
| & | \\
\text{H}-\text{C}-\text{OH} & \text{H}-\text{C}-\text{OH} \\
| & | \\
\text{H}-\text{C}-\text{OH} & \text{H}-\text{C}-\text{OH} \\
| & | \\
\text{CH}_2\text{OH} & \text{CH}_2\text{OPO}_3^{2-}
\end{array}
$$

2-Carboxyarabinitol-1-phosphate (CA1P)   2-Carboxyarabinitol-1,5-bisphosphate (CABP)

provide strong evidence for the existence of this intermediate.

The driving force for the overall reaction, which is highly exergonic ($\Delta G°' = -35.1 \ kJ \cdot mol^{-1}$), *is provided by the cleavage of the β-keto acid intermediate to yield an additional resonance-stabilized carboxylate group.*

RuBP carboxylase activity requires a bound divalent metal ion, physiologically Mg^{2+}, which probably acts to stabilize developing negative charges during catalysis. The Mg^{2+} is, in part, bound to the enzyme by a catalytically essential carbamate group that is generated by the reaction of a nonsubstrate CO_2 with the ε-amino group of Lys 201. Although the *in vitro* activation reaction occurs spontaneously in the presence of Mg^{2+} and HCO_3^-, it is catalyzed *in vivo* by the enzyme **RuBP carboxylase activase** in an ATP-driven process.

GAP Is the Precursor of Glucose-1-phosphate and Other Biosynthetic Products

The overall stoichiometry of the Calvin cycle is

$$3CO_2 + 9ATP + 6NADPH \longrightarrow$$
$$GAP + 9ADP + 8P_i + 6NADP^+$$

GAP, the primary product of photosynthesis, is used in a variety of biosynthetic pathways, both inside and outside the chloroplast. For example, it can be converted to fructose-6-phosphate by the further action of Calvin cycle enzymes and then to glucose-1-phosphate (G1P) by phosphoglucose isomerase and phosphoglucomutase (Section 17-1B). *G1P is the precursor of the higher carbohydrates characteristic of plants.* These most notably include sucrose (Section 10-2B), their major transport sugar for delivering carbohydrates to nonphotosynthesizing cells; starch (Section 10-2D), their chief storage polysaccharide; and cellulose (Section 10-2C), the primary structural component of

their cell walls. In the synthesis of all these substances, G1P is activated by the formation of either ADP-, CDP-, GDP-, or UDP-glucose (Section 17-2), depending on the species and the pathway. Its glucose unit is then transferred to the nonreducing end of a growing polysaccharide chain much as occurs in the synthesis of glycogen (Section 17-2B). In the case of sucrose synthesis, the acceptor is the reducing end of F6P with the resulting **sucrose-6-phosphate** being hydrolyzed to sucrose by a phosphatase. Fatty acids and amino acids are synthesized from GAP as is described, respectively, in Sections 23-4 and 24-5.

B. Control of the Calvin Cycle

During the day, plants satisfy their energy needs via the light and dark reactions of photosynthesis. At night, however, like other organisms, they must use their nutritional reserves to generate their required ATP and NADPH through glycolysis, oxidative phosphorylation, and the pentose phosphate pathway. Since the stroma contains the enzymes of glycolysis and the pentose phosphate pathway as well as those of the Calvin cycle, *plants must have a light-sensitive control mechanism to prevent the Calvin cycle from consuming this catabolically produced ATP and NADPH in a wasteful futile cycle.*

As we saw in Section 16-4A, the control of flux in a metabolic pathway occurs at enzymatic steps that are far from equilibrium; that is, those that have a large negative value of ΔG. Inspection of Table 22-1 indicates that the three best candidates for flux control in the Calvin cycle are the reactions catalyzed by RuBP carboxylase, FBPase, and SBPase (Reactions 2, 7, and 10, Fig. 22-25). In fact, the catalytic efficiencies of these three enzymes all vary, *in vivo*, with the level of illumination.

The activity of RuBP carboxylase responds to three light-dependent factors:

1. It varies with pH. Upon illumination, the pH of the stroma increases from around 7.0 to about 8.0 as protons are pumped from the stroma into the thylakoid space. RuBP carboxylase has a sharp pH optimum near pH 8.0.

2. It is stimulated by Mg^{2+}. Recall that the light-induced influx of protons to the thylakoid space is accompanied by the efflux of Mg^{2+} to the stroma (Section 22-2D).

3. It is strongly inhibited by its transition state analog 2-carboxyarabinitol-1-phosphate (CA1P; Section 22-3A), which many plants synthesize only in the dark. **RuBP carboxylase activase** facilitates the release of the tight-binding CA1P from RuBP carboxylase as well as catalyzing its carbamoylation (Section 22-3A).

FBPase and SBPase are also activated by increased pH and Mg^{2+}, and by NADPH as well. The action of these factors is complemented by a second regulatory system that

TABLE 22-1. STANDARD AND PHYSIOLOGICAL FREE ENERGY CHANGES FOR THE REACTIONS OF THE CALVIN CYCLE

Step[a]	Enzyme	$\Delta G^{\circ\prime}$ (kJ · mol^{-1})	ΔG (kJ · mol^{-1})
1	Phosphoribulokinase	−21.8	−15.9
2	Ribulose bisphosphate carboxylase	−35.1	−41.0
3 + 4	Phosphoglycerate kinase + glyceraldehyde-3-phosphate dehydrogenase	+18.0	−6.7
5	Triose phosphate isomerase	−7.5	−0.8
6	Aldolase	−21.8	−1.7
7	Fructose bisphosphatase	−14.2	−27.2
8	Transketolase	+6.3	−3.8
9	Aldolase	−23.4	−0.8
10	Sedoheptulose bisphosphatase	−14.2	−29.7
11	Transketolase	+0.4	−5.9
12	Phosphopentose isomerase	+0.8	−0.4
13	Ribose phosphate isomerase	+2.1	−0.4

[a] Refer to Fig. 22-25.

Source: Bassham, J.A. and Buchanan, B.B., *in* Govindjee (Ed.), *Photosynthesis* Vol. II, *p*. 155, Academic Press (1982).

responds to the redox potential of the stroma. **Thioredoxin,** a 12-kD protein that occurs in many types of cells, contains a reversibly reducible cystine disulfide group. Reduced thioredoxin activates both FBPase and SBPase by a disulfide interchange reaction (Fig. 22-29). This explains why these Calvin cycle enzymes are activated by reduced disulfide reagents such as dithiothreitol. The redox level of thioredoxin is maintained by a second disulfide-containing enzyme, ferredoxin–thioredoxin reductase, which directly responds to the redox state of the soluble ferredoxin in the stroma. This in turn varies with the illumination level. The thioredoxin system also deactivates phosphofructokinase (PFK), the main flux-generating enzyme of glycolysis (Section 16-4B). Thus in plants, *light stimulates the Calvin cycle while deactivating glycolysis, whereas darkness has the opposite effect* (that is, the so-called dark reactions do not occur in the dark).

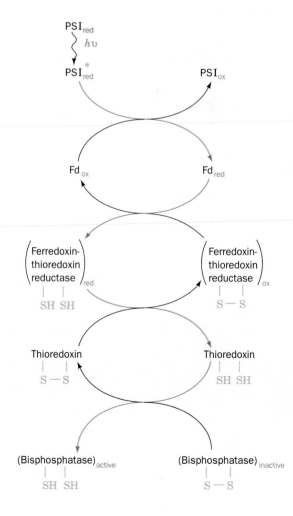

FIGURE 22-29. The light-activation mechanism of FBPase and SBPase. Photoactivated PSI reduces soluble ferredoxin (Fd), which reduces ferredoxin–thioredoxin reductase which, in turn, reduces the disulfide linkage of thioredoxin. Reduced thioredoxin reacts with the inactive bisphosphatases by disulfide interchange, thereby activating these flux-generating Calvin cycle enzymes.

C. Photorespiration and the C₄ Cycle

It has been known since the 1960s that *illuminated plants consume O_2 and evolve CO_2 in a pathway distinct from oxidative phosphorylation. In fact, at low CO_2 and high O_2 levels, this* **photorespiration** *process can outstrip photosynthetic CO_2 fixation.* The basis of photorespiration was unexpected: *O_2 competes with CO_2 as a substrate for RuBP carboxylase* (RuBP carboxylase is therefore also called **RuBP carboxylase–oxygenase** or **rubisco**). In the oxygenase reaction, O_2 reacts with rubisco's second substrate, RuBP, to form 3PG and **2-phosphoglycolate** (Fig. 22-30). The 2-phosphoglycolate is hydrolyzed to **glycolate** by **glycolate phosphatase** and, as described below, is partially oxidized to yield CO_2 by a series of enzymatic reactions that occur in the peroxisome and the mitochondrion. Thus photorespiration is a seemingly wasteful process that undoes some of the work of photosynthesis. In the following subsections we discuss the biochemical basis of photorespiration, its significance, and how certain plants manage to evade its deleterious effects.

Photorespiration Dissipates ATP and NADPH

The photorespiration pathway is outlined in Fig. 22-31. Glycolate is exported from the chloroplast to the peroxisome (also called the glyoxisome, Section 1-2A), where it is oxidized by **glycolate oxidase** to **glyoxylate** and H_2O_2. The H_2O_2, a powerful and potentially harmful oxidizing agent, is disproportionated to H_2O and O_2 in the peroxisome by the heme-containing enzyme **catalase.** Some of the glyoxylate is further oxidized by glycolate oxidase to oxalate. The remainder is converted to glycine in a **transamination reaction,** as discussed in Section 24-1A, and exported to the mitochondrion. There, two molecules of glycine are converted to one molecule of serine and one of CO_2 by a reaction described in Section 24-1B. *This is the origin of the CO_2 generated by photorespiration.* The serine is transported back to the peroxisome where a transamination reaction converts it to **hydroxypyruvate.** This substance is reduced to **glycerate** and phosphorylated in the cytosol to 3PG, which reenters the chloroplast where it is reconverted to RuBP in the Calvin cycle. *The net result of this complex photorespiration cycle is that some of the ATP and NADPH generated by the light reactions is uselessly dissipated.*

Although photorespiration has no known metabolic function, the rubiscos from the great variety of photosynthetic organisms so far tested all exhibit oxygenase activity. Yet, over the eons, the forces of evolution must have optimized the function of this important enzyme. It is thought that photosynthesis evolved at a time when the Earth's atmosphere contained large quantities of CO_2 and very little O_2 so that photorespiration was of no consequence. It has therefore been suggested that the rubisco reaction has an obligate intermediate that is inherently autooxidizable. Another possibility is that photorespiration protects the photosynthetic apparatus from photooxidative damage when insufficient CO_2 is available to otherwise dissipate its absorbed light energy. This hypothesis is supported by the

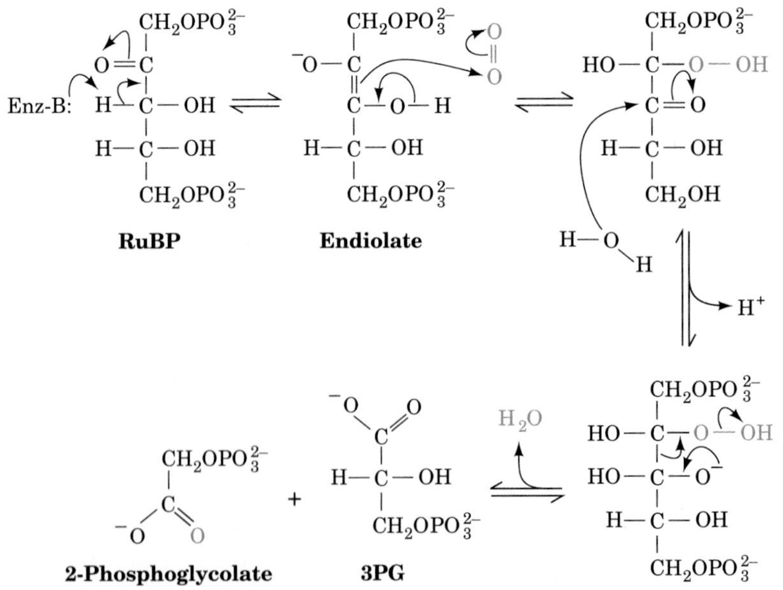

FIGURE 22-30. The probable mechanism of the oxygenase reaction catalyzed by RuBP carboxylase–oxygenase. Note the similarity of this mechanism to that of the carboxylase reaction catalyzed by the same enzyme (Fig. 22-28).

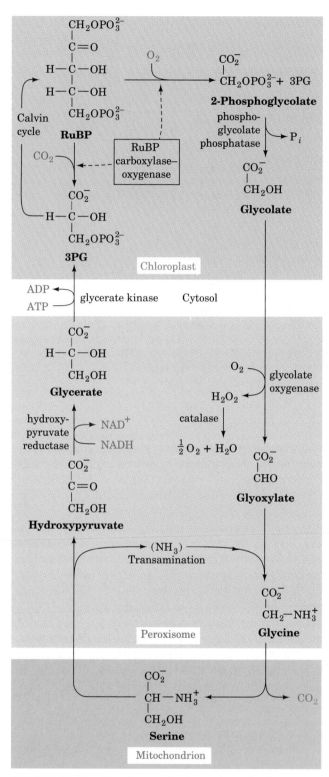

FIGURE 22-31. The photorespiration pathway for the metabolism of the phosphoglycolate produced by the RuBP carboxylase–catalyzed oxidation of RuBP. The reactions occur, as indicated, in the chloroplast, the peroxisome, the mitochondrion, and the cytosol. Note that two glycines are required to form serine + CO_2 (Section 24-3B).

observation that when chloroplasts or leaf cells are brightly illuminated in the absence of both CO_2 and O_2, their photosynthetic capacity is rapidly and irreversibly lost.

Photorespiration Limits the Growth Rates of Plants

The steady state CO_2 concentration attained when a photosynthetic organism is illuminated in a sealed system is named its **CO_2 compensation point.** For healthy plants, this is the CO_2 concentration at which the rates of photosynthesis and photorespiration are equal. For many species it is ~40 to 70 ppm (parts per million) CO_2 (the normal atmospheric concentration of CO_2 is 330 ppm), so their photosynthetic CO_2 fixation usually dominates their photorespiratory CO_2 release. However, the CO_2 compensation point increases with temperature because the oxygenase activity of rubisco increases more rapidly with temperature than its carboxylase activity. Thus, *on a hot bright day, when photosynthesis has depleted the level of CO_2 at the chloroplast and raised that of O_2, the rate of photorespiration may approach that of photosynthesis. This phenomenon is, in fact, a major limiting factor in the growth of many plants.* Indeed, plants possessing a rubisco with significantly less oxygenase activity would not only have increased photosynthetic efficiency but would need less water because they could spend less time with their **stomata** (the pores leading to their internal leaf spaces) open acquiring CO_2 and would have a reduced need for fertilizer because they would require less rubisco. The control of photorespiration is therefore an important unsolved agricultural problem that is presently being attacked through genetic engineering studies (Section 28-8).

C₄ Plants Concentrate CO₂

Certain species of plants, such as sugar cane, corn, and most important weeds, have a metabolic cycle that concentrates CO_2 in their photosynthetic cells, thereby almost totally preventing photorespiration (their CO_2 compensation points are in the range 2 to 5 ppm). The leaves of plants that have this so-called **C₄ cycle** have a characteristic anatomy. Their fine veins are concentrically surrounded by a single layer of so-called **bundle-sheath cells,** which in turn are surrounded by a layer of **mesophyll cells.**

The C₄ cycle (Fig. 22-32) was elucidated in the 1960s by Marshall Hatch and Rodger Slack. It begins with the uptake of atmospheric CO_2 by the mesophyll cells which, lacking rubisco in their chloroplasts, do so by condensing it as HCO_3^- with phosphoenolpyruvate (PEP) to yield oxaloacetate. The oxaloacetate is reduced by NADPH to **malate,** which is exported to the bundle-sheath cells (the name C₄ refers to these four-carbon acids). There the malate is oxidatively decarboxylated by $NADP^+$ to form CO_2, pyruvate, and NADPH. The CO_2, which has been concentrated by this process, enters the Calvin cycle. The pyruvate is returned to the mesophyll cells, where it is phosphorylated to again form PEP. The enzyme that mediates this reaction, **pyruvate-phosphate dikinase,** has the unusual action of ac-

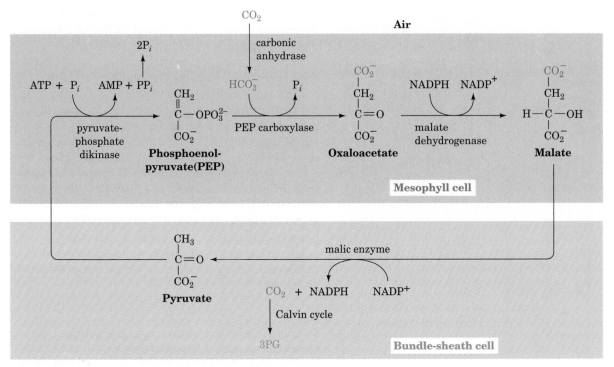

FIGURE 22-32. The C_4 pathway for concentrating CO_2 in the mesophyll cells and transporting it to the bundle-sheath cells for entry into the Calvin cycle.

tivating a phosphate group through the hydrolysis of ATP to AMP + PP$_i$. This PP$_i$ is further hydrolyzed to two P$_i$, which is tantamount to the consumption of a second ATP. *CO_2 is thereby concentrated in the bundle-sheath cells at the expense of two ATPs per CO_2. Photosynthesis in C_4 plants therefore consumes a total of five ATPs per CO_2 fixed versus the three ATPs required by the Calvin cycle alone.*

C_4 **plants** occur largely in tropical regions because they grow faster under hot and sunny conditions than other, so called C_3 **plants** (so named because they initially fix CO_2 in the form of three-carbon acids). In cooler climates, where photorespiration is less of a burden, C_3 plants have the advantage because they require less energy to fix CO_2.

CAM Plants Store CO_2 through a Variant of the C_4 Cycle

A variant of the C_4 cycle that separates CO_2 acquisition and the Calvin cycle in time rather than in space occurs in

many desert-dwelling succulent plants. If, as most plants, they opened their stomata by day to acquire CO_2, they would simultaneously transpire (lose by evaporation) what for them would be unacceptable amounts of water. To minimize this loss, these succulents only absorb CO_2 at night when the temperature is relatively cool. They store this CO_2, in a process known as **Crassulacean acid metabolism** (**CAM;** so named because it was first discovered in plants of the family Crassulaceae), by the synthesis of malate through the reactions of the C_4 pathway (Fig. 22-32). The large amount of PEP necessary to store a day's supply of CO_2 is obtained by the breakdown of starch via glycolysis. During the course of the day, this malate is broken down to CO_2, which enters the Calvin cycle, and pyruvate, which is used to resynthesize starch. CAM plants are able, in this way, to carry out photosynthesis with minimal water loss.

CHAPTER SUMMARY

Photosynthesis is the light-driven fixation of CO_2 to form carbohydrates and other biological molecules. In plants, photosynthesis takes place in the chloroplast, which consists of an inner and an outer membrane surrounding the stroma, a concentrated enzyme solution, in which the thylakoid membrane system is immersed.

Photosynthesis occurs in two stages, the so-called light reactions in which light energy is harnessed to synthesize ATP and NADPH, and the dark reactions in which these products are used to drive the synthesis of carbohydrates from CO_2 and H_2O. The thylakoid membrane is the site of the photosynthetic light reactions, whereas the dark reactions take place in the stroma. The counterpart of the thylakoid in photosynthetic bacteria is a portion of the plasma membrane termed the chromatophore.

Chlorophyll is the principal photoreceptor of photosynthesis. Light is absorbed initially by a light-harvesting antenna system consisting of chlorophyll and accessory pigments. The resulting excitation then migrates via exciton transfer until it reaches the reaction center chlorophyll, where it is trapped.

In purple photosynthetic bacteria, the reaction center is a particle that consists of three subunits and several redox-active small molecules. The primary photon absorbing species of the bacterial reaction center is a "special pair" of BChl a molecules known as P870. By rapid measurement techniques it has been determined that the electron ejected by P870* passes by a third BChl a to a BPheo a molecule and then sequentially to a menaquinone (Q_A) and a ubiquinone (Q_B). The resulting Q_B^- is subsequently further reduced in a second one-electron transfer process and then takes up two protons from the cytosol to form Q_BH_2. The electrons taken up by this species are returned to P870 via a cytochrome bc_1 complex, cytochrome c_2, and, in some bacteria, a four-heme c-type cytochrome associated with the photosynthetic reaction center. This cyclic electron-transport process functions to translocate protons, in part through a Q cycle, from the cytoplasm to the outside of the cell. The resulting proton gradient, in a process known as photophosphorylation, drives the synthesis of ATP. Since bacterial photosynthesis does not generate the reducing equivalents needed in many biosynthetic processes, photosynthetic bacteria require an outside source of reducing agents such as H_2S.

In plants and cyanobacteria, the light reactions occur in two reaction centers, PSI and PSII, that are electrically "connected" in series. This enables the system to generate sufficient electromotive force to form NADPH by oxidizing H_2O in a noncyclic pathway known as the Z-scheme. PSII contains an Mn complex that oxidizes two H_2Os to four H^+ and O_2 in four one-electron steps. The electrons are passed singly, through a Tyr-containing carrier named Z, to photooxidized P680, the reaction center's photon-absorbing species, which consists of one or two Chl a molecules. The electron previously ejected from P680* passes through a series of carriers similar in character to those of the bacterial reaction center

to a pool of plastoquinone molecules. The electrons then enter the cytochrome $b_6 f$ complex, which transports protons, via a Q cycle, from the stroma to the thylakoid space. These electrons are transferred individually, by a plastocyanin carrier, directly to PSI's photooxidized photon-absorbing pigment, P700, which is a single molecule of Chl a. The electron that had been previously released by P700* migrates through a chain of Chl a molecules and then through a chain of ferredoxin molecules. The electron may be returned cyclically, via cytochrome b_6, to the plastoquinone pool so as to translocate protons across the thylakoid membrane. Alternatively, it may act to reduce $NADP^+$ in a noncyclic process mediated by ferredoxin–$NADP^+$ reductase. ATP is synthesized by the CF_1CF_0–ATP synthase, which closely resembles the analogous mitochondrial complex, in a reaction driven by the dissipation of the proton gradient across the thylakoid membrane.

CO_2 is fixed in the photosynthetic dark reactions of plants and cyanobacteria by reactions of the Calvin cycle. The first stage of the Calvin cycle, in sum, mediates the reaction $3RuBP + 3CO_2 \rightarrow 6GAP$ with the consumption of nine ATP and six NADPHs. The second stage reshuffles the atoms of five GAPs to reform the three RuBPs with which the cycle began, a process that requires no further input of free energy or reduction equivalents. The sixth GAP, the product of the Calvin cycle, is used to synthesize carbohydrates, amino acids, and fatty acids. The flux-controlling enzymes of the Calvin cycle are activated in the light through variations in the pH and the Mg^{2+} and NADPH concentrations, and by the redox level of thioredoxin. The central enzyme of the Calvin cycle, RuBP carboxylase, catalyzes both a carboxylase and an oxygenase reaction with RuBP. The latter reaction is the first step in the photorespiration cycle that liberates CO_2. The rate of photorespiration increases with temperature and decreases with CO_2 concentration, so photorespiration constitutes a significant energetic drain on most plants on hot bright days. C_4 plants, which are most common in the tropics, have a system for concentrating CO_2 in their photosynthetic cells so as to minimize the effects of photorespiration but at the cost of two ATPs per CO_2 fixed. Certain desert plants conserve water by absorbing CO_2 at night and releasing it to the Calvin cycle by day. This Crassulacean acid metabolism occurs through a process similar to the C_4 cycle.

REFERENCES

General

Danks, S.M., Evans, E.H., and Whittaker, P.A., *Photosynthetic Systems,* Wiley (1983).

Deisenhofer, J. and Norris, J.R. (Eds.), *The Photosynthetic Reaction Center,* Vols. I and II, Academic Press (1993). [Vol. I covers chemical and biochemical aspects of photosynthesis and Vol. II is oriented towards its physical principles.]

Foyer, C.H., *Photosynthesis,* Wiley (1984).

Nicholls, D.G. and Ferguson, S.J., *Bioenergetics 2,* Chapter 6, Academic Press (1992).

Chloroplasts

Bogorad, L. and Vasil, I.K. (Eds.), *The Molecular Biology of Plastids,* Academic Press (1991).

Hoober, J.K., *Chloroplasts,* Plenum Press (1984).

Light Reactions

Allen, J.F., How does protein phosphorylation regulate photosynthesis? *Trends Biochem. Sci.* **17,** 12–17 (1992).

Anderson, J.M., Photoregulation of the composition, function and structure of thylakoid membranes, *Annu. Rev. Plant Physiol.* **37,** 93–136 (1986).

Andréasson, L.-E. and Vänngard, T., Electron transport in photosystems I and II, *Annu. Rev. Plant Physiol. Plant Mol. Biol.* **39,** 379–411 (1988).

Barber, J. and Anderson, B., Revealing the blueprint of photosynthesis, *Nature* **370,** 31–34 (1994).

Barber, J. (Ed.), *The Photosystems: Structure Function and Molecular Biology,* Elsevier (1992).

Beck, W.F. and dePaula, J.C., Mechanism of photosynthetic water oxidation, *Annu. Rev. Biophys. Biophys. Chem.* **18** 25–46 (1989).

Bogorad, L. and Vasil, I.K. (Eds.), T*he Photosynthetic Apparatus,* Academic Press (1991).

Brudvig, G.W., Probing the mechanism of of water oxidation in photosystem II, *Acc. Chem. Res.* **24,** 311–316 (1991); *and* Brudvig, G.W., Beck, W.F., and de Paula, J.C., Mechanism of photosynthetic water oxidation, *Annu. Rev. Biophys. Biophys. Chem.* **18,** 25–46 (1989).

Cramer, W.A., Martinez, S.E., Furbacher, P.N., Huang, D., and Smith, J.L., The cytochrome $b_6 f$ complex, *Curr. Opin. Struct. Biol.* **4,** 536–544 (1994).

Curr. Top. Bioenerg. **16** (1991). [Contains comprehensive reviews on photosynthetic electron transfer and ATP synthesis.]

Debus, R.J., The manganese and calcium ions of photosynthetic oxygen evolution, *Biochim. Biophys. Acta* **1102,** 269–352 (1992).

Deisenhofer, J. and Michel, H., High-resolution structures of photosynthetic reaction centers, *Annu. Rev. Biophys. Biophys. Chem.* **20,** 247–266 (1991); *and* Structures of bacterial photosynthetic reaction centers, *Annu. Rev. Cell Biol.* **7,** 1–23 (1991).

DiMagno, T.J., Wang, Z., and Norris, J.R., Initial electron-transfer events in photosynthetic bacteria, *Curr. Opin. Struct. Biol.* **2,** 836–842 (1992).

El-Kabbani, O., Chang, C.-H., Tiede, D., Norris, J., and Schiffer, M., Comparison of reaction centers from *Rhodobacter sphaeroides* and *Rhodopseudomonas viridis:* Overall architecture and protein-pigment interactions, *Biochemistry* **30,** 5361–5369 (1991). [A detailed comparison of the two known X-ray structures of bacterial photosynthetic reaction centers.]

Feher, G., Allen, J.P., Okamura, M.Y., and Rees, D.C., Structure and function of bacterial photosynthetic reaction centres, *Nature* **339,** 111–116 (1989).

Gennis, R.B., Barquera, B., Hacker, B., Van Doren, S.R., Arnaud, S., Crofts, A.R., Davidson, E., Gray, K.A., and Daldal, F., The bc_1 complexes of *Rhodobacter sphaeroides* and *Rhodobacter capsulatus, J. Bioenerg. Biomembr.* **25,** 195–209 (1993).

Ghanotakis, D.F. and Yocum, C.F., Photosystem II and the oxygen-evolving complex, *Annu. Rev. Plant Physiol. Plant Mol. Biol.* **41,** 255–276 (1990).

Golbeck, J.H., Structure and function of photosystem I, *Annu. Rev. Plant Physiol. Plant Mol. Biol.* **43,** 293–324 (1992); The structure of photosystem I, *Curr. Opin. Struct. Biol.* **3,** 508–514 (1993); *and* Shared thematic elements in photochemical reaction centers, *Proc. Natl. Acad. Sci.* **90,** 1642–1646 (1993).

Govindjee and Coleman, W.J., How plants make oxygen, *Sci. Am.* **262**(2): 50–59 (1990).

Gutteridge, S., Limitations of the primary events of CO_2 fixation in photosynthetic organisms: the structure and mechanism of rubisco, *Biochim. Biophys. Acta* **1015,** 1–14 (1990).

Hunter, C.N., van Grondelle, R., and Olsen, J.D., Photosynthetic antenna proteins: 100 ps before photochemistry starts, *Trends Biochem. Sci.* **14,** 72–76 (1989).

Karplus, P.A., Daniels, M.J., and Herriott, J.R., Atomic structure of ferredoxin-NADP$^+$ reductase: Prototype for a structurally novel flavoenzyme family, *Science* **251,** 60–66 (1991).

Knaff, D.B., The cytochrome bc_1 complex of photosynthetic bacteria, *Trends Biochem. Sci.* **15,** 289–291 (1990).

Knaff, D.B. and Hirasawa, M., Ferredoxin-dependent chloroplast enzymes, *Biochim. Biophys. Acta* **1056,** 93–125 (1991).

Krauss, N., Hinrichs, W., Witt, I., Fromme, P., Pritzkow, W., Dauter, Z., Betzel, C., Wilson, K.S., Witt, H.T., and Saenger, W., Three-dimensional structure of system I of photosynthesis at 6 Å resolution, *Nature* **361,** 326–331 (1993).

Kühlbrandt, W., Wang, D.N., and Fujiyoshi, Y., Atomic model of plant light-harvesting complex by electron crystallography, *Nature* **367,** 614–621 (1994); *and* Kühlbrandt, W., Structure and function of the plant light-harvesting complex, LHC-II, *Curr. Opin. Struct. Biol.* **4,** 519–528 (1994).

Martinez, S.E., Huang, D., Szczepaniak, A., Cramer, W.A., and Smith, J.L., Crystal structure of chloroplast cytochrome *f* reveals a novel cytochrome fold and unexpected heme ligation. *Structure* **2,** 95–105 (1994).

Okamura, M.Y. and Feher, G., Proton transfer in reaction centers from photosynthetic bacteria, *Annu. Rev. Biochem.* **61,** 861–896 (1992).

Rutherford, A.W., Photosystem II, the water-splitting enzyme, *Trends Biochem. Sci.* **14,** 227–232 (1989).

Staehelin, J.K. and Arntzen, C.J. (Eds.), *Encyclopedia of Plant Physiology,* Vol. 19, Photosynthesis III, Springer–Verlag (1986). [Authoritative articles on the major aspects of light reactions.]

Stanier, R.Y., Ingraham, J., Wheelis, M.L., and Painter, P.R., *The Microbial World* (5th ed.), Chapter 15, Prentice–Hall (1986). [The biology of photosynthetic eubacteria.]

Strotmann, H. and Bickel-Sandkötter, S., Structure, function, and regulation of chloroplast ATPase, *Annu. Rev. Plant Physiol.* **35,** 97–120 (1984).

Youvain, D.C. and Marrs, B.L., Molecular mechanisms of photosynthesis, *Sci. Am.* **256**(6): 42–48 (1987).

Dark Reactions

Brändén, C.-I., Lindqvist, Y., and Schneider, G., Protein engineering of rubisco, *Acta Cryst.* **B47,** 824–835 (1991); *and* Schneider, G., Lindqvist, Y., and Brändén, C.-I., RUBISCO: Structure and mechanism, *Annu. Rev. Biophys. Biomol. Struct.* **21,** 119–143 (1992).

Edwards, G. and Walker, D., C$_3$, C$_4$.: *mechanisms, and cellular and environmental regulation, of photosynthesis,* University of Californis Press (1983).

Hartman, F.C. and Harpel, M.R., Chemical and genetic probes of the active site of D-ribulose-1,5,-bisphosphate carboxylase/oxygenase: A retrospective based on the three-dimensional structure, *Adv. Enzymol. Relat. Areas Mol. Biol.* **67,** 1–75 (1993).

Hatch, M.D., C$_4$ photosynthesis: a unique blend of modified biochemistry, anatomy, and ultrastructure, *Biochem. Biophys. Acta* **895,** 81–106 (1987).

Miziorko, H.M. and Lorimer, G.H., Ribulose-1,5-bisphosphate carboxylase–oxygenase, *Annu. Rev. Biochem.* **52,** 507–535 (1983).

Ogren, W.L., Photorespiration: pathways, regulation, and modification, *Annu. Rev. Plant Physiol.* **35**, 415–442 (1984).

Portis, A.R., Jr., Regulation of ribulose 1,5-bisphosphate carboxylase/oxygenase activity, *Annu. Rev. Plant Physiol. Plant Mol. Biol.* **43**, 415–437 (1992); *and* Rubisco activase, *Biochim. Biophys. Acta* **1015**, 15–28 (1990).

Schreuder, H.A., Knight, S., Curmi, P.M.G., Andersson, I., Cascio, D., Sweet, R.M., Brändén, C.-I., and Eisenberg, D., Crystal

structure of activated tobacco rubisco complexed with the reaction-intermediate analogue 2-carboxy-arabinitol 1,5-bisphosphate, *Protein Sci.* **2**, 1136–1146 (1993).

Spreitzer, R.J., Genetic dissection of rubisco structure and function, *Annu. Rev. Plant Physiol. Plant Mol. Biol.* **44**, 411–434 (1993).

Ting, I.P., Crassulacean acid metabolism, *Annu. Rev. Plant Physiol.* **36**, 595–622 (1985).

PROBLEMS

1. Why is chlorophyll green in color when it absorbs in the red and the blue regions of the spectrum (Fig. 22-5)?

2. Indicate, where appropriate, the analogous components in the photosynthetic electron-transport chains of purple photosynthetic bacteria and chloroplasts.

3. Antimycin A inhibits photosynthesis in chloroplasts. Indicate its most likely site of action and explain your reasoning.

4. Calculate the energy efficiency of cyclic and noncyclic photosynthesis in chloroplasts using 680-nm light. What would this efficiency be with 500-nm light? Assume that ATP formation requires 59 kJ·mol^{-1} under physiological conditions.

*5. What is the minimum pH gradient required to synthesize ATP from ADP + P$_i$? Assume [ATP]/([ADP] [P$_i$]) = 10^3, $T =$ 25°C, and that three protons must be translocated per ATP generated. (See Table 15-3 for useful thermodynamic information.)

6. Indicate the average Calvin cycle labeling pattern in ribulose-5-phosphate after two rounds of exposure to $^{14}CO_2$.

7. Chloroplasts are illuminated until the levels of their Calvin cycle intermediates reach a steady state. The light is then turned off. How do the levels of RuBP and 3PG vary after this time?

8. What is the energy efficiency of the Calvin cycle combined with glycolysis and oxidative phosphorylation; that is, what percentage of the input energy can be metabolically recovered in synthesizing starch from CO_2 using photosynthetically produced NADPH and ATP rather than somehow directly storing these "high-energy" intermediates? Assume that each NADPH is energetically equivalent to three ATPs and that starch synthesis and breakdown are energetically equivalent to glycogen synthesis and breakdown.

9. If a C$_3$ plant and a C$_4$ plant are placed together in a sealed illuminated box with sufficient moisture, the C$_4$ plant thrives while the C$_3$ plant sickens and eventually dies. Explain.

10. The leaves of some species of desert plants taste sour in the early morning but, as the day wears on, they become tasteless and then bitter. Explain.

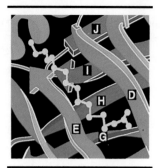

C H A P T E R

23

Lipid Metabolism

Lipids play indispensable roles in cell structure and metabolism. For example, triacylglycerols are the major storage form of metabolic energy in animals; cholesterol is a vital component of cell membranes and a precursor of the steroid hormones and bile acids; arachidonate is an unsaturated fatty acid that serves as the precursor of the prostaglandins, prostacyclins, thromboxanes, and leukotrienes, potent intercellular mediators that control a variety of complex processes; and complex glycolipids and phospholipids are major components of biological membranes. We discussed the structures of simple and complex lipids in Section 11-1. In the first half of this chapter, we consider the metabolism of fatty acids and triacylglycerols, including their digestion, oxidation, and biosynthesis. We then consider how cholesterol is synthesized and utilized, and how arachidonate is converted to prostaglandins, prostacyclins, thromboxanes, and leukotrienes. We end by studying how complex glycolipids and phospholipids are synthesized from their simpler lipid and carbohydrate components.

1. LIPID DIGESTION, ABSORPTION, AND TRANSPORT

***Triacylglycerols** (also called **fats** or **triglycerides**) constitute both ~90% of the dietary lipid and the major form of metabolic energy storage in humans.* Triacylglycerols consist of glycerol triesters of fatty acids such as palmitic and oleic acids

1-Palmitoyl-2,3-dioleoyl-glycerol

(the names and structural formulas of some biologically common fatty acids are listed in Table 11-1). Like glucose, they are metabolically oxidized to CO_2 and H_2O. Yet, since

TABLE 23-1. ENERGY CONTENT OF FOOD CONSTITUENTS

Constituent	$\Delta H(\text{kJ} \cdot \text{g}^{-1}$ dry weight)
Carbohydrate	16
Fat	37
Protein	17

Source: Newsholme, E.A. and Leech, A.R., *Biochemistry for the Medical Sciences*, p. 16, Wiley (1983).

most carbon atoms of triacylglycerols have lower oxidation states than those of glucose, *the oxidative metabolism of fats yields over twice the energy of an equal weight of dry carbohydrate or protein (Table 23-1).* Moreover, fats, being nonpolar, are stored in an anhydrous state, whereas glycogen, the storage form of glucose, is polar and is consequently stored in a hydrated form that contains about twice its dry weight of water. Fats therefore provide up to six times the metabolic energy of an equal weight of hydrated glycogen.

Lipid Digestion Occurs at Lipid–Water Interfaces

Since triacylglycerols are water insoluble, whereas digestive enzymes are water soluble, *triacylglycerol digestion takes place at lipid–water interfaces.* The rate of triacylglycerol digestion therefore depends on the surface area of the interface, a quantity that is greatly increased by the churning peristaltic movements of the intestine combined with the emulsifying action of **bile acids.** The bile acids are powerful digestive detergents that, as we shall see in Section 23-6C, are synthesized by the liver and secreted via the gallbladder into the small intestine where lipid digestion and absorption mainly take place.

Pancreatic Lipase Requires Interfacial Activation and Has a Catalytic Triad

Pancreatic **lipase (triacylglycerol lipase)** catalyzes the hydrolysis of triacylglycerols at their 1 and 3 positions to form sequentially **1,2-diacylglycerols** and **2-acylglycerols,** together with the Na^+ and K^+ salts of fatty acids (soaps). These soaps, being amphipathic, aid in the lipid emulsification process.

The enzymatic activity of pancreatic lipase greatly increases when it contacts the lipid–water interface, a phenomenon known as **interfacial activation.** However, the enzyme does not bind to the lipid–water interface unless it is in complex with pancreatic **colipase,** a protein that forms a 1 : 1 complex with lipase. The X-ray structures, determined by Christian Cambillau, of pancreatic lipase–procolipase complexes, alone and cocrystallized with mixed micelles of phosphatidylcholine (Fig. 11-4) and bile acid, have revealed the structural basis of the interfacial activation of lipase as well as how colipase aids lipase in binding to the lipid–water interface (Fig. 23-1).

The active site of the 449-residue pancreatic lipase, which is contained in the enzyme's N-terminal domain (residues

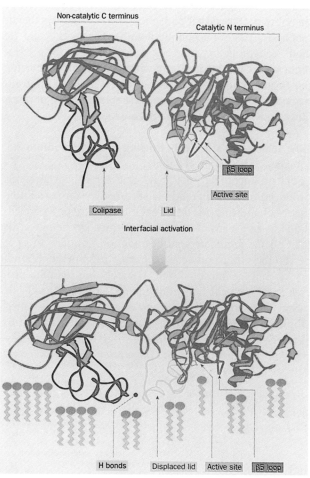

FIGURE 23-1. The mechanism of interfacial activation of triacylglycerol lipase in complex with procolipase (*purple*). Upon binding to a phospholipid micelle (*green*), the 25-residue lid (*yellow*) covering the enzyme's active site (*pink*) changes conformation so as to expose its hydrophobic residues, thereby uncovering the active site. This causes the 10-residue β5 loop (*red*) to move aside in a way that forms the enzyme's oxyanion hole. The procolipase also changes its conformation so as to hydrogen bond to the "open" lid, thereby stabilizing it in this conformation and, together with lipase, forming an extended hydrophobic surface. [From *Nature*, **362**, 793 (1993). Reproduced with permission.]

1–336), contains a catalytic triad that closely resembles the one in the serine proteases (Section 14-3B; recall that ester hydrolysis is mechanistically similar to peptide hydrolysis). In the absence of mixed micelles, lipase's active site is covered by a 25-residue helical lid. However, in the presence of the mixed micelles, the lid undergoes a complex structural reorganization that exposes the active site; causes a contacting 10-residue loop, the β5 loop, to change conformation in a way that forms the active enzyme's oxyanion hole; and generates a hydrophobic surface about the entrance to the active site. Indeed, the active site of the mixed micelle–containing complex contains a long rod of electron density that contacts the catalytic triad's Ser residue and appears to be a phosphatidylcholine molecule.

Procolipase binds to the C-terminal domain of lipase (residues 337-449) such that the hydrophobic tips of the three loops that comprise much of this 90-residue protein extend from the complex on the same face as lipase's active site. A continuous hydrophobic plateau is thereby created that extends over a distance of > 50 Å past the active site and that, presumably, helps bind the complex to the lipid surface. The procolipase also forms three hydrogen bonds to the opened lid, thereby stabilizing it in this conformation.

Pancreatic Phospholipase A₂ Has a Catalytic Diad

Phospholipids are degraded by pancreatic **phospholipase A₂**, which hydrolytically excises the fatty acid residue at C2 to yield the corresponding **lysophospholipids** (Fig. 23-2), which are also powerful detergents. Indeed, the phospholipid lecithin (phosphatidylcholine) is secreted in the bile, presumably to aid in lipid digestion.

Phospholipase A₂, as does triacylglycerol lipase, preferentially catalyzes reactions at interfaces. However, as Paul Sigler's determinations of the X-ray structures of the phospholipases A₂ from cobra venom and bee venom revealed, its mechanism of interfacial activation differs from that of triacylglycerol lipase in that it does not change its confor-

mation. Instead, phospholipase A₂ contains a hydrophobic channel that provides the substrate with direct access from the phospholipid aggregate (micelle or membrane) surface to the bound enzyme's active site. Hence, on leaving its micelle to bind to the enzyme, the substrate need not become solvated and then desolvated (Fig. 23-3). In contrast, soluble and dispersed phospholipids must first surmount these significant kinetic barriers in order to bind to the enzyme.

The catalytic mechanism of phospholipase A₂ also differs substantially from that of triacylglycerol lipase. Although the phospholipase A₂ active site contains the His and Asp components of a catalytic triad, an enzyme-bound water molecule occupies the position expected for an active site Ser. Moreover, the active site contains a bound Ca^{2+} ion and does not form an acyl–enzyme intermediate. Sigler therefore proposed that phospholipase A₂ catalyzes the direct hydrolysis of phospholipid with a His–Asp "catalytic diad" activating an active site water molecule for nucleophilic attack on the ester, and with the Ca^{2+} ion stabilizing the oxyanion transition state. Figure. 23-3*b* shows this transition state along with a schematic representation of interfacial activation via a hydrophobic channel.

Bile Acids and Fatty Acid–Binding Protein Facilitate the Intestinal Absorption of Lipids

The mixture of fatty acids and mono- and diacylglycerols produced by lipid digestion is absorbed by the cells lining the small intestine (the intestinal mucosa) in a process facilitated by bile acids. The micelles formed by the bile acids take up the nonpolar lipid degradation products so as to permit their transport across the unstirred aqueous boundary layer at the intestinal wall. The importance of this process is demonstrated in individuals with obstructed bile ducts: They absorb little of their dietary lipids but, rather, eliminate them in hydrolyzed form in their feces (**steatorrhea**). Evidently, *bile acids are not only an aid to lipid digestion but are essential for the absorption of lipid digestion*

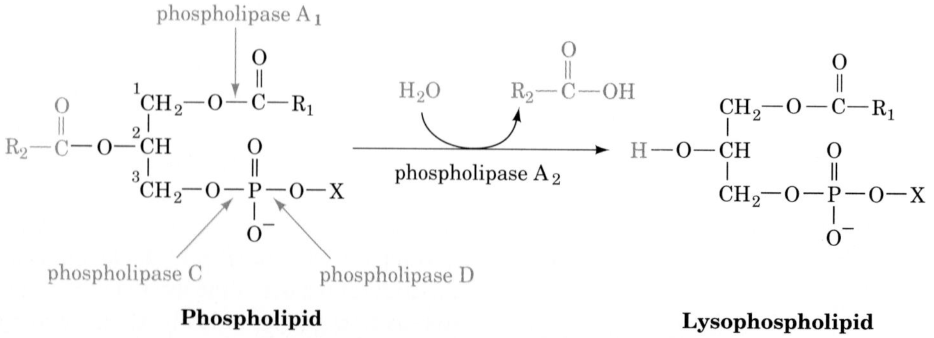

FIGURE 23-2. Phospholipase A₂ hydrolytically excises the C2 fatty acid residue from a triacylglycerol to yield the corresponding lysophospholipid. The bonds hydrolyzed by other types of phospholipases, which are named according to their specificities, are also indicated.

products. Bile acids are likewise required for the efficient intestinal absorption of the lipid-soluble vitamins A, D, E, and K.

Inside the intestinal cells, fatty acids form complexes with **intestinal fatty acid–binding protein (I-FABP)**, a cytoplasmic protein, which serves to increase the effective solubility of these water-insoluble substances and also to protect the cell from their detergent-like effects (recall that soaps are fatty acid salts). The X-ray structures of rat I-FABP, both alone and in complex with a single molecule of palmitate, were determined by James Sacchettini. This monomeric, 131-residue protein consists largely of 10 antiparallel

(a)

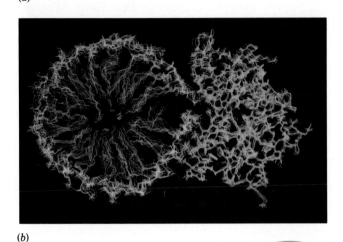

FIGURE 23-3. (*a*) A hypothetical model of phospholipase A$_2$ in complex with a micelle of lysophosphatidylethanolamine as shown in cross-section. The protein is drawn in light blue, the phospholipid head groups are yellow, and their hydrocarbon tails are blue. The calculated atomic motions of the assembly are indicated through a series of superimposed images taken at 5-ps intervals. [Courtesy of Raymond Salemme, E.I. du Pont de Nemours & Company.] (*b*) Schematic diagram of a productive interaction between phospholipase A$_2$ and a phosphatidylethanolamine contained in a micelle. The mechanism of action of phospholipase A$_2$ involves a His 48 – Asp 99 "catalytic diad" (*red*) that activates a water molecule (*blue*) for nucleophilic attack on the ester while Ca^{2+} stabilizes the oxyanion transition state. The fatty acid leaving group is shown in green and EA represents the ethanolamine head group. [After Scott, D. L., White, S. P., Otwinowski, Z., Yuan, W., Gelb, M. H., and Sigler, P. B., *Science* **250**, 1545 (1990).]

(b)

Phospholipase A$_2$

Lipid micelle

Interfacial surface

Phospholipase A$_2$

Asp 99

O — EA$^+$

His 48

Gly 30

Asp 49

Ca^{2+}

Hydrophobic channel

Interfacial surface

β strands organized into a stack of 2 approximately orthogonal β sheets (Fig. 23-4). The palmitate occupies a gap between two of the β strands such that it lies between the β sheets with an orientation that, over much of its length, is more or less parallel to the gapped β strands (this structure has been described as forming a "β-clam"). The palmitate's carboxyl group interacts with Arg 106, Gln 115, and two bound water molecules, whereas the methylene chain is encased by the side chains of several hydrophobic, mostly aromatic, residues.

Lipids Are Transported in Lipoprotein Complexes

The lipid digestion products absorbed by the intestinal mucosa are converted by these tissues to triacylglycerols (Section 23-4F) and then packaged into lipoprotein particles called **chylomicrons**. These, in turn, are released into the bloodstream via the lymph system for delivery to the tissues. Similarly, triacylglycerols synthesized by the liver are packaged into **very low density lipoproteins (VLDL)** and released directly into the blood. These lipoproteins, whose origins, structures, and functions are discussed in Section 11-4, maintain their otherwise insoluble lipid components in aqueous solution.

The triacylglycerol components of chylomicrons and VLDL are hydrolyzed to free fatty acids and glycerol in the capillaries of adipose tissue and skeletal muscle by **lipoprotein lipase** (Section 11-5B). The resulting free fatty acids are taken up by these tissues while the glycerol is transported to the liver or kidneys. There it is converted to the glycolytic intermediate dihydroxyacetone phosphate by the sequential actions of **glycerol kinase** and **glycerol-3-phosphate dehydrogenase** (Fig. 23-5).

Mobilization of triacylglycerols stored in adipose tissue involves their hydrolysis to glycerol and free fatty acids by **hormone-sensitive triacylglycerol lipase** (Section 23-5). The free fatty acids are released into the bloodstream, where they bind to **albumin,** a soluble 66.5-kD monomeric protein that comprises about half of the blood serum protein. In the absence of albumin, the maximum solubility of free fatty acids is $\sim 10^{-6}M$. Above this concentration, free fatty acids form micelles that act as detergents to disrupt protein and membrane structure and would therefore be toxic. However, the effective solubility of fatty acids in fatty acid–albumin complexes is as much as 2 mM. Nevertheless,

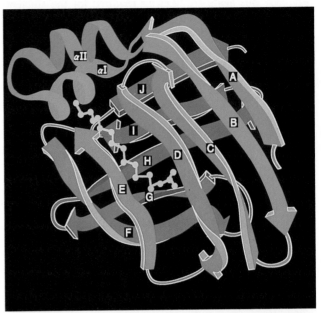

FIGURE 23-4. The X-ray structure of rat intestinal fatty acid–binding protein, shown in ribbon form (*blue*), in complex with palmitate, shown in ball-and-stick form (*yellow*). [Courtesy of James Sacchettini, Albert Einstein College of Medicine.]

those rare individuals with **analbuminemia** (severely depressed levels of albumin) suffer no apparent adverse symptoms; evidently, their fatty acids are transported in complex with other serum proteins.

2. FATTY ACID OXIDATION

The biochemical strategy of fatty acid oxidation was understood long before the advent of modern biochemical techniques involving enzyme purification or the use of radioactive tracers. In 1904, Franz Knoop, in the first use of chemical labels to trace metabolic pathways, fed dogs fatty acids labeled at their ω (last) carbon atom by a benzene ring and isolated the phenyl-containing metabolic products from their urine. Dogs fed labeled odd-chain fatty acids excreted **hippuric acid,** the glycine amide of **benzoic acid,**

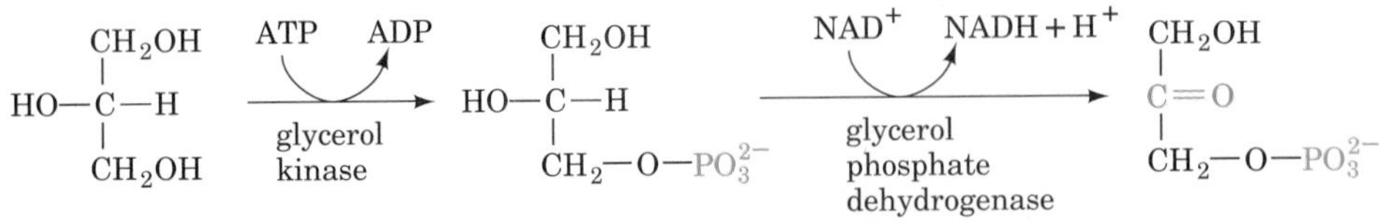

FIGURE 23-5. The conversion of glycerol to the glycolytic intermediate dihydroxyacetone phosphate.

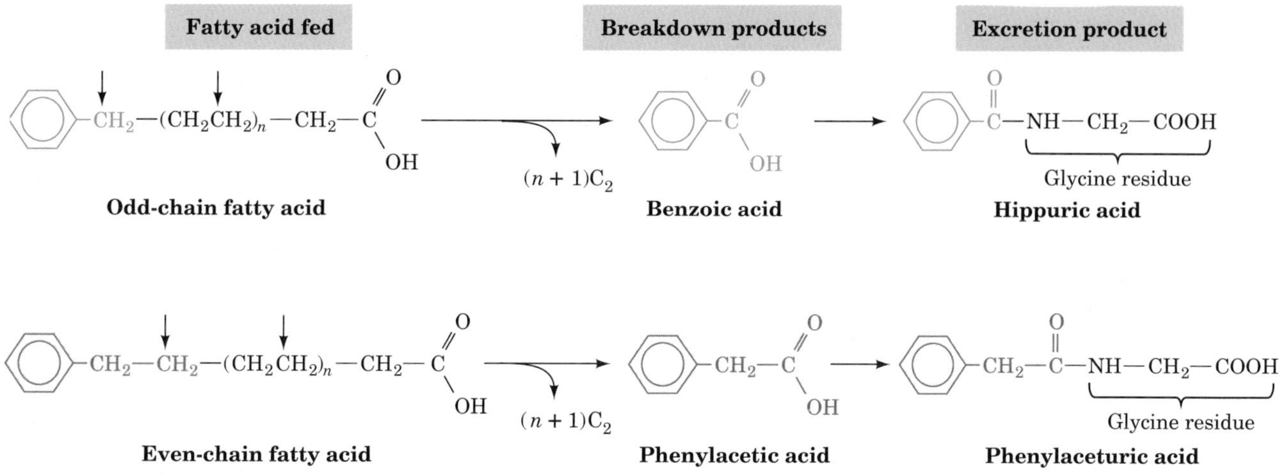

FIGURE 23-6. Franz Knoop's classic experiment indicating that fatty acids are metabolically oxidized at their β-carbon atom. ω-Phenyl-labeled fatty acids containing an odd number of carbon atoms are oxidized to the phenyl-labeled C_1 product, benzoic acid, whereas those with an even number of carbon atoms are oxidized to the phenyl-labeled C_2 product, phenylacetic acid. These products are excreted as their respective glycine amides, hippuric and phenylaceturic acids. The vertical arrows indicate the deduced sites of carbon oxidation. The intermediate C_2 products are oxidized to CO_2 and H_2O and were therefore not isolated.

whereas those fed labeled even-chain fatty acids excreted **phenylaceturic acid,** the glycine amide of **phenylacetic acid** (Fig. 23-6). Knoop therefore deduced that the oxidation of the carbon atom β to the carboxyl group is involved in fatty acid breakdown. Otherwise, the phenylacetic acid would be further oxidized to benzoic acid. Knoop proposed that this breakdown occurs by a mechanism known as **β oxidation** in which the fatty acid's C_β atom is oxidized. It was not until after 1950, following the discovery of coenzyme A, that the enzymes of fatty acid oxidation were isolated and their reaction mechanisms elucidated. This work confirmed Knoop's hypothesis.

A. Fatty Acid Activation

Before fatty acids can be oxidized, they must be "primed" for reaction in an ATP-dependent acylation reaction to form fatty acyl-CoA. This "activation" process is catalyzed by a family of at least three **acyl-CoA synthetases** (also called **thiokinases**) that differ according to their chain-length specificities. These enzymes, which are associated with either the endoplasmic reticulum or the outer mitochondrial membrane, all catalyze the reaction

Fatty acid + CoA + ATP $\rightleftharpoons$ acyl-CoA + AMP + PP_i

In the activation of ^{18}O-labeled palmitate by a long-chain acyl-CoA synthetase, both the AMP and the acyl-CoA products become ^{18}O labeled. This observation indicates that the reaction has an acyladenylate mixed anhydride intermediate that is attacked by the sulfhydryl group of CoA to form the thioester product (Fig. 23-7). The reaction involves both the cleavage and the synthesis of bonds with large negative free energies of hydrolysis so that the free energy change associated with the overall reaction is close to zero. The reaction is driven to completion in the cell by the highly exergonic hydrolysis of the product pyrophosphate (PP_i) catalyzed by the ubiquitous **inorganic pyro-**

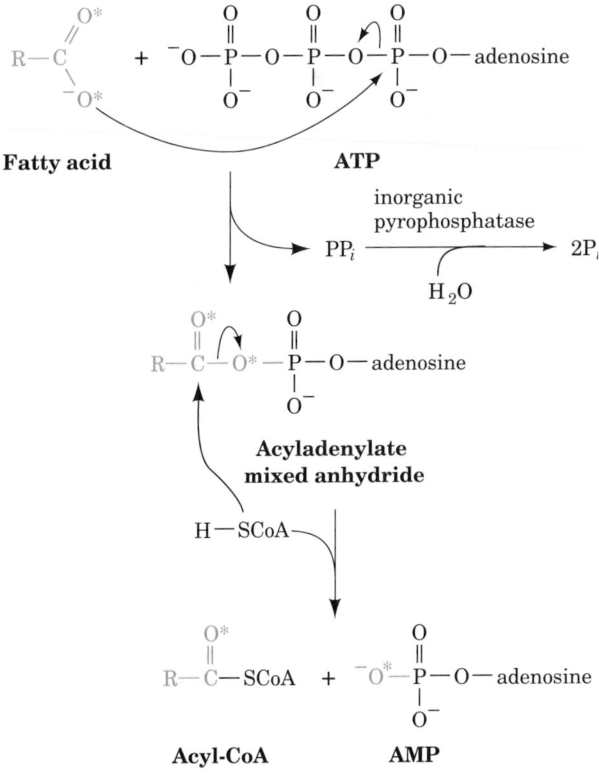

FIGURE 23-7. The mechanism of fatty acid "activation" catalyzed by acyl-CoA synthase. Experiments utilizing ^{18}O-labeled fatty acids (*) demonstrate that the formation of acyl-CoA involves an intermediate acyladenylate mixed anhydride.

phosphatase. Thus, as commonly occurs in metabolic pathways, *a reaction forming a "high-energy" bond through the hydrolysis of one of ATP's phosphoanhydride bonds is driven to completion by the hydrolysis of its second such bond.*

B. Transport across the Mitochondrial Membrane

Although fatty acids are activated for oxidation in the cytosol, they are oxidized in the mitochondrion as Eugene Kennedy and Albert Lehninger established in 1950. We must therefore consider how fatty acyl-CoA is transported across the inner mitochondrial membrane. A long-chain fatty acyl-CoA cannot directly cross the inner mitochondrial membrane. Rather, its acyl portion is first transferred to **carnitine** (Fig. 23-8), a compound that occurs in both plant and animal tissues. This transesterification reaction has an equilibrium constant close to 1, which indicates that the *O*-acyl bond of **acyl-carnitine** has a free energy of hydrolysis similar to that of the thioester. **Carnitine palmitoyl transferases I** and **II,** which can transfer a variety of acyl groups, are located, respectively, on the external and internal surfaces of the inner mitochondrial membrane. The translocation process itself is mediated by a specific carrier protein that transports acyl-carnitine into the mitochondrion while transporting free carnitine in the opposite direction. Acyl-CoA transport therefore occurs via four reactions (Fig. 23-9):

1. The acyl group of a cytosolic acyl-CoA is transferred to carnitine, thereby releasing the CoA to its cytosolic pool.

2. The resulting acyl-carnitine is transported into the mitochondrial matrix by the transport system.

3. The acyl group is transferred to a CoA molecule from the mitochondrial pool.

4. The product carnitine is returned to the cytosol.

The cell thereby maintains separate cytosolic and mitochondrial pools of CoA. The mitochondrial pool functions

in the oxidative degradation of pyruvate (Section 19-2A) and certain amino acids (Sections 24-3E – G) as well as fatty acids, whereas the cytosolic pool supplies fatty acid biosynthesis (Section 23-4). The cell similarly maintains separate cytosolic and mitochondrial pools of ATP and NAD^+.

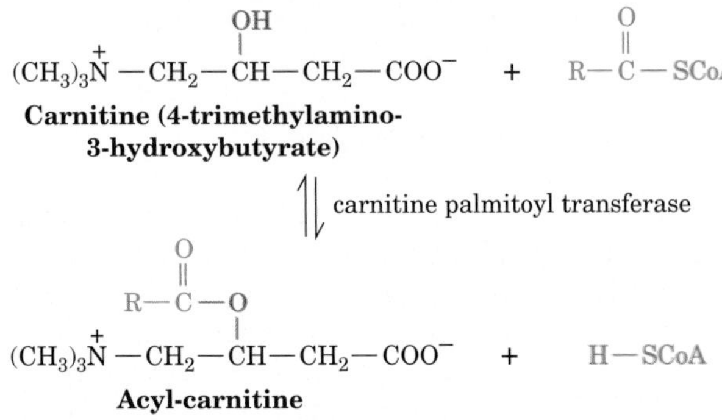

C. β Oxidation

Fatty acids are dismembered through the β oxidation of fatty acyl-CoA, a process that occurs in four reactions (Fig. 23-10):

1. Formation of a *trans-α,β* double bond through dehydrogenation by the flavoenzyme **acyl-CoA dehydrogenase.**

2. Hydration of the double bond by **enoyl-CoA hydratase** to form a **3-L-hydroxyacyl-CoA.**

3. NAD^+-dependent dehydrogenation of this β-hydroxyacyl-CoA by **3-L-hydroxyacyl-CoA dehydrogenase** to form the corresponding β-ketoacyl-CoA.

4. C_α—C_β cleavage in a thiolysis reaction with CoA as catalyzed by **β-ketoacyl-CoA thiolase** (also called just **thiolase**) to form acetyl-CoA and a new acyl-CoA containing two less C atoms than the original one.

The first three steps of this process chemically resemble the citric acid cycle reactions that convert succinate to oxaloacetate (Sections 19-3F – H).

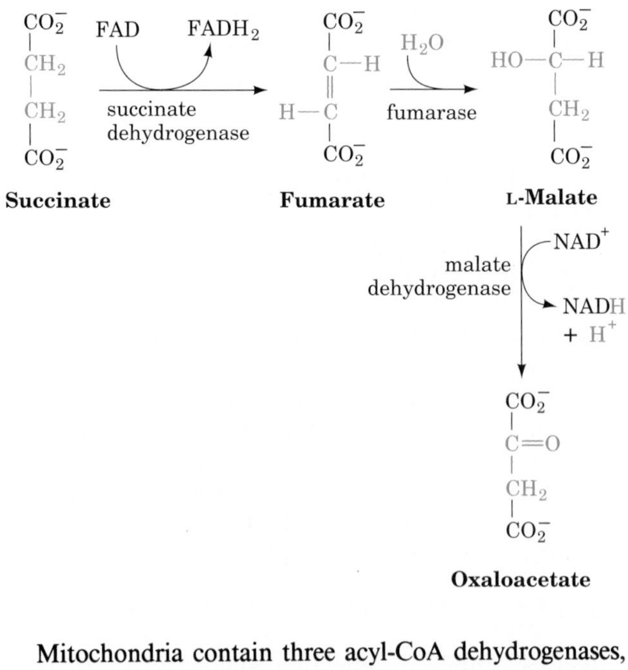

Mitochondria contain three acyl-CoA dehydrogenases, with specificities for short-, medium-, and long-chain fatty acyl-CoAs. The reaction catalyzed by these enzymes is thought to involve removal of a proton at C_α and transfer of a hydride ion equivalent from C_β to FAD (Fig. 23-10, Reaction 1). The X-ray structure of the **medium-chain acyl-CoA dehydrogenase (MCAD)** in complex with **octanoyl-CoA,** determined by Jung-Ja Kim, clearly shows how the enzyme

FIGURE 23-8. The acylation of carnitine is catalyzed by carnitine – palmitoyl transferase.

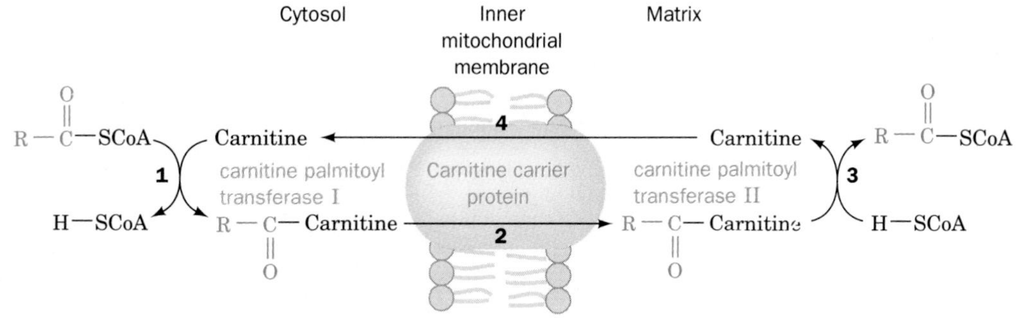

FIGURE 23-9. The transport of fatty acids into the mitochondrion.

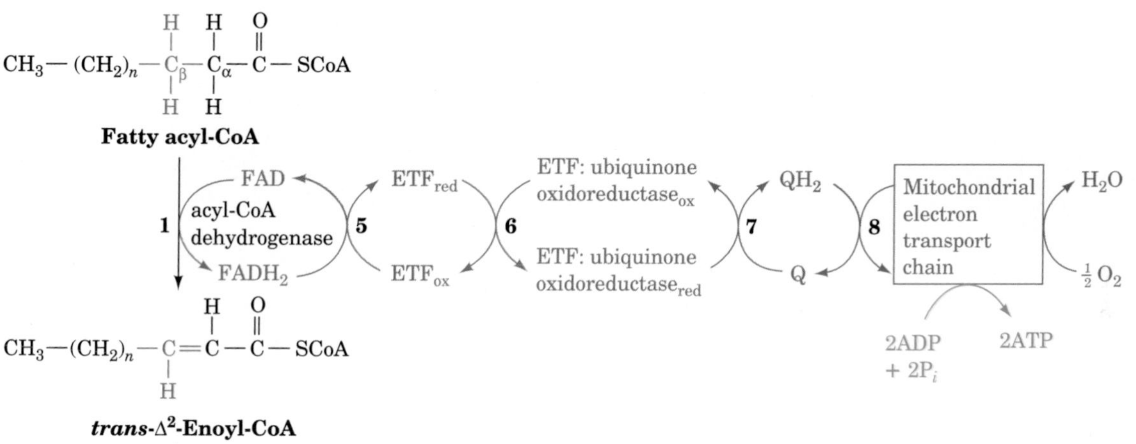

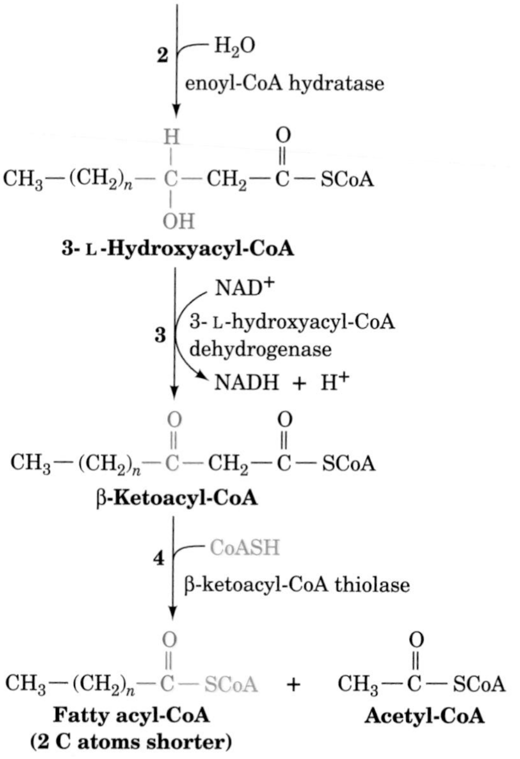

FIGURE 23-10. The β-oxidation pathway of fatty acyl-CoA.

orients the enzyme's base (Glu 376), the substrate C_α—C_β bond, and the FAD prosthetic group for reaction (Fig. 23-11).

Acyl-CoA Dehydrogenase Is Reoxidized via the Electron-Transport Chain

The $FADH_2$ resulting from the oxidation of the fatty acyl-CoA substrate is reoxidized by the mitochondrial electron-transport chain through the intermediacy of a series of electron-transfer reactions. **Electron-transfer flavoprotein (ETF)** transfers an electron pair from $FADH_2$ to the flavo-iron–sulfur protein **ETF:ubiquinone oxidoreductase**, which in turn transfers an electron pair to the mitochondrial electron-transport chain by reducing coenzyme Q (CoQ; Fig. 23-10, Reactions 5–8). Reduction of O_2 to H_2O by the electron-transport chain beginning at the CoQ stage results in the synthesis of two ATPs per electron pair transferred (Section 20-2B).

Acyl-CoA Dehydrogenase Deficiency Has Fatal Consequences

The unexpected death of an apparently healthy infant, often overnight, has been, for lack of any real explanation, termed **sudden infant death syndrome (SIDS)**. MCAD has been shown to be deficient in up to 10% of these infants, making this genetic disease more prevalent than **phenylketonuria (PKU)** (Section 24-3H), a genetic defect in phenylalanine degradation for which babies born in the United States are routinely tested. Glucose is the principal energy

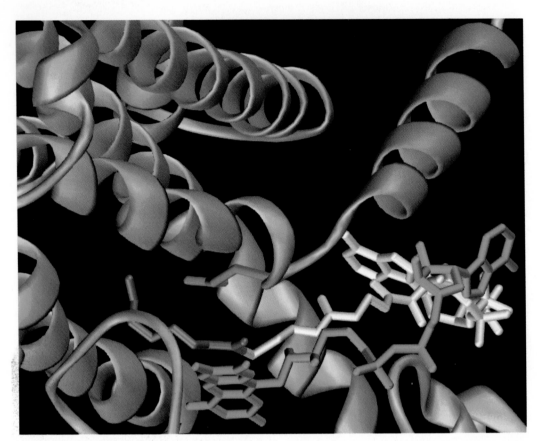

FIGURE 23-11. A ribbon diagram of the active site region in a subunit of medium-chain acyl-CoA dehydrogenase from pig liver mitochondria in complex with octanoyl-CoA. The enzyme is a tetramer of identical 385-residue subunits, each of which binds an FAD prosthetic group (*green*) and its octanoyl-CoA substrate (whose octanoyl and CoA moieties are light blue and white) in largely extended conformations. The octanoyl-CoA binds such that it's C_α—C_β bond is sandwiched between the carboxylate group of Glu 376 (*red*) and the flavin ring (*green*), consistent with the proposal that Glu 376 is the general base that abtracts the α proton in the α,β dehydrogenation reaction catalyzed by the enzyme. [Based on an X-ray structure by Jung-Ja Kim, Medical College of Wisconsin.]

metabolism substrate just after eating, but when the glucose level later decreases, the rate of fatty acid oxidation must correspondingly increase. The sudden death in infants lacking MCAD may be caused by the imbalance between glucose and fatty acid oxidation.

Lys 304, which becomes Glu in the most prevalent mutation among individuals with MCAD deficiency, is ~20 Å distant from the enzyme's active site and hence cannot participate in binding substrate or FAD. However, since the side chains of Asp 300 and Asp 346 lie within 6 Å of Glu 304, near a subunit–subunit interface, it seems likely that the high concentration of negative charges resulting from the Lys 304 → Glu mutation structurally destabilizes the enzyme.

Deficiency of acyl-CoA dehydrogenase has also been implicated in **Jamaican vomiting sickness,** whose victims suffer violent vomiting followed by convulsions, coma, and death. Severe hypoglycemia is observed in most cases. This condition results from eating unripe **ackee fruit,** which contains **hypoglycin A,** an unusual amino acid, which is metab-

olized to **methylenecyclopropylacetyl-CoA (MCPA-CoA;** Fig. 23-12). MCPA-CoA, a substrate for acyl-CoA dehydrogenase, is thought to undergo the first step of the reaction that this enzyme catalyzes, removal of a proton from C_α, to form a reactive intermediate that covalently modifies the enzyme's FAD prosthetic group (Fig. 23-12). Since a normal step in the enzyme's reaction mechanism generates the reactive intermediate, MCPA-CoA is said to be a **mechanism-based inhibitor.**

The Thiolase Reaction Occurs via Claisen Ester Cleavage

Following the dehydrogenation of the fatty acyl-CoA starting material of fatty acid oxidation to an enoyl-CoA, enoyl-CoA hydratase catalyzes stereospecific addition of H_2O to its substrate's trans-α,β double bond to form 3-L-(S)-hydroxyacyl-CoA. 3-L-Hydroxyacyl-CoA dehydrogenase oxidizes this secondary alcohol to a ketone utilizing NAD^+ as its oxidizing agent.

The final stage of the fatty acid β-oxidation process, the thiolase reaction, forms acetyl-CoA and a new acyl-CoA,

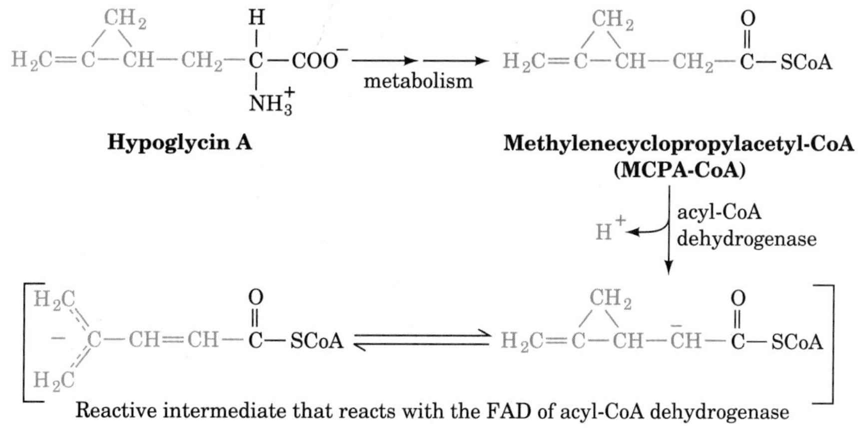

Hypoglycin A

Methylenecyclopropylacetyl-CoA (MCPA-CoA)

Reactive intermediate that reacts with the FAD of acyl-CoA dehydrogenase

FIGURE 23-12. Metabolic conversions of hypoglycin A to yield a product that inactivates acyl-CoA dehydrogenase. Spectral changes suggest that the enzyme's FAD prosthetic group has been modified, although the purported adduct has not yet been characterized.

β-Ketoacyl-CoA

Acetyl-CoA

Enzyme–thioester intermediate

Acyl-CoA

which is two carbon atoms shorter than the one that began the cycle (Fig. 23-13):

1. The first step of the thiolase reaction involves formation of a thioester bond to the substrate by an active site thiol group, a deduction based on the observation that [^{14}C]acetyl-CoA labels a specific enzyme Cys residue.

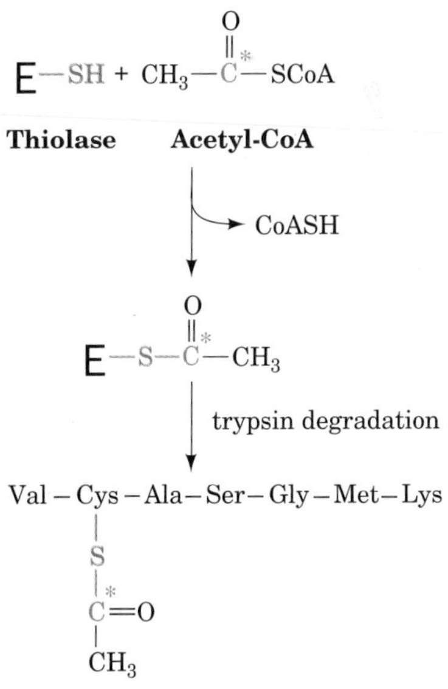

Thiolase   **Acetyl-CoA**

Val – Cys – Ala – Ser – Gly – Met – Lys

2. The second step involves carbon–carbon bond cleavage to form an acetyl-CoA carbanion intermediate that is stabilized by electron withdrawal into this thioester's carbonyl group. This type of reaction is known as a Clai-

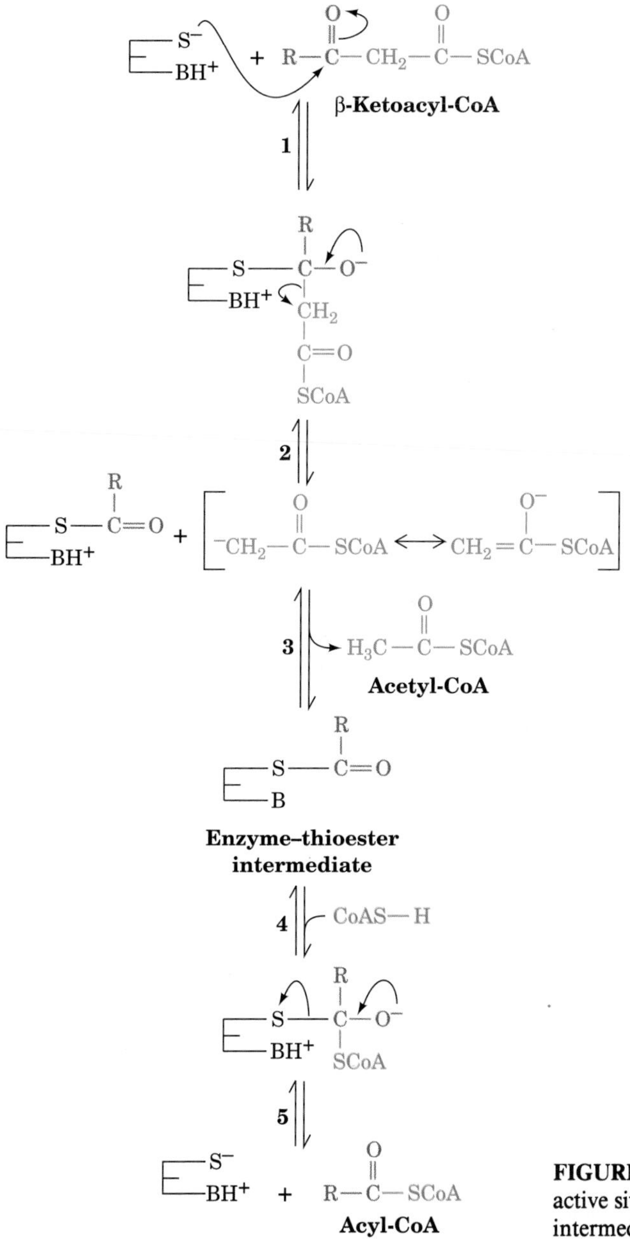

FIGURE 23-13. The mechanism of action of β-ketoacyl-CoA thiolase. An active site Cys residue participates in the formation of an enzyme thioester intermediate.

sen ester cleavage (the reverse of a Claisen condensation). The citric acid cycle enzyme citrate synthase also catalyzes a reaction that involves a stabilized acetyl-CoA carbanion intermediate (Section 19-3A).

3. The acetyl-CoA carbanion intermediate is protonated by an enzyme acid group, yielding acetyl-CoA.
4. Finally, CoA displaces the enzyme thiol group from the enzyme–thioester intermediate, yielding acyl-CoA.

Fatty Acid Oxidation Is Highly Exergonic

The function of fatty acid oxidation is, of course, to generate metabolic energy. Each round of β oxidation produces one NADH, one $FADH_2$, and one acetyl-CoA. Oxidation of acetyl-CoA via the citric acid cycle generates additional $FADH_2$ and NADH, which are reoxidized through oxidative phosphorylation to form ATP. Complete oxidation of a fatty acid molecule is therefore a highly exergonic process, which yields numerous ATPs. For example, oxidation of palmitoyl-CoA (which has a C_{16} fatty acyl group) involves seven rounds of β oxidation, yielding $7FADH_2$, 7NADH, and 8acetyl-CoA. Oxidation of the 8acetyl-CoA, in turn, yields 8GTP, 24NADH, and $8FADH_2$. Since oxidative phosphorylation of the 31NADH molecules yields 93ATP and that of the $15FADH_2$ yields 30ATPs, subtracting the 2ATP equivalents required for fatty acyl-CoA formation (Section 23-2A), *the oxidation of one palmitate molecule has a net yield of 129ATP.*

D. Oxidation of Unsaturated Fatty Acids

Almost all unsaturated fatty acids of biological origin (Section 11-1A) contain only cis double bonds, which most often begin between C9 and C10 (referred to as a Δ^9 or 9-double bond; Table 11-1). Additional double bonds, if any, occur at three-carbon intervals and are therefore never conjugated. Two examples of unsaturated fatty acids are oleic acid and linoleic acid (Fig. 23-14). Note that one of the double bonds in linoleic acid is at an odd-numbered carbon atom and the other is at an even-numbered carbon atom. Double bonds at these positions in fatty acids pose two problems for the β-oxidation pathway that are solved through the actions of three additional enzymes (Fig. 23-15):

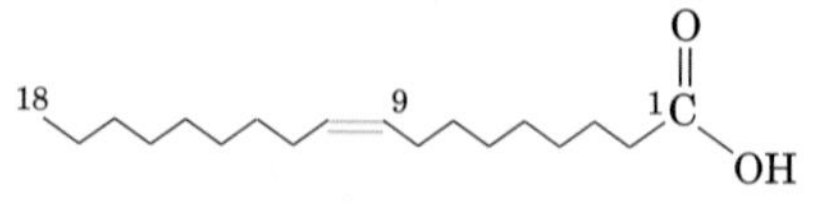

Oleic acid
(9-*cis*-Octadecenoic acid)

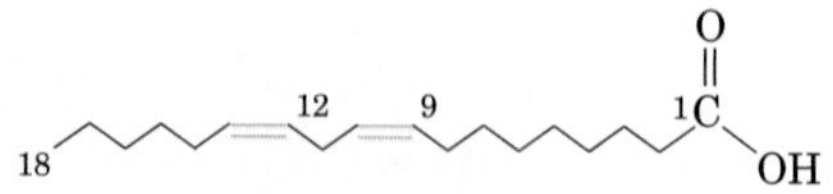

Linoleic acid
(9,12-*cis*-Octadecadienoic acid)

FIGURE 23-14. The structures of two common unsaturated fatty acids. Most unsaturated fatty acids contain unconjugated cis double bonds.

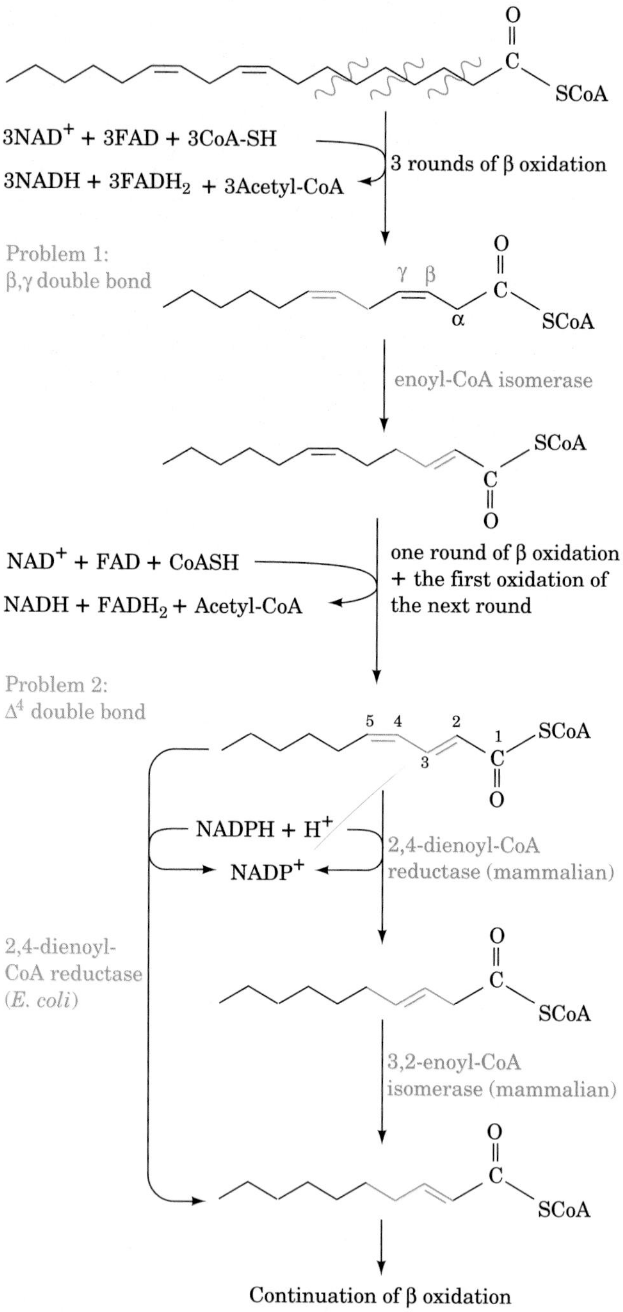

FIGURE 23-15. Problems in the oxidation of unsaturated fatty acids and their solutions. Linoleic acid is used as an example. The first problem, the presence of a β,γ double bond, is solved by the bond's enoyl-CoA isomerase-catalyzed conversion to a trans-α,β double bond. The second problem, that a 2,4-dienoyl-CoA is not a substrate for enoyl-CoA hydratase, is eliminated by the NADPH-dependent reduction of the 4-double bond by 2,4-dienoyl-CoA-reductase to yield the β-oxidation substrate *trans*-2-enoyl-CoA in *E. coli* but *trans*-3-enoyl-CoA in mammals. Mammals therefore also have 3,2-enoyl-CoA isomerase, which converts the *trans*-3-enoyl-CoA to *trans*-2-enoyl-CoA.

Problem 1: A β, γ Double Bond

The first enzymatic difficulty occurs after the third round of β oxidation: The resulting cis-β,γ double bond-containing enoyl-CoA is not a substrate for enoyl-CoA hydratase. **Enoyl-CoA isomerase**, however, mediates conversion of the cis-Δ^3 double bond to the more stable, ester-conjugated trans-Δ^2 form:

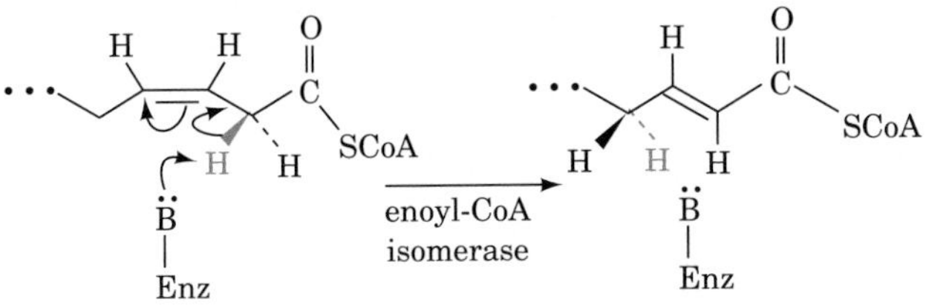

Such compounds are normal substrates of enoyl-CoA hydratase so that β oxidation can then continue.

Problem 2: A Δ^4 Double Bond Inhibits Hydratase Action

The next difficulty arises in the fifth round of β oxidation. Presence of a double bond at an even-numbered carbon atom results in the formation of 2,4-dienoyl-CoA, which is a poor substrate for enoyl-CoA hydratase. However, NADPH-dependent **2,4-dienoyl-CoA reductase** reduces the Δ^4 double bond. The *E. coli* reductase produces *trans*-2-enoyl-CoA, a normal substrate of β oxidation. The mammalian reductase, however, yields *trans*-3-enoyl-CoA, which, to proceed along the β-oxidation pathway, must first be isomerized to *trans*-2-enoyl-CoA by **3,2-enoyl-CoA isomerase**.

> Until recently, it was generally accepted that 2,4-dienoyl-CoA was a substrate for enoyl-CoA hydratase, yielding, after a further cycle of β oxidation, *cis*-Δ^2-enoyl-CoA, which in turn is converted by enoyl-CoA hydratase to 3-D-hydroxyacyl-CoA. This D isomer is not a β-oxidation substrate but was thought to be converted to its normally metaboliz-able L stereoisomer by **3-hydroxyacyl-CoA epimerase.** The finding that this epimerase is, in fact, a per-oxisomal rather than a mitochondrial enzyme, combined with the discovery of 2,4-dienoyl-CoA re-ductase, has led to the formulation of the revised pathway diagrammed at the bottom of Fig. 23-15.

E. Oxidation of Odd-Chain Fatty Acids

Most fatty acids have even numbers of carbon atoms and are therefore completely converted to acetyl-CoA. Some plants and marine organisms, however, synthesize fatty acids with an odd number of carbon atoms. *The final round of β oxidation of these fatty acids forms propionyl-CoA, which, as we shall see, is converted to succinyl-CoA for entry into the citric acid cycle.* Propionate or propionyl-CoA is also produced by oxidation of the amino acids isoleucine, valine, and methionine (Section 24-3E). Furthermore, ru-minant animals such as cattle derive most of their caloric intake from the acetate and propionate produced in their rumen (stomach) by bacterial fermentation of carbohy-drates. These products are absorbed by the animal and me-tabolized after conversion to the corresponding acyl-CoA.

Propionyl-CoA Carboxylase Has a Biotin Prosthetic Group

The conversion of propionyl-CoA to succinyl-CoA in-volves three enzymes (Fig. 23-16). The first reaction is that of **propionyl-CoA carboxylase**, a tetrameric enzyme that contains a biotin prosthetic group (Section 21-1A). The reaction occurs in two steps (Fig. 23-17):

1. Carboxylation of biotin at N1′ by bicarbonate ion as in the reaction catalyzed by pyruvate carboxylase (Fig. 21-4). This step, which is driven by the concomitant hydrol-ysis of ATP to ADP and P_i, activates the resulting car-boxyl group for transfer without further free energy input.

2. Stereospecific transfer of the activated carboxyl group from carboxybiotin to propionyl-CoA to form **(S)-methylmalonyl-CoA.** This step occurs via nucleophilic attack on carboxybiotin by a carbanion at C2 of pro-pionyl-CoA (see below).

These two reaction steps occur at different catalytic sites on propionyl-CoA carboxylase. It has therefore been proposed that the biotinyllysine linkage attaching the biotin ring to the enzyme forms a flexible tether that permits the efficient transfer of the biotin ring between these two active sites as postulated for the biotin enzyme pyruvate carboxylase (Section 21-1A).

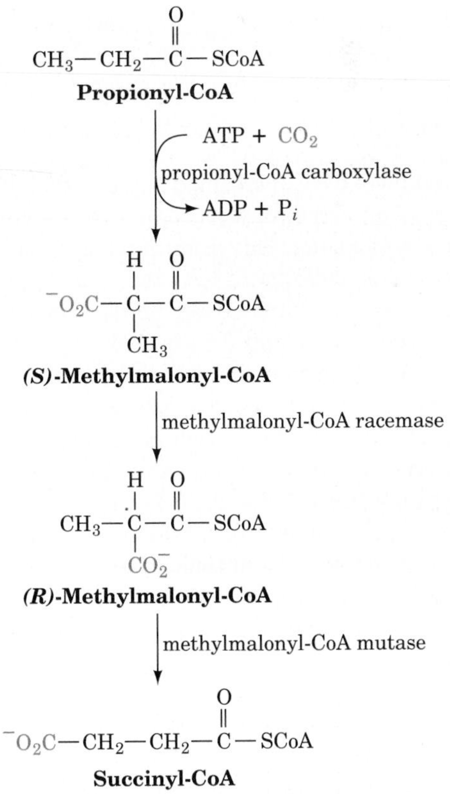

FIGURE 23-16. The conversion of propionyl-CoA to succinyl-CoA.

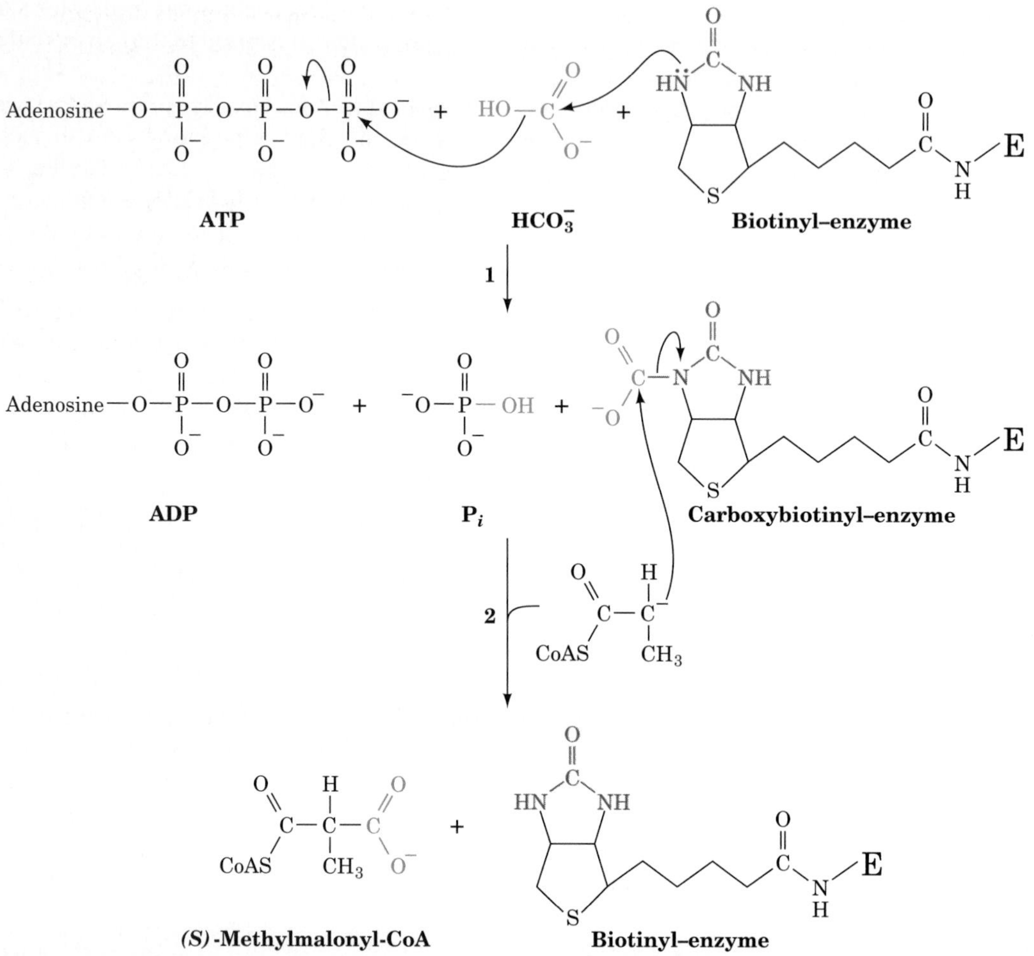

FIGURE 23-17 . The propionyl-CoA carboxylase reaction involves: **(1)** the carboxylation of biotin with the concomitant hydrolysis of ATP; followed by **(2)** the carboxylation of a propionyl-CoA carbanion by its attack on carboxybiotin. Each reaction step probably involves the intermediate formation of CO_2 as occurs in the pyruvate carboxylase reaction (Fig. 21-4).

Formation of the C2 carbanion in the second stage of the propionyl-CoA carboxylase reaction involves removal of a proton α to a thioester. This proton is relatively acidic since, as we have seen in Section 23-2C, the negative charge on a carbanion α to a thioester can be delocalized into the thioester's carbonyl group. This may explain the relatively convoluted path taken in the conversion of propionyl-CoA to succinyl-CoA (Fig. 23-16). It would seem simpler, at least on paper, for this process to occur in one step, with carboxylation occurring on C3 of propionyl-CoA so as to form succinyl-CoA directly. Yet, the C3 carbanion required for such a carboxylation would be extremely unstable. Nature has instead chosen a more facile, albeit less direct route, which carboxylates propionyl-CoA at a more reactive position and then rearranges the C_4 skeleton to form the desired product.

Methylmalonyl-CoA Mutase Contains a Coenzyme B_{12} Prosthetic Group

Methylmalonyl-CoA mutase, which catalyzes the third reaction of the propionyl-CoA to succinyl-CoA conversion (Fig. 23-16), is specific for *(R)*-methylmalonyl-CoA even though propionyl-CoA carboxylase stereospecifically synthesizes *(S)*-methylmalonyl-CoA. This diversion is rectified by **methylmalonyl-CoA racemase,** which interconverts the *(R)* and *(S)* configurations of methylmalonyl-CoA, presumably by promoting the reversible dissociation of its acidic α-H via formation of a resonance-stabilized carbanion intermediate.

Resonance-stabilized carbanion intermediate

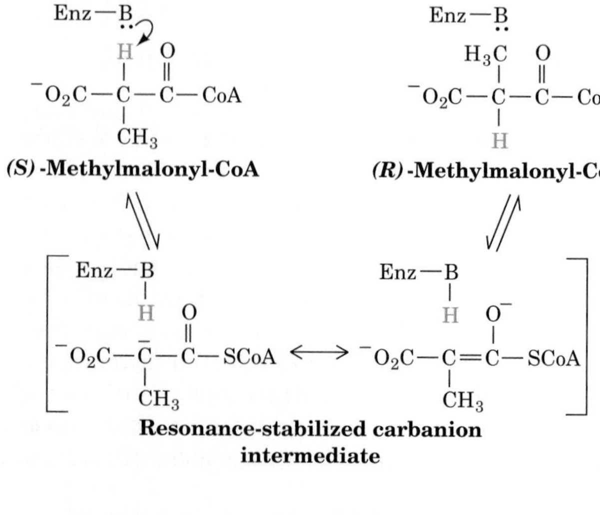

Methylmalonyl-CoA mutase, which catalyzes an unusual carbon skeleton rearrangement (Fig. 23-18), utilizes a **5′-deoxyadenosylcobalamin** prosthetic group (also called **coenzyme B_{12}**). Dorothy Hodgkin determined the structure of this complex molecule (Fig. 23-19) in 1956, a landmark achievement, through X-ray crystallographic analysis combined with chemical degradation studies. 5′-Deoxyadenosylcobalamin contains a hemelike **corrin** ring whose four pyrrole N atoms each ligand a 6-coordinate Co ion. The fifth Co ligand is an N atom of a **5,6-dimethylbenzimidazole (DMB)** nucleotide that is covalently linked to the corrin D ring. The sixth ligand is a 5′-deoxyadenosyl group in which the deoxyribose C5′ atom forms a covalent C—Co bond, *the only carbon–metal bond known in biology.* In some enzymes, the sixth ligand instead is a CH_3 group that likewise forms a C—Co bond.

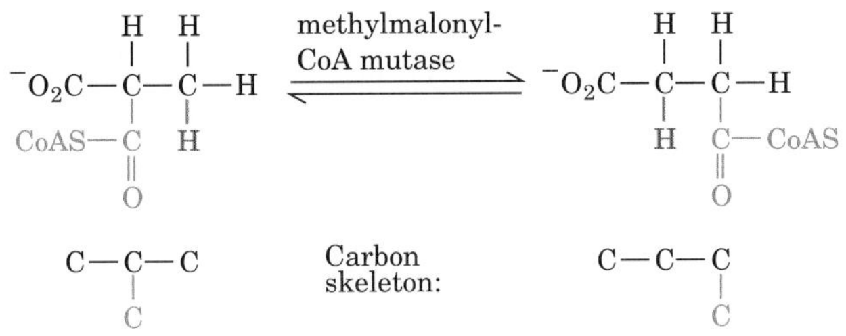

***(R)*-Methylmalonyl-CoA** **Succinyl-CoA**

FIGURE 23-18. The rearrangement catalyzed by methylmalonyl-CoA mutase.

Coenzyme B_{12}'s reactive C—Co bond participates in two types of enzyme-catalyzed reactions:

1. Rearrangements in which a hydrogen atom is directly transferred between two adjacent carbon atoms with concomitant exchange of the second substituent, X:

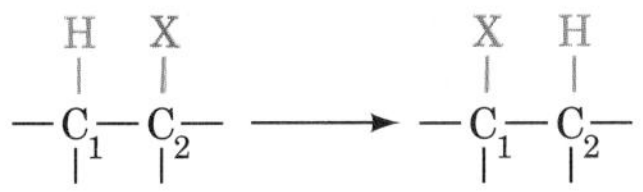

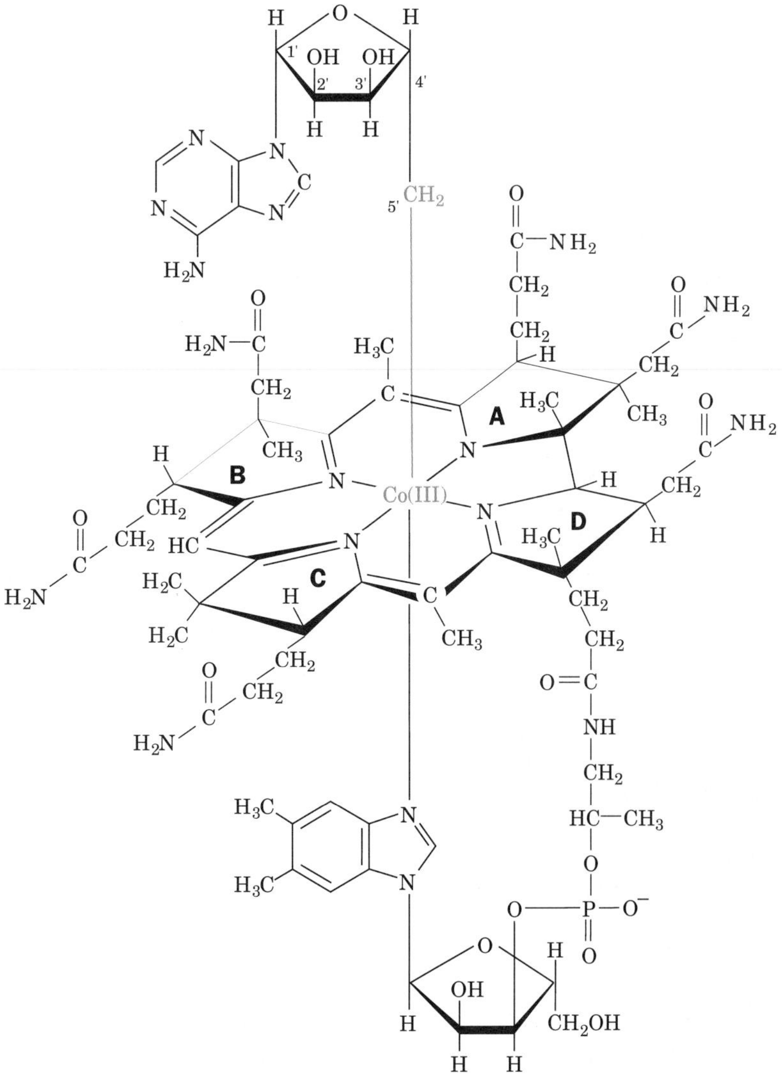

5'-Deoxyadenosylcobalamin (coenzyme B_{12})

FIGURE 23-19. The structure of 5′-deoxyadenosylcobalamin (coenzyme B_{12}).

where X may be a carbon atom with substituents, an oxygen atom of an alcohol, or an amine.

2. Methyl group transfers between two molecules.

There are about a dozen known cobalamin-dependent enzymes. Only two occur in mammalian systems: methyl-malonyl-CoA mutase, which catalyzes a carbon skeleton rearrangement (the X group in the rearrangement is —COSCoA; Fig. 23-18) and **homocysteine methyltransferase,** a methyl transfer enzyme that participates in methionine biosynthesis (Sections 24-3E and 5B).

The proposed methylmalonyl-CoA mutase reaction mechanism (Fig. 23-20) begins with **homolytic cleavage** of the cobalamin C—Co(III) bond (the C and Co atoms each acquire one of the electrons that formed the cleaved electron pair bond). The Co ion therefore fluctuates between its Co(III) and Co(II) oxidation states [the two states are spectroscopically distinguishable: Co(III) is red and diamagnetic (no unpaired electrons), whereas Co(II) is yellow and paramagnetic (unpaired electrons)]. Hence, *the role of coenzyme B_{12} in the catalytic process is that of a reversible free radical generator.* The C—Co(III) bond is well suited to this function because it is inherently weak (dissociation energy = 109 kJ · mol^{-1}) and appears to be further weakened through steric interactions with the enzyme. Note that a homolytic cleavage reaction is unusual in biology; most other biological bond cleavage reactions occur via **heterolytic cleavage** (in which the electron pair forming the cleaved bond is fully acquired by one of the separating atoms).

Succinyl-CoA Cannot Be Directly Consumed by the Citric Acid Cycle

Methylmalonyl-CoA mutase catalyzes the conversion of a metabolite to a citric acid cycle intermediate other than acetyl-CoA. You should note, however, that the route of succinyl-CoA oxidation is not as simple as it may first ap-

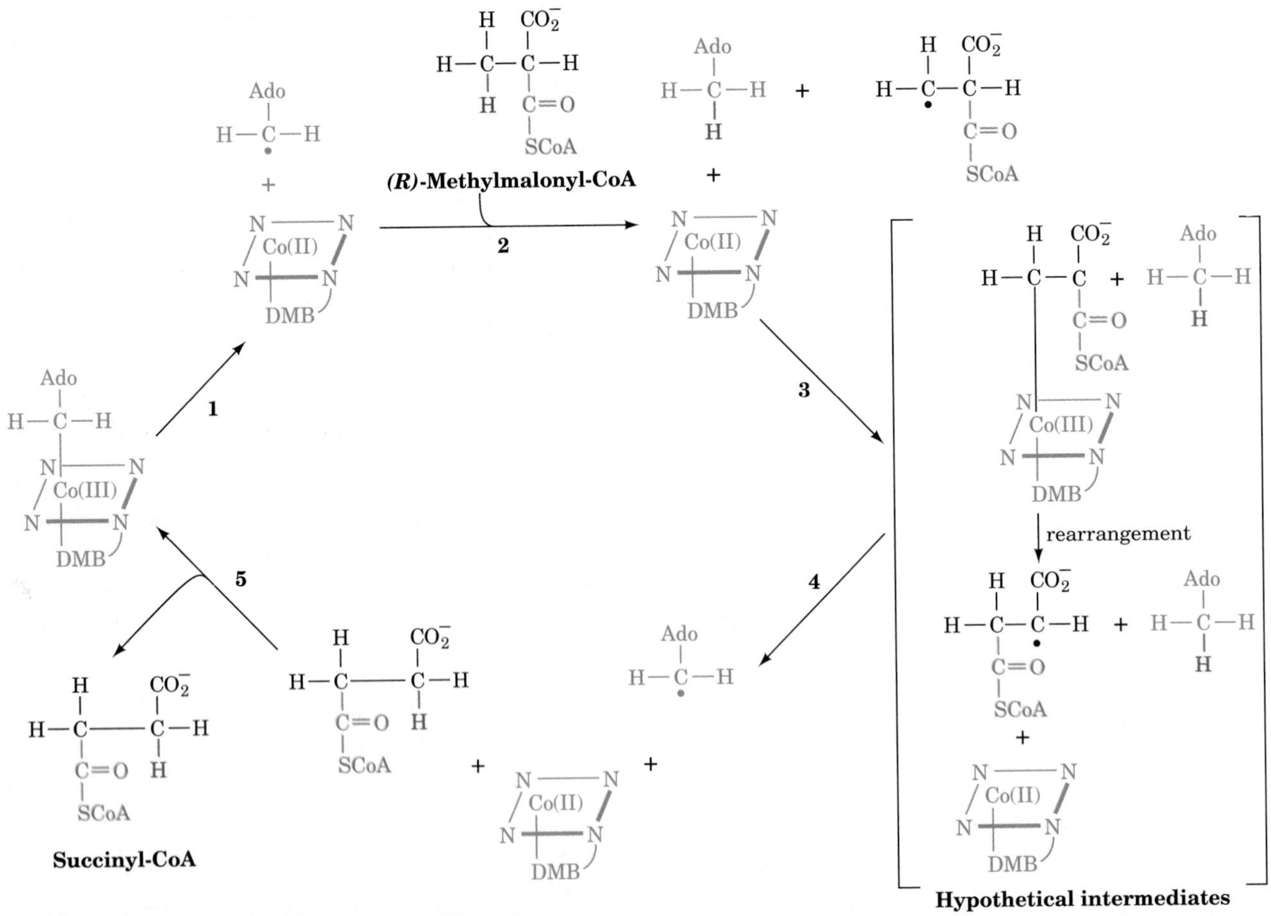

FIGURE 23-20. The proposed mechanism of methylmalonyl-CoA mutase: **(1)** The homolytic cleavage of the C—Co(III) bond yielding a 5′-deoxyadenosyl radical and cobalamin in its Co(II) oxidation state. **(2)** Abstraction of a hydrogen atom from the methylmalonyl-CoA by the 5′-deoxyadenosyl radical, thereby generating a methylmalonyl-CoA radical. **(3)** Hypothetical formation of a C—Co bond between the methylmalonyl-CoA radical and the coenzyme, followed by carbon skeleton rearrangement to form a succinyl-CoA radical. **(4)** Abstraction of a hydrogen atom from 5′-deoxyadenosine by the succinyl-CoA radical to regenerate the 5′-deoxyadenosyl radical. **(5)** Release of succinyl-CoA and reformation of the coenzyme.

pear. The citric acid cycle regenerates all of its C_4 intermediates so that these compounds are really catalysts, not substrates. Consequently, succinyl-CoA cannot undergo net degradation by citric acid cycle enzymes alone. Rather, *in order for a metabolite to undergo net oxidation by the citric acid cycle, it must first be converted either to pyruvate or directly to acetyl-CoA.* Net degradation of succinyl-CoA begins with its conversion, via the citric acid cycle, to malate. At high concentrations, malate is transported, by a specific transport protein, to the cytosol, where it may be oxidatively decarboxylated to pyruvate and CO_2 by **malic enzyme (malate dehydrogenase, decarboxylating);**

Malate   **Pyruvate**

(an enzyme we previously encountered in the C_4 cycle of photosynthesis; Fig. 22-32). Pyruvate is then completely oxidized via pyruvate dehydrogenase and the citric acid cycle.

Pernicious Anemia Results from Vitamin B_{12} Deficiency

The existence of **vitamin B_{12}** came to light in 1926 when George Minot and William Murphy discovered that **pernicious anemia,** an often fatal disease of the elderly characterized by decreased numbers of red blood cells, low hemoglobin levels, and progressive neurological deterioration, can be treated by the daily consumption of large amounts of raw liver (a treatment that some patients considered worse than the disease). It was not until 1948, however, after a bacterial assay for antipernicious anemia factor had been developed, that vitamin B_{12} was isolated.

Vitamin B_{12} is synthesized by neither plants nor animals but only by a few species of bacteria. Herbivores obtain their vitamin B_{12} from the bacteria that inhabit their gut (in fact, some animals, such as rabbits, must periodically eat some of their feces to obtain sufficient amounts of this essential substance). Humans, however, obtain almost all their vitamin B_{12} directly from their diet, particularly from meat. The vitamin is specifically bound in the intestine by the glycoprotein **intrinsic factor** that is secreted by the stomach. This complex is absorbed by a specific receptor in the intestinal mucosa, where the complex is dissociated and the liberated vitamin B_{12} transported to the bloodstream. There it is bound by at least three different plasma globulins, called **transcobalamins,** which facilitate its uptake by the tissues.

Pernicious anemia is not usually a dietary deficiency disease but, rather, results from insufficient secretion of intrinsic factor. The normal human requirement for cobalamin is very small, ~ 3 μg $\cdot$ day^{-1}, and the liver stores a 3- to 5-year

supply of this vitamin. This accounts for the insidious onset of pernicious anemia and the fact that true dietary deficiency of vitamin B_{12}, even among strict vegetarians, is extremely rare.

F. Peroxisomal β Oxidation

The β oxidation of fatty acids occurs in the peroxisome as well as in the mitochondrion. Peroxisomal β oxidation in animals functions to shorten very long chain fatty acids (> 22 C atoms) so as to facilitate their degradation by the mitochondrial β-oxidation system. In plants, fatty acid oxidation occurs exclusively in the peroxisomes and glyoxysomes (specialized peroxisomes, Sections 21-2 and 1-2A).

The peroxisomal pathway results in the same chemical changes to fatty acids as does the mitochondrial pathway, although the enzymes in these two organelles are different. There is no carnitine requirement for transport of fatty acyl-CoA into the peroxisome. Rather, very long chain fatty acids diffuse into this compartment, are activated by a peroxisomal very long chain acyl-CoA synthetase to form their CoA esters, and are oxidized directly. The shorter chain acyl products of this β-oxidation process are then linked to carnitine for transport to mitochondria for further oxidation.

X-Adrenoleukodystrophy May Be Caused by a Deficiency of Peroxisomal Acyl-CoA Synthetase

X-Adrenoleukodystrophy (X-ALD), a rare X-linked inherited disease (made famous by the recent film, "Lorenzo's Oil") causes very long chain saturated fatty acids to accumulate in the blood and destroy myelin, the insulating sheath surrounding the axons of many neurons (Section 34-4C). This disease may be caused by a deficiency of peroxisomal very long chain acyl-CoA synthetase. Thus **lignoceric acid** (24:0; recall that the symbol $n:m$ indicates a C_n fatty acid with m double bonds) is transported normally into peroxisomes from X-ALD patients but is converted to lignoceroyl-CoA at only 13% of the normal rate (although once activated, it undergoes β-oxidation at the normal rate).

Peroxisomal β Oxidation Differs in Detail from Mitochondrial β Oxidation

The β-oxidation process in peroxisomes involves three enzymatic reactions:

1. The **acyl-CoA oxidase** reaction:

Fatty acyl-CoA + O_2 $\longrightarrow$ *trans*-Δ^2-enoyl-CoA + H_2O_2

This reaction involves participation of an FAD cofactor but differs from its mitochondrial counterpart in that the abstracted electrons are transferred directly to O_2 rather than passing through the electron-transport chain with its concomitant oxidative phosphorylation (Fig. 23-10). Peroxisomal fatty acid oxidation is therefore less efficient than the mitochondrial process by two ATPs for each C_2 cycle. The H_2O_2 produced disproportionates to H_2O and O_2 through the action of peroxisomal catalase (Section 1-2A).

2. Peroxisomal enoyl-CoA hydratase and 3-L-hydroxy-acyl-CoA dehydrogenase are activities that occur on a single polypeptide and therefore join the growing list of multifunctional enzymes. The reactions catalyzed are identical to those of the mitochondrial system (Fig. 23-10).

3. Peroxisomal thiolase, which has a different chain-length specificity than its mitochondrial counterpart. It is almost inactive with acyl-CoAs of length C_8 or less so that fatty acids are incompletely oxidized by peroxisomes.

Although peroxisomal β oxidation is not dependent on the transport of acyl groups into the peroxisome as their carnitine esters, the peroxisome contains both a carnitine–acetyl transferase and a transferase specific for longer chain acyl groups. Acyl-CoAs that have been chain-shortened by peroxisomal β oxidation are thereby converted to their carnitine esters. These substances, for the most part, passively diffuse out of the peroxisome to the mitochondrion, where they are oxidized further.

G. Minor Pathways of Fatty Acid Oxidation

β Oxidation is blocked by an alkyl group at the C_β of a fatty acid, and thus at any odd-numbered carbon atom. One such branched-chain fatty acid, a common dietary component, is **phytanic acid**. This metabolic breakdown product of chlorophyll's phytyl side chain (Section 22-2A) is present in dairy products and ruminant fats, although, surprisingly, chlorophyll itself is but a poor dietary source of phytanic acid for humans. The oxidation of branched-chain fatty acids such as phytanic acid is facilitated by α **oxidation** (Fig. 23-21). In this process, the fatty acid C_α is hydroxylated and the resulting product is oxidatively decarboxylated to yield a new fatty acid with an unsubstituted C_β. Further degradation of the molecule can then continue via six cycles of normal β oxidation to yield three propionyl-CoAs, three acetyl-CoAs, and one 2-methylpropionyl-CoA (which is converted to succinyl-CoA).

A rare genetic defect, **Refsum's disease** or **phytanic acid storage syndrome**, results from the accumulation of this metabolite throughout the body. The disease, which is characterized by progressive neurological difficulties such as tremors, unsteady gait, and poor night vision, results from a greatly reduced α-hydroxylation activity. Its symp-

toms can therefore be attenuated by a diet that restricts the intake of phytanic acid–containing foods.

Medium- and long-chain fatty acids are converted to dicarboxylic acids through ω **oxidation** (oxidation of the last carbon atom). This process, which is catalyzed by enzymes of the endoplasmic reticulum (microsomes), involves hydroxylation of a fatty acid's C_ω atom by **cytochrome P$_{450}$**, a monooxygenase that utilizes NADPH and O_2. The OH group is then oxidized to a carboxyl group, converted to a CoA derivative at either end, and oxidized via the β-oxidation pathway. ω Oxidation is probably of only minor significance in fatty acid oxidation.

3. KETONE BODIES

Acetyl-CoA produced by oxidation of fatty acids in liver mitochondria can be further oxidized via the citric acid cycle as is discussed in Chapter 19. A significant fraction of

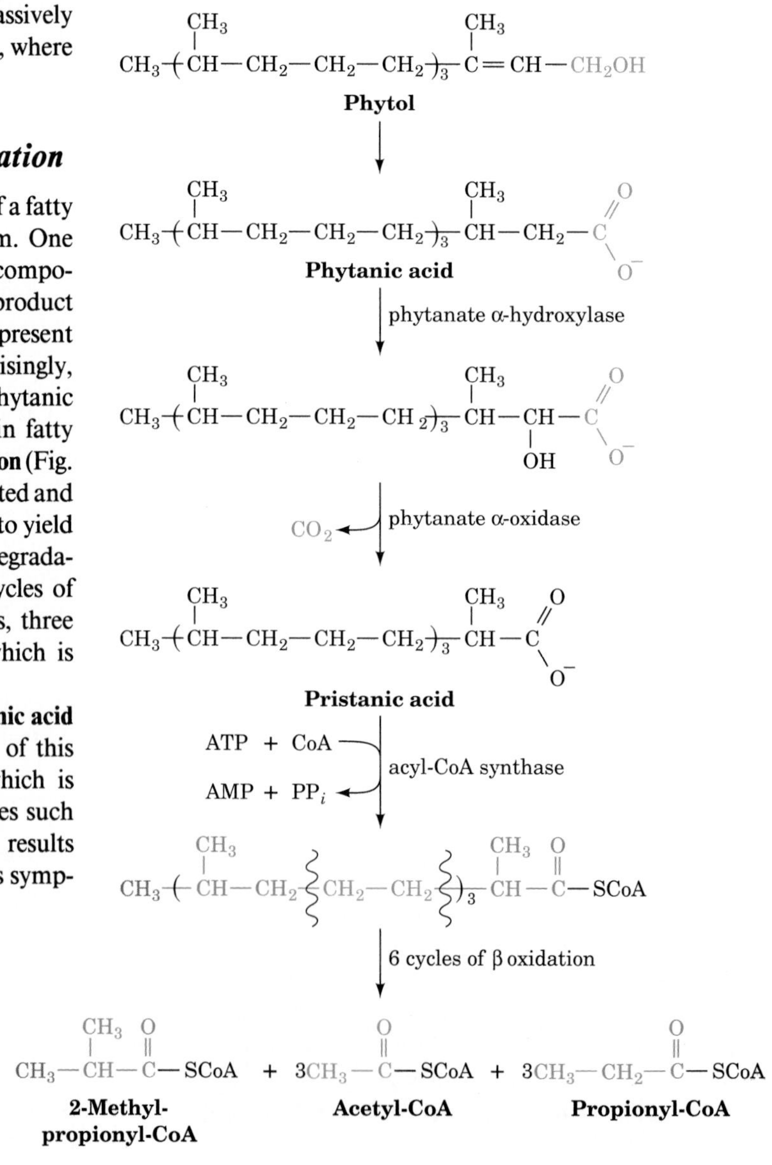

FIGURE 23-21. Phytanic acid, a degradation product of the phytol side chain of chlorophyll, is metabolized through α oxidation to **pristanic acid** followed by β oxidation.

this acetyl-CoA has another fate, however. *By a process known as **ketogenesis**, which occurs primarily in liver mitochondria, acetyl-CoA is converted to **acetoacetate** or **D-β-hydroxybutyrate**. These compounds, which together with **acetone** are somewhat inaccurately referred to as **ketone bodies**,*

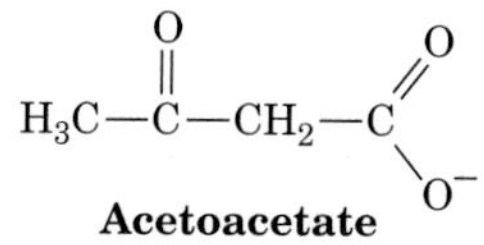

Acetoacetate 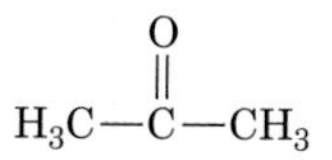   **Acetone**

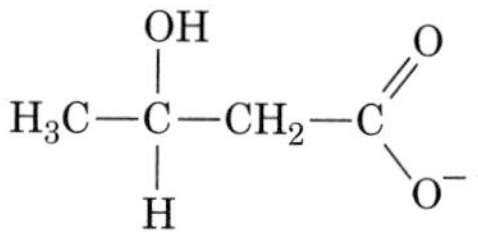

D-β-Hydroxybutyrate

serve as important metabolic fuels for many peripheral tissues, particularly heart and skeletal muscle. The brain, under normal circumstances, uses only glucose as its energy source (fatty acids are unable to pass the blood–brain barrier) but during starvation, ketone bodies become the brain's major fuel source (Section 25-3A). *Ketone bodies are water-soluble equivalents of fatty acids.*

Acetoacetate formation occurs in three reactions (Fig. 23-22):

1. Two molecules of acetyl-CoA are condensed to **acetoacetyl-CoA** by thiolase (also called **acetyl-CoA acetyltransferase**) working in the reverse direction from the way it does in the final step of β oxidation (Section 23-2C).

2. Condensation of the acetoacetyl-CoA with a third acetyl-CoA by **HMG-CoA synthase** forms **β-hydroxy-β-methylglutaryl-CoA (HMG-CoA)**. The mechanism of this reaction resembles the reverse of the thiolase reaction (Fig. 23-13) in that an active site thiol group forms an acyl–thioester intermediate.

3. Degradation of HMG-CoA to acetoacetate and acetyl-CoA in a mixed aldol–Claisen ester cleavage by **HMG-CoA lyase**. The mechanism of this reaction is analogous to the reverse of the citrate synthase reaction (Section 19-3A). (HMG-CoA is also a precursor in cholesterol biosynthesis and hence may be diverted to this purpose as is discussed in Section 23-6A.)

The overall reaction catalyzed by HMG-CoA synthase and HMG-CoA lyase is

$$\text{Acetoacetyl-CoA} + H_2O \longrightarrow \text{acetoacetate} + \text{CoA}$$

One may well ask why this apparently simple hydrolysis reaction occurs in such an indirect manner. The answer is unclear but may lie in the regulation of the process.

Acetoacetate may be reduced to D-β-hydroxybutyrate by **β-hydroxybutyrate dehydrogenase:**

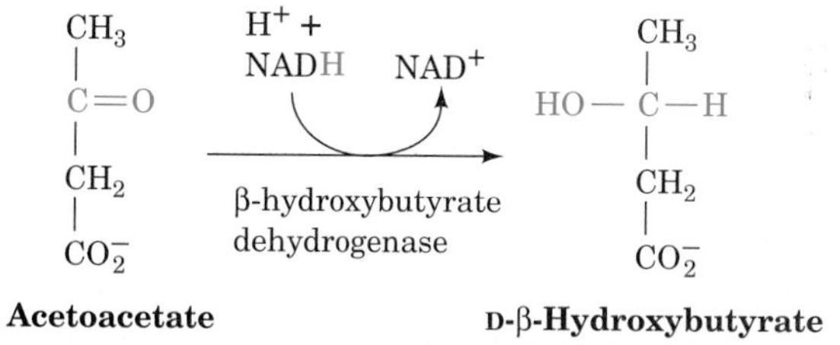

Acetoacetate   **D-β-Hydroxybutyrate**

Note that this product is the stereoisomer of the L-β-hydroxyacyl-CoA that occurs in the β-oxidation pathway. Acetoacetate, being a β-keto acid, also undergoes relatively facile nonenzymatic decarboxylation to acetone and CO_2. Indeed, the breath of individuals with **ketosis**, a pathological condition in which acetoacetate is produced faster than it can be metabolized (a symptom of diabetes; Section 25-3B), has the characteristic sweet smell of acetone.

The liver releases acetoacetate and β-hydroxybutyrate, which are carried by the bloodstream to the peripheral tissues for use as alternative fuels. There, these products are

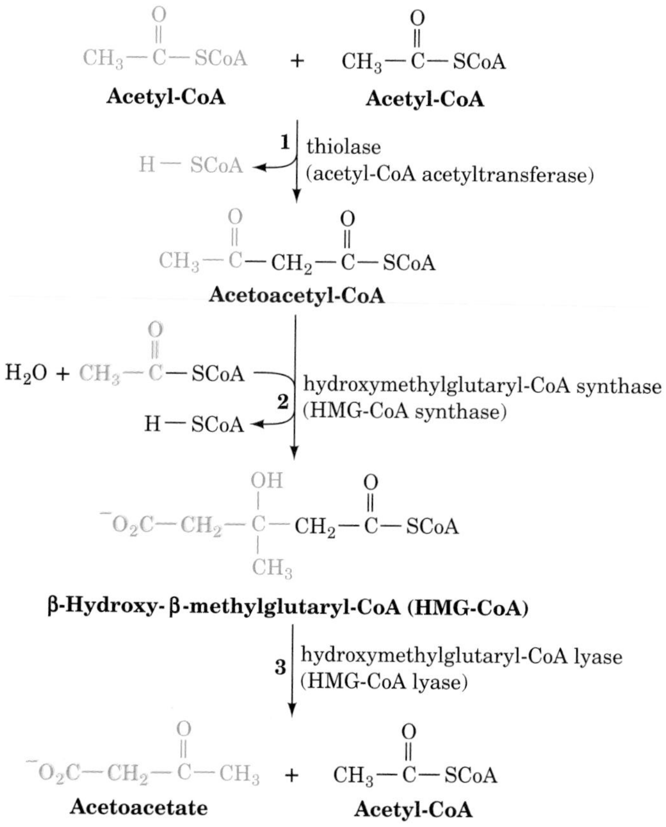

FIGURE 23-22. Ketogenesis: the enzymatic reactions forming acetoacetate from acetyl-CoA. **(1)** Two molecules of acetyl-CoA condense to form acetoacetyl-CoA in a thiolase-catalyzed reaction. **(2)** A Claisen ester condensation of the acetoacetyl-CoA with a third acetyl-CoA to form β-hydroxy-β-methylglutaryl-CoA (HMG-CoA) as catalyzed by HMG-CoA synthase. **(3)** The degradation of HMG-CoA to acetoacetate and acetyl-CoA in a mixed aldol–Claisen ester cleavage catalyzed by HMG-CoA lyase.

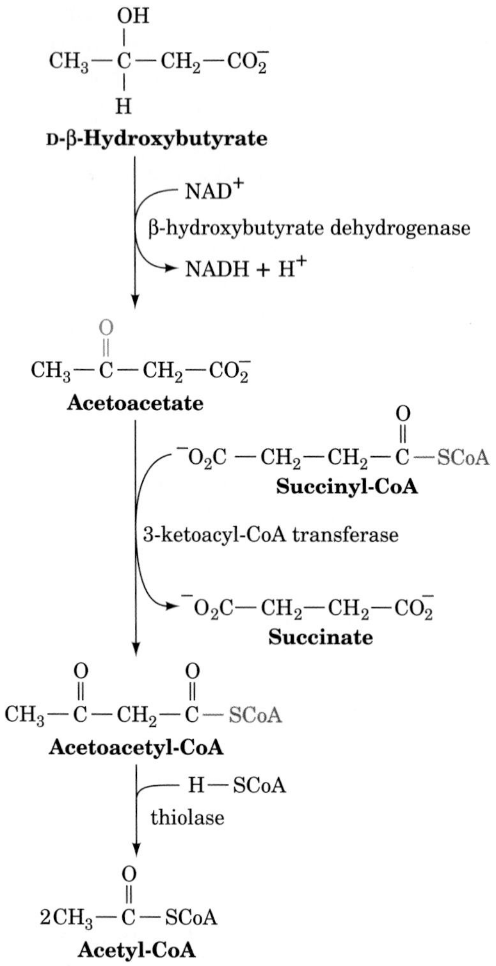

$$CH_3 - \underset{\underset{H}{|}}{\overset{\overset{OH}{|}}{C}} - CH_2 - CO_2^-$$

D-β-Hydroxybutyrate

NAD⁺

β-hydroxybutyrate dehydrogenase

NADH + H⁺

$$CH_3 - \overset{\overset{O}{\|}}{C} - CH_2 - CO_2^-$$

Acetoacetate

$$^-O_2C - CH_2 - CH_2 - \overset{\overset{O}{\|}}{C} - SCoA$$

Succinyl-CoA

3-ketoacyl-CoA transferase

$$^-O_2C - CH_2 - CH_2 - CO_2^-$$

Succinate

$$CH_3 - \overset{\overset{O}{\|}}{C} - CH_2 - \overset{\overset{O}{\|}}{C} - SCoA$$

Acetoacetyl-CoA

H — SCoA

thiolase

$$2CH_3 - \overset{\overset{O}{\|}}{C} - SCoA$$

Acetyl-CoA

FIGURE 23-23. The metabolic conversion of ketone bodies to acetyl-CoA.

converted to acetyl-CoA as is diagrammed in Fig. 23-23. The proposed reaction mechanism of **3-ketoacyl-CoA-transferase** (Fig. 23-24), which catalyzes this pathway's second step, involves the participation of an active site carboxyl group both in an enzyme–CoA thioester intermediate and in an unstable anhydride. Succinyl-CoA, which acts as the CoA donor in this reaction, can also be converted to succinate with the coupled synthesis of GTP in the succinyl-CoA synthase reaction of the citric acid cycle (Section 19-3E). The "activation" of acetoacetate bypasses this step and therefore "costs" the free energy of GTP hydrolysis. The liver lacks 3-ketoacyl-CoA transferase, which permits it to supply ketone bodies to other tissues.

4. FATTY ACID BIOSYNTHESIS

Fatty acid biosynthesis occurs through condensation of C₂ units, the reverse of the β-oxidation process. Through isotopic labeling techniques, David Rittenberg and Konrad Bloch demonstrated, in 1945, that these condensation units

are derived from acetic acid. Acetyl-CoA was soon proven to be a precursor of the condensation reaction but its mechanism remained obscure until the late 1950s when Salih Wakil discovered a requirement for bicarbonate in fatty acid biosynthesis and malonyl-CoA was shown to be an intermediate. In this section, we discuss the reactions of fatty acid biosynthesis.

A. Pathway Overview

The pathway of fatty acid synthesis differs from that of fatty acid oxidation. This situation, as we saw in Section 17-1D, is typically the case of opposing biosynthetic and degradative pathways because it permits them both to be thermodynamically favorable and independently regulated under similar physiological conditions. Figure 23-25 outlines fatty acid oxidation and synthesis with emphasis on the

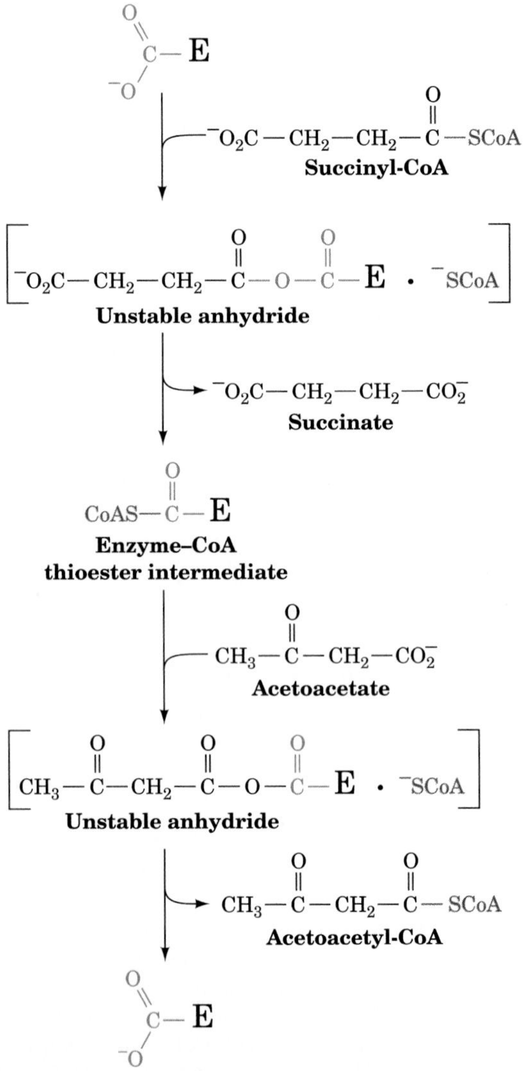

$$\underset{-O}{\overset{O}{\underset{\|}{C}}} - E$$

$$^-O_2C - CH_2 - CH_2 - \overset{\overset{O}{\|}}{C} - SCoA$$

Succinyl-CoA

$$\left[^-O_2C - CH_2 - CH_2 - \overset{\overset{O}{\|}}{C} - O - \overset{\overset{O}{\|}}{C} - E \cdot {}^-SCoA \right]$$

Unstable anhydride

$$^-O_2C - CH_2 - CH_2 - CO_2^-$$

Succinate

$$CoAS - \overset{\overset{O}{\|}}{C} - E$$

Enzyme–CoA thioester intermediate

$$CH_3 - \overset{\overset{O}{\|}}{C} - CH_2 - CO_2^-$$

Acetoacetate

$$\left[CH_3 - \overset{\overset{O}{\|}}{C} - CH_2 - \overset{\overset{O}{\|}}{C} - O - \overset{\overset{O}{\|}}{C} - E \cdot {}^-SCoA \right]$$

Unstable anhydride

$$CH_3 - \overset{\overset{O}{\|}}{C} - CH_2 - \overset{\overset{O}{\|}}{C} - SCoA$$

Acetoacetyl-CoA

$$\underset{-O}{\overset{O}{\underset{\|}{C}}} - E$$

FIGURE 23-24. The proposed mechanism of 3-ketoacyl-CoA transferase involves an enzyme-CoA thioester intermediate.

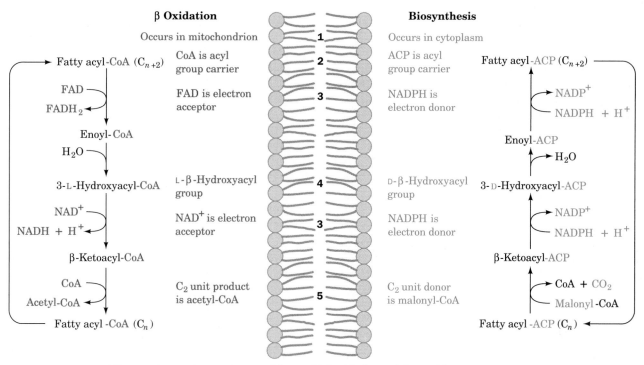

FIGURE 23-25. Differences between the pathways of fatty acid β oxidation and fatty acid biosynthesis with respect to: **(1)** cellular location; **(2)** acyl group carrier; **(3)** electron acceptor/donor; **(4)** stereochemistry of the hydration/dehydration reaction; and **(5)** the form in which C_2 units are produced/donated.

differences between these pathways. Whereas fatty acid oxidation occurs in the mitochondrion and utilizes fatty acyl-CoA esters, fatty acid biosynthesis occurs in the cytosol, as Roy Vagelos discovered, with the growing fatty acids esterified to **acyl-carrier protein (ACP;** Fig. 23-26). ACP, like CoA, contains a phosphopantetheine group that forms thioesters with acyl groups. The phosphopantetheine phosphoryl group is esterified to a Ser OH group of ACP, whereas in CoA it is esterified to AMP. In *E. coli*, ACP is a 10-kD polypeptide, whereas in animals it is part of a large multifunctional protein.

The redox coenzymes of the fatty acid oxidative and biosynthetic pathways differ (NAD^+ and FAD for oxidation; NADPH for biosynthesis) as does the stereochemistry of their intermediate steps, but their main difference is the manner in which C_2 units are removed from or added to the fatty acyl thioester chain. In the oxidative pathway, β-ketothiolase catalyzes the cleavage of the C_α—C_β bond of β-ketoacyl-CoA so as to produce acetyl-CoA and a new fatty acyl-CoA, which is shorter by a C_2 unit. The $\Delta G°'$ of this reaction is very close to zero so it can also function in the reverse direction (ketone body formation). In the bio-

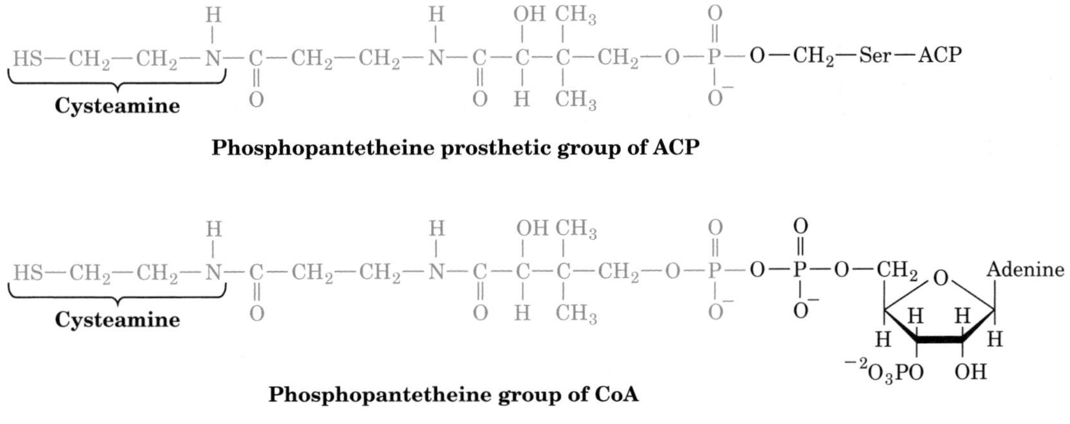

Phosphopantetheine prosthetic group of ACP

Phosphopantetheine group of CoA

FIGURE 23-26. The phosphopantetheine group in acyl-carrier protein (ACP) and in CoA.

synthetic pathway, the condensation reaction is coupled to the hydrolysis of ATP, thereby driving the reaction to completion. This process involves two steps: (1) the ATP-dependent carboxylation of acetyl-CoA by **acetyl-CoA carboxylase** to form **malonyl-CoA,** and (2) the exergonic decarboxylation of the malonyl group in the condensation reaction catalyzed by **fatty acid synthase.** The mechanisms of these enzymes are described in the next section.

B. Acetyl-CoA Carboxylase

Acetyl-CoA carboxylase catalyzes the first committed step of fatty acid biosynthesis and one of its rate-controlling steps. The mechanism of this biotin-dependent enzyme is very similar to those of propionyl-CoA carboxylase (Section 23-2E) and pyruvate carboxylase (Fig. 21-4) in that it occurs in two steps, a CO_2 activation and a carboxylation:

$$E\text{—biotin}$$
Biotinyl-enzyme

$$^-O_2C\text{—}CH_2\text{—}\overset{\overset{O}{\|}}{C}\text{—}SCoA + E\text{—biotin}$$
Malonyl-CoA

$$\begin{array}{c} HCO_3^- \\ + \text{ATP} \\ \\ ADP + P_i \end{array}$$

$$CH_3\text{—}\overset{\overset{O}{\|}}{C}\text{—}SCoA$$
Acetyl-CoA

$$E\text{—biotin—}CO_2^-$$
Carboxybiotinyl-enzyme

In *E. coli,* these steps are catalyzed by separate subunits, known as **biotin carboxylase** and **transcarboxylase,** respectively. In addition, the biotin is bound as a biocytin residue to a third subunit, termed **biotin carboxyl-carrier protein.** The mammalian and avian enzymes contain both enzymatic activities as well as the biotin carboxyl carrier on a single 230-kD polypeptide chain.

Avian and Mammalian Acetyl-CoA Carboxylase Are Regulated through Enzyme Polymerization

Electron microscopy reveals that the flat rectangular protomers of both avian and mammalian acetyl-CoA carboxylases associate to form long filaments with molecular masses in the range 4000 to 8000 kD (Fig. 23-27). *This polymeric form of the enzyme is catalytically active but the protomer is not.* The rate of fatty acid biosynthesis is therefore controlled by the position of the equilibrium between these forms:

Protomer *(inactive)* $\rightleftharpoons$ polymer *(active)*

Metabolites that most affect the position of this equilibrium are citrate, which shifts the equilibrium towards polymer formation, and palmitoyl-CoA, which promotes polymer disaggregation. Thus, cytosolic citrate, whose concentration increases when the acetyl-CoA concentration builds up in the mitochondrion (Section 23-4D), activates fatty acid biosynthesis, whereas palmitoyl-CoA, the pathway product, is a feedback inhibitor.

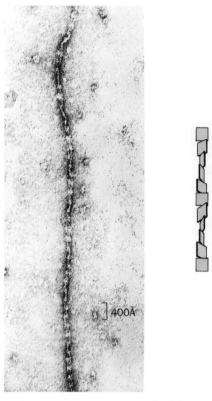

FIGURE 23-27. An electron micrograph with an accompanying interpretive drawing indicating that filaments of avian liver acetyl-CoA carboxylase consist of linear chains of flat rectangular protomers. [Courtesy of Malcolm Lane, The Johns Hopkins University School of Medicine.]

Acetyl-CoA carboxylase is also subject to hormonal regulation. Glucagon as well as epinephrine and norepinephrine (adrenalin and noradrenalin; Section 17-3E) trigger the enzyme's cAMP-dependent increase in phosphorylation, which shifts the polymerization equilibrium in favor of the inactive protomer. Insulin, on the other hand, stimulates dephosphorylation, promoting the formation of the active polymer.

The mechanism by which cAMP causes an increase in the phosphorylation state of acetyl-CoA carboxylase is interesting. Acetyl-CoA carboxylase is phosphorylated, *in vitro,* by two different kinases, cAMP-dependent protein kinase (cAPK; Section 17-3C) at Ser 77 and **AMP-dependent protein kinase (AMPK)** (which is cAMP independent) at Ser 79. Yet, when liver cells are incubated with cAMP-elevating hormones in the presence of ^{32}P-ATP, only Ser 79 is found to be labeled. Evidently, a [cAMP] increase results in a phosphorylation increase at sites modified by AMPK rather than by cAPK. How can this be? It appears that, *in vivo,* the cAMP-dependent increase in phosphorylation occurs not through the phosphorylation of new sites but, rather, through the inhibition of dephosphorylation of previously phosphorylated positions. We have already seen such a mechanism in operation in the control of glycogen

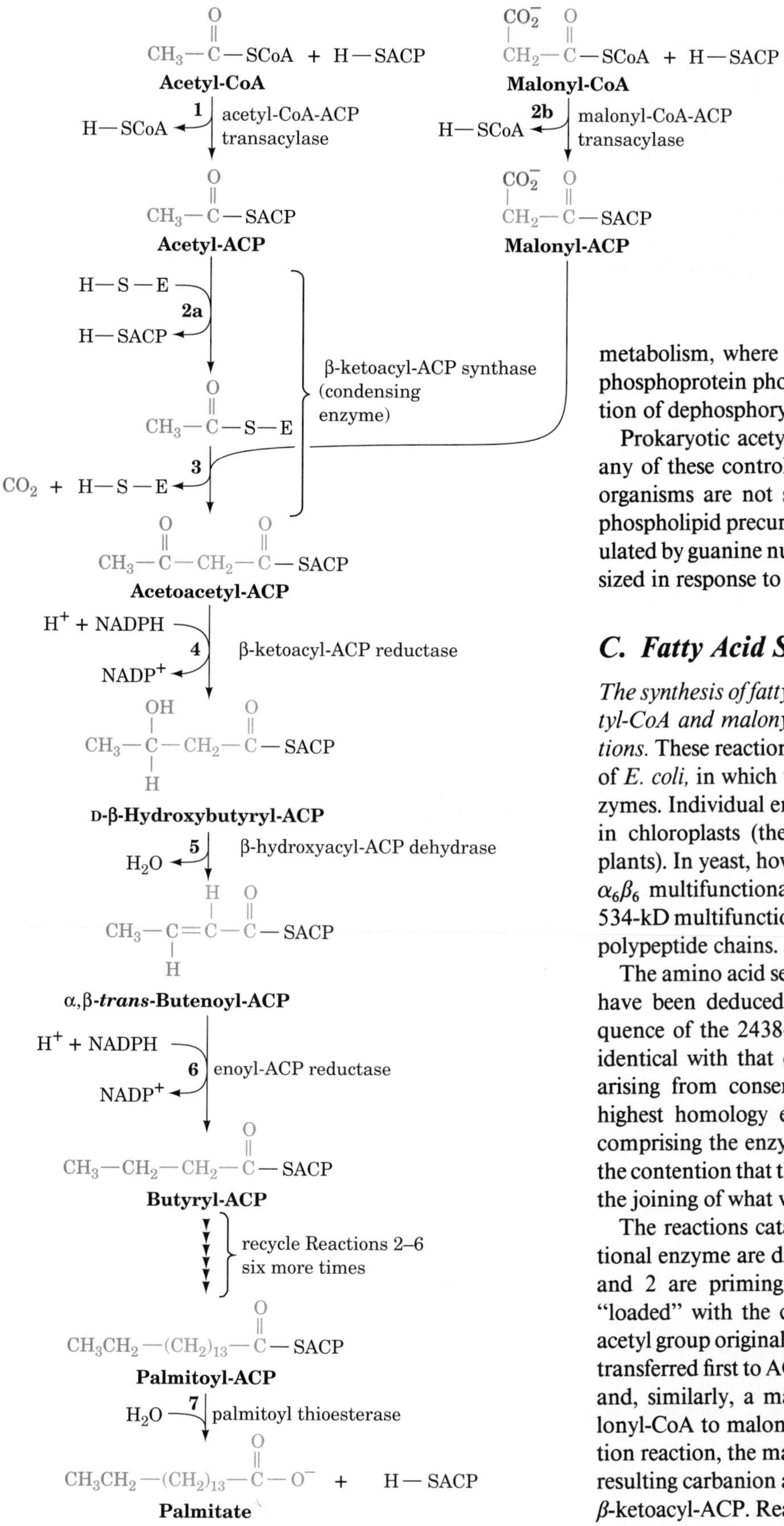

$$\underset{\textbf{Acetyl-CoA}}{CH_3-\overset{\overset{\textstyle O}{\|}}{C}-SCoA} \;+\; H-SACP$$

$$\underset{\textbf{Malonyl-CoA}}{\overset{CO_2^-}{\underset{|}{CH_2}}-\overset{\overset{\textstyle O}{\|}}{C}-SCoA} \;+\; H-SACP$$

$$H-SCoA \xleftarrow{\;\textbf{1}\;} \begin{array}{l}\text{acetyl-CoA-ACP}\\ \text{transacylase}\end{array}$$

$$H-SCoA \xleftarrow{\;\textbf{2b}\;} \begin{array}{l}\text{malonyl-CoA-ACP}\\ \text{transacylase}\end{array}$$

$$\underset{\textbf{Acetyl-ACP}}{CH_3-\overset{\overset{\textstyle O}{\|}}{C}-SACP}$$

$$\underset{\textbf{Malonyl-ACP}}{\overset{CO_2^-}{\underset{|}{CH_2}}-\overset{\overset{\textstyle O}{\|}}{C}-SACP}$$

$$H-S-E \;\diagdown$$
$$\xleftarrow{\;\textbf{2a}\;}$$
$$H-SACP \;\diagup$$

β-ketoacyl-ACP synthase (condensing enzyme)

$$\underset{}{CH_3-\overset{\overset{\textstyle O}{\|}}{C}-S-E}$$

$$CO_2 \;+\; H-S-E \xleftarrow{\;\textbf{3}\;}$$

$$\underset{\textbf{Acetoacetyl-ACP}}{CH_3-\overset{\overset{\textstyle O}{\|}}{C}-CH_2-\overset{\overset{\textstyle O}{\|}}{C}-SACP}$$

$$H^+ + NADPH \diagdown$$
$$\xrightarrow{\;\textbf{4}\;}\; \text{β-ketoacyl-ACP reductase}$$
$$NADP^+ \diagup$$

$$\underset{\textbf{D-β-Hydroxybutyryl-ACP}}{CH_3-\overset{\overset{\textstyle OH}{|}}{\underset{\underset{\textstyle H}{|}}{C}}-CH_2-\overset{\overset{\textstyle O}{\|}}{C}-SACP}$$

$$H_2O \xleftarrow{\;\textbf{5}\;} \text{β-hydroxyacyl-ACP dehydrase}$$

$$\underset{\textbf{α,β-\textit{trans}-Butenoyl-ACP}}{CH_3-\overset{\overset{\textstyle H}{|}}{C}=\overset{\overset{\textstyle}{}}{\underset{\underset{\textstyle H}{|}}{C}}-\overset{\overset{\textstyle O}{\|}}{C}-SACP}$$

$$H^+ + NADPH \diagdown$$
$$\xrightarrow{\;\textbf{6}\;}\; \text{enoyl-ACP reductase}$$
$$NADP^+ \diagup$$

$$\underset{\textbf{Butyryl-ACP}}{CH_3-CH_2-CH_2-\overset{\overset{\textstyle O}{\|}}{C}-SACP}$$

recycle Reactions 2–6 six more times

$$\underset{\textbf{Palmitoyl-ACP}}{CH_3CH_2-(CH_2)_{13}-\overset{\overset{\textstyle O}{\|}}{C}-SACP}$$

$$H_2O \xleftarrow{\;\textbf{7}\;} \text{palmitoyl thioesterase}$$

$$\underset{\textbf{Palmitate}}{CH_3CH_2-(CH_2)_{13}-\overset{\overset{\textstyle O}{\|}}{C}-O^-} \;+\; H-SACP$$

FIGURE 23-28. The reaction sequence for the biosynthesis of fatty acids. In forming palmitate, the pathway is repeated for seven cycles of C_2 elongation followed by a final hydrolysis step.

metabolism, where cAMP-dependent phosphorylation of phosphoprotein phosphatase inhibitor-1 causes the inhibition of dephosphorylation (Section 17-3C).

Prokaryotic acetyl-CoA carboxylases are not subject to any of these controls. This is because fatty acids in these organisms are not stored as fats but function largely as phospholipid precursors. The *E. coli* enzyme is instead regulated by guanine nucleotides so that fatty acids are synthesized in response to the cell's growth requirements.

C. Fatty Acid Synthase

The synthesis of fatty acids, mainly palmitic acid, from acetyl-CoA and malonyl-CoA involves seven enzymatic reactions. These reactions were first studied in cell-free extracts of *E. coli*, in which they are catalyzed by independent enzymes. Individual enzymes with these activities also occur in chloroplasts (the only site of fatty acid synthesis in plants). In yeast, however, fatty acid synthase is a 2500-kD $\alpha_6\beta_6$ multifunctional enzyme, whereas in animals it is a 534-kD multifunctional enzyme consisting of two identical polypeptide chains.

The amino acid sequences of several fatty acid synthases have been deduced from their gene sequences. The sequence of the 2438-residue chicken liver enzyme is 67% identical with that of rat, with many of the mismatches arising from conservative substitutions. The regions of highest homology encompass the polypeptide segments comprising the enzymatic active sites, thereby supporting the contention that this multifunctional enzyme evolved by the joining of what were previously independent enzymes.

The reactions catalyzed by the mammalian multifunctional enzyme are diagrammed in Fig. 23-28. Reactions 1 and 2 are priming reactions in which the synthase is "loaded" with the condensation reaction precursors: An acetyl group originally linked as a thioester in acetyl-CoA is transferred first to ACP and then to an enzyme Cys residue; and, similarly, a malonyl group is transferred from malonyl-CoA to malonyl-ACP. In Reaction 3, the condensation reaction, the malonyl-ACP is decarboxylated with the resulting carbanion attacking the acetyl-thioester to form a β-ketoacyl-ACP. Reactions 4–6 are the reductions and dehydration that convert this ketone to an alkyl group. The coenzyme in both reductive steps is NADPH, whereas in β oxidation, the analogs of Reactions 4 and 6, respectively, use NAD$^+$ and FAD (Fig. 23-25). Moreover, Reaction 5 requires a D-β-hydroxyacyl substrate, whereas the analogous reaction in β oxidation forms the corresponding L isomer.

All of the enzyme activities but those catalyzing Reactions 2a and 4 remain functional when the native dimeric enzyme is dissociated into monomers. Electron microscopy of these monomers indicates that they consist of a linear chain of at least four 50-Å-diameter lobes. Moreover, fragments resulting from the limited proteolysis of fatty acid synthase exhibit many of the enzymatic activities of the intact protein. Thus, *contiguous stretches of its polypeptide chain fold to form a series of autonomous domains, each with a specific but different catalytic activity.* Several other enzymes, such as mammalian acetyl-CoA carboxylase (Section 23-4B), exhibit similar multifunctionality but none has as many separate catalytic activities as does animal fatty acid synthase.

Since the condensation reaction requires the juxtaposition of the sulfhydryl groups of the ACP phosphopantetheine and an enzyme Cys residue, it has been proposed that these groups are on separate subunits that interact in a head-to-tail manner (Fig. 23-29). The mechanism of palmitate synthesis by this dimer is therefore thought to occur as is diagrammed in Fig. 23-30 with the long flexible phosphopantetheine chain of ACP (Fig. 23-26) functioning to transport the substrate between the enzyme's various catalytic sites:

1. The "priming" of the condensing enzyme (Fig. 23-28, Reaction 1) followed by formation of an acetyl-Cys residue (Fig. 23-28, Reaction 2a).

2. The "loading" of the condensing enzyme by the formation of malonyl–ACP (Fig. 23-28, Reaction 2b).

3. The coupling of the acetyl group to the C$_\beta$ of the malonyl group on the other subunit with the latter's accompanying decarboxylation so as to form acetoacetyl-ACP and free the active site Cys-SH group (Fig. 23-28, Reaction 3). Consequently, the CO$_2$ taken up in the acetyl-CoA carboxylase reaction (Section 23-4B) does not appear in the product fatty acid. Rather, the decarboxylation functions to drive the condensa-

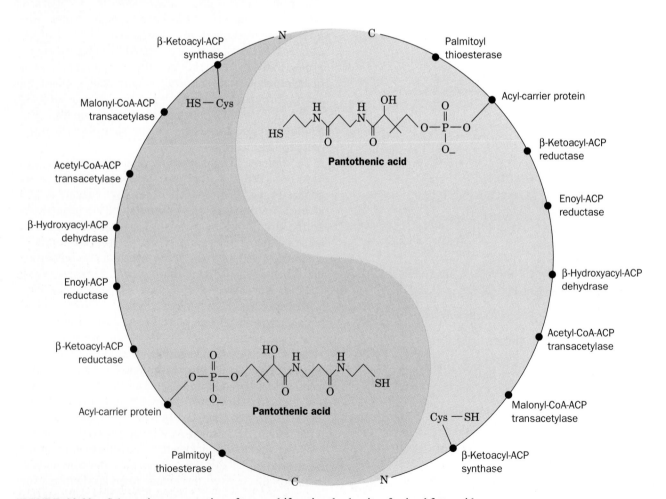

FIGURE 23-29. Schematic representation of two multifunctional subunits of animal fatty acid synthase in head-to-tail association to form the active dimer. [After Wakil, S.J., Stoops, J.K., and Joshi, V.C., *Annu. Rev. Biochem.* **52,** 556 (1983).]

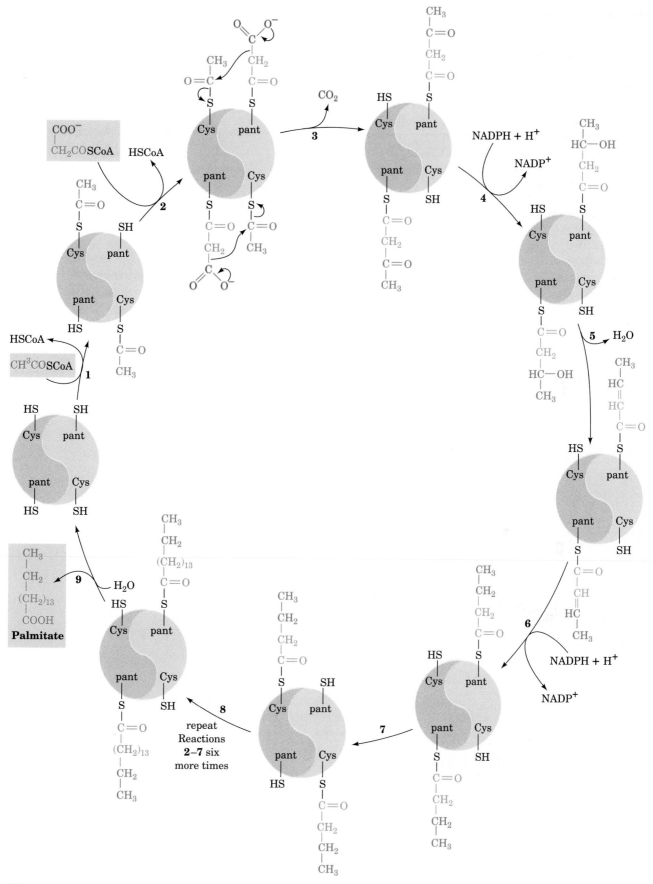

FIGURE 23-30. Proposed mechanism of palmitate synthesis as mediated by animal fatty acid synthase. The circles represent the multifunctional dimer of fatty acid synthase. The Cys—SH represents the active cysteine sulfhydryl of the β-ketoacyl-ACP synthase and pant—SH represents the pantetheine sulfhydryl of the acyl-carrier protein. The other catalytic domains indicated in Fig. 23-29 are not shown but are present in both subunits of the dimer. [After Wakil, S.J., Stoops, J.K., and Joshi, V.C., *Annu. Rev. Biochem.* **52**, 568 (1983).]

tion reaction which, through the acetyl-CoA carboxylase reaction, is coupled to ATP hydrolysis.

4–6. The reduction, dehydration, and further reduction form butyryl-ACP followed by the transfer of this group to the Cys-SH of the first subunit. Thus the acetyl group with which the system was initially primed has been elongated by a C_2 unit.

7. The transfer of the butyryl group to the Cys-SH of the first subunit (Fig. 23-28, repeat of Reaction 2a).

The ACP group is "reloaded" with a malonyl group, and another cycle of C_2 elongation occurs. This process occurs altogether seven times to form palmitoyl-ACP (Fig. 23-30, Reaction 8). Its thioester bond is then hydrolyzed by **palmitoyl thioesterase** (Fig. 23-28, Reaction 7), yielding palmitate, the normal product of the fatty acid synthase pathway,

and regenerating the enzyme for a new round of synthesis. The stoichiometry of palmitate synthesis therefore is

$$\text{Acetyl-CoA} + 7\text{malonyl-CoA} + 14\text{NADPH} + 7\text{H}^+ \longrightarrow$$
$$\text{palmitate} + 7\text{CO}_2 + 14\text{NADP}^+ + 8\text{CoA} + 6\text{H}_2\text{O}$$

Since the 7malonyl-CoA are derived from acetyl-CoA as follows:

$$7\text{Acetyl-CoA} + 7\text{CO}_2 + 7\text{ATP} \longrightarrow$$
$$7\text{malonyl-CoA} + 7\text{ADP} + 7\text{P}_i + 7\text{H}^+$$

the overall stoichiometry for palmitate biosynthesis is

$$8\text{Acetyl-CoA} + 14\text{NADPH} + 7\text{ATP} \longrightarrow$$
$$\text{palmitate} + 14\text{NADP}^+ + 8\text{CoA} + 6\text{H}_2\text{O} + 7\text{ADP} + 7\text{P}_i$$

We next consider the means of transport of mitochondrial acetyl-CoA to the cytosol, the site of fatty acid synthe-

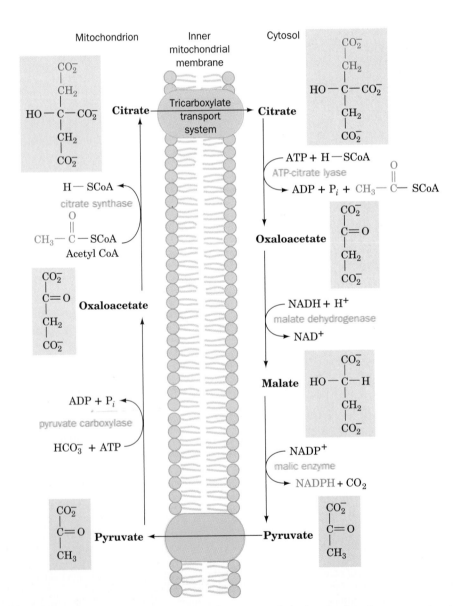

FIGURE 23-31. The transfer of acetyl-CoA from mitochondrion to cytosol via the tricarboxylate transport system.

sis. Following that, we examine the reactions by which fatty acids are elongated and desaturated.

D. Transport of Mitochondrial Acetyl-CoA into the Cytosol

Acetyl-CoA is generated in the mitochondrion by the oxidative decarboxylation of pyruvate as catalyzed by pyruvate dehydrogenase (Section 19-2A) as well as by the oxidation of fatty acids. When the need for ATP synthesis is low, so that the oxidation of acetyl-CoA via the citric acid cycle and oxidative phosphorylation is minimal, this mitochondrial acetyl-CoA may be stored for future use as fat. Fatty acid biosynthesis occurs in the cytosol but the mitochondrial membrane is essentially impermeable to acetyl-CoA. *Acetyl-CoA enters the cytosol in the form of citrate via the* **tricarboxylate transport system** *(Fig. 23-31)*. Cytosolic **ATP-citrate lyase** then catalyzes the reaction

Citrate + CoA + ATP $\rightleftharpoons$

acetyl-CoA + oxaloacetate + ADP + P_i

which resembles the reverse of the citrate synthase reaction (Section 19-3A) except that ATP hydrolysis is required to drive the intermediate synthesis of the "high-energy" citryl-CoA (whose hydrolysis drives the citrate synthase reaction to completion). ATP hydrolysis is therefore required in the ATP–citrate lyase reaction to power the resynthesis of this thioester bond. Oxaloacetate is reduced to malate by **malate dehydrogenase**. Malate may be oxidatively decarboxylated to pyruvate by **malic enzyme** and be returned in this form to the mitochondrion. The malic enzyme reaction resembles that of isocitrate dehydrogenase in which a β-hydroxy acid is oxidized to a β-keto acid, whose decarboxylation is strongly favored (Section 19-3C). Malic enzyme's coenzyme is $NADP^+$, so when this route is used NADPH is produced for use in the reductive reactions of fatty acid biosynthesis.

Citrate transport out of the mitochondrion must be balanced by anion transport into the mitochondrion. Malate, pyruvate, and P_i can act in this capacity. Malate may therefore also be transported directly back to the mitochondrion without generating NADPH. As we have seen in Section 23-4C, synthesis of each palmitate ion requires 8 molecules of acetyl-CoA and 14 molecules of NADPH. As many as 8 of these NADPH molecules may be supplied with the 8 molecules of acetyl-CoA if all the malate produced in the cytosol is oxidatively decarboxylated. The remaining NADPH is provided through the pentose phosphate pathway (Section 21-4).

E. Elongases and Desaturases

Palmitate (16:0), the normal product of the fatty acid synthase pathway, is the precursor of longer chain saturated and unsaturated fatty acids through the actions of **elongases** *and* **desaturases**. Elongases are present in both the mitochon-

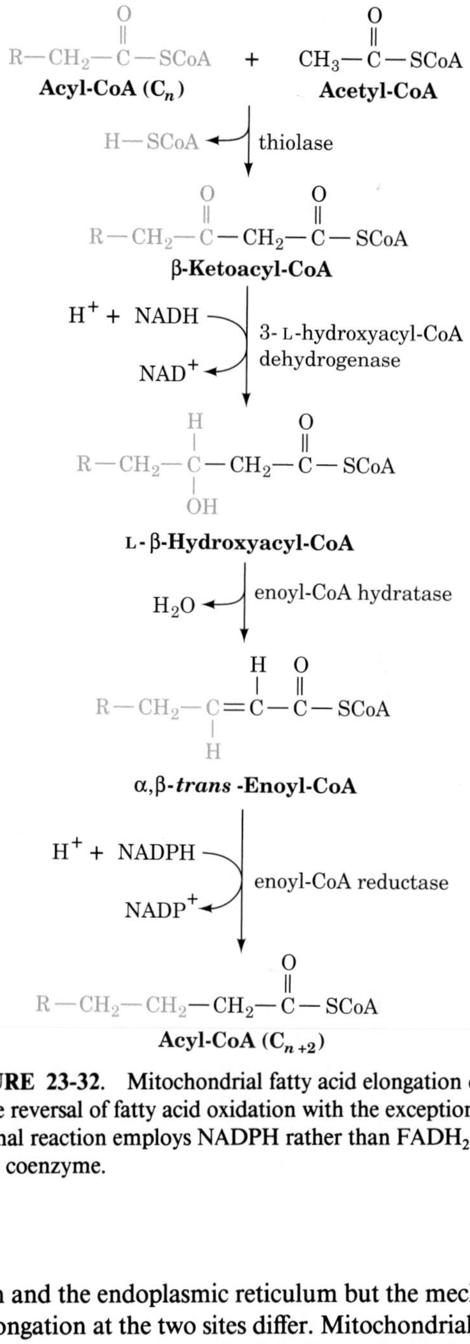

FIGURE 23-32. Mitochondrial fatty acid elongation occurs by the reversal of fatty acid oxidation with the exception that the final reaction employs NADPH rather than $FADH_2$ as its redox coenzyme.

drion and the endoplasmic reticulum but the mechanisms of elongation at the two sites differ. Mitochondrial elongation (a process independent of the fatty acid synthase pathway) occurs by successive addition and reduction of acetyl units in a reversal of fatty acid oxidation; the only chemical difference between these two pathways occurs in the final reduction step in which NADPH takes the place of $FADH_2$ as the terminal redox coenzyme (Fig. 23-32). Elongation in the endoplasmic reticulum involves the successive condensations of malonyl-CoA with acyl-CoA. These reactions are each followed by NADPH-associated reductions similar to those catalyzed by fatty acid synthase, the only difference being that the fatty acid is elongated as its CoA derivative rather than as its ACP derivative.

Unsaturated fatty acids are produced by **terminal desaturases**. Mammalian systems contain four terminal desaturases of broad chain-length specificities designated Δ^9-, Δ^6-, Δ^5-, and Δ^4-fatty acyl-CoA desaturases. These non-heme iron–containing enzymes catalyze the general reaction:

$$CH_3-(CH_2)_x-\overset{\overset{\displaystyle H}{|}}{\underset{\underset{\displaystyle H}{|}}{C}}-\overset{\overset{\displaystyle H}{|}}{\underset{\underset{\displaystyle H}{|}}{C}}-(CH_2)_y-\overset{\overset{\displaystyle O}{\|}}{C}-SCoA + NADH + H^+ + O_2$$

$$\downarrow$$

$$CH_3-(CH_2)_x-\overset{\overset{\displaystyle H}{|}}{C}=\overset{\overset{\displaystyle H}{|}}{C}-(CH_2)_y-\overset{\overset{\displaystyle O}{\|}}{C}-SCoA + 2H_2O + NAD^+$$

where x is at least five and where $(CH_2)_x$ can contain one or more double bonds. The $(CH_2)_y$ portion of the substrate is always saturated. Double bonds are inserted between existing double bonds in the $(CH_2)_x$ portion of the substrate and the CoA group such that the new double bond is three carbon atoms closer to the CoA group than the next double bond (not conjugated to an existing double bond) and, in animals, never at positions beyond C9.

A variety of unsaturated fatty acids may be synthesized by combinations of elongation and desaturation reactions. However, since palmitic acid is the shortest available fatty acid in animals, the above rules preclude the formation of the Δ^{12} double bond of linoleic acid ($\Delta^{9,12}$-octadecadienoic acid), a required precursor of **prostaglandins** (Section 23-7). *Linoleic acid must consequently be obtained in the diet (ultimately from plants that have Δ^{12}- and Δ^{15}-desaturases) and is therefore an* **essential fatty acid.** Indeed, animals maintained on a fat-free diet develop an ultimately fatal condition that is initially characterized by poor growth, poor wound healing, and dermatitis. Linoleic acid is also an important constituent of epidermal sphingolipids that function as the skin's water-permeability barrier.

Mammalian terminal desaturases are components of mini-electron-transport systems that contain two other proteins; **cytochrome b_5** and **NADH–cytochrome b_5 reductase.** The electron-transfer reactions mediated by these complexes occur at the inner surface of the endoplasmic reticulum membrane (Fig. 23-33) and are therefore not associated with oxidative phosphorylation.

F. Synthesis of Triacylglycerols

Triacylglycerols are synthesized from fatty acyl-CoA esters and glycerol-3-phosphate or dihydroxyacetone phosphate (Fig. 23-34). The initial step in this process is catalyzed either by **glycerol-3-phosphate acyltransferase** in mitochondria and endoplasmic reticulum, or by **dihydroxyacetone phosphate acyltransferase** in endoplasmic reticulum or peroxisomes. In the latter case, the product acyl-dihydroxyacetone phosphate is reduced to the corresponding **lysophosphatidic acid** by an NADPH-dependent reductase. The lysophosphatidic acid is converted to a triacylglycerol by the successive actions of **1-acylglycerol-3-phosphate acyltransferase, phosphatidic acid phosphatase,** and **diacylglycerol acyltransferase.** The intermediate phosphatidic acid and diacylglycerol can also be converted to phospholipids by the pathways described in Section 23-8. The acyltransferases are not completely specific for particular fatty acyl-CoAs, either in chain length or in degree of unsaturation, but in human adipose tissue triacylglycerols, palmitate tends to be concentrated at position 1 and oleate at position 2.

5. REGULATION OF FATTY ACID METABOLISM

Discussions of metabolic control are usually concerned with the regulation of metabolite flow through a pathway in response to the differing energy needs and dietary states of an organism. For example, the difference in the energy requirement of muscle between rest and vigorous exertion may be as much as 100-fold. Such varying demands may be placed on the body when it is in either a fed or a fasted state. For instance, Eric Newsholme, an authority on the biochemistry of exercise, enjoys a 2-hour run before breakfast. Others might wish for no greater exertion than the motion of hand to mouth. In both individuals, glycogen and triacylglycerols serve as primary fuels for energy-requiring processes and are synthesized in times of quiet plenty for future use.

Synthesis and breakdown of glycogen and triacylglycerols, as detailed in Chapter 17 and above, are processes that

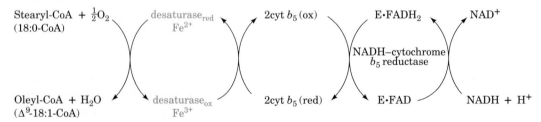

FIGURE 23-33. The electron-transfer reactions mediated by the Δ^9-fatty acyl-CoA desaturase complex. Its three proteins, desaturase, cytochrome b_5, and NADH–cytochrome b_5 reductase, are situated in the endoplasmic reticulum membrane. [After Jeffcoat, R., *Essays Biochem.* **15,** 19 (1979).]

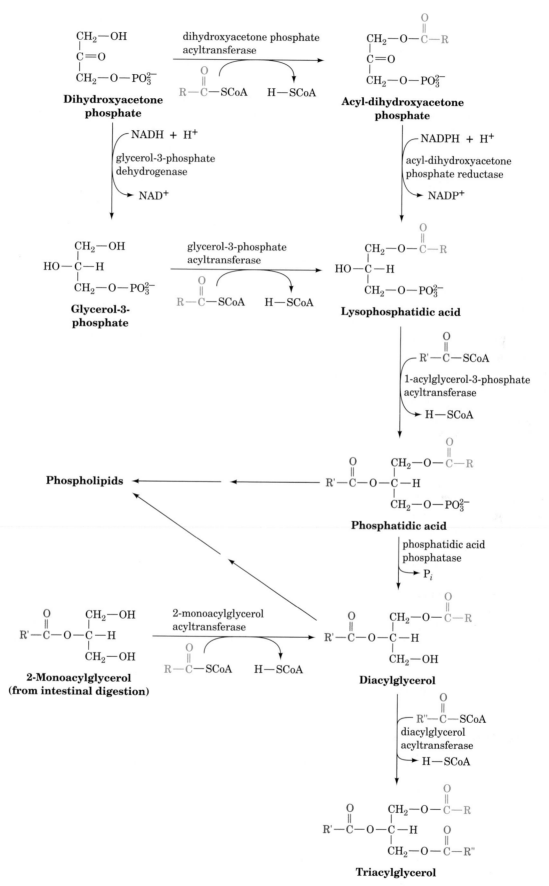

FIGURE 23-34. The reactions of triacylglycerol biosynthesis.

concern the whole organism, with its organs and tissues forming an interdependent network connected by the bloodstream. The blood carries the metabolites responsible for energy production: triacylglycerols in the form of chylomicrons and VLDL (Section 11-4A), fatty acids as their albumin complexes (Section 23-1), ketone bodies, amino acids, lactate, and glucose. The pancreatic α and β cells sense the organism's dietary and energetic state mainly through the glucose concentration in the blood. The α cells respond to the low blood glucose concentration of the fasting and energy-demanding states by secreting glucagon. The β cells respond to the high blood glucose concentration of the fed and resting states by secreting insulin. We have previously discussed (Sections 17-3E and F) how these hormones are involved in glycogen metabolism. *They also regulate the rates of the opposing pathways of lipid metabolism and therefore control whether fatty acids will be oxidized or synthesized.* Their targets are the regulatory (flux-generating) enzymes of fatty acid synthesis and breakdown in specific tissues (Fig. 23-35).

We are already familiar with most of the mechanisms by which the catalytic activities of regulatory enzymes may be controlled: substrate availability, allosteric interactions, and covalent modification (phosphorylation). These are examples of **short-term regulation,** regulation that occurs with a response time of minutes or less. *Fatty acid synthesis is controlled, in part, by short-term regulation.* Acetyl-CoA carboxylase, which catalyzes the first committed step of this pathway, is inhibited by palmitoyl-CoA and by the glucagon-stimulated cAMP-dependent increase in phosphorylation, and is activated by citrate and by insulin-stimulated dephosphorylation (Section 23-4B).

Another mechanism exists for controlling a pathway's regulatory enzymes: alteration of the amount of enzyme present by changes in the rates of protein synthesis and/or breakdown. This process requires hours or days and is therefore called **long-term regulation** (the control of protein synthesis and breakdown is discussed in Chapters 29 and 30). *Lipid biosynthesis is also controlled by long-term regulation,* with insulin stimulating and starvation inhibiting the synthesis of acetyl-CoA carboxylase and fatty acid synthase. The presence in the diet of polyunsaturated fatty acids also decreases the concentrations of these enzymes. The amount of adipose tissue lipoprotein lipase, the enzyme that initiates the entry of lipoprotein-packaged fatty acids into adipose tissue for storage (Section 11-4B), is also increased by insulin and decreased by starvation. In contrast, the concentration of heart lipoprotein lipase, which controls the entry of fatty acids from lipoproteins into heart tissue for oxidation rather than storage, is decreased by insulin and increased by starvation. *Starvation and/or regular exercise, by decreasing the glucose concentration in the blood, change the body's hormone balance. This situation results in long-term increases in the levels of fatty acid oxidation enzymes accompanied by long-term decreases in those of lipid biosynthesis.*

Hormones Regulate Fatty Acid Metabolism

*Fatty acid oxidation is regulated largely by the concentration of fatty acids in the blood, which is, in turn, controlled by the hydrolysis rate of triacylglycerols in adipose tissue by **hormone-sensitive triacylglycerol lipase.*** This enzyme is so named because it is susceptible to regulation by phosphorylation and dephosphorylation in response to hormonally controlled cAMP levels. Epinephrine and norepinephrine, as does glucagon, act to increase adipose tissue cAMP concentrations. cAMP allosterically activates cAMP-dependent protein kinase (cAPK) which, in turn, increases the phosphorylation levels of susceptible enzymes. Phosphorylation activates hormone-sensitive lipase, thereby stimulating lipolysis in adipose tissue, raising blood fatty acid levels, and ultimately activating the β-oxidation pathway in other tissues such as liver and muscle. In liver, this process leads to the production of ketone bodies that are secreted into the bloodstream for use as an alternative fuel to glucose by peripheral tissues. cAMP-dependent protein kinase, acting in concert with AMP-dependent protein kinase (AMPK), also causes the inactivation of acetyl-CoA carboxylase (Section 23-4B), one of the rate-determining enzymes of fatty acid synthesis, so that *cAMP-dependent phosphorylation simultaneously stimulates fatty acid oxidation and inhibits fatty acid synthesis.*

Insulin has the opposite effect of glucagon and epinephrine: It stimulates the formation of glycogen and triacylglycerols. This protein hormone, which is secreted in response to high blood glucose concentrations, decreases cAMP levels. This situation leads to the dephosphorylation and thus the inactivation of hormone-sensitive lipase, thereby reducing the amount of fatty acid available for oxidation. Insulin also stimulates the dephosphorylation of acetyl-CoA carboxylase, thereby activating this enzyme (Section 23-4B). *The glucagon–insulin ratio is therefore of prime importance in determining the rate and direction of fatty acid metabolism.*

Another control point that inhibits fatty acid oxidation when fatty acid synthesis is stimulated is the inhibition of carnitine palmitoyl transferase I by malonyl-CoA. This inhibition keeps the newly synthesized fatty acids out of the mitochondrion (Section 23-1) and thus away from the β-oxidation system.

6. CHOLESTEROL METABOLISM

Cholesterol is a vital constituent of cell membranes and the precursor of steroid hormones and bile acids. It is clearly essential to life, yet its deposition in arteries has been associated with cardiovascular disease and stroke, two leading causes of death in humans. In a healthy organism, an intricate balance is maintained between the biosynthesis, utili-

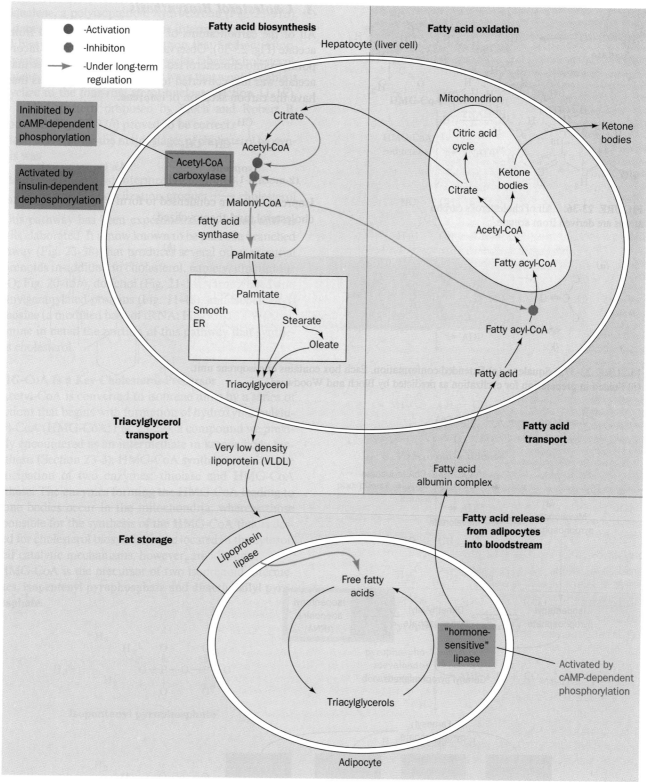

FIGURE 23-35. Sites of regulation of fatty acid metabolism.

zation, and transport of cholesterol, keeping its harmful deposition to a minimum. In this section, we study the pathways of cholesterol biosynthesis and transport and how they are controlled. We also examine how cholesterol is utilized in the biosynthesis of steroid hormones and bile acids.

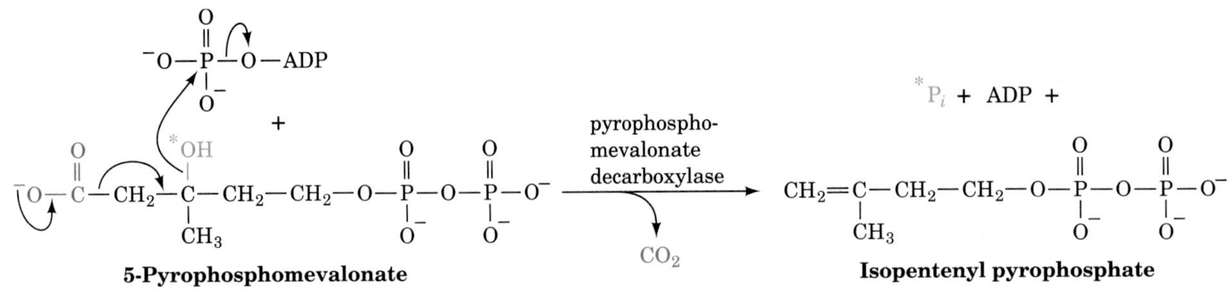

FIGURE 23-40. Pyrophosphomevalonate decarboxylase catalyzes an ATP-dependent concerted dehydration–decarboxylation of pyrophosphomevalonate, yielding isopentenyl pyrophosphate.

2. The new OH group is phosphorylated by **mevalonate-5-phosphotransferase.**

3. The phosphate group is converted to a pyrophosphate by **phosphomevalonate kinase.**

4. The molecule is decarboxylated and the resulting alcohol dehydrated by **pyrophosphomevalonate decarboxylase.**

HMG-CoA reductase mediates the rate-determining step of cholesterol biosynthesis and is the most elaborately regulated enzyme of this pathway. This 97-kD membrane-bound enzyme of the endoplasmic reticulum is regulated, as we shall see in Section 23-6B, by competitive and allosteric mechanisms, phosphorylation/dephosphorylation, and long-term regulation. Cholesterol itself is an important feedback regulator of the enzyme.

Pyrophosphomevalonate Decarboxylase and Isopentenyl Pyrophosphate Isomerase Both Catalyze Apparently Concerted Reactions

5-Pyrophosphomevalonate is converted to isopentenyl pyrophosphate by an ATP-dependent dehydration–decarboxylation reaction catalyzed by **pyrophosphomevalonate decarboxylase** (Fig. 23-40). When [3-^{18}O]-5-pyrophosphomevalonate (*O in Fig. 23-40) is used as a substrate, the labeled oxygen appears in P_i. This observation suggests that 3-phospho-5-pyrophosphomevalonate is a reaction intermediate. Since all attempts to isolate this intermediate have

failed, however, it has been proposed that phosphorylation, the α,β elimination of CO_2, and the elimination of P_i occur in a concerted reaction.

The equilibration between isopentenyl pyrophosphate and dimethylallyl pyrophosphate is catalyzed by **isopentenyl pyrophosphate isomerase.** The reaction is thought to occur via a concerted protonation/deprotonation reaction (Fig. 23-41).

Squalene Is Formed by the Condensation of Six Isoprene Units

Four isopentenyl pyrophosphates and two dimethylallyl pyrophosphates condense to form the C_{30} cholesterol precursor squalene in three reactions catalyzed by two enzymes (Fig. 23-42):

1. **Prenyl transferase (farnesyl pyrophosphate synthase)** catalyzes the head-to-tail (1′–4) condensation of dimethylallyl pyrophosphate and isopentenyl pyrophosphate to yield **geranyl pyrophosphate.**

2. Prenyl transferase catalyzes a second head-to-tail condensation of geranyl pyrophosphate and isopentenyl pyrophosphate to yield **farnesyl pyrophosphate.**

3. **Squalene synthase** then catalyzes the head-to-head (1–1′) condensation of two farnesyl pyrophosphate molecules to form squalene. Farnesyl pyrophosphate is also a precursor to dolichol, farnesylated and geranylgeranylated proteins, and ubiquinone (Fig. 23-38).

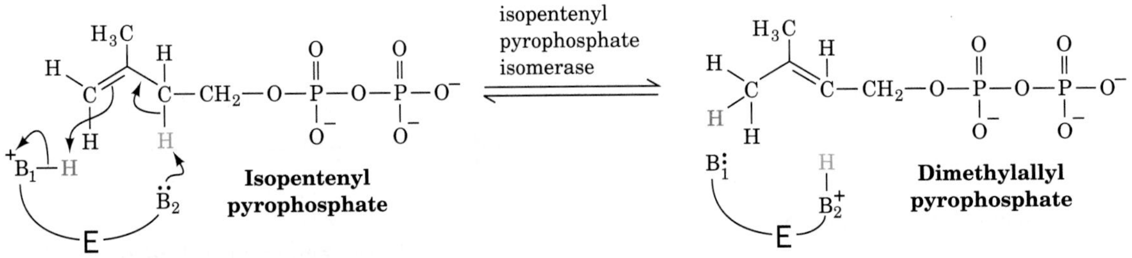

FIGURE 23-41. Isopentenyl pyrophosphate isomerase interconverts isopentenyl pyrophosphate and dimethylallyl pyrophosphate by a concerted protonation/deprotonation reaction.

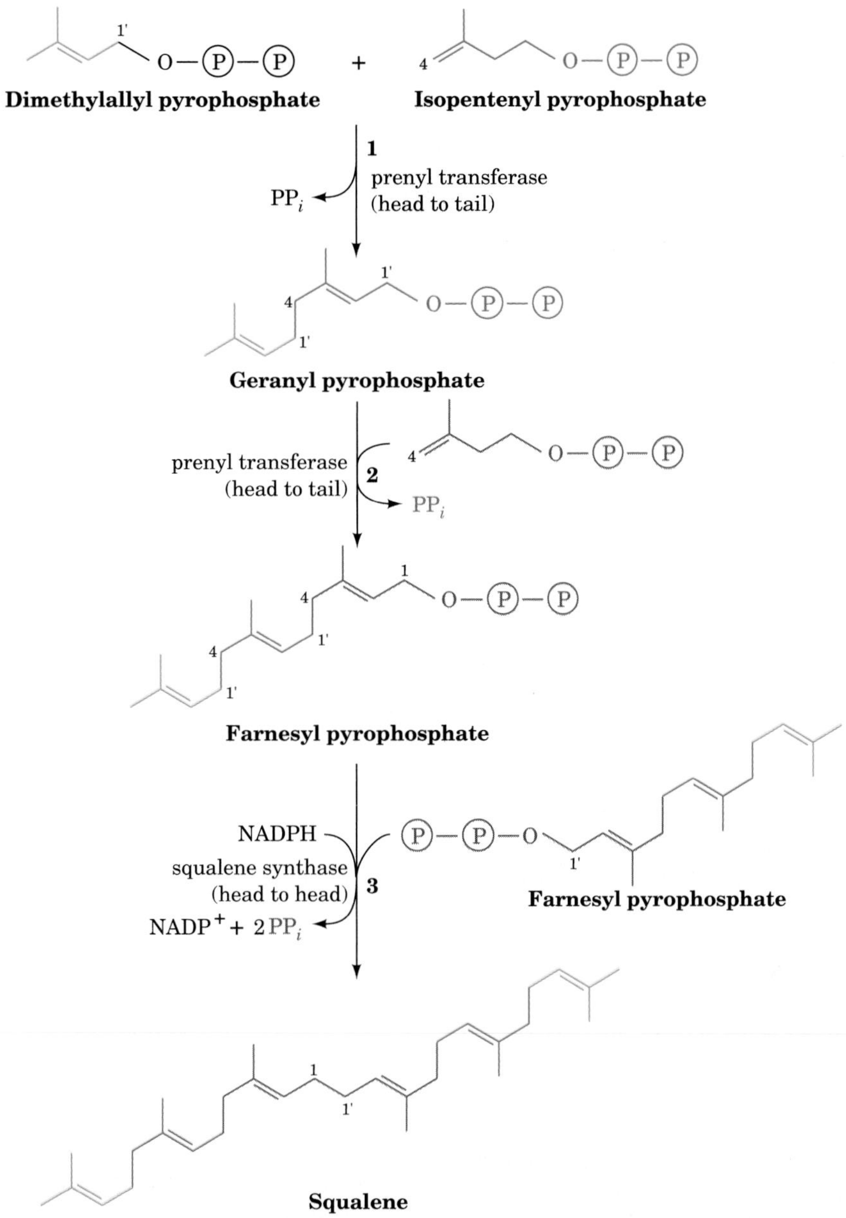

FIGURE 23-42. The formation of squalene from isopentenyl pyrophosphate and dimethylallyl pyrophosphate. The pathway involves two head-to-tail condensations catalyzed by prenyl transferase and a head-to-head condensation catalyzed by squalene synthase.

Prenyl transferase catalyzes the condensation of isopentenyl pyrophosphate with an allylic pyrophosphate. It is specific for isopentenyl pyrophosphate but can use either the 5-carbon dimethylallyl pyrophosphate or the 10-carbon **geranyl pyrophosphate** as its allylic substrate. The prenyl transferase–catalyzed condensation mechanism is particularly interesting since it is one of the very few known enzyme-catalyzed reactions that proceed via a carbocation intermediate. Two possible condensation mechanisms can be envisioned (Fig. 23-43):

Scheme I An S_N1 mechanism in which an allylic carbocation forms by the elimination of PP_i. Isopentenyl pyrophosphate then condenses with this carbocation, forming a new carbocation that eliminates a proton to form product.

Scheme II An S_N2 reaction in which the allylic PP_i is displaced in a concerted manner. In this case, an enzyme nucleophile, X, assists in the reaction. This group is eliminated in the second step with the loss of a proton to form product.

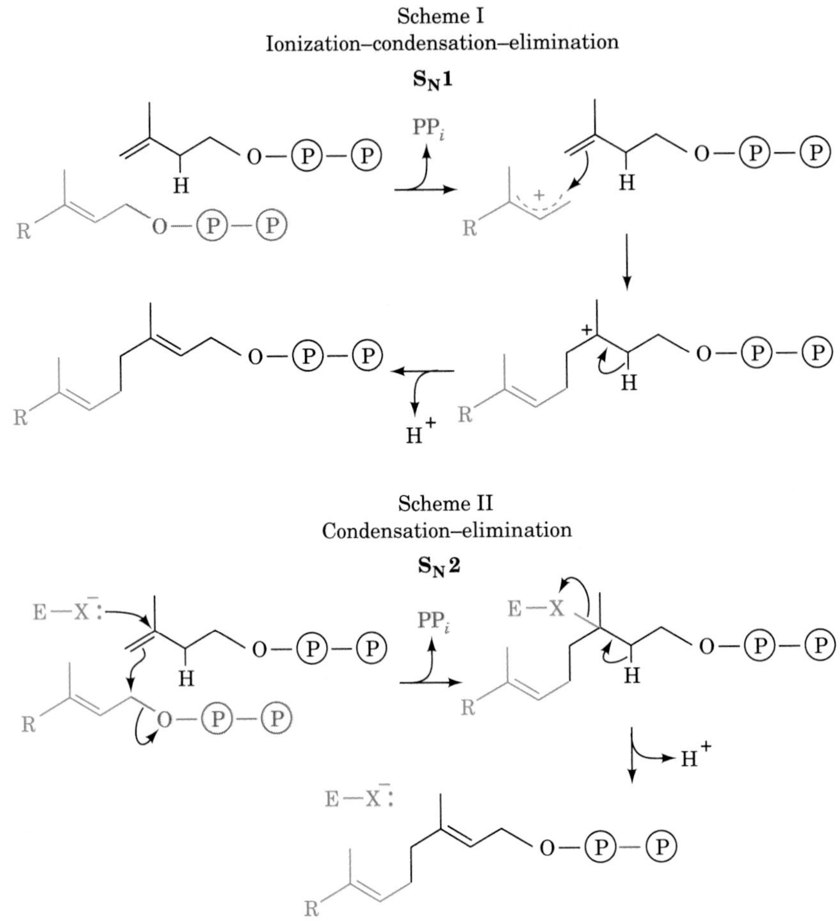

Scheme I
Ionization–condensation–elimination

$S_N 1$

Scheme II
Condensation–elimination

$S_N 2$

FIGURE 23-43. Two possible mechanisms for the prenyl transferase reaction. Scheme I involves the formation of a carbocation intermediate, whereas Scheme II involves the participation of an enzyme nucleophile, X.

C. Dale Poulter and Hans Rilling used chemical logic to differentiate between these two mechanisms. Capitalizing on the observation that $S_N 1$ reactions are much more sensitive to electron-withdrawing groups than $S_N 2$ reactions, they synthesized a geranyl pyrophosphate derivative in which the H at C3 is replaced by the electron-withdrawing group F. This allylic substrate for the second $(1'-4)$ condensation catalyzed by prenyl transferase, not surprisingly, has the same K_M as the natural substrate (F and H have similar atomic radii).

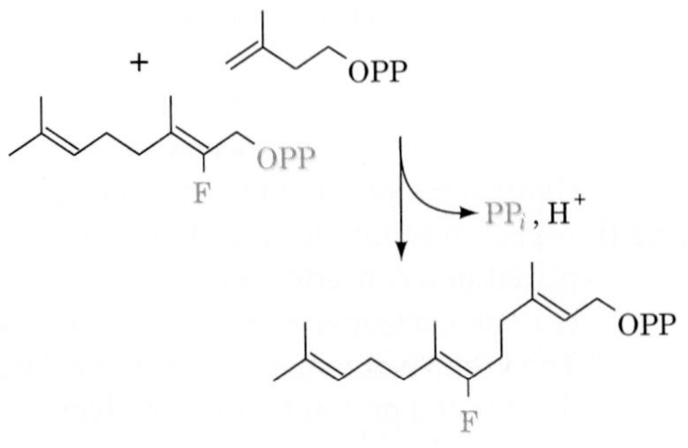

It is, however, the V_{max} of this reaction that tells the story. If the reaction is an $S_N 2$ displacement, the fluoro derivative should react at a rate similar to that of the natural substrate. If, instead, the reaction has an $S_N 1$ mechanism, the fluoro derivative should react orders of magnitude more slowly than the natural substrate. In fact, 3-fluorogeranyl pyrophosphate forms product at $< 1\%$ of the rate of the natural substrate, strongly supporting an $S_N 1$ mechanism with a carbocation intermediate.

Squalene, the immediate sterol precursor, is formed by the head-to-head condensation of two farnesyl pyrophosphate molecules by **squalene synthase**. The reaction is not a simple head-to-tail condensation, as might be expected, but, rather, proceeds via a complex two-step reaction (Fig. 23-44):

Step I The insertion of C1 of one farnesyl pyrophosphate molecule into the C2—C3 double bond of the second molecule, eliminating PP$_i$ and forming **presqualene pyrophosphate**, a cyclopropylcarbinyl pyrophosphate.

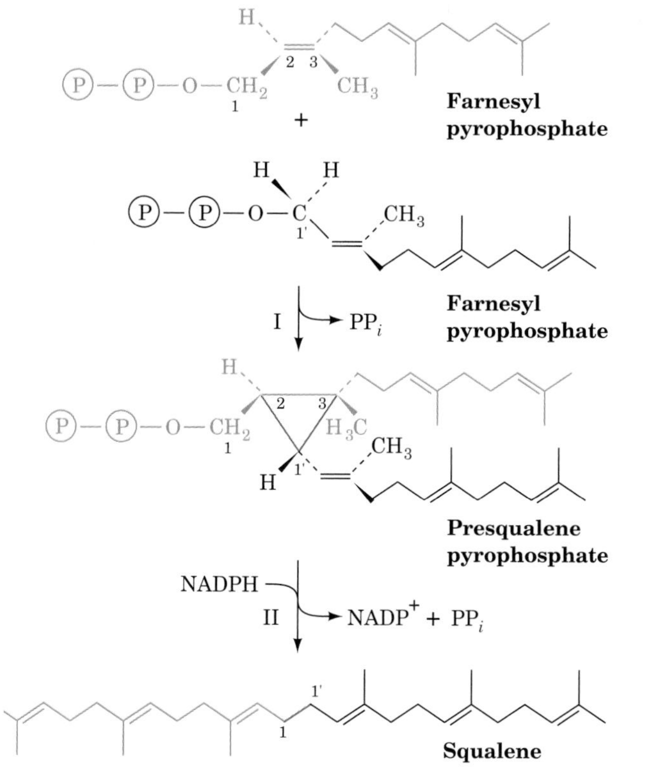

FIGURE 23-44. Squalene synthase catalyzes the head-to-head condensation of two farnesyl pyrophosphate molecules to form squalene.

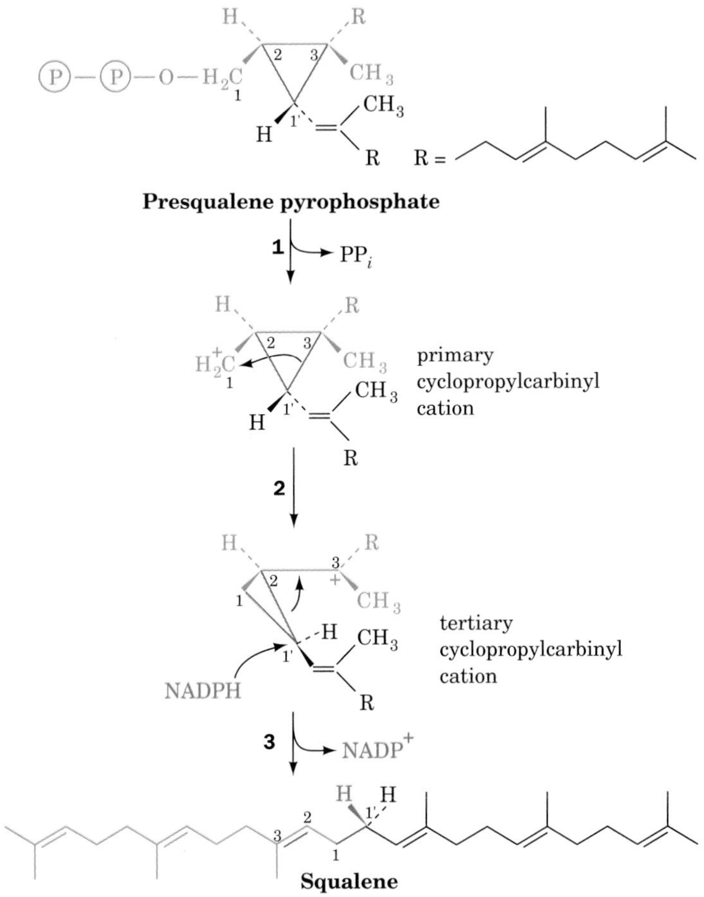

Step II Rearrangement and reduction of presqualene pyrophosphate by NADPH to form squalene. This reaction involves the formation and rearrangement of a cyclopropylcarbinyl cation in a complex reaction sequence called a **1′–2–3 process** (Fig. 23-45).

Lanosterol Is Produced by Squalene Cyclization

Squalene, an open-chain C_{30} hydrocarbon, is cyclized to form the tetracyclic steroid skeleton in two steps. **Squalene epoxidase** catalyzes oxidation of squalene to form **2,3-oxidosqualene** (Fig. 23-46). **Squalene oxidocyclase** converts this epoxide to **lanosterol,** the sterol precursor of cholesterol. The reaction is a complex process involving cycliza-

FIGURE 23-45. The mechanism of rearrangement and reduction of presqualene pyrophosphate to squalene as catalyzed by squalene synthase: **(1)** Presqualene's pyrophosphate group leaves, yielding a primary carbocation at C1. **(2)** The electrons forming the C1′—C3 bond migrate to C1, forming squalene's C1—C1′ bond and a tertiary carbocation at C3. **(3)** The process is completed by the addition of an NADPH-supplied hydride ion to C1′ and the formation of the C2═C3 double bond.

tion of 2,3-oxidosqualene to a **protosterol** cation and rearrangement of this cation to lanosterol by a series of 1,2 hydride and methyl shifts (Fig. 23-47).

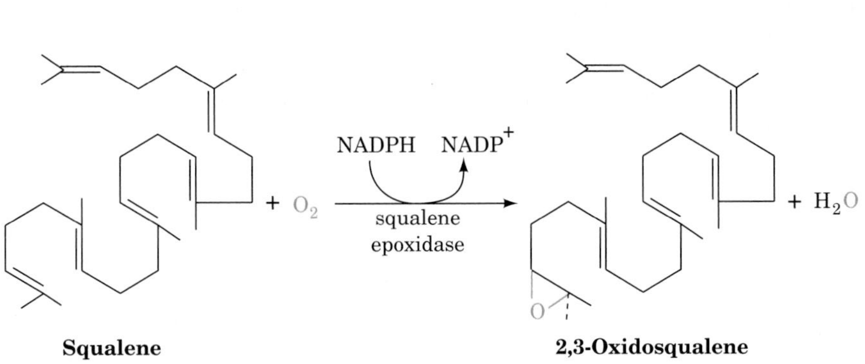

FIGURE 23-46. The squalene epoxidase reaction.

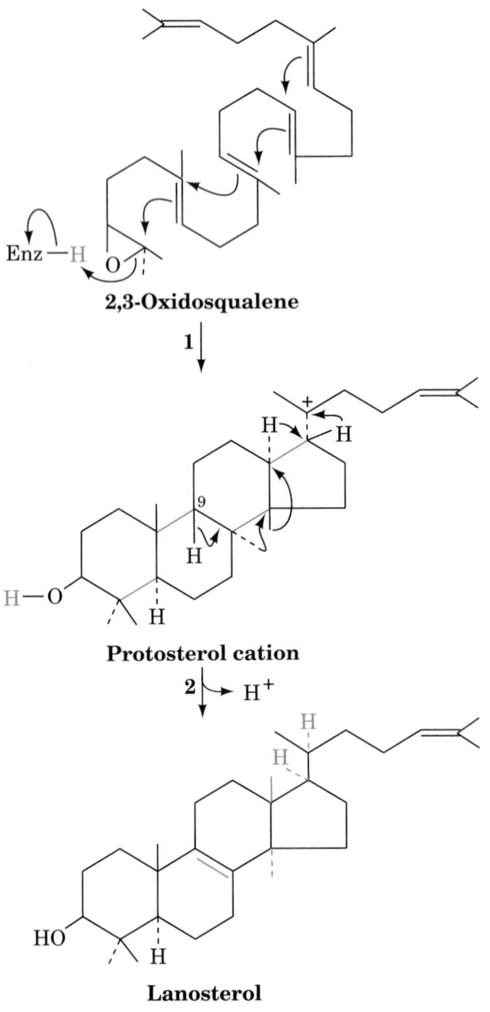

2,3-Oxidosqualene

Protosterol cation

Lanosterol

FIGURE 23-47. The squalene oxidocyclase reaction: **(1)** 2,3-Oxidosqualene is cyclized to the protosterol cation in a process that is initiated by the enzyme-mediated protonation of the squalene epoxide oxygen while this extended molecule is folded in the manner predicted by Bloch and Woodward. The opening of the epoxide leaves an electron-deficient center whose migration drives the series of cyclizations that form the protosterol cation. **(2)** The elimination of a proton from C9 of the sterol to form a double bond initiates a series of methyl and hydride migrations that ultimately yields neutral lanosterol.

Cholesterol Is Synthesized from Lanosterol

Conversion of lanosterol to cholesterol (Fig. 23-48) is a 19-step process that we shall not explore in detail. It involves an oxidation and loss of three methyl groups. The first methyl group is removed as formate and the other two are eliminated as CO_2 in reactions that all require NADPH and O_2. The enzymes involved in this process are embedded in the endoplasmic reticulum membrane.

Cholesterol Is Transported in the Blood and Taken Up by Cells in Lipoprotein Complexes

Transport and cellular uptake of cholesterol is described in Section 11-5. To recapitulate, cholesterol synthesized by

the liver is either converted to bile acids for use in the digestive process (Section 23-1) or esterified by **acyl-CoA:cholesterol acyl transferase (ACAT)** to form **cholesteryl esters**

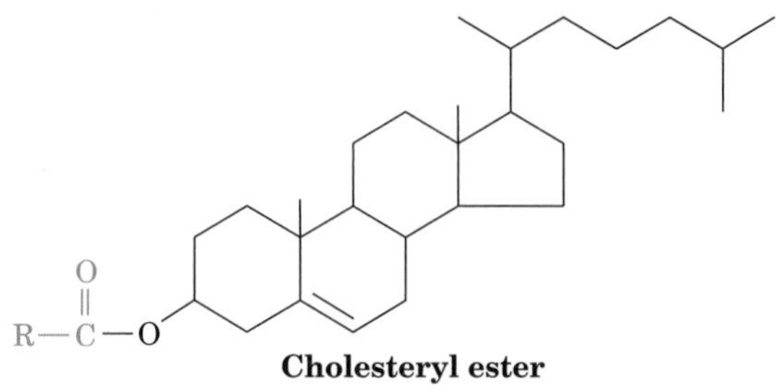

Cholesteryl ester

which are secreted into the bloodstream as part of the lipoprotein complexes called **very low density lipoproteins (VLDL).** As the VLDL circulate, their component triacylglycerols and most types of their **apolipoproteins** (Table 11-6) are removed in the capillaries of muscle and adipose tissues, sequentially converting the VLDL to **intermediate-density lipoproteins (IDL)** and then to **low-density lipoproteins (LDL).** Peripheral tissues normally obtain most of their exogenous cholesterol from LDL by receptor-mediated endocytosis (Fig. 23-49; Section 11-4B). Inside the cell, cholesteryl esters are hydrolyzed by a lysosomal lipase to free cholesterol, which is either incorporated into cell membranes or reesterified by ACAT for storage as cholesteryl ester droplets.

Dietary cholesterol, cholesteryl esters, and triacylglycerols are transported in the blood by intestinally synthesized lipoprotein complexes called **chylomicrons.** After removal of their triacylglycerols at the peripheral tissues, the resulting **chylomicron remnants** bind to specific liver cell remnant receptors and are taken up by receptor-mediated endocytosis in a manner similar to that of LDL. In the liver, dietary cholesterol is either used in bile acid biosynthesis (Section 23-6C) or packaged into VLDL for export. *Liver and peripheral tissues therefore have two ways of obtaining cholesterol: They may either synthesize it from acetyl-CoA by the de novo pathway we have just discussed, or they may obtain it from the bloodstream by receptor-mediated endocytosis.* A small amount of cholesterol also enters cells by a non-receptor-mediated pathway.

Cholesterol actually circulates back and forth between the liver and peripheral tissues. While LDL transports cholesterol from the liver, cholesterol is transported back to the liver by **high-density lipoproteins (HDL).** Surplus cholesterol is disposed of by the liver as bile acids, thereby protecting the body from an overaccumulation of this water-insoluble substance.

B. Control of Cholesterol Biosynthesis and Transport

Cholesterol biosynthesis and transport must be tightly regulated. There are three ways in which the cellular cholesterol supply is maintained:

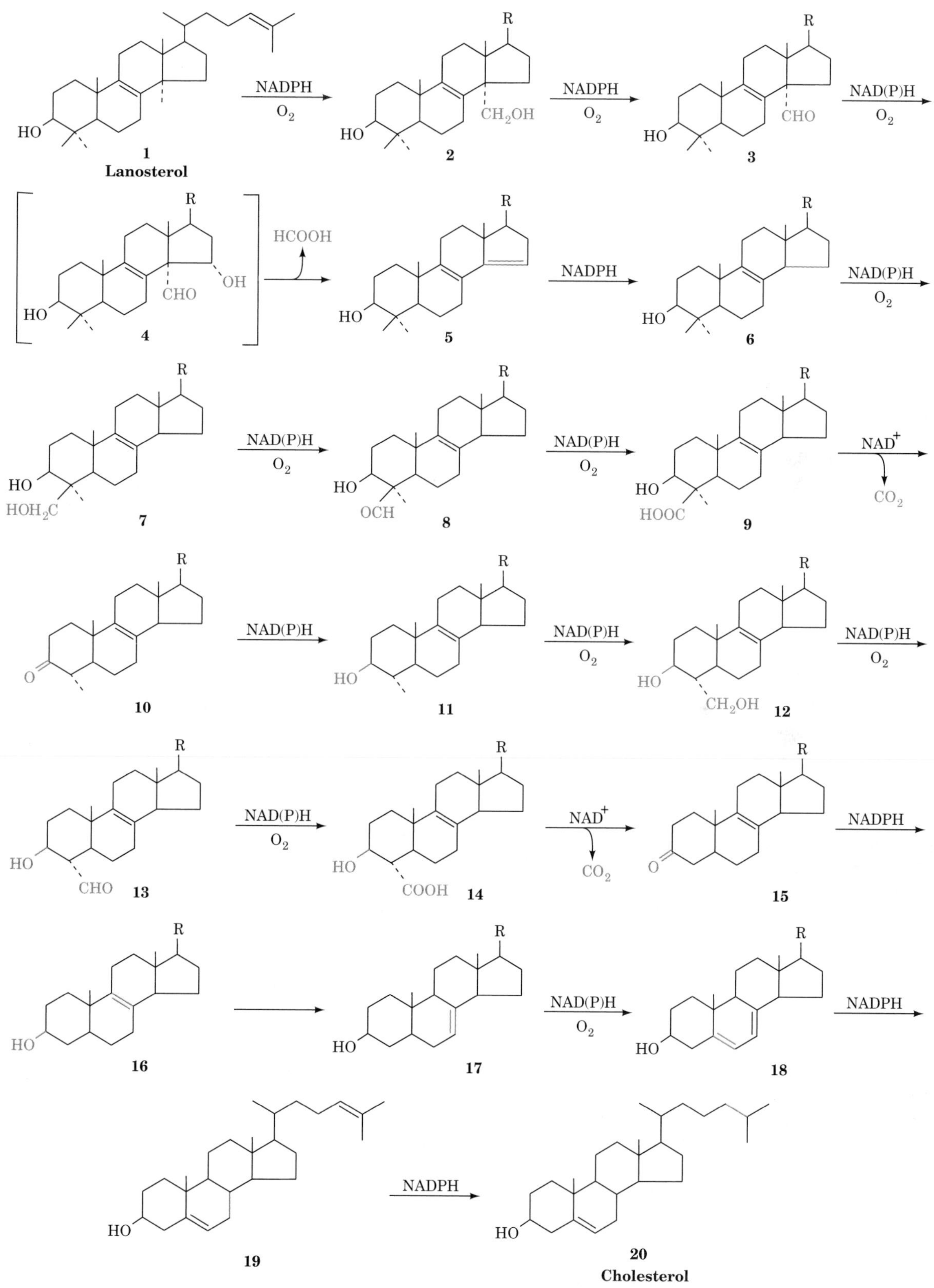

FIGURE 23-48. The 19-reaction conversion of lanosterol to cholesterol. [After Rilling, H.C. and Chayet, L.T., *in* Danielsson, H. and Sjövall, J. (Eds.), *Sterols and Bile Acids, p.* 33, Elsevier (1985).]

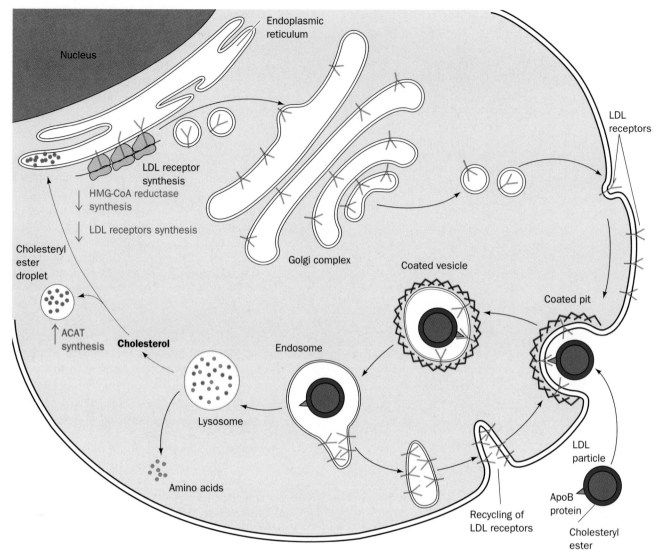

FIGURE 23-49. LDL receptor-mediated endocytosis in mammalian cells. LDL receptor is synthesized on the endoplasmic reticulum, processed in the Golgi complex, and inserted into the plasma membrane as a component of coated pits. LDL is specifically bound by the receptor on the coated pit and brought into the cell in endosomes that deliver LDL to lysosomes while recycling LDL receptor to the plasma membrane (Section 11-5C). Lysosomal degradation of LDL releases cholesterol, whose presence decreases the rate of synthesis of HMG-CoA reductase and LDL receptors (*down arrows*) while increasing that of acyl-CoA:cholesterol acyltransferase (ACAT; *up arrow*). [After Brown, M.S. and Goldstein, J.L., *Curr. Top. Cell. Reg.* **26,** 7 (1985).]

1. By regulating the activity of HMG-CoA reductase, the enzyme catalyzing the rate-limiting step in the *de novo* pathway. This is accomplished in two ways:

   (i) Short-term regulation of the enzyme's catalytic activity by (a) competitive inhibition, (b) allosteric effects, and (c) covalent modification involving reversible phosphorylation.

   (ii) Long-term regulation of the enzyme's concentration by modulating its rates of synthesis and degradation.

2. By regulating the rate of LDL receptor synthesis. High intracellular concentrations of cholesterol suppress LDL receptor synthesis, whereas low cholesterol concentrations stimulate it.

3. By regulating the rate of esterification and hence the removal of free cholesterol. ACAT, the enzyme that catalyzes intracellular cholesterol esterification, is regulated by reversible phosphorylation and by long-term control.

HMG-CoA Reductase Is the Primary Control Site for Cholesterol Biosynthesis

HMG-CoA reductase is the rate-limiting enzyme in cholesterol biosynthesis and, as therefore might be expected, constitutes the pathway's main regulatory site. The pathway branches after this reaction, however (Fig. 23-38); ubiquinone, dolichol, farnesylated and geranylgeranylated proteins, and isopentenyl adenosine are also essential, albeit minor products. HMG-CoA is therefore subject to "multivalent" control, both long-term and short-term, in order to coordinate the synthesis of all of these products.

Long-Term Feedback Regulation of HMG-CoA Reductase Is Its Primary Means of Control

The main way in which HMG-CoA reductase is controlled is by long-term feedback control of the amount of enzyme present in the cell. When either LDL–cholesterol or mevalonate levels fall, the amount of HMG-CoA reductase present in the cell can rise as much as 200-fold, due to an increase in enzyme synthesis combined with a decrease in its degradation. When LDL–cholesterol or **mevalonolactone** (an internal ester of mevalonate that is hydrolyzed to mevalonate and metabolized in the cell)

$$\begin{array}{c} H_3C \quad CH_2{-}CH_2 \\ \diagdown \quad \diagup \qquad \diagdown \\ C \qquad\qquad O \\ \diagup \quad \diagdown \qquad \diagup \\ HO \quad CH_2{-}C \\ \qquad\qquad \diagdown \\ \qquad\qquad O \end{array}$$

Mevalonolactone

are added back to a cell, these effects are reversed.

Short-Term Feedback Regulation of HMG-CoA Reductase Constitutes a Means of Cellular Energy Conservation

HMG-CoA reductase exists in interconvertible more active and less active forms, as do glycogen phosphorylase (Section 17-3C), glycogen synthase (Section 17-3D), pyruvate dehydrogenase (Section 19-2B), and acetyl-CoA carboxylase (Section 23-4B), among others. The unmodified form of HMG-CoA reductase is more active; the phosphorylated form is less active. HMG-CoA reductase is phosphorylated (inactivated) at its Ser 871 in a bicyclic cascade system by the covalently modifiable enzyme **HMG-CoA reductase kinase (RK)**. This enzyme has recently been found to be identical with **AMP-dependent protein kinase (AMPK)** which, as we saw in Section 23-4B, also acts on acetyl-CoA carboxylase. It appears that this control is exerted to conserve energy when ATP levels fall and AMP levels rise, by inhibiting biosynthetic pathways. This hypothesis was tested by Michael Brown and Joseph Goldstein, who used genetic engineering techniques to produce hamster cells containing a mutant HMG-CoA reductase with Ala replacing Ser 871 and therefore incapable of phosphorylation control. These cells respond normally to feedback regulation of cholesterol biosynthesis by LDL–cholesterol and mevalonate but, unlike normal cells, do not decrease their synthesis of cholesterol on ATP depletion, supporting the idea that control of HMG-CoA reductase by phosphorylation is involved in energy conservation.

LDL Receptor Activity Controls Cholesterol Homeostasis

LDL receptors clearly play an important role in the maintenance of plasma LDL–cholesterol levels. In normal individuals, about one half of the IDL formed from the VLDL reenters the liver through LDL receptor–mediated endocytosis (IDL and LDL both contain apolipoproteins that specifically bind to the LDL receptor; Section 11-5C). The remaining IDL are converted to LDL (Fig. 23-50*a*). *The serum concentration of LDL therefore depends on the rate that liver removes IDL from the circulation, which, in turn, depends on the number of functioning LDL receptors on the liver cell surface.*

High blood cholesterol **(hypercholesterolemia),** which results from the overproduction and/or underutilization of LDL, is known to be caused by two metabolic irregularities: (1) the genetic disease **familial hypercholesterolemia (FH),** or (2) the consumption of a high-cholesterol diet. FH is a dominant genetic defect that results in a deficiency of functional LDL receptors (Section 11-5D). Homozygotes for this disorder lack functional LDL receptors, so their cells can absorb neither IDL nor LDL, by receptor-mediated endocytosis. The increased concentration of IDL in the bloodstream leads to a corresponding increase in LDL, which is, of course, underutilized since it cannot be taken up by the cells (Fig. 23-50*b*). FH homozygotes therefore have plasma LDL–cholesterol levels three to five times higher than average. FH heterozygotes, which are far more common, have about one half of the normal number of functional LDL receptors and plasma LDL–cholesterol levels of about twice the average.

The ingestion of a high-cholesterol diet has an effect similar, although not as extreme, as FH (Fig. 23-50c). Excessive dietary cholesterol enters the liver cells in chylomicron remnants and represses the synthesis of LDL–receptor protein. The resulting insufficiency of LDL receptors on the liver cell surface has consequences similar to those of FH.

LDL receptor deficiency, whether of genetic or dietary origin, raises the LDL level by two mechanisms: (1) increased LDL production resulting from decreased IDL uptake; and (2) decreased LDL uptake. Two strategies for reversing these conditions (besides maintaining a low cholesterol diet) are being used in humans:

1. *Ingestion of resins that bind bile acids, thereby preventing their intestinal absorption.* Bile acids, which are derived from cholesterol, are normally efficiently recycled by the liver (Section 23-6C). Elimination of resin-bound cholesterol in the feces forces the liver to convert more cholesterol to bile acids than normal. The consequent decrease in the serum cholesterol concentration induces synthesis of LDL receptors (of course, not in FH homozygotes). Unfortunately, the decreased serum cholesterol level also induces the synthesis of HMG-CoA reductase, which increases the rate of cholesterol biosynthesis. Ingestion of bile acid–binding resins therefore provides only a 15 to 20% drop in serum cholesterol levels.

2. *Treatment with competitive inhibitors of HMG-CoA reductase, notably the fungal products **compactin** and **lo-***

FIGURE 23-50. Liver LDL receptors control plasma LDL production and uptake. (*a*) In normal human subjects, VLDL is secreted by the liver and converted to IDL in the capillaries of the peripheral tissues. About half of the plasma IDL particles bind to the LDL receptor and are taken up by the liver. The remainder are converted to LDL at the peripheral tissues. (*b*) In individuals with familial hypercholesterolemia (FH), liver LDL receptors are diminished or eliminated because of a genetic defect. (*c*) In normal individuals who ingest a high-cholesterol diet, the liver is filled with cholesterol, which represses the rate of LDL receptor production. Receptor deficiency, whether of genetic or dietary cause, raises the plasma LDL level by increasing the rate of LDL production and decreasing the rate of LDL uptake. [After Goldstein, J.L. and Brown, M.S., *J. Lipid Res.* **25,** 1457 (1984).]

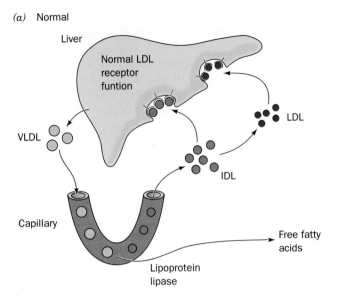

(*a*) Normal

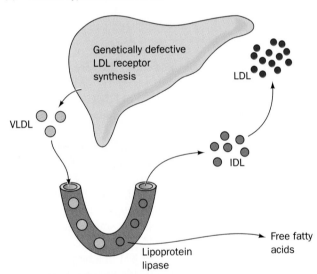

(*b*) Familial hypercholesterolemia

*vastatin (also called **mevinolin**; Fig. 23-51), so as to decrease the rate of cholesterol biosynthesis.* Indeed, lovastatin is now in routine use for the treatment of hypercholesterolemia. The resulting decreased cholesterol supply is again met by induction of LDL receptors and HMG-CoA reductase. Lovastatin-treated FH heterozygotes nevertheless routinely show a serum cholesterol decrease of 30%.

The combined use of these agents, moreover, results in a clinically dramatic 50 to 60% decrease in serum cholesterol levels.

Overexpression of LDL Receptor Prevents Diet-Induced Hypercholesterolemia

Experiments are well underway towards the treatment of hypercholesterolemic individuals by **gene therapy** (Section 28-8F). A line of transgenic mice (mice that express a foreign gene; Section 27-2A) has been developed that overproduce the human LDL receptor. When fed a diet high in cholesterol, fat, and bile acids, these transgenic animals did not develop a detectable increase in plasma LDL. In contrast, normal mice fed the same diet exhibited large increases in plasma LDL levels. Evidently, the unregulated

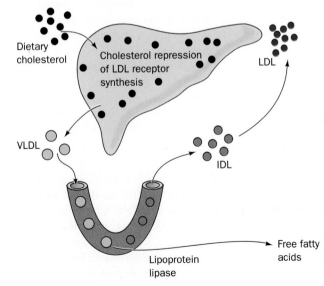

(*c*) High cholesterol diet

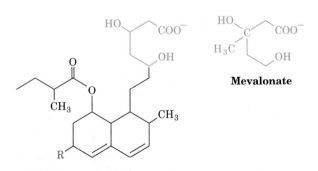

R = H   **Compactin**
R = CH₃   **Lovastatin (mevinolin)**

FIGURE 23-51. Compactin and lovastatin, two potent inhibitors of HMG-CoA reductase. The structure of mevalonate is shown for comparison.

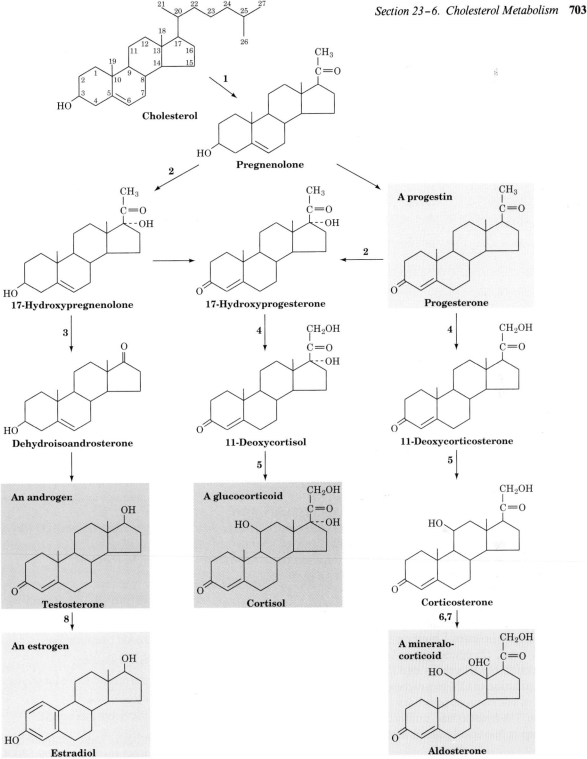

FIGURE 23-52. A simplified scheme of steroid biosynthesis. The enzymes involved are **(1)** the cholesterol side chain cleavage enzyme; **(2)** steroid C17 hydroxylase; **(3)** steroid C17,C20 lyase; **(4)** steroid C21 hydroxylase; **(5)** steroid 11β-hydroxylase; **(6)** steroid C18 hydroxylase; **(7)** 18-hydroxysteroid oxidase; and **(8)** aromatase.

overexpression of LDL receptors can prevent diet-induced hypercholesterolemia, at least in mice.

C. Cholesterol Utilization

*Cholesterol is the precursor of **steroid hormones** and bile acids.* Steroid hormones are grouped into five categories: ***progestins, glucocorticoids, mineralocorticoids, androgens, and estrogens.*** These hormones, as described in Section 34-4A, mediate a wide variety of vital physiological functions. All contain the four-ring structure of the sterol nucleus and are remarkably similar in structure, considering the enormous differences in their physiological effects. A simplified biosynthetic scheme (Fig. 23-52) indicates their structural

FIGURE 23-53. Structures of the major bile acids and their glycine and taurine conjugates.

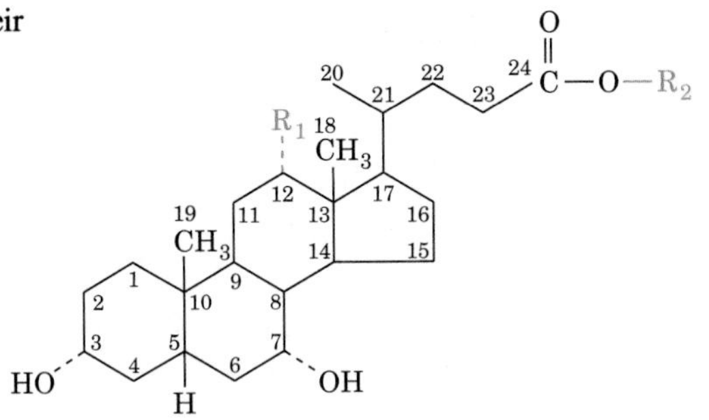

	$R_1 = OH$	$R_1 = H$
$R_2 = H$	**Cholic acid**	**Chenodeoxycholic acid**
$R_2 = NH-CH_2-COOH$	**Glycocholic acid**	**Glycochenodeoxycholic acid**
$R_2 = NH-CH_2-CH_2-SO_3H$	**Taurocholic acid**	**Taurochenodeoxycholic acid**

similarities and differences. We shall not discuss the details of these pathways.

*The quantitatively most important pathway for the excretion of cholesterol in mammals is the formation of bile acids (also called **bile salts**).* The major bile acids, **cholic acid** and **chenodeoxycholic acid,** are synthesized in the liver and secreted as glycine or **taurine** conjugates (Fig. 23-53) into the gallbladder. From there, they are secreted into the small intestine, where they act as emulsifying agents in the digestion and absorption of fats and fat-soluble vitamins (Section 23-1). An efficient recycling system allows the bile acids to reenter the bloodstream and return to the liver for reuse several times each day. The < 1 g $\cdot$ day^{-1} of bile acids that normally escape this recycling system are further metabolized by microorganisms in the large intestine and excreted. *This is the body's only route for cholesterol excretion.*

Comparison of the structures of cholesterol and the bile acids (Figs. 23-36 and 23-53) indicates that biosynthesis of bile acids from cholesterol involves (1) saturation of the 5,6-double bond, (2) epimerization of the 3β-OH group, (3) introduction of OH groups into the 7α and 12α positions, (4) oxidation of C24 to a carboxylate, and (5) conjugation of this side chain carboxylate with glycine or taurine. **Cholesterol 7-α-hydroxylase** catalyzes the first and rate-limiting step in bile acid synthesis and is closely regulated.

7. ARACHIDONATE METABOLISM: PROSTAGLANDINS, PROSTACYCLINS, THROMBOXANES, AND LEUKOTRIENES

Prostaglandins (PGs) were first identified in human semen by Ulf von Euler in the early 1930s through their ability to stimulate uterine contractions and lower blood pressure.

von Euler thought that these compounds originated in the prostate gland (hence their name) but they were later shown to be synthesized in the seminal vesicles. By the time the mistake was realized, the name was firmly entrenched. In the mid-1950s, crystalline materials were isolated from biological fluids and called PGE (*e*ther-soluble) and PGF (phosphate buffer–soluble; *f*osfat in Swedish). This began an explosion of work on these potent substances.

*Almost all mammalian cells except red blood cells produce prostaglandins and their related compounds, the **prostacyclins, thromboxanes,** and **leukotrienes** (known collectively as **eicosanoids** since they are all C_{20} compounds; Greek: eikosi, twenty). The eicosanoids, like hormones, have profound physiological effects at extremely low concentrations.* For example, they mediate:

1. The inflammatory response, notably as it involves the joints (rheumatoid arthritis), skin (psoriasis), and eyes.
2. The production of pain and fever.
3. The regulation of blood pressure.
4. The induction of blood clotting.
5. The control of several reproductive functions such as the induction of labor.
6. The regulation of the sleep/wake cycle.

The enzymes that synthesize these compounds and the receptors to which they bind are therefore the targets of intensive pharmacological research.

The eicosanoids are also hormonelike in that many of their effects are intracellularly mediated by cAMP. Unlike hormones, however, they are not transported in the bloodstream to their sites of action. Rather, these chemically and biologically unstable substances (some decompose within minutes or less *in vitro*) are **local mediators;** that is, *they act in the same environment in which they are synthesized.*

In this section, we discuss the structures of the eicosanoids and outline their biosynthetic pathways and modes of action. As we do so, note the great diversity of their structures and functions, a phenomenon that makes the elucidation of the physiological roles of these potent substances a challenging research area.

A. Background

Prostaglandins are all derivatives of the hypothetical C_{20} fatty acid **prostanoic acid** *in which carbon atoms 8 to 12 comprise a cyclopentane ring (Fig. 23-54a).* Prostaglandins A through I differ in the substituents on the cyclopentane ring (Fig. 23-54b): **PGAs** are α,β-unsaturated ketones, **PGEs** are β-hydroxy ketones, **PGFs** are 1,3-diols, etc. In **PGF$_\alpha$**, the C9—OH group is on the same side of the ring as R_1; it is on the opposite side in **PGF$_\beta$**. The numerical subscript in the name refers to the number of double bonds contained on the side chains of the cyclopentane ring (Fig. 23-54c).

In humans, the most important prostaglandin precursor is **arachidonic acid (5,8,11,14-eicosatetraenoic acid),** *a C_{20} polyunsaturated fatty acid that has four nonconjugated double bonds.* The double bond at C14 is six carbon atoms from the terminal carbon atom (the ω-carbon atom), making arachidonic acid an $\omega - 6$ fatty acid. Arachidonic acid is synthesized from linoleic acid (also an $\omega - 6$ fatty acid) by elongation and desaturation (Fig. 23-55; Section 23-4E). Prostaglandins with the subscript 1 (the "series-1" prostaglandins) are synthesized from **8,11,14-eicosatrienoic acid,** whereas "series-2" prostaglandins are synthesized from arachidonic acid. **α-Linolenic acid** is a precursor of **5,8,11,14,17-eicosapentaenoic acid (EPA)** and the "series-3" prostaglandins. Since arachidonate is the primary prostaglandin precursor in humans, we shall mostly refer to the series-2 prostaglandins in our examples.

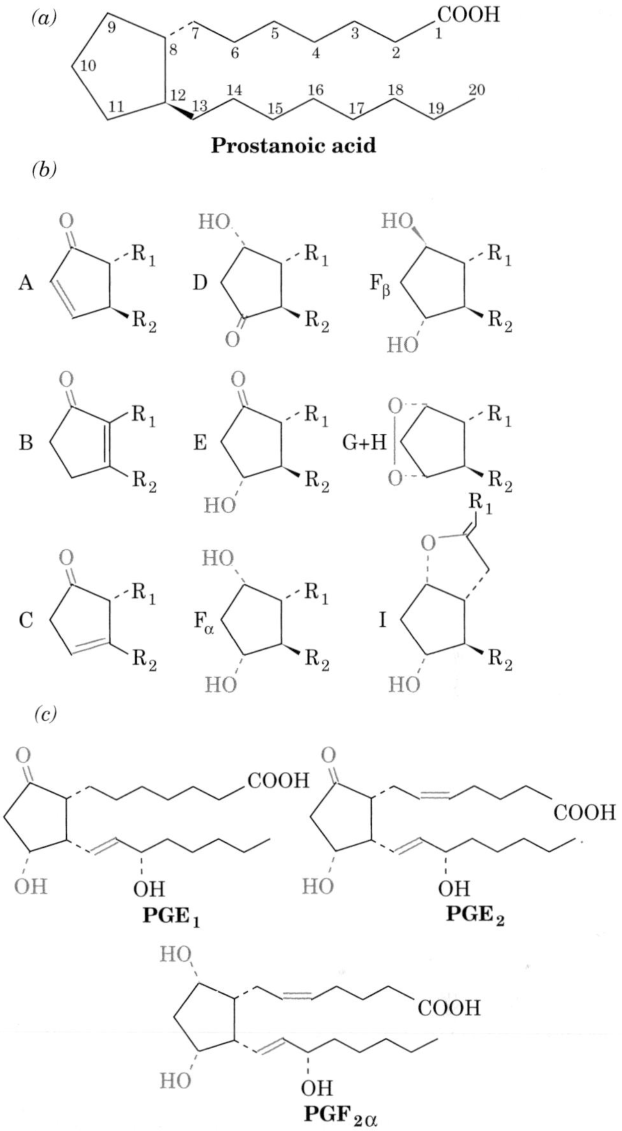

Arachidonate Is Generated by Phospholipid Hydrolysis

Arachidonate is stored in cell membranes esterified at glycerol C2 of phosphatidylinositol and other phospholipids. The production of arachidonate metabolites is controlled by the rate of arachidonate release from these phospholipids through three alternative pathways (Fig. 23-56):

1. **Phospholipase A$_2$** hydrolyzes acyl groups at C2 of phospholipids (Fig. 23-56b, left).

2. **Phospholipase C** specifically hydrolyzes the phosphatidylinositol head group to yield a 1,2-diacylglycerol, which is phosphorylated by **diglycerol kinase** to phosphatidic acid, a phospholipase A$_2$ substrate (Fig. 23-56b, center).

3. The 1,2-diacylglycerol also may be hydrolyzed directly by **diacylglycerol lipase** (Fig. 23-56b, right).

Corticosteroids are used as anti-inflammatory agents because they inhibit phospholipase A$_2$, reducing the rate of arachidonate production.

Aspirin Inhibits Prostaglandin Synthesis

The use of **aspirin** as an analgesic (pain-relieving), antipyretic (fever-reducing), and anti-inflammatory agent has

FIGURE 23-54. Prostaglandin structures: (*a*) the carbon skeleton of prostanoic acid, the prostaglandin parent compound. (*b*) Structures of prostaglandins A through I. (*c*) Structures of prostaglandins E$_1$, E$_2$, and F$_{2\alpha}$ (the first prostaglandins to be identified).

been widespread since the nineteenth century. Yet, it was not until 1971 that John Vane discovered its mechanism of action. *Aspirin, as well as other* **nonsteroidal anti-inflammatory drugs (NSAIDs),** *inhibits the synthesis of prostaglandins from arachidonic acid (Section 23-7B).* These inhibitors have therefore proved to be valuable tools in the elucidation of prostaglandin biosynthesis pathways and have provided a starting point for the rational synthesis of new anti-inflammatory drugs.

Arachidonic Acid Is a Precursor of Leukotrienes, Thromboxanes, and Prostacyclins

Arachidonic acid also serves as a precursor to compounds whose synthesis is not inhibited by aspirin. In fact,

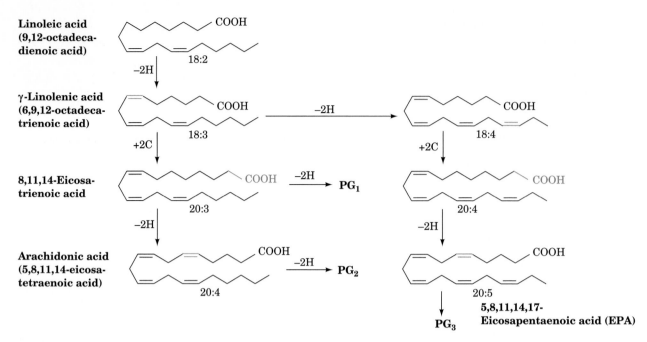

FIGURE 23-55. The synthesis of prostaglandin precursors. The linoleic acid derivatives 8,11,14-eicosatrienoic acid and arachidonic acid are the respective precursors of the series-1 and series-2 prostaglandins. The γ-linolenic acid derivative 5,8,11,14-eicosapentaenoic acid is the series-3 prostaglandin precursor.

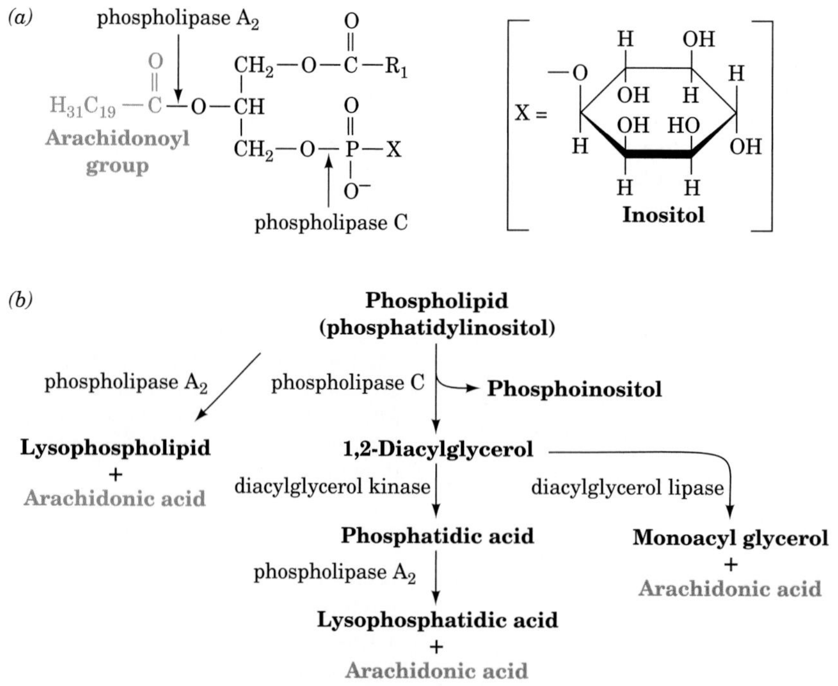

FIGURE 23-56. (*a*) The sites of hydrolytic cleavage mediated by phospholipases A_2 and C. The polar head group, X, is often inositol. (*b*) Pathways of arachidonic acid liberation from phospholipids.

there are two main pathways of arachidonate metabolism. The so-called "cyclic pathway," which is inhibited by NSAIDs, forms prostaglandin's characteristic cyclopentane ring, whereas the so-called "linear pathway," which is not inhibited by these agents, leads to the formation of the **leukotrienes** and **HPETEs** (Fig. 23-57; Section 23-7C).

Studies using NSAID's helped demonstrate that two structurally related and highly short-lived classes of compounds, the prostacyclins and the thromboxanes (Fig. 23-58), are also products of the cyclic pathway of arachidonic acid metabolism. The specific products produced by this branched pathway depend on the tissue involved. For ex-

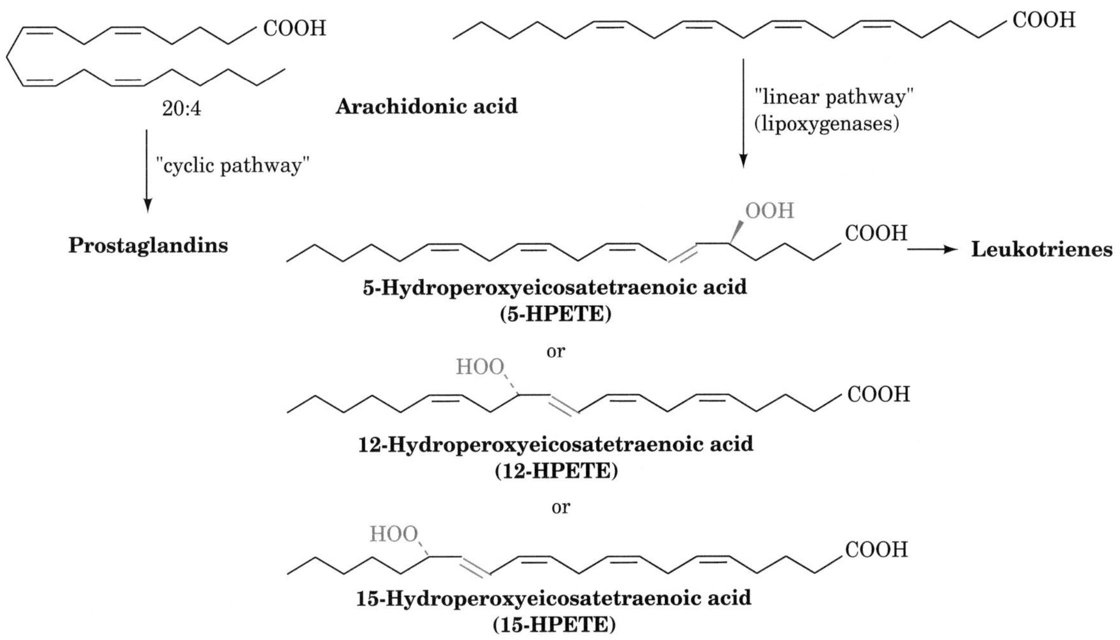

FIGURE 23-57. The cyclic and linear pathways of arachidonic acid metabolism.

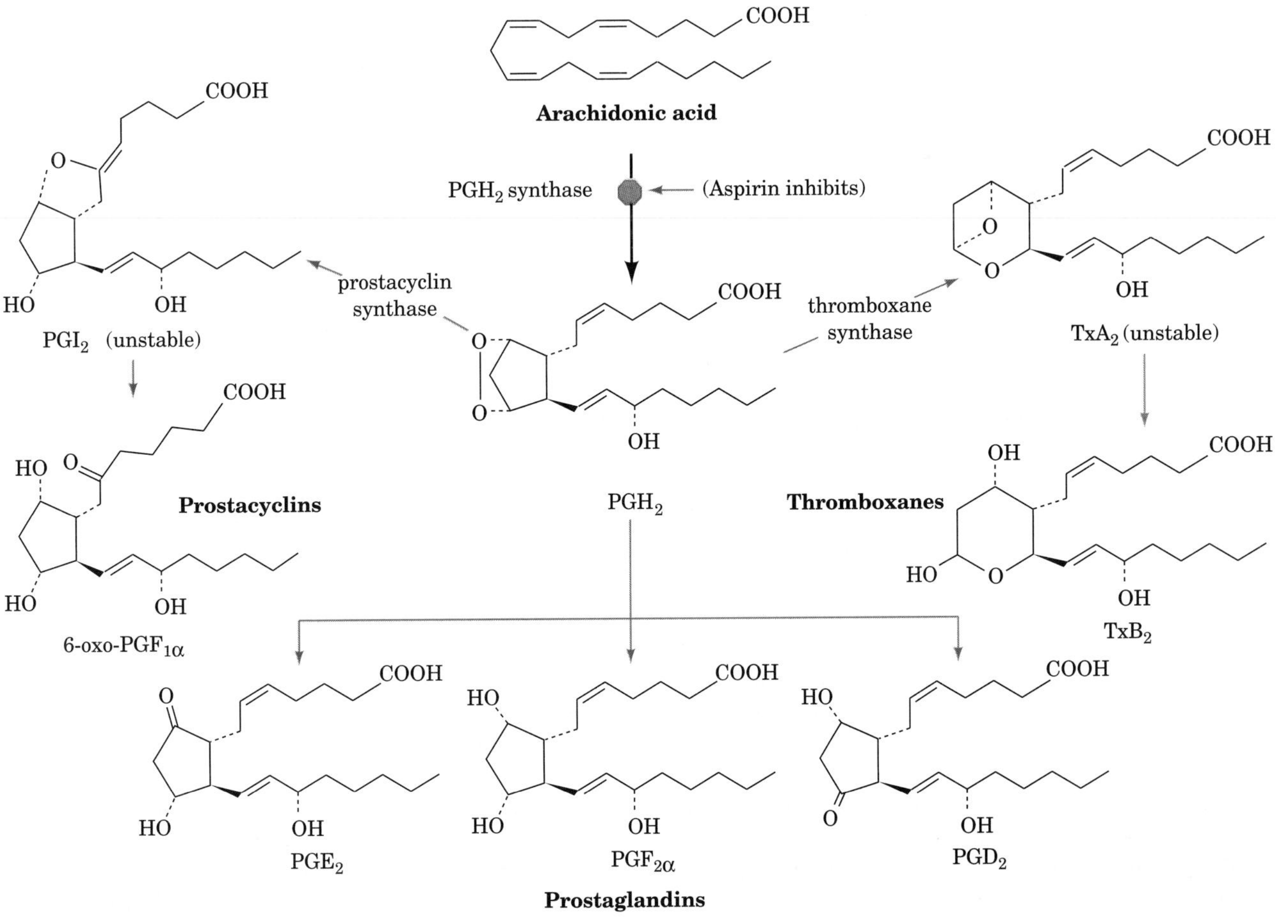

FIGURE 23-58. The cyclic pathway of arachidonic acid metabolism is branched, leading to prostaglandins, prostacyclins, and thromboxanes.

ample, blood platelets (thrombocytes) produce thromboxanes almost exclusively; vascular endothelial cells, which make up the walls of veins and arteries, predominantly synthesize the prostacyclins; and heart muscle makes PGI_2, PGE_2, and $PGF_{2\alpha}$ in more or less equal quantities. In the remainder of this section, we study the cyclic and the linear pathways of arachidonate metabolism.

B. The Cyclic Pathway of Arachidonate Metabolism: Prostaglandins, Prostacyclins, and Thromboxanes

The first step in the cyclic pathway of arachidonic acid metabolism is catalyzed by **PGH$_2$ synthase (prostaglandin H$_2$ synthase; prostaglandin endoperoxide synthase;** Fig. 23-59). This heme-containing enzyme contains two catalytic activities: a cyclooxygenase activity and a peroxidase activity. The former catalyzes the addition of two molecules of O_2 to arachidonic acid, forming **PGG$_2$**. The latter converts the hydroperoxy function of PGG$_2$ to an OH group **(PGH$_2$)**. *PGH$_2$ is the immediate precursor of all series-2 prostaglandins, prostacyclins, and thromboxanes (Fig. 23-58).*

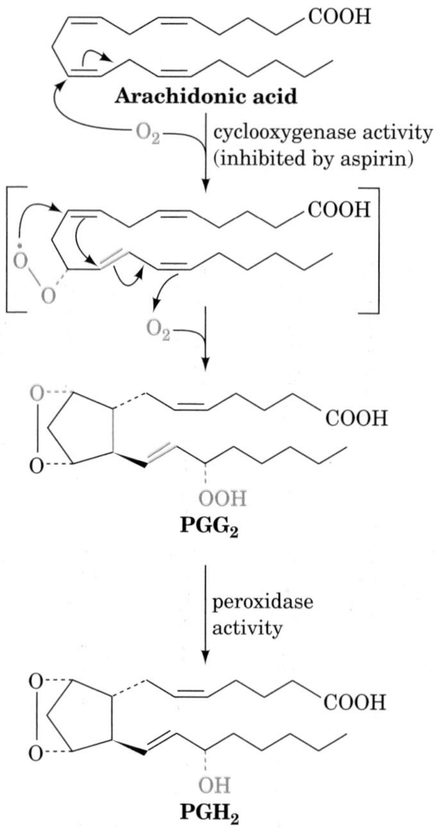

PGH$_2$ synthase, a dimeric glycoprotein of identical 576-residue subunits, is an integral membrane protein that extends into the lumen of the endoplasmic reticulum. The X-ray structure of PGH$_2$ synthase, determined by Michael Garavito, reveals that each of its subunits folds into three domains (Fig. 23-60*a*): an N-terminal module that structurally resembles **epidermal growth factor (EGF;** a hormonally active polypeptide that stimulates cell proliferation; Section 33-4C); a central membrane-binding motif; and a C-terminal enzymatic domain. The 44-residue membrane-binding motif has a hydrophobic surface that faces away from the body of the protein but is of insufficient depth to penetrate more than one leaflet of a lipid bilayer. Hence, *PGH$_2$ synthase constitutes the first known structure of an integral membrane protein that is not a transmembrane protein.*

The peroxidase active site of PGH$_2$ synthase occurs at the interface between the large and small lobes of the catalytic domain, in a shallow cleft that contains the enzyme's Fe(III)–heme prosthetic group. The cleft exposes a large portion of the heme to solvent and is therefore thought to comprise the substrate binding site. The structure of the region containing the peroxidase active site resembles that of cytochrome *c* peroxidase (CCP; Section 20-2C), even though there is no obvious sequence homology between these two heme-containing enzymes.

The cyclooxygenase active site lies on the opposite side of the heme at the end of a long narrow channel (~8 × 25Å) extending from the outer surface of the membrane-binding motif to the center of each subunit (Fig. 23-60*b*). Tyr 385, which lies near the top of the channel, just beneath the heme, has been shown to form a transient radical during the

FIGURE 23-59. The reactions catalyzed by PGH$_2$ synthase. The enzyme, which is inhibited by aspirin, contains two activities: a cyclooxygenase and a peroxidase.

cyclooxgenase reaction as does Trp 191 in the CCP reaction (Section 20-2C). Indeed, the mutagenic replacement of PGH$_2$ synthase's Tyr 385 by Phe abolishes its cyclooxgenase activity.

The fate of PGH$_2$ depends on the relative activities of the enzymes catalyzing the specific interconversions (Fig. 23-58). Platelets contain **thromboxane synthase,** which mediates formation of **thromboxane A$_2$ (TxA$_2$),** a vasoconstrictor and stimulator of platelet aggregation (an initial step in blood clotting; Section 34-1). Vascular endothelial cells contain **prostacyclin synthase,** which catalyzes the synthesis of **prostacyclin I$_2$ (PGI$_2$),** a vasodilator and inhibitor of platelet aggregation. These two substances act in opposition, maintaining a balance in the cardiovascular system.

NSAIDs Inhibit PGH$_2$ Synthase

Nonsteroidal anti-inflammatory drugs (NSAIDs; Fig. 23-61) inhibit the synthesis of the prostaglandins, prostacyclins, and thromboxanes by inhibiting or inactivating the cyclooxygenase activity of PGH$_2$ synthase. Aspirin **(acetylsalicylic acid),** for example, acetylates this enzyme: If [^{14}C-acetyl] salicylic acid is incubated with the enzyme, radioactivity becomes irreversibly associated with the inactive

(b)

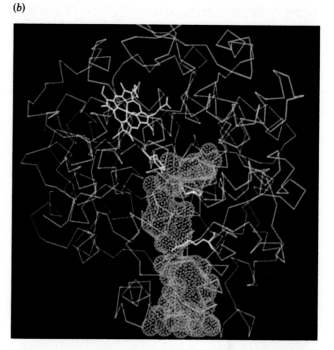

(a)

FIGURE 23-60. The X-ray structure of PGH_2 synthase from sheep seminal vesicles. (*a*) A diagram of the dimer as viewed along the dimer interface with its twofold axis of symmetry vertical. The EGF module, the membrane-binding motif, and the catalytic domain are colored green, tan, and blue, respectively, whereas the heme is red and the five disulfide bonds in each subunit are yellow. (*b*) A C_α diagram of a PGH_2 subunit (*green*), the left subunit in Part *a* as viewed from 30° to the left. The peroxidase active site is located above the heme (*pink*). The hydrophobic channel, which penetrates the subunit from the membrane-binding motif at the bottom of the figure to the cyclooxygenase active site below the heme, is represented by its van der Waals surface (*blue dots*). The three residues in the channel that are shown in yellow are, from top to bottom: Tyr 385, which forms a transient radical during the cyclooxygenase reaction; Ser 530, which is acetylated by aspirin; and Arg 120, which forms an ion pair with the NSAID flurbiprofen when it binds in the channel. [Courtesy of Michael Garavito, University of Chicago.]

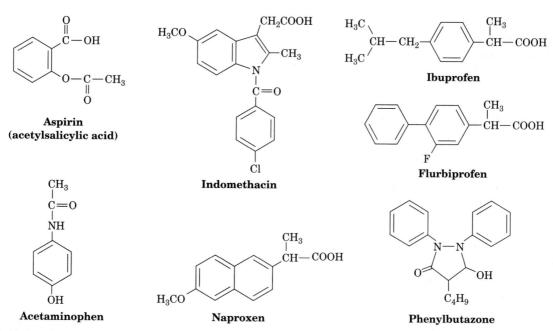

FIGURE 23-61. Some nonsteroidal anti-inflammatory drugs.

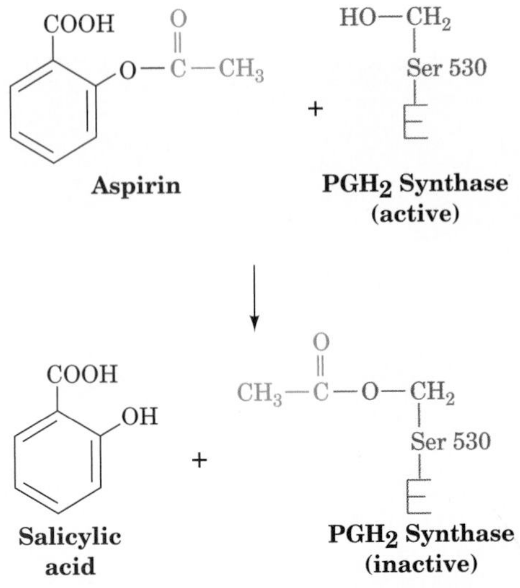

FIGURE 23-62. Aspirin acetylates Ser 530 of PGH$_2$ synthase, thereby blocking the enzyme's cyclooxygenase activity.

enzyme as Ser 530 becomes acetylated (Fig. 23-62). The X-ray structure of PGH$_2$ synthase reveals that Ser 530, which is not implicated in catalysis, extends into the cyclooxygenase channel just below Tyr 385 such that its acetylation would block arachidonic acid's access to the active site (Fig. 23-60b). The structure of PGH$_2$ synthase, which was crystallized with the NSAID **flurbiprofen** (Fig. 23-61), indicates that it binds in the cycloxygenase channel, with its carboxyl group forming an ion pair with Arg 120. Evidently, flurbiprofen, and by implication other NSAIDs, inhibits the cyclooxygenase activity of PGH$_2$ synthase by blocking its active site channel.

Low doses of aspirin, ~75 mg every 2 days, significantly reduce the long-term incidence of heart attacks and strokes. Such low doses selectively inhibit platelet aggregation and thus blood clot formation because these enucleated cells, which have a lifetime in the circulation of ~10 days, cannot resynthesize their inactivated enzymes. Vascular endothelial cells are not as drastically affected since, for the most part, they are far from the site where aspirin is absorbed, are exposed to lesser concentrations of aspirin and, in any case, can synthesize additional PGH$_2$ synthase.

C. The Linear Pathway of Arachidonate Metabolism: Leukotrienes

Arachidonic acid can be converted by a linear pathway to several different **hydroperoxyeicosatetraenoic acids** **(HPETEs)** by 5-, 12-, and 15-lipoxygenases (Fig. 23-57). The products derived from the 12- and 15-lipoxygenase pathways, **hepoxins** and **lipoxins**, show biological activities in various systems, although little is yet known about them. Leukotrienes, derived from the 5-lipoxygenase reaction, are synthesized by a variety of white blood cells, mast cells (connective tissue cells derived from the blood-forming tis-

sues that secrete substances that mediate inflammatory and allergic reactions) as well as lung, spleen, brain, and heart. **Peptidoleukotrienes (LTC$_4$, LTD$_4$, and LTE$_4$)** are now recognized to be the components of the **slow reacting substances of anaphylaxis (SRS-A;** anaphalaxis is a violent and potentially fatal allergic reaction) released from sensitized lung after immunological challenge. These substances act at very low concentrations (as little as $10^{-10}M$) to contract vascular, respiratory, and intestinal smooth muscle. Peptidoleukotrienes, for example, are ~10,000-fold more potent than histamine, a well-known stimulant of allergic reactions. In the respiratory system, they constrict bronchi, especially the smaller airways; increase mucus secretion; and are thought to be the mediators in asthma. They are also implicated in immediate hypersensitivity (allergic) reactions, inflammatory reactions, and heart attacks.

The first reaction in the conversion of arachidonate to leukotrienes is its **5-lipoxygenase (5-LO)**-catalyzed oxidation to form **5-hydroperoxyeicosatetraenoic acid (5-HPETE;** Fig. 23-57), a substances that, in itself, is not a physiological mediator. 5-LO contains a nonheme, non-[Fe – S] cluster iron atom that must be in its Fe(III) state to be active. The 5-LO reaction is thought to proceed as follows (Fig. 23-63):

1. The active site iron atom, in its active Fe(III) state, abstracts an electron from the central methylene group of the 5,8-pentadiene moiety of arachidonate and the resulting free radical loses a proton to an enzymatic base.

2. The free radical rearranges and adds O$_2$ to form a hydroperoxide radical.

3. The hydroperoxide radical reacts with the active site iron, now in its Fe(II) form, to yield the hydroperoxide in its anionic form, which the enzyme then protonates to yield the hydroperoxide product, regenerating the active Fe(III) enzyme.

The X-ray structure of the monomeric 839-residue soybean lipoxygenase-1, a homolog of 5-LO, was determined by Mario Amzel. Its active site Fe atom is coordinated by the invariant His residues 499, 504, and 690 and by a C-terminal carboxylate oxygen at Ile 839. This liganding arrangement is best described as a distorted octahedron with only four of its six vertices occupied. The Fe, which is well below the protein surface, faces two large internal cavities that open onto the surface of the enzyme (Fig. 23-64). Cavity I, an 18-Å-long tunnel that is 8 Å-wide along most of its length but that narrows to 2.5 Å at the Fe and is lined by hydrophobic residues, provides what appears to be an ideal path to conduct O$_2$ from the outside the enzyme to one of the Fe's unoccupied liganding sites. Cavity II, which is 40 Å long and < 3.5 Å wide in some places and which is mostly lined by conserved hydrophobic and neutral residues, follows an irregular pathway past the Fe atom. Model building indicates that an arachidonic acid molecule can fit snugly into this tunnel such that its reacting diene system approaches the Fe atom opposite His 690.

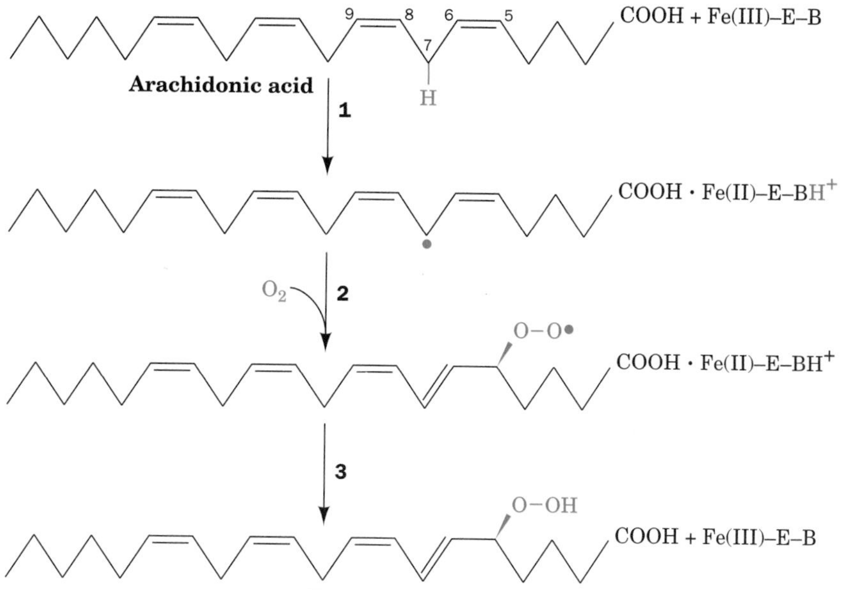

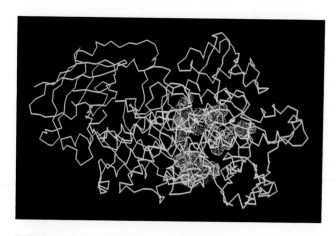

FIGURE 23-63. The lipoxygenase-catalyzed oxidation of arachidonate to 5-HPETE.

FIGURE 23-64. The X-ray structure of soybean lipoxygenase-1. The protein is represented by its C_α diagram (*yellow*). Its internal cavities are outlined by dot surfaces with cavity I in green and cavity II in pink. The active site Fe atom is represented by an orange sphere. [Courtesy of Mario Amzel, The Johns Hopkins University.]

Intriguingly, lipoxygenase-1 contains two rarely observed π helices (Fig. 7-14c). The first, which consists of 13 residues, the longest π helix yet observed, contains the Fe ligands His 499 and His 504, whereas the second, which consists of 6 residues, contains a third Fe ligand, His 690. Each of these π helices is embedded in a longer helix rather than being at the end of a helix as is the case for all previously observed π helices.

5-HPETE, the product of the 5-LO-catalyzed oxidation of arachidonic acid (Fig. 23-57), is converted to peptidoleukotrienes by first forming an unstable epoxide, **leukotriene A₄ (LTA₄**; Fig. 23-65; the subscript indicates the number of

carbon–carbon double bonds in the molecule as well as the series to which the leukotriene belongs). **Glutathione-*S*-transferase** then catalyzes the addition of the glutathione sulfhydryl group to the epoxide, forming the first of the peptidoleukotrienes, **leukotriene C₄ (LTC₄)**. **γ-Glutamyl-transferase** removes glutamic acid, converting LTC₄ to **leukotriene D₄ (LTD₄)**. LTD₄ is converted to **leukotriene E₄ (LTE₄)** by a dipeptidase that removes glycine. LTA₄ can also be converted to **leukotriene B₄ (LTB₄)**, a potent chemotactic agent (a substance that attracts motile cells) involved in attracting certain types of white blood cells to fight infection.

Various inflammatory and hypersensitivity disorders (such as asthma) are associated with elevated levels of leukotrienes. The development of drugs that inhibit leukotriene synthesis has therefore been an active field of research. 5-Lipoxygenase requires the presence of **5-lipoxygenase-activating protein (FLAP)**, an 18-kD integral membrane protein, for activity. FLAP binds the arachidonic acid substrate of 5-LO and facilitates enzyme-substrate binding as well as 5-LO's interaction with the membrane. Several inhibitors of leukotriene synthesis, such as **MK0886**,

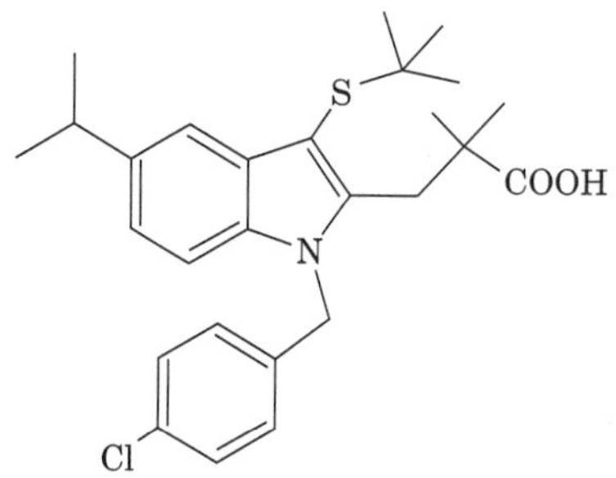

MK0886

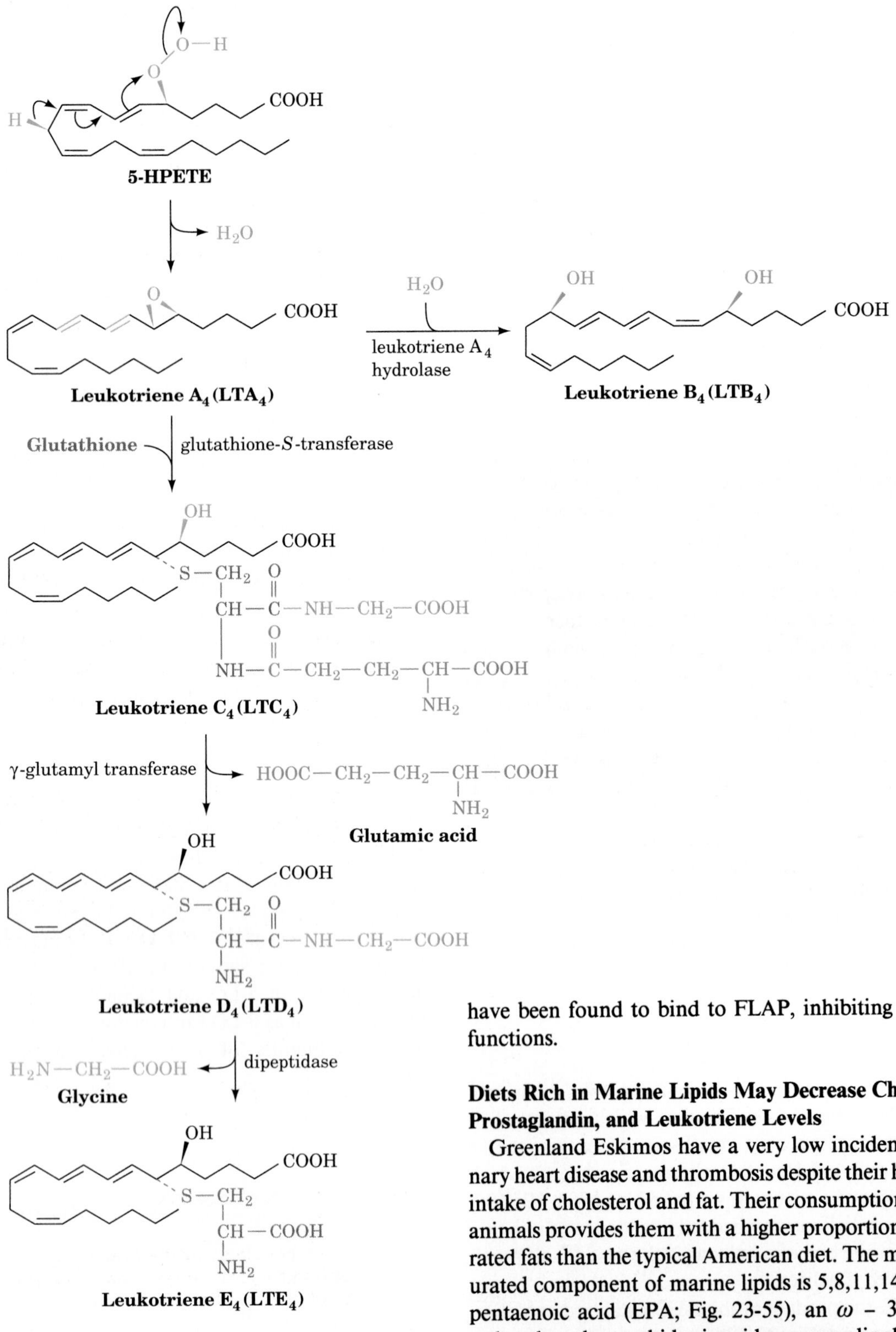

FIGURE 23-65. The formation of the leukotrienes from 5-HPETE via the unstable epoxide leukotriene A_4.

have been found to bind to FLAP, inhibiting both of its functions.

Diets Rich in Marine Lipids May Decrease Cholesterol, Prostaglandin, and Leukotriene Levels

Greenland Eskimos have a very low incidence of coronary heart disease and thrombosis despite their high dietary intake of cholesterol and fat. Their consumption of marine animals provides them with a higher proportion of unsaturated fats than the typical American diet. The major unsaturated component of marine lipids is 5,8,11,14,17-eicosapentaenoic acid (EPA; Fig. 23-55), an $\omega - 3$ fatty acid, rather than the arachidonic acid precursor linoleic acid, an $\omega - 6$ fatty acid. EPA inhibits formation of TxA_2 (Fig.

23-58) and is a precursor of the **series-5 leukotrienes**, compounds with substantially lower physiological activities than their arachidonate-derived (series-4) counterparts. This suggests that a diet containing marine lipids should decrease the extent of prostaglandin- and leukotriene-mediated inflammatory responses. Indeed, dietary enrichment with EPA inhibits the *in vitro* chemotactic and aggregating activities of neutrophils (a type of white blood cell). Moreover, an EPA-rich diet decreases the cholesterol and triacylglycerol levels in the plasma of hypertriacylglycerolemic patients.

These are indeed exciting times in the study of arachidonate metabolism and its physiological manifestations. As the mechanisms of action of the prostaglandins, prostacyclins, thromboxanes, and leukotrienes are becoming better understood, they are providing the insights required for the development of new and improved therapeutic agents.

8. PHOSPHOLIPID AND GLYCOLIPID METABOLISM

The "complex lipids" are dual-tailed amphipathic molecules composed of either 1,2-diacyl-sn-glycerol, or N-acylsphingosine (ceramide) linked to a polar head group that is either a carbohydrate or a phosphate ester (Fig. 23-66; Sections 11-1C and 1D; sn stands for stereospecific numbering, which assigns the 1 position to the group occupying the pro-S position of a prochiral center). Hence, there are two categories of phospholipids, **glycerophospholipids** and **sphingophospholipids,** and two categories of glycolipids, **glyceroglycolipids** and **sphingoglycolipids.** In this section we describe the biosynthesis of the complex lipids from their simpler components. We shall see that the great variety of these substances is matched by the numerous enzymes required for their specific syntheses. Note also that these substances are synthesized in membranes, mostly on the cytosolic face of the endoplasmic reticulum, and from there are transported to their final cellular destinations as indicated in Section 11-4B.

A. Glycerophospholipids

Glycerophospholipids have significant asymmetry in their C1- and C2-linked fatty acyl groups: C1 substituents are mostly saturated fatty acids, whereas those at C2 are by and large unsaturated fatty acids. We shall examine the major pathways of biosynthesis and metabolism of the glycerophospholipids with an eye towards understanding the origin of this asymmetry.

Biosynthesis of Diacylglycerophospholipids

The triacyglycerol precursors 1,2-diacyl-sn-glycerol and phosphatidic acid are also the precursors of certain glycerophospholipids (Figs. 23-34 and 23-66). Activated phosphate esters of the polar head groups (Table 11-2) react with the C3—OH group of 1,2-diacyl-sn-glycerol to form the phospholipid's phosphodiester bond. In some cases the phosphoryl group of phosphatidic acid is activated and reacts with the unactivated polar head group.

The mechanism of activated phosphate ester formation is the same for both the polar head groups **ethanolamine** and **choline** (Fig. 23-67):

1. ATP first phosphorylates the OH group of choline or ethanolamine.

2. The phosphoryl group of the resulting **phosphoethanolamine** or **phosphocholine** then attacks CTP, displacing PP_i, to form the corresponding CDP derivatives, which are activated phosphate esters of the polar head group.

3. The C3—OH group of 1,2-diacyl-sn-glycerol attacks the phosphoryl group of the activated CDP–ethanolamine or CDP–choline, displacing CMP to yield the corresponding glycerophospholipid.

The liver also converts phosphatidylethanolamine to phosphatidylcholine by trimethylating its amino group, using **S-adenosylmethionine** (Section 24-3E) as the methyl donor.

Phosphatidylserine is synthesized from phosphatidylethanolamine by a head group exchange reaction catalyzed by **phosphatidylethanolamine:serine transferase** in which

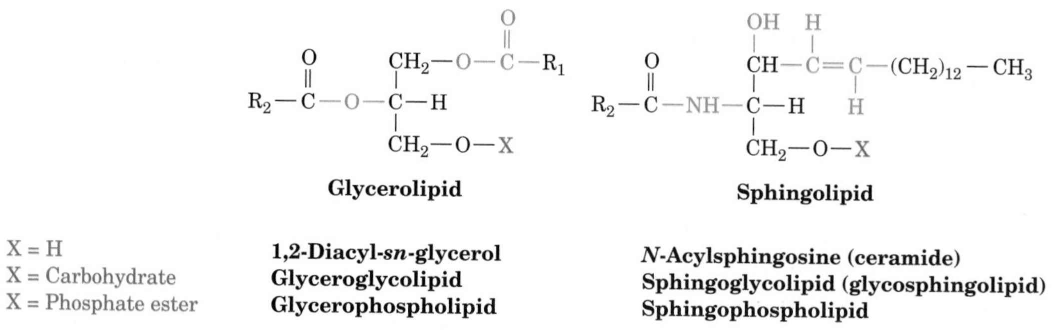

Glycerolipid   **Sphingolipid**

X = H	**1,2-Diacyl-sn-glycerol**	**N-Acylsphingosine (ceramide)**
X = Carbohydrate	**Glyceroglycolipid**	**Sphingoglycolipid (glycosphingolipid)**
X = Phosphate ester	**Glycerophospholipid**	**Sphingophospholipid**

FIGURE 23-66. The glycerolipids and sphingolipids. The structures of the common head groups, X, are presented in Table 11-2.

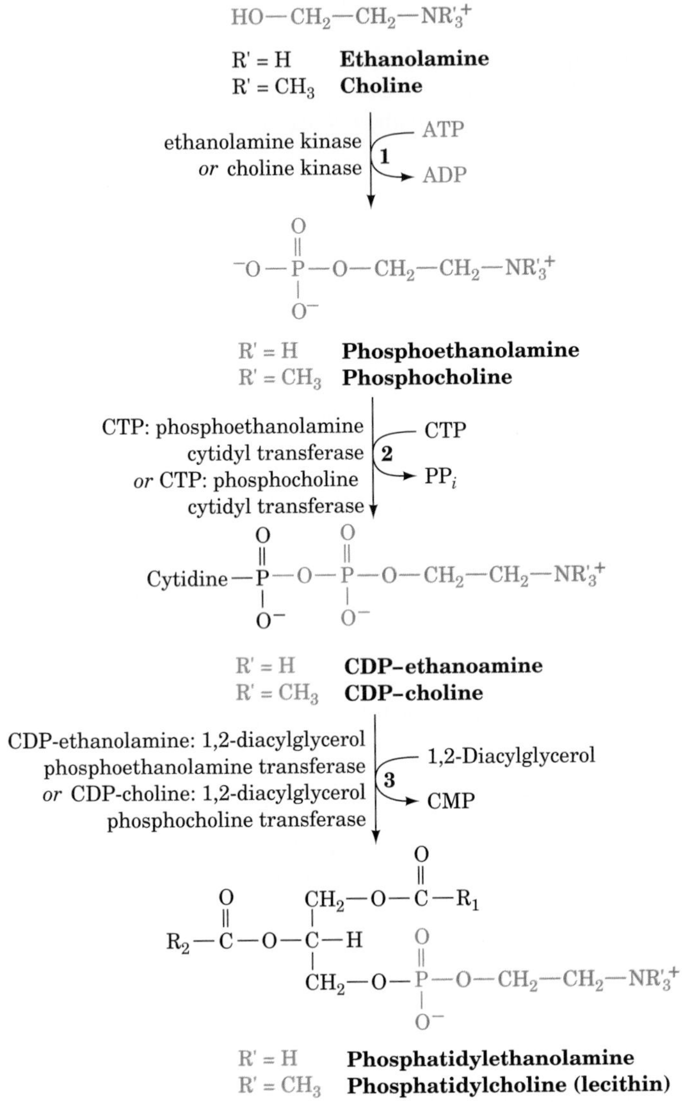

$$HO-CH_2-CH_2-NR_3'^+$$

R' = H   **Ethanolamine**
R' = CH_3 **Choline**

ethanolamine kinase
or choline kinase **1** ⟵ ATP
⟶ ADP

$$^-O-\overset{\displaystyle O}{\overset{\|}{P}}-O-CH_2-CH_2-NR_3'^+$$
$$\underset{O^-}{|}$$

R' = H   **Phosphoethanolamine**
R' = CH_3 **Phosphocholine**

CTP: phosphoethanolamine
cytidyl transferase **2** ⟵ CTP
or CTP: phosphocholine
cytidyl transferase ⟶ PP$_i$

$$Cytidine-\overset{O}{\overset{\|}{P}}-O-\overset{O}{\overset{\|}{P}}-O-CH_2-CH_2-NR_3'^+$$
$$\underset{O^-}{|}\qquad\underset{O^-}{|}$$

R' = H   **CDP–ethanoamine**
R' = CH_3 **CDP–choline**

CDP-ethanolamine: 1,2-diacylglycerol
phosphoethanolamine transferase **3** ⟵ 1,2-Diacylglycerol
or CDP-choline: 1,2-diacylglycerol
phosphocholine transferase ⟶ CMP

$$R_2-\overset{O}{\overset{\|}{C}}-O-\overset{\displaystyle CH_2-O-\overset{O}{\overset{\|}{C}}-R_1}{\underset{\displaystyle CH_2-O-\overset{}{\underset{O^-}{\overset{\|}{\underset{}{P}}}}-O-CH_2-CH_2-NR_3'^+}{\overset{|}{\underset{|}{C}}-H}}$$

R' = H   **Phosphatidylethanolamine**
R' = CH_3 **Phosphatidylcholine (lecithin)**

FIGURE 23-67. The biosynthesis of phosphatidyl-ethanolamine and phosphatidylcholine involves CDP–ethanolamine and CDP–choline.

serine's OH group attacks the donor's phosphoryl group (Fig. 23-68). The original head group is then eliminated, forming phosphatidylserine.

In the synthesis of **phosphatidylinositol** and **phosphatidylglycerol,** the hydrophobic tail is activated rather than the polar head group. Phosphatidic acid, the precursor of 1,2-diacyl-sn-glycerol (Fig. 23-34), attacks the α-phosphoryl group of CTP to form the activated **CDP–diacylglycerol** and PP$_i$ (Fig. 23-69). Phosphatidylinositol results from the attack of inositol on CDP–diacylglycerol. Phosphatidylglycerol is formed in two reactions: (1) attack of the C1—OH group of sn-glycerol-3-phosphate on CDP–diacylglycerol, yielding **phosphatidylglycerol phosphate;** and (2) hydrolysis of the phosphoryl group to form phosphatidylglycerol.

Cardiolipin, an important phospholipid first isolated from heart tissue, is synthesized from two molecules of phosphatidylglycerol (Fig. 23-70). The reaction occurs by the attack of the C1—OH group of one of the phosphatidylglycerol molecules on the phosphoryl group of the other, displacing a molecule of glycerol.

Enzymes that synthesize phosphatidic acid have a general preference for saturated fatty acids at C1 and for unsaturated fatty acids at C2. Yet, this general preference cannot account, for example, for the observations that ~80% of brain phosphatidylinositol has a stearoyl group (18:0) at C1 and an arachidonoyl group (20:4) at C2, and that ~40% of lung phosphatidylcholine has palmitoyl groups (16:0) at both positions (this latter substance is the major component of the surfactant that prevents the lung from collapsing when air is expelled; its deficiency is responsible for **respiratory distress syndrome** in premature infants). William Lands showed that *such side chain specificity results from "remodeling" reactions in which specific acyl groups of individual glycerophospholipids are exchanged by specific phospholipases and acyl transferases.*

Biosynthesis of Plasmalogens and Alkylacylglycerophospholipids

Eukaryotic membranes contain significant amounts of two other types of glycerophospholipids:

1. **Plasmalogens,** which contain a hydrocarbon chain linked to glycerol C1 via a vinyl ether linkage.

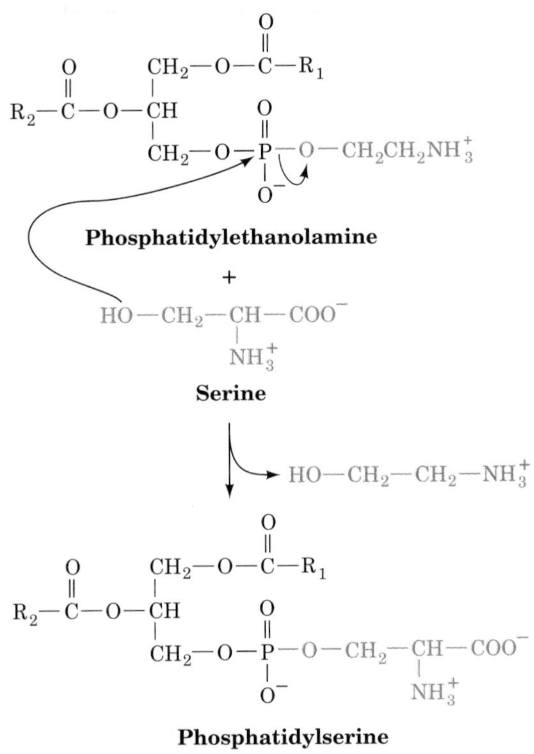

$$R_2-\overset{O}{\overset{\|}{C}}-O-\overset{\displaystyle CH_2-O-\overset{O}{\overset{\|}{C}}-R_1}{\underset{\displaystyle CH_2-O-\overset{}{\underset{O^-}{\overset{\|}{\underset{}{P}}}}-O-CH_2CH_2NH_3^+}{\overset{|}{\underset{|}{CH}}}}$$

Phosphatidylethanolamine

+

$$HO-CH_2-\underset{\underset{NH_3^+}{|}}{CH}-COO^-$$

Serine

⟶ $HO-CH_2-CH_2-NH_3^+$

$$R_2-\overset{O}{\overset{\|}{C}}-O-\overset{\displaystyle CH_2-O-\overset{O}{\overset{\|}{C}}-R_1}{\underset{\displaystyle CH_2-O-\overset{}{\underset{O^-}{\overset{\|}{\underset{}{P}}}}-O-CH_2-\underset{\underset{NH_3^+}{|}}{CH}-COO^-}{\overset{|}{\underset{|}{CH}}}}$$

Phosphatidylserine

FIGURE 23-68. Phosphatidylserine synthesis from phosphatidylethanolamine occurs by a head group exchange reaction.

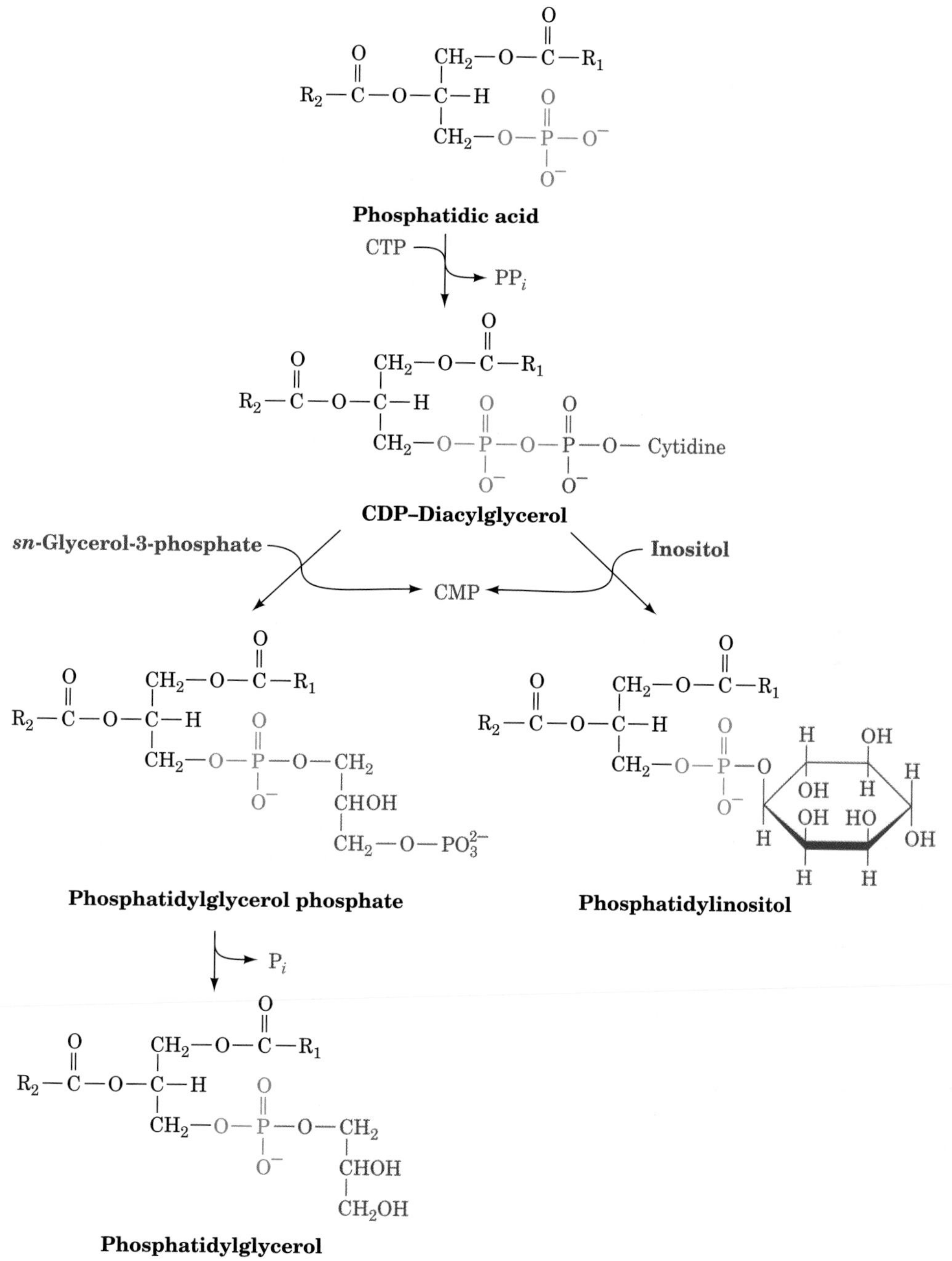

FIGURE 23-69. The biosynthesis of phosphatidylinositol and phosphatidyglycerol involves a CDP–diacylglycerol intermediate.

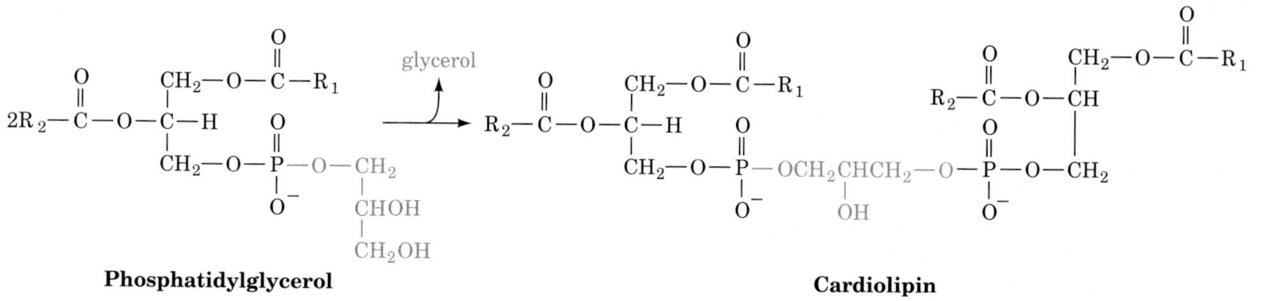

FIGURE 23-70. The formation of cardiolipin.

2. Alkylacylglycerophospholipids, in which the alkyl substituent at glycerol C1 is attached via an ether linkage.

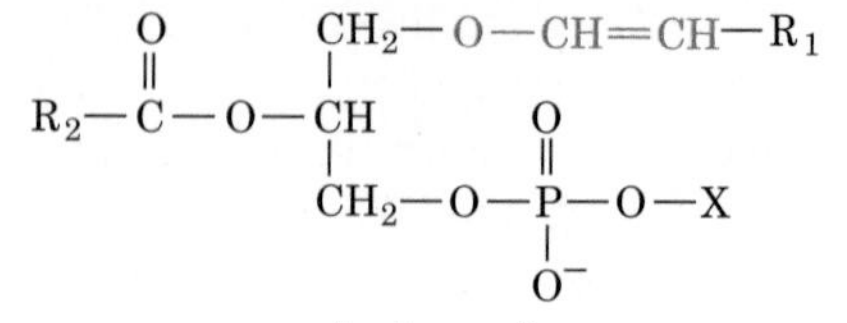

A plasmalogen

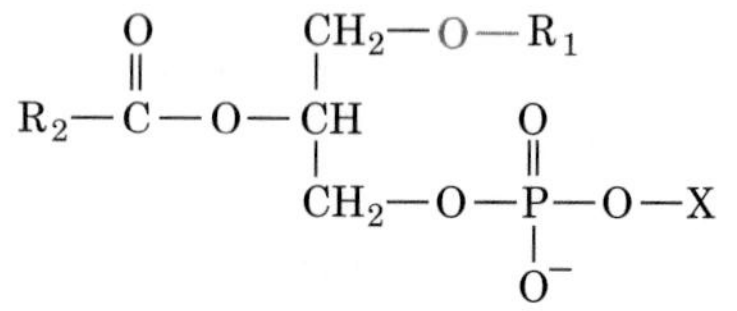

**An alkylacyl
glycerophospholipid**

About 20% of mammalian glycerophospholipids are plasmalogens. The exact percentage varies both from spe-

cies to species and from tissue to tissue within a given organism. While plasmalogens comprise only 0.8% of the phospholipids in human liver, they account for 23% of those in human nervous tissue. The alkylacylglycerophospholipids are less abundant than the plasmalogens; for instance, 59% of the ethanolamine glycerophospholipids of human heart are plasmalogens, whereas only 3.6% are alkylacylglycerophospholipids. However, in bovine erythrocytes, 75% of the ethanolamine glycerophospholipids are of the alkylacyl type.

The pathway forming ethanolamine plasmalogens and alkylacylglycerophospholipids involves several reactions (Fig. 23-71):

1. Exchange of the acyl group of **1-acyldihydroxyacetone phosphate** for an alcohol.

2. Reduction of the ketone to **1-alkyl-*sn*-glycerol-3-phosphate.**

3. Acylation of the resulting C2—OH group by acyl-CoA.

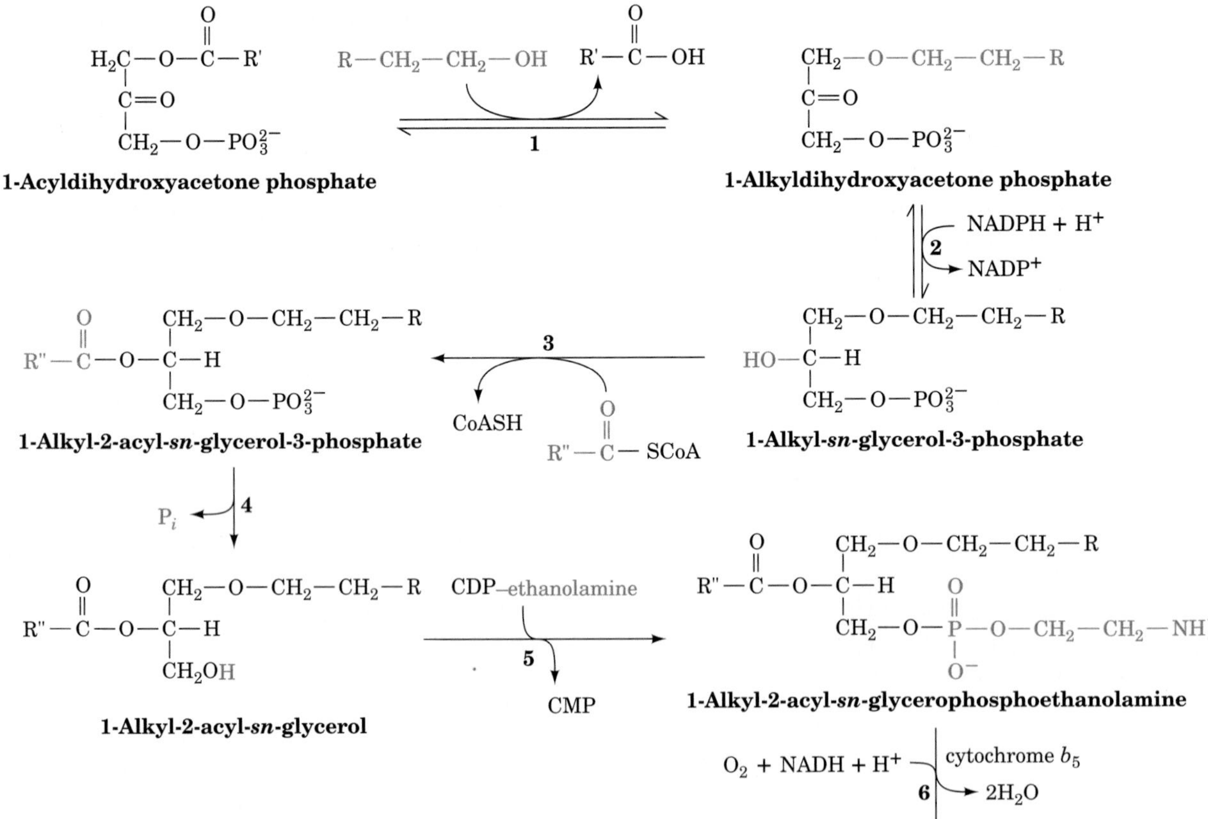

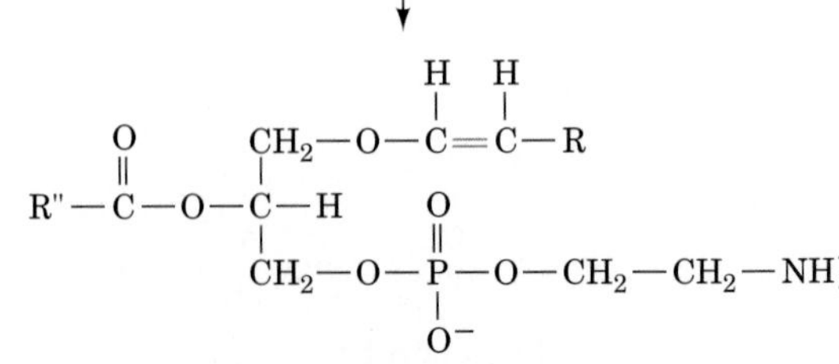

FIGURE 23-71. The biosynthesis of ethanolamine plasmalogen via a pathway in which 1-alkyl-2-acyl-*sn*-glycerolphosphoethanolamine is an intermediate. The participating enzymes are (**1**) alkyl-DHAP synthase; (**2**) 1-alkyl-*sn*-glycerol-3-phosphate dehydrogenase; (**3**) acyl-CoA:1-alkyl-*sn*-glycerol-3-phosphate acyl transferase; (**4**) 1-alkyl-2-acyl-*sn*-glycerol-3-phosphate phosphatase; (**5**) CDP–ethanolamine:1-alkyl-2-acyl-*sn*-glycerophosphoethanolamine transferase; and (**6**) 1-alkyl-2-acyl-*sn*-glycerophosphoethanolamine desaturase.

4. Hydrolysis of the phosphoryl group to yield an alkylacylglycerol.

5. Attack by the new OH group of alkylacylglycerol on CDP–ethanolamine to yield **1-alkyl-2-acyl-*sn*-glycerophosphoethanolamine.**

6. Introduction of a double bond into the alkyl group to form the plasmalogen by a desaturase having the same cofactor requirements as the fatty acid desaturases (Section 23-4E).

Recall that the precursor–product relationship between the alkylacylglycerophospholipid and the plasmalogen was established through studies using [^{14}C]ethanolamine (Section 15-3B).

The plasmalogen with an acetyl group at R_2 and a choline polar head group (X), **1-*O*-hexadec-1'-enyl-2-acetyl-*sn*-glycero-3-phosphocholine,** is known as **platelet-activating factor (PAF).** This molecule has diverse functions and acts at very low concentrations ($10^{-10}M$) to lower blood pressure and to cause blood platelets to aggregate.

B. *Sphingophospholipids*

Only one major phospholipid contains ceramide (*N*-acylsphingosine) as its hydrophobic tail: **sphingomyelin (*N*-acylsphingosine phosphocholine;** Section 11-D), an important structural lipid of nerve cell membranes. The molecule was once thought to be synthesized from *N*-acylsphingosine and CDP–choline. Recent evidence has shown, however, that the main route of sphingomyelin synthesis occurs through donation of the phosphocholine group of

phosphatidylcholine to *N*-acylsphingosine (Fig. 23-72). These pathways were differentiated by establishing the precursor-product relationships between CDP-choline, phosphatidylcholine and sphingomyelin (Section 15-3B). Mouse liver microsomes were isolated and incubated for a short time with [^{3}H]choline. Radioactivity appeared in sphingomyelin only after first appearing in both CDP–choline and phosphatidylcholine, ruling out the direct transfer of phosphocholine from CDP–choline to *N*-acylsphingosine.

The most prevalent acyl groups of sphingomyelin are palmitoyl (16:0) and stearoyl (18:0) groups. Longer-chain fatty acids such as nervonic acid (24:1) and behenic acid (22:0) occur with lesser frequency in sphingomyelins.

C. *Sphingoglycolipids*

Most sphingolipids are sphingoglycolipids (alternatively, **glycosphingolipids**); that is, their polar head groups consist of carbohydrate units (Section 11-1D). The principal classes of sphingoglycolipids, as indicated in Fig. 23-73, are **cerebrosides** (ceramide monosaccharides), **sulfatides** (ceramide monosaccharide sulfates), **globosides** (neutral ceramide oligosaccharides), and **gangliosides** (acidic, sialic acid–containing ceramide oligosaccharides). The carbohydrate unit is glycosidically attached to the *N*-acylsphingosine at its C1—OH group (Fig. 23-66). In the following subsections, we discuss the biosynthesis and breakdown of *N*-acylsphingosine and sphingoglycolipids and consider the diseases caused by deficiencies in their degradative enzymes.

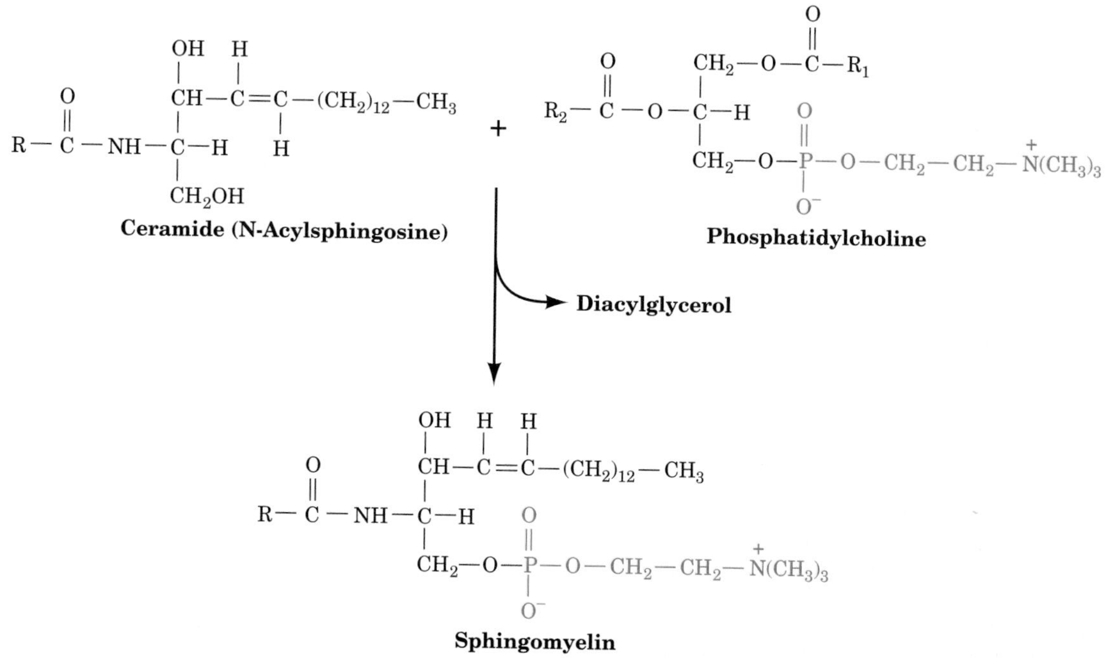

FIGURE 23-72. The synthesis of sphingomyelin from *N*-acylsphingosine and phosphatidylcholine.

Cerebrosides

Glucocerebroside

Galactocerebroside

Sulfatide

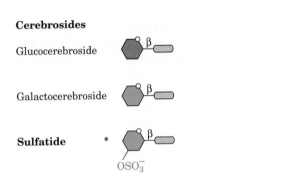

OSO_3^-

Globosides

Lactosyl ceramide

Trihexosyl ceramide

Globoside

Gangliosides

G_{M3}

G_{M2}

G_{M1}

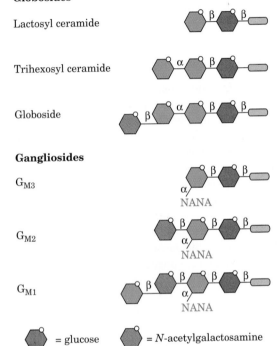

NANA

= glucose   = *N*-acetylgalactosamine

= galactose   = ceramide

NANA = *N*-acetylneuraminic acid (sialic acid)

FIGURE 23-73. Diagrammatic representation of the principal classes of sphingoglycolipids. The G_M ganglioside structures are presented in greater detail in Fig. 11-7.

Biosynthesis of Ceramide (*N*-Acylsphingosine)

Biosynthesis of *N*-acylsphingosine occurs in four reactions from the precursors palmitoyl-CoA and serine (Fig. 23-74):

1. **3-Ketosphinganine synthase,** a pyridoxal phosphate–dependent enzyme, catalyzes condensation of palmitoyl-CoA with serine yielding **3-ketosphinganine** (pyridoxal phosphate–dependent reactions are discussed in Section 24-1A).

2. **3-Ketosphinganine reductase** catalyzes the NADPH-dependent reduction of 3-ketosphinganine's keto group to form **sphinganine (dihydrosphingosine).**

$$CoA-S-\overset{\overset{O}{\|}}{C}-CH_2-CH_2-(CH_2)_{12}-CH_3 \quad + \quad \underset{\underset{CH_2OH}{|}}{\overset{\overset{CO_2^-}{|}}{H_2N-\overset{|}{C}-H}}$$

Palmitoyl-CoA   **Serine**

1 | 3-ketosphinganine synthase
→ CO_2^- + CoASH

$$\underset{\underset{CH_2OH}{|}}{\overset{\overset{\overset{O}{\|}}{C}-CH_2-CH_2-(CH_2)_{12}-CH_3}{H_2N-\overset{|}{C}-H}}$$

**3-Ketosphinganine
(3-ketodihydrosphingosine)**

2 | — NADPH + H⁺
3-ketosphinganine reductase
→ NADP⁺

$$\underset{\underset{CH_2OH}{|}}{\overset{\overset{\overset{OH}{|}}{CH-CH_2-CH_2-(CH_2)_{12}-CH_3}}{H_2N-\overset{|}{C}-H}}$$

**Sphinganine
(dihydrosphingosine)**

3 | — $R-\overset{\overset{O}{\|}}{C}-SCoA$
acyl-CoA transferase
→ CoASH

$$\underset{\underset{CH_2OH}{|}}{\overset{\overset{\overset{OH}{|}}{CH-CH_2-CH_2-(CH_2)_{12}-CH_3}}{R-\overset{\overset{O}{\|}}{C}-NH-\overset{|}{C}-H}}$$

**Dihydroceramide
(*N*-acylsphinganine)**

4 | — FAD
dihydroceramide reductase
→ FADH₂

$$\underset{\underset{CH_2OH}{|}}{\overset{\overset{\overset{OH \quad H}{|}}{CH-C=C-(CH_2)_{12}-CH_3}}{R-\overset{\overset{O}{\|}}{C}-NH-\overset{|}{C}-H \quad\quad H}}$$

**Ceramide
(*N*-acylsphingosine)**

FIGURE 23-74. The biosynthesis of ceramide (*N*-acylsphingosine).

3. **Dihydroceramide** is formed by transfer of an acyl group from an acyl-CoA to the sphinganine's 2-amino group, forming an amide bond.

4. **Dihydroceramide reductase** converts dihydroceramide to ceramide by an FAD-dependent oxidation reaction.

Biosynthesis of Cerebrosides

Galactocerebroside (**1-β-galactoceramide**) and **glucocerebroside (1-β-glucoceramide)** are the two most common cerebrosides. In fact, the term cerebroside is often used synonymously with galactocerebroside. Both are synthesized from ceramide by addition of a glycosyl unit from the corresponding UDP–hexose (Fig. 23-75). Galactocerebroside is a common component of brain lipids. Glucocerebroside, although relatively uncommon, is the precursor of globosides and gangliosides.

Biosynthesis of Sulfatides

Sulfatides (galactocerebroside-3-sulfate) account for 15% of the lipids of white matter in the brain. They are formed by transfer of an "activated" sulfate group from **3′-phosphoadenosine-5′-phosphosulfate (PAPS)** to the C3—OH group of galactose in galactocerebroside (Fig. 23-76).

Biosynthesis of Globosides and Gangliosides

Biosynthesis of both globosides (neutral ceramide oligosaccharides) and gangliosides (acidic, sialic acid-containing ceramide oligosaccharides) is catalyzed by a series of **glycosyl transferases.** While the reactions are chemically similar, each is catalyzed by a specific enzyme. The pathways begin with transfer of a galactosyl unit from UDP–Gal to glucocerebroside to form a $\beta(1 \rightarrow 4)$ linkage (Fig. 23-77). Since this bond is the same as that linking glucose and galactose in lactose, this glycolipid is often referred to as **lactosyl ceramide.** Lactosyl ceramide is the precursor of both globosides and gangliosides. To form a globoside, one galactosyl and one N-acetylgalactosaminyl unit are sequentially added to lactosyl ceramide from UDP–Gal and UDP–GalNAc, respectively. The G_M gangliosides are formed by addition of **N-acetylneuraminic acid (NANA, sialic acid)**

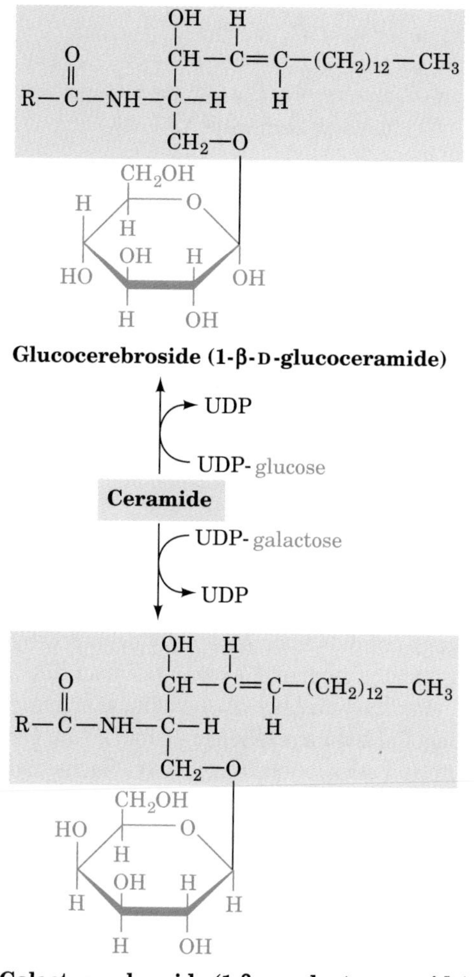

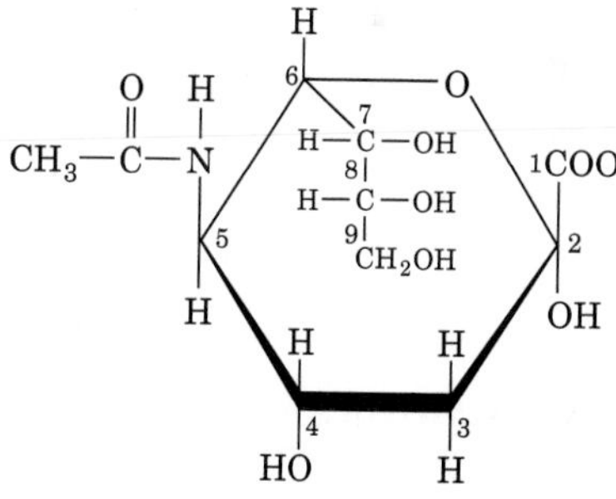

FIGURE 23-75. The biosynthesis of cerebrosides.

N-Acetylneuraminic acid
(NANA, sialic acid)

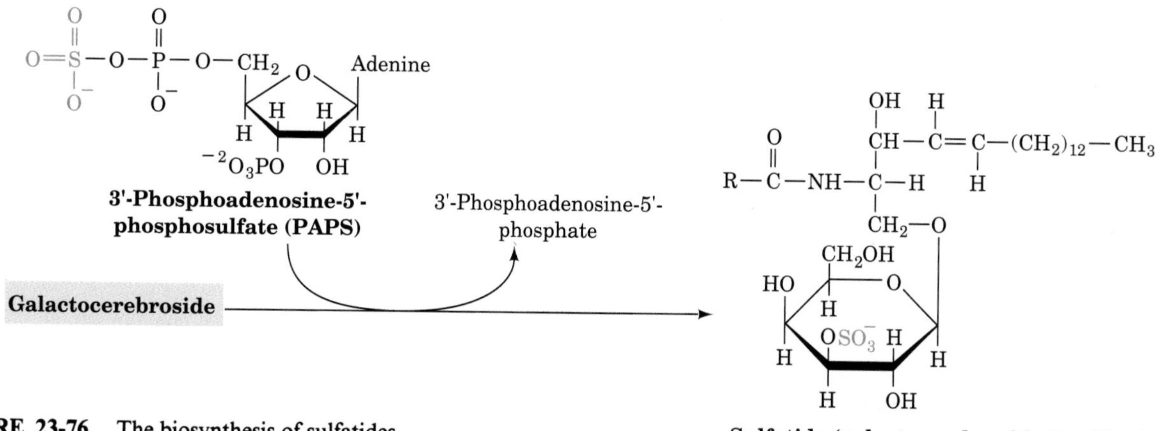

FIGURE 23-76. The biosynthesis of sulfatides.

Sulfatide (galactocerebroside-3-sulfate)

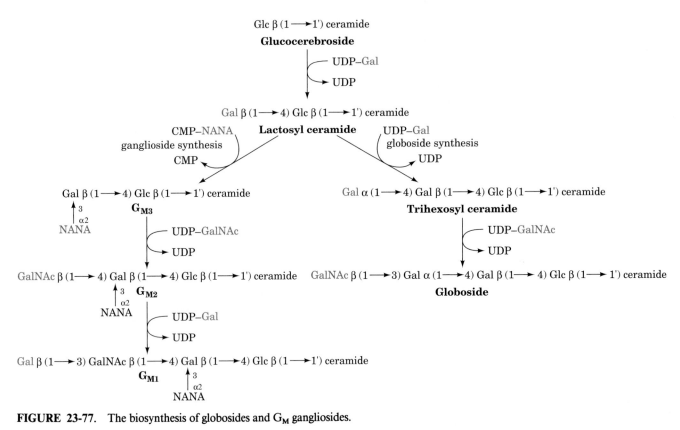

FIGURE 23-77. The biosynthesis of globosides and G_M gangliosides.

from CMP–NANA to lactosyl ceramide in $\alpha(2 \rightarrow 3)$ linkage yielding G_{M3}. The sequential addition to G_{M3} of the *N*-acetylgalactosamine and galactose units from UDP–GalNAc and UDP–Gal yield gangliosides G_{M2} and G_{M1}. Other gangliosides are formed by adding a second NANA group to G_{M3}, forming G_{D3}, or by adding an *N*-acetylglucosamine unit to lactosyl ceramide before NANA addition, forming G_{A2}. Over 60 different gangliosides are known.

Sphingoglycolipid Degradation and Lipid Storage Diseases

Sphingoglycolipids are lysosomally degraded by a series of enzymatically mediated hydrolytic reactions (Fig. 23-78). The hereditary absence of one of these enzymes results in a **sphingolipid storage disease** (Table 23-2). One of the most common such condition is **Tay–Sachs disease,** an autosomal recessive deficiency in **hexosaminidase A,** which hydrolyzes *N*-acetylgalactosamine from ganglioside G_{M2}.

TABLE 23-2. Sᴘʜɪɴɢᴏʟɪᴘɪᴅ Sᴛᴏʀᴀɢᴇ Dɪsᴇᴀsᴇs

Disease	Enzyme Deficiency	Principal Storage Substance	Major Symptoms
G_{M1} Gangliosidosis	G_{M1} β-Galactosidase	Ganglioside G_{M1}	Mental retardation, liver enlargement, skeletal involvement, death by age 2
Tay–Sachs disease	Hexosaminidase A	Ganglioside G_{M2}	Mental retardation, blindness, death by age 3
Fabry's disease	α-Galactosidase A	Trihexosylceramide	Skin rash, kidney failure, pain in lower extremities
Sandhoff's disease	Hexosaminidases A and B	Ganglioside G_{M2} and globoside	Similar to Tay–Sachs disease but more rapidly progressing
Gaucher's disease	Glucocerebrosidase	Glucocerebroside	Liver and spleen enlargement, erosion of long bones, mental retardation in infantile form only
Niemann–Pick disease	Sphingomyelinase	Sphingomyelin	Liver and spleen enlargement, mental retardation
Farber's lipogranulomatosis	Ceramidase	Ceramide	Painful and progressively deformed joints, skin nodules, death within a few years
Krabbe's disease	Galactocerebrosidase	Deacylated galactocerebroside	Loss of myelin, mental retardation, death by age 2
Sulfatide lipidosis	Arylsulfatase A	Sulfatide	Mental retardation, death in first decade

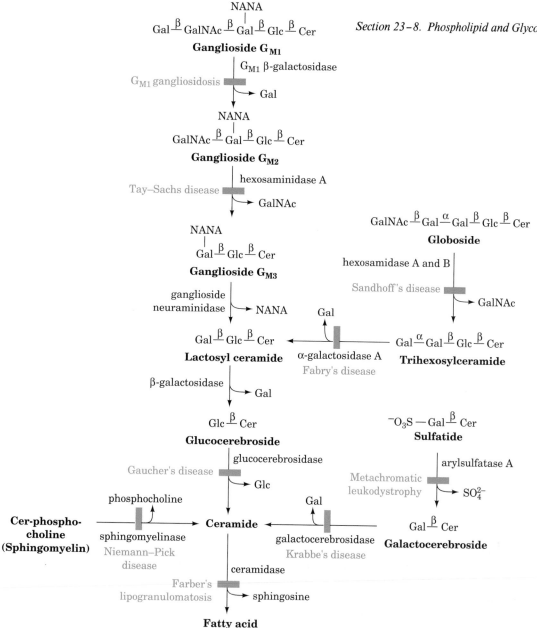

FIGURE 23-78. The breakdown of sphingolipids by lysosomal enzymes. The genetic diseases caused by the corresponding enzyme deficiencies are noted in red.

The absence of hexosaminidase A activity results in the neuronal accumulation of G_{M2} as shell-like inclusions (Fig.23-79).

Although infants born with Tay–Sachs disease at first appear normal, by ~1 year of age, when sufficient G_{M2} has accumulated to interfere with neuronal function, they become progressively weaker, retarded, and blinded until they die, usually by the age of 3 years. It is possible, however, to screen potential carriers of this disease by a simple serum

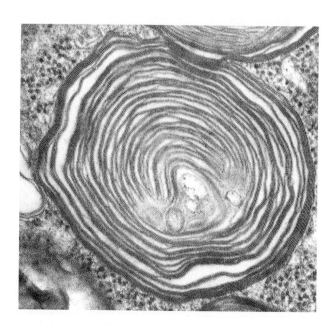

FIGURE 23-79. Cytoplasmic membranous body in a neuron affected by Tay–Sachs disease. [Courtesy of John S. O'Brien, University of California at San Diego Medical School.]

assay. It is also possible to detect the disease *in utero* by assay of amniotic fluid or amniotic cells obtained by amniocentesis. The assay involves use of an artificial hexosaminidase substrate, **4-methylumbelliferyl-β-D-N-acetylglucosamine,** which yields a fluorescent product on hydrolysis.

4-Methylumbelliferyl-β-D-N-acetylglucosamine

4-Methylumbelliferone
(fluorescent in alkaline medium)

Since this substrate is also recognized by **hexosaminidase B,** which is unaffected in Tay–Sachs disease, the hexosaminidase B is first heat inactivated since it is more heat labile than hexosaminidase A. As a result of mass screening efforts, the tragic consequences of this genetic enzyme deficiency are being averted. The other sphingolipid storage diseases, although less common, have similar consequences (Table 23-2).

CHAPTER SUMMARY

Triacylglycerols, the storage form of metabolic energy in animals, provide up to six times the metabolic energy of an equal weight of hydrated glycogen. Dietary lipids are digested by pancreatic digestive enzymes such as lipase and phospholipase A_2 that are active at the lipid–water interface of bile acid–stabilized emulsions. Bile acids are also essential for the intestinal absorption of dietary lipids, as is fatty acid–binding protein. Dietary triacylglycerols and those synthesized by the liver are transported in the blood as chylomicrons and VLDL, respectively. Triacylglycerols present in these lipoproteins are hydrolyzed by lipoprotein lipase outside the cells and enter them as free fatty acids. Fatty acids resulting from hydrolysis of adipose tissue triacylglycerols by hormone-sensitive lipase are transported in the bloodstream as fatty acid–albumin complexes.

Before fatty acids are oxidized, they are converted to their acyl-CoA derivatives by acyl-CoA synthase in an ATP-requiring process, transported into mitochondria as carnitine esters, and reconverted inside the mitochondrial matrix to acyl-CoA. β Oxidation of fatty acyl-CoA occurs in 2-carbon increments so as to convert even-chain fatty acyl-CoAs completely to acetyl-CoA. The pathway involves FAD-dependent dehydrogenation of an alkyl group, hydration of the resulting double bond, NAD^+-dependent oxidation of this alcohol to a ketone, and C—C bond cleavage to form acetyl-CoA and a new fatty acyl-CoA with two fewer carbon atoms. The process then repeats itself. Complete oxidation of the acetyl-CoA, NADH, and $FADH_2$ is achieved by the citric acid cycle and oxidative phosphorylation. Oxidation of unsaturated fatty acids and odd-chain fatty acids also occur by β oxidation but

require the participation of additional enzymes. Odd-chain fatty acid oxidation generates propionyl-CoA, whose further metabolism requires the participation of (1) propionyl-CoA carboxylase, which has a biotin prosthetic group, (2) methylmalonyl-CoA racemase, and (3) methylmalonyl-CoA mutase, which contains coenzyme B_{12}. β Oxidation of fatty acids takes place in the peroxisomes in addition to the mitochondrion. The peroxisomal pathway differs from the mitochondrial pathway in that the $FADH_2$ produced in the first step, rather than generating ATP by oxidative phosphorylation, is directly oxidized by O_2 to produce H_2O_2. Peroxisomal enzymes are specific for long-chain fatty acids and are thought to function in a chain-shortening process. The resultant intermediate chain-length products are transferred to the mitochondrion for complete oxidation.

A significant fraction of the acetyl-CoA produced by fatty acid oxidation in the liver is converted to acetoacetate and D-β-hydroxybutyrate, which, together with acetone, are referred to as ketone bodies. The first two compounds serve as important fuels for the peripheral tissues.

Fatty acid biosynthesis differs from fatty acid oxidation in several respects. Whereas fatty acid oxidation occurs in the mitochondrion utilizing fatty acyl-CoA esters, fatty acid biosynthesis occurs in the cytosol with the growing fatty acids esterified to acyl-carrier protein (ACP). The redox coenzymes differ (FAD and NAD^+ for oxidation; NADPH for biosynthesis), as does the stereochemistry of the pathway's intermediate steps. Oxidation produces acetyl-CoA, whereas malonyl-CoA is the immediate precursor in biosynthesis. HCO_3^- is required for biosynthesis but is not a product of oxidation. Acetyl-CoA is transferred from the mitochondrion to the cytosol as citrate via the tricarboxylate transport system and citrate cleavage, a process that also generates some of the NADPH required for biosynthesis. Palmitate is the primary product of fatty acid biosynthesis in animals. Longer chain fatty acids and unsaturated fatty acids are synthesized from palmitate by elongation and desaturation reactions. Triacylglycerols are synthesized from fatty acyl-CoA esters and glycerol-3-phosphate.

Fatty acid metabolism is regulated through the allosteric control of hormone-sensitive triacylglycerol lipase and acetyl-CoA carboxylase, phosphorylation/dephosphorylation, and/or changes in the rates of protein synthesis and breakdown. This regulation is mediated by the hormones glucagon, epinephrine, and norepinephrine, which activate degradation, and by insulin, which activates biosynthesis. These hormones interact to control the cAMP concentration, which in turn controls phosphorylation/dephosphorylation ratios.

Cholesterol is a vital constituent of cell membranes and is the precursor of the steroid hormones and bile acids. Its biosynthesis, transport, and utilization are rigidly controlled. Cholesterol is synthesized in the liver from acetate in a pathway that involves formation of HMG-CoA from three molecules of acetate followed by reduction, phosphorylation, decarboxylation, and dehydration to the isoprene units isopentenyl pyrophosphate and dimethylallyl pyrophosphate. These isoprene units are then condensed to form squalene, which, in turn, undergoes a cyclization reaction to form lanosterol, the sterol precursor to cholesterol. The pathway's major control point is at HMG-CoA reductase. This enzyme is regulated by competitive and allosteric mechanisms, by phosphorylation/dephosphorylation, and, most importantly, by long-term control of the rates of enzyme synthesis and degradation. The liver secretes cholesterol into the bloodstream in esterified form as part of the VLDL. This complex is sequentially converted to IDL and then to LDL. LDL, which is brought into the cells by receptor-mediated endocytosis, carries the major portion of cholesterol to peripheral tissues for utilization. Excess cholesterol is returned to the liver from peripheral tissues by HDL. The cellular supply of cholesterol is controlled by three mechanisms: (1) long- and short-term regulation of HMG-CoA reductase; (2) control of LDL receptor synthesis by cholesterol concentration; and (3) long- and short-term regulation of acyl-CoA:cholesterol acyl transferase (ACAT), which mediates cholesterol esterification. Cholesterol is the precursor to the steroid hormones, which are classified as progestins, glucocorticoids, mineralocorticoids, androgens, and estrogens. The quantitatively most important pathway for the excretion of cholesterol in mammals is the formation of bile acids.

Prostaglandins, prostacyclins, thromboxanes, and leukotrienes are products of arachidonate metabolism. These highly unstable compounds have profound physiological effects at extremely low concentrations. They are involved in the inflammatory response, the production of pain and fever, the regulation of blood pressure, and many other important physiological processes. Arachidonate is synthesized from linoleic acid, an essential fatty acid, and stored as phosphatidylinositol and other phospholipids. Prostaglandins, prostacyclins, and thromboxanes are synthesized via the "cyclic pathway," whereas leukotrienes are synthesized via the "linear pathway." Aspirin and other nonsteroidal anti-inflammatory drugs (NSAIDs) inhibit the cyclic pathway but not the linear pathway. Peptidoleukotrienes have been identified as the slow reacting substances of anaphylaxis (SRS-A) released from sensitized lung after immunological challenge.

Complex lipids have either a phosphate ester or a carbohydrate as their polar head group and either 1,2-diacyl-*sn*-glycerol or ceramide (*N*-acylsphingosine) as their hydrophobic tail. Phospholipids are either glycerophospholipids or sphingophospholipids, whereas glycolipids are either glyceroglycolipids or sphingoglycolipids. The polar head groups of glycerophospholipids, which are phosphate esters of either ethanolamine, serine, choline, inositol, or glycerol, are attached to 1,2-diacyl-*sn*-glycerol's C3—OH group by means of CTP-linked transferase reactions. The specific long-chain fatty acids found at the C1 and C2 positions are incorporated by "remodeling reactions" after the addition of the polar head group. Plasmalogens and alkylacylglycerophospholipids, respectively, contain a long-chain alkyl group in a vinyl–ether linkage or an ether linkage to glycerol's C1—OH group. Platelet-activating factor (PAF) is an important plasmalogen. The only major sphingophospholipid is sphingomyelin (*N*-acylsphingosine phosphocholine), an important structural lipid of nerve cell membranes. Most sphingolipids contain polar head groups composed of carbohydrate units and are therefore referred to as sphingoglycolipids. The principal classes of sphingoglycolipids are cerebrosides, sulfatides, globosides, and gangliosides. Their carbohydrate units, which are attached to *N*-acylsphingosine's C1—OH group by glycosidic linkages, are formed by stepwise addition of activated monosaccharide units. Several lysosomal sphingolipid storage diseases, including Tay–Sachs disease, result from deficiencies in the enzymes that degrade sphingoglycolipids.

REFERENCES

General

Boyer, P.D. (Ed.), *The Enzymes* (3rd ed.), Vol. 16, Academic Press (1983). [An excellent collection of reviews on lipid enzymology. Section 1 deals with fatty acid biosynthesis; Section 2 covers glyceride synthesis and degradation; Sections 3–5 review phospholipid, sphingolipid, and glycolipid metabolism; and Section 6 deals with aspects of cholesterol metabolism.]

Newsholme, E.A. and Leech, A.R., *Biochemistry for the Medical Sciences,* Wiley (1983). [Chapters 6–8 contain a wealth of information on the control of fatty acid metabolism and its integration into the overall scheme of metabolism.]

Numa, S. (Ed.), *Fatty Acid Metabolism and Its Regulation,* Elsevier (1984).

Scriver, C.R., Beaudet, A.C., Sly, W.S., and Valle, D. (Eds.), *The Metabolic Basis of Inherited Disease* (6th ed.), McGraw–Hill (1989). [Contains numerous chapters on defects in lipid metabolism.]

Thompson, G.A., *The Regulation of Membrane Lipid Metabolism* (2nd ed.), CRC Press (1992).

Vance, D.E. and Vance, J.E. (Eds.), *Biochemistry of Lipids, Lipoproteins and Membranes,* Elsevier (1991).

Lipid Digestion

Borgström, B., Barrowman, J.A., and Lindström, M., Roles of bile acids in intestinal lipid digestion and absorption, *in* Danielsson, H.

PROBLEMS

1. The venoms of many poisonous snakes, including rattlesnakes, contain a phospholipase A_2 that causes tissue damage that is seemingly far out of proportion to the small amount of enzyme injected. Explain.

2. Why are the livers of Jamaican vomiting sickness victims usually depleted of glycogen?

3. Compare the metabolic efficiencies, in moles of ATP produced per gram, of completely oxidized fat (tripalmitoyl glycerol) versus glucose derived from glycogen. Assume that the fat is anhydrous and the glycogen is stored with twice its weight in water.

4. Methylmalonyl mutase is incubated with deuterated methylmalonyl-CoA. The coenzyme B_{12} extracted from this mutase is found to contain deuterium at its 5'-methylene group. Account for the transfer of label from substrate to coenzyme.

5. What is the energetic price, in ATP units, of converting acetoacetyl-CoA to acetoacetate and then resynthesizing acetoacetyl-CoA?

6. A fasting animal is fed palmitic acid that has a ^{14}C-labeled carboxyl group. (a) After allowing sufficient time for fatty acid breakdown and resynthesis, what would be the ^{14}C-labeling pattern in the animal's palmitic acid residues? (b) The animal's liver glycogen becomes ^{14}C labeled although there is no net increase in the amount of this substance present. Indicate the sequence of reactions whereby the glycogen becomes labeled. Why is there no net glycogen synthesis?

7. What is the ATP yield from the complete oxidation of a molecule of (a) α-linolenic acid (9,12,15-octadecatrienoic acid, 18:3), and (b) **margaric acid** (heptadecanoic acid, 17:0)? Which has the greater amount of available biological energy on a per carbon basis?

*8. The role of coenzyme B_{12} in mediating hydrogen transfer was established using the coenzyme B_{12}-dependent bacterial enzyme **dioldehydrase,** which catalyzes the reaction:

$$CH_3-\underset{\underset{OH}{|}}{CH}-\underset{\underset{H}{|}}{CH}-OH \longrightarrow CH_3-\underset{\underset{H}{|}}{CH}-\underset{\underset{OH}{|}}{CH}-OH$$

$$\downarrow H_2O$$

$$CH_3-CH_2-\overset{\overset{O}{\|}}{CH}$$

Propionaldehyde

The enzyme converts $[1-^3H_2]1,2$-propanediol to $[1,2-^3H]$propionaldehyde with the incorporation of tritium into both C5' positions of 5'-deoxyadenosylcobalamin's 5'-deoxyadenosyl residue. Suggest the mechanism of this reaction. What would be the products of the dioldehydrase reaction if the enzyme was supplied with $[5'-^3H]$deoxyadenosylcobalamin and unlabeled 1,2-propanediol?

9. What is the energetic price, in ATP equivalents, of breaking down palmitic acid to acetyl-CoA and then resynthesizing it?

10. What is the energetic price, in ATP equivalents, of synthesizing cholesterol from acetyl-CoA?

11. What would be the ^{14}C-labeling pattern in cholesterol if it was synthesized from HMG-CoA that was ^{14}C labeled (a) at C5, its carboxyl carbon atom or (b) C1, its thioester carbon atom?

*12. A child suffering from severe abdominal pain is admitted to the hospital several hours after eating a meal consisting of hamburgers, fried potatoes, and ice cream. Her blood has the appearance of "creamed tomato soup" and upon analysis is found to contain massive quantities of chylomicrons. As attending physician, what is your diagnosis of the patient's difficulty (the cause of the abdominal pain is unclear)? What treatment would you prescribe to alleviate the symptoms of this inherited disease?

13. Although linoleic acid is an essential fatty acid in animals, it is not required by animal cells in tissue culture. Explain.

24

Amino Acid Metabolism

1. Amino Acid Deamination

   A. Transamination

   B. Oxidative Deamination: Glutamate Dehydrogenase

   C. Other Deamination Mechanisms

2. The Urea Cycle

   A. Carbamoyl Phosphate Synthetase: Acquisition of the First Urea Nitrogen Atom

   B. Ornithine Transcarbamoylase

   C. Argininosuccinate Synthetase: Acquisition of the Second Urea Nitrogen Atom

   D. Argininosuccinase

   E. Arginase

   F. Regulation of the Urea Cycle

3. Metabolic Breakdown of Individual Amino Acids

   A. Amino Acids Can Be Glucogenic, Ketogenic, or Both

   B. Alanine, Cysteine, Glycine, Serine, and Threonine Are Degraded to Pyruvate

   C. Asparagine and Aspartate Are Degraded to Oxaloacetate

   D. Arginine, Glutamate, Glutamine, Histidine, and Proline Are Degraded to α-Ketoglutarate

   E. Isoleucine, Methionine, and Valine Are Degraded to Succinyl-CoA

   F. Leucine and Lysine Are Degraded to Acetoacetate and/or Acetyl-CoA

   G. Tryptophan Is Degraded to Alanine and Acetyl-CoA

   H. Phenylalanine and Tyrosine Are Degraded to Fumarate and Acetoacetate

4. Amino Acids as Biosynthetic Precursors

   A. Heme Biosynthesis and Degradation

   B. Biosynthesis of Physiologically Active Amines

   C. Glutathione

   D. Tetrahydrofolate Cofactors: The Metabolism of C_1 Units

5. Amino Acid Biosynthesis

   A. Biosynthesis of the Nonessential Amino Acids

   B. Biosynthesis of the Essential Amino Acids

6. Nitrogen Fixation

α-Amino acids, in addition to their role as protein monomeric units, are energy metabolites and precursors of many biologically important nitrogen-containing compounds, notably heme, physiologically active amines, glutathione, nucleotides, and nucleotide coenzymes. Amino acids are classified into two groups: **essential** and **nonessential.** Mammals synthesize the nonessential amino acids from metabolic precursors but must obtain the essential amino acids from their diet. Excess dietary amino acids are neither stored for future use nor excreted. Rather, they are converted to common metabolic intermediates such as pyruvate, oxaloacetate, and α-ketoglutarate. Consequently, *amino acids are also precursors of glucose, fatty acids, and ketone bodies and are therefore metabolic fuels.*

In this chapter, we consider the pathways of amino acid breakdown, synthesis, and utilization. We begin by examining the three common stages of amino acid breakdown:

1. Deamination (amino group removal), whereby amino groups are converted either to ammonia or to the amino group of aspartate.

2. Incorporation of ammonia and aspartate nitrogen atoms into urea for excretion.

3. Conversion of amino acid carbon skeletons (the α-keto acids produced by deamination) to common metabolic intermediates.

Many of these reactions are similar to those we have considered in other pathways. Others employ enzyme cofactors we have not previously encountered. One of our goals in studying amino acid metabolism is to understand the mechanisms of action of these cofactors.

After our discussion of amino acid breakdown, we examine the pathways by which amino acids are utilized in the biosynthesis of heme, physiologically active amines, and glutathione (the synthesis of nucleotides and nucleotide coenzymes is the subject of Chapter 26). Next, we study amino acid biosynthesis pathways. The chapter ends with a discussion of nitrogen fixation, a process that converts atmospheric N_2 to ammonia and is therefore the ultimate biological source of metabolically useful nitrogen.

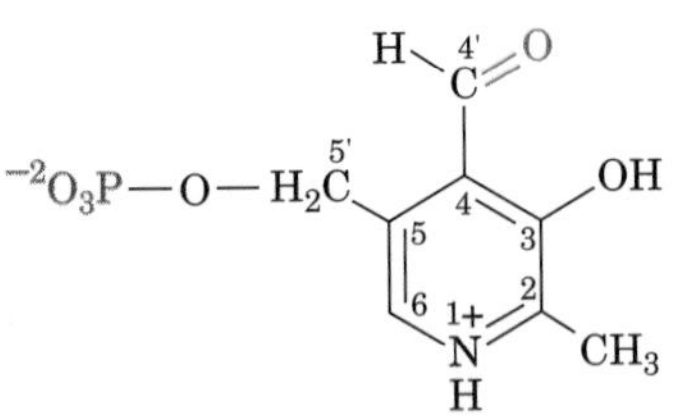

1. AMINO ACID DEAMINATION

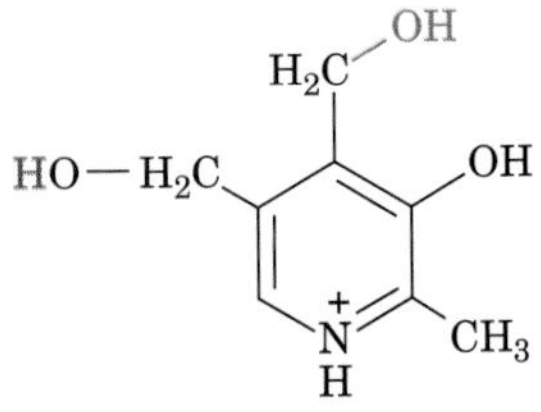

The first reaction in the breakdown of an amino acid is almost always removal of its α-amino group with the object of excreting excess nitrogen and degrading the remaining carbon skeleton. Urea, the predominant nitrogen excretion product in terrestrial mammals, is synthesized from ammonia and aspartate. Both of these latter substances are derived mainly from glutamate, a product of most deamination reactions. In this section we examine the routes by which α-amino groups are incorporated into glutamate and then into aspartate and ammonia. In Section 24-2, we discuss urea biosynthesis from these precursors.

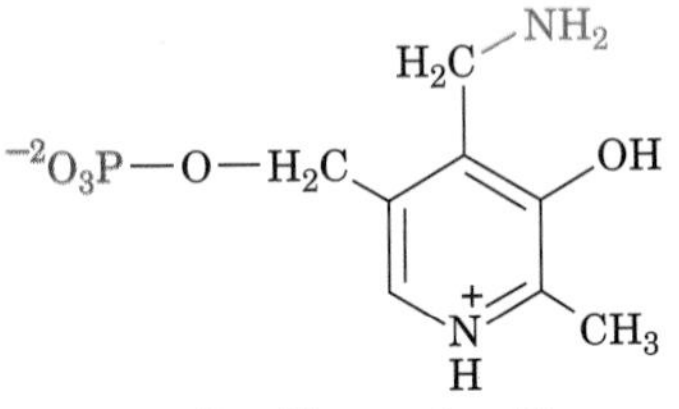

Most amino acids are deaminated by **transamination,** the transfer of their amino group to an α-keto acid to yield the α-keto acid of the original amino acid and a new amino acid, in reactions catalyzed by **aminotransferases** (alternatively, **transaminases**). The predominant amino group acceptor is α-ketoglutarate, producing glutamate as the new amino acid:

Amino acid + α-ketoglutarate ⇌

α-keto acid + glutamate

Glutamate's amino group, in turn, is transferred to oxaloacetate in a second transamination reaction, yielding aspartate:

Glutamate + oxaloacetate ⇌

α-ketoglutarate + aspartate

Transamination, of course, does not result in any net deamination. Deamination occurs largely through the oxidative deamination of glutamate by **glutamate dehydrogenase,** yielding ammonia. The reaction requires NAD^+ or $NADP^+$ as an oxidizing agent and regenerates α-ketoglutarate for use in additional transamination reactions:

Glutamate + $NAD(P)^+$ + H_2O ⇌

α-ketoglutarate + NH_4^+ + $NAD(P)H$

FIGURE 24-1.   The coenzymes pyridoxal-5′-phosphate (PLP) and pyridoxamine-5′-phosphate (PMP) are derived from pyridoxine (vitamin B_6).

The mechanisms of transamination and oxidative deamination are the subjects of this section. We also consider other means of amino group removal from specific amino acids.

A. Transamination

Aminotransferase reactions occur in two stages:

1. The amino group of an amino acid is transferred to the enzyme, producing the corresponding keto acid and the aminated enzyme.

   Amino acid + enzyme ⇌

   α-keto acid + enzyme—NH_2

2. The amino group is transferred to the keto acid acceptor (e.g., α-ketoglutarate), forming the amino acid product (e.g., glutamate) and regenerating the enzyme.

   α-Ketoglutarate + enzyme — NH_2 ⇌

   enzyme + glutamate

*To carry the amino group, aminotransferases require participation of an aldehyde-containing coenzyme, **pyridoxal-5′-phosphate (PLP),** a derivative of **pyridoxine (vitamin B_6;** Fig. 24-1).* The amino group is accommodated by conversion of this coenzyme to **pyridoxamine-5′-phosphate (PMP;** Fig. 24-1). PLP is covalently attached to the enzyme via a Schiff base (imine) linkage formed by the condensation of its aldehyde group with the ε-amino group of an enzymatic Lys residue.

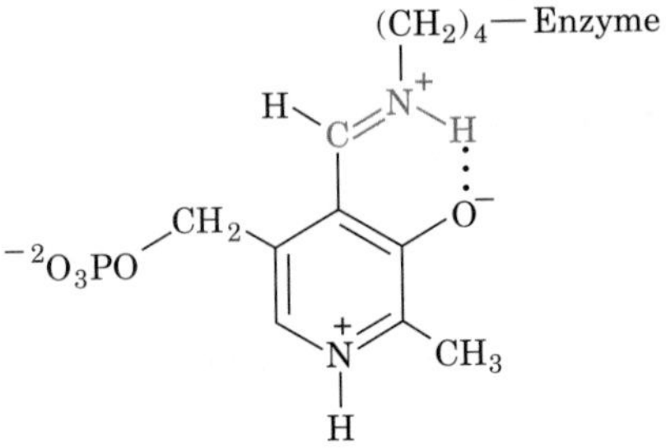

Steps 1 & 1': Transimination:

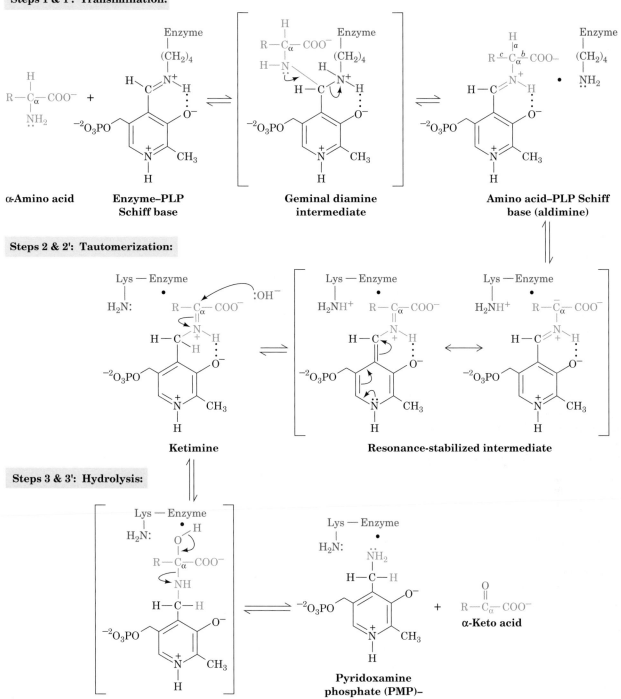

α-Amino acid   **Enzyme–PLP Schiff base**   **Geminal diamine intermediate**   **Amino acid–PLP Schiff base (aldimine)**

Steps 2 & 2': Tautomerization:

Ketimine   **Resonance-stabilized intermediate**

Steps 3 & 3': Hydrolysis:

Carbinolamine   **Pyridoxamine phosphate (PMP)– enzyme**   **α-Keto acid**

FIGURE 24-2. The mechanism of PLP-dependent enzyme-catalyzed transamination. The first stage of the reaction, in which the α-amino group of an amino acid is transferred to PLP yielding an α-keto acid and PMP, consists of three steps: **(1)** transimination; **(2)** tautomerization, in which the Lys released during the transimination reaction acts as a general acid–base catalyst; and **(3)** hydrolysis. The second stage of the reaction, in which the amino group of PMP is transferred to a different α-keto acid to yield a new α-amino acid and PLP, is essentially the reverse of the first stage: Steps 3', 2', and 1' are, respectively, the reverse of Steps 3, 2, and 1.

This Schiff base, which is conjugated to the coenzyme's pyridinium ring, is the focus of the coenzyme's activity.

Esmond Snell, Alexander Braunstein, and David Metzler demonstrated that the aminotransferase reaction occurs via a Ping Pong Bi Bi mechanism whose two stages consist of three steps each (Fig. 24-2):

Stage I: Conversion of an Amino Acid to a Keto Acid

1. The amino acid's nucleophilic amino group attacks the enzyme–PLP Schiff base carbon atom in a **transimination (trans-Schiffization)** reaction to form an amino acid–PLP Schiff base (aldimine), with concomitant release of the enzyme's Lys amino group. This Lys is then free to act as a general base at the active site.

2. The amino acid–PLP Schiff base tautomerizes to an α-keto acid–PMP Schiff base by the active-site Lys–catalyzed removal of the amino acid α hydrogen and protonation of PLP atom C4' via a resonance-stabilized carbanion intermediate. This resonance stabilization facilitates the cleavage of the C_α—H bond.

3. The α-keto acid–PMP Schiff base is hydrolyzed to PMP and an α-keto acid.

Stage II: Conversion of an α-Keto Acid to an Amino Acid

To complete the aminotransferase's catalytic cycle, the coenzyme must be converted from PMP back to the enzyme–PLP Schiff base. This involves the same three steps as above, but in reverse order:

3'. PMP reacts with an α-keto acid to form a Schiff base.

2'. The α-keto acid–PMP Schiff base tautomerizes to form an amino acid–PLP Schiff base.

1'. The ε-amino group of the active-site Lys residue attacks the amino acid–PLP Schiff base in a transimination reaction to regenerate the active enzyme–PLP Schiff base, with release of the newly formed amino acid.

The reaction's overall stoichiometry therefore is

Amino acid 1 + α-keto acid 2 $\rightleftharpoons$
$$\alpha\text{-keto acid 1} + \text{amino acid 2}$$

Examination of the amino acid–PLP Schiff base's structure (Fig. 24-2, Step 1) reveals why this system is called "an electron-pusher's delight." *Cleavage of any of the amino acid C_α atom's three bonds (labeled a, b, and c) produces a resonance-stabilized C_α carbanion whose electrons are delocalized all the way to the coenzyme's protonated pyridinium nitrogen atom; that is, PLP functions as an electron sink.* For transamination reactions, this electron-withdrawing capacity facilitates removal of the α proton (a bond cleavage) in the tautomerization of the Schiff base. PLP-dependent reactions involving b bond cleavage (amino acid decarboxylation) and c bond labilization are discussed in Sections 24-4B and 24-3B and G, respectively.

Aminotransferases differ in their specificity for amino acid substrates in the first stage of the transamination reaction, thereby producing the correspondingly different α-keto acid products. Most aminotransferases, however, accept only α-ketoglutarate or (to a lesser extent) oxaloacetate as the α-keto acid substrate in the second stage of the reaction, thereby yielding glutamate or aspartate as their only amino acid products. *The amino groups of most amino acids are consequently funneled into the formation of glutamate or aspartate, which are themselves interconverted by glutamate–aspartate aminotransferase:*

Glutamate + oxaloacetate $\rightleftharpoons$
$$\alpha\text{-ketoglutarate} + \text{aspartate}$$

Oxidative deamination of glutamate (Section 24-1B) yields ammonia and regenerates α-ketoglutarate for another round of transamination reactions. Ammonia and aspartate are the two amino group donors in the synthesis of urea.

The Glucose–Alanine Cycle Transports Nitrogen to the Liver

An important exception to the foregoing is a group of muscle aminotransferases that accept pyruvate as their α-keto acid substrate. The product amino acid, alanine, is released into the bloodstream and transported to the liver, where it undergoes transamination to yield pyruvate for use in gluconeogenesis (Section 21-1A). The resulting glucose is returned to the muscles, where it is glycolytically degraded to pyruvate. This is the **glucose–alanine cycle.** The amino group ends up in either ammonia or aspartate for urea biosynthesis. Evidently, the glucose–alanine cycle functions to transport nitrogen from muscle to liver.

During starvation the glucose formed in the liver by this route is also used by the other peripheral tissues, breaking the cycle. Under these conditions both the amino group and the pyruvate originate from muscle protein degradation, providing a pathway yielding glucose for other tissue use (recall that muscle is not a gluconeogenic organ; Section 21-1).

Nitrogen is also transported to the liver in the form of glutamine, synthesized from glutamate and ammonia in a reaction catalyzed by **glutamine synthetase** (Section 24-5A). The ammonia is released through the action of **glutaminase** (Section 24-3D).

B. Oxidative Deamination: Glutamate Dehydrogenase

Glutamate is oxidatively deaminated in the mitochondrion by glutamate dehydrogenase, the only known enzyme that, in at least some organisms, can accept either NAD^+ or $NADP^+$ as its redox coenzyme. Oxidation is thought to occur with transfer of a hydride ion from glutamate's C_α to $NAD(P)^+$, thereby forming α-iminoglutarate, which is hydrolyzed to α-ketoglutarate and ammonia (Fig. 24-3).

Glutamate dehydrogenase is inhibited by GTP and activated by ADP *in vitro*, suggesting that these nucleotides regulate the enzyme *in vivo*. Studies of the cellular substrate and product concentrations nevertheless indicate that the enzyme functions close to equilibrium ($\Delta G \approx 0$) *in vivo*. Changes in glutamate dehydrogenase activity resulting from allosteric interactions are therefore unlikely to result

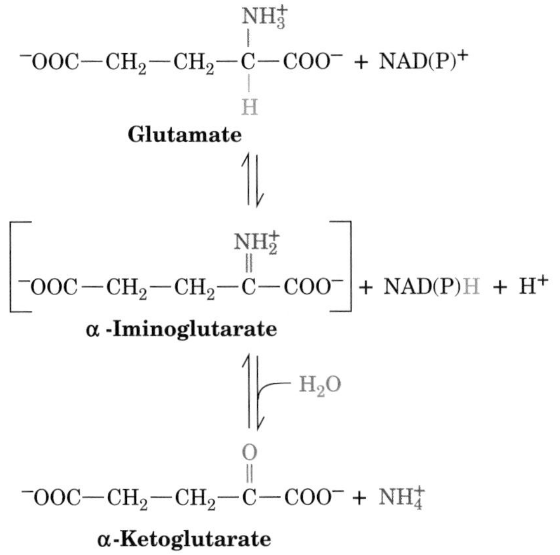

$$\underset{\text{Glutamate}}{{}^-OOC-CH_2-CH_2-\overset{\overset{\displaystyle NH_3^+}{|}}{\underset{\underset{\displaystyle H}{|}}{C}}-COO^-} + NAD(P)^+$$

$$\Updownarrow$$

$$\underset{\alpha\text{-Iminoglutarate}}{\left[{}^-OOC-CH_2-CH_2-\overset{\overset{\displaystyle NH_2^+}{\|}}{C}-COO^-\right]} + NAD(P)H + H^+$$

$$\Updownarrow{\small-H_2O}$$

$$\underset{\alpha\text{-Ketoglutarate}}{{}^-OOC-CH_2-CH_2-\overset{\overset{\displaystyle O}{\|}}{C}-COO^-} + NH_4^+$$

FIGURE 24-3. The oxidation of glutamate by glutamate dehydrogenase involves the intermediate formation of α-iminoglutarate.

in flux changes. Most probably the flux is controlled by the concentrations of substrates and products (Section 16-4A). The equilibrium position greatly favors glutamate formation over ammonia formation ($\Delta G^{\circ\prime} \approx 30$ kJ $\cdot$ mol^{-1} for the reaction as written in Fig. 24-3). Because high concentrations of ammonia are toxic, this equilibrium position is physiologically important; it helps maintain low ammonia concentrations. The ammonia produced is converted to urea (Section 24-2).

C. Other Deamination Mechanisms

Two nonspecific amino acid oxidases, **L-amino acid oxidase** and **D-amino acid oxidase,** catalyze the oxidation of L- and D-amino acids, utilizing FAD as their redox coenzyme [rather than NAD(P)$^+$]. The resulting FADH$_2$ is reoxidized by O$_2$.

$$\text{Amino acid} + FAD + H_2O \longrightarrow$$
$$\alpha\text{-keto acid} + NH_3 + FADH_2$$

$$FADH_2 + O_2 \longrightarrow FAD + H_2O_2$$

D-Amino acid oxidase occurs mainly in kidney. Its function is an enigma since D-amino acids are associated mostly with bacterial cell walls (Section 10-3B). A few amino acids, such as serine and histidine, are deaminated nonoxidatively (Sections 24-3B and D).

2. THE UREA CYCLE

Living organisms excrete the excess nitrogen resulting from the metabolic breakdown of amino acids in one of three ways. Many aquatic animals simply excrete ammonia.

Where water is less plentiful, however, processes have evolved that convert ammonia to less toxic waste products that therefore require less water for excretion. One such product is urea, which is excreted by most terrestrial vertebrates; another is **uric acid,** which is excreted by birds and terrestrial reptiles.

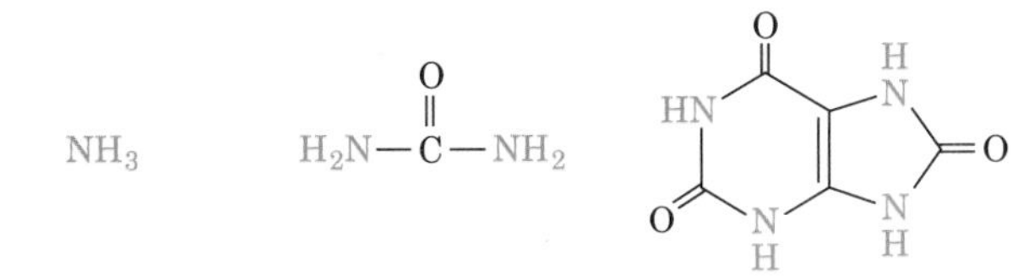

$$\underset{\text{Ammonia}}{NH_3} \qquad \underset{\text{Urea}}{H_2N-\overset{\overset{\displaystyle O}{\|}}{C}-NH_2} \qquad \underset{\text{Uric acid}}{}$$

Accordingly, living organisms are classified as being either **ammonotelic** (ammonia excreting), **ureotelic** (urea excreting), or **uricotelic** (uric acid excreting). Some animals can shift from ammonotelism to ureotelism or uricotelism if their water supply becomes restricted. Here we focus our attention on urea formation. Uric acid biosynthesis is discussed in Section 26-5A.

Urea is synthesized in the liver by the enzymes of the urea cycle. It is then secreted into the bloodstream and sequestered by the kidneys for excretion in the urine. The urea cycle was elucidated in outline in 1932 by Hans Krebs and Kurt Henseleit (the first known metabolic cycle; Krebs did not elucidate the citric acid cycle until 1937). Its individual reactions were later described in detail by Sarah Ratner and Philip Cohen. The overall urea cycle reaction is

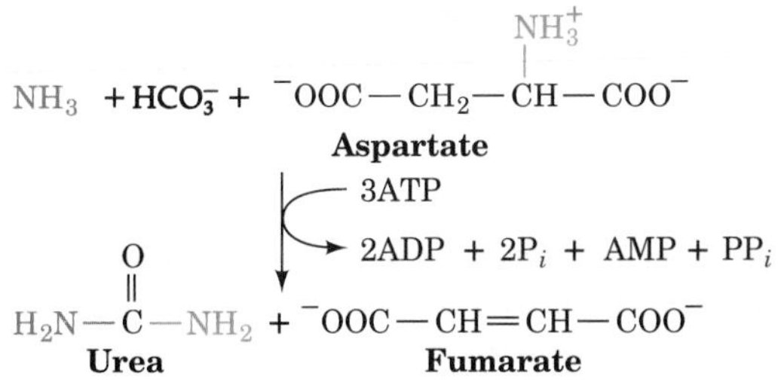

$$NH_3 + HCO_3^- + \underset{\text{Aspartate}}{{}^-OOC-CH_2-\overset{\overset{\displaystyle NH_3^+}{|}}{CH}-COO^-}$$

$$\Bigg\downarrow \begin{array}{l} \nearrow 3ATP \\ \searrow 2ADP + 2P_i + AMP + PP_i \end{array}$$

$$\underset{\text{Urea}}{H_2N-\overset{\overset{\displaystyle O}{\|}}{C}-NH_2} + \underset{\text{Fumarate}}{{}^-OOC-CH=CH-COO^-}$$

Thus, the two urea nitrogen atoms are contributed by ammonia and aspartate, whereas the carbon atom comes from HCO$_3^-$. Five enzymatic reactions are involved in the urea cycle, two of which are mitochondrial and three cytosolic (Fig. 24-4). In this section, we examine the mechanisms of these reactions and their regulation.

A. Carbamoyl Phosphate Synthetase: Acquisition of the First Urea Nitrogen Atom

Carbamoyl phosphate synthetase (CPS) is technically not a member of the urea cycle. It catalyzes the condensation and activation of NH$_4^+$ and HCO$_3^-$ to form **carbamoyl phosphate,** the first of the cycle's two nitrogen-containing substrates, with the concomitant hydrolysis of two ATPs. Eukaryotes have two forms of CPS:

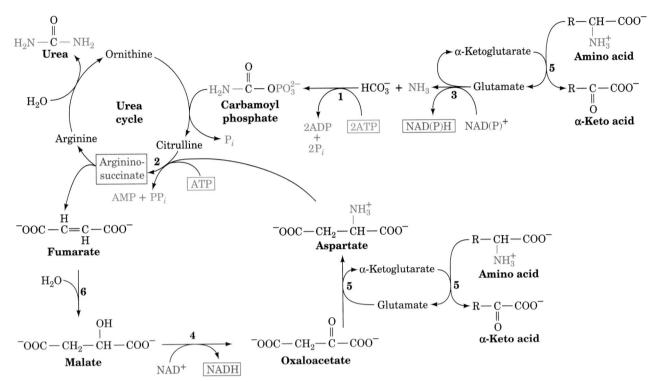

FIGURE 24-7. Energy sources and requirements for the urea cycle. ATP is hydrolyzed in the reactions catalyzed by (1) carbamoyl phosphate synthetase I and (2) argininosuccinate synthetase. The ATP is regenerated by oxidative phosphorylation from the NAD(P)H produced in (3) the glutamate dehydrogenase reaction and (4) the malate dehydrogenase reaction. The remaining reactions shown are catalyzed by (5) aminotransferases and (6) fumarase.

cycle substrates. The ammonia released by the glutamate dehydrogenase reaction is accompanied by NADH formation, as is the reconversion of fumarate through oxaloacetate to aspartate (Fig. 24-7). Mitochondrial reoxidation of this NADH yields six ATPs.

F. Regulation of the Urea Cycle

Carbamoyl phosphate synthetase I, the mitochondrial enzyme that catalyzes the first committed step of the urea cycle, is allosterically activated by *N*-acetylglutamate.

$$
\begin{array}{c}
\text{COO}^- \\
| \\
(\text{CH}_2)_2 \quad \text{O} \\
| \qquad \quad \| \\
\text{H}-\text{C}-\text{N}-\text{C}-\text{CH}_3 \\
| \quad \ | \\
{}^-\text{OOC} \quad \text{H}
\end{array}
$$

***N*-Acetylglutamate**

This metabolite is synthesized from glutamate and acetyl-CoA by *N*-acetylglutamate synthase and hydrolyzed by a specific hydrolase. The rate of urea production by the liver is, in fact, correlated with the *N*-acetylglutamate concentra-

tion. Increased urea synthesis is required when amino acid breakdown rates increase, generating excess nitrogen that must be excreted. Increases in these breakdown rates are signaled by an increase in glutamate concentration through transamination reactions (Section 24-1). This situation, in turn, causes an increase in *N*-acetylglutamate synthesis, stimulating carbamoyl phosphate synthetase and thus the entire urea cycle.

The remaining enzymes of the urea cycle are controlled by the concentrations of their substrates. Thus, inherited deficiencies in urea cycle enzymes other than arginase do not result in significant decreases in urea production (the total lack of any urea cycle enzyme results in death shortly after birth). Rather, the deficient enzyme's substrate builds up, increasing the rate of the deficient reaction to normal. The anomalous substrate buildup is not without cost, however. The substrate concentrations become elevated all the way back up the cycle to ammonia, resulting in **hyperammonemia** (elevated levels of ammonia in the blood). Although the root cause of ammonia toxicity is not completely understood, a high ammonia concentration puts an enormous strain on the ammonia-clearing system, especially in the brain (symptoms of urea cycle enzyme deficiencies include mental retardation and lethargy). This clearing system involves glutamate dehydrogenase (working in reverse) and **glutamine synthetase,** which decrease the α-ketoglutarate and glutamate pools (Sections 24-1 and 24-5A). The brain is most sensitive to the depletion of these pools. Depletion of α-ketoglutarate decreases the rate of the energy-generating citric acid cycle, whereas glutamate is both a neurotransmitter and a precursor to γ-**aminobutyrate (GABA),** another neurotransmitter (Section 34-4C).

3. METABOLIC BREAKDOWN OF INDIVIDUAL AMINO ACIDS

The degradation of amino acids converts them to citric acid cycle intermediates or their precursors so that they can be metabolized to CO_2 and H_2O or used in gluconeogenesis. Indeed, oxidative breakdown of amino acids typically accounts for 10 to 15% of the metabolic energy generated by animals. In this section we consider how amino acid carbon skeletons are catabolized. The 20 "standard" amino acids (the amino acids of proteins) have widely differing carbon skeletons, so their conversions to citric acid cycle intermediates follow correspondingly diverse pathways. We shall not describe all of the many reactions involved in detail. Rather, we shall consider how these pathways are organized and focus on a few reactions of chemical and/or medical interest.

A. Amino Acids Can Be Glucogenic, Ketogenic, or Both

"Standard" amino acids are degraded to one of seven metabolic intermediates: pyruvate, α-ketoglutarate, succinyl-CoA, fumarate, oxaloacetate, acetyl-CoA, or acetoacetate (Fig. 24-8). The amino acids may therefore be divided into two groups based on their catabolic pathways (Fig. 24-8):

1. **Glucogenic amino acids,** whose carbon skeletons are degraded to pyruvate, α-ketoglutarate, succinyl-CoA, fumarate, or oxaloacetate and are therefore glucose precursors (Section 21-1A).

2. **Ketogenic amino acids,** whose carbon skeletons are broken down to acetyl-CoA or acetoacetate and can thus be converted to fatty acids or ketone bodies (Section 23-3).

For example, alanine is glucogenic because its transamination product, pyruvate (Section 24-1A), can be converted to glucose via gluconeogenesis (Section 21-1A). Leucine, on the other hand, is ketogenic; its carbon skeleton is converted to acetyl-CoA and acetoacetate (Section 24-3F). Since animals lack any metabolic pathway for the net conversion of acetyl-CoA or acetoacetate to gluconeogenic precursors, no net synthesis of carbohydrates is possible from

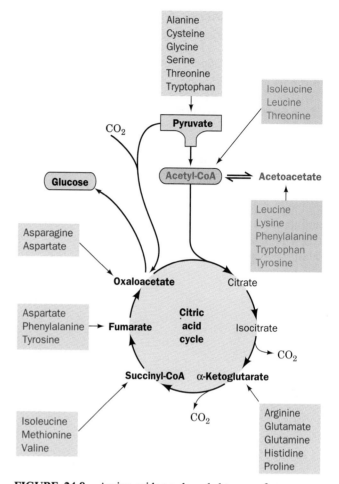

FIGURE 24-8. Amino acids are degraded to one of seven common metabolic intermediates. Glucogenic and ketogenic degradations are indicated in green and red, respectively.

leucine, or from lysine, the only other purely ketogenic amino acid. Isoleucine, phenylalanine, threonine, tryptophan, and tyrosine, however, are both glucogenic and ketogenic; isoleucine, for example, is broken down to succinyl-CoA and acetyl-CoA and hence is a precursor of both carbohydrates and ketone bodies (Section 24-3E). The remaining 13 amino acids are purely glucogenic.

In studying the specific pathways of amino acid breakdown, we shall organize the amino acids into groups that are degraded into each of the seven metabolic intermediates mentioned above: pyruvate, oxaloacetate, α-ketoglutarate, succinyl-CoA, fumarate, acetyl-CoA, and acetoacetate. When acetoacetyl-CoA is a product in amino acid degradation, it can, of course, be directly converted to acetyl-CoA (Section 23-2). We also discuss the pathway by which, in liver, it is converted instead to acetoacetate for use an an alternative fuel source in peripheral tissues (Section 23-3).

B. Alanine, Cysteine, Glycine, Serine, and Threonine Are Degraded to Pyruvate

Five amino acids, alanine, cysteine, glycine, serine, and threonine, are broken down to yield pyruvate (Fig. 24-9). Tryptophan should also be included in this group since one of its breakdown products is alanine (Section 24-3G), which, as we have seen (Section 24-1A), is transaminated to pyruvate.

Serine is converted to pyruvate through dehydration by **serine dehydratase**. This PLP–enzyme, like the aminotransferases (Section 24-1), functions by forming a PLP–amino acid Schiff base so as to facilitate the removal of the amino acid's α-hydrogen atom. In the serine dehydratase reaction, however, the C_α carbanion breaks down with the elimination of the amino acid's C_β—OH, rather than with

tautomerization (Fig. 24-2, Step 2), so that the substrate undergoes α,β elimination of H_2O rather than deamination (Fig. 24-10). The product of the dehydration, the enamine **aminoacrylate,** tautomerizes nonenzymatically to the corresponding imine, which spontaneously hydrolyzes to pyruvate and ammonia.

Cysteine may be converted to pyruvate via several routes in which the sulfhydryl group is released as H_2S, SO_3^{2-}, or SCN^-.

Glycine is converted to serine by the enzyme **serine hydroxymethyl transferase,** another PLP-containing enzyme (Fig. 24-9, Reaction 4). This enzyme utilizes N^5,N^{10}-**methylene-tetrahydrofolate (N^5,N^{10}-methylene-THF)** as a cofactor to provide the C_1 unit necessary for this conversion. We shall defer discussion of reactions involving THF cofactors until Section 24-4D.

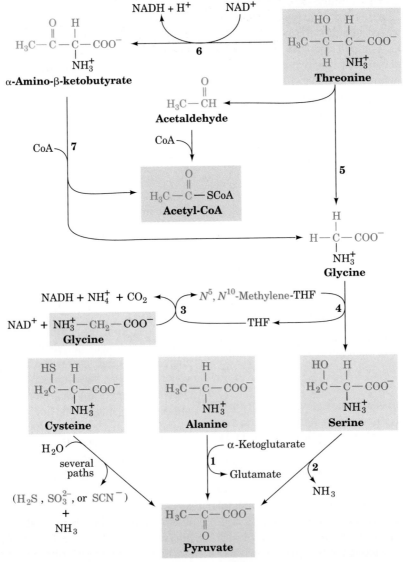

FIGURE 24-9. The pathways converting alanine, cysteine, glycine, serine, and threonine to pyruvate. The enzymes involved are **(1)** alanine aminotransferase, **(2)** serine dehydratase, **(3)** glycine cleavage system, **(4)** and **(5)** serine hydroxymethyl transferase, **(6)** threonine dehydrogenase, and **(7)** α-amino-β-ketobutyrate lyase.

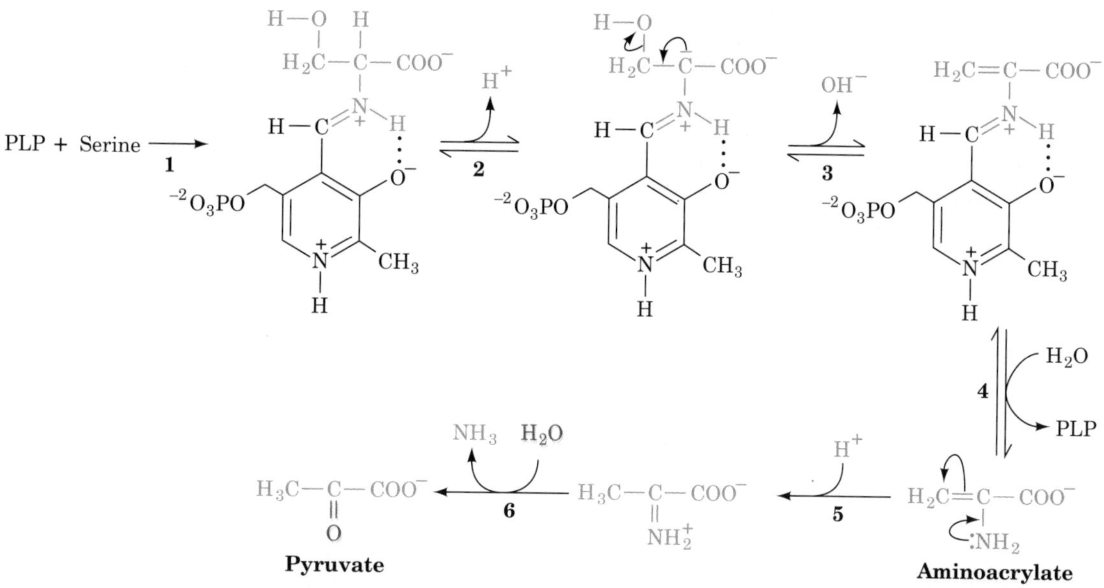

FIGURE 24-10. Serine dehydratase, a PLP-dependent enzyme, catalyzes the elimination of water from serine. The steps in the reaction are (1) formation of a serine–PLP Schiff base, (2) removal of the α-H atom of serine to form a resonance-stabilized carbanion, (3) β elimination of OH$^-$, (4) hydrolysis of the Schiff base to yield the PLP–enzyme and aminoacrylate, (5) nonenzymatic tautomerization to the imine, and (6) nonenzymatic hydrolysis to form pyruvate and ammonia.

The methylene group of the N^5,N^{10}-methylene-THF utilized in conversion of glycine to serine is obtained through a second glycine degradation (Fig. 24-9, Reaction 3) catalyzed by the **glycine cleavage system** (also called **glycine synthase** when acting in the reverse direction; Section 24-5A). The glycine cleavage system, a multienzyme complex that resembles pyruvate dehydrogenase (Section 19-2A), contains four protein components (Fig. 24-11):

1. A PLP-dependent glycine decarboxylase (P protein).

2. A lipoamide-containing aminomethyltransferase (H protein), which carries the aminomethyl group remaining after glycine decarboxylation.

3. An N^5,N^{10}-methylene-THF synthesizing enzyme (T protein), which accepts a methylene group from the aminomethyltransferase (the amino group is released as ammonia).

4. An NAD$^+$-dependent, FAD-requiring lipoamide dehydrogenase (L protein).

Two observations indicate that this pathway is the major route of glycine degradation in mammalian tissues:

1. The serine isolated from an animal that has been fed [2-^{14}C]glycine is ^{14}C labeled at both C2 and C3. This observation indicates that the methylene group of the N^5,N^{10}-methylene-THF utilized by serine hydroxymethyl transferase is derived from glycine C2.

2. The inherited human disease **nonketotic hyperglycinemia,** which is characterized by mental retardation and

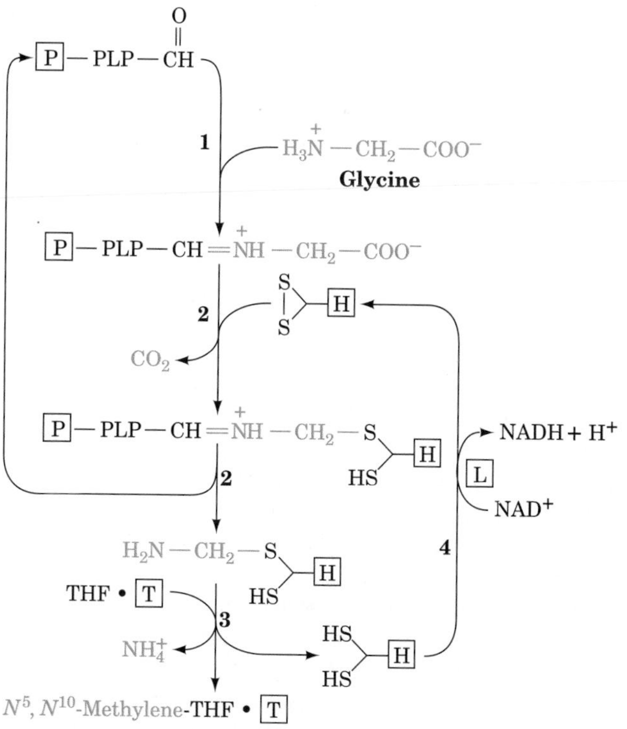

FIGURE 24-11. The reactions catalyzed by the glycine cleavage system, a multienzyme complex (also called glycine synthase). The enzymes involved are (1) a PLP-dependent glycine decarboxylase (P protein), (2) a lipoamide-containing protein (H protein), (3) a THF-requiring enzyme (T protein), and (4) an NAD$^+$-dependent, FAD-requiring lipoamide dehydrogenase (L protein).

accumulation of large amounts of glycine in body fluids, results from the absence of the glycine cleavage system.

Threonine is both glucogenic and ketogenic, since one of its degradation routes produces both pyruvate and acetyl-CoA (Fig. 24-9, Reactions 6 and 7). Its major route of breakdown is through **threonine dehydrogenase**, producing α-amino-β-ketobutyrate, which is converted to acetyl-CoA and glycine by α-amino-β-ketobutyrate lyase. The glycine may be converted, through serine, to pyruvate.

Serine Hydroxymethyl Transferase Catalyzes PLP-Dependent C_α—C_β Bond Cleavage

Threonine may also be converted directly to glycine and acetaldehyde (that latter being subsequently oxidized to acetyl-CoA), at least *in vitro*, via Reaction 5 of Fig. 24-9. Surprisingly, this reaction is catalyzed by serine hydroxymethyl transferase. We have heretofore considered PLP-catalyzed reactions that begin with the cleavage of an amino acid's C_α—H bond (Fig. 24-2). Degradation of threonine to glycine and acetaldehyde by serine hydroxymethyl transferase demonstrates that PLP also facilitates cleavage of an amino acid's C_α—C_β bond by delocalizing the electrons of the resulting carbanion into the conjugated PLP ring.

How can the same amino acid–PLP Schiff base be involved in the cleavage of the different bonds to an amino acid C_α in different enzymes? The answer to this conundrum was suggested by Harmon Dunathan. For electrons to be withdrawn into the conjugated ring system of PLP, the π-orbital system of PLP must overlap with the bonding orbital containing the electron pair being delocalized. This is possible only if the bond being broken lies in the plane perpendicular to the plane of the PLP π-orbital system (Fig. 24-12a). Different bonds to C_α can be placed in this plane by rotation about the C_α—N bond. Indeed, the X-ray structure of aspartate aminotransferase reveals that the C_α—H of its aspartate substrate assumes just this conformation (Fig. 24-12b). Evidently, *each enzyme specifically cleaves its corresponding bond because the enzyme binds the amino acid–PLP Schiff base adduct with this bond in the plane perpendicular to that of the PLP ring.* This is an example of

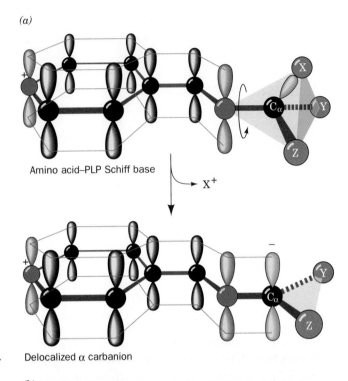

(a)

Amino acid–PLP Schiff base

Delocalized α carbanion

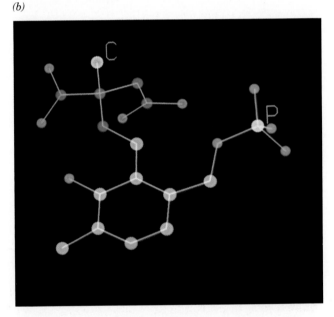

(b)

FIGURE 24-12. (*a*) The π-orbital framework of a PLP–amino acid Schiff base. The bond to C_α in the plane perpendicular to the PLP π-orbital system (from X in the illustration) is labile as a consequence of its overlap with the π system, which permits the broken bond's electron pair to be delocalized over the conjugated molecule. (*b*) The Schiff base complex of the inhibitor α-methylaspartate with pyridoxal phosphate in the X-ray structure of porcine aspartate aminotransferase. Here, the methyl C atom (*yellow ball marked C*) occupies the position of the H atom that the enzyme normally excises from aspartate. Note that the bond linking the methyl C atom to aspartate is in the plane perpendicular to the pyridoxal ring and is thus ideally oriented for bond cleavage. [Courtesy of Craig Hyde, National Institutes of Health.]

stereoelectronic assistance (Section 14-1E): *The enzyme binds substrate in a conformation that minimizes the electronic energy of the transition state.*

C. Asparagine and Aspartate Are Degraded to Oxaloacetate

Transamination of aspartate leads directly to oxaloacetate:

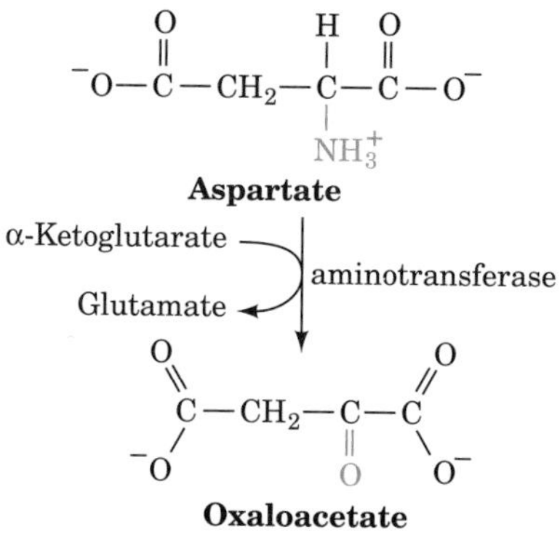

Aspartate

α-Ketoglutarate ⟶
Glutamate ⟵ | aminotransferase

Oxaloacetate

Asparagine is also converted to oxaloacetate in this manner after its hydrolysis to aspartate by **L-asparaginase**:

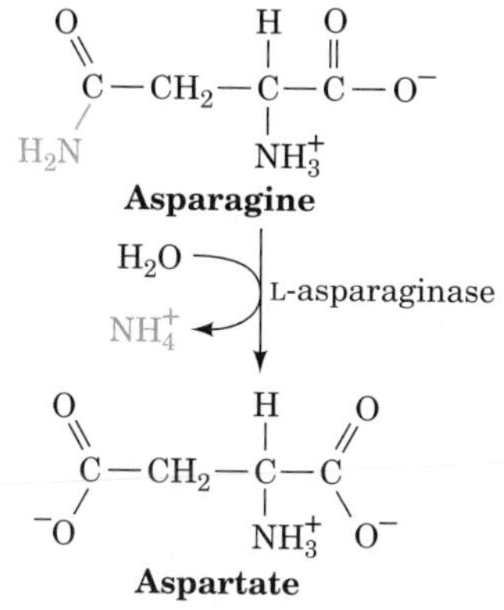

Asparagine

H_2O ⟶
NH_4^+ ⟵ | L-asparaginase

Aspartate

Interestingly, L-asparaginase is an effective chemotherapeutic agent in the treatment of cancers that must obtain asparagine from the blood, particularly **acute lymphoblastic leukemia**. It is uncertain, however, whether cell death results from the depletion of asparagine levels in the blood or from some other metabolite of the L-asparaginase reaction.

D. Arginine, Glutamate, Glutamine, Histidine, and Proline Are Degraded to α-Ketoglutarate

Arginine, glutamine, histidine, and proline are all degraded by conversion to glutamate (Fig. 24-13), which in turn is oxidized to α-ketoglutarate by glutamate dehydrogenase (Section 24-1). Conversion of glutamine to glutamate involves only one reaction: hydrolysis by **glutaminase**. Histidine's conversion to glutamate is more complicated: It is nonoxidatively deaminated, then it is hydrated, and its im-

idazole ring is cleaved to form ***N*-formiminoglutamate**. The formimino group is then transferred to tetrahydrofolate forming glutamate and ***N*^5^-formiminotetrahydrofolate** (Section 24-4D). Both arginine and proline are converted to glutamate through the intermediate formation of **glutamate-5-semialdehyde**.

E. Isoleucine, Methionine, and Valine Are Degraded to Succinyl-CoA

Isoleucine, methionine, and valine have complex degradative pathways that all yield propionyl-CoA. Propionyl-CoA, which is also a product of odd-chain fatty acid degradation, is converted, as we have seen, to succinyl-CoA by a series of reactions involving the participation of biotin and coenzyme B_{12} (Section 23-2E).

Methionine Breakdown Involves Synthesis of *S*-Adenosylmethionine and Cysteine

Methionine degradation (Fig. 24-14) begins with its reaction with ATP to form ***S*-adenosylmethionine (SAM;** alternatively **AdoMet).** *This sulfonium ion's highly reactive methyl group makes it an important biological methylating agent.* For instance, we have already seen that SAM is the methyl donor in the synthesis of phosphatidylcholine from phosphatidylethanolamine (Section 23-8A). It is also the methyl donor in the conversion of norepinephrine to epinephrine (Section 24-4B).

Methylation reactions involving SAM yield ***S*-adenosylhomocysteine** in addition to the methylated acceptor. The former product is hydrolyzed to adenosine and **homocysteine** in the next reaction of the methionine degradation pathway. The homocysteine may be methylated to form methionine via a reaction in which N^5-**methyl-THF** is the methyl donor. Alternatively, the homocysteine may combine with serine to yield **cystathionine,** which subsequently forms cysteine (cysteine biosynthesis) and **α-ketobutyrate.** The α-ketobutyrate continues along the degradative pathway to propionyl-CoA and then succinyl-CoA.

Branched-Chain Amino Acid Degradation Pathways Contain Themes Common to All Acyl-CoA Oxidations

Degradation of the branched-chain amino acids isoleucine, leucine, and valine begins with three reactions that employ common enzymes (Fig. 24-15, *top*): (1) transamination to the corresponding α-keto acid, (2) oxidative decarboxylation to the corresponding acyl-CoA, and (3) dehydrogenation by FAD to form a double bond.

The remainder of the isoleucine degradation pathway (Fig. 24-15, *left*) is identical to that of fatty acid oxidation (Section 23-2C): (4) double-bond hydration, (5) dehydrogenation by NAD^+, and (6) thiolytic cleavage yielding acetyl-CoA and propionyl-CoA, which is subsequently converted to succinyl-CoA. Valine degradation is a variation on this theme (Fig. 24-15, *center*): Following (7) double-bond hydration, (8) the CoA thioester bond is hydrolyzed before (9) the second dehydrogenation reaction. The thioester bond is then regenerated as propionyl-CoA in the sequence's last reaction (10), an oxidative decarboxylation rather than a thiolytic cleavage.

Maple Syrup Urine Disease Results from a Defect in Branched-Chain Amino Acid Degradation

Branched chain α-keto acid dehydrogenase (BCKDH; also known as α-ketoisovalerate dehydrogenase), which catalyzes Reaction 2 of branched-chain amino acid degradation (Fig. 24-15), is a multienzyme complex that closely resembles the pyruvate dehydrogenase and α-ketoglutarate

dehydrogenase multienzyme complexes (Sections 19-2A and 19-3D). Indeed, all three of these multienzyme complexes share a common protein component, E$_3$ (dihydrolipoamide dehydrogenase), and employ the coenzymes TPP, lipoamide, and FAD in addition to their terminal oxidizing agent, NAD$^+$.

A genetic deficiency in BCKDH causes **maple syrup urine disease,** so named because the consequent buildup of branched-chain α-keto acids imparts the urine with the characteristic odor of maple syrup. Unless promptly treated by a diet low in branched-chain amino acids, maple syrup urine disease is rapidly fatal.

FIGURE 24-13. Degradation pathways of arginine, glutamate, glutamine, histidine, and proline to α-ketoglutarate. The enzymes catalyzing the reactions are (1) glutamate dehydrogenase, (2) glutaminase, (3) arginase, (4) ornithine-δ-aminotransferase, (5) glutamate semialdehyde dehydrogenase, (6) proline oxidase, (7) spontaneous, (8) histidine ammonia lyase, (9) urocanate hydratase, (10) imidazolone propionase, and (11) glutamate formimino transferase.

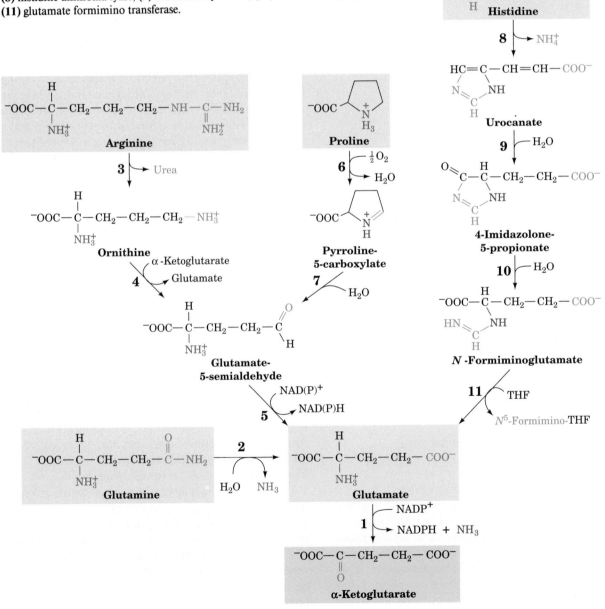

F. Leucine and Lysine Are Degraded to Acetoacetate and/or Acetyl-CoA

Leucine is oxidized by a combination of reactions used in β oxidation and ketone body synthesis (Fig. 24-15, *right*). The first dehydrogenation and the hydration reactions are interspersed by (11) a carboxylation reaction catalyzed by a biotin-containing enzyme. The hydration reaction (12) then produces β-**hydroxy-β-methylglutaryl-CoA (HMG-**

CoA), which is cleaved by HMG-CoA lyase to form acetyl-CoA and the ketone body acetoacetate (13) (which, in turn, may be converted to 2 acetyl-CoA; Section 23-3).

Although there are several pathways for lysine degradation, the one that proceeds via formation of the α-ketoglutarate–lysine adduct **saccharopine** predominates in

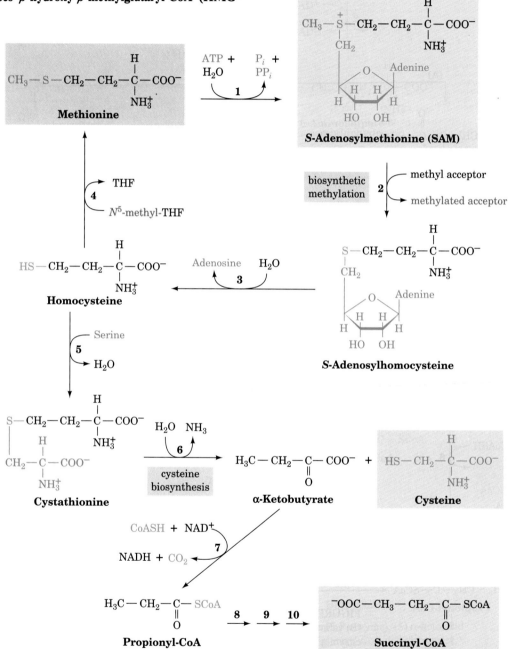

FIGURE 24-14. The pathway of methionine degradation, yielding cysteine and succinyl-CoA as products. The enzymes involved in the pathway are **(1)** methionine adenosyl transferase in a reaction that yields the biological methylating agent *S*-adenosylmethionine (SAM), **(2)** methylase, **(3)** adenosylhomocysteinase, **(4)** homocysteine methyltransferase (a coenzyme B_{12}-dependent enzyme), **(5)** cystathionine β-synthase (a PLP-dependent enzyme), **(6)** cystathionine γ-lyase (a PLP-dependent enzyme), **(7)** α-keto acid dehydrogenase, **(8)** propionyl-CoA carboxylase, **(9)** methylmalonyl-CoA racemase, and **(10)** methylmalonyl-CoA mutase (a coenzyme B_{12}-dependent enzyme; Reactions 8–10 are discussed in Section 23-2E).

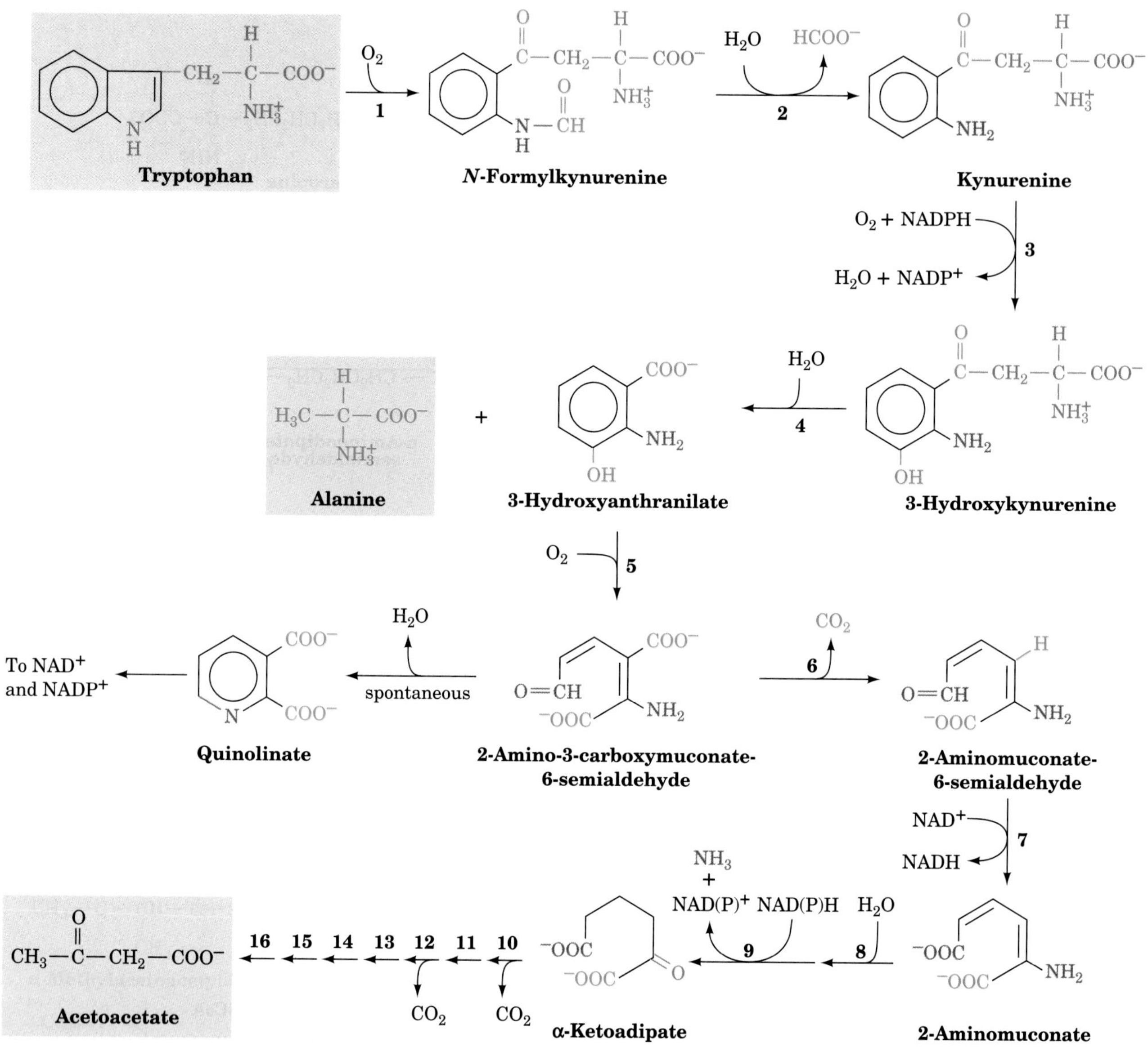

FIGURE 24-17. The pathway of tryptophan degradation. The enzymes involved are **(1)** tryptophan-2,3-dioxygenase, **(2)** formamidase, **(3)** kynurenine-3-monooxygenase, **(4)** kynureninase (PLP dependent), **(5)** 3-hydroxyanthranilate-3,4-dioxygenase, **(6)** amino carboxymuconate semialdehyde decarboxylase, **(7)** aminomuconate semialdehyde dehydrogenase, **(8)** hydratase, **(9)** dehydrogenase, **(10–16)** Reactions 5 through 11 in lysine degradation (Fig. 24-16). 2-Amino-3-carboxymuconate-6-semialdehyde, in addition to undergoing Reaction 6, spontaneously forms **quinolinate**, an NAD^+ and $NADP^+$ precursor (Section 26-6A).

19-2A and 19-3D). Reactions 6, 8, and 9 are standard reactions of fatty acyl-CoA oxidation: dehydrogenation by FAD, hydration, and dehydrogenation by NAD^+. Reactions 10 and 11 are standard reactions in ketone body formation. Two moles of CO_2 are produced at Reactions 5 and 7 of the pathway.

The saccharopine pathway is thought to predominate in mammals because a genetic defect in the enzyme that catalyzes Reaction 1 in the sequence results in **hyperlysinemia** and **hyperlysinuria** (elevated levels of lysine in the blood and urine, respectively) along with mental and physical retardation. This is yet another example of how the study of rare inherited disorders has helped to trace metabolic pathways.

Leucine's carbon skeleton, as we have seen, is converted to one molecule each of acetoacetate and acetyl-CoA, whereas that of lysine is converted to one molecule of acetoacetate and two of CO_2. Since neither acetoacetate nor acetyl-CoA can be converted to glucose in animals, leucine and lysine are purely ketogenic amino acids.

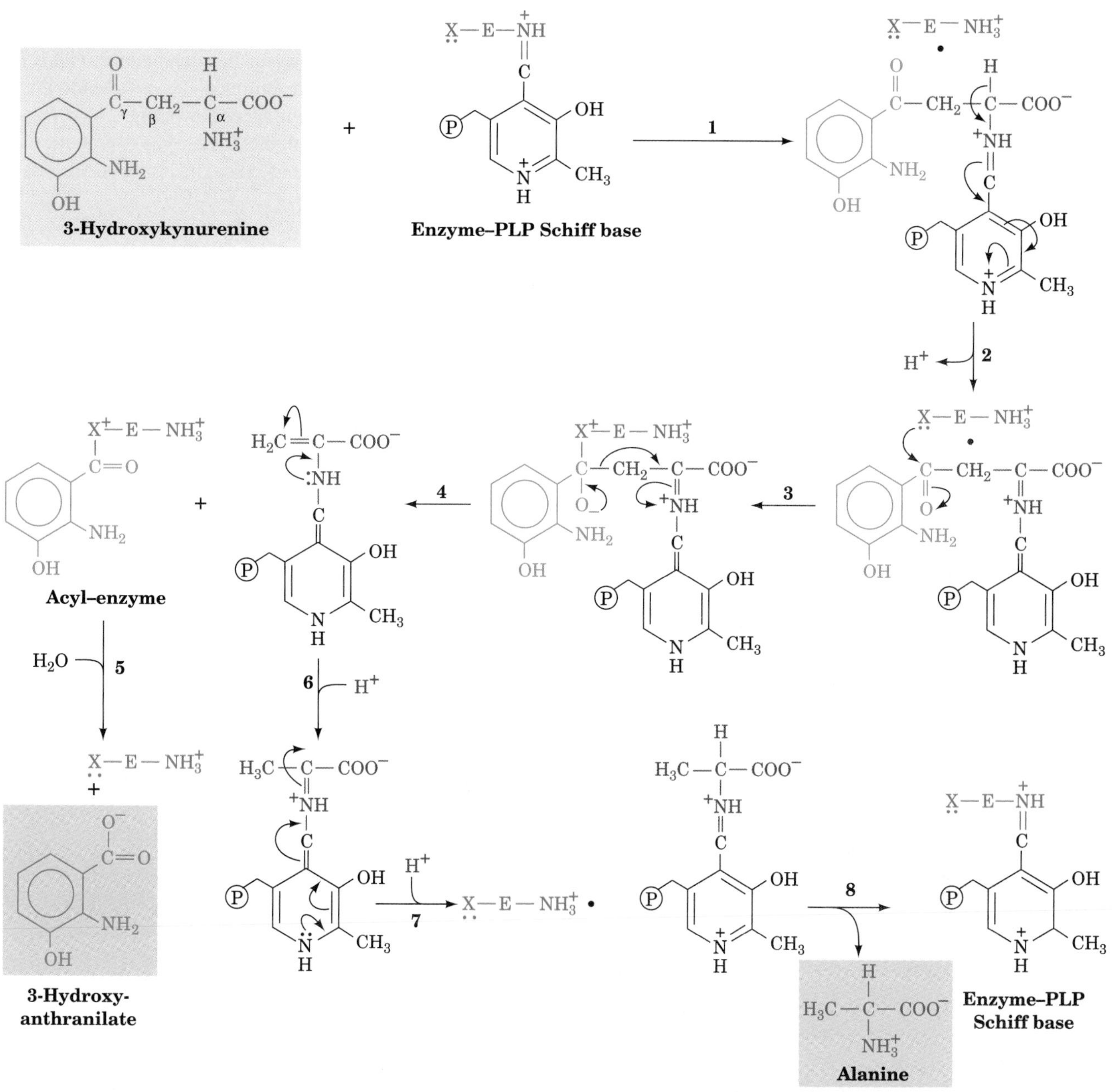

FIGURE 24-18. The proposed mechanism for the PLP-dependent kynureninase-catalyzed C_β—C_γ bond cleavage of 3-hydroxykynurenine occurs in eight steps: **(1)** transimination, **(2)** tautomerization, **(3)** attack of an enzyme nucleophile,

(4) C_β—C_γ bond cleavage with formation of an acyl–enzyme intermediate, **(5)** acyl–enzyme hydrolysis, **(6)** and **(7)** tautomerization, and **(8)** transimination.

G. Tryptophan Is Degraded to Alanine and Acetoacetate

The complexity of the major tryptophan degradation pathway (Fig. 24-17) precludes the detailed discussion of all of its reactions. However, one reaction in the pathway is of particular interest. Reaction 4, cleavage of **3-hydroxykynurenine** to alanine and **3-hydroxyanthranilate,** is catalyzed by **kynureninase,** a PLP-dependent enzyme. The reaction further demonstrates the enormous versatility of PLP. We

have seen how PLP can labilize an α-amino acid's C_α—H bond. Here we see the facilitation of C_β—C_γ bond cleavage. The reaction follows the same steps as transimination reactions but does not hydrolyze the tautomerized Schiff base (Fig. 24-18). The proposed reaction mechanism involves an attack of an enzyme nucleophile on the carbonyl carbon (C_γ) of the tautomerized 3-hydroxykynurenine–PLP Schiff base (Fig. 24-18, Step 3). This is followed by C_β—C_γ bond cleavage to generate an acyl–enzyme intermediate together with a tautomerized alanine–PLP adduct (Fig. 24-

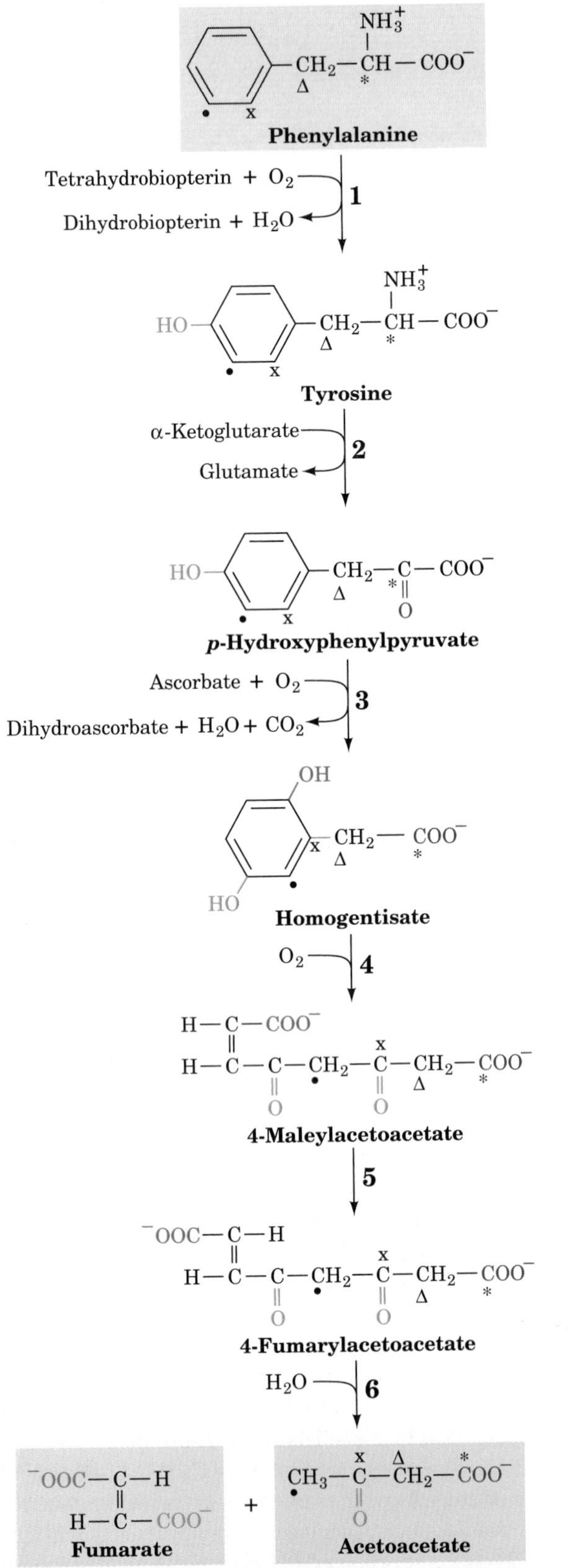

NH$_3^+$

CH$_2$—CH—COO$^-$

Phenylalanine

Tetrahydrobiopterin + O$_2$ ⟶ **1**

Dihydrobiopterin + H$_2$O ⟵

NH$_3^+$

HO— CH$_2$—CH—COO$^-$

Tyrosine

α-Ketoglutarate ⟶ **2**

Glutamate ⟵

HO— CH$_2$—C—COO$^-$ ‖ O

***p*-Hydroxyphenylpyruvate**

Ascorbate + O$_2$ ⟶ **3**

Dihydroascorbate + H$_2$O + CO$_2$ ⟵

OH

CH$_2$—COO$^-$

HO

Homogentisate

O$_2$ ⟶ **4**

$$H-C-COO^-$$
$$\|$$
$$H-C-C-CH_2-C-CH_2-COO^-$$
$$\| \quad\quad \|$$
$$O \quad\quad O$$

4-Maleylacetoacetate

5

$$^-OOC-C-H$$
$$\|$$
$$H-C-C-CH_2-C-CH_2-COO^-$$
$$\| \quad\quad \|$$
$$O \quad\quad O$$

4-Fumarylacetoacetate

H$_2$O ⟶ **6**

$$^-OOC-C-H$$
$$\|$$
$$H-C-COO^-$$
Fumarate

+

$$CH_3-C-CH_2-COO^-$$
$$\|$$
$$O$$
Acetoacetate

FIGURE 24-19. The pathway of phenylalanine degradation. The enzymes involved are **(1)** phenylalanine hydroxylase, **(2)** aminotransferase, **(3)** *p*-hydroxyphenylpyruvate dioxygenase, **(4)** homogentisate dioxygenase, **(5)** maleylacetoacetate isomerase, and **(6)** fumarylacetoacetase. The symbols labeling the various carbon atoms serve to indicate the group migration that occurs in Reaction 3 of the pathway (see Fig. 24-23).

18, Step 4). Hydrolysis of the acyl–enzyme then yields 3-hydroxyanthranilate, whose further degradation yields **α-ketoadipate** (Fig. 24-17, Reactions 4–9). α-Ketoadipate is also an intermediate in lysine breakdown (Fig. 24-16, Reaction 4) so that the last seven reactions in the degradation of both these amino acids are identical, forming acetoacetate and two molecules of CO$_2$.

H. Phenylalanine and Tyrosine Are Degraded to Fumarate and Acetoacetate

Since the first reaction in phenylalanine degradation is its hydroxylation to tyrosine, a single pathway (Fig. 24-19) is responsible for the breakdown of both of these amino acids. The final products of the six-reaction degradation are fumarate, a citric acid cycle intermediate, and acetoacetate, a ketone body.

Pterins Are Redox Cofactors

The hydroxylation of phenylalanine by the Fe(III)-containing enzyme **phenylalanine hydroxylase** requires the participation of a cofactor we have not yet encountered: **biopterin**, a **pterin** derivative. Pterins are compounds that contain the **pteridine** ring (Fig. 24-20). Note the resemblance between the pteridine ring and the isoalloxazine ring of the flavin coenzymes; the positions of the nitrogen atoms

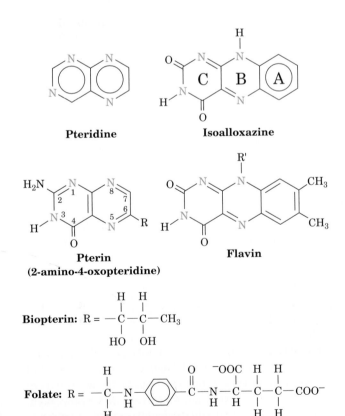

Pteridine

Isoalloxazine

Pterin (2-amino-4-oxopteridine)

Flavin

Biopterin: R = —C—C—CH$_3$ (with H, H above and HO, OH below)

Folate: R = —C—N— ... —C—N—C—C—C—COO$^-$

FIGURE 24-20. The pteridine ring is the nucleus of biopterin and folate. Note the similar structures of pteridine and the isoalloxazine ring of flavin coenzymes.

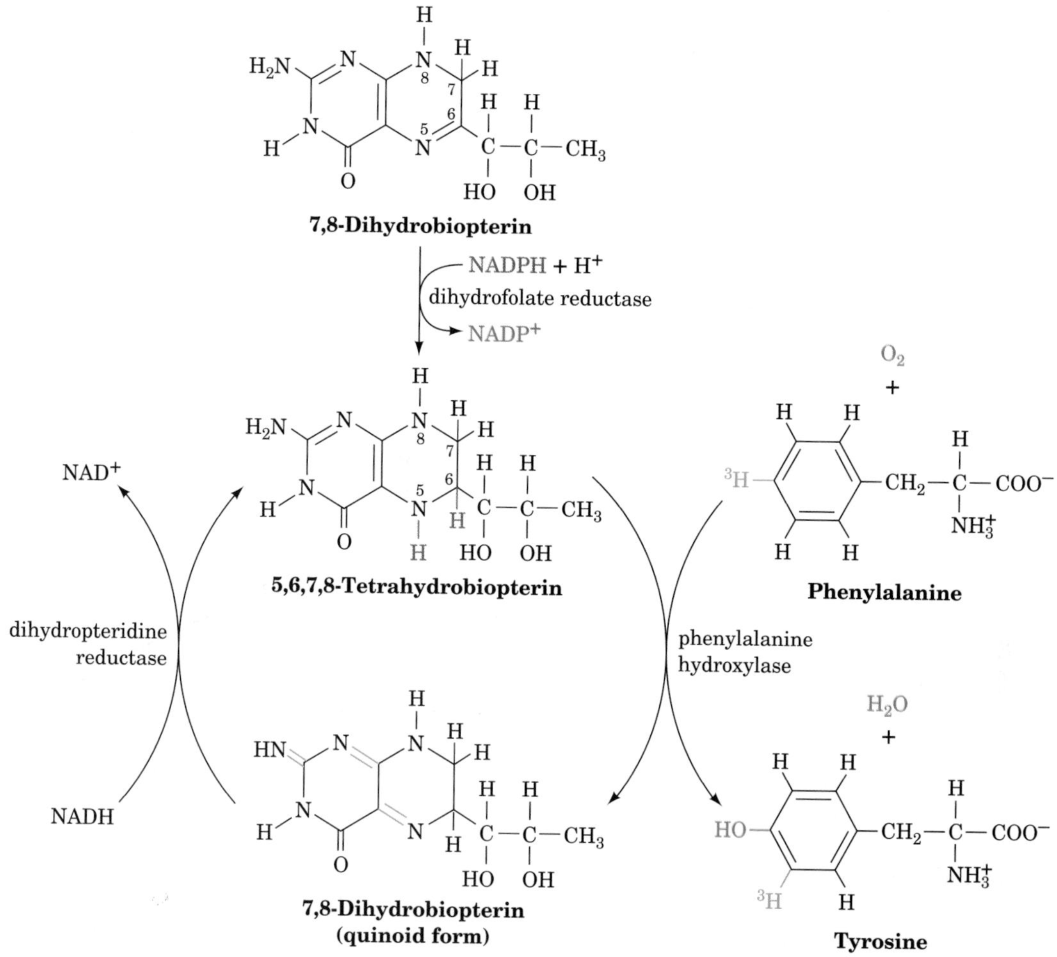

FIGURE 24-21. The formation, utilization, and regeneration of 5,6,7,8-tetrahydrobiopterin in the phenylalanine hydroxylase reaction.

in pteridine are identical with those of the B and C rings of isoalloxazine. Folate derivatives also contain the pterin ring (Section 24-4D).

Pterins, like flavins, participate in biological oxidations. The active form of biopterin is the fully reduced form, **5,6,7,8-tetrahydrobiopterin.** It is produced from **7,8-dihydrobiopterin** and NADPH, in what may be considered a priming reaction, by **dihydrofolate reductase** (Fig. 24-21). In the phenylalanine hydroxylase reaction, 5,6,7,8-tetrahydrobiopterin is oxidized to **7,8-dihydrobiopterin (quinoid form).** This quinoid is subsequently reduced by the NADH-requiring enzyme **dihydropteridine reductase** to regenerate the active cofactor. Note that although dihydrofolate reductase and dihydropteridine reductase produce the same product, they utilize different tautomers of the substrate. Although this suggests that these enzymes may be evolutionarily related, the comparison of their X-ray structures indicates that they are unrelated. Instead, dihydropteridine reductase resembles nicotinamide coenzyme-requiring flavin-dependent enzymes such as glutathione reductase (Section 14-4).

The NIH Shift

An unexpected aspect of the phenylalanine hydroxylase reaction is that a ³H atom, which begins on C4 of phenylalanine's phenyl ring, ends up on C3 of this ring in tyrosine (Fig. 24-21, *right*) rather than being lost to the solvent by replacement with the OH group. The mechanism postulated to account for this **NIH shift** (so-called because it was first characterized by chemists at the National Institutes of Health) involves the activation of oxygen to form an epoxide across the phenyl ring's 3,4 bond, followed by epoxide opening to form a carbocation at C3 (Fig. 24-22). Migration of a hydride from C4 to C3 forms a more stable carbocation (an oxonium ion). This migration is followed by ring aromatization to form tyrosine.

Reaction 3 (Fig. 24-19) in the phenylalanine degradation pathway provides a second example of an NIH shift. This reaction, which is catalyzed by the Cu-containing *p*-hydroxyphenylpyruvate dioxygenase, involves the oxidative decarboxylation of an α-keto acid as well as ring hydroxylation. In this case, the NIH shift involves migration of an alkyl group rather than of a hydride ion to form a more

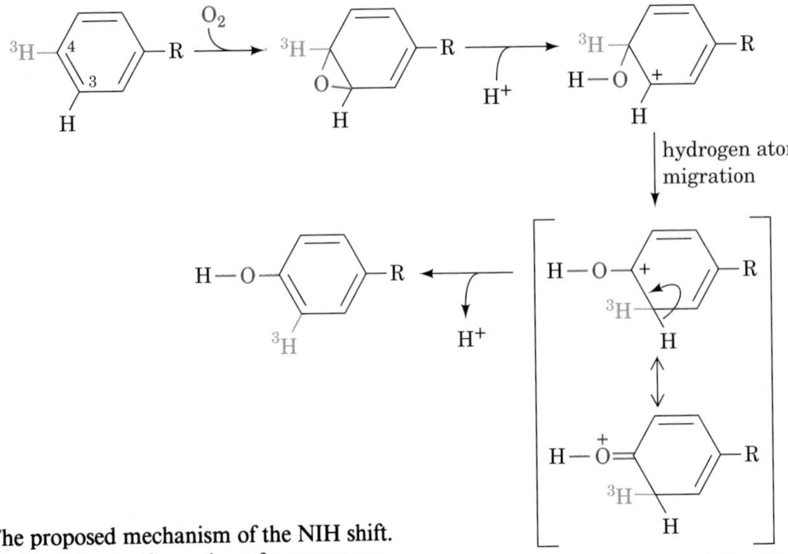

FIGURE 24-22. The proposed mechanism of the NIH shift. The rearrangement is driven by the formation of a resonance-stabilized oxonium ion.

stable carbocation (Fig. 24-23). This shift, which has been demonstrated through isotope-labeling studies (represented by the different symbols in Figs. 24-19 and 24-23), accounts for the observation that C3 is bonded to C4 in *p*-hydroxyphenylpyruvate but to C5 in **homogentisate.**

Alkaptonuria and Phenylketonuria Result from Defects in Phenylalanine Degradation

Archibald Garrod realized in the early 1900s that human genetic diseases result from specific enzyme deficiencies. We have repeatedly seen how this realization has contributed to the elucidation of metabolic pathways. The first such disease to be recognized was **alkaptonuria,** which, Garrod observed, resulted in the excretion of large quantities of homogentisic acid. This condition results from deficiency of **homogentisate dioxygenase** (Fig. 24-19, Reaction 4). Alkaptonurics suffer no ill effects other than arthritis later in life (although their urine darkens alarmingly because of the rapid air oxidation of the homogentisate they excrete).

Individuals suffering from **phenylketonuria (PKU)** are not so fortunate. Severe mental retardation occurs within a few months of birth if the disease is not detected and treated immediately (see below). Indeed, ~1% of the patients in mental institutions were, at one time (before routine screening), phenylketonurics. PKU is caused by the inability to hydroxylate phenylalanine (Fig. 24-19, Reaction 1) and therefore results in increased blood levels of phenylalanine **(hyperphenylalaninemia).** The excess phenylalanine is transaminated to **phenylpyruvate**

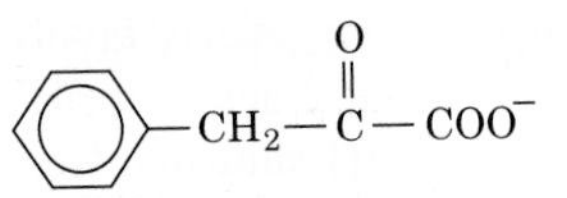

by an otherwise minor pathway. The "spillover" of phenylpyruvate (a phenylketone) into the urine was the first observation connected with the disease and gave the disease its name. All babies born in the United States are now screened for PKU immediately after birth by testing for elevated levels of phenylalanine in the blood.

Classical PKU results from a deficiency in phenylalanine hydroxylase. When this was established in 1947, it was the first human inborn error of metabolism whose basic biochemical defect had been identified. Since all of the tyrosine breakdown enzymes are normal, treatment consists in providing the patient with a low-phenylalanine diet and monitoring the blood level of phenylalanine to ensure that it remains within normal limits for the first 5 to 10 years of life (the adverse effects of hyperphenylalaninemia seem to disappear after that age). Phenylalanine hydroxylase deficiency also accounts for another common symptom of PKU: Its victims have lighter hair and skin color than their siblings. This is because tyrosine hydroxylation, the first reaction in the formation of the black skin pigment **melanin** (Section 24-4B), is inhibited by elevated phenylalanine levels.

Other causes of hyperphenylalaninemia have been discovered since the introduction of infant screening techniques. These are caused by deficiencies in the enzymes catalyzing the formation or regeneration of 5,6,7,8-tetrahydrobiopterin, the phenylalanine hydroxylase cofactor (Fig. 24-21). In such cases, patients must also be supplied with L-3,4-dihydroxyphenylalanine (L-DOPA) and 5-hydroxytryptophan, metabolic precursors of the neurotransmitters **norepinephrine** and **serotonin,** respectively, since the enzymes that produce these physiologically active amines also require 5,6,7,8-tetrahydrobiopterin (Section 24-4B). Simply adding 5,6,7,8-tetrahydrobiopterin to the diet of affected individuals is not sufficient because this compound is unstable and cannot cross the blood–brain barrier.

Phenylpyruvate

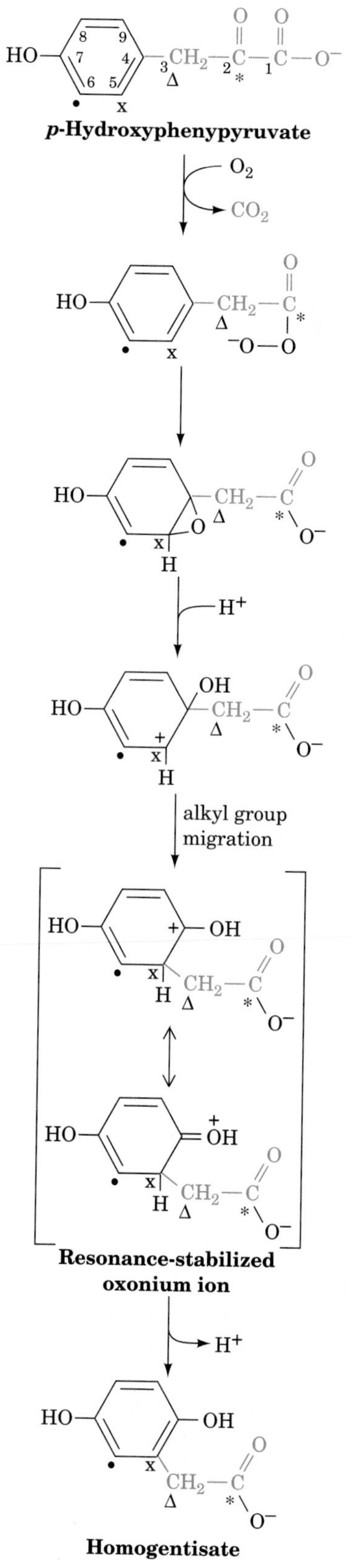

FIGURE 24-23. The NIH shift in the *p*-hydroxyphenyl-pyruvate dioxygenase reaction. Carbon atoms are labeled as an aid to following the group migration constituting the shift.

4. AMINO ACIDS AS BIOSYNTHETIC PRECURSORS

Certain amino acids, in addition to their major function as protein building blocks, are essential precursors of a variety of important biomolecules, including nucleotides and nucleotide coenzymes, heme, various hormones and neurotransmitters, and glutathione. In this section, we therefore consider the pathways producing some of these substances. We begin by discussing the biosynthesis of heme from glycine and succinyl-CoA. We then examine the pathways by which tyrosine, tryptophan, glutamate, and histidine are converted to various neurotransmitters and study certain aspects of glutathione biosynthesis and the involvement of this tripeptide in amino acid transport and other processes. Finally, we consider the role of folate derivatives in the biosynthetic transfer of C_1 units. The biosynthesis of nucleotides and nucleotide coenzymes is the subject of Chapter 26.

A. Heme Biosynthesis and Degradation

Heme (Fig. 24-24), as we have seen, is an Fe-containing prosthetic group that is an essential component of many proteins, notably hemoglobin, myoglobin, and the cytochromes. The initial reactions of heme biosynthesis are common to the formation of other tetrapyrroles including

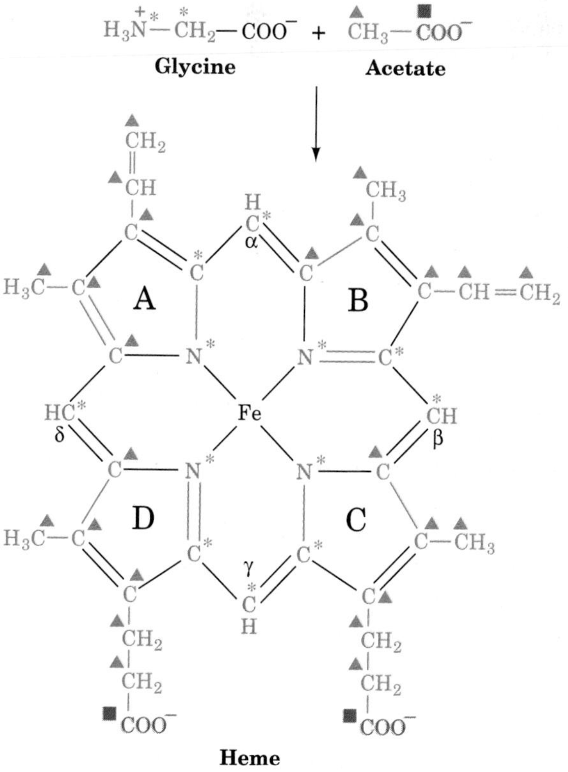

FIGURE 24-24. Heme's C and N atoms are derived from those of glycine and acetate.

chlorophyll in plants and bacteria (Section 22-1A) and coenzyme B_{12} in bacteria (Section 23-2E).

Porphyrins Are Derived from Succinyl-CoA and Glycine

Elucidation of the heme biosynthesis pathway involved some interesting detective work. David Shemin and David Rittenberg, who were among the first to use isotopic tracers in the elucidation of metabolic pathways, demonstrated, in 1945, that *all of heme's C and N atoms can be derived from acetate and glycine*. Only glycine, out of a variety of ^{15}N-labeled metabolites they tested (including ammonia, glutamate, leucine, and proline) yielded ^{15}N-labeled heme in the hemoglobin of experimental subjects to whom these metabolites were administered. Similar experiments, using acetate labeled with ^{14}C in its methyl or carboxyl groups, or [$^{14}C_\alpha$]glycine, demonstrated that 24 of heme's 34 carbon atoms are derived from acetate's methyl carbon, 2 from acetate's carboxyl carbon, and 8 from glycine's C_α atom (Fig. 24-24). None of the heme atoms is derived from glycine's carboxyl carbon atom.

Figure 24-24 indicates that heme C atoms derived from acetate methyl groups occur in groups of three linked atoms. Evidently, acetate is first converted to some other metabolite that has this labeling pattern. Shemin and Rittenbenberg postulated that this metabolite is succinyl-CoA based on the following reasoning (Fig. 24-25):

1. Acetate is metabolized via the citric acid cycle (Section 19-3 I).

2. Labeling studies indicate that atom C3 of the citric acid cycle intermediate succinyl-CoA is derived from acetate's methyl C atom, whereas atom C4 comes from acetate's carboxyl C atom.

3. After many turns of the citric acid cycle, C1 and C2 of succinyl-CoA likewise become fully derived from acetate's methyl C atom.

We shall see that this labeling pattern indeed leads to that of heme.

The first phase of heme biosynthesis is a condensation of succinyl-CoA with glycine followed by decarboxylation to form **δ-aminolevulinic acid (ALA)** as catalyzed by the PLP-dependent enzyme **δ-aminolevulinate synthase** (Fig. 24-26). The carboxyl group lost in the decarboxylation (Fig. 24-26, Reaction 5) originates in glycine, which is why heme contains no label from this group.

The Pyrrole Ring Is the Product of Two ALA Molecules

The pyrrole ring is formed in the next phase of the pathway through linkage of two molecules of ALA to yield **porphobilinogen (PBG)**. The reaction is catalyzed by the Zn-requiring enzyme **porphobilinogen synthase** (alternatively, **δ-aminolevulinic acid dehydratase**) and involves Schiff base formation of one of the substrate molecules with an enzyme amine group. One possible mechanism of this condensation–elimination reaction involves formation of

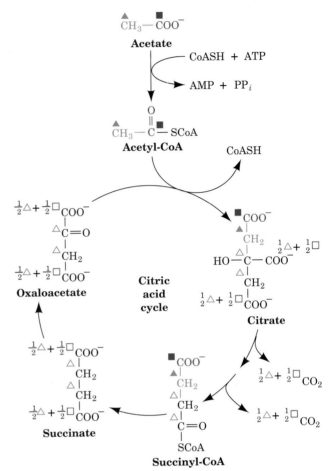

FIGURE 24-25. The origin of the C atoms of succinyl-CoA as derived from acetate via the citric acid cycle. C atoms labeled with triangles and squares are derived, respectively, from acetate's methyl and carboxyl C atoms. Filled symbols label atoms derived from acetate in the present round of the citric acid cycle, whereas open symbols label atoms derived from acetate in previous rounds of the citric acid cycle. Note that the C1 and C4 atoms of succinyl-CoA are scrambled on forming the twofold symmetric succinate.

a second Schiff base between the ALA–enzyme Schiff base and the second ALA molecule (Fig. 24-27). At this point, if we continue tracing the acetate and glycine labels through the PBG synthase reaction (Fig. 24-27), we can begin to see how heme's labeling pattern arises.

Inhibition of PBG synthase by lead is one of the major manifestations of acute lead poisoning. Indeed, it has been suggested that the accumulation, in the blood, of ALA, which resembles the neurotransmitter γ-aminobutyric acid (Section 24-4B), is responsible for the psychosis that often accompanies lead poisoning.

The Porphyrin Ring Is Formed from Four PBG Molecules

The next phase of heme biosynthesis is the condensation of four PBG molecules to form **uroporphyrinogen III**, the porphyrin nucleus, in a series of reactions catalyzed by **por-**

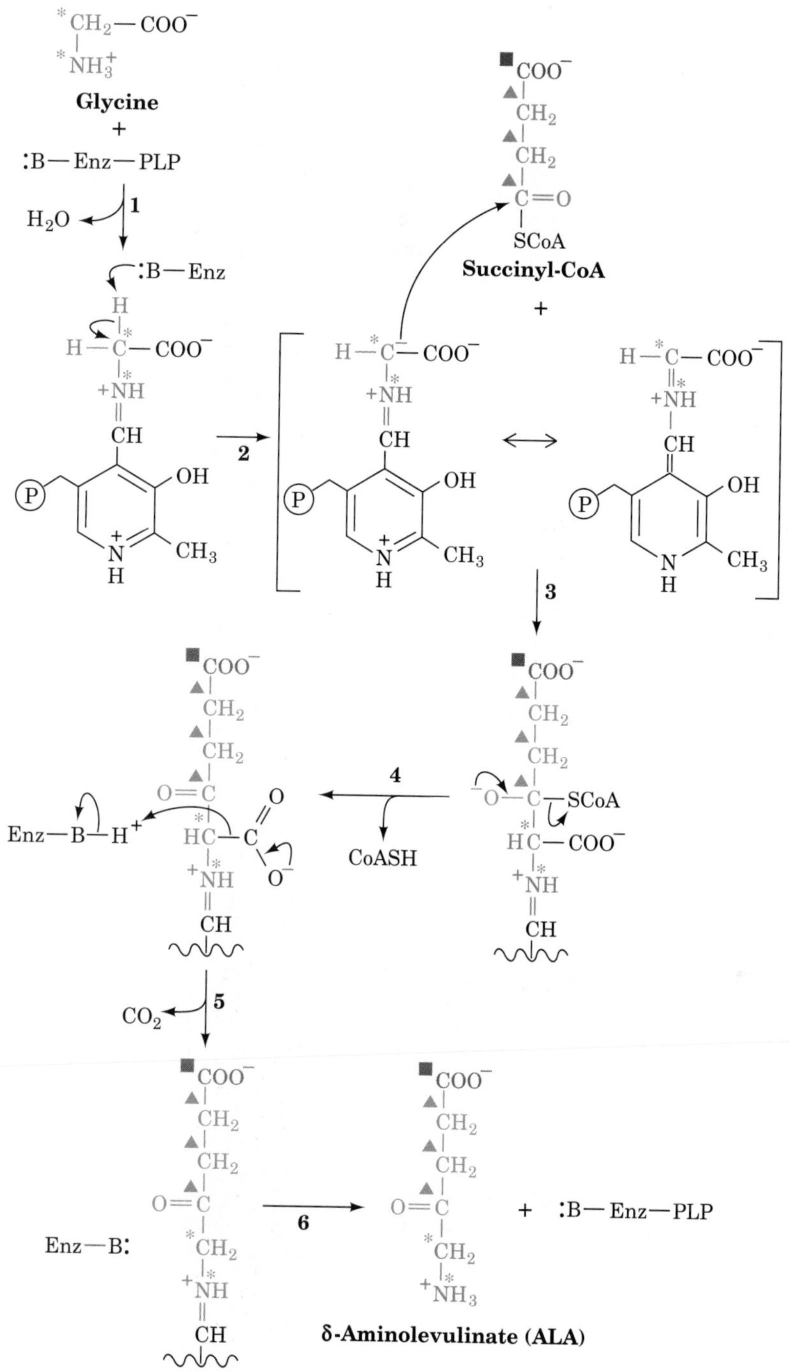

FIGURE 24-26. The mechanism of action of the PLP-dependent enzyme, δ-aminolevulinate synthase. The reaction steps are **(1)** transimination, **(2)** PLP-stabilized carbanion formation, **(3)** C—C bond formation, **(4)** CoA elimination, **(5)** decarboxylation facilitated by the PLP–Schiff base, and **(6)** transimination yielding ALA and regenerating the PLP–enzyme.

phobilinogen deaminase (alternatively, **hydroxymethyl-bilane synthase** or **uroporphyrinogen synthase**) and **uropor-phyrinogen III cosynthase.** The reaction (Fig. 24-28) begins with the enzyme's displacement of the amino group in PBG to form a covalent adduct. A second, third, and fourth PBG then sequentially add through the displacement of the primary amino group on one PBG by a carbon atom on the pyrrole ring of the succeeding PBG to yield the linear tetra-pyrrole **hydroxymethylbilane** (also called **preuroporphyrin-ogen**).

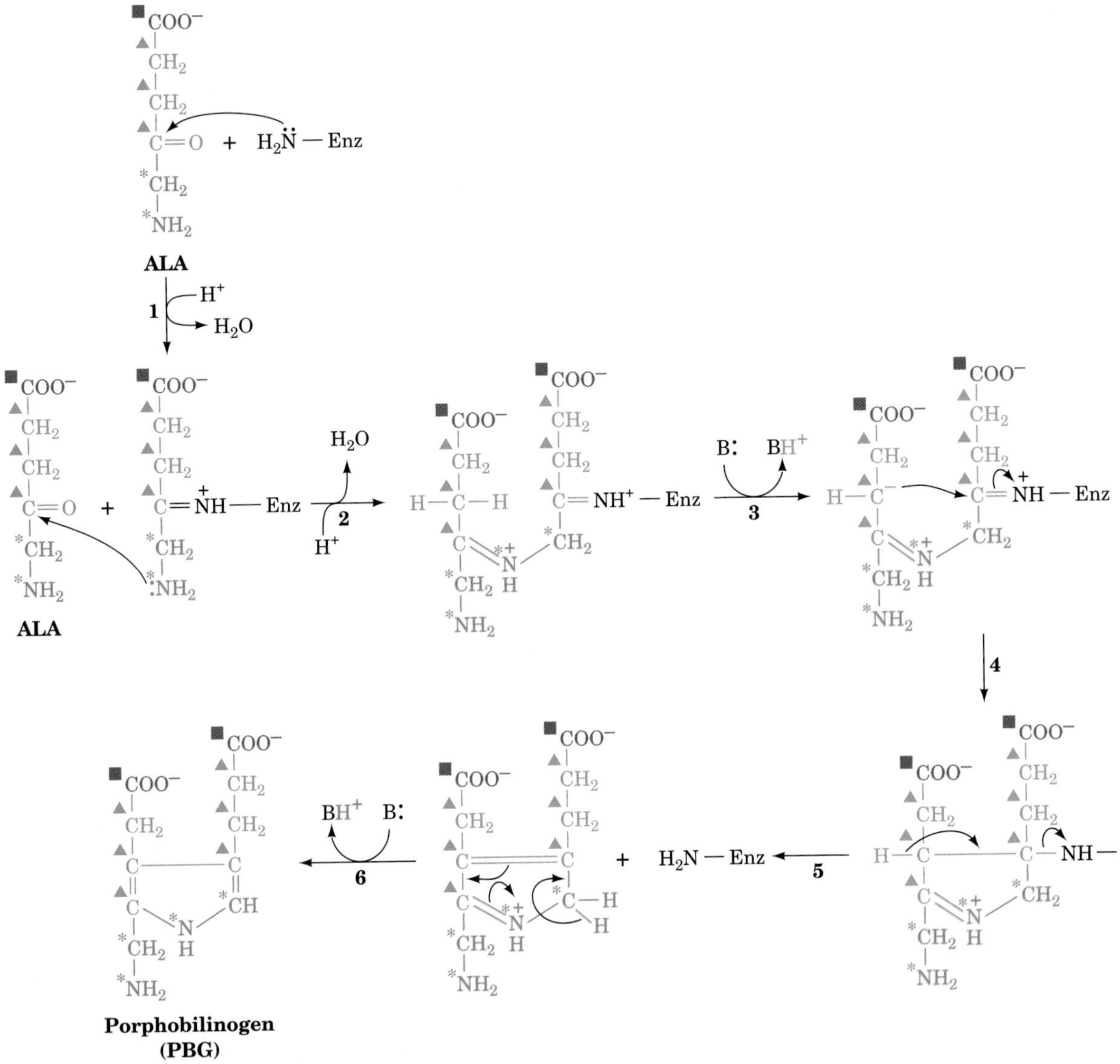

FIGURE 24-27. A possible mechanism for porphobilinogen synthase. The reaction involves:
(1) Schiff base formation, **(2)** second Schiff base formation, **(3)** formation of a carbanion α to
a Schiff base, **(4)** cyclization by an aldol-type condensation, **(5)** elimination of the enzyme NH_2
group, and **(6)** tautomerization.

Porphobilinogen Deaminase Has a Dipyrromethane Cofactor

Porphobilinogen deaminase contains a unique **dipyrromethane** cofactor (two pyrroles linked by a methylene bridge; rings C_1 and C_2 in Figure 24-28), which is covalently linked to the enzyme via a C—S bond to an enzyme Cys residue. Thus, the methylbilane–enzyme complex really contains a linear hexapyrrole. The subsequent reaction step, also catalyzed by porphobilinogen deaminase (Step 5 in Figure 24-28), is the hydrolysis of the bond linking the second and third pyrrole units of the hexapyrrole to yield methylbilane and the dipyrromethane cofactor. This latter

is still linked to the enzyme, which is therefore ready to catalyze a new round of methylbilane synthesis.

How is the dipyrromethane cofactor synthesized? It turns out that porphobilinogen deaminase synthesizes its own cofactor from two PBG units using, it appears, the same catalytic machinery with which it synthesizes methylbilane. The cofactor remains associated with the enzyme rather than being incorporated into product as ^{14}C-labeling studies have shown.

The X-ray structure of *E. coli* porphobilinogen deaminase (whose sequence is >45% identical to those of mammalian enzymes) in covalent complex with its dipyrro-

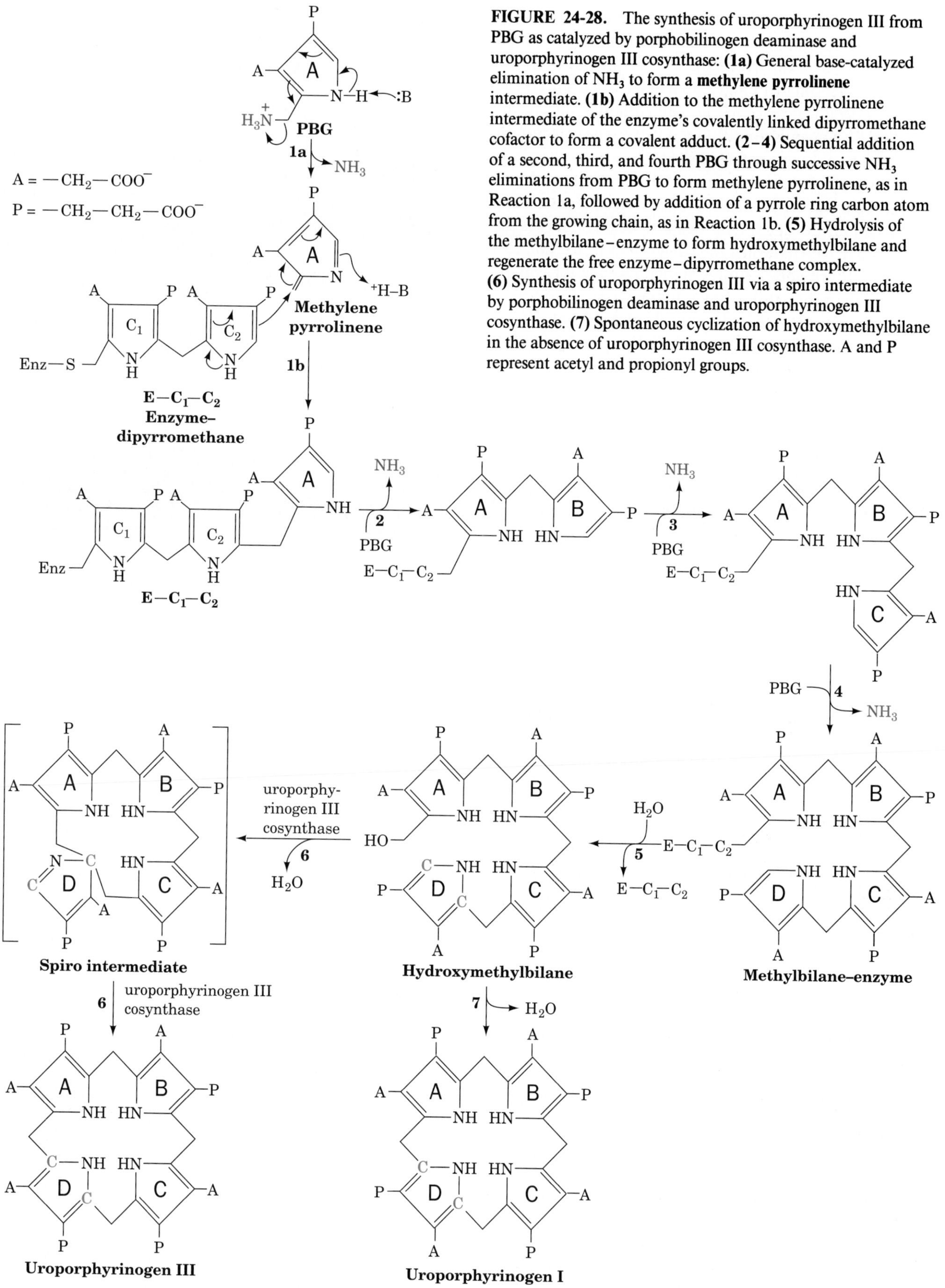

FIGURE 24-28. The synthesis of uroporphyrinogen III from PBG as catalyzed by porphobilinogen deaminase and uroporphyrinogen III cosynthase: **(1a)** General base-catalyzed elimination of NH₃ to form a **methylene pyrrolinene** intermediate. **(1b)** Addition to the methylene pyrrolinene intermediate of the enzyme's covalently linked dipyrromethane cofactor to form a covalent adduct. **(2–4)** Sequential addition of a second, third, and fourth PBG through successive NH₃ eliminations from PBG to form methylene pyrrolinene, as in Reaction 1a, followed by addition of a pyrrole ring carbon atom from the growing chain, as in Reaction 1b. **(5)** Hydrolysis of the methylbilane–enzyme to form hydroxymethylbilane and regenerate the free enzyme–dipyrromethane complex. **(6)** Synthesis of uroporphyrinogen III via a spiro intermediate by porphobilinogen deaminase and uroporphyrinogen III cosynthase. **(7)** Spontaneous cyclization of hydroxymethylbilane in the absence of uroporphyrinogen III cosynthase. A and P represent acetyl and propionyl groups.

methane cofactor, indicates that this monomeric, 307-residue protein folds into three nearly equal sized domains (Fig. 24-29). The dipyrromethane cofactor lies deep in a cleft between domains 1 and 2 such that there is still considerable unoccupied space in the cleft. Although the enzyme sequentially appends four PBG residues to the cofactor, the enzyme has only one catalytic site.

If the enzyme has only one catalytic site, how does it reposition the polypyrrole chain after each catalytic cycle so that it can further extend this chain? One possibility is that the polypyrrole chain fills the cavity next to the cofactor. This model provides a simple steric rationale for why the length of the polypyrrole chain is limited to six residues (the final four of which are hydrolytically cleaved away by the enzyme to yield the hydroxymethylbilane product and regenerate the dipyrromethane cofactor).

Cyclization of the hydroxymethylbilane product requires the participation of uroporphyrinogen III cosynthase (Fig. 24-28). In the absence of the cosynthase, hydroxymethylbilane is released from the synthase and rapidly cyclizes nonenzymatically to the symmetric **uroporphyrinogen I.** Heme, however, is an asymmetric molecule; the methyl substituent of pyrrole ring D has an inverted placement compared to those of rings A, B, and C (Fig. 24-24). This ring reversal to yield uroporphyrinogen III is thought to proceed through attachment of the methylenes from rings A and C to the same carbon of ring D so as to form a spiro compound (a bicyclic compound with a carbon atom common to both rings; Fig. 24-28).

Heme biosynthesis takes place partly in the mitochondrion and partly in the cytosol (Fig. 24-30). ALA is mitochondrially synthesized and is transported to the cytosol for conversion to PBG and then to uroporphyrinogen III. **Protoporphyrin IX,** to which Fe is added to form heme, is produced from uroporphyrinogen III in a series of reactions catalyzed by (1) **uroporphyrinogen decarboxylase,** which decarboxylates all four acetate side chains (A) to form methyl groups (M); (2) **coproporphyrinogen oxidase,** which oxidatively decarboxylates two of the propionate side chains (P) to vinyl groups (V); and (3) **protoporphyrinogen oxidase,** which oxidizes the methylene groups linking the pyrrole rings to methenyl groups. Altogether, six carboxyl groups originally from carboxyl-labeled acetate are lost as CO_2. The only remaining C atoms from carboxyl-labeled acetate are the carboxyl groups of heme's two propionate side chains (P). During the coproporphyrinogen oxidase reaction, the macrocycle is transported back into the mitochondrion for the pathway's final reactions. Protoporphyrin IX is converted to heme by the insertion of Fe(II) into the tetrapyrrole nucleus by **ferrochelatase.**

Heme Biosynthesis Is Regulated Differently in Erythroid and Liver Cells

The two major sites of heme biosynthesis are erythroid cells, which synthesize ~85% of the body's heme groups, and the liver, which synthesizes most of the remainder. An

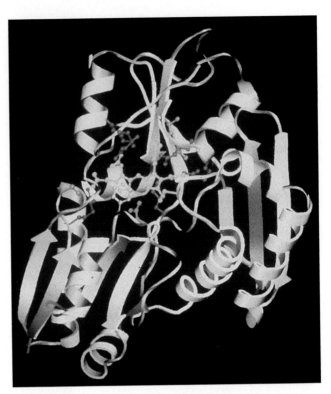

FIGURE 24-29. The X-ray structure of *E. coli* porphobilinogen deaminase in covalent complex with its dipyrromethane cofactor. The protein is shown in ribbon form and the dipyrromethane cofactor (*yellow*) together with the side chains that it contacts are shown in ball-and-stick form. [Courtesy of Gordon Louie, Stephan Wood, Peter Jordan, and Tom Blundell, Birbeck College, U.K.]

important function of heme in liver is as the prosthetic group of **cytochrome P_{450},** an oxidative enzyme involved in detoxification, which is required throughout the liver cell's lifetime in amounts that vary with conditions. In contrast, erythroid cells, in which heme is, of course, a hemoglobin component, engage in heme synthesis only upon differentiation when they synthesize hemoglobin in vast quantities. This is a one-time synthesis; the heme must last the erythrocyte's lifetime (normally 120 days) since heme and hemoglobin synthesis stop upon red cell maturation (protein synthesis stops upon the loss of nuclei and ribosomes). The different ways that heme biosynthesis is regulated in liver and erythroid cells reflect these different demands: In liver, heme biosynthesis must really be "controlled," whereas in erythroid cells, the process is more like breaking a dam.

In liver, the main control target in heme biosynthesis is ALA-synthase, the enzyme catalyzing the pathway's first committed step. Heme, or its Fe(III) oxidation product **hemin,** controls this enzyme's activity through three mechanisms: (1) feedback inhibition, (2) inhibition of the transport of ALA-synthase from its site of synthesis in the cytosol to its reaction site in the mitochondrion (Fig. 24-30), and (3) repression of ALA-synthase synthesis.

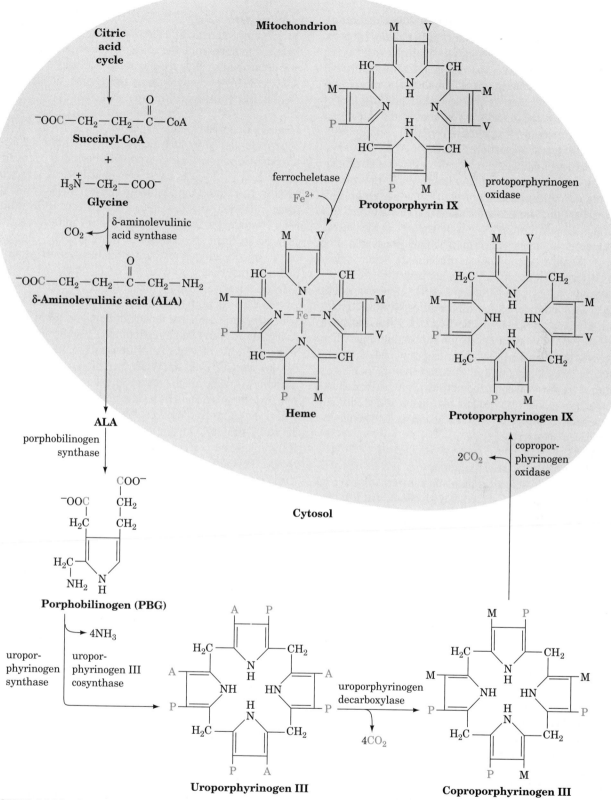

FIGURE 24-30. The overall pathway of heme biosynthesis. δ-Aminolevulinic acid (ALA) is synthesized in the mitochondrion by ALA-synthase. ALA (*left*) leaves the mitochondrion and is converted to PBG, four molecules of which condense to form a porphyrin ring. The next three reactions involve oxidation of the pyrrole ring substituents yielding protoporphyrinogen IX whose formation is accompanied by its transport back into the mitochondrion. After oxidation of the methylene groups linking the pyrroles to yield protoporphyrin IX, ferrochelatase catalyzes the insertion of Fe^{2+} to yield heme. A, P, M, and V, respectively, represent acetyl, propionyl, methyl, and vinyl ($-CH_2=CH_2$) groups. C atoms originating as the carboxyl group of acetate are red.

In erythroid cells, heme exerts quite a different effect on its biosynthesis. Heme stimulates, rather than represses, protein synthesis in reticulocytes (immature erythrocytes; Section 30-4A). Although the vast majority of the protein synthesized by reticulocytes is globin, there is evidence that heme also induces these cells to synthesize the enzymes of the heme biosynthesis pathway. Moreover, the rate-determining step of heme biosynthesis in erythroid cells may not be the ALA-synthase reaction. Experiments on various systems of differentiating erythroid cells implicate ferrochelatase, the enzyme catalyzing iron insertion, and porphobilinogen deaminase in the control of heme biosynthesis in these cells. There are also indications that cellular uptake of iron may be rate limiting. Iron is transported in the plasma complexed with the iron transport protein **transferrin**. The rate at which the iron–transferrin complex enters most cells, including those of liver, is controlled by receptor-mediated endocytosis (Section 11-4B). However, lipid-soluble iron complexes that diffuse directly into reticulo-cytes stimulate *in vitro* heme biosynthesis. The existence of several control points supports the supposition that when erythroid heme biosynthesis is "switched on," all of its steps function at their maximal rates rather than any one step limiting the flow through the pathway. Heme-stimulated synthesis of globin also ensures that heme and globin are synthesized in the correct ratio for assembly into hemoglobin (Section 30-4A).

Porphyrias Have Bizarre Symptoms

Several genetic defects in heme biosynthesis, in liver or erythroid cells, are recognized. All involve the accumulation of prophyrin and/or its precursors and are therefore known as **porphyrias.** Two such defects are known to affect erythroid cells: uroporphyrinogen III cosynthase deficiency **(congenital erythropoietic porphyria)** and ferrochelatase deficiency **(erythropoietic protoporphyria).** The former results in accumulation of uroporphyrinogen I and its decarboxylation product **coproporphyrinogen I.** Excretion of these compounds colors the urine red, their deposition in the teeth turns them a fluorescent reddish brown, and their accumulation in the skin renders it extremely photosensitive such that it ulcerates and forms disfiguring scars. Increased hair growth is also observed in afflicted individuals such that fine hair may cover much of their faces and extremities. These symptoms have prompted speculation that the werewolf legend has a biochemical basis.

The most common porphyria that primarily affects liver is porphobilinogen deaminase deficiency **(acute intermittent porphyria).** This disease is marked by intermittent attacks of abdominal pain and neurological dysfunction. Excessive amounts of ALA and PBG are excreted in the urine during and after such attacks. The urine may become red resulting from the excretion of excess porphyrins synthesized from PBG in nonhepatic cells although the skin does not become unusually photosentitive. King George III, who ruled England during the American Revolution, and

who has been widely portrayed as being mad, in fact had attacks characteristic of acute intermittent porphyria, was reported to have urine the color of port wine, and had several descendants who were diagnosed as having this disease. American history might have been quite different had George III not inherited this metabolic defect.

Heme Is Degraded to Bile Pigments

At the end of their lifetime, red cells are removed from the circulation and their components degraded. Heme catabolism (Fig. 24-31) begins with oxidative cleavage, by **heme oxygenase,** of the porphyrin between rings A and B to form **biliverdin,** a green linear tetrapyrrole. Biliverdin's central methenyl bridge (between rings C and D) is then reduced to form the red-orange **bilirubin.** The changing colors of a healing bruise are a visible manifestation of heme degradation.

In the reaction forming biliverdin, the methenyl bridge carbon between porphyrin rings A and B is released as CO, which, we have seen, is a tenacious heme ligand (with 200-fold greater affinity for hemoglobin than O_2; Section 9-1A). Consequently, ~1% of hemoglobin's O_2-binding sites are blocked by CO, even in the absence of air pollution. This amount would be much greater were it not for the presence of hemoglobin's distal His residue (E7, the His residue that hydrogen bonds to bound O_2; Section 9-2). The distal His sterically strains CO from its preferred linear geometry in liganding heme Fe(II) towards the bent geometry favored by O_2. This reduces heme's affinity for CO over 100-fold, thus permitting the CO to be exhaled slowly. Indeed, in individuals with the mutant **Hb Zurich** [His E7(63)$\beta \rightarrow$ Arg], around 10% of the hemes carry CO.

The highly lipophilic bilirubin is insoluble in aqueous solutions. Like other lipophilic metabolites, such as free fatty acids, it is transported in the blood in complex with serum albumin. In the liver, its aqueous solubility is increased by esterification of its two propionate side groups with glucuronic acid, yielding **bilirubin diglucuronide,** which is secreted into the bile. Bacterial enzymes in the large intestine hydrolyze the glucuronic acid groups and, in a multistep process, convert bilirubin to several products, most notably **urobilinogen.** Some urobilinogen is reabsorbed and transported via the bloodstream to the kidney, where it is converted to the yellow **urobilin** and excreted, thus giving urine its characteristic color. Most of the urobilinogen, however, is microbially converted to the deeply red-brown **stercobilin,** the major pigment of feces.

When the blood contains excessive amounts of bilirubin, the deposition of this highly insoluble substance colors the skin and the whites of the eyes yellow. This condition, called **jaundice** (French: *jaune,* yellow), signals either an abnormally high rate of red cell destruction, liver dysfunction, or bile duct obstruction. Newborn infants, particularly when premature, often become jaundiced because their livers do not yet make sufficient **bilirubin glucuronyl transferase** to glucuronidate the incoming bilirubin. Jaun-

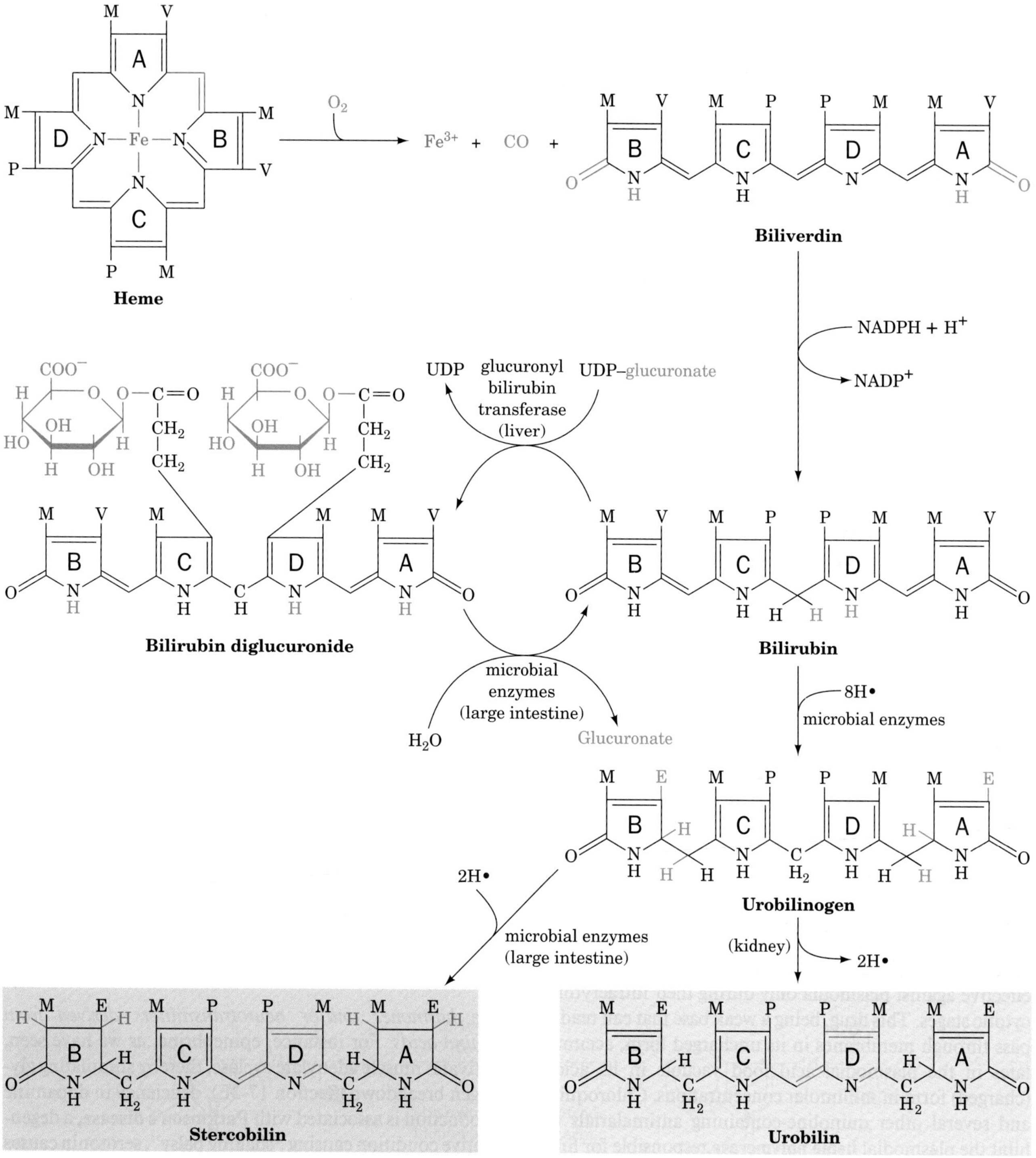

FIGURE 24-31. The heme degradation pathway. M, V, P, and E, respectively, represent methyl, vinyl, propionyl, and ethyl groups.

diced infants are treated by bathing them with light from a fluorescent lamp; this photochemically converts bilirubin to more soluble isomers that the infant can degrade and excrete.

Chloroquine Prevents Malaria by Inhibiting Plasmodial Heme Sequestration

Malaria is caused by the mosquito-borne parasite *Plasmodium falciparum* (Section 6-3A), which multiplies

3. A second hydroxylation yields norepinephrine.

4. Methylation of norepinephrine's amino group by *S*-adenosylmethionine (SAM; Section 24-3E) produces epinephrine.

The specific catecholamine that a cell produces depends on which enzymes of the pathway are present. In adrenal medulla, which functions to produce hormones (Section 34-4A), epinephrine is the predominant product. In some

areas of the brain, norepinephrine is more common. In other areas, most prominently the **substantia nigra,** the pathway stops at dopamine synthesis. Indeed, Parkinson's disease, which is caused by degeneration of the substantia nigra, has been treated with some success by the administration of L-DOPA, dopamine's immediate precursor. Dopamine itself is ineffective because it cannot cross the blood–brain barrier. L-DOPA, however, is able to get to its sites of action where it is decarboxylated to dopamine. The enzyme catalyzing this reaction, **aromatic amino acid decarboxylase,** decarboxylates all aromatic amino acids and is therefore also responsible for serotonin formation. In a recently developed approach to the treatment of Parkinsonism, a portion of the patient's adrenal medulla is surgically transplanted to his or her brain. Presumably, the dopamine and L-DOPA released by this tissue serve to replace that lost via degeneration of the substantia nigra. L-DOPA is also a precursor of the black skin pigment melanin.

C. Glutathione

Glutathione (GSH; γ-glutamylcysteinylglycine),

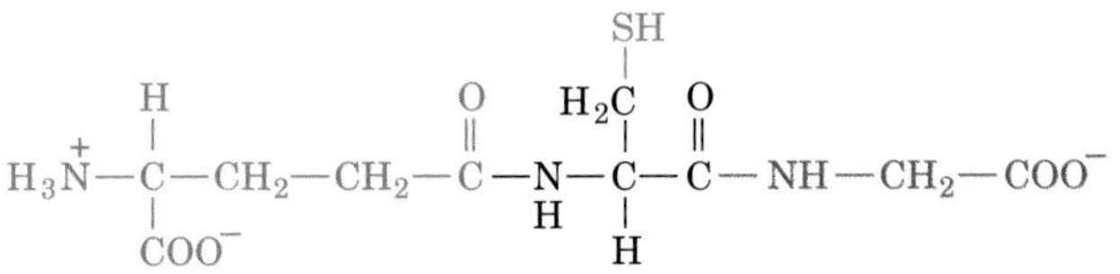

γ-Glutamylcysteinylglycine,
glutathione (GSH)

a tripeptide that contains an unusual γ-amide bond, participates in a variety of detoxification, transport, and metabolic processes (Fig. 24-35). For instance, it is a substrate for peroxidase reactions, helping to destroy peroxides generated by oxidases; it is involved in leukotriene biosynthesis (Section 23-7C); and the balance between its reduced (GSH) and oxidized (GSSG) forms maintains the sulfhy-

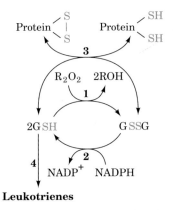

FIGURE 24-35. Some reactions involving glutathione: **(1)** peroxide detoxification by **glutathione peroxidase,** **(2)** regeneration of GSH from GSSG by glutathione reductase (Section 14-4), **(3)** thiol transferase modulation of protein thiol–disulfide balance, and **(4)** leukotriene biosynthesis by glutathione-*S*-transferase.

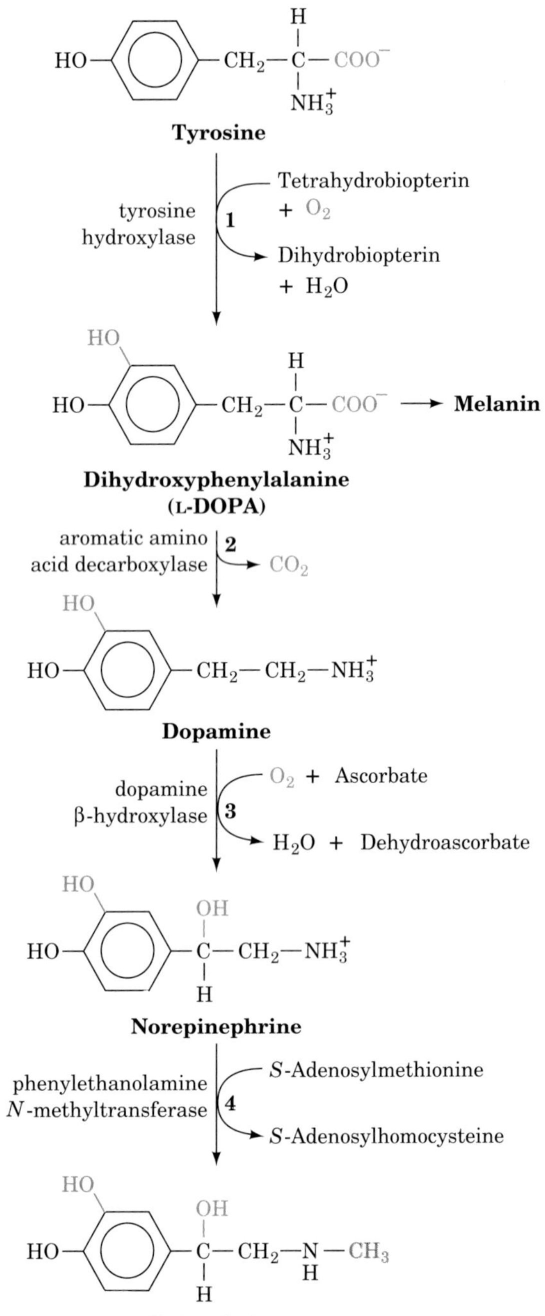

FIGURE 24-34. The sequential synthesis of L-DOPA, dopamine, norepinephrine, and epinephrine from tyrosine. L-DOPA is also the precursor of the black skin pigment melanin, an oxidized polymeric material.

dryl groups of intracellular proteins in their correct oxidation states.

The *γ-glutamyl cycle*, which was elucidated by Alton Meister, *provides a vehicle for the energy-driven transport of amino acids into cells through the synthesis and breakdown of GSH (Fig. 24-36).* GSH is synthesized from glutamate, cysteine, and glycine by the consecutive action of *γ-glutamylcysteine synthetase* and **GSH synthetase** (Fig. 24-36, Reactions 1 and 2). ATP hydrolysis provides the free energy for each reaction. The carboxyl group is activated for peptide bond synthesis by formation of an acyl phosphate intermediate.

$$
\begin{array}{c}
\overset{O}{\overset{\|}{R-C-O^-}} + ATP \xrightarrow{\quad\nearrow^{ADP}\quad} \overset{O}{\overset{\|}{R-C-OPO_3^{2-}}} \\
\\
\overset{\downarrow\; \nwarrow P_i \;\;\; \nwarrow NH_2-R'}{} \\
\\
\underset{\underset{H}{|}}{\overset{O}{\overset{\|}{R-C-N-R'}}}
\end{array}
$$

The breakdown of GSH is catalyzed by *γ-glutamyl transpeptidase*, *γ-glutamyl cyclotransferase*, **5-oxoprolinase**, and an intracellular protease (Fig. 24-36, Reactions 3–6).

Amino acid transport occurs because, whereas GSH is synthesized intracellularly and is located largely within the cell, *γ*-glutamyl transpeptidase, which catalyzes GSH breakdown (Fig. 24-36, Reaction 3), is situated on the cell membrane's external surface and accepts amino acids, notably cysteine and methionine. GSH is first transported to the external surface of the cell membrane, where the transfer of the *γ*-glutamyl group from GSH to an external amino acid occurs. The *γ*-glutamyl amino acid is then transported back into the cell and converted to glutamate by a two-step process in which the transported amino acid is released and **5-oxoproline** is formed as an intermediate. The last step in the cycle, the hydrolysis of 5-oxoproline, requires ATP hydrolysis. This surprising observation (amide bond hydrolysis is almost always an exergonic process) is a consequence of 5-oxoproline's unusually stable internal amide bond.

D. Tetrahydrofolate Cofactors: The Metabolism of C_1 Units

Many biosynthetic processes involve the addition of a C_1 unit to a metabolic precursor. A familiar example is car-

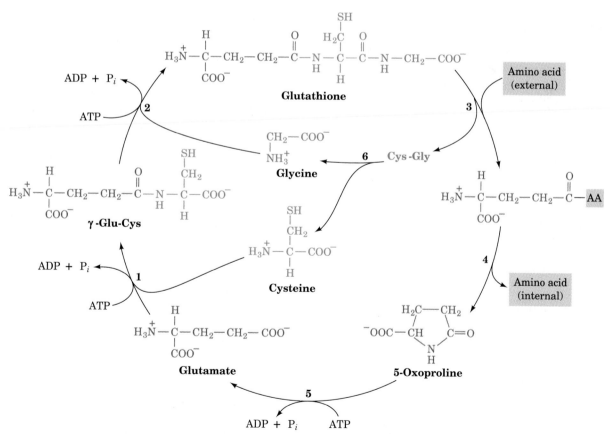

FIGURE 24-36. Glutathione synthesis as part of the *γ*-glutamyl cycle of glutathione metabolism. The cycle's reactions are catalyzed by: **(1)** *γ*-glutamylcysteine synthetase, **(2)** glutathione synthetase, **(3)** *γ*-glutamyl transpeptidase, **(4)** *γ*-glutamyl cyclotransferase, **(5)** 5-oxoprolinase, and **(6)** an intracellular protease.

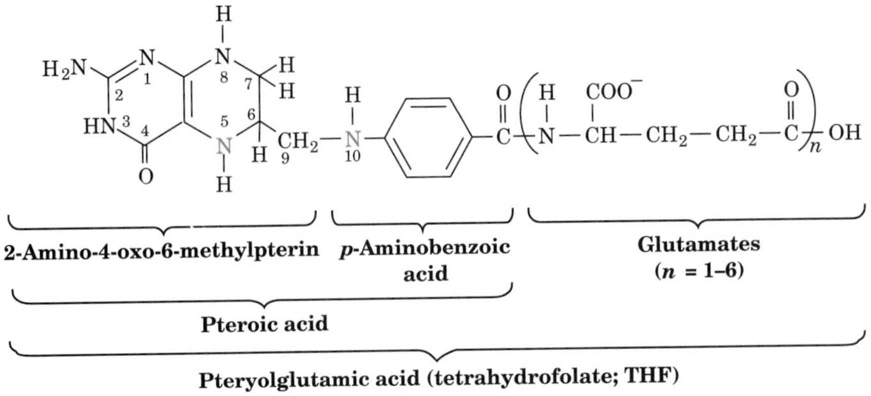

FIGURE 24-37.   Tetrahydrofolate (THF).

boxylation. For instance, gluconeogenesis from pyruvate begins with the addition of a carboxyl group to form oxaloacetate (Section 21-1A). The coenzyme involved in this and most other carboxylation reactions is biotin (Section 21-1A). In contrast, *S*-adenosylmethionine functions as a methylating agent (Section 24-2E).

Tetrahydrofolate (THF) is more versatile than the above cofactors in that it functions to transfer C_1 units in several oxidation states. THF is a 6-methylpterin derivative linked in sequence to *p*-aminobenzoic acid and Glu residues (Fig. 24-37). Up to five additional Glu residues may be linked to the first glutamate via isopeptide bonds to form a polyglutamyl tail.

THF is derived from **folic acid** (Latin: *folium,* leaf), a doubly oxidized form of THF that must be enzymatically reduced before it becomes an active coenzyme (Fig. 24-38). Both reductions are catalyzed by **dihydrofolate reductase (DHFR).** Mammals cannot synthesize folic acid so it must be provided in the diet or by intestinal microorganisms.

C_1 units are covalently attached to THF at its positions N5, N10, or both N5 and N10. These C_1 units, which may be at the oxidation levels of formate, formaldehyde, or methanol (Table 24-1), are all interconvertible by enzymatic redox reactions (Fig. 24-39).

The main entry of C_1 units into the THF pool is as N^5,N^{10}-methylene-THF through the conversion of serine to glycine by serine hydroxymethyl transferase (Sections 24-3B and 24-5A) and the cleavage of glycine by glycine synthase (the glycine cleavage system; Section 24-3B, Fig. 24-11). Histidine also contributes C_1 units through its degradation with formation of N^5-**formimino-THF** (Fig. 24-13, Reaction 11).

A C_1 unit in the THF pool can have several fates (Fig. 24-40):

1. It may be used directly as N^5,N^{10}-methylene-THF in the conversion of the deoxynucleotide dUMP to dTMP by **thymidylate synthase** (Section 26-4B).

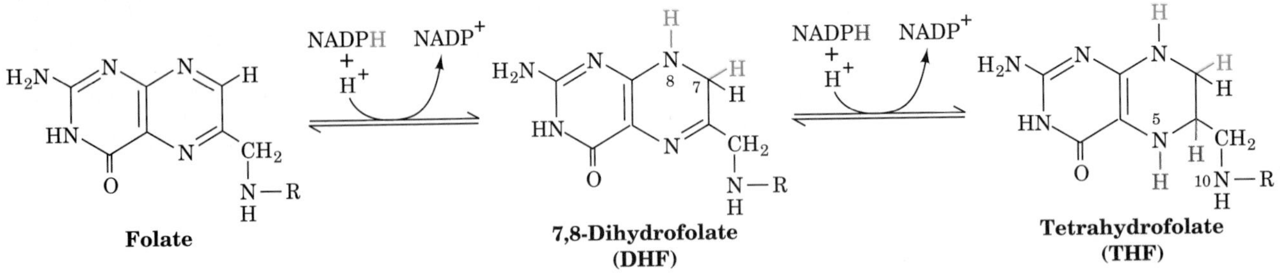

FIGURE 24-38.   The two-stage reduction of folate to THF. Both reactions are catalyzed by dihydrofolate reductase (DHFR).

TABLE 24-1.   OXIDATION LEVELS OF C_1 GROUPS CARRIED BY THF

Oxidation Level	Group Carried	THF Derivative(s)
Methanol	Methyl (—CH_3)	N^5-Methyl-THF
Formaldehyde	Methylene (—CH_2—)	N^5,N^{10}-Methylene-THF
Formate	Formyl (—CH=O)	N^5-Formyl-THF, N^{10}-formyl-THF
	Formimino (—CH=NH)	N^5-Formimino-THF
	Methenyl (—CH=)	N^5,N^{10}-Methenyl-THF

Serine
Glycine
serine hydroxymethyl transferase
THF
glycine synthase
Glycine + NAD+
CO2 + NH4+ + NADH
N5-Methyl-THF
N5,N10-methylene-THF reductase
NAD+
NADH + H+
N5,N10-Methylene-THF
N5,N10-methenyl-THF reductase
NADP+
NADPH + H+
N5,N10-Methenyl-THF
N5,N10-methenyl-THF cyclohydrolase
H2O
N10-formyl-THF synthetase
HCO2− + ATP
Formate
ADP + Pi
N10-Formyl-THF
THF
Histidine
Glutamate + NH4+
N5-formimino-THF cyclodeaminase
NH3
N5-Formimino-THF
ADP + Pi
ATP
N10-formyl-THF isomerase
N5-Formyl-THF

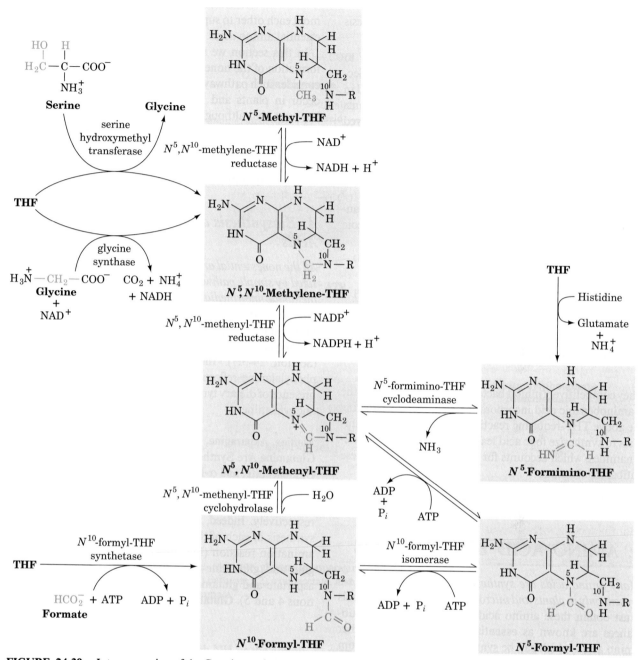

FIGURE 24-39. Interconversion of the C_1 units carried by THF.

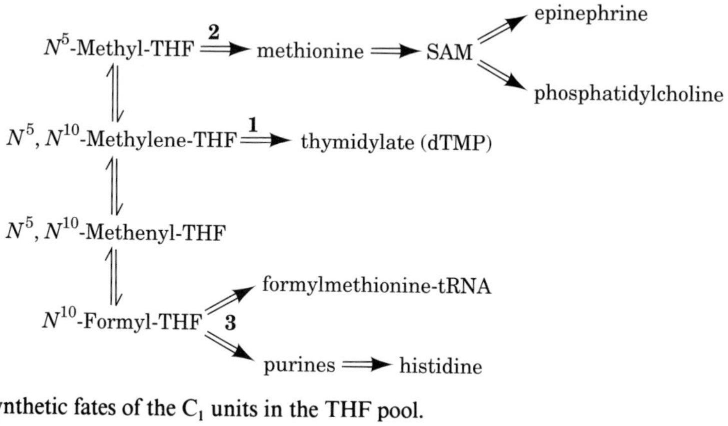

N5-Methyl-THF
2
methionine
SAM
epinephrine
phosphatidylcholine
N5, N10-Methylene-THF
1
thymidylate (dTMP)
N5, N10-Methenyl-THF
N10-Formyl-THF
3
formylmethionine-tRNA
purines
histidine

FIGURE 24-40. The biosynthetic fates of the C_1 units in the THF pool.

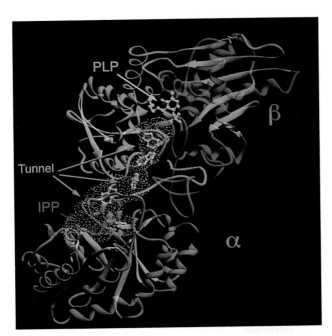

FIGURE 24-51. A ribbon diagram of the bifunctional enzyme tryptophan synthase from *S. typhimurium*. Only one $\alpha\beta$ unit of this twofold symmetric $\alpha\beta\beta\alpha$ heterotetramer is shown. The 268-residue α subunit is blue, the 397-residue β subunit's N-terminal domain is orange, its C-terminal domain is red, and all β sheets are tan. The active site of the α subunit is located at its bound competitive inhibitor, **indolpropanol phosphate** (IPP; *red ball-and-stick model*), whereas that of the β subunit is marked by its PLP coenzyme *(tan ball-and-stick model)*. The solvent-accessible surface of the ~25-Å-long "tunnel" connecting the active sites of the α and β subunits is outlined by yellow dots. Several indole molecules *(green ball-and-stick models)* have been modeled into the tunnel in head to tail fashion, thereby demonstrating that the tunnel has sufficient width to permit the indole product of the α subunit to pass through the tunnel to the β subunit's active site. [Courtesy of Craig Hyde, National Institutes of Health.]

Histidine Biosynthesis

Five of histidine's six C atoms are derived from **5-phosphoribosyl-α-pyrophosphate (PRPP;** Fig. 24-52), an intermediate also involved in the biosynthesis of tryptophan, (Fig. 24-50, Reaction 2), purine nucleotides (Section 26-2A), and pyrimidine nucleotides (Section 26-3A). The histidine's sixth carbon originates from ATP. The ATP atoms that are not incorporated into histidine are eliminated as **5-aminoimidazole-4-carboxamide ribonucleotide** (Fig. 24-52, Reaction 5), which is also an intermediate in purine biosynthesis (Section 26-2A).

The unusual biosynthesis of histidine from a purine has been cited as evidence supporting the hypothesis that life was originally RNA based (Section 1-4C). His residues, as we have seen, are often components of enzyme active sites where they act as nucleophiles and/or general acid–base catalysts. The discovery that RNA can have catalytic properties (Section 29-4C) therefore suggests that the imidazole

FIGURE 24-52. *(Opposite)* The biosynthesis of histidine. The pathway enzymes are **(1)** ATP phosphoribosyl transferase, **(2)** pyrophosphohydrolase, **(3)** phosphoribosyl-AMP cyclohydrolase, **(4)** phosphoribosylformimino-5-aminoimidazole carboxamide ribonucleotide isomerase, **(5)** glutamine amidotransferase, **(6)** imidazole glycerol phosphate dehydratase, **(7)** L-histidinol phosphate aminotransferase, **(8)** histidinol phosphate phosphatase, and **(9)** histidinol dehydrogenase.

moiety of purines plays a similar role in these RNA enzymes (**ribozymes**). This further suggests that the histidine biosynthesis pathway is a "fossil" of the transition to more efficient protein-based life forms.

6. NITROGEN FIXATION

The most prominent chemical elements in living systems are O, H, C, N, and P. The elements O, H, and P occur widely in metabolically available forms (H_2O, O_2, and P_i). However, the major available forms of C and N, CO_2 and N_2, are extremely stable (unreactive); for example, the N—N triple bond has a bond energy of 945 kJ · mol^{-1} (vs 351 kJ · mol^{-1} for a C—O single bond). CO_2, with only minor exceptions, is metabolized (fixed) only by photosynthetic organisms (Chapter 22). *N_2 fixation is even less common; this element is converted to metabolically useful forms by only a few strains of bacteria, named diazatrophs.*

Diazatrophs of the genus *Rhizobium* live in symbiotic relationship with root nodule cells of legumes (plants belonging to the pea family, including beans, clover, and alfalfa; Fig. 24-53) where they convert N_2 to NH_3.

$$N_2 + 8H^+ + 8e^- + 16ATP + 16H_2O \longrightarrow$$
$$2NH_3 + H_2 + 16ADP + 16P_i$$

The NH_3 thus formed can be incorporated either into glutamate by glutamate dehydrogenase (Section 24-1) or into glutamine by glutamine synthetase (Section 24-5A). This nitrogen-fixing system produces more metabolically useful nitrogen than the legume needs; the excess is excreted into the soil, enriching it. It is therefore common agricultural practice to plant a field with alfalfa every few years to build up the supply of usable nitrogen in the soil for later use in growing other crops.

Nitrogenase Contains Several Novel Redox Centers

Nitrogenase, which catalyzes the reduction of N_2 to NH_3, consists of two proteins:

1. The **Fe-protein,** an ~64-kD dimer of identical subunits that contains one [4Fe–4S] cluster and two ATP binding sites.

2. The **MoFe-protein,** an ~220-kD protein of subunit structure $\alpha_2\beta_2$ that contains Fe and Mo.

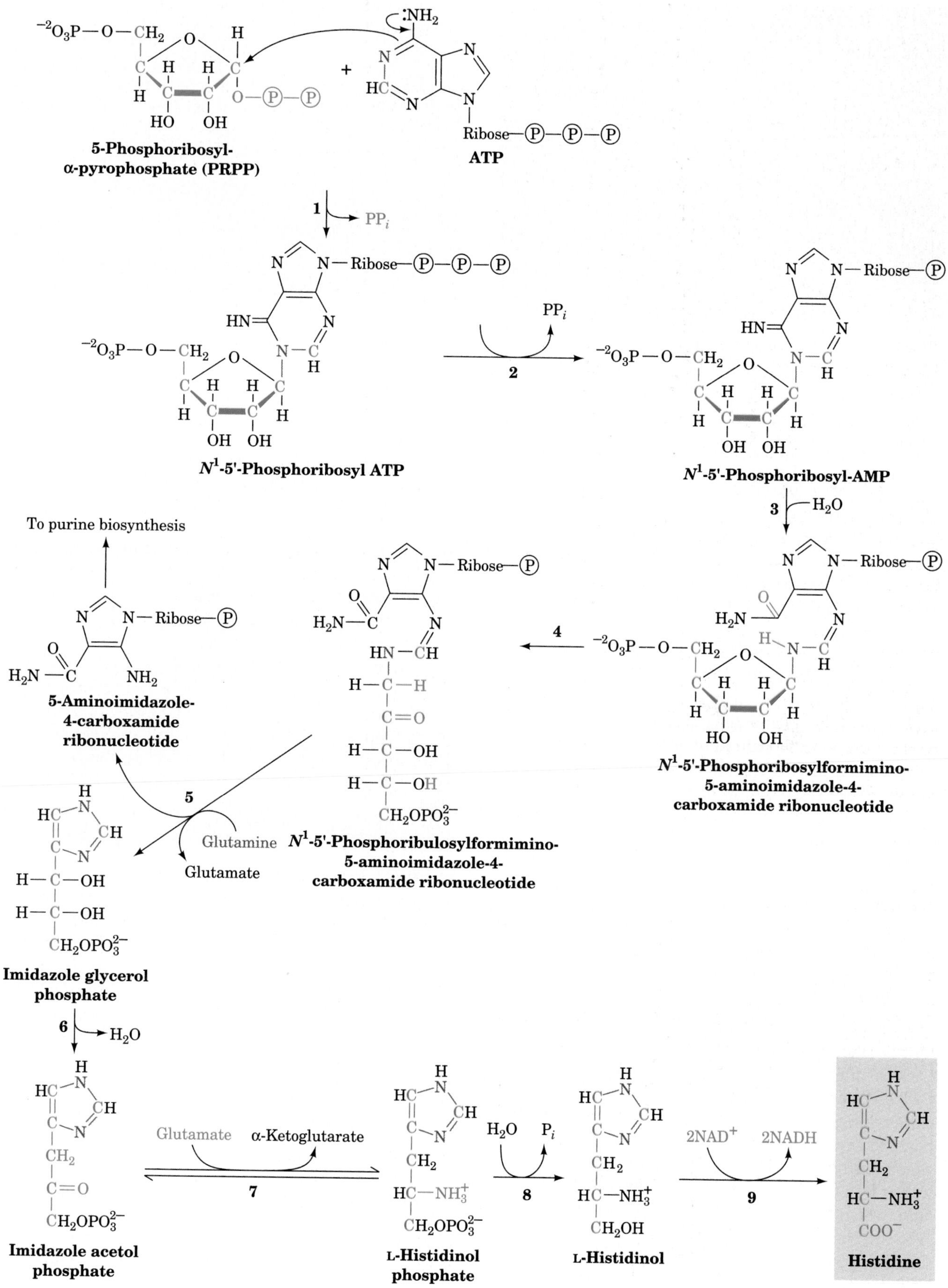
5-Phosphoribosyl-α-pyrophosphate (PRPP)
ATP
1
PP_i
N^1-5'-Phosphoribosyl ATP
2
PP_i
N^1-5'-Phosphoribosyl-AMP
3
H_2O
N^1-5'-Phosphoribosylformimino-5-aminoimidazole-4-carboxamide ribonucleotide
4
N^1-5'-Phosphoribulosylformimino-5-aminoimidazole-4-carboxamide ribonucleotide
To purine biosynthesis
5-Aminoimidazole-4-carboxamide ribonucleotide
5
Glutamine
Glutamate
Imidazole glycerol phosphate
6
H_2O
Imidazole acetol phosphate
Glutamate
α-Ketoglutarate
7
L-Histidinol phosphate
H_2O
P_i
8
L-Histidinol
2NAD^+
2NADH
9
Histidine

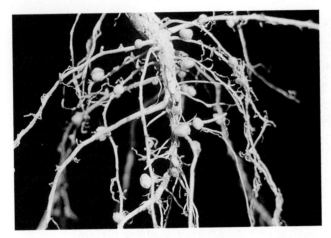

FIGURE 24-53. A photograph showing the root nodules of the legume bird's foot trefoil. [Vu/Cabisco/Visuals Unlimited]

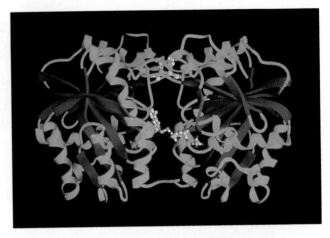

FIGURE 24-54. The X-ray structure of the *Azotobacter vinelandii* nitrogenase Fe-protein dimer as viewed with its twofold molecular axis vertical in the plane of the paper. The protein is drawn in ribbon form with its helices blue, its β sheets red, and its other segments green. The [4Fe-4S] cluster *(upper center)* and a bound ADP *(center)* and MoO_4^{2-} ion *(right of center)* are shown in ball-and-stick form (the MoO_4^{2-} ion presumably occupies the phosphate binding site). [Courtesy of Douglas Rees, California Institute of Technology.]

The X-ray structure of *Azotobacter vinelandii* Fe-protein has been determined by Douglas Rees, whereas those of the MoFe-protein from *A. vinelandii* and *Clostridium pasteurianum* were independently determined by Rees and Jeffrey Bolin. Each 289-residue subunit of the Fe-protein folds into a single domain (Fig. 24-54). The Fe-protein dimer's single [4Fe-4S] cluster is located in a solvent-exposed cleft between the two subunits and is symmetrically linked to Cys 97 and Cys 132 from both subunits. The structure resembles an "iron butterfly" with the [4Fe-4S] cluster at its head. One of the ATP binding sites is occupied by a single bound nucleotide, thought to be ADP, located at the interface between the two subunits (no nucleotide had been added to the crystallization preparation so this molecule must have copurified with the protein). Since the ADP's phosphate groups and the [4Fe-4S] cluster are separated by ~20 Å, a distance too large for direct coupling between electron transfer and ATP hydrolysis, it appears that these processes are indirectly coupled through allosteric changes at the subunit interface. These allosteric changes are thought to drive the electron transfer process.

The MoFe-protein's 491-residue α subunit and 522-residue β subunit assume similar polypeptide folds and extensively associate to form an $\alpha\beta$ dimer, two of which more loosely associate to form the $\alpha_2\beta_2$ tetramer (Fig. 24-55). Each $\alpha\beta$ dimer contains two bound redox centers: (1) The **P-cluster pair** (Fig. 24-56*a*), which consists of two [4Fe-4S] clusters bridged by two Cys thiol ligands and a disulfide bond between an S atom in each of the two clusters and in which one of the Fe atoms is liganded by a Ser oxygen as well as by a Cys thiol; and (2) the **FeMo-cofactor** (Fig. 24-56*b*), which consists of a [4Fe-3S] cluster and a [1Mo-3Fe-3S] cluster bridged by three nonprotein ligands: two sulfide ions, and an unidentified third ligand (most probably S, but possibly N or O) designated Y. The FeMo-cofactor's Mo atom is approximately octahedrally coordinated

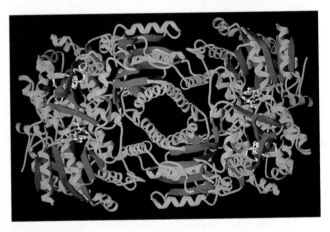

FIGURE 24-55. The X-ray structure of the *A. vinelandii* nitrogenase MoFe-protein $\alpha_2\beta_2$ tetramer as viewed down its molecular twofold axis. The protein is drawn in ribbon form and is colored as is Fig. 24-54. The P-cluster pair *(upper left and lower right)* and the FeMo-cofactor *(lower left and upper right)* are shown in ball-and-stick form. [Courtesy of Douglas Rees, California Institute of Technology.]

FIGURE 24-56 *(Opposite).* The prosthetic groups of the *A. vinelandii* nitrogenase MoFe-protein. *(a)* The P-cluster pair, which consists of two linked [4Fe-4S] complexes; *(b)* the FeMo-cofactor; and *(c)* the predicted interaction of N_2 with the FeMo-cofactor. The atoms are colored according to type with C green, N blue, O red, S yellow, Fe brown, and Mo silver. [Based on an X-ray structure by Douglas Rees, California Institute of Technology.]

(a)

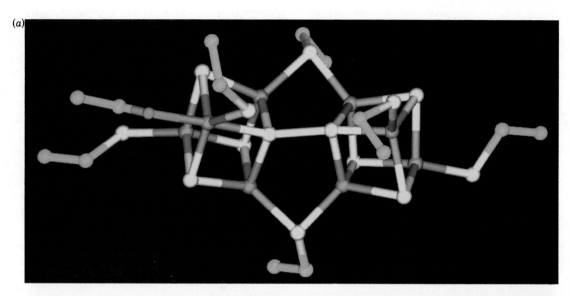

(b)

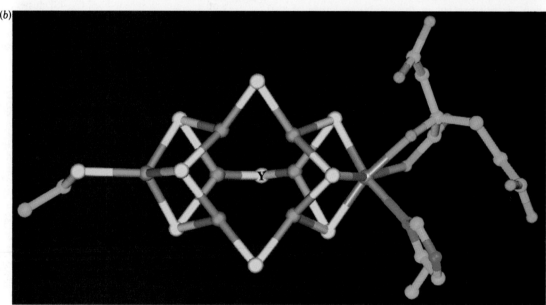

(c)

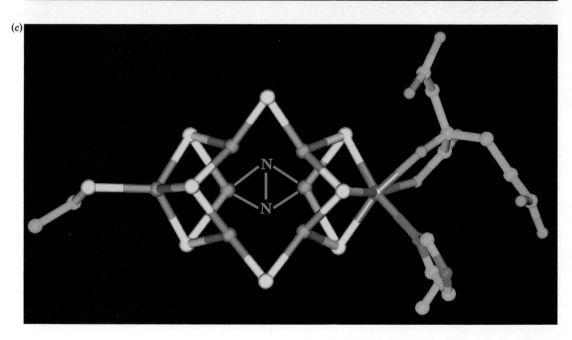

by three cofactor sulfurs, a His imidazole nitrogen, and two oxygens from a bound **homocitrate** ion:

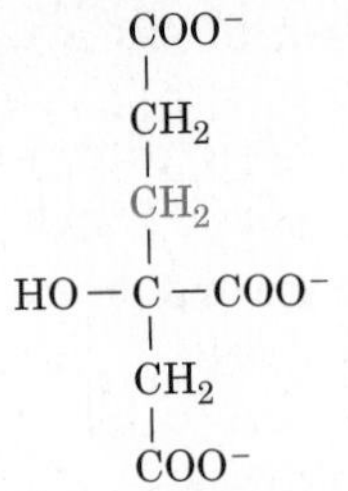

$$
\begin{array}{c}
COO^- \\
| \\
CH_2 \\
| \\
CH_2 \\
| \\
HO-C-COO^- \\
| \\
CH_2 \\
| \\
COO^-
\end{array}
$$

Homocitrate

(an essential component of the FeMo-cofactor). In contrast, six of the FeMo-cofactor's seven Fe atoms have but three ligands; only the apical Fe is tetrahedally liganded, as are the Fe atoms of most other iron–sulfur complexes (Section 20-2C). The FeMo-cofactor contains a central cavity, which is postulated to form the N_2 binding site.

Although N_2 is not present in the crystal structure of MoFe-protein, molecular orbital calculations by Roald Hoffman suggest that N_2 replaces the Y ligand and is oriented such that its N≡N bond is perpendicular to the axis between the two Fe atoms coordinating it (Fig. 24-56c). The FeMo-cofactor is located ~10 Å below the α subunit surface, so the N_2 is thought to gain access to its binding site through conformational fluctuations of the protein (recall that myoglobin and hemoglobin likewise have no clear path for O_2 to approach its heme-binding sites in these proteins; Section 8-2). The P-cluster pair, which is also ~10 Å below the protein surface, is at the interface between the α and β subunits on the pseudo-twofold axis that roughly relates these two subunits.

The Fe- and MoFe-proteins have been shown to associate. Hence, the sequence of the electron transfer steps in the nitrogenase reaction is thought to be

$$\text{Fe-protein} \rightarrow \text{P-cluster pair} \rightarrow \text{FeMo-cofactor} \rightarrow N_2$$

The Fe-protein residues that have been shown to interact with the MoFe-protein are located on the same side of the Fe-protein as its [4Fe–4S] cluster and hence this surface very probably contains part of the the two protein's interaction region. The twofold axis of the Fe-protein and the pseudo-twofold axis of the MoFe-proteins presumably coincide in the complex.

Leghemoglobin Protects Nitrogenase from Oxygen Inactivation

Nitrogenase is rapidly inactivated by O_2, so the enzyme must be protected from this reactive substance. Cyanobacteria (photosynthetic oxygen-evolving bacteria; Section 1-1A) provide protection by carrying out nitrogen fixation in specialized nonphotosynthetic cells called **heterocysts,** which have Photosystem I but lack Photosytem II (Section 22-2C). In the root nodules of legumes (Fig. 24-53), however, protection is afforded by the symbiotic synthesis of **leghemoglobin.** The globin portion of this monomeric oxygen-binding protein is synthesized by the plant (an evolutionary curiosity since globins are otherwise known to occur only in animals), whereas the heme is synthesized by the *Rhizobium*. Leghemoglobin has a very high O_2 affinity, thus keeping the pO_2 low enough to protect the nitrogenase while providing passive O_2 transport for the aerobic bacterium.

N_2 Reduction Is Energetically Costly

Nitrogen fixation requires two participants in addition to N_2 and nitrogenase: (1) a source of electrons and (2) ATP. Electrons are generated either oxidatively or photosynthetically, depending on the organism. These electrons are transferred to ferredoxin, a [4Fe–4S]-containing electron carrier that transfers an electron to the Fe-protein of nitrogenase, beginning the nitrogen fixation process (Fig. 24-57). Two molecules of ATP bind to the reduced Fe-protein and are hydrolyzed as the electron is passed from the Fe-protein to the MoFe-protein. ATP hydrolysis is thought to cause a conformational change in the Fe-protein that alters its redox potential from –0.29 to –0.40 V, making the electron capable of N_2 reduction ($\mathscr{E}°'$ for the half-cell $N_2 + 6H^+ + 6e^- \rightleftharpoons 2NH_3$ is –0.34 V).

The actual reduction of N_2 occurs on the MoFe-protein in three discrete steps, each involving an electron pair:

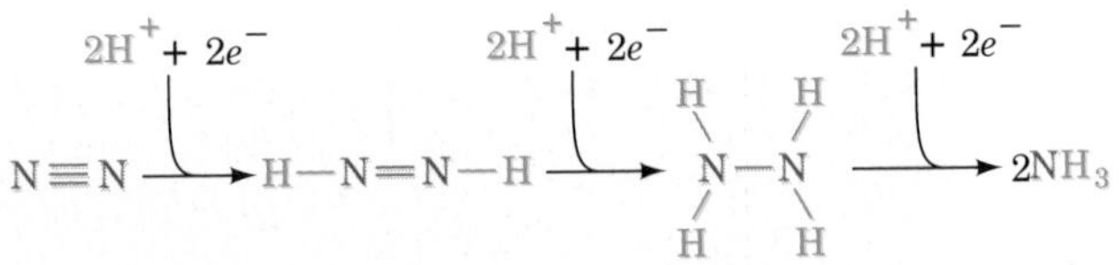

$$
N\equiv N \xrightarrow{2H^+ + 2e^-} H-N=N-H \xrightarrow{2H^+ + 2e^-} \underset{\substack{H \quad H}}{\overset{\substack{H \quad H}}{N-N}} \xrightarrow{2H^+ + 2e^-} 2NH_3
$$

Diimine   **Hydrazine**

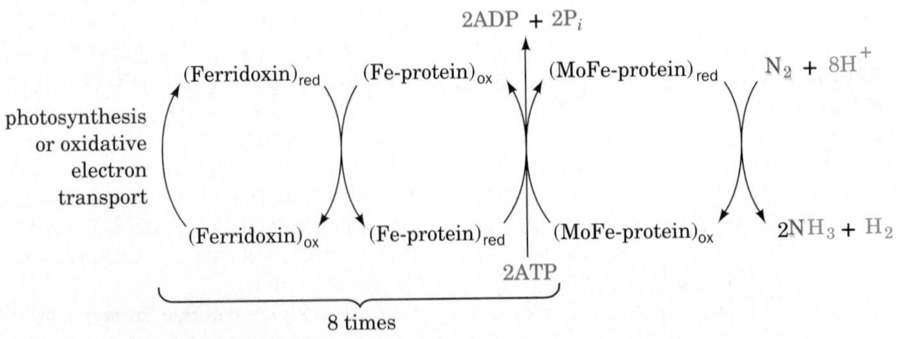

FIGURE 24-57. The flow of electrons in the nitrogenase-catalyzed reduction of N_2.

An electron transfer must occur six times per N_2 molecule fixed so that a total of 12 ATPs are required to fix one N_2 molecule. However, nitrogenase also reduces H_2O to H_2, which in turn reacts with **diimine** to reform N_2.

$$HN{=}NH + H_2 \longrightarrow N_2 + 2H_2$$

This futile cycle is favored when the ATP level is low and/or the reduction of the Fe-protein is sluggish. Even when ATP is plentiful, however, the cycle cannot be suppressed beyond about one H_2 molecule produced per N_2 reduced and hence appears to be a requirement of the nitrogenase reaction. The total cost of N_2 reduction is therefore 8 electrons transferred and 16 ATPs hydrolyzed (physiologically, 20–30 ATPs). Hence nitrogen fixation is an energetically expensive process; indeed, the nitrogen fixing bacteria in the root nodules of pea plants consume nearly 20% of the ATP the plant produces.

Although atmospheric N_2 is the ultimate nitrogen source for all living things, most plants do not support the symbiotic growth of nitrogen-fixing bacteria. They must therefore depend on a source of "prefixed" nitrogen such as nitrate or ammonia. These nutrients come from lightening discharges (the source of ~10% of naturally fixed N_2), decaying organic matter in the soil, or from fertilizer applied to it. The Haber process, which was invented by Fritz Haber in 1910, is a chemical process for N_2 fixation that is still widely used in fertilizer manufacture. This direct reduction of N_2 by H_2 to form NH_3 requires temperatures of 300 to 500°C, pressures of >300 atm, and an iron catalyst. One of the major long-term goals of genetic engineering (Section 28-8) is to induce agriculturally useful nonleguminous plants to fix their own nitrogen, a complex undertaking in which the plant must be made to provide a hospitable environment for nitrogen fixation as well as to acquire the enzymatic machinery to do so. This would free farmers, particularly those in developing countries, from the need for either purchasing fertilizers, periodically letting their fields lie fallow (giving legumes the opportunity to grow), or following the slash-and-burn techniques that are rapidly destroying the world's tropical forests and contributing significantly to the greenhouse effect (atmospheric CO_2 pollution causing long-term global warming).

CHAPTER SUMMARY

Amino acids are the precursors for numerous nitrogen-containing compounds such as heme, physiologically active amines, and glutathione. Excess amino acids are converted to common metabolic intermediates for use as fuels. The first step in amino acid breakdown is removal of the α-amino group by transamination. Transaminases require pyridoxal phosphate (PLP) and convert amino acids to their corresponding α-keto acids. The amino group is transferred to α-ketoglutarate to form glutamate, oxaloacetate to form aspartate, or pyruvate to form alanine. Glutamate is subsequently oxidatively deaminated to form ammonia and regenerate α-ketoglutarate.

In the urea cycle, amino groups from ammonia and aspartate combine with HCO_3^- to form urea. This pathway takes place in the liver, partially in the mitochondrion, and partially in the cytosol. It begins with the ATP-dependent condensation of NH_4^+ and HCO_3^- by carbamoyl phosphate synthetase. The resulting carbamoyl phosphate then combines with ornithine to yield citrulline, which combines with aspartate to form argininosuccinate, which in turn is cleaved to fumarate and arginine. The arginine is then hydrolyzed to urea, which is excreted, and ornithine, which reenters the urea cycle. *N*-Acetylglutamate regulates the urea cycle by activating carbamoyl phosphate synthetase allosterically.

The α-keto acid products of transamination reactions are degraded to citric acid cycle intermediates or their precursors. The amino acids leucine and lysine are ketogenic in that they are converted only to the ketone body precursors acetyl-CoA and acetoacetate. The remaining amino acids are, at least in part, glucogenic in that they are converted to the glucose precursors pyruvate, oxaloacetate, α-ketoglutarate, succinyl-CoA, and fumarate. Alanine, cysteine, glycine, serine, and threonine are converted to pyruvate. Serine hydroxymethyl transferase catalyzes the PLP-dependent C_α—C_β bond cleavage of serine to form glycine. The reaction requires tetrahydrofolate (THF) to act as a C_1 unit acceptor. Asparagine and aspartate are converted to oxaloacetate. α-Ketoglutarate is a product of arginine, glutamate, glutamine, histidine, and proline degradation. Methionine, isoleucine, and valine are degraded to succinyl-CoA. Methionine breakdown involves the synthesis of *S*-adenosylmethionine (SAM), a sulfonium ion that acts as a methyl donor in many biosynthetic reactions. Maple syrup urine disease is caused by an inherited defect in branched-chain amino acid degradation. Branched-chain amino acid degradation pathways contain reactions common to all acyl-CoA oxidations. Tryptophan is degraded to alanine and acetoacetate. Phenylalanine and tyrosine are degraded to fumarate and acetoacetate. Most individuals with the hereditary disease phenylketonuria lack phenylalanine hydroxylase, which converts phenylalanine to tyrosine.

Heme is synthesized from glycine and succinyl-CoA. These precursors condense to form δ-aminolevulinic acid (ALA), which cyclizes to form the pyrrole, porphobilinogen (PBG). Four molecules of PBG condense to form uroporphyrinogen III, which then goes on to form heme. Heme is degraded to form linear tetrapyrroles, which are subsequently excreted as bile pigments. The hormones and neurotransmitters L-DOPA, epinephrine, norepinephrine, serotonin, γ-aminobutyric acid (GABA), and histamine are all synthesized from amino acid precursors. Glutathione, a tripeptide that is synthesized from glutamate, cysteine, and glycine, is involved in a variety of protective, transport, and metabolic processes. Tetrahydrofolate is a coenzyme that participates in the transfer of C_1 units.

Amino acids are required for many vital functions of an organism. Those that mammals can synthesize are known as nonessential amino acids; those that mammals must obtain from their diet are called essential amino acids. The biosynthesis of nonessential amino acids involves relatively simple pathways, whereas those forming the essential amino acids are generally more complex.

Although the ultimate source of nitrogen for amino acid biosynthesis is atmospheric N_2, this nearly inert gas must first be reduced to a metabolically useful form, NH_3, by nitrogen fixation. This process occurs only in certain types of bacteria, one genus of which occurs in symbiotic relationship with legumes. N_2 is fixed in these organisms by an oxygen-sensitive enzyme, nitrogenase, that consists of two proteins: the Fe-protein, which contains a [4Fe–4S] cluster, and the MoFe-protein, which contains two P-cluster pairs (each consisting of two linked [4Fe–4S] clusters) and two FeMo-cofactors ([1Mo–3Fe–3S] clusters). These cofactors function as the two electron carrier for the ATP-driven reduction of N_2 to NH_3.

REFERENCES

General

Bender, David A., *Amino Acid Metabolism,* Wiley (1985).

Scriver, C.R., Beaudet, A.L., Sly, W.S., and Valle, D. (Eds.), *The Metabolic Basis of Inherited Disease* (6th ed.), Chapters 15–28, 31, 52, and 53, McGraw-Hill (1989).

Walsh, C., *Enzymatic Reaction Mechanisms,* Chapters 24 and 25, Freeman (1979). [Discusses reactions involving PLP, THF, and SAM cofactors.]

Amino Acid Deamination and the Urea Cycle

Baker, P.J., Britton, K.L., Engel, P.C., Farrants, G.W., Lilley, K.S., Rice, D.W., and Stillman, T.J., Subunit assembly and active site location in the structure of glutamate dehydrogenase, *Proteins* **12,** 75–86 (1992). [The X-ray structure of the *Clostridium symbiosum* enzyme.]

Braunstein, A.E., Amino group transfer, *in* Boyer, P.D. (Ed.), *The Enzymes* (3rd ed.), Vol. 9, *pp.* 379–482, Academic Press (1973).

Cohen, P.P., The ornithine–urea cycle: biosynthesis and regulation of carbamyl phosphate synthetase I and ornithine transcarbamylase, *Curr. Top. Cell. Reg.* **18,** 1–19 (1981). [An interesting historical review of the discovery of urea and the urea cycle, as well as a discussion of the cycle's regulation.]

Hayashi, H., Wada, H., Yoshimura, T., Esaki, N., and Soda, K., Recent topics in pyridoxal 5′-phosphate enzyme studies, *Annu. Rev. Biochem.* **59,** 87–110 (1990).

Martell, A.E., Vitamin B_6 catalyzed reactions of α-amino and α-keto acids: model systems, *Acc. Chem. Res.* **22,** 115–124 (1989).

Meijer, A.J., Lamers, W.H., and Chamuleau, R.A.F.M., Nitrogen metabolism and ornithine cycle function, *Physiol. Rev.* **70,** 701–748 (1990).

Smith, E.L., Austen, B.M., Blumenthal, K.M., and Nyc, J.F., Glutamate dehydrogenases, *in* Boyer, P.D. (Ed.), *The Enzymes* (3rd ed.), Vol. 11, *pp.* 293–367, Academic Press (1975).

Torchinsky, Yu.M., Transamination: its discovery, biological and chemical aspects (1937–1987), *Trends Biochem. Sci.* **12,** 115–117 (1987).

Degradation and Biosynthesis of Amino Acids

Adams, E. and Frand, L., Metabolism of proline and the hydroxyprolines, *Annu. Rev. Biochem.* **49,** 1005–1061 (1980).

Almassey, R.J., Janson, C.A., Hamlin, R., Xuong, N.-H., and Eisenberg, D., Novel subunit–subunit interactions in the structure of glutamine synthetase, *Nature* **323,** 304–309 (1986); Yamashita, M., Almassy, R.J., Janson, C.A., Cascio, D., and Eisenberg, D., Refined atomic model of glutamine synthetase at 3.5 Å resolution, *J. Biol. Chem.* **264,** 17681–17690 (1989); *and* Liaw, S.-H. and Eisenberg, D., Structural model for the reaction mechanism of glutamine synthetase, based on five crystal structures of enzyme–substrate complexes, *Biochemistry* **33,** 675–681 (1994).

Cooper, A.J.L., Biochemistry of sulfur-containing amino acids, *Annu. Rev. Biochem.* **52,** 187–222 (1983).

Herrmann, K.M. and Somerville, R.L. (Eds.), *Amino Acids: Biosynthesis and Genetic Regulation,* Addison–Wesley (1983).

Hyde, C.C. and Miles, E.W., The tryptophan synthase multienzyme complex: Exploring structure-function relationships with X-ray crystallography and mutagenesis, *Biotechnology* **8,** 27-32 (1990).

Jansonius, J.N. and Vincent, M.G., Structural basis for catalysis by aspartate aminotransferase, *in* Jurnak, F.A. and McPherson, A. (Eds.), *Biological Macromolecules and Assemblies,* Vol. 3, *pp.* 187–285, Wiley (1987).

Kishore, G.M. and Shah, D.M., Amino acid biosynthesis inhibitors as herbicides, *Annu. Rev. Biochem.* **57,** 627–663 (1988). [Discusses the biosynthesis of the essential amino acids.]

Miles, E.W., Structural basis for catalysis by tryptophan synthase, *Adv. Enzymol.* **64,** 93–172 (1991); *and* Hyde, C.C., Ahmed, S.A., Padlan, E.A., Miles, E.W., and Davies, D.R., Three-dimensional structure of the tryptophan synthase $\alpha_2\beta_2$ multienzyme complex from *Salmonella typhimurium, J. Biol. Chem.* **263,** 17857–17871 (1988).

Nichol, C.A., Smith, G.K., and Duch, D.S., Biosynthesis and metabolism of tetrahydrobiopterin and molybdopterin, *Annu. Rev. Biochem.* **54,** 729–764 (1985).

Stadtman, E.R. and Ginsburg, A., The glutamine synthetase of *E. coli:* structure and control, *in* Boyer, P.D. (Ed.), *The Enzymes* (3rd ed.), Vol. 10, *pp.* 755–807, Academic Press (1974).

Stallings, W.C., Abdel-Meguid, S.S., Lim, L.W., Shieh, H.-S., Dayringer, H.E., Leimgruber, N.K., Stegeman, R.A., Anderson, K.S., Sikorski, J.A., Padgette, S.R., and Kishore, G.M., Structure and topological symmetry of the glyphosate target 5-*enol*-pyruvylshikimate-3-phosphate synthase: A distinctive protein fold, *Proc. Natl. Acad. Sci.* **88,** 5046–5050 (1991). [The enzyme that catalyzes Reaction 6 of Fig. 24-49 in complex with glyphosate, an inhibitor that is a broad-spectrum herbicide.]

Swain, A.L., Jaskólski, M., Housset, D., Rao, J.K.M., and Wladower, A., Crystal structure of *Eschericia coli* L-asparaginase, an enzyme used in cancer therapy, *Proc. Natl. Acad. Sci.* **90**, 1474–1478 (1993).

Tyler, B., Regulation of the assimilation of nitrogen compounds, *Annu. Rev. Biochem.* **47**, 1127–1162 (1978).

Umbarger, H.E., Amino acid biosynthesis and its regulation, *Annu. Rev. Biochem.* **47**, 533–606 (1978).

Varughese, K.I., Skinner, M.M., Whiteley, J.M., Matthews, D.A., and Xuong, N.H., Crystal structure of rat liver dihydropteridine reductase, *Proc. Natl. Acad. Sci.* **89**, 6080–6084 (1992).

Wellner, D. and Meister, A., A survey of inborn errors of amino acid metabolism and transport in man, *Annu. Rev. Biochem.* **50**, 911–968 (1981).

Wilmanns, M., Priestle, J.P., Niermann, T., and Jansonius, J.N., Three-dimensional structure of the bifunctional enzyme phosphoribosylanthranilate isomerase:indoleglycerolphosphate synthase from *Escherichia coli* refined at 2.0 Å resolution, *J. Mol. Biol.* **223**, 477–507 (1992). [The enzyme that catalyzes Reactions 3 and 4 in Fig. 24-50.]

Amino Acids as Biosynthetic Precursors

Battersby, A.R., Fookes, C.J.R., Matcham, G.W.J., and McDonald, E., Biosynthesis of the pigments of life: formation of the macrocycle, *Nature* **285**, 17–21 (1980).

Beru, N. and Goldwasser, E., The regulation of heme biosynthesis during erythropoietin-induced erythroid differentiation, *J. Biol. Chem.* **260**, 9251–9257 (1985).

Cooper, J.R., Bloom, F.E., and Roth, R.H., *The Biochemical Basis of Neuropharmacology* (4th ed.), Chapters 6–9 and 12, Oxford University Press (1982).

Grandchamp, B., Beaumont, C., de Verneuil, H., and Nordmann, Y., Accumulation of porphobilinogen deaminase, uroporphyrinogen decarboxylase, and α- and β-globin mRNAs during differentiation of mouse erythroleukemic cells: effects of succinylacetone, *J. Biol. Chem.* **260**, 9630–9635 (1985).

Jordan, P.M. and Gibbs, P.N.B., Mechanism of action of 5-aminolevulinate dehydratase from human erythrocytes, *Biochem. J.* **227**, 1015–1021 (1985).

Louie, G.V., Brownlie, P.D., Lambert, R., Cooper, J.B., Blundell, T.L., Wood, S.P., Warren, M.J., Woodcock, S.C., and Jordan, P.M., Structure of porphobilinogen deaminase reveals a flexible multidomain polymerase with a single catalytic site, *Nature* **359**, 33–39 (1992).

Macalpine, I. and Hunter, R., Porphyria and King George III, *Sci. Am.* **221**(1): 38–46 (1969).

Meister, A., Glutathione metabolism and its selective modification, *J. Biol. Chem.* **263**, 17205–17208 (1988); *and* Meister, A. and Anderson, M.E., Glutathione, *Annu. Rev. Biochem.* **52**, 711–760 (1983).

Meister, A., Glutamine synthetase of mammals, *in* Boyer, P.D. (Ed.), *The Enzymes* (3rd ed.), Vol. 10, *pp.* 699–754, Academic Press (1974).

Padmanaban, G., Venkateswar, V., and Rangarajan, P.N., Haem as a multifunctional regulator, *Trends Biochem. Sci.* **14**, 492–496 (1989).

Perutz, M.F., Myoglobin and hemoglobin: role of distal residues in reactions with haem ligands, *Trends Biochem. Sci.* **14**, 42–44 (1989).

Ponka, P. and Schulman, H.M., Acquisition of iron from transferrin regulated reticulocyte heme synthesis, *J. Biol. Chem.* **260**, 14717–14721 (1985).

Rutherford, T., Thompson, G.G., and Moore, M.R., Heme biosynthesis in Friend erythroleukemia cells: control by ferrochelatase, *Proc. Natl. Acad. Sci.* **76**, 833–836 (1979).

Slater, A.F.G., and Cerami, A., Inhibition by chloroquine of a novel haem polymerase enzyme activity in malaria trophozoites, *Nature* **355**, 167–169 (1992).

Warren, M.J. and Scott, A.I., Tetrapyrrole assembly and modification into the ligands of biologically functional cofactors, *Trends Biochem. Sci.* **15**, 486–491 (1990).

Wellems, T.E., How chloroquine works, *Nature* **355**, 108–109 (1992).

Nitrogen Fixation

Bolin, J.T., Campobasso, N., Muchmore, S.W., Morgan, T.V., and Mortenson, L.E., Structure and environment of metal clusters in nitrogenase molybdenum–iron protein from *Clostridium pasteurianum, in* Stiefel, E.I., Coucouvanis, D., and Newton, W.E. (Eds.), *Molybdenum Enzymes, Cofactors, and Model Systems, pp.* 186–195, American Chemical Society (1993).

Burris, R.H., Nitrogenases, *J. Biol. Chem.* **266**, 9339–9342 (1991).

Chan, M.K., Kim, J., and Rees, D.C., The nitrogenase FeMo-cofactor and P-cluster pair: 2.2 Å resolution structure, *Science* **260**, 792–794 (1993).

Deng, H. and Hoffman, R., How N_2 might be activated by the FeMo-cofactor in nitrogenase, *Angew. Chem. Int. Ed. Engl.* **32**, 1062–1065 (1993).

Dilworth, M. and Glenn, A., How does a legume nodule work? *Trends Biochem. Sci.* **9**, 519–523 (1984).

Fisher, R.F. and Long, S.R., *Rhizobium*–plant signal exchange, *Nature* **357**, 655–660 (1992). [Discusses the signals through which Rhizobiaceae and legumes communicate to symbiotically generate the root nodules in which nitrogen fixation occurs.]

Georgiadis, M.M., Kormia, H., Chakrabarti, P., Woo, D., Kornuc, J.J., and Rees, D.C., Crystallographic structure of the nitrogenase iron protein from *Azotobacter vinelandii, Science* **257**, 1653–1659 (1993).

Haaker, H. and Veeger, C., Enzymology of nitrogen fixation, *Trends Biochem. Sci.* **9**, 188–192 (1984).

Kim, J. and Rees, D.C., Nitrogenase and biological nitrogen fixation, *Biochemistry* **33**, 389–397 (1994); *and* Rees, D. C., Dinitrogen reduction by nitrogenase: if N_2 isn't broken, it can't be fixed, *Curr. Opin. Struct. Biol.* **3**, 921–928 (1993).

Kim, J. and Rees, D.C., Crystallographic structure and functional implications of the nitrogenase molybdenum–iron protein from *Azotobacter vinelandii, Nature* **360**, 553–560 (1992); *and* Structural models for the metal centers in the nitrogenase molybdenum–iron protein, *Science* **257**, 1677–1682 (1992).

Orme-Johnson, W.H., Molecular basis of nitrogen fixation, *Annu. Rev. Biophys. Biophys. Chem.* **14**, 419–459 (1985).

PROBLEMS

1. The symptoms of the partial deficiency of a urea cycle enzyme may be attenuated by a low-protein diet. Explain.

2. Why are people on a high-protein diet instructed to drink lots of water?

3. A student on a particular diet expends $10,000 \text{ kJ} \cdot \text{day}^{-1}$ while excreting 40 g of urea. Assuming that protein is 16% N by weight and that its metabolism yields $18 \text{ kJ} \cdot \text{g}^{-1}$, what percentage of the student's energy requirement is met by protein?

*4. Among the many eat-all-you-want-and-lose-weight diets that have been popular for a time is one that eliminates all carbohydrates but permits the consumption of all the protein and fat desired. Would such a diet be effective? (*Hint:* Individuals on such a diet often complain that they have bad breath.)

5. Why are phenylketonurics warned against eating products containing the artificial sweetener **aspartame (NutraSweet®**; chemical name L-aspartyl-L-phenylalanine methyl ester)?

6. Demonstrate that the synthesis of heme from PBG as labeled in Fig. 24-27 results in the heme-labeling pattern given in Fig. 24-24.

7. Explain why certain drugs and other chemicals can precipitate an attack of acute intermittent porphyria.

8. One of the symptoms of **kwashiorkor,** the dietary protein deficiency disease in children, is the depigmentation of the skin and hair. Explain the biochemical basis of this symptom.

9. What are the metabolic consequences of a defective uridylyl-removing enzyme in *E. coli*?

10. Figure 24-47, Reaction 9, indicates that methionine is synthesized in microorganisms by the methylation of homocysteine in a reaction in which N^5-methyl-THF is the methyl donor. Yet, in the breakdown of methionine (Fig. 24-14), its demethylation occurs in three steps in which SAM is an intermediate. Discuss why this reaction does not occur via the simpler one-step reverse of the methylation reaction.

*11. In the glucose–alanine cycle, glycolytically derived pyruvate is transaminated to alanine and exported to the liver for conversion to glucose and return to the cell. Explain how a muscle cell is able to participate in this cycle under anaerobic (vigorously contracting) conditions. (*Hint:* The breakdown of many amino acids yields NH_3.)

12. Suggest a reason why the nitrogen fixing heterocysts of cyanobacteria have lost Photosystem II but retain Photosystem I.

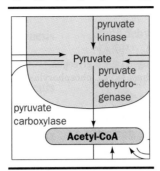

25

Energy Metabolism: Integration and Organ Specialization

At this point in our narrative we have studied all of the major pathways of energy metabolism. Consequently, we are now in a position to consider how organisms, mammals in particular, orchestrate the metabolic symphony to meet their energy needs. This chapter therefore begins with a recapitulation of the major metabolic pathways and their control systems, then considers how these processes are apportioned among the various organs of the body, and ends with a discussion of how the body deals with the metabolic challenge of starvation and how it responds to the loss of control resulting from diabetes mellitus.

1. MAJOR PATHWAYS AND STRATEGIES OF ENERGY METABOLISM: A SUMMARY

Figure 25-1 indicates the interrelationships among the major pathways involved in energy metabolism. Let us review these pathways and their control mechanisms.

1. Glycolysis (Chapter 16)

The metabolic degradation of glucose begins with its conversion to two molecules of pyruvate with the net generation of two molecules each of ATP and NADH. Under anaerobic conditions, pyruvate is converted to lactate (or, in yeast, to ethanol) so as to recycle the NADH. Under aerobic conditions, however, when glycolysis serves to prepare glucose for further oxidation, the NAD^+ is regenerated through oxidative phosphorylation (see below). The flow of metabolites through the glycolytic pathway is largely controlled by the activity of phosphofructokinase (PFK). This enzyme is activated by AMP and ADP, whose concentrations rise as the need for energy metabolism increases, and is inhibited by ATP and citrate, whose concentrations increase when the demand for energy metabolism has slackened. PFK is also activated by fructose-2,6-bisphosphate, whose concentration is regulated by the levels of glucagon, epinephrine, and norepinephrine through the intermediacy

tion of biologically active substances such as drugs, poisons, and hormones by a variety of oxidation, reduction, hydrolysis, conjugation, and methylation reactions.

3. METABOLIC ADAPTATION

In this section we consider the body's responses to two metabolically abnormal situations: (1) starvation and (2) the disease diabetes mellitus.

A. Starvation

Glucose is the metabolite of choice of both brain and working muscle. Yet, the body stores less than a day's supply of carbohydrate (Table 25-1). Thus, the low blood sugar resulting from even an overnight fast results, through an increase in glucagon secretion and a decrease in insulin secretion, in the mobilization of fatty acids from adipose tissue (Section 23-5). The diminished insulin level also inhibits glucose uptake by muscle tissue. Muscles therefore switch from glucose to fatty acid metabolism for energy production. The brain, however, still remains heavily dependent on glucose.

In animals, glucose cannot be synthesized from fatty acids. This is because neither pyruvate nor oxaloacetate, the precursors of glucose in gluconeogenesis (Section 21-1), can be synthesized from acetyl-CoA (the oxaloacetate in the citric acid cycle is derived from acetyl-CoA but the cyclic nature of this process requires that the oxaloacetate be consumed as fast as it is synthesized; Section 19-1A). During starvation, glucose must therefore be synthesized from the glycerol product of triacylglycerol breakdown and, more importantly, from the amino acids derived from the proteolytic degradation of proteins, the major source of which is muscle. Yet, the continued breakdown of muscle during prolonged starvation would ensure that this process became irreversible since a large muscle mass is essential for an animal to move about in search of food. The organism must therefore make alternate metabolic arrangements.

After several days of starvation, gluconeogenesis has so depleted the liver's oxaloacetate supply that this organ's ability to metabolize acetyl-CoA via the citric acid cycle is greatly diminished. Rather, the liver converts the acetyl-CoA to ketone bodies (Section 23-3), which it releases into the blood. The brain gradually adapts to using ketone bodies as fuel through the synthesis of the appropriate enzymes: After a 3-day fast, only about one third of the brain's energy requirements are satisfied by ketone bodies but after 40 days of starvation, $\sim$70% of its energy needs are so met. The rate of muscle breakdown during prolonged starvation consequently decreases to $\sim$25% of its rate after a several-day fast. The survival time of a starving individual is therefore much more dependent on the size of his or her fat reserves than it is on his or her muscle mass. Indeed, highly obese individuals can survive for over a year without eating (and have occasionally done so in clinically supervised weight reduction programs).

B. Diabetes Mellitus

The polypeptide hormone insulin acts mainly on muscle, liver, and adipose tissue cells to stimulate the synthesis of glycogen, fats, and proteins while inhibiting the breakdown of these metabolic fuels. In addition, insulin stimulates the uptake of glucose by most cells, with the notable exception of brain and liver cells. Together with glucagon, which has largely opposite effects, insulin acts to maintain the proper level of blood glucose.

In the disease **diabetes mellitus,** which is the third leading cause of death in the United States after heart disease and cancer, insulin either is not secreted in sufficient amounts or does not efficiently stimulate its target cells. As a consequence, blood glucose levels become so elevated that the glucose "spills over" into the urine, providing a convenient diagnostic test for the disease. Yet, despite these high blood glucose levels, cells "starve" since insulin-stimulated glucose entry into cells is impaired. Triacylglycerol hydrolysis, fatty acid oxidation, gluconeogenesis, and ketone body formation are accelerated and, in a condition termed **ketosis,** ketone body levels in the blood become abnormally high. Since ketone bodies are acids, their high concentration puts a strain on the buffering capacity of the blood and on the kidney, which controls blood pH by excreting the excess H^+ into the urine. This H^+ excretion is accompanied by Na^+, K^+, P_i, and H_2O excretion, causing severe dehydration (which compounds the dehydration resulting from the osmotic effect of the high glucose concentration in the blood; excessive thirst is a classical symptom of diabetes) and a decrease in blood volume—ultimately life-threatening situations.

Their are two major forms of diabetes mellitus:

1. Insulin-dependent or **juvenile-onset diabetes mellitus,** which most often strikes suddenly in childhood.

TABLE 25-1.   FUEL RESERVES FOR A NORMAL 70-KG MAN

Fuel	Mass (kg)	Calories[a]
Tissues		
Fat (adipose triacyglycerols)	15	141,000
Protein (mainly muscle)	6	24,000
Glycogen (muscle)	0.150	600
Glycogen (liver)	0.075	300
Circulating fuels		
Glucose (extracellular fluid)	0.020	80
Free fatty acids (plasma)	0.0003	3
Triacylglycerols (plasma)	0.003	30
Total		166,000

[a] 1 (dieter's) Calorie = 1 kcal = 4.184 kJ.

Source: Cahill, G.F., Jr., *New Engl. J. Med.* **282,** 669 (1970).

2. Noninsulin-dependent or **maturity-onset diabetes mellitus,** which usually develops rather gradually after the age of 40.

Insulin-Dependent Diabetes Is Caused by a Deficiency of Pancreatic β Cells

In insulin-dependent (type I) diabetes mellitus, insulin is absent or nearly so because the pancreas lacks or has defective β cells. This condition results, in genetically susceptible individuals (see below), from an autoimmune response that selectively destroys their β cells. Individuals with insulin-dependent diabetes, as Frederick Banting and George Best first demonstrated in 1921, require daily insulin injections to survive and must follow carefully balanced diet and exercise regimens. Their lifespans are, nevertheless, reduced by up to one third as a result of degenerative complications such as kidney malfunction, nerve impairment, and cardiovascular disease, as well as blindness, that apparently arise from the imprecise metabolic control provided by periodic insulin injections. Perhaps newly developed systems that monitor blood glucose levels and continuously deliver insulin in the required amounts will rectify this situation.

The usually rapid onset of the symptoms of insulin-dependent diabetes had suggested that the autoimmune attack on the pancreatic β cells responsible for this disease is one of short duration. Typically, however, the disease "brews" for several years as the aberrantly aroused immune system slowly destroys the β cells. Only when >80% of these cells have been eliminated do the classic symptoms of diabetes suddenly emerge.

Why does the immune system attack the pancreatic β cells? It has long been known that certain alleles (genetic variants) of the **Class II major histocompatability complex (MHC) proteins** are particularly common in insulin-dependent diabetics [MHC proteins are highly polymorphic (variable within a species) immune system components to which cell-generated antigens such as viral proteins must bind in order to be recognized as foreign; Sections 34-2A and E]. It is thought that autoimmunity against β cells is induced in a susceptible individual by a foreign antigen, perhaps a virus, that immunologically resembles some β cell component. The Class II MHC protein that binds this antigen does so with such tenacity that it stimulates the immune sytem to launch an unusually vigorous and prolonged attack on the antigen. Some of the activated immune system cells eventually make their way to the pancreas, where they initiate an attack on the β cells due to the close resemblance of the β cell component to the foreign antigen.

Noninsulin-Dependent Diabetes Is Probably Caused by a Deficiency of Insulin Receptors

Non-insulin-dependent (type II) diabetes mellitus (NIDDM), which accounts for over 90% of the diagnosed cases of diabetes and affects 18% of the population over 65 years of age, usually occurs in obese individuals with a genetic predisposition for this condition (although one that differs from that associated with insulin-dependent diabetes). These individuals have normal or even greatly elevated insulin levels. Their symptoms arise from an apparent paucity of **insulin receptors** on normally insulin-responsive cells. Perhaps the increased insulin production resulting from overeating (obesity is almost always the consequence of overeating) eventually suppresses the synthesis of insulin receptor (a plasma membrane-bound glycoprotein; Section 34-4B). This hypothesis accounts for the observation that diet alone is often sufficient to control this type of diabetes.

CHAPTER SUMMARY

The complex network of processes involved in energy metabolism are distributed among different compartments within cells and in different organs of the body. These processes function to generate ATP "on demand," to generate and store glucose, triacylglycerols, and proteins in times of plenty for use when needed, and to keep the concentration of glucose in the blood at the proper level for use by organs such as the brain whose sole fuel source, under normal conditions, is glucose. The major energy metabolism pathways include glycolysis, glycogen degradation and synthesis, gluconeogenesis, the pentose phosphate pathway, and triacylglycerol and fatty acid synthesis, which are cytosolically based, in addition to fatty acid oxidation, the citric acid cycle, and oxidative phosphorylation, which are confined to the mitochondrion. Amino acid degradation occurs, in part, in both compartments. The mediated membrane transport of metabolites therefore also plays an essential metabolic role.

The main organs involved in energy metabolism are brain, muscle, adipose tissue, and liver. The brain normally consumes large amounts of glucose. Muscle, under intense ATP demand such as in sprinting, degrades glucose and glycogen anaerobically, thereby producing lactate, which is exported via the blood to the liver for reconversion to glucose through gluconeogenesis. During moderate activity, muscle generates ATP by oxidizing glucose from glycogen, fatty acids, and ketone bodies completely to CO_2 and H_2O via the citric acid cycle and oxidative phosphorylation. Adipose tissue stores triacylglycerols and releases fatty acids into the bloodstream in response to the organism's metabolic needs. These metabolic needs are communicated to adipose tissue by means of the hormones insulin, which indicates a fed state in which storage is appropriate, and glucagon, epinephrine, and norepinephrine, which signal a need for fatty acid release to provide fuel for other tissues. The liver, the body's central metabolic clearing house, maintains blood glucose concentrations by storing glucose as glycogen in times of plenty and releasing glucose in times of need both by glycogen breakdown and gluconeogenesis. It also converts fatty acids to ketone bodies for use by peripheral tissues. During a fast, it

relatively insensitive to FdUMP (some exceptions are the bone marrow cells that comprise the blood-forming tissues and much of the immune system, the intestinal mucosa, and hair follicles). **5-Fluorouracil** and **5-fluorodeoxyuridine** are also effective antitumor agents since they are converted to FdUMP through salvage reactions.

N^5, N^{10}-Methylene-THF Is Regenerated in Two Reactions

The thymidylate synthase reaction is biochemically unique in that it oxidizes THF to DHF; no other enzymatic reaction employing a THF cofactor alters this coenzyme's net oxidation state. The DHF product of the thymidylate synthase reaction is recycled to the enzyme's N^5, N^{10}-meth-

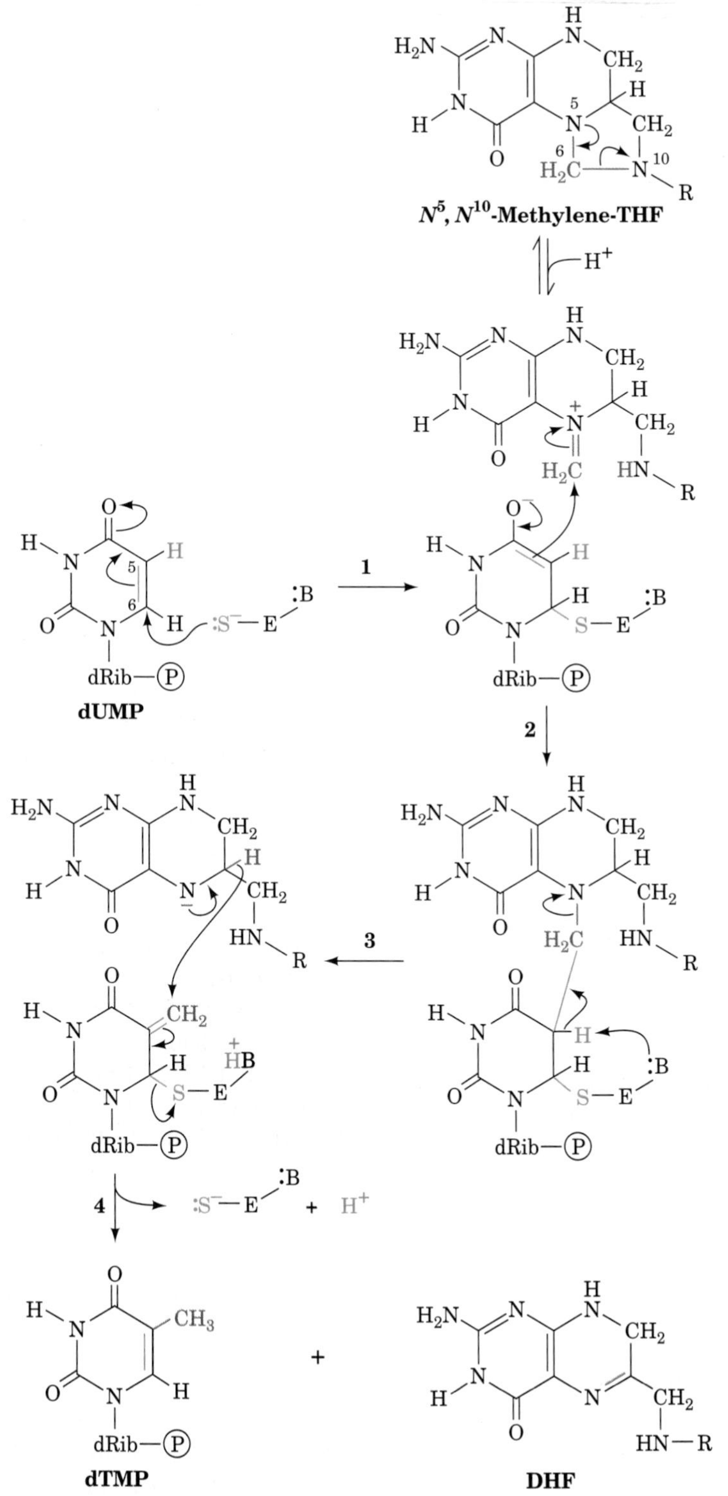

FIGURE 26-18. The catalytic mechanism of thymidylate synthase. The methyl group is supplied by N^5, N^{10}-methylene-THF, which is concomitantly oxidized to dihydrofolate.

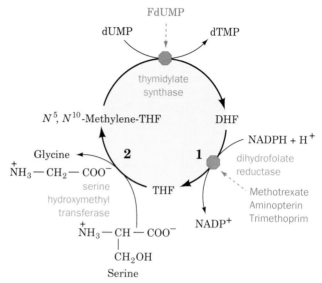

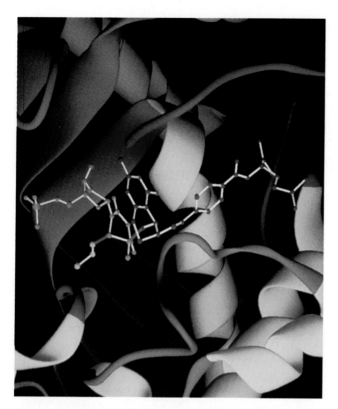

FIGURE 26-19. A ribbon diagram of the active site region of one subunit of the dimeric *E. coli* enzyme thymidylate synthase in covalent complex with FdUMP (*green spheres*) and N^5, N^{10}-methylene-THF (*blue spheres*). The C5 and C6 atoms of FdUMP form covalent bonds (*red*) with the CH_2 group substituent to N5 of THF and the S atom of Cys 146 (*yellow spheres*). The polypeptide backbone is colored such that its helices are yellow, its β sheets are brown, and its remaining segments are blue. [Courtesy of Jesus Villafranca and David Matthews, Agouron Pharmaceuticals, La Jolla, California.]

FIGURE 26-20. The N^5, N^{10}-methylenetetrahydrofolate that is converted to DHF in the thymidylate synthase reaction is regenerated by the sequential actions of (1) dihydrofolate reductase and (2) serine hydroxymethyl transferase. Thymidylate synthase is inhibited by FdUMP, whereas dihydrofolate reductase is inhibited by the antifolates methotrexate, aminopterin, and trimethoprim.

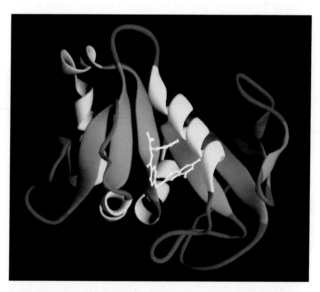

FIGURE 26-21. A ribbon diagram of human dihydrofolate reductase in complex with folate. The helices of this monomeric enzyme are drawn in yellow, the β sheets in brown, and the other polypeptide segments in blue. [Courtesy of Jay F. Davies, II, and Joseph Kraut, University of California at San Diego.]

ylene-THF cofactor through two sequential reactions (Fig. 26-20):

1. DHF is reduced to THF by NADPH as catalyzed by **dihydrofolate reductase (DHFR;** Section 24-4D). Although, in most organisms, DHFR is a monomeric monofunctional enzyme, in protozoa and at least some plants, DHFR and TS occur on the same polypeptide chain to form a bifunctional enzyme that has been shown to channel DHF from its TS to its DHFR active sites.

2. Serine hydroxymethyl transferase (Section 24-3B) transfers the hydroxymethyl group of serine to THF yielding N^5, N^{10}-methylene-THF and glycine.

Antifolates Are Anticancer Agents

Inhibition of DHFR quickly results in all of a cell's limited supply of THF being converted to DHF by the thymidylate synthase reaction. Inhibition of DHFR therefore not only prevents dTMP synthesis (Fig. 26-20), but also blocks all other THF-dependent biological reactions such as the synthesis of purines (Section 26-2A), histidine, and methionine (Section 24-5B). DHFR (Fig. 26-21) therefore offers an attractive target for chemotherapy.

Methotrexate (amethopterin), aminopterin, and **trimethoprim**

R = H   **Aminopterin**
R = CH$_3$ **Methotrexate (amethopterin)**

Trimethoprim

are DHF analogs that competitively although nearly irreversibly bind to DHFR with an ~1000-fold greater affinity than does DHF. These **antifolates** (substances that interfere with the action of folate cofactors) are effective anticancer agents, particularly against childhood leukemias. In fact, a successful chemotherapeutic strategy is to treat a cancer victim with a lethal dose of methotrexate and some hours later "rescue" the patient (but hopefully not the cancer) by administering massive doses of 5-formyl-THF and/or thymidine. Trimethoprim, which was discovered by George Hitchings and Gertrude Elion, binds much more tightly to bacterial DHFRs than to those of mammals and is therefore a clinically useful antibacterial agent.

5. NUCLEOTIDE DEGRADATION

Most foodstuffs, being of cellular origin, contain nucleic acids. Dietary nucleic acids survive the acid medium of the stomach; they are degraded to their component nucleotides, mainly in the duodenum, by pancreatic nucleases and intestinal phosphodiesterases. These ionic compounds, which cannot pass through cell membranes, are then hydrolyzed to nucleosides by a variety of group-specific nucleotidases and nonspecific phosphatases. Nucleosides may be directly absorbed by the intestinal mucosa or first undergo further degradation to free bases and ribose or ribose-1-phosphate through the action of **nucleosidases** and **nucleoside phosphorylases:**

$$\text{Nucleoside} + \text{H}_2\text{O} \xrightarrow{\text{nucleosidase}} \text{base} + \text{ribose}$$

$$\text{Nucleoside} + \text{P}_i \xrightarrow[\text{phosphorylase}]{\text{nucleoside}} \text{base} + \text{ribose-1-P}$$

Radioactive labeling experiments have demonstrated that only a small fraction of the bases of ingested nucleic acids are incorporated into tissue nucleic acids. Evidently, the *de novo* pathways of nucleotide biosynthesis largely satisfy an organism's need for nucleotides. Consequently, ingested bases, for the most part, are degraded and excreted. Cellular nucleic acids are also subject to degradation as part of the continual turnover of nearly all cellular components. In this section we outline these catabolic pathways and discuss the consequences of several of their inherited defects.

A. Catabolism of Purines

The major pathways of purine nucleotide and deoxynucleotide catabolism in animals are diagrammed in Fig. 26-22. Other organisms may have somewhat different pathways among these various intermediates (including adenine) but all of these pathways lead to uric acid. Of course, the intermediates in these processes may instead be reused to form nucleotides via salvage reactions. In addition, ribose-1-phosphate, a product of the reaction catalyzed by **purine nucleoside phosphorylase (PNP;** Fig. 26-23), is isomerized by **phosphoribomutase** to the PRPP precursor ribose-5-phosphate.

Adenosine and deoxyadenosine are not degraded by mammalian PNP. Rather, adenine nucleosides and nucleotides are deaminated by **adenosine deaminase (ADA)** and **AMP deaminase** to their corresponding inosine derivatives, which, in turn, may be further degraded. The X-ray structure of murine ADA that was crystallized in the presence of its inhibitor **purine ribonucleoside** was determined by Florante Quiocho (Fig. 26-24a). The enzyme forms an eight-stranded α/β barrel (Fig. 7-19b) with its active site in a pocket at the C-terminal end of the β barrel, as occurs in all known α/β barrel enzymes (Section 16-2E). Purine ribonucleoside binds to ADA in a normally rare hydrated form, **6-hydroxyl-1,6-dihydropurine ribonucleoside (HDPR),**

Purine ribonucleoside

6-Hydroxyl-1,6-dihydropurine ribonucleoside (HDPR)

a nearly ideal transition state analog of the ADA reaction. Although it had been previously reported that ADA does not require a cofactor, its X-ray structure clearly reveals that a zinc ion is bound in the deepest part of the active site pocket, where it is pentacoordinated by three His side chains, a carboxyl oxygen of Asp 295, and the O6 atom of HDPR. ADA's active site complex suggests a catalytic mechanism (Fig. 26-24b) reminiscent of that of carbonic anhydrase (Section 14-1C): His 238, which is properly posi-

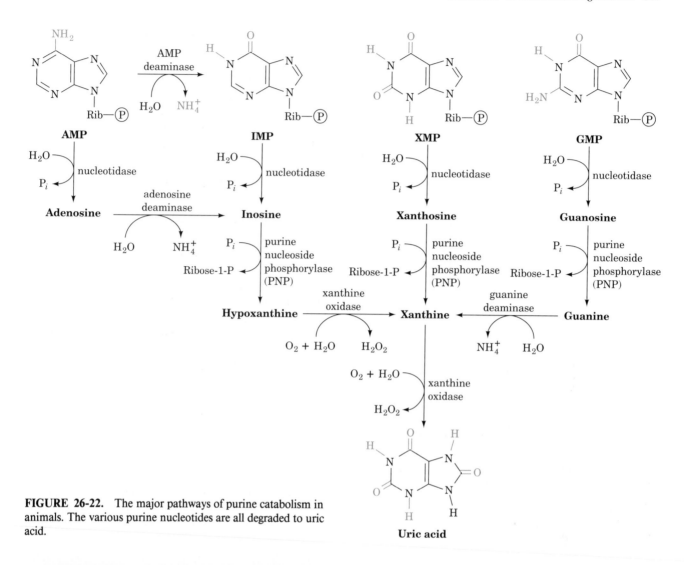

FIGURE 26-22. The major pathways of purine catabolism in animals. The various purine nucleotides are all degraded to uric acid.

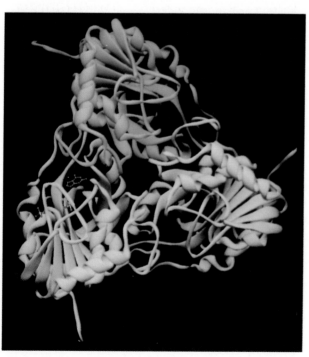

tioned to act as a general base, abstracts a proton from a bound Zn^{2+}-activated water molecule, which nucleophilically attacks the adenine C6 atom to form a tetrahedral intermediate. Products are then formed by the elimination of ammonia.

Genetic Defects in ADA Result in Severe Combined Immunodeficiency Disease

Abnormalities in purine nucleoside metabolism arising from rare genetic defects in ADA selectively kill **lymphocytes** (a type of white blood cell). Since lymphocytes mediate much of the immune response (Section 34-2A), ADA

FIGURE 26-23. The X-ray structure of human erythrocyte purine nucleoside phosphorylase as viewed along this trimeric enzyme's threefold axis. Each identical subunit is differently colored with that in yellow shown in complex with a guanine molecule and two phosphate ions. [Courtesy of Mike Carson, University of Alabama at Birmingham; X-ray structure determined by Stephen Ealick and Charles Bugg, University of Alabama at Birmingham.]

(a)

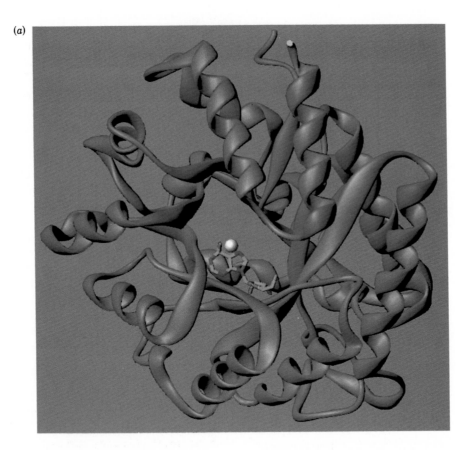

FIGURE 26-24. (*a*) A ribbon diagram of murine adenosine deaminase in complex with its transition state analog HDPR as viewed approximately down the axis of the enzyme's α/β barrel from the N-terminal ends of its β strands. The HDPR is shown in skeletal form with its C, N, and O atoms green, blue and red. The enzyme-bound Zn^{2+} ion, which is coordinated by HDPR's 6-hydroxyl group, is represented by a silver sphere. [Based on an X-ray structure by Florante Quiocho, Baylor College of Medicine.] (*b*) *(Opposite)* The proposed catalytic mechanism of adenosine deaminase. A Zn^{2+}-polarized H_2O molecule (Section 14-1C) nucleophilically attacks C6 of the enzyme-bound adenosine molecule in a process that is facilitated by His 238 acting as a general base, Glu 217 acting as a general acid, and Asp 295 acting to orient the water molecule via hydrogen bonding. The resulting tetrahedral intermediate decomposes by the elimination of ammonia in a reaction that is aided by the now imidazolium and carboxyl side chains of His 238 and Glu 217 acting as a general acid and a general base, respectively. This yields inosine in its enol tautomeric form which, upon its release from the enzyme, largely assumes its dominant keto form. The Zn^{2+} is coordinated by three His side chains that are not shown. [After Wilson, D.K. and Quiocho, F.A., *Biochemistry* **32**, 1692 (1993).]

deficiency results in **severe combined immunodeficiency disease (SCID)** that, without special protective measures, is invariably fatal in infancy because of overwhelming infection. The mutations in all eight known ADA variants obtained from SCID patients appear to structurally perturb the active site of ADA.

Biochemical considerations provide a plausible explanation of SCID's etiology (causes). In the absence of active ADA, deoxyadenosine is phosphorylated to yield levels of dATP that are 50-fold greater than normal. This high concentration of dATP inhibits ribonucleotide reductase (Section 26-4A), thereby preventing the synthesis of the other dNTPs, choking off DNA synthesis, and thus cell proliferation. The tissue-specific effect of ADA deficiency on the immune system may be explained by the observation that lymphoid tissue is particularly active in deoxyadenosine phosphorylation.

SCID caused by ADA defects does not respond to treatment by the intravenous injection of ADA because the liver clears this enzyme from the bloodstream within minutes. If, however, several molecules of the biologically inert polymer **polyethylene glycol (PEG)**

$$HO-[-CH_2-CH_2-O-]_n-H$$

Polyethylene glycol

are covalently linked to surface groups on ADA, the resulting **PEG–ADA** remains in the blood for 1 to 2 weeks, thereby largely resuscitating the SCID victim's immune

(b)

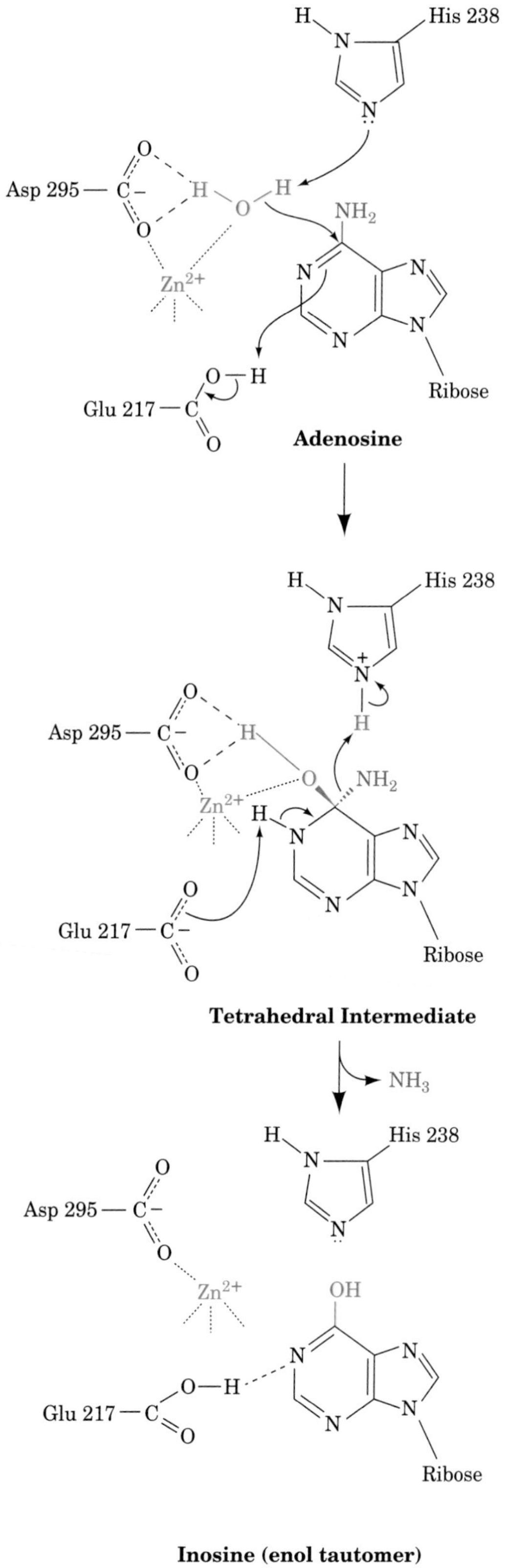

Adenosine

↓

Tetrahedral Intermediate

↓ NH₃

Inosine (enol tautomer)

system. The protein-linked PEG only reduces the catalytic activity of ADA by ~40% but, evidently, masks it from the receptors that filter it out of the blood. SCID can therefore be treated effectively by PEG–ADA. This treatment, however, is expensive and not entirely satisfactory. Consequently, ADA deficiency has been selected as one of the first genetic diseases to be treated by **gene therapy**: Lymphocytes are extracted from the blood of an ADA-deficient child, grown in the laboratory, have a normal ADA gene inserted into them via genetic engineering techniques (Section 28-8F), and are then returned to the child. Highly encouraging preliminary results indicate that ADA deficiency in lymphocytes is reversed by this procedure and that these gene-corrected cells proliferate in the body and/or that their lifespan is over 6 months.

PNP Is a Target of Structure-Based Drug Design

PNP deficiency kills the so-called *T* **lymphocytes** but not the *B* **lymphocytes** and therefore causes an immunodeficiency syndrome of lesser severity than SCID (the *T* and *B* lymphocytes mediate different aspects of the immune response known as **cellular** and **humoral immunity;** Section 34-2A). This observation suggests that the selective inhibition of PNP may suppress the excess *T* lymphocyte activity associated with such autoimmune diseases as rheumatoid arthritis, psoriasis, and insulin-dependent diabetes and impede the growth of cancers such as *T* cell lymphomas and leukemias. In the recent past, the design of an otherwise nontoxic drug that inhibits a particular enzyme was largely carried out by trial-and-error procedures: A very large number of diverse substances (typically ~10,000) are screened for possible inhibitory effects and, upon finding a suitable so-called **lead compound,** a large number of its variants are synthesized and likewise tested for drug efficacy. Although nearly all of the therapeutic drugs in use today were discovered in this way, it is an expensive and time-consuming method.

The use of recently developed **structure-based drug design** techniques promises to yield more efficacious drugs and to increase the rate at which they are developed. Technological and theoretical advances in macromolecular X-ray crystallography have greatly reduced the time and effort required to determine the structure of a protein (which is why the number of known protein X-ray structures has so enormously increased in recent years). Thus, the X-ray structure of an enzyme such as PNP that is a potential drug target can be determined in a relatively short time and its active site geometry used to design inhibitors of the enzyme via computerized molecular modeling techniques. The X-ray structure of such an inhibitor in complex with the enzyme can then be determined and correlated with the observed inhibitory properties of the designed molecule, and the results can be used to design yet more effective inhibitors. In this way, a PNP inhibitor that shows great promise in treating both psoriasis and a form of *T* cell lymphoma has been designed in ~3 years, rather than the ~10 years

thymidylate synthase based on the X-ray structure of the covalent inhibitory ternary complex with 5-fluoro-2′-deoxyuridylate and 5,10-methylenetetrahydrofolate, *J. Mol. Biol.* **214**, 937–948 (1990).

Meyer, E., Leonard, N.J., Bhat, B., Stubbe, J., and Smith, J.M., Purification and characterization of the *pur*E, *pur*K, and *pur*C gene products: Identification of a previously unrecognized energy requirement in the purine biosynthetic pathway, *Biochemistry* **31**, 5022–5032 (1992).

Nordlund, P. and Eklund, H., Structure and function of the *Eschericia coli* ribnucleotide reductase protein R2, *J. Mol. Biol.* **232**, 123–164 (1993).

Scriver, C.R., Beaudet, A.L., Sly, W.S., and Valle, D. (Eds.), *The Metabolic Basis of Inherited Disease* (6th ed.), Chapters 37–43, McGraw-Hill (1989).

Reichard, P., From RNA to DNA, why so many ribonucleotide reductases, *Science* **260**, 1773–1777 (1993).

Reichard, P., Interactions between deoxyribonucleotide and DNA synthesis, *Annu. Rev. Biochem.* **57**, 349–374 (1988).

Uhlin, U. and Eklund, H., Structure of ribonucleotide reductase protein R1, *Nature* **370**, 533–539 (1994); *and* Sjöberg, B.M., The ribonucleotide reductase jigsaw puzzle: a large piece falls into place, *Structure* **2**, 793–796 (1994).

Wilson, D.K., Rudolph, F.B., and Quiocho, F.A., Atomic structure of adenosine deaminase complexed with a transition-state analog: Understanding catalysis and immunodeficiency mutations, *Science* **252**, 1278–1284 (1991); Wilson, D.K. and Quiocho, F.A., A pre-transition-state mimic of an enzyme: X-ray structure of adenosine deaminase with bound 1-deazaadenosine and zinc-activated water, *Biochemistry* **32**, 1689-1694 (1993); *and* Crystallographic observation of a trapped tetrahedral intermediate in a metalloenzyme, *Nature Struct. Biol.* **1**, 691–694 (1994).

Zalkin, H. and Dixon, J.E., *De novo* purine nucleotide biosynthesis, *Prog. Nucl. Acid Res. Mol. Biol.* **42**, 259–285 (1992).

PROBLEMS

1. Azaserine (*O*-diazoacetyl-L-serine) and **6-diazo-5-oxo-L-norleucine (DON)**

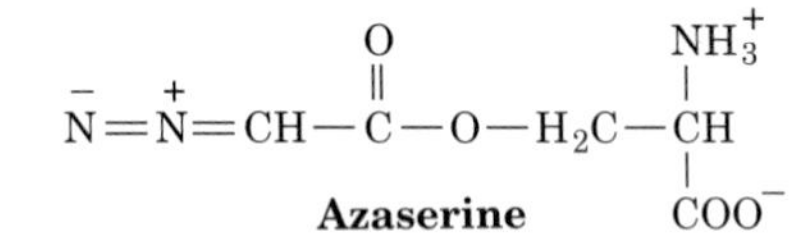

$$\overset{-}{N}=\overset{+}{N}=CH-\overset{\overset{O}{\|}}{C}-O-H_2C-\overset{\overset{NH_3^+}{|}}{\underset{\underset{COO^-}{|}}{CH}}$$

Azaserine

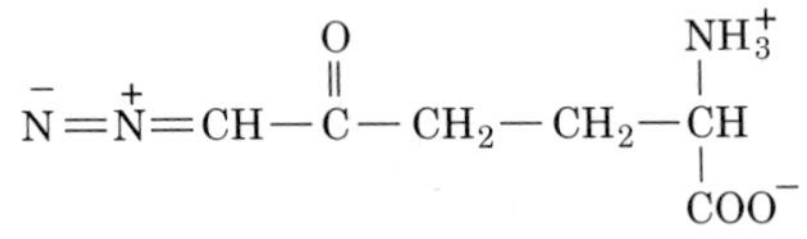

$$\overset{-}{N}=\overset{+}{N}=CH-\overset{\overset{O}{\|}}{C}-CH_2-CH_2-\overset{\overset{NH_3^+}{|}}{\underset{\underset{COO^-}{|}}{CH}}$$

6-Diazo-5-oxo-L-norleucine (DON)

are glutamine analogs. They form covalent bonds to nucleophiles at the active sites of enzymes that bind glutamine, thereby irreversibly inactivating these enzymes. Identify the nucleotide biosynthesis intermediates that accumulate in the presence of either of these glutamine antagonists.

2. Suggest a mechanism for the AIR synthetase reaction (Fig. 26-3, Reaction 6).

3. Although the product of the ATP-driven PurK reaction in the synthesis of CAIR (Fig. 26-3, Reaction 7) has not been identified, describe some reasonable possibilities for this intermediate given that AIR is not carboxylated by PurK alone.

*4. What is the energetic price, in ATPs, of synthesizing the hypoxanthine residue of IMP from CO_2 and NH_4^+?

5. Assuming that Classes I, II, and III of ribonucleotide reductase evolved from a common ancestor that existed before the development of O_2-generating photosynthesis, describe a plausible scenario for the order in which these different enzyme classes arose.

6. Why is deoxyadenosine toxic to mammalian cells?

7. Indicate which of the following substances are mechanism-based inhibitors and explain your reasoning. (a) Tosyl-L-phenylanine chloromethylketone with chymotrypsin (Section 14-3A). (b) Trimethoprim with bacterial dihydrofolate reductase. (c) The δ-lactone analog of $(NAG)_4$ with lysozyme (Section 14-2C). (d) Allopurinol with xanthine oxidase.

8. Why do individuals who are undergoing chemotherapy with cytotoxic (cell killing) agents such as FdUMP or methotrexate temporarily go bald?

9. Normal cells die in a nutrient medium containing thymidine and methotrexate that supports the growth of mutant cells defective in thymidylate synthase. Explain.

10. FdUMP and methotrexate, when taken together, are less effective chemotherapeutic agents than when either drug is taken alone. Explain.

11. Why is gout more prevalent in populations that eat relatively large amounts of meat?

12. Gout resulting from the *de novo* overproduction of purines can be distinguished from gout caused by impaired excretion of uric acid by feeding a patient ^{15}N-labeled glycine and determining the distribution of ^{15}N in his or her excreted uric acid. What isotopic distributions are expected for each type of defect?

13. 6-Mercaptopurine,

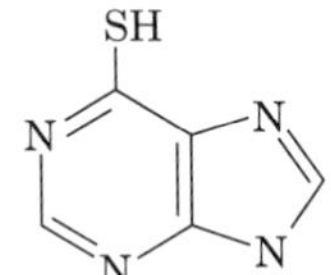

6-Mercaptopurine

after conversion to the corresponding nucleotide through salvage reactions, is a potent competitive inhibitor of IMP in the pathways for AMP and GMP biosynthesis. It is therefore a clinically useful anticancer agent. The chemotherapeutic effectiveness of 6-mercaptopurine is enhanced when it is administered with allopurinol. Explain the mechanism of this enhancement.

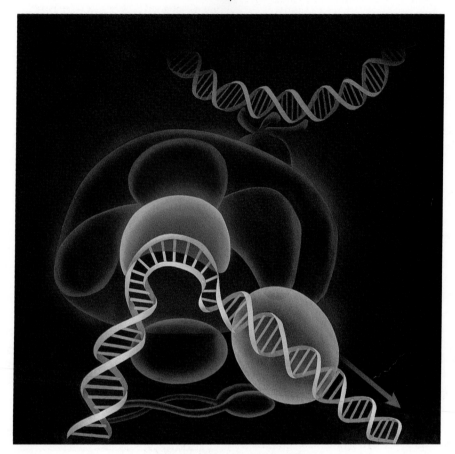

THE EXPRESSION AND TRANSMISSION OF GENETIC INFORMATION

*Schematic diagram of the eukaryotic
preinitiation complex that is required for the
transcription of DNA to messenger RNA.
The TATA-box binding protein is shown in orange.*

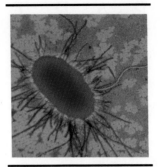

27

DNA: The Vehicle of Inheritance

DNA, as is now common knowledge, is the carrier of genetic information in all cellular life as well as in many viruses. Yet, a period of over 75 years passed from the time the laws of inheritance were discovered until the biological role of DNA was elucidated. Even now, many details of how genetic information is expressed and transmitted to future generations are still unclear, particularly in eukaryotes, and are subjects of intense research. Thus, our ideas of how genetic information is expressed and transmitted have been attained through a slow evolutionary process that has been only occasionally punctuated by incisive experiments or brilliant insights.

In this chapter, we commence our study of **molecular genetics;** that is, how genetic information is transmitted and expressed on the molecular level. We begin by reviewing "classical" genetics, whose understanding is prerequisite for assimilating molecular genetics. We then consider

how we have come to know that DNA is the carrier of genetic information. This chapter has been written with a historical perspective in order to illustrate how these ideas developed, in particular, and how scientific concepts evolve, in general. The major aspects of molecular genetics are considered in detail in subsequent chapters.

1. GENETICS: A REVIEW

One has only to note the resemblance between parent and child to realize that physical traits are inherited. Yet, the mechanism of inheritance has, until recent decades, been unknown. The theory of **pangenesis,** which originated with the ancient Greeks, held that semen, which clearly has something to do with procreation, consists of representative particles from all over the body **(pangenes).** This idea was extended in the late eighteenth century by Jean Baptiste de Lamarck who, in a theory known as **Lamarckism,** hypothesized that an individual's acquired characteristics, such as large muscles resulting from exercise, would be transmitted to his/her offspring. Pangenesis and at least some aspects of Lamarckism were accepted by most nineteenth century biologists, including Charles Darwin.

The realization, in the mid-nineteenth century, that all organisms are derived from single cells set the stage for the development of modern biology. In his **germ plasm theory,** August Weismann pointed out that sperm and ova, the **germ cells** (whose primordia are set aside early in embryonic development), are directly descended from the germ cells of the previous generation and that other cells of the body, the **somatic cells,** although derived from germ cells, do not give rise to them. He refuted the ideas of pangenesis and Lamarckism by demonstrating that the progeny of many successive generations of mice whose tails had been cut off had tails of normal length.

A. Chromosomes

In the 1860s, eukaryotic cell nuclei were observed to contain linear bodies that were named **chromosomes** (Greek: *chromos,* color; *soma,* body) because they are strongly stained by certain basic dyes (Fig. 27-1). There are normally two copies of each chromosome (**homologous pairs**) present in every somatic cell. The number of unique chromosomes (*N*) in such a cell is known as its **haploid number** and the total number of chromosomes (2*N*) is its **diploid number.** Different species differ in their haploid number of chromosomes (Table 27-1).

Somatic Cells Divide by Mitosis

The division of somatic cells, a process known as **mitosis** (Fig. 27-2), is preceded by the duplication of each chromosome to form a cell with 4*N* chromosomes. During cell division, each chromosome attaches by its **centromere** to the **mitotic spindle** such that the members of each duplicate pair line up across the equatorial plane of the cell. The members of each duplicate pair are then pulled to opposite poles of the dividing cell by the action of the spindle to yield

TABLE 27-1. NUMBER OF CHROMOSOMES (2*N*) IN SOME EUKARYOTES

Organism	Chromosomes
Humans	46
Dog	78
Rat	42
Turkey	82
Frog	26
Fruit fly	8
Hermit crab	~254
Garden pea	14
Potato	48
Yeast	34
Green alga	~20

Source: Ayala, F.J. and Kiger, J.A., Jr., *Modern Genetics* (2nd ed.), *p.* 9, Benjamin/Cummings (1984).

diploid daughter cells that each have the same 2*N* chromosomes as the parent cell.

Germ Cells Are Formed by Meiosis

The formation of germ cells, a process known as **meiosis** (Fig. 27-3), requires two consecutive cell divisions. Before the first meiotic division each chromosome replicates, but the resulting sister **chromatids** remain attached at their centromere. The homologous pairs of the doubled chromosomes then line up across the equatorial plane of the cell in zipperlike fashion, which permits an exchange of the corresponding sections of homologous chromosomes in a process known as **crossing over** (Section 27-1B). The spindle then moves the members of each homologous pair to opposite poles of the cell so that, after the first meiotic division, each daughter cell contains *N* doubled chromosomes. In the second meiotic division, the sister chromatids separate to form chromosomes and move to opposite poles of the dividing cell to yield a total of four haploid cells that are known as **gametes.** Fertilization consists of the fusion of a male gamete (sperm) with a female gamete (ovum) to yield a diploid cell known as a **zygote** that has received *N* chromosomes from each of its parents.

B. Mendelian Inheritance

The basic laws of inheritance were reported in 1866 by Gregor Mendel. They were elucidated by the analysis of a series of **genetic crosses** between true-breeding strains (producing progeny that have the same characteristics as the parents) of garden peas, *Pisum sativum,* that differ in certain well-defined traits such as seed shape (round vs wrinkled), seed color (yellow vs green), or flower color (purple vs white). Mendel found that in crossing parents (*P*) that differ in a single trait, say seed shape, the progeny (*F*₁; first filial generation) all have the trait of only one of the parents, in

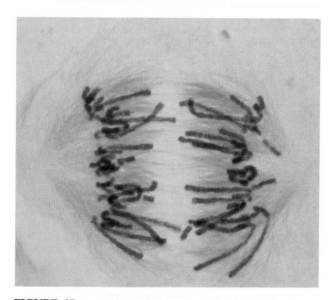

FIGURE 27-1. A photomicrograph of a plant cell (*Scadoxus katherinae* Bak.) during anaphase of mitosis showing its chromosomes being pulled to opposite poles of the cell by the mitotic spindle. The microtubules forming the mitotic spindle are stained red and the chromosomes are blue. [Courtesy of Andrew S. Bajer, University of Oregon.]

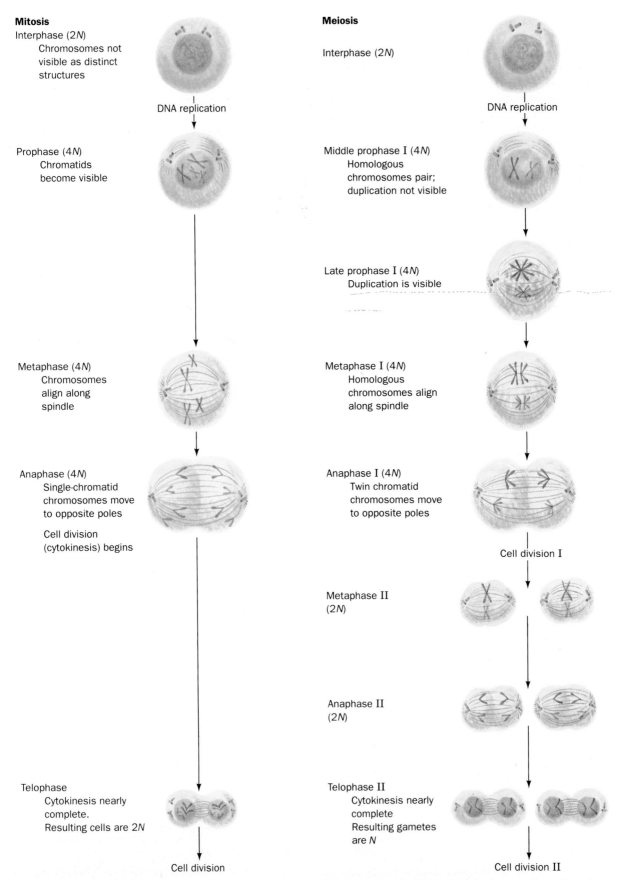

Mitosis

Interphase (2N)
Chromosomes not
visible as distinct
structures

DNA replication

Prophase (4N)
Chromatids
become visible

Metaphase (4N)
Chromosomes
align along
spindle

Anaphase (4N)
Single-chromatid
chromosomes move
to opposite poles

Cell division
(cytokinesis) begins

Telophase
Cytokinesis nearly
complete.
Resulting cells are 2N

Cell division

Meiosis

Interphase (2N)

DNA replication

Middle prophase I (4N)
Homologous
chromosomes pair;
duplication not visible

Late prophase I (4N)
Duplication is visible

Metaphase I (4N)
Homologous
chromosomes align
along spindle

Anaphase I (4N)
Twin chromatid
chromosomes move
to opposite poles

Cell division I

Metaphase II
(2N)

Anaphase II
(2N)

Telophase II
Cytokinesis nearly
complete
Resulting gametes
are N

Cell division II

FIGURE 27-2. Mitosis, the usual form of cell division in eukaryotes, yields two daughter cells, each with the same chromosomal complement as the parental cell.

FIGURE 27-3. Meiosis, which leads to the formation of gametes (sex cells), comprises two consecutive cell divisions to yield four daughter cells, each with half of the chromosomal complement of the parental cell.

this case round seeds (Fig. 27-4). The trait appearing in F_1 is said to be **dominant**, whereas the alternative trait is called **recessive**. In F_2, the progeny of F_1, three quarters have the dominant trait and one quarter have the recessive trait. Those peas with the recessive trait breed true; that is, self-crossing recessive F_2's results in progeny (F_3) that also have the recessive trait. The F_2's exhibiting the dominant trait, however, fall into two categories: one third of them breed true, whereas the remainder have progeny with the same 3:1 ratio of dominant to recessive traits as do the members of F_2.

Mendel accounted for his observations by hypothesizing that *the various pairs of contrasting traits each result from a factor (now called a* **gene***) that has alternative forms (**alleles***). Every plant contains a pair of genes governing a particular trait, one inherited from each of its parents.* The alleles for seed shape are symbolized R for round seeds and r for wrinkled seeds (gene symbols are generally given in italics). The pure breeding plants with round and wrinkled seeds, respectively, have RR and rr **genotypes** (genetic com-

position) and are both said to be **homozygous** in seed shape. Plants with the Rr genotype are **heterozygous** in seed shape and have the round seed **phenotype** (appearance or character) because R is dominant over r. *The two alleles do not blend or mix in any way in the plant and are independently transmitted through gametes to progeny (Fig. 27-5).*

Mendel also found that *different traits are independently inherited.* For example, crossing peas that have round yellow seeds ($RRYY$) with peas that have wrinkled green seeds ($rryy$) results in F_1 progeny ($RrYy$) that have round yellow seeds (yellow seeds are dominant over green seeds). The F_2 phenotypes appear in the ratio 9 round yellow : 3 round green : 3 wrinkled yellow : 1 wrinkled green. This result indicates that there is no tendency for the genes from any parent to assort together (Fig. 27-6). It was later shown, however, that *only genes that occur on different chromosomes exhibit such independence.*

The dominance of one trait over another is a common but not universal phenomenon. For example, crossing a pure-breeding red variety of the snapdragon *Antirrhinum* with a pure-breeding white variety results in pink colored F_1 progeny. The F_2 progeny have red, pink, and white flowers in 1:2:1 ratio because the flowers of homozygotes for the red color (AA) contain more red pigment than do the heterozy-

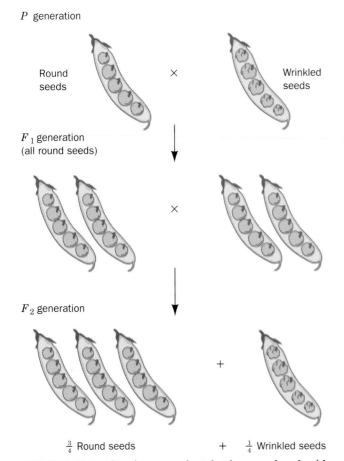

FIGURE 27-4. Crossing a pea plant that has round seeds with one that has wrinkled seeds yields F_1 progeny that all have round seeds. Crossing these F_1 peas yields an F_2 generation, of which three quarters have round seeds and one quarter have wrinkled seeds.

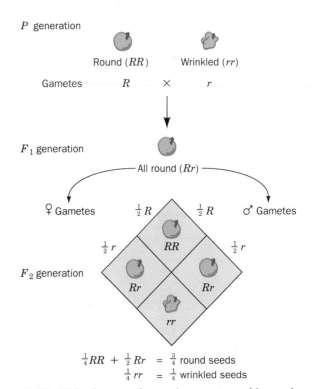

FIGURE 27-5. In a genetic cross between peas with round seeds and peas with wrinkled seeds the F_1 generation has the round seed phenotype because round seeds are dominant over wrinkled seeds. Three fourths of the F_2 generation's seeds are round and one fourth are wrinkled because the genes for these alleles are independently transmitted by haploid gametes.

gotes (*Aa*; Fig. 27-7). The red and white traits are therefore said to be **codominant.** In the case of codominance, the phenotype reveals the genotype.

A given gene may have multiple alleles. A familiar example is the human ABO blood group system (Section 11-3D). The A and B antigens are specified by the codominant *I*^A and *I*^B alleles, respectively, and the O phenotype is homozygous for the recessive *i* allele. The different blood types, it will be recalled, arise from the action of a glycosyltransferase that, if specified by an *I*^A, *I*^B, or *i* allele, respectively, transfers an *N*-acetylgalactosamine residue, transfers a galactose residue, or is inactive.

C. Chromosomal Theory of Inheritance

Mendel's theory of inheritance was almost universally ignored by his contemporaries. This was partially because in analyzing his data he used probability theory, an alien subject to most biologists of the time. The major reason his theory was ignored, however, is that it was ahead of its time: Contemporary knowledge of anatomy and physiology provided no basis for its understanding. For instance, mitosis and meiosis had yet to be discovered. Yet, after Mendel's work was rediscovered in 1900, it was shown that his principles explained inheritance in animals as well as in plants. In 1903, as a result of the realization that chromosomes and genes behave in a parallel fashion, Walter Sutton formulated the **chromosomal theory of inheritance** in which he hypothesized that genes are parts of chromosomes.

The first trait to be assigned a chromosomal location was that of sex. *In most eukaryotes, the cells of females each contain two copies of the* **X chromosome** *(XX), whereas male cells contain one copy of X and a morphologically distinct* **Y chromosome** *(XY; Fig. 27-8).* Ova must therefore contain a single X chromosome and sperm contain either an X or a Y

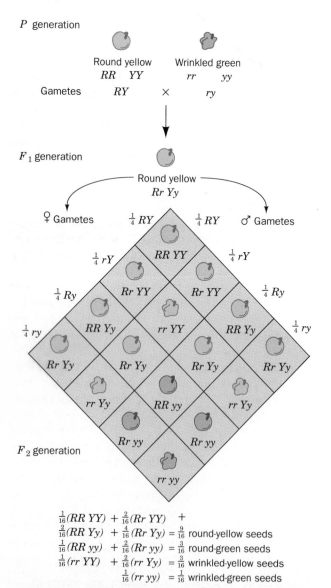

P generation

Round yellow   Wrinkled green
RR YY rr yy

Gametes RY × ry

F_1 generation

Round yellow
$Rr\,Yy$

♀ Gametes $\frac{1}{4}RY$ $\frac{1}{4}RY$ ♂ Gametes

F_2 generation

$\frac{1}{16}(RR\ YY) + \frac{2}{16}(Rr\ YY)\ +$
$\frac{2}{16}(RR\ Yy) + \frac{4}{16}(Rr\ Yy) = \frac{9}{16}$ round-yellow seeds
$\frac{1}{16}(RR\ yy) + \frac{2}{16}(Rr\ yy) = \frac{3}{16}$ round-green seeds
$\frac{1}{16}(rr\ YY) + \frac{2}{16}(rr\ Yy) = \frac{3}{16}$ wrinkled-yellow seeds
$\frac{1}{16}(rr\ yy) = \frac{1}{16}$ wrinkled-green seeds

FIGURE 27-6. The genes for round (*R*) versus wrinkled (*r*) and yellow (*Y*) versus green (*y*) pea seeds assort independently. The F_2 progeny consist of nine genotypes comprising the four possible phenotypes.

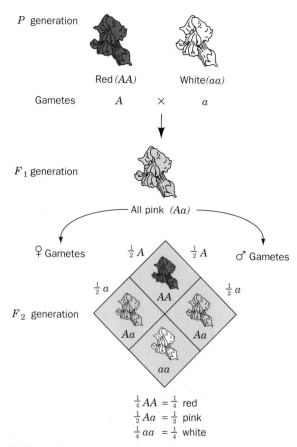

P generation

Red *(AA)*   White *(aa)*

Gametes A × a

F_1 generation

All pink *(Aa)*

♀ Gametes $\frac{1}{2}A$ $\frac{1}{2}A$ ♂ Gametes

F_2 generation

$\frac{1}{4}AA = \frac{1}{4}$ red
$\frac{1}{2}Aa = \frac{1}{2}$ pink
$\frac{1}{4}aa = \frac{1}{4}$ white

FIGURE 27-7. In a cross between snapdragons with red (*AA*) and white (*aa*) flowers, the F_1 generation is pink (*Aa*), which demonstrates that the two alleles, *A* and *a*, are codominant. The F_2 flowers are red, pink, and white in 1:2:1 ratio.

P generation

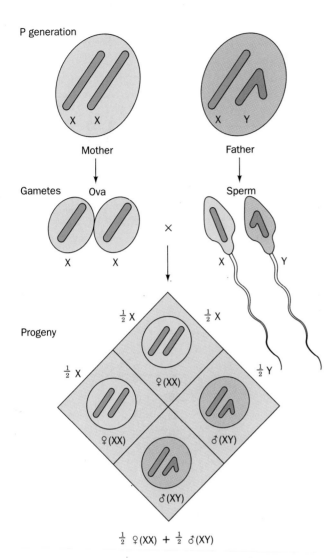

Mother   Father

Gametes   Ova   Sperm

Progeny

$$\tfrac{1}{2}\ ♀(XX) + \tfrac{1}{2}\ ♂(XY)$$

FIGURE 27-8. The independent segregation of the sex chromosomes, X and Y, results in a 1:1 ratio of males to females.

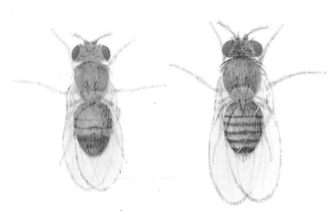

FIGURE 27-9. The fruit fly *Drosophila melanogaster*. The male (*left*) and the female (*right*) are shown in their relative sizes; they are actually ~2 mm long and weigh ~1 mg.

chromosome (Fig. 27-8). Fertilization by an X-bearing sperm therefore results in a female zygote and by a Y-bearing sperm yields a male zygote. This explains the observed 1:1 ratio of males to females in most species. The X and Y chromosomes are referred to as **sex chromosomes;** the others are known as **autosomes.**

Fruit Flys Are Favorite Genetic Subjects

The pace of genetic research greatly accelerated after Thomas Hunt Morgan began using the fruit fly *Drosophila melanogaster* as an experimental subject. This small prolific insect (Fig. 27-9), which is often seen hovering around ripe fruit in summer and fall, is easily maintained in the laboratory where it produces a new generation every 14 days. With *Drosophila,* the results of genetic crosses can be determined some 25 times faster than they can with peas. *Drosophila* is presently the genetically best characterized higher organism.

The first known mutant strain of *Drosophila* had white eyes rather than the red eyes of the **wild-type** (occurring in nature). Through genetic crosses of the white eye strain with the wild type, Morgan showed that the distribution of the white eye gene (*wh*) parallels that of the X chromosome. This indicates that the *wh* gene is located on the X chromosome and that the Y chromosome does not contain it. The *wh* gene is therefore said to be **sex linked.**

Genetic Maps Can Be Constructed from an Analysis of Crossover Rates

In succeeding years, the chromosomal locations of many *Drosophila* genes were determined. Those genes that reside on the same chromosome do not assort independently. However, any pair of such **linked** genes **recombine** (exchange relative positions with their allelic counterparts on the homologous chromosome) with a characteristic frequency. The cytological basis of this phenomenon was found to occur at the start of meiosis when the homologous doubled chromosomes line up in parallel (Metaphase I; Fig. 27-3). Homologous chromatids are observed to exchange equivalent sections in a process known as **crossing over** (Fig. 27-10). The chromosomal location of the crossover point varies nearly randomly from event to event. Consequently, *the crossover frequency of a pair of linked genes varies directly with their physical separation along the chromosome.* Morgan and Alfred Sturtevant made use of this phenomenon to **map** (locate) the relative positions of genes on *Drosophila*'s four unique chromosomes. Such studies have demonstrated that *chromosomes are linear unbranched structures.* We now know that such **genetic maps** (Fig. 27-11) parallel the corresponding base sequences of the DNA within the chromosomes.

Nonallelic Genes Complement One Another

Whether or not two recessive traits that affect similar functions are allelic (different forms of the same gene) can be determined by a **complementation test.** In this test, a

(a)

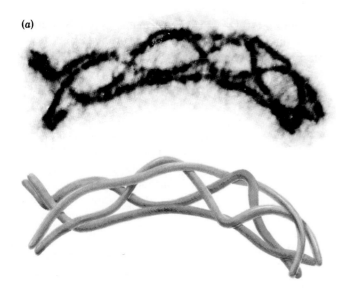

(b)

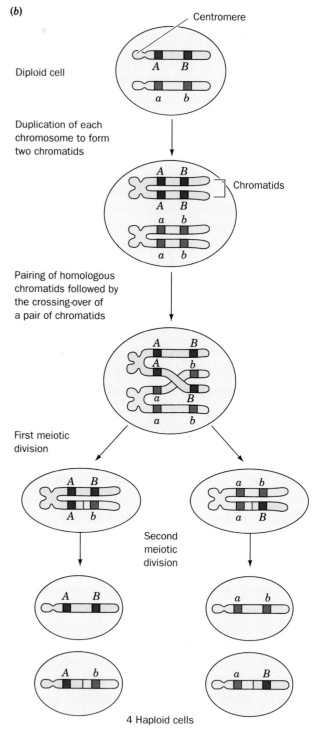

FIGURE 27-10. Crossing over. (*a*) An electron micrograph, together with an interpretive drawing, of two homologous pairs of chromatids during meiosis in the grasshopper *Chorthippus parallelus*. Nonsister chromatids (*different colors*) may recombine at any of the points where they cross over. [Courtesy of Bernard John, The Australian National University.] (*b*) A diagram showing the recombination of pairs of allelic genes (*A, B*) and (*a, b*) during crossover.

homozygote for one of the traits is crossed with a homozygote for the other. If the two traits are nonallelic, the progeny will have the wild-type phenotype because each of the homologous chromosomes supplies the wild-type function that the other lacks; that is, they complement each other. For example, crossing a *Drosophila* that is homozygous for an eye color mutation known as purple (*pr*) with a homozygote for another eye color mutation known as brown (*bw*) yields progeny with wild-type eye color, thereby demonstrating that these two genes are not allelic (Fig. 27-12a). In contrast, in crossing a female *Drosophila* that is homozygous for the sex-linked white eye color allele (*wh*) with a male carrying the sex-linked coffee eye color allele (*cf*), the female progeny do not have wild-type eye color (Fig. 27-12b). The *wh* and *cf* genes must therefore be allelic.

Genes Direct Protein Expression

The question of how genes control the characteristics of organisms took some time to be answered. Archibald Garrod was the first to suggest a specific connection between genes and enzymes. Individuals with **alkaptonuria** produce urine that darkens alarmingly on exposure to air, a consequence of the oxidation of the **homogentisate** they excrete (Section 15-3A). In 1902, Garrod showed that this rather benign metabolic disorder (its only adverse effect is arthritis in later life) results from a recessive trait that is inherited in a Mendelian fashion. He further demonstrated that alkaptonurics are unable to metabolize the homogentisate fed to them and therefore concluded that *they lack an enzyme that metabolizes this substance.* This enzyme is

now known to be homogentisate dioxygenase, which is involved in the degradation of phenylalanine and tyrosine (Section 24-3H). Garrod described alkaptonuria and several other inherited human diseases he had studied as **inborn errors of metabolism.**

Beginning in 1940, George Beadle and Edward Tatum, in a series of investigations that mark the beginning of biochemical genetics, showed that *there is a one-to-one corre-*

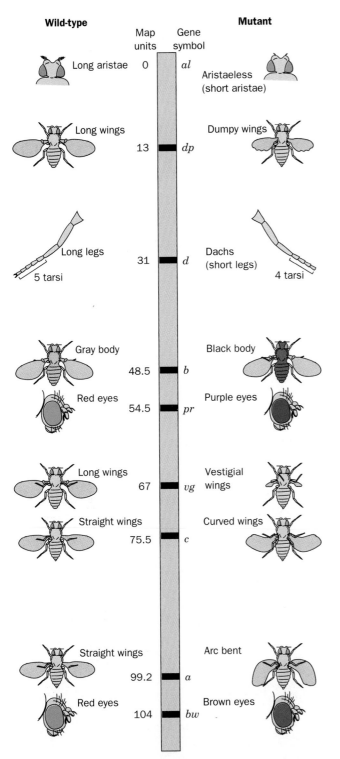

FIGURE 27-11. A portion of the genetic map of chromosome 2 of *Drosophila*. The positions of the genes are given in map units. Two genes separated by *m* map units recombine with a frequency of *m%*.

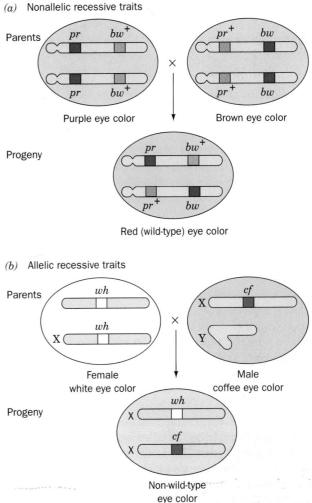

FIGURE 27-12. The complementation test indicates whether two recessive traits are allelic. Two examples in *Drosophila* are (*a*) Crossing a homozygote for purple eye color (*pr*) with a homozygote for brown eye color (*bw*) yields progeny with wild-type eye color. This indicates that *pr* and *bw* are nonallelic. Here the superscript "+" indicates the wild-type allele. (*b*) In crossing a female that is homozygous for the sex-linked white eye color gene *wh* with a male bearing the sex-linked coffee eye color gene *cf*, the female progeny do not have the wild-type eye color. The *wh* and *cf* genes must therefore be allelic.

spondence between a mutation and the lack of a specific enzyme. The wild-type mold *Neurospora* grows on a "minimal medium" in which the only sources of carbon and nitrogen are glucose and NH_3. Certain mutant varieties of *Neurospora* that were generated by means of irradiation with X-rays, however, require an additional substance, such as arginine or thiamine, in order to grow. Beadle and Tatum demonstrated, in several cases, that the mutants lack a normally present enzyme that participates in the biosynthesis of the required substance (Section 15-3A). This resulted in their famous maxim **one gene–one enzyme.** Today we know this principle to be only partially true since many genes specify proteins that are not enzymes and many proteins consist of several independently specified subunits (in humans, e.g., the α and β subunits of hemoglobin are specified by genes that reside on different chromosomes). A more accurate dictum might be **one gene–one**

polypeptide. Yet, even this is not completely correct because RNAs with structural and functional roles, such as transfer RNAs and ribosomal RNAs, are also genetically specified.

D. *Bacterial Genetics*

Bacteria offer several advantages for genetic study. Foremost of these is that *under favorable conditions, many have generation times of under 20 min. Consequently, the results of a genetic experiment with bacteria can be ascertained in a matter of hours rather than the weeks or years required for an analogous study with higher organisms. The tremendous number of bacteria that can be quickly grown (~10^{10} mL^{-1}) permits the observation of extremely rare biological events.* For example, an event that occurs with a frequency of 1 per million can be readily detected in bacteria, with only a few minutes work. To do so in *Drosophila* would be an enormous and probably futile effort. Moreover, bacteria are usually haploid, so their phenotype indicates their genotype. Nevertheless, the basic principles of genetics were elucidated from the study of higher plants and animals. This is because bacteria do not reproduce sexually in the manner of higher organisms, so the basic technique of classical genetics, the genetic cross, is not normally applicable to bacteria (but see below). In fact, before it was shown that DNA is the carrier of hereditary information, it was not altogether clear that bacteria had chromosomes.

The study of bacterial genetics effectively began in the 1940s when procedures were developed for isolating bacterial mutants. Since bacteria have few easily recognized morphological features, *their mutants are usually detected (selected for) by their ability or inability to grow under certain conditions.* For example, wild-type *E. coli* can grow on a medium in which glucose is the only carbon source. Mutants that are unable to synthesize methionine, for instance, require the presence of methionine in their growth media. Mutants that are resistant to an antibiotic, say ampicillin, can grow in the presence of this antibiotic, whereas the wild type cannot. Mutants in which an essential protein has become temperature sensitive grow at 30 but not at 42°C, whereas the wild type grows at either temperature. By using a suitable screening protocol, a bacterial colony containing a particular mutation or combination of mutations can be selected. This is conveniently done by the method of **replica plating** (Fig. 27-13).

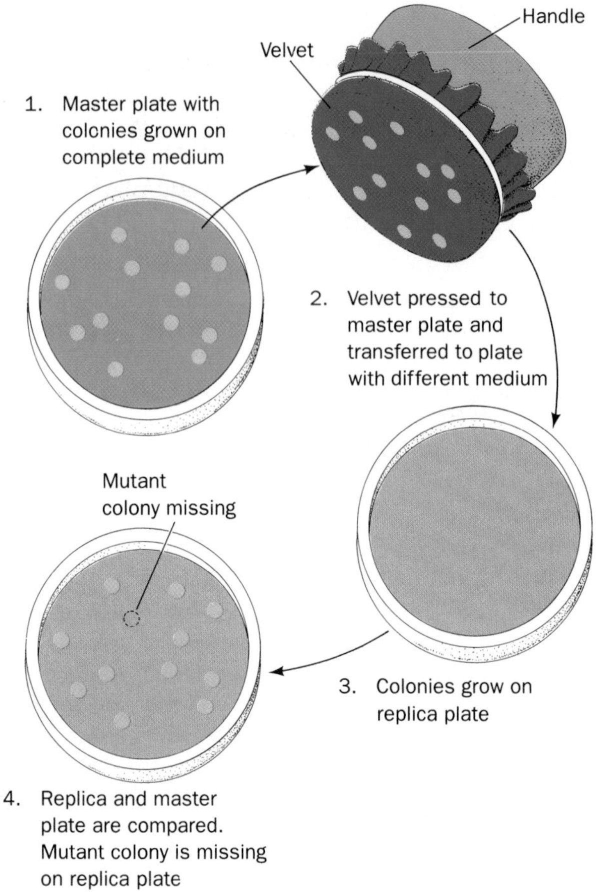

Bacterial Chromosomes Have Been Mapped through Interrupted Mating

In 1946, Joshua Lederberg and Tatum discovered *that some bacteria can transfer genetic information to others through a process known as **conjugation**.* The ability to conjugate ("mate") is conferred on otherwise indifferent bacteria by a **plasmid** (a DNA molecule distinct from the bacterial chromosome that is replicated by the cell; Section 28-8A) called an **F factor** (for fertility). Bacteria that possess an F factor (designated F⁺ or male) are covered by hairlike

1. Master plate with colonies grown on complete medium

Velvet

Handle

2. Velvet pressed to master plate and transferred to plate with different medium

Mutant colony missing

3. Colonies grow on replica plate

4. Replica and master plate are compared. Mutant colony is missing on replica plate

FIGURE 27-13. Replica plating is a technique for rapidly and conveniently transferring colonies from a "master" culture plate (Petri dish) to a different medium on another culture plate. Since the colonies on the master plate and on the replicas should have the same spatial distribution, it is easy to identify the desired mutants.

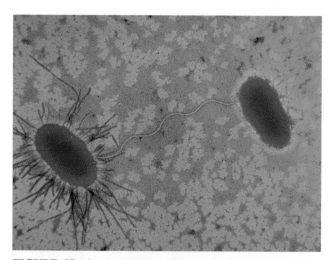

FIGURE 27-14. An electron micrograph of an F⁺ (*left*) and an F⁻ (*right*) *E. coli* engaged in sexual conjugation. [Dennis Kunkel/Phototake.]

projections known as **F pili.** These bind to cell-surface receptors on bacteria that lack the F factor (F⁻ or female), which leads to the formation of a cytoplasmic bridge between the cells (Fig. 27-14). The F factor then replicates and, as the newly replicated single strand is formed, it passes through the cytoplasmic bridge to the F⁻ cell where the complementary strand is synthesized (Fig. 27-15). This converts the F⁻ cell to F⁺ so that the F factor is an infectious agent (a bacterial venereal disease?).

On very rare occasions, the F factor spontaneously integrates into the chromosome of the F⁺ cell (plasmids with this capability are termed **episomes**). In the resulting **Hfr** (for *high frequency of recombination*) cells, the F factor behaves much as it does in the autonomous state. Its replication commences at a specific internal point in the F factor, and the replicated section passes through a cytoplasmic bridge to the F⁻ cell, where its complementary strand is synthesized. In this case, however, *the replicated chromosome of the Hfr cell is also transmitted to the F⁻ cell (Fig. 27-16).* Usually, only part of the Hfr bacterial chromosome

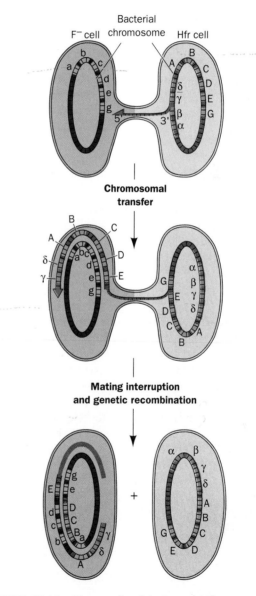

FIGURE 27-16. The transfer of the bacterial chromosome from an Hfr cell to an F⁻ cell and its subsequent recombination with the F⁻ chromosome. Here, Greek letters represent F factor genes, upper case Roman letters represent bacterial genes from the Hfr cell, and lower case Roman letters represent the corresponding alleles in the F⁻ cell. Since chromosomal transfer, which begins within the F factor, is rarely complete, the entire F factor is seldom transferred. Hence the recipient cell usually remains F⁻.

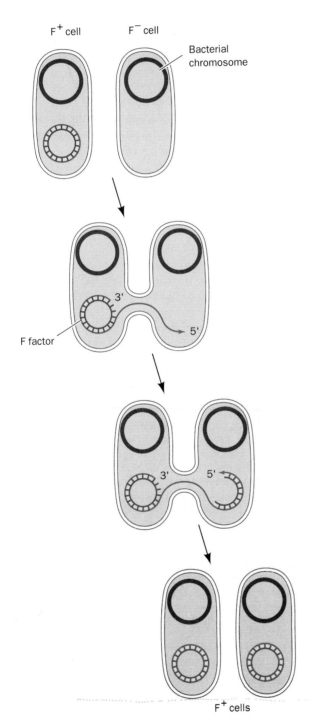

FIGURE 27-15. A diagram showing how an F⁻ cell acquires an F factor from an F⁺ cell. A single strand of the F factor is replicated, via the rolling circle mode (Section 31-3B), and is transferred to the F⁻ cell where its complementary strand is synthesized to form a new F factor.

is transferred during sexual conjugation because the cytoplasmic bridge almost always breaks off sometime during the ~90 min required to complete the transfer process. In the resulting **merozygote** (a partially diploid bacterium), the chromosomal fragment, which lacks a complete F factor, neither transforms the F⁻ cell to Hfr nor is subsequently

replicated. However, *the transferred chromosomal fragment recombines with the chromosome of the F⁻ cell in a manner similar to chromosomal crossing over in eukaryotes, thereby permanently endowing the F⁻ cell with some of the traits of the Hfr strain.*

Bacterial genes are transferred from the Hfr cell to the F⁻ cell in fixed order. This is because the F factor in a given Hfr strain is integrated into the bacterial chromosome at a specific site and because only a particular strand of the Hfr chromosomal DNA is replicated and transferred to the F⁻ cell. The bacterial chromosome can therefore be mapped by the following procedure. Hfr and F⁻ strains containing different alleles of the genes to be mapped are mixed, permitted to conjugate, and, after a certain time, the conjugation is interrupted by violent agitation in a kitchen blender so as to break the cytoplasmic bridges between conjugating cells (**interrupted mating;** Fig. 27-17). Subsequent screening procedures reveal which allelic genes have entered the F⁻ cell and recombined with its chromosome. *By interrupting the conjugation after various times, the order of the allelic genes can be determined* (Fig. 27-18). The difficulty in determining the order of genes located towards the end of the rarely completely transferred chromosome is circumvented by using several Hfr strains that differ according to the point at which the F factor is integrated into the chromosome. This mapping procedure has demonstrated that the *E. coli* chromosome is circular, as is its DNA (Fig. 27-18). Other bacterial chromosomes have been similarly mapped.

The integrated F factor in an Hfr cell occasionally undergoes spontaneous excision to yield an F⁺ cell. In rare in-

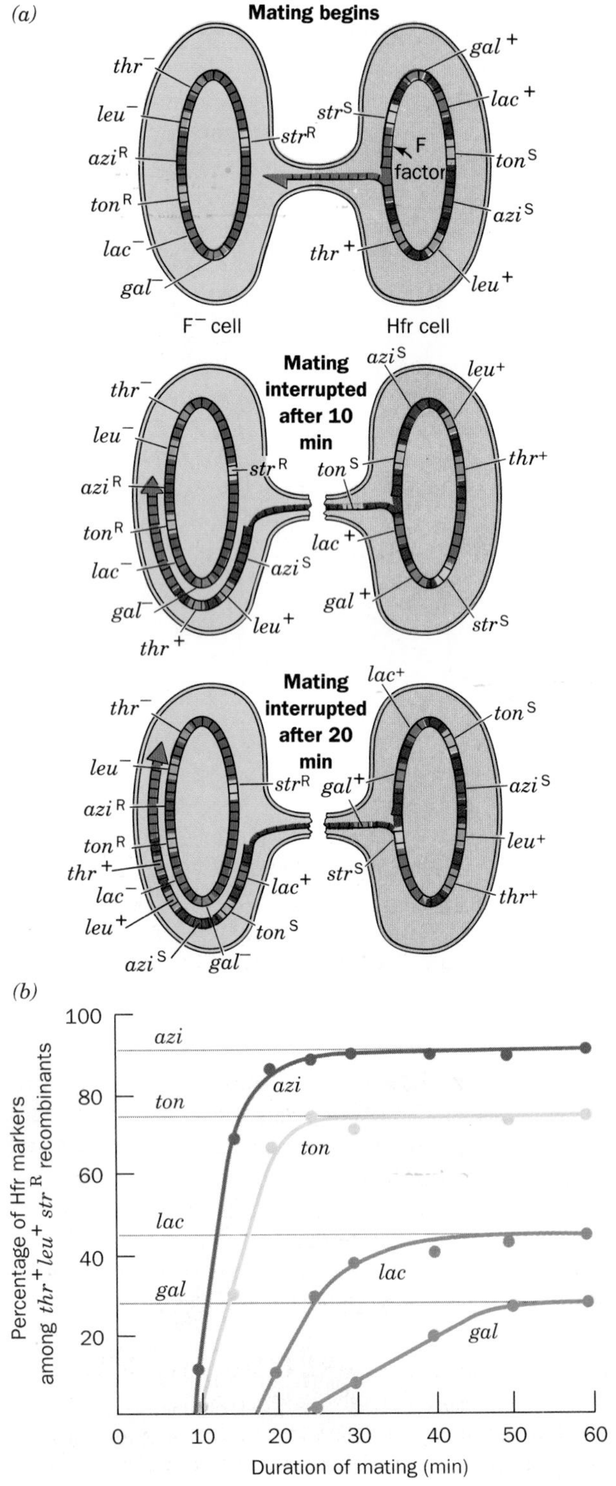

FIGURE 27-17. An example of the mapping of a bacterial chromosome by interrupted mating. (*a*) The ordered transfer of an Hfr chromosome to an F⁻ cell. Bacterial mating is interrupted at various times by agitation in a blender. Bacteria carrying the mutant alleles *thr⁻* and *leu⁻*, respectively, require the amino acids threonine and leucine in their growth media; those with *gal⁻* and *lac⁻* are, respectively, unable to grow on media containing the sugars lactose or galactose as their only carbon source; *azi*ᴿ confers resistance to azide; *ton*ᴿ confers resistance to **bacteriophage T1;** and *str*ᴿ confers resistance to the antibiotic streptomycin. The superscripts +, R, and S indicate wild type, "resistance," and "sensitivity," respectively. (*b*) The frequencies of occurrence of the genetic markers *azi*, *ton*, *lac*, and *gal* in recombinants as a function of mating time. After their mating was interrupted, the bacteria were grown on a medium containing streptomycin with glucose as the only carbon source to select for recombinants containing the *thr⁺*, *leu⁺*, and *str*ᴿ alleles. These recombinants were scored for their sensitivity to azide or bacteriophage T1 or for their ability to grow on lactose or galactose as their only carbon source. Extrapolation to zero of the frequency of bacterial colonies on the various restrictive media indicates the earliest times that the corresponding alleles became available for recombination with the F⁻ chromosome. [After Jacob, F. and Wollman, E.L., *Sexuality and the Genetics of Bacteria*, p. 135, Academic Press (1961).]

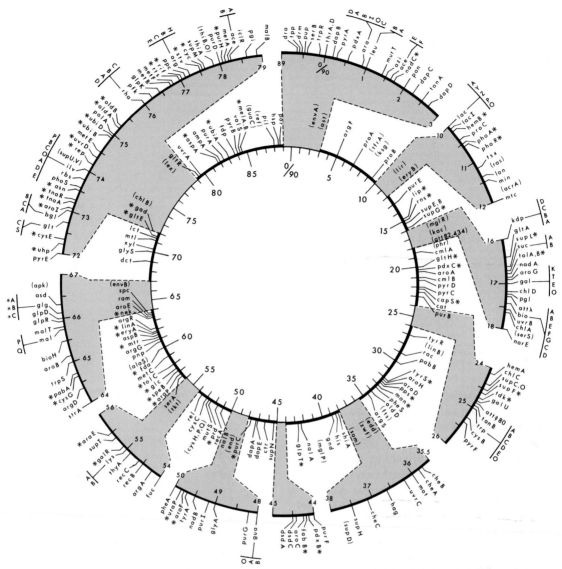

FIGURE 27-18. The genetic map of *E. coli* based on the results of interrupted mating experiments with the finer details determined by transductional mapping procedures. The inner circle indicates gene transfer time, in minutes, with the *thrA* locus arbitrarily placed at 0. The outer circle displays crowded sections of the map in expanded form. The genetic symbols are defined in the reference. The 310 genes mapped account for ~10% of the potential information content of the *E. coli* genome. Since the time this map was made (1970), many hundreds more *E. coli* genes have been discovered and mapped. [Courtesy of Austin L. Taylor, University of Colorado.]

stances, the F factor is aberrantly excised such that a portion of the adjacent bacterial chromosome is incorporated in the subsequently autonomously replicating F factor. Bacteria carrying such a so-called **F′ factor** are permanently diploid for its bacterial genes.

E. *Viral Genetics*

*Viruses are infectious particles consisting of a nucleic acid molecule enclosed by a protective **capsid** (coat) that consists largely or entirely of protein.* A virus specifically adsorbs to a susceptible cell into which it insinuates its nucleic acid. Over the course of the infection (Fig. 27-19), the viral chro-

mosome redirects the cell's metabolism so as to produce new viruses. A viral infection usually culminates in the lysis of the host cell, thereby releasing large numbers (tens to thousands) of mature virus particles that can each initiate a new round of infection. Viruses, having no metabolism of their own, are the ultimate parasites. They are not living organisms since, in the absence of their host, they are as biologically inert as any other large molecule.

The Fine Details of a Bacterial Genetic Map Can Be Elucidated through Transductional Mapping

Certain species of **bacteriophages** (viruses infecting bacteria, **phages** for short; Greek: *phagein,* to eat) have been

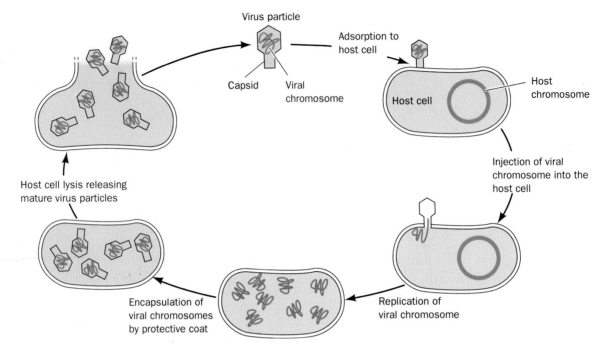

FIGURE 27-19. The life cycle of a virus.

useful in elucidating bacterial genetics. In an infection by such bacteriophages, about one in every thousand progeny particles contains a segment of the bacterial chromosome in place of the viral chromosome. These defective phage particles can inject their DNA into another bacterial cell but this does not kill the bacterium since the viral genome is absent. The transferred chromosomal segment can, however, recombine with homologous portions of the bacterium's chromosome. This phage-mediated recombinational process is known as **transduction.**

A transducing phage can contain no more than a capsid full of DNA [typically 50,000 base pairs **(bp)**] so that it can only transduce genes that are no further than this distance apart (maximally 2 min) on a bacterial chromosome. Consequently, *the relative frequency with which closely linked bacterial genes are cotransduced accurately reflects their separation on the bacterial chromosome.* The finer details of the *E. coli* genetic map shown in Fig. 27-18 were elucidated by such **transductional mapping** using **bacteriophage P1.**

Viruses Are Subject to Complementation and Recombination

The genetics of viruses can be studied in much the same way as that of cellular organisms. Since viruses have no metabolism, however, their presence is usually detected by their ability to kill their host. The presence of viable bacteriophages is conveniently indicated by **plaques** (clear spots) on a "lawn" of bacteria on a culture plate (Fig. 27-20). Plaques mark the spots where single phage particles had multiplied with the resulting lysis of the bacteria in the area. A mutant phage, which can produce progeny under certain

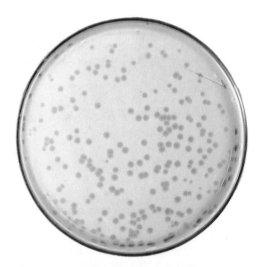

FIGURE 27-20. A culture plate covered with a lawn of *E. coli* on which bacteriophage T4 has formed plaques. [Bruce Iverson.]

permissive conditions, is detected by its inability to do so under other **restrictive conditions** in which the wild-type phage is viable. These conditions usually involve differences in the strain of the bacterial host employed or in the temperature.

Viruses are subject to complementation. Simultaneous infection of a bacterium by two different mutant varieties of a phage may yield progeny under conditions in which neither variety by itself can reproduce. If this occurs, then each mutant phage must have supplied a function that could not be supplied by the other. Each such mutation is said to

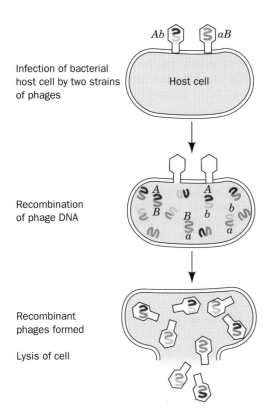

FIGURE 27-21. Recombination of bacteriophage chromosomes occurs on simultaneous infection of a bacterial host by two phage strains carrying the genes *Ab* and *aB*.

belong to a different **complementation group,** a term synonymous for gene.

Viral chromosomes are also subject to recombination. This occurs when a single cell is simultaneously infected by two mutant strains of a virus (Fig. 27-21). The dynamics of viral recombination differ from those in eukaryotes or bacteria because the viral chromosome undergoes recombination throughout the several rounds of DNA replication that occur during the viral life cycle. Recombinant viral progeny therefore consist of many if not all of the possible recombinant types.

The Recombinational Unit Is a Base Pair

The enormous rate at which bacteriophages reproduce permits the detection of recombinational events that occur with a frequency of as little as one in 10^8. In the 1950s, Seymour Benzer carried out high-resolution genetic studies of the *rII* region of the **bacteriophage T4** chromosome. This ~4000 bp region, which represents ~2% of the T4 chromosome, consists of two adjacent complementation groups designated *rIIA* and *rIIB*. In a permissive host, *E. coli* B, a mutation that inactivates the product of either gene causes the formation of plaques that are easily identified because they are much larger than those of the wild-type phage (the designation *r* stands for rapid lysis). However, only the wild-type will lyse the restrictive host, *E. coli* K12(λ). The

presence of plaques in an *E. coli* K12(λ) culture plate that had been simultaneously infected with two different *rII* mutants in the same complementation group demonstrated that *recombination can take place within a gene.* This refuted a then widely held model of the chromosome in which genes were thought to be discrete entities, rather like beads on a string, such that recombination could take place only between intact genes. The genetic mapping of mutations at over 300 distinguishable sites in the *rIIA* and *rIIB* regions indicated that *genes, as are chromosomes, are linear unbranched structures.*

Benzer also demonstrated that a complementation test between two mutations on the same complementation group yields progeny in the restrictive host when the two mutations are in the **cis** configuration (on the same chromosome; Fig. 27-22a), but fails to do so when they are in the **trans** configuration (on physically different chromosomes; Fig. 27-22b). This is because only when both mutations physically occur in the same gene will the other gene be functionally intact. The term **cistron** was coined to mean a functional genetic unit defined according to this **cis–trans test.** This word has since become synonomous with gene or complementation group.

The recombination of pairs of *rII* mutants was observed to occur at frequencies as low as 0.01% (although frequencies as low as 0.0001% could, in principle, have been detected). Since a recombination frequency in T4 of 1% corresponds to a 240-bp separation of mutation sites, the unit of recombination can be no larger than 0.01 × 240 = 2.4 bp. For reasons having to do with the mechanism of recombination, this is an upper limit estimate. On the basis of high-resolution genetic mapping, it was therefore concluded that *the unit of recombination is about the size of a single base pair.*

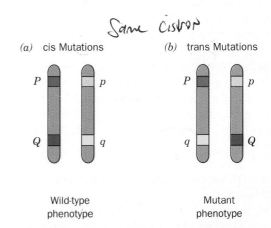

FIGURE 27-22. The cis–trans test. Consider a chromosome that is present in two copies in which two positions on the same gene, *P* and *Q*, have defective recessive mutants, *p* and *q*, respectively. (a) If the two mutations are cis (physically on the same chromosome), one gene will be wild type, so the organism will have a wild-type phenotype. (b) If the mutations are trans (on physically different chromosomes), both genes will be defective and the organism will have a mutant phenotype.

2. DNA IS THE CARRIER OF GENETIC INFORMATION

Nucleic acids were first isolated in 1869 by Friedrich Miescher and so named because he found them in the nuclei of **leukocytes** (pus cells) from discarded surgical bandages. The presence of nucleic acids in other cells was demonstrated within a few years but it was not until some 75 years after their discovery that their biological function was elucidated. Indeed, in the 1930s and 1940s, it was widely held, in what was termed the **tetranucleotide hypothesis,** that nucleic acids have a monotonously repeating sequence of all four bases so that they were not suspected of having a genetic function. Rather, it was generally assumed that genes were proteins since proteins were the only biochemical entities that, at that time, seemed capable of the required specificity. In this section, we outline the experiments that established DNA's genetic role.

A. Transforming Principle Is DNA

The virulent form of pneumococcus *(Diplococcus pneumoniae),* a bacterium that causes pneumonia, is encapsulated by a gelatinous polysaccharide coating that contains the O antigens through which it recognizes the cells it infects (Section 10-3B). Mutant pneumococci that lack this coating, because of a defect in an enzyme involved in its formation, are not pathogenic. The virulent and nonpathogenic pneumococci are known as the S and R forms, respectively, because of the smooth and rough appearances of their colonies in culture (Fig. 27-23).

In 1928, Frederick Griffith made a startling discovery. He injected mice with a mixture of live R and heat-killed S pneumococci. This experiment resulted in the death of most of the mice. More surprising yet was that the blood of the dead mice contained live S pneumococci. The dead S pneumococci initially injected into the mice had somehow **transformed** the otherwise innocuous R pneumococci to the virulent S form. Furthermore, the progeny of the transformed pneumococci were also S; the transformation was permanent. Eventually, it was shown that the transformation could also be made *in vitro* by mixing R cells with a cell-free extract of S cells. The question remained: What is the nature of the **transforming principle?**

In 1944, Oswald Avery, Colin MacLeod, and Maclyn McCarty, after a 10-year investigation, reported that *transforming principle is DNA.* The conclusion was based on the observations that the laboriously purified (few modern fractionation techniques were then available) transforming principle had all the physical and chemical properties of DNA; contained no detectable protein; was unaffected by trypsin, chymotrypsin, or ribonuclease; and was totally inactivated by treatment with DNase. *DNA must therefore be the carrier of genetic information.*

Avery's discovery was another idea whose time had not yet come. This seminal advance was initially greeted with skepticism and then largely ignored. Indeed, even Avery did not directly state that DNA is the hereditary material but merely that it has "biological specificity." His work, however, influenced several biochemists, including Erwin Chargaff, whose subsequent accurate determination of DNA base ratios (Section 28-1) refuted the tetranucleotide

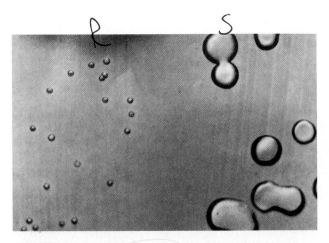

FIGURE 27-23. The large glistening colonies are virulent S-type pneumococci that resulted from the transformation of nonpathogenic R-type pneumococci (smaller colonies) by DNA from heat-killed S pneumococci. [From Avery, O.T., MacLeod, C.M., and McCarty, M., *J. Exp. Med.* **79,** 153 (1944). Copyright © 1944 by Rockefeller University Press.]

FIGURE 27-24. The gigantic mouse *(left)* grew from a fertilized ovum that had been microinjected with DNA bearing the rat growth hormone gene. His normal littermate *(right)* is shown for comparison. [Courtesy of Ralph Brinster, University of Pennsylvania.]

hypothesis and thereby indicated that DNA could be a complex molecule.

It was eventually demonstrated that eukaryotes are also subject to transformation by DNA. Thus DNA, which cytological studies had shown resides in the chromosomes, must also be the hereditary material of eukaryotes. In a spectacular demonstration of eukaryotic transformation, Ralph Brinster, in 1982, microinjected DNA bearing the gene for rat **growth hormone** (a polypeptide) into the nuclei of fertilized mouse eggs (a technique discussed in Section 28-8F) and implanted these eggs into the uteri of foster mothers. The resulting "supermice" (Fig. 27-24), which had high levels of rat growth hormone in their serum, grew to nearly twice the weight of their normal litter mates. Such genetically altered animals are said to be **transgenic.**

B. The Hereditary Molecule of Many Bacteriophages Is DNA

Electron micrographs of phage-infected bacteria show empty-headed phage "ghosts" attached to the bacterial surface (Fig. 27-25). This observation led Roger Herriott to suggest "that the virus may act like a little hypodermic needle full of transforming principle," which it injects into the bacterial host (Fig. 27-26). This proposal was tested in 1952 by Alfred Hershey and Martha Chase as is diagrammed in Fig. 27-27. **Bacteriophage T2** was grown on *E. coli* in a medium containing the radioactive isotopes ^{32}P and ^{35}S. This labeled the phage capsid, which contains no P, with ^{35}S, and its DNA, which contains no S, with ^{32}P. These

phages were added to an unlabeled culture of *E. coli* and, after sufficient time was allowed for the phages to infect the bacterial cells, the culture was agitated in a kitchen blender so as to shear the phage ghosts from the bacterial cells. This rough treatment neither injured the bacteria nor altered the course of the phage infection. When the phage ghosts were separated from the bacteria by centrifugation, the ghosts were found to contain most of the ^{35}S, whereas the bacteria contained most of the ^{32}P. Furthermore, 30% of the ^{32}P appeared in the progeny phages but only 1% of the ^{35}S did so. Hershey and Chase therefore concluded that only the phage DNA was essential for the production of progeny. *DNA therefore must be the hereditary material.* In later years it was shown that, in a process known as **transfection,**

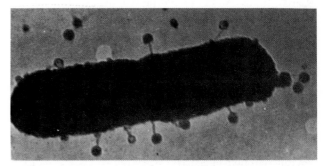

FIGURE 27-25. An early electron micrograph of an *E. coli* cell to which **bacteriophage T5** are adsorbed by their tails. [Courtesy of Thomas F. Anderson, Fox Chase Cancer Center.]

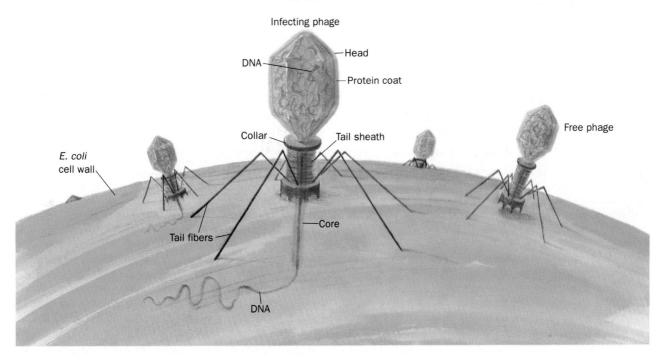

FIGURE 27-26. A diagram of T2 bacteriophage injecting its DNA into an *E. coli* cell.

purified phage DNA can, by itself, induce a normal phage infection in a properly treated bacterial host (transfection differs from transformation in that the latter results from the recombination of the bacterial chromosome with a fragment of homologous DNA).

In 1952, the state of knowledge of biochemistry was such that Hershey's discovery was much more readily accepted than Avery's identification of the transforming principle had been some 8 years earlier. Within a few months, the first speculations arose as to the nature of the genetic code, and James Watson and Frances Crick were inspired to investigate the structure of DNA. In 1955, it was shown that the somatic cells of eukaryotes have twice the DNA of the corresponding germ cells. When this observation was proposed to be a further indicator of DNA's genetic role, there was little comment even though the same could be said of any other chromosomal component.

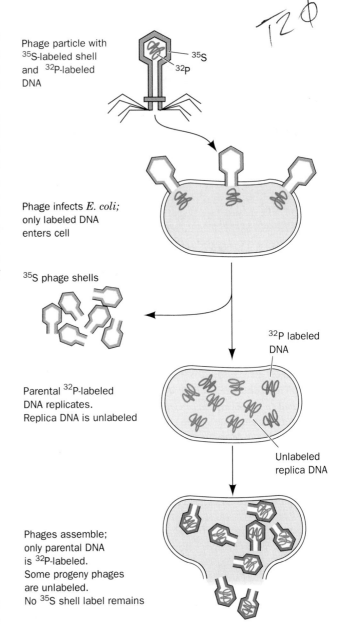

Phage particle with ^{35}S-labeled shell and ^{32}P-labeled DNA

^{35}S

^{32}P

Phage infects *E. coli;* only labeled DNA enters cell

^{35}S phage shells

^{32}P labeled DNA

Parental ^{32}P-labeled DNA replicates. Replica DNA is unlabeled

Unlabeled replica DNA

Phages assemble; only parental DNA is ^{32}P-labeled. Some progeny phages are unlabeled. No ^{35}S shell label remains

FIGURE 27-27. A diagram of the Hershey–Chase experiment demonstrating that only the nucleic acid component of bacteriophages enters the bacterial host during phage infection.

CHAPTER SUMMARY

Eukaryotic cells contain a characteristic number of homologous pairs of chromosomes. In mitosis, each daughter cell receives a copy of each of these chromosomes but in meiosis each resulting gamete receives only one member of each homologous pair. Fertilization is the fusion of two haploid gametes to form a diploid zygote.

The Mendelian laws of inheritance state that alternative forms of true-breeding traits are specified by different alleles of the same gene. Alleles may be dominant, codominant, or recessive depending on the phenotype of the heterozygote. Different genes assort independently unless they are on the same chromosome. The linkage between genes on the same chromosome, however, is never complete because of crossing over among homologous chromosomes during meiosis. The rate that genes recombine varies with

their physical separation because crossing over occurs essentially at random. This permits the construction of genetic maps. Whether two recessive traits are allelic may be determined by the complementation test. The nature of genes is largely defined by the dictum "one gene–one polypeptide."

The rapid reproduction rate of bacteria and bacteriophages permits the detection of extremely rare genetic events. F$^+$ strains of bacteria transfer a copy of their F factor to F$^-$ cells through conjugation. In Hfr cells, the F factor is integrated into the bacterial chromosome. Hfr cells transfer the leading section of their F factor together with a portion of the attached bacterial chromosome, in fixed order, to the F$^-$ cell where the chromosomal fragment recombines with the F$^-$ chromosome. This permits the genetic mapping of bacterial chromosomes by interrupting the mating process.

The fine structures of these maps may be deduced through transductional mapping.

Mutant varieties of bacteriophages are detected by their ability to kill their host under various restrictive conditions. The fine structure analysis of the *rII* region of the bacteriophage T4 chromosome has revealed that recombination may take place within a gene, that genes are linear, unbranched structures, and that the unit of mutation is ~1 bp.

Extracts of virulent S-type pneumococci transform nonpathogenic R-type pneumococci to the S form. The transforming principle is DNA. Similarly, radioactive labeling has demonstrated that the genetically active substance of bacteriophage T2 is its DNA. The viral capsid serves only to protect its enclosed DNA and to inject it into the bacterial host. This demonstrates that DNA is the hereditary molecule.

REFERENCES

Genetics

Benzer, S., The fine structure of the gene, *Sci. Am.* **206**(1): 70–84 (1962).

Cairns, J., Stent, G.S., and Watson, J. (Eds.), *Phage and the Origins of Molecular Biology,* (expanded ed.), Cold Spring Harbor Laboratory (1992). [A series of scientific memoirs by many of the pioneers of molecular biology.]

Griffiths, A.J.F., Miller, J.H., Suzuki, D.T., Lewontin, R.C., and Gelbart, W.N., *An Introduction to Genetic Analysis* (5th ed.), Freeman (1993).

Russell, P.J., *Genetics* (3rd ed.), Harper Collins (1992).

Stent, G.S. and Calender, R., *Molecular Genetics* (2nd ed.), Freeman (1978).

The Role of DNA

Avery, O.T., MacLeod, C.M., and McCarty, M., Studies on the chemical nature of the substance inducing transformation of pneumococcal types, *J. Exp. Med.* **79**, 137–158 (1944). [The milestone report identifying transforming principle as DNA.]

Hershey, A.D. and Chase, M., Independent functions of viral proteins and nucleic acid in growth of bacteriophage, *J. Gen. Physiol.* **36**, 39–56 (1952).

McCarty, M., *The Transforming Principle,* Norton (1985). [A chronicle of the discovery that genes are DNA.]

Palmiter, R.D., Brinster, R.L., Hammer, R.E., Trumbauer, M.E., Rosenfeld, M.G., Birmberg, N.C., and Evans, R.M., Dramatic growth of mice that develop from eggs microinjected with metallothionein-growth hormone fusion genes, *Nature* **300**, 611–615 (1982).

Stent, G.S., Prematurity and uniqueness in scientific discovery, *Sci. Am.* **227**(6): 84–93 (1972). [A fascinating philosophical discourse on what it means for discoveries such as Avery's to be "ahead of their time" and on the nature of creativity in science.]

PROBLEMS

1. One method that Mendel used to test his laws is known as a **testcross**. In it, F_1 hybrids are crossed with their recessive parent. What is the expected distribution of progeny and what are their phenotypes in a testcross involving peas with different color seeds? What is it for snapdragons with different flower colors (use the white parent in this testcross)?

2. The disputed paternity of a child can often be decided on the basis of blood tests. The M, N, and MN blood groups (Section 11-3D) result from two alleles, L^M and L^N; the Rh$^+$ blood group arises from a dominant allele, R. Both sets of alleles occur on a different chromosome from each other and from the alleles responsible for the ABO blood groups. The following table gives the blood types of three children, their mother, and the two possible fathers. Indicate, where possible, each child's paternity and justify your answer.

Child 1	B	M	Rh$^-$
Child 2	B	MN	Rh$^+$
Child 3	AB	MN	Rh$^+$
Mother	B	M	Rh$^+$
Male 1	B	MN	Rh$^+$
Male 2	AB	N	Rh$^+$

3. The most common form of color blindness, red–green color blindness, afflicts almost only males. What are the genotypes and phenotypes of the children and grandchildren of a red–green color blind man and a woman with no genetic history of color blindness? Assume the children mate with individuals who also have no history of color blindness.

4. How might F$'$ cells be useful in the genetic analysis of bacteria?

5. Hfr strains of *E coli* differ both in their origins of replication and in their directions of chromosomal transfer to F$^-$ recipients. The following table presents the transfer order of genes near the replication origin, reading left to right, in several Hfr strains. Use this data to construct a genetic map of the bacterial chromosome.

Hfr Strain	Order of Gene Transfer
1	*met-thi-thr-leu-azi-ton-pro*
2	*thi-met-ile-mtl-xyl-mal-str-his*
3	*ton-pro-lac-ade-gal-trp*
4	*xyl-mal-str-his-trp-gal-ade-lac*
5	*thi-thr-leu-azi-ton*

C H A P T E R

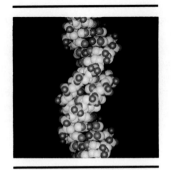

28

Nucleic Acid Structures And Manipulation

*There are two classes of nucleic acids, **deoxyribonucleic acid (DNA)** and **ribonucleic acid (RNA)**. DNA is the hereditary molecule in all cellular life forms, as well as in many viruses.* It has but two functions:

1. To direct its own replication during cell division.

2. To direct the **transcription** of complementary molecules of RNA.

RNA, in contrast, has more varied biological functions:

1. The RNA transcripts of DNA sequences that specify polypeptides, **messenger RNA (mRNA),** direct the ribosomal synthesis of these polypeptides in a process known as **translation.**

2. The RNAs of ribosomes, which are about two thirds RNA and one third protein, almost certainly have functional as well as structural roles.

3. During protein synthesis, amino acids are delivered to the ribosome by molecules of **transfer RNA (tRNA).**

4. Certain RNAs are associated with specific proteins to form **ribonucleoproteins** that participate in the post-transcriptional processing of other RNAs.

5. In many viruses, RNA, not DNA, is the carrier of hereditary information.

In this chapter we examine the structures of nucleic acids with emphasis on DNA (the structures of RNAs are detailed in Section 30-2A), and discuss methods of purifying, sequencing, and chemically synthesizing nucleic acids. We end by outlining how recombinant DNA technology, which has revolutionized the study of biochemistry, is used to manipulate, synthesize, and express DNA.

1. CHEMICAL STRUCTURE AND BASE COMPOSITION

The chemical structures of the nucleic acids were elucidated by the early 1950s largely through the efforts of Phoebus Levine, followed by the work of Alexander Todd. *Nucleic acids are, with few exceptions, linear polymers of* nucleotides whose phosphates bridge the 3' and 5' positions of successive sugar residues (e.g., Fig. 28-1). The phosphates of these **polynucleotides,** the **phosphodiester** groups, are acidic so that, *at physiological pH's, nucleic acids are polyanions.*

DNA's Base Composition Is Governed by Chargaff's Rules

DNA has equal numbers of adenine and thymine residues (A = T) and equal numbers of guanine and cytosine residues (G = C). These relationships, known as **Chargaff's rules,** were discovered in the late 1940s by Erwin Chargaff who first devised reliable quantitative methods for the separation (by paper chromatography) and analysis of DNA hydrolysates. Chargaff also found that the base composition of DNA from a given organism is characteristic of that organism; that is, it is independent of the tissue from which the DNA is taken as well as the age of the organism, its nutri-

(a)

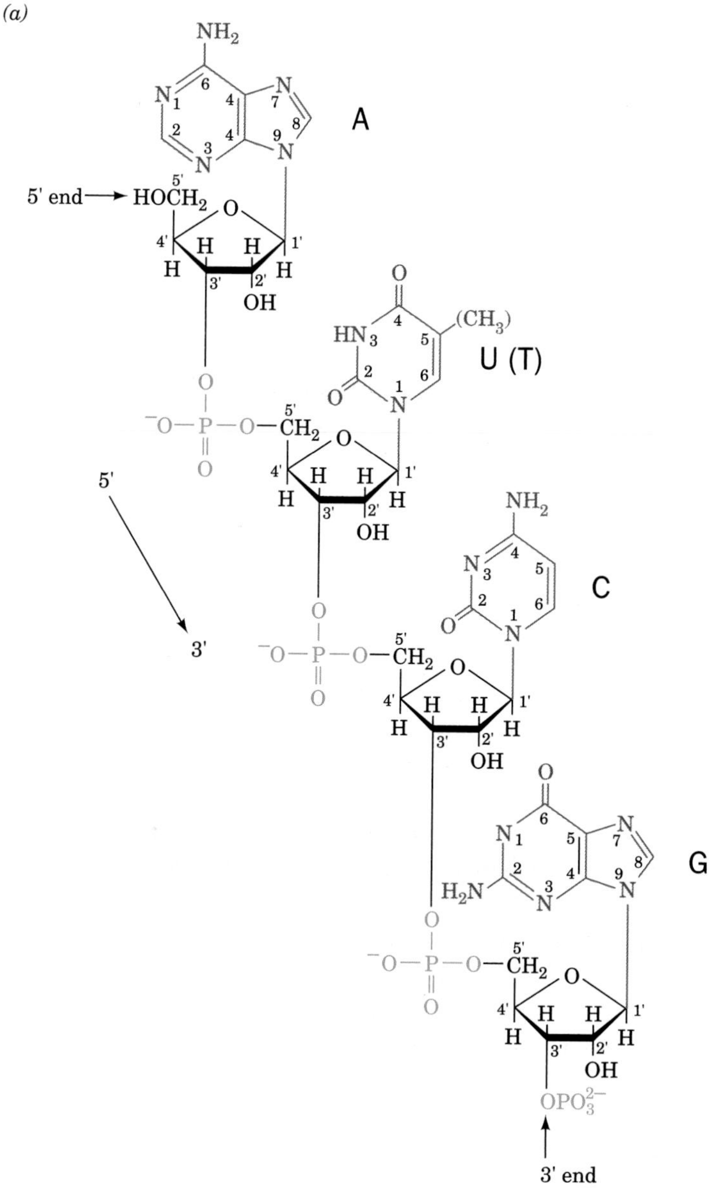

(b)

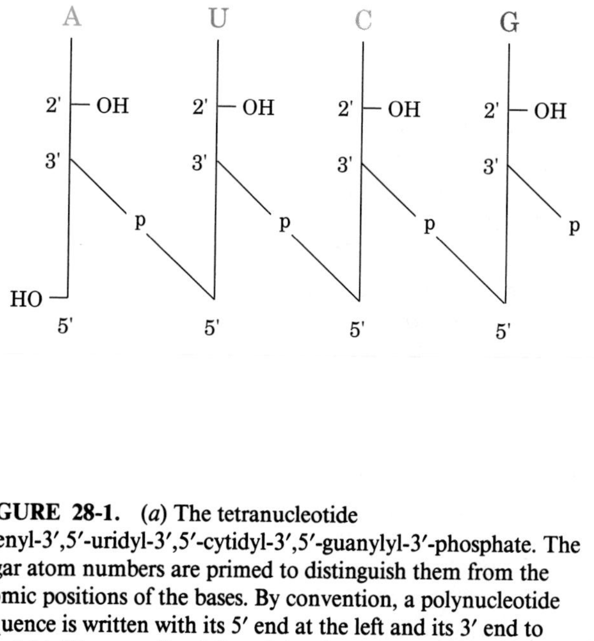

FIGURE 28-1. (*a*) The tetranucleotide adenyl-3′,5′-uridyl-3′,5′-cytidyl-3′,5′-guanylyl-3′-phosphate. The sugar atom numbers are primed to distinguish them from the atomic positions of the bases. By convention, a polynucleotide sequence is written with its 5′ end at the left and its 3′ end to the right. Thus, reading left to right, the phosphodiester bond links neighboring ribose residues in the 5′ → 3′ direction. The above sequence may be abbreviated ApUpCpGp or just AUCGp (where a "p" to the left and/or right of a nucleoside symbol indicates a 5′ and/or a 3′ phosphoryl bond, respectively; see Table 26-1 for other symbol definitions). The corresponding deoxytetranucleotide, in which the 2′-OH groups are all replaced by H and the uracil (U) base is replaced by thymine (5-methyluracil; T), is abbreviated d(ApTpCpGp) or d(ATCGp). (*b*) A schematic representation of AUCGp. Here a vertical line denotes a ribose residue, its attached base is indicated by the corresponding one-letter abbreviation, and a diagonal line flanking an optional "p" represents a phosphodiester bond. The atom numbering of the ribose residues, which is indicated here, is usually omitted. The equivalent representation of deoxypolynucleotides differ only by the absence of the 2′-OH groups and the replacement of U by T.

tional state, or any other environmental factor. The structural basis of Chargaff's rules, as we shall see, derives from DNA's double-stranded character (Section 28-2A).

DNA's base composition varies widely among different organisms. It ranges from ~25 to 75% G + C in different species of bacteria. It is, however, more or less constant among related species; for example, in mammals G + C ranges from 39 to 46%.

RNA, which usually occurs as a single-stranded molecule, has no apparent constraints on its base composition. However, double-stranded RNA, which comprises the genetic material of several viruses, obeys Chargaff's rules. Conversely, single-stranded DNA, which occurs in certain viruses, does not obey Chargaff's rules. Upon entering its

host organism, however, such DNA is replicated to form a double-stranded molecule, which then obeys Chargaff's rules.

Nucleic Acid Bases May Be Modified

Some DNAs contain bases that are chemical derivatives of the standard set. For example, dA and dC in the DNAs of many organisms are partially replaced by N^6-methyl-dA and **5-methyl-dC**, respectively.

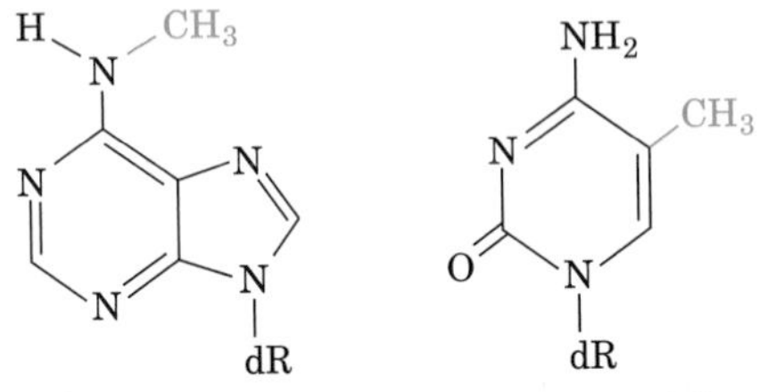

N^6-Methyl-dA   5-Methyl-dC

The altered bases are generated by the sequence-specific enzymatic modification of normal DNA (Sections 28-6A and 31-7). The modified DNAs obey Chargaff's rules if the derivatized bases are taken as equivalent to their parent bases. Likewise, many bases in RNA and, in particular, in tRNA (Section 30-2), are derivatized.

RNA but Not DNA Is Susceptible to Base-Catalyzed Hydrolysis

RNA is highly susceptible to base-catalyzed hydrolysis by the reaction mechanism diagrammed in Fig. 28-2 so as to yield a mixture of 2' and 3' nucleotides. In contrast, DNA, which lacks 2'-OH groups, is resistant to base-catalyzed hydrolysis and is therefore much more chemically stable than RNA. This is probably why DNA rather than RNA evolved to be the cellular genetic archive.

2. DOUBLE HELICAL STRUCTURES

The determination of the structure of DNA by James Watson and Francis Crick in 1953 is often said to mark the birth of modern molecular biology. The **Watson–Crick structure** of DNA is of such importance because, in addition to providing the structure of what is arguably the central molecule of life, it suggested the molecular mechanism of heredity. Watson and Crick's accomplishment, which is ranked as one of science's major intellectual achievements, tied together the less than universally accepted results of several diverse studies:

1. Chargaff's rules. At the time, these relationships were quite obscure because their significance was not apparent. In fact, even Chargaff did not emphasize them.

2. The correct tautomeric forms of the bases. X-Ray, NMR, and spectroscopic investigations have firmly established that the nucleic acid bases are overwhelmingly in the keto tautomeric forms shown in Fig. 28-1. In

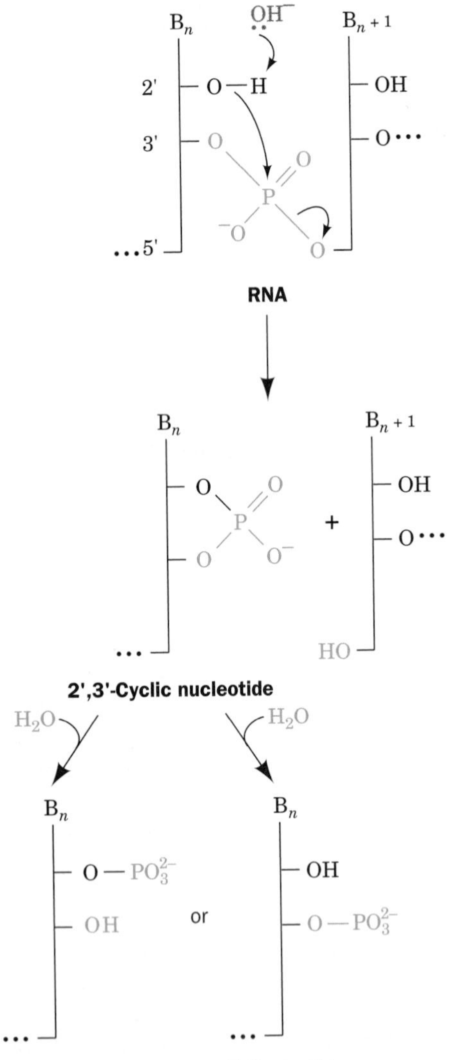

FIGURE 28-2. The mechanism of base-catalyzed RNA hydrolysis. The base-induced deprotonation of the 2'-OH group facilitates its nucleophilic attack on the adjacent phosphorus atom thereby cleaving the RNA backbone. The resultant 2',3'-cyclic phosphate group subsequently hydrolyzes to either the 2' or the 3' phosphate. Note that the RNase-catalyzed hydrolysis of RNA follows a nearly identical reaction sequence (Section 14-1A).

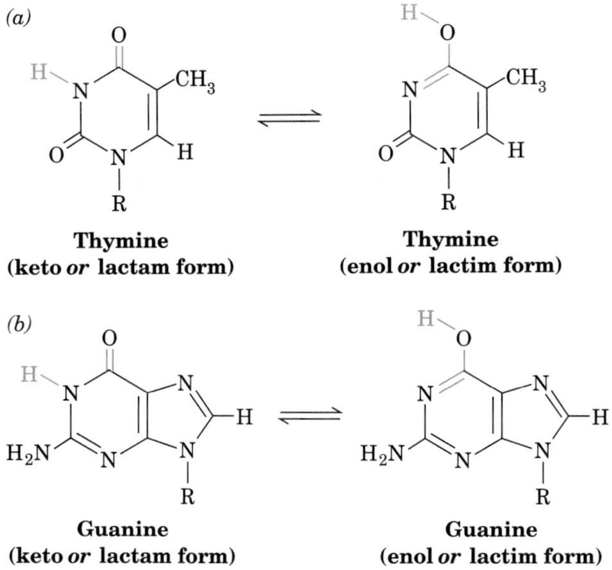

(a)

Thymine
(keto *or* lactam form)

Thymine
(enol *or* lactim form)

(b)

Guanine
(keto *or* lactam form)

Guanine
(enol *or* lactim form)

FIGURE 28-3. Some possible tautomeric conversions for (a) thymine and (b) guanine residues. Cytosine and adenine residues can undergo similar proton shifts.

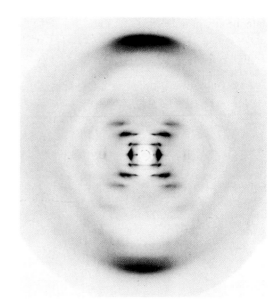

FIGURE 28-4. An X-ray diffraction photograph of a vertically oriented Na⁺ DNA fiber in the B conformation taken by Rosalind Franklin. This is the photograph that provided key information for the elucidation of the Watson–Crick structure. The central X-shaped pattern of spots is indicative of a helix, whereas the heavy black arcs on the top and bottom of the diffraction pattern correspond to a distance of 3.4 Å and indicate that the DNA structure largely repeats every 3.4 Å along the fiber axis. [Courtesy of Maurice Wilkins, King's College, London.]

1953, however, this was not generally appreciated. Indeed, guanine and thymine were widely believed to be in their enol forms (Fig. 28-3) because it was thought that the resonance stability of these aromatic molecules would thereby be maximized. Knowledge of the dominant tautomeric forms, which was prerequisite for the prediction of the correct hydrogen bonding associations of the bases, was provided by Jerry Donohue, an office mate of Watson and Crick and an expert on the X-ray structures of small organic molecules.

3. Information that DNA is a helical molecule. This was provided by an X-ray diffraction photograph of a DNA fiber taken by Rosalind Franklin (Fig. 28-4; DNA, being a threadlike molecule, does not crystallize but, rather, can be drawn out in fibers consisting of parallel bundles of molecules; Section 7-2). A description of the photograph enabled Crick, an X-ray crystallographer by training who had earlier derived the equations describing diffraction by helical molecules, to deduce that DNA is (a) a helical molecule, and (b) that its planar aromatic bases form a stack of parallel rings that is parallel to the fiber axis.

This information only provided a few crude landmarks that guided the elucidation of the DNA structure; it mostly sprang from Watson and Crick's imaginations through model building studies. Once the Watson–Crick model had been published, however, its basic simplicity combined with its obvious biological relevance led to its rapid acceptance. Later investigations have confirmed the essential correctness of the Watson–Crick model although its details have been modified.

It is now realized that double helical DNA and RNA can assume several distinct structures that vary with such factors as the humidity and the identities of the cations present, as well as with base sequence. In this section, we describe these various structures.

A. The Watson–Crick Structure: B-DNA

Fibers of DNA assume the so-called B conformation, as indicated by their X-ray diffraction patterns, when the counterion is an alkali metal such as Na⁺ and the relative humidity is 92%. **B-DNA** *is regarded as the native form because its X-ray pattern resembles that of the DNA in intact sperm heads.*

The Watson-Crick structure of B-DNA has the following major features (Table 28-1):

1. *It consists of two polynucleotide strands that wind about a common axis with a right-handed twist to form an ~20-Å-diameter double helix (Fig. 28-5). The two strands are antiparallel (run in opposite directions) and* wrap around each other such that they cannot be separated without unwinding the helix (a phenomenon known as **plectonemic coiling**). The bases occupy the core of the helix and sugar–phosphate chains are coiled about its periphery, thereby minimizing the repulsions between charged phosphate groups.

2. The planes of the bases are nearly perpendicular to the helix axis. Each base is hydrogen bonded to a base on the opposite strand to form a planar *base pair (Fig. 28-5).* It

(a)

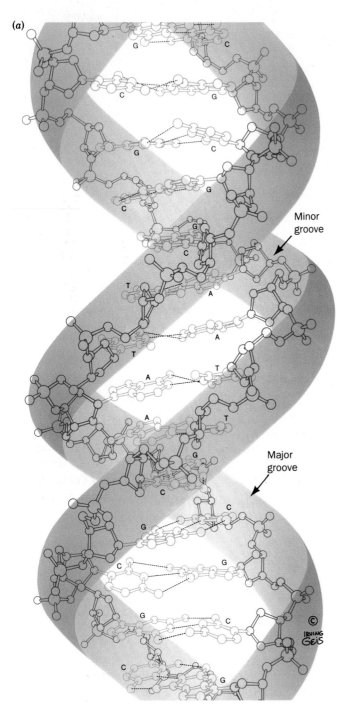

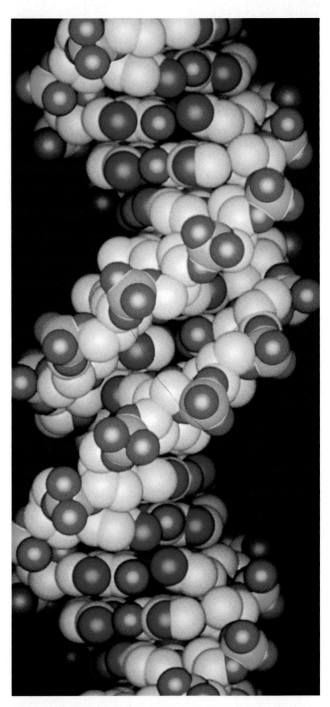

FIGURE 28-5. The structure of B-DNA as represented by ball-and-stick drawings and the corresponding computer-generated space-filling models. The repeating helix is based on the X-ray structure of the self-complementary dodecamer d(CGCGAATTCGCG) determined by Richard Dickerson and Horace Drew. (*a*) View perpendicular to the helix axis. In the drawing, the sugar–phosphate backbones, which wind about the periphery of the molecule, are blue, and the bases, which occupy its core, are red. In the space-filling model, C, N, O, and P atoms are white, blue, red, and green, respectively. H atoms have been omitted for clarity in both drawings. Note that the two sugar–phosphate chains run in opposite directions. (*b*) (*Opposite*) View down the helix axis. In the drawing, the ribose ring O atoms are red and the nearest base pair is white. Note that the helix axis passes through the base pairs so that the helix has a solid core. [Drawings copyrighted © by Irving Geis.]

is these hydrogen bonding interactions, a phenomenon known as **complementary base pairing,** that result in the specific association of the two chains of the double helix.

3. The "ideal" B-DNA helix has 10 base pairs (bp) per turn (a helical twist of 36° per bp) and, since the aromatic bases have van der Waals thicknesses of 3.4 Å and are

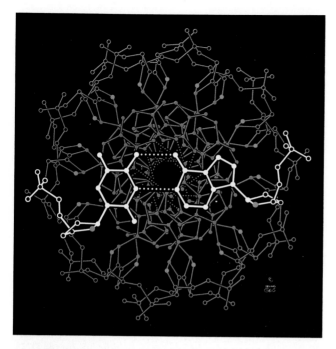

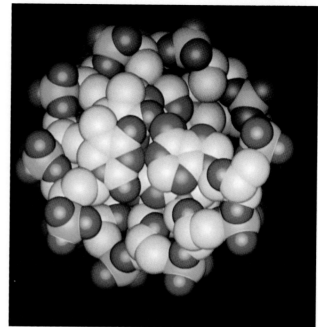

FIGURE 28-5. *(b)*

partially stacked on each other (**base stacking;** Fig. 28-5b), the helix has a pitch (rise per turn) of 34 Å.

The most remarkable feature of the Watson–Crick structure is that *it can accommodate only two types of base pairs: Each adenine residue must pair with a thymine residue and vice versa, and each guanine residue must pair with a cytosine residue and vice versa.* The geometries of these A·T and G·C base pairs, the so-called **Watson–Crick base pairs,** are shown in Fig. 28-6. It can be seen that *both of these base pairs are interchangeable in that they can replace each other in the double helix without altering the positions of the sugar–phosphate backbone's C1' atoms. Likewise, the double helix is undisturbed by exchanging the partners of a Watson–Crick base pair, that is, by changing a G·C to a C·G or a A·T to a T·A.* In contrast, any other combina-

tion of bases would significantly distort the double helix since the formation of a non-Watson–Crick base pair would require considerable reorientation of the sugar–phosphate chain.

B-DNA has two deep exterior grooves that wind between its sugar–phosphate chains as a consequence of the helix axis passing through the approximate center of each base pair (Fig 28-5b). The grooves are of unequal size (Fig. 28-5a) because: (1) the top edge of each base pair, as drawn in Fig. 28-6, is structurally distinct from the bottom edge; and (2) the deoxyribose residues are asymmetric. The **minor groove** is that in which the C1'–helix axis–C1' angle is <180° (opening towards the bottom in Fig. 28-6; the helix axis passes through the middle of each base pair in B-DNA), whereas the **major groove** opens towards the opposite edge of each base pair (Fig. 28-6).

TABLE 28-1. STRUCTURAL FEATURES OF IDEAL A-, B-, AND Z-DNA

	A	B	Z
Helical sense	Right handed	Right handed	Left handed
Diameter	~26 Å	~20 Å	~18 Å
Base pairs per helical turn	11	10	12 (6 dimers)
Helical twist per base pair	33°	36°	60° (per dimer)
Helix pitch (rise per turn)	28Å	34Å	45Å
Helix rise per base pair	2.6 Å	3.4 Å	3.7 Å
Base tilt normal to the helix axis	20°	6°	7°
Major groove	Narrow and deep	Wide and Deep	Flat
Minor groove	Wide and shallow	Narrow and deep	Narrow and deep
Sugar pucker	C3'-*endo*	C2'-*endo*	C2'-*endo* for pyrimidines; C3'-*endo* for purines
Glycosidic bond	Anti	Anti	Anti for pyrimidines; syn for purines

The Watson–Crick structure can accommodate any sequence of bases on one polynucleotide strand if the opposite strand has the complementary base sequence. This immediately accounts for Chargaff's rules. More importantly, *it suggests that hereditary information is encoded in the sequence of bases on either strand.*

Real DNA Deviates from the Ideal Watson–Crick Structure

By the late 1970s, advances in nucleic acid chemistry permitted the synthesis and crystallization of ever longer oligonucleotides of defined sequences (Section 28-7). Consequently, some 25 years after the Watson–Crick structure had been formulated, the X-ray crystal structures of DNA fragments were clearly visualized for the first time (fiber diffraction studies provide only crude, low-resolution images in which the base pair electron density is the average electron density of all the base pairs in the fiber). Richard

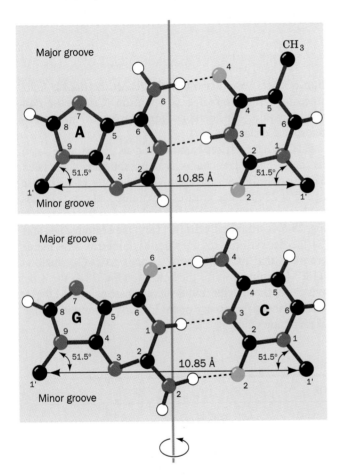

FIGURE 28-6. The Watson–Crick base pairs. The line joining the C1′ atoms is the same length in both base pairs and makes equal angles with the glycosidic bonds to the bases. This gives DNA a series of pseudo-twofold symmetry axes (often referred to as **dyad axes**) that pass through the center of each base pair (*red line*) and are perpendicular to the helix axis. Note that A·T base pairs associate via two hydrogen bonds, whereas C·G base pairs are joined by three hydrogen bonds. [After Arnott, S., Dover, S.D., and Wonacott, A.J., *Acta Cryst.* **B25,** 2196 (1969).]

Dickerson and Horace Drew have shown that the self-complementary dodecamer d(CGCGAATTCGCG) crystallizes in the B conformation. The molecule has an average rise per residue of 3.4 Å and has 10.1 bp per turn (a helical twist of 35.6° per bp), which is nearly equal to that of ideal B-DNA. Nevertheless, *individual residues significantly depart from this average conformation in a manner that appears to be sequence dependent (Fig. 28-5).* For example, the helical twist per base pair in this dodecamer ranges from 28 to 42°. Each base pair further deviates from its ideal conformation by such distortions as propeller twisting (the opposite rotation of paired bases about the base pair's long axis; in the above dodecamer these values range from 10 to 20°) and base pair roll (the tilting of a base pair as a whole about its long axis). Indeed, X-ray and NMR studies of numerous other double helical DNA oligomers have amply demonstrated that *the conformation of DNA is irregular in a sequence-specific manner,* although the rules specifying how sequence governs conformation have proved to be surprisingly elusive. *This phenomenon, as we shall see (Section 29-3), is important for the sequence-specific binding to DNA of proteins that process genetic information.*

DNA Is Semiconservatively Replicated

The Watson–Crick structure also suggests how DNA can direct its own replication. Each polynucleotide strand can act as a template for the formation of its complementary strand through base pairing interactions. The two strands of the parent molecule must therefore separate so that a complementary daughter strand may be enzymatically synthesized on the surface of each parent strand. This results in two molecules of **duplex** (double stranded) DNA, each consisting of one polynucleotide strand from the parent molecule and a newly synthesized complementary strand (Fig. 1-16). Such a mode of replication is termed **semiconservative** in contrast with **conservative** replication which, if it occurred, would result in a newly synthesized duplex copy of the original DNA molecule with the parent DNA molecule remaining intact. The mechanism of DNA replication is the main subject of Chapter 31.

The semiconservative nature of DNA replication was elegantly demonstrated in 1958 by Matthew Meselson and Franklin Stahl. The density of DNA was increased by labeling it with ^{15}N, a heavy isotope of nitrogen (^{14}N is the naturally abundant isotope). This was accomplished by growing *E. coli* for 14 generations in a medium that contained $^{15}NH_4Cl$ as its only nitrogen source. The labeled bacteria were then abruptly transferred to an ^{14}N-containing medium and the density of their DNA was monitored as a function of bacterial growth by equilibrium density gradient ultracentrifugation (Section 5-5B; a technique Meselson, Stahl, and Jerome Vinograd had developed for the purpose of distinguishing ^{15}N-labeled DNA from unlabeled DNA).

The results of the Meselson–Stahl experiment are displayed in Fig. 28-7. After one generation (doubling of the cell population), all of the DNA had a density exactly half-

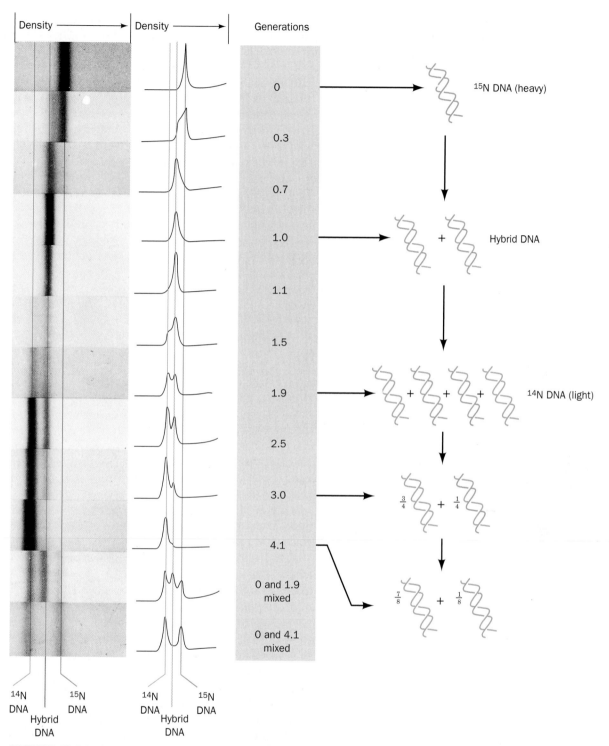

FIGURE 28-7. The demonstration of the semiconservative nature of DNA replication in *E. coli.* DNA in a CsCl solution of density 1.71 g·cm⁻³ was subjected to equilibrium density gradient ultracentrifugation at 140,000g in an analytical ultracentrifuge (a device in which the spinning sample can be optically observed). The enormous centrifugal acceleration caused the CsCl to form a density gradient in which DNA migrated to its position of buoyant density. The left panels are UV absorption photographs of ultracentrifuge cells (DNA strongly absorbs UV light) and are arranged such that regions of equal density have the same horizontal positions. The middle panels are microdensitometer traces of the corresponding photographs in which the vertical displacement is proportional to the DNA concentration. The buoyant density of DNA increases with its ¹⁵N content. The bands furthest to the right (greatest radius and density) arise from DNA that is fully ¹⁵N labeled, whereas unlabeled DNA, which is 0.014 g · cm⁻³ less dense, forms the leftmost bands. The bands in the intermediate position result from duplex DNA in which one strand is ¹⁵N labeled and the other strand is unlabeled. The accompanying interpretive drawings (*right*) indicate the relative numbers of DNA strands at each generation donated by the original parents (*blue,* ¹⁵N labeled) and synthesized by succeeding generations (*red,* unlabeled). [From Meselson, M. and Stahl, F.W., *Proc. Natl. Acad. Sci.* **44,** 674 (1958).]

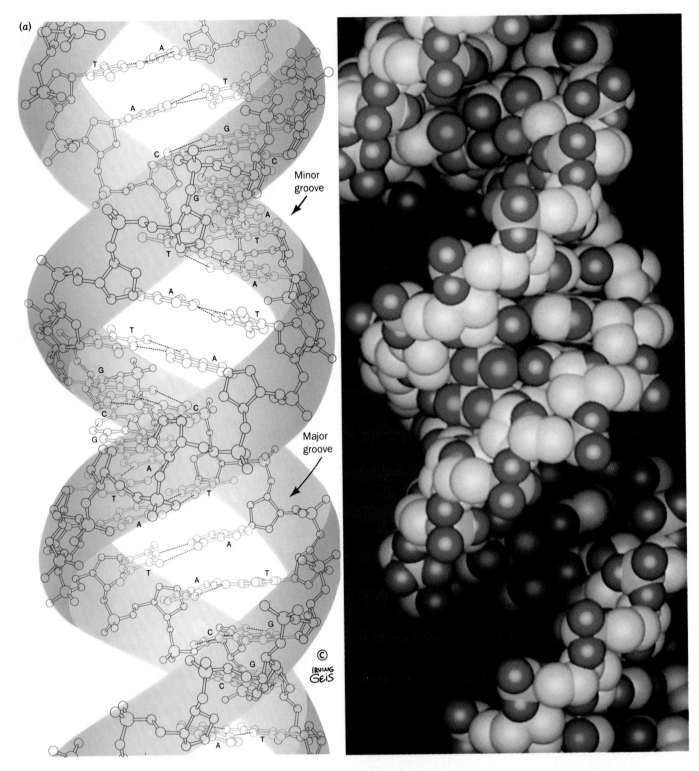

FIGURE 28-8. Ball-and-stick drawings and the corresponding space-filling models of A-DNA as viewed (a) perpendicular to the helix axis, and (b) (*Opposite*) down the helix axis. The color codes are given in the legend to Fig. 28-5. The repeating helix was generated by Richard Dickerson based on the X-ray structure of the self-complementary octamer d(GGTATACC) determined by Olga Kennard, Dov Rabinovitch, Zippora Shakked, and Mysore Viswamitra. Note that the base pairs are inclined to the helix axis and that the helix has a hollow core. Compare this figure with Fig. 28-5. [Drawings copyrighted © by Irving Geis.]

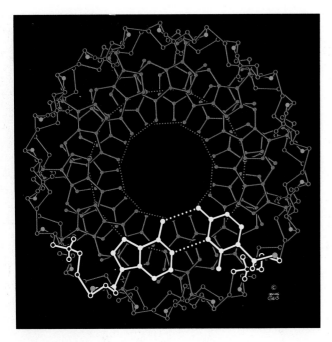

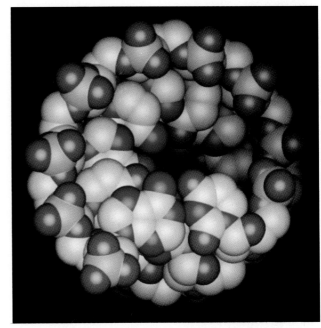

FIGURE 28-8. *(b)*

way between the densities of fully ^{15}N-labeled DNA and unlabeled DNA. This DNA must therefore contain equal amounts of ^{14}N and ^{15}N as is expected after one generation of semiconservative replication. Conservative DNA replication, in contrast, would result in the preservation of the parental DNA, so that it maintained its original density, and the generation of an equal amount of unlabeled DNA. After two generations, half of the DNA molecules were unlabeled and the remainder were ^{14}N – ^{15}N hybrids. This is also in accord with the predictions of the semiconservative replication model and in disagreement with the conservative replication model. In succeeding generations, the amount of unlabeled DNA increased relative to the amount of hybrid DNA although the hybrid never totally disappeared. This is again in harmony with semiconservative replication but at odds with conservative replication, which predicts that the fully labeled parental DNA will always be present and that hybrid DNA never forms.

Meselson and Stahl also demonstrated that DNA is double stranded. DNA from ^{15}N-labeled *E. coli* that were grown for one generation in an ^{14}N medium was heat denatured at 100°C (which causes strand separation; Section 28-3A) and then subjected to density gradient ultracentrifugation. Two bands were observed; one at the density of fully ^{15}N-labeled DNA and the other at the density of unlabeled DNA. Moreover the molecular masses of the DNA in these bands, as estimated from their peak shapes, was half that of undenatured DNA (the peak width varies with molecular mass). Native DNA must therefore be composed of two equal-sized strands that separate upon heat denaturation.

B. Other Nucleic Acid Helices

Double-stranded DNA is a conformationally variable molecule. In the following subsections we discuss its major conformational states besides B-DNA and also those of double-stranded RNA.

A-DNA's Base Pairs Are Inclined to the Helix Axis

When the relative humidity is reduced to 75%, B-DNA undergoes a reversible conformational change to the so-called A form. Fiber X-ray studies indicate that *A-DNA forms a wider and flatter right-handed helix than does B-DNA* (Fig. 28-8; Table 28-1). A-DNA has 11 bp per turn and a pitch of 28 Å, which gives A-DNA an axial hole (Fig. 28-8b). A-DNA's most striking feature, however, is that the planes of its base pairs are tilted 20° with respect to the helix axis. Since its helix axis does not pass through its base pairs (Fig. 28-8b), A-DNA has a deep major groove and a very shallow minor groove; it can be described as a flat ribbon wound around a 6-Å-diameter cylindrical hole. Most self-complementary oligonucleotides of <10 base pairs, for example, d(GGCCGGCC) and d(GGTATACC), crystallize in the A-DNA conformation. Like B-DNA, these molecules exhibit considerable sequence-specific conformational variation.

A-DNA has, so far, been observed in only one biological context. Gram-positive bacteria undergoing **sporulation** (the formation, under enviromental stress, of resistant although dormant cell types known as **spores**; a sort of biological lifeboat) contain a high proportion (20%) of **small acid-soluble spore proteins (SASPs).** Some of these SASPs

(a)

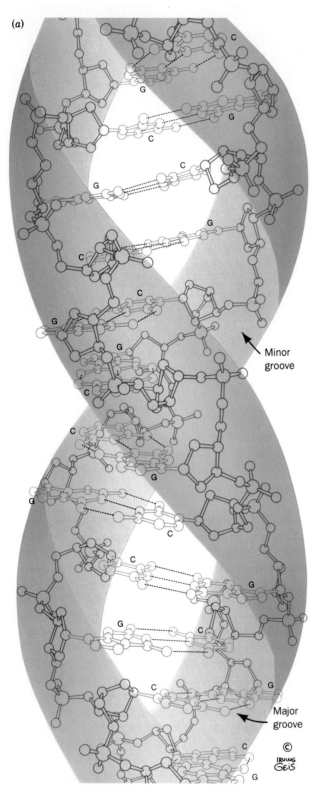

Minor groove

Major groove

© IRVING GEIS

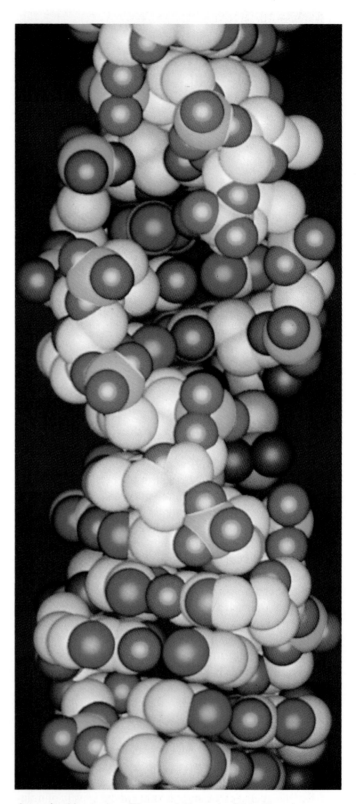

FIGURE 28-9. Ball-and-stick drawings and the corresponding space-filling models of Z-DNA as viewed (*a*) perpendicular to the helix axis and (*b*) (*opposite*) down the helix axis. The color codes are given in the legend to Fig. 28-5. The repeating helix was generated by Richard Dickerson based on the X-ray structure of the self-complementary hexamer d(CGCGCG) determined by Andrew Wang and Alexander Rich. Note that the helix is left handed and that the sugar–phosphate chains follow a zigzag course (alternate ribose residues lie at different radii in Part *b*) indicating that the Z-DNA's repeating motif is a dinucleotide. Compare this figure with Figs. 28-5 and 28-8. [Drawings copyrighted © by Irving Geis.]

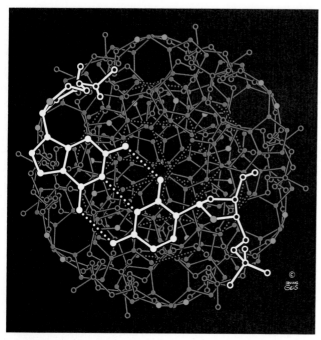

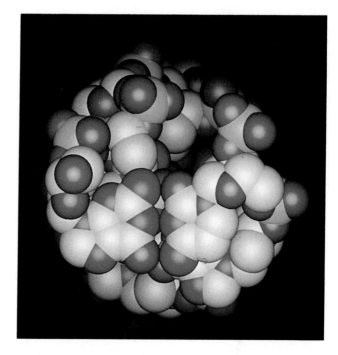

FIGURE 28-9. *(b)*

induce B-DNA to assume the A form, at least *in vitro*. The DNA in bacterial spores exhibits a resistance to UV-induced damage that is abolished in mutants that lack these SASPs. This occurs because the B→A conformation change inhibits the UV-induced covalent cross-linking of pyrimidine bases (Section 31-5A), in part by increasing the distance between successive pyrimidines.

Z-DNA Forms a Left-Handed Helix

Occasionally, a seemingly well understood or at least familiar system exhibits quite unexpected properties. Over 25 years after the discovery of the Watson–Crick structure, the crystal structure determination of d(CGCGCG) by Andrew Wang and Alexander Rich revealed, quite surprisingly, *a left-handed double helix (Fig. 28-9; Table 28-1). A* similar helix is formed by d(CGCATGCG). *This helix, which has been dubbed* **Z-DNA,** *has 12 Watson–Crick base pairs per turn, a pitch of 45 Å, and, in contrast to A-DNA, a deep minor groove and no discernible major groove.* Z-DNA therefore resembles a left-handed drill bit in appearance. The base pairs in Z-DNA are flipped 180° relative to those in B-DNA (Fig. 28-10) through conformational changes discussed in Section 28-3B. As a consequence, the repeating unit of Z-DNA is a dinucleotide, d(XpYp), rather than a single nucleotide as it is in the other DNA helices. The line joining successive phosphate groups on a polynucleotide strand of Z-DNA therefore follows a zigzag path around the helix (Fig. 28-9a; hence the name Z-DNA) rather than a smooth curve as it does in A- and B-DNAs (Figs. 28-5a and 28-8a).

Fiber diffraction and NMR studies have shown that complementary polynucleotides with alternating purines and pyrimidines, such as poly d(GC)·poly d(GC) or poly d(AC)·poly d(GT), take up the Z-DNA conformation at high salt concentrations. Evidently, *the Z-DNA conformation is most readily assumed by DNA segments with alternating purine–pyrimidine base sequences (for structural reasons explained in Section 28-3B).* A high salt concentration stabilizes Z-DNA relative to B-DNA by reducing the otherwise increased electrostatic repulsions between closest approaching phosphate groups on opposite strands (8 Å in Z-DNA vs 12 Å in B-DNA). The methylation of cytosine residues at C5, a common biological modification (Section 31-7), also promotes Z-DNA formation since a hydrophobic methyl group in this position is less exposed to solvent in Z-DNA than it is in B-DNA.

Does Z-DNA have any biological significance? Rich has proposed that the reversible conversion of specific segments of B-DNA to Z-DNA under appropriate circumstances acts as a kind of switch in regulating genetic expression. Yet, the *in vivo* existence of Z-DNA has been difficult to prove. A major problem is demonstrating that a particular probe for detecting Z-DNA, a Z-DNA-specific antibody, for example, does not in itself cause what would otherwise be B-DNA to assume the Z conformation—a kind of biological uncertainty principle (the act of measurement inevitably disturbs the system being measured). Recently, however, Z-DNA has been shown to be present in *E. coli* by employing an *E. coli* enzyme that methylates a specific base sequence *in vitro* when the DNA is in the B form but not when

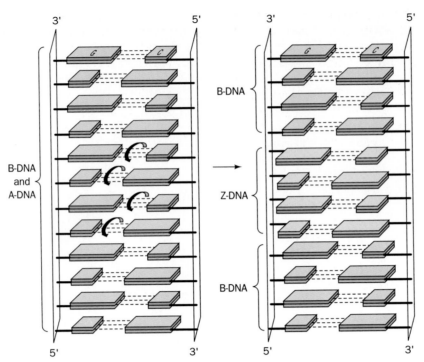

FIGURE 28-10. The conversion of B-DNA to Z-DNA, here represented by a 4-bp segment, involves a 180° flip of each base pair (*curved arrows*) relative to the sugar–phosphate chains. Here, the different faces of the base pairs are colored red and green. Note that if the drawing on the left is taken as looking into the minor groove of unwound A- or B-DNA, then in the drawing on the right, we are looking into the major groove of the unwound Z-DNA segment. [After Rich, A., Nordheim, A., and Wang, A.H.-J., *Annu. Rev. Biochem.* **53**, 799 (1984).]

it is in the Z form. The *in vivo* methylation of this base sequence is inhibited when it is cloned in *E. coli* (by techniques discussed in Section 28-8) within or adjacent to a DNA segment that can form Z-DNA. Moreover, there is a balance between the *in vivo* B and Z forms of these DNAs that is thought to be influenced by environmental factors such as salt concentration and protein binding. Nevertheless, the biological function of Z-DNA, if any, remains unknown.

RNA-11 and RNA–DNA Hybrids Have an A-DNA-Like Conformation

Double helical RNA is unable to assume a B-DNA-like conformation because of steric clashes involving its 2'-OH groups. Rather, it usually assumes a conformation resembling A-DNA (Fig. 28-8), known as **A-RNA** or **RNA-11**, which has 11 bp per helical turn, a pitch of 30 Å, and its base pairs inclined to the helix axis by ~14°. Many RNAs, for example, transfer and ribosomal RNAs (whose structures are detailed in Sections 30-2A and 30-3A), contain complementary sequences that form double helical stems.

Hybrid double helices, which consist of one strand each of RNA and DNA, are thought to also have an A-DNA-like conformation. Indeed, the X-ray structure, by Rich, of a 10-bp complex between the RNA–DNA oligonucleotide r(GCG)d(TATACCC) and the complementary DNA oligonucleotide d(GGGTATACGC) reveals that it forms an overall A-type double helix that has 11 bp per turn. Although the base pairing geometry, particularly that in the central TATA segment, is distorted, there is no indication of a transition from an A- to a B- type helix at the junction between the RNA–DNA hybrid and the DNA duplex (although recall that duplex DNAs of <10 bp tend to crystallize in the A form). Small segments of RNA·DNA hybrid helices must occur in both the transcription of RNA on DNA templates (Section 29-2D) and in the initiation of DNA replication by short lengths of RNA (Section 31-1D).

C. The Size of DNA

DNA molecules are generally enormous (Fig. 28-11). The molecular mass of DNA has been determined by a variety of techniques including hydrodynamic methods (Section 5-5), length measurements by electron microscopy, and autoradiography [Fig. 28-12; a base pair of Na⁺ B-DNA has an average molecular mass of 660 D and a length (thickness) of 3.4 Å]. The number of base pairs and the **contour lengths** (the end-to-end lengths of the stretched out native molecules) of the DNAs from a selection of organisms of increasing complexity are presented in Table 28-2. Not surprisingly, an organism's haploid quantity (unique amount) of DNA varies more or less with its complexity (although there are notable exceptions to this generalization such as the last entry in Table 28-2).

The visualization of DNAs from prokaryotes has demonstrated that their entire **genome** (complement of genetic information) is contained on a single, usually circular, length of DNA. Similarly, Bruno Zimm demonstrated that the *largest chromosome of the fruit fly Drosophila melanogaster contains a single molecule of DNA* by comparing the molecular mass of this DNA with the cytologically measured amount of DNA contained in the chromosome. Presumably other eukaryotic chromosomes also contain only single molecules of DNA.

The highly elongated shape of duplex DNA (recall B-DNA is only 20 Å in diameter), together with its stiffness, make it extremely susceptible to mechanical damage outside the cell's protective environment (for instance, if the *Drosophila* DNA of Fig. 28-12 were expanded by a factor of 500,000, it would have the shape and some of the mechanical properties of a 6-km long strand of uncooked spaghetti). The hydrodynamic shearing forces generated by such ordinary laboratory manipulations as stirring, shaking, and pipetting break DNA into relatively small pieces so that the

isolation of an intact molecule of DNA requires extremely gentle handling. Before 1960, when this was first realized, the measured molecular masses of DNA were no higher than 10 million D. DNA fragments of uniform molecular mass and as small as a few hundred base pairs may be generated by **shear degrading** DNA in a controlled manner; for instance, by pipetting, through the use of a high-speed blender, or by **sonication** (exposure to intense high-frequency sound waves).

TABLE 28-2. SIZES OF SOME DNA MOLECULES

Organism	Number of base pairs (kb)[a]	Contour length (μm)
Viruses		
Polyoma, SV40	5.1	1.7
λ Bacteriophage	48.6	17
T2, T4, T6 bacteriophage	166	55
Fowlpox	280	193
Bacteria		
Mycoplasma hominis	760	260
Eschericia coli	4,700	1,600
Eukaryotes		
Yeast (in 17 haploid chromosomes)	13,500	4,600
Drosophila (in 4 haploid chromosomes)	165,000	56,000
Human (in 23 haploid chromosomes)	2,900,000	990,000
Lungfish (in 19 haploid chromosomes)	102,000,000	34,700,000

[a] kb = kilobase pair = 1000 base pairs (bp).

Source: Kornberg, A. and Baker, T.A., DNA Replication (2nd ed.), p. 20, Freeman (1992).

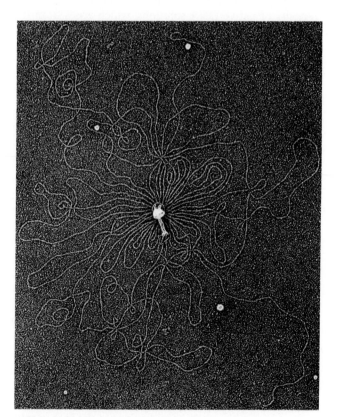

FIGURE 28-11. An electron micrograph of a T2 bacteriophage that had been osmotically lysed in distilled water so that its DNA spilled out. Without special treatment, duplex DNA, which is only 20 Å in diameter, is difficult to visualize in the electron microscope. In the **Kleinschmidt procedure,** DNA is fattened to ~200 Å in diameter by coating it with denatured cytochrome *c* or some other basic protein. The preparation is rendered visible in the electron microscope by shadowing it with platinum. [From Kleinschmidt, A.K., Lang, D., Jacherts, D., and Zahn, R.K., *Biochim. Biophys. Acta* **61,** 861 (1962).]

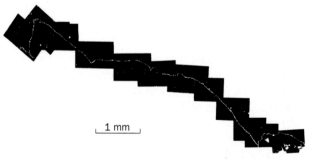

1 mm

FIGURE 28-12. An autoradiograph of *Drosophila melanogaster* DNA. Lysates of *D. melanogaster* cells that had been cultured with [³H]thymidine were spread on a glass slide and covered with a photographic emulsion that was developed after a 5-month exposure. The measured contour length of the DNA is 1.2 cm. [From Kavenoff, R., Klotz, L.C., and Zimm, B.H., *Cold Spring Harbor Symp. Quant. Biol.* **38,** 4 (1973). Copyright © 1973 by Cold Spring Harbor Laboratory.]

3. FORCES STABILIZING NUCLEIC ACID STRUCTURES

DNA does not exhibit the structural complexity of proteins because it has only a limited repertoire of secondary struc-

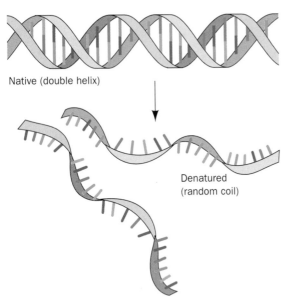

Native (double helix)

Denatured
(random coil)

FIGURE 28-13. A schematic representation of DNA denaturation.

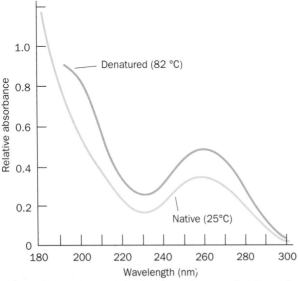

FIGURE 28-14. The UV absorbance spectra of native and heat-denatured *E. coli* DNA. Note that denaturation does not change the general shape of the absorbance curve but only increases its intensity. [After Voet, D., Gratzer, W.B., Cox, R.A., and Doty, P., *Biopolymers* **1**, 205 (1963).]

tures and no comparable tertiary or quaternary structures. This is perhaps to be expected since there is a far greater range of chemical and physical properties among the 20 amino acid residues of proteins than there is among the four DNA bases. As we discuss in Sections 30-2B and 3A, however, many RNAs have well-defined tertiary structures.

In this section we examine the forces that give rise to the structures of nucleic acids. These forces are, of course, much the same as those that are responsible for the structures of proteins (Section 7-4) but, as we shall see, the way they combine gives nucleic acids properties that are quite different from those of proteins.

A. Denaturation and Renaturation

When a solution of duplex DNA is heated above a characteristic temperature, its native structure collapses and its two complementary strands separate and assume the random coil conformation (Fig. 28-13). This denaturation process is accompanied by a qualitative change in the DNA's physical properties. For instance, the characteristic high viscosity of native DNA solutions, which arises from the resistance to deformation of its rigid and rodlike duplex molecules, drastically decreases when the DNA decomposes to relatively freely jointed single strands.

DNA Denaturation Is a Cooperative Process

The most convenient way of monitoring the native state of DNA is by its ultraviolet (UV) absorbance spectrum. When DNA denatures, its UV absorbance, which is almost entirely due to its aromatic bases, increases by ~40% at all wavelengths (Fig. 28-14). This phenomenon, which is known as the **hyperchromic effect** (Greek: *hyper,* above; *chroma,* color), results from the disruption of the electronic interactions among nearby bases. DNA's hyperchromic shift, as monitored at a particular wavelength (usually 260 nm), occurs over a narrow temperature range (Fig. 28-15). This indicates that the denaturation of DNA is a cooperative phenomenon in which the collapse of one part of the structure destabilizes the remainder. The denaturation of DNA may be described as the melting of a one-dimensional solid, so Fig. 28-15 is referred to as a **melting curve** and the temperature at its midpoint is known as its **melting temperature, T_m.**

The stability of the DNA double helix, and hence its T_m, depends on several factors, including the nature of the solvent, the identities and concentrations of the ions in solution, and the pH. T_m also increases linearly with the mole fraction of $G \cdot C$ base pairs (Fig. 28-16), which indicates that triply hydrogen bonded $G \cdot C$ base pairs are more stable than doubly hydrogen bonded $A \cdot T$ base pairs.

Denatured DNA Can Be Renatured

If a solution of denatured DNA is rapidly cooled below its T_m, the resulting DNA will be only partially base paired

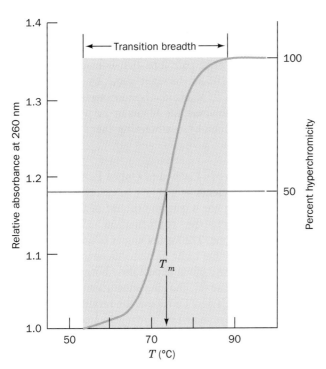

FIGURE 28-15. An example of a DNA melting curve. The relative absorbance is the ratio of the absorbance (customarily measured at 260 nm) at the indicated temperature to that at 25°C. The melting temperature, T_m, is the temperature at which half of the maximum absorbance increase is attained.

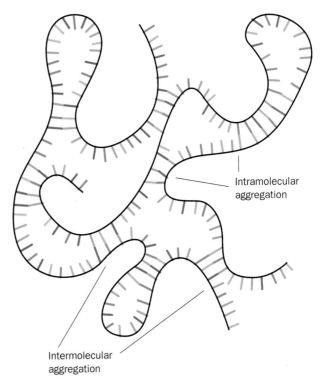

FIGURE 28-17. A schematic representation of the imperfectly base paired structures assumed by DNA that has been heat denatured and then rapidly cooled. Note that both intramolecular and intermolecular aggregation may occur.

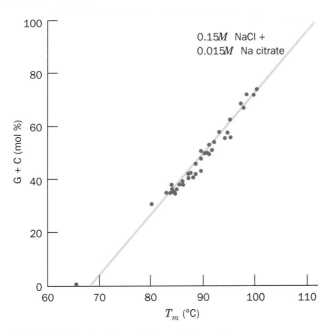

FIGURE 28-16. The variation of the melting temperatures, T_m, of various DNAs with their G + C content. The DNAs were dissolved in a solution containing 0.15M NaCl and 0.015M Na citrate. [After Marmur, J. and Doty, P., *J. Mol. Biol.* **5,** 113 (1962).]

(Fig. 28-17) because the complementary strands will not have had sufficient time to find each other before the partially base paired structures become effectively "frozen in." If, however, the temperature is maintained ~25°C below the T_m, enough thermal energy is available for short base paired regions to rearrange by melting and reforming but not so much as to melt out long complementary stretches. Under such **annealing conditions,** as Julius Marmur discovered in 1960, denatured DNA eventually completely renatures. Likewise, complementary strands of RNA and DNA, in a process known as **hybridization,** form RNA–DNA hybrid double helices that are only slightly less stable than the corresponding DNA double helices.

B. Sugar–Phosphate Chain Conformations

The conformation of a nucleotide unit, as Fig. 28-18 indicates, is specified by the six torsion angles of the sugar–phosphate backbone and the torsion angle describing the orientation of the base about the glycosidic bond (the bond joining C1′ to the base). It would seem that these seven degrees of freedom per nucleotide would render polynucleotides highly flexible. Yet, as we shall see, these torsion angles are subject to a variety of internal constraints that greatly restrict their conformational freedom.

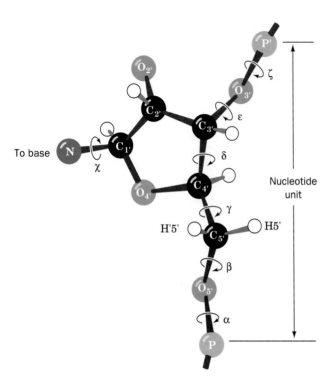

FIGURE 28-18.   The conformation of a nucleotide unit is determined by the seven indicated torsion angles.

Torsion Angles about Glycosidic Bonds Have One or Two Stable Positions

The rotation of a base about its glycosidic bond is greatly hindered, as is best seen by the manipulation of a space-filling molecular model. Purine residues have two sterically permissible orientations relative to the sugar known as the **syn** (Greek: *with*) and **anti** (Greek: *against*) conformations (Fig. 28-19). For pyrimidines, only the anti conformation is easily formed because, in the syn conformation, the sugar residue sterically interferes with the pyrimidine's C2 substituent. In most double helical nucleic acids, all bases are in the anti conformation (e.g., Figs. 28-5*b* and 8*b*). The excep-

tion is Z-DNA (Section 28-2B), in which the alternating pyrimidine and purine residues are anti and syn (Fig. 28-9*b*). *This explains Z-DNA's pyrimidine–purine alternation.* Indeed, the base pair flips that convert B-DNA to Z-DNA (Fig. 28-10) are brought about by rotating each purine base about its glycosidic bond from the anti to syn conformations. However, the sugars in the pyrimidine nucleotides rotate, thereby maintaining them in their anti conformations.

Sugar Ring Pucker Is Limited to Only a Few of Its Possible Arrangements

The ribose ring has a certain amount of flexibility that significantly affects the conformation of the sugar–phosphate backbone. The vertex angles of a regular pentagon are 108°, a value quite close to the tetrahedral angle (109.5°), so that one might expect the ribofuranose ring to be nearly flat. However, the ring substituents are eclipsed when the ring is planar. To relieve the resultant crowding, which even occurs between hydrogen atoms, the ring **puckers;** that is, it becomes slightly nonplanar, so as to reorient the ring substituents (Fig. 28-20; this is readily observed by the manipulation of a skeletal molecular model).

One would, in general, expect only three of a ribose ring's five atoms to be coplanar since three points define a plane. Nevertheless, in the great majority of the >50 nucleoside and nucleotide crystal structures that have been reported, four of the ring atoms are coplanar to within a few hundredths of an Ångstrom and the remaining atom is out of this plane by several tenths of an Ångstrom (the **half-chair** conformation). If the out-of-plane atom is displaced to the same side of the ring as atom C5′, it is said to have the **endo** conformation (Greek: *endon,* within), whereas displacement to the opposite side of the ring from C5′ is known as the **exo** conformation (Greek: *exo,* out of). In the great majority of known nucleoside and nucleotide structures, the out-of-plane atom is either C2′ or C3′ (Fig. 28-21). C2′-endo is the most frequently occurring ribose pucker with C3′-endo and -exo also being common. Other ribose conformations are rare.

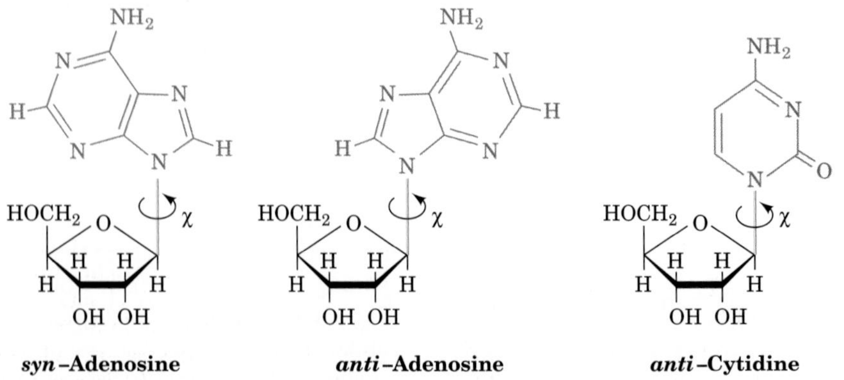

---

syn–Adenosine   *anti*–Adenosine   *anti*–Cytidine

FIGURE 28-19.   The sterically allowed orientations of purine and pyrimidine bases with respect to their attached ribose units.

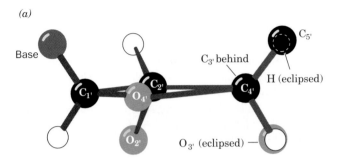

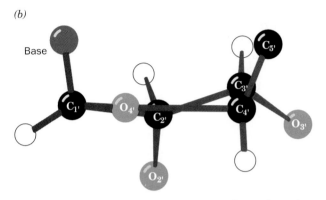

FIGURE 28-20. The substituents to (*a*) a planar ribose ring (here viewed down the C3'—C4' bond) are all eclipsed. The resulting steric strain is partially relieved by ring puckering such as in (*b*), a half-chair conformation in which C3' is the out-of-plane atom.

The ribose pucker is conformationally important in nucleic acids because it governs the relative orientations of the phosphate substituents to each ribose residue. For instance, it is difficult to build a model of a double helical nucleic acid unless the sugars are either C2'-*endo* or C3'-*endo*. In fact, B-DNA has the C2'-*endo* conformation, whereas A-DNA and RNA-11 are C3'-*endo*. In Z-DNA, the purine nucleotides are all C3'-*endo* and the pyrimidine nucleotides are C2'-*endo*, which is another reason that the repeating unit of Z-DNA is a dinucleotide. Note that the most common sugar puckers of independent nucleosides and nucleotides, molecules that are subject to few of the conformational constraints of double helices, are the same as those of double helices.

The Sugar–Phosphate Backbone Is Conformationally Constrained

If the torsion angles of the sugar–phosphate chain (Fig. 28-18) were completely free to rotate, there could probably be no stable nucleic acid structure. However, the comparison, by Muttaiya Sundaralingam, of some 40 nucleoside and nucleotide crystal structures has revealed that these angles are really quite restricted. For example, the torsion angle about the C4'—C5' bond (γ in Fig. 28-18) is rather narrowly distributed such that O4' usually has a gauche

conformation with respect to O5' (Fig. 28-22). This is because the presence of the ribose ring together with certain noncovalent interactions of the phosphate group stiffens the sugar–phosphate chain by restricting its range of torsion angles. These restrictions are even greater in polynucleotides because of steric interference between residues.

The sugar–phosphate conformational angles of the various double helices are all reasonably strain free. *Double helices are therefore conformationally relaxed arrangements of the sugar–phosphate backbone.* Nevertheless, the sugar–phosphate backbone is by no means a rigid structure so, upon strand separation, it assumes a random coil conformation.

C. Base Pairing

Base pairing is apparently a "glue" that holds together double-stranded nucleic acids. Only Watson–Crick pairs occur in the crystal structures of self-complementary oligonucleotides. It is therefore important to understand how

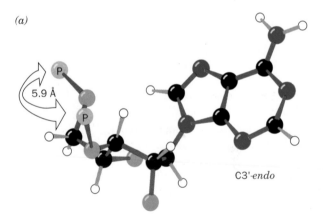

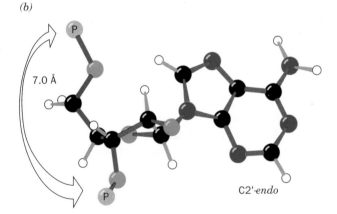

FIGURE 28-21. Nucleotides in (*a*) the C3'-*endo* conformation (on the same side of the sugar ring as C5'), which occurs in A-RNA and RNA-11; and (*b*) the C2'-*endo* conformation, which occurs in B-DNA. The distances between adjacent P atoms in the sugar–phosphate backbone are indicated. [After Saenger, W., *Principles of Nucleic Acid Structure*, p. 237, Springer–Verlag (1983).]

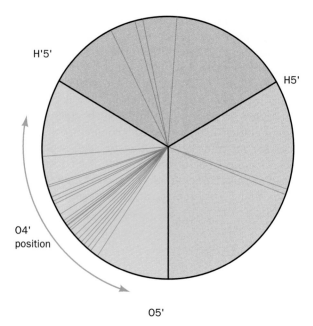

FIGURE 28-22. A conformational wheel showing the distribution of the torsion angle about the C4′—C5′ bond (ψ in Fig. 28-18) in 33 X-ray structures of nucleosides, nucleotides, and polynucleotides. Each radial line represents the position of the C4′—O4′ bond in a single structure relative to the substituents of C5′ as viewed from C5′ to C4′. Note that most of the observed torsion angles fall within a relatively narrow range. [After Sundaralingam, M., *Biopolymers* **7**, 838 (1969).]

Watson–Crick base pairs differ from other doubly hydrogen bonded arrangements of the bases that have reasonable geometries (e.g., Fig. 28-23).

Unconstrained A·T Base Pairs Assume Hoogsteen Geometry

When monomeric adenine and thymine derivatives are cocrystallized, the A·T base pairs that form invariably have adenine N7 as the hydrogen bonding acceptor (**Hoogsteen geometry;** Fig. 28-23*b*) rather than N1 (Watson–Crick geometry; Fig. 28-6). This suggests that Hoogsteen geometry is inherently more stable for A·T pairs than is Watson–Crick geometry. Apparently steric and other environmental influences make Watson–Crick geometry the preferred mode of base pairing in double helices. A·T pairs with Hoogsteen geometry are nevertheless of biological importance; for example, they help stabilize the tertiary structures of tRNAs (Section 30-2B). In contrast, monomeric G·C pairs always cocrystallize with Watson–Crick geometry as a consequence of their triply hydrogen bonded structures.

Non-Watson–Crick Base Pairs Are of Low Stability

The bases of a double helix, as we have seen (Section 28-2A), associate such that any base pair position may interchangeably be A·T, T·A, G·C, or C·G without affecting the conformations of the sugar–phosphate chains. One

might reasonably suppose that this requirement of **geometric complementarity** of the Watson–Crick base pairs, A with T and G with C, is the only reason that other base pairs do not occur in a double helical environment. In fact, this was precisely what was believed for many years after the DNA double helix was discovered.

Eventually, the failure to detect pairs of different bases in nonhelical environments other than A with T (or U) and G with C led Richard Lord and Rich to demonstrate, through spectroscopic studies, that *only the bases of Watson–Crick pairs have a high mutual affinity.* Figure 28-24*a* shows the infrared (IR) spectrum in the N—H stretch region of guanine and cytosine derivatives, both separately and in a mixture. The band in the spectrum of the G + C mixture that is

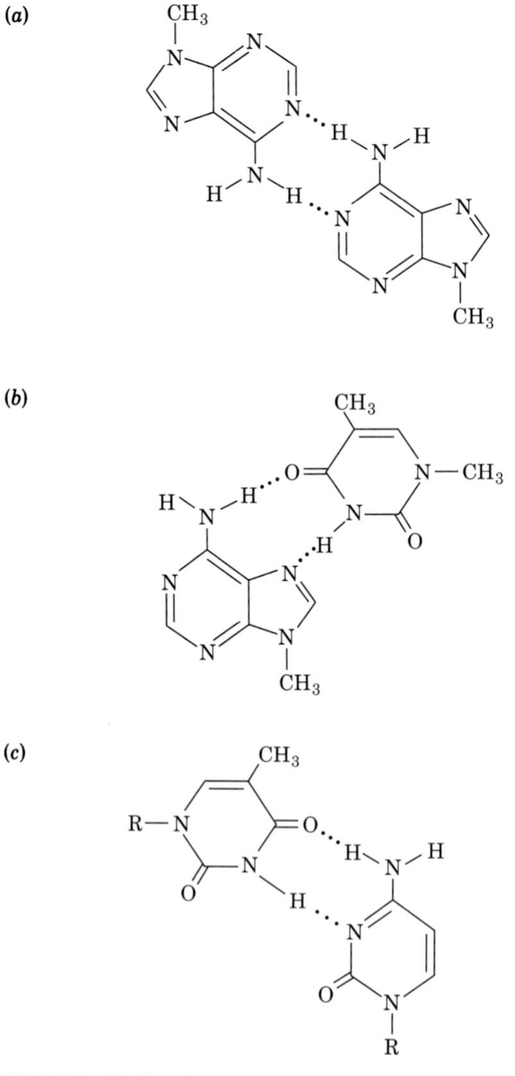

(a)

(b)

(c)

FIGURE 28-23. Some non-Watson–Crick base pairs. (*a*) The pairing of adenine residues in the crystal structure of 9-methyladenine. (*b*) Hoogsteen pairing between adenine and thymine residues in the crystal structure of 9-methyladenine · 1-methylthymine. (*c*) A hypothetical pairing between cytosine and thymine residues.

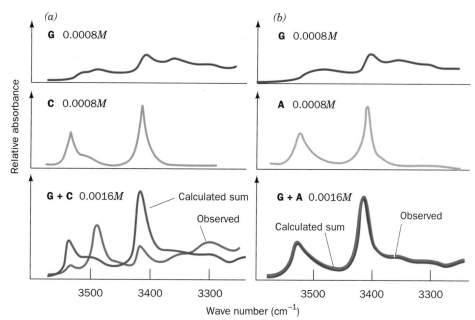

FIGURE 28-24. The IR spectra, in the N—H stretch region, of guanine, cytosine, and adenine derivatives, both separately and in the indicated mixtures. The solvent, CDCl₃, does not hydrogen bond with the bases and is relatively transparent in the frequency range of interest. (*a*) G + C. The brown curve in the lower panel, which is the sum of the spectra in the two upper panels, is the calculated spectrum of G + C for noninteracting molecules. The band near 3500 cm⁻¹ in the observed G + C spectrum is indicative of a specific hydrogen bonding association between G and C. (*b*) G + A. The close match between the calculated and observed spectra of the G + A mixture indicates that G and A do not significantly interact. [After Kyogoku, Y., Lord, R.C., and Rich, A., *Science* **154,** 5109 (1966).]

not present in the spectra of either of its components is indicative of a specific hydrogen bonding interaction between G and C. Such an association, which can occur between like as well as unlike molecules, may be described by ordinary mass action equations.

$$B_1 + B_2 \rightleftharpoons B_1 \cdot B_2 \qquad K = \frac{[B_1 \cdot B_2]}{[B_1][B_2]} \quad [28.1]$$

From analyses of IR spectra such as Fig. 28-24, the values of K for the various base pairs have been determined. The self-association constants of the Watson–Crick bases are given in the top of Table 28-3 (the hydrogen bonded association of like molecules is indicated by the appearance of new IR bands as the concentration of the molecule is increased). The bottom of Table 28-3 lists the association constants of the Watson–Crick pairs. Note that each of these latter quantities is larger than the self-association constants of both their component bases so that Watson–Crick base pairs preferentially form from their constituents. In contrast, the non-Watson–Crick base pairs, A·C, A·G, C·U, and G·U, whatever their geometries, have association constants that are negligible compared with the self-pairing association constants of their constituents (e.g., Fig. 28-24*b*). *Evidently, a second reason that non-Watson–Crick base pairs do not occur in DNA double helices is that they have relatively little stability.* Conversely, the exclusive presence of Watson–Crick base pairs in DNA results, in

part, from an **electronic complementarity** matching A to T and G to C. The theoretical basis of this electronic complementarity, which is an experimental observation, is obscure. This is because the approximations inherent in present day theoretical treatments make them unable to accurately account for the few kJ · mol⁻¹ energy differences between specific and nonspecific hydrogen bonding associations. The double helical segments of many RNAs, however, contain occasional non-Watson–Crick base pairs,

TABLE 28-3. **ASSOCIATION CONSTANTS FOR BASE PAIR FORMATION**

Base Pair	$K(M^{-1})^a$
Self-Association	
A·A	3.1
U·U	6.1
C·C	28
G·G	10^3–10^4
Watson-Crick Base Pairs	
A·U	100
G·C	10^4–10^5

^{a} Data measured in deuterochloroform at 25°C.

Source: Kyogoku, Y., Lord, R.C., and Rich, A., *Biochim. Biophys. Acta* **179,** 10 (1969).

most often G · U, which have functional as well as structural significance (e.g., Sections 30-2B and D).

Hydrogen Bonds Do Not Stabilize DNA

It is clear that hydrogen bonding is required for the specificity of base pairing in DNA that is ultimately responsible for the enormous fidelity required to replicate DNA with almost no error (Section 31-3D). Yet, as is also true for proteins (Section 7-4B), *hydrogen bonding contributes little to the stability of the double helix.* For instance, adding the relatively nonpolar ethanol to an aqueous DNA solution, which strengthens hydrogen bonds, destabilizes the double helix as is indicated by its decreased T_m. This is because hydrophobic forces, which are largely responsible for DNA's stability (see Section 28-3D), are disrupted by nonpolar solvents. In contrast, *the hydrogen bonds between the base pairs of native DNA are replaced in denatured DNA by energetically more or less equivalent hydrogen bonds between the bases and water.*

D. Base Stacking and Hydrophobic Interactions

Purines and pyrimidines tend to form extended stacks of planar parallel molecules. This has been observed in the structures of nucleic acids (Figs. 28-5, 8, and 9) and in the several hundred reported X-ray crystal structures that contain nucleic acid bases. The bases in these structures are usually partially overlapped (e.g., Fig. 28-25). In fact, crystal structures of chemically related bases often exhibit similar stacking patterns. Apparently stacking interactions, which in the solid state are a form of van der Waals interaction (Section 7-4A), have some specificity, although certainly not as much as base pairing.

Nucleic Acid Bases Stack in Aqueous Solution

Bases aggregate in aqueous solution, as has been demonstrated by the variation of osmotic pressure with concentration. The van't Hoff law of osmotic pressure is

$$\pi = RTm \qquad [28.2]$$

where π is the osmotic pressure, m is the molality of the solute (mol solute/kg solvent), R is the gas constant, and T is the temperature. The molecular mass, M, of an ideal solute can be determined from its osmotic pressure since $M = c/m$, where $c = $ g solute/kg solvent.

If the species under investigation is of known molecular mass but aggregates in solution, Eq. [28.2] must be rewritten:

$$\pi = \phi RTm \qquad [28.3]$$

where ϕ, the **osmotic coefficient,** indicates the solute's degree of association. ϕ varies from 1 (no association) to 0 (infinite association). The variation of ϕ with m for nucleic acid bases in aqueous solution (e.g., Fig. 28-26) is consistent with a model in which the bases aggregate in successive steps:

$$A + A \rightleftharpoons A_2 + A \rightleftharpoons A_3 + A \rightleftharpoons \cdots \rightleftharpoons A_n$$

where n is at least 5 (if the reaction goes to completion, $\phi = 1/n$). This association cannot be a result of hydrogen bonding since N^6,N^6**-dimethyladenosine,**

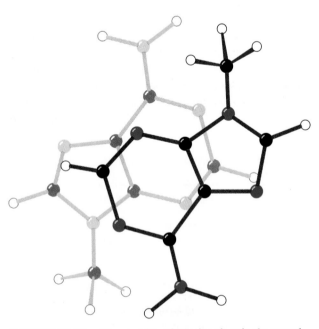

N^6,N^6**-Dimethyladenosine**

which cannot form interbase hydrogen bonds, has a greater degree of association than does adenosine (Fig. 28-26). Apparently *the aggregation arises from the formation of stacks of planar molecules.* This model is corroborated by proton NMR studies: The directions of the aggregates' chemical shifts are compatible with a stacked but not a hydrogen bonded model. The stacking associations of monomeric bases are not observed in nonaqueous solutions.

Single-stranded polynucleotides also exhibit stacking interactions. For example, poly(A) shows a broad increase of UV absorbance with temperature (Fig. 28-27a). This hyperchromism is independent of poly(A) concentration so that it cannot be a consequence of intermolecular aggrega-

FIGURE 28-25. The stacking of adenine rings in the crystal structure of 9-methyladenine. The partial overlap of the rings is typical of the association between bases in crystal structures and in double helical nucleic acids. [After Stewart, R.F. and Jensen, L.H., *J. Chem. Phys.* **40**, 2071 (1964).]

tion. Likewise, it is not due to intramolecular hydrogen bonding because poly(N^6,N^6-dimethyl A) has a greater degree of hyperchromism than does poly(A). The hyperchromism must therefore arise from some sort of stacking associations within a single strand that melt out with increasing temperature. This is not a very cooperative process, as is indicated by the broadness of the melting curve and the observation that short polynucleotides, including dinucleo-

side phosphates such as ApA, exhibit similar melting curves (Fig. 28-27*b*).

Nucleic Acid Structures Are Stabilized by Hydrophobic Forces

Stacking associations in aqueous solutions are largely stabilized by hydrophobic forces. One might reasonably suppose that hydrophobic interactions in nucleic acids are similar in character to those that stabilize protein structures. However, closer examination reveals that these two types of interactions are qualitatively different in character. Thermodynamic analysis of dinucleoside phosphate melting curves in terms of the reaction

Dinucleoside phosphate *(unstacked)* $\rightleftharpoons$
dinucleoside phosphate *(stacked)*

(Table 28-4) indicates that *base stacking is enthalpically driven and entropically opposed.* Thus the hydrophobic interactions responsible for the stability of base stacking associations in nucleic acids are diametrically opposite in character to those that stabilize protein structures (which are enthalpically opposed and entropically driven; Section 7-4C). This is reflected in the differing structural properties of these interactions. For example, the aromatic side chains of proteins are almost never stacked and the crystal structures of aromatic hydrocarbons such as benzene, which resemble these side chains, are characteristically devoid of stacking interactions.

Hydrophobic forces in nucleic acids are but poorly understood. The observation that they are different in character from the hydrophobic forces that stabilize proteins is nevertheless not surprising because the nitrogenous bases are considerably more polar than the hydrocarbon residues of proteins that participate in hydrophobic bonding. There is, however, no theory available that adequately explains the nature of hydrophobic forces in nucleic acids (our understanding of hydrophobic forces in proteins, it will be recalled, is similarly incomplete). They are complex interactions of which base stacking is probably a significant

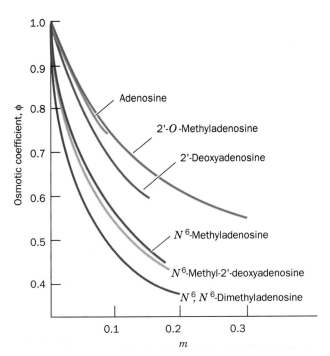

FIGURE 28-26. The variation of the osmotic coefficient ϕ with the molal concentrations m of adenosine derivatives in H₂O. The decrease of ϕ with increasing m indicates that these derivatives aggregate in solution. [After Broom, A.D., Schweizer, M.P., and Ts'o, P.O.P., *J. Am. Chem. Soc.* **89**, 3613 (1967).]

TABLE 28-4. **THERMODYNAMIC PARAMETERS FOR THE REACTION**

Dinucleoside phosphate $\rightleftharpoons$ dinucleoside phosphate
(unstacked)   *(stacked)*

Dinucleoside Phosphate	$\Delta H_{stacking}$ (kJ · mol^{-1})	$-T\Delta S_{stacking}$ (kJ · mol^{-1} at 25°C)
ApA	−22.2	24.9
ApU	−35.1	39.9
GpC	−32.6	34.9
CpG	−20.1	21.2
UpU	−32.6	36.2

Source: Davis, R.C. and Tinoco, I., Jr., *Biopolymers* **6**, 230 (1968).

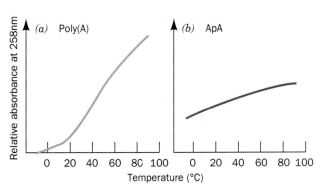

FIGURE 28-27. The broad temperature range of hyperchromic shifts at 258 nm of (*a*) poly(A) and (*b*) ApA is indicative of noncooperative conformational changes in these substances. Compare this figure with Fig. 28-15. [After Leng, M. and Felsenfeld, G., *J. Mol. Biol.* **15**, 457 (1966).]

component. Whatever their origins, hydrophobic forces are of central importance in determining nucleic acid structures.

E. Ionic Interactions

Any theory of the stability of nucleic acid structures must take into account the electrostatic interactions of their charged phosphate groups. Unfortunately, the theory of polyelectrolytes is, as yet, incapable of making reliable predictions of molecular conformations. We can, however, make experimental observations.

The melting temperature of duplex DNA increases with the cation concentration because these ions electrostatically shield the anionic phosphate groups from each other. The observed relationship for Na^+ is

$$T_m = 41.1X_{G+C} + 16.6 \log [Na^+] + 81.5 \quad [28.4]$$

where X_{G+C} is the mole fraction of $G \cdot C$ base pairs (recall that T_m increases with the G + C content); the equation is valid in the ranges $0.3 < X_{G+C} < 0.7$ and $10^{-3}M < [Na^+] < 1.0M$. Other monovalent cations such as Li^+ and K^+ have similar nonspecific interactions with phosphate groups. Divalent cations, such as Mg^{2+}, Mn^{2+}, and Co^{2+}, in contrast, specifically bind to phosphate groups so that *divalent cations are far more effective shielding agents for nucleic acids than are monovalent cations*. For example, an Mg^{2+} ion has an influence on the DNA double helix comparable to that of 100 to 1000 Na^+ ions. Indeed, enzymes that mediate reactions with nucleic acids or just nucleotides (e.g., ATP) usually require Mg^{2+} for activity. Moreover, Mg^{2+} ions play an essential role in stabilizing the complex structures assumed by many RNAs such as transfer RNAs (tRNAs; Section 30-2B) and ribosomal RNAs (Section 30-3A).

4. NUCLEIC ACID FRACTIONATION

In Chapter 5 we considered the most commonly used procedures for isolating and, to some extent, characterizing proteins. Most of these methods, often with some modification, are also regularly used to fractionate nucleic acids according to size, composition, and sequence. There are also many techniques that are applicable only to nucleic acids. In this section we shall outline some of the most useful of the separation procedures that are specific for nucleic acids.

A. Solution Methods

Nucleic acids are invariably associated with proteins. Once cells have been broken open (Section 5-1B), the nucleic acids must be deproteinized. This may be accomplished by shaking (very gently if high molecular mass DNA is being isolated) the protein–nucleic acid mixture with a phenol

solution and/or a $CHCl_3$–isoamyl alcohol mixture so that the protein precipitates and can be removed by centrifugation. Alternatively, the protein can be dissociated from the nucleic acids by detergents, guanidinium chloride, or high salt concentrations, or it can be enzymatically degraded by proteases. In all cases, the nucleic acids, a mixture of RNA and DNA, can then be isolated by precipitation with ethanol. The RNA can be recovered from such precipitates by treating them with pancreatic DNase to eliminate the DNA. Conversely, the DNA can be freed of RNA by treatment with RNase. Alternatively, RNA and DNA may be separated by ultracentrifugation (Section 28-4D).

In all these and subsequent manipulations, the nucleic acids must be protected from degradation by nucleases that occur both in the experimental materials and on human hands. Nucleases may be inhibited by the presence of chelating agents such as EDTA, which sequester the divalent metal ions that nucleases require for activity. In cases where no nuclease activity can be tolerated, all glassware must be autoclaved to heat denature the nucleases and the experimenter should wear plastic gloves. Nevertheless, nucleic acids are generally easier to handle than proteins because their lack, in most cases, of a complex tertiary structure makes them relatively tolerant of extreme conditions.

B. Chromatography

Many of the chromatographic techniques that are used to separate proteins (Section 5-3) are also applicable to nucleic acids. Paper chromatography and thin layer chromatography are useful in fractionating oligonucleotides. They have been largely replaced, however, by the more powerful techniques of HPLC, particularly those using reverse-phase chromatography. Larger nucleic acids are often separated by procedures that include ion exchange chromatography and gel filtration chromatography.

Hydroxyapatite Binds Double-Stranded DNA More Tightly Than Single-Stranded DNA

Hydroxyapatite (a form of calcium phosphate; Section 5-3E) is particularly useful in the chromatographic purification and fractionation of DNA. Double-stranded DNA binds to hydroxyapatite more tightly than do most other molecules. Consequently, DNA can be rapidly isolated by passing a cell lysate through a hydroxyapatite column, washing the column with a phosphate buffer of concentration low enough to release only the RNA and proteins, and then eluting the DNA with a concentrated phosphate solution.

Single-stranded DNA elutes from hydroxyapatite at a lower phosphate concentration than does double-stranded DNA (Fig. 28-28). This phenomenon forms the basis of a technique, known as **thermal chromatography,** for separating DNA according to its base composition. A hydroxyapatite column to which double-stranded DNA is bound is eluted with a phosphate buffer that releases only single-stranded DNA while the temperature of the column is grad-

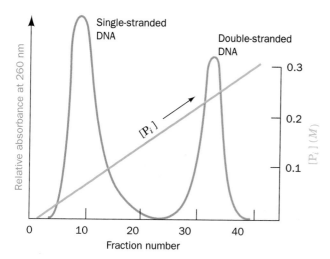

FIGURE 28-28. The chromatographic separation of single-stranded and duplex DNAs on hydroxyapatite by elution with a solution of increasing phosphate concentration.

ually increased. As the DNA melts and is converted to the single-stranded form it is eluted from the column. Since the T_m of a duplex DNA varies with its G + C content (Eq. [28.4]), thermal chromatography permits the fractionation of double-stranded DNA according to its base composition.

Messenger RNAs Can Be Isolated by Affinity Chromatography

Affinity chromatography is useful in isolating specific nucleic acids. For example, most eukaryotic messenger RNAs (mRNAs) have a poly(A) sequence at their 3′ ends (Section 29-4A). They can be isolated on agarose or cellulose to which poly(U) is covalently attached. The poly(A) sequences specifically bind to the complementary poly(U) in high salt and at low temperatures and can later be released by altering these conditions. Moreover, if the (partial) sequence of an mRNA is known (e.g., as deduced from the corresponding protein's amino acid sequence), the complementary DNA strand may be synthesized (via methods discussed in Section 28-7) and used to isolate that particular mRNA.

C. Electrophoresis

Nucleic acids of a given type may be separated by polyacrylamide gel electrophoresis (Sections 5-4B and C) because their electrophoretic mobilities in such gels vary inversely with their molecular masses. However, DNAs of more than a few thousand base pairs cannot penetrate even a weakly cross-linked polyacrylamide gel. This difficulty is partially overcome through the use of agarose gels. By using gels with an appropriately low agarose content, relatively large DNAs in various size ranges may be fractionated. In this manner, **plasmids** (small, autonomously replicating DNA molecules that occur in bacteria and yeast), for exam-

ple, may be separated from the larger chromosomal DNA of bacteria.

Very Large DNAs Are Separated by Pulsed-Field Gel Electrophoresis

The sizes of the DNAs that can be separated by conventional gel electrophoresis are limited to ~100,000 bp, even when gels containing as little as 0.1% agarose (which makes an extremely fragile gel) are used. However, the development of **pulsed-field gel electrophoresis (PFG)** by Charles Cantor and Cassandra Smith extended this limit to DNAs with up to 10 million bp (6.6 million kD). The electrophoresis apparatus used in PFG has two or more pairs of electrodes arrayed around the periphery of an agarose slab gel. The different electrode pairs are sequentially pulsed for times varying from 0.1 to 1000 s depending on the sizes of the DNAs being separated. Gel electrophoresis of DNA requires that these elongated molecules worm their way through the gel's labyrinthine channels more or less in the direction from the cathode to the anode. If the direction of the electric field abruptly changes, these DNAs must reorient their long axes along the new direction of the field before they can continue their passage through the gel. The time required to reorient very long gel-embedded DNA molecules evidently increases with their size. Consequently, a judicious choice of electrode distribution and pulse lengths causes shorter DNAs to migrate through the gel faster than longer DNAs, thereby effecting their separation.

Duplex DNA Is Detected by Selectively Staining It with Intercalation Agents

The various DNA bands in a gel must be detected if they are to be isolated. Double-stranded DNA is readily stained by planar aromatic cations such as **ethidium ion, acridine orange,** or **proflavin.**

Ethidium

Acridine orange

Proflavin

These dyes bind to duplex DNA by **intercalation** (slipping in between the stacked base pairs), where they exhibit a fluorescence under UV light that is far more intense than that of the free dye. As little as 50 ng of DNA may be detected in a gel by staining it with ethidium bromide (Fig. 28-29). Single-stranded DNA and RNA also stimulate the fluorescence of ethidium but to a lesser extent than does duplex DNA.

Southern Blotting Identifies DNAs with Specific Sequences

DNA with a specific base sequence may be identified through a procedure developed by Edwin Southern known as the **Southern transfer technique** or more colloquially as **Southern blotting** (Fig. 28-30). This procedure takes advantage of the valuable property of nitrocellulose that it tenaciously binds single-stranded (but not duplex) DNA. Following the gel electrophoresis of double-stranded DNA, the gel is soaked in $0.5M$ NaOH solution, which converts the DNA to the single-stranded form. The gel is then overlaid by a sheet of nitrocellulose paper which, in turn, is covered by a thick layer of paper towels and the entire assembly is compressed by a heavy plate. The liquid in the gel is thereby forced (blotted) through the nitrocellulose so that the single-stranded DNA binds to it at the same position it had in the gel (the transfer to nitrocellulose can alternatively be accomplished by an electrophoretic process named **electroblotting**). After vacuum drying the nitrocellulose at 80°C, which permanently fixes the DNA in place, the nitrocellulose sheet is moistened with a minimal quantity of solution containing ^{32}P-labeled single-stranded DNA or RNA that is complementary in sequence to the DNA of interest (the "probe"). The moistened filter is held at a suitable renaturation temperature for several hours to permit

FIGURE 28-29.   An agarose gel electrophoretogram of double helical DNA. After electrophoresis, the gel was soaked in a solution of ethidium bromide, washed, and photographed under UV light. The fluorescence of the ethidium cation is strongly enhanced by binding to DNA, so each fluorescent band marks a different sized DNA fragment. The three parallel lanes contain identical DNA samples so as to demonstrate the technique's reproducibility. [Photo by Elizabeth Levine. From Freifelder, D., *Biophysical Chemistry. Applications to Biochemistry and Molecular Biology* (2nd ed.), p. 292, W.H. Freeman (1982). Used by permission.]

the probe to hybridize to its target sequence(s), washed to remove the unbound radioactive probe, dried, and then autoradiographed by placing it for a time over a sheet of X-ray film. The positions of the molecules that are complementary to the radioactive sequences are indicated by a blackening of the developed film. A DNA segment contain-

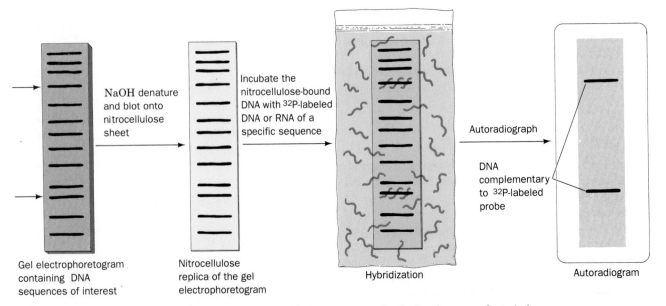

FIGURE 28-30.   The detection of DNAs containing specific base sequences by the Southern transfer technique.

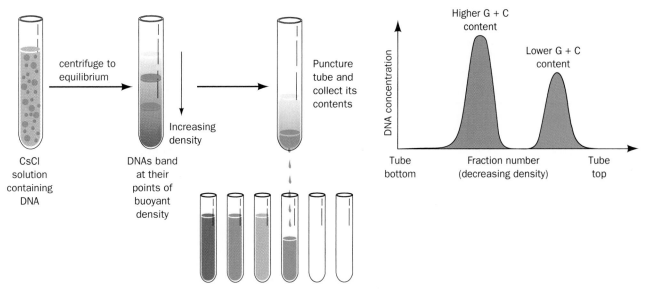

FIGURE 28-31. The separation of DNAs according to base composition by equilibrium density gradient ultracentrifugation in CsCl solution. An initially 8M CsCl solution forms a density gradient that varies linearly from ~1.80 g·cm^{-3} at the bottom of the centrifuge tube to ~1.55 g·cm^{-3} at the top. The amount of DNA in each fraction is estimated from its UV absorbance, usually at 260 nm.

ing a particular base sequence (e.g., a gene specifying a certain protein) may, in this manner, be detected and isolated. Specific DNAs may likewise be detected by linking the probe to an enzyme that generates a colored or fluorescent deposit on the blot. Such nonradioactive detection techniques are desirable in a clinical setting because of health hazards, disposal problems, and the more cumbersome nature of autoradiographic methods.

Specific RNA sequences may be detected through a variation of the Southern transfer, punningly named a **Northern transfer (Northern blot),** in which the RNA is immobilized on nitrocellulose paper and detected through the use of complementary radiolabeled RNA or DNA probes. A specific protein may be analogously detected, in a procedure named **immunoblotting** or **Western blotting,** through the use of antibodies directed against the protein in a manner discussed in Section 5-4B.

D. Ultracentrifugation

Equilibrium density gradient ultracentrifugation (Fig. 28-31; Section 5-5B) in CsCl constitutes one of the most commonly used DNA separation procedures. The bouyant density, ρ, of double-stranded Cs$^+$ DNA depends on its base composition:

$$\rho = 1.660 + 0.098\,X_{G+C} \qquad [28.5]$$

so that a CsCl density gradient fractionates DNA according to its base composition. For example, eukaryotic DNAs often contain minor fractions that band separately from the major species. Some of these **satellite bands** consist of mito-chondrial and chloroplast DNAs. Another important class of satellite DNA is composed of **repetitive sequences** that are short segments of DNA tandemly (one behind the other) repeated hundreds, thousands, and in some cases, millions of times in a chromosome (Section 33-2B). Likewise, plasmids may be separated from bacterial chromosomal DNA by equilibrium density gradient ultracentrifugation.

Single-stranded DNA is ~0.015 g·cm^{-3} denser than the corresponding double-stranded DNA so that the two may be separated by equilibrium density gradient ultracentrifugation. RNA is too dense to band in CsCl but does so in Cs$_2$SO$_4$ solutions. RNA–DNA hybrids will band in CsCl but at a higher density than the corresponding duplex DNA.

RNA may be fractionated by zonal ultracentrifugation through a sucrose gradient (Fig. 28-32; Section 5-5B). RNAs are separated by this technique largely on the basis of their size. In fact, ribosomal RNA, which constitutes the major portion of cellular RNA, is classified according to its sedimentation rate; for example, the RNA of the *E. coli* small ribosomal subunit is known as 16S RNA (Section 30-3A).

5. SUPERCOILED DNA

The circular genetic maps of viruses and bacteria implies that their chromosomes are likewise circular. This conclusion has been confirmed by electron micrographs in which

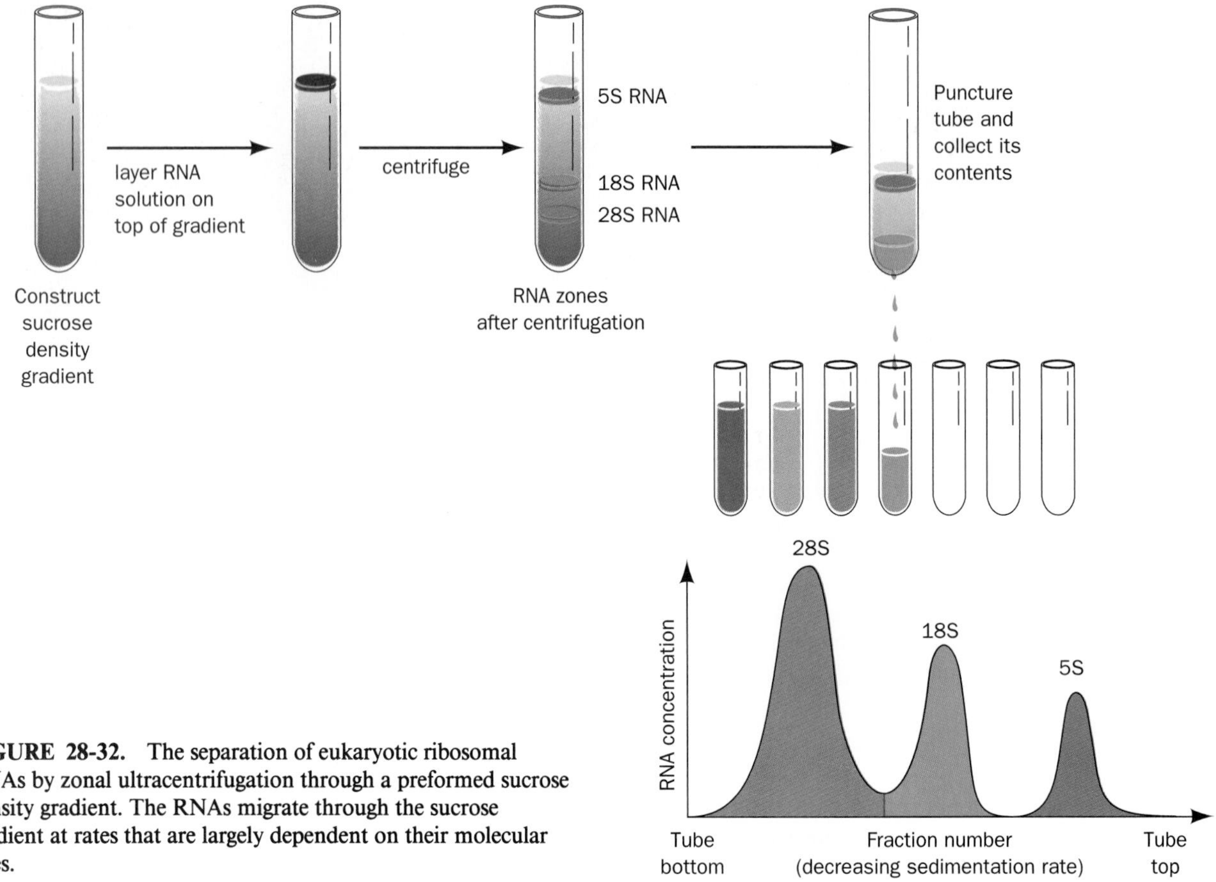

FIGURE 28-32. The separation of eukaryotic ribosomal RNAs by zonal ultracentrifugation through a preformed sucrose density gradient. The RNAs migrate through the sucrose gradient at rates that are largely dependent on their molecular sizes.

circular DNAs are seen (Fig. 28-33). Some of these circular DNAs have a peculiar twisted appearance, a phenomenon that is known equivalently as **supercoiling, supertwisting,** or **superhelicity.** Supercoiling arises from a biologically important topological property of covalently closed circular duplex DNA that is the subject of this section. It is occasionally referred to as DNA's tertiary structure.

A. Superhelix Topology

Consider a double helical DNA molecule in which both strands are covalently joined to form a circular duplex molecule as is diagrammed in Fig. 28-34 (each strand can be joined only to itself because the strands are antiparallel). *A geometric property of such an assembly is that its number of coils cannot be altered without first cleaving at least one of its polynucleotide strands.* You can easily demonstrate this to yourself with a buckled belt in which each edge of the belt represents a strand of DNA. The number of times the belt is twisted before it is buckled cannot be changed without unbuckling or cutting the belt (cutting a polynucleotide strand).

This phenomenon is mathematically expressed

$$L = T + W \qquad [28.6]$$

in which:

1. L, the **linking number,** is the number of times that one DNA strand winds about the other. This integer quantity is most easily counted when the molecule's duplex axis is constrained to lie in a plane (see below). However, *the linking number is invariant no matter how the circular molecule is twisted or distorted so long as both its polynucleotide strands remain covalently intact; the linking number is therefore a topological property of the molecule.*

2. T, the **twist,** is the number of complete revolutions that one polynucleotide strand makes about the duplex axis in the particular conformation under consideration. By convention, T is positive for right-handed duplex turns so that, for B-DNA in solution, the twist is normally the number of base pairs divided by 10.4 (the number of base pairs per turn of the B-DNA double helix under physiological conditions; see Section 28-5B).

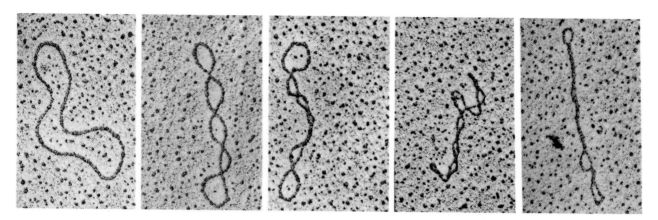

FIGURE 28-33. Electron micrographs of circular duplex DNAs that vary in their conformations from no supercoiling (*left*) to tightly supercoiled (*right*). [Electron micrographs by Laurien Polder. From Kornberg, A. and Baker, T.A., *DNA Replication* (2nd ed.), *p.* 36, W.H. Freeman (1992). Used by permission.]

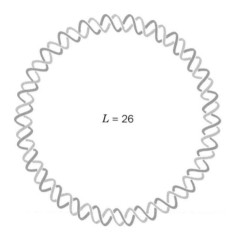

$L = 26$

FIGURE 28-34. A schematic diagram of covalently closed circular duplex DNA that has 26 double helical turns. Its two polynucleotide strands are said to be **topologically bonded** to each other because, although they are not covalently linked, they cannot be separated without breaking covalent bonds.

Large writhing number, small twist

Small writhing number, large twist

FIGURE 28-35. The difference between writhing and twist as demonstrated by a coiled telephone cord. In its relaxed state (*left*), the cord is in a helical form that has a large writhing number and a small twist. As the coil is pulled out (*middle*) until it is nearly straight (*right*), its writhing number becomes small as its twist becomes large.

3. *W*, the **writhing number,** is the number of turns that the duplex axis makes about the superhelix axis in the conformation of interest. *It is a measure of the DNA's superhelicity.* The difference between writhing and twisting is illustrated by the familiar example in Fig. 28-35. $W = 0$ when the DNA's duplex axis is constrained to lie in a plane (e.g., Fig. 28-34); then $L = T$, so L may be evaluated by counting the DNA's duplex turns.

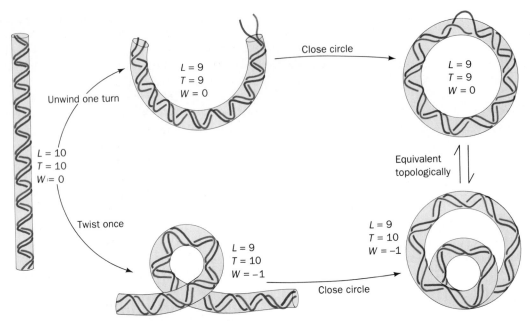

FIGURE 28-36. Two ways of introducing one supercoil into a DNA with 10 duplex turns. The two closed circular forms shown (*right*) are topologically equivalent; that is, they are interconvertible without breaking any covalent bonds. The linking number *L*, twist *T*, and writhing number *W* are indicated for each form. Strictly speaking, the linking number is only defined for a covalently closed circle.

The two DNA conformations diagrammed on the right of Fig. 28-36 are topologically equivalent; that is, they have the same linking number, *L*, but differ in their twists and writhing numbers. Note that *T* and *W* need not be integers, only *L*.

Since L is constant in an intact duplex DNA circle, for every new double helical twist, ΔT, there must be an equal and opposite superhelical twist; that is, $\Delta W = -\Delta T$. For example, a closed circular DNA without supercoils (Fig. 28-36, *upper right*) can be converted to a negatively supercoiled conformation (Fig. 28-36, *lower right*) by winding the duplex helix the same number of positive (right-handed) turns.

Supercoils May Be Toroidal or Interwound

A supercoiled duplex may assume two topologically equivalent forms:

1. **A toroidal helix** in which the duplex axis is wound as if about a cylinder (Fig. 28-37a).

2. **An interwound helix** in which the duplex axis is twisted around itself (Fig. 28-37b).

Note that these two interconvertible superhelical forms have opposite handedness. Since left-handed toroidal turns may be converted to left-handed duplex turns (see Fig. 28-35), left-handed toroidal turns and right-handed interwound turns both have negative writhing numbers. Thus an underwound duplex ($T <$ number of bp/10.4), for example, will tend to develop right-handed interwound or left-handed toroidal superhelical turns when the constraints causing it to be underwound are released (the molecular forces in a DNA double helix promote its winding to its normal number of helical turns).

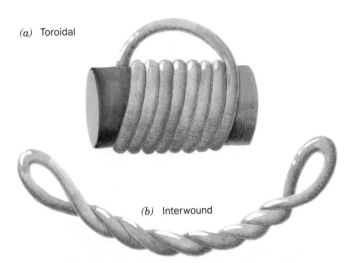

(a) Toroidal

(b) Interwound

FIGURE 28-37. A rubber tube that has been (*a*) toroidally coiled in a left-handed helix around a cylinder with its ends joined such that it has no twist, jumps to (*b*) an interwound helix with the opposite handedness when the cylinder is removed. Neither the linking number, the twist, nor the writhing number are changed in this transformation.

Supercoiled DNA Is Relaxed by Nicking One Strand

Supercoiled DNA may be converted to **relaxed circles** (as appears in the leftmost panel of Fig. 28-33) by treatment with **pancreatic DNase I,** an **endonuclease** (an enzyme that cleaves phosphodiester bonds within a polynucleotide strand), which cleaves only one strand of a duplex DNA. *One single-strand nick is sufficient to relax a supercoiled DNA.* This is because the sugar–phosphate chain opposite the nick is free to swivel about its backbone bonds (Fig. 28-18) so as to change the molecule's linking number and thereby alter its superhelicity. Supercoiling builds up elastic strain in a DNA circle, much as it does in a rubber band. This is why the relaxed state of a DNA circle is not supercoiled.

B. Measurements of Supercoiling

Supercoiled DNA, far from being just a mathematical curiosity, has been widely observed in nature. In fact, its discovery in polyoma virus DNA by Jerome Vinograd stimulated the elucidation of the topological properties of superhelices rather than *vice versa.*

Intercalating Agents Control Supercoiling by Unwinding DNA

All naturally occurring DNA circles are underwound; that is, their linking numbers are less than those of their corresponding relaxed circles. This phenomenon has been established by observing the effect of ethidium binding on the sedimentation rate of circular DNA (Fig. 28-38). Intercalating agents such as ethidium alter a circular DNA's degree of superhelicity because they cause the DNA double helix to unwind by ~26° at the site of the intercalated molecule (Fig. 28-39). $W < 0$ in an unconstrained underwound circle because of the tendency of a duplex DNA to maintain its normal twist of 1 turn/10.4 bp. The titration of a DNA circle by ethidium unwinds the duplex (decreases T), which must be accompanied by a compensating increase in W. This, at first, lessens the superhelicity of an underwound circle. However, as the circle binds more and more ethidium, its value of W passes through zero (relaxed circles) and then becomes positive so that the circle again becomes superhelical. Thus the sedimentation rate of underwound DNAs, which is a measure of their compactness and therefore their superhelicity, passes through a minimum as the ethidium concentration increases. This is what is observed with native DNAs (Fig. 28-38). In contrast, the sedimentation rate of an overwound circle would only increase with increasing ethidium concentration.

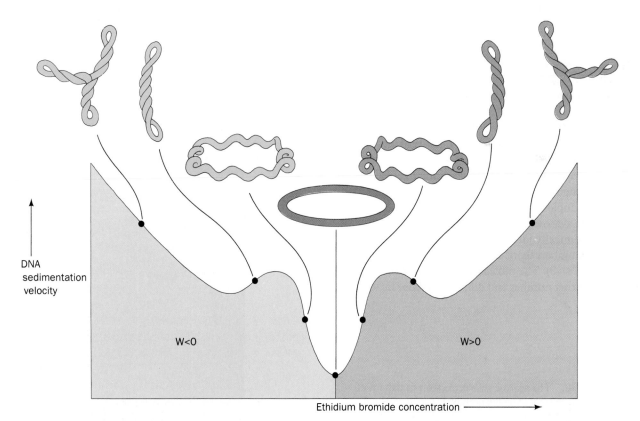

DNA sedimentation velocity

W<0

W>0

Ethidium bromide concentration ⟶

FIGURE 28-38. The sedimentation rate of closed circular duplex DNA as a function of ethidium bromide concentration. The intercalation of ethidium between the base pairs locally unwinds the double helix (Fig. 28-39) which, since the linking number of the circle is constant, is accompanied by an equivalent increase in the writhing number. As the negatively coiled superhelix unwinds, it becomes less compact and sediments more slowly. At the low point on the curve, the DNA circles have bound sufficient ethidium to become fully relaxed. As the ethidium concentration is further increased, the DNA supercoils in the opposite direction yielding a positively coiled superhelix. The supertwisted appearances of the depicted DNAs have been verified by electron microscopy. [After Bauer, W.R., Crick, F.H.C., and White, J.H., *Sci. Am.* **243**(1): 129 (1980). Copyright © 1981 by Scientific American, Inc.]

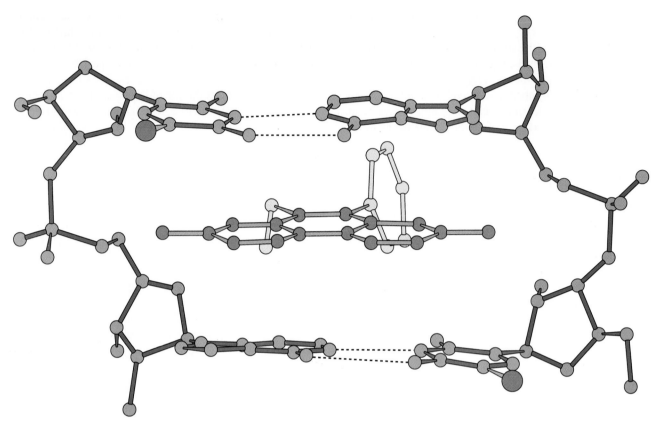

FIGURE 28-39.   The X-ray structure of a complex of ethidium with 5-iodo-UpA. Ethidium (*red*) intercalates between the base pairs of the double helically paired dinucleoside phosphate and thereby provides a model for the binding of ethidium to duplex DNA. [After Tsai, C.-C., Jain, S.C., and Sobell, H.M., *Proc. Natl. Acad. Sci.* **72**, 629 (1975).]

DNAs Are Separated According to Their Linking Number by Gel Electrophoresis

Gel electrophoresis also separates similar molecules on the basis of their compactness so that the rate of migration of a circular duplex DNA increases with its degree of super-helicity. The agarose gel electrophoresis pattern of a population of chemically identical DNA molecules with different linking numbers therefore consists of a series of discrete bands (Fig. 28-40). The molecules in a given band all have the same linking number and differ from those in adjacent bands by $\Delta L \pm 1$.

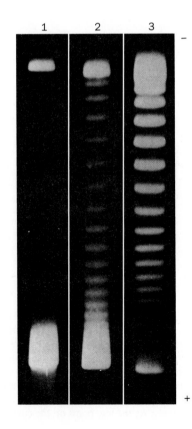

FIGURE 28-40.   The agarose gel electrophoresis pattern of SV40 DNA. Lane 1 contains the negatively supercoiled native DNA *(lower band;* the DNA was applied to the top of the gel). In lanes 2 and 3, the DNA has been exposed for 5 and 30 min, respectively, to an enzyme, known as a Type I topoisomerase (Section 28-5C), that relaxes negative supercoils one at a time by increasing the linking number (L). The DNAs in consecutively higher bands of a given gel have successively increasing linking numbers ($\Delta L = +1$). [From Keller, W., *Proc. Natl. Acad. Sci.* **72**, 2553 (1975).]

Comparison of the electrophoretic band patterns of **simian virus 40 (SV40)** DNA that had been enzymatically relaxed to varying degrees and then resealed (Fig. 28-40) reveals that 26 bands separate native from fully relaxed SV40 DNAs. Native SV40 DNA therefore has $W = -26$ (although it is somewhat heterogeneous in this quantity). Since SV40 DNA consists of 5243 bp, it has 1 superhelical turn per ~19 duplex turns. Such a **superhelix density** is typical of circular DNAs from various biological sources.

DNA in Physiological Solution Has 10.4 Base Pairs Per Turn

The insertion, using genetic engineering techniques (Section 28-8B), of an additional x base pairs into a superhelical DNA with a given linking number will increase the DNA's twist and hence decrease its writhing number by $x/h°$, where $h°$ is the number of base pairs per duplex turn. Such an insertion shifts the position of each band in the DNA's gel electrophoretic pattern by $x/h°$ of the spacing between bands. By measuring the effects of several such insertions James Wang established that $h° = 10.4 \pm 0.1$ bp for B-DNA in solution under physiological conditions.

C. Topoisomerases

The normal biological functioning of DNA occurs only if it is in the proper topological state. In such basic biological processes as RNA transcription and DNA replication, the recognition of a base sequence requires the local separation of complementary polynucleotide strands. The negative supercoiling of naturally occurring DNAs results in a torsional strain that promotes such separations since it tends to unwind the duplex helix (an increase in W must be accompanied by a decrease in T). *If DNA lacks the proper superhelical tension, the above vital processes (which themselves supercoil DNA; Sections 29-2D and 31-2C) occur quite slowly, if at all.*

The supercoiling of DNA is controlled by a remarkable group of enzymes known as **topoisomerases.** They are so named because they alter the topological state (linking number) of circular DNA but not its covalent structure. There are two classes of topoisomerases:

1. **Type I topoisomerases** act by creating transient single-strand breaks in DNA.

2. **Type II topoisomerases** act by making transient double-strand breaks in DNA.

Type I Topoisomerases Incrementally Relax Supercoiled DNA

Type I topoisomerases, which are also known as **nicking–closing enzymes,** are monomeric ~100-kD proteins that are widespread in both prokaryotes and eukaryotes. *They catalyze the relaxation of negative supercoils in*

DNA by increasing its linking number in increments of one turn. The exposure of a negatively supercoiled DNA to nicking–closing enzyme sequentially increases its linking number until the supercoil is entirely relaxed.

A clue to the mechanism of action of this enzyme was provided by the observation that it reversibly **catenates** (interlinks) single-stranded circles (Fig. 28-41a). Apparently the enzyme operates by cutting a single strand, passing a single-strand loop through the resulting gap, and then resealing the break (Fig. 28-41b), thereby twisting double helical DNA by one turn. In support of this hypothesis, the denaturation of prokaryotic nicking–closing enzyme that has been incubated with single-stranded circular DNA yields a linear DNA that has its 5′-terminal phosphoryl

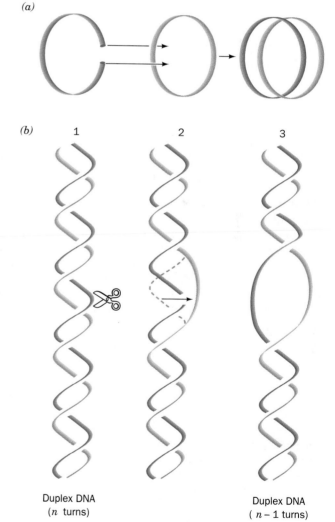

(a)

(b)

| 1 | 2 | 3 |

Duplex DNA (n turns)　　Duplex DNA ($n-1$ turns)

FIGURE 28-41. By cutting a single-stranded DNA, passing a loop of it through the break and then resealing the break, Type I topoisomerase can (a) catenate two single-stranded circles or (b) unwind duplex DNA by one turn.

group linked to the enzyme via a phosphotyrosine diester linkage.

Denatured eukaryotic nicking–closing enzymes are instead linked to the 3′ end of DNA in a like manner. *By forming such covalent enzyme–DNA intermediates, the free energy of the cleaved phosphodiester bond is preserved so that no energy input is required to reseal the nick.*

The 67-kD N-terminal fragment of *E. coli* topoisomerase I cannot relax supercoiled DNA but does cleave single-stranded oligonucleotides. Its X-ray structure (Fig. 28-42*a*), determined by Wang and Alfonso Mondragón, reveals that this protein folds into four domains which enclose a 27.5-Å-diameter hole that is large enough to loosely contain a double-stranded DNA. The hole's inner surface is lined with an excess of positively charged residues (18 Arg and Lys residues but only 9 Asp and Glu residues) as is expected for a DNA-binding surface. The active site tyrosyl residue, Tyr 319, is buried between two domains, where it is inaccessible to single-stranded DNA. Thus, to cleave single-stranded DNA, the protein must undergo a conformational change that exposes Tyr 319. This has led to the formulation of a model for the strand passage reaction that is diagrammed in Fig. 28-42*b*.

Type II Topoisomerases Supercoil DNA at the Expense of ATP Hydrolysis

Prokaryotic Type II topoisomerases, which are also known as **DNA gyrases**, are ~375-kD proteins that consist of two pairs of subunits designated *A* and *B* (875 and 804 residues in *E. coli*). *These enzymes catalyze the stepwise negative supercoiling of DNA with the concomitant hydrolysis of an ATP to ADP + Pi.* In the absence of ATP, DNA

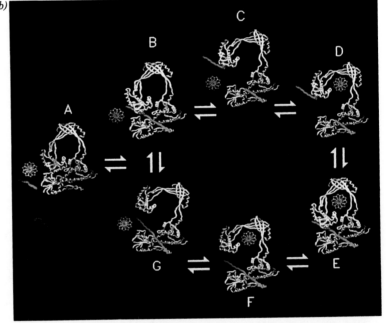

FIGURE 28-42. (*a*) The X-ray structure of the 67-kD N-terminal fragment of *E. coli* DNA topoisomerase I. The protein consists of four domains surrounding a hole large enough to enclose B-DNA, which is shown in hypothetical complex with the protein as viewed down the DNA's helix axis. (*b*) The proposed mechanism of the strand passage reaction catalyzed by DNA topoisomerase I. The protein (A) binds a single-stranded DNA (*green tube*) in a cleft that adjoins the catalytic tyrosine, Tyr 319 (*red dot;* B). The single-stranded DNA is then cleaved (C), the protein opens up to permit a duplex DNA (or, alternatively, a different single-stranded DNA) to pass through the break (D), and the two segments of single-stranded DNA are rejoined (E). The protein again opens up (F), which allows the duplex DNA to escape (G). The complex can then return to state B to initiate another strand passage reaction or, alternatively, it can decompose to its component macromolecules (A). Although this model shows duplex DNA being passed through a transiently cleaved single strand, the passing segment could just as well be single-stranded DNA. [Courtesy of Alfonso Mondragón, Northwestern University.]

gyrase relaxes negatively supercoiled DNA but at a relatively slow rate. It can also tie knots in double-stranded circles as well as catenate them. Eukaryotic Type II topoisomerases only relax supercoils; they neither generate them nor hydrolyze ATP. DNA supercoiling in eukaryotes is generated somewhat differently (Section 33-1B).

Prokaryotic DNA gyrases are specifically inhibited by two classes of antibiotics. One of these classes includes the *Streptomyces*-derived **novobiocin** and the other contains the clinically useful synthetic antibacterial agent **oxolinic acid.**

Novobiocin

Oxolinic acid

Both classes of antibiotics profoundly inhibit bacterial DNA replication and RNA transcription, thereby demonstrating the importance of properly supercoiled DNA in these processes. Studies using antibiotic-resistant *E. coli* mutants have demonstrated that oxolinic acid associates with DNA gyrase's *A* subunit and novobiocin binds to its *B* subunit.

The gel electrophoretic pattern of duplex circles that have been exposed to DNA gyrase, with or without ATP, show a band pattern in which the linking numbers differ by incre-

ments of two rather than one as occurs with nicking–closing enzymes. *Evidently, DNA gyrase acts by cutting both strands of a duplex, passing the duplex through the break, and resealing it* (Fig. 28-43). This hypothesis is corroborated by the observation that when DNA gyrase is incubated with DNA and oxolinic acid, and subsequently denatured with guanidinium chloride, its *A* subunits remain covalently linked to the 5′ ends of both cut strands through phosphotyrosine linkages. Apparently oxolinic acid interferes with gyrase action by blocking the strand breaking–rejoining process. Novobiocin, on the other hand, prevents ATP from binding to the enzyme.

Electron microscopic studies indicate that gyrase is a "heart-shaped" protein in which the *A* subunits comprise the upper and larger lobes of the structure. The exposure of a gyrase–DNA complex to staphylococcal nuclease reveals that gyrase protects an ~140-bp DNA fragment that is roughly centered on the gyrase cleavage site from nucleolytic degradation. The observation that this length of DNA can form a single turn around the protein led Nicholas Cozzarelli to propose the **sign inversion** mechanism for DNA gyrase (Fig. 28-44), so called because it converts a right-handed toroidal supercoil to a left-handed toroidal supercoil.

The X-ray structure of a fragment of the *B* subunit comprising residues 2 to 393 of the intact subunit, in complex with the nonhydrolyzable ATP analog **adenosine-5′-(β,γ-imido)triphosphate (ADPNP),**

Adenosine-5'-(β,γ-imido)triphosphate (ADPNP)

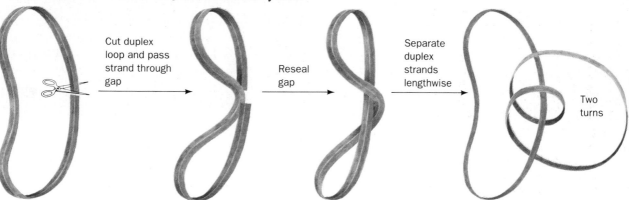

FIGURE 28-43. A demonstration, in which DNA is represented by a ribbon, that cutting a duplex circle, passing the strand through the resulting gap, and then resealing the break

changes the linking number by two. Separating the resulting strands (slitting the ribbon along its length; *right*), indicates that one strand makes two complete revolutions about the other.

was determined by Guy Dodson and Eleanor Dodson (Fig. 28-45). The protein fragment, which dimerizes in solution in the presence of ADPNP, consist of two domains. The N-terminal domain, which has been implicated in ATP hydrolysis, binds Mg^{2+}–ADPNP. The C-terminal domains form the walls of a remarkable 20-Å-diameter hole through

the dimer, the same diameter as that of the B-DNA double helix. All of this domain's numerous Arg residues line the walls of the cavity, as might be expected for a DNA-binding surface. Since the mechanism of supercoiling by gyrase almost certainly involves the passage of a DNA strand though a double-stranded break that is held open by the protein (Figure 28-44), it seems likely that the foregoing hole somehow participates in the strand passage process, perhaps in a way that resembles the postulated mechanism of topoisomerase I (Fig. 28-42*b*).

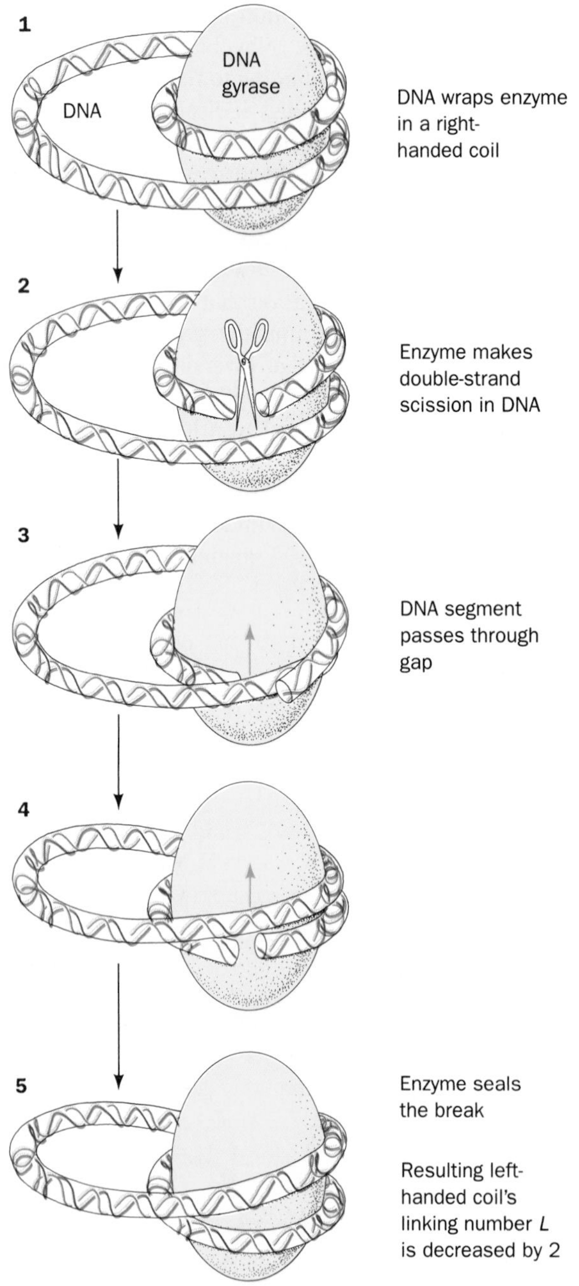

6. NUCLEIC ACID SEQUENCING

The basic strategy of nucleic acid sequencing is identical to that of protein sequencing (Section 6-1). It involves:

1. The specific degradation and fractionation of the polynucleotide of interest to fragments small enough to be fully sequenced.

2. The sequencing of the individual fragments.

FIGURE 28-44. The sign inversion mechanism of DNA gyrase action. The duplex DNA is initially wrapped about the enzyme in a right-handed toroidal coil (*1*). The enzyme then makes a double-strand scission in the DNA (*2*), passes a DNA segment through the gap (*3,4*), and reseals the break (*5*). This changes the handedness of the coil to the left-handed form so that the DNA's linking number *L* is decreased by 2.

1 — DNA gyrase / DNA — DNA wraps enzyme in a right-handed coil

2 — Enzyme makes double-strand scission in DNA

3 — DNA segment passes through gap

4

5 — Enzyme seals the break — Resulting left-handed coil's linking number *L* is decreased by 2

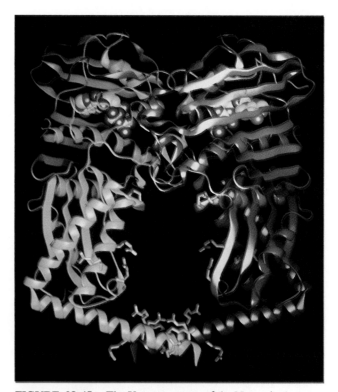

FIGURE 28-45. The X-ray structure of the N-terminal fragment of *E. coli* DNA gyrase's *B* subunit in complex with ADPNP as viewed with the two-fold axis of this dimeric protein vertical. The two identical subunits, which are colored red and green, each fold into two domains, which are represented by lighter and darker shades of color. The side chains of the Arg residues lining the 20-Å-diameter hole through the protein are shown in stick form (*blue*) and the bound ADPNP molecules are shown in space-filling form. [Courtesy of Eleanor Dodson and Guy Dodson, York University, U.K.]

```
G C A C U U G A
            | snake venom
            | phosphodiesterase
            ↓
G C A C U U G A
G C A C U U G
G C A C U U
G C A C U
G C A C
G C A
G C    + Mononucleotides
```

FIGURE 28-46. The sequence determination of an oligonucleotide by partial digestion with snake venom phosphodiesterase. This enzyme sequentially cleaves the nucleotides from the 3′ end of a polynucleotide that has a free 3′-OH group. Partial digestion of an oligonucleotide with snake venom phosphodiesterase yields a mixture of fragments of all lengths, as indicated, that may be chromatographically separated. Comparison of the base compositions of pairs of fragments that differ in length by one nucleotide establishes the identity of the 3′-terminal nucleotide of the larger fragment. In this way the base sequence of the oligonucleotide may be elucidated.

3. The ordering of the fragments by repeating the preceding steps using a degradation procedure that yields a set of polynucleotide fragments that overlap the cleavage points in the first such set.

Before about 1975, however, nucleic acid sequencing techniques lagged far behind those of protein sequencing largely because there were no available endonucleases that were specific for sequences greater than a nucleotide. Rather, nucleic acids were cleaved into relatively short fragments by partial digestion with enzymes such as **ribonuclease T1** (from *Aspergillus oryzae),* which cleaves RNA after guanine residues, or **pancreatic ribonuclease A,** which does so after pyrimidine residues. Moreover, there is no reliable polynucleotide reaction analogous to the Edman degradation for proteins (Section 6-1A). Consequently, the polynucleotide fragments were sequenced by their partial digestion with either of two **exonucleases** (enzymes that sequentially cleave nucleotides from the end of a polynucleotide strand): **snake venom phosphodiesterase,** which removes residues from the 3′ end of polynucleotides (Fig. 28-46), or **spleen phosphodiesterase,** which does so from the 5′ end. The resulting oligonucleotide fragments were identified from their chromatographic and electrophoretic mobilities. Sequencing RNA in this manner is a lengthy and painstaking procedure.

The first biologically significant nucleic acid to be sequenced was that of yeast **alanine tRNA** (Section 30-2A). The sequencing of this 76-nucleotide molecule by Robert Holley, a labor of 7 years, was completed in 1965, some 12 years after Frederick Sanger had determined the amino acid sequence of insulin. This was followed, at an accelerating pace, by the sequencing of numerous species of tRNAs and

the **5S ribosomal RNAs** (Section 30-3A) from several organisms. The art of RNA sequencing by these techniques reached its zenith in 1976 with the sequencing, by Walter Fiers, of the entire 3569-nucleotide genome of the **bacteriophage MS2.** In comparison, DNA sequencing was in a far more primitive state because of the lack of available DNA endonucleases with any sequence specificity.

After 1975, dramatic progress was made in nucleic acid sequencing technology. Three advances made this possible:

1. The discovery of **restriction endonucleases,** enzymes that cleave duplex DNA at specific sequences.

2. The development of DNA sequencing techniques.

3. The development of **molecular cloning** techniques (Section 28-8), which permit the acquisition of any identifiable DNA segment in the amounts required for sequencing. Their use is necessary because most specific DNA sequences are normally present in a genome in only a single copy.

These procedures are largely responsible for the enormous advances in our understanding of molecular biology that have been made over the past two decades and which we discuss in succeeding chapters. The use of restriction endonucleases and DNA sequencing techniques is the subject of this section.

The pace of nucleic acid sequencing has become so rapid that directly determining a protein's amino acid sequence is far more difficult than determining the base sequence of its corresponding gene (although amino acid and base sequences provide complementary information; Section 6-1K). There has been such a flood of new DNA sequences — 130 million bases as of March 1994, and doubling every 18 months — that only computers can keep track of them. Recent high points in the sequencer's art include the determination of the entire 315-, 562-, and 666-kilobase pair **(kb)** sequences of chromosomes III, VIII, and XI from *Saccharomyces cerevisiae* (baker's yeast) and that of a contiguous 2.2-megabase pair **(Mb)** segment of chromosome III from the nematode worm *Caenorhabditis elegans.* Indeed, preparations are well along to sequence the 2.9 billion-bp human genome (although the magnitude of this project is such that if the DNA sequencing rate can be increased, as it is hoped, to 1 million bases/day, the project will still take nearly 10 years to complete).

A. Restriction Endonucleases

Bacteriophages that propagate efficiently on one bacterial strain, such as *E. coli* K12, have a very low rate of infection (~0.001%) in a related bacterial strain such as *E. coli* B. However, the few viral progeny of this latter infection propagate efficiently in the new host but only poorly in the original host. What is the molecular basis of this **host-specific modification** system? Werner Arber showed that it results from a **restriction–modification system** in the bacterial host that consists of a **restriction endonuclease** and a matched **modification methylase.** *The restriction endonuclease recognizes a specific base sequence of four to eight bases in double-stranded DNA and cleaves both strands of the duplex.* The modification methylase methylates a specific base (at the amino group of an adenine or either the 5 position or the amino group of a cytosine) in the same base sequence recognized by the restriction enzyme. The restriction enzyme does not cleave such a modified DNA. A newly replicated strand of bacterial DNA, which is protected from degradation by the methylated parent strand with which it forms a duplex, is modified before the next cycle of replication. The restriction–modification system is therefore thought to protect the bacterium against invasion by foreign (usually viral) DNAs which, once they have been cleaved by a restriction endonuclease, are further degraded by bacterial exonucleases. Invading DNAs are only rarely modified before being attacked by restriction enzymes. Once a viral genome becomes modified, however, it is able to reproduce in its new host. Its progeny, however, are no longer modified in the way that permits them to propagate in the original host.

There are three known types of restriction endonucleases. **Type I** and **Type III** restriction enzymes each carry

both the endonuclease and the methylase activity on a single protein molecule. Type I restriction enzymes cleave the DNA at a possibly random site located at least 1000 bp from the recognition sequence, whereas Type III enzymes do so 24 to 26 bp distant from the recognition sequence. However, **Type II** restriction enzymes, which were discovered and characterized by Hamilton Smith and Daniel Nathans in the late 1960s, are separate entities from their corresponding modification methylases. *They cleave DNAs at specific sites within the recognition sequence, a property that makes Type II restriction enzymes indispensible biochemical tools for DNA manipulation.* In the remainder of this section we discuss only Type II restriction enzymes.

Around 2500 species of Type II restriction enzymes with nearly 200 differing sequence specificities and from a variety of bacteria have been characterized. Surprisingly, their polypeptide sequences, for the most part, appear to be unrelated. Several of the more widely used species are listed in Table 28-5. A restriction endonuclease is named by the first letter of the genus of the bacterium that produced it and the first two letters of its species, followed by its serotype or strain designation, if any, and a roman numeral if the bacterium contains more than one type of restriction enzyme. For example, *Eco*RI is produced by *E. coli* strain RY13.

Most Restriction Endonucleases Recognize Palindromic DNA Sequences

Most restriction enzyme recognition sites possess exact twofold rotational symmetry as is diagrammed in Fig. 28-47. Such sequences are known as **palindromes.**

> A palindrome is a word, verse, or sentence that reads the same backwards or forwards. Two examples are "Madam, I'm Adam" and "Sex at noon taxes."

Many restriction enzymes, such as *Eco*RI (Fig. 28-47*a*), catalyze the cleavage of the two DNA strands at positions that are symmetrically staggered about the center of the palindromic recognition sequence. This yields restriction

TABLE 28-5. RECOGNITION AND CLEAVAGE SITES OF SOME TYPE II RESTRICTION ENZYMES

Enzyme	Recognition Sequence[a]	Microorganism
*Alu*I	AG↓C*T	*Arthrobacter luteus*
*Bam*HI	G↓GATC*C	*Bacillus amyloliquefaciens* H
*Bgl*I	GCCNNNN↓NGCC	*Bacillus globigii*
*Bgl*II	A↓GATCT	*Bacillus globigii*
*Eco*RI	G↓AA*TTC	*Escherichia coli* RY13
*Eco*RII	↓CC*(^A_T)GG	*Escherichia coli* R245
*Eco*RV	GA*T↓ATC	*Eschericia coli* J62P7G74
*Hae*II	RGCGC↓Y	*Haemophilus aegyptius*
*Hae*III	GG↓C*C	*Haemophilus aegyptius*
*Hind*III	A*↓AGCTT	*Haemophilus influenzae* R$_d$
*Hpa*II	C↓C*GG	*Haemophilus parainfluenzae*
*Msp*I	C*↓CGG	*Moraxella* species
*Pst*I	CTGCA*↓G	*Providencia stuartii* 164
*Pvu*II	CAG↓C*TG	*Proteus vulgaris*
*Sal*I	G↓TCGAC	*Streptomyces albus* G
*Taq*I	T↓CGA*	*Thermus acuaticus*
*Xho*I	C↓TCGAG	*Xanthomonas holcicola*

[a] The recognition sequence is abbreviated so that only one strand, reading 5′ to 3′, is given. The cleavage site is represented by an arrow (↓) and the modified base, where it is known, is indicated by an asterisk (A* is N^6-methyladenine and C* is 5-methylcytosine). R, Y, and N represent purine nucleotide, pyrimidine nucleotide, and any nucleotide, respectively.

Source: Roberts, R.J. and Macelis, D., REBASE — restriction enzymes and methylases, *Nucl. Acids Res.* **21,** 3125–3127 (1993).]

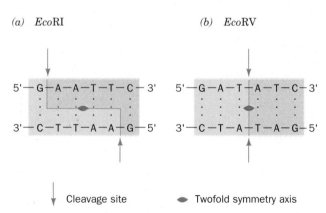

↓ Cleavage site   ⬥ Twofold symmetry axis

FIGURE 28-47. The recognition sequences of the restriction endonucleases (*a*) *Eco*RI and (*b*) *Eco*RV showing their twofold (palindromic) symmetry and indicating their cleavage sites.

fragments with complementary single-stranded ends that are from one to four nucleotides in length. Restriction fragments with such **cohesive** or **sticky ends** can associate by complementary base pairing with other restriction fragments generated by the same restriction enzyme. Some restriction cuts, such as that of *Eco*RV (Fig. 28-47*b*), pass through the twofold axis of the palindrome to yield restriction fragments with fully base paired **blunt ends.** Since a given base has a one fourth probability of occurring at any nucleotide position (assuming the DNA has equal proportions of all bases), a restriction enzyme with an *n*-base pair recognition site produces restriction fragments that are, on average, 4^n base pairs long. Thus *Alu*I (4-bp recognition sequence) and *Eco*RI (6-bp recognition sequence) restriction fragments should average $4^4 = 256$ and $4^6 = 4096$ bp in length, respectively.

The X-Ray Structures of the *Eco*RI · DNA, *Eco*RV · DNA, and *Pvu*II · DNA Complexes Reveal How These Enzymes Recognize DNA

The X-ray structure of *Eco*RI endonuclease in complex with a segment of B-DNA containing the enzyme's recognition sequence was determined by John Rosenberg. The DNA binds in the twofold symmetric cleft between the two identical 276-residue subunits of the dimeric enzyme (Fig. 28-48), thereby accounting for the DNA's palindromic recognition sequence. Protein binding causes the dihedral angle between the recognition sequence's central two base pairs to open up by ~50° towards the minor groove and to thereby become unstacked (the DNA is nevertheless nearly straight due to compensating bends at the adjacent base pair steps). This unwinds the DNA by 28° and widens the major groove by 3.5 Å at the recognition site. Recognition occurs through an extensive and apparently cooperative hydrogen bonded network involving the bases of the recognition sequence and protein residues at or near the N-terminal ends of a pair of parallel helices from each subunit, which are inserted into the widened major groove.

The X-ray structure of *Eco*RV, a dimer of identical 244-residue subunits, in complex with a 10-bp DNA containing the enzyme's 6-bp recognition sequence (Fig. 28-47*b*), was determined by Fritz Winkler (Fig. 28-49). *Eco*RV and *Eco*RI have <20% sequence identity and little structural similarity. The DNA binds to *Eco*RV with its minor groove facing the floor of the binding cleft (vs the major groove for *Eco*RI) such that the DNA is kinked by ~50°

(*a*)

(*b*)

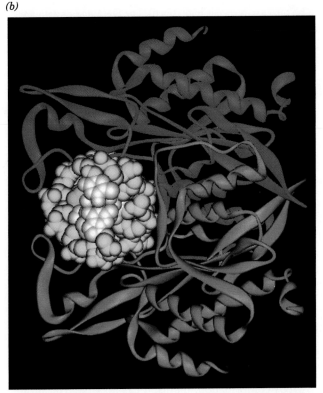

FIGURE 28-48. The X-ray structure of *Eco*RI endonuclease in complex with a segment of duplex DNA that has the self-complementary sequence TCGCGAATTCGCG (12 bp with an overhanging T at both 5′ ends; the enzyme's 6-bp target sequence is underlined). The DNA is drawn as a space-filling model with its sugar–phosphate chains yellow, its recognition sequence bases light blue, and its other bases white. The protein is drawn in ribbon form with its two identical subunits red and blue. The complex is shown with (*a*) its DNA helix axis vertical, and (*b*) in end view. The complex's two-fold axis is horizontal and the DNA's major groove faces right (towards the protein) in both views. [Based on an X-ray structure by John Rosenberg, University of Pittsburgh.]

(a)

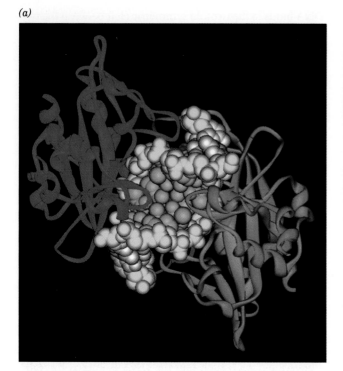

(b)

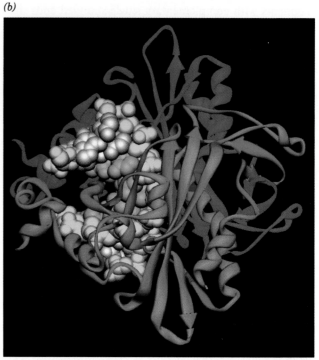

FIGURE 28-49.   The X-ray structure of *Eco*RV endonuclease in complex with a 10-bp segment of DNA of self-complementary sequence GGG<u>GATATC</u>CC (the enzyme's 6-bp target sequence is underlined). The DNA and protein are colored as is described in the legend to Figure 28-48. (*a*) The complex as viewed along its two-fold axis, facing the DNA's major groove. The two symmetry-related protein loops that overlie the major groove (comprised of residues 182 to 186) are the only parts of the enzyme that make base-specific contacts with the DNA. (*b*) The complex as viewed from the right in *a* (the DNA's major groove faces left). Note how the protein kinks the DNA towards its major groove. [Based on an X-ray structure by Fritz Winkler, Hoffman-LaRoche Ltd., Switzerland.]

towards its major groove (Fig 28-49*b*; opposite the direction in *Eco*RI). Consequently, the two central base pairs, which span the cleavage site, are unstacked, the major groove in this region is narrowed, and the minor groove is widened. Base-specific hydrogen bonds between *Eco*RV and the DNA's recognition sequence occur exclusively in the DNA's major groove through a short protein surface loop from each subunit, which extends over the top of the binding cleft (Fig. 28-49*a*). These interactions appear to be highly cooperative, which accounts for the observation that a change of even one base pair in the recognition sequence reduces *Eco*RV's k_{cat}/K_M (catalytic efficiency; Section 13-2B) for DNA cleavage by ~10^6-fold. Indeed, the X-ray structure of *Eco*RV in complex with a noncognate self-complementary DNA of sequence CGAGCTCG reveals that two of these 8-bp DNAs, in end-to-end contact and related by the 2-fold symmetry of the complex, bind to the protein as a pseudocontinuous 16-mer that has a typical B-DNA conformation; that is, it has no kink. The sugar–phosphate backbones, in both the cognate and the noncognate DNA complexes, participate in numerous non-sequence-specific interactions with the protein.

The X-ray structure of **PvuII endonuclease,** a dimer of identical 157-residue subunits (the smallest known restriction endonuclease), was determined, by Xiaodong Cheng, in complex with a DNA segment containing its 6-bp target sequence (Table 28-5). *Pvu*II which, like *Eco*RV, makes a blunt-ended cleavage and binds to its target DNA from its minor groove side, shows some structural similarity to *Eco*RV despite their lack of sequence identity. Nevertheless, *Pvu*II does not significantly bend its bound target B-DNA, which indicates that DNA distortion is not a prerequisite for the formation of a sequence-specific protein–DNA complex.

Restriction Maps Provide a Means of Characterizing a DNA Molecule

The treatment of a DNA molecule with a restriction endonuclease produces a series of precisely defined fragments that can be separated according to size by gel electrophoresis (Fig. 28-50). Complementary single strands can be separated either by melting the DNA and subjecting it to gel electrophoresis, or by density gradient ultracentrifugation in alkaline CsCl. The single strands can be sequenced by one of the methods described below. If a DNA segment is too long to sequence, it may be further fragmented with a second, *etc.,* restriction enzyme before its strands are separated.

A diagram of a DNA molecule showing the relative positions of the cleavage sites of various restriction enzymes is

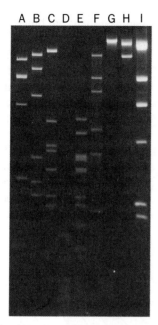

FIGURE 28-50. Agarose gel electrophoretograms of restriction digests of *Agrobacterium radiobacter* plasmid pAgK84 with (A) *Bam*HI, (B) *Pst*I, (C) *Bgl*II, (D) *Hae*III, (E) *Hinc*II, (F) *Sac*I, (G) *Xba*I, and (H) *Hpa*I. Lane (I) contains λ phage DNA digested with *Hin*dIII as a standard since these fragments have known sizes. [From Slota, J. E. and Farrand, S.F., *Plasmid* **8**, 180 (1982). Copyright © 1982 by Academic Press.]

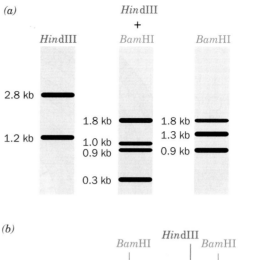

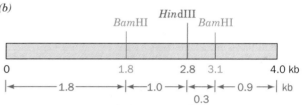

FIGURE 28-51. (*a*) The gel electrophoretic patterns of digests of a hypothetical DNA molecule with *Hin*dIII, *Bam*HI, and their mixture. The lengths of the various fragments are indicated. (*b*) The restriction map of the DNA resulting from the information in Part *a*. This map is equivalent to one that has been reversed, right to left.

known as its **restriction map.** Such a map is generated by subjecting the DNA to digestion with two or more restriction enzymes, both individually and in mixtures. By comparing the lengths of the fragments in the various digests, as determined, for instance, by their electrophoretic mobilities, a restriction map can be constructed. For example, consider the 4-kb linear DNA molecule that *Bam*HI, *Hin*dIII, and their mixture cut to fragments of the lengths indicated in Fig. 28-51*a*. This information is sufficient to deduce the positions of the restriction sites in the intact DNA and hence to construct the restriction map diagrammed in Fig. 28-51*b*. The restriction map of the SV40 chromosome is shown in Fig. 28-52. The restriction sites are physical reference points on a DNA molecule that are easily located. *Restriction maps therefore constitute a convenient framework for locating particular base sequences on a chromosome and for estimating the degree of difference between related chromosomes.*

Restriction-Fragment Length Polymorphisms Provide Markers for Characterizing Genes

Individuality in humans and other species derives from their high degree of genetic polymorphism; homologous human chromosomes differ in sequence, on average, every 200 to 500 bp. These genetic differences create or eliminate restriction sites. Restriction enzyme digests of the corresponding segments from homologous chromosomes there-

fore contain fragments with different lengths; that is, these DNAs exhibit **restriction-fragment length polymorphisms (RFLPs;** Fig. 28-53).

RFLPs are useful markers for identifying chromosomal differences (Fig. 28-54). They are particularly valuable for diagnosing inherited diseases for which the molecular defect is unknown. If a particular RFLP is so closely linked to a defective gene that there is little chance the two will recombine from generation to generation (recall that the probability of recombination between two genes increases with their physical separation on a chromosome; Section 27-1C), then the detection of that RFLP in an individual is indicative that the individual has also inherited the defective gene. For example, **Huntington's disease,** a progressive and invariably fatal neurological deterioration, whose symptoms first appear around age 40, is caused by a dominant but until recently unknown genetic defect (Section 31-7). The identification of an RFLP that is closely linked to the defective Huntington's gene has permitted the children of Huntington's disease victims (50% of whom inherit this devastating condition) to make informed decisions in ordering their lives. (Note that the availability of fetal testing has actually increased the number of births because many couples who knew they had a high risk of conceiving a genetically defective child previously chose not to have children.)

RFLPs are valuable markers for isolating and thus sequencing their closely linked but unknown genes. Indeed,

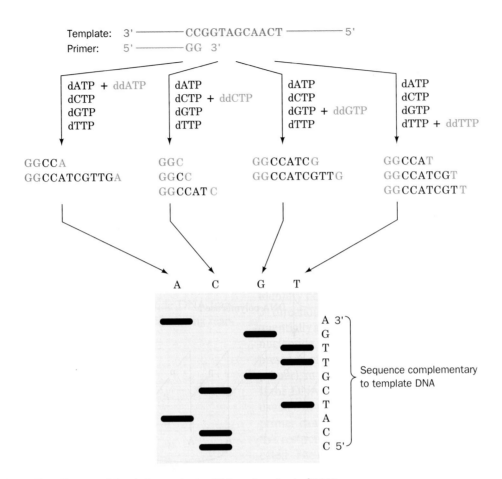

FIGURE 28-58. A flow diagram of the chain-terminator (dideoxy) method of DNA sequencing. The symbol ddATP represents dideoxyadenosine triphosphate, *etc.*

This is because the β particles emitted by ³⁵S nuclei have less energy and hence shorter path lengths than those of ³²P, thereby yielding sharper gel bands. More readily interpretable gels may also be obtained by the replacement of Klenow fragment with DNA polymerases either from **bacteriophage T7 (T7 DNA polymerase,** which is less sensitive to the presence of ddNTPs than is Klenow fragment and hence yields gel bands of more even intensities), or from the thermophilic bacteria ***Thermus aquaticus*** or ***Thermococcus litoralis* (*Taq* polymerase** or Vent_R™ DNA polymerase, respectively, which are stable above 90°C and hence can be

FIGURE 28-59. An autoradiograph of a sequencing gel containing DNA fragments produced by the chain-terminator method of DNA sequencing. A second loading of the gel (*right*) was made 90 min after the initial loading. The deduced sequence of 140 nucleotides is written along side. [From Hindley, J., DNA Sequencing, *In* Work, T.S. and Burdon, R.H. (Eds.), *Laboratory Techniques in Biochemistry and Molecular Biology*, Vol. 10, *p.* 82, Elsevier (1983). Used by permission.]

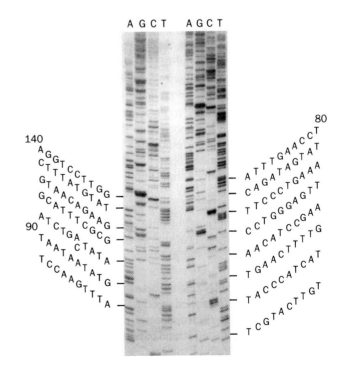

used at the temperatures required to denature particularly stable segments of DNA secondary structure that would otherwise interfere with chain extension; Vent$_R$™ DNA polymerase is so-named because *Thermococcus litoralis* inhabits the areas around deep submarine volcanic vents).

Both the chain-terminator and the chemical cleavage procedures are widely used for DNA sequencing. With a few hours effort by a skilled operator, either method can sequence a DNA segment of up to 800 nucleotides. Indeed, the major obstacle to sequencing a very long DNA molecule is ensuring that all of its fragments are cloned (by methods discussed in Section 28-8C) rather than sequencing them once they have been obtained. The chemical cleavage method is somewhat easier to set up for occasional use, whereas the chain-terminator method is generally chosen for routine use. The chemical cleavage method is also the basis for **dimethyl sulfate footprinting,** a widely used technique for identifying the segments of a DNA that specifically bind to a particular protein (Section 33-3B). Note that the sequence obtained by the chain-terminator method is complementary to the DNA strand being sequenced, whereas the sequence obtained by the chemical cleavage method is that of the original DNA strand.

The Chain-Terminator Method Is Readily Automated

If large DNA segments such as entire chromosomes are to be sequenced, then existing sequencing methods must be greatly accelerated, that is, automated. The chain-terminator method has been adapted to computerized procedures. Three types of automated systems are in use in commercially available DNA sequencers:

1. Four Reaction/Four Gel Systems
The primer is linked at its 5′ end to a highly fluorescent dye (which avoids the use of radiolabeled nucleotides with their inherent health hazards and storage problems), and the chain extension reactions are carried out in four separate vessels as described above. The reaction products are then subject to sequencing gel electrophoresis in four parallel lanes and the order in which the fluorescent fragments pass through the gel is recorded by a laser-activated fluoresence detection system.

2. Four Reaction/One Gel Systems
The primers used in each of the four chain extension reactions are each 5′ linked to a differently fluorescing dye. The separately reacted mixtures are combined, subjected to sequencing gel electrophoresis in a single lane, and the terminal base on each fragment identified according to its characteristic fluorescence spectrum (Fig. 28-60).

3. One Reaction/One Gel Systems
Each of the four ddNTPs used to terminate chain extension is covalently linked to a differently fluorescing dye, the chain-extension reaction is carried out in a single vessel, the resulting fragment mixture is subjected to sequencing gel electrophoresis in a single lane, and the terminal base on each fragment is identified according to its characteristic fluoresence spectrum.

The fluorescence detectors used in all these devices are computer controlled and hence data acquisition is automated. Such systems can identify ~10,000 bases per day, in contrast to the ~50,000 bases per year that a skilled operator can identify using the above-described manual methods (note that with the use of only one such device, it would still take nearly 1000 years to sequence the human genome). Sequencing rates have been further increased by automating the setup of DNA sequencing reactions through the use of robotics and/or by separating the DNA fragments by capillary electrophoresis (Section 5-4E).

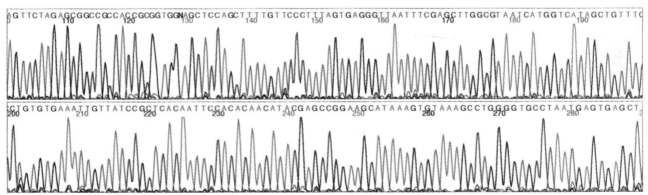

FIGURE 28-60. A portion of the output of a four reaction/one gel sequencing system. Each of the four differently colored curves indicates the fluoresence intensity of a particular dye that is linked to the primer used with a specific ddNTP in terminating the primer extension reaction (green, red, black, and blue with ddATP, ddTTP, ddGTP, and ddCTP, respectively). The 3′-terminal base of each terminated oligonucleotide, which the gel separates according to size, is identified by the fluoresence of its gel band (letters above the bands; the numbers indicate the positions of the bases in the DNA segment being sequenced). [Courtesy of Mark Adams, The Institute for Genomic Research, Gaithersburg, Maryland.]

D. RNA Sequencing

RNA may be rapidly sequenced by only a slight modification of DNA sequencing procedures. The RNA to be sequenced is transcribed into a complementary strand of DNA (**cDNA**) through the action of **RNA-directed DNA polymerase** (also known as **reverse transcriptase**). This enzyme, which is produced by certain RNA-containing viruses (Section 31-4C), uses an RNA template but is otherwise similar in its action to DNA polymerase I. The resulting cDNA may then be sequenced by either the chemical cleavage or the chain-terminator method.

7. CHEMICAL SYNTHESIS OF OLIGONUCLEOTIDES

Molecular cloning techniques (Section 28-8) have permitted the genetic manipulation of organisms in order to investigate their cellular machinery, change their characteristics, and produce scarce or specifically altered proteins in large quantities. *The ability to chemically synthesize DNA oligonucleotides of specified base sequences is an indispensable part of this powerful technology.* For example, suppose we wished to obtain the gene specifying a protein whose amino acid sequence is at least partially known. Reference to the **genetic code** (the correspondence between an amino acid sequence and the base sequence of the gene specifying it; Section 30-1) permits the synthesis of a short (~15-nucleotide) ^{32}P-labeled oligonucleotide that is complementary to a segment of the gene of interest. The oligonucleotide is used as a probe in the Southern transfer procedure (Section 28-4C) on restriction enzyme-digested DNA from the organism that produced the protein. The probe specifically labels the required gene and thereby permits its isolation.

Synthetic oligonucleotides are also required to specifically alter genes through **site-directed mutagenesis**, a technique pioneered by Michael Smith. An oligonucleotide containing a short gene segment with the desired altered base sequence is used as a primer in the DNA polymerase I replication of the gene of interest. Such a primer will hybridize to the corresponding wild-type sequence if there are only a few mismatched base pairs, and its extension, by DNA polymerase I (Section 28-6C), yields the desired altered gene (Fig. 28-61). The altered gene can then be inserted in a suitable organism via techniques discussed in Section 28-8 and grown (cloned) in quantity.

Oligonucleotides Are Valuable Diagnostic Tools

Synthetic oligonucleotides are widely used as probes in Southern transfer analysis for the diagnosis and prenatal detection of genetic diseases. These diseases often result from a specific change in a single gene such as a base substitution, deletion, or insertion. The temperature at which probe hybridization is carried out may be adjusted so that

only an oligonucleotide that is perfectly complementary to a length of DNA will hybridize to it. Even a single base mismatch, under appropriate conditions, will result in a failure to hybridize. For example, sickle-cell anemia arises from a single base change that causes the amino acid substitution Glu β6 → Val in hemoglobin (Section 9-3B). A 19-residue oligonucleotide that is complementary to the sickle-cell gene's mutated segment hybridizes, at the proper temperature, to DNA from homozygotes for the sickle-cell gene but not to DNA from normal individuals. An oligonucleotide that is complementary to the normal Hb β gene gives opposite results. DNA from sickle-cell heterozygotes hybridizes to both probes but in reduced amounts relative to the DNAs from homozygotes. The oligonucleotides may consequently be used in the prenatal diagnosis of sickle-cell disease. DNA probes are also rapidly replacing the much slower and less accurate culturing techniques for the identification of pathogenic bacteria.

Oligonucleotides Are Synthesized in a Stepwise Manner

The basic strategy of oligonucleotide synthesis is analogous to that of polypeptide synthesis (Section 6-4): *A suitably protected nucleotide is coupled to the growing end of the*

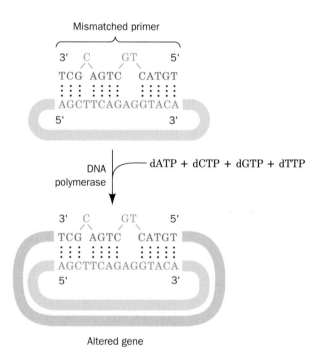

FIGURE 28-61. Site-directed mutagenesis. A chemically synthesized oligonucleotide incorporating the desired base changes is hybridized to the DNA encoding the gene to be altered. The mismatched primer is then extended by DNA polymerase I, thereby generating the mutated gene. The mutated gene can subsequently be inserted into a suitable host organism so as to yield the mutant DNA, or the corresponding RNA, in quantity, produce a specifically altered protein, and/or generate a mutant organism.

oligonucleotide chain, the protecting group is removed, and the process is repeated until the desired oligonucleotide has been synthesized. The first practical technique for DNA synthesis, the **phosphodiester method,** which was developed by H. Gobind Khorana in the 1960s, is a laborious process in which all reactions are carried out in solution and the products must be isolated at each stage of the multistep synthesis. Khorana, nevertheless, used this method, in combination with enzymatic techniques, to synthesize a 126-nucleotide tRNA gene, a project that required several years of intense effort by numerous skilled chemists.

The Phosphoramidite Method

By the early 1980s, these difficult and time-consuming processes had been supplanted by much faster solid phase methodologies that permitted oligonucleotide synthesis to be automated. The presently most widely used chemistry, which was formulated by Robert Letsinger and further developed by Marvin Caruthers, is known as the **phosphoramidite method.** This nonaqueous reaction sequence adds a single nucleotide to a growing oligonucleotide chain as follows (Fig. 28-62):

1. The **dimethoxytrityl (DMTr)** protecting group at the 5′ end of the growing oligonucleotide chain (which is anchored via a linking group at its 3′ end to a solid support, S) is removed by treatment with acid.

2. The newly liberated 5′ end of the oligonucleotide is coupled to the 3′-phosphoramidite derivative of the next deoxynucleoside to be added to the chain. The coupling agent in this reaction is **tetrazole.**

3. Any unreacted 5′ end (the coupling reaction has a yield of over 99%) is capped by acetylation so as to block its extension in subsequent coupling reactions. This prevents the extension of erroneous oligonucleotides.

4. The phosphite triester group resulting from the coupling step is oxidized to the phosphotriester, thereby yielding a chain that has been lengthened by one nucleotide.

This reaction sequence, in commercially available automated synthesizers, can be repeated up to ~150 times with a cycle time of 40 min or less. Once an oligonucleotide of desired sequence has been synthesized, it is released from its support and its various blocking groups, including those on the bases, are removed. The product can then be purified by HPLC and/or gel electrophoresis.

8. MOLECULAR CLONING

A major problem in almost every area of biochemical research is obtaining sufficient quantities of the substance of interest. For example, a 10-L culture of *E. coli* grown to its maximum titer of ~10^{10} cells·mL^{-1} contains, at most, 7 mg of DNA polymerase I, and many of its proteins are present in far lesser amounts. Yet, it is rare that as much as half of any protein originally present in an organism can be recovered in pure form. Eukaryotic proteins may be even more difficult to obtain because many eukaryotic tissues, whether acquired from an intact organism or grown in tissue culture, are available in only small quantities. As far as the amount of DNA is concerned, our 10-L *E. coli* culture would contain ~0.1 mg of any 1000-bp length of chromosomal DNA (a length sufficient to contain most prokaryotic genes) but its purification in the presence of the rest of the chromosomal DNA would be an all but impossible task. These difficulties have been largely eliminated in recent years through the development of **molecular cloning** techniques (a **clone** is a collection of identical organisms that are derived from a single ancestor). These methods, which are also referred to as **genetic engineering** and **recombinant DNA** technology, deserve much of the credit for the enormous progress in biochemistry and the dramatic rise of the biotechnology industry since the mid-1970s.

The main idea of molecular cloning is to insert a DNA segment of interest into an autonomously replicating DNA molecule, a so-called **cloning vector** *or* **vehicle,** *so that the DNA segment is replicated with the vector.* Cloning such a **chimeric vector** (*chimera:* a monster in Greek mythology that has a lion's head, a goat's body, and a serpent's tail) in a suitable **host organism** such as *E. coli* or yeast results in the production of large amounts of the inserted DNA segment. If a cloned gene is flanked by the properly positioned control sequences for RNA and protein synthesis (Chapters 29 and 30), the host may also produce large quantities of the RNA and protein specified by that gene. The techniques of genetic engineering are outlined in this section.

A. Cloning Vectors

Both plasmids, bacteriophages, and **yeast artificial chromosomes** are used as cloning vectors in genetic engineering.

Plasmid-Based Cloning Vectors

Plasmids are circular DNA duplexes of 1 to 200 kb that contain the requisite genetic machinery, such as a **replication origin** (a site at which DNA replication is initiated; Section 31-2), to permit their autonomous propagation in a bacterial host or in yeast. Plasmids may be considered molecular parasites but in many instances they benefit their host by providing functions, such as resistance to antibiotics, that the host lacks. Indeed, the widespread and alarming appearance, since antibiotics came into use, of antibiotic-resistant pathogens is a result of the rapid proliferation among these organisms of plasmids containing genes that confer resistance to antibiotics.

Some types of plasmids, which are present in one or a few copies per cell, replicate once per cell division as does the bacterial chromosome; their replication is said to be under **stringent control.** The plasmids used in molecular cloning,

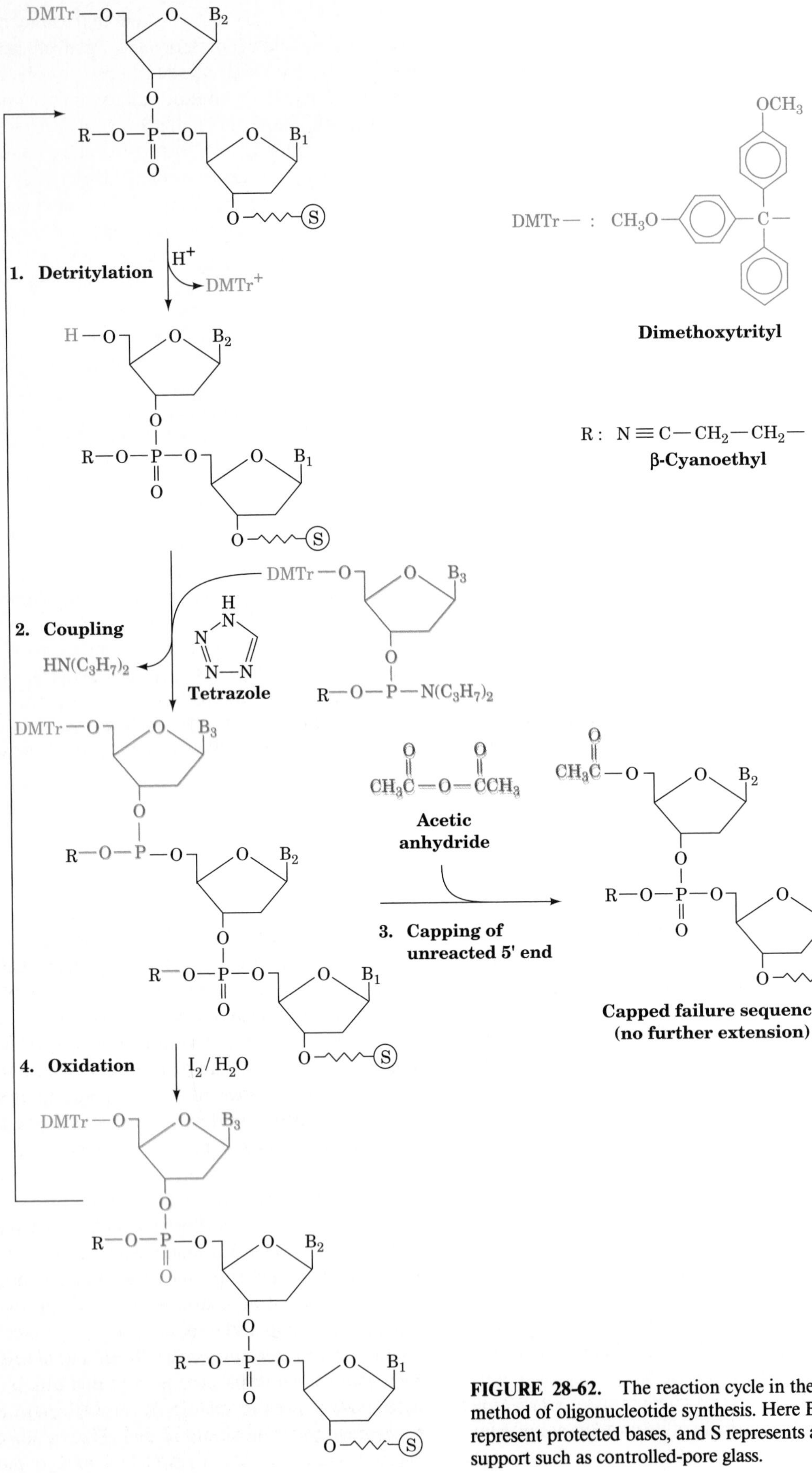

FIGURE 28-62. The reaction cycle in the phosphite-triester method of oligonucleotide synthesis. Here B_1, B_2, and B_3 represent protected bases, and S represents an inert solid phase support such as controlled-pore glass.

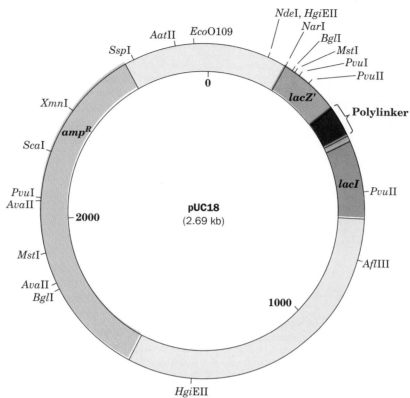

FIGURE 28-63. A restriction map of the plasmid pUC18 indicating the positions of its **amp^R**, **lacZ'**, and **lacI** genes. The *amp^R* gene confers resistance to the antibiotic **ampicillin** (a penicillin derivative); *lacZ'* is a modified form of the **lacZ** gene, which encodes the enzyme β-galactosidase (Section 10-1C); and *lacI* encodes the **lac repressor,** a protein that controls the

transcription of *lacZ* as is discussed in Sections 29-1A and 3B. The polylinker, which encodes an 18-residue polypeptide segment inserted near the N-terminus of β-galactosidase, incorporates 13 different restriction sites that do not occur elsewhere in the plasmid.

however, are under **relaxed control;** they are normally present in 10 to as many as 700 copies per cell. Moreover, if protein synthesis in the bacterial host is inhibited, for example, by the antibiotic **chloramphenicol** (Section 30-3G), thereby preventing cell division, these plasmids continue to replicate until two or three thousand copies have accumulated per cell (which represents about half of the cell's total DNA). The plasmids that have been constructed (by genetic engineering techniques) for use in molecular cloning are relatively small, replicate under relaxed control, carry genes specifying resistance to one or more antibiotics, and contain a number of conveniently located restriction endonuclease sites into which the DNA to be cloned may be inserted (via techniques described in Section 28-8B). Indeed, many plasmid vectors contain a strategically located short (<100-bp) segment of DNA known as a **polylinker** that has been synthesized to contain a variety of restriction sites that are not present elsewhere in the plasmid. The *E. coli* plasmid designated **pUC18** (Fig. 28-63) is representative of the cloning vectors presently in use.

The expression of a chimeric plasmid in a bacterial host was first demonstrated in 1973 by Herbert Boyer and Stanley Cohen. The host bacterium takes up a plasmid when the

two are mixed together in a process that is greatly enhanced by the presence of divalent cations such as Ca^{2+} (which increase membrane permeability to DNA). An absorbed plasmid vector becomes permanently established in its bacterial host (transformation) with an efficiency of ~0.1%.

Plasmid vectors cannot be used to clone DNAs of more than ~10 kb. This is because the time required for plasmid replication increases with plasmid size. Hence intact plasmids with large unessential (to them) inserts are lost through the faster proliferation of plasmids that have eliminated these inserts by random deletions.

Bacteriophage-Based Cloning Vectors

Bacteriophage λ (Fig. 28-64) is an alternative cloning vehicle that can be used to clone DNAs of up to 16 kb. The central third of this virus' 48.5-kb genome is not required for phage infection (Section 32-3A) and can therefore be replaced by foreign DNAs of up to slightly greater size using techniques discussed in Section 28-8B. The chimeric phage DNA can then be introduced into the host cells by infecting them with phages formed from the DNA by an *in vitro* packaging system (Section 32-3B). The use of phages as cloning vectors has the additional advantage that the chi-

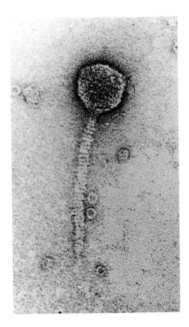

FIGURE 28-64. An electron micrograph of bacteriophage λ. [Courtesy of A.F. Howatson. From Lewin, B., *Gene Expression*, Vol. 3, Fig. 5.23, John Wiley & Sons Inc. (1977).]

meric DNA is produced in large amounts and in easily purified form.

λ Phages can be used to clone even longer DNA inserts. The viral apparatus that packages DNA into phage heads requires only that the DNA have a specific 14-bp sequence known as a *cos* site located at each end and that these ends be 36- to 51-kb apart (Section 32-3B). Placing two *cos* sites the proper distance apart on a plasmid vector yields, via an *in vitro* packaging system, a so-called **cosmid** vector, which can contain foreign DNA of up to ~49 kb. Cosmids have no phage genes and hence, upon introduction into a host cell via phage infection, reproduce as plasmids.

The **filamentous bacteriophage M13** (Fig. 28-65) is also a useful cloning vector. It has a single-stranded circular DNA that is contained in a protein tube composed of ~2700 helically arranged identical protein subunits. This number is controlled, however, by the length of the phage DNA being coated; insertion of foreign DNA in a nonessential region of the M13 chromosome results in the production of longer phage particles. Although M13 cloning vectors cannot stably maintain DNA inserts of >1 kb, they are widely used in the production of DNA for sequence analysis by the chain-terminator method (Section 28-6C) because these phages directly produce the single-stranded DNA that this technique requires. Furthermore, since the DNA to be sequenced is always inserted at the same point in the viral chromosome (a restriction site; Section 28-8B), an ~15-base synthetic oligonucleotide that is complementary to the viral DNA on the 3' side of the cloning site (the so-called "universal primer") may be used as the primer for any DNA segment sequenced by this method.

YAC Vectors

DNA segments larger than those that can be carried by cosmids may be cloned in **yeast artificial chromosomes (YACs).** YACs are linear DNA segments that contain all the molecular paraphernalia required for replication in yeast: A replication origin [known as an **autonomously replicating sequence (ARS)**], a centromere (the chromosomal segment attached to the spindle during mitosis and meiosis), and telomeres (the ends of linear chromosomes that permit their replication; Section 31-4D). DNAs of several hundred kb have been spliced into YACs and successfully cloned.

B. Gene Splicing

A DNA to be cloned is, in many cases, obtained as a defined fragment through the application of restriction endonucleases (for M13 vectors, the restriction enzymes' requirement of duplex DNA necessitates converting this phage DNA to its double-stranded form). Recall that most restriction endonucleases cleave duplex DNA at specific palindromic sites so as to yield single-stranded ends that are complementary to each other (cohesive or "sticky" ends; Section 28-6A). Therefore, as Janet Mertz and Ron Davis first demonstrated in 1972, *a restriction fragment may be inserted into a cut made in a cloning vector by the same restriction enzyme (Fig. 28-66). The complimentary (cohesive) ends of the two DNAs specifically associate under annealing conditions and are covalently joined (spliced) through the action of an enzyme named **DNA ligase*** (Section 31-2C; DNA ligase produced by **bacteriophage T4** must be used for blunt-ended restriction cuts such as those generated by *Alu*I, *Eco*RV, or *Hae*III; Table 28-5). *A great*

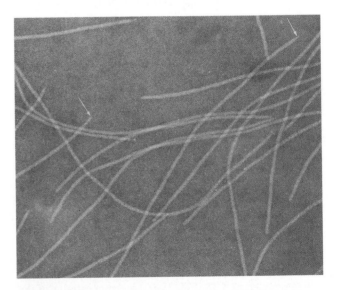

FIGURE 28-65. An electron micrograph of the filamentous bacteriophage M13. Note that some filaments appear to be pointed at one end (*arrows*). [Courtesy of Robley Williams, Stanford University, Emeritus and Harold Fisher, University of Rhode Island.]

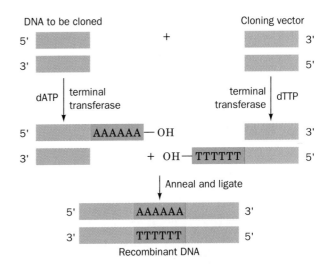

FIGURE 28-67. Two DNA fragments may be joined through the generation of complementary homopolymer tails. The poly(dA) and poly(dT) tails shown in this example may be replaced by poly(dC) and poly(dG) tails.

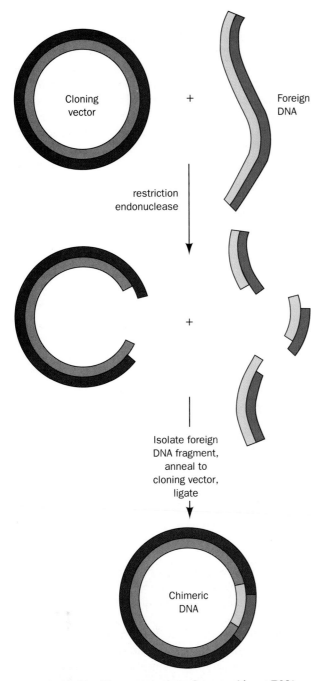

FIGURE 28-66. The construction of a recombinant DNA molecule by the insertion of a restriction fragment in a cloning vector's corresponding restriction cut.

advantage of using a restriction enzyme to construct a chimeric vector is that the DNA insert can be precisely excised from the cloned vector by cleaving it with the same restriction enzyme.

If the foreign DNA and cloning vector have no common restriction sites at innocuous positions, they may still be spliced, using a procedure pioneered by Dale Kaiser and Paul Berg, through the use of **terminal deoxynucleotidyl transferase (terminal transferase).** This mammalian en-

zyme, which has been implicated in the generation of antibody diversity (Section 34-2C), adds nucleotides to the 3′-terminal OH group of a DNA chain; it is the only known DNA polymerase that does not require a template. Terminal transferase and dTTP, for example, can build up poly(dT) tails of ~100 residues on the 3′ ends of the DNA segment to be cloned (Fig. 28-67). The cloning vector is enzymatically cleaved at a specific site and the 3′ ends of the cleavage site are similarly extended with poly(dA) tails. The complimentary homopolymer tails are annealed, any gaps resulting from differences in their lengths filled in by DNA polymerase I, and the strands joined by DNA ligase.

A disadvantage of the above technique is that it eliminates the restriction sites that were used to generate the foreign DNA insert and to cleave the vector. It may therefore be difficult to recover the insert from the cloned vector. This difficulty can be circumvented by appending to both ends of the foreign DNA a chemically synthesized palindromic "linker" which has a restriction site matching that of the cloning vector. The linker is attached to the foreign DNA by blunt end ligation with T4 ligase and then cleaved with the appropriate restriction enzyme to yield the correct cohesive ends for ligation to the vector (Fig. 28-68).

Properly Transformed Cells Must Be Selected

How can one select only those host organisms that contain a properly constructed vector? In the case of plasmid transformation, this is usually done through the use of antibiotics and/or chromogenic (color-producing) substrates. For example, *E. coli* transformed by a pUC18 plasmid (Fig. 28-63) containing a foreign DNA insert in its polylinker region lack β-galactosidase activity because the insert interrupts the protein-encoding sequence of the *lacZ′* gene. Thus, when grown in the presence of **5-bromo-4-chloro-3-indolyl-β-D-galactoside** (commonly known as **X-gal**), a col-

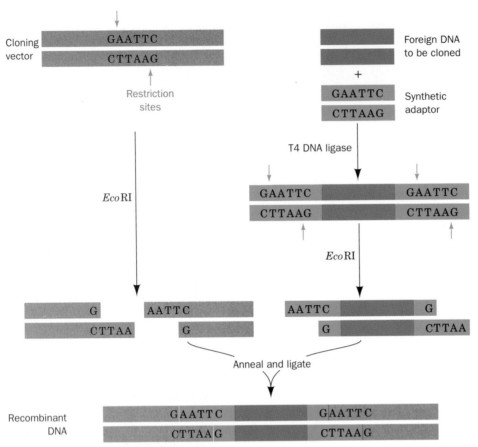

FIGURE 28-68. The construction of a recombinant DNA molecule through the use of synthetic oligonucleotide adaptors. In this example, the adaptor and the cloning vector have *Eco*RI restriction sites (*red arrows*).

orless substance which when hydrolyzed by β-galactosidase yields a blue product,

5-Bromo-4-chloro-3-indolyl-β-D-galactose (X-gal)
(colorless)

β-D-Galactose   **5-Bromo-4-chloro-3-hydroxyindole**
                              (blue)

bacterial colonies that have an insert in their polylinker site form colorless colonies, whereas bacteria containing only plasmids that lack such an insert form blue colonies. Bacteria that have failed to take up any plasmid, which would otherwise also form colorless colonies in the presence of X-gal, are excluded by adding the antibiotic ampicillin to the growth medium. Bacteria that do not contain the plasmid are sensitive to ampicillin, whereas bacteria containing the plasmid will grow, because the plasmid's intact *amp*R gene confers ampicillin resistance. Genes such as *amp*R are therefore known as **selectable markers.**

Genetically engineered λ phage variants contain restriction sites that flank the dispensable central third of the phage genome (Section 28-8A). This segment may therefore be replaced, as is described above, by a foreign DNA insert (Fig. 28-69). DNA is only packaged in λ phage heads if its length is from 75 to 105% of the 48.5-kb wild-type λ genome. Consequently, λ phage vectors that have failed to acquire a foreign DNA insert are unable to propagate because they are too short to form infectious phage particles. Cosmid vectors are subject to the same limitation. Moreover, cloned cosmids are harvested by repackaging them into phage particles. Hence, any cosmids that have lost sufficient DNA through random deletion to make them

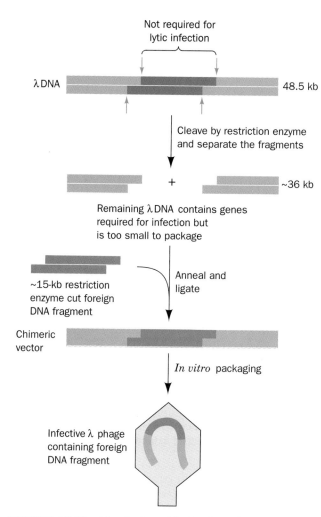

FIGURE 28-69. The cloning of foreign DNA in λ phages.

be complementary to a segment of the gene's inferred base sequence (Section 30-1E).

In practice, it is usually more difficult to identify a particular gene from an organism and then clone it than it is to clone the organism's entire genome as DNA fragments and then identify the clone(s) containing the sequences(s) of interest. Such a set of cloned fragments is known as a **genomic library.** A genomic library of a particular organism need only be made once since it can be perpetuated for use whenever a new probe becomes available.

Genomic libraries are generated according to a procedure known as **shotgun cloning.** The chromosomal DNA of the organism of interest is isolated, cleaved to fragments of clonable size, and inserted in a cloning vector by the methods described in Section 28-8B. The DNA is fragmented by partial rather than exhaustive restriction digestion so that the genomic library contains intact representatives of all the organism's genes, including those whose sequences contain restriction sites. Shear fragmentation by rapid stirring of a DNA solution can also be used but requires further treatment of the fragments to insert them into cloning vectors. Genomic libraries have been established for numerous organisms including yeast, *Drosophila,* and humans.

Many Clones Must Be Screened to Obtain a Gene of Interest

The number of random cleavage fragments that must be cloned to ensure a high probability that a given sequence is represented at least once in the genomic library is calculated as follows: The probability P that a set of N clones contains a fragment that constitutes a fraction f, in bp, of the organism's genome is

$$P = 1 - (1 - f)^N \qquad [28.7]$$

Consequently,

$$N = \log(1 - P)/\log(1 - f) \qquad [28.8]$$

Thus, in order for $P = 0.99$ for fragments averaging 10 kb in length, $N = 2162$ for the 4700-kb *E. coli* chromosome and 76,000 for the 165,000-kb *Drosophila* genome. The use of YAC-based genomic libraries therefore greatly reduces the effort necessary to obtain a given gene segment from a large genome.

Since a genomic library lacks an index, it must be screened for the presence of a particular gene. This is done by a process known as **colony** or *in situ* hybridization (Fig. 28-70; Latin: *in situ,* in position). The cloned yeast colonies, bacterial colonies, or phage plaques to be tested are transferred, by replica plating, from a master plate, to a nitrocellulose filter. The filter is treated with NaOH, which lyses the cells or phages and denatures the DNA so that it binds to the nitrocellulose (recall that single-stranded DNA is preferentially bound to nitrocellulose). The filter is then dried to fix the DNA in place, treated under annealing conditions with a radioactive probe for the gene of interest, washed, and autoradiographed. *Only those colonies or*

shorter than the above limit are not recovered. This is why cosmids can support the proliferation of large DNA inserts, whereas other types of plasmids cannot.

C. Genomic Libraries

In order to clone a particular DNA fragment, it must first be obtained in relatively pure form. The magnitude of this task may be appreciated when it is realized that, for example, a 1-kb fragment of human DNA represents only 0.000035% of the 2.9 billion-bp human genome. A DNA fragment may be identified by Southern blotting of a restriction digest of the genomic DNA under investigation (Section 28-4C). The radioactive probe used in this procedure can be the corresponding mRNA if it is produced in sufficient quantity to be isolated (e.g., reticulocytes, which produce little protein besides hemoglobin, are rich in globin mRNAs). Alternatively, in cases where the amino acid sequence of the protein encoded by the gene is known, the probe may be a mixture of the various synthetic oligonucleotides that may

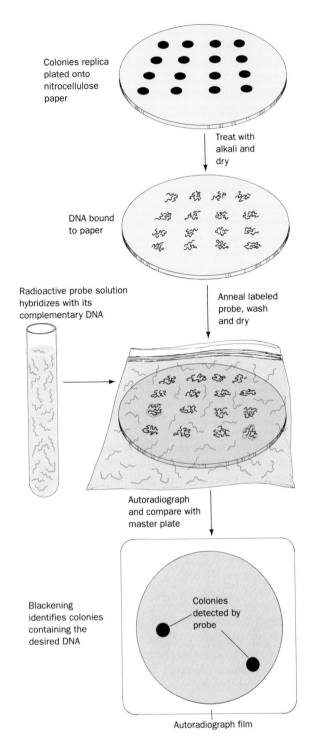

FIGURE 28-70.   Colony (*in situ*) hybridization identifies the clones containing a DNA of interest.

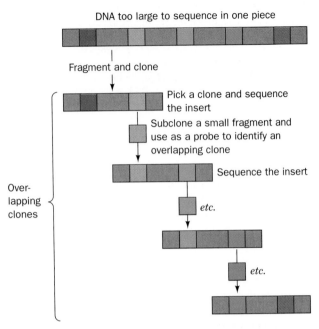

FIGURE 28-71.   Chromosome walking. A DNA segment too large to sequence in one piece is fragmented and cloned. A clone is picked and the DNA insert it contains is sequenced. A small fragment of the insert near one end is subcloned (cloned from a clone) and used as a probe to select a clone containing an overlapping insert, which, in turn, is sequenced. The process is repeated so as to "walk" down the chromosome. Chromosome walking can, of course, extend in both directions.

Many eukaryotic genes and gene clusters span enormous tracts of DNA (Section 33-2); some consist of >1000 kb. With the use of plasmid-, phage-, or cosmid-based genomic libraries, such long DNAs can only be obtained as a series of overlapping fragments (Fig. 28-71): Each gene fragment that has been isolated is, in turn, used as a probe to identify a successive but partially overlapping fragment of that gene, a process called **chromosome walking.** The use of YACs, however, greatly reduces the need for this laborious and error-prone process.

D. DNA Amplification by the Polymerase Chain Reaction

Although molecular cloning techniques are indispensible to modern biochemical research, the use of the **polymerase chain reaction (PCR)** offers a faster and more convenient method of amplifying a specific DNA segment of up to 6 kb. In this technique (Fig. 28-72), which was formulated by Kerry Mullis, a denatured (strand-separated) DNA sample is incubated with DNA polymerase, dNTPs, and two oligonucleotide primers whose sequences flank the DNA segment of interest so that they direct the DNA polymerase to synthesize new complementary strands. Multiple cycles of this process, each doubling the amount of DNA present, geometrically amplify the DNA starting from as little as a

plaques containing the sought-after gene will bind the probe and thereby blacken the film. The corresponding clones can then be retrieved from the master plate. Using this technique, even an ~1 million clone human genomic library can be readily screened for the presence of one particular DNA segment.

FIGURE 28-72. The polymerase chain reaction (PCR). In each cycle of the reaction, the strands of the duplex DNA are separated by heat denaturation, the preparation is cooled such that synthetic DNA primers anneal to a complementary segment on each strand, and the primers are extended by DNA polymerase. The process is then repeated for numerous cycles. The number of "unit-length" strands doubles with every cycle after the second cycle.

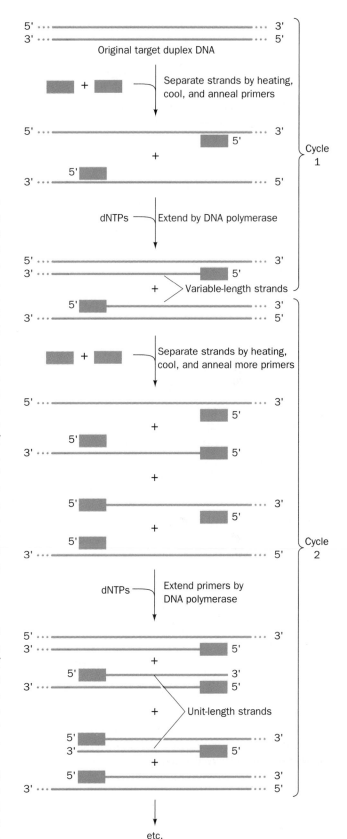

single gene copy. In each cycle, the two strands of the duplex DNA are separated by heat denaturation, the primers are annealed to their complementary segments on the DNA, and the DNA polymerase directs the synthesis of the complementary strands (Section 28-6C). The use of a heat-stable DNA polymerase, such as *Taq* polymerse (Section 28-6C), eliminates the need to add fresh enzyme after each heat denaturation step. Hence, in the presence of sufficient quantities of primers and dNTPs, the PCR is carried out simply by cyclically varying the temperature.

Twenty cycles of PCR amplification increase the amount of the target sequence around one-millionfold with high specificity. Indeed, the method has been shown to amplify a target DNA present only once in a sample of 10^5 cells, thereby demonstrating that the method can be used without prior DNA purification (although, as a consequence of this enormous amplification, particular care must be taken that the DNA sample of interest is not contaminated by extraneous DNA). The amplified DNA can be characterized by the various techniques we have discussed: Southern blotting, RFLP analysis, and direct sequencing. PCR amplification is therefore a form of "cell-free molecular cloning" that can accomplish in an automated 3 to 4 h *in vitro* reaction what would otherwise take days or weeks via the cloning techniques discussed above.

PCR Has Many Uses

PCR amplification has become an indispensible tool in a great variety of applications. Clinically, it is used for the rapid diagnosis of infectious diseases and the detection of rare pathological events such as mutations leading to cancer (Section 33-4C). Forensically, the DNA from a single hair or sperm can be used to identify the donor unambiguously. RNA may be amplified by the PCR method by first converting it to cDNA through the use of reverse transcriptase (Section 28-6D).

Variations on the theme of PCR have found numerous applications. For instance, the single-stranded DNA required for sequencing by the chain-terminator method (Section 28-6C) can be rapidly generated via **asymmetric PCR,** in which such small amounts of one primer are used that it is exhausted after several PCR cycles. In subsequent cycles, only the strand extended from the other primer, which is present in excess, is synthesized (note that PCR amplification becomes linear rather than geometric after one primer is used up). In another example, PCR may be used as a vehicle for site-directed mutagenesis simply by using a mutagenized primer in amplifying a gene of interest so that the resulting DNA contains the altered sequence.

PCR is also largely responsible for the newly emerging science of **molecular paleontology** (paleontology is the study of ancient life forms from their fossil remains). Thus, DNA that was extracted from a 120 to 135 million year old amber-entombed weevil (amber is fossil pine resin) was sufficiently preserved to permit its amplification by PCR and its subsequent sequencing. The comparison of the sequences of segments of this weevil's ribosomal RNA genes with those of several living coleopterans (beetles) and dipterans (two-winged flies) indicated that the DNA came from an extinct nemonychid weevil. Since the DNA is ~80 million years older than any that had been previously sequenced, this result suggests that many of the animal remains preserved in amber contain DNA that can be sequenced and hence that amber is a DNA repository of enormous paleontological potential.

E. Production of Proteins

One of the greatest uses of recombinant DNA technology is in the production of large quantities of scarce and/or novel proteins. This is a relatively straightforward procedure for bacterial proteins: A cloned structural gene must be inserted into an **expression vector,** a plasmid that contains the properly positioned transcriptional and translational control sequences for the protein's expression. With the use of a relaxed control plasmid and an efficient **promoter** (a type of transcriptional control element; Section 29-3A), the production of a protein of interest may reach 30% of the host's total cellular protein. Such genetically engineered organisms are called **overproducers.** Bacterial cells often sequester such large amounts of useless and possibly toxic (to the bacterium) protein as insoluble and denatured inclusions. Protein extracted from these inclusions must therefore be renatured, usually by dissolving it in a guanidinium chloride or urea solution (Section 8-1A) and then dialyzing away the denaturant. A strategy for circumventing this difficulty is to precede the protein of interest with the signal sequence of a secreted bacterial protein. Such a protein is secreted into the bacterial periplasm (the compartment between the plasma membrane and the cell wall of gram-negative bacteria such as *E. coli;* Fig. 10-23) with the concomitant removal of its signal sequence by a bacterial protease. Secreted proteins, which are relatively few in number, can be released into the medium by hypotonic disruption of the bacterial outer membrane (the bacterial cell wall is porous), so their purification is greatly simplified relative to that of intracellular proteins.

Eukaryotic Proteins Can Be Produced in Bacteria

The synthesis of a eukaryotic protein in a prokaryotic host presents several problems not encountered with prokaryotic proteins:

1. The eukaryotic control elements for RNA and protein synthesis are not recognized by bacterial hosts.

2. Most eukaryotic genes contain one or more internal unexpressed sequences called **introns,** which are specifically excised from the gene's RNA transcript to form the mature mRNA (Section 29-4A). Bacterial genes lack introns and hence bacteria are unable to excise them.

3. Bacteria lack the enzyme systems to carry out the specific posttranslational processing that many eukaryotic proteins require for biological activity (Section 30-5). Most conspicuously, bacteria do not glycosylate proteins (although, in many cases, glycosylation does not seem to affect protein function).

4. Eukaryotic proteins may be preferentially degraded by bacterial proteases (Section 30-6A).

The problem of nonrecognition of eukaryotic control elements can be eliminated by inserting the protein-encoding portion of a eukaryotic gene into a vector containing correctly placed bacterial control elements. The need to excise introns can be circumvented by cloning the cDNA of the protein's mRNA. Alternatively, genes encoding small proteins of known sequence can be chemically synthesized (Section 28-7). Neither of these strategies is universally applicable, however, because few mRNAs are sufficiently abundant to be isolated and many eukaryotic proteins are large (although the maximum available size of synthetic polynucleotides is increasing rapidly). Likewise, no general approach has been developed for the posttranslational modification of eukaryotic proteins, although polypeptide cleavage by treatment with trypsin or cyanogen bromide (Section 6-1E) has been successfully employed in the *in vitro* activation of some eukaryotic proenzymes. Lastly, the preferential bacterial proteolysis of certain eukaryotic proteins has been prevented by inserting the eukaryotic gene within a bacterial gene. The resulting hybrid protein has an N-terminal polypeptide of bacterial origin that, in some cases, prevents bacterial proteases from recognizing the eukaryotic segment as being foreign. The two polypeptide segments can later be separated by treatment with a protease that specifically cleaves a susceptible site that had been placed at the boundary between the segments. However, the development of cloning vectors that propagate in eukaryotic hosts, such as yeast or cultured animal cells, has led to the elimination of many of these problems (although posttranslational processing may vary among different eukaryotes). Indeed, **shuttle vectors** are available that can propagate in both yeast and *E. coli* and thus transfer (shuttle) genes between these two types of cells.

Recombinant Protein Production Has Important Practical Consequences

The ability to synthesize a given protein in large quantities is already having enormous medical, agricultural, and industrial impact. Those that are in routine clinical use include human insulin, human growth hormone, **erythro-**

poietin (a protein growth factor secreted by the kidney that stimulates the production of red blood cells and is used in the treatment of anemia arising from kidney disease), and several types of **colony-stimulating factors** (protein growth factors that stimulate the production and activation of white blood cells and are used clinically to counter the white cell–killing effects of chemotherapy and to facilitate bone marrow transplantation). Synthetic vaccines consisting of harmless but immunogenic components of pathogens are eliminating the risks attendant in using killed or attenuated viruses or bacteria in vaccines as well as making possible new strategies of vaccine development. The use of recombinant **blood clotting factors** that are defective in individuals with the inherited disease **hemophilia** (Section 34-1C), has replaced the need to extract these scarce proteins from large quantities of human blood and has thereby eliminated the high risk that hemophiliacs faced of contracting such blood-borne diseases as hepatitis and AIDS. Bovine growth hormone has long been known to stimulate milk production in dairy cows by ~15%. However, its use has been made cost effective by the advent of recombinant DNA technology since bovine growth hormone could previously only be obtained in small quantities from cow pituitaries.

Site-Directed Mutagenesis Tailors Proteins to Specific Applications

Of equal importance is the ability to tailor proteins to specific applications through site-directed mutagenesis (Section 28-7). Thus, the development of the subtilisin variant Met 222 → Ala has permitted the use of this bacterial protease in laundry detergent that contains bleach (which largely inactivates wild-type subtilisin by oxidizing Met 222). Monoclonal antibodies (Sections 5-1D and 34-2B) can be targeted against specific proteins and hence, it is hoped, be used as antitumor agents. However, since monoclonal antibodies, as presently made, are mouse proteins, they are ineffective as therapeutic agents in humans because humans mount an immune response against mouse proteins. These difficulties may be rectified by "humanizing" monoclonal antibodies by replacing their mouse-specific sequences with those of humans (which the human immune system ignores).

We have seen numerous instances throughout this textbook of protein function being characterized by the replacement of a specific residue or polypeptide segment suspected of having an important mechanistic or structural role through site-directed mutagensis. Indeed, site-directed mutagenesis has become an indispensible tool in the practice of enzymology.

F. Transgenic Organisms and Gene Therapy

For many purposes it is preferable to tailor an intact organism rather than just a protein — true genetic engineering. Multicellular organisms expressing a foreign (from another organism) gene are said to be **transgenic** and their transplanted foreign genes are often referred to as **transgenes**.

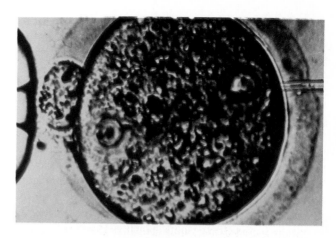

FIGURE 28-73. The microinjection of DNA into the pronucleus of a fertilized mouse ovum. The ovum is being held in place by gentle suction from the pipette on the left. [Science Vu/Visuals Unlimited]

For the change to be permanent, that is, heritable, a transgene must be stably integrated into the organism's germ cells. For mice, this is accomplished by microinjecting cloned DNA encoding the desired altered characteristics into a **pronucleus** of a fertilized ovum (Fig. 28-73; a fertilized ovum contains two pronuclei, one from the sperm and the other from the egg, which eventually fuse to form the nucleus of the one-celled embryo), and implanting it into the uterus of a foster mother. The use of transgenic mice has greatly enhanced our understanding of vertebrate gene expression (Section 33-4C).

Procedures are being developed to generate transgenic farm animals such as cows, pigs, and sheep. Animals may thus be induced to grow larger on lesser amounts of feed and/or to be resistant to particular diseases, although this will require a greater understanding of the genes involved than is presently available. An intriguing application of transgenic farm animals is for them to secrete medicinally useful proteins, such as human growth hormone and blood clotting factors, into their milk. Such a transgenic cow, it is expected, will yield several grams of a foreign protein per liter of milk (tens of kilograms per year), which can thereby be produced far more economically than it can by bacteria. Thus, a small herd of such cows would be able to satisfy the world's need for a particular medicinally useful protein.

Transgenic plants are also starting to become available, promising a significant extension of the "green revolution" that has changed the face of agriculture throughout the world over the past three decades. For example, during sporulation, various strains of the soil microbe *Bacillus thuringiensis* (**Bt**) express proteins that, upon activation, specifically bind to the brush border cells of the intestinal epitheliun in certain insects. There, they form transmembrane pores, thereby lysing the cells and killing the insect through starvation and infection. These so-called **δ-endotoxins** (also known as **crystal proteins** because Bt spores contain them in microcrystalline form) are innocuous to vertebrates and, hence, Bt spores have been used to control such pests as the **gypsy moth.** Unfortunately, Bt's protective effect is short lived. However, recently the gene for a δ-endotoxin has been cloned into maize and shown, in field testing, to confer protection against the **European corn borer** (a commercially significant pest that, for much of its life cycle, lives inside the maize plant, where it is largely inaccessible to chemical insecticides). In a more speculative example, if nitrogen-fixing bacteria (Section 24-6) can be persuaded to associate with agriculturally important plants besides legumes (a complicated process whose requirements are not well understood), the need for nitrogenous fertilizers to grow these plants in high yield will perhaps be entirely eliminated, thereby also reducing a major source of environmental pollution.

Gene therapy, the transfer of new genetic material to the cells of an individual resulting in therapeutic benefit to that individual, has been a clinical reality since 1990 when French Anderson and Michael Blaese successfully employed this technology in a child to alleviate severe combined immunodeficiency disease (SCID) resulting from adenosine deaminase (ADA) deficiency (Section 26-5A). Around 4000 genetic diseases are presently known and are thereby potential targets of gene therapy. Among the gene therapy protocols that are presently under development are those aiming to insert normal LDL genes into the liver cells of individuals suffering from familial hypercholesterolemia (Sections 11-4D and 23-6B) and those designed to treat cancer by genetically modifying tumor cells so as to render them more susceptible to attack by the immune system.

Although several types of gene-transfer protocols are currently under development for use in gene therapy, those which have been most widely employed utilize retroviral vectors. **Retroviruses** are RNA-containing viruses that, upon entering the host cell, use virally encoded reverse transcriptase to transcribe the viral RNA to its complementary DNA, thereby forming an RNA–DNA hybrid helix. The enzyme then uses the newly synthesized DNA as a template to synthesize the complementary DNA while degrading the original RNA (Section 31-4C). The resulting duplex DNA is then integrated into a host chromosome, a characteristic that makes retroviral vectors of great value in gene transfer. The retroviral RNAs that are used in these procedures have been engineered so as to replace the genes encoding essential viral proteins with therapeutic genes. Hence, cells that have been infected by these "viruses" contain the therapeutic genes in their chromosomes but they lack the genetic information to replicate the virus.

G. Social Considerations

In the early 1970s, when strategies for genetic engineering were first being discussed, it was realized that little was known about the safety of the proposed experiments. Certainly it would be foolhardy to attempt experiments such as introducing the gene for **diphtheria toxin** (Section 30-3G) into *E. coli* so as to convert this human symbiont into a deadly pathogen. But what biological hazards would result, for example, from cloning tumor virus genes in *E. coli* (a useful technique for analyzing these viruses)? Consequently, in 1975, molecular biologists declared a voluntary moratorium on molecular cloning experiments until these risks could be assessed. There ensued a spirited debate, at first among molecular biologists and later in the public arena, between two camps: those who thought that the enormous potential benefits of recombinant DNA research warranted its continuation once adequate safety precautions had been instituted, and those who felt that its potential dangers were so great that it should not be pursued under any circumstances.

The former viewpoint eventually prevailed with the promulgation, in 1976, of a set of United States' government regulations for recombinant DNA research. Experiments that are obviously dangerous were forbidden. In other experiments, the escape of laboratory organisms was to be prevented by both physical and biological containment. By biological containment it is meant that vectors will only be cloned in host organisms with biological defects that prevent their survival outside the laboratory. For example, χ1776, the first approved "safe" strain of *E. coli,* has among its several defects the requirement for diaminopimelic acid, an intermediate in lysine biosynthesis (Fig. 24-47), which is neither present in human intestines nor commonly available in the environment.

As experience with recombinant DNA research accumulated, it became evident that the foregoing reservations were largely groundless. No genetically altered organism yet reported has caused an unexpected health hazard. Indeed, recombinant DNA techniques have, in many cases, eliminated the health hazards of studying dangerous pathogens such as the virus causing AIDS. Consequently, since 1979, the regulations governing recombinant DNA research have been gradually relaxed.

There are other social, ethical, and legal considerations that will have to be faced as new genetic engineering techniques become available (Fig. 28-74). Bacterially produced human growth hormone is now routinely prescribed to increase the stature of abnormally short children. However, should athletes be permitted to use this protein, as some reportedly have, to increase their size and strength? Few would dispute the use of gene therapy, if it can be developed, to cure such genetic defects as sickle-cell anemia (Section 9-3B) and Lesch–Nyhan syndrome (Section 26-2D). If, however, it becomes possible to alter complex (i.e., multigene) traits such as athletic ability or intelligence, which changes would be considered desirable, under what circumstances would they be made, and who would decide whether to make them? Should gene therapy be used on individuals with inheirited diseases only to correct defects in their somatic cells or should it also be used to alter genes in their germ cells which could then be transmitted to suc-

FIGURE 28-74. [Drawing by T.A. Bramley, *in* Andersen, K., Shanmugam, K.T., Lim, S.T., Csonka, L.N., Tait, R., Hennecke, H., Scott, D.B., Hom, S.S.M., Haury, J.F., Valentine, A., and Valentine, R.C., *Trends Biochem. Sci.* **5**, 35 (1980). Copyright © Elsevier Biomedical Press, 1980. Used by permission.]

ceeding generations? If it becomes easy to determine an individual's genetic makeup, should this information be used, for example, in evaluating applications for educational and employment opportunities, or in assessing a person's eligibility for health insurance? Under present United States' laws, novel life forms developed in the laboratory and perhaps even newly sequenced human genes of unknown function may be patented. But to what extent will such proprietary rights impede the free exchange of ideas, information, and material that has heretofore permitted the rapid development of recombinant DNA technology?

CHAPTER SUMMARY

Nucleic acids are linear polymers of nucleotides containing either ribose residues in RNA or deoxyribose residues in DNA that are linked by $3' \rightarrow 5'$ phosphodiester bonds. In double helical DNAs and RNAs, the base compositions obey Chargaff's rules: A = T and G = C. RNA, but not DNA, is susceptible to base-catalyzed hydrolysis.

B-DNA consists of a right-handed double helix of antiparallel sugar–phosphate chains with ~10 bp per turn of 34 Å and with the bases all perpendicular to the helix axis. Bases on opposite strands hydrogen bond in a geometrically complementary manner to form A·T and G·C Watson–Crick base pairs. DNA replicates in a semiconservative manner as has been demonstrated by the Meselson–Stahl experiment. At low humidity, B-DNA undergoes a reversible transformation to a wider, flatter right-handed double helix known as A-DNA. Z-DNA, which is formed at high salt concentrations by polynucleotides of alternating purine and pyrimidine base sequences, is a left-handed double helix. Double-helical RNA and RNA·DNA hybrids have A-DNA-like structures.

DNA occurs in nature as molecules of enormous lengths which, because they are also quite stiff, are easily mechanically cleaved by laboratory manipulations.

When heated past its melting temperature, T_m, DNA denatures and undergoes strand separation. This process may be monitored by the hyperchromism of the DNA's UV spectrum. The orientations about the glycosidic bond and the various torsion angles in the sugar–phosphate chain are sterically constrained in nucleic acids. Likewise, only a few of the possible sugar pucker conformations are commonly observed. Watson–Crick base pairing is both geometrically and electronically complementary. Yet, hydrogen bonding interactions do not significantly stabilize nucleic acid structures. Rather, they are largely stabilized by hydrophobic interactions. Nevertheless, the hydrophobic forces in nucleic acids are qualitatively different in character from those that stabilize proteins. Electrostatic interactions between charged phosphate groups are also important structural determinants of nucleic acids.

Nucleic acids are fractionated by many of the techniques that are used to separate proteins. Hydroxyapatite chromatography separates single-stranded from double-stranded DNA. Polyacrylamide or agarose gel electrophoresis separates DNA largely on the basis of size. Very large DNAs can be separated by pulsed-field gel electrophoresis on agarose gels. Specific base sequences may be detected in DNA with the Southern transfer technique and in RNA by the similar Northern transfer technique. DNA may be fractionated according to base composition by CsCl density gradient ultracentrifugation. Different species of RNA are separated by zonal ultracentrifugation through a sucrose gradient.

The linking number of a covalently closed circular DNA is topologically invariant. Consequently, any change in the twist of a circular duplex must be balanced by an equal and opposite change in its writhing number, which indicates its degree of supercoiling. Supercoiling can be induced by intercalation agents. The gel electrophoretic mobility of DNA increases with its degree of superhelicity. Naturally occurring DNAs are all negatively supercoiled and must be so in order to participate in DNA replication, RNA transcription, and genetic recombination. Type I topoisomerases (nicking–closing enzymes) relax negatively supercoiled DNAs, one supertwist at a time, by creating a single-strand break, passing a single-strand loop through the gap, and resealing it. Type II topoisomerases (gyrases) generate negative supertwists at the expense of ATP hydrolysis. They do so, two supertwists at a time, by making a double-strand scisson in the DNA, passing the duplex through the break, and resealing it. The X-ray structures of both *E. coli* enzymes reveal that they each contain a hole large enough to enclose B-DNA, which therefore probably play important roles in the strand passage reaction that these enzymes catalyze.

Nucleic acids may be sequenced by the same basic strategy used to sequence proteins. Defined DNA fragments are generated by Type II restriction endonucleases, which cleave DNA at specific and usually palindromic sequences of four to six bases. Restriction maps provide easily located physical reference points on a DNA molecule. In the chemical cleavage method of DNA sequencing, a defined fragment of DNA is ^{32}P-labeled at one end and subjected to a chemical cleavage process that randomly cleaves it after a particular type of base. The electrophoresis of the four differently cleaved DNA samples in parallel lanes of a sequencing gel resolves fragments that differ in size by one nucleotide. The base sequence of the DNA can be directly read from an autoradiogram of the gel. In the chain-terminator method, the DNA to be sequenced is replicated by DNA polymerase I in the presence of an $[\alpha\text{-}^{32}P]$-labeled deoxynucleoside triphosphate and a small amount of the dideoxy analog of one of the nucleoside triphosphates. This results in a series of ^{32}P-labeled chains that are terminated after the various positions occupied by the corresponding base. An autoradiograph of the sequencing gel containing the four sets of fragments, each terminated after a different type of base, indicates the DNA's base sequence. RNA may be sequenced by determining the sequence of its corresponding cDNA. Automated methods, which are greatly speeding up DNA sequence determinations, are making feasible the sequencing of very large tracts of DNA such as the human genome.

Oligonucleotides are indispensible to recombinant DNA technology; they are used to identify normal and mutated genes and to alter specific genes through site-directed mutagenesis. Oligonucleotides of defined sequence are efficiently synthesized by the phosphoramadite method, a cyclic, nonaqueous, solid phase process that has been automated.

A DNA fragment may be produced in large quantities by inserting it, using recombinant DNA techniques, into a suitable cloning vector. These may be genetically engineered plasmids, bacteriophages, cosmids, or yeast artificial chromosomes (YACs). The DNA to be cloned is usually obtained as a restriction fragment so that it can be specifically ligated into a corresponding restriction cut in the cloning vector. Gene splicing may also occur through the generation of complementary homopolymer tails on the DNA fragment and the cloning vector or through the use of synthetic palindromic linkers containing restriction sequences. Introduction of a recombinant cloning vector into a suitable host organism permits the foreign DNA segment to be produced in nearly unlimited quantities. A particular gene may be isolated through the screening of a genomic library of the organism producing the gene. The polymerase chain reaction (PCR) is a particularly fast and convenient method of identifying and obtaining specific sequences of DNA. Genetic engineering techniques may be used to produce otherwise scarce or specifically altered proteins in large quantities. They are also used to produce transgenic plants and animals and in gene therapy.

REFERENCES

General

Bloomfield, V.A., Crothers, D.M., and Tinoco, I., Jr., *Physical Chemistry of Nucleic Acids,* Harper & Row (1974).

Calladine, C.R. and Drew, H.R., *Understanding DNA,* Academic Press (1992). [The molecule and how it works.]

Cantor, C.R. and Schimmel, P.R., *Biophysical Chemistry,* Chapters 3, 5, 22–24, Freeman (1980).

Saenger, W., *Principles of Nucleic Acid Structure,* Springer–Verlag (1984). [A detailed and authoritative exposition.]

Structures and Stabilities of Nucleic Acids

Dickerson, R.E., DNA structures from A to Z, *Methods Enzymol.* **211**, 67–111 (1992).

Egli, M., Usman, N., Zhang, S., and Rich, A., Crystal structure of an Okazaki fragment at 2.0-Å resolution, *Proc. Natl. Acad. Sci.* **89**, 534–538 (1992). [The structure of an RNA–DNA oligonucleotide in complex with its complementary DNA.]

Fairhead, H., Setlow, B., and Setlow, P., Prevention of DNA damage in spores and in vitro by small, acid-soluble proteins from *Bacillus* species, *J. Bacteriol.* **175**, 1367–1374 (1993).

Jaworski, A., Hsieh, W.-T., Blaho, J.A., Larson, J.E., and Wells, R.D., Left-handed DNA in vivo, *Science* **238**, 773–777 (1988).

Joshua-Tor, L. and Sussman, J.L., The coming of age of DNA crystallography, *Curr. Opin. Struct. Biol.* **3**, 323–335 (1993).

Meselson, M. and Stahl, F.W., The replication of DNA in Escherichia coli, *Proc. Natl. Acad. Sci.* **44**, 671–682 (1958). [The classic paper establishing the semiconservative nature of DNA replication.]

Rich, A., Nordheim, A., and Wang, A.H.-J., The chemistry and biology of left-handed Z-DNA, *Annu. Rev. Biochem.* **53**, 791–846 (1984).

Sundaralingam, M., Stereochemistry of nucleic acids and their constituents. IV. Allowed and preferred conformations of nucleosides, nucleoside mono-, di-, tri-, and tetraphosphates, nucleic acids and polynucleotides, *Biopolymers* **7**, 821–860 (1969).

Voet, D. and Rich, A., The crystal structures of purines, pyrimidines and their intermolecular structures, *Prog. Nucleic Acid Res. Mol. Biol.* **10**, 183–265 (1970).

Watson, J.D. and Crick, F.H.C., Molecular structure of nucleic acids, *Nature* **171**, 737–738 (1953); *and* Genetical implications of the structure of deoxyribonucleic acid, *Nature* **171**, 964–967 (1953). [The seminal papers that are widely held to mark the origin of modern molecular biology.]

Wing, R., Drew, H., Takano, T., Broka, C., Tanaka, S., Itakura, K., and Dickerson, R.E., Crystal structure analysis of a complete turn of B-DNA, *Nature* **287**, 755–758 (1980).

Fractionation of Nucleic Acids

Birren, B. and Lai, E. (Eds.), *Pulsed-Field Gel Electrophoresis, Methods* **1**(2) (1990).

Cantor, C.R., Smith, C.L., and Mathew, M.K., Pulsed-field gel electrophoresis of very large molecules, *Annu. Rev. Biophys. Biophys. Chem.* **17**, 287–304 (1988).

Freifelder, D., *Physical Biochemistry. Applications to Biochemistry and Molecular Biology* (2nd ed.), Freeman (1982).

Rickwood, D. and Hames, B.D. (Eds.), *Gel Electrophoresis of Nucleic Acids. A Practical Approach,* IRL Press (1982).

Schleif, R.F. and Wensink, P.C., *Practical Methods in Molecular Biology,* Chapter 5, Springer–Verlag (1981).

Walker, J.M. (Ed.), *Methods in Molecular Biology,* Vol. 2, *Nucleic Acids,* Humana Press (1984).

Supercoiled DNA

Bates, A.D. and Maxwell, A. *DNA Topology,* IRL Press (1993). [A monograph.]

Bauer, W.R., Crick, F.H.C., and White, J.H., Supercoiled DNA, *Sci. Am.* **243**(1): 118–133 (1980). [The topology of supercoiling.]

Cozarelli, N.R. and Wang, J.C. (Eds.), *DNA Topology and Its Biological Effects,* Cold Spring Harbor Laboratory Press (1990).

Gellert, M., Mechanistic aspects of DNA topoisomerases, *Adv. Protein Chem.* **38**, 69–107 (1986).

Kanaar, R. and Cozarelli, N.R., Roles of supercoiled DNA structure in DNA transactions, *Curr. Opin. Struct. Biol.* **2**, 369–379 (1992).

Lima, C.D., Wang, J.C., and Mondragón, A., Three-dimensional structure of the 67K N-terminal fragment of *E. coli* DNA topoisomerase I, *Nature* **367**, 138–146 (1994).

Reese, R.J. and Maxwell, A., DNA gyrase: Structure and function, *Crit. Rev. Biochem. Mol. Biol.* **26**, 335–375 (1991). [A detailed review.]

Wigley, D.B., Davies, G.J., Dodson, E.J., Maxwell, A., and Dodson, G., Crystal structure of an N-terminal fragment of DNA gyrase B, *Nature* **351**, 624–629 (1991).

Nucleic Acid Sequencing

Cheng, X., Balendiran, K., Schildkraut, I., and Anderson, J.E., Structure of *Pvu*II endonuclease with cognate DNA, *EMBO J.* **13**, 3927–3935 (1994).

Cooperative Human Linkage Center, A comprehensive human linkage map with centimorgan density, *Science* **265**, 2049–2054 (1994); Cohen, D., Chumakov, I., and Weissenbach, J., A first-generation physical map of the human genome, *Nature* **366**, 698–701 (1993); *and* NIH/CEPH Collaborative Mapping Group, A comprehensive genetic linkage map of the human genome, *Science* **258**, 67–86 (1992).

Database issue, *Nucleic Acids Res.* **21**(3) (1993). [Annually updated descriptions of databases containing nucleic acid sequences of various types.]

Gusella, J.F., DNA polymorphism and human disease, *Annu. Rev. Biochem.* **55**, 831–854 (1986).

Hindley, J., DNA sequencing, *in* Work, T.S. and Burdon, R.S. (Eds.), *Laboratory Techniques in Molecular Biology,* Vol. 10, North–Holland (1983).

Hunkapiller, T., Kaiser, R.J., Koop, B.F., and Hood, L., Large-scale and automated DNA determination, *Science* **254**, 59–67 (1991).

Lipschutz, R.J. and Fodor, S.P.A., Advanced DNA sequencing technologies, *Curr. Opin. Struct. Biol.* **4**, 376–380 (1994).

Maxam, A.M. and Gilbert, W., Sequencing end-labeled DNA with base-specific chemical cleavages, *Methods Enzymol.* **65**, 499–560 (1980).

Newman, M., Strzelecka, T., Dorner, L.F., Schildkraut, I., and Aggarwal, A.K., Structure of restriction endonuclease *Bam*HI and its releationship to *Eco*RI, *Nature* **368**, 660–664 (1994).

Nathans, D. and Smith, M.O., Restriction endonucleases in the analysis and restructuring of DNA, *Annu. Rev. Biochem.* **44,** 273–293 (1975).

Olson, M.V., The human genome project, *Proc. Natl. Acad. Sci.* **90,** 4338–4344 (1993).

Prober, J.M., Trainor, G.L., Dam, R.J., Hobbs, F.W., Robertson, C.W., Zagursky, R.J., Cocuzza, A.J., Jensen, M.A., and Baumeister, K., A system for rapid DNA sequencing with fluorescent chain-terminating dideoxynucleotides, *Science* **238,** 336–341 (1987).

Roe, B.A. (Ed.), *DNA Sequencing, Methods* **3**(1) (1991).

Rosenberg, J.M., Structure and function of restriction endonucleases, *Curr. Opin. Struct. Biol.* **1,** 104–113 (1991).

Wells, R.D., Klein, R.D., and Singleton, C.K., Type II restriction enzymes, *in* Boyer, P.D. (Ed.), *The Enzymes* (3rd ed.), Vol. 14, *pp.* 137–156, Academic Press (1981).

White, R. and Lalouel, J.-M., Chromosome mapping with DNA markers, *Sci. Am.* **258**(2): 40–48 (1988). [Describes the use of RFLPs.]

Wilson, R., et al., 2.2 Mb of contiguous nucleotide sequence from chromosome III of *C. elegans, Nature* **368,** 32–38 (1994); Johnston, M. et al., Complete nucleotide sequence of Saccharomyces cerevisiae chromosome VIII, *Science* **265,** 2077–2082 (1994); Dujon, B. et al., Complete DNA sequence of yeast chromosome XI, *Nature* **369,** 371–378 (1994); *and* Oliver, S.G., et al., The complete DNA sequence of yeast chromosome III, *Nature* **357,** 36–46 (1992).

Winkler, F.K., Banner, D.W., Oefner, C., Tsernoglu, D., Brown, R.S., Heathman, S.P., Bryan, R.K., Martin, P.D., Petratos, K., and Wilson, K.S., The crystal structure of EcoRV endonuclease and of its complexes with cognate and non-cognate DNA fragments, *EMBO J.* **12,** 1781–1795 (1993); *and* Winkler, F. K., Structure and function of restriction endonucleases, *Curr. Opin. Struct. Biol.* **2,** 93–99 (1992).

Chemical Synthesis of Oligonucleotides

Caruthers, M.H., Beaton, G., Wu, J.V., and Wiesler, W., Chemical synthesis of deoxynucleotides and deoxynucleotide analogs, *Methods Enzymol.* **211,** 3–20 (1992); *and* Caruthers, M.H., Chemical synthesis of DNA and DNA analogues, *Acc. Chem. Res.* **24,** 278–284 (1991).

Conner, B.J., Reyes, A.A., Morin, C., Itakura, K., Teplitz, R.L., and Wallace, R.B., Detection of sickle cell β^S-globin allele by hybridization with synthetic oligonucleotides, *Proc. Natl. Acad. Sci.* **80,** 278–282 (1983).

Gait, M.J. (Ed.), *Oligonucleotide Synthesis. A Practical Approach,* IRL Press (1984).

Molecular Cloning

Arnheim, N. (Ed.), *Polymerase Chain Reaction, Methods* **2**(1) (1990).

Berger, S.L. and Kimmel, A.R. (Eds.), Guide to Molecular Cloning Techniques, *Methods Enzymol.* **152** (1987). [A "cookbook" describing the basic techniques of molecular biology.]

Burke, D.T., Carle, G.F., and Olso, M.V., Cloning of large segments of exogenous DNA into yeast by means of artificial chromosome vectors, *Science* **236,** 806–812 (1987).

Cano, R.J., Poinar, H.N., Pieniazek, N.J., Acra, A., and Poinar, G.O., Jr., Amplification and sequencing of DNA from a 120–135 million-year-old weevil, *Nature* **363,** 536–538 (1993).

Erlich, H.A., Gelfand, D., and Sninsky, J.J., Recent advances in the polymerase chain reaction, *Science* **252,** 1643–1650 (1991); *and* Erlich, H.A. and Arnheim, N., Genetic analysis using the polymerase chain reaction, *Annu. Rev. Genet.* **26,** 479–506 (1992).

Fersht, A. and Winter, G., Protein engineering, *Trends Biochem. Sci.* **17,** 292–294 (1992).

Gadowski, P.J. and Henner, D. (Eds.), *Protein Overproduction in Heterologous Systems, Methods* **4**(2) (1992).

Glover, D.M. (Eds.), *DNA Cloning. A Practical Approach,* Vols. 1 and 2, IRL Press (1985).

Goeddel, D.V. (Ed.), *Gene Expression Technology, Methods Enzymol.* **185** (1990).

Morgan, R.A. and Anderson, W.F., Human gene therapy, *Annu. Rev. Biochem.* **62,** 191–217 (1993).

Mullis, K.B., The unusual origin of the polymerase chain reaction. *Sci. Am.* **262**(4): 56–65 (1990).

Rees, A.R., Sternberg, M.J.E., and Wetzel, R. (Eds.), *Protein Engineering. A Practical Approach,* IRL Press (1992).

Saiki, R.K., Gelfand, D.H., Stoffel, S., Scharf, S.J., Higuchi, R., Horn, G.T., Mullis, K.B., and and Erlich, H.A., Primer-directed enzymatic amplification of DNA with a thermostable DNA polymerase, *Science* **239,** 487–494 (1988).

Sambrook, J., Fritsch, E.F., and Maniatis, T., *Molecular Cloning* (2nd ed.), Cold Spring Harbor Laboratory (1989). [A three-volume "bible" of laboratory protocols with accompanying background explanations.]

Watson, J.D., Gilman, M., Witkowski, J., and Zoller, M., *Recombinant DNA* (2nd ed.), Freeman (1992). [A detailed exposition of the methods, findings, and results of recombinant DNA technology and research.]

Wu, R., Grossman, L., and Moldave, K. (Eds.), *Recombinant DNA,* Parts A–I, *Methods Enzymol.* **68, 100, 101, 153–155,** and **216–218** (1979, 1983, 1987, 1992, and 1993).

Historical Aspects

Crick, F., *What Mad Pursuit,* Basic Books (1988). [A scientific autobiography.]

Judson, H.F., *The Eighth Day of Creation,* Part I, Simon & Schuster (1979). [A fascinating narration of the discovery of the DNA double helix.]

Lebowitz, J., Through the looking glass: The discovery of supercoiled DNA, *Trends Biochem. Sci.* **15,** 202–207 (1990). [An informative eyewitness account of how DNA supercoiling was discovered.]

Olby, R., *The Path to the Double Helix,* Macmillan (1974).

Portugal, F.H. and Cohen, J.S., *A Century of DNA,* MIT Press (1977).

Sanger, F., Sequences, sequences, and sequences, *Annu. Rev. Biochem.* **57,** 1–28 (1988). [A scientific memoir.]

Sayre, A., *Rosalind Franklin and DNA,* Norton (1975). [A biographical work which argues that Rosalind Franklin, who died in 1958, deserves far more credit than is usually accorded her for the discovery of the structure of DNA.]

Schlenk, F., Early nucleic acid chemistry, *Trends Biochem. Sci.* **13,** 67–68 (1988).

Watson, J.D., *The Double Helix,* Atheneum (1968). [A provocative autobiographical account of the discovery of the DNA structure.]

Watson, J.D. and Tooze, J., *The DNA Story,* Freeman (1981). [A scrapbooklike account of the recombinant DNA debate of the 1970s.]

PROBLEMS

1. Non-Watson–Crick base pairs are of biological importance. For example: (a) **Hypoxanthine** (6-oxopurine) is often one of the bases of the anticodon of tRNA (the three consecutive nucleotides that base pair with mRNA). With what base on mRNA is hypoxanthine likely to pair? Draw the structure of this base pair. (b) tRNA often makes a G·U base pair with mRNA. Draw a plausible structure for such a base pair. (c) Many species of tRNA contain a hydrogen bonded U·A·U assembly. Draw two plausible structures for this assembly in which each U forms at least two hydrogen bonds with the A. (d) Mutations may arise during DNA replication when mispairing occurs as a result of the transient formation of a rare tautomeric form of a base. Draw the structure of a base pair with proper Watson–Crick geometry that contains a rare tautomeric form of adenine. What base sequence change would be caused by such mispairing?

2. What is the molecular mass and contour length of a section of B-DNA that specifies a 40-kD protein? Each amino acid is specified by three contiguous bases on a single strand of DNA (Section 30-1).

*3. The antiparallel orientation of complementary strands in duplex DNA was elegantly demonstrated in 1960 by Arthur Kornberg by **nearest-neighbor analysis.** In this technique, DNA is synthesized by DNA polymerase I from one [α-^{32}P]-labeled and three unlabeled deoxynucleoside triphosphates. The resulting product is hydrolyzed by a DNase that cleaves phosphodiester bonds on the 3′ sides of all deoxynucleotides.

$$ppp^{*}A + pppC + pppG + pppT$$

$$PP_i \longleftarrow \Big\downarrow \text{DNA polymerase}$$

$$\cdots pCpTp^{*}ApCpCp^{*}ApGp^{*}Ap^{*}ApTp\cdots$$

$$H_2O \longleftarrow \Big\downarrow \text{DNase I}$$

$$\cdots + Cp + Tp^{*} + Ap + Cp + Cp^{*} + Ap + Gp^{*} + Ap^{*} + Ap + Tp + \cdots$$

In this example, the relative frequencies of occurrence of ApA, CpA, GpA, and TpA in the DNA can be determined by measuring the relative amounts of Ap*, Cp*, Gp*, and Tp*, respectively, in the product. The relative frequencies with which the other 12 dinucleotides occur may likewise be determined by labeling, in turn, the other 3 nucleoside triphosphates in the above reactions. There are equivalencies between the amounts of certain pairs of dinucleotides. However, the identities of these equivalencies depend on whether the DNA consists of parallel or antiparallel strands. What are these equivalences in both cases?

4. What would be the effect of the following treatments on the melting curve of an aqueous solution of duplex DNA? Explain. (a) Decreasing the ionic strength of the solution. (b) Squirting the DNA solution, at high pressure, through a very narrow orifice. (c) Bringing the solution to $0.1M$ adenine. (d) Heating the solution to 25°C above the DNA's melting point and then rapidly cooling it to 35°C below the DNA's melting point. (e) Adding a small amount of ethanol to the DNA solution.

5. What is the mechanism of alkaline denaturation of DNA?

*6. At Na$^+$ concentrations >5M, the T_m of DNA decreases with increasing [Na$^+$]. Explain this behavior. (*Hint:* Consider the solvation requirements of Na$^+$.)

*7. Why are the most commonly observed conformations of the ribose ring those in which either atom C2′ or atom C3′ is out of the plane of the other four ring atoms? (*Hint:* In puckering a planar ring such that one atom is out of the plane of the other four, the substituents about the bond opposite the out-of-plane atom remain eclipsed. This is best observed with a ball-and-stick model.)

8. Polyoma virus DNA can be separated by sedimentation at neutral pH into three components that have sedimentation coefficients of 20, 16, and 14.5S and which are known as Types I, II, and III DNAs, respectively. These DNAs all have identical base compositions and molecular masses. In 0.15M NaCl, both Types II and III DNA have melting curves of normal cooperativity and a T_m of 88°C. Type I DNA, however, exhibits a very broad melting curve and a T_m of 107°C. At pH 13, Types I and III DNAs have sedimentation coefficients of 53 and 16S, respectively, and Type II separates into two components with sedimentation coefficients of 16S and 18S. How do Types I, II, and III DNAs differ from one another? Explain their different physical properties.

9. A closed circular duplex DNA has a 100-bp segment of alternating C and G residues. Upon transfer to a solution containing a high salt concentration, this segment undergoes a transition from the B conformation to the Z conformation. What is the accompanying change in its linking number, writhing number, and twist?

We begin this chapter by discussing experiments that led to the elucidation of mRNA's central role in protein synthesis. We then study the mechanism of transcription and its control in prokaryotes. Finally, in the last section, we consider posttranscriptional processing of RNA in both prokaryotes and eukaryotes. Translation is the subject of Chapter 30.

1. THE ROLE OF RNA IN PROTEIN SYNTHESIS

Proteins are specified by mRNA and synthesized on ribosomes. This idea arose from the study of **enzyme induction,** a phenomenon in which bacteria vary the synthesis rates of specific enzymes in response to environmental changes. We shall see below that *enzyme induction occurs as a consequence of the regulation of mRNA synthesis by proteins that specifically bind to the mRNA's DNA templates.*

A. Enzyme Induction   1000X

E. coli can synthesize an estimated 3000 different polypeptides (Section 27-1D). There is, however, enormous variation in the amounts of these different polypeptides that are produced. For instance, the various ribosomal proteins may each be present in over 10,000 copies per cell, whereas certain regulatory proteins (see below) normally occur in < 10 copies per cell. Many enzymes, particularly those involved in basic cellular "housekeeping" functions, are synthesized at a more or less constant rate; they are called **constitutive enzymes.** Other enzymes, termed **adaptive** or **inducible enzymes,** are synthesized at rates that vary with the cell's circumstances.

Lactose-Metabolizing Enzymes Are Inducible

Bacteria, as has been recognized since 1900, adapt to their environments by producing enzymes that metabolize certain nutrients, for example, lactose, only when those substances are available. *E. coli* grown in the absence of lactose are initially unable to metabolize this disaccharide. To do so they require the presence of two proteins: **β-galactosidase,** which catalyzes the hydrolysis of lactose to its component monosaccharides;

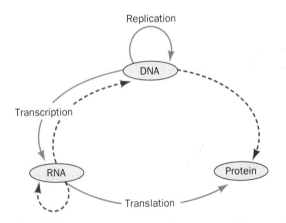

and **galactoside permease** (also known as **lactose permease;** Section 18-4B), which transports lactose into the cell. *E. coli* grown in the absence of lactose contain only a few molecules of these proteins. Yet, a few minutes after lactose is introduced into their medium, *E. coli* increase the rate at which they synthesize these proteins by ~1000-fold and maintain this pace until lactose is no longer available. The synthesis rate then returns to its original miniscule level (Fig. 29-2). *This ability to produce a series of proteins only when the substances they metabolize are present permits bacteria to adapt to their environment without the debilitating need to continuously synthesize large quantities of otherwise unnecessary substances.*

Lactose or one of its metabolic products must somehow trigger the synthesis of the above proteins. Such a substance is known as an **inducer.** The physiological inducer of the lactose system, the lactose isomer **1,6-allolactose,**

1,6-Allolactose

arise's from lactose's occasional transglycosylation by β-ga-

FIGURE 29-1. The central dogma of molecular biology. Solid arrows indicate the types of genetic information transfers that occur in all cells. Special transfers are indicated by the dashed arrows: RNA-directed RNA polymerase occurs both in certain RNA viruses and in some plants (where it is of unknown function); RNA-directed DNA polymerase (reverse transcriptase) occurs in other RNA viruses; and DNA directly specifying a protein is unknown but does not seem beyond the realm of possibility. However, the missing arrows are information transfers the central dogma postulates never occur: protein specifying either DNA, RNA, or protein. In other words, *proteins can only be recipients of genetic information.* [After Crick, F., *Nature* **227,** 562 (1970).]

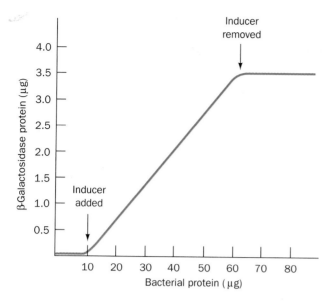

FIGURE 29-2. The induction kinetics of β-galactosidase in *E. coli.* [After Cohn, M., *Bacteriol. Rev.* **21,** 156 (1957).]

lactosidase. Most studies of the lactose system use **isopropylthiogalactoside (IPTG),**

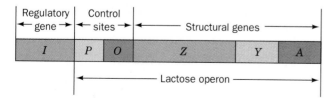

a potent inducer that structurally resembles allolactose but that is not degraded by β-galactosidase.

Lactose system inducers also stimulate the synthesis of **thiogalactoside transacetylase,** an enzyme that, *in vitro,* transfers an acetyl group from acetyl-CoA to the C6—OH group of a β-thiogalactoside such as IPTG. Since lactose fermentation proceeds normally in the absence of thiogalactoside transacetylase, however, this enzyme's physiological role is unknown.

lac System Genes Form an Operon

The genes specifying wild-type β-galactosidase, galactoside permease, and thiogalactoside transacetylase are desig-

nated Z^+, Y^+, and A^+, respectively. Genetic mapping of the defective mutants Z^-, Y^-, and A^- indicated that these *lac* **structural genes** (genes that specify polypeptides) are contiguously arranged on the *E. coli* chromosome (Fig. 29-3; genetic mapping is reviewed in Section 27-1). *These genes, together with the control elements P and O, form a genetic unit called an* **operon,** *specifically the* **lac** *operon.* The nature of the control elements is discussed below. The role of operons in prokaryotic gene expression is examined in Section 29-3.

lac Repressor Inhibits the Synthesis of *lac* Operon Proteins

An important clue as to how *E. coli* synthesizes protein was provided by a mutation that causes the proteins of the *lac* operon to be synthesized in large amounts in the absence of inducer. This so-called **constitutive mutation** occurs in a gene, designated *I,* that is distinct from although closely linked to the genes specifying the *lac* enzymes (Fig. 29-3). What is the nature of the *I* gene product? This riddle was solved through an ingeneous experiment performed by Arthur Pardee, Francois Jacob, and Jacques Monod. Hfr bacteria of genotype I^+Z^+ were mated to an F^- strain of genotype I^-Z^- in the absence of inducer while the β-galactosidase activity of the culture was monitored (Fig. 29-4; bacterial mating is described in Section 27-1D). At first, as

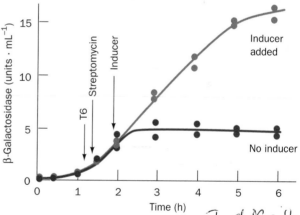

FIGURE 29-4. The demonstration of the existence of the *lac* repressor through the appearance of β-galactosidase in the transient merozygotes (partial diploids) formed by mating I^+Z^+ Hfr donors with $I^-Z^-F^-$ recipients. The F^- strain was also resistant to both **bacteriophage T6** and **streptomycin,** whereas the Hfr strain was sensitive to these agents. Both types of cells were grown and mated in the absence of inducer. After sufficient time had passed for the transfer of the *lac* genes, the Hfr cells were selectively killed by the addition of T6 phage and streptomycin. In the absence of inducer (*lower curve*), β-galactosidase synthesis commenced at around the time that the *lac* genes had entered the F^- cells but stopped after ~1 h. If inducer was added shortly after the Hfr donors had been killed (*upper curve*), enzyme synthesis continued unabated. This demonstrates that the cessation of β-galactosidase synthesis in uninduced cells is not due to the intrinsic loss of the ability to synthesize this enzyme but to the production of a repressor specified by the I^+ gene. [After Pardee, A.B., Jacob, F., and Monod, J., *J. Mol. Biol.* **1,** 173 (1959).]

FIGURE 29-3. A genetic map of the *E. coli lac* operon, that is, the genes encoding the proteins mediating lactose metabolism and the genetic sites that control their expression. The *Z, Y,* and *A* genes, respectively, specify β-galactosidase, galactoside permease, and thiogalactoside transacetylase.

expected, there was no β-galactosidase activity because the Hfr donors lacked inducer and the F⁻ recipients were unable to produce active enzyme (only DNA passes through the cytoplasmic bridge connecting mating bacteria). About 1 h after conjugation began, however, when the I^+Z^+ genes had just entered the F⁻ cells, β-galactosidase synthesis began and only ceased after about another hour. The explanation for these observations is that the donated Z^+ gene, upon entering the cytoplasm of the I^- cell, directs the synthesis of β-galactosidase in a constitutive manner. Only after the donated I^+ gene has had sufficient time to be expressed is it able to repress β-galactosidase synthesis. *The I^+ gene must therefore give rise to a diffusible product, the **lac repressor,** which inhibits the synthesis of β-galactosidase (and the other lac proteins).* Inducers such as IPTG temporarily inactivate *lac* repressor, whereas I^- cells constitutively synthesize *lac* enzymes because they lack a functional repressor. *Lac* repressor, as we shall see in Section 29-3B, is a protein.

[handwritten: diffusible inhibitor]
[handwritten: Inducers inactivate it]

B. *Messenger RNA*

The nature of the *lac* repressor's target molecule was deduced in 1961 through a penetrating genetic analysis by Jacob and Monod. A second type of constitutive mutation in the lactose system, designated O^c (for **operator constitutive**), which complementation analysis has indicated to be independent of the I gene, maps between the I and Z genes (Fig. 29-3). In the partially diploid F' strain O^cZ^-/F O^+Z^+, β-galactosidase activity is inducible by IPTG whereas the strain O^cZ^+/F O^+Z^- constitutively synthesizes this enzyme (in F' bacteria, the F factor plasmid contains a segment of the bacterial chromosome, in this case a portion of the *lac* operon; Section 27-1D). *An O^+ gene can therefore only control the expression of a Z gene on the same chromosome.* The same is true with the Y^+ and A^+ genes.

Jacob and Monod's observations led them to conclude the proteins are synthesized in a two-stage process:

1. The structural genes on DNA are transcribed onto complementary strands of **messenger RNA (mRNA).**

2. The mRNAs transiently associate with ribosomes, which they direct in polypeptide synthesis.

This hypothesis explains the behavior of the *lac* system *(Fig. 29-5). In the absence of inducer, the lac repressor specifically binds to the O gene (the **operator**) so as to physically block the enzymatic transcription of mRNA. Upon binding inducer, the repressor dissociates from the operator, thereby permitting the transcription and subsequent translation of the lac enzymes.* The operator–repressor–inducer system thereby acts as a molecular switch so that the *lac* operator can only control the expression of *lac* enzymes on the same chromosome. The O^c mutants constitutively synthesize *lac* enzymes because they are unable to bind repressor. The **coordinate** (simultaneous) expression of all three *lac* en-

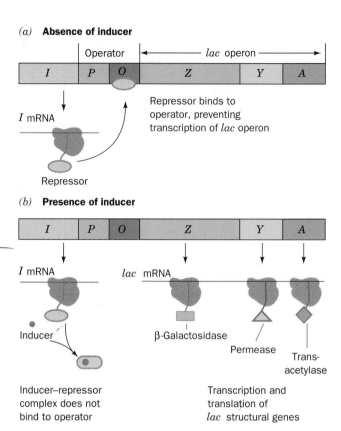

FIGURE 29-5. The expression of the *lac* operon. (*a*) In the absence of inducer, the repressor, the product of the I gene, binds to the operator, thereby preventing transcription of the *lac* operon. (*b*) Upon binding inducer, the repressor dissociates from the operator, which permits the transcription and subsequent translation of the *lac* structural genes to proceed.

zymes under the control of a single operator site arises, as Jacob and Monod theorized, from the transcription of the *lac* operon as a single **polycistronic mRNA** which directs the ribosomal synthesis of each of these proteins. This transcriptional control mechanism is further discussed in Section 29-3. [DNA sequences, which are on the same DNA molecule, are said to be in cis (Latin: on this side), whereas those on different DNA molecules are said to be in trans (Latin: across). Control sequences such as the O gene, which are only active on the same DNA molecule as the genes they control, are called **cis-acting elements.** Those such as *lacI*, which specify the synthesis of diffusible products and can therefore be located on a different DNA molecule from the genes they control, are said to direct the synthesis of **trans-acting factors.**]

mRNAs Have Their Predicted Properties

The kinetics of enzyme induction, as indicated, for example, in Figs. 29-2 and 29-4, requires that the postulated mRNA be both rapidly synthesized and rapidly degraded. An RNA with such quick turnover had, in fact, been observed in T2-infected *E. coli*. Moreover, the base composition of this RNA fraction resembles that of the viral DNA

[handwritten at bottom: Coordinate control via one-site (cis)]

N^{15}, C^{13} heavy ribosomes

rather than that of the bacterial RNA. Ribosomal RNA, which comprises up to 90% of a cell's RNA, turns over much more slowly than mRNA. Ribosomes are therefore not permanently committed to the synthesis of a particular protein (a once popular hypothesis). Rather, *ribosomes are nonspecific protein synthesizers that produce the polypeptide specified by the mRNA with which they are transiently associated.* A bacterium can therefore respond within a few minutes to changes in its environment.

Evidence favoring the Jacob and Monod model rapidly accumulated. Sydney Brenner, Jacob, and Matthew Meselson carried out experiments designed to characterize the RNA that *E. coli* synthesized after T4 phage infection. *E. coli* were grown in a medium containing ^{15}N and ^{13}C so as to label all cell constituents with these heavy isotopes. The cells were then infected with T4 phages and immediately transferred to an unlabeled medium (which contained only the light isotopes ^{14}N and ^{12}C) so that cell components synthesized before and after phage infection could be separated by equilibrium density gradient ultracentrifugation in CsCl solution. No "light" ribosomes were observed, which indicates, in agreement with the above-mentioned T2 phage results, that no new ribosomes are synthesized after phage infection.

^{32}P RNA
^{35}S prot.
on heavy ribos

The growth medium also contained either ^{32}P or ^{35}S so as to radioactively label the newly synthesized and presumably phage-specific RNA and protein, respectively. Much of the ^{32}P-labeled RNA was associated, as was postulated for mRNA, with the preexisting "heavy" ribosomes (Fig. 29-6). Likewise, the ^{35}S-labeled proteins were transiently associated with, and therefore synthesized by, these ribosomes.

hbdztn

Sol Spiegelman developed the RNA–DNA hybridization technique (Section 28-3A) in 1961 to characterize the RNA synthesized by T2-infected *E. coli*. He found that this phage-derived RNA hybridizes with T2 DNA (Fig. 29-7) but neither does so with DNAs from unrelated phage nor with the DNA from uninfected *E. coli*. This RNA must therefore be complementary to T2 DNA in agreement with Jacob and Monod's prediction; that is, the phage-specific RNA is a messenger RNA. Hybridization studies have likewise shown that mRNAs from uninfected *E. coli* are complementary to portions of *E. coli* DNA. In fact, other RNAs, such as transfer RNA and ribosomal RNA, have corresponding complementary sequences on DNA from the same organism. Thus, *all cellular RNAs are transcribed from DNA templates.*

2. RNA POLYMERASE

RNA polymerase, the enzyme responsible for the DNA-directed synthesis of RNA, was discovered independently in 1960 by Samuel Weiss and Jerard Hurwitz. *The enzyme couples together the ribonucleoside triphosphates ATP,*

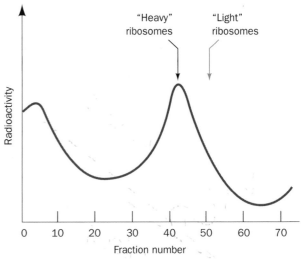

FIGURE 29-6. The distribution, in a CsCl density gradient, of ^{32}P-labeled RNA that had been synthesized by *E. coli* after T4 phage infection. Free RNA, being relatively dense, bands at the bottom of the centrifugation cell (*left*). Much of the RNA, however, is associated with the ^{15}N- and ^{13}C-labeled "heavy" ribosomes that had been synthesized before the phage infection. The predicted position of unlabeled "light" ribosomes, which are not synthesized by phage-infected cells, is also indicated. [After Brenner, S., Jacob, F., and Meselson, M., *Nature* **190,** 579 (1961).]

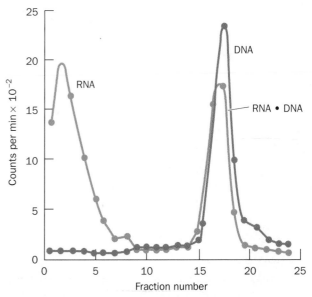

FIGURE 29-7. The hybridization of ^{32}P-labeled RNA produced by T2-infected *E. coli* with ^{3}H-labeled T2 DNA. Upon radioactive decay, ^{32}P and ^{3}H emit β particles with characteristically different energies so that these isotopes can be independently detected. Although free RNA (*left*) in a CsCl density gradient is denser than DNA, much of the RNA bands with the DNA (*right*). This indicates that the two polynucleotides have hybridized and are therefore complementary in sequence. [After Hall, B.D. and Spiegelman, S., *Proc. Natl. Acad. Sci.* **47,** 141 (1961).]

CTP, GTP, and UTP, on DNA templates in a reaction that is driven by the release and subsequent hydrolysis of PP$_i$:

$$(RNA)_{n \text{ residues}} + \underset{\substack{\text{Nucleoside} \\ \text{triphosphate}}}{\text{NTP}} \rightleftharpoons (RNA)_{n+1 \text{ residues}} + PP_i$$

All cells contain RNA polymerase. In bacteria, one species of this enzyme synthesizes all of the cell's RNA except the short RNA primers employed in DNA replication (Section 31-1D). Various bacteriophages generate RNA polymerases that synthesize only phage-specific RNAs. Eukaryotic cells contain four or five RNA polymerases that each synthesize a different class of RNA. In this section we first concentrate on the properties of the *E. coli* enzyme because it is the best characterized RNA polymerase; other bacterial RNA polymerases have similar properties. We then consider the eukaryotic enzymes.

A. Enzyme Structure

E. coli RNA polymerase's so-called **holoenzyme** is an ~449-kD protein with subunit composition $\alpha_2\beta\beta'\sigma$. Once RNA synthesis has been initiated, however, the σ subunit (also called **σ factor** or σ^{70} since its molecular mass is 70 kD) dissociates from the **core enzyme**, $\alpha_2\beta\beta'$, which carries out the actual polymerization process (see below). The β' subunit contains two Zn^{2+} ions, which are thought to participate in the enzyme's catalytic function. The active enzyme also requires the presence of Mg^{2+}.

Electron micrographs (Fig. 29-8) clearly indicate that RNA polymerase, which has a characteristic large size,

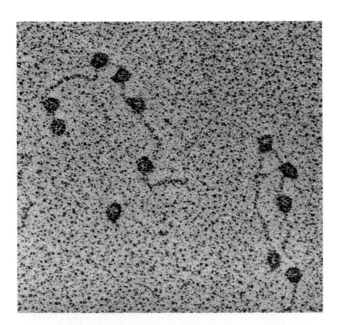

FIGURE 29-8. An electron micrograph of *E. coli* RNA polymerase holoenzyme, one of the largest known soluble enzymes, attached to various promoter sites on bacteriophage T7 DNA. [From Williams, R. C., *Proc. Natl. Acad. Sci.* **74**, 2313 (1977).]

binds to DNA as a protomer. This large size is presumably a consequence of the holoenzyme's several complex functions including (1) template binding, (2) RNA chain initiation, (3) chain elongation, and (4) chain termination. We discuss these various functions below.

RNA Polymerases Contain a Large DNA-Binding Channel Similar to Those in DNA Polymerases

Although the *E. coli* enzyme has not been crystallized in a manner suitable for X-ray analysis, its low-resolution structure (Fig. 29-9a) was determined by Roger Kornberg via electron crystallography (Section 11-3A). The enzyme's most striking features are a thumblike projection which flanks a cylindrical channel that is ~25 Å in diameter and 55 Å in length and is therefore of appropriate dimensions to bind ~16 bp of B-DNA.

The RNA polymerase encoded by **bacteriophage T7** is an 882-residue (99-kD) monomeric enzyme and hence is considerably smaller than that from *E. coli*. Nevertheless, the two enzymes catalyze a remarkably similar series of reaction steps (see below). The X-ray structure of **T7 RNA polymerase** (Fig. 29-9b), determined by Bi-Cheng Wang, reveals a channel that closely resembles that of the *E. coli* holoenzyme. Moreover, the location of several catalytically critical residues along one wall of the T7 RNA polymerase channel clearly indicates that DNA in fact binds in this channel. *E. coli* DNA polymerase I Klenow fragment (Section 28-6C) and reverse transcriptase (Section 31-8F) from **HIV-1** (an AIDS virus), which catalyze reactions similar to those of RNA polymerase, both have a structurally homologous channel (Sections 31-2A and 4A).

B. Template Binding *specific sites*
polar

RNA synthesis is normally initiated only at specific sites on the DNA template. This was first demonstrated through hybridization studies of **bacteriophage ϕX174** DNA with the RNA produced by ϕX174-infected *E. coli*. Bacteriophage ϕX174 carries a single strand of DNA known as the "plus" strand. Upon its injection into *E. coli*, the plus strand directs the synthesis of the complementary "minus" strand with which it combines to form a circular duplex DNA known as the **replicative form** (Section 31-3B). The RNA produced by ϕX174-infected *E. coli* does not hybridize with DNA from intact phages but does so with the replicative form. Thus only the minus strand of ϕX174 DNA, the so-called **antisense strand,** is transcribed, that is, acts as a template; the plus strand, the **sense strand** (so called because it has the same sequence as the transcribed RNA), does not do so. Similar studies indicate that in larger phages, such as T4 and λ, the two viral DNA strands are the antisense (template) strands for different sets of genes. The same appears to be true of cellular organisms.

Holoenzyme Specifically Binds to Promoters

*RNA polymerase binds to its initiation sites through base sequences known as **promoters** that are recognized by the*

(b)

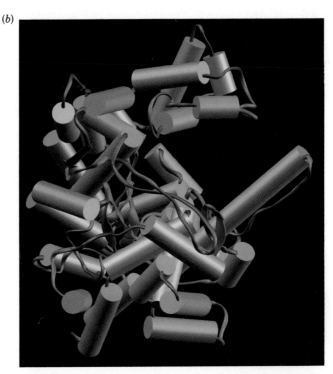

(a)

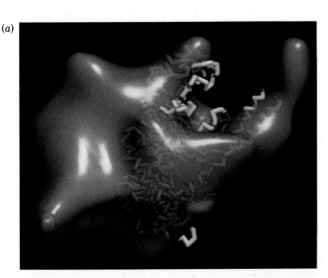

FIGURE 29-9. (*a*) The structure of *E. coli* RNA polymerase at ~27-Å resolution as determined by electron crystallography (*blue*). The irregularly shaped enzyme is ~100 × 100 × 160 Å in size and is viewed along its ~25-Å-wide and 55-Å-long cylindrical channel. Note the thumblike projection flanking this putative DNA-binding channel. The similarly oriented C_α backbone of DNA polymerase I Klenow fragment (Sections 28-6C and 31-2A) is superimposed in yellow with the C_α atoms on either end of an unobserved gap in this structure in red

(under the "thumb"). [Courtesy of Roger Kornberg, Stanford University.] (*b*) The X-ray structure of T7 RNA polymerase oriented so as to match the low resolution image of *E. coli* RNA polymerase shown in Part *a*. Helices are yellow cylinders, β-sheets are green arrows, and other segments of the protein are blue. Portions of the monomeric protein are not visible in the electron density map and hence the polypeptide chain appears to be discontinuous. [Based on an X-ray structure by Rui Sousa and Bi-Cheng Wang, University of Pittsburgh.]

corresponding σ factor. The existence of promoters was first recognized through mutations that enhance or diminish the transcription rates of certain genes, including those of the *lac* operon. *Genetic mapping of such mutations indicated that the promoter consists of an ~40-bp sequence that is located on the 5′ side of the transcription start site.* [By convention, the sequence of template DNA is represented by its sense (nontemplate) strand so that it will have the same directionality as the transcribed RNA. A base pair in a promoter region is assigned a negative or positive number that indicates its position, upstream or downstream in the direction of RNA polymerase travel, from the first nucleotide that is transcribed to RNA; this start site is +1 and there is no 0.] RNA, as we shall see, is synthesized in the 5′ → 3′ direction (Section 29-2D). Consequently, the promoter lies on the "upstream" side of the RNA's starting nucleotide. Sequencing studies indicate that the *lac* promoter *(lacP)* overlaps the *lac* operator (Fig. 29-3).

The holoenzyme forms tight complexes with promoters (dissociation constant $K \approx 10^{-14}M$) and thereby protects the bound DNA segments from digestion by DNase I. The region from about −20 to +20 is protected against exhaus-

tive DNase I degradation. The region extending upstream to about −60 is also protected but to a lesser extent, presumably because it binds holoenzyme less tightly.

Sequence determinations of the protected regions from numerous *E. coli* and phage genes have revealed the "consensus" sequence of *E. coli* promoters (Fig. 29-10). *Their most conserved sequence is a hexamer centered at about the −10 position* (sometimes called the **Pribnow box** after David Pribnow, who pointed out its existence in 1975). It has a consensus sequence of TATAAT in which the leading TA and final T are highly conserved. *Upstream sequences around position −35 also have a region of sequence similarity,* TTGACA, which is most evident in efficient promoters. The initiating (+1) nucleotide, which is nearly always A or G, is centered in a poorly conserved CAT or CGT sequence. Most promoter sequences vary considerably from the consensus sequence (Fig. 29-10). Nevertheless, a mutation in one of the partially conserved regions can greatly increase or decrease a promoter's initiation efficiency. *The rates at which genes are transcribed, which span a range of at least 1000, vary directly with the rate that their promoters form stable initiation complexes with the holoenzyme.*

Operon	−35 region	−10 region (Pribnow box)	Initiation site (+1)
lac	ACCCCAGGCTTTACACTTTATGCTTCCGGCTCG	TATGTTGTGTGG	AATTGTGAGCGG
lacI	CCATCGAATGGCGCAAAACCTTTCGCGGTATGG	CATGATAGCGCCC	GGAAGAGAGTC
galP2	ATTTATTCCATGTCACACTTTTCGCATCTTTGT	TATGCTATGG	TTATTTCATACCAT
araBAD	GGATCCTACCTGACGCTTTTTATCGCAACTCTC	TACTGTTTCTCCATACCGTTTT
araC	GCCGTGATTATAGACACTTTTGTTACGCGTTTT	TGTCATGGCGATTGGGTCCCGCTTTG
trp	AAATGAGCTGTTGACAATTAATCATCGAACTAG	TTAACTAGTACGCAAGTTCACGTA
bioA	TTCCAAAACGTGTTTTTTGTTGTTAATTCGGTG	TAGACTTGTAAACCTAAATCTTTT
bioB	CATAATCGACTTGTAAACCAAATTGAAAAGATTT	AGGTTTACAAGTCTACACCGAAT
*t*RNA^Tyr	CAACGTAACACTTTACAGCGGCGCGTCATTTGA	TATGATGCGCCCCGCTTCCCGATA
rrnD1	CAAAAAAATACTTGTGCAAAAAATTGGGATCCC	TATAATGCGCCTCCGTTGAGACGA
rrnE1	CAATTTTTCTATTGCGGCCTGCGGAGAACTCCC	TATAATGCGCCTCCATCGACACGG
rrnA1	AAAATAAATGCTTGACTCTGTAGCGGGAAGGCG	TATTATGCACACCCCGCGCCGCTG

	−35 region						−10 region						Initiation site
Consensus sequence:	T	T	G	A	C	A	... 16–19 bp ...	T	A	T	A	A	T	... 5–8 bp ...	A 51 / C 55 / G 42 / T 48
	69	79	61	56	54	54		77	76	60	61	56	82	

FIGURE 29-10. The sense (noncoding) strand sequences of selected *E. coli* promoters. A 6-bp region centered around the −10 position (*red shading*) and a 6-bp sequence around the −35 region (*blue shading*) are both conserved. The transcription initiation sites (+1), which in most promoters occurs at a single purine nucleotide, are shaded in green. The bottom row shows the consensus sequence of 298 *E. coli* promoters with the number below each base indicating its percentage occurrence. [After Rosenberg, M. and Court, D., *Annu. Rev. Genet.* **13**, 321–323 (1979). Consensus sequence from Lisser, S. and Margalit, H., *Nucleic Acids Res.* **21**, 1512 (1993).]

Initiation Requires the Formation of an Open Complex

The promoter regions in contact with the holoenzyme have been identified by determining where the enzyme alters the susceptibility of the DNA to alkylation by agents such as dimethyl sulfate (DMS), a procedure named **footprinting** (Section 33-3B). These experiments demonstrated that the holoenzyme contacts the promoter only around its −10 and −35 regions. These protected sites are both on the same side of the B-DNA double helix as the initiation site, which suggests that RNA polymerase binds to only one face of the promoter.

DMS, in addition to methylating G residues at N7 and A residues at N3 (Section 28-6B), methylates N1 of A and N3 of C. Since these latter positions participate in base pairing interactions, however, they can only react with DMS in single-stranded DNA. This differential methylation of single- and double-stranded DNAs provides a sensitive test for DNA strand separation or "melting." Footprinting studies indicate that the binding of holoenzyme "melts out" the promoter in a region of at least 11 bp extending from the middle of the −10 region to just past the initiation site (−9 to +2). The need to form this "open complex" explains why promoter efficiency tends to decrease with the number of G·C base pairs in the −10 region; this presumably increases the difficulty in opening the double helix as is required for chain initiation (G·C pairs, it will be recalled, are more stable than A·T pairs).

Core enzyme, which does not specifically bind promoter, tightly binds duplex DNA (the complex's dissociation constant is $K \approx 5 \times 10^{-12}M$ and its half-life is ~60 min). Holoenzyme, in contrast, binds to nonpromoter DNA comparatively loosely ($K \approx 10^{-7}M$ and a half-life > 1 s). Evidently, the σ subunit allows holoenzyme to move rapidly along a DNA strand in search of the σ subunit's corresponding promoter. Once transcription has been initiated and the σ subunit jettisoned, the tight binding of core enzyme to DNA apparently stabilizes the ternary enzyme-DNA-RNA complex.

C. Chain Initiation

The 5′-terminal base of prokaryotic RNAs is almost always a purine with A occurring more often than G. The initiating reaction of transcription is the coupling of two nucleoside triphosphates in the reaction

$$pppA + pppN \rightleftharpoons pppApN + PP_i$$

Bacterial RNAs therefore have 5′-triphosphate groups as was demonstrated by the incorporation of radioactive label into RNA when it was synthesized with [γ-^{32}P]ATP. Only the 5′ terminus of the RNA can retain the label because the internal phosphodiester groups of RNA are derived from the α-phosphate groups of nucleoside triphosphates.

The difficulty in forming an open complex is reflected in the observation that RNA synthesis, at least *in vitro*, is frequently aborted after usually 2 or 3 but up to 9 nucleotides

$\wedge\digamma \rightarrow \beta$

have been joined. However, the holoenzyme does not release the promoter, but rather, reinitiates transcription. Eventually, the open complex forms and **processive** (continuous) RNA synthesis commences. At this point, σ factor dissociates from the core–DNA–RNA complex and can join with another core to form a new initiation complex. This is demonstrated by a burst of RNA synthesis upon the addition of core enzyme to a transcribing reaction mixture that initially contained holoenzyme.

Rifamycins Inhibit Prokaryotic Transcription Initiation

Two related antibiotics, **rifamycin B,** which is produced by *Streptomyces mediterranei,* and its semisynthetic derivative **rifampicin,**

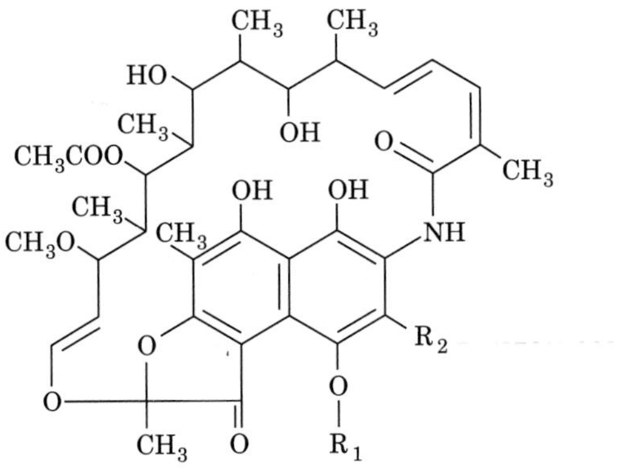

Rifamycin B $R_1 = CH_2COO^-$; $R_2 = H$

Rifampicin $R_1 = H$; $R_2 = CH = N^+\!\!\!\diagdown\!\!\!N - CH_3$

specifically inhibit transcription by prokaryotic, but not eukaryotic, RNA polymerases. This selectivity and their high potency (bacterial RNA polymerase is 50% inhibited by $2 \times 10^{-8} M$ rifampicin) has made them medically useful bacteriocidal agents against gram-positive bacteria and tuberculosis. The isolation of rifamycin-resistant mutants whose β subunits have altered electrophoretic mobilities indicates that this subunit contains the rifamycin-binding site. Rifamycins inhibit neither the binding of RNA polymerase to the promoter nor the formation of the first phosphodiester bond, but they prevent further chain elongation. The inactivated RNA polymerase remains bound to the promoter, thereby blocking its initiation by uninhibited enzyme. Once RNA chain initiation has occurred, however, rifamycins have no effect on the subsequent elongation process. The rifamycins are useful research tools because they permit the transcription process to be dissected into its initiation and its elongation phases.

D. Chain Elongation

What is the direction of RNA chain elongation; that is, does it occur by the addition of incoming nucleotides to the 3' end of the nascent (growing) RNA chain ($5' \rightarrow 3'$ growth; Fig. 29-11a), or by their addition to its 5' terminus ($3' \rightarrow 5'$ growth; Fig. 29-11b)? This question was answered by determining the rate that the radioactive label from [γ-^{32}P]GTP is incorporated into RNA. For $5' \rightarrow 3'$ elongation, the 5' γ-P is permanently labeled and hence, the chain's level of radioactivity would not change upon re-

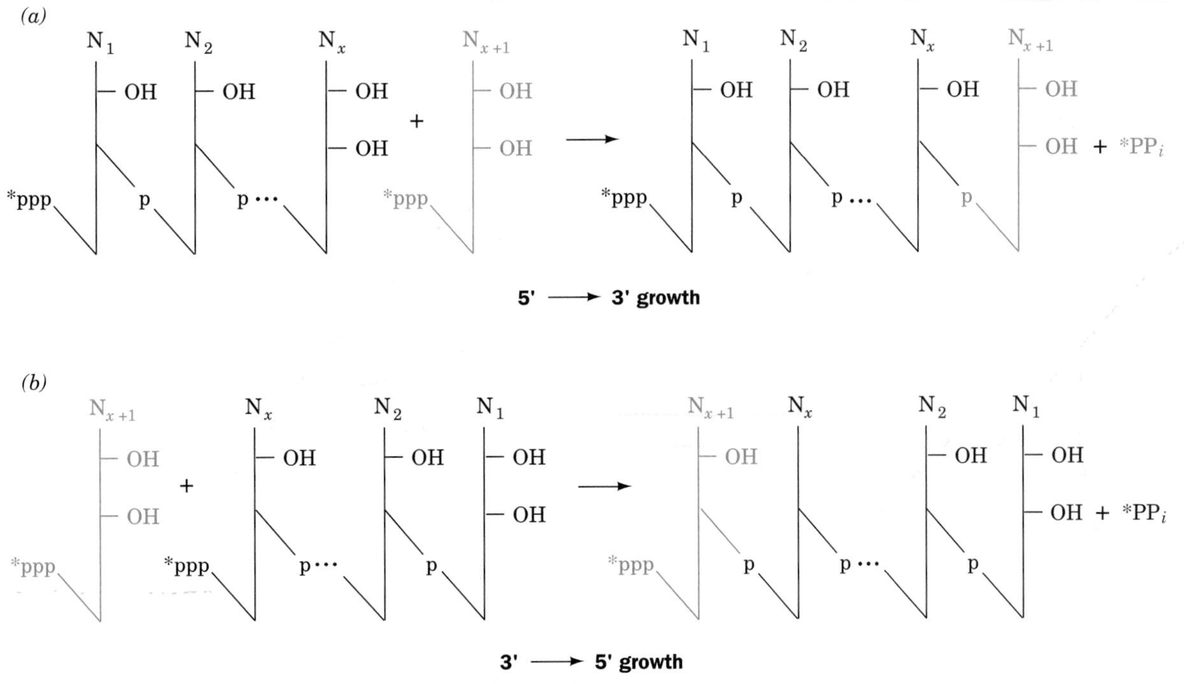

FIGURE 29-11.   The two possible modes of RNA chain growth: (*a*) by the addition of nucleotides to the 3' end, and (*b*) by the addition of nucleotides to the 5' end. RNA polymerase catalyzes the former reaction.

placement of the labeled GTP with unlabeled GTP. However, for 3′ → 5′ elongation, the 5′ γ-P is replaced with the addition of every new nucleotide so that, upon replacement of labeled with unlabeled GTP, the nascent RNA chains would loose their radioactivity. The former was observed. *Chain growth must therefore occur in the 5′ → 3′ direction (Fig. 29-11a).* This conclusion is corroborated by the observation that the antibiotic **cordycepin,**

NH₂

(purine structure)

HOCH₂ O

H   H

H   H

H   OH

**Cordycepin
(3′-deoxyadenosine)**

an adenosine analog that lacks a 3′-OH group, inhibits bacterial RNA synthesis. Its addition to the 3′ end of RNA, as is expected for 5′ → 3′ growth, prevents the RNA chain's further elongation. Cordycepin would not have this effect if chain growth occurred in the opposite direction because it could not be appended to an RNA's 5′ end.

Transcription Probably Supercoils DNA

RNA chain elongation requires that the double-stranded DNA template be opened up at the point of RNA synthesis so that the antisense strand can be transcribed onto its complementary RNA strand. In doing so, the RNA chain only transiently forms a short length of RNA–DNA hybrid duplex, as is indicated by the observation that transcription leaves the template duplex intact and yields single-stranded RNA. The unpaired "bubble" of DNA in the open initiation complex apparently travels along the DNA with the RNA polymerase. There are two ways this might occur (Fig. 29-12):

1. If the RNA polymerase followed the template strand in its helical path around the DNA, the DNA would build up little supercoiling because the DNA duplex would never be unwound by more than about a turn. However, the RNA transcript would wrap around the DNA, once per duplex turn. This model is implausible since it is unlikely that its DNA and RNA could be readily untangled: The RNA would not spontaneously unwind from the long and often circular DNA in any reasonable time,

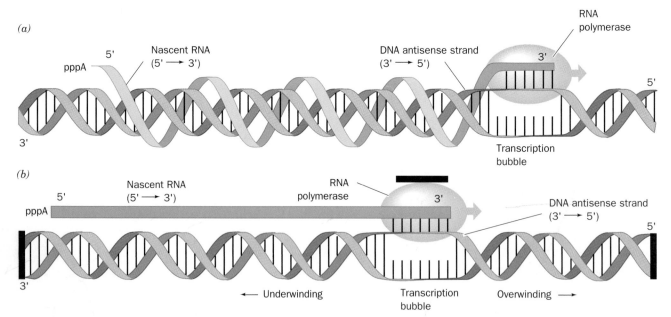

FIGURE 29-12. RNA chain elongation by RNA polymerase. In the region being transcribed, the DNA double helix is unwound by about a turn to permit the DNA's sense strand to form a short segment of DNA–RNA hybrid double helix with the RNA's 3′ end. As the RNA polymerase advances along the DNA template (here to the right), the DNA unwinds ahead of the RNA's growing 3′ end and rewinds behind it, thereby stripping the newly synthesized RNA from the template (antisense) strand. (*a*) One way this might occur is by the RNA polymerase following the path of the template strand about the DNA double helix, in which case the transcript would become wrapped about the DNA once per duplex turn. (*b*) A second, and more plausible possibility, is that the RNA moves in a straight line while the DNA rotates beneath it. In this case the RNA would not wrap around the DNA but the DNA would become overwound ahead of the advancing transcription bubble and unwound behind it (consider the consequences of placing your finger between the twisted DNA strands in this model and pushing towards the right). The model presumes that the ends of the DNA as well as the RNA polymerase, are prevented from rotating by attachments within the cell (*black bars*). [After Futcher, B., *Trends Genet.* **4,** 271, 272 (1988).]

and no topoisomerase is known to accelerate this process.

2. If the RNA polymerase moves in a straight line while the DNA rotates, the RNA and DNA will not become entangled. Rather, the DNA's helical turns are pushed ahead of the advancing transcription bubble so as to more tightly wind the DNA ahead of the bubble (which promotes positive supercoiling) and the DNA behind the bubble becomes equivalently unwound (which promotes negative supercoiling, although note that the linking number of the entire DNA remains unchanged). This model is supported by the observations that the transcription of plasmids in *E. coli* causes their positive supercoiling in gyrase mutants (which cannot relax positive supercoils; Section 28-5C) and their negative supercoiling in topoisomerase I mutants (which cannot relax negative supercoils).

Whatever the case, recall that inappropriate superhelicity halts transcription (Section 28-5C). Perhaps the torsional tension in the DNA generated by negative superhelicity behind the transcription bubble is required to help drive the transcriptional process, whereas too much such tension prevents the opening and maintenance of the transcription bubble.

Transcription Occurs Rapidly and Accurately

The *in vivo* rate of transcription is 20 to 50 nucleotides per second at 37°C as indicated by the rate that *E. coli* incorporate ^{3}H-labeled nucleosides into RNA (cells cannot take up nucleoside triphosphates from the medium). Once an RNA polymerase molecule has initiated transcription and moved away from the promoter, another RNA polymerase can follow suit. The synthesis of RNAs that are needed in large quantities, ribosomal RNAs, for example, is initiated as often as is sterically possible, about once per second (Fig. 29-13).

The error frequency in RNA synthesis, as estimated from the analysis of transcripts of simple templates such as poly[d(AT)]·poly[d(AT)], is one wrong base incorporated for every ~10^4 transcribed. This rate is tolerable because of the repeated transcription of most genes, because the genetic code contains numerous synonyms (Section 30-1E), and because amino acid substitutions in proteins are often functionally innocuous.

Intercalating Agents Inhibit Both RNA and DNA Polymerases

Actinomycin D,

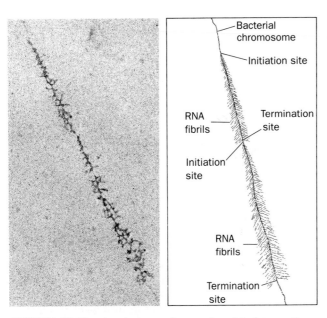

Actinomycin D

FIGURE 29-13. An electron micrograph and its interpretive drawing of two contiguous *E. coli* ribosomal genes undergoing transcription. The "arrowhead" structures result from the increasing lengths of the nascent RNA chains as the RNA polymerase molecules synthesizing them move from the initiation site on the DNA to the termination site. [Courtesy of Oscar L. Miller, Jr., University of Virginia.]

a useful antineoplastic (anticancer) agent produced by *Streptomyces antibioticus,* tightly binds to duplex DNA and, in doing so, strongly inhibits both transcription and DNA replication, presumably by interfering with the passage of RNA polymerase and DNA polymerse. The X-ray structure of actinomycin D in complex with a duplex DNA composed of two strands of the self-complementary octamer d(GAAGCTTC) reveals that the DNA assumes a B-like conformation in which the actinomycin's **phenoxazone ring system,** as had previously been shown, is intercalated between the DNA's central G · C base pairs (Fig. 29-14). Consequently, the DNA helix is unwound by 23° at the intercalation site and the central G · C base pairs are separated by 7.0 Å. The DNA helix is severely distorted from the normal B-DNA conformation such that its minor groove is wide and shallow in a manner resembling that of A-DNA. Actinomycin D's two chemically identical cyclic **depsipeptides** (having both peptide bonds and ester linkages), which assume different conformations, extend in opposite directions from the intercalation site along the minor groove of the DNA. The complex is stabilized through the formation of base–peptide and phenoxazone–sugar–phosphate backbone hydrogen bonds, as well as by hydrophobic interactions, in a way that explains the preference of actinomycin D to bind to DNA with its phenoxazone ring intercalated between the base pairs of a 5'-GC-3' sequence.

Numerous other intercalation agents, including ethidium and proflavin (Section 28-4C), also inhibit nucleic acid synthesis, presumably by similar mechanisms. Many of these substances are valuble antibiotic and/or antineoplastic agents.

E. Chain Termination

Electron micrographs such as Fig. 29-13 suggest that DNA contains specific sites at which transcription is terminated. The transcriptional termination sequences of many *E. coli* genes share two common features (Fig. 29-15a):

1. A series of 4 to 10 consecutive A · T's with the A's on the template strand. The transcribed RNA is terminated in or just past this sequence.

2. A G + C-rich region with a palindromic (twofold symmetric) sequence that immediately precedes the series of A · T's.

The RNA transcript of this region can therefore form a self-complementary "hairpin" structure that is terminated by several U residues (Fig. 29-15b).

The stability of a terminator's G + C-rich hairpin and the weak base pairing of its oligo(U) tail to template DNA appear to be important factors in ensuring proper chain termination. Indeed, model studies have shown that oligo(dA · rU) forms a particularly unstable hybrid helix although oligo(dA · dT) forms a helix of normal stability. The formation of the G + C-rich hairpin causes RNA polymerase to pause for several seconds at the termination site. This, it has been proposed, induces a conformational change in the RNA polymerase, which permits the noncoding DNA strand to displace the weakly bound oligo(U) tail from the template strand, thereby terminating transcription. Consistent with this notion is the observation that

mutations that alter the strengths of these associations reduce the efficiency of chain termination and often eliminate it. Termination is similarly diminished when *in vitro* transcription is carried out with GTP replaced by **inosine triphosphate (ITP).**

Inosine triphosphate (ITP)

I · C pairs are weaker than those of G · C because the hypoxanthine base of I, which lacks the 2-amino group of G, can only make two hydrogen bonds to C, thereby decreasing the hairpin's stability.

Despite the foregoing, experiments by Michael Chamberlin in which segments of highly efficient terminators were swapped via recombinant DNA techniques indicate that the RNA-terminator hairpin and U-rich 3' tail do not function independently of their upstream and downstream flanking regions. Indeed, terminators that lack a U-rich segment can be highly efficient when joined to the appropriate sequence immediately downstream from the termination site. Termination efficiency also varies with the concentrations of nucleoside triphosphates, with the level of supercoiling in the DNA template, and with changes in the concentrations of salts in ways that do not affect the relative stabilities of DNA, RNA, and hybrid double helices. Moreover, the extent to which these quantities affect termination efficiency varies with the sequence of the terminator. These results suggest that termination is a complex multistep process in which RNA polymerase is a key player. Indeed, mutations in the β subunit can both increase and decrease termination efficiency.

Termination Often Requires the Assistance of Rho Factor

The above termination sequences induce the spontaneous termination of transcription. Other termination sites, however, lack any obvious similarities and are unable to form strong hairpins; *they require the participation of a protein known as rho factor to terminate transcription.* The existence of rho factor was suggested by the observation that *in vivo* transcripts are often shorter than the corresponding *in vitro* transcripts. Rho factor, a hexamer of identical 419-residue subunits, enhances the termination efficiency of spontaneously terminating transcripts as well as inducing the termination of nonspontaneously terminating transcripts.

Several key observations have led to a model of rho-dependent termination:

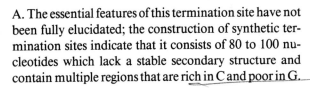

1. Rho factor is an enzyme that catalyzes the unwinding of RNA–DNA and RNA–RNA double helices. This process is powered by the hydrolysis of nucleoside triphosphates (NTPs) to nucleoside diphosphates + P_i with little preference for the identity of the base. NTPase activity is required for rho-dependent termination as is demonstrated by its *in vitro* inhibition when the NTPs are replaced by their β,γ-imido analogs,

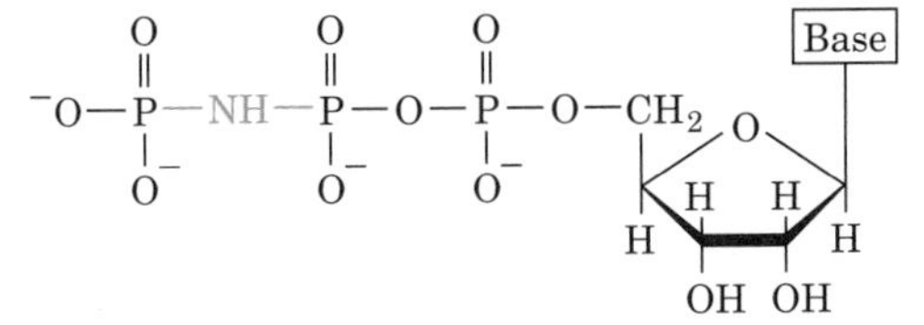

$$^-O-\overset{\overset{\displaystyle O}{\|}}{\underset{\underset{\displaystyle O^-}{|}}{P}}-NH-\overset{\overset{\displaystyle O}{\|}}{\underset{\underset{\displaystyle O^-}{|}}{P}}-O-\overset{\overset{\displaystyle O}{\|}}{\underset{\underset{\displaystyle O^-}{|}}{P}}-O-CH_2$$

Base

OH OH

β,γ**-Imido nucleoside triphosphate**

substances that are RNA polymerase substrates but cannot be hydrolyzed by rho factor.

2. Genetic manipulations indicate that rho-dependent termination requires the presence of a specific recognition sequence upstream of the termination site. The recognition sequence must be on the nascent RNA rather than the DNA as is demonstrated by rho's inability to terminate transcription in the presence of pancreatic RNase

A. The essential features of this termination site have not been fully elucidated; the construction of synthetic termination sites indicate that it consists of 80 to 100 nucleotides which lack a stable secondary structure and contain multiple regions that are rich in C and poor in G.

These observations suggest that rho factor attaches to nascent RNA at its recognition sequence and then migrates

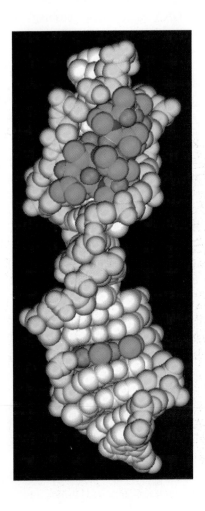

FIGURE 29-14. The X-ray structure of actinomycin D in complex with a duplex DNA of self-complementary sequence d(GAAGCTTC). The complex is shown in space-filling form in which the DNA's sugar–phosphate backbone is yellow, its bases are white, and the actinomycin D is colored according to atom type with C green, N blue, and O red. The two symmetry-related DNA molecules that are shown stack vertically to form a pseudo-continuous helix. The upper DNA is viewed towards its minor groove into which its bound actinomycin D's two cyclic depsipeptides are tightly wedged. The lower DNA (which is turned 180° about its helix axis relative to the upper DNA) is viewed towards its major groove, into which its bound actinomycin D's intercalated phenoxazone ring system projects from the minor groove side. [Based on an X-ray structure by Fusao Takusagawa, University of Kansas.]

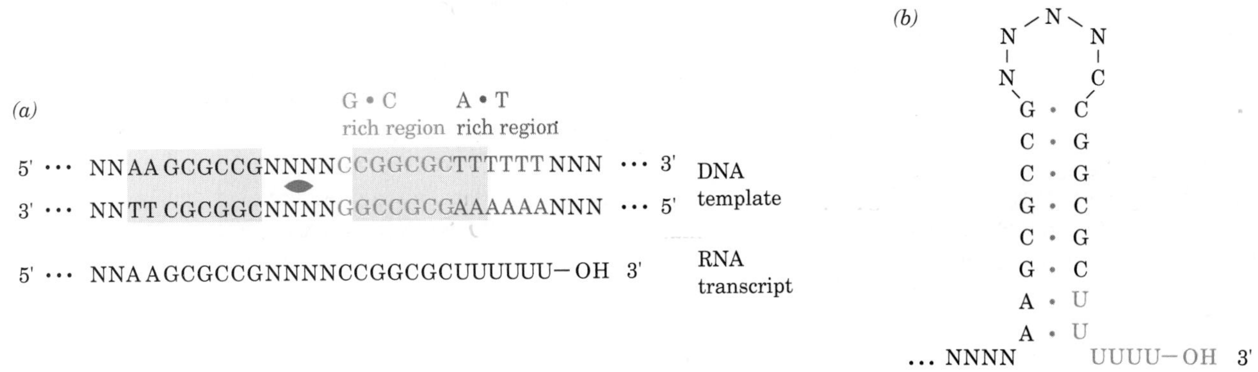

(a)

G · C   A • T
rich region rich region

5′ ··· NN AA GCGCCG NNNN CCGGCGC TTTTTT NNN ··· 3′   DNA
3′ ··· NN TT CGCGGC NNNN GGCCGCG AAAAAA NNN ··· 5′   template

5′ ··· NNA AGCGCCG NNNN CCGGCGC UUUUUU– OH 3′   RNA transcript

(b)

```
        N
      /   \
     N     N
     |     |
     N     C
      \   /
    G · C
    C · G
    C · G
    G · C
    C · G
    G · C
    A · U
    A · U
··· NNNN   UUUU– OH  3′
```

FIGURE 29-15. The base sequence of a hypothetical strong (efficient) *E. coli* terminator as deduced from the sequences of several transcripts. (*a*) The DNA sequence together with its corresponding RNA. The A · T-rich and G · C-rich sequences are shown in blue and red, respectively. The twofold symmetry axis (*lenticular symbol*) relates the flanking shaded segments that form an inverted repeat. (*b*) The RNA hairpin structure and poly(U) tail that triggers transcription termination. [After Pribnow, D., *in* Goldberger, R.F. (Ed.), *Biological Regulation and Development,* Vol. 1, *p.* 253, Plenum Press (1979).]

along the RNA in the 5′ → 3′ direction until it encounters an RNA polymerase paused at the termination site (without the pause, rho might not be able to overtake the RNA polymerase). There, rho unwinds the RNA–DNA duplex forming the transcription bubble, thereby releasing the RNA transcript. Rho-terminated transcripts have 3′ ends that typically vary over a range of ~50 nucleotides. This suggests that rho somehow pries the RNA away from its template DNA rather than "pushing" an RNA release "button."

F. Eukaryotic RNA Polymerases

Eukaryotic nuclei contain three distinct types of RNA polymerases that differ in the RNAs they synthesize:

1. **RNA polymerase I,** which is located in the nucleoli (dense granular bodies in the nuclei that contain the ribosomal genes; Section 29-4B), synthesizes precursors of most ribosomal RNAs (rRNAs).

2. **RNA polymerase II,** which occurs in the nucleoplasm, synthesizes mRNA precursors.

3. **RNA polymerase III,** which also occurs in the nucleoplasm, synthesizes the precursors of 5S ribosomal RNA, the tRNAs, and a variety of other small nuclear and cytosolic RNAs.

In addition to these nuclear enzymes, eukaryotic cells contain separate mitochondrial and (in plants) chloroplast RNA polymerases.

Eukaryotic RNA polymerases, whose molecular masses vary between 500 and 700 kD, are characterized by subunit compositions of Byzantine complexity. Each type of enzyme contains two nonidentical "large" (> 100 kD) subunits and an array of up to 12 different "small" (< 50 kD) subunits. For example, yeast RNA polymerase II, which has a molecular mass of ~550 kD, contains 10 different subunits. The three largest subunits, which are homologs of the prokaryotic RNA polymerase subunits α, β, and β′, are therefore thought to constitute the structural and functional core of the enzyme. Three other subunits are also components of RNA polymerases I and III, whereas three of the four remaining subunits are not essential for yeast viability, although they are thought to fine tune the transcription apparatus. Interestingly, one of these nonessential subunits has a 102-residue segment that is 30% identical to σ⁷⁰, the predominant prokaryotic σ factor. The sequences of RNA polymerase II subunits from other eukaryotic species reveal that almost 40% of their amino acid residues are invariant.

Although no eukaryotic RNA polymerase has yet been crystallized in a form suitable for X-ray analysis, the low-resolution structures of yeast RNA polymerases I and II have been determined by electron crystallography (Fig. 29-16). Both proteins have irregular shapes which resemble each other and that of *E. coli* RNA polymerase (Fig. 29-9*a*) in that their most prominent feature is a protein arm flanking a ~25-Å-wide channel that curves around the protein's surface and presumably functions to bind B-DNA. In the RNA polymerase II structure (Fig. 29-16*b*), which is of

(a)

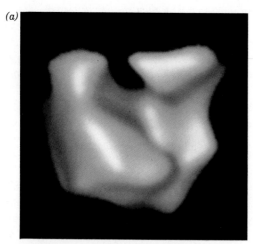

(b)

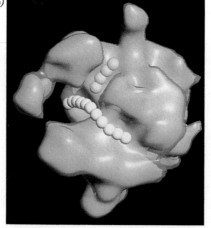

FIGURE 29-16. The structures of eukaryotic RNA polymerases as determined by electron crystallography. (*a*) Yeast RNA polymerase I at ~30-Å resolution. The irregularly shaped protein is around 150 × 110 × 110 Å in size. The 30-Å wide and 100-Å-long groove that extends from the upper middle to the lower right of the image probably constitutes the enzyme's DNA binding site. [Courtesy of Patrick Schultz and Pierre Oudet, Laboratoire de Genétique Moléculaire des Eucaryotes, Strasbourg, France.] (*b*) Yeast RNA polymerase II at ~16-Å resolution. The enzyme is about 140 × 136 × 110 Å in size. The beads are 8 Å in diameter and are placed every 6.8 Å. The chain of pink beads marks the putative DNA-binding channel, which is 25 Å wide and 70 to 80 Å long. The chain of light green beads follows a 12 to 15-Å-wide and 30-Å-long channel that branches off from the DNA-binding channel and that probably functions to bind single-stranded RNA. The entrance to an 8-Å-diameter and 35-Å-long tunnel that passes through the protein lies to the lower right of the lowest two green beads. [Courtesy of Roger Kornberg, Stanford University.]

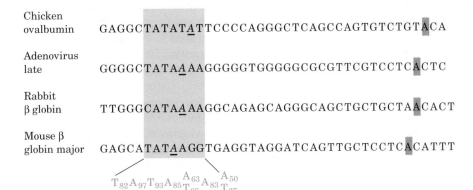

FIGURE 29-17. The promoter sequences of selected eukaryotic structural genes. The homologous segment, the TATA box, is shaded in red with the base at position −27 underlined and the initial nucleotide to be transcribed (+1) shaded in green. The bottom row indicates the consensus sequence of several such promoters with the subscripts indicating the percent occurrence of the corresponding base. [After Gannon, F., O'Hare, K., Perrin, F., Le Pennec, J.P., Benoist, C., Cochet, M., Breathnach, R., Royal, A., Garapin, A., Cami, B., and Chambon, P., *Nature* **278,** 433 (1978).]

higher resolution than the other two structures, a channel wide enough to accomodate a single strand of RNA branches off from the "DNA binding" channel in a nearly perpendicular direction. Intriguingly, an 8-Å-diameter tunnel that extends for ~35 Å completely through the enzyme begins near the branch point of the two channels. The tunnel's role is unclear. Two possibilities are that it provides substrate NTPs access to the enzyme's catalytic site or that it simply is a solvent-accessible region at the contact points of several subunits.

Mammalian RNA Polymerase I Has a Bipartite Promoter

Since, as we shall see in Section 29-4B, the numerous rRNA genes in a given eukaryotic cell have essentially identical sequences, its RNA polymerase I only recognizes one promoter. Yet, in constrast to the case for RNA polymerases II and III, RNA polymerase I promoters are species specific, that is, an RNA polymerase I only recognizes its own promoter and those of closely related species. This is because only closely related species exhibit recognizable sequence identities near the transcriptional start sites of their rRNA genes. RNA polymerse I promoters were therefore identified by determining how the transcription rate of an rRNA gene is affected by a series of increasingly longer deletions approaching its start site from either its upstream or its downstream sides. Such studies have indicated, for example, that mammalian RNA polymerases I require the presence of a so-called **core promoter element,** which spans positions −31 to +6 and hence overlaps the transcribed region. However, efficent transcription additionally requires an **upstream promoter element,** which is located between residues −187 and −107.

RNA Polymerase II Promoters Are Complex and Diverse

The promoters recognized by RNA polymerase II, which are considerably longer and more diverse than those of pro-

karyotic genes, have as yet been described only superficially. The structural genes expressed in all tissues, the so-called "housekeeping" genes, which are thought to be constituitively transcribed, have one or more copies of the sequence GGGCGG or its complement (the **GC box**) located upstream from their transcription start sites. The analysis of deletion and point mutations in eukaryotic viruses such as SV40 indicates that GC boxes function analogously to prokaryotic promoters. On the other hand, structural genes that are selectively expressed in one or a few types of cells often lack these GC-rich sequences. Rather, *they contain a conserved AT-rich sequence located 25 to 30 bp upstream from their transcription start sites (Fig. 29-17).* Note that this so-called **TATA box** resembles the −10 region of prokaryotic promoters (TATAAT) although they differ in their locations relative to the transcription start site (−27 vs −10). The functions of these two promoter elements are not strictly analogous, however, since the deletion of the TATA box does not necessarily eliminate transcription. Rather, TATA box deletion or mutation generates heterogeneities in the transcriptional start site, thereby indicating that the TATA box participates in selecting this site.

The gene region extending between about −50 and −110 also contains promoter elements. For instance, many eukaryotic structural genes, including those encoding the various globins, have a conserved sequence of consensus CCAAT (the **CCAAT box**) located between about −70 and −90 whose alteration greatly reduces the gene's transcription rate. Globin genes have, in addition, a conserved **CACCC box** upstream from the CCAAT box that has also been implicated in transcriptional initiation. Evidently, the promoter sequences upstream of the TATA box form the initial DNA-binding sites for RNA polymerase II and the other proteins involved in transcriptional initiation (see below).

Enhancers Are Transcriptional Activators That Can Have Variable Positions and Orientations

Perhaps the most suprising aspect of eukaryotic transcriptional control elements is that some of them need not have fixed positions and orientations relative to their corresponding transcribed sequences. For example, the SV40 genome, in which such elements were first discovered, contains two repeated sequences of 72 bp each that are located upstream from the promoter for early gene expression. Transcription is unaffected if one of these repeats is deleted but is nearly eliminated when both are absent. The analysis of a series of SV40 mutants containing only one of these repeats demonstrated that its ability to stimulate transcription from its corresponding promoter is all but independent of its position and orientation. Indeed, transcription is unimpaired when this segment is several thousand base pairs upstream or downstream from the transcription start site. Gene segments with such properties are named **enhancers** to indicate that they differ from promoters, with which they must be associated in order to trigger site-specific and strand-specific transcription initiation (although the characterization of numerous promoters and enhancers indicates that their functional properties are similar). Enhancers occur in both eukaryotic viruses and cellular genes.

Enhancers are required for the full activities of their cognate promoters. But how do they act? Two not mutually exclusive possibilities are given the most credence:

1. Enhancers are "entry points" on DNA for RNA polymerase II, perhaps through a lack of binding affinity for the histones that normally coat eukaryotic DNA, so as to (as seems likely) block RNA polymerase II binding (Section 33-1A). Alternatively, enhancers may alter DNA's local conformation in a way that favors RNA polymerase II binding. In fact, some enhancers contain a segment of alternating purines and pyrimidines which, we have seen, is just the type of sequence most likely to form Z-DNA (Section 28-2B).

2. Enhancers are recognized by specific proteins called **transcription factors** that stimulate RNA polymerase II to bind to a nearby promoter.

All cellular enhancers that have yet been identified are associated with genes that are selectively expressed in specific tissues. It therefore seems, as we discuss in Section 33-3B, that *enhancers mediate much of the selective gene expression in eukaryotes.*

RNA Polymerase III Promoters Can Be Located Downstream from Their Transcription Start Sites

The promoters of genes transcribed by RNA polymerase III can be located entirely within the genes' transcribed regions. Donald Brown established this through the construction of a series of deletion mutants of a *Xenopus borealis* 5S RNA gene. Deletions of base sequences that start from outside one or the other end of the transcribed portion of the 5S gene only prevent transcription if they extend into the segment between nucleotides $+40$ and $+80$. Indeed, a fragment of the 5S RNA gene consisting of only nucleotides 41 to 87, when cloned in a bacterial plasmid, is sufficient to direct specific initiation by RNA polymerase III at an upstream site. This is because, as was subsequently demonstrated, the sequence contains the binding site for a transcription factor that stimulates the upstream binding of RNA polymerase III. Further studies have shown, however, that the promoters of other RNA polymerase III-transcribed genes may lie partially or even entirely upstream of their start sites.

Amatoxins Specifically Inhibit RNA Polymerases II and III

The poisonous mushroom ***Amanita phalloides* (death cap),** which is responsible for the majority of fatal mushroom poisonings, contains several types of toxic substances, including a series of unusual bicyclic octapeptides known as **amatoxins.** α-Amanitin,

α-Amanitin

which is representative of the amatoxins, forms a tight $1:1$ complex with RNA polymerase II ($K = 10^{-8}M$) and a looser one with RNA polymerase III ($K = 10^{-6}M$), so as to specifically block their elongation steps. α-Amanitin is therefore a useful tool for mechanistic studies of these enzymes. RNA polymerase I as well as mitochondrial, chloroplast, and prokaryotic RNA polymerases are insensitive to α-amanitin.

Despite the amatoxins' high toxicity (5–6 mg, which occur in ~40 g of fresh mushrooms, are sufficient to kill a human adult), they act slowly. Death, usually from liver dysfunction, occurs no earlier than several days after mushroom ingestion (and after recovery from the effects of other mushroom toxins). This, in part, reflects the slow turnover rate of eukaryotic mRNAs and proteins.

3. CONTROL OF TRANSCRIPTION IN PROKARYOTES

Prokaryotes respond to sudden environmental changes, such as the influx of nutrients, by inducing the synthesis of the appropriate proteins. This process takes only minutes

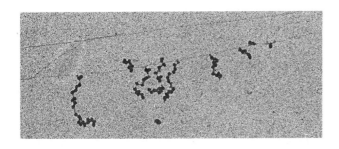

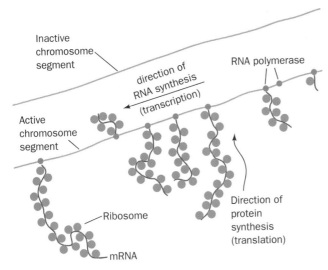

FIGURE 29-18. An electron micrograph and its interpretive drawing showing the simultaneous transcription and translation of an *E. coli* gene. RNA polymerase molecules are transcribing the DNA from right to left while ribosomes are translating the nascent RNAs (mostly from bottom to top). [Courtesy of Oscar L. Miller, Jr., University of Virginia.]

because transcription and translation in prokaryotes are closely coupled: *Ribosomes commence translation near the 5′ end of a nascent mRNA soon after it is extruded from RNA polymerase (Fig. 29-18).* Moreover, *most prokaryotic mRNAs are enzymatically degraded within 1 to 3 min of their synthesis,* thereby eliminating the wasteful synthesis of unneeded proteins after a change in conditions (protein degradation is discussed in Section 30-6). In fact, the 5′ ends of some mRNAs are degraded before their 3′ ends have been synthesized.

In contrast, the induction of new proteins in eukaryotic cells frequently takes hours or days, in part, because transcription takes place in the nucleus and the resulting mRNAs must be transported to the cytoplasm, where translation occurs. However, eukaryotic cells, particularly those of multicellular organisms, have relatively stable environments; changes in their transcriptional patterns usually occur only during cell differentiation.

In this section we examine some of the ways in which prokaryotic gene expression is regulated through transcriptional control. Eukaryotes, being vastly more complex creatures than are prokaryotes, have a correspondingly more complicated transcriptional control system whose

general outlines are beginning to come into focus. We therefore defer discussion of eukaryotic transcriptional control until Section 33-3, where it can be considered in light of what we know about the structure and organization of the eukaryotic chromosome.

A. Promoters

In the presence of high concentrations of inducer, the *lac* operon is rapidly transcribed. In contrast, the *lacI* gene is transcribed at such a low rate that a typical *E. coli* cell contains < 10 molecules of the *lac* repressor. Yet, the *I* gene has no repressor. Rather, it has such an inefficient promoter (Fig. 29-10) that it is transcribed an average of about once per bacterial generation. *Genes that are transcribed at high rates have efficient promoters.* In general, the more efficient a promoter, the more closely its sequence resembles that of the corresponding consensus sequence.

Gene Expression Can Be Controlled by a Succession of σ Factors

The processes of development and differentiation involve the temporally ordered expression of sets of genes according to genetically specified programs. Phage infections are among the simplest examples of developmental processes. Typically, only a subset of the phage genome, often referred to as *early* genes, are expressed in the host immediately after phage infection. As time passes, *middle* genes start to be expressed and the *early* genes as well as the bacterial genes are turned off. In the final stages of phage infection, the *middle* genes give way to the *late* genes. Of course some phage types express more than three sets of genes and some genes may be expressed in more than one stage of an infection.

One way in which families of genes are sequentially expressed is through "cascades" of σ factors. In the infection of *Bacillus subtilis* by **bacteriophage SP01,** for example, the *early* gene promoters are recognized by the bacterial RNA polymerase holoenzyme. Among these *early* genes is gene 28, whose gene product is a new σ subunit, designated σ^{gp28}, that displaces the bacterial σ subunit from the core enzyme. This reconstituted holoenzyme recognizes only the phage *middle* gene promoters, which all have similar −35 and −10 regions, but bear little resemblance to the corresponding regions of bacterial and phage *early* genes. The *early* genes therefore become inactive once their corresponding mRNAs have been degraded. The phage *middle* genes include genes 33 and 34, which together specify yet another σ factor, $\sigma^{gp33/34}$ which, in turn, permits the transcription of only *late* phage genes.

Several bacteria, including *E. coli* and *B. subtilis,* likewise have several different σ factors. These are not necessarily utilized in a sequential manner. Rather, those that differ from the predominant or primary σ factor control the transcription of coordinately expressed groups of special purpose genes, whose promoters are quite different from those recognized by the primary σ factor. For example, sporulation in *B. subtilis,* a process in which the bacterial cell is asymmetrically partitioned into two compartments, the **forespore** (which becomes the **spore,** a germline cell from which subsequent progeny arise) and the **mother cell** (which synthesizes the spore's protective cell wall and is eventually discarded), is governed by five σ factors in addition to that of the **vegatative** (nonsporulating) cell: one that is active before cell partition occurs, two that are sequentially active in the forespore, and two that are sequentially active in the mother cell. Cross regulation of the compartmentalized σ factors permits the forespore and mother cell to tightly coordinate this differentiation process.

B. lac Repressor

In 1966, Beno Müller-Hill and Walter Gilbert isolated *lac* repressor on the basis of its ability to bind ^{14}C-labeled IPTG and demonstrated that it is a protein. This was an exceedingly difficult task because *lac* repressor comprises only ~0.002% of the protein in wild-type *E. coli*. Now, however, *lac* repressor is available in quantity through the application of molecular cloning techniques (Section 28-8D).

lac Repressor Finds Its Operator by Sliding Along DNA

The *lac* repressor is a tetramer of identical 360-residue subunits arranged with three mutually perpendicular twofold axes (D_2 symmetry; Section 7-5B). Each subunit is capable of binding one IPTG molecule with a dissociation constant of $K = 10^{-6}M$. In the absence of inducer, the repressor tetramer nonspecifically binds duplex DNA with a dissociation constant of $K \approx 10^{-4}M$. However, it specifically binds to the *lac* operator with far greater affinity: $K \approx 10^{-13}M$. Limited proteolysis of *lac* repressor with trypsin reveals that each subunit consists of two functional domains: Its 58-residue N-terminal peptide binds DNA but not IPTG, whereas the remaining "core tetramer" binds only IPTG.

The observed rate constant for the binding of *lac* repressor to *lac* operator is $k_f \approx 10^{10}M^{-1}$ s^{-1}. This "on" rate is much greater than that calculated for the diffusion-controlled process in solution: $k_f = 10^7M^{-1}$ s^{-1} for molecules the size of *lac* repressor. Since it is impossible for a reaction to proceed faster than its diffusion-controlled rate, the *lac* repressor must not encounter operator from solution in a random three-dimensional search. Rather, *it appears that lac repressor finds operator by nonspecifically binding to DNA and diffusing along it in a far more efficient one-dimensional search.*

lac Operator Has a Nearly Palindromic Sequence

The availability of large quantities of *lac* repressor made it possible to characterize the *lac* operator. *E. coli* DNA that had been sonicated to small fragments was mixed with *lac* repressor and passed through a nitrocellulose filter. Protein, with or without bound DNA, sticks to nitrocellulose,

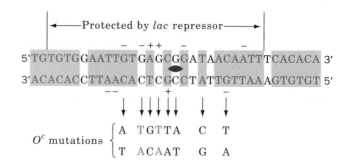

FIGURE 29-19. The base sequence of the *lac* operator. The symmetry related regions (*red*), comprise 28 of its 35 bp. A "+" denotes positions at which repressor binding enhances methylation by dimethyl sulfate (which methylates G at N7 and A at N3) and a "−" indicates where this footprinting reaction is inhibited. The bottom row indicates the positions and identities of different point mutations that prevent *lac* repressor binding (O^c mutants). Those in red increase the operator's symmetry. [After Sobell, H.M., *in* Goldberger, R.F. (Ed.), *Biological Regulation and Development*, Vol. 1, *p.* 193, Plenum Press (1979).]

whereas duplex DNA, by itself, does not. The DNA was released from the filter-bound protein by washing it with IPTG solution, recombined with *lac* repressor, and the resulting complex treated with DNase I. The DNA fragment that *lac* repressor protects from nuclease degradation consists of a run of 26 bp that is embedded in a nearly twofold symmetric sequence of 35 bp (Fig. 29-19, *top*). *Such palindromic symmetry is a common feature of DNAs that are specifically bound by proteins;* recall that restriction endonuclease recognition sites are also palindromic (Section 28-6A).

It has been suggested that the *lac* operator's symmetry matches that of its repressor; that is, operator binds to repressor in a twofold symmetric cleft between two subunits, much in the same way as *Eco*RI and *Eco*RV restriction endonucleases bind to their recognition sites (Section 28-6A). Methylation protection experiments, however, do not support this contention. There is an asymmetric pattern of differences between free and repressor-bound operator in the susceptibility of its bases to reaction with DMS (Fig. 29-19). Furthermore, point mutations in the operator that render it operator constitutive (O^c), and that invariably weaken the binding of repressor to operator, may increase as well as decrease the operator's twofold symmetry (Fig. 29-19). We discuss further aspects of *lac* operator organization in Section 29-3E.

lac Repressor Prevents RNA Polymerase from Forming a Productive Initiation Complex

Operator occupies positions −7 through +28 of the *lac* operon relative to the transcription start site (Fig. 29-20). Nuclease protection studies, it will be recalled, indicate that, in the initiation complex, RNA polymerase tightly

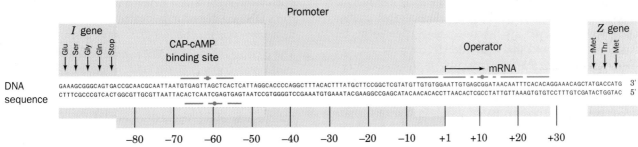

FIGURE 29-20. The nucleotide sequence of the *E. coli lac* promoter–operator region extending from the C-terminal region of *lacI* (*left*) to the N-terminal region of *lacZ* (*right*). The palindromic sequences of the operator and the CAP-binding site (Section 29-3C) are overscored or underscored. [After Dickson, R.C., Abelson, J., Barnes, W.M., and Reznikoff, W.A., *Science* **187**, 32 (1975).]

binds to the DNA between positions -20 and $+20$ (Section 29-2B). Thus, *the lac operator and promoter sites overlap.* It was therefore widely assumed for many years that *lac* repressor simply physically obstructs the binding of RNA polymerase to the *lac* promoter. However, the observation that *lac* repressor and RNA polymerase can simultaneously bind to the *lac* operon indicates that *lac* repressor must act by somehow interfering with the initiation process. Closer investigation of this phenomenon has revealed that, in the presence of bound *lac* repressor, RNA polymerase holoenzyme still abortively synthesizes oligonucleotides, although they tend to be shorter than those made in the absence of repressor. Evidently, *lac repressor acts by somehow increasing the normally high kinetic barrier for RNA polymerase to generate the open complex and commence processive elongation.*

C. Catabolite Repression: An Example of Gene Activation

Glucose is E. coli's metabolite of choice; the availability of adequate amounts of glucose prevents the full expression of genes specifying proteins involved in the fermentation of numerous other catabolites, including lactose (Fig. 29-21), arabinose, and galactose, even when they are present in high concentrations. This phenomenon, which is known as **catabolite repression,** prevents the wasteful duplication of energy-producing enzyme systems.

cAMP Signals the Lack of Glucose

The first indication of the mechanism of catabolite repression was the observation that, in *E. coli,* the level of cAMP, which was known to be a second messenger in animal cells (Section 17-3E), is greatly diminished in the presence of glucose. This observation led to the finding that the addition of cAMP to *E. coli* cultures overcame catabolite repression by glucose. Recall that, in *E. coli,* adenylate cyclase is activated by a phosphorylated enzyme (E III^{glc}), which is dephosphorylated upon the transport of glucose across the cell membrane (Section 18-3D). *The presence of* *glucose, therefore, normally lowers the cAMP level in E. coli.*

CAP–cAMP Complex Stimulates the Transcription of Catabolite Repressed Operons

Certain *E. coli* mutants, in which the absence of glucose does not relieve catabolite repression, are missing a cAMP-binding protein that is synonymously named **catabolite gene activator protein (CAP)** or **cAMP receptor protein**

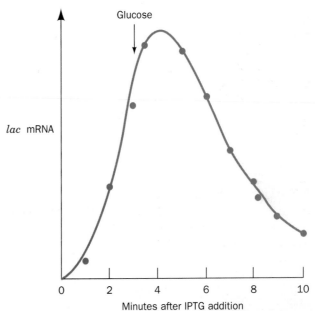

FIGURE 29-21. The kinetics of *lac* operon mRNA synthesis following its induction with IPTG, and of its degradation after glucose addition. *E. coli* were grown on a medium containing glycerol as their only carbon-energy source and ^{3}H-labeled uridine. IPTG was added to the medium at the beginning of the experiment to induce the synthesis of the *lac* enzymes. After 3 min, glucose was added to stop the synthesis. The amount of ^{3}H-labeled *lac* RNA was determined by hybridization with DNA containing the *lacZ* and *lacY* genes. [After Adesnik, M. and Levinthal, C., *Cold Spring Harbor Symp. Quant. Biol.* **35**, 457 (1970).]

(CRP). CAP is a dimeric protein of identical 210-residue subunits that undergoes a large conformational change upon binding cAMP. Its function was elucidated by Ira Pastan, who showed that *CAP–cAMP complex, but not CAP itself, binds to the lac operon (among others) and stimulates transcription from its otherwise low-efficiency promoter in the absence of repressor.* CAP is therefore a **positive regulator** (turns on transcription), in contrast to *lac* repressor, which is a **negative regulator** (turns off transcription).

Why is CAP–cAMP complex necessary to stimulate the transcription of its target operons? And how does it do so? The *lac* operon has a weak (low-efficiency) promoter; its −10 and −35 sequences (TATGTT and TTTACA; Figs. 29-10 and 29-20) differ significantly from the corresponding consensus sequences of strong (high-efficiency) promoters (TATAAT and TTGACA; Fig. 29-10). Such weak promoters evidently require some sort of help for efficient transcriptional initiation. There are two plausible (and not mutually exclusive) ways that CAP–cAMP might provide such help:

1. CAP–cAMP may stimulate transcriptional initiation through direct interaction with RNA polymerase. This hypothesis is supported by the observation that the *lac* operon fragment which CAP–cAMP complex protects from DNase I digestion is located in the *lac* promoter's upstream segment (Fig. 29-20).

2. The binding of CAP–cAMP complex to promoter may conformationally alter this DNA in a way that facilitates transcriptional initiation. Indeed, the X-ray structure, by Thomas Steitz, of CAP–cAMP in complex with a palindromic 30-bp segment of duplex DNA whose sequence resembles that of the CAP binding sequence (Fig. 29-20) reveals that the DNA is bent by ~90° around the protein (Fig. 29-22a). The bend arises from two ~45° kinks in the DNA between the 5th and 6th bases out from the complex's twofold axis in both directions. This distortion results in the closing of the major groove and an enormous widening of the minor groove at each kink.

It seems almost certain that CAP-induced DNA bending plays an important role in activating RNA polymerase. Since DNA bending changes the relative orientation of bound CAP and RNA polymerase compared to what it would be on straight DNA, it may well be that, on the *lac* operon, the cAMP-binding domain of CAP rather than its DNA-binding domain is adjacent to RNA polymerase. In fact, RNA polymerase holoenzyme binds CAP–cAMP in

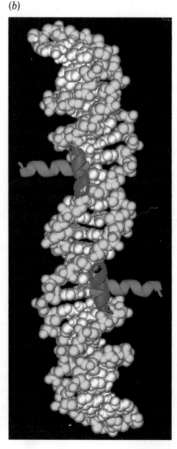

(b)

(a)

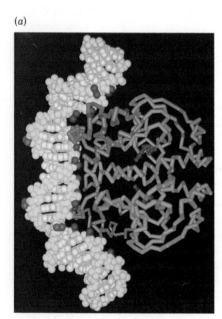

FIGURE 29-22. The X-ray structure of the CAP–cAMP dimer in complex with a self-complementary 30-bp duplex DNA. (*a*) The complex as viewed with its molecular two-fold axis horizontal and in the plane of the paper. The protein is represented by its C$_\alpha$ backbone with its N-terminal cAMP-binding domain blue and its C-terminal DNA-binding domain purple. The DNA is shown in space-filling form with its sugar–phosphate backbone yellow and its bases white (the atoms are drawn with slightly less than van der Waals radii). The DNA phosphates whose ethylation interferes with CAP binding are red. Those in the complex that are hypersensitive to DNase I are blue (these latter phosphates bridge the CAP-induced kinks and hence occur where the minor groove has been dramatically widened, which apparently increases their susceptability to DNase I digestion). The bound cAMPs are shown in ball-and-stick form in red. (*b*) The binding of the CAP dimer's two helix–turn–helix (HTH) motifs in successive major grooves of the DNA. The HTH motif's N-terminal helix is blue and its C-terminal recognition helix is red. The view is along the molecular two-fold axis and is related to that in Part *a* by a 90° rotation about its vertical axis. [Part *a* courtesy of and part *b* based on an X-ray structure by Thomas Steitz, Yale University.]

solution with a dissociation constant of ~1 μM. Moreover, several point mutants of CAP have been characterized that reduce transcriptional activation without altering DNA binding.

Another possibility is that CAP-induced DNA bending forms a loop in the DNA that brings a segment of upstream DNA into contact with the RNA polymerase in a way that facilitates initiation (see Section 29-3E). This latter idea is supported by experiments using a DNA sequence consisting of four segments of 5 or 6 consecutive A·T's, which each bend the DNA by ~18° (as indicated by its reduced electrophoretic mobility on polyacrylamide gels although the reason for this bending is obscure), interspersed by 5-bp spacers so as to put consecutive bends on the same side of the DNA. Hence the entire sequence is bent by ~70° (recall that B-DNA normally has ~10.5 bp/turn). When this so-called A-tract DNA is positioned so as to bend the DNA around RNA polymerase in the same direction as would bound CAP–cAMP, the rate of transcriptional initiation from the *lac* promoter is increased by ~10-fold relative to DNA constructs in which the DNA bends in the opposite direction.

D. Sequence-Specific Protein–DNA Interactions

Since genetic expression is controlled by proteins such as CAP and *lac* repressor, an important issue in the study of gene regulation is how these proteins recognize their target base sequences on DNA. Sequence-specific DNA-binding proteins generally do not disrupt the base pairs of the duplex DNA to which they bind. Consequently, these proteins can only discriminate among the four base pairs (A·T, T·A, G·C, and C·G) according to the functional groups of these base pairs that project into DNA's major and minor grooves. An inspection of Fig. 28-6 reveals that the groups exposed in the major groove have a greater variation in their types and arrangements than do those that are exposed in the minor groove. Moreover, the ~5-Å-wide and ~8-Å-deep minor groove of canonical (ideal) B-DNA is too narrow to admit protein structural elements such as an α helix, whereas its ~12-Å-wide and ~8-Å-deep major groove can do so. Thus, in the absence of major conformational changes to B-DNA, it would be expected that proteins could more readily differentiate base sequences from its major groove than from its minor groove. We shall see below that this is, in fact, the case.

The Helix–Turn–Helix Motif Is a Common DNA Recognition Element in Prokaryotes

The CAP dimer's two symmetrically disposed F helices protrude from the protein surface in such a way that they fit into successive major grooves of B-DNA (Fig. 29-22). *CAP's E and F helices form a* **helix–turn–helix (HTH)** *motif (supersecondary structure) that conformationally resembles analogous HTH motifs in several other prokaryotic*

repressors of known X-ray and NMR structure, including the *lac* repressor, the *E. coli* **trp** repressor (Section 29-3F) and the **cI repressors** and **Cro proteins** from **bacteriophages** λ and **434** (Section 32-3D). HTH motifs are ~20-residue polypeptide segments that form two α helices which cross at ~120° (Fig. 29-22b). They occur as components of domains that otherwise have widely varying structures, although all of them bind DNA. Note that HTH motifs are structurally stable only when they are components of larger proteins.

The X-ray and NMR structures of a number of protein–DNA complexes (see below) indicates that *DNA-binding proteins containing an HTH motif associate with their target base pairs mainly via the side chains extending from the second helix of the HTH motif, the so-called* **recognition helix** *(helix F in CAP, E in trp repressor, and $\alpha 3$ in the phage proteins).* Indeed, replacing the outward-facing residues of the 434 repressor's recognition helix with the corresponding residues of the related **bacteriophage P22** (using the genetic engineering techniques described in Section 28-8) yields a hybrid repressor that binds to P22 operators but not to those of 434. Moreover, the HTH motifs in all these proteins have amino acid sequences that are similar to each other and to polypeptide segments in numerous other prokaryotic DNA-binding proteins, including *lac* repressor. Evidently, *these proteins are evolutionarily related and bind their target DNAs in a similar manner* (but in a way that differs from those of *Eco*RI, *Eco*RV, and *Pvu*II restriction endonucleases; Section 28-6A).

How does the recognition helix recognize its target sequence? Since each base pair presents a different and presumably readily differentiated constellation of hydrogen bonding groups in DNA's major groove, it seemed likely that there would be a simple correspondence, analogous to Watson–Crick base pairing, between the amino acid residues of the recognition helix and the bases they contact in forming sequence-specific associations. The above X-ray structures, however, indicate this idea to be incorrect. Rather, base sequence recognition arises from complex structural interactions. For instance:

1. The X-ray structures of the closely similar N-terminal domain of 434 repressor (residues 1–69) and the entire 71-residue 434 Cro protein in complex with the identical 20-bp target DNA (434 phage expression is regulated through the differential binding of these proteins to the same DNA segments; Section 32-3D) were both determined by Stephen Harrison. Both dimeric proteins, as seen for CAP (Fig. 29-22), associate with the DNA in a twofold symmetric manner with their recognition helices bound in successive turns of the DNA's major groove (Figs. 29-23 and 29-24). In both complexes, the protein closely conforms to the DNA surface and interacts with its paired bases and sugar–phosphate chains through elaborate networks of hydrogen bonds, salt bridges, and van der Waals contacts. Nevertheless, the

(a) *(b)* *(c)*

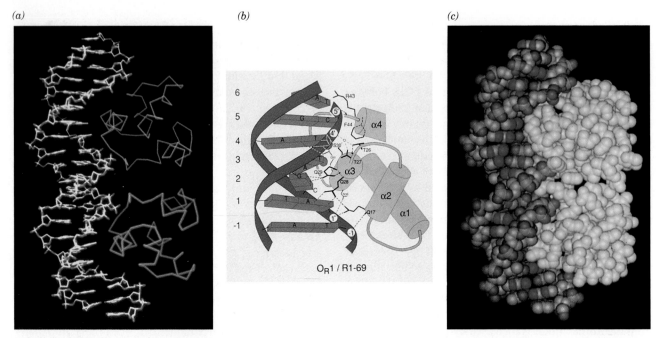

FIGURE 29-23. The X-ray structure of 434 phage repressor (actually only the repressor's 69-residue N-terminal domain) in complex with a 20-bp fragment of its target sequence [one strand of which has the sequence d(TATACAAGAAAGTTTGTACT)] as viewed perpendicularly to the complex's twofold axis of symmetry. (*a*) A skeletal model with the DNA on the left and with the protein's two identical subunits (C$_\alpha$ backbone only) shown in red and blue. Only the first 63 residues of the protein are visible. (*b*) A schematic drawing indicating how the helix–turn–helix motif, which encompasses helices $\alpha 2$ and $\alpha 3$, interacts with its target DNA. Residues are identified by their single letter code (Table 4-1). Short bars emanating from the polypeptide chain represent peptide NH groups, hydrogen bonds are represented by dashed lines, and DNA phosphates are represented by numbered circles. The small circle is a water molecule. (*c*) A space-filling model corresponding to Part *a*. All of the protein's non-H atoms are drawn in yellow. [Courtesy of Aneel Aggarwal, John Anderson, and Stephen Harrison, Harvard University.]

(a) *(b)* *(c)*

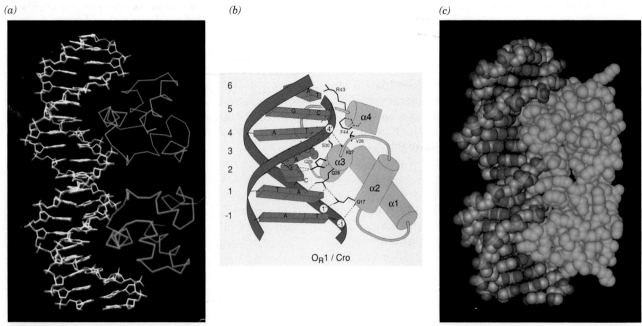

FIGURE 29-24. The X-ray structure of the 72-residue 434 Cro protein in complex with the same 20-bp DNA shown in Fig. 29-23 as viewed perpendicular to the complex's two-fold axis of symmetry. Only the first 64 residues of the protein are visible. Parts *a*, *b*, and *c* correspond to those in Fig. 29-23 with the protein in Part *c* shown in light blue. Note the close but not identical correspondence between the two structures. [Courtesy of Alfonso Mondragón, Cynthia Wolberger, and Stephen Harrison, Harvard University.]

detailed geometries of these associations are significantly different. In the repressor–DNA complex (Fig. 29-23), the DNA bends around the protein in an arc of radius ~65 Å which compresses the minor groove by ~2.5 Å near its center (between the two protein monomers) and widens it by ~2.5 Å towards its ends. In contrast, the DNA in complex with Cro (Fig. 29-24), although also bent, is nearly straight at its center and has a less compressed minor groove (compare Figs. 29-23*a* and 29-24*a*). This explains why the simultaneous replacement of three residues in the repressor's recognition helix with those occurring in Cro does not cause the resulting hybrid protein to bind DNA with Cro-like affinity: *The different conformations of the DNA in the repressor and Cro complexes prevents any particular side chain from interacting identically with the DNA in the two complexes.*

2. Paul Sigler determined the X-ray structure of *E. coli trp* repressor in complex with a DNA containing an 18-bp palindrome (TGT<u>ACTAGT</u>TA<u>ACTAGT</u>AC, where the *trp* repressor's target sequence is underlined) that closely resembles *trp* operator (Section 29-3F). The dimeric protein's recognition helices bind, as expected, in successive major grooves of the DNA, each in contact with an operator half-site (<u>ACTAGT</u>; Fig. 29-25). There are numerous hydrogen bonding contacts between the *trp* repressor and its bound DNA's nonesterified phosphate oxygens. Astoundingly, however, *there are no direct hydrogen bonds or nonpolar contacts that can explain the repressor's specificity for its operator. Rather, all but one of the side chain–base hydrogen bonding interactions are mediated by bridging water molecules* (the one direct such interaction involves a base that can be mutated without greatly affecting repressor binding affinity). Such buried water molecules have therefore been described as "honorary" protein side chains. In addition, the operator contains several base pairs that are not in contact with the repressor but whose mutation nevertheless greatly decreases repressor binding affinity. This suggests that the operator assumes a sequence-specific conformation that makes favorable contacts with the repressor. Indeed, comparison of the X-ray structure of an uncomplexed 10-bp self-complementary DNA containing the *trp* operator's half-site (CC<u>ACTAGT</u>GG) with that of the DNA in the *trp* repressor–operator complex reveals that the ACTAGT half-site assumes nearly identical idiosyncratic conformations and patterns of hydration in both structures. However, the B-DNA helix, which is straight in the DNA 10-mer, is bent by 15° towards the major groove in each operator half-site of the repressor–operator complex. Other DNA sequences could conceivably assume the repressor-bound operator's conformation but at too high an energy cost to form a stable complex with repressor (*trp* repressor's measured 10⁴-fold preference for its operator over other DNAs implies an ~23 kJ · mol⁻¹ difference in their binding free energies). This phenome-

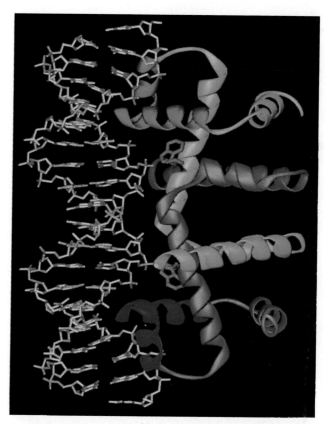

FIGURE 29-25. The X-ray structure of an *E. coli trp* repressor–operator complex as viewed with its molecular two-fold axis horizontal and in the plane of the paper. The protein's two identical subunits are shown in ribbon form in green and blue with the HTH motifs (helices D and E) more deeply colored. The 18-bp-containing self-complementary DNA is yellow. *trp* repressor binds its operator only when L-tryptophan (*red*) is simultaneously bound. Note that the protein's recognition helices (E) bind, as expected, in successive major grooves of the DNA but extend approximately perpendicular to the DNA duplex axis. In contrast, the recognition helices of 434 repressor and Cro proteins are nearly parallel to the major grooves of their bound DNAs (Figs. 29-23 and 29-24), whereas those of CAP assume an intermediate orientation (Fig. 29-22). [Based on an X-ray structure by Paul Sigler, Yale University.]

non, in which a protein senses the base sequence of DNA through the DNA's backbone conformation and/or flexibility, is referred to as **"indirect readout."** 434 repressor apparently also employs "indirect readout": Replacing the central A · T base pair of the operator shown in Fig. 29-23 with G · C reduces repressor binding affinity by 50-fold even though 434 repressor does not contact this region of the DNA.

It therefore appears that *there are no simple rules governing how particular amino acid residues interact with bases. Rather, sequence specificity results from an ensemble of mutually favorable interactions between a protein and its target DNA.*

met Repressor Contains a Two-Stranded Antiparallel β Sheet That Binds in Its Target DNA's Major Groove

The *E. coli* **met repressor,** when complexed with *S*-adenosylmethionine (SAM; Figure 24-14), represses the transcription of its own gene and those encoding enzymes involved in the synthesis of methionine (Figure 24-45) and SAM. The X-ray structure of the *met* repressor–SAM–operator complex (Fig. 29-26), determined by Simon Phillips, reveals a symmetric dimer of intertwined monomers that lacks an HTH motif. Rather, *met* repressor binds to its palindromic target DNA sequence through a symmetry-related pair of symmetrical two-stranded antiparallel β sheets (called β **ribbons**) that are inserted in successive major grooves of the DNA. Each β ribbon makes sequence-specific contacts with its target DNA sequence via hydrogen bonding and, probably, "indirect readout." The X-ray structure of the ***arc* repressor** of bacteriophage P22 reveals that this protein binds to its target DNA in a similar manner.

Phillips first determined the X-ray structure of *met* repressor in the absence of DNA. Model building studies aimed at elucidating how *met* repressor binds to its palindromic target DNA assumed that the two-fold rotation axes of both molecules would be coincident, as they are in all prokaryotic protein–DNA complexes of known structure. There were, consequently, two reasonable choices: (1) The protein could dock to the DNA with the above pairs of β ribbons entering successive major grooves; or (2) a symmetry-related pair of protruding α helices on the opposite face of the protein could do so in a manner resembling the way the recognition helices of HTH motifs interact with DNA. A variety of structural criteria suggested that the α helices make significantly better contacts with the DNA than do the β ribbons. Thus, the observation that it is, in fact, the β ribbons that bind to the DNA provides an important lesson: *The results of model building studies must be treated with utmost caution.* This is because our imprecise understanding of the energetics of intermolecular interactions (Sections 7-4 and 28-3) prevents us from reliably predicting how associating macromolecules conform to one another.

Prokaryotic repressors of known structures contain either an HTH motif or resemble the *met* repressor. However, eukaryotic transcription factors, as we shall see in Section 33-3B, employ a much wider variety of structural motifs to bind their target DNAs.

E. araBAD Operon: Positive and Negative Control by the Same Protein

Humans neither metabolize nor intestinally absorb the plant sugar L-arabinose. Hence, the *E. coli* that normally inhabit the human gut are periodically presented with a banquet of this pentose. Three of the five *E. coli* enzymes that metabolize L-arabinose are products of the catabolite repressible ***araBAD* operon** (Fig. 29-27).

The *araBAD* operon, as Robert Schleif has shown, contains, moving upstream from its transcriptional start site, the *araI*, *araO*$_1$, and *araO*$_2$ control sites (Fig. 29-28). The *araI* site (*I* for inducer) consists of two identical 17-bp sub-

(a)

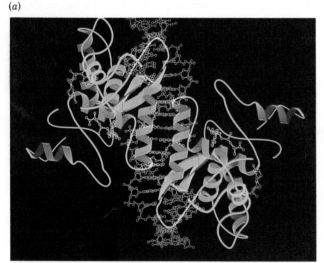

(b)

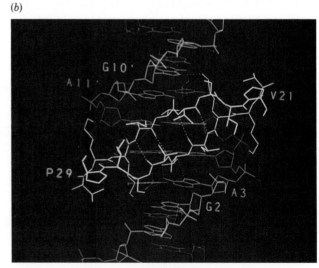

FIGURE 29-26. The X-ray structure of the *E. coli met* repressor–SAM–operator complex. (*a*) The overall structure of the complex as viewed along its 2-fold axis of symmetry. The 104-residue repressor subunits are shown in gold. The self-complementary 19-bp DNA and SAM, which must be bound to the repressor for it to also bind DNA, are shown in ball-and-stick form with the DNA blue and SAM green. Note that the DNA has four bound repressor subunits: Pairs of subunits form symmetric dimers in which each subunit donates one strand of the 2-stranded antiparallel β ribbon that is inserted in the DNA's major groove (*upper left and lower right*). Two such dimers pair across the complex's 2-fold axis via their antiparallel N-terminal helices which contact one another over the DNA's minor groove. (*b*) Detailed view of the 2-stranded antiparallel β ribbon (*yellow,* residues 21–29) inserted into the DNA's major groove (*blue*). Hydrogen bonds (*dashed lines*) indicate how Lys 23 and Thr 25 make sequence-specific contacts with bases G2 and A3. [Courtesy of Simon Phillips, University of Leeds, Leeds, U.K.]

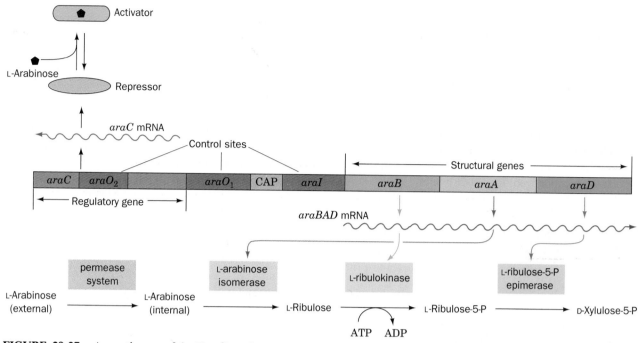

FIGURE 29-27. A genetic map of the *E. coli araC* and *araBAD* operons indicating the proteins they encode and the reactions in which these proteins participate. The permease system, which transports arabinose into the cell, is the product of the *araE* and *araF* genes, which occur in two independent operons. The pathway product, xylulose-5-phosphate, is converted, via the transketolase reaction, to the glycolytic intermediate fructose-6-phosphate (Section 21-4C). [After Lee, N., *in* Miller. J.H. and Rezinkoff, W.S. (Eds.), *The Operon, pp.* 390, Cold Spring Harbor Laboratory (1979).]

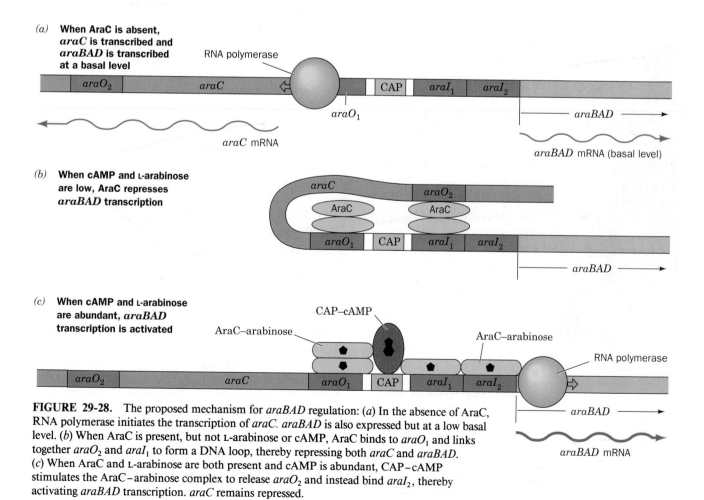

FIGURE 29-28. The proposed mechanism for *araBAD* regulation: (*a*) In the absence of AraC, RNA polymerase initiates the transcription of *araC*. *araBAD* is also expressed but at a low basal level. (*b*) When AraC is present, but not L-arabinose or cAMP, AraC binds to *araO₁* and links together *araO₂* and *araI₁* to form a DNA loop, thereby repressing both *araC* and *araBAD*. (*c*) When AraC and L-arabinose are both present and cAMP is abundant, CAP–cAMP stimulates the AraC–arabinose complex to release *araO₂* and instead bind *araI₂*, thereby activating *araBAD* transcription. *araC* remains repressed.

sites, *araI*$_1$ and *araI*$_2$, which are direct repeats separated by 4 bp, and are oriented such that *araI*$_2$, which overlaps the -35 region of the *araBAD* promoter, is downstream of *araI*$_1$. Intriguingly, *araO*$_2$, is located in a noncoding upstream region of the *araC* gene, at position -270 relative to the *araBAD* start site.

The transcription of the araBAD operon is regulated by both CAP–cAMP and the L-arabinose-binding protein, **AraC** *(the araC gene product; proteins may be assigned the name of the gene specifying them but in roman letters with the first letter capitalized; Fig. 29-28):*

1. In the absence of AraC, RNA polymerase initiates transcription of the *araC* gene in the direction away from its upstream neighbor, *araBAD*. The *araBAD* operon is expressed at a low basal level.

2. When AraC is present, but neither L-arabinose nor CAP–cAMP (high glucose), AraC binds to *araO*$_1$, *araO*$_2$, and *araI*$_1$. *araO*$_1$ is the operator for the *araC* gene; its association with AraC blocks *araC* transcription so that this process is autoregulatory, although this requires high levels of AraC. The binding of AraC to *araI*$_1$ represses the expression of *araBAD* (negative control). A series of deletion mutations indicate that the presence of *araO*$_2$ is also required for the repression of *araBAD*. The remarkably large 211-bp separation between *araO*$_2$ and *araI*$_1$ therefore suggests that the DNA between them is looped such that a dimeric molecule of AraC protein simultaneously binds to both *araO*$_2$ and *araI*$_1$. This is corroborated by the observation that the level of repression is greatly diminished by the insertion of 5 bp (half a turn) of DNA between these two sites, thereby transferring *araO*$_2$ to the opposite face of the DNA relative to *araI*$_1$ in the putative loop. Yet, the insertion of 11 bp (one turn) of DNA has no such affect.

3. When L-arabinose is present, it allosterically induces the AraC subunits bound to *araO*$_2$ to instead bind *araI*$_2$. This activates RNA polymerase to transcribe the *araBAD* genes (positive control). When the cAMP level is high (low glucose), CAP–cAMP binds to a site between *araO*$_1$ and *araI*$_1$ (although in most other CAP-binding operons, CAP–cAMP binds adjacent to RNA polymerase), where it functions both to help break the loop between *araO*$_2$ and *araI*$_1$ and to increase the affinity of AraC for *araI*$_2$. *araC* remains repressed by AraC–L-arabinose at *araO*$_1$.

If the *araI*$_2$ subsite is mutated so as to increase AraC's affinity for it, L-arabinose is no longer required for transcriptional activation. This suggests that L-arabinose does not conformationally transform AraC to an activator but, rather, weakens its binding affinity for *araO*$_2$. If the *araI* site is turned around or if it is moved upstream so that *araI*$_2$ does not overlap the *araBAD* promoter, AraC cannot stimulate transcription. Evidently, *AraC activates RNA polymerase through specific and relatively inflexible protein–protein interactions.*

Loop Formation Is Also Important in the Expression of the *lac* Operon

The function of DNA loop formation is obscure, although it has been demonstrated to occur in numerous bacterial and eukaryotic systems. It probably permits several regulatory proteins and/or regulatory sites on one protein to simultaneously influence transcription initiation by RNA polymerase. In fact, *the lac repressor has three binding sites on the lac operon:* the primary operator, now known as O_1, and two so-called pseudo-operators (previously thought to be nonfunctional evolutionary fossils), O_2 and O_3, which are located 401 bp downstream and 92 bp upstream of O_1 (within the *lacZ* gene and overlapping the CAP binding site). Müller-Hill determined the relative contributions of these various operators to the repression of the *lac* operon through the construction a set of eight plasmids: Each contained the *lacZ* gene under the control of the natural *lac* promoter as well as the three *lac* operators (O_1, O_2, and O_3), which were either active or mutagenically inactive in all possible combinations. When all three operators are active, *lacZ* expression is repressed 1300-fold relative to when all three operators are inactive. The inactivation of only O_1 results in almost complete loss of repression whereas, the inactivation of only O_2 or O_3 causes only a ~2-fold loss in repression. However, when O_2 and O_3 are both inactive, repression is decreased ~70-fold. These results suggest that efficient repression requires the formation of a DNA loop between O_1 and either O_2 or O_3. Indeed, such loop formation, and/or the cooperativity of repressor binding arising from it, appears to be a greater contributor to repression than repressor binding to O_1 alone, which provides only 19-fold repression.

Helix–turn–helix proteins, such as *lac* repressor, must have at least dimeric symmetry to specifically bind to their palindromic operator DNA. Thus, *lac* repressor's tetrameric character (Section 29-3B) permits it to bind two operators simultaneously, thereby forming a DNA loop.

F. *trp* Operon: Attenuation

In the following paragraphs we discuss a sophisticated transcriptional control mechanism named **attenuation** through which bacteria regulate the expression of certain operons involved in amino acid biosynthesis. This mechanism was discovered through the study of the *E. coli* **trp operon** (Fig. 29-29), which encodes five polypeptides comprising three enzymes that mediate the synthesis of tryptophan from chorismate (Section 24-5B). Charles Yanofsky established that the *trp* operon genes are coordinately expressed under the control of **trp repressor,** a dimeric protein of identical 107-residue subunits that is the product of the *trpR* gene (which forms an independent operon). *The trp repressor binds L-tryptophan, the pathway's end product, to form a complex that specifically binds to trp operator (trpO; Fig. 29-30) so as to reduce the rate of trp operon transcription 70-fold.* The X-ray structure of the *trp* repressor–operator

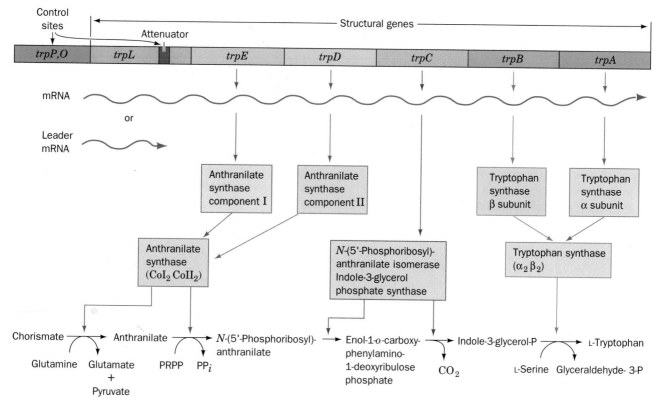

FIGURE 29-29. A genetic map of the *E. coli trp* operon indicating the enzymes it specifies and the reactions they catalyze. The gene product of *trpC* catalyzes two sequential reactions in the synthesis of tryptophan. [After Yanofsky, C., *J. Am. Med. Assoc.* **218**, 1027 (1971).]

complex (Section 29-3D) indicates that tryptophan binding allosterically orients *trp* repressor's two symmetry related helix–turn–helix "DNA reading heads" so that they can simultaneously bind to *trpO* (Fig. 29-25). Moreover, the bound tryptophan forms a hydrogen bond to a DNA phosphate group, thereby strengthening the repressor–operator association. Tryptophan therefore acts as a **corepressor;** its presence prevents what is then superfluous tryptophan biosynthesis (SAM similarly functions as a corepressor with the *met* repressor; Fig. 29-26*a*). The *trp* repressor also controls the synthesis of at least two other operons: the **trpR operon** and the **aroH operon** (which encodes one of three isozymes that catalyze the initial reaction of chorismate biosynthesis: Section 24-5B).

Tryptophan Biosynthesis Is Also Regulated by Attenuation

The *trp* repressor–operator system was at first thought to fully account for the regulation of tryptophan biosynthesis in *E. coli*. However, the discovery of *trp* deletion mutants located downstream from *trpO* that increase *trp* operon expression sixfold indicated the existence of an additional transcriptional control element. Sequence analysis established that *trpE*, the *trp* operon's leading structural gene, is preceded by a 162-nucleotide **leader sequence** *(trpL)*. Genetic analysis indicated that the new control element is lo-

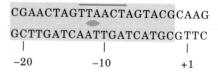

FIGURE 29-30. The base sequence of the *trp* operator. The nearly palindromic sequence is boxed and its −10 region is overscored.

cated in *trpL*, ~30 to 60 nucleotides upstream of *trpE* (Fig. 29-29).

When tryptophan is scarce, the entire 6720-nucleotide polycistronic *trp* mRNA, including the *trpL* sequence, is synthesized. As the tryptophan concentration increases, the rate of *trp* transcription decreases as a result of the *trp* repressor–corepressor complex's consequent greater abundance. Of the *trp* mRNA that is transcribed, however, an increasing proportion consists of only a 140-nucleotide segment corresponding to the 5′ end of *trpL*. *The availability of tryptophan therefore results in the premature termination of trp operon transcription.* The control element responsible for this effect is consequently termed an **attenuator.**

The *trp* Attenuator's Transcription Terminator Is Masked When Tryptophan Is Scarce

What is the mechanism of attenuation? The attenuator transcript contains four complementary segments that can form one of two sets of mutually exclusive base paired hairpins (Fig. 29-31). *Segments 3 and 4 together with the succeeding residues comprise a normal transcription terminator (Section 29-2E):* a G + C-rich sequence that can form a self-complementary hairpin structure followed by several sequential U's (compare with Fig. 29-15). *Transcription rarely proceeds beyond this termination site unless tryptophan is in short supply.*

A section of the leader sequence, which includes segment 1 of the attenuator, is translated to form a 14-residue polypeptide that contains two consecutive Trp residues (Fig. 29-31, *left*). The position of this particularly rare dipeptide segment (~1% of the residues in *E. coli* proteins are Trp) provided an important clue to the mechanism of attenuation. An additional essential aspect of this mechanism is that ribosomes commence the translation of a prokaryotic mRNA shortly after its 5′ end has been synthesized.

The above considerations led Yanofsky to propose the following model of attenuation (Fig. 29-32). An RNA

polymerase that has escaped repression initiates *trp* operon transcription. Soon after the ribosomal initiation site of the *trpL* gene has been transcribed, a ribosome attaches to it and begins translation of the leader peptide. When tryptophan is abundant, so that there is a plentiful supply of **tryptophanyl–tRNA^Trp** (the transfer RNA specific for Trp with an attached Trp residue; Section 30-2C), the ribosome follows closely behind the transcribing RNA polymerase so as to sterically block the formation of the 2·3 hairpin. Indeed, RNA polymerase pauses past position 92 of the transcript and only continues transcription upon the approach of a ribosome, thereby ensuring the proximity of these two entities at this critical position. The prevention of 2·3 hairpin formation permits the formation of the 3·4 hairpin, the transcription terminator pause site, which results in the termination of transcription (Fig. 29-32*a*). When tryptophan is scarce, however, the ribosome stalls at the tandem UGG codons (the three sequential nucleotides specifying Trp; Section 30-1E) because of the lack of tryptophanyl–tRNA^Trp. As transcription continues, the newly synthesized segments 2 and 3 form a hairpin because the stalled ribosome prevents the otherwise competitive formation of the 1·2 hairpin (Fig. 29-32*b*). The formation of the transcrip-

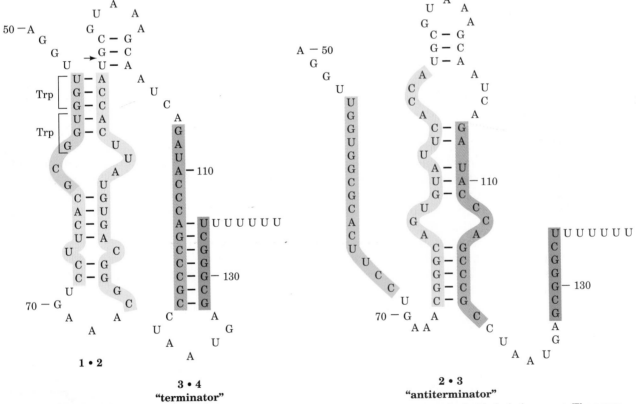

FIGURE 29-31. The alternative secondary structures of *trpL* mRNA. The formation of the base paired 2·3 (antiterminator) hairpin (*right*) precludes the formation of the 1·2 and 3·4 (terminator) hairpins (*left*) and *vice versa*. Attenuation results in the premature termination of transcription immediately after nucleotide 140 when the 3·4 hairpin is present. The arrow indicates the mRNA site past which RNA polymerase pauses until approached by an active ribosome. [After Fisher, R.F. and Yanofsky, C., *J. Biol. Chem.* **258,** 8147 (1983).]

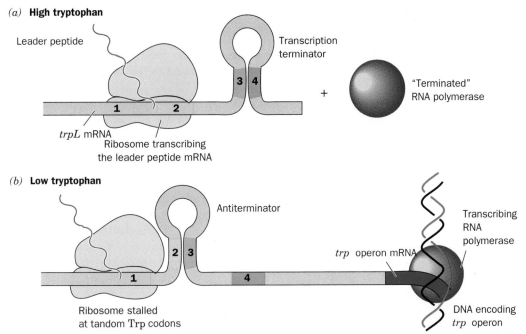

FIGURE 29-32. Attenuation in the *trp* operon. (*a*) When tryptophanyl–tRNA^Trp is abundant, the ribosome translates *trpL* mRNA. The presence of the ribosome on segment 2 prevents the formation of the base paired 2·3 hairpin. The 3·4 hairpin, an essential component of the transcriptional terminator, can thereby form thus aborting transcription.

(*b*) When tryptophanyl–tRNA^Trp is scarce, the ribosome stalls on the tandem Trp codons of segment 1. This situation permits the formation of the 2·3 hairpin which, in turn, precludes the formation of the 3·4 hairpin. RNA polymerase therefore transcribes through this unformed terminator and continues *trp* operon transcription.

tional terminator's 3·4 hairpin is thereby preempted for sufficient time for RNA polymerase to transcribe through it and consequently through the remainder of the *trp* operon. The cell is thus provided with a regulatory mechanism that is responsive to the tryptophanyl–tRNA^Trp level, which, in turn, depends on the protein synthesis rate as well as on the tryptophan supply.

There is considerable evidence supporting this model of attenuation. The *trpL* transcript is resistant to limited RNase T1 digestion indicating that it has extensive second-

ary structure. The significance of the tandem Trp codons in the *trpL* transcript is corroborated by their presence in *trp* leader regions of several other bacterial species. Moreover, the leader peptides of the five other amino acid–biosynthesizing operons known to be regulated by attenuation (most exclusively so) are all rich in their corresponding amino acid residues (Table 29-1). For example, the *E. coli* **his operon,** which specifies enzymes synthesizing histidine (Fig. 24-52), has seven tandem His residues in its leader peptide whereas the **ilv operon,** which specifies enzymes

TABLE 29-1. AMINO ACID SEQUENCES OF SOME LEADER PEPTIDES IN OPERONS SUBJECT TO ATTENUATION

Operon	Amino Acid Sequence[a]
trp	Met-Lys-Ala-Ile-Phe-Val-Leu-Lys-Gly-TRP-TRP-Arg-Thr-Ser
pheA	Met-Lys-His-Ile-Pro-PHE-PHE-PHE-Ala-PHE-PHE-PHE-Thr-PHE-Pro
his	Met-Thr-Arg-Val-Gln-Phe-Lys-HIS-HIS-HIS-HIS-HIS-HIS-HIS-Pro-Asp
leu	Met-Ser-His-Ile-Val-Arg-Phe-Thr-Gly-LEU-LEU-LEU-LEU-Asn-Ala-Phe-Ile-Val-Arg-Gly-Arg-Pro-Val-Gly-Gly-Ile-Gln-His
thr	Met-Lys-Arg-ILE-Ser-THR-THE-ILE-THR-THR-THR-ILE-THR-ILE-THR-THR-Gln-Asn-Gly-Ala-Gly
ilv	Met-Thr-Ala-LEU-LEU-Arg-VAL-ILE-Ser-LEU-VAL-VAL-ILE-Ser-VAL-VAL-VAL-ILE-ILE-ILE-Pro-Pro-Cys-Gly-Ala-Ala-Leu-Gly-Arg-Gly-Lys-Ala

[a] Upper case residues are synthesized in the pathway catalyzed by the operon's gene products.

Source: Yanofsky, C., *Nature* **289,** 753 (1981).

participating in isoleucine, leucine, and valine biosynthesis (Fig. 24-46), has five Ile's, three Leu's, and six Val's in its leader peptide. Finally, the leader transcripts of these operons resemble that of the *trp* operon in their capacity to form two alternative secondary structures, one of which contains a trailing termination structure.

G. Regulation of Ribosomal RNA Synthesis: The Stringent Response

E. coli cells growing under optimal conditions divide every 20 min. Such cells contain nearly 10,000 ribosomes. Yet, RNA polymerase can initiate the transcription of an rRNA gene no faster than about once every second. If the *E. coli* genome contained only one copy of each of the three types of rRNA genes (those specifying the so-called 23S, 16S, and 5S rRNAs; Section 30-3A), fast-growing cells could contain no more than ~1200 ribosomes. However, *the E. coli chromosome contains seven separately located rRNA operons, all of which contain one nearly identical copy of each type of rRNA gene,* thereby accounting for the observed rRNA synthesis rate.

Cells have the remarkable ability to coordinate the rates at which their thousands of components are synthesized. For example, *E. coli* adjust their ribosome content to match the rate at which they can synthesize proteins under the prevailing growth conditions. The rate of rRNA synthesis is therefore proportional to the rate of protein synthesis. One mechanism by which this occurs is known as the **stringent response:** *A shortage of any species of amino acid–charged tRNA (usually a result of "stringent" or poor growth conditions) that limits the rate of protein synthesis triggers a sweeping metabolic readjustment.* A major facet of this change is an abrupt 10- to 20-fold reduction in the rate of rRNA and tRNA synthesis. This **stringent control,** moreover, depresses numerous metabolic processes (including DNA replication and the biosynthesis of carbohydrates, lipids, nucleotides, proteoglycans, and glycolytic intermediates) while stimulating others (such as amino acid biosynthesis). The cell is thereby prepared to withstand nutritional deprivation.

ppGpp Mediates the Stringent Response

*The stringent response is correlated with a rapid intracellular accumulation of the unusual nucleotide **ppGpp** and its prompt decay when amino acids become available.* The observation that mutants, designated *relA⁻*, which do not exhibit the stringent response (they are said to have **relaxed control**), lack ppGpp suggests that this substance mediates the stringent response. This idea was corroborated by *in vitro* studies demonstrating, for example, that ppGpp inhibits the transcription of rRNA genes but stimulates the transcription of the *trp* and *lac* operons as does the stringent response *in vivo*. It therefore seems that ppGpp acts by somehow altering RNA polymerase's promoter specificity at stringently controlled operons, a hypothesis that is supported by the isolation of RNA polymerase mutants that exhibit reduced responses to ppGpp.

Experiments with cell-free *E. coli* extracts have established that the protein encoded by wild-type *relA* gene, named **stringent factor,** catalyzes the reaction

$$ATP + GDP \rightleftharpoons AMP + ppGpp$$

Stringent factor is only active in association with a ribosome that is actively engaged in translation. ppGpp synthesis occurs at a maximal rate when a ribosome binds its mRNA-specified but uncharged (lacking an amino acid residue) tRNA. The binding of a specified and charged tRNA greatly reduces the rate of ppGpp synthesis. *The ribosome apparently signals the shortage of an amino acid by stimulating the synthesis of ppGpp which, acting as an intracellular messenger, influences the rates at which a great variety of operons are transcribed.*

ppGpp degradation is catalyzed by the *spoT* gene product. The *spoT⁻* mutants show a normal increase in ppGpp level upon amino acid starvation but an abnormally slow decay of ppGpp to basal levels when amino acids again become available. The *spoT⁻* mutants therefore exhibit a sluggish recovery from the stringent response. *The ppGpp level is apparently regulated by the countervailing activities of stringent factor and the spoT gene product.*

4. POSTTRANSCRIPTIONAL PROCESSING

The immediate products of transcription, the **primary transcripts,** are not necessarily functional entities. In order to acquire biological activity, many of them must be specifically altered in several ways: (1) by the exo and endonucleolytic removal of polynucleotide segments; (2) by appending nucleotide sequences to their 3' and 5' ends; and (3) by the modification of specific nucleosides. The three major classes of RNAs, mRNA, rRNA, and tRNA, are altered in different ways in prokaryotes and in eukaryotes. In this section we shall outline these **posttranscriptional modification** processes.

A. Messenger RNA Processing

In prokaryotes, most primary mRNA transcripts function in translation without further modification. Indeed, as we have seen, ribosomes in prokaryotes usually commence translation on nascent mRNAs. In eukaryotes, however, mRNAs are synthesized in the cell nucleus whereas translation occurs in the cytosol. Eukaryotic mRNA transcripts can therefore undergo extensive posttranscriptional processing while still in the nucleus.

Eukaryotic mRNAs Are Capped

*Eukaryotic mRNAs have a peculiar enzymatically appended **cap structure** consisting of a 7-methylguanosine res-*

idue joined to the transcript's initial (5') nucleoside via a 5'–5' triphosphate bridge (Fig. 29-33). The cap, which a specific guanylyltransferase adds to the growing transcript before it is >20-nucleotides long, defines the eukaryotic translational start site (Section 30-3C). A cap may be $O^{2'}$-methylated at the transcript's leading nucleoside (**cap-1**, the predominant cap in multicellular organisms), at its first two nucleosides (**cap-2**), or at neither of these positions (**cap-0**, the predominant cap in unicellular eukaryotes). If the leading nucleoside is adenosine (it is usually a purine), it may also be N^6-methylated.

Eukaryotic mRNAs Have Poly(A) Tails

Eukaryotic mRNAs, in contrast to those of prokaryotes, are invariably monocistronic. Yet, the sequences signaling transcriptional termination in eukaryotes have not been identified. This is largely because the termination process is imprecise; that is, the primary transcripts of a given structural gene have heterogeneous 3' sequences. Nevertheless,

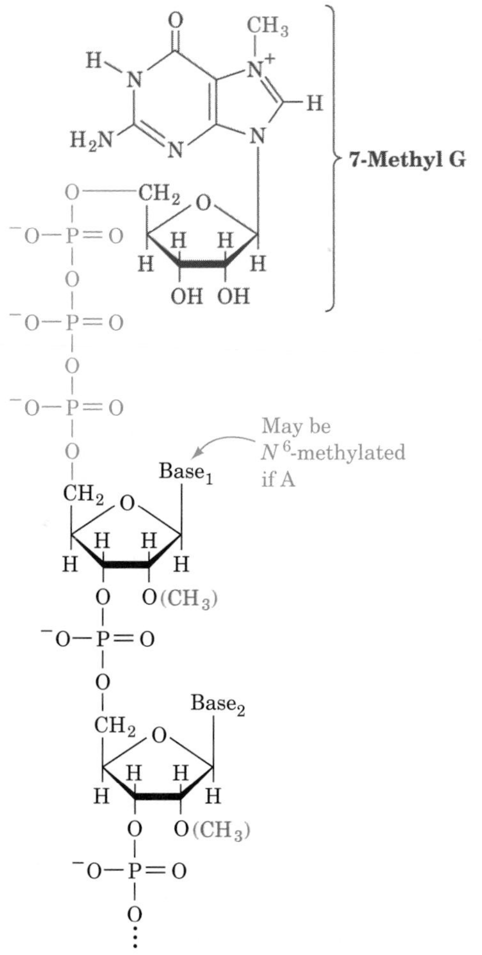

7-Methyl G

May be N^6-methylated if A

Base$_1$

O(CH$_3$)

Base$_2$

O(CH$_3$)

FIGURE 29-33. The structure of the 5' cap of eukaryotic mRNAs. It is known as cap-0, cap-1, or cap-2, respectively, if it has no further modifications, if the leading nucleoside of the transcript is $O^{2'}$-methylated, or if its first two nucleosides are $O^{2'}$-methylated.

mature eukaryotic mRNAs have well-defined 3' ends; almost all of them have 3'-poly(A) tails of 20 to 50 nucleotides. The poly(A) tails are enzymatically appended to the primary transcripts in two reactions:

1. A transcript is cleaved 15 to 25 nucleotides past a highly conserved AAUAAA sequence, whose mutation abolishes cleavage and polyadenylation, and within 50 nucleotides before a less conserved U-rich or GU-rich sequence. The precision of the cleavage reaction has apparently eliminated the need for accurate transcriptional termination; to put things another way, all's well that ends well.

2. The poly(A) tail is subsequently generated from ATP through the stepwise action of **poly(A) polymerase**. This enzyme is activated by **cleavage and polyadenylation specificity factor (CPSF)** upon this latter protein's recognition of the AUUAAA sequence, which it does with almost no tolerance for sequence variation. Once the poly(A) tail has grown to ~10 residues, the AUUAAA sequence is no longer required for further chain elongation. This suggests that CPSF becomes disengaged from its recognition site in a manner reminiscent of the way σ factor is released from the transcriptional initiation site once the elongation of prokaryotic mRNA is under way (Section 29-2C). It is unclear what limits the length of the poly(A) tails although it seems unlikely that it is an intrinsic property of poly(A) polymerase. The polyadenylation machinery is part of a 500- to 1000-kD complex that also contains at least three proteins that are required for mRNA cleavage.

In vitro studies indicate that a poly(A) tail is not required for mRNA translation. Rather, the observations that an mRNA's poly(A) tail shortens as it ages in the cytosol and that unadenylated mRNAs have abbreviated cytosolic lifetimes suggest that poly(A) tails have a protective role. In fact, the only mature mRNAs that generally lack poly(A) tails, those of histones (which, with few exceptions, lack the AAUAAA cleavage–polyadenylation signal), have lifetimes of <30 min in the cytosol, whereas most other mRNAs last hours or days. The poly(A) tails are specifically complexed in the cytosol by **poly(A) binding protein (PABP)**, which organizes the mRNAs into ribonucleoprotein particles. PABP is thought to protect mRNA from degradation as is suggested, for example, by the observation that the addition of PABP to a cell-free system containing mRNA and mRNA-degrading nucleases greatly reduces the rate at which the mRNAs are degraded and the rate at which their poly(A) tails are shortened.

Eukaryotic Genes Consist of Alternating Expressed and Unexpressed Sequences

The most striking difference between eukaryotic and prokaryotic structural genes is that the coding sequences of most eukaryotic genes are interspersed with unexpressed regions. Early investigations of eukaryotic structural gene transcription found, quite surprisingly, that primary transcripts are highly heterogeneous in length (from ~2000 to well over 20,000 nucleotides) and are much larger than is expected from the known sizes of eukaryotic proteins. Rapid labeling experiments demonstrated that little of this so-called **heterogeneous nuclear RNA (hnRNA)** is ever transported to the cytosol; most of it is quickly turned over (degraded) in the nucleus. Yet, the hnRNA's 5′ caps and 3′ tails eventually appear in cytosolic mRNAs. *The straightforward explanation of these observations, that **pre-mRNAs** are processed by the excision of internal sequences, seemed so bizarre that it came as a great suprise in 1977 when Phillip Sharp and Richard Roberts independently demonstrated that this is actually the case.* In fact, pre-mRNAs typically contain eight noncoding **intervening sequences (IVSs** or **introns)** whose aggregate length averages four to ten times that of its flanking **expressed sequences (exons).** This situation is graphically illustrated in Fig. 29-34, which is an electron micrograph of chicken **ovalbumin** mRNA hybridized to the antisense strand of the ovalbumin gene (ovalbumin is the major protein component of egg white). The lengths of introns in vertebrate genes ranges from ~65 to ~200,000 nucleotides with no obvious periodicity. Indeed, the corresponding introns from genes in two vertebrate species can vary extensively in both length and sequence so as to bear little resemblance to one another.

Further investigations established that the formation of eukaryotic mRNA begins with the transcription of an entire structural gene, including its introns, to form pre-mRNA (Fig. 29-35). Then, following capping and perhaps polyadenylation, the introns are excised and their flanking

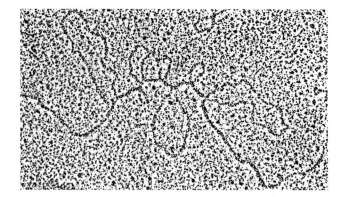

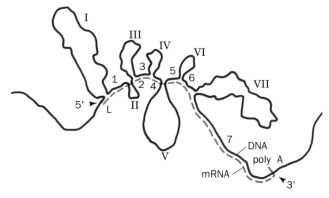

FIGURE 29-34. An electron micrograph and its interpretive drawing of a hybrid between the antisense strand of the chicken ovalbumin gene (as obtained by molecular cloning methods; Section 28-8) and its corresponding mRNA. The complementary segments of the DNA (*purple line in drawing*) and mRNA (*red line*) have annealed to reveal the exon positions (*L*, 1–7). The looped-out segments (I–VII), which have no complementary sequences in the mRNA, are the introns. [From Chambon, P., *Sci. Am.* **244**(5): 61 (1981)].

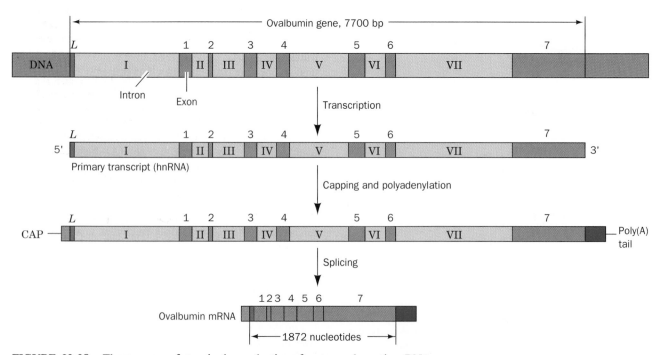

FIGURE 29-35. The sequence of steps in the production of mature eukaryotic mRNA as shown for the chicken ovalbumin gene. Following transcription, the primary transcript is capped and polyadenylated. The introns are then excised and the exons spliced together to form the mature mRNA.

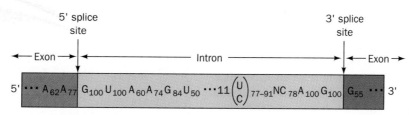

FIGURE 29-36. The consensus sequence at the exon–intron junctions of eukaryotic pre-mRNAs. The subscripts indicate the percentage of pre-mRNAs in which the specified base(s) occurs. Note that the 3′ splice site is preceded by a tract of 11 predominantly pyrimidine nucleotides. [Based on data from Padgett, R.A., Grabowski, P.J., Konarska, M.M., Seiler, S.S., and Sharp, P.A., *Annu. Rev. Biochem.* **55**, 1123 (1986).]

exons are connected, a process called **gene splicing**, to yield the mature mRNA. *The most striking aspect of gene splicing is its precision; if one base too few or too many were excised, the resulting mRNA could not be translated properly (Section 30-1B). Moreover, exons are never shuffled; their order in the mature mRNA is exactly the same as that in the gene from which it is derived.* In the following subsections we discuss the mechanism of this remarkable splicing process.

Exons Are Spliced in a Two-Stage Reaction

Sequence comparisons of exon–intron junctions from a diverse group of eukaryotes indicate that they have a high degree of homology (Fig. 29-36), including, as Richard Breathnach and Pierre Chambon first pointed out, *an invariant GU at the intron's 5′ boundary and an invariant AG*

at its 3′ boundary. These sequences are necessary and sufficient to define a splice junction: Mutations that alter the sequences interfere with splicing, whereas mutations that change a nonjunction to a consensus-like sequence can generate a new splice junction.

Investigations of both cell free and *in vivo* splicing systems by Argiris Efstradiadis, Tom Maniatis, Michael Rosbash, and Sharp established that intron excision occurs via two transesterification reactions (Fig. 29-37):

1. The formation of a 2′,5′-phosphodiester bond between an intron adenosine residue and its 5′-terminal phosphate group with the concomitant release of the 5′ exon. *The intron thereby assumes a novel **lariat structure.*** The adenosine residue at the lariat branch has been identified as the *A* in the sequence CUR*A*Y [where R represents

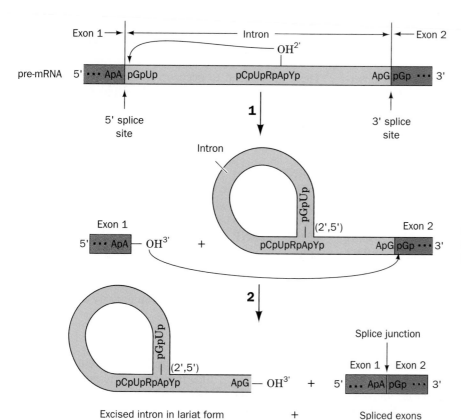

FIGURE 29-37. The sequence of transesterification reactions that splice together the exons of eukaryotic pre-mRNAs (the exons and introns are drawn in blue and orange; R and Y represent purine and pyrimidine residues): **(1)** The 2′-OH group of a specific intron A residue nucleophilically attacks the 5′-phosphate at the 5′ intron boundary to yield an unusual 2′,5′-phosphodiester bond and thus form a lariat structure. **(2)** The liberated 3′-OH group forms a 3′,5′-phosphodiester bond with the 5′ terminal residue of the 3′ exon, thereby splicing the two exons together and releasing the intron in lariat form.

purines (A or G) and Y represents pyrimidines (C or U)], which is highly conserved in vertebrate mRNAs and is typically located 20 to 50 residues upstream of the 3′ splice site (yeast have a similar UACUAAC sequence that occurs ~50-residues upstream from all its 3′ splice sites). Mutations that change this branch point A residue abolish splicing at that site.

2. The now free 3′-OH group of the 5′ exon forms a phosphodiester bond with the 5′-terminal phosphate of the 3′ exon yielding the spliced product. The intron is thereby eliminated in its lariat form and, *in vivo,* is rapidly degraded. Mutations that alter the conserved AG at the 3′ splice junction block this second step, although they do not interfere with lariat formation.

Note that the splicing process proceeds without free energy input; its transesterification reactions preserve the free energy of each cleaved phosphodiester bond through the concomitant formation of a new one.

Splicing Is Mediated by snRNPs

How are splice junctions recognized and how are the two exons to be joined brought together in the splicing process? Part of the answer to this question was established by Joan Steitz going on the assumption that one nucleic acid is best recognized by another. The eukaryotic nucleus, as has been known since the 1960s, contains numerous copies of several highly conserved 60- to 300-nucleotide RNAs called **small nuclear RNAs (snRNAs),** which form protein complexes termed **small nuclear ribonucleoproteins (snRNPs;** pronounced "snurps"). Steitz recognized that the 5′ end of one of these snRNAs, **U1-snRNA** (so called because it is a member of a U-rich subfamily of snRNAs), is partially complementary to the consensus sequence of 5′ splice junctions. The consequent hypothesis, that *U1-snRNA recognizes the 5′ splice junction,* was corroborated by the observations that splicing is inhibited by the selective destruction of the U1-snRNA sequences that are complementary to the 5′ splice junction and by the presence of anti-U1-snRNP antibodies (produced by patients suffering from **systemic lupus erythematosus,** an often fatal autoimmune disease). Other snRNPs that have been implicated in splicing are **U2-snRNP, U4–U6-snRNP** (in which the U4- and U6-snRNAs associate via base pairing), and **U5-snRNP.** Although the splicing reaction sequence is far from being understood, it appears that U1-snRNA recognizes the 5′ splice junction in the first transesterification reaction, **U2-snRNA** recognizes the intron region that forms the lariat branch point, **U5-snRNA** aligns the two exons for the second transesterification reaction, and **U6-snRNA** pairs with the 5′ splice junction in the second transesterification reaction (the same region recognized by U1 in the first transesterification reaction). Altogether, ~65 pre-mRNA nucleotides participate in this recognition process, which rationalizes why introns are minimally ~65 nucleotides in length.

Splicing takes place in an as yet poorly characterized 50S to 60S particle dubbed the **spliceosome** *(Fig. 29-38). The*

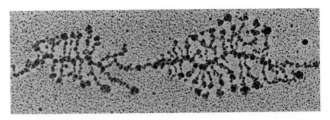

FIGURE 29-38. An electron micrograph of spliceosomes in action. The spliceosomes are the large beads on the pre-mRNAs extending above and below the horizontal DNA. [From Steitz, J. A., *Sci. Am.* **258**(6): 59 (1988). Electron micrograph by Yvonne N. Osheim.]

spliceosome brings together a pre-mRNA, the foregoing snRNPs, and a variety of pre-mRNA binding proteins. Note that the spliceosome, which consists of 5 RNAs and at least 50 polypeptides, is comparable in size and complexity to the *E. coli* ribosome's large (50S) subunit (which consists of 2 RNAs and 31 polypeptides; Section 30-3A).

A simplistic interpretation of Fig. 29-37 suggests that any 5′ splice site could be joined with any following 3′ splice site, thereby eliminating all the intervening exons together with the introns joining them. This does not occur. Rather, all of a pre-mRNA's introns are individually excised in what appears to be a largely fixed order that more or less proceeds in the 5′ → 3′ direction. The spliceosome, no doubt, plays a major role in ensuring the proper excision of all introns. The biochemical significance of splicing is discussed in Section 33-2F.

mRNA Is Methylated at Certain Adenylate Residues

During or shortly after the synthesis of vertebrate pre-mRNAs, ~0.1% of their A residues are methylated at N6. These m^6A's tend to occur in the sequence RRm^6ACX, where X is rarely G. Although the functional significance of these methylated A's is unknown, it should be noted that a large fraction of them are components of the corresponding mature mRNAs.

hnRNP Proteins

Throughout their residency in the nucleus, hnRNAs (pre-mRNAs) are closely associated with a great variety of proteins, thereby forming **hnRNPs.** Although the functions of these **hnRNP proteins** are largely unknown, it can be anticipated that they facilitate the processing of the hnRNAs and their transport to different regions of the nucleus and, ultimately, to the cytoplasm, where the resulting mature mRNAs associate with an entirely different set of proteins.

RNA Editing

Certain mRNAs from a variety of eukaryotic organisms have been found to differ from their corresponding genes in several unexpected ways, including C → U and U → C changes, the insertion or deletion of U residues, and the

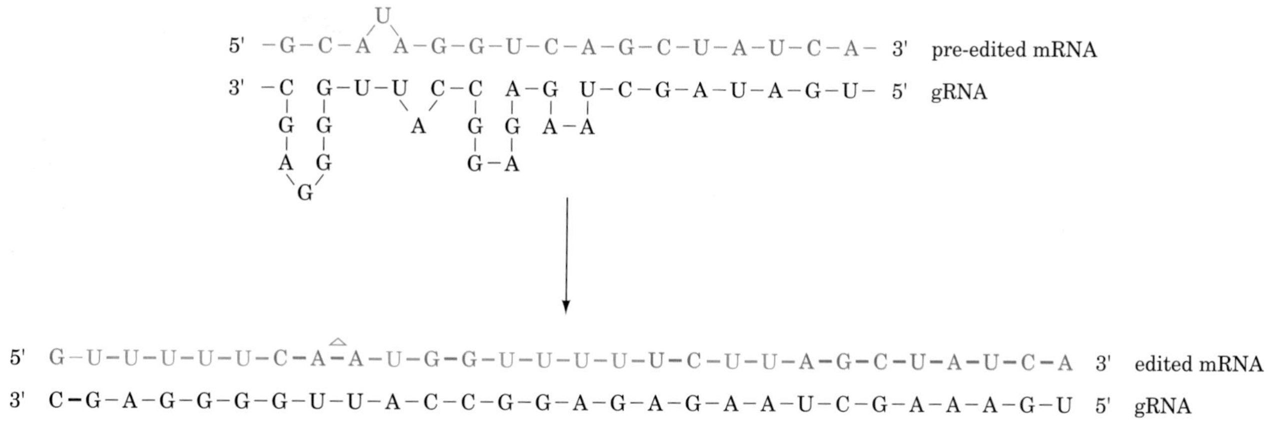

```
                       U
 5' –G–C–A   A–G–G–U–C–A–G–C–U–A–U–C–A–  3'   pre-edited mRNA
 3' –C  G–U–U  C–C  A–G  U–C–G–A–U–A–G–U– 5'   gRNA
        G  G      A    G  G  A–A
        A  G         G–A
         \G/
```

```
 5'  G–U–U–U–U–U–C–A–Â–A–U–G–G–U–U–U–U–U–C–U–U–A–G–C–U–A–U–C–A  3'   edited mRNA
 3'  C–G–A–G–G–G–G–U–U–A–C–C–G–G–A–G–A–G–A–A–U–C–G–A–A–A–G–U  5'   gRNA
```

FIGURE 29-39. A schematic diagram indicating how gRNAs direct the editing of kinetoplastid pre-edited-mRNAs. Several gRNAs may be necessary to direct the editing of consecutive segments of a pre-edited mRNA. [After Bass, B.L., *in* Gesteland, R.F. and Atkins, J.F., *The RNA World*, p. 387, Cold Spring Harbor Laboratory Press (1993).]

insertion of multiple G or C residues. The most extreme examples of this phenomenon, which occur in the mitochondria of **kinetoplastid protozoa (trypanosomes),** involve the addition and removal of up to hundreds of U's to and from otherwise untranslatable mRNAs. The process whereby a transcript is altered in this manner is called **RNA editing** because it originally seemed that the required enzymatic reactions occurred without the direction of a nucleic acid template and hence violated the central dogma of molecular biology (Figure 29-1). Eventually, however, a new class of kinetoplastid mitochondrial transcripts called **guide RNAs (gRNAs)** was identified. gRNAs, which consist of 60 to 80 nucleotides, have 3' oligo(U) tails, an internal segment that is precisely complementary to the edited portion of the mRNA (if G·U pairs, which are common in RNAs, are taken to be complementary), and a short so-called anchor sequence near the 5' end that is largely complementary in the Watson–Crick sense to a segment of the mRNA that is not edited. An unedited transcript presumably associates with the corresponding gRNA via its anchor sequence (Fig. 29-39). Then, in a process mediated by the appropriate enzymatic machinery in what appears to be a high molecular mass particle named the **editosome,** the gRNA's internal segment is used as a template to "correct" the transcript, thereby yielding the edited mRNA. Interestingly, the gRNA's 3' oligo(U) tail is the source of the added U's. Thus, the genetic information specifying an edited mRNA is derived from at least two different genes. The functional advantage of this complicated process, either presently or more likely in some ancestral organism, is obscure.

Editing of Apolipoprotein B mRNA

Humans express two forms of **apolipoprotein B (apoB):** **apoB-48,** which is made in the small intestine and functions in chylomicrons to transport triacylglycerols from the intestine to the liver and peripheral tissues; and **apoB-100,** which is made in the liver and functions in VLDL, IDL, and LDL to transport cholesterol from the liver to the periph-

eral tissues (Sections 11-5B and C). ApoB-100 is an enormous 4536-residue protein, whereas apoB-48 consists of apoB-100's N-terminal 2152 residues and therefore lacks the C-terminal domain of apoB-100 that mediates LDL receptor binding.

Despite their differences, both apoB-48 and apoB-100 are expressed from the same gene. How does this occur? Comparison of the mRNAs encoding the two proteins indicates that they differ by a single C → U change: The triplet of nucleotides (codon) that codes for Gln 2153 (CAA) in apoB-100 mRNA is, in apoB-48 mRNA, a triplet that terminates ribosomal polypeptide synthesis (a **stop codon,** UAA; we shall see in Section 30-1 that the amino acid residues in proteins are specified by codons on mRNA). The activity that catalyzes this conversion is a protein: It is destroyed by proteases and protein-specific reagents but not by nucleases. When apoB mRNA is synthesized with [α-^{32}P]CTP, *in vitro* editing yields a [^{32}P]UMP residue solely at the editing site. Therefore, the editing activity is most probably a site-specific **cytidine deaminase.** This type of RNA editing differs in character from that in trypanosomal mitochondria, which insert and delete multiple U's into mRNAs under the direction of guide RNAs. ApoB mRNA editing therefore falls into a new class of RNA editing, **substitutional editing,** which resembles the type of posttranscriptional modifications that yield the modified nucleosides in tRNAs (Section 30-2A). As yet, only one other example of this type of RNA editing is known: Rat brain **glutamate receptor** mRNA undergoes an A → G change, which transforms a Gln codon (CAG) to that of a functionally important Arg (CGG).

B. Ribosomal RNA Processing

The seven *E. coli* rRNA operons all contain one (nearly identical) copy of each of the three types of rRNA genes (Section 29-3F). Their polycistronic primary transcripts, which are > 5500 nucleotides in length, contain 16S rRNA

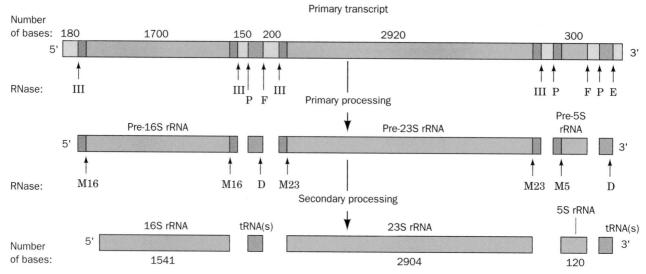

FIGURE 29-40. The posttranscriptional processing of *E. coli* rRNA. The transcriptional map is shown approximately to scale. The labeled arrows indicate the positions of the various nucleolytic cuts and the nucleases that generate them. [After Apiron, D., Ghora, B.K., Plantz, G., Misra, T.K., and Gegenheimer, P., *in* Söll, D., Abelson, J.N., and Schimmel P.R. (Eds.), *Transfer RNA: Biological Aspects, p.* 148, Cold Spring Harbor Laboratory (1980).]

at their 5′ ends followed by the transcripts for 1 or 2 tRNAs, 23S rRNA, 5S rRNA and, in some rRNA operons, 1 or 2 more tRNAs at the 3′ end (Fig. 29-40). The steps in processing these primary transcripts to mature rRNAs (Fig. 29-40) were elucidated with the aid of mutants defective in one or more of the processing enzymes.

The initial processing, which yields products known as **pre-rRNAs,** commences while the primary transcript is still being synthesized. It consists of specific endonucleolytic cleavages by **RNase III, RNase P, RNase E,** and **RNase F** at the sites indicated in Fig. 29-40. The base sequence of the primary transcript suggests the existence of several base paired stems. The RNase III cleavages occur in a stem consisting of complementary sequences flanking the 5′ and 3′ ends of the 23S segment (Fig. 29-41) as well as that of the 16S segment. Presumably certain features of these stems constitute the RNase III recognition site.

The 5′ and 3′ ends of the pre-rRNA's are trimmed away in secondary processing steps (Fig. 29-40) through the action of **RNASes M16, M23,** and **M5** to produce the mature rRNAs. These final cleavages only occur after the pre-rRNAs become associated with ribosomal proteins.

Ribosomal RNAs Are Methylated

During ribosomal assembly, the 16S and 23S rRNAs are methylated at a total of 24 specific nucleosides. The methylation reactions, which employ *S*-adenosylmethionine (Section 24-3E) as a methyl donor, yield N^6, N^6-dimethyl-adenine and $O^{2'}$-methylribose residues. $O^{2'}$-methyl groups are thought to protect adjacent phosphodiester bonds from degradation by intracellular RNases (the mechanism of RNase hydrolysis involves utilization of the free 2′-OH

group of ribose to eliminate the substituent on the 3′-phosphoryl group via the formation of a 2′,3′-cyclic phosphate intermediate; Section 28-1). However, the function of base methylation is unknown.

Eukaryotic rRNA Processing Resembles That of Prokaryotes

The eukaryotic genome typically has several hundred tandemly repeated copies of rRNA genes that are contained in small dark-staining nuclear bodies known as nucleoli (the site of rRNA transcription and processing and ribosomal subunit assembly; Fig. 1-5). The primary rRNA transcript is an ~7500-nucleotide 45S RNA that contains, starting from its 5′ end, the 18S, 5.8S, and 28S rRNAs separated by spacer sequences (Fig. 29-42). In the first stage of its processing, 45S RNA is specifically methylated at ~110 sites that occur mostly in its rRNA sequences. About 80% of these modifications yield $O^{2'}$-methylribose residues and the remainder form methylated bases such as N^6, N^6-dimethyladenine and 2-methylguanine. The subsequent cleavage and trimming of the 45S RNA superficially resembles that of prokaryotic rRNAs. In fact, enzymes exhibiting RNase III- and RNase P-like activities occur in eukaryotes. The 5S eukaryotic rRNA is separately processed in a manner resembling that of tRNA (Section 29-4C).

Some Eukaryotic rRNA Genes Are Self-Splicing

Only a few eukaryotic rRNA genes contain introns. Nevertheless, Thomas Cech's study of how such genes are spliced in the ciliated protozoan *Tetrahymena thermophila* led to an astonishing discovery: *RNA can act as an enzyme. When the isolated pre-rRNA of this organism is incubated*

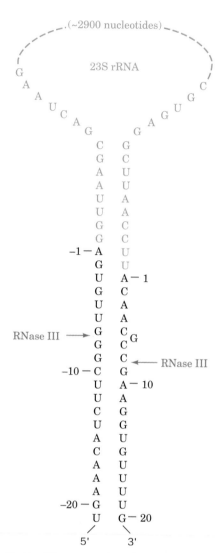

FIGURE 29-41. The proposed stem-and-giant-loop secondary structure in the 23S region of the *E. coli* primary rRNA transcript. The RNase III cleavage sites are indicated. [After Young. R.R., Bram, R.J., and Steitz, J.A., *in* Söll, D., Abelson, J.N., and Schimmel, P.R. (Eds.), *Transfer RNA: Biological Aspects*, p. 102, Cold Spring Harbor Laboratory (1980).]

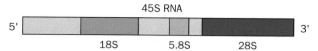

FIGURE 29-42. The organization of the 45S primary transcript of eukaryotic rRNA.

3. The 3′-terminal OH group of the intron forms a phosphodiester bond with the phosphate of the nucleotide 15 residues from the intron's 5′ end, yielding the 5′-terminal fragment with the remainder of the intron in cyclic form.

This self-splicing process consists of a series of transesterifications and therefore does not require free energy input. Cech further established the enzymatic properties of the *Tetrahymena* intron [a member of the so-called **group I introns,** which occur in the nuclei, mitochondria and chloroplasts of diverse eukaryotes (although not vertebrates), and even in some bacteria], which presumably stem from its three-dimensional structure, by demonstrating that it catalyzes the *in vitro* cleavage of poly(C) with an enhancement factor of 10^{10} over the rate of spontaneous hydrolysis. Indeed, this RNA catalyst even exhibits Michaelis–Menten kinetics ($K_M = 42$ μM and $k_{cat} = 0.033$ s^{-1} for C_5). Such RNA enzymes have been named **ribozymes.**

Several unrelated types of naturally occurring ribozymes have been characterized. Among them are the self-splicing pre-rRNAs known as **group II introns,** which occur in the mitochondria of fungi and plants and comprise the majority of the introns in chloroplasts. They react via a lariat intermediate and do not utilize an external nucleotide, a process that resembles the splicing of nuclear pre-mRNAs (Fig. 29-37).

Although the idea that an RNA can have enzymatic properties may be unorthodox, *there is no fundamental reason that an RNA, or any other macromolecule, cannot have catalytic activity* (recall that it was likewise once generally accepted that nucleic acids lack the complexity to carry hereditary information; Section 27-2). Of course, in order to be an efficient catalyst, a macromolecule must be able to assume a stable structure but, as we shall see in Sections 30-2B and 3A, RNAs such as tRNAs and rRNAs do just that. The chemical similarities of the nuclear mRNA and group II intron splicing reactions therefore suggest that *spliceosomes are ribozymal systems whose snRNA components have evolved from primordial self-splicing RNAs and that their protein components serve mainly to fine tune ribozymal structure and function* (e.g., promote snRNA–snRNA, snRNA–intron, and snRNA–exon interactions). Similarly, the RNA components of ribosomes, which are more than half RNA and the rest protein, almost certainly have catalytic functions in addition to the structural and recognition roles traditionally attributed to them (Section 30-3).

with guanosine or a free guanine nucleotide (GMP, GDP, or GTP), but in the absence of protein, its single 413-nucleotide intron excises itself and splices together its flanking exons; that is, this pre-rRNA is self-splicing. The three-step reaction sequence of this process (Fig. 29-43) resembles that of mRNA splicing:

1. The 3′-OH group of the guanosine forms a phosphodiester bond with the intron's 5′ end.

2. The 3′-terminal OH group of the newly liberated 5′ exon forms a phosphodiester bond with the 5′-terminal phosphate of the 3′ exon, thereby splicing together the two exons and releasing the intron.

FIGURE 29-43. The sequence of reactions in the self-splicing of *Tetrahymena* pre-rRNA: **(1)** The 3'-OH group of a guanine nucleotide attacks the intron's 5'-terminal phosphate so as to form a phosphodiester bond and release the 5' exon. **(2)** The newly generated 3'-OH group of the 5' exon attacks the 5'-terminal phosphate of the 3' exon, thereby splicing the two exons and releasing the intron. **(3)** The 3'-OH group of the intron attacks the phosphate of the nucleotide that is 15 residues from the 5' end so as to cyclize the intron and release its 5'-terminal fragment. Throughout this process, the RNA maintains a folded, internally hydrogen bonded conformation that permits the precise excision of the intron.

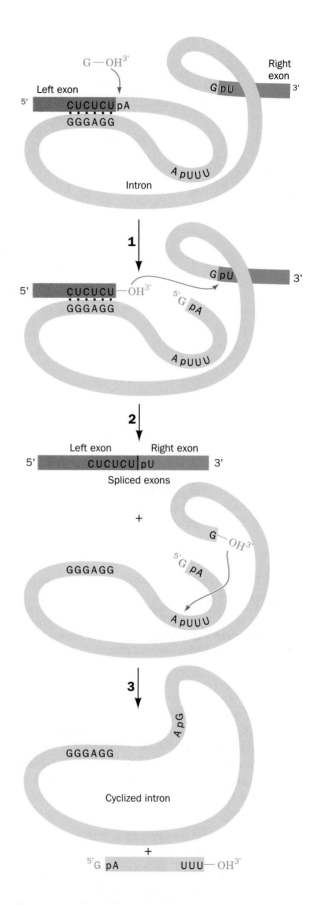

Thus, the observations that nucleic acids but not proteins can direct their own synthesis, that cells contain batteries of protein-based enzymes for manipulating DNA but few for processing RNA, and that many coenzymes are ribonucleotides (e.g., ATP, NAD$^+$, and CoA), led to the hypothesis that *RNAs were the original biological catalysts in precellular times (the so-called* **RNA world)** *and that the chemically more versatile proteins were relative latecomers in macromolecular evolution (Section 1-4C).*

The Structure of a Hammerhead Ribozyme

The simplest known ribozymes, which are embedded in the RNAs of certain plant viruses, are named **hammerhead ribozymes** due to the superficial resemblance of their secondary structures, as customarily laid out, to a carpenter's hammer (Fig. 29-44*a*). Hammerhead ribozymes catalyze a transesterification reaction in which the 3',5'-phosphodiester bond between nucleotides 17 and 1.1 is cut so as to yield a cyclic 2',3'-phosphodiester on nucleotide 17 and a free 5'-OH on nucleotide 1.1, much like the intermediate product in the RNA hydrolysis reaction catalyzed by RNase A (Section 14-1A). Consequently, a free 2'-OH on nucleotide 17 is essential for this reaction: Hammerhead ribozymes do not cleave DNA, although they will cleave a DNA strand that has a single ribonucleotide at the cleavage site. The reaction requires the presence of a divalent cation, preferentially Mg^{2+} or Mn^{2+}, in roughly millimolar concentrations. Hammerhead ribozymes have three duplex stems and a conserved core of two nonhelical segments.

David McKay has determined the X-ray structure of a hammerhead ribozyme in which the RNA strand that is normally cleaved has been replaced by the analogous DNA strand (Fig. 29-44*b*; note that this is the first known structure of a complex RNA molecule besides those of tRNAs—which are discussed in Section 30-2B). The ribozyme has the expected secondary structure of three A-form helical segments although its overall shape more closely resembles a wishbone than a hammerhead. The nucleotides in the helical stems form normal Watson–Crick base pairs, whereas nucleotides U7 through A9 form non-Watson–Crick base pairs with nucleotides G12 through A14 (Fig. 29-44*c*) in which ribose oxygen atoms participate as both hydrogen bond donors and acceptors. This explains the observations that most helical positions can be occupied by

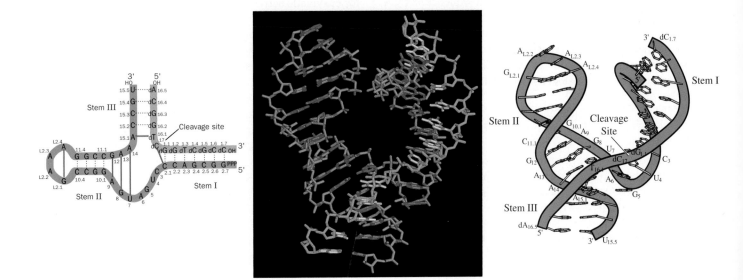

FIGURE 29-44. The X-ray structure of a hammerhead ribozyme. (*a*) The secondary structure and the sequence numbering of the RNA–DNA complex. Normal Watson–Crick base pairing interactions are represented by dashed lines and non-Watson–Crick base pairing interactions are represented by solid blue lines. Note that in normal hammerhead ribozymes the DNA strand is replaced by the analogous RNA strand.

(*b*) A stick diagram of the ribozyme in which the RNA's sugar-phosphate backbone is red, its bases are blue, the DNA's sugar-phosphate backbone is green, and its bases are yellow. (*c*) A schematic diagram of the ribozyme in which the RNA and DNA backbones are drawn as red and green ribbons. [Part *b* based on an X-ray structure by and Part *c* courtesy of David McKay, Stanford University.]

any Watson–Crick base pair but that few core bases can be changed without reducing ribozymal activity. Although biochemical evidence indicates that the catalytically essential divalent metal ion is bound to an oxygen of the scissile phosphate, no metal ion has yet been observed in the crystal structure.

The hammerhead-catalyzed RNA cleavage reaction, like that catalyzed by RNase A, proceeds via inversion of configuration about the P atom. This suggests an "in line" mechanism such as that diagrammed in Fig. 15-7*b* in which the transition state forms a trigonal bipyramidal intermediate in which the attacking nucleophile, the 2'-OH group (Y in Fig. 15-7*b*), and the leaving group, which forms the free 5'-OH group (X in Fig. 15-7*b*), occupy the axial positions. In the hammerhead structure, however, the inferred and observed positions of the 2' and 5' oxygens are quite different from that required for an in line mechanism. This, quite possibly, is due to the absence of the 2'-OH group in the structure. In fact, it would require only a modest conformational reorientation of the structure to bring these groups into the reactive configuration.

C. Transfer RNA Processing

tRNAs, as discussed in Section 30-2A, consist of ~80 nucleotides that assume a secondary structure with four base paired stems known as the **cloverleaf structure** (Fig. 29-45). All tRNAs have a large fraction of modified bases (whose

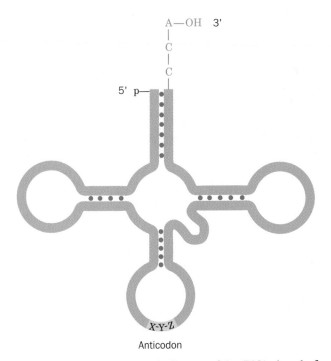

FIGURE 29-45. A schematic diagram of the tRNA cloverleaf secondary structure. Each dot indicates a base pair in the hydrogen bonded stems. The position of the anticodon triplet and the 3'-terminal —CCA are indicated.

structure, function, and synthesis is also discussed in Section 30-2A) and each has the 3′-terminal sequence —CCA to which the corresponding amino acid is appended in the amino acid–charged tRNA. The **anticodon** (which is complementary to the codon specifying the tRNA's corresponding amino acid) occurs in the loop of the cloverleaf structure opposite the stem containing the terminal nucleotides.

The *E. coli* chromosome contains ~60 tRNA genes. Some of them are components of rRNA operons (Section 29-4A); the others are distributed, often in clusters, throughout the chromosome. The primary tRNA transcripts, which contain from one to as many as four or five identical tRNA species, have extra nucleotides at the 3′ and 5′ ends of each tRNA sequence. The excision and trimming of these tRNA sequences resembles that for *E. coli* rRNAs (Section 29-4B) in that both processes employ some of the same nucleases.

RNase P Is a Ribozyme

RNase P, which generates the 5′ ends of tRNAs (Fig. 29-40), is a particularly interesting enzyme because it has, in *E. coli*, a 377-nucleotide RNA component (~125 kD vs

14 kD for its 119-residue protein subunit) that is essential for enzymatic activity. The enzyme's RNA was, quite understandably, first proposed to function in recognizing the substrate RNA through base pairing and to thereby guide the protein subunit, which was presumed to be the actual nuclease, to the cleavage site. However, Sidney Altman has shown that *the RNA component of RNase P is, in fact, the enzyme's catalytic subunit* by demonstrating that protein-free RNase P RNA catalyzes the cleavage of substrate RNA at high salt concentrations. RNase P protein, which is basic, evidently functions at physiological salt concentrations to electrostatically reduce the repulsions between the polyanionic ribozyme and substrate RNAs. The argument that trace quantities of RNase P protein are really responsible for the RNase P reaction was disposed of by showing that catalytic activity is exhibited by RNase P RNA that has been transcribed in a cell-free system. RNase P activity occurs in eukaryotes (nuclei, mitochondria, and chloroplasts) as well as in prokaryotes.

Many Eukaryotic Pre-tRNAs Have Introns

Eukaryotic genomes contain from several hundred to several thousand tRNA genes. Many eukaryotic primary

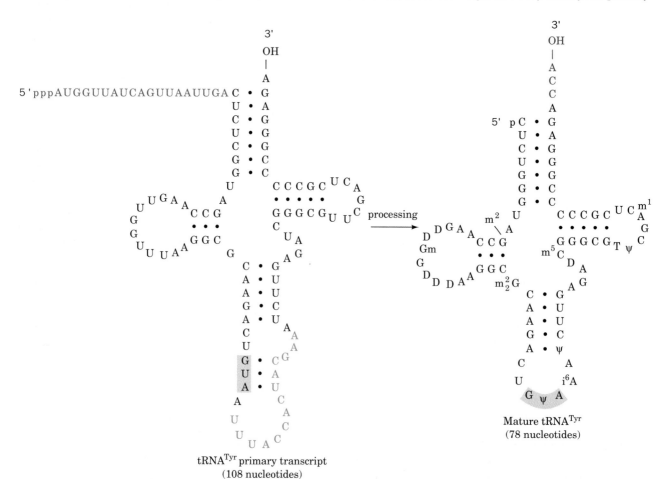

FIGURE 29-46. The posttranscriptional processing of yeast tRNA^Tyr. A 14-nucleotide intervening sequence and a 19-nucleotide 5′-terminal sequence are excised from the primary transcript, a —CCA is appended to the 3′ end and several of the bases are modified (their symbols are defined in Fig. 30-13) to form the mature tRNA. The anticodon is shaded. [After DeRobertis, E.M. and Olsen, M.V., *Nature* **278**, 142 (1989).]

tRNA transcripts, for example, yeast tRNA^{Tyr} (Fig. 29-46), contain a small intron adjacent to their anticodons as well as extra nucleotides at their 5' and 3' ends. Note that this intron is unlikely to disrupt the tRNA's cloverleaf structure.

Eukaryotic tRNA transcripts lack the obligatory —CCA sequence at their 3' end. This is appended to the immature tRNAs by the enzyme **tRNA nucleotidyltransferase,** which sequentially adds two C's and an A to tRNA using CTP and ATP as substrates. This enzyme also occurs in prokaryotes, although, at least in *E. coli,* the tRNA genes all encode a —CCA terminus. The *E. coli* tRNA nucleotidyltransferase is therefore thought to function in the repair of degraded tRNAs.

CHAPTER SUMMARY

The central dogma of molecular biology states that "DNA makes RNA makes protein" (although RNA can also "make" DNA). There is, however, enormous variation among the rates that the various proteins are made. Certain enzymes, such as those of the *lac* operon, are synthesized only when the substances they metabolize are present. The *lac* operon consists of the control sequences *lacP* and *lacO* followed by the tandemly arranged genes for β-galactosidase *(lacZ),* galactoside permease *(lacY),* and thiogalactoside transacetylase *(lacA).* In the absence of inducer, physiologically allolactose, the *lac* repressor, the product of the *lacI* gene, binds to operator *(lacO)* so as to prevent the transcription of the *lac* operon by RNA polymerase. The binding of inducer causes the repressor to release the operator that allows the *lac* structural genes to be transcribed onto a single polycistronic mRNA. The mRNAs transiently associate with ribosomes so as to direct them to synthesize the encoded polypeptides.

The holoenzyme of *E. coli* RNA polymerase has the subunit structure $\alpha_2\beta\beta'\sigma$. It initiates transcription on the antisense strand of a gene at a position designated by its promoter. The most conserved region of the promoter is centered at about the -10 position and has the consensus sequence TATAAT. The -35 region is also conserved in efficient promoters. Methylation protection studies indicate that holoenzyme forms an "open" initiation complex with the promoter. After the initiation of RNA synthesis, the σ subunit dissociates from the core enzyme, which then autonomously catalyzes chain elongation in the $5' \rightarrow 3'$ direction. RNA synthesis is terminated by a segment of the transcript that forms a G + C-rich hairpin with an oligo(U) tail that spontaneously dissociates from the DNA. Termination sites that lack these sequences require the assistance of rho factor for proper chain termination. In the nuclei of eukaryotic cells, RNA polymerases I, II, and III, respectively, synthesize rRNA precursors, hnRNA, and tRNAs + 5S RNA. The minimal RNA polymerase I promoter extends between nucleotides -7 and $+6$. Many RNA polymerase II promoters contain a conserved TATAAAA sequence, the TATA box, located around position -27. Enhancers are transcriptional activators that can have variable positions and orientations relative to the transcription start site. RNA polymerase III promoters are located within the transcribed regions of their gene between positions $+40$ and $+80$.

Prokaryotes can respond rapidly to environmental changes, in part because the translation of mRNAs commences during their transcription and because most mRNAs are degraded within 1 to 3 min of their synthesis. The ordered expression of sets of genes in some bacteriophages and bacteria is controlled by cascades of σ factors. The *lac* repressor is a tetrameric protein of identical subunits that, in the absence of inducer, nonspecifically binds to duplex DNA but binds much more tightly to *lac* promoter. The promoter sequence that *lac* repressor protects from nuclease digestion has nearly palindromic symmetry. Yet, methylation protection and mutational studies indicate that repressor is not symmetrically bound to promoter. *lac* repressor prevents RNA polymerase from properly initiating transcription at the *lac* promoter.

The presence of glucose represses the transcription of operons specifying certain catabolic enzymes through the mediation of cAMP. Upon binding cAMP, which is formed only in the absence of glucose, catabolite gene activator protein (CAP) binds to the promoters of certain operons, such as the *lac* operon, thereby activating their transcription. CAP bends its target DNA by ~90°, thereby activating an adjacently bound RNA polymerase. CAP's two symmetry equivalent DNA-binding domains each bind in the major groove of their target DNA via a helix–turn–helix (HTH) motif that occurs in numerous prokaryotic repressors. The binding between these repressors and their target DNAs is mediated by mutually favorable associations between these macromolecules rather than any specific interactions between particular base pairs and amino acid side chains analogous to Watson–Crick base pairing. Sequence-specific interactions between the *met* repressor and its target DNA occur through a two-fold symmetric antiparallel β ribbon that this protein inserts into the DNA's major groove. *araBAD* transcription is controlled by the levels of L-arabinose and CAP–cAMP through a remarkable complex of AraC to two binding sites, *araO$_2$* and *araI$_1$*, that forms a DNA loop. Upon binding L-arabinose and when CAP–cAMP is adjacently bound, AraC releases *araO$_2$* and instead binds *araI$_2$* thereby activating RNA polymerase to transcribe the *araBAD* operon. The expression of the *lac* operon is also in part controlled by DNA loop formation. The expression of the *E. coli trp* operon is regulated by both attenuation and repression. Upon binding tryptophan, its corepressor, *trp* repressor binds to the *trp* operator, thereby blocking *trp* operon transcription. When tryptophan is available, much of the *trp* transcript that has escaped repression is prematurely terminated in the *trpL* sequence because its transcript contains a segment that forms a normal terminator structure. When tryptophanyl–tRNA^{Trp} is scarce, ribosomes stall at the transcript's two tandem Trp codons. This permits the newly synthesized RNA to form a base paired stem and loop that prevents the formation of the terminator structure. Several other operons are similarly regulated by attenuation. The stringent response is another mechanism by which *E. coli* match the rate of transcription to charged tRNA availability. When a specified charged tRNA is scarce, stringent factor on active ribosomes synthesizes ppGpp, which inhibits the transcription of rRNA and some mRNAs while stimulating the transcription of other mRNAs.

Prokaryotic mRNA transcripts require no additional processing. However, eukaryotic mRNAs have an enzymatically appended 5' cap and, in most cases, an enzymatically generated poly(A) tail. Moreover, the introns of eukaryotic mRNA primary transcripts (hnRNAs) are precisely excised and their flanking exons are spliced together to form mature mRNAs in a snRNP-mediated process that takes place in splicosomes. Certain mRNAs are subject to RNA editing. The primary transcript of *E. coli* rRNAs contains all three rRNAs together with some tRNAs. These are excised and trimmed by specific endonucleases and exonucleases. The rRNAs are also modified by the methylation of specific nucleosides. The eukaryotic 18S, 5.8S, and 28S rRNAs are similarly transcribed as a 45S precursor, which is processed in a manner resembling that of *E. coli* rRNAs. The intron of *Tetrahymena* pre-rRNA is removed in an RNA-catalyzed self-splicing reaction. Prokaryotic tRNAs are excised from their primary transcripts and trimmed in much the same manner as rRNAs. In RNase P, one of the enzymes mediating this process, the catalytic subunit is an RNA. Eukaryotic tRNA transcripts also require the excision of a short intron and the enzymatic addition of a 3'-terminal —CCA to form the mature tRNA.

REFERENCES

General

Adams, R.L.P., Knowler, J.T., and Leader, D.P., *The Biochemistry of the Nucleic Acids* (11th ed.), Chapters 9–11, Chapman and Hall (1992).

Gesteland, R.F. and Atkins, J.F. (Eds.), *The RNA World*, Cold Spring Harbor Laboratory (1993). [A series of authoritative articles on the nature of the prebiotic "RNA world" as revealed by the RNA "relics" in modern organisms. See, in particular, Chapters 11, 13, 14, and 15.]

Lewin, B., *Genes* V, Chapters 14–16, 29, 30, and 32, Oxford (1994).

Schleif, R., *Genetics and Molecular Biology* (2nd ed.), Chapters 4, 5, 11–13, The Johns Hopkins University Press (1993).

Watson, J.D., Hopkins, N.H., Roberts, J.W., Steitz, J.A., and Weiner, A.M., *Molecular Biology of the Gene* (4th ed.), Chapters 13, 16, and 20, Benjamin/Cummings (1987).

The Genetic Role of RNA

Brachet, J., Reminiscences about nucleic acid cytochemistry and biochemistry, *Trends Biochem. Sci.* **12**, 244–246 (1987).

Brenner, S., Jacob, F., and Meselson, M., An unstable intermediate carrying information from genes to ribosomes for protein synthesis, *Nature* **190**, 576–581 (1960). [The experimental verification of mRNA's existence.]

Crick, F., Central dogma of molecular biology, *Nature* **227**, 561–563 (1970).

Hall, B.D. and Spiegelman, S., Sequence complementarity of T2-DNA and T2-specific RNA, *Proc. Natl. Acad. Sci.* **47**, 137–146 (1964). [The first use of RNA–DNA hybridization.]

Jacob, F. and Monod, J., Genetic regulatory mechanisms in the synthesis of proteins, *J. Mol. Biol.* **3**, 318–356 (1961). [The classic paper postulating the existence of mRNA and operons and explaining how the transcription of operons is regulated.]

RNA Polymerase and mRNA

Cheng, S.-W.C., Lynch, E.C., Leason, K.R., Court, D.L., Shapiro, D.A., and Friedman, D.I., Functional importance of sequence in the stem-loop of a transcription terminator, *Science* **254**, 1205–1207 (1991).

Darst, S.A., Edwards, A.M., Kubalek, E.W., and Kornberg, R.D., Three-dimensional structure of yeast RNA polymerase II at 16 Å resolution, *Cell* **66**, 121–128 (1991); *and* Darst, S.A., Kubalek, E.W., and Kornberg, R.D., Three-dimensional structure of Escherischia coli RNA polymerase holoenzyme determined by electron crystallography, *Nature* **340**, 731–732 (1989).

Das, A., Control of transcription termination by RNA-binding proteins, *Annu. Rev. Biochem.* **62**, 893–930 (1993).

Erie, D.A., Yager, T.D., and von Hippel, P.H., The single nucleotide addition cycle in transcription, *Annu. Rev. Biophys. Biomol. Struct.* **21**, 379–415 (1992).

Futcher, B., Supercoiling and transcription, or vice versa? *Trends Genet.* **4**, 271–272 (1988).

Gale, E.F., Cundliffe, E., Reynolds, P.E., Richmond, M.H., and Waring, M.J., *The Molecular Basis of Antibiotic Action* (2nd ed.), Chapter 5, Wiley (1981).

Gannan, F., O'Hare, K., Perrin, F., LePennec, J.P., Benoist, C., Cochet, M., Breathnach, R., Royal, A., Garapin, A., Cami, B., and Chambon, P., Organization and sequences of the 5' end of a cloned complete ovalbumin gene, *Nature* **278**, 428–434 (1979).

Geiduschek, E.P. and Tocchini-Valentini, G.P., Transcription by RNA polymerase III, *Annu. Rev. Biochem.* **57**, 873–914 (1988).

Kamitori, S. and Takusagawa, F., Crystal structure of the 2:1 complex between d(GAAGCTTC) and the anticancer drug actinomycin D, *J. Mol. Biol.* **225**, 445–456 (1992).

Khoury, G. and Gruss, P., Enhancer elements, *Cell* **33**, 313–314 (1983).

Lewis, M.K. and Burgess, R.R., Eukaryotic RNA polymerases, *in* Boyer, P.D. (Ed.), *The Enzymes* (3rd ed.), Vol. 15, *pp.* 110–153, Academic Press (1982).

Reynolds, R., Bermúdez-Cruz, R.M., and Chamberlin, M.J., Parameters affecting transcription termination by *Eschericia coli* RNA. I. Analysis of 13 rho-independent terminators, *J. Mol. Biol.* **224**, 31–51 (1992); *and* Reynolds, R. and Chamberlin, M. J., Parameters affecting transcription termination by *Eschericia coli* RNA. II. Construction of hybrid terminators, *J. Mol. Biol.* **224**, 53–63 (1992).

Richardson, J.P., Transcription termination, *Crit. Rev. Biochem. Mol. Biol.* **28**, 1–30 (1993).

Sawadogo, M. and Sentenac, A., RNA polymerase B (II) and general transcription factors. *Annu. Rev. Biochem.* **59**, 711–754 (1990).

Schultz, P., Célia, H., Riva, M., Sentenac, A., and Oudet, P., Three-dimensional model of yeast RNA polymerase I determined by electron microscopy of two-dimensional crystals, *EMBO J.* **12**, 2601–2607 (1993).

Sentenac, A., Eukaryotic RNA polymerases, *CRC Crit. Rev. Biochem.* **18**, 31–90 (1985).

Sousa, R., Chung, Y.J., Rose, J.P., and Wang, B.-C., Crystal structure of bacteriophage T7 RNA polymerase at 3.3 Å resolution, *Nature* **364**, 593–599 (1993).

Willis, I. M., RNA polymerase III, *Eur. J. Biochem.* **212**, 1–11 (1993).

Young, R.A., RNA polymerase II, *Annu. Rev. Biochem.* **60**, 689–715 (1991).

Control of Transcription

Anderson, J.E., Ptashne, M., and Harrison, S.C., The structure of the repressor–operator complex of bacteriophage 434, *Nature* **326**, 846–852 (1987).

Chuprina, V.P., Rullmann, J.A.C., Lamerichs, R.M.J.N., van Boom, J.H., Boelins, R., and Kaptein, R., Structure of the complex of *lac* repressor headpiece and an 11 base pair half–operator determined by nuclear magnetic resonance spectroscopy and restrained molecular dynamics, *J. Mol. Biol.* **234**, 446–462 (1993).

Freemont, P.S., Lane, A.N., and Sanderson, M.R., Structural aspects of protein–DNA recognition, *Biochem. J.* **278**, 1–23 (1991).

Friedman, D.I., Imperiale, M.J., and Adhya, S.L., RNA 3′ end formation in the control of gene expression, *Annu. Rev. Genet.* **21**, 453–488 (1987).

Gallant, J.A., Stringent control in *E. coli*, *Annu. Rev. Genet.* **13**, 393–415 (1979).

Gartenberg, M.R. and Crothers, D.M., Synthetic DNA bending sequences increase the rate of *in vitro* transcription initiation at the *Eschericia coli lac* promoter, *J. Mol. Biol.* **219**, 217–230 (1991).

Gilbert, W. and Müller-Hill, B., Isolation of the lac repressor, *Proc. Natl. Acad. Sci.* **56**, 1891–1898 (1966).

Gralla, J.D., Specific repression in the *lac* repressor—the 1988 version, *in* Gralla, J.D. (Ed.), *DNA–Protein Interactions in Transcription, pp.* 3–10, Liss (1989).

Harrison, S.C., A structural taxonomy of DNA-binding domains, *Nature* **353**, 715–719 (1991).

Helmann, J.D. and Chamberlin, M.J., Structure and function of bacterial sigma factors, *Annu. Rev. Biochem.* **57**, 839–872 (1988).

Kolb, A., Busby, S., Buc, H., Garges, S., and Adhya, S., Transcriptional regulation by cAMP and its receptor protein, *Annu. Rev. Biochem.* **62**, 749–795 (1993).

Kolter, R. and Yanofsky, C., Attenuation in amino acid biosynthetic operons, *Annu. Rev. Genet.* **16**, 113–134 (1982).

Lamond, A.I. and Travers, A.A., Stringent control of bacterial transcription, *Cell* **41**, 6–8 (1985).

Lee, J. and Goldfarb, A., *lac* repressor acts by modifying the initial transcribing complex so that it cannot leave the promoter, *Cell* **66**, 793–798 (1991).

Lobel, R.B. and Schleif, R.F., DNA looping and unlooping by AraC protein, *Science* **250**, 528–532 (1990).

Losick, R. and Stragier, P., Crisscross regulation of cell-type-specific gene expression during development in *B. subtilis, Nature* **355**, 601–604 (1992).

Luisi, B.F. and Sigler, P.B., The stereochemistry and biochemistry of the *trp* repressor-operator complex, *Biochim. Biophys. Acta* **1048**, 113–126 (1990).

McClure, W.R., Mechanism and control of transcription initiation in prokaryotes, *Annu. Rev. Biochem.* **54**, 171–204 (1985).

McKnight, S.L. and Yamamoto, K.R. (Eds.), *Transcriptional Regulation,* Cold Spring Harbor Laboratory Press (1992). [A two-volume compendium that contains authoritative articles on many aspects of prokaryotic transcriptional control.]

Mondragón, A. and Harrison, S.C., The phage 434 Cro/O_R1 complex at 2.5 Å resolution, *J. Mol. Biol.* **219**, 321–334 (1991); *and* Wolberger, C., Dong, Y., Ptashne, M., and Harrison, S.C., Structure of phage 434 Cro/DNA complex, *Nature* **335**, 789–795 (1988).

Oehler, S., Eismann, E.R., Krämer, H., and Müller-Hill, B. The three operators of the *lac* operon cooperate in repression, *EMBO J.* **9**, 973–979 (1990).

Raumann, B.E., Rould, M.A., Pabo, C.O., and Sauer, R.T., DNA recognition by β-sheets in the Arc repressor–operator crystal system, *Nature* **367**, 754–757 (1994); *and* Raumann, B.E., Brown, B.M., and Sauer, R.T., Major groove DNA recognition by β-sheets: the ribbon-helix-helix family of gene regulatory proteins, *Curr. Opin. Struct. Biol.* **4**, 36–43 (1994).

Reznikoff, W.S., Siegele, D.A., Cowing, D.W., and Gross, C.A., The regulation of transcription initiation in bacteria, *Annu. Rev. Genet.* **19**, 355–387 (1985).

Reeder, T. and Schleif, R., AraC protein can activate transcription from only one position and when pointed in only one direction, *J. Mol. Biol.* **231**, 205–218 (1993).

Rogers, D.W. and Harrison, S.C., The complex between phage 434 repressor DNA-binding domain and operator site O_R3: structural differences between consensus and non-consensus half-sites, *Structure* **1**, 227–240 (1993).

Schleif, R., DNA looping, *Annu. Rev. Biochem.* **61**, 199–223 (1992).

Schultz, S.C., Shields, G.C., and Steitz, T.A., Crystal structure of a CAP-DNA complex: The DNA is bent by 90°, *Science* **253**, 1001–1007 (1991).

Shakked, Z., Guzikevich–Guerstein, G., Frolow, F., Rabinovich, D., Joachimiak, A., and Sigler, P.B., Determinants of repressor/operator recognition from the structure of the *trp* operator binding site, *Nature* **368**, 469–473 (1994).

Somers, W.S. and Phillips, S.E.V., Crystal structure of the *met* repressor-operator complex at 2.8 Å resolution reveals DNA recognition by β-strands, *Nature* **359**, 387–393 (1992).

Steitz, T.A., Structural studies of protein–nucleic acid interaction: the sources of sequence-specific binding, *Quart. Rev. Biophys.* **23**, 205–280 (1990). [Also published as a book of the same title by Cambridge University Press (1993).]

Yanofsky, C., Transcription attenutation, *J. Biol. Chem.* **263**, 609–612 (1988); *and* Attenuation in the control of expression of bacterial operons, *Nature* **289**, 751–758 (1981).

Posttranscriptional Processing

Altman, S., Kirsebom, L., and Talbot, S., Recent studies of ribonuclease P, *FASEB J.* **7**, 7–14 (1993).

acid sequence of a protein must somehow be specified by the base sequence of the corresponding segment of DNA.

A DNA base sequence might specify an amino acid sequence in many conceivable ways. With only 4 bases to code for 20 amino acids, a group of several bases, termed a **codon,** is necessary to specify a single amino acid. A triplet code, that is, one with 3 bases per codon, is minimally required since there are $4^3 = 64$ different triplets of bases, whereas there can be only $4^2 = 16$ different doublets, which is insufficient to specify all the amino acids. In a triplet code, as many as 44 codons might not code for amino acids. On the other hand, many amino acids could be specified by more than one codon. Such a code, in a term borrowed from mathematics, is said to be **degenerate.**

Another mystery was, how does the polypeptide synthesizing apparatus group DNA's continuous sequence of bases into codons? For example, the code might be overlapping; that is, in the sequence

ABCDEFGHIJ · · ·

ABC might code for one amino acid, BCD for a second, CDE for a third, *etc.* Alternatively, the code might be nonoverlapping, so that ABC specifies one amino acid, DEF a second, GHI a third, *etc.* The code might also contain internal "punctuation" such as in the nonoverlapping triplet code

ABC,DEF,GHI, · · ·

in which the commas represent particular bases or base sequences. A related question is, how does the genetic code specify the beginning and the end of a polypeptide chain?

The genetic code is, in fact, a nonoverlapping, comma-free, degenerate, triplet code. How this was determined and how the genetic code dictionary was elucidated are the subject of this section.

A. Chemical Mutagenesis

The triplet character of the genetic code, as we shall see below, was established through the use of **chemical mutagens,** substances that induce mutations. We therefore precede our study of the genetic code with a discussion of these substances. There are two major classes of mutations:

1. **Point mutations,** in which one base pair replaces another. These are subclassified as:

   (a). **Transitions,** in which one purine (or pyrimidine) is replaced by another.

   (b). **Transversions,** in which a purine is replaced by a pyrimidine or vice versa.

2. **Insertion/deletion mutations,** in which one or more nucleotide pairs are inserted in or deleted from DNA.

A mutation in any of these three categories may be reversed by a subsequent mutation of the same but not another category.

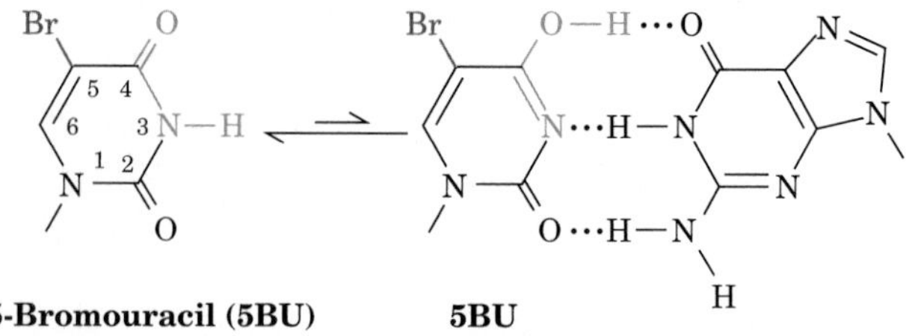

5-Bromouracil (5BU)   5BU
(keto tautomer)   (enol tautomer)   Guanine

FIGURE 30-1. The keto form of 5-bromouracil (*left*) is its most common tautomer. However, it frequently assumes the enol form (*right*), which base pairs with guanine.

Point Mutations Are Generated by Altered Bases

Point mutations can result from the treatment of an organism with base analogs or substances that chemically alter bases. For example, the base analog **5-bromouracil (5BU)** sterically resembles thymine (5-methyluracil) but, through the influence of its electronegative Br atom, frequently assumes a tautomeric form that base pairs with guanine instead of adenine (Fig. 30-1). Consequently, when 5BU is incorporated into DNA in place of thymine, as it usually is, it occasionally induces an $A \cdot T \rightarrow G \cdot C$ transition in subsequent rounds of DNA replication. Occasionally, 5BU is also incorporated into DNA in place of cytosine, which instead generates a $G \cdot C \rightarrow A \cdot T$ transition.

The adenine analog **2-aminopurine (2AP),** normally base pairs with thymine (Fig. 30-2a) but occasionally forms an undistorted but singly hydrogen bonded base pair with cytosine (Fig. 30-2b). Thus 2AP also generates $A \cdot T \rightarrow G \cdot C$ and $G \cdot C \rightarrow A \cdot T$ transitions.

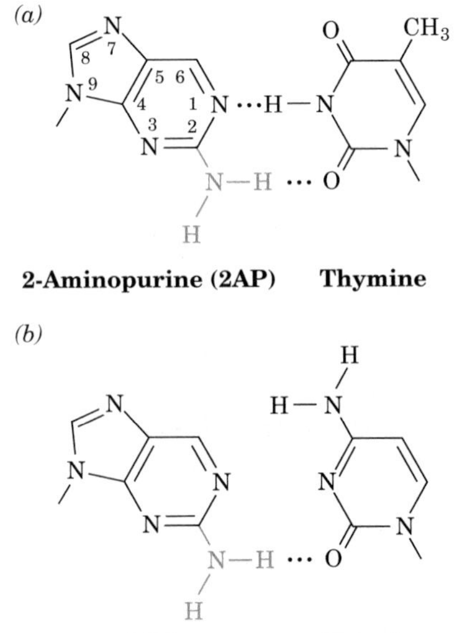

2-Aminopurine (2AP)   Thymine

2AP   Cytosine

FIGURE 30-2. The adenine analog 2-aminopurine normally base pairs with (*a*) thymine but occasionally also does so with (*b*) cytosine.

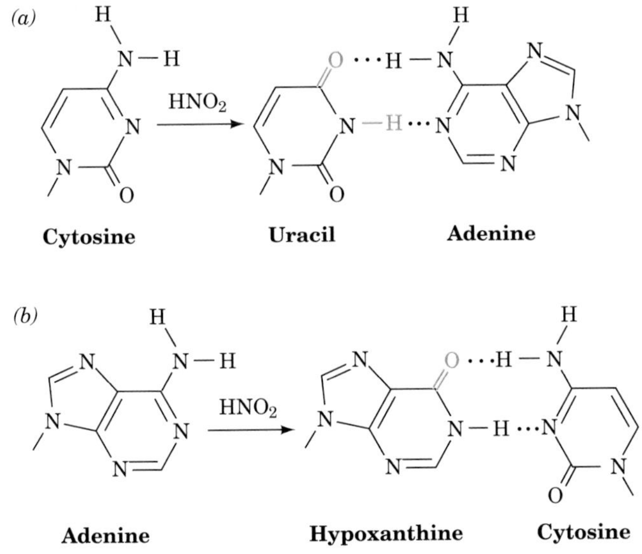

FIGURE 30-3. Reaction with nitrous acid converts (*a*) cytosine to uracil, which base pairs with adenine; and (*b*) adenine to hypoxanthine, a guanine derivative (it lacks guanine's 2-amino group) that base pairs with cytosine.

In aqueous solutions, **nitrous acid** (HNO$_2$) oxidatively deaminates aromatic primary amines so that it converts cytosine to uracil (Fig. 30-3*a*) and adenine to the guanine-like **hypoxanthine** (which forms two of guanine's three hydrogen bonds with cytosine; Fig. 30-3*b*). Hence, treatment of DNA with nitrous acid, or compounds such as **nitrosamines**

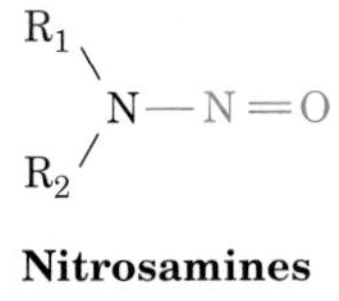

Nitrosamines

that react to form nitrous acid, results in both A·T → G·C and G·C → A·T transitions.

> **Nitrite,** the conjugate base of nitrous acid, has long been used as a preservative of prepared meats such as frankfurters. However, the observation that many mutagens are also carcinogens (Section 31-5E) suggests that the consumption of nitrite-containing meat is harmful to humans. Proponents of nitrite preservation nevertheless argue that to stop it would result in far more fatalities. This is because lack of such treatment would greatly increase the incidence of **botulism,** an often fatal form of food poisoning caused by the ingestion of protein neurotoxins secreted by the anaerobic bacterium *Clostridium botulinum* (Section 34-4C).

Hydroxylamine (NH$_2$OH) also induces G·C → A·T transitions by specifically reacting with cytosine to convert it to a compound that base pairs with adenine (Fig. 30-4). The use

of alkylating agents such as dimethyl sulfate, **nitrogen mustard,** and **ethylnitrosourea**

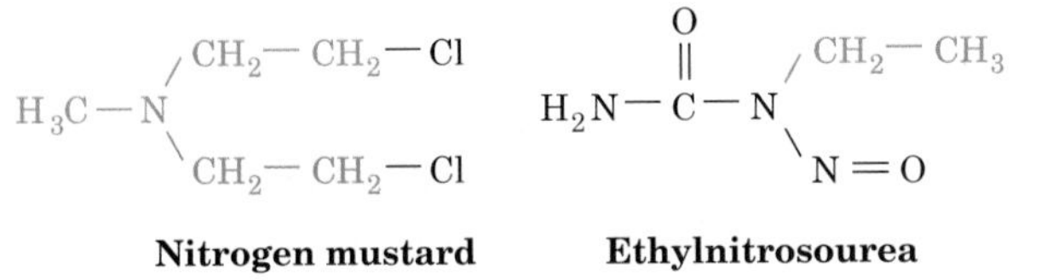

Nitrogen mustard   **Ethylnitrosourea**

often generates transversions. The alkylation of the N7 position of a purine nucleotide causes its subsequent depurination in a reaction similar to that diagrammed in Fig. 28-55*a*. The resulting gap in the sequence is filled in by an error-prone enzymatic repair system (Section 31-5B). Transversions arise when the missing purine is replaced by a pyrimidine. The enzymatic repair of DNA that has been damaged by UV radiation may also generate transversions.

Insertion/Deletion Mutations Are Generated by Intercalating Agents

Insertion/deletion mutations may arise from the treatment of DNA with intercalating agents such as acridine orange or proflavin (Section 28-4C). The distance between two consecutive base pairs is doubled by the intercalation of such a molecule between them. The replication of such a distorted DNA occasionally results in the insertion or deletion of one or more nucleotides in the newly synthesized polynucleotide. (Insertions and deletions of large DNA segments generally arise from aberrant crossover events; Section 33-2C.)

B. Codons Are Triplets

In 1961, Francis Crick and Sydney Brenner, through genetic investigations into the previously unknown character of proflavin-induced mutations, determined the triplet character of the genetic code. In bacteriophage T4, a particular proflavin-induced mutation, designated *FC*0, maps in the *rIIB* cistron (Section 27-1E). The growth of this mutant phage on a permissive host (*E. coli* B) resulted in the occasional spontaneous appearance of phenotypically wild-type phages as was demonstrated by their ability to grow on a restrictive host [*E. coli* K12(λ); recall that *rIIB* mutants form characteristically large plaques on *E. coli* B but cannot

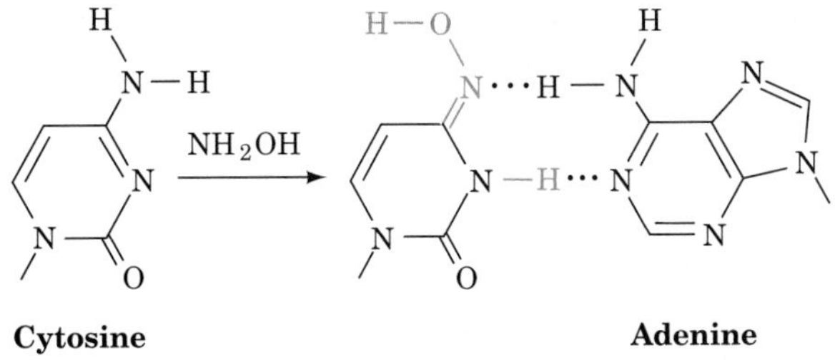

Cytosine   **Adenine**

FIGURE 30-4. Reaction with hydroxylamine converts cytosine to a derivative that base pairs with adenine.

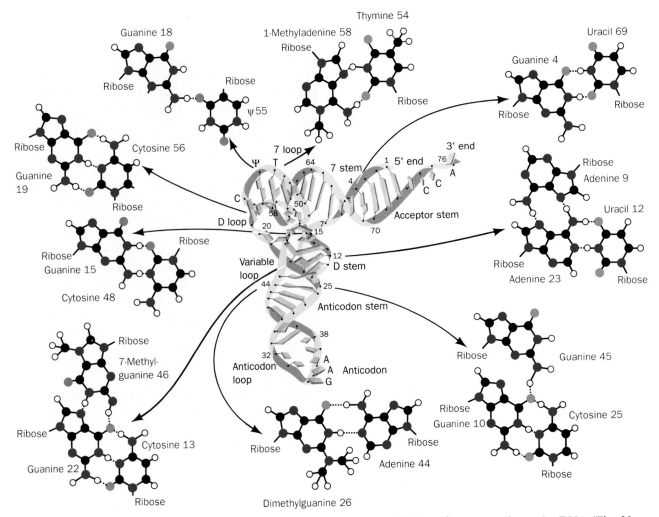

FIGURE 30-15. The nine tertiary base pairing interactions in yeast tRNA^Phe. Note that all but one involve non-Watson–Crick pairs and that they are all located near the corner of the L. [After Kim, S.H., *in* Schimmel, P.R., Söll, D., and Abelson, J.N. (Eds.), *Transfer RNA: Structure, Properties and Recognition, p.* 87, Cold Spring Harbor Laboratory (1979). Drawing of tRNA copyrighted © by Irving Geis.]

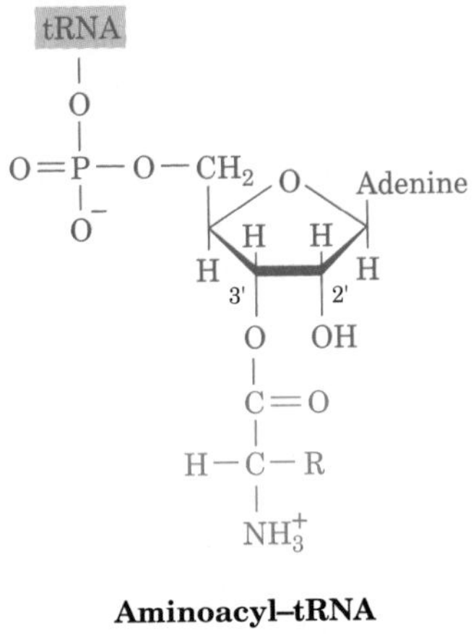

cognate tRNA to form an **aminoacyl–tRNA** (Fig. 30-16). This otherwise unfavorable process is driven by the hydrolysis of ATP in two sequential reactions that are catalyzed by a single enzyme.

1. The amino acid is first "activated" by reaction with ATP to form an **aminoacyl-adenylate:**

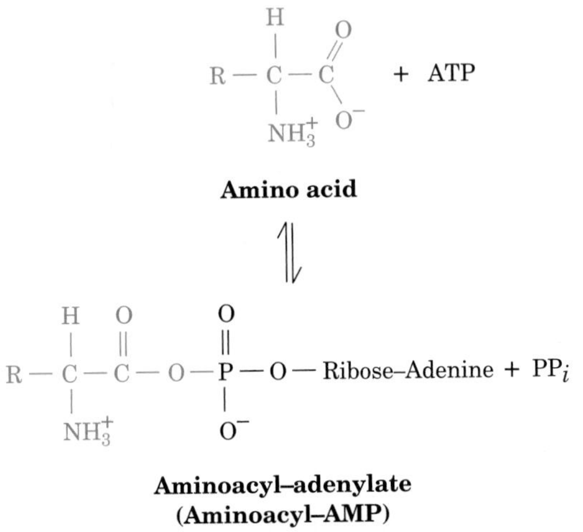

Aminoacyl–tRNA

FIGURE 30-16. In aminoacyl-tRNAs, the amino acid residue is esterified to the tRNA's 3'-terminal nucleoside at either its 3'-OH group, as shown here, or its 2'-OH group.

Amino acid

Aminoacyl–adenylate
(Aminoacyl–AMP)

which, with all but three aaRSs, can occur in the absence of tRNA. Indeed, this intermediate may be isolated although it normally remains tightly bound to the enzyme.

2. This mixed anhydride then reacts with tRNA to form the aminoacyl-tRNA:

$$\text{Aminoacyl--AMP + tRNA} \rightleftharpoons$$
$$\text{aminoacyl--tRNA + AMP}$$

Some aaRSs exclusively append an amino acid to the terminal 2'-OH group of their cognate tRNAs, and others do so at the 3'-OH group. This selectivity was established with the use of chemically modified tRNAs that lack either the 2'- or 3'-OH group of their 3'-terminal ribose residue. The use of these derivatives was necessary because, in solution, the aminoacyl group rapidly equilibrates between the 2' and 3' positions.

The overall aminoacylation reaction is

$$\text{Amino acid + tRNA + ATP} \rightleftharpoons$$
$$\text{aminoacyl--tRNA + AMP + PP}_i$$

These reaction steps are readily reversible because the free energies of hydrolysis of the bonds formed in both the aminoacyl-adenylate and the aminoacyl–tRNA are comparable to that of ATP hydrolysis. The overall reaction is driven to completion by the inorganic pyrophosphatase-catalyzed hydrolysis of the PP_i generated in the first reaction step. Amino acid activation therefore chemically resembles fatty acid activation (Section 23-2A); the major difference between these two processes, which were both elucidated by Paul Berg, is that tRNA is the acyl acceptor in amino acid activation, whereas CoA performs this function in fatty acid activation.

There are Two Classes of Aminoacyl–tRNA Synthetases

Cells must have at least one aaRS for each of the 20 amino acids. The similarity of the reactions catalyzed by these enzymes and the structural resemblance of all tRNAs suggests that all aaRSs evolved from a common ancestor and should therefore be structurally related. This is not the case. In fact, *the aaRSs form a diverse group of enzymes.* The over 100 such enzymes that have been characterized each have one of four different types of subunit structures, α, α_2 (the predominant form), α_4, and $\alpha_2\beta_2$, with known subunit sizes ranging from 334 to 1112 residues. Moreover, there is little sequence similarity among synthetases specific for different amino acids. Quite possibly, aminoacyl–tRNA synthetases arose very early in evolution, before the development of the modern protein synthesis apparatus other than tRNAs.

Detailed sequence and structural comparisons of aminoacyl–tRNA synthetases by Dino Moras indicates that these enzymes form two unrelated families, termed **Class I** and **Class II aaRSs,** that each have the same 10 members in all organisms. The Class I enzymes, although of largely dissimilar sequences, share two homologous polypeptide segments, not present in other proteins, that have the consensus sequences His-Ile-Gly-His (HIGH), and Lys-Met-Ser-Lys-Ser (KMSKS). In the Class I enzymes of known X-ray structure [**glutaminyl–tRNA synthetase (GlnRS), MetRS, TyrRS**; see below], both of these segments are components of a dinucleotide-binding fold (Rossmann fold, which is also possessed by many NAD^+- and ATP-binding proteins; Section 7-3B) in which they participate in ATP binding and are implicated in catalysis. The Class II synthetases lack the foregoing sequences but have three other sequences in common. The known X-ray structures of Class II synthetases (**AspRS and SerRS**) reveal that these sequences occur in a so-called signature motif, a fold found only in Class II enzymes that consists of a 7-stranded antiparallel β sheet with 3 flanking helices and that forms the core of their catalytic domains.

Many Class I aaRSs require anticodon recognition to aminoacylate their cognate tRNAs. In contrast, several Class II enzymes, including SerRS, do not interact with their bound tRNA's anticodon. Indeed, several class II aaRSs accurately aminoacylate "microhelices" derived from only the acceptor stems of their cognate tRNAs. Another difference between Class I and Class II synthetases is that all Class I enzymes aminoacylate their bound tRNA's 3'-terminal 2'-OH group, whereas Class II enzymes charge the 3'-OH group. The amino acids for which the Class I synthetases are specific tend to be larger and more hydrophobic than those used by Class II synthetases.

Prokaryotic aaRSs occur as individual protein molecules. However, in many higher eukaryotes (e.g., *Drosophila* and mammals), 9 aaRSs, some of each class, associate to form a multienzyme particle in which the glutamyl and prolyl synthetase functions are fused into a single polypeptide named **GluProRS.** The advantages of this system are unknown.

The Structural Features Recognized by Aminoacyl–tRNA Synthetases May Be Quite Simple

As we shall see in Section 30-2D, ribosomes select aminoacyl–tRNAs only via codon–anticodon interactions, not according to the identities of their aminoacyl groups. *Accurate translation therefore requires not only that each tRNA be aminoacylated by its cognate aaRS but that it not be aminoacylated by any of its 19 noncognate aaRSs.* Considerable effort has therefore been expended, notably by LaDonne Schulman, Paul Schimmel, Olke Uhlenbeck, and John Abelson, in elucidating how aaRSs manage this feat, despite the close structural similarities of nearly all tRNAs. The methods used involved the use of specific tRNA fragments, mutationally altered tRNAs, chemical cross-linking agents, computerized sequence comparisons, and, more recently, X-ray crystallography. The most common synthetase contact sites on tRNA occur on the inner (concave) face of the L. Other than that, there appears to be little regularity in how the various tRNAs are recognized by their cognate synthetases. Indeed, as we shall see, some aaRSs recognize only their cognate tRNA's acceptor stem, whereas others also interact with its anticodon region. Additional tRNA regions may also be recognized.

Genetic manipulations by Schimmel revealed that the tRNA features recognized by at least one type of aaRS are surprisingly simple. Numerous sequence alterations of *E. coli* tRNA^{Ala} do not appreciably affect its capacity to be aminoacylated with alanine. Yet, most base substitutions in the G3·U70 base pair located in the tRNA's acceptor stem (Fig. 30-17a) greatly diminish this reaction. Moreover, the introduction of a G·U base pair into the analogous position of **tRNA^{Cys}** and **tRNA^{Phe}** causes them to be aminoacylated with alanine even though there are few other sequence identities between these mutant tRNAs and tRNA^{Ala} (e.g., Fig. 30-18). In fact, *E. coli* AlaRS even efficiently aminoacylates a 24-nucleotide "microhelix" derived from only the G3·U70-containing acceptor stem of *E. coli* tRNA^{Ala}. Since the only known *E. coli* tRNAs that normally have a G3·U70 base pair are the tRNA^{Ala}, and this base pair is also present in the tRNA^{Ala} from many organisms including yeast (Fig. 30-11), the foregoing observations strongly suggest that *the G3·U70 base pair is a major feature recognized by AlaRSs*. These enzymes presumably recognize the distorted shape of the G·U base pair (Fig. 30-15), an idea corroborated by the observation that base changes at G3·U70, which least affect the acceptor identity of tRNA^{Ala}, yield base pairs that structurally resemble G·U.

The elements of three other tRNAs, which are recognized by their cognate tRNA synthetases, are indicated in Fig. 30-17. As with tRNA^{Ala}, these identifiers appear to comprise only a few bases. Note that the anticodon is an identifier in two of these tRNAs. In another example of an anticodon identifier, the *E. coli* **tRNA^{Ile}** specific for the codon AUA has the anticodon LAU, where L is **lysidine,** a modified cytosine whose 2-keto group is replaced by the amino acid lysine (Fig. 30-13). The L in this context pairs with A rather than G, a unique case of base modification altering base pairing specificity. The replacement of this L with unmodified C, as expected, yields a tRNA that recognizes the Met codon AUG (codons bind anticodons in an antiparallel fashion). Surprisingly, however, this altered tRNA^{Ile} is also a much better substrate for MetRS than it is for IleRS. Thus, both the codon and the amino acid specificity of this tRNA are changed by a single posttranscriptional modification.

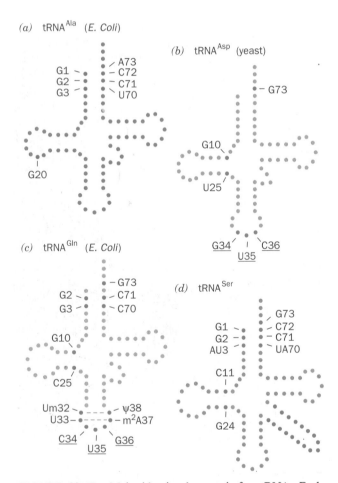

FIGURE 30-17. Major identity elements in four tRNAs. Each base in the tRNA is represented by a filled circle. Red circles indicate positions that have been shown to be identity elements for the recognition of the tRNA by its cognate aminoacyl–tRNA synthetase. The anticodon bases that are identity elements are underlined. In each case, additional identity elements may yet be discovered.

FIGURE 30-18. A three-dimensional model of *E. coli* tRNA^{Ala} based on the X-ray structure of yeast tRNA^{Phe} (Fig. 30-14) in which the nucleotides that are different in *E. coli* tRNA^{Cys} are highlighted in light blue and the G3·U70 base pair is highlighted in ivory. [Courtesy of Ya-Ming Hou, MIT.]

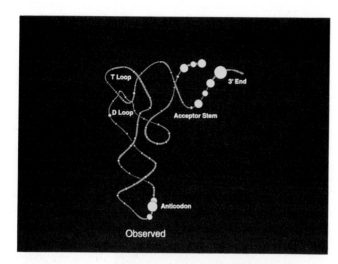

FIGURE 30-19. The experimentally observed identity elements of tRNAs. The tRNA backbone is blue and each of its nucleotides is represented by a yellow circle whose diameter is proportional to fraction of the 20 tRNA acceptor types for which the nucleotide is an observed determinant. [Courtesy of William McClain, University of Wisconsin.]

The available experimental evidence has largely located the various tRNA identifiers in the acceptor stem and the anticodon loop (Fig 30-19). The recently determined X-ray structures of several aaRS·tRNA complexes, which we consider next, have structurally rationalized some of these observations.

The X-Ray Structure of GlnRS·tRNA^{Gln}, a Class I Complex

The X-ray structure of *E. coli* GlnRS, a Class I synthetase, in complex with **tRNA^{Gln}** and ATP (Fig. 30-20), which was determined by Thomas Steitz, was the first known structure of an aaRS·tRNA complex. The tRNA^{Gln} assumes an L-shaped conformation that resembles those of tRNAs of known structures (e.g., Fig. 30-14*b*). GlnRS, a 553-residue monomeric protein that consists of four domains arranged to form an elongated molecule, interacts with the tRNA along the entire inside face of the L such that

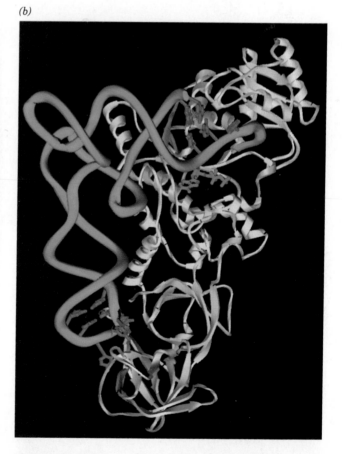

FIGURE 30-20. The X-ray structure of *E. coli* GlnRS·tRNA^{Gln}·ATP. (*a*) The tRNA and ATP are shown in skeletal form with the tRNA sugar–phosphate backbone green, its bases purple, and the ATP red. The protein is represented by a translucent blue-white space-filling model that reveals the buried portions of the tRNA and ATP. Note that both the 3′ end of the tRNA (*top right*) and its anticodon bases (*bottom*) are inserted into deep pockets in the protein. (*b*) A ribbon drawing of the complex viewed as in Part *a*. The tRNA's sugar–phosphate backbone is represented by a green rod and the bases forming its identity elements (Fig. 30-17*c*) are purple. The protein's four domains are differently colored with the dinucleotide binding fold gold and the remainder of the catalytic domain that contains it, yellow. The ATP is shown in skeletal form (*red*). [Based on an X-ray structure by Thomas Steitz, Yale University.]

the anticodon is bound near one end of the protein and the acceptor stem is bound near its other end.

Genetic and biochemical data indicate that the identity elements of tRNA^{Gln} are largely clustered in its anticodon loop and acceptor stem (Fig. 30-17c). The anticodon stem of tRNA^{Gln} is extended by two novel non-Watson–Crick base pairs (2'-O-methyl-U32·ψ38 and U33·m^{2}A37), thereby causing the bases of the anticodon to unstack and splay outward in different directions so as to bind in separate recognition pockets of GlnRS. These structural features suggest that GlnRS uses all seven bases of the anti-codon loop to discriminate among tRNAs. Indeed, changes to any one of the bases of residues C34 through ψ38 yield tRNAs with decreases in k_{cat}/K_M for aminoacylation by GlnRS by factors ranging from 70 to 28,000.

The GCCA at the 3' end of the tRNA^{Gln} makes a hairpin turn towards the inside of the L rather than continuing helically onward (as does the ACCA at the 3' end in the X-ray structure of tRNA^{Phe}; Fig. 30-14b). This conformation change is facilitated by the insinuation of a Leu side chain between the 5' and 3' ends of the tRNA so as to disrupt the first base pair of the acceptor stem (U1·A72).

(a)

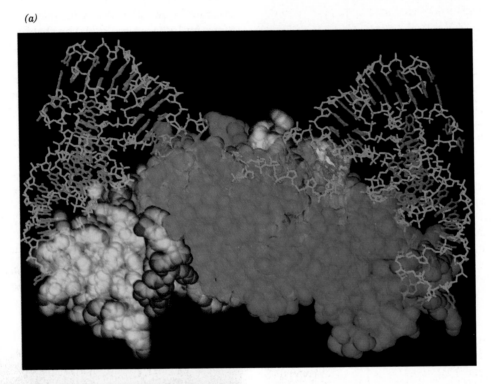

(b)

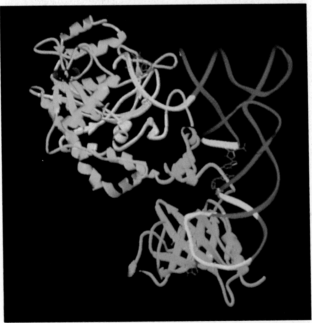

FIGURE 30-21. The X-ray structure of yeast AspRS·tRNA^{Asp}·ATP. (*a*) The dimeric enzyme with its two symmetrically bound tRNAs viewed with its two-fold axis approximately vertical. The tRNAs are shown in skeletal form with their sugar–phosphate backbones green and their bases purple. The two protein subunits are represented by translucent yellow and blue space-filling models that reveal buried portions of the tRNAs. (*b*) A ribbon diagram of the AspRS·tRNA^{Asp}·ATP monomer. The tRNA's contact regions with the protein are yellow and its identity elements are shown in stick form in red as is the ATP. The protein's N-terminal domain is green, the central domain is blue, and the C-terminal catalytic domain is orange with its component signature motif (the 7-stranded antiparallel β sheet with 3 flanking helices characteristic of Type II aaRSs) in white. [Part *a* based on an X-ray structure by and Part *b* courtesy of Dino Moras, Institut de Biologie Moléculaire et Cellulaire du CNRS, Strasbourg, France.]

The GlnRS reaction is therefore relatively insensitive to base changes in these latter two positions except when base pairing is strengthened by their conversion to G1·C72. The GCCA end of the tRNA^{Gln} plunges deeply into a protein pocket that also binds the enzyme's ATP and glutamine substrates. Three protein "fingers" are inserted into the minor groove of the acceptor stem to make sequence-specific interactions with base pairs G2·C71 and G3·C70 [recall that double helical RNA has an A-DNA-like structure (Section 28-2B) whose wide minor groove readily admits protein but whose major groove is normally too narrow to do so].

The GlnRS domain that binds glutamine, ATP, and the GCCA end of tRNA^{Gln}, the so-called catalytic domain, contains, as we previously discussed, a dinucleotide-binding fold. Much of this domain is nearly superimposable with and thus evolutionarily related to the corresponding domains of TyrRS and MetRS whose X-ray structures were determined in the absence of tRNA.

The X-Ray Structure of AspRS·tRNA^{Asp}, a Class II Complex

Yeast AspRS, a Class II synthetase, is an α_2 dimer of 557-residue subunits. Its X-ray structure in complex with tRNA^{Asp}, determined by Moras, reveals that the protein symmetrically binds two tRNA molecules (Fig. 30-21). Like GlnRS, AspRS principally contacts its bound tRNA at both the end of its acceptor stem and in its anticodon region. The contacts in these two enzymes are, nevertheless, quite different in character (Fig. 30-22): Although both

tRNAs approach their cognate synthetases along the inside of their L shapes, tRNA^{Gln} does so from the direction of the minor groove of its acceptor stem, whereas tRNA^{Asp} does so from the direction of its major groove. The GCCA at the 3' end of tRNA^{Asp} thereby continues its helical track as it plunges into AspRS's catalytic site, whereas, as we saw, the GCCA end of tRNA^{Gln} bends backwards into a hairpin turn that opens up the first base pair (U1·A72) of its acceptor stem. Although the deep major groove of an A-RNA helix is normally too narrow to admit groups larger than water molecules (Section 28-2B), the major groove at the end of the acceptor stem in AspRS·tRNA^{Asp} is sufficiently widened for its base pairs to interact with a protein loop.

The anticodon arm of tRNA^{Asp} is bent by as much as 20 Å towards the inside of the L relative to that in the X-ray structure of uncomplexed tRNA^{Asp} and its anticodon bases are unstacked. The hinge point for this bend is a G30·U40 base pair in the anticodon stem which, in nearly all other species of tRNA, is a Watson–Crick base pair. The anticodon bases of tRNA^{Gln} are also unstacked in contacting GlnRS but with a backbone conformation that differs from that in tRNA^{Asp}. Evidently, the conformation of a tRNA in complex with its cognate synthetase appears to be dictated more by its interactions with the protein (induced fit) than by its sequence.

Structural analyses of complexes of AspRS·tRNA^{Asp} with ATP and aspartic acid, and of GlnRS·tRNA^{Gln} with ATP have permitted models of the aminoacyl–AMP complexes of these enzymes to be independently formulated. Comparison of these models reveals that the 3'-terminal A

(a)

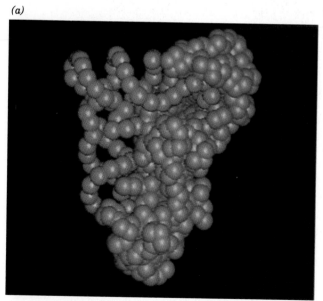

(b)

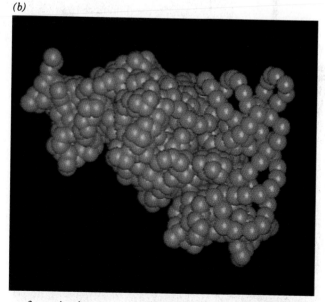

FIGURE 30-22. Comparison of the modes by which (*a*) GlnRS and (*b*) AspRS bind their cognate tRNAs. The proteins and tRNAs are represented by blue and red spheres marking the C$_\alpha$ and P atom positions. Note how GlnRS, a Class I synthetase, binds tRNA^{Gln} from the minor groove side of its acceptor stem so as to bend its 3' end into the hairpin

conformation it assumes on binding to the enzyme's active site. In contrast, AspRS, a Class II synthetase, binds tRNA^{Asp} from the major groove side of its acceptor stem so that its 3' end continues its helical path on entering the active site. [Courtesy of Dino Moras, Institut de Biologie Moléculaire et Cellulaire du CNRS, Strasbourg, France.]

residues of tRNA^{Gln} and tRNA^{Asp} (to which the aminoacyl groups are appended; Fig. 30-16) are positioned on opposite sides of the enzyme-bound aminoacyl–AMP intermediate (Fig. 30-23). The 3'-terminal ribose residues are puckered C2'-*endo* for tRNA^{Asp} and C3'-*endo* for tRNA^{Gln}; see Fig. 28-21) such that the 2'-hydroxyl group of tRNA^{Gln} (Class I) is stereochemically positioned to attack the aminoacyl–AMP's carboxyl group, whereas for tRNA^{Asp} (Class II), only the 3' hydroxyl group is situated to do so. This clearly explains the different aminoacylation specificities of the Class I and Class II aaRSs.

Proofreading Enhances the Fidelity of Amino Acid Attachment to tRNA

The charging of a tRNA with its cognate amino acid is a remarkably accurate process. Experimental measurements indicate, for example, that, at equal concentrations of isoleucine and valine, IleRS transfers ~50,000 isoleucines to tRNA^{Ile} for every valine it so transfers. Yet, *there are insufficient structural differences between Val and Ile to permit such a high degree of accuracy in the direct generation of aminoacyl–tRNAs.* It seems likely that IleRS has a binding site of sufficient size to admit isoleucine but that excludes larger amino acids. On the other hand, valine, which differs from isoleucine by only the lack of a single methylene group, fits into the isoleucine-binding site. The binding free energy of a methylene group is estimated to be ~12 kJ·mol^{-1}. Equation [3.16] indicates that the ratio f of the equilibrium constants, K_1 and K_2, with which two substances bind to a given binding site is given by

$$f = \frac{K_1}{K_2} = \frac{e^{-\Delta G_1^{\circ\prime}/RT}}{e^{-\Delta G_2^{\circ\prime}/RT}} = e^{-\Delta\Delta G^{\circ\prime}/RT} \qquad [30.1]$$

where $\Delta\Delta G^{\circ\prime} = \Delta G_1^{\circ\prime} - \Delta G_2^{\circ\prime}$ is the difference between the free energies of binding of the two substances. It is therefore estimated that isoleucyl–tRNA synthetase could discriminate between isoleucine and valine by no more than a factor of ~100.

Berg resolved this apparent paradox by demonstrating that, in the presence of tRNA^{Ile}, IleRS catalyzes the quantitative hydrolysis of valine–adenylate to valine + AMP rather than forming Val-tRNA^{Ile}. Thus, *isoleucyl–tRNA synthetase subjects aminoacyl–adenylates to a **proofreading** or editing step that has been shown to occur at a separate catalytic site.* This site presumably binds Val residues but excludes the larger Ile residues. *The enzymes's overall selectivity is therefore the product of the selectivities of its adenylation and proofreading steps, thereby accounting for the high fidelity of translation.* Many other synthetases discriminate against noncognate amino acids in a similar fashion. However, synthetases that have adequate selectivity for their corresponding amino acid (e.g., TyrRS discriminates between tyrosine and phenylalanine through hydrogen bonding with the tyrosine —OH group) lack editing functions. Note that *editing occurs at the expense of ATP hydrolysis, the thermodynamic price of high fidelity (increased order).*

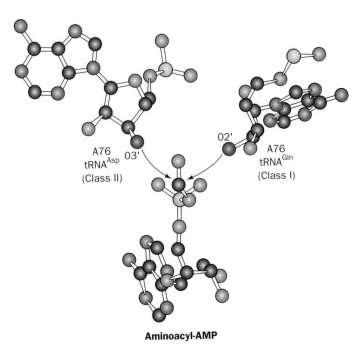

Aminoacyl-AMP

FIGURE 30-23. Comparison of the positions of the the 3' terminal adenosine residues (A76) of AspRS (Class II, *left*) and GlnRS (Class I, *right*) relative to that of the enzyme-bound aminoacyl–AMP (*below*; only the carbonyl group of its aminoacyl residue is shown). Note how only atoms O3' of tRNA^{Gln} and O2' of tRNA^{Asp} are suitably positioned to attack the aminoacyl residue's carbonyl group and thereby transfer the aminoacyl residue to the tRNA. [After Cavarelli J., Eriani, G., Rees, B., Ruff, M., Boeglin, M., Mitschler, A., Martin, F., Gangloff, J., Thierry, J.-C., and Moras, D., *EMBO J.* **13**, 335 (1994).]

D. Codon–Anticodon Interactions

In protein synthesis, the proper tRNA is selected only through codon–anticodon interactions; the aminoacyl group does not participate in this process. This phenomenon was demonstrated as follows. Cys–tRNA^{Cys}, in which the Cys residue was ^{14}C labeled, was reductively desulfurized with Raney nickel so as to convert the Cys residue to Ala:

$$HS-CH_2-\underset{\underset{NH_3^+}{|}}{\overset{\overset{H}{|}}{C}}-\overset{\overset{O}{\|}}{C}-O-tRNA^{Cys} \quad + \quad Ni(H)_x$$

Cys-tRNA^{Cys}   **Raney nickel**

$$H-CH_2-\underset{\underset{NH_3^+}{|}}{\overset{\overset{H}{|}}{C}}-\overset{\overset{O}{\|}}{C}-O-tRNA^{Cys} \quad + \quad H_2S + Ni$$

Ala–tRNA^{Cys}

The resulting ^{14}C-labeled hybrid, Ala-tRNA[Cys], was added to a cell-free protein synthesizing system extracted from rabbit reticulocytes. The product hemoglobin α chain's only radioactive tryptic peptide was the one that normally contains the subunit's only Cys. No radioactivity was found in the peptides that normally contain Ala but no Cys. Evidently, only the anticodons of aminoacyl–tRNAs are involved in codon recognition.

Genetic Code Degeneracy Is Largely Mediated by Variable Third Position Codon–Anticodon Interactions

One might naively guess that each of the 61 codons specifying an amino acid would be read by a different tRNA. Yet, even though most cells contain several groups of **isoaccepting tRNAs** (different tRNAs that are specific for the same amino acid), *many tRNAs bind to two or three of the codons specifying their cognate amino acids.* For example, yeast tRNA[Phe], which has the anticodon GmAA, recognizes the codons UUC and UUU (remember that the anticodon pairs with the codon in an antiparallel fashion),

$$
\begin{matrix}
& 3' & & 5' & & 3' & & & 5' \\
\text{Anticodon:} & -A-A-Gm- & & & & -A-A-Gm- \\
& \cdot & \cdot & \cdot & & & \cdot & \cdot & \cdot \\
& 5' & \cdot & \cdot & \cdot & 3' & 5' & \cdot & \cdot & \cdot & 3' \\
\text{Codon:} & -U-U-C- & & & & -U-U-U-
\end{matrix}
$$

and yeast tRNA[Ala], which has the anticodon IGC, recognizes the codons GCU, GCC, and GCA.

$$
\begin{matrix}
& 3' & & 5' & & 3' & & & 5' \\
\text{Anticodon:} & -C-G-I- & & & & -C-G-I- \\
& \cdot & \cdot & \cdot & & & \cdot & \cdot & \cdot \\
& 5' & \cdot & \cdot & \cdot & 3' & 5' & \cdot & \cdot & \cdot & 3' \\
\text{Codon:} & -G-C-U- & & & & -G-C-C-
\end{matrix}
$$

$$
\begin{matrix}
& 3' & & 5' \\
\text{Anticodon:} & -C-G-I- \\
& \cdot & \cdot & \cdot \\
& 5' & \cdot & \cdot & \cdot & 3' \\
\text{Codon:} & -G-C-A-
\end{matrix}
$$

It therefore seems that non-Watson–Crick base pairing can occur at the third codon–anticodon position (the anticodon's first position is defined as its 3' nucleotide), the site of most codon degeneracy (Table 30-2). Note also that the third (5') anticodon position commonly contains a modified base such as Gm or I.

The Wobble Hypothesis Structurally Accounts for Codon Degeneracy

By combining structural insight with logical deduction, Crick proposed, in what he named the **wobble hypothesis,** how a tRNA can recognize several degenerate codons. He assumed that the first two codon–anticodon pairings have normal Watson–Crick geometry. The structural constraints that this places on the third codon–anticodon pairing ensure that its conformation does not drastically differ from that of a Watson–Crick pair. Crick then proposed that there could be a small amount of play or "wobble" in

the third codon position which allows limited conformational adjustments in its pairing geometry. This permits the formation of several non-Watson–Crick pairs such as U·G and I·A (Fig. 30-24a). The allowed "wobble" pairings are indicated in Fig. 30-24b. Then, by analyzing the known pattern of codon–anticodon pairing, Crick deduced the most plausible sets of pairing combinations in the third codon–anticodon position (Table 30-4). Thus, an anticodon with C or A in its third position can only pair with its Watson–Crick complementary codon. If U, G, or I occupies the third anticodon position, two, two, or three codons are recognized, respectively.

No prokaryotic or eukaryotic cytoplasmic tRNA is known to participate in a nonwobble pairing combination. There is,

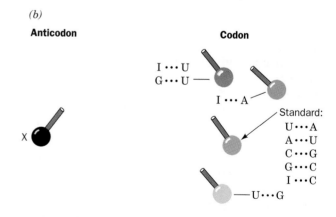

FIGURE 30-24. Wobble pairing. (a) U·G and I·A wobble pairs. Both have been observed in X-ray structures. (b) The geometry of wobble pairing. The spheres and their attached bonds represent the positions of ribose C1' atoms with their accompanying glycosidic bonds. X (*left*) designates the nucleoside at the 5' end of the anticodon (tRNA). The positions on the right are those of the 3' nucleoside of the codon (mRNA) in the indicated wobble pairings. [After Crick, F.H.C., *J Mol. Biol.* **19,** 552 (1966).]

TABLE 30-4. ALLOWED WOBBLE PAIRING COMBINATIONS IN THE THIRD CODON–ANTICODON POSITION

5'-Anticodon Base	3'-Codon Base
C	G
A	U
U	A or G
G	U or C
I	U, C, or A

however, no known instance of such a tRNA with an A in its third anticodon position, which suggests that the consequent A·U pair is not permitted. The structural basis of wobble pairing is poorly understood, although it is clear that it is influenced by base modifications.

A consideration of the various wobble pairings indicates that at least 31 tRNAs are required to translate all 61 coding triplets of the genetic code (there are 32 tRNAs in the minimal set because translational initiation requires a separate tRNA; Section 30-3C). Most cells have >32 tRNAs, some of which have identical anticodons. In fact, mammalian cells have >150 tRNAs. Nevertheless, *all isoaccepting tRNAs in a cell are recognized by a single aminoacyl–tRNA synthetase.*

Some Mitochondrial tRNAs Have More Permissive Wobble Pairings Than Other tRNAs

The codon recognition properties of mitochondrial tRNAs must reflect the fact that mitochondrial genetic codes are variants of the "standard" genetic code (Table 30-3). For instance, the human mitochondrial genome, which consists of only 16,569 bp, encodes 22 tRNAs (together with 2 ribosomal RNAs and 13 proteins). Fourteen of these tRNAs each read one of the synonymous pairs of codons indicated in Tables 30-2 and 30-3 (MNX, where X is either C or U or else A or G) according to normal G·U wobble rules: The tRNAs have either a G or a modified U in their third anticodon position that, respectively, permits them to pair with codons having X = C or U or else X = A or G. The remaining 8 tRNAs, which, contrary to wobble rules, each recognize 1 of the groups of 4 synonymous codons (MNY, where Y = A, C, G, or U), all have anticodons with a U in their third position. Either this U can somehow pair with any of the 4 bases or these tRNAs read only the first two codon positions and ignore the third. Thus, not surprisingly, many mitochondrial tRNAs have unusual structures in which, for example, the GTψCRA sequence (Fig. 30-12) is missing, or, in the most bizarre case, a tRNA^{Ser} lacks the entire D arm.

Frequently Used Codons Are Complementary to the Most Abundant tRNA Species

The analysis of the base sequences of several highly expressed structural genes of baker's yeast, *Saccharomyces*

cerevisiae, has revealed a remarkable bias in their codon usage. Only 25 of the 61 coding triplets are commonly used. *The preferred codons are those that are most nearly complementary, in the Watson–Crick sense, to the anticodons in the most abundant species in each set of isoaccepting tRNAs.* Furthermore, codons that bind anticodons with two consecutive G·C pairs or three A·U pairs are avoided so that the preferred codon–anticodon complexes all have approximately the same binding free energies. A similar phenomenon occurs in *E. coli,* although several of its 22 preferred codons differ from those in yeast. The degree with which the preferred codons occur in a given gene is strongly correlated, in both organisms, with the gene's level of expression (the measured rates of aminoacyl–tRNA selection in *E. coli* span a 25-fold range). This, it has been proposed, permits the mRNAs of proteins that are required in high abundance to be rapidly and smoothly translated.

Selenocysteine Is Specified by a tRNA

Although it is widely stated, even in this text, that proteins are synthesized from the 20 "standard" amino acids, that is, those specified by the "standard" genetic code, some organisms, in fact, use a 21st amino acid, **selenocysteine (SeCys),** in synthesizing a few of their proteins.

$$
\begin{array}{c}
| \\
NH \\
| \\
CH - CH_2 - Se - H \\
| \\
C = O \\
|
\end{array}
$$

The selenocysteine residue

Selenium, a biologically essential trace element, is a component of several enzymes in both prokaryotes and eukaryotes. *E. coli* contains three selenoproteins, all **formate dehydrogenases,** which each contain an SeCys residue. The SeCys residues are ribosomally incorporated into these proteins by a unique tRNA, **tRNA^{Sec},** bearing a UCA anticodon that is specified by a particular (in the mRNA) UGA codon (normally the *opal* stop codon). The SeCys–tRNA^{Sec} is synthesized by the aminoacylation of tRNA^{Sec} with L-serine by the same SerRS that charges tRNA^{Ser}, followed by the enzymatic selenylation of the resulting Ser residue.

How does the ribosomal system differentiate SeCys-specifying UGA codons from normal *opal* stop codons? Clearly, mRNA context effects must be important. Thus, in constructs that fuse the N-terminal segment of the *E. coli* formate dehydrogenase selenopeptide gene upstream of the β-galactosidase gene, selenium is required for the translational readthrough of the mRNA and hence the synthesis of β-galactosidase. However, changing the SeCys-specifying TGA codon to TGC, TGT (Cys codons), or TCA (a Ser codon) abolishes this selenium requirement.

E. Nonsense Suppression

Nonsense mutations are usually lethal when they prematurely terminate the synthesis of an essential protein. An organism with such a mutation may nevertheless be "rescued" by a second mutation on another part of the genome. For many years after their discovery, the existence of such **intergenic suppressors** was quite puzzling. It is now known, however, that they usually arise from mutations in a tRNA gene that cause the tRNA to recognize a nonsense codon. Such a **nonsense suppressor** tRNA appends its amino acid (which is the same as that carried by the corresponding wild-type tRNA) to a growing polypeptide in response to the recognized stop codon, thereby preventing chain termination. For example, the *E. coli amber* suppressor known as *su*3 is a tRNA^{Tyr} whose anticodon has mutated from the wild-type GUA (which reads the Tyr codons UAU and UAC) to CUA (which recognizes the *amber* stop codon UAG). An *su*3$^+$ *E. coli* with an otherwise lethal *amber* mutation in a gene coding for an essential protein would be viable if the replacement of the wild-type amino acid residue by Tyr does not inactivate the protein.

There are several well-characterized examples of *amber* (UAG), *ochre* (UAA), and *opal* (UGA) suppressors in *E. coli* (Table 30-5). Most of them, as expected, have mutated anticodons. UGA-1 tRNA, however, differs from the wild-type only by a G $\rightarrow$ A mutation in its D stem, which changes a G·U pair to a stronger A·U pair. This mutation apparently alters the conformation of the tRNA's CCA anticodon so that it can form an unusual wobble pairing with UGA as well as with its normal codon, UGG. Nonsense suppressors also occur in yeast.

Suppressor tRNAs Are Mutants of Minor tRNAs

How do cells tolerate a mutation that both eliminates a normal tRNA and prevents the termination of polypeptide synthesis? They survive because the mutated tRNA is usually a minor member of a set of isoaccepting tRNAs and because nonsense suppressor tRNAs must compete for stop codons with the protein factors that mediate the termination of polypeptide synthesis (Section 30-3E). Consequently, the rate of suppressor-mediated synthesis of active proteins with either UAG or UGA nonsense mutations rarely exceeds 50% of the wild-type rate, whereas mutants with UAA, the most common termination codon, have suppression efficiencies of <5%. Many mRNAs, moreover, have two tandem stop codons so that even if their first stop codon was suppressed, termination could occur at the second. Nevertheless, many suppressor-rescued mutants grow relatively slowly because they cannot make an otherwise prematurely terminated protein as efficiently as do wild-type cells.

Other types of suppressor tRNAs are also known. **Missense suppressors** act similarly to nonsense suppressors but substitute one amino acid in place of another. **Frameshift suppressors** have eight nucleotides in their anticodon loops rather than the normal seven. They read a four base codon beyond a base insertion thereby restoring the wild-type reading frame.

3. RIBOSOMES

Ribosomes were first seen in cellular homogenates by dark field microscopy in the late 1930s by Albert Claude who referred to them as "microsomes." It was not until the mid-1950s, however, that George Palade observed them in cells by electron microscopy, thereby disposing of the contention that they were merely artifacts of cell disruption. The name ribosome derives from the fact that these particles in *E. coli* consist of ~2/3 RNA and 1/3 protein. (**Microsomes** are now defined as the artifactual vesicles formed by the endoplasmic reticulum upon cell disruption. They are easily isolated by differential centrifugation and are rich in ribosomes.) The correlation between the amount of RNA in a cell and the rate at which it synthesizes protein led to the suspicion that ribosomes are the site of protein synthesis. This hypothesis was confirmed in 1955 by Paul Zamecnik, who demonstrated that ^{14}C-labeled amino acids are transiently associated with ribosomes before they appear in free proteins. Further research showed that ribosomal polypeptide synthesis has three distinct phases: (1) chain initiation, (2) chain elongation, and (3) chain termination.

In this section we examine the structure of the ribosome, insofar as it is known, and then outline the ribosomal mechanism of polypeptide synthesis. In doing so we shall compare the properties of ribosomes from prokaryotes (mostly *E. coli*) with those of eukaryotes (mostly rat liver cytoplasm).

A. Ribosome Structure

The *E. coli* ribosome, which has a particle mass of ~2.5 × 10^6 D and a sedimentation coefficient of 70S, is a spheroidal particle that is ~250 Å across in its largest dimension. It may be dissociated, as James Watson discovered, into two

TABLE 30-5. SOME *E. COLI* NONSENSE SUPPRESSORS

Name	Codon Suppressed	Amino Acid Inserted
*su*1	UAG	Ser
*su*2	UAG	Gln
*su*3	UAG	Tyr
*su*4	UAA, UAG	Tyr
*su*5	UAA, UAG	Lys
*su*6	UAA	Leu
*su*7	UAA	Gln
UGA-1	UGA	Trp
UGA-2	UGA	Trp

Source: Körner, A.M., Feinstein, S.I., and Altman, S., *in* Altman, S. (Ed.), *Transfer RNA, p.* 109, MIT Press (1978).

TABLE 30-6. **COMPONENTS OF *E. COLI* RIBOSOMES**

	Ribosome	Small Subunit	Large Subunit
Sedimentation coefficient	70S	30S	50S
Mass (kD)	2520	930	1590
RNA			
Major		16S, 1542 nucleotides	23S, 2904 nucleotides
Minor			5S, 120 nucleotides
RNA mass (kD)	1664	560	1104
Proportion of mass	66%	60%	70%
Proteins		21 polypeptides	31 polypeptides
Protein mass (kD)	857	370	487
Proportion of mass	34%	40%	30%

unequal subunits (Table 30-6). The small (30S) subunit consists of a 16S rRNA molecule and 21 different polypeptides, whereas the large (50S) subunit contains a 5S and a 23S rRNA together with 31 different polypeptides. The up to 20,000 ribosomes in an *E. coli* cell account for ~80% of its RNA content and 10% of its protein.

Although the ribosome has been crystallized by Ada Yonath, it is such a complex entity that it will be many years before its structure is known in molecular detail. However, the low-resolution structures of the ribosome and its subunits have been determined through image reconstruction techniques, pioneered by Aaron Klug, in which electron

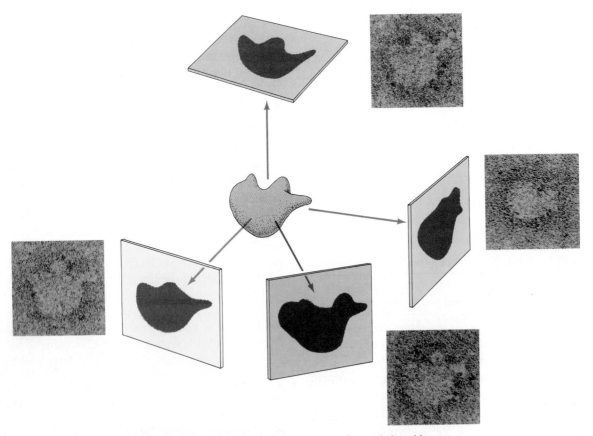

FIGURE 30-25. The three-dimensional model of the large ribosomal subunit was deduced by mathematically combining its two-dimensional electron microscope images as viewed from different directions. The model of the small subunit was similarly determined. [Courtesy of James Lake, UCLA.]

(a)

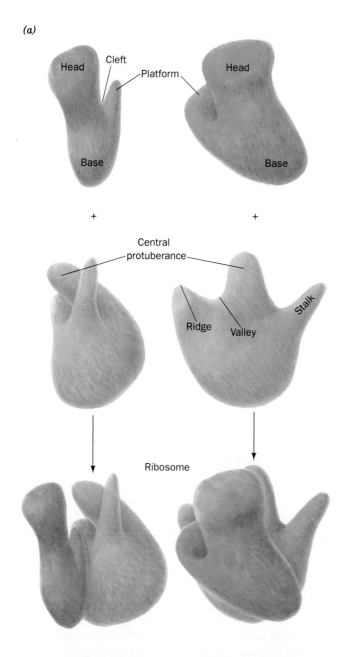

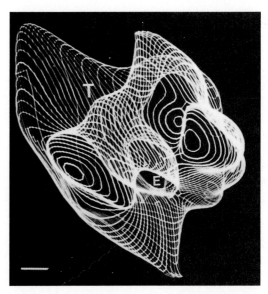

FIGURE 30-27. A computer-generated image of the large ribosomal subunit from *Bacillus stearothermophilus* as determined by electron micrographic image reconstruction of oriented two-dimensional arrays of particles. An ~25-Å-diameter tunnel extends ~100 Å from the cleft between the subunit's three protrusions (T) to the nascent polypeptide's probable exit site (E). The bar is 20 Å long. [Courtesy of Ada Yonath, Weizmann Institute of Science, Israel.]

micrographs of a single particle or ordered sheets of particles taken from several directions are combined to yield its three-dimensional image (Fig. 30-25). The small subunit is a roughly mitten-shaped particle, whereas the large subunit is spheroidal with three protuberances on one side (Fig. 30-26). The large subunit also contains a tunnel, up to 25 Å in diameter and 100 to 120 Å long, that extends from a cleft between the subunit's three protuberances and is postulated to provide the nascent polypeptide's exit path (Fig. 30-27).

Ribosomal RNAs Have Evolutionarily Conserved Secondary Structures

The *E. coli* 16S rRNA, which was sequenced by Harry Noller, consists of 1542 nucleotides. A computerized search of this sequence for stable double helical segments yielded many plausible but often mutually exclusive secondary structures. However, the comparison of the sequences of 16S rRNAs from several prokaryotes, under the assumption that their structures have been evolutionarily conserved, led to the flowerlike secondary structure for 16S rRNA proposed in Fig. 30-28. This four-domain structure, which is 46% base paired, is reasonably consistent with the results of nuclease digestion and chemical modification studies. Its double helical stems tend to be short (<8 bp) and many of them are imperfect. Intriguingly, electron micrographs of the 16S rRNA resemble those of the complete 30S subunit, thereby suggesting that the 30S subunit's overall shape is largely determined by the 16S rRNA.

(b)

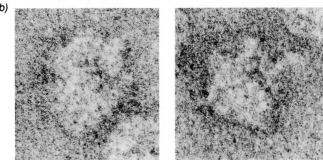

FIGURE 30-26. (*a*) A three-dimensional model of the *E. coli* ribosome deduced as indicated in Fig. 30-25. The small subunit (*top*) combines with the large subunit (*middle*) to form the complete ribosome (*bottom*). The two views of the ribosome match those seen in (*b*) the electron micrographs. [Courtesy of James Lake, UCLA.]

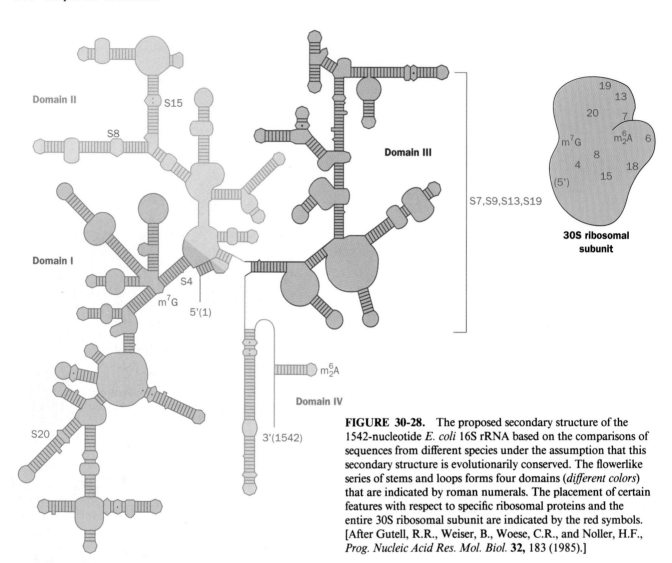

FIGURE 30-28. The proposed secondary structure of the 1542-nucleotide *E. coli* 16S rRNA based on the comparisons of sequences from different species under the assumption that this secondary structure is evolutionarily conserved. The flowerlike series of stems and loops forms four domains (*different colors*) that are indicated by roman numerals. The placement of certain features with respect to specific ribosomal proteins and the entire 30S ribosomal subunit are indicated by the red symbols. [After Gutell, R.R., Weiser, B., Woese, C.R., and Noller, H.F., *Prog. Nucleic Acid Res. Mol. Biol.* **32,** 183 (1985).]

The large ribosomal subunit's 5S and 23S rRNAs, which consist of 120 and 2904 nucleotides, respectively, have also been sequenced. As with the 16S rRNA, they appear to have extensive secondary structures. That proposed for 5S rRNA is shown in Fig. 30-29a. The NMR structure of its Helix I (Fig. 30-29*b*), determined by Peter Moore, indicates that this molecular fragment assumes an A-RNA conformation but with a distortion at its $G9 \cdot U111$ base pair arising from its meager stacking on the adjacent $G8 \cdot C112$ base pair. However, the structure of the entire ~40-kD 5S RNA is too large to determine by known NMR techniques (Section 7-3A).

Ribosomal Proteins Have Been Partially Characterized

Ribosomal proteins are difficult to separate because most of them are insoluble in ordinary buffers. By convention, ribosomal proteins from the small and large subunits are designated with the prefixes S and L, respectively, followed

by a number indicating their position, from upper left to lower right, on a two-dimensional gel electrophoretogram (roughly in order of decreasing molecular mass; Fig. 30-30). Only protein S20/L26 is common to both subunits; it is apparently located at their interface. One of the large subunit proteins is partially acetylated at its N-terminus so that it gives rise to two electrophoretic spots (L7/L12). Four copies of this protein are present in the large subunit. Moreover, these four copies of L7/L12 aggregate with L10 to form a stable complex that was initially thought to be a unique protein, "L8." All the other ribosomal proteins occur in only one copy per subunit.

The amino acid sequences of all 52 *E. coli* ribosomal proteins have been elucidated, mainly by Heinz-Günter Wittmann and Brigitte Wittmann-Liebold. They range in size from 46 residues for L34 to 557 residues for S1. Most of these proteins, which exhibit little sequence similarity with one another, are rich in the basic amino acids Lys and Arg

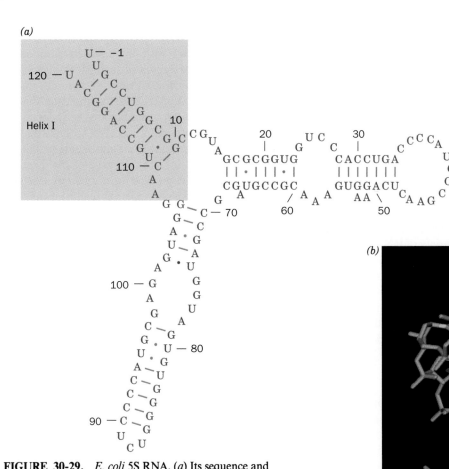

(a)

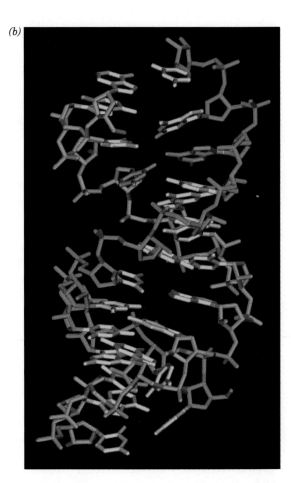

(b)

FIGURE 30-29. *E. coli* 5S RNA. (*a*) Its sequence and probable secondary structure. [After Noller, H.F., *Annu. Rev. Biochem.* **53**, 134 (1984).] (*b*) The NMR structure of its Helix I (the shaded area in Part *a*). The sugar–phosphate chain of residues –1 to 11 is green, that of residues 108 to 120 is light blue, and the bases are yellow except for the poorly stacked G9·U111 base pair, which is red. [Based on an NMR structure by Peter Moore, Yale University.]

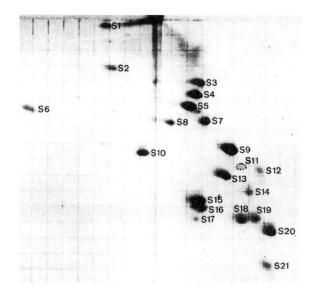

FIGURE 30-30. A two-dimensional gel electrophoretogram of *E. coli* small ribosomal subunit proteins. First dimension (*vertical*): 8% acrylamide, pH 8.6; second dimension (*horizontal*): 18% acrylamide; pH 4.6. [From Kaltschmidt, E. and Wittmann, H.G., *Proc. Natl. Acad. Sci.* **67**, 1277 (1970).]

and contain few aromatic residues as expected for proteins that are closely associated with polyanionic RNA molecules.

The X-ray or NMR structures of seven ribosomal proteins, S5, S6, S17, L6, L7/L12, L9 and L30, have been determined by Stephen White and by Anders Liljas (Fig. 30-31). Intriguingly, five of them (all but S5 and S17) contain homologous structural motifs that consist of a 3-stranded antiparallel β sheet with the latter two strands connected by an α helix (Fig. 30-31). This structural motif has also been observed in the X-ray structure of **U1-snRNP A protein** (Section 29-4A) and, through sequence homology, is implicated as a component of a variety of RNA-binding proteins including rho protein (the transcriptional termination factor, which contains four such motifs; Section 29-2E), poly(A) binding protein (PABP; Section 29-4A), and the translation factor **eIF-4B** (Section 30-3C). This protein fold has therefore been dubbed the **RNA-recognition motif (RRM)**. All of these proteins presumably evolved from an ancient RNA-binding protein.

Ribosomal Subunits Are Self-Assembling

Ribosomal subunits form, under proper conditions, from mixtures of their numerous macromolecular components. *Ribosomal subunits are therefore self-assembling entities.* Masayasu Nomura determined how this occurs through partial reconstitution experiments. If one macromolecular component is left out of an otherwise self-assembling mixture of proteins and RNA, the other components that fail to bind to the resulting partially assembled subunit

must somehow interact with the omitted component. Through the analysis of a series of such partial reconstitution experiments, Nomura constructed an assembly map of the small subunit (Fig. 30-32). This map indicates that the first steps in small subunit assembly are the independent binding of certain proteins to 16S rRNA. The resulting assembly intermediates provide the molecular scaffolding for binding other proteins. At one stage of the assembly process, an intermediate particle must undergo a marked conformational change before assembly can continue. The large subunit self-assembles in a similar manner. The observation that similar assembly intermediates occur *in vivo* and *in vitro* suggests that *in vivo* and *in vitro* assembly processes are much alike.

Ribosomal Architecture Has Been Deduced through Immune Electron Microscopy and Neutron Diffraction Studies

The positions of most ribosomal components have been determined through a variety of physical and chemical techniques. Many proteins have been located by James Lake and by Georg Stöffler through **immune electron mi-**

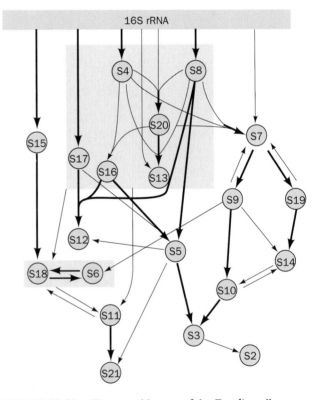

FIGURE 30-32. The assembly map of the *E. coli* small subunit. Thick and thin arrows between components indicate strong and weak facilitation of binding. For example, the thick arrow from 16S rRNA to S4 indicates that S4 binds directly to 16S rRNA in the absence of other proteins, whereas the thin arrows from 16S rRNA, S4, S8, S9, S19, and S20 to S7 indicate that the former components all participate in binding S7. [After Held, W.A., Ballou, B., Mizushima, S., and Nomura, M., *J. Biol. Chem.* **249**, 3109 (1974).]

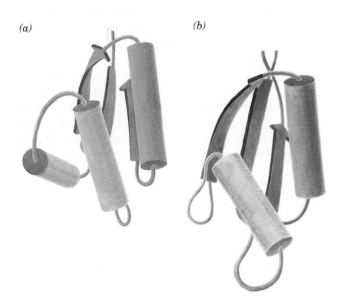

(a)   *(b)*

FIGURE 30-31. The X-ray structures of two ribosomal proteins: (*a*) The 74-residue C-terminal fragment of *E. coli* L7/L12. (*b*) *Bacillus stearothermophilus* L30 (61 residues). The two protein molecules are oriented so as to show their closely similar RNA-recognition motifs (RRMs, *darker shading*). [After Leijonmarck, M., Appelt, K., Badger, J., Liljas, A., Wilson, K.S., and White, S.W., *Proteins* **3**, 244 (1988).]

croscopy. Rabbit antibodies [**immunoglobulin G (IgG)**; Section 34-2A] raised against a specific ribosomal protein bind to this protein where it is exposed on the surface of its subunit. Electron microscopy of the ribosomal sub-

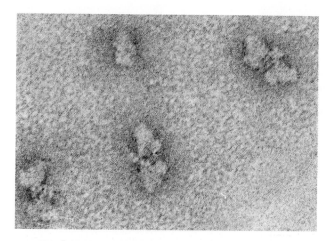

FIGURE 30-33. Immune electron microscopy reveals the positions of ribosomal proteins. Immunoglobin G (IgG) raised against a particular ribosomal protein, here S6, is mixed with ribosomes. The IgG, which is a Y-shaped protein (Section 34-2B), binds to its corresponding antigen at the ends of the two short prongs of the Y, thereby binding together two ribosomes. The position of the protein on the surface of the ribosome is indicated by the point of attachment of the IgG. [Courtesy of James Lake, UCLA.]

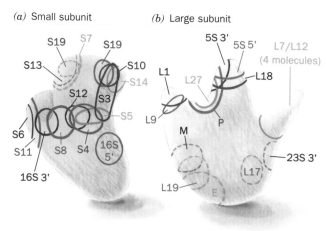

FIGURE 30-34. Maps of the *E. coli* ribosomal (*a*) small and (*b*) large subunits indicating the locations of some of their component proteins as determined by immune electron microscopy. Sites that are dashed are located on the back side of the subunit. On the small subunit, the symbols 16S 3′ and 16S 5′ mark the ends of the 16S RNA. On the large subunit, P indicates the peptidyl transferase site, E marks the site where the nascent polypeptide emerges from the ribosome (the end of the tunnel in Fig. 30-27), M specifies the ribosome's membrane anchor site, and 5S 3′ and 5S 5′ mark the ends of the 5S rRNA. [After Lake, J.A., *Annu. Rev. Biochem.* **54**, 512 (1985).]

unit · IgG complex indicates the point of attachment of the IgG and hence the site of the ribosomal protein to which it binds (Figs. 30-33 and 30-34). These results have been confirmed and extended through neutron diffraction measurements of 30S subunits conducted by Donald Engleman and Peter Moore (Fig. 30-35*a*) and by similar studies on 50S subunits by Knud Nierhaus. The protein positions indicated in Figs. 30-34 and 30-35*a* are consistent with the subunit assembly map shown in Fig. 30-32 in that pairs of proteins that must interact for proper subunit assembly (although not necessarily by direct contact) are in close proximity.

Many of the secondary structural elements of the 16S rRNA have been located on the small subunit (see Figs. 30-28 and 30-35*b*). Their positions were indirectly established from the known positions of proteins that nuclease protection and RNA–protein cross-linking experiments indicate bind to these elements. Thus, we now have a complete, albeit crude, model of the *E. coli* 30S ribosomal subunit.

Affinity Labeling Has Helped Identify the Ribosome's Functional Components

Considerable effort has gone into identifying the ribosome's functional components such as the peptidyl transferase center that catalyzes peptide bond formation (Section 30-3D). Many of these investigations have involved **affinity labeling,** a technique in which a reactive group is attached to a natural ligand of the system of interest such as an antibiotic that binds to a specific site on the ribosome (Section 30-3G). The reactive group, which may be spontaneously reactive or photolabile so that it only reacts upon UV illumination (**photoaffinity labeling**), is carried to the ligand-binding site, where it reacts to cross-link the ligand to the surrounding groups. Dissociation of the resulting particle permits the identification of the components with which the usually radioactive affinity label has reacted.

The results of affinity labeling the ribosome have often been difficult to interpret because its various functions each appear to involve several ribosomal components. For example, mRNA binding apparently involves proteins S1, S3, S4, S5, S9, S12, and S18 as well as the 16S rRNA, whereas proteins L2, L11, L15, L16, L18, L23, and L27, and the 23S RNA are implicated in the peptidyl transferase function. To further confuse matters, the omission of single proteins from 30S particles rarely fully inactivates the ribosome (most exceptions are proteins whose omission causes assembly defects). Nevertheless, the following functionalities have been located (Figs. 30-26 and 30-34):

1. The 3′ end of the 16S rRNA, which is known to participate in mRNA binding (Section 30-3C), is located on the small subunit's "platform." The locations of the proteins implicated in ribosomal mRNA binding, together with the observation that the ribosome protects an ~40-nucleotide mRNA segment from RNase digestion, indi-

(a)

(i)

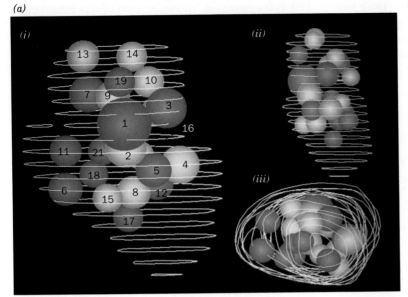

(ii)

(iii)

(b)

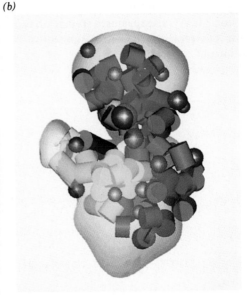

FIGURE 30-35. The structure of the 30S ribosomal subunit. (*a*) The relative positions of all 21 proteins of the 30S ribosomal subunit superimposed on its surface outline. Calling Part (*i*) the front view, then Parts (*ii*) and (*iii*) are the left side and bottom views, respectively. The proteins are assigned their standard numbers in Part (*i*) in which S20 is directly behind S3 (the different colors of spheres are only a viewing aid). The distances between pairs of these proteins were determined from the neutron scattering of concentrated solutions of reconstituted 30S subunits in which the two proteins of interest were heavily deuterated while all other subunit components were normally protonated (deuterons scatter neutrons quite differently from protons). Such measurements on many different pairs of proteins permitted the construction of this map in which the volume of each sphere is proportional to the corresponding protein's mass and its position marks the protein's center of mass. Compare this map with Fig. 30-34*a*. [Courtesy of Peter Moore, Yale University and Malcolm Capel, Brookhaven National Laboratory.] (*b*) A model indicating the locations of the double helical elements of the 16S RNA (*cylinders* colored by domain as in Fig. 30-28 except that domains III and IV are both blue) relative to the 30S subunit proteins (*silver spheres shown at half their size in Part a*) and the electron microscopy–based outline of the 30S subunit (*translucent gray surface*). The model was generated by a computerized energy minimization algorithm constrained by a variety of experimental distance measurements including RNA–RNA and protein–RNA cross-linking studies. The view is the same as in Part *a* (*i*) and Fig. 30-34*a*. The uncertainty in the model is estimated to average ~15 Å. [Courtesy of Stephen Harvey, University of Alabama at Birmingham.]

cates that mRNA binds to the small subunit across the region connecting its "head" to its "base."

2. The anticodon-binding sites occur in the small subunit's "cleft" region.

3. The four L7/L12 subunits forming the large subunit's "stalk" participate in the ribosome's various GTPase reactions.

4. The peptidyl transferase function (P) occupies the "valley" between the large subunit's other two protuberances.

5. The site that binds ribosomes to membranes (M; Section 11-4B), occurs on the large subunit adjacent to the polypeptide exit tunnel, E.

Thus, *the large subunit appears to be mainly involved in mediating biochemical tasks such as catalyzing the reactions of polypeptide elongation, whereas the small subunit is the major actor in ribosomal recognition processes such as mRNA and tRNA binding (although the large subunit is also implicated in tRNA binding)*. We shall see (Section 30-3D) that rRNA almost certainly has a major functional role in ribosomal processes (recall that RNA has demonstrated catalytic properties; Sections 29-4A and C).

Eukaryotic Ribosomes Are Larger and More Complex Than Prokaryotic Ribosomes

Although eukaryotic and prokaryotic ribosomes resemble each other in both structure and function, they differ in nearly all details. Eukaryotic ribosomes have particle masses in the range 3.9 to 4.5×10^6 D and have a nominal sedimentation coefficient of 80S. They dissociate into two unequal subunits that have compositions that are distinctly different from those of prokaryotes (Table 30-7; compare with Table 30-6). The small (40S) subunit of the rat liver cytoplasmic ribosome, the most well-characterized eukaryotic ribosome, consists of 33 unique polypeptides and an **18S rRNA**. Its large (60S) subunit contains 49 different polypeptides and three rRNAs of 28S, 5.8S, and 5S. Electron microscopy indicates that these subunits, as well as the

TABLE 30-7. COMPONENTS OF RAT LIVER CYTOPLASMIC RIBOSOMES

	Ribosome	Small Subunit	Large Subunit
Sedimentation coefficient	80S	40S	60S
Mass (kD)	4220	1400	2820
RNA			
Major		18S, 1874 nucleotides	28S, 4718 nucleotides
Minor			5.8S, 160 nucleotides
			5S, 120 nucleotides
RNA mass (kD)	2520	700	1820
Proportion of mass	60%	50%	65%
Proteins		33 polypeptides	49 polypeptides
Protein mass (kD)	1700	700	1000
Proportion of mass	40%	50%	35%

intact ribosome, have shapes that are similar to those of their prokaryotic counterparts.

Sequence comparisons of the corresponding rRNAs from various species indicates that evolution has conserved their secondary structures rather than their base sequences (Figs. 30-28 and 30-36). For example, a G·C in a base paired stem of *E. coli* 16S rRNA has been replaced by an A·U in the analogous stem of yeast 18S rRNA. The **5.8S rRNA,** which occurs in the large eukaryotic subunit in base paired complex with **28S rRNA,** is homologous in sequence to the 5′ end of prokaryotic 23S rRNA. Apparently 5.8S RNA arose through mutations that altered rRNA's posttranscriptional processing producing a fourth rRNA.

B. Polypeptide Synthesis: An Overview

Before we commence our detailed discussion of polypeptide synthesis, it will be helpful to outline some of its major features.

Polypeptide Synthesis Proceeds from N Terminus to C Terminus

The direction of ribosomal polypeptide synthesis was established, in 1961, by Howard Dintzis through radioactive labeling experiments. He exposed reticulocytes that were actively synthesizing hemoglobin to ^{3}H-labeled leucine for times less than that required to make an entire polypeptide.

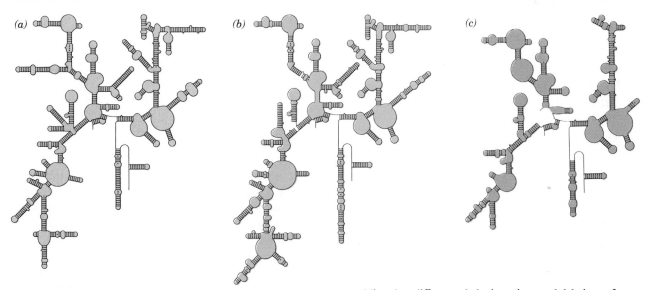

FIGURE 30-36. The predicted secondary structures of evolutionarily distant 16S-like rRNAs from (*a*) archaebacteria (*Halobacterium volcanii*), (*b*) eukaryotes (baker's yeast), and (*c*) mammalian mitochondria (bovine). Compare them with Fig. 30-28, the predicted secondary structure of 16S RNA from eubacteria (*E. coli*). Note the close similarities of these assemblies; they differ mostly by insertions and deletions of stem-and-loop structures. The 23S-like rRNAs from a variety of species likewise have similar secondary structures. [After Gutell, R.R., Weiser, B., Woese, C.R., and Noller, H.F., *Prog. Nucleic Acid Res. Mol. Biol.* **32,** 183 (1985).]

The extent that the tryptic peptides from the soluble (completed) hemoglobin molecules were labeled increased with their proximity to the C terminus (Fig. 30-37). Incoming amino acids must therefore be appended to a growing polypeptide's C terminus; that is, *polypeptide synthesis proceeds from N terminus to C terminus.*

Ribosomes Read mRNA in the 5′ → 3′ Direction

The direction that the ribosome reads mRNAs was determined through the use of a cell-free protein synthesizing system in which the mRNA was poly(A) with a 3′-terminal C.

$$5'\ A-A-A-\cdots-A-A-A-C\ \ 3'$$

Such a system synthesizes a poly(Lys) that has a C-terminal Asn.

$$H_3\overset{+}{N}-Lys-Lys-Lys-\cdots-Lys-Lys-Asn-COO^-$$

This, together with the knowledge that AAA and AAC code for Lys and Asn and the polarity of polypeptide synthesis, indicates that *the ribosome reads mRNA in the 5′ → 3′ direction.* Since mRNA is synthesized in the 5′ → 3′ direction, this accounts for the observation that, in prokaryotes, ribosomes initiate translation on nascent mRNAs (Section 29-3).

Active Translation Occurs on Polyribosomes

Electron micrographs reveal that ribosomes engaged in protein synthesis are tandemly arranged on mRNAs like

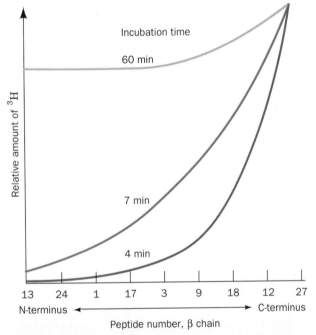

FIGURE 30-37. Distribution of [³H]Leu among the tryptic peptides from the *β* subunit of soluble rabbit hemoglobin after the incubation of rabbit reticulocytes with [³H]leucine for the indicated times. [After Dintzis, H.M., *Proc. Natl. Acad. Sci.* **47,** 255 (1961).]

beads on a string (Figs. 30-38 and 29-18). The individual ribosomes in these **polyribosomes (polysomes)** are separated by gaps of 50 to 150 Å so they have a maximum density on mRNA of ~1 ribosome per 80 nucleotides. Polysomes arise because once an active ribosome has cleared its initiation site, a second ribosome can initiate at that site.

Chain Elongation Occurs by the Linkage of the Growing Polypeptide to the Incoming tRNA's Amino Acid Residue

During polypeptide synthesis, amino acid residues are sequentially added to the C terminus of the nascent, ribosomally bound polypeptide chain. If the growing polypeptide is released from the ribosome by treatment with high salt concentrations, its C-terminal residue is invariably esterified to a tRNA molecule as a **peptidyl–tRNA.**

Peptidyl–tRNA

The nascent polypeptide must therefore grow by being transferred from the peptidyl–tRNA to the incoming aminoacyl–tRNA to form a peptidyl–tRNA with one more residue (Fig. 30-39). Apparently, the ribosome has at least two tRNA-binding sites: The so-called **P site,** which binds the *p*eptidyl–tRNA, and the **A site,** which binds the incoming *a*minoacyl–tRNA (Fig. 30-39). Consequently, after the formation of a peptide bond, the newly deacylated P-site tRNA must be released and replaced by the newly formed peptidyl–tRNA from the A site, thereby permitting a new round of peptide bond formation. The finding that each ribosome can bind up to three deacylated tRNAs but only two aminoacyl–tRNAs indicates, however, that the

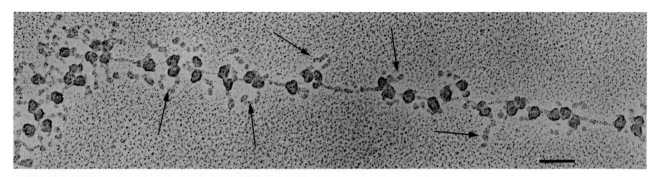

FIGURE 30-38. Electron micrographs of polysomes from silk gland cells of the silkworm *Bombyx mori.* The 3′ end of the mRNA is on the right. Arrows point to the silk fibroin polypeptides. The bar represents 0.1 μm. [Courtesy of Oscar L. Miller, Jr., University of Virginia.]

ribosome has a third tRNA-binding site: the **exit** or **E site**, which transiently binds the outgoing tRNA.

The details of the chain elongation process are discussed in Section 30-3D. Chain initiation and chain termination, which are special processes, are examined in Sections 30-3C and 30-3E, respectively. In all of these sections we shall first consider the process of interest in *E. coli* and then compare it with the analogous eukaryotic activity.

C. Chain Initiation

fMet Is the N-Terminal Residue of Prokaryotic Polypeptides

The first indication that the initiation of translation requires a special codon, since identified as AUG (and, in prokaryotes, occasionally GUG), was the observation that almost half of the *E. coli* proteins begin with the otherwise uncommon amino acid Met. This was followed by the discovery of a peculiar form of Met–tRNA^{Met} in which the Met residue is *N*-formylated.

$$
\begin{array}{c}
S-CH_3 \\
| \\
CH_2 \\
| \\
O \qquad\quad CH_2 \quad O \\
\| \qquad\qquad | \qquad\; \| \\
HC-NH-CH-C-O-tRNA_f^{Met}
\end{array}
$$

N–Formylmethionine-tRNA$_f^{Met}$
(**fMet-tRNA$_f^{Met}$**)

The *N*-formylmethionine residue (**fMet**), which already has an amide bond, can therefore only be the N-terminal residue of a polypeptide. In fact, polypeptides synthesized in an *E. coli*-derived cell-free protein synthesizing system always

FIGURE 30-39. The ribosomal peptidyl transferase reaction forming a peptide bond. The amino group of the aminoacyl–tRNA in the A site nucleophilically displaces the tRNA of the peptidyl–tRNA in the P site, thereby transferring the nascent polypeptide to the A site tRNA.

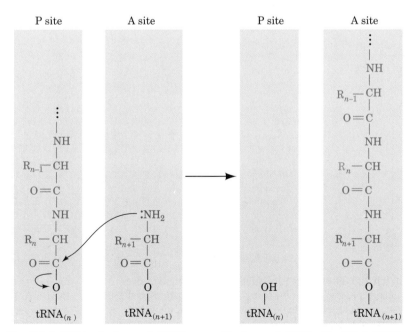

have a leading fMet residue. *fMet must therefore be E. coli's initiating residue.*

The tRNA that recognizes the initiation codon, tRNA$_f^{Met}$ (Fig. 30-40), differs from the tRNA that carries internal Met residues, tRNA$_m^{Met}$, although they both recognize the same codon. In *E. coli,* uncharged (deacylated) tRNA$_f^{Met}$ is first aminoacylated with Met by the same MetRS that charges

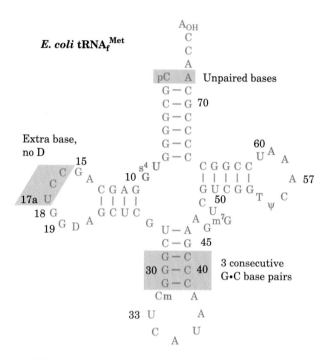

FIGURE 30-40. The nucleotide sequence of *E. coli* tRNA$_f^{Met}$ shown in cloverleaf form. The shaded boxes indicate the significant differences between this initiator tRNA and noninitiator tRNAs such as yeast tRNA^{Ala} (Fig. 30-11). [After Woo, N. M., Roe, B. A., and Rich, A., *Nature* **286,** 346 (1980).]

tRNA$_m^{Met}$. The resulting Met–tRNA$_f^{Met}$ is specifically *N*-formylated to yield fMet–tRNA$_f^{Met}$ in an enzymatic reaction that employs N^{10}-formyltetrahydrofolate (Section 24-4D) as its formyl donor. The formylation enzyme does not recognize Met–tRNA$_m^{Met}$. The X-ray structures of *E. coli* tRNA$_f^{Met}$ and yeast tRNA^{Phe} (Fig. 30-14) are largely similar but differ conformationally in their acceptor stems and anticodon loops. Perhaps these structural differences permit tRNA$_f^{Met}$ to be distinguished from tRNA$_m^{Met}$ in the reactions of chain initiation and elongation (see Section 30-3D).

E. coli proteins are posttranslationally modified by deformylation of their fMet residue and, in many proteins, by the subsequent removal of the resulting N-terminal Met. This processing usually occurs on the nascent polypeptide, which accounts for the observation that mature *E. coli* proteins all lack fMet.

Base Pairing between mRNA and the 16S rRNA Helps Select the Translational Initiation Site

AUG codes for internal Met residues as well as the initiating Met residue of a polypeptide. Moreover, mRNAs usually contain many AUGs (and GUGs) in different reading frames. Clearly, *a translational initiation site must be specified by more than just an initiation codon.*

In *E. coli,* the 16S rRNA contains a pyrimidine-rich sequence at its 3′ end. This sequence, as John Shine and Lynn Dalgarno pointed out in 1974, is partially complementary to a purine-rich tract of 3 to 10 nucleotides, the **Shine–Dalgarno sequence,** that is centered ~10 nucleotides upstream from the start codon of nearly all known prokaryotic mRNAs (Fig. 30-41). *Base pairing interactions between an mRNA's Shine–Dalgarno sequence and the 16S rRNA apparently permit the ribosome to select the proper initiation codon.* Thus ribosomes with mutationally altered anti-Shine–Dalgarno sequences often have greatly reduced

Initiation
codon

araB	– U U U G G A U G G A G U G A A A C G A U G G C G A U U –
galE	– A G C C U A A U G G A G C G A A U U A U G A G A G U U –
LacI	– C A A U U C A G G G U G G U G A U U G U G A A A C C A –
lacZ	– U U C A C A C A G G A A A C A G C U A U G A C C A U G –
Qβ phage replicase	– U A A C U A A G G A U G A A A U G C A U G U C U A A G –
φX174 phage A protein	– A A U C U U G G A G G C U U U U U U A U G G U U C G U –
R17 phage coat protein	– U C A A C C G G G G U U U G A A G C A U G G C U U C U –
Ribosomal S12	– A A A A C C A G G A G C U A U U U A A U G G C A A C A –
Ribosomal L10	– C U A C C A G G A G C A A A G C U A A U G G C U U U A –
trpE	– C A A A A U U A G A G A A U A A C A A U G C A A A C A –
trp leader	– G U A A A A A G G G U A U C G A C A A U G A A A G C A –

3′ end of 16S rRNA	$3'_{HO}$ A U U C C U C C A C U A G – 5′

FIGURE 30-41. Some translational initiation sequences recognized by *E. coli* ribosomes. The mRNAs are aligned according to their initiation codons (*blue shading*). Their Shine–Dalgarno sequences (*red shading*) are complementary, counting G·U pairs, to a portion of the 16S rRNA's 3′ end

(*below*). [After Steitz, J.A., in Chambliss, G., Craven, G.R., Davies, J., Davis, K., Kahan, L., and Nomura, M. (Eds.), *Ribosomes. Structure, Function and Genetics,* pp. 481–482, University Park Press (1979).]

ability to recognize natural mRNAs, although they efficiently translate mRNAs whose Shine–Dalgarno sequences have been made complementary to the altered anti-Shine–Dalgarno sequences. Moreover, treatment of ribosomes with the bacteriocidal protein **colicin E3** (produced by *E. coli* strains carrying the E3 plasmid), which specifically cleaves a 49-nucleotide fragment from the 3′ terminus of 16S rRNA, yields ribosomes that cannot initiate new polypeptide synthesis but can complete the synthesis of a previously initiated chain. In fact, when ribosomes that have bound a fragment of **R17 phage** mRNA containing the initiation sequence for its so-called A protein are treated with colicin E3 and then dissociated in 1% SDS, the mRNA fragment is released in complex with the 49-nucleotide rRNA fragment (Fig. 30-42).

Initiation Is a Three-Stage Process that Requires the Participation of Soluble Protein Initiation Factors

Intact ribosomes do not directly bind mRNA so as to initiate polypeptide synthesis. Rather, *initiation is a complex process in which the two ribosomal subunits and fMet–tRNA$_f^{Met}$ assemble on a properly aligned mRNA to form a complex that is competent to commence chain elongation. This assembly process also requires the participation of protein* **initiation factors** *that are not permanently associated with the ribosome.* Initiation in *E. coli* involves three initiation factors designated **IF-1, IF-2,** and **IF-3** (Table 30-8). Their existence was discovered when it was found that washing small ribosomal subunits with 1*M* ammonium chloride solution, which removes the initiation factors but not the "permanent" ribosomal proteins, prevents initiation.

The initiation sequence in *E. coli* ribosomes has three stages (Fig. 30-43):

1. Upon completing a cycle of polypeptide synthesis, the 30S and 50S subunits remain associated as inactive 70S ribosomes. IF-3 binds to the 30S subunit so as to promote the dissociation of this complex. IF-1 increases this dissociation rate, perhaps by assisting the binding of IF-3.

2. mRNA and IF-? in a ternary complex with GTP and fMet–tRNA$_f^{Met}$ subsequently bind to the 30S subunit in either order. Hence, fMet–tRNA$_f^{Met}$ recognition must not be mediated by a codon–anticodon interaction; it is the only tRNA–ribosome association not to require one. This interaction, nevertheless, helps bind fMet–tRNA$_f^{Met}$ to the ribosome. IF-3 also functions in this

TABLE 30-8. THE SOLUBLE PROTEIN FACTORS OF *E. COLI* PROTEIN SYNTHESIS

Factor	Mass (kD)	Function
Initiation Factors		
IF-1	9	Assists IF-3 binding.
IF-2	97	Binds initiator tRNA and GTP.
IF-3	22	Releases 30S subunit from inactive ribosome and aids mRNA binding.
Elongation Factors		
EF-Tu	43	Binds aminoacyl-tRNA and GTP.
EF-Ts	74	Displaces GDP from EF-Tu.
EF-G	77	Promotes translocation by binding GTP to the ribosome.
Release Factors		
RF-1	36	Recognizes UAA and UAG Stop codons.
RF-2	38	Recognizes UAA and UGA Stop codons.
RF-3	46	Binds GTP and stimulates RF-1 and RF-2 binding.

stage of the initiation process: It assists the 30S subunit in binding the mRNA although it does not appear to influence the selection of the initiation codon.

3. Last, in a process that is preceded by IF-3 release, the 50S subunit joins the 30S initiation complex in a manner that stimulates IF-2 to hydrolyze its bound GTP to GDP + P$_i$. This irreversible reaction conformationally rearranges the 30S subunit and releases IF-1 and IF-2 for participation in further initiation reactions.

Initiation results in the formation of an fMet–tRNA$_f^{Met}$ · mRNA · ribosome complex in which the fMet–tRNA$_f^{Met}$ occupies the ribosome's P site while its A site is poised to accept an incoming aminoacyl–tRNA (an arrangement analogous to that at the conclusion of a round of elongation: Section 30-3D). This arrangement was established through the use of the antibiotic **puromycin** as is discussed in Section 30-3D. Note that tRNA$_f^{Met}$ is the only tRNA that directly enters the P site. All other tRNAs must do so via the A site during chain elongation (Section 30-3D).

fMet–Arg–Ala–

R17 phage A protein mRNA   –AUUCCUAGGAGGUUUGACCUAUG CGAGCU–

3′ end of 16S rRNA
(colicin E3 fragment)   3′ $_{HO}$AU^{UCCUCCA} CCACUAG– 5′

FIGURE 30-42. Base pairing interactions between the colicin E3 fragment of *E. coli* 16S rRNA and the R17 phage A protein initiator region. [After Steitz, J.A. and Jakes, K., *Proc. Natl. Acad. Sci.* **72**, 4735 (1975).]

Eukaryotic Initiation Resembles that of Prokaryotes

Eukaryotic initiation resembles the overall prokaryotic process but differs from it in detail. Eukaryotes have a far more extensive "zoo" of initiation factors (designated eIF-*n*; "e" for eukaryotic) than do prokaryotes. Over 10 such factors, many with multiple subunits, occur in some eukaryotic systems, although they are more difficult to distinguish from ribosomal proteins than are prokaryotic initiation factors.

The most striking difference between eukaryotic and prokaryotic ribosomal initiation occurs in the second stage of the process, the binding of mRNA and a complex of **eIF-2**, GTP, and **Met-tRNA$_i^{Met}$** to the 40S ribosomal subunit (here the subscript "i" distinguishes eukaryotic initiator tRNA, whose appended Met residue is never *N*-formylated, from that of prokaryotes; both species are, nevertheless, readily interchangeable *in vitro*). Eukaryotic mRNAs lack the complementary sequences to bind 18S rRNA in the Shine–Dalgarno manner. Rather, *translation of eukaryotic mRNAs, which are invariably monocistronic, almost always starts at their first AUG.* This, together with the observations that (1) prokaryotic but not eukaryotic ribosomes can initiate on circular RNAs, and (2) **eIF-4E** is a **cap-binding protein,** suggests that the 40S subunit binds at or near eukaryotic mRNA's 5′ cap (Section 29-4A) and migrates downstream until it encounters the first AUG. This hypothesis explains the greatly reduced initiation rates of improperly capped mRNAs.

D. Chain Elongation

Ribosomes elongate polypeptide chains in a three-stage reaction cycle that adds amino acid residues to a growing polypeptide's C-terminus (Fig. 30-44). This process, which occurs at a rate of up to 40 residues/s, involves the participation of several nonribosomal proteins known as **elongation factors** (Table 30-8).

Aminoacyl–tRNA Binding

In the "binding" stage of the *E. coli* elongation cycle, a binary complex of GTP with the elongation factor **EF-Tu** combines with an aminoacyl–tRNA. The resulting ternary complex binds to the ribosome and, in a reaction that hydrolyzes the GTP to GDP + P$_i$, the aminoacyl–tRNA is bound in a codon–anticodon complex to the ribosomal A site and EF-Tu · GDP + P$_i$ is released. In the remainder of this stage, which serves to regenerate the EF-Tu · GTP complex, GDP is displaced from EF-Tu · GDP by the elongation factor **EF-Ts** which, in turn, is displaced by GTP.

Aminoacyl–tRNAs can bind to the ribosomal A site without the mediation of EF-Tu but at a rate too slow to support cell growth. The importance of EF-Tu is indicated by the fact that it is the most abundant *E. coli* protein; it is present in ~100,000 copies per cell (>5% of the cell's protein), which is approximately the number of tRNA molecules in the cell. Consequently, *the cell's entire complement of aminoacyl–tRNAs is essentially sequestered by EF-Tu.*

EF-Tu binds neither formylated nor unformylated Met–tRNA$_f^{Met}$, which is why the initiator tRNA never reads internal AUG or GUG codons. What is the structural basis of this discrimination? *E. coli* tRNA$_f^{Met}$ differs from other *E. coli* tRNAs by the absence of a base pair at the end of its amino acid stem (Fig. 30-40). The conversion of its 5′-ter-

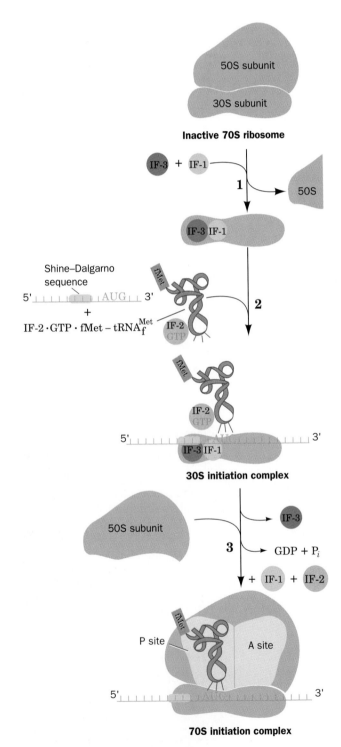

FIGURE 30-43.   The initiation pathway in *E. coli* ribosomes.

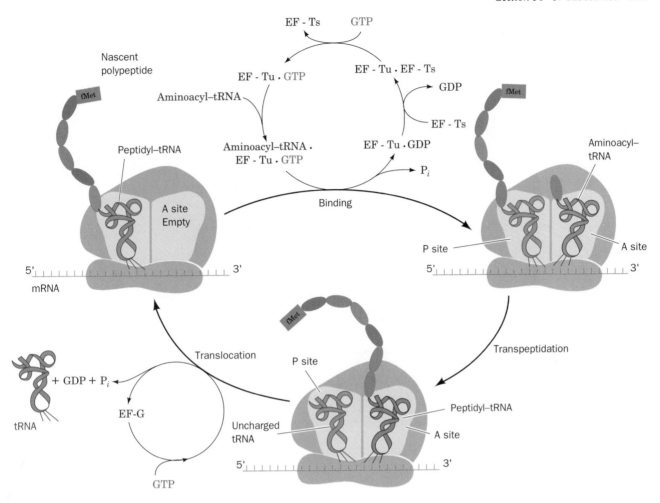

FIGURE 30-44. The elongation cycle in *E. coli* ribosomes. The E site, to which discharged tRNAs are transferred before being released to solution, is not shown. Eukaryotic elongation follows a similar cycle but EF-Tu and EF-Ts are replaced by a single multisubunit protein, eEF-1, and EF-G is replaced by eEF-2.

minal C residue to U by bisulfite treatment, which reestablishes the missing base pair as U · A, allows EF-Tu binding. Evidently, EF-Tu recognizes the amino acid stem of noninitiator tRNAs. However, the initiator tRNAs from several other sources have fully base paired amino acid stems, and recall that the U1 · A72 base pair of tRNA^{Gln} is opened up upon binding to GlnRS (Section 30-2C).

EF-Tu Undergoes a Major Conformational Change on Hydrolyzing GTP

EF-Tu is a member of the **GTP-binding protein** family, many of whose members are also soluble ribosomal factors (e.g., IF-2) or essential participants in intracellular signaling pathways (Sections 33-4C and 34-4B). All of these proteins share a common structural motif that binds guanine nucleotides (GDP and GTP) and catalyzes the hydrolysis of GTP to GDP + P$_i$. They are often accompanied by two other proteins: a **GTPase activating protein (GAP),** which as its name implies, stimulates the GTP-binding protein to hydrolyze its bound GTP; and a **guanine nucleotide releas-**

ing factor [GRF; alternatively **guanine nucleotide release protein (GNRP)** or **guanine nucleotide exchange factor (GEF)],** which induces the GTP-binding protein to exchange its bound GDP for GTP. In the case of EF-Tu, the ribosome is its GAP and EF-Ts is its GRF.

Morten Kjeldgaard and Jens Nyborg determined the X-ray structures of the 393-residue EF-Tu from *E. coli* in complex with GDP and the 70% homologous 405-residue EF-Tu from *Thermus aquaticus* (a thermophilic bacterium) in complex with the slowly hydrolyzing GTP analog **guanosine-5′-(β,γ-imido)triphosphate (GDPNP).**

Guanosine-5′-(β,γ-imido)triphosphate (GDPNP)

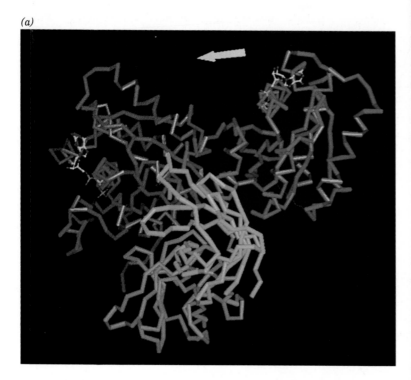

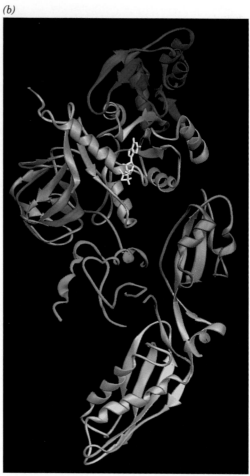

FIGURE 30-45.   X-Ray structures of ribosomal elongation factors. (*a*) Comparison of the X-ray structures of EF-Tu in its complexes with GDP and GDPNP. The protein is represented by its C$_\alpha$ backbone with domain 1, its GTP-binding domain, purple in the GDP complex and red in the GDPNP complex. Domains 2 and 3, which have the same orientation in both complexes, are green and blue. The bound GDP and GDPNP are shown in skeletal form with C yellow, N blue, O red, and P green. [Courtesy of Morten Kjeldgaard and Jens Nyborg, Aarhus University, Århus, Denmark.] (*b*) Ribbon diagram of EF-G in complex with GDP. Domain 1, its GTP-binding domain, is light purple, its insert is dark purple, and domains 2, 3, 4, and 5 are blue, green, yellow, and red, respectively. The bound ADP is shown in stick form in white. A 25-residue segment in domain 1 is not seen in the X-ray structure and domain 3 is poorly defined, thereby accounting for the several polypeptide segments seen in this model. [Based on an X-ray structure by John Czworkowski, Jimin Wang, Joan Steitz, and Peter Moore, Yale University.]

EF-Tu folds into three distinct domains that are connected by flexible peptides, rather like beads on string. The N-terminal domain 1, which binds GTP/GDP and catalyzes GTP hydrolysis, structurally resembles other known GTP-binding proteins. Comparison of the GDPNP and GDP complexes (Fig. 30-45*a*) indicates that, on hydrolyzing GTP, EF-Tu undergoes a major structural reorganization: Helix B of domain 1, which is in proximity to GTP's γ-phosphate group, shifts its direction by 42°, thereby inducing domain 1 to rigidly change its orientation with respect to domains 2 and 3 by ~91°.

EF-Tu · GTP, but not EF-Tu · GDP, binds aminoacyl–tRNA. EF-Tu · GDPNP has a solvent-filled cleft formed by domains 1 and 2 that is absent in EF-Tu · GDP. This cleft is lined with several positively charged side chains that could bind to tRNA's phosphate groups and is therefore likely to form EF-Tu's aminoacyl–tRNA binding site. This suggestion is supported by the observations that the formation of the putative aminoacyl–tRNA binding cleft requires a conformational change at Gly 84 and that the replacement of Gly 84 with Ala, which sterically prevents cleft formation, yields a mutant EF-Tu that does not bind aminoacyl–tRNA.

Transpeptidation

Note that in Fig. 30-44, *the peptide bond is formed in the second stage of the elongation cycle through the nucleophilic displacement of the P site tRNA by the amino group of the*

3′-linked aminoacyl–tRNA in the A site (Fig. 30-39). The nascent polypeptide chain is thereby lengthened at its C terminus by one residue and transferred to the A-site tRNA, a process called **transpeptidation.** The reaction occurs without the need of activating cofactors such as ATP because the ester linkage between the nascent polypeptide and the P-site tRNA is a "high-energy" bond. The peptidyl transferase center that catalyzes peptide bond formation is located entirely on the large subunit as is demonstrated by the observation that in high concentrations of organic solvents such as ethanol, the large subunit alone catalyzes peptide bond formation. The organic solvent apparently distorts the large subunit in a way that mimics the effect of small subunit binding. Peptidyl transferase activity appears to be resident on the 23S RNA and is probably facilitated by several polypeptide chains in the large subunit (see below).

Translocation

*In the final stage of the elongation cycle, the now uncharged P site tRNA (at first tRNA$_f^{Met}$ but subsequently a noninitiator tRNA) is transferred to the E site (not shown in Fig. 30-44), its former occupant having been previously expelled (see below). Simultaneously, in a process known as **translocation**, the peptidyl–tRNA in the A site, together with its bound mRNA, is moved to the P site.* This prepares the ribosome for the next elongation cycle. The maintenance of the peptidyl–tRNA's codon–anticodon association is no longer necessary for amino acid specification. Rather, it probably acts as a place-keeper that permits the ribosome to precisely step off the three nucleotides along the mRNA required to preserve the reading frame. Indeed, the observation that frameshift suppressor tRNAs induce a four-nucleotide translocation (Section 30-2E) indicates that mRNA movement is directly coupled to tRNA movement.

The translocation process requires the participation of an elongation factor, **EF-G,** that binds to the ribosome together with GTP and is only released upon hydrolysis of the GTP to GDP + P$_i$. EF-G release is prerequiste for beginning the next elongation cycle because the ribosomal binding sites of EF-G and EF-Tu partially overlap and hence their ribosomal binding is mutually exclusive.

The X-ray structures of EF-G from *Thermus thermophilus,* both alone and in complex with GDP, were respectively determined by Liljas and by Joan Steitz and Moore. EF-G, which has a closely similar structure in the two complexes, is a tadpole-shaped monomeric protein of 691 residues that consists of five domains (Fig. 30-45*b*). Its first two domains, starting at its N-terminus, form the head of the tadpole and closely resemble EF-Tu's first two domains (Fig. 30-45*a*). Interestingly, however, the relative disposition of these two domains in EF-G·GDP resembles that in EF-Tu·GTP rather than that in EF-Tu·GDP. Perhaps this is indicative of the reciprocal functions of these two elongation factors: EF-Tu·GTP facilitates the conversion of the ribosome from its pre- to its post-translocational state,

whereas EF-G·GTP promotes the reverse transition.

EF-G is unusual among GTP-binding proteins in that it lacks a corresponding guanine nucleotide releasing factor (GRF). However, its N-terminal GTP-binding domain contains a unique subdomain that contacts the domain's conserved core at sites analogous to those in EF-Tu that interact with EF-Ts. This suggests that this subdomain acts as an internal GRF.

EF-G's three C-terminal domains, which comprise ~500 residues, have no counterparts in EF-Tu. However, they structurally resemble ribosomal proteins of known structures (Fig. 30-31), which suggests that these EF-G domains interact with RNA.

Translocation Occurs Via Intermediate States

Chemical footprinting studies (Sections 29-2B and 33-3B) by Noller reveal that certain bases in 16S rRNA are protected by tRNAs bound in the ribosomal A and P sites and that certain bases in the 23S rRNA are protected by tRNAs in the A, P, and E sites. Almost all of these protected bases are absolutely conserved in evolution and many of them have been implicated in ribosomal function through biochemical or genetic studies.

Variations in chemical footprinting patterns during the elongation cycle indicate that the translocation of tRNA occurs in two discrete steps (Fig. 30-46, steps 1 and 2):

1. Peptide bond formation causes the acceptor end of the new peptidyl tRNA to shift from the A site to the P site of the large ribosomal subunit while its anticodon end remains associated with the A site of the small subunit (yielding a so-called A/P hybrid binding state). The acceptor end of the newly deacylated tRNA simultaneously moves from the P site to the E site of the large subunit while its anticodon end remains associated with the P site of the small subunit (the P/E binding state).

2. The ribosomal binding of the EF-G·GTP complex causes the anticodon ends of these tRNAs, together with their bound mRNA, to move relative to the small ribosomal subunit such that the peptidyl–tRNA occupies the P site of both ribosomal subunits (the P/P binding state) and the deacylated tRNA occupies the E site of the large ribosomal subunit (a tRNA in the E site does not protect the 16S rRNA and hence its state of occupancy cannot be determined by chemical footprinting).

The elongation cycle is then completed by the release of the tRNA in the E site and the binding of a new aminoacyl–tRNA in the A site (Fig. 30-46, *top*).

The binding of tRNA to the A and E sites exhibits negative allosteric cooperativity. At the end of an elongation cycle, the E site binds the newly deacylated tRNA with high affinity, whereas the now empty A site has but low affinity for aminoacyl–tRNA (the posttranslational state). Upon binding an incoming aminoacyl–tRNA, the ribosome undergoes a conformational change that converts the A site to a high-affinity state and the E site to a low-affinity state, which consequently releases the deacylated tRNA (the pre-

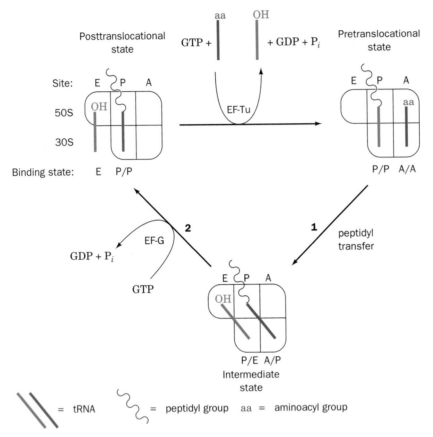

FIGURE 30-46. The ribosomal binding states in the elongation cycle. Note how this scheme elaborates the classical elongation cycle diagrammed in Fig. 30-44. [In part after Moazed, D. and Noller, H.F., *Nature* **342**, 147 (1989).]

translocational state). Evidently, GTP hydrolysis by the elongation factors EF-Tu and EF-G as well as the transpeptidation reaction function to reduce the activation barriers between these conformational states. The unidirectional A → P → E flow of tRNAs through the ribosome is thereby facilitated.

Puromycin Is an Aminoacyl–tRNA Analog

The ribosomal elongation cycle was originally characterized through the use of the antibiotic **puromycin** (Fig. 30-47). *This substance, which resembles the 3′ end of Tyr–tRNA, causes the premature termination of polypeptide chain synthesis.* Puromycin, in competition with the mRNA- specified aminoacyl–tRNA but without the need of elongation factors, binds to the ribosomal A site which, in turn, catalyzes a normal transpeptidation reaction to form peptidyl–puromycin. Yet, the ribosome cannot catalyze the transpeptidation reaction in the next elongation cycle because puromycin's "amino acid residue" is linked to its "tRNA" via an amide rather than an ester bond. Polypeptide synthesis is therefore aborted and the peptidyl–puromycin is released.

In the absence of EF-G and GTP, an active ribosome cannot bind puromycin because its A site is at least partially occupied by a peptidyl–tRNA. A newly initiated ribosome, however, violates this rule; it catalyzes fMet–puromycin formation. *These observations demonstrated the functional existence of the ribosomal P and A sites and established that fMet–tRNA$_f^{Met}$ binds directly to the P site, whereas other aminoacyl–tRNAs must first enter the A site.*

23S rRNA Almost Certainly Participates in Catalyzing Peptide Bond Formation

The idea that rRNA functions catalytically in polypeptide synthesis arose from the following observations:

1. RNA can act as a catalyst (Section 29-4B).

2. Ribosomes consist mostly of RNA.

3. RNAs are more highly conserved throughout evolution than are ribosomal proteins.

4. The absence of any of numerous ribosomal proteins does not abolish ribosomal function.

5. Most mutations that confer resistance to antibiotics

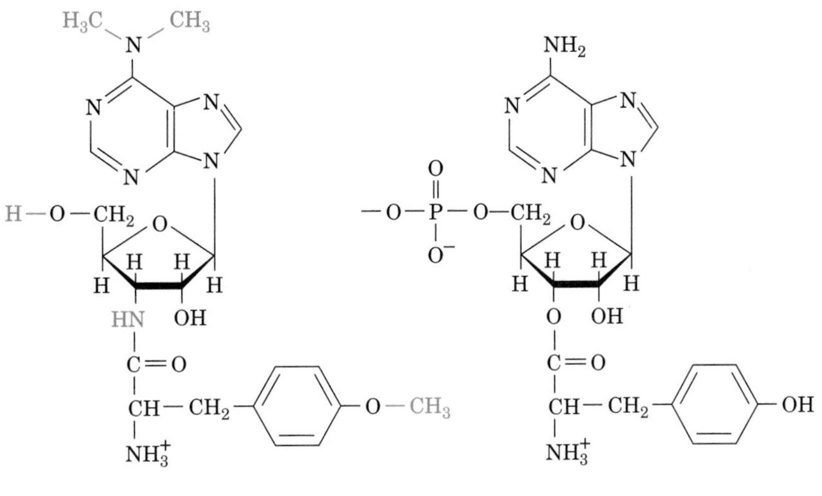

Puromycin   **Tyrosyl–tRNA**

FIGURE 30-47.   Puromycin (*left*) resembles the 3′ terminus of tyrosyl–tRNA (*right*).

which inhibit protein synthesis mostly occur in genes encoding rRNAs rather than ribosomal proteins.

However, with the exception that the anti-Shine–Dalgarno sequence at the 3′ end of 16S RNA has been shown to participate in mRNA selection (Section 30-3C), it has not been conclusively demonstrated that rRNA has any specific function.

The ribosomal peptidyl transferase function can be monitored by the so-called "fragment reaction," in which ^{35}S-labeled fMet is transferred from CAACCA–fMet (the 3′-terminal segment of fMet–tRNA$_f^{Met}$, which presumably binds in the ribosomal P site as does intact fMet–tRNA$_f^{Met}$) to the amino group of puromycin (bound in the A site) to form a model peptide bond (CAACCA–fMet + puromycin → CAACCA—OH + fMet–puromycin). Under appropriate conditions (which include a solution that is 33% methanol or ethanol), the fragment reaction is catalyzed by the large ribosomal subunit alone. The biological authenticity of this process is demonstrated by its stereochemical specificity for its substrates and its inhibition by antibiotics such as **chloramphenicol,** which are known to inhibit the ribosomal peptidyl transferase function (Section 30-3G).

Noller has shown that the *E. coli* large ribosomal subunit retains 20 to 40% of its peptidyl transferase activity, as monitored by the fragment reaction, after treatment with **proteinase K** and extraction with SDS, but loses all of this activity when further extracted by phenol (which denatures proteins). However, the more stable large ribosomal subunit from the thermophilic bacterium *T. aquaticus* retains >80% activity after such treatments. Yet, this activity is abolished by treatment with **ribonuclease T1,** which cleaves RNA only after G residues and hence does not alter the CAACCA substrate.

The foregoing observations, together with a variety of circumstantial evidence, strongly implicate the 23S RNA of the large ribosomal subunit in the peptidyl transferase function. However, since ~5% of the protein initially present in the large ribosomal subunit was not extracted by the above treatments, which 2-dimensional gel electrophoresis indicates is due to the presence of 2 to 8 intact ribosomal proteins, the catalytic function of rRNA has not been unequivocally established. Of course, considering that ribosomal proteins and RNAs have coevolved for over 3.5 billion years, it would hardly be surprising if it turns out that the peptidyl transferase function presently arises from the synergistic interactions of certain ribosomal proteins with the 23S RNA.

The Eukaryotic Elongation Cycle Resembles that of Prokaryotes

The eukaryotic elongation cycle closely resembles that of prokaryotes. In eukaryotes, the functions of EF-Tu and EF-Ts are assumed by two different subunits of the eukaryotic elongation factor **eEF-1.** Likewise, **eEF-2** functions in a manner analogous to EF-G. However, the corresponding eukaryotic and prokaryotic elongation factors are not interchangable.

E. Chain Termination

Polypeptide synthesis under the direction of synthetic mRNAs such as poly(U) terminates with a peptidyl–tRNA in association with the ribosome. However, *the translation of natural mRNAs, which contain the termination codons UAA, UGA, or UAG, results in the production of free poly-*

peptides (Fig. 30-48). In *E. coli,* the termination codons, the only codons that normally have no corresponding tRNAs, are recognized by protein **release factors** (Table 30-8): **RF-1** recognizes UAA and UAG, whereas **RF-2** recognizes UAA and UGA. Neither of these release factors can bind to the ribosome simultaneously with EF-G. A third release factor, **RF-3,** a GTP-binding protein, in complex with GTP, stimulates the ribosomal binding of RF-1 and RF-2. The release factors act at the ribosomal A site as is indicated by the observation that they compete with suppressor tRNAs for termination codons.

The binding of a release factor to the appropriate termination codon induces the ribosomal peptidyl transferase to transfer the peptidyl group to water rather than to an aminoacyl–tRNA (Fig. 30-49). The consequent uncharged tRNA subsequently dissociates from the ribosome and the release factors are expelled with the concomitant hydrolysis of the RF-3–bound GTP to GDP + P_i. The resulting inactive ribosome then releases its bound mRNA preparatory to a new round of polypeptide synthesis.

Termination in eukaryotes resembles that in prokaryotes but requires only a single release factor, **eRF,** that binds to the ribosome together with GTP. This GTP is hydrolyzed to GDP + P_i in a reaction that is thought to trigger eRF's dissociation from the ribosome.

GTP Hydrolysis Speeds Up Ribosomal Processes

What is the role of the GTP hydrolysis reactions mediated by the various GTP-binding factors (IF-2, EF-Tu, EF-G, and RF-3 in *E. coli*) and which are essential for normal ribosomal function? Translation occurs in the absence of GTP, albeit extremely slowly, so that the free energy of the transpeptidation reaction is sufficient to drive the entire translational process. Moreover, none of the GTP hydrolysis reactions yields a "high-energy" covalent intermediate as does, say ATP hydrolysis in numerous biosynthetic reactions. It is therefore thought that the ribosomal binding of a GTP-binding factor in complex with GTP allosterically causes ribosomal components to change their conformations in a way that facilitates a particular process such as translocation (e.g., Fig. 30-45*a*). This conformational change also catalyzes GTP hydrolysis, which, in turn, permits the ribosome to relax to its initial conformation with the concomitant release of products including GDP + P_i. *The high rate and irreversibility of the GTP hydrolysis reaction therefore ensures that the various complex ribosomal processes to which it is coupled, initiation, elongation, and termination, will themselves be fast and irreversible.* GTP hydrolysis also facilitates translational accuracy (see below).

F. Translational Accuracy

The genetic code is normally translated with remarkable fidelity. We have already seen that transcription and tRNA aminoacylation both proceed with high accuracy (Sections

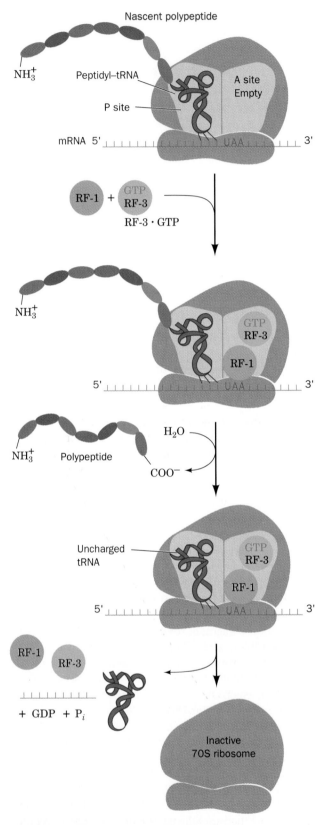

FIGURE 30-48. The termination pathway in *E. coli* ribosomes. RF-1 recognizes the termination codons UAA and UAG, whereas RF-2 recognizes UAA and UGA. Eukaryotic termination follows an analogous pathway but requires only a single release factor, eRF, that recognizes all three termination codons.

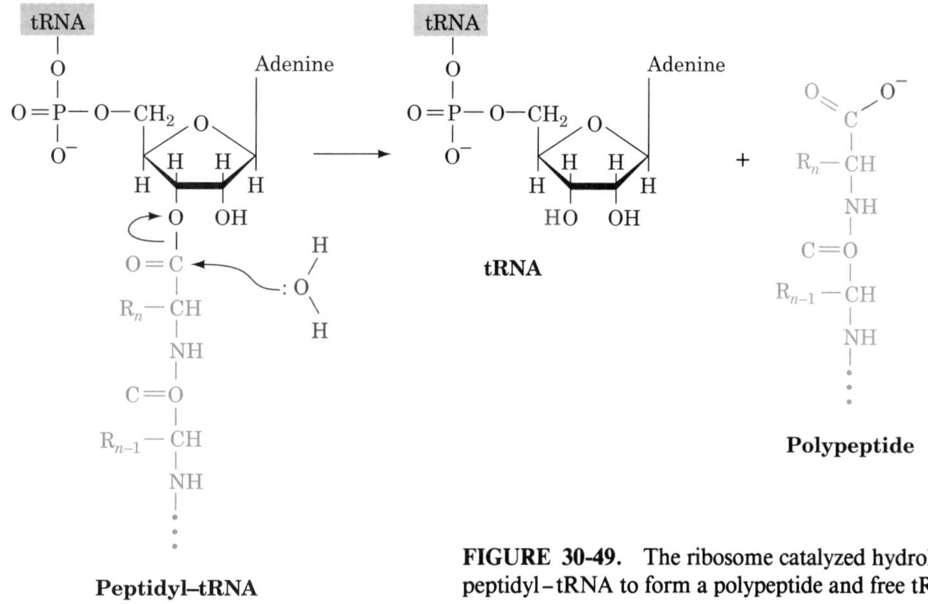

Peptidyl–tRNA

FIGURE 30-49. The ribosome catalyzed hydrolysis of peptidyl–tRNA to form a polypeptide and free tRNA.

29-2D and 30-2C). The accuracy of ribosomal mRNA decoding was estimated from the rate of misincorporation of ^{35}S-Cys into highly purified **flagellin,** an *E. coli* protein (Section 34-3G) that normally lacks Cys. These measurements indicated that the mistranslation rate is $\sim 10^{-4}$ errors per codon. This rate is greatly increased in the presence of **streptomycin,** an antibiotic that increases the rate of ribosomal misreading (Section 30-3G). From the types of reading errors that streptomycin is known to induce, it was concluded that the mistranslation arose almost entirely from the confusion of the Arg codons CGU and CGC for the Cys codons UGU and UGC. The above error rate is therefore largely caused by mistakes in ribosomal decoding.

Aminoacyl–tRNAs are selected by the ribosome only according to their anticodon. Yet, the binding energy loss arising from a single base mismatch in a codon–anticodon interaction is estimated to be ~ 12 kJ·mol^{-1}, which, according to Eq. [30.1], cannot account for a ribosomal decoding accuracy of less than $\sim 10^{-2}$ errors per codon. Evidently, the ribosome has some sort of proofreading mechanism that increases its overall decoding accuracy.

How might a ribosome proofread a codon–anticodon interaction? Two types of mechanisms can be envisaged: (1) A selective binding mechanism, such as those of aminoacyl–tRNA synthetases (Section 30-2C); and (2) a kinetic mechanism. The problem with a ribosomal selective binding mechanism is that there is little evidence indicating the existence of a second aminoacyl–tRNA binding site that functions to exclude improper codon–anticodon interactions. Evidence is accumulating, however, that is consistent with a **kinetic proofreading** mechanism.

Kinetic Proofreading Requires a Branched Reaction Path

Kinetic proofreading models of tRNA selection require only one binding site. John Hopfield theorized that such a process can occur via a branched reaction mechanism of polypeptide chain elongation such as is diagrammed in Fig. 30-50:

1. The initial binding reaction discriminates, as we discussed, between cognate (specified) and noncognate

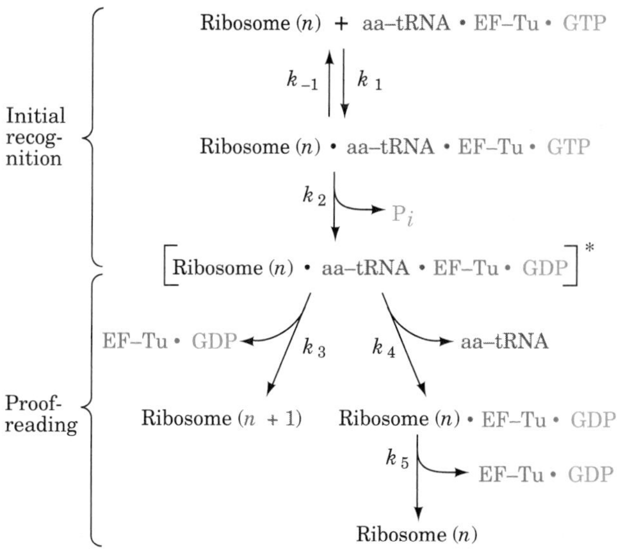

FIGURE 30-50. A kinetic proofreading mechanism for selecting a correct codon–anticodon interaction. The initial recognition reaction screens the aminoacyl–tRNA (aa–tRNA) for the correct codon–anticodon interaction. The resulting complex converts, in a GTP hydrolysis-driven process, to a "high-energy" intermediate (*) which, in turn, either releases EF-Tu·GDP preparatory to forming a peptide bond or releases aminoacyl–tRNA before EF-Tu·GDP is released. If k_4/k_3 is greater for a codon–anticodon mismatch than it is for a match, then these latter steps constitute a proofreading mechanism for proper tRNA binding.

There are three families of interferons: **type α** or **leukocyte interferon** (leukocytes are white blood cells), the related **type β** or **fibroblast interferon** (fibroblasts are connective tissue cells), and **type γ** or **lymphocyte interferon** (lymphocytes are immune system cells). *Interferon synthesis is induced by double-stranded RNA (dsRNA), which is probably generated during infection by both DNA and RNA viruses, as well as by the synthetic dsRNA poly(I)·poly(C).* Interferons are effective antiviral agents in concentrations as low as $3 \times 10^{-14} M$, which makes them among the most potent biological substances. Moreover, they have far wider specificities than antibodies raised against a particular virus.

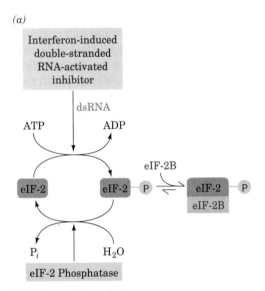

(a)

Inhibition of Translation

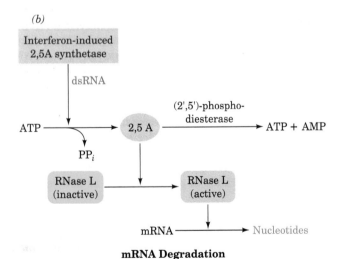

(b)

mRNA Degradation

FIGURE 30-54. In interferon-treated cells, the presence of dsRNA, which normally results from a viral infection, causes (*a*) the inhibition of translational initiation, and (*b*) the degradation of mRNA, thereby blocking translation and preventing virus replication.

They have therefore elicited great medical interest, particularly since some cancers are virally induced (Section 33-4C). Indeed, they are in clinical use against certain tumors and viral infections. These treatments are made possible by the production of large quantities of these otherwise quite scarce proteins through molecular cloning techniques (Section 28-8).

Interferons prevent viral proliferation largely by inhibiting protein synthesis in infected cells (lymphocyte interferon also modulates the immune response). They do so in two independent ways (Fig. 30-54):

1. Interferons induce the production of a protein kinase, **double-stranded RNA-activated inhibitor (DAI)**, that, in the presence of dsRNA, phosphorylates the eIF-2 α subunit identically to the action of HCR in reticulocytes, thereby inhibiting ribosomal initiation. This observation suggests that eIF-2α phosphorylation may be a general mechanism of eukaryotic translational control. Indeed, increased eIF-2α phosphorylation results from a variety of physiological challenges that continuing protein synthesis would aggravate, including amino acid deprivation and the presence of heavy metals. (It should be noted that the specific phosphorylation of the mRNA cap-binding protein eIF-4E at its Ser 53 stimulates translation and is therefore also likely to function in translational control. However, the responsible kinases are, as yet, poorly characterized.).

2. Interferons also induce the synthesis **(2′,5′)-oligoadenylate synthetase**. In the presence of dsRNA, this enzyme catalyzes the synthesis from ATP of the unusual oligonucleotide **pppA(2′p5′A)$_n$** *where n = 1 to 10. This compound, **2,5-A**, activates a preexisting endonuclease, **RNase L**, to degrade mRNA, thereby inhibiting protein synthesis.* 2,5-A is itself rapidly degraded by an enzyme named **(2′,5′)-phosphodiesterase** so that it must be continually synthesized to maintain its effect.

The independence of the 2,5-A and DAI systems is demonstrated by the observation that the effect of 2,5-A on protein synthesis is reversed by added mRNA but not by eIF-2.

C. mRNA Masking

It has been known since the 19th century that early embryonic development in organisms such as sea urchins, insects, and frogs is governed almost entirely by information present in the egg before fertilization. Indeed, sea urchin embryos exposed to sufficient actinomycin D (Section 29-2D) to inhibit RNA synthesis without blocking DNA synthesis develop normally through their early stages without a change in their protein synthesis program. This is because *an unfertilized egg contains large quantities of mRNA that is "masked" by associated proteins to form ribonucleoprotein particles, thereby preventing the mRNAs' association with the ribosomes that are also present. Upon fertilization, this mRNA is "unmasked" in a controlled fashion, quite possibly by the dephosphorylation of the associated proteins, and commences directing protein synthesis.* Development of the embryo can therefore start immediately upon fertilization rather than waiting for the generation of paternally specified mRNAs.

The cytoplasms of many eukaryotic cells contain large amounts of protein-complexed mRNAs that are not associated with ribosomes. It remains to be seen, however, whether mRNA masking is used for translational control in nonembryonic tissues.

D. Antisense RNA

Since ribosomes cannot translate double-stranded RNA, the translation of a given mRNA can be inhibited by a segment of its complementary sequence, a so-called **antisense RNA**. The expression of antisense RNA in a selected tissue or organism therefore has enormous biomedical and biotechnological potential. For example, fruit ripening is controlled by the plant hormone **ethylene** (C_2H_4), the product of a metabolic pathway whose rate-determining step is catalyzed by **1-aminocyclopropane-1-carboxylate synthase (ACC synthase).** The constitutive expression of the antisense RNA of the ACC synthase gene in fruit would therefore be expected to inhibit fruit ripening. In fact, the transgenic expression of this antisense RNA in tomato plants prevents tomato ripening, an effect that can be reversed by the administration of ethylene. This effect raises the prospect that fruits and vegetables can be ripened on demand, thereby preventing spoilage during transportation or because of lack of refrigeration.

Antisense RNA Inhibits the Blockage of Injured Arteries

The difficulty of delivering antisense RNA molecules in the required amounts to the proper cells has heretofore prevented their therapeutic use. Recently, however, antisense RNA has been used *in vivo* to control the expression of a normal gene product.

A relatively nontraumatic method for widening an artery that is blocked by an atherosclerotic plaque (Section 11-5D) is to inflate a narrow cylindrical balloon inside it so as to compress the plaque. Although this procedure, which is named **balloon angioplasty** (Greek: *angeion,* vessel + *plasso,* to fashion), is effective in the short term, the artery frequently becomes reblocked because the injury caused by the procedure stimulates the proliferation of smooth muscle cells in the artery's inner wall. Smooth muscle cell proliferation is induced by a protein growth factor encoded by the **c-*myb*** **proto-oncogene** (protein growth factors and oncogenes are discussed in Section 33-4C). Thus, if the expression of the c-*myb* gene in an injured artery could somehow be inhibited, the artery's eventual blockage might be prevented.

A nontoxic gel containing an 18-nucleotide RNA complementary to a segment of c-*myb* mRNA was prepared.

When this gel was applied to the inner wall of a rat's artery immediately after the artery had been subjected to balloon angioplasty, smooth muscle cells did not proliferate over at least the following 2 weeks. In contrast, significant blockage occurred in untreated segments of the injured artery as well as in injured arteries that had been coated with gel that contained either the corresponding sense RNA segment or no RNA at all. The gel presumably retains the antisense RNA at a high enough concentration and for a sufficient time to inhibit c-*myb* expression in the coated region. The use of antisense RNA directed against c-*myb* mRNA therefore holds great promise in the treatment of blood vessel injuries.

5. POSTTRANSLATIONAL MODIFICATION

To become mature proteins, polypeptides must fold to their native conformations, their disulfide bonds, if any, must form, and, in the case of multisubunit proteins, the subunits must properly combine. Moreover, as we have seen throughout this text, many proteins are modified in enzymatic reactions that proteolytically cleave certain peptide bonds and/or derivatize specific residues. In this section we shall review some of these **posttranslational modifications.**

A. Proteolytic Cleavage

Proteolytic cleavage is the most common type of posttranslational modification. Probably all mature proteins have been so modified, if by nothing else than the proteolytic removal of their leading Met (or fMet) residue shortly after it emerges from the ribosome. Many proteins, which are involved in a wide variety of biological processes, are synthesized as inactive precursors that are activated under proper conditions by limited proteolysis. Some examples of this phenomenon that we have encountered are the conversion of trypsinogen and chymotrypsinogen to their active forms by tryptic cleavages of specific peptide bonds (Section 14-3E), and the formation of active insulin from the 84-residue proinsulin by the excision of its internal 33-residue C chain (Section 8-1A). Inactive proteins that are activated by removal of polypeptides are called **proproteins,** whereas the excised polypeptides are termed **propeptides.**

Propeptides Direct Collagen Assembly

Collagen biosynthesis is illustrative of many facets of posttranslational modification. Recall that collagen, a major extracellular component of connective tissue, is a fibrous triple helical protein whose polypeptides each have the amino acid sequence $(Gly-X-Y)_n$ where X is often Pro, Y is often 4-hydroxyproline (Hyp), and $n \approx 340$ (Section

7-2C). The polypeptides of **procollagen** (Fig. 30-55) differ from those of the mature protein by the presence of both N-terminal and C-terminal propeptides of ~100 residues

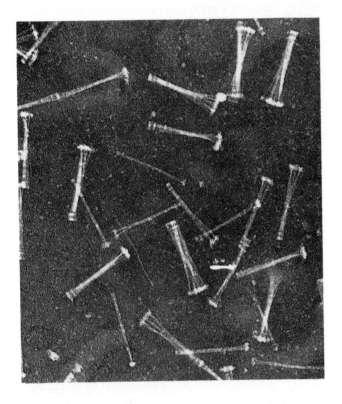

FIGURE 30-55. An electron micrograph of procollagen aggregates that have been secreted into the extracellular medium. [Courtesy of Jerome Gross, Harvard Medical School.]

whose sequences, for the most part, are unlike those of mature collagen. The procollagen polypeptides rapidly assemble, *in vitro* as well as *in vivo,* to form a collagen triple helix. In contrast, polypeptides extracted from mature collagen will reassemble only over a period of days, if at all. *The collagen propeptides are apparently necessary for proper procollagen folding.*

The N- and C-terminal propeptides of procollagen are respectively removed by **amino-** and **carboxylprocollagen peptidases** (Fig. 30-56), which may also be specific for the different collagen types. An inherited defect of aminoprocollagen peptidase in cattle and sheep results in a bizarre condition, **dermatosparaxis,** that is characterized by extremely fragile skin. An analogous disease in man, **Ehlers–Danlos syndrome VII,** is caused by a mutation in one of the procollagen polypeptides that inhibits the enzymatic removal of its aminopropeptide. Collagen molecules normally spontaneously aggregate to form collagen fibrils (Fig. 7-33 and 7-34). However, electron micrographs of dermatosparaxic skin show sparse and disorganized collagen fibrils. *The retention of collagen's aminopropeptides apparently interferes with proper fibril formation.* (The dermatosparaxis gene was bred into some cattle herds because heterozygotes produce tender meat.)

Signal Peptides Are Removed from Nascent Proteins by a Signal Peptidase

Many transmembrane proteins or proteins that are destined to be secreted are synthesized with an N-terminal **signal peptide** of 13 to 36 predominantly hydrophobic residues. According to the **signal hypothesis** (Section 11-4B), a

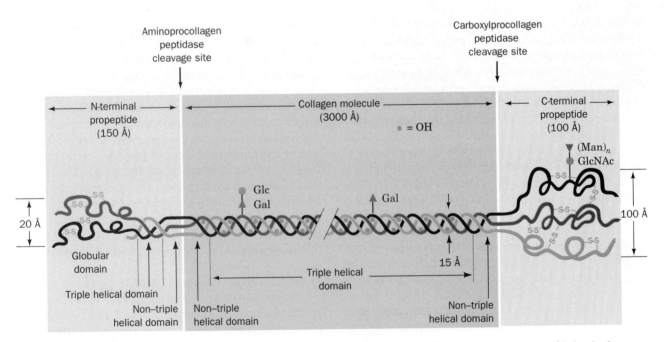

FIGURE 30-56. A schematic representation of the procollagen molecule. Gal, Glc, GlcNAc, and Man, respectively, denote galactose, glucose, *N*-acetylglucosamine, and mannose residues. Note that the N-terminal propeptide has intrachain disulfide bonds while the C-terminal propeptide has both intrachain and interchain disulfide bonds. [After Prockop, D.J., Kivirikko, K.I., Tuderman, L., and Guzman, N.A., *New Engl. J. Med.* **301,** 16 (1979).]

signal peptide is recognized by a **signal recognition particle (SRP).** The SRP binds a ribosome synthesizing a signal peptide to a receptor on the membrane [the rough endoplasmic reticulum (RER) in eukaryotes and the plasma membrane in bacteria] and conducts the signal peptide and its following nascent polypeptide through it.

Proteins bearing a signal peptide are known as **proproteins** or, if they also contain propeptides, as **preproproteins.** Once the signal peptide has passed through the membrane, it is specifically cleaved from the nascent polypeptide by a membrane-bound **signal peptidase.** Both insulin and collagen are secreted proteins and are therefore synthesized with leading signal peptides in the form of **preproinsulin** and **preprocollagen.** These and many other proteins are therefore subject to three sets of sequential proteolytic cleavages: (1) the deletion of their initiating Met residue, (2) the removal of their signal peptides, and (3) the excision of their propeptides.

Polyproteins

Some proteins are synthesized as segments of **polyproteins,** polypeptides that contain the sequences of two or more proteins. Examples include most polypeptide hormones (Section 33-3C); the proteins synthesized by many viruses, including those causing polio (Section 32-2C) and AIDS, and **ubiquitin,** a highly conserved eukaryotic protein involved in protein degradation (Section 30-6B). Specific proteases posttranslationally cleave polyproteins to their component proteins, presumably through the recognition of the cleavage site sequences. Some of these proteases are conserved over remarkable evolutionary distances. For instance, ubiquitin is synthesized as several tandem repeats **(polyubiquitin)** that *E. coli* properly cleave even though prokaryotes lack ubiquitin. Other proteases have more idiosyncratic cleavage sequences. Thus, medicinal chemists have designed and synthesized numerous inhibitors of **HIV**

protease (which catalyzes an essential step in the viral life cycle) in an effort to slow the progress of, if not cure, AIDS.

B. Covalent Modification

Proteins are subject to specific chemical derivatizations, both at the functional groups of their side chains and at their terminal amino and carboxyl groups. Over 150 different types of side chain modifications, involving all side chains but those of Ala, Gly, Ile, Leu, Met, and Val, are known (Section 4-3A). These include acetylations, glycosylations, hydroxylations, methylations, nucleotidylations, phosphorylations, and ADP-ribosylations as well as numerous "miscellaneous" modifications.

Some protein modifications, such as the phosphorylation of glycogen phosphorylase (Section 17-1A) and the ADP-ribosylation of eEF-2 (Section 30-3G), modulate protein activity. Several side chain modifications covalently bond cofactors to enzymes, presumably to increase their catalytic efficiency. Examples of linked cofactors that we have encountered are N^ε-lipoyllysine in dihydrolipoyl transacetylase (Section 19-2A) and 8α-histidylflavin in succinate dehydrogenase (Section 19-3F). The attachment of complex carbohydrates, which occur in almost infinite variety, alter the structural properties of proteins and form recognition markers in various types of targeting and cell–cell interactions (Sections 10-3C, 11-3D, and 21-3B). Modifications that cross-link proteins, such as occur in collagen and elastin (Sections 7-2C and D), stabilize supramolecular aggregates. The functions of most side chain modifications, however, remain enigmatic.

Collagen Assembly Requires Chemical Modification

Collagen biosynthesis (Fig. 30-57) is illustrative of protein maturation through chemical modification. As the nascent procollagen polypeptides pass into the RER of the

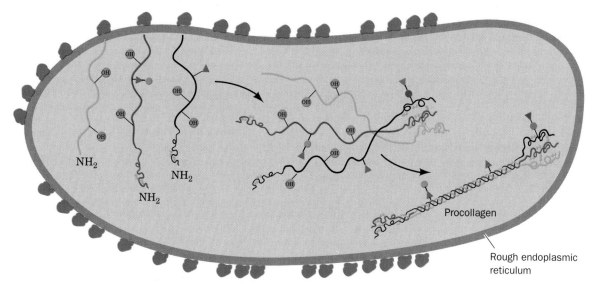

FIGURE 30-57. A schematic representation of procollagen biosynthesis. The diagram does not indicate the removal of signal peptides. [After Prockop, D.J., Kivirikko, K.I., Tuderman, L., and Guzman, N.A., *New Engl. J. Med.* **301,** 18 (1979).]

fibroblasts that synthesized them, the Pro and Lys residues are hydroxylated to Hyp, 3-hydroxy-Pro, and 5-hydroxy-Lys. The enzymes that do so are sequence specific: **Prolyl 4-hydroxylase** and **lysyl hydroxylase** act only on the Y residues of the Gly-X-Y sequences, whereas **prolyl 3-hydroxylase** acts on the X residues but only if Y is Hyp. Glycosylation, which also occurs in the RER, subsequently attaches sugar residues to 5-hydroxy-Lys residues (Section 7-2C). The folding of three polypeptides into the collagen triple helix must follow hydroxylation and glycosylation because the hydroxylases and glycosyl transferases do not act on helical substrates. Moreover, the collagen triple helix denatures below physiological temperatures unless stabilized by hydrogen bonding interactions involving Hyp residues (Section 7-2C). Folding is also preceded by the formation of specific interchain disulfide bonds between the carboxyl-propeptides. This observation bolsters the previously discussed conclusion that collagen propeptides help select and align the three collagen polypeptides for proper folding.

The procollagen molecules pass into the Golgi apparatus where they are packaged into secretory granules (Sections 11-4B and 21-3B) and secreted into the extracellular spaces of connective tissue. The aminopropeptides are excised just after procollagen leaves the cell and the carboxylpropeptides are removed sometime later. The collagen molecules then spontaneously assemble into fibrils, which suggests that an important propeptide function is to prevent intracellular fibril formation. Finally, after the action of the extracellular enzyme lysyl oxidase, the collagen molecules in the fibrils spontaneously cross-link (Fig. 7-35).

6. PROTEIN DEGRADATION

The pioneering work of Henry Borsook and Rudolf Schoenheimer around 1940 demonstrated that the components of living cells are constantly turning over. Proteins have lifetimes that range from as short as a few minutes to weeks or more. In any case, *cells continuously synthesize proteins from and degrade them to their component amino acids.* The function of this seemingly wasteful process is twofold: (1) to eliminate abnormal proteins whose accumulation would be harmful to the cell, and (2) to permit the regulation of cellular metabolism by eliminating superfluous enzymes and regulatory proteins. Indeed, since the level of an enzyme depends on its rate of degradation as well as its rate of synthesis, *controlling a protein's rate of degradation is as important to the cellular economy as is controlling its rate of synthesis.* In this section we consider the processes of intracellular protein degradation and their consequences.

A. Degradation Specificity

Cells selectively degrade abnormal proteins. For example, hemoglobin that has been synthesized with the valine ana-

log **α-amino-β-chlorobutyrate**

α-Amino-β-Chorobutyrate   Valine

has a half-life in reticulocytes of ~10 min, whereas normal hemoglobin lasts the 120-day lifetime of the red cell (which makes it perhaps the longest lived cytoplasmic protein). Likewise, unstable mutant hemoglobins are degraded soon after their synthesis, which, for reasons explained in Section 9-3A, results in the hemolytic anemia characteristic of these molecular disease agents. Bacteria also selectively degrade abnormal proteins. For instance, *amber* and *ochre* mutants of β-galactosidase have half-lives in *E. coli* of only a few minutes, whereas the wild-type enzyme is almost indefinitely stable. Most abnormal proteins, however, probably arise from the chemical modification and/or spontaneous denaturation of these fragile molecules in the cell's reactive environment rather than by mutations or the rare errors in transcription or translation. *The ability to eliminate damaged proteins selectively is therefore an essential recycling mechanism that prevents the buildup of substances that would otherwise interfere with cellular processes.*

Normal intracellular proteins are eliminated at rates that depend on their identities. A given protein is eliminated with first-order kinetics indicating that the molecules being degraded are chosen at random rather than according to their age. The half-lives of different enzymes in a given tissue vary substantially as is indicated for rat liver in Table 30-10. Remarkably, *the most rapidly degraded enzymes all occupy important metabolic control points, whereas the rel-*

TABLE 30-10. Half-Lives of Some Rat Liver Enzymes

Enzyme	Half-Life (h)
Short-Lived Enzymes	
Ornithine decarboxylase	0.2
RNA polymerase I	1.3
Tyrosine aminotransferase	2.0
Serine dehydratase	4.0
PEP carboxylase	5.0
Long-Lived Enzymes	
Aldolase	118
GAPDH	130
Cytochrome *b*	130
LDH	130
Cytochrome *c*	150

Source: Dice, J.F. and Goldberg, A.L., *Arch. Biochem. Biophys.* **170,** 214 (1975).

atively stable enzymes have nearly constant catalytic activities under all physiological conditions. The susceptibilities of enzymes to degradation have evidently evolved together with their catalytic and allosteric properties so that cells can efficiently respond to environmental changes and metabolic requirements. The criteria through which native proteins are selected for degradation are considered in Section 30-6B.

The rate of protein degradation in a cell also varies with its nutritional and hormonal state. Under conditions of nutritional deprivation, cells increase their rate of protein degradation so as to provide the necessary nutrients for indispensible metabolic processes. The mechanism that increases degradative rates in *E. coli* is the stringent response (Section 29-3G). A similar mechanism may be operative in eukaryotes since, as happens in *E. coli,* increased rates of degradation are prevented by antibiotics that block protein synthesis.

B. Degradation Mechanisms

Eukaryotic cells have dual systems for protein degradation: lysosomal mechanisms and ATP-dependent cytosolically based mechanisms. We consider both mechanisms below.

Lysosomes Mostly Degrade Proteins Nonselectively

Lysosomes are membrane-encapsulated organelles (Section 1-2A) that contain ~50 hydrolytic enzymes, including a variety of proteases known as **cathepsins.** The lysosome maintains an internal pH of ~5 and its enzymes have acidic pH optima. This situation presumably protects the cell against accidental lysosomal leakage since lysosomal enzymes are largely inactive at cytosolic pH's.

Lysosomes recycle intracellular constituents by fusing with membrane-enclosed bits of cytoplasm known as **autophagic vacuoles** and subsequently breaking down their contents. They similarly degrade substances that the cell takes up via endocytosis (Section 11-5C). The existence of these processes has been demonstrated through the use of lysosomal inhibitors. For example, the antimalarial drug **chloroquine**

Chloroquine

is a weak base that, in uncharged form, freely penetrates the lysosome where it accumulates in charged form, thereby increasing the intralysosomal pH and inhibiting lysosomal function. The treatment of cells with chloroquine reduces their rate of protein degradation. Similar effects arise from treatment of cells with cathepsin inhibitors such as the polypeptide antibiotic **antipain.**

Lysosomal protein degradation in well nourished cells appears to be nonselective. Lysosomal inhibitors do not affect the rapid degradation of abnormal proteins or short-lived enzymes. Rather, they prevent the acceleration of nonselective protein breakdown upon starvation. However, the continued nonselective degradation of proteins in starving cells would rapidly lead to an intolerable depletion of essential enzymes and regulatory proteins. Lysosomes therefore also have a selective pathway, which is activated only after a prolonged fast, that takes up and degrades proteins containing the pentapeptide Lys-Phe-Glu-Arg-Gln (KFERQ) or a closely related sequence. Such KFERQ proteins are selectively lost in fasting animals from tissues that atrophy in response to fasting (e.g., liver and kidney) but not from tissues that do not do so (e.g., brain and testes). KFERQ proteins are specifically bound in the cytosol and delivered to the lysosome by a 73-kD **peptide recognition protein (prp73),** a member of the 70-kD heat shock protein (hsp70) family (Section 8-1C).

Many normal and pathological processes are associated with increased lysosomal activity. **Diabetes mellitus** (Section 25-3B) stimulates the lysosomal breakdown of proteins. Similarly, muscle wastage caused by disuse, denervation, or traumatic injury arises from increased lysosomal activity. The regression of the uterus after childbirth, in which this muscular organ reduces its mass from 2 kg to 50 g in 9 days, is a striking example of this process. Many chronic inflammatory diseases, such as **rheumatoid arthritis,** involve the extracellular release of lysosomal enzymes which break down the surrounding tissues.

Ubiquitin Marks Proteins Selected for Degradation

It was initially assumed that protein degradation in eukaryotic cells is primarily a lysosomal process. Yet, reticulocytes, which lack lysosomes, selectively degrade abnormal proteins. The observation that protein breakdown is inhibited under anaerobic conditions led to the discovery of a cytosolically based ATP-dependent proteolytic system that is independent of the lysosomal system. This phenomenon was thermodynamically unexpected since peptide hydrolysis is an exergonic process.

Analysis of a cell-free rabbit reticulocyte system has demonstrated that **ubiquitin** (Fig. 30-58) is required for ATP-dependent protein degradation. *This 76-residue monomeric protein, so named because it is ubiquitous as well as abundant in eukaryotes, is the most highly conserved protein known:* It is identical in such diverse organisms as humans, toad, trout, and *Drosophila* and differs in only three resi-

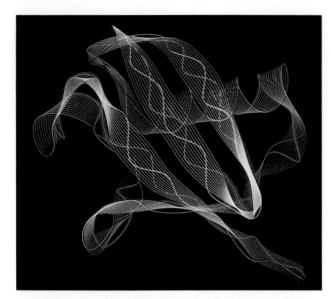

FIGURE 30-58. The X-ray structure of ubiquitin. The white ribbon represents the polypeptide backbone and the red and blue curves, respectively, indicate the directions of the carbonyl and amide groups. [Courtesy of Michael Carson, University of Alabama at Birmingham. X-ray structure determined by Charles Bugg.]

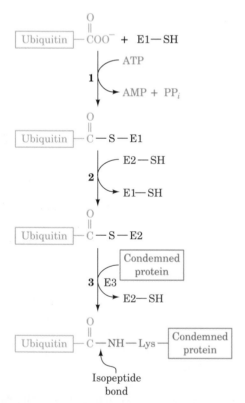

FIGURE 30-59. The reactions involved in the attachment of ubiquitin to a protein. In the first part of the process, ubiquitin's terminal carboxyl group is joined, via a thioester linkage, to E1 in a reaction driven by ATP hydrolysis. The activated ubiquitin is subsequently transferred to a sulfhydryl group of E2 and then, in a reaction catalyzed by E3, to a Lys ε-amino group on a condemned protein, thereby flagging the protein for proteolytic degradation by the 26S proteosome.

dues between humans and yeast. Evidently, ubiquitin is all but uniquely suited for some essential cellular process.

Proteins that are selected for degradation are so marked by covalently linking them to ubiquitin. This process, which is reminiscent of amino acid activation (Section 30-2C), occurs in a three-step pathway elucidated notably by Avram Hershko (Fig. 30-59):

1. In an ATP-requiring reaction, ubiquitin's terminal carboxyl group is conjugated, via a thioester bond, to **ubiquitin-activating enzyme (E1),** a 105-kD dimer of identical subunits.

2. The ubiquitin is then transferred to a specific sulfhydryl group on one of numerous proteins named **ubiquitin-conjugating enzymes (E2's).** The various E2's are characterized by ~150-residue cores containing the active site Cys that exhibit at least 25% sequence identities and which, in part, vary by the presence or absence of N- and/or C-terminal extensions that exhibit little sequence identity to each other. Comparison of the X-ray structures of E2's from the plant *Arabidopsis thaliana* (Fig. 30-60) and from baker's yeast, both determined by William Cook, reveals that most of the identical residues in these 42% identical and structurally nearly superimposable proteins are clustered on one surface near the ubiquitin-accepting Cys residue. Cook has therefore postulated that this conserved surface area comprises the binding site for ubiquitin and/or E1.

3. **Ubiquitin-protein ligase (E3; ~180 kD)** transfers the activated ubiquitin from E2 to a Lys ε-amino group of a previously bound protein, thereby forming an **isopeptide bond.** E3 therefore appears to have a key role in selecting the protein to be degraded. However, the large number of different E2's in a cell (>10 in yeast and >20 in *Arabidopsis*) suggests that these proteins also function in target protein selection. Indeed, some E2's transfer ubiquitin directly to target proteins.

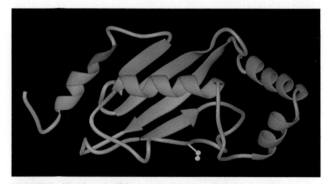

FIGURE 30-60. A ribbon drawing showing the X-ray structure of an *Arabidopsis* E2 protein. α Helices are blue, the 3₁₀ helical segment is purple, β strands are green, and the remainder of the molecule is silver. The side chain of Cys 88, to which ubiquitin is covalently linked, is shown in ball-and-stick form in yellow. [Courtesy of William Cook, University of Alabama at Birmingham.]

Usually, several ubiquitin molecules are linked to a condemned protein. In addition, as many as 50 or more ubiquitin molecules may be tandemly linked to a target protein to form a multiubiquitin chain in which Lys 48 of each ubiquitin forms an isopeptide bond with the C-terminal carboxyl group of the following ubiquitin (Fig. 30-61). In fact, the attachment of multiubiquitin chains, which are synthesized by several species of E2, appears to be essential for the degradation of at least some proteins. Ubiquinated proteins are dynamic entities, with ubiquitin molecules being rapidly attached and removed (the latter by **ubiquitin isopeptidases**).

*The ubiquitinated protein is proteolytically degraded in an ATP-dependent process mediated by a large (2000 kD, 26S) multiprotein complex, named the **26S proteosome**, that electron micrographic studies indicate has the shape of a bicapped hollow barrel (Fig. 30-62).* The 26S proteosome,

(a)

(b)

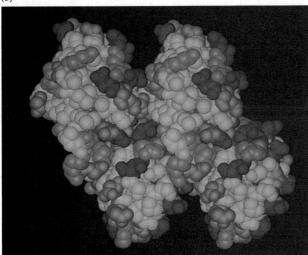

FIGURE 30-61. The X-ray structure of **tetraubiquitin.** (*a*) A ribbon drawing in which the isopeptide bonds connecting successive ubiquitin molecules, together with the Lys side chains making them, are orange. However, since the isopeptide bond connecting ubiquitins 2 and 3 is not visible in the X-ray structure, it is represented by a stick bond (this isopeptide bond nevertheless exists as was demonstrated by the SDS–PAGE of dissolved crystals). It seems likely that the monomer units in a multiubiquitin chain of any length would be arranged with the repeating symmetry of the tetraubiquitin structure. (*b*) A space-filling model, viewed as in Part *a*, in which basic residues (Arg, Lys, His) are blue, acidic residues (Asp, Glu) are red, uncharged polar residues (Gly, Ser, Thr, Asn, Gln) are purple, and hydrophobic residues (Ile, Leu, Val, Ala, Met, Phe, Tyr, Pro) are green. Note the unusually large solvent-exposed surface occupied by hydrophobic residues. [Courtesy of William Cook, University of Alabama at Birmingham.]

(a)

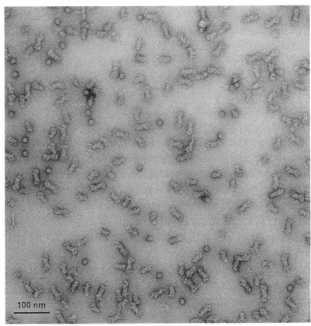

(b)

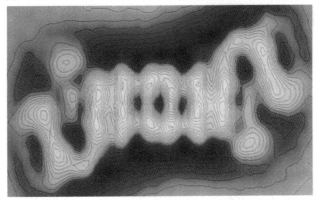

FIGURE 30-62. The structure of the 26S proteosome from *Xenopus* oocytes. (*a*) Electron micrograph and (*b*) image reconstruction based on the averaging of 527 images. The complex is around 450×190 Å and is seen with a resolution estimated to be ~25 Å. The central portion of this 2-fold symmetric multiprotein complex consists of 4 stacked 6- or 7-membered rings of 20 to 32 kD subunits that form a hollow barrel in which, it is thought, the proteolysis of ubiquitin-linked proteins occurs. [Courtesy of Wolfgang Baumeister, Max-Planck Institute for Biochemistry, Martinsried, Germany.]

Khorana, H.G., Nucleic acid synthesis in the study of the genetic code, *Nobel Lectures in Molecular Biology, 1933–1975, pp.* 303–331, Elsevier (1977).

Nirenberg, M., The genetic code, *Nobel Lectures in Molecular Biology, 1933–1975, pp.* 335–360, Elsevier (1977).

Nirenberg, M.W., The genetic code: II, *Sci. Am.* **208:** 80–94 (1963). [Discusses the use of synthetic mRNAs to analyze the genetic code.]

Nirenberg, M. and Leder, P., RNA code words and protein synthesis, *Science* **145,** 1399–1407 (1964). [The determination of the genetic code by the ribosomal binding of tRNAs using specific trinucleotides.]

Nirenberg, M.W. and Matthaei, J.H., The dependence of cell-free protein synthesis in *E. coli* upon naturally occurring or synthetic polyribonucleotides, *Proc. Natl. Acad. Sci.* **47,** 1588–1602 (1961). [The landmark paper reporting the finding that poly(U) stimulates the synthesis of poly(Phe).]

Singer, B. and Kuśmierek, J.T., Chemical mutagenesis, *Annu. Rev. Biochem.* **51,** 655–693 (1982).

Yanofsky, C., Gene structure and protein structure, *Sci. Am.* **216**(5): 80–94 (1967).

Yanofsky, C., Carlton, B.C., Guest, J.R., Helinski, D.R., and Henning, U., On the colinearity of gene structure and protein structure, *Proc. Natl. Acad. Sci.* **51,** 266–272 (1964).

Transfer RNA and Its Aminoacylation

Biou, V., Yaremchuk, A., Tukalo, M., and Cusack, S., The 2.9 Å crystal structure of *T. thermophilus* seryl-tRNA synthetase complexed with tRNA^Ser, *Science* **263,** 1404–1410 (1994).

Björk, G.R., Ericson, J.U., Gustafsson, C.E.D., Hagervall, T.G., Jösson, Y.H., and Wikström, P.M., Transfer RNA modification, *Annu. Rev. Biochem.* **56,** 263–287 (1987).

Böck, A., Forschhammer, K., Heider, J., and Baron, C., Selenoprotein synthesis: an expansion of the gentic code, *Trends Biochem. Sci.* **16,** 463–467 (1991).

Carter, C.W., Jr., Cognition, mechanism, and evolutionary relationships in aminoacyl-tRNA synthetases, *Annu. Rev. Biochem.* **62,** 715–748 (1993).

Crick, F.H.C., Codon–anticodon pairing: the wobble hypothesis, *J. Mol. Biol.* **19,** 548–555 (1966).

Freist, W., Isoleucyl-tRNA synthetase: an enzyme with several catalytic cycles displaying variation in specificity and energy consumption, *Angew. Chem. Int. Ed. Engl.* **27,** 773–788 (1988).

Gesteland, R.F., Weiss, R.B., and Atkins, J.F., Recoding: Reprogrammed genetic coding, *Science* **257,** 1640–1641 (1992). [Discusses contextual signals in mRNA that alter the way the ribosome reads certain codons.]

Geigé, R. Puglisi, J.D., and Florentz, C., tRNA structure and aminoacylation efficiency, *Prog. Nucleic Acid Res. Mol. Biol.* **45,** 129–206 (1993). [A detailed review.]

Hatfield, D.L., Lee, B.J., and Pirtle, R.M. (Eds.), *Transfer RNA in Protein Synthesis,* CRC Press (1992). [Contains articles on such subjects as the role of modified nucleosides in tRNAs, variations in reading the genetic code, patterns of codon usage, and tRNA identity elements.]

Kersten, H., On the biological significance of modified nucleosides in tRNA, *Prog. Nucleic Acid Res. Mol. Biol.* **31,** 59–114 (1984).

Kim, S.H., Suddath, F.L., Quigley, G.J., McPherson, A., Sussman, J.L., Wang, A.M.J., Seeman, N.C., and Rich, A., Three-dimensional tertiary structure of yeast phenylalanine transfer RNA, *Science* **185,** 435–440 (1974); *and* Robertus, J.D., Ladner, J.E., Finch, J.T., Rhodes, D., Brown, R.S., Clark, B.F.C., and Klug, A., Structure of yeast phenylalanine tRNA at 3 Å resolution, *Nature* **250,** 546–551 (1974). [The landmark papers describing the high-resolution structure of a tRNA.]

Kline, L.K. and Söll, D., Nucleotide modifications in RNA, *in* Boyer, P.D. (Ed.), *The Enzymes* (3rd ed.), Vol. 15, *pp.* 567–582, Academic Press (1982).

Leinfelder, W., Zehelein, E., Mandrand-Berthelot, M.-A., and Böck, A., Gene for a novel tRNA species that accepts L-serine and cotranslationally inserts selenocysteine, *Nature* **331,** 723–725 (1988).

McClain, W.H., Rules that govern tRNA identity in protein synthesis, *J. Mol. Biol.* **234,** 257–280 (1993).

Mirande, M., Aminoacyl-tRNA synthetase family from prokaryotes and eukaryotes: Structural domains and their implications, *Prog. Nucleic Acid Res. Mol. Biol.* **40,** 95–142 (1991).

Moras, D. Transfer RNA, *Curr. Opin. Struct. Biol.* **1,** 410–415 (1991).

Moras, D., Structural and functional relationships between aminoacyl-tRNA synthetases, *Trends Biochem. Sci.* **17,** 159–169 (1992). [Discussess Class I and Class II enzymes.]

Muramatsu, T., Nishakawa, K., Nemoto, F., Kuchino, Y., Nishamura, S., Miyazawa, T., and Yokoyama, S., Codon and amino-acid specificities of a transfer RNA are both converted by a single posttranscriptional modification, *Nature* **336,** 179–181 (1988).

Rich, A. and Kim, S.H., The three-dimensional structure of transfer RNA, *Sci. Am.* **238**(1): 52–62 (1978).

Rould, M.A., Perona, J.J., Söll, D., and Steitz, T.A., Structure of *E. coli* glutaminyl-tRNA synthetase complexed with tRNA^Gln and ATP at 2.8 Å resolution, *Science* **246,** 1135–1142 (1989); Rould, M.A., Perona, J.J., and Steitz, T.A., Structural basis of anticodon loop recognition by glutaminyl-tRNA synthetase, *Nature* **352,** 213–218 (1991); *and* Perona, J.J., Rould, M.A., and Steitz, T.A., Structural basis for transfer RNA aminoacylation by *Eschericia coli* glutaminyl-tRNA synthetase, *Biochemistry* **32,** 8758–8771 (1993).

Ruff, M., Krishnaswamy, S., Boeglin, M., Poterszman, A., Mitschler, A., Podjarny, A., Rees, B., Thierry, J.C., and Moras, D., Class II aminoacyl transfer RNA synthetases: Crystal structure of yeast aspartyl-tRNA synthetase complexed with tRNA^Asp, *Science* **252,** 1682–1689 (1991); Cavarelli, J., Rees, B., Ruff, M., Thierry, J.-C., and Moras, D., Yeast tRNA^Asp recognition by its cognate class II aminoacyl-tRNA synthetase, *Nature* **362,** 181–184 (1993); *and* Cavarelli, J., Eriani, G., Rees, B., Ruff, M., Boeglin, M., Mitschler, A., Martin, F., Gangloff, J., Thierry, J.-C., and Moras, D., The active site of yeast aspartyl-tRNA synthetase: structural and functional aspects of the aminoacylation reaction, *EMBO J.* **13,** 327–337 (1994).

Saks, M.E., Sampson, J.R., and Abelson, J.N., The transfer identity problem: A search for rules, *Science* **263,** 191–197 (1994).

Schimmel, P., Giegé, R., Moras, D., and Yokoyama, S., An operational RNA code for amino acids and possible relationship to genetic code, *Proc. Natl. Acad. Sci.* **90**, 8763–8768 (1993).

Schulman, L.H., Recognition of tRNAs by aminoacyl-tRNA synthetases, *Prog. Nucl. Acid Res. Mol. Biol.* **41**, 23–87 (1991).

Steege, D.A. and Söll, D.G., Suppression, *in* Goldberger, R.F. (Ed.), *Biological Regulation and Development*, Vol. 1, *pp.* 433–485, Plenum Press (1979).

Ribosomes

Basavappa, R. and Sigler, P.B., The 3 Å crystal structure of yeast initiator tRNA: functional implications in initiator/elongator discrimination, *EMBO J.* **10**, 3105–3111 (1991).

Berkovitch-Yellin, Z., Bennett, W.S., and Yonath, A., Aspects in structural studies on ribosomes, *Crit. Rev. Biochem. Mol. Biol.* **27**, 403–444 (1992).

Brinacombe, R., The emerging three-dimensional structure and function of 16S ribosomal RNA, *Biochemistry* **27**, 4207–4214 (1988).

Burgess, S.M. and Guthrie, C., Beat the clock: paradigms for NTPases in the maintenance of biological fidelity, *Trends Biochem. Sci.* **18**, 381–390 (1993). [Discusses kinetic proofreading.]

Capel, M.S., Engelman, D.M., Freeborn, B.R., Kjeldgaard, M., Langer, J.A., Ramakrishnan, V., Schindler, D.G., Schneider, D.K., Schoenborn, B.P., Siller, I.-Y., Yabuki, S., and Moore, P.B., A complete mapping of the proteins in the small ribosomal subunit of *Escherichia coli, Science* **238**, 1403–1406 (1987).

Choe, S., Bennett, M.J., Fuji, G., Curmi, P.M.G., Kantardjieff, K.A., Collier, R.J., and Eisenberg, D., The crystal structure of diphtheria toxin, *Nature* **357**, 216–222 (1992).

Czworkowski, J., Wang, J., Steitz, J.A., and Moore, P.B., The crystal structures of elongation factor G complexed with GDP, at 2.7 Å resolution; *and* Ævarsson, A., Brazhnikov, E., Garber, M., Zheltonosova, J., Chirgadze, Yu., Al-Karadaghi, S., Svensson, L.A., and Liljas, A., Three-dimensional structure of the ribosomal translocase: elongation factor G from *Thermus thermophilus, EMBO J.* **13**, 3661–3668 *and* 3669–3677 (1994).

Dintzis, H.M., Assembly of the peptide chains of hemoglobin, *Proc. Natl. Acad. Sci.* **47**, 247–261 (1961). [The determination of the direction of polypeptide biosynthesis.]

Edelmann, P. and Gallant, J., Mistranslation in E. coli, *Cell* **10**, 131–137 (1977).

Fersht. A., *Enzyme Structure and Mechanism* (2nd ed.), Chapter 13, Freeman (1985). [A discussion of enzymatic specificity and editing mechanisms.]

Gale, E.F., Cundliffe, E., Reynolds, P.E., Richmond, M.H., and Waring, M.J., *The Molecular Basis of Antibiotic Action*, Chapter 6, Wiley (1981).

Gluick, T.C. and Draper, D.E., Tertiary structure of ribosomal RNA, *Curr. Opin. Struct. Biol.* **2**, 338–344 (1992).

Gualerzi, C.O. and Pon, C.L., Initiation of mRNA translation in prokaryotes, *Biochemistry* **29**, 5881–5889 (1990).

Gutell, R.R., Weiser, B., Woese, C.R., and Noller, H. F., Comparitive anatomy of 16-S-like ribosomal RNA, *Prog. Nucleic Acid Res. Mol. Biol.* **32**, 155–216 (1985).

Held, W.A., Ballou, B., Mizushima, S., and Nomura, M., Assem-bly mapping of 30S ribosomal proteins from *Escherichia coli, J. Biol. Chem.* **249**, 3103–3111 (1974).

Hoffman, D.W., Davies, C., Gerchman, S.E., Kycia, J.H., Porter, S.J., White, S.W., and Ramakrishnan, V., Crystal structure of the prokaryotic ribosomal protein L9: a bi-lobed RNA-binding protein, *EMBO J.* **13**, 205–212 (1994); *and* Golden, B.L., Ramakrishnan, V., and White, S.W., Ribosomal protein L6: structural evidence of gene duplication from a primative RNA binding protein, *EMBO J.* **12**, 4901–4908 (1993).

Kjeldgaard, M., and Nyborg, J., Refined structure of elongation factor EF-Tu from *Escherichia coli, J. Mol. Biol.* **223**, 721–742 (1992); Kjeldgaard, M., Nissen, P., Thirup, S., and Nyborg, J., The crystal structure of elongation factor EF-Tu from *Thermus aquaticus* in the GTP conformation, *Structure* **1**, 35–50 (1993); *and* Berchtold, H., Reshetnikova, L., Reiser, C.O.A., Schirmer, N.K., Sprinzl, M., and Hilgenfeld, R., Crystal structure of active elongation factor Tu reveals major domain rearrangements, *Nature* **365**, 126–132 (1993). [The latter paper reports the structure of the GDPNP complex of *T. thermophilus* EF-Tu, which is 97.5% identical in sequence to *T. aquaticus* EF-Tu and essentially identical in structure.]

Kurland, C.G. and Ehrenberg, M., Optimization of translation, *Prog. Nucleic Acid Res. Mol. Biol.* **31**, 191–219 (1984).

Lake, J.A., Evolving ribosome structure: domains in archaebacteria, eubacteria, eocytes and eukaryotes, *Annu. Rev. Biochem.* **54**, 507–530 (1985).

Lake, J.A., The ribosome, *Sci. Am.* **245**(2): 84–97 (1981).

Maden, B.E.H., The numerous modified nucelotides in eukaryotic ribosomal RNA, *Prog. Nucleic Acid Mol. Biol.* **39**, 241–303 (1990).

Moazed, D. and Noller, H.F., Intermediate states in the movement of tranfer RNA in the ribosome, *Nature* **342**, 142–148 (1989).

Moore, P.B. and Capel, M.S., Structure–function correlations in the small ribosomal subunit from *Escherichia coli, Annu. Rev. Biophys. Biophys. Chem.* **17**, 349–367 (1988).

Nierhaus, K.H., The allosteric three-site model for the ribosomal elongation cycle: Features and future, *Biochemistry* **29**, 4997-5008 (1990).

Nierhaus, K.H., Franceschi, F., Subramanian, A.R., Erdmann, V.A., and Wittmann-Liebold, B. (Eds.), *The Translational Apparatus. Structure, Function, Regulation, Evolution,* Plenum Press (1993); *and* Hill, W.E., Dahlberg, A., Garrett, R.A., Moore, P.B., Schlessinger, D., and Warner, J.R. (Eds.), *The Ribosome. Structure, Function, & Evolution,* American Society for Microbiology (1990). [The 1993 and 1990 "bibles" of ribosomology.]

Noller, H.F., Ribosomal RNA and translation, *Annu. Rev. Biochem.* **60**, 191–227 (1991). [An authoritative review.]

Noller, H.F., Jr., and Moldave, K., (Eds.), *Ribosomes, Methods Enzymol.* **164** (1988). [Contains numerous articles on ribosome methodology.]

Noller, H.F., Structure of ribosomal RNA, *Annu. Rev. Biochem.* **53**, 119–162 (1984).

Noller, H.F., Hoffarth, V., and Zimniak, L., Unusual resistance of peptidyl transferase to protein extraction procedures, *Science* **256**, 1416–1419 (1992); *and* Noller, H.F., Peptidyl transferase: Protein, ribonucleoprotein, or RNA? *J. Bacteriol.* **175**, 5297–5300 (1993).

Pappenheimer, A.M., Jr., Diphtheria toxin, *Annu. Rev. Biochem.* **46,** 69–94 (1977).

Rané, H.A., Klootwijk, J., and Musters, W., Evolutionary conservation of structure and function of high molecular weight ribosomal RNA, *Prog. Biophys. Mol. Biol.* **51,** 77–129 (1988).

Rhoads, R.E., Cap recognition and the entry of mRNA into the protein initiation cycle, *Trends Biochem. Sci.* **13,** 52–56 (1988).

Riis, B., Rattan, S.I.S., Clark, B.F.C., and Merrick, W.C., Eukaryotic protein elongation factors, *Trends Biochem. Sci.* **15,** 420–424 (1990).

Schüler, D. and Brinacombe, R., The *Escherichia coli* 30S ribosomal subunit; an optimized three-dimensional fit between the ribosomal proteins and the 16S RNA, *EMBO J.* **7,** 1509–1513 (1988).

Shatkin, A.J., mRNA cap binding proteins: essential factors for initiating translation, *Cell* **40,** 223–224 (1985).

Shine, J. and Dalgarno, L., The 3'-terminal sequence of *Escherichia coli* 16S ribosomal RNA: complementarity to nonsense triplets and ribosome binding sites, *Proc. Natl. Acad. Sci.* **71,** 1342–1346 (1974).

Steitz, J.A. and Jakes, K., How ribosomes select initiator regions in mRNA: base pair formation between the 3' terminus of 16S RNA and the mRNA during initiation of protein synthesis in *Escherichia coli*, *Proc. Natl. Acad. Sci.* **72,** 4734–4738 (1975).

Stern, S., Powers, T., Changchien, L.-I., and Noller, H.F., RNA–protein interactions in 30S ribosomal subunits: folding and function of 16S rRNA, *Science* **244,** 783–790 (1989).

Stöffler, G. and Stöffler-Meilicke, M., Immunoelectron microscopy of ribosomes, *Annu. Rev. Biophys. Bioeng.* **13,** 303–330 (1984).

Thompson, R.C., EFTu provides an internal kinetic standard for translational accuracy, *Trends Biochem. Sci.* **13,** 91–93 (1988).

Trachsel, E. (Ed.), *Translation in Eukaryotes,* CRC Press (1991). [Contains numerous useful articles on various aspects of eukaryotic translation.]

Yonath, A. and Franceschi, F., Structural aspects of ribonuceoprotein interactions in ribosomes; *Curr. Opin. Struct. Biol.* **3,** 45–49 (1993); *and* Yonath, A. and Berkovitch-Yellin, Z., Hollows, voids, gaps and tunnels in the ribosome, *Curr. Opin. Struct. Biol.* **3,** 175–181 (1993).

Control of Translation

Eguchi, Y., Itoh, T., and Tomizawa, J., Antisense RNA, *Annu. Rev. Biochem.* **60,** 631–652 (1991).

Hershey, J.W.B., Translational control in mammals, *Annu. Rev. Biochem.* **60,** 717–755 (1991).

Kozak, M., Regulation of translation in eukaryotic systems, *Annu. Rev. Cell Biol.* **8,** 197–225 (1992).

Oeller, P.W., Min-Wong, L., Taylor, L.P., Pike, D.A., and Theologis, A., Reversible inhibition of tomato fruit senescence by antisense RNA, *Science* **254,** 437–439 (1991).

Pestka, S., Langer, J.A., Zoon, K.C., and Samuel, C.E., Interferons and their actions, *Annu. Rev. Biochem.* **56,** 757–777 (1987).

Rhoads, R.E., Regulation of eukaryotic protein synthesis by initiation factors, *J. Biol. Chem.* **268,** 3017–3020 (1993).

Samuel, C.E., The eIF-2α protein kinases, regulators of translation in eukaryotes from yeasts to humans, *J. Biol. Chem.* **268,** 7603–7606 (1993).

Sen, G.C. and Lengyel, P., The interferon system, *J. Biol. Chem.* **267,** 5017–5020 (1992).

Simons, M., Edelman, E.R., DeKeyser, J.-L., Langer, R., and Rosenberg, R.D., Antisense c-*myb* oligonucleotides inhibit intimal arterial smooth muscle accumulation *in vivo. Nature* **359,** 67–70 (1992).

Tafuri, S.R. and Wolffe, A.P., Dual roles for transcription and translation factors in the RNA storage particles of *Xenopus* oocytes, *Trends Cell. Biol.* **3,** 94–98 (1993).

Thach, R.E., Cap recap: The involvement of eIF-4F in regulating gene expression, *Cell* **68,** 177–180 (1992).

Weintraub, H.M., Antisense RNA and DNA, *Sci. Am.* **262**(1): 40–46 (1990).

Posttranslational Modification

Fessler, J.H. and Fessler, L.I., Biosynthesis of procollagen, *Annu. Rev. Biochem.* **47,** 129–162 (1978).

Harding, J.J., and Crabbe, M.J.C. (Eds.), *Post-Translational Modifications of Proteins,* CRC Press (1992).

Peters, J.-M., Proteosomes: protein degradation machines of the cell, *Trends Biochem. Sci.* **19,** 377–382 (1994).

Prockop, D.J., Kivirikko, K.I., Tuderma, L., and Guzman, N.A., The biosynthesis of collagen and its disorders, *New Engl. J. Med.* **301,** 13–23 (1979).

Wold, F., In vivo chemical modification of proteins, *Annu. Rev. Biochem.* **50,** 783–814 (1981).

Wold, F. and Moldave, K. (Eds.), Posttranslational Modifications, Parts A and B, *Methods Enzymol.* **106** and **107** (1984). [Contains extensive descriptions of the amino acid "zoo."]

Protein Degradation

Cook, W.J., Jeffrey, L.C., Kasperek, E., and Pickart, C.M., Structure of tetraubiquitin shows how multiubiquitin chains can be formed, *J. Mol. Biol.* **236,** 601–609 (1994).

Cook, W.J., Jeffrey, L.C., Sullivan, M.L., and Vierstra, R.D., Three-dimensional structure of a ubiquitin-conjugating enzyme (E2), *J. Biol. Chem.* **267,** 15116–15121; *and* Cook, W.J., Jeffrey, L.C., Carson, M., Chen, Z., and Pickart, C.M., Structure of a diubiquitin conjugate and a model for interaction with ubiquitin conjugating enzyme (E2), *J. Biol. Chem.* **267,** 16467–16471 (1992).

Dice, F., Peptide sequences that target cytosolic proteins for lysosomal proteolysis, *Trends Biochem Sci.* **15,** 305–309 (1990).

Goldberg, A.L., and Rock, K.L., Proteolysis, proteasomes and antigen presentation, *Nature* **357,** 375–379 (1992).

Hershko, A. and Ciechanover, A., The ubiquitin system for protein degradation, *Annu. Rev. Biochem.* **61,** 761–807 (1992); *and* Ciechanover, A., The ubiquitin–proteasome proteolytic pathway, *Cell* **79,** 13–21 (1994).

Hochstrasser, M., Ubiquitin and intracellular protein degradation, *Curr. Opin. Cell Biol.* **4,** 1024–1031 (1992).

Jentsch, S., The ubiquitin-conjugation system, *Annu. Rev. Genet.* **26,** 179–207 (1992).

Peters, J.-M., Cejka, Z., Harris, J.R., Kleinschmidt, J.A., and Baumeister, W., Structural features of the 26 S proteasome complex, *J. Mol. Biol.* **234,** 932–937 (1993).

Rechsteiner, M., Hoffman, L., and Dubiel, W., The multicatalytic and 26 S proteases, *J. Biol. Chem.* **268,** 6065–6068 (1993).

Rivett, A.J., Proteasomes: multicatalytic proteinase complexes, *Biochem. J.* **291,** 1–10 (1993).

Senahdi, V.-J., Bugg, C.E., Wilkinson, K.D., and Cook, W.J., Three-dimensional structure of ubiquitin at 2.8 Å resolution, *Proc. Natl. Acad. Sci.* **82,** 3582–3585 (1985).

Varshavsky, A., The N-end rule, *Cell* **69,** 725–735 (1992).

PROBLEMS

1. What is the product of reacting guanine with nitrous acid? Is the reaction mutagenic? Explain.

2. What is the polypeptide specified by the following DNA antisense strand? Assume translation starts after the first initiation codon.

   5′-TCTGACTATTGAGCTCTCTGGCACATAGCA-3′

*3. The fingerprint of a protein from a phenotypically revertant mutant of bacteriophage T4 indicates the presence of an altered tryptic peptide with respect to the wild-type. The wild-type and mutant peptides have the following sequences:

   Wild-type   Cys-Glu-Asp-His-Val-Pro-Gln-Tyr-Arg
   Mutant   Cys-Glu-Thr-Met-Ser-His-Ser-Tyr-Arg

   Indicate how the mutant could have arisen and give the base sequences, as far as possible, of the mRNAs specifying the two peptides. Comment on the function of the peptide in the protein.

4. Explain why the various classes of mutations can reverse a mutation of the same class but not a different class.

5. Which amino acids are specified by codons that can be changed to an *amber* codon by a single point mutation?

6. The mRNA specifying the α chain of human hemoglobin contains the base sequence

   · · ·UCCAAAUACCGUUAAGCUGGA· · ·

   The C-terminal tetrapeptide of the normal α chain, which is specified by part of this sequence, is

   -Ser-Lys-Tyr-Arg

   In hemoglobin Constant Spring, the corresponding region of the α chain has the sequence

   -Ser-Lys-Tyr-Arg-Gln-Ala-Gly-· · ·

   Specify the mutation that causes hemoglobin Constant Spring.

7. Explain why a minimum of 32 tRNAs are required to translate the "standard" genetic code.

8. Draw the wobble pairings not in Fig. 30-24*a*.

9. A colleague of yours claims that by exposing *E. coli* to HNO₂ she has mutated a tRNA^Gly to an *amber* suppressor. Do you believe this claim? Explain.

*10. Deduce the anticodon sequences of all suppressors listed in Table 30-5 except UGA-1 and indicate the mutations that caused them.

11. How many different types of macromolecules must be minimally contained in a cell-free protein synthesizing system from *E. coli*? Count each type of ribosomal component as a different macromolecule.

12. Why do oligonucleotides containing Shine–Dalgarno sequences inhibit translation in prokaryotes? Why don't they do so in eukaryotes?

13. Why does m⁷GTP inhibit translation in eukaryotes? Why doesn't it do so in prokaryotes?

14. What would be the distribution of radioactivity in the completed hemoglobin chains upon exposing reticulocytes to ³H-labeled leucine for a short time followed by a chase with unlabeled leucine?

15. Design an mRNA with the necessary prokaryotic control sites that codes for the octapeptide Lys-Pro-Ala-Gly-Thr-Glu-Asn-Ser.

16. Indicate the translational control sites in and the amino acid sequence specified by the following prokaryotic mRNA.

   5′-CUGAUAAGGAUUUAAAUUAUGUGUCAAUCACGA-
   AUGCUAAUCGAGGCUCCAUAAUAACACUUCGAC-3′

17. What is the energetic cost, in ATPs, for the *E. coli* synthesis of a polypeptide chain of 100 residues starting from amino acids and mRNA? Assume that no losses are incurred as a result of proofreading.

*18. It has been suggested that Gly–tRNA synthetase does not require an editing mechanism. Why?

19. An antibiotic named fixmycin, which you have isolated from a fungus growing on ripe passion fruit, is effective in curing many types of venereal disease. In characterizing fixmycin's mode of action, you have found that it is a bacterial translational inhibitor that binds exclusively to the large subunit of *E. coli* ribosomes. The initiation of protein synthesis in the presence of fixmycin results in the generation of dipeptides that remain associated with the ribosome. Suggest a mechanism of fixmycin action.

20. Heme inhibits protein degradation in reticulocytes by allosterically regulating ubiquitin-activating enzyme (E1). What physiological function might this serve?

21. Genbux Inc., a genetic engineering firm, has cloned the gene encoding an industrially valuable enzyme into *E. coli* such that the enzyme is produced in large quantities. However, since the firm wishes to produce the enzyme in ton quantities, the expense of isolating it would be greatly reduced if the bacterium could be made to secrete it. As a high-priced consultant, what general advice would you offer to solve this problem?

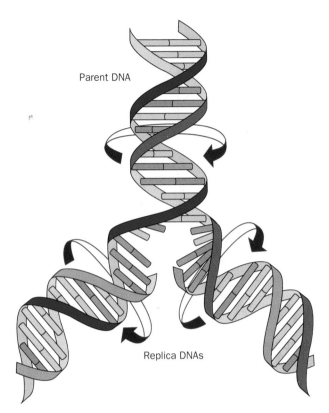

FIGURE 31-3. The replication of DNA.

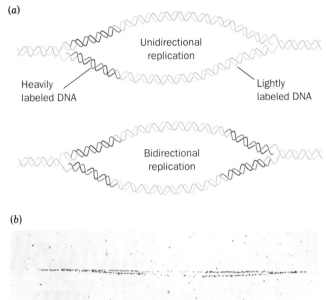

(a)

Unidirectional replication

Heavily labeled DNA

Lightly labeled DNA

Bidirectional replication

(b)

FIGURE 31-4. The autoradiographic differentiation of unidirectional and bidirectional θ replication of DNA. (*a*) An organism is grown for several generations in a medium that is lightly labeled with [³H]thymine so that all of its DNA will be visible in an autoradiogram. A large amount of [³H]thymine is then added to the medium for a few seconds before the DNA is isolated (**pulse labeling**) in order to label only those bases near the replication fork(s). Unidirectional DNA replication will exhibit only one heavily labeled branch point (*above*), whereas bidirectional DNA replication will exhibit two such branch points (*below*). (*b*) An autoradiogram of *E. coli* DNA so treated, demonstrating that it is bidirectionally replicated. [Courtesy of David M. Prescott, University of Colorado.]

strands (*Fig. 31-3*). DNA replication involving θ structures is known as θ **replication.**

A branch point in a replication eye at which DNA synthesis occurs is called a **replication fork.** A replication bubble may contain one or two replication forks (**unidirectional** or **bidirectional replication**). Autoradiographic studies have demonstrated that θ replication is almost always bidirectional (Fig. 31-4). Moreover, such experiments, together with genetic evidence, have established that prokaryotic and bacteriophage DNAs have but one **replication origin** (point where DNA synthesis is initiated).

B. Role of DNA Gyrase

The requirement that the parent DNA unwind at the replication fork (Fig. 31-3) presents a formidable topological obstacle. For instance, *E. coli* DNA is replicated at a rate of ~1000 nucleotides/s. If its 1300-μm-long chromosome were linear, it would have to flail around within the confines of a 3-μm-long *E. coli* cell at ~100 revolutions/s (recall that B-DNA has ~10 bp per turn). But since the *E. coli* chromosome is, in fact, circular, even this could not occur. Rather, the DNA molecule would accumulate +100 supercoils/s (see Section 28-5A for a discussion of supercoiling) until it became too tightly coiled to permit further unwinding. Naturally occurring DNA's negative supercoiling promotes DNA unwinding but only to the extent of ~5% of its duplex turns (recall that naturally occurring DNAs are typ-

ically underwound by one supercoil per ~20 duplex turns; Section 28-5B). In prokaryotes, however, negative supercoils may be introduced into DNA through the action of a Type II topoisomerase (DNA gyrase; Section 28-5C) at the expense of ATP hydrolysis. This process is essential for prokaryotic DNA replication as is demonstrated by the observation that DNA gyrase inhibitors, such as novobiocin and oxolinic acid, arrest DNA replication except in mutants whose DNA gyrase does not bind these antibiotics.

C. Semidiscontinuous Replication

The low-resolution images provided by autoradiograms such as Figs. 31-2 and 31-4*b* suggest that duplex DNA's two antiparallel strands are simultaneously replicated at an advancing replication fork. Yet, all known DNA polymerases can only extend DNA strands in the $5' \rightarrow 3'$ direction. How, then, does DNA polymerase copy the parent strand that extends in the $5' \rightarrow 3'$ direction past the replication fork? This question was answered in 1968 by Reiji Okazaki through the following experiments. If a growing *E. coli* culture is pulse labeled for 30 s with [³H]thymidine, much of the radioactive and hence newly synthesized DNA has a

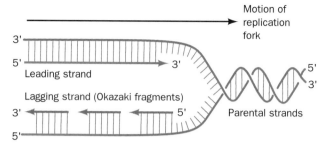

FIGURE 31-5. In DNA replication, both daughter strands (*red*) are synthesized in their $5' \rightarrow 3'$ direction. The leading strand is synthesized continuously, whereas the lagging strand is synthesized discontinuously.

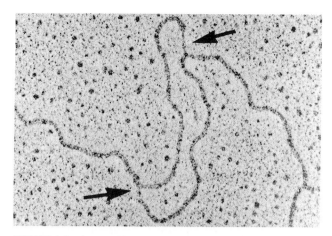

FIGURE 31-6. An electron micrograph of a replication eye in *Drosophila melanogaster* DNA. Note that the single-stranded regions (*arrows*) near the replication forks have the trans configuration consistent with the semidiscontinuous model of DNA replication. [From Kreigstein, H.J. and Hogness, D.S., *Proc. Natl. Acad. Sci.* **71,** 173 (1974).]

sedimentation coefficient in alkali of 7S to 11S. These so-called **Okazaki fragments** evidently consist of only 1000 to 2000 nucleotides (**nt;** 100-200 nt in eukaryotes). If, however, following the 30 s [³H]thymidine pulse, the *E. coli* are transferred to an unlabeled medium (a **pulse-chase** experiment), the resulting radioactively labeled DNA sediments at a rate that increases with the time that the cells had grown in the unlabeled medium. The Okazaki fragments must therefore become covalently incorporated into larger DNA molecules.

Okazaki interpreted his experimental results in terms of the **semidiscontinuous replication** model (Fig. 31-5). The two parent strands are replicated in different ways. *The newly synthesized DNA strand that extends $5' \rightarrow 3'$ in the direction of replication fork movement, the so-called **leading strand,** is essentially continuously synthesized in its $5' \rightarrow 3'$ direction as the replication fork advances. The other newly synthesized strand, the **lagging strand,** is also synthesized in its $5' \rightarrow 3'$ direction but discontinuously as Okazaki fragments. The Okazaki fragments are only covalently joined together sometime after their synthesis in a reaction catalyzed by the enzyme **DNA ligase** (Section 31-2C).*

The semidiscontinuous model of DNA replication is corroborated by electron micrographs of replicating DNA showing single-stranded regions on one side of the replication fork (Fig. 31-6). In bidirectionally replicating DNA, moreover, the two single-stranded regions occur, as expected, on diagonally opposite sides of the replication bubble.

D. RNA Primers

DNA polymerases' all but universal requirement for a free 3'-OH group to extend a DNA chain poses a question that was emphasized by the establishment of the semidiscontinuous model of DNA replication: How is DNA synthesis initiated? Careful analysis of Okazaki fragments revealed that *their 5' ends consist of RNA segments of 1 to 60 nt (a length that is species dependent) that are complementary to the template DNA chain* (Fig. 31-7). *E. coli* has two en-

zymes that can catalyze the formation of these **RNA primers:** RNA polymerase, the enzyme that mediates transcription (Section 29-2), and the much smaller **primase** (60 kD), the monomeric product of the *dnaG* gene.

Primase is insensitive to the RNA polymerase inhibitor rifampicin (Section 29-2C). The observation that rifampicin inhibits only leading strand synthesis therefore indicates that *primase initiates the Okazaki fragment primers.* The initiation of leading strand synthesis in *E. coli,* a much rarer event than that of Okazaki fragments, can be mediated *in vitro* by either RNA polymerase or primase alone but is greatly stimulated when both enzymes are present. It is therefore thought that these enzymes act synergistically *in vivo* to prime leading strand synthesis.

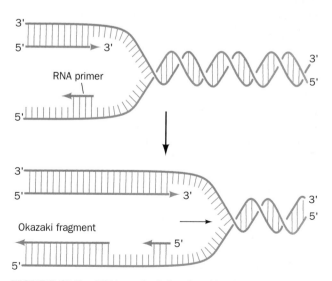

FIGURE 31-7. DNA synthesis is primed by short RNA segments.

Mature DNA does not contain RNA. The RNA primers are eventually removed and the resulting single-strand gaps are filled in with DNA by a mechanism described in Section 31-2A.

2. ENZYMES OF REPLICATION

DNA replication is a complex process involving a great variety of enzymes. It requires, to list only its major actors in their order of appearance: (1) a DNA gyrase, (2) enzymes known as **helicases** that separate the DNA strands at the replication fork, (3) proteins that prevent them from reannealing before they are replicated, (4) enzymes that synthesize RNA primers, (5) a DNA polymerase, (6) an enzyme to remove the RNA primers, and (7) an enzyme to covalently link successive Okazaki fragments. In this section, we describe the properties and functions of many of these proteins.

A. DNA Polymerase I

In 1957, Arthur Kornberg reported that he had discovered an enzyme that catalyzes the synthesis of DNA in extracts of *E. coli* through its ability to incorporate the radioactive label from [¹⁴C]thymidine triphosphate into DNA. This enzyme, which has since become known as **DNA polymerase I** or **Pol I**, consists of a single 928-residue polypeptide.

Pol I couples deoxynucleoside triphosphates on DNA templates (Fig. 31-1) in a reaction that occurs through the nucleophilic attack of the growing DNA chain's 3'-OH group on the α-phosphoryl of an incoming nucleoside triphosphate. The reaction is driven by the resulting elimination of PP_i and its subsequent hydrolysis by inorganic pyrophosphatase. The overall reaction resembles that catalyzed by RNA polymerase (Section 29-2) but differs from it by the strict requirement that the incoming nucleoside be linked to a free 3'-OH group of a polynucleoside that is base paired to the template (recall that RNA polymerase initiates transcription by linking together two ribonucleoside triphosphates on a DNA template). The complementarity between the product DNA and the template was at first inferred through base composition and hybridization studies but was eventually directly established by base sequence determinations. The error rate of Pol I in copying the template is quite low as was demonstrated by its *in vitro* replication of ϕX174 DNA to yield fully infective phage DNA.

The specificity of Pol I for an incoming base is thought to arise from the requirement that it form a Watson–Crick base pair with the template rather than direct recognition of the incoming base (recall that the four base pairs, A·T, T·A, G·C, and C·G, have nearly identical shapes; Section 28-2A). This accounts for the observation that Pol I may substitute 5-bromouracil only for thymine and hypoxanthine only for guanine. Pol I is said to be **processive** in that it catalyzes a series of successive polymerization steps, typi-

cally 20 or more, without releasing the template. Pol I can, of course, work in reverse by degrading DNA through pyrophosphorolysis. This reverse reaction, however, probably has no physiological significance because of the low *in vivo* concentration of PP_i resulting from the action of inorganic pyrophosphatase.

Pol I Can Edit Its Mistakes

In addition to its polymerase activity, Pol I has two independent hydrolytic activities:

1. It can act as a $3' \rightarrow 5'$ exonuclease.
2. It can act as a $5' \rightarrow 3'$ exonuclease.

The $3' \rightarrow 5'$ exonuclease reaction differs chemically from the pyrophosphorolysis reaction only in that H_2O rather than PP_i is the nucleotide acceptor. Kinetic and crystallographic studies (see below), however, indicate that these two catalytic activities occupy separate active sites. The $3' \rightarrow 5'$ exonuclease function is activated by an unpaired 3'-terminal nucleotide with a free OH group. If Pol I erroneously incorporates a wrong (unpaired) nucleotide at the end of a growing DNA chain, the polymerase activity is inhibited and the $3' \rightarrow 5'$ exonuclease excises the offending nucleotide (Fig. 31-8). The polymerase activity then resumes DNA replication. *Pol I therefore has the ability to proofread a DNA chain as it is synthesized so as to correct its mistakes.* This explains the great fidelity of DNA replication by Pol I despite the relatively low free energy of a single base pairing interaction (the energetics of binding fidelity is discussed in Section 30-2C). The price of this high fidelity is that ~ 3% of correctly incorporated nucleotides are also excised.

The Pol I $5' \rightarrow 3'$ exonuclease binds to duplex DNA at single-strand nicks with little regard to the character of the 5' nucleotide (5'-OH or phosphate group; base paired or not). It cleaves the DNA in a base paired region beyond the nick such that the DNA is excised as either mononucleotides or oligonucleotides of up to 10 residues (Fig. 31-9). In

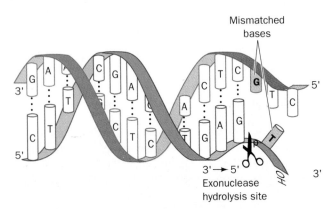

FIGURE 31-8. The $3' \rightarrow 5'$ exonuclease function of DNA polymerase I excises mispaired nucleotides from the 3' end of the growing DNA strand.

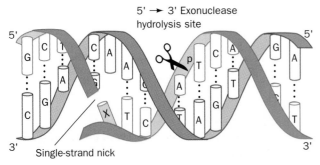

FIGURE 31-9. The $5' \rightarrow 3'$ exonuclease function of DNA polymerase I excises up to 10 nucleotides from the 5' end of a single-strand nick. The nucleotide immediately past the nick (X) may or may not be paired.

contrast, the $3' \rightarrow 5'$ exonuclease removes only unpaired mononucleotides with 3'-OH groups.

Pol I's Polymerase and Two Exonuclease Functions Each Occupy Separate Active Sites

The $5' \rightarrow 3'$ exonuclease activity of Pol I is independent of both its $3' \rightarrow 5'$ exonuclease and its polymerase activities. In fact, as we saw in Section 28-6C, proteases such as subtilisin or trypsin cleave Pol I into two fragments: the large or "Klenow" fragment (**KF;** residues 324–928), which contains both the polymerase and the $3' \rightarrow 5'$ exonuclease activities; and a smaller fragment (residues 1–

323), which contains the $5' \rightarrow 3'$ exonuclease activity. Thus Pol I contains three active sites on a single polypeptide chain.

The X-Ray Structure of Klenow Fragment Indicates How It Binds DNA

The X-ray structure of KF, determined by Thomas Steitz, reveals that this protein consists of two domains (Fig. 31-10). The smaller domain (residues 324–517) contains the $3' \rightarrow 5'$ exonuclease site, as was demonstrated by the absence of this function but not polymerase activity in a genetically engineered Klenow fragment mutant that lacks the divalent metal ion–binding sites known to be essential for $3' \rightarrow 5'$ exonuclease activity but which otherwise has a normal structure. The larger domain (residues 521–928; helix G and beyond in Fig. 31-10*b*) contains the polymerase active site at the bottom of a prominent cleft, a surprisingly large distance (~25 Å) from the $3' \rightarrow 5'$ exonuclease site. The cleft, which is lined with positively charged residues, has the appropriate size (~22-Å wide × ~30-Å deep) and shape to bind a B-DNA molecule in a manner resembling a right hand grasping a rod (in which the "thumb" consists of helices H–I, the "fingers" of helices L–P, and the remainder of the larger domain, the "palm," includes a 6-stranded antiparallel β sheet that forms the floor of the cleft and contains the polymerase function's active site residues). Indeed, the active sites of all DNA and RNA polymerases of known structure are located at the bottoms of similarly shaped clefts (Sections 29-2A, 31-4B, and 31-4C).

(a)

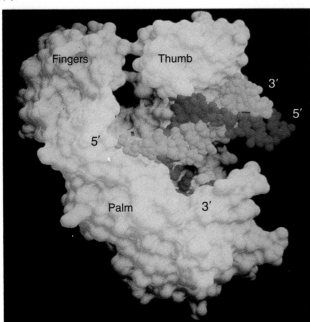

(b)

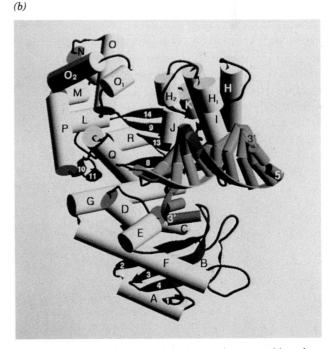

FIGURE 31-10. The X-ray structure of *E. coli* DNA polymerase I Klenow fragment (KF) in complex with a double helical DNA (whose probable sequence is given in Fig. 31-11). (*a*) The solvent accessible surface of KF (*yellow*) with the 12-nt template strand in light blue and the 14-nt primer strand in red. (*b*) A tube-and-arrow representation of the complex in the same orientation as Part *a* in which the template strand is blue and the primer strand is purple. [Courtesy of Thomas Steitz, Yale University.]

Steitz cocrystallized KF with an 11-nt DNA "template" strand (5'-TGCCTCGCGGCC-3'), a 7-nt "primer" strand (3'-GCGCCGG-5') that is complementary to the the 3' end of the template strand, and 2',3'-epoxy-ATP,

$$
\begin{array}{c}
O \quad\quad O \quad\quad O \\
\| \quad\quad\quad \| \quad\quad\quad \| \\
{}^-O-P-O-P-O-P-O-CH_2 \quad O \quad\quad A
\end{array}
$$

2',3'-Epoxy-ATP

which promotes the tight binding of DNA to the polymerase site. The X-ray structure of the resulting complex shows that the primer strand base pairs, as expected, to the 3' end of the template strand to form a distorted segment of B-DNA, and that the polymerase has apparently appended an epoxy-A residue to the primer's 3' end (where it base pairs to a T on the template strand; Fig. 31-11). In addition, a second primer strand, whose 3' G residue has apparently been removed, continues the 3' end of the primer (after a break in the sugar–phosphate chain) by base pairing, via its 5'-terminal three nt, to the template strand. Thus, this complex contains an 11-bp duplex DNA, which has a single-nt overhang at the 5' end of its template strand and a 3-nt overhang at the 3' end of its primer strand. The 3'-terminal nucleotide of the primer strand (the last one that an active polymerase would have added) is bound at the 3' → 5' exonuclease active site whereas the 5'-terminal nucleotide of the template strand is bound at the entrance of the polymerase active site. Evidently, KF has bound the DNA in an "editing" complex rather than in the polymerase cleft between the thumb and fingers as had previously been antici-

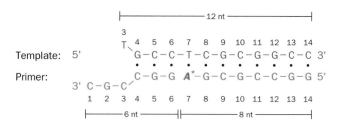

FIGURE 31-11. The probable sequence of the double-stranded DNA seen in the X-ray structure of KF (Fig. 31-10). *A** represents a 2',3'-epoxy-A residue that KF has appended to the 3' end of the 7-nt DNA with which the KF crystal was incubated (together with the 12-nt DNA). The 2-part primer strand's 3' segment appears to be a second copy of the 7-nt DNA with its 3' nt removed.

pated. KF only contacts the phosphate backbone of the DNA, consistent with Pol I's lack of sequence specificity in binding DNA.

In forming the DNA complex, KF's thumb, which directly contacts the duplex portion of the DNA, undergoes a conformational change involving atomic shifts of up to 12 Å relative to the X-ray structure of KF alone. In addition, the 50 residues connecting helices H and I, which are unobserved and presumably disordered in the structure of KF alone, become ordered, thus forming helices H_1 and H_2. The protein thereby forms a second cleft in which the duplex portion of the DNA binds and which joins with, but is nearly perpendicular to, the cleft that contains the polymerase site (Fig. 31-10). This suggests a model for DNA polymerase I in which the DNA lies in the polymerase cleft such that the primer strand extends 5' → 3' from the exonuclease site towards the polymerase site. Model building

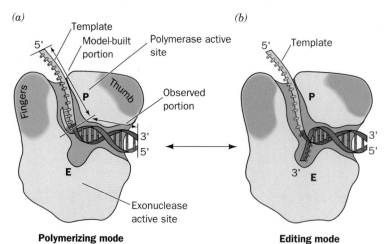

FIGURE 31-12. A schematic model of Klenow fragment with DNA binding to (*a*) the polymerase active site; and (*b*) the 3' → 5' exonuclease active site. The template strand is blue and the primer strand is red with that portion of the DNA that is observed in the crystal structure in darker shades than the portion that was modeled. The 3' terminus of the primer strand can shuttle between the polymerase active site (P) and the

3' → 5' exonuclease site (E) without dissociating from the enzyme. Any factor that destabilizes double-stranded DNA, such as a mismatched base pair, favors the editing complex and hence promotes excision of the primer strand's 3' nucleotide. [After Beese, L.S., Derbeyshire, V., and Steitz, T.A., *Science* **260**, 354 (1993).]

suggests that the nascent 3′ end of the primer strand shuttles between the polymerase and exonuclease active sites (Fig. 31-12). This, however, requires that, when the 3′ end of the primer occupies the polymerase site, the preceding duplex DNA undergoes an ~80° bend. Protein-induced bending of duplex DNA of this magnitude has, in fact, been observed in the X-ray structure of a CAP–cAMP·DNA complex (Section 29-3C). Such DNA bending would probably make the equilibrium between single- and double-stranded DNAs at the primer terminus more sensitive to mismatched base pairs.

Pol I Functions Physiologically to Repair DNA

For some 13 years after Pol I's discovery, it was generally assumed that this enzyme was *E. coli*'s DNA replicase because no other DNA polymerase activity had been detected in *E. coli*. This assumption was made untenable by Cairns and Paula DeLucia's isolation, in 1969, of a mutant *E. coli* whose extracts exhibit <1% of the normal Pol I activity (although it has nearly normal levels of the 5′ → 3′ exonuclease activity) but which nevertheless reproduce at the normal rate. This mutant strain, however, is highly susceptible to the damaging effects of UV radiation and chemical mutagens. *Pol I evidently plays a central role in the repair of damaged (chemically altered) DNA.*

Damaged DNA, as we discuss in Section 31-5, is detected by a variety of DNA repair systems. Many of them endonucleolytically cleave the damaged DNA on the 5′ side of the lesion, thereby activating Pol I's 5′ → 3′ exonuclease. While excising this damaged DNA, Pol I simultaneously fills in the resulting single-strand gap through its polymerase activity. In fact, its 5′ → 3′ exonuclease activity increases 10-fold when the polymerase function is active. Perhaps the simultaneous excision and polymerization activities of Pol I protects DNA from the action of cellular nucleases that would further damage the otherwise gapped DNA.

Pol I Catalyzes Nick Translation

Pol I's combined 5′ → 3′ exonuclease and polymerase activities can replace the nucleotides on the 5′ side of a single-strand nick on otherwise undamaged DNA. These reactions, in effect, translate (move) the nick towards the DNA strand's 3′ end without otherwise changing the molecule (Fig. 31-13). This **nick translation** process, in the presence of labeled deoxynucleoside triphosphates, is synthetically employed to prepare highly radioactive DNA (the required nicks may be generated by treating the DNA with a small amount of pancreatic DNase I).

Pol I's 5′ → 3′ Exonuclease Functions Physiologically to Excise RNA Primers

The 5′ → 3′ exonuclease also removes the RNA primers at the 5′ ends of newly synthesized DNA and fills in the resulting gaps. The importance of this function was demonstrated by the isolation of temperature-sensitive *E. coli*

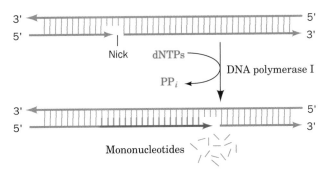

FIGURE 31-13. Nick translation as catalyzed by Pol I.

mutants that neither are viable nor exhibit any 5′ → 3′ exonuclease activity at the restrictive temperature of ~43°C (the low level of polymerase activity in the Pol I mutant isolated by Cairns and DeLucia is apparently sufficient to carry out this essential gap-filling process during chromosome replication). Thus Pol I has an indispensable role in *E. coli* DNA replication although a different one than was first supposed.

B. DNA Polymerase III

The discovery of normally growing *E. coli* mutants that have very little Pol I activity stimulated the search for an additional DNA polymerizing activity. This effort was rewarded by the discovery of two more enzymes, designated, in the order they were discovered, **DNA polymerase II (Pol II)** and **DNA polymerase III (Pol III)**. The properties of these enzymes are compared with that of Pol I in Table 31-1. Pol II and Pol III had not previously been detected because their combined activities in the assays used are normally <5% that of Pol I.

A mutant *E. coli* lacking measurable Pol II activity grows normally. However, Pol II has been implicated as a participant in repairing DNA damage via the **SOS response** (Section 31-5D).

TABLE 31-1. Properties of *E. Coli* DNA Polymerases

	Pol I	Pol II	Pol III
Mass (kD)	103	90	130
Molecules/cell	400	?	10–20
Turnover number[a]	600	30	9000
Structural gene	*polA*	*polB*	*polC*
Conditionally lethal mutant	+	−	+
Polymerization: 5′ → 3′	+	+	+
Exonuclease: 3′ → 5′	+	+	+
Exonuclease: 5′ → 3′	+	−	−

[a] Nucleotides polymerized min^{-1}·molecule^{-1} at 37°C.

Source: Kornberg, A. and Baker, T.A., *DNA Replication* (2nd ed.), p. 167, Freeman (1992).

Pol III Is *E. coli's* DNA Replicase

The cessation of DNA replication in temperature-sensitive *polC* mutants above the restrictive (high) temperature demonstrates that Pol III is *E. coli's DNA replicase*. Its **Pol III core** has the subunit composition $\alpha\varepsilon\theta$ where α, the *polC* gene product (Table 31-2), contains the polymerase function. The catalytic properties of Pol III core resemble those of Pol I (Table 31-1) except for Pol III core's inability to replicate primed single-stranded or nicked duplex DNA. Rather, Pol III core acts *in vitro* at single-strand gaps of <100 nucleotides, a situation that probably resembles the state of DNA at the replication fork. The Pol III $3' \rightarrow 5'$ exonuclease function, which resides on the enzyme's ε subunit, is DNA's primary editor during replication; it enhances the enzyme's replication fidelity by up to 200-fold. However, the Pol III $5' \rightarrow 3'$ exonuclease acts only on single-stranded DNA, so it cannot catalyze nick translation.

Pol III core functions in vivo as part of a complex and labile multisubunit enzyme, the **Pol III holoenzyme,** *which consists of at least 10 types of subunits (Table 31-2).* The latter 7 subunits in Table 31-2 act to modulate Pol III core's activity. For example, Pol III core has a processivity of 10 to 15 residues; it can only fill in short single-stranded regions of DNA. However, Pol III core is rendered processive by association with the β **subunit** in the presence of the 5-subunit γ **complex** ($\gamma\delta\delta'\chi\psi$). Assembly of the processive enzyme is a two-stage process in which the γ complex transfers the β subunit to the primed template in an ATP-dependent reaction followed by the assembly of Pol III core with the β

TABLE 31-2. COMPONENTS OF DNA POLYMERASE III HOLOENZYME

Subunit	Mass (kD)	Structural Gene
α[a]	130	*polC (dnaE)*
ϵ[a]	27.5	*dnaQ*
θ[a]	10	*holE*
τ	71	*dnaX*[c]
γ[b]	45.5	*dnaX*[c]
δ[b]	35	*holA*
δ'[b]	33	*holB*
χ[b]	15	*holC*
ψ[b]	12	*holD*
β	40.6	*dnaN*

[a] Components of Pol III.
[b] Components of the γ complex.
[c] The γ and τ subunits are encoded by the same gene sequence; the γ subunit comprises the N-terminal end of the τ subunit.

Sources: Kornberg, A. and Baker, T.A., *DNA Replication* (2nd ed.), p 169, Freeman (1992); *and* Baker, T.A. and Wickner, S.H., *Annu. Rev. Genet.* **26,** 450 (1992).

subunit on the DNA. The β subunit confers essentially unlimited processivity (>5000 residues) on the core enzyme even if the γ complex is subsequently removed. In fact, the β subunit is very strongly bound to the DNA, although it can freely slide along it.

(a)

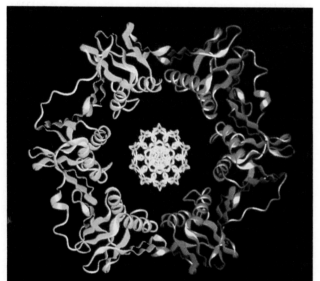

(b)

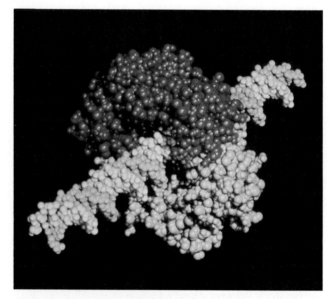

FIGURE 31-14. The X-ray structure of the β subunit of *E. coli* Pol III holoenzyme. (a) A ribbon drawing showing the two monomeric units of the dimeric protein in yellow and red as viewed along the dimer's two-fold axis. A stick model of B-DNA is placed with its helix axis coincident with the protein dimer's two-fold axis. (b) A space-filling model of the protein, colored as in Part *a*, in the hypothetical complex with the B-DNA shown in Part *a* in light blue. [Courtesy of John Kuriyan, The Rockefeller University.]

The β Subunit Forms a Ringlike Structure

The observation that a β subunit clamped to a cut circular DNA slides to the break and falls off suggests that the β subunit forms a closed ring around the DNA, thereby preventing its escape, and that the γ complex functions to open and close this ring. The X-ray structure of the β subunit (Fig. 31-14*a*), determined by John Kuriyan, reveals that it is a dimer of C-shaped, 366-residue monomer units which associate to form an ~80-Å-diameter doughnut-shaped structure whose central hole is ~35 Å in diameter, larger than the 20- and 26-Å diameters of B- and A-DNAs (recall that the hybrid helices which RNA primers make with DNA have A-DNA-like conformations; Section 28-2B). Each monomer consists of six βαββ motifs of identical topology, which associate in pairs to form three pseudotwofold symmetric domains of very similar structures (although with <20% sequence identity). The dimeric ring therefore has the shape of a 6-pointed star in which the 12 helices line the central hole and the β strands associate in six β sheets that form the protein's outer surface. Electrostatic calculations indicate that the interior surface of the ring is positively charged, whereas its outer surface is negatively charged.

Model building studies in which a B-DNA helix is threaded through the central hole (Fig. 31-14) indicate that the helices are all oriented such that they are perpendicular to their radially adjacent segments of sugar–phosphate backbone. These helices therefore span the major and minor grooves of the DNA rather than entering into them as do, for example, the recognition helices of helix–turn–helix motifs (Section 29-3C). Since A- and B-DNAs have 11 and 10 bp per turn, whereas the β subunit has a pseudo-12-fold symmetry, it appears that the β subunit is designed to minimize its associations with its threaded DNA. This presumably permits the protein to freely slide along the DNA helix. Indeed, the radius of the β subunit's central hole is at least 3.5 Å larger than that of DNA, so any interactions between them are probably mediated by intervening water molecules.

TABLE 31-3. Unwinding and Binding Proteins of *E. coli* DNA Replication

Protein	Subunit Structure	Subunit Mass (kD)
DnaB protein	hexamer	50
SSB	tetramer	19
Rep protein	monomer	68
PriA protein	monomer	76

Source: Kornberg, A. and Baker, T.A., *DNA Replication* (2nd ed.), *p.* 366, Freeman (1992).

cooperatively coat single-stranded DNA, thereby maintaining it in an unpaired state. Note, however, that DNA must be stripped of SSB before it can be replicated by Pol III holoenzyme.

Two other helicases, **Rep protein** and **PriA protein,** have been implicated in the replication of various *E. coli* phage DNAs (Section 31-3B) and also participate in certain aspects of *E. coli* DNA replication (Section 31-3C). Both proteins translocate along DNA in the 3′ → 5′ direction while consuming ATP. Rep helicase, which is a monomer in solution but dimerizes as it assembles onto a DNA strand, is not essential for *E. coli* DNA replication but the rate that *E. coli* replication forks propagate is reduced ~2-fold in *rep⁻* mutants.

DNA Ligase Seals Single-Strand Nicks

Pol I, as we saw in Section 31-1D, replaces the Okazaki fragments' RNA primers with DNA through nick translation. *The resulting single-strand nicks between adjacent Okazaki fragments, as well as the nick on circular DNA after leading strand synthesis, are sealed in a reaction catalyzed by **DNA ligase.*** The free energy required by this reaction is obtained, in a species-dependent manner, through the coupled hydrolysis of either NAD⁺ to NMN⁺ + AMP or ATP to PPᵢ + AMP. The *E. coli* enzyme, a 77-kD mono-

C. Helicases, Binding Proteins, and DNA Ligases

Pol III holoenzyme, unlike Pol I, cannot unwind duplex DNA. Rather, *two proteins, **DnaB protein** (the product of the **dnaB** gene) and **single-strand binding protein (SSB)** (Table 31-3), work in concert to unwind the DNA before an advancing replication fork (Fig. 31-15) in a process that is driven by ATP hydrolysis.* The hexameric DnaB protein, a **helicase,** separates the duplex DNA strands by translocating along the lagging strand template in the 5′ → 3′ direction while hydrolyzing ATP (it can also use GTP and CTP but not UTP). The separated DNA strands behind the advancing helicase are prevented from reannealing by the binding of SSB. Numerous copies of this tetrameric protein

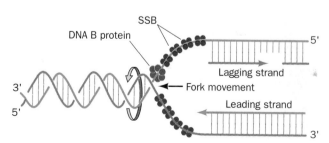

FIGURE 31-15. The unwinding of DNA by the combined action of DnaB and SSB proteins. The hexameric DnaB protein moves along the lagging strand template in the 5′ → 3′ direction. The resulting separated DNA strands are prevented from reannealing by SSB binding.

1. The reaction sequence begins in the same way as that for M13: The (+) strand is coated with SSB except for a 44-nt hairpin near position 2300 (between genes F and G; Fig. 30-9). A 70-nt sequence containing this hairpin known as **pas** (for *primosome assembly site*), is then recognized and bound by the PriA, **PriB**, and **PriC** proteins (formerly named **n′**, **n**, and **n″**).

2. DnaB, and **DnaC** proteins in the form of a $DnaB_6 \cdot DnaC_6$ complex add to the DNA with the help of **DnaT protein** (formerly **i**) in an ATP-requiring process. DnaC protein is then released yielding the **preprimosome**. The preprimosome, in turn, binds primase yielding the primosome.

3. The primosome is propelled in the $5' \rightarrow 3'$ direction along the (+) strand by PriA- and DnaB-catalyzed ATP hydrolysis. This motion, which displaces the SSB in its path, is opposite in direction to that of template reading during DNA chain propagation.

4. At randomly selected sites, the primosome reverses its migration while primase synthesizes an RNA primer. The initiation of primer synthesis requires the participation of DnaB protein which, through concomitant ATP hydrolysis, is thought to alter template DNA conformation in a manner required by primase.

5. Pol III holoenzyme extends the primers to form Okazaki fragments.

6. Pol I excises the primers and replaces them by DNA. The fragments are then joined by DNA ligase and supercoiled by DNA gyrase to form the ϕX174 RF I.

The primosome remains complexed with the DNA (Fig. 31-19) where it participates in (+) strand synthesis (see below).

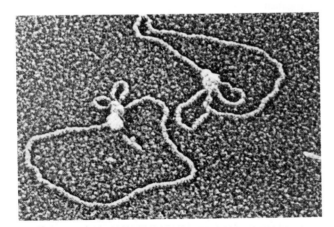

FIGURE 31-19. Electron micrograph of a primosome bound to a ϕX174 RFI DNA. Such complexes always contain a single primosome with one or two associated small DNA loops. [Courtesy of Jack Griffith, Lineberger Cancer Research Center, University of North Carolina.]

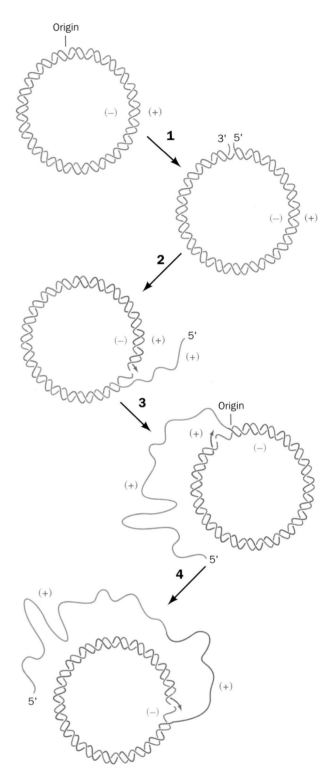

FIGURE 31-20. The rolling circle mode of DNA replication. The (+) strand being synthesized is extended from a specific cut made at the replication origin (1) so as to strip away the old (+) strand (2 and 3). The continuous synthesis of the (+) strand on a circular (−) strand template produces a series of tandemly linked (+) strands (4), which may later be separated by a specific exonuclease.

φX174 (+) Strand Replication Serves as a Model for Leading Strand Synthesis

One strand of a circular duplex DNA may be synthesized via the **rolling circle** or **σ-replication** mode (so-called because of the resemblance of the replicating structure to the Greek letter sigma; Fig. 31-20). *The φX174 (+) strand is synthesized on an RF I template by a variation on this process, the* **looped rolling circle mode** *(Fig. 31-21):*

1. (+) strand synthesis begins with the primosome-aided binding of the phage-encoded enzyme **gene A protein** (60 kD) to its ~30-bp recognition site. There, gene A protein specifically cleaves the phosphodiester bond preceding (+) strand nucleotide 4306 (near the beginning of gene A; Fig. 30-9) by forming a covalent bond between a Tyr residue and the DNA's 5'-phosphoryl group, thereby conserving the cleaved bond's energy.

2. Rep protein (Section 31-2C) subsequently attaches to the (−) strand at the gene A protein and, with the aid of the primosome still associated with the (+) strand, commences unwinding the duplex DNA from the (+) strand's 5' end. The displaced (+) strand is coated with SSB, which prevents it from reannealing to the (−) strand. Rep protein is essential for the replication of φX174 DNA, but not for the *E. coli* chromosome, as is demonstrated by the inability of φX174 to multiply in *rep⁻ E. coli*. Pol III holoenzyme extends the (+) strand from its free 3'-OH group.

3. The extension process generates a **looped rolling circle** structure in which the 5' end of the old (+) strand remains linked to the gene A protein at the replication fork. It is thought that as the old (+) strand is peeled off the RF, the primosome synthesizes the primers required for the later generation of a new (−) strand.

4. When it has come full circle around the (−) strand, the gene A protein again makes a specific cut at the replication origin so as to form a covalent linkage with the new (+) strand's 5' end. Simultaneously, the newly formed 3'-terminal OH group of the old, looped-out (+) strand nucleophilically attacks its 5'-phosphoryl attachment to the gene A protein, thereby liberating a covalently closed (+) strand. This is possible because the gene A protein has two closely spaced Tyr residues that alternate in their attachment to the 5' ends of successively synthesized (+) strands. The replication fork continues its progress about the duplex circle, producing new (+) strands in a manner reminiscent of linked sausages being pulled off a reel.

In the intermediate stages of a φX174 infection, each newly synthesized (+) strand directs the synthesis of the (−) strand to form RF I as described above. In the latter stages of infection, however, the newly formed (+) strands are packaged into phage particles.

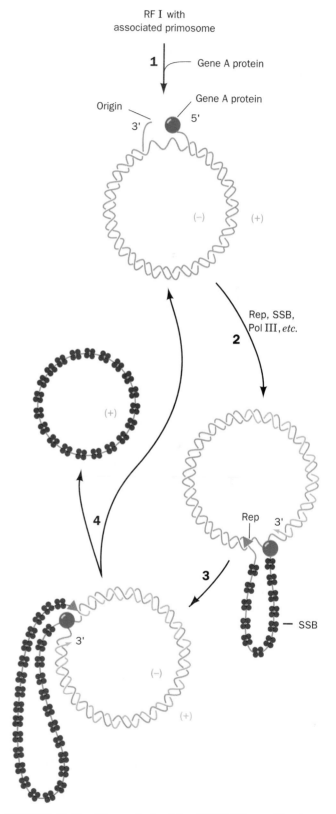

FIGURE 31-21. The synthesis of the φX174 (+) strand by the looped rolling circle mode. The numbered steps are described in the text.

The Initiation of *E. coli* DNA Replication Is Strictly Regulated

Chromosome replication in E. coli occurs only once per cell division so this process must be tightly controlled. The doubling time of *E. coli* at 37°C varies with growth conditions from <20 min to ~10 h. Yet the constant ~1000 nt/s rate of movement of each replication fork fixes the 4.7×10^6-bp *E. coli* chromosome's replication time, *C*, at ~40 min. Moreover, the segregation of cellular components and the formation of a septum between them, which must precede cell division, requires a constant time, $D = 20$ min, after the completion of the corresponding round of chromosome replication. *Cells with doubling times* $< C + D = 60$ *min must consequently initiate chromosome replication before the end of the preceding cell division cycle.* This results in the formation of **multiforked chromosomes** as is indicated in Fig. 31-24 for a cell division time of 35 min.

The above considerations indicate that there must be a signal that triggers each cycle of chromosome replication. Two observations suggest roles for DnaA and/or DNA methylation in generating this signal:

1. Genetically engineed variations in the intracellular concentration of DnaA protein reveal that at low DnaA levels, the ratio of DNA to cell mass is quite low, indicating infrequent initiation, whereas at high levels of DnaA, the DNA to cell mass ratio is greater than normal, indicating an increased rate of initiation. This suggests that initiation is triggered by the accumulation of DnaA protein.

2. The sequence most commonly methylated in *E. coli* is GATC (Section 31-7), which occurs 14 times in *oriC* including at the beginning of all four of its 13-bp repeats (see above). Thus, the observation that *E. coli* defective in the GATC methylation enzyme are very inefficiently transformed by *oriC*-containing plasmids, suggests that the DNA replication trigger in *E. coli* also responds to the level of *oriC* methylation.

There is extensive morphological evidence, such as Fig. 31-25, that the *E. coli* chromosome is associated with the cell membrane. This association presumably permits the segregation of replicated chromosomes into different cells during cell division. There is, nevertheless, no direct evidence that any membrane component is required for DNA replication.

Replication Termination

The *E. coli* replication terminus is a large (350-kb) region flanked by six nearly identical nonpalindromic ~23-bp terminator sites, **TerE, TerD,** and **TerA** on one side and **TerF, TerB,** and **TerC** on the other (Fig. 31-26; note that *oriC* is directly opposite the terminus region on the *E. coli* chromosome). A replication fork traveling counterclockwise as drawn in Fig. 31-26 passes through *TerF, TerB,* and *TerC* but stops upon encountering either *TerA, TerD,* or *TerE* (*TerD* and *TerE* are presumably backup sites for *TerA*).

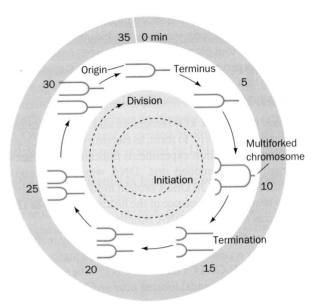

FIGURE 31-24. In cells that are dividing every 35 min, the fixed 60-min interval between the initiation of replication and cell division results in the production of multiforked chromosomes. [After Lewin, B., *Genes V, p.* 554, Oxford Univ. Press (1994).

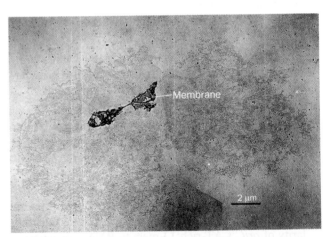

FIGURE 31-25. An electron micrograph of an intact and supercoiled *E. coli* chromosome attached to two fragments of the cell membrane. [From Delius H. and Worcel, A., *J. Mol. Biol.* **82,** 108 (1974).]

Similarly, a clockwise-traveling replication fork transits *TerE, TerD* and *TerA* but halts at *TerC* or, failing that, *TerB* or *TerF*. Thus, these termination sites are polar; they act as one-way valves that allow replication forks to enter the terminus region but not to leave it. This arrangement guarantees that the two replication forks generated by bidirectional initiation at *oriC* will meet in the replication terminus even if one of them arrives there well ahead of its counterpart.

The arrest of replication fork motion at *Ter* sites requires the action of **Tus protein**, a 309-residue monomer that is the

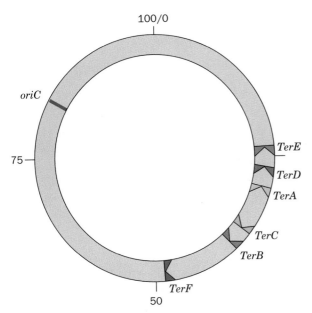

FIGURE 31-26. A map of the *E. coli* chromosome, in the same orientation as Fig. 27-18, showing the positions of the *Ter* sites (*right*) and the *oriC* site (*left*). The *TerC, TerB,* and *TerF* sites, in combination with Tus protein, allow a counter-clockwise-moving replisome to pass but not a clockwise-moving replisome. The opposite is true of the *TerA, TerD,* and *TerE* sites. Consequently, two replication forks that initiate bidirectional DNA replication at *oriC* will meet between the oppositely facing *Ter* sites.

product of the **tus** gene (for *t*erminator *u*tilization *s*ubstance). Tus protein specifically binds to a *Ter* site, where it prevents strand displacement by DnaB helicase, thereby arresting replication fork motion. When Tus protein is fused to another DNA-binding protein, replication is inhibited at the other protein's binding site, which suggests that Tus protein does not act as a simple clamp, but interacts with DnaB protein to inhibit its helicase action. Curiously, however, this termination system is not essential for termination. When the replication terminus is deleted, replication simply stops, apparently through the collision of opposing replication forks. Nevertheless, this termination system is highly conserved in gram-negative bacteria.

The final step in *E. coli* DNA replication is the topological unlinking of the catenated parental DNA strands, thereby permitting the separation of the two replication products. This reaction, no doubt, is catalyzed by one or more topoisomerases such as those discussed in Section 28-5C, although they have not yet been unambiguously identified.

D. Fidelity of Replication

Since a single polypeptide as small as the Pol I Klenow fragment can replicate DNA by itself, why does *E. coli* maintain a battery of > 20 intricately coordinated proteins

to replicate its chromosome? The answer apparently is *to ensure the nearly perfect fidelity of DNA replication required to preserve the genetic message's integrity from generation to generation.*

The rates of reversion of mutant *E. coli* or T4 phage to the wild-type indicates that only one mispairing occurs per 10^8 to 10^{10} base pairs replicated. This corresponds to ~1 error per 1000 bacteria per generation. Such high replication accuracy arises from four sources:

1. Cells maintain balanced levels of dNTPs through the mechanism discussed Section 26-4A. This is an important aspect of replication fidelity because a dNTP present at aberrantly high levels is more likely to be misincorporated and, conversely, one present at low levels is more likely to be replaced by the dNTPs present at higher levels.

2. The polymerase reaction itself has extraordinary fidelity. This is because, as enzymological studies indicate, the polymerase reaction occurs in two stages: (1) a binding step in which the incoming dNTP base pairs with the template while the enzyme is in an open conformation that cannot catalyze the polymerase reaction; and (2) a catalysis step in which the polymerase has formed a closed conformation about the newly formed base pair, which properly positions its catalytic residues (induced fit). Since the formation of the closed conformation requires that the incoming dNTP be Watson–Crick base paired to the template, the conformation change constitutes a double check for correct base pairing reminiscent of the kinetic proof reading mechanism in ribosomes (Section 30-3F).

3. The $3' \rightarrow 5'$ exonuclease functions of Pol I and Pol III detect and eliminate the occasional errors made by their polymerase functions. In fact, mutations that increase a DNA polymerase's proofreading exonuclease activity decrease the rates of mutation of other genes.

4. A remarkable battery of enzyme systems, contained in all cells, function to repair residual errors in the newly synthesized DNA as well as any damage that it may incur after its synthesis through chemical and/or physical insults. We discuss these DNA repair systems in Section 31-5.

In addition, *the inability of a DNA polymerase to initiate chain elongation without a primer is a feature that increases DNA replication fidelity.* The first few nucleotides of a chain to be coupled together are those most likely to be mispaired because of the cooperative nature of base pairing interactions (Section 28-3). The editing of a short duplex oligonucleotide is similarly an error-prone process. The use of RNA primers eliminates this source of error since the RNA is eventually replaced by DNA under conditions that permit accurate base pairing to be achieved.

One might wonder why cells have evolved the complex system of discontinuous lagging strand synthesis rather

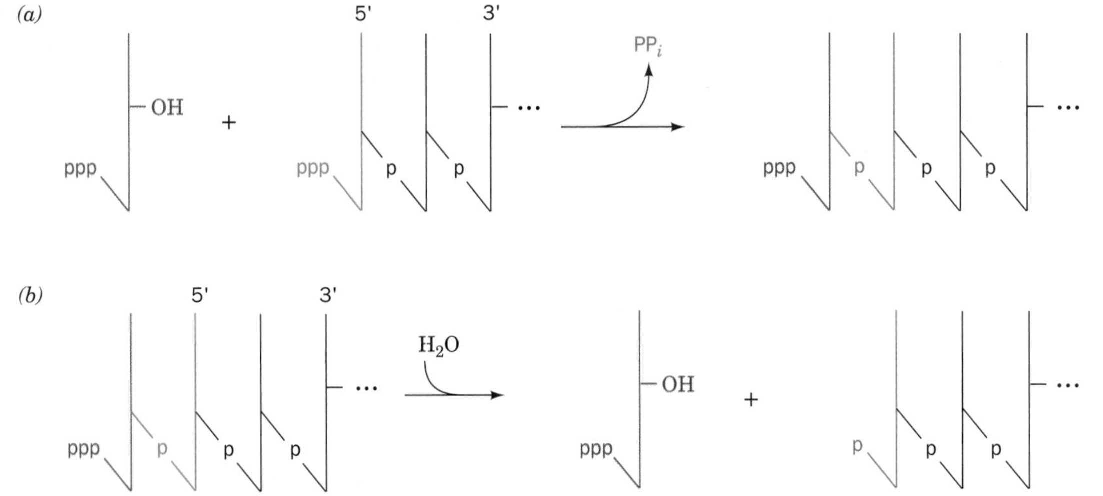

FIGURE 31-27. If a DNA polymerase could synthesize DNA in its $3' \rightarrow 5'$ direction: (*a*) The coupling of each nucleoside triphosphate to the growing chain would be driven by the hydrolysis of the previously appended nucleoside triphosphate.

(*b*) The editorial removal of an incorrect $5'$-terminal nucleoside triphosphate would render the DNA chain incapable of further extension.

than a DNA polymerase that could simply extend DNA chains in their $3' \rightarrow 5'$ direction. Consideration of the chemistry of DNA chain extension also leads to the conclusion that this system promotes high-fidelity replication. The linking of $5'$-deoxynucleotide triphosphates in the $3' \rightarrow 5'$ direction would require the retention of the growing chain's $5'$-terminal triphosphate group to drive the next coupling step (Fig. 31-27*a*). Upon editing a mispaired $5'$-terminal nucleotide (Fig. 31-27*b*), this putative polymerase would—in analogy with Pol I, for example—excise the offending nucleotide, leaving either a $5'$-OH or a $5'$-phosphate group. Neither of these terminal groups is capable of energizing further chain extension. A proofreading $3' \rightarrow 5'$ DNA polymerase would therefore have to be capable of reactivating its edited product. The inherent complexity of such a system has presumably selected against its evolution.

4. EUKARYOTIC DNA REPLICATION

It is becoming increasingly evident that *there is a remarkable degree of similarity between eukaryotic and prokaryotic DNA replication mechanisms.* There are, nevertheless, distinct differences between these two replication systems as a consequence of the vastly greater complexity of eukaryotes in comparison to prokaryotes. We consider these differences in this section. We also discuss two DNA polymerases that are peculiar to eukaryotic systems; reverse transcriptase and **telomerase.**

A. The Cell Cycle

The **cell cycle,** the general sequence of events that occur during the lifetime of a eukaryotic cell, is divided into four distinct phases (Fig. 31-28):

1. Mitosis and cell division occur during the relatively brief **M phase** (for mitosis).

2. This is followed by the G_1 **phase** (for gap), which covers the longest part of the cell cycle.

3. G_1 gives way to the **S phase** (for synthesis) which, in contrast to events in prokaryotes, *is the only period in the cell cycle when DNA is synthesized.*

4. During the relatively short G_2 **phase,** the now tetraploid cell prepares for mitosis. It then enters M phase once again and thereby commences a new round of the cell cycle.

The cell cycle for cells in culture typically occupies a 16- to 24-h period. In contrast, cell cycle times for the different types of cells of a multicellular organism may vary from as little as 8 h to > 100 days. Most of this variation occurs in the G_1 phase. Moreover, many terminally differentiated cells, such as neurons or muscle cells, never divide; they assume a quiescent state known as the G_0 **phase.**

A cell's irreversible "decision" to proliferate is made during G_1. Quiescence is maintained if, for example, nutrients are in short supply or the cell is in contact with other cells **(contact inhibition).** Conversely, DNA synthesis may be induced by various agents such as carcinogens or tumor viruses, which trigger uncontrolled cell proliferation (cancer; Section 33-4C); by the surgical removal of a tissue, which results in its rapid regeneration; or by proteins known as **mitogens,** which bind to cell surface receptors and induce cell division (Section 33-4C). Growing cells contain cytoplasmic factors that stimulate DNA replication. For example, extracts of frog eggs induce DNA synthesis in frog spleen cells. The mode of action of these factors, however, is unknown (although see below).

B. Eukaryotic DNA Polymerases

Animal cells contain at least five distinct types of DNA polymerases, designated, in the order of their discovery, **DNA polymerases** α, β, γ, δ, and ε (Table 31-5). Their functions were largely elucidated by their different responses to inhibitors (Table 31-5) because mutant forms of these enzymes have not, until recently, been available.

DNA polymerase α, *which occurs only in the cell nucleus, participates in the replication of chromosomal DNA.* This function was largely established through the use of its specific inhibitor **aphidicolin**

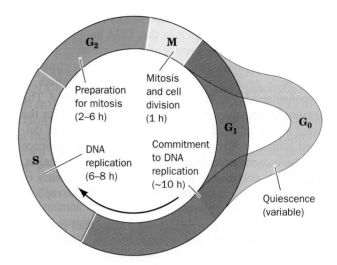

Aphidicolin

and by the observation that DNA polymerase α activity varies with the rate of cellular proliferation. This multisubunit protein (four types of subunit in *Drosophila*; five in rat liver), as do all DNA polymerases, replicates DNA by extending a primer in the $5' \rightarrow 3'$ direction under the direction of a single-stranded DNA template. DNA polymerase α has a tightly associated primase activity. However, it lacks exonuclease activity so that the DNA it replicates must be proofread by some other means.

DNA polymerase δ, a nuclear enzyme with inhibitor sensitivities similar to those of DNA polymerase α (Table 31-5), lacks an associated primase but exhibits a proofreading $3' \rightarrow 5'$ exonuclease activity. Moreover, whereas DNA

FIGURE 31-28. The eukaryotic cell cycle. Cells in G_1 may enter a quiescent phase (G_0) rather than continuing about the cycle.

polymerase α exhibits only moderate processivity (~100 nucleotides), that of DNA polymerase δ is essentially unlimited (replicates the entire length of a template), but only when it is in complex with a protein known as **proliferating cell nuclear antigen (PCNA**; so named because it occurs only in the nuclei of proliferating cells and reacts with the sera from a subset of patients with the autoimmune disease systemic lupus erythrematosus). *It is therefore likely that a complex between DNA polymerase* δ *and PCNA is the eukaryotic leading strand replicase (which requires high processivity but only occasional need of a primer), whereas DNA polymerase* α *is the lagging strand replicase (which requires frequent priming but a processivity of only 100–200 nucleotides).* The amino acid sequence of the 261-residue PCNA can be weakly aligned with that of the 366-residue β subunit of *E. coli* Pol III holoenzyme (Section 31-2B). This suggests that PCNA forms a trimeric, rather than a dimeric, ring about DNA similar to that postulated for the β subunit (Fig. 31-14) in which each PCNA subunit consists of 4 rather than 6 of the structurally similar $\beta\alpha\beta\beta$ motifs of which β subunit is comprised.

TABLE 31-5. PROPERTIES OF ANIMAL DNA POLYMERASES

	α	β	γ	δ	ϵ
Location	nucleus	nucleus	mitochondrion	nucleus	nucleus
Mass (kD)	>250	36–38	160–300	170	256
Inhibitors:					
Aphidicolin	strong	none	none	strong	strong
Dideoxy NTPs	none	strong	strong	weak	weak
N-Ethylmaleimide (NEM)[a]	strong	none	strong	strong	strong

[a] A cysteine reagent (Table 6-3).

Source: Kornberg, A. and Baker, T.A., *DNA Replication* (2nd ed.), *p.* 199, Freeman (1992).

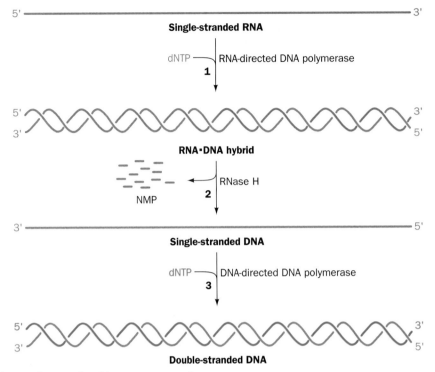

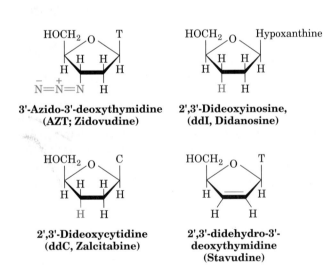

FIGURE 31-32.   The reactions catalyzed by reverse transcriptase.

dine), 2',3'-dideoxyinosine (ddI; didanosine), 2',3'-dideoxycytosine (ddC; zalcitabine), and 2',3'-didehydro-3'-deoxythymidine (stavudine),

3'-Azido-3'-deoxythymidine (AZT; Zidovudine)

2',3'-Dideoxyinosine, (ddI, Didanosine)

2',3'-Dideoxycytidine (ddC, Zalcitabine)

2',3'-didehydro-3'-deoxythymidine (Stavudine)

are RT inhibitors. Unfortunately, resistant strains of HIV-1 arise quite rapidly because RT lacks a proofreading exonuclease function and hence is highly error prone. Indeed, it is HIV's capacity to rapidly evolve, even within a single patient, that has heretofore prevented the formulation of an effective long-term anti-HIV therapy.

The X-ray structure of RT complexed to an 18-bp DNA with a 1-nt overhang at the 5' end of one strand has been determined by Edward Arnold (Fig. 31-33). This complex also contains a monoclonal **Fab fragment** (the antigen-binding segment of an immunoglobulin; Section 34-2B) that specifically binds to RT and presumably facilitates the crystallization of the complex. Steitz has independently determined the X-ray structure of RT in the absence of DNA.

The two RT structures are closely similar, although there appear to be shifts in some secondary structural elements, particularly those that contact the DNA and the Fab fragment. The polymerase domains of p66 and p51 each contain four subdomains, which, because of their collective resemblance in p66 to KF, are named, from N- to C-terminus, "fingers," "palm," "thumb," and "connection." In p66, the RNase H domain follows the connection.

p51 has undergone a remarkable conformational change relative to p66: The connection has rotated by 155° and translated by 17 Å to bring it from a position in p66 in which it contacts the RNase H domain (Fig. 31-33a), to one in p51 in which it contacts all three other polymerase subdomains (Fig. 31-33b). This permits p66 and p51 to bring different surfaces of their connections into juxtaposition to form, in part, RT's DNA-binding groove. Thus, the chemically identical polymerase domains of p66 and p51 are not related by twofold molecular symmetry (a rare but not unprecedented phenomenon), but, rather, associate in a sort of head-to-tail arrangement. Consequently, RT has only one active site. This is an example of viral genetic economy: HIV-1, with its limited genome size, has succeeded in using a single polypeptide for what are essentially two different functions.

The DNA assumes a conformation that, near the polym-

(a)

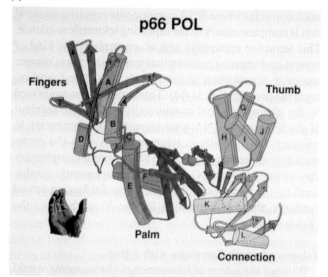

(b)

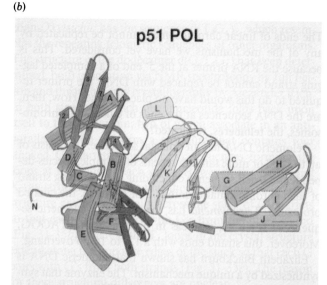

(c)

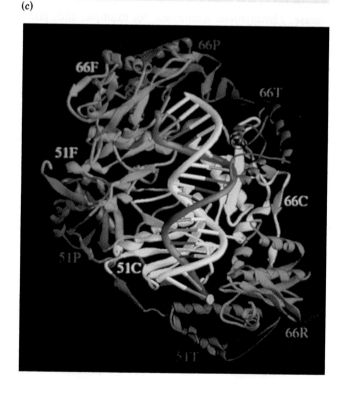

FIGURE 31-33. The X-ray structure of HIV-1 reverse transcriptase. (*a*) A tube-and-arrow representation of the p66 subunit's polymerase domain in which the N-terminal finger subdomain is light blue, the palm is pink, the thumb is green, and the connection is yellow. The RNase H domain (not shown) follows the connection. (*b*) The p51 subunit with its pink palm subunit oriented identically to that in p66. Note the different relative orientations of the four subdomains in the two subunits. The G helix is shown in dashed outline because its electron density is weak and ambiguous. (*c*) A ribbon diagram of the HIV-1 RT p66/p51 heterodimer in complex with DNA. The subdomains of p66 and p51 are colored as in Parts *a* and *b* and the RNase H subdomain of p66 is orange [the labels indicate subunit and (sub)domain; e.g., 51F and 66R denote the p51 finger subdomain and the p66 RNase H domain]. The DNA is shown in ladder representation with the 18-nt primer strand white and the 19-nt template strand blue. The complex is oriented with its p66 polymerase domain towards the top of the figure and viewed from above the protein's template–primer binding cleft (whose floor is largely comprised of the connection subdomains of p66 and p51) so as to show the bend in the DNA. [Parts *a* and *b* courtesy of Thomas Steitz, Yale University. Part *c* courtesy of Edward Arnold, Rutgers University.]

erase active site, resembles A-DNA but, near the RNase H domain, more closely resembles B-DNA (Fig. 31-33*c*). The A- and B-form regions of the DNA are separated by a 40-45° bend. Because RT must also bind DNA–RNA hybrids, which have an A-DNA-like conformation (Section 28-2B), it is not surprising that RT also induces this conformation in duplex DNA. The 3'-OH group at the end of the 18-nt DNA strand, the so-called primer strand, is near p66's three catalytically essential Asp side chains, where it is properly positioned to nucleophilically attack the α phosphate of an incoming dNTP that had been shown to bind near this site.

Most of the protein–DNA interactions involve the DNA's sugar–phosphate backbone and the residues of p66's palm, thumb, and fingers.

The RT active site region contains the few sequence motifs that are conserved among the various polymerases. Indeed, this region of p66 has a striking structural resemblance to KF (Fig. 31-10), T7 RNA polymerase (Fig. 29-9*b*), and even DNA polymerase β (Fig. 31-29; which, recall, has a different folding topology from these other polymerases). This suggests that other polymerases are likely to bind DNA in a similar manner.

5. REPAIR OF DNA

DNA is by no means the inert substance that might be supposed from naive consideration of the genome's stability. Rather, the reactive environment of the cell, the presence of a variety of toxic substances, and exposure to UV or ionizing radiation subjects it to numerous chemical insults that excise or modify bases and alter sugar–phosphate groups. Indeed, some of these reactions occur at surprisingly high rates. For example, under normal physiological conditions, the glycosidic bonds of some 10,000 purine nucleotides in the genome of each human cell hydrolyze spontaneously each day.

Any DNA damage must be repaired if the genetic message is to maintain its integrity. Such repair is possible because of duplex DNA's inherent information redundancy. The biological importance of **DNA repair** is indicated by the great variety of such pathways possessed by even relatively simple organisms such as *E. coli*. In fact, *the major DNA repair processes in E. coli and mammalian cells are chemically quite similar.* These processes are outlined in this section.

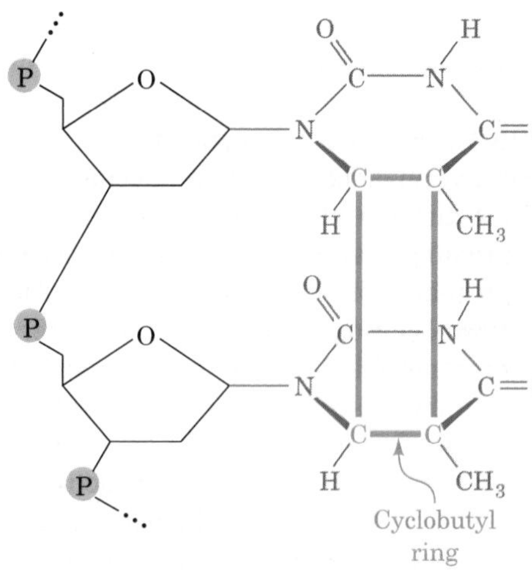

A. Direct Reversal of Damage

Pyrimidine Dimers Are Split by Photolyase

UV radiation (200–300 nm) promotes the formation of a cyclobutyl ring between adjacent thymine residues on the same DNA strand to form an intrastrand **thymine dimer** (Fig. 31-36). Similar cytosine and thymine–cytosine dimers are likewise formed but at lesser rates. Such **pyrimidine dimers** locally distort DNA's base paired structure so that it can form neither a proper transcriptional nor replicational template.

Pyrimidine dimers may be restored to their monomeric forms through the action of light-absorbing enzymes present in all life forms named **photoreactivating enzymes** or **DNA photolyases.** These enzymes are 55- to 65-kD monomers that bind to a pyrimidine dimer in DNA, a process that can occur in the dark. A noncovalently bound chromophore, in some species an N^5, N^{10}-methenyltetrahydrofolate (Fig. 24-39) and in others a **5-deazaflavin,**

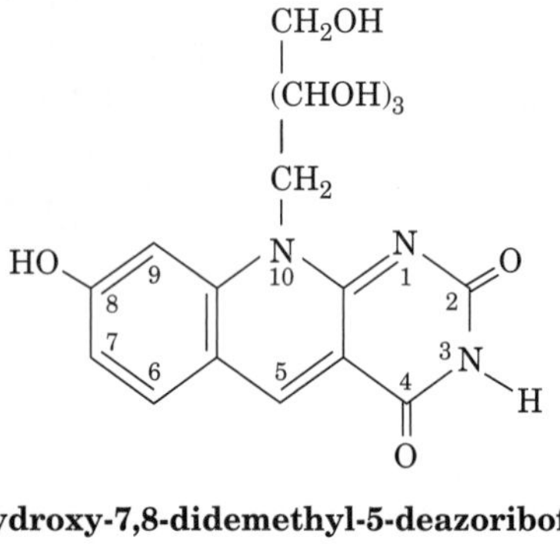

CH$_2$OH
|
(CHOH)$_3$
|
CH$_2$

8-Hydroxy-7,8-didemethyl-5-deazoriboflavin

then absorbs 300- to 500-nm light and transfers the excitation energy to a noncovalently bound FADH$^-$, which in

FIGURE 31-36. The cyclobutylthymine dimer that forms upon UV irradiation of two adjacent thymine residues on a DNA strand. The ~1.6-Å-long covalent bonds joining the thymine rings (*red*) are much shorter than the normal 3.4-Å spacing between stacked rings in B-DNA, thereby locally distorting the DNA.

turn transfers an electron to the pyrimidine dimer, thereby splitting it. Finally, the resulting pyrimidine anion re-reduces the FADH· and the now unblemished DNA is released, thereby completing the catalytic cycle.

Alkyltransferases Dealkylate Alkylated Nucleotides

The exposure of DNA to alkylating agents such as *N*-methyl-*N'*-nitro-*N*-nitrosoguanidine (MNNG)

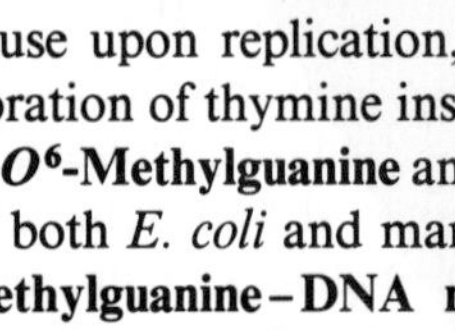

N-**Methyl-***N'***-nitro-***N*-
nitrosoguanidine (MNNG)

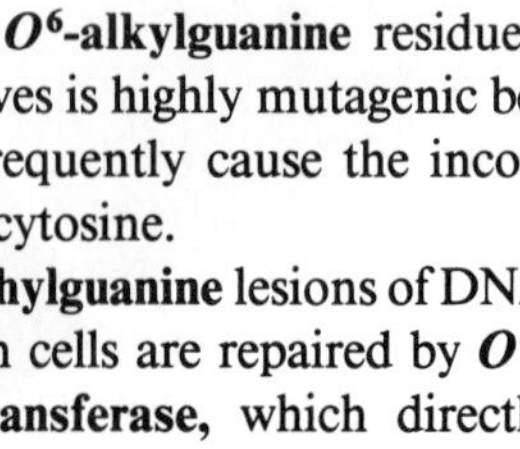

O^6-**Methylguanine residue**

yields, among other products, O^6-**alkylguanine** residues. The formation of these derivatives is highly mutagenic because upon replication, they frequently cause the incorporation of thymine instead of cytosine.

O^6-**Methylguanine** and O^6-**ethylguanine** lesions of DNA in both *E. coli* and mammalian cells are repaired by O^6-**methylguanine–DNA methyltransferase,** which directly transfers the offending alkyl group to one of its own Cys residues. The reaction inactivates this protein, which therefore cannot be strictly classified as an enzyme. The alkyltransferase reaction has elicited considerable attention because carcinogenesis induced by methylating and ethylating agents is correlated with deficient repair of O^6-alkylguanine lesions.

(a)

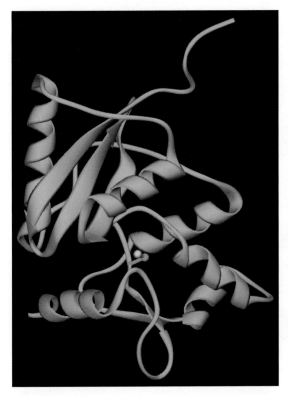

(b)

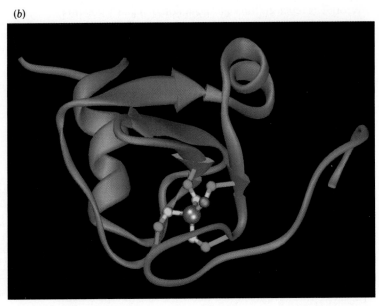

FIGURE 31-37. The structure of *E. coli* Ada protein. (*a*) The X-ray structure of Ada's 178-residue C-terminal segment, which contains its O^6-methylguanine–DNA methyltransferase function. The side chain of Cys 146 (Cys 321 in the intact protein), to which the methyl group is irreversibly transferred, is shown in ball-and-stick form with C green and S yellow. Note that this residue is almost entirely buried within the protein. [Based on an X-ray structure determined by Eleanor Dodson and Peter Moody, University of York, York, U.K.] (*b*) The NMR structure of Ada's 92-residue, N-terminal segment, which mediates its methyl phosphotriester repair function. The protein's bound Zn^{2+} ion is represented by a gray sphere and its four tetrahedrally coordinating Cys side chains are shown in ball-and-stick form, with C green and S yellow except for the orange S atom of Cys 69, which becomes irreversibly methylated when the protein encounters a methylated phosphate group on DNA. [Based on an NMR structure determined by Gregory Verdine and Gerhard Wagner, Harvard University.]

The *E. coli* O^6-methylguanine–DNA methyltransferase activity occurs on the 178-residue C-terminal segment of the 354-residue **Ada protein** (the product of the ***ada*** gene). Its X-ray structure (Fig. 31-37*a*), determined by Eleanor Dodson and Peter Moody, reveals, unexpectedly, that its active site Cys residue, Cys 321, is buried inside the protein. Apparently, the protein must undergo a significant conformation change upon DNA binding in order to effect the methyl transfer reaction.

Ada protein's 92-residue N-terminal segment has an independent function: It repairs methyl phosphotriesters in DNA (methylated phosphate groups) by irreversibly transferring the offending methyl group to its Cys 69. The NMR structure of Ada's N-terminal domain (Fig. 31-37*b*), determined by Gregory Verdine and Gerhard Wagner, reveals that Cys 69, together with three other Cys residues, tetrahedrally coordinates a Zn^{2+} ion. This presumably stabilizes

the thiolate form of Cys 69 over its thiol form, thereby facilitating its nucleophilic attack on the methyl group.

Intact Ada protein that is methylated at its Cys 69 binds to a specific DNA sequence, which is located upstream of the *ada* gene and several other genes encoding DNA repair proteins, thereby inducing their transcription. Evidently, Ada also functions as a chemosensor of methylation damage.

B. Nucleotide Excision Repair

Pyrimidine dimers may also be mended by a process known as **nucleotide excision repair (NER).** *In such repair pathways, an oligonucleotide containing the lesion is excised from the DNA and the resulting single-strand gap filled in.* In *E. coli,* pyrimidine dimers are recognized by a multisubunit enzyme, the product of the ***uvrA, uvrB,*** and ***uvrC*** genes. This **UvrABC endonuclease,** in an ATP-dependent reaction, cleaves the dimer-containing DNA strand at the seventh and fourth phosphodiester bonds on the dimer's 5' and 3' sides, respectively (Fig. 31-38). The excised oligonucleotide is replaced through the action of a DNA polymerase, most probably Pol I, followed by that of DNA ligase.

UvrABC endonuclease excises other types of DNA lesions besides pyrimidine dimers. These lesions are characterized by the displacement of bases from their normal positions, as with pyrimidine dimers, or by the addition of a bulky substituent to a base. Evidently, UvrABC endonuclease is activated by a helix distortion rather than by the recognition of any particular group.

Xeroderma Pigmentosum and Cockayne Syndrome Are Caused by Genetically Defective NER

In humans, the rare inherited disease **xeroderma pigmentosum** (**XP;** Greek: *xeros,* dry + *derma,* skin) is mainly characterized by the inability of skin cells to repair UV-induced DNA lesions. Individuals suffering from this autosomal recessive condition are extremely sensitive to sunlight. During infancy they develop marked skin changes such as dryness, excessive freckling, and keratoses (a type of skin tumor; the skin of these children is described as resembling that of farmers with many years of sun exposure), together with eye damage, such as opacification and ulceration of the cornea. Moreover, they develop often fatal skin cancers at a 2000-fold greater rate than normal. Curiously, many individuals with XP also have a bewildering variety of seemingly unrelated symptoms including progressive neurological degeneration and developmental deficits.

Cultured skin fibroblasts from individuals with xeroderma pigmentosum are defective in the NER of pyrimidine dimers. Cell-fusion experiments with cultured cells taken from various patients have demonstrated that this disease results from defects in any of 7 complementation groups, indicating that there must be at least 7 gene products, XP-A through XP-G, involved in this clearly important UV damage repair pathway. **Cockayne syndrome (CS),** an inherited disease, which is likewise associated with defective NER, arises from defects in XP-B, XP-D, and XP-G. as well as in two additional complementation groups that are distinct from those of XP. Individuals with CS are hypersensitive to UV radiation and exhibit stunted growth as well as neurological dysfunction due to neuron demyelination, but, intriguingly, have a normal incidence of skin cancer. What is the biochemical basis for the diverse group of symptoms associated with impaired NER?

The retarded development typical of XP-B and perhaps the demyelination that occurs in CS appear to be due more to impaired transcription than to defective NER. Moreover, pyrimidine dimers are more efficiently removed from transcribed genes than from unexpressed sequences. These observations are explained by the discovery that the gene implicated in the XP-B complementation group, *ERCC3,*

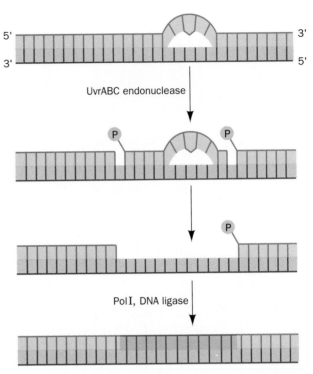

FIGURE 31-38. The mechanism of nucleotide excision repair (NER) of pyrimidine photodimers.

encodes a subunit of the human transcription factor **TFIIH,** a helicase that participates in the initiation of mRNA transcription by RNA polymerase II (Section 33-3B).

Glycosylases Remove Altered Bases

DNA bases are modified by reactions that occur under normal physiological conditions as well as through the action of environmental agents. For example, adenine and cytosine residues spontaneously deaminate at finite rates to yield hypoxanthine and uracil residues, respectively. *S*-Adenosylmethionine (SAM), a common metabolic methylating agent, occasionally nonenzymatically methylates a base to form derivatives such as 3-methyladenine and 7-methylguanine residues. Ionizing radiation can promote ring opening reactions in bases. Such changes modify or eliminate base pairing properties.

DNA containing a damaged base may be restored to its native state through a form of excision repair. Cells contain a variety of **DNA glycosylases** that each cleave the glycosidic bond of a corresponding specific type of altered nucleotide (Fig. 31-39), thereby leaving a deoxyribose residue in the backbone. Such **apurinic** or **apyrimidinic (AP)** sites are also generated under normal physiological conditions by the spontaneous hydrolysis of a glycosidic bond. The deoxyribose residue is then cleaved on one side by an **AP endonuclease,** the deoxyribose and several adjacent residues are removed by the action of DNA polymerase or

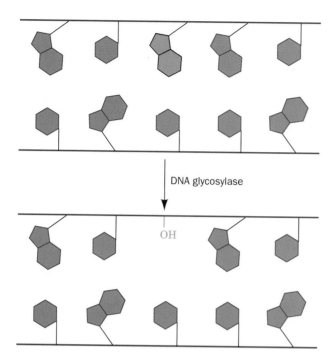

FIGURE 31-39. DNA glycosylases hydrolyze the glycosidic bond of their corresponding altered base (*red*) to yield an AP site.

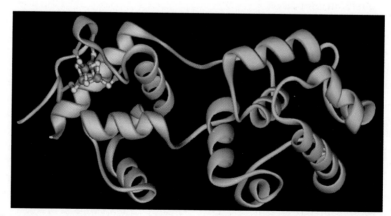

some other cellular exonuclease, and the gap is filled in and sealed by DNA polymerase and DNA ligase.

E. coli **endonuclease III,** an enzyme that participates in the NER of DNA, has two enzymatic functions: It is a DNA glycosidase that removes oxidized pyrimidines from DNA and it is an AP endonuclease that introduces a single strand nick at the resulting AP site. This 211-residue monomeric enzyme, which contains a covalently bound [4Fe–4S] cluster, is conserved in bacteria, yeast, and mammals. Its X-ray

structure (Fig. 31-40), determined by John Tainer, reveals that it consists of two domains separated by a deep cleft: a novel bundle of 6 antiparallel helices (residues 22-132), and a domain that binds the [4Fe–4S] cluster (residues 1-21 and 133-211). The observations that the surface facing the viewer in Fig. 31-40 is populated with conserved residues and has a positive electrostatic potential, whereas that of the opposite surface is negative, suggests that B-DNA binds to the former surface. The observation that endonuclease III protects a 46-Å-long segment of DNA from DNase I digestion further suggests that the DNA binds with its helix axis parallel to the long axis of the protein. The enzyme's [4Fe–4S] cluster is resistant to oxidation and reduction and exhibits no spectroscopic changes on base or oligonucleotide binding. It apparently functions structurally, probably to position the protein's conserved basic residues to interact with the DNA's phosphate backbone, rather than as a redox agent as it does in other enzymes that contain this prosthetic group (Section 20-2C).

Uracil in DNA Would Be Highly Mutagenic

For some time after the basic functions of nucleic acids had been elucidated there seemed no apparent reason for nature to go to the considerable metabolic effort of using thymine in DNA and uracil in RNA when these substances have virtually identical base pairing properties. This enigma was solved by the discovery of cytosine's penchant for conversion to uracil by deamination, either spontaneously or by reaction with nitrites (Section 30-1A). If U were a normal DNA base, the deamination of C would be highly mutagenic because there would be no indication of whether the resulting mismatched G·U base pair had initially been G·C or A·U. *Since T is DNA's normal base, however, any U in DNA is almost certainly a deaminated C.* U's that occur in DNA are efficiently excised by the DNA glycosy-

FIGURE 31-40. The X-ray structure of *E. coli* endonuclease III. The 211-residue enzyme is shown in ribbon form with the domain consisting of the 6-helix bundle in light blue and that binding the [4Fe–4S] cluster in pink. The [4Fe–4S] cluster, together with its four liganding Cys side chains, are shown in ball-and-stick form with C green, S yellow, and Fe orange. The DNA is thought to bind to the surface facing the viewer. [Based on an X-ray structure determined by John Tainer, The Scripps Research Institute, La Jolla, California.]

lase **uracil *N*-glycosylase** and then replaced by C through nucleotide excision repair.

Uracil *N*-glycosylase also has an important function in DNA replication. dUTP, an intermediate in dTTP synthesis, is present in all cells in small amounts (Section 26-4B). DNA polymerases do not discriminate well between dUTP and dTTP (recall that DNA polymerases select a base for incorporation into DNA according to its ability to base pair with the template) so that, despite the low dUTP level that cells maintain (Section 26-4B), newly synthesized DNA contains an occasional U. These U's are rapidly replaced by T through excision repair. However, since excision occurs more rapidly than repair, all newly synthesized DNA is fragmented. When Okazaki fragments were first discovered (Section 31-1C), it therefore seemed that all DNA was synthesized discontinuously. This ambiguity was resolved with the discovery of *E. coli* defective in uracil *N*-glycosylase. In these so-called ***ung*⁻** mutants, only about half of the newly synthesized DNA is fragmented, strongly suggesting that DNA's leading strand is synthesized continuously.

C. Recombination Repair

Damaged DNA may undergo replication before the lesion can be eliminated by the previously described repair systems. The replication of DNA containing a pyrimidine dimer is interrupted by this template distortion and is only reinitiated at some point past the dimer site. The resulting daughter strand has a gap opposite the pyrimidine dimer (Fig. 31-41). This genetic lesion cannot be eliminated by excision repair, which requires an intact complementary strand. Yet, such an intact strand occurs in the sister duplex that was formed at the same replication fork. The lesion can therefore be corrected through a process that is alternatively known as **recombination** or **postreplication repair** (Fig. 31-41). *This pathway exchanges the corresponding segments of sister DNA strands, thereby placing the gapped DNA segment in apposition to the undamaged strand, where the gap can be filled in and sealed.* The pyrimidine dimer, which is likewise associated with its intact complementary strand, can then be eliminated by NER or photoreactivation. Re-

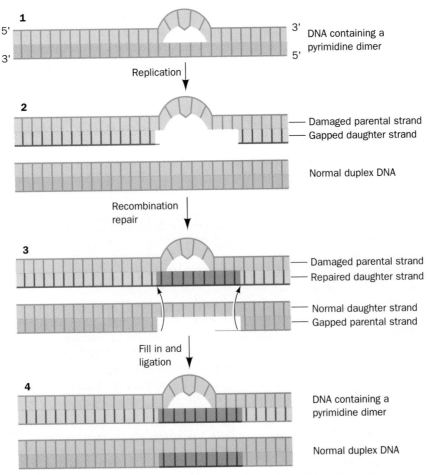

FIGURE 31-41. In recombination repair, the gap in a newly synthesized DNA strand opposite a damage site is filled by the corresponding segment from its sister duplex.

combination repair has been directly detected by the observation that segments of isotopically labeled parental DNA are transferred into daughter strands.

Recombination repair closely resembles genetic recombination. Indeed, both processes are mediated in *E. coli* by **RecA protein,** a 38-kD nuclease, which promotes sister strand exchange between homologous DNA segments. *E. coli* with a mutant *recA* gene are therefore deficient in both recombination repair (which makes them extremely sensitive to UV radiation) and genetic recombination. We consider the mechanism of recombination in Section 31-6A.

D. The SOS Response

Agents that damage DNA, such as UV radiation, alkylating agents, and cross-linking agents, induce a complex system of cellular changes in *E. coli* known as the **SOS response.** *E. coli so treated cease dividing and increase their capacity to repair damaged DNA.*

LexA Protein Represses the SOS Response

Clues as to the nature of the SOS response were provided by the observations that *E. coli* with mutant *recA* or **lexA** genes have their SOS response permanently switched on. Moreover, when wild-type *E. coli* are exposed to agents that damage DNA or inhibit DNA replication, their RecA specifically mediates the proteolytic cleavage of **LexA** protein (22 kD) at an Ala–Gly bond. RecA is activated to do so, at least *in vitro,* when it binds to single-stranded DNA (it was initially assumed that RecA directly proteolyzes LexA but subsequent experiments by John Little indicate that activated RecA stimulates LexA to cleave itself). Further genetic analysis indicated that LexA functions as a repressor of a number of operons including those of *recA* and *lexA*. Cynthia Kenyon identified these other LexA-controlled operons by inserting the *lacZ* gene (which codes for β-galactosidase; Section 29-1A) at random positions in the *E. coli* chromosome and examining the clones that had increased β-galactosidase activity in the presence of DNA-damaging agents. Altogether, 11 genes, including the excision repair genes uvrA and uvrB, lack normal function in these clones (the genes are inactivated by *lacZ* insertion). DNA sequence analyses of the LexA-repressible genes revealed that they are all preceded by a homologous 20-nt sequence, the so-called **SOS box,** that has the palindromic symmetry characteristic of operators (Section 29-3B). Indeed, LexA has been shown to directly bind the SOS boxes of *recA* and *lexA*.

The preceding information suggests a model for the regulation of the SOS response (Fig. 31-42). During normal growth, LexA largely represses SOS gene expression. When DNA damage has been sufficient to produce postreplication gaps, however, this single-stranded DNA binds to RecA so as to stimulate LexA cleavage. The LexA-repressible genes are consequently released from repression and direct the synthesis of SOS proteins including that of LexA

(although this repressor continues to be cleaved through the influence of RecA). When the DNA lesions have been eliminated, RecA ceases stimulating LexA's autoproteolysis. The newly synthesized LexA can then function as a repressor, which permits the cell to return to normality.

SOS Repair Is Error Prone

SOS repair is an error prone and therefore mutagenic process. Yet, DNA damage that normally activates the SOS response is nonmutagenic in the $recA^-$ *E. coli* that survive. This is because the intact SOS repair system will replace the bases at a DNA lesion even when there is no information as to which bases were originally present (via a poorly characterized process that involves the products of the SOS genes **umuC** and **umuD** together with RecA). The SOS repair system is therefore a testimonial to the proposition that survival with a chance of loss of function (and the possible gain of new ones) is advantageous, in the Darwinian sense, over death.

E. Identification of Carcinogens

Many forms of cancer are known to be caused by exposure to certain chemical agents that are therefore known as **carcinogens.** It has been estimated that as much as 80% of human cancer arises in this fashion. There is considerable evidence that the primary event in carcinogenesis is often damage to DNA (carcinogenesis is discussed in Section 33-4C). Carcinogens are consequently also likely to induce the SOS response in bacteria and thus act as indirect mutagenic agents. In fact, there is a high correlation between carcinogenesis and mutagenesis (recall, e.g., the progress of xeroderma pigmentosum; Section 31-5B).

There are presently over 60,000 man-made chemicals of commercial importance and ~1000 new ones are introduced each year. The standard animal tests for carcinogenesis, exposing rats or mice to high levels of the suspected carcinogen and checking for cancer, are expensive and require ~3 years to complete. Thus relatively few substances have been tested in this manner.

The Ames Test Assays for Probable Carcinogenicity

Bruce Ames devised a rapid and effective bacterial assay for carcinogenicity that is based on the high correlation between carcinogenesis and mutagenesis. He constructed special tester strains of *Salmonella typhimurium* that are **his⁻** (cannot synthesize histidine so that they are unable to grow in its absence), have cell envelopes that lack the lipopolysaccharide coating which renders normal *Salmonella* impermeable to many substances (Section 10-3B), and have an inactivated excision repair system. Mutagenesis in these tester strains is indicated by their reversion to the *his⁺* phenotype.

In the **Ames test,** $\sim 10^9$ tester strain bacteria are spread on a culture plate that lacks histidine. Usually a mixture of several *his⁻* strains is used so that both point and frameshift

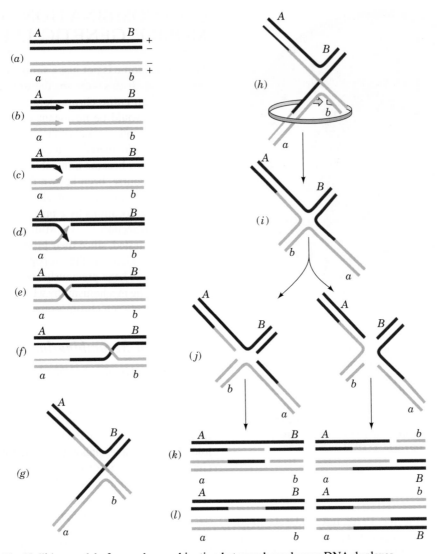

FIGURE 31-44.   The Holliday model of general recombination between homologous DNA duplexes.

nick sealing, the traditional recombinant DNA molecule (right branch of Fig. 31-44*j–l*).

2. The cleavage of the strands that crossed over exchanges a pair of homologous single-stranded segments (left branch of Fig. 34-44*j–l*).

The recombination of circular duplex DNAs results in the types of structures diagrammed in Fig. 31-46. Electron microscopic evidence for the existence of the postulated "figure-8" structures are shown in Fig. 31-47*a*. These figure-8 structures were shown not to be just twisted circles by cutting them with a restriction enzyme to yield **chi structures** (after their resemblance to the Greek letter χ) such as that pictured in Fig. 31-47*b*.

General Recombination in *E. coli* Is Catalyzed by RecA

The observation that *recA⁻ E. coli* have a 10⁴-fold lower recombination rate than the wild-type indicates that *RecA*

protein has an important function in recombination. Indeed, RecA greatly increases the rate at which complementary strands renature *in vitro*. This versatile 352-residue protein (recall it also stimulates the autoproteolysis of LexA to trigger the SOS response; Section 31-5D) polymerizes cooperatively without regard to base sequence on single-stranded DNA or on duplex DNA that has a single-stranded gap. The resulting filaments specifically bind the homologous duplex DNA, and, in an ATP-dependent reaction, catalyze strand exchange. Electron microscopy (EM; Figure 31-48) reveals that RecA filaments bound to single- or double-stranded DNA form a right handed helix with ~6.2 RecA monomers per turn and a pitch (rise per turn) of 95 Å. The DNA in these filaments, which binds to the protein with 3 nt (or bp) per RecA monomer unit and hence has ~18.6 nt (or bp) per turn, is so extended (having a rise of 5.1 Å/bp vs 3.4 Å/bp in B-DNA) that it must lie near the center of the helical filament as Fig. 31-48 indicates.

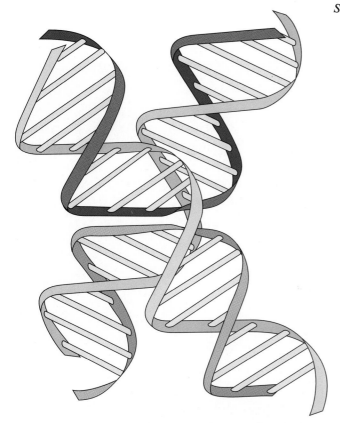

FIGURE 31-45. A proposed model of the Holliday junction. The most energetically favorable conformation, which does not violate stereochemical principles, appears to be that in which the exchanging strands enter and leave the junction on the same side of the X. [After Murchie, A.I.H., Clegg, R.M., von Kitzing, E., Duckett, D.R., Diekman, S., and Lilley, D.M.J., *Nature* **341**, 765 (1989).]

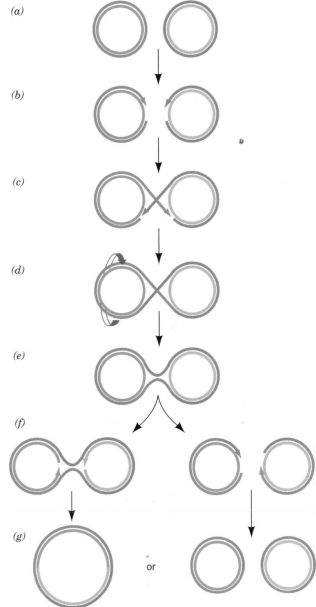

FIGURE 31-46. General recombination between two circular DNA duplexes. This process can result in the production of either two circles of the original sizes or in a single composite circle.

(a)

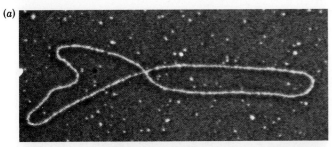

(b)

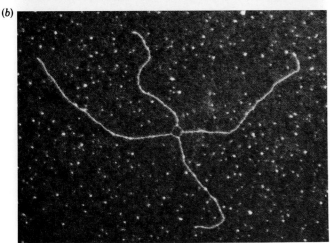

FIGURE 31-47. Electron micrographs of intermediates in the general recombination of two plasmids. (*a*) A figure-8 structure. This corresponds to Fig. 31-46*d*. (*b*) A chi structure that results from the treatment of a figure-8 structure with a restriction endonuclease. Note the single-stranded connections in the crossover region. [Courtesy of David Dressler and Huntington Potter, Harvard Medical School.]

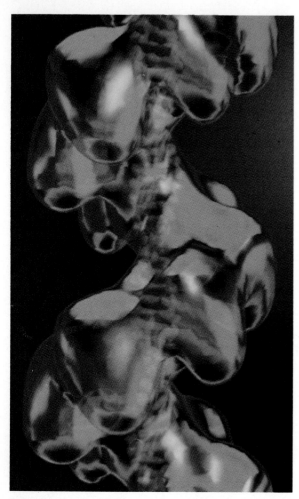

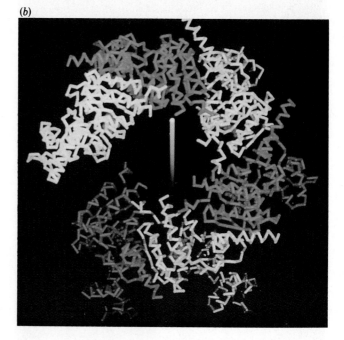

(a)

(b)

FIGURE 31-48.   An electron microscopy–based image (*transparent surface*) of an *E. coli* RecA–DNA–ADP filament. Its extended and untwisted duplex DNA (*red*) has been modeled into this image. [Courtesy of Edward Egelman, University of Minnesota Medical School.]

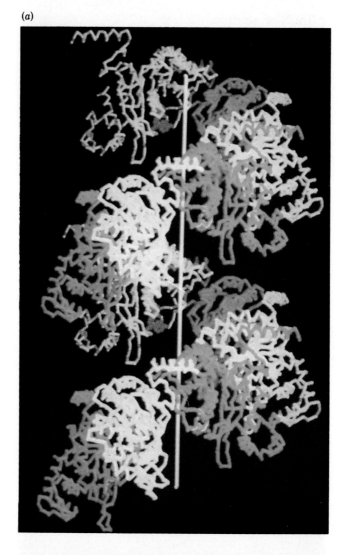

FIGURE 31-49.   The X-ray structure of *E. coli* RecA protein, in which the RecA monomers are represented by their C_α chains. Alternate RecA monomers are yellow and blue, and their bound ADPs are red. (*a*) A view perpendicular to the protein filament's helix axis (*light blue rod*) showing 12 monomers constituting two turns of the helix in the same orientation as in Fig. 31-48. (*b*) View nearly parallel to the helix axis showing one turn of the helix. [Courtesy of Thomas Steitz, Yale University.]

The X-ray structure of RecA (Fig. 31-49), determined by Thomas Steitz, reveals that the protein consists of a major central domain which is flanked by smaller N- and C-terminal domains. The monomers associate to form a ~120-Å-wide helical filament with 6 monomer units per turn and a pitch of 82.7 Å. The helical filament is remarkably open, so much so that there are gaps between the monomer units in successive turns. This arrangement results in a large helical groove running the length of the filament that, when viewed down the helix axis, forms a 25-Å-wide central hole. This helical filament is strikingly similar to that of a RecA–duplex DNA filament as visualized at lower resolution by electron microscopy (Fig 31-48; EM studies have also shown that LexA protein binds within this helical groove such that it spans the two RecA subunits on successive turns of the helix).

Comparison of the sequences of 16 bacterial RecA proteins indicates that 105 of their residues (~30%) are invariant. Not surprisingly, these residues cluster on the inner surface of the filament's helical groove where they presumably contact the DNA that normally binds there. Two loops comprising residues 157 to 164 and 195 to 209, that are not visualized in the X-ray structure and therefore presumably disordered, are even more highly conserved (44% invariant) than the rest of the protein. These loops are located close to the filament axis, and are therefore implicated in DNA binding.

How does RecA mediate DNA strand exchange between single-stranded and duplex DNAs? Upon encountering a duplex DNA with a strand that is complementary to its bound single-stranded DNA, RecA partially unwinds the duplex and, in a reaction driven by RecA-catalyzed ATP hydrolysis, exchanges the single-stranded DNA with the corresponding strand on the duplex. A model of how RecA might do so is diagrammed in Fig. 31-50a. *This process tolerates only a limited degree of mispairing and requires that one of the participating DNA strands have a free end.* The assimilation (exchange) of a single-stranded circle with a strand on a linear duplex (Fig. 31-51) cannot proceed past the 3′ end of a highly mismatched segment in the complementary strand. *The invasion of the single strand must therefore begin with its 5′ end.* A model for the consequent branch migration process is diagrammed in Fig. 31-52a. Of course, two such strand exchange processes must simulta-

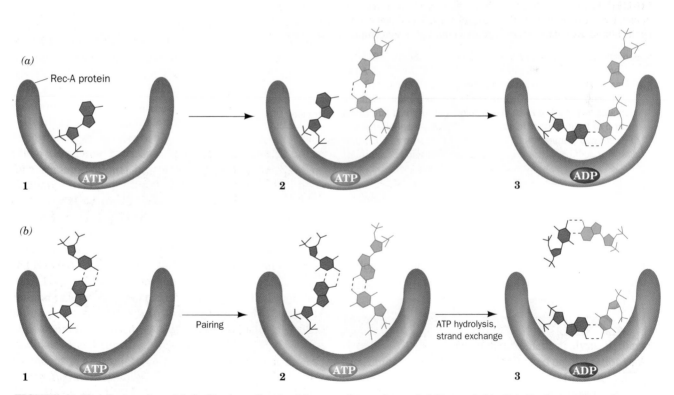

FIGURE 31-50. Proposed models for RecA-mediated pairing and strand exchange between (*a*) a single-stranded and a duplex DNA and (*b*) two duplex DNAs: (1) A single-stranded (duplex in *b*) DNA binds to RecA to form an initiation complex; (2) duplex DNA binds to the initiation complex so as to transiently form a 3-stranded (4-stranded in *b*) helix that mediates the correct pairing of the homologous strands; and (3) RecA rotates the bases of the aligned homologous strands to effect strand exchange in an ATP-driven process. [After West, S.C., *Annu. Rev. Biochem.* **61,** 618 (1992).]

3. **Spinal and bulbar muscular atrophy (Kennedy's disease)** is an X-linked adult onset form of motor neuron disease associated with **androgen** (male sex hormone) insensitivity. It is caused by a polymorphic (CAG)$_n$ repeat in the **androgen receptor** gene although the disease is not associated with genetic anticipation.

4. **Spinocerebellar ataxia type 1** is a progressive neurodegenerative disease whose age of onset is typically in the third or fourth decade, although it exhibits genetic anticipation. Like DM, it is caused by selective neuronal loss and is associated with an an expansion of a CAG repeat, in this case from ~28 to between 43 and 81 copies

Since 13 other known genes have at least six-fold triplet repeats, it seems likely that other genetic diseases will be found to arise from the expansion of such repeats.

CHAPTER SUMMARY

DNA is replicated in the $5' \rightarrow 3'$ direction by the assembly of deoxynucleoside triphosphates on complementary DNA templates. Replication is initiated by the generation of short RNA primers, as mediated in *E. coli* by primase and RNA polymerase. The DNA is then extended from the $3'$ ends of the primers through the action of DNA polymerase (Pol III in *E. coli*). The leading strand at a replication fork is synthesized essentially continuously, whereas the lagging strand is synthesized discontinuously by the formation of Okazaki fragments. RNA primers on newly synthesized DNA are excised and replaced by DNA through Pol I-catalyzed (in *E. coli*) nick translation. The single-strand nicks are then sealed by DNA ligase. Mispairing errors during DNA synthesis are corrected by the $3' \rightarrow 5'$ exonuclease functions of both Pol I and Pol III. DNA synthesis in *E. coli* requires the participation of many auxilliary proteins including DnaB protein, SSB, and DNA gyrase.

DNA synthesis commences from specific sites known as replication origins. In the synthesis of the bacteriophage M13 (−) strand on the (+) strand template, the origin is recognized and primer synthesis is initiated by RNA polymerase. The analogous process in bacteriophage ϕX174, as well as in *E. coli*, is mediated by a complex primase-containing particle known as the primosome. ϕX174 (+) strands are synthesized according to the looped rolling circle mode of DNA replication on (−) strand templates of the replicative form in a process that is directed by the virus-specific gene *A* protein. The *E. coli* chromosome is bidirectionally replicated in the θ mode from a single origin, *oriC*, which is recognized by DnaA protein. Leading strand synthesis is probably primed by RNA polymerase and primase working together, whereas Okazaki fragments are primed by primase in the primosome. Replication termination is facilitated by Tus protein, which upon binding to an appropriately oriented *Ter* site, arrests the motion of a replication fork by binding to DnaB helicase. The great complexity of the DNA replication process apparently ensures the enormous fidelity necessary to maintain genome integrity.

In eukaryotes, DNA is synthesized during the S phase of the cell cycle. In animal cells, chromosomal DNA is bidirectionally replicated from multiple origins by DNA polymerases α and δ, which probably synthesize the lagging and leading strands, respectively. Mitochondrial DNA is replicated in the D-loop mode by DNA polymerase γ. Retroviruses produce DNA on RNA templates in a reaction sequence catalyzed by reverse transcriptase. Telomeric DNA is synthesized by the RNA-containing enzyme telomerase, which is active in germ cells but not somatic cells, a phenomenon that may, in part, be responsible for cellular senescence and aging.

Cells have a great variety of DNA repair mechanisms. DNA damage may be directly reversed such as in the photoreactivation of UV-induced pyrimidine dimers or in the repair of O^6-methylguanine lesions. Pyrimidine dimers, as well as many other types of lesions, may also be removed by excision repair. DNA glycosylases specifically remove the corresponding chemically altered bases, including uracil, to form AP sites that are eliminated by nucleotide excision repair. Xeroderma pigmentosum is an inherited human disease characterized by defects in any of seven complementation groups that participate in nucleotide excision repair. A lesion in a DNA strand resulting from its synthesis on a damaged template may be corrected through recombination repair. Large amounts of DNA damage induce the SOS response, which involves an error-prone DNA repair system. The high correlation between mutagenesis and carcinogenesis permits the detection of carcinogens by the Ames test.

Genetic information may be exchanged between homologous DNA sequences through general recombination. This process, which occurs according to the Holliday model, is mediated in *E. coli* by RecA together with numerous other enzymes. DNA may also be rearranged through the action of transposons. These DNA segments carry the genes coding for the proteins that mediate the transposition process as well as other genes. Transposition may be important in chromosomal and plasmid evolution and has been implicated in the control of phenotypic expression such as phase alternation in *Salmonella*, a process that is catalyzed by the Hin DNA invertase. Transposons in eukaryotes appear to be degenerate retroviruses.

Prokaryotic DNA may be methylated at its A or C bases. This prevents the action of restriction endonucleases and permits the correct mismatch repair of newly replicated DNA. In many but not all eukaryotes, DNA methylation, which occurs through the formation of m^{5}C, has been implicated in the control of gene expression and, via maintenance methylation, in genomic imprinting. Several inherited neurological diseases, including fragile X syndrome, myotonic dystrophy, and Huntington's disease, are characterized by the expansion of segments of repeating GC-rich triplets.

REFERENCES

General

Adams, R.L.P., Knowler, J.T., and Leader, D.P., *The Biochemistry of the Nucleic Acids* (11th ed.), Chapters 6 and 7, Chapman & Hall (1992).

Kornberg, A., *For Love of Enzymes: The Odyssey of a Biochemist,* Harvard University Press (1989). [A scientific autobiography.]

Kornberg, A. and Baker, T.A., *DNA Replication* (2nd ed.), Freeman (1992). [A compendium of information about DNA replication whose first author is the founder of the field. If you only read one work on DNA replication, this should be it.]

Lewin, B., *Genes V*, Chapters 18–20 and 33–36, Oxford University Press (1994).

Watson, J.D., Hopkins, N.H., Roberts, J.W., Steitz, J.A., and Weiner, A.M., *Molecular Biology of the Gene* (4th ed.), Chapters 10–12, Benjamin/Cummings (1987).

DNA Replication

Allsopp, R.C., Vaziri, H., Patterson, C., Goldstein, S., Younglai, E.V., Futcher, A.B., Greider, C.W., and Harley, C.B., Telomere length predicts replicative capacity of human fibroblasts, *Proc. Natl. Acad. Sci.* **89**, 10114–10118 (1992).

Baker, T.A. and Wickner, S.H., Genetics and enzymology of DNA replication in *Escherichia coli, Annu. Rev. Genet.* **26**, 447–477 (1992).

Bambara, R.A. and Jessee, C.B., Properties of DNA polymerases δ and ε, and their roles in eukaryotic DNA replication, *Biochim. Biophys. Acta* **1088**, 11–24 (1991).

Beese, L.S., Derbyshire, V., and Steitz, T.A., Structure of DNA polymerase I Klenow fragment bound to duplex DNA, *Science* **260**, 352–355 (1993).

Blackburn, E.H., Telomerases, *Annu. Rev. Biochem.* **61**, 113–129 (1992); *and* Telomeres and their synthesis, *Harvey Lectures* **86**, 1–18 (1992).

Burgers, P.M.J., Eukaryotic polymerases α and δ: Conserved properties and interactions, from yeast to mammalian cells, *Prog. Nucleic Acid Res. Mol. Biol.* **37**, 235–280 (1989).

Clayton, D.A., Replication and transcription of vertebrate mitochondrial DNA, *Annu. Rev. Cell Biol.* **7**, 453–478 (1991).

Counter, C.M., Hirte, H.W., Baccetti, S., and Harley, C.B., Telomerase activity in human ovarian carcinoma, *Proc. Natl. Acad. Sci.* **91**, 2900–2904 (1994).

Davies, J.F., II, Almassey, R.J., Hostomska, Z., Ferre, R.A., and Hostomsky, Z., 2.3 Å crystal structure of the catalytic domain of DNA polymerase β, *Cell* **76**, 1123–1133 (1994).

DePamphillus, M.L., Eukaryotic DNA replication, *Annu. Rev. Biochem.* **62**, 29–63 (1993); *and* Origins of DNA replication in metazoan chromosomes, *J. Biol. Chem.* **268**, 1–4 (1993).

Goodman, M.F., Creighton, S., Bloom, L.B., and Petruska, J., Biochemical basis of DNA replication fidelity, *Crit. Rev. Biochem. Mol. Biol.* **28**, 83–126 (1993).

Hill, T.M., Arrest of bacterial DNA replication, *Annu. Rev. Microbiol.* **46**, 603–633 (1992).

Hübscher, U. and Thömmes, P., DNA polymerase ε: in search of a function, *Trends Biochem. Sci.* **17**, 55–58 (1992). [Describes a model for the possible function of DNA polymerase ε.]

Jacobo-Molina, A., Ding, J., Nanni, R.G., Clark, A.D., Jr., Lu, X., Tantillo, C., Williams, R.L., Kamer, G., Ferris, A.L., Clark, P., Hizi, A., Hughes, S.H., and Arnold, E., Crystal structure of human immunodeficiency virus type 1 reverse transcriptase complexed with double-stranded DNA at 3.0 Å resolution shows bent DNA, *Proc. Natl. Acad. Sci.* **90**, 6320–6324 (1993).

Johnson, K.A., Conformational coupling in DNA polymerase fidelity, *Annu. Rev. Biochem.* **62**, 685–713 (1993).

Kang, C.H., Zhang, X., Ratliff, R., Moyzis, R., and Rich, A., Crystal structure of four-stranded *Oxytrichia* telomeric DNA, *Nature* **356**, 126–131 (1992).

Kohlstaedt, L.A., Wang, J., Friedman, J.M., Rice, P.A., and Steitz, T.A., Crystal structure at 3.5 Å resolution of HIV-1 reverse transcriptase complexed with an inhibitor, *Science* **256**, 1783–1790 (1992).

Kong, X.-P., Onrust, R., O'Donnell, M., and Kuriyan, J., Three-dimensional structure of the β subunit of E. coli DNA polymerase III holoenzyme: A sliding DNA clamp, *Cell* **69**, 425–437 (1992).

Lehman, I.R. and Kagun, L.S. DNA polymerase α, *J. Biol. Chem.* **264**, 4265 – 4268 (1989).

Lohman, T.M., Helicase-catalyzed DNA unwinding, *J. Biol. Chem.* **268**, 2269–2272 (1993).

Lohman, T.M., Bujalowski, W., and Overman, L.B., *E. coli* single strand binding protein, *Trends Biochem. Sci.* **13**, 250–255 (1988).

Matson, S.W., DNA helicases of *Escherichia coli, Prog. Nucleic Acid Res. Mol. Biol.* **40**, 289–326 (1991); *and* Matson, S.W. and Kaiser-Rogers, K.A., DNA helicases, *Annu. Rev. Biochem.* **59**, 289–329 (1990).

Marians, K.J., Prokaryotic DNA replication, *Annu. Rev. Biochem.* **61**, 673–719 (1992).

McHenry, C.S., DNA polymerase III holoenzyme, *J. Biol. Chem.* **266**, 19127–19130 (1991).

Schultze, P., Smith, F.W., and Feigon, J., Refined solution structure of the dimeric quadruplex formed from the *Oxytrichia* telomeric oligonucleotide d(GGGGTTTTGGGG), *Structure* **2**, 221–233 (1994).

So, A.G. and Downey, K.M., Eukaryotic DNA replication, *Crit. Rev. Biochem. Mol. Biol.* **27**, 129–155 (1992).

Steitz, T.A., DNA- and RNA-dependent DNA polymerases, *Curr. Opin. Struct. Biol.* **3**, 31–38 (1993).

Varmus, H., Reverse transcription, *Sci. Am.* **257**(3): 56–64 (1987).

Watson, J.D. and Crick, F.H.C., Genetical implications of the structure of deoxyribonucleic acid, *Nature* **171**, 964–967 (1953). [The paper in which semiconservative DNA replication was first postulated.]

Zyskind, J.W. and Smith, D.W., DNA replication, the bacterial cell cycle, and cell growth, *Cell* **69**, 5–8 (1992). [Hypothesizes that DNA replication and cell division in *E. coli* are coordinated by the methylation of the GATC sequences in *oriC,* which, in turn, controls the availability of these sites to the DnaA protein binding that initiates the DNA replication process.]

Repair of DNA

Ames, B.N., Identifying environmental chemicals causing mutations and cancer, *Science* **204,** 587–593 (1979).

Cleaver, J.E. and Kraemer, K.H, Xeroderma pigmentosum, *in* Scriver, C.R., Beaudet, A.L., Sly, W.S., and Valle, D. (Eds.), *The Metabolic Basis of Inherited Disease* (6th ed.), pp. 2949–2971, McGraw-Hill (1986).

Devoret, R., Bacterial tests for potential carcinogens, *Sci. Am.* **241**(2): 40–49 (1979).

Friedberg, E.C., *DNA Repair,* Freeman (1985). [An authoritative treatise.]

Grossman, L. and Thiagalingam, S., Nucleotide excision repair, a tracking mechanism in search of damage, *J. Biol. Chem.* **268,** 16871–16874 (1993).

Hoejimakers, J.H.J., Nucleotide excision repair I: from *E. coli* to yeast; *and* II: from yeast to mammals, *Trends Genet.* **9,** 173–177; *and* 211–216 (1993).

Howard-Flanders, P., Inducible repair of DNA, *Sci. Am.* **245**(5): 72–80 (1981).

Kenyon, C.J., The bacterial response to DNA damage, *Trends Biochem. Sci.* **8,** 84–87 (1983).

Kuo, C.-F., McRee, D.E., Fisher, C.L., O'Handley, S.F., Cunningham, R.P., and Tainer, J.A., Atomic structure of the DNA repair [4Fe–4S] enzyme endonuclease III, *Science* **258,** 434–440 (1992).

Lindahl, T., Instability and decay of the primary structure of DNA, *Nature* **363,** 709–715 (1993).

Mitra, S. and Kaina, B., Regulation of repair of alkylation damage in mammalian genomes, *Prog. Nucleic Acid Res. Mol. Biol.* **44,** 109–142 (1993).

Moore, M.H., Gulbis, J.M., Dodson, E.J., Demple, B., and Moody, P.C.E., Crystal structure of a suicidal DNA repair protein: Ada O^6-methylguanine-DNA methyltransferase from *E. coli, EMBO J.* **13,** 1495–1501 (1994).

Morikawa, K., DNA repair enzymes, *Curr. Opin. Struct. Biol.* **3,** 17–23 (1993).

Myers, L.C. and Verdine, G.L., DNA repair proteins, *Curr. Opin. Struct. Biol.* **5,** 51–59 (1994).

Myers, L.C., Verdine, G.L., and Wagner, G., Solution structure of the DNA methyl triester repair domain of *Eschericia coli* Ada, *Biochemistry* **32,** 14089–14094 (1993).

Radman, M. and Wagner, R., The high fidelity of DNA replication, *Sci. Am.* **259**(2): 40–46 (1988).

Sancar, A., Structure and function of DNA photolyase, *Biochemistry* **33,** 2–9 (1994).

Tanaka, K. and Wood, R.D., Xeroderma pigmentosum and nucleotide excision repair of DNA, *Trends Biochem. Sci.* **19,** 83–86 (1994).

Van Houtten, B. and Snowden, A., Mechanism of action of the

Eschericia coli UvrABC nuclease: Clues to the damage recognition problem, *BioEssays* **15,** 51–59 (1993).

Recombination and Mobile Genetic Elements

Cohen, S.N. and Shapiro, J.A., Transposable genetic elements, *Sci. Am.* **242**(2): 40–49 (1980).

Cox, M.M., Why does RecA protein hydolyse ATP? *Trends Biochem. Sci.* **19,** 217–222 (1994).

Egelman, E.H., What do X-ray crystallographic and electron microscopic structural studies of RecA protein tell us about recombination? *Curr. Opin. Struct. Biol.* **3,** 189–197 (1993).

Feng, J.-A, Dickerson, R.E., and Johnson, R.C., Proteins that promote DNA inversion and deletion, *Curr. Opin. Struct. Biol.* **4,** 60–66 (1994).

Haber, J.E., Exploring the pathways of homologous recombination, *Curr. Opin. Cell Biol.* **4,** 401–412 (1992).

Haselkorn, R., Developmentally regulated gene rearrangements in prokaryotes, *Annu. Rev. Genet.* **26,** 113–130 (1992).

Howard-Flanders, P., West, S.C., and Stasiak, A., Role of RecA protein in genetic recombination, *Nature* **309,** 215–220 (1984).

Kowalczykowski, S.C., Biochemistry of genetic recombination: Energetics and mechanism of DNA strand exchange, *Annu. Rev. Biophys. Biophys. Chem.* **20,** 539–575 (1991).

Mizuuchi, K., Polynucleotidyl transfer reactions in transpositional DNA recombination, *J. Biol. Chem.* **267,** 21273–21276 (1992); *and* Transpositional recombination, *Annu. Rev. Biochem.* **61,** 1011–1051 (1992).

Radding, C.M., Helical interactions in homologous pairing and strand exchange driven by RecA protein, *J. Biol. Chem.* **266,** 5355–5358 (1991).

Schleif, R., *Genetics and Molecular Biology* (2nd ed.), Chapter 19, The Johns Hopkins University Press (1993).

Simon, M., Zieg, J., Silverman, M., Mandel, G., and Doolittle, R., Phase variation: evolution of a controlling element, *Science* **209,** 1370–1374 (1980).

Stahl, F.W., Genetic recombination, *Sci. Am.* **256**(2): 90–101 (1987).

Story, R.M., Weber, I.T., and Steitz, T.A., The structure of the *E. coli recA* protein monomer and polymer, *Nature* **355,** 318–325 (1992); *and* the erratum for this paper, *Nature* **355,** 367 (1992). [These two papers should be read together.]

Taylor, A.F., Movement and resolution of Holliday junctions by enzymes from E. coli, *Cell* **69,** 1063–1065 (1992).

West, S.C., The processing of recombination intermediates: Mechanistic insights from studies of bacterial proteins, *Cell* **76,** 9–15 (1994); *and* Enzymes and molecular mechanisms of genetic recombination, *Annu. Rev. Biochem.* **61,** 603–640 (1992).

DNA Methylation and Trinucleotide Repeat Expansions

Adams, R.L.P., DNA methylation, *Biochem. J.,* **265,** 309–320 (1990).

Cedar, H. and Razin, A., DNA methylation and development, *Biochim. Biophys. Acta* **1049**, 1–8 (1990).

Holliday, R., A different kind of inheritance, *Sci. Am.* **260**(6): 60–73 (1989). [Discusses how DNA methylation may control gene activity patterns and how these patterns may be passed from one cell generation to another.]

Huntington's Disease Collaborative Research Group, A novel gene containing a trinucleotide repeat that is expanded and unstable on Huntington's disease chromosomes, *Cell* **72**, 971–983 (1993).

Klimasauskas, S., Kumar, S., Roberts, R.J., and Cheng, X., HhaI methyltransferase flips its target base out of the DNA helix, *Cell* **76**, 357–369 (1994).

Li, E., Beard, C., and Jaenisch, R., Role for DNA methylation in genomic imprinting, *Nature* **366**, 362–365 (1993).

Marinus, M.G., DNA methylation in *Escherichia coli, Annu. Rev. Genet.* **21**, 113–131 (1987).

Messer, W. and Noyer-Weidner, W., Timing and targeting: the biological function of Dam methylation in E. coli, *Cell* **54**, 735–737 (1988).

Modrich, P., Methyl-directed DNA mismatch correction, *J. Biol. Chem.* **264**, 6597–6600 (1989).

Modrich, P., DNA mismatch correction, *Annu. Rev. Biochem.* **56**, 435–466 (1987).

Richards, R.I. and Sutherland, G.R., Dynamic mutations: A new class of mutations causing human disease, *Cell* **70**, 709–712 (1992); *and* Ross, C.A., McInnis, M.G., Margolis, R.L., and Li, S.-H., Genes with triplet repeats: candidate mediators of neuropsychiatric disorders, *Trends Neurobiol.* **16**, 254–260 (1993).

Verdine, G.L., The flip side of DNA methylation, *Cell* **76**, 197–200 (1994).

Warren, S.T. and Nelson, D.L., Trinucleotide repeat expansions in neurological disease, *Curr. Opin. Neurobiol.* **3**, 752–759 (1993).

PROBLEMS

1. Explain how certain mutant varieties of Pol I can be nearly devoid of DNA polymerase activity but retain almost normal levels of $5' \rightarrow 3'$ exonuclease activity.

2. Why haven't Pol I mutants been found that completely lack $5' \rightarrow 3'$ activity at all temperatures?

3. Why aren't Type I topoisomerases necessary in DNA replication?

*4. The $3' \rightarrow 5'$ exonuclease activity of Pol I excises only unpaired 3'-terminal nucleotides from DNA, whereas this enzyme's pyrophosphorolysis activity removes only properly paired 3'-terminal nucleotides. Discuss the mechanistic significance of this phenomenon in terms of the polymerase reaction.

5. You have isolated *E. coli* with temperature sensitive mutations in the following genes. What are their phenotypes above their restrictive temperatures? Be specific. (a) *dnaB*, (b) *dnaE*, (c) *dnaG*, (d) *lig*, (e) *polA*, (f) *rep*, (g) *ssb*, and (h) *recA*.

6. About how many Okazaki fragments are synthesized in the replication of an *E. coli* chromosome?

*7. What are the minimum and maximum number of replication forks that occur in a contiguous chromosome of an *E. coli* that is dividing every 25 min; every 80 min?

8. Why can't linear duplex DNAs, such as occur in bacteriophage T7, be fully replicated by only *E. coli*-encoded proteins?

*9. What is the half-life of a particular purine base in the human genome assuming that it is subject only to spontaneous depurination? What fraction of the purine bases in a human genome will have depurinated in the course of a single generation (assume 25 years)? The DNAs of ~4000-year-old Egyptian mummies have been sequenced. Assuming that mummification did not slow the rate of DNA depurination, what fraction of the purine bases originally present in the mummy would still be intact today.

10. Why is the methylation of DNA to form O^6-methylguanine mutagenic?

11. There are certain sites in the *E. coli* chromosome known as **hot spots** that have unusually high rates of point mutation. Many of these sites contain a 5-methylcytosine residue. Explain the existence of such hot spots.

12. Explain why the brief exposure of a cultured eukaryotic cell line to 5-azacytosine results in permanent phenotypic changes to the cells.

13. Explain why chi structures, such as that shown in Fig. 31-47*b*, have two pairs of equal length arms.

*14. Single-stranded circular DNAs containing a transposon have a characteristic stem-and-double-loop structure such as that shown in Fig. 31-68. What is the physical basis of this structure?

15. A composite transposon integrated in a circular plasmid occasionally transposes the DNA comprising the original plasmid rather than the transposon's central region. Explain how this is possible.

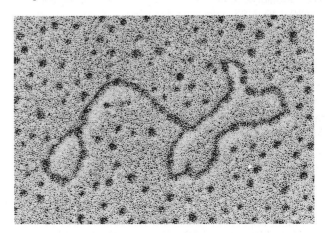

FIGURE 31-68. An electron micrograph of a single circular DNA containing a transposon. [Courtesy of Stanley Cohen, Stanford University School of Medicine.]

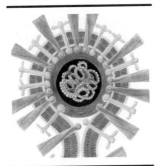

C H A P T E R

32

Viruses: Paradigms For Cellular Functions

1. **Tobacco Mosaic Virus**
   A. Structure
   B. Assembly

2. **Spherical Viruses**
   A. Virus Architecture
   B. Tomato Bushy Stunt Virus
   C. Picornaviruses
   D. Simian Virus 40 (SV40)
   E. Bacteriophage MS2

3. **Bacteriophage λ**
   A. The Lytic Pathway
   B. Virus Assembly
   C. The Lysogenic Mode
   D. Mechanism of the λ Switch

4. **Influenza Virus**
   A. Virus Structure and Life Cycle
   B. Mechanism of Antigenic Variation
   C. Mechanism of Membrane Fusion

5. **Subviral Pathogens**
   A. Viroids
   B. Prions

***Viruses** are parasitic entities, consisting of nucleic acid molecules with protective coats, which are replicated by the enzymatic machinery of suitable host cells.* Since they lack metabolic apparatus, viruses are not considered to be alive (although this is a semantic rather than a scientific distinction). They range in complexity from **satellite tobacco necrosis virus (STNV),** whose genome has only one gene, to the **pox viruses,** which code for ~240 genes.

Viruses were originally characterized at the end of the nineteenth century as infectious agents that could pass through filters that held back bacteria. Yet viral diseases, varying in severity from small pox and rabies to the common cold, have no doubt plagued mankind since before the dawn of history. It is now known that viruses can infect plants and bacteria as well as animals. Each viral species has a very limited **host range;** that is, it can reproduce in only a small group of closely related species.

An intact virus particle, which is referred to as a **virion,** consists of a nucleic acid molecule encased by a protein **capsid.** In some of the more complex virions, the capsid is surrounded by a lipid bilayer and glycoprotein-containing **envelope,** which is derived from a host cell membrane. Since the small size of a viral nucleic acid severely limits the number of proteins that can be encoded by its genome, its capsid, as Francis Crick and James Watson pointed out in 1957, must be built up of one or a few kinds of protein subunits that are arranged in a symmetrical or nearly symmetrical fashion. There are two ways that this can occur:

1. In the **helical viruses** (Section 32-1), the coat protein subunits associate to form helical tubes.

2. In the **spherical viruses** (Section 32-2), coat proteins aggregate as closed polyhedral shells.

In both cases, the viral nucleic acid occupies the capsid's central region. In many viruses, the coat protein subunits may be "decorated" by other proteins so that the capsid exhibits spikes and, in larger bacteriophages, a complex tail. These assemblies are involved in recognizing the host cell and delivering the viral nucleic acid into its interior. Figure 32-1 is a "rogues gallery" of viruses of varying sizes and morphologies.

1074

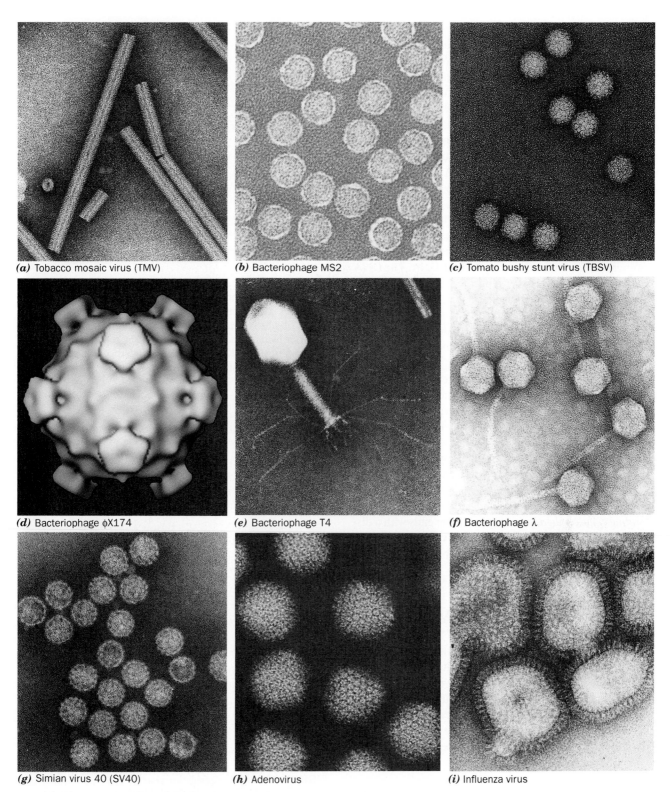

(a) Tobacco mosaic virus (TMV)

(b) Bacteriophage MS2

(c) Tomato bushy stunt virus (TBSV)

(d) Bacteriophage φX174

(e) Bacteriophage T4

(f) Bacteriophage λ

(g) Simian virus 40 (SV40)

(h) Adenovirus

(i) Influenza virus

FIGURE 32-1. Electron micrographs of a selection of viruses. TMV, MS2, TBSV, and influenza virus are single-stranded RNA viruses; φX174 is a single-stranded DNA virus; and λ, T4, SV40, and adenovirus are double-stranded DNA viruses. Bacteriophage M13, a filamentous, single-stranded DNA coliphage, is shown in Fig. 28-65. [Parts *a–c* and *f–i* courtesy of Robley Williams, University of California at Berkeley and Harold Fisher, University of Rhode Island; Part *d* courtesy of Michael Rossmann, Purdue University; and Part *e* courtesy of John Finch, Cambridge University.]

The great simplicity of viruses in comparison to cells makes them invaluable tools in the elucidation of gene structure and function, as well as our best characterized models for the assembly of biological structures. Although all viruses use ribosomes and other host factors for the RNA-instructed synthesis of proteins, their modes of genome replication are far more varied than that of cellular life. In contrast to cells, in which the hereditary molecules are invariably double-stranded DNA, viruses contain either single- or double-stranded DNA or RNA. In RNA viruses, the viral RNA may be directly replicated or act as a template in the synthesis of DNA. The RNA of single-stranded RNA viruses may be the positive strand (the mRNA) or the negative strand (complementary to the mRNA). Viral DNA may replicate autonomously or be inserted in the host chromosome for replication with the host DNA. The DNA of eukaryotic viruses is either replicated and transcribed in the cell nucleus by cellular enzymes or in the cytoplasm by virally specified enzymes. In fact, in the case of negative strand RNA viruses, enzymes that mediate viral RNA transcription must be carried by the virion because most cells lack the ability to transcribe RNA.

This chapter is a discussion of the structures and biology of a variety of viruses. In it, we examine mainly **tobacco mosaic virus (TMV),** a helical RNA virus; **tomato bushy stunt virus (TBSV),** a spherical RNA virus; **bacteriophage λ,** a tailed DNA bacteriophage; and **influenza virus,** an enveloped RNA virus. These examples have been chosen to illustrate important aspects of viral structure, assembly, molecular genetics, and evolutionary strategy. *Much of this information is relevant to the understanding of the corresponding cellular phenomena.* The chapter ends with a discussion of **subviral pathogens,** relatively recently discovered disease agents that are even simpler than viruses.

1. TOBACCO MOSAIC VIRUS

Tobacco mosaic virus causes leaf mottling and discoloration in tobacco and many other plants. It was the first virus to be discovered (by Dmitri Iwanowsky in 1892), the first virus to be isolated (by Wendell Stanley in 1935), and even now is among the most extensively investigated and well-understood viruses from the standpoint of structure and assembly. In this section, we discuss these aspects of TMV.

A. Structure

TMV is a rod-shaped particle (Fig. 32-1*a*) that is ~3000 Å long, 180 Å in diameter, and has a particle mass of 40 million D. Its ~2130 identical copies of coat protein subunits (158 amino acid residues; 17.5 kD) are arranged in a hollow right-handed helix that has $16\frac{1}{3}$ subunits/turn, a pitch (rise per turn) of 23 Å, and a 40-Å-diameter central cavity (Fig. 32-2). TMV's single RNA strand (~6400 nt; 2

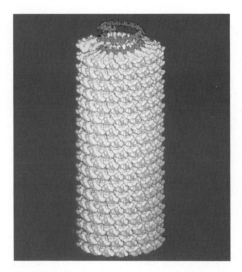

FIGURE 32-2. A model of TMV illustrating the helical arrangement of its coat protein subunits and RNA molecule. The RNA is represented by the red chain exposed at the top of the viral helix. Only 18 turns (415 Å) of the TMV helix are shown, which represent ~14% of the TMV rod. [Courtesy of Gerald Stubbs and Keiichi Namba, Vanderbilt University; and Donald Caspar, Brandeis University.]

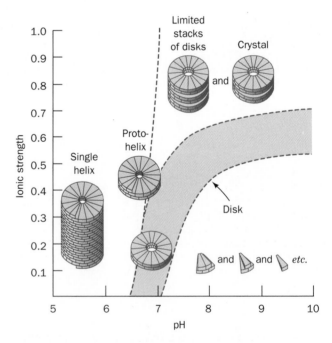

FIGURE 32-3. The aggregation state of TMV coat protein as a function of pH and ionic strength. Under basic conditions, the subunits aggregate into small clusters. Around neutrality and at high ionic strengths, the protein forms a 34-subunit double-layered disk. Under acidic conditions and at low ionic strengths, the subunits form protohelices that stack to form long helices. At neutral pH and low ionic strength, which resembles physiological conditions, the protein forms helices only in the presence of TMV RNA. [After Durham, A.C.H., Finch, J.T., and Klug, A., *Nature New Biol.* **229,** 38 (1971).]

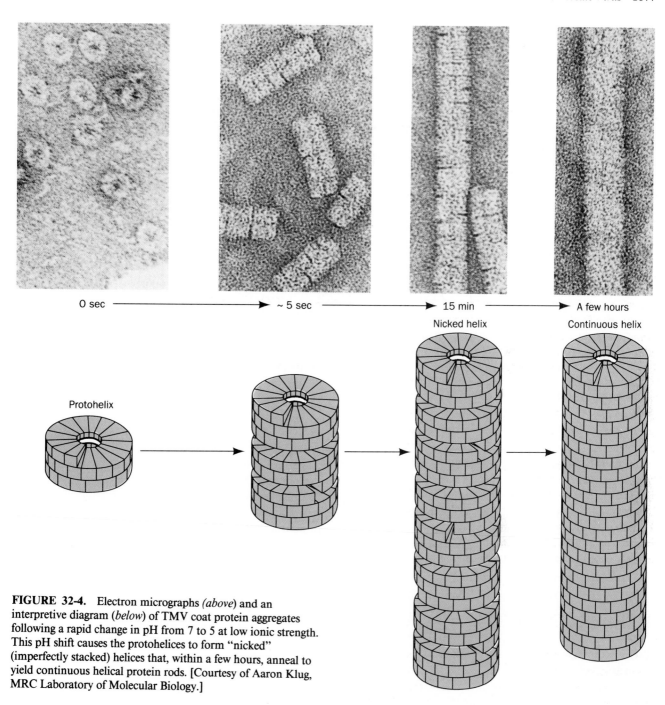

FIGURE 32-4. Electron micrographs *(above)* and an interpretive diagram *(below)* of TMV coat protein aggregates following a rapid change in pH from 7 to 5 at low ionic strength. This pH shift causes the protohelices to form "nicked" (imperfectly stacked) helices that, within a few hours, anneal to yield continuous helical protein rods. [Courtesy of Aaron Klug, MRC Laboratory of Molecular Biology.]

million D) is coaxially wound within the turns of the coat protein helix such that 3 nt are bound to each protein subunit (Fig. 32-2).

TMV Coat Protein Aggregates To Form Viruslike Helical Rods

The aggregation state of TMV coat protein is both pH and ionic strength dependent (Fig. 32-3). At slightly alkaline pH's and low ionic strengths, the coat protein forms complexes of only a few subunits. At higher ionic strengths,

however, the subunits associate to form a double-layered disk of 17 subunits/layer, a number that is nearly equal to the number of subunits per turn in the intact virion. At neutral pH and low ionic strengths, the subunits form short helices of slightly more than two turns (39 ± 2 subunits) termed "protohelices" (also known as "lockwashers"). If the pH of these protohelices is shifted to ~5, they stack in imperfect register and eventually anneal to form indefinitely long helical rods that, although they lack RNA, resemble intact virions (Fig. 32-4). These observations, as we

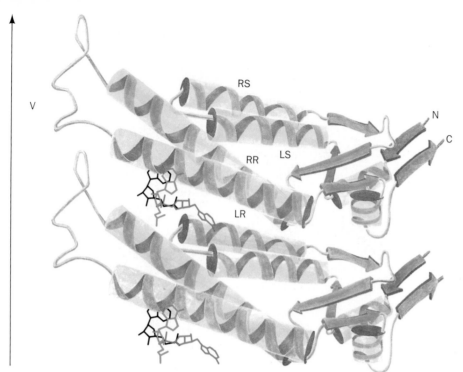

FIGURE 32-5. A ribbon diagram of two vertically stacked TMV subunits as viewed perpendicular to the virus helix axis (*vertical arrow on the left*). Each subunit has four more or less radially extending helices (LR, RR, LS, and RS), as well as a short vertical segment (V), which comprises part of the flexible loop in the disk structure (dashed lines in Fig. 32-7). Two successive turns of RNA are shown passing through their binding sites. Each subunit binds three nucleotides, here represented by GAA with each of its nucleotides differently colored, such that their three bases lie flat against the LR helix so as to grasp it in a clawlike manner. [After Namba, K., Pattanayek, R., and Stubbs, G., *J. Mol. Biol.* **208**, 314 (1989).]

shall see below, lead to the explanation of how TMV assembles.

TMV Coat Protein Interacts Flexibly with Viral RNA

X-Ray studies of TMV have been pursued on two fronts. The virus itself does not crystallize but forms a highly oriented gel of parallel viral rods. The X-ray analysis of this gel by Kenneth Holmes and Gerald Stubbs yielded a structure of sufficient resolution (2.9 Å) to reveal the folding of the protein and the RNA (Figs. 32-5 and 32-6). This study is complemented by Aaron Klug's X-ray crystal structure determination, at 2.8-Å resolution, of the 34-subunit coat protein disk (Fig. 32-7).

A major portion of each subunit consists of a bundle of four alternately parallel and antiparallel α helices that project more or less radially from the virus axis (Figs. 32-5–32-7). In the disk, one of the inner connections between these α helices, a 24-residue loop (residues 90–113; dashed line in Fig. 32-7), is not visible, apparently because it is highly mobile. This disordered loop is also present in the protohelix as shown by NMR studies. In the virus, however, the loop adopts a definite conformation containing of a series of reverse turns arranged such that the overall direction of

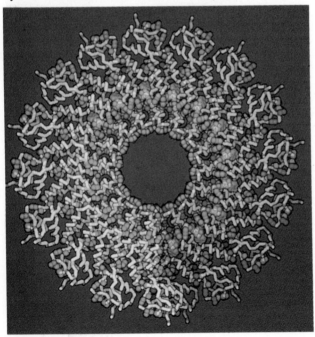

FIGURE 32-6. Top view of 17 TMV coat protein subunits comprising slightly more than one helical turn in complex with a 33-nucleotide RNA segment. The protein is represented by its C_α atoms, shown as connected 2.5 Å diameter helical rods, together with its acidic side chains (Asp and Glu) in red and its basic side chains (Arg and Lys) in blue. The RNA's phosphate atoms are green and its bases are purple. Note that the acidic side chains form a 25-Å-radius helix that lines the virion's inner cavity and the basic side chains form a 40-Å-radius helix that interacts with the RNA's anionic sugar–phosphate chain. [Courtesy of Gerald Stubbs and Keiichi Namba, Vanderbilt University; and Donald Caspar, Brandeis University.]

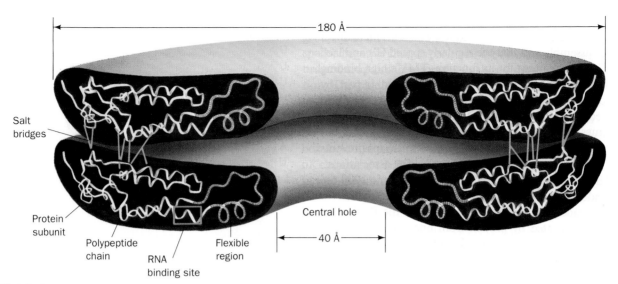

FIGURE 32-7. The structure of the TMV protein disk in cross-section showing its polypeptide chains as ribbon diagrams. The dashed lines represent disordered loops of polypeptide chain that are therefore not visible in the disk X-ray structure. The stacked protein rings interact along their outer rims through a system of salt bridges (*red lines*). [After Butler, P.J.G. and Klug, A., *Sci. Am.* **239**(5): 67 (1978). Copyright © 1978 by Scientific American, Inc.]

this polypeptide segment is approximately parallel to the virus axis (V in Fig. 32-5). This conformational change, as we shall see, is an important aspect of virus assembly.

In the virus, the RNA is helically wrapped between the coat protein subunits at a radius of ~40 Å. The triplet of bases binding to each subunit forms a clawlike structure around one of the radial helices (LR in Fig. 32-5) with each base occupying a hydrophobic pocket in which it lies flat against LR. Arg residues 90 and 92, which are invariant in the several known TMV strains and which are part of the disk and protohelix's disordered loop, as well as Arg 41, form salt bridges to the RNA phosphate groups.

B. Assembly

How is the TMV virion assembled from its component RNA and coat protein subunits? *The assembly of any large molecular aggregate, such as a crystal or a virus, generally occurs in two stages: (1) **nucleation**, the largely random aggregation of subunits to form a quasi-stable nucleation complex, which is almost always the rate-determining step of the assembly process; followed by (2) **growth**, the cooperative addition of subunits to the nucleation complex in an orderly arrangement that usually proceeds relatively rapidly.* For TMV, it might reasonably be expected that the nucleation complex minimally consists of the viral RNA in association with the 17 or 18 subunits necessary to form a stable helical turn, which could then grow by the accumulation of subunits at one or both ends of the helix. The low probability for the formation of such a complicated nucleation complex from disaggregated subunits accounts for the observed 6-h time necessary to complete this *in vitro* assembly process. Yet, the *in vivo* assembly of TMV probably occurs

much faster. A clue as to the nature of this *in vivo* process was provided by the observation that if protohelices rather than disaggregated subunits are mixed with TMV RNA, complete virus particles are formed in 10 min. Other RNAs do not have this effect. Evidently, *the in vivo nucleation complex in TMV assembly is the association of a protohelix with a specific segment of TMV RNA.* (Although it was originally assumed that the double-layered disk rather than the protohelix formed the nucleating complex, experimental evidence indicates that the disk does not form under physiological conditions and that its rate of conversion to the protohelix under these conditions is too slow to account for the rate of TMV assembly. Other experiments, however, suggest that it is the disk that predominates at pH 7.0, the pH at which TMV most rapidly assembles from its component protein and RNA. Thus, keep in mind that the question as to whether TMV assembles from protohelices, as we state here, or from double-layered disks, has not been fully resolved.)

TMV Assembly Proceeds by the Sequential Addition of Protohelices

The specific region of the TMV RNA responsible for initiating the virus particle's growth was isolated using the now classical nuclease protection technique. The RNA is mixed with a small amount of coat protein so as to form a nucleation complex that cannot grow because of the lack of coat protein. The RNA that is not protected by coat protein is then digested away by RNase, leaving intact only the initiation sequence. This RNA fragment forms a hairpin loop whose 18-nucleotide apical sequence, AGAAGAAGUUGUUGAUGA has a G at every third residue (recall that each coat protein subunit binds three nucleotides) but

(a)

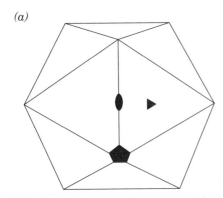

(b)

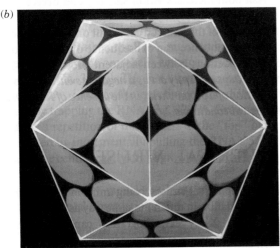

FIGURE 32-11. An icosahedron. (*a*) This regular polyhedron has 12 vertices, 20 equilateral triangular faces of identical size, and 30 edges. It has a fivefold axis of symmetry through each vertex, a threefold axis through the center of each face, and a twofold axis through the center of each edge (also see Fig. 7-58). (*b*) A drawing of 60 identical subunits (*lobes*) arranged with icosahedral symmetry. [Drawing copyrighted © by Irving Geis.]

hedron (Fig. 32-11*a*) has 20 triangular faces, each with threefold symmetry, for a total of $20 \times 3 = 60$ equivalent positions (each represented by a lobe in Fig. 32-11*b*). Of these polyhedra, the icosahedron encloses the greatest volume per subunit. Indeed, electron microscopy of the so-called spherical viruses (such as Fig. 32-1*b–h*) has revealed that *all of them have icosahedral symmetry.*

Viral Capsids Resemble Geodesic Domes

A viral nucleic acid, if it is to be protected effectively against a hostile environment, must be completely covered by coat protein. Yet, many viral nucleic acids occupy so large a volume that their coat protein subunits would have to be prohibitively large if their capsids were limited to the 60 subunits required by exact icosahedral symmetry. In fact, nearly all viral capsids have considerably more than 60 chemically identical subunits. How is this possible?

Donald Caspar and Klug pointed out the solution to this dilemma. *The triangular faces of an icosahedron can be*

subdivided into integral numbers of equal sized equilateral triangles (e.g., Fig. 32-12*a*). The resulting polyhedron, an **icosadeltahedron,** has "local" symmetry elements relating its subunits (lobes in Fig. 32-12*b*) in addition to its exact icosahedral symmetry. By local symmetry, we mean that the symmetry is only approximate so that, in contrast to the case for exact symmetry, it breaks down over larger distances. For instance, the subunits (lobes) in Fig. 32-12*b* that are distributed about each exact triangular vertex form clusters whose members are related by a local sixfold axis of

(a)

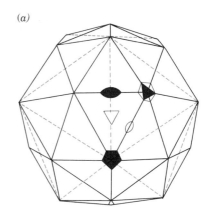

(b)

FIGURE 32-12. A $T = 3$ icosadeltahedron. (*a*) This polyhedron has the exact rotational symmetry of an icosahedron (*solid symbols*) together with local sixfold, threefold, and twofold rotational axes (*hollow symbols*). Note that the edges of the underlying icosahedron (*dashed red lines*), are not edges of this polyhedron and that its local sixfold axes are coincident with its exact threefold axes. (*b*) A drawing of a $T = 3$ icosadeltahedron showing its arrangement of 3 quasi-equivalent sets of 60 icosahedrally related subunits (*lobes*). The A lobes (*orange*) pack about the icosadeltahedron's exact fivefold axes, whereas the B and C lobes (*blue and green*) alternate about its local sixfold axes. TBSV's chemically identical coat protein subunits are arranged in this manner. [Drawing copyrighted © by Irving Geis.]

symmetry. *Adjacent subunits in these clusters are not exactly equivalent; they are quasi-equivalent.* In contrast, the subunits clustered about the 12 fivefold axes of icosahedral symmetry are exactly equivalent. The interactions between the subunits clustered about the local sixfold axes are therefore essentially distorted versions of those about the exact fivefold axes. Consequently, *the coat protein subunits of any viral capsid with icosadeltahedral symmetry must make alternative sets of intersubunit associations and/or have sufficient conformational flexibility to accommodate these distortions.*

Icosadeltahedra are actually familiar figures. The faceted surface of a soccer ball is an icosadeltahedron. Likewise, **geodesic domes** (Fig. 32-13), which were originally designed by Buckminster Fuller, are portions of icosadeltahedra. It was, in fact, Fuller's designs that inspired Caspar and Klug. *Geodesic domes are inherently rigid shell-like structures that are constructed from a few standard parts, make particularly efficient use of structural materials, and can be rapidly and easily assembled. Presumably the evolution of spherical virus capsids was guided by these very principles.*

The number of subunits in an icosadeltahedron is 60T, where T is called the **triangulation number** (it can be shown that the permissible values of T are given by $T = h^2 + hk + k^2$, where h and k are positive integers). An icosahedron, the simplest icosadeltahedron, has $T = 1$ ($h = 1$, $k = 0$) and therefore 60 subunits. The icosadeltahedron with the next level of complexity has a triangulation number of $T = 3$ ($h = 1$, $k = 1$) and hence 180 subunits (Fig. 32-12). A capsid with this geometry has three different sets of icosahedrally related subunits that are quasi-equivalent to each other

(lobes A, B, and C in Fig. 32-12*b*). Viruses with capsids consisting of $T = 1$, 3, and 4 icosadeltahedra have been identified. Some of the larger polyhedral viruses may form icosadeltahedra with even greater triangulation numbers (although several of them have been shown to be based on somewhat different assembly principles as we shall see in Section 32-2D). The T value for any particular capsid, presumably, depends on its subunit's innate curvature.

B. Tomato Bushy Stunt Virus

TBSV (Fig. 32-1*c*) is a $T = 3$ spherical virus that is ~175 Å in radius. It consists of 180 identical coat protein subunits, each of 386 residues (43 kD), encapsulating a single-stranded RNA molecule of ~4800 nt (1500 kD; the positive or message strand) and a single copy of an ~85-kD protein. The X-ray crystal structure of TBSV, the first of a virus to be determined at high resolution, was reported in 1978 by Stephen Harrison. TBSV's coat protein subunits have three domains (Fig. 32-14): P, the C-terminal domain, which projects outward from the virus; S, which forms the protein shell; and R, the protein's inwardly extending N-terminal

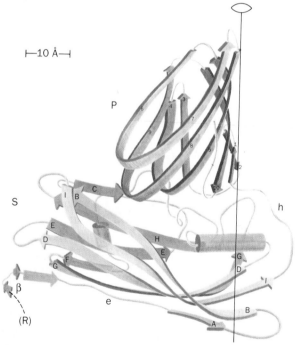

FIGURE 32-14. The TBSV coat protein subunit consists of three domains: P, which projects from the virion's surface; S, which forms the capsid; and R, which extends below the capsid surface where it participates in binding the viral RNA. The S domain is largely comprised of an 8-stranded antiparallel β barrel that has been dubbed a **jellyroll** or **Swiss roll β-barrel** due to its topological resemblance to these pastries. The S domain is also composed largely of an antiparallel β sheet, whereas the R domain is not visible in the viral X-ray structure so its tertiary structure is unknown. [After Olsen, A.J., Bricogne, G., and Harrison, S.C., *J. Mol. Biol.* **171**, 78 (1983).]

FIGURE 32-13. A geodesic dome built on the plan of a $T = 36$ icosadeltahedron. Two of its pentagonal vertices are visible in this photograph. [Stanley Schoenberger/Grant Heilman.]

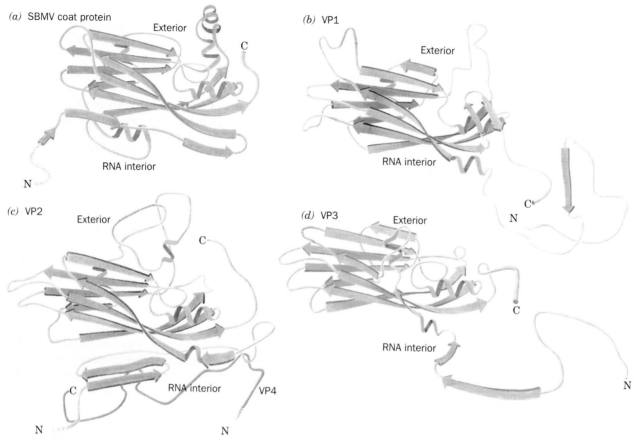

(a) SBMV coat protein

Exterior

C

RNA interior

N

(b) VP1

Exterior

RNA interior

C

N

(c) VP2

Exterior

C

RNA interior

VP4

C

N

N

(d) VP3

Exterior

C

RNA interior

N

FIGURE 32-18. The structures of *(a)* SBMV coat protein, and the *(b)* VP1, *(c)* VP2 (together with VP4), and *(d)* VP3 proteins of human rhinovirus. Note the close structural similarities of their 8-stranded β-barrel cores and that of TBSV's S domain (Fig. 32-14). The VP1, VP2, and VP3 proteins of poliovirus likewise have this fold. [After Rossmann, M.G., Arnold, E., Erickson, J.W., Frankenberger, E.A., Griffith, J.P., Hecht, H.-J., Johnson, J.E., Kamer, G., Luo, M., Mosser, A.G., Rueckert, R.R., Sherry, B., and Vriend, G., *Nature* **317,** 148 (1985).]

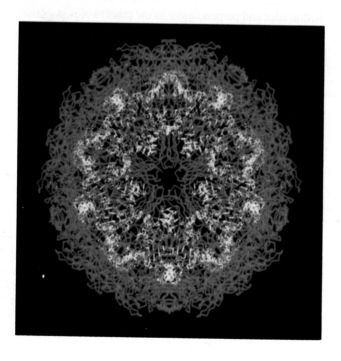

FIGURE 32-19. The X-ray structure of STMV in which the protein capsid *(blue)*, which forms a shell from radii 57 to 86 Å, is associated with 30 double helical segments of RNA *(yellow)* that lie between radii 52 and 64 Å. The icosahedral *(T* = 1) particle is viewed along one of its 5-fold axes. [Courtesy of Alexander McPherson, University of California at Riverside.]

C. Picornaviruses

The X-ray structures of two viral pathogens of humans have been elucidated: that of **poliovirus,** the cause of **poliomyelitis,** by James Hogle; and that of **rhinovirus,** the cause of **infectious rhinitis** (the common cold), by Rossmann. Both pathogens are **picornaviruses,** a large family of animal viruses that also includes the agents causing human **hepatitis A** and **foot-and-mouth disease.** Picornaviruses (*pico,* small + *rna*) are among the smallest RNA-containing animal viruses: They have a particle mass of ~8.5 × 10⁶ D of which ~30% is a single-stranded RNA of ~7500 nucleotides. Their icosahedral protein shell, which is ~300 Å in

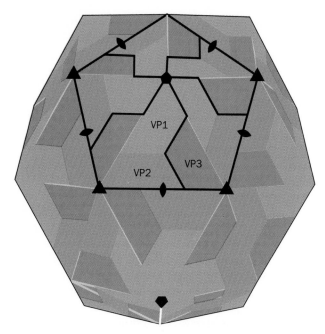

FIGURE 32-20. The arrangement of the 60 trimers *(triangles)* of pseudo-equivalent VP1, VP2, and VP3 subunits on human rhinovirus's icosahedral capsid. This arrangement resembles that of TBSV in which 180 chemically identical subunits are quasi-symmetrically related to form a $T = 3$ icosadeltahedron (Figs. 32-12 and 32-15). The positions of the icosahedron's exact fivefold, threefold, and twofold axes are marked. [After Rossmann, M.G., Arnold, E., Erickson, J.W., Frankenberger, E.A., Griffith, J.P., Hecht, H.-J., Johnson, J.E., Kamer, G., Luo, M., Mosser, A.G., Rueckert, R.R., Sherry, B., and Vriend, G., *Nature* **317**, 147 (1985).]

diameter, contains 60 protomers, each consisting of 4 structural proteins, **VP1, VP2, VP3**, and **VP4**. These 4 proteins are synthesized by an infected cell as a single **polyprotein,** which is cleaved to the individual subunits during virion assembly. Picornaviruses can be highly specific as to the cells they infect; for example, poliovirus binds to receptors that occur only on certain types of primate cells.

The structures of poliovirus, rhinovirus, and **foot-and-mouth disease virus (FMDV;** determined by David Stuart) are remarkably alike, both to each other and to TBSV and SBMV. Although VP1, VP2, and VP3 of picornaviruses have no apparent sequence similarities with each other or with the coat proteins of TBSV and SBMV, these proteins all exhibit striking structural similarities (Figs. 32-14 and 32-18; VP4, which is much smaller than the other subunits, forms, in effect, an N-terminal extension of VP2). Indeed, the picornaviruses' chemically distinct VP1, VP2, and VP3 subunits are pseudosymmetrically related by pseudo-threefold axes passing through the center of each triangular face of the icosahedral ($T = 1$) virion, which therefore has pseudo-$T = 3$ symmetry (Fig. 32-20). The chemically identical but conformationally distinct A, B, and C subunits of the $T = 3$ plant viruses are likewise quasi-symmetrically related by analogously located local threefold axes (Fig. 32-15). These structural similarities strongly suggest that the picornaviruses and the spherical plant viruses all diverged from a common ancestor.

The protein capsids of poliovirus, rhinovirus, and FMDV form a hollow shell enclosing a disordered core composed of the viral RNA and some protein, much as in the spherical plant viruses. This arrangement is vividly illustrated in Fig. 32-21, which shows both the inner and outer views of the poliovirus capsid. Note that VP4 largely lines the inside of the capsid. Also note the rugged topogra-

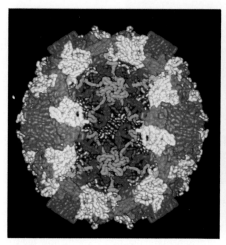
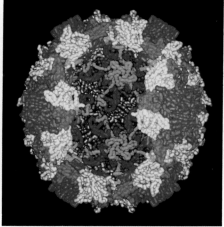

FIGURE 32-21. A stereo diagram of the poliovirus capsid in which the inner surface is revealed by the removal of two pentagonal faces. Here, the polypeptide chain is represented by a folded tube that approximates the volume of the protein and which is blue in VP1, yellow in VP2, red in VP3, and green in VP4. The VP4 subunits, which line the capsid's inner surface,

associate about its fivefold axes of symmetry to form a framework similar to although geometrically distinct from that formed by the C subunit arms in TBSV (Fig. 32-16). [Courtesy of Arthur Olson, The Scripps Research Institute, La Jolla, California.]

phy of the capsid's outer surface. Some of its crevices form the receptor-binding site through which the virus is targeted to specific cells.

D. *Simian Virus 40 (SV40)*

Simian virus 40 (SV40) is a **polyomavirus**, the simplest class of viruses containing double-stranded DNA. This ~500-Å external diameter spherical virus (Figure 32-1g) functions to transfer a 5243-bp circular "minichromosome" (DNA in complex with histone-containing particles known as **nucleosomes;** Section 33-3B) from the nucleus of one cell to that of another. The viral capsid consists of 360 copies of a 361-residue protein, VP1, that are arranged with icosahedral symmetry. However, this number of particles cannot be arranged with the icosadeltahedral symmetry characteristic of TBSV, for example, because $T = 360/60 = 6$ is a forbidden value for icosadeltahedra (for which $T = h^2 + hk + k^2$). Rather, as Caspar demonstrated through low-resolution X-ray studies of **polyomaviruses,** VP1 exclusively forms pentamers that take up two nonequivalent positions (Fig. 32-22a). Twelve of the pentamers lie on the icosahedron's 12 five-fold rotation axes, each surrounded by 5 pentamers of a different class. This latter class of 60 pentamers, which do not lie on icosahedral symmetry axes, are each surrounded by 6 pentamers, 5 of its own class and one of the former class. As a consequence, each capsid contains 6 symmetry-inequivalent classes of the chemically identical VP1 subunits. What conformational adjustments must the

subunits make to form such a structure and, in particular, how does a pentameric structure coordinate with 6 other such pentamers?

The X-ray structure of SV40, determined by Harrison, indicates that VP1 consists of three modules: (1) an N-terminal arm that extends across the inside of the pentamer beneath the clockwise neighboring subunit (looking from the outside in) and whose first 15 residues are not visible in the structure (they probably extend inwards to interact with the minichromosome which is likewise not visible); (2) an antiparallel β barrel with the same topology as that in RNA plant viruses and picornaviruses (Figures 32-14 and 32-18), although oriented more or less radially with respect to the capsid rather than tangentially; and (3) a long C-terminal arm, the site of the only major conformational variation among the 6 symmetry-inequivalent sets of VP1 subunits. The C-terminal arms form the principal interpentamer contacts by extending from their pentamer of origin so as to invade a neighboring pentagon (Fig. 32-22b and c). Each pentamer thereby receives five invading arms from adjacent pentamers as well as donating five such arms. *It is the differing patterns of C-terminal arm exchange among the various pentamers that determines how they associate in forming the capsid.* Since these C-terminal arms are probably flexible and unstructured on a free pentamer, the capsid's pentameric building blocks probably behave, so to speak, as if they are tied together with ropes rather than being cemented together across extended complementary surfaces. Indeed, deletion of the C-terminal arms from re-

(a)   *(b)*   *(c)*

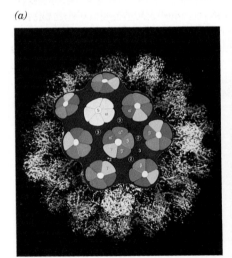 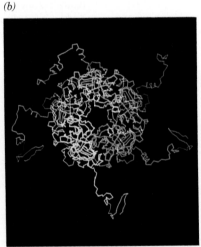 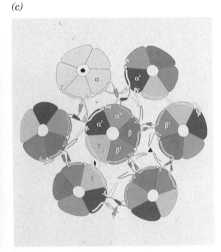

FIGURE 32-22. The X-ray structure of SV40. (*a*) The SV40 virion consists of 360 copies of VP1 that are organized into 72 pentamers of which 12 (*white*) are 5-coordinated and 60 (*colored*) are 6-coordinated. Three types of interpentamer clustering are indicated on the schematic part of the drawing: The white (α), purple (α'), and green (α'') subunits form a 3-fold interaction ③; the red (β) and blue (β') form one type of 2-fold interaction ②; and the yellow subunits (γ) form a second type of 2-fold interaction (2). The icosahedral axes of symmetry are indicated by the numerals 5, 3, and 2. (*b*) A 6-coordinated

pentamer as viewed from outside the virion. The VP1 subunits, which are represented by their C_α chains, are colored as in Part *a*. Note the C-terminal arms extending out from each subunit. (*c*) Schematic diagram showing how the C-terminal arms tie the pentamers together. The C-terminal arms are represented by lines and small cylinders (helices). The icosahedral particle's exact 5-, 3-, and 2-fold axes are represented by the conventional symbols, whereas the asterisk indicates a local 2-fold axis relating 5- and 6-coordinated pentamers. [Courtesy of Stephen Harrison, Harvard University.]

combinant VP1 subunits does not prevent their associating into pentamers but precludes these pentamers from assembling into the viruslike shells that they would otherwise form.

E. Bacteriophage MS2

The RNA bacteriophage MS2 infects only F⁺ (male) *E. coli* (Section 27-1D) because infection is initiated by viral attachment to bacterial F pili. The 275-Å diameter MS2 virion consists of 180 identical 129-residue coat protein subunits arranged with $T = 3$ icosadeltahedral symmetry encapsidating a 3569-nt single-stranded RNA molecule. The virion also contains a single copy of the 44-kD A-protein, which is thought to be responsible for viral attachment to the F pili and must therefore be exposed on the phage surface.

The X-ray structure of MS2, determined by Karin Valegård and Lars Liljas, reveals that its protein shell is formed by 60 icosahedrally related triangular protomers, each of which consists of three chemically identical subunits with slightly different conformations, much as in TBSV (Figure 33-15). However, *the MS2 coat protein does not contain the 8-stranded antiparallel β barrel present in all other spherical viruses of known structure*. Rather, each subunit consists of a 5-stranded antiparallel β sheet facing the interior of the particle overlaid with a short β hairpin and two α helices facing the viral exterior (Fig. 32-23). This protein fold is unlike that in any other known virus.

3. BACTERIOPHAGE λ

Bacteriophage λ (Figs. 32-1*f* and 32-24), a midsized (58 million D) coliphage, has a 55 nm diameter icosahedral head and a flexible 15- to 135-nm long tail that bears a single thin fiber at its end. The virion contains a 48,502-bp linear double-stranded B-DNA molecule of known sequence. Phage λ is, at present, the most extensively characterized complex virus with respect to its molecular biology. Indeed, as we shall see in this section, *its genetic regulatory mechanisms form our best paradigm for the control of development in higher organisms and its assembly is among our best characterized examples of the morphogenesis of biological structures*.

Bacteriophage λ adsorbs to *E. coli* through a specific interaction between the viral tail fiber and a maltose transport protein (the product of the *E. coli* **lamB** gene) that is a component of the bacterium's outer membrane. This interaction initiates a complex and poorly understood process in which the phage DNA is injected through the viral tail into the host cell. Soon after entering the host, the λ DNA, which has complementary single-stranded ends of 12 nucleotides (cohesive ends), circularizes and is covalently closed and supertwisted by the host DNA ligase and DNA gyrase (Fig. 32-25, Stages 1–4).

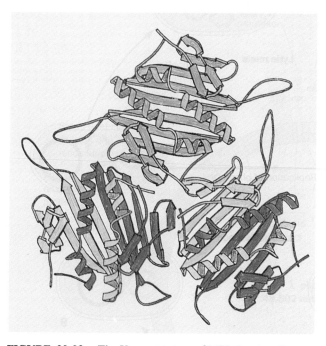

FIGURE 32-23. The X-ray structure of MS2 showing three dimers related by a quasi-threefold axis of the $T = 3$ icosadeltahedral particle. The A, B, and C subunits, as defined in Fig. 32-12*b*, are, respectively, yellow, red, and orange (those in Fig. 32-12*b* are differently colored). The two C monomers shown are related by the particle's exact twofold axis, whereas closely asociated A and B monomers are related by quasi-twofold axes. In all cases, each monomer's five-stranded antiparallel β sheet is extended across the twofold axis and its helices interlock with those of its dimeric mate. Note the lack of structural resemblance between the MS2 subunits and the eight-stranded antiparallel β barrels that form the coat proteins of nearly all other spherical viruses with known structures (Figs. 32-14 and 32-18). [Courtesy of Karin Valegård, Uppsala University, Uppsala, Sweden.]

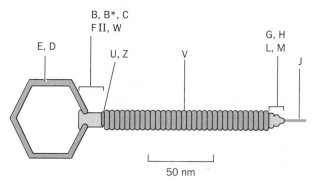

FIGURE 32-24. A sketch of bacteriophage λ indicating the locations of its protein components, The letters refer to specific proteins (gene products; see text). The bar represents 50 nm. [After Eiserling, F.A. *in* Fraenkel-Conrat, H. and Wagner, R.R. (Eds.), *Comparative Virology,* Vol. 13, p. 550, Plenum (1979).]

tion. Its genes are clustered according to function. For example, the genes concerned with the synthesis of phage tail proteins are tandemly arranged on the bottom of Fig. 32-26. This organization, as we shall see, enables these genes to be transcribed together, that is, as an operon. The functions of many of the λ genes and control sites, together with those of the host that are important in phage function, are tabulated in Table 32-1.

In the lytic replication of phage λ, as in love and war, proper timing is essential. This is because the DNA must be replicated in sufficient quantity before it is made unavailable by packaging into phage particles and because packaging must be completed before the host cell is enzymatically lysed. The transcription of the λ genome, which is carried out by host RNA polymerase, is controlled in both the lytic and the lysogenic programs by the regulatory genes that are shaded in red in Fig. 32-26.

The Lytic Mode Has Early, Delayed-Early, and Late Phases

The lytic transcriptional program has three phases (Fig. 32-27):

1. **Early transcription**

   Soon after phage infection or induction, E. coli RNA polymerase commences "leftward" transcription of the phage DNA starting at the promoter p_L and "rightward" transcription (and thus from the opposite DNA strand) from the promoters p_R and p'_R (Fig. 32-27a):

   (i). The "leftward" transcript, L1, which terminates at termination site t_{L1}, encodes the N gene.

   (ii). "Rightward" transcription from p_R terminates with ~50% efficiency at t_{R1}, to yield transcript R1, and otherwise at t_{R2} to yield transcript R2. R1 contains only the *cro* gene transcript, whereas R2 also contains the *cII, O,* and *P* gene transcripts.

   (iii). "Rightward" transcription from p'_R terminating at t'_R yields a short transcript, R4, that specifies no protein.

   L1, R1, and R2 are translated by host ribosomes to yield proteins whose functions are described below.

2. **Delayed-early transcription**

   The second transcriptional phase commences as soon as a significant quantity of the protein **gp$N$** (gp for gene product) accumulates. *This protein, through a mechanism considered below, acts as a **transcriptional antiterminator** at termination sites t_{L1}, t_{R1}, and t_{R2} (Fig. 32-27b):*

   (i). Leftward transcript L1 is extended to form L2, which additionally contains the transcripts of the *cIII, xis,* and *int* genes (which encode proteins involved in switching between the lytic and lysogenic modes; Sections 32-3C and D) together with the **b region** gene transcripts (which specify the so-called

TABLE 32-1.   IMPORTANT GENES AND GENETIC SITES FOR BACTERIOPHAGE

Phage genes

cI	λ Repressor; establishment and maintenance of lysogeny
cII,cIII	Establishment of lysogeny
cro	Repressor of *cI* and early genes
N,Q	Antiterminators for early and delayed early genes
O,P	Origin recognition in DNA replication
γ	Inhibits RecBCD
int	Prophage integration and excision
xis	Prophage excision
B,C,D,E,W,Nu3,FI,FII	Head assembly
G,H,I,J,K,L,M,U,V,Z	Tail assembly
A,Nu1	DNA packaging
R,R$_z$,S	Host lysis
b	Accessory gene region

Phage sites

*att*P	Attachment site for prophage integration
*att*L,*att*R	Prophage excision sites
cos	Cohesive end sites in linear duplex DNA
o_L, o_R	Operators
p_1,p_L,p_R,p_{RM},p_{RE},p'_R	Promoters
t_{L1},t_{R1},t_{R2},t_{R3},t'_R	Transcriptional termination sites
*nut*L,*nut*R	N utilization sites
qut	Q utilization site
ori	DNA replication origin

Host genes

lamB	Host recognition protein
lig	DNA ligase
gyrA,gyrB	DNA gyrase
rpoA,rpoB,rpoC	RNA polymerase core enzyme
rho	Transcription termination factor
nusA,nusB,nusE	Necessary for gp$N$ function
groEL,groES	Head assembly
himA,himD	Integration host factor
hflA,hflB	Degrades gp*cII*
cap,cya	Catabolite repressor system
attB	Prophage integration site
recA	Induction of lytic growth

   accessory proteins which, although not essential for lytic growth, increase its efficiency).

   (ii). Transcript R3, which includes R1 and R2, also encodes a second antiterminator, **gp$Q$**, whose function is discussed below. The continuing trans-

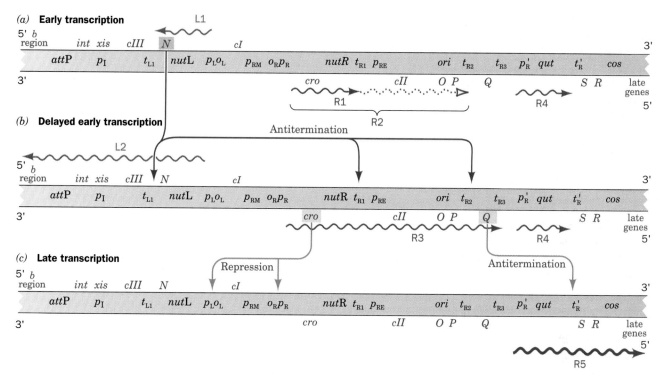

FIGURE 32-27. Gene expression in the lytic pathway of phage λ. Genes specifying proteins that are transcribed to the "left" and "right" are shown above and below the phage chromosome. Control sites are indicated between the DNA strands. The genetic map is not drawn to scale and not all of the genes or control sites are indicated. Transcripts are represented by wiggly arrows pointing in the direction of mRNA elongation; the actions of regulatory proteins are denoted by arrows pointing from each regulatory protein to the site(s) it controls. The lytic pathway has three transcriptional phases: (*a*) early transcription, (*b*) delayed early transcription, and (*c*) late transcription. Gene expression in each of the latter two phases is regulated by proteins synthesized in the preceding phase as is explained in the text. [After Arber, W., *in* Hendrix, R.W., Roberts, J.W., Stahl, F.W., and Weisberg, R.A. (Eds.), *Lambda II, p.* 389, Cold Spring Harbor Laboratory (1983).]

lation of R2 and later R3 to yield **gpO** and **gpP**, proteins that are both required for λ DNA replication, stimulates viral DNA production. Similarly, the translation of R1 and later R3 yields **Cro protein (gpcro),** a repressor of both the "rightward" and "leftward" genes (see below; *cro* stands for *c*ontrol of *r*epressor and *o*ther things).

At this stage, ~15-min postinfection, Cro protein has accumulated in sufficient quantity to bind to operators o_L and o_R, thereby shutting off transcription from p_L and p_R. This is more than just efficient use of resources; the overexpression of the early genes, as occurs in λcro⁻ phage, poisons the lytic cycle's late phase.

3. Late transcription

In the final transcriptional phase (Fig. 32-27c), *the antiterminator gpQ acts to extend the R4 transcript through t'_R to form the R5 transcript.* The "gene dosage" effect of the ~30 copies of phage DNA that have accumulated by the beginning of this stage results in the rapid synthesis of the capsid-forming proteins (which are all encoded by late genes; their assembly to form mature phage particles is described in Section 32-3B), as well as **gpR, gpR$_z$,** and

gpS, which catalyze host cell lysis [gpR is a transglycosidase that cleaves the bond between NAG and NAM in the host cell wall peptidoglycan (Section 10-3B); gpR$_z$ is an endopeptidase that hydrolyzes a peptidoglycan peptide bond; and gpS forms pores in the cell membrane, thereby providing gpR and gpR$_z$ with access to their peptidoglycan substrate]. The first phage particle is completed ~22-min postinfection.

Antitermination Requires the Action of Several Proteins

Transcriptional control in the λ lytic phase is exerted by gpN- and gpQ-mediated antitermination rather than by repressor binding at an operator site through which, for example, *lac* operon expression (Section 29-1B) is regulated. gpN (12 kD) acts at both rho-dependent and rho-independent termination sites (t_{L1} and t_{R1} are rho dependent, whereas t_{R2} is rho independent; transcriptional termination is discussed in Section 29-2E). Yet, gpN does not act at just any transcriptional termination site. Rather, genetic analysis of mutant phage defective for antitermination has established the existence of two so-called *nut* (for *N ut*ilization) sites that are required for antitermination: **nutL,** which is located between p_L and *N,* and **nutR,** which occurs between

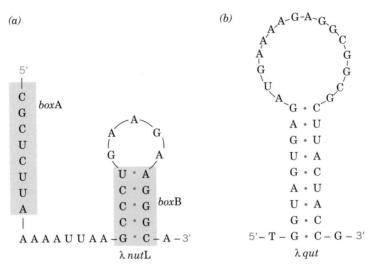

FIGURE 32-28. The RNA sequences of the phage λ control sites (*a*) *nut*L, which closely resembles *nut*R, and (*b*) *qut*. Each of these control sites is thought to form a base paired hairpin.

cro and t_{R1} (Fig. 32-27). These sites have closely similar sequences consisting of two elements, *box*B, whose transcripts can form hydrogen bonded hairpin loops, and *box*A (Fig. 32-28*a*).

What is the mechanism of gp*N*-mediated antitermination? The observation that some *E. coli* defective in antitermination have mutations that map in the *rpoB* gene (which encodes the RNA polymerase β subunit), suggests that gp*N* acts at *nut* sites to render RNA polymerase resistant to termination. Indeed, gp*N*-modulated RNA polymerase will pass over many different terminators that it encounters either naturally or by experimental design. A variety of evidence, including the observation that covering *nut* RNA with ribosomes prevents antitermination, indicates that gp*N* recognizes this site on RNA, not DNA.

Genetic analyses have revealed that antitermination requires several other host factors termed **Nus** (for *N u*tilization *s*ubstance) **proteins** (Fig. 32-29): **NusA**, which specifically binds to both gp*N* and RNA polymerase; **NusE** (which, interestingly, is ribosomal protein S10) and **NusG**, which both bind to RNA polymerase; and **NusB**, which binds to S10. Upon encountering a *nut* site, gp*N* forms a

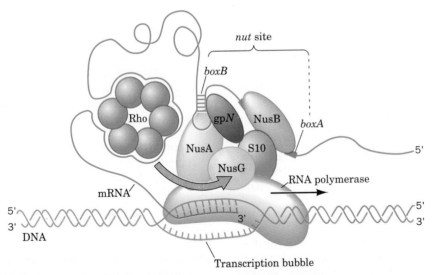

FIGURE 32-29. The proposed antitermination complex between transcribing RNA polymerase, gp*N*, and Nus proteins. gp*N* and the Nus proteins form a complex on a *nut* site of the nascent RNA that binds to the transcribing RNA polymerase further along the looped-out RNA. This complex inhibits RNA polymerase from pausing at a transcriptional termination site, which may prevent rho factor from overtaking the RNA polymerase so as to release the transcript. Another possibility is that transcript release may be inhibited by a gp*N*-modulated direct interaction between NusG and rho factor (*curved arrow*). [After Greenblatt, J., Nodwell, J.R., and Mason, S.W., *Nature* **364**, 402 (1993).]

complex with the Nus proteins and RNA polymerase that travels with this enzyme during elongation and inhibits it from pausing at termination sites. At rho-independent terminators, this deters the release of the transcript at the terminator's weakly bound poly(U) segment, whereas at rho-dependent terminators, it may prevent rho factor from overtaking RNA polymerase, thereby stopping it from unwinding and thus releasing the transcript at the transcription bubble. Alternatively, since it has been shown that NusG binds directly to rho, this interaction, as modulated by gp*N*, may inhibit rho from releasing the nascent transcipt.

Transcriptional antitermination is not limited only to certain bacteriophage. Indeed, the 7 ribosomal RNA (*rrn*) operons of λ's host organism, *E. coli* (which encode its 5S, 16S, and 23S RNAs; Section 29-4B), each contain a *box*A-like element which, together with the Nus proteins, mediates antitermination at *rrn* (which probably explains the function of S10 as a Nus protein). This suggests that λ *box*A is a defective form of *rrn box*A that requires the presence of gp*N* bound to *box*B in addition to the Nus proteins to inhibit termination.

gp*Q*, which overrides t'_R to permit late transcription, acts at a ***qut*** site (analogous to the *nut* sites) that is located some 20 bp downstream from p'_R and that can form an RNA hairpin similar to those of the *nut* sites (Fig. 32-28b). Curiously, however, gp*Q*-mediated antitermination occurs via a mechanism that is quite different from that mediated by gp*N*. In fact, gp*Q* binds specifically to *qut* DNA, not to RNA, where together with NusA it binds to RNA polymerase that is paused at p'_R, thereby accelerating it out of the pause site and somehow inducing it not to terminate transcription at t'_R.

gp*O* and gp*P* Participate in λ DNA Replication

The course of DNA replication in phage λ is diagrammed in Fig. 32-30. Electron microscopy indicates that in the early stages of lytic infection, λ DNA replication occurs both by the bidirectional θ mode (Section 31-1A) from a single replication origin **(*ori*)**, and by the rolling circle (σ) mode (Section 31-3B). By the late stage of the lytic program, however, DNA replication has completely switched, via an unknown mechanism, to the rolling circle mode (with the accompanying synthesis of the complementary strand). The host RecBCD protein (Section 31-6A), a nuclease that would rapidly fragment the resulting concatemeric (consisting of tandemly linked identical units) linear duplex DNA, is inactivated by the phage γ **protein**.

In the process of phage assembly (Section 32-3B), the concatemeric DNA is specifically cleaved in its *cos* (for *co*hesive-end *s*ite) site to yield the linear duplex DNA with complementary 12-nt single-stranded ends that are contained by mature phage particles. The staggered double-stranded scission is made by the so-called **terminase**, which is a complex of the phage proteins **gp*A*** and **gp*Nu1*.**

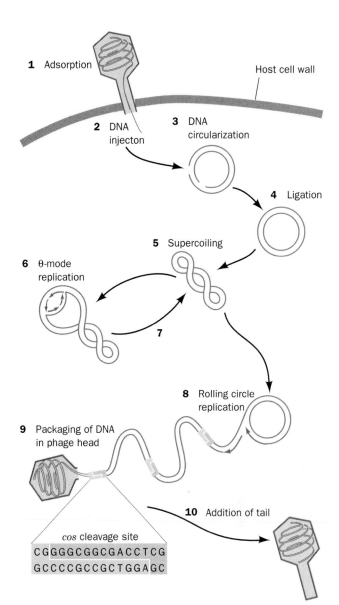

FIGURE 32-30. DNA replication in the lytic mode of bacteriophage λ. The phage particle adsorbs to the host cell (**1**) and injects its linear duplex DNA chromosome (**2**). The DNA circularizes by base pairing at its complementary single-stranded ends (**3**), and the resulting nicked circle is covalently closed (**4**) and supercoiled (**5**) by the sequential actions of host DNA ligase and host DNA gyrase. DNA replication commences according to both the bidirectional θ mode (**6** and **7**) and the rolling circle mode (**8**) but in the later stages of infection occurs exclusively by the rolling circle mode. Here blue arrows indicate the most recently synthesized DNA at the replication forks and the arrowheads represent the 3′ ends of the growing DNA chains. The concatemeric DNA produced by the rolling circle mode is specifically cleaved at its *cos* sites (*shaded boxes*) and is packaged into phage heads (**9**). The addition of tails (**10**) completes the assembly of the mature phage particles, which are each capable of initiating a new round of infection. [After Furth, M.E. and Wickner, S.H., *in* Hendrix, R.W., Roberts, J.W., Stahl, F.W., and Weisberg, R.A. (Eds.), *Lambda II, p.* 146, Cold Spring Harbor Laboratory (1983).]

Phage λ is replicated by the host DNA replication machinery (Sections 31-1, -2, and -3) with the participation of only two phage proteins, gpO and gpP. gpO specifically binds to four repeated palindromic segments within the phage DNA *ori* region, whereas gpP interacts with both gpO and the DnaB protein of the host primosome. gpO and gpP, it is thought, act analogously to host DnaA and DnaC proteins, which are required for the initiation of replication of *E. coli* DNA (Section 31-3C) but not of λ DNA. Evidently, gpO and gpP function to recognize the λ *ori* site, which, curiously, lies within the *O* gene.

B. Virus Assembly

The mature λ phage head contains two major proteins: **gpE** (38 kD), which forms its polyhedral shell, and **gpD** (12 kD), which "decorates" its surface. Electron microscopy indicates that these proteins, which are present in equal numbers, are arranged on the surface of a $T = 7$ icosadeltahedron. However, the λ head also contains four major proteins, **gpB, gpC, gpFII**, and **gpW**, which form a cylindrical structure that attaches the tail to the head. This connector occurs at one of the head's fivefold vertices and thereby breaks its icosahedral symmetry. Hence, gpE and gpD are present in somewhat fewer than the 420 copies/phage in a perfect $T = 7$ icosadeltahedron.

The tail is a tubular entity that consists of 32 stacked hexagonal rings of **gpV** (31 kD) for a total of 192 subunits. The tail begins with a complex adsorption organelle composed of 5 different proteins, **gpG, gpH, gpL, gpM,** and **gpJ,** and ends with an assembly of **gpU** and **gpZ** (Fig. 32-24).

The study of complex virus assembly has been motivated by the conviction that it will provide a foundation for understanding the assembly of cellular organelles. Phage assembly is studied through a procedure developed by Robert Edgar and William Wood that combines genetics, biochemistry, and electron microscopy. Conditionally lethal mutations (either temperature-sensitive mutants, which appear normal at low temperatures but exhibit a mutant phenotype at higher temperatures; or supressor-sensitive *amber* mutants, Section 30-2E) are generated that, under nonpermissive conditions, block phage assembly at various stages. This process results in the accumulation of intermediate assemblies or side products that can be isolated and structurally characterized through electron microscopy. The mutant protein can be identified, through a process known as *in vitro* **complementation** (in analogy with *in vivo* genetic complementation; Section 27-1C), by mixing cell-free extracts containing these structural intermediates with the corresponding normal protein to yield infectious phage particles.

The assembly of bacteriophage λ occurs through a branched pathway in which the phage heads and tails are formed separately and then join to yield mature virions.

Phage Head Assembly

λ Phage head assembly occurs in five stages (Fig. 32-31, *right*):

1. Two phage proteins, gpB and **gpNu3** (19 kD), together with two host-supplied chaperonin proteins, GroEL and GroES, interact to form an "initiator" that consists of 12 copies of gpB arranged in a ring with a central orifice. This precursor of the mature phage head–tail connector (Fig. 32-24) apparently organizes the phage head's subsequent formation. GroEL and GroES, it will be recalled, provide a protected environment that facilitates the proper folding and assembly of proteins and protein complexes such as the connector precursor (Section 8-1C). In fact, these chaperonins were discovered through their role in λ assembly. gpNu3, as we shall see, also functions as a molecular chaperone in that it has but a transient role in phage head assembly.

2. gpE and additional gpNu3 associate to form a structure called an immature **prohead.** If gpB, GroEL, or GroES is defective or absent, some gpE assembles into spiral or tubular structures, which indicates that the missing proteins guide the formation of a proper shell. The absence of gpNu3 results in the formation of but a few shells that contain only gpE. gpNu3 evidently facilitates proper shell construction and promotes the association of gpE with gpB.

3. In the formation of the mature prohead, ~75% of the gpB (61 kD) is cleaved to form **gpB*** (56 kD); the gpNu3 is degraded and lost from the structure; and 10 copies of gpC participate in a fusion–cleavage reaction with 10 additional copies of gpE to yield the hybrid proteins **pX1** and **pX2** (p for protein), which form the collar that apparently holds the connector in place. This maturation process, which involves only phage gene products that are part of the immature prohead, requires that all of the prohead components be present and functional; that is, that the immature prohead be correctly assembled to start with. The enzyme(s) that catalyze this process have, nevertheless, not been identified.

4. The concatemeric viral DNA is packaged in the phage head and cleaved by mechanisms discussed below. During this process, the capsid proteins undergo a conformational change that results in an expansion of the phage head to twice its original volume (a process that occurs in $4M$ urea in the absence of DNA). gpD then attaches to newly exposed binding sites on gpE, thereby partially stabilizing the capsid's expanded structure.

5. In the final stage of phage head assembly, gpW and gpFII add in that order to stabilize the head and form the tail-binding site.

These stages of phage head assembly, as well as some of their component reactions, must proceed in an obligatory order for proper assembly to occur. Of particular interest is that *the components of the mature phage head are not en-*

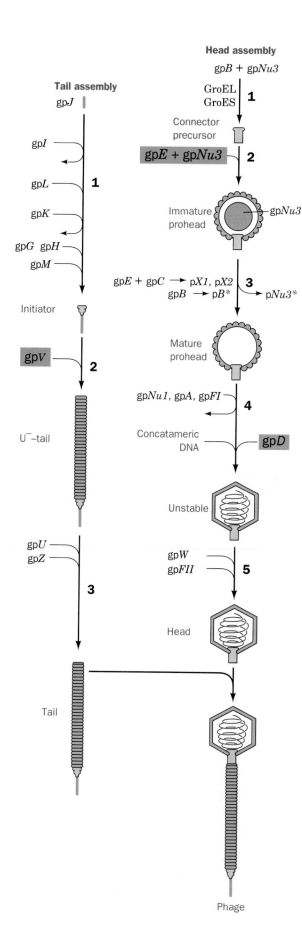

Tail assembly

gp*J*

gp*I*

gp*L* **1**

gp*K*

gp*G* gp*H*

gp*M*

Initiator

gp*V* **2**

U⁻–tail

gp*U*
gp*Z*

3

Tail

Head assembly

gp*B* + gp*Nu3*

GroEL
GroES **1**

Connector
precursor

gp*E* + gp*Nu3* **2**

Immature
prohead ——gp*Nu3*

gp*E* + gp*C* ⟶ p*X1*, p*X2* **3**
gp*B* ⟶ p*B** ——p*Nu3**

Mature
prohead

gp*Nu1*, gp*A*, gp*FI* **4**

Concatameric
DNA —— gp*D*

Unstable

gp*W*
gp*FII* **5**

Head

Phage

FIGURE 32-31. The assembly of bacteriophage λ. The heads and tails are assembled in separate pathways before joining to form the mature phage particle. Within each pathway the order of the various reactions is obligatory for proper assembly to occur. gp*E*, gp*Nu3*, gp*D*, and gp*V* are highlighted in red boxes to indicate that relatively large numbers of these proteins are required for phage assembly. The numbered steps are described in the text.

tirely self-assembling as are, for example, TMV (Section 32-1B) and ribosomes (Section 30-3A). Rather, the *E. coli* proteins groEL and groES, as we saw, facilitate head–tail connector assembly. Moreover, gp*Nu3*, which occurs in ~200 copies inside the immature prohead but is absent from the mature prohead, evidently acts as a "scaffolding" protein that organizes gp*E* to form a properly assembled phage head. Finally, *since phage assembly involves several proteolytic reactions, it must also be considered to occur via enzyme-directed processes.*

DNA Is Tightly Packed in the Phage Head

An intriguing question of λ phage assembly is, How does a 55-nm-diameter phage head package a 16,500-nm-long, stiff duplex DNA molecule? Two models have been proposed:

1. Electron microscopy of gently disrupted phages and X-ray scattering from phage solutions both suggest that the DNA is tightly wound in a spool-like structure (Fig. 32-32a). Since the DNA linearly enters the phage prohead through the head–tail connector (see below), it has been

(a)   *(b)*

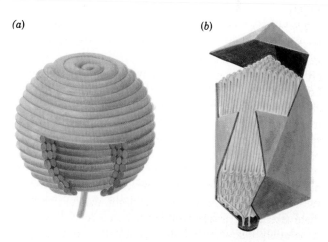

FIGURE 32-32. Models for the packing of double-stranded DNA inside a phage head: *(a)* The concentric shell model in which the DNA is wound inward like a spool of twine about the phage's long axis. [After Harrison, S.C., *J. Mol. Biol.* **171**, 579 (1983).] *(b)* The spiral-fold model in which the DNA strands run parallel to the phage's long axis with sharp 180° bends at the ends of the capsid. The folds themselves are radially arranged about the phage's long axis in spirally organized shells. [After Black, L.W., Newcomb, W.W., Boring, J.W., and Brown, J.C., *Proc. Natl. Acad. Sci.* **82**, 7963 (1985).]

proposed that its stiffness would cause it to first coil against the inner wall of the rigid protein shell and then to wind concentrically inward, much like a spool of twine.

2. Ion etching of phages (a process in which frozen phages are progressively worn away by bombardment with a beam of Ar^+ ions) in which only one end of the phage DNA is radioactively labeled indicates that the first DNA to enter the prohead is the most shielded from the ion beam. This observation suggests that the leading DNA segment is condensed in the center of the capsid, thereby supporting the "spiral-fold" model of DNA packaging (Fig. 32-32*b*).

In both models, the DNA's detailed winding path varies randomly from particle to particle as is indicated by the observation that packaged DNA can be cross-linked to the capsid along its entire length. Both DNA packing models also predict that the injection of phage DNA into the host bacterium proceeds by a reversal of the packaging process.

DNA Is "Pumped" into the Phage Head by an ATP-Driven Process

The packaging of λ DNA begins when terminase (gp$A$ + gp$Nu1$) binds to its recognition sequences on a randomly selected ~200-bp *cos* site. The resulting complex then binds to the prohead so as to introduce the DNA into it through a 20-Å-diameter orifice in its head–tail connector. The "left" end of the DNA chromosome enters the prohead first as is indicated by the observation that only this end of the chromosome is packaged by an *in vitro* system when λ DNA restriction fragments are used. Whether the cutting of the initial *cos* site precedes or follows the initiation of packaging is unknown. However, at least *in vitro,* this process requires the binding to *cos* of either of the *E. coli* histone-like proteins known as **integration host factor (IHF)** and **termination host factor (THF)**. IHF binds specific sequences of duplex DNA, which it wraps around its surface, thereby inducing a sharp bend.

The packing of DNA inside a phage head must be an enthalpically as well as entropically unfavorable process because of duplex DNA's stiffness and its intramolecular charge repulsions. The observation that DNA packaging requires the presence of ATP therefore strongly suggests that DNA is actively "pumped" into the phage head by an ATP-driven process. In a particularly intriguing model of a DNA pump, for which there is no proof, ATP hydrolysis drives the rotation of the connector with respect to the prohead so as to screw the helically grooved DNA into the phage head much like a threaded rod in a nut. The injection of λ DNA into a host bacterium by a mature phage is presumably a spontaneous process that, once it has been triggered, is driven by the free energy stored in the compacted DNA.

The final step in the DNA packaging process is the recognition and cleavage of the next *cos* site (Fig. 32-30) on the concatemeric DNA by terminase, possibly with the participation of **gpFI**. Phage λ therefore contains a unique segment of DNA (in contrast to some phages in which the amount of DNA packaged is limited by a "headful" mechanism that results in their containing somewhat more DNA than an entire chromosome). Indeed, the λ packaging system will efficiently package a DNA that is 75 to 105% the length of the wild-type λ DNA as long as it is flanked by *cos* sites (the central third of the phage DNA, which encodes the dispensable accessory genes, can be replaced by other sequences, thereby making phage λ a useful cloning vector; Section 28-8A).

Tail Asssembly

Tail assembly, which occurs independently of head assembly, proceeds, as a comparison of Figs. 32-24 and 32-31 indicates, from the 200-Å-long tail fiber towards the head-binding end. This strictly ordered series of reactions can be considered to have three stages (Fig. 32-31, *left*):

1. The formation of the "initiator," which ultimately becomes the adsorption organelle, requires the sequential actions on gp$J$ (the tail fiber protein) of the products of phage genes *I, L, K, G, H,* and *M,* respectively. Of these, only **gp$I$** and **gp$K$** are not components of the mature tail.

2. The initiator forms the nucleus for the polymerization of gp$V,$ the major tail protein, to form a stack of 32 hexameric rings. The length of this stack is thought to be regulated by gp$H,$ which, the available evidence suggests, becomes extended along the length of the growing tail and somehow limits its growth. λ tail length is apparently specified in much the same way that the helical length of TMV is governed (Section 32-1B), although in TMV, the regulating template is an RNA molecule rather than a protein.

3. In the termination and maturation stage of tail assembly, **gp$U$** attaches to the growing tail, thereby preventing its further elongation. The resultant immature tail has the same shape as the mature tail and can attach to the head. In order to form an infectious phage particle, however, the immature tail must be activated by the action of **gp$Z$** before joining the head.

The completed tail then spontaneously attaches to a mature phage head to form an infectious λ phage particle (Fig. 32-31, *bottom*).

The Assembly of Other Double-Stranded DNA Phages Resembles That of λ

The assembly of several other double-stranded DNA bacteriophages have been studied in detail, notably those of **coliophages T4 and T7** and the lambdoid (λ-like) phage **P22** (which grows on *Salmonella typhimurium*). All of them are formed in assembly processes that closely resemble that of phage λ. For example, their head assembly processes proceed in obligatory reaction sequences through an initiation stage; the scaffolded assembly of a prohead; an ATP-driven DNA packaging process, in which the DNA assumes a tightly packed conformation and the prohead undergoes an expansion; and a final stabilization. The mature phages then form by the attachment of separately assembled tails to the completed and DNA-filled heads.

C. The Lysogenic Mode

Lysogeny is established by the integration of viral DNA into the host chromosome accompanied by the shutdown of all lytic gene expression. With phage λ, integration takes place through a **site-specific recombination** process that differs from general recombination (Section 31-6A) in that it occurs only between the chromosomal sites designated *att*P on the phage and *att*B on the bacterial host (Fig. 32-33).

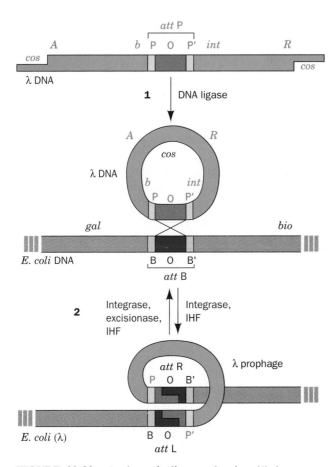

These two *att*achment sites have a 15-bp homology (Fig. 32-34), so they can be represented as having the sequences POP' for *att*P and BOB' for *att*B, where O denotes their common sequence. Phage integration occurs through a process that yields the inserted phage chromosome flanked by the sequence BOP' on the "left" (the *att*L site) and POB' on the "right" (the *att*R site; Fig. 32-33). The nature of the crossover site was determined through the use of ³²P-labeled bacterial DNA and unlabeled phage DNA. The crossover site occurs at a unique position on each strand that is displaced with respect to its complementary strand so as to form a staggered recombination joint (Fig. 32-34).

Integrase Mediates λ DNA Integration, whereas Excisionase Is Additionally Required for λ DNA Excision

Phage integration is mediated by a phage-specific **integrase**, the **λ*int*** gene product, acting in concert with IHF. Integrase, which specifically binds the region common to the crossover sites (O), acts *in vitro* as a Type I topoisomerase (which nicks and reseals only one strand of double helical DNA; Section 28-5C) and has been shown to resolve synthetic Holliday structures (crossover structures; Section 31-6A). It therefore seems likely that the breaking and re-

FIGURE 32-33. A schematic diagram showing: **(1)** the circularization of the linear phage λ DNA through base pairing between its complementary ends to form the *cos* site; and **(2)** the integration/excision of this DNA into/from the *E. coli* chromosome through site-specific recombination between the phage *att*P and host *att*B sites. The darker colored regions in the *att* sites represent the homologous 15-bp crossover sequences (O), whereas the lighter colored regions symbolize the unique sequences of bacterial (B and B') and phage (P and P') origin. [After Landy, A. and Weisberg, R.A., in Hendrix, R.W., Roberts, J.W., Stahl, F.W., and Weisberg, R.A. (Eds.), *Lambda II, p.* 212, Cold Spring Harbor Laboratory (1983).]

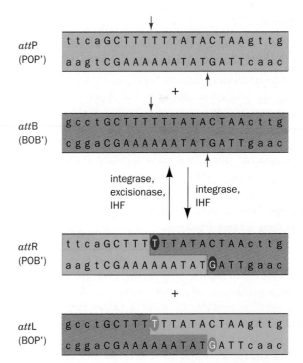

FIGURE 32-34. The site-specific recombination process that inserts/excises phage λ DNA into/from the chromosome of its *E. coli* host. Exchange occurs between the phage *att*P site (*red*) and the bacterial *att*B site (*blue*), and the prophage *att*L and *att*R sites. The strand breaks occur at the approximate positions indicated by the short blue arrows. The sources of the more darkly shaded bases in *att*R and *att*L are uncertain. The upper case letters represent bases in the O region common to the phage and bacterial DNA's, whereas lower case letters symbolize bases in the flanking B, B', P, and P' sites.

sealing of DNA strands, which constitutes the crossover event, is carried out by integrase. IHF has no demonstrable endonuclease or topoisomerase activity but specifically binds to DNAs bearing various *att* sequences.

Since viral integration is not an energy-consuming process, why is phage integration not readily reversible? The answer is that the prophage excision requires the participation of an **excisionase**, the *λxis* gene product, in concert with integrase, IHF, and Fis (a DNA-binding protein that also stimulates Hin-mediated gene inversion; Section 31-6B). Apparently the *λ* recombination system has an inherent asymmetry that ensures the kinetic stability of the lysogenic integration product. The mechanism by which excisionase reverses the integration process in unknown, although it has been shown that this protein specifically binds to POB′, where it induces a sharp bend in this DNA.

The Relative Levels of Cro Protein and cI Repressor Determine the λ Phage Life Cycle

The establishment of lysogeny in phage λ is triggered by high concentrations of **gp*cII*** *(see below).* This early gene product stimulates "leftward" transcription from two promoters, p_I (I for *int*egrase) and p_{RE} (RE for *r*epressor *e*stablishment; Fig. 32-35a):

1. Transcription initiated from p_I, which is located within the *xis* gene, results in the production of integrase but not excisionase. λ DNA is consequently integrated into the host chromosome to form the prophage.

2. The transcript initiated from p_{RE} encodes the *cI* gene whose product is called the *λ* or **cI repressor**. The *λ* repressor, as does Cro protein (Section 32-3A), binds to the o_L and o_R operators, thereby blocking transcription from p_L and p_R, respectively (note that these operators are upstream from their corresponding promoters rather than downstream as in the *lac* operon; Fig. 29-3). *Both repressors therefore act to shut down the synthesis of early gene products, including Cro protein and gpcII.*

gp*cII* is metabolically unstable with a half-life of ~1 min (see below) so that *cI* transcription from p_{RE} soon ceases. λ Repressor bound at o_R, but not Cro protein, however, stimulates "leftward" transcription of *cI* from p_{RM} (RM for *r*epressor *m*aintenance; Fig. 32-35b). In other words, *Cro protein represses all mRNA synthesis, whereas λ repressor stimulates transcription of its own gene while repressing all other mRNA synthesis. This conceptually simple difference between the actions of λ repressor and Cro protein forms the basis of a genetic switch that stably maintains phage λ in either the lytic or the lysogenic state.* The molecular mechanism of this switch is described in Section 32-3D. In the following subsection we discuss how this switch is "thrown" from one state to another. You should recognize, however, that, *once the switch is thrown in favor of the lytic cycle, that is, when Cro protein occupies o_L and o_R, the phage is irrevocably committed to at least one generation of lytic growth.*

gpcII Is Activated when Phage Multiplicity Is High or Nutritional Conditions Are Poor

The reason why a high gp*cII* concentration is required to establish lysogeny is that this early gene product can stimulate transcription from p_I and p_{RE} only when it is in oligomeric form. This phenomenon accounts for the observation that lysogeny is induced when the **multiplicity of infection** (ratio of infecting phages to bacteria) is large (≥ 10) since this gene dosage effect results in gp*cII* being synthesized at a high rate.

gp*cII* is metabolically unstable because it is preferentially proteolyzed by host proteins, notably **gp*hflA*** and **gp*hflB***. However, gp*cIII* somehow protects gp*cII* from the action of gp*hflA*, which is why its presence enhances lysogenation

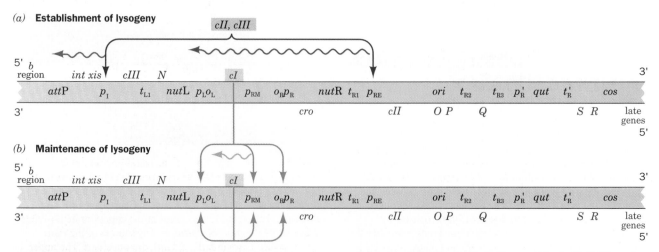

FIGURE 32-35. Gene expression in (*a*) the establishment and (*b*) the maintenance of lysogeny by bacteriophage λ. The symbols used are described in the legend of Fig. 32-27. [After

Arber, W., *in* Hendrix, R.W., Roberts, J.W., Stahl, F.W., and Weisberg, R.A. (Eds.), *Lambda II, p.* 389, Cold Spring Harbor Laboratory (1983).]

(Fig. 32-35*a*). The activity of gp*hflA* is dependent on the host cAMP-activated catabolite repression system (Section 29-3C) as is indicated by the observation that *E. coli* mutants defective in this system lysogenize with less than normal frequency. Yet, if these mutant strains are also *hflA⁻*, they lysogenize with greater than normal frequency. Apparently the *E. coli* catabolite repression system, which is known to regulate the transcription of many bacterial genes, controls *hflA* activity, perhaps by directly repressing this protein's synthesis at high cAMP concentrations. *This explains why poor host nutrition, which results in elevated cAMP concentrations, stimulates lysogenation.*

Once a prophage has been integrated in the host chromosome, lysogeny is stably maintained from generation to generation by λ repressor. This is because λ repressor stimulates its own synthesis at a rate sufficient to maintain lysogeny in the progeny while repressing the transcription of all other phage genes. In fact, *λ repressor is synthesized in sufficient excess to also repress transcription from superinfecting λ phage, thereby accounting for the phenomenon of immunity.* We shall see below how induction occurs.

D. Mechanism of the λ Switch

The lysogenic cycle is a highly stable mode of phage λ replication; under normal conditions lysogens spontaneously induce only about once per 10⁵ cell divisions. Yet, transient exposure to inducing conditions triggers lytic growth in almost every cell of a lysogenic bacterial culture. In this section, we consider how this genetic switch, whose mechanism was largely elucidated by Mark Ptashne, can so tightly repress lytic growth and yet remain poised to turn it on efficiently.

o_R Consists of Three Homologous Palindromic Subsites

Both of the operators to which λ repressor and Cro protein bind, o_L and o_R, consist of three subsites (Fig. 32-36). These are designated o_{L1}, o_{L2}, and o_{L3} for o_L, and o_{R1}, o_{R2}, and o_{R3} for o_R. *Each of these subsites consists of a homologous 17-bp segment that has approximate palindromic symmetry.* Nevertheless, *only the elements of o_R form components of the λ switch.*

λ Repressor and Cro Protein Structurally Resemble Other Repressors

λ repressor binds to DNA as a dimer so that its twofold symmetry matches those of the operator subsites to which it binds. The monomer's 236-residue polypeptide chain is folded into two roughly equal sized domains connected by an ~30-residue segment that is readily cleaved by proteolytic enzymes. The isolated N-terminal domains retain their ability to bind specifically to operators (although with only one half of the binding energy of the intact repressor) but cannot dimerize. The C-terminal domains can still dimerize but lack the capacity to bind DNA. Evidently, *repressor's N-terminal domain binds operator, whereas its C-terminal domain provides the contacts for dimer formation.*

Although the λ repressor has not been crystallized, its N-terminal domain comprising residues 1 to 92, as excised by treatment with the papaya protease **papain,** does crystallize. The X-ray structure of this protein, both alone and in complex with a 20-bp DNA containing the o_{L1} sequence, has been determined by Carl Pabo. The N-terminal domain crystallizes as a symmetric dimer with each subunit containing an N-terminal arm and five α helices (Fig. 32-37). Two of these helices, α2 and α3, form a helix–turn–helix (HTH) motif, much like those in other prokaryotic repressors of known structure (Sections 29-3D). The α3 helix, the recognition helix, protrudes from the protein surface such that the two α3 helices of the dimeric protein fit into successive major grooves of the operator DNA. Similar associations are observed in the X-ray structures of the closely related **bacteriophage 434 repressor** N-terminal fragment in complex with a 20-bp DNA containing its operator sequence (Fig. 29-23).

Cro protein also forms dimers. In contrast to λ or 434 repressor, however, this 66-residue polypeptide forms but

(*a*)

```
5'... TGTGCTCAGTATCACCGCCAGTGGTATTTATGTCAACACCGCCAGAGATAATTTATCACCGCAGATGGTTATCTGTAT... 3'
3'... ACACGAGTCATAGTGGCGGTCACCATAAATACAGTTGTGGCGGTCTCTATTAAATAGTGGCGTCTACCAATAGACATA... 5'
```

p_L o_{L1} o_{L2} o_{L3}

(*b*)

                                                                p_R

```
5'... TACGTTAAATCTATCACCGCAAGGGATAAATATCTAACACCGTGCGTGTTGACTATTTTACCTCTGGCGGTGATAATGGTTGCA...3'
3'... ATGCAATTTAGATAGTGGCGTTCCCTATTTATAGATTGTGGCACGCACAACTGATAAAATGGAGACCGCCACTATTACCAACGT...5'
```

p_{RM} o_{R3} o_{R2} o_{R1}

FIGURE 32-36
The base sequences of (*a*) the o_L and (*b*) the o_R regions of the phage λ chromosome. Each of these operators consists of three homologous 17-bp subsites separated by short AT-rich spacers. Each subsite has approximate palindromic (twofold) symmetry as is demonstrated by the comparison of the two sets of red letters in each subsite. The wiggly arrows mark the transcriptional start sites and directions at the indicated promoters.

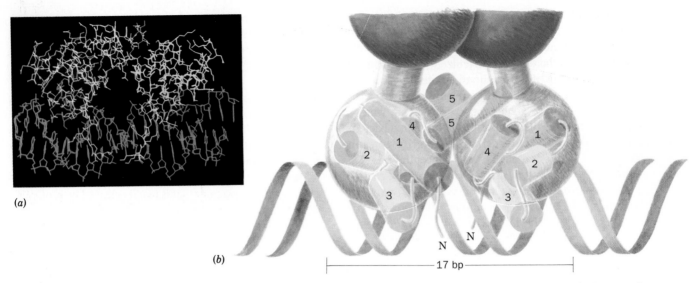

(a)

(b)

17 bp

FIGURE 32-37. The X-ray structure of a dimer of λ repressor N-terminal domains in complex with B-DNA. (*a*) Stick-form representation of the complex in which the DNA is blue, the two repressor N-terminal domains are yellow and white, and their recognition helices are red. Note that the protein's N-terminal arms wrap around the DNA. This accounts for the observation that the G residues in the major groove on the repressor–operator complex's "back side" are protected from methylation only when these N-terminal arms are intact. [Courtesy of Carl Pabo, The Johns Hopkins University.]

(*b*) An interpretive drawing indicating how contacts between the repressor's C-terminal domains (*upper lobes; not part of the X-ray structure*) maintain the intact protein's dimeric character. The λ repressor binds to the 17-bp operator subsites of o_L and o_R as symmetric dimers with the N-terminal domain of each subunit specifically binding to a half-subsite. Note how the α3 recognition helices of the symmetry related α2–α3 HTH units (*light yellow*) fit into successive turns of the DNA's major groove. [After Ptashne, M., *A Genetic Switch* (2nd ed.), *p.* 38, Cell Press & Blackwell Scientific Publications (1992).]

one domain that contains both its operator recognition site and its dimerization contacts. The X-ray structure of Cro in complex with a 17-bp tight binding operator DNA, determined by Brian Matthews, reveals that this dimer likewise contains a pair of HTH units (Fig. 32-38), but which bind to the DNA such that they induce it to bend about the protein by 40°. The sequence-specific binding predicted by this

structure is supported by Robert Sauer's genetic studies indicating that mutant varieties of Cro, in which the proposed DNA-contacting residues have been changed, are defective in operator binding. Moreover, this structure closely resembles that of the related **phage 434 Cro protein** in complex with a 20-bp DNA containing its operator sequence (Fig. 29-24).

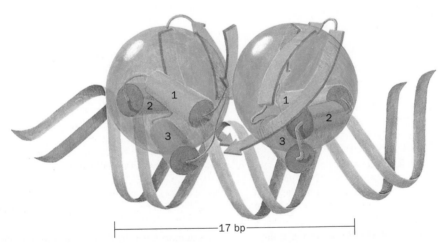

17 bp

FIGURE 32-38. The X-ray structure of the Cro protein dimer shown in its complex with B-DNA. Note that the λ repressor (Fig. 32-37), although otherwise dissimilar, contains HTH units

that also bind in successive turns of the DNA's major groove. [After Ptashne, M., *A Genetic Switch* (2nd ed.), *p.* 40, Cell Press & Blackwell Scientific Publications (1992).]

Repressor Stimulates Its Own Synthesis While Repressing All Other λ Genes

Chemical and nuclease protection experiments have indicated that λ repressor has the following order of intrinsic affinities for the subsites of o_R (Fig. 32-39):

$$o_{R1} > o_{R2} > o_{R3}$$

Despite this order, o_{R1} and o_{R2} are filled nearly together. This is because *λ repressor bound at o_{R1} cooperatively binds repressor at o_{R2} through associations between their C-terminal domains (Fig. 32-39c).* o_{R1} and o_{R2} are therefore both occupied at low λ repressor concentrations, whereas o_{R3} becomes occupied only at higher repressor concentrations.

The binding of λ repressor to o_R, as we previously mentioned, abolishes transcription from p_R and stimulates it from p_{RM} (Fig. 32-39c). At high concentrations of λ repressor, however, transcription from p_{RM} is also repressed (Fig. 32-39d). These phenomena have been clearly demonstrated through the construction of a series of hybrid operons that permit the effect of λ repressor on a promoter to be studied in a controlled manner. The system has two elements (Fig. 32-40):

1. A plasmid bearing the *lacI* gene (which encodes *lac* repressor; Section 29-1A) and the *lac* operator–promoter sequence fused to the *cI* gene. This construct permits the amount of λ repressor produced to be directly controlled by varying the concentration of the *lac* inducer IPTG (Section 29-1A).

2. A prophage containing o_R and either p_{RM}, as Fig. 32-40 indicates, or p_R, fused to the *lacZ* gene. The amount of the *lacZ* gene product, β-galactosidase, produced, which can be readily assayed, reflects the activity of p_{RM} (or p_R).

The manipulation of these systems has demonstrated that at intermediate λ repressor concentrations (when o_{R1} and o_{R2} are occupied), transcription from p_R is indeed repressed,

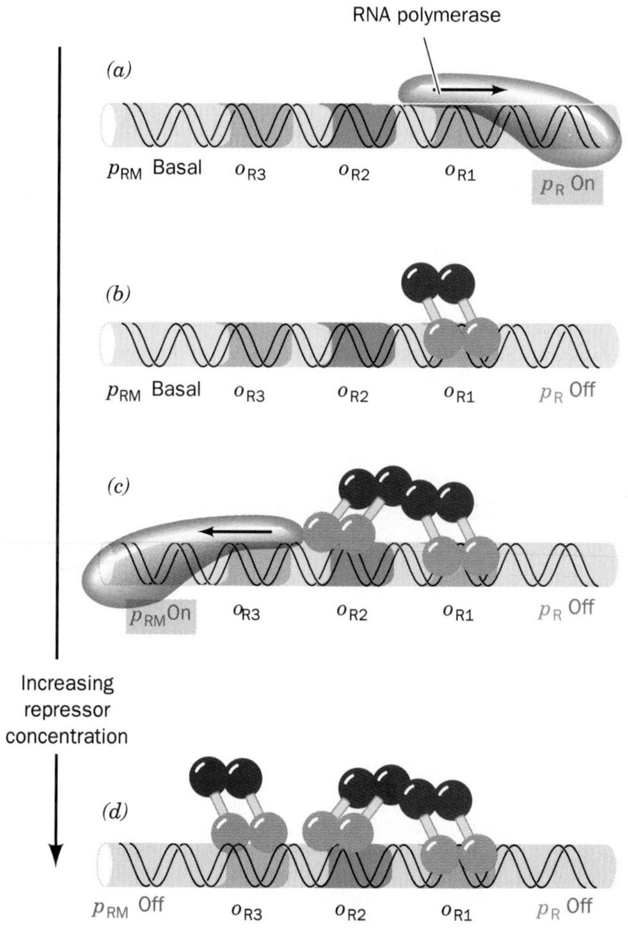

RNA polymerase

(a)

p_{RM} Basal o_{R3} o_{R2} o_{R1} p_R On

(b)

p_{RM} Basal o_{R3} o_{R2} o_{R1} p_R Off

(c)

p_{RM} On o_{R3} o_{R2} o_{R1} p_R Off

Increasing repressor concentration

(d)

p_{RM} Off o_{R3} o_{R2} o_{R1} p_R Off

FIGURE 32-39. The binding of λ repressor to the three subsites of o_R. (*a*) In the absence of repressor, RNA polymerase initiates transcription at a high level from p_R (*right*) and at a basal level from p_{RM}. (*b*) Repressor has ~10 times higher affinity for o_{R1} than it does for o_{R2} or o_{R3}. Repressor dimer therefore first binds to o_{R1} so as to block transcription from p_R. (*c*) A second repressor dimer binds to o_{R2} at only slightly higher repressor concentrations due to specific binding between the C-terminal domains of neighboring repressors. In doing so, it stimulates RNA polymerase to initiate transcription from p_{RM} at a high level (*left*). (*d*) At high repressor concentrations, repressor binds to o_{R3} so as to block transcription from p_{RM}. [After Ptashne, M., *A Genetic Switch* (2nd ed.), p. 23, Cell Press & Blackwell Scientific Publications (1992).]

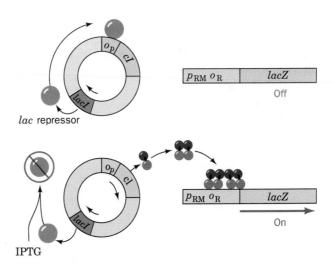

o_P *cI*

lacI

lac repressor

p_{RM} o_R | *lacZ*

Off

o_P *cI*

lacI

IPTG

p_{RM} o_R | *lacZ*

On

FIGURE 32-40. The genetic system used to study the effect of λ repressor on p_{RM}. The bacterium contains two hybrid operons. The first (*left*) is a plasmid bearing the *lac* operator–promoter (*Op*) fused to the λ*cI* gene so as to provide a source of repressor. The *lacI* gene, which encodes *lac* repressor, is also incorporated in the plasmid so that the level of λ repressor in the bacterium may be controlled by the concentration of the *lac* inducer IPTG. The second operon (*right*) is carried on a prophage that contains the promoter p_{RM} fused to the *lacZ* gene. The level of β-galactosidase (gp*lacZ*) in these cells therefore reflects the activity of p_{RM}. In similar experiments, the *cro* gene was substituted for λ*cI* and/or p_{RM} was replaced by p_R. [After Ptashne, M., *A Genetic Switch* (2nd ed.), p. 89, Cell Press & Blackwell Scientific Publications (1992).]

whereas that from p_{RM} is stimulated (Fig. 32-41). Transcription from p_{RM} only becomes repressed at high levels of λ repressor (when o_{R3} is also occupied). The stimulation of transcription from p_{RM} is abolished by mutations in o_{R2} that prevent repressor binding, whereas its repression at high repressor concentrations is relieved by mutations in o_{R3}.

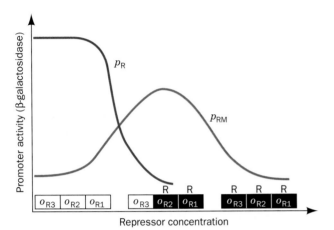

FIGURE 32-41. The response of p_{RM} and p_R to the λ repressor level. The p_{RM} curve was derived using the system diagrammed in Fig. 32-40, whereas the p_R curve was obtained using a similar system but with p_R rather than p_{RM} fused to *lacZ*. The amount of λ repressor that maximally stimulates p_{RM} is approximately that which occurs in a λ lysogen. At least fivefold more repressor is required to half-maximally repress p_{RM}. The boxes indicate the states of each o_R subsite at the various repressor concentrations; black represents repressor occupancy. [After Ptashne, M., *A Genetic Switch* (2nd ed.), *p.* 90, Cell Press & Blackwell Scientific Publications (1992).]

Thus, *occupancy of o_{R2} by λ repressor stimulates transcription from p_{RM}, whereas occupancy of o_{R3} prevents it (Fig. 32-39c and d). By the same token, occupancy of o_{R1} and/or o_{R2} prevents transcription from p_R.* In this way, λ repressor prevents the synthesis of all phage gene products but itself. Yet, at high repressor concentrations, its synthesis is also repressed, thereby maintaining the repressor concentration within reasonable limits.

What is the basis of λ repressor's remarkable property of inhibiting transcription from one promoter while stimulating it from another? Knowledge of the sizes and shapes of repressor and RNA polymerase, as well as their positions on the DNA as demonstrated by chemical protection experiments, indicate that repressor at o_{R2} and RNA polymerase at p_{RM} are in contact (Fig. 32-42). Evidently, *repressor stimulates RNA polymerase activity through their cooperative binding to DNA.* This model was corroborated by the analysis of repressor mutants that bind normally (or nearly so) to operators but fail to stimulate the binding of RNA polymerase: All of the mutated residues occur either in helix $\alpha 2$ or in the link connecting it to helix $\alpha 3$ and lie on the surface of the protein that is thought to face the RNA polymerase-binding site (Fig. 32-42).

Cro Protein Binding to o_R Represses All λ Genes

Cro protein binds to the subsites of o_R in an order opposite to that of λ repressor (Fig. 32-43):

$$o_{R3} > o_{R2} \approx o_{R1}$$

This binding is noncooperative. Through experiments similar to that diagrammed in Fig. 32-40, but with *cro* in place of *cI*, the binding of Cro protein to o_{R3} was shown to abolish

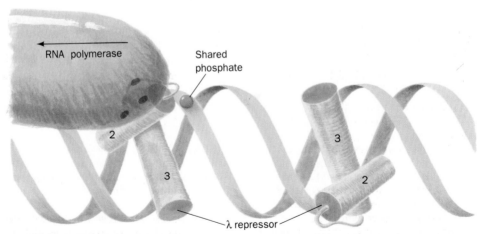

FIGURE 32-42. Repressor bound at o_{R2} is proposed to stimulate transcription at p_{RM} through a specific association with RNA polymerase that helps the polymerase bind to the promoter. This model is supported by the locations of the altered residues (*blue dots*) in three mutant repressors that bind normaly to o_{R2} but fail to stimulate transcription at p_{RM}. The relative positions of repressor and RNA polymerase are established by the location of a phosphate group (*orange sphere*) whose ethylation interferes with the binding of both proteins to the DNA. For the sake of clarity, only the $\alpha_2 - \alpha_3$ helix – turn – helix units of the repressor dimer are shown.

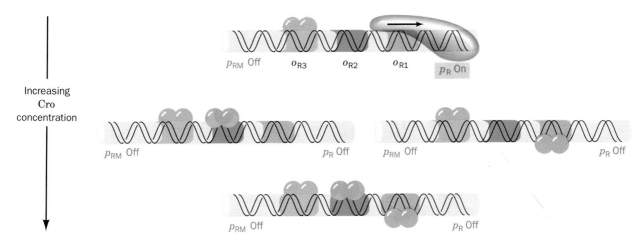

Increasing
Cro
concentration

FIGURE 32-43. The binding of Cro protein to the three o_R subsites. o_{R3} binds Cro ~10 times more tightly than does o_{R1} or o_{R2}. Cro dimer therefore first binds to o_{R3}. A second dimer then binds to either o_{R1} or o_{R2} and in each case blocks transcription from p_R. At high Cro concentrations, all three operator subsites are occupied. Compare this binding sequence with that of λ repressor (Fig. 32-39). [After Ptashne, M., *A Genetic Switch* (2nd ed.), p. 27, Cell Press & Blackwell Scientific Publications (1992).]

transcription from p_{RM}. Additional Cro binding to o_{R2} and/or o_{R1} turns off transcription from p_R.

The SOS Response Induces the RecA-Mediated Cleavage of λ Repressor

A final piece of information allows us to understand the workings of the λ switch. *The lytic phase is induced by agents that damage host DNA or inhibit its replication.* These are just the conditions that induce *E. coli*'s SOS response: The resulting fragments of single-stranded DNA activate RecA protein to stimulate the self-cleavage of LexA protein, the SOS gene repressor, at an Ala—Gly bond (Section 31-5D). *Activated RecA protein likewise stimulates the autocatalytic cleavage of λ repressor monomer's Ala 111—Gly 112 bond (which occurs in the polypeptide segment linking the repressor's two domains).* Repressor's ability to cooperatively bind to o_{R2} is thereby abolished (Fig. 32-44a and b; the C-terminal domains can still dimerize but they no longer link the DNA-binding N-terminal domains). The consequent reduction in concentration of intact free monomers shifts the monomer–dimer equilibrium such that the operator-bound dimers dissociate to form monomers, which are then cleaved through the influence of activated RecA before they can rebind to their target DNA.

In the absence of repressor at o_R, the λ early genes, including *cro,* are transcribed (Fig. 32-44c). As Cro accumulates, it first binds to o_{R3} so as to block even basal levels of λ repressor synthesis (Fig. 32-44d). Thus, *there being no mechanism for selectively inactivating Cro, the phage irreversibly enters the lytic mode:* The λ switch, once thrown, cannot be reset. The prophage is subsequently excised from the host chromosome by the integrase and excisionase that are produced in the delayed early phase.

The λ Switch's Responsiveness to Conditions Arises from Cooperative Interactions among Its Components

The complexity of the above switch mechanism endows it with a sensitivity that is not possible in simpler systems. The degree of repression at p_R is a steep function of repressor concentration (Fig. 32-45, *right*): The repression of p_R in a lysogen is normally 99.7% complete but drops to one half this level upon inactivation of 90% of the repressor. This steep sigmoid binding curve arises from the much greater operator affinity of repressor dimers compared to monomers. This situation, in turn, results from the cooperative linking of the monomer–dimer equilibrium, the binding of dimer to operator, and the association of dimers bound at o_{R1} and o_{R2}. In contrast, a 99.7% repressed promoter controlled by a stably oligomeric repressor binding to a single operator site, such as occurs in the *lac* system, requires 99% repressor inactivation for 50% expression (Fig. 32-45, *left*). *The cooperativity of λ repressor oligomerization and multiple operator site binding are therefore responsible for the remarkable responsiveness of the λ switch to the health of its host.*

4. INFLUENZA VIRUS

Influenza is one of the few common infectious diseases that is poorly controlled by modern medicine. Its annual epidemics, one of which was recorded by Hippocrates in 412 B.C., are occasionally punctuated by devastating pandemics. For example, the influenza pandemic of 1918, which killed over 20 million people (almost 1% of the world's population at the time) and affected perhaps 50 times that

transport of the virus to and from the infection site by permitting its passage through mucin (mucus) and preventing viral self-aggregation. Each virion incorporates ~100 copies of NA.

Just beneath the viral membrane is a 6-nm-thick protein shell composed of ~3000 copies of **matrix protein (M_1)**, the virion's most abundant protein.

The influenza virus genome is unusual in that it consists of 8 different sized segments of single-stranded RNA. These RNA molecules are negative strands; that is, they are complementary to the viral mRNAs. In the viral core, these RNAs occur in complex with 4 different proteins: **nucleocapsid protein (NP)**, which occurs in ~1000 copies, and 3 proteins, **PA, PB1,** and **PB2**, present in 30 to 60 copies each. The resulting **nucleocapsids** have the appearance of flexible rods.

The 8 viral RNAs, which vary in length from 890 to 2341 nucleotides, have all been sequenced. They code for the virus' 7 structural proteins (HA, NA, M_1, NP, PA, PB1, and PB2) and 3 nonstructural proteins that occur only in infected cells (NS_1, NS_2, and M_2). The sizes of the RNAs and the proteins they encode are listed in Table 32-2.

Virus Life Cycle

The influenza infection of a susceptible cell begins with the HA-mediated adsorption of the virus to specific cell-surface receptors. This is followed by uptake of the virus via an endocytotic mechanism (Section 11-5C) in which the viral and endosome membranes fuse through a process discussed in Section 32-4C. The viral contents are thereby introduced into the cell. By ~20-min postinfection, the still intact nucleocapsids have been transported to the cell nucleus, where they commence transcription of the viral

TABLE 32-2.   THE INFLUENZA VIRUS GENOME

RNA Segment	Length (nt)	Polypeptide(s) Encoded
1	2341	PB2
2	2341	PB1
3	2233	PA
4	1778	HA
5	1565	NP
6	1413	NA
7	1027	M_1,M_2
8	890	NS_1,NS_2

Source: Lamb, R.A. and Choppin, P.W., *Annu. Rev. Biochem.* **52**, 473 (1983).

RNAs (**vRNAs**). Cellular enzyme systems are incapable of mediating such RNA-directed RNA synthesis. Rather, it is carried out by a viral RNA transcriptase system that consists of the nucleocapsid proteins.

The transcription of the influenza virus genome is terminated if infected cells are treated with inhibitors of RNA polymerase II (which synthesizes cellular mRNA precursors; Section 29-2F) such as actinomycin D or α-amanitin. Yet, none of these agents affects the viral transcriptase's *in vitro* activity. The resolution of this seeming paradox is that *in vivo* viral mRNA synthesis is primed by newly synthesized cellular mRNA fragments consisting of a 7-methyl-G cap (Section 29-4A) followed by a 10- to 13-nt chain ending in A or G (Fig. 32-48, *top*). Viral mRNAs, as do most mature cellular mRNAs, have poly(A) tails appended to their 3′ ends (Section 29-4A).

The viral mRNAs lack 15- to 22-nt segments that are complementary to the 5′ ends of their parental vRNAs.

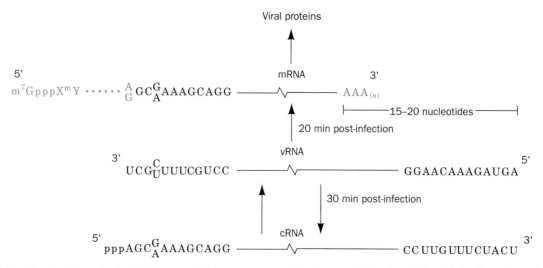

FIGURE 32-48. The biosynthesis of influenza vRNA, mRNA, and cRNA. The conserved nucleotides at the ends of the RNA segments are indicated. The viral mRNA's host-derived capped 5′ head and 3′ poly(A) tail are shown in color. [After Lamb, R.A. and Choppin, P.W., *Annu. Rev. Biochem.* **52**, 490 (1983).]

They therefore cannot act as templates in vRNA replication. Rather, in an alternative transcription process that begins some 30-min postinfection, complete vRNA complements are synthesized. These so-called **cRNAs** (Fig. 32-48, *bottom*), whose synthesis does not require a primer, begin with pppA at their 5′ ends and lack poly(A) tails (Fig. 32-48). Hence cRNAs, unlike viral mRNAs, do not associate with polysomes in infected cells. The synthesis of cRNA, in contrast to that of viral mRNA, requires the continuing production of viral proteins. This observation suggests that the viral nonstructural proteins, NS_1, NS_2, and/or M_2, modify the viral transcriptase so as to render it primer-independent, thereby permitting it to fully transcribe the vRNAs. The cRNAs are the templates for vRNA synthesis in a process whose mechanism has not yet been elucidated.

Cross-linking experiments, which indicate that PB2 and PB1, respectively, bind to the capped primers and the priming chains of viral mRNAs, suggest that PB2 functions to recognize the 5′ cap and that PB1 is the endonuclease that cleaves primers from the cellular mRNAs. Mutational experiments indicate that PA plays an important role in vRNA but not mRNA synthesis, whereas mutations in PB1 or PB2 inhibit mRNA synthesis. It therefore appears that it is the complex of the three P proteins rather than any one of them that catalyzes viral RNA synthesis. The abundance of NP suggests that it has a structural role in the nucleocapsid, although it has also been implicated in the antitermination required to synthesize cRNAs rather than vRNAs.

The mechanism of influenza virus assembly is not well characterized. The viral spike glycoproteins, HA and NA, are ribosomally synthesized on the rough endoplasmic reticulum, further processed in the Golgi apparatus (Section 11-4B), and then transported, presumably in clathrin-coated vesicles, to specific areas of the plasma membrane. There, they aggregate in sufficient numbers to exclude host proteins (Fig. 32-49a and b). The M_1 protein is thought to form a nucleocapsid-enclosing shell that binds to HA and NA on the inside of the plasma membrane (Fig. 32-49b). This binding process causes the entire assembly to bud from the cell surface, thereby forming the mature virion (Fig. 32-49c). The complete infection cycle occupies ~8 to 12 h.

One of the mysteries of influenza virus assembly is how each virion acquires a complete set of the eight vRNAs. There is no evidence that the newly formed nucleocapsids are physically linked. On the contrary, in mixed infections with various influenza strains, the reassortment of their genomic segments occurs with high frequency. It has therefore been suggested that the nucleocapsids are randomly selected but that each virion contains sufficient numbers of vRNAs to ensure a reasonable probability that a given particle be infectious. This proposal is in agreement with the observation that aggregates of influenza virus have enhanced infectivity, a process that presumably occurs through the complementation of their vRNAs. Alternatively, the eight vRNAs may be selected by an ordered pro-

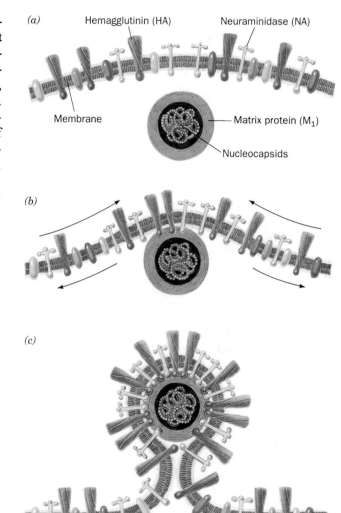

FIGURE 32-49. The budding of influenza virus from the host cell membrane. *(a)* The viral glycoproteins, HA and NA, are inserted into the plasma membrane of the host cell and the matrix protein, M_1, forms the nucleocapsid-containing shell. *(b)* The binding of the matrix protein to the cytoplasmic domains of HA and NA results in the aggregation of these glycoproteins so as to exclude host cell membrane proteins. *(c)* This binding process induces the membrane to envelop the matrix protein shell such that the mature virion buds from the host cell surface. [After Wiley, D.C., Wilson, I.A., and Skehel, J.J., *in* Jurnak, F.A. and McPherson, A. (Eds.), *Biological Macromolecules and Assemblies,* Vol. 1: *Virus Structures,* Wiley (1984).]

cess, a hypothesis that is supported by the observation that mature viruses, but not infected cells, contain roughly equimolar amounts of the vRNAs.

B. Mechanism of Antigenic Variation

Influenza virus infects a wide variety of mammalian and avian species in addition to humans. Indeed, it is thought that migratory birds are the major vectors that transport influenza viruses around the world. The species specificity of a particular viral strain presumably arises from the binding specificity of its HA for cell surface glycolipids. Influenza viruses are classified into three immunological types, A, B, and C, depending on the antigenic properties of their nucleoproteins and matrix proteins. The A virus has caused all of the major pandemics in humans and is the only influenza virus known to infect animals. It has therefore been more extensively investigated than the B and C viruses.

HA Residue Changes Are Responsible for Most of the Antigenic Variation in Influenza Viruses

HA, being the influenza virus' major surface protein, is largely responsible for stimulating the production of the antibodies that neutralize the virus. Consequently, the different influenza virus subtypes arise mainly through the variation of HA. Antigenic variation in NA, the virus' only other surface protein, also occurs but this has lesser immunological consequences.

Two distinct mechanisms of antigenic variation have been observed in influenza-A viruses:

1. **Antigenic shift,** in which the gene encoding one HA species is replaced by an entirely new one. This change may or may not be accompanied by a replacement of NA. It is thought that these new viral strains arise from the reassortment of genes among animal and human flu viruses. *Antigenic shift is responsible for influenza pandemics because the human population's immunity against previously existing viral strains is ineffective against the newly generated strain.* Evidently, these viruses had retained the (unknown) genetic traits responsible for their virulence in humans.

2. **Antigenic drift,** which occurs through a succession of point mutations in the HA gene, resulting in an accumulation of amino acid residue changes that attenuate the host's immunity. This process occurs in response to the selective pressure brought about by the buildup in the human population of immunity to the extant viral strains.

In early 1976, at Fort Dix, New Jersey, there was an outbreak of an influenza strain that carried an HA type that occurs in swine flu virus. This viral subtype is thought to have caused the great pandemic of 1918 (although influenza virus was not isolated until 1933, individuals who had contracted influenza during this pandemic have antibodies against swine flu virus in their serum). If this new strain had been virulent, no one under the age of 50 at the time would have been immune to it. There was, consequently, grave concern that a deadly influenza pandemic would ensue. This situation led to a crash program in which well over a million people deemed to be at high risk (such as pregnant women and the elderly) were vaccinated against swine flu. Fortunately, the 1976 swine flu was not virulent; it did not spread beyond Fort Dix.

HA Is an Elongated Trimeric Transmembrane Glycoprotein

Influenza virus hemagglutinin plays a central role in both the viral infection process and in the immunological measures and countermeasures taken in the continuing biological contest between host and parasite. This has motivated considerable efforts to elucidate the structural basis of these properties. HA is a trimer of 550-residue identical subunits that is 19% carbohydrate by weight. The protein has three domains (Fig. 32-50):

1. A large hydrophilic, carbohydrate-containing domain that occupies the viral membrane's external surface and that contains its sialic acid–binding site.

2. A hydrophobic 24 to 28 residue membrane-spanning domain that is located near the polypeptide's C-terminus.

3. A hydrophilic domain that occurs on the membrane's inner side and that consists of the protein's 10 C-terminal residues.

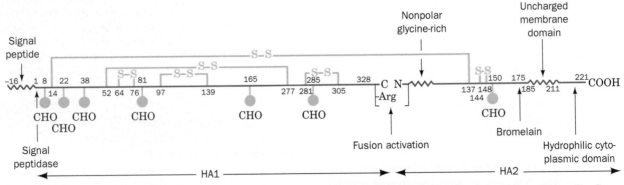

FIGURE 32-50. The 1968 Hong Kong influenza virus hemagglutinin amino acid sequence, indicating its external domain (all of HA₁ and HA₂ through 185), its membrane anchoring domain (185–211 of HA₂), and its cytoplasmic domain (212–221 of HA₂). The positions of the signal sequence directing the protein's insertion into the membrane, S—S bridges, carbohydrate (CHO) attachment sites, fusion-activation site, and bromelain cleavage site are indicated. [After Wilson, I.A., Skehel, J.J., and Wiley, D.C., *Nature* **289**, 367 (1981).]

HA is posttranslationally cleaved by the excision of Arg 329, thereby yielding two chains, HA$_1$ and HA$_2$, that are linked by a disulfide bond. This cleavage, which does not affect HA's receptor-binding affinity, is required for the fusion of the virus with the host cell and therefore activates viral infectivity (see below).

HA can be removed from the virion by treatment with detergent but the resulting solubilized protein has not been made to crystallize. However, treatment of HA from a Hong Kong-type virus (influenza virus subtypes are named according to their site of discovery) with the pineapple protease **bromelain,** which cleaves the polypeptide just before the membrane-spanning segment, yields a water-soluble protein named **BHA** that has been crystallized. X-Ray analysis of these crystals by Don Wiley revealed an unusual structure (Fig. 32-51). The monomer consists of a long fibrous stalk extending from the membrane surface upon

which is perched a globular region. The fibrous stalk consists of segments from HA$_1$ and HA$_2$ and includes a remarkable 76-Å long (53 residues in 14 turns) α helix. The globular region, which is comprised of only HA$_1$ residues, contains an eight-stranded antiparallel β-sheet structure that forms the sialic acid–binding pocket.

The dominant interaction stabilizing BHA's trimeric structure is a triple-stranded coiled coil consisting of the 76-Å α helices from each of its subunits (Fig. 32-51c). The BHA trimer is therefore an elongated molecule, some 135 Å in length, with a triangular cross-section that varies in radius from 15 to 40 Å. The carbohydrate chains, which are attached to the protein via N-glycosidic linkages at each of its seven Asn-X-Thr/Ser sequences (Section 10-3C), are located almost entirely along the trimer's lateral surfaces. The role of the carbohydrates is unclear despite the fact that they cover some 20% of the protein's surface. However, the ob-

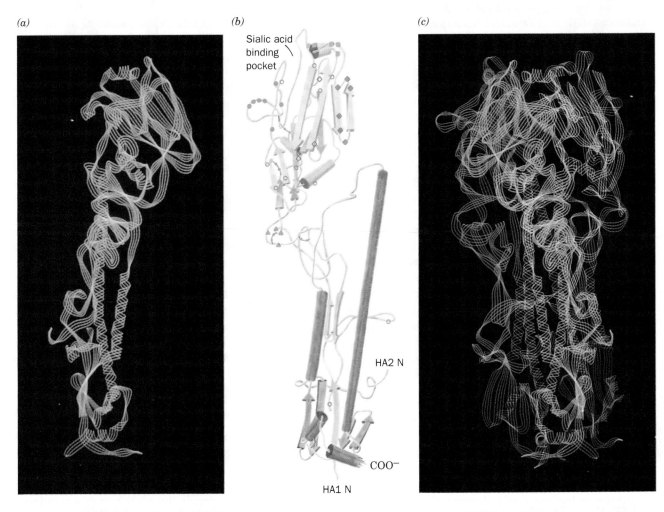

FIGURE 32-51. The X-ray structure of the influenza hemagglutinin monomer. *(a)* The polypeptide backbone drawn as a ribbon. HA$_1$ is green and HA$_2$ is blue. *(b)* A cartoon diagram from a somewhat different point of view as Part *a* but similarly colored. The pairs of linked, small, filled circles represent disulfide groups. The positions of the mutant residues at the four antigenic sites are indicated by filled circles, squares, triangles, and diamonds. Open symbols represent antigenically neutral residues. Note the position of the sialic acid–binding pocket. *(c)* A ribbon diagram of the HA trimer. Each HA$_1$ and HA$_2$ chain is drawn in a different color. [Parts *a* and *c* courtesy of Michael Carson, University of Alabama at Birmingham; Part *b* after a drawing by Hidde Ploegh, *in* Wilson, I.A., Skehel, J.J., and Wiley, D.C., *Nature* **289**, 366 (1981).]

(a)

(b)

(c)

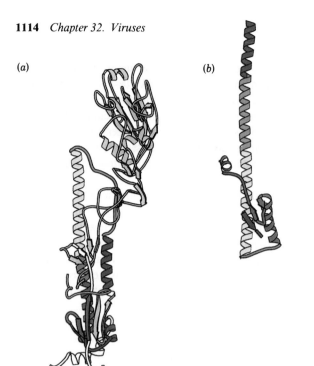

Receptor binding sites

B

C

A

D

E

F

HA trimer

Low–pH fragment

A

B

F

C

E

D

FIGURE 32-53. Comparison of the X-ray structures of BHA and TBHA$_2$. (*a*) Ribbon diagram of BHA in which the structural elements of the HA$_2$ chain in TBHA$_2$ are colored in rainbow order (red, orange, yellow, green, blue, violet) from N- to C-terminus and the HA$_1$ segment of TBHA$_2$ which is disulfide-linked to HA$_2$, is blue. Regions of BHA that are proteolytically excised to form TBHA$_2$ are gray and those that are apparently disordered in TBHA$_2$ are white. (*b*) Ribbon diagram of TBHA$_2$ colored as in Part *a*. The heights of the various structural element relative to the yellow helix segment,

which is common to both BHA and TBHA$_2$, are indicated. (*c*) Schematic diagram showing the positions and heights above the viral membrane surface of TBHA$_2$'s various structural elements in the HA trimer and in the low-pH fragment. The structural elements are color coded as in Parts *a* and *b*. In the low-pH fragment, the fusion peptide (*not shown*) would protrude well above the receptor-binding heads where it would presumably insert itself into the cellular membrane. [Parts *a* and *b* courtesy of Don Wiley, Harvard University.]

vince the proponents of the by then predominant microorganism theory of disease that these subcellular entities could also be pathogens. Presently, our ideas as to the nature of pathogenic agents are again evolving. It has become evident that *subviral agents can also cause infectious diseases.* Two types of these substances have been discovered:

1. **Viroids,** which are small single-stranded RNA molecules.

2. **Prions,** which appear to be only protein molecules.

In this section we outline what is known about these surprising entities. In doing so, we shall see that some of these new discoveries appear to be challenging the basic principles of molecular biology and thereby leading to a deeper understanding of biological processes.

A. *Viroids*

The **potato spindle tuber disease,** which was first described in the 1920s, causes the production of gnarled elongated potatoes. It was soon established that the disease was contagious but that no microorganisms were associated with it. It was therefore assumed to be a viral disease. The fact that

the putative virus could not be isolated was no doubt frustrating but not surprising; many viruses have been difficult to isolate.

Viroids Are Single-Stranded Circular RNAs

In the 1960s, it was discovered that the infectious agent of potato spindle tuber disease would grow in tomato plants so as to yield much more highly infectious extracts than were obtainable from infected potato plants. It was expected that the causative virus could be easily purified from these extracts by differential centrifugation. This was not the case; the infectious agent would not form a pellet even with a centrifugal acceleration of 100,000g for 4 h. The infectious agent must therefore be remarkably small.

Further investigations, largely by Theodor Diener, demonstrated, in 1971, that the potato spindle tuber agent is not only smaller than any known virus but is even smaller than any viral nucleic acid. The infectivity of these agents is exquisitely sensitive to RNase but insensitive to DNase or proteases. *The infectious agent evidently consists of only RNA.* This observation, by itself, is not unprecedented; the naked nucleic acids of many viruses are infectious, albeit at very low efficiency. Yet, the RNA's very small size (~120 kD) together with the observation that it was not associated with protein, was indeed unique. Diener therefore named these infectious agents viroids to distinguish them from viruses.

Once **potato spindle tuber viroid (PSTV)** had been isolated, it became possible to characterize it. Electron micrographs (Fig. 32-54) indicate that it has an average length of 50 nm, much shorter than the nucleic acids of viruses, although its width is similar to that of the double-stranded DNAs in the same micrograph. Electron micrographs of denatured PSTV, however, exhibit circular structures so that *the viroid must be a heretofore unobserved species: single-stranded covalently closed RNA circles.* This observa-

tion was eventually confirmed by sequence analysis that established PSTV to be a highly self-complementary single-stranded circle of 359 nucleotides (Fig. 32-55). Its most probable secondary structure consists of short double-stranded regions alternating with shorter unpaired regions so as to form the apparently double-stranded linear structures observed in electron micrographs of the native viroid.

Viroids Are Replicated by Host Enzymes via the Intermediacy of cRNA

In the 1970s, shortly after the nature of PSTV had been established, several economically significant plant diseases previously assumed to be of viral origin were shown to be caused by viroids (Table 32-3). One of them, **coconut cadang-cadang viroid (CCCV),** has killed ~30 million coconut palms in the Philippine Islands and has therefore seriously affected an entire nation's economy. All these viroids, of which 15 are known, are highly self-complementary, single-stranded, covalently closed RNA circles that are, in most cases, >50% homologous with PSTV. Most likely, other plant diseases presently attributed to viruses will also be found to be caused by viroids.

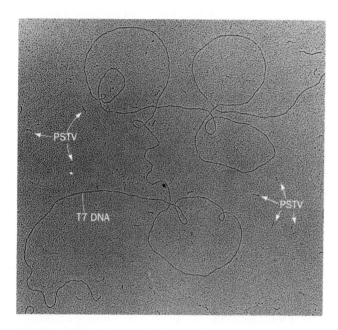

FIGURE 32-54. An electron micrograph of a mixture of PSTV (scattered short rods) and bacteriophage T7 DNA. The 359-nucleotide viroid is ~50 nm long, whereas the nearly 40,000-bp double-stranded T7 DNA, which is shown for comparison, is some 1400 nm long. [Courtesy of Jose Sogo, The Scripps Research Institute, La Jolla, California.]

TABLE 32-3. SOME VIROIDS CAUSING ECONOMICALLY IMPORTANT DISEASES

Viroid	Number of Nucleotides
Avocado sun blotch viroid (ASBV)	247
Citrus exocortis viroid (CEV)	371
Chrysanthemum stunt viroid (CSV)	354
Coconut cadang-cadang viroid (CCCV)	246
Cucumber pale fruit viroid (CPFV)	303
Hop stunt viroid (HSV)	297
Potato spindle tuber viroid (PSTV)	359

Source: Riesner, D. and Gross, H.J., *Annu. Rev. Biochem.* **54,** 542–543 (1985).

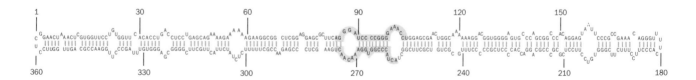

FIGURE 32-55. The base sequence of PSTV. Its most probable secondary structure is indicated above (*a*) and the corresponding three-dimensional structure is sketched below (*b*). The shaded regions in the sequence are highly conserved in all

but three known viroids, including CCCV, which is only 11% homologous to PSTV. [After Gross, H.J. and Riesner, D., *Angew. Chem. Int. Ed. Engl.* **19,** 237 (1980).]

How do viroids replicate? A reasonable hypothesis is that they are "defective" viruses that have lost the ability to specify coat protein. There is, however, no evidence that viroids can specify any proteins that do not occur in uninfected plants. Indeed, several viroids and their complements, including PSTV, are devoid of the translation initiation codon AUG, and none have any of mRNA's other characteristic control sequences.

An alternative hypothesis is that viroids somehow trigger the transcription of preexisting viroidal DNA sequences in susceptible plants. Yet, hybridization studies indicate that uninfected plants lack viroidal sequences. *Viroids must therefore be directly replicated by plant enzymes.* An obvious candidate for the viroidal replicase is the **RNA-directed RNA polymerase** that occurs in several higher plants (but not in animals; its normal function is unknown). Another possibility is RNA polymerase II since this enzyme, from certain plants, transcribes viroids *in vitro* with reasonable efficiency. The latter possibility is favored by the observations that viroid replication is inhibited by the RNA polymerase II inhibitors actinomycin D and α-amanitin which do not affect RNA-directed RNA polymerase. Whatever the enzyme(s) responsible, the isolation, from viroid-infected plants, of RNAs complementary to these viroids indicates that they are replicated through the intermediate synthesis of cRNAs. The finding, in infected plants, of concatemers of these cRNAs that are themselves infectious suggests that RNA replication occurs via a rolling circle mechanism (Section 31-3B), which must eventually be followed by site-specific cleavage of the concatemeric product and its religation to form covalently closed circles. *Viroids are therefore the only known autonomously replicating entities that do not specify at least one subunit of their replicating enzymes.*

Viroids May Be Molecular Fossils

The mechanism of viroidal pathogenesis is by no means obvious. In fact, viroids that cause disease in certain plant species replicate harmlessly in others. It has been variously suggested that the replication of viroids in susceptible species somehow interferes with gene regulation, with the normal RNA maturation process, or perhaps with cellular differentiation. It has been demonstrated that a 68-kD plant protein exhibiting kinase activity is more heavily phosphorylated in PSTV-infected plants than in uninfected plants. This phosphoprotein resembles the protein kinase from interferon-treated cells that acts to inhibit ribosomal initiation in the presence of double-stranded RNA (Section 30-4B). Perhaps the viroid-affected plant protein adversely influences protein synthesis in susceptible plants. In any case, it is clear that viroids are pathogenic agents that are distinctly different from viruses.

The realization that all known viroidal diseases were only first detected after 1920, in contrast to most viral plant diseases, which were described in the nineteenth century or earlier, suggests that viroids are of recent origin. Indeed, the observation that only cultivated plants are known to be adversely affected by viroids additionally suggests that modern agricultural techniques are responsible for spreading these disease agents. How, then, did viroids originate? Sequence comparisons indicate that viroids are of similar size to, and share many features with, the self-splicing group I introns that occur in mitochondrial and ribosomal RNA genes (Section 29-4B). In fact, these introns contain the shaded sequences in Fig. 32-55 that are common to nearly all viroids, suggesting that these two types of entities have similar secondary and tertiary structures (viroids *in vivo* are complexed with proteins so they are unlikely to assume the rodlike conformation they exhibit in the pure state). Group I introns are normally sequestered in the nucleus or the mitochondrion so they probably never enter the cytoplasm. Moreover, at least one viroid, **avocado sun blotch viroid (ABSV)**, is, in fact, self-cleaving. It has therefore been proposed that viroids are "escaped introns." An alternative hypothesis, however, is that viroids are molecular fossils of precellular self-replicating RNAs (the "RNA world"; Section 1-4C) and hence represent an even more primitive stage of evolution than do introns.

The Causative Agent of Hepatitis δ Contains a Viroid-like RNA Segment

Human **hepatitis δ virus (HDV)** replicates only in hepatocytes (liver cells) that are coinfected with **hepatitis B virus (HBV)**, thereby exacerbating the already occasionally fatal liver damage caused by HBV. HDV is an enveloped virus whose lipid bilayer-containing envelope is similar to that of HBV in that it contains HBV's three surface antigens (proteins). The HDV core contains a single-stranded circular RNA of ~1700 nt, its so-called **genomic RNA**, together with the so-called **δ antigen**, a 195-residue RNA-binding protein (which may be extended to 214 residues), which is HDV's only protein product.

HDV is unique among animal viruses in that it is the smallest such virus known, its RNA is circular, and it is ~70% base paired to form an unbranched rodlike structure. Intriguingly, *a 379-nt segment that only slightly overlaps the genomic RNA's coding sequence is similar to viroid RNA, including its highly conserved region (Fig. 32-55).* Indeed, HDV's genomic RNA, as does viroid RNA, replicates through the formation of its cRNA, its so-called **antigenomic RNA**, via a rolling circle mechanism. Moreover, both the genomic and the antigenomic RNAs each undergo self-splicing at specific sites as well as the reverse of these reactions, **self-ligation,** to form RNA circles. This suggests that the viroidlike and protein-coding regions of HDV RNA arose separately and were somehow joined together. Thus, at least one viroidlike agent is implicated in human disease.

B. Prions

Certain transmissible diseases that affect the mammalian central nervous systems were originally classified as being caused by "slow viruses" because they take months, years, or even decades to develop. Among them are **scrapie,** a neurological disorder of sheep and goats, so named for the tendency of infected sheep to scrape off their wool; **bovine spongiform encephalopathy (BSE** or **mad cow disease),** which similarly afflicts cattle; and **kuru,** a degenerative brain disease that occurred among the Fore people of Papua New Guinea, and which was transmitted by ritual cannibalism. There is also a sporadically occurring human disease with similar symptoms, **Creutzfeldt–Jakob disease (CJD),** a rare, progressive, cerebellar disorder, which resembles and may be identical to kuru. These diseases, all of which are ultimately fatal, have similar symptoms, which suggests that they are closely related. None of them exhibit any sign of an inflammatory process or fever, which indicates that the immune system, which is not impaired by the disease, is not activated by it.

The classical technique for isolating an unknown disease agent involves the fractionation of diseased tissue as monitored by assays for the disease. The long incubation time for scrapie, the most extensively studied "slow virus" disease, has enormously hampered efforts to characterize its disease agent. Indeed, in the early work on scrapie, an entire herd of sheep and several years of observation were necessary to evaluate the results of a single fractionation. Assays for scrapie were greatly accelerated, however, by the discovery that hamsters develop the disease in a time, minimally 60 days, that decreases as the dose given is increased. Using a hamster assay, Stanley Prusiner has purified the scrapie agent to a high degree and has been instrumental in characterizing it.

Scrapie Appears To Be Caused by Prion Protein

The scrapie agent apparently is a single species of protein. This astonishing conclusion was established by the observations that the scrapie agent is inactivated by substances that modify proteins, such as proteases, detergents, phenol, urea, and protein-specific reagents, whereas it is unaffected by agents that alter nucleic acids, such as nucleases, UV irradiation, and substances that specifically react with nucleic acids. For instance, scrapie agent is inactivated by treatment with diethylpyrocarbonate, which carboxyethylates the His residues of proteins (Table 6-3), but is unaltered by the cytosine-specific reagent hydroxylamine (Section 30-1A). In fact, the infectivity of diethylpyrocarbonate-inactivated scrapie agent is restored by treatment with hydroxylamine, presumably by the following known reaction:

$$CH_3CH_2OC\text{—}N \quad N \quad + \quad NH_2OH$$

Ethylcarboxamido-His   **Hydroxylamine**

$$CH_3CH_2OC\text{—}NHOH \quad + \quad$$

His

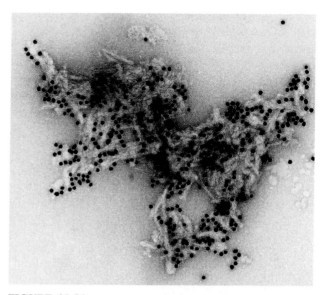

FIGURE 32-56
An electron micrograph of a cluster of partially proteolyzed prion rods. The black dots are colloidal gold beads that are coupled to anti-PrP antibodies adhering to the PrP. [Courtesy of Stanley Prusiner, University of California at San Francisco Medical Center.]

The novel properties of the scrapie agent, which distinguish it from viruses, plasmids, and viroids, has resulted in its being termed a **prion** (for *pro*teinaceous *in*fectious particle). The scrapie protein, which is named **PrP** (for *Pr*ion *P*rotein), is a 208-residue hydrophobic glycoprotein. This hydrophobicity, as we shall see below, causes partially proteolyzed PrP to aggregate as clusters of rodlike particles (Fig. 32-56). There is a close resemblance between these clusters and the so-called **amyloid plaques** that are seen on microscopic examination of prion-infected brain tissue. In fact, brain tissue from CJD victims contains protease-resistant protein that cross-reacts with antibodies raised against scrapie PrP.

PrP Is a Widely Expressed Product of a Normal Cellular Gene That Has No Apparent Function

The bizarre composition of prions immediately raises the question: How are they synthesized? Three possibilities have been suggested:

1. Despite all evidence to the contrary, prions contain a nucleic acid genome that is somehow shielded from detection; that is, prions are conventional viruses. The rapidly growing body of information concerning the nature of prions, however, makes this notion increasingly untenable.

2. Prions might somehow specify their own amino acid sequence by "reverse translation" to yield a nucleic acid that is normally translated by the cellular system. Such a process, of course, would directly contravene the "central dogma" of molecular biology (Section 29-1), which states that genetic information flows undirectionally

from nucleic acids to proteins. Alternatively, prions might directly catalyze their own synthesis. Such protein-directed protein synthesis is likewise unknown (although many small bacterial polypeptides are enzymatically rather than ribosomally synthesized).

3. Susceptible cells carry a gene that codes for the corresponding PrP. Infection of such cells by prions activates this gene and/or alters its protein product in some autocatalytic way.

The latter hypothesis seems to be the most plausible mechanism of prion replication. Indeed, the use of oligonucleotide probes complementary to the PrP gene (which is named **Prn-p** for *prion protein*), as inferred from the amino acid sequence of PrP's N-terminus (Section 28-7), established that the brains of both scrapie-infected and normal mice contain *Prn-p*. The most surprising discovery, however, is that *Prn-p is transcribed at similar levels in both normal and scrapie-infected brain tissue.* Moreover, the use of the above probes has revealed that *Prn-p genes occur in all vertebrates so far tested, including humans, as well as in invertebrates such as Drosophila..* This evolutionary conservation suggests that PrP, a GPI-anchored protein that occurs mainly on the surfaces of neurons, has an important function. Thus it came as a further surprise that transgenic mice in which both *Prn-p* genes have been disrupted through genetic engineering techniques appear to be entirely normal and that mating two such *Prn-p$^{0/0}$* mice gives rise to normal *Prn-p$^{0/0}$* progeny.

Scrapie Disease Requires the Expression of the Corresponding PrP^C Protein

Prn-p$^{0/0}$ mice remain completely free of scrapie symptoms after innoculation with a dose of mouse scrapie PrP (PrP^{Sc}; Sc for *scrapie*) that causes wild-type (*Prn-p$^{+/+}$*) mice to die of scrapie within 6 months after innoculation. Evidently, *PrP^{Sc} induces the conversion of normal PrP (PrP^C; C for cellular) to PrP^{Sc}.* This unorthodox hypothesis is supported by the observation that when wild-type mice are innoculated with PrP^{Sc} that has been continuously passaged (incubated) in hamsters, the incubation time for developing disease symptoms is, at first, 500 days but then, in all further passages in mice, diminishes to 140 days. Conversely, when PrP^{Sc} that has been passaged in mice is innoculated into hamsters, the incubation time is first 400 days but subsequently shortens to 75 days. This suggests that the conversion of host PrP^C (whose sequence in mice differs from that in hamsters) to PrP^{Sc} by a foreign PrP^{Sc} is a rare event but, once it has occurred, the newly formed host PrP^{Sc} catalyzes the conversion much more efficiently. Indeed, after innoculation with hamster PrP^{Sc}, transgenic mice expressing hamster PrP have incubation times that are reduced to between 48 and 250 days, depending on the transgenic line.

Mutant *Prn-p* Genes Give Rise to Prion Diseases

Three dominantly inherited neurodegenerative disorders in humans have been traced to mutations in the *Prn-p* gene. These are **familial CJD, Gerstmann–Stråussler-Scheinker syndrome (GSS),** and **fatal familial insomnia (FFI).** All of them are extremely rare. In fact, FFI has been found in only five families.

PrP^{Sc} Is Apparently a Stable Conformational Variant of PrP^C

How does PrP^{Sc} differ from PrP^C? The direct sequencing of PrP^{Sc} (Section 6-1) indicates that its amino acid sequence is identical to that deduced from the *Prn-p* gene sequence, thereby eliminating both RNA editing (Section 29-4A) and alternative site gene splicing (Section 33-3C) as possible causes for the pathogenic properties of PrP^{Sc}. Furthermore, mass spectrometric studies on PrP^{Sc} designed to reveal previously uncharacterized posttranslational chemical modifications indicated that, in fact, PrP^{Sc} and PrP^C are chemically identical. Thus, although the possibility that only a small fraction of PrP^{Sc} is chemically modified has not been eliminated, it seems more likely that PrP^{Sc} and PrP^C differ in their secondary and/or tertiary structures. Indeed, spectroscopic measurements indicate that PrP^C has no β sheet (3%) but a high α helix content (42%), whereas PrP^{Sc} has an even higher β sheet content (54%) but a lower α helix content (21%). *Evidently, this conformation change is autocatalytic; that is, PrP^{Sc} induces PrP^C to convert to PrP^{Sc}.* In fact, PrP^{Sc} in a cell-free system has been shown to catalyze the conversion of PrP^C from an uninfected source to PrP^{Sc}.

In cells, PrP^{Sc} is deposited in cytosolic vesicles rather than being GPI-anchored to the cell surface as is PrP^C. Both PrP^C and PrP^{Sc} are subject to eventual proteolytic degradation in the cell (Section 30-6). However, although PrP^C is completely degraded, PrP^{Sc} only looses its N-terminal 67 residues to form a 27- to 30-kD protease-resistant core, known as **PrP 26-30,** which still exhibits a high β sheet content. *The PrP 26-30 then aggregates to form the amyloid plaques that, it is thought, are directly responsible for the neuronal degeneration characteristic of prion diseases.*

According to this model, sporadically occurring prion diseases such as CJD arise from the spontaneous although infrequent conversion of sufficient quantities of PrP^C to PrP^{Sc} to support the autocatalytic conformational isomerization reaction, a model that is corroborated by the observation that transgenic mice that overexpress wild-type *Prn-p* invariably develop scrapie late in life. The model similarly explains inherited prion diseases such as FFI as arising from a lower kinetic barrier for the conversion of the mutant PrP^C to PrP^{Sc} relative to that of normal PrP^C.

There are numerous chronic neurodegenerative diseases of humans whose causes are unknown. Pruisner has hypothesized that some of these diseases, such as **Alzheimer's disease** and the paralytic disease **amyotrophic lateral sclerosis,** may have similar causes to those of prion diseases. Indeed, Alzeimer's disease is also characterized by the deposition of amyloid plaques in the brain, although they consist of different types of proteins.

CHAPTER SUMMARY

Viruses are complex molecular aggregates that exhibit many attributes of living systems. Their structural and genetic properties have therefore served as valuable paradigms for the analogous cellular functions. The TMV virion consists of a helix of identical and therefore largely quasi-equivalent coat protein subunits containing a coaxially wound single strand of RNA. X-Ray studies of TMV gels reveal that this RNA is bound, with three nucleotides per subunit, between the subunits of the protein helix. In the absence of TMV RNA, the subunits aggregate at high ionic strengths to form double-layered disks, and at low ionic strengths to form protohelices, which stack to form helical rods under acidic conditions. The virus' innermost polypeptide loop is disordered in both the disk and the protohelix. Virus assembly is initiated when a protohelix (or possibly a double-layered disk) binds to the initiation sequence of TMV RNA, which is located ~1000 nucleotides from the RNA's 3' end. Interactions between the RNA and protohelix trigger the ordering of the disordered loop, thereby converting the protohelix to the helical form. Elongation of the virus particle then proceeds by the sequential addition of protohelices (disks) to the "top" of the assembly so as to pull the 5' end of the RNA up through the center of the growing viral helix.

Viral capsids are formed from one or a few types of coat protein subunits. These must be either helically arranged as in TMV, or quasi-equivalently arranged in a polyhedral shell so as to enclose the viral nucleic acid. The coat proteins of many spherical viruses are arranged in icosadeltahedra consisting of $60T$ subunits. The coat protein of TBSV is arranged in a $T = 3$ icosadeltahedron so that TBSV subunits occupy three symmetrically distinct positions. The subunits must therefore associate through several sets of nonidentical intersubunit contacts. Some of the R domains form a structurally disordered inner protein shell. The viral RNA together with the remaining R domains are tightly packed in the space between the inner and outer protein shells. Other spherical plant viruses, SBMV and STMV, have tertiary and quaternary structures that are clearly related to those of TBSV. The structurally similar VP1, VP2, and VP3 coat proteins of poliovirus, rhinovirus, and FMDV are likewise icosahedrally arranged. However, the SV40 capsid consists of 72 pentagons of identical subunits in two different environments that are linked together by differering arrangements of their C-terminal arms in a nonicosadeltahedral arrangement. Although the coat proteins of most spherical viruses consist mainly of structurally similar 8-stranded antiparallel β barrels, that of bacteric phage MS2, a $T = 3$ virion, has an unrelated fold.

Lytic growth of bacteriophage λ in *E. coli* is controlled by the sequential syntheses of antiterminators, which inhibit both rho-independent and rho-dependent transcriptional terminators. Thus gp$N$, which is synthesized in the early stage of growth, permits the synthesis of gp$Q$ in the delayed early stage which, in turn, permits the synthesis of the capsid proteins in the late stage. Early gene transcription is repressed in the delayed early stage by Cro protein. DNA replication, which commences in the early stage, is mediated by the host DNA replication machinery with the aid of the phage proteins gp$O$ and gp$P$. DNA synthesis initially occurs by both the θ and rolling circle (σ) modes but eventually switches entirely to the rolling circle mode.

The λ virion heads and tails are separately assembled. Head assembly is a complex process involving the participation of many phage gene products, not all of which are part of the mature virion.

Phage heads are not self-assembling in that their formation is guided by host chaperonins and a viral scaffolding protein and requires several enzymatically catalyzed protein modification reactions. The mature phage head is a $T = 7$ icosadeltahedron of gp$E$, which is decorated by an equal number of gp$D$ subunits. Just before the final stage of its assembly, the phage head is filled with a linear double strand of DNA in a process that is driven by ATP hydrolysis. The packaged DNA is thought to be either wound in a spool or folded back and forth in a spirally arranged manner. Tail assembly occurs in a stepwise process from the tail fiber to the head-binding end. The body of the tail consists of a stack of hexameric rings of gp$V$. The completed heads and tails spontaneously join to form the mature virion.

Lysogeny is established by site-specific recombination between the phage *att*P site and the bacterial *att*B sites in a process mediated by phage integrase (gp*int*) and host IHF. Induction, in which this process is reversed, requires the additional action of phage excisionase (gp*xis*). Lysogeny is established by a high level of gp*cII*, which stimulates the transcription of *int* and the λ repressor gene, *cI*. Repressor, as does Cro, binds to the o_L and o_R operators to shut down early gene transcription, including that of *cro* and *cII*. Each of these dimeric proteins, like other repressors of known structure, contains two symmetrically related HTH units that bind in successive turns of B-DNA's major groove. However, repressor, but not Cro, induces its own synthesis from the promoter p_{RM} by binding to o_{R2} so as to interact with RNA polymerase. The induction of repressor synthesis therefore throws the genetic switch that stably maintains the phage in the lysogenic state from generation to generation. Damage to host DNA, nevertheless, stimulates host RecA protein to mediate λ repressor cleavage so as to release repressor from o_L and o_R. This initiates the synthesis of early gene products, including gp*int* and gp*xis*, from p_L and p_R and thus triggers induction. If sufficient Cro protein is then synthesized to repress the synthesis of repressor, the phage becomes irrevocably committed to at least one generation of lytic growth. The tripartite character of o_R, the site of the λ switch, together with the cooperative nature of repressor binding to o_R, confers the λ switch with a remarkable sensitivity to the health of its host.

The influenza virion's enveloping membrane is studded with protein spikes consisting of hemagglutinin (HA), which mediates host recognition, and neuraminidase (NA), which facilitates the passage of the virus to and from the infection site. Inside the membrane is a shell of matrix protein that contains the virus' genome of 8 single-stranded RNAs, each in a separate protein complex known as a nucleocapsid. These vRNAs are templates for the transcription of mRNAs as catalyzed by the nucleocapsid proteins. This process is primed by 7-methyl-G-capped host mRNA fragments. The viral mRNAs, which have poly(A) tails, lack the sequences complementary to the vRNA's 5' ends. The vRNAs, however, also act as templates for the transcription of the corresponding cRNAs which, in turn, are the templates for vRNA synthesis. The virus is assembled in and near the plasma membrane and forms by budding from the cell surface. Influenza viruses infect a variety of mammals besides humans as well as many birds. Variation in the antigenic character of HA has been mainly responsible for the different influenza subtypes. Antigenic variation in HA occurs by either antigenic shift, in which the HA gene from an animal virus replaces that from a human virus, or antigenic drift, which occurs by a succession of point mutations in the

such intense scrutiny, however, that significant advances in its understanding are made almost daily. Thus, perhaps more so than for other subject matter considered in this text, it is important that the reader supplement the material in this chapter with that in the recent biochemical literature.

1. CHROMOSOME STRUCTURE

Eukaryotic chromosomes, which consist of a complex of DNA, RNA, and protein called **chromatin,** are dynamic entities whose appearance varies dramatically with the stage of the cell cycle. The individual chromosomes assume their familiar condensed forms (Figs. 27–1 and 33–1) only during cell division (M phase of the cell cycle; Section 31-4A). During interphase, the remainder of the cell cycle, when the chromosomal DNA is transcribed and replicated, the chromosomes of most cells become so highly dispersed that they cannot be individually distinguished (Fig. 33-2). Cytologists have long recognized that there are two types of this dispersed chromatin: a less densely packed variety named **euchromatin** and a more densely packed variety termed **heterochromatin** (Fig. 33-2). These two types of chromatin differ, as we shall see, in that euchromatin is genetically expressed, whereas heterochromatin is not expressed.

The 46 chromosomes in a human cell each contain between 48 and 240 million bp, so their DNAs, which are most probably continuous (Section 28-2C), have contour lengths between 1.6 and 8.2 cm (3.4 Å/bp). Yet, in metaphase, their most condensed state (Fig. 33-1), these chromosomes range in length from 1.3 to 10 μm. *Chromosomal DNA therefore has a **packing ratio** (ratio of its contour length to the length of its container) of >8000.* How does the DNA in chromatin attain such a high degree of condensation? Structural studies have revealed that this results from three levels of folding. We discuss these levels below, starting with the lowest level. We begin, however, by studying the proteins responsible for much of this folding.

A. Histones

*The protein component of chromatin, which comprises somewhat more than half its mass, consists mostly of **histones.*** There are five major classes of these proteins, **histones H1, H2A, H2B, H3,** and **H4,** all of which have a large proportion of positively charged residues (Arg and Lys; Table 33–1). These proteins therefore ionically bind DNA's negatively charged phosphate groups. Indeed, histones may be extracted from chromatin by 0.5*M* NaCl, a salt solution of sufficient concentration to interfere with these electrostatic interactions.

Histones Are Evolutionarily Conserved

The amino acid sequences of histones H2A, H2B, H3, and H4 have remarkably high evolutionary stability (Table 33–1). For example, histones H4 from cows and peas, spe-

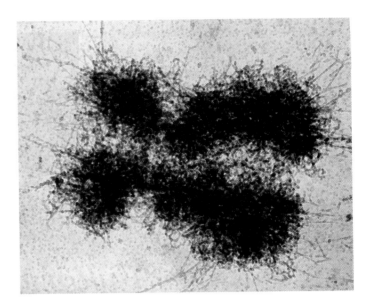

FIGURE 33-1. Electron micrograph of a human metaphase chromosome. [Courtesy of Gunther Bahr, Armed Forces Institute of Pathology.]

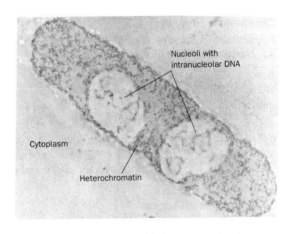

FIGURE 33-2. A thin section through a cell nucleus treated with **Feulgen reagent** (which reacts with DNA to form an intense red stain). Heterochromatin appears as dark-staining regions near the nucleolus and the nuclear membrane. The less darkly staining material is euchromatin. [Courtesy Edmund Puvion, CNRS, France.]

TABLE 33-1. CALF THYMUS HISTONES

Histone	Number of Residues	Mass (kD)	% Arg	% Lys	UEP[a] (×10⁻⁶ year)
H1	215	23.0	1	29	8
H2A	129	14.0	9	11	60
H2B	125	13.8	6	16	60
H3	135	15.3	13	10	330
H4	102	11.3	14	11	600

[a] Unit evolutionary period: The time for a protein's amino acid sequence to change by 1% after two species have diverged (Section 6-3B).

Ac — Ser —	Gly —	Arg —	Gly —	Lys —	Gly —	Gly —	Lys —	Gly — Leu —	10
Gly —	Lys —	Gly —	Gly —	Ala —	Lys —	Arg —	His —	Arg — Lys —	20
Val —	Leu —	Arg —	Asp —	Asn —	Ile —	Gln —	Gly —	Ile — Thr —	30
Lys —	Pro —	Ala —	Ile —	Arg —	Arg —	Leu —	Ala —	Arg — Arg —	40
Gly —	Gly —	Val —	Lys —	Arg —	Ile —	Ser —	Gly —	Leu — Ile —	50
Tyr —	Glu —	Glu —	Thr —	Arg —	Gly —	Val —	Leu —	Lys — Val —	60
Phe —	Leu —	Glu —	Asn —	Val —	Ile —	Arg —	Asp —	Ala — Val —	70
Thr —	Tyr —	Thr —	Glu —	His —	Ala —	Lys —	Arg —	Lys — Thr —	80
Val —	Thr —	Ala —	Met —	Asp —	Val —	Val —	Tyr —	Ala — Leu —	90
Lys —	Arg —	Gln —	Gly —	Arg —	Thr —	Leu —	Tyr —	Gly — Phe —	100
Gly —	Gly								102

FIGURE 33-3. The amino acid sequence of calf thymus histone H4. This 102-residue protein's 25 Arg and Lys residues are indicated in red. Pea seedling H4 differs from that of calf thymus by conservative changes at the two shaded residues: Val 60 → Ile and Lys 77 → Arg. The underlined residues are subject to posttranslational modification: Ser 1 is invariably *N*-acetylated and may also be *O*-phosphorylated; Lys residues 5, 8, 12, and 16 may be *N*-acetylated; and Lys 20 may be mono- or di-*N*-methylated. [After DeLange, R.J., Fambrough, D.M., Smith, E.L., and Bonner, J., *J. Biol. Chem.* **244,** 5678 (1969).]

cies that diverged 1.2 billion years ago, differ by only two conservative residue changes (Fig. 33-3) which makes histone H4, the most invariant histone, among the most evolutionarily conserved proteins known (Section 6-3B). *Such rigid evolutionary stability implies that the above four histones have critical functions to which their structures are so well tuned that they are all but intolerant to change.* The fifth histone, histone H1, is more variable than the other histones; we shall see below that its role differs from that of the other histones.

Histones May Be Modified

Histones are subject to posttranslational modifications that include methylations, acetylations, and phosphorylations of specific Arg, His, Lys, Ser, and Thr residues. These modifications, many of which are reversible, all decrease the histones' positive charges, thereby significantly altering histone–DNA interactions. Yet, despite the histones' great evolutionary stability, their degree of modification varies enormously with the species, tissue, and the stage of the cell cycle. A particularly intriguing modification is that 10% of the H2As have an isopeptide bond between the ε-amino group of their Lys 119 and the terminal carboxyl group of the protein ubiquitin. Although such ubiquitination marks cytosolic proteins for degradation by cellular proteases (Section 30-6B), it is not known whether this is the case with H2A. It would be most surprising, however, if this ubiquitination, as well as the other histone modifications, do not somehow serve to modulate eukaryotic gene expression.

Many, if not all, eukaryotes have genetically distinct subtypes of histones H1, H2A, H2B, and H3 whose syntheses are switched on or off during specific stages of embryogenesis and in the development of certain cell types. The sequence variations of these subtypes are limited to only a few residues in H2A, H2B, and H3 but are much more extensive in H1. Indeed, the erythroid cells of chick embryos contain an H1 variant that differs so greatly from other H1s that it is named **histone H5** (avian erythrocytes, unlike those of mammals, have nuclei). Histone switching seems to be related to cell differentiation but the nature of this relationship is unknown.

B. Nucleosomes: The First Level of Chromatin Organization

The first level of chromatin organization was pointed out by Roger Kornberg in 1974 through the synthesis of several lines of evidence:

1. Chromatin contains roughly equal numbers of molecules of histones H2A, H2B, H3, and H4, and no more than half that number of histone H1 molecules.

2. X-Ray diffraction studies indicate that chromatin fibers have a regular structure that repeats about every 100 Å along the fiber direction. This same X-ray pattern is observed when purified DNA is mixed with equimolar amounts of all the histones except histone H1.

3. Electron micrographs of chromatin (Fig. 33-4) reveal that it consists of ~100-Å-diameter particles connected by thin strands of apparently naked DNA, rather like

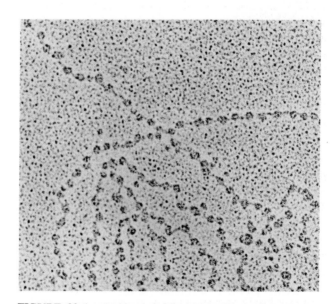

FIGURE 33-4. An electron micrograph of *D. melanogaster* chromatin showing that its 100-Å fibers are strings of closely spaced nucleosomes. [Courtesy of Oscar L. Miller, Jr., University of Virginia.]

(a)

(b)

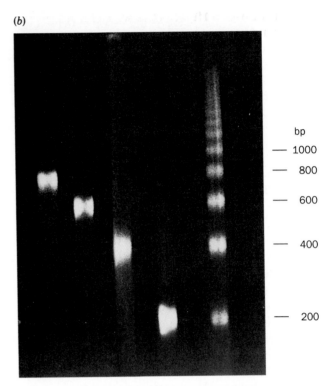

FIGURE 33-5. Defined lengths of calf thymus chromatin are obtained by the sucrose density gradient ultracentrifugation of chromatin that has been partially digested by micrococcal nuclease. (*a*) Electron micrographs of sucrose density gradient fractions containing, from top to bottom, nucleosome monomers, dimers, trimers, and tetramers. (*b*) Gel electrophoresis of DNA extracted from the nucleosome multimers indicates that they are the corresponding multiples of ~200 bp. The rightmost lane contains DNA from the unfractionated nuclease digest. [Courtesy of Roger Kornberg, MRC Laboratory of Molecular Biology, U.K.]

beads on a string. These particles are presumably responsible for the foregoing X-ray pattern.

4. Brief digestion of chromatin by **micrococcal nuclease** (which cleaves double-stranded DNA) cleaves the DNA between some of the above particles (Fig. 33-5*a*); apparently the particles protect the DNA closely associated with them from nuclease digestion. Gel electrophoresis indicates that each particle *n*-mer contains ~200*n* bp of DNA (Fig. 33-5*b*).

5. Chemical crosslinking experiments, such as are described in Section 7-5C, indicate that histones H3 and H4 associate to form the tetramer $(H3)_2(H4)_2$ (Fig. 33-6).

These observations led Kornberg to propose that *the chromatin particles, which are called* **nucleosomes,** *consist of the octamer* $(H2A)_2(H2B)_2(H3)_2(H4)_2$ *in association with ~200 bp of DNA.* The fifth histone, H1, was postulated to be associated in some manner with the outside of the nucleosome (see below).

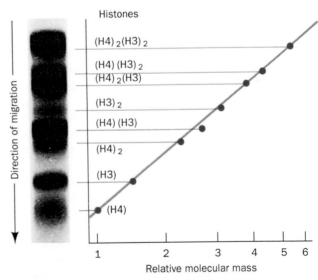

FIGURE 33-6. The SDS–gel electrophoresis of a mixture of calf thymus histones H3 and H4 that had been crosslinked by dimethylsuberimidate contains all the bands expected from an $(H3)_2(H4)_2$ tetramer. [Courtesy of Roger Kornberg, MRC Laboratory of Molecular Biology, U.K.]

DNA Coils around a Histone Octamer To Form the Nucleosome Core Particle

Micrococcal nuclease, as described above, initially degrades chromatin to single nucleosomes in complex with histone H1 (particles called **chromatosomes**). Upon further digestion, some of the chromatosomes' DNA is trimmed away in a process that releases histone H1. This yields the so-called **nucleosome core particle,** which consists of a 146-bp strand of DNA in association with the above histone octamer. The DNA removed by this digestion, which had previously joined neighboring nucleosomes, is named **linker DNA.** Its length has been found to vary between 8 and 114 bp from organism to organism and tissue to tissue although it is usually ~55 bp.

The X-ray structure of the histone octamer, determined by Evangelos Moudrianakis, reveals that it has a twofold symmetric structure comprised of two H2A–H2B dimers flanking a centrally located (H3–H4)₂ tetramer (Fig. 33-7a). The surface of the octamer forms a left-handed helical ramp of ~1⅔ turns whose helix axis is perpendicular to the octamer's molecular twofold axis such that the order of the histones along it are (H2A–H2B)-(H3–H4)-(H3–H4)-(H2A–H2B). The DNA is apparently bound along this ramp (see below), which is is lined by numerous positively charged side chains (Arg and Lys) and the positive dipole (N-terminal) ends of several helices. The octamer's overall shape is that of a 65-Å-diameter wedge-shaped disk, whose thickness varies from 10 Å on the side that the DNA presumably enters and exits the particle (the facing side in Fig. 33-7a) to 60 Å on the opposite side.

Despite only weak sequence similarity, all four histones contain a common fold in which a long central helix is flanked on each side by a loop and a shorter helix (Fig. 33-7b). The members of each H2A–H2B dimer and each H3–H4 half-tetramer interdigitate in a sort of "molecular handshake" to form extensive protein–protein interfaces. Quite possibly these histones are all evolutionarily related.

The X-ray structure of the nucleosome core particle had been previously determined by Gerard Bunick and Edward Uberbacher (Fig. 33-8), and independently by Aaron Klug, John Finch, and Timothy Richmond to a resolution of ~8 Å (vs 3.1-Å resolution for the histone octamer structure). At this low resolution, gross molecular features such as α helices just begin to be discernible. Nevertheless, the structural features of the nucleosome core particle are consistent with those of the histone octamer.

(a)

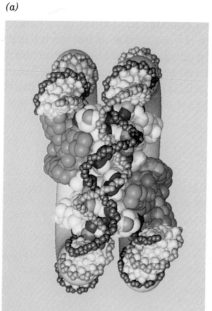

(b)

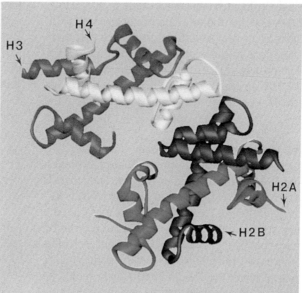

FIGURE 33-7. The X-ray structure of the nucleosomal core histone octamer. (*a*) The entire octamer as represented by spheres centered on its Cα atoms and viewed along its twofold axis with the DNA's superhelix axis horizontal. The two H2A–H2B dimers are blue and the (H3–H4)₂ tetramer is white except for the positively charged residues (Arg and Lys), which are red, and the residues at the positive dipole (N-terminal) ends of helices, which are yellow. The model-built path of the B-DNA, which is wrapped around the octamer in a left-handed helix, is represented at its ends by a space-filling model in which the two sugar–phosphate chains (drawn undersized for clarity) are different shades of grey and the base pairs are white, in its central region by only its backbone atoms, and in the intervening regions at the back of the particle by a grey tube. Note how the negatively charged DNA backbones apparently track along the the histone octamer's positive charges. (*b*) The unique half of the histone octamer shown in ribbon form and viewed approximately down the superhelix axis with H2A light blue, H2B dark blue, H3 green. and H4 white. The N-terminus of each chain is marked by an arrow. [Courtesy of Evangelos Moudrianakis, The Johns Hopkins University.]

(a)

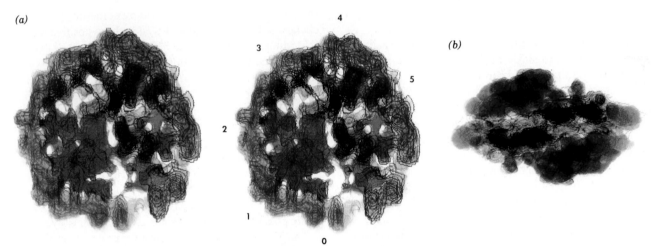

(b)

FIGURE 33-8. The X-ray structure of the nucleosome core particle at 8-Å resolution. (*a*) A stereo image of the electron density map viewed along the DNA supercoiling axis and showing slightly more than the upper half of the particle. The upper turn of the DNA (*orange-brown*) winds about the periphery of the particle, starting from position 0, such that the minor groove faces outwards at the positions marked by successive numbers. The histones that can be seen here comprise almost all of the upper H3 (*blue*) and H4 (*green*), small portions of the lower H3 and H4, and small portions of H2A (*violet*) and H2B (*dark brown*). Instructions for viewing stereo diagrams are given in the appendix to Chapter 7. (*b*) A portion of the electron density map as viewed along the core particle's twofold axis from +4 in Part *a*. [Courtesy of Gerard Bunick and Edward Uberbacher, University of Tennessee and Oak Ridge National Laboratory.]

B-DNA is wrapped around the histone octamer in 1.8 turns of a flat left-handed superhelix of pitch 28 Å. However, the DNA does not follow a smooth superhelical path; it is bent fairly sharply at several locations such that there are large variations in the widths of its major and minor grooves. The protein–DNA interactions occur on the inside of the DNA superhelix; no histone protein appears to surround the DNA or to protrude between the turns of the superhelix.

Histone H1 "Seals Off" the Nucleosome

In the micrococcal nuclease digestion of chromatosomes, the ~200-bp DNA is first degraded to 166 bp. Then there is a pause before histone H1 is released and the DNA is further shortened to 146 bp. The twofold symmetry of the core particle suggests that the reduction in length of the 166-bp DNA comes about by the removal of 10 bp from each of its two ends. Since the 146-bp DNA of the core particle makes 1.8 superhelical turns, the 166-bp intermediate should be able to make two full superhelical turns, which would bring its two ends as close together as possible. Klug has therefore proposed that histone H1 binds to nucleosomal DNA in a cavity formed by the central segment of its DNA and the segments that enter and leave the core particle (Fig. 33-9). This model is supported by the observation that in chromatin filaments containing H1, the DNA enters and leaves the nucleosome on the same side (Fig. 33-10*a*), whereas in H1-depleted chromatin, the entry and exit points are more randomly distributed and tend to occur on opposite sides of the nucleosome (Fig. 33-10*b*). The model also suggests that the

length of the linker DNA is controlled by the subspecies of histone H1 bound to it.

Histone H5 is a variant of histone H1 that has several Lys → Arg substitutions and binds chromatin more tightly. The observations that the expression of histone H5 in rat sarcoma cells inhibits DNA replication, thereby arresting cells in the G1 phase of the cell cycle, and that histone H5

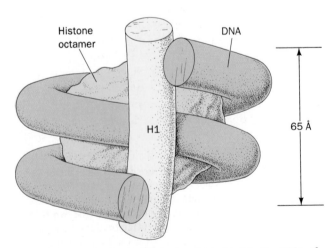

FIGURE 33-9. Histone H1 is thought to bind to the DNA of the 166-bp nucleosome. The DNA's two complete superhelical turns enable H1 to bind to the DNA's two ends and its middle. Here the histone octamer is represented by the central spheroid and the H1 molecule is represented by the cylinder.

(a)   (b)

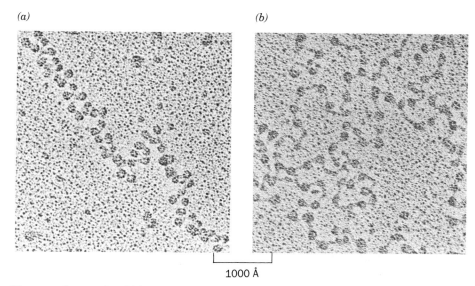

1000 Å

FIGURE 33-10. Electron micrographs of (a) H1-containing chromatin and (b) H1-depleted chromatin, both in 5 to 15 m*M* salt. [Courtesy of Fritz Thoma, Eidgenössische Technische Hochschule, Switzerland.]

more closely resembles **histone H1°** (a histone H1 variant that occurs in terminally differentiated cells) than does histone H1 itself, suggests that histone H5 is associated with replicationally and transcriptionally inactive chromatin.

Histones H1 and H5 both consist of a highly conserved globular, trypsin-resistant domain that is flanked by extended N- and C-terminal arms that are rich in basic residues. These basic arms, which comprise more than half of the intact protein, are therefore thought to interact with the linker DNA connecting adjacent nucleosomes even though it is the globular domain that is required for the binding of histone H1 to the nucleosome.

The Globular Domain of Histone H5 Structurally Resembles CAP Protein

V. Ramakrishnan has determined the X-ray structure of **GH5,** an 89-residue polypeptide that contains the 81-residue globular domain of histone H5 (although its five N-terminal and eleven C-terminal residues were not observed and, hence, are probably disordered). The polypeptide chain folds into a 3-helix bundle with a 2-stranded β sheet at its C-terminus (Fig. 33-11). This structure and, in particular, its 3-helix bundle, is strikingly similar in conformation to that of the helix–turn–helix (HTH) motif-containing DNA binding domain of *E. coli* catabolite activator protein (CAP; Fig. 29-22). Thus, even though there is little sequence identity between GH5 and CAP, their similar structures suggest that that GH5 binds DNA in a manner analogous to CAP. Indeed, a model of the GH5–DNA complex based on the known X-ray structure of the CAP–DNA complex (Section 29-3C) positions GH5's highly conserved Lys 69, Arg 73, and Lys 85 side chains to interact with the DNA (Fig. 33-11). These residues, which all have counterparts in CAP, are protected against chemical modification in chromatin. Moreover, GH5 contains a cluster of four

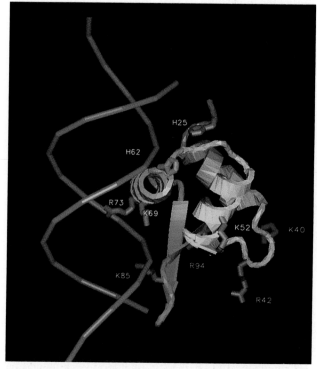

FIGURE 33-11. The X-ray structure of GH5 shown in hypothetical complex with DNA. This model was constructed by superimposing the structure of GH5 on that of CAP in the CAP–DNA structure (Fig. 29-22a). However, to avoid any presumptions about the nature of the DNA, that in the CAP structure, which is bent, was replaced by ideal B-DNA, which is represented here by its phosphate backbone (*red*). GH5 is shown in ribbon form and is color-ramped from red to blue going from its N- to its C-terminus. Conserved basic residues, as well as two His residues that have been crosslinked to DNA, are shown in stick form (*blue*). Residues are identified by their one letter codes (Table 4-1). [Courtesy V. Ramakrishnan, Brookhaven National Laboratory, NY.]

conserved basic residues on the opposite face of the protein from its "recognition helix," which could interact with a second segment of duplex DNA in agreement with the experimental evidence that GH5 simultaneously binds two DNA duplexes.

Parental Nucleosomes Are Transferred to Daughter Duplexes upon DNA Replication

The *in vivo* replication of eukaryotic DNA is accompanied by its packaging into chromatin; that is, it is the chromatin that actually is replicated. What, then, is the fate of the histone octamers originally associated with the parental DNA? There are several possibilities: The "parental" octamers may remain associated with either the leading strand or the lagging strand, or they may be partitioned between the two daughter DNA duplexes, either at random or in some systematic way. Attempts to resolve this issue have yielded contradictory results. However, the weight of the evidence now indicates that parental octamers are distributed at random between the daughter duplexes. Moreover, the parental octamers remain associated with DNA during the replication process instead of dissociating from the parental DNA and later rebinding the daughter duplexes. Thus, nucleosomes either open up to permit the passage of a replication fork or parental histone octamers immediately in front of an advancing replication fork are somehow transferred to the daughter duplexes immediately behind the replication fork.

Nucleosome Assembly Is Facilitated by Molecular Chaperones

How are nucleosomes formed *in vivo? In vitro,* at high salt concentrations, nucleosomes self-assemble from the proper mixture of DNA and histones. In fact, when only H3, H4, and DNA are present, the mixture forms nucleosome-like particles that each contain an $(H3)_2(H4)_2$ tetramer. Presumably, nucleosome cores are formed by the addition of H2A–H2B dimers to these particles.

At physiological salt concentrations, *in vitro* nucleosome assembly occurs much more slowly than at high salt concentrations and, unless the histone concentrations are carefully controlled, is accompanied by considerable histone precipitation. However, in the presence of **nucleoplasmin,** an acidic protein that has been isolated from *Xenopus laevis* oocyte nuclei, and DNA topoisomerase I (nicking-closing enzyme; Section 28-5C), nucleosome assembly proceeds rapidly without histone precipitation. Nucleoplasmin binds to histones but neither to DNA nor to nucleosomes. Evidently, *nucleoplasmin functions as a molecular chaperone (Section 8-1C) to bring histones and DNA together in a controlled fashion, thereby preventing their nonspecific aggregation through their otherwise strong electrostatic interactions.* The nicking–closing enzyme, no doubt, acts to provide the nucleosome with its preferred level of supercoiling.

C. 300-Å Filaments: The Second Level of Chromatin Organization

The 166-bp nucleosomal DNA has a packing ratio of ~7 (its 560-Å contour length is wound into a ~80-Å-high supercoil). Clearly, the 100-Å filament of nucleosomes, which occurs at low ionic strengths, represents only the first level of chromosomal DNA compaction. Only at physiological ionic strengths does the next level of chromosomal organization become apparent.

As the salt concentration is raised, the H1-containing nucleosome filament initially folds to a zigzag conformation (Fig. 33-10a) whose appearance suggests that nucleosomes interact through contacts between their H1 molecules. Then, as the salt concentration approaches the physiological range, chromatin forms a 300-Å thick filament in which the nucleosomes are visible (Fig. 33-12). Klug proposed that the 300-Å filament is constructed by winding the 100-Å nucleosome filament into a solenoid with ~6 nucleosomes per turn and a pitch of 110 Å (the diameter of a nucleosome; Fig. 33-13). The solenoid is stabilized by H1 molecules whose relatively variable, extended N-terminal and C-terminal arms (which are absent in GH5; Fig. 33-11) are thought to contact adjacent nucleosomes, at least in part by interacting with neighboring H1s in a head-to-tail fashion. This model, which is consistent with the X-ray diffraction pattern of the 300-Å filaments, has a packing ratio of ~40 (6 nucleosomes, each with ~200 bp DNA, rising a total of 110 Å). Note, however, that several other plausible models for the 300-Å chromatin filament have also been formulated.

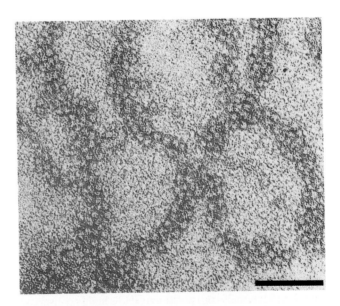

FIGURE 33-12. Electron micrograph of the 300-Å chromatin filaments. Note that the filaments are two to three nucleosomes across. The bar represents 1000 Å. [Courtesy of Jerome B. Rattner, University of Calgary, Canada.]

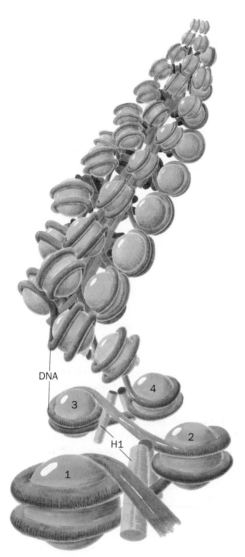

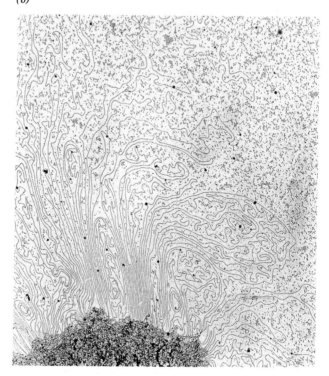

FIGURE 33-13. A proposed model of the 300-Å chromatin filament. The filament is represented (*bottom to top*) as it might form with increasing salt concentrations. The zigzag pattern of nucleosomes (*1, 2, 3, 4*) closes up to form a solenoid with ~6 nucleosomes per turn. The H1 molecules (yellow cylinders), which stabilize the structure, are thought to form a helical polymer running along the center of the solenoid.

D. Radial Loops: The Third Level of Chromatin Organization

Histone-depleted metaphase chromosomes exhibit a central fibrous protein "scaffold" surrounded by an extensive halo of DNA (Fig. 33-14*a*). The strands of DNA that can be followed are observed to form loops that enter and exit the scaffold at nearly the same point (Fig. 33-14*b*). Most of these loops have lengths in the range 15 to 30 μm (which corresponds to 45–90 kb), so that when condensed as 300-Å filaments they would be ~0.6 μm long. Electron micrographs of chromosomes in cross-section, such as Fig. 33-

FIGURE 33-14. Electron micrographs of a histone-depleted metaphase human chromosome. (*a*) The central protein matrix (scaffold) serves to anchor the surrounding DNA, (*b*) At higher magnification it can be seen that the DNA is attached to the scaffold in loops. [Courtesy of Ulrich Laemmli, University of Geneva, Switzerland.]

15*a*, strongly suggest that the chromatin fibers of metaphase chromosomes are radially arranged. If the observed loops correspond to these radial fibers, they would each contribute 0.3 μm to the diameter of the chromosome (a fiber must double back on itself to form a loop). Taking into account the 0.4-μm width of the scaffold, this model predicts the diameter of the metaphase chromosome to be 1.0 μm, in agreement with observation (Fig. 33-15*b*). A typical human chromosome, which contains ~140 million bp, would therefore have ~2000 of these ~70-kb radial loops. The 0.4-μm-diameter scaffold of such a chromosome has sufficient surface area along its 6-μm length to bind this number of radial loops. The radial loop model therefore accounts for DNA's observed packing ratio in metaphase chromosomes.

Almost nothing is known about how the 300-Å filaments are organized to form radial loops or about how metaphase chromosomes and the far more dispersed interphase chromosomes interconvert. Certainly, **nonhistone proteins,** whose hundreds or even thousands of varieties constitute ~10% of chromosomal proteins, must be involved in these processes. Moreover, there are intriguing indications that the radial loops are the chromosomal transcriptional units.

E. *Polytene Chromosomes*

The diffuse structure of most interphase chromosomes (Fig. 33-2) makes it all but impossible to characterize them at the level of individual genes. Nature, however, has greatly ameliorated this predicament through the production of "giant" banded chromosomes in certain nondividing secretory cells of dipteran (two-winged) flies (Fig. 33-16). These chromosomes, of which those from the salivary glands of *Drosophila melanogaster* larvae are the most extensively studied, are produced by multiple replications of a synapsed (joined in parallel) diploid pair in which the replicas remain attached to one another and in register. Each diploid pair may replicate in this manner as many as nine times so that the final **polytene chromosome** contains up to $2 \times 2^9 = 1024$ DNA strands.

Drosophila's 4 giant chromosomes have an aggregate length of ~2 mm so that its haploid genome of 1.65×10^8 bp has an average packing ratio in these chromosomes of almost 30. About 95% of this DNA is concentrated in chromosomal bands (Fig. 33-17). These bands (more properly, **chromomeres**), as microscopically visualized through staining, form a pattern that is characteristic of each *Drosophila* strain. Indeed, chromosomal rearrangements such as duplications, deletions, and inversions result in a corresponding change in the banding pattern. *A polytene chromosome's banding pattern therefore forms a cytological map that parallels its genetic map.*

Drosophila chromosomes exhibit ~5000 bands, more or less matching the estimated number of proteins that *Drosophila* produces. This correlation suggests that each chromosomal band corresponds to a single structural gene, a hypothesis corroborated by the application of *in situ* (on site) **hybridization.** In this technique, which Mary Lou Pardue and Joseph Gall have developed, an immobilized chromosome preparation is treated with NaOH to denature its DNA; it is then hybridized with a purified species of radioactively labeled mRNA (or its corresponding cDNA), and the chromosomal binding site of the radioactive probe is determined by autoradiography. A given mRNA hybridizes with one or, at most, a few chromosomal bands (Fig.

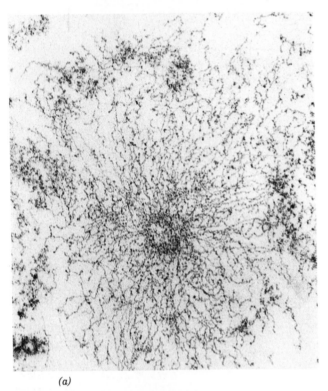

(a)

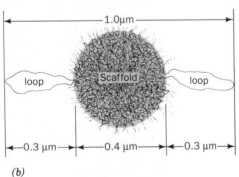

(b)

FIGURE 33-15. (*a*) Electron micrograph of a metaphase human chromosome in cross-section. Note the mass of chromatin fibers radially projecting from the central scaffold. [Courtesy of Ulrich Laemmli, University of Geneva, Switzerland.] (*b*) Interpretive diagram indicating how the 0.3-μm-long radial loops are thought to combine with the 0.4-μm-wide scaffold to form the 1.0-μm-diameter metaphase chromosome.

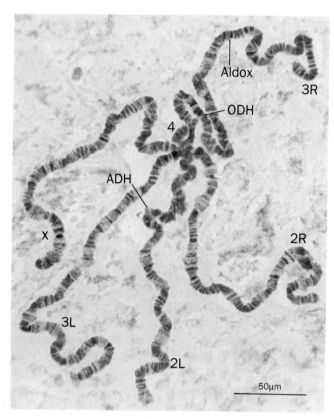

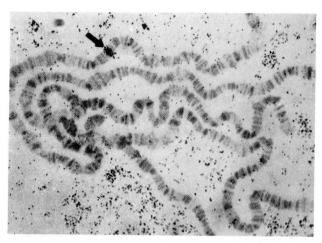

FIGURE 33-18. Autoradiograph of a *D. melanogaster* polytene chromosome that has been *in situ* hybridized with yolk protein cDNA. The dark grains (*arrow*) identify the chromosomal location of the yolk protein gene. [From Barnett, T., Pachl, C., Gergen, J.P., and Wensink, P.C., *Cell* **21**, 735 (1980). Copyright © 1980 by Cell Press.]

FIGURE 33-16. Photomicrograph of the stained polytene chromosomes from the *D. melanogaster* salivary gland. Such chromosomes consist of darkly staining bands interspersed with light-staining interband regions. All four chromosomes in a single cell are held together by their centromeres. The chromosomal positions for the genes specifying alcohol dehydrogenase (ADH), aldehyde oxidase (Aldox), and octanol dehydrogenase (ODH) are indicated. [Courtesy of B.P. Kaufmann, University of Michigan.]

33-18). We shall see that these bands, which probably correspond to the radial loops of metaphase chromosomes, are the chromosome's transcriptional units (Section 33-3).

2. GENOMIC ORGANIZATION

Higher organisms contain a great variety of cells that differ not only in their appearances (e.g., Fig. 1-10) but in the proteins they synthesize. Pancreatic acinar cells, for example, synthesize copious amounts of digestive enzymes, including trypsin and chymotrypsin, but no insulin, whereas the neighboring pancreatic β cells produce large quantities of insulin but no digestive enzymes. Clearly, each of these different types of cells expresses different genes. Yet, most of a multicellular organism's somatic cells contain the same genetic information as the fertilized ovum from which they are descended (a phenomenon described as **totipotency**). This was demonstrated, for instance, by John Gurdon, who raised a normal adult frog from a fertilized frog egg whose nucleus he had replaced with the nucleus of a tadpole intestinal cell. In this section we describe the genetic organization of the eukaryotic chromosome, which permits its enormous expressional flexibility. How this genetic expression is controlled is the subject of Section 33-3.

A. The C-Value Paradox

One might reasonably expect the morphological complexity of an organism to be roughly correlated with its **C value,** the amount of DNA in its haploid genome. After all, the morphological complexity of an organism must reflect an underlying genetic complexity. Nevertheless, in what is

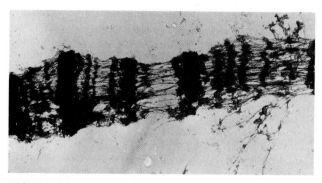

FIGURE 33-17. An electron micrograph of a segment of polytene chromosome from *D. melanogaster.* Note that its interband regions consist of chromatin fibers that are more or less parallel to the long axis of the chromosome, whereas its bands, which contain ~95% of the chromosome's DNA, are much more highly condensed. [Courtesy of Gary Burkholder, University of Saskatechewan, Canada.]

known as the **C-value paradox,** many organisms have unexpectedly large C values (Fig. 33-19). For instance, the genomes of lungfish are 10 to 15 times larger than of those of mammals and those of some salamanders are yet larger. Moreover, the C-value paradox even applies to closely related species; for example, the C values for several species of *Drosophila* have a 2.5-fold spread. Does the "extra" DNA in the larger genomes have a function, and if not, why is it preserved from generation to generation?

The 4.7 million-bp *E. coli* genome is thought to code for ~3000 gene products (about half of which have been characterized). In contrast, the 2.9 billion-bp haploid human genome, which is >600 times larger than that of *E. coli,* is estimated to code for 30 to 40 thousand proteins; that is, humans have only 10 to 13 times as many structural genes as do *E. coli.* Certainly the control of genetic expression in eukaryotes must be a far more elaborate process than it is in prokaryotes. Yet, does all the unexpressed DNA in the human genome, at least 98% of the total, function in the control of genetic expression?

At present, we are unable to answer the foregoing questions satisfactorily. As we shall see below, however, we are learning a considerable amount about the detailed genetic organization of the eukaryotic chromosome. This information, no doubt, will contribute heavily to finding the answers to these questions.

C_0t Curve Analysis Indicates DNA Complexity

The rate at which DNA renatures is indicative of the lengths of its unique sequences. If DNA is sheared into uniform fragments of 300 to 10,000 bp (Section 28-2C), denatured, and kept at a low concentration so that the effects of mechanical entanglement are small, the rate-determining step in renaturation is the collision of complementary sequences. Once the complementary sequences have found each other through random diffusion, they rapidly zip up to form duplex molecules. The rate of renaturation of denatured DNA is therefore expressed

$$\frac{d[A]}{dt} = -k[A][B] \qquad [33.1]$$

where A and B represent complementary single-stranded sequences and k is a second-order rate constant (Section

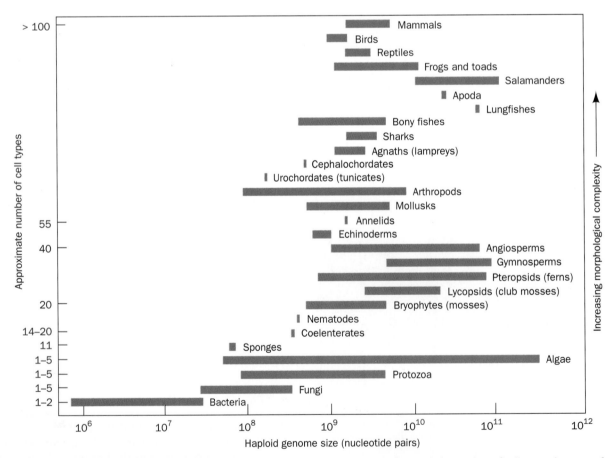

FIGURE 33-19. The range of haploid genome DNA contents in various categories of organisms indicating the C-value paradox. The morphological complexity of the organisms, as estimated according to their number of cell types, increases from bottom to top. [After Raff, R.A. and Kaufman, T.C., *Embryos, Genes, and Evolution, p.* 314, Macmillan (1983).]

13-1B). Since [A] = [B] for duplex DNA, Eq. [33.1] integrates to

$$\frac{1}{[A]} = \frac{1}{[A]_0} + kt \qquad [33.2]$$

where $[A]_0$ is the initial concentration of A.

It is convenient to measure the fraction f of unpaired strands:

$$f = \frac{[A]}{[A]_0} \qquad [33.3]$$

Combining Eqs. [33.2] and [33.3] yields

$$f = \frac{1}{1 + [A]_0 kt} \qquad [33.4]$$

The concentration terms in these equations refer to unique sequences since the collision of noncomplementary sequences does not lead to renaturation. Hence, if C_0 is the initial concentration of base pairs in solution, then

$$[A]_0 = \frac{C_0}{x} \qquad [33.5]$$

where x is the number of base pairs in each unique sequence and is known as the DNA's **complexity**. For example, the repeating sequence $(AGCT)_n$ has a complexity of 4, whereas an *E. coli* chromosome, which consists of 4.7 million bp of unrepeated sequence, has a complexity of 4.7 million. Combining Eqs. [33.4] and [33.5] yields

$$f = \frac{1}{1 + C_0 kt/x} \qquad [33.6]$$

When one half of the molecules in the sample have renatured, $f = 0.5$ so that

$$C_0 t_{1/2} = \frac{x}{k} \qquad [33.7]$$

where $t_{1/2}$ is the time for this to occur. The rate constant k is characteristic of the rate at which single strands collide in solution under the conditions employed, so it is independent of the complexity of the DNA and, for reasonably short DNA fragments, the length of a strand. Consequently, *for a given set of conditions, the value of $C_0 t_{1/2}$ depends only on the complexity x of the DNA.* This situation is indicated in Fig. 33-20, which is a series of plots of f versus $C_0 t$ for various DNAs. Such plots are referred to as $C_0 t$ (pronounced "cot") curves. The complexities of the DNAs in Fig. 33-20 vary from 1 for the synthetic duplex poly(A) · poly(U) to $\sim 3 \times 10^9$ for some fractions of mammalian DNAs. Their corresponding values of $C_0 t_{1/2}$ vary accordingly.

The speed and sensitivity of $C_0 t$ curve analysis is greatly enhanced through the hydroxyapatite fractionation of the renaturing DNA. Hydroxyapatite, it will be recalled (Section 28-4B), binds double-stranded DNA at a higher phosphate concentration than it binds single-stranded DNA. The single- and double-stranded DNAs in a solution of renaturing DNA may therefore be separated by hydroxyapatite chromatography and the amounts of each measured. The single-stranded DNA can then be further renatured and the process repeated. If the renaturing DNA is radioactively labeled, much smaller quantities of it can be detected than is possible by spectroscopic means. Thus, through the

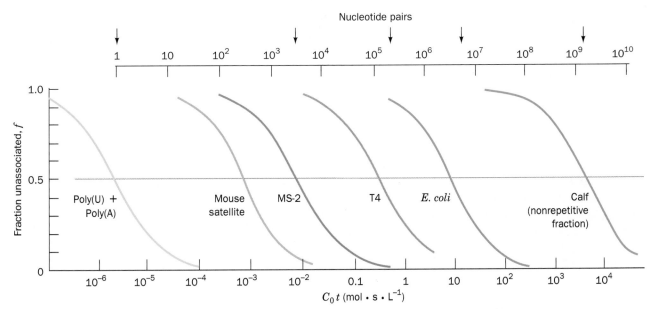

FIGURE 33-20. The reassociation ($C_0 t$) curves of duplex DNAs from the indicated sources. The DNA was dissolved in a solution containing $0.18M$ Na$^+$ and sheared to an average length of 400 bp. The upper scale indicates the genome sizes of some of the DNA's (**MS2** and **T4** are bacteriophages). [After Britten, R.J. and Kohne, D.E., *Science* **161,** 530 (1968).]

hydroxyapatite chromatography of radioactively labeled DNA, the C_0t curve analysis of a DNA of such a high complexity that its $t_{1/2}$ is days or weeks can be conveniently measured in a small fraction of that time.

B. Repetitive Sequences

Consider a sample of DNA that consists of sequences with varying degrees of complexity. Its C_0t curve, Fig. 33-21 for example, is the sum of the individual C_0t curves for each complexity class of DNA. *C_0t curve analysis has demonstrated that viral and prokaryotic DNAs have few, if any, repeated sequences (e.g., Fig. 33-20 for MS2, T4, and E. coli). In contrast, eukaryotic DNAs exhibit complicated C_0t curves (e.g., Fig. 33-22) that must arise from the presence of DNA segments of several different complexities.*

Kinetic analyses indicate that eukaryotic C_0t curves may be attributed to the presence of four somewhat arbitrarily defined classes of DNAs: (1) **unique sequences** (~1 copy per haploid genome), (2) **moderately repetitive sequences** (<10^6 copies per haploid genome), (3) **highly repetitive sequences** (>10^6 copies per haploid genome), and (4) **inverted repeats**. The sequences and chromosomal distributions of these DNA segments vary with the species, so a unifying description of their arrangements cannot be made. Nevertheless, several broad generalizations are possible as we shall see below.

Inverted Repeats Form Foldback Structures

The most rapidly reassociating eukaryotic DNA, which represents as much as 10% of some genomes, renatures with first-order kinetics. Evidently, this DNA contains inverted (self-complementary) sequences in close proximity, which can fold back on themselves to form hairpinlike **foldback structures** (Fig. 33-23a). Inverted sequences may be isolated by adsorbing the duplex DNA formed at very low C_0t values to hydroxyapatite and subsequently degrading its single-stranded loop and tails with **S1 nuclease** (an endonuclease from *Aspergillus oryzae* that preferentially cleaves single strands). The resulting inverted repeats range in length from 100 to 1000 bp, sizes much too large to have evolved at random. *In situ* hybridization studies on metaphase chromosomes using these inverted repeats as probes indicate that they are distributed at many chromosomal sites.

The function of inverted repeats, some 2 million of which occur in the human genome, is unknown. However, since the cruciform structures formed by paired foldback structures (Fig. 33-23b) are only slightly less stable than the corresponding normal duplex DNA, it has been suggested that the inverted repeats function in chromatin as some sort of molecular switch.

Highly Repetitive DNA Is Clustered at Centromeres

Highly repetitive DNA consists of clusters of nearly identical sequences up to 10 bp long that are tandemly repeated

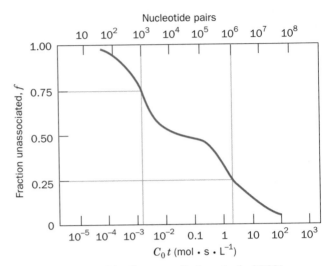

FIGURE 33-21. The C_0t curve of a hypothetical DNA molecule that, before fragmentation, was 2 million bp in length and consisted of a unique sequence of 1 million bp and 1000 copies of a 1000-bp sequence. Note the curve's biphasic nature.

thousands of times. Such **simple sequence DNAs** can often be separated from the bulk of the chromosomal DNA by shear degradation followed by density gradient ultracentrifugation in CsCl since their distinctive base compositions cause them to form "satellites" to the main DNA band (Fig. 33-24; recall that the buoyant density of DNA in CsCl increases with its G + C content; Section 28-4D). The sequences of these DNAs, which are also known as **satellite DNAs**, are species specific. For example, the crab *Cancer borealis* has a simple sequence DNA comprising 30% of its

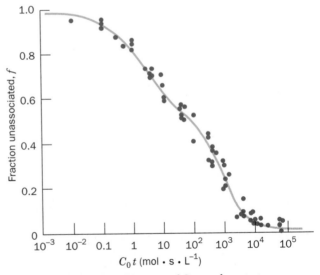

FIGURE 33-22. The C_0t curve of *Strongylocentrotus purpuratus* (a sea urchin) DNA. [After Galau, G.A., Britten, R.J., and Davidson, E.H., *Cell* **2**, 11 (1974).]

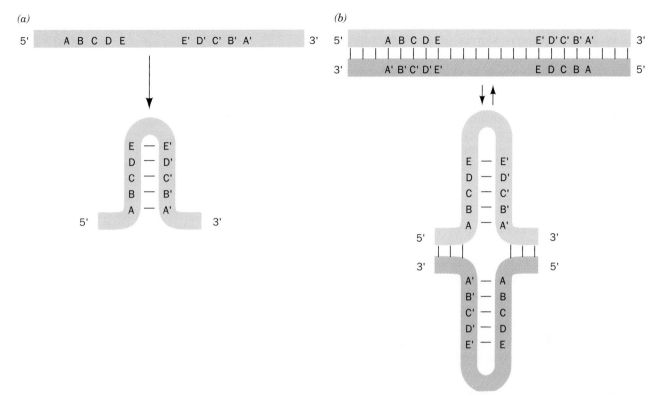

(a) *(b)*

FIGURE 33-23. Foldback structures in DNA. *(a)* Single-stranded DNA containing an inverted repeat will, under renaturing conditions, form a base paired loop known as a foldback structure. Here A is complementary to A′, B is complementary to B′, etc. *(b)* An inverted repeat in duplex

DNA could assume a cruciform conformation consisting of two opposing foldback structures. The stability of this structure would be less than that of the corresponding duplex but only by the loss of the base pairing energy in the unpaired loops.

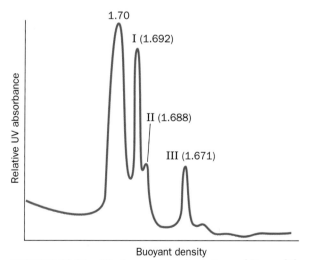

FIGURE 33-24. The buoyant density pattern of *Drosophila virilus* DNA centrifuged to equilibrium in neutral CsCl. Three prominent bands of satellite DNA (ρ = 1.692, 1.688, and 1.671) are present, in addition to the main DNA band (ρ = 1.70). [After Gall, J.G., Cohen, E.H., and Atherton, D.D., *Cold Spring Harbor Symp. Quant. Biol.* **38**, 417 (1973).]

genome in which the repeating unit is the dinucleotide AT. The DNA of *Drosophila virilus* exhibits three satellite bands (Fig. 33-24), which each consist of a different although closely related repeating heptanucleotide sequence:

$$5'-\text{ACAAACT}-3'$$
$$3'-\text{TGTTTGA}-5'$$
Satellite I

$$5'-\text{ATAAACT}-3'$$
$$3'-\text{TATTTGA}-5'$$
Satellite II

$$5'-\text{ACAAATT}-3'$$
$$3'-\text{TGTTTAA}-5'$$
Satellite III

These comprise 25, 8, and 8% of the 4.4×10^7-bp *D. virilus* genome, so that these sequences are repeated 1.6, 0.5, and 0.5 million times, respectively.

The *in situ* hybridization of mouse chromosomes with ^{3}H-labeled RNA synthesized on mouse simple sequence DNA templates established that simple sequence DNA is concentrated in the heterochromatic region associated with the chromosomal centromere (Fig. 33-25). This observation suggests that simple sequence DNA, which is not transcribed *in vivo,* functions to align homologous chromosomes during meiosis and/or to facilitate their recombination. This hypothesis is supported by the observations that satellite DNAs are largely or entirely eliminated in the somatic cells of a variety of eukaryotes (which are consequently no longer totipotent) but not in their germ cells. The putative chromosomal proteins that specifically bind simple sequence DNAs have not been detected, however.

Moderately Repetitive DNAs Are Arranged in Dispersed Repeats

Moderately repetitive DNAs occur in segments of 100 to several thousand bp that are interspersed with larger blocks of unique DNA. Some of this repetitive DNA consists of tandemly repeated groups of genes that specify products that cells require in large quantities, such as ribosomal RNAs, tRNAs, and histones. The organization of these repeated genes is discussed in Section 33-2C. However, most moderately repetitive DNAs, although they may be transcribed, do not specify RNAs of known function. The best characterized such DNA is known as the *Alu* family because most of its ~300 bp segments contain a cleavage site for the restriction endonuclease *Alu*I (Table 28-5). The *Alu* family is the human genome's most abundant moderately repetitive DNA; the genome contains 300 to 500 thousand widely distributed *Alu* sequences that are, on average, 80 to 90% homologous with their consensus sequence. *Alu* DNA also occurs in monkeys and rodents, and *Alu*-like sequences occur in such distantly related organisms as slime molds, echinoderms, amphibians, and birds. Although the *Alu* family is the most prominent moderately repetitive DNA in many organisms, it is by no means the only one. Indeed, vertebrate genomes, as sequence analyses have shown, generally contain several different varieties of moderately repetitive DNAs.

Moderately Repetitive DNAs Have Unknown Functions

It seems likely, considering their ranges of segment lengths and copy numbers, that nonexpressed, moderately repetitive DNAs have several different functions. There is, however, little experimental evidence in support of any of the various proposals that have been put forward in this regard. The proposal that is usually given the most credence is that moderately repetitive DNAs function as control sequences that participate in coordinately activating nearby genes. Another possibility, which is based on the observation that *Alu* DNA contains a segment that is homologous to the **papovavirus** replication origin, is that certain families of moderately repetitive DNAs act as DNA replication origins. A third class of proposed functions for moderately repetitive DNAs is that they increase the evolutionary versatility of eukaryotic genomes by facilitating chromosomal rearrangements and/or forming reservoirs from which new functional sequences can be recruited. Genetic evidence

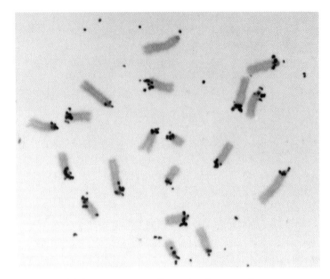

FIGURE 33-25. An autoradiograph of mouse chromosomes showing the location of their satellite DNA through *in situ* hybridization. [Courtesy of Joseph Gall, Carnegie Institution of Washington.]

indicates that retrotransposons (Section 31-6B), which, for example, comprise ~3% of the *Drosophila* genome, indeed promote chromosomal rearrangements.

Considering both the enormous amount of repetitive DNA in most eukaryotic genomes and the dearth of confirmatory evidence for any of the above proposals, a possibility that must be seriously entertained is that much repetitive DNA serves no useful purpose whatever for its host. Rather, it is **selfish** or **junk DNA,** a molecular parasite that, over many generations, has disseminated itself throughout the genome through some sort of transpositional process. The theory of natural selection indicates that the increased metabolic burden imposed by the replication of an otherwise harmless selfish DNA would eventually lead to its elimination. Yet, for slowly growing eukaryotes, the relative disadvantage of replicating say an additional 1000 bp of selfish DNA in an ~1 billion-bp genome would be so slight that its rate of elimination would be balanced by its rate of propagation. The C-value paradox may therefore simply indicate that a significant fraction, if not the great majority, of each eukaryotic genome is selfish DNA.

C. Tandem Gene Clusters

Most genes occur but once in an organism's haploid genome. This situation is feasible, even for genes specifying proteins required in large amounts, through the accumulation of their corresponding mRNAs. However, the great cellular demand for rRNAs (which comprise ~80% of a cell's RNA) and tRNAs, which are all transcription products, can only be satisfied through the expression of multiple copies of the genes specifying them. In the following subsections we discuss the organization of the genes coding

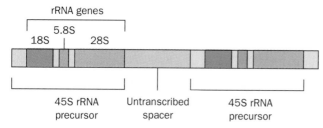

FIGURE 33-26. The 18S, 5.8S, and 28S rRNA genes are organized in tandem repeats in which sequences coding for the 45S rRNA precursor are interspersed by untranscribed spacers.

for rRNAs and tRNAs. We shall also consider the organization of histone genes, the only protein-coding genes that occur in multiple identical copies.

rRNA Genes Are Organized into Repeating Sets

We have seen in Sections 29-4B and C that even the *E. coli* genome, which otherwise consists of unique sequences, contains multiple copies of rRNA and tRNA genes. In eukaryotes, the genes specifying the 18S, 5.8S, and 28S rRNAs are invariably arranged in this order, reading $5' \to 3'$ on the RNA strand, and separated by short transcribed spacers to form a single transcription unit of ~7500 bp (Fig. 33-26). (Recall that the primary transcript of this gene cluster is a 45S RNA from which the mature rRNAs are derived by posttranscriptional cleavage; Section 29-4B.) *Indeed, this rRNA gene arrangement is universal since the 5' end of prokaryotic 23S rRNA is homologous to eukaryotic 5.8S rRNA (Section 30-3A).*

Electron micrographs, such as Fig. 33-27, indicate that *the blocks of transcribed eukaryotic rRNA genes are arranged in tandem repeats that are separated by untranscribed spacers (Fig. 33-26).* These tandem repeats are typically ~12,000 bp in length, although the untranscribed spacer varies in length between species and, to a lesser extent, from gene to gene. Quantitative measurements of the amounts of radioactively labeled rRNAs that can hybridize with the corresponding nuclear DNA **(rDNA)** indicate that these rRNA genes, which may be distributed among several chromosomes, vary in haploid number from less than 50 to over 10,000, depending on the species. Humans, for example, have 50 to 200 blocks of rDNA spread over 5 chromosomes.

The Nucleolus Is the Site of rRNA Synthesis and Ribosome Assembly

In a typical interphase cell nucleus, the rDNA condenses to form a single nucleolus (Fig. 1-5). There, as Fig. 33-27 suggests, these genes are rapidly and continuously transcribed by RNA polymerase I (Section 29-2F). The nucleolus, as demonstrated by radioactive labeling experiments, is also the site where these rRNAs are posttranscriptionally processed and assembled with cytoplasmically synthesized ribosomal proteins into immature ribosomal subunits.

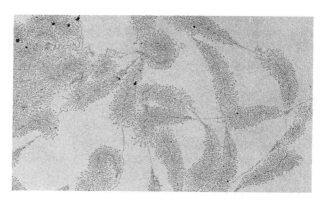

FIGURE 33-27. An electron micrograph of tandem arrays of actively transcribing 18S, 5.8S, and 28S rRNA genes from the nucleoli of the newt *Notophthalmus viridescens.* The axial fibers are DNA. The fibrillar "Christmas tree" matrices, which consist of newly synthesized RNA strands in complex with proteins, outline each transcriptional unit. Note that the longest ribonucleoprotein branches of each "Christmas tree" are only ~10% the length of their corresponding DNA stem. Apparently, the RNA strands are compacted through secondary structure interactions and/or protein associations. The matrix-free segments of DNA are the untranscribed spacers. [Courtesy of Oscar L. Miller, Jr., University of Virginia.]

Final assembly of the ribosomal subunits only occurs as they are being transferred to the cytoplasm, which presumably prevents the premature translation of partially processed mRNAs (hnRNAs) in the nucleus.

5S rRNA Is Synthesized Separately from Other rRNAs

The genes coding for the 120-nucleotide 5S rRNAs, much like the other rRNA genes, are arranged in clusters that contain a total of several hundred to several hundred thousand tandem repeats distributed among one or more chromosomes. In *Xenopus laevis,* the organism whose 5S rRNA genes are best characterized, the repeating unit consists of the 5S rRNA gene, a nearby **pseudogene** (a 101-bp segment of the 5S rRNA gene that, curiously, is not transcribed), and an untranscribed spacer of variable length but averaging ~400 bp (Fig. 33-28). The 5S rRNA genes are transcribed outside of the nucleolus by RNA polymerase III (Section 29-2F), an enzyme distinct from RNA polymerase I. 5S rRNA must therefore be transported into the nucleo-

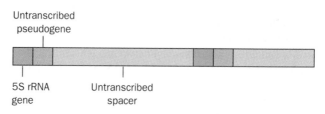

FIGURE 33-28. The organization of the 5S RNA genes in *Xenopus laevis.* Each of the ~750-nt tandemly repeated units consists of a 5S rRNA gene trailed by an untranscribed spacer in which a pseudogene closely follows the 5S gene.

lus for incorporation into the large ribosomal subunit. The tRNA genes, which are likewise transcribed by RNA polymerase III, are also multiply reiterated and clustered, but the organization of these ~60 different gene types is largely unknown.

Histone Genes Are Reiterated

Histone mRNAs have relatively short cytoplasmic lifetimes because of their lack of the poly(A) tails that are appended to other eukaryotic mRNAs (Section 29-4A). Yet, histones must be synthesized in large amounts during S phase of the cell cycle (when DNA is synthesized). *This process is made possible through the multiple reiteration of histone genes, which in most organisms are the only identically repeated genes that code for proteins.* This organization, it is thought, permits the sensitive control of histone synthesis through the coordinate transcription of sets of histone genes. Histone genes also differ from nearly all other eukaryotic genes in that almost all histone sequences lack introns (noncoding intervening sequences; Section 29-4A). The significance of this observation is unknown.

There is little relationship between a genome's size and its total number of histone genes. For example, birds and mammals have 10 to 20 copies of each of the 5 histone genes, *Drosophila* has ~100, and sea urchins have several hundred. This suggests that the efficiency of histone gene expression varies with species. In many organisms, as sequencing studies of cloned genes have shown, the histone genes are organized into tandemly repeated quintets consisting of a gene coding for each of the 5 different histones interspersed by untranscribed spacers (Fig. 33-29). The gene order and the direction of transcription in these quintets is preserved over large evolutionary distances. Corresponding spacer sequences vary widely among species and, to a limited extent, among the repeating quintets within a genome. In birds and mammals, this repetitious organization has broken down; their histone genes occur in clusters but in no particular order.

Reiterated Sequences May Be Generated and Maintained by Unequal Crossovers and/or Gene Conversion

How do reiterated genes maintain their identity? The usual mechanism of Darwinian selection would seem ineffective in accomplishing this since deleterious mutations in a few members of a multiply repeated set of identical genes would have little phenotypic effect. Indeed, many mutations do not affect the function of a gene product and are therefore selectively neutral. Reiterated gene sets must therefore maintain their homogeneity through some additional mechanism. Two such mechanisms seem plausible:

1. In the **unequal crossover** mechanism (Fig. 33-30*a*), recombination occurs between homologous segments of misaligned chromosomes, thereby excising a segment from one of the chromosomes and adding it to the other. Computer simulations indicate that such repeated expansions and contractions of a chromosome will, by

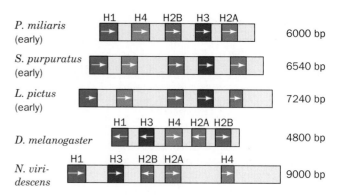

FIGURE 33-29. The organization and lengths of the histone gene cluster repeating units in a variety of organisms (the top three organisms are distantly related sea urchins). Coding regions are indicated in color and spacers are gray. The arrows denote the directions of transcription.

random processes, generate a cluster of reiterated sequences that have been derived from a much smaller ancestral cluster.

2. In the **gene conversion** mechanism (Fig. 33-30*b*), one member of a reiterated gene set "corrects" a nearby variant through a process resembling recombination repair (Section 31-5C).

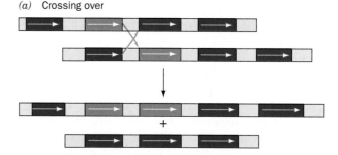

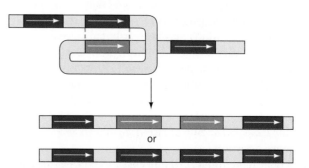

FIGURE 33-30. Two possible mechanisms for maintaining the homogeneity of a tandem multigene family. (*a*) Unequal crossing over between mispaired but similar genes results in an unpaired DNA segment being deleted from one chromosome and added to the other. (*b*) Gene conversion "corrects" one member of a tandem array with respect to the other via a recombination repair mechanism. Repeated cycles of either process may either eliminate a variant gene or spread it throughout the entire tandem array.

Since point mutations are rare events compared to crossovers, either mechanism would eventually result in a newly arisen variant copy of a repeated sequence either being eliminated or taking over the entire cluster. If a mutation that has been so concentrated is deleterious, it will be eliminated by Darwinian selection. In contrast, variant spacers, which are not as subject to selective pressure, would be eliminated at a slower rate. The existence of reiterated sets of identical genes separated by somewhat heterogeneous spacers may therefore be reasonably attributed to either homogenization model.

D. Gene Amplification

The selective replication of a particular set of genes, a process known as **gene amplification,** normally occurs only at specific stages of the life cycle of certain organisms. In the following subsections, we outline what is known about this phenomenon.

rRNA Genes Are Amplified During Oogenesis

The rate of protein synthesis during the early stages of embryonic growth is so great that in some species the normal genomic complement of rRNA genes cannot satisfy the demand for rRNA. In these species, notably certain insects, fish, and amphibians, the rDNA is differentially replicated in developing oocytes (immature egg cells). In one of the most spectacular examples of this process, the rDNA in *Xenopus laevis* oocytes is amplified by ~1500 times its amount in somatic cells to yield some 2 million sets of rRNA genes comprising nearly 75% of the total cellular DNA. The amplified rDNA occurs as extrachromosomal circles, each containing one or two transcription units, that are organized into hundreds of nucleoli (Fig. 33-31). Mature *Xenopus* oocytes therefore contain ~10^{12} ribosomes, 200,000 times the number in most larval cells. This is so many that mutant zygotes (fertilized ova), which lack nucleoli (and thus cannot synthesize new ribosomes; the oocyte's extra nucleoli are destroyed during its first meiotic division) survive to the swimming tadpole stage with only their maternally supplied ribosomes.

What is the mechanism of rDNA amplification? An important clue is that the untranscribed spacers from a given extrachromosomal nucleolus all have the same length, whereas we have seen that the corresponding chromosomal spacers exhibit marked length heterogeneities. This observation suggests that the rDNA circles in a single nucleolus are all descended from a single chromosomal gene. Gene amplification has been shown to occur in two stages: A low level of amplification in the first stage followed by massive amplification in the second stage. It therefore seems likely that, in the first stage, no more than a few chromosomal rRNA genes are replicated by an unknown mechanism and the daughter strands released as extrachromosomal circles. Then, in the second stage, these circles are multiply replicated by the rolling circle mechanism (Section 31-3B). In support of this hypothesis are electron micrographs of am-

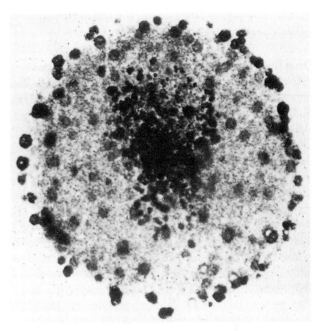

FIGURE 33-31. A photomicrograph of an isolated oocyte nucleus from *X. laevis.* Its several hundred nucleoli, which contain amplified rRNA genes, appear as darkly staining spots. [Courtesy of Donald Brown, Carnegie Institution of Washington.]

plified genes showing the "lariat" structures postulated to be rolling circle intermediates (Fig. 31-20).

Chorion Genes Are Amplified

The only other known example of programmed gene amplification is that of the *Drosophila* ovarian follicle cell genes that code for **chorion** (egg shell) **proteins** (ovarian follicle cells surround and nourish the maturing egg). Prior to chorion synthesis, the entire haploid genome of each ovarian follicle cell is replicated 16-fold. This process is followed by an ~10-fold selective replication of only the chorion genes to form a multiply branched (partially polytene) structure in which the amplified chorion genes remain part of the chromosome (Fig. 33-32). Interestingly, chorion gene amplification does not occur in silk moth oocytes.

FIGURE 33-32. An electron micrograph of a chorion gene–containing chromatin strand from an oocyte follicle cell of *D. melanogaster.* The strand has undergone several rounds of partial replication (*arrows at replication forks*) to yield a multiforked structure containing several parallel copies of chorion genes. [Courtesy of Oscar L. Miller, Jr., University of Virginia.]

Rather, this organism's genome has multiple copies of chorion genes.

Drug Resistance Can Result from Gene Amplification

In cancer chemotherapy, a common observation is that the continued administration of a cytotoxic drug causes an initially sensitive tumor to become increasingly drug resistant to the point that the drug loses its therapeutic efficacy. One mechanism by which a cell line can acquire such drug resistance is through the overproduction of the drug's target enzyme. Such a process can be observed, for example, by exposing cultured animal cells to the dihydrofolate analog methotrexate. This substance, it will be recalled, all but irreversibly binds to dihydrofolate reductase (DHFR), thereby inhibiting DNA synthesis (Section 26-4B). Slowly increasing the methotrexate dose yields surviving cells that ultimately contain up to 1000 copies of the DHFR gene and are thereby capable of tremendous overproduction of this enzyme—a clear laboratory demonstration of Darwinian selection. Members of some of these cell lines contain extrachromosomal elements known as **double minute chromosomes** that each bear one or more copies of the DHFR gene, whereas in other cell lines the additional DHFR genes are chromosomally integrated. The mechanism of gene amplification in either cell type is not well understood, although it is worth noting that this phenomenon is only known to occur in cancer cells. Both types of amplified genes are genetically unstable; further cell growth in the absence of methotrexate results in the gradual loss of the extra DHFR genes.

E. Clustered Gene Families: Hemoglobin Gene Organization

Few proteins in a given organism are really unique. Rather, like the digestive enzymes trypsin, chymotrypsin, and elastase (Section 14-3), or the various collagens (Section 7-2C), they are usually members of families of structurally and functionally related proteins. In many cases, the family of genes specifying such proteins are clustered together in a single chromosomal region. In the following subsections, we consider the organization of two of the best characterized clustered gene families, those coding for the two types of human hemoglobin subunits. The clustered gene families that encode immune system proteins are discussed in Section 34-2C.

Human Hemoglobin Genes Are Arranged in Two Developmentally Ordered Clusters

Human adult hemoglobin (HbA) consists of $\alpha_2\beta_2$ tetramers in which the α and β subunits are structurally related. The first hemoglobin made by the human embryo, however, is a $\zeta_2\varepsilon_2$ tetramer (**Hb Gower 1**) in which ζ and ε are α- and β-like subunits, respectively (Fig. 33-33). By around 8-weeks postconception, the embryonic subunits have been supplanted (in newly formed erythrocytes) by the α subunit

and the β-like γ subunit to form fetal hemoglobin (HbF), $\alpha_2\gamma_2$ (the hemoglobins present during the changeover period, $\alpha_2\varepsilon_2$ and $\zeta_2\gamma_2$, are named **Hb Gower 2** and **Hb Portland,** respectively). The γ subunit is gradually superseded by β starting a few weeks before birth. Adult blood normally contains ~97% HbA, 2% **HbA$_2$** ($\alpha_2\delta_2$ in which δ is a β variant), and 1% HbF.

In mammals, the genes specifying the α- and β-like hemoglobin subunits form two different gene clusters that occur on separate chromosomes. This distribution was largely determined through the sequence analysis of cloned hemoglobin genes, which were identified in genomic libraries (Section 28-8C) by Southern blotting (Section 28-4C). The probes used in this process were derived from hemoglobin mRNAs, which, being the major mRNA products of reticulocytes, are readily isolated. In humans and many other mammals, the genes in each globin cluster are arranged, $5' \rightarrow 3'$ on the coding strands, in the order of their developmental expression (Fig. 33-34). This ordering is common in mammals but not universal; in the mouse β gene cluster, for instance, the adult genes precede the embryonic genes.

The β-globin gene cluster (Fig. 33-34), which spans >60 kb, contains five functional genes: the embryonic ε gene, two fetal genes, $^G\gamma$ and $^A\gamma$ (duplicated genes that encode polypeptides differing only by having either Gly or Ala at their positions 136), and the two adult genes, δ and β. The β-globin cluster also contains one **pseudogene,** $\psi\beta$ (an un-

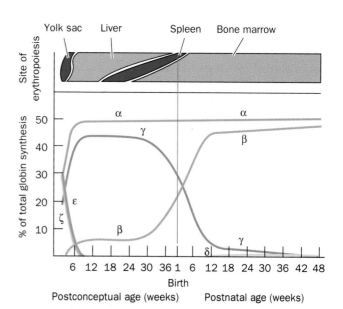

FIGURE 33-33. The progression of human globin chain synthesis with embryonic and fetal development. Note that any red blood cell contains only one type each of α- and β-like subunits. The progression in the sites of **erythropoiesis** (red cell formation), which is indicated in the upper panel, corresponds roughly to the major switches in hemoglobin types. [After Weatherall, D.J. and Clegg, J.B., *The Thalassaemia Syndromes* (3rd ed.), p. 64, Blackwell Scientific Publications (1981).]

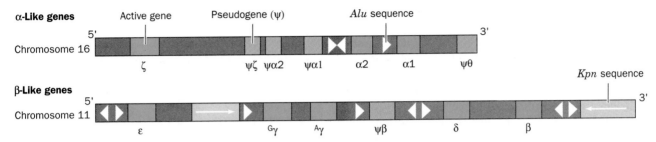

FIGURE 33-34. The organization of human globin genes on their respective coding strands. Red boxes represent active genes; green boxes represent pseudogenes; yellow boxes represent *Kpn* sequences, with the arrows indicating their

relative orientations; and triangles represent *Alu* sequences in their relative orientations. [After Karlsson, S. and Nienhuis, A. W., *Annu. Rev. Biochem.* **54,** 1074 (1985).]

transcribed relic of an ancient gene duplication that is ~75% homologous with the β gene), eight copies of the *Alu* family sequence, and two copies of the ***Kpn* family** (a 6.0-kb moderately repetitive DNA, so named because most of its ~10^4 members in the primate haploid genome have a cleavage site for the restriction endonuclease ***Kpn*I**).

The α-globin gene cluster (Fig. 33-34), which spans 28 kb, contains three functional genes: the embryonic ζ gene and two slightly different α genes, α1 and α2, which encode identical polypeptides. The α cluster also contains four pseudogenes, ψζ, ψα2, ψα1, and ψθ, and three *Alu* sequences.

Hemoglobin Genes All Have the Same Exon–Intron Structure

Protein-coding sequences represent <5% of either globin gene cluster. This situation is largely a consequence of the heterogeneous collection of untranscribed spacers separating the genes in each cluster. In addition, *all known vertebrate globin genes, including that of myoglobin and most hemoglobin pseudogenes, consist of three nearly identically*

placed coding sequences (exons) separated by two somewhat variable unexpressed intervening sequences (introns; Fig. 33-35). This gene structure apparently arose quite early in vertebrate history, well over 500 million years ago. Indeed, much of this structure even predates the divergence of plants and animals. The structure of the gene encoding leghemoglobin (a plant globin that functions in legumes to protect nitrogenase from O_2 poisoning; Section 24-6) differs from that of vertebrates only in that the central exon of vertebrate globins is split by a third intron in the leghemoglobin gene. Quite possibly the central exon in vertebrate globins arose through the fusion of the two interior exons in a leghemoglobin-like ancestral gene.

DNA Polymorphisms Can Establish Genealogies

Unexpressed sequences, which are subject to little selective pressure, evolve so much faster than expressed sequences that they even accumulate significant numbers of sequence **polymorphisms** (variations) within a single species. Consequently, the evolutionary relationships among populations within a species can be established by deter-

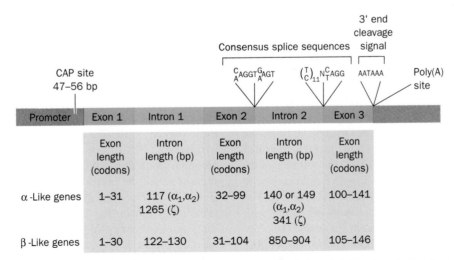

FIGURE 33-35. The structure of the prototypical hemoglobin gene, indicating the conserved sequences at the exon–intron boundaries (splice sequences) and at the 3' end of the gene

(polyadenylation site). The length of each exon (in codons) and each intron (in base pairs) is given. [After Karlsson, S. and Nienhuis, A.W., *Annu. Rev. Biochem.* **54,** 1079 (1985).]

mining how a series of polymorphic DNA sequences are distributed among them. For example, the genealogy of several diverse human populations has been inferred from the presence or absence of certain restriction sites [restriction-site length polymorphisms (RFLPs); Section 28-6A] in five segments of their β-globin gene clusters. This study has led to the construction of a "family tree" (Fig. 33-36), which indicates that non-African (Eurasian) populations are much more closely related to each other than they are to African populations. Fossil evidence indicates that anatomically modern man arose in Africa about 100,000 years ago and rapidly spread throughout that continent. This family tree therefore suggests that all Eurasian populations are descended from a surprisingly small "founder population" (perhaps only a few hundred individuals) that left Africa ~50,000 years ago. A similar analysis indicates that the sickle-cell variant of the β gene arose on at least three separate occasions in geographically distinct regions of Africa.

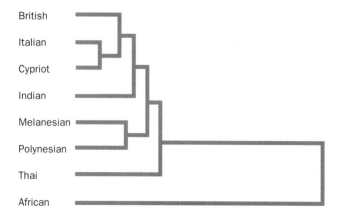

FIGURE 33-36. A family tree showing the lines of descent among eight human population groups as determined from the distribution of five restriction fragment-length polymorphisms in their β-globin gene clusters. The horizontal axis is indicative of the genetic distances between related populations and therefore of the times between their divergence. [After Wainscoat, J.S., Hill, A.V.S., Boyce, A.L., Flint, J., Hernandez, M., Thein, S.L., Old, J.M., Lynch, J.R., Falusi, A.G., Weatherall, D.J., and Clegg, J.B., *Nature* **319**, 493 (1986).]

F. Significance of Introns

The rapidly growing body of known DNA sequences reveals that introns are rare in prokaryotic structural genes, uncommon in lower eukaryotes such as yeast, and abundant in higher eukaryotes (the only known vertebrate structural genes lacking introns are those encoding histones and interferons). The exons in most interrupted genes have sizes in the range 100 to 250 bp. In contrast, intron sizes are broadly distributed from <50 to >200,000 bp. Moreover, the number of introns in a given gene can be surprisingly large; the most observed so far are the 70 or more introns in the ~2500-kb **dystrophin** gene (which encodes a 427-kD muscle protein whose defects are responsible for Duchenne/Becker muscular dystrophy; Section 34-3A). Unexpressed sequences thereby constitute ~80% of a typical vertebrate structural gene (but ~99.5% of the dystrophin gene, which is the largest known gene by a factor of ~10!).

What are the functions of introns? The argument that all introns are simply selfish DNA seems untenable since it would otherwise be difficult to rationalize why the evolution of splicing machinery offered any selective advantage over the simple elimination of the split genes. Yet, in most genes, introns have no obvious function [although introns may protect the integrity of the genes in gene families (see below) and act as regulatory elements in the transcription of certain genes (Section 33-3C)]. In fact, the number of introns in the gene encoding a given protein is not necessarily the same in all organisms or even within one organism. For instance, the rat has two functional insulin genes, one with two introns, as do most rodent insulin genes, and the other in which one of these introns has been lost. It has therefore been proposed that introns had an essential function at an earlier stage of genetic evolution which is no longer important; that is, *introns may be genetic fossils.*

Introns May Be the Products of Modular Gene Assembly

Walter Gilbert proposed that *primordial protein-coding genes arose as collections of exons that were assembled by recombination between intron sequences. Modern introns, according to this hypothesis, are therefore the remnants of a process that facilitated protein evolution.* Considerable experimental evidence has accumulated that supports this supposition. For example, the triple helical region of chicken α2(I) collagen, which consists of a 332-fold repeated triplet Gly-X-Y (Section 7-2C), is encoded by 42 of its gene's 52 exons. All of these exons are integral multiples of 9 bp (23 of them are 54 bp long with the rest consisting of either 45, 99, 108, or 162 bp) with each exon beginning with a Gly codon and ending with a Y codon. This distribution suggests that the gene segment specifying collagen's triple helical region evolved through multiple duplications of its repeating intron-flanked genetic element.

Exons Often Encode Discrete Structural Elements

The structural analysis of pyruvate kinase in terms of the chicken muscle gene's base sequence suggests that its exon–intron boundaries are functionally positioned. Each of this gene's 10 exons encodes a discrete element of protein secondary structure, with most of the introns marking positions at which the polypeptide chain makes a reverse turn (Fig. 33-37). *The pyruvate kinase gene was apparently assembled by combining a series of smaller protein-coding units and exploiting RNA splicing to express them as a single polypeptide.*

The exon boundaries in vertebrate globins (Fig. 33-35) also appear to have functional significance. They occur at precisely the same sites in all vertebrate globin chains: be-

(a)

(b)

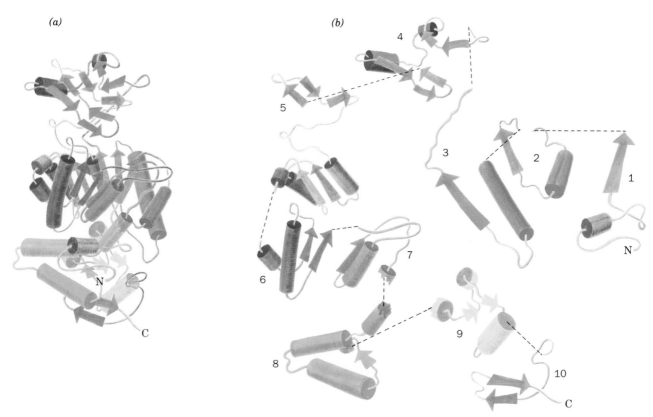

FIGURE 33-37. The structure of pyruvate kinase. *(a)* A single subunit of the tetrameric enzyme from cat muscle colored according to the exons encoding it. *(b)* An exploded view of the cat muscle pyruvate kinase subunit in which the structural segments have been separated at the exon boundaries in the gene specifying the 88% homologous chicken muscle enzyme. The exons are numbered in the order they occur, $5' \rightarrow 3'$, on the mRNA. [After Lonberg, N. and Gilbert, W., *Cell* **40**, 84 (1985).]

tween residues B12 and B13 and between residues G6 and G7 (Recall that in globin nomenclature residue B12 is the 12th residue in helix B; Section 9-2A). Thus, as Fig. 33-38 indicates: Exon 1 encodes the A and B helices, which form a scaffolding for the heme pocket (Fig. 9-11); exon 2 encodes the heme pocket (helices E and F) and the $\alpha_1\beta_2$ contacts (helix C and the FG corner), which assume alternative stable positions in the T and R states (Fig. 9-17); and exon 3 encodes the G and H helices, which provide most of the $\alpha_1\beta_1$ contacts and which are unchanged by the T→R shift (Section 9-2B).

Rat fatty acid synthase provides a third example of the functional significance of exons. Recall that this mutifunctional enzyme's seven catalytic activities occur on independent enzymes in bacteria (Section 23-4C). In the gene encoding rat fatty acid synthase (which has a total of 42 introns), the boundaries between its constituent enzyme functionalities, with but one exception, coincide with the locations of introns, thereby supporting the hypothesis that this multifunctional enzyme evolved via exon shuffling.

The Exons of LDL Receptor Occur in Other Proteins

The gene sequence of the **LDL receptor** provides what is perhaps the most convincing evidence favoring Gilbert's

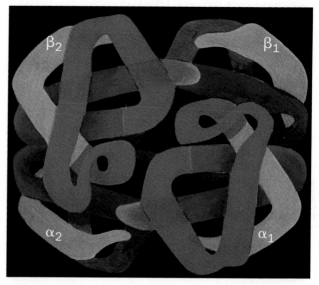

FIGURE 33-38. The structure of vertebrate hemoglobin colored according to the exons encoding it (Fig. 33-35): Exon 1 is yellow, exon 2 is red, and exon 3 is blue in both the α and the β chains. The tetrameric protein is viewed from the bottom of the page relative to Fig. 9-13. [Figure copyrighted © by Irving Geis.]

hypothesis. This 839-residue plasma membrane protein functions to bind low-density lipoprotein (LDL) to coated pits for transport into the cell via endocytosis (Section 11-5C). LDL receptor's 45-kb gene contains 18 exons, most of which encode specific functional domains of the protein. *The most intriguing aspect of this sequence, however, is that 13 of its exons specify polypeptide segments that are homologous to segments in other proteins:*

1. Five exons encode a sevenfold repeat of a 40-residue sequence that occurs once in **complement C9** (an immune system protein; Section 34-2F).

2. Three exons each encode a 40-residue repeat similar to that occurring four times in **epidermal growth factor (EGF) precursor** (EGF is a hormonally active polypeptide that stimulates cell proliferation) and once each in three blood clotting system proteins: **factor IX, factor X,** and **protein C** (Section 34-1).

3. Five exons encode a 400-residue sequence that is 33% homologous with a polypeptide segment that is shared only with EGF precursor.

Evidently, the LDL receptor gene is modularly constructed from exons that also encode portions of other proteins. Numerous other eukaryotic proteins are similarly constituted.

How Many Exons Exist?

Sequence comparisons indicate that the intron/exon structures of genes are very ancient, certainly predating the divergence of plants and animals 1 billion years ago and possibly even being components of the first protogenes when life arose. If exon shuffling is the basis of protein diversity, how many different exons have contributed to forming current proteins? Gilbert answered this question by identifying the exons in the >2500 known sequences of eukaryotic genes and eliminating the duplicates through statistical considerations. These computations suggest that a surprisingly small number of exons, only 1000 to 7000, were needed to construct all current proteins. Since the number of possible 40-residue polypeptide segments, $20^{40} = 10^{52}$, is a far larger number than the 10^3 to 10^4 that this treatment predicts really exist, modern proteins probably employ only a small fraction of the theoretically possible structural and functional motifs. Apparently the advantages of an increased protein assembly rate through exon shuffling evolutionarily outweigh the need for structural variety.

Introns May Have Been Selectively Eliminated from the Genes of Lower Organisms

If introns are the remnants of a primordial gene shuffling process, why are they absent or nearly so in the "lower" forms of life from which the "higher" forms have evolved (e.g., the gene for yeast pyruvate kinase, whose amino acid sequence is 45% homologous with that from chicken, has

no introns)? A plausible explanation of this observation, in light of the foregoing data, is that in lower forms of life, whose life styles place a premium upon efficiency (Section 1-2), introns have been selectively eliminated. In contrast, higher organisms, which are adapted to stable environments, have had much less selective pressure for intron elimination (although the rat insulin genes discussed previously provide a clear example of this). Indeed, the large sizes of many vertebrate introns suggests that they have been invaded by selfish DNA.

Introns May Genetically Stabilize Gene Families

In addition to their putative role in facilitating the evolution of new proteins, *the introns in gene families may function to protect their neighboring exons from elimination via unequal crossing over (Fig. 33-30a).* Duplicated genes are particularly susceptible to this form of degradation because their similar base sequences promote their mispairing. Their alternation with the much more variable and therefore less readily mispaired introns, however, inhibits this process. Prokaryotes and yeast, which have few gene families, have little need of such protection.

Introns May Have Arisen in Eukaryotic Nuclear Genes

Despite the foregoing, the hypothesis that exons correspond to functional units of protein structure in primordial protein-encoding genes (the so-called introns-early theory) is by no means proven. For example, the analysis of 62 intron positions in four ancient proteins with known X-ray structures revealed no significant tendency of introns to avoid interrupting the secondary structural elements of globular domains. Thus, the origins and functions of introns remains enigmatic. Quite possibly split genes arose not through the joining of primordial minigenes (exons) separated by spacers (introns) but, rather, through the insertion of introns into previously unsplit eukaryotic nuclear genes (the introns-late theory).

G. *The Thalassemias: Genetic Disorders of Hemoglobin Synthesis*

The study of mutant hemoglobins (Section 9-3) has provided invaluable insights into structure–function relationships in proteins. Likewise, the study of defects in hemoglobin expression has greatly facilitated our understanding of eukaryotic gene expression.

The most common class of inherited human disease results from the impaired synthesis of hemoglobin subunits. These anemias are named **thalassemias** (Greek: *thalassa,* sea) because they commonly occur in the region surrounding the Mediterranean Sea (although they are also prevalent in Central Africa, India, and the Far East). The observation that malaria is or was endemic in these same areas (Fig. 6-13) led to the realization that heterozygotes for thalassemic genes (who appear normal or are only mildly anemic; a condition known as **thalassemia minor**) are resistant to malaria. Thus, as we have seen in our study of sickle-cell anemia (Section 9-3B), mutations that are seriously debilitating or even lethal in homozygotes (who are said to suffer from **thalassemia major**) may offer sufficient selective advantage to heterozygotes to ensure the propagation of the mutant gene.

Thalassemia can arise from many different mutations, each of which causes a disease state of characteristic severity. In α^0- and β^0-thalassemias, the indicated globin chain is absent, whereas in α^+- and β^+-thalassemias, the normal globin subunit is synthesized in reduced amounts. In what follows, we shall consider thalassemias that are illustrative of several different types of genetic lesions.

α-Thalassemias

Most α-thalassemias are caused by the deletion of one or both of the α-globin genes in an α gene cluster (Fig. 33-34). A variety of such mutations have been cataloged. In the absence of equivalent numbers of α chains, the fetal γ chains and the adult β chains form homotetramers: **Hb Bart's** (γ_4) and **HbH** (β_4). Neither of these tetramers exhibits any cooperativity or Bohr effect (Sections 9-1C and D), which makes their oxygen affinities so high that they cannot release oxygen under physiological conditions. Consequently, α^0-thalassemia occurs with four degrees of severity depending on whether an individual has 1, 2, 3, or 4 missing α-globin genes:

1. **Silent-carrier state:** The loss of one α gene is an asymptomatic condition. The rate of expression of the remaining α genes largely compensates for the less than normal α gene dosage so that, at birth, the blood contains only ~1 to 2% Hb Bart's.

2. **α-Thalassemia trait:** With two missing α genes (either one each deleted from both α gene clusters or both deleted from one cluster), only minor anemic symptoms occur. The blood contains ~5% Hb Bart's at birth.

3. **Hemoglobin H disease:** Three missing α genes results in a mild to moderate anemia. Affected individuals can usually lead normal or nearly normal lives.

4. **Hydrops fetalis:** The lack of all four α genes is invariably lethal. Unfortunately, the synthesis of the embryonic ζ-chain continues well past the 8 weeks postconception when it normally ceases (Fig. 33-33), so the fetus usually survives until around birth.

α-Thalassemias caused by nondeletion mutations are relatively uncommon. One of the best characterized such lesions changes the UAA stop codon of the α2-globin gene to CAA (a Gln codon) so that protein synthesis continues for the 31 codons beyond this site to the next UAA. The resultant **Hb Constant Spring** is produced in only small amounts because, for unknown reasons, its mRNA is rapidly degraded in the cytosol. Another point mutation in the α2 gene changes Leu H8(125)α to Pro, which no doubt disrupts the H helix. The consequent α^+-thalassemia results from the rapid degradation of this abnormal **Hb Quong Sze**.

β-Thalassemias

Heterozygotes of β-thalassemias are usually asymptomatic. Homozygotes become so severely anemic, however, that once their HbF production has diminished, many require frequent blood transfusions to sustain life and all require them to prevent the severe skeletal deformities caused by bone marrow expansion. The anemia results not only from the lack of β-chains but also from the surplus of α chains. The latter form insoluble membrane-damaging precipitates that cause premature red cell destruction (Section 9-3A). The coinheritance of α-thalassemia therefore tends to lessen the severity of β-thalassemia major.

In β-thalassemia, there may be an increased production of the δ- and γ-chains so that the consequent extra HbA$_2$ and HbF can compensate for some of the missing HbA. In $\delta\beta$-**thalassemia,** the neighboring δ and β genes have both been deleted so that only increased production of the γ chain is possible. Yet many adult $\delta\beta$-thalassemics, for reasons that are not understood, produce so much HbF that they are asymptomatic. Such individuals are said to have **hereditary persistence of fetal hemoglobin (HPFH)**. This condition is therefore of medical interest because it could also alleviate the symptoms of β-thalassemia and sickle-cell anemia.

The so-called Greek form of HPFH is associated with a G $\rightarrow$ A mutation at position -117 of the γ-globin gene (its promoter region). In an effort to establish whether this mutation does, in fact, cause HPFH, the mutated γ-globin gene was introduced into mice. The resulting fetal and adult transgenic animals synthesized γ-globin at a high level, with a concomitant decrease in the synthesis of the β-globin gene. These changes in gene expression correlate with the loss of binding of the transcription factor **GATA-1** to the γ-globin promoter, thereby suggesting that this protein is a negative regulator of the γ-globin gene expression in normal human adults (transcription factors are discussed in Section 33-3B).

β^0-Thalassemias caused by deletions are rare compared to those causing α^0-thalassemias. This is probably because the long repeated sequences in which the α-globin genes are embedded make them more prone to unequal crossing over than the β-globin gene. Nevertheless, a β-thalassemic lesion causing the production of **Hb Lepore** is a particularly clear instance of this deletion mechanism. This lesion, the consequence of a deletion extending from within the δ gene to the corresponding position of its neighboring β gene, yields a δ/β hybrid subunit. Such deletions almost certainly arose through unequal crossovers between the β gene on one

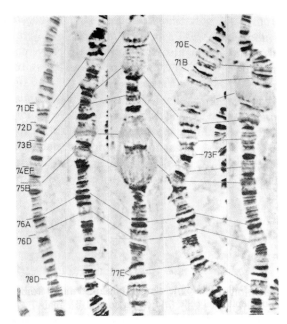

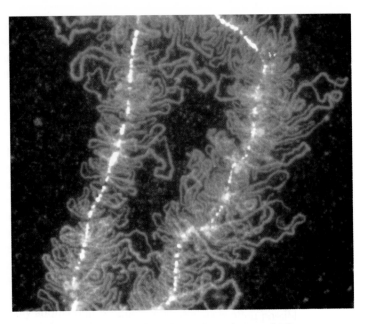

FIGURE 33-41. A series of photomicrographs showing the formation and regression of chromosome puffs (*lines*) in a *D. melanogaster* polytene chromosome over a 22-h period of larval development. Very large puffs are also known as **Balbiani rings.** [Courtesy of Michael Ashburner, Cambridge University.]

FIGURE 33-42. An immunofluorescence micrograph of lampbrush chromosome from an oocyte nucleus of the newt *Notophthalmus viridescens.* The chromosome's numerous transcriptionally active loops give rise to the name "lampbrush" (an obsolete implement for cleaning kerosene lamps). [From Roth, M.B. and Gall, J.G., *J. Cell Biol.* **105,** 1049 (1987). Copyright © 1987 by Rockefeller University Press.]

bands of giant polytene chromosomes (Fig. 33-41). These puffs reproducibly form and regress as part of the normal larval development program (Fig. 33-41) and in response to such physiological stimuli as hormones and heat. Autoradiography studies with ^{3}H-labeled uridine and immunofluorescence studies using antibodies against RNA polymerase II clearly demonstrate that *puffs are the major sites of RNA synthesis in polytene chromosomes.*

The analogous decondensation of nonpolytene chromosomes occurs most conspicuously in the so-called **lampbrush chromosomes** of amphibian oocytes (Fig. 33-42). During their prolonged meiotic prophase I (Fig. 27-3), these previously condensed chromosomes loop out segments of transcriptionally active DNA that electron micrographs such as Fig. 33-43 indicate are usually single transcription units.

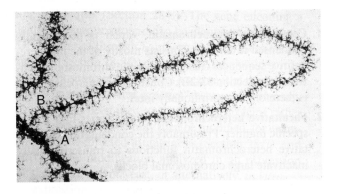

FIGURE 33-43. An electron micrograph of a single loop of a lampbrush chromosome. The ribonucleoprotein matrix coating the loop increases in thickness from one end of the loop (A) to the other (B), which indicates that the loop comprises a single transcriptional unit. [Courtesy of Oscar L. Miller, Jr., University of Virginia.]

B. Regulation of Transcriptional Initiation

The foregoing observations suggest that selective transcription is mainly responsible for the differential protein synthesis among the various types of cells in the same organism. It was not until 1981, however, that James Darnell actually demonstrated this to be the case, as follows. Experimentally useful amounts of mouse liver genes were obtained by inserting the cDNAs of mouse liver mRNAs (some 95% of which are cytosolic) into plasmids and replicating them in *E. coli* (Section 28-8A). By hybridizing the resulting cloned cDNAs with radioactively labeled mRNAs from various mouse cell types, the *E. coli* colonies containing liver-specific genes were distinguished from colonies containing genes common to most mouse cells. In this way, 12 liver-specific cDNA clones and three common cDNA clones were obtained. The question was then asked, does a eukaryotic cell transcribe only the genes encoding the proteins it synthesizes, or does it transcribe all of its genes but only process properly the transcripts it translates? This

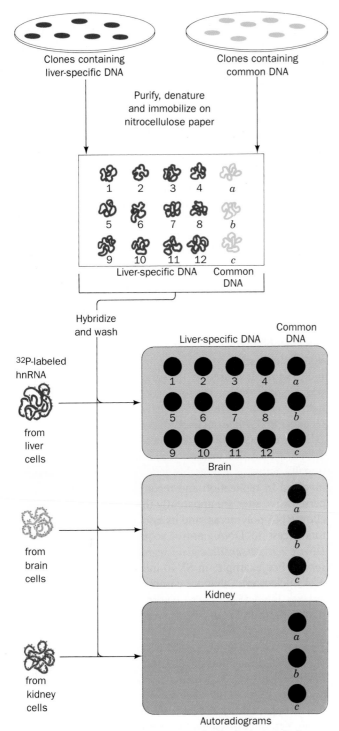

Clones containing
liver-specific DNA

Clones containing
common DNA

Purify, denature
and immobilize on
nitrocellulose paper

1 2 3 4 *a*

5 6 7 8 *b*

9 10 11 12 *c*

Liver-specific DNA   Common
DNA

Hybridize
and wash

Liver-specific DNA   Common
DNA

^{32}P-labeled
hnRNA

from
liver
cells

1 2 3 4 *a*

5 6 7 8 *b*

9 10 11 12 *c*

Brain

from
brain
cells

a

b

c

Kidney

from
kidney
cells

a

b

c

Autoradiograms

FIGURE 33-44. The primary role of selective transcription in the control of eukaryotic gene expression was established through gene cloning and hybridization techniques. Cloned cDNAs encoding 12 different mouse liver–specific proteins (*1–12*) and 3 different proteins common to most mouse cells (*a–c*) were purified, denatured, and spotted onto filter paper (*upper left*). The DNAs were hybridized with newly formed and therefore unprocessed radioactively labeled RNAs produced by either mouse liver, kidney, or brain nuclei (*lower left*). Autoradiography showed that the liver RNAs hybridized with all 12 liver-specific cDNAs and all 3 common cDNAs but that the kidney and brain RNAs only hybridized with the common cDNAs (*right*).

question was answered by hybridizing the cloned mouse genes with freshly synthesized and therefore unprocessed RNAs (hnRNAs) obtained from the nuclei of mouse liver, kidney, and brain cells (Fig. 33-44). Only the RNAs extracted from liver nuclei hybridized with the 12 liver-specific cDNAs that were probed. The RNAs from all three cell types, however, hybridized with the DNA from the three clones containing the common mouse genes. Evidently, *liver-specific genes are not transcribed by brain or kidney cells. This strongly suggests that the control of genetic expression in eukaryotes is primarily exerted at the level of transcription.*

Transcriptionally Active Chromatin Is Sensitive to Nuclease Digestion

The open structure of transcriptionally active chromatin presumably gives the transcriptional machinery access to the active genes. This hypothesis is corroborated by Harold Weintraub's demonstration that *transcriptionally active chromatin is more susceptible to digestion by pancreatic DNase I (a relatively nonspecific endonuclease) than is transcriptionally inactive chromatin.* For example, globin genes from chicken erythrocytes (avian red cells are nucleated) are more sensitive to DNase I digestion than are those from chicken oviduct (where eggs are made) as is indicated by the loss of the abilities of these genes to hybridize with a complementary DNA probe after DNase I treatment. Conversely, the gene coding **ovalbumin** (the major egg white protein) from oviduct is more sensitive to DNase I than is that from erythrocytes. Yet, nuclease sensitivity apparently reflects a gene's potential for transcription rather than transcription itself: The DNase I sensitivity of oviduct ovalbumin gene is independent of whether or not the oviduct has been hormonally stimulated to produce ovalbumin.

Nonhistone Proteins Confer Nuclease Sensitivity

The variation of a given gene's transcriptional activity with the cell in which it is located indicates that chromosomal proteins participate in the gene activation process. Histones' chromosomal abundance and lack of variety, however, make it highly unlikely that they have the specificity required for this role. Among the most conspicuous nonhistone proteins are the members of the **high mobility group (HMG),** so named because of their high electrophoretic mobilities in polyacrylamide gels. These highly conserved, low molecular mass (<30 kD) proteins have the unusual amino acid composition of ~25% basic side chains and 30% acidic side chains. The major HMG proteins constitute two pairs of homologous proteins: **HMG1** and **HMG2** (~25 kD), which bind to both single- and double-stranded DNAs, and **HMG14** and **HMG17** (~10 kD), which have a higher affinity for nucleosomes than for DNA. The HMG proteins can be eluted from chick erythrocyte chromatin by 0.35*M* NaCl without gross structural changes to the nucleosomes. This treatment eliminates the preferential nuclease sensitivity of the erythrocyte globin genes. Their nuclease sensitivity can be restored, however, by adding HMG14 and HMG17, either individually or together, to the salt-extracted chromatin.

The HMGs are not tissue specific: HMG14 and HMG17 eluted from brain nuclei can also restore nuclease sensitivity to globin genes in HMG-depleted erythrocyte chromatin. Yet, the reverse process, adding HMG14 and HMG17 from erythrocytes to HMG-depleted chromatin from brain, induces neither nuclease sensitivity in the brain globin genes nor their selective transcription. These HMGs apparently do not recognize specific DNA sequences; rather, they must bind to tissue-specific chromatin components.

HMG14 and HMG17 bind directly to nucleosomes, possibly by displacing histone H1. Indeed, both have basic regions in their N-terminal segments that are identical from chickens to humans. Nucleosomal core particles in association with HMG14 and/or HMG17 exhibit exactly the same nuclease sensitivity as does intact chromatin. *This observation indicates that genes need not be stripped of nucleosomes to be at least potentially transcriptionally active.*

Nucleosome Cores Are Transferred out of the Path of an Advancing RNA Polymerase

Since nucleosomes bind their component DNA tightly and quite stably, how does an actively transcribing RNA polymerase, which is roughly the size of a nucleosome and must separate the strands of duplex DNA to transcribe it, get access to the DNA? Two classes of models have been proposed: The advancing RNA polymerase either (1) induces a conformational change in the nucleosome that permits its DNA to be transcribed while still associated with the nucleosome, or (2) displaces the nucleosome from the DNA. These models were differentiated by Gary Felsenfeld as follows: A single nucleosome core was assembled onto a short DNA segment of defined sequence. Then, under conditions that nucleosome cores are stable (don't decompose or move) in the absence of transcription, the resulting assembly was ligated into a plasmid between a promoter and terminators for the RNA polymerase from **bacteriophage SP6** and the DNA between these two sites was transcribed by this enzyme. This treatment caused the nucleosome to move to a different site on the same plasmid, with a small preference for the untranscribed region preceding the promoter. However, the use of a very short (227-bp) DNA template containing the SP6 promoter and a bound nucleosome revealed that nucleosome transfer occurred only to the same template molecule, 40 to 95 bp upstream of its original site, even in the presence of a large excess of competitor DNA. Evidently, the histone octamer somehow steps around a transcribing RNA polymerase so as to transfer to a nearby segment of the same DNA. Felsenfeld has proposed that this occurs via a DNA looping mechanism in which the histone octamer incrementally spools onto its new position behind the advancing RNA polymer-

ase as the polymerase peels the octamer away from its original position (Fig. 33-45).

How does RNA polymerase displace nucleosomes from DNA? SP6 RNA polymerase, being a phage enzyme, cannot have evolved to interact with histones but, nevertheless, appears to do so. Other prokaryotic RNA polymerases can likewise transcribe through nucleosomes. A plausible mechanism for this phenomenon is that it is promoted by the transcriptionally induced supercoiling of DNA. A moving transcription bubble, it will be recalled (Section 29-2D), generates positive supercoils in the the DNA ahead of it and negative supercoils behind it. However, nucleosomal DNA is wound around its histone core in a left-handed toroidal coil and is therefore negatively supercoiled (Section 28-5A). Consequently, an advancing RNA polymerase molecule should destabilize the nucleosomes ahead of it while facilitating nucleosome assembly in its wake, precisely what is observed.

Active Genes Have Nuclease Hypersensitive Control Sites

RNA polymerase together with its associated transcription factors (see below) must gain access to a gene's promoter in order to initiate the gene's transcription. How is this possible on nucleosome-associated genes? The very light digestion of transcriptionally active chromatin with DNase I and other nucleases has revealed the presence of **DNase I hypersensitive sites.** These specific DNA segments are mostly located in the 5'-flanking regions of transcriptionally active or activatable genes as well as in sequences involved in replication and recombination. Nuclease hypersensitive sites are apparently the "open windows" that allow RNA polymerase and its associated transcription factors access to DNA control sequences. This is because *DNase I hypersensitive gene segments are free of nucleosomes.* For example, in SV40-infected cells, none of the ~24 nucleosomes that are complexed to the virus' 5.2-kb circular DNA (Fig. 33-46) incorporate the ~250-bp viral transcription initiation site, thereby rendering that site nuclease hypersensitive.

Felsenfeld has similarly shown that the 5'-flanking region of the β^A-globin gene from chicken erythrocytes contains a 114-bp DNase I hypersensitive segment, that can be excised by the restriction endonuclease *Msp*I. The accessibility of such a long fragment indicates that it is not part of a nucleosome. Yet, since naked DNA is not nuclease hypersensitive, the special properties of nuclease hypersensitive chromatin must arise from the sequence-specific binding of proteins so as to exclude nucleosomes. In fact, two proteins, present in chicken erythrocytes but not oviducts, specifically bind to the β^A-globin gene so as to confer on it, when inserted in a plasmid and complexed with histones, the same nuclease hypersensitivity pattern it exhibits when isolated from erythrocytes. These proteins apparently prevent the binding of histones to the hypersensitive site, thereby facilitating the transcriptional initiation of the associated gene.

The exclusive expression of globin genes in erythroid cells

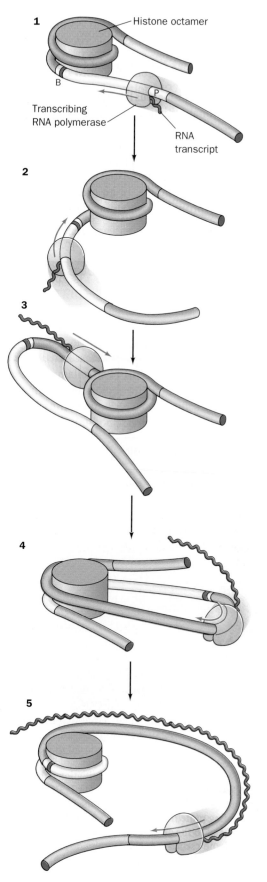

1

Histone octamer

B

Transcribing
RNA polymerase

P

RNA
transcript

2

3

4

5

FIGURE 33-45. A spooling model for transcription through a nucleosome. (**1**) RNA polymerase commences transcription at a promoter, P; the border of the nucleosome is indicated by B. (**2**) As the RNA polymerase approaches the nucleosome, it induces the dissociation of the proximal (nearest) DNA, thereby exposing part of the histone octamer surface. (**3**) The exposed histone surface binds to the DNA behind the RNA polymerase, thus forming a loop. Note that this loop is topologically isolated from the rest of the DNA and, consequently, is subject to the superhelical stress that the advancing RNA polymerase generates (see the text). (**4**) As the RNA polymerase continues to advance, the DNA ahead of it peels off the histone octamer while the trailing DNA spools onto it. (**5**) The nucleosome is thereby reformed behind the RNA polymerase, thus permitting the transcript to be completed. [After Studitsky, V.M., Clark, D.J., and Felsenfeld, G., *Cell* **76**, 379 (1994).]

appears to be largely a consequence of local chromatin structure. The human β cluster has five super-hypersensitive (SH) sites (so named because they are cleaved earlier than the normal hypersensitive sites during DNase I digestion) in a region 6 to 25 kb on the 5′ side of the ε, gene as well as one SH site ~20 kb on the 3′ side of the β gene (see Fig. 33-34). These SH sites appear to demarcate the boundaries of a large segment of transcriptionally active chromatin. Thus, an individual with an extensive upstream deletion that eliminates the SH sites, but with normal β genes, has **$\gamma\delta\beta$-thalassemia** (reduced synthesis of γ, δ, and β globins). Moreover, mice that are transgenic for the entire region of the human β cluster between its SH sites express high levels of human β-globin in a tissue-specific manner, whereas mice transgenic for only the β-gene together with its local regulatory sites either fail to express or express very low levels of human β-globin. A plausible explanation for the latter observations is that a DNA segment that is randomly inserted into a genome will most often occupy a position in

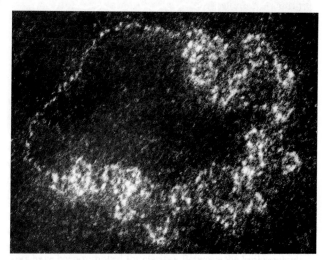

FIGURE 33-46. An electron micrograph of an SV40 minichromosome that has a nucleosome-free DNA segment. [Courtesy of Moshe Yaniv, Institut Pasteur, France.]

ilar and topologically identical domains, each composed of one of the direct repeats. These are arranged with pseudo-twofold symmetry such that the protein consists of a 10-stranded antiparallel β sheet, 5 strands from each domain, flanked at each end by two α helices and a loop that is reminiscent of a stirrup hanging from the protein saddle. The curvature of the β pleated sheet saddle is such that it appears that TBP, in agreement with biochemical and genetic evidence, could fit snugly astride the DNA. However, the X-ray structures of the DNA complexes tell quite a different story.

Two closely similar X-ray structures of TBP–DNA complexes have been determined: one by Paul Sigler of yeast TBP in complex with a 27-nt DNA that forms an 11-bp TATA box-containing stem whose ends are joined by a 5-nt loop; and one by Burley of *Arabidopsis* TBP in complex with a 14-bp TATA box-containing duplex DNA. The DNA indeed binds to the concave surface of TBP but with its duplex axis nearly perpendicular rather than parallel to the saddle's "cylindrical" axis (Fig. 33-48*b*). The DNA is kinked by ~45° between the first two and the last two base pairs of its 8-bp TATA element. Between these kinks, the DNA is severely, although smoothly, bent with a radius of curvature of ~25 Å and unwound by ~1/3 of a turn. This permits the protein's antiparallel β sheet to bind in the DNA's greatly widened and more shallow minor groove through hydrogen bonding and van der Waals interactions (the protein does not contact the DNA's major groove). A noteworthy aspect of this remarkable structure is that each kink in the DNA is stabilized by a wedge of two Phe side chains extending from the adjacent stirrup that pries apart the base pairs flanking the kink from their minor groove side and severely buckles the interior base pair. As a result of these unprecedented distortions to the DNA (the protein undergoes only slight conformational adjustments on binding DNA), there is a ~100° angle and a lateral 18-Å displacement between the helix axes of the B-form DNA entering and leaving TBP's binding site, thereby giving the DNA a cranklike shape. The DNA, nevertheless, maintains normal Watson–Crick pairing throughout the distorted region.

Some Class II Promoters Lack a TATA Box

Several class II genes have promoters that lack TATA boxes. They are mostly "housekeeping" genes; that is, genes that are expressed in all cells and at relatively low rates. How can RNAPII properly initiate transcription at these TATA-less promoters? Investigations have shown that TATA-less promoters contain an ~11-bp element surrounding the transcription start site, a so-called **initiator (Inr)**, whose presence is sufficient to direct RNAPII to the correct start site. These systems require the participation of many of the same GTFs that initiate transcription from TATA box–containing promoters. Most unexpectedly, however, they also require TBP. How TBP binds to a TATA-less promoter is unknown, although its seems un-

likely that it would do so in the same way that it binds to a TATA box. Indeed, it may well be that TBP only indirectly contacts TATA-less promoters, via other GTFs.

Class I and Class III Genes Also Require TBP for Transcriptional Initiation

RNA polymerase I **(RNAPI,** which synthesizes most rRNAs) and RNA polymerase III **(RNAPIII,** which synthesizes 5S rRNA and tRNAs) require different sets of GTFs from each other and from RNAPII to initiate transcription at their respective promoters. This is not unexpected considering the very different organizations of these three classes of promoters (Section 29-2F). Indeed, the promoters recognized by RNAPI (class I promoters) and nearly all those recognized by RNAPIII (class III promoters) lack TATA boxes. Thus, it came as a great surprise when it was demonstrated that TBP is required for initiation by both RNAPI and RNAPIII, as well as by RNAPII. It participates by combining with different sets of TAFs to form the GTFs **SL1** (with class I promoters) and **TFIIIB** (with class III promoters).

As with class II TATA-less promoters, its seems unlikely that TBP interacts directly with DNA when taking part in initiation at classses I and III promoters. In fact, a TBP mutant that is defective for TATA-box binding can still support *in vitro* transcriptional initiation by both RNAPI and RNAPIII. Clearly, TBP, the only known universal transcription factor, is an unusually versatile protein.

Transcriptional Initiation of Class II Genes Is Mediated by Cell-Specific Upstream Transcription Factors Bound to Promoter and Enhancer Elements

The use of molecular cloning procedures has permitted the demonstration that *eukaryotic promoter and enhancer elements mediate the expression of cell-specific genes* (recall that an enhancer is a gene sequence that is required for the full activity of its associated promoter but which may have a variable position and orientation with respect to that promoter; Section 29-2F). For example, William Rutter has linked the 5′-flanking sequences of either the insulin or the chymotrypsin gene to the sequence encoding **chloramphenicol acetyltransferase (CAT),** an easily assayed enzyme not normally present in eukaryotic cells. A plasmid containing the insulin gene recombinant elicits expression of the CAT gene only when introduced into cultured cells that normally produce insulin. Likewise, the chymotrypsin recombinants are only active in chymotrypsin-producing cells. Dissection of the insulin control sequence indicates that the segment between its positions −103 and −333 contains an enhancer: In insulin producing cells only, it stimulates the transcription of the CAT gene with little regard to the enhancer's position and orientation relative to its promoter.

The foregoing indicates that cells contain specific transcription factors, the upstream transcription factors, that recognize the promoters and enhancers in the genes they transcribe. For instance, Robert Tjian has isolated a pro-

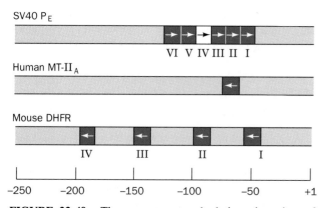

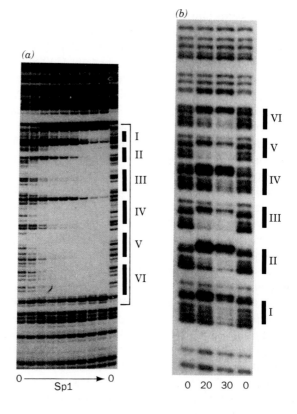

FIGURE 33-49. The arrangement and relative orientations of the GC boxes in the indicated promoters (each arrow represents the GC box sequence NGGGCGGNNN). The blue boxes represent Sp1-binding sites, whereas SV40 GC box IV is shown as a white box because Sp1 bound at GC box V prevents this transcription factor from efficiently binding to GC box IV. The transcription start site is designated by +1. DHFR = dihydrofolate reductase; MT = metallothionein. [After Kadonaga, J.T., Jones, K.A., and Tjian, R., *Trends Biochem. Sci.* **11**, 21 (1986).]

tein, **Sp1,** from cultured human cells that stimulates, by factors of 10 to 50, the transcription of cellular and viral genes containing at least one properly positioned GC box [GGGCGG (Section 29-2F); Fig. 33-49]. This protein binds, for example, to the 5′-flanking region of the SV40 virus early genes so as to protect its GC boxes from DNase I digestion (Fig. 33-50a; DNase I footprinting) and from methylation by dimethyl sulfate (Fig. 33-50b; DMS footprinting). Likewise, Sp1 specifically interacts with the four GC boxes in the upstream region of the mouse dihydrofolate reductase gene and with the single GC boxes in the human **metallothionein I_A** and **II_A** promoters (metallothioneins are metal ion–binding proteins that are implicated in heavy metal ion detoxification processes and whose synthesis is triggered by heavy metal ions).

Upstream transcription factors are essential participants in controlling the differential expression of the various globin genes in the human embryo, fetus, and adult (Section 33-2E). A typical β-globin gene promoter, in addition to its TATA box, has two positive-acting promoter elements: a CCAAT box near the −70 to −90 region and a CACCC motif at variable sites but often near positions −95 to −120 (Section 29-2F). Their importance is demonstrated by the observations that individuals with point mutations in their TATA or CACCC elements have reduced β-globin levels. These promoter elements are specifically bound by upstream transcription factors. The CCAAT box is bound by the ubiquitous transcription factor **CP1** (a mutation in the human γ-globin promoter that increases the similarity of its CCAAT box to CP1's consensus binding sequence, yields a stronger binding site for CP1, and results in increased γ-globin expression). The CACCC element is bound by Sp1,

FIGURE 33-50. The identification of the Sp1-binding sites on the SV40 early promoter. (*a*) In a DNase I footprinting assay, a DNA segment that is ^{32}P end labeled on one strand is incubated with a binding protein and then lightly digested with DNase I such that, on average, each labeled DNA strand is nicked only once. The DNA is then denatured, the resulting labeled fragments separated according to size by electrophoresis on a sequencing gel, and detected by autoradiography. Unprotected DNA is cleaved more or less at random and therefore appears as a "ladder" of bands, each representing an additional nucleotide (as in a sequencing ladder; Sections 28-6B and C). In contrast, the DNA sequences that the protein protects from DNase I cleavage have no corresponding bands. In the above footprint, the lanes labeled "0" are the DNase I digestion pattern in the absence of Sp1 and in the other lanes the amount of Sp1 increases from left to right. The footprint boundary is delineated by the bracket and the positions of SV40 GC boxes I to VI are indicated. [From Kadonaga, J.T., Jones, K.A., and Tjian, R., *Trends Biochem. Sci.* **11**, 21 (1986). Copyright © 1986 by Elsevier Biomedical Press.] (*b*) In dimethyl sulfate (DMS) footprinting, a protein-complexed end-labeled DNA segment is treated with DMS and cleaved at its G residues; the resulting fragments are electrophoretically separated as in the chemical cleavage (Maxam–Gilbert) method of DNA sequence analysis (Section 28-6B). The DNA regions that the protein protects from methylation are not cleaved by this procedure and therefore are not represented in the resulting G residue "ladder." In the above autoradiogram, the number below each lane indicates the amount, in μL, of an Sp1 fraction added to a given quantity of SV40 early promoter DNA. The positions of its GC boxes are indicated. [From Gidoni, D., Katonaga, J.T., Barrera-Saldana, H., Takahashi, K., Chambon, P., and Tjian, R., *Science* **230**, 516 (1985). Copyright © 1985 by the American Society for the Advancement of Science.]

which also binds to other globin promoter sequences that resemble Sp1's consensus binding sequence. Four erythroid-specific upstream transcription factors have also been implicated in globin gene expression: **GATA-1** (so named because it binds to sequences that contain the conserved core GATA), **NF-E2** (NF-E for *n*uclear *f*actor-*e*rythroid), **NF-E3**, and **NF-E4** (GATA-1 was previously named NF-E1).

Analysis of hereditary persistence of fetal hemoglobin (HPFH), a syndrome characterized by the inappropriate expression of γ-genes in human adults (Section 33-2G), has provided valuable insights into the basis of stage-specific globin expression. There are several HPFH variants that differ from normal only by a point mutation in the γ-gene promoter. Such mutations might result in either tighter binding of a positive transcription factor or looser binding of a negative regulator. Thus, an HPFH mutation at position −117, which is located in the more upstream of the γ-gene's two CCAAT boxes, increases the resemblance of this site to CP1's consensus binding sequence and results in a twofold tighter binding of CP1 to the mutant site. Similarly, HPFH mutations in a GC-rich region close to position −200 result in tighter Sp1 binding.

Upstream Transcription Factors Interact Cooperatively with Each Other and the PIC

How do upstream transcription factors stimulate (or inhibit) transcription? Evidently, when these proteins bind to their target DNA sites in the vicinity of a PIC (in some cases, many thousands of base pairs distant), they somehow activate (or repress) its component RNAPII to initiate transcription. Transcription factors may bind cooperatively to each other and/or to RNAPII in a manner resembling the binding of two λ repressor dimers and RNA polymerase to the o_R operator of bacteriophage λ (Section 32-3D), thereby synergistically stimulating (or repressing) transcriptional initiation. Indeed, molecular cloning experiments indicate that many enhancers consist of segments (modules) whose

individual deletion reduces but does not eliminate enhancer activity. Such complex arrangements presumably permit transcriptional control systems to respond to a variety of stimuli in a graded manner.

The functional properties of many upstream transcription factors are surprisingly simple. They appear to have two domains:

1. A DNA-binding domain that binds to the protein's target DNA sequence (and whose structural properties are discussed below).

2. A domain containing the transcription factor's activation function. Sequence analysis indicates that many of these activation domains have conspicuously acidic surface regions whose negative charges, if mutationally increased/decreased, respectively raise or lower the transcription factor's activity. Evidently, the associations between these transcription factors and a PIC are mediated by relatively nonspecific electrostatic interactions rather than by conformationally more demanding hydrogen bonds. Other types of activation domains have also been characterized, including those with Gln-rich regions, such as Sp1, and those with Pro-rich regions.

The DNA-binding and activation functions of eukaryotic transcription factors can be physically separated (which is why they are thought to occur on different domains). Thus, a genetically engineered hybrid protein containing the DNA-binding domain of one transcription factor and the activation domain of a second, activates the same genes as the first transcription factor. Indeed, it makes little functional difference as to whether the activation domain is placed on the N-terminal side of the DNA-binding domain or on its C-terminal side. This geometric permissiveness in the binding between the activation domain and its target protein is also indicated by the observation that transcription factors are largely insensitive to the orientations and positions of their corresponding enhancers rela-

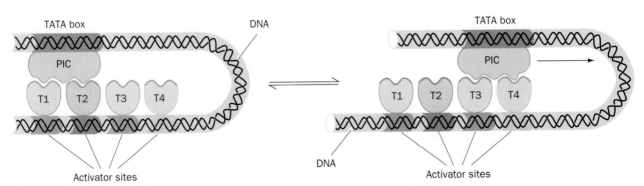

FIGURE 33-51. A model for the action of transcription factors. Here, four transcription factors, T1, T2, T3, and T4, are shown bound to their target DNA sequences (activator sites) and simultaneously, in groups of two via relatively nonspecific electrostatic interactions (*red*), to the preinitiation complex

(PIC) bound at the TATA box. The transcription factor–PIC association is probably rather weak but it is conjectured that continual sampling maintains this interaction so that RNA polymerase can attach to the PIC and initiate transcription. [After Ptashne, M., *Nature* **335**, 687 (1988).]

tive to the transcriptional start site (Section 29-2F). Of course, the DNA between an enhancer and its distant transcriptional start site must be looped out for the transcription factor to interact with a PIC (Fig. 33-51).

The nonspecific nature of transcriptional activators has led Mark Ptashne to formulate a model which explains the obervation that the rate of transcription of a given gene tends to increase with the number of upstream transcription factors that can interact with its bound PIC (Fig. 33-51). A PIC presumably has a limited number of activation sites at which it can interact simultaneously with transcriptional activators: for example, one at TFIIB and another at TFIID (which appear to be the most common activation targets). If, however, many transcriptional activators are available for binding, any activation site from which a relatively weakly bound activator has dissociated will rapidly rebind another, thereby maintaining the PIC in an activated state. Transcriptional activation, according to this model, is essentially a mass action effect. This explains the observation that unphysiologically high concentrations of a transcription factor that is not bound to DNA or even lacks its DNA-binding domain can nevertheless stimulate transcriptional initiation.

Steroid Receptors Are Examples of Inducible Transcription Factors

Eukaryotic cells express many cell-specific proteins in response to hormonal stimuli. Many of these hormones are **steroids,** cholesterol derivatives that mediate a wide variety of physiological and developmental responses (Section 34-4A). For example, the administration of **estrogens** (female sex hormones) such as *β*-**estradiol**

β-Estradiol

Ecdysone

causes chicken oviducts to increase their ovalbumin mRNA level from ~10 to ~50,000 molecules per cell, and the amount of ovalbumin they produce rises from undetectable levels to a majority of their newly synthesized protein.

Similarly, the insect steroid hormone **ecdysone** mediates several aspects of larval development (the temporal sequence of chromosome puffing shown in Fig. 33-41 can be induced by ecdysone administration).

Steroid hormones, which are nonpolar molecules, spontaneously pass through the plasma membranes of their target cells to the cytosol, where they bind to their cognate receptors. The steroid–receptor complexes, in turn, enter the nucleus, where they bind to specific segments of chromosomal enhancers known as **response elements** *so as to induce, or in some cases repress, the transcription of their associated genes.* For example, receptors for **glucocorticoids** (a class of steroids that affect carbohydrate metabolism; Section 34-4A) bind to specific 15-bp **glucocorticoid response elements (GREs)** in the upstream regions of many genes, including those of metallothioneins. Thus, eukaryotic steroid receptors are inducible transcription factors: Their actions resemble those of prokaryotic transcriptional regulators such as the *E. coli* CAP–cAMP complex (Section 29-3C). However, eukaryotic systems are much more complex. For instance, different cell types may have the same receptor for a given steroid hormone and yet synthesize different proteins in response to the hormone. Apparently, only some of the genes inducible by a given steroid are made available for activation in each type of cell responsive to that steroid. The structures of steroid receptors are discussed below.

Eukaryotic Transcription Factors Have a Great Variety of DNA-Binding Motifs

How do DNA-binding transcription factors recognize their target DNA sequences? In prokaryotes, as we have seen (Section 29-3D), most repressors and activators do so via helix–turn–helix (HTH) motifs and, in a few cases, via *β* ribbon motifs. Eukaryotes, as we shall see, employ a far greater variety of DNA-binding motifs in their transcription factors. In the remainder or this section we discuss the structures of several of the more common of these motifs and how they bind their target DNAs.

Zinc Finger DNA-Binding Motifs

The first of the predominantly eukaryotic DNA-binding motifs was discovered by Aaron Klug in *Xenopus* **transcription factor IIIA (TFIIIA),** a protein that binds to the internal control sequence of the 5S rRNA gene (Section 29-2F). This complex then sequentially binds TFIIIB (which contains TBP), **TFIIIC,** and RNA polymerase III which, in turn, initiates transcription of the 5S rRNA gene. The 344-residue TFIIIA contains 9 similar, tandemly repeated, ~30-residue modules, each of which contains two invariant Cys residues, two invariant His residues, and several conserved hydrophobic residues (Fig. 33-52*a*). Each of these units binds a Zn^{2+} ion, which X-ray absorption measurements indicate is tetrahedrally liganded by the invariant Cys and His residues. Sequence analyses have since revealed that these so-called **zinc fingers** occur 2 to at least 37 times each in a variety of eukaryotic transcription factors,

(a)

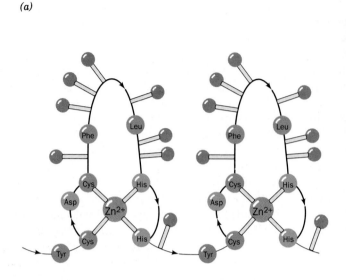

(b)

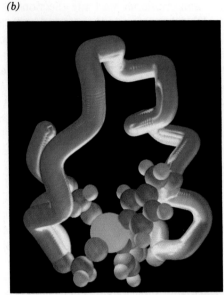

FIGURE 33-52. Zinc fingers. (*a*) A schematic diagram of tandemly repeated zinc finger motifs indicating their tetrahedrally liganded Zn²⁺ ions. Conserved amino acid residues are labeled. Gray balls represent the most probable DNA-binding side chains. [After Klug, A. and Rhodes, D., *Trends Biochem. Sci.* **12,** 465 (1988).] (*b*) The NMR structure of a single zinc finger from the *Xenopus* protein Xfin. The Zn²⁺ ion together with the atoms of its His and Cys ligands are represented as spheres with Zn light blue, C gray, N blue, S yellow, and H white. [Courtesy of Michael Pique, The Scripps Research Institute , La Jolla, California. Based on an NMR structure by Peter E. Wright, The Scripps Research Institute.]

including Sp1, several *Drosophila* developmental regulators (Section 33-4B), and certain proto-oncogene proteins (proteins whose mutant forms induce cancerous growth; Section 33-4C), as well as the *E. coli* UvrA protein (Section 31-5B). In some zinc fingers, the two Zn²⁺-liganding His residues are replaced by two additional Cys residues, whereas others have six Cys residues liganding two Zn²⁺ ions. Indeed, as we shall see, structural diversity is a hallmark of zinc finger proteins. In all cases, however, the Zn²⁺ ions appear to knit together relatively small globular domains, thereby eliminating the need for much larger hydrophobic cores.

Cys₂–His₂ Zinc Fingers: Xfin and Zif268 Proteins

The first reported zinc finger structure, an NMR structure by Peter Wright of a single zinc finger from the *Xenopus* **Xfin protein,** revealed that it forms a compact globule containing a 2-stranded antiparallel β sheet and one α helix (a ββα unit) that are held together by the tetrahedrally liganded Zn²⁺ ion (Fig. 33-52*b*). This was followed by Carl Pabo's X-ray structure of a 72-residue segment of the mouse protein **Zif268** that incorporates the protein's three zinc fingers in complex with a DNA segment containing the protein's 9-bp consensus binding sequence. The structures of Zif268's three zinc finger motifs (Fig. 33-53*a*) are closely superimposable and are nearly identical to that of the Xfin zinc finger (Fig. 33-52*b*). The three Zif268 zinc fingers are arranged as separate domains in a C-shaped structure that fits snugly into the DNA's major groove (Fig. 33-53*b*). Each

zinc finger interacts in a conformationally identical manner with successive 3-bp segments of the DNA, predominantly through hydrogen bonding interactions between the zinc finger's α helix and one strand of the DNA (here, a G-rich strand). Each zinc finger makes specific hydrogen bonding contacts with two bases in the major groove. Interestingly, five of these six associations involve interactions between Arg and G residues. In addition to these sequence-specific interactions, each zinc finger hydrogen bonds with the DNA's phosphate groups via conserved Arg and His residues.

The Cys₂–His₂ zinc finger broadly resembles the prokaryotic HTH motif as well as most other DNA-binding motifs we shall encounter (including other types of zinc finger modules) in that *all of these DNA-binding motifs provide a platform for inserting an α helix into the major groove of B-DNA.* However, Cys₂–His₂ zinc finger proteins, unlike those containing other DNA-binding motifs, possess repeated protein modules that each contact successive DNA segments. Such a modular system can recognize extended asymmetric base sequences.

Cys₂–Cys₂ Zinc Fingers: The Glucocorticoid Receptor and Estrogen Receptor DNA-Binding Domains

The **glucocorticoid receptor (GR)** is a member of the **nuclear receptor superfamily,** which includes the receptors for various steroid hormones, **vitamin D, retinoic acid,** and **thyroid hormones** (Section 34-4A). These receptor proteins, many of which activate distinct but overlapping sets of

(a)

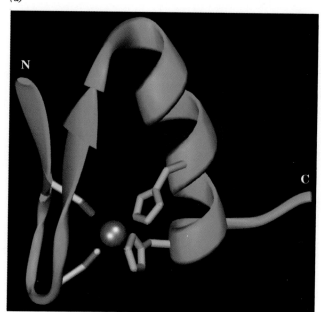

(b)

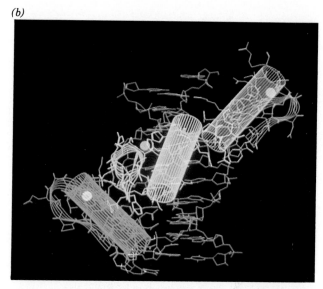

FIGURE 33-53. The X-ray structure of a three-zinc finger segment of Zif268 in complex with a 10-bp DNA that has a single nucleotide overhang at each end. (*a*) A ribbon diagram of a single zinc finger motif (finger 1) with its Zn^{2+} ion's tetrahedrally liganding His (*light blue*) and Cys (*yellow*) side chains shown in stick form and its Zn^{2+} ion represented by a silver sphere. (*b*) The complex of the entire protein segment with DNA. The protein and DNA are shown in stick form, with superimposed cylinders and ribbons marking the protein's α helices and β sheets. Finger 1 is orange, finger 2 is yellow, finger 3 is purple, the DNA is blue, and the Zn^{2+} ions are represented by light blue spheres. Note how the N-terminal end of each zinc finger's helix extends into the DNA's major groove to contact three base pairs. [Part *a* based on an X-ray structure by and Part *b* courtesy of Carl Pabo, MIT.]

genes, have a common organization of discrete functional domains for ligand binding, DNA binding, and transcriptional regulation. The DNA-binding domains of nuclear receptors contain 8 Cys residues that, in groups of four, tetrahedrally coordinate two Zn^{2+} ions. Members of this receptor family often recognize DNA response elements with similar or even identical half-site sequences with different numbers of base pairs acting as spacers between them.

The X-ray structures of two related DNA segments complexed with the 86-residue DNA-binding domain of rat GR have been determined by Paul Sigler and Keith Yamamoto. One segment, designated GRE_{4S}, contains two ideal 6-bp glucocorticoid response element (GRE) half-sites symmetrically disposed about a 4-bp (non-native) spacer, whereas the other DNA, GRE_{3S}, differs from GRE_{4S} in that its spacer has the naturally occurring length of 3 bp. In both complexes, the protein forms a symmetric dimer involving protein–protein contacts even though it exhibits no tendency to dimerize in the absence of DNA (NMR measurements indicate that the contact region is flexible in solution).

The X-ray structure of the DNA-binding domain of the GR subunit complexed to DNA resembles that of its NMR structure in the absence of DNA: It consists of two structur-ally distinct modules, each nucleated by a Zn^{2+} coordination center, that closely associate to form a compact globular fold (Fig. 33-54*a*). The C-terminal module provides the entire dimerization interface as well as making several contacts with the phosphate groups of the DNA backbone. The N-terminal module, which is also anchored to the phosphate backbone, makes all of the GR's sequence-specific interactions with the GRE via three side chains extending from the N-terminal α helix, which is inserted into the GRE's major groove.

In the GRE_{3S} complex, a subunit of the GR DNA-binding domain binds to each GRE half-site in a structurally identical manner, making sequence-specific contacts even though the odd number of base pairs in its spacer, which the protein does not contact, renders the DNA sequence nonpalindromic. However, in the GRE_{4S} complex (Fig. 33-54*b*), the protein dimer maintains a structure that is essentially identical to that in the GRE_{3S} complex so that only one of its subunits can bind to a GRE half site in a manner resembling that in the GRE_{3S} complex. The other subunit is shifted out of register with the GRE sequence by 1 bp and hence can only make nonspecific contacts with the DNA. The dimer interactions are apparently stronger than the protein–DNA interactions, a surprising finding in light of the protein's failure to dimerize in the absence of DNA.

(a)

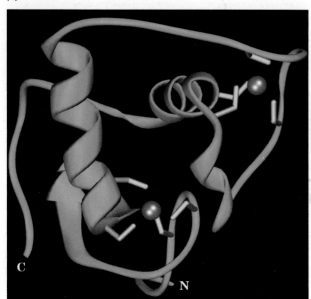

(b)

FIGURE 33-54. The X-ray structure of the dimeric glucocorticoid receptor (GR) DNA-binding domain in complex with an 18-bp DNA that has a single nucleotide overhang at both of its 5′ ends. The DNA contains two symmetrically disposed 6-bp glucocorticoid response element half sites (5′-AGAACA-3′) separated by a 4-bp spacer (GRE$_{4S}$). *(a)* A ribbon diagram of a single subunit of the GR with its two Zn^{2+} ions represented by silver spheres and their tetrahedrally liganding Cys side chains shown in stick form. Compare this structure with Fig. 33-53*a*. *(b)* The complex of the dimeric protein with GRE$_{4S}$ DNA as viewed with its approximate twofold molecular axis horizontal and in the plane of the paper. The protein is shown in ribbon form with its two subunits differently colored and its bound Zn^{2+} ions represented by silver spheres. The DNA is drawn in stick form with its two 6-bp GRE half sites colored purple and the remainder light blue. Note how the GR's two N-terminal helices are inserted into adjacent major grooves of the DNA. However, only the upper (*green*) subunit binds to the DNA in a sequence-specific manner; the lower (*yellow*) subunit binds to the palindromic DNA one base pair closer to the center of the DNA molecule than does the upper subunit and hence does not make sequence-specific contacts with the DNA. [Based on an X-ray structure by Paul Sigler, Yale University.]

Thus, the two subunits and the DNA associate in a cooperative fashion that favors the binding of the glucocorticoid receptor to targets with properly spaced half sites.

The **estrogen response element (ERE),** the DNA segment to which the **estrogen receptor (ER)** specifically binds, differs from the GRE only by changes in the central two base pairs in their identical 6-bp half-sites. The X-ray structure of the ER DNA-binding domain in complex with an ERE-containing DNA segment, determined by Daniela Rhodes, closely resembles that of the GR–GRE complex. However, the side chains that make base-specific contacts with each ERE half-site are quite differently arranged from those contacting the GRE$_{4S}$ half-sites. Evidently, the discrimination of a half site sequence is not simply a matter of substituting one or more different amino acid residues into

a common framework but, rather, involves considerable side chain rearrangement.

Members of the nuclear receptor family often recognize response elements with similar or even identical half-site sequences but with different orientations and/or spacings. The foregoing observations provide a structural basis for the graded affinities of these receptors towards their various target genes.

Binuclear Cys$_6$ Zinc Fingers: The GAL4 DNA-Binding Domain

The yeast protein **GAL4** is a transcriptional activator of several genes that encode galactose-metabolizing proteins. This 881-residue protein binds to a 17-bp DNA segment as a homodimer. Residues 1 to 65, which contain six Cys resi-

(a)

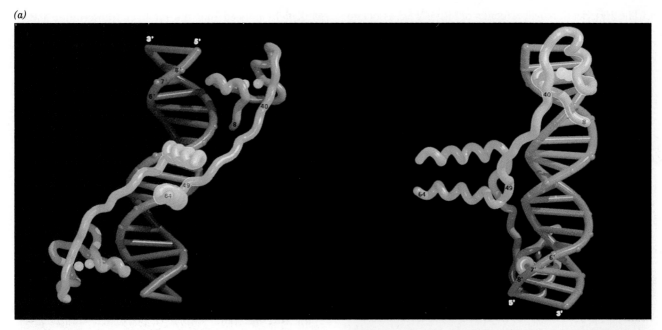

FIGURE 33-55. The X-ray structure of the GAL4 DNA-binding domain in complex with a palindromic 19-bp DNA (except for the central base pair) containing the protein's consensus binding sequence. (*a*) The complex of the dimeric protein with the DNA as shown in tube form and with the DNA red, the protein backbone light blue, and the Zn^{2+} represented by yellow spheres. The views are along the complex's twofold axis (*left*) and turned 90° with the twofold axis horizontal (*right*). Note how the C-terminal end of each subunit's N-terminal helix extends into the DNA's major groove. (*b*) A ribbon diagram of the protein's zinc finger domain (residues 8-40) with the Cys side chains of its $Zn_2^{2+}Cys_6$ complex shown in stick form (*yellow*) and its Zn^{2+} ions shown as silver spheres. Compare this structure with Figs. 33-53*a* and 33-54*a*. [Part *a* courtesy of and Part *b* based on an X-ray structure by Stephen Harrison and Ronen Mamorstein, Harvard University.]

(b)

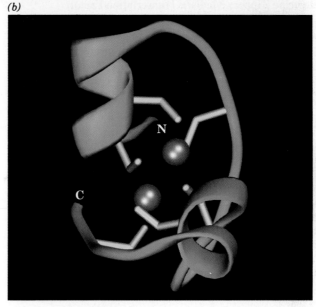

dues that collectively ligand two Zn^{2+} ions (Fig. 33-55), have been implicated in DNA binding; residues 65 to 94 participate in dimerization (although, as we shall see, residues 50-64 also have a weak dimerization function); and residues 148 to 196 and 768 to 881 function as acidic transcriptional activating regions. The X-ray crystal structure of the 65-residue N-terminal fragment of GAL4 in complex with a symmetrical 19-bp DNA containing GAL4's palindromic 17-bp consensus sequence has been determined by Ptashne, Ronen Mamorstein, and Stephen Harrison.

The protein binds to the DNA as a symmetric dimer (Fig. 33-55*a*) although in the absence of DNA it is only monomeric. Each subunit folds into three distinct modules: a compact Zn^{2+}-liganding domain that binds specific se-

quences of DNA (residues 8–40), an extended linker (residues 41–49), and a short α helical dimerization element (residues 50–64). In the Zn^{2+}-liganding module (Fig. 33-55*b* and top and bottom of Fig. 33-55*a*), the two Zn^{2+} ions are each tetrahedrally coordinated by four of the six Cys residues, with two of these residues ligating both metal ions so as to form a binuclear cluster. This module's polypeptide chain forms two short α helices connected by a loop such that the module, together with its bound Zn^{2+} ions, has pseudo-twofold symmetry. The N-terminal helix is inserted into the DNA's major groove, thereby making sequence-specific contacts with a highly conserved CCG sequence at each end of the consensus sequence. The DNA's conformation deviates little from that of ideal B-DNA.

The dimerization helices (center of Fig. 33-55*a*) associate to form a short segment of parallel coiled coil in which the contact region between the coiled coil's component helices is hydrophobically stabilized by three pairs of Leu residues and a pair of Val residues (an arrangement similar to that in the so-called **leucine zipper** described below). The coiled coil is positioned over the minor groove of the DNA such that its superhelix axis coincides with the DNA's twofold axis. The linkers connecting the coiled coil to the DNA-binding modules wrap around the DNA, largely following its minor groove, until, upon reaching the DNA-binding module, they shift over into the DNA's major groove. The two symmetrically related DNA-binding modules thereby approach the major groove from opposite sides of the DNA, ~1.5 helical turns apart, rather than from the same side of the DNA, ~1 helical turn apart, as do, for example, HTH motifs and the glucocorticoid receptor. The resulting relatively open structure could permit some other proteins to bind simultaneously to the DNA.

Leucine Zippers Mediate Transcription Factor Dimerization

Transcriptional activation requires, as we have seen, the cooperative association of several proteins that bind to specific sequences on DNA. Steven McKnight discovered one way in which such associations occur. We have seen (Section 7-2A) that α helices with the 7-residue pseudorepeating sequence $(a\text{-}b\text{-}c\text{-}d\text{-}e\text{-}f\text{-}g)_n$, in which the a and d residues are hydrophobic, have a hydrophobic strip along one side, which induces them to dimerize so as to form a coiled coil.

McKnight noticed that the rat liver transcription factor named **C/EBP** (for *CCAAT/enhancer binding protein*), which specifically binds to the CCAAT box (Section 29-2F), has a Leu at every seventh position of a 28-residue segment in its DNA-binding domain. Similar heptad repeats occur in a number of known dimeric DNA-binding proteins, including the yeast transcriptional activator **GCN4** and several DNA-binding proteins encoded by proto-oncogenes (Section 33-4C). McKnight suggested that these proteins form coiled coils in which the Leu side chains are interdigitated, much like the teeth of a zipper. He therefore named this motif the **leucine zipper**. The leucine zipper, as we shall see, mediates both the homodimerization and the heterodimerization of DNA-binding proteins (but note that it is not, in itself, a DNA-binding motif).

The X-ray structure of the 33-residue polypeptide corresponding to the leucine zipper of the 281-residue GCN4 was determined by Peter Kim and Thomas Alber. Its first 30 residues, which contain ~3.6 heptad repeats (Fig. 33-56*a*), coil into an ~8-turn α helix that dimerizes as McKnight predicted to form ~1/4 turn of a parallel left-handed coiled coil (Fig. 33-56*b*). The dimer can be envisioned as a twisted ladder whose sides consist of the helix backbones and whose rungs are formed by the interacting hydrophobic side chains. The conserved Leu residues at heptad position *d*, which comprise every second rung, are not interdigitated as McKnight originally suggested but, instead, make side-to-side contacts. The alternate rungs are likewise formed by the *a* residues of the heptad repeat (which are mostly Val) in side-to-side contact. Each Leu

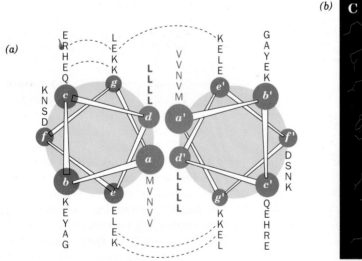
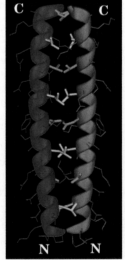

FIGURE 33-56. The GCN4 leucine zipper motif. (*a*) A helical wheel representation of two such motifs as viewed from their N-termini. The sequences of residues at each position are indicated by the adjacent column of one-letter codes. Residues that form ion pairs in the crystal structure are connected by dashed lines. Note that all residues at positions *d* and *d'* are Leu (L), those at positions *a* and *a'* are mostly Val (V), and those at other positions are mostly polar. [After O'Shea, E. K.,

Klemm, J. D., Kim, P. S., and Alber, T., *Science* **254,** 540 (1991).] (*b*) The X-ray structure, in side view, in which the helices are shown in ribbon form. Side chains are shown in stick form with the contacting Leu residues at positions *d* and *d'* yellow and residues at positions *a* and *a'* green. [Based on an X-ray structure by Peter Kim, MIT, and Tom Alber, University of Utah School of Medicine.]

side chain at position *d,* in addition to packing against the symmetry-related Leu side chain, *d',* from the other polypeptide, packs against the side chain of the succeeding residue, *e'.* Similarly, each side chain at position *a* packs between its symmetry mate, *a',* and the preceding residue, *g'.* These two sets of alternating layers thereby form an extensive hydrophobic interface between the coiled coil's component helices.

bZIP Motifs: The GCN4 DNA-Binding Domain

In many but not all leucine zipper proteins, a DNA-binding region, which is rich in basic residues, is immediately N-terminal to the leucine zipper. Sequence comparisons among 11 of these so-called **basic region leucine zipper (bZIP) proteins** revealed that the 16-residue basic sequence invariably ends 7 residues before the leucine zipper's N-terminal Leu residue. Moreover, all of these basic regions, as well as the 6-residue segment linking them to the leucine

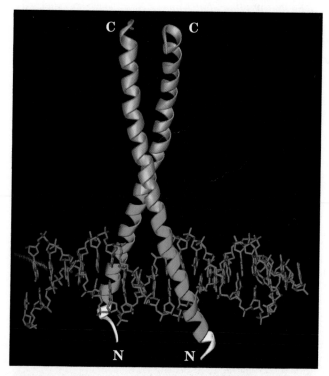

FIGURE 33-57. The X-ray structure of the GCN4 bZIP region in complex with its target DNA viewed with its molecular twofold axis vertical. The DNA (*red*), which is represented in stick form, consists of a 19-bp segment with a single nucleotide overhang at each each end and contains the protein's palindromic (except for the central base pair) 7-bp target sequence. The two identical subunits, shown in ribbon form, each contain a continuous 52-residue α helix. At their C-terminal ends (*yellow*), the two subunits associate in a parallel coiled coil (a leucine zipper), and at their basic regions (*green*), they smoothly diverge to each engage the DNA in its major groove at the target sequence. The N-terminal ends are white. [Based on an X-ray structure by Stephen Harrison, Harvard University.]

zipper, are devoid of the two strongest helix-destabilizing residues, Pro and Gly (Section 8-1D), thereby suggesting that each bZIP polypeptide is entirely α helical.

The C-terminal 56 residues of GCN4 constitute its bZIP element. Harrison and Kevin Struhl determined the X-ray structure of this polypeptide segment in complex with a 19-bp-containing duplex DNA whose central 9 bp consist of GCN4's symmetrized target sequence (Fig. 33-57). The bZIP element forms a symmetric dimer in which each subunit consists, almost entirely, of a continuous α helix. The C-terminal 25 residues of two such helices associate via a leucine zipper whose geometry closely resembles that of the 33-residue GCN4 leucine zipper element alone. Past this point, the two α helices smoothly diverge to bind in the DNA's major groove on opposite sides of the helix, thereby clasping the DNA in a sort of scissors grip. The DNA, whose helix axis is nearly perpendicular to that of the coiled coil, maintains what is essentially a straight and undistorted B-form conformation. The basic region residues that are conserved in bZIP proteins thereby make numerous contacts with both the bases and with phosphate oxygens of the DNA target sequence.

b/HLH Motifs: The Max DNA-Binding Domain

The **basic/helix–loop–helix (b/HLH) motif,** which occurs in a variety of eukaryotic transcription factors, contains a conserved DNA-binding basic region. This is immediately followed by two amphipathic helices connected by a loop that mediate the protein's dimerization. The b/HLH motif in many proteins is followed by a conserved leucine zipper (Z) motif that presumably augments protein dimerization. The transcription factor **Max** is such a **b/HLH/Z** protein, which, *in vivo,* forms a heterodimer with the proto-oncogene protein **Myc** and is required for both its normal and cancer-inducing activities. Max, by itself, readily homodimerizes and binds DNA with high affinity but Myc does not do so.

The X-ray structure of a truncated version of the 160-residue Max, Max(22-113), which contains the parent protein's b/HLH and leucine zipper elements, was determined, by Edward Ziff and Burley, in complex with a 22-bp quasipalindromic DNA containing Max's 6-bp central recognition element. Each subunit of this homodimeric protein consists of two long α helices connected by a loop to form a novel protein fold (Fig. 33-58). The N-terminal α helix (b/H1) contains residues from the protein's basic region (b) followed, without interruption, by those of the HLH motif's leading helix (H1). The C-terminal α helix (H2/Z), which is composed of the second HLH helix (H2) and the leucine zipper (Z), mediates the protein's homodimerization through the formation of a parallel left-handed coiled coil similar to that in GCN4 (Fig 33-57). Each of the dimer's two b/H1 helices projects from the resulting parallel 4-helix bundle to engage the DNA in a manner reminiscent of a pair of forceps by binding in its major groove on opposite sides of the helix (much like the way GCN4 grips its target

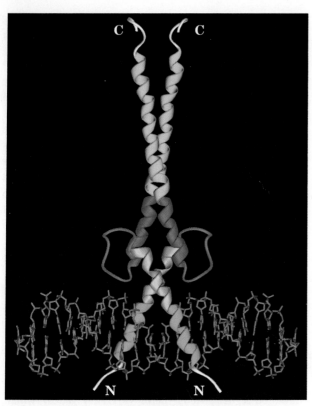

FIGURE 33-58. The X-ray structure of the Max(22-113) dimer in complex with a 22-bp DNA containing the protein's palindromic 6-bp target sequence. The DNA (*red*) is shown in stick form and the homodimeric protein is shown in ribbon form. The protein's N-terminal basic region (*green*) forms an α helix that engages its target sequence in the DNA's major groove and then merges smoothly with the H1 helix (*yellow*) of the helix–loop–helix (HLH) motif. Following the loop (*pink*), the protein's two H2 helices (*purple*) of the HLH motif form a parallel left-handed four-helix bundle with the two H1 helices. Each H2 helix then merges smoothly with the leucine zipper (Z) motif (*light blue*) to form a parallel coiled coil. The protein's N- and C-terminal ends are white. [Based on an X-ray structure by Stephen Burley, The Rockefeller University.]

1. Selection of Alternative Initiation Sites

The expression of several eukaryotic genes is controlled, in part, through the selection of alternative transcriptional initiation sites. For example, identical molecules of α-amylase are produced by mouse liver and salivary gland but the corresponding mRNAs synthesized by these two organs differ at their 5′ ends. Comparison of the sequences of these mRNAs with that of their corresponding genomic DNA indicates that the different mRNAs arise from separate initiation sites that are ~2.8-kb apart (Fig. 33-59). Thus, after being spliced, the liver and salivary gland α-amylase mRNAs have different untranslated 5′ leaders but the same coding sequences. The two initiation sites, it is thought, support different rates of initiation. This hypothesis accounts for the observation that α-amylase mRNA comprises 2% of the polyadenylated mRNA in salivary gland but only 0.02% of that in liver.

2. Selection of Alternative Splice Sites

The expression of numerous cellular genes is modulated by the selection of alternative splice sites. Thus, certain exons in one type of cell may be introns in another. For example, a single rat gene encodes seven tissue-specific variants of the muscle protein α-**tropomyosin** (Section 34-3B) through the selection of alternative splice sites (Fig. 33-60).

The phenomenon of alternative splicing can have far reaching physiological consequences. Thus, sex determination in *Drosophila* is principally governed through splice site selection as mediated by the products of the ***sex lethal, transformer,*** and ***transformer-2*** genes. Similarly, in SV40-infected cells, the so-called **large T antigen**, which participates in SV40 DNA replication, is encoded by an mRNA produced by selecting the first of two alternative 5′ splice sites in its corresponding pre-mRNA. The selection of the second 5′ splice site but the same 3′ splice site yields the **small t antigen**, which con-

DNA, although GCN4's bZIP element consists of only two α helices rather than the four of Max). The DNA helix is essentially straight with only small deviations from the ideal B-DNA structure. Each basic region makes several sequence-specific interactions with the bases of the DNA's 6-bp recognition element as well as numerous contacts with its phosphate groups. Side chains of both the loop and the N-terminal end of the H2 helix also contact DNA phosphate groups.

C. Other Expressional Control Mechanisms

Most eukaryotic genes are specifically regulated only by the control of transcriptional initiation. Many viral genes and cellular genes, however, additionally respond to other types of control processes. The various mechanisms employed by these secondary systems are outlined below.

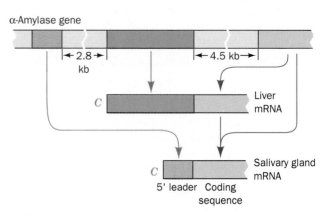

FIGURE 33-59. The transcription start site of the mouse α-amylase gene is subject to tissue-specific selection so as to yield mRNAs with different cap (*C*) and leader segments but the same coding sequences. [After Young, R. A., Hagenbüchle, O., and Schibler, U., *Cell* **23**, 454 (1981).]

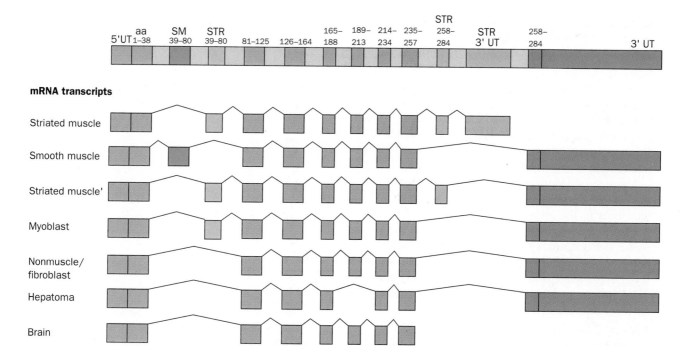

mRNA transcripts

FIGURE 33-60. The organization of the rat α-tropomyosin gene and the seven alternative splicing pathways that give rise to cell-specific α-tropomyosin variants. The thin kinked lines indicate the positions occupied by the introns before they are spliced out to form the mature mRNAs. Tissue-specific exons are indicated together with the amino acid (aa) residues they encode: "constitutive" exons (those expressed in all tissues) are green; those expressed only in smooth muscle (SM) are brown; those expressed only in striated muscle (STR) are purple; and those variably expressed are yellow. Note that the smooth and striated muscle exons encoding amino acid residues 39 to 80 are mutually exclusive and, likewise, there are alternative 3'-untranslated (UT) exons. [After Breitbart, R.E., Andreadis, A., and Nadal-Ginard, B., *Annu. Rev. Biochem.* **56,** 481 (1987).]

tains the same N-terminal sequences as the large T antigen but, past the position of the splice, different C-terminal sequences as a consequence of the shift in reading frame arising from the alternative splicing events

The observation that the ratio of small t to large T mRNAs varies with the nature of the infected cells suggests that cells contain a factor that regulates splice site selection. This led to the isolation of a host-encoded 248-residue protein, named **alternative splicing factor 1 (ASF-1).** The N-terminal 80 residues of ASF-1 are homologous to the RNA-recognition motifs (RRMs) that occur in many RNA-binding proteins (Section 30-3A), including those of several spliceosomal snRNPs (Section 29-4A). It also contains a 50-residue segment rich in Arg–Ser and Ser–Arg dipeptides that is similar to polypeptide sequences encoded by the *Drosophila transformer* and *transformer-2* genes. The use of bacterially expressed ASF-1 in *in vitro* splicing systems has clearly demonstrated that the selection of alternative splice sites can be influenced by the action of a general splicing factor.

3. Translocational Control

The observation that only ~5% of nuclear RNA ever makes its way to the cytosol, probably less than can be accounted for by gene splicing, suggests that differential mRNA translocation to the cytosol may be an impor-

tant expressional control mechanism in eukaryotes. Evidence is accumulating that this is, in fact, the case. Cellular RNA is never "naked" but rather is always in complex with a variety of conserved proteins. Intriguingly, nuclear and cytosolic mRNAs are associated with different sets of proteins, indicating that there is protein exchange on translocating mRNA out of the nucleus.

4. Control of mRNA Degradation

The rates at which eukaryotic mRNAs are degraded in the cytosol vary widely. Whereas most have half-lives of hours or days, some are degraded within 30 min of entering the cytosol. A given mRNA may also be subject to differential degradation. For example, the major egg yolk protein **vitellogenin** is synthesized in chicken liver in response to estrogens (in roosters as well as in hens) and transported via the bloodstream to the oviduct. Radioactive-labeling experiments established that estrogen stimulation increases the rate of vitellogenin mRNA transcription by several hundredfold and that this mRNA has a cytosolic half-life of 480 h. When estrogen is withdrawn, the synthesis of vitellogenin mRNA returns to its basal rate and its cytosolic half-life falls to 16 h.

The poly(A) tails appended to nearly all eukaryotic mRNAs apparently help protect them from degradation (Section 29-4A). For example, histone mRNAs, which lack poly(A) tails, have much shorter half-lives than most other mRNAs. Histones, in contrast to most other cellular proteins, are largely synthesized during the relatively short S phase of the cell cycle, when they are required in massive amounts for chromatin replication (the small amounts of histones synthesized during the rest of the cell cycle are thought to be used for repair purposes). The short half-lives of histone mRNAs ensure that the rate of histone synthesis closely parallels the rate of histone gene transcription.

A structural feature that increases the rate at which mRNAs are degraded is the presence of certain AU-rich sequences in the untranslated 3′ segments. These sequences, when grafted to mRNAs that lack them, decrease the mRNAs cytosolic lifetimes. By and large, however, the nature of the signals through which mRNAs are selected for degradation are poorly understood, in part, no doubt, because the nucleases that do so have not been identified.

5. Control of Translational Initiation Rates

The rates of translational initiation of eukaryotic mRNAs, as we have seen (Section 30-4), are responsive to the presence of certain substances, including heme (in reticulocytes) and interferon, as well as to mRNA masking.

6. Selection of Alternative Posttranslational Processing Pathways

Polypeptides synthesized in both prokaryotes and eukaryotes are subject to proteolytic cleavage and covalent modification (Section 30-5). These posttranslational processing steps are important regulators of enzyme activity (e.g., see Section 14-3E) and, in the case of glycosylations, are major determinants of a protein's final cellular destination (Sections 11-4C and 21-3B). The selective degradation of proteins (Section 30-6) is also a significant factor in eukaryotic gene expression.

In addition to the foregoing, most eukaryotic polypeptide hormones (whose functions are discussed in Section 34-4A) are synthesized as segments of large precursor polypeptides known as **polyproteins**. These are posttranslationally cleaved to yield several, not necessarily different, polypeptide hormones. *The cleavage pattern of a particular polyprotein may vary among different tissues so that the same gene product can yield different sets of polypeptide hormones.* For example, the polyprotein **pro-opiomelanocortin (POMC)**, which, in the rat, is synthesized in both the anterior and intermediate lobes of the pituitary, contains seven different polypeptide hormones (Fig. 33-61). In both of these lobes, which are functionally separate glands, posttranslational processing of POMC yields an N-terminal fragment, **ACTH** and **β-LPH**. Processing in the anterior lobe ceases at this point. In the intermediate lobe, however, the N-terminal fragment is further cleaved to yield **γ-MSH**, ACTH is converted to **α-MSH** and CLIP, and **β-LPH** is split to **γ-LPH** and **β-END** (Fig. 33-61). These various hor-

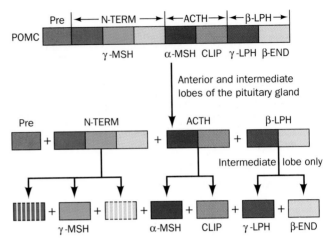

FIGURE 33-61. The tissue-specific posttranscriptional processing of POMC yields two different sets of polypeptide hormones. In both the anterior and intermediate lobes of the pituitary gland, POMC is proteolytically cleaved to yield its N-terminal fragment (N-TERM), **adrenocorticotropic hormone (ACTH)** and **β-lipotropin (β-LPH)**. In the intermediate lobe only, these polypeptide hormones are further cleaved to yield **γ-melanocyte stimulating hormone (γ-MSH)**, **α-MSH**, **corticotropin-like intermediate lobe peptide (CLIP)**, **γ-LPH**, and **β-endorphin (β-END)**. [After Douglass, J., Civelli, O., and Herbert, E., *Annu. Rev. Biochem.* **53**, 698 (1984).]

mones have different activities so that the products of the anterior and intermediate lobes of the pituitary are physiologically distinct.

Most of the cleavage sites in POMC and other polyproteins consist of pairs of basic amino acid residues, Lys–Arg, for example, which suggests that cleavage is mediated by enzymes with trypsin-like activity. Indeed, the enzymes that process POMC also activate other prohormones such as proinsulin. Moreover, the observation that a yeast protease that normally functions to activate a yeast prohormone, also properly processes POMC, suggests that prohormone processing enzymes are evolutionarily conserved.

4. CELL DIFFERENTIATION

Perhaps the most awe inspiring event in biology is the growth and development of a fertilized ovum to form an extensively differentiated multicellular organism. No outside instruction is required to do so; *fertilized ova contain all the information necessary to form complex multicellular organisms such as human beings.* Since, contrary to the beliefs of the earliest microscopists, zygotes do not contain miniature adult structures, these structures must somehow be generated through genetic specification. In this section we discuss the little we presently know about this astounding process. We end by considering the genetic basis of cancer, a group of diseases caused by the proliferation of cells that have lost some of their developmental constraints.

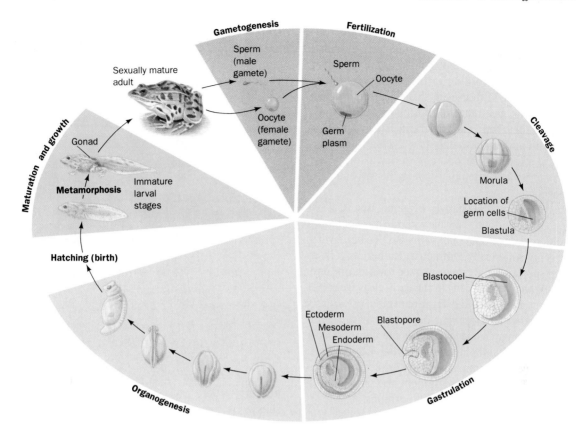

FIGURE 33-62. Embryogenesis in a representative animal, the frog.

A. Embryological Development

The formation of multicellular animals can be considered as occurring in four somewhat overlapping stages (Fig. 33-62):

1. **Cleavage,** in which the zygote undergoes a series of rapid mitotic divisions to yield many smaller cells arranged in a hollow ball known as a **blastula.**

2. **Gastrulation,** whereby the blastula, through a structural reorganization that includes the blastula's invagination, forms a triple-layered bilaterally symmetric structure called a **gastrula.** Cleavage and gastrulation together take from a few hours to several days depending on the organism.

3. **Organogenesis,** in which the body structures are formed in a process requiring various groups of proliferating cells to migrate from one part of the embryo to another in a complicated but reproducible choreography. Organogenesis occupies hours to weeks.

4. **Maturation and growth,** whereby the embryonic structures achieve their final sizes and functional capacities. This stage stretches into and sometimes throughout adulthood.

Cell Differentiation Is Mediated by Developmental Signals
As an embryo develops, its cells become progressively and

irreversibly committed to specific lines of development. What this means is that these cells undergo sequences of self-perpetuating internal changes that distinguish them and their progeny from other cells. A cell and its descendents therefore "remember" their developmental changes even when placed in a new environment. For example, the dorsal (upper) ectoderm (outer layer) of an amphibian embryo (Fig. 33-63) is normally fated to give rise to brain

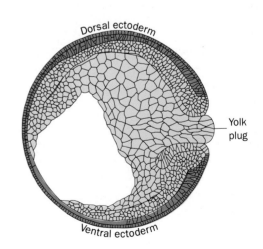

FIGURE 33-63. The dorsal and ventral ectoderm of an amphibian embryo.

tissue, whereas its ventral (lower) ectoderm becomes epidermis. If a block of an early gastrula's dorsal ectoderm is cut out and exchanged with a block of its ventral ectoderm, both blocks develop according to their new locations to yield a normal adult. If, however, this experiment is performed on the late gastrula, the transplanted tissues will differentiate as they had originally been fated, that is, as misplaced brain and epidermal tissues. Evidently, the dorsal and ventral ectoderms become committed to form brain and epidermal tissues sometime between the early and late gastrula stages.

How are developmental changes triggered; that is, What are the signals that induce two cells with identical genomes to follow different developmental pathways? To begin with, the zygote is not spherically symmetric. Rather, its yolk, as well as other substances, are concentrated towards one end. Consequently, the various cells in the early cleavage stages inherit different cytoplasmic determinants that apparently govern their further development. Even as early as an embryo's eight-cell stage, some of its cells are demonstrably different in their developmental potential from others. However, as the above transplantation experiments indicate, cells in later stages of development also obtain developmental cues from their embryonic positions.

Cells may obtain spatial information in two ways:

1. Through direct intercellular interactions.

2. From the gradients of diffusible substances called **morphogens** released by other cells.

For most developmental programs, the interacting tissues must be in direct contact, but this is not always the case. For example, mouse ectoderm fated to become eye lens will only do so in the presence of mesenchyme (embryonic tissue that gives rise to the muscle, skeleton, and connective tissue) but this process still occurs if the interacting tissues are separated by a porous filter. Lens development must therefore be mediated by diffusible substances.

Developmental signals may be recognized over great evolutionary distances. For instance, the epidermis from the back of a chick embryo, through interactions with the underlying dermis, forms feather buds that are arrayed in a characteristic hexagonal pattern. If embryonic chick epidermis is instead combined with dermis from the whiskered region of mouse embryo snout, the chick epidermis still forms feather buds but arranged in the pattern of mouse whiskers.

Even though mammals and birds diverged ~300 million years ago, mouse inducers can still activate the appropriate chicken genes, although, of course, they cannot alter the products these genes specify. In an intriguing example of this phenomenon, combining epithelium from the jaw-forming region of a chick embryo with molar mesenchyme from mouse embryo, induces the chick tissue to grow teeth that are unlike those of mammals (Fig. 33-64). Apparently chickens, whose ancestors have been toothless for ~60 mil-

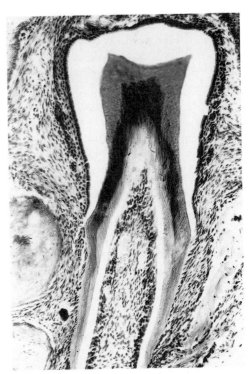

FIGURE 33-64. The proverbial "hen's tooth" forms in chick embryo jaw-forming epithelium under the influence of mouse embryo molar mesenchyme tissue. [Courtesy of Edward Kollar, University of Connecticut Health Center.]

lion years (the original bird, *Archaeopterix,* had teeth), retain the genetic potential to grow teeth even though they lack the developmental capacity to activate these genes. This observation corroborates the hypothesis that organismal evolution proceeds largely via mutations that alter developmental programs rather than the structural genes whose expression they control (Section 6-3B).

Developmental Signals Act in Combination

An additional developmental stimulus to a previously determined cell will modulate, but not reverse, its developmental state. Consider, for example, what happens in a chicken embryo if undifferentiated tissue from the base of a leg bud, which normally gives rise to part of the thigh, is transplanted beneath the end of a wing bud, which normally develops into the handlike wing tip. The transplant does not become a wing tip or even misplaced thigh tissue; instead it forms a foot (Fig. 33-65). Apparently the same stimulus that causes the end of a wing bud to form a wing tip causes tissue that is already committed to be part of a leg to form a leg's morphological equivalent to a wing tip, a foot. Evidently, the many different tissues of a higher organism do not each form in response to a tissue-specific developmental stimulus. Rather, *a given tissue results from the effects of a particular combination of relatively nonspecific developmental stimuli.* This situation, of course, greatly reduces the number of different developmental stimuli neces-

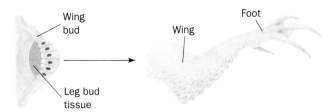

FIGURE 33-65. Presumptive thigh tissue from a chicken leg bud develops into a misplaced foot when implanted beneath the tip of a chicken wing bud.

sary to form a complex organism and therefore simplifies the regulation of the developmental process.

B. The Molecular Basis of Development

The study of the molecular basis of cell differentiation has only become possible in recent years with the advent of modern methods of molecular genetics. Much of what we know about this subject is based on studies of the fruit fly *Drosophila melanogaster*. We therefore begin this section with a synopsis of embryogenesis in this genetically best characterized multicellular organism.

Drosophila Development

Almost immediately after the *Drosophila* egg (Fig. 33-66*a*) is laid (which, rather than the earlier fertilization, triggers development), it commences a series of synchronized nuclear divisions, one every 6 to 10 min. The DNA must therefore be replicated at a furious rate, among the fastest known for eukaryotes. Most probably each of its replicons (Section 31-4B) are simultaneously active. The nuclear division process is unusual in that it is not accompanied by the formation of new cell membranes; the nuclei continue sharing their common cytoplasm to form a so-called **syncytium** (Fig. 33-66*b*). After the 8th round of nuclear division, the ~256 nuclei begin to migrate towards the cortex (outer layer) of the egg where, by around the 11th nuclear division, they have formed a single layer surrounding a yolk-rich core (Fig. 33-66*c*). At this stage, the mitotic cycle time begins to lengthen while the nuclear genes, which have heretofore been fully engaged in DNA replication, become transcriptionally active (a freshly laid egg contains an enormous store of mRNA that has been contributed by the developing oocyte's surrounding "nurse" cells). In the 14th nuclear division cycle, which lasts ~60 min, the egg's

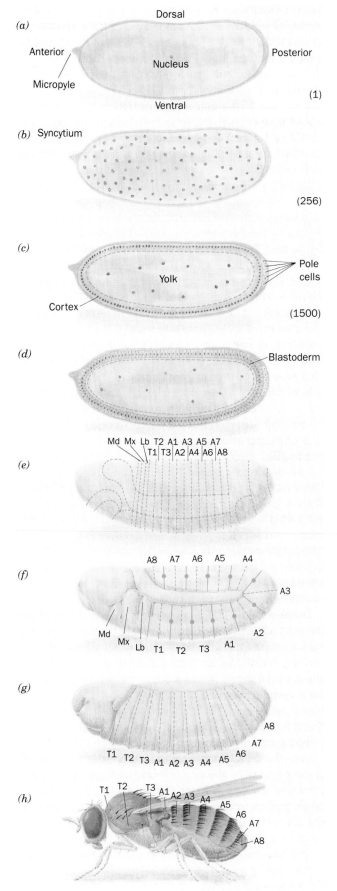

FIGURE 33-66. Development in *Drosophila*. The various stages are explained in the text. Note that the embryos and newly hatched larva are all the same size, ~0.5 mm long. The adult is, of course, much larger. The approximate number of cells in the early stages of development are given in parentheses.

plasma membrane invaginates around each of the ~6000 nuclei to yield a cellular monolayer called a **blastoderm** (Fig. 33-66*d*). At this point, after ~2.5 h of development, genomic transcriptional activity reaches its maximum in the embryo, mitotic synchrony is lost, the cells become motile, and gastrulation begins.

Until the blastoderm is formed, most of the embryo's nuclei maintain the ability to colonize any portion of the cortical cytoplasm and hence to form any part of the larva or adult except its germ cells [the germ cell progenitors, the **pole cells** (Fig. 33-66*c*), are set aside after the 9th nuclear division]. *Once the blastoderm has formed, however, its cells become progressively committed to ever narrower lines of development.* This has been demonstrated, for example, by tracing the developmental fates of small clumps of cells by excising them or ablating (destroying) them with a laser microbeam and characterizing the resultant deformity.

During the embryo's next few hours, it undergoes gastrulation and organogenesis. A striking aspect of this remarkable process, in *Drosophila* as well as in higher animals, is the division of the embryo into a series of segments corresponding to the adult organism's organization (Fig. 33-66*e*). The *Drosophila* embryo has at least three segments that eventually merge to form its head (Md, Mx, and Lb for mandibulary, maxillary, and labial), three thoracic segments (T1–T3), and eight abdominal segments (A1–A8). As development continues, the embryo elongates and several of its abdominal segments fold over its thoracic segments (Fig. 35-66*f*). At this stage, the segments become subdivided into anterior (forward) and posterior (rear) compartments. The embryo then shortens and unfolds to form a larva that hatches 1 day after beginning development (Fig. 33-66*g*). Over the next 5 days, the larva feeds, grows, molts twice, pupates, and commences metamorphosis to form an adult (**imago;** Fig. 33-66*h*). In this latter process, the larval epidermis is almost entirely replaced by the outgrowth of apparently undifferentiated patches of larval epithelium known as **imaginal disks** that are committed to their developmental fates as early as the blastoderm stage. These structures, which maintain the larva's segmental boundaries, form the adult's legs, wings, antennae, eyes, etc. (Fig. 33-67). About 10 days after commencing development, the adult emerges and, within a few hours, initiates a new reproductive cycle.

Developmental Patterns Are Genetically Mediated

What is the mechanism of embryonic pattern formation? Much of what we know about this process stems from genetic analyses of a series of bizarre mutations in three classes of *Drosophila* genes that normally specify progressively finer regions of cellular specialization in the developing embryo:

1. *Maternal-effect genes,* which define the embryo's polarity, that is, its anteroposterior (head to tail) and dorsoventral (back to belly) axes. Mutations of these genes

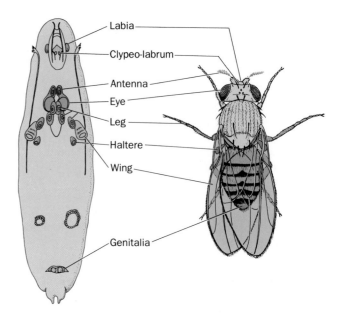

FIGURE 33-67. The locations and developmental fates, in *Drosophila,* of the imaginal disks (*left*), pouches of larval tissue that form the adult's outer structures. [After Fristrom, J.W., Raikow, R., Petri, W., and Stewart, D., *in* Hanly, E.W. (Ed.), *Problems in Biology: RNA in Development,* p. 382, University of Utah Press (1970).]

globally alter the embryonic body pattern regardless of the paternal genotype. For instance, females homozygous for the *dicephalic* (two-headed) **mutation** lay eggs that develop into nonviable two-headed monsters. These are embryos with two anterior ends pointing in opposite directions and completely lacking posterior structures. Similarly, the *bicaudal* (two-tailed) and *snake* **mutations** give rise to mirror-symmetric embryos with two abdomens (Fig. 33-68*a*).

2. *Segmentation genes, which specify the correct number and polarity of embryonic body segments.* These are subclassified as follows:

   a. **Gap genes,** the first of a developing embryo's to be transcribed, are so named because their mutations result in gaps in the embryo's segmentation pattern. Embryos with defective *hunchback (hb)* genes, for example, lack mouthparts and thorax structures.

   b. **Pair-rule genes** specify the division of the embryo's broad gap domains into segments. These genes are so named because their mutations usually delete portions of every second segment. This occurs, for example, in embryos that are homozygous for mutations in the *fushi tarazu (ftz;* Japanese for not enough segments) gene (Fig. 33-68*b*).

   c. **Segment polarity genes** specify the polarities of the developing segments. Thus, homozygous *engrailed* (*en;* indented with curved notches) mutants lack the posterior compartment of each segment.

(a) Wild-type embryo

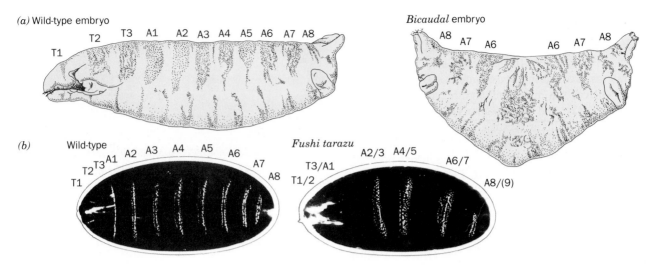

Bicaudal embryo

(b)

(c) *Antennapedia* mutation

(e) Bithorax mutations

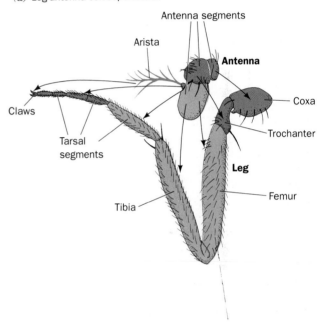

(d) Leg-antenna correspondence

FIGURE 33-68. Developmental mutants of *Drosophila.* (a) The cuticle patterns of wild-type embryos (*left*) exhibit 11 body segments, T1 to T3 and A1 to A8 (the head segments have retracted into the body and hence are not visible here). In contrast, the nonviable "monsters" produced by homozygous *bicaudal* mutant females (*right*) develop only abdominal segments arranged with mirror symmetry. [After Gergen, P.J., Coulter, D., and Weischaus, E., *in* Gall, J.G., *Gametogenesis and the Early Embryo, p.* 200, Liss (1986).] (b) In the wild-type embryo (*left*), the anterior edge of each of the 11 abdominal and thoracic segments has a belt of tiny projections known as denticles (which help larvae crawl) that appear in these

photomicrographs as white stripes. *Fushi tarazu* mutants (*right*) lack portions of alternate segments and the remaining segments are fused together (e.g., A2/3), yielding a nonviable embryo with only half of the normal number of denticle belts. [Courtesy of Walter Gehring, University of Basel, Switzerland.] (c) Head of an adult fly that is homozygous for the homeotic *Antennapedia* mutation. Absence of the *Antp* gene product causes the imaginal disks that normally form antennae to develop as the legs that normally occur only on segment T2. [Courtesy of Walter Gehring, University of Basel, Switzerland.] (d) The corresondence (*arrows*) between antennae and the legs to which the *Antp* mutation transforms them. [After Postlethwait, J.H. and Schneiderman, H.A., *Devel. Biol.* **25,** 622 (1971).] (e) A four-winged *Drosophila* (it normally has two wings; Fig. 33-67) that results from the presence of three mutations in the bithorax complex, *abx, bx,* and *pbx.* These mutations cause the nomally haltere-bearing segment T3 to develop as if it were the wing-bearing segment T2. This striking architectural change may reflect evolutionary history: *Drosophilia* evolved from more primitive insects that had four wings. [Courtesy of Edward B. Lewis, Caltech.]

3. **Homeotic selector genes,** *which specify segmental iden-tity;* their mutations transform one body part into an-other. For instance, ***Antennapedia* (*antp,* antenna-foot)** mutants have legs in place of antennae (Fig. 33-68*c* and *d*), whereas the mutations ***bithorax (bx), anteriorbi-thorax (abx),*** and ***postbithorax (pbx)*** each transform sections of halteres (vestigial wings that function as balancers), which normally occur only on segment T3, to the corresponding sections of wings, which normally occur only on segment T2 (Fig. 33-68*e*).

Maternal-Effect Gene Products Specify the Egg's Directionality through Gradient Formation

The properties of maternal-effect gene mutants suggest that maternal-effect genes specify morphogens whose distri-butions in the egg cytoplasm define the future embryo's spa-tial coordinate system. Indeed, immunofluoresence studies by Christiane Nüsslein-Volhard have demonstrated that the product of the ***bicoid (bcd)*** gene is distributed in a gra-dient that decreases towards the posterior end of the normal embryo (Figs. 33-69 and 33-70*a*), whereas embryos with *bcd*-deficient mothers lack this gradient. The gradient,

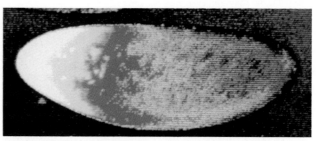

FIGURE 33-69. The distribution of Bicoid protein in a *Drosophila* syncytial blastoderm as revealed by immunofluorescence. High concentrations of the protein are yellow, lower concentrations are red, and its absence is black. [Courtesy of Christiane Nüsslein-Volhard, Max-Planck-Institut für Entwicklungsbiologie, Germany.]

which is facilitated by the syncytium's lack of cellular boundaries, arises through the secretion, by ovarian nurse cells, of *bcd* mRNA into the anterior end of the oocyte during oogenesis and its translation in the early embryo. The ***nanos*** gene mRNA is similarly deposited in the egg but

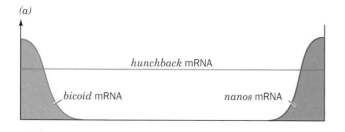

(*a*)

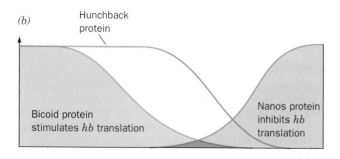

(*b*)

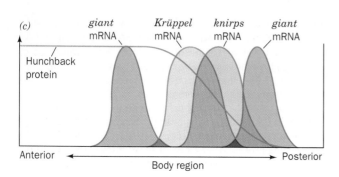

(*c*)

FIGURE 33-70. The formation and effects of the Hunchback protein gradient in *Drosophila* embryos. (*a*) The unfertilized egg contains maternally supplied *bicoid* and *nanos* mRNAs placed at its anterior and posterior poles, together with a uniform distribution of *hunchback* mRNA. (*b*) Upon fertilization, the three mRNAs are translated. Bicoid and Nanos proteins are not bound in place as are their mRNAs and hence their gradients are broader than those of the mRNAs. Bicoid protein stimulates the translation of *hunchback* mRNA, whereas Nanos protein inhibits it, resulting in a gradient of Hunchback protein that decreases nonlinearly from anterior to posterior. (*c*) Specific concentrations of Hunchback protein induce the transcription of the *giant, Krüppel,* and *knirps* genes. The gradient of Hunchback protein thereby specifies the positions at which these latter mRNAs are synthesized. (*d*) A photo-micrograph of a *Drosophila* embryo (*anterior end left*) that has been immunofluorescently stained for both Hunchback (*green*) and Krüppel proteins (*red*). The region where these proteins overlap is yellow. [Parts *a, b,* and *c* after Gilbert, S.F., *Developmental Biology* (4th ed.), *p.* 543, Sinauer Associates (1994); Part *d* courtesy of Jim Langeland, Steve Paddock, and Sean Carroll, Howard Hughes Medical Institute, University of Wisconsin-Madison.]

(*d*)

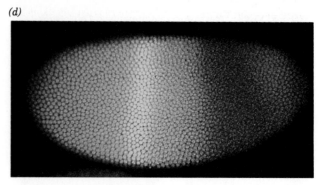

it is localized near the egg's posterior pole (Fig. 33-70*a*). The *bcd* and *nanos* gene products are both transcription factors that, as we shall see, regulate the expression of specific gap genes. Other maternal-effect genes that participate in anterior–posterior axis formation specify proteins that function to trap the localized mRNAs in their area of deposition. This explains why early embryos produced by females homozygous for maternal-effect mutations can often be "rescued" by the injection of cytoplasm, or sometimes just the mRNA, from early wild-type embryos. With some of these mutations, the polarity of the rescued embryo is determined by the site of the injection.

Gap Genes Are Expressed in Specific Regions

The mRNA of the gap gene *hunchback (hb)* is maternally deposited in a uniform distribution throughout the unfertilized egg (Fig. 33-70*a*). However, **Bicoid protein** activates the transcription of the embryonic *hb* gene, whereas **Nanos protein** inhibits the translation of *hb* mRNA. Consequently, **Hunchback protein** becomes distributed in a gradient that decreases from anterior to posterior (Fig. 33-70*b*).

DNase I footprinting studies have demonstrated that Bicoid protein binds to five homologous sites (consensus sequence TCTAATCCC) in the *hb* gene's upstream promoter region. Nüsslein-Volhard demonstrated the ability of Bicoid protein to activate the *hb* gene by fusing the *hb* promoter upstream of the CAT reporter gene (Section 33-3B) and injecting the resulting construct into early *Drosophila* embryos. CAT was produced in wild-type but not in *bcd*-deficient embryos. Moreover, by using progressively shorter segments of the *hb*-derived promoter region it was shown that at least three of the five Bicoid protein-binding sites must be present to obtain full CAT expression.

Hunchback protein, in turn, controls the expression of several other gap genes (Figs. 33-70*c* and *d*): High levels of Hunchback protein induce *giant* expression, *Krüppel* (German: cripple) is expressed where the level of Hunchback protein begins to decline, *knirps* (German: pigmy) is expressed at even lower levels of Hunchback protein, and *giant* is again activated in regions where Hunchback protein is undetectable.

Although the original positions of the proteins encoded by these latter gap genes are elicited by the approriate concentrations of Hunchback protein, these positions are stabilized and maintained through their mutual interactions. Thus **Krüppel protein** binds to the promoters of the *hb* gene, which it activates, and the *knirps* gene, which it represses. Conversely, **Knirps protein** represses the *Krüppel* gene. This mutual repression is thought to be responsible for the sharp boundaries between the various gap domains.

Pair-Rule Genes Are Expressed in "Zebra Stripes"

Pair-rule genes are expressed in sets of 7 stripes, each just a few nuclei wide, along the embryo's anterior–posterior axis. The embryo (which, at this stage, is just beginning to

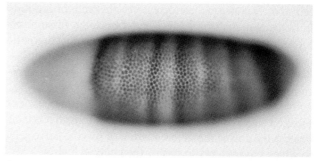

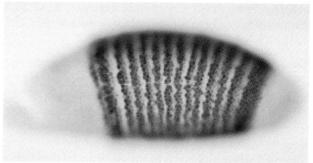

FIGURE 33-71. *Drosophila* embryos stained for Ftz (*brown*) and Eve (*gray*) proteins. These proteins are each expressed in seven stripes which, at first, are relatively blurred (*above*) but within a short time become sharply defined (*below*). [Courtesy of Peter Lawrence, MRC Laboratory of Molecular Biology, U.K.]

cellularize) is thereby divided into 15 domains (Fig. 33-71). These "zebra stripe" expression patterns for the various pair-rule genes are offset relative to one another.

The gap gene products directly control three **primary pair-rule genes**: *hairy, even-skipped (eve),* and *runt.* The striped pattern of expression arises because the promoters of most primary pair-rule genes are comprised of a series of modules, each of which induce their gene's expression in a particular stripe. Thus, the transformation of an embryo by a *lacZ* reporter gene (whose β-galactosidase product can be readily detected by the blue color produced upon administration of the β-galactosidase substrate X-Gal; Section 28-8B) preceded by a particular fragment of the *hairy* promoter led to β-galactosidase production in only stripe 6, whereas a different promoter fragment did so in only stripe 7. Each of these promoter modules contains a particular arrangement of activating and inhibitory binding sites for the various gap gene proteins so as to enable the expression of the associated pair-rule gene under the particular combination of gap gene proteins present in the corresponding stripe. As with the gap genes, the patterns of expression of the primary pair-rule genes become stabilized through interactions among themselves.

The primary pair-rule gene products also induce or inhibit the expression of five **secondary pair-rule genes** including *ftz.* Thus, as Walter Gehring demonstrated, *ftz* transcripts first appear in the nuclei lining the cortical cytoplasm during the embryo's 10th nuclear division cycle. The rate of *ftz* expression then increases as the embryo develops until the 14th division cycle, when the cellular blastoderm forms. At this stage, as immunochemical staining dramatically shows, *ftz* is expressed in a pattern of 7 belts around the blastoderm, each 3 or 4 cells wide (Fig. 33-71), which correspond precisely to the missing regions in homozygous *ftz⁻* embryos. Then, as the embryonic segments form, *ftz* expression subsides to undetectable levels (although it is later reactivated during the differentiation of specific nerve cells in which it is required to specify their correct "wiring" pattern). Evidently, the *ftz* gene must be expressed in alternate sections of the embryo for normal segmentation to occur.

Segment Polarity Genes Define Parasegment Boundaries

The expression of eight known segment polarity genes is initiated by pair-rule gene products. For example, by the 13th nuclear division cycle, as Thomas Kornberg demonstrated, *engrailed (en)* transcripts become detectable but are more or less evenly distributed throughout the embryonic cortex. However, since *en* is expressed in nuclei containing high concentrations of either **Eve** or **Ftz** proteins (Fig. 33-71), by the 14th cycle they form a striking pattern of 14 stripes around the blastoderm (half the spacing of *ftz* expression). Continuing development reveals that these stripes are localized in the primordial posterior compartment of every segment (Fig. 33-72), just those compartments that are missing in homozygous *en⁻* embryos. Thus, much like we saw for *ftz,* the *en* gene product induces the posterior half of each segment to develop in a different fashion from its anterior half.

Another segment polarity gene, **wingless (wg),** is expressed simultaneously with *en* but in narrow bands on the

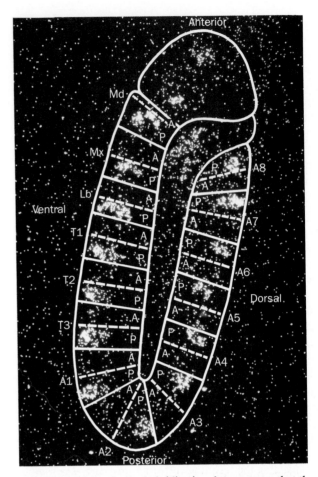

FIGURE 33-72. *In situ* hybridization demonstrates that the *Drosophila engrailed* gene is expressed in the posterior compartment of every embryonic segment. [Courtesy of Walter Gehring, University of Basel, Switzerland.]

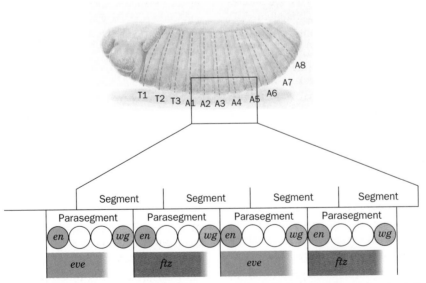

FIGURE 33-73. The pair-rule proteins Eve and Ftz regulate the expression of the segment polarity genes *engrailed (en)* and *wingless (wg).* When either Eve or Ftz is present, *en* is expressed, whereas when both proteins are absent, *wg* is expressed. The parasegment boundaries are thereby defined. Other pair-rule proteins are thought to inhibit *en* and *wg* expression in nuclei not at the parasegment boundaries.

anterior side of most *en* bands (Fig. 33-73). Cells expressing *en* and *wg* genes thereby define the boundaries of the so-called **parasegments,** embryonic regions that consist of the posterior portion of one segment and the anterior portion of the segment behind it. Parasegments do not become morphological units in the larva or adult but, nevertheless, are thought to be the embryo's actual developmental units.

Homeotic Selector Genes Direct the Development of the Individual Body Segments

The structural components of developmentally analogous body parts, say *Drosophila* antennae and legs, are nearly identical; only their organizations differ (Fig. 33-68*d*). *Consequently, developmental genes must function to control the pattern of structural gene expression rather than simply turning these genes on or off.* Thus, as we saw for the segmentation genes, the expression of the structural genes charactistic of any given tissue must be controlled by a complex network of regulatory genes. The homeotic selector genes, as we shall see, are the "master" genes in the control networks governing segmental differentiation.

Most homeotic mutations in *Drosophila* (which were first described in 1894 by William Bateson who coined the name "homeotic") map into two large gene families: the **bithorax complex (*BX-C*),** which controls the development of the thoracic and abdominal segments, and the **antennapedia complex (*ANT-C*),** which primarily affects head and thoracic segments. *Recessive mutations in BX-C, when homozygous, cause one or more segments to develop as if they were more anterior segments.* Thus, the combined ***bx, abx,*** and ***pbx*** mutations cause segment T3 to develop as if it was segment T2 (Fig. 33-68*e*). Similarly, the entire deletion of *BX-C* causes all segments posterior to T2 to resemble T2; apparently T2 is the developmental "ground state" of these 10 segments. The evolution of such gene families, it is thought, permitted arthropods (the phylum containing insects) to arise from the more primitive annelids (segmented worms) in which all segments are nearly alike.

Detailed genetic analysis of *BX-C* led Edward B. Lewis to formulate a model for segmental differentiation (Fig. 33-74). *BX-C*, Lewis proposed, contains at least one gene for each segment from T3 to A8, which for simplicity are numbered 0 to 8 in Fig. 33-74. These genes, for reasons that are not understood, are arranged in the same order, from "left" to "right," as the segments whose development they influence. Starting with segment T3, progressively more posterior segments express successively more *BX-C* genes until, in segment A8, all of these genes are expressed. The developmental fate of a segment is thereby determined by its position in the embryo.

Sequence analysis of the *BX-C* region led to a difficulty with Lewis' model: The *BX-C* contains only three protein-encoding genes, ***Ultrabithorax (Ubx), Abdominal-A (Abd-A),*** and ***Abdominal-B (Abd-B).*** However, further analysis indicated, for example, that mutations such as *bx, abx,* and *pdx*, which were previously assumed to occur on separate

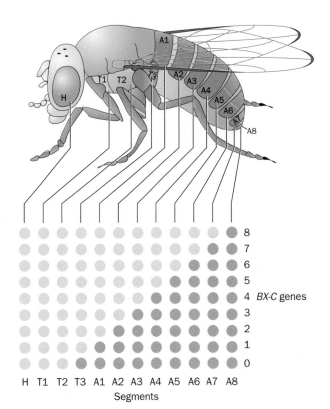

FIGURE 33-74. A model for the differentiation of embryonic segments in *Drosophila* as directed by the genes of the bithorax complex *(BX-C)*. Segments T2, T3, and A1-8 in the embryo, as the lower drawing indicates, are each characterized by a unique combination of active *(purple circles)* and inactive *(yellow circles)* *BX-C* "genes." These "genes" (which sequencing studies later demonstrated are really enhancer elements), here numbered 0 to 8, are thought to be sequentially activated from anterior to posterior in the embryo so that segment T2, the developmentally most primitive segment, has no active *BX-C* genes, while in segment A8, all of them are active. Such a pattern of gene expression may result from a gradient in the concentration of a *BX-C* repressor that decreases from the anterior to the posterior of the embryo. [After Ingham, P., *Trends Genet.* **1,** 113 (1985).]

genes, are actually mutations of enhancer elements that enable the position-specific expression of the *Ubx* gene. Thus, the nine "genes" in Lewis' model have turned out to be enhancer elements on the three *BX-C* genes. The targets of the homeotic selector genes, the genes that presumably function to form the specified tissues and structures, are presently the objects of intensive research.

Developmental Genes Have Common Sequences

In characterizing the ***Antennapedia (Antp)*** gene, Gehring and Matthew Scott independently discovered that *Antp* cDNA hybridizes to both the *Antp* and the *ftz* gene and that, therefore, *these genes share a common base sequence.* This startling observation rapidly led to the discovery that *the Drosophila genome contains numerous such sequences, many of which occur in the homeotic gene complexes*

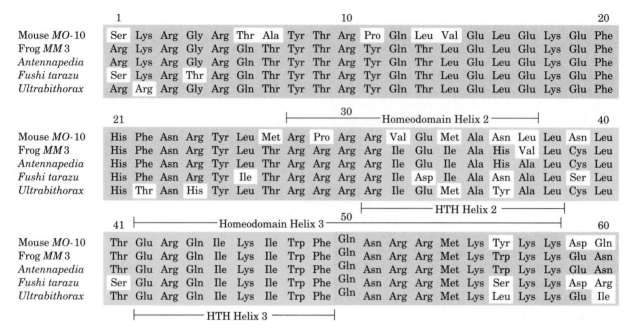

FIGURE 33-75.   The amino acid sequences of the polypeptides encoded by the homeodomains of 5 genes from mouse, *Xenopus,* and *Drosophila* (*Ultrabithorax* is a *BX-C* gene). Discrepancies between the polypeptide specified by the *Antp* homeobox and those of the other genes lack shading. Each polypeptide has a 19-residue segment (*red shading*), which is homologous to the DNA-binding HTH fold of prokaryotic repressors. The positions of the helices observed in the X-ray and NMR structures of homeodomains, together with the corresponding positions of HTH helices in prokaryotic proteins, are indicated.

ANT-C and BX-C. DNA sequencing studies of these genes revealed that each contains a 180-bp sequence, the so-called **homeodomain** or **homeobox,** which are 70 to 90% homologous to one another and which encode even more homologous 60-residue polypeptide segments (Fig. 33-75).

Further hybridization studies using homeodomain probes led to the truly astonishing finding that *multiple copies of the homeodomain are also present in the genomes of segmented animals ranging from annelids to vertebrates such as Xenopus, mice, and humans.* In some of these sequences the degree of homology is remarkably high; for example, the homeodomains of the *Drosophila Antp* gene and the *Xenopus MM3* **gene** encode polypeptides that have 59 of their 60 amino acids in common (Fig. 33-75).

The Homeodomain's DNA-Binding Motif Resembles a Helix–Turn–Helix Motif

Since vertebrates and invertebrates diverged over 600 million years ago, this strongly suggests that the gene product of the homeodomain has an essential function. What might this function be? The ~30% Arg + Lys content of homeodomain polypeptides suggest that they bind DNA. Sequence comparisons and NMR studies further suggest that these polypeptide segments form helix–turn–helix (HTH) motifs resembling those of prokaryotic gene regulators such as the *E. coli trp* repressor (Section 29-3C) and the λCro protein (Section 32-3D). Indeed, the polypeptide encoded by the homeodomain of the *Drosophila engrailed*

gene specifically binds to the DNA sequences just upstream from the transcription start sites of both the *en* and the *ftz* genes. Moreover, fusing the *ftz* gene's upstream sequence to other genes imposes *ftz*'s pattern of stripes (Fig. 33-71) on the expression of these genes in *Drosophila* embryos. *These observations suggest, in agreement with the idea that the products of developmental genes act to regulate the expression of other genes, that homeodomain-containing genes encode transcription factors.* In fact, not all homeodomain-encoded proteins are involved in regulating development. The homeodomain is apparently a widespread genetic motif that specifies the DNA-binding segments of a variety of proteins.

Thomas Kornberg and Pabo have determined the X-ray structure of the 61-residue homeodomain from the *Drosophila engrailed* protein in complex with a 21-bp DNA (Fig. 33-76). Two copies of the protein bind to the DNA, one near the center of the DNA and the other near one end, where it also contacts a second DNA molecule that, in the crystal, forms a pseudo-continuous helix with the first. The conformations of the two protein molecules, and the contacts they make with the DNA, are nearly identical. The two homeodomains are not in contact so, in contrast to other DNA-binding motifs of known structure, *they bind to their target DNAs as monomers.* The X-ray structure is largely consistent with the NMR structure of the *Antennapedia* homeodomain in complex with a 14-bp DNA determined by Gehring and Kurt Wüthrich.

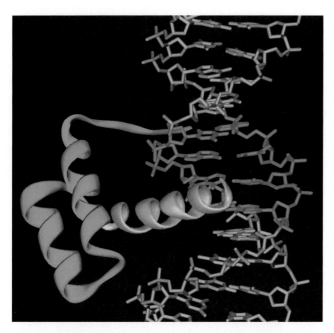

FIGURE 33-76.   The X-ray structure of the Engrailed protein homeodomain in complex with a 21-bp DNA containing its target sequence. The 60-residue protein is shown in ribbon form (*green*) with its recognition helix (Helix 3, residues 42–58), which is bound in the DNA's major groove, highlighted in yellow. The DNA is shown in stick form (*light blue*) with the base pairs comprising its TAAT subsite highlighted in purple. A second homeodomain that binds to the lower end of the DNA in a nearly identical manner but does not contact the homeodomain shown has been omitted for clarity. Note how the N-terminal segment (*red*, residues 3–5; residues 1–2 are disordered) binds in the minor groove of the DNA. [Based on an X-ray structure by Carl Pabo, The Johns Hopkins University.]

The homeodomain consists largely of three α helices, the last two of which, as sequence comparisons had previously suggested, form an HTH motif that is closely superimposable with the HTH motifs of prokaryotic repressors such as that of the λ repressor (Fig. 32-37). However, although helix 3, the HTH motif's recognition helix, fits into the major groove of its corresponding DNA, it does so quite differently in the two complexes. In the λ repressor complex, for example, the N-terminal end of the recognition helix is inserted into the DNA's major groove, whereas in the homeodomain complex the DNA is shifted towards the C-terminal end of the helix, which is longer than that of the λ repressor (it extends from residues 42 to 58 in Fig. 33-75). As a consequence, the way in which the first helix of the HTH motif (helix 2; residues 28 to 37 in Fig. 33-75) contacts the DNA also differs between the two complexes.

Most homeodomain binding sites have the subsequence TAAT. The recognition helix in the X-ray structure makes base-specific hydrogen bonding contacts with this subsequence in the major groove through residues that are highly conserved in higher eukaryotic homeodomains. It

therefore appears that these interactions function to align the homeodomain with the other bases that it contacts. In addition, two conserved Arg residues located in the N-terminal tail of the homeodomain make base-specific hydrogen bonding contacts with the TAAT subsequence in the minor groove of the DNA. The protein thereby grips the TAAT subsequence from two sides. Note that few other sequence-specific DNA-binding proteins contact bases in the minor groove. Finally, the homeodomain makes extensive contacts with the DNA backbone that, it is presumed, also play an important part in binding and recognition.

Homeodomain Genes Function Analogously in Vertebrates and *Drosophila*

Homeodomain-encoding genes have collectively become known as ***Hox*** genes. In vertebrates, they are organized in four clusters of 9 to 11 genes, each located on a separate chromosome and spanning more than 100 kb. In contrast, *Drosophila*, as we saw, have two *Hox* clusters, whereas nematodes (roundworms), which are evolutionarily more primitive than insects, have only one *Hox* cluster. As in *Drosophila*, the genes in each vertebrate *Hox* cluster are activated in the same order, left to right, as they are expressed from the anterior end of the embryo to its posterior end. Perhaps this arrangement is necessary for the homeodomain genes to be activated in the proper order, although, at least in *Drosophila*, gap and pair-rule proteins can still act on *Hox* control regions that have been transplanted to other parts of the genome. Whatever the case, the various *Hox* clusters, as well as their component genes, almost certainly arose through a series of gene duplications and diversifications starting with a single *Hox* gene in a primitive ancestral organism.

Vertebrate *Hox* genes, like those of *Drosophila*, are expressed in specific patterns and at particular stages during embryogenesis. Most *Hox* genes are expressed at a gestational time when organogenesis prevails. That the *Hox* genes directly specify the identities and fates of embryonic cells, that is, are homeotic in character, was shown, for example by the following experiment. Mouse embryos were made transgenic for the *Hox-1.1* gene that had been placed under the control of a promoter that is active throughout the body even though *Hox-1.1* is normally expressed only below the neck. The resulting mice had severe craniofacial abnormalities such as a cleft palate and an extra vertebra and an intervertebral disk at the base of the skull. Some also had an extra pair of ribs in the neck region. Thus, this *Hox* gene's "gain of function" induced a homeotic mutation, that is, a change in the development pattern, analogous to those observed in *Drosophila*.

Homozygotic mice resulting from the replacement of their *Hox-3.1* gene coding sequence in embryonic stem cells of mice with that of the *lacZ* are born alive but usually die within a few days. They exhibit skeletal deformities in their trunk regions in which several skeletal segments are transformed into the likenesses of more anterior segments.

The pattern of *β*-galactosidase activity (Fig. 33-77) as color-imetrically detected through the use of X-Gal; Section 28-8B), in both homozygotes and heterozygotes, indicates that *Hox-3.1* deletion modifies the identities but not the positions of the embryonic cells that normally express *Hox-3.1*.

Retinoic Acid Is a Vertebrate Morphogen

Retinoic acid (RA), a derivative of **vitamin A (retinol)**,

X=COOH: **Retinoic acid (RA)**

X=CHO: **Retinal (Vitamin A)**

has been found to have a graded distribution in developing chick limbs and is therefore thought to be a morphogen. The systematic administration of RA during mouse embryogenesis results in severe malformations, notably skeletal deformities that appear to arise from anterior or posterior shifts of their normal characteristics. A variety of evidence suggests that the expression of *Hox* genes mediates the positional information that RA disrupts. The *Hox* genes are differentially activated by RA according to their positions in their Hox clusters: Those towards the 3′ end of a cluster are maximally induced by as little as $10^{-8}M$ RA, those towards the 5′ end of the cluster require $10^{-5}M$ RA to do so, and those at the 5′ ends are insensitive to RA. Moreover, $10^{-5}M$ RA sequentially activates the *Hox* genes from the 3′ to the 5′ end of a cluster, the same order as their expression patterns in developing axial systems such as the skeleton and the central nervous system.

The foregoing explains why the RA analog **13-*cis*-retinoic acid,** which taken orally, has been invaluable in the treatment of severe **cystic acne,** induces birth defects if used by pregnant women. The characteristic pattern of cranial deformities in the resulting infants, whose analog is induced in mouse embryos that had been exposed to low concentrations ($2 \times 10^{-6}M$) of this drug, indicates that its presence alters the expression of *Hox* genes early in gestation (~1 month postfertilization in humans, ~9 days in mice).

C. The Molecular Basis of Cancer: Oncogenes

The cells of the body normally remain under strict developmental control. Thus, during embryogenesis, cells must differentiate, proliferate, migrate, and even die in the correct spatial arrangement and temporal sequence to yield a normally functioning organism. In the adult, the cells of certain tissues, such as the intestinal epithelium and the blood-forming tissues of the bone marrow, continue to proliferate. Most adult body cells, however, remain quiescent, that is, in the G_0 phase of the cell cycle.

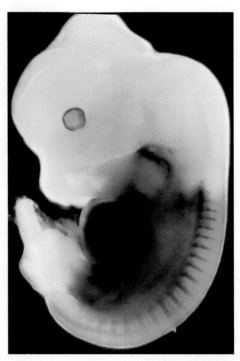

FIGURE 33-77. The pattern of expression of the *Hox 3.1* gene in a 12.5 day postconception mouse embryo. The protein-encoding portion of the embryo's *Hox-3.1* gene was replaced by the *lacZ* gene. The regions of this transgenic embryo in which *Hox-3.1* is expressed are revealed by the blue color that develops on soaking the embryo in X-Gal-containing buffer. [Courtesy of Phillipe Brûlet, Collège de France and the Pasteur Institute, France.]

Cells occasionally lose their developmental controls and commence excessive proliferation. The resulting tumors can be of two types:

1. **Benign tumors,** such as warts and moles, grow by simple expansion and often remain encapsulated by a layer of connective tissue. Benign tumors are rarely life threatening, although if they occur in an enclosed space such as in the brain or secrete large amounts of certain hormones, they can prove fatal.

2. **Malignant tumors** or **cancers** grow in an invasive manner and shed cells that, in a process known as **metastasis,** colonize new sites in the body. Malignant tumors are almost invariably life threatening; they are responsible for 20% of the mortalities in the United States.

Cancer, being one of the major human health problems, has received enormous biomedical attention over the past few decades. Around 100 different types of human cancers are recognized, methods of cancer detection and treatment are highly developed, and cancer epidemiology has been extensively characterized. We are, nevertheless, just beginning to understand the biochemical basis of this collection of diseases. In this section we outline what is known about this rapidly developing field of knowledge. However, since cancer can be considered as resulting from an aberrant cell cycle, we first discuss how the cell cycle is controlled.

The Regulation of the Cell Cycle

What are the molecular events that control the initiation of mitosis (cell division)? The first clues to this process came from studies of marine invertebrate embryos, revealing that a class of proteins named **cyclins** accumulate steadily throughout the cell cycle (Section 31-4A) and then abruptly disappear just before the anaphase portion of mitosis (Fig. 27-2). Clam embryos have two types of cyclins, **cyclin A** and **cyclin B,** which are distinguished by their differing amino acid sequences and their precise patterns of accumulation and destruction during the cell cycle. Homologs of these proteins have since been discovered in many eukaryotes, from sea urchins to humans, which is, perhaps, not surprising since all eukaryotic cells follow some version of the cell cycle (yeasts also have cyclin-like proteins, although they do not closely resemble cyclins A or B).

Cyclin B combines with a 34-kD protein named **p34**^{cdc2} (*cdc* for *cell division cycle*), whose sequence clearly indicates that it is a member of the Ser/Thr protein kinase family and which is highly conserved from yeasts to humans. *It is this latter protein that is the central cell cycle regulator in species ranging from yeasts to humans. It does so by phosphorylating a variety of nuclear proteins, among them histone H1, several oncogene proteins (see below), and proteins involved in nuclear disassembly and cytoskeletal rearrangement. This presumably initiates a cascade of cellular events that culminates in mitosis.*

The binding of cyclin B to p34^{cdc2} forms an activated complex that is alternatively called **cyclin-dependent protein kinase (Cdk)** or **maturation promoting factor (MPF).** This, however, is by no means the entire activation story. p34^{cdc2} is a phosphoprotein itself, which can be phosphorylated on its Thr 14, Tyr 15, and Thr 161 residues: Cdk is active only when both Thr 14 and Tyr 15, which occupy the region of the ATP-binding site, are dephosphorylated and when Thr 161 is phosphorylated (the significance of Tyr phosphorylation is discussed below). Moreover, the phosphorylation of Tyr 15 requires the presence of cyclin B. Thus, a reasonable although much oversimplified scenario for the regulation of the cell cycle is as follows (Fig. 33-78):

1. The cell enters the G1 phase of the cell cycle with cyclin B absent and with p34^{cdc2} dephosphorylated. There, an as of yet unknown enzyme phosphorylates Thr 161.

2. Newly synthesized cyclin B then binds to p34^{cdc2}, while Thr 14 and Tyr 15 become phosphorylated. The resulting triply phosphorylated cyclin B–p34^{cdc2} complex is

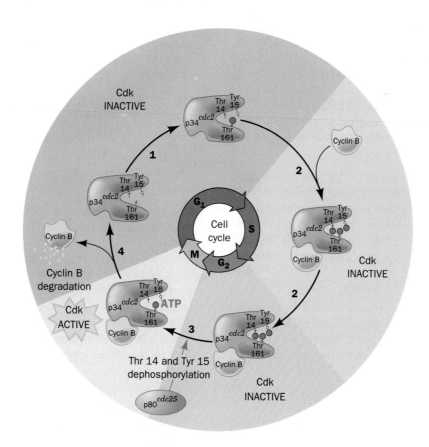

FIGURE 33-78. The regulation of the cyclin *B*-dependent protein kinase (Cdk) activity of p34^{cdc2} in the animal cell cycle. Details are described in the text. [After Norbury, C. and Nurse, P., *Annu. Rev. Biochem.* **61,** 451 (1992).]

enzymatically inactive because Thr 14 and Tyr 15 prevent p34*cdc2* from binding ATP. Thus, the entire system appears designed to maintain p34*cdc2* in an inactive state while cyclin B gradually accumulates during the S phase of the cell cycle.

3. The rapid and specific dephosphorylation of Thr 14 and Tyr 15 at the cell cycle's G2/M boundary by **p80*cdc25*** (the 80-kD protein product of the **cdc25** gene), a phosphatase, activates Cdk, which in turn triggers mitosis (M phase).

4. Cyclin B is quickly proteolyzed in a ubiquitin-mediated pathway (Section 30-6B), followed by the rapid dephosphorylation of Thr 161 on p34*cdc2*. This inactivates the Cdk function, thereby returning the now divided cell to the G1 phase.

The X-Ray Structure of Cyclin-Dependent Kinase Resembles That of cAPK

The cell cycle in human cells is governed by several Cdks, the most extensively studied of which are the closely related ~34-kD proteins known as **CDC2,** which participates in the control of mitosis, and **CDK2,** which binds cyclin A and is implicated in the control of G1 and S events. Both enzymes are activated by the binding of a cyclin followed by phosphorylation at Thr 160 in CDK2 and Thr 161 in CDC2 by a specific protein kinase named **Cdk-activating kinase.** Both enzymes are also negatively regulated by the phosphorylation of Tyr 15 and, to a lesser extent, the adjacent Thr 14.

The X-ray structure of CDK2 in complex with MgATP (Fig. 33-79), by Sung-Hou Kim, indicates that this Ser/Thr protein kinase closely resembles the catalytic subunit of cAMP-dependent protein kinase (cAPK; Section 17-3C), a protein whose sequence is 24% identical to that of CDK2. However, there are functionally significant structural differences between these two enzymes:

1. The relative arrangement of ATP's β- and γ-phosphate groups in CDK2 is expected to greatly reduce the reactivity of the γ-phosphate relative to that in the cAPK–ATP complex (stereoelectronic control), thereby rationalizing, in part, why CDK2 is catalytically inactive, whereas the catalytic subunit of cAPK alone is catalytically active.

2. Access to the γ-phosphate of CDK2's bound ATP by its protein substrates appears to be blocked by an apparently flexible 19-residue protein loop (residues 152–170) that has been named the "T loop" because it contains Thr 160. Mutational analysis has shown that cyclin A binding to CDK2 is inhibited by changes in charged residues in the protein's N-terminal lobe near its active site cleft. Thus, cyclin A binding to this site may well result in a conformational reorientation that induces the ATP to take up a more reactive conformation and moves the T loop out of the active site cleft. The phosphorylation of Thr 160 may contribute to this activa-

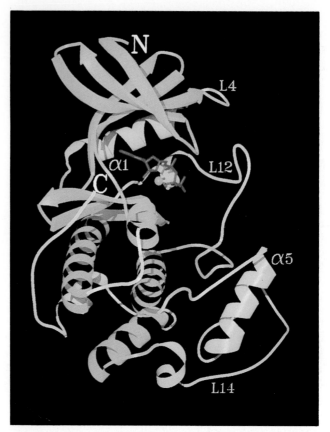

FIGURE 33-79. The X-ray structure human cyclin-dependent kinase 2 (CDK2) in complex with MgATP. The protein is represented by a ribbon diagram that is blue in regions that are conserved between CDK2 and cAPK (Fig. 17-14) and yellow in divergent regions. The ATP is shown in stick form in red and the Mg²⁺ ion is represented by a green ball. [Courtesy of Sung-Hou Kim, University of California at Berkeley.]

tion process by stabilizing the T loop in the active conformation.

The X-ray structure also explains why the phosphorylation of Thr 14 inactivates CDK2: The hydroxyl group of this side chain is close to the ATP's γ-phosphate so that phosphorylation of Thr 14 would probably disrupt the conformation of the ATP's phosphate groups. It is unclear, however, how the phosphorylation of Tyr 15 affects CDK2 activity.

Cancer Cells Differ in Many Ways from Normal Cells

The most obvious and the medically most significant property of cancer cells is that they proliferate uncontrollably. For instance, when grown in a tissue culture dish, normal cells form a monocellular layer on the bottom of the dish and then, through a process termed **contact inhibition,** cease dividing (Fig. 33-80a). In contrast, the growth of malignant cells is unhampered by intercellular contacts; in culture they form multicellular layers (Fig. 33-80b). More-

(a) **Normal cells**

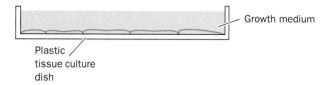

Growth medium

Plastic
tissue culture
dish

(b) **Transformed cells**

FIGURE 33-80. The growth pattern of vertebrate cells in culture: (*a*) Normal cells stop growing through contact inhibition once they have formed a confluent monolayer. (*b*) In contrast, transformed cells lack contact inhibition; they pile up to form a multilayer.

over, even in the absence of contact inhibition, normal cells are far more limited in their capacity to reproduce than are cancer cells. Normal cells, depending on the species and age of the animal from which they were taken, will only divide in culture 20 to 60 times before they reach senescence and die (a phenomenon that, no doubt, is at the heart of the aging process; cell senescence may be caused by telomeric erosion as is discussed in Section 31–4D). *Cancer cells, on the other hand, are immortal; there is no limit to the number of times they can divide.* In fact, some cancer cell lines have been maintained in culture through thousands of divisions spanning several decades. Immortal cells, however, are not necessarily malignant: *The hallmark of cancer is immortal-*

ity combined with uncontrolled growth.
ity combined with uncontrolled growth.

The properties of cancer cells differ from those of the normal cells from which they are derived. The plasma membranes of malignant cells have a more fluid character than those of normal cells and have altered ratios of many of their cell surface components such as glycoproteins and glycolipids. Internally, the cytoskeletons of cancer cells are less organized than those of normal cells. This, presumably, is why cancer cells have a more rounded appearance than the corresponding normal cells (Fig. 33-81). Metabolically, cancer cells have a high rate of glycolysis that results in a debilitating energy drain on the host. The conversion of a normal to a cancerous cell is therefore accompanied by a complex series of structural, biochemical, and, as we shall see, genetic changes.

Cancer Is Caused by Carcinogens, Radiation, and Viruses

Most cancers are caused by agents that damage DNA or interfere with its replication or repair. These include a great variety of man made and naturally occurring substances known as chemical carcinogens (Section 31-5E), as well as radiation, both electromagnetic and particulate, with sufficient energy to break chemical bonds. In addition, *certain viruses induce the formation of malignant tumors in their hosts (see below).*

Almost all malignant tumors result from the **transformation** of a single cell (conversion to the cancerous state; this term should not be confused with the acquisition of genetic information from exogenously supplied DNA), which, being free of its normal developmental constraints, proliferates. Yet, considering, for example, that the human body

(a)   *(b)*

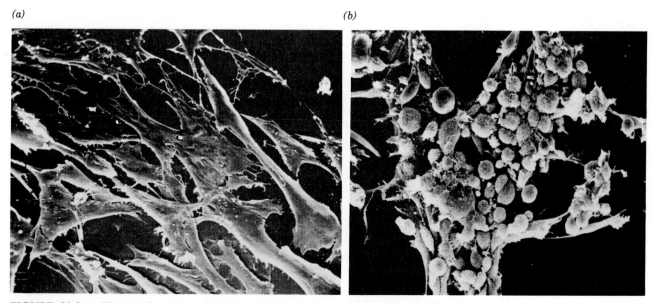

FIGURE 33-81. The transformation of cultured chicken fibroblasts by Rous sarcoma virus: (*a*) Normal cells adhere to the surface of the culture dish, where they assume a flat extended conformation. (*b*) Upon infection with RSV, these cells become rounded and cluster together in piles. [Courtesy of G. Steven Martin, University of California at Berkeley.]

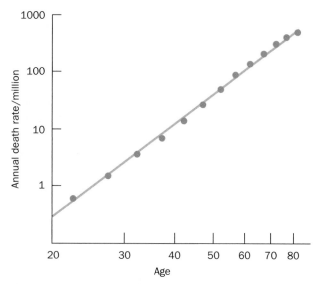

FIGURE 33-82. The variation of the cancer death rate in humans with age. The linearity of this log–log plot can be explained by the hypothesis that several randomly occurring mutations are required to generate a malignancy. The slope of the line suggests that, on the average, five such mutations are required for a malignant transformation.

consists of around 10^{14} cells, transformation must be a very rare event. One of the major reasons for this, as the age distribution of the cancer death rate indicates (Fig. 33-82), is that *transformation requires a cell or its ancestors to have undergone several independent and presumably improbable carcinogenic changes.* Consequently, exposure to a carcinogen may prime many cells for transformation but a malignant tumor may not form until decades later when one of these cells suffers a final transforming event.

Certain Retroviruses Carry Oncogenes

The viral induction of cancer was first observed in 1911 by Peyton Rous, who demonstrated that cell-free filtrates from certain chicken **sarcomas** (malignant tumors arising from connective tissues) promote new sarcomas in chickens. Although decades were to pass before the significance of this work was appreciated (Rous was awarded the Nobel prize in 1966 at the age of 85), many other such **tumor viruses** have since been characterized.

The **Rous sarcoma virus (RSV),** as are all known RNA tumor viruses, is a **retrovirus** [an RNA virus that replicates its chromosome by copying it to DNA in a reaction mediated by virally specified reverse transcriptase (Section 31-4C), inserting the DNA into the host cell's genome, and then transcribing this DNA]. The base sequence of the RSV chromosome reveals that it contains four genes (Fig. 33-83), of which only three are essential for viral replication: *gag,* which encodes an inner capsid protein; *pol,* which specifies reverse transcriptase; and *env,* which codes for the outer envelope protein. The fourth gene, **v-src** ("v" for *v*iral, "*src*" for *s*arcoma), encodes a protein known as **pp60ᵛ⁻ˢʳᶜ**

("pp60" signifies that it is a 60-kD *p*hospho*p*rotein), which mediates host cell transformation. v-*src* has therefore been termed an **oncogene** (Greek: *onkos,* mass or tumor).

What is the origin of v-*src* and what is its viral function? Hybridization studies by Michael Bishop and Harold Varmus in 1976 led to the remarkable discovery that *uninfected chicken cells contain a gene,* **c-src** *("c" for cellular), that is homologous to* v-*src* (the two genes differ mainly in that c-*src* is interrupted by six introns, whereas v-*src* is uninterrupted). Moreover, c-*src* is highly conserved in a wide variety of eukaryotes that span the evolutionary scale from *Drosophila* to humans. This observation strongly suggests that c-*src,* which antibodies directed against pp60ᵛ⁻ˢʳᶜ indicate is expressed in normal cells, is an essential cellular gene. In fact, as we shall see below, *both* pp60ᵛ⁻ˢʳᶜ *and its normal cellular analog,* **pp60ᶜ⁻ˢʳᶜ**, *function to stimulate cell proliferation.* Apparently, v-*src* was originally acquired from a cellular source by an initially nontransforming ancestor of RSV. By maintaining the host cell in a proliferative state (cells are usually not killed by RSV infection), pp60ᵛ⁻ˢʳᶜ presumably enhances the viral replication rate.

Viral Oncogene Products Mimic the Effects of Polypeptide Growth Factors and Hormones

In order to understand how oncogenes subvert the normal processes of cell division, we must first understand these processes. *Cell proliferation is stimulated by hormonelike polypeptide growth factors.* These **mitogens** (substances that induce mitosis), such as **epidermal growth factor (EGF)** and **platelet-derived growth factor (PDGF),** bind with high affinity to the extracellular domains of specific protein receptors that span the plasma membranes of certain types of cells (Fig. 33-84). In doing so, they activate the receptors' cytoplasmic domains to phosphorylate their target proteins which, in turn, are thought to act as intracellular messengers that stimulate cell division by mechanisms discussed in Section 34-4B. For example, **transforming growth factor α (TGF-α),** a protein synthesized by and required for the growth of epithelial cells, is produced in excessive amounts in the skin of individuals with **psoriasis,** a common skin disease characterized by epidermal hyperproliferation.

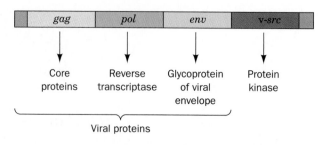

FIGURE 33-83. The genetic map of RSV. Its 9-kb genome encodes only four genes: *gag, pol,* and *env* are essential for viral replication, whereas v-*src* is an oncogene. The segments at the ends of the genome represent terminally redundant sequences.

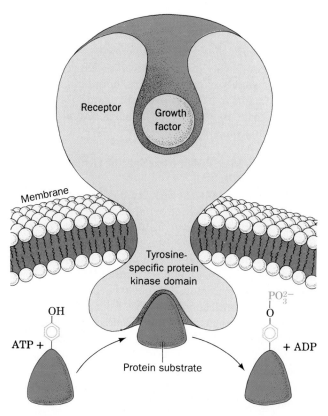

FIGURE 33-84. The binding of growth factor to the external domain of its corresponding receptor activates the receptor's cytoplasmic tyrosine-specific protein kinase domain to phosphorylate specific Tyr residues of its protein substrates.

Many growth factor receptors are **tyrosine-specific protein kinases**; *that is, they phosphorylate specific Tyr OH groups in their target proteins* (Fig. 33-84). Most protein kinases, it should be noted, specifically phosphorylate Ser or Thr residues (recall, e.g., that phosphorylase kinase phosphorylates Ser 14 of glycogen phosphorylase; Section 17-3C); only about one phosphorylated amino acid residue in 2000 is Tyr. Nevertheless, as we discuss in Section 34-4B, tyrosine phosphorylation of central importance in regulating a variety of essential cellular processes, notably growth and differentiation. Interestingly, many activated tyrosine kinases phosphorylate themselves. This **autophosphorylation** itself serves as a signal that the receptor has bound ligand.

Hormones such as epinephrine and glucagon also profoundly affect the physiology of their target cells (Sections 17-3E–G and 34-4A and B). These hormones bind to specific receptors, thereby stimulating adenylate cyclase to catalyze the formation of cAMP, the second messenger that actually triggers the cellular response to the hormone. Hormone receptors, which face out from the plasma membrane, and adenylate cyclase, which is located on the membrane's cytoplasmic surface, are separate proteins that do not physically interact. Rather, they are functionally coupled by **G-proteins** (Fig. 33-85), so-called because they specifically bind GTP and GDP. Adenylate cyclase is activated by G-protein but only when the G-protein is complexed

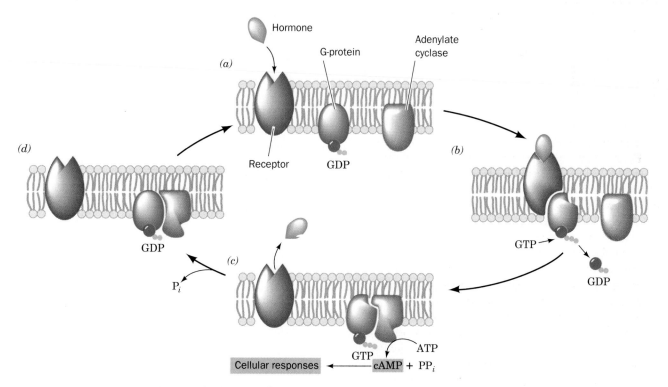

FIGURE 33-85. The activation/deactivation cycle for hormonally stimulated adenylate cyclase. (*a*) In the absence of hormone, G-protein binds GDP and adenylate cyclase is catalytically inactive. (*b*) The hormone–receptor complex stimulates G-protein to exchange its bound GDP for GTP.

(*c*) The G-protein–GTP complex, in turn, binds to and thereby activates adenylate cyclase to produce cAMP. (*d*) The eventual G-protein–catalyzed hydrolysis of its bound GTP to GDP causes G-protein to dissociate from and hence deactivate adenylate cyclase.

with GTP. However, G-protein slowly hydrolyzes GTP to GDP + P_i and thereby deactivates itself. G-protein is reactivated by a GDP–GTP exchange reaction that is catalyzed by hormone–receptor complex but not by unoccupied receptor. *G-protein therefore mediates the hormonal signal.* We discuss this process in greater detail in Section 34-4B.

The proteins encoded by many viral oncogenes are analogs of various growth factor and hormone system components. For instance:

1. The **v-*sis*** oncogene of **simian sarcoma virus** encodes a protein secreted by infected cells that is nearly identical with one of the two subunits of PDGF. Hence, the uncontrolled growth of simian sarcoma virus–infected cells apparently results from the continuous and inappropriate presence of this PDGF homolog.

2. Nearly half of the more than 20 known retrovirus oncogenes, including v-*src,* encode tyrosine-specific protein kinases. Indeed, the **v-*erbB*** oncogene specifies a truncated version of the **EGF receptor** that lacks the EGF-binding domain but retains its transmembrane segment and its protein kinase domain. *Evidently, oncogene-encoded protein kinases inappropriately phosphorylate the target proteins normally recognized by growth factor receptors, thereby driving the afflicted cells to a state of unrestrained proliferation.*

3. The **v-*ras*** oncogene encodes a protein, $p21^{v\text{-}ras}$, that functionally resembles G-proteins in that it is localized on the cytoplasmic side of the mammalian plasma membrane where, when binding GTP, it activates a variety of cellular processes by stimulating the phosphorylation of numerous proteins at specific Ser and Thr residues (Section 34-4B). Although $p21^{v\text{-}ras}$ hydrolyzes GTP to GDP, it does so much more slowly than its normal cellular analog. The restraint to protein phosphorylation that GTP hydrolysis would normally impose is therefore greatly reduced in $p21^{v\text{-}ras}$, thereby transforming the cell.

4. Several retroviral oncogenes, including **v-*jun*** and **v-*fos,*** encode nuclear proteins whose corresponding normal cellular analogs are synthesized in response to mitogenic signals. Many such proteins, including the v-*jun* and v-*fos* gene products, bind to DNA, strongly suggesting that they influence its transcription and/or replication. Indeed, v-*jun* is 80% homologous to the **proto-oncogene** (normal cellular analog of an oncogene) c-*jun,* which encodes a transcription factor named **Jun** (also called **AP-1**). Moreover, Jun/AP-1 forms a tight complex with the protein product of the proto-oncogene c-*fos,* which sequence analysis and mutational alteration studies indicates is mediated by a leucine zipper (Section 33-3B). This association greatly increases the ability of Jun/AP-1 to stimulate transcription from Jun-responsive genes.

Oncogene products therefore appear to be functionally modified or inappropriately expressed components of elaborate control networks that regulate cell growth and differentiation. The complexity of these networks (cells generally respond to a variety of growth factors, hormones, and transcription factors in partially overlapping ways) is probably why malignant transformation requires several independent carcinogenic events.

Malignancies May Result from Specific Genetic Alterations

Although much of what we know concerning oncogenes stems from the study of retroviral oncogenes, few human cancers are caused by retroviruses. Nevertheless, *it seems likely that all cancers are caused by genetic alterations.* Robert Weinberg demonstrated this to be the case for mouse fibroblasts that had been transformed by a known carcinogen: Normal mouse fibroblasts in culture are transformed upon transfection with DNA from the transformed cells. Moreover, these newly transformed cells, when innoculated into mice, form tumors. Similar investigations indicate that DNAs from a wide variety of malignant tumors likewise have transforming activity.

What sorts of genetic changes can give rise to cancer? Several types of changes have been observed:

1. **Altered Proteins**
An oncogene, as we have seen, may give rise to a protein product with an anomalous activity relative to that of the corresponding proto-oncogene. This may even result from a simple point mutation. For example, Weinberg, Michael Wigler, and Mariano Barbacid showed that the *ras* oncogene isolated from a human bladder **carcinoma** (a malignant tumor arising from epithelial tissue) differs from its corresponding proto-oncogene by the mutation of the Gly 12 codon (GGC) to a Val codon (GTC). The resulting amino acid change attenuates the GTPase activity of $p21^{c\text{-}ras}$, evidently without affecting its ability to stimulate protein phosphorylation, thereby prolonging the time this G-protein remains in the "on" state. Indeed, Sung-Hou Kim's X-ray structure determinations of normal human $p21^{c\text{-}ras}$ (Fig. 33-86) and its oncogenic counterpart (Gly 12→Val), both in complex with GDP, indicates that the mutation mainly alters the normal protein structure in the vicinity of its presumed GTPase function. Most other *ras* oncogene–activating mutations also change residues close to this site. Interestingly, c-*ras* is one the most commonly implicated proto-oncogenes in human cancers.

2. **Altered Regulatory Sequences**
Malignant transformation can result from the inappropriately high expression of a normal cellular protein. For example, the retroviral oncogene v-*fos* and its corresponding proto-oncogene c-*fos* encode similar proteins. These genes differ mainly in their regulatory sequences: v-*fos* has an efficient enhancer, whereas c-*fos* has a 67-nucleotide AT-rich segment in its unexpressed 3′-terminal end that, when transcribed, promotes rapid mRNA

FIGURE 33-86. The X-ray structure of the human proto-oncogene protein, p21$^{c\text{-}ras}$ in complex with GDP (*yellow*). The protein's helical and β sheet regions are drawn in red and green, respectively. The native protein's 18 C-terminal residues, which have no known biochemical function and which are thought to be flexible, were removed to facilitate the protein's crystallization. p21$^{c\text{-}ras}$ structurally resembles the GTP-binding domain of the *E. coli* ribosomal elongation factors EF-Tu and EF-G (Section 30-3D). [Courtesy of Sung-Hou Kim, University of California at Berkeley.]

degradation (Section 33-3C). Thus, c-*fos* can be converted to an oncogene by deleting its 3′ end and adding the v-*fos* enhancer.

3. Loss of Degradation Signals

An oncogene protein that is degraded more slowly than the corresponding normal cellular protein may cause malignant transformation through its consequent inappropriately high concentration in the cell. Thus c-Jun protein but not v-Jun is efficiently multiubiquinated and hence proteolytically degraded by the cell (Section 30-6B). This is because v-Jun lacks a 27-residue segment present in c-Jun that mutagenesis experiments indicate is essential for the efficient ubiquination of c-Jun even though this segment does not contain the protein's principal ubiquitin attachment sites.

4. Chromosomal Rearrangements

An oncogene may be inappropriately transcribed when brought under the control of a foreign regulatory sequence through chromosomal rearrangement. For example, Carlo Croce found that the human cancer **Burkitt's lymphoma** (a lymphoma is an immune system cell malignancy) is characterized by an exchange of chromosomal segments in which the proto-oncogene **c-myc** is translocated from its normal position at one end of chromosome 8 to the end of chromosome 14 adjacent to

certain immunoglobulin genes. The misplaced c-*myc* gene is thereby brought under the transcriptional control of the highly active (in immune system cells) immunoglobulin regulatory sequences. The consequent overproduction of the normal c-*myc* gene product (which encodes a DNA-binding protein whose transient increase is normally correlated with the onset of cell division) or alternatively, its production at the wrong time in the cell cycle, is apparently a major factor in cell transformation.

5. Gene Amplification

Oncogene overexpression can also occur when the oncogene is replicated multiple times, either as sequentially repeated chromosomal copies or as extra-chromosomal particles. The amplification of the c-*myc* gene, for example, has been observed in several types of human cancers. Gene amplification is usually an unstable genetic condition that can only be maintained under strong selective pressure such as that conferred by cytotoxic drugs (Section 33-2D). It is not known how oncogene amplification is stably maintained.

6. Viral Insertion into a Chromosome

Inappropriate oncogene expression may result from the insertion of a viral genome into a cellular chromosome such that the proto-oncogene is brought under the transcriptional control of a viral regulatory sequence. For instance, **avian leukosis virus,** a retrovirus that lacks an oncogene but that nevertheless induces lymphomas in chickens, has a chromosomal insertion site near c-*myc*. Some DNA tumor viruses also transform cells in this manner.

7. Loss or Inactivation of Tumor Suppressor Genes

The high incidence of particular cancers in certain families suggests that there are genetic predispositions towards these diseases. A particularly clearcut example of this phenomenon occurs in **retinoblastoma,** a cancer of the developing retina that therefore afflicts only infants and young children. The offspring of surviving retinoblastoma victims also have a high incidence of this disease, as well as several other types of malignancies. In fact, retinoblastoma is associated with the inheritance of a copy of chromosome 13 from which a particular segment has been deleted. Retinoblastoma develops, as Alfred Knudson first explained, through a somatic mutation in a **retinoblast** (a retinal precursor cell) that alters the same segment of the second, heretofore normal copy of chromosome 13. This is because *the affected chromosomal segment contains a gene, the **Rb gene,** which specifies a factor that somehow restrains uninhibited cell proliferation; that is, the Rb gene product, **pRb,** is a **tumor suppressor** (alternatively, an **anti-oncogene** protein).*

Mutations altering normal gene products, causing chromosomal rearrangements and deletions, and perhaps gene amplification can all result from the actions of carcinogens on cellular DNA. Thus, normal cells bear the seeds of their own cancers. To date, ~40 viral and cellular oncogenes have been identified. It therefore seems likely that only a relatively small number of genes and perhaps an even smaller number of general molecular mechanisms are involved in oncogenesis.

pRb Functions by Binding to Certain Transcription Factors

pRb is a 105-kD DNA-binding protein that is localized in the nucleus of normal retinal cells but is absent in retinoblastoma cells. It is a phosphoprotein that is phosphorylated in a cell cycle–dependent manner by the protein kinase CDC2 (see above). Hypophosphorylated forms of pRb form complexes with certain transcription factors, including **E2F**, which regulates the expression of several cellular and viral genes. E2F was first identified as a cellular factor involved in the regulation of the **adenovirus** early *E2* gene by the adenovirus oncogene product **E1A**, although further investigation revealed that the adenovirus *E4* gene product also participates in this process. E1A protein, which does not bind DNA, promotes the dissociation of pRB from E2F by complexing pRB. It thereby frees E2F protein to combine with **E4 protein** on the adenovirus *E2* promoter so as to stimulate the transcription of the *E2* gene. These observations suggest that *the interaction of pRB with E2F and other transcription factors to which it binds plays an important role in the suppression of cellular proliferation* and that the dissociation of this complex is, at least in part, the means by which E1A inactivates pRB function. Thus, *an additional way that oncogenes can cause cancer is by inactivating the products of normal cellular tumor supressor genes.*

What is the role of pRb in normal development? To answer this question strains of mice that are heterozygous for an inactive *Rb* gene were generated. Cross-breeding such mice yields embryos with two normal *Rb* genes as well as both heterozygotes and homozygotes of the inactive *Rb* gene. The heterozygotes grow normally and, curiously, do not develop retinoblastomas (but they do develop pituitary tumors, which suggests that pRb functions in the development of different tissues in different organisms). The homozygotes for the inactive *Rb* gene develop in synchrony with their normal littermates until until at least 11.5 days of gestation but then die, *in utero,* around their 12th or 13th embryonic day due to improper differentiation of their central nervous system cells and erythrocytes. Other tissues, however, appear normal. Moreover, by the 12th embryonic day, millions of cell divisions have already occurred and numerous developmental decisions have occurred. Evidently, the *Rb* gene only plays an essential role in the development of a few cell types.

p53 Is a DNA-Binding Transcriptional Activator That Monitors Genome Integrity

Several tumor suppressor genes in addition to *Rb* have

now been characterized, including the ***p53* gene,** the **von Recklinghausen neurofibromatosis (*NF-1*) gene** (whose defect causes benign tumors of the peripheral nerves such as those of the famous "Elephant Man" of Victorian England), the ***DCC* gene** (whose defect promotes colon carcinoma), and the **thyroid hormone receptor (c-*erbA*) gene.** *The p53 gene, which encodes a 53-kD nuclear phosphoprotein, has emerged as the most commonly altered gene in human cancers: Around 50% of them contain a mutation in p53.*

The idea that *p53* encodes a tumor suppressor first arose from the discovery that germ line mutations in *p53* invariably occur in individuals with the rare inherited condition, known as **Li–Fraumeni syndrome,** that renders them highly susceptible to a variety of malignant tumors, particularly breast cancer, which they often develop before their 30th birthdays. That *p53* is indeed a tumor supresor gene has been clearly demonstrated by inactivating the *p53* gene of murine embryonic stem cells. Mice homozygous for this null allele appear to be developmentally normal but are prone to the spontaneous development of a variety of cancers by the age of 6 months.

The normal function of **p53 protein** has only recently begun to come to light. It has been shown that p53 specifically binds to the human analog of the mouse **MDM2 protein.** The ***MDM2* gene** is the dominant transforming oncogene present on *m*ouse *d*ouble *m*inute chromosomes (amplified extrachromosomal segments of DNA; Section 33-2D). Indeed, 17 out of 47 human sarcomas tested show amplification of the *MDM2* locus. Yet none of these tumors has a mutated *p53* gene. It therefore apppears that the high levels of MDM2 protein resulting from amplified *MDM2* genes inactivate p53 by binding to it, as do the DNA tumor virus oncoproteins SV40 large T antigen, adenovirus **E1B protein,** and **papilloma virus** (which causes warts) **E6 protein.**

It had been previously shown that p53 binds to specific sequences on double-stranded DNA. Further experiments have demonstrated that p53 is, in fact, a powerful transcriptional activator. Indeed, all point mutated forms of p53 that are implicated in cancer have lost their sequence-specific DNA-binding properties. But then, why is p53 a tumor suppressor? A clue to this riddle comes from observations that treatment of cells with radiation or chemotherapeutic drugs that damage DNA induce the accumulation of normal p53. Evidently, p53 acts as a "molecular policeman" in monitoring genome integrity. If the genome is damaged, p53 accumulates, thereby activating the transcription of a of gene, *Pic1,* encoding a 21-kD protein that binds to and inhibits various Cdks. This arrests the cell cycle in G1, thus allowing time for DNA repair. If DNA repair is unsuccessful, p53 may trigger cell suicide, a process named **apoptosis,** in order to prevent the proliferation of the genetically damaged and hence cancer-prone cell. In fact, the presence of p53 is required for radiation-induced apoptosis in the immune system cells known as **thymocytes.**

The X-Ray Structure of p53 Explains Its Oncogenic Mutations

p53 is a tetramer of identical 413-residue subunits. Each subunit has been shown to consist of four domains: an N-terminal transcriptional activation (transactivation) domain, a core DNA-binding domain, an oligimerization domain, and a basic C-terminal nuclear localization domain. Although the entire protein has so far resisted crystallization, Nikola Pavletich has determined the X-ray structure of the DNA binding core (residues 102–292) in complex with a DNA segment containing its 5-bp target sequence. *The vast majority of the >1000 p53 mutations that have been found in human tumors occur in this core.*

The structure of the p53 DNA-binding core domain (Fig. 33-87) contains a sandwich of two antiparallel β pleated sheets, one with four strands and the other with five, and a loop–sheet–helix motif that packs against one edge of the β sandwich. This edge of the β sandwich also contains two large loops running between the two β sheets that are held together through their tetrahedral coordination of a Zn^{2+} ion via one His and three Cys side chains.

The p53 DNA binding motif does not resemble any other that has previously been characterized. The helix and loop from the loop–sheet–helix motif are inserted in the DNA's major groove where they make sequence-specific contacts with the bases (lower right of Fig. 33-87). One of the large loops provides a side chain (Arg 248) that fits in the minor groove (upper right of Fig. 33-87). The protein also contacts the DNA backbone between the major and minor grooves in this region (notably with Arg 273).

The structure's most striking feature is that *its DNA-binding motif consists of conserved regions comprised of the most frequently mutated residues in the p53 variants found in tumors.* Among them are one Gly and five Arg residues (highlighted in yellow in Fig. 33-87) whose mutations collectively account for over 40% of the *p53* variants in

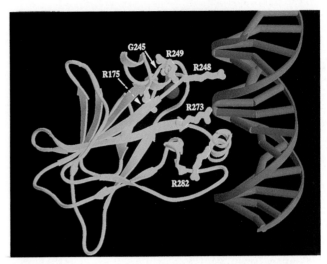

FIGURE 33-87. The X-ray structure of the DNA-binding domain of human p53 in complex with its target DNA. The protein is shown in ribbon form (*light blue*), the DNA in ladder form with its bases represented by cylinders (*dark blue*), the tetrahedrally liganded Zn^{2+} ion is shown as a red sphere, and the side chains of the six most frequently mutated side chains in human tumors are shown in stick form (*yellow*) and identified with their one-letter codes. [Courtesy of Nikola Pavletich, Memorial Sloan-Kettering Cancer Center, New York.]

tumors. The two most frequently mutated residues, Arg 248 and Arg 273, as we saw, directly contact the DNA. The other four "mutational hotspot" residues appear to play a critical role in structurally stabilizing p53's DNA-binding surface. The relatively sparse secondary structure in the polypeptide segments forming this surface (one helix and three loops) accounts for this high mutational sensitivity: Its structural integrity mostly relies on specific side chain–side chain and side chain--backbone interactions.

CHAPTER SUMMARY

Eukaryotic chromatin consists of DNA, RNA, and proteins, the majority of which are the highly conserved histones. Chromatin is structurally organized in a hierarchial manner. In the first level of chromatin organization, ~200 bp of DNA are doubly wrapped around a histone octamer, $(H2A)_2(H2B)_2(H3)_2(H4)_2$, to form a nucleosome. Each nucleosome is associated with one molecule of histone H1. The passage of transcribing RNA polymerase causes nucleosomes to dissociate from the DNA and then to rebind it in a process that appears to be driven by supercoiling. DNA replication causes the parental nucleosomes to be randomly distributed between the daughter duplexes. The assembly of nucleosomes from their components is mediated by the molecular chaperone nucleoplasmin. In the second level of chromatin organization, the nucleosome filaments coil into 300-Å-thick filaments that probably contain six nucleosomes per turn. Then, in the third and final level

of chromatin organization, the 300-Å-thick filaments form 15- to 30-μm-long radial loops that project from the axis of the metaphase chromosome. This accounts for DNA's packing ratio of <8000 in the metaphase chromosome. The larvae of certain dipteran flies, including *Drosophila,* contain banded polytene chromosomes, which consist of up to 1024 identical DNA strands in parallel register. The bands, as shown by *in situ* hybridization, correspond to the chromosomes' transcription units.

The complexity of a DNA sample can be determined from its renaturation rate through C_0t curve analysis. Eukaryotic DNAs have complex C_0t curves that arise from the presence of unique, moderately repetitive and highly repetitive sequences, as well as from inverted repeats. The function of inverted repeats, which form foldback structures, is unknown. Highly repetitive sequences, which occur in the heterochromatic regions near the chromosomal centromeres, probably function to align homologous chromosomes during meiosis and/or to facilitate their recombination. Moderately repetitive DNAs, for the most part, have unknown functions; many of them may simply be selfish DNA. The genes specifying rRNAs and tRNAs are organized into tandemly repeated clusters. The rDNA condenses to form nucleoli, the sites of rRNA transcription by RNA polymerase I and of partial ribosomal assembly. The 5S RNA and tRNAs are transcribed outside the nucleoli by RNA polymerase III. The genes specifying histones, which are required in large quantities only during S phase of the cell cycle, are the only repeated protein-encoding genes. The identity of a series of repeated genes is probably maintained through unequal crossing over and/or gene conversion. Many families of genes specifying related proteins are clustered into gene families. In mammals, the gene clusters encoding the α- and β-like hemoglobin subunits occur on separate chromosomes. Nevertheless, all vertebrate globin genes have the same exon–intron structure: three exons separated by two introns. The genes encoding modern proteins appear to have arisen through the recombinational assembly of exons so that modern introns are the remnants of a process that facilitated protein evolution. This can be seen in the structure of chicken muscle pyruvate kinase, whose 10 exons each encode discrete elements of protein secondary structure, as well as in the structure of hemoglobin. Likewise, many of the polypeptide segments encoded by the 18 exons of the LDL receptor gene are homologous to segments in several other proteins. The scarcity of introns in the genes of lower organisms probably reflects the greater selective advantage that metabolic efficiency confers on them relative to higher organisms. In higher organisms, introns may function to protect the exons of gene families from elimination through unequal crossing over. The thalassemias are inherited diseases caused by the genetic impairment of hemoglobin synthesis. Most α-thalassemias are caused by the deletion of one or more of the α-globin genes, whereas most β-thalassemias arise from point mutations that affect the transcription or the posttranscriptional processing of the β-globin mRNAs.

Heterochromatin may be subclassified as constitutive heterochromatin, which is never transcriptionally active, and facultative heterochromatin, whose activity varies in a tissue-specific manner. The Barr bodies in the cells of female mammals constitute a common form of facultative heterochromatin: One of each cell's two X chromosomes is permanently condensed and confers its state of inactivity on its progeny, possibly through maintenance methylation. Active chromatin, in contrast, has a relatively open structure that makes it available to the transcriptional machinery. Two well-characterized examples of transcriptionally active chromatin are the chromosome puffs that emanate from single bands in polytene chromosomes, and the lampbrush chromosomes of amphibian oocytes. The differential protein synthesis characteristic of the cells in a multicellular organism largely stems from the selective transcription of the expressed genes. Transcriptionally active chromatin is more susceptible to nuclease digestion than is nontranscribing chromatin. This nuclease sensitivity appears to be promoted by the presence of nonhistone proteins such as HMGs 14 and 17, although these proteins are not tissue specific. RNA polymerase transcribes through nucleosomes by inducing the histone octamers it encounters to step around it. Active genes have nuclease-hypersensitive sites that occur in nucleosome-free regions of DNA. Nuclease hypersensitivity is conferred on DNA by the binding of specific proteins that presumably make the genes accessible to the proteins mediating transcriptional initiation. The cell-specific expression of genes is mediated by the genes' promoter and enhancer elements. Consequently, cells must contain specific transcription factors that recognize these genetic elements. For example, Sp1 binds to the GC box that precedes many genes. Likewise, steroid hormones bind to their cognate receptors, which in turn, bind to specific enhancers so as to modulate their transcriptional activities. The first step in the transcriptional initiation of RNAPII-transcribed genes is the the binding TBP to the promoter's TATA box, followed by the sequential addition of TAFs and GTFs, together with RNAPII to form the PIC, which is capable of a basal rate of transcription. TBP, together with other GTFs, is also required for the transcriptional initiation of class I and III genes. The cooperative binding of several transcriptional factors to their target promoter and enhancer sites stimulates an associated PIC to increase its rate of transcriptional initiation. Many transcription factors have two domains, a DNA-binding domain targeted to a specific sequence, and an activation domain, which interacts with the PIC in a largely nonspecific manner via a negatively charged surface region. Eukaryotic transcription factors have a great variety of DNA-binding motifs, including several types of zinc fingers, the b/ZIP motif, and the bHLH/Z motif. Many transcription factors, including those with the latter two types of motif, dimerize through the formation of a leucine zipper. Other forms of selective gene expression involve the use of alternative initiation sites in a single gene, the selection of alternative splice sites, the possible regulation of mRNA translocation across the nuclear membrane, the control of mRNA degradation, the control of translational initiation rates, and the selection of alternative posttranslational processing pathways.

Embryogenesis occurs in four stages: cleavage, gastrulation, organogenesis, and maturation and growth. One of the most striking characteristics of embryological development is that cells become progressively and irreversibly committed to specific lines of development. The signals that trigger developmental changes, which are recognized over great evolutionary distances, may be transmitted through direct intercellular contacts or from the gradients of substance, known as morophogens, released by other embryonic cells. Developmental signals act combinatorially; that is, the developmental fate of a specific tissue is determined by several not necessarily unique developmental stimuli. In *Drosophila*, early embryonic development is governed by maternal-effect genes whose distribution imposes the embryo's spatial coordinate system. These encode transcription factors that regulate the expression of gap genes, which in turn regulate the expression of pair-rule genes, which in turn regulate the expression of segment polarity genes. Sequentially finer domains of the the embronic body are thereby defined in a way that specifies the number and polarity of the larval and adult body segments. Homeotic selector or *Hox* genes, whose mutations transform one body part into another, then regulate the differentiation of the individual segments. These regulatory genes, which occur in two gene clusters, are, as the preceding genes, selectively expressed in the embryonic tissues whose development they control. They have closely related base sequences that encode ~60-residue polypeptide segments known as homeodomains, which bind their target DNA sequences in a manner similar to but distinct from that of the homologous HTH module. In vertebrates, *Hox* genes occur in four clusters and likewise control development. Cancer cells differ in a variety of structural, functional, and metabolic ways from the normal cells from which they are derived. Their medically most significant properties, their immortality and their uncontrolled proliferation, endows them with the capacity to form invasive and metastatic tumors. Malignant tumors are caused by agents that alter DNA sequences: carcinogens, radiation, and viruses. However, several such changes must occur before a cell becomes transformed. Rous sarcoma virus, a retrovirus causing sarcomas in chickens, carries an oncogene, v-*src*, whose protein product, pp60$^{v\text{-}src}$, mediates the transformation of the host cell. Uninfected chicken cells contain a gene, c-*src*, that is homologous to v-*src*. Both genes encode tyrosine-specific protein kinases whose action is thought to stimulate cell division. The ~40 viral and cellular oncogenes that have been characterized stimulate cell division by mimicking the effects of growth factors and certain hormones. The oncogene products include analogs of growth factors, growth factor receptors, nuclear proteins that stimulate transcription and/or cell division, and G-proteins. The types of genetic changes that distinguish oncogenes from their cellular homologs are point mutations, chromosomal rearrangements that bring oncogenes under the influence of inappropriate regulatory sequences, oncogene amplification, and the insertion of a viral genome in a position that places the cellular oncogene under the transcriptional control of viral regulatory sequences. The loss or inactivation of tumor supressor genes may also cause cancer.

REFERENCES

General

Alberts, B., Bray, D., Lewis, J., Raff, M., Roberts, K., and Watson, J.D., *Molecular Biology of the Cell* (3rd ed.), Chapters 8, 9, 17, 21, and 24, Garland Publishing (1994).

Darnell, J., Lodish, H., and Baltimore, D., *Molecular Cell Biology* (2nd ed.), Chapters 8–11 and 24, Scientific American Books (1990).

De Pomerai, D., *From Gene to Animal,* Cambridge University Press (1985).

Felsenfeld, G., DNA, *Sci. Am.* **253**(4): 58–67 (1985).

Lewin, B., *Genes V,* Chapters 22–24, 26–30, 38, and 39, Oxford (1994).

Watson, J.D., Hopkins, N.H., Roberts, J.W., Steitz, J.A., and Weiner, A.M., *Molecular Biology of the Gene* (4th ed.), Chapters 20–22 and 25–27, Benjamin/Cummings (1987).

Chromosome Structure

Adolph, K.W., (Ed.), *Chromosomes and Chromatin,* Vols. I–III, CRC Press (1988).

Arents, G., Burlingame, R.W., Wang, B.-C., Love, W.E., and Moudrianakis, E.N., The nucleosomal core histone octamer at 3.1 Å resolution: A tripartite protein assembly and a left-handed superhelix, *Proc. Natl. Acad. Sci.* **88**, 10148–10152 (1991); *and* Arents, G. and Moudrianakis, E.N., Topography of the histone octamer surface: Repeating structural motifs utilized in the docking of nucleosomal DNA, *Proc. Natl. Acad. Sci.* **90**, 10489–10493 (1993).

Berry, M., Grosveld, F., and Dillon, N., A single point mutation is the cause of the Greek form of hereditary persistance of fetal hemoglobin, *Nature* **358**, 499–502 (1992).

Bradbury, E.M., Reversible histone modifications and the cell cycle, *BioEssays* **14**, 9–16 (1992).

Earnshaw, W.C., Large scale chromosome structure and organization, *Curr. Opin. Struct. Biol.* **1**, 237–244 (1991).

Felsenfeld, G. and McGhee, J.D., Structure of the 30 nm chromatin fiber, *Cell* **44**, 375–377 (1986).

Gruss, C. and Sogo, J.M., Chromatin replication, *BioEssays* **14**, 1–8 (1992).

Kornberg, R.D., Chromatin structure: a repeating unit of histones and DNA, *Science* **184**, 868–871 (1974). [The classic paper first indicating the constitution of nucleosomes.]

Kornberg, R.D. and Klug, A., The nucleosome, *Sci. Am.* **244**(2): 52–64 (1981).

Kornberg, R.D. and Lorch, Y., Chromatin structure and transcription, *Annu. Rev. Cell Biol.* **8**, 563–587 (1992).

Paranjape, S.M., Kamakaka, R.T., and Kadonaga, J.T., Role of chromatin structure in the regulation of transcription by RNA polymerase II, *Annu. Rev. Biochem.* **63**, 265–297 (1994).

Pienta, K.J., Getzenberg, R.H., and Coffey, D.S., Cell structure and DNA organization, *CR Eukar. Gene Express* **1**, 355–385 (1991).

Ramakrishnan, V., Histone structure, *Curr. Opin. Struct. Biol.* **4**, 44–50 (1994).

Ramakrishnan, V., Finch, J.T., Graziano, V., Lee, P.L., and Sweet, R.M., Crystal structure of globular domain of histone H5 and its implications for nucleosome binding, *Nature* **362**, 219–223 (1993).

Richmond, T.J., Finch, J.T., Rushton, B., Rhodes, D., and Klug, A., Structure of the nucleosome core particle at 7 Å resolution, *Nature* **311**, 532–537 (1984).

Turner, B.M., Decoding the nucleosome, *Cell* **75**, 5–8 (1993). [Discusses the functions of histone modifications.]

Uberbacher, E.C. and Bunick, G.J., Structure of the nucleosome core particle at 8 Å resolution, *J. Biomol. Struct. Dynam.* **7**, 1–18 (1989).

van Holde, K. E., *Chromatin,* Springer–Verlag (1989).

Widom, J., Toward a unified model of chromatin folding, *Annu. Rev. Biophys. Biophys. Chem.* **18**, 365–395 (1989).

Widom, J. and Klug, A., Structure of the 300 Å chromatin filament: X-ray diffraction from oriented samples, *Cell* **43**, 207–213 (1985).

Wu, R.S., Panusz, H.T., Hatch, C.L., and Bonner, W.M., Histones and their modifications, *CRC Crit. Rev. Biochem.* **20**, 201–263 (1986).

Genomic Organization

Amy, C.M., Williams-Ahlf, B., Naggert, J., and Smith, S., Intron-exon organization of the gene for the multifunctional animal fatty acid synthase, *Proc. Natl. Acad. Sci.* **89**, 1105–1108 (1992).

Breathnach, R. and Chambon, P., Organization and expression of eucaryotic split genes coding for proteins, *Annu. Rev. Biochem.* **50,** 349–383 (1981).

Britten, R.J. and Kohne, D., Repeated sequences in DNA, *Science,* **161,** 529–540 (1968).

Brutlag, D.L., Molecular arrangement and evolution of heterochromatic DNA, *Annu. Rev. Genet.* **14,** 121–144 (1980). [Deals with highly repetitive sequences.]

Deininger, P.L., Batzer, M.A., Hutchinson, C.A., III, and Edgell, M.H., Master genes in mammalian repetitive DNA amplification, *Trends Genet.* **8,** 307–311 (1992).

Doolittle, R.F., The genealogy of some recently evolved vertebrate proteins, *Trends Biochem. Sci.* **10,** 233–237 (1985).

Doolittle, W.F. and Sapienza, C., Selfish genes, the phenotype paradigm and genome evolution, *Nature* **284,** 601–603 (1980).

Dorit, R.L., Schoenbach, L., and Gilbert, W., How big is the universe of exons? *Science* **250,** 1377–1382 (1990); *and* Dorit, R.L. and Gilbert, W., The limited universe of exons, *Curr. Opin. Struct. Biol.* **1,** 973–977 (1991).

Gilbert, G., Marchionni, M., and McKnight, G., On the antiquity of introns, *Cell* **46,** 151–154 (1986).

Hamlin, J.L., Leu, T.-H., Vaughn, J.P., Ma, C., and Dijkwel, P.A., Amplification of DNA sequences in mammalian cells, *Prog. Nucleic Acid Res. Mol. Biol.* **41,** 203–239 (1991).

Jelinek, W.R. and Schmid, C.W., Repetitive sequences in eukaryotic DNA and their expression, *Annu. Rev. Biochem.* **51,** 813–844 (1982). [Concentrates on moderately repetitive sequences.]

Johnson, P.F. and McKnight, S.L., Eukaryotic transcriptional regulatory proteins, *Annu. Rev. Biochem.* **58,** 799–839 (1989).

Kafatos, F.C., Orr, W., and Delidakis, C., Developmentally regulated gene amplification, *Trends Genet.* **1,** 301–306 (1985).

Lonberg, N. and Gilbert, W., Intron/exon structure of the chicken pyruvate kinase gene, *Cell* **40,** 81–90 (1985).

Long, E.O. and Dawid, I.B., Repeated genes in eukaryotes, *Annu. Rev. Biochem.* **49,** 727–764 (1980). [Discusses structural genes that occur in multiple copies.]

Mandal, R.K., The organization and transcription of eukaryotic ribosomal RNA genes, *Prog. Nucleic Acid Res. Mol. Biol.* **31,** 115–160 (1984).

Maxson, R., Cohn, R., and Kedes, L., Expression and organization of histone genes. *Annu. Rev. Genet.* **17,** 239–277 (1983).

Orgel, L.E. and Crick, F.H.C., Selfish DNA: the ultimate parasite, *Nature* **284,** 604–607 (1980).

Orkin, S.H. and Kazazian, H.H., Jr., The mutation and polymorphism of the human β-globin gene and its surrounding DNA, *Annu. Rev. Genet.* **18,** 131–171 (1984).

Orr-Weaver, T.L., *Drosophila* chorion genes: Cracking the eggshell's secrets, *BioEssays* **13,** 97–105 (1991).

Schimke, R.T., Gene amplification in cultured cells, *J. Biol. Chem.* **263,** 5989–5992 (1988).

Stamatoyannopoulos, G. and Nienhuis, A.W., Therapeutic approaches to hemoglobin switching in treatment of hemoglobinapathies, *Annu. Rev. Med.* **43,** 497–521 (1992); *and* Stamatoyan-

nopoulos, G., Nienhuis, A.W., Leder, P., and Majerus, P.W. (Eds.), *The Molecular Basis of Blood Diseases,* Chapters 2–4, Saunders (1987). [Discusses hemoglobin genes and their normal and thalassemic expression.]

Stark, G.R. and Wahl, G.M., Gene amplification, *Annu. Rev. Biochem.* **53,** 447–491 (1984).

Südhof, T.C., Goldstein, J.L., Brown, M.S., and Russell, D.W., The LDL receptor gene: a mosaic of exons shared with different proteins, *Science* **228,** 815–828 (1985).

Ulla, E., The human *Alu* family of repeated DNA sequences, *Trends Biochem. Sci.* **7,** 216–219 (1982).

Wainscoat, J.S., Hill, A.V.S., Boyce, A.L., Flint, J., Hernandez, M., Thein, S.L., Old, J.M., Lynch, J.R., Falusi, A.G., Weatherall, D.J., and Clegg, J.B., Evolutionary relationships of human populations from an analysis of nuclear DNA polymorphisms, *Nature* **319,** 491–493 (1986).

Weatherall, D.J., Clegg, J.B., Higgs, D.R., and Wood, W.G., The hemoglobinopathies, *in* Scriver, C.R., Beaudet, A.L., Sly, W.S., and Valle, D. (Eds.), *The Metabolic Basis of Inherited Disease* (6th ed.), *pp.* 2281–2339, McGraw–Hill (1989). [Contains a section on the thalassemias.]

Control of Expression

Adams, C.C. and Workman, J.L., Nucleosome displacement in transcription, *Cell* **72,** 305–308 (1993).

Andres, A.J. and Thummel, C.S., Hormones, puffs and flies: the molecular control of metamorphosis by ecdysone, *Trends Genet.* **8,** 132–138 (1992).

Ashburner, M., Puffs, genes, and hormones revisited, *Cell* **61,** 1–3 (1990).

Atchison, M.L., Enhancers: mechanisms of action and cell specificity, *Annu. Rev. Cell Biol.* **4,** 127–153 (1988).

Breitbart, R.E., Andreadis, A., and Nadal-Ginard, B., Alternative splicing: a ubiquitous mechanism for the generation of multiple protein isoforms from single genes, *Annu. Rev. Biochem.* **56,** 467–495 (1987).

Buratowski, S., The basics of basal transcription by RNA polymerase II, *Cell* **77,** 1–3 (1994).

Bustin, M., Lehn, D.A., and Landsman, D., Structural features of the HMG chromosomal proteins and their genes, *Biochim. Biophys. Acta* **1049,** 231–243 (1990).

Chasman, D.I., Flaherty, K.M., Sharp, P.A., and Kornberg, R.D., Crystal structure of yeast TATA-binding protein and model for interaction with DNA, *Proc. Natl. Acad. Sci.* **90,** 8174–8178 (1993); *and* Nikolov, D.B., Hu, S.-H., Lin, J., Gasch, A., Hoffmann, A., Horikoshi, M., Chua, N.-H., Roeder, R.G., and Burley, S.K., Crystal structure of TFIID TATA-box binding protein, *Nature* **360,** 40–46 (1992).

Conway, R.C. and Conway, J.W., General initiation factors for RNA polymerase II, *Annu. Rev. Biochem.* **62,** 161–190 (1993).

Douglass, J., Civelli, O., and Herbert, E., Polyprotein gene expression, *Annu. Rev. Biochem.* **53,** 665–715 (1984).

Elgin, S.C.R., The formation and function of DNase I hypersensitive sites in the process of gene activation. *J. Biol. Chem.* **263**, 19259–19262 (1988).

Ellenberger, T.E., Getting a grip on DNA recognition: structures of the basic region leucine zipper, and the basic region helix-loop-helix DNA-binding domains, *Curr. Opin. Struct. Biol.* **4**, 12–21 (1994).

Ellenberger, T.E., Brandl, C.J., Struhl, K., and Harrison, S.C., The GCN4 basic region leucine zipper binds DNA as a dimer of uninterrupted α helices: Crystal structure of the protein–DNA complex, *Cell* **71**, 1223–1237 (1992).

Evans, R.M., The steroid and thyroid hormone receptor superfamily, *Science* **240**, 889–895 (1988).

Evans, T., Felsenfeld, G., and Reitman, M., Control of globin gene transcription, *Annu. Rev. Cell Biol.* **6**, 95–124 (1990).

Felsenfeld, G., Chromatin as an essential part of the transcriptional mechanism, *Nature* **355**, 219–224 (1992).

Ferré-d'Amaré, A.R., Prendergast, G.C., Ziff, E.B., and Burley, S.K., Recognition by Max of its cognate DNA through a dimeric b/HLH/Z domain, *Nature* **363**, 38–45 (1993).

Frohman, M.A. and Martin, G.R., Cut, paste, and save: new approaches to alternating specific genes in mice, *Cell* **56**, 145–147 (1989).

Gross, D.S. and Garrard, W.T., Nuclease hypersensitive sites in chromatin, *Annu. Rev. Biochem.* **57**, 159–197 (1988).

Heintz, N. (Ed.), Transcription – A special issue, *Trends Biochem Sci.* **16**, 393–447 (1991). [Contains numerous articles on transcriptional regulation.]

Hurst, H.C., Transcription factors 1: bZIP proteins, *Protein Profiles* **1**, 123–168 (1994).

Jeang, K.-T. and Khoury, G., The mechanistic role of enhancer elements in eukaryotic transcription, *BioEssays* **8**, 104–107 (1988).

Kim, Y., Geiger, J.H., Hahn, S., and Sigler, P.B., Crystal structure of a yeast TBP/TATA-box complex; Kim, J.L., Nikolov, D.B., and Burley, S.K., Co-crystal structure of TBP recognizing the minor groove of a TATA element, *Nature* **365**, 512–520 *and* 520–527 (1993); *and* Nikolov, D.B. and Burley, S.K., 2.1 Å resolution refined structure of TATA box-binding protein (TBP), *Nature Struct. Biol.* **1**, 621–637 (1994).

Klug, A. and Rhodes, D., 'Zinc fingers': a novel protein motif for nucleic acid recognition, *Trends Biochem. Sci.* **12**, 464–469 (1987).

Lamond, A.I., ASF/SF2: a splice site selector, *Trends Biochem. Sci.* **16**, 452–453 (1991).

Landschultz, W.H., Johnson, P.F., and McKnight, S.L., The leucine zipper: a hypothetical structure common to a new class of DNA binding proteins, *Science* **240**, 1759–1764 (1988).

Luisi, B.F., Xu, W.X., Otwinowski, Z., Freedamn, L.P., Yamamoto, K.R., and Sigler, P.B., Crystallographic analysis of the interaction of the glucocorticoid receptor with DNA, *Nature* **352**, 497–505 (1991).

Marmorstein, R., Carey, M., Ptashne, M., and Harrison, S.C., DNA recognition by GAL4: structure of a protein–DNA complex, *Nature* **356**, 408–414 (1992).

Martin, G.M., X-Chromosome inactivation in mammals, *Cell* **29**, 721–724 (1982).

McKnight, S.L. and Yamamoto, K.R. (Eds.), *Transcriptional Regulation,* Cold Spring Harbor Laboratory Press (1992). [A two-volume compendium that contains authoritative articles on many aspects of eukaryotic transcriptional control.]

O'Shea, E.K., Klemm, J.D., Kim, P.S., and Alber, T., X-ray structure of the GCN4 leucine zipper, a two-stranded, parallel coiled coil, *Science* **254**, 539–544 (1991).

Pavletich, N.P. and Pabo, C.O., Zinc finger–DNA recognition: Crystal structure of a Zif268-DNA complex at 2.1 Å, *Science* **252**, 809–817 (1991).

Pelham, H., Activation of heat-shock genes in eukaryotes, *Trends Genet.* **1**, 31–35 (1985).

Pelz, S.W., Brewer, G., Bernstein, P., Hart, P.A., and Ross, J., Regulation of mRNA turnover in eukaryotic cells, *CR Eukar. Gene Express.* **1**, 99–126 (1991); *and* Atwater, J.A., Wisdom, R., and Verma, I.M., Regulated mRNA stability, *Annu. Rev. Genet.* **24**, 519–541 (1990).

Ptashne, M., *A Genetic Switch* (2nd ed.), Cell Press & Blackwell Scientific Publications (1992). [Chapters 5 and 6 discuss eukaryotic gene regulation.]

Ptashne, M., How gene activators work, *Sci. Am.* **260**(1): 41–47 (1989); *and* How eukaryotic transcriptional activators work, *Nature* **335**, 683–689 (1988).

Raghow, R., Regulation of messenger RNA turnover in eukaryotes, *Trends Biochem. Sci.* **12**, 358–360 (1987).

Riggs, A.D. and Pfeifer, G.P., X-chromosome inactivation and cell memory, *Trends Genet.* **8**, 169–174 (1992).

Schmiedeskamp, M. and Klevit, R.E., Zinc finger diversity, *Curr. Opin. Struct. Biol.* **4**, 28–35 (1994).

Schröder, H.C., Bachmann, M., Diehl-Siefert, B., and Müller, W.E.G., Transport of mRNA from nucleus to cytoplasm, *Prog. Nucleic Acid Res. Mol. Biol.* **34**, 89–142 (1987).

Schwabe, J.W.R., Chapman, L., Finch, J.T., and Rhodes, D., The crystal structure of the estrogen receptor DNA-binding domain bound to DNA: How receptors discriminate between their response elements, *Cell* **75**, 567–578 (1993).

Stamatoyannopoulos, G. and Nienhuis, A.W. (Eds.), *The Regulation of Hemoglobin Switching,* The Johns Hopkins University Press (1991).

Stolzfus, A., Spencer, D.F., Zuker, M., Logsdon, J.M., Jr., and Doolittle, W.F., Testing the exon theory of genes: The evidence from protein structure, *Science* **265**, 202–207 (1994). [An analysis of the boundary positions of exons relative to those of secondary structural elements.]

Struhl, K., Duality of TBP, the universal transcription factor, *Science* **263**, 1103–1104 (1994); Rigby, P. W. J., Three in one and one in three: It all depends on TBP, *Cell* **72**, 7–10 (1993); *and* White, R.J. and Jackson, S.P., The TATA-binding protein: a central role in transcription by RNA polymerases I, II, and III, *Trends Genet.* **8**, 284–288 (1992).

Studitsky, V.M., Clark, D.J., and Felsenfeld, G., A histone octamer can step around a transcribing polymerase without leaving the template, *Cell* **76**, 371–382 (1994); *and* Clark, D.J. and Fel-

TABLE 34-1. HUMAN BLOOD COAGULATION FACTORS

Factor Number	Common Name	Molecular Mass (kD)
I	Fibrinogen	330
II	Prothrombin	66
III	Tissue factor *or* thromboplastin	30
IV	Ca^{2+}	
V$_a$	Proaccelerin	249
VII	Proconvertin	46
VIII	Antihemophilic factor	265
IX	Christmas factor	47
X	Stuart factor	50
XI	Plasma thromboplastin antecedent (PTA)	136
XII	Hageman factor	67
XIII	Fibrin-stabilizing factor (FSF)	301
	Prekallikrein	69
	High molecular weight kininogen (HMK)	70

^{a} Factor V$_a$ was once called factor VI; consequently there is no factor VI.

Source: Halkier, T., *Mechanisms in Blood Coagulation, Fibrinolysis, and the Complement System,* pp. 4–5, Cambridge University Press (1991).

many of its components are listed in Table 34-1. All but two of these factors are designated by both a roman numeral and a common name, although, unfortunately, the order of the roman numerals has historical rather than mechanistic significance. Seven of the clotting factors are zymogens (inactive forms) of serine proteases that are proteolytically activated by serine proteases further up the cascade. Other clotting proteins, termed **accessory factors,** which are also activated by these serine proteases, enhance the rate of activation of some of the zymogens. In both cases, the active form of a factor is designated by the subscript *a.*

The blood clotting system, which occurs in recognizable form in all vertebrates, contains a number of homologous serine proteases and therefore appears to have arisen through a series of gene duplications. The C-terminal ~250 residues of these proteases, which comprise their catalytically active domains, are also homologous to the pancreatic serine proteases trypsin, chymotrypsin, and elastase (Section 14-3). Like these digestive enzymes, the blood clotting proteases are activated by proteolytic cleavages that precede their C-terminal segments (Section 14-3E). However, the clotting proteases differ from the digestive enzymes in that the zymogen-to-protease conversion only takes place in the presence of Ca^{2+} and on an appropriate phospholipid membrane (see below); the resulting N-terminal fragments are quite large (150–582 residues) and, with the exception of prothrombin (Section 34-1B), are linked to their C-terminal segments via disulfide bonds so that these segments do not separate upon activation. These N-terminal segments are thought to be responsible, at least in part, for the exquisite specificities of the proteolytic blood clotting factors:

Their substrates are limited, as we shall see, to the few inactive factors they function to activate.

In what follows, we describe the mechanism of blood clotting in humans from the bottom up, that is, starting with the formation of the clot itself and working backwards through the sequence of activation steps leading to this process.

A. Fibrinogen and Its Conversion to Fibrin

*Blood clots consist of arrays of cross-linked **fibrin** that form an insoluble fibrous network (Fig. 34-2). Fibrin is made from the soluble plasma protein **fibrinogen (factor I)** through a proteolytic reaction catalyzed by the serine protease **thrombin.*** Fibrinogen comprises 2 to 3% of plasma protein. A molecule of fibrinogen consists of three pairs of nonidentical but homologous polypeptide chains, Aα (610 residues), Bβ (461 residues) and γ (411 residues), and two pairs of *N*-linked oligosaccharides of ~2.5 kD each. Here A and B represent the 16- and 14-residue N-terminal **fibrinopeptides** that thrombin cleaves from fibrinogen, so a fibrin monomer is designated $\alpha_2\beta_2\gamma_2$. The reaction forming a blood clot from fibrinogen may therefore be represented

$$n(A\alpha)_2(B\beta)_2\gamma_2 \xrightarrow{\;2nA + 2nB\;} n\alpha_2\beta_2\gamma_2 \longrightarrow (\alpha_2\beta_2\gamma_2)_n$$

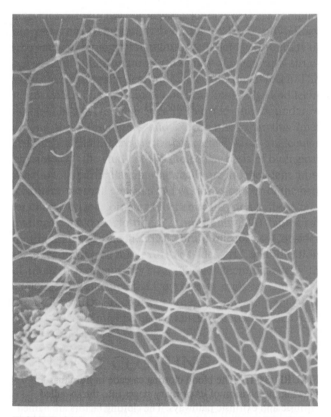

FIGURE 34-2. Scanning electron micrograph of a blood clot showing a red cell enmeshed in a fibrin network. [Manfred Kaga/Peter Arnold Inc.]

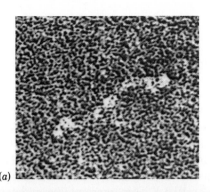

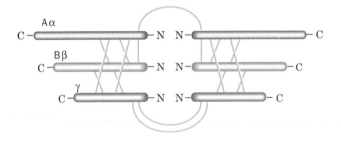

FIGURE 34-3. Fibrinogen. (*a*) An electron micrograph. (*b*) A 30-Å resolution model viewed along the molecule's twofold rotation axis. The light colored balls represent protuberances that are seen in combined electron microscopic and low resolution X-ray diffraction studies of fibrinogen. [Courtesy of John Weisel, University of Pennsylvania, and Carolyn Cohen, Brandeis University.]

FIGURE 34-4. A schematic diagram of the structure of fibrinogen, $(A\alpha)_2(B\beta)_2\gamma_2$, in which the interchain disulfide bonds are represented by yellow lines.

Combined electron microscopic and low-resolution X-ray crystallographic studies indicate that fibrinogen is a twofold symmetric elongated molecule, ~450 Å in length, that has two nodules at each end and one in the middle (Fig. 34-3). Its 6 polypeptide chains are joined by 17 disulfide bonds, 7 within each half of the dimer and 3 linking these two protomers (Fig. 34-4). Structural indications together with sequence-based conformation predictions suggest that the central region of each protomer consists mainly of a three-stranded coiled coil of α helices in which the α, β, and γ chains each contribute a strand (Fig. 34-5). The peripheral nodules diagrammed in Fig. 34-3 are formed by the C-terminal domains of the β and γ chains. The C-terminal segment of the α chain apparently lacks a definite conformation and is therefore not represented in Fig. 34-3.

How does fibrin polymerize to form a clot? Thrombin specifically cleaves the Arg—X peptide bond (where X is Gly in most species) joining each fibrinopeptide to fibrin. Fibrin then spontaneously aggregates to form fibers that electron micrographs indicate have a banded structure that repeats every 225 Å (Fig. 34-6). This repeat distance is exactly half the 450-Å length of a fibrin monomer, suggesting that fibrin monomers associate as a half-staggered array (Fig. 34-7). But why do fibrin monomers aggregate while fibrinogen, which has an all but identical structure, remains in solution? The main reason is that the loss of the fibrinopeptides exposes otherwise masked sites that mediate intermolecular association (Fig. 34-7). In addition, fibrinogen aggregation is inhibited by charge–charge repulsions: The fibrinopeptides are highly anionic, so much so that fibrinogen's central region, where the fibrinopeptides reside, has a charge of -8, whereas that of fibrin is $+5$. Fibrinogen's end segments each have a similar charge of -4 but maintain that charge in fibrin. The repulsions between fibrinogen's like-charged segments helps prevent this protein from ag-

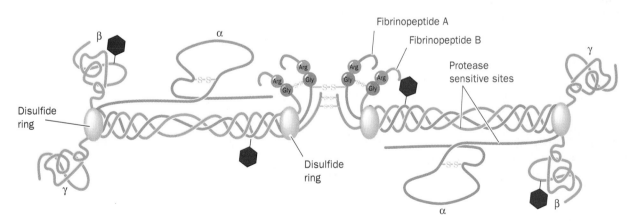

FIGURE 34-5. The proposed structure of fibrinogen based on low-resolution structural studies, primary structure determinations, and chain-folding predictions. The so-called disulfide rings are regions containing three disulfide bonds cyclically linking homologous segments of the α, β, and γ chains. *N*-linked polysaccharides are represented by filled hexagons. The Arg–Gly bonds that are cleaved by thrombin in fibrin activation are indicated. [After the cover illustration, *Annu. N.Y. Acad. Sci.* **408** (1983).]

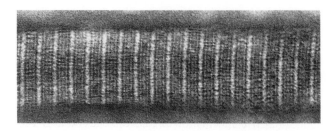

FIGURE 34-6. An electron micrograph of a fibrin fiber. The striations repeat every 225 Å, exactly half the length of a fibrin molecule. [Courtesy of John Weisel, University of Pennsylvania, and Carolyn Cohen, Brandeis University.]

gregating, whereas the attractions between fibrin's central and end segments promotes its specific association.

The diameters of fibrin fibers, which are fairly uniform (maximally ~50 nm), are important determinants of a clot's physical properties. What controls this fiber diameter? John Weisel and Lee Makowski have shown through electron microscopy studies that fibrin fibers are uniformly twisted. Consequently, the in-register molecules near the periphery of a twisted fiber must traverse a longer path than molecules near the fiber's center. The degree to which a molecule can stretch therefore limits the diameter of the fiber: Molecules add to the outside of a growing fiber until the energy to stretch an added molecule exceeds its energy of binding. A similar mechanism may limit the diameters of

other biological fibers such as those of collagen (Section 7-2C).

Fibrin-Stabilizing Factor Cross-Links Fibrin Clots

The above "soft clot," as this name implies, is rather fragile. It is rapidly converted to a more stable "hard clot," however, by the covalent cross-linking of neighboring fibrin molecules in a reaction catalyzed by **fibrin-stabilizing factor** (**FSF** or **XIII$_a$**). This transamidase initially joins the C-terminal segments of adjacent γ chains by forming isopeptide bonds between the side chains of a Gln residue on one γ chain and a Lys residue on another (Fig. 34-8). Two such symmetrically equivalent bonds are rapidly formed between each neighboring pair of γ chains (Fig. 34-7). The α chains are similarly cross-linked to one another but at a slower rate. The physiological importance of fibrin cross-linking is demonstrated by the observation that individuals deficient in FSF have a pronounced tendency to bleed.

FSF is present in both platelets and plasma. Platelet FSF consists of two 75-kD a chains, whereas plasma FSF additionally has two 88-kD b chains. Both species of FSF occur as zymogens that undergo thrombin-catalyzed cleavage of a specific Arg—Gly bond near the N-terminus of each a chain with the consequent release of a 37-residue propeptide. This treatment activates platelet FSF, designated a'_2, but plasma FSF, a'_2b_2 remains inactive until its b chains dissociate, a process that is triggered by the binding of Ca^{2+} to the a' subunits. We shall see below that *Ca^{2+} is an essential factor in most stages of the blood clotting cascade.* The b

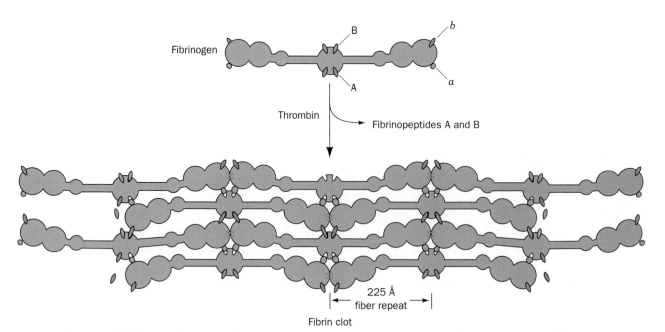

FIGURE 34-7. A model of the conversion of fibrinogen to a fibrin blood clot showing the half-staggered arrangement of its fibrin monomers. In "soft" clots, the fibrin monomers associate only by noncovalent interactions between knob*s* *a* and *b* that protrude from fibrin's γ chain and their complementary binding sites that are exposed by the excision of fibrinopeptides A and B. In "hard" clots, however, fibrin monomers are also covalently cross-linked by isopeptide bonds between the C-terminal segments of the γ chains in neighboring molecules (*red lines*) as well as between the protruding segments of α chains (*not shown*). [After Weisel, J.W., *Biophys. J.* **50,** 1080 (1986).]

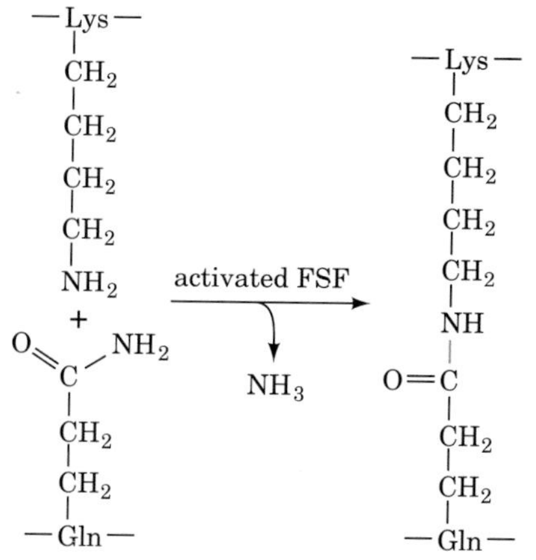

FIGURE 34-8. The transamidation reaction forming the isopeptide bonds cross-linking fibrin monomers in "hard" clots as catalyzed by activated fibrin-stabilizing factor (FSF, XIII$_a$).

subunits are thought to prolong the survival of plasma FSF in the circulation.

B. Thrombin Activation and the Function of Vitamin K

Thrombin is a serine protease that consists of two disulfide-linked polypeptide chains: in humans, a 36-residue A chain and a 259-residue B chain. The thrombin B chain is homologous to trypsin and has similar specificity but is far more selective: It cleaves only certain Arg—X and, less frequently, Lys—X bonds with a clear preference for a Pro preceeding the Arg or Lys.

Human thrombin is synthesized as a 579-residue zymogen, **prothrombin (II),** *which is activated by two proteolytic cleavages catalyzed by activated* **Stuart factor (X$_a$),** *the product of the preceding step of the clotting cascade.* The cleavage of prothrombin's Arg 271—Thr 272 and Arg 320—Ile 321 bonds releases its N-terminal propeptide and separates the A and B chains (Fig. 34-9). The latter cleavage, which yields active enzyme, results in the formation of an ion pair between Ile 321 and Asp 524, much like that formed between chymotrypsinogen's homologous

Ile 16 and Asp 194 in the activation of this zymogen (Section 14-3E). Thrombin then autolytically cleaves its Arg 285—Thr 286 bond, thereby trimming away the N-terminal 13 residues of the A chain to yield α-thrombin, hereinafter referred to as simply thrombin (recall that chymotrypsin also undergoes an autolytic cleavage that does not affect its catalytic activity; Section 14-3E).

Prothrombin's propeptide consists of three domains (Fig. 34-9; *left*): an N-terminal 40-residue Gla domain (so named for reasons indicated below) followed by two 40% identical ~115-residue **kringle** domains. Kringles are crosslinked by three characteristically located disulfide bonds that gives these triple-looped sequence motifs a folded appearance as drawn in Fig. 34-9 reminiscent of a Scandinavian pastry of the same name. Gla domains and kringles occur in several of the proteins involved in the formation and the breakdown of blood clots.

Vitamin K Is an Essential Cofactor in the Synthesis of γ-Carboxyglutamate

Prothrombin, as well as the homologous factors VII, IX, and X, is synthesized in the liver in a process that requires an adequate dietary intake of **vitamin K** *(Fig. 34-10; K for the Danish koagulation).* Lack of vitamin K or the presence of a competitive inhibitor such as **dicoumarol** (which was discovered in spoiled sweet clover because it causes fatal hemorrhaging in cattle) or **warfarin** (a rat poison) causes the production of an abnormal prothrombin that is activated by factor X$_a$ at only 1 to 2% of the normal rate. This observation was, at first, quite puzzling because normal and abnormal prothrombins seemed to have identical amino acid compositions. NMR studies eventually established, however, that normal prothrombin contains **γ-carboxyglutamate (Gla)** residues,

γ-**Carboxyglutamate (Gla)**

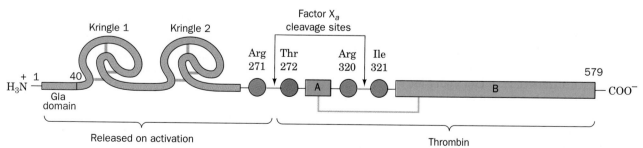

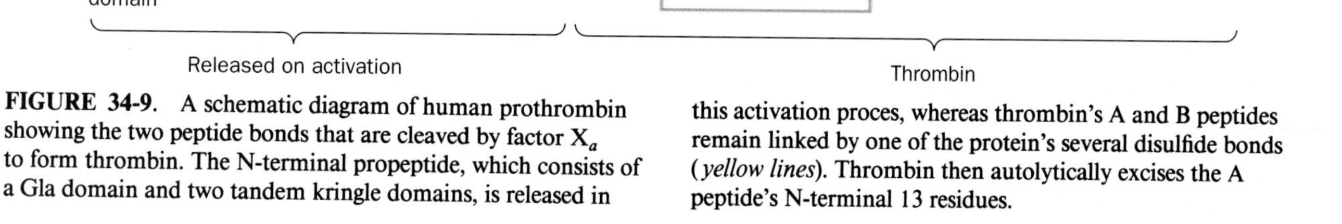

FIGURE 34-9. A schematic diagram of human prothrombin showing the two peptide bonds that are cleaved by factor X$_a$ to form thrombin. The N-terminal propeptide, which consists of a Gla domain and two tandem kringle domains, is released in this activation proces, whereas thrombin's A and B peptides remain linked by one of the protein's several disulfide bonds (*yellow lines*). Thrombin then autolytically excises the A peptide's N-terminal 13 residues.

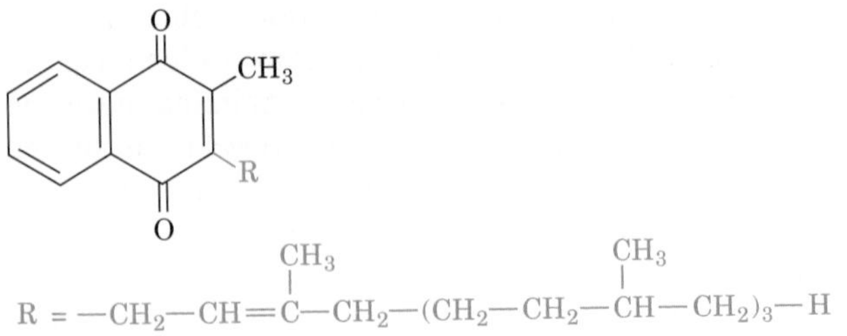

R = —CH₂—CH═C—CH₂—(CH₂—CH₂—CH—CH₂)₃—H

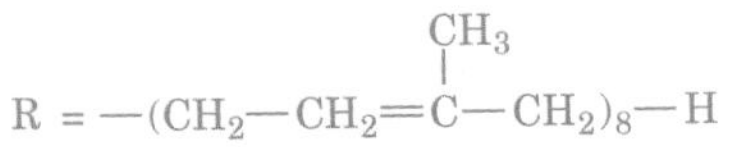

Vitamin K₁ (Phylloquinone)

R = —(CH₂—CH₂═C—CH₂)₈—H

Vitamin K₂ (Menaquinone)

R = —H

Vitamin K₃ (Menadione)

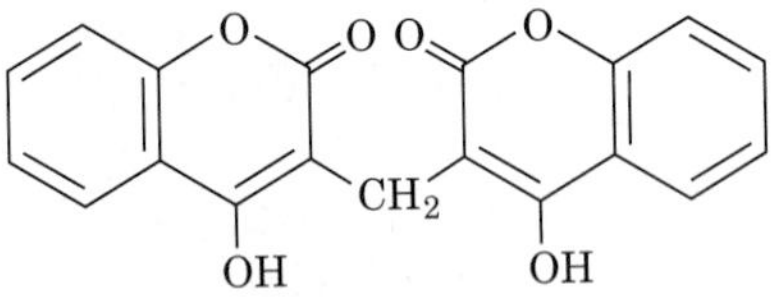

Dicoumarol

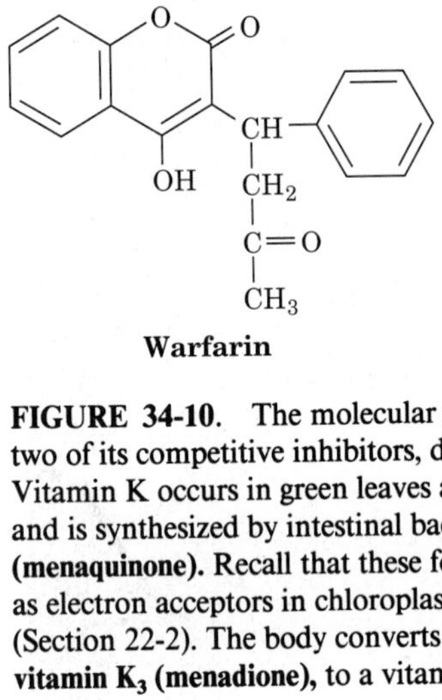

Warfarin

FIGURE 34-10. The molecular formulas of vitamin K and two of its competitive inhibitors, dicoumarol and warfarin. Vitamin K occurs in green leaves as **vitamin K₁ (phylloquinone)**, and is synthesized by intestinal bacteria as **vitamin K₂ (menaquinone)**. Recall that these forms of the vitamin function as electron acceptors in chloroplast and bacterial photosynthesis (Section 22-2). The body converts the parent compound, **vitamin K₃ (menadione)**, to a vitamin-active form.

10 of which occur between residues 6 and 32 in human prothrombin (in its Gla domain; Fig. 34-11). Abnormal prothrombin, in contrast, contains Glu in place of these Gla residues. *Vitamin K must therefore be a cofactor in the post-translational conversion of Glu to Gla.* The reason why prothrombin's Gla residues were not initially detected is because they decarboxylate to Glu under the conditions of acid hydrolysis normally used in amino acid composition determinations (Section 6-1D).

The liver reaction cycle that synthesizes Gla from Glu and then regenerates the vitamin K cofactor may be considered to occur in four reactions (Fig. 34-12):

1.   Vitamin K, in its active hydroquinone form, abstracts a γ proton from Glu in an O_2-consuming

reaction that yields the γ-carbanion of Glu and the 2,3-epoxide of vitamin K. The nature of the oxygenated vitamin K intermediate that abstracts Glu's γ proton has not been definitively established. It nevertheless appears that this intermediate has the high basicity required to abstract the proton that is lacking in the vitamin K forms shown in Fig. 34-12.

2.   The Glu carbanion then reacts with CO_2 to yield Gla.

3. & 4. Vitamin K hydroquinone is regenerated in two sequential reactions, both apparently catalyzed by the same enzyme, that employ thiols such as lipoic acid. Dicoumarol and warfarin act by blocking both these steps. Reaction 4 may also be catalyzed by certain NADH- or NADPH-dependent reductases.

The discovery of Gla residues in clotting factors led to their discovery in other tissues, which must therefore also contain vitamin K-dependent carboxylases. We shall see below that the Gla residues of clotting factors function to bind Ca^{2+}. Presumably they have similar roles in other tissues.

Prothrombin Activation Is Accelerated in the Presence of Factor V_a, Ca^{2+}, and Phospholipid

Factor X_a, by itself, is an extremely sluggish prothrombin activator. Yet, in the presence of activated **proaccelerin (V_a)**, Ca^{2+}, and phospholipid membrane, its activity is enhanced 20,000-fold. The membrane surface in contact with the activation complex must contain negatively charged phospholipids such as phosphatidylserine in order to stimulate this rate enhancement. Such phospholipids occur almost exclusively on the cytoplasmic sides of cell membranes (Section 11-3B) which, of course, are normally not in contact with the blood plasma. Moreover, ~20% of the total factor V in blood is stored in the platelets and released only upon platelet activation. Consequently, *physiological prothrombin activation normally takes place at a significant rate only in the vicinity of an injury.*

Ca^{2+} is required for either prothrombin or factor X_a to bind to phospholipid membranes; these proteins are anchored to the membrane via Ca^{2+} bridges. Prothrombin and factor X_a from vitamin K-deficient animals have greatly reduced membrane-binding affinities compared to the corresponding normal proteins. Evidently, *the Gla side chains, which are much stronger Ca^{2+} chelators than Glu, form the proteins' Ca^{2+}-binding sites.* In fact, the 9 to 12 conserved Gla residues that occur in each of the vitamin

H₃N⁺—¹Ala – Asn – Thr – Phe – Leu – Gla – Gla – Val – Arg – Lys –¹⁰
    ¹¹Gly – Asn – Leu – Gla – Arg – Gla – Cys – Val – Gla – Gla –²⁰
    ²¹Thr – Cys – Ser – Tyr – Gla – Gla – Ala – Phe – Gla – Ala –³⁰
    ³¹Leu – Gla – Ser – Ser – Thr – Ala – Thr – Asp – Val – Phe –⁴⁰

FIGURE 34-11. The sequence of the Gla domain of human prothrombin showing its 10 Gla residues.

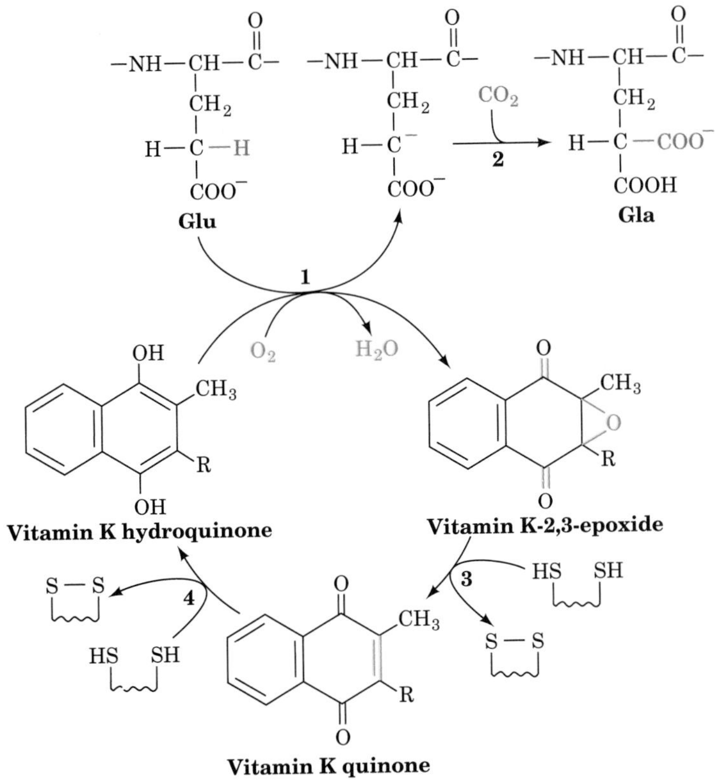

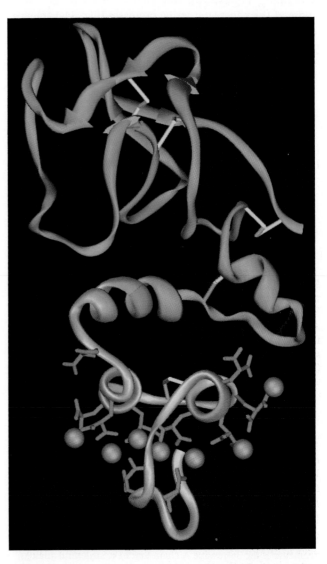

Glu ⟶ **Gla**

Vitamin K hydroquinone

Vitamin K-2,3-epoxide

Vitamin K quinone

FIGURE 34-12. The vitamin K metabolism cycle in liver. Reactions 3 and 4, which are both inhibited by dicoumarol and warfarin, are thought to be catalyzed by the same enzyme. The R group is indicated in Fig. 34-10. [After Suttie, J.W., *Annu. Rev. Biochem.* **54**, 472 (1985).]

K-dependent clotting zymogens — prothrombin and factors VII, IX, and X — are contained in these proteins' highly homologous N-terminal segments (Fig. 34-11 for prothrombin). The excision of prothrombin's N-terminal propeptide releases the resulting thrombin from the phospholipid membrane so that it can activate fibrinogen in the plasma. Thrombin differs in this respect from the other vitamin K–dependent zymogens, which remain bound to the phospholipid membrane after their activation.

Active thrombin specifically cleaves prothrombin's propeptide at its Arg 155–Ser 156 bond to yield its 155-residue N-terminal segment, the so called **fragment 1**, which consists of prothrombin's Gla domain and its kringle 1. The X-ray structure of bovine fragment 1 was determined by Alexander Tulinsky in both the presence and absence of Ca^{2+} ion. In the absence of Ca^{2+}, the Gla domain is disordered. However, when Ca^{2+} is present, the Gla domain folds to form two internal carboxylate surfaces that coordinate seven Ca^{2+} ions in a nearly linear array (Fig. 34-13). Kringle 1, which assumes the same conformation in the presence and absence of Ca^{2+}, folds into a compact globule that contains three short two-stranded antiparallel β sheets.

Activated proaccelerin (V_a), the accessory factor in prothrombin activation, is activated by a thrombin-catalyzed proteolytic cleavage. Prothrombin activation, in this indi-

rect way, is thereby autocatalytic (thrombin, *in vitro*, can also directly activate prothrombin by cleaving its Arg 283–Thr 284 bond but this reaction has been shown to be physiologically insignificant). V_a, however, is subject to further thrombin-catalyzed proteolysis, which inactivates it. Moreover, thrombin can proteolytically inactivate other thrombin molecules. *Clot formation is therefore self-limiting, a safeguard that helps prevent blood clots from propagating away from the site of an injury.*

Thrombin Structrally Resembles Trypsin

The X-ray structure of human thrombin as inactivated by **D-Phe-Pro-Arg chloromethylketone [PPACK;** which specifically alkylates thrombin's active site His (Section 14-

FIGURE 34-13. The X-ray structure of bovine fragment 1 in complex with Ca^{2+}. The protein is shown in ribbon form with its Gla domain gold and its kringle 1 green. The protein's 10 Gla side chains (*red*) and 5 disulfide linkages (*yellow*) are drawn in stick form and its 7 bound Ca^{2+} ions are represented by light blue spheres. [Based on an X-ray structure by Alexander Tulinsky, Michigan State University.]

3A) as well as forming a hemiketal (a tetrahedral interme- diate analog) with its active site Ser], was independently determined by Alexander Tulinsky and by Wolfram Bode and Robert Huber (Fig. 34-14). The A chain and B chains, which are linked only by a disulfide bond (Fig. 34-9), are not organized into separate domains but, rather, form a nearly spherical molecule. The boomerang-shaped A chain, which is analogous to the propeptide of chymotrypsinogen (Section 14-3E), is nestled against the B chain globule op- posite the substrate binding cleft, not in close association with it as had been predicted.

The structure of the B chain closely resembles that of the the pancreatic serine proteases (compare Figs. 14-20 and 34-14a) as had been expected from the high degree of se- quence homology among these various proteins. However, thrombin's substrate-binding cleft is much deeper than those of the pancreatic serine proteases (Fig. 34-14b) as a consequence of several elongated and exposed loops that are located around thrombin's substrate-binding cleft and which are not present in the pancreatic enzymes. Steric hindrance by these loops greatly restricts access to the active

site and presumably contributes to thrombin's high speci- ficity and its poor binding of most natural serine protease inhibitors. Indeed, when a model of bovine pancreatic tryp- sin inhibitor (BPTI) is docked to that of thrombin in the same position and orientation that BPTI assumes in its complex with trypsin (Fig. 14-24), there is a significant steric clash between BPTI and one of the loops. Neverthe- less, the specificity of thrombin for fibrinogen is largely attributable to its so called anion-binding exosite, an exten- sion of thrombin's substrate-binding cleft (on the right in Fig. 34-14), which is lined with positively charged side chains and which binds the highly anionic fibrinopeptides.

C. The Intrinsic Pathway

Factor X may be activated by two different proteases (Fig. 34-1):

1. By **factor IX$_a$**, the product of the **intrinsic pathway** (so- named because all of its protein components are con- tained in the blood).

2. By **factor VII$_a$**, the product of the **extrinsic pathway**

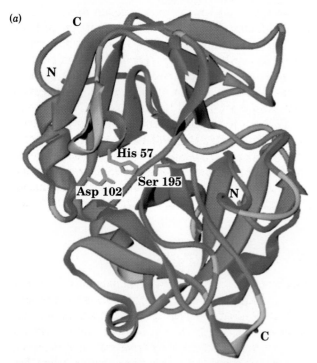

(a)

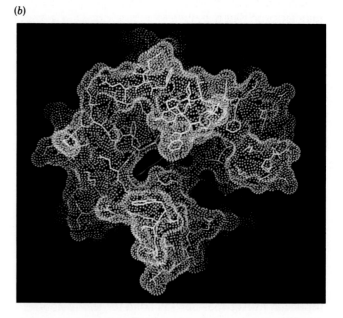

(b)

FIGURE 34-14. The X-ray structure of human thrombin inactivated by D-Phe-Pro-Arg chloromethylketone (PPACK). (a) A ribbon diagram oriented similarly to the structurally homologous trypsin as drawn in Fig. 14-20. Those segments of the 259-residue B chain that have a structural analog in trypsin are purple, whereas the remaining loops are light blue. Similarly, segments of the 36-residue A chain that have structural analogs in chymotrypsinogen are red and other segments are dark blue. The side chains of the catalytic triad (His 57, Asp 102, and Ser 195 using chymotrypsinogen numbering) are drawn in stick form in green. The PPACK chain which, like normal substrates, runs left to right, has been omitted for clarity. Compare this diagram with Fig. 14-20b. (b) The protein (*C yellow, N blue, O*

red) and its bound PPACK residue (*C purple*) shown in stick form superimposed on the complex's solvent-accessible surface (*light blue dots*) in approximately the same orientation as in Part a. For clarity, only the front part of the protein is shown (and hence, what appears to be a slot opening to the right is really part of thrombin's substrate-binding cleft, which extends from left to right across the front surface of the enzyme). The surface "hole" below the PPACK Phe and Pro residues, which is partially occluded by a loop not present in trypsin, marks the entrance to the specificity pocket (Fig. 14-20a). [Part a based on an X-ray structure by and Part b courtesy of Wolfram Bode, Max-Planck-Institut für Biochemie, Germany.]

(so-called because one of its important components occurs in the tissues).

We shall discuss these two pathways separately beginning with the intrinsic pathway.

Clotting May Be Initiated by the Contact System

It has long been known that bringing blood into contact with negatively charged surfaces, such as those of glass or kaolin (a clay used to make porcelain), initiates clotting. *In vivo,* collagen and platelet membranes are thought to have the same effect. The nature of this so-called **contact system** has only been worked out in outline and its physiological significance, if any, is still unclear.

The contact system consists of four glycoproteins: the serine protease zymogens named **Hageman factor (XII), prekallikrein, plasma thromboplastin antecedent (PTA or XI),** and **high molecular weight kininogen (HMK),** an accessory factor that is also a precursor of the nonapeptide hormone **bradykinin** (a potent vasodilator and diuretic factor). Adsorption to a suitable surface is thought to somehow activate Hageman factor which, in the presence of HMK, proteolyzes prekallikrein to form the active protease **kallikrein.** Kallikrein, in turn, proteolytically activates Hageman factor so that these two proteins reciprocally activate each other.

The nature of contact-activated Hageman factor is enigmatic; it is by no means certain that physical adsorption to a surface cleaves the same bond as does kallikrein or, for that matter, cleaves any bond at all. Much of the experimental difficulty in resolving this issue is a consequence of the contact-activation process's autocatalytic nature: Prekallikrein, contact-activated Hageman factor's substrate, is the zymogen of the protease that activates Hageman factor. Consequently, in any measurement of its activity, the nature of contact-activated Hageman factor is immediately obscured by large amounts of rapidly generated kallikrein-activated Hageman factor.

The final reaction mediated by the contact system is the proteolytic activation of factor XI by activated Hageman factor in a process that also uses HMK as an accessory factor. Although the contact system is clearly effective in initiating *in vitro* clot formation, its *in vivo* importance is in doubt because individuals deficient in Hageman factor, prekallikrein, or HMK do not suffer from bleeding problems.

The Last Two Steps of the Intrinsic Pathway Are Similar

The intrinsic pathway has two remaining steps leading to the activation of Stuart factor (X, Fig. 34-1). Factor XI_a catalyzes the proteolytic activation of **Christmas factor (IX),** a Gla-containing glycoprotein, in a Ca^{2+}-requiring reaction that takes place on a phospholipid membrane surface. No accessory factor is known for this reaction. Christmas factor may also be activated by activated **proconvertin (VII$_a$),** a product of the extrinsic pathway (Section 34-1D).

In the final step of the intrinsic pathway, factor X is proteolytically cleaved by activated Christmas factor (IX_a) on a phospholipid membrane surface in a reaction involving Ca^{2+} and the accessory factor activated **antihemophilic factor (VIII$_a$).** Antihemophilic factor, as is proaccelerin (V), is proteolytically activated by thrombin in a second autocatalytic process leading to prothrombin activation (Fig. 34-1). Not surprisingly, proaccelerin and antihemophilic factor are homologous proteins. Antihemophilic factor circulates in the plasma in complex with von Willebrandt factor; in fact, the activities of these two substances were initially attributed to a single protein.

Hemophilias Result from Clotting Factor Deficiencies

The discovery of antihemophilic factor came about through its deficiency in individuals with the most common clotting disorder, **hemophilia A,** a sex-linked inherited deficiency (~1 per 10,000 male births). Indeed, several of the clotting factors were discovered through the diagnosis of their deficiencies in various clotting disorders (the existence of Christmas factor was discovered through its absence in Stephen Christmas, a hemophiliac whose deficiency, **hemophilia B,** is the second most common form of hemophilia). Hemophiliacs may lose large amounts of blood from even the smallest injury and frequently hemorrhage without any apparent cause. However, the symptoms of their diseases may be alleviated by the intravenous administration of the deficient factor. In the past, this treatment was expensive and not without risk because large amounts of blood must be fractionated to obtain therapeutic doses of most clotting factors. Hemophiliacs were therefore inordinately subject to a variety of dangerous bloodborne viral diseases including hepatitis and AIDS. These difficulties have now been largely eliminated through the production of the required clotting factors by recombinant DNA techniques.

D. The Extrinsic Pathway

The extrinsic pathway (Fig. 34-1), the alternative arm of the clotting cascade, is initiated by the proteolysis of proconvertin (VII), a process that can be catalyzed by activated Hageman factor (XII_a) as well as by thrombin. Activated proconvertin, in turn, mediates the activation of factor X in a process analogous to that of prothrombin activation (Section 34-1B) in that its rate is enhanced 16,000-fold by the presence of phospholipid membrane, Ca^{2+}, and an accessory factor named **tissue factor (III).** Intact proconvertin can also catalyze factor X activation but at only 2% the rate of activated proconvertin. Apparently this rate is so low that, in the absence of tissue factor, unactivated proconvertin cannot initiate *in vivo* clot formation.

Tissue factor is an integral membrane glycoprotein that occurs in many tissues and is particularly abundant in brain, lung, blood vessel walls, and placenta. Consequently, an injury that exposes blood to tissue rapidly initiates the extrinsic pathway. In fact, the addition of tissue factor to the extrinsic system causes clot formation in ~12 s, whereas the intrinsic system requires several minutes to do so. These observations suggest that the intrinsic pathway is normally of little significance. However, the severity of the hemophilias resulting from intrinsic pathway clotting factor deficiencies clearly establishes the importance of the intrinsic pathway in blood clotting. Of course the two pathways are not really independent since they are coupled through a number of reactions (Fig. 34-1).

E. Control of Clotting

The multilevel cascade of the blood clotting system permits enormous amplification of its triggering signals. Moving down the extrinsic pathway, for example, proconvertin (VII), Stuart factor (X), prothrombin, and fibrinogen are present in plasma in concentrations of <1, 8, 150, and up to 4000 $\mu g \cdot mL^{-1}$, respectively. Yet, clotting must be very strictly regulated since even one inappropriate clot can have fatal consequences. Indeed, blood clots are the leading cause of strokes and heart attacks, the two major causes of human death in developed countries.

A Variety of Factors Limit Clot Growth

There are numerous physiological mechanisms that limit clot formation. We have seen that there are several interactions among the various clotting factors that inhibit blood coagulation (Fig. 34-1). The blood flow dilution of active clotting factors also does so as does their selective removal from the circulation by the liver. In addition, plasma contains several serine protease inhibitors whose presence prevents clots from spreading beyond the vicinity of an injury. For example, **antithrombin** (58 kD) inhibits all active proteases of the clotting system except VII$_a$ by binding to them in tight 1:1 complex (much like BPTI binds trypsin; Section 14-3D). The presence of **heparin**, a sulfated glycosaminoglycan (Section 10-2E), enhances the activity of antithrombin by several hundredfold. Heparin occurs almost exclusively in the intercellular granules of the mast cells that line certain blood vessels. Its release, presumably by injury, activates antithrombin, thereby preventing runaway clot growth.

Protein C is another plasma protein that limits clotting. This Gla residue-containing 62-kD zymogen is activated by thrombin to proteolytically inactivate proaccelerin (V) and antihemophilic factor (VIII). Activated protein C attacks the activated forms of these accessory factors more readily than their nonactive forms. The importance of protein C is demonstrated by the observation that individuals who lack it often die in infancy of massive thrombotic complications.

Despite the foregoing, blood clotting *in vitro* is not self-limiting. This observation, which suggests the existence of additional *in vivo* clot-limiting factors, led to the discovery of **thrombomodulin,** a 74-kD glycoprotein that projects from the cell surface membranes of the vascular endothe-

lium (inner lining). Thrombomodulin specifically binds thrombin so as to convert it to a form with decreased ability to catalyze clot formation but with a >1000-fold increased capacity to activate protein C.

The control of clotting is a major medical concern. Heparin, the most frequently used anticoagulant, is administered before and after surgery to retard clot formation. For long-term control of hemostasis, dicoumarol is often employed. In the design of an artificial heart, the elimination of mechanically induced clots remains the major unsolved problem (the construction of an adequate pump is a relatively simple task). The prevention of clotting is also a concern of blood-sucking organisms. The leech *Hirudo medicinalis* solves this problem by secreting **hirudin,** a 65-residue protein, in its saliva. Hirudin, the most powerful naturally occurring anticoagulant known, specifically binds to thrombin with an association constant of ~5 × $10^{13} M^{-1}$ (greater than that of trypsin with BPTI; Section 14-3D), thereby inactivating it. Consequently a leech bite, although a minor wound, bleeds quite freely.

The X-ray structure of the complex of hirudin with human thrombin, which was determined by Tulinsky, Bode, and Huber, reveals that hirudin binds to thrombin in a manner not previously observed for protease inhibitors (Fig. 34-15). Hirudin has a globular N-terminal domain and an extended C-terminal domain. The N-terminal do-

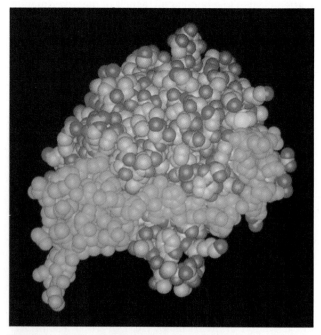

FIGURE 34-15. The X-ray structure of human thrombin in complex with hirudin as shown in space-filling form. The hirudin is green and the thrombin, which is oriented similarly to that in Fig. 34-14, is colored according to atom type with C gray, N blue, O red, and S yellow. The hirudin's N-terminal domain is on the left and its C-terminal helix is on the right. [Courtesy of Alexander Tulinsky, Michigan State University.]

main binds in the region of thrombin's active site such that hirudin's N-terminal amino group forms a hydrogen bond with the active site Ser of thrombin. However, hirudin does not occupy thrombin's specificity pocket. The three N-terminal residues of hirudin form a parallel β strand with Ser 214 to Glu 217 of thrombin (using chymotrypsinogen numbering; Section 14-3B). Interestingly, this segment of thrombin forms an antiparallel β sheet in all other known natural serine protease inhibitor complexes (with which hirudin has no sequence homology) as well as in the above described D-Phe-Pro-Arg chloromethylketone (PPACK) complex. Thus, the hirudin chain runs in the direction opposite to that expected for fibrinogen. Hirudin's highly anionic and extended 39-Å-long C-terminal tail, which comprises its residues 48 to 65, wraps around thrombin in its anion-binding exosite (where the highly anionic fibrinopeptides also bind) making extensive ionic and hydrophobic contacts. These numerous interactions are almost certainly responsible for hirudin's tight binding to thrombin.

F. Clot Lysis

Blood clots are only temporary patches; they must be eliminated as wound repair progresses. This is a particularly urgent need when a clot has inappropriately formed or has broken free into the general circulation. Fibrin is a molecule that is "designed" to be easily dismantled in a process termed **fibrinolysis.** The demolition agent is a plasma serine protease named **plasmin,** an enzyme that specifically cleaves fibrin's triple-stranded coiled coil segment and cuts away its covalently cross-linked α chain protuberances (Fig. 34-5). The rather open meshlike structure of a blood clot (Fig. 34-2) gives plasmin relatively free access to the polymerized fibrin molecules thereby facilitating clot lysis.

Plasmin is formed through the proteolytic cleavage of the 86-kD zymogen **plasminogen,** a protein that is homologous to the zymogens of the blood clotting cascade. There are several serine proteases that activate plasminogen, most notably the 54-kD enzyme **urokinase,** which is synthesized by the kidney and occurs, as its name implies, in the urine, and the homologous 70-kD enzyme **tissue-type plasminogen activator (t-PA),** which occurs in vascular tissues. In addition, activated Hageman factor, in the presence of prekallikrein and HMK (the contact activation system) activates plasminogen although the physiological significance of the contact activation of the fibrinolytic system has not been determined. Nevertheless, the fibrinolysis system, as our experience might have led us to expect, is not so simple as just a zymogen and its activators. It also incorporates several inhibitors, principally the 70-kD glycoprotein α_2-**antiplasmin,** which forms an irreversible equimolar complex with plasmin that prevents it from binding to fibrin. The α_2-antiplasmin cross-links to fibrin α chains through the action of activated FSF (XIII$_a$, the enzyme that also cross-links fibrin), thereby making "hard" clots less susceptible to fibrinolysis than "soft" clots. The importance of this serine protease inhibitor (it also inhibits chymotrypsin) is indicated by the observation that homozygotes for a defective α_2-antiplasmin have a serious tendency to bleed.

Plasminogen activators have received considerable medical attention aimed at rapidly dissolving the blood clots responsible for heart attacks and strokes. **Streptokinase,** a 45-kD protein produced by certain streptococci, has shown considerable utility in this regard, particularly when administered together with aspirin (which inhibits platelet aggregation; Section 23-7B). Despite its name, streptokinase exhibits no enzymatic activity. Rather, it acts by forming a tight 1:1 complex with plasminogen that proteolytically activates other plasminogen molecules. The use of streptokinase to dissolve clots has the apparent disadvantage that it activates plasmin to degrade fibrinogen as well as fibrin thereby increasing the risk of bleeding problems, particularly strokes. The therapeutic use of t-PA, which has been synthesized by recombinant DNA techniques, is thought to eliminate these problems because this enzyme activates plasminogen only in the presence of a blood clot (although the medical significance of these problems appears to be minimal).

2. IMMUNITY

All organisms are continually subject to attack by other organisms. In response to predators, animals have developed an enormous variety of defensive strategies. An even more insidious threat, however, is attack by disease-causing microorganisms and viruses (pathogens). In order to deal with them, animals have evolved an elaborate protective array known as the **immune system** (Latin: *immunis,* exempt). Pathogens that manage to breach the physical barrier presented by the skin and mucous membranes (a vital first line of defense) are identified as foreign invaders and destroyed. In this section we discuss how the immune system recognizes foreign invaders, how they are distinguished from normal components of self, and how they are destroyed. As we do so, keep in mind that the immune system exhibits many of the qualities that are characteristic of the nervous system such as the ability to detect and react to stimuli and to remember. Indeed, the size and complexity of the vertebrate immune system rivals that of the vertebrate nervous system.

A. The Immune Response

Immunity in vertebrates is conferred by certain types of white blood cells collectively known as **lymphocytes.** They arise, as do all blood cells, from common precursor cells **(stem cells)** in the bone marrow. Lymphocytes, however, in contrast to red blood cells, can leave the blood vessels and patrol the intercellular spaces for foreign intruders. They eventually return to the blood via the lymphatic vessels but not before interacting with specialized **lymphoid tissues** such as the thymus, the lymph nodes, and the spleen, the sites where much of the immune response occurs.

Two types of immunity have been distinguished:

1. **Cellular immunity,** which guards against virally infected cells, fungi, parasites, and foreign tissue, is mediated by *T* **lymphocytes** or *T* **cells,** so called because their development occurs in the *t*hymus.

2. **Humoral immunity** (*humor* is an archaic term for fluid), which is most effective against bacterial infections and the extracellular phases of viral infections, is mediated by an enormously diverse collection of related proteins known as **antibodies** or **immunoglobulins.** Antibodies are produced by *B* **lymphocytes** or *B* **cells** which, in mammals, mature in the *bone* marrow.

We shall outline the operations and interactions of these systems as a prelude to discussing their biochemistry.

The Cellular Immune System

*The immune response is triggered by the presence of foreign macromolecules, normally proteins, carbohydrates, and nucleic acids, known as **antigens**.* This process occurs through a complex series of interactions among various types of *T* cells and *B* cells that specifically bind a particular antigen (Fig. 34-16). In the following paragraphs, italic numerals and letters refer to the corresponding drawing in Fig. 34-16.

The cellular immune response leads to the destruction of the offending cells. It begins when a **macrophage** (a type of white blood cell) engulfs (*1a, 1b*) and partially digests (*2a, 2b*) a foreign antigen and then displays the resulting antigenic fragments on its surface (*3a, 3b*). There, it is thought, these fragments bind to one of two types of cell-surface proteins known as **major histocompatibility complex (MHC) proteins** (so called because they are transcribed from a closely linked series of genes called the MHC; Section 34-2E). The MHC is remarkably polymorphic (has numerous alleles); so much so that any two unrelated individuals of the same species are highly unlikely to have an identical set of MHC proteins. *MHC proteins are therefore markers of individuality.*

Class I MHC proteins are displayed on the surfaces of nearly all nucleated vertebrate cells. Macrophages exhibiting Class I MHC proteins are recognized by proteins known as *T* **cell receptors,** which occur on the surfaces of immature **cytotoxic** *T* **cells.** *In order to bind to the antigen-displaying macrophage, however, these receptors must specifically complex the antigen together with the Class I MHC protein (4a);* neither molecule alone can do the job. In the same way, macrophages displaying antigenic fragments complexed to **Class II MHC proteins** are bound by immature **helper** *T* **(T_H) cells** bearing the cognate receptor (*4b*). This elaborate recognition system, as we shall see, focuses the attention of *T* cells on cell surfaces and thereby prevents the resources of the cellular immune system from being futilely squandered on noncellular targets.

T cells that bind to a macrophage-displayed antigen–MHC protein complex are induced to propagate, a process known as **clonal selection,** which was first recognized in the 1950s by Niels Kaj Jerne, Macfarlane Burnet, Joshua Lederberg, and David Talmadge. Consequently, *only those T cells that specifically recognize the intruding antigen are produced in quantity.* Clonal selection occurs because a macrophage bound to a *T* cell releases a protein growth factor named **interleukin-1** *(5a, 5b), which specifically stimulates T cells to proliferate and differentiate (6a, 6b).* This process is enhanced by the *T* cells' autostimulatory secretion of **interleukin-2.** *T cells only make **interleukin-2 receptor** so long as they remain bound to a macrophage (7a, 7b),* thereby preventing unlimited *T* cell proliferation. Nevertheless, a large number of mature cytotoxic *T* cells *(8a),* which through their receptors are specifically targeted for host cells displaying both the foreign antigen and Class I MHC proteins, are generated starting a few days after the antigen is first encountered. The cytotoxic *T* cells, which are also known as **killer** *T* **cells,** live up to their name: *They bind to antigen-bearing host cells (9) and, at the point of contact, release a 70-kD protein, **perforin,** that lyses these target cells (10) by aggregating to form pores in their plasma membranes (Section 34-2F).*

The cellular immune system functions mainly to prevent the spread of a viral infection by killing virus-infected host cells (viral coat proteins are generally displayed on the surface of an animal cell during the latter stages of its viral infection; e.g., Section 32-4A). It is also effective against fungal infections, parasites, and certain types of cancers. Indeed, the cellular immune system's vital function has become painfully evident in recent years through the tragic spread of **acquired immune deficiency syndrome (AIDS),** whose causative agent, **human immunodeficiency virus (HIV),** acts by specifically attacking helper *T* cells. The cellular immune system is also responsible for various difficulties elicited by modern medicine that do not occur in nature such as the rejection of tissue and organ grafts from foreign donors. Such grafts, which are recognized as foreign because they almost always bear MHC proteins that differ from those of the host, have only been made possible by the development of drugs known as **immunosuppressants,** such as cyclosporin A and FK506 (Section 8-1C), that suppress the immune response (but not so much as to leave the body defenseless against pathogens).

FIGURE 34-16 (*Opposite*).   An outline of the immune response. See the text for an explanation. [After Marrack, P. and Kappler, J., *Sci. Am.* **254**(2): 38–39 (1986). Copyright © 1986 by Scientific American, Inc.]

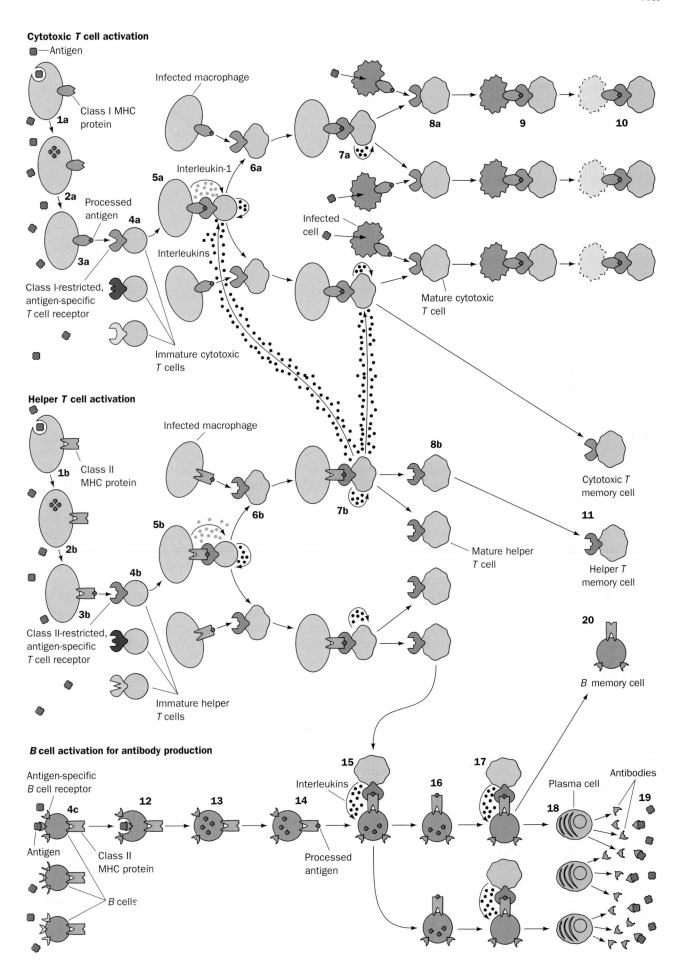

Cytotoxic *T* cell activation

Antigen

Class I MHC protein

1a

2a Processed antigen

3a **4a**

Class I-restricted, antigen-specific *T* cell receptor

Immature cytotoxic *T* cells

Infected macrophage

5a Interleukin-1

6a

Interleukins

7a

8a **9** **10**

Infected cell

Mature cytotoxic *T* cell

Helper *T* cell activation

Class II MHC protein

1b

2b

3b **4b**

Class II-restricted, antigen-specific *T* cell receptor

Immature helper *T* cells

Infected macrophage

5b **6b** **7b** **8b**

Mature helper *T* cell

Cytotoxic *T* memory cell

11 Helper *T* memory cell

20 *B* memory cell

B cell activation for antibody production

Antigen-specific *B* cell receptor

4c **12** **13** **14** **15** **16** **17** **18** **19**

Antigen

Class II MHC protein

B cells

Interleukins

Processed antigen

Plasma cell

Antibodies

The Humoral Immune System

B cells display both immunoglobulins and Class II MHC proteins *(4c)* on their surfaces. If a *B* cell encounters an antigen that binds to its particular immunoglobulin, it engulfs the complex *(12)*, partially digests the antigen *(13)*, and displays the fragments on its surface in complex with the Class II MHC protein *(14)*. Mature helper *T* cells *(8b)* bearing receptors specific for this complex bind to the *B* cell *(15)* and, in response, release interleukins that stimulate the *B* cell to proliferate and differentiate *(16)*. Cell division continues so long as the *B* cells are stimulated by the helper *T* cells *(17)* which, in turn, depends on the continuing presence of antigen *(1b–8b)*. *Most of the B cell progeny are* **plasma cells** *(18) that are specialized to secrete large amounts of the antigen-specific antibody. The antibodies bind to the available antigen (19), thereby marking it for destruction either by* **phagocytosis** *(ingestion by white cells known as* **phagocytes***) or by activating the* **complement system** *(a series of interacting proteins that lyse cells and trigger local inflammatory reactions; Section 34-2F).*

Most *T* cells and *B* cells live only a few days unless stimulated by their corresponding antigen. Moreover, the proliferation of *B* cells is limited by their interactions with **suppressor *T* (*T*$_S$) cells,** an additional type of *T* lymphocyte progeny which have essentially the opposite function of helper *T* cells. Yet, one of the hallmarks of the immune system is that an animal is rarely infected twice by exactly the same type of pathogen; that is, *recovery from an infection by a pathogen renders an animal immune from that pathogen.* This so-called **secondary immune response** is mediated by long lived **memory *T* cells** *(11)* and **memory *B* cells** *(20)* which, upon reencountering their cognate antigen, perhaps decades after its previous appearance, proliferate much faster and more massively than do "virgin" *T* and *B* cells (those that have never encountered their corresponding antigen) as is indicated in Fig. 34-17. This characteristic of the immune system has been recognized since ancient times: The Greek historian Thucydides noted over 2400 years ago that the sick could be treated by those who had recovered for a man was never attacked twice by the same disease.

The Immune System Is Self-Tolerant

Nearly all biological macromolecules are antigenic. To prevent self-destruction, an animal's immune system must therefore discriminate between self-antigens and foreign antigens. Such a process must be exquisitely selective. After all, a vertebrate, for example, has tens of thousands of different macromolecules, each with numerous distinctive antigenic sites.

What is the mechanism of immunological **self-tolerance?** The immune system in mammals becomes active around the time of birth. If a foreign antigen is implanted in an embryo before this time, the resulting animal is unable to mount an immune attack against that antigen. Apparently, the immune system eliminates the clones of *B* and *T*

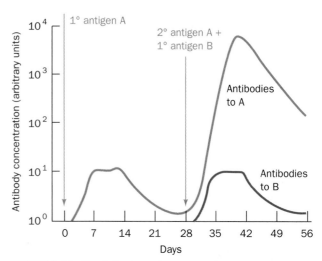

FIGURE 34-17. Primary and secondary immune responses: The rates of appearance of antibodies in the blood serum following primary (1°) immunization on day 0 with antigen A and secondary (2°) immunization on day 28 with antigens A and B. Antigen B is included in the secondary immunization to demonstrate the specificity of immunological memory for antigen A. Note that the secondary response to antigen A is both faster and greater than the primary response. *T* cell-mediated responses exhibit similar immunological memory.

cells that recognize the antigens that were present during the critical period when the immune system became active (clonal deletion). Yet, because new clones of lymphocytes, each with a nearly unique set of antigenic determinants, arise throughout an animal's lifetime (Section 34-2C; antibodies and *T* cell receptors are themselves antigenic), self-toleration must be an ongoing process. This process, although of dimly perceived mechanism, has been shown to occur in the thymus where, it appears, *only those virgin T cells displaying receptors that bind MHC proteins but which have no affinity for self-antigens are selected for further propagation.* Indeed, only a small fraction of the lymphocytes that are processed by the thymus ever leave that organ.

Occasionally, the immune system loses tolerance to some of its self-antigens, resulting in an **autoimmune disease.** For example, **myasthenia gravis,** an autoimmune disease in which individuals make antibodies against the **acetylcholine receptors** of their own skeletal muscles (acetylcholine is a neurotransmitter that triggers muscle contraction; Section 34-4C), results in a progressive and often fatal muscular weakness. Similarly, individuals with **systemic lupus erythematosis,** an often fatal inflammatory disease, produce antibodies against many of their own cellular components including certain ribonuclear proteins (Section 29-4A) and DNA. Other common autoimmune diseases are **rheumatoid arthritis** (in which the immune system attacks the connective tissue in the joints), **insulin-dependent diabetes mellitus** (Section 25-3B), and **multiple sclerosis** [in which the immune system destroys the myelin insulation surrounding nerve fibers in the brain and spinal cord (Section 34-4C), thereby causing paralysis].

B. Antibody Structures

The immunoglobulins form a related but yet enormously diverse group of proteins. In this section, we consider the structures of these essential molecules. How their diversity is generated is the subject of the following section.

There Are Five Classes of Immunoglobulins

Most immunoglobulins, and the basic building blocks of all of them, consist, as Gerald Edelman and Rodney Porter showed, of four subunits: two identical ~23-kD **light chains (L)** and two identical 53- to 75-kD **heavy chains (H)**. These subunits associate via disulfide bonds as well as by noncovalent interactions to form, as electron micrographs indicate, a Y-shaped symmetric dimer, $(L-H)_2$ (Fig. 34-18). Immunoglobulins are glycoproteins; each heavy chain has an *N*-linked oligosaccharide.

*Humans have five classes of secreted immunoglobulins, designated **IgA** (for **immunoglobulin A**), **IgD, IgE, IgG,** and **IgM,** which differ in their corresponding types of heavy chains, designated α, δ, ε, γ, and μ, respectively (Table 34-2). There are also two types of light chain, κ and λ but these*

occur in immunoglobulins of all classes. IgD, IgE, and IgG exist only as $(L-H)_2$ dimers. IgM, however, consists of pentamers of its respective dimers and IgA occurs as monomers, dimers, and trimers of its corresponding dimers (Fig.

TABLE 34-2. CLASSES OF HUMAN IMMUNOGLOBULINS

Class	Heavy Chain	Light Chain	Subunit Structure	Molecular Mass (kD)
IgA	α	κ or λ	$(\alpha_2\kappa_2)_n J^a$ $(\alpha_2\lambda_2)_n J^a$	360–720
IgD	δ	κ or λ	$\delta_2\kappa_2$ $\delta_2\lambda_2$	160
IgE	ε	κ or λ	$\varepsilon_2\kappa_2$ $\varepsilon_2\lambda_2$	190
IgG[b]	γ	κ or λ	$\gamma_2\kappa_2$ $\gamma_2\lambda_2$	150
IgM	μ	κ or λ	$(\mu_2\kappa_2)_5 J$ $(\mu_2\lambda_2)_5 J$	950

[a] $n = 1$, 2, or 3.
[b] IgG has four subclasses, IgG1, IgG2, IgG3, and IgG4, which differ in their γ chains.

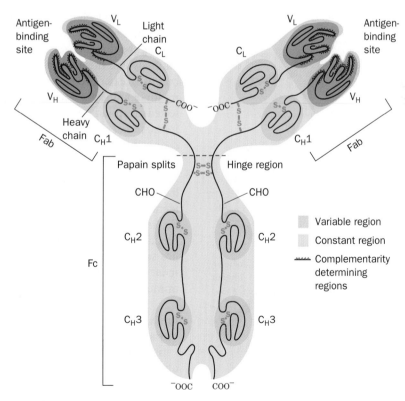

FIGURE 34-18. A diagram of the human IgG molecule. Each light (L) chain consists of two homologous units, V_L and C_L, where V and C indicate the polypeptide chain's variable and constant regions. Each heavy (H) chain is composed of four such units, V_H, C_H1, C_H2, and C_H3. Treatment of IgG by the proteolytic enzyme papain results in the cleavage of this immunoglobulin molecule in its "hinge" region yielding two Fab fragments and one Fc fragment. CHO represents carbohydrate chains. [Drawing copyrighted © by Irving Geis.]

34-19). The dimeric units of these multimers are linked by disulfide bonds to each other and to an ~20-kD protein termed the **joining chain (J)**. IgM also occurs in a *B* cell–displayed monomeric membrane-bound form. It is antigen binding by this latter form of IgM that triggers the humoral immune response.

The various classes of secreted immunoglobulins have different physiological functions. IgM, which is largely confined to the blood, is most effective against invading microorganisms. It is the first immunoglobulin to be secreted in response to an antigen; its production begins 2 to 3 days after antigen is first encountered. IgG, the most common immunoglobulin, is equally distributed between the blood

and the interstitial fluid. It is the only antibody that can cross the placenta (via receptor-mediated endocytosis) and thus provide the fetus with immunity. IgG production begins 2 to 3 days after IgM first appears. IgA occurs predominantly in the intestinal tract and in such secretions as saliva, sweat, and tears; it defends against invading pathogens by binding to their antigenic sites so as to block their attachment to epithelial (outer) surfaces. IgA is also the major antibody of milk and colostrum (the first milk secreted after pregnancy) and thereby protects nursing infants from gastrointestinal invasion by pathogens. IgE, which is normally present in the blood in minute concentrations, protects against parasites and has been implicated in allergic reactions. IgD, which is also present in blood in very small amounts, is of unknown function.

Immunoglobulin's Functional Segments May Be Proteolytically Separated

In 1959, Porter showed that IgG, the most common class of immunoglobulin, is cleaved, through limited proteolysis with papain, into three ~50-kD fragments: two identical **Fab fragments** and one **Fc fragment.** The Fab fragments, which form the arms of the Y-shaped IgG molecule and which each consist of an entire L chain and the N-terminal half of an H chain (Fig. 34-18), contain IgG's antigen-binding sites ("ab" stands for *a*ntigen *b*inding). Immunoglobulin's consequent divalent (or, for IgA and IgM, multivalent) antigen-binding character forms the basis of the **precipitin reaction,** a sensitive test that has long been used for determining the presence of antibody or antigen: A mixture of antibody and the antigen against which it is directed combine as an extended cross-linked lattice (most antigens have multiple antigenic determinants) that yields an easily detected precipitate (Fig. 34-20). *The formation of these cross-linked lattices enhances antibody–antigen binding through cooperative interactions and is required to trigger B cell proliferation.*

The Fc fragment (so named because it is readily *c*rystallized), derives from the stem of the Y and consists of the identical C-terminal segments of two H chains (Fig. 34-18). Fc fragments contain the effector sites that mediate the functions common to a particular class of immunoglobulins such as inducing phagocytosis, triggering the complement system (Section 34-2F), and directing the transport of immunoglobulins to their sites of action.

IgG's Heavy and Light Chains Both Have Constant and Variable Regions

In order to characterize a molecule, it is necessary to obtain it in reasonably pure form. This requirement, at first, presented immunologists with a seemingly insurmountable obstacle. Exposing an animal to a particular antigen elicits the formation of numerous clones of plasma cells, each of which synthesizes a slightly different immunoglobulin molecule that binds the antigen. The resulting

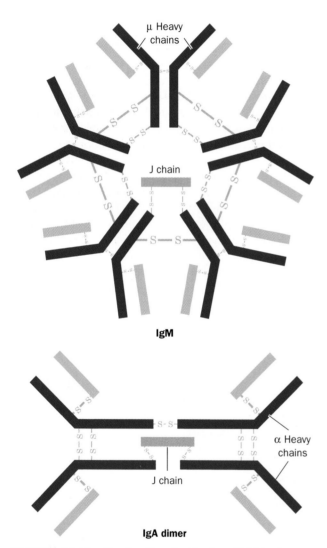

IgM

IgA dimer

FIGURE 34-19. The five dimeric subunits of IgM (*top*) are held together by disulfide bonds. A single J chain joins two of the pentamer's μ heavy chains and is therefore thought to initiate assembly of this immunoglobulin. The J chain also participates in joining IgA chains to form dimers (*bottom*) and trimers.

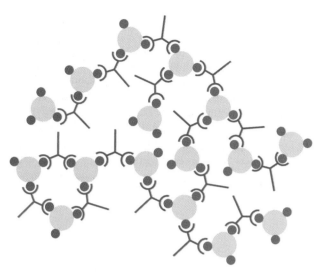

FIGURE 34-20. Divalent antibody (*green*) can cross-link its corresponding multivalent antigen (*red*) so as to form an extended lattice. Although the antigen is shown here as having three copies of only one type of antigenic determinant, most naturally occurring antigens, for example, a bacterium, have multiple copies of several different antigenic determinants. Such complex antigens are efficiently cross-linked by the mixtures of antibodies directed against their various antigenic sites.

antibodies are therefore quite heterogeneous. This obstacle was largely removed by the demonstration, in the early 1960s, that an individual with **multiple myeloma,** a plasma cell cancer, synthesizes large amounts of a single species of immunoglobulin termed a **myeloma protein.** Some myelomas make excess light chains, which, when they are ex-

creted in the urine, are known as **Bence Jones proteins** (after Henry Bence Jones who first described them in 1847).

The amino acid sequences of several different Bence Jones proteins, which each have 214 residues, revealed that *the sequence differences among light chains are largely confined to their N-terminal halves.* Light chains are therefore said to have a **variable region, V_L,** spanning residues 1 to 108, and a **constant region, C_L,** comprising residues 109 to 214 (Fig. 34-18). Similarly, comparisons of myeloma heavy chains, which have 446 residues, revealed that all the sequence differences among them occur between residues 1 to 125. *Thus heavy chains also have a variable region, V_H, and a constant region, C_H* (Fig. 34-18).

Additional sequence comparisons indicated that the C_H region consists of three ~110-residue segments, **C_H1, C_H2,** and **C_H3,** which are homologous to each other and to C_L. In fact, even the constant and variable sequences are related albeit not as closely as the members of each of these groups are related to each other. These homologies, together with the observation that each homology unit is cross-linked by a disulfide bond, correctly suggest (see below) that *an immunoglobulin molecule's 12 homology units each fold into an independent domain.* Apparently, modern light chain and heavy chain genes evolved through duplications of a primordial gene encoding an ~110-residue protein.

The V_L and V_H regions are not uniformly variable. Rather, most of their amino acid variations are concentrated into three short **hypervariable sequences** (Fig. 34-21). Elvin Kabat therefore predicted that *the hypervariable sequences line the immunoglobulin's antigen-binding site and that their amino acids determine its binding specificity.*

Kabat's hypothesis was supported by affinity labeling experiments. Molecules of <5 kD are rarely antigenic. Yet, when small organic groups termed **haptens,** such as the

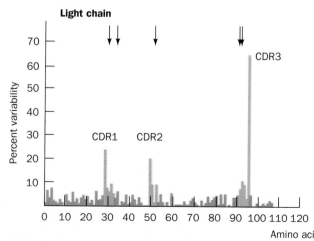

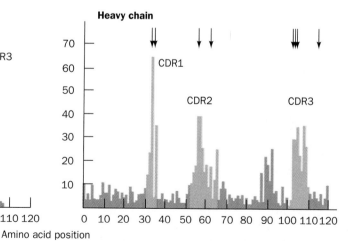

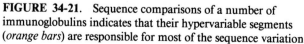

FIGURE 34-21. Sequence comparisons of a number of immunoglobulins indicates that their hypervariable segments (*orange bars*) are responsible for most of the sequence variation in the variable regions of both the light and heavy chains. The arrows mark the sites on anti-DNP antibodies that are derivatized by the affinity label *p*-nitrophenyldiazonium.

2,4-dinitrophenyl (DNP) group, are covalently attached to a carrier protein such as bovine serum albumin (by reaction of fluorodinitrobenzene with its Lys residues)

DNP-hapten

and then injected into an animal, the animal produces antibodies that bind to the hapten in the absence of the carrier. If the DNP analog *p*-**nitrophenyldiazonium**

p-**Nitrophenyldiazonium**

is combined with anti-DNP antibodies, the hapten's highly reactive diazonium group will form diazo bonds with the His, Lys, and Tyr side chains in the vicinity of the antibodies' DNP-binding sites (affinity labeling; Section 30-3A). Most of the side chains so derivatized are, in fact, members of the antibodies' hypervariable sequences (Fig. 34-21), thereby indicating that *antigen-binding sites are lined with hypervariable residues.* Immunoglobulin's hypervariable segments are therefore also called **complementarity-determining regions (CDRs).**

FIGURE 34-22. Procedure for producing monoclonal antibodies against an antigen, X. **HAT medium,** so called because it contains *h*ypoxanthine, *a*methopterin (methotrexate, an antifolate; Section 26-4B), and *t*hymine, prevents the growth of mutant cell lines lacking hypoxanthine-guanine phosphoribosyl transferase (HGPRT, a purine salvage enzyme that catalyzes the formation of the AMP and GMP precursor IMP; Section 26-2D). The amethopterin blocks the *de novo* synthesis of purines, which *HGPRT⁻* cells cannot replace through salvage pathways. Thymine, whose synthesis is also inhibited by the amethopterin, is available from the HAT medium. *HGPRT⁻* myeloma cells are fused with spleen-derived lymphocytes from a mouse immunized against X and the resultant preparation is transferred to a HAT medium. This treatment selects for fused cells (hybridomas): The *HGPRT⁻* myeloma cells cannot grow in HAT medium; lymphocytes, which make HGPRT, do not grow in culture; but the hybridoma cells, which have the lymphocytes' HGPRT and the myeloma cells' immortality, proliferate. Individual hybridoma cells are then cloned and screened for the production of anti-X antibody. A satisfactory clone can be grown in virtually unlimited quantities, either in culture or as a mouse tumor, so as to synthesize the desired amounts of monoclonal antibody.

Monoclonal Antibodies Are Indispensable Biomedical Tools

One might expect that homogeneous immunoglobulins could be obtained in quantity by simply cloning a single lymphocyte and harvesting the immunoglobulin the clone produced. Unfortunately, lymphocytes do not grow continuously in culture. In the late 1970s, however, Cesar Milstein and Georges Köhler developed a technique for immortalizing such clones (Fig. 34-22). *Monoclonal antibodies can now be obtained in virtually unlimited quantities and specific for almost any antigen by fusing myeloma cells*

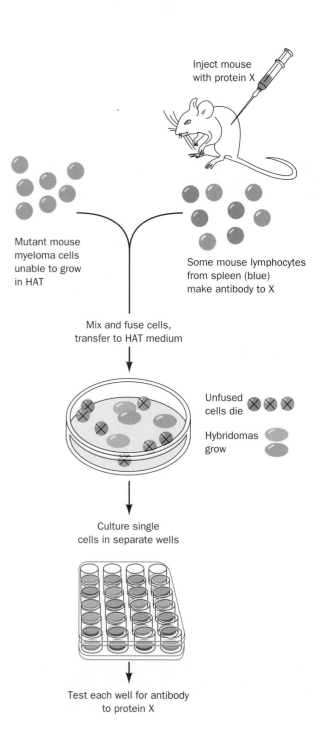

with lymphocytes raised against that antigen (that is, isolated from an animal that has been immunized with the antigen). A clone of the resulting **hybridoma** (hybrid myeloma) cell synthesizes the lymphocyte's immunoglobulin but has the myeloma cell's immortality. Monoclonal antibodies have become indispensable biomedical tools; they can be used to assay for and to isolate extremely small amounts of nearly any specific biological substance. For example, they have made possible the routine testing of blood for the presence of HIV (AIDS virus), thereby protecting the public blood supply.

Immunoglobulin Homology Units Are Similarly Folded

The proposed immunoglobulin structure was confirmed and extended by X-ray structure determinations of Fab and Fc fragments and entire myeloma proteins variously carried out by David Davies, Allan Edmondson, Robert Huber, Roberto Poljak, and many others. IgG is a molecule with twofold chemical equivalence whose homology units form separate domains (Fig. 34-23). Each of these domains is closely associated with a domain from another polypeptide chain so that the entire molecule may be considered to consist of six globular modules, two that form the stem of the Y (Fc region) and two that form each of its two arms (Fab regions; Fig. 34-18).

The immunoglobulin homology units all have the same characteristic ***immunoglobulin fold:*** A barrel composed of a three- and a four-stranded antiparallel β sheet that are linked by a disulfide bond (Fig. 34-24). The V domains differ from the C domains mainly by an additional polypeptide loop flanking each V domain's three-stranded β sheet.

Immunoglobulins, as both physicochemical studies and X-ray structural analyses indicate, exhibit considerable intersegmental flexibility (Fig. 34-23). This is particularly evident in the protein's so-called **"hinge" region,** the polypep-

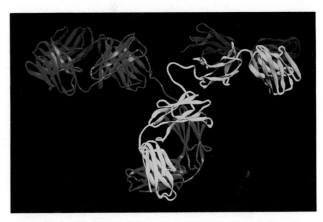

FIGURE 34-23. X-Ray structure of a murine antibody against canine lymphoma (a type of cancer against which this antibody is therapeutically useful) shown in ribbon form with its two heavy chains yellow and blue and its two light chains both red. The antigen-combining sites are located at the ends of the two approximately horizontal Fab arms formed by the association of the light chains with the heavy chains. Compare this figure with Fig. 34-18. [Courtesy of Alexander McPherson, University of California at Riverside.]

tide segment joining each Fab region to its Fc region (Fig. 34-18), although these regions are more like tethers than hinges as Fig. 34-23 (in which the two hinge angles are ~65° and ~115°) strikingly indicates. In fact, Fig. 34-23 represents the only high resolution X-ray structure of an intact antibody yet reported; the many other IgGs whose crystallization has been attempted apparently take up multiple conformations, thereby forming poorly ordered crystals, if they crystallize at all.

Since the basic immunoglobulin structure must accommodate an enormous variety of antigens, its flexibility presumably facilitates antigen binding by permitting an opti-

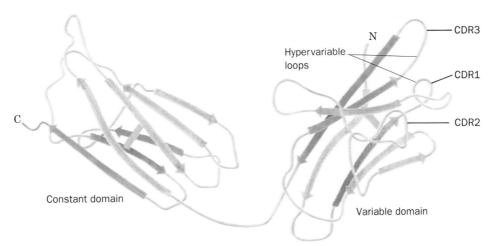

FIGURE 34-24. The chain folding of a myeloma protein light chain. Both its constant and variable domains assume the immunoglobin fold: a barrel-shaped sandwich of a four-stranded antiparallel β sheet (*blue arrows*) and a three-stranded antiparallel β sheet (*brown arrows*) that are linked by a disulfide bond (*yellow*). The positions of its three CDRs are indicated. [After Schiffer, M., Girling, R.L., Ely, K.R., and Edmundson, A.B., *Biochemistry* **12,** 4628 (1973).]

mal fit between the antigen and its combining site. The immunoglobulin's carbohydrate moiety is wedged between the C_H1 and C_H2 homology units and therefore also modulates the interactions between the Fab and Fc regions.

Antigen-Binding Sites Are Complementary to Their Corresponding Antigens

Myeloma proteins, upon which much of our structural knowledge of immunoglobulins is based, are produced by cancer cells that originally proliferated in response to unknown, if any, antigens. Nevertheless, haptens that bind to particular myeloma proteins have been identified by screening many different compounds.

The X-ray structures of several hapten–myeloma protein complexes indicate that an immunoglobulin's antigen-binding site is located at the tip of each Fab region in a crevice between its V_L and V_H units (Fig. 34-18). *The size and shape of this crevice depends on the amino acid sequences of the V_L and V_H units and its walls are formed, as predicted, by the six hypervariable segments (CDRs; Fig. 34-25).* Antibody–hapten complexes, not surprisingly, resemble enzyme–substrate complexes; both types of associ-

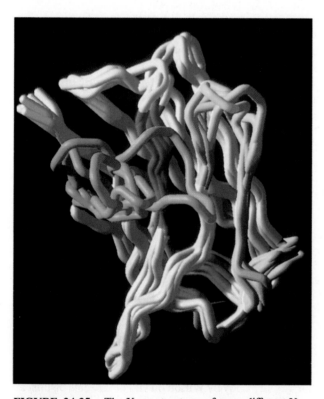

FIGURE 34-25. The X-ray structures of seven different V_H units superimposed on their conserved framework residues. The C_α backbone is colored light gray for framework (non-CDR) residues, red for CDR1, yellow for CDR2, and green for CDR3. Note that the conformational variation among these structures resides almost entirely in their CDRs. The CDRs of the V_L units are similarly varied. [Courtesy of Elizabeth Getzoff, Victoria Roberts, Michael Pique, and John Tainer, The Scripps Research Institute, La Jolla, California.]

ations involve van der Waals interactions, hydrophobic forces, hydrogen bonding, and ionic interactions. Indeed, antibody–hapten complexes and enzyme–substrate complexes have similar ranges of dissociation constants, from 10^{-4} to $10^{-10}M$, which correspond to binding energies of 25 to 65 kJ · mol^{-1}.

Antibody–hapten complexes are imperfect models of antibody–antigen complexes because a hapten only partially fills its corresponding antigen-binding site. However, the advent of monoclonal antibodies made it possible to determine the X-ray structures of protein antigens in complex with Fab's derived from the monoclonal antibodies raised against them. For example, the X-ray structures of three complexes of hen egg white (HEW) lysozyme with Fab's derived from different anti-HEW lysozyme monoclonal antibodies have been reported. Each of these Fab's binds to a largely independent, irregularly shaped, ~700-Å^2 surface patch on lysozyme such that one molecule's protruding side chains fit neatly into depressions on its mate (Fig. 34-26a). In each of these associations, all six Fab CDRs participate in lysozyme binding. These complexes, much like other known protein–protein associations, are cemented by highly complementary and thus solvent-excluding sets of van der Waals interactions, salt bridges, and hydrogen bonds. In several of these interactions, the lysozyme backbone and side chains maintain conformations identical to those in isolated lysozyme (Section 14-2A), but in others, there are significant local conformational variations (this comparison cannot be extended to the lysozyme-binding Fab's because they have not crystallized by themselves).

The exquisite specificity of anti-lysozyme antibodies for their antigenic sites is demonstrated by the effect of a single amino acid change on the lysozyme contact surface. The dissociation constant of the anti-HEW lysozyme immunoglobulin named D1.3 with HEW lysozyme is $2.2 \times 10^{-8}M$. Yet, the dissociation constant of this monoclonal antibody with those of the nearly identical egg white lysozymes from partridge, California quail, and turkey are all $> 10^{-5}M$. In all these latter lysozymes, Gln 121, which conspicuously protrudes from the HEW lysozyme surface into its Fab antigen-binding site (Fig. 34-26b), is replaced by His.

What are the special characteristics, if any, of the **epitopes** (antigenic sites) to which antibodies bind? All of the above lysozyme epitopes consist of 14 to 16 surface residues from two or more polypeptide segments. Some of these residues exhibit high mobility (Section 8-2) but others do not. Thus, the observation that our sample of only three antibody–lysozyme complexes cover around half of lysozyme's surface strongly suggests that *a protein's entire accessible surface is potentially antigenic.*

How do Fab's respond to the binding of antigen? The answer, determined in part through X-ray studies by Ian Wilson, is that it depends on the Fab and antigen in question. Comparison of the X-ray structures of an Fab from a monoclonal antibody directed against a 36-residue segment

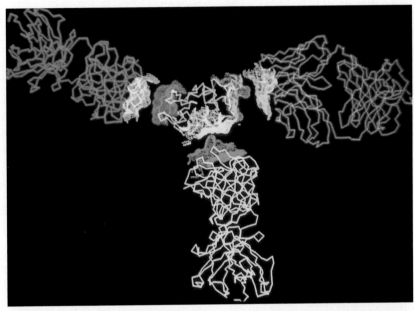

(a)

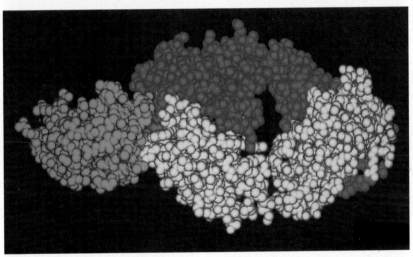

(b)

FIGURE 34-26. Hen egg white lysozyme in complex with the Fab fragments of monoclonal antibodies raised against it. (a) An exploded-view collage indicating how three different anti-lysozyme Fab's interact with lysozyme (*center*) in the X-ray structures of their respective complexes. The proteins are represented by their C_α chains and their interacting surfaces are outlined by juxtaposed dot surfaces. Note that the three crystal structures on which this diagram is based each contain only one Fab species; the Fab's do not crystallize together. [Courtesy of Steven Sheriff and David Davies, NIH.] (b) The X-ray structure of HEW lysozyme in complex with the anti-lysozyme Fab named D1.3 (Part *a, upper right*). In this space-filling representation, the Fab's L chain is yellow, its H chain is blue, the lysozyme molecule is green, and lysozyme Gln 121 is red. [From Amit, A.G., Mariuzza, R.A., Phillips, S.E.V., and Poljak, R.J., *Science* **233**, 749 (1986).]

of influenza hemagglutinin (Section 32-4B), alone and in complex with a 9-residue fragment of this antigen that the Fab binds with high affinity (Fig. 34-27), reveals that antigen binding causes the Fab to undergo a major structural rearrangement, most notably through a conformational shift of its heavy chain CDR3. Indeed, had this conformational change not occurred, the Fab would have been sterically unable to bind the antigen, at least in the conformation it assumes in the complex. On the other hand, in a

similar X-ray study by Wilson on an Fab directed against a 19-residue segment of **myohemerythrin** (a nonheme Fe-containing oxygen-binding protein from a marine worm), both alone and in complex with this antigen, the Fab underwent only a few small main chain and side chain conformational adjustments on binding this antigen.

In both of the foregoing complexes, the conformation of Fab-bound antigen is substantially different from its conformation in its intact parent protein. This raises the in-

(a)

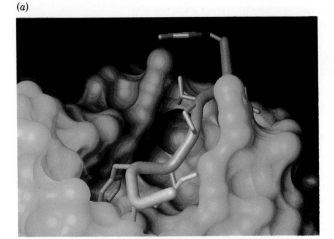

(b)

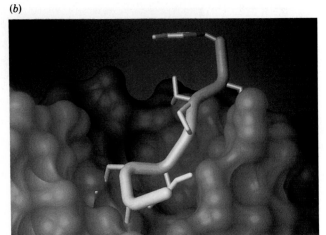

FIGURE 34-27. Comparison of the shapes of the antigen binding pockets in the X-ray structures of the liganded and unliganded forms of an Fab derived from a monoclonal antibody directed against a 36-residue segment of influenza hemagglutinin. (*a*) The solvent-accessible surface of the Fab antigen-binding pocket (*green*) with its bound 9-residue peptide fragment of the antigen shown in stick form (*pink*; sequence YDVPDYASL–amide). (*b*) The solvent accessible surface of the Fab antigen-binding pocket in the absence of the peptide ligand (*blue*) with the peptide positioned as in the complex to illustrate the extensive conformational changes that the Fab makes on peptide binding. Note that the peptide's Tyr 6 side chain (lower left of Part *a*) would collide with a portion of the binding pocket of the unliganded Fab (Part *b*) formed by its heavy chain CDR3. Consequently, the complex could not form without these conformational changes. [Courtesy of Ian Wilson, The Scripps Research Institute, La Jolla, California.]

triguing and not yet answered question of how an antibody can bind to both a peptide and to the protein from which the peptide is derived. The answer to this question will almost certainly have important consequences for the generation of peptide-based vaccines.

C. Generation of Antibody Diversity

The immune system has the capacity to generate antibodies against almost any antigen that it encounters; it can produce a virtually unlimited variety of antigen-binding sites. What is the origin of this enormous diversity? One might reasonably expect that immunoglobulin gene expression resembles that of other proteins in that every distinct H and L chain is encoded by a separate germline gene. If this were true, then to encode the billions of different antibodies each vertebrate appears capable of producing would require huge numbers of these genes. For example, it would require 10^3 H and L chain genes each to encode $10^3 \times 10^3 = 10^6$ different immunoglobulins. However, hybridization studies using radioactive cDNA probes transcribed from immunoglobulin mRNAs indicate that the mouse embryo genome, for example, contains far too few immunoglobulin genes to account for the mouse's observed level of antibody diversity. Consequently, this so-called **germline hypothesis** must be rejected.

Two other models for the origin of antibody diversity have been seriously considered:

1. The **somatic recombination hypothesis,** which was originally formulated in 1965 by William Dreyer and Claude Bennett, proposes that *antibody diversity is generated by genetic recombination among a relatively few gene segments encoding the variable region of an immunoglobulin chain.* This process occurs via intrachromosomal recombination during *B* cell differentiation so that each *B* cell clone expresses an all but unique immunoglobulin.

2. The **somatic mutation hypothesis** proposes that *antibody diversity arises through an extraordinarily high rate of immunoglobulin gene mutation during B cell differentiation.*

We shall see below that both of these mechanisms contribute to antibody diversity.

κ Light Chain Genes Are Assembled from Three Sets of Gene Segments

DNA sequencing studies by Leroy Hood, Philip Leder, and Susumu Tonegawa have revealed that κ light chains are each encoded by four exons (Fig. 34-28):

1. A **leader** or L_κ **segment,** which encodes a 17 to 20-residue hydrophobic signal peptide. This polypeptide directs newly synthesized κ chains to the endoplasmic reticulum and is then excised (Section 11-4B).

2. A V_κ **segment,** which encodes the first 95 residues of the κ chain's 108-residue variable region.

3. A **joining** or J_κ **segment** (not to be confused with the J chain of IgA and IgM), which encodes the variable region's remaining 13 residues.

4. The C_κ **segment,** which encodes the κ chain's constant region.

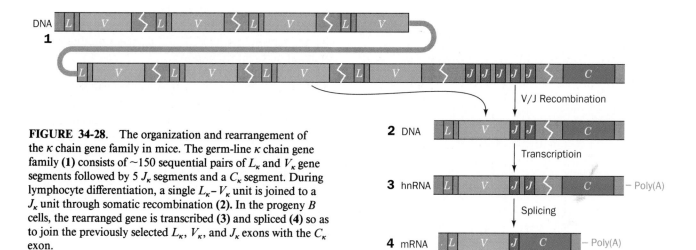

V/J Recombination

FIGURE 34-28. The organization and rearrangement of the κ chain gene family in mice. The germ-line κ chain gene family (**1**) consists of ~150 sequential pairs of L_κ and V_κ gene segments followed by 5 J_κ segments and a C_κ segment. During lymphocyte differentiation, a single L_κ–V_κ unit is joined to a J_κ unit through somatic recombination (**2**). In the progeny B cells, the rearranged gene is transcribed (**3**) and spliced (**4**) so as to join the previously selected L_κ, V_κ, and J_κ exons with the C_κ exon.

The arrangement of these exons in human embryonic tissues (which do not make antibodies) differs strikingly from those in gene families we have previously encountered. The L_κ and V_κ segments are separated by an intron as occurs in other split genes. However, the κ chain gene family contains an array of ~150 of these ~400-bp L_κ–V_κ units separated from each other by ~7-kb spacers. This sequence of exon pairs is followed, well downstream, by 5 J_κ segments at intervals of ~300 bp, a 2.4-kb spacer, and a single C_κ segment.

The assembly of a κ chain mRNA is a complex process involving both somatic recombination and selective gene splicing (Fig. 34-28). The first step of this process in mice, which occurs in a progenitor of each B cell clone, is an intrachromosomal recombination that joins an L_κ–V_κ unit to a J_κ segment and deletes the intervening sequences. Then, in later cell generations, the entire modified gene is transcribed and selectively spliced so as to join the L_κ–V_κ–J_κ unit to the C_κ segment. The L_κ and V_κ segments are also spliced together in this step, yielding an mRNA that encodes one of each of the elements of a κ chain gene.

Highly conserved sequences on the 3' side of each V_κ segment and on the 5' side of each J_κ segment suggest how the somatic recombination sites are selected. The V_κ sequence is immediately followed by the heptameric sequence CACAGTG, a 12 ± 1 nucleotide spacer, and an AT-rich nonamer. The J_κ chain is preceded by the complementary heptamer, a 23 ± 1 nucleotide spacer, and the complementary AT-rich nonamer. These sequences combine under the influence of a system of recombinatory enzymes, discussed below, to form a stem-and-loop structure that acts as a recombination signal (Fig. 34-29).

Recombinational Flexibility Contributes to Antibody Diversity

The joining of 1 of 150 V_κ segments to 1 of 5 J_κ segments can generate only $150 \times 5 = 750$ different κ chains, far less

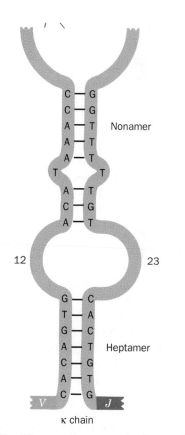

FIGURE 34-29. The germ-line κ gene family contains complementary heptamers and nonamers succeeding each V_κ segment and preceding each J_κ segment. These sequences are thought to mediate somatic recombination by forming the indicated stem-and-loop structure.

than the number observed. However, studies of many joining events involving the same V_κ and J_κ segments revealed that *the V/J recombination site is not precisely defined; these two gene segments can join at different crossover*

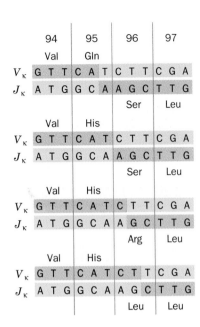

FIGURE 34-30.   The cross-over point at which the V_κ and J_κ sequences somatically recombine varies by several nucleotides, thereby giving rise to different nucleotide sequences (*brown bands*) in the active κ gene. For example, as is indicated here, amino acid 96, which occurs in the κ chain's third hypervariable region, can be Ser, Arg, or Leu.

points (Fig. 34-30). Consequently, the amino acids specified by the codons in the vicinity of the *V/J* recombination site depend on what part of the sequence is supplied by the germline V_κ segment and what part is supplied by the germline J_κ segment. Indeed, the amino acids specified by the codons surrounding the recombination junction form the light chain hypervariable region in the vicinity of residue 96 (CDR3; Fig. 34-21). Assuming that this recombinational flexibility increases the possible κ chain diversity 10-fold, the expected number of possible different κ chains is increased to $150 \times 5 \times 10 = 7500$.

The imprecision of the *V/J* joining often results in the random loss of a few nucleotides from the ends of the V_κ and J_κ segments. Consequently, up to two thirds of the recombination products have an out-of-phase reading frame downstream from the recombination joint, so the resulting gene encodes a nonsense protein. Such proteins are not expressed. *A cell in which a nonproductive recombination event has occurred will attempt further κ gene rearrangements between its remaining L_κ–V_κ and J_κ units and, if all of these fail, will rearrange its λ genes (see below).* This phenomenon accounts for the observation that κ express-

ing cells rarely have their λ genes rearranged, whereas λ expressing cells invariably have their κ genes rearranged. The mechanism by which the cell detects a productive recombination event is unknown.

λ Light Chains Derive from Multiple Constant Regions

The κ and λ chain gene families, which occur on different chromosomes, have different germline arrangements of their *L, V, J,* and *C* segments. Mice have only two L_λ–V_λ segments, each followed by a pair of J_λ–C_λ units (Fig. 34-31). Mice therefore have relatively little λ chain diversity compared to that of their κ chains. This is probably why murine light chains are 95% κ and only 5% λ. Humans, in contrast, have many more L_λ–V_λ and J_λ–C_λ units than mice; human immunoglobulins contain approximately equal amounts of κ and λ chains.

Heavy Chain Genes Are Assembled from Four Sets of Gene Segments

*Heavy chain genes are assembled in much the same way as are light chain genes but with the additional inclusion of an ~13-bp **diversity** or **D segment** between their V_H and J_H segments.* The human heavy chain gene family, which occurs on a different chromosome from either of the light chain gene families, consists of clusters of ~250 different L_H–V_H units, perhaps 10 D segments, 6 J_H segments, and 8 C_H segments, 1 for each of the 8 immunoglobulin classes and subclasses (Fig. 34-32). The D segments encode the core of the heavy chain's third hypervariable region (Fig. 34-21). Germline V_H, D, and J_H segments are flanked by heptamer–nonamer recombination signals similar to those that occur in light chain genes (Fig. 34-33). Moreover, heavy chain *V/D* and *D/J* joining sites are subject to the same recombinational flexibility as are light chain *V/J* sites. This ***V(D)J* joining** process is tightly regulated in that it occurs in a particular temporal order (e.g., D_H is joined to J_H before V_H is joined to $D_H J_H$) and in a lineage-specific manner [e.g., *T* cell receptor loci, which are also subject to *V(D)J* joining (Section 34-2D), are never fully rearranged in *B* cells].

Assuming that recombinational flexibility contributes a factor of 100 towards heavy chain diversity, somatic recombination can generate some $250 \times 10 \times 6 \times 100 = 1.5 \times 10^6$ different heavy chains of a given class. Then, taking into account κ chain diversity (and neglecting that of λ chains), *there can be as many as $7500 \times 1.5 \times 10^6 = 11$ billion different types of immunoglobulins of each class formed by somatic recombination among ~400 different gene segments.*

FIGURE 34-31.   The germline organization of the λ gene family in mice. $J_{\lambda 4}$ is a pseudogene segment.

Germ line heavy chain DNA

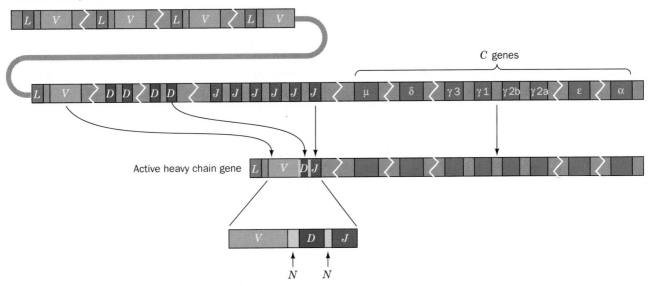

FIGURE 34-32. The organization and rearrangement of the heavy chain gene family in humans. This gene family consists of ~250 sequential pairs of L_H and V_H gene segments followed by ~10 D segments, 6 J_H segments and 8 C_H segments (one for each class or subclass of heavy chains). During lymphocyte differentiation, an $L_H–V_H$ unit is recombinationally joined to a D segment and a J_H segment. In this process, the D segment becomes flanked by short segments of random sequence called N regions. In the B cell and its progeny, transcription and splicing joins the $L_H–V_H–N–D–N–J_H$ unit to one of the 8 C_H gene segments.

RAG1 and RAG2 Proteins Probably Comprise the *V(D)J* Recombinase

The recombination signals that mediate *V(D)J* joining (Figures 34-29 and 34-33), which are necessary and sufficient to direct this process, are conserved among the different loci and species that carry it out and are functionally interchangeable. This latter observation suggests that all *V(D)J* joining reactions are catalyzed by a single evolutionarily conserved ***V(D)J* recombinase.** Through the use of recombinant DNA techniques, David Baltimore identified **a recombination activating gene, *RAG1,*** which from human and mouse, encodes putative 1043-and 1040-residue proteins that have 90% sequence identity. RAG1 protein appears to be partially homologous to the glucocorticoid receptor, including its metal-binding Cys residues (Section 33-3B) and hence, not surprisingly, appears to be a DNA-binding protein.

Although the presence of RAG1 promotes *V(D)J* recombination, it does so quite inefficiently. This led to the discovery of a second evolutionarily conserved gene, ***RAG2,*** whose protein product increases the *V(D)J* recombination rate 1000-fold over that of RAG1 alone. *RAG2,* which is chromosomally adjacent to *RAG1,* encodes a 527-residue protein that is unrelated to RAG1 or any other protein of

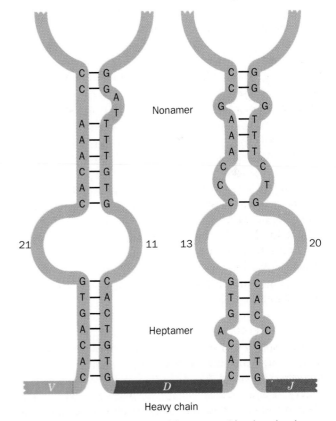

FIGURE 34-33. The stem-and-loop recombination sites in the germline heavy chain gene family that mediate somatic recombination between its V_H and D segments (*left*) and between its D and J_H segments (*right*). Compare them to the κ chain recombination signal (Fig. 34-29). The recombination system's requirement for both the 20/21 and the 11/13-bp spacers prevents it from inadvertently skipping the D segment by directly joining the V_H and J_H segments.

known sequence. Evidently, *RAG1* and *RAG2* encode two synergistically acting components of the *V(D)J* recombinase, although the hypothesis that these proteins instead regulate the expression of the *V(D)J* recombinase has not been eliminated. Note that recombination systems requiring the participation of two proteins are not unknown; for instance, the products of the *int* and *xis* genes are both required to excise bacteriophage λ DNA from the *E. coli* host chromosome upon lytic induction (Section 32-3C).

Somatic Mutation Is a Further Source of Antibody Diversity

Despite the enormous antibody diversity generated by somatic recombination, *immunoglobulins are subject to even more variation through somatic mutations of two types:*

1. During V_H/D and D/J_H joining, a few nucleotides may be added or removed from the recombination joints. The added nucleotides, which form so-called *N* **regions,** yield *NDN* units of up to 30 bp that encode enormously variable heavy chain segments of 0 to 10 amino acid residues (Fig. 34-32). Baltimore hypothesized that the *N* regions arise through the action of **terminal deoxynucleotidyl transferase,** a template-independent DNA polymerase present in the *B* cell progenitors that make the heavy chain joints but is probably absent in later cell generations when the light chain joints are formed.

2. The variable regions of both heavy and light chains are more diverse than is expected on the basis of comparisons of their amino acid sequences with their corresponding germline nucleotide sequences. Indeed, these regions mutate at rates of up to 10^{-3} base changes per nucleotide per cell generation, rates that are at least a millionfold higher than the rates of spontaneous mutation in other genes. *B* cells and/or their progenitors apparently possess enzymes that mediate this **somatic hypermutation** of immunoglobulin variable gene segments. Since the rate at which memory *B* cells are activated for proliferation increases with the affinity of their surface-displayed antibodies for antigen, *somatic hypermutation is thought to act, over many cell generations, to tailor antibodies to a particular antigen.*

These somatic mutation processes increase the possible number of different antibodies that humans can produce by many orders of magnitude beyond the 11 billion we estimated on the basis of somatic recombination alone. The final number is so large, probably > 10^{10}, that an individual synthesizes only a small fraction of its potential immunoglobulin repertoire. Somatic diversification arising from both recombination and mutation thereby permits an individual organism to cope, in a kind of Darwinian struggle, with the rapid mutational rates of pathogenic microorganisms.

Allelic Exclusion Ensures that Antibodies Are Monospecific

The immunoglobulins synthesized by a given *B* cell, as we have seen, consist of two identical heavy chains and two identical light chains. Such homogeneity is essential for the immune system's proper functioning because immunoglobulins consisting of two types of heavy and/or light chains would have two different antigen-combining sites and therefore could not form lattices of cross-linked antigens. Yet *B* cells, which like other somatic cells are diploid, contain two gene families specifying heavy chains (one maternal allele and one paternal allele) and four gene families encoding light chains (two κ's and two λ's). Apparently *B* cells are able to suppress the expression of all but one heavy chain allele and one light chain allele, a process known as **allelic exclusion,** by inhibiting further somatic recombination of heavy and light chain genes after a productive recombination has occurred. Allelic exclusion was experimentally demonstrated by microinjecting plasmids containing already recombined κ chain genes into fertilized mouse ova. The resulting transgenic mice suppress the somatic recombination of their endogenous κ chain genes. Analogous results were obtained for heavy chain genes. Although the mechanism of allelic exclusion is unknown, an intriguing possibility is that the protein products of a successful recombination event inhibit all further analogous recombinations.

The Switch from the Membrane Bound to the Secreted Form of an Antibody Involves a Change in Its Heavy Chain Transcript

The clonal selection model of antibody generation requires that the antibody displayed on the surface of a virgin *B* cell have the same specificity for antigen as the antibody secreted by its mature *B* cell progeny. Membrane-bound IgM (the antibody synthesized by virgin *B* cells) is anchored to the plasma membrane by a 41-residue hydrophobic polypeptide forming the C-terminus of its heavy chain (μ_m). In the secreted form of IgM (the first antibody secreted by mature *B* cells), the heavy chain (μ_s) has a different C-terminal segment but is otherwise identical. How does the *B* cell alter the synthesis of this heavy chain?

Somatically recombined heavy chain genes consist of eight exons (Fig. 34-34): an *L* segment that encodes a signal peptide leader; a *VDJ* unit that encodes the V_H domain; four exons that encode the C_H1 domain, the hinge region, the C_H2 domain, and the C_H3 domain (a further example of exons each specifying a structurally significant polypeptide unit); and two exons that collectively encode the transmembrane tail of μ_m. In forming the mRNA specifying μ_m, the splicing system excludes the segment at the end of the C_H3 exon that specifies the μ_s tail and the entire transcript is terminated, as usual, by poly(A). In forming μ_s mRNA, however, the splicing system retains the μ_s segment and the transcript is polyadenylated after this point, thereby elimi-

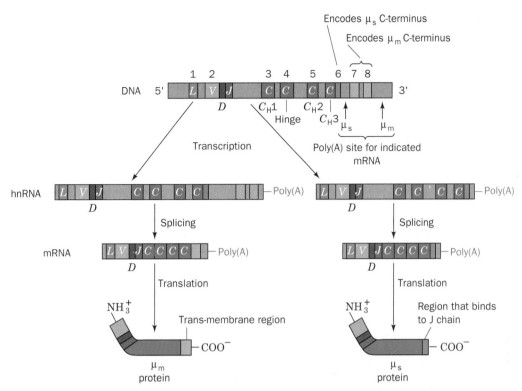

FIGURE 34-34. The C_μ gene specifies both the μ_m and the μ_s proteins through the selection of alternative splice and polyadenylation sites. In μ_m mRNA (*left*), the segment at the end of the C_H3 exon (6), which specifies the μ_s tail, has been spliced away and the transcript has been polyadenylated after the two exons specifying its transmembrane segment (7 + 8). μ_s mRNA (*right*), on the other hand, is polyadenylated just past its retained μ_s tail segment.

nating the transmembrane tail. How antigen-stimulated *B* cells switch between these alternative splice and polyadenylation sites is unknown.

B Cells Can Switch the Class of Immunoglobulin They Synthesize

Virgin *B* cells mostly synthesize membrane-bound IgM. Yet, the progeny of *B* cells that have been stimulated to proliferate may eventually synthesize immunoglobulins of different classes that have the same variable regions as the original IgM (recall that these different immunoglobulin classes have distinct physiological roles). The nucleic acid sequences specifying the variable region of the heavy chain must therefore become juxtaposed with the sequences specifying the constant regions of various types of heavy chains. What is the mechanism of this **class switching**?

The downstream regions of the human heavy chain gene family consist, as we have discussed, of eight segments encoding the constant regions for the various immunoglobulin classes and subclasses (Fig. 34-35). Class switching might occur either through RNA processing or through DNA processing. In fact, both mechanisms occur. In the RNA processing mechanism, it is uncertain whether the switching event is a change in transcriptional termination, polyadenylation, or splicing but, in any case, the result is the synthesis of heavy chain mRNAs with identical variable regions but different constant regions. The cell can therefore simultaneously synthesize two or more classes of immunoglobulins with identical antigen-binding sites.

The DNA processing mechanism of class switching occurs through somatic recombination between the *VDJ* unit and *C* region of choice. In doing so the intervening segment of DNA is deleted so that this mechanism is progressive and irreversible. For example, in recombinational switching from making IgM to making IgG1 (Fig. 34-35), a *B* cell loses its C_μ, C_δ, and $C_{\gamma3}$ segments so that its progeny cannot synthesize IgM, IgD, or IgG3. Yet, the progeny still have the potential to switch to IgG2, IgE, and IgA synthesis since the recombination does not disturb the $C_{\gamma2}$, C_ε, and C_α segments. Each of the C_H segments, with the exception of C_δ, is preceded by a **switch** or **S region** that consists of multiply repeated short complementary elements (C_δ is only expressed through RNA processing). These *S* regions are therefore thought to form the recombination signals used in class switching.

D. T Cell Receptors

T cell receptors, as we have seen, are in many ways the cellular immunity system's analog of immunoglobulins.

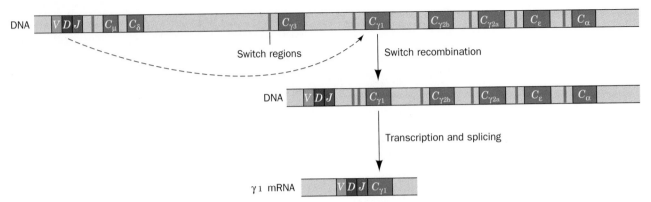

FIGURE 34-35. DNA-mediated class switching. An irreversible change in heavy chain synthesis from μ to a downstream constant region (shown here as $\gamma 1$) occurs through somatic recombination between the switch regions located upstream of all but one of the constant regions. Each constant region consists of multiple exons and encodes alternative secreted and membrane-binding C-termini (see Fig. 34-34).

Like immunoglobulins, *T* cell receptors exhibit enormous specificity in binding antigens. Yet, *T* cell receptors have proved quite difficult to characterize because they occur only as cell surface proteins and are therefore present in small quantities. They were finally isolated in 1983 through the use of monoclonal antibodies directed against them and characterized through reverse genetics (Section 33-3).

The *T* cell receptor consists of two glycosylated polypeptide chains, α and β, which in humans have respective molecular masses of 50 and 39 kD (there is also a related class of *T* cell receptors that consist of γ and δ subunits). Each of these chains has a constant domain and a variable domain of approximately equal size as well as a C-terminal transmembrane segment (Fig. 34-36a). Not surprisingly, considering their similar functions, the *T* cell receptor subunits are sufficiently homologous to the immunoglobulin subunits (Fig. 34-36b) that their constant and variable domains are each predicted to assume the immunoglobulin fold. Moreover, the α and β subunit gene families are each organized into clusters of *V* and *J* regions, with the α family having an additional *D* region. The somatic recombination of these regions, as mediated by heptamer–nonamer sequences similar to those guiding the analogous process in immunoglobulin genes (Figs. 34-29 and 34-33), is a major source of *T* cell receptor diversity. The homology between the *T* cell receptor and immunoglobulin subunits indicates that these proteins all have a common ancestor. Nevertheless, the gene families encoding the α and β chains are distinct from the immunoglobulin gene families and, like them, reside on different chromosomes.

E. The Major Histocompatibility Complex

The membrane-bound proteins encoded by the major histocompatibility complex (MHC; *histo* refers to tissue), as we have seen, are the antigen-presenting markers through which the immune system distinguishes body cells from invading antigens (Class I MHC proteins) and immune system cells from other cells (Class II MHC proteins). In the following paragraphs, we outline the structures and genetic properties of these essential proteins.

MHC Proteins Are Highly Polymorphic

The MHC has been extensively studied in both humans and mice. In humans, the Class I MHC proteins are encoded by three separate although homologous genetic loci, **HLA-A, HLA-B,** and **HLA-C** (Fig. 34-37; *HLA* stands for *h*uman-*l*eukocyte-*a*ssociated antigen since these proteins were first observed on leukocytes), so each individual synthesizes up to six different Class I MHC proteins (see below). There are also three human Class II MHC proteins whose α and β chains are encoded by genes designated DP_α, DP_β, DQ_α, DQ_β, DR_α, and DR_β (Fig. 34-37). Mouse MHC genes, which occupy the **H-2** loci, are similarly arranged.

The most striking feature of the Class I and Class II MHC genes is their high level of polymorphism among individuals of the same species; *they are, in fact, the most polymorphic genes known in higher vertebrates.* For example, in humans, 23 *A* alleles, 49 *B* alleles, and 12 *C* alleles have been characterized, and no doubt more will be discovered. Likewise, there are > 50 known alleles for each of the mouse Class I MHC genes. *Two unrelated individuals are therefore highly unlikely to have the same set of MHC genes.*

Class I MHC Proteins

Tissues can be readily transplanted from one part of an individual's body to another or between genetically identical individuals (e.g., identical twins). Yet, when tissues are transplanted between even closely related individuals, the graft is generally destroyed by the recipient's immune system (a phenomenon that is a major impediment to the transplantation of organs such as hearts and kidneys). Studies of such **graft rejection** led, nearly 50 years ago, to the discovery of the Class I MHC proteins, which are therefore also known as **transplantation antigens.**

The Class I MHC proteins are ~44-kD transmembrane

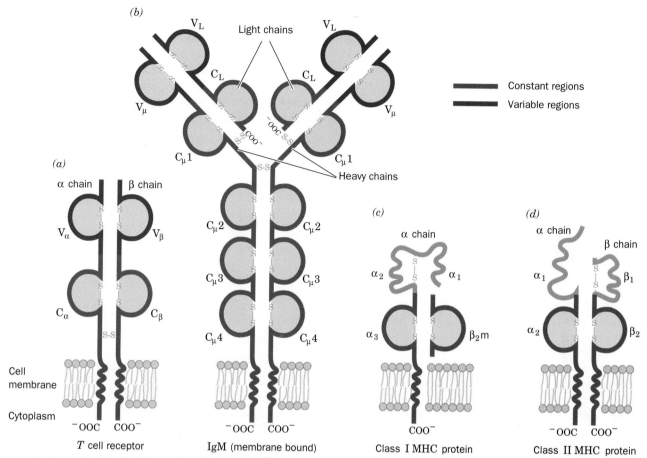

FIGURE 34-36. The members of the immunoglobulin gene superfamily, such as (*a*) the *T* cell receptor, (*b*) membrane-bound IgM, (*c*) Class I MHC protein, and (*d*) Class II MHC protein, all have similar domain structures. Each of these proteins contains multiple immunoglobulin homology units (*brown and purple regions*). Note that the IgM heavy chain has four constant regions compared with three in IgG (Fig. 34-18). Other members of the immunoglobulin gene superfamily include the **CD4 cell surface glycoprotein** of helper *T* cells (a monomeric transmembrane protein that contains four tandem immunoglobulin homology units), and the **CD8 cell surface glycoprotein** of cytotoxic and suppressor *T* cells (which occurs as both a transmembrane α_2 homodimer and a transmembrane

$\alpha\beta$ heterodimer, each of whose subunits contains one immunoglobulin homology unit). These proteins help their cells to recognize the Class II and Class I MHC proteins on other cells (Section 34-2A), presumably by binding to nonpolymorphic regions of these MHC proteins [CD4 is also the HIV (AIDS virus) receptor, which is why this virus specifically infects helper *T* cells]. Evidently, the members of the immunoglobulin gene superfamily are specialized for molecular recognition rather than limited to the immune system. Thus, the immunoglobulin fold-containing protein named **neuronal cell-adhesion molecule (N-CAM)** participates in cell–cell recognition but is not an immune system component.

glycoproteins that are displayed on the surfaces of nearly all nucleated vertebrate cells. These proteins' amino acid sequences suggest that they are folded into five domains that are, from C-terminus to N-terminus, an ~30-residue cytoplasmic domain, a transmembrane segment of ~40 resi-

dues, and three external domains of ~90 residues each designated α_3, α_2, and α_1 (Fig. 34-36c). The Class I MHC proteins are invariably noncovalently associated in a 1 : 1 ratio with β_2**-microglobulin (β_2m;** Fig. 34-36c), a 12-kD protein. The X-ray structure of the extracellular portion of a

Chromosome 6

Marker loci		DP_β	DP_α		DQ_β	DQ_α	DR_β	DR_α		$C4B$	$C4A$	Bf	$C2$		B		C		A	
Class		II	II		II	II	II	II		III	III	III	III		I		I		I	

FIGURE 34-37. The genetic map of the MHC which, in humans, encodes the HLA proteins. The Class III genes encode several complement system proteins (Section 34-2F).

Class I MHC protein (Fig. 34-38a), HLA-A2, which was elucidated by Don Wiley and Jack Strominger, indicates that its α_3 domain as well as β_2-microglobulin, both of which are homologous with immunoglobulins, assume the immunoglobulin fold. *Evidently, all of these proteins, together with T cell receptors, are evolutionarily related and therefore form a **gene superfamily** (a set of evolutionarily related genes with divergent functions).*

The homologous α_1 and α_2 domains of HLA-A2 form a relatively flat eight-stranded antiparallel β sheet that is parallel to the cell membrane and which is flanked by two α helices (Fig. 34-38a and b). *The resulting deep groove or cleft, which is of sufficient size and convoluted shape to bind a largely extended 8- to 10-residue polypeptide, forms the binding site of a cell-processed antigen fragment that, together with the Class I MHC protein itself, is recognized by*

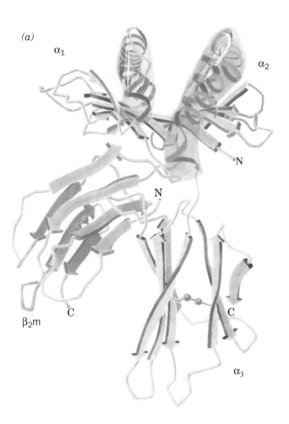

(a)

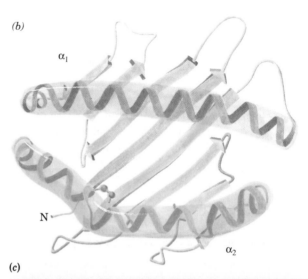

(b)

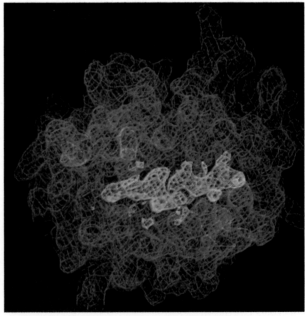

(c)

FIGURE 34-38. The X-ray structure of a human Class I MHC protein, HLA-A2. The protein was obtained by the papain digestion of plasma membranes from cultured human cells, which cleaves the protein at residue 271, 13 residues before its transmembrane segment. (a) A ribbon diagram of the protein indicating the relationships among its immunoglobulin-like domains, β_2m and α_3 (*bottom*), and its polymorphic α_1 and α_2 domains (*top*). The protein is oriented such that the plasma membrane from which it normally projects would be horizontal at the bottom of the drawing. The protein's antigen recognition site, which located in the groove between the α_1 and α_2 helices, is apparently readily accessible from outside the cell. Disulfide bonds are represented by connected yellow spheres. (b) Top view of the α_1 and α_2 domains showing the surface that presumably contacts the *T* cell receptor (viewed from above in Part a). The domains, which each consist of four antiparallel β strands followed by a long helix, pair with pseudo-twofold symmetry to form a relatively flat eight-stranded antiparallel β sheet flanked by the two helices. The resulting groove forms the binding site for the ~9 residue antigen fragment that the Class I MHC protein presents to the *T* cell receptor. Most of the protein's polymorphic residues line the surface of this groove as do many residues critical for *T* cell receptor recognition. (c) The protein's van der Waals surface (*blue*), viewed as in Part b, showing the extra electron density (*red*) that is observed to occupy the protein's binding site. This extra electron density represents a collection of unidentified peptides that the protein had bound before its isolation. [Courtesy of Don Wiley, Harvard University.]

(a)

(b)

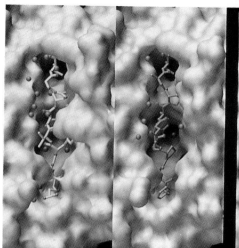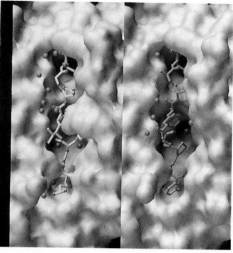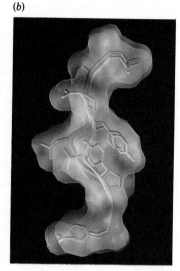

FIGURE 34-39. The X-ray structures of the complexes of murine H-2K^b Class I MHC protein with a nonapeptide fragment of **Sendai virus** nucleoprotein (SEV-9; sequence FAPGNYPAL) and with an octapeptide fragment of vesicular stomatitis virus nucleoprotein (VSV-8; sequence RGYVYQCL) showing their similarity. (*a*) A double stereo drawing of H-2K^b displayed as its solvent-accessible surface and viewed into its antigen-binding cleft with the SEV-9 complex (*blue*) on the right and the VSV-8 complex (*yellow*) on the left. The peptides are shown in ball-and-stick form with N blue and O red, and with bound water molecules represented by blue spheres. Note the depth of the cleft and the relative inaccessibility of the peptides. Directions for viewing stereo drawings are given in the appendix to Chapter 7. (*b*) The SEV-9 (*blue*) and the VSV-8 (*yellow*) polypeptides shown superimposed in their H-2K^b binding clefts. The polypeptide backbones are each represented by a rod with their side chains shown in skeletal form. The view is from the right in Part *a* such that the top of the cleft is on the left. [Courtesy of Ian Wilson, The Scripps Research Institute, La Jolla, California.]

a T cell receptor (Section 34-2A). Indeed, the HLA-A2 structure contains an unknown "antigen" bound in the cleft (Fig. 34-38*c*) that apparently copurified and cocrystallized with this Class I MHC protein (which was synthesized in a cultured human cell line). Moreover, the amino acid residues that differ among HLA-A2 and the two other human Class I MHC proteins whose X-ray structures have been determined (HLA-Aw68 and HLA-B27) are concentrated in and around this antigen-binding cleft (as are the variable residues in other Class I MHC proteins when modeled on the HLA-A2 structure).

The X-ray structures of both human and murine Class I MHC proteins complexed to either endogenous peptides or to specific exogenous octa- and nonapeptides have revealed how these proteins bind their cognate peptides and present them to *T* cell receptors. Peptides associate with Class I MHC proteins, mostly via hydrogen bonds involving the peptide's backbone, such that the peptides assume nearly extended but twisted conformations resembling that of the polyproline II helix (Section 7-1B). As a consequence, consecutive peptide side chains protrude in largely opposite directions, somewhat like the side chains of a strand of β sheet (Figure 7-17). In the complex of murine H-2K^b protein with a viral nonapeptide, for example (Fig. 34-39), the side chains of residues P2, P3, P6, and P9 (where P*n* represents the *n*th residue of the peptide) face inward to contact the protein in pockets that appear designed to bind them.

The remaining side chains are at least partially in contact with solvent and presumably could interact with a *T* cell receptor. In addition, the peptide's N- and C-termini bind to deep and conserved pockets at each end of the MHC protein's binding cleft through hydrogen bonding contacts to conserved residues that thereby dictate the peptide's orientation. Thus, a viral octapeptide bound to H-2K^b (Fig. 34-39) maintains essentially the same contacts as does the nonapeptide because the nonapeptide's P5 residue is accommodated through the formation of a central bulge (i.e., residues P6 to P9 in the nonamer correspond to residues P5 to P8 in the octamer).

The sizes and amino acid compositions of the pockets containing the occluded side chains suggest that any particular Class I MHC protein can only bind a limited selection of peptides. For example, in the complex of HLA-B27 with endogenous peptides, the P2 side chain binds in a hydrophobic pocket that ends near Cys 67 and the negatively charged Glu 45, which suggests that this site preferentially binds a long positively charged side chain. In fact, all 11 peptides that were eluted from HLA-B27 contain Arg at P2. The identities of its other occluded side chains, those at P3, P7, and P9, although not as restricted as that of P2, are more or less consistent with the properties of the pockets in which they bind, whereas the solvent-exposed side chains have the expected wide range of identities.

The peptides eluted from other class I MHC proteins also

exhibit distinct allele-specific sequence motifs. In particular, each sequence motif contains two so-called anchor positions that are occupied by only one or, at most, a few residues with closely related side chains. The anchor positions vary with the MHC protein. In contrast to the anchor positions, side chains that are solvent-exposed or bound in pockets that are tolerant of sequence variation show considerable diversity. This system presumably permits T cell receptors to distinguish between self and foreign antigens.

In an effort to elucidate the roles of the conserved residues on HLA-A2 implicated in interacting with the N- and C-termini of peptides to which HLA-A2 binds, Strominger investigated the effects of mutating these residues on an HLA-A2–peptide complex's ability to activate killer T cells. Either of two Tyr residues implicated in binding the N-terminus of associated influenza virus–derived nonapeptides were mutated to Phe. This reduced, by up to two orders of magnitude, the ability of killer T cells to lyse cells bearing either of these HLA-A2 mutants and which had been exposed either to one of the nonapeptides or to intact influenza virus. Evidently, the hydrogen bonds that the Tyr side chains make with the N-terminal amine group of these and probably other HLA-A2-bound nonapeptides are critical for activating killer T cells. Nevertheless, the mutation of Tyr and Thr residues that form hydrogen bonds to the C-terminal carboxylate group of nonapeptides to Phe and Val did not affect killer T cell activation.

Class II MHC Proteins

The discovery of Class II MHC proteins came about through Baruj Benacerraf's observation that certain T cell–dependent immune responses are mediated by gene products that are not antibodies. For example, when guinea pigs are inoculated with a simple antigen such as polylysine, some individuals mount a vigorous immunological response to the antigen, whereas others fail to respond. Immunological responsiveness to a given simple antigen is a dominant genetic trait. *A small number of so-called immune response (Ir) genes apparently govern how an individual responds to all simple antigens.* [A naturally occurring antigen, such as a protein, is complex, that is, it has numerous different epitopes (antigenic determinants; Section 34-2B). Consequently, an individual will almost always be able to mount an immune response against a naturally occurring antigen.]

The *Ir* genes map into the MHC so that they are now known as Class II MHC genes. They encode the two subunits of a heterodimeric transmembrane glycoprotein composed of a 33-kD α chain and a 28-kD β chain, each of which consists of two domains (Fig. 34-36d). The amino acid sequences of these subunits indicate that the C-terminal α_2 and β_2 domains are members of the immunoglobulin gene superfamily. Moreover, their α_1 and β_1 domains can be convincingly aligned on the known structures of the Class I MHC proteins' α_1 and α_2 domains, respectively. This indicates that Class I and Class II MHC proteins are structurally as well as functionally similar.

This prediction has been largely confirmed by the X-ray structures of the extracellular portion of a Class II MHC protein, HLA-DR1, in its complexes with both a mixture of endogenous peptides and with a 13-residue fragment of influenza virus hemagglutinin protein (HA), both determined by Strominger and Wiley. However, the peptide-binding site of HLA-DR1 is an open-ended groove (Fig. 34-40a and b), whereas those of Class I MHC proteins are elongated but closed-ended clefts (Figs. 34-38c and 34-39). This explains why Class II MHC proteins bind peptides of arbitrary length, whereas Class I MHC proteins bind mostly extended but bulged-out nonapeptides. Indeed, the X-ray structure of HLA-DR1 in complex with the 13-residue HA peptide reveals that the peptide extends out from both ends of its binding groove (Fig. 34-40b).

Although Class I and Class II MHC proteins are both $\alpha\beta$ dimers, HLA-DR1 crystallizes as a dimer of $\alpha\beta$ dimers in which all four C-termeni face in one direction and both peptide-binding clefts face in the opposite direction (Fig. 34-40c). This is the way in which Class II MHC protein $\alpha\beta$ dimers would be expected to associate, if they do so, on the cell-surface membrane. Since the ligand-induced dimerization of cell-surface receptors is a known mechanism of signal transduction (Section 34-4B), it may well be that the dimerization of Class II MHC proteins induces the cooperative dimerization of the cognate T cell receptors in a way that triggers T cell activation (although it cannot be ruled out that HLA-DR1 dimerization is simply an artifact of crystallization). This would explain why T cell receptors are activated by being crosslinked with a divalent antibody but not by the corresponding monovalent Fab fragment.

MHC Polymorphism Has An Important Protective Function

Most of the polymorphic residues in MHC proteins are clustered, as we saw, in their antigen-building grooves so that each polymorph binds a given antigenic fragment with characteristic affinity (e.g., it is estimated that any Class I MHC polymorph can bind < 1% of the octa- and nonapeptides it encounters). The above-described observations on the variation of immunological responses with Class II MHC *(Ir)* genes therefore suggest that some Class II MHC protein polymorphs are less effective than others in associating with a given epitope. Indeed, *epidemiological studies indicate that certain polymorphs of MHC genes are associated with increased or decreased susceptibilities to particular infectious and/or autoimmune diseases.* For example, 95% of individuals with insulin-dependent diabetes mellitus (Section 25-3B) carry at least one *DR2* or *DR3* allele of the *DR* gene in comparison to 50% of normal individuals, whereas **celiac disease** (a violent intestinal upset resulting from eating wheat gluten) is 100% linked to the *DQw2* allele of the *DQ* gene. Conversely, a study of the distribution of MHC alleles in West African children with severe (probably fatal in the absence of treatment) malaria compared to that in infected but largely unaffected children (only a small fraction of those infected with malaria have life-threatening

(a)

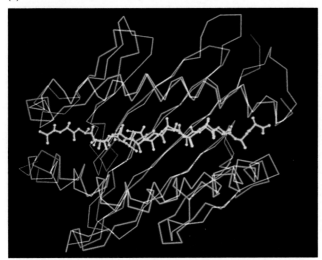

(b)

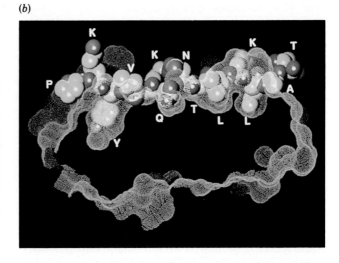

(c)

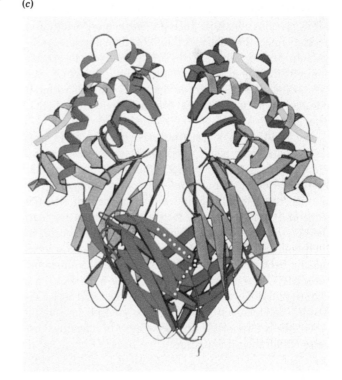

FIGURE 34-40. The X-ray structure of the human Class II MHC protein HLA-DR1. (*a*) Superposition of the peptide-binding sites, as seen in "top" view, of HLA-DR1 in complex with a 13-residue HA fragment (*orange;* peptide sequence PKYVKQNTLKLAT) and the Class I MHC protein HLA-B27 in complex with an endogenous mixture of nonapeptides (*light blue*). The proteins are represented by their C_α backbones and their bound peptides are shown in ball-and-stick form with their N-termeni to the left and their side chains removed for clarity. Note the close structural similarity of these two classes of MHC proteins and that the HLA-DR1-bound peptide extends beyond the edges of its binding site, whereas peptides bound to Class I MHC proteins are entirely embedded in protein (Figs. 34-38*c* and 34-39). (*b*) A "vertical" section through HLA-DR1's molecular surface (*light blue*) with its bound 13-residue HA peptide shown in space-filling form (C yellow, N dark blue, and O red). The peptide's side chains are identified by their one-letter codes. (*c*) A ribbon diagram indicating how two HLA-DR1 $\alpha\beta$ heterodimers associate in the crystal to form a dimer of dimers. One $\alpha\beta$ heterodimer is blue and the other is red with the α-chains shown in lighter shades than the β-chains. The bound peptides are shown in yellow. The twofold axis relating the two $\alpha\beta$ heterodimers is vertical. [Courtesy of Don Wiley, Harvard University.]

illness), showed that the Class I MHC protein HLA-Bw53 and the Class II MHC protein DRB1*1302-DQB1*0501 independently associated with protection from severe malaria. These alleles are quite common in West African populations but rare or even nonexistent in other areas. Indeed, in West African populations (in which ~1% of the children under 5 years of age die from malaria), these MHC alleles provide greater protection against malaria than does the sickle-cell trait (Section 6-3A).

What is the function of MHC protein polymorphism? It seems unlikely that it evolved only to prevent tissue grafts. Recall, however, that *T* cell receptors only recognize antigens when they are presented together with MHC proteins (Section 34-2A). If every member of a single species had an identical set of MHC proteins, a pathogen whose epitopes interacted poorly with these MHC proteins would obliterate that species. MHC gene polymorphism is thought to prevent pathogens from evolving the capacity to do so. Natural selection therefore tends to maintain a large variety of MHC proteins in a population.

Why Did So Many Amerinds Die of Introduced Infectious Diseases?

The majority of the 56 million Amerinds (indigenous peoples of North and South America) estimated to have died as a consequence of the European exploration of the New World succumbed to a variety of common infectious diseases that originated in the Old World (e.g., measles and small pox). Moreover, this excess mortality, which constituted as much as 90% of some populations, has persisted into modern times through continued episodes of the same diseases, even with medical treatment. Yet, Amerinds do not have unusual genetic susceptibilities for these diseases or immune systems that are somehow deficient. Rather, it appears that the biochemical basis of this calamity stems from the fact that Amerind populations are unusually homogeneous.

A clue to this enigma came from the observation that children who catch measles from a family member have twice the fatality rate of children who catch this disease from an unrelated individual. The measles virus, as is true of many other RNA viruses, replicates with low fidelity (because reverse transcriptase lacks a proofreading exonuclease function; Section 31-4C) and hence rapidly evolves to better adapt to its particular host. Thus, a virus that has adapted its peptides to bind less effectively to the particular MHC proteins of its host will, on average, more readily and severely infect the relatives of this host than unrelated individuals, because the relatives are more likely to express some of the same MHC protein alleles. Conversely, the greater the allelic diversity of the MHC proteins within a population, the less readily a virus can spread through that population.

The MHC genes of Amerinds have been found to have less than half the allelic variation at their *HLA–A* and *HLA-B* loci (and, by inference, other MHC loci) than do most other ethnic populations. This effect is magnified by the fact that the probability of a virus encountering the same MHC allele in two successive hosts is proportional to the square of the frequency with which that allele occurs in the population. Thus, there is a 32% chance that a virus passing between two South Amerinds will encounter the same MHC type at either the *A* or *B* locus but only a 0.5% chance of it doing so when passing between Africans. Evidently, the relatively low MHC polymorphism among Amerinds is responsible for the severity of infectious diseases within their populations.

F. The Complement System

Antibodies, for all their complications, only serve to identify foreign antigens. Other biological systems must then inactivate and dispose of the intruders. The **complement system,** a complex series of interacting plasma proteins, is one of these essential defensive systems. Indeed, it was named to indicate that it "complements" the function of antibodies in eliminating antigens. It does so in three ways:

1. It kills foreign cells by binding to and lysing their cell membranes, a process known as **complement fixation.**

2. It stimulates the phagocytosis of foreign particles, a process named **opsonization.**

3. It triggers a local acute inflammatory reaction that walls off the area and attracts phagocytotic cells.

In the following paragraphs, we describe the organization and function of the complement system.

The complement system consists of ~20 plasma proteins (Table 34-3) that interact in two related sets of reactions (Fig. 34-41): the antibody-dependent **classical pathway** and the antibody-independent **alternative pathway.** *Both pathways largely consist of the sequential activation of a series of serine proteases, much like the blood clotting pathway (Section 34-1).*

The complement system has its own peculiar nomenclature. Most complement protein names consist of the upper case letter "C" followed by a component number and, if the protein is either a subunit or fragment of a larger protein, a lower case letter. Active proteases are indicated by a bar over the component identifier. For example, $\overline{C4b}$ is a protease that has been activated by the proteolysis of C4.

TABLE 34-3. PROTEIN COMPONENTS OF THE COMPLEMENT SYSTEM

Protein	Subunit Structure	Molecular Mass (kD)
Recognition Unit (C1)		
C1q	$A_6B_6C_6$	460
C1r	α_2	157
C1s	α_2	150
Activation Unit		
C2	Monomer	81
C3	$\alpha\beta$	174
C4	$\alpha\beta\gamma$	187
Membrane Attack Unit		
C5	$\alpha\beta$	190
C6	Monomer	102
C7	Monomer	91
C8	$\alpha\beta\gamma$	142
C9	Monomer	61
Alternative Pathway		
Factor B	Monomer	83
Factor $\overline{D}$	Monomer	24
Properdin (P)	α_4	224
Regulatory Proteins		
Factor H	Monomer	137
Factor I	$\alpha\beta$	63
C4b Binding protein	α_7	570
$\overline{C1}$ Inhibitor	Monomer	53
S Protein	Monomer	52

Source: Halkier, T., *Mechanisms in Blood Coagulation, Fibrinolysis, and the Complement System, pp.* 164–165, Cambridge University Press (1991).

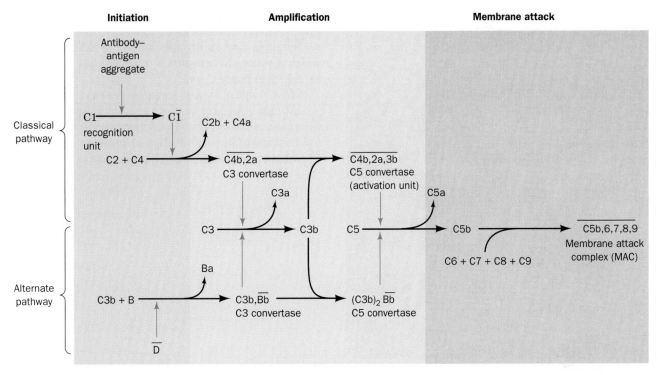

FIGURE 34-41. A schematic diagram of the complement system activation pathways. Colored arrows indicate proteolytic activations. Active proteases are indicated by a bar over the component number.

The Classical Pathway Is Triggered by Antibody–Antigen Complexes

In the classical pathway, the complement proteins form three sequentially activated membrane-bound complexes (Fig. 34-41), *top*):

1. The **recognition unit,** which assembles on cell surface–bound antibody–antigen complexes.

2. The **activation unit,** which amplifies the recognition event through a proteolytic cascade.

3. The **membrane attack complex (MAC),** which punctures the antibody-marked cell's plasma membrane causing cell lysis and death.

The Recognition Unit

*The classical pathway is initiated when **C1**, the recognition unit, specifically binds to a cell-surface antigen–antibody aggregate.* C1 occurs in the plasma as a loosely bound complex of **C1q, C1r,** and **C1s.** C1q is a remarkable 18-polypeptide chain protein, $A_6B_6C_6$, in which the ~80 N-terminal residues of each chain has the repeating sequence Gly-X-Y characteristic of collagen, where X is often Pro and Y is often 4-hydroxyproline or 5-hydroxylysine (Section 7-2C). C1q is therefore a bundle of six collagen-like triple helices that each end in a C-terminal globular domain so as to form an assembly that resembles a bunch of six tulips (Fig. 34-42). It is these globular domains that bind antigen-bound antibody through their recognition of the Fc

regions of IgM and several subclasses of IgG (although how or even if the Fc region in an antibody–antigen complex conformationally differs from that in the free antibody is unknown). Moreover, C1 is only activated if at least two of its C1q heads are simultaneously bound to antibody, a process that requires the participation of at least two IgG's but only one IgM (recall that IgM is pentameric). IgM is therefore far more effective in activating the complement system than is IgG. A variety of foreign substances, including bacterial lipopolysaccharides and viral membranes, can also activate C1.

The remaining C1 components, C1r and C1s, are homologous serine protease zymogens which, like most of the blood clotting zymogens, are each activated by a single proteolytic cleavage that yields two disulfide-linked chains. The binding of antibody–antigen complex stimulates C1q to bind two subunits each of C1r and C1s more tightly which, in a Ca^{2+}-dependent process, results in the autoactivation of C1r through the cleavage of an Arg–Ile bond. C̄1r, in turn, specifically cleaves C1s, its only known substrate, also at an Arg–Ile bond, to yield C̄1s.

The Activation Unit

*The activation unit consists of components derived from **C2, C3,** and **C4.*** In the initial step forming the activation unit, C̄1s cleaves C4 at an Arg–Ala bond and the larger of the resulting fragments, C4b, covalently binds to the cell membrane (as described below) in the vicinity of the recog-

(a)

(b)

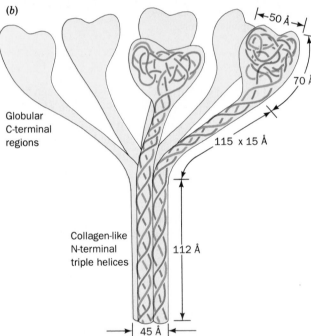

Globular
C-terminal
regions

Collagen-like
N-terminal
triple helices

← 50 Å →

70 Å

115 x 15 Å

112 Å

← 45 Å →

FIGURE 34-42. The structure of the complement protein C1q. (*a*) An electron micrograph of C1q seen in "side" view showing its 6 C-terminal domains attached to a central stalk. [Courtesy of Tibor Borso, USPHS–NIH.] (*b*) A schematic diagram of C1q. Its central stalk consists of a bundle of 6 collagen-like triple helices. The 80-residue triple helices are $80 \times 2.9 \approx 227$ Å in length. [After Porter, R.R. and Reid, K.B.M., *Nature* **275**, 701 (1978).]

nition unit. Membrane-bound C4b, in association with $\overline{\text{C1s}}$, specifically cleaves C2. C2a, the resulting larger fragment, combines with C4b to yield $\overline{\text{C4b,2a}}$, a protease

named **C3 convertase,** that cleaves C3 to C3a and C3b. Finally, C3b combines with C3 convertase to yield the activation unit, $\overline{\text{C4b,2a,3b}}$, also known as **C5 convertase,** which functions to activate **C5** proteolytically by cleaving an Arg–Leu bond.

Both C4 and C3 have buried hyper-reactive thioester groups that, when exposed, can covalently link these proteins to the cell membrane. In C3, the thioester consists of a Cys thiol and a Glu γ-carboxyl group forming a macrocyclic ring of sequence Gly-Cys-Gly-Glu-Glu-Asn:

Upon the cleavage of C3, the product C3b undergoes a conformational rearrangement that exposes its thioester group. The thioester then reacts with a nearby cell-surface amine or OH group to yield the corresponding amide or ester bond together with a free Cys sulfhydryl group. The function of this process is further discussed below. C4 is thought to behave in a similar manner on activation.

The activation of C3, C4, and C5 also triggers other immune system functions. C3b and, to a lesser extent, C4b are **opsonins,** substances that stimulate phagocytosis (opsonization), whereas C3a, C4a, and C5a (a product of the C5 convertase reaction) are **anaphylatoxins,** substances that trigger local acute inflammatory reactions and smooth muscle contraction.

The Membrane Attack Complex

C5b, the other product of the C5 convertase reaction, exhibits no proteolytic activity. Rather, it sequentially binds **C6** and **C7** to form a complex that spontaneously inserts into cell membranes. *This C5b,6,7 complex then binds one molecule of **C8** followed by anywhere between 1 and 18 molecules of **C9** to form the MAC. In this latter process, the C9 molecules polymerize to form a tubular membrane-embedded structure to which the C5b,6,7,8 complex is firmly attached (Figs. 34-43 and 34-44).* Cell

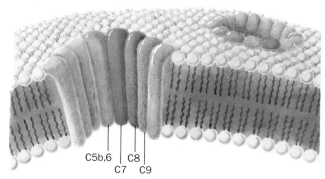

FIGURE 34-43. The membrane attack complex (MAC) is a tubular structure that forms a transmembrane pore in the target cell's plasma membrane. The perforin pores generated by cytotoxic *T* lymphocytes (Section 34-2A) have similar structures (perforin and C9 are homologous) but lack an analog of the C5b,6,7,8 complex.

lysis ensues both because the MAC forms a 30- to 100-Å-diameter aqueous channel that pierces the membrane (Fig. 34-43) and because it perturbs the surrounding membrane structure so as to increase its permeability. Both mechanisms permit the cell's small molecules, but not its macromolecules, to exchange with the surrounding medium. Water is therefore osmotically drawn into the cell, causing it to swell and burst. MACs are efficient cell killers; very few, possibly only one, can lyse a cell.

The Alternative Pathway Is Antibody Independent

The alternative pathway of complement fixation (Fig. 34-41, bottom) uses many of the same components as the classical pathway and likewise causes formation of a C5 conver-

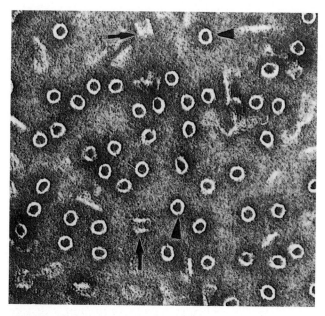

FIGURE 34-44. Electron micrograph of C9 ring complexes seen in side view (*arrows*) and top view (*arrow heads*). [Courtesy of Zanvil Cohn, The Rockefeller University.]

tase that triggers MAC assembly. The two pathways differ in that the alternative pathway is antibody independent. It is therefore thought that the alternative pathway functions to defend against invading microorganisms before an immune response against them can be mounted (although the classical pathway may also do so). Once sufficient antibody has been synthesized, the alternative pathway assumes a secondary role relative to that of the classical pathway.

The alternative pathway is thought to always operate at a low level (see below) so as to continually produce small amounts of C3b, the same molecule produced by C3 convertase of the classical pathway. In the alternative pathway, however, C3b combines with the plasma protein **factor B** in a Mg^{2+}-dependent reaction. The resultant complex, C3b,B, is the only known substrate for the active plasma serine protease, **factor $\overline{D}$,** which cleaves the B subunit of C3b,B to yield C3b,$\overline{Bb}$. This latter complex is a C3 convertase that is equivalent to but distinct from that of the classical pathway. It cleaves C3 to C3b, which participates in the formation of more C3 convertase in a cyclically amplified process. The additional C3b also binds to C3 convertase to yield $(C3b)_2\overline{Bb}$, a C5 convertase distinct from that of the classical pathway but which likewise catalyzes the formation of the MAC.

What is the origin of the C3b that initiates the alternative pathway? It may, of course, be generated by the classical pathway in which case the alternative pathway acts as an amplification mechanism for antibody-induced complement activation. In the absence of this process, however, it is thought that the reactive but unexposed thioester bond of native C3 undergoes slow spontaneous hydrolysis to yield a C3b-like protein, **C3i**, in that it binds factor B and mediates its $\overline{D}$-catalyzed activation. The resulting C3 convertase, in turn, generates authentic C3b.

How does the alternative pathway target invading microorganisms? The C3b concentration in solution is limited by a plasma protein named **factor I** which, together with a second protein, **factor H**, forms a complex (I,H) that proteolytically degrades C3b in solution. When C3b is covalently bound to a surface, however, its degradation rate is greatly reduced. Moreover, the surface-bound C3 convertase complex is stabilized by the binding of the plasma protein **properdin (P)**, which further protects C3b from I,H-mediated degradation as well as retards the dissociation of $\overline{Bb}$ from C3 convertase. Consequently, the faster C3b covalently attaches to a surface, the more slowly it is degraded. *Substances to which C3b efficiently attaches are therefore alternative pathway activators.* These include polymers of microbial origin such as the lipopolysaccharides of gram-negative bacteria known as **endotoxins** and cell wall teichoic acids from gram-positive bacteria (Section 10-3B), certain whole bacteria, fungi, and cells infected by certain viruses. The alternative pathway therefore provides an effective defense against invading microorganisms. Indeed, individuals who are genetically deficient in certain complement components are highly susceptible to various infections.

The Complement System Is Strictly Regulated

The inability of many complement components to discriminate between normal tissues and foreign substances requires that the complement system be maintained under tight control. Otherwise the complement system would destroy host cells. Indeed, the actual damage in many autoimmune diseases is caused by the complement system.

The complement system is regulated by the inactivation of its activated components. This occurs in three ways:

1. Complement components are inactivated through their spontaneous decay. For example, the hyperreactive thioesters of newly activated C3b and C4b react with water with half-lives of ~60 μs. These proteins are therefore lost to the classical pathway unless they attach to a membrane in the immediate vicinity of their activating recognition unit, that is, to the membranes of the invading microorganisms that triggered their activation (rather than to those of host cells). Similarly, classical pathway C3 convertase, $\overline{\text{C4b,2a}}$, is but transiently active; its C2a component readily dissociates with the consequent loss of enzymatic activity.

2. Complement components are inactivated through their degradation by specific proteases. For instance, **C4b-binding protein** forms a complex with factor I that proteolytically inactivates C4b, much like, as we have seen, the I,H complex degrades C3b. Apparently, C4b-binding protein and factor H act as cofactors that target factor I for C4b and C3b. From this point of view, C4b-binding protein limits the activities of classical pathway C3 convertase ($\overline{\text{C4b,2a}}$) and C5 convertase ($\overline{\text{C4b,2a,3b}}$), whereas, as we have seen, factor H does so for proteases containing C3b in both the classical and alternative pathways.

3. Complement components are inactivated through their association with specific binding proteins. For example, **$\overline{\text{C1}}$ inhibitor** tightly binds $\overline{\text{C1r}}$ and $\overline{\text{C1s}}$ to form a complex that dissociates from, and hence inactivates, the recognition unit. Similarly, **S protein** attaches to MACs assembling in the plasma so as to prevent their later attachment to cell membranes. Such attachment is consequently limited to the site of complement activation.

The regulation of the complement system therefore functions to target foreign invaders while minimizing host cell damage.

3. MOTILITY: MUSCLES, CILIA, AND FLAGELLA

Perhaps the most striking characteristic of living things is their capacity for organized movement. Such phenomena occur at all structural levels and include such diverse vec-

FIGURE 34-45.   A photomicrograph of a muscle fiber in longitudinal section. Its transverse dark A bands and light I bands are clearly visible. [J.C. Révy/CNRI.]

torial processes as active transport through membranes, the translocation of DNA polymerase along DNA, the separation of replicated chromosomes during cell division, the beating of flagella and cilia, and, most obviously, the contraction of muscles. In this section we consider the structural and chemical basis of biological motility. In doing so we shall be mainly concerned with **striated muscle** since it is the most familiar and best understood motility system. We shall also briefly discuss smooth muscle and two entirely different types of biological motility systems: those of eukaryotic cilia and prokaryotic flagella.

A. Structure of Striated Muscle

The voluntary muscles, which include the skeletal muscles, have a striated appearance when viewed under the light microscope (Fig. 34-45). Such muscles consist of long parallel bundles of 20 to 100-μm-diameter **muscle fibers** (Fig. 34-46), which may span the entire length of the muscle and which are actually giant multinucleated cells that arise during muscle development by the end-to-end fusion of numerous precursor cells. Muscle fibers are, in turn, composed of parallel bundles of around one thousand 1- to 2-μm-diameter **myofibrils** (Greek: *myos,* muscle), which may extend the full length of a fiber.

Electron micrographs show that muscle fiber striations arise from an underlying banded structure of multiple in-register myofibrils (Fig. 34-47). The bands are formed by alternating regions of greater and lesser electron density, respectively, named **A bands** and **I bands** (Fig. 34-48). The myofibril's repeating unit, the **sarcomere** (Greek: *sarkos,* flesh), which is 2.5 to 3.0 μm long in relaxed muscle but progressively shortens as the muscle contracts, is bounded by dark (electron dense) **Z disks** or **lines** at the center of each I band. The A band is centered on the lighter **H zone** which, in turn, is centered on the dark **M disk** or **line**.

Cross-sections through the sarcomere reveal the origin of its banded pattern. The H zones contain an array of paral-

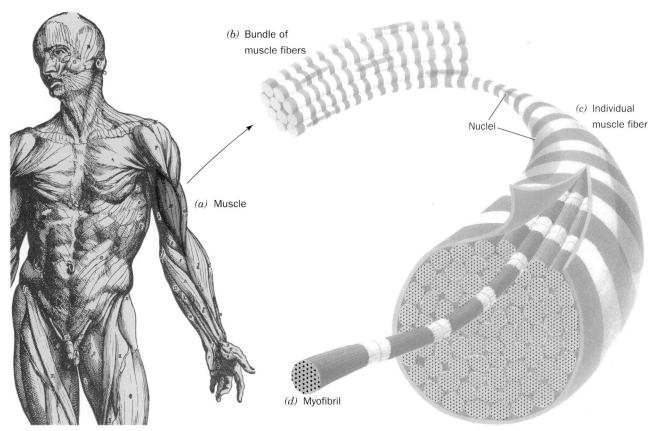

(b) Bundle of
muscle fibers

(c) Individual
muscle fiber

Nuclei

(a) Muscle

(d) Myofibril

FIGURE 34-46.   Skeletal muscle organization. A muscle (*a*),
consists of bundles of muscle fibers (*b*), each of which is a long
thin multinucleated cell (*c*), that may run the length of the
muscle. Muscle fibers contain bundles of laterally aligned
myofibrils (*d*), which consist of bundles of alternating thick and
thin filaments.

lel, hexagonally packed 150-Å-diameter **thick filaments,**
whereas the lighter I bands consist of twice as many hexagonally arranged 70-Å-diameter **thin filaments** that are anchored to the Z disk. The darker areas at the ends of each A
band mark the regions where the two sets of fibers interdigitate. The thick and thin filaments associate in this region by
means of regularly spaced **cross-bridges** (Fig. 34-49). We
shall see below that it is these associations that are responsible for the generation of muscular tension.

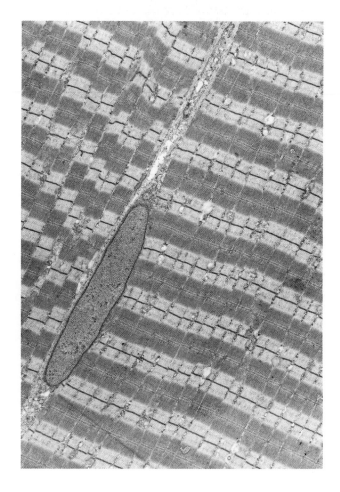

FIGURE 34-47.   An electron micrograph of skeletal muscle
showing that the myofibrils in muscle fiber are in register (the
out-of-register portion of the myofibril on the upper left is
probably an artifact of preparation). The ovoid object near the
center is a nucleus. Note its peripheral location in its fiber.
[Courtesy of Don Fawcett, Harvard University.]

Thick Filaments Consist of Myosin

The major protein components of striated muscle are listed in Table 34-4. Vertebrate thick filaments are composed almost entirely of a single type of protein, **myosin**, which occurs in virtually every vertebrate cell. *Myosin molecules consist of six highly conserved polypeptide chains: two 220-kD **heavy chains** and two pairs of different **light chains**, the so called **essential and regulatory light chains (ELC and RLC)**, that vary in size between 15 and 22 kD depending on their source.* Myosin is an unusual protein in that it has both fibrous and globular properties (Fig. 34-50). The N-terminal half of its heavy chain folds into an elongated globular head, around 55 × 200 Å, whereas its C-terminal half forms a long fibrous α-helical tail. Two of these α-helical tails associate to form a left-handed parallel coiled

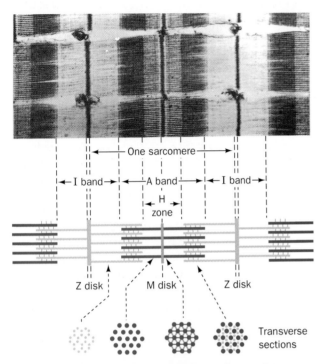

FIGURE 34-48. An electron micrograph of parts of three myofibrils in longitudinal section. The myofibrils are separated by horizontal gaps. A myofibril's major features, as indicated in the accompanying interpretive drawings, are the light I band, which contains only hexagonally arranged thin filaments; the A band, whose dark H zone contains only hexagonally packed thick filaments, and whose even darker outer segments contain overlapping thick and thin filaments; the Z disk, to which the thin filaments are anchored; and the M disk, which arises from a bulge at the center of each thick filament. The myofibril's functional unit, the sarcomere, is the region between two successive Z disks. [Courtesy of Hugh Huxley, Brandeis University.]

TABLE 34-4. PROTEINS OF STRIATED MUSCLE

Protein	Molecular Mass (kD)
Myosin	540
Heavy chain	230
Essential light chain (ELC)	~20
Regulatory light chain (RLC)	~20
G-Actin	42
Tropomyosin	33
Troponin	72
TnC	18
TnI	23
TnT	31
α-Actinin	200
Desmin	50
Vimentin	52
Titin	~3600
Nebulin	~800
Dystrophin	427
C-Protein	150
M-Protein	100

FIGURE 34-49. Electron micrograph of deep-etched, freeze-fractured myofibril showing its alternating thick and thin filaments. The knobs (cross-bridges) projecting from the thick filaments are helically arrayed. [Courtesy of John Heuser, Washington University School of Medicine.]

(a)

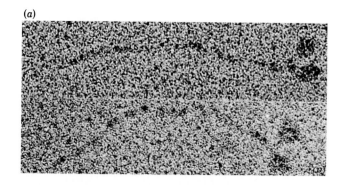

(b)

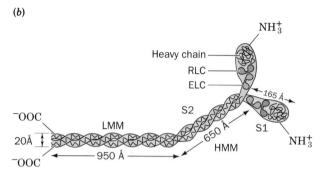

FIGURE 34-50. The myosin molecule. (*a*) Electron micrograph showing that the myosin molecule is a fibrous entity with two globular heads. [Courtesy of Henry Slayter, Harvard Medical School.] (*b*) Its rod-shaped tail is formed by the two extended α helices, one from each of its two identical heavy chains, that wrap around each other to form a parallel coiled coil. One of each type of myosin light chain, an essential light chain (ELC) and a regulatory light chain (RLC), is associated with each of myosin's identical globular heads.

coil, yielding an ~1600-Å-long rodlike segment with two globular heads. The amino acid sequence of myosin's α-helical tail is characteristic of coiled coils: It has a seven-residue pseudorepeat, *a-b-c-d-e-f-g*, with nonpolar residues concentrated at positions *a* and *d*. Thus, much like in the coiled coils of the fibrous protein keratin (Section 7-2A) and leucine zippers (Section 33-3B), the myosin helix has a hydrophobic strip along one side that promotes its lengthwise association with another such helix. One of each type of light chain is associated with each of the heavy chain dimer's globular heads.

Myosin only exists as single molecules at low ionic strengths. However, under physiological conditions, these proteins form aggregates that resemble thick filaments. *Natural thick filaments consist of several hundred myosin molecules with their rodlike tails packed end-to-end in a regular staggered array (Fig. 34-51).* The thick filament is therefore a bipolar entity in which the globular myosin heads project from either end, leaving a bare central region. It is these myosin heads that form the cross-bridges that interact with the thin filaments in intact myofibrils.

In addition to its structural function, the myosin heavy chain is an ATPase: It hydrolyzes ATP to ADP and P_i in a reaction that powers muscle contraction. Muscle is therefore a device for transducing the chemical free energy of ATP hydrolysis to mechanical energy. The myosin light chains, through their level of phosphorylation, are thought to modulate the ATPase activity of their associated heavy chains.

(a)

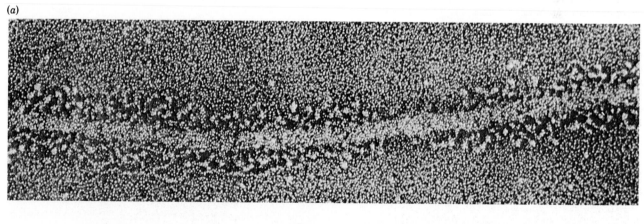

(b)

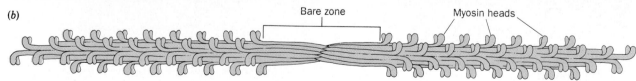

FIGURE 34-51. The thick filament of striated muscle. (*a*) An electron micrograph showing the myosin heads projecting from the thick filament's outer segments and its bare central zone. [From Trinick, J. and Elliott, A., *J. Mol. Biol.* **131,** 135 (1977).]

(*b*) A thick filament typically contains several hundred myosin molecules organized in a repeating staggered array such that the myosin molecules are oriented with their globular heads pointing away from the filament's center.

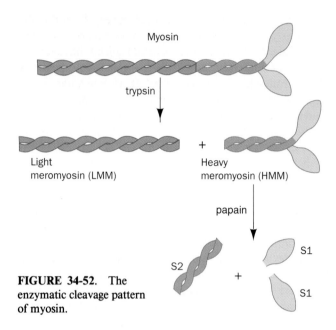

FIGURE 34-52. The enzymatic cleavage pattern of myosin.

FIGURE 34-53. The X-ray structure of chicken muscle myosin subfragment-1 (S1). (*a*) A ribbon diagram in which the heavy chain's 25-, 50-, and 20-kD segments are green, red, and blue, respectively, and its essential and regulatory light chains, RLC and ELC, are purple and yellow. Residue numbers are indicated at various positions, with 2000 and 3000 being added to those of the RLC and ELC to distinguish them from the heavy chain. A sulfate ion, shown in space-filling form (*red*), is bound near the confluence of the three heavy chain fragments, where it is thought to occupy the binding site of ATP's β-phosphate group (sulfate is a competitive inhibitor of myosin's ATPase function). An RLC-bound Ca^{2+} ion (*lower left*) is represented by a gray ball. (*b*) A space-filling representation of S1, colored and oriented similarly to that in Part *a*. Note the prominent vertical cleft that divides the 50-kD fragment. [Courtesy of Ivan Rayment and Hazel Holden, University of Wisconsin.]

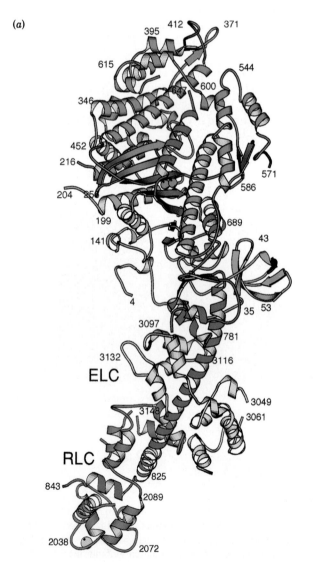

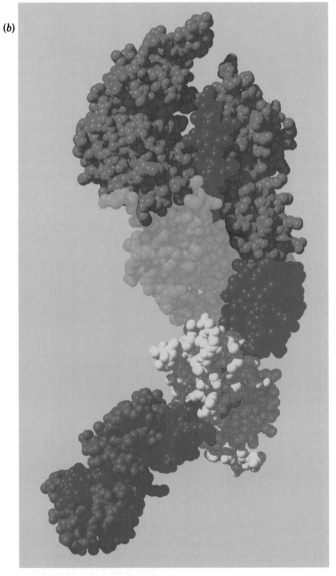

In 1953, Andrew Szent-Györgi demonstrated that limited trypsin digestion cleaves myosin into two fragments (Fig. 34-52):

1. **Light meromyosin (LMM)**, a 950-Å long α-helical rod that aggregates to form filaments but lacks both ATPase activity and the ability to associate with light chains.

2. **Heavy meromyosin (HMM)**, which has a rodlike tail and two globular heads, does not aggregate but has ATPase activity and binds to light chains.

HMM can be further split by treatment with papain to yield two identical molecules of **subfragment-1 (S1)** and one of the rod-shaped **subfragment-2 (S2)**. The 130-kD S1, which contains myosin's ATPase activity and its thin filament–binding site, consists of a 95-kD heavy chain fragment and one molecule each of ELC and RLC. S1 is a pear-shaped molecule that electron microscopy studies indicate is 190 Å long and 50 Å wide at its widest point. Further tryptic digestion of S1 yields three fragments: a 25-kD N-terminal segment that binds nucleotide (ATP or ADP), a central 50-kD segment, and a 20-kD C-terminal segment.

X-Ray Structure of Myosin Subfragment-1

The X-ray structure of chicken muscle S1, determined by Ivan Rayment and Hazel Holden, reveals it to be nearly 50% α helical (Fig. 34-53). However, the core of the molecule consists of a mostly parallel seven-strand β sheet whose strands are contributed by all three tryptic segments. Perhaps the most conspicuous structural feature of S1 is an ~85-Å-long helix that extends from the thick part of the head, the so-called motor domain, to the protein's C-terminal region at its narrow end. The motor domain contains the binding site for ATP as well as that for the protein **actin**, the thin filament's major component (see below).

The ATP-binding site, which is located in a 13-Å-deep V-shaped pocket at the point where the three tryptic segments come together, was identified from its resemblance to such sites in certain other nucleotide-binding proteins and from the positions of amino acids previously shown to be at myosin's ATP-binding site. The tight binding of ATP to myosin (association constant, $K = 3 \times 10^{11}\ M^{-1}$) suggests that this pocket must close around the ATP. This notion is supported by the observation that the particularly reactive Cys 697 and Cys 707 residues can be cross-linked by bifunctional sulfhydryl reagents whose reactive groups are 3 to 14 Å apart only when nucleotide is bound to S1. These residues, which lie near the C-terminal ends of two consecutive and highly conserved α helices, are 18 Å apart in the nucleotide-free X-ray structure of S1. The site that is implicated in binding actin is located on the opposite side of the S1 globule from the nucleotide-binding site and is formed by the 50-kD tryptic fragment. The nucleotide and actin binding sites are connected by a deep cleft that divides the 50-kD fragment into two domains, the so called actin cleft. The way in which myosin and actin interact is considered below.

The two light chains share both sequence and structural homology with calmodulin (CaM; Section 17-3C). However, RLC contains only one of CaM's four Ca^{2+}-binding motifs and ELC has none. ELC embraces the middle region of myosin's long helix (Fig. 34-53a) in a manner resembling the way that CaM interacts with the CaM-binding helix of myosin light chain kinase (Fig. 17-17). RLC also clasps the long helix, but near its C-terminus and in a manner different from that of ELC (Fig. 34-53a).

Thin Filaments Consist of Actin, Tropomyosin, and Troponin

Actin, a ubiquitous and highly abundant eukaryotic protein, is the major constituent of thin filaments. At low ionic strengths, actin occurs as 375-residue bilobal globular monomers called **G-actin** (G for globular) that normally bind one molecule of ATP each. Under physiological conditions, however, G-actin polymerizes to form fibers known as **F-actin** (Fig. 34-54; F for fibrous), a process that hydrolyzes the ATP to ADP, which remains bound to the F-actin monomer unit. *F-actin forms the core of the thin filament.*

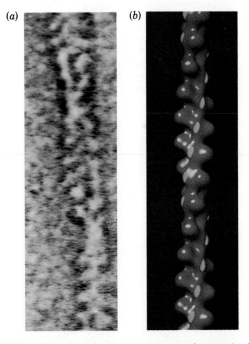

FIGURE 34-54. F-actin. (a) An electron micrograph of a thin filament from striated muscle. [Courtesy of Hugh Huxley, Brandeis University.] (b) An actin fiber (*red*) as visualized through image analysis of cryoelectron micrographs. Note the bilobal appearance of each monomeric (repeating) unit. The tropomyosin binding sites (see text) are blue. The F-actin helix has a maximum diameter of ~100 Å, 2.17 actin monomers per left-handed helical turn (13 subunits in 6 turns), and a rise per turn of ~60 Å. [Alternatively, F-actin may be described as a double (two-start) helix with 13 subunits per right-handed turn of each strand and a pitch of 720 Å.] [Courtesy of Daniel Safer, University of Pennsylvania, and Ronald Milligan, The Scripps Research Institute.]

Each of F-actin's monomeric units is capable of binding a single myosin S1 head. Electron micrographs of S1-decorated F-actin have the appearance of a series of head-to-tail arrowheads (Fig. 34-55a). F-actin must therefore be a polar entity; that is, all of its monomer units have the same orientation with respect to the fiber axis (Fig. 34-55b). The "arrowheads" in S1-decorated thin filaments that are still attached to their Z disk all point away from the Z disk indicating that *the thin filament bundles extending from the two sides of the Z disk have opposite orientations.*

Myosin and actin, the major components of muscle, account for 60 to 70% and 20 to 25% of total muscle protein, respectively. Of the remainder, two proteins that are asso-

(a)

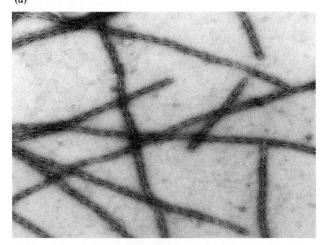

(b)

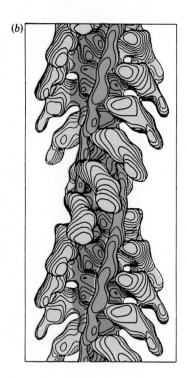

FIGURE 34-55. (a) An electron micrograph of a thin filament decorated with myosin S1 fragments. Note its resemblance to a series of arrowheads all pointing in the same direction along the filament. [Courtesy of Hugh Huxley, Brandeis University.] (b) Image reconstruction of S1-decorated actin filaments at a resolution of ~30 Å. The actin is colored green, the S1 fragments are pink, and the bound tropomyosin (see text) is orange. The helical filament has a pitch of 370 Å. [After a drawing provided by Ronald Milligan, The Scripps Research Institute, and Paula Flicker, University of California at San Francisco.]

FIGURE 34-56. A model of the striated muscle thin filament based on the 15-Å resolution X-ray structure of tropomyosin and electron micrographic studies of F-actin. Tropomyosin, a coiled coil of two α-helical subunits, wraps in the groove of the F-actin helix (*large blue spheres*) in a head-to-tail manner such that one tropomyosin molecule contacts seven consecutive bilobal actin monomer units (the red and blue regions of tropomyosin identify the seven homologous segments that are presumed to form its actin-binding sites). Each tropomyosin molecule binds a single troponin molecule at its head-to-tail joint (*left,* the small white spheres represent bound Ca²⁺ ions). The tropomyosin chain winding about the opposite side of the F-actin helix from the tropomyosin shown has been omitted for clarity. [Courtesy of George N. Phillips, Jr., Rice University.]

FIGURE 34-57. The low-resolution X-ray structure of rabbit cardiac muscle tropomyosin. Its two identical 284-residue polypeptides (*gold and light blue*) form a parallel coiled coil of α helices. Only the residues at the ends of the chains are not in α helices. [Based on an X-ray structure by George N. Phillips, Jr. and Carolyn Cohen, Brandeis University.]

ciated with the thin filaments are particularly prominent (Fig. 34-56):

1. **Tropomyosin,** a homodimer whose two 284-residue α helical subunits wrap around each other to form a parallel coiled coil (Fig. 34-57). These 400-Å long rod-shaped molecules are joined head-to-tail to form cables wound in the grooves of the F-actin helix such that each tropomyosin molecule contacts seven consecutive actin monomers in a quasi-equivalent manner.

2. **Troponin,** which consists of three subunits: **TnC,** a Ca^{2+}-binding protein (Fig. 34-58) that is 70% homologous to CaM; **TnI,** which binds to actin; and **TnT,** an elongated molecule, which binds to tropomyosin at its head-to-tail junction.

The tropomyosin–troponin complex, as we shall see, regulates muscle contraction by controlling the access of the myosin S1 cross-bridges to their actin-binding sites.

X-Ray Structure of Actin

The tendency of G-actin to polymerize has thwarted its crystallization in a manner suitable for X-ray crystallographic analysis. Curiously, however, it binds pancreatic deoxyribonuclease I (DNase I) in a 1:1 complex that, together with a Ca^{2+} ion and a molecule of either ATP or ADP, form crystals whose X-ray structures were determined by Wolfgang Kabsch and Kenneth Holmes (Fig. 34-59). The actin monomer consists of two domains that, for historical reasons, are referred to as the small and large domains, even though the former is only slightly smaller than the latter. Each domain has extensive secondary structure and is divided into two subdomains. The ATP and ADP bind in a cleft between the two domains, where the Ca^{2+} ion is liganded by the protein as well as by the β-phosphate of ADP or the β and γ-phosphates of ATP. The two actin structures are otherwise very similar. Two of the subdomains, one on each domain, contain a five-stranded β sheet consisting of a β hairpin motif followed by a right handed βαβ motif (Fig. 7-48*a*), suggesting that these subdomains arose by gene duplication even though their amino acid sequences exhibit no significant similarity.

The atomic model of ADP–actin has been used to generate a model of F-actin that fits the observed X-ray fiber

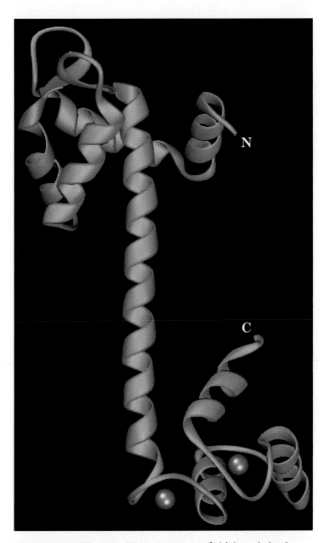

FIGURE 34-58. The X-ray structure of chicken skeletal muscle TnC. Its two globular domains, which are connected by an ~9-turn α helix, can each bind two Ca^{2+} ions. Under the conditions of the crystal structure determination, however, only the Ca^{2+}-binding sites of the C-terminal (*lower*) domain are occupied (*silver spheres*). These latter sites remain occupied at the lowest physiological Ca^{2+} concentrations in muscle so that the regulatory effects exerted by TnC must arise from the binding of Ca^{2+} to the N-terminal (*upper*) domain. Note the resemblance of the TnC structure to that of the homologous Ca^{2+}-binding regulatory protein calmodulin (Fig. 17-15). [Based on an X-ray structure by Muttaiya Sundaralingam, University of Wisconsin.]

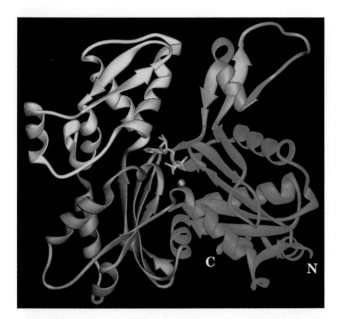

FIGURE 34-59. The X-ray structure of rabbit skeletal muscle actin·Ca^{2+}–ATP in its complex with bovine pancreatic DNase I. The protein is shown in ribbon form with subdomains 1 and 2, which together form the small domain, purple and light blue, and subdomains 3 and 4, which together form the large domain, orange and yellow. The ATP is shown in stick form (C green, N dark blue, O red, and P gold) and the bound Ca^{2+} ion is represented by a silver sphere. Note the structural similarity between subdomains 1 and 3. [Based on an X-ray structure by Wolfgang Kabsch and Kenneth Holmes, Max-Planck-Institute für medizinische Forschung, Germany.]

largest polypeptide known (it was only recently discovered because it does not enter polyacrylamide gels), extends from the thick filament to the Z disk, where it is thought to act as a spring to keep the thick filament centered in the sarcomere. **Nebulin,** which is also extemely large (~800 kD, ~7000 residues), consists, over at least 80% of its length, of a repeating 35-residue actin-binding motif that is predicted to be α helical (although it lacks the heptad repeat of α helices that form coiled coils). The observations that the

diagram from oriented gels of F-actin (Fig. 34-60). The monomer has a specific orientation with its small domain at a high radius from the F-actin helix axis and with all four subdomains making contacts with neighboring monomers. This model is in excellent agreement with three-dimensional maps of vertebrate muscle thin filaments obtained by cryoelectron microscopy and image reconstruction (Fig. 34-54): The two studies agree in terms of filament polarity, monomer orientation, and the three-dimensional location of the C-terminus of the actin monomer.

Minor Muscle Proteins Control Myofibril Assembly

The Z disk, which anchors two sets of oppositely oriented thin filaments (Fig. 34-48), is an amorphous entity that contains several fibrous proteins. For instance, **α-actinin,** which binds to the ends of F-actin filaments *in vitro,* is localized in the Z disk's interior (Fig. 34-61a). α-Actinin is therefore thought to attach thin filaments to the Z disk. Two other proteins, **desmin** and **vimentin,** largely occur at the Z disk periphery (Fig. 34-61b), where they apparently act to keep adjacent myofibrils in lateral register. **Titin,** whose ~3600-kD molecular mass (~33,000 residues!) makes it the

FIGURE 34-60. A model of the actin filament based on fitting the known X-ray structure of the actin monomer to the X-ray fiber diffraction pattern of F-actin. Actin monomers are shown in space-filling representation, in alternating blue, red, and white, with each amino acid residue represented by a sphere. The lowest monomer shown is oriented identically to that in Fig. 34-59. The residues which cross-linking studies indicate form the myosin binding site are green. [Courtesy of Wolfgang Kabsch and Kenneth Holmes, Max-Planck-Institute für medizinische Forschung, Germany.]

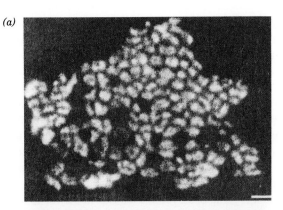

(a)

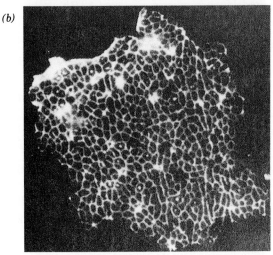

(b)

FIGURE 34-61. Indirect immunofluorescence micrographs of isolated sheets of skeletal muscle Z disks: (*a*) Using antibodies to α-actinin, indicating that α-actinin occurs at the interior of the Z disk. The bar represents 2.5 μm. (*b*) Using antibodies to desmin, showing that desmin is distributed about the Z-disk periphery. Antibodies to vimentin exhibit the same distribution. In the indirect immunofluorescence technique, proteins are labeled with rabbit antibodies raised against them. The bound rabbit antibodies are subsequently labeled with goat anti-rabbit immunoglobulin antibodies to which fluorescent molecules such as fluorescein are covalently linked. The proteins are then observed under UV light. This indirect approach of using two types of antibodies rather than directly using fluorescently tagged rabbit antibodies increases the sensitivity of the method because several fluorescently labeled goat antibodies can bind to each rabbit antibody. [From Lazarides, E., *Nature* **283**, 251 (1980).]

aggregate length of these putative α helices is approximately that of a thin filament, that nebulin antibodies label the thin filament, and that this labeling pattern is fixed with respect to the Z disk suggest that nebulin winds along the ~1 μm full length of a thin filament, thereby controlling this length. Titin may similarly control the length of thick filaments.

The M disk (Fig. 34-48) arises from the local enlargement of in-register thick filaments. The two proteins that are associated with this structure, **C-protein** and **M-protein,** probably participate in thick filament assembly. Invertebrate thick filaments contain a core of **paramyosin** which, in some muscles, is the dominant component.

Duchenne muscular dystrophy (DMD) and the less severe **Becker muscular dystrophy (BMD)** are both sex-linked muscle-wasting diseases. In DMD, which has an onset age of 2 to 5 years, muscle degeneration exceeds muscle regeneration causing progressive muscle weakness and ultimately death, usually at around age 20. In BMD, the onset age is 5 to 10 years and there is an overall less progressive course of muscle degeneration and a longer (sometimes normal) lifespan than in individuals with DMD.

The ~2500-kb gene responsible for DMD/BMD, which contains at least 70 introns (the most known in a single gene; Section 33-2F), encodes a 3685-residue protein named **dystrophin.** However, dystrophin has numerous isoforms that differ at their C-termeni through alternative mRNA splicing, as well as at their N-termeni through alternative transcriptional initiation (Section 34-3C). Dystrophin appears to be a member of the family of flexible rod-shaped proteins that includes the actin-binding cytoskeletal components spectrin (Section 11-3C) and α-actinin, each of which contains segments homologous to portions of dystrophin. Subcellular fractionation and immunofluorescence studies reveal that dystrophin, which has a normal abundance in muscle tissue of 0.002%, is associated with the inner surface of the muscle plasma membrane where it probably functions to anchor specific membrane glycoproteins, much as do spectrin and ankyrin in the erythrocyte (Fig. 11-35*d*).

The dystrophin gene in most individuals with DMD/BMD contains deletions or, less frequently, duplications of one or more exons. Individuals with DMD usually have no detectable dystrophin in their muscles, whereas those with BMD mostly have dystrophins of altered sizes. Evidently, the dystrophins of individuals with DMD are rapidly degraded, whereas those of individuals with BMD are semifunctional.

B. Mechanism of Muscle Contraction

So far we have simply described the components of striated muscle. Now, like good engineers, we must ask how do these components fit together and how do they interact? In other words, how does muscle work?

Thick and Thin Filaments Slide Past Each Other during Muscle Contraction

Physiologists have long known that a contracted muscle is as much as one third shorter than its fully extended length. Electron micrographs have demonstrated that this shortening is a consequence of a decrease in the length of the sarcomere (Fig. 34-62). Yet, during muscle contraction, the thick and the thin filaments maintain constant lengths as is indicated by the observations that the width of the A band as well as the distance between the Z disk and the edge of the adjacent H zone do not change. Rather, sarcomere contraction is accompanied by equal reductions in the widths of the I band and the H zone. These observations were independently explained by Hugh Huxley and Jean Hanson and by Andrew Huxley and R. Niedergerke who, in 1954, proposed the **sliding filament model:** *The force of muscle contraction is generated by a process in which interdigitated sets of thick and thin filaments slide past each other (Fig. 34-62).*

Actin Stimulates Myosin's ATPase Activity

The sliding filament model partially explains the mechanics of muscle contraction but not the origin of the contractile force. Albert Szent-Györgi's work in the 1940s pointed the way towards the elucidation of the contraction mechanism. The mixing of solutions of actin and myosin to form a complex known as **actomyosin** is accompanied by a large increase in the solution's viscosity. This viscosity increase is reversed, however, when ATP is added to the actomyosin solution. *Evidently, ATP reduces myosin's affinity for actin.*

Further insight into the role of ATP in muscle contraction was provided by kinetic studies. Isolated myosin's ATPase function has a turnover number of ~0.05 s^{-1}, far less than that in contracting muscle. Paradoxically, however, the presence of actin increases myosin's ATP hydrolysis rate to the physiologically more realistic turnover number of ~10 s^{-1}, a rate enhancement of ~200 (indeed, actin was so named because it *acti*vates myosin). This is because isolated myosin rapidly hydrolyzes ATP

$$\text{ATP}^{4-} + \text{H}_2\text{O} \rightleftharpoons \text{ADP}^{3-} + \text{HPO}_4^{2-} + \text{H}^+$$

but only slowly releases the products ADP + P$_i$ as is indicated by the observation that myosin-catalyzed ATP hydrolysis begins with a rapid burst of H$^+$, whereas free ADP and P$_i$ appear much more slowly. Actin enhances myosin's ATPase activity by binding to the myosin – ADP – P$_i$ complex and stimulating it to sequentially release P$_i$ followed by ADP. The myosin – ADP – P$_i$ complex cannot be formed by simply mixing myosin, ADP, and P$_i$, which suggests that this complex is a "high-energy" intermediate in which the free energy of ATP hydrolysis has somehow been conserved.

The foregoing observations led Edwin Taylor to formulate a model for actomyosin-mediated ATP hydrolysis (Fig. 34-63):

Step 1 ATP binding to the myosin component of actomyosin results in the dissociation of actin and myosin.

Step 2 The myosin-bound ATP is rapidly hydrolyzed to form a stable "high-energy" myosin – ADP – P$_i$ complex.

Step 3 Actin binds to the myosin – ADP – P$_i$ complex.

Step 4 In a process accompanied by a conformational relaxation to its resting state, the actin – myosin – ADP – P$_i$ complex sequentially releases P$_i$ followed by ADP yielding actomyosin that can undergo another round of ATP hydrolysis.

This ATP-driven alternate binding and release of actin by myosin provides, as we shall see, the vectorial force of muscle contraction.

A Structure-Based Model for the Interaction of Actin and Myosin

Rayment, Holden, and Ronald Milligan have formulated a model for the so called **rigor complex** of the myosin

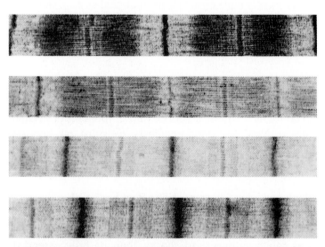

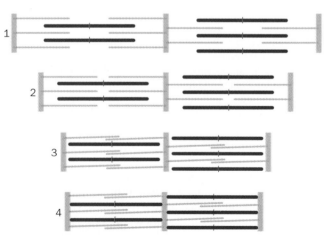

FIGURE 34-62. Electron micrographs with accompanying interpretive drawings of myofibrils in progressively more contracted states (1-4). Note that the widths of the I band and H zone decrease upon contraction, whereas the lengths of the thick zone decrease upon contraction, whereas the lengths of the thick and thin filaments remain constant upon myofibril contraction. The interpenetrating sets of thick and thin filaments must therefore slide past each other, as drawn. [Courtesy of Hugh Huxley, Brandeis University.]

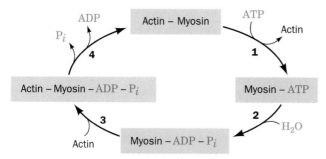

FIGURE 34-63. The reaction sequence in actomyosin-catalyzed ATP hydrolysis.

S1 head and F-actin (Fig. 34-64; the rigor complex is that taken up by ATP-deprived muscle, which occurs in **rigor mortis,** the temporary rigidity of muscles after death). This was done, with an estimated accuracy of 5 to 8 Å, by fitting the X-ray structure of the myosin S1 head (Fig. 34-53) and the X-ray fiber structure of F-actin as derived from the X-ray structure of G-actin (Fig. 34-60) to the electron density map obtained from the electron microscopy–based image of this complex (Fig. 34-55). The bulky motor domain of S1 binds tangentially to the actin filament at a ~45° angle to the filament axis. Its extension, the narrow S1 tail,

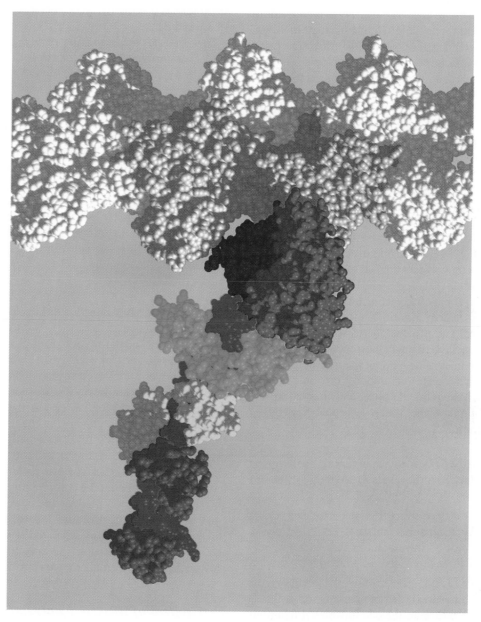

FIGURE 34-64. A space-filling atomic model of the myosin–actin interaction constructed by fitting the atomic models of myosin S1 (Fig. 34-53) and the actin filament (Fig. 34-60) to the electron microsopy–based image of S1-decorated actin filaments (Fig. 34-55). The myosin S1 is rotated about its long axis relative to the view in Fig. 34-53; its tryptic fragments and light chains are similarly colored to those in that figure. [Courtesy of Ivan Rayment and Hazel Holden, University of Wisconsin.]

which contains the two light chain binding regions, projects tangentially away from the filament axis at ~90°, an orientation that permits S1 to impose tension on the rodlike myosin tail that associates with other such tails to form the thick filament. The myosin head appears to interact with actin via ion pairing involving several Lys residues on myosin and several Asp and Glu residues on actin. These interactions are bolstered by what appears to be a stereospecific association between juxtaposed surface-exposed patches of hydrophobic residues on actin and myosin.

Myosin Heads "Walk" Along Actin Filaments

In order to complete our description of muscle contraction we must determine how ATP hydrolysis is coupled to the sliding filament model. If the sliding filament model is correct then it would be impossible for a myosin cross-bridge to remain attached to the same point on a thin filament during muscle contraction. Rather, it must repeatedly detach and then reattach itself at a new site further along the thin filament towards the Z disk. This, in turn, suggests that muscular tension is generated *through the interaction of myosin cross-bridges with thin filaments.*

The X-ray structure of myosin S1 (Fig. 34-53) together with the model of its rigor complex with actin (Fig. 34-64) suggests how ATP hydrolysis is coupled to myosin's conformational change. Despite the excellent fit of the X-ray structures of actin and myosin S1 to the image of their rigor complex (Fig. 34-55), the resulting atomic model of the rigor complex (Fig. 34-64) contains a steric clash between residues at the actin–myosin contact region. However, it seems quite plausible that, upon ATP binding, this clash is relieved by the opening of the actin cleft, the cleft that divides the 50-kD segment into two domains (Fig. 34-53b). Rayment, Holden, and Milligan therefore postulated that the closure of the actin cleft under the impetus of the release of ADP from the nucleotide binding site is responsible for the conformational change that produces myosin's "power stroke" in muscle's contractile cycle. This has led to the following variation of the widely accepted "rowboat" model for the contractile cycle (Fig. 34-65):

1. ATP binds to the S1 head in a manner that opens up the actin cleft. This, in turn, causes S1 to release its bound actin.

2. The active site cleft (distinct from the actin cleft) closes about the ATP in a manner that catalyzes its hydrolysis. This process "cocks" the myosin molecule; that is, puts it into its "high energy" state in which its S1 head is approximately perpendicular to the thick filament.

3. The S1 head binds weakly to an actin monomer that is closer to the Z disk than the one to which it had been bound previously.

4. S1 releases P_i, which causes its actin cleft to close, thereby increasing S1's binding affinity for actin.

5. The resulting transient state is immediately followed by the power stroke, a conformational shift that sweeps S1's

C-terminal tail by an estimated ~60 Å towards the Z disk relative to the motor domain, thus translating the attached thin filament by this distance towards the M-disk. It seems likely that the ~85-Å-long helix that connects S1's motor domain to its C-terminal tail is the conformational coupler in this energy transduction step.

6. ADP is released, thereby completing the cycle.

The cyclic nature of this process is necessary to prevent this molecular motor from reversing its power stroke while still bound to the actin which, if it occurred, would result in no net movement of the myosin head relative to the actin filament. The ATP-driven active transport of ions accross a membrane is a similarly cyclic vectorial process (Sections 18-3A and B).

The ~500 S1 heads on every thick filament asynchronously cycle through this reaction sequence about five times per second each during a strong muscular contraction. The S1 heads thereby "walk" or "row" up adjacent thin filaments towards the Z disk with the concomitant contraction of the muscle.

A remarkably similar model of actomyosin to that in Fig. 34-64 has been derived by fitting the X-ray structures of myosin S1 and F-actin to the image reconstruction of rabbit F-actin decorated with *Dictyostelium discoideum* (slime mold) myosin S1. The observation that *Dictyostelium* myosin binds to rabbit F-actin with the same affinity as does rabbit muscle myosin indicates a high degree of conservation at the interface between these two proteins and thus supports the forgoing mechanistic model for the contractile cycle.

Myosin Light Chains Function to Increase the Rate of Muscle Contraction

The function of the myosin light chains in vertebrate skeletal muscle had, for many years, been enigmatic. Recently, however, a motility assay has demonstrated that the rate at which myosin heavy chains slide along actin filaments is reduced 10-fold when the light chains are removed, even though removing the chains does not significantly reduce myosin's ATPase activity. It is therefore suggested that the 85-Å-long α helix to which both light chains bind (Fig. 34-53) and which is thought to act as the lever arm that amplifies the conformation change in myosin's motor domain to produce the power stroke, is stabilized by its bound light chains. Indeed, a bare α helix is rarely observed in proteins in aqueous solution. Thus, in the absence of light chains, the long α helix is likely to collapse leading to a smaller power stroke.

C. Control of Muscle Contraction

Striated muscles are, for the most part, under voluntary control; that is, their contraction is triggered by motor nerve impulses. How do these nerve impulses trigger muscle contraction? To answer this question, let us begin at the level of the myofibril and work up.

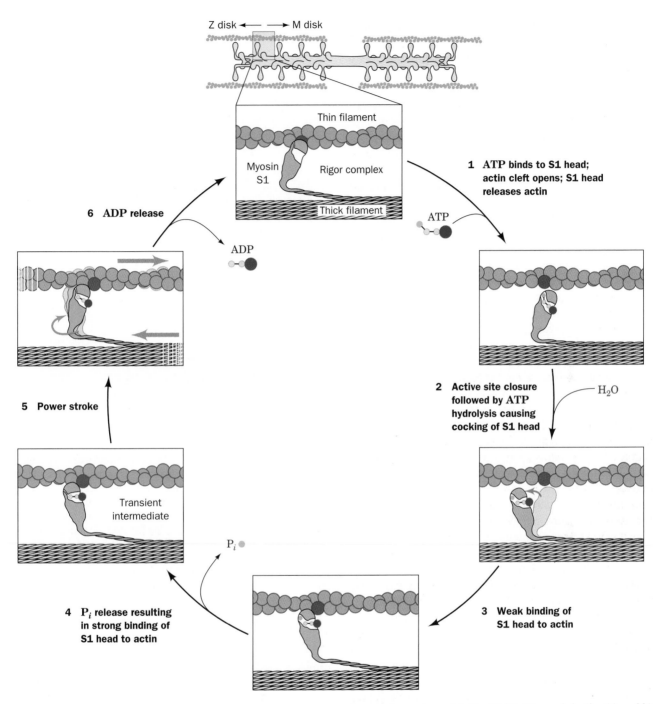

FIGURE 34-65. The mechanism of force generation in muscle. The myosin head "walks" up the actin thin filament through a cyclic vectorial process that is driven by ATP hydrolysis. Only one of myosin's two independent S1 heads is shown. The narrow cleft that splits the 50-kD segment of the S1 head into two domains (Fig. 34-53*b*) is represented by a horizontal gap perpendicular to the thin filament (although in the actual atomic model, Fig. 34-64, this gap is inclined by ~30° to the thin filament, thereby obscuring it in that figure). The actin monomer to which S1 was bound at the beginning of the cycle is more darkly colored for reference. [After Rayment, I., and Holden, M., *Curr. Opin. Struct. Biol.* **3**, 949 (1993).]

Ca²⁺ Regulates Muscle Contraction in a Process Mediated by Troponin and Tropomyosin

It has been known since the 1940s that Ca^{2+} is somehow involved in controlling muscle contraction. It was not until the early 1960s, however, that Setsuro Ebashi demonstrated that the effect of Ca^{2+} is mediated by troponin and tropomyosin. He did so by showing that actomyosin extracted directly from muscle, and therefore bound to troponin and tropomyosin, contracts in the presence of ATP only when Ca^{2+} is also present, whereas actomyosin prepared from purified actin and myosin contracts in the presence of ATP regardless of the Ca^{2+} concentration. The addition of tropomyosin and troponin to the purified actomyosin system restored its sensitivity to Ca^{2+}. Indeed, it was through these experiments that troponin was discovered.

The TnC subunit of troponin (Fig. 34-58) is the only Ca^{2+}-binding component of the tropomyosin–troponin complex. Tropomyosin, as we saw, binds along the thin filament groove in relaxed muscle (Fig. 34-56), where it apparently blocks the attachment of S1 myosin heads to seven consecutive actin units. *X-Ray diffraction studies indicate that when the [Ca^{2+}] reaches a critical level, an allosteric interaction between Ca^{2+}–troponin and tropomyosin causes tropomyosin to move ~10-Å deeper into the thin filament groove (Fig. 34-66). This movement, it is thought, uncovers the actins' myosin-binding sites, thereby switching on muscle contraction.* This switching mechanism may well be cooperative, in that the binding of a myosin head to an actin subunit might push tropomyosin away from neighboring myosin-binding sites in a conformational change that could also increase the Ca^{2+}-binding affinity of its associated TnC subunit.

Nerve Impulses Release Ca^{2+} from the Sarcoplasmic Reticulum

In order to understand how a nerve impulse affects the [Ca^{2+}] in a myofibril we must further consider the anatomy of striated muscle fibers. A nerve impulse arriving at a **neuromuscular junction** is transmitted directly to each sarcomere by a system of **transverse** or **T tubules,** nervelike infoldings of the muscle fiber's plasma membrane that surround each myofibril at its Z disk (Fig. 34-67; nerve impulse transmission is the subject of Section 34-4C). All of a muscle's sarcomeres therefore receive the signal to contract within a few milliseconds of each other so that the muscle contracts as a unit. The electrical signal is transferred, in a poorly understood manner, to the **sarcoplasmic reticulum (SR),** a system of flattened membranous vesicles derived from the endoplasmic reticulum that surround each myofibril rather like a net stocking. The SR membrane, which is normally impermeable to Ca^{2+}, contains a transmembrane Ca^{2+}–ATPase (Section 18-3B) that pumps Ca^{2+} into the SR so as to maintain the cytosolic [Ca^{2+}] of resting muscle below $10^{-7}M$, whereas that in the SR is over $10^{-3}M$. The SR's ability to store Ca^{2+} is enhanced by the presence of a highly acidic (37% Asp + Glu) 55-kD protein named **calsequestrin,** which has >40 Ca^{2+}-binding sites. *The arrival of a nerve impulse renders the SR permeable to Ca^{2+} which, in a few milliseconds, diffuses through specific Ca^{2+} channels into the myofibril so as to raise its internal [Ca^{2+}] to ~$10^{-5}M$. This Ca^{2+} concentration is sufficient to trigger the conformational change in troponin–tropomyosin that permits muscle contraction.* Once nerve excitation has subsided, the SR membrane again becomes impermeable to Ca^{2+}, so the Ca^{2+} inside the myofibril is pumped back into the SR. Tropomyosin therefore resumes its resting conformation causing the muscle to relax.

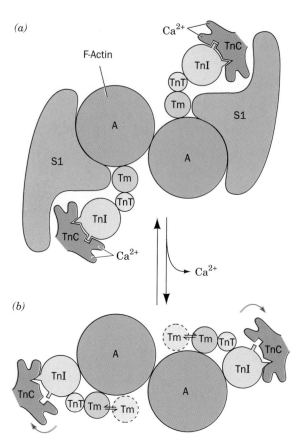

FIGURE 34-66. The control of skeletal muscle contraction by troponin and tropomyosin. (*a*) In contracting muscle, here diagrammed in cross-section, the myosin S1 heads freely interact with and thereby "walk" up the F-actin filaments (A). (*b*) Muscle relaxes when Ca^{2+} dissociates from troponin's TnC subunit thereby allosterically moving the tropomyosin (Tm) molecules to positions which sterically block myosin-actin interactions. [After Zot, A.S. and Potter, J.D., *Annu. Rev. Biophys. Biophys. Chem.* **16,** 555 (1987).]

D. Smooth Muscle

Vertebrates have two major types of muscle besides skeletal muscle: **cardiac muscle** and **smooth muscle.** Cardiac muscle, which is responsible for the heart's pumping action, is striated, indicating the similarity of its organization to that of skeletal muscle. Cardiac and skeletal muscle differ mainly in their metabolism, with cardiac muscle, which must function continuously for a lifetime, being much more dependent on aerobic metabolism than is skeletal muscle. Vertebrate heart muscle contraction is also spontaneously initiated by the heart muscle itself rather than through external nervous stimuli, although the nervous system can influence this contractile response. Smooth muscle, which is responsible for the slow, long-lasting, and involuntary contractions of such tissues as the intestinal walls, uterus, and large blood vessels, has a quite different organization from that of striated muscle. Smooth muscle consists of spindle-shaped, mononucleated cells whose

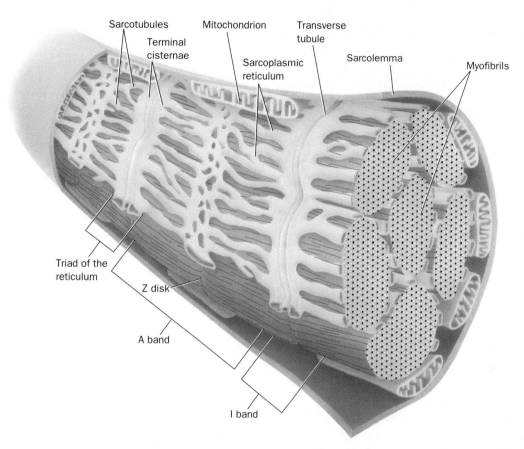

Sarcotubules
Terminal cisternae
Mitochondrion
Transverse tubule
Sarcoplasmic reticulum
Sarcolemma
Myofibrils
Triad of the reticulum
Z disk
A band
I band

FIGURE 34-67. A diagram of myofibrils showing the sarcoplasmic reticulum (*yellow*), the system of connected membranous vesicles that relay contractile signals from nerve to myofibril.

thick and thin filaments are more or less aligned along the cells' long axes but which do not form myofibrils.

Smooth muscle myosin, a genetically distinct protein, is functionally distinct from striated muscle myosin in several ways:

1. Its maximum ATPase activity is only ~10% of that of striated muscle.

2. It interacts with actin only when one of its light chains is phosphorylated at a specific Ser residue.

3. It forms thick filaments whose cross-bridges lack the regular repeating pattern of striated muscle and are distributed along the thick filament's entire length.

Smooth Muscle Contraction Is Triggered by Ca^{2+}

The thin filaments of smooth muscle contain actin and tropomyosin but lack troponin. *Smooth muscle contraction is nevertheless triggered by Ca^{2+} because* **myosin light chain kinase (MLCK),** *an enzyme that phosphorylates myosin light chains and thereby stimulates smooth muscle to contract, is enzymatically active only when it is associated with* Ca^{2+}–*calmodulin* (Fig. 34-68, *bottom;* myosin light chain phosphorylation in skeletal muscle appears to modulate the degree of tension produced by contraction). The mecha-

nism whereby Ca^{2+}–CaM activatates MLCK is discussed in Section 17-3C.

The intracellular $[Ca^{2+}]$ varies with the permeability of the smooth muscle cell plasma membrane to Ca^{2+} which, in turn, is under the control of the autonomic (involuntary) nervous system. When the $[Ca^{2+}]$ rises to ~$10^{-5}M$, smooth muscle contraction is initiated as described. When the $[Ca^{2+}]$ falls to ~$10^{-7}M$ through the action of the plasma membrane's Ca^{2+}-ATPase, the MLCK is deactivated, the myosin light chain is dephosphorylated by **myosin light chain phosphatase,** and muscle relaxation ensues. *Thus, Ca^{2+}, like cAMP, is a second messenger that transmits extracellular signals within the interior of a cell. In the many situations in which Ca^{2+} is a second messenger, calmodulin or a calmodulin-like protein is invariably the intracellular signal receiver.*

Smooth Muscle Activity Is Hormonally Modulated

Smooth muscles also respond to hormones such as epinephrine (Fig. 34-68, top). The binding of epinephrine to its plasma membrane–bound receptor activates adenylate cyclase. The cytosolic cAMP that is thereby generated binds to and causes the dissociation of the regulatory dimer, R_2, of an inactive protein kinase, R_2C_2, yielding active catalytic

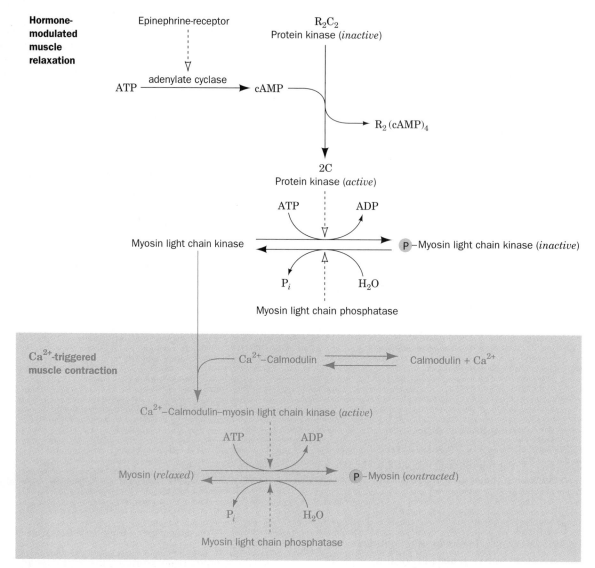

FIGURE 34-68. The control of smooth muscle contraction. Dashed arrows indicate stimulation or catalysis of a modification or demodification reaction. The lower part of the diagram (*shading*) indicates how Ca^{2+}, whose intracellular concentration increases in response to nerve impulses, triggers muscle contraction. The upper part of the diagram indicates how hormones such as epinephrine inhibit the contractile response causing smooth muscle relaxation.

subunits, C, that phosphorylate MLCK. Phosphorylated MLCK binds Ca^{2+}–calmodulin only weakly, so the extracellular presence of epinephrine causes smooth muscles to relax. Note the resemblance of this system to that controlling glycogen metabolism in skeletal muscle (Section 17-3). **Asthma,** a breathing disorder caused by the inappropriate contraction of bronchial smooth muscle, is often treated by the inhalation of an aerosol containing epinephrine, thereby relaxing the contracted bronchi.

The sequence of events culminating in smooth muscle contraction are inherently much slower than those leading to skeletal muscle contraction. Indeed, the structure and regulatory apparatus of smooth muscle suits it to its function: the maintenance of tension for prolonged periods

while consuming ATP at a much lower rate than skeletal muscle performing the same task. The structural and functional resemblance of TnC to CaM therefore suggests that TnC is a CaM variant that has evolved in skeletal muscle to provide a rapid response to the presence of Ca^{2+}.

E. Actin and Myosin in Nonmuscle Cells

Although actin and myosin are most prominent in muscle, they also occur in other tissues. In fact, actin is ubiquitous and is usually the most abundant cytoplasmic protein in eukaryotic cells, typically comprising 5 to 10% of their total protein. Myosin, in contrast, is usually present in only about one tenth the quantity of actin. This ratio reflects the fact that actin, in addition to its role in actomyosin-based contractile systems, participates in several myosin-independent motility systems as well as being a principal cytoskeleton component.

Actin Forms Microfilaments

Actin in muscles is entirely in the form of thin filaments. Nonmuscle actin, however, is about equally partitioned between soluble G-actin and F-actin fibers known as **microfilaments.** The actin content of microfilaments was established both through the immunofluorescence microscopy of living cells (e.g., Fig. 34-69) and because microfilaments can be decorated with S1 myosin heads to form arrowhead structures that are visually indistinguishable from those formed by muscle thin filaments (Fig. 34-55a). Such decoration is possible because actin is highly conserved throughout the eukaryotic kingdom. For instance, slime mold and rabbit muscle actins differ at only 17 of their 375 residues.

Actin *in vitro* is monomeric at low temperatures, low ionic strengths, and alkaline pH's. *Under physiological conditions, G-actin polymerizes in a process that is accelerated by the presence of ATP. In vivo,* microfilament assembly and disassembly is also influenced by numerous actin-binding proteins. For example, **profilin,** a 16-kD protein, binds G-actin in a 1:1 **profilactin** complex so as to prevent actin polymerization. A dramatic example of the effect of profilin occurs in many invertebrates upon the encounter of sperm and egg. A sea urchin sperm, for example, contains a reservoir of profilactin in its **acrosome,** a vesicle that lies just beneath the front of the sperm's head. Contact with the egg's jellylike coat triggers a reaction that dissociates the profilactin by raising the acrosomal pH. The newly liber-

ated G-actin undergoes "explosive" polymerization so as to erect, in a matter of seconds, a 90-μm long bundle of F-actin filaments, the **acrosomal process,** that is projected outwards from the sperm head (Fig. 34-70). It is the acrosomal process that penetrates the egg's jelly coat to initiate the fusion of sperm and egg.

There is clear evidence that the assembly and disassembly of actin filaments plays an important role in such cellular motility processes as ameboid locomotion, phagocytosis, **cytokinesis** (the separation of daughter cells in the last stage of mitosis), and the extension and retraction of various cellular protuberances, such as **microvilli** (fingerlike projections of cell surfaces) and neuronal axons. This evidence was obtained largely through the use of drugs that interfere with actin aggregation. For example, the fungal alkaloid **cytochalasin B**

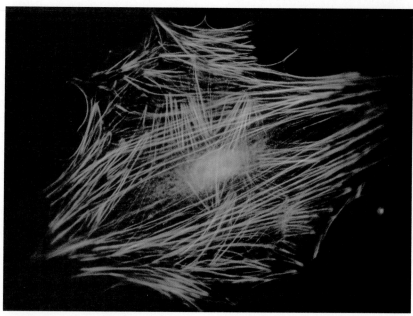

Cytochalasin B

FIGURE 34-69. The microfilaments in a fibroblast resting on the surface of a culture dish as revealed by indirect immunofluorescence microscopy using anti-actin antibody. When the cell begins to move, the filaments disassemble to form a diffuse mesh, thereby suggesting that actin plays a central role in cellular movement. [Courtesy of Elias Lazarides, California Institute of Technology.]

FIGURE 34-70. A series of light micrographs showing the elongation of the acrosomal process in a sea urchin sperm. The photographs were taken at 0.75-s intervals beginning 2 s after the sperm was artificially stimulated to begin the acrosomal reaction. The arc to the right of the sperm head is the sperm tail that has curved around outside the field of the micrograph. The acrosomal process' final length (*bottom frame*) is ~70 μm. [From Tilney, L.G. and Inoué, S., *J. Cell Biol.* **93**, 822 (1982).]

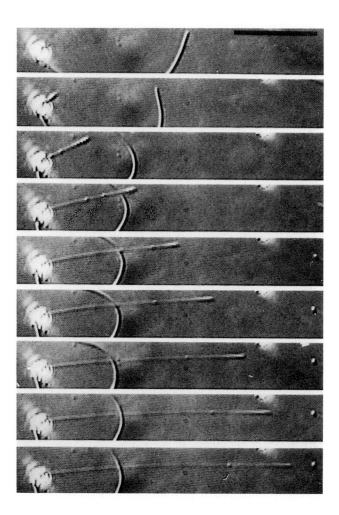

(which we have seen inhibits Na$^+$-independent glucose transport; Section 18-4A) blocks actin polymerization by specifically binding to the end of a growing F-actin filament (the "barbed" end of S1-decorated filaments) so as to inhibit actin polymerization from that end. In contrast, **phalloidin,**

Phalloidin

a bicyclic heptapeptide produced by the poisonous mushroom *Amanita phalloides* (which also synthesizes the chemically similar eukaryotic RNA polymerase inhibitor α-amanitin; Section 29-2F), blocks microfilament depolymerization by specifically binding to its actin units.

ATP accelerates actin polymerization by activating G-actin to add preferentially to a particular end of a growing actin filament. The ATP is eventually hydrolyzed, but not until after polymerization has occurred. This vectorial process has the consequence that there is a certain "critical" G-actin concentration at which activated monomer units add predominantly (although not exclusively) to their preferred end of an actin filament at the same rate that monomer units dissociate predominantly from the opposite end of the filament. Under these conditions, the actin filament neither grows nor shrinks; rather, it assumes a steady state in which, through this **treadmilling** process, actin monomer units are continually translocated from one end of the actin filament to the other (Fig. 34-71).

Actomyosin Has Contractile Functions in Nonmuscle Cells

Myosin is not so well conserved a protein as is actin. Nevertheless, nonmuscle myosin forms thick filaments that participate in contractile processes with microfilaments. One of the best characterized such processes occurs

during cytokinesis (cytoplasmic division) in animal cells and protozoa. In the final stages of mitosis, a **cleavage furrow** forms around the equator of the dividing cell in the plane perpendicular to the long axis of the mitotic spindle. Immunofluorescence microscopy demonstrates that the cleavage furrow is lined with an actomyosin belt (Fig. 34-72). Cytokinesis is accomplished through the tightening of this so called **contractile ring,** which disperses once cleavage has occurred. Blood platelets also contain actomyosin, which, upon blood clot formation (Section 34-1), contracts so as to strengthen the clot. The contraction is initiated by the Ca^{2+}–CaM activation of MLCK as occurs in smooth muscle.

F. Ciliary Motion and Vesicle Transport

Eukaryotes have two nearly ubiquitous but unrelated types of motility systems:

1. **Microfilament**-based systems, such as muscle, which contain actin.

2. **Microtubule**-based systems, such as cilia (see below), which contain the protein **tubulin.**

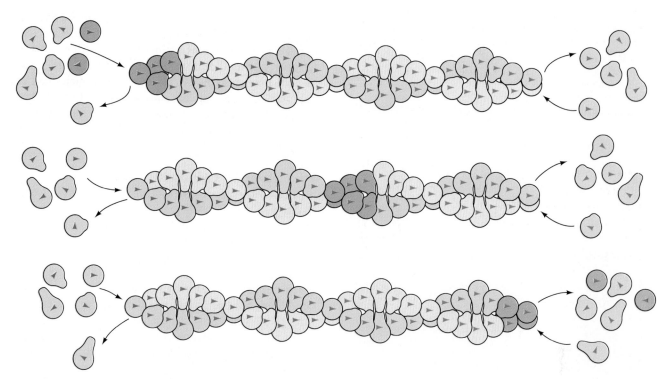

FIGURE 34-71. The "treadmilling" of actin monomer units along an actin filament. Actin monomers continually add to the left end of the filament, with eventual ATP hydrolysis, but dissociate at the same rate from the right end, so the filament maintains a constant length while its component monomer units translocate from left to right.

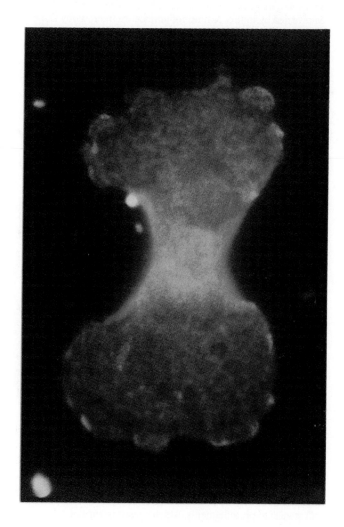

FIGURE 34-72. An indirect immunofluorescence micrograph, using anti-myosin II antibodies (*red*), of a dividing *Dictyostelium* (slime mold) amoeba showing that its contractile ring contains **myosin II** (a double-headed protein such as muscle myosin). The cell has also been treated with anti-myosin I antibodies (*green;* **myosin I** is a single-headed protein that has ~40% sequence identity with myosin II), thereby demonstrating that myosin I is localized at the leading edges of the daughter cell's **lamellipodia** (sheetlike extensions of the cell surface that participate in cell locomotion). [Courtesy of Edward Korn, National Institutes of Health, and Yoshio Fukui, Northwestern University Medical School.]

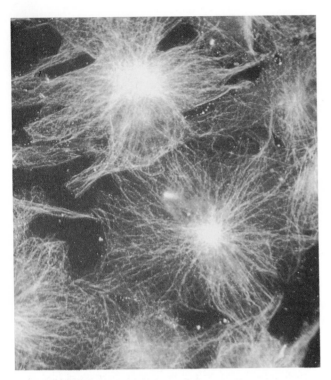

FIGURE 34-73. The networks of microtubules in fibroblasts as revealed by indirect immunofluorescence microscopy using anti-tubulin antibodies. [K.G. Marti/Visuals Unlimited.]

Microtubules (Fig. 34-73), as their name implies, are tubular structures, ~300 Å in diameter, which form a class of cytoskeletal components distinct from the ~70 Å in diameter microfilaments and the 100- to 150-Å-diameter **intermediate filaments** (the cytoskeleton's third major component, which apparently has only a structural role; Section 1-2A). *Microtubules comprise the major components of such cellular organelles as the mitotic spindle and cilia, and are thought to form the framework that organizes the cell.*

Microtubules Are Composed of Tubulin

Microtubules are polymers of tubulin, a dimer of globular α- and β-tubulin subunits (55 kD each). Each of these types of subunits are highly homologous throughout the eukaryotic kingdom and, to a lesser extent, with each other. At low temperatures, and in the presence of Ca^{2+}, tubulin assumes a soluble protomeric form (the $\alpha\beta$ dimers dissociate only in the presence of denaturing agents). Under physiological conditions, however, tubulin polymerizes to microtubules through a process in which each tubulin molecule binds 2 GTPs and hydrolyzes one of them to GDP + P_i during or shortly after the incorporation of the $\alpha\beta$ dimer into a microtubule. Electron microscopy and X-ray studies indicate that microtubules consist of 13 parallel but staggered protofilaments arranged about a hollow core (Fig. 34-74). The protofilaments, which consist of alternating head-to-tail–

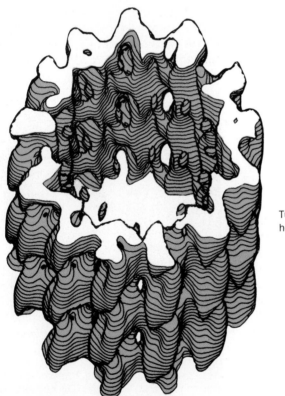

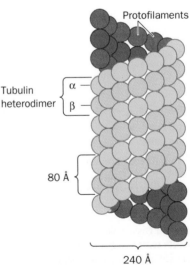

FIGURE 34-74. The 18-A resolution X-ray structure of a microtubule, together with an interpretive drawing. Microtubules may be considered to be composed of 13 parallel but staggered protofilaments, which consist in turn of

alternating α- and β-tubulin subunits linked head to tail. [Courtesy of Gerald Stubbs, Vanderbilt University, and Lorena Beese and Carolyn Cohen, Brandeis University.]

linked α- and β-tubulin subunits, all run in the same direction. Consequently, microtubules, like microfilaments, are polar entities: The end that grows most rapidly is called the plus end, whereas the other end is called the minus end. Microtubules are generally oriented in a cell with their minus ends towards a **centrosome** (the organizing centers from which they eminate; Fig. 34-73), and their plus ends towards the cell periphery.

Microtubules undergo continuous assembly and disassembly. Indeed, in a given population of microtubules, some may grow while others simultaneously shrink. This **dynamic instability** occurs because if the second GTP in a tubulin subunit at the microtubule's plus end becomes hydrolyzed to GDP before it is "capped" by another tubulin subunit, the resulting GDP–subunit rapidly dissociates from the microtubule. The balance between net microtubule growth or shrinkage in a cell therefore depends on the rate that tubulin hydrolyzes its second bound GTP together with the availability of GTP–tubulin subunits. Even when microtubules maintain a constant length, they are by no means static: They undergo GTP-driven treadmilling in which tubulin subunits add to the plus end at the same rate that they dissociate from the minus end. By regulating the rate of tubulin polymerization, cells presumably vary their shapes and induce the formation and dissolution of such cellular apparatus as the mitotic spindle.

Antimitotic Drugs Inhibit Microtubule Formation

Colchicine,

![Colchicine structure]

Colchicine

an alkaloid produced by the meadow saffron, inhibits microtubule-dependent cellular processes by inhibiting the polymerization of tubulin protomers. For example, colchicine arrests mitosis in both plant and animal cells at metaphase (when the condensed and replicated chromosomes line up on the cell's equator; Section 27-1A) by preventing the formation of the mitotic spindle. It also inhibits cell motility.

> Colchicine has been used for centuries to treat acute attacks of gout (which result from elevated uric acid levels in body fluids; Section 26-5B). The lysosomes of the white cells that engulf urate microcrystals are ruptured by these needle-shaped crystals, causing cell lysis and triggering the local acute inflammatory reaction responsible for the exquisite pain characteristic of gout attacks. Colchicine, it is thought, slows the

ameboid movements of white cells by inhibiting their microtubule-based systems.

The **vinca alkaloids, vinblastine** and **vincristine,**

![Vinblastine/vincristine structure]

Vinblastine: R = CH₃
Vincristine: R = CHO

products of the Madagascan periwinkle *Vinca rosea*, also inhibit microtubule polymerization by binding to tubulin. These substances are widely used in cancer chemotherapy since blocking mitosis preferentially kills fast growing cells. Curiously, colchicine is not selectively toxic to cancer cells.

Cilia and Eukaryotic Flagella Contain Organized Sheaves of Microtubules

Cilia are the hairlike organelles on the surfaces of many animal and lower plant cells that function to move fluid over the cell's surface or to "row" single cells through a fluid. In humans, for example, epithelial cells lining the respiratory tract each bear ~200 cilia that beat in synchrony to sweep mucus-entrained foreign particles towards the throat for elimination (Fig. 34-75; individuals with the in-

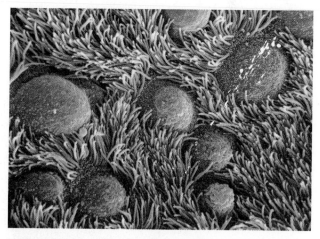

FIGURE 34-75. Scanning electron micrograph showing cilia lining the epithelial surface of a human bronchial tube as well as the rounded surfaces of a number of mucus-secreting **goblet cells.** Individuals with hereditary ciliary defects suffer from recurrent respiratory tract infections resulting from their reduced ability to clear away foreign particles. [From Kessel, R.G. and Kardon, R.H., *Tissues and Organs: A Text-Atlas of Scanning Electron Microscopy, p. 210,* Freeman (1979). Copyright ©1979 W.H. Freeman and Company. Reproduced by permission.]

herited recessive disease **immotile-cilia syndrome** suffer from chronic respiratory disorders). Cilia are relatively short, operate with a whiplike motion, and occur in large numbers on a single cell. **Eukaryotic flagella** (as distinct from prokaryotic flagella; Section 34-3G), which occur on certain protozoa and comprise sperm tails, are much longer by comparison, carry out their propulsive function via undulatory motions, and occur in quantities of only one or a few per cell. Nevertheless, both types of organelles have the same basic architecture (males with immotile-cilia syndrome are usually sterile because their sperm are also immotile).

A cilium or flagellum consists of a plasma membrane–coated bundle of microtubules called an **axoneme.** Electron micrographs indicate that *an axoneme contains a ring of 9*

*double microtubules surrounding 2 single microtubules to form a common biological motif known as a **9 + 2 array** (Fig. 34-76).* Each outer doublet consists of a ring of 13-protofilaments, **subfiber A,** fused to a C-shaped assembly of 10, or in some cases 11, protofilaments, **subfiber B** (Fig. 34-77). The 11 microtubules forming an axoneme are held together by three types of connectors (Fig. 34-77):

1. Subfibers A are joined to the central microtubules by radial **spokes,** which each terminate in a knoblike feature termed a **spoke head.**

2. Adjacent outer doublets are joined by circumferential linkers that, in part, consist of a highly elastic protein named **nexin.**

3. The central microtubules are joined by a connecting bridge.

Each type of connector is repeated along the length of the axoneme with its own characteristic periodicity. Finally, every subfiber A bears two arms, an **inner dynein arm** and an **outer dynein arm** (Fig. 34-78), which both point clockwise when viewed from the base of the cilium (Fig. 34-77).

Ciliary Motion Results from the ATP-Powered "Walking" of Dynein Arms Along an Adjacent Subfiber B

An isolated flagellum (excised by a laser microbeam) whose plasma membrane has been removed by treatment with a nonionic detergent will continue to beat when supplied with ATP. Evidently, the eukaryotic flagellar "motor" is contained in the axoneme itself rather than at its base, as occurs in bacterial flagella (Section 34-3G). What is the site of the eukaryotic flagellar motor? Several observations point to the dynein arms:

1. The dynein arms can be selectively extracted from naked axonemes by solutions containing high salt con-

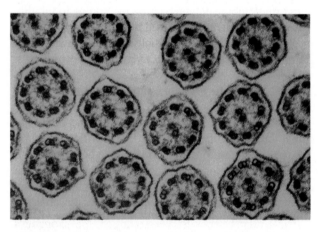

FIGURE 34-76. An electron micrograph of hamster oviduct cilia in cross-section. Two single microtubules are surrounded by 9 doublets to form a 9 + 2 array. [David M. Phillips/Visuals Unlimited.]

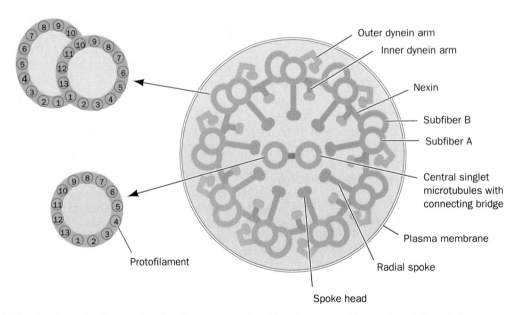

FIGURE 34-77. A schematic diagram showing the structure of a cilium in cross-section as viewed from its base.

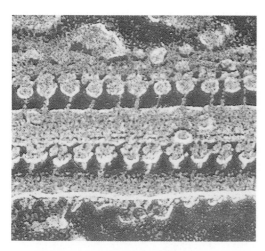

FIGURE 34-78. A freeze-etch electron micrograph of a flagellar microtubule from the unicellular algae *Chlamydomonas reinhardtii,* in transverse view, showing its dynein arms projecting like "lollipops." The outer dynein arms, which are spaced every 240 Å along the microtubule, have 100-Å-diameter heads attached to stalks that are <30 Å wide (an arrangement reminiscent of myosin's S1 heads). The spacing of the inner dynein arms is less regular. [Courtesy of John Heuser and Ursala W. Goodenough, Washington University School of Medicine.]

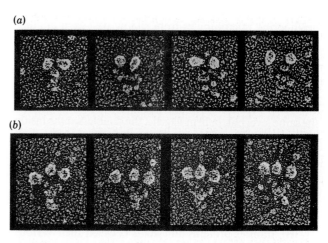

FIGURE 34-79. Electron micrographs of axonemal outer dynein arms. (*a*) Two-headed dyneins from sperm of the sea urchin *Strongylocentrotus purpuratus.* (*b*) Three-headed dyneins from *Chlamydomonus.* [Courtesy of Ursula Goodenough, Washington University School of Medicine.]

centrations. This treatment immobilizes the axonemes while it solubilizes their ATPase activity (although it is much lower in solution than in intact axonemes). The addition of purified dynein to the salt-extracted axonemes restores their ability to beat.

2. In the absence of ATP, flagella become rigid. Electron micrographs indicate that the dynein arms in such ATP-deprived flagella are attached to their adjacent subfiber B.

3. Dynein resembles the S1 heads of myosin in both appearance and function. The outer dynein arms consist of either two-headed (~1200-kD) or three-headed (~1900-kD) entities, depending on the species, in which the globular heads are joined to a common base by flexible stems (Fig. 34-79; the inner dynein arms, which are not well characterized, probably have one or two globular heads). Dynein's ATPase functions are located in these heads.

4. Brief trypsin treatment, which selectively cleaves the radial spokes and nexin circumferential linkers, followed by the addition of ATP, causes axonemes to elongate up to nine times their original length (Fig. 34-80). The elongation results from the telescoping of the axoneme's component microtubules out of the disrupted structure; the individual microtubules do not change in length.

These observations indicate that ciliary motion results from an ATP-driven process reminiscent of the sliding filament model of muscle contraction: *The dynein arms on one microtubule "walk" up the neighboring subfiber B so that these two microfilaments slide past each other.* However, the cross-links between microtubules in an intact cilium prevent

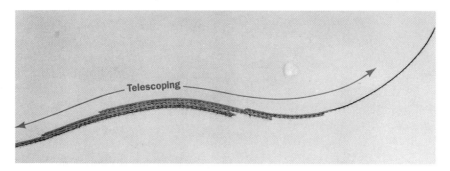

FIGURE 34-80. An electron micrograph of an isolated axoneme from a *Tetrahymena* cilium that has been briefly treated with trypsin to degrade its protein connectors and then exposed to ATP. The individual microtubule doublets telescope out from each other so that the axoneme elongates by up to a factor of 9. [From Warner, F.D. and Mitchell, D.R., *J. Cell Biol.* **89,** 36 (1981).]

neighboring microtubules from sliding past each another by more than a short distance. *These cross-links therefore convert the dynein-induced sliding motion to a bending motion of the entire axoneme.* This model is supported by electron microscopy studies showing that in straight flagella all the outer doublets have the same length and terminate at the same level but, in bent flagella, the doublets at the inside of the bend extend further than those on the outside of the bend (Fig. 34-81).

Dynein and Kinesin Motivate the Intracellular Transport of Vesicles and Organelles along Microtubule Tracks

Eukaryotic cells, as we have seen (Sections 11-4B and 21-3B), transfer proteins and lipids between their various organelles via membranous vesicles. But how do these vesicles find their way to their proper destinations at a reasonable rate? The answer to this question was determined, in part, through the study of vesicle transport in the **axons** of neurons (cellular projections that extend from the cell body by up to 1 m; Fig. 1-10*d*). The use of **video-enhanced contrast microscopy** (subcellular components are generally smaller than the resolution limit of light) revealed that vesicles and even entire organelles such as mitochondria are unidirectionally transported within axons at rates of 1 to 5 μm·s^{-1}, so they can traverse the length of even the longest

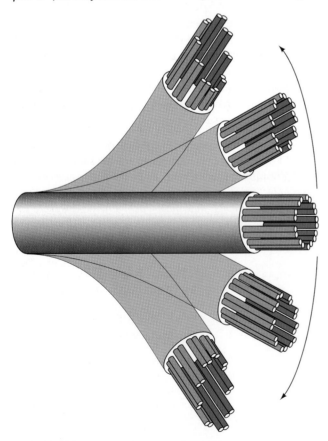

FIGURE 34-81.   Diagram of sliding outer microtubules in a beating cilium. When the cilium is straight, all the outer doublets end at the same level (*center*). Cilium bending occurs when the doublets on the inner side of the bend slide beyond those on the outer side (*top* and *bottom*).

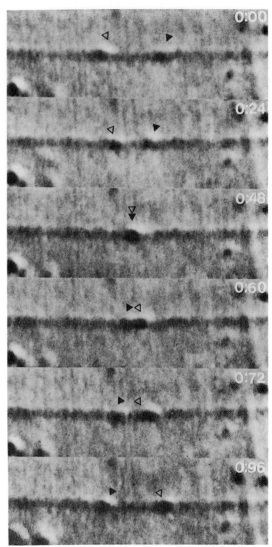

FIGURE 34-82.   A series of successive video-enhanced contrast micrographs showing two organelles (*triangles*) moving in opposite directions along a microtubule and passing one another without colliding. The number in the upper right corner of each frame is the elapsed time in seconds from the top frame. [Courtesy of Bruce Schnapp, Boston Medical Center, and Thomas S. Reese, NIH.]

axon in ~2 days. This apparently purposeful traffic, which simultaneously occurs in both directions (Fig. 34-82), moves along filamentous tracks that have been identified as microtubules through their binding of specific antibodies.

What are the "motors" that drive vesicle and organelle transport? Two types have been identified:

1. Cytosolic dyneins, which resemble axonemal two-headed dyneins in appearance (Fig. 34-79*a*), transport vesicles and organelles in the plus to minus direction along microtubules (towards the cell center). This is corroborated by the observation that dynein that is immobilized by adsorption to a glass surface transports free microtubules in the direction of their plus end when supplied with ATP.

(a)

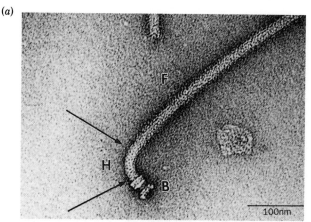

(b)

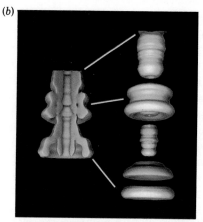

FIGURE 34-83. Electron micrographs of *Salmonella* flagella. (*a*) An intact flagellum. Its helical filament (F), which extends to the upper right, is connected via its hook (H) to its four-ringed basal body (B). [Courtesy of Robert McNab, Yale University.] (*b*) A closeup of the basal body, as generated by image reconstruction from electron micrographs, showing it in cutaway (*left*) and exploded (*right*) views. The top line points at the hook–filament junction, the middle line points at the L–P ring complex, and the bottom line points at the M ring in the center of the motor. [Courtesy of Noreen Francis and David DeRosier, Brandeis University.]

2. Kinesin transports vesicles and organelles in the minus to plus direction along microtubules (away from the cell center) and similarly, upon adsorption to a glass surface, transports free microtubules in the direction of their minus end. This ~600-kD, elongated (~1000 Å in length) protein resembles myosin and dynein in that it has twin globular ATPase-containing heads which power its motion.

Thus, eukaryotes have three classes of force-generating ATPases: myosins, dyneins, and kinesins.

G. Bacterial Flagella

The final aspect of biological motility that we shall consider is the nature of the bacterial flagellum. This remarkable propulsive organelle generates true rotary motion; *it is a propeller rather than a bending or a contractile device.*

Many species of bacteria, *E. coli,* for example (Fig. 1-3*b*), "swim" through solution via the action of a few flagella. Bacterial flagella, however, are entirely different from those of eukaryotes in both structure and chemistry. To begin with, bacterial flagella are only ~200 Å in diameter, less than the width of a single microtubule, and contain no tubulin. Electron micrographs (Fig. 34-83) indicate that bacterial flagella consist of three major segments (Fig. 34-84):

1. The **flagellar filament,** its most prominant portion, is a tightly coiled helix up to 10 μm long that consists only of ~490-residue subunits of the protein **flagellin.**

2. The **flagellar hook,** which is assembled from subunits of the 42-kD **hook protein,** forms a short, curved structure to which the flagellar filament is attached.

3. The **basal body,** which in gram-negative bacteria penetrates the outer membrane, the peptidoglycan cell wall,

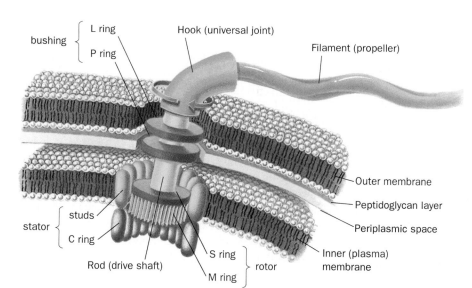

FIGURE 34-84. The structure of the gram-negative bacterial flagellum. The rotational "motor" consists of the basal body's studs and the C, M, and S rings. The studs and C ring are anchored to the plasma membrane, whereas the M and S rings, which are attached via the rod and hook to the flagellar filament, rotate freely in the plasma membrane. Torque is generated through the interaction of this "motor" with the electrochemical gradient across the plasma membrane. Gram-positive bacteria, which lack the outer membrane, also lack the "bushing" formed by the L and P rings. [After Schuster, S.C. and Khan, S., *Annu. Rev. Biophys. Biomol. Struct.* **23,** 524 (1994).]

and the inner (plasma) membrane, consists of a rod connecting several ringlike structures that anchor the flagellar hook to the bacterium.

The Flagellar Filament Is Hollow

The structure of the *Salmonella typhimurium* flagellar filament, as determined by Keiichi Namba to a resolution of 12.5 Å using cryoelectron microscopy, reveals that it is 240 Å in diameter and has ~5.5 subunits per turn. A section of the filament that contains 11 subunits in two helical turns, has, when viewed end on, the appearance of an 11-bladed propeller with a ~25-Å-diameter central channel (Fig. 34-85). Flagella have been shown to grow by adding flagellin subunits at the end away from the cell body and it has therefore been postulated that these subunits diffuse through the flagllum's hollow core, thereby reaching the tip without being lost to the medium. The structure reveals that the flagellum's hollow core may have sufficient width to accommodate a folded flagellin subunit.

Bacterial Flagella Rotate

Microscopic observations of swimming bacteria indicate that they are driven by what appear to be flagellar undulations. These undulations are an illusion; one cannot, by only watching through a light microscope, distinguish wave propagation along a helical flagellum from its rigid rotation. That bacterial flagella can rotate freely, a previously unknown and unexpected biological phenomenon, was established by the observation that when a bacterial flagellum is immobilized by "gluing" it to a microscope slide with anti-flagellin antibody, its attached bacterial cell slowly rotates. Similarly, if small latex beads are so glued to a bacterial flagellum, they can be seen to rotate about the flagellum.

What is the nature of the flagellar "motor"? It cannot be located in either the flagellar filament or the flagellar hook since both flagellin and hook protein have no demonstrable enzymatic activity. The basal body must therefore form the rotary element. If this is so, then it must have the same mechanical elements as other rotary devices: a rotor (the rotating element) and a stator (the stationary element). Indeed, as Figs. 34-83*b* and 34-84 indicate, the ringlike structures of the basal body form just such elements. The M and S rings are the rotor, the C ring and studs, which are anchored to the plasma membrane, form the stator, and the L and P rings form a bushing (sleeve) through which the rotating rod penetrates the bacterial outer membrane. The flagellar hook and filament are therefore chemically passive elements that mechanically convert rotary motion to linear thrust in the manner of a propeller.

A final question we shall ask is what makes the flagellar rotor rotate? One might guess, in analogy with muscles and cilia, that the M and S rings form an ATPase that acts as a mechanochemical transducer. However, the observation that bacterial swimming is unaffected by drastic reductions of the bacterial ATP pool requires that this hypothesis be

FIGURE 34-85
An electron density map of the *Salmonella typhimurium* flagellar filament showing a 52.5-Å-thick section, which contains 11 subunits of the 489-residue flagellar protein in 2 helical turns. Note the ~25-Å-diameter central hole through the filament. [Courtesy of Keiichi Namba, Matsushita Electrical Industrial Co., Ltd., Japan.]

abandoned. Rather, *the driving force behind flagellar rotation is the electrochemical proton gradient across the plasma membrane;* the same proton gradient that powers oxidative phosphorylation. This phenomenon was first demonstrated in *E. coli* mutants that lack an active F_1F_0-ATPase (the enzyme that mediates the proton gradient-driven synthesis of ATP; Section 20-3C) and are therefore unable to generate a proton gradient under anaerobic conditions (when an active F_1F_0-ATPase works in reverse). These mutant bacteria can only swim under aerobic conditions, whereas normal bacteria can also swim under anaerobic conditions. Such proton gradient–driven rotational motors may be far more common in biological systems than has previously been suspected: The F_1F_0-ATPase appears to contain just such a motor to drive ATP synthesis (Section 20-3C).

The rotation of a flagellum requires the downhill transport of at least 1000 protons/turn for each of its ~10 rotations/s in metabolizing tethered *E. coli* (a small fraction of the bacterium's total proton circulation). The transduction of the proton gradient's free energy to mechanical work is thought to occur via the passage of protons across the M ring in a way that allows the protons to interact with suitably disposed charges fixed to the surface of the studs. The elucidation of a detailed model of this process is a matter of active research.

4. BIOCHEMICAL COMMUNICATIONS: HORMONES AND NEUROTRANSMISSION

Living things coordinate their activities at every level of their organization through complex chemical signaling systems. Intracellular communications are maintained by the synthesis or alteration of a great variety of different substances that are often integral components of the processes they control; for example, metabolic pathways are regulated via the feedback control of allosteric enzymes by metabolites in those pathways or by the covalent modification of macromolecules. Intercellular signals occur through the mediation of chemical messengers known as **hormones** and, in higher animals, via neuronally transmitted electrochemical impulses. In this section we outline the operations of both these complex signaling systems.

A. The Endocrine System

Hormones are classified according to the distance over which they act (Fig. 34-86):

1. **Autocrine hormones** act on the same cell that released them. Interleukin-2, which stimulates T cell proliferation (Section 34-2A), is an autocrine hormone.

2. **Paracrine hormones** (alternatively, **local mediators**) act only on cells close to the cell that released them. Prostaglandins (Section 23-7) and many polypeptide growth factors (Section 33-4C) are examples of paracrine hormones.

3. **Endocrine hormones** act on cells distant from the site of their release. Endocrine hormones, for example, insulin

and epinephrine, are synthesized and released in the bloodstream by specialized ductless **endocrine glands.**

We are already familiar with many aspects of hormonal control. For instance, we have considered how epinephrine, insulin, and glucagon regulate energy metabolism through the intermediacy of cAMP (Sections 17-3E and G), how steroid hormones activate transcription factors and thus influence protein synthesis (Section 33-3B), and how epinephrine relaxes smooth muscle through the intermediacy of Ca^{2+} (Section 34-3D). In this section we shall extend and systematize this information. Before we do so, it should be noted that biochemical communications are not limited to intracellular and intercellular signals. Many organisms release substances called **pheromones** that alter the behavior of other organisms of the same species in much the same way as hormones. Pheromones are commonly sexual attractants but some have other functions in species, such as ants, that have complex social interactions.

The human endocrine system (Fig. 34-87) secretes a wide variety of hormones (Table 34-5) that enable the body to:

1. Maintain homeostasis (e.g., insulin and glucagon maintain the blood glucose level within rigid limits during feast or famine).

2. Respond to a wide variety of external stimuli (such as the preparation for "fight or flight" engendered by epinephrine and norepinephrine).

3. Follow various cyclic and developmental programs (for instance, sex hormones regulate sexual differentiation, maturation, the menstrual cycle, and pregnancy).

Most hormones are either polypeptides, amino acid derivatives, or steroids, although there are important exceptions to this generalization. In any case, only those cells with a

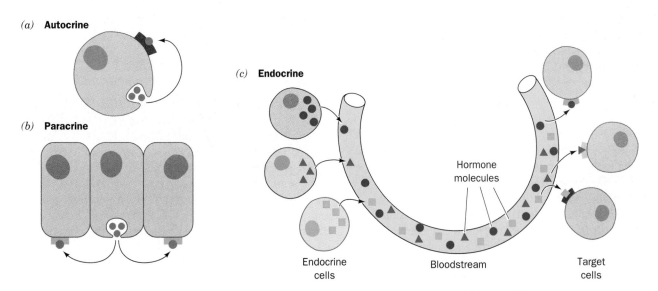

FIGURE 34-86. Hormonal communications are classified according to the distance over which the signal acts: (*a*) autocrine signals are directed at the cell that produced them, (*b*) paracrine signals are directed at nearby cells, and (*c*) endocrine signals are directed at distant cells through the intermediacy of the bloodstream.

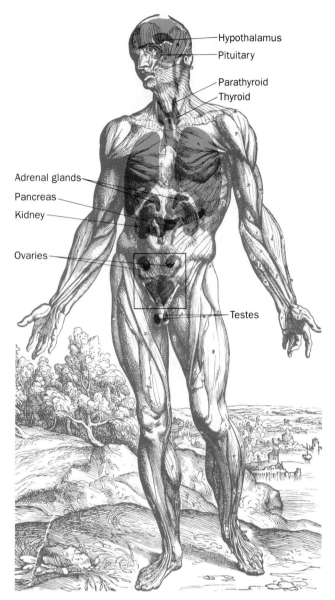

—— Hypothalamus
—— Pituitary
—— Parathyroid
—— Thyroid
Adrenal glands ——
Pancreas ——
Kidney ——
Ovaries ——
—— Testes

FIGURE 34-87. The major glands of the human endocrine system. Other tissues, the intestines, for example, also secrete endocrine hormones.

specific receptor for a given hormone will respond to its presence even though nearly all cells in the body may be exposed to the hormone. Hormonal messages are therefore quite specifically addressed.

In what follows, we outline the hormonal functions of the various endocrine glands. Throughout this discussion keep in mind that these glands are not just a collection of independent secretory organs but form a complex and highly interdependent control system. Indeed, as we shall see, the secretion of many hormones is under feedback control through the secretion of other hormones to which the original hormone-secreting gland responds. Much of our understanding of hormonal function has come from careful measurements of hormone concentrations and the effects of

changes of these concentrations on physiological functions. We begin, therefore, with a consideration of how physiological hormone concentrations are measured.

Hormone Concentrations Are Measured by Radioimmunoassays

The serum concentrations of hormones are extremely small, generally between 10^{-12} and $10^{-7}M$, so they usually must be measured by indirect means. Biological assays have traditionally been employed for this purpose but they are generally slow, cumbersome, and imprecise. Such assays have therefore been largely supplanted by **radioimmunoassays.** In this technique, which was developed by Rosalyn Yalow, the unknown concentration of a hormone, H, is determined by measuring how much of a known amount of the radioactively labeled hormone, H*, binds to a fixed quantity of anti-H antibody in the presence of H. This competition reaction is easily calibrated by constructing a standard curve indicating how much H* binds to the antibody as a function of [H]. The high ligand affinity and specificity that antibodies possess gives radioimmunoassays the advantages of great sensitivity and specificity.

Pancreatic Islet Hormones Regulate the Storage and Release of Glucose and Fatty Acids

The pancreas is a large glandular organ, the bulk of which is an **exocrine gland** dedicated to producing digestive enzymes such as trypsin, RNase A, α-amylase, and phospholipase A_2 that it secretes via the pancreatic duct into the small intestine. However, ~1 to 2% of pancreatic tissue consists of scattered clumps of cells known as **islets of Langerhans,** which comprise an endocrine gland that functions to maintain energy metabolite homeostasis. Pancreatic islets contain three types of cells, each of which secretes a characteristic polypeptide hormone:

1. The α cells secrete glucagon (29 residues; Section 17-3E).
2. The β cells secrete insulin (51 residues; Fig. 8-4).
3. The δ cells secrete **somatostatin** (14 residues).

Insulin, which is secreted in response to high blood glucose levels, primarily functions, as we have seen (Section 25-2), to stimulate muscle, liver, and adipose cells to store glucose for later use by synthesizing glycogen, protein, and fat. Glucagon, which is secreted in response to low blood glucose, has essentially the opposite effects: It stimulates liver to release glucose through glycogenolysis and gluconeogenesis and it stimulates adipose tissue to release fatty acids through lipolysis. Somatostatin, which is also secreted by the hypothalamus (see below), inhibits the release of insulin and glucagon from their islet cells and is therefore thought to have a paracrine function in the pancreas.

Polypeptide hormones, as are other proteins destined for secretion, are ribosomally synthesized as preprohormones, processed in the rough endoplasmic reticulum and Golgi apparatus to form the mature hormone, and then packaged

TABLE 34-5. SOME HUMAN HORMONES

Hormone	Origin	Major Effects
Polypeptides		
Corticotropin-releasing factor (CRF)	Hypothalamus	Stimulates ACTH release
Gonadotropin-releasing factor (GnRF)	Hypothalamus	Stimulates FSH and LH release
Thyrotropin-releasing factor (TRF)	Hypothalamus	Stimulates TSH release
Growth hormone-releasing factor (GRF)	Hypothalamus	Stimulates growth hormone release
Somatostatin	Hypothalamus	Inhibits growth hormone release
Adrenocorticotropic hormone (ACTH)	Adenohypophysis	Stimulates the release of adrenocorticosteroids
Follicle-stimulating hormone (FSH)	Adenohypophysis	In ovaries, stimulates follicular development, ovulation, and estrogen synthesis; in testes, stimulates spermatogenesis
Lutinizing hormone (LH)	Adenohypophysis	In ovaries, stimulates oocyte maturation and follicular synthesis of estrogens and progesterone; in testes, stimulates androgen synthesis
Chorionic gonadotropin (CG)	Placenta	Stimulates progesterone release from the corpus luteum
Thyrotropin (TSH)	Adenohypophysis	Stimulates T_3 and T_4 release
Somatotropin (growth hormone)	Adenohypophysis	Stimulates growth and synthesis of somatomedins
Met-enkephalin	Adenohypophysis	Opioid effects on central nervous system
Leu-enkephalin	Adenohypophysis	Opioid effects on central nervous system
β-Endorphin	Adenohypophysis	Opioid effects on central nervous system
Vasopressin	Neurohypophysis	Stimulates water resorption by kidney and increases blood pressure
Oxytocin	Neurohypophysis	Stimulates uterine contractions
Glucagon	Pancreas	Stimulates glucose release through glycogenolysis and stimulates lipolysis
Insulin	Pancreas	Stimulates glucose uptake through gluconeogenesis, protein synthesis, and lipogenesis
Gastrin	Stomach	Stimulates gastric acid and pepsinogen secretion
Secretin	Intestine	Stimulates pancreatic secretion of HCO_3^-
Cholecystokinin (CCK)	Intestine	Stimulates gallbladder emptying and pancreatic secretion of digestive enzymes and HCO_3^-
Gastric inhibitory peptide (GIP)	Intestine	Inhibits gastric acid secretion and gastric emptying; stimulates pancreatic insulin release
Parathyroid hormone	Parathyroid	Stimulates Ca^{2+} uptake from bone, kidney, and intestine
Calcitonin	Thyroid	Inhibits Ca^{2+} uptake from bone and kidney
Somatomedins	Liver	Stimulate cartilage growth; have insulin-like activity
Steroids		
Glucocorticoids	Adrenal cortex	Affect metabolism in diverse ways, decrease inflammation, increase resistance to stress
Mineralocorticoids	Adrenal cortex	Maintain salt and water balance
Estrogens	Gonads	Maturation and function of secondary sex organs, particularly in females
Androgens	Gonads	Maturation and function of secondary sex organs, particularly in males; male sexual differentiation
Progestins	Ovaries and placenta	Mediate menstrual cycle and maintains pregnancy
Vitamin D	Diet and sun	Stimulates Ca^{2+} absorption from intestine, kidney and bone
Amino Acid Derivatives		
Epinephrine	Adrenal medulla	Stimulates contraction of some smooth muscles and relaxes others, increases heart rate and blood pressure, stimulates glycogenolysis in liver and muscle, stimulates lipolysis in adipose tissue
Norepinephrine	Adrenal medulla	Stimulates arteriole contraction, decreases peripheral circulation, stimulates lipolysis in adipose tissue
Triiodothyronine (T_3)	Thyroid	General metabolic stimulation
Thyroxine (T_4)	Thyroid	General metabolic stimulation

in secretory granules to await the signal for their release by exocytosis (Section 11-3F). The most potent physiological stimuli for the release of insulin and glucagon are, respectively, high and low blood glucose concentrations so that islet cells act as the body's primary glucose sensors. However, the release of these hormones is also influenced by the autonomic (involuntary) nervous system and by hormones secreted by the gastrointestinal tract (see below).

Gastrointestinal Hormones Regulate Digestion

The digestion and absorption of nutrients is a complicated process that is regulated by the autonomic nervous system in concert with a complex system of polypeptide hormones. Indeed, gastrointestinal hormones are secreted into the bloodstream by a system of specialized cells lining the gastrointestinal tract whose aggregate mass is greater than that of the rest of the endocrine system. Four gastrointestinal hormones have been well characterized:

1. **Gastrin** (17 residues), which is produced by the gastric mucosa, stimulates the gastric secretion of HCl and **pepsinogen** (the zymogen of the digestive protease pepsin). Gastrin release is stimulated by amino acids and partially digested protein as well as by the vagus nerve (which innervates the stomach) in response to stomach distension. Gastrin release is inhibited by HCl and by other gastrointestinal hormones (see below).

2. **Secretin** (27 residues), which is produced by the upper small intestinal mucosa in response to acidification by gastric HCl, stimulates the pancreatic secretion of HCO_3^- so as to neutralize this acid.

3. **Cholecystokinin** (**CCK;** 33 residues), which is produced by the upper small intestine, stimulates gallbladder emptying, the pancreatic secretion of digestive enzymes and HCO_3^- (and thus enhances the effect of secretin), and inhibits gastric emptying. CCK is released in response to the products of lipid and protein digestion, that is, fatty acids, monoacylglycerols, amino acids, and peptides.

4. **Gastric inhibitory peptide** (**GIP;** 43 residues), which is produced by specialized cells lining the small intestine, is a potent inhibitor of gastric acid secretion, gastric mobility, and gastric emptying. However, GIP's major physiological function is to stimulate pancreatic insulin release. Indeed, the release of GIP is stimulated by the presence of glucose in the gut which accounts for the observation that, after a meal, the blood insulin level increases before the blood glucose level does.

These gastrointestinal hormones form families of related polypeptides: The C-terminal pentapeptides of gastrin and CCK are identical; secretin, GIP, and glucagon, are closely similar.

Several other polypeptides that affect gastrointestinal function have been isolated from the gut. However, the physiological roles of these so-called **candidate hormones** is unclear. Much of this difficulty stems from the diffuse distribution of gastrointestinal hormone-secreting cells that precludes their excision, a procedure that is commonly used in controlled studies of the effects of other endocrine hormones.

Thyroid Hormones Are Metabolic Regulators

The thyroid gland produces two related hormones, triiodothyronine (T_3) and thyroxine (T_4),

X = H   **Triiodothyronine (T_3)**
X = I   **Thyroxine (T_4)**

that stimulate metabolism in most tissues (adult brain is a conspicuous exception). The production of these unusual iodinated amino acids begins with the synthesis of **thyroglobin,** a 660-kD globular protein. This protein is post-translationally modified in a series of biochemically unique reactions (Fig. 34-88):

1. Around 20% of thyroglobin's 140 Tyr residues are iodinated in an **iodoperoxidase**-catalyzed reaction forming **2,5-diiodo-Tyr** residues.

2. Two such residues are coupled to yield T_3 and T_4 residues. Mature thyroglobin itself is hormonally inactive; some five or six molecules of the active hormones, T_3 and T_4, are produced by the lysosomal proteolysis of thyroglobin upon hormonal stimulation of the thyroid (see below).

How do thyroid hormones work? T_3 and T_4, being nonpolar substances, are transported by the blood in complex with plasma carrier proteins, primarily **thyroxine-binding globin,** but also **prealbumin** and **albumin.** The hormones then pass through the cell membranes of their target cells into the cytosol where they bind to a specific protein. Since the resulting hormone–protein complex does not enter the nucleus, it is thought that this complex acts to maintain an intracellular reservoir of thyroid hormones. The true **thyroid hormone receptor** is a nonhistone chromosomal protein and therefore does not leave the nucleus. *The binding of T_3, and to a lesser extent T_4, activates this receptor as a transcription factor, resulting in increased rates of synthesis of numerous metabolic enzymes.* Indeed, thyroid hormone receptor is homologous to steroid hormone receptors (Section 33-3B). High affinity thyroid hormone-binding sites also occur on the inner mitochondrial membrane, suggesting that these receptors may directly regulate O_2 consumption and ATP production.

Abnormal levels of thyroid hormones are common human afflictions. **Hypothyroidism** is characterized by lethargy, obesity, and cold dry skin, whereas **hyperthyroidism** has the opposite effects. The inhabitants of areas in which the soil has a low iodine content often develop hypo-

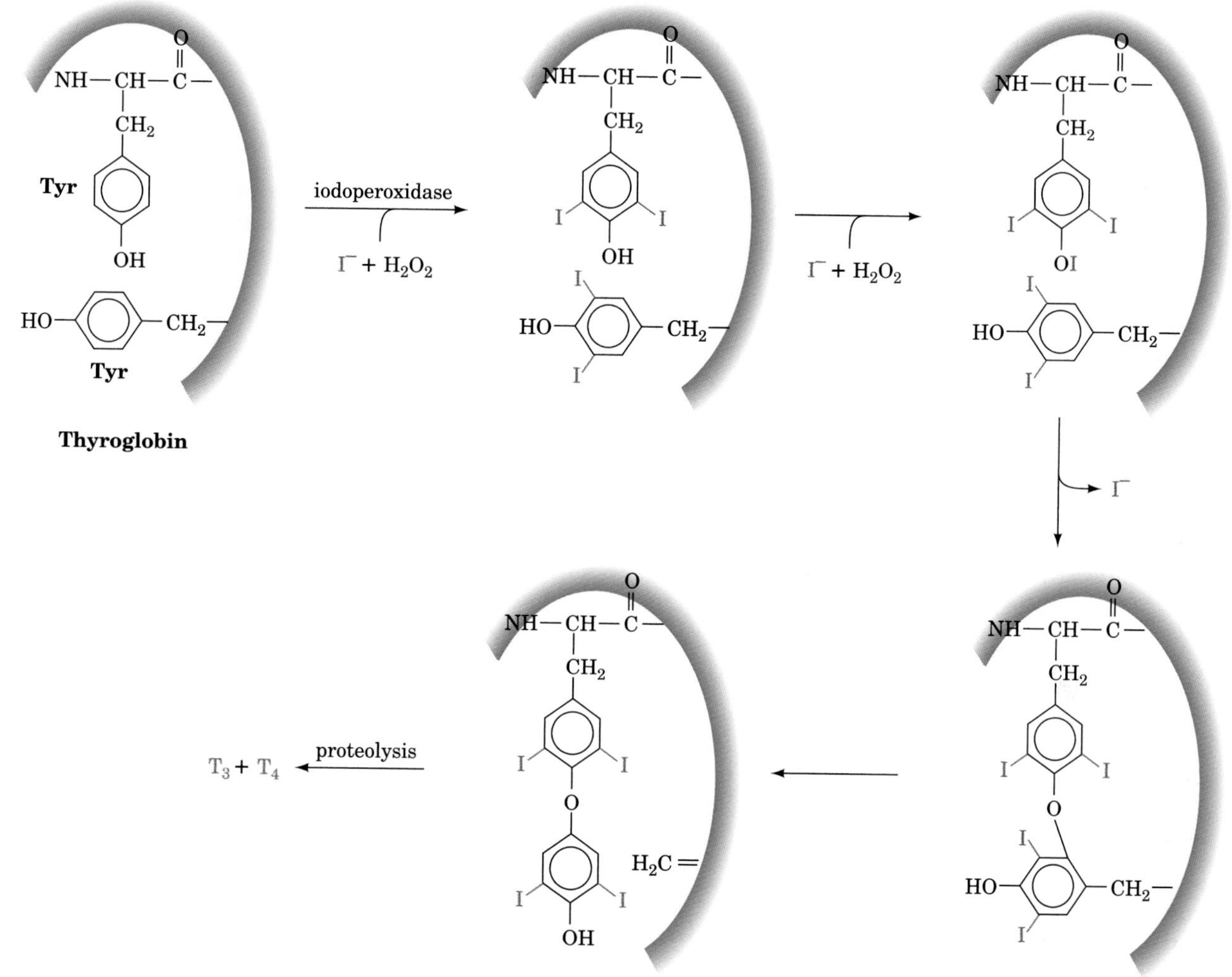

FIGURE 34-88. The biosynthesis of T_3 and T_4 in the thyroid gland through the iodination, rearrangement, and hydrolysis (proteolysis) of thyroglobin Tyr residues. The relatively scarce I^- is actively sequestered by the thyroid gland.

thyroidism accompanied by an enlarged thyroid gland, a condition known as **goiter.** The small amount of NaI often added to commercially available table salt ("iodized" salt) easily prevents this iodine deficiency disease. Young mammals require thyroid hormone for normal growth and development: Hypothyroidism during the fetal and immediate postnatal periods results in irreversible physical and mental retardation, a syndrome named **cretinism.**

Calcium Metabolism Is Regulated by Parathyroid Hormone, Vitamin D, and Calcitonin

Ca^{2+} forms **hydroxyapatite,** $Ca_5(PO_4)_3OH$, the major mineral constituent of bone, and is an essential element in many biological processes including the mediation of hormonal signals as a second messenger, the triggering of muscle contraction, the transmission of nerve impulses, and blood clotting. The extracellular $[Ca^{2+}]$ must therefore be closely regulated to keep it at its normal level of ~1.2 mM. Three hormones have been implicated in maintaining Ca^{2+} homeostasis (Fig. 34-89):

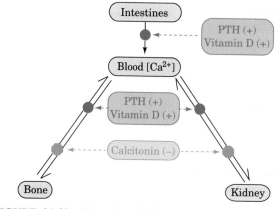

FIGURE 34-89. The roles of PTH, vitamin D, and calcitonin in controlling Ca^{2+} metabolism.

1. **Parathyroid hormone (PTH),** an 84-residue polypeptide secreted by the parathyroid gland, increases serum $[Ca^{2+}]$ by stimulating its resorption from bone and kid-

ney and by increasing the dietary absorption of Ca^{2+} from the intestine.

2. **Vitamin D,** a group of steroid-like substances that act in a synergistic manner with PTH to increase serum $[Ca^{2+}]$.

3. **Calcitonin,** a 33-residue polypeptide synthesized by specialized thyroid gland cells, decreases serum $[Ca^{2+}]$ by inhibiting the resorption of Ca^{2+} from bone and kidney.

We shall briefly discuss the functions of these hormones.

The bones, the body's main Ca^{2+} reservoir, are by no means metabolically inert. They are continually "remodeled" through the action of two types of bone cells: **osteoblasts,** which synthesize the collagen fibrils that form the bulk of bone's organic matrix, the scaffolding upon which its $Ca_5(PO_4)_3OH$ mineral phase is laid down; and **osteoclasts,** which participate in bone resorption. *PTH inhibits collagen synthesis by osteoblasts and stimulates bone resorption by osteoclasts. The main effect of PTH, however, is to increase the rate that the kidneys excrete phosphate, the counterion of Ca^{2+} in bone.* The consequent decreased serum $[P_i]$ causes $Ca_5(PO_4)_3OH$ to leach out of bone through mass action and thus increase serum $[Ca^{2+}]$. Finally, PTH stimulates the production of the active form of vitamin D by kidney which, in turn, enhances the transfer of intestinal Ca^{2+} to the blood (see below).

Vitamin D is a group of dietary substances that prevent **rickets,** a disease of children characterized by stunted growth and deformed bones stemming from insufficient bone mineralization (vitamin D deficiency in adults is known as **osteomalacia,** a condition characterized by weakened, demineralized bones). Although rickets was first described in 1645, it was not until the early twentieth century that it was discovered that animal fats, particularly fish liver oils, are effective in preventing this deficiency disease. Moreover, rickets can also be prevented by exposing children to sunlight or just UV light in the wavelength range 230 to 313 nm, regardless of their diets.

The D vitamins, which we shall see are really hormones, are sterol derivatives in which the steroid B ring is disrupted at its 9,10 position. The natural form of the vitamin, **vita-**

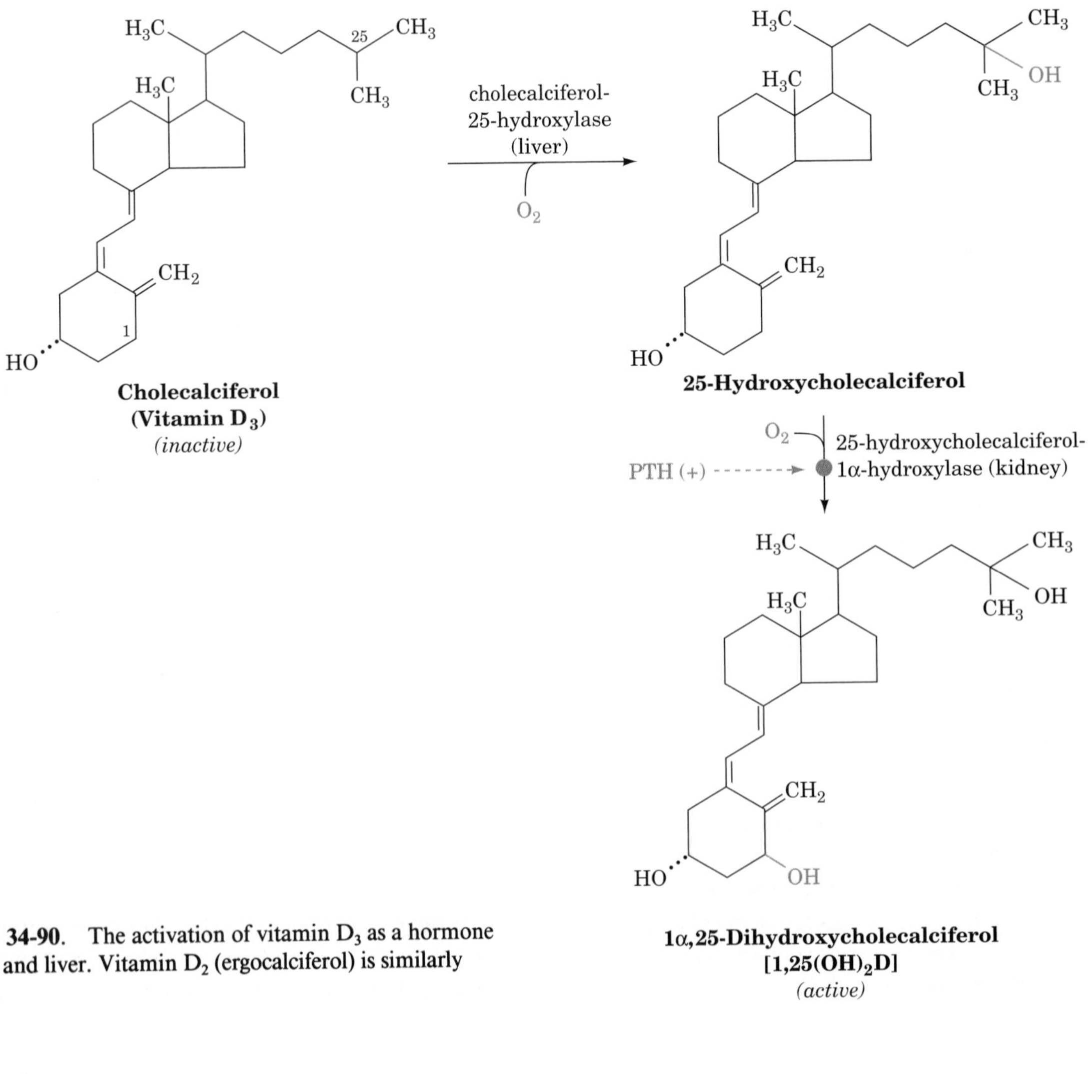

FIGURE 34-90. The activation of vitamin D_3 as a hormone in kidney and liver. Vitamin D_2 (ergocalciferol) is similarly activated.

min D$_3$ (cholecalciferol), is nonenzymatically formed in the skin of animals through the photolytic action of UV light on **7-dehydrocholesterol.**

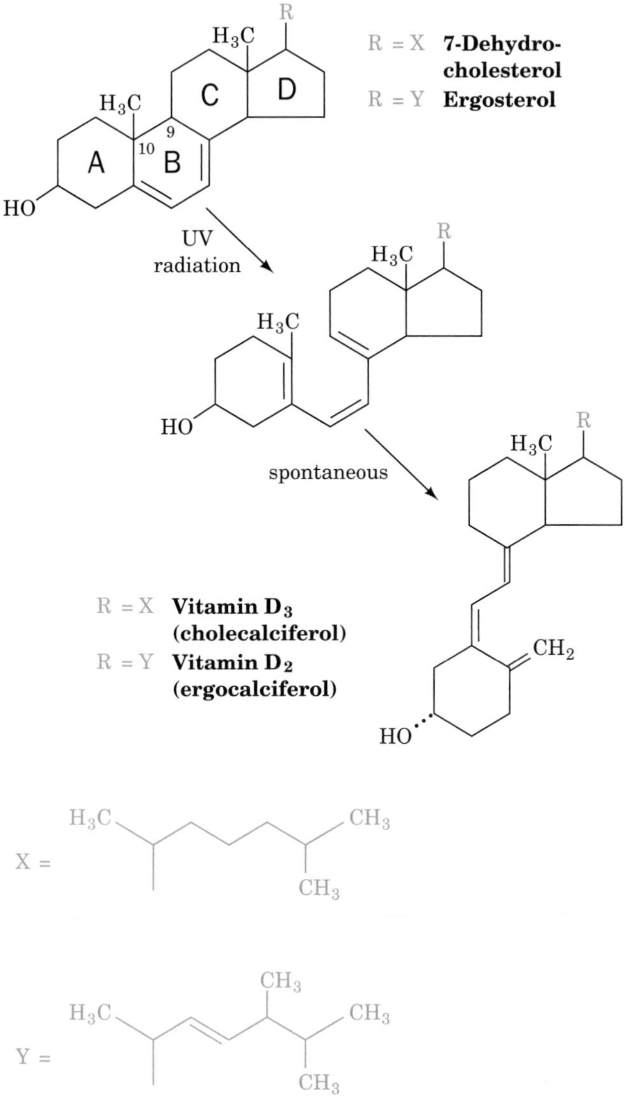

R = X **7-Dehydro-
cholesterol**

R = Y **Ergosterol**

R = X **Vitamin D$_3$
(cholecalciferol)**

R = Y **Vitamin D$_2$
(ergocalciferol)**

X =

Y =

Vitamin D$_2$ (ergocalciferol), which differs from vitamin D$_3$ only by a side chain double bond and methyl group, is formed by the UV irradiation of the plant sterol **ergosterol.** Since vitamins D$_2$ and D$_3$ have essentially identical biological activities, vitamin D$_2$ is commonly used as a vitamin supplement, particularly in milk.

Vitamins D$_2$ and D$_3$ are hormonally inactive as such; they gain biological activity through further metabolic processing, first in the liver and then in the kidney (Fig. 34-90):

1. In the liver, vitamin D$_3$ is hydroxylated to form **25-hydroxycholecalciferol** in an O$_2$-requiring reaction catalyzed by **cholecalciferol-25-hydroxylase.**

2. The 25-hydroxycholecalciferol is transported to the kidney, where it is further hydroxylated by a mitochondrial oxygenase, **25-hydroxycholecalciferol-1α-hydroxylase,**

to yield the active hormone **1α,25-dihydroxycholecalciferol [1,25(OH)$_2$D].** *The activity of 25-hydroxycholecalciferol-1α-hydroxylase is regulated by PTH, so this reaction is an important control point in Ca^{2+} homeostasis.*

1,25(OH)$_2$D acts to increase serum [Ca^{2+}] by promoting the intestinal absorption of dietary Ca^{2+} and by stimulating Ca^{2+} release from bone. Intestinal Ca^{2+} absorption is stimulated through increased synthesis of a **Ca^{2+}-binding protein,** which functions to transport Ca^{2+} across the intestinal mucosa. 1,25(OH)$_2$D binds to cytoplasmic receptors in intestinal epithelial cells that, upon transport to the nucleus, function as transcription factors for the Ca^{2+}-binding protein. The maintenance of electroneutrality requires that Ca^{2+} transport be accompanied by that of counterions, mostly P$_i$, so that 1,25(OH)$_2$D also stimulates the intestinal absorption of P$_i$. The observation that 1,25(OH)$_2$D, like PTH, stimulates the release of Ca^{2+} and P$_i$ from bone seems paradoxical in view of the fact that low levels of 1,25(OH)$_2$D result in subnormal bone mineralization. Presumably the increased serum [Ca^{2+}] resulting from 1,25(OH)$_2$D-stimulated intestinal uptake of Ca^{2+} causes bone to take up more Ca^{2+} than it loses through direct hormonal stimulation.

Vitamin D, unlike the water-soluble vitamins, is retained by the body so that excessive intake of vitamin D over long periods causes **vitamin D intoxication.** The consequent high serum [Ca^{2+}] results in aberrant calcification of a wide variety of soft tissues. The kidneys are particularly prone to calcification, a process that can lead to the formation of kidney stones and ultimately kidney failure. In addition, vitamin D intoxication promotes bone demineralization to the extent that bones are easily fractured. The observation that the level of skin pigmentation in indigenous human populations tends to increase with their proximity to the equator is explained by the hypothesis that skin pigmentation functions to prevent vitamin D intoxication by filtering out excessive solar radiation.

Calcitonin has essentially the opposite effect of PTH; it lowers serum [Ca^{2+}]. It does so primarily by inhibiting osteoclastic resorption of bone. Since PTH and calcitonin both stimulate the synthesis of cAMP in their target cells (see below), it is unclear how these hormones can oppositely affect osteoclasts. Calcitonin also inhibits kidney from resorbing Ca^{2+} but in this case the kidney cells that calcitonin influences differ from those that PTH stimulates to resorb Ca^{2+}.

The Adrenals Secrete Steroids and Catecholamines

The adrenal glands consist of two distinct types of tissue: the **medulla** (core), which is really an extension of the sympathetic nervous system (a part of the autonomic nervous system), and the more typically glandular **cortex** (outer layer). We shall first consider the hormones of the adrenal medulla and then those of the cortex.

The Adrenal Medulla Synthesizes Catecholamines

The adrenal medulla synthesizes two hormonally active **catecholamines** *(amine-containing derivatives of* **catechol,** *1,2-dihydroxybenzene), norepinephrine and its methyl derivative epinephrine.*

HO—, HO— (benzene ring)—CH—CH_2—$\overset{+}{NH_2}$—R
|
OH

R = H **Norepinephrine (noradrenalin)**
R = CH_3 **Epinephrine (adrenalin)**

These hormones are synthesized from tyrosine as is described in Section 24-4B and stored in granules to await their exocytotic release under the control of the sympathetic nervous system.

The biological effects of catecholamines are mediated by two classes of plasma transmembrane receptors, the α- and the **β-adrenoreceptors** (also known as **adrenergic receptors**). These glycoproteins were originally identified on the basis of their varying responses to certain **agonists** (substances that bind to a hormone receptor so as to evoke a hormonal response) and **antagonists** (substances that bind to a hormone receptor but fail to elicit a hormonal response, thereby blocking agonist action). The β- but not the α-adrenoreceptors, for example, are stimulated by **isopro-**

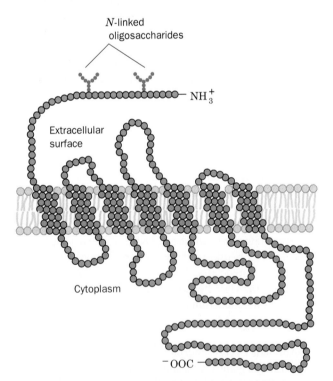

FIGURE 34-91. The amino acid sequence of human β-adrenoreceptor has 7 segments of ~24 hydrophobic residues (*brown circles*), suggesting that this glycoprotein has 7 membrane-spanning helices as diagrammed. [After Dohlman, H.G., Caron, M.G., and Lefkowitz, R.J., *Biochemistry* **26,** 2660 (1987).]

terenol but blocked by **propranolol,** whereas α- but not β-adrenoreceptors are blocked by **phentolamine.**

Isoproterenol

Propranolol

Phentolamine

The α- and β-adrenoreceptors, which occur on separate tissues in mammals, generally respond differently and often oppositely to catecholamines. For instance, β-adrenoreceptors, which activate adenylate cyclase, stimulate glycogenolysis and gluconeogenesis in liver and skeletal muscle (Sections 17-3E and G), lipolysis in adipose tissue, smooth muscle relaxation in the bronchi and the blood vessels supplying the skeletal muscles, and increased heart action. In contrast, α-adrenoreceptors, whose intracellular effects are mediated either by the inhibition of adenylate cyclase (**α₂ receptors;** Section 34-4B) or via the phosphoinositide cascade (**α₁ receptors;** Section 34-4B), stimulate smooth muscle contraction in blood vessels supplying peripheral organs such as skin and kidney, smooth muscle relaxation in the gastrointestinal tract, and blood platelet aggregation. *Most of these diverse effects are directed towards a common end: the mobilization of energy resources and their shunting to where they are most needed to prepare the body for sudden action.*

The varying responses and tissue distributions of the α- and β-adrenoreceptors and their subtypes to different agonists and antagonists have important therapeutic consequences. For example, propranolol is widely used for the treatment of high blood pressure and protects heart attack victims from further heart attacks, whereas epinephrine's bronchodilator effects make it clinically useful in asthma treatment (Section 34-3D).

β-Adrenoreceptors are transmembrane glycoproteins that all contain 7 stretches of 20 to 28 hydrophobic amino acids that are each thought to form a membrane-spanning helix (Fig. 34-91). *This 7-helix bundle is a common receptor motif.* Sequencing studies indicate that it also occurs in α₂-adrenoreceptors, **rhodopsin** (the photoreceptor protein of retinal rod cells), and the muscarinic acetylcholine receptors of nerve synapses (Section 34-4C). We have previously encountered such a 7-helix bundle in our study of bacteriorhodopsin (Section 11-3A; Fig. 11-26).

The Adrenal Cortex Synthesizes a Variety of Steroids

*The adrenal cortex produces at least 50 different **adrenocortical steroids** (whose synthesis is outlined in Section 23-6C).* These have been classified according to the physiological responses they evoke:

1. The **glucocorticoids** affect carbohydrate, protein, and lipid metabolism in a manner nearly opposite to that of insulin and influence a wide variety of other vital functions, including inflammatory reactions and the capacity to cope with stress.
2. The **mineralocorticoids** largely function to regulate the excretion of salt and water by kidney.
3. The **androgens** and **estrogens** affect sexual development and function. They are made in larger quantities by the gonads and are therefore further discussed below.

Glucocorticoids, the most common of which are **cortisol** (also known as **hydrocortisone**) and **corticosterone**, and the mineralocorticoids, the most common of which is **aldosterone**, are all C_{21} compounds.

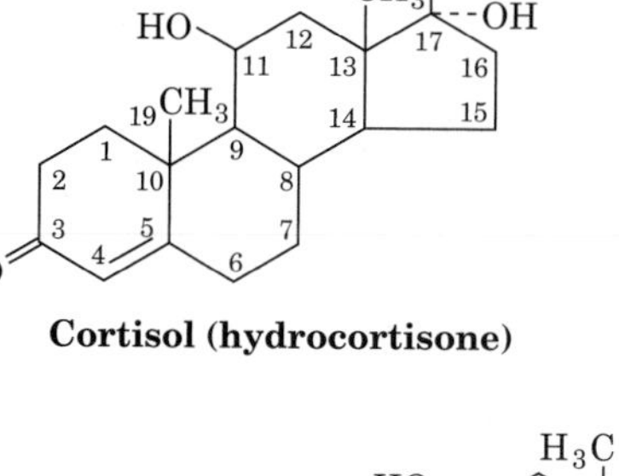

Cortisol (hydrocortisone)

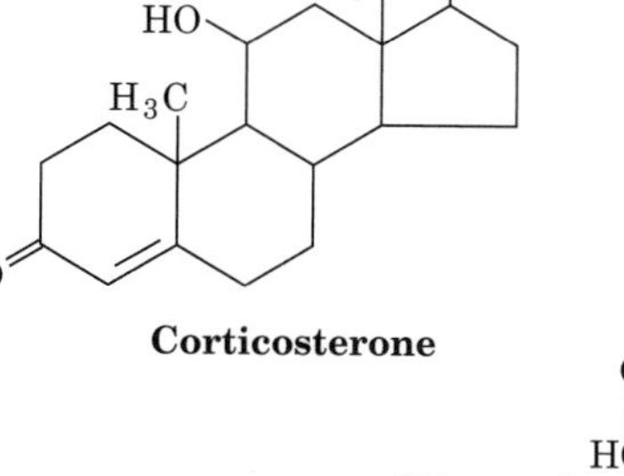

Corticosterone

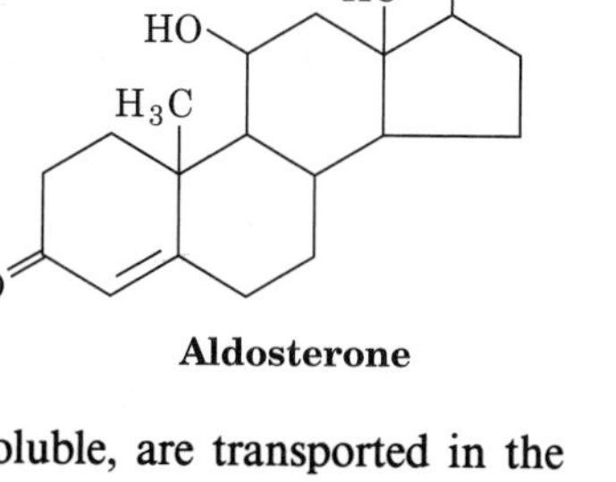

Aldosterone

Steroids, being water insoluble, are transported in the blood in complex with the glycoprotein **transcortin** and, to a lesser extent, by albumin. The steroids enter their target cells, apparently spontaneously, where they bind to their cytoplasmic receptors so as to activate them as transcription factors for specific proteins (Section 33-3B). Indeed, the glucocorticoids and the mineralocorticoids induce the synthesis of numerous metabolic enzymes in their respective target tissues.

Impaired adrenocortical function, either through disease or trauma, results in a condition known as **Addison's disease,** which is characterized by hypoglycemia, muscle weakness, Na^+ loss, K^+ retention, impaired cardiac function, loss of appetite, and a greatly increased susceptibility to stress. The victim, unless treated by the administration of glucocorticoids and mineralocorticoids, slowly languishes and dies without any particular pain or distress. The opposite problem, adrenocortical hyperfunction, which is usually caused by a tumor of the adrenal cortex or the pituitary gland (see below), results in **Cushing's syndrome,** which is characterized by fatigue, hyperglycemia, edema (water retention), and a redistribution of body fat to yield a characteristic "moon face." Long-term treatments of various diseases with synthetic glucocorticoids result in similar symptoms.

Gonadal Steroids Mediate Sexual Development and Function

*The **gonads** (testes in males, ovaries in females), in addition to producing sperm or ova, secrete steroid hormones (androgens and estrogens) that regulate sexual differentiation, the expression of secondary sex characteristics, and sexual behavior patterns.* Although testes and ovaries both synthesize androgens and estrogens, the testes predominantly secrete androgens, which are therefore known as male sex hormones, whereas ovaries produce mostly estrogens, which are consequently termed female sex hormones.

Androgens, of which **testosterone** is prototypic,

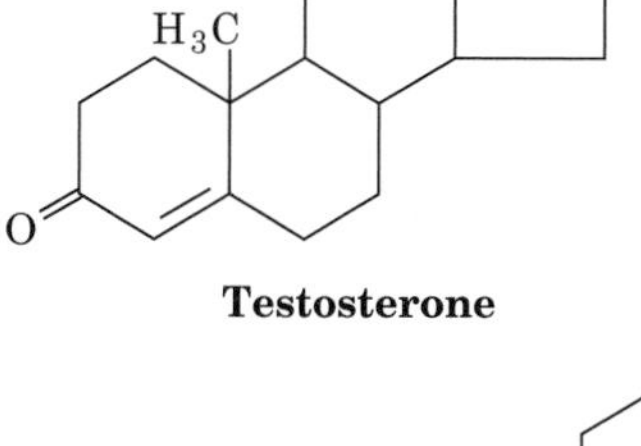

Testosterone

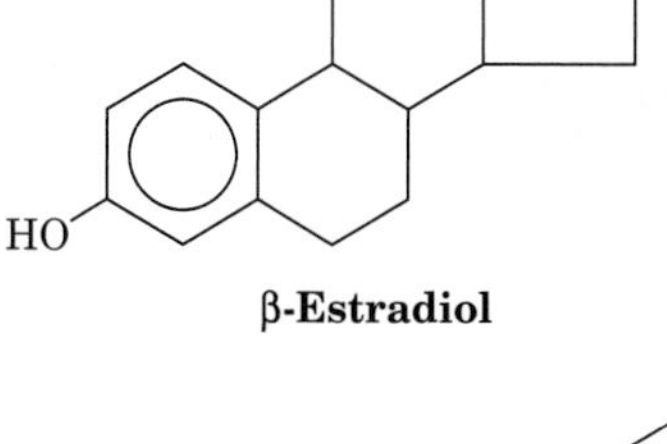

β-Estradiol

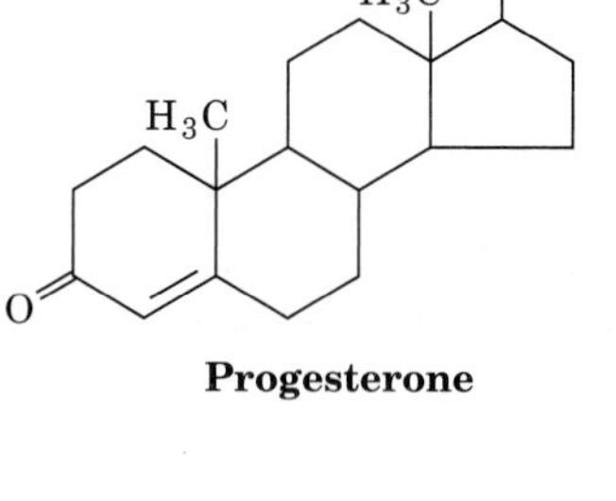

Progesterone

lack the C_2 substituent at C17 occurring in glucocorticoids and are therefore C_{19} compounds. Estrogens, such as **β-estradiol,** resemble androgens but lack a C10 methyl group because they have an aromatic A ring and are therefore C_{18} compounds. Interestingly, testosterone is an intermediate in estrogen biosynthesis (Fig. 23-52). A second class of ovarian steroids, C_{21} compounds called **progestins,** help mediate the menstrual cycle and pregnancy (see below). **Progesterone,** the most abundant progestin, is, in fact, a precursor of glucocorticoids, mineralocorticoids, and testosterone.

What factors control sexual differentiation? If the gonads of an embryonic male animal are surgically removed, that individual will become a phenotypic female. Evidently, *embryonic mammals are programmed to develop as females unless subjected to the influence of testicular hormones.* Indeed, genetic males with absent or nonfunctional cytosolic androgen receptors are phenotypic females, a condition named **testicular feminization.** Curiously, estrogens appear to play no part in embryonic female sexual development, although they are essential for female sexual maturation and function.

Normal individuals have either the XY (male) or the XX (female) genotypes. However, those with the abnormal genotypes XXY **(Klinefelter's syndrome)** and X0 (only one sex chromosome; **Turner's syndrome)** are, respectively, phenotypic males and phenotypic females, although both are sterile. Apparently, *the normal Y chromosome confers the male phenotype, whereas its absence results in the female phenotype.* There are, however, rare (1 in 20,000) XX males and XY females. DNA hybridization studies have revealed that these XX males (who are sterile and have therefore been identified through infertility clinics) have a small segment of a normal Y chromosome translocated onto one of their X chromosomes, whereas XY females are missing this segment.

Early male and female embryos—through the sixth week of development in humans—have identical undifferentiated genitalia. Evidently, the Y chromosome contains a gene, **testes-determining factor (*TDF*),** that induces the differentiation of testes, whose hormonal secretions, in turn, promote male development. The misplaced chromosomal segments in XY females and XX males have a common 140-kb sequence, which must therefore contain *TDF.* This segment contains a gene that appears to encode a zinc finger protein and hence this *ZFY* gene (for *zinc finger Y*) was proposed as a candidate for *TDF.* However, the discovery of three XX males who lack *ZFY* clearly indicates that *ZFY* is not *TDF.* Further winnowing of the *TDF*-containing region narrowed it down to 35 kb, which was found to contain a structural gene dubbed ***SRY*** (for *sex-determining region of Y*) that encodes an 80-residue DNA-binding motif. This motif also occurs in several eukaryotic regulatory proteins, as well as in the high mobility group proteins HMG1 and HMG2 (Section 33-3B), and is therefore named the **HMG domain.** The HMG domain encoded by ***Sry,*** the mouse

analog of *SRY,* binds in the minor groove of duplex DNA containing the sequence 5′-CCATTGTTCT-3′, which it bends by ~85°.

It is now all but certain that *SRY* is, in fact, *TDF.* Several sex-reversed XY women have been shown to have a mutation in the HMG domain-encoding region of their *SRY* gene not present in their father's gene, which eliminates their HMG domain's DNA binding. *Sry* is expressed in embryonic gonadal somatic cells previously shown to be responsible for testis determination. Moreover, of eleven XX mice that were made transgenic for *Sry,* three were males. Thus, *TDF/SRY* is the first clear example of a mammalian gene that controls the development of an entire organ system.

The Hypothalamus Controls Pituitary Secretions That Regulate Other Endocrine Glands

The anterior lobe of the **pituitary gland** (the **adenohypophysis**) and the **hypothalamus,** a nearby portion of the brain, constitute a functional unit that hormonally controls much of the endocrine system. *The neurons of the hypothalamus synthesize a series of polypeptide hormones known as **releasing factors** and **release-inhibiting factors** which, upon delivery to the adenohypophysis via a direct circulatory connection (their half lives are on the order of a few minutes), stimulate or inhibit the release of the corresponding **trophic hormones** into the blood stream. Trophic hormones, by definition, stimulate their target endocrine tissues to secrete the hormones they synthesize.* Since releasing and release-inhibiting factors, trophic hormones, and endocrine hormones are largely secreted in nanogram, microgram, and milligram quantities, respectively, and tend to have progressively longer half lives, these hormonal systems can be said to form amplifying cascades. Four such systems are prominent in humans (Fig. 34-92; *left*):

1. **Corticotropin-releasing factor (CRF; 41 residues)** causes the adenohypophysis to release **adrenocorticotropic hormone (ACTH; 39 residues),** which stimulates the release of adrenocortical steroids.

> Ser - Gln - Glu - Pro - Pro - Ile - Ser - Leu - Asp - Leu -
> Thr - Phe - His - Leu - Leu - Arg - Glu - Val - Leu - Glu -
> Met - Thr - Lys - Ala - Asp - Gln - Leu - Ala - Gln - Gln -
> Ala - His - Ser - Asn - Arg - Lys - Leu - Leu - Asp - Ile -
> Ala - NH_2

Sheep corticotropin-releasing factor (CRF)

> Ser - Tyr - Ser - Met - Glu - His - Phe - Arg - Trp - Gly -
> Lys - Pro - Val - Gly - Lys - Lys - Arg - Arg - Pro - Val -
> Lys - Val - Tyr - Pro - Asn - Gly - Ala - Glu - Asp - Glu -
> Ser - Ala - Glu - Ala - Phe - Pro - Leu - Glu - Phe

Human adrenocorticotropic hormone (ACTH)

The entire system is under feedback control: ACTH inhibits the release of CRF and the adrenocortical steroids

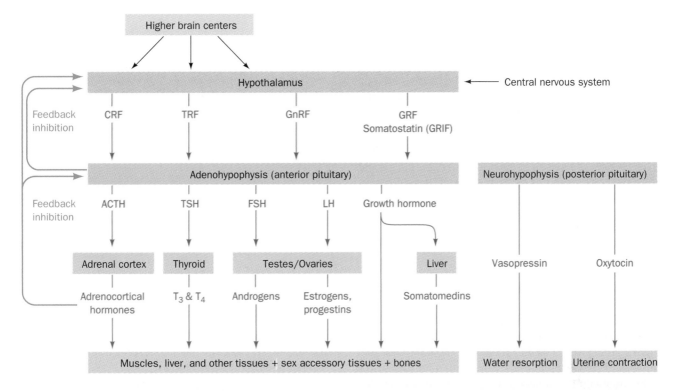

FIGURE 34-92. Hormonal control circuits, indicating the relationships between the hypothalamus, the pituitary, and the target tissues. Releasing factors and release-inhibiting factors secreted by the hypothalamus signal the adenohypophysis to secrete or stop secreting the corresponding trophic hormones, which, for the most part, stimulate the corresponding endocrine gland(s) to secrete their respective endrocrine hormones. The endrocrine hormones, in addition to controlling the growth, differentiation, and metabolism of their corresponding target tissues, influence the secretion of releasing factors and trophic hormones through feedback inhibition. The levels of trophic hormones likewise influence the levels of their corresponding releasing factors.

inhibit the release of both CRF and ACTH. Moreover, the hypothalamus, being part of the brain, is also subject to neuronal control, so the hypothalamus forms the interface between the nervous system and the endocrine system. Recall that ACTH is synthesized together with several other hormones as a single polyprotein, proopiomelanocortin, that is posttranslationally cleaved to yield the individual polypeptides (Section 33-3C).

2. **Thyrotropin-releasing factor (TRF),** a tripeptide with an N-terminal **pyroGlu** residue (a Glu derivative in which the side chain carboxyl group forms an amide bond with its amino group),

$$\text{pyroGlu}\text{—}\overset{3}{\text{His—Pro}}\text{—}NH_2$$

Thyrotropin-releasing factor (TRF)

stimulates the adenohypophysis to release the trophic hormone **thyrotropin (thyroid-stimulating hormone; TSH)** which, in turn, stimulates the thyroid to synthesize and release T_3 and T_4. TRF, as are other releasing factors, is present in the hypothalamus in only vanish-

ingly small quantities. It was independently characterized in 1969 by Roger Guillemin and Andrew Schally using extracts of the hypothalami from over 2 million sheep and 1 million pigs.

3. **Gonadotropin-releasing factor (GnRF; 10 residues)** stimulates the adenohypophysis to release **lutinizing hormone (LH)** and **follicle-stimulating hormone (FSH),** which are collectively known as **gonadotropins.** In males, LH stimulates the testes to secrete androgens while FSH promotes spermatogenesis. In females, FSH stimulates the development of ovarian follicles (which contain the immature ova), whereas LH triggers ovulation.

$$\overset{1}{\text{pyroGlu}}\text{-His-Trp-Ser-Tyr-Gly-Leu-Arg-Pro-}\overset{10}{\text{Gly}}\text{-}NH_2$$
Gonadotropin-releasing factor (GnRF)

4. **Growth hormone-releasing factor (GRF; 44 residues)** and **somatostatin** [14 residues; also known as **growth hormone release-inhibiting factor (GRIF)**], stimulate/ inhibit the release of **growth hormone (GH)** from the adenohypophysis. GH (also called **somatotropin**), in turn, stimulates generalized growth (see Fig. 27-24 for a striking example of its effect). GH directly accelerates

the growth of a variety of tissues (in contrast to TSH, LH, and FSH, which act only indirectly by activating endocrine glands) and induces the liver to synthesize a series of polypeptide growth factors termed **somatomedins** that stimulate cartilage growth and have insulin-like activities.

TSH, LH, and FSH are all $\alpha\beta$-glycoproteins which, in a given species, all have the same α subunit (92 residues) and a homologous β subunit (114, 114, and 118 residues, respectively, in humans). Human GH (**hGH**) consists of a single 191-residue polypeptide chain, which is unrelated to TSH, LH, or FSH.

The Menstrual Cycle and Pregnancy Are Governed by Several Interacting Hormones

The menstrual cycle and pregnancy are particularly illustrative of the interactions among hormonal systems. The ~28 day human menstrual cycle (Fig. 34-93) begins during menstruation with a slight increase in the FSH level that initiates the development of a new ovarian follicle. As the follicle matures, it secretes estrogens that act to sensitize the adenohypophysis to GnRF. This process culminates in a surge of LH and FSH, which triggers ovulation. The ruptured ovarian follicle, the **corpus luteum,** secretes progesterone and estrogens, which inhibit further gonadotropin secretion by the adenohypophysis and stimulate the uterine lining to prepare for the implantation of a fertilized ovum. If fertilization does not occur, the corpus luteum regresses, progesterone and estrogen levels fall, and menstruation (the sloughing off of the uterine lining) ensues. The reduced steroid levels also permit a slight increase in the FSH level, which initiates a new menstrual cycle.

A fertilized ovum that has implanted into the hormonally prepared uterine lining soon commences synthesizing **chorionic gonadotropin (CG).** This $\alpha\beta$-glycoprotein hormone contains a 145-residue β subunit that has a high degree of sequence identity to those of LH (85%), FSH (45%), and TSH (36%) over their first 114 residues and the same α subunit. CG stimulates the corpus luteum to continue secreting progesterone rather than regressing and thus prevents menstruation. Present-day pregnancy tests utilize immunoassays that can detect CG in blood or urine within a few days after embryo implantation. Most female oral contraceptives (birth control pills) contain progesterone derivatives, whose ingestion induces a state of pseudopregnancy in that they inhibit the midcycle surge of FSH and LH so as to prevent ovulation.

The X-ray structure of human CG (hCG), which was independently determined by Wayne Hendrickson and Neil Isaacs, reveals that this strikingly elongated protein's closely associated α and β subunits have similar folds (Fig. 34-94*a*) despite their lack of sequence homology. Much of each subunit consists of three looped-out pairs of β strands that extend from a so called **cysteine knot.** In this motif, a

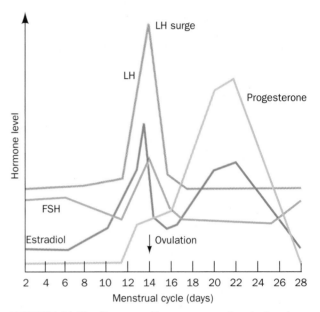

FIGURE 34-93.   Patterns of hormone secretion during the menstrual cycle in the human female.

disulfide bridge connecting two β strands passes through an eight-residue circle formed by a pair of disulfide bonds joining two other β strands, thereby forming two catenated (interlocked) rings. Cysteine knots are also present in several

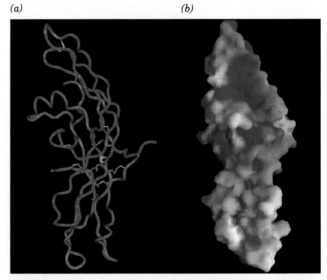

FIGURE 34-94.   The X-ray structure of human chorionic gonadotropin (hCG). (*a*) A C_α diagram showing the α subunit in red and the β subunit in blue with disulfide linkages in yellow. The pseudo-twofold axis relating the two subunits is horizontal. (*b*) The surface electrostatic potential of hCG as viewed towards its receptor-binding surface. The surface is colored according to its electrostatic potential, with the most positive regions dark blue and the most negative regions dark red. The view is related to that in Part *a* by a 60° rotation about the vertical axis. [Courtesy of Hao Wu and Wayne Hendrickson, Columbia University.]

other protein growth factors of known structure, including platelet-derived growth factor (PDGF; Section 33-4C).

The biological effects of hCG are elicited through its binding to the G-protein-coupled **hCG receptor** (which is also the receptor for the closely similar LH, whose binding elicits identical responses to those of hCG; G-proteins are discussed in Sections 33-4C and 34-4B). The hCG residues that have been implicated, through genetic engineering experiments, in binding to the hCG receptor lie in close proximity on one face of hCG. Calculations reveal that this face has a markedly positive electrostatic potential (Fig. 33-94*b*). Thus, the observation that the extracellular portion of hCG receptor (a transmembrane protein) is rich in acidic residues suggests that hCG and its receptor have electrostatically complementary surfaces.

Human Growth Hormone Causes Its Receptor to Dimerize

The binding of growth hormone activates its receptor to stimulate growth and metabolism in muscle, bone, and cartilage cells. This 620-residue receptor is a member of a rapidly growing list of structurally related protein growth factor receptors, which includes those for various interleukins. All of these receptors consist of an N-terminal extracellular ligand-binding domain, a single transmembrane segment that is probably helical, and a C-terminal cytoplasmic domain that is not homologous within the superfamily but in many cases contains a tyrosine kinase function (Section 34-4B).

The X-ray structure of the 191-residue human growth hormone (hGH) in complex with the 238-residue extracellular domain of its binding protein (**hGHbp**), determined by Abraham de Vos and Anthony Kossiakoff, revealed that this complex consists of two molecules of hGHbp bound to a single hGH molecule (Fig. 34-95). hGH consists largely of a four-helix bundle, which closely resembles that in the previously determined X-ray structure of porcine GH, although with significant differences that may be attributable to the binding of hGH to its receptor. A variety of other protein growth factors with known structures, including several interleukins, contain similar four-helix bundles. Each hGHbp molecule consists of two structurally homologous domains, each of which forms a topologically identical sandwich of a three- and a four-stranded antiparallel β sheet that resembles the immunoglobulin fold (Figure 34-24).

The two hGHbp molecules bind to hGH with near twofold symmetry about an axis that is roughly perpendicular to the helical axes of the hGH four-helix bundle and, presumably, to the plane of the cell membrane to which the intact hGH receptor is anchored (Fig. 34-95). The C-terminal domains of the two hGHbp molecules are almost parallel and in contact with one another. Intriguingly, the two hGHbp molecules use essentially the same residues to bind to sites that are on opposite sides of hGH's four-helix bundle and which have no structural similarity. The X-ray

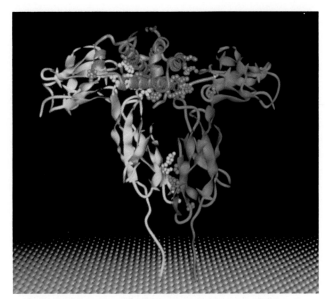

FIGURE 34-95. The X-ray structure of human growth hormone (hGH) in complex with two molecules of its receptor's extracellular domain (hGHbp). The proteins are shown in ribbon form, with the two hGHbp molecules, which together bind one molecule of hGH, green and blue and with the hGH red. The side chains involved in intersubunit interactions are shown in space-filling form. The yellow pebbled surface represents the cell membrane, through which the C-terminal ends of the hGHbp molecules are shown penetrating as they do in the intact hGH receptor. [Courtesy of Abraham de Vos and Anthony Kossiakoff, Genentech Inc., South San Franciso, California.]

structure is largely consistent with the results of mutational studies designed to identify the hGH and hGHbp residues important for receptor binding.

The ligand-induced dimerization of hGHbp has important implications for the mechanism of signal transduction. The dimerization, which does not occur in the absence of hGH, apparently brings together the intact receptors' intracellular domains in a way that activates an effector protein such as a tyrosine kinase (Section 34-4B). Indeed, hGH mutants that cannot induce receptor dimerization are biologically inactive. Recent investigtions have shown that numerous other protein growth factors also induce the dimerization of their receptors.

Overproduction of GH, usually a consequence of a pituitary tumor, results in excessive growth. If this condition commences while the skeleton is still growing, that is, before its growth plates have ossified, then this excessive growth is of normal proportions over the entire body resulting in **gigantism.** Moreover, since excessive GH inhibits the testosterone production necessary for growth plate ossification, such "giants" continue growing throughout their abnormally short lives. If, however, the skeleton has already matured, GH stimulates only the growth of soft tissues, resulting in enlarged hands and feet and thickened facial features, a condition named **acromegaly** (Fig. 34-96). The opposite problem, GH deficiency, which results in insufficient growth (**dwarfism**) can be treated before skeletal maturity by regular injections of hGH (animal GH is ineffective in humans). Since hGH was, at first, available only from the pituitaries of cadavers, it was in very short supply. Now, however, hGH can be synthesized in virtually unlimited amounts by recombinant DNA techniques. Indeed, there is concern that GH will be taken in an uncontrolled manner by individuals wishing to increase their athletic prowess (it would be very difficult to prove that an individual had used exogenously supplied hGH because it is normally present in the human body, even in adults, and is rapidly degraded).

Opioid Peptides Are Naturally Occurring Opiates

Among the most intriguing hormones secreted by the adenohypophysis are polypeptides that have opiate-like effects on the central nervous system. These include the 31-residue **β-endorphin,** its N-terminal pentapeptide, termed **methionine-enkephalin,** and the closely similar **leucine-enkephalin** (although the enkaphalins are independently expressed).

$$\overset{1}{\text{Tyr}} - \text{Gly} - \text{Gly} - \text{Phe} - \overset{5}{\text{Met}} - \text{Thr} - \text{Ser} - \text{Glu} - \text{Lys} - \overset{10}{\text{Ser}}$$
$$\overset{11}{\text{Gln}} - \text{Thr} - \text{Pro} - \text{Leu} - \overset{15}{\text{Val}} - \text{Thr} - \text{Leu} - \text{Phe} - \text{Lys} - \overset{20}{\text{Asn}}$$
$$\overset{21}{\text{Ala}} - \text{Ile} - \text{Val} - \text{Lys} - \overset{25}{\text{Asn}} - \text{Ala} - \text{His} - \text{Lys} - \text{Lys} - \overset{30}{\text{Gly}}$$
$$\overset{31}{\text{Gln}} -$$

β-Endorphin

Tyr-Gly-Gly-Phe-Met

Methionine-enkephalin (Met-enkephalin)

Tyr-Gly-Gly-Phe-Leu

Leucine-enkephalin (Leu-enkephalin)

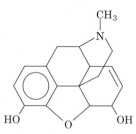

Morphine (an opiate)

These substances bind to **opiate receptors** in the brain and have been shown to be their physiological agonists. The role of these so-called **opioid peptides** has yet to be definitively established but it appears they are important in the control of pain and emotional states. Pain relief through the use of acupuncture and placebos as well as such phenomena as "runner's high" may be mediated by opioid peptides.

The Neurohypophysis Secretes Vasopressin and Oxytocin

The posterior lobe of the pituitary, the **neurohypophysis,** which is anatomically distinct from the adenohypophysis,

FIGURE 34-96. The characteristic enlarged features of Akhenaton, the Pharaoh who ruled Egypt in the years 1379–1362 B.C., strongly suggest that he suffered from acromegaly. [Agytisches Museum. Staadtliche Museen Preussicher Kulturbesitz, Berlin, Germany. Photo by Margarete Busing.]

secretes two homologous nonapeptide hormones (Fig. 34-92, *right*); **vasopressin** [also known as **antidiuretic hormone (ADH)**], which increases blood pressure and stimulates the kidneys to retain water; and **oxytocin,** which causes contraction of uterine smooth muscle and therefore induces labor.

$$\overset{1}{\text{Cys}}-\text{Tyr}-\text{Phe}-\text{Gln}-\text{Asn}-\text{Cys}-\text{Pro}-\text{Arg}-\overset{9}{\text{Gly}}-\text{NH}_2$$
$$\underset{\text{S---S}}{\rule{4cm}{0.4pt}}$$

Human vasopressin

$$\overset{1}{\text{Cys}}-\text{Tyr}-\text{Ile}-\text{Gln}-\text{Asn}-\text{Cys}-\text{Pro}-\text{Leu}-\overset{9}{\text{Gly}}-\text{NH}_2$$
$$\underset{\text{S---S}}{\rule{4cm}{0.4pt}}$$

Human oxytocin

The rate of vasopressin release is largely controlled by osmoreceptors, which monitor the osmotic pressure of the blood.

B. Second Messengers

Since hormones are chemical signals, their actual chemical identities are themselves of little real significance. What is of importance is how the messages they carry are interpreted; that is, how hormone receptors receive hormonal messages and how they transmit these messages to the cellular machinery. Four classes of receptors are known:

1. Receptors epitomized by steroid receptors: cytoplasmic and nuclear proteins that upon binding their corresponding hormones are activated as transcription factors for specific proteins (Section 33-3B).

2. Receptors represented by growth factor receptors such as the EGF receptor (Section 33-4C): membrane-spanning proteins whose cytoplasmic domains are either activated as tyrosine-specific protein kinases or mediate the activation of independent tyrosine-specific protein kinases when hormone binds to their external domains. The tyrosine-specific protein kinases, in turn, modulate the activities of specific cytoplasmic proteins as we discuss below.

3. Receptors exemplified by adrenoreceptors: transmembrane proteins that upon hormone binding stimulate the synthesis and/or release of second messengers (alternatively, **intracellular mediators**) such as cAMP and Ca^{2+} that mediate the hormonal signal within the cell.

4. Receptors typified by the **acetylcholine receptor**: transmembrane ion channels that open or close upon binding ligand, thereby initiating or stopping the flow of specific ions such as Ca^{2+} or K^+ into or out of a cell or organelle. Such ligand-gated ion channels are most conspicuously associated with nerve impulse transmission and are therefore largely discussed in Section 34-4C.

In the following subsections we consider the formation and actions of second messengers in greater detail than we have before. We shall see that, *although cells have numerous seemingly independent signaling pathways, these pathways are, in fact, extensively interconnected on many levels, thereby forming a signaling network of exquisite sensitivity and enormous flexibility.* We begin, however, with a discussion of how receptor–ligand interactions are quantified.

Receptor Binding

Receptors, as do other proteins, bind their corresponding ligands (agonists and antagonists) according to the laws of mass action:

$$R + L \rightleftharpoons R \cdot L$$

Here R and L represent receptor and ligand, and the reaction's dissociation constant is expressed:

$$K_L = \frac{[R][L]}{[R \cdot L]} = \frac{([R]_T - [R \cdot L])[L]}{[R \cdot L]} \quad [34.1]$$

where the total receptor concentration, $[R]_T = [R] + [R \cdot L]$. Equation [34.1] may be rearranged to a form analogous to the Michaelis–Menten equation of enzyme kinetics (Section 13-2A):

$$Y = \frac{[R \cdot L]}{[R]_T} = \frac{[L]}{K_L + [L]} \quad [34.2]$$

where Y is the fractional occupation of the ligand-binding

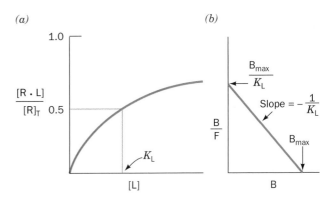

FIGURE 34-97. The binding of ligand to receptor: (*a*) A hyperbolic plot. (*b*) A Scatchard plot. Here, $B \equiv [R \cdot L]$, $F \equiv [L]$, and $B_{max} \equiv [R]_T$.

sites. Equation [34.2] represents a hyperbolic curve (Fig. 34-97*a*) in which K_L may be operationally defined as the ligand concentration at which the receptor is half-maximally occupied by ligand.

Although K_L and $[R]_T$ may, in principle, be determined from an analysis of a hyperbolic plot such as Fig. 34-97*a*, the analysis of a linear form of the equation is a more accurate procedure. Equation [34.1] may be rearranged to:

$$\frac{[R \cdot L]}{[L]} = \frac{([R]_T - [R \cdot L])}{K_L} \quad [34.3]$$

Now, in keeping with customary receptor-binding nomenclature, let us redefine [R · L] as B (for bound ligand), [L] as F (for free ligand), and $[R]_T$ as B_{max}. Then Eq. [34.3] becomes:

$$\frac{B}{F} = \frac{(B_{max} - B)}{K_L} \quad [34.4]$$

A plot of B/F versus B, which is known as a **Scatchard plot** (after George Scatchard, its originator), therefore yields a straight line of slope $-1/K_L$ whose intercept on the B axis is B_{max} (Fig. 34-97*b*). Here, both B and F may be determined by filter-binding assays as follows. Most receptors are insoluble membrane-bound proteins and may therefore be separated from soluble free ligand by filtration (receptors that have been solubilized may be separated from free ligand by filtration, for example, through nitrocellulose; recall that proteins nonspecifically bind to nitrocellulose). Hence, through the use of radioactively labeled ligand, the values of B and F ([R · L] and [L]) may be determined, respectively, from the radioactivity on the filter and that remaining in solution. The rate of R · L dissociation is generally so slow (half-times of minutes to hours) as to cause insignificant errors when the filter is washed to remove residual free ligand.

Competitive-Binding Studies

Once the receptor-binding parameters for one ligand have been determined, the dissociation constant of other ligands for the same ligand-binding site may be determined through competitive-binding studies. The model describing this competitive binding is analogous to the competitive inhibition of a Michaelis–Menten enzyme (Section 13-3A):

$$R + L \; \underset{\longleftarrow}{\overset{\longrightarrow}{\rightleftharpoons}} \; R \cdot L$$
$$+$$
$$I$$
$$\Big\updownarrow$$
$$R \cdot I + L \longrightarrow \text{No reaction}$$

where I is the competing ligand whose dissociation constant with the receptor is expressed:

$$K_I = \frac{[R][I]}{[R \cdot I]} \qquad [34.5]$$

Thus, in direct analogy with the derivation of the equation describing competitive inhibition:

$$[R \cdot L] = \frac{[R]_T \, [L]}{K_L \left(1 + \dfrac{[I]}{K_I}\right) + [L]} \qquad [34.6]$$

The relative affinities of a ligand and an inhibitor may therefore be determined by dividing Eq. [34.6] in the presence of inhibitor with that in the absence of inhibitor:

$$\frac{[R \cdot L]_I}{[R \cdot L]_0} = \frac{K_L + [L]}{K_L \left(1 + \dfrac{[I]}{K_I}\right) + [L]} \qquad [34.7]$$

When this ratio is 0.5 (50% inhibition), the competitor concentration is referred to as $[I_{50}]$. Thus, solving Eq. [34.7] for K_I at 50% inhibition:

$$K_I = \frac{[I_{50}]}{1 + \dfrac{[L]}{K_L}} \qquad [34.8]$$

Through these procedures it has been shown, for example, that epinephrine tightly binds to β-adrenoreceptors ($K_L = 5 \times 10^{-6}M$), the agonist isoproterenol binds more tightly ($K_L = 0.4 \times 10^{-6}M$), whereas the antagonist propranolol binds yet more tightly ($K_L = 0.0034 \times 10^{-6}M$) even though it fails to activate adenylate cyclase (Section 34-4A). Further studies with a variety of epinephrine analogs indicate that epinephrine's amine function is essential for its tight binding to β-adrenoreceptor, whereas its catechol moiety is required for adenylate cyclase activation.

G-Proteins Mediate Adenylate Cyclase Activation and Inhibition

Many adrenoreceptors and polypeptide hormone receptors mediate the formation of cAMP as a second messenger (intracellular mediator). cAMP, as we have seen for the case of glycogen metabolism (Section 17-3C), activates or

inhibits various enzymes or cascades of enzymes by promoting their phosphorylation or dephosphorylation. In Section 33-4C, we saw that when the hormone receptor for such a system binds its cognate hormone, it activates a membrane-bound G-protein·GDP complex by stimulating it to exchange its GDP for GTP. The resulting G-protein·GTP complex, in turn, activates adenylate cyclase, but this activation is short lived because G protein hydrolyzes GTP to GDP + P_i at a rate of 2 to 3 min^{-1}, and, upon doing so, reverts to its inactive state. Nevertheless, *this system amplifies the hormonal signal because each hormone–receptor complex activates many G-proteins before it becomes inactive and, during its lifetime, each G-protein·GTP·adenylate cyclase complex catalyzes the formation of much cAMP.*

G-protein, as shown by Alfred Gilman and Martin Rodbell, is actually more complex than we previously indicated: It consists of three different subunits, α, β, and γ (45, 37, and 9 kD, respectively), of which it is G_α, a member of the GTP-binding protein family (Section 30-3D), that binds GDP or GTP (Fig. 34-98). The binding of $G_\alpha \cdot GDP \cdot G_\beta G_\gamma$ to hormone–receptor complex induces G_α to exchange its bound GDP for GTP and, in so doing, to dissociate from $G_\beta G_\gamma$. GTP binding also decreases G_α's affinity for activated hormone receptor while increasing its affinity for adenylate cyclase. Thus, *it is the binding of $G_\alpha \cdot GTP$ that activates adenylate cyclase.* Upon the eventual G_α-catalyzed hydrolysis of GTP, the resulting $G_\alpha \cdot GDP$ complex dissociates from adenylate cyclase and reassociates with $G_\beta G_\gamma$ to reform inactive G-protein. Both G_α and $G_\beta G_\gamma$ are membrane-anchored proteins: G_α through its myristoylation and, in some cases, also palmitoylation; and $G_\beta G_\gamma$ through the prenylation of G_γ (Section 11-5A).

Several types of hormone receptors may activate the same G protein. This occurs, for example, in liver cells in response to the binding of the corresponding hormone to glucagon receptors and to β-adrenoreceptors. In such cases, the amount of cAMP produced is the sum of that induced by the individual hormones. G-proteins may also act in other ways than by activating adenylate cyclase: They are known, for example, to stimulate the opening of K$^+$ channels in heart cells and to participate in the phosphoinositide signaling system (see below). Nevertheless, all known G-protein-coupled receptors are 7-helix transmembrane proteins such as the β-adrenoreceptor (Fig. 34-91).

Some hormone receptors inhibit rather than stimulate adenylate cyclase. These include the α_2-adrenoreceptor and receptors for somatostatin and opioids. The inhibitory effect is mediated by "inhibitory" G-protein, G_i, which probably has the same β and γ subunits as does "stimulatory" G protein, G_s, but has a different α subunit, $G_{i\alpha}$ (41 kD). G_i acts analogously to G_s in that upon binding to its corresponding hormone–receptor complex, its $G_{i\alpha}$ subunit exchanges bound GDP for GTP and dissociates from $G_\beta G_\gamma$ (Fig. 34-98). $G_{i\alpha}$, however, inhibits rather than activates adenylate kinase, through direct interactions and possibly

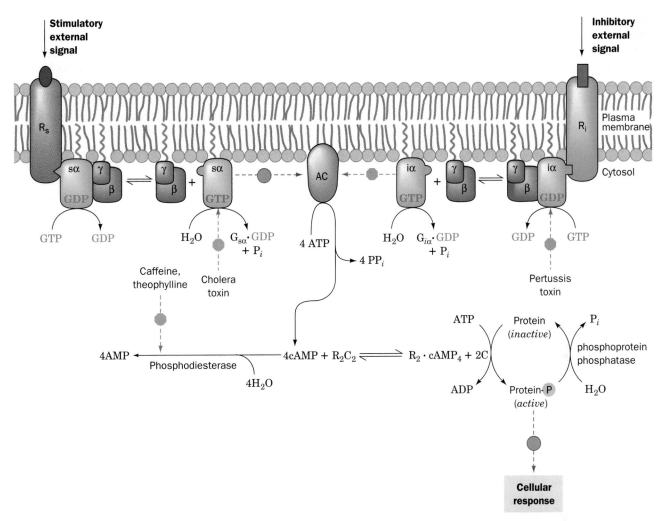

FIGURE 34-98. The mechanism of receptor-mediated activation/inhibition of adenylate cyclase (AC). The binding of hormone to a stimulatory receptor, R_s (*left*), induces it to bind G_s protein which, in turn, stimulates the $G_{s\alpha}$ subunit of this $G_{s\alpha}G_{\beta}G_{\gamma}$ heterotrimer to exchange its bound GDP for GTP. The $G_{s\alpha} \cdot$ GTP complex then dissociates from $G_{\beta}G_{\gamma}$ and, until it catalyzes the hydrolysis of its bound GTP to GDP, stimulates adenylate cyclase to convert ATP to cAMP. The binding of

hormone to the inhibitory receptor, R_i (*right*), triggers an almost identical chain of events except that the presence of $G_{i\alpha} \cdot$ GTP complex inhibits AC from synthesizing cAMP. R_2C_2 represents cAMP-dependent protein kinase whose catalytic subunit, C, when activated by the dissociation of the regulatory dimer as $R_2 \cdot$ cAMP$_4$, activates various cellular proteins by catalyzing their phosphorylation. The sites of action of cholera and pertussis toxins are indicated.

because the liberated $G_{\beta}G_{\gamma}$ binds to and sequesters $G_{s\alpha}$. The latter mechanism is supported by the observation that liver cell membranes contain far more G_i than G_s. The activation of G_i in such cells would therefore release enough $G_{\beta}G_{\gamma}$ to bind the available $G_{s\alpha}$. Moreover, in some systems, $G_{\beta}G_{\gamma}$ itself acts as an intracellular mediator.

G_s and G_i are members of a large family of related signal-transducing G-proteins. This family also includes:

1. **G_q**, which forms a link in the phosphoinositide signaling system (see below).

2. **Transducin (G_t),** which transduces visual stimuli by coupling the light-induced conformational change of the visual pigment rhodopsin to the activation of a specific phosphodiesterase that hydrolyzes **3′,5′-cyclic GMP**

(cGMP) to GMP. This heterotetrameric ($\alpha\beta\gamma_2$) phosphodiesterase is activated by the displacement of its inhibitory γ subunits by $G_{t\alpha} \cdot$ GTP.

3. **G_{olf},** which is expressed only in olfactory sensory neurons and is involved in odorant signal transduction.

This heterogeneity in G-proteins occurs in the β and γ subunits as well as in the α subunits. Indeed, at least 15 different α subunits, 5 different β subunits, and 5 different γ subunits have been identified. Thus, a cell may contain several closely related G-proteins of a given type that interact with varying specificities with receptors and effectors. This complex signaling system presumably permits cells to respond in a graded manner to a variety of stimuli.

The X-Ray Structures of G_α Proteins Rationalize Their Functions

The X-ray structures of the C-terminal 325 residues of the 350-residue bovine transducin-α ($G_{t\alpha}$) in its complexes with GDP (Figs. 34-99a and b) and with the nonhydrolyzable GTP analog **GTPγS**

GTPγS

(Figs. 34-99c and d) were determined by Paul Sigler. $G_{t\alpha}$ consists of two clearly delineated domains connected by two polypeptide linkers: (1) a highly conserved "GTPase" domain that is structurally similar to the proto-oncogene product **Ras** (p21^{c-ras}; Section 33-4C) and to the N-terminal domains of the translational elongation factors EF-Tu and EF-G (Section 30-3D), even though these GTPases have little (~17%) sequence identity; and (2) a helical domain that is unique to G-proteins. Guanine nucleotides bind to $G_{t\alpha}$ in a deep cleft that is flanked by these domains. The structure of the $G_{t\alpha}\cdot$GTPγS complex closely resembles that of a $G_{i\alpha}\cdot$GTPγS complex determined by Gilson and Stephan Sprang.

Comparison of the structures of the $G_{t\alpha}\cdot$GDP and $G_{t\alpha}\cdot$GTPγS complexes reveals that GTP's γ-phosphate group promotes significant conformational shifts in only three so-called switch regions, all of which are located on the facing side of $G_{t\alpha}$ in Fig. 34-99. The γ-phosphate hydrogen bonds to side chains on switches I and II, thereby pulling these polypeptide segments in towards it and causing switch II to contact switch III in a way that pulls it to the right (Fig. 34-99). These concerted conformational shifts cause an extensive cavity over the GDP-binding site to largely fill in in the GTPγS complex.

Switches I and II have counterparts in Ras and EF-Tu. Portions of these polypeptide segments have been implicated in the interactions of $G_{t\alpha}$ with the phosphodiesterase it functions to activate and in the interactions between the closely related $G_{s\alpha}$ with its target adenylate cyclase. Note, however, that EF-Tu undergoes a quite different conformational change on replacing its bound GDP with GTP (Fig. 30-45). A question that remains unanswered is how a liganded receptor induces its target G_α subunit to exchange its bound GDP for GTP.

Receptors Are Subject to Desensitization

One of the hallmarks of biological signaling systems is that they adapt to long-term stimuli by reducing their response to them, a process named **desensitization.** *These signaling systems therefore respond to changes in stimulation levels rather than to their absolute values.* What is the mechanism of desensitization? In the case of β-adrenore-

ceptors, continuous exposure to epinephrine leads to the phosphorylation of one or more of the receptor's Ser residues. This phosphorylation, which is catalyzed by a specific kinase that acts on the hormone–receptor complex but not on the receptor alone, decreases the influence of hormone on G-protein, at least in part, by reducing the receptor's epinephrine-binding affinity. The phosphorylated receptors, moreover, are endocytotically sequestered in specialized vesicles that are devoid of both G-protein and adenylate cyclase, thereby further attenuating the cell's response to epinephrine. If the epinephrine level is reduced, the receptor is slowly dephosphorylated by a phosphorylase and returned to the cell surface, thereby restoring the cell's epinephrine sensitivity.

Cholera Toxin Stimulates Adenylate Cyclase by Permanently Activating G_s Protein

The major symptom of **cholera,** an intestinal disorder caused by the bacterium *Vibrio cholerae,* is massive diarrhea that, if untreated, often results in death from dehydration. This dreaded disease is not an infection in the usual sense since the vibrio neither invades nor damages tissues but merely colonizes the intestine, as does *E. coli.* Rather, the catastrophic fluid loss that cholera induces (over a liter per hour) occurs in response to a bacterial toxin. Indeed, merely replacing cholera victims' lost water and salts enables them to survive the few days necessary to immunologically eliminate the bacterial infestation.

Cholera toxin (CT) is an 87-kD protein of subunit composition AB$_5$ in which the B subunits (103 residues each) form a pentagonal ring surrounding the A subunit (240 residues; see below). Upon binding its cell surface receptor, ganglioside G_{M1} (Sections 11-1D and 23-8C), CT is taken into the cell, possibly via receptor-mediated endocytosis. This process is accompanied by cholera toxin activation through the proteolytic cleavage and disulfide bond reduction of the A subunit to two fragments, A1 (~195 residues) and A2 (~45 residues), whereupon A1 is released into the cytosol (the B subunits do not enter the cell). Another possibility, which may accompany endocytosis, is that A1 is injected into the cytoplasm by direct insertion through the cell membrane.

Once inside the cell, A1 catalyzes the transfer of the ADP–ribose unit from NAD$^+$ to an Arg side chain of $G_{s\alpha}$ (Fig. 34-100; recall that diphtheria toxin similarly ADP-ribosylates eukaryotic elongation factor eEF-2, which is also a GTP-binding protein; Section 30-3G). *ADP-ribosylated $G_{s\alpha}\cdot$GTP can activate adenylate cyclase but is incapable of hydrolyzing its bound GTP.* As a consequence, the adenylate cyclase remains "locked" in its active state. The epithelial cells of the small intestine normally secrete digestive fluid (an HCO$_3^-$-rich salt solution) in response to small [cAMP] increases. The ~100-fold rise in intracellular [cAMP] induced by cholera toxin causes the symptoms of cholera by inducing these epithelial cells to pour out enormous quantities of digestive fluid. Cholera toxin also affects

(a)

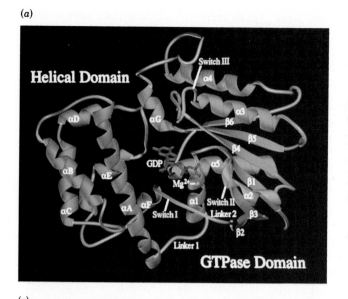

(b)

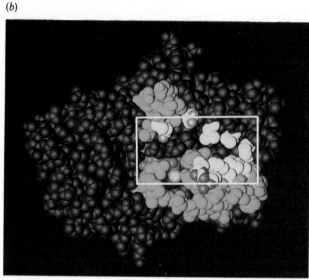

(c)

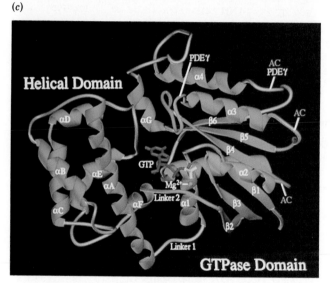

(d)

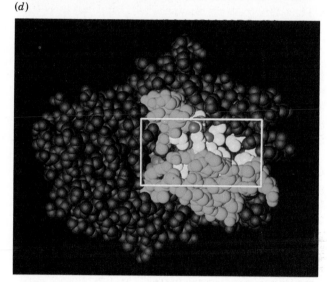

FIGURE 34-99. Structural differences between the inactive and active forms of $G_{t\alpha}$ (transducin) as indicated by the comparison of the X-ray structure of $G_{t\alpha} \cdot GDP$ in its (a) ribbon and (b) spacing-filling forms with that of $G_{t\alpha} \cdot GTP\gamma S$ in its (c) ribbon and (d) spacing-filling forms, all viewed from the same direction. In the ribbon drawings, helices and sheets are green; the segments linking them are gold; the guanine nucleotides are purple, except for the γ-phosphate of $GTP\gamma S$ which is yellow; and the bound Mg^{2+} ion is represented by a blue ball. The protein's three switch regions (I, II, and III) are highlighted in blue. In Part c, the the two loop regions of the protein that are implicated in its interaction with the phosphodiesterase it activates ($PDE\gamma$) are pointed out with yellow labels, whereas the three loop regions that are implicated in the interaction of the homologous $G_{s\alpha}$ with adenylate cyclase (AC) are indicated with purple labels. The space-filling models are colored similarly to the ribbon diagrams except for the gold residues, which here represent those that appear to propagate or stabilize the structural transitions induced by the binding of the γ-phosphate group. The box in the space-filling models outlines the cavity in $G_{t\alpha} \cdot GDP$ that closes when the GDP is replaced by $GTP\gamma S$ and that has been implicated in modulating the affinity of $G_{t\alpha}$ for $G_{t\beta\gamma}$ and for the receptor. [Courtesy of Paul Sigler, Yale University.]

other tissues *in vitro* but does not do so *in vivo* because cholera toxin is not absorbed from the gut into the bloodstream.

Certain strains of *E. coli* cause a diarrheal disease similar to, although less serious than cholera through their production of **heat-labile enterotoxin (LT)**, a protein that is closely similar to CT (their A and B subunits are >80% identical) and has the same mechanism of action. LT's remarkable X-ray structure, both alone and in complex with lactose (which has a galactose residue at its nonreducing end, as does its target ligand, ganglioside G_{M1}), was determined by Wim Hol. The C-terminus of the A2 (C-terminal) fragment

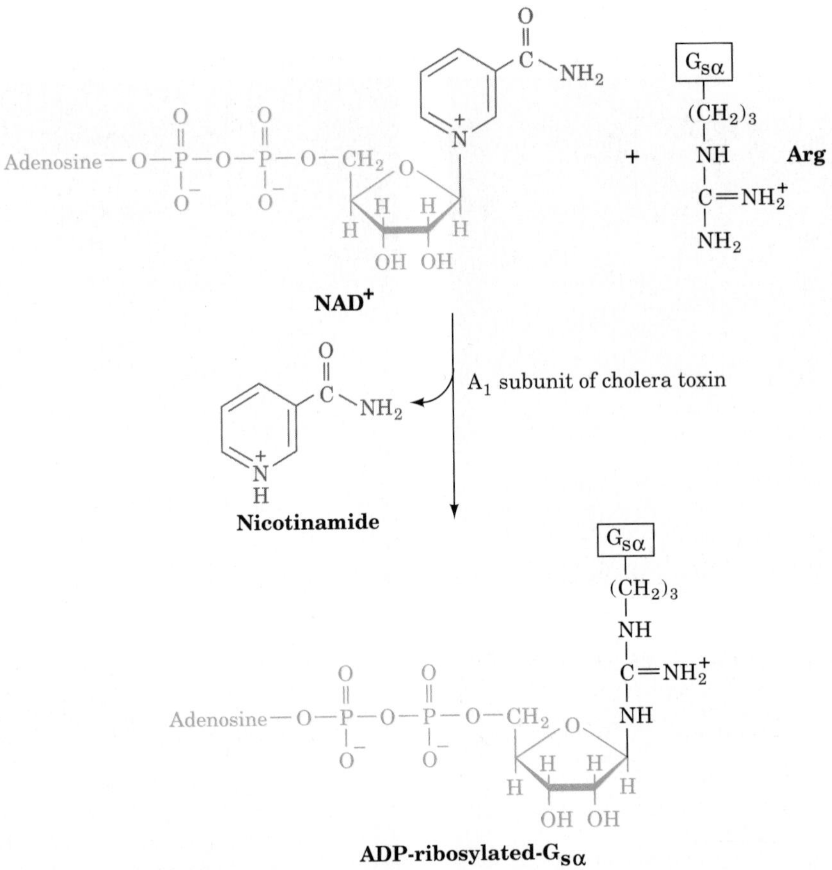

FIGURE 34-100.   The cholera toxin's A_1 fragment catalyzes the ADP-ribosylation of a specific Arg residue on G_s protein's α subunit by NAD^+, thereby rendering this subunit incapable of hydrolyzing GTP.

forms an unusually extended segment that inserts into the B_5 pentamer's solvent-filled central pore (Fig. 34-101a), where it is noncovalently anchored. The N terminal segment of A2 forms an α helix that extends beyond the B_5 pentamer so as to tether the wedge-shaped A1 fragment to B_5, much like a balloon on a string (Fig. 34-101b). In the lactose complex, which is conformationally similar to the uncomplexed LT, a galactose residue binds through an extensively hydrogen bonded network to each B subunit on the face opposite the B_5 pentamer's A-binding face. These binding sites are at least 8 Å away from the sides of the B_5 pentamer and 25 Å away from its relatively flat A-binding face, which is thought to contact the cell surface. There is no evidence that B_5 undergoes a large conformational change on binding G_{M1}, even though the oligosaccharide portion of ganglioside G_{M1} can extend no more than 25 Å from the surface of the cell membrane. Hol has therefore proposed that the LT protein associates with the cell membrane in a way that inserts the hydrophilic B_5 pentamer into the hydrophobic part of the membrane by a distance of at least 8 Å, thereby forcing the A subunit into the membrane.

A1 contains an elongated crevice in the vicinity of a catalytically implicated residue, Glu 112. Indeed, this putative active site region structurally resembles that of diphtheria toxin, which catalyzes a similar reaction, although these two proteins are otherwise dissimilar (recall that diphtheria toxin is a monomer that is proteolytically activated to yield

an A fragment, which also catalyzes the ADP-ribosylation of eEF-2, and a B fragment, which binds to cell-surface receptors and facilitates the transfer of A to the cytoplasm; Section 30-3G).

Pertussis Toxin ADP-Ribosylates $G_{i\alpha}$

Bordetella pertussis, the bacterium that causes **pertussis** (whooping cough; a disease that is still responsible for ~400,000 infant deaths per year worldwide), produces a 76-kD AB_5 protein, **pertussis toxin (PT),** that ADP-ribosylates $G_{i\alpha}$. In doing so, it prevents $G_{i\alpha}$ from exchanging its bound GDP for GTP and therefore from inhibiting adenylate cyclase. PT's X-ray structure, determined by Randy Read, reveals that its A and B subunits are structurally homologous to those of LT, although the A subunit of PT projects from the opposite face of its B_5 pentamer relative to that in LT. Several other bacterial toxins are also known to be AB_5 proteins.

Receptor Tyrosine Kinases Mediate Growth and Differentiation

Many protein growth factors stimulate the proliferation and differentiation of their target cells by binding to their cognate **receptor tyrosine kinases (RTKs).** The RTKs form a diverse family of transmembrane proteins (Fig. 34-102) that each have a C-terminal cytoplasmic tyrosine kinase

(b)

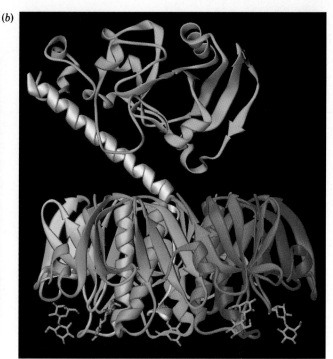

(a)

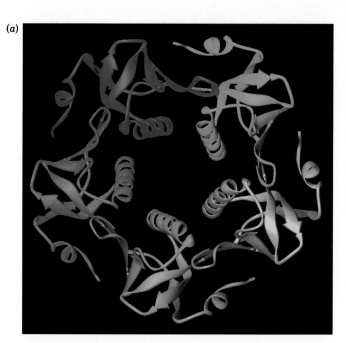

FIGURE 34-101. The X-ray structure of *E. coli* heat-labile enterotoxin (LT) in complex with lactose. (*a*) A ribbon diagram of only the B$_5$ pentamer as viewed along its fivefold axis of symmetry from the side containing the lactose binding sites. Each subunit has a different color. Note the pentamer's large central pore. (*b*) A ribbon diagram of the entire complex as viewed perpendicular to Part *a*. The A1 segment is white, the A2 segment is light blue, and the B subunits are colored as in Part *a*. The lactose molecules, whose galactose residues bind

to the B subunit, are shown in stick form with C green and O red. Although the A1 and A2 segments in this structure form a continuous polypeptide chain, residues 189–195, which encompass the site that is cleaved on toxin activation, are disordered and hence not visible here (*upper left*). The C-terminal end of the A2 segment binds in the pentamer's central pore with the remaining space occupied by at least 66 water molecules. [Based on an X-ray structure by Wim Hol, University of Groningen, The Netherlands.]

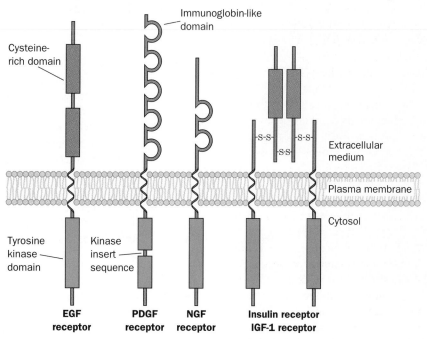

FIGURE 34-102. A selection of receptor tyrosine kinase (RTK) subfamilies as represented by one or two members of each subfamily. The significance of the extracellular cysteine-rich and immunoglobulin-like domains is unknown (compare with Fig. 34-36). Note that the tyrosine kinase domain of the

PDGF receptor is interrupted by a kinase insert sequence and that the **insulin receptor** and the **insulin-like growth factor-1 (IGF-1) receptor** are $\alpha_2\beta_2$ heterotetramers. NGF stands for **nerve growth factor.**

domain and a single-pass transmembrane segment that is presumably an α helix.

Certain protein growth factors, as we saw for human growth hormone (hGH; Fig. 34-95), bind to their cognate receptors in a way that induces their dimerization (although keep in mind that the hGH receptor is not an RTK). In the RTKs, this brings the cytoplasmically located tyrosine kinase activities into proximity, thereby inducing them to cross-phosphorylate each other on several Tyr residues each (Fig. 34-103). This **autophosphorylation** activates the tyrosine kinase to bind to and/or phosphorylate specific Tyr residues on other cytoplasmic signaling proteins, many of which are protein kinases themselves. The latter, as we discuss below, phosphorylate the proteins that participate in executing the instructions implied by the extracellular presence of the protein growth factor and/or transmit this signal to other parts of the cell.

SH2 and SH3 Domains Mediate Signal Transduction

How do autophosphorylated RTKs relay the signal that they have bound their cognate ligand; that is, how do they recognize and activate their target proteins? Many of the diverse cytoplasmic proteins that bind to autophosphorylated receptors, for example the proto-oncogene product **Src** (pp60^{c-src}; Section 33-4C), **GTPase activating protein (GAP)**, and **phospholipase C**γ (see below), contain one or two conserved ~100-residue modules known as **Src homology 2 (SH2) domains** (so named because they were first noticed in tyrosine kinases related to Src). *SH2 domains specifically bind phospho-Tyr residues with high affinity.* The sequences of a variety of SH2-containing proteins reveal no apparent preference for the location of this domain in a protein.

The X-ray and NMR structures of SH2 domains from several proteins, both alone and in complex with phospho-

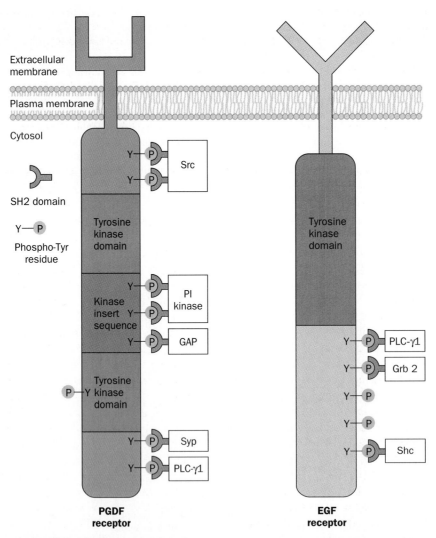

FIGURE 34-103. Schematic diagrams of the PGDF and EGF RTKs, indicating their autophosphorylation sites and the proteins that are activated by binding to these sites via their SH2 domains (most of which are discussed in the text). Note that almost all of the autophosphoryated Tyr residues lie outside the tyrosine kinase domains. [After Pawson, T. and Schlessinger, J., *Curr. Biol.* **3,** 435 (1993).]

(a)

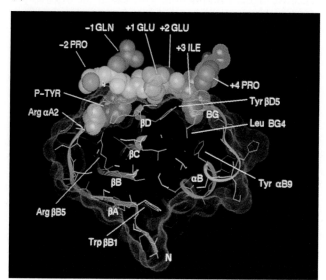

(b)

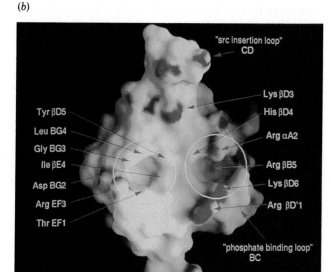

FIGURE 34-104. X-Ray structure of the 104-residue Src SH2 domain in complex with an 11-residue polypeptide (EPQpYEEIPIYL) containing the protein's pYEEI target tetrapeptide. (a) A cutaway view of the complex, in which its solvent-accessible surface is represented by red dots, the protein (*purple*) is shown in ribbon form with its side chains in stick form, and the N terminal 8-residue segment of the bound polypeptide is shown in space-filling form with its backbone yellow, its side chains green, and its phosphate group white [the N-terminal Pro side chain (*left*) is largely obscured in this view and the C-terminal three residues are disordered]. (b) The molecular surface of the protein only, as viewed towards the peptide binding site and colored according to its local electrostatic potential with the most positive regions deep blue and the most negative regions deep red. The binding pockets for the phospho-Tyr (*right*) and Ile (*left*) side chains are circled in yellow and important residues are identified by red arrows. [Courtesy of John Kuriyan, The Rockefeller University.]

Tyr–containing polypeptides, have been determined. SH2 is a hemispherically shaped protein, which contains a central 5-stranded antiparallel β sheet that is sandwiched between two nearly parallel α helices. The N- and C-terminal residues of SH2 are in close proximity on the surface opposite the peptide binding site, which suggests that this domain can be inserted between any two surface residues on a protein without disturbing its fold or function.

John Kuriyan determined the X-ray structure of the SH2 domain of Src in complex with an 11-residue polypeptide containing the sequence pYEEI (phospho-Tyr-Glu-Glu-Ile), a tetrapeptide segment that binds to this SH2 domain with high affinity. The 11-residue peptide binds to the SH2 domain in an extended conformation with contact being made primarily by the pYEEI tetrapeptide (Fig. 34-104a). The phospho-Tyr side chain inserts into a small cleft formed, in part, by three highly conserved positively charged residues with which it makes specific contacts. The Ile side chain is similarly inserted into a nearby hydrophobic pocket and the entire tetrapeptide segment interacts very tightly with SH2, although the side chains of the peptide's two central Glu residues do not project towards SH2. Thus, the peptide resembles a two-pronged plug that is inserted into a two-holed socket on SH2 (Fig. 34-104b). Comparison of this structure with that of uncomplexed Src SH2 indicates that, upon binding peptide, SH2 undergoes only small conformational changes that are localized at its pep-tide-binding site. These structures provide a simple explanation for why SH2 does not bind the far more abundant phospho-Ser– and phospho-Thr–containing peptides: The side chains of the these residues are too short to interact with the Arg side chain at the bottom of the phospho-Tyr binding pocket.

Many of the RTKs that contain SH2 domains also have one or more 50- to 75-residue **SH3 domains.** Moreover, SH3 is contained in several membrane-associated proteins that lack an SH2. The SH3 domain, which is unrelated to SH2, binds Pro-rich sequence motifs of 9 or 10 residues. The physiological function of SH3 is less apparent than that of SH2 because SH3 occurs in a greater variety of proteins, including receptor and nonreceptor tyrosine kinases, adaptor proteins such as **sem-5** (see below), and structural proteins such as spectrin and myosin. However, the observation that the deletion of the SH3 domain–encoding segments from the proto-oncogenes *Src* and *Abl* (which both encode tyrosine kinases) converts them to oncogenes suggests that SH3, much like SH2, functions to mediate the interactions between kinases and regulatory proteins. Thus, both SH2 and SH3 have been called "molecular velcro."

The X-ray and NMR structures of SH3 domains from several proteins indicate that the SH3 core consists of two three-stranded, antiparallel β sheets that pack against each other with their strands nearly perpendicular. As with SH2, the close proximity of SH3's N- and C-termini suggests that this domain could be modularly inserted between two residues on the surface of another protein without greatly perturbing either structure. The X-ray structures of the SH3 domains from the tyrosine kinases Abl and **Fyn** in complex with different 10-residue Pro-rich polypeptides to which they tightly bind was determined by Andrea Musacchio and Matti Sareste. Both decapeptides assume nearly identical conformations with their C-terminal seven residues in the polyproline II helix conformation (Section 7-1B). The peptides bind to SH3 over their entire length in three geometrically complementary cavities (Fig. 34-105), which are mostly occupied by Pro side chains.

Ras Is Activated by Receptor Tyrosine Kinases Via a Grb2–Sos Adaptor Complex

Ras protein (p21^{c-ras}), *a proto-oncogene product (Section 33-4C), is a membrane anchored (by prenylation) GTP-binding protein (Section 30-3D) that lies at the center of an intracellular signaling system that regulates such essential cellular functions as growth and differentiation through the phosphorylation and hence activation of a variety of proteins.* In the pathway described here (Fig. 34-106), the binding of ligand to RTKs activates a **guanine-nucleotide-releasing factor (GRF)** to exchange a Ras-bound GDP for GTP. Only Ras·GTP is capable of further relaying the signal. However, as do G-proteins, Ras hydrolyzes its bound GTP to GDP, thereby halting further signal transduction and limiting the magnitude of the signal generated by the binding of ligand to the receptor. Ras by itself hydrolyzes bound GTP with a half-life of 1 to 5 hours, too slowly for

effective signal transduction. This led to the discovery of **GTPase activating protein (GAP),** a 120-kD protein that, on binding Ras·GTP, accelerates the rate of GTP hydrolysis by up to five orders of magnitude. GAP's physiological importance as a negative regulator of Ras-mediated signal transduction is demonstrated by the observation that the relative biological activity of Ras mutants is better correlated with their resistance to regulation by GAP than by their intrinsic GTPase activity. [Recall that members of GTP-binding family of proteins often have associated GAPs and guanine nucleotide releasing factors (GRFs); Section 30-3E.]

Molecular genetic analyses of signaling in a variety of distantly related organisms (notably humans, mice, *Xenopus, Drosophila,* and the round worm *Caenorhabditis elegans*) had revealed a remarkably conserved pathway in which *RTKs funnel the signal that they have bound ligand to Ras, which, in turn, relays the signal, via a "kinase cascade", to the transcriptional apparatus in the nucleus.* Nevertheless, the way in which messages are passed between the RTKs and Ras had remained enigmatic until recently, when investigations in numerous laboratories revealed their major details. In particular, these studies have demonstrated that *two previously characterized proteins,* **Grb2** *and* **Sos,** *form a complex that bridges activated RTKs and Ras in a way that induces Ras to exchange its bound GDP for GTP, thereby activating it (i.e., they act as a GRF).*

The mammalian protein Grb2, a 217-residue homolog of **drk** in *Drosophila* and **Sem-5** in *Caenorhabditis,* consists almost entirely of an SH2 domain flanked by two SH3 domains. Sos protein (the 1596-residue product of the *Son of Sevenless* gene, so named because Sos interacts with the *Sevenless* gene product, an RTK that regulates the development of the R7 photoreceptor cell in the *Drosophila* compound eye), which is required for Ras-mediated signaling, contains a central domain homologous to known Ras-GRFs and a Pro-rich sequence in its C-terminal segment similar to known SH3-binding motifs. Moreover, mammalian homologs of Sos (**mSos**) have been shown to specifically stimulate guanine nucleotide exchange by mammalian Ras proteins. Western blotting techniques (Section 5-4B) using anti-Grb2 and anti-mSos antibodies indicate that Grb2 binds to the C-terminal segment of mSos but not when one of the Pro residues in Sos' SH3-binding motif has been replaced by Leu or in the presence of synthetic polypeptides that have these Pro-rich sequences. Similar studies indicate that in the presence of epidermal growth factor (EGF), the EGF receptor (an RTK; Section 33-4C) specifically binds the Grb2–mSos complex, an interaction that is blocked by the presence of a phosphopeptide with the sequence of the peptide segment containing one of the activated EGF receptor's phospho-Tyr residues. Evidently, Grb2's SH2 domain binds a phospho-Tyr–containing peptide segment in an activated RTK while its two SH3 domains bind the Pro-rich sequences of Sos. The GRF function of Sos is thereby stimulated to activate Ras.

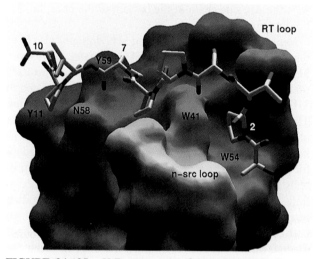

FIGURE 34-105. X-Ray structure of the SH3 domain from Abl protein in complex with its 10-residue target Pro-rich polypeptide (APTMPPPLPP). The protein is represented by its surface diagram and the peptide is drawn in stick form with C white, N blue, O red, and S green. Residues contacting the polypeptide are identified by their one letter codes. [Courtesy of Andrea Musacchio, European Molecular Biology Laboratory, Germany.]

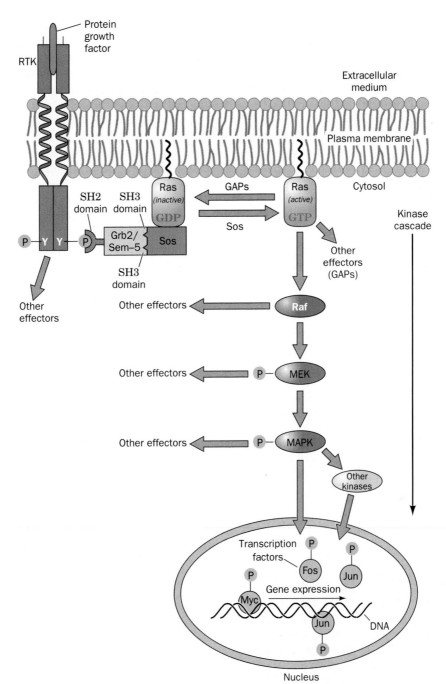

FIGURE 34-106. The Ras signaling cascade begins when an RTK binds its cognate growth factor, thereby inducing the autophosphorylation of this RTK's cytosolic domain. Grb2/Sem-5 binds to the resulting phospho-Tyr–containing peptide segment via its SH2 domain and simultaneously binds to Pro-rich segments on Sos via its two SH3 domains. This activates Sos as a guanine nucleotide releasing factor (GRF) to exchange Ras' bound GDP for GTP, which activates Ras to bind to Raf. Then, in a so-called kinase cascade, Raf, a Ser/Thr

kinase, phosphorylates MEK, which in turn phosphorylates MAPK, which then migrates to the nucleus, where it phosphorylates transcription factors such as Fos, Jun, and Myc, thereby modulating gene expression. The kinase cascade eventually returns to its resting state through the actions of phosphoprotein phosphatases after a GTPase activating protein (GAP) deactivates Ras by inducing it to hydrolyze its bound GTP to GDP. [After Egan, S.E. and Weinberg, R.A., *Nature* **365,** 782 (1993).]

Ras·GTP Activates a Cascade of Ser/Thr Kinases

The signaling pathway downstream of Ras consists of a linear cascade of Ser/Thr kinases (Fig. 34-106), many of which are the products of proto-oncogenes:

1. Raf, a Ser/Thr protein kinase, is activated by direct interaction with Ras·GTP. Evidence for this direct interaction includes the following: Ras that has been permanently activated by mutation (Gly 12 → Val) or that is in complex with the nonhydrolyzable GTP analog GDPNP (Section 30-3D) binds to **Raf** protein with high affinity, whereas mutant forms of Ras that lack transforming activity (e.g., Ile 36 → Ala or Asp 38 → Glu) fail to do so. (Other signaling pathways may instead activate Raf by phosphorylating it at multiple Ser and Thr residues, as we discuss below).

2. Activated Raf phosphorylates a protein alternatively known as **MEK** or **MAP kinase kinase (MAPKK)** at specific Ser and Thr residues, thereby activating it as a Ser/Thr kinase. [Raf is therefore a **MAP kinase kinase kinase (MAPKKK)**].

3. Activated MEK phosphorylates a family of proteins variously named **mitogen-activated protein (MAP) kinases** or **extracellular-signal-regulated kinases (ERKs)**. The more than marginal activation of a MAP kinase requires that it be phosphorylated at both its Thr and Tyr residues in the sequence TEY (Thr-Glu-Tyr). MEK (which stands for *M*AP kinase/*E*RK-activating *k*inase) catalyzes both phosphorylations and thus has dual specificity for Ser/Thr and Tyr. The X-ray structure of the unphosphorylated MAP kinase ERK2, determined by Elizabeth Goldsmith, reveals that this protein structurally resembles the cAMP-dependent protein kinase C subunit (Section 17-3C) and that its Tyr residue which becomes phosphorylated blocks the peptide-binding site in its unphosphorylated form.

4. The MAP kinases migrate from the cytosol to the nucleus, where they phosphorylate a variety of transcription factors such as **Jun/AP-1**, **Fos**, and **Myc** (Section 33-4C), thereby activating them in various ways to produce the effects commissioned by the extracellular presence of the protein growth factor that initiated the signaling cascade.

This kinase cascade can be activated in other ways besides by liganded RTKs. For example, Raf may also be activated via its Ser/Thr phosphorylation by **protein kinase C,** which is activated via the phosphoinositide signaling system described below. Alternatively, Ras may be activated by subunits of certain heterotrimeric G-proteins. Thus, the kinase cascade serves to integrate a variety of extracellular signals.

Tyrosine Kinase – Associated Receptors

A variety of cell surface receptors are not members of the receptor families that we have discussed and do not respond to agonist binding by autophosphorylation. These include the receptors for growth hormone (Fig. 34-95), the **cytokines** (local mediator proteins that regulate the differentiation, proliferation, and activities of the various types of blood cells; interleukins are cytokines), the interferons (Section 30-4B), and *T* cell receptors (Section 34-2D). Ligand binding induces these **tyrosine kinase – associated receptors** to dimerize (and, in some cases, to trimerize), often with different types of subunits, in a way that activates an associated **nonreceptor tyrosine kinase.**

Many of the nonreceptor tyrosine kinases that are activated by tyrosine kinase – associated receptors are members of the **Src family.** These ~500-residue membrane-anchored (by myristoylation) proteins, which include Src, Abl, and Fyn, contain SH2 and SH3 domains as well a kinase domain. Hence, a Src-related kinase may also be activated by asssociation with an autophosphorylated RTK. Moreover, Src-related kinases may respectively be activated or deactivated by the phosphorylation of Tyr residues in its kinase domain or in its C-terminal tail. Finally, although Src-related kinases are each associated with different receptors, they phosphorylate overlapping sets of target proteins. This complex web of interactions explains why different effectors often activate some of the same signaling pathways.

The activated receptors for growth hormone and a variety of cytokines form complexes with members of the recently discovered **Janus kinase (JAK)** family of nonreceptor tyrosine kinases. These ~130-kD proteins, which include **JAK1, JAK2,** and **Tyk2,** have both an active tyrosine kinase domain and a tyrosine kinase – like domain, but lack SH2 and SH3 domains. Liganded and thus dimerized **growth hormone receptor (GHR)** stimulates JAK2 to phosphorylate tyrosine residues on both GHR and itself, thereby mimicking the autophosphorylation of dimerized RTKs. The succeeding steps of this signaling pathway are as yet unknown, although there are indications that it activates MAP kinases.

Protein Tyrosine Phosphatases Also Mediate Signal Transduction

In order to be an effective second messenger, a substance must be rapidly eliminated once it has delivered its message. Otherwise the signaling sustem would become stuck in the "on" position. cAMP, for example, is hydrolyzed to AMP by a specific phosphodiesterase (Section 17-3C). Indeed, the methylated purine derivatives **caffeine,** an ingredient of coffee, and **theophylline,** which occurs in tea,

Caffeine
(1,3,7-trimethylxanthine)

Theophylline
(1,3-dimethylxanthine)

are stimulants, in part, because they inhibit this phosphodiesterase.

The enzymes that dephosphorylate Tyr residues, the **protein tyrosine phosphatases (PTPs),** are not just simple housekeeping enzymes but are signal transducers in their own right. The PTPs form a large family of diverse proteins that are present in all eukaryotes. Each PTP contains at least one conserved ~250-residue phosphatase domain that

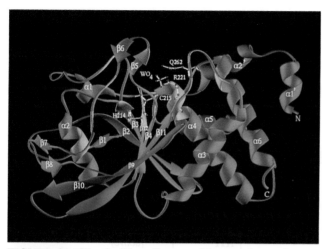

FIGURE 34-107. A ribbon diagram of the protein tyrosine phosphatase PTP1B from human placenta. This 321-residue monomeric protein is shown in ribbon form with its catalytic site and invariant residues in yellow and its four active site residues as well as its active site–bound tungstate ion (WO_4^-; a phosphate analog that inhibits the enzyme) in ball-and-stick form with C green, N blue, O red, S yellow, W purple, and all bonds white. [Courtesy of David Barford, Cold Spring Harbor Laboratory, New York.]

incorporates an 11-residue sequence motif, which contains its catalytically implicated Arg and Cys residues. PTPs have no sequence similarities with other types of phosphatases such as Ser/Thr phosphatases.

The diversity among PTPs largely arises from sequences that flank their phosphatase domains. Thus, some PTPs have receptor-like transmembrane segments and two cytoplasmic phosphatase domains, which suggests that they dephosphorylate phosphotyrosine proteins in a ligand-modulated fashion. In cytosolic PTPs, the catalytic domains are flanked by regulatory sequences including SH2 domains. Indeed, the X-ray structure of an SH2 domain from the cytoplasmic PTP **Syp,** determined by Kuriyan, is remarkably similar to that of Src (Fig. 34-104), despite their low (~30%) sequence identity.

David Barford and Nicholas Tonks determined the X-ray structure of the 321-residue human placental enzyme **PTP1B** (the first PTP to be isolated and sequenced, also by Tonks). PTP1B consists of a single domain in which the catalytic site is located at the base of a shallow cleft (Fig. 34-107). The phosphate recognition site is formed by a loop that contains the 11-residue sequence motif characteristic of PTPs. The position of the essential Cys residue is consistent with its observed role as a nucleophile in the catalytic reaction.

The Phosphoinositide Cascade: Ca²⁺, Inositol Triphosphate, and Diacylglycerol Are Intracellular Second Messengers

Extracellular signals often cause a transient rise in the cytosolic [Ca²⁺], which, in turn, activates a great variety of enzymes through the intermediacy of calmodulin and its

homologs. We have seen, for example, that an increase in cytosolic [Ca²⁺] triggers such diverse cellular processes as glycogenolysis and muscle contraction. What is the source of this Ca²⁺ and how does it enter the cytosol? In certain types of cells, neurons, for example (Section 34-4C), the Ca²⁺ originates in the extracellular fluid. However, the observation that the absence of extracellular Ca²⁺ does not inhibit certain Ca²⁺-mediated processes led to the discovery that, in these cases, cytosolic Ca²⁺ is obtained from intracellular reservoirs, mostly the endoplasmic reticulum (the sarcoplasmic reticulum in muscle). Extracellular stimuli leading to Ca²⁺ release must therefore be mediated by an intracellular signal.

The first clue as to the nature of this signal came from observations that the intracellular mobilization of Ca²⁺ and the turnover of **phosphatidylinositol-4,5-bisphosphate [PIP₂ or PtdIns(4,5)P₂],** a minor component of the plasma membrane's inner leaflet, are strongly correlated. This information led Robert Michell to propose, in 1975, that PIP₂ hydrolysis is somehow associated with Ca²⁺ release.

Investigations, notably by Mabel and Lowell Hokin, Michael Berridge, and Michell, have revealed that *PIP₂ is part of an important second messenger system, the phosphoinositide cascade, that mediates the transmission of numerous hormonal signals* including those of vasopressin, CRF, TRF, acetylcholine (a neurotransmitter; Section 34-4C), epinephrine (with α₁-adrenoreceptors), EGF, and PDGF. Remarkably, this system yields up to three separate types of second messengers through the following sequence of events (Fig. 34-108):

1.–3. The agonist–receptor interactions described below activate **phospholipase C (PLC)** to hydrolyze PIP₂ to **inositol-1,4,5-trisphosphate (IP₃ or InsP₃)** and *sn*-**1,2-diacylglycerol (DG)** as indicated in the top line of Fig. 34-109.

4. The water-soluble IP₃, acting as a second messenger, diffuses through the cytoplasm to the ER from which it stimulates the release of Ca²⁺ into the cytoplasm by binding to and thereby opening an ER-bound transmembrane Ca²⁺ transport channel known as the **IP₃ receptor.**

5. The Ca²⁺, in turn, stimulates a variety of cellular processes through the intermediacy of calmodulin and its homologs.

6. The nonpolar DG is constrained to remain in the plane of the plasma membrane where it nevertheless also acts as a second messenger by activating **protein kinase C (PKC)** in the presence of phosphatidylserine (PS) and Ca²⁺. This membrane-bound enzyme (actually a large family of enzymes), in turn, phosphorylates and thereby modulates the activities of several different proteins including glycogen synthase (Section 17-3D) and smooth muscle myosin light chains (which, it will be recalled, are also phosphorylated by myosin light chain kinase; Section 34-

$G_\beta G_\gamma$ functions to activate certain PLC isoforms, notably **PLC-β2.**

In contrast to G-protein-linked receptors, certain ligand-activated and thus autophosphorylated RTKs, including those for PDGF and EGF (Fig. 34-103), directly activate **PLC-γ1** by binding to this isoform via its two SH2 domains (it also has an SH3 domain) and phosphorylating at least three of its Tyr residues. PLC-γ1's association with the membrane-bound RTK, besides facilitating its phosphorylation, brings this otherwise cytosolic enzyme into contact with its substrate, PIP_3, in the inner leaflet of the plasma membrane. In *T* cells, PLC-γ1 is likewise phosphorylated through the action of activated *T* cell receptors, but here these tyrosine kinase–associated receptors recruit members of the Src family such as Fyn and **Lck** to carry out the phosphorylation.

IP_3 and DG are rapidly recycled to form PIP_2 through the bicyclic metabolic pathway diagrammed in Fig. 34-109. Some of these inositol phosphates, as well as many not appearing in Fig. 34-109, also act as signal molecules in certain cells, thereby increasing an organism's ability to respond to complex stimuli. Intriguingly, the enzyme that catalyzes the hydrolysis of **inositol-1-phosphate (IP_1), IP_1 phosphatase,** is inhibited by Li^+. The therapeutic efficacy of Li^+ in controlling the incapacitating mood swings of manic-depressive individuals therefore suggests that this mental illness is caused by an aberration in a phosphoinositide signaling system in the brain, possibly causing abnormal activation of Ca^{2+}-mobilizing receptors.

The activating effects of IP_3 and DG explain many cellular phenomena. In skeletal muscle, for example, IP_3 mobilizes Ca^{2+} from the sarcoplasmic reticulum. Since Ca^{2+} triggers muscle contraction (Section 34-3C), this observation suggests that nerve impulses mobilize Ca^{2+} by releasing neurotransmitters in the myofibril's T tubules (Fig. 34-67), which then bind to PLC-activating receptors. In a second example, several polypeptide growth factors, including PDGF, act to mobilize IP_3 and DG, which in turn stimulate cell proliferation. The v-*sis* oncogene, which it will be recalled specifies an analog of PDGF (Section 33-4C), may therefore act to permanently switch on PIP_2 degradation, thus forcing the cell into a state of continuous proliferation. Several other oncogene products, including v-Src, are thought to aberrantly activate the synthesis of PIP_2 from its precursors (Fig. 34-109). Similarly, **phorbol esters** such as **12-*O*-tetra-decanoylphorbol-13-acetate,**

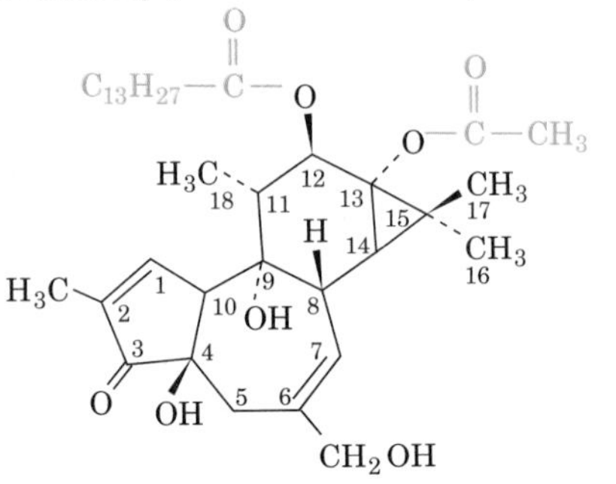

12-*O*-Tetradecanoylphorbol-13-acetate

which are potent activators of protein kinase C (they structurally resemble DG), are the most effective known **tumor promotors** (substances that are not in themselves carcinogenic but increase the potency of known carcinogens; phorbol esters induce the synthesis of the transcription factor AP-1, the product of the *c-jun* proto-oncogene; Section 33-4C). Clearly, the phosphoinositide cascade plays a central role in the control of cellular metabolism.

NO and Possibly CO Are Biological Messengers

Nitric oxide (NO) is a reactive and toxic free radical gas. Thus, it came as a great surprise that *this molecule functions as an intercellular signal in regulating blood vessel dilation and serves as a neurotransmitter. It also functions in the immune response.* The role of NO in vasodilation was discovered through the observation that substances such as acetylcholine (Section 34-4C) and bradykinin (Section 6-4B), which act through the phosphoinositide signaling system to increase the flow through blood vessels by eliciting smooth muscle relaxation, require an intact endothelium overlying the smooth muscle. Evidently, endothelial cells respond to the presence of these vasodilation agents by releasing a diffusible and highly labile substance (half-life ~5 s) that induces the relaxation of smooth muscle cells. This substance was identified as NO, in part, through parallel studies identifying NO as the active metabolite that mediates the well-known vasodilating effects of antianginal organic nitrates such as **nitroglycerin**

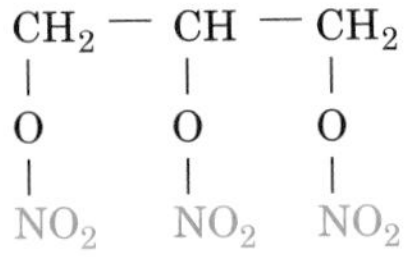

$$\begin{array}{ccccc}
CH_2 & - & CH & - & CH_2 \\
| & & | & & | \\
O & & O & & O \\
| & & | & & | \\
NO_2 & & NO_2 & & NO_2
\end{array}$$

Nitroglycerin

(**angina pectoris** is a disease caused by insufficient blood flow to the heart muscle, leading to severe chest pain).

NO is synthesized by **NO synthase (NOS),** which catalyzes the NADPH-dependent reaction of L-arginine with O_2 to yield NO and the amino acid citrulline (Fig. 34-110). NOS is a homodimeric protein of 125- to 160-kD subunits that are homologous to **cytochrome P_{450} reductase,** an enzyme involved in detoxification processes. Each NOS subunit contains one FMN, one FAD, one tetrahydobiopterin (Fig. 24-21), and one Fe(III)-heme, cofactors that presumably facilitate the 5-electron oxidation of L-arginine to yield NO. The enzyme is activated by Ca^{2+} through its interaction with Ca^{2+}–calmodulin. Hence, the stimulatory action of vasodilatants on the phosphoinositide signaling system in endothelial cells to produce an influx of Ca^{2+} results in the synthesis of NO.

NO rapidly diffuses across cell membranes, although its high reactivity prevents it from getting >1 mm from its site of synthesis (in particular, it efficiently reacts with both oxyhemoglobin and deoxyhemoglobin: $NO + HbO_2 \rightarrow NO_3^- + Hb$; and $NO + Hb \rightarrow HbNO$). *The physiological target of NO in smooth muscle cells is **guanylate cyclase,***

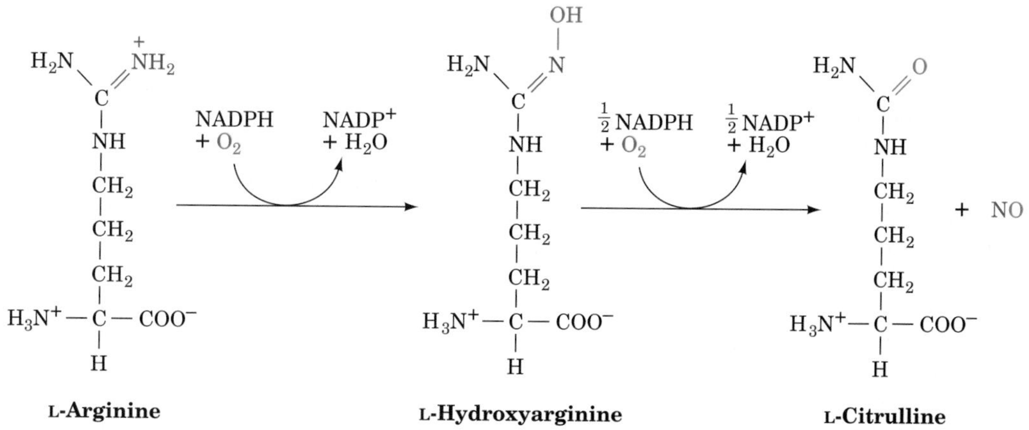

FIGURE 34-110. The NO synthase (NOS) reaction. The L-hydroxyarginine intermediate is tightly bound to the enzyme.

which catalyzes the reaction of GTP to yield **3′,5′-cyclic GMP (cGMP),** an intracellular second messenger that resembles cAMP. cGMP causes smooth muscle relaxation through its stimulation of protein phosphorylation by **cGMP-dependent protein kinase.** NO reacts with guanylate cyclase's heme prosthetic group to yield **nitrosoheme,** whose presence increases the enzyme's activity 50-fold, presumably via a conformation change resembling that in hemoglobin upon binding O_2 (Section 9-2B). Thus, *NO functions to transduce hormonally induced increases in intracellular [Ca^{2+}] in endothelial cells to increased rates of production of cGMP in neighboring smooth muscle cells.*

NO also mediates vasodilation through endothelium-independent neural stimulation of smooth muscle. In this signal transduction pathway, which is responsible for the dilation of cerebral and other arteries as well as penile erection, nerve impulses cause an increased [Ca^{2+}] in nerve terminals, thereby stimulating neuronal NOS (which is ~55% homologous to endothelial NOS). The resultant NO diffuses to nearby smooth muscle cells, where it binds to guanylate cyclase and activates it to synthesize cGMP as described above.

A third type of NOS, which is calmodulin- independent, is transcriptionally induced in macrophages and **neutrophils** (white blood cells that function to ingest and kill bacteria), as well as in endothial and smooth muscle cells. Several hours after exposure to cytokines and/or endotoxins (bacterial cell wall lipopolysaccharides that elicit inflammatory responses; Section 34-2F), these cells begin to produce large quantities of NO and continue doing so for many hours. Activated macrophages and neutrophils also produce superoxide radical ($O_2^- \cdot$), which chemically combines with NO to form the even more toxic **peroxynitrite** (OONO⁻, which rapidly decomposes to the highly reactive **hydroxide radical, OH·,** and NO_2) that they use to kill ingested bacteria. Indeed, NOS inhibitors block the cytotoxic actions of macrophages. In endothelial and smooth muscle cells, cytokines and endotoxins induce a long-lasting and profound vasodilation and a poor response to vaso-

constrictors such as epinephrine. The sustained release of NO has been implicated in **endotoxic shock** (an often fatal immune system overreaction to bacterial infection), inflammation-related tissue damage, and in the damage to neurons in the vicinity of but not directly killed by a stroke (which often does greater harm than the stroke itself).

The physical resemblance of CO and NO (both are highly toxic diatomic gases) and the ability of CO to activate guanylate cyclase (presumably by binding to its heme group as does NO) suggests that CO may also act as an intercellular messenger. CO is physiologically generated through the oxidative breakdown of heme to form biliverdin, Fe^{3+}, and CO in a reaction catalyzed by **heme oxygenase (HO;** Section 24-4A). Two forms of HO are known: **HO-1,** which is abundant in spleen and other tissues involved in red cell breakdown; and **HO-2,** which is prevalent in brain tissue. Intriguingly, cytochrome P_{450} reductase, the only mammalian protein known to be homologous to NOS, is the electron donor in the HO reaction. *In situ* hybridization studies indicate that HO-2 mRNA is present in discrete neuronal locations throughout the brain that closely overlap those for guanylate cyclase mRNA but differ somewhat from those of NOS mRNA. The presence of **zinc protoporphyrin IX** (the Zn^{2+} analog of heme), a potent selective inhibitor of HO, results in a large decrease in the concentration of endogenous cGMP. All of these observations constitute strong circumstantial evidence that CO is, in fact, a neurotransmitter.

C. Neurotransmission

In higher animals, the most rapid and complex intercellular communications are mediated by nerve impulses. Neurons (nerve cells; e.g., Fig. 1-10d) electrically transmit these signals along their highly extended lengths (commonly over 1 m in larger animals) as traveling waves of ionic currents. Signal transmission between neurons as well as between neurons and muscles or glands, is usually chemically mediated by neurotransmitters. In the remainder of this section, we discuss both the electrical and chemical aspects of nerve impulse transmission.

Nerve Impulses Are Propagated by Action Potentials

Neurons, like other cells, generate ionic gradients across their plasma membranes through the actions of the corresponding ion-specific pumps. In particular, a $(Na^+–K^+)$-ATPase (Section 18-3A) pumps K^+ into and Na^+ out of the neuron to yield intracellular and extracellular concentrations of these ions similar to those listed in Table 34-6. The consequent membrane potential, $\Delta\Psi$, across a cell membrane is described by the **Goldman equation,** an extension of Eq. [18.3] that explicitly takes into account the various ions' different membrane permeabilities:

$$\Delta\Psi = \frac{RT}{\mathscr{F}} \ln \frac{\sum P_c[C(out)] + \sum P_a[A(in)]}{\sum P_c[C(in)] + \sum P_a[A(out)]} \quad [34.9]$$

Here, C and A represent cations and anions, respectively, and, for the sake of simplicity, we have made the physiologically reasonable assumption that only monovalent ions have significant concentrations. The quantities P_c and P_a, the respective **permeability coefficients** for the various cations and anions, are indicative of how readily the corresponding ions traverse the membrane (each is equal to the corresponding ion's diffusion coefficient through the membrane divided by the membrane's thickness; Section 18-2A). Note that Eq. [34.9] reduces to Eq. [18.3] if the permeability coefficients of all mobile ions are assumed to be equal.

Applying Eq. [34.9] to the data in Table 34-6 and assuming a temperature of 25°C yields $\Delta\Psi = -83$ mV (negative inside), which is in good agreement with experimentally measured membrane potentials for mammalian cells. This value is somewhat greater than the K^+ equilibrium potential, the value of $\Delta\Psi = -91$ mV obtained assuming the membrane is permeable to only K^+ ions ($P_{Na^+} = P_{Cl^-} = 0$). The membrane potential is generated by a surprisingly small imbalance in the ionic distribution across the membrane; only ~1 ion pair/per million is separated by the membrane with the anion going to the cytoplasmic side and

TABLE 34-6. IONIC CONCENTRATIONS AND MEMBRANE PERMEABILITY COEFFICIENTS IN MAMMALS

Ion	Cell (mM)	Blood (mM)	Permeability Coefficient (cm·s^{-1})
K^+	139	4	5×10^{-7}
Na^+	12	145	5×10^{-9}
Cl^-	4	116	1×10^{-8}
X^{-a}	138	9	0

[a] X^- represents macromolecules that are negatively charged under physiological conditions.

Source: Darnell, J., Lodish, H., and Baltimore, D., *Molecular Cell Biology,* pp. 618 and 725, Scientific American Books (1986).

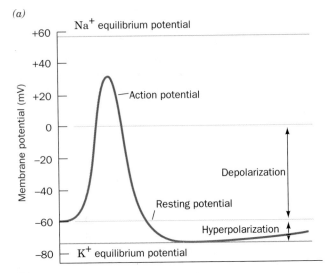

(a)

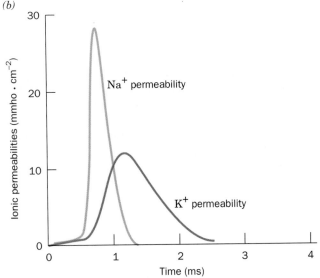

(b)

FIGURE 34-111. The time course of an action potential. (*a*) The axon membrane undergoes rapid depolarization, followed by a nearly as rapid hyperpolarization and then a slow recovery to its resting potential. (*b*) The depolarization is caused by a transient increase in Na^+ permeability (conductance), whereas the hyperpolarization results from a more prolonged increase in K^+ permeability that begins a fraction of a millisecond later. [After Hodgkin, A.L. and Huxley, A.F., *J. Physiol.* **117**, 530 (1952).]

the cation going to the external side. The resulting electric field is, nevertheless, enormous by macroscopic standards: Assuming a typical membrane thickness of 50 Å, it is nearly 170,000 V·cm^{-1}.

*A nerve impulse consists of a wave of transient membrane depolarization known as an **action potential** that passes along a nerve cell.* A microelectrode implanted in an **axon** (the long process emanating from the nerve cell body) will record that during the first ~0.5 ms of an action potential, $\Delta\Psi$ increases from its resting potential of around -60 mV to about ~30 mV (Fig. 34-111*a*). This depolarization is followed by a nearly as rapid repolarization past the resting potential to the K^+ equilibrium potential (hyperpolarization) and then a slower recovery to the resting potential. What is the origin of this complicated electrical behavior? In 1953, Alan Hodgkin and Andrew Huxley demonstrated that the action potential results from a transient increase in the membrane's permeability to Na^+ followed, within a fraction of a millisecond, by a transient increase in its K^+ permeability (Fig. 34-111b).

The ion-specific permeability changes that characterize an action potential result from the presence of trans-axonal membrane proteins that function as Na^+- and K^+-specific **voltage-gated channels** *(also known as* **voltage-sensitive channels**). As a nerve impulse reaches a given patch of nerve cell membrane, the increased membrane potential induces the transient opening of the Na^+ channels so that Na^+ ions diffuse into the nerve cell at the rate of ~6000 ions · ms^{-1} per channel. This increase in P_{Na^+} causes $\Delta\Psi$ to increase (Eq. [34.9]) which, in turn, induces more Na^+ channels to open, *etc.*, leading to an explosive entry of Na^+ into the cell. Yet, before this process can equilibrate at its Na^+ equilibrium potential of around ~60 mV, the K^+ channels open (P_{K^+} increases) while the Na^+ channels close (P_{Na^+} returns to its resting value). $\Delta\Psi$ therefore reverses sign and overshoots its resting potential to approach its K^+ equilibrium value. Eventually the K^+ channels also close and the membrane patch regains its resting potential. The Na^+ channels, which remain open only 0.5 to 1.0 ms, will not reopen until the

membrane has returned to its resting state, thereby limiting the axon's firing rate.

An action potential is triggered by an ~20-mV rise in $\Delta\Psi$ to about −40 mV. Action potentials therefore propagate along an axon because the initially rising value of $\Delta\Psi$ in a given patch of axonal membrane triggers the action potential in an adjacent membrane patch that does so in an adjacent membrane patch, etc. (Fig. 34-112). The nerve impulse is thereby continuously amplified so that its signal amplitude remains constant along the length of the axon (in contrast, an electrical impulse traveling down a wire dissipates as a consequence of resistive and capacitive effects). Note, however, that since the relative ion imbalance responsible for the resting membrane potential is small, only a tiny fraction of a nerve cell's Na^+–K^+ gradient is discharged by a single nerve impulse. An axon can therefore transmit a nerve impulse every few milliseconds without letup. This capacity to fire rapidly is an essential feature of neuronal communications: *Since nerve impulses all have the same amplitude, the magnitude of a stimulus is conveyed by the rate that a nerve fires.*

The Voltage-Gated Na^+ Channel Is the Target of Numerous Neurotoxins

Neurotoxins have proved to be invaluable tools for dissecting the various mechanistic aspects of neurotransmission. Many neurotoxins, as we shall see, interfere with the action of neuronal voltage-gated Na^+ channels but,

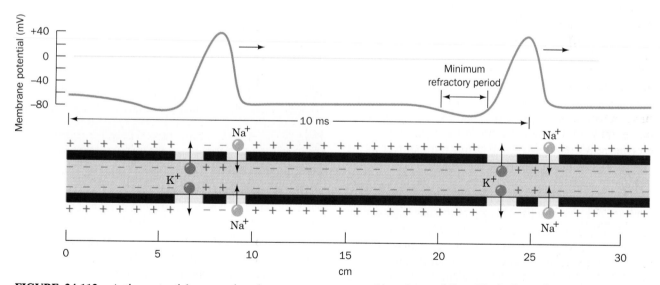

FIGURE 34-112. Action potential propagation along an axon. Membrane depolarization at the leading edge of an action potential triggers an action potential at the immediately downstream portion of the axon membrane by inducing the opening of its voltage-gated Na^+ channels. As the depolarization wave moves further downstream, the Na^+ channels close and the K^+ channels open to hyperpolarize the membrane. After a brief refractory period, during which the K^+ channels close and the hyperpolarized membrane recovers its resting potential, a

second impulse can follow. The indicated impulse propagation speed is that measured in the giant axon of the squid which, because of its extraordinary width (~1 mm) is a favorite experimental subject of neurophysiologists. Note that the action potential in this figure appears backwards from that in Fig. 34-111 because this figure shows the distribution of the membrane potential along an axon at an instant in time, whereas Fig. 34-111 shows the membrane potential's variation with time at a fixed point on the axon.

curiously, few are known that affect K$^+$ channels. **Tetrodotoxin,**

Tetrodotoxin

a paralytic poison of enormous potency, which occurs mainly in the skin, ovaries, liver, and intestines of the puffer fish (known as fugu in Japan, where it is a delicacy that may be prepared only by chefs certified for their knowledge of puffer fish anatomy), acts by specifically blocking the Na$^+$ channel. The Na$^+$ channel is similarly blocked by **saxitoxin,**

Saxitoxin

a product of marine dinoflagellates (a type of plankton known as the "red tide") that is concentrated by filter-feeding shell fish to such an extent that a small mussel can contain sufficient saxitoxin to kill 50 people. Both of these neurotoxins have a cationic guanidino group, and both are effective only when applied to the external surface of a neuron (their injection into the cytoplasm elicits no response). It is therefore thought that these toxins specifically interact with an anionic carboxylate group located at the mouth of the Na$^+$ channel on its extracellular side.

The specific binding of radioactive tetrodotoxin to the detergent-solubilized Na$^+$ channel greatly aided in the channel's purification. The Na$^+$ channel from mammalian brain is a heterotrimeric complex of α (270 kD), β_1 (36 kD), and β_2 (33 kD) subunits. This glycoprotein's large size no doubt reflects the complexity of the tasks it performs: It forms an ion-selective transmembrane pore that contains two voltage-sensitive "gates," one to open the channel upon membrane depolarization and one to later close it. In fact, minute "gating currents" arising from the movements of these positively charged gates in opening and closing the Na$^+$ channel can be detected (electrical current is the movement of charge) if the much larger ionic currents through the membrane are first blocked by plugging the Na$^+$ and K$^+$ channels with tetrodotoxin and Cs$^+$ (the K$^+$ channel can be blocked from its cytoplasmic side by high concentrations of Cs$^+$ or tetraethylammonium ion).

Batrachotoxin,

Batrachotoxin

a steroidal alkaloid secreted by the skin of a Columbian arrow-poison frog, *Phyllobates aurotaenia,* is the most potent known venom (2 μg/kg body weight is 50% lethal in mice). This substance also specifically binds to the voltage-gated Na$^+$ channel but, in contrast to the actions of tetrodotoxin and saxitoxin, renders the axonal membrane highly permeable to Na$^+$. Indeed, batrachotoxin-induced axonal depolarization is reversed by tetrodotoxin. The observation that the repeated electrical stimulation of a neuron enhances the action of batrachotoxin indicates that this toxin binds to the Na$^+$ channel in its open state.

FIGURE 34-113. The X-ray structure of the 65-residue variant-3 toxin from the Southwestern American scorpion *Centruroides sculpturatus* Ewing. The residues shown in color are conserved among the various scorpion neurotoxins. These residues are clustered on the protein's near surface, which is therefore thought to form the site that binds to the voltage-gated Na$^+$ channel. The color code used for the conserved residues only is: Pro and Gly side chains and the main chain are light blue; Asp and Glu side chains and the C-terminus are red; Arg and Lys side chains and the N-terminus are dark blue; uncharged polar side chains are purple; aliphatic side chains are green; and Cys side chains are yellow. [Courtesy of Mike Carson, University of Alabama at Birmingham; X-ray structure determined by Charles Bugg, University of Alabama at Birmingham.]

Venoms from American scorpions contain families of 60- to 70-residue protein neurotoxins that also act to depolarize neurons by binding to their Na^+ channels (Fig. 34-113; the different neurotoxins in the same venom appear to be specialized for binding to the Na^+ channels in the various species the scorpion is likely to encounter). Scorpion toxins and tetrodotoxin do not, however, compete with each other for binding to the Na^+ channel and therefore must bind at separate sites.

Nerve Impulse Velocity Is Increased by Myelination

The axons of the larger vertebrate neurons are sheathed with **myelin,** a biological "electrical insulating tape" that is wrapped about the axon (Fig. 34-114*a*) so as to electrically isolate it from the extracellular medium. Impulses in myelinated nerves propagate with velocities of up to 100 m · s^{-1}, whereas those in unmyelinated nerves are no faster than 10 m · s^{-1} (imagine the coordination difficulties that, say, a giraffe would have if it had to rely on only unmyelinated nerves). How does myelination increase the velocity of nerve impulses?

Myelin sheaths are interrupted every millimeter or so along the axon by narrow unmyelinated gaps known as **nodes of Ranvier** (Fig. 34-114*b*) where the axon contacts the extracellular medium. Binding studies using radioactive tetrodotoxin indicate that the voltage-gated Na^+ channels of unmyelinated axons have rather sparse although uniform, distributions in the axonal membrane of ~20 channels · μm^{-2}. In contrast, the Na^+ channels of myelinated axons occur only at the nodes of Ranvier, where they are concentrated with a density of ~$10^4 · \mu m^{-2}$. The action potential of a myelinated axon evidently hops between these nodes, a process named **saltatory conduction** (Latin: *saltare,* to jump). Nerve impulse transmission between the nodes must therefore occur by the passive conduction of an ionic current, a mechanism that is inherently much faster than the continuous propagation of an action potential but that is also dissipative. The nodes act as amplification stations to maintain the intensity of the electrical impulse as it travels down the axon. Without the myelin insulation, the electrical impulse would become too attenuated through transmembrane ion leakage and capacitive effects to trigger

(a)

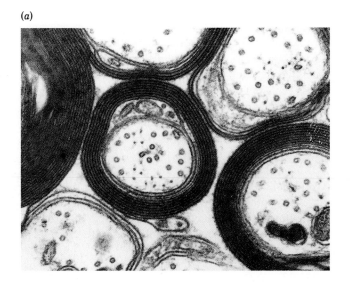

FIGURE 34-114. (*a*) An electron micrograph of myelinated nerve fibers in cross-section. The myelin sheath surrounding an axon is the plasma membrane of a **Schwann cell,** which, as it spirally grows around an axon, extrudes its cytoplasm from between the layers. The resulting double bilayer, which makes between 10 and 150 turns about the axon, is a good electrical insulator because of its particularly high (79%) lipid content. [Courtesy of Cedric Raine, Albert Einstein College of Medicine of Yeshiva University.] (*b*) A schematic diagram of a myelinated axon in longitudinal cross-section, indicating that in the nodes of Ranvier (the gaps between adjacent Schwann cells), the axonal membrane is in contact with the extracellular medium. A depolarization generated by an action potential at one node hops, via ionic conduction, down the myelinated axon (*red arrows*), to the neighboring node where it induces a new action potential. Nerve impulses in myelinated axons are therefore transmitted by saltatory conduction.

(b)

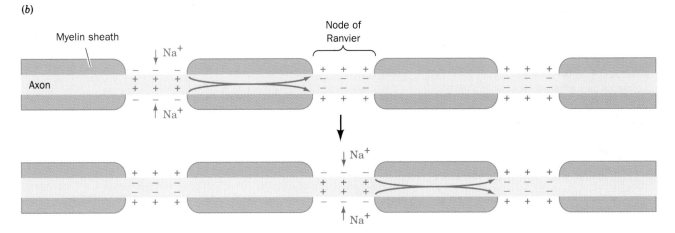

an action potential at the next node. In fact, **multiple sclerosis,** an autoimmune disease that demyelinates nerve fibers in the brain and spinal cord, results in serious and often fatal neurological deficiencies.

Neurotransmitters Relay Nerve Impulses across Synapses

The junctions at which neurons pass signals to other neurons, muscles, or glands are called **synapses.** In **electrical synapses,** which are specialized for rapid signal transmission, the cells are separated by a gap, the **synaptic cleft,** of only 20 Å, so an action potential arriving at the presynaptic side of the cleft can sufficiently depolarize the postsynaptic membrane to trigger its action potential directly. However, the >200-Å gap of most synapses is too great a distance for such direct electrical coupling. In these **chemical synapses,** the arriving action potential triggers the release from the presynaptic neuron of a specific substance known as a **neurotransmitter,** which diffuses across the cleft and binds to its corresponding receptors on the postsynaptic membrane. In **excitatory synapses,** neurotransmitter binding stimulates membrane depolarization, thereby triggering an action potential on the postsynaptic membrane. Conversely, neurotransmitter binding in **inhibitory synapses** alters postsynaptic membrane permeability so as to inhibit an action potential and thus attenuate excitatory signals. What is the mechanism through which an arriving action potential stimulates the release of a neurotransmitter, and by what means does its binding to a receptor alter the postsynaptic membrane's permeability? To answer these questions let us consider the workings of **cholinergic synapses;** that is, synapses that use **acetylcholine (ACh)** as a neurotransmitter.

Acetylcholine (ACh)

Nicotine

Muscarine

Two types of cholinergic synapses are known:

1. Those containing **nicotinic receptors** (receptors that respond to **nicotine**).

2. Those containing **muscarinic receptors** (receptors that respond to **muscarine,** an alkaloid produced by the poisonous mushroom *Amanita muscaria).*

In what follows, we shall focus on cholinergic synapes containing nicotinic receptors since this best characterized type

of synapse occurs at all excitatory neuromuscular junctions in vertebrates and at numerous sites in the nervous system.

Electric Organs of Electric Fish Are Rich Sources of Cholinergic Synapses

The study of synaptic function has been greatly facilitated by the discovery that the homogenization of nerve tissue causes its presynaptic endings to pinch off and reseal to form **synaptosomes.** The use of synaptosomes, which can be readily isolated by density gradient ultracentrifugation, has the advantage that they can be manipulated and analyzed without interference from other neuronal components.

The richest known source of cholinergic synapses is the electric organs of the freshwater electric eel *Electrophorus electricus* and saltwater electric fish of the genus *Torpedo.* Electric organs, which these organisms use to stun or kill their prey, consist of stacks of ~5000 thin flat cells called **electroplaques** that begin their development as muscle cells but ultimately lose their contractile apparatus. One side of an electroplaque is richly innervated and has high electrical resistance, whereas its opposite side lacks innervation and has low electrical resistance. Both sides maintain a resting membrane potential of around −90 mV. Upon neuronal stimulation, all the innervated membranes in a stack of electroplaques simultaneously depolarize to a membrane potential of around ~40 mV, yielding a potential difference across each cell of 130 mV (Fig. 34-115). Since the 5000 electroplaques in a stack are "wired" in series like the batteries in a flashlight, the total potential difference across the stack is ~5000 × 0.130 V = 650 V, enough to kill a human being.

Acetylcholine Release Is Triggered by Ca²⁺

ACh is synthesized near the presynaptic end of a neuron by the transfer of an acetyl group from acetyl-CoA to **choline** in a reaction catalyzed by **choline acetyltransferase.**

Acetyl-CoA   **Choline**

choline acetyltransferase

Acetylcholine

Much of this ACh is sequestered in ~400-Å-diameter **synaptic vesicles,** which typically contain ~10^4 ACh molecules each.

*The arrival of an action potential at the presynaptic membrane triggers the opening of **voltage-gated Ca²⁺ channels**. The resulting influx of extracellular Ca²⁺ into the presynaptic terminal, in turn, stimulates the exocytosis of its synaptic vesicles so as to release their packets of ACh into the synaptic cleft (Fig. 34-116).* The black widow spider takes advan-

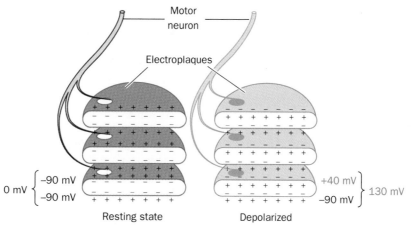

FIGURE 34-115. The simultaneous depolarization (*red*, right) of the innervated membranes in a stack of electroplaques "wired" in series results in a large voltage difference between the two ends of the stack. This is because the total voltage is the sum of the voltages generated by each electroplaque.

(*a*)

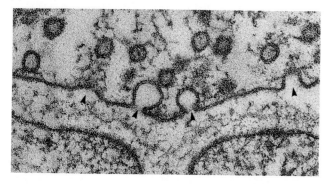

(*b*)

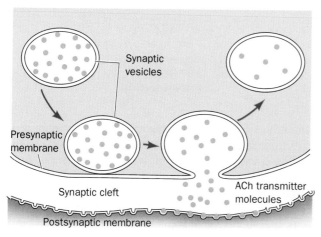

FIGURE 34-116. The transmission of nerve impulses across a synaptic cleft. (*a*) An electron micrograph of a frog neuromuscular junction in which synaptic vesicles are undergoing exocytosis (*arrows*) with the presynaptic membrane (*top*). [Courtesy of John Heuser, Washington University School of Medicine, St. Louis, Missouri.] (*b*) This process discharges the ACh contents of the synaptic vesicles into the synaptic cleft, where it diffuses in <100 μs to the postsynaptic membrane and binds to its receptor. The synaptic vesicle membranes are later reclaimed by endocytosis and refilled with ACh.

tage of this system: Its highly neurotoxic venom protein, **α-latrotoxin** (130 kD), causes massive release of ACh at the neuromuscular junction, possibly by acting as a Ca^{2+} ionophore. In contrast, ACh release is inhibited by **botulinus toxin,** a mixture of eight 135- to 170-kD proteins produced by the anaerobic bacterium *Clostridium botulinum* and the agent responsible for the deadly food poisoning syndrome **botulism.** However, carefully controlled quantities of botulinus toxin are medically useful in relieving the symptoms of certain types of chronic muscle spasms.

The hypothesis that ACh is released by exocytosis of synaptic vesicles was proposed in 1952 by Bernard Katz on the basis of his observation that unstimulated neuromuscular junctions exhibit small (0.1-3.0 mV) randomly occurring depolarizing pulses known as **miniature end plate potentials** (the **end plate** is the neuromuscular junctions's postsynaptic membrane). The intensities of these pulses are consistent with the release of ~10^4 molecules of ACh each, which corresponds to the exocytosis of a single synaptic vesicle. A full end plate potential presumably results from the simultaneous release of as many as 400 such ACh packets. Although the mechanism of Ca^{2+}-stimulated synaptic vesicle exocytosis is unknown, it is likely to involve **synapsin I,** a 75-kD synaptic vesicle membrane protein that, in brain, is a major substrate for Ca^{2+}-calmodulin and cAMP-dependent protein kinases. Activated synapsin I presumably promotes fusion of the synaptic vesicle membrane with the plasma membrane.

Acetylcholine Receptor Is a Ligand-Gated Cation Channel

The **acetylcholine receptor** is an ~250-kD $\alpha_2\beta\gamma\delta$ transmembrane glycoprotein whose four different subunits have similar sequences. Electron crystallography studies, by Nigel Unwin, of the ACh receptor in its closed (unliganded) form indicate that it is an 80-Å-diameter by 125-Å-long cylinder that protrudes ~60 Å into the synaptic space and

~20 Å into the cytoplasm (Fig. 34-117). Its five rodlike subunits are arranged with quasi-fivefold symmetry over much of their length. The ACh receptor's most striking structural feature is a ~20-Å-diameter by 60-Å-long water-filled central channel that extends from the receptor's synaptic entrance to the level of the lipid bilayer, where it forms a more constricted ~30-Å-long pore that appears to be blocked near the middle of the bilayer. The cytoplasmic end of the ACh receptor forms a second ~20-Å-diameter central channel that is ~20-Å long.

The binding of two ACh molecules, one per α subunit, allosterically induces the opening of the channel through the bilayer to permit Na⁺ and K⁺ ions to diffuse, respectively, in and out of the cell at rates of ~20,000 of each type of ion per millisecond. The resulting depolarization of the postsynaptic membrane initiates a new action potential. After 1 to 2 ms, the ACh spontaneously dissociates from the receptor and the channel closes.

The ACh receptor is the target of some of the most deadly known neurotoxins (death occurs through respiratory arrest) whose use has greatly aided in the elucidation of receptor function. **Histrionicatoxin,** an alkaloid secreted by the skin of the Columbian arrow-poison frog *Dendrobates histrionicus,* and ***d*-tubocurarine,** the active ingredient of the Amazonian arrow poison **curare** as well as a medically useful paralytic agent, are both ACh antagonists that prevent ACh receptor channel opening.

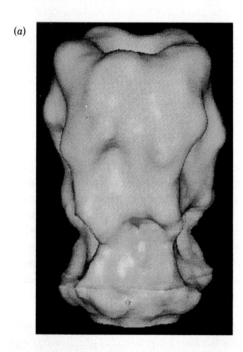

(a)

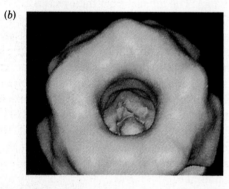

(b)

FIGURE 34-117. The electron crystal structure of the nicotinic acetylcholine receptor from the electric fish *Torpedo marmorata.* (*a*) Side view with the synaptic side up. The band across the structure marks the position of membrane bilayer in which the ACh receptor is embedded and separates its large extracellular portion from its smaller cytoplasmic portion. (*b*) View into the synaptic entrance of the channel. The channel narrows quite abruptly at the level of the lipid bilayer, ~60 Å below this entrance. [Courtesy of Nigel Unwin, MRC Laboratory of Molecular Biology, Cambridge, U.K.]

Histrionicatoxin

***d*-Tubocurarine**

Similarly, a family of homologous 7 to 8-kD venom proteins from some of the worlds most poisonous snakes, including **α-bungarotoxin** from snakes of the genus *Bungarus,* **erabutoxin** from sea snakes, and **cobratoxin** from cobras (Fig. 34-118) prevent ACh receptor channel opening by binding specifically and all but irreversibly to its α subunits. Indeed, detergent-solubilized ACh receptor has been purified by affinity chromatography on a column containing covalently attached cobratoxin.

Acetylcholine Is Rapidly Degraded by Acetylcholinesterase

*An ACh molecule that participates in the transmission of a given nerve impulse must be degraded in the few milliseconds before the potential arrival of the next nerve impulse. This essential task is accomplished by **acetylcholinesterase (AChE),** a 75-kD fast-acting enzyme that is GPI anchored*

(Section 11-5A) to the surface of the postsynaptic membrane

$$H_3C-\overset{\overset{\displaystyle O}{\|}}{C}-O-CH_2-CH_2-\overset{+}{N}(CH_3)_3 \; + \; H_2O$$
Acetylcholine

$$\downarrow \text{acetylcholinesterase}$$

$$H_3C-\overset{\overset{\displaystyle O}{\|}}{C}-O^- + \; HO-CH_2-CH_2-\overset{+}{N}(CH_3)_3 \; + \; H^+$$
Acetate   **Choline**

(the turnover number of AChE is $k_{cat} = 14{,}000 \text{ s}^{-1}$; the enzyme's catalytic efficiency, $k_{cat}/K_M = 1.5 \times 10^8 M^{-1} \cdot \text{s}^{-1}$, is close to the diffusion-controlled limit so that it is a nearly perfect catalyst; Section 13-2B). The products, acetate and choline, are transported back into the presynaptic terminal for recycling to ACh.

AChE is a serine esterase; that is, its catalytic mechanism resembles that of serine proteases such as trypsin. These enzymes, as we have seen in Section 14-3A, are irreversibly inhibited by alkylphosphofluoridates such as diisopropylphosphofluoridate (DIPF). Indeed, related compounds such as **tabun** and **sarin**

Tabun   **Sarin**

are military nerve gases because their efficient inactivation of human AChE causes paralysis stemming from cholinergic nerve impulse blockade and thus death by suffocation. **Succinylcholine,**

$$\begin{array}{l} H_2C-\overset{\overset{\displaystyle O}{\|}}{C}-O-CH_2-CH_2-\overset{+}{N}(CH_3)_3 \\ \;\;\;| \\ H_2C-\underset{\underset{\displaystyle O}{\|}}{C}-O-CH_2-CH_2-\overset{+}{N}(CH_3)_3 \end{array}$$
Succinylcholine

which is used as a muscle relaxant during surgery, is an ACh agonist that although rapidly released by ACh receptor, is but slowly hydrolyzed by AChE. Succinylcholine therefore produces persistent end plate depolarization. Its effects are short-lived, however, because it is rapidly hydrolyzed by the relatively nonspecific liver and plasma enzyme **butyrylcholinesterase.** Certain snake venoms, such as that of the green mamba snake, inactivate AChE, although they do so by binding to a site on AChE distinct from its active site.

X-Ray Structure of Acetylcholinesterase

The X-ray structure of the 537-residue AChE from the electric fish *Torpedo californica,* determined by Joel Sussman, Israel Silman, and Michal Harel, confirms that the previously identified Ser 200 and His 440 are members of AChE's catalytic triad. The structure further reveals that the third member of AChE's catalytic triad is Glu 237 rather than an Asp residue, only the second instance of a Glu in this position among the many serine proteases, li-

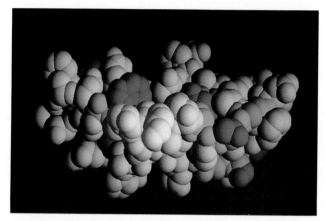

FIGURE 34-118. The X-ray structure of α-cobratoxin, a 71-residue neurotoxin from the venom of the cobra *Naja naja siamensis,* which specifically binds to the ACh receptor so as to inhibit its opening. The related snake venom neurotoxins α-bungarotoxin and erabutoxin have similar X-ray structures. The conserved residues thought to be essential for the toxicity of these neurotoxins are colored as is indicated in the legend to Fig. 34-113. [Courtesy of Mike Carson, University of Alabama at Birmingham; X-ray structure determined by Wolfram Saenger, Freie Universität Berlin, Germany.]

pases, and esterases of known structure. AChE's catalytic triad is arranged in what appears to be the mirror image of the catalytic triads in trypsin and subtilisin, for example (Figure 14-21), although, of course, this is not actually the case since all proteins consist of L-amino acid residues.

AChE's catalytic site is near the bottom of a narrow and 20-Å-deep gorge that extends halfway through the protein and widens out near its base (Fig. 34-119). The sides of this

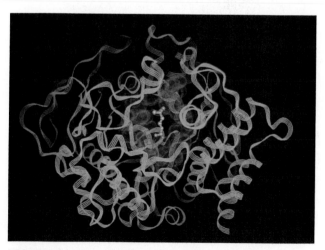

FIGURE 34-119. The X-ray structure of acetylcholinesterase (AChE) as represented in ribbon form. The aromatic side chains lining its active site gorge (*purple*) are shown in stick form surrounded by their van der Waals dot surface. The ACh substrate, which was modeled into the active site (the enzyme was crystallized in its absence), is shown in ball-and-stick form with its atoms gold and its bonds light blue. The entrance to the gorge is at the top of the figure. [Courtesy of Joel Sussman, The Weizmann Institute of Science, Israel.]

so-called active site gorge are lined with the side chains of 14 aromatic residues that comprise 40% of its surface area. Since the side chain O atom of the active site Ser is only 4 Å from the bottom of the gorge, ACh must bind in the gorge with its positively charged trimethylammonium group surrounded by aromatic side chains. This conclusion came as a surprise since it had been understandably expected that the trimethylammonium group would be bound at an anionic site. Perhaps the weak binding provided by the interactions of the trimethylammonium group with the π electrons of the aromatic rings facilitates the rapid diffusion of ACh to the bottom of the gorge, thereby accounting for the enzyme's high turnover number. In fact, model aromatic compounds have been synthesized that also bind quaternary ammonium compounds.

Amino Acids and Their Derivatives Function as Neurotransmitters

The mammalian nervous system employs well over 30 substances as neurotransmitters. Some of these substances, such as glycine and glutamate, are amino acids; many others are amino acid decarboxylation products or their derivatives (often referred to as **biogenic amines**). Thus, as we saw in Section 24-4B, the catecholamines dopamine, norepinephrine, and epinephrine are sequentially synthesized from tyrosine, whereas γ-aminobutyric acid (GABA), histamine, and serotonin are derived from glutamate, histidine, and tryptophan, respectively (Fig. 34-120). You will recognize many of these compounds as hormonally active substances that are present in the bloodstream. However, since the brain is largely isolated from the general circulation by a selective filtration system known as the **blood–brain barrier,** the presence of these substances in the blood

has no direct effect on the brain. The use of the same compounds as hormones and neurotransmitters apparently has no physiological significance but, rather, is thought to reflect evolutionary opportunism in adapting already available systems to new roles.

The use of selective staining techniques has established that each of the different neurotransmitters is used in discrete and often highly localized regions of the nervous system. The various neurotransmitters are, nevertheless, not simply functional equivalents of acetylcholine. Rather, many of them have distinctive physiological roles. For example, both GABA and glycine are inhibitory rather than excitatory neurotransmitters. The receptors for these substances are ligand-gated channels that are selectively permeable to Cl$^-$. Hence, their opening tends to hyperpolarize the membrane (make its membrane potential more negative) rather than depolarize it. A neuron inhibited in this manner must therefore be more intensely depolarized than otherwise to trigger an action potential (note that these neurons respond to more than one type of neurotransmitter). Thus, anion channels are inhibitory, whereas cation channels are excitatory. Ethanol, the oldest and most widely used psychoactive drug, is thought to act by inducing GABA receptors in the brain to open their Cl$^-$ channels.

The subunits of the various neurotransmitter-gated cation channels have 20 to 40% sequence identity, as do those of the anion channels. However, the two families of channel proteins appear to be unrelated. Despite this lack of homology, the sequences of the two types of channels suggest that they have considerable structural similarity.

The actual nature of a neuron's response to a neurotransmitter depends more on the characteristics of the corresponding receptor than on the neurotransmitter's identity.

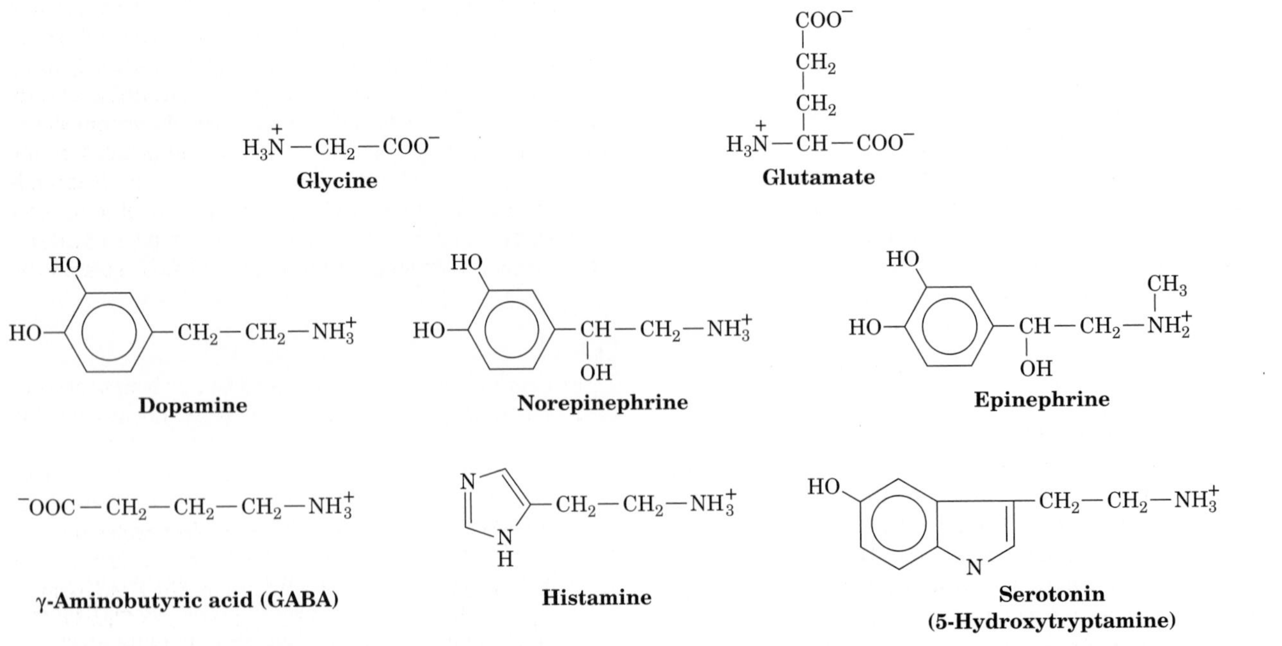

FIGURE 34-120. A selection of neurotransmitters.

Thus, as we have seen, nicotinic ACh receptors, which trigger the rapid contraction of skeletal muscles, respond to ACh within a few milliseconds by depolarizing their postsynaptic membrane. In contrast, the binding of ACh to muscarinic ACh receptors in heart muscle inhibits muscle contraction over a period of several seconds (several heart beats). This is accomplished by hyperpolarizing the postsynaptic membrane through the closure of otherwise open K^+ channels. Slow-acting neurotransmitters may act by inducing the formation of a second messenger such as cAMP. In fact, the brain has the highest concentration of cAMP-dependent kinases in the body. The binding of catecholamines to their respective neuronal receptors, through the intermediacy of adenylate cyclase and cAMP, evidently activates protein kinases to phosphorylate ion channels so as to alter the neuron's electrical properties. The ultimate effect of this process can be either excitatory or inhibitory. Note that catecholamines, whether acting as hormones or neurotransmitters, have similar mechanisms of receptor activation.

Neuropeptides Are Neurotransmitters

A large and growing list of hormonally active polypeptides also act as neurotransmitters. Not surprisingly, perhaps, the opioid peptides β-endorphin, met-enkephalin, and leu-enkephalin, as well as the hypothalamic releasing factors TRF, GnRF, and somatostatin (Section 34-4A), are in this category. What is less expected is that several gastrointestinal polypeptides, including the hormones gastrin, secretin, and cholecystokinin (CCK), may also act as neurotransmitters in discrete regions of the brain as do the pituitary hormones oxytocin, and vasopressin. Such **neuropeptides** differ from the simpler neurotransmitters in that they seem to elicit complex behavior patterns. For example, intracranially injecting rats with a nanogram of vasopressin greatly enhances their ability to learn and remember new tasks. Similarly, injecting a male or a female rat with GnRF evokes the respective postures they require for copulation. Just how these neuropeptides operate is but one of the many enigmas of brain function and organization.

CHAPTER SUMMARY

Blood Clotting

Blood coagulation occurs through a double cascade of proteolytic reactions, which result in the formation of cross-linked fibrin clots. Fibrin, which is formed from fibrinogen by the thrombin-catalyzed proteolytic excision of this soluble precursor's fibrinopeptides, self-associates to form a fibrous meshwork of half-staggered elongated molecules. The clot is later cross-linked by the formation of intermolecular Gln-Lys isopeptide bonds in a reaction catalyzed by the thrombin-activated fibrin-stabilizing factor (FSF or XIII_a).

Thrombin, which like all the proteolytic clotting factors is a serine protease, is synthesized as a zymogen that is proteolytically activated by Stuart factor (X_a). Thrombin, as do several other clotting factors, contains a number of γ-carboxylglutamate (Gla) residues whose posttranslational synthesis from Glu residues requires a vitamin K cofactor. The Gla residues chelate Ca^{2+} whose presence is required for prothrombin activation together with thrombin-activated proaccelerin (V_a) and a phosphatidylserine-containing membrane surface. The Gla residues apparently function to anchor X_a to the membrane via Ca^{2+} bridges. Thrombin is a trypsin-like enzyme but has a much deeper active site cleft, thereby accounting for its high substrate specificity.

Factor X may be activated through either the intrinsic pathway or the extrinsic pathway. In the intrinsic pathway, clotting is initiated by the contact system in which Hageman factor (XII), in the presence of high molecular weight kininogen (HMK), is activated by adsorption to a negatively charged surface such as glass to proteolyze prekallikrein to kallikrein. Kallikrein then reciprocally proteolyzes XII which, in the presence of HMK, proteolytically activates plasma thromboplastin antecedent (PTA or XI). In the final two steps of the intrinsic pathway, XI_a proteolytically activates Christmas factor (IX) which, in turn, proteolytically activates X in the presence of thrombin-activated antihemophilic factor

(VIII_a). Both of these latter reactions must also take place on a phospholipid membrane surface in the presence of Ca^{2+}. The extrinsic pathway begins with the proteolytic activation of proconvertin (VII) by either XII_a or thrombin. VII_a, in turn, mediates the activation of X in the presence of Ca^{2+}, phospholipid membrane and tissue factor (III), a membrane glycoprotein that occurs in many tissues.

Clot formation is inhibited by numerous physiological mechanisms, including the inactivation of all of the clotting system proteases but VII_a by the binding of antithrombin III. The presence of heparin, a component of many tissues, greatly stimulates this reaction. Similarly, the thrombin-activated protein C proteolytically inactivates V and VIII. Thrombin, in turn, is activated to activate protein C by binding to the cell surface protein thrombomodulin. Once wound healing is underway, clots are dismantled by the serine protease plasmin, a process termed fibrinolysis. Plasmin is formed by the proteolysis of plasminogen by several serine proteases, notably urokinase and tissue-type plasminogen activator (t-PA). This process is limited through the inhibition of plasmin by $α_2$-antiplasmin.

Immunity

The immune response, which is conferred by lymphocytes in association with lymphoid tissues such as the thymus and the lymph nodes, results from cellular and humoral immunity. Cellular immunity, which guards against parasites, virally infected cells, cancers, and foreign tissue, is mediated by *T* cells, whereas humoral immunity, which is most effective against bacterial infections and extracellular viruses, is mediated by antibodies produced by *B* cells. A *T* cell is selected for proliferation if its *T* cell receptor simultaneously binds a foreign antigen and the host's Class I MHC protein. Some of the progeny, cytotoxic *T* cells, bind to antigen-bearing host cells and kill them by inserting pore-forming proteins in their plasma membranes. A *B* cell displaying an antibody that binds to a foreign antigen is similarly selected for proliferation and its plasma cell progeny secrete large amounts of that antibody. The immune system must be self-tolerant; failure to prevent the proliferation of *T* and *B* cells bearing antibodies against self-antigens results in autoimmune diseases.

Antibodies (immunoglobulins) are glycoproteins that consist of two identical light (L) chains and two identical heavy (H) chains which, in turn, each have a constant (C) region and a variable (V) region. The five classes of secreted immunoglobulins, IgA, IgD, IgE, IgG, and IgM, have different physiological functions and vary only in the identities of their H chains, whose differences mainly affect the antibodies' Fc segments. The L chain in all of these immunoglobulins is either a κ or a λ chain. An immunoglobulin's two identical Fab segments, which each consist of one C and one V domain from both the H and the L chains, contain the antibody's antigen-binding sites. The antigen-binding specificity of an immunoglobulin is largely dependent on the sequences of the hypervariable segments from both its H and its L chains that line its antigen-binding site, thereby forming its six complementarity determining regions (CDRs). X-Ray studies show that L and H chains are folded, respectively, into two and four domains that each have the characteristic immunoglobulin fold.

The immune system generates a virtually unlimited variety of antigen-binding sites through both somatic recombination and somatic mutation. The κ light chain is encoded by four exons known as the leader (L_κ), variable (V_κ), joining (J_κ), and constant (C_κ) segments. During *B* cell differentiation, one of the ~300 $L_\kappa + V_\kappa$ units contained in the embryonic human genome somatically recombines with one of the genome's five J_κ segments in a reaction that is probably mediated by the *RAG1* and *RAG2* gene products. Then, in later cell generations, the resulting $L_\kappa - V_\kappa - J_\kappa$ unit is transcribed and selectively spliced to the genome's single C_κ segment yielding κ chain mRNA. The *V/J* recombination joint in this process is not precisely defined leading to further variation in the κ chain. Heavy chain genes are similarly assembled but, in addition, have a *D* segment between their V_H and J_H segments that leads to even greater heavy chain diversity. Somatic mutation provides yet more diversity: Nucleotides may be added at random at the V_H/D and D/J_H joints through the action of terminal deoxynucleotidyl transferase and, furthermore, both heavy and light chain genes in *B* cell progenitors are subject to somatic hypermutation. *B* cells express only one heavy chain and one light chain allele, a phenomenon named allelic exclusion, thereby ensuring that each cell expresses only one species of immunoglobulin. Differentiating *B* cells switch from the synthesis of membrane-bound IgM, whose heavy chain has a hydrophobic transmembrane tail, to a secreted IgM with the same antigen specificity but lacking this C-terminal polypeptide. This switch occurs via the selection of alternative polyadenylation sites located before or after the exons specifying the transmembrane tail. *B* cell progeny also progressively switch from the synthesis of IgM to other classes of immunoglobulins, a process known as class switching, through either selective splicing or somatic recombination leading to the expression of alternative C_H segments.

T cell receptors resemble immunoglobulins in that they consist of α and β chains that have constant and variable domains, each of which assumes the immunoglobulin fold. *T* cell receptor diversity is generated by the somatic selection of different *V*, *J*, and for α chains, *D* regions.

The major histocompatibility complex encodes a highly polymorphic group of membrane-bound proteins that act as individuality markers among members of the same species (Class I MHC proteins) and differentiate immune system cells from other body cells (Class II MHC proteins). MHC proteins have domains that structurally resemble those in immunoglobulins and *T* cell receptors and, therefore, the genes encoding all these proteins form a gene superfamily. An 8- or 9-residue polypeptide fragment of the processed antigen is presented to *T* cell receptors in complex with the Class I MHC protein through its binding in the cleft formed by the Class I MHC protein's α_1 and α_2 domains. Class II MHC proteins have very similar structures but their bound peptides extend out from both ends of the binding groove. The MHC gene's polymorphism is thought to prevent pathogens from evolving antigens that interact poorly with a particular MHC protein during the antigen recognition process.

The complement system defends against foreign invaders by killing foreign cells through complement fixation, inducing the phagocytosis of foreign particles (opsonization), and triggering local acute inflammatory reactions. The complement system consists of ~20 proteins that interact in the antibody-dependent classical pathway and in the antibody-independent alternative pathway. The classical pathway contains three sequentially activated protein complexes: The recognition unit, which assembles on cell surface–bound antibody–antigen complexes; the activation unit, which amplifies the recognition process through a proteolytic cascade involving a series of serine proteases; and the membrane attack complex (MAC), which punctures the antibody-marked cell's plasma membrane, causing cell lysis and death. The alternative pathway, which is thought to defend against invading microorganisms before an effective immune response can be mounted, also leads to the assembly of the MAC but in a series of reactions that are activated by the presence of certain bacterially synthesized polymers, whole bacteria, and host cells that are infected by certain viruses. The complement system is tightly regulated by the structural instabilities of certain activated complement proteins, by the degradation of complement components through the actions of specific proteases such as C4b binding protein, and by the sequestering of complement components by their specific binding of proteins such as C1 inhibitor and S protein.

Motility

Skeletal muscle fibers consist of banded myofibrils which are, in turn, comprised of interdigitated thick and thin filaments. The thick filaments are composed almost entirely of myosin, a dimeric protein with two globular heads and an elongated rodlike segment comprised of two α helices in a coiled coil. The myosin molecules aggregate end to end in a regular staggered array to form the bipolar thick filament. The myosin elongated S1 head contains an ATPase. The thin filaments consist mainly of actin, a globular protein (G-actin) that polymerizes to form a helical filament (F-actin) in which each monomer unit is capable of binding a single myosin head. The thin filament also contains two other major proteins, tropomyosin and troponin. Tropomyosin is a heterodimeric protein that consists mainly of a coiled coil of two α helices that is wound in the grooves of the F-actin helix. Troponin consists of three subunits: TnC, a Ca^{2+}-binding calmodulin homolog; TnI, which binds actin; and TnT, which binds tropomyosin. The troponin–tropomysin complex regulates muscle contraction by varying the access of myosin heads to their actin-binding sites in response to the concentration of Ca^{2+}. Other, less abundant proteins function in muscle assembly and stabilization including α-actinin, desmin, vimentin, nebulin, titin, and dystrophin.

Structural studies indicate that the thick and thin filaments slide past each other during muscle contraction. Tension is generated through a six-part ATP-driven reaction cycle in which myosin heads on thick filaments attach to actin monomers on thin filaments, change conformation, and then release the thin filament. Repeated such cycles cause the myosin heads to "walk" up the adjacent thin filaments, resulting in muscle contraction.

Muscle contraction is triggered by an increase in $[Ca^{2+}]$. The Ca^{2+} binds to the TnC subunit of troponin, with the resulting conformational change causing tropomyosin to move deeper into the thin filament groove, thereby exposing actin's myosin head-binding sites, and thus switching on muscle contraction. The Ca^{2+} is released from the sarcoplasmic reticulum (SR) in response to nerve impulses that render the SR membrane permeable to Ca^{2+}. The cytosolic $[Ca^{2+}]$ is otherwise maintained at a very low level through the action of SR membrane–bound Ca^{2+}–ATPases, which pump the Ca^{2+} into the SR thus terminating muscle contraction.

Smooth muscle, which is responsible for long-lasting and involuntary contractions, lacks the banded pattern of skeletal muscles. Its myosin heads only interact with actin when one of their light chains is phosphorylated at a specific Ser residue. Smooth muscle contraction is nevertheless triggered by Ca^{2+} because myosin light chain kinase, the enzyme that catalyzes the phosphorylation of myosin light chains, is active only when associated with Ca^{2+}–calmodulin. Myosin light chain phosphatase hydrolyzes myosin's activating phosphate group, so in the absence of active myosin light chain kinase, smooth muscle relaxation ensues. Nerve impulses increase the permeability of the smooth muscle cell plasma membrane to Ca^{2+}, which acts as an intracellular second messenger in stimulating smooth muscle contraction. Smooth muscles also respond to hormones such as epinephrine through the intermediacy of cAMP, whose presence activates a protein kinase to phosphorylate myosin light chains.

Actin and myosin are also prominent in nonmuscle cells, where they have both structural and functional roles. Nonmuscle actin is generally in a state of equilibrium between its monomeric G-actin form and polymeric F-actin microfilaments. The assembly and disassembly of microfilaments, as influenced by the presence of actin-binding proteins such as profilin, has an important role in cellular motility. Nonmuscle myosin also forms thick filaments that, in concert with microfilaments, participate in intracellular contractile processes such as the tightening of the contractile ring during cytokinesis.

Ciliary motion is a microtubule-based phenomenon. Microtubules are formed from the protein tubulin, an $\alpha\beta$ dimer, that polymerizes with the concomitant hydrolysis of GTP. In cilia and eukaryotic flagella, the microtubules are arranged in a 9 + 2 array in which 9 double microtubules surround 2 single microtubules in an assembly that is cross-linked by three types of proteins. Subfibers A of the outer fibers each bear two dynein arms that "walk" up neighboring subfibers B in an ATP-powered process. However, the cross-links between neighboring fibers prevent these fibers from sliding past each other; rather, the cilia bend, which accounts for their oarlike motion. Cytosolic dyneins and kinesin oppositely motivate the transport of vesicles along microtubule tracks. Bacterial flagella, which are responsible for bacterial propulsion, are entirely different from eukaryotic flagella. Bacterial flagella consist of a flagellin filament, a flagellar hook made of hook protein, and a complex basal body that is embedded in the bacterial plasma membrane. The basal body is a true rotary motor. Thus, the flagellar filament and hook are passive elements that, like a propeller, convert the rotary motion of the basal body to linear thrust. The basal body's rotary motion is directly powered by the discharge of the metabolically generated electrochemical proton gradient across the plasma membrane.

Biochemical Communications

Chemical messengers are classified as autocrine, paracrine, or endocrine hormones if they act on the same cell, cells that are nearby, or cells that are distant from the cell that secreted them, respectively. The body contains a complex endocrine system that controls many aspects of its metabolism. The pancreatic islet cells secrete insulin and glucagon, polypeptide hormones that induce liver and adipose tissue to store or release glucose and fat, respectively. Gastrointestinal polypeptide hormones coordinate various aspects of digestion. The thyroid hormones, T_3 and T_4, are iodinated amino acid derivatives that generally stimulate metabolism by activating cellular transcription factors. Ca^{2+} metabolism is regulated by the levels of PTH, vitamin D, and calcitonin. PTH and vitamin D induce an increase in blood $[Ca^{2+}]$ by stimulating Ca^{2+} release from bone and its adsorption from kidney and intestine, whereas calcitonin has the opposite effects. Vitamin D is a steroid derivative that must be obtained in the diet or by exposure to UV radiation. Vitamin D, after being sequentially processed in the liver and kidney to $1,25(OH)_2D$, stimulates the synthesis of a Ca^{2+}-binding protein in the intestinal epithelium. The adrenal medulla secretes the catecholamines epinephrine and norepinephrine, which bind to α- and β-adrenoreceptors on a great variety of cells so as to prepare the body for "fight or flight." The adrenal cortex secretes glucocorticoid and mineralocorticoid steroids. Glucocorticoids affect metabolism in a manner opposite to that of insulin as well as mediating a wide variety of other vital functions. Mineralocorticoids regulate the excretion of salt and water by kidney. The gonads secrete steroid sex hormones, the androgens (male hormones), and estrogens (female hormones), which regulate sexual differentiation, the development of secondary sex characteristics, and sexual behavior patterns. Ovaries, in addition, secrete progestins that help mediate the menstrual cycle and pregnancy. Mammalian embryos will develop as females unless subjected to the influence of the androgen testosterone. *TDF*, a gene that encodes a DNA-binding protein and that is normally located on the Y chromosome, induces the development of testes, which in turn secrete testosterone.

The hypothalamus secretes a series of polypeptide releasing factors and release-inhibiting factors such as CRF, TRF, GnRF, and somatostatin that control the secretion of the corresponding trophic hormones from the adenohypophysis. Most of these trophic hormones, such a ACTH, TSH, LH, and FSH, stimulate their target endocrine glands to secrete the corresponding hormones. Growth hormone acts directly on tissues as well as stimulating liver to synthesize growth factors known as somatomedins. The menstrual cycle results from a complex interplay of hypothalamic, adenohypophyseal, and steroid sex hormones. A fertilized and implanted ovum secretes CG, which binds to the same receptor and has similar affects as LH, thus preventing menstruation. The binding of hGH to its receptor causes the receptor to dimerize, thereby providing the intracellular signal that the receptor has bound hCG. Many other hormonal signals are similarly mediated. The adenohypophysis also secretes opioid peptides that have opiate-like effects on the central nervous system. The neurohypophysis secretes the polypeptides vasopressin, which stimulates the kidneys to retain water, and oxytocin, which stimulates uterine contraction.

Receptors are membrane-bound proteins that bind their ligands according to the laws of mass action. The parameters describing the binding of a radiolabeled ligand to its receptor can be determined from Scatchard plots. The dissociation constants of additional ligands for the same receptor-binding site can then be determined through competitive binding studies. The intracellular effects of most polypeptide and catecholamine hormones are mediated by second messengers such as cAMP. Hormone binding to certain 7-transmembrane receptors activates the $G_{s\alpha}$ subunit of a stimulatory G-protein to replace its bound GDP with GTP, release its associated $G_\beta G_\gamma$ subunits, and activate adenylate cyclase to synthesize cAMP. Activation continues until $G_{s\alpha}$ spontaneously hydrolyzes its bound GTP to GDP and recombines with $G_\beta G_\gamma$. Several types of activated hormone receptors in a cell may stimulate the same G_s protein. There are also inhibitory G-proteins, which have the same G_β and G_γ subunits as does G_s, but which have an inhibitory $G_{i\alpha}$ subunit that deactivates adenylate cyclase. Biological signaling systems are subject to desensitization through the phosphorylation and endocytotic sequestering of the cell-surface receptors. Cholera toxin and heat-labile enterotoxin (LT), related bacterial AB_5 proteins, induce uncontrolled cAMP production by ADP-ribosylating $G_{s\alpha}$ so as to render it incapable of hydrolyzing GTP. Pertussis toxin, also an AB_5 protein, similarly ADP-ribosylates $G_{i\alpha}$.

Agonist binding activates receptor tyrosine kinases (RTKs) by inducing them to dimerize and thus autophosphorylate specific Tyr residues in their cytoplasmic domains. An autophosphorylated RTK may activate other proteins by Tyr-phosphorylating them. It can also modulate the activities of specific proteins through the binding of an RTK phospho-Tyr–containing peptide segment to SH2 domains on these proteins. Grb2/Sem-5 protein binds to an RTK in this way and simultaneously to Sos protein via its SH3 domains. The bound Sos, in turn, induces Ras to exchange its bound GDP for GTP. This causes Ras to activate Raf, a Ser/Thr kinase, to initiate a kinase cascade, which utimately phosphorylates and thus modulates the activities of transcription factors in the cell nucleus. Tyrosine kinase–associated receptors transduce the signal that they have bound effector by activating associated nonreceptor tyrosine kinases, many of which are members of the Src family. Protein tyrosine phosphatases (PTPs) deactivate Tyr-phosphorylated proteins. Some do so in a receptor-activated way.

PIP_2, a minor phospholipid component of the plasma membrane's inner leaflet, can yield up to three types of second messengers. Agonist–receptor interactions, through the intermediacy of a G-protein or an RTK, stimulate the corresponding phospholipases C (PLCs) to hydrolyze PIP_2 to IP_3 and DG. The IP_3 stimulates the release of Ca^{2+} from the endoplasmic reticulum through ligand-gated channels. The Ca^{2+} binds to calmodulin, which in turn activates a variety of cellular processes. The membrane-bound DG activates protein kinase C (PKC) to phosphorylate and thereby modulate the activities of numerous cellular proteins. DG may also be degraded to yield arachidonate, an obligate intermediate in the biosynthesis of prostaglandins and related compounds. NO is a local mediator that regulates vasodilation, serves as neurotrans-

mitter, and functions in the immune response.

Nerve impulses are transmitted electrically along a neuron and, in most cases, chemically between neurons and from neurons to muscle or gland cells. Neurons, as do most cells, actively pump K^+ into and Na^+ out of the cell, so as to generate a membrane potential of around -60 mV. A nerve impulse consists of a wave of transient membrane depolarization known as an action potential that passes along an axon. An action potential begins with the opening of transmembrane Na^+-specific voltage-gated channels followed, a fraction of a millisecond later, by the opening of K^+-specific voltage-gated channels and the closing of the Na^+ channels. As a consequence, the membrane potential increases from its resting potential to about $+30$ mV in a fraction of a millisecond and then, almost as rapidly, decreases past the resting potential to the membrane's K^+ equilibrium potential. As the K^+ channels close over the next few milliseconds, the resting potential is restored. The action potential is triggered by adjacent membrane depolarization, so a nerve impulse is a self-amplifying phenomenon that maintains its intensity along the length of an axon. The voltage-gated Na^+ channel is the target of many potent neurotoxins including tetrodotoxin, which blocks the channel, and batrachotoxin, which locks it open. Nerve impulses in myelinated neurons occurs by saltatory conduction in which the impulse is rapidly transmitted by ionic currents through the myelinated (electrically insulated) segments of the axon, and is renewed by the generation of an action potential at the unmyelinated nodes of Ranvier, which contain high concentrations of Na^+ channels.

Nerve impulses are chemically transmitted across most synapses by the release of neurotransmitters. Acetylcholine (ACh), the best characterized neurotransmitter, is packaged in synaptic vesicles that are exocytotically released into the synaptic cleft. This process is triggered by an increase in cytosolic $[Ca^{2+}]$ resulting from the arriving action potential's opening of voltage-gated Ca^{2+} channels. The ACh diffuses across the synaptic cleft, where it binds to the ACh receptor, a transmembrane ion channel that opens in response to ACh binding. The resultant flow of Na^+ into and K^+ out of the postsynaptic cell depolarizes the postsynaptic membrane, which, if sufficient neurotransmitter has been released, triggers a postsynaptic action potential. The ACh receptor is the target of numerous deadly neurotoxins, including histrionicatoxin, *d*-tubocurarine, and cobratoxin, which all bind to the ACh receptor so as to prevent its opening. The ACh is rapidly degraded, before the possible arrival of the next nerve impulse, through the action of acetylcholinesterase, a fast-acting serine esterase that has an unusual aromatic side chain–lined active site gorge. Nerve gases and succinylcholine inhibit acetylcholinesterase and therefore block nerve impulse transmission at cholinergic synapses. Many specific regions of the nervous system employ neurotransmitters other than ACh. Most of these neurotransmitters are amino acids, such as glycine and glutamate, or their decarboxylation products and their derivatives, including catecholamines, GABA, histamine, and serotonin. Many of these compounds are also hormonally active, but they are excluded from the brain by the blood–brain barrier. Although many neurotransmitters, such as ACh, are excitatory, others are inhibitory. The latter stimulate the opening of anion (Cl^-) channels, thereby causing the postsynaptic membrane to become hyperpolarized so that it must be more highly depolarized than otherwise to trigger an outgoing action potential. There is also a growing list of polypeptide neurotransmitters, many of which are also polypeptide hormones. These polypeptides apparently elicit complex behavior patterns.

REFERENCES

Blood Clotting

Arni, R.K., Padmanabhan, K., Padmanabhan, K.P., Wu, T.-P., and Tulinsky, A., Structures of the noncovalent complexes of human and bovine prothrombin fragment 2 with human PPACK–thrombin, *Biochemistry* **32**, 4727–4737 (1993).

Bode, W., Turk, D., and Karshikov, A., The refined 1.9-Å crystal structure of D-Phe-Pro-Arg chloromethylketone-inhibited human α-thrombin: Structure analysis, overall structure, electrostatic properties, detailed active site geometry, and structure-function relationships, *Protein Sci.* **1**, 426–471 (1992); *and* Stubbs, M.T. and Bode, W., A player of many parts: The spotlight falls on thrombin's structure, *Thrombosis Res.* **69**, 1–58 (1993).

Berliner, L.J., *Thrombin, Structure and Function,* Plenum Press (1992).

Colman, R.W., Hirsh, J., Marder, V.J., and Salzman, E.W. (Eds,), *Hemostasis and Thrombosis* (3rd ed.), Lipincott (1994).

Doolittle, R.F., Fibrinogen and fibrin, *Annu. Rev. Biochem.* **53**, 195–229 (1984); *and* Doolittle, R.F., Fibrinogen and fibrin, *Sci. Am.* **245**(6): 126–135 (1981).

Esmon. C.T., Cell mediated events that control blood coagulation and vascular injury, *Annu. Rev. Cell Biol.* **9**, 1–26 (1993).

Francis, C.W. and Marder, V.J., Concepts of clot lysis, *Annu. Rev. Med.* **37**, 187–204 (1986).

Furie, B. and Furie, B.C., Molecular and cellular biology of blood coagulation, *New Engl. J. Med.* **326**, 800–806 (1992).

Furie, B. and Furie, B.C., Molecular basis of vitamin K-dependent γ-carboxylation, *Blood* **75**, 1753–1762 (1990).

Halkier, T., *Mechanisms in Blood Coagulation, Fibrinolysis and the Complement System,* Cambridge University Press (1991).

Lawn, R.M. and Vehar, G.A., The molecular genetics of hemophilia, *Sci. Am.* **254**(3): 48–54 (1986).

Mann, K.G., Jenny, R.J., and Krishnaswamy, S., Cofactor proteins in the assembly and expression of blood clotting enzyme complexes, *Annu. Rev. Biochem.* **57**, 915–956 (1988).

Rao, S.P.S., Poojary, M.D., Elliott, B.W., Jr., Melanson, L.A., Oriel, B., and Cohen, C., Fibrinogen structure in projection at 18 Å resolution, *J. Mol. Biol.* **222**, 89–98 (1991).

Rydel, T.J., Ravichandran, K.G., Tulinsky, A., Bode, W., Huber, R., Roitsch, C., and Fenton, J.W., II, The structure of a complex of recombinant hirudin and human α-thrombin, *Science* **249**, 277–283 (1990); *and* Rydel, T.J., Tulinsky, A., Bode, W., and Huber, R., Refined structure of the hirudin–thrombin complex, *J. Mol. Biol.* **221**, 583–604 (1991).

Scriver, C.R., Beaudet, A.L., Sly, W.S., and Valle, D. (Eds.), *The Metabolic Basis of Inherited Disease* (6th ed.), Chapters 84–90, McGraw-Hill (1989).

Seshadri, T.P., Tulinsky, A., Skrzypczak-Jankun, E., and Park, C.H., Structure of bovine prothrombin fragment I refined at 2.25 Å resolution, *J. Mol. Biol.* **220**, 481–494 (1991); *and* Soriano-Garcia, M., Padmanabhan, K., de Vos, A.M., and Tulinsky, A., The Ca^{2+} ion and membrane binding structure of the Gla domain of Ca-prothrombin fragment 1, *Biochemistry* **31**, 2554–2566 (1992).

Stamatoyannopoulos, G., Nienhuis, A.W., Leder, P., and Majerus, P.W. (Eds.), *The Molecular Basis of Blood Diseases,* Chapters 15–18, Saunders (1987). [Authoritative discussions of hemostasis, hemophilia, fibrinogen, and fibrinolysis.]

Weisel, J.W., Nagaswami, C., and Makowski, L., Twisting of fibrin fibrils limits their radial growth, *Proc. Natl. Acad. Sci.* **84**, 8991–8995 (1987).

Zwaal, R.F.A. and Hemker, H.C. (Eds.), *Blood Coagulation,* Elsevier (1986).

Immunity

Ada, G.L. and Nossal, G., The clonal-selection theory, *Sci. Am.* **257**(2): 62–69 (1987).

Alberts, B., Bray, D., Lewis, J., Raff, M., Roberts, K., and Watson, J.D., *Molecular Biology of the Cell* (3rd ed.), Chapter 23, Garland Publishing (1994).

Alt, F.W., Blackwell, T.K., and Yancopoulos, G.D., Development of the primary antibody repertoire, *Science* **238**, 1079–1087 (1987).

Amit, A.G., Mariuzza, R.A., Phillips, S.E.V., and Poljak, R.J., Three-dimensional structure of an antigen-antibody complex at 2.8 Å resolution, *Science* **233**, 747–753 (1986); *and* Mariuzza, R.A., Phillips, S.E.V., and Poljak, R.J., The structural basis of antigen-antibody recognition, *Annu. Rev. Biophys. Biophys. Chem.* **16**, 139–159 (1987). [The structure of a lysozyme–Fab complex.]

Barber, L.D. and Parham, P., Peptide binding to major histocompatibity complex molecules, *Annu. Rev. Cell Biol.* **9**, 163–206 (1993).

Bjorkman, P.J. and Parham, P., Structure, function, and diversity of Class I major histocompatibility complex molecules, *Annu. Rev. Biochem.* **59**, 253–288 (1990).

Bjorkman, P.J., Saper, M.A., Samraoui, B., Bennett, W.S., Stro-minger, J.L., and Wiley, D.C., Structure of human class I histo-compatibility antigen, HLA-A2; *and* The foreign antigen binding site and T cell recognition regions of class I histocompatibility antigens, *Nature* **329**, 506–512 *and* 512–518 (1987).

Black, F.L., Why did they die? *Science* **258**, 1739–1740 (1992). [Discusses why so many Amerinds died of infectious diseases.]

Blackwell, T.K. and Alt, F.W., Molecular characterization of the lymphoid V(D)J recombination activity, *J. Biol. Chem.* **264**, 10327–10330 (1989).

Brady, R.L., Dodson, E.J., Dodson, G.G., Lange, G., Davis, S.J., Williams, A.F., and Barclay, A.N., Crystal structure of domains 3 and 4 of rat CD4: Relation to the NH_2-terminal domains, *Science* **260**, 979–983 (1993); Ryu, S.-E., Kwong, P.D., Truneh, A., Porter, T.G., Arthos, J., Rosenberg, M., Dai, X., Xuong, N., Axel, R., Sweet, R.W., and Hendrickson, W.A., Crystal structure of an HIV-binding recombinant fragment of human CD4, *Nature* **348**, 419–426 (1990); *and* Wang, J., Yan, Y., Garrett, T.P.J., Liu, J.,

Rodgers, D.W., Garlick, R.L., Tarr, G.E., Husain, Y., Reinherz, E.L., and Harrison, S.C., Atomic structure of a fragment of human CD4 containing two immunoglobulin-like domains, *Nature* **348**, 411–418 (1990). [The latter two papers discuss the X-ray structure of domains 1 and 2.]

Capra, J.D. and Edmundson, A.B., The antibody combining site, *Sci. Am.* **236**(1): 50–59 (1977).

Cohen, I.R., The self, the world and autoimmunity, *Sci. Am.* **258**(4): 52–60 (1988).

Darnell, J., Lodish, H., and Baltimore, D., *Molecular Cell Biology* (2nd ed.), Chapter 25, Scientific American Books (1990).

Davies, D.R. and Chacko, S., Antibody structure, *Acc. Chem. Res.* **26**, 421–427 (1993); *and* Davies, D.R., Padlan, E.A., and Sheriff, S., Antibody-antigen complexes, *Annu. Rev. Biochem.* **59**, 439–473 (1990).

Davis, M.M., T cell receptor gene diversity and selection, *Annu. Rev. Biochem.* **59**, 475–496 (1990).

Englehard, V.H., Structure of peptides associated with Class I and Class II MHC molecules, *Annu. Rev. Immunol.* **12**, 181–207 (1994).

French, D.L., Laskov, R., and Scharff, M.D., The role of somatic hypermutation in the generation of antibody diversity, *Science* **244**, 1152–1157 (1989).

Harris, L.J., Larson, S.B., Hasel, K.W., Day, J., Greenwood, A., and McPherson, A., The three-dimensional structure of an intact monoclonal antibody for canine lymphoma, *Nature* **360**, 369–372 (1992). [The first high-resolution X-ray structure of an intact IgG.]

Harrison, S.C., CD4: Structure and interactions of an immunoglobulin superfamily adhesion molecule, *Acc. Chem. Res.* **26**, 449–453 (1993).

Hill, A.V.S., Allsopp, C.E.M., Kwiatkowski, D., Anstey, N.M., Twumasi, P., Rowe, P.A., Bennett, S., Brewster, D., McMichael, A.J., and Greenwood, B.M., Common West African HLA antigens associated with protection from severe malaria, *Nature* **352**, 595–600 (1991).

Honjo, T. and Habu, S., Origin of immune diversity: genetic variation and selection, *Annu. Rev. Biochem.* **54**, 803–830 (1985).

Hunkapiller, T. and Hood, L., Diversity of the immunoglobulin gene superfamily, *Adv. Immunol.* **44**, 1–63 (1989).

Kappes, D. and Strominger, J.L., Human class II major histocompatibility genes and proteins, *Annu. Rev. Biochem.* **57**, 991–1028 (1988).

Kuby, J., *Immunology* (2nd ed.), Freeman (1994).

Kuma, K., Iwabe, N., and Miyata, T., The immunoglobulin family, *Curr. Opin. Struct. Biol.* **1**, 384–393 (1991). [Discusses the phylogenetics of the immunoglobulin family.]

Latron, F., Pazmany, L., Morrison, J., Moots, R., Saper, M.A., McMichael, A., and Strominger, J.L., A critical role for conserved residues in the cleft of HLA-A2 in presentation of a nonapeptide to T cells, *Science* **257**, 964–967 (1992).

Leahy, D.J., Axel, R., and Hendrikson, W.A., Crystal structure of a soluble form of the human T cell coreceptor CD8 at 2.6 Å resolution, *Cell* **68**, 1145–1162 (1992).

Lewis, S.M., The mechanism of V(D)J joining: Lessons from molecular, immunological, and comparative analyses, *Adv. Immunol.* **56**, 27–150 (1994).

MacDonald, H.R. and Nabholz, M., T-cell activation, *Annu. Rev. Cell Biol.* **2**, 231–253 (1986).

Madden, D.R., Gorga, J.C., Strominger, J.L., and Wiley, D.C., The structure of HLA-B27 reveals nonamer self-peptides bound in an extended conformation, *Nature* **353**, 326–329 (1991); Peptide binding to the major histocompatibility complex molecules, *Curr. Opin. Struct. Biol.* **2**, 300–304 (1992); *and* Jardetzky, T.S., Lane, W.S., Robinson, R.A., Madden, D.R., and Wiley, D.C., Identification of self peptides bound to purified HLA-B27, *Nature* **353**, 321–325 (1991).

Marrack, P. and Kappler, J., The T cell receptor, *Science* **238**, 1073–1079 (1987).

Matsumura, M., Fremont, D.H., Peterson, P.A., and Wilson, I.A., Emerging principles for the recognition of peptide antigens by MHC Class I molecules, *Science* **257**, 927–934 (1992); *and* Stanfield, R.L., Fieser, T.M., Lerner, R.A., and Wilson, I.A., Crystal structure of an antibody to a peptide and its complex with peptide antigen at 2.8 Å, *Science* **248**, 712–719 (1990).

Müller-Eberhard, H.J., Molecular organization and function of the complement system, *Annu. Rev. Biochem.* **57**, 321–337 (1988).

Müller-Eberhard, H.J., The membrane attack complex of complement, *Annu. Rev. Immunol.* **4**, 503–528 (1986).

Podack, E.R. and Kupfer, A., T-Cell effector functions: Mechanisms for the delivery of cytotoxicity and help, *Annu. Rev. Cell Biol.* **7**, 479–304 (1991).

Reid, K.B.M., Activation and control of the complement system, *Essays Biochem.* **22**, 27–68 (1986).

Ross, G.D. (Ed.), *Immunobiology of the Complement System*, Academic Press (1986).

Schatz, D.G., Oettinger, M.A., and Schissel, M.S., V(D)J Recombination: Molecular biology and regulation, *Annu. Rev. Immunol.* **10**, 359–383 (1992); *and* Oettinger, M.A., Activation of V(D)J recombination by RAG1 and RAG2, *Trends Genet.* **8**, 408–412 *and* 413–416 (1992).

Sci. Am. **269** (3): (1993). [A special issue on the immune system.]

Stern, L.J., Brown, J.H., Jardetzky, T.S., Gorga, J.C., Urban, R.G., Strominger, J.L., and Wiley, D.C., Crystal structure of the human class II MHC protein HLA-DR1 complexed with an influenza virus peptide, *Nature* **368**, 215–221 (1994); *and* Brown, J.H., Jardetzky, T.S., Gorga, J.C., Stern, L.J., Urban, R.G., Strominger, J. L., and Wiley, D. C., Three-dimensional structure of the human class II histocompatibility antigen HLA-DR1, *Nature* **364**, 33–39 (1993).

Thomson, G., HLA disease associations, *Annu. Rev. Genet.* **22**, 31–50 (1988).

Todd, J.A., Acha-Orbea, H., Bell, J.I., Chao, N., Fronek, Z., Jacob, C.O., McDermott, M., Sinha, A.A., Timmerman, L., Steinman, L., and McDevitt, H.O., A molecular basis for MHC Class II-associated autoimmunity, *Science* **240**, 1003–1009 (1988).

Tonegawa, S., Somatic generation of immune diversity, *Angew. Chem. Int. Ed. Engl.* **27**, 1028–1039 (1988). [A Nobel lecture.]

Tonegawa, S., Somatic generation of antibody diversity, *Nature* **302**, 575–581 (1983).

Watson, J.D., Hopkins, N.H., Roberts, J.W., Steitz, J.A., and Weiner, A.M., *Molecular Biology of the Gene* (4th ed.), Chapter 25, Benjamin/Cummings (1987).

Young, J.D.-E. and Cohn, Z.A., How killer cells kill, *Sci. Am.* **258**(1): 38–44 (1988).

Motility

Alberts, B., Bray, D., Lewis, J., Raff, M., Roberts, K., and Watson, J.D., *Molecular Biology of the Cell* (3rd ed.), Chapter 16, Garland Publishing (1994).

Allen, B.D. and Walsh, M.P., The biochemical basis of the regulation of smooth-muscle contraction, *Trends Biochem. Sci.* **19**, 362–368 (1994).

Allen, R.D., The microtubule as an intracellular engine, *Sci. Am.* **256**(2): 42–49 (1987).

Amos, L.A., Structure of muscle filaments studied by electron microscopy, *Annu. Rev. Biophys. Biophys. Chem.* **14**, 291–313 (1985).

Anderson, M.S. and Kunkel, L.M., The molecular and biochemical basis of Duchenne muscular dystrophy, *Trends Biochem. Sci.* **17**, 289–292 (1992); *and* Ervasti, J.M. and Campbell, K.P., Dystrophin and the membrane skeleton, *Curr. Opin. Cell Biol.* **5**, 82–87 (1993).

Darnell, J., Lodish, H., and Baltimore, D., *Molecular Cell Biology* (2nd ed.), Chapters 21 and 22, Scientific American Books (1990).

Dustin, P., *Microtubules* (2nd ed.), Springer–Verlag (1984).

Fleischer, S. and Inui, M., Biochemistry and biophysics of excitation-contraction coupling, *Annu. Rev. Biophys. Biophys. Chem.* **18**, 333–364 (1989).

Gibbons, I.R., Dynein ATPases as microtubule motors, *J. Biol. Chem.* **263**, 15837–15840 (1988).

Harrington, W.F. and Rodgers, M.E., Myosin, *Annu. Rev. Biochem.* **53**, 35–73 (1984).

Hibberd, M.G. and Trentham, D.R., Relationships between chemical and mechanical events during muscular contraction, *Annu. Rev. Biophys. Biophys. Chem.* **15**, 119–161 (1986).

Hyams, J.S. and Lloyd, C.W., *Microtubules,* Wiley-Liss (1994).

Johnson, K.A., Pathway of the microtubule-dynein ATPase and the structure of dynein: a comparison with actomyosin, *Annu. Rev. Biophys. Biophys. Chem.* **14**, 161–188 (1985).

Kabsch, W., Mannherz, H.G., Suck, D., Pai, E.F., and Holmes, K.C., Atomic structure of actin:DNase I complex, *Nature* **347**, 37–44 (1990); Holmes, K.C., Popp, D., Gebhard, W., and Kabsch, W., Atomic model of the actin filament, *Nature* **347**, 44–49 (1990); *and* Holmes, K.C. and Kabsch, W., Muscle proteins: actin, *Curr. Opin. Struct. Biol.* **1**, 270–280 (1991).

Korn, E.D. and Hammer, J.A., III, Myosins of nonmuscle cells, *Annu. Rev. Biophys. Biophys. Chem.* **17**, 23–45 (1988).

Lowey, S., Waller, G.S., and Trybus, K.M., Skeletal muscle myosin light chains are essential for physiological speeds of shortening, *Nature* **365**, 454–456 (1993).

Mandelkow, E. and Mandlekow, E.-M., Microtubule structure; *and* Murray, J.M., Eukaryotic flagella, *Curr. Opin. Struct. Biol.* **4**, 171–179; *and* 180–186 (1994).

McIntosh, J.R. and Porter, M.E., Enzymes for microtubule-dependent motility, *J. Biol. Chem.* **264**, 6001–6004 (1989).

Milligan, R.A., Whittaker, M., and Safer, D., Molecular structure of F-actin and the location of surface binding sites, *Nature* **348**, 217–221 (1990).

Namba, K., Yamashita, I., and Vonderviszt, F., Structure of the core and central channel of bacterial flagella, *Nature* **342**, 648–653 (1989).

Ohtsuki, I., Maruyama, K., and Ebashi, S., Regulatory and cyto-skeletal proteins of vertebrate skeletal muscle, *Adv. Protein Chem.* **38**, 1–67 (1986).

Rayment, I., Rypniewski, W.R., Schmidt-Bäse, K., Smith, R., Tomchick, D.R., Benning, M.M., Winkelmann, D.A., Wesenberg, G., and Holden, H.M., Three-dimensional structure of myosin subfragment-1: A molecular motor, *Science* **261**, 50–58 (1993); Rayment, I., Holden, H.M., Whittaker, M., Yohn, C.B., Lorenz, M., Holmes, K.C., and Milligan, R.A., Structure of the actin-myosin complex and its implications for muscle contraction, *Science* **261**, 58–65 (1993); *and* Rayment, I. and Holden, H.M., Myosin subfragment-1: structure and function of a molecular motor, *Curr. Opin. Struct. Biol.* **3**, 944–952 (1993) *and* The three-dimensional structure of a molecular motor, *Trends Biochem. Sci.* **19**, 129–134 (1994).

Satyshur, K.A., Pyzalska, D., Greaser, M., Rao, S.T., and Sundaralingam, M., Structure of chicken skeletal troponin C at 1.78 Å resolution, *Acta. Cryst.* **D50**, 40–49 (1994).

Schröder, D.J., Jahn, W., Holden, H., Rayment, I., Holmes, K.C., and Spudich, J.A., Three-dimensional atomic model of F-actin decorated with *Dictyostelium* myosin S1, *Nature* **364**, 171–174 (1993).

Schuster, S.C. and Khan, S., The bacterial flagellar motor, *Annu. Rev. Biophys. Biomol. Struct.* **23**, 509–539 (1994).

Schutt, C.E., Myslik, J.C., Rozycki, M.D., Goonesekere, N.C.W., and Lindberg, U., The structure of crystalline profilin–β-actin, *Nature* **365**, 810–816 (1993).

Sellers, J.R. and Adelstein, R.S., Regulation of contractile activity, *in* Boyer, P.D. and Krebs, E.G. (Eds.), *The Enzymes* (3rd ed.), Vol. 18, *pp.* 381–418, Academic Press (1987).

Sheterline, P. and Sparrow, J.C., Actin, *Protein Profiles* **1**, 1–121 (1994).

Sosinsky, G. E., Francis, N. R., Stallmeyer, M.J.B., and DeRosier, D.J., Substructure of the flagellar basal body of *Salmonella typhimuriun, J. Mol. Biol.* **223**, 171–184 (1992); *and* Francis, N.R., Sosinsky, G., Thomas, G., and DeRosier, D.J., Isolation, characterization, and structure of bacterial flagellar motors containing the switch complex, *J. Mol. Biol.* **235**, 1261–1270 (1994).

Tan, J.L., Ravid, S., and Spudich, J.A., Control of nonmuscle myosins by phosphorylation, *Annu. Rev. Biochem.* **61**, 721–759 (1992).

Trinick, J. Molecular rulers in muscle, *Curr. Biol.* **2**, 75–77 (1992); *and* Titin and nebulin: Protein rulers in muscle? *Trends Biochem. Sci.* **19**, 405–409 (1994).

Vale, R.D., Intracellular transport using microtubule-based motors, *Annu. Rev. Cell Biol.* **3**, 347–378 (1987).

Warrick, H.M. and Spudich, J.A., Myosin structure and function in cell motility, *Annu. Rev. Cell Biol.* **3**, 379–421 (1987).

Whitby, F.G., Kent, H., Stewart, F., Stewart, M., Xie, X., Hatch, V., Cohen, C., and Phillips, G.N., Jr., Structure of tropomyosin at 9 angstroms resolution, *J. Biol. Chem.* **227**, 441–452 (1992).

Xie, X., Harrison, D.H., Schlichting, I., Sweet, R.M., Kalabokis, V.N., Szent–Györgyi, A.G., and Cohen, C., Structure of the regulatory domain of scallop myosin at 2.8 Å resolution, *Nature* **368**, 306–312 (1994). [The regulatory portion of a myosin that is directly regulated by Ca²⁺ binding and which consists of the homolog of skeletal muscle myosin S1's C-terminal (long) helix together with an ELC and an RLC.]

Zot, A.S. and Potter, J.D., Structural aspects of troponin–tropomyosin regulation of skeletal muscle contraction, *Annu. Rev. Biophys. Biophys. Chem.* **16**, 535–559 (1987).

Communications

Alberts, B., Bray, D., Lewis, J., Raff, M., Roberts, K., and Watson, J.D., *Molecular Biology of the Cell* (3rd ed.) Chapters 11 and 15, Garland Publishing (1994).

Almassy, R.J., Fontecilla-Camps, J.C., Suddath, F.L., and Bugg, C.E., Structure of variant 3 scorpion neurotoxin from *Centruoides sculpturates* Ewing refined at 1.8 Å resolution, *J. Mol. Biol.* **170**, 497–527 (1983).

Barford, D., Flint, A.J., and Tonks, N.K., Crystal structure of human protein tyrosine kinase phosphatase 1B, *Science* **263**, 1397–1404 (1994).

Benovic, J.L., Bouvier, M., Caron, M.G., and Lefkowitz, R.J., Regulation of adenylyl cyclase-coupled β-adrenergic receptors, *Annu. Rev. Cell Biol.* **4**, 405–428 (1988).

Berridge, M.J., Inositol triphosphate and calcium signalling, *Nature* **361**, 315–325 (1993).

Blumer, K.J. and Johnson, G.L., Diversity in function and regulation of MAP kinase pathways, *Trends Biochem. Sci.* **19**, 236–240 (1994).

Bogusky, M.S. and McCormick, F., Proteins regulating Ras and its relatives, *Nature* **366**, 643–654 (1993).

Bradford, H.F., *Chemical Neurobiology,* Freeman (1986).

Bredt, D.S. and Snyder, S.H., Nitric oxide: A physiologic messenger molecule, *Annu. Rev. Biochem.* **63**, 175–195 (1994); Snyder, S.H., Nitric oxide: First in a new class of neurotransmitters, *Science* **257**, 494–496 (1992); *and* Lowenstein, C.J. and Snyder, S.H., Nitric oxide, a novel biologic messenger, *Cell* **70**, 705–707 (1992).

Burnette, W.N., AB₅ ADP-ribosylating toxins: comparative anatomy and physiology, *Structure* **2**, 151–158 (1994). [Compares the structures of CT, LT, and PT.]

Catt, K.J. and Balla, T., Phosphoinositide metabolism and hormone action, *Annu. Rev. Med.* **40**, 487–509 (1989).

Catterall, W.A., Structure and function of voltage-sensitive ion channels, *Science* **242**, 50–61 (1988).

Charbonneau, H. and Tonks, N.K., 1002 protein phosphatases? *Annu. Rev. Cell Biol.* **8**, 463–493 (1992).

Chard, T., An introduction to radioimmunoassay and related techniques, *in* Work, T.S. and Work, E. (Eds.), *Laboratory Techniques in Biochemistry and Molecular Biology,* Vol. 6, Part II, North–Holland (1978).

Coleman, D.E., Berghuis, A.M., Lee, E., Linder, M.E., Gilman, A.G., and Sprang, S.R., Structures of active conformations of $G_{i\alpha 1}$ and the mechanism of GTP hydrolysis, *Science* **265**, 1405–1412 (1994).

Crapo, L., *Hormones,* Freeman (1985). [A highly readable introductory work.]

Darnell, J., Lodish, H., and Baltimore, D., *Molecular Cell Biology* (2nd ed.), Chapters 19 and 20, Scientific American Books (1990).

Dekker, L.V. and Parker, P.J., Protein kinase – a question of specificity, *Trends Biochem. Sci.* **19**, 73–77 (1994).

Eck, M.J., Atwell, S.K., Shoelson, S.E., and Harrison, S.C., Structure of the regulatory domains of the Src-family tyrosine kinase Lck, *Nature* **368**, 764–769 (1994). [The X-ray structure of tandemly linked SH3 and SH2 domains.]

Egan, S.E. and Weinberg, R.A., The pathway to signal achievement, *Nature* **365**, 781–783 (1993); Blenis, J., Signal transduction via the MAP kinases: Proceed at your own RSK, *Proc. Natl. Acad. Sci.* **90**, 5889–5892 (1993); *and* Avruch, J., Zhang, X., Kyriakis, J.M., Raf meets Ras: completing the framework of a signal transduction pathway, *Trends Biochem. Sci.* **19**, 279–283 (1994). [Reviews of the Ras-mediated signaling pathway.]

Fantl, W.J., Johnson, D.E., and Williams, L.T., Signaling by receptor tyrosine kinases, *Annu. Rev. Biochem.* **62**, 453–481 (1993).

Goodfellow, P.N. and Lovell-Badge, R., *SRY* and sex determination in mammals, *Annu. Rev. Genet.* **27**, 71–92 (1993); *and* Harley, V.R., Jackson, D.I., Hextall, P.J., Hawkins, J.R., Berkovitz, G.D., Sockanathan, S., Lovell-Badge, R., and Goodfellow, P.N., DNA binding activity of recombinant SRY from normal males and XY females, *Science* **255**, 453–456 (1992).

Hadley, M., *Endocrinology* (2nd ed.), Prentice–Hall (1988).

Jan, L.Y. and Jan, Y.N., Voltage-sensitive channels, *Cell* **56**, 13–25 (1989).

Jessell, T.M. and Kandel, E.R., Synaptic transmission: A directional and self-modifiable form of cell-cell communication, *Cell* **72** (Suppl.), 1–30 (1993). [A wide-ranging review.]

Kaziro, Y., Itoh, H., Kozasa, T., Nakafuku, M., and Satoh, T., Structure and function of signal-transducing GTP-binding proteins, *Annu. Rev. Biochem.* **60**, 349–400 (1991).

Kikkawa, U., Kishimoto, A., and Nishizuku, Y., The protein kinase C family: heterogeneity and its implications, *Annu. Rev. Biochem.* **58**, 31–44 (1989).

Koopman, P., Gubbay, J., Vivian, N., Goodfellow, P., and Lovell-Badge, R., Male development of chromosomally female mice transgenic for *Sry, Nature* **351**, 117–121 (1991).

Kuriyan, J. and Cowburn D., Structures of SH2 and SH3 domains, *Curr. Opin. Struct. Biol.* **3**, 828–837 (1993).

Lambright, D.G., Noel, J.P., Hamm, H.E., and Sigler, P.B., Structure determinants for activation of the α-subunit of a heterotrimeric G protein, *Nature* **369**, 621–628 (1994); *and* Noel, J. P.,

Hamm, H.E., and Sigler, P.B., The 2.2 Å crystal structure of trans-ducin-α complexed with GTPγS, *Nature* **366**, 654–663 (1993). [The first paper describes the GDP complex and the second compares it with the GTPγS complex.]

Lee, C.-H., Kominos, D., Jacques, S., Margolis, B., Schlessinger, J., Shoelson, S. E., and Kuriyan, J., Crystal structures of peptide complexes of the amino-terminal SH2 domain of the Syp tyrosine phosphatase, *Structure* **2**, 423–438 (1994).

Lefkowitz, R.J. and Caron, M.G., Adrenergic receptors, *J. Biol. Chem.* **263**, 4993–4996 (1988).

Levitzki, A., From epinephrine to cyclic AMP, *Science* **241**, 800–806 (1988). [Discusses the mechanism of signal transduction via the β-adrenergic receptor.]

Lynch, D.R. and Snyder, S.H., Neuropeptides: multiple molecular forms, metabolic pathways and receptors, *Annu. Rev. Biochem.* **55**, 773–799 (1986).

Majerus, P.W., Inositol phosphate biochemistry, *Annu. Rev. Biochem.* **61**, 225–250 (1992).

Musacchio, A., Sareste, M., and Wilmanns, M., High-resolution crystal structures of tyrosine kinase SH3 domains complexed with proline-rich peptides, *Nature Struct. Biol.* **1**, 546–551 (1994).

Neer, E.J. and Clapham, D.E., Roles of G protein subunits in transmembrane signaling, *Nature* **333**, 129–134 (1988).

Nishida, E. and Gotoh, Y., The MAP kinase cascade is essential for diverse signal transduction pathways, *Trends Biochem. Sci.* **18**, 128–131 (1993).

Norman, A.W. and Litwack, G., *Hormones,* Academic Press (1987).

Pawson, T. and Schlessinger, J., SH2 and SH3 domains, *Curr. Biol.* **3**, 434–432 (1993); *and* Mayer, B.J. and Baltimore, D., Signalling through SH2 and SH3 domains, *Trends Cell Biol.* **3**, 8–13 (1993).

Pelech, S.L. and Vance, D.E., Signal transduction via phosphatidylcholine cycles, *Trends Biochem. Sci.* **14**, 28–30 (1989).

Pennington, S.R., GTP-binding proteins 1: Heterotrimeric G proteins, *Protein Profiles* **1**, 169–342 (1994).

Rasmussen, H., The cycling of calcium as an intracellular messenger, *Sci. Am.* **261**(4): 66–73 (1989).

Rhee, S. G., Inositol phospholipid-specific phospholipase C: interaction of the γ₁ isoform with tyrosine kinase, *Trends Biochem. Sci.* **16**, 297–301 (1991).

Sinclair, A.H., Berta, P., Palmer, M.S., Hawkins, J.R., Griffiths, B.L., Smith, M.J., Foster, J. W., Frischauf, A.-M., Lovell-Badge, R., and Goodfellow, P.N., A gene from the human sex-determining region encodes a protein with homology to a conserved DNA-binding motif, *Nature* **346**, 240–244 (1990).

Sixma, T.K., Pronk, S.E., Kalk, K.H., van Zanten, B.A.M., Berhuis, A.M., and Hol, W.G.J., Lactose binding to heat-labile enterotoxin revealed by X-ray crystallography, *Nature* **355**, 561–566 (1992); *and* Sixma, T.K., Pronk, S.E., Kalk, K.H., Wartna, E.S., van Zanten, B.A.M., Witholt, B., and Hol, W.G.J., Crystal structure of a cholera toxin-related heat-labile enterotoxin from *E. coli, Nature* **351**, 371–377 (1991).

Stahl, N. and Yancopoulos, G.D., The alphas, betas, and kinases of cytokine receptor complexes, *Cell* **74**, 587–590 (1993).

Stamler, J.S., and Singel, D.J., and Loscalzo, J., Biochemistry of nitric oxide and its redox-activated forms, *Science* **258**, 1898–1902 (1992).

Stein, P.E., Boodhoo, A., Armstrong, G.D., Cockle, S.A., Klein, M.H., and Read, R.J., The crystal structure of pertussis toxin, *Structure* **2**, 45–57 (1994); *and* Stein, P.E., Boodhoo, A., Armstrong, G.D., Heerze, L.D., Cockle, S.A., Klein, M.H., and Read, R.J., Structure of pertussis toxin–sugar complex as a model for receptor binding, *Nature Struct. Biol.* **1**, 591–596 (1994).

Sternweis, P.C. and Smrcka, A.V., Regulation of phospholipase C by G proteins, *Trends Biochem. Sci.* **17**, 502–506 (1992).

Strader, C.D., Fong, T.M., Tota, M.R., Underwood, D., and Dixon, R.A.F., Structure and function of G protein–coupled receptors, *Annu. Rev. Biochem.* **63**, 101–132 (1994).

Sussman, J.L., Harel, M., Frolow, F., Oefner, C., Goldman, A., Toker, L., and Silman, I., Atomic structure of acetylcholinesterase from *Torpedo californica:* A prototypic acetylcholine-binding protein, *Science* **253**, 872–879 (1991); *and* Sussman, J.L. and Silman, I., Acteylcholinesterase: structure and use as model for specific cation–protein interactions, *Curr. Opin. Struct. Biol.* **2**, 721–729 (1992).

Taylor, W., The role of G proteins in transmembrane signalling, *Biochem. J.* **272**, 1–13 (1990).

Unwin, N., Neurotransmitter action: Opening of ligand-gated ion channels, *Cell* **72** (Suppl.), 31–41 (1993).

Unwin, N., Nicotinic acetylcholine receptor at 9 Å resolution, *J. Mol. Biol.* **229**, 1101–1124 (1993).

Verma, A., Hirsch, D.J., Glatt, C.E., Ronnett, G.V., and Snyder, S.H., Carbon monoxide: A putative neural messenger, *Science* **259**, 381–384 (1993).

de Vos, A.M., Ultsch, M., and Kossiakoff, A.A., Human growth hormone and the extracellular domain of its receptor: Crystal structure of the complex, *Science* **255**, 306–312 (1992); *and* De Meyts, P., Structure of growth hormone and its receptor: an unexpected stoichiometry, *Trends Biochem. Sci.* **17**, 169–170 (1992).

Wachtel, S.S. (Ed.), *Molecular Genetics of Sex Determination,* Academic Press (1994).

Waksman, G., Shoelson, S.E., Pant, N., Cowburn, D., and Kuriyan, J., Binding of high affinity phosphotyrosyl peptide to the Src SH2 domain: Crystal structures of the complexed and peptide-free forms, *Cell* **72**, 779–790 (1993).

Walkinshaw, M.D., Saenger, W., and Maelicke, A., Three-dimensional structure of the "long" neurotoxin from cobra venom, *Proc. Natl. Acad. Sci.* **77**, 2400–2404 (1980).

Walton, K.M. and Dixon, J.E., Protein tyrosine phosphatases, *Annu. Rev. Biochem.* **62**, 101–120 (1993).

Wu, H., Lustbader, J.W., Liu, Y., Canfield, R.E., and Hendrickson, W.A., Structure of human chorionic gonadotropin at 2.6 Å resolution from MAD analysis of the selenomethionyl protein, *Structure* **2**, 545–558 (1994); *and* Lapthorn, A.J., Harris, D.C., Littlejohn, A., Lustbader, J.W., Canfield, R.E., Machin, K.J., Morgan, F.J., and Isaacs, N.W., Crystal structure of human chorionic gonadotropin, *Nature* **369**, 455–461 (1994).

Zhang, F., Strand, A., Robbins, D., Cobb, M.H., and Goldsmith, E., Atomic structure of the MAP kinase ERK2 at 2.3 Å resolution, *Nature* **367**, 704–711 (1994).

PROBLEMS

1. There is only one known symptom, in humans, of vitamin K deficiency. What is it?

2. Clotting is prevented in stored whole blood by mixing it with citrate. What is the function of the citrate?

3. Blood samples taken from patients with either hemophilia A or hemophilia B exhibit little tendency to clot. Yet, when these two types of blood are mixed, the resulting mixture has nearly normal clotting properties. Explain.

4. Explain why the following mixtures do not form a precipitate. (a) An Fab fragment with its corresponding antigen. (b) A hapten with the antibodies raised against it. (c) An antigen and its corresponding antibody when either component is in great excess.

5. Why do antibodies raised against a native protein rarely bind to the corresponding denatured protein?

6. Explain why: (a) Antibodies to antibodies raised against a particular enzyme occasionally specifically bind the enzyme's substrate. (b) Antibodies raised against a transition state analog of a particular reaction occasionally catalyze that reaction (such antibodies have been named **abzymes**).

*7. Although only ~0.1% of an individual's *T* cells will respond to a given antigen, as many as 10% of them respond to the MHC proteins of another individual. Suggest the reason(s) for this observation.

8. The injection of bacterial cell wall constituents into an animal can trigger many of the symptoms caused by an infection including fever and inflammation. Explain how these symptoms are elicited.

*9. When you hold a weight at arm's length, you are not doing any thermodynamic work but the muscles supporting the weight are nevertheless consuming energy. Describe, on the molecular level, how muscles might maintain such state of constant tension without contracting. Why does this state consume ATP?

10. When deprived of ATP, muscles assume a rigid and inextensible form known as the **rigor state** (after **rigor mortis**, the stiffening of the body after death). What is the molecular basis of rigor?

11. In nonmuscle cells, tropomyosin is associated with microfilaments that have a structural function but not with those that participate in contractile processes. Rationalize this observation.

12. Explain why the treatment of cells with colchicine results in the disappearance of their previously existing microtubules, even though colchicine does not cause microtubule dissociation *in vitro*.

13. Explain the following observations: (a) Thyroidectomized rats, when deprived of food, survive for 20 days while normal rats starve to death within 7 days. (b) Cushing's syndrome, which results from excessive secretion of adrenocortical steroids, can be caused by a pituitary tumor. (c) **Diabetes insipidus,** which is characterized by unceasing urination and unquenchable thirst, results from an injury to the pituitary. (d) The growth of malignant tumors derived from sex organs may be slowed or even reversed by the surgical removal of the gonads and the adrenal glands.

14. How does the presence of the nonhydrolyzable GTP analog

$$GMP-O-\overset{\overset{\displaystyle O}{\|}}{\underset{\underset{\displaystyle O^-}{|}}{P}}-NH-\overset{\overset{\displaystyle O}{\|}}{\underset{\underset{\displaystyle O}{|}}{P}}-O^-$$

Guanosine-5′-(β,γ-imido)-triphosphate (GDPNP)

affect cAMP-dependent receptor systems?

15. What is the resting membrane potential across an axonic membrane at 25°C (a) in the presence of tetrodotoxin or (b) with a high concentration of Cs^+ inside the axon (use the data in Table 34-6)? How do these substances affect the axon's action potential?

16. Why don't nerve impulses propagate in the reverse direction?

17. **Decamethonium ion** is a synthetic muscle relaxant.

$$(H_3C)_3\overset{+}{N}-(CH_2)_{10}-\overset{+}{N}(CH_3)_3$$

Decamethonium

What is its mechanism of action?

Index

Page numbers in **bold face** refer to a major discussion of the entry. F after a page number refers to a figure and/or its legend or a structural formula. T after a page number refers to a table. Positional and configurational designations in chemical names (e.g., 3-, α-, *N*-, *p*-, *trans*-, D-, *sn*-) are ignored in alphabetizing. Numbers and Greek letters are otherwise alphabetized as if they were spelled out.

SOME COMMON BIOCHEMICAL ABBREVIATIONS[a]

A	adenine		ER	endoplasmic reticulum
aa	amino acid		FAD	flavin adenine dinucleotide, oxidized form
aaRS	amino-acyl tRNA synthetase		FADH·	flavin adenine dinucleotide, radical form
ACAT	acyl-CoA:cholesterol acyl transferase		$FADH_2$	flavin adenine dinucleotide, reduced form
ACh	acetylcholine		FBP	fructose-1,6-bisphosphate
ACP	acyl carrier protein		FBPase	fructose-1,6-biphosphatase
ADA	adenosine deaminase		Fd	ferredoxin
ADH	alcohol dehydrogenase		FH	familial hypercholesterolemia
ADP	adenosine diphosphate		fMet	N-formylmethionine
AIDS	acquired immunodeficiency syndrome		FMN	flavin mononucleotide
AMP	adenosine monophosphate		F1P	fructose-1-phosphate
AMPK	AMP-dependent protein kinase		F6P	fructose-6-phosphate
ALA	δ-aminolevulinic acid		G	guanine
ATCase	aspartate transcarbamoylase		GABA	γ-aminobutyric acid
ATP	adenosine triphosphate		Gal	galactose
BChl	bacteriochlorophyll		GalNAc	N-acetylgalactosamine
bp	base pair		GAP	glyceraldehyde-3-phosphate
BPG	D-2,3-bisphosphogylcerate		GAPDH	glyceraldehyde-3-phosphate dehydrogenase
BPheo	bacteriopheophytin		GC	gas chromatography
BPTI	bovine pancreatic trypsin inhibitor		GDP	guanosine diphosphate
C	cytosine		Glc	glucose
CaM	calmodulin		GMP	guanosine monophosphate
cAMP	cyclic AMP		G1P	glucose-1-phosphate
CAP	catabolite gene activating protein		G6P	glucose-6-phosphate
cAPK	cAMP-dependent protein kinase		GPI	glycosylphosphatidyl inositol
cDNA	complimentary DNA		GSH	glutathione
CDP	cytidine diphosphate		GSSG	glutathione disulfide
CDR	complimentarity determining region		GTP	guanosine triphosphate
CE	capillary electrophoresis		HA	hemagglutinin
Chl	chlorophyll		Hb	hemoglobin
CM	carboxymethyl		HDL	high density lipoprotein
CMP	cytidine monophosphate		HGPRT	hypoxanthine–guanine phosphoribosyl transferase
CoA or CoASH	coenzyme A		HIV	human immunodeficiency virus
CoQ	coenzyme Q (ubiquinone)		HMG-CoA	β-hydroxy-β-methylglutaryl-CoA
CTP	cytidine triphosphate		hnRNA	heterogeneous nuclear RNA
D	dalton		HPLC	high-performance liquid chromatography
d	deoxy		hsp	heat shock protein
dd	dideoxy		Hyl	5-hydroxylysine
DEAE	diethylaminoethyl		Hyp	4-hydroxyproline
DG	sn-1,2-diacylglycerol		IDL	intermediate density lipoprotein
DHAP	dihydroxyacetone phosphate		IF	initiation factor
DHF	dihydrofolate		IgG	immunoglobulin G
DHFR	dihydrofolate reductase		IHP	inositol hexaphosphate
DMF	N,N-dimethylformamide		IMP	inosine monophosphate
DMS	dimethyl sulfate		IP_1	inositol-1-phosphate
DNP	2,4-dinitrophenol		IP_3	inositol 1, 4, 5-triphosphate
DNA	deoxyribonucleic acid		IPTG	isopropylthiogalactoside
Dol	dolichol		IR	infrared
L-DOPA	L-3,4-dihydroxyphenylalanine		IS	insertion sequence
EF	elongation factor		ITP	inosine triphosphate
EGF	epidermal growth factor		K_M	Michaelis constant
EPR	electron paramagnetic resonance		kb	kilo base pair

[a] The three-letter and one-letter abbreviations for the "standard" amino acid residues are given in Table 4-1.